Antimicrobials in Food

Antimicrobials in Food

Fourth Edition

Edited by

P. Michael Davidson

T. Matthew Taylor

Jairus R. D. David

CRC Press
Taylor & Francis Group
Boca Raton London New York

CRC Press is an imprint of the
Taylor & Francis Group, an **informa** business

Fourth edition published 2021
by CRC Press
6000 Broken Sound Parkway NW, Suite 300, Boca Raton, FL 33487-2742

and by CRC Press
2 Park Square, Milton Park, Abingdon, Oxon, OX14 4RN

First edition published by Marcel Dekker 1983

Third edition published by CRC Press 2005

CRC Press is an imprint of Taylor & Francis Group, LLC

Library of Congress Cataloging-in-Publication Data

Names: Davidson, P. Michael, 1950- editor. | Taylor, T. M. (T. Matthew), editor. | David, Jairus R. D., editor.
Title: Antimicrobials in food / edited by P. Michael Davidson, T. Matthew Taylor, Jairus R. D. David.
Description: Fourth edition. | Boca Raton : CRC Press, 2020. | Includes bibliographical references and index.
Identifiers: LCCN 2020020320 | ISBN 9780367178789 (hardback) | ISBN 9780429058196 (ebook)
Subjects: LCSH: Food additives. | Anti-infective agents.
Classification: LCC TX553.A3 A57 2020 | DDC 664/.06--dc23
LC record available at https://lccn.loc.gov/2020020320

ISBN: 978-0-367-17878-9 (hbk)
ISBN: 978-0-429-05819-6 (ebk)

Typeset in Times
by Deanta Global Publishing Services, Chennai, India

Contents

Preface

For the sake of maintaining the health and wellbeing of all, protection of the microbiological quality and safety of food is of utmost importance. The presence of contamination, and the unchecked growth of microbes in foods, is well-known to open the door both to the transmission of human pathogens to consumers and potential disease, and the loss of a food's usefulness as a nutrient via spoilage. Foods may be prevented from becoming vehicles of disease transmission and protected from spoilage by application of one or more physical processing measures, such as heat, dehydration, or high-pressure processing, or by the use of chemical antimicrobial preservatives. Chemical antimicrobials, which act by a wide variety of mechanisms to inhibit microbial growth or inactivate microbes, may be derived from both natural and synthetic sources. Their activity is variable as to types of microorganisms affected, and their activity is often highly influenced by the environment and the physical and chemical composition of foods. They are subject to a diverse and complex set of regulatory requirements from nations around the globe based upon their usefulness and toxicology.

It has been nearly 15 years since the publication of the 3rd edition of *Antimicrobials in Food*, an excellent reference for the food microbiologist, food safety specialist, industrial researcher, and student alike. This 4th edition comes about in a time of great change within the food industry, driven in large part by consumer purchasing trends demanding foods with fewer traditional additives, including antimicrobials, as well as foods processed so as to retain a greater fresh-like character during the food's shelf-life. Compared to 2005, consumer purchase trends are apparently more influenced by numerous considerations, including budgetary restrictions, labeling claims regarding whether a food is "Clean," "Natural," "Organic," "Raised Without Antibiotics," or any of a number of others. The popularity of differing social media applications and networks also contributes to a consumer's decision on whether to purchase and consume a particular food product or brand versus another.

The 4th edition of *Antimicrobials in Food* addresses these and a number of other concerns in the following ways. First, a comprehensive revision has been completed of many chapters that appeared in previous editions of the text. Recent research has been integrated with important historical work that is still relevant. Highlights include chapters on individual or groups of compounds that generally detail current knowledge on the spectrum of activity of the antimicrobials, their mechanisms of activity, influences of food process measures on antimicrobial activity, regulatory status, and toxicology, among others. Traditional antimicrobials are covered, but new chapters have been added on lauric arginate, bacteriophages, and delivery systems for antimicrobials. A new comprehensive revision of Chapter 1 provides discussion on the key market trends which have influenced the use of food antimicrobials in recent years, including the clean label trend, the voice of the consumer and its ever-evolving role in the manufacture of foods, and the greater awareness of the need to reduce food insecurity through enhanced preservation technologies. There are new chapters on processing aides (Chapter 19), an evaluation of the hurdle concept (Chapter 21), and applications of antimicrobial food preservatives, including a discussion of the challenges and opportunities for industry members (Chapter 22). The final chapter (Chapter 23) is a comprehensive discussion of the mechanisms of action of antimicrobial preservatives as well as resistance development. Last, while P. M. Davidson, the heart and soul of this text, has returned as editor for this 4th edition, the current text brings on new editors from both academia and industry. This change from the last edition brings a broader perspective by blending both academic research and knowledge and experience from food industry applications.

As with past editions, the editors are delighted to report a diverse and highly qualified set of contributing authors, assembled from universities and corporations around the U.S. and a host of other nations. An outline of content, albeit slightly different from chapter to chapter, has been retained from previous editions in an effort to improve the reader's ability to utilize individual chapters for easy reference, wherein an antimicrobial's spectrum of activity, uses within foods, mechanisms of action, regulations surrounding its use, and its toxicology are described. Finally, the editors wish to express their deep gratitude to the contributing authors whose efforts at assembling, reviewing, and integrating a large amount of recently published refereed research into existing knowledge, or the development of wholly new content, made this book possible. We do hope this book will be a useful reference and resource for those needing insight as to the use of antimicrobial food preservatives for protecting the quality and safety of foods from microorganisms.

T. Matthew Taylor
College Station, Texas

P. Michael Davidson
Coeur d'Alene, Idaho

Jairus R. D. David
Omaha, Nebraska

Acknowledgments

To my family, friends, and co-workers at Texas A&M University whose desire for my success is constant – thank you. To Mike Davidson, to whom, like my fellow editor Dr. Jairus David, I must say congratulations for his contributions to the field of food safety microbiology and the use of food preservatives. In addition, I thank Dr. Davidson for his mentorship during my time at University of Tennessee. I thank also my students, both past and present, who have shouldered many burdens I've laid on them, because without them my research is not completed, my classes are not taught as effectively, and my time spent on this book would not be possible. Thank you all.

T. Matthew Taylor

First, I extend many thanks to the authors of the chapters of this 4th edition of *Antimicrobials in Foods*. I believe this to be the best of all the editions based on its breadth, depth, and appeal to academic, industrial, and governmental food microbiologists alike. Thanks too to my fellow co-editors, Drs. Matt Taylor and Jairus David, for, as Jairus often calls it, the "heavy lifting," and to my previous co-editors, Dr. A. Larry Branen and Dr. John N. Sofos, both titans in the food science field, whom I both admire and respect.

Few people advance science without support from others. If others judge that I have contributed positively to the field of food microbiology, I could not have done so without the cooperation and collaboration of outstanding professional colleagues, post-docs, and graduate and undergraduate students. Thus, I must thank them all for the opportunity to have interacted with them in an area that I have truly enjoyed.

And finally I would like to dedicate the book to my spouse, Linda, for her encouragement and support during what is supposed to be our retirement, and my daughters, Laurel and Holly, and their families for making our life fun and livable.

P. Michael Davidson

I would like to express my thanks to the following: my wife, Shelley Zylstra-David, for her loving encouragement of this my work, and to our three millennial children, Adriana, Brennan, and Blake, for "daring" me to write "another book."

Dr. P. Michael Davidson for his pioneering contributions to the advances in antimicrobials and food safety, and to my professional development in the field covered.

Dr. Larry Steenson, John Wyatt, Jerry Erdmann (Danisco); Dr. Athula Ekanayake (P&G); Weylan Bosse, Saurabh Kumar (Corbion-Purac); Michael Matthews, Emma Cahill (Kerry); Dr. Kathy Glass, Dr. Chuck Czuprynski, Max Golden, Dr. Wendy Bedale (Food Research Institute, University of Wisconsin); Dr. Emilia Rico (BCN Mold Research Labs) for introducing me to the world of natural and clean label antimicrobials, delivery systems, and microbial challenge studies.

Dr. Al Bolles, Senior Executive Vice President, R&D, Conagra Brands, Dr. Richard McArdle, Vice President, Innovation, and Dr. Corey Berends, Vice President, R&D for their visionary leadership and entrusting me to lead, scout, and manage a program on the use of natural and clean label antimicrobials and novel processing technologies to provide competitive advantage and food protection. And to my peers – Kevin Gatlin, Susan Janssen, John Hinchik, Kathy Landis, Christie Hancock, Amy Kerby, Denise Becker, Edith Zambrana, Dr. Fred Cook, Dr. Lalit Bohra, Dr. Shecoya White, Dr. Lenora Howard, Larry Quint, Cort Ballard, Indarpal Singh, Dr. Eric Brown, Dr. Ric Gonzalez, Dr. Brian Degner, Dr. Curtis Stowe, Dr. Rodney Green, and Dr. Andrew Wassinger.

My fellow editors – Dr. T. Matthew Taylor and Dr. P. Michael Davidson – for their visionary leadership and this our collaboration, and my learning.

Jairus R. D. David

Editors

Jairus R. D. David – Biographical Sketch

Jairus R. D. David, Ph.D., is a thought leader in the agro-food industry and an authority on food preservation science, food safety, and quality. Jairus has worked with the food industry on the use of natural and clean-label ingredients, especially antimicrobial preservatives for designing premium foods, and in developing microbiology and thermal-processing food safety objectives and compliance for over 35 years, balancing applied research with process and quality optimization, and the launch of innovative products. His industrial experience spans working for family-owned small companies and large corporations, on highly regulated baby foods, infant formula, and consumer-packaged goods. Jairus earned his Ph.D. in microbiology with emphasis in thermal processing from the University of California, Davis, under the tutelage of Dr. Larry Merson. Earlier, he earned his M.Sc. (Food Technology) from FAO Central Food Technological Institute (CFTRI), Mysore, India, and his B.Sc. (Agriculture) from the University of Agricultural Sciences, Bangalore, India. He is a Certified Quality Manager (CQM) and Certified Quality Engineer (CQE), from American Society for Quality. He has participated in the leadership program at the Kellogg School of Management, Center for Creative Leadership, Stephen Covey Leadership Forum, and the Massachusetts Institute of Technology. A Fellow of the Institute of Food Technologists (IFT), he is the recipient of IFT's prestigious Industrial Scientist Award (2006). He is recognized for developing and influencing public health food safety policy on the use of honey in cereals and bakery products for the prevention of infant botulism in infants under 12 months of age. Currently, all honey and honey-containing food products in commerce carry a warning label: "Do not feed honey to infants less than one year of age." Jairus has authored or coauthored several refereed papers, book chapters, books on aseptic technology, abstracts, and patents. Jairus is currently Principal Consultant at JRD Food Technology Consulting, LLC, Omaha, Nebraska. www.JRDFoodTech.com. Open for both domestic and international assignments, able and willing to do *pro bono* consulting for projects in developing "Third-World" and "Fourth-World" countries, rural communities, and non-profit organizations.

P. Michael Davidson – Biographical Sketch

Dr. P. Michael Davidson is a University of Tennessee (UT) Institute of Agriculture Chancellor's Professor Emeritus and former Head (2005–2013) of the Department of Food Science & Technology at UT. Prior to retirement in 2016, he served on the faculty at UT for 30 years and was Professor in Food Science and Toxicology at the University of Idaho for eight years. He earned a Ph.D. in food science at Washington State University in 1979, an M.S. in food science from the University of Minnesota, and a B.S. in microbiology from the University of Idaho. Dr. Davidson's research program involves microbiological food safety. His primary research area in food safety has been characterizing regulatory-approved and naturally occurring antimicrobial food preservatives. He is co-editor of the book *Antimicrobials in Foods*, 3rd Edition along with John Sofos and Larry Branen. A secondary research area has been the development and characterization of thermal and novel non-thermal processes to control pathogenic and spoilage microorganisms in foods. Dr. Davidson has authored or co-authored over 200 refereed journal articles, book chapters, and books and given over 300 scientific presentations at national and international meetings, industry workshops, and universities. He previously served as a Co-Scientific Editor for the *Journal of Food Protection* and on the Board of Directors of the Institute of Food Technologists (IFT). Davidson was presented with Honorary Life Membership in the International Association of Food Protection (IAFP) in 2018, the inaugural IFT Gerhardt Haas Award in 2017 for outstanding contribution to food safety, the Frozen Food Research Award from the Frozen Food Foundation in 2016, the IFT IAFP President's Recognition Award in 2005, and the IFT Food Microbiology Division Distinguished Service Award in 2000. He was elected Chair of the IFT Food Microbiology Division in 1996 and Chair of the Food Microbiology Division of the American Society for Microbiology in 1993. For his contributions to microbiology, food science and technology, and food safety, Dr. Davidson was elected a Fellow of the American Academy of Microbiology, the Institute of Food Technologists, and the International Association of Food Protection, respectively.

T. Matthew Taylor – Biographical Sketch

Dr. Matthew Taylor is an associate professor of food microbiology in the Department of Animal Science at Texas A&M University, College Station, TX. He received the B.S. in Food Science and B.A. in Sociology in 2000 from North Carolina State University. He then obtained the M.S. degree in Food Science from North Carolina State University in 2003, and earned his Ph.D. in Food Science and Technology from the University of Tennessee, Knoxville in 2006. He joined Texas A&M University in June 2007. Dr. Taylor's primary research interests are in the utilization and mechanisms of food antimicrobials to inhibit bacterial foodborne pathogens. Specifically, research is conducted to investigate and determine the manner by which food antimicrobials inhibit microbial pathogens. Additionally, research is conducted that seeks to overcome obstacles to the use of food antimicrobials in some product by the encapsulation of food antimicrobials. Dr. Taylor functions as lead instructor for undergraduate and graduate courses elaborating the microbiology of human foods. Dr. Taylor also regularly provides guest lectures for Texas A&M University courses, as well as teaching in food safety microbiology for members of the food processing industry and regulatory agency officers. Dr. Taylor is an active member of the Institute of Food Technologists and was previously recognized for excellence in service to the IFT Food Microbiology Division. He currently chairs and co-chairs various committees for the International Association for Food Protection, Phi Tau Sigma Honorary Food Science Society, and Gamma Sigma Delta Society. He sits on the editorial boards of multiple refereed journals. He also provides expert reviews in food safety microbiology for multiple journals publishing refereed food microbiology research. He has participated in over $33 million dollars in competitively funded and contracted research throughout his career.

List of Contributors

Stella M. Alzamora
Departamento de Industrias
Ciudad Autónoma de Buenos Aires
Buenos Aires, Argentina

Wendy Bedale
Food Research Institute
University of Wisconsin
Madison, Wisconsin

Adriano Brandelli
Department of Food Science
Federal University of Rio Grande do Sul
Porto Alegre, Brazil

Haiqiang Chen
Department of Animal and Food Sciences
University of Delaware
Newark, Delaware

Rebecca M. Cheng
Department of Food Science
Cornell University
Geneva, New York

Nour-Eddine Chihib
UMET-PIHM Laboratory
University of Lille
Villeneuve d'Ascq, France

John R. Chipley
Food Microbiologist (Retired)
Watkinsville, Georgia

María B. Coronel
Departamento de Industrias
Ciudad Autónoma de Buenos Aires
Buenos Aires, Argentina

Jairus R. D. David
JRD Food Technology Consulting, LLC
Omaha, Nebraska

P. Michael Davidson
Department of Food Science (Retired)
University of Tennessee
Knoxville, Tennessee
Current Address: Coeur d'Alene, Idaho

Craig Doan
Impact Washington
Bothell, Washington

Stephanie Doores
Department of Food Science (Retired)
Pennsylvania State University
University Park, Pennsylvania

Noushin Eghbal
Department of Food Science, Engineering and Technology
University of Tehran
Karaj, Iran

Leandra Filiaci
Schmid College of Science and Technology
Chapman University
Orange, California

Adem Gharsallaoui
LAGEPP Laboratory
University of Lyon
Villeurbanne, France

Kathleen Glass
Food Research Institute
University of Wisconsin
Madison, Wisconsin

Carmen L. Gomes
Department of Mechanical Engineering
Iowa State University
Ames, Iowa

Paula L. Gómez
Departamento de Industrias
Ciudad Autónoma de Buenos Aires
Buenos Aires, Argentina

Dallas G. Hoover
Department of Animal and Food Sciences
University of Delaware
Newark, Delaware

Morten Hyldgaard
DuPont Nutrition and Biosciences
Brabrand, Denmark

Eric A. Johnson
Food Research Institute
University of Wisconsin
Madison, Wisconsin

Jon J. Kabara (Deceased)
Med-Chem Labs
Longboat Key, Florida

Kalmia E. Kniel
Department of Animal and Food Sciences
University of Delaware
Newark, Delaware

Travis Kudron
Nobilus, LLC
Seattle, Washington

Ann E. Larson
Department of Environment,
Health, and Safety
University of Wisconsin
Madison, Wisconsin

Leonor Lastra-Vargas
Departamento de Ingenieria Química y
Alimentos
Universidad de las Américas Puebla
Cholula, Puebla, Mexico

J. David Legan
Eurofins Microbiology Laboratories
Madison, Wisconsin

Pushpinder K. Litt
Department of Animal and Food Sciences
University of Delaware
Newark, Delaware

Aurelio López-Malo
Departamento de Ingenieria Química y
Alimentos
Universidad de las Américas Puebla
Cholula, Puebla, Mexico

Emma Mani-López
Departamento de Ingenieria Química y
Alimentos
Universidad de las Américas Puebla
Cholula, Puebla, Mexico

Douglas Marshall
Eurofins Scientific Inc.
Fort Collins, CO

Andrew Milkowski
Department of Animal Sciences
University of Wisconsin
Madison, Wisconsin

Emefa A. Monu
Department of Poultry Science
Auburn University
Auburn, Alabama

Cornelius S. Ough (Deceased)
University of California
Davis, California

Enrique Palou
Departamento de Ingenieria Química y
Alimentos
Universidad de las Américas Puebla
Cholula, Puebla, Mexico

Maria J. Paris
Departamento de Ingenieria Química y
Alimentos
Universidad de las Américas Puebla
Cholula, Puebla, Mexico

Cicero C. Pola
Department of Mechanical Engineering
Iowa State University
Ames, Iowa

Manan Sharma
Agricultural Research Service
U.S. Department of Agriculture
Beltsville, Maryland

Cangliang Shen
Davis College of Agriculture, Natural Resources
and Design
West Virginia University
Morgantown, West Virginia

Jeffrey J. Sindelar
Department of Animal Sciences
University of Wisconsin
Madison, Wisconsin

Jarret Stopforth
Nobilus, LLC
Seattle, Washington

Peter J. Taormina
Etna Consulting Group
Jacksonville, Florida

T. Matthew Taylor
Department of Animal Science
Texas A&M University
College Station, Texas

R. Bruce Tompkin
ConAgra Refrigerated Foods (Retired)
Downers Grove, Illinois

Lilian Were
Schmid College of Science and Technology
Chapman University
Orange, California

Randy W. Worobo
Department of Food Science
Cornell University
Geneva, New York

1

Food Antimicrobials – An Introduction

T. Matthew Taylor, P. Michael Davidson, and Jairus R. D. David

CONTENTS

1.1 Role of Additives in Food

Food additives may be classified by one of six primary functions they serve: preservation, improvement in nutritional value, addition or replacement of color, addition or replacement of flavor, improvement in texture, or processing aids (Branen and Haggerty, 2002). The focus of this book is food additives that contribute to preservation, although some food antimicrobials also contribute to color stability (sulfites, nitrites) or flavor (nitrites and certain organic acids). Despite the recognized requirement for food additives, their toxicological safety continues to be evaluated and is often questioned. Few new additives have been approved by regulatory agencies in recent years, and it is doubtful that many additional ones will be approved in the future. Although there is no question that the risk from an additive must be minimal, it is apparent that such risk must be balanced against the benefits of the use of such additives. For example, reduction in the number of foodborne illness cases or reduced food waste via spoilage losses are benefits provided by antimicrobial additives. Balancing the risks vs. the benefits is not easy and requires extensive research about the usefulness and toxicological safety of the additives in question. In very few cases are additives totally devoid of risk, and thus an assessment of the degree of acceptable risk is often required. Responsibility for determining if risks outweigh benefits for any particular additive is with scientists, legislators, regulatory personnel, food processors, and consumers. It is essential that all involved in the decision process be acutely aware of the risks and benefits of all additives.

1.2 Definition and Function of Chemical Food Preservatives

Drying, cold, fermentation, and heating have always been the primary methods used to prolong the shelf-life of food products. However, since prehistoric times, some chemicals, such as salt, have been added to preserve freshly harvested foods for later use. While certain chemical food preservatives, such as salt, nitrites, and sulfites, have been in use for many years, most others have seen extensive use only recently. One of the reasons for increased use of chemical preservatives has been the change in the ways foods are produced and marketed. Today, consumers expect foods to be readily available year-round, to be "free" of foodborne pathogens, and to have a reasonably long shelf-life. While many improvements have been made using packaging and processing systems to preserve foods, antimicrobials play a significant role in protecting consumers from premature food loss and potential episodes of foodborne illness. Because of changes in the marketing for foods to a more global system, products are seldom grown and sold locally as in the past. Today, foods produced in one geographic area are often shipped to another area for processing and to several other areas for distribution. Several days, months, or even years may elapse from the time food is produced until it is consumed. To accomplish the long-term shelf-life necessary for such a system, multiple effective means of preservation are often required including the use of antimicrobials. It is important to note that food antimicrobials are not designed to be able to conceal spoilage of a food product. Rather they actually prevent or slow the growth of foodborne microorganisms which means the food remains wholesome during its extended shelf-life. Also, because most food antimicrobials are generally only bacteriostatic or fungistatic, they will not preserve a food indefinitely.

Regulatory agencies generally classify food antimicrobials as "preservatives." Chemical preservatives are defined by the U.S. Food and Drug Administration (FDA; 21CFR 101.22(a)(5)) as

> any chemical that, when added to food, tends to prevent or retard deterioration thereof, but does not include common salt, sugars, vinegars, spices, or oils extracted from spices, substances added to food by direct exposure thereof to wood smoke, or chemicals applied for their insecticidal or herbicidal properties.

Therefore, preservatives are used to prevent or retard both chemical and biological deterioration of foods. Those preservatives used to prevent or delay chemical deterioration include antioxidants used against autoxidation of pigments, flavors, lipids, and vitamins, anti-browning compounds used against enzymatic and nonenzymatic browning, and anti-staling compounds used against deleterious texture changes. Those additives used to prevent or delay biological deterioration are termed "antimicrobials."

FDA defines antimicrobial agents (21CFR 170.3(o)(2)) as "substances used to preserve food by preventing growth of microorganisms and subsequent spoilage, including fungistats, mold and rope inhibitors." The European Union definition for "preservatives" is the same as the US definition of antimicrobials, i.e., those compounds that "extend the shelf-life of foods by protecting against loss of quality caused by microorganisms or protecting against the growth of illness-causing microorganisms" (European Union, 2019).

Historically, the primary function of food antimicrobials has been to prolong shelf-life and preserve quality through the inhibition of spoilage microorganisms. Surprisingly, few food antimicrobials have been used exclusively to control the growth of specific foodborne pathogens. The one example of long-term usage is that of nitrite to inhibit *Clostridium botulinum* in cured meats. More recently, a number of compounds have been used against pathogens including organic acids as spray sanitizers against pathogens on beef carcasses, nisin and lysozyme against *Clostridium botulinum* in pasteurized process cheese, and lactate and diacetate to inactivate *Listeria monocytogenes* in processed meats. Thus, antimicrobials are being increasingly looked at as primary interventions for the inhibition or inactivation of pathogenic microorganisms in foods.

Antimicrobials may be classified as traditional or naturally occurring (Davidson et al., 2013a). The former is approved for use in foods by many international regulatory agencies (Tables 1.1 and 1.2). Naturally occurring antimicrobials include compounds from microbial, plant, and animal sources. They are, for the most part, only proposed for use in foods. A few, such as nisin, natamycin, lactoferrin, lauric arginate, and lysozyme, are approved by regulatory agencies in some countries for application to foods (Tables 1.1 and 1.2). Extensive reviews on natural antimicrobials may be found in Dillon and Board (1994), Sofos et al. (1998), Naidu (2000), Burt (2004), Davidson et al. (2013a,b), and Davidson et al. (2014).

TABLE 1.1

Traditional or Regulatory-Approved (U.S. Food and Drug Administration, 2018) Food Antimicrobials

Compound(s)	Microbial Target	Primary Food Applications	21 CFR[1] Section
Acetic acid, acetates, diacetates, dehydroacetic acid	Yeasts, bacteria	Baked goods, condiments, confections, dairy products, fats/oils, meats,[a] sauces	172.130, 182.6197, 184.1005, 184.1185, 184.1721, 184.1754
Benzoic acid, benzoates	Yeasts, molds	Beverages, fruit products, margarine	184.1021, 184.1733
Dimethyl dicarbonate	Yeasts	Beverages	172.133
Lactic acid, lactates	Bacteria	Meats,[a] fermented foods	184.1061, 184.1207, 184,1639, 184.1768
Lactoferrin	Bacteria	Meats[a]	GRAS Notices 000067 and 000130
Lauric arginate	Bacteria	Meats[a]	USDA-FSIS
Lysozyme	*Clostridium botulinum*, other bacteria	Cheese, casings for frankfurters,[a] cooked meat, and poultry products	GRAS Notice No. 000064
Natamycin	Molds	Cheese	172.155
Nisin	*Clostridium botulinum*, other bacteria	Cheese, casings for frankfurters, cooked meat, and poultry products[a]	184.1538
Nitrite, nitrate	*Clostridium botulinum*	Cured meats[a]	172.160, 172.170, 172.175, 172.177
Parabens (alkyl esters (propyl, methyl, heptyl) of *p*-hydroxybenzoic acid)	Yeasts, molds, bacteria (Gram-positive)	Beverages, baked goods, syrups, dry sausage	172.145, 184.1490, 184.1670
Propionic acid, propionates	Molds	Bakery products, dairy products	184.1081, 184.1221, 184.1784
Sorbic acid, sorbates	Yeasts, molds, bacteria	Most foods, beverages, wines	182.3089, 182.3225, 182.3640, 182.3795
Sulfites	Yeasts, molds	Fruits, fruit products, potato products, wines	182.3616, 182.3637, 182.3739, 182.3766, 182.3798, 182.3862

[a] For meat products, antimicrobials permitted by USDA Food Safety and Inspection Service are listed in USDA-FSIS (2020).

1.3 Selection of Antimicrobials

It is not an easy process to select the appropriate preservation system for a particular food product. The target pathogen or spoilage microorganisms must be identified and then the possible preservation systems must be evaluated via model studies and studies in the food product in question. Generally, a combination of chemical preservatives and other preservation methods is needed (Leistner, 2000).

Selection of the proper antimicrobial is dependent upon several primary factors, including the spectrum of antimicrobial activity, chemical properties of the antimicrobial, physicochemical properties and composition of the food product in question, and the type of preservation or processing and storage systems utilized. The individual chapters in this book examine these factors for all of the various natural and synthetic food antimicrobials. This discussion gives a brief overview of those factors and their importance.

TABLE 1.2

Regulatory-Approved Food Antimicrobial Designations in the European Union (E Numbers) and in *Codex Alimentarius* (INS Numbers)

Compound	E or INS Number	Compound	E or INS Number
Sorbic acid	200	DMDC[a]	242
K sorbate	202	Na lauroyl arginate	243
Ca sorbate	203	K nitrite	249
Benzoic acid	210	Na nitrite	250
Na benzoate	211	Na nitrate	251
K benzoate	212	K nitrate	252
Ca benzoate	213	Acetic acid	260
Ethyl paraben	214	K acetate	261
Na ethyl paraben	215	Na acetate	262
Methyl paraben	218	Ca acetate	263
Na methyl paraben	219	Lactic acid	270
Sulfur dioxide	220	Propionic acid	280
Na sulfite	221	Na propionate	281
Na hydrogen sulfite	222	Ca propionate	282
Na metabisulfite	223	K propionate	283
K metabisulfite	224	Boric acid	284
Ca sulfite	226	Na tetraborate	285
Ca hydrogen sulfite	227	Na lactate	325
K hydrogen sulfite	228	K lactate	326
Nisin	234	Ca lactate	327
Natamycin	235		

Source: European Union, 2008.
[a] Dimethyl Dicarbonate.

1.3.1 Antimicrobial Spectrum

The initial selection of the antimicrobial is normally based on an assessment of the overall microbial spectrum of the chemical in question. Antimicrobial spectrum may be defined as the efficacy of a compound against various types of microorganisms (e.g., bacteria, yeasts, molds) and forms of those microorganisms (vegetative cells vs. spores). Even species, strain, and Gram reaction (positive vs. negative) can have dramatic influences on apparent activity. For example, the activity of very hydrophobic antimicrobials may be limited against Gram-negative bacteria which have an outer membrane lipopolysaccharide layer which helps to screen out certain antimicrobials. Quite often, a broad spectrum of activity, although desired, is not easy to achieve. Few chemicals have the ability to inhibit several different types, species, or strains of microorganism.

The antimicrobial spectrum of a compound is generally determined by following the growth of organisms in the presence of various concentrations of the antimicrobial. Appropriate methods for *in vitro* evaluation of the activity of food antimicrobials are described in Chapter 2, "Methods for Activity Assay and Evaluation of Results." Seldom, however, does growth in a synthetic microbiological medium parallel that in a food product. Thus, one must be wary of an antimicrobial spectrum determined only in synthetic medium. The final confirmatory test for the antimicrobial spectrum and activity must be carried out in a food product since food components and properties can dramatically alter the overall spectrum and activity of the antimicrobial. As a rule, it is important to know the conditions under which any antimicrobial spectrum was determined before projections are made regarding the usefulness of an antimicrobial.

Probably the best method for determining what type of food antimicrobial to use would be based upon its mechanism of action and/or target in the cell. However, the exact mechanisms through which antimicrobials affect microbial growth are complex and difficult to determine. Mechanisms of action of food antimicrobials generally are classified as reaction with the cell membrane causing permeability changes or interference with uptake and transport, inactivation of essential enzymes, interference with genetic mechanisms, or inhibition of protein synthesis. These are discussed within individual chapters and collectively in Chapter 23, "Mechanisms of Action, Resistance, and Stress Adaptation." If the mechanism of the compound is known, combinations of antimicrobials with different mechanisms could be utilized against the microorganisms in the food product (Branen and Davidson, 2004; Davidson et al., 2013a).

1.3.2 Physico-Chemical Properties of the Antimicrobial

The overall microbial spectrum, the mode of action, and the efficacy of compounds are largely dependent on the chemical and physical properties of the antimicrobial. The polarity of a compound is probably the most important physical property. Water solubility or hydrophilic properties appear to be necessary to assure that the antimicrobial is soluble in the water phase, where microbial growth occurs. At the same time, however, antimicrobials acting on the hydrophobic cell membrane most likely require some lipophilic properties. Thus, the most effective food antimicrobials are amphiphilic and require a specific hydrophile–lipophile balance for optimal activity. The balance needed in a synthetic medium, however, may differ significantly from that needed for a food product owing to polarity of the food components.

1.3.3 Food-Related Factors

The chemical reactivity of the antimicrobial with other food components can significantly affect activity. Chemical reactions, in addition to decreasing antimicrobial activity, can also result in the formation of off-flavors, -odors, and -colors. Sorbic acid, for example, can be degraded by certain *Penicillium* species isolated from cheese to produce 1,3 pentadiene which has a kerosene off-odor. A sensory evaluation is often needed to assure that antimicrobials do not directly or indirectly through chemical reaction alter the color, flavor, or texture of a food product.

Another primary factor leading to reduced effectiveness among food antimicrobials is food component interactions (Davidson et al., 2013a). As stated above, since most food antimicrobials are amphiphilic, they can solubilize in or be bound by lipids or hydrophobic proteins in foods, making them less available to inhibit microorganisms in the food product. Interaction with lipids probably results in the greatest interference with antimicrobial activity. Highly active antimicrobial compounds that are hydrophobic tend to partition into the lipid areas of the food and away from the water phase, where microbial growth occurs (Rico-Muñoz Davidson, 1983).

The pH of a food can alter the effectiveness of an antimicrobial. For example, organic acids are most effective in their undissociated form. The concentration of the undissociated acid is dictated by the food pH and pK_a of the acid. In high-acid foods (generally less than pH 4.5–4.6), this is not a problem because all regulatory-approved organic acids have pK_a values less than 5.0. For food products with a pH of 5.5 or greater, there are very few organic acid-based compounds that are effective at low concentrations.

In the undissociated form, organic acids can penetrate the cell membrane lipid bilayer more easily. Once inside the cell, the acid dissociates because the cell interior has a higher pH than the exterior (Davidson, 2013b). Bacteria maintain internal pH near neutrality to prevent conformational changes to the cell structural proteins, enzymes, nucleic acids, and phospholipids. Protons generated from intracellular dissociation of the organic acid acidify the cytoplasm and must be extruded to the exterior. Since protons generated by the organic acid inside the cell must be extruded using energy in the form of ATP, the constant influx of these protons will eventually deplete cellular energy.

Chelating compounds such as ethylenediaminetetraacetic acid (EDTA) or its salts have a potentiating effect on some antimicrobials. They expand the activity of certain antimicrobials (e.g., nisin, lysozyme) to include Gram-negative bacteria which are not normally inhibited by the compounds alone (Cutter and Siragusa, 1995; Branen and Davidson, 2004). In addition, some Gram-positive bacteria are more susceptible to certain antimicrobials in the presence of chelators including EDTA. It is theorized that chelators may destabilize the LPS layer of the outer cell membrane of the Gram-negative bacteria allowing the antimicrobials access to the inner cell membrane.

1.3.4 Process Factors

The type of preservation process used in conjunction with antimicrobials has a significant influence on the type and level of antimicrobial needed. Certain preservation processes may result in the need to control spore-formers that have the ability to survive the heating process. A lowering of the water activity can select for those organisms that have the ability to survive and/or grow at lower water activity. Generally, molds survive and yeasts can grow at a lower water activity than bacteria, thus indicating the need for a different antimicrobial. Refrigeration procedures generally select for psychrotrophic Gram-negative microorganisms, thus requiring an antimicrobial capable of limiting the growth and activity of these organisms.

Packaging can directly alter the environment of the food and thus influence the overall growth pattern and type of organisms in the food. Vacuum or modified-atmosphere packaging results in low oxygen tension which inhibits the growth of molds and several bacteria but allows certain facultative anaerobic microorganisms such as lactic acid bacteria to grow. Much research has been done on the addition or incorporation of food antimicrobials to packaging materials, which themselves may be antimicrobials (e.g., chitosan). Also, active packaging systems which incorporate compounds that continuously alter the atmosphere are available. This could allow for inhibition of spoilage or pathogenic microorganisms in the packaged product.

Food antimicrobial compounds are often primary contributors to a combination of inhibitors and inhibitory conditions (e.g., low pH, low temperature). This is sometimes termed "hurdle technology" (Leistner and Gorris, 1995; Leistner, 2000). The concept of the use of multiple factors is discussed in detail in Chapter 21, "Hurdle Technology – or Is It? Multifactorial Food Preservation for the Twenty-First Century."

1.3.5 Resistance Development

Since the activity spectra are often different for each antimicrobial, the microflora contaminating a food product significantly influences the choice of the antimicrobial needed. One should be cautious, however, not to select an antimicrobial solely according to its ability to control the predominant microorganism present. Because of their specificity, selecting antimicrobials that control some genera but not others may result in selecting for and creating favorable conditions for the growth of other organisms. For example, phenolic compounds may inhibit certain Gram-positive food-poisoning bacteria, but because of the reduced activity against Gram-negative bacteria, favorable conditions can be created for growth and spoilage by these latter organisms.

Potential food antimicrobials should not contribute to the development of resistant strains nor alter the environment of the food in such a way that growth of another pathogen is selected. This subject is discussed in detail in Chapter 23, "Mechanisms of Action, Resistance, and Stress Adaptation." Microorganisms exposed to a stress factors (e.g., heat, cold, starvation, low pH/organic acids) may become more resistant and have enhanced survival to subsequent stresses. For example, it has been demonstrated that some bacterial pathogens may develop a tolerance or adaptation to organic acids following prior exposure to low pH. While this increased resistance may be a problem in the application of organic acids for controlling pathogens, it has not been shown to occur in an actual food-processing system (Davidson and Harrison, 2002).

1.4 Considerations in the Applications of Food Antimicrobials

Detailed considerations in the application of food antimicrobials may be found in Chapter 22, "Applications of Antimicrobials to Foods – A Food Industry Perspective." Subjects covered include research and development perspectives, how to determine efficacy with single and multiple antimicrobials, and the steps used in the food industry for identifying and determining the feasibility for application of antimicrobials to a food product. The technology development steps involve determinations of efficacy, cost-in-use, capital expenditure, minimum order quantity, patents, regulatory aspects, challenge studies, sensory and shelf-life studies, and production impacts. Scale-up and commercialization for various types of food products are also detailed.

Other considerations for application include toxicological safety, and labeling.

1.4.1 Toxicological Safety

Perhaps the most important aspect of any compound proposed for use as a food additive would be its toxicological characteristics. It is obviously essential that an additive for use as a food antimicrobial be safe for human consumption. The safety of some food antimicrobials has been questioned which has led to limitations on their use. The stringent requirements for toxicological safety testing can result in costs in the millions of dollars before approval for a new additive can be obtained. Ultimately, of course, an antimicrobial must be nontoxic to test-animals and humans, based on several studies. It is also important that the antimicrobial be metabolized and excreted by the body. The compound or its breakdown products should also not result in buildup of residues in body tissues. Valid assay methods for the antimicrobial are also a necessity so that levels can be easily followed. The majority of the antimicrobials now being used in food products have been extensively tested for toxicological safety, and although questions will continue to be asked, an evaluation of the risks vs. the benefits indicates that these antimicrobials are acceptable.

A possible total or partial shift to naturally occurring antimicrobials is being investigated by many researchers at the present time. Because they occur in nature, it is often thought that naturally occurring antimicrobials are less toxic than synthetic compounds. Obviously, this is not always true. A naturally occurring antimicrobial must be shown to be non-toxic either by animal testing or by its continuous consumption by consumers as a food over a long period of time. The latter may be problematic even for some common potential natural antimicrobials such as spice extracts. This is because, while spices have been consumed for centuries, they are not normally consumed in the concentrations necessary to achieve antimicrobial activity. In addition to lack of toxicity, naturally occurring compounds must be able to be metabolized and excreted so as to not lead to residue build-up. In addition, they should be non-allergenic and should not bind or destroy important nutrients in a food product (Harlander, 1993).

1.4.2 Labeling with Respect to Antimicrobials

One of the alleged attractions of naturally occurring antimicrobials is their reduced negative impact on the labeling of foods. Consumers are reportedly concerned about the presence of synthetic chemicals in their foods and would prefer natural compounds. This has led to the so-called "clean-label" movement which has its roots in the health and wellness movement of the early 2000s. Clean labels are not regulatory body-driven but rather appear to be consumer-driven (Table 1.3). To date there is not an established, objective, and common definition of what a clean label is, but rather several definitions or interpretations,

TABLE 1.3

Regulatory Definitions for Use of Food Antimicrobials

Processing Aids	21 CFR 178 Indirect Food Additives
Preservatives	21 CFR 101.22 (a) (5)
Antimicrobials:	
-FDA	21 CFR 170.3 (o) (2)
-USDA FSIS	9 CFR 430.1
Natural Antimicrobials:	
-FDA	No definition currently, proposed rule and comment period
-USDA	21 CFR 101.22 (1982) – Vinegar and Lemon Juice
Clean Label:	
FDA and USDA	Not defined

TABLE 1.4

Clean Label "Interpretation"—Based on Consumer, Retail, and Restaurant Perspective (David, 2016)

Consumer Requirements for "Clean Label"
Fewer ingredients (<5?)
Common/pantry or kitchen names
Natural Ingredients (not chemical-sounding names)
The three no's
• No synthetic colors
• No artificial flavors
• No artificial preservatives
Minimally processed
"Free-from"—allergens, antibiotics, GMO
Uncluttered label

TABLE 1.5

Consumer Perceptions of Natural vs. Artificial Ingredients

Ingredients	Natural	Not Sure	Artificial
Chicory root fiber	56%	6%	38%
Inulin (chemical name for chicory)	14%	22%	64%
Protein from soy	63%	12%	25%
Soy protein isolate (ISP)	36%	20%	44%
Xanthan gum	17%	39%	44%
Carrageenan	15%	25%	60%
Citric acid	59%	17%	24%

Survey of 1,100 diverse and stratified sample of US customers; sample demographics: Millennials (18–34 years)—31%, Gen X (35–49 years)—27%, Baby Boomers (50–64 years)—25%, Seniors (65+ years)—17%.

Source: Michael Eliasen, 2017. Personal Communication.

often provided by market-trend reports but not backed up by consumer behavior research or sales data (Table 1.4).

It has been reported by some that consumers expect a clean label to have simple (less than five) common name ingredients and be free of the "no" list (no artificial flavors, no artificial preservatives, no synthetic colors). Further, they expect products to be nutritious (reduced sugar, added protein, "good" fat) and sustainable (organic, non-GMO, ethical product and business practices, reduction in waste and packaging) (FONA International, 2016; Kerry, 2017; Mallardi, 2018). There are several potential negative impacts of the clean-label movement on the microbiological status of foods. Wholesale removal of antimicrobial food preservatives would lead to reduced shelf-life of products with a resultant increase in food waste. Food waste is a major global problem affecting food security and results in 27 million tons of food loss in the US alone (McHugh, 2018). Additionally, removal of food antimicrobials could potentially increase the chances for foodborne illness outbreaks, particularly in foods that are potentially hazardous (Doyle, 2016). Outbreaks of foodborne illness have occurred associated with foods in which antimicrobial compounds were absent or reduced including baby food, yogurt, UHT-packaged juice, processed cheese food, and turkey bacon (Doyle, 2016). In addition, trying to achieve a clean label can be difficult for food companies as illustrated by consumer confusion shown in Table 1.5. Consumers want to purchase foods and beverages that contain no artificial ingredients but often lack the knowledge to know what is and is not "natural" or "artificial." Achieving a clean label can offer a marketing opportunity for food manufacturers, however a lack of a regulatory definition and lack of specific understanding

TABLE 1.6

Clean Label Project Management Tool (David, 2016)

Criteria	Current	Clean	Cleaner	Cleanest
Basis and process—dictated by customer segmentation, and company success criteria	Current formula or "*Legacy*"	Remove and replace one "questionable" ingredient in legacy formula	Renovation—remove and replace two or more "questionable" ingredients in legacy formula	Innovation—new product development; clean-label product formulated from scratch
Budget and resources needed—FTE,[a] pilot plant, validation, sensory, shelf-life studies, opportunity cost[b]		+$	+ $$	+$$$$
Cycle time—product development, sensory equivalency, microbiological validation study, shelf-life, label development, claims		6–9 months	6–18 months	12–18+ months
On-package communication and claims—subject to regulatory and labeling constraints		Call out removal of a negative ingredient	Call out removal of a negative ingredients	Multiple claims with substantiation and science data facts from studies
Total quality—value, taste, sensory, nutrition, life cycle stability	Datum[c]	Should be equal to or greater than legacy	Should be equal to or greater than legacy	Should be greater than legacy or new
Final cost-in-use and upcharge per case	Datum[c]	+ $	+ $$	+ $$$$

[a] FTE—Full Time Equivalent; full-time employment status at 40 h per week; considered "direct cost" to a project or product development activity.

[b] Opportunity cost the loss of potential gain from other alternatives when one alternative is chosen.

[c] Datum—reference to "Current Formula" used for comparison to understand improvements to product, taste, intellectual property, and cost-in-use.

means it can be difficult to navigate company brands through this trend and determine how to meet consumer needs. Based on industrial experience (David, 2016), a project management tool for putting into operation a clean-label product is proposed (Table 1.6). The basis, process, complexity, and resources are shown stepwise. The proposed project management process can be applied for "renovation" of existing (legacy) brands and/or for "innovation" of new premium-products. The tool can be used to align front-end innovation objectives and back-end success criteria among diverse multiple stakeholders in product development, research and development, procurement in supply chain, regulatory, marketing claims, messaging, and advertisements. Inputs of budget, human capital, resources, and the cycle time needed for proper claims and target messaging are also shown. Figure 1.1 shows X Y Z coordinates representing the three major constraints of cost-in-use, efficacy and regulatory, total quality and sensory, respectively. The cost-in-use and associated upcharge per case for final product will depend on whether the new product is clean, cleaner, or cleanest compared to current product formulation as discussed in the project management tool (Table 1.6).

Traditional antimicrobials such as sorbate, benzoate, propionate, nitrite, and lactate can sometimes be replaced with new "clean-label" natural fermentation-based antimicrobials, i.e., cultured wheat starch, cultured dextrose, cultured corn solids, cultured brown rice, cultured sugar, cultured cane sugar, cultured celery juice powder, and vinegar, etc. These products must be tested however because they may not give equivalent product protection compared to the synthetic alternatives. Additionally, the approximate market price may differ by a factor of 10 to 25 between existing and natural clean-label ingredients as shown in Table 1.7. Like traditional antimicrobials, clean-label antimicrobial preservatives will need to be labeled along with other ingredients based on regulatory and labeling guidelines.

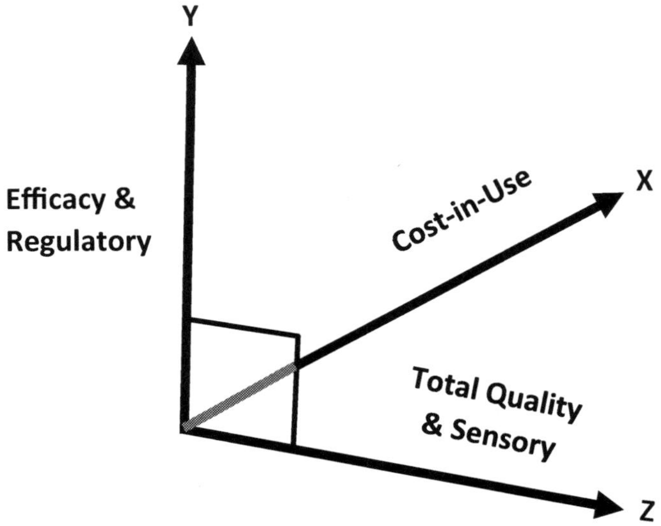

FIGURE 1.1 X Y Z coordinates representing the three major constraints for pairing a food product with a potential antimicrobial system—cost-in-use, efficacy and regulatory, and total quality and sensory, respectively.

TABLE 1.7

Approximate Market Price for "Synthetic" Vs. "Natural" Antimicrobials

Traditional Preservative (Legacy)	Approximate cost (US$)/ Pound (kg)	Clean-Label Preservative ("Natural")	Approximate cost (US$)/ Pound (kg)
Sodium benzoate	$1.27 ($2.79)	Cultured sugar	$5.10 ($11.22)
Potassium sorbate	$2.06 ($4.53)	Cultured sugar	$5.10 ($11.22)
Calcium propionate	$1.30 ($2.86)	Cultured wheat starch	$2.17 ($4.77)
Sodium nitrite	$1.11 ($2.44)	Cultured celery juice powder	$25.00 ($55.00)
Lactate-diacetate	$0.75 ($1.65)	Dry vinegar	$3.89 ($8.56)

As an alternative, it may be possible to eliminate the need for certain traditional synthetic antimicrobials via enhanced sanitation, and/or alternate processing, packaging, or distribution technologies. For example, unlike ingredient solutions, process-package solution need not be labeled. However, while ingredient solutions may persist beyond opening of a package by a consumer, the hurdle effect of process-package solutions does not and any unused product will likely need to be refrigerated for further use.

1.5 Future of Antimicrobials

Antimicrobials will undoubtedly continue to be needed to provide the food supply that will be needed in the future. There exists food insecurity throughout the globe, and antimicrobials can contribute to reducing food shortages through reducing food waste (CAST, 2016). It is hoped, through the information presented in the following chapters, that food scientists, regulators, and consumers will be better able to determine whether the benefits outweigh the risks for selected additives now available, as well as for some compounds of potential future value. The information should also serve as a basis for the selection of any new antimicrobial developed in the future.

The future of research in the area of food antimicrobials should be on several fronts. First is the expansion of information on the antimicrobial spectrum of natural antimicrobials. This research should be focused on the appropriate use of natural antimicrobials or the utilization of compounds in situations in which they are compatible with the sensory properties of the food product. Appropriate or compatible use would involve using these compounds in foods in which they add to the positive sensory characteristics of the product in addition to improving food safety or increasing shelf-life. Also needed are antimicrobial compounds or combinations of compounds that function in potentially hazardous foods with pHs above 5.5–6.0. Most commercially available natural antimicrobials are not generally useful at higher pHs. A second major area of research involves the use of antimicrobials in combinations with each other and with traditional or novel processing methods to leverage multifactorial preservation for premium-quality foods (see Chapter 21, "Hurdle Technology – or Is It? Multifactorial Food Preservation for the Twenty-First Century"). Using antimicrobials in combination with each other, so that synergistic activity is possible, requires knowledge of the mechanisms of action of the compounds. It is logical to propose that to attain synergistic activity with antimicrobial combinations requires that the individual antimicrobial components have different mechanisms. While we have made some strides to identify mechanisms of action, there is still much that is not known about specific targets of synthetic or natural antimicrobials (see Chapter 23, "Mechanisms of Action, Resistance, and Stress Adaptation").

It is likely that natural antimicrobials will be increasingly looked upon as stand-alone methods for microbial control as well as adjuncts in hurdle technology and/or used with milder non-sterilizing non-thermal processing methods (e.g., high hydrostatic pressure). It is imperative that research be undertaken on two fronts associated with reducing the use of synthetic antimicrobials. First, it must be known what the consequences of wholesale removal are on the spoilage characteristics, i.e., shelf-life as well as the safety of these products if challenged with foodborne pathogens. Second, commercial natural antimicrobial products must be evaluated by unbiased third-party researchers to determine the limits of their efficacy as to spectrum and factors affecting activity, including food composition. Finally, microbial modeling can and should be used to reduce the amount of experimentation needed to determine the potential effectiveness of natural antimicrobials against specific target microorganisms in food products under conditions of process and storage (ComBase, 2019; Halder et al., 2011; Stewart et al., 2002; USDA ARS, 2019).

REFERENCES

Branen, A.L. and Haggerty, R.J. 2002. Introduction to food additives, p. 1–9, in Branen, A.L., Davidson, P.M., Salminen, S. and Thorngate, J.H. (eds.), *Food Additives*, 2nd ed. Marcel Dekker, New York.

Branen, J.K. and Davidson, P.M. 2004. Enhancement of nisin, lysozyme, and monolaurin antimicrobial activities by ethylenediaminetetraacetic acid and lactoferrin. *Intl. J. Food Microbiol.* 90:63–74.

Burt, S. 2004. Essential oils: Their antibacterial properties and potential applications in foods—A review. *Intl. J. Food Microbiol.* 94:223–253.

CAST (Council for Agriculture and Technology). 2016. *Food Waste Across the Supply Chain: US Perspective on a Global Problem.* CAST, Ames, IA.

Combase. 2019. A web resource for quantitative and predictive food microbiology. https://www.combase.cc/index.php/en/

Cutter, C.N. and Siragusa, G.R. 1995. Population reductions of gram-negative pathogens following treatments with nisin and chelators under various conditions. *J. Food Prot.* 58:977–983.

David, J.R.D. 2016. What is clean label? A food industry perspective. Institute of Food Technologists Annual Meeting, Chicago, IL.

Davidson, P.M. and Harrison, M.A. 2002. Resistance and adaptation to food antimicrobials, sanitizers, and other process controls. *Food Technol.* 56:69–78.

Davidson, P.M., Critzer, F.J. and Taylor, T.M. 2013a. Naturally occurring antimicrobials for minimally processed foods. *Ann. Rev. Food Sci. Technol.* 4:163–190.

Davidson, P.M., Taylor, T.M. and Schmidt, S.E. 2013b. Chemical preservatives and natural antimicrobial compounds, pp. 765–801, in Doyle, M.P. and Buchanan, R.L. (eds.), *Food Microbiology: Fundamentals and Frontiers*, 4th ed. American Society for Microbiology, Washington, DC.

Davidson, P.M., Cekmer Bozkurt, H., Monu, E. and Techathuvanan, C. 2014. The use of natural antimicrobials in food: An overview, pp. 1–26, in Taylor, T.M. (ed.), *Handbook of Natural Antimicrobials for Food Safety and Quality.* Woodhead Publishing/Elsevier, Cambridge, UK.

Dillon, V.M. and Board, R.G. 1994. *Natural Antimicrobial Systems and Food Preservation*. CAB International, Wallingford, Oxon, UK.
Doyle, M.P. 2016. Unprecedented challenges in producing and serving safe foods. William C. Frazier Memorial Award Lecture, Food Research Institute Annual Meeting, University of Wisconsin, Madison.
European Union. 2008. Regulation (EC) No 1333/2008 of The European Parliament and of the Council of 16 December 2008 on Food Additives.
European Union. 2019. https://www.eufic.org/en/whats-in-food/article/food-additives-used-in-the-eu
FDA. 2018. Code of Federal Regulations Title 21 Part 101. Department of Health and Human Services.
FONA International. 2016. CLEAN: The Opportunity Report. FONA International Inc. http://www.fona.com/wp-content/uploads/2018/09/FONA_Clean-The-Opportunity-Report.pdf
Halder, A., Dhall A., Datta, A.K., Black, D.G., Davidson, P.M., Li, J. and Zivanovic, S. 2011. A user-friendly general-purpose predictive software package for food safety. *J. Food Egr.* 104:173–185.
Harlander, S.K. 1993. Regulatory aspects of bacteriocin use, in Hoover, D.G. and Steenson, L.R. (eds.), *Bacteriocins of Lactic Acid Bacteria*. Academic Press, San Diego, CA.
Kerry, Inc. 2017. Beyond the label–The clean food revolution. Natural Products Expo West, Anaheim, CA.
Leistner, L. and Gorris, L.G.M. 1995. Food preservation by hurdle technology. *Trends Food Sci. Technol.* 6:41–46.
Leistner, L. 2000. Basic aspects of food preservation by hurdle technology. *Intl. J. Food Microbiol.* 55:181–186.
Mallardi, M. 2018. NEXT Trend Guide Book. Natural Products Expo West, Anaheim, CA. https://www.newhope.com/clean-label-essentials-guide?
McHugh, T. 2018. Solving the food waste disgrace. *Food Technol.* 72(8):76–78.
Naidu, A.S. (ed.) 2000. *Natural Food Antimicrobial Systems*. CRC Press, LLC, Boca Raton, FL.
Rico-Muñoz, E. and Davidson, P.M. 1983. The effect of corn oil and casein on the antimicrobial activity of phenolic antioxidants. *J. Food Sci.* 48:1284–1288.
Sofos, J.N., Beuchat, L.R., Davidson, P.M. and Johnson, E.A. 1998. Naturally occurring antimicrobials in food. Task Force Report No. 132, Council for Agricultural Science and Technology, Ames, IA. 103 pp.
Stewart, C.M., Cole, M.B., Legan, D.J., Slade, L., Vandeven, M.H. and Schaffner, D.W. 2002. Staphylococcus aureus growth boundaries: Moving towards mechanistic predictive models based on solute-specific effects. *Appl. Environ. Microbiol.* 68:1864–1871.
USDA ARS. 2019. Pathogen modeling program. https://www.ars.usda.gov/northeast-area/wyndmoor-pa/eastern-regional-research-center/residue-chemistry-and-predictive-microbiology-research/docs/pathogen-modeling-program/pathogen-modeling-program-models/
USDA FSIS. 2013. FSIS Directive 7120.1 Revision 15. Safe and suitable ingredients used in the production of meat, poultry, and egg products. https://www.fsis.usda.gov/wps/wcm/connect/7f981741-94f1-468c-b60d-b428c971152d/7120_68.pdf?MOD=AJPERES

2

Methods for Activity Assay and Evaluation of Results

Aurelio López-Malo, Emma Mani-López, P. Michael Davidson and Enrique Palou

CONTENTS

2.1 Introduction

Hundreds of compounds that are proven or potential antimicrobials for use in foods are discussed in this book. Unfortunately, there have been nearly as many methods used for determining antimicrobial effectiveness as there are compounds. This situation leads to several difficulties in comparing results from different laboratories, determining antimicrobial effectiveness, establishing minimal inhibitory concentrations and evaluating antimicrobial spectrum, among others. In addition, it is often difficult to determine the potential success of an antimicrobial in a food or foods because of the use of methods that are inappropriate or lack relevance. Davidson and Parish (1989) reviewed the status of, and made recommendations concerning, methods for testing the effectiveness of food antimicrobials. They suggested that more uniformity and standardization should be instituted in testing food antimicrobials. Since that time, researchers have made some improvements in testing methods for food antimicrobials. The purpose of this chapter is to complement information presented in this book on antimicrobial effectiveness by suggesting methods that could be used in future studies. The subjects presented include factors affecting the activity of antimicrobials, methods for evaluating the effectiveness of single and multiple antimicrobials and techniques for interpreting data gathered in testing. It is hoped that this information will be of benefit to academic, industrial and regulatory researchers in producing data that are comparable among laboratories and that have relevance to applications of antimicrobials to foods.

2.2 Factors That Affect Antimicrobial Activity

In considering the testing of antimicrobials, the first issue which reasonably presents itself is: what is the purpose of such testing? For antimicrobial agents in foods it is difficult to answer at least in a simple and direct form, because of the many and diverse foods to which they are added and the several possible antimicrobial agents that are being evaluated. The principle of all antimicrobial testing is to assess the degree of efficiency inhibiting or inactivating a selected range of organisms in specified conditions. All methods used to evaluate antimicrobial efficacy are influenced by factors that can affect the apparent activity of a compound. Factors inherent in the test microorganism, the antimicrobial agent, the test medium and/or the test procedure itself may alter results such that an efficacious compound is judged ineffective, or vice versa. Additionally, a compound could be efficacious under the test conditions used but not provide adequate activity in the target food item under practical conditions. It is highly desirable, but often impractical or logistically impossible, to control all variables that could influence test results. It is therefore necessary to identify quantitative factors capable of influencing results. Table 2.1 highlights several factors that may influence the activity of antimicrobial agents.

One of the more important factors is the inherent nature of the specific test organism. Individual strains of a species often demonstrate genotypes and phenotypes different from the type strain. For instance, the development and detection of antibiotic resistance by specific bacterial strains are well-documented (Tenover, 1986). The microorganism(s) selected for use in the antimicrobial test method should adequately reflect the microbiota of the target food item. In deciding upon the type of microorganism to be used in any antimicrobial test, due attention is always given to the purpose for which the antimicrobial is required. Having selected a culture for a particular purpose, it should be immediately preserved in order to have always available a standard culture organism. If kept in continuous culture, it is not unknown for a microorganism to change its characteristics somewhat and for its resistance to vary. This is likely to occur even under the most carefully controlled cultural conditions, and in such circumstances the only solution is to revert to the original culture. All microorganisms have an optimum growth temperature, and they will grow over a reasonably wide range either side of this temperature. But it is self-evident that changes in temperature must affect the metabolic and growth rate of the organism and might even cause some change in the metabolic activity. This would inevitably lead to an alteration in the resistance of the culture and so affect the levels of response in a test. Sudden changes in the incubation temperature can also dramatically affect microbial viability.

The size and physiological state of the inoculum are critical and must be well controlled. Piddock (1990) emphasized the critical nature of inoculum size in antibiotic susceptibility testing for the broth dilution and agar dilution techniques. Inocula that are too concentrated or too dilute can result in false-resistant or false-sensitive results, respectively. Research is needed to determine appropriate cell concentrations

TABLE 2.1

Selected Factors That May Affect Antimicrobial Activity

- Test microorganism
    - Strain
    - Inoculum size
    - Cell physiology
    - Culture medium in which the organism is grown
- Antimicrobial agent
    - Interaction of the test compound with components of the medium
    - Partial partition coefficient
- Test medium
    - pH
    - Water activity
    - Redox potential
- Test procedure
    - Incubation conditions
        - Oxygen pressure
        - Atmospheric CO_2 concentration
        - Incubation temperature
    - Equipment variability

for use in testing food-related compounds. The physicochemical configuration of the target food and proposed shelf life impact inoculum preparation procedures and cell concentration used in antimicrobial testing. Cell physiology can also affect antimicrobial testing results. It is often assumed that actively growing vegetative cells (log-phase growth) are more sensitive to environmental stress caused by antimicrobial factors (chemicals, heating, water activity and freezing, among others) than are stationary-phase vegetative cells or spores. Small variations in the age of the culture can give rise to large variations in its resistance during antimicrobial agent testing. Physiology may indeed play a role in determining cell sensitivity and resistance, but the effect of microbial growth phase on susceptibility to food antimicrobials is not well-defined. The culture medium in which the organism is grown is also an important factor in influencing the characteristics of the microorganism. In several antimicrobial tests, it plays a dual role, first in maintaining the culture prior to its use in the test and second in providing the means of recovery and growth after treatment in the test.

Inherent factors of the test medium and test method have definite effects on cell sensitivity or resistance response. It is well-documented that the pH of the medium affects the antimicrobial activity of such compounds as benzoic and sorbic acids. The antimicrobial action of these compounds is due to the concentrations of the undissociated acid form, which is dependent upon the pH of the medium. Water activity, partial oxygen pressure, atmospheric CO_2 concentration, redox potential, incubation temperature and partition coefficient can also affect the physicochemical properties of specific antimicrobial compounds.

Other less obvious factors that can influence test results include interaction of the test compound with components of the medium, combined effects of factors that give the appearance of increased or decreased susceptibility, equipment variability, limitations of the plating method used for enumeration, purity of water and analyst errors. For example, in agar diffusion techniques, the overall negative charge of agar due to sulfate groups may slow the diffusion rate of cationic antimicrobial agents, thereby affecting expected activity patterns (Piddock, 1990).

2.3 Current Methodology

One of the first methods for evaluating the activity of disinfectants was developed by Robert Koch in 1881 (Block, 1983; Crémieux and Fleurette, 1983). He immersed silk threads in culture fluid containing *Bacillus anthracis* spores and exposed the dried threads to disinfectant for varying periods. He demonstrated that mercuric chloride was highly effective against the spores but that neither carbolic acid (phenol) nor ethanol had significant antimicrobial effects. Koch was probably the first to distinguish between bacteriostasis and lethality (Block, 1983). In 1897, Kroning and Paul noted that the antimicrobial activity of disinfectants could be compared only under controlled conditions and that kill rate was dependent upon concentration and temperature (Block, 1983). Rideal and Walker used the principles of Kronig and Paul to develop the phenol coefficient method in 1903 (Crémieux and Fleurette, 1983). The novelty of their method was standardization of components, including microorganisms, recovery medium, disinfectant stock solution and test methods. Although this method had many deficiencies, it continues to be used today with modifications (Beloian, 1990).

With the discovery of penicillin by Fleming in 1929, methods for determining the antimicrobial activity of antibiotics were developed. Assay methods were developed in the 1940s that involved the diffusion of compounds in agar form wells or paper disks (Piddock, 1990). At the same time, methods for the determination of minimum inhibitory concentrations (MIC) using broth or agar dilution were developed. In 1966, Bauer et al. proposed a disk susceptibility method that had standardized procedures. Today, in the United States, the National Center for Clinical Laboratory Standards (NCCLS) publishes standardized procedures for agar diffusion and dilution assays for the activity of antibiotics and the bactericidal activity of antimicrobial agents (NCCLS, 1999, 2002, 2018). The available documents published by NCCLS include "Performance standards for antimicrobial disk susceptibility test" and "Methods for dilution antimicrobial susceptibility test for bacteria that grow aerobically", among others.

Methods for evaluating the effectiveness of food-related antimicrobials have been in existence nearly as long as those for disinfectants and antibiotics. For example, Walton et al. (1936) determined the antimicrobial effect of vapors of crushed garlic against *Mycobacterium* sp., *Escherichia coli*, *Serratia marcescens* and *Bacillus subtilus*. In their procedure, crushed garlic was placed on the lid of a Petri dish and the bottom of the dish containing nutrient medium was inverted over the top. The vapors of the garlic were allowed to penetrate into the agar for varying periods. Following exposure, the medium with the test microorganism was incubated to determine inhibition. Antimicrobial effectiveness was related to time of exposure of the medium to the garlic vapors. This particular procedure was unique; most methods for evaluating the activity of food antimicrobials have been adopted directly or with modifications for other disciplines.

The classification of some of the most used methods to evaluate the antimicrobial efficiency in foods is shown in Table 2.2. Screening or *in vitro* tests (generally made in laboratory media or in model systems) provide preliminary information about the potential antimicrobial activity: the endpoint tests, in which the microorganisms are challenged for an arbitrary period of incubation time, give qualitative information about approximate effective concentration, and the descriptive screening methods, in which the microorganism growth is analyzed over time, give quantitative information about the growth dynamics. Finally, in applied tests, the antimicrobial is applied to the actual food and the results allow the evaluation of some factors that can affect the efficacy of the natural antimicrobial (Davidson and Parish, 1989; López-Malo et al., 2000a). Current methods used to evaluate the efficacy of food antimicrobials may be divided into *in vitro* and application methods. *In vitro* methods include any test in which the compound is not applied to a product under use conditions. These methods can provide only preliminary information to determine the potential usefulness of the test compound in a food. Application methods are those in which an antimicrobial is applied directly to a food product to determine its effects on natural microbiota or an inoculated microorganism.

In vitro methods include end-point and descriptive analyses. End-point methods are those in which a microorganism is challenged for an arbitrary period and the results reflect the inhibitory power of a compound only for the time specified. The end-point methods section of the third edition of this book focused on direct contact between the tested antimicrobial and the indicator microorganism studied; due to increased research in alternative antimicrobials and food active package development, the end-point methods of this edition include assays that evaluate vapors of antimicrobials as well as their volatile compounds. In descriptive methods, periodic sampling is carried out to determine changes in viable cell numbers over time. Both these methods may be used to determine the effectiveness of antimicrobial combinations.

TABLE 2.2

Methods for Testing the Efficacy of Food Antimicrobials

In vitro, screening or exploratory tests
- Endpoint methods
    - * Agar diffusion
    - * Agar and broth dilution
    - * Gradient plates
    - * Spiral plating
    - * Sanitizer and disinfectant tests
    - * Vapor-phase and volatile compounds methods
- Descriptive methods
    - * Turbidimetric assays
    - * Inhibition curves

Applied tests
- Endpoint methods
- Inhibition curves

Combined antimicrobials tests
- Agar diffusion
- Dilution assays
- Inhibition curves

Adapted from Davidson and Parish (1989) and López-Malo et al. (2000).

2.4 *In Vitro* Methods

2.4.1 Agar Diffusion

The agar diffusion method (Acar and Goldstein, 1986; Piddock, 1990; NCCLS, 1999, 2002; CLSI, 2018a) has probably been the most widely used method for the determination of antimicrobial activity throughout history. In this test, an antimicrobial compound is added to an agar plate on a paper disk or in a well (Figures 2.1 and 2.2). The compound diffuses through the agar, resulting in a concentration gradient that is inversely proportional to the distance from the disk or well. The degree of inhibition, which is indicated by a zone of no growth around the disk or well, is dependent upon the rate of diffusion of the compound and cell growth (Barry, 1986). Therefore, the antimicrobial evaluated should not be highly hydrophobic, because the compound will not diffuse and little or no inhibition will be detected. The test microorganism selected must also grow rapidly and uniformly. Slow-growing strains produce

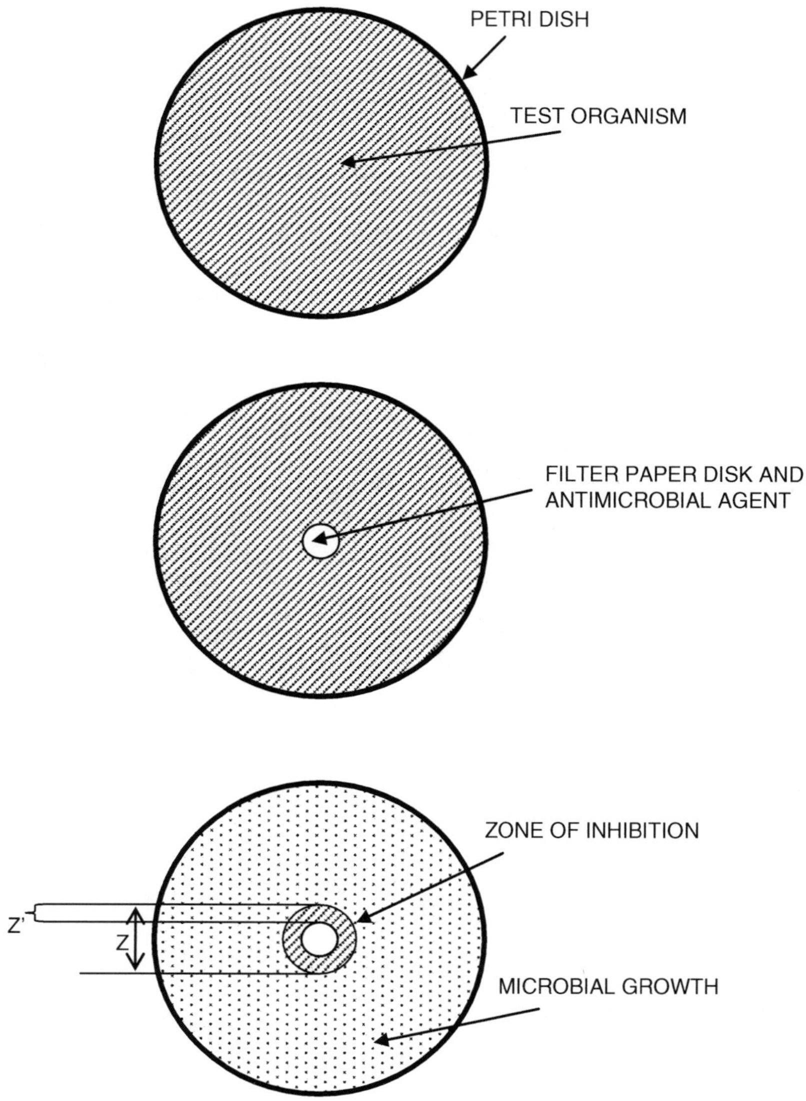

FIGURE 2.1 Determination of the "zone of inhibition" by the filter paper disk diffusion method.

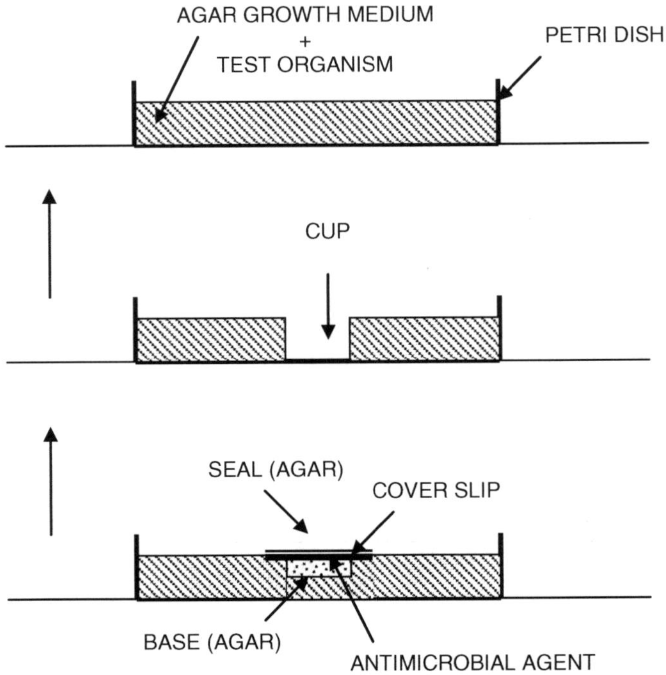

FIGURE 2.2 Agar cup method for the determination of antimicrobial activity of antimicrobial agents.

large zones of inhibition and vice versa (Piddock, 1990). This method should not be used for anaerobic microorganisms.

To run the test, Petri dishes are prepared to contain a nonselective medium at a depth of approximately 4 mm. The medium is surface-inoculated with a suspension containing approximately log 6.0 colony-forming units (CFU)/ml of the test microorganism (Piddock, 1990). Known concentrations of antimicrobial are the added to filter paper disks (6 mm diameter), which are dried and placed on the surface of the previously inoculated plate (Figure 2.1). Alternatively, the antimicrobial may be added to wells cut in the agar with a sterile cork borer (Figure 2.2). Plates are incubated under optimum conditions for the test microorganism for 16–24 h (NCCLS, 1999, 2002; CLSI, 2018a). Following incubation, plates are examined for zones of no growth indicated by halos around the disks (Figure 2.1). A control test microorganism with known susceptibility to the antimicrobial should be included.

The results of the agar diffusion test are generally qualitative; since microbial inhibition does not mean the microorganism is dead, this method cannot distinguish between biocidal or biostatic effects. The susceptibility of the test microorganism is related to the inhibition zone size in millimeters. Microorganisms are termed susceptible when the zone is >30–35 mm in diameter, intermediate with a zone of 20–30 mm, or resistant with a zone <15–20 mm (Piddock, 1990). Inhibition zones can be measured by automatic scanner Protocol 3 or Protocol 3 HD which is able to read multi-well or single radial immunodiffusion with a detection limit of 0.1 mm (Synbiosis, 2019). Determination of minimum inhibitory concentration is possible with this technique, but other methods are better. Quantification of the antimicrobial agent diffused into the agar medium is impossible in this technique. However, because the method is versatile and straightforward, several variations have been proposed.

A considerable amount of research has been done on the effectiveness of biologically derived antimicrobials, primarily bacteriocins produced by the lactic acid bacteria. Several methods related to the agar diffusion method are available for determining the activity of these compounds, including the flip-streak and flip-spot methods (Kekessy and Piguet, 1970) and the spot-on-lawn method. In studies by Harris et al. (1989) and Lewus and Montville (1991), the agar diffusion assay using wells was found to give a large number of false negative results. Nisin activity is determined following the well diffusion assay described by Tramer and Fowler (1964) using an indicator strain (*Micrococcus luteus* ATCC 10420).

Bacteriocin activity can be estimated from the size of the inhibition zones produced by the diffusion of the bacteriocin in the confluent lawns of sensitive bacteria (Montville et al., 2001).

The overlay method (Pereira et al., 2013) has been proposed for antifungal activity in co-culture (filamentous fungi and yeast), where the filamentous fungi are first inoculated and, after 5–6 days, it is covered with the indicator culture (yeast). This assay is less time-consuming than isolating and purifying the anti-yeast compound from filamentous fungi. This method is very useful when performing large screenings of fungal antimicrobial activities.

An alternative method based on the agar diffusion test is the agar plug diffusion. This procedure begins by growing the microorganism of interest on its appropriate agar medium using tight streaks on the plate surface; during their growth, microbial cells secrete molecules which diffuse into the medium. After adequate incubation, an agar-plug or cylinder is aseptically cut with a sterile cork borer and deposited on the agar surface previously inoculated with the target microorganism (Jiménez-Esquilín and Roane, 2005; Elleuch et al., 2010; Balouiri et al., 2016). The substances diffuse from the plug to the agar medium, and inhibition zones around the agar plugs can be observed if antimicrobial activity is present. This method allows screening for antimicrobial compounds produced by one microorganism without previous purification.

The cross-streak method proposed by Lertcanawanichakul and Sawangnop (2008) is another variant of agar diffusion method; the strain with antimicrobial activity is inoculated by a single streak in the center of the agar plate and incubated for an established period. Then, the plate is seeded with the tested indicator microorganisms by a single streak perpendicular to the central streak, and after incubation, inhibition zones can be observed.

Antimicrobial agents can be tested after separation by a chromatogram. After separation, they are allowed to diffuse from the chromatography paper or thin layer chromatogram to a previous inoculated agar plate with the target microorganism (Marston, 2011; Balouiri et al., 2016). Diffusion takes place in minutes or hours, and after that, the chromatogram is removed and then the agar plate is incubated. The growth inhibition zones are located where the compounds contact the agar layer. No quantification of the antimicrobial agent can be determined. For plant extracts antimicrobial screening, direct bioautography is preferred. This technique was first used by Goodall and Levi in 1946 to estimate the purity of penicillin (Dewanjee et al., 2015). Bioautography is a microbial screening method commonly utilized for the detection of antimicrobial activity (antibacterial, antifungal and antiprotozoae) of organic extracts (generally plant extracts) which previously are separated by thin-chromatography layer (TLC). This technique is simple, cheap and time-saving (Choma and Grzelak, 2011). The developed TLC plate is sprayed with or dipped into a microbial suspension (fungal or bacterial). Then, the bioautogram is incubated at 25°C for 48 h under humid conditions. After that, a tetrazolium salt has to be sprayed onto the bioautogram to visually assess microbial growth, since these salts form intensely colored formazan by means of the dehydrogenases of living cells (Balouiri et al., 2016). The TLC plate is re-incubated at 25°C for 24 h or at 37°C for 3–4 h; *p*-iodonitrotetrazolium violet is the most suitable detection reagent. Schematic representations of this method can be consulted in Choma and Grzelak (2011) and Dewanjee et al. (2015).

2.4.1.1 Dilution Methods

Agar and broth dilution assays are generally used when quantitative data are desired, to determine whether an antimicrobial is lethal to a test microorganism, for microorganisms with a variable growth rate and for anaerobic or microaerophilic microorganisms (Barry, 1986). If a microaerophilic microorganism is tested, adequate CO_2 atmosphere should be provided (CLSI, 2012a). These methods offer the food microbiologist a great deal of flexibility in determining the activity of a novel antimicrobial and/or the effect of environmental conditions on the growth of a microorganism in the presence of a known antimicrobial. In both methods a single statistic known as the minimum inhibitory concentration (MIC) is generated. The MIC is generally defined as the lowest concentration of an antimicrobial that prevents growth of a microorganism after a specified incubation period.

In the agar dilution method, a number of containers of molten nonselective agar are prepared. A single concentration of antimicrobial is then added to the agar in each container. The range of concentrations used is normally determined by trial and error. In clinical microbiology, doubling dilutions are used (e.g.,

512, 256, 128, 64 and 32 μg/ml); however, any interval may be used as long as they are narrow enough to provide a realistic MIC. After addition of the antimicrobial to the agar, plates are poured and the agar allowed to solidify. The test microorganism is diluted to around log 7.0 CFU/ml and added to the plates in 1–2 μl spots (around log 4.0 CFU per spot) (NCCLS, 1999, 2002; CLSI, 2018b). More than one species or strain may be used on one plate. A control plate, without added antimicrobial, should be prepared and inoculated to ensure adequate growth of the test microorganism. In addition, a control microorganism with known susceptibility should be included. Plates are incubated at the optimum temperature of the test microorganism for 16–24 h (NCCLS, 1999, 2002; CLSI, 2018b). For some food-borne microorganisms, it may be desirable to incubate for longer periods under environmental stress conditions (Moir and Eyles, 1992). The MIC is considered the lowest concentration that completely inhibits growth (ignoring faint haze or a single colony). There are several advantages of the agar dilution method, including the ability to test a large number of strains at once, contamination is easily detected and the medium can contain opaque materials (Barry, 1986).

An agar dilution method termed the "poisoned food method" is an alternative technique designed to evaluate potential antifungal compounds against molds (Mukherjee and Raghu, 1997; Ali-Shtayeh and Abu Ghdeib, 1999; Balouiri et al., 2016). The antifungal agent is added to molten agar at desired concentrations, mixed well and poured into Petri dishes. Inoculation can be performed by a mycelia disc between 2 to 5 mm, which is set on the center of the plate. After the incubation period under suitable conditions, diameters of fungal growth in control and tested plates are measured. The antifungal effect can be estimated by the following formula:

$$\text{Antifungal activity } (\%) = \left(\left(\text{Dc} - \text{Ds}\right)/\text{Dc}\right) \times 100$$

Where Dc is the diameter of growth in the control plate and Ds is the diameter of growth in the plate containing the antifungal agent. Furthermore, sporulation can be also compared between these two plates.

In the broth dilution assay, an antimicrobial is serially diluted and a single concentration added to a culture tube of nonselective broth medium. The concentrations are selected in the same manner as for the agar dilution method. Tubes (usually 1–10 ml total volume) are inoculated to contain approximately log 5.7 CFU/ml of the test microorganism (Thrupp, 1986; Piddock, 1990; NCCLS, 1999; CLSI, 2018b). As with the agar dilution assay, a control tube without antimicrobial should be prepared and inoculated and a strain with known susceptibility should be included. Tubes are incubated for 16–20 h at the optimum temperature of the test microorganism. As before, the incubation time may have to be adjusted depending upon the environmental conditions being evaluated. The MIC is defined as the lowest concentration at which no growth occurs (absence of turbidity) in a nutrient medium following incubation.

The broth dilution assay may also be carried out on a micro scale. In the micro dilution broth assay, 50–100 μl broth plus a single concentration of antimicrobial is added to the well of a microtiter plate (commonly contains 96 wells with approximate capacity of 300 μl). The inoculum is added at a level of approximately 4.7 CFU per well. The remainder of the assay is the same as the standard broth dilution assay. Micro dilution assays have recently been expanded using the Multidrop Combi (Thermo Fisher Scientific) microtiter plate with 1536 wells for bacterial cultures and a process called "ultra-high-throughput bacterial growth assay" (HIGA) (Sun et al., 2016). Broth volumes are 2.5 μl in order to inoculate with 2.5 μl of bacterial culture and utilize 23 nl of the test antimicrobial compounds allowing final dilutions of 1:500.

As mentioned, broth dilution assays may be used to determine the lethality of the test microorganism. To establish lethality, the standard or micro dilution broth assay must be carried out as described. Following incubation of the tubes or micro titer plates, 10–100 μl medium from the last tubes or wells in which the microorganism exhibits growth and all tubes or wells showing no growth (no turbidity) are plated on a nonselective agar using the spread-plate method. The lowest concentration of antimicrobial that produces ≥99.9% kill of the test microorganism is termed the minimum bactericidal concentration (MBC) or minimum lethal concentration (MLC) (NCCLS, 2002; CLSI, 2018b). Figure 2.3 presents, as an example, the tube dilution method determining the minimal inhibitory concentration as well as the minimal bactericidal or lethal concentration.

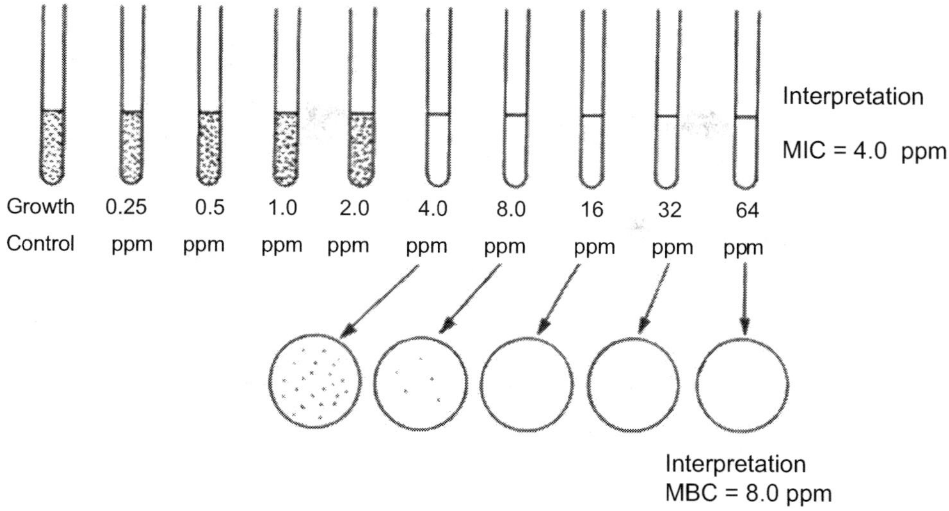

FIGURE 2.3 Determination of minimal inhibitory concentrations (MIC) and minimal bactericidal (MBC) or lethal concentrations (MLC). The MIC is defined as the lowest concentration of antimicrobial concentration that results in no visible growth (turbidity). The MLC is the lowest concentration of antimicrobial that produces a 99.9% (10^3) reduction in organisms from the original inoculum of approximately 10^5. (Adapted from Conte and Barriere, 1992.)

For MIC endpoint determinations, viewing devices can facilitate reading microdilution tests. Common measurement of turbidity is through optical density (OD; 600 or 450 nm) which reflects the viability of cells. Recently, several colorimetric methods based on the use of dye reagents have been developed. Tetrazolium salts are often utilized for MIC endpoint determination of antifungal and antibacterial microdilution assays (Balouiri et al., 2016). Resazurin or Alamar blue dye which detect metabolically active cells through fluorescence at 560 and 590 nm (Foerster et al., 2017) are also effective growth indicators. For certain fastidious microorganisms these alternatives are rapid, objective, high-throughput, quantitative and cost-effective tools.

2.4.1.2 Gradient Plates

Wedge plating is a subjective method of observing relative sensitivity to antimicrobial agents among several microorganisms on an agar medium. It is similar to the agar diffusion methods since the concentration of antimicrobial agent to which a microorganism is exposed depends on the ability of the agent to diffuse through agar. In this technique, a known concentration of antimicrobial is aseptically added to a nonselective base agar medium, which is poured into a Petri dish with one side raised to create a slanted surface. Square Petri plates are often used for this method to create a uniform gradient. After hardening, the dish is placed on a flat surface and the base medium without added antimicrobial is poured over the surface of the slanted agar (Figure 2.4). The gradient is established depending upon the rate at which the antimicrobial diffuses through the agar. Cells of known concentration from a broth culture are then streaked across the gradient surface. Control strains with known susceptibility or resistance to the test compound are streaked next to the test organism using broth cultures of the same cell concentration. Plates are incubated with an appropriate temperature and time regimen for the test microorganism. Growth suppression of the test organism is observed in relation to the control strains.

This method of antimicrobial testing can be semi-quantitative if proper control strains of known sensitivity are selected for use. Such factors as incubation (exposure) time and inoculum preparation and size must be consistent for reproducible results.

The spiral plating system is used for the enumeration of cells in microbial suspensions (Spiral Biotech, Inc., Norwood, MA); however, a spiral gradient end-point (SGE) test for the determination of MIC has been developed for antibiotics (Paton et al., 1990; Spiral Biotech, 2001; Hill, 1991) and could potentially be used to determine MIC of antimicrobial compounds used in food systems (Schalkowsky, 1986). The spiral

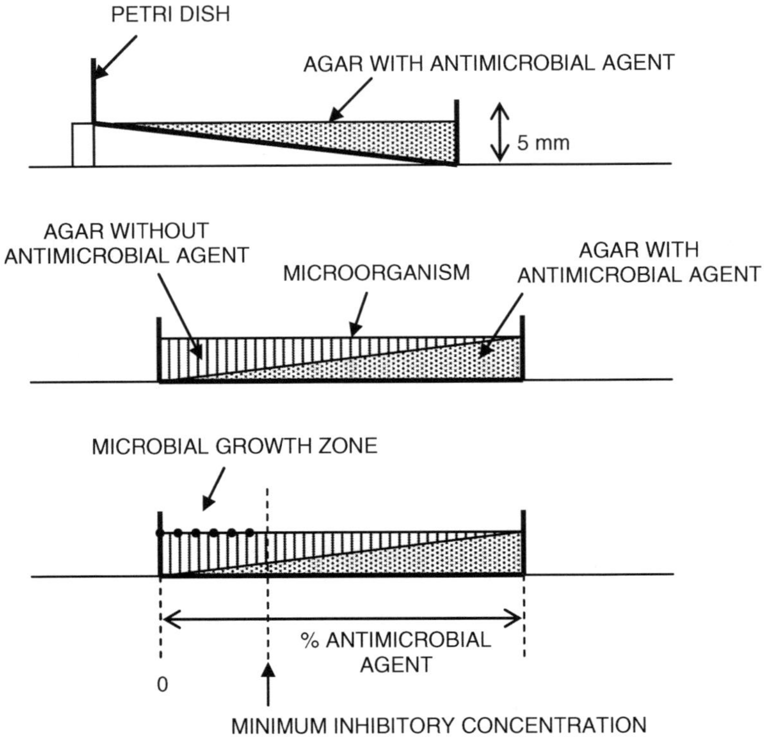

FIGURE 2.4 Gradient plate method for the determination of antimicrobial activity of antimicrobial agents.

plater dispenses a continuously decreasing volume of a known concentration of antimicrobial onto an agar medium surface in an Archimedes spiral pattern. This results in a gradient of the test compound with a high concentration near the center of the Petri plate. Cell suspensions of control and test organisms are radially streaked onto the surface of the plate from the outside edge to near the plate center. After appropriate incubation, the distance from the end of growth to where the antimicrobial deposition began (about 13 mm from the plate center) is measured and recorded. The distance is used to calculate a discrete MIC, as opposed to an MIC based on traditional incremental dilution methods (Paton et al., 1990). Observation of the growth to no growth region provides a basis for measuring the MIC. Figure 2.5 presents the template used to read the end growth point and shows an example of how microbial growth is observed when using this technique. Using this methodology Astorga-Solari et al. (2001) reported that it was possible to evaluate the antimicrobial effectiveness of thymol, eugenol, potassium sorbate and sodium benzoate to inhibit the growth of different bacteria (*Listeria innocua*, *Escherichia coli* and *Lactobacillus* spp.). The SGE test correlates well with conventional dilution methods; however, there are few data regarding its use with food antimicrobials.

2.4.1.3 Automated Methods

Automated and semi-automated methods have been developed to provide rapid enumeration and increased end-point accuracy with reduced technical error. Some systems were originally developed for rapid enumeration and identification but can also provide information regarding antimicrobial susceptibility testing. Thornsberry (1986) and Piddock (1990) reviewed the use of automated methods for antimicrobial susceptibility testing. These systems correlate well with conventional susceptibility testing; however, their use has been directed at clinical testing. Methods involving impedance, bioluminescence, pH changes, radiometry and micro calorimetry have also been investigated (Piddock, 1990). As the field of rapid methods and automation develops, the boundaries between instrumentation and diagnostic tests began to merge (Fung, 2002). Instruments are needed to monitor changes in a population such as ATP levels, specific enzymes, pH, electrical impedance, conductance and capacitance, generation of heat,

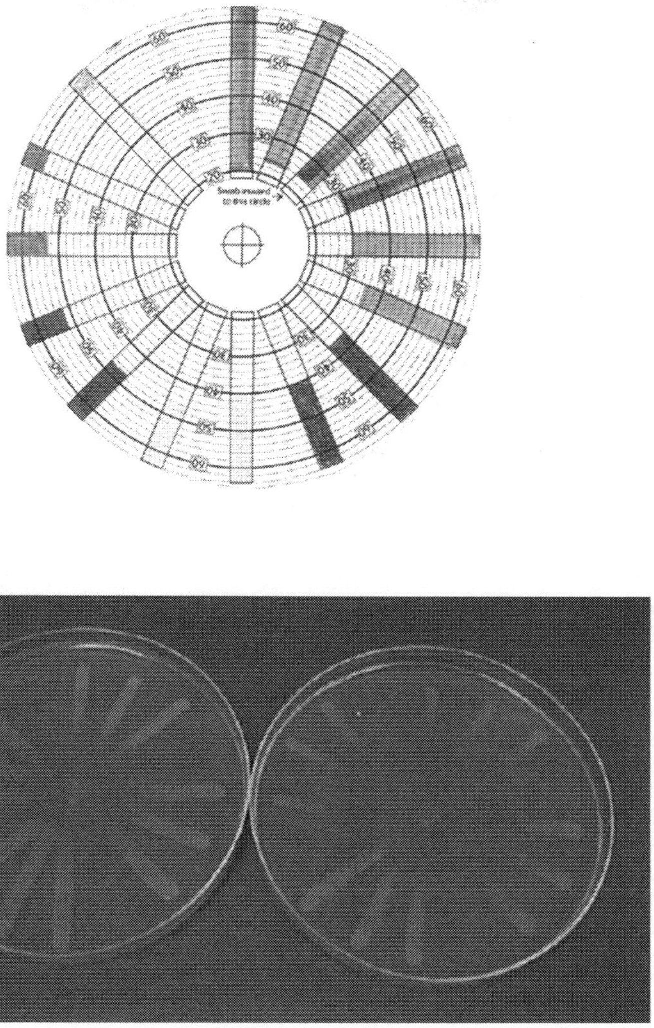

FIGURE 2.5 Spiral gradient endpoint template and *Listeria innocua* growth response to sodium benzoate (left) and potassium sorbate (right) deposited using this technique over solidified SPC agar.

radioactive carbon dioxide and so on. It is important to note that for the information to be useful; these parameters must be related to viable cell count of the same sample series.

In general, the larger the number of viable cells in the sample, the shorter the detection time of these systems. The assumption is that as the number of microorganisms increases in the sample, these physical, biophysical and biochemical events will also increase accordingly. When a sample has 5 to 6 log microorganisms/ml, detection time can be achieved in about a few hours. The following ATP instruments are available: EnSURE™, WaterShot™, SnapShot™ (Hygiena, LLC, Camarillo, CA, USA), Celsis Advance II™ system (Charles River, Hollister, CA, USA), 3M™ Clean-Trace™ (Thermo Fisher Scientific Inc., Waltham, MA, USA), Lightning MVP ICON™ (BioControl Systems, MVPICON™, Merck KGaA, Darmstadt, Germany), HY-LiTE® (Merck, KGaA, Darmstadt, Germany), PocketSwab Plus® and FieldSwab® (Charm Sciences, Inc., Lawrence, MA, USA) and others. Investigations of biocide efficacy by automated methods involving optical density measurements are complicated by the fact that a poor correlation often exists between the log reductions obtained using the automated method and those obtained by the traditional plate count methods (Lambert and van der Ouderaa, 1999; Rodrigues et al., 2018). It was assumed that the differences observed between the two methods were due to the level of cell injury, which was masked by the optical density methods but which was recognized by the plate counts.

As microorganisms grow and metabolize nutrients, large molecules change to smaller molecules in a liquid system and cause a change in electrical conductivity and resistance in the liquid as well as at the interface of electrodes. These changes can be expressed as impedance, conductance and capacitance changes. When a population of cells reaches approximately log 5 cells per ml, it will cause a change of these parameters. Thus, when a food has a large initial population, the time to make this change will be shorter than in a food that has a smaller initial population (Fung, 2002). The time it takes for the curve to change from the baseline and accelerate upward is the detection time of the test sample, which is inversely proportional to the initial concentration of microorganisms in the food. In order to use these methods a series of standard curves must be constructed by making viable cell counts of a series of food with different initial concentrations of cells and then measuring the resultant detection time.

2.4.1.4 Vapor-Phase or Volatile Compounds Methods

These assays are designed for antimicrobial substances with volatile compounds capable of exerting antimicrobial activity, such as essential oils (EOs) or ethanol. These antimicrobials have high vapor pressures, so they are suitable for exhibiting antimicrobial activity at a distance that is mediated by their vapor phase. The rapid volatilization of compounds allows them to reach the target microorganism in order to avoid its reproduction or growth in as little as one hour. Essential oils have great impact on flavor properties when they are incorporated directly in foods as antimicrobials. Thus, researchers have proposed generating an environment or atmosphere containing the volatile compounds in order to reduce sensory impacts on foods. Therefore, traditional antimicrobial assays are not useful for screening this type of application of EOs. Most vapor-phase assays are difficult and time-consuming since only one antimicrobial is tested at a time. However, chambers, boxes or large containers of glass, polycarbonate or polypropylene can support using one antimicrobial compound on test different microorganisms and/or different growth media, including real foods, if incubation conditions are similar. The antimicrobial activity of tested EOs depends on their volatile compounds' activity and volatility. To date, no standard assay has been recommended for vapor phase (Reyes-Jurado et al., 2019b) or volatile compounds from antimicrobial agents; however various methods are already reported and commonly utilized for this purpose, especially for *in vitro* studies.

For one method, the vapor-phase assay using an inverted Petri dish, a nonselective solidified medium is prepared in a Petri dish and inoculated with the target microorganism. Then, a known volume of the antimicrobial compound (or diluted using different solvents such as ethyl acetate) is added to a sterile, blank paper disk placed on the medium-free cover of the Petri dish (Figure 2.6). Afterwards, the Petri

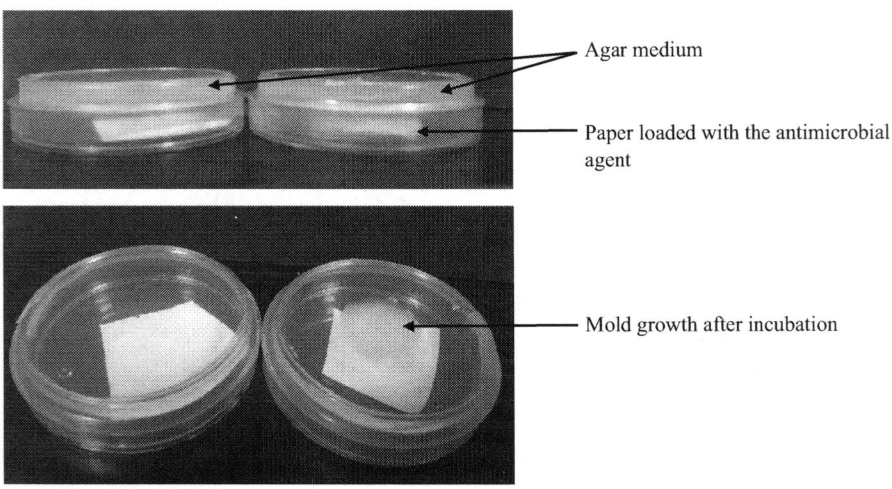

FIGURE 2.6 Inverted Petri dish assay for evaluating the vapor phase of antimicrobial agents (top). After suitable incubation period, mold growth is evident when less than minimal inhibition concentration is applied (bottom).

dish is closed and sealed using sterile adhesive tape or Parafilm™, inverted and incubated. After the incubation period under controlled temperature conditions, without eliminating the antimicrobial atmosphere, growth is observed, or survival counts are determined. The concentration of the antimicrobial compound in the atmosphere is determined as the volume between the agar surface and the lid and could be expressed per liter of air. This technique is mainly utilized for MIC determination on model systems.

In the disc volatilization technique, microbiological medium is added into a Petri dish and its cover; after solidification, an indicator microorganism is inoculated on the bottom Petri dish medium. In the medium of the cover, a sterile paper disc is placed with the antimicrobial agent to be tested. The dish is then closed and the medium in the cover of the Petri dish serves as a seal to avoid the loss of volatile compounds. Plates are incubated and observed for visible inhibition zones. This technique can be used for simple screening of antimicrobial activity or to determine the MIC of an antimicrobial agent (Mousavi Khaneghah et al., 2018). The concentration of the antimicrobial compound can be defined as the volume between the two agar surfaces and could be expressed per liter of air.

The agar plug vapor-phase assay utilizes a microtube cap containing a specific concentration of antimicrobial agent (~100 µl) which is embedded in an agar plate. Over the cap is placed an agar plug inoculated with the target microorganism (~log 7.0). The system is incubated at controlled temperature and visually examined for growth (or inhibition). Biocidal effect can be assessed by removing the agar plug and putting it in fresh broth (5 ml) that is vortexed; after that, the broth is plated to recover viable cells and counting. More details about the system are available in Amat et al. (2017). In this technique, the concentration of antimicrobial agent is difficult to measure as a head space; however, the use of small volumes and growth medium recovery for counting could be advantageous.

Seo et al. (2015) designed and constructed an apparatus consisting of an upper chamber with seven wells, in which each well contains solid medium on which the indicator microorganism(s) could be uniformly distributed. There is a lower chamber with seven wells containing the liquid EO(s) that generate the volatiles to reach the inoculated microorganism(s). Between the upper and lower parts, O-rings are positioned at the junctions to prevent leakage of vapors; the complete system design is shown in Seo et al. (2015). Their chamber was constructed in polycarbonate and thus is autoclavable. This apparatus allows testing of seven concentrations or different microorganisms at the same time. Deriving from the original design, Lee et al. (2018) designed an autoclavable polycarbonate vial with upper and lower wells.

Vapor-phase-mediated patch assay and vapor-phase-mediated susceptibility assay are two methods proposed by Feyaerts et al. (2017, 2018). Their assays were based on the Clinical and Laboratory Standards Institute protocol for the broth microdilution assay using standard 96-well plates (CLSI, 2008, 2012b). The vapor phase is assessed for its biological activity in a liquid culture. The authors hypothesized that, under ideal conditions, volatile substance added to the four central wells of the plate would spread circularly, thus establishing a concentration gradient across the plate. For the patch assay, the same assumption was made if a patch was located in the center of the microplate cover. Wells around the central wells (volatility-center) or patch are equidistant and can be categorized and ranked ordinally from the central point. Therefore, the content of every well belonging to one category would be affected equally, due to radial symmetry, and therefore could be considered as replicates. As expected, growth inhibition increases if wells are closer to the volatility-center (or patch) which could be confirmed spectrophotometrically. Inhibition doses can be established resulting from the comparison of the absorbances between the control growth and corresponding exposed wells. These two techniques can be useful for a quick screening using very low volumes of an antimicrobial agent against various microorganisms of the same nature. However, the inhibitory concentration of the antimicrobial agent is difficult to determine and the biocidal effect needs to be confirmed in solid medium.

2.5 Descriptive Methods

In end-point methods, the concentration of an antimicrobial required to totally inhibit a microorganism is determined. These methods provide little information concerning the effect of an antimicrobial that is below the MIC that may still cause inhibition in the form of increased lag phase, reduced growth rate or even initial lethality followed by growth. In food products, total inhibition of a pathogen of spoilage microorganism is not always required. An increased lag phase, especially under conditions of severe abuse, is often sufficient to protect the consumer or increase shelf life. Therefore, to determine the effect of a compound on the growth (or death) kinetics of a microorganism, a descriptive method is required.

The most commonly used descriptive technique involves producing an inhibition curve with the colony count procedure. In clinical microbiology, these inhibition curves are known as "time-kill curves" (Schoenknecht et al., 1985; NCCLS, 2002). The procedure utilizes a nonselective broth medium (10–100 ml) to which is added a single concentration of antimicrobial. The concentration may be determined from the end-point methods previously described. The test microorganism is diluted to achieve a final cell concentration of approximately log 5.7 CFU/ml in the test medium (NCCLS, 2002). If the effect of the compound on the lag phase is to be evaluated, a lower number (e.g., log 3–4 CFU/ml) can be used. The NCCLS (2002) recommends against inoculum levels > log 6 CFU/ml because selection of resistant mutants may occur. A tube or flask without added antimicrobial should be prepared as a control. The medium is incubated at the optimum temperature of the test microorganism for up to 48 h. The medium is sampled at appropriate times (e.g., 0, 2, 4, 8, 12, 24 and 48 h) and the number of viable microorganisms determined by the spread or pour plate method. Several responses of the test microorganism may be encountered in this type of analysis, including stationary-phase growth level suppression, lag-phase increase, decrease in the growth rate during log phase and lethality (Davidson and Parish, 1989). In practice a distinction is frequently drawn between a fungistatic or bactericidal action and a fungicidal or bactericidal action. This difference may not be completely justified since the former two differ from the latter in the death rate of the microorganisms. In the long term, the effect of added antimicrobial in food is either to inactivate (kill) the microorganisms or to eventually allow them to recover and grow. The governing factor here is the dosage of antimicrobial (Figure 2.7).

Inhibition curves can be used to evaluate the antimicrobial action of several agents against bacteria, yeast and molds. However, for molds their radial growth rate as well as the germination time (time to form a visible colony) are often used as measures of antimicrobial effectiveness (López-Malo et al., 1995, 1997, 1998; Lahlali et al., 2005; Baert et al., 2007; Gómez-Ramírez et al., 2013; Mani-López et al., 2018). The inhibition curves methodology is the only technique that demonstrates lethality. This method is versatile but has some disadvantages, including that no single statistic is produced to compare such treatments as MIC. It is also labor-intensive and expensive.

Modeling the growth kinetics of microorganisms (Zwietering et al., 1990, 1991; McMeekin et al., 1993; Whiting, 1993, 1995, 1997; Velázquez-Nuñez et al., 2013; Ozcakmak and Gul, 2017; Mani-López et al., 2018; Ro et al., 2018) allows improved statistical analysis of growth-inhibition curves in the presence of food antimicrobials. Secondary models also can describe the individual or combined effects of antimicrobials under specific growth conditions (pH, aw, temperature) using parameters estimated from primary models (germination time, radial growth rate, lag phase duration and decimal reduction rate). In addition, secondary empirical models describe only the effect of antimicrobials under specific tested conditions on microorganism growth. Zhang et al. (2015) described the antimicrobial interaction effects of apple skin polyphenols, acetic acid, oregano essential oil and carvacrol on *Salmonella* in

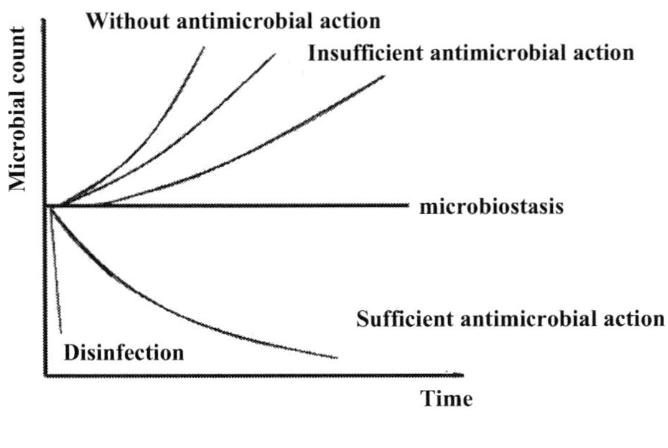

FIGURE 2.7 Effect of antimicrobial agent dose on microbial response. (Adapted from Lück and Jager, 1997.)

sliced cooked ham through a secondary empirical model. A secondary model obtained from Gompertz parameters was used to describe the growth rate of *Aspergillus flavus* under different values of cinnamon essential oil and pH, aw, casein or fat, to identify the effects of their interactions on the mold growth (Kosegarten et al., 2017). Probabilistic modeling has been used to determine the growth/no-growth interface. Probabilistic models based on logistic regression analysis provide a useful way to describe the growth/no growth boundary (Ratkowsky and Ross, 1995; Presser et al., 1998; López-Malo, et al., 2000; Gómez-Ramírez et al., 2013; Jamshidi et al., 2014; Kosegarten et al., 2017), including the effect of selected antimicrobial concentrations. This methodology allows the definition of the antimicrobial concentration range necessary to inhibit growth under selected environmental conditions (Alzamora and López-Malo, 2002; Gómez-Ramírez et al., 2013; Jamshidi et al., 2014; Kosegarten et al., 2017).

A second method for determining antimicrobial effectiveness over time is to measure turbidity increases with a spectrophotometer. A major disadvantage of this type of analysis is the sensitivity of the instrument. Spectrophotometers generally require log 6.0–7.0 CFU/ml for detection (Brock et al., 1984; Piddock, 1990). This may create a situation in which no growth (i.e., no absorbance increase) is observed when, in fact, undetectable growth is occurring at levels below log 5.0 CFU/ml. An erroneous interpretation of "lethality" could result (Davidson and Parish, 1989). A method of using turbidimetric analysis to determine infinite inhibitory concentrations (IIC, analogous to MIC) was developed by Marwan and Nagel (1986). In their procedure, the time for a microbial population to reach a specific turbidity in the presence of an antimicrobial was divided by the time for the same population to reach the turbidity in the absence of the antimicrobial. This was termed the relative effectiveness (RE). When 1/RE was plotted versus concentration of the inhibitor, the response was linear. Marwan and Nagel determined that at 1/RE = 0 (IIC), the concentration of antimicrobial would totally inhibit the test microorganism. As an example, the inverse of lag time of *Penicillium glabrum* inoculated in PDA formulated with selected concentrations of vanillin and potassium sorbate is shown in Figure 2.8. The response was linear and the calculated inhibitory concentrations, infinite lag time, correspond to MIC determined using the agar dilution method. Lambert and Pearson (2000) described a test to obtain MIC using optical density (OD) determinations. The basis of the technique is the comparison of the area under the optical density versus the time curve of the control (microorganism without antimicrobial) with the areas of the microbial curve with the test antimicrobial in several dilutions. The authors use 100-well plates and an OD automatic reader capable of giving an output signal based on a microbial growth criterion. As the amount of antimicrobial agent in a well increases, the effect on the growth is noticed by a reduction in the area under the curve. In general, plotting the inhibitor concentration on a logarithmic scale gives a characteristic sigmoid-shaped curve. The curve can be split into three principle regions: a region where the presence of the antimicrobial has no effect on the organisms relative to the control growth, a region where there is increasing inhibition of growth and a region where there is no measurable growth relative to the

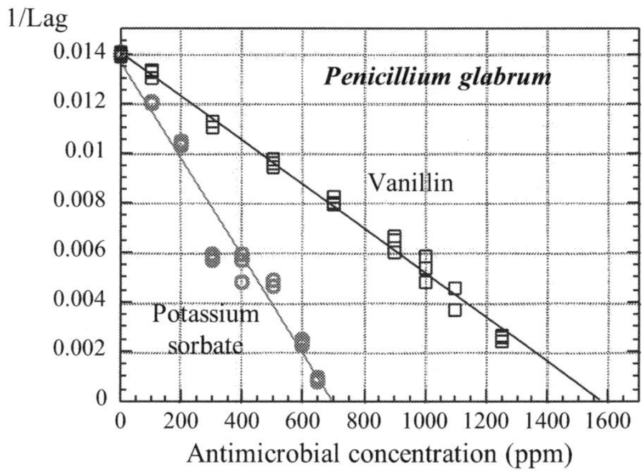

FIGURE 2.8 Relative effectiveness (inverse of lag time) of selected concentrations of vanillin and potassium sorbate on *Penicillium glabrum* inoculated in PDA.

control. This approach permits the calculation of the MIC as the concentration above which no growth is observed relative to the control.

To summarize the use of *in vitro* methods, they should be used together, one end-point and one descriptive method. The end-point method helps to determine the approximate effective concentration, and the descriptive method evaluates the effect of a compound on growth over time.

2.6 Combined Antimicrobial Systems

Traditionally, it was common to use only one chemical antimicrobial agent in a food product for preservation purposes (Busta and Foegeding, 1983); however, in recent years, the use of combined agents in a single food system has become more frequent. The use of combined antimicrobial agents theoretically provides a greater spectrum of activity, with increased antimicrobial action against the pathogenic or spoilage organisms. It is thought that combined agents act on different species of a mixed microbiota or act on different metabolic elements within similar species or strains. This theoretically results in improved microbial control over the use of one antimicrobial agent alone; however, actual proof of improved efficacy requires subjective interpretations. Although testing of combined antibiotics for clinical use is well-studied and relatively well-standardized (Barry, 1976; Krogstad and Moellering, 1986; Eliopoulos and Moellering, 1991), application of such methodology to antimicrobials in food is much less so. However, such systems have gained attention more recently because of the desire to minimize the sensory effects of high concentrations of certain antimicrobials or to potentiate the antimicrobial effect of single antimicrobials. There have been several studies on the antimicrobial combinations to preserve or extend the shelf life of foods including nisin and bovine lactoferrin (Turkish-style meatball, Colak et al., 2008), reuterin and nisin (milk, Arqués et al., 2011), cumin essential oil and nisin (commercial barley soup, Pajohi et al., 2011), chitosan and oregano essential oil (red porgy, Vatavali et al., 2013), dimethyl decarbonate and nisin (litchi juice, Yu et al., 2013), clove and mustard essential oil both in vapor phase (strawberries, Aguilar-González et al., 2015), Chinese cinnamon and cinnamon essential oils (ready-to-cook meat; Ghabraie et al., 2016), oregano and rosemary essential oils (iceberg lettuce and chard, de Medeiros Barbosa et al., 2016), *Blepharis cuspidata* and *Boswella agadensis* essential oils (Gadisa et al., 2019) and thyme and Mexican oregano essential oils in vapor phase (Reyes-Jurado et al., 2019a).

Methods of testing combined antimicrobials usually involve agar diffusion, agar or broth dilution, time-kill curves and vapor-phase methods. Agar diffusion is a simple qualitative method that often requires subjective interpretation of inhibition zone shapes to determine the efficacy of combined antimicrobials (Barry, 1976). This method provides a measure of microbial growth inhibition but does not address biocidal activity. Dilution and vapor-phase methods yield quantitative data and are most often conducted with various combined concentrations of two antimicrobials arranged in a "checkerboard" array. The checkerboard method is the technique used most frequently to assess antimicrobial combinations *in vitro*, presumable because (a) its rationale is easy to understand, (b) the mathematics necessary to calculate and interpret the results are simple, (c) it can readily be performed in the laboratory using microdilution systems and (d) it has been the technique most frequently used in studies that have suggested an advantage of synergistic interactions of antibiotics in clinical treatments (Eliopoulos and Moellering, 1991). The term checkerboard refers to the pattern (of tubes or microtiter wells or closed systems) formed by multiple dilutions of the two antimicrobials being tested in concentrations equal to, above and below their MIC (Table 2.3). Traditional clinical dilution testing uses two-fold dilutions of test compounds, but testing of food antimicrobials is often conducted with alternate dilution schemes (Rehm, 1959; Davidson and Parish, 1989; Eliopoulos and Moellering, 1991). Inhibition or time-kill curves can also be used for the evaluation of combined antimicrobials (Krogstad and Moellering, 1986; Eliopoulos and Moellering, 1991). The advantage of this test is that rates of lethality can be determined; however, repetitive sampling limits the number of combinations that can be assayed at any one time.

Combined studies are conducted to determine if specific types of interactions occur between the two combined antimicrobials. Traditionally, the terms "additive", "antagonistic" and "synergistic" were used to describe possible antimicrobial interactions. Garreti (1958) suggested the terms "additivity" and

TABLE 2.3

Checkerboard Array in Which Serial Dilutions of Two Antimicrobials Are Performed Using Concentrations Proportional to the MICs of the Antimicrobials Being Tested

Antimicrobial B	2.0	2.0/0	2.0/0.06	2.0/0.12	2.0/0.25	2.0/0.50	2.0/1.0	2.0/0
	1.0	1.0/0	1/0.06	1.0/0.12	1.0/0.25	1.0/0.50	1.0/1.0	1.0/0
	0.50	0.50/0	0.50/0.06	0.50/0.12	0.50/0.25	0.50/0.50	0.50/1.0	0.50/0
	0.25	0.25/0	0.25/0.06	0.25/0.12	0.25/0.25	0.25/0.50	0.25/1.0	0.25/0
	0.12	0.12/0	0.12/0.06	0.12/0.12	0.12/0.25	0.12/0.50	0.12/1.0	0.12/0
	0.06	0.06/0	0.06/0	0.06/0.12	0.06/0.25	0.06/0.50	0.06/1.0	0.06/0
	0	0/0	0/0.06	0/0.12	0/0.25	0/0.50	0/1.0	0/2.0
		0	0.06	0.12	0.25	0.50	1.0	2.0
					Antimicrobial A			

"equivalence" to describe combination interactions; however, our discussion is limited to the more traditional descriptors.

Additivity occurs when two combined antimicrobials give results that are equivalent to the sum of each antimicrobial acting independently. There is no enhancement or reduction in overall efficacy for the combined antimicrobials compared to the individual results. This is also sometimes referred to as "indifference" (Krogstand and Moellering, 1986; Eliopoulos and Moellering, 1991). Antagonism refers to reduced efficacy of the combined agents compared to the sum of the individual results. Synergism is an increase or enhancement of overall antimicrobial activity when two agents are combined compared to the sum of individual results.

A conclusion of synergism must be approached with caution since it implies that a reduction in the overall antimicrobial concentrations might be achieved in a food system without a reduction in efficacy. Gardner (1977) stated that true synergism is quite rare in relation to combined antibiotics. Other concerns about the misuse of the term "synergism" in relation to antimicrobials have been previously cited (Davidson and Parish, 1989; Garrett, 1958). Most commonly, additive interactions are misidentified as synergistic. A case in which an increase in antimicrobial activity is observed upon the addition of a second compound to a food system does not necessarily constitute synergy. A conclusion of synergism requires that the overall efficacy of the combination be significantly greater than the sum of the efficacies of the individual compounds.

From a practical standpoint, processors must beware of the temptation to utilize combined antimicrobials without thorough scientific scrutiny. This is mainly because of possible antagonistic interactions that can lower overall antimicrobial efficacy compared to the individual compounds used singly. Some food-related antimicrobial combinations that have been interpreted as antagonistic include sulfite and butylated hydroxyanisole, ethanol and butyl paraben, SO_2 and butyl paraben, SO_2 and benzoate and SO_2 and sorbate (Parish and Carroll, 1988b; Davidson and Parish 1989).

2.7 Application Studies

The primary purpose of *in vitro* methods is to determine the concentration of an antimicrobial necessary to inhibit a microorganism in a laboratory medium. However, once it has been determined that an antimicrobial is effective using *in vitro* methods, it is still necessary to apply the compound to a food system. Often, an antimicrobial that performed well in a microbiological medium is shown to have little or no effect in a food. This is the result of many interacting factors in a food, such as proteins, lipids, cations, binding to food components, inactivation by other additives, pH effects on antimicrobial stability and activity, uneven distribution in the food matrix and poor solubility, among others.

Before actual application of an antimicrobial to a food, it may be useful to evaluate the effect of some components in a food system that may influence the effectiveness of the compounds, such as lipids, proteins, additives and divalent cations. Lipids may case a decrease in activity of lipophilic compounds, and since many food antimicrobials have a hydrophobic character, there is invariably some reduction (Rico-Muñoz and Davidson, 1983; García-Díez et al., 2017; Kosegarten et al., 2017). Proteins may cause binding of some compounds and reduce activity (Rico-Muñoz and Davidson, 1983) or not influence the antimicrobial effect (Kosegarten et al., 2017; García-Díez et al., 2017). López-Malo et al. (1995) prepared fruit-based agars containing mango, papaya, pineapple, apple and banana with up to 2000 ppm vanillin to assess the effect of fruit composition on the MIC to inhibit *A. flavus*, *A. niger*, *A. parasiticus* and *A. ochraceus*. The small differences in protein and lipid composition between banana or mango and the other fruits increased vanillin MIC; the effect is attributed to binding of the phenolic vanillin by protein. Divalent cations may affect the activity of some compounds by affecting the microorganism itself or by interacting with the antimicrobial (Rico-Muñoz and Davidson, 1984). Further information can be gained by using model systems that include a percentage of a food in a buffer or microbiological medium. For example, Dje et al. (1989) evaluated antimicrobials in model systems containing a 10% (wt/vol) suspension of raw chicken or frankfurters. They prepared the suspensions by homogenizing 25 g product with 225 ml 0.1% peptone diluent. A homogeneous suspension of frankfurter was facilitated by adding 0.25% Tween 80 (sorbitan monooleate polyoxyethylene). In addition to determining the effect of food components on activity, this procedure allows better control of environmental conditions and natural microbiota. In both preceding model systems, a modified broth dilution assay could be used to determine an MIC (end-point method), or periodic sampling (descriptive method) could be run.

In applying an antimicrobial to a food, the microorganisms should include natural contaminants (bioburden) and the pathogen of interest. The incubation conditions should reflect normal storage and abuse. The success of the test may be through increased shelf life or reduced growth rate of a food-borne pathogen. Few food antimicrobials actually cause lethality to food-borne microorganisms at normal use concentrations. No standardized methods are available or application methods; however, a modification of the end-point method or time-kill curve is possible. Although it is impractical to design uniform procedures for all products, as is done for screening methods, a good approach is to develop a uniform in-house procedure based upon standardized procedures. Microbial challenge testing can be used to study the fate of pertinent microorganisms in foods formulated with antimicrobial agents as a way to assess product stability and safety (Tapia et al., 1995). The microbial challenge test has become an established technique within the food industry in order to simulate what can happen to a product during processing, distribution and subsequent handling, following inoculation with relevant microorganisms and further holding under controlled conditions (Notermans et al., 1993). This procedure can be used to verify antimicrobial agent effectiveness in foods.

For foods, the common application method of antimicrobial agents in vapor-phase is through the vapor-phase assay in which the system is scaled to a closed "big box" instead of a Petri dish. However, for certain foods such as bread slices, Petri dishes can be utilized; the antimicrobial agent is loaded in a paper disc placed in the plate lid (Krisch et al., 2013). Hermetic glass jars, hermetic polypropylene boxes or polyethylene terephthalate (PET) jars are commonly utilized to create an atmosphere of vapors of the antimicrobial agent which are thus in contact with the tested food (Aguilar-González et al., 2015; Mani-López et al., 2018). A known concentration of antimicrobial agent is placed inside the container separately from food. If the amount to be tested is small (μl range), the compound may be added to a piece of paper, or if a larger amount is necessary (ml range) a glass container may be used (see Figures 2.9 and 2.10). The jars (or boxes) are kept at a controlled temperature under standard conditions depending on the target microorganism and observed after the corresponding incubation period to determine the antimicrobial affect. The MIC is defined as the lower concentration in which microbial growth is inhibited and is expressed as μl (or ml) of antimicrobial agent per liter of air.

2.7.1 Interpretation of Results

The most difficult aspect of antimicrobial testing is appropriate data analysis and interpretation. Erroneous research conclusions from improper data interpretation can ultimately lead to industrial losses when practiced commercially. It is therefore necessary for data analysis to be conducted with great care.

Agar diffusion techniques produce quantitative zone diameter data. These results indicate relative microbial inhibition among a group of antimicrobials or test microorganisms. In antibiotic disk testing,

FIGURE 2.9 Vapor-phase test application for apple or strawberry in polyethylene terephthalate (PET) jars.

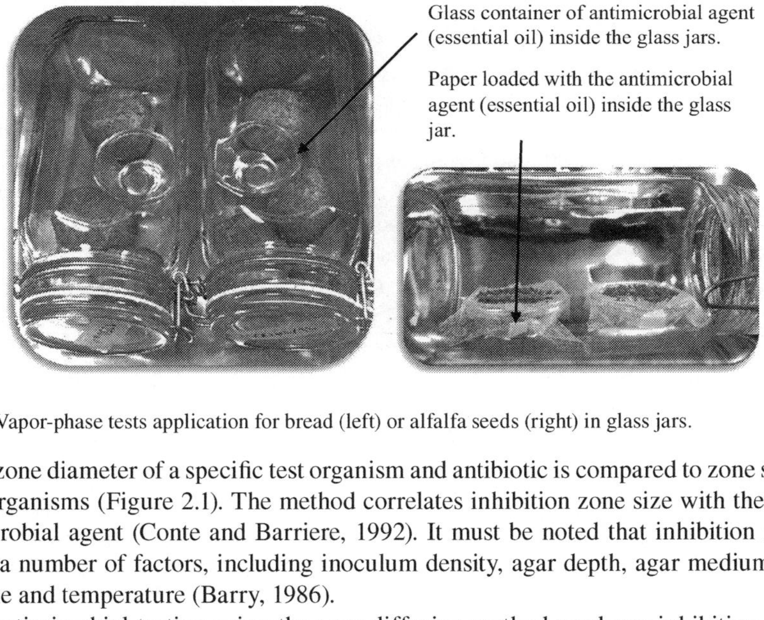

FIGURE 2.10 Vapor-phase tests application for bread (left) or alfalfa seeds (right) in glass jars.

the inhibition zone diameter of a specific test organism and antibiotic is compared to zone sizes produced with control organisms (Figure 2.1). The method correlates inhibition zone size with the concentration of the antimicrobial agent (Conte and Barriere, 1992). It must be noted that inhibition zone sizes are influenced by a number of factors, including inoculum density, agar depth, agar medium composition, incubation time and temperature (Barry, 1986).

Combined antimicrobial testing using the agar diffusion method produces inhibition zones that are interpreted depending upon their shape. This technique makes use of paper strips or disks impregnated with a specific quantity of antimicrobial gent and placed on an agar surface seeded with the test microorganism. This method is more qualitative than dilution testing but is useful as an end-point method for large numbers of preservative combinations.

As previously discussed, dilution testing results in an MIC. This reflects the minimum concentration of a compound necessary to inhibit the growth of a microorganism under specific conditions including time and temperature. It should be noted that MIC could vary depending on factors mentioned earlier.

Interpretation of dilution results for combined testing is usually conducted with isobolograms. An isobologram may be thought of as an array of differing concentrations of two compounds, where one compound ranges from lowest to highest concentration on the x-axis and the other on the y-axis (Figure 2.11). All possible permutations of combined concentrations are reflected within the array. If those concentrations that inhibit the growth of the test organism fall on an approximately straight line that connects the

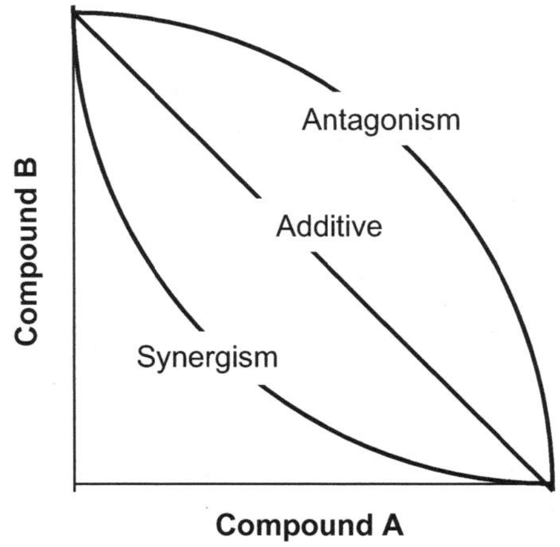

FIGURE 2.11 Interpretation of time-kill curves for combined antimicrobials.

individual MIC on the *x*- and *y*-axes, the combined effect is additive. Deviation of linearity to the left or right of the additive line is interpreted as synergism or antagonism, respectively.

Isobologram construction can be simplified using fractional inhibitory concentrations (FIC), which are MIC normalized to unit. The FIC is the concentration of a compound needed to inhibit growth (expressed as a fraction of its MIC) when combined with a known amount of a second antimicrobial compound. It is calculated as the ratio of the MIC of a compound when combined with a second compound divided by the MIC of the first compound alone. Additive, synergistic or antagonistic interactions are interpreted as with an MIC isobologram. For example, if the MIC for sorbate (under specific test conditions) is determined to be 200 ppm and the MIC for sorbate when combined with x ppm of propylparaben is 120 ppm, then the FIC for sorbate in the presence of x ppm propylparaben is 120/200 = 0.6.

A modified method for determining the fractional inhibitory concentration of two or more compounds in combination was described by Techathuvanan et al. (2014). The method essentially selects combinations of antimicrobials to be tested that lie in the three potential areas of an isobologram, antagonism, additive or synergism. By starting with MIC and using fractions thereof for each antimicrobial, one can interpret responses of growth or no growth as antagonism, additive or synergism (Table 2.4). This technique has potential to be used to rapidly screen multiple combinations only for synergistic effects if desired.

The FIC of two compounds in an inhibitory combination may be added to give a total FIC_{index}. If the FIC of propylparaben in the previous example was determined to be 0.5, then FIC_{index} would be FIC (sorbate) + FIC (propylparaben) = 0.6 + 0.5 = 1.1. An FIC_{index} near 1 indicates additivity, whereas <1 indicates synergy and >1 indicates antagonism (Table 2.5). The degree to which a result must be less than or greater than 1 to indicate synergism or antagonism is a matter of interpretation. Squires and Cleeland (1985) proposed that additive results are indicated by FIC_{index} between 0.5 and 2.0 for antibiotic testing. Synergism and antagonism are indicated by results <0.5 and >2.0, respectively. Research is needed to provide a database for proper interpretation of FIC and of the FIC_{index} in relation to food antimicrobial systems. Data interpretation must be conducted conditionally and will depend upon a number of variables, such as specific test conditions, microbial strain and target food system. It should be noted that interpretations might also vary depending upon the specific concentrations of each antimicrobial used in combination. Parish and Carroll (1988a) observed additivity between SO_2 and either sorbate or butylparaben when the inhibitory concentration contained less than 0.25 FIC of SO_2; however, the same combination at higher SO_2 FIC indicated antagonistic results. Rehm (1959) observed similar anomalies when sodium sulfite was combined with formate or borate.

Fractional lethal concentrations (FLC) can be determined when a broth dilution assay is used. This is the concentration of single or combined agents that kills the population of test organism to a non-recoverable level under the test conditions. The FLC is determined by inoculating broth media without antimicrobial agents with aliquots from those samples in the MIC test that inhibited growth. The concentrations

TABLE 2.4

Use of Modified Fractional Antimicrobial Concentrations for Antimicrobial Combination Testing of Two or More Antimicrobials

Antimicrobial (AM) (x MIC)			Fractional Inhibitory Concentration Index	Bacterial Growth	Descriptor of Results
AM_1[a]	AM_2[b]	AM_3[c]			
Binary antimicrobial combinations					
0	0	NA[d]	0	Growth	Control
1.0	0	NA	1	No growth	Minimum Inhibitory Conc.
0	1.0	NA	1	No growth	Minimum Inhibitory Conc. Additive[e] or antagonistic
0.75	0.25	NA	1	Growth No growth	Additive effect
0.50	0.50	NA	1	Growth No growth	Additive or antagonistic Additive effect
0.25	0.75	NA	1	Growth No growth	Additive or antagonistic Additive effect
0.25	0.25	NA	0.5	Growth No growth	Additive or antagonistic **Synergistic** effect
0.75	0.75	NA	1.5	Growth No growth	Antagonistic effect Additive or synergistic
Triple antimicrobial combinations					
0	0	0	0	Growth	Control
1	0	0	1	No growth	Minimum Inhibitory Conc.
0	1	0	1	No growth	Minimum Inhibitory Conc.
0	0	1	1	No growth	Minimum Inhibitory Conc.
0.33	0.33	0.33	1	Growth No growth	Additive or antagonistic Additive effect
0.50	0.25	0.25	1	Growth No growth	Additive or antagonistic Additive effect
0.25	0.50	0.25	1	Growth No growth	Additive or antagonistic Additive effect
0.25	0.25	0.50	1	Growth No growth	Additive or antagonistic Additive effect
0.167	0.167	0.167	0.5	Growth No growth	Additive or antagonistic **Synergistic effect**
0.50	0.50	0.50	1.5	Growth No growth	Antagonistic effect Additive or synergistic

[a] AM_1 = first antimicrobial, [b]AM_2 = second antimicrobial, [c]AM_3 = third antimicrobial.

[d] NA = not applicable.

[e] Synergistic, additive and antagonistic effects of combined antimicrobials correspond to FICI of <0.5, 1.0 and >1.5, respectively.

Adapted from: Techathuvanan et al., 2016.

that had a lethal effect remain negative for growth after appropriate incubation, but growth is observed in the concentrations that had only an inhibitory effect.

Inhibition or time-kill curves (log CFU/ml versus time) can be graphically interpreted for combination interaction (Barry, 1976; Krogstad and Moellering, 1986; Eliopoulos and Moellering, 1991). Since this technique uses a fixed concentration of each antimicrobial and must be repeated when a concentration is altered, it is logistically difficult and is most often used to enhance results observed with dilution methods. Survivor curves are generated with four treatments: (1) compound A alone, (2) compound B alone, (3) compound A and B together and (4) control A without A or B (Figure 2.12). Additivity is observed when the time-kill curve for the combined antimicrobials is similar to the curve for the more active

TABLE 2.5
Quantitative Definitions of Results with Antimicrobial Combinations

Additivity	
The result with two antimicrobials is equal to the sum of the results for each of the antimicrobials used separately	$FIC_A + FIC_B = 1.0$
Antagonism	
The result with two antimicrobials is significantly less than the additive response	$FIC_A + FIC_B > 1.0$
Synergism	
The result with two antimicrobials is significantly greater than the additive response	$FIC_A + FIC_B < 1.0$

FIC_A and FIC_B are the fractional inhibitory concentrations of antimicrobial A and B, respectively.

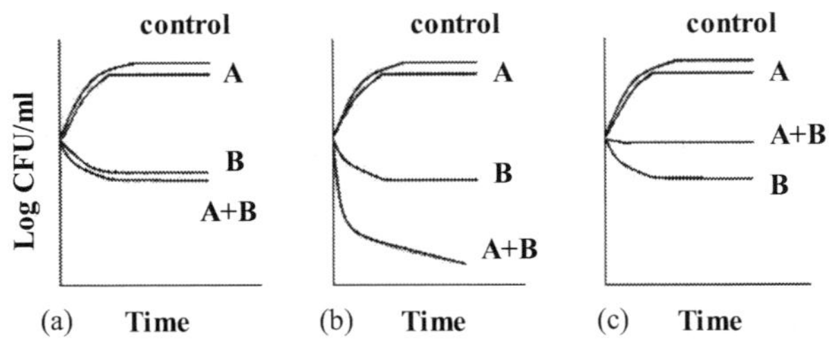

FIGURE 2.12 An isobologram than enables interpretation of dilution results when testing combined antimicrobials: (a) additive, (b) synergistic, or (c) antagonistic results.

compound alone. Antagonism results if the combined effect is significantly less than that observed for either compound alone. If the time-kill curve for the combined compounds is significantly reduced from both individual curves, the result is interpreted as synergistic.

2.8 Acknowledgments

Authors López-Malo, Mani-López and Palou gratefully acknowledge financial support from the National Council for Science and Technology (CONACyT) of Mexico (Project CB-2016-01-283636) and Universidad de las Américas Puebla (UDLAP Projects 2409 and 3555).

REFERENCES

Acar, J. F. and Goldstein, F. W. 1986. Disk susceptibility test. p. 27. In: *Antibiotics in Laboratory Medicine*, 2nd ed., edited by V. Lorian, Williams & Wilkins, Baltimore, MD.

Aguilar-González, A. E., Palou, E. and López-Malo, A. 2015. Antifungal activity of essential oils of clove (*Syzygium aromaticum*) and/or mustard (*Brassica nigra*) in vapor phase against gray mold (*Botrytis cinerea*) in strawberries. *Innov. Food Sci. Emerg. Technol.* 32:181–185.

Ali-Shtayeh, M. S. and Abu Ghdeib, S. I. 1999. Antifungal activity of plant extracts against dermatophytes. *Mycoses* 42:665–672.

Alzamora, S. M. and López-Malo, A. 2002. Microbial behavior modeling as a tool in the design and control of minimally processed foods. p. 631. In: *Engineering and Food for the 21st Century*, edited by J. Welti-chanes, G. V. Barbosa-Cánovas and J. M. Aguilera, CRC Press LLC, Boca Raton, FL.

Amat, S., Baines, D. and Alexander, T. W. 2017. A vapour phase assay for evaluating the antimicrobial activities of essential oils against bovine respiratory bacterial pathogens. *Lett. Appl. Microbiol.* 65:489–495.

Arqués, J. L., Rodríguez, E., Nuñez, M. and Medina, M. Combined effect of reuterin and lactic acid bacteriocins on the inactivation of food-borne pathogens in milk. *Food Control* 22:457–461.

Astorga-Solari, R., Santiesteban-López, A., Palou, E. and López-Malo, A. 2001. Antimicrobial spectrum of thymol, eugenol, potassium sorbate and sodium benzoate at selected pHs. Presented at 2001 International Association for Food Protection (IAFP) Annual Meeting, Minneapolis, MN.

Baert, K., Valero, A., De Meulenaer, B., Samapundo, S., Ahmed, M. M., Bo, L., Debevere, J. and Devlieghere, F. 2007. Modeling the effect of temperature on the growth rate and lag phase of *Penicillium expansum* in apples. *Int. J. Food Microbiol.* 118:139–150.

Balouiri, M., Sadiki, M. and Ibnsouda, S. K. 2016. Methods for in vitro evaluating antimicrobial activity: A review. *J. Pharm. Anal.* 6:71–79.

Barry, A. L. 1976. *The Antimicrobic Susceptibility Test: Principles and Practices*, Lea & Febiger, Philadelphia, PA.

Barry, A. L. 1986. Procedure for testing antimicrobial agents in agar media: Theoretical considerations. p. 1. In: *Antibiotics in Laboratory Medicine*, 2nd ed., edited by V. Lorian, Williams & Wilkins, Baltimore, MD.

Bauer, A. W., Kirby, W. M. M., Sherris, J. C. and Turck, M. 1966. Antibiotic susceptibility testing by a standardized single disk method. *Am. J. Clin. Pathol.* 45:493–496.

Beloian, A., editor. 1990. Disinfectants. p. 133. In: *Official Methods of Analysis*, 15th ed., Vol. I, edited by K. Helrich, Association of Official Analysis Chemists, Inc., Arlington, VA.

Block, S. S. 1983. Historical review. p. 868. In: *Disinfection, Sterilization and Preservation*, 3rd ed., edited by S. S. Block, Lea & Febiger, Philadelphia, PA.

Brock, T. D., Smith, D. W. and Madigan, M. T. 1984. *Biology of Microorganisms*, 4th ed., Prentice-Hall, Englewood Cliffs, NJ.

Busta, F. F. and Foegeding, P. M. 1983. Chemical food preservatives. p. 656. In: *Disinfection, Sterilization and Preservation*, 3rd ed., edited by S. S. Block, Lea & Febiger, Philadelphia, PA.

Choma, I. M. and Grzelak, E. M. 2011. Bioautography detection in thin-layer chromatography. *J. Chromatogr. A.* 1218:2684–2691.

CLSI, Clinical and Laboratory Standards Institute. 2008. Reference method for broth dilution antifungal susceptibility testing of yeast; approved standard, 3rd ed., CLSI document M27-A3, Clinical and Laboratory Standards Institute, Wayne, PA.

CLSI, Clinical and Laboratory Standards Institute. 2012a. Methods for dilution antimicrobial susceptibility test for bacteria that grow aerobically; approved standard, 9th ed., CLSI document M07-A9, Clinical and Laboratory Standards Institute, Wayne, PA.

CLSI, Clinical and Laboratory Standards Institute. 2012b. Reference method for broth dilution antifungal susceptibility testing of yeasts. Fourth Informational Supplement, CLSI document M27-S4, Clinical and Laboratory Standards Institute, Wayne, PA.

CLSI, Clinical and Laboratory Standards Institute. 2018a. Performance standards for antimicrobial disk susceptibility tests, 13th ed., CLSI standard M02 (ISBN 1-56238-835-5), Clinical and Laboratory Standards Institute, Wayne, PA.

CLSI, Clinical and Laboratory Standards Institute. 2018b. Methods for dilution antimicrobial susceptibility tests for bacteria that grow aerobically, 11th ed., CLSI standard M07 (ISBN 1-56238-837-1), Clinical and Laboratory Standards Institute, Wayne, PA.

Colak, H., Hampikyan, H., Bingol, E. B. and Aksu, H. 2008. The effect of nisin and bovine lactoferrin on the microbiological quality of Turkish-style meatball (tekirdag köfte). *J. Food Safety* 28:355–375.

Conte, J. E. and Barriere, S. L. 1992. *Manual of Antibiotics and Infectious Diseases*, 7th ed., Lea & Febiger, Philadelphia, PA.

Crémieux, A. and Fleurette, J. 1983. Methods of testing disinfectants. p. 918. In: *Disinfection, Sterilization and Preservation*, 3rd ed., edited by S. S. Block, Lea & Febiger, Philadelphia, PA.

Davidson, P. M. and Parish, M. E. 1989. Methods for testing the efficacy of food antimicrobials. *Food Technol.* 43:148–155.

De Medeiros Barbosa, I., Medeiros, J. A. C., de Oliveira, K. A. R., Gomes-Neto, N. J., Tavares, J. F., Magnani, M. and de Souza, E. L. 2016. Efficacy of the combined application of oregano and Rosemary essential oils for the control of *Escherichia coli*, *Listeria monocytogenes* and *Salmonella* Enteritidis in leafy vegetables. *Food Control* 59:468–477.

Dewanjee, S., Gangopadhyay, M., Bhattacharya, N., Khanra, R. and Dua, T. K. 2015. Bioautography and its scope in the field of natural product chemistry. *J. Pharm. Anal.* 5:75–84.

Dje, Y., Payne, K. D. and Davidson, P. M. 1989. Inhibition of *Listeria monocytogenes* Scott A by propyl paraben in non-fat milk and chicken and hot dog suspensions. *Inst. Food Techno. Ann. Mtg.*, Chicago.

Eliopoulos, G. M. and Moellering, R. C. 1991. Antimicrobial combinations. p. 432. In: *Antibiotics in Laboratory Medicine*, 3rd ed., edited by V. Lorian, Williams & Wilkins, Baltimore, MD.

Elleuch, L., Shaaban, M., Smaoui, S., Mellouli, L., Karray-Rebai, I., Fguira, L. F.-B., Shaaban, K. A. and Laatsch, H. 2010. Bioactive secondary metabolites from a new terrestrial *Streptomyces* sp. TN262. *Appl. Biochem. Biotechnol.* 162:579–593.

Feyaerts, A. F., Mathé, L., Luyten, W., Tournu, H., Van Dyck, K., Broekx, L. and Van Dijck, P. 2017. Assay and recommendations for the detection of vapour-phase-mediated antimicrobial activities. *Flavour Fragr. J.* 32:347–353.

Feyaerts, A. F., Mathé, L., Luyten, W., De Graeve, S., Van Dyck, K., Broekx, L. and Van Dijck, P. 2018. Essential oils and their components are a class of antifungals with potent vapour-phase-mediated anti-*Candida* activity. *Nature Sci. Rep.* 8:3958 (10 p).

Foerster, S., Desilvestro, V., Hathaway, L. J., Althaus, C. L. and Unemo, M. 2017. A new resazurin-based microdilution assay for antimicrobial susceptibility testing of Neisseria gonorrhoeae. *J. Antimicrob. Chemother.* 72:1961–1968.

Fung, D. Y. C. 2002. Rapid methods and automation in microbiology. *Compr. Rev. Food Sci. Food Safety* 1: 3–22.

Gadisa, E., Weldearegay, G., Desta, K., Tsegaye, G., Hailu, S., Jote, K. and Takele, A. 2019. Combined antibacterial effect of essential oils from three most commonly used Ethiopian traditional medicinal plants on multidrug resistant bacteria. *BMC Complement. Altern. Med.* 19: 9 pp. https://doi.org/10.1186/s12906-019-2429-4

García-Díez, J., Alheiro, J., Pinto, A. L., Soares, L., Falco, V., Fraqueza, M. J. and Patarata, L. 2017. Influence of food characteristics and food additives on the antimicrobial effect of garlic and oregano essential oils. *Foods.* 6:10 pp.

Gardner, J. R. 1977. Principles of antimicrobial activity. In: *Disinfection, Sterilization and Preservation*, 2nd ed., edited by S. S. Block, Lea & Febiger, Philadelphia, PA.

Garrett, E. R. 1958. Classification and evaluation of combined antibiotic activity. *Antibiot. Chemother.* 8:8.

Ghabraie, M., Vu, K. D., Tata, L., Salmieri, S. and Lacroix, M. 2016. Antimicrobial effect of essential oils in combinations against five bacteria and their effect on sensorial quality of ground meat. *LWT-Food Sci. Technol.* 66:332–339.

Gómez-Ramírez, C., Sosa-Morales, M. E., Palou, E. and López-Malo, A. 2013. *Aspergillus niger* time to growth in dried tomatoes. *Int. J. Food Microbiol.* 164:23–25.

Harris, L. J., Daeschel, M. A., Stiles, M. E. and Klaenhammer, T. R. 1989. Antimicrobial activity of lactic acid bacteria against *Listeria monocytogenes. J. Food. Prot.* 52:384–387.

Hill, G. B. 1991. Spiral gradient endpoint method compared to standard agar dilution for susceptibility testing of anaerobic Gram-negative bacilli. *J. Clin. Microbial.* 29:975–979.

Jamshidi, A., Khanzadi, S., Azizi, M., Azizzadeh, M. and Hashemi, M. 2014. Modeling the growth of Staphylococcus aureus as affected by black zira (*Bunium persicum*) essential oil, temperature, pH and inoculum levels. *Vet. Res. Forum.* 5:107–114.

Jiménez-Esquilín, A. E. and Roane, T. M. 2005. Antifungal activities of actinomycete strains associated with high-altitude sagebrush rhizosphere. *J. Ind. Microbiol. Biotechnol.* 32:378–381.

Kekessy, D. A. and Piguet, J. D. 1970. New method for detecting bacteriocin production. *Appl. Microbiol.* 20:282–383.

Kosegarten, C. E., Ramírez-Corona, N., Mani-López, E., Palou, E. and López-Malo, A. 2017. Description of *Aspergillus flavus* growth under the influence of different factors (water activity, incubation temperature, protein and fat concentration, pH and cinnamon essential oil concentration) by kinetic, probability of growth and time-to-detection models. *Int. J. Food Microbiol.* 240:115–123.

Krisch, J., Rentskenhand, T., Horváth, G. and Vágvölgyi, C. 2013. Activity of essential oils in vapor phase against bread spoilage fungi. *Acta Biol. Szeged.* 57:9–12.

Krogstad, D. J. and Moellering, R. C., Jr. 1986. Combinations of antibiotics, mechanisms of interaction against bacteria. p. 537. In: *Antibiotics in Laboratory Medicine*, 2nd ed., edited by V. Lorian, Williams & Wilkins, Baltimore, MD.

Lahlali, R., Serrhini, M. N. and Jijakli, M. H. 2005. Studying and modelling the combined effect of temperature and water activity on the growth rate of *P. expansum*. *Int. J. Food Microbiol.* 103:315–322.

Lambert, R. J. W. and Pearson, J. 2000. Susceptibility testing: Accurate and reproducible minimum inhibitory concentration (MIC) and non-inhibitory concentration (NIC) values. *J. Appl. Microbiol.* 88:784–790.

Lambert, R.J.W. and van der Ouderaa, M.-L.H. 1999. An Investigation into the differences between the Bioscreen and the traditional plate count disinfectant test methods. *J. Appl. Microbiol.* 86:6899–694.

Lee, G., Kim, Y., Kim, H., Beuchat, L. and Ryu, J. 2018. Antimicrobial activities of gaseous essential oils against *Listeria monocytogenes* on a laboratory medium and radish sprouts. *Int. J. Food Microbiol.* 265:49–54.

Lertcanawanichakul, M. and Sawangnop, S. 2008. A comparison of two methods used for measuring the antagonistic activity of *Bacillus* species. *Walailak J. Sci. & Tech.* 5:161–171.

Lewus, C. B. and Montville, T. J. 1991. Detection of bacteriocins produced by lactic acid bacteria. *J. Microbiol. Methods* 13:145–150.

López-Malo, A., Alzamora, S. M. and Argaiz, A. 1995. Effect of natural vanillin on germination time and radial growth of moulds in fruit-based agar systems. *Food Microbiol.* 12:213–219.

López-Malo, A, Alzamora, S. M. and Argaiz, A. 1997. Effect of vanillin concentration, pH and incubation temperature on *Aspergillus flavus, A. niger, A. ochraceus* and *A. parasiticus* growth. *Food Microbiol.* 14:117–124.

López-Malo, A., Alzamora, S. M. and Argaiz, A. 1998. Vanillin and pH synergistic effects on mold growth. *J. Food Sci.* 63:143–146.

López-Malo, A., Alzamora, S. M. and Guerrero, S. 2000a. Natural Antimicrobials from Plants. p. 237. In: *Minimally Processed Fruits and Vegetables: Fundamentals Aspects and Applications*, edited by S. M. Alzamora, M. S. Tapia and A. López-Malo, Aspen Publishers, Inc. Gaithersburg, MD.

López-Malo, A., Guerrero, S. and Alzamora, S. M. 2000b. Probabilistic modeling of *Saccharomyces cerevisiae* inhibition under the effects of water activity, pH and potassium sorbate. *J. Food Prot.* 63:91–95.

Lück, E. and Jager, M. 1997. *Antimicrobial Food Additives: Characteristics, Uses, Effects*, 2nd ed, Springer Verlag, Germany.

Mani-López, E., Valle-Vargas, G. P., Palou, E. and López-Malo, A. 2018. *Penicillium expansum* inhibition on bread by lemongrass essential oil in vapor phase. *J. Food Prot.* 81:467–471.

Marston, A. 2011. Thin-layer chromatography with biological detection in phytochemistry. *J. Chromatogr. A.* 1218: 2676–2683.

Marwan, A. G. and Nagel, C. W. 1986. Quantitative determination of infinite inhibition concentrations of antimicrobial agents. *Appl. Environ. Microbial.* 51:559–561.

McMeekin, T. A., Olley, J. N., Ross, T. and Ratkowsky, D. A. 1993. *Predictive Microbiology: Theory and Application*, Research Studies Press, Ltd., Taunton.

Moir, C. J. and Eyles, M. J. 1992. Inhibition, injury and inactivation of four psychotropic food borne bacteria by the preservatives methyl p-hydroxybenzoate and potassium sorbate. *J. Food Prot.* 55:360–366.

Montville, T. J., Winkowski, K. and Chikindas, M. L. 2001. Biologically based preservation systems. p. 629. In: *Food Microbiology: Fundamentals and Frontiers*, 2nd ed., edited by M. P. Doyle, L. R. Beuchat and T. J. Montville, ASM Press, Washington, DC.

Mousavi Khaneghah, A., Bagher Hashemi, S. M., Eş, I., Fracassetti, D. and Limbo, S. 2018. Efficacy of antimicrobial agents for food contact applications: biological activity, incorporation into packaging and assessment methods: a review. *J. Food Prot.* 81:1142–1156.

Mukherjee, P. K. and Raghu, K. 1997. Effect of temperature on antagonistic and biocontrol potential of *Trichoderma* sp. on *Sclerotium rolfsii. Mycopathologia.* 139:151–155.

NCCLS. 1999. Methods for determining bactericidal activity of antimicrobial agents, Approved Guideline, National Committee for Clinical Laboratory Standards, Wayne, PA.

NCCLS. 2002. Methods for Dilution Antimicrobial Susceptibility Tests for Bacteria That Grow Aerobically; Approved Standard, 6th ed., National Committee for Clinical Laboratory Standards, Wayne, PA.

Notermans, S., Veld, P., Wijtzes, T. and Mead, G. C. 1993. A user's guide to microbial challenge testing for ensuring the safety and stability of food products. *Food Microbiol.* 10: 145–157.

Ozcakmak, S. and Gul, O. 2017. Inhibition kinetics of Penicillium verrucosum using different essential oils and application of predictive inactivation models. *Int. J. Food Prop.* 20:5684–5692.

Pajohi, M. R., Tajik, H., Farshid, A. A. and Hadian, M. 2011. Synergistic antibacterial activity of the essential oil of *Cuminum cyminum* L. seed and nisin in a food model. *J. Appl. Microbiol.* 110:943–951.

Parish, M. E. and Carroll, D. E. 1988a. Minimum inhibitory concentration studies of antimicrobial combination against *Saccharomyces cerevisae* in a model broth system. *J. Food Sci.* 53:237–239.

Parish, M. E. and Carroll, D. E. 1988b. Effects of combined antimicrobial agents on fermentation initiation by *Saccharomyces cerevisae* in a model broth system. *J. Food Sci.* 53:240–242.

Paton, J. H., Holt, H. A. and Bywater, M. J. 1990. Measurement of MICs of antibacterial agents by spiral gradient endpoint compared with conventional dilution method. *Int. J. Exp. Clin. Chemotherapy.* 3:31–38.

Pereira, E., Santos, A., Reis, F., Tavares, R. M., Baptista, P., Lino-Neto, T. and Almeida-Aguilar, C. 2013. A new effective assay to detect antimicrobial activity of filamentous fungi. *Microbiol. Res.* 168:1–5.

Piddock, L. J. 1990. Techniques used for determination of antimicrobial resistance and sensitivity in bacteria. *J. Appl. Bacteriol.* 68:307–318.

Presser, K. A., Ross, T. and Ratkowsky, D. A. 1998. Modelling the growth limits (growth/no growth interface) of *Escherichia coli* as a function of temperature, pH, lactic acid concentration and water activity. *Appl. Environ. Microbiol.* 64:1773–1779.

Ratkowsky, D. A. and Ross, T. 1995. Modeling the bacterial growth/no growth interface. *Lett. Appl. Microbiol.* 20:29–33.

Rehm, H. 1959. Untersuchung zur Wirkung von Konservierungsmittel kombinationen I. *Z. Lebesm. Untersuch. Forsch.* 110:283–293.

Reyes-Jurado, F., Cervantes-Rincón, T., Bach, H., López-Malo, A. and Palou, E. 2019a. Antimicrobial activity of Mexicano regano (*Lippia berlandieri*), thyme (*Thymus vulgaris*) and mustard (*Brassica nigra*) essential oils in gaseous phase. *Ind. Crops Prod.* 131:90–95.

Reyes-Jurado, F., Navarro-Cruz, A. R., Ochoa-Velasco, C. E., Palou, E., López-Malo, A. and Ávila-Sosa, R. 2019b. Essential oils in vapor phase as alternative antimicrobials: A review. *Crit. Rev. Food Sci. Nutr.* 18:1–10.

Rico-Muñoz, E. and Davidson, P. M. 1983. Effect of corn oil and casein on the antimicrobial activity of phenolic antioxidants. *J. Food Sci.* 48:1284–1288.

Rico-Muñoz, E. and Davidson, P. M. 1984. Effect of calcium and magnesium on the antibacterial activity of phenolic antioxidants against *Staphylococcus aureus* A100. *J. Food Sci.* 49:282–283.

Ro, E. Y., Kim, G. S., Kwon, D. Y., Park, Y. M., Cho, S. W., Lee, S. Y., Yeo, I. H. and Yoon. K. S. 2018. Effect of natural antimicrobials with modified atmosphere packaging on the growth kinetics of *Listeria monocytogenes* in ravioli at various temperatures. *J. Food Safety* 38:e12392 (8 pp).

Rodrigues, L. B., dos Santos, L. R., Rizzo, N. N., Ferreira, D., de Oliveira, A. P., Levandowski, R., Webber, B. and do Nascimento, V. P. 2018. ATP-bioluminescence and conventional microbiology for hygiene evaluation of cutting room surfaces in poultry slaughterhouse. *Acta Sci. Vet.* 46:1534 (6 pp).

Schalkowsky, S. 1986. Plating systems. In: *Food Borne Microorganisms and Their Toxins: Developing Methodology*, edited by M. Pierson and N. Stern, Marcel Dekker, New York.

Schoenknecht, F. D., Sabath, L. D. and Thornsberry, C. 1985. Susceptibility tests: Special test. p. 1000. In: *Manual of Clinical Microbiology*, 4th ed., edited by E. Lennette, American Society for Microbiology, Washington, DC.

Seo, H. S., Beuchat, L. R., Kim, H. and Ryu, J. H. 2015. Development of an experimental apparatus and protocol for determining antimicrobial activities of gaseous plant essential oils. *Int. J. Food Microbiol.* 215:95–100.

Spiral Biothec, Inc. 2001. *SGE, Spiral Gradient Endpoint, User Guide*, Spiral Biothec, Norwood, MA.

Squires, E. and Cleeland, R. 1985. *Methods of Testing Combinations of Antimicrobial Agents*, Hoffman-LaRoche, Inc., Nutley, NJ.

Sun, W., Weingarten, R. A., Xu, M., Southall, N., Dai, S., Shinn, P., Sanderson, P. E., Williamson, P. R., Frank, K. M. and Zheng, W. 2016. Rapid antimicrobial susceptibility test for identification of new therapeutics and drug combinations against multidrug-resistant bacteria. *Emerg. Microbes Infect.* 5:11 p.

Synbiosis, A Division of Synoptics Ltd. 2019. Protocol 3 – Automatic colony counting and zone measuring, Product description, Cambridge, UK.

Tapia de Daza, M. S., Argaiz, A., López-Malo, A. and Díaz, R. V. 1995. Microbial stability assessment in high and intermediate moisture foods: special emphasis on fruit products. p. 575. In: *Food Preservation by Moisture Control, Fundamentals and Applications*, edited by G. V. Barbosa-Cánovas and J. Welti-Chanes, Technomic Publishing, Inc, Lancaster, PA.

Techathuvanan, C., Reyes, F., David, J. R. D. and Davidson, P. M. 2016. Efficacy of commercial natural antimicrobials alone and in combinations against pathogenic and spoilage microorganisms. *J. Food Prot.* 77:269–275.

Tenover, F. C. 1986. Studies of antimicrobial resistance genes using DNA probes. *Antimicrob. Agents Chemother.* 29:721–725.

Thornsberry, C. 1986. Automation in antibiotic susceptibility testing. In: *Antibiotics in Laboratory Medicine*, 2nd ed., edited by V. Lorian, Williams & Wilkins, Baltimore, MD.

Thrupp, L. D. 1986. Susceptibility testing of antibiotics in liquid media. p. 93. In: *Antibiotics in Laboratory Medicine*, 2nd ed., edited by V. Lorian, Williams & Wilkins, Baltimore, MD.

Tramer, J. and Fowler, G. C. 1964. Estimation of nisin in foods. *J. Sci. Food Agric.* 15:522–528.

Vatavali, K., Karakosta, L., Nathanailides, C., Georgantelis, D. and Kontominas, M. G. 2013. Combined effect of chitosan and oregano essential oil dip on the microbiological, chemical and sensory attributes of red porgy (*Pagrus pagrus*) stored in ice. *Food Bioprocess Technol.* 6:3510–3521.

Velázquez-Nuñez, M. J., Ávila-Sosa, R., Palou, E. and López-Malo, A. 2013. Antifungal activity of orange (*Citrus sinensis* var. Valencia) peel essential oil applied by direct addition or vapor contact. *Food Control* 31:1–4.

Walton, L., Herbold, M. and Lindergren, C. C. 1936. Bactericidal effects of vapors from crushed garlic. *J. Food Sci.* 1:163–169.

Whiting, R. C. 1993. Modeling bacterial survival in unfavorable environments. *J. Ind. Microbiol.* 12:240–246.

Whiting, R. C. 1995. Microbial modeling in foods. *Crit. Rev. Food Sci. Nutr.* 35:464–494.

Whiting, R. C. 1997. Microbial database building—What have we learned? (In Schaffner, D., ed., *Predictive Food Microbiology: Where Do We Go From Here?*). *Food Technol.* 51:82.

Yu, Y., Wu, J., Xiao, G., Xu, Y., Tang, D., Chen, Y. and Zhang, Y. 2013. Combined effect of dimethyl dicarbonate (DMDC) and nisin on indigenous microorganisms of litchi juice and its microbial shelf life. *J. Food Sci.* 78:M1236–M1241.

Zhang, Q., Mukhopadhyay, S., Hwang, C. A., Xu, X. and Juneja, V. K. 2015. Modelling the survival of *Salmonella* on slice cooked ham as a function of apple skin polyphenols, acetic acid, oregano essential oil and carvacrol. *J. Food Process. Preserv.* 39:2371–2378.

Zwietering, M. H., Jongenburger, I., Rombouts, F. M. and Van't Riet, K. 1990. Modeling of the bacterial growth curve. *Appl. Environ. Microbiol.* 56:1875–1881.

Zwietering, M. H., de Koos, J. T., Hasenack, B. E., de Wit, J. C. and Van't Riet, K. 1991. Modeling of bacterial growth as a function of temperature. *Appl. Environ. Microbiol.* 57:1094–1101.

3

Sodium Benzoate and Benzoic Acid

John R. Chipley

CONTENTS

3.1 Introduction and Historical Background

Benzoic acid is one of the oldest chemical preservatives used in the cosmetic, drug, and food industries. Sodium benzoate was the first chemical preservative approved for use in foods by the U.S. Food and Drug Administration (FDA) (Jay, 2000). Its preservative action appears to have been first described in 1875 when researchers discovered that sodium benzoate had strong antifungal properties, and was relatively tasteless in foods, easy to make, and inexpensive. It was not introduced for commercial food preservation until around 1900 because benzoic acid could not be produced synthetically in large quantities (Lueck, 1980).

An informative historical narrative published recently (Blum, 2018) details the dangers of consuming processed food in the U.S. at the end of the nineteenth century. In most cases, substandard dairy and meat products, alcoholic beverages, and canned vegetables which should have been discarded were knowingly sold to customers for consumption after the addition of chemicals, often at lethal concentrations. For example, milk adulterated with formaldehyde was estimated to have killed thousands of children in New York City alone. In the production of ketchup, tomato trimmings were "thickened" with various non-tomato byproducts, processed under non-sanitary conditions, and a large dose of preservative added to extend shelf life. The preservative of choice was sodium benzoate. Food companies found it profitable to use rotten tomato waste and scraps in their production ketchups and supplement them with up to four times the proposed government standard of 0.10% for this preservative (Blum, 2019).

Currently, the advantages of low cost, ease of incorporation into products, lack of color, and relatively low toxicity have subsequently caused benzoic acid (and its salts) to become one of the most widely used preservatives in the world (Davidson, 1997) when used at legally prescribed levels.

During the last 15 years, several articles and references concerning various aspects of food additives and preservatives (e.g., benzoic acid) have been published. General evaluations of the use of preservatives in foods (Vogel, 1992), in prevention of microbial spoilage (Giese, 1994), in meat products (Gerhardt, 1995), in beverage manufacture (Giese, 1995), and in consumer attitudes toward the use of preservatives (Jager, 1994) serve as examples. In addition, references detailing characteristics of food spoilage yeasts (Deak, 2007), food preservation (Rahman, 2007), fungal food spoilage (Pitt and Hocking, 2009), and chemical preservatives and natural antimicrobial compounds (Davidson and Taylor, 2007; Davidson et al., 2013) provide excellent background information.

Recently, an extensive review of benzoic acid as a naturally occurring compound in foods and as an antimicrobial additive was published by del Olmo et al. (2017). Detailed coverage of its uses in foods and non-food products, toxicology in humans and animals, current controversy involving long-term exposure, and analytical methods of detection were also presented.

3.2 Physical and Chemical Properties and Natural Occurrence

Benzoic acid (C_6H_5COOH) and sodium benzoate (C_6H_5COONa) have the structural formulas shown in Figure 3.1. Benzoic acid (molecular weight 122.1), also called phenylformic acid or benzenecarboxylic acid, occurs in pure form as colorless or white needles or leaflets. It is soluble to a limited extent in water (0.18, 0.27, and 2.2 g dissolves in 100 ml water at 4°C, 18°C, and 75°C, respectively.)

Sodium benzoate (molecular weight 144.1) is a white granular or crystalline powder. It is much more soluble in water than benzoic acid (62.8, 66.0, and 74.2 g dissolves in 100 ml water at 0°C, 20°C, and 100°C, respectively). For this reason, it is preferred for use in many cases. Potassium benzoate and calcium benzoate have also been approved for use, although their solubility in water is less than that of the sodium salt.

Benzoic acid occurs naturally in several foods and commodities (Table 3.1). It accounted for approximately 16% of the growth inhibition of *Saccharomyces bayanus* and *Pseudomonas fluorescens* from ethanolic extracts of cranberries (Marwan and Nagel, 1986a). Benzoic acid has also been identified as a major constituent in extracts of blackberries (Humpf and Schreier, 1991); of mushrooms, depending on the variety (Abdullah et al., 1994); and of fresh tomatoes (Marlatt et al., 1992).

Yogurts have been found to contain natural levels of benzoic acid (Stijve and Hischenhuber, 1984; Teuber, 1995). In an extensive review, Sieber et al. (1995) surveyed and analyzed many types of cultured dairy products and cheeses for the natural occurrence of benzoic acid (Table 3.2). Its presence appears to occur as a by-product of the microbial degradation of either hippuric acid or phenylalanine in these products (Figure 3.2). Oxidation of benzaldehyde may also contribute to benzoic acid generation. A fourth reaction involves hydrolysis of hippuric acid to yield benzoic acid and glycine, e.g., during the fermentation of goat's and sheep's milk (Hornickova et al., 2014). Both of these acids were detected at higher levels in milk from sheep (Table 3.3). This may be of significance in countries like Switzerland, where benzoic acid has not been approved as a food additive (Sieber et al., 1989, 1990).

Benzoic acid has also been identified as a natural by-product in culture filtrates of *Lactobacillus plantarum* (Niku-Paavola et al., 1999). Several new types of low-molecular-mass compounds were

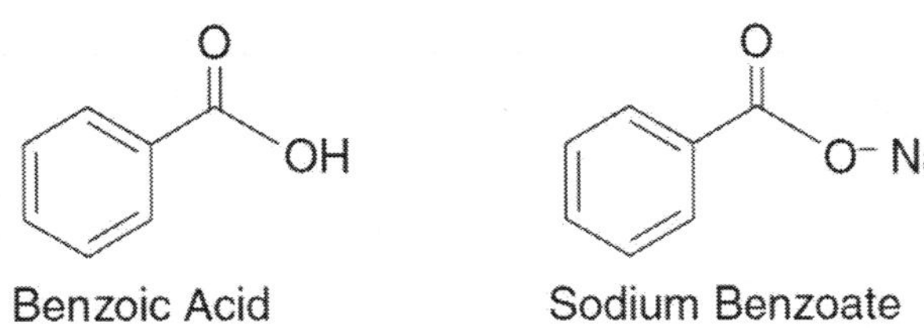

FIGURE 3.1 Structures of benzoic acid and sodium benzoate.

TABLE 3.1

Natural Occurrence of Benzoic Acid

Category	Product
Fruits and berries	Apples
	Apricots
	Berries (i.e., blackberries)
	Blueberries
	Cherries
	Cranberries
	Grapes
	Plums, greengage
	Prunes
	Strawberries
	Tomatoes
Fermented products	Beers
	Dairy, cultured
	Teas, black
	Wines
Spices and flavors	Anise
	Cinnamon
	Chocolate
	Cloves, ripe
	Licorice
	Vanilla
Others	Brewer's yeast
	Coffee beans
	Honey
	Milk
	Mushrooms
	Teas, green
	Tobacco

Note: Levels of benzoic acid vary (10 to 1000 mg/kg) depending on the food.
Source: Adapted from Fenaroli (2002); Chipley (2005); Qi et al. (2009).

TABLE 3.2

Benzoic Acid Content of Cultured Dairy Products

Product	Range of Concentration (mg/kg)
Yogurts	9–56
Yogurt, fruit	5–39
Sour cream	10–18
Buttermilk	10–19
Cottage cheese	2–18
Cheese	0–200
Cheese, nonsmear, ripened	0–41
Cheese, smear, ripened	0–622

Source: Adapted from Sieber et al. (1995).

also identified, and a mixture of these, along with benzoic acid, was found to have antimicrobial properties.

Naturally occurring non-flavonoid phenolic compounds, including derivatives of benzoic acid, were used to characterize commercial fruit juices (Fernandez de Simon et al., 1992). Orange, apple, pineapple, peach, apricot, pear, and grape juices were analyzed to establish phenolic profiles unique to each type of juice. In four of the seven juices, p-hydroxybenzoic acid was detected at 0.2 to 2.6 mg/l.

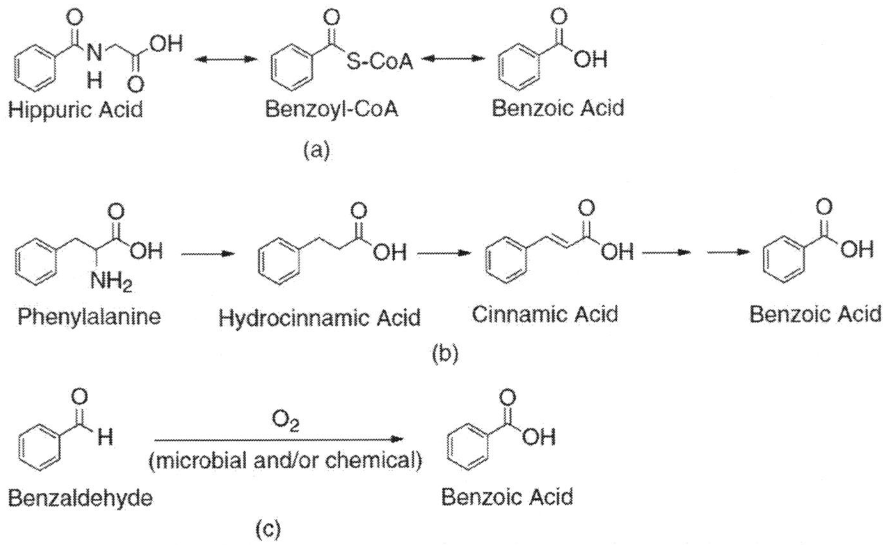

FIGURE 3.2 Biosynthesis of benzoic acid by lactic acid bacteria in fermented dairy products. [*Source*: Adapted from Sieber et al. (1995).]

TABLE 3.3

Content of Hippuric Acid in Raw Milk and Benzoic Acid in Fermented Milk

(A) Hippuric *Acid (mg/kg)*	*Raw Goat's Milk*	*Raw Sheep's Milk*
Average	15.5 ± 8.3	43.3 ± 12.3
Range	5.8–25.1	20.3–55.6
(B) Benzoic *Acid (mg/kg)*	*Ferm. Goat's Milk*	*Ferm. Sheep's Milk*
Average	20.3 ± 13.9	29.5 ± 16.1
Range	5.0–52.8	4.9–77.8

Source: Adapted from Hornickova et al. (2014).

3.3 Antimicrobial Activity

3.3.1 Spectrum of Action

Because the quantity of undissociated acid decreases with increasing pH (Table 3.4), the use of benzoic acid or sodium benzoate as a food preservative has been limited to those products that are acid in nature.

TABLE 3.4

Effect of pH on the Dissociation of Benzoic Acid

pH	Undissociated Acid (%)
3	93.5
4	59.3
5	12.8
6	1.44
7	0.144
pK	4.19

Source: From Baird-Parker (1980).

Currently, these compounds are used primarily as antimycotic agents, and most yeasts and fungi are inhibited by 0.05% to 0.1% of the undissociated acid. Food-poisoning and spore-forming bacteria are generally inhibited by 0.01% to 0.02% undissociated acid, but many spoilage bacteria are much more resistant. Therefore, benzoic acid cannot be relied on to effectively preserve foods capable of supporting bacterial growth (Chichester and Tanner, 1972; Baird-Parker, 1980). MICs for some of the bacteria, yeasts, and fungi involved in food poisoning and food spoilage are given in Tables 3.5–3.6. Several factors interact to determine the MIC, including pH, temperature, genus and species of the microorganism in question, composition of the growth medium, prior exposure to the preservative, and environment from which the microorganism was originally isolated.

The sensitivity of 42 yeast cultures to sorbic and benzoic acids, potassium sorbate, and sodium benzoate was determined (Manganelli and Casolari, 1983). The scattering of MIC values was lower with potassium sorbate and benzoic acid and higher with sorbic acid and sodium benzoate. The addition of benzoate to chemostat cultures of *S. cerevisiae* decreased the biomass and increased the specific oxygen uptake rate of cells (Verduyn et al., 1992).

The effects of several preservatives and antimicrobial agents on aflatoxin B_1 production by *A. flavus* have been reported (Bauer et al., 1981). In liquid media, all treated cultures produced measurable levels of toxin 3 to 7 days later than controls. In benzoic acid-supplemented cultures, aflatoxin B_1 production was higher than in controls. The authors concluded that subinhibitory concentrations of these compounds may stimulate toxin production in some cases.

Sodium benzoate was found to control both growth and aflatoxin production by *Aspergillus parasiticus* in liquid media (El-Gazzar and Marth, 1987). Increasing the concentration of sodium benzoate increased the percentage of inhibition at the end of incubation (10 days). The average accumulation of mycelial dry weight was greater in the presence of benzoate than in its absence, however, with the greatest increase occurring when the medium contained 0.3% sodium benzoate.

Different concentrations of benzoic acid were tested to determine the effective levels capable of reducing the mycelial growth of six *Fusarium* and eight *Penicillium* species by 50% (Thompson, 1997). In general, *Fusarium* species were more sensitive to benzoic acid (210 to 420 μg/ml) than were *Penicillium* species (250 to 3000 μg/ml).

Growth and aflatoxin production by toxigenic strains of *Aspergillus* were partially or completely inhibited by the undissociated form of six organic acid preservatives, including benzoic (Rusul and Marth, 1988). Salts, such as sodium and potassium chlorides and sodium nitrate, enhanced aflatoxin production when present at low levels but became inhibitory at higher levels.

Garri, produced from fermented cassava roots, spoils quickly due to the germination and growth of *Aspergillus*. Effects of sodium benzoate on survival, growth, and aflatoxin production in packaged garri have been reported (Ogiehor and Ikenebomeh, 2004). Three *Aspergillus* species were inoculated into garri with or without benzoate and stored at ambient temperature. Decreased growth and no aflatoxin production occurred in benzoate-treated samples at the end of storage.

Neosartorya fischeri is one of the most frequently isolated heat-resistant fungi causing spoilage of fruit juices and other heat-processed fruit-based products (Nielsen et al., 1989). Growth of this fungus was accompanied by production of fumitremorgin mycotoxins. Fungal growth was reduced by lowering the pH of laboratory media from 7.0 to 2.5; selected organic acids promoted growth and toxin production when added to the media. Small amounts (75 mg/L) of potassium sorbate or sodium benzoate completely inhibited the germination of ascospores and subsequent outgrowth.

Both fungistatic and fungicidal properties have been attributed to benzoic acid, according to the results of a study involving several strains of *Trichophyton* and *Microsporum* (Pelayo, 1979). Sodium benzoate has been suggested as an inhibitor of cellulose-decomposing bacteria and fungi (Sauer, 1977).

Under appropriate conditions, bacteriostatic and bactericidal properties of benzoic acid can also be demonstrated. Beuchat (1980) reported that sodium benzoate (300 μg/ml) inhibited the growth of *Vibrio parahaemolyticus* in laboratory media and enhanced the rate of thermal inactivation of this organism at slightly higher concentrations. Ten generally regarded as safe (GRAS) substances, including benzoic acid, were tested against both the opaque and translucent morphotypes of *Vibrio vulnificus* (Sun and Oliver, 1995). Eight of these had a lethal effect on both morphotypes of this bacterium.

TABLE 3.5

Antimicrobial Spectrum of Benzoic Acid against Selected Bacteria, Yeasts, and Fungi

Microorganisms	pH	MIC*
Bacteria		
Bacillus cereus	6.3	500
Escherichia coli	5.2–5.6	50–120
Escherichia coli	6.0	100–200
Klebsiella pneumoniae	6.0	100–200
Lactobacillus sp.	4.3–6.0	300–1800
Listeria monocytogenes	5.6 (21°C)	3000
	5.6 (4°C)	2000
Micrococcus sp.	5.5–5.6	50–100
Pseudomonas sp.	6.0	200–480
Pseudomonas aeruginosa	6.0	200–500
Staphylococcus aureus	6.0	50–100
Streptococcus sp.	5.2–5.6	200–400
Yeasts		
Sporogenic yeasts	2.6–4.5	20–200
Asporogenic yeasts	4.0–5.0	70–150
Candida krusei		300–700
Debaryomyces hansenii	4.8	500
Hansenula sp.	4.0	180
Hansenula subpelliculosa		200–300
Oospora lactis		300
Pichia membranefaciens		700
P. pastori		300
Rhodotorula sp.		100–200
Saccharomyces bayanus	4.0	330
S. cerevisiae	4.0	600
Torulopsis sp.		200–500
Zygosaccharomyces bailii	4.8	4500
	4.0	1200
Z. lentus	4.0	500–1100
Z. rouxii	4.8	1000
Fungi		
Alternaria solani	5.0	1500
Aspergillus sp.	3.0–5.0	20–300
Aspergillus parasiticus	5.5	>4000
Aspergillus niger	5.0	2000
Byssochlamys nivea	3.3	500
Chaetomonium globosum	5.0	1000
Cladosporium herbarum	5.1	100
Mucor racemosus	5.0	30–120
Penicillium sp.	2.6--5.0	30–280
Penicillium citrinum	5.0	2000
Penicillium glaucum	5.0	400–500
Rhizopus nigricans	5.0	30–120

* Minimum inhibitory concentration in μg/ml (ppm).

Source: Adapted from Chipley (1983, 1993, 2005); Davidson and Juneja (1990); Russell (1991); and Steels et al. (1999).

TABLE 3.6

Minimum Inhibitory Concentration of Benzoic Acid for Food Spoilage Yeasts

Isolate*	MIC (ppm)
Kluveromyces fragilis	173
Kloeckera apiculata	188
Pichia ohmeri	200
Hansenula anomala	223
Saccharomyces cerevisiae	170–450
Zygosaccharomyces rouxii	242–330
Zygosaccharomyces bisporus	200–350
Candida krusei	440
Saccharomycodes ludwigii	500–600
Schizosaccharomyces pombe	500–567
Zygosaccharomyces bailii	600–1300

* A total of 23 isolates were tested. Most were isolated from spoiled foods that had contained preservative. Isolates were grown at 25°C in yeast extract medium containing 5% glucose (pH 3.5) without addition of benzoic acid.

Source: Adapted from Warth (1989c).

In a series of studies involving *L. monocytogenes* (El-Shenawy and Marth, 1988; Yousef et al., 1989), it was found that benzoic acid at concentrations of approximately 1000 to 3000 ppm had strong bacteriostatic, but relatively modest bactericidal, activities against cells in a liquid minimal medium. Incubation of cells in minimal media caused injury that depended on the temperature of incubation but not on the presence of benzoic acid. The authors questioned the suitability of benzoic acid alone to control this pathogen in foods. This organism was isolated from milk, and its survival and growth were determined in media supplemented with organic acids and sodium benzoate (El-Shenawy and Marth, 1989). In general, inactivation or inhibition of growth occurred in inoculated media when a lower incubation temperature (13°C) and pH (5.0) were used and required lower levels of sodium benzoate. Combining glycerol monolaurate with benzoic acid gave greater inhibition of *L. monocytogenes* than when each preservative was tested alone (Oh and Marshall, 1994).

The effects of temperature, pH, sodium chloride, and three preservatives on the growth of three foodborne bacterial pathogens were studied using gradient gel plates (Thomas et al., 1993). Potassium sorbate was completely effective against Verocytotoxigenic *E. coli* at all temperature/pH/NaCl combinations. It was also the most effective against B_1 *cereus*. At <25°C, sorbate was more effective than sodium benzoate against *S. aureus* when used with higher concentrations of NaCl. At 35°C, benzoate was the most effective preservative against *S. aureus*, especially when used at pH <6. Sodium nitrite was the least effective preservative tested. Increasing the acidity and/or NaCl generally improved the effect of all the preservatives.

Potassium sorbate and sodium benzoate did not have any significant effects on the heat resistance of four strains of *B. stearothermophilus* (now *Geobacillus thermophilus*) spores (Lopez et al., 1996). However, when these preservatives were added to recovery media, they were very effective inhibitory agents for heat-damaged spores (Lopez et al., 1998). Effectiveness increased as the pH of the recovery media decreased to 5.0.

The mycoflora of two types of Portuguese dry-smoked sausages have been isolated and identified (Matos et al., 2007). Three preservatives were tested for inhibitory effects on isolates from seven fungal genera. Methyl paraben was more effective than potassium sorbate, which was more effective than sodium benzoate. In Portugal, as in the EU, these preservatives can only be used for surface treatments of dry sausage casings.

Statistical modeling for the heat resistance of *E. coli* O157:H7 in apple cider has been reported (Steenstrup and Floros, 2002). The addition of sorbate, benzoate, and malic acid significantly decreased heat resistance. A 5-log reduction of viable cells could be obtained at 50°C in a few minutes and at 20°C

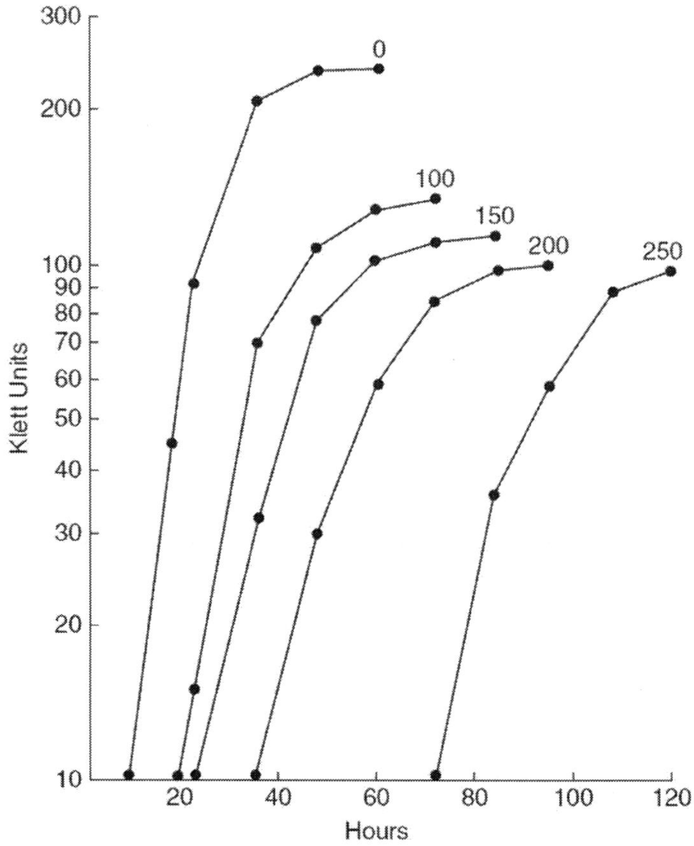

FIGURE 3.3 Inhibition of *Saccharomyces bayanus* by benzoic acid in Trypticase soy broth at 30°C and pH 4.0. The concentrations of benzoic acid tested (in parts per million) are indicated. [*Source*: Marwan and Nagel (1986b).]

in about 90 minutes. In apple cider to which fumaric acid and sodium benzoate were added, pH and storage temperatures had important effects (Chikthimmah et al., 2003). At 5°C, destruction increased with time. At 15°C to 25°C, the rate of destruction was significantly increased in cider with the preservatives. Increasing the pH of cider (from 3.2 to 4.7) decreased the rate of destruction.

Predictive microbiology (modeling) has become a useful tool for microbial risk assessment in foods (Devlieghere et al., 2004). Modeling of microbial growth vs. temperature, influence of intrinsic and extrinsic parameters, and the effects of chemicals on microbial inactivation/survival are reviewed. Within each section, research involving benzoate or benzoic acid is cited.

Z. rouxii is an osmophilic yeast responsible for significant spoilage and economic losses in high-sugar food production (Rojo et al., 2015). Statistical modeling was used to determine the effects of seven chemical preservatives in controlling the growth of five strains of this yeast. None of the preservatives tested completely inhibited growth. Resistance to preservatives was enhanced by growth in a high-sugar culture medium. A maximum growth reduction (≈40%) was obtained for four of the preservatives, including benzoate.

The growth patterns of *S. bayanus* in the presence of different concentrations of benzoic acid are shown in Figure 3.3. A typical dose-response effect may be observed.

3.3.2 Influence of Other Chemicals and Physical Environment

No antimicrobial is completely effective against all microorganisms present in a given foodstuff. In theory, one should be able to combine various antimicrobials having different modes of action to compensate for this deficiency. It should then be possible to achieve (1) a broader spectrum of action or (2) an increased antimicrobial action by using this combination (Lueck, 1980). For example, combinations of sorbic acid and benzoic acid inhibit several bacterial strains better than either antimicrobial alone (Rushing and Senn, 1963). Synergistic effects have also been reported using combinations of benzoate and sulfur dioxide, carbon dioxide, sodium chloride, boric acid, or sucrose (Rehm, 1960; Chichester and Tanner, 1972). Additive, synergistic, and antagonistic effects reported in several early studies have been extensively reviewed by Lueck (1980) for benzoic acid as well as for other preservatives. The influence of combinations of chemical antimicrobials on the growth of food-poisoning and spoilage microorganisms has been reported (Banks et al., 1988). This will be discussed in more detail in subsequent sections of the present chapter.

Individual and synergistic effects of pH, benzoic acid, and sorbic acid on the growth rate of the yeast *Z. bailii* were determined and expressed in polynomial equations (Cole and Keenan, 1986). Synergism between benzoic and sorbic acids was pH-dependent. The anionic form of benzoic acid was shown to have a distinct effect on doubling time.

Antagonistic as well as synergistic effects were observed when a wine yeast strain of *S. cerevisiae* was inoculated in a laboratory medium supplemented with several combinations of preservatives (Parish and Carroll, 1988). Sulfur dioxide was apparently antagonistic when used in combination with any of the other antimicrobial agents (including benzoate) tested in this study.

In addition to combining several preservatives together, it may also be advantageous to combine preservatives with physical methods of food preservation, such as heating, refrigeration, irradiation, or drying. Synergistic effects on yeasts and fungi using a combination of benzoic acid and heat have been reviewed by Lueck (1980) and have also been reported by Beuchat and Jones (1979) and Beuchat (1981a, b). The combined effects of preservatives and incubation temperature on biomass and patulin production by *Byssochlamys nivea* in apple juice have been determined (Roland and Beuchat, 1984). Growth at 21°C, 30°C, and 37°C over a 25-day incubation period was significantly retarded by 75 ppm sulfur dioxide, 150 ppm potassium sorbate, and 500 ppm sodium benzoate. Sulfur dioxide had the most significant effect on the rate of biomass and patulin production by this organism, followed by potassium sorbate and sodium benzoate, respectively.

The effects of pH and benzoate on the growth of yeasts in an experimental grape drink containing 1.5 volumes of CO_2 have been reported (Kelly, 1975). Likewise, an experimental bottled beverage made from papaw fruit was reported to have a shelf life of 80 weeks at 10°C and 30°C when pasteurized and supplemented with sodium benzoate (1250 ppm and pH 3.9) (Okoli and Ezenweke, 1990). A combination of benzoic acid and low temperatures was used effectively to reduce the numbers of yeasts in grape juice (Pederson et al., 1961). Sodium benzoate prevented the growth of *Staphylococcus aureus* in an intermediate-moisture food (Boylan et al., 1976).

The presence of other chemicals may be beneficial or detrimental to the action of the preservative, depending on the chemical in question. For example, the antimicrobial activity of benzoic acid is reduced by nonionic surfactants (de Navarre and Bailey, 1956). However, benzoic acid can be solubilized with a surfactant to hold it in solution in its undissociated form (Lewis and Payne, 1958).

The effects of quinic acid on yeasts and molds were investigated (Kallio et al., 1985). Growth of yeasts was unaltered, but mold growth was accelerated in the presence of quinic acid. This acid was antagonistic to the antifungal effects of both potassium sorbate and sodium benzoate. Adding 1% quinic acid to media almost completely eliminated the inhibitory effect of 0.01% to 0.02% sodium benzoate.

In a study to evaluate synergism, several combinations of sodium benzoate, potassium sorbate, and sodium nitrite were tested with food spoilage and pathogenic bacteria and fungi (Stanojevic et al., 2009). Overall, the most effective antimicrobial was sodium nitrite. Synergism was found for the nitrite–benzoate combination (40% of organisms tested) and for the nitrite–sorbate combination (30% of organisms tested). *Escherichia coli* was most sensitive and *Aspergillus flavus* most resistant to the combinations tested.

UV pasteurization of apple juice and the effect of UV exposure on the stability of ascorbic acid, sodium benzoate, potassium sorbate, and sulfur dioxide, when previously added to juice, were evaluated (Usaga et al., 2017). UV treatment had no significant effect on total vitamin C and benzoate, but decreased sulfur dioxide, ascorbic acid, and sorbate levels. Increases in ascorbic acid decreased inactivation of *E. coli* in juice. The authors suggested that additives that could either adversely affect UV efficiency or be degraded due to UV exposure should be added after UV treatment.

A chemical preservative–sodium chloride–moisture interaction was evaluated in ready-to-eat meat products which affected the growth of *Listeria monocytogenes* (Seman et al., 2008). Sodium benzoate was more effective when used with increasing levels of sodium diacetate and sodium chloride and decreasing finished product moisture. High-moisture products in which sodium benzoate is limited to 0.1% (current limits) may need additional ingredients to effectively inhibit growth of *L. monocytogenes*.

A combination of the polysaccharide chitosan (0.005%) and 0.025% sodium benzoate acted synergistically against three food spoilage yeasts suspended in saline solutions at pH 6.2 and 4.5 (Sagoo et al., 2002). Chitosan (0.005%) almost doubled the extent of death caused by 0.025% benzoate alone, therefore demonstrating significant synergism. At these pH values, the authors proposed that chitosan is highly polycationic and could react with the anionic components of the microbial surface; this might enhance the antimicrobial effects of benzoate. The authors suggested that chitosan might be useful as an adjunct to potentiate the efficacy of antimicrobial compounds such as benzoates.

Glass et al. (2007a, b, 2013) conducted a series of studies with combinations of benzoate, propionate, and sorbate to prevent the growth of *L. monocytogenes* on sliced, cooked, cured or uncured turkey and ham. Data showed that 0.1% benzoate prevented growth of this organism on cured ham stored at 4°C for 12 weeks. Inhibition was less pronounced when formulations were stored at abuse temperatures. These antimycotic agents inhibited the growth of *L. monocytogenes* on ready-to-eat meats but were more effective when used with nitrite and proper storage temperature. Treatment with propionate and benzoate limited listerial growth (<1 log through 9 weeks at 4°C) on cured turkey. Bologna and ham slices were inoculated with *L. monocytogenes*, then immersed in 5% potassium benzoate and vacuum-packaged. This treatment was lethal to microbial cells, but lowered the sensory scores (Geornaras et al., 2005).

Slices of chicken luncheon meat or surfaces of turkey frankfurters were treated by spraying or dipping with a solution of one of four preservatives (Islam et al., 2002a, b). Each product was then surface-inoculated with a five-strain mixture of *L. monocytogenes* and incubated at three different temperatures. In both products, sodium diacetate was the most effective, followed by sodium benzoate. For frankfurters, growth inhibition was directly proportional to the concentration of the preservative.

Combinations of sodium chloride and organic acids (e.g., benzoic) are frequently used to inhibit growth of spoilage and pathogenic bacteria in food (Jordan and Davies, 2001). To reduce the risk of foodborne illness, many fermented and minimally processed foods rely on the "hurdle effect", i.e., a combination of two or more inhibitory agents being more inhibitory than any of the agents alone (Casey and Condon, 2002). However, in both studies, the presence of sodium chloride promoted survival and/or growth of two strains of *E. coli* O157:H7 under stressful conditions of low pH. The authors suggested that *E. coli* cells could use sodium chloride to counteract acidification of its cytoplasm by organic acids.

Sodium chloride also appears to affect survival of *E. coli* O157:H7 at selected concentrations of acetic acid (Hosein et al., 2011). Modeling predictions showed that decreases in intracellular pH were significantly correlated with times required to achieve a 5-log reduction in the number of viable bacteria.

Protective effects of organic acids on survival of *E. coli* O157:H7 in acidic environments have been reported (Bjornsdottir et al., 2006). Acid-adapted strains were reduced by 4-log cycles in the absence of added organic acids (acetic, citric, malic, or lactic). Only a 2–3-log reduction was observed when low concentrations of fully protonated organic acids were added. Would this affect the antibacterial action of benzoic acid if it were subsequently added to these cultures?

The survival of *E. coli* O157:H7, *Salmonella*, and *L. monocytogenes* was evaluated in cranberry juice concentrates with different Brix levels (Enache and Chen, 2007). All three pathogens in the stationary phase survived longer than in the acid-adapted phase under the same conditions. The authors suggested that concentrated acid (e.g., benzoic) and/or some intrinsic compounds may play an important role in the bactericidal effects of cranberry juice concentrates.

The growth of four spoilage yeasts was inhibited with three-component combinations of preservatives, including sodium benzoate (Dai et al., 2010). These combinations were highly effective against all four yeasts at practical pH levels. Concentrations of each antimicrobial in the three-component combinations that were required to inhibit the growth of yeasts were lower than those in single or binary combinations. Synergistic, additive, and some antagonistic effects were noted, depending on the concentrations of each component (Dai et al., 2010). Synergism was also noted when combinations of food antimicrobials at low levels inhibited the growth of *S.* Typhimurium (Nazer et al., 2005).

Mayonnaise and organic acid solutions with or without chemical preservatives (benzoate et al.) were used to determine the survival of three phage and three serotypes of *Salmonella* (Zhu et al., 2012). Survival depended on the type of *Salmonella*, the medium (mayonnaise or acid solution), and the presence or absence of preservatives. The authors suggested that the inhibition of *Salmonella* by vinegar and lemon juice was due to the hurdle effect of organic acids and chemical preservatives.

A model was established for the growth of *L. monocytogenes* in mayonnaise-based salads (Vermeulen et al., 2007). The influence of water activity, pH, acetic acid levels, and microbial strain variability was determined. Maximum-allowed levels of benzoic and sorbic acids in mayonnaise-based salads, according to EU legislation, eliminated the growth of this pathogen. Levels of 600 and 300 ppm of a 1:1 mixture of these preservatives were sufficient to inhibit growth at 7 and 4°C, respectively.

Fruit juices have not been normally identified in foodborne outbreaks since low pH and high levels of organic acids were believed to provide adequate protection for these products. However, several outbreaks linked to unpasteurized fruit juices have been reported, raising concern over the safety of these products. A study was conducted to determine if *E. coli* O157:H7, *L. monocytogenes*, and *Salmonella* could survive in cranberry, lemon, and lime juices at temperatures at or above those commonly used for transportation or storage of concentrates (Nogueira et al., 2003). The results indicated that concentrates contained intrinsic antimicrobial properties that would eliminate pathogens if juices were contaminated after concentration. Microbial inactivation was demonstrated regardless of the genus or storage temperature. However, in four different juice concentrates and one puree, these same pathogens were recovered from all five products through 12 weeks of storage at –10°F (Oyarzabal et al., 2003). Survival of stationary phase (non-adapted) and acid-adapted *E. coli* O157:H7 in single-strength lemon and lime juices was evaluated at room temperature (22°C) (Enache et al., 2009). Lemon juice was more inhibitory than lime juice. A greater than 5-log reduction of stationary-phase *E. coli* cells occurred in single-strength juices made from concentrates (no preservatives) or in commercially produced shelf-stable juices (containing sodium bisulfite and sodium benzoate) after storage at 22°C for 3 days. The authors proposed that 22°C storage for 3 days might be an alternative to heat treatments to achieve the required 5-log reduction of vegetative pathogens. Six other strains of Enterohemorrhagic *E. coli* were also inactivated in shelf-stable lemon and lime juices following storage at 22°C for 3 days (Kataoka et al., 2011). Storage conditions that inactivated *E. coli* O157:H7 similarly inactivated non-O157 serotypes.

Selected combinations of cinnamon extract, sodium benzoate, and pH were evaluated for their effects on growth of *Aspergillus flavus* on laboratory media (Lopez-Malo et al., 2007). Depending upon the pH, mixtures exhibited either additive or synergic effects and appeared to be promising antifungal agents. The combined effects of water activity, chemical preservatives, and natural antimicrobial compounds on this fungus have also been reported (Lopez-Malo et al., 2005). In general, natural antimicrobials were less pH-dependent than were chemical preservatives (e.g., potassium sorbate and sodium benzoate) for growth inhibition.

Cinnamon extract was also used along with sodium benzoate, and potassium sorbate (alone or in combinations) against the growth of *E. coli* O157:H7 in apple juice at 8°C and 25°C (Ceylan et al., 2004). Combinations of 0.3% cinnamon extract with either 0.1% sodium benzoate or 0.1% potassium sorbate were most effective. A synergistic effect was observed between the extract and both preservatives. A 5.3-log reduction in viability occurred in 11–14 days at 8°C, but in 3 days at 25°C. A significant interaction was observed between strains of *E. coli*, storage temperature, and the physiological state of microbial cells during survival studies in apple juice (Usaga et al., 2014).

Components used in the formulation of soft drinks can act synergistically with preservatives to prevent spoilage (Ndagigimana et al., 2004). Two types of orange-based soft drinks were evaluated. Sodium benzoate was added to one type at a concentration of 160 ppm (maximum level allowed in these products in Italy). The second type contained sodium benzoate (160 ppm) and orange essential oil. This latter type did not spoil for 45 days of storage at room temperature. Terpenes in the essential oil apparently acted synergistically with the sodium benzoate according to the authors.

In several of the articles cited above, combinations of weak acid preservatives have been used to demonstrate enhanced preservative effects (synergism) compared to the effects of each preservative used alone. Combining this with hurdle technology (use of one or more preservation methods) may prove most effective for protecting foods from food spoilage and pathogenic microorganisms.

Basic aspects of food preservation through the use of hurdle technology have been reviewed by Leistner (2000). This author discusses the principles of major preservative factors for foods and the effects of food preservation methods on microorganisms in foods. Emphasis is placed on the effects of hurdle technology on these organisms. Prospects for a multitargeted preservation system(s) for foods are also discussed.

3.3.3 Mechanism of Action

It is the undissociated molecule of benzoic acid that is responsible for antimicrobial activity (Table 3.4). Gabel (1921) was one of the first to demonstrate that benzoic acid was effective against bacteria in acid media at a level of 0.1% and in neutral media at 0.2% but inactive in alkaline media. Similar results were reported for fungi and yeasts (Cruess and Richert, 1929).

Rahn and Conn (1944) reported that the antimicrobial effect of benzoic acid was nearly 100 times as efficient in strongly acidic solutions as in neutral solutions, that only the undissociated acid was antimicrobial, and that the toxicity of sodium benzoate in solution was a result of the undissociated benzoic acid molecule.

Marcris (1975) observed a rapid uptake of benzoic acid in *Saccharomyces cerevisiae*. Saturation was reached in about 2 minutes and remained constant thereafter. The strong dependence of uptake on pH was because of the relative distribution of undissociated and dissociated forms in solution, not just the pH itself. The undissociated form was the only one taken up by cells. The effect of temperature on the uptake was similar to that observed in enzymatic reactions. Proteinaceous material was apparently involved in the uptake of this preservative.

The effects of pH, NaCl, sucrose, and sorbic and benzoic acids on the growth of 30 strains of food spoilage yeasts have been reported (Praphailong and Fleet, 1997). *Zygosaccharomyces bailii* and *Yarrowia lipolytica* were the most resistant to the two preservatives at pH 5.0. Sorbate was more inhibitory than benzoate on the strains that were tested. The combined effects of pH and benzoate on the growth of these strains are presented in Table 3.7. As expected, increasing the pH of the culture medium decreased the effectiveness of this preservative.

The effects of benzoic acid were studied in the preservative-resistant yeast *Z. bailii* (Warth, 1991a). At concentrations up to 4 mM, fermentation was stimulated and low levels of benzoate were accumulated. Near the minimum inhibitory concentration (MIC) (10mM), fermentation was inhibited, adenosine triphosphate (ATP) levels declined, and benzoate was accumulated to higher levels. The author concluded

TABLE 3.7

Effect of Benzoic Acid on the Growth of Some Important Food Spoilage Yeasts at Different pH Values

Yeast	No. of Strains Tested	Maximum Concentration (mg/L) Allowing Growth			
		pH 2	pH 3	pH 5	pH 7
Debaromyces hansenii	3	NG	–	500	1200
	1	–	–	250	1200
Yarrowia lipolytica	5	–	250	1200	1200
Pichia anomala	1	–	–	500	1200
P. membranaefaciens	2	–	–	750	1200
Saccharomyces cerevisiae	2	NG	–	750	1200
Kluyveromyces marxianus	4	NG	–	500	1200
Kloeckera apiculata	4	–	–	750	1200
Zygosaccharomyces bailii	6	NG	250	1200	NG
Z. rouxii	2	NG	–	750	1200

Note: NG, no growth in the absence of benzoic acid; –, no growth in the presence of 250 mg of benzoic acid/L.
Source: Adapted from Praphailong and Fleet (1997).

that the primary action of benzoic acid in *Z. bailii* was to cause a general energy loss (ATP depletion). Fermentation was inhibited as a result of high ATP usage rather than of lowered intracellular pH (Warth, 1991b). Inhibition of several glycolytic enzymes was not observed.

Burlini et al. (1993) preincubated cells of *S. cerevisiae* with benzoate or sorbate at an extracellular pH of 6.8, and then added glucose. The resulting metabolic effects included reduced glucose consumption and suppression or reduction of glucose-triggered reactions within cells. The authors stated that, in general, the glucose-induced switch from gluconeogenesis to glycolysis was prevented. Growth rates were substantially reduced by both preservatives even at an extracellular pH of 6.8. The authors concluded that the metabolic effects of benzoate or sorbate were only marginally pH-dependent.

Sorbic and benzoic acids both affected the heat shock response and thermotolerance of *S. cerevisiae* (Cheng and Piper, 1994). These effects were dependent on the pH of the culture medium. The authors reported that these preservatives were among the first compounds shown to act as selective inhibitors of heat-induced protein expression in yeasts.

Two modeling studies to predict the inhibitory effects of organic acid preservatives have been reported. In yeast (*S. cerevisiae*), Lambert and Stratford (1999) demonstrated the following:

1. Inhibition depends more on the preservative concentration within cells rather than on undissociated acid concentration per se.
2. Microbial cells can pump protons out during extended lag phase and raise internal pH despite further influx of preservative.
3. Duration of the lag phase can be predicted from the model.
4. Theoretical ATP consumption for proton pumping can be directly correlated with reduction in cell yield.

Hsiao and Siebert (1999) constructed a model based on physical and chemical properties of organic acids and principal components analysis. Acid-susceptible bacteria (*Bacillus* and *Alicyclobacillus*) and acid-resistant bacteria (*Lactobacillus* and *Escherichia coli*) were grown in media containing organic acids. MICs were determined for each acid and used with principal components to produce models with good correlations.

Only those organic acids that are lipophilic, such as benzoic acid, show antimicrobial activity. One hypothesis is that they inhibit or kill microorganisms by interfering with the permeability of the microbial cell membrane, causing uncoupling of both substrate transport and oxidative phosphorylation from the electron transport system (Freese et al., 1973). In *Bacillus subtilis* (Freese, 1978), membrane transport of amino acids is inhibited, resulting in nutritional starvation of cells (Gould et al., 1983; Gould, 1989). Inhibition of transport in turn results from the destruction of the proton-motive force caused by the continuous shuttle of protons into cells by benzoic acid (Davidson, 1997). The inhibitory potency of this acid can be determined by its lipid–water partition coefficient, its pK value (Table 3.4), and the ability of benzoic acid molecules to delocalize the negative charge of ions within the interior of cells and thus increase their membrane mobility.

More than 99% of the viable cells of *Listeria monocytogenes* were injured after exposure to a solution of 8.5% sodium benzoate (pH 7.0) for 1 hour (Buazzi and Marth, 1992). Metabolic injury was evident by the inability of cells to tolerate 6% NaCl in tryptose agar and the ability to grow on this medium with no added salt. Synthesis of messenger ribonucleic acid (mRNA) was critical for the restoration of salt tolerance. Increased permeability of cell membranes during the course of benzoate injury was not observed.

The effects of organic acid preservatives on the growth and intracellular pH of *E. coli* have been reported (Salmond et al., 1984). The authors suggested that the potency of weak acids, such as benzoic, as food preservatives is related to their capacity specifically to reduce the intracellular pH. They also stated that although both the undissociated and dissociated forms of these acids cause the intracellular pH to fall, growth inhibition is the result predominantly of the undissociated acid. A pH-related increase in protein production was observed when cultures of *E. coli* were grown in liquid media with or without 20 mM benzoate (Lambert et al., 1997). At pH 6.5, benzoate addition resulted in an increased production of 33 proteins. Production of 12 of these was induced at pH 8.0 as well as pH 6.5.

Benzoate also inhibits amino acid uptake in *Penicillium chrysogenum* (Hunter and Seel, 1973), *B. subtilis*, *E. coli*, and *Pseudomonas aeruginosa* (Freese et al., 1973; Sheu et al., 1975; Eklund, 1980). In these studies it was suggested that the undissociated form of benzoic acid may diffuse freely through the cell membrane and then ionize in the cell, yielding protons that acidify the alkaline interior of the cell. Similar effects have been found using a variety of microorganisms, and the effectiveness of undissociated acids as preservatives has been noted for many years (Ingram et al., 1956; Corlett and Brown, 1980; Eklund, 1989; Davidson, 1997).

In addition to these mechanisms, benzoic acid or sodium benzoate can also inhibit specific enzyme systems within cells (Webb, 1966). For example, in many bacteria and yeasts, enzymes involved in acetic acid metabolism and oxidative phosphorylation can be inhibited (Bosund, 1960, 1962). In the citric acid cycle, alpha-ketoglutarate and succinate dehydrogenases appear to be quite sensitive (Bosund, 1962). The production of lipase by *P. fluorescens* was completely inhibited by sodium benzoate (Anderson et al., 1980). In resting cells of *E. coli*, 50% of the trimethylamine-N-oxide reductase activity was inhibited by 1.2 mM benzoic acid (Kruk and Lee, 1982). In yeast cells, phosphofructokinase activity was inhibited (Francois et al., 1986). Aflatoxin production by a toxigenic strain of *Aspergillus flavus* was greatly reduced by the presence of these compounds in synthetic media (Uraih and Chipley, 1976; Uraih et al., 1977; Chipley and Uraih, 1980). Reduction was accompanied by the appearance of a yellow pigment that could be converted into aflatoxin B_1 by cell-free extracts. Conversion could be prevented by the addition of benzoic acid or sodium benzoate. The results suggested that these preservatives blocked an enzymatic step late in the biosynthetic pathway of aflatoxin B_1.

With the previously mentioned results in mind, benzoic acid might also change the permeability of the microbial cell membranes by possibly causing a conformational change in the lipid moieties of these structures. This might result in inhibition of membrane-bound enzymes as a secondary effect. Similar results were observed with 2, 4-dinitrophenol and cell envelope-associated enzymes of *E. coli* and *Salmonella* Enteritidis (Chipley, 1974).

A microarray study was conducted to determine the transcriptional response of *E. coli* O157:H7 to 0.5% sodium benzoate in laboratory media (Critzer et al., 2010). The phosphate-specific transport system in benzoate-treated cells was found to be highly upregulated. Increased expression of several efflux systems was also observed. Benzoate treatment may also affect gene-regulating pathogen virulence and cell survival (Critzer et al., 2008).

The safety of acidified foods relies on the thermal process(es) that will assure the destruction of acid-resistant vegetative pathogenic and spoilage bacteria. Breidt et al. (2013) determined the times needed at 10°C to achieve a 5-log reduction of *E. coli* O157:H7, *Salmonella enterica*, and *Listeria monocytogenes* in pickling brines. A variety of acetic and benzoic acid combinations at pH 3.5 and 3.8 were tested. Strains of *E. coli* were more acid-resistant compared to the other pathogens. For holding times of 4 days or less, a 5-log reduction was achieved at pH 3.5 with 2.5% acetic acid or at pH 3.8 with 2.5% acetic acid + 0.1% benzoic acid. Benzoic acid was effective in reducing the time required to achieve a 5-log reduction in target pathogens.

Hazan et al. (2004) identified a novel synergistic effect from combining benzoic acid with nitrogen starvation in cells of *Saccharomyces cerevisiae*. In yeast cells, survival under nitrogen starvation conditions requires a cellular response in which cytosolic material is targeted for degradation and then recycled into essential molecular building blocks (e.g., amino acids). This allows cellular physiology to continue in the absence of external resources. Benzoic acid specifically inhibited this cellular response. Synergism between benzoic acid and inhibition of this cellular response during nitrogen starvation could be beneficial in the development of novel food preservation technologies.

Cunningham et al. (2009) investigated the extent to which *Listeria monocytogenes* was killed by several weak acid food preservatives, benzoate, propionate, acetate, and sorbate. The inhibition by the preservatives was greatest at pH 4.0. At this pH, sodium benzoate was the most effective preservative tested. Potassium sorbate was effective at 50 mM and 37°C, but was not effective at 4°C or 21°C. Cells grown with sorbate at pH 4.0 entered a viable but nonculturable (VBNC) state, but became nonviable by 24 hours. Cells incubated without sorbate entered the VBNC state, but maintained viability throughout the 24-hour study period.

The ingestion of cranberry juice prevents bacterial adhesion which is crucial in urinary tract infections (Johnson-White et al., 2006). Prevention of adhesion of bacteria responsible for oral infections and stomach ulcers has also been reported.

Gram-negative bacteria have a unique outer membrane which makes them resistant to many antimicrobial agents (Alakomi et al., 2007). However, with permeabilizers, outer membranes can be disintegrated and weakened. Using a fluorescent probe uptake technique and a lipopolysaccharide release assay, the authors found that benzoic acid was an efficient permeabilizer for *Salmonella*.

In *E. coli* O157:H7 cells, bacteriophages may be the principal means for the dissemination and release of Shiga toxins (Subils et al., 2012). Three food preservatives, including sodium benzoate, were tested for their effects on phage production and Shiga toxin activity. Inhibition was more effective on lysogenic phage release than on free phages. A reduction in Shiga toxin titer and an inactivation of toxin activity were also noted.

Uropathogenic strains of *E. coli* are involved in urinary tract infections. An *in vitro* assay for the determination of these strains and for the isolation of compounds from cranberry juice active in inhibiting adherence of these strains to a human cell line has been reported (Turner et al., 2005). Dose-dependent inhibition was demonstrated.

In a separate study, uropathogenic *E. coli* was grown in liquid media with or without the addition of cranberry material and cranberry proanthocyanidins (Hidalgo et al., 2011). In test cultures, expression of the flagellin gene was inhibited, fewer flagella were seen, and swimming and swarming motilities were decreased. The authors proposed that the inhibition of flagellum-mediated motility by cranberry materials might be key to preventing urinary tract infections.

Growth inhibition by weak organic acids was determined in cells of *Saccharomyces cerevisiae* (Ullah et al., 2012). The cell's ability to restore the intracellular pH was involved in the inhibition of growth. A proposed anion export pump was shown to play an important role in restoration following lipophilic (sorbic and benzoic) acid stress.

Several acids used in food preservation were compared for their effects against strains of *E. coli* O157:H7 (Lu et al., 2011). Tests were conducted under conditions representing those of non-heat-processed acidified vegetables in hermetically sealed jars. Jars were incubated under anaerobic conditions, with the pH (3.2) and the ionic strength (0.342) of the contents being held constant. Sulfite was most effective, followed by benzoic acid and sorbic acid, in the present study. The authors also reported that most of the acids tested were more effective for killing *E. coli* O157:H7 strains than acetic acid.

3.3.4 Acquired Resistance to Benzoic Acid

Inhibition or control of metabolic and/or reproductive activities in microorganisms is one of the most basic goals in food preservation. Microorganisms adapt for existence under environments and nutritional conditions which may be unique for a particular food. Any significant changes will require microbial cells to respond by adapting to the new conditions, e.g., stress responses.

Research involving the adaptation of food spoilage and pathogenic microorganisms to changing environmental and/or nutritional conditions has flourished during the past 15 years. This is being driven by greater demand for more unique foods from worldwide sources, and less processing and less addition of chemical preservatives while maintaining safety, nutrition, and consumer appeal.

A comprehensive review of microbial adaptation (stress response) has been published and includes a discussion of response following addition of weak acid preservative such as benzoates (Beales, 2004). This author proposed that prevention of acid adaption might be obtained by using different acid types such as lactic and acetic acids, which are more antimicrobial than other weak acid preservatives. The storage temperature of food products supplemented with any of these acids is also important. Room temperature storage decreases survival more than storage at chill temperatures (Skandamis and Nychas, 2000). This might be related to lipid phase transition occurring in the membranes of microbial cells during storage.

More detailed information on resistance and stress adaptation in general is presented in another chapter of the present reference. The following citations focus mainly on the use of benzoate.

An isolate of *Pichia membranifaciens* from spoiled mango fruit was resistant to sodium benzoate (1500 and 3000 ppm) when grown in laboratory media at pH 4.0 and 4.5, respectively (Ethiraj and Suresh, 1988). However, this isolate was less resistant to potassium sorbate. The authors observed that preservative resistance may be strain specific. The heat-resistant yeast *Talaromyces flavus* was isolated from fruit juice concentrates involved in spoilage of packaged reconstituted fruit juice (King and Halbrook, 1987). Potassium sorbate and sodium benzoate prevented outgrowth, with the required concentrations lower when the pH of the growth media was reduced from 5.4 to 3.5.

Three strains of *T. flavus*, isolated from spoiled fruit products, were tested for tolerance to several organic acids during and after heating (Beuchat, 1988). Fumaric, sorbic, and benzoic acids were the most effective, and lethality was enhanced as the pH of the heating medium was reduced. The effects of sorbic and benzoic acids on the viability of ascospores varied depending on the strain and were influenced by other constituents in the heating medium. Heat resistance increased with age, and in some cases, heat tolerance appeared to be related to the morphology of the ascospores.

A strain of *S. cerevisiae* was found to have multiple resistances to both physical treatment and chemical preservatives (Guerzoni et al., 1990). In an inoculated peach-based system, the effectiveness of benzoic acid on this strain decreased as the water activity decreased and only a slight synergistic action with thermal treatments was observed.

Warth (1977) reported that cells of *Z. bailii* grew in the presence of high concentrations (600 mg/L) of benzoic acid at pH values below the pK of this acid (4.19). Resistance resulted primarily from an inducible energy-requiring system that transported benzoate from the cell. Because this system required energy, cells were resistant only when a sufficient energy source, such as glucose, was present. In a later study (Warth, 1988) it was determined that energy was required for the reduction in cytoplasmic benzoate concentration and for the maintenance of internal pH. Thus, the energy source was unavailable for growth, resulting in reduced growth yields and rates of cell production by yeasts. Several species of yeasts grown in the presence of benzoic acid tolerated 40% to 100% higher benzoic acid concentrations than yeasts grown in its absence (Table 3.8).

The genus *Zygosaccharomyces* is often regarded as being synonymous with food spoilage as a result of the remarkable resistance of its species to commonly used food preservatives such as sorbic, benzoic, and acetic acids and ethanol (Thomas and Davenport, 1985). *Z. bailii* can grow in concentrations of food preservatives in great excess of those normally, or legally, allowed in beverages and can adapt to concentrations even higher. The *Zygosaccharomyces* genus also contains some of the most osmotolerant species known, and most can grow at very low pH. Foods at particular risk are acidic and contain relatively high levels of sugar.

TABLE 3.8

Resistance of Yeast Species to Benzoic Acid

	MIC of Benzoic Acid (mg/L)	
Isolate	**Cells Grown without Benzoic Acid**	**Cells Grown with Benzoic Acid***
Saccharomyces cerevisiae	100	170
Kluveromyces fragilis	125	173
Kloeckera apiculata	125	188
Hansenula anomala	140	223
Saccharomyces cerevisiae	175	285
Candida krusei	300	440
Saccharomycodes ludwigii	300	650
Schizosaccharomyces pombe	325	567
Zygosaccharomyces bailii	600	1250

* Adapted to benzoic acid by overnight growth in the presence of 0.25 mM benzoic acid (first five species) or 2 mM benzoic acid (last four species).

Source: Adapted from Warth (1988).

Uptake rates of benzoic and propanoic acids in cells of *Z. bailii* were proportional to the concentration of undissociated acid (Warth, 1989a). Benzoic acid was taken up 27 times faster by cells than propanoic acid. Glucose stimulated the rates of uptake and was also required for the continuous elimination of benzoate from cells. The principal mechanism of uptake appeared to be diffusion of undissociated acid. Growth of yeasts in the presence of benzoic acid reduced their subsequent permeability to this acid. In another study (Warth, 1989b), the growth of 22 isolates of 11 yeast species in the presence of benzoic acid enhanced their subsequent resistance to benzoic acid and other weak acid preservatives. The minimum pH for growth was not related to this acquired resistance. Growth in the presence of weak acid preservatives appeared to involve a common resistance mechanism.

Three selective media were compared for detecting and enumerating nine strains of *Z. bailii* grown in six commercial food products (Makdesi and Beuchat, 1996). These media were also evaluated for enumerating two of these strains grown in blueberry syrup with or without sodium benzoate (0, 300, and 600 μg/ml). The nonselective medium supported higher recovery of all nine strains from the six food products compared to the three selective media. Cells grown in syrup without sodium benzoate were more sensitive to selective media than were cells grown in syrup containing the preservative.

Mustard spoilage by *Z. bailii* has been described (Buchta et al., 1996). Eight tons of table mustard were exported to the United States. It was preserved using benzoic acid (1.5 g/kg) and packaged in glass jars. Spoilage resulted from a *Z. bailii* strain that was resistant to both benzoic and sorbic acids. A new osmophilic, preservative-resistant spoilage yeast, *Zygosaccharomyces lentus*, has been described (Steels et al., 1999). Five strains of this yeast grew over a wide range of temperatures (4°C to 25°C) and pH (2.2 to 7.0). Growth at 4°C was significant, and all strains were osmotolerant. They were resistant to sorbic acid (up to 400 mg/l) and benzoic acid (up to 900 mg/l) at pH 4.0. Adaptation to higher preservative concentrations was demonstrated with benzoic acid. Resistance to dimethyldicarbonate was greater than that of *Z. bailii* and *S. cerevisiae*. The authors concluded that *Z. lentus* was likely to be more significant than *Z. bailii* in the spoilage of refrigerated foods.

Differences in resistance among yeast species could be the result of differences in the rate of penetration of the cell by benzoic acid, the capacity of the cell to remove it, or an intrinsic sensitivity to benzoic acid or its anion (Warth, 1989c). Resistant species may be less permeable than sensitive ones to undissociated benzoic acid (Warth, 1988). However, results also suggested that the resistance mechanism, in which benzoic acid is continuously removed from the cell, is a common and major determinant of the preservative tolerance of yeast species (Table 3.6) (Warth, 1989c).

Some yeasts are intrinsically resistant to relatively high concentrations of preservatives (Stratford and Lambert, 1999). *Zygosaccharomyces* and *Saccharomyces* are able to adapt to changing levels of preservatives such as benzoate. Adaptation can result in rapid growth in moderate levels or slow growth in previously inhibitory levels. The authors stated that microorganisms resist or adapt to weak acid preservatives (e.g., benzoate) by one of the following three methods:

1. Removal or metabolism. Uptake of weak acid preservatives by passive diffusion is quite rapid, requiring only 5 to 15 minutes to reach equilibrium. It is likely that removal of weak acids by metabolism could occur fast enough to alter the initial impact of these substances on spoilage yeasts.
2. Prevention of uptake. This appears unlikely because uptake occurs rapidly by passive diffusion. Whether efflux pumps are involved is open to debate.
3. Adaptation to or repair of damage. Several causes of damage by weak acid preservatives have been proposed, and adaptive mechanisms have been suggested. For example, if damage is caused by low intracellular pH, adaptation could involve enhanced removal of protons using H^+ATPase. The authors proposed that some yeast adaptation to weak acid preservatives might simply be the result of accumulation of anion. However, earlier studies by Henriques et al. (1997) found that cells of *S. cerevisiae* were able to extrude benzoic acid by an energy-dependent mechanism. Piper et al. (1998) demonstrated the existence of a multidrug-resistance pump that actively extruded preservative anions from cells of this yeast.

In an extensive review, Brul and Coote (1999) described the modes of action of several types of preservative agents for foods and the microbial resistance mechanisms induced by these compounds. Weak organic acid preservatives have maximum inhibitory activity at low pH. This favors the uncharged, undissociated state of the molecule, which is freely permeable across the plasma membrane and is thus able to enter the cell. The inhibitory action is classically believed to occur when the molecule encounters the higher pH inside the cell and then dissociates, resulting in the release of charged anions and protons, which cannot cross the plasma membrane. The net result is that the preservative diffuses into the cell until equilibrium is reached in accordance with the pH gradient across the membrane. Anions and protons accumulate inside the cell. When this happens, inhibition of microbial cell growth occurs because various key metabolic reactions are inhibited, as cited previously (see "Mechanism of Action"). In yeasts, several workers have proposed that the actual inhibitory action of weak acid preservatives (benzoate and sorbate) could be the result of the induction of a stress response that attempts to restore homeostasis but results in the reduction of available energy pools needed for growth and other essential metabolic reactions (Warth 1991a, b; Holyoak et al., 1996; Henriques et al., 1997; Bracey et al., 1998).

Several microbial-resistance mechanisms induced by weak organic acid preservatives were also reviewed by Brul and Coote (1999). They included the acid tolerance responses of *E. coli* O157:H7 and *L. monocytogenes* and the role of membrane ATPase in the resistance of food spoilage yeasts. The authors proposed that in yeasts, simply pumping preservative anions out of cells could create a so-called "futile cycle" where the anions would reassociate at the lower external pH and reenter the cell. They assumed that any rate of diffusion across the plasma membrane remained unchanged and that cells would not alter their membrane composition or structure to reduce uptake of the toxic compound. However, they cited reports that adapted yeasts reduced the diffusion coefficient of preservatives across their plasma membrane so that transport of weak acids into cells was indeed reduced. Therefore, efflux of protons and anions by cells would not create a "futile cycle" if there was a concurrent reduction in the ability of weak acids to diffuse across the cell membrane and enter the cytosol.

Acquired resistance to benzoic acid has been reported for other microorganisms. Benzoic acid occurs naturally as a phytoalexin of immature apples (Seng et al., 1985). Mutants of the fungus *Nectria galligena*, which can infect apple tissues, were isolated and found to be resistant to benzoic acid. The pathogenicity of this fungus could be increased by growth on a medium supplemented with benzoic acid before inoculation into fruit.

MICs of sorbic and benzoic acids for *Gluconobacter oxydans* were 1000 and 900 mg/L, respectively, in media at pH 3.8. Reducing the pH of the media to 3.3 reduced the MIC of both antimicrobials by about 300 mg/L (Eyles and Warth, 1989). When *G. oxydans* was grown in the presence of sublethal concentrations of either compound before the MIC was determined, however, the MIC of both antimicrobials increased substantially within 1 hour.

Cells of *E. coli* initially grown in media at pH 5.0 survived exposures to inorganic acid or to acid pH plus organic acid that prevented subsequent growth of cells grown at pH 7.0 (Goodson and Rowbury, 1989). Six organic acids, including benzoic, were evaluated; none of these acids inhibited the subsequent growth of acid-adapted cells. The authors proposed that acid-adapted organisms might survive in acid foods.

Eleven *E. coli* O157:H7 strains and four commensal strains of *E. coli* were tested for their ability to survive extreme acid exposures (Lin et al., 1996). Three previously reported acid-resistance systems were tested. They included an acid-induced oxidative system, an acid-induced arginine-dependent system, and a glutamate-dependent system. When challenged at pH 2.0, the arginine-dependent system provided more protection to the O157:H7 strains than to commensal strains. However, the glutamate-dependent system provided better protection than the arginine system and appeared equally effective in all strains. Both of these systems were effective in protecting *E. coli* against the bactericidal effects of a variety of weak acids including benzoic acid. The authors suggested that several acid-resistance systems might contribute to the survival of pathogenic *E. coli* in the gastrointestinal tract. They also found that once induced, acid-resistance systems remained active for prolonged periods of cold storage at 4°C.

Russell (1991) examined mechanisms whereby bacteria resist the effects of preservatives and non-antibiotic antibacterial agents. He stated that there were two types of bacterial resistance to preservatives. First, **intrinsic resistance** could result from a natural chromosomally controlled property of a bacterial cell. Examples of this type would include basic cell wall structure/organization and the possession of constitutive enzymes that enable bacteria to metabolize preservatives. Second, **acquired resistance** could result from genetic changes in a cell either by mutation or by acquisition of genetic material from another cell (usually by a plasmid). Resistance to preservatives caused by chromosomal gene mutation is an example. Others include the induction of acid resistance in *E. coli* O157:H7 (Lin et al., 1996) and protection of *L. monocytogenes* against lethal chemicals after adaptation to sublethal environmental stress (Lou and Yousef, 1997). When considering ways of overcoming resistance or of preventing it, Russell (1991) proposed that choosing a suitable preservative or a combination of preservatives (for synergy) might be the most effective.

Acquired resistance in cells of *E. coli* was demonstrated following cultivation in increasing levels of either potassium sorbate or sodium benzoate (Santiesteban-Lopez et al., 2009). Bacterial growth was observed in media containing up to 7000 ppm of either preservative. The author suggested that *E. coli* could be adapted to high levels of these compounds if exposure was performed under sublethal conditions. Inhibition of phosphofructokinase and succinic dehydrogenase by sodium benzoate and potassium sorbate, respectively, was demonstrated. Adaptation was not related to the presence of plasmids. Adaptation and resistance are not new phenomena in bacteria (Russell, 2004), but are in response to several environmental factors which may affect survival (Beales, 2004).

Pathogenic and non-pathogenic surrogate strains of *Escherichia coli* were inoculated into strawberry juice with or without sodium benzoate, potassium sorbate, and citric acid (Gurtler et al., 2011). The juices were then treated with a laboratory-scale pulsed electric field unit. Although some variation occurred between pathogenic and surrogate strains in juice with or without antimicrobials and citric acid, in general, the surrogate provided a margin of safety for studies involving inoculated juice. Although *E. coli* O157:H7 has at least three distinct mechanisms of acid resistance, it is not exceptionally acid-resistant compared to other pathogenic and commensal strains (Large et al., 2005).

Several spoilage yeasts isolated from high sugar/salt environments were characterized by their physiological traits (Martorell et al., 2007). Most strains adapted to high levels of sugar which conferred some protection against pasteurization. Isolates of *Zygosaccharomyces bailii* were very resistant to preservatives (e.g., sorbic and benzoic acids) and to heat. These isolates were not exceptionally resistant to biocides such as peracetic acid or hypochlorite. The authors suggested that spoilage by yeasts, i.e., *Z. bailii*, might be better prevented by use of biocidal cleaning agents in the processing facility, rather than treating the food with preservatives.

Most of the bacterial research involving resistance to organic acids (i.e., benzoic acid) has been conducted with *E. coli*, *Salmonella*, and *L. monocytogenes*. Systems of bacterial (and microbial) resistance are discussed in another chapter in the present reference.

3.4 Microbial Metabolism of Benzoic Acid

Johnson and Stanier (1971) stated that aerobic metabolism of benzoate by bacteria frequently proceeded through the ß-ketoadipate pathway (Figure 3.4). The enzymes of the pathway are inducible, and their synthesis is elicited by primary substrates or metabolic intermediates in the pathway.

In a taxonomic study, Moscoso-Vizcarra and Popoff (1977) found that Enterobacteriaceae could be divided into two groups with respect to benzoic acid metabolism. In the first group, benzoate could be metabolized through the ß-ketoadipate pathway; the second group consisted of those organisms that did not metabolize benzoate.

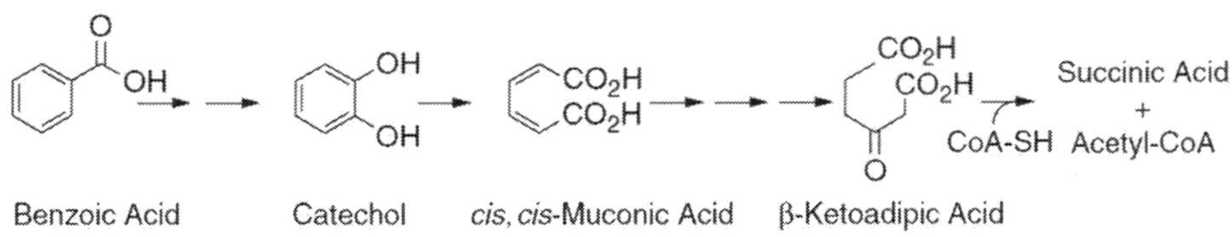

FIGURE 3.4 Proposed pathway for the aerobic microbial metabolism of benzoic acid. [*Source*: Adapted from Zeyer and Kearney (1984).]

Thayer and Wheelis (1976) characterized an inducible benzoate permease system involved in the uptake of benzoate by cells of *Pseudomonas putida*. Metabolism of benzoate occurred by the ß-ketoadipate pathway. A mutant incapable of growth on benzoate was isolated and characterized. Active transport of benzoate may occur in this organism (Thayer and Wheelis, 1982).

Both chromosomal and plasmid-encoded pathways have been described for the catabolism of benzoate in *P. putida* (Jeffrey et al., 1992). In *Alcaligenes eutrophus*, a commonly occurring soil bacterium, the rate of benzoate metabolism may be regulated by the presence of succinate, end product of the ß-ketoadipate pathway (Figure 3.4) (Ampe et al., 1997). A permease-like protein involved in benzoate transport and degradation has now been isolated from cells of *Acinetobacter* (Collier et al., 1997).

Pseudomonas solanacearum, one of the causal agents of wilt in banana plants, was also capable of metabolizing benzoate (Kumari and Mahadevan, 1984). Catechol, *cis*, *cis*-muconate, and ß-ketoadipate were identified as intermediates of benzoate metabolism (Figure 3.4). These results were confirmed by assaying key enzymes from benzoate-grown cells. Kinetic models have been derived for the metabolism of [^{14}C]-benzoate by *Pseudomonas* species (Simkins and Alexander, 1984). Similar results have been reported with *Geobacillus stearothermophilus* (Adams and Ribbons, 1988), with *Rhodococcus* (Janke et al., 1988), and with *Micrococcus* (Haribabu et al., 1984).

Hugo (1991) described the ability of 9 species of *Pseudomonas* and *Vibrio* to degrade 18 aromatic compounds used as preservatives in medicines, foods, and cosmetics. He reported that benzoic acid was easily metabolized by all nine bacterial species tested. Seventeen strains of aerobic bacteria were tested for their ability to degrade 15 aromatic substrates (Lang, 1996). Benzoate was metabolized by 16 of these strains when added as the sole carbon source to liquid mineral media.

A variant of aerobic benzoate degradation has been found in a denitrifying bacterium (*Pseudomonas*) in which benzoyl-coenzyme A (CoA) is the first intermediate (Altenschmidt et al., 1993; Niemetz et al., 1995). This isolate was able to oxidize benzoate completely to carbon dioxide, both aerobically and anaerobically. Outlines of the aerobic pathway were deduced. The proposed intermediates included benzoyl-CoA, 3-hydroxybenzoyl-CoA, gentisate, maleylpyruvate, fumarylpyruvate, and fumarate + pyruvate.

Harwood and Gibson (1986) stated that aerobic degradation of benzoate by various Gram-negative bacteria, including *Pseudomonas*, involves monooxygenase and dioxygenase enzymes that catalyze the incorporation of molecular oxygen into the aromatic nucleus to form dehydroxylated intermediates (see also Heider and Fuchs, 1997; Heider et al., 1999). Anaerobic metabolism can also occur but involves reductive saturation of the nucleus.

Compared to relatively well-characterized aerobic mechanisms for the metabolism of benzoate and other aromatic compounds, anaerobic pathways have been far less studied and reported. One pathway proposed for the anaerobic microbial metabolism of benzoate is shown in Figure 3.5.

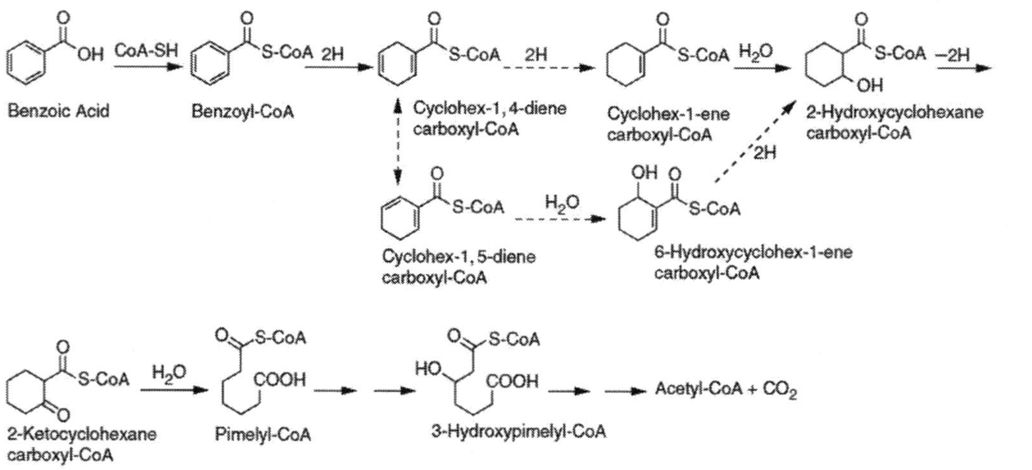

FIGURE 3.5 Proposed pathway for the anaerobic microbial metabolism of benzoic acid. **Note**: This sequence of reaction is compiled from several studies. Solid arrows indicate intermediates detected in benzoate-grown cultures. Dashed arrows indicate proposed intermediates. [*Source*: Adapted from Koch et al. (1993); Harwood and Gibson (1997).]

Harwood and Gibson (1997) stated that this pathway could be divided into several phases: (1) formation of CoA thioesters, which are central intermediates in the anaerobic metabolism of many aromatic compounds (Elder and Kelly, 1994; Fuchs, 2008); (2) ring reduction; (3) introduction of a carbonyl group; (4) ring opening; and (5) a ß-oxidation sequence. This is followed by conversion of the remainder of the molecule to acetyl-CoA. Nitrate, iron, and sulfate can be used as electron acceptors during metabolism (Colberg and Young, 1995).

Several actinomycetes were tested for the degradation of benzoate by growth in a liquid medium containing sodium benzoate (1 to 3 g/L) as the principal carbon source (Grund et al., 1990). Benzoate was converted to catechol by the strains able to grow on this substrate. Further metabolism proceeded via the catechol branch of the ß-ketoadipate pathway with the occurrence of high specific activities of catechol 1, 2-dioxygenase.

In *Aspergillus niger*, benzoate and salicylate are metabolized via the protocatechuate and ß-ketoadipate pathways, respectively (Shailubhai et al., 1982). The enzymes of these pathways appeared to be under coordinate control and were repressed in the presence of a primary substrate, such as glucose. The enzymes responsible for ring cleavage were induced depending on the substrate used for growth. In the presence of benzoate, protocatechuate 3, 4-dioxygenase was induced, whereas salicylate induced catechol 1, 2-dioxygenase for ring fission.

Microbial reduction of benzoate to benzyl alcohol under aerobic conditions has been reported (Kato et al., 1988). *Nocardia asteroides* and several fungi, including species of *Rhizopus* and *Mucor*, were found to have this capability. Although this appears to be a unique reaction occurring in microbial cells under aerobic conditions, the authors concluded that it may be relatively common. However, Arfmann and Abraham (1993) found that benzoic acid as well as 3 hydroxyl derivatives could not be reduced to their respective alcohols when screened with a set of 20 microorganisms, most of which were fungi.

Progress in more thorough understanding of pathways involved in benzoate metabolism has recently been reported (Rather et al., 2010; Valderrama et al., 2012). Normally, benzoate is converted by dioxygenase enzymes to catechol (bacteria) or protocatechuate (mainly fungi), followed by ring cleavage catalyzed by central ring-cleaving dioxygenases. The ß-ketoadipate pathway is central to this scheme (Rather et al., 2010). *Three* major pathways depending on the presence or absence of oxygen are now known (Valderrama et al., 2012). *Aerobically*, molecular oxygen is involved in the activation and cleavage of the aromatic ring. *Anaerobically*, metabolism relies on a strategy based upon coenzyme A-dependent activation of the ring, reductive dearomatization, and then cleavage. The latter authors also reported a *third pathway*, in two bacterial strains, that cross-regulated between anaerobic and aerobic pathways for metabolism of aromatic compounds (benzoate). They stated that this had never been shown before and may be a strategy for organisms to survive in environments with changing oxygen levels.

In the genus *Campylobacter*, one species, *C. jejuni*, has become recognized as a leading cause of human gastroenteritis (Hani and Chan, 1995). The N-benzoyl-glycine amidohydrolase (hippuricase) test is of paramount importance to differentiate *C. jejuni* (hippuricase positive) from other species. Currently, the hippuricase assay is the only reliable biochemical assay that can distinguish *C. jejuni* from *C. coli*. This enzyme cleaves hippuric acid into the constituent products glycine and benzoic acid. The authors isolated the gene responsible for the expression of hippuricase from 12 strains to *C. jejuni* and demonstrated that this gene was not present in 17 isolates of *C. coli*. They also showed that the gene was absent in other *Campylobacter* species. Based on these results, the authors proposed that a species-specific diagnostic DNA probe for human gastrointestinal infection caused by *C. jejuni* could be developed.

3.5 Reaction with Food Constituents

The variability of the levels of benzoic acid in samples of commercial food products may indicate a preferential binding of this compound by certain food constituents. For example, Ganzfried and McFeeters (1976) developed a model system to evaluate binding of benzoic acid by lipids and proteins. Both bound benzoic acid to approximately the same extent. An equation was obtained to estimate the ratio of bound to unbound acid in food products based on their lipid and protein content. Heintze (1978) reported that benzoic acid accumulated in the lipid phase of herring salad, although stability was not adversely affected with time. Sodium benzoate has been shown to inhibit the growth of some of the microorganisms used for analyzing the vitamin content of foods (Voigt et al., 1979).

Models were tested in several food systems to determine the inhibition of 18 strains of *L. monocytogenes* by benzoic acid and other compounds (Ramos-Nino et al., 1996). In foods with high protein or lipid content, antilisterial activity was much lower than predicted, making the models unacceptable under these conditions.

In the early 1990s it was discovered that the combination of ascorbic acid and sodium or potassium benzoate in beverages could, under certain conditions, generate detectable levels of benzene. In one case (a fruit juice–mineral water combination), benzene levels higher than the permissible levels for mineral waters prompted a recall in one state. Subsequent studies indicated that soft drinks containing both ascorbic acid and sodium benzoate could produce very low levels of detectable benzene. Gardner and Lawrence (1993) demonstrated that low levels of detectable benzene could be produced in aqueous solutions of sodium benzoate and ascorbic acid in the presence of the transition metal catalysts iron (III) or copper (II). Although the detectable yield was extremely low (<1 ppb), the authors recommended that the combination of ascorbic acid and sodium benzoate in foods and beverages be evaluated more carefully. Similar results were reported by McNeal et al. (1993) when they analyzed several foods and fruit-flavored soft drinks for benzene.

An automated method was developed for benzene detection in soft drinks (Liang Cao et al., 2007). This method was used to assess benzene levels in 124 soft drinks and beverages, but benzene was not detected in 60% of these products. Levels in six products exceeded Canadian limits (5 µg/L), while two of the six products exceeded World Health Organization limits (10 µg/L). The highest level of benzene, 23 µg/L, was found in a children's soft drink. In a subsequent U.S. survey (Nyman et al., 2008) as well as in a Belgian survey (Van Poucke et al., 2008), the vast majority of soft drinks and beverages either contained no detectable levels of benzene or levels below the U.S. limits (5 µg/L) or the Belgian limits (1 µg/L) for drinking water. Several soft drink companies have reformulated their products to meet or exceed federal standards for benzene.

Studies using model systems have revealed that the high antioxidant activity of some foods (e.g., pickles) may have a strong inhibitory effect on benzene formation (Kharat et al., 2016). However, adding potassium benzoate to turkey breast rolls, followed by exposure of rolls to electron beam irradiation, greatly increased the formation of benzene during irradiation (Zhu et al., 2004).

3.6 Regulatory Status

In the United States, benzoic acid and sodium benzoate are GRAS preservatives (Code of Federal Regulations, 2019; Title 21, Secs. 184.1021 and 184.1733) up to a maximum permitted level of 0.1%. They may also be used in certain standardized foods. Information provided in this code is still current as of April, 2018. In most other countries, the maximum permissible quantities generally range between 0.15% and 0.25% (Tables 3.9 and 3.10; del Olmo et al., 2017). They may also include foods from which benzoate is excluded in the United States. An average daily intake (ADI) value of 0 to 5 mg/kg of body weight has been specified for benzoic acid (Pollard, 1990).

A *Food Technology* special report (1986) reviewed the characteristics, applications, and limitations of preservatives and antimicrobial agents, including benzoic acid. Potassium benzoate is commercially available and apparently evolved in response to consumers' interest in reduced sodium intake. Regulatory agencies consider it to be GRAS to the same extent that sodium benzoate is GRAS as a preservative with the limitation of 0.1%. The potassium salt is less water soluble (42.4 g per 100 ml solution) than the sodium salt. Approximately 1.1 times more potassium salt than sodium salt is required to obtain the same level of benzoic acid in solution and the same level of antimicrobial effect. Product information for both salts is available (Miles Laboratories, Inc.). Calcium benzoate has also been approved for use, although its solubility in water is much less than the sodium salt (Pollard, 1990).

Three food preservatives, previously prohibited for use in meat and poultry, have been approved effective May, 2013 (Federal Register, 2013). Benzoic acid, sodium benzoate, and sodium propionate were found to be safe and suitable for use as antimicrobial agents in ready-to-eat (RTE) meat and poultry

TABLE 3.9

Maximum Permitted Levels (ppm) for Benzoic Acid and Its Salts in Foods in Selected Countries

Food	Canada	Denmark	Norway	Sweden	U.K.	U.S.*
Fish products	1000	50			200	
Fish, semi-preserved	1000	1000	5000	2000	1000	1000
Fruit juice	1000	200	1500	1000	800	1000
Fruit pulp	1000	600	1500	1000	800	1000
Jam, jellies	1000	200	1500	1000	250	1000
Liquid egg white, yolk, or whole		500	5000	10,000	2000	
Margarine, salted	1000			2000		1000
Mayonnaise		1000	3000	2000	500	1000
Mustard	1000	1000	3000	2000	250	1000
Pickles, relishes	1000	250	1500	2000	250	1000
Salads, salad dressings	1000	1000	1500	2000	250	1000
Sauces, ketchup	1000	750	1500	2000	250	1000
Soft drinks containing fruit juice	1000	120	1500	1000	150	1000
Soft drinks, Carbonated	1000	120	500	1000	150	GMP[b]

* In the United States, a maximum level of 1000 ppm benzoic acid or its salts may be used for all products listed. A maximum of 2000 ppm may be used in orange juice for manufacturing.

Orange juice not for manufacturing may not contain a preservative (federal standards of identity).

[b] Good manufacturing practice.

Source: Adapted from Chipley (2005); del Olmo et al. (2017).

products. A large meat products company submitted data showing that these agents did not conceal damage or inferiority or make products appear better or of greater value than they were under the proposed conditions of use. These agents were also effective in controlling the growth of *Listeria* in RTE meat and poultry products.

3.7 Applications

Benzoic acid and sodium benzoate are most suitable for foods and beverages that naturally are in a pH range below 4.5 or that can be brought into this range by acidification. As food preservatives, their main advantages include low price, ease of incorporation into products, and lack of color. However, the narrow pH range in which they are effective, the off-flavor they may impart to the foods they are to preserve, and their toxicologic properties compared to some other preservatives have all contributed to recent efforts by food processors to replace benzoic acid and sodium benzoate with other preservatives with better characteristics. Benzoate does not control the growth of high levels of microorganisms and therefore cannot be used to compensate for poor ingredients or processing. An excellent review of the uses of benzoic acid, as well as human and animal exposure and the resulting controversies, has recently been published (del Olmo et al., 2017).

3.7.1 Use in Various Food Systems

Benzoic acid and sodium benzoate have been widely used to preserve beverages, fruit products, bakery products, and other food in several countries. Applications of these preservatives to foods have been reviewed by Chichester and Tanner (1972), Lueck (1980), Kimble (1977), and Chipley (1983, 1993, 2005). In addition, the use of benzoic acid and other chemicals in the processing and freezing of shrimp has been reviewed (Chandrasekaran, 1994).

Reports indicate that these preservatives continue to be of benefit in a variety of products. Fresh catfish were dipped in various concentrations of sodium benzoate or potassium sorbate and smoked, then

TABLE 3.10

Maximum Permitted Levels (ppm) for Benzoic Acid and Its Salts in Foods in the European Union*

Commodity	Food	Level for Use
Beverages	Nonalcoholic flavored drinks (excluding dairy-based drinks)	600–1000
	Spirits with <15% alcohol by volume	1000
	Alcohol-free beer in keg	1000
	Grape juice, unfermented	1000
	Liquid tea concentrates	1000
Fish	Semi-preserved fish products (including fish roe)	2000–2500
	Salted, dried fish	200
	Shrimp, cooked	6000
Fruits and vegetables	Low-sugar jams, jellies, marmalades, and other fruit-based spreads	1000
	Candied fruits and vegetables	1000
	Vegetables in vinegar, brine, or oil	2000
	Prepared salads	1500
	Olives and olive-based preparations	2000
	Aspic	1500
Sauces	Emulsified sauces with fat content >60%	1000
	Emulsified sauces with fat content <60%	1000
	Non-emulsified sauces	1000
Other	Non-heat-treated dairy-based desserts	300
	Confectionery (excluding chocolate)	1500
	Chewing gum	1500
	Seasonings, condiments, mustard	1000
	Liquid egg white, yolk, or whole egg	5000
	Liquid dietary food supplements	1500
	Dietetic foods (excluding foods for infants/young children)	2000

* The European Union (EU) is currently composed of 27 member nations. In the EU, benzoic acid and its sodium, potassium, and calcium salts are all permitted for use (E210-E 213) (Pollard, 1990).

Source: Adapted from Anon, (1995); Chipley, (2005); del Olmo et al. (2017).

evaluated during shelf life (Efiuvwevwere and Ajiboye, 1996). Sorbate treatment was more effective than benzoate and increased the shelf life by 8 days.

Halotolerant fungi were isolated from salted dried fish (Chakrabarti and Varma, 2000). *Aspergillus flavus*, *A. niger*, and *Penicillium* sp. were dominant. *Penicillium* sp. was most sensitive to propionic acid and sodium benzoate, but *A. flavus* was most sensitive to potassium sorbate. Propionic acid (0.06%) or potassium sorbate (0.02%) or sodium benzoate (0.04%) inhibited the growth of all three fungi.

Several additives and preservatives were evaluated as inhibitors of melanosis (black spot) and microbial spoilage in prawns in chilled storage (Montero et al., 2001). A combination of sodium benzoate and kojic acid was effective in inhibiting melanosis. However, 4-hexylresorcinol was the most effective inhibitor of both melanosis and microbial spoilage in prawns. The shelf life of sardines stored in an acid brine containing 0.3% sodium benzoate was three times longer than that of the controls (Ponce de Leon et al., 1994).

The effectiveness of sodium benzoate and potassium sorbate, separately or combined, was evaluated against the growth of *Z. bailii* in an inoculated salsa mayonnaise at room temperature (Wind and Restaino, 1995). Potassium sorbate was more effective than sodium benzoate. However, neither preservative, separately or combined, was able to prevent spoilage of the product by *Z. bailii.*

The microbial stability and quality of tomato juice and fermented cucumbers and carrots were improved by the addition of a combination of sorbate and benzoate or benzoate alone (Bizri and Wahem, 1994; Fleming et al., 1996; Montano et al., 1997). The levels of some nutrients were reduced when tomato juice was treated with dimethyl dicarbonate. Tomato concentrate could be kept for up to 12 months at room temperature with the addition of salt, acetic acid, sodium benzoate, and potassium sorbate (Uboldi Eiroa et al., 1995). No viable microorganisms, including *Z. bailii*, were observed in any samples containing preservatives.

The effects of sodium benzoate and potassium sorbate on the thermal death rates of ascospores from the heat-resistant mold *N. fischeri* were evaluated in fruit juices (Rajashekhara et al., 1998). Comparable rates were noted when each preservative or the combination of both was used in mango juice heated at 85°C. In grape juice, potassium sorbate was more effective than sodium benzoate or their combination.

The growth and control of four *Salmonella* serotypes in a soft Hispanic type cheese were evaluated (Kasrazadeh and Genigeorgis, 1994). The minimum temperature that allowed growth was 8°C. The growth and control of two strains of *E. coli* O157:H7 were also evaluated in this cheese (Kasrazadeh and Genigeorgis, 1995). The minimum temperature that allowed growth of the *E. coli* strains was 10°C. In both studies, models were developed relating lag time and specific growth rate to temperature. Growth was either prevented or delayed by adding sodium benzoate (0.3%) to cheese (pH 6.6) or adding potassium sorbate (0.3%) to cheese (pH 6.0) made from milk acidified to pH 5.9 with propionic acid.

Chicken skin was inoculated with *Salmonella* spp., *L. monocytogenes*, *C. jejuni*, or *S. aureus*, and wash solutions were evaluated for effectiveness in decontaminating the skin (Hwang and Beuchat, 1995). Washing the skin with solutions of either 0.3% or 0.5% lactic acid combined with 0.05% sodium benzoate reduced the numbers of these pathogens compared to washing with water. No viable cells from any of the four pathogens were detected on skin washed with lactic acid/benzoate solutions and stored for 8 days at 4°C. These solutions could be used for dips to sanitize chickens intended for frying before presentation to consumers (Hathcox et al., 1995).

Listeria monocytogenes is recognized as a foodborne pathogen of major concern to humans. Large outbreaks of infection have been linked to consumption of contaminated coleslaw, cheeses, milk, hot dogs, and luncheon meats. Contamination of many of these foods frequently occurs after processing. The application of antimicrobials as sprays or mists to meats following processing may be more effective than their addition in the meat formulation. The antimicrobial can be applied directly onto the product surface where cells of *L. monocytogenes* usually attach following cooking and during slicing and packaging (Farber and Peterkin, 1999; Tompkin et al., 1999). A study was conducted to evaluate aqueous dipping solutions of organic acids or salts to control *L. monocytogenes* on sliced, vacuum-packaged bologna stored at 4°C for up to 120 days. No significant increase in the numbers of this pathogen occurred on bologna slices treated with 2.5% or 5% acetic acid, 5% sodium diacetate, or 5% potassium benzoate for 120 days (Samelis et al., 2001). Sorbates and benzoates are currently approved in various countries for use as dipping solutions to prevent fungal growth in dry sausages (Sofos, 1989).

Combinations of organic acids, low pH, and ethanol were very effective bactericidal treatments for *L. monocytogenes* (Barker and Park, 2001). Benzoate was one of the most effective compounds tested in the presence or absence of ethanol. Ethanol-enhanced killing correlated with damage to the bacterial cytoplasmic membrane.

Unpasteurized apple cider has been implicated in several foodborne disease outbreaks caused by pathogenic microorganisms. Most research efforts have been directed toward *E. coli* O157:H7 in cider because of the severity of illness this pathogen causes, especially in younger children and the elderly. Survival of *E. coli* O157:H7 in apple juice and apple cider with or without potassium sorbate or sodium benzoate has been investigated. Both preservatives reduced the heat resistance of *E. coli* O157:H7, but benzoate was about two to eight times more effective than sorbate (Gurtler et al., 2019; Splittstoesser et al., 1996; Dock et al., 2000). At 8°C, the combination of sorbate and benzoate was more effective than either preservative alone (Zhao et al., 1993). Resistance to these preservatives was greater at 4°C than at 25°C (Fisher and Golden, 1998). These strains grew well in unpasteurized and pasteurized apple juice; growth was inhibited by benzoate or sorbate (Koodie and Dhople, 2001). Control strains of *E. coli* failed to grow in either type of apple juice. However, the growth of some O157:H7 strains in apple cider was not affected by the presence of either preservative (Miller and Kaspar, 1994).

A preservative treatment was developed that was capable of achieving the FDA mandate for a 5-log reduction of *E. coli* O157:H7 in apple cider (Comes and Beelman, 2002). The treatment that was successful included the addition of 0.15% fumaric acid and 0.05% sodium benzoate followed by holding at

25°C for 6 hours and at 4°C for 24 hours. The same preservatives added to cider resulted in a greater than 5-log reduction in less than 5 hours and 2 hours when held at 25°C and 35°C, respectively. Increasing storage temperatures increased the rate of microbial destruction, while increasing pH decreased the rate (Chikthimmah et al., 2003). Blends of apple and apricot juices with 0.1% sodium benzoate maintained good overall acceptability during storage (Hussain et al., 2011).

3.7.2 Use as a Postharvest Fungicide

Because of the long and successful use of benzoic acid in the processed food industry, this antimicrobial was evaluated for control of postharvest diseases of various fruits and vegetables. However, several factors may influence the effectiveness of benzoic acid in the treatment of fresh fruits and vegetables, with perhaps the most pertinent consideration the pH of the superficial tissues (Eckert, 1967).

A benzoic acid-based polymer coating for apples has been developed (Ivanov et al., 1989). A mixture of fruit-coating polymers and potassium sorbate or sodium benzoate completely inhibited postharvest fungal growth on bananas (Al Zaemey et al., 1993). The shelf life of sliced apples and potatoes was extended by 1 week when they were dipped in a polysaccharide/protein edible coating containing sodium benzoate or potassium sorbate (Baldwin et al., 1996). Methylcellulose was mixed with chitosan and 4% sodium benzoate or potassium sorbate to produce a food-grade film (Chen et al., 1996). Significant antifungal properties were demonstrated against food spoilage fungi.

Benzoic acid and its derivatives have been proposed for use as fungicides, especially against toxigenic strains of *A. flavus*, in peanuts (Uraih and Offonry, 1981). It is currently used in animal feeds and in some tobacco products as a fungicide at levels of up to 0.1% and 0.025%, respectively.

3.7.3 Other Applications

Benzoic acid is one of the oldest preservatives in the cosmetic and pharmaceutical industries. Generally, concentrations of 0.1% to 0.5%, incorporated as sodium benzoate, have been used to preserve cosmetic formulations. Currently, the cosmetic applications for benzoic acid have largely been taken over by more potent antimicrobial agents and compounds that are active over a wider pH range (Manowitz, 1968). A summary of the antimicrobial activities of 30 preservatives, including benzoic acid, used in cosmetics has been published (Bach et al., 1990). In the pharmaceutical industry, benzoic acid was used for oral preparations in concentrations of 0.05% to 0.1%. Although its use in pharmaceuticals has largely been displaced by other compounds (Grundy, 1968), possible areas for specialized applications still exist – for example, as a disinfectant for artificial kidneys (Kolmos, 1976), as a control for denture-borne *Candida albicans* (Lambert and Kolstad, 1986), and as a preservative in certain brands of cough syrup (Chen et al., 1988; Hewala, 1994).

The minimum inhibitory concentrations of sodium benzoate and dichlorobenzyl alcohol for 115 strains of dental plaque microorganisms were determined (Ostergaard, 1994). Sodium benzoate did not inhibit the growth of any Gram-positive cocci. However, saliva samples from volunteers who used a dentifrice containing these antimicrobials indicated that for 5 to 10 minutes after toothbrushing, their concentrations were high enough to inhibit the growth of periodontal pathogens. A single oral rinse with a mouth rinse containing these antimicrobials did not affect plaque removal (Danielsen et al., 1996).

A sodium benzoate-sorbic acid combination used in a pharmaceutical product for treating ulcers was effective against a wild strain of *Pseudomonas cepacia*, following official methods of analysis (Zani et al., 1997). However, this preservative system was ineffective against an adaptive-resistant strain. A p-hydroxybenzoate-based system was effective in protecting the product against a variety of strains of *P. cepacia* grown under different conditions.

Sodium benzoate has also been used to inhibit postharvest changes in fruits and vegetables. For example, Wang and Baker (1979) found that chilling injury to cucumber and sweet pepper fruits could be reduced by the addition of 10 mM sodium benzoate as a 5-minute dip before chilling. The chilling-induced production of ethylene in cucumbers was inhibited by the addition of sodium benzoate (Wang and Adams, 1980).

Several food preservatives were incorporated into synthetic plastics to produce antimicrobial packaging material for foodstuffs (Vartiainen et al., 2003). No antimicrobial activity was observed for any of the films. In addition, chitosan-coated plastic films incorporating five antimicrobials (including sodium benzoate) were evaluated for control of *L. monocytogenes* on ham steaks and salmon (Hudaa Neetoo and Chen, 2008a, b). Film containing sodium lactate was the most effective for both food products. Film containing benzoate was not as effective, possibly due to the low levels of benzoate in the film.

Sodium benzoate is one of the most commonly used antimicrobials in edible films because it is soluble in most film solutions and remains active after film preparation (Cagri et al., 2004). Methylcellulose, chitosan, and collagen films, all with a relatively low pH, work effectively with benzoate. Edible films containing benzoic acid or sodium benzoate are best suited for acidic foods such as cheeses and fermented meat products.

Polylactic acid films containing up to four antimicrobials (alone or in combinations) were developed and tested for their effects on *E. coli* O157:H7 and background microflora in strawberry puree (Jin et al., 2010). The most effective treatment was polymer film containing all four antimicrobials, followed by film containing sodium benzoate + potassium sorbate. The authors suggested using antimicrobial packaging and combinations of preservatives for control of *E. coli* O157:H7 and background microflora in this product.

Technological applications for meat safety are being developed at a rapid pace. For example, nanoparticles are being used to deliver benzoic acid (Hettiarachchy and Reavichandran, 2012). This proved to be an efficient system against *Salmonella* Typhimurium and *Listeria monocytogenes* for up to 14 days of storage in model studies. Its use in meats remains to be evaluated according to the authors.

Interest in the utilization of antimicrobial active packaging for food products has increased in recent years (Khaneghah et al., 2018). The authors reviewed the different types of antimicrobial agents used in active food packaging materials. These included weak organic acids, such as sorbic and benzoic, either alone or in mixtures. In other cases, the incorporation of these acids into biopolymers, such as chitosan, has been reported.

The effects of an additive (sodium benzoate, potassium sorbate, and sodium nitrite) on the quality of fermented silages have been reported (Knicky and Sporndly, 2011). All treated silages from all crops were aerobically stable during the examination time. The additive reduced the growth of undesirable microflora, reduced silage losses, and prolonged the aerobic stability.

For more than 60 years, enteric disease suppression and livestock growth promotion have been effectively achieved by the inclusion of various antibiotics at subtherapeutic doses into diets. Although growth rates and feed efficiency are improved, evidence of a relationship between feed medication and pathogen resistance in human therapy has been established (Mroz, 2005). As a result, all in-feed antibiotics were banned in the European Union in 2006. As a substitute for antibiotics, short chain fatty acids as well as carboxylic acids (e.g., benzoic) and their salts are being evaluated (Table 3.11). Another alternative for the preservation of livestock food and feed systems involves the use of natural preservatives from animal-derived systems, plant-derived products, and microbial metabolites (Lee and Paik, 2016), some of which may contain benzoic acid or its salts.

Worldwide patents have also been issued for the use of benzoic acid or sodium benzoate in many products. These include several categories of foods, cosmetic, medical, and commercial products (Table 3.12). Benzoate may also serve as a scavenger for free radicals (Harvath, 1979), as an inhibitor of D-amino acid oxidases (Quay and Massay, 1977; Yagi et al., 1980), as a weak inhibitor of poly (ADP ribose) polymerase (Oikawa et al., 1980), as an inhibitor of passive anion transport (Lucas-Heron and Fontenaille, 1979), and as an inhibitor of nitrosamine formation (Sung et al., 1988). It has been reported to have a flower-inducing effect in plants (Watanabe and Takimoto, 1979).

3.7.4 Storage and Handling

Sodium benzoate should be stored in a cool, dry place. Containers should be kept closed as much as possible. This product is not corrosive. There is little danger of its being a toxicant or fire hazard under normal conditions of use. However, both benzoic acid and sodium benzoate are moderately toxic by ingestive, intramuscular, and intraperitoneal routes (Sax and Lewis, 1989, 1996).

TABLE 3.11

Health Effects Reported from Experimental Feeding of Livestock with Rations Supplemented with Benzoic Acid

Observation	Animal	BA concentration	Reference
-Lowers pH of diet -Increases gastro-intestinal acidity -Promotes beneficial bacteria -Improves health and disease resistance -Faster growth -Increases efficiency of diet util. -Better carcass quality -Prevents bacteriuria of urinary tract	Pigs	Various	Mroz (2005)
-Increases feed intake -Improves feed conversion -Antimicrobial effects in GI tract	Piglets	5 + 10g/Kg	Kluge et al. (2006)
-Improves average daily gain and feed intake -Most BA excreted as HAc in urine -Reduced urinary pH	Piglets	3.5 and 5 g/Kg	Graber et al. (2012)
-Improves intestinal antioxidant capacity -Improves average daily feed intake & average daily weight gain	Young pigs	5000 mg/Kg	Diao et al. (2016)
-Improves feed conversion ratio -Enhances growth performance	Piglets; grower-finisher pigs	0.3–0.5%	Zhai et al. (2017)
-Piglet growth same with BA as with AB growth promoter	Weaned pigs	0.5%	Voth et al. (2018)
-Feed flavoring agent -No risk to consumers -Low deposition, if any, in edible tissue	Piglets, pigs, sows, chickens, hens, turkeys	Various	EFSA (2018)

BA, benzoic acid; HAc, hippuric acid; AB, antibiotic.

3.8 Toxicology

Several toxicologic studies involving different animal species have been conducted using both benzoic acid and sodium benzoate. These have been summarized by exposure category in Table 3.13. In addition, Sax and Lewis (1989, 1996) reported data for several animal species based on the route of administration (Table 3.14). Extensive human feeding trials conducted early in the twentieth century led Chittenden et al. (1909) and Dakin (1909) to conclude that sodium benzoate was not deleterious to human health.

Benzoate does not appear to be accumulated in the body. The apparent reason for this involves a detoxifying mechanism whereby benzoate is absorbed from the intestine and "activated" by linkage with CoA to yield benzoyl coenzyme A. The overall reaction sequence proceeds as shown in Figure 3.6 (White et al., 1964).

The first reaction is catalyzed by a synthetase enzyme; the second is catalyzed by an acyltransferase enzyme. Hippuric acid, synthesized in the liver, is then excreted in the urine (White et al., 1964). Its formation can be increased greatly by the administration of benzoic acid. Excretion of benzoic acid by this particular pathway was first reported by Dakin (1909). Griffith (1929) found that this mechanism accounted for 66% to 95% of benzoic acid fed in studies in which the quantities were far in excess of those that might be ingested from foods preserved with permitted levels of benzoate. He also suggested that the remaining portion of the benzoate not excreted as hippuric acid may have been detoxified by conjugation with glucuronic acid and excreted in the urine by this route.

Akira et al. (1994) used deuterated benzoic acid and nuclear magnetic resonance spectroscopy to monitor metabolism in a healthy male volunteer. A single oral dose of 250 mg of benzoic acid was quantitatively metabolized to hippuric acid and excreted in the urine within 4 hours. The amount of sodium

TABLE 3.12

Foods, Personal Care Items, and Household Products Supplemented with Benzoate for Sale in Various Countries

Foods	Personal Care Items	Household Products
Bakery products	Acne treatments	Animal repellants (goose, deer, rabbit, snake)
Canned fruits	Antifungal treatments (topical)	
Carbonated beverages	Baby wipes	Antifreeze (corrosion inhibitor)
Casseroles	Coated pills	
Cereal bars	Cough syrups	
	Dental plaque	Anti-odor sprays and powders
	Deodorants	
Ciders	Depilatory creams	Caulks
Condiments (ketchup, BBQ sauce, marinades)	Diaper rash creams	Furniture polishes
Fast foods	Hair dyes	Glass and surface cleaners
Frozen dinners	Hair sprays	Home main. lubricants
Fruit juices	Insect repellents	Molding and wood fillers
Margarine	Lip moisturizers	Paints and enamels
Meat and poultry Pastries	Lotions	Pet products (anti-chew aids, mouth hygiene)
Pickled products	Makeup (foundation, lipstick, eye shadow, blush, etc.)	Printer inks
	Mouthwashes	Plasticizers
Prepacked meats	Nail polish	
Protein bars	Perfumes and colognes	
Salad dressings	Scar treatments	
Syrups	Shampoos and conditioners	
	Shaving creams and gels	
	Styling gels	
	Sunscreens	
	Toothpastes	

Source: Adapted from McGowan and Jacob (2016); Shahmohammadi et al. (2016).

benzoate absorbed by human subjects may be estimated by determining the amount of hippuric acid excreted in the urine (Fujii et al., 1991).

Since the 1970s, benzoic acid has been used in the treatment of patients with congenital errors of urea synthesis to develop an alternative pathway of nitrogen waste excretion (Kubota and Ishizaki, 1991). These authors analyzed plasma concentrations of benzoic and hippuric acids and urinary levels of hippuric acid after oral doses of sodium benzoate (40, 80, or 160 mg/kg) were given at least 1 week apart to six healthy male volunteers. The authors reported that both the biotransformation of benzoic acid to hippuric acid and the urinary excretion of hippuric acid followed first-order reaction rates.

These results have been confirmed in several animal studies. Different animal species differ in their ability to metabolize benzoic acid. This may be related to differences in the ability of liver and kidney cells to carry out glycine and glucuronic acid conjugation (for review, see Chipley, 1993).

Sodium benzoate inhibited the synthesis of glucose from lactate and the generation of urea from ammonia when added to suspensions of rat hepatocytes (Cyr et al., 1991). Inhibition was caused by the accumulation of benzoyl CoA with a resultant depletion of free CoA and acetyl CoA. Acceleration of the conversion of benzoyl CoA to hippurate by the addition of glycine restored the levels of free CoA and acetyl CoA and the rates of gluconeogenesis and ureagenesis.

In guinea pigs, the route of administration significantly affected the efficiency of benzoic acid metabolism (Nathan et al., 1990). When applied topically, only a small portion (6.9%) of the absorbed benzoic

TABLE 3.13

Toxicity of Benzoic Acid and Sodium Benzoate by Exposure Category[a]

Category	Species Tested	Time Period	Level	Results
Acute	Rat		1.7–4.0 g/kg body weight	50% mortality
	Guinea pig[b]		1.4–2.0 g/kg body weight	100% mortality
	Rabbit, cat, dog			
Subchronic	Mouse	3 months	80 mg/kg body weight	Mortality rate increase
	Mouse	5 days	3% of diet	50% mortality
	Mouse	3 months	4% of diet (as sodium benzoate)	No effect
	Human	3 months	1 g/day	No effect
	Human	14 days	12 g per 14 days	No effect
	Human	60–100 days	0.3–4.0 g per 60–100 days	No effect
	Human	Several days	5–10 g for several days (as sodium benzoate)	No effect
Chronic	Mouse	17 months	40 mg/kg body weight per day	Growth disturbance
	Rat	18 months	40 mg/kg body weight per day	Growth disturbance
	Rat	2 weeks	5% of diet (as sodium benzoate)	100% mortality
	Rat		1.5% of diet	Decreased growth rate
	Rat	2 weeks	1% of diet (as sodium benzoate)	No effect

[a] Route of administration was peroral.
[b] Route of administration was per os.
Source: Adapted from Lueck (1980).

TABLE 3.14

Toxicity of Benzoic Acid and Sodium Benzoate by Route of Administration

Benzoic Acid				Sodium Benzoate			
Species Tested	Route of Administration[a]	Toxicity Results[b]	Level mg/kg	Species Tested	Route of Administration[a]	Toxicity Results[b]	Level mg/kg
Human	Skin	TDLo	6	Rabbit	Oral	LDLo	1994
Human	Oral	LDLo	500	Rabbit	Scu	LDLo	2000
Rabbit	Skin	MLD	500 mg[c]	Dog	Oral	LDLo	2018
Rabbit	Eye	SEV	100 mg[c]	Rat	Oral	TDLo	44,000[d]
Rabbit	Oral	LDLo	2000	Rat	Oral	LD50	4070
Rabbit	Scu	LDLo	2000	Rat	Ipr	TDLo	3000[e]
Dog	Oral	LD50	2000	Mouse	Ims	LD50	2306
Cat	Oral	LD50	2000	Guinea pig	Ipr	LDLo	1400
Rat	Oral	LD50	2530				
Mouse	Oral	LD50	1940				
Mouse	Ipr	LD50	1460				
Guinea pig	Oral	LDLo	2000				
Guinea pig	Ipr	LDLo	1400				

[a] Scu, subcutaneous; Ipr, intraperitoneal; Ims, intramuscular
[b] TDLo, lowest published toxic dose; LDLo, lowest published lethal dose; LD50, lethal dose 50% kill; MLD, mild irritation effects; SEV, severe irritation effects.
[c] Total dosage.
[d] Total dosage for 22 days to pregnant rats.
[e] Total dosage for 3 days to pregnant rats.
Source: Adapted from Sax and Lewis (1989, 1996).

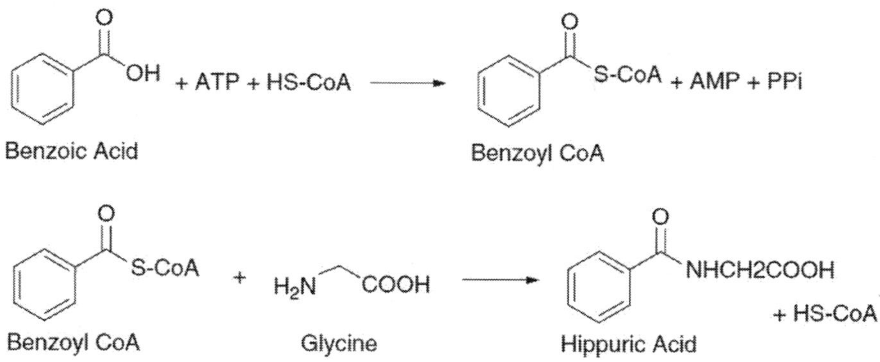

FIGURE 3.6 Metabolism of benzoic acid by mammals. [*Source*: White et al. (1964).]

acid was conjugated with glycine to form hippuric acid. However, it was excreted almost completely as hippuric acid after systemic administration.

S. cerevisiae was used as a model system for testing antioxidant or prooxidant properties of benzoic acid and other weak organic acid food preservatives (Piper, 1999). Cell sensitivity to these acids was enhanced by aerobic rather than anaerobic growth conditions. The food preservatives were shown to have a strong prooxidant effect on aerobically grown yeast cells. They were also mutagenic to the yeast mitochondrial genome but only in the presence of oxygen. The author proposed that large-scale consumption of these preservatives in the human diet might generate oxidative stress within the epithelia of the gastrointestinal tract.

Some adverse reactions to benzoate ingestion have been reported (Schaubschlager et al., 1991). In an oral provocation test with 29 patients, the release of histamine and prostaglandin from mucosa was significantly increased by sodium benzoate exposure. In addition to occasional hypersensitivity reactions, severe anaphylaxis may occur (Michils et al., 1991). Sax and Lewis (1989, 1996) list benzoic acid as a human skin and eye irritant. They also cited data that indicated that sodium benzoate may be both an experimental teratogen and a mutagen.

The concentrations of several preservatives were determined and daily intakes were estimated in a survey of Japanese foods (Ishiwata et al., 1997). The estimated daily intake of benzoic acid was 11 mg/person based on the consumption of foods analyzed in this survey. Consumption of nonalcoholic beverages containing benzoic acid accounted for about 87% of the daily intake. Benzoic acid and its salts have been given a recommended average daily intake level of 0 to 5 mg/kg of body weight (Pollard, 1990).

A medical study was conducted to determine the effect of regular intake of cranberry juice, a natural source of benzoic acid (Table 3.1), on bacteriuria and pyuria in elderly women (Avorn et al., 1994). Subjects were randomly assigned to consume 300 ml/day of a standard cranberry beverage or a placebo drink for 6 months. At the end of the study, bacteriuria with pyuria was found in 28% of urine samples in the placebo group but in only 15% of the group randomized to the cranberry beverage. The findings suggested that use of a cranberry beverage reduced the frequency of bacteriuria with pyuria in older women. The authors stated that beliefs about the effects of cranberry juice on the urinary tract might have microbiological justification.

Sodium benzoate therapy is known to increase ammonia nitrogen elimination by conjugation with glycine and excretion as urinary hippurate (Feoli-Fonseca et al., 1996). In 16 children with inborn errors of urea synthesis, the authors found that therapy was not associated with a constant level of hippurate elimination. This could result in irregular removal of ammonia nitrogen.

Benzyl alcohol is used in a wide variety of cosmetics. It is metabolized to benzoic acid, which reacts with glycine and is excreted as hippuric acid in humans (Nair, 2001). Acceptable daily intakes are 5 mg/kg for benzyl alcohol, benzoic acid, and sodium benzoate as established by the WHO. In chronic-exposure animal studies, reduced feed intake and reduced growth, but no reproductive or developmental toxicity was noted in mice and rats. Carcinogenicity studies were negative. However, allergic reactions (e.g., asthma, allergies and skin rashes) have been reported following the ingestion of foods preserved with benzoate in susceptible individuals (Petrus et al., 1996; Sharma, 2015; Piper, 2018). This has helped to fuel public demand for the reduced use of chemical food preservatives. In addition, recent studies have reported adverse effects in children and adults from the consumption of foods and beverages containing these preservatives (Table 3.15). Summaries of acute and chronic exposures, toxicity, and genetic studies have also been reviewed by the World Health Organization (Anon., 2005).

Several excellent reviews of the toxicologic and safety aspects of food additives used as antimicrobials or antioxidants have been written. For example, Parke and Lewis (1992) stated that the main toxic effects of benzoic and sorbic acids were various allergic responses in humans. In their opinion, steady increases in the incidences of bacterial food poisoning in Europe during the previous decade along with the emergence of "new" bacterial pathogens indicated that misplaced concern for potential toxicity of preservatives had resulted in increasing rather than decreasing health risks to consumers. These authors also described a procedure known as computer-optimized molecular parametric analysis of chemical toxicity (COMPACT). This procedure predicted if food additives were potential substrates of the cytochrome P450 family of enzymes and therefore susceptible to becoming potential carcinogens. Both benzoic and sorbic acids had COMPACT ratios that were moderate. For further information, the reader should also consult the proceedings of two World Health Organization conferences (Anon., 1996, 2005), and three reviews of benzoic and sorbic acids (Frias et al., 1996; del Olmo et al., 2017; Piper, 2018).

3.9 Assay

Because benzoic acid is volatile in steam, it can be quantitatively isolated by acid steam distillation from foods. Occasionally, extraction with organic solvents may be advisable (Lueck 1980; Nishiyama et al., 1995).

Two early collaborative studies should be noted. The first involved spectrophotometric determination of benzoate in ground beef (Maxstadt and Pollman, 1980). Reversed-phase, high-performance liquid chromatography was used in the second study to quantitate sodium benzoate in three types of soda beverages (Woodward et al., 1979). Both methods were adopted as official first action. Benzoic acid in human plasma and urine has been quantitated using a gas-liquid chromatographic technique (Sioufi and Pommier, 1980). Concentrations of 10 ng/ml could be detected in both substances. Microbiological assays for benzoic acid have also been reported. One procedure used two different genera of yeasts, and the minimum detectable concentration of benzoic acid was approximately 100 ppm (Ellerman, 1977).

TABLE 3.15

Recent Toxicology Studies Involving Benzoic Acid Consumption by Humans

Evaluation	Results	Reference
Glucose homeostasis and metabolic profiles	No significant impact on glucose, insulin, or glucagon from acute SB dose.	Lennerz et al. (2015)
Immune response in blood mononuclear cells	Response was suppressed in dose-dependent manner.	Maier et al. (2010)
Genotoxicity of SB and PB in lymphocytes	Both preserv. were clastogenic, mutagenic, and cytotoxic.	Zengin et al. (2011)
ADHD symptoms in college students	High intake may add to ADHD-related symptoms.	Beezhold et al. (2014)
Hyperactivity in three- and eight-/nine-yr.-old children	SB in diet increased hyper-activity in both groups.	McCann et al. (2007)
Allergy in children and adults	SB allergy possibly linked to colas and personal hygiene products.	Jacob et al. (2016); McGowan and Jacob (2016)
Asthma, allergy, ADHD, cognitive functioning (mood, learning, personality) in children	High dietary intake may be associated with these disorders.	Piper (2018)

SB, sodium benzoate; PB, potassium benzoate.

TABLE 3.16

Recent Developments in Detection and Quantitation Methodologies for Benzoic Acid, Including Sorbic Acid, Parabens, and Hippuric Acid

Method[a]	Analyte[b]	Foods	Reference
HPLC	BA, SA, MP and PP	Various	Saad et al. (2005)
Solid phase ext. with TD-GC	BA, SA, MP, EP and PP	Soft drinks, yogurt, sauces	Wang et al. (2006)
HPLC	BA and SA	Quince jam	Ferreira et al. (2000)
Micellar LC	BA and various parabens	Various foods; cosmetics	Memon et al. (2005)
HPLC/UV diode	SB and PS; BA and SA	Various foods; beverages	Pylypiw and Grether (2000); Ristovska et al. (2012)
Liq. microext. + HPLC	BA and SA	Various	Javanmardi et al. (2015)
Microext. + GC	BA and SA	Milk products	Abedi et al. (2014)
GC–MS	BA and SA	Various	El-Ziney (2009)
Rev. phase HPLC	SB	Soft drinks, jams	Altiokka et al. (2007)
Capill. electrophoresis	Hippuric acid	Goat's milk	Carpio et al. (2010)

[a] HPLC, high-performance liquid chromatography; TD–GC, thermal desorption–gas chromatography; LC, liquid chromatography; GC–MS, gas chromatography–mass spectrometry.

[b] BA, benzoic acid; SA, sorbic acid; MP, EP, PP, methyl, ethyl and propyl parabens; SB, sodium benzoate; PS, potassium sorbate.

A novel impedance-splitting method was evaluated for the microbial analysis of benzoic and sorbic acids in selected foods (Kroyer and Futschik, 1996). *E. coli* was used as the test organism, parameters were optimized, and quantitative analysis of these preservatives in various food extracts was performed. Advantages of the method were noted, and it was recommended for screening purposes.

The use of impedance microbiology to study effects of environmental factors on fungi has been reported by Nielsen (1991). Three isolates of the heat-resistant fungus *Neosartorya fischeri* were tested for their resistance to sodium benzoate. Inhibitory effects were strongly affected by pH, inoculum amount, and temperature. In addition, a biosensor prepared from tyrosinase-containing mushroom tissues has been developed and used to assay for the presence of benzoic acid (Wang et al., 1996).

Procedures for both qualitative and quantitative determination of benzoic acid and sodium benzoate can also be found in the Association of Official Analytical Chemists' *Official Methods of Analysis* (AOAC, 2002). Recent developments in detection and quantitation methodologies for several food products are outlined in Table 3.16.

3.10 Acknowledgments

Sincere appreciation is given to Miriam Chipley for compiling the reference citations, and to Julie Ryan for her patience in typing the manuscript. I dedicate this review to my sons, James and John.

REFERENCES

Abdullah, I., Young, J.C. and Games, D.E. 1994. Supercritical fluid extraction of carboxylic and fatty acids from *Agaricus* spp. Mushrooms. *J. Agric. Food Chem.* 42:718.

Abedi, A.S., Mohammadi, A., Azadniya, E., Mortazavian, A.M. and Khaksar, R. 2014. Simultaneous determination of sorbic and benzoic acids in milk products using an optimised microextraction technique followed by gas chromatography. *Food Add. Contamin: Part* A 31:30–36.

Adams, D. and Ribbons, D.W. 1988. The metabolism of aromatic compounds by thermophilic bacilli. *Appl. Biochem. Biotechnol.* 17:231–44.

Akira, K., Farrant, R.D., Lindon, J.C., Caddick, S.T., Nicholls, A.W. and Nicholson, J.K. 1994. High-field deuterium nuclear magnetic resonance spectroscopic monitoring of the pharmacokinetics of selectively deuterated benzoic acid in man. *Anal. Biochem.* 221:297–302.

Al Zaemey, A.B., Magan, N. and Thompson, A.K. 1993. Studies on the effect of fruit-coating polymers and organic acids on growth of *Collectotrichem musae* in vitro and on post-harvest control of anthracnose of bananas. *Mycolog. Res.* 97:1463–68.

Alakomi, H.L., Puupponen-Pimiä, R., Aura, A.M., Helander, I.M., Nohynek, L., Oksman-Caldentey, K.M. and Saarela, M. 2007. Weakening of *Salmonella* with selected microbial metabolites of berry-derived phenolic compounds and organic acids. *J. Agric. Food Chem.* 55:3905–12.

Altenschmidt, U., Oswald, B., Steiner E., Hermann, H. and Fuchs, G. 1993. New aerobic benzoate oxidation pathway via benzoyl-coenzyme A and 3-hydroxybenzoyl-coenzyme A in a denitrifying *Pseudomonas* sp. *J. Bacteriol.* 175:4851–58.

Altiokka, G., Ergun, B., Can, N.O. and Aboul-Enein, H.Y. 2007. Validated reversed phase HPLC method for the analysis of the food additive, sodium benzoate in soft drinks and jams. *J. Liq. Chromatog.& Related Technol.* 30:40–45.

Ampe, F., Uribelarrea, J.L., Aragao, G.M.F. and Lindley, N.D. 1997. Benzoate degradation via the ortho pathway in *Alcaligenes eutrophus* is perturbed by succinate. *Appl. Environ. Microbiol.* 63:2765.

Andersson, R.E., Bodin, H.G. and Snygg, B.G. 1980. The effect of some preservatives on growth, lipase production and lipase activity of *Pseudomonas fluorescens. Chem. Mikrobiol. Technol. Lebensm.* 6:161.

Anonymous. 1995. Food additives other than colours and sweeteners. *Off. J. Eur. Comm.* L61:48.

Anonymous. 1996. Toxicologic evaluation of certain food additives. *WHO Food Addit. Ser.* 37:155.

Anonymous. 2005. Concise international chemical assessment document. Benzoic acid and sodium benzoate. World Health Organization. p. 1–46.

AOAC (Association of Analytical Chemists International). 2002. *Official Methods of Analysis of AOAC International*, William Horwitz, Ed. AOAC International, Gaithersburg, MD.

Arfmann, H.A. and Abraham, W.R. 1993. Microbial reduction of aromatic carboxylic qacids. *Naturfosch* 48:52.

Avorn, J., Monane, M., Gurwitz, J.H., Glynn, R.J., Choodnovsky, I. and Lipsitz, L.A. 1994. Reduction of bacteriuria and pyuria after ingestion of cranberry juice. *J. Amer. Med. Assoc.* 271:751.

Bach, M., Domsch, A., Hoelzl, M., Klein, B., Pohl, L., Scholtyssek, R., Siemer, E. and Waider, H.G. 1990. Preservatives and their practical application in cosmetics. *Seifen. Oele Fette Wachse* 116:345.

Baird-Parker, A.C. 1980. Organic acids. In *Microbial Ecology of Foods.* Vol. 1. Silliker, J.H., Ed. New York: Academic Press. p. 126.

Baldwin, E.A., Nisperos, M.O., Chen, X. and Hagenmaier, R.D. 1996. Improving storage life of cut apple and potato with edible coating. *Postharvest Biol. Technol.* 9:151.

Banks, J.G., Morgan, S. and Stringer, M.F. 1988. The influence of combinations of chemical preservatives on the growth of food poisoning and spoilage microorganisms. *Campden Food Preserv. Res. Assoc.* Technical Memorandum No. 471.

Barker, C. and Park, S.F. 2001. Sensitization of *Listeria monocytogenes* to low pH, organic acids, and osmotic stress by ethanol. *Appl. Environ. Microbiol.* 67:1594.

Bauer, J., Montgelas, A.F. and Gedek, B. 1981. Aflatoxin B1 Production in the Presence of Preservatives and Antimicrobial Agents. In *Proceedings of the International Symposium on Myotoxin*, Naguib K., Park, D.L., Nagui M.M. and Pohland A.E., Eds. Cairo: National Research Center. p. 249.

Beales, N. 2004. Adaptation of microorganisms to cold temperatures, weak acid preservatives, low pH, and osmotic stress: A review. *Comp. Rev. Food Sci. Food Safety* 3:1–20.

Beezhold, B.L., Johnston, C.S. and Nochta, K.A. 2014. Sodium benzoate-rich beverage consumption is associated with increased reporting of ADHD symptoms in college students: A pilot investigation. *J. Atten. Disord.* 18:236–41.

Beuchat, L.R. 1980. Comparison of anti-vibrio activities of potassium sorbate, sodium benzoate, and glycerol and sucrose esters of fatty acids. *Appl. Environ. Microbiol.* 39:1178.

Beuchat, L.R. 1981a. Combined effects of solutes and food preservatives on rates of inactivation of colony formation by heated spores and vegetative cells of molds. *Appl. Environ. Microbiol.* 41:472.

Beuchat, L.R. 1981b. Synergistic effects of potassium sorbate and sodium benzoate on thermal inactivation of yeasts. *J. Food Sci.* 46:771.

Beuchat, L.R. 1988. Influence of organic acids on heat resistance characteristics of *Talaromyces flavus* ascospores. *Intl. J. Food Microbiol.* 6:97.

Beuchat, L.R. and Jones, W.K. 1979. Effects of food preservatives and antioxidants on colony formation by heated conidia of *Aspergillus flavus. Acta Aliment Acad. Sci. Hung.* 7:373.

Bizri, J.N. and Wahem, I.A. 1994. Citric acid and antimicrobials affect microbiological stability and quality of tomato juice. *J. Food Sci.* 59:130.

Bjornsclottin, K., Breidt, Jr., F. and McFeeters, R.F. 2006. Protective effects of organic acids on survival of *Escherichia coli* O157:H7 in acidic environments. *Appl. Environ. Microbiol* 72:660–64.

Blum, D. 2018. *The Poison Squad*. New York: Penguin Press. 330 pp.

Blum, D. 2019. How ketchup made food safer. *National Geographic*, 235:17.

Bosund, I. 1960. The bacteriostatic action of benzoic and salicylic acids. V. Influence of pH on the total uptake of benzoic acid by *Proteus vulgaris* and baker's yeast. *Physiol. Plant.* 13:793.

Bosund, I. 1962. The action of benzoic and salicylic acids on metabolism of microorganisms. *Adv. Food Res.* 11:331.

Boylan, S.L., Acott, K.A. and Labuza, T.P. 1976. *Staphylococcus aureus* challenge study in an intermediate moisture food. *J. Food Sci.* 41:918.

Bracey, D., Holyoak, C.D., Nebe-von Caron, G. and Coote, P.J. 1998. Determination of the intracellular pH (pH1) of growing cells of *Saccharomyces cerevisiae*: The effect of reduced-expression of the membrane H+ATPase. *J. Microbiol. Meth.* 31:113.

Breidt, Jr., F., Kay, K., Cook, J., Osborne, J., Ingham, B. and Arritt, F. 2013. Determination of 5-log reduction times for *Escherichia coli* O157:H7, *Salmonella enterica*, or *Listeria monocytogenes* in acidified foods with pH 3.5 or 3.8. *J. Food Prot* 76:1245–49.

Brul, S. and Coote, P. 1999. Preservative agents in foods. Mode of action and microbial resistance mechanisms. *Intl. J. Food Microbiol.* 50:1.

Buazzi, M.M. and Marth, E.H. 1992. Characteristics of sodium benzoate injury of *Listeria monocytogenes. Microbios* 70:199.

Buchta, V., Slavikova, E., Vadkartiova, R., Alt, S. and Jilek, P. 1996. *Zygosaccharomyces bailii* as a potential spoiler of mustard. *Food Microbiol.* 13:133.

Burlini, N., Pellegrini, R., Facheris, P., Tortora, P. and Guerritore, A. 1993. Metabolic effects of benzoate and sorbate in the yeast *Saccharomyces cerevisiae* at neutral pH. *Arch. Microbiol.* 159:220.

Cagri, A., Ustunol, Z. and Ryser, E.T. 2004. Antimicrobial edible films and coatings. *J. Food Prot.* 67:833–48.

Carpio, A., Rodriquez-Estevez, V., Sanchez-Rodriquez, M., Arce, L. and Valcarcel, M. 2010. Differentation of organic goat's milk based on its hippuric acid content as determined by capillary electrophoresis. *Electrophoresis* 31:2211–17.

Casey, P.G. and Condon, S. 2002. Sodium chloride decreases the bactericidal effect of acid pH on *Escherichia coli* O157:H7. *Inter. J. Food Microbiol.* 76:199–206.

Ceylan, E., Fung, D.Y.C. and Sabah, J.R. 2004. Antimicrobial activity and synergistic effect of cinnamon with sodium benzoate or potassium sorbate in controlling *Escherichia coli* O157:H7 in apple juice. *J. Food Sci.* 69:FMS 102–FMS 106.

Chakrabarti, R. and Varma, P.G.R. 2000. The sensitivity of halotolerant *Aspergillus flavus*, *Aspergillus niger* and *Penicillium* sp. to propionate, sorbate and benzoate. *J. Food Sci. Technol.* 37:72.

Chandrasekaran, M. 1994. Methods for preprocessing and freezing of shrimp: A critical evaluation. *J. Food Sci. Technol. India* 31:441.

Chen, M.-C., Yeh, H.-C. and Chiang B.-H. 1996. Antimicrobial and physicochemical properties of methylcellulose and chitosan films containing a preservative. *J. Food Process. Preserv.* 20:379.

Chen, T.M., Pacifieo, J.R. and Daly R.E. 1988. High-pressure liquid chromatographic assay of dextromethorphan hydrobromide, guaifenesin, and sodium benzoate in an expectorant syrup. *J. Chromatogr. Sci.* 26:636.

Cheng, L. and Piper, P.W. 1994. Weak acid preservatives block the heat shock response and heat-shock-element-directed lacZ expression of low pH *Saccharomyces cerevisiae* cultures, an inhibitory action partially relieved by respiratory deficiency. *Microbiol.* 140:1085.

Chichester, D.F. and Tanner, F.W. 1972. Antimicrobial Food Additives. In *Handbook of Food Additives.* 2nd ed., Furia, T.E., Ed. Chemical Rubber Publishing, Cleveland, OH.

Chikthimmah, N., LaBoard, L.F. and Beelman, R.B. 2003. Critical factors affecting the destruction of *Escherichia coli* O157:H7 in apple cider treated with fumaric acid and sodium benzoate. *J. Food Sci.* 68:1438–41.

Chipley, J.R. 1974. Effects of 2,4-dinitrophenol and N, N+dicyclohexylcarbodiimide on cell envelope-associated enzymes of *Escherichia coli* and *Salmonella entertidis. Microbios* 10:115.

Chipley, J.R. 1983. Sodium benzoate and benzoic acid. In *Antimicrobials in Foods*, Branen A., and Davidson P.M., Eds. New York: Marcel Dekker.

Chipley, J.R. 1993. Sodium benzoate and benzoic acid. In *Antimicrobials in Foods*, Davidson P.M., and Branen, A., Eds. New York: Marcel Dekker.

Chipley, J.R. 2005. Sodium benzoate and benzoic acid. In *Antimicrobials in Foods*, Davidson P.M., Sofos, J.N. and Branen, A, Eds. New York: Marcel Dekker.

Chipley, J.R. and Uraih, N. 1980. Inhibition of *Aspergillus* growth and aflatoxin release by dreivatives of benzoic acid. *Appl. Environ. Microbiol.* 40:352.

Chittenden, R.H., Long, J.H. and Herter, C.A. 1909. *U.S. Dept. of Agric. Bull.* 88.

Code of Federal Regulations. 2019. *Title 21: Food and Drugs, Parts 100–199.* Washington, DC: U.S. Government Printing Office.

Colberg, P.J.S. and Young, L.Y. 1995. Anaerobic degradation of nonhalogenated homocyclic aromatic compounds coupled with nitrate, iron, or sulfate reduction. In *Microbial Transformation and Degradation of Toxic Chemicals*, Young L.Y. and Cerniglia, C.E., Eds. New York: Wiley-Liss. p. 307.

Cole, M.B. and Keenan, M.H.J. 1986. Synergistic effects of weak-acid preservatives and PH on the growth of *Zygosaccharomyces bailii*. *Yeast.* 2:93.

Collier, L.S., Nichols, N.N. and Neidle, E.L. 1997. BenK encodes a hydrophobic permease-like protein involved in benzoate degradation by *Acinetobacter* sp. strain ADPI. *J. Bacteriol.* 179:5943–46.

Comes, J.E. and Beelman, R.B. 2002. Addition of fumaric acid and sodium benzoate as an alternative method to achieve a 5-log reduction of *Escherichia coli* O157:H7 populations in apple cider. *J. Food Prot.* 65:476.

Corlett, D.A., Jr. and Brown, M.H. 1980. pH and acidity. In *Microbial Ecology of Foods*, vol. 1, International Commission on Microbiological Specifications for Foods, Eds. New York: Academic Press. P. 92.

Critzer, F.J., D'Souza, D.H. and Golden, D.A. 2008. Transcription analysis of stx1, marA, and eaeA genes in *Escherichia coli* O157:H7 treated with sodium benzoate. *J. Food Prot* 71:1469–74.

Critzer, F.J., D'Souza, D.H., Saxton, A.M. and Golden, D.A. 2010. Increased transcription of the phosphate - specific transport system of *Escherichia coli* O157:H7 after exposure to sodium benzoate. *J. Food Prot.* 73:819–24.

Cruess, W.V. and Richert, P.H. 1929. Effect of hydrogen ion concentration on the toxicity of sodium benzoate to microorganisms. *J. Bacteriol.* 17:363.

Cunningham, E., O'Byrne, C. and Oliver, J.D. 2009. Effect of weak acids on *Listeria monocytogenes* survival: Evidence for a viable but nonculturable state in response to low pH. *Food Control* 20:1141–44.

Cyr, D.M., Egan, S.G., Brini, C.M. and Tremblay, G.C. 1991. On the mechanisms of inhibition of gluconeogenesis and ureagenesis by sodium benzoate. *Biochem. Pharmacol.* 42:3.

Dai, Y., Normand, M.D., Weiss, J. and Peleg, M. 2010. Modeling the efficacy of triplet antimicrobial combinations: Yeast suppression by lauric arginate, cinnamic acid, and sodium benzoate or potassium sorbate as a case study. *J. Food Prot.* 73:515–23.

Dakin, J.D. 1909. The Fate of sodium benzoate in the human organism. *J. Biol. Chem.* 7:103.

Danielsen, B., Fejerskkov, O. and Scheie, A.A. 1996. The effects of a mouth rinse containing sodium benzoate and alcohol on the quantity and metabolism of dental plaque. *Intl. Dent. J.* 46:91.

Davidson, P.M. 1997. Chemical preservatives and natural antimicrobial compounds. In *Food Microbiology Fundamentals and Frontiers*. 2nd ed., Doyle M.P., Beuchat L.R. and Montville T.J., Eds. Washington, DC: ASM Press. p. 520.

Davidson, P.M. and Juneja, V.K. 1990. Antimicrobial agents. In *Food Additives*, Branen A.L., Davidson P.M. and Salminen S., Eds. New York: Marcel Dekker. p. 83.

Davidson, P.M. and Taylor, T.M. 2007. Chemical Preservatives and Natural Antimicrobial Compounds. In *Food Microbiology Fundamentals and Frontiers*. 3rd ed. Doyle, M.P. and Buchanan, R.L., Eds. Washington, DC: ASM Press.

Davidson, P.M., Taylor, T.M. and Schmidt, S.E. 2013. Chemical Preservatives and Natural Antimicrobial Compounds. In *Food Microbiology Fundamentals and Frontiers*. 4th ed., Doyle, M.P. and Buchanan, R.L., Eds. Washington, DC: ASM Press.

De Navarre, M.G. and Bailey, H.E. 1956. The interference of non-ionic emulsifiers with preservatives. II. *J. Soc. Cosmet. Chem.* 7:427.

Deak, T. 2007. *Handbook of Food Spoilage Yeasts*. 2nd ed. Boca Raton, FL: CRC Press.

Devlieghere, F. 2004. Effect of chemicals on the microbial evolution in foods. *J. Food Prot.* 67:1977–90.

Diao, H., Gao, Z., Yu, B., Zheng, P., He, J., Yu, J., Huang, Z., Chen, D. and Mao, X. 2016. Effects of benzoic acid (VevoVitall®) on the performance and jejunal digestive physiology in young pigs. *J. Animal Sci. Biotechnol.* 7:32–36.

Dock, L.L., Floros, J.D. and Linton, R.H. 2000. Heat inactivation of *Escherichia coli* O157:H7 in apple cider containing malic acid, sodium benzoate, and potassium sorbate. *J. Food Prot.* 63:1026.

Eckert, J.W. 1967. Application and use of post harvest fungicides. In *Fungicides, Vol. 1*, Torgeson, D.C., Ed. New York: Academic Press. p. 287.

Efiuvwevwere, B.J. and Ajiboye, M.O. 1996. Control of microbiological quality and shelf-life of catfish (*Clarias gariepinus*) by chemical preservatives and smoking. *J. Appl. Bacteriol.* 80:465.

EFSA Panel on Additives and Products. 2018. Safety and efficacy of benzoic acid for pigs and poultry. *EFSA J.* 16:5210–30.

Eklund, T. 1980. Inhibition of growth and uptake processes in bacteria by some chemical food preservatives. *J. Appl. Bacteriol.* 48:423.

Eklund, T. 1989. Organic acids and esters. In *Mechanisms of Action of Food Preservation Procedures*, Gould G.S., Ed. London: Elsevier Applied Science. p. 161.

Elder, D.J.E. and Kelly, J. 1994. The bacterial degradation of benzoic acid and benzenoid compounds under anaerobic conditions: unifying trends and new perspectives. *FEMS Michobiol. Rev.* 13:441.

El-Gazzar, F.E. and Marth, E.H. 1987. Sodium benzoate in the control of growth and aflatoxin production by *Aspergillus parasiticus. J. Food Prot.* 50:305.

Ellerman, J.A.G. 1977. Assay of sorbate and benzoate by the disc plate method. *Aust. J. Dairy Technol.* 32:13.

El-Shenawy, M.A. and Marth, E.H. 1988. Sodium benzoate inhibits growth of or inactivates *Listeria monocytogenes. J. Food Prot.* 51:525.

El-Shenawy, M.A. and Marth, E.H. 1989. Inhibition or inactivation of *Listeria monocytogenes* by sodium benzoate together with some organic acids. *J. Food Prot.* 52:771.

El-Ziney, M.G. 2009. GC-MS analysis of benzoate and sorbate in Saudi dairy and food products with estimation of daily exposure. *J. Food Technol.* 7:127–34.

Enache, E. and Chen, Y. 2007. Survival of *Escherichia coli* O157:H7, *Salmonella*, and *Listeria monocytogenes* in cranberry juice concentrates at different brix levels. *J. Food Prot.* 70:2072–77.

Enache, E., Chen, Y. and Elliott, P.H. 2009. Inactivation of *Escherichia coli* O157:H7 in single-strength lemon and lime juices. *J. Food Prot.* 72:235–40.

Ethiraj, S. and Suresh, E.R. 1988. *Pichia membranaefaciens*; A benzoate-resistant yeast from spoiled mango pulp. *J. Food Sci. Technol. India* 25:63.

Eyles, M.J. and Warth, A.D. 1989. The response of *Gluconobacter oxydans* to sorbic acid and benzoic acids. *Intl. J. Food Microbiol.* 8:335.

Farber, J.M. and Petgerkin, P.I. 1999. Incidence and behavior of *Listeria monocytogenes* in meat products. In *Listeria, Listeriosis and Food Safety*, Marth E.H., Ed. New York: Marcel Dekker. p. 505.

Federal Register. 2013. 78:14636–14640.

Fenaroli, B. 2002. *Fenaroli's Handbook of Flavor Ingredients.* 4th ed., Burdock, G.A., Ed. Boca Raton, FL: CRC Press.

Feoli-Fonseca, J.C., Lambert, M., Mitchell, G., Melançon, S.B., Dallaire, D., Millington, D. and Qureshi, I.A. 1996. Chronic sodium benzoate therapy in children with inborn errors of urea synthesis : Effect on carnitine metabolism and ammonia nitrogen removal. *Biochem. Mol. Med.* 57:31–36.

Fernandez de Simon, B., Perez-Ilzarbe, J., Gomez-Cordoves, C. and Estrella, I. 1992. Importance of phenolic compounds for the characterization of fruit juices. *J. Agric. Food Chem.* 40:1531.

Ferreira, I., Mendes, E., Brito, P. and Ferreira, M.A. 2000. Simultaneous determination of benzoic acid and sorbic acids in quince jam by HPLC. *Food Res. Intl.* 33:113–17.

Fisher, T.L. and Golden, D.A. 1998. Survival of *Escherichia coli* O157:H7 in apple cider as affected by dimethyl dicarbonate, sodium bisulfite, and sodium benzoate. *J. Food Sci.* 63:904.

Fleming, H.P., Thompson, R.L. and McFeeters, R.F. 1996. Assuring microbial and textural stability of fermented cucumbers by pH adjustment and sodium benzoate addition. *J. Food Sci.* 61:832.

Food Technology. 1986. Special Report. Antimicrobial agents, a means toward product stability. *Food Technol.* 40:104.

Francois, J., Van Scjaftingen, E. and Hers, H.G. 1986. Effect of benzoate on the metabolism of fructose 2,6-biphosphate in yeast. *Euro. J. Biochem.* 154:141.

Freese, E. 1978. Mechanism of growth inhibition by lipophilic acids. In *The Pharmacological Effect of Lipids*, Kabara J.J., Ed., Champaign: American Oil Chemists' Society. p. 123.

Freese, E., Sheu, C.W. and Galliers, E. 1973. Function of lipophilic acids as antimicrobial food additives. *Nature* 241:321.

Frias, I., Alvarez, R., Sierra, A. and Hardisson, A. 1996. Bromatological and toxicological aspects of the preservatives benzoic and sorbic acids. *Alimentaria* 27:109.

Fuchs, G. 2008. Anaerobic metabolism of aromatic compounds. *Ann. N.Y. Acad. Sci.* 1125:82–99.

Fujii, T., Omori, T., Taguchi, T. and Ogata, M. 1991. Urinary excretion of hippuric acid after administration of sodium benzoate: (Biological Monitoring I). *J. Food Hyg. Soc. Jpn.* 32:177.

Gabel, L.F. 1921. The relative action of preservatives in pharmaceutical preparations. *J. Am. Pharm. Assoc.* 10:767.

Ganzfried, M.C. and McFeeters, R.F. 1976. Binding of benzoic acid by lipids and proteins in acidic foods. *Lebensm. Wiss. Technol.* 9:171.

Gardner, L.K. and Lawrence, G.D. 1993. Benzene production from decarboxylation of benzoic acid in the presence of ascorbic acid and a transition-metal catalyst. *J. Agric. Food Chem.* 41:693.

Geornaras, I., Belk, K.E., Scanga, J.A., Kendall, P.A., Smith, G.C. and Sofos, J.N. 2005. Postprocessing antimicrobial treatments to control *Listeria monocytogenes* in commercial vacuum-packaged bologna and ham stored at 10C. *J. Food Prot.* 68:991–98.

Gerhardt, U. 1995. Preserving agents in meat products: A real necessity? *Fleischerei* 46:6.

Giese, J. 1994. Antimicrobials: Assuring food safety. *Food Technol.* 48:101.

Giese, J. 1995. Developments in beverage additives. *Food Technol.* 49:63.

Glass, K., Preston, D. and Veesenmeyer, J. 2007a. Inhibition of *Listeria monocytogenes* in turkey and pork-beef bologna by combinations of sorbate, benzoate, and propionate. *J. Food Prot.* 70:214–17.

Glass, K.A., McDonnell, L.M., Rassel, R.C. and Zierke, K.L. 2007b. Controlling *Listeria monocytogenes* on sliced ham and turkey products using benzoate, propionate, and sorbate. *J. Food Prot.* 70:2306–12.

Glass, K.A., McDonnell, L., VonTayron, M.R., Wanless, B. and Badvela, M. 2013. Inhibition of *Listeria monocytogenes* by propionic acid-based ingredients in cured deli-style turkey. *J. Food Prot.* 76:2074–78.

Goodson, M. and Rowbury, R.J. 1989. Resistance of acid-habituated Escherichia coli to organic acids and its medical and applied significance. *Lett. Appl. Microbiol.* 8:211.

Gould, G.W. 1989. *Mechanisms of Action of Food Preservation Procedures.* London: Elsevier Applied Science.

Gould, G.W., Brown, M.H. and Fletcher, B.C. 1983. Mechanisms of action of food preservation procedures. In *Food Microbiology: Advances and Prospects*, Roberts T.A. and Skinner F.A., Eds. New York and London: Academic Press. p. 67.

Graber, T., Kluge, H., Hirche, F., Broz, J. and Stangl, G.I. 2012. Effects of dietary benzoic acid and sodium benzoate on performance, nitrogen and mineral balance and hippuric acid excretion of piglets. *Arch. Animal Nutr.* 66:227–36.

Griffith, W.H. 1929. Benzoylated amino acids in the animal organism. IV. A method for the investigation of the origin of glycine. *J. Biol. Chem.* 82:415.

Grund, E., Knorr, C. and Eichenlauv, R. 1990. Catabolism of benzoate and monohydroxylated benzoates by *Amycolatopsis* and *Streptomyces* spp. *Appl. Environ. Microbiol.* 56:1459.

Grundy, W.E. 1968. Antimicrobial preservatives in pharmaceuticals. In *Disinfection, Sterilization, and Preservation*, Lawrence, C.A. and Block, S.S., Eds. Philadelphia: Lea and Febiger. p. 566.

Guerzoni, M.E., Gardini, F. and Duan, J. 1990. Interacton between inhibition factors on microbial stability of fruit-based systems. *Intl. J. Food Microbiol.* 10:1.

Gurtler, J.B., Bailey, R.B., Geveke, D.J. and Zhang, H.Q. 2011. Pulsed electric field inactivation of *E. coli* O157:H7 and non-pathogenic *E. coli* in strawberry juice as influenced by sodium benzoate, potassium sorbate, and citric acid. *Food Control* 22:1689–94.

Gurtler, J.B., Fan, X., Jin, T. and Niemira, B.A. 2019. Influence of antimicrobial agents on the thermal sensitivity of foodborne pathogens: A review. *J. Food Prot.* 82:628–44.

Hani, E.K. and Chan, V.L. 1995. Expression and characterization of *Campylobacter jejuni* benzoylglycine amidohydrolase (Hippuricase) gene in *Escherichia coli*. *J. Bacteriol.* 177:2396.

Haribabu, B., Kamath, A.V. and Vaidyanathan, C.S. 1984. Degradation of substituted benzoic acids by a *Micrococcus* species. *FEMS Microbiol. Lett.* 21:97.

Harvath, L. 1979. Enhancement of granulocyte chemiluminescence with hydroxyl radical scavengers. *Infect. Immun.* 25:473.

Harwood, C.S. and Gibson, J. 1986. Uptake of benzoate by *Rhodopseudomonas palustris* grown anaerobically in light. *J. Bacteriol.* 165:504.

Harwood, C.S. and Gibson, J. 1997. Shedding light on anaerobic benzene ring degradation: A process unique to prokaryotes? *J. Bacteriol.* 179:301–9.

Hathcox, A.K., Hwang, C.A., Resurreccion, A.V.A. and Beuchat, L.R. 1995. Consumer evaluation of raw fried chicken after washing in trisodium phosphate or lactic acid/sodium benzoate solutions. *J. Food Sci.* 60:604.

Hazan, R., Levine, A. and Abeliovich, H. 2004. Benzoic acid, a weak organic acid food preservative, exerts specific effects on intracellular membrane trafficking pathways in *Saccharomyces cerevisiae*. *Appl. Environ. Microbiol* 70:4449–57.

Heider, J. and Fuchs, G. 1997. Microbial anaerobic aromatic metabolism. *Anaerobe* 3:1.

Heider, J., Spormann, A.M., Beller, H.R. and Widdel, F. 1999. Anaerobic bacterial metabolism of hydrocarbons. *FEMS Microbiol. Rev.* 22:459–73.

Heintze, K. 1978. Preservative effect and stability of benzoic acid in herring salad. *Lebensm. Wiss. Technol.* 11:147.

Henriques, M., Quintas, C. and Loureiro-Dias, M.C. 1997. Extrusion of benzoic acid in *Saccharomyces cerevisiae* by an energy-dependent mechanism. *Microbiol.* 143:1877.

Hettiarachcy, N. and Ravichandran, M. 2012. Potential use of edible nanoscale coatings for meat. *Food Safety* 18:54–57.

Hewala, I.I. 1994. Stability-indicating HPLC assay for paracetamol, guaiphenesin, sodium benzoate and oxomezanine in cough syrup. *Anal. Lett.* 27:71–93.

Hidalgo, G., Chan, M. and Tufenkji, N. 2011. Inhibition of *Escherichia coli* CFT073 fliC expression and motility by cranberry materials. *Appl. Environ. Microbiol* 77:6852–57.

Holyoak, C.D., Stratford, M., McMullin, A., Cole, M.B., Crimmins, K., Brown, A.J.P. and Coote, P. 1996. Activity of the plasma-membrane H+-ATPase, and optimal glycolytic flux are required for rapid adaptation and growth in the presence of the weak acid preservative sorbic acid. *Appl. Environ. Microbiol.* 62:3158.

Hornickova, S., Dragounova, H., Hejtmankova, K., Michlova, T. and Hejtmankova, A. 2014. Production of benzoic acid in fermented goat's and sheep's milk. *Sci. Agric. Bohemica* 45:247–53.

Hosein, A.M., Breidt, Jr., F. and Smith, C.E. 2011. Modeling the effects of sodium chloride, acetic acid, and intracellular pH on survival of *Escherichia coli* O157:H7. *Appl. Environ. Microbiol* 77:889–95.

Hsiao, C.P. and Siebert, K.J. 1999. Modeling the inhibitory effects of organic acids on bacteria. *Intl. J. Food Microbiol.* 47:189.

Hudaa Neetoo, M.Y. and Chen, H. 2008a. Control of *Listeria monocytogenes* on ham steaks by antimicrobials incorporated into chitosan-coated plastic films. *Food Microbiol* 25:260–68.

Hudaa Neetoo, M.Y. and Chen, H. 2008b. Effectiveness of chitosan-coated plastic films incorporating antimicrobials in inhibition of *Listeria monocytogenes* on cold-smoked salmon. *J. Food Microbiol.*127:235–40.

Hugo, W.B. 1991. The degradation of preservatives by microorganisms. *Inter. Biodeterior.* 27:185.

Humpf, J.U. and Schreier, P. 1991. Bound aroma compounds from the fruit and the leaves of blackberry (*Rubus laciniata* L.). *J. Agric. Food Chem.* 39:1830.

Hunter, D.R. and Seel, I.H. 1973. Effect of weak acids on amino acid transport by *Penicillium chrysogenum*: evidence of a proton or charge gradient as the driving force. *J. Bacteriol.* 113:1184.

Hussain, I., Zeb, A. and Ayub, M. 2011. Evaluation of apple and apricot blend juice preserved with sodium benzoate at refrigeration temperature. *World J. Dairy and Food Sci.* 6:79–85.

Hwang, C.A. and Beuchat, L.R. 1995. Efficacy of selected chemicals for killing pathogenic and spoilage microorganisms on chicken skin. *J. Food Prot.* 58:19.

Ingram, M., Ottoway, F.J.H. and Coppock, J.B.M. 1956. The preservative action of acid substances in food. *Chem. Ind. (Lond.)* 42:1154.

Ishiwata, H., Nishijima, M., Fukusawa, Y., Ito, Y. and Yamada, T. 1997. Evaluation of preservatives contents in foods and the daily intake deduced from the results of the official inspection in Japan in F.Y. 1994. *J. Food Hyg. Soc. Japan* 38:145–154.

Islam, M., Chen, J., Doyle, M.P. and Chinnan, M. 2002a. Effect of selected generally-recognized-as safe preservative sprays on growth of *Listeria monocytogenes* on chicken luncheon meat. *J. Food Prot.* 65:794–98.

Islam, M., Chen, J., Doyle, M.P. and Chinnan, M. 2002b. Control of *Listeria monocytogenes* on turkey frankfurters by generally-recognized-as safe preservatives. *J. Food Prot.* 65:1411–16.

Ivanov, L. I., Dorofeev, A.L. and Rogozhnik, F.A. 1989. Storage of polymer-coated apples. *Pischevaya Promyshlenost USSR* 3:66.

Jacob, S.E., Hill, H., Lucero, H. and Nedorost, S. 2016. Benzoate allergy in children—From foods to personal hygiene. *Pediatr. Dermatol.* 33:213–15.

Jager, M. 1994. Are preservatives still necessary? *Food Market. Technol.* 8:4.

Janke, D., Al-Mofarji, T., Straube, G., Schumann, P. and Prauser, H. 1988. Critical steps in degradation of chloroaromatics by rhodococci. I. initial enzyme reactions involved in catabolism of aniline, phenol and benzoate by *Rhodococcus* Sp. An 117 and An 213. *J. Basic Microbiol.* 28:509.

Javanmardi, F., Nemati, M., Ansarin, M. and Arefhosseini, S.R. 2015. Benzoic and sorbic acid in soft drinks, milk, ketchup sauce and bread by dispersive liquid-liquid microextraction coupled with HPLC. *Food Add. Contamin:Part* B 8:32–39.

Jay, J.M. 2000. *Modern Food Microbiology.* 6th ed. Gaithersburg: Aspen Publishers, Inc.

Jeffrey, W.H., Cuskey, S.M., Chapman, P.J., Resnick, S. and Olsen, R.H. 1992. Characterization of *Pseudomonas putida* mutants unable to catabolize benzoate: cloning and characterization of *Pseudomonas* genes involved in benzoate catabolism and isolation of a chromosomal DNA fragment able to substitute for XylS in activation of the TOL lower-pathway promoter. *J. Bacteriol.* 174:4986.

Jin, T., Zhang, H. and Boyd, G. 2010. Incorporation of preservatives in polylactic acid films for inactivating *Escherichia coli* O157:H7 and extending microbiological shelf life of strawberry puree. *J. Food Prot.* 73:812–18.

Johnson, B.F. and Stanier, R.Y. 1971. Regulation of the B-ketoadipate pathway in *Alcaligenes eutrophus. J. Bacteriol.* 107:476.

Johnson-White, B., Buquo, L., Zeinali, M. and Ligler, F.S. 2006. Prevention of nonspecific bacterial cell adhesion in immunoassays by use of cranberry juice. *Anal. Chem.* 78:853–57.

Jordan, K.N. and Davies, K.W. 2001. Sodium chloride enhances recovery and growth of acid-stressed *E. coli* O157:H7. *Lett. Appl. Microbiol.* 32:312–15.

Kallio, J., Ahtonen, S. and Sarimo, S.S. 1985. Effect of quinic acid on the growth of wild yeasts and molds. *J. Food Prot.* 48:327.

Kasrazadeh, M. and Genigeorgis, C. 1994. Potential growth and control of *Salmonella* in hispanic type soft cheese. *Intl. J. Food Microbiol.* 22:127.

Kasrazadeh, M. and Genigeorgis, C. 1995. Potential growth of *Escherichia coli* O157:H7 in soft hispanic type cheese. *Intl. J. Food Microbiol.* 25:289.

Kataoka, A., Enache, E., Sohail, M., Elliott, P. and Black, D.G. 2011. Inactivation of shiga toxin-producing *Escherichia coli* in single-strength lemon and lime juices containing preservatives. *J. Food Prot.* 74:1746–50.

Kato, N., Konishi, H., Uda, K., Massayuki, S. and Sakazawa, C 1988. Microbial reduction of benzoate to benzyl alcohol. *Agric. Biol. Chem.* 52:1855.

Kelly, S.F. 1975. The Effect of sodium benzoate as a preservative for carbonated beverages. presented at the Proceedings of the 22nd Annual Meeting of Soft Drink Technologists, Washington, DC.

Khaneghah, A.M., Hashemi, S.M.B. and Limbo, S. 2018. Antimicrobial agents and packaging systems in antimicrobial active food packaging: An overview of approaches and interactions. *Food Bioprod. Proc.* 111:1–19.

Kharat, M.M. 2016. Antioxidant compounds in traditional Indian pickles may prevent the process-induced formation of benzene. *J. Food Prot.* 79:123–31.

Kimble, C.E. 1977. Chemical Food Preservatives. In *Disinfection, Sterilization, and Preservation.* 2nd ed., Block S.S., Ed. Philadelphia: Lea and Febiger. p. 834.

King, Jr., A.D. and Halbrook, W.U. 1987. Ascospore heat resistant and control measures for *Talaromyces flavus* isolated from fruit juice concentrate. *J. Food Sci.* 52:1252.

Kluge, H., Broz, J. and Eder, K. 2006. Effect of benzoic acid on growth performance, nutrient digestibility, nitrogen balance, gastrointestinal microflora and parameters of microbial metabolism in piglets. *J. Animal Physiol. Animal Nutr.* 90:316–24.

Knicky, M. and Sporndly, R. 2011. The ensiling capacity of a mixture of sodium benzoate, potassium sorbate, and sodium nitrite. *J. Dairy Sci.* 94:824–31.

Koch, J., Eisenreich, W., Bacher, A. and Puchs, G. 1993. Products of enzymatic reduction of benzoyl-CoA, a key reaction in anaerobic aromatic metabolism. *Eur. J. Biochem.* 211:649.

Kolmos, H.J. 1976. The bactericidal action of benzoic acid and sodium acetate on the gram-negative flora of dialysis fluid. *Acta. Pathol. Microbiol. Scand. B* 84:259.

Koodie, L. and Dhople, A.M. 2001. Acid tolerance of *Escherichia coli* O157:H7 and its survival in apple juice. *Microbios* 104:167.

Kroyer, G.T. and Futschik, K. 1996. Application of a new impedance-splitting method for the microbiological analysis of food preservatives. IFT annual meeting: Book of Abstracts. p. 26.

Kruk, M. and Lee, J.S. 1982. Inhibition of *Escherichia coli* trimethylamine-n-oxide reductase by food preservatives. *J. Food Prot.* 45:241.

Kubota, K. and Ishizaki, T. 1991. Dose-dependent pharmacokinetics of benzoic acid following oral administration of sodium benzoate to humans. *Eur. J. Clin. Pharmacol.* 41:363.

Kumari, A.A. and Mahadevan, A. 1984. Degradation of benzoate by *Pseudomonas solanacearum. A. Pfanzenkr. Pfanschutz* 91:383.

Lambert, J.P. and Kolstad, R. 1986. Effect of a benzoic acid-detergent germicide on denture-born candida albicans. *J. Prosthet. Dent.* 55:699.

Lambert, L.A., Abshire, K., Blankenhorn, D. and Slonczewski, J.L. 1997. Protein induced in *Escherichia coli* by benzoic acid. *J. Bacteriol.* 179:7975.

Lambert, R.J. and Stratford, M. 1999. Weak-acid preservatives: Modelling microbial inhibition and response. *J. Appl. Microbiol.* 86:57.

Lang, E. 1996. Diversity of bacterial capabilities in utilizing alkylated benzenes and other aromatic compounds. *Lett. Appl. Microbiol.* 23:257.

Large, T.M., Walk, S.T. and Whittam, T.S. 2005. Variation in acid resistance among shiga toxin-producing clones of pathogenic *Escherichia coli. Appl. Environ. Microbiol.* 71:2493–500.

Lee, N.K. and Paik, H.D. 2016. Status, antimicrobial mechanism, and regulation of natural preservatives in livestock food systems. *Kor. J. Food Sci. An.* 36:547–57.

Leistner, L. 2000. Basic aspects of food preservation by hurdle technology. *Intl. J. Food Microbiol.* 55:181–86.

Lennerz, B., Vafai, S.B. Delaney, N.F., Clish, C.B., Deik, A.A., Pierce, K.A., Ludwig, D.S. and Mootha, V.K. 2015. Effects of sodium benzoate, a widely used food preservative, on glucose homeostasis and metabolic profiles in humans. *Mol. Genet. Metab.* 114:73–79.

Lewis, D.R. and Payne, J. 1958. Benzoic acid as a fungicide, particularly for citrus fruit. *British Patent*, 794, 769.

Liang Cao, X., Carey, V., Seaman, S., Tague, B. and Becalski, A. 2007. Determination of benzene in soft drinks and other beverages by isotope dilution headspace gas chromatography/mass spectrometry. *J. AOAC Inter.* 90:479–84.

Lin, J., Smith, M.P., Chapin, K.C., Baik, H.S., Bennett, G.N. and Foster, J.W. 1996. Mechanisms of acid resistance in enterohemorrhagic Escherichia coli. *Appl. Environ. Microbiol.* 62:3094.

Lopez, M., Martinez, S., Gonzalez, J., Martin, R. and Bernardo, A. 1998. Sensitization of thermally injured spores of *Bacillus stearothermophilus* to sodium benzoate and potassium sorbate. *Appl. Microbiol.* 27:331.

Lopez, M., Mazas, M., Gonzalez, I., Gonzalez, J. and Bernardo, A. 1996. Thermal resistance of *Bacillus stearothermophilus* spores in different heating systems containing some approved food additives. *Lett. Appl. Microbiol.* 23:187.

Lopez-Malo, A., Alzamora, S.M. and Palou, E. 2005. *Aspergillus flavus* growth in the presence of chemical preservatives and naturally occurring antimicrobial compounds. *Intl. J. Food Microbiol.* 99:119–28.

Lopez-Malo, A., Barreto-Valdivieso, J., Palou, E. and San Martin, F. 2007. *Aspergillus flavus* growth response to cinnamon extract and sodium benzoate mixtures. *Food Control* 18:1658–1362.

Lou, Y. and Yousef, A.E. 1997. Adaptation to sublethal environmental stresses protects *Listeria monocytogenes* against lethal preservation factors. *Appl. Environ. Microbiol.* 63:1252.

Lu, H.J., Breidt, Jr., F., Perez-Diaz, I.M. and Osborne, J.A. 2011. Antimicrobial effects of weak acids on the survival of *Escherichia coli* O157:H7 under anacrobic conditions. *J. Food Prot.* 74:893–98.

Lucas-Heron, B. and Fontenaille, C. 1979. Urate transport in human blood cells: activation by ATP. *Biochim. Biophys. Acta.* 553:284.

Lueck, E. 1980. *Antimicrobial Food Additives*. New York: Springer-Verlag.

Macris, B.J. 1975. Mechanism of benzoic acid uptake by *Saccharomyces cerevisiae*. *Appl. Microbiol.* 30:503.

Maier, E., Kurz, K., Jenny, M., Schennach, H., Ueberall, F. and Fuchs, D. 2010. Food preservatives sodium benzoate and propioic acid and colorant curcumin suppress Th1-type immune response *in vitro*. *Food Chem. Toxicol.* 48:1950–56.

Makdesi, A.K. and Beuchat, L.R. 1996. Performance of selective media for enumerating *Zygosaccharomyces bailii* in acid foods and beverages. *J. Food Prot.* 59:652.

Manganelli, E. and Casolari, A. 1983. Sensitivity of yeasts to sorbic and benzoic acids and their salts. *Ind. Conserve* 58:23.

Manowitz, M. 1968. Cosmetic preservatives. In *Disinfection, Sterilization, and Preservation*, Lawrence, C.A. and Block, S.S., Eds. Philadelphia: Lea and Febiger. p. 555.

Marlatt, C., Hoct, C. and Chien, M.J. 1992. Studies of aroma constituents bound as glycosides in tomato. *J. Agr. Food Chem.* 40:249.

Martorell, P., Stratford, M., Steels, H., Fernandez-Espinar, M.T. and Querol, A. 2007. Physiological characterization of spoilage strains of *Zygosaccharomyces bailii* and *Zygosaccharomyces rouxii* isolated from high sugar environments. *Intl. J. Food Microbiol.* 114:234–42.

Marwan, A.G. and Nagel, C.W. 1986a. Microbial inhibitors of cranberries. *J. Food Sci.* 51:1009.

Marwan, A. G., and Nagel. C. W. 1986b. Quantitative determination of infinite inhibition concentrations of antimicrobial agents. *Appl. Environ. Microbiol.* 51:559.

Matos, T.J.S., Jensen, B.B., Bernardo, F.M.A., Barreto, A.H.S. and Hojberg, O. 2007. Mycoflora of two types of Portuguese dry-smoked sausages and inhibitory effect of sodium benzoate, potassium sorbate, and methyl p-hydroxy benzoate on mold growth rate. *J. Food Prot.* 70:1468–74.

Maxstadt, J.J. and Pollman, R.M. 1980. Spectrophotometric determination of sulfites, benzoates, sorbates, ascorbates and added blood in ground beef: Collaborative study. *J. Assoc. Off Anal. Chem.* 63:667.

McCann, D., Barrett, A., Cooper, A., Crumpler, D., Dalen, L., Grimshaw, K., Kitchin, E., Lok, K., Porteous, L., Prince, E., Sonuga-Barke E., Warner, J.O. and Stevenson, J. 2007. Food additives and hyperactive behavior in 3-year-old and 8/9-year-old children in the community: A randomised, double-blinded, placebo-controlled trial. *Lancet* 370:1560–67.

McGowan, M. and Jacob, S.E. 2016. An update on benzoates. *Dermatologist* 24:360–63.

McNeal, T.P., Nyman, P.J., Diachenko, W.G. and Hollifield, H.C. 1993. Survey of benzene in foods by using headspace concentration techniques and capillary gas chromatography. *J. Assoc. Off. Anal. Chem. Intl.* 76:1213.

Memon, N., Bhanger, M.I. and Khuhawer, N.Y. 2005. Determination of preservatives in cosmetics and food samples by micellar liquid chromatography. *J. Sep. Sci.* 28:635–38.

Michils, A., Vandermoten, G., Duchateau, J. and Yernault, J.C. 1991. Anaphylaxis with sodium benzoate. *Lancet* 337:1424.

Miller, L.G. and Kasper, C.W. 1994. *Escherichia coli* O157:H7 acid tolerance and survival in apple cider. *J. Food Prot.* 57:460.

Montano, A., Sanchez, A.H., Rejano, L. and deCastro, A. 1997. Processing and storage of lye-treated carrots fermented by a mixed starter culture. *Intl. J. Food Microbiol.* 35:83.

Montero, P., Lopez-Caballero, M.E. and Perez-Mateos, M. 2001. The effect of inhibitors and high pressure treatment to prevent melanosis and microbial growth on chilled prawns (*Penaeus japonicus*). *J. Food Sci.* 66:1201.

Moscoso-Vizcarra, M. and Poppoff, M. 1977. Pathways of benzoic acid dissimilation in *Enterobacteriaceae*. *Ann. Microbiol. (Paris)* 128A:199.

Mroz, Z. 2005. Organic acids as potential alternatives to antibiotic growth promoters for pigs. *Adv. Pork Prod.* 16:169–82.

Nair, B. 2001. Final report on the safety assessment of benzyl alcohol, benzoic acid and sodium benzoate. *Inter. J. Toxicol. 20* Suppl. 3:23–50.

Nathan, D., Sakr, A., Lichtin J.L. and Bronaugh, R.L. 1990. *In vitro* skin absorption and metabolism of benzoic acid, p-aminobenzoic acid and benzocaine in the hairless guinea pig. *Pharmaceut. Res.* 7:1147.

Nazer, A.I., Kobilinsky, A., Tholozan J.L. and Dubois-Brissonnet, F. 2005. Combinations of food antimicrobials at low levels to inhibit the growth of *Salmonella* sv. Typhimurium: A synergistic effect? *Food Microbiol.* 22:391–98.

Ndagijimana, M., Belletti, N., Lanciotti, R., Guerzoni M.E. and Gardini, F. 2004. Effect of aroma compounds on the microbial stabilization of orange-based soft drinks. *J. Food Sci.* 69:20–24.

Nielsen, P.V. 1991. Preservative and temperature effect on growth of three varieties of the heat-resistant mold, *Neosartorya fischeri* as measured by an impedimetric method. *J. Food Sci.* 56:1735.

Nielson, P.V., Beuchat, L.R. and Frisvad, J.C. 1989. Growth and fumitremorgin production by *Neosartorya fischeri* as affected by food preservatives and organic acids. *J. Appl. Bacteriol.* 66:197.

Niemetz, R., Altenschmidt, U., Brucker A. and Fuchs, G. 1995. Benzoyl-coenzyme-a-monooxygenase, a flavin-dependent hydrolyase. purification, some properties and its role in aerobic benzoate oxidation via gentisate in a denitrifying bacterium. *Eur. J. Biochem.* 227:161.

Niku-Paavola, M.L., Laitila, A., Mattila-Sandholm T. and Haikaara, A. 1999. New types of antimicrobial compounds produced by *Lactobacillus plantarum*. *J. Appl. Microbiol.* 86:29.

Nishiyama, O., Kurita, R., Kurozuka, M., Matsumura, M., Oukuni, N., Awano, K., Morioka, S. and Kaneda, Y. 1995. Analysis of preservatives in foods by using steam distillation with ammonium sulfate saturated solution. *J. Food Hyg. Soc. Jpn.* 46:495.

Nogueira, M.C.L., Oyarzabal O.A. and Gombas, D.E. 2003. Inactivation of *Escherichia coli* O157:H7, *Listeria monocytogenes*, and *Salmonella* in cranberry, lemon, and lime juice concentrates. *J. Food Prot.* 66:1637–41.

Nyman, P.J., Diachenko, G.W., Perfetti, G.A., McNeal, T.P., Hiatt, M.H. and Morehouse, K.M. 2008. Survey results of benzene in soft drinks and other beverages by headspace gas chromatography/mass spectrometry. *J. Agric. Food Chem.* 56:571–76.

Ogiehor, I.S. and Ikenebomeh, M.J. 2004. Antimicrobial effects of sodium benzoate on the growth, survival and aflatoxin production potential of some species of *Aspergillus* in garri during storage. *Pak J. Nutr.* 3:300–3.

Oh, D.-H. and Marshall, D.L. 1994. Enhanced inhibition of *Listeria monocytogenes* by glycerol monolaurate with organic acids. *J. Food Sci.* 59:1258.

Oikawa, A., Tohda, H., Kanai, M., Miwa, M. and Sugmura, T. 1980. Inhibitors of poly (ADP Ribose) polymerase induce sister chromatid Exchanges. *Biochem. Biophys. Res. Commun.* 97:1311.

Okoli, E.C. and Ezenweke, L.O. 1990. Formulation and shelf life of a bottled pawpaw juice beverage. *J. Food Sci. Technol.* 25:706.

del Olmo, A., Calzada J. and Nunez, M. 2017. Benzoic acid and its derivatives as naturally occurring compounds in foods and as additives : uses, exposure, and controversy. *Crit. Rev. Food Sci. Nutr.* 57:3084–103.

Ostergaard, E. 1994. Evaluation of the antimicrobial effects of sodium benzoate and dichlorobenzyl alcohol against dental plaque microorganisms. An in vitro study. *Acta. Odontol. Scand.* 52:335.

Oyarzabal, O.A., Nagueira, M.C.L. and Gombas, D.E. 2003. Survival of *Escherichia coli* O157:H7, *Listeria monocytogenes*, and *Salmonella* in juice concentrates. *J. Food Prot.* 66:1595–98.

Parish, M.E. and Carrol, D.E. 1988. Effects of combined antimicrobial agents on fermentation initiation by *Saccharomyces cerevisiae* in a model broth system. *J. Food Sci.* 53:240.

Parke, D.V. and Levis, D.F.V. 1992. Safety aspects of food preservatives. *Food Addit. Contamin.* 9:561.

Pederson, C.S., Albury, N.N. and Christensen, M.D. 1961. The growth of yeasts in grape juice stored at low temperature. IV. Fungistatic effect of organic acids. *Appl. Microbiol.* 9:162.

Pelayo, S.U. 1979. The in vitro efficacy of salicylic acid, benzoic acid, boric acid and diiodate thymol solutions on several species of dermatophytes. *Rev. Cubana Med. Trop.* 31:121.

Petrus, M., Bonaz, S. and Causse, E. 1996. Asthma induced by benzoate-containing foods and antiallergic drugs. *Arch. de Pediatrie*, 3:984.

Piper, P., Mahe, Y., Thompson, S., Pandjaitan, R., Holyoak, C., Egner, R., Muhbauer, M, Coote, M. and Kuchler, K. 1998. The Pdr12 ATP-binding cassette ABC is required for the development of weak acid resistance in *Saccharomyces cerevisiae*. *EMBO J.* 17:4257.

Piper, P.W. 1999. Yeast superoxide dismutase mutants reveal a pro-oxidant action of weak organic acid food preservatives. *Free Radic. Biol. Medic.* 27:1219.

Piper, P.W. 2018. Potential safety issues surrounding the use of benzoate preservatives. *Bev.* 4:33–39.

Pitt, J.L. and Hocking, A.D. 2009. *Fungi and Food Spoilage*. 4th ed. Gaithersburg: Aspen Publishers.

Pollard, J.A. 1990. Legislative aspects. In *Food Preservatives*, Russell, N.J. and Gould, G.W., Eds., New York: Van Nostrand Reinhold. p. 235.

Ponce de Leon, S., Kimura, T., Inoue N. and Shinano, H. 1994. Sardine storage at 15 deg. C in brine containing sodium benzoate. *Bul. Faculty Fisheries, Hokkaido Univ.* 45:19.

Poucke, C.V., Detavernier, C., Van Bocxlaer, J.F., Vermelen R. and Peteghem, C.V. 2008. Monitoring the benzene contents in soft drinks using headspace gas chromatography-mass spectrometry: A survey of the situation on the Belgian market. *J. Agric. Food Chem.* 56:4504–10.

Praphailong, W. and Fleet, G.H. 1997. The effect of pH, sodium chloride, sucrose, sorbate, and benzoate on the growth of food spoilage yeasts. *Food Microbiol.* 14:459.

Pylypiw, H.M. and Grether, M.T. 2000. Rapid high-performance liquid chromatography method for the analysis of sodium benzoate and potassium sorbate in foods. *J. Chromato. A* 883:299–304.

Qi, P., Hong, H., Liang X. and Liu, D. 2009. Assessment of benzoic acid levels in milk in China. *Food Control* 20:414–18.

Quay, S. and Massay, V. 1977. Effect of pH on the interaction of benzoate and D-amino acid oxidase. *Biochemistry* 16:3348.

Rahman, M.S. 2007. *Handbook of Food Preservation*. 2nd ed. Boca Raton, FL: CRC Press.

Rahn, O. and Conn, J.C. 1944. Effect of increase in acidity on antiseptic efficiency. *Ind. Eng. Chem.* 36:185.

Rajashekhara, E.R., Suresh E.R. and Ethiraj, S. 1998. Thermal death rate of ascospores of *Neosartorya fischeri* ATCC 200957 in the presence of organic acids and preservatives in fruit juices. *J. Food Prot.* 61:1358.

Ramos-Nino, M.E., Clifford M.N. and Adams, M.R. 1996. Quantitative structure activity relationship for the effect of benzoic acid, cinnamic acid and benzaldehyde on *Listeria monocytogenes*. *J. Appl. Bacteriol.* 80:303.

Rather, L.J., Knapp, B., Haehnel, W. and Fuch, G. 2010. Coenzyme A-dependent aerobic metabolism of benzoate via epoxide formation. *J. Biol. Chem.* 285:20615–24.

Rehm, H.J. 1960. A study on the antimicrobial activity of mixtures of agents used for the preservation of liquids. *Ann. Inst. Pasteur* 11:217.

Ristovska, G., Dimitrovska, M. and Najdenkoska, A. 2012. Safety issues associated with nonalcoholic beverages. *Food Safety* 18:58–61.

Rojo, M.C., Arroyo Lopez, F.N., Lerena, M.C., Mercado, L., Torres, A. and Combina, M. 2015. Evaluation of different chemical preservatives to control *Zygosaccharomyces rouxii* growth in high sugar culture media. *Food Control* 50:349–55.

Roland, J.O. and Beuchat, L.R. 1984. Biomass and patulin production by *Byssochlamys nivea* in apple juice as affected by sorbate, benzoate, SO2, and temperature. *J. Food Sci.* 49:202.

Rushing, N.B. and Senn, B.J. 1963. The effect of benzoic, sorbic, and dehydroacetic acids on the growth of citrus products spoilage organisms. *Proc. Fla. State Hort. Soc.* 76:271.

Russell, A.D. 1991. Mechanisms of bacterial resistance to non-antibiotics: food additives and food pharmaceutical preservatives. *J. Appl. Bacteriol.* 71:191.

Russell, A.D. 2004. Bacterial adaptation and resistance to antiseptics, disinfectants and preservatives is not a new phenomenon. *J. Hosp. Infect.* 57:97–104.

Rusul, G. and Marth, E.H. 1988. Food additives and plant components control growth and aflatoxin production by toxigenic aspergilli: A review. *Mycopathologia* 101:13.

Saad, B., Baari, M.F., Saleh, M.I., Ahmad K. and Khairuddin, M. 2005. Simultaneous determination of preservatives (benzoic acid, sorbic acid, methyl paraben and propyl paraben) in foodstuffs using high-performance liquid chromatography. *J. Chromato. A* 1073:393–97.

Sagoo, S.K., Board, R. and Roller, S. 2002. Chitosan potentiates the antimicrobial action of sodium benzoate on spoilage yeasts. *Lett. Appl. Microbiol.* 34:168–72.

Salmond, C.V., Kroll, R.G. and Booth, I.R. 1984. The effect of food preservatives on pH homoeostasis in *Escherichia coli*. *J. Gen. Microbiol.* 130:2845.

Samelis, J., Sofos, J.N., Kain, M.L., Scanga, J.A., Belk, K.E. and Smith, G.C. 2001. Organic acids and their salts as dipping solutions to control *Listeria monocytogenes* inoculated following processing of sliced pork bologna stored at 4°C in vacuum packages. *J. Food Prot.* 11:1772.

Santiesteban-Lopez, N.A., Rosales, M., Palou E. and Lopez-Malo, A.L. 2009. Growth response of *Escherichia coli* ATCC 35218 adapted to several concentrations of sodium benzoate and potassium sorbate. *J. Food Prot.* 72:2301–7.

Sauer, F. 1977. Control of yeasts and molds with preservatives. *Food Technol.* 31:66.

Sax, N.I. and Lewis, Sr., R.J. 1996. *Dangerous Properties of Industrial Materials*. 8th ed. Vol. 1. New York: Van Nostrand Reinhold.

Sax, N. I., and Lewis R. J., Sr. 1989. Dangerous Properties of Industrial Materials, 7th ed., Vol. 2, Van Nostrand Reinhold, New York.

Schaubschlager, W.W., Becker, W.M., Schade, U., Zabel P. and Schlaak, M. 1991. Release of mediators from human gastric mucosa and blood in adverse reactions to benzoate. *Intl. Arch. Allergy Appl. Immunol.* 96:97.

Seman, D.L., Quickert, S.C., Borger A.C. and Meyer, J.D. 2008. Inhibition of *Listeria monocytogenes* growth in cured ready-to-eat meat products by use of sodium benzoate and sodium diacetate. *J. Food Prot.* 71:1386–92.

Seng, J.-M., Saindrenan, P. and Bompeix, G. 1985. Induction of *Nectria galligena* mutants resistant to benzoic acid and study of their aggressiveness towards immature apples. *J. Gen. Microbiol.* 131:1863.

Shahmohammadi, M., Javadi, M. and Nassiri-Asl, M. 2016. An overview on the effects of sodium benzoate as a preservative in food products. *Biotech. Health Sci.* 3:1–5.

Shailubhai, K., Rao, N.N. and Modi, V.V. 1982. Degradation of benzoate and salicylate by *Aspergillus niger. Indian J. Exp. Biol.* 20:166.

Sharma, S. 2015. Food preservatives and their harmful effects. *Intl. J. Sci. Res. Pub.* 5:1–2.

Sheu, C.W., Salomon, D., Simmons, J.L., Sreevalsan, T. and Freese, E. 1975. Inhibitory effects of lipophilic acids and related compounds on bacteria and mammalian cells. *Antimicrob. Agents Chemother.* 7:349.

Sieber, R., Butikofer, U., Baumann, E. and Bosset, J.O. 1990. Occurrence of benzoic acid in fermented milk products and cheese. *Mitteil. Geb. Lebensm. Hyg.* 81:484.

Sieber, R., Butikofer, U. and Bosset, J.O. 1995. Benzoic acid as a natural compound in cultured dairy products and cheese. *Intl. Dairy J.* 5:227.

Sieber, R., Butikofer, U., Bosset, J.O. and Ruegg, M. 1989. Benzoic acid as a natural compound of foods-review. *Mitteil. Geb. Lebensm Hyg.* 80:345.

Simpkins, S. and Alexander, M. 1984. Models for mineralization kinetics with the variables of substrate concentration and population density. *Appl. Environ. Microbiol.* 47:1299.

Sioufi, A. and Pommier, F. 1980. Gas chromatographic determination of low concentrations of benzoic acid in human plasma and urine. *J. Chromatogr. Biomed. Appl.* 181:161.

Skandamis, P.N. and Nychas, G.J.E. 2000. Development and evaluation of a model predicting the survival of *Escherichia coli* O157:H7 NCTC 12900 in homemade eggplant salad at various temperatures, pHs and oregano essential oil concentrations. *Appl. Environ. Microbiol* 66:1646–53.

Sofos, J.N. 1989. *Sorbate Food Preservatives.* Boca Raton, FL: CRC Press.

Splittstoesser, D.F., Mclellan, M.R. and Churey, J.J. 1996. Heat resistance of *Escherichia coli* O157:H7 in apple juice. *J. Food Prot.* 59:226.

Stanojevic, D., Comic, L., Stefanovic O. and Solujic-Sukdolak, S. 2009. Antimicrobial effects of sodium benzoate, sodium nitrite and potassium sorbate and their synergistic action in vitro. *Bulg. J. Agric. Sci.* 15:307–11.

Steels, H., James, S.A., Roberts, I.N. and Stratford, M. 1999. *Zygosaccharomyces lentus*: a significant new osmophilic, preservative-resistant spoilage yeast, capable of growth at low temperature. *J. Appl. Microbiol.* 87:520.

Steenstrup, L.L. and Floros, J.D. 2002. Statistical modeling of D- and z- value of *E. coli* O157:H7 and pH in apple cider containing preservatives. *J. Food Sci.* 67:793–96.

Stijve, T. and Hischenguber, C. 1984. High performance liquid chromatographic determination of low levels of benzoic acid and sorbic acid in yogurts. *Dtsch. Lebensm. Runschau.* 80:81.

Stratford, M. and Lambert, R.J.W. 1999. Weak-acid preservatives: Mechanisms of adaptation and resistance by yeasts. *Food Australia* 51:26.

Subils, T., Aquili, V., Ebner, G. and Balague, C. 2012. Effect of preservatives on shiga toxigenic phages and shiga toxin of *Escherichia coli* O157:H7. *J. Food Prot.* 75:959–65.

Sun, Y and Oliver, J.D. 1995. Antimicrobial action of some GRAS compounds against *Vibrio vulnificus. Food Additives Contamin.* 11:549.

Sung, N.J., Hwang, O.J. and Lee, E.H. 1988. N-Nitrosamines in Korean ordinary soy sauce. *J. Korean Soc. Food Nutr.* 17:125.

Teuber, M. 1995. The influence of fermentation on the nutritional quality of dairy products. Facts and fiction. *World Ingred.* Jan./Feb:43.

Thayer, R.R. and Wheelis, M.L. 1976. Characteristics of a benzoate permease mutant of *Pseudomonas putida. Arch. Microbiol.* 110:37.

Thayer, R.R. and Wheelis, M.L. 1982. Active transport of benzoate in *Pseudomonas putida. J. Gen. Microbiol.* 28:1749.

Thomas, D.S. and Davenport, R.R. 1985. *Zygosaccharomyces baillii*–a profile of characteristics and spoilage activities. *Food Microbiol.* 2:157.

Thomas, L.V., Wimpenny, J.W.T. and Davis, G. 1993. Effect of three preservatives on the growth of *Bacillus cereus*, Vero cytotoxigenic *Escherichia coli* and *Staphylococcus aureus*, on plates with gradients of pH and sodium chloride concentration. *Intl. J. Food Microbiol.* 17:289.

Thompson, D.P. 1997. Effect of phenolic compounds on mycelial growth of *Fusarium* and *Penicillium* species. *J. Food Prot.* 60:1262.

Tompkin, R. B., Scott, V.N., Bernard, W.H., Sveum, W.H. and Gombas, K.S. 1999. Guidelines to prevent post-processing contamination from *Listeria monocytogenes. J. Food Environ. Sanit.* 19:551.

Turner, A., Chen, S.-N., Joike, M.K., Pendland, S.L., Pauli, G.F. and Farnsworth, N.R. 2005. Inhibition of uropathogenic *Escherichia coli* by cranberry juice: A new antiadherence assay. *J. Ag. Food Chem.* 53:8940–47.

Ubodi Eiroa, M.N., Jardim, D.P. and Rossi, R.A. 1995. Microbial stability of tomato concentrate preserved by hurdle technic and stored at room temperature. *Ciencia e Technologia de Alimentos* 15:128.

Ullah, A., Orij, R., Brul, S. and Smits, G.J. 2012. Quantitative analysis of the modes of growth inhibition by weak organic acids in *Saccharomyces cerevisiae. Appl. Environ. Microbiol* 78:8377–87.

Uraih, N., Cassity, T.R. and Chipley, J.R. 1977. Partial characterization of the mode of action of benzoic acid on aflatoxin biosynthesis. *Can. J Microbiol.* 23:1580.

Uraih, N. and Chipley, J.R. 1976. Effects of various acids and salts on growth and aflatoxin production by *Aspergillus flavus. Microbios* 17:51.

Uraih, N. and Offonry, S. 1981. Inhibition of aflatoxin production in groundnut with benzoic acid derivatives and possible toxic effects of their aromatic residues. *Microbios* 31:93.

Usaga, J., Manns, D.C., Moraru, C.I., Worobo R.W. and Padilla-Zakour, O.I. 2017. Ascorbic acid and selected preservatives influence effectiveness of UV treatment of apple juice. *LWT-Food Sci. Technol.* 75:9–16.

Valderrama, J.A., Durante-Rodriguez, G., Blazquez, B., Luis Garcia, J., Carmona, M. and Diaz, E. 2012. Bacterial degradation of benzoate. Cross-regulation between aerobic and anaerobic pathways. *J. Biol. Chem.* 287:10494–508.

Vartiainen, J., Skytta, E., Enqvist, J. and Ahvenainen, R. 2003. Properties of antimicrobial plastics containing traditional food preservatives. *Pack. Technol. Sci.* 16:223–29.

Verduyn, C., Postma, E., Scheffers, W.A. and Van Dijken, J.P. 1992. Effect of benzoic acid on metabolic fluxes in yeasts: a continuous-culture study on the regulation of respiration and alcoholic fermentation. *Yeast* 8:501.

Vermeulen, A., Smigic, N. Rajkovic, A., Gysemans, K., Bernaerts, K., Geeraerd, A., Van Impe, J., Debevere, J. and Devlieghere, F. 2007. Performance of a growth - no growth model for *Listeria monocytogenes* developed for mayonnaise-based salads : influence of strain variability, food matrix, inoculation level, and presence of sorbic and benzoic acid. *J. Food Prot.* 70:2118–26.

Vogel, R. 1992. Nutritional and practical significance of food additives. *Food Market. Technol.* 6:6.

Voigt, M.N., Eitenmller, R.R. and Ware, G.O. 1979. Vitamin Analysis by microbial and protozoan organisms: response to food preservatives and neutralization salts. *J. Food Sci.* 44:723.

Voth, C., Zhu, C., Wey, D., Vingerhoeds, P., Borucki, S. and Kiarie, E. 2018. Evaluation of benzoic acid as an alternative to antibiotic growth promoter in weaned pigs. *J. Animal Sci.* 96(Issue Suppl. 2):111–12.

Wang, C.Y. and Adams, D.O. 1980. Ethylene production by chilled cucumbers (*Cucumis sativus*). *Plant Physiol.* 66:841.

Wang, C.Y. and Baker, J.E. 1979. Effect of two free radical scavengers and intermittent warming on chilling injury and polar lipid composition of cucumber and sweet pepper fruits. *Plant Cell Physiol.* 20:243.

Wang, J., Kane, S.A., Liu, J., Smyth, M.R. and Rogers, K. 1996. Mushroom tissue-based biosensor for inhibitor monitoring. *Food Technol. Biotechnol.* 34:51.

Wang, L., Zhang, X., Wang, Y. and Wang, W. 2006. Simultaneous determination of preservatives in soft drinks, yogurts and sauces by a novel solid-phase extraction element and thermal desorption—Gas chromatography. *Anal. Chim. Acta.* 577:62–67.

Warth, A.D. 1977. Mechanism of resistance of *Saccharomyes bailii* to benzoic, sorbic and other weak acids used as food preservatives. *J. Appl. Bacteriol.* 43:215–230.

Warth, A.D. 1988. Effect of benzoic acid on growth yield of yeasts differing in their resistance to preservatives. *Appl. Environ. Microbiol.* 54:2091–2095.

Warth, A.D. 1991a. Mechanism of action of benzoic acid on *Zygosaccharomyces bailli*: effects of glycolytic metabolite levels, energy reduction, and intracellular pH. *Appl. Environ. Microbiol.* 57:3410.

Warth, A.D. 1991b. Effect of benzoic acid on glycolytic metabolite levels and intracellular pH in *Saccharomyces cerevisiae. Appl. Environ. Microbiol.* 57:3415.

Warth, A.D. 1989a. Transport of benzoic and propanoic acids by *Zygosaccharomyces bailli. J. Gen. Microbiol.* 135:1383.

Warth, A.D. 1989b. Relationships between the resistance of yeasts to acetic, propanoic and benzoic acids and to methyl paraben and pH. *Intl. J. Food Microbiol.* 8:343.

Warth, A.D. 1989c. Relationships among cell size, membrane permeability, and preservative resistance in yeast species. *Appl. Environ. Microbiol.* 55:2995.

Watanabe, K. and Takimoto, A. 1979. Flower-inducing effects of benzoic acid and some related compounds in Lemna Paucicostata. *Plant Cell Physiol.* 20:847.

Webb, J.L. 1966. *Enzyme and Metabolic Inhibitors.* Vol. 2. New York: Academic press. p. 245.

White, A., Handler, P. and Smith, E.L. 1964. *Principles of Biochemistry.* 3rd ed. New York: McGraw-Hill.

Wing, C.E. and Restaino, L. 1995. Antimicrobial effectiveness of potassium sorbate and sodium benzoate against *Zygosaccharomyces bailii* in a salsa mayonnaise. *J. Food Prot.* 58:1257–59.

Woodward, B.B., Heffelfinger, G.P. and Ruggles, D.I. 1979. High pressure liquid chromatographic determination of sodium saccharin, sodium benzoate and caffeine in soda beverages: collaborative study. *J. Assoc. Off. Anal. Chem.* 62:1011.

Yagi, K., Lang, R. and Douzou, P. 1980. Spectroscopic demonstration of an initial stage of the complex of D-amino acid oxidase and its substrate D-a-aminobutyric acid. *Biochem. Biophys. Res. Commun.* 97:370.

Yousef, A.E., El-Shenawy, M.A. and Marth, E.H. 1989. Inactivation and injury *of Listeria monocytogenes* in a minimal medium as affected by benzoic acid and incubation temperature. *J. Food Sci.* 54:560.

Zani, F., Minutello, A., Maggi, L., Santi, P. and Mazza, P. 1997. Evaluation of preservative effectiveness in pharmaceutical products; the use of a wild strain of *Pseudomonas cepacia. J. Appl. Microbiol.* 83:322.

Zengin, N., Yuzbasioglu, D., Unal, F., Yilmaz, S. and Aksoy, H. 2011. The evaluation of the genotoxicity of two food preservatives: Sodium benzoate and potassium benzoate. *Food Chem. Toxicol.* 49:763–69.

Zeyer, J. and Kearney, P.C. 1984. Degradation of o-nitrophenol and m-nitrophenol by *Pseudomonas putida. J. Agric. Food Chem.* 32:238.

Zhai, H., Ren, W., Wang, S., Wu, J., Guggenbuhl, P. and Kluenter, A.-M. 2017. Growth performance of nursery and grower-finisher pigs fed diets supplemented with benzoic acid. *Animal Nutr.* 3:232–35.

Zhao, T., Doyle, M.P. and Besser, R. 1993. Fate of enterohemorrhagic Escherichia coli O157:H7 in apple cider with and without preservatives. *Appl. Environ. Microbiol.* 59:2526.

Zhu, J., Li, J. and Chen, J. 2012. Survival of *3* in home-style mayonnaise and acid solutions as affected by acidulant type and preservatives. *J. Food Prot.* 75:465–71.

Zhu, M.J., Mendonca, A., Min, B., Lee, E.J., Nam, K.C., Park, K., Du, M., Ismail, H.A. and Ahn, D.U. 2004. Effects of electron beam, irradiation and antimicrobials on the volatiles, color, and texture of ready-to-eat turkey breast roll. *J. Food Sci.* 69:C382–87.

4

Sorbic Acid and Sorbates

Jarret Stopforth and Travis Kudron

CONTENTS

4.1 Introduction

The antimicrobial and preservative properties of sorbic acid were first discovered, both in Germany and in the United States by E. Miller and C. M. Gooding, in the late 1930s and early 1940s, respectively (Lück, 1976, 1980; Sofos and Busta, 1981). The compound was first isolated, as naturally occurring from the oil of unripened rowanberries (sorbapple or mountain ash tree), by the German chemist A. W. von Hoffmann in 1859 in London (Sofos, 1989). The first U.S. patent on sorbic acid was awarded in 1945 to C. M. Gooding and Best Foods, Inc. for their discovery that the compound was a good fungistatic agent for application to foods and food wrappers (Gooding, 1945).

Sorbic acid became available commercially during the late 1940s and early 1950s when testing as a preservative agent increased. Studies of that period also examined the biological properties and established guidelines for the safety of the compound. These developments resulted in the extensive use of sorbic acid and its salts as preservatives of foods and other materials throughout the world (Lück, 1976, 1980; Sofos et al., 1979a; Sofos and Busta, 1981; Sofos, 1989). Research efforts during the late 1950s and in the 1960s dealt mostly with the mechanism of sorbic acid activity against microbial growth and the potential application of the compound in additional food products (Sofos and Busta, 1981).

During the 1970s extensive research was performed on the potential for using sorbic acid and its salts as antibotulinal agents in meat products, especially in combination with reduced nitrite levels for a reduction in nitrosamine formation (Robach and Sofos, 1982; Sofos et al., 1979a; Sofos, 1981), while the health effects of sorbate were reexamined (Robach and Sofos, 1982; Sofos, 1981, 1989).

Currently, sorbic acid and its more water-soluble salts, especially potassium sorbate, are collectively known as sorbates and are used widely throughout the world as preservatives for various foods, animal feeds, pharmaceuticals and cosmetics, and in other industrial applications (Sofos, 1989). The extensive use of sorbates as preservatives is based on their ability to inhibit or delay growth of numerous microorganisms, including yeasts, molds, and bacteria. There is variation, however, in inhibition of microorganisms by sorbates, depending on differences in microbial types, species and strains, substrate properties, and environmental factors. Under certain conditions, some microbial strains are resistant to inhibition by sorbate or even metabolize the compound. In general, however, sorbates are considered effective food preservatives when used under sanitary conditions and in products processed using good manufacturing practices.

The mechanisms of antimicrobial activity of sorbates are not fully defined (Sofos and Busta, 1981; Sofos et al., 1986; Sofos, 1989). Inhibition of bacterial spore germination (Sofos et al., 1979d) by sorbate is believed to occur at the connecting reactions of the germination process, probably through action on spore membranes or protease enzymes involved in germination (Blocher and Busta, 1985; Sofos et al., 1986). Inhibition of microbial metabolic function is probably due to morphological alterations in the structure of the cells, changes in the genetic material, alterations in cell membranes, and inhibition of enzymes or transport functions (Sofos, 1989).

4.2 Chemistry

Sorbic acid is a straight-chain, trans-trans unsaturated fatty acid (2,4-hexadienoic acid; CH_3–CH=CH–CH=CH–COOH), with a molecular weight of 112.13 (Table 4.1). The carboxyl group of sorbic acid is highly reactive and results in the formation of various salts and esters. The conjugated double bonds of

TABLE 4.1

Physical and Chemical Properties of Sorbic Acid and Potassium Sorbate

Property	Sorbic Acid	Potassium Sorbate
Chemical name	2,4-Hexadienoic acid	2,4-Hexadienoic acid potassium salt
Molecular formula	CH_3–CH=CH–CH=CH–COOH	CH_3–CH=CH–CH=CH=COOK
Molecular weight	112.13	150.22
Flash point, °C (COC ASTM D-92)	126	—
Ionization constant at 25°C	1.73×10^{-5}	—
Density at 20°C g/ml	—	1.36
Melting range, °C	132-137	Decomposes above 270°C
Heat of Combustion at 25°C, Btu/ lb	11,927	—
Alkalinity/acidity	—	1.1 ml 0.1 *N* NaOH to 0.8 ml 0.1 *N* HCl per 1.1 g
Vapor pressure, mm Hg, at		
20°C	<0.001	—
120°C	10	—
140°C	43	—
Boiling point, °C		
760 mm Hg	Decomposes	—
50 mm Hg	143	—
10 mm Hg	119	—
Purity, %	>98	>98
Water content, maximum %	0.5	1.0
Heavy metal content, maximum ppm	10	10
Arsenic content, maximum %	3	3
Ash, maximum %	0.2	—

Source: From Sofos (1989).

sorbic acid are also reactive and can be influential in its antimicrobial activity as well as on the quality and safety of food products (Sofos, 1989; Wedzicha and Brook, 1989). The volatility of the compound in steam is useful in its isolation for quantitative detection in foods or other materials.

Sorbic acid forms colorless flakes or needles when crystallized, it is available as a white free-flowing powder or as granules, and it has a weak but characteristic acrid odor and acid taste. Commercially available salts include calcium, sodium, and potassium sorbates. Potassium sorbate may be manufactured as a powder or granules and, on an equivalent weight basis with the acid, has an antimicrobial potency of 74%. Sodium sorbate is a white fluffy powder and is commercially available as an aqueous solution that is sensitive to oxidation and stable for only a few weeks (Lück, 1980). The calcium salt is formed as a white odorless and tasteless powder.

The solubility of sorbic acid (Table 4.2) in water at room temperature is only 0.15 g per 100 ml, but it increases with temperature or the pH of the solution, or both. The solubility of sorbic acid is also higher in alcohol, especially ethanol, and in glacial acetic acid. The salts (e.g., potassium sorbate) of sorbic acid may find more frequent applications in foods because of their greater solubility in water. Calcium sorbate has a water solubility of 1.2% and is insoluble in fats, which makes it valuable as a delayed-release form of sorbic acid (Sofos, 1989, 1992). The sodium salt has a water solubility of 32% (wt/vol). The molecular weight of potassium sorbate is 150.22, and it constitutes the most soluble form of sorbate (Table 4.2). In food systems the solubility of the compound is estimated as above 50%. Good solubility (58.2 % w/v at 20°C), stability, and ease of manufacture make potassium sorbate the most widely used form in food systems (Sofos, 1989, 1992).

The amount of sorbic acid in the aqueous phase is lowered in food systems of higher lipid content because the solubility of sorbic acid in fat is approximately three times that in water (Gooding et al., 1955). The partition quotient is increased from 3.0 to 7.0 with increased levels (50%) of soluble food components, such as sugars and salts (Sofos and Busta, 1981). In general, the partitioning of sorbic acid between the water and fat phases of foods depends on the pH of the food, the type and amount of fat, and other ingredients present (Sofos, 1989). Aqueous solutions of sorbates are unstable and degrade through oxidation, while in the pure dry form they are stable. Oxidation of sorbic acid yields various carbonyl compounds, such as crotonaldehyde, malonaldehyde, acetaldehyde, acrolein, formic acid, malonic acid, and ß-carboxylactolein (Arya and Thakur, 1988; Sofos, 1989). The rate of oxidation in aqueous solutions is increased at lower pH values by the presence of light and acids or increased temperatures. Oxidation and losses of sorbic acid are inhibited by antioxidants, appropriate packaging materials, and anaerobic conditions (Sofos, 1992). Ethylenediamine tetraacetic acid (EDTA) enhances sorbate degradation probably as a result of iron scavenging from the packaging material (glass or polypropylene) through EDTA complexation. It is proposed that EDTA–Fe^{2+} complexes catalyze sorbic acid autoxidation, increasing the production of carbonyls which take part in nonenzymatic browning. Thus, the higher iron content in the EDTA-containing systems might be responsible for the higher rates of sorbic acid degradation as well as for the increase in nonenzymatic browning observed (Campos et al., 1996).

Losses of sorbic acid during storage of foods depend on sorbate levels, the nature and pH of the food, amino acids, metal ions, light, antioxidants, types of food materials present, moisture content, processing conditions, other additives present, packaging material, and storage temperature and time (Bolin et al., 1980; Gerschenson et al., 1986a, b, 1987; Sofos, 1989; 1992; 1994; Vidyasagar and Arya, 1983, 1984). Sodium chloride at 3.5 and 8.0% enhanced sorbic acid destruction, but at 13.0% it had a protective effect (Guerrero et al., 1990). In general, oxidative degradation of sorbic acid in foods is less extensive than in aqueous solutions, and it is influenced by the nature of the food and its components, and by processing, handling, and storage conditions (Sofos, 1994).

Uptake and diffusion of sorbate into foods should be influenced by properties of the substrate, such as composition, physical state, structure, moisture, and water activity (Giannakopoulos and Guilbert, 1986a, b; Vojdani and Torres, 1990). In general, it is important to be able to predict or to control migration of sorbic acid during processing and storage of foods because it can influence product preservation (Sofos, 1989). When wheat flour doughs are treated with sorbic acid and heated, a significant amount of sorbic acid is not recovered by extraction with methanol. This suggests that sorbic acid is indeed reacting with thiol groups in the flour protein, and the acid-labile nature of this binding is due to the instability of the thiol-sorbic acid monoadducts under acid conditions (Khandelwal et al., 1992). It is suggested that

TABLE 4.2
Solubility (%) of Sorbic Acid and Potassium Sorbate

Solvent	Sorbic Acid	Potassium Sorbate
Water, 20°C (pH 3.1)	0.15	58.20
Water, 20°C (pH 4.4)	0.22	—
Water, 20°C (pH 5.9)	1.02	—
Water, 50°C	0.55	61.00
Water, 100°C	4.00	64.00
Corn oil, 20°C	0.08	0.01
Corn oil, 50°C	2.00	0.03
Soybean oil, 20°C	0.52	—
Cottonseed oil, 20°C	1.00	0.01
Sucrose, 10% solution	0.15	58.00
Sucrose, 40% solution	0.10	45.00
Sucrose, 60% solution	0.08	28.00
Sodium chloride, 5% solution	0.11	47.00
Sodium chloride, 10% solution	0.07	34.00
Sodium chloride, 15% solution	0.04	12.00–15.00
Acetic acid, glacial	11.50–12.30	—
Lactic acid, 85.5%	2.25	—
Citric acid, 50%	0.26	—
Phosphoric acid, 85%	0.12	—
Ethanol, 5%	0.16	57.40
Ethanol, 20%	0.29	54.60
Ethanol, 50%	—	45.30
Ethanol, 95%	12.60–14.50	6.50
Ethanol, 100%	12.90–14.80	2.00
Propylene glycol, 20°C	0.02	55.00
Propylene glycol, 50°C	0.05	48.00
Propylene glycol, 100°C	5.50	20.00
Pentane, 25°C	0.15	—
Pentane, 50°C	0.60	—
Pentane, 75°C	1.80	—
Benzene, 25°C	2.34	<0.01
Benzene, 50°C	8.14	—
Benzene, 75°C	24.00	—
Ethyl ether, 20°C	5.00–5.30	0.10
Glycerol, 20°C	0.31	0.20
Acetone, 20°C	9.20	0.10
Methanol, 20°C	12.90	16.00
Cyclohexane, 20°C	0.28	—
Carbon tetrachloride, 20°C	1.30	<0.01

Source: From Sofos (1989).

decomposition of the adducts requires the sulfur atom at position 5 of the substituted 3-hexanoic acid to be protonated and that a proportion of sorbic acid present in flour-based products may be reversibly bound to components of the food (Khandelwal et al., 1992). Jideani and Wedzicha (1994) showed that when sorbic acid is incorporated into millet dough a significant proportion (approximately 40%) of the additive becomes unavailable on extraction with methanol, and a further loss is apparent after the dough is cooked to give the traditional West African food, fura. The lipid and protein fractions which appear to bind sorbic acid or its degradation products were identified with the help of ^{14}C-sorbic acid. Fractionation of the lipids and proteins from fura showed the ^{14}C to be associated with the petroleum spirit crude fat extract (28 ± 5%), the water-soluble (6.2 ± 0.7%), the ethanol-soluble (4.5 ± 0.7%), the salt-soluble (3.5 ± 0.6%), and the acetic acid-soluble components (0.9 ± 0.3%). Dialysis of the ethanol-, salt- and acetic acid-soluble fractions suggested that ^{14}C activity tends to be associated with the low-molecular-weight components of these fractions. The results show that sorbic acid binds with the lipid components, the water-, salt-, and ethanol-soluble components of fura from pearl millet. A proportion of the binding is with the low-molecular-weight components of these fractions (Jideani and Wedzicha, 1995).

Even though the salts of sorbic acid have been commercially developed, broadly tested, and widely used as antimicrobial agents, other derivatives of sorbic acid have also been examined. These include esters, alcohols, aldehydes, amine salts, and amide derivatives. Several other compounds that have previously been evaluated include sorboyl palmitate, sorbamide, methyl sorbate, ethyl sorbate, sorbohydroxamic acid, and sorbic anhydride. A mixed anhydride of sorbic and palmitic acids called sorboyl palmitate has found some application in yeast-leavened bakery products. Sorbamide was 1000 times more inhibitory than sorbic acid against yeast alcohol dehydrogenase (Martoadiprawito and Whitaker, 1963); sorbohydroxamic acid, with a dissociation constant (pK_a value) of 8.8, was an effective mold inhibitor over a wide pH range (3.6–9.2) (Dudman, 1963; Troller and Olsen, 1967); and sorbic aldehydes were also effective antimicrobials (Troller and Olsen, 1967). Commercial application of such derivatives, however, is hindered by disadvantages, such as low solubility in water, strong off-flavors and odors, and potential health problems (Lück, 1980; Sofos, 1989; Sofos and Busta, 1981).

The development of new sorbic acid amide derivatives is important as resistance and adaptation to sorbic acid and its existing derivatives have been found in various microbes over time. Figure 4.1 presents several amide compounds that were recently evaluated against a variety of organisms (Wei et al., 2018). The compounds evaluated included: Etheyl N-[1-oxo-2, 4-hexadien-1-yl] glyciante (a1), propyl N-[1-oxo-2, 4-hexadien-1-yl] glyciante (a2), isopropyl N-[1-oxo-2, 4-hexadien-1-yl] glycinate (a3), methyl N-[1-oxo-2, 4-hexadien-1-yl]-L-phenylalaninate (a4), ethyl N-[1-oxo-2, 4hexadien-1-yl]-L-phenylalaninate (a5), propyl N-[1-oxo-2, 4-hexadien-1-yl]-L-phenylalaninate (a6), isopropyl N-[1-oxo-2, 4hexadien-1-yl]-L-phenylalaninate (a7).

New controlled-release, water-soluble formulations of sorbic acid in the form of epoxidized polymers of polyvinylpyrrolidone (PVP) containing covalently bonded sorbic acid (polymeric esters of sorbic acid) and complexes of PVP with hydrogen-bonded sorbic acid have been shown to be more effective fungicidal agents than sorbic acid polymeric esters (Charvalos et al., 2001; Tzatzarakis et al., 2000). In addition, these products have low toxicity and are effective in a wide range of concentrations and thus may be considered for food application or feed protection.

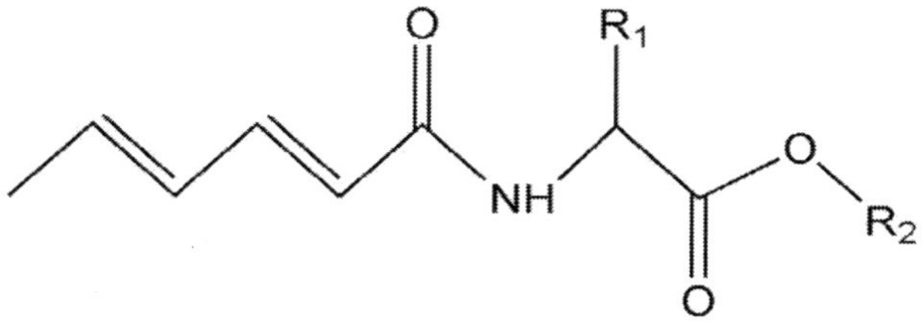

$\mathbf{a_1}$: $R_1=H$, $R_2=CH_2CH_3$;
$\mathbf{a_2}$: $R_1=H$, $R_2=CH_2CH_2CH_3$;
$\mathbf{a_3}$: $R_1=H$, $R_2=CH(CH_3)_2$;
$\mathbf{a_4}$: $R_1=CH_2C_6H_5$, $R_2=CH_3$;
$\mathbf{a_5}$: $R_1=CH_2C_6H_5$, $R_2=CH_2CH_3$;
$\mathbf{a_6}$: $R_1=CH_2C_6H_5$, $R_2=CH_2CH_2CH_3$;
$\mathbf{a_7}$: $R_1=CH_2C_6H_5$, $R_2=CH(CH_3)_2$

FIGURE 4.1 Structures of sorbic acid amides.

Commercial applications of sorbates include use of the free acid or its potassium and, to a lesser extent, calcium or sodium, salts, which are available as powders, granules, suspensions, or solutions. Since the compounds are sensitive to heat, moisture, and light, they are shipped and stored in fiber drums with a moisture barrier in the carton wall. These cartons should be kept closed in cool, dry, and dark storage areas.

Several procedures have been patented for the manufacture of sorbates (Sofos, 1989), but commercial production of sorbic acid has mostly involved the methods of oxidation of 2,4-hexadienal and the condensation reaction of ketene and crotonaldehyde. Numerous patents describing catalysts, reaction conditions, and purification procedures follow the second method, which is most widely used. Alkaline salts of sorbic acid are formed by its neutralization with alkali metal carbonates, hydroxides, and other suitable salts in the presence of necessary additives.

4.3 Antimicrobial Activity

4.3.1 Inhibition

Sorbic acid is active against yeasts and molds, as well as against many bacteria (Sofos, 2000). Extensive research during the 1950s demonstrated the impressive effectiveness of sorbates against yeasts and molds and resulted in the extensive use of the compounds as fungistatic agents in many foods. Effective antimicrobial concentrations of sorbates in most foods are in the range 0.02–0.30%.

The inhibitory action of sorbate against yeasts was first documented in the 1950s in fermented vegetable products. The effectiveness of sorbates against yeasts has been documented by numerous studies (Bracey, et al., 1998; El Halouat et al., 1998; Emard and Vaughn, 1952; Ferguson and Powrie, 1957; Geminder, 1959; Huang and Armstrong, 1970; Pederson et al., 1961; Piper et al., 1998). Yeasts inhibited by sorbates include species of the genera *Brettanomyces*, *Candida*, *Cryptococccus*, *Debaryomyces*, *Endomycopsis*, *Hansenula*, *Kloeckera*, *Pichia*, *Rhodotorula*, *Saccharomyces*, *Sporobolomyces*, *Torulaspora*, *Torulopsis*, and *Zygosaccharomyces* (Sofos, 1989). In addition to their effectiveness in fermented vegetables, sorbates inhibit yeasts in fruit juices, wines, cottage cheese, dried fruits, and meat and fish products. The use of sorbates for the inhibition of yeasts is especially important in low-pH and/or intermediate water activity (a_w) products, such as carbonated beverages, salad dressings, syrups, tomato products, jams, candy, jellies, and chocolate syrup (Liewen and Marth, 1985a; Restaino et al., 1982). Although *Zygosaccharomyces rouxii* is considered resistant to sorbate treatments, the use of a hurdle approach resulted in a synergistic effect (El Halouat and Debevere, 1996). Hurdles used in combination included a_w, CO_2-atmospheres, and sorbic acid. Under an 80% CO_2 atmosphere, the inhibitory amounts of sorbic acid were reduced by 40 to 50%. Under this atmosphere and at a_w values in the studied range (0.80–0.90), potassium sorbate used at 150 ppm resulted in final counts, after 21 days, lower than the inoculum level (10^3 CFU/g), while no growth was observed when the preservative concentration was increased to 220 ppm (El Halouat and Debevere, 1996).

Numerous studies have also documented the effectiveness of sorbates against molds (Aly, 1996; Baldock et al., 1979; Deuel et al., 1954a, b; Emard and Vaughn, 1952; Fan and Chen, 1999; Garza et al., 1993; Huang and Armstrong, 1970Kaul et al., 1979; Kivanç, 1992; Melnick and Luckmann, 1954a, b; Matamoros-León et al., 1999; Melnick et al., 1954a, b; Skirdal and Eklund, 1993; Smith and Rollin, 1954a, b). Mold species inhibited by sorbates belong to the genera *Alternaria*, *Ascochyta*, *Ascosphaera*, *Aspergillus*, *Botrytis*, *Cephalosporium*, *Chaetomium*, *Cladosporium*, *Colletotrichum*, *Cunninghamella*, *Curvularia*, *Fusarium*, *Geotrichum*, *Gliocladium*, *Helminthosporium*, *Heterosporium*, *Humicola*, *Monilia*, *Mucor*, *Penicillium*, *Phoma*, *Pepularia*, *Pestalotiopsis*, *Pullularia*, *Rhizoctonia*, *Rhizopus*, *Rosellinia*, *Sporotrichum*, *Trichoderma*, *Truncatella*, *Ulocladium*, and others. A major application of sorbates in food products is their use for inhibition of molds in cheeses (Chichester and Tanner, 1972). Sorbates also inhibit molds in butter, sausages, fruits and juices, cakes, grains, bread, and smoked fish (Liewen and Marth, 1985a; Sofos, 1989).

Sorbates inhibit the formation of mycotoxins by various molds in culture media and in foods (Bhattacharya and Majumdar, 1984; Bullerman, 1983, 1984, 1985; Gourama and Bullerman, 1988;

Lennox and McElory, 1984; Liewen and Marth, 1983, 1984; Marshall and Bullerman, 1986; Roland and Beuchat, 1984; Tong and Draughon, 1985; Tsai et al., 1984). However, neither sorbate nor propionate, applied at 10 mg/ml, were able to depress mycelial weight and aflatoxin production by *Aspergillus flavus* or *A. parasiticus* cultured in yeast extract-sucrose (YES) broth (Fan and Chen, 1999). In fact, under certain conditions, subinhibitory levels of sorbate may simulate the production of mycotoxins (Bullerman and Olivigni, 1974; Gareis et al., 1984; Liewen and Marth, 1985a, b; Monnet et al., 1988; Rusul and Marth, 1987; Sanchis et al., 1988). Stimulation of mycotoxin formation by low levels of sorbate depends on species and strains of molds, storage temperature, and other factors (Sofos, 1989).

Bacterial species inhibited by sorbate belong to the genera *Acetobacter, Achromobacter, Acinetobacter, Enterobacter, Aeromonas, Alcaligenes, Alteromonas, Arthrobacter, Bacillus, Campylobacter, Clostridium, Escherichia, Klebsiella, Lactobacillus, Micrococcus, Moraxella, Mycobacterium, Pediococcus, Proteus, Pseudomonas, Salmonella, Serratia, Staphylococcus, Vibrio, Yersinia*, and others (El-Shenawy and Marth, 1988; Koodie and Dhople, 2001; Kouassi and Shelef, 1995a, b; Sofos, 2000; Zhao et al., 1993). Depending on pH and concentration, sorbate inhibited or inactivated *Listeria monocytogenes* in a broth substrate (El-Shenawy and Marth, 1988) and in a cold-pack cheese food (Ryser and Marth, 1988). Potassium sorbate sensitized cells of *L. monocytogenes* and *Zygosaccharomyces bailii* to inactivation by high hydrostatic pressure (Mackey et al., 1995; Palou et al., 1997). In other studies (Kouassi and Shelef, 1995a, b), although sorbate did not affect growth of *L. monocytogenes* in broth, it suppressed cysteine activation of listeriolysin. It was concluded that combinations of sorbate with propionate or lactate, which inhibited growth, could extend shelf-life and increase safety (Kouassi and Shelef, 1995a, b; Sofos, 2000). At the relatively high level of 1%, potassium sorbate slightly decreased *L. monocytogenes* presence in two commercial cheese brines and thus could be used as an antilisterial agent in commercial brines, but the cost effectiveness of adding high levels (1%) of substrate is questionable, and their long-term stability in the high-salt and low-pH environment of brines is not well-documented (Larson et al., 1999). Recently studies have been done evaluating the effect of sorbate on the lag phase and growth rates of *Listeria monocytogenes* on meat surfaces and are shown in Table 4.3. Response surface models were developed to predict the effect of sorbate solutions on the surfaces of meat. These models have reasonable predictive capabilities with the predicted lag phase durations (LPD) and growth rates (GR) with R^2 values of 0.953 and 0.926 respectively (Hwang et al., 2015).

Staphylococcus aureus producing staphylococcal thermonuclease (TNase) retained its full activity and was not inhibited even after exposure to chilling and refrigeration temperatures when sorbic acid was applied at 0.04 to 0.5% (Kumar et al., 2000). A hydroxypropyl methylcellulose (HPMC) coating significantly reduced the number of viable *Salmonella* Montevideo cells on the surface of tomatoes; however, the addition of sorbic acid (0.2 to 0.4%) to HPMC did not substantially enhance bactericidal activity (Zhuang et al., 1996). The combination of 0.1% potassium sorbate and 0.1% sodium benzoate substantially decreased *E. coli* O157:H7 contamination and suppressed the growth of yeasts and molds which could have a protective effect on *E. coli* O157:H7, increasing its survival, and result in spoilage of the product (Zhao et al., 1993). The addition of 0.05% sorbic acid was found to inhibit growth of *E.*

TABLE 4.3

Observed Lag Phase (LPD) and Growth Rate (GR) from Validation Experiments Compared with Predicted Values from the LPD and GR Models

Temp (°C)	Sorbate Solution (% [wt/vol])	pH	LPD (day)		GR (log CFU/day)	
			Observed	Predicted	Observed	Predicted
4	2.75	4.5	37.3 ± 1.4	39.1	0.15 ± 0.08	0.16
4	2.0	4.5	17.3 ± 1.4	15.8	0.18 ± 0.08	0.2
6	3.0	4.5	28.3 ± 1.8	27.2	0.15 ± 0.1	0.16
6	2.75	4.0	32.3 ± 1.4	33.2	0.14 ± 0.1	0.12
8	3.0	4.0	37.3 ± 1.4	32.6	0.14 ± 0.1	0.14
8	2.0	4.0	14.3 ± 1.4	16.2	0.18 ± 0.09	0.20

Source: Hwang et al., 2015.

coli O157:H7 in apple cider (Koodie and Dhople, 2001). However, it has also been shown that potassium sorbate or sodium benzoate did not affect survival of *E. coli* O157:H7 during storage of apple cider (Miller and Kaspar, 1994). Certain bacterial strains are not inhibited by sorbate, however, and some may even metabolize the compound (Sofos, 1989). Overall, however, sorbates can inhibit Gram-positive and Gram-negative, catalase-positive and catalase-negative, aerobic and anaerobic, and mesophilic and psychrotrophic microorganisms, as well as spoilage and pathogenic bacteria. Inhibition of bacteria by sorbate appears to effect an extension of the lag phase, with a lesser influence on rate and extent of growth (Chung and Lee, 1981, 1982; Greer, 1982; Larocco and Martin, 1981; Tsay and Chou, 1989; Zamora and Zaritzky, 1987a, b).

The effect of sorbate on spore-forming bacteria may be exerted on spore germination, outgrowth, and/or vegetative cell division (Blocher and Busta, 1985; Lund et al., 1987; Sofos et al., 1979a-d, 1980a, 1986; Smoot and Pierson, 1981; Seward et al., 1982). Overall, however, sorbic acid is considered as a more effective inhibitor of yeasts and molds than bacteria (Skirdal and Eklund, 1993; Sofos, 1989).

4.3.2 Selective Action

Microbial inhibition by sorbate is variable and depends on species, strains, composition of substrate, pH, a_w, additives present, food processing treatments, temperature of storage, gas atmosphere, type of packaging, and concentration of sorbate. Variations and resistance to inhibition by sorbate may lead to failures in preservation and defective food products (Sofos, 1989).

Early studies indicated that sorbate could be used as a selective agent for catalase-negative lactic acid-producing bacteria and clostridia because it was highly inhibitory against catalase-positive organisms (Emard and Vaughn, 1952; Phillips and Mundt, 1950; Vaughn and Emard, 1951; York and Vaughn, 1954, 1955). In contrast, other studies have reported either no effect or inhibition of lactics and clostridia by sorbate (Costilow et al., 1955; Hamdan et al., 1971; Hansen and Appleman, 1955). Overall, however, the inhibitory action against lactics by sorbate is less than that against yeasts, which explains the usefulness of the compound as a preservative in vegetable fermentations. Another bacterium that appears to be more resistant to inhibition by sorbate than other spore formers is *Sporolactobacillus* (Botha and Holzapfel, 1987). Growth of *Gluconobacter oxydans* in the presence of sublethal concentrations of sorbic acid before determination of the minimal inhibitory concentration (MIC) resulted in a substantial increase in the MIC within 1 h (Eyles and Warth, 1989). In general, various species and strains of microorganisms exhibit different sensitivities to inhibition by sorbate. Varying sensitivities of bacterial species and strains to sorbate may lead to shifts in the microbial flora during storage of foods (Blocher et al., 1982; Blocher and Busta, 1983, 1985; Chung and Lee, 1981, 1982; Kondaiah et al., 1985; Lahellec et al., 1981; Lynch and Potter, 1982; McMeekin et al., 1984).

In addition to bacteria, under certain conditions some species and strains of yeasts and molds are resistant to inhibition by sorbate. Yeast strains resistant to sorbate belong to the genera *Zygosaccharomyces*, *Saccharomyces*, *Torulopsis*, *Brettanomyces*, *Candida*, and *Triganopsis* (Bills et al., 1982; Cole et al., 1987; Lenovich et al., 1988; Mihyar et al., 1997; Restaino et al., 1982, 1983; Splittstoesser et al., 1978; Warth, 1977, 1985). Of 100 yeast strains isolated from spoiled foods and beverages, most tolerated 150 ppm sorbic acid, 40% tolerated 500 ppm, and two strains of *Z. bailii* tolerated 800 ppm of sorbic acid (Neves et al., 1994). The resistance of yeasts to inhibition by sorbate depends on species and strains, sorbate concentration, pH, inoculum level, storage temperature, and previous exposure of the organism to low levels of sorbate (Sofos, 1989). When the yeast cells have been previously adapted to sorbate in media containing glucose or sucrose, subsequent exposure of the cells shows little effect of solute type on sorbate resistance (Lenovich et al., 1988). However, potassium sorbate suppressed the growth of *Z. bailii* in salsa mayonnaise more than sodium benzoate (Wind and Restaino, 1995). Potassium sorbate or sodium benzoate resulted in complete inhibition of *Z. rouxii* in high-moisture prunes (El Halouat et al., 1998).

The resistance of osmotolerant yeasts to inhibition by sorbate was acquired by preconditioning the yeast to sorbate (Bills et al., 1982). One proposed mechanism of resistance of osmotolerant yeasts has involved an inducible, energy-requiring system that transports the preservative out of the cell (Warth, 1977). Other proposed mechanisms of yeast resistance to sorbate at reduced a_w have been related to yeast cell shrinkage and decreases in membrane pore size, retarding the flow of sorbate into the cell (Restaino et al., 1983), or protection of enzyme systems from inhibition by sorbate through production of compatible solutes, such as polyols (Bills et al., 1982). Exposure of *S. cerevisiae* to sorbic acid caused strong induction of two plasma membrane proteins, one of which was identified as ATP-binding cassette transporter (Pdr12), which is essential for the adaptation of yeast to growth under weak acid stress and confers weak acid resistance by mediating energy-dependent extrusion of water soluble carboxylate ions (Holyoak et al., 1996; Piper et al., 1998). Exposure of *S. cerevisiae* to 0.9 mM sorbic acid at pH 4.5 resulted in the increased transcription and translation (upregulation) of genes encoding ten different proteins and the downregulation of three proteins (de Nobel et al., 2001). Functional categories of genes that are induced by sorbic acid stress included cell stress (particularly oxidative stress), transposon function, mating response, and energy generation. The induction of Hsp26, a heat shock protein of *S. cerevisiae*, which occurs during adaptation to sorbic acid, confers resistance to the inhibitory effects of sorbic acid (de Nobel et al., 2001).

4.3.3 Degradation

Animals and certain microorganisms can metabolize sorbate, under certain conditions, as a fatty acid through ω-oxidation. When sorbate levels are high, there is also evidence of some ω-oxidation (Deuel et al., 1954b; Lück, 1980). Like caproic and butyric acids, under normal conditions of alimentation, sorbate is completely oxidized to carbon dioxide and water. Because it is metabolized like other fatty acids, sorbate yields 6.6 kcal/g, of which 50% is biologically utilizable.

Some mold strains can grow and metabolize sorbate under certain conditions as detected in cheeses and fruit products (Melnick et al., 1954b; Sofos, 1989). Mold strains of the genus *Penicillium* isolated from cheese treated with sorbate were able to grow and metabolize high (0.18–1.20%) sorbate levels (Bullerman, 1977; Finol et al., 1982; Marth et al., 1966). It should be noted that 0.1% sorbate is usually sufficient to inhibit sensitive molds (Liewen and Marth, 1985a, b). It appears that selection may occur in sorbate-treated cheeses for certain molds tolerant to the compound (Schroeder and Bullerman, 1985). Products of sorbate metabolism by molds include 1,3-pentadiene, which is a volatile compound formed through a decarboxylation reaction and has a kerosene-like, plastic paint, or hydrocarbon-like odor (Liewen and Marth, 1985a-c; Marth et al., 1966). Other strains of molds that may degrade sorbate belong to the genera *Aspergillus*, *Fusarium*, *Mucor*, and *Geotrichum* (Sofos, 1989). It appears, however, that there is no apparent relationship between sorbate resistance and the toxigenic properties of molds (Tsai et al., 1988). In general, although many molds are sensitive to inhibition by sorbate, certain strains are resistant and can metabolize the compound, using it as a carbon source. Degradation of sorbate by molds depends on species and strains, prior exposure to subinhibitory levels of sorbate, level of inoculum, amount of sorbate present, and type of substrate (Sofos, 1989).

In addition to certain molds, some bacterial strains may also degrade sorbate under appropriate conditions. This metabolism is mostly associated with lactic acid-producing bacterial strains present as high inocula in sublethal concentrations of sorbate (Crowell and Guymon, 1975; Horwood et al., 1981; Liewen and Marth, 1985a). Degradation of sorbate by lactic acid bacteria has been associated with geranium-type off-odors in wines and fermented vegetables, caused by ethyl sorbate, 4-hexenoic acid, 1 -ethoxyhexa-2,4-diene, and 2-ethoxyhexa-3,5-diene (Edinger and Splittstoesser, 1986a, b; Sofos, 1989). In general, a geranium-like odor is usually associated with wines treated with sorbate and contaminated with high microbial loads.

4.3.4 Interactions

The antimicrobial activity of sorbates is influenced by compositional, processing, and environmental factors, such as concentration, other ingredients, pH, a_w, temperature, gas atmosphere, packaging, microbial flora, inoculum size, and other additives (Sofos and Busta, 1981; Sofos, 1989; Steels et al., 2000). These factors can act synergistically or be antagonistic and either enhance or negate the antimicrobial activity of sorbate (Sofos, 1989, 1992).

The minimum inhibitory concentration (MIC) of sorbic acid increases with the size of the inoculum; large inocula at high cell concentrations therefore require considerably higher concentrations of the inhibitor to prevent growth than do dilute cell suspensions. A study found a pronounced positive inoculum effect of *Z. bailii* resistance to sorbic acid activity, that was not an artifact caused by insufficient growth time, dehydration of cultures, substantial metabolism of sorbic acid, or binding of sorbic acid to dead cells, but an inoculum effect which may be caused by the diversity of the cells in inocula or initial contamination as regards to sorbic acid resistance (Steels et al., 2000). The study showed that the resistance of *Z. bailii* to sorbic acid was largely due to the presence of a small fraction of resistant cells and was not heritable or due to the existence of a mixed culture (Steels et al., 2000). Increasing the concentration of sorbic acid from 200 to 1000 mg/l in apple cider decreased the $D_{50°C}$ value of *E. coli* O157:H7 from 36 to 5.2 min, about a 7-fold increase in lethality (Splittstoesser et al., 1996). Sorbic acid (0.1%) reduced the $D_{50°C}$ of *E. coli* O157:H7 during storage of apple juice from 18 to 5.2 min, while benzoic acid reduced it to 0.64 min (Splittstoesser et al., 1996).

The antimicrobial action of sorbate is pH-dependent and increases as the pH of the substrate decreases, approaching its dissociation constant ($pK_a = 4.76$) (Cerruti et al., 1990; Cowles, 1941; Hoffman et al., 1944; Lück, 1976, 1980; Rahn and Conn, 1944; Sofos and Busta, 1981). Although activity is greater at low pH values, sorbates have the advantage of being effective at pH values as high as 6.5 (Bell et al., 1959, Lück, 1976; Sofos et al., 1979a; Sofos and Busta, 1980, 1981), while certain studies have indicated antimicrobial activity by sorbate at pH values as high as 7.0 (Chung and Lee, 1982; Raevuori, 1976; Statham and McMeekin, 1988). In contrast, the maximum pH for antimicrobial activity by most other common food preservatives is lower, for example, 5.0–5.5 and 4.0–4.5 for propionate and benzoate, respectively (Sofos and Busta, 1981).

The increased activity of sorbates at pH values above 5.5 is advantageous because it allows for their use in foods of higher pH values in which preservatives, such as parabens, might not be effective owing to their increased solubility in fat. In certain instances, sorbates can partially or totally replace benzoate even in foods of lower pH in order to avoid possible off-flavors caused by the higher benzoate levels needed for inhibition and to extend the range of microbial groups inhibited compared to benzoate or propionate used singly (Gooding et al., 1955; Melnick et al., 1954a; Sofos and Busta, 1981; 1982; 1993).

The increased antimicrobial activity of sorbate at lower pH values has been attributed to the increased amount of undissociated acid present, which is believed to be the effective antimicrobial form (Lück, 1980; Lund et al., 1987; Skirdal and Eklund, 1993; Sofos and Busta, 1981; Sofos, 1989). This popular theory has been questioned, however, (Sofos, 1989). Studies have indicated that the dissociated sorbic acid also had antimicrobial activity, but it was 10–600 times less inhibitory than the undissociated acid (Eklund, 1983; Statham and McMeekin, 1988). In environments of pH above 6.0, however, more than 50% of the inhibition was due to dissociated sorbic acid (Eklund, 1983). Nevertheless, it is believed that both undissociated and dissociated sorbic acid have antimicrobial activity (Skirdal and Eklund, 1993). Use of artificial saltwater prevented dissociation of the 1% sorbic acid which exhibited favorable antimicrobial properties against *Vibrio vulnificus*, supporting previous evidence that sorbic acid is effective in its undissociated form (Sun and Oliver, 1994). Sorbic acid (0.1%) enhanced the destruction of *E. coli* O157:H7 cells at pH 3.4 but not at pH 6.4 (Liu et al., 1997). Increased amounts of fat in a product reduce the concentration of sorbate in the water phase, where it is needed for microbial control (Oka, 1960). Other food ingredients (e.g., salt and sugars) also reduce the concentration of sorbate in the aqueous phase (Gooding et al., 1955; Liewen and Marth, 1985a). Sugar and salt, however, act synergistically to enhance the antimicrobial activity of sorbate (Acott et al., 1976; Beuchat, 1981c; Costilow et al., 1955, 1956, 1957; Robach and Stateler, 1980; Sheneman and Costilow, 1955). In general, solutes should increase the inhibitory activity of sorbate by reducing the a_w of the substrate (Cerruti et al., 1990; Sofos, 1989). Sucrose, glucose, and sodium chloride, however, have reduced the synergistic effect of sorbate and heat on thermal inactivation of microorganisms (Beuchat, 1981a–c; Cerruti et al., 1988). Sodium chloride (1.25 and 2.5%) reduced the inhibition of *Clostridium botulinum* by sorbate in a nutrient broth (Wagner and Busta, 1984, 1985a, b). Preconditioning of *Saccharomyces rouxii* cells in 60% sucrose + 0.1% sorbate rendered the cells more sensitive to inhibition by sorbate than preconditioning in 0% sucrose + 0.1% sorbate (Bills et al., 1982). Lowering the a_w enhanced the resistance of the same organism to increasing concentrations of sorbate (Restaino et al., 1981, 1983). Not only are certain strains of microorganisms resistant to inhibition by sorbate, but increased levels of microbial contamination reduce antimicrobial activity. Thus, sorbates should be used to preserve foods processed using good manufacturing practices, not as a substitute for appropriate sanitation and hygienic practices.

Interactions of sorbate with heat may affect the rate and extent of microbial destruction during heating, as well as dormancy and recovery of heated microorganisms (Sofos, 1989). Sorbate may enhance heat activation and destruction of spores, and it may inhibit the repair and growth of thermally injured organisms (Banks et al., 1988; Beuchat, 1980, 1981a–d, 1982; Lopez et al., 1996; Lusher et al., 1984; Oloyede and Scholefield, 1994; Splittstoesser et al., 1996), but the effect of sorbate on thermal inactivation and recovery of injured microorganisms is variable among species and strains. In more recent studies the effects of sorbic acid, heat, and the combination of both have been evaluated for *B. subtilis* spores and their germination. Sorbic acid mainly affected the generation time and outgrowth while heat mainly affected the germination process with the combination effectively increasing the mean and variation of germination time and burst time pointing to a synergistic effect. While sorbic acid alone did not affect germination and outgrowth efficiency, heat-treated spores germinating with sorbic acid present showed lower germination than without sorbic acid. As spore-forming bacteria are a serious issue for the food industry, a combination of sorbic acid and heat may be developed to better prevent potential spoilage (Pandey et al., 2015).

Low concentrations of sorbic and fumaric acids in the heating medium had little effect on the heat resistance of *Eurotium herbariorum*, a true aerophilic mold involved in the spoilage of grape preserves (Splittstoesser et al., 1989). Concentrations of sorbate as high as 0.1% had little effect on the thermal resistance of ascospores of *Neosartorya fischeri*, but growth of surviving spores that had been exposed to high temperatures was greatly inhibited by sorbate concentrations as low as 0.007% (Splittstoesser and Churey, 1989). Another report also indicated that potassium sorbate inhibited the heat-resistant *N. fischeri* (Nielsen et al., 1989). Sorbate may also eliminate the protective effect of sucrose against the thermal inactivation of yeasts and molds (Sofos, 1989). Sorbic and benzoic acids affected the thermotolerance and heat shock response of *S. cerevisiae* depending on pH (Cheng and Piper, 1994; Sofos, 2000). At low pH, sorbate inhibited the induction of thermotolerance by sublethal heat shock, but at pH 5.5 it acted as a powerful inducer of thermotolerance in the absence of sublethal heat treatment. Sorbic acid was found to induce thermotolerance without inducing the heat shock response through accumulation of trehalose in *S. cerevisiae* (Cheng et al., 1999).

It is of particular interest that 0.1% potassium sorbate did not have any significant effects on the heat resistance of *Bacillus* (now *Geobacillus*) *stearothermophilus* spores in distilled water, and as such would therefore be important only in improving microbiological stability through inhibitory effects on germination and/or outgrowth of heat-damaged spores (López et al., 1996). The tolerance of *Z. rouxii* to sorbate was enhanced in cells preconditioned to elevated sorbate concentrations, especially in media containing 0.1% sorbate. Cells grown in sucrose-supplemented media tolerated higher concentrations of sorbate than did those grown in glucose-supplemented media. Heat resistance increased substantially when cells were grown in media containing sorbate, and particularly in the presence of sucrose as opposed to glucose. The finding that plasmolysis increased in cells grown in sucrose- as opposed to glucose-containing media provides evidence that cell morphology influences heat resistance (Golden and Beuchat, 1992).

Inhibition of microbial growth by sorbate is more effective as storage temperature decreases, indicating that the compound should be more useful as a preservative in refrigerated foods (McMeekin et al., 1984; Pederson et al., 1961; Park et al., 1970; Park and Marth, 1972; Robach, 1980; Roberts et al., 1982; Roland and Beuchat, 1984; Roland et al., 1984; Tuncan and Martin, 1985). Combinations of acidification and treatment with sorbate may enhance the storage stability of fruit juices even at temperatures higher than normal refrigeration (5°C) (Alli and Kermasha, 1989). Reductions of 5–10 log-units of *E. coli* O157:H7 or *S.* Typhimurium DT104 in apple cider were achieved through freeze-thaw treatments in the presence of sorbic acid (Uljas and Ingham, 1999).

The antimicrobial activity of sorbate is usually enhanced under vacuum or modified gas atmosphere storage conditions, as indicated with several food items, including meat (McMeekin et al., 1984; Myers et al., 1983; Wagner et al., 1982) and fish products (Bremmer and Statham, 1983; Statham et al., 1985). Combinations of carbon dioxide and sorbate have also been reported as effective inhibitors of microbial growth (Danzinger et al., 1973; Elliot and Gray, 1981; Elliot et al., 1982, 1985; Gray et al., 1984). Sorbic acid (0.2–0.4%), however, did not enhance the bactericidal activity of cellulose-based edible films against *Salmonella* Montevideo (Zhuang et al., 1996).

Although food acids may reduce the water solubility of sorbate, they can enhance its antimicrobial activity by increasing the concentration of undissociated sorbic acid. In addition, the specific anion itself may contribute antimicrobial activity (Huhtanen and Feinberg, 1980; Huhtanen et al., 1981, 1983; Juven, 1976). Specific effects, however, vary with substrates, microorganisms, and types of acids (Restaino et al., 1981, 1982; Sofos, 1989).

Combinations of sorbate with antioxidants, such as butylated hydroxyanisole (BHA), butylated hydroxytoluene (BHT), tertiary butyl hydroquinone (TBHQ), and propyl gallate (PG), had increased antimicrobial activity compared to individual components (Davidson et al., 1981; Klindworth et al., 1979; Lahellec et al., 1981; Morad et al., 1982; Poerschke and Cunningham, 1985). Variations existed, however, with types of microorganisms, antioxidants, and substrates. These combinations offer the advantage of simultaneous inhibition of microbial growth and development of rancidity. Potassium sorbate was found more inhibitory against yeasts than hydroxycinnamic acid (Stead, 1995).

Several studies have also indicated increased antimicrobial effects when sorbate was combined with various phosphates (Ivey and Robach, 1978; Mendonca et al., 1989; Robach, 1979b; Seward et al., 1982; Nelson et al., 1983; Sofos, 1986a; Thomas and Wagner, 1987; Wagner and Busta, 1983). Combinations of sorbate with benzoate or propionate may be used to expand the range of microorganisms inhibited with reduced concentrations of each preservative.

The synergistic effect observed when vanillin and potassium sorbate were used in combination could be considered for use in order to reduce the amounts needed to inhibit mold growth (Matamoros-León et al., 1999). *Penicillium glabrum* was considered as the most resistant *Penicillium* mold tested and found to be inhibited by the combination of 500 ppm of vanillin and 300 ppm of potassium sorbate for at least 1 month at 25°C, pH 3.5, and a_w 0.98; these concentrations of chemicals represented a greater than 50% reduction of the amounts needed to suppress growth when used individually (Matamoros-León et al., 1999). In addition, mixtures of sorbate with various antibiotics have demonstrated increased antimicrobial activity (Amano et al., 1968; Gourama and Bullerman, 1988; Miller and Brown, 1984; Schmidt, 1960b). Interactions and increased antimicrobial effects have also been observed in combinations of sorbates with propionate, ascorbate, certain amino acids, fatty acids, sucrose fatty acid esters, sulfur dioxide, propylene glycol, glucose oxidase, and other compounds (Bell and De Lacy, 1987; Ferguson and Powrie, 1957; Kabara, 1984; Kadam et al., 1985; Marshall and Bullerman, 1986; Ough and Ingraham, 1960; Robach et al., 1981; Tellez-Giron et al., 1988).

Numerous publications and patents describe interactions of sorbate with many other factors, as well as multifactorial interactions (Deak and Beuchat, 1994; Guerrrero et al., 1994; Sofos, 1989; Thomas et al., 1993). The microbial stability of shelf-stable banana puree was enhanced through the inhibition of inoculated osmophilic and non-osmophilic yeasts, molds, *B. coagulans*, *C. pasteurianum*, and *C. butyricum* by the adjustment of a_w to 0.97, pH to 3.4, and the addition of 250 ppm ascorbic acid, 100 ppm potassium sorbate, 400 ppm sodium bisulfite, and mild heat (Guerrero et al., 1994). The effect of temperature, pH, sodium chloride concentration, and potassium sorbate was evaluated against growth of three foodborne bacterial pathogens (*Bacillus cereus*, verocytotoxigenic *E. coli*, and *S. aureus*) using gradient gel plates (Thomas et al., 1993). Sorbate was completely effective against *E. coli* at all temperature/pH/NaCl combinations, and was the most effective preservative tested against *B. cereus*. Increase in the acidity and/or the NaCl concentration improved the effect of all the preservatives, except nitrite when used against *S. aureus*. At <25°C, sorbate was more effective than benzoate against *S. aureus* when used with higher concentrations of NaCl (Thomas et al., 1993). As indicated, *Z. bailii* is highly resistant to individual inhibitory factors, including sorbate. A study found that growth was not inhibited at pH 3.8, in a medium containing 0.06% potassium sorbate, at a_w of 0.93 or at 10°C (Deak and Beuchat, 1994). However, interactions between two or more of these factors resulted in marked inhibition with the inhibitory effect of potassium sorbate occurring at the highest concentration (0.06%) combined with a_w values lower than 0.95 (Deak and Beuchat, 1994). The ability of *C. botulinum* non-proteolytic type B to grow in the presence of sorbic acid was at least as great at 20 as at 30°C (Lund et al., 1990). After 14 d at 30°C in the presence of 280 mg undissociated sorbic acid/l, the log probability of growth of a single vegetative cell is approximately –4. At temperatures of 12°C and below, sorbic acid resulted in marked inhibition, the log probability of growth of a single bacterium at 12°C after 60 d at pH 5.5 in the absence acid being 0 and –1 under the same conditions, but in the presence of a calculated undissociated sorbic acid concentration of 200 mg/l, equivalent to a total sorbic acid concentration of 1300 mg/l, the probability of growth was approximately –6 (Lund et al., 1990).

Not all microbes respond well to sorbic acid; *C. perfringens A* was sensitive to sorbate in a nutrient-rich medium; however, the effect was negligible when applied to cooked chicken meat under abusive conditions (Alnoman et al., 2014).

In general, the extent of the antimicrobial activity of sorbate under commercial conditions is difficult to predict because it is influenced by a variety of factors. These factors include product pH and composition (fat and moisture), presence of other ingredients and preservatives, product processing, sanitation and contamination, packaging, gas atmosphere, and storage temperature (Sofos and Busta, 1981; Sofos, 1989).

4.3.5 Mechanisms of Action

Sorbate concentrations used in foods (<0.3%) usually inhibit microorganisms, while higher amounts may result in death. Consequently, when the sorbate hurdle is reduced or removed, the surviving microorganisms may resume growth and spoil the food. Sorbate inhibits cell growth and multiplication as well as germination and outgrowth of spore-forming bacteria, but the exact mechanisms of antimicrobial activity are not well-defined (Freese et al., 1973; Freese and Levin, 1978; Sofos et al., 1986; Sofos, 1989).

Several studies have indicated that sorbate inhibits bacterial spore germination (Blocher and Busta, 1985; Seward et al., 1982; Smoot and Pierson, 1981; Sofos et al., 1979d, 1986). Inhibition has involved various species of bacteria in laboratory culture media and in foods and has been influenced by species, strains, pH, and sorbate concentration (Sofos, 1989). Published data have suggested that sorbate acts as a competitive and reversible inhibitor of amino acid-induced germination (Smoot and Pierson, 1981). Another study reported that sorbate inhibited spores triggered to germinate or after germinant binding (Blocher and Busta, 1985). It was also postulated that sorbate probably inhibits a postgerminant binding step in the process of germination, and it was thus concluded that it inhibits spore commitment to germination, and not triggering of germination. Inhibition of the not well-defined connecting reactions of spore germination may be taking place through the interaction of sorbate with spore membranes and through increases in their fluidity (Blocher and Busta, 1985; Sofos et al., 1986; Sofos, 1989).

Several mechanisms of inhibition of metabolic function by sorbate have been proposed, and it may be possible that several of them may be functional under various conditions, including types and species of microorganisms, type of substrate, environmental conditions, and the type of food processing. Under certain conditions sorbates have changed the morphology and appearance of microbial cells (Ronning and Frank, 1989; Statham and McMeekin, 1988; Sofos, 1989). Such changes have been observed in yeast cells as dense phosphoprotein granules, irregular nuclei, increased numbers and variable sizes of mitochondria, and vacuoles. Cells of *C. botulinum* were long, with bulbous formation and defective division (Seward et al., 1982; Wagner and Busta, 1985a, b). Sorbate-treated cells of *C. sporogenes* were usually filamentous and nonseptate, but with distorted shapes characterized by numerous bends and bulges. Septation, when present, resulted in minicells, and the inner cell wall appeared to be thickened; the outer cell wall was absent in many areas (Ronning and Frank, 1989). Treatment of *Alteromonas putrefaciens* with sorbate at pH 7.0 increased cell hydrophobicity and cell wall lysis upon exposure to lysozyme, which could be overcome to some extent by the addition of magnesium ions. There was also evidence of other membrane damage in sorbate-treated cells (Statham and McMeekin, 1988). Although the significance of such alterations is unknown, they may be due to the incorporation of sorbate into specific cell structures and alteration of biosynthetic processes in the cell (Sofos et al., 1986).

Other proposed mechanisms of inhibition of microbial growth by sorbate include alterations in the morphology, integrity, and function of cell membranes and inhibition of transport functions and metabolic activity (Sofos, 1989). The death of microorganisms exposed to high concentrations of preservatives, such as sorbate, has been attributed to the generation of holes in the cell membrane (Freese and Levin, 1978). Sorbate has decreased the assimilation of carbon from several substrates, including glucose, acetate, succinate, pyruvate, lactate, oxaloacetate, α-ketoglutarate, ethanol, and acetaldehyde (Harada et al., 1968; Sofos and Busta, 1981; York and Vaughn, 1964). Inhibition of cell metabolism by sorbate in these studies may have been due to the inhibition of enzymes, nutrient uptake, or various transport systems.

Sorbate inhibits the activity of several enzyme systems, which may lead to the disruption of vital processes involved in transport functions, cell metabolism, growth, and replication. Enzymes inhibited by sorbate include alcohol dehydrogenase, fumarase, enolase, aspartase, catalase, malate dehydrogenase, α -ketoglutarate dehydrogenase, succinic dehydrogenase, and ficin (Azukas et al., 1961; Martoadiprawito and Whitaker, 1963; Melnick et al., 1954b; Troller, 1965; Whitaker, 1959; York and Vaughn, 1964). Some reports, however, have indicated no inhibition of enzymatic activity by sorbate (Sofos, 1989).

Sorbate is known to inhibit the in vitro activity of many enzymes, especially sulfhydryl-containing enzymes (Kouassi and Shelef, 1995a, b). Inhibition of sulfhydryl enzymes by sorbate has been attributed to binding the compound with sulfhydryl groups and decreasing the number of such active sites on the enzyme. Inhibition of yeast alcohol dehydrogenase has been attributed to the formation of a covalent bond with the sulfhydryl or ZnOH group of the enzyme and the α- and/or β-carbon of sorbate (Martoadiprawito and Whitaker, 1963). It was also proposed (Whitaker, 1959) that sorbate inhibits the sulfhydryl enzymes through the formation of a thiohexenoic acid derivative (CH_3–CH=CH=RSCH–CH_2CO_2H). Inhibition of catalase by sorbate was attributed to the formation of sorbyl peroxides through the autoxidation of sorbic acid. These peroxides then would inactivate catalase (Troller, 1965). Another postulation has indicated that sorbate may act competitively with acetate at the site of acetyl coenzyme A formation (Harada et al., 1968; Sofos, 1992; Wakil and Hubscher, 1960). Binding and inhibition of coenzyme A should result in the inhibition of oxygen uptake and microbial growth (Sofos, 1992).

Lipophilic acid food preservatives, such as sorbate, may interfere with substrate and electron transport mechanisms. Inhibition of substrate transport into the cell by uncoupling it from the electron transport system results in cell starvation (Deak and Novak, 1970; Freese and Levin, 1978; Sheu and Freese, 1972, 1973; Sheu et al., 1972, 1975). Sorbic acid has inhibited the uptake of glucose (Yousef and Marth, 1983) and amino acids (Hunter and Segel, 1973; Tuncan and Martin, 1985), as well as the electron transport system (Anderson and Costilow, 1963; Freese et al., 1973).

Inhibition of nutrient uptake may be due to neutralization of the proton-motive force needed for substrate uptake, inhibition of the electron transport system, inhibition of synthesis or depletion of ATP, inhibition of transport enzymes, and inhibition of metabolic energy utilization by the amino acid transport systems (Sofos, 1989). This effect may be occurring through incorporation of sorbic acid, or unsaturated fatty acid, into the cell membrane, where it may cause steric disorganization of active membrane transport proteins (Sofos, 1989; 1992). The highly reproducible and characteristic thermograms of *S. aureus* metabolism were significantly affected by sorbic acid in a concentration-dependent manner (Sayeed and Sankaran, 1991). Both the peak heat and total heat dissipation profiles were affected by 50% of the common concentration (0.2%) for use in foods. Additionally, the suggestion that inhibition of the membrane function is the primary site of action for sorbic acid is strengthened by an observation on damage to the outer cell membranes by sorbate ions at pH 7.0 in *Alteromonas putrefaciens* (Sayeed and Sankaran, 1991). Thus, microcalorimetric measurement of the effects of sorbic acid on *S. aureus* cells has clearly shown that the acid acts in its dissociated anionic form primarily on the substrate transport systems, creating a starvation condition in the cells. Secondary events such as partial inhibition of the electron transport and inhibition of the activity or synthesis of catalase enzyme may have resulted in cell death (Sayeed and Sankaran, 1991). Sorbic acid could conceivably inhibit microorganisms as a weak-acid preservative, a membrane-active compound, or as a specific inhibitor of metabolism (Azukas et al., 1961). There is evidence to suggest that protein denaturation or enhanced protein turnover is a further consequence of weak acid stress due to the expression of many proteins which act as molecular chaperones (de Nobel et al., 2001). Conversely, it has been shown that sorbic acid acts at high pH where weak-acid preservatives are not expected to be active (Stratford and Anslow, 1996). Sorbic acid in its MIC releases far fewer protons than other weak-acid preservatives. Similar degrees of inhibition by sorbic alcohol and sorbic aldehyde suggest that sorbic acid does not act as a weak-acid preservative. Specific inhibition of metabolism is also unlikely as inhibitory activity is shared by several small carbon compounds. Correlation of sorbic acid resistance with ethanol tolerance and the partition coefficient strongly suggest an inhibitory role for sorbic acid as a membrane-active compound (Stratford and Anslow, 1998).

Cells of *Z. rouxii* cultured in media supplemented with sorbate contained higher percentages of C18:1 fatty acids than cells cultured in media without sorbate (Golden et al., 1994). Since inhibition of microbial growth by sorbate has been associated with decreased levels of ATP (Harada et al., 1968; Hunter and Segal, 1973), a proposed mechanism of ATP depletion includes hydrolysis of ATP by the primary sodium/hydrogen pump in an attempt to maintain ion balance in the cell (Przybylski and Bullerman, 1980). As the hydrogen influx exceeds the pumped efflux, a shift in charge may potentially take place and lead to a decrease in the net negative intercellular charge. This could then discharge the pH gradient required for ATP formation according to the chemostatic theory of oxidative phosphorylation (Sofos, 2000). In another study using a novel method to measure pH in growing cells, little correlation was found between reduced growth upon exposure to sorbic acid and reduction of intracellular pH (pH_i) (Bracey et al., 1998). In fact, growth inhibition correlated with an increase in the intracellular ADP/ATP ratio due to increased ATP consumption by the cells. This was partly attributed to the activation of protective mechanisms, such as increased proton pumping by the membrane H^+-ATPase, which ensured that pH_i did not decline when cells were exposed to sorbic acid (Henriques et al., 1997). Therefore, the available evidence suggested that the inhibitory action of sorbic acid was due to the induction of an energetically expensive protective mechanism that compensated for any disruption of pH_i homeostasis but resulted in less available energy for normal growth. Thus, it is believed that the inhibitory action of sorbic acid is due to excessive consumption of cellular energy that occurs as a consequence of the cell eliciting a stress response that attempts to maintain pH_i homeostasis such that the available energy for growth and cell division is drastically reduced (Bracey et al., 1998). Evidence also suggests that this energy-demanding stress could also be due to the induction of ATP-driven plasma membrane pumps for active extrusion of weak acids from the cell (Henriques et al., 1997). Indeed, as indicated, a pump recently identified as the Pdr12 protein is induced by sorbic acid in *S. cerevisiae* (Piper et al., 1998). Previous studies with bacteria, however, found no decreases in ATP in the presence of sorbic acid (Sofos, 1989, 1992, 2000). Inhibition of microbial growth by sorbate may be based on neutralization of the proton-motive force (PMF) that exists across the cell membranes by lipophilic, weak acid preservatives such as sorbic acid, and it may lead to starvation of cells from compounds that are transported actively by the PMF (Eklund, 1980, 1983, 1985; Ronning and Frank, 1987, 1988; Salmond et al., 1984; Sofos, 1989). In lower pH environments, the difference across the cell membrane is large, and the amount of acid entering and dissociating in the cytoplasm is higher. This accumulation of hydrogen ions inside the cell acts as an inhibitor by interfering with metabolic processes and causing a dissipation of the transmembrane proton gradient, which is one of the components of the PMF (Sofos, 2000). According to this theory, undissociated sorbic acid acts as a protonophore, which decreases the intracellular pH and dissipates the PMF of the membrane that energizes the transport of compounds such as amino and keto acids. Inhibition of uptake of such components is believed to induce a stringent-type regulatory response in the cells, resulting in inhibition of growth but in maintenance of cell viability (Sofos, 2000). A stringent response involves readjustments occurring in bacteria when amino acids become limiting or their specific ratios are disturbed (Sofos, 2000). Preincubation of *S. cerevisiae* yeast cells in the presence of benzoate or sorbate at an extracellular pH value of 6.8 elicited a set of metabolic effects on sugar metabolism, which became apparent after the subsequent addition of glucose. These effects can be summarized as follows: (1) reduced glucose consumption; (2) inhibition of glucose- and fructose-phosphorylating activities; (3) suppression of glucose-triggered peak of hexose monophosphates; (4) substantial reduction of glucose-triggered peak of fructose 2,6-bisphosphate; (5) block of catabolite inactivation of fructose-1,6-bisphosphatase and phosphoenolpyruvate carboxy-kinase, but not of cytoplasmic malate dehydrogenase (Burlini et al., 1993). On the whole, this pattern resulted in prevention of glucose-induced switch of metabolism from a gluconeogenic to a glycolytic state. Thus, unlike former assumptions, intracellular acidification is not likely to mediate the bulk of metabolic effects of benzoate and sorbate, since under the described working conditions intracellular pH was kept close to neutrality (Burlini et al., 1993). Overall, it is believed that neutralization of the pH difference across the cell membrane is not the only mechanism of microbial inhibition by sorbate (Bracey et al., 1998; Stratford and Anslow, 1998; Sofos, 2000). Reasons for this conclusion include: (1) although dissociation of the acid inside the cell may eliminate the pH difference across the membrane, its effect on the other component of the PMF, which is the difference in electrical potential, is smaller, but it is believed to be adequate to energize the uptake of substances needed for cell maintenance and growth; (2) sorbic acid has been shown to inhibit microbial growth, although less efficiently, even at pH values near neutrality; in such situations, uptake of the compound based on pH difference across the cell membrane would not be adequate to explain the observed antimicrobial activity; (3) neutralization of the PMF is probably not involved in growth inhibition in the presence of carbohydrates, because they do not depend on this force for their transport (Sofos, 2000). Thus, the mechanisms involved in inhibition of microbial metabolism and proliferation by sorbic and other similar lipophilic acid preservatives appear to be different, depending on the type of microorganism, substrates, and environmental conditions (Sofos, 1989, 1992).

4.4 Applications

Sorbates are widely used food preservatives throughout the world. The most commonly used forms include sorbic acid and the potassium salt. The compounds find application in human foods of all types, animal feeds, pharmaceuticals, cosmetic products, packaging materials, and technical preparations that come in contact with foods or the human body. In the United States, sorbates may be used in any food product that allows generally recognized as safe (GRAS) food additives and in approximately 80 additional food products that have federal standards of identity (Sofos, 1989; 1992).

As yeast and mold inhibitors, sorbates have found wide application in various foods, including dairy products, bakery items, fruit and vegetable products, edible fat emulsion products, certain meat and fish products, and sugar and confectionery items (Sofos, 1989) (Table 4.4). However, as is the case with all food preservatives, sorbates should not be considered as substitutes for good sanitation or agents to be used to improve the quality of partially spoiled or degraded foods. They should be considered as adjuncts to proper sanitation and hygiene.

Application methods for sorbates include direct addition in the formulation, spraying or immersing the food material in a solution, dusting with a powder, or addition in a coating or packaging material (Table 4.5). The inclusion of sorbates into edible films and coatings is a particularly unique application for the control of microorganisms on food surfaces. The diffusion coefficient representing the speed with which a molecule moves through the diffusion matrix is important, thus the film composition used to carry sorbate is the primary factor determining its effect to diffuse. Researchers (Redl et al., 1996) have explored a number of different materials to include sorbate into an edible film and found that films containing lipids had lower diffusion coefficients than those without a lipid component. Other researchers demonstrated that increasing the relative amounts of protein decreased potassium sorbate diffusivity, while increasing relative amounts of sorbitol and initial potassium sorbate in the films increased the diffusion of potassium sorbate (Ozdemir and Floros, 2003).

TABLE 4.4

Common Applications of Sorbates as Antimicrobial Agents

Products	Levels (%)
Dairy products: aged cheeses, processed cheeses, cottage cheese, cheese spreads, cheese dips, sour cream, yogurt	0.05–0.03
Bakery products: cakes, cake mixes, pies, fillings, mixes, icings, fudges, toppings, doughnuts	0.03–0.30
Vegetable products: fermented, pickles, olives, relishes, fresh salads	0.02–0.20
Fruit products: dried fruit, jams, jellies, juices, fruit salads, syrups, purees, concentrates	0.02–0.25
Beverages: still wines, carbonated and non-carbonated beverages, fruit drinks, low-calorie drinks	0.02–-0.10
Food emulsions: mayonnaise, margarine, salad dressings	0.05–0.10
Meat and fish products: smoked and salted fish, dry sausages	0.05–0.30
Miscellaneous: dry sausage casings, semimoist pet foods, confectionary	0.05–0.30

Source: From Sofos (1989).

TABLE 4.5

Methods of Application of Sorbates

Methods	Products	Solution (%)	Application Level (%)
Immersion	Dairy: provolone, Siciliano, Caciocarallo, pasta filata, Edam, Muenster, Gouda	20–40	0.1–0.3
	Dry sausage or casings	2.5–10	0.05–0.3
Spraying	Dairy: Swiss, Gruyere, Emmentaler	20–40	0.1–0.3
Immersion or spraying	Dairy: Cheddar, Colby, Monterey Jack, blue, mozzarella	20–40	0.01–0.3
	Dried fruits: prunes	2.5–5	0.02–0.06
	Raisins	5	0.25
	Figs	2.5–5	0.05–0.1
	Smoked and salted fish	5	0.05–0.15
Dry mixing	Bakery: cheesecake		0.1–0.3
	Angel food cake		0.03–0.06
	Fruitcake		0.1–0.4
	Cake mixes		0.05–0.1
	Doughnut mixes		0.05–0.1
	Pie crust dough		0.05–0.1
Direct addition	Dairy: processed cheese, spreads		0.1–0.2
	Cottage cheese		0.05–0.075
	Bakery: icings, fillings, toppings, fudges		0.05–0.1
	Vegetables: fresh salads (potato, vegetable, slaw)		0.05–0.15
	Pickles, relishes, olives, sauerkraut		0.05–0.1
	Fruits: syrups, beverages, drinks		0.025–0.075
	Still wines		0.02–0.04
	Other: margarine		0.05–0.1
	Semimoist pet food		0.1–0.3
Dry mixing or direct addition	Bakery: pound cake		0.075–0.15
	Devil's food cake		0.3
	Chocolate cake		0.1–0.3

Source: From Sofos (1989).

Barzegar et al. (2014) evaluated the effect of including potassium sorbate in active starch-clay nanocomposite films as an antimicrobial film for fresh food packaging. Inclusion rates of 0.5, 7.5, and 10% potassium sorbate in the film were evaluated for their physical, mechanical, and antimicrobial properties. The inclusion rate of at least 7.5% demonstrates favorable water vapor permeability, tensile strength, and the ability to inhibit *A. niger*, making it a viable packaging option for preserving foods, especially bakery products. Hauser and Wunderlich (2011) demonstrated the effect of packaging films coated with a lacquer containing sorbic acid to inhibit *E. coli* on Gouda cheese and pork loins; the results suggest that packaging films with a sorbic acid coating are promising for the quality and safety of packaged goods. Numerous studies have been conducted in the development of antimicrobial packaging films that delay the growth and spoilage of food products, especially refrigerated meats. The key considerations in developing such a film include: (1) the loss of antimicrobial agents through volatilization and poor thermal stability (Kayaci et al., 2013) and (2) rapid release of antimicrobial agents for long-term effect (Helmroth et al., 2002). To solve these issues, Hu et al. (2017) adopted the controlled release systems used for conventional agents (Hosseini et al., 2016) for sorbic acid. Sorbic acid-chitosan microcapsules (S-MPs) were designed to serve as an antibacterial agent to be included in packaging film material (active ethylene vinyl alcohol copolymer and polyethylene terephthalate [EVOH/PET]) and tested for inhibitory capacity against *E. coli*, *L. monocytogenes*, and *S.* Enteritidis. The sorbic acid-chitosan microcapsule treated film was able to extend the shelf-life of fish fillets by 4 days at refrigeration when compared to film not using S-MPs.

Selection of the most appropriate method depends on processing procedures, type of food, objectives to be accomplished, equipment available, and convenience (Sofos, 1989). Amounts of sorbate/sorbic acid used in foods are in the range of 0.02–0.3%. Although these concentrations have no major impact on food quality, higher levels may cause undesirable changes in the taste of most foods. In general, amounts of 0.1–0.3% are tolerated, but levels as low as 0.1% may be detectable in some foods. Furthermore, it is important to consider the intrinsic and extrinsic parameters of the matrix when introducing sorbic acid or its salts; Soljic et al. (2018) demonstrated clearly the impact of both pH and fat phase (solid or liquid) on the antimicrobial activity in an aqueous model. The presence of liquid oil caused partitioning of sorbic acid into the lipid phase, reducing the effect of sorbic acid in inhibiting *Candida guilliermondii* introduced into the food model system. This information may help product developers and production facilities control the solid fat content of the lipid phases in their recipes to ensure better retention of sorbic acid in the aqueous phase, thereby reducing potential spoilage of their products.

4.4.1 Dairy Products

The use of potassium sorbate alone and in combination was evaluated on vegetative cell growth and spore germination of *Bacillus sporothermodurans* in milk (Aouadhi et al., 2015). Results indicated that potassium sorbate used at 0.2% had a bacteriostatic effect on vegetative cells followed by slower regrowth as well as inhibition of germination and outgrowth of *B. sporothermodurans*. Furthermore, the combination of potassium sorbate and nisin at 50 IU/ml demonstrated a synergistic effect that resulted in total inhibition of cell growth after 5 days.

In labneh (a yogurt product) produced under strict hygienic processing measures and effectively chilled, initial numbers of yeasts were low, and relatively low concentrations of potassium sorbate and sodium benzoate (<150 and <250 mg/kg, respectively) extended the shelf-life by 14 d. Labneh with higher initial yeast counts needed higher concentrations of preservative (>300 and >400 mg/kg for K-sorbate and Na-benzoate, respectively) to maintain stability when kept at 5°C for 7 or 14 d. These concentrations are in agreement with the use of these preservatives in other dairy products, such as cottage cheese and yogurt (Mihyar et al., 1997).

Cheese preservation is one of the most common applications of sorbates. In the United States, federal standards of identity permit the use of sorbates as preservatives in more than 40 types of cheese, cheese spreads, and cheese food [21 CFR Part 133].

Kivanç (1992) found that *Penicillium* was the major genus found on Cheddar cheese, Swiss cheese, domestic, and imported cheeses. Sorbates are widely used in preserving cheese owing to the susceptibility of all types of cheese to microbial deterioration, especially surface mold growth during storage, aging, and distribution; the high activity of sorbates against molds; and the improved action of sorbates at high pH values compared to other common preservatives (Lück, 1976, 1980; Sofos and Busta, 1981). In these applications sorbates not only prevent cheese spoilage by mold, but they also protect human health by preventing the formation of toxic mold metabolites (mycotoxins). It is obvious however, that sorbates should not be used in mold-ripened cheeses, such as Roquefort or blue cheese and Camembert cheese, in which they can interfere with the desirable molds involved in the manufacture of these cheeses.

Sorbate application, in general, is limited to the surface of cheese, where mold development usually occurs. In certain products, however (e.g., cottage cheese), sorbic acid may be incorporated into the cheese curd. The method of application varies with the type of cheese, preservation objectives, available facilities and equipment, and regulations, which may vary among countries (Lück, 1976, 1980). Addition of sorbate does not significantly affect the organoleptic properties of mozzarella cheese; however, depending on the technique used to treat the cheese (i.e., dipping or in brine salting), there may develop a slight objectionable flavor (Aly, 1996). For application on the surface of the cheese, one of the following procedures may be followed: (1) potassium sorbate in solution or in a brine, which can be used to dip, spray, or wash; (2) sorbic acid powder for dusting; and (3) sorbic acid, potassium, or calcium sorbate impregnated into the packaging material or other coatings (e.g., wax) used in cheese storage and marketing (Han and Floros, 1998; Sofos, 1989; Weng and Chen, 1997). Treatment of mozzarella cheese with potassium sorbate in the kneading water or brine may be more effective than dipping (Aly, 1996). Amounts of sorbic acid permitted and applied in cheese preservation range from 0.05 to 0.30%

(Lück, 1980; Sofos and Busta, 1981). Levels of 0.05–0.07% sorbic acid are generally used in direct addition to cheese. Concentrations applied to cheese surfaces range from 0.1 to 0.3 g/dm^2, while when applied to packaging films, amounts of 2–4 g/m^2 are used (Lück, 1980). The appearance of *Penicillium verrucosum* var. *cyclopium* mycelia and spores isolated from cheeses and inoculated as suspensions into YES broth containing 1000 µg of potassium sorbate/ml was not detected at 5, 15, and 25°C after 30 d (Kivanç, 1992). The addition of potassium sorbate ($\leq$6%) to mozzarella cheese inhibited growth of *Streptococcus salivarius* var. *thermophilus* and *Lactobacillus delbrueckii* var. *bulgaricus*, and was especially effective against mold and yeast contamination by *Penicillium roqueforti* and *Mucor miehi*, respectively (Aly, 1996). Inclusion of sorbic acid in cottage cheese was effective as a preservative but not against *L. monocytogenes* (Piccinin and Shelef, 1995); however, at the relatively high level of 1%, potassium sorbate inhibited *L. monocytogenes* survival in two commercial cheese brines (Larson et al., 1999). In another study, potassium sorbate (0.3%) or sodium benzoate (0.3%) showed a significant effect in delaying growth of *E. coli* O157:H7 in Queso Fresco, soft cheese (Kasrazadeh and Genigeorgis, 1995). Milk acidification to pH 6.0 with hydrochloric, acetic, or propionic acid significantly increased the effectiveness of potassium sorbate, especially at lower ($\leq$6.6°C) temperatures (Kasrazadeh and Genigeorgis, 1994). Growth of *Salmonella* in soft cheese was inhibited if the cheese was prepared from milk which was acidified with propionic acid to pH 5.9, potassium sorbate was added to the cheese (pH 6.0) at levels of $\geq$0.3%, and the cheese was stored at $\leq$30°C (Kasrazadeh and Genigeorgis, 1994).

The effect of sorbic acid on storage quality was also explored with *Kaladhi*, a traditional acid-coagulated cheese from India (Ahmad et al., 2014). Sorbic acid inclusion of 0.1, 0.2, and 0.3% was tested on the shelf-life and storage quality for physicochemical and sensory properties over 35 days at ambient temperature. The results indicated that up to the 0.3% level of sorbic acid, the sensory and physicochemical qualities were superior to that of the cheeses treated at lower levels or without sorbic acid.

Absorption of the compound from the surface into the cheese is dependent on the porosity of the surface and the amount and distribution of fat in the cheese. Calcium sorbate may be preferred in cheese preservation over other forms of sorbate owing to its solubility properties. The compound is insoluble in fat and only slightly soluble in water. Therefore, when applied to the surface of hard, ripened cheeses, most of the preservative remains on the surface of the cheese without migrating inside the cheese or dissolving in the water on the shelves during the long maturation period of these cheeses (Lück, 1976). In addition to reducing cheese spoilage and the possibility of aflatoxin formation, use of sorbates in aged cheeses also reduces labor and other costs through reduction of the need for cheese washing and trimming. Furthermore, the addition of potassium sorbate may increase the meltability and improve the fat leakage of certain cheeses (Aly, 1996).

4.4.2 Vegetable Products

The water-soluble salts of sorbic acid are used widely in the preservation of a variety of plant products, including fresh, fermented, and pickled vegetables (Sofos, 1989). The use of sorbates in the processing of vegetables was one of the first applications, and it was initiated through extensive research performed during the early stages of sorbate testing. This research demonstrated the effectiveness of sorbates against undesirable organisms without interfering with the desirable fermentations. Use of sorbates at levels between 0.05 and 0.20% inhibits the growth of organisms not desired in vegetable fermentations, such as yeasts, molds, and putrefactive bacteria. At these concentrations the selective action of sorbic acid allows growth of the desirable lactic acid-producing organisms. The selection of precise levels of sorbate, however, depends upon products and specific fermentation conditions (e.g., salt concentration). Inclusion of sorbate at the initial stage of the process results in a "clean" fermentation and prevents turbidity development in the final product (Costilow et al., 1955; Jones and Harper, 1952; Lück, 1980). Mold and yeast spoilage is also prevented by sorbates in pickled vegetable products (e.g., cucumbers and olives). The addition of 0.02–0.05% sorbate to a 15–20% brine retards yeast scum formation. Levels of 0.02–0.10% sorbate reduce the number of bloaters and poor seed cavities. In sweet cucumber packs, 0.1% sorbate and 0.5% acetic acid completely inhibit microbial spoilage at sucrose concentrations ranging from 2 to 40%. In sweet relishes, sorbate is more effective than benzoate owing to the pH of these products. Also, the pH values of fermented green olives favor the use of 0.02–0.05% sorbates in the fermenting and holding tanks. Sorbate levels in the range of 0.05–0.10% may be used to preserve refrigerated fresh salads. Other vegetable products preserved with sorbates include tomato products and fermented Asian sauces (Chichester and Tanner, 1972; Lück, 1980; Sofos, 1989).

4.4.3 Fruit Products

Fruit products preserved with sorbates throughout the world include dried fruits, fruit juices and syrups, fruit cocktails, jams, jellies, preserves, beverages, and wines. Potassium sorbate is seldom used to treat or control the growth of molds on fruit surfaces such as citrus given the irregular disposition of microorganisms on the surface of fruit and the brief persistence of potassium sorbate residues on the surface (D'Aquino et al., 2013) – instead a more successful application is to treat further processed fruit. Sorbate concentrations as low as 0.02–0.05% are adequate for preserving high-moisture dried fruits (e.g., prunes, raisins, figs). The higher moisture content in these products is preferred by the consumers, but it makes them susceptible to mold and yeast spoilage. Depending on the particular product, sorbate acts as an inhibitor of yeasts, molds, or bacteria. The lower the moisture content of the product (e.g., dried fruits), the lower is the sorbate level needed for preservation. Alagoz et al. (2015) indeed demonstrated that regardless of the moisture content of dried apricots, 0.05% of sorbic acid was sufficient in preventing brown color formation and microbial growth. Also, in high-sugar products (e.g., jams and jellies), smaller quantities of sorbic acid are adequate for preservation because of the synergistic action of sorbate with sugar (Lück, 1980). In these products, sorbate is either added directly into the product or applied to the surface of the product or packaging material. Potassium sorbate was found more effective than chitosan in inhibiting growth of *A. niger* on low-sugar candied "kumquat" (Fang et al., 1994). Other fruit products that may be preserved with sorbates include maraschino cherries and strawberry puree.

In fruit juice and drink processing, sorbates are used mostly at the preprocessing stage, together with sulfur dioxide and pasteurization, to inhibit chemical, enzymatic, or microbiological deterioration (e.g., fermentation). Concentrations as low as 0.02–0.10% are adequate in improving the preservation of soft drinks. Combinations of sorbic acid and sulfur dioxide are also used in the preservation of high-pulp fruit juices. In these products sorbic acid acts as the microbial inhibitor, and sulfur dioxide prevents oxidation and enzymatic spoilage (Lück, 1980). Potassium sorbate and sodium benzoate were highly effective in reducing yeast and mold populations in tomato juice (Bizri and Wahem, 1994). Use of the combination of 0.1% sodium benzoate and 0.1% potassium sorbate substantially inhibited *E. coli* O157:H7 in apple cider, while additionally suppressing the growth of yeasts and molds which may influence the survival of *E. coli* O157:H7 and result in spoilage of the product (Zhao et al., 1993). Another study showed that *E. coli* O157:H7 growth was inhibited in both tryptic soy broth (TSB) and in apple cider by sorbic acid applied at 0.05% and as such may be an option which processors have to reduce contamination in their products (Koodie and Dhople, 2001).

In wine processing, sorbates are used to prevent refermentation (Parish and Carroll, 1988a, b). Potassium sorbate is preferred because of its solubility. Since potassium, however, may result in potassium bitartrate precipitation in some wines, sodium sorbate may be used to avoid the problem. Good wine preservation is achieved with concentrations of 0.02% sorbic acid in combination with 0.002–0.004% sulfur dioxide. Sorbate acts as an inhibitor of yeast refermentation during storage, and sulfur dioxide protects against enzymatic and oxidative changes and bacterial fermentation. Sorbate levels above 0.03% are not recommended in wine preservation because they may impart an off-flavor to the delicate taste of wine. Sorbate can also be added before bottling the wine and is very valuable in the preservation of sweet wines. It has also been used in champagne preservation (Sofos, 1989).

Sorbates, in general, may be more suitable for use in fruit products than other common preservatives because of their milder organoleptic properties and their neutral taste, which is especially important in the preservation of juices, soft drinks, and other beverages. Often, however, sorbates are used in combination with benzoate in preserving fruit juices and drinks. In fruit preserves (e.g., jams, jellies, and marmalades), mold inhibition may be achieved by sprinkling sorbate on the surface of the product. In dried products with an irregular surface conformation (e.g., figs), the dipping method of application is more effective (Sofos, 1989).

4.4.4 Bakery Products

Even though propionic acid-based preservatives are the most widely used compounds in preserving bakery goods, sorbates are also valuable because they are active at higher pH values and effective against molds (Chichester and Tanner, 1972; Lück, 1980). In addition to preventing mold spoilage, sorbates may also be useful in preventing aflatoxin formation (Lück, 1980). Concentrations of 0.03–0.30% sorbic acid are adequate for preserving bakery products. Sorbic acid and calcium propionate have been shown to be significantly more effective preservatives than salt and sugar when used to prolong the shelf-life of bread (Doulia et al., 2000). In addition to undesirable molds, sorbate also inhibits rope-forming bacteria and pathogens, such as *S. aureus*, in cream pies (Schmidt et al., 1969; Sofos, 1989). Potassium sorbate increased the shelf-life of acidified tortillas, especially when used in combination with calcium propionate (Tellez-Giron et al., 1988).

Use of sorbates must be avoided in products raised by yeasts, unless the amount of yeast and the leavening time are increased. Yeast-leavened products may be protected with sorbate spraying after baking. In such applications sorbate may be used in the coating oil or the pan grease (e.g., brown-and-serve products). To avoid the effect of sorbates on desirable yeasts in yeast-leavened products, the derivative sorboyl palmitate, which is a mixed anhydride of sorbic and palmitic acids that has no direct antimicrobial activity, may also be used; upon heating in the baking process, the compound is chemically degraded and releases sorbic acid after the yeast-leavened dough is raised. Use of this compound has the advantage of preserving yeast-leavened bread without interfering with the fermentation process (Lück, 1976, 1980).

Sorbates should be valuable in baking powder-raised products, including cakes and confectionery. Concentrations of 0.05–0.10% sorbate inhibit the spoilage of cake icings, toppings, and fillings. Sorbates are also used in the preservation of pie crusts and fillings, refrigerated dough products, pizza shells and toppings, muffins, and doughnut mixes (Sofos, 1989; 1992). Researchers surveyed the concentration of potassium sorbate in baked goods in Persian markets and found levels did not exceed 0.0415% and 0.07% in bread and cakes, respectively (Chaleshtori et al., 2018).

4.4.5 Meat Products

The only approved use of sorbates in meat products in the United States is to suppress mold growth on the surface of dry sausages during the drying period. For this purpose a solution of up to 10% potassium sorbate may be used to protect unrefrigerated dry sausages. However, sorbates are more common meat preservatives in other countries, such as Japan and Korea (Sofos, 1989).

In 1974, Tompkin et al., considering the permitted use of potassium sorbate as a mold inhibitor in dry sausages and reports suggesting that sorbic acid was promoting growth of clostridia (Emard and Vaughn, 1952; York and Vaughn, 1954), examined its activity against various pathogenic bacteria in an uncured, cooked sausage product. Wang et al. (2018) demonstrated the antimicrobial and antioxidant effects of sorbic acid-loaded chitosan/tripolyphosphate nanoparticles (SAN) in Chinese sausage stored at ambient for 72 days. The results showed that sausage treated with SAN had significantly lower levels of surviving bacteria, molds, and yeasts than a control without SAN.

A few sporadic studies in the 1950s and 1960s also indicated its potential as a preservative for meat products (Sofos, 1989). Tompkin et al. (1974) showed that sorbate retarded growth of *Salmonella* spp. and *S. aureus* and delayed growth and toxin production by *C. botulinum*. This work (Tompkin et al., 1974) was followed by extensive testing that examined the action of sorbates against clostridia and other pathogens in a variety of fresh, uncured, and processed cured meat products (National Academy of Sciences, 1982; Robach and Sofos, 1982; Sofos, 1981, 1989). The reason for extensive testing of the antimicrobial properties of sorbate in meat products was the perceived need for an alternative to sodium nitrite as an inhibitor of *C. botulinum* in cured meat products. Nitrite was reported to be a potential carcinogen, and it can react with amines to form carcinogenic nitrosamines during meat product processing and cooking (National Academy of Sciences, 1981; Sofos et al., 1979a).

Extensive studies published in the 1970s and 1980s established the antibotulinal and overall antimicrobial activity of sorbates in various cured and uncured meat and poultry products. In addition, the effect of sorbates on overall product shelf-life, eating quality, nitrosamine formation, and safety were examined. These studies have been thoroughly reviewed by the National Academy of Sciences (1981, 1982), Robach and Sofos (1982), Sofos and Busta (1980), Sofos et al. (1979a), and Sofos (1981, 1989). The antimicrobial activity of sorbate was demonstrated in bacon (Ivey et al., 1978; Sofos et. al., 1980b; U.S. Department of Agriculture, 1979a, b), comminuted pork products (Ivey and Robach, 1978; Robach et al., 1978; Robach, 1979b), beef and pork frankfurters (Sofos et. al., 1979b; Wagner and Busta, 1983), pork slurries (Draughon et al., 1982; Robets et al., 1982), uncured cooked sausage (Tompkin et al., 1974), sliced bologna (Chang et al., 1983), raw (Mendonca et al., 1989) and cooked pork chops (Prabhu et al., 1988), beef steaks (Unda et al., 1990), poultry products (Hall and Maurer, 1980; Huhtanen and Feinberg, 1980; Nelson et al., 1983; Robach et al., 1980c; Sofos et al., 1979b, 1980a), and hamburger beef (Yamamura et al., 2000). In addition to *C. botulinum*, other pathogenic and spoilage bacteria inhibited by sorbate in various meat products include *B. cereus*, *B. licheniformis*, *Brochothrix thermosphacta*, *C. perfringens*, *C. sporogenes*, *E. coli*, *Lactobacillus*, *Pseudomonas*, *Salmonella*, *S. aureus*, *Serratia liquefaciens*, *Y. enterocolitica*, mesophiles, psychrotrophs, and lipolytic organisms (Bushway et al., 1982; Chang et al., 1983, Cunningham, 1979; Gray et al., 1984; Greer, 1982; Elliot et al., 1985; Hallerbach and Potter, 1981; Kaloyereas et al., 1961; McMeekin et al., 1984; Morad et al., 1982; Myers et al., 1983; Perry et al., 1964; Petaja et al., 1979; Pierson et al., 1979; Rice and Pierson, 1982; Robach, 1979a; Robach and Ivey, 1978; Sofos et al., 1980b; Sofos, 1986b, c; Tompkin et al., 1974; To and Robach, 1980a, b; Wagner et al., 1982; Zamora and Zaritzky, 1987a, b). Products in which these microorganisms are inhibited include bacon and other cured meats, cured poultry products, raw beef and pork, and uncured poultry meat and carcasses. The antimicrobial activity of sorbate in these products has been enhanced when combined with nitrite, sodium chloride, phosphates, antioxidants, acids, low pH, low storage temperature, low oxygen, and increased carbon dioxide atmospheres (Sofos, 1989).

Sorbate in meat products (<0.3%) had no major adverse effects on sensory qualities, such as color and flavor (Amundson et al., 1981a, b; Bauermann, 1979; Berry and Blumer, 1981; Berry et al., 1981; Chambers et al., 1981; Frank, 1977; Huhtanen et al., 1981, 1983; Ivey et al., 1978; Kemp et al., 1979, 1983; Paquette et al., 1980; Vareltzis et al., 1984; Vareltzis and Buck, 1984; Wagner and Gehrke, 1982). In addition, the inclusion of sorbate and reduction of nitrite levels in bacon-pumping brine solutions resulted in reduced levels of nitrosamines in the fried product (Ivey et al., 1978; Robach et al., 1980b; U.S. Department of Agriculture, 1979b). Based on the results of the various studies, sorbate was proposed as a means of reducing nitrite and nitrosamine levels in bacon while maintaining antimicrobial activity and inhibition of *C. botulinum* spores during temperature abuse of the product. This proposal, however, did not take effect following a report that consumption of experimental bacon resulted in allergic-type reactions in certain individuals (Berry and Blumer, 1981). It should be noted, however, that no other studies reported undesirable flavors or allergic reactions (Sofos, 1989) and that a National Academy of Sciences committee concluded that although the flavor of bacon treated with combinations of sorbate and nitrite was not the same as the flavor of bacon treated only with nitrite, the flavors of both types of bacon were equally desirable (National Academy of Sciences, 1982).

4.4.6 Miscellaneous Food Products

Sorbates are commonly used in the preservation of oil-in-water or water-in-oil emulsion-based products, such as margarine, mayonnaise, salad dressings, delicatessen, and similar products. A survey of products in Persian markets revealed that levels of potassium sorbate in mayonnaise typically did not exceed 0.033% (Chaleshtori et al., 2018). In certain instances and for better preservation, sorbic acid may be mixed with its potassium salt or with other preservatives, such as benzoates. Other food-related applications of sorbates include sugar-based and confectionery products at concentrations of 0.05–0.20%, at which they prevent the growth of molds and osmophilic yeasts in items such as fillings for chocolate and pralines (Chichester and Tanner, 1972; Lück, 1976, 1980; Sofos and Busta, 1981). Sorbates are also used to prevent mold growth on the surface of dried and smoked fish (e.g., dried cod). In Asian countries sorbates are commonly used in combination with other preservatives to preserve fish sausages and similarly processed fish and meat products, as well as fermented plant foods (Chichester and Tanner, 1972; Lück,

1976, 1980; Robach and Sofos, 1982; Sofos, 1989). Several studies have reported improved preservation of various types of fish with sorbate (Bremner et al., 1978; Bremner and Statham, 1983; Chung and Lee, 1981; Debevere and Voets, 1972; Fey and Regenstein, 1982; Geminder, 1959; Robach and Hickey, 1978; Lynch and Potter, 1982; Regenstein, 1983; Robach and Shaw et al., 1983; Statham and Bremner, 1983; Sharp et al., 1986). One study, however, indicated that toxin production by *C. botulinum* in shucked scallops developed slightly more rapidly at 27°C in vacuum packages when sorbate (0.1%) was used (Fletcher et al., 1988). A sorbate (0.1%)–benzoate (0.1%) solution preserved brined shrimp for 59 days, which is longer than the preservation (31 days) achieved by the most effective bacteriocin (nisin) tested (Einarsson and Lauzon, 1995). Application of potassium sorbate with bifidobacteria extended the shelf-life of whole and peeled shrimp by three days (Al-Dagal and Bazaraa, 1999). Garcia-Soto et al. (2015) demonstrated the effect of combining lyophilized algae (*Fucus spiralis*) (8%) and sorbic acid (1%) into polylactic acid (PLA) biodegradable film was able to maintain freshness of refrigerated fish up to 11 days while fish stored in conventional polyethylene film were spoiled by the same time period.

Sorbic acid at a concentration of 0.5% did not adversely affect the germination energy and germination capacities of sorghum grain and did not significantly affect malting loss, diastatic powers, and cold and hot water extracts of the malt samples. This optimal concentration of sorbic acid coincided with the lethal dosage for most of the associated fungal microorganisms during malting, inhibiting their growth and, thus, it may be desirable to apply sorbic acid during malting (Ogundiwin et al., 1991).

The literature contains reports on several other food products that may be preserved with sorbates (Sofos, 1989). They include high-moisture marshmallow candy, sour cream, tangerine sherbet base, orange peels, strawberry puree, pineapples, tomato juice, Mexican-type hot sauces, pre-peeled carrots, potato chips, cut squash, shredded cabbage, cucumbers, garlic oil, and pancake batter (Sofos, 1989). A multistate outbreak of listeriosis in commercially produced and prepackaged caramel apples in 2014–2015 that resulted in 35 people becoming ill and 7 fatalities drove research into controlling this pathogen. Carstens et al. (2018) conducted a study to assess the effect of pre-soaking the sticks used in the manufacture of caramel apples with up to a 40% potassium sorbate solution to control post-process *L. monocytogenes* populations. The results demonstrated that a 40% potassium sorbate solution could drop *L. monocytogenes* populations by up to 3 logs post-process thereby proving it to be a viable preventive measure.

4.4.7 Other Applications

Other items that may be preserved by sorbic acid and its salts include pharmaceuticals, cosmetic products, and animal feeds (Sofos, 1989). In addition, the esters of sorbic acid (e.g., ethyl and propyl) may find applications in cosmetics and pharmaceuticals, especially those of the emulsion type, in which the emulsifiers may be inactivated by nonionic compounds (Chichester and Tanner, 1972; Lück, 1976). Sorbates may be particularly important in intermediate-moisture (25% moisture content) pet foods, in which levels of up to 0.30% can be effective mold inhibitors. Sorbic acid levels in the range of 0.03–0.10%, depending on moisture content, are also effective in preventing mold growth on animal feeds during storage. This should assure mycotoxin-free animal feeds, and, thus, eliminate any hazards associated with mycotoxin-contaminated feed. Mayer and Hillebrandt (1997) established that chemical preservation with sorbic acid was a possible way of preventing potato pulp from spoiling for subsequent practical application (feed). Samples treated with sorbic acid lost biological activity within a few days of aerobic storage (confirmed by aerobic degradation of starch and triglycerides). Although the presence of enzyme activities was shown, there was no detection of pectinolytic activities (Mayer and Hillebrandt, 1997). Blanching (80°C) and the application of 0.1% sorbic acid to citrus pulp were effective in inhibiting yeasts, molds, enterobacteria, and clostridia common in the spoilage of silage, and, therefore, reduced ensiling losses and increased the feed value characteristics of citrus pulp silages (Filya et al., 2001). The preservatives sorbic acid and methyl paraben, when used individually or in combination as preservatives of an artificial diet for leaf-cutting bees, did not influence the susceptibility of the bees to the fungus *Ascosphaera aggregate* (Goettel and Duke, 1996). Other reported uses of sorbates include preservation of tobacco products, adhesives, and glues (Lück, 1976; Sofos, 1989).

4.5 Toxicology and Safety

In the United States, sorbate is considered a generally recognized as safe compound, and since it is a metabolizable fatty acid, the World Health Organization has set the acceptable daily intake for sorbate at 25 mg/kg body weight per day. The relative nontoxicity of sorbates was established early in their testing as food preservatives. The compounds were fed to various animal species for determination of acute toxicity, as well as influence on metabolism, carcinogenicity, and teratogenicity after short- or long-term exposure. Overall, these studies demonstrated the relative harmlessness of sorbates and their relative superiority in safety compared to other chemical additives.

Acute toxicity studies with rats have determined mean lethal dose (LD_{50}) values for sorbates in the range of 4.2–10.5 g/kg body weight (Deuel et al., 1954a; Smyth and Carpenter, 1948). For comparison, the LD_{50} for common salt (NaCl) is 5 g/kg body weight. Thus, based on acute toxicity, sorbate is considered one of the least harmful preservatives in use.

Rat feeding studies have indicated that a dosage of 10% sorbic acid in the diet could be tolerated for 40 days (Lück, 1980). Upon extending the feeding period to 120 days, however, the growth rate and liver weight of the animals increased (Demaree et al., 1955). Additional studies with rats and dogs have shown no damage with feeding 5% sorbic acid for 90 days, but 8% sorbic acid resulted in an increase in liver weight (Demaree et al., 1955; Lück, 1976; Deuel et al., 1954a). The increase in liver weight has not been associated with histopathological changes and has been interpreted as functional hypertrophy due to the caloric utilization of sorbic acid (Deuel et al., 1954a; Food and Drug Research Laboratories, 1973; Lück, 1980; Sofos, 1989).

Chronic toxicity studies have involved feeding rats and mice through the life span of one or two generations with sorbic acid concentrations as high as 90 mg/kg body weight or diets containing 10% sorbic acid. No abnormalities have been observed, and in some instances the growth rate of sorbate-fed animals increased significantly, apparently due to an increased caloric intake resulting from the metabolizable sorbic acid. No carcinogenic or mutagenic effects have been observed with sorbate alone (Dickens et al., 1968; Food and Drug Research Laboratories, 1975; Gaunt et al., 1975; Hendy et al., 1976; Litton Biometrics, 1974, 1977; Lück, 1980; Shtenberg and Ignat'ev, 1970; Sofos and Busta, 1981, 1993; Sofos, 1994).

An allergic-type response has been reported for sorbic acid in one study in which bacon samples with sorbate–nitrite combinations were tested against *C. botulinum*. Some taste panelists reported certain "allergic-type" symptoms after tasting uninoculated experimental bacon (Berry and Blumer, 1981; U.S. Department of Agriculture, 1979b). Even though it was implied that sorbate might have been involved in producing those symptoms, no direct relation could be proven between symptoms and specific ingredients used in formulating the bacon (U.S. Department of Agriculture, 1979b). In addition, no such symptoms were observed by other individuals who tasted sorbate–nitrate experimental bacon from other studies, and Robach and Adam (1980) were able to induce such symptoms from panelists consuming commercial bacon manufactured without sorbate. This, of course, does not rule out the possibility that high concentrations of sorbate may act as irritants to certain susceptible individuals (Sofos, 1989).

As preservatives for cosmetics and pharmaceutical products, sorbates have been extensively examined for skin tolerance (Lück, 1976; Patrizi et al., 1999). The literature is contradictory, but it indicates that most people are not affected by sorbic acid applied to the skin. Some sensitive individuals, however, show skin irritations when exposed to sorbic acid (Dejobert et al., 2001; National Academy of Sciences, 1982). Average sorbic acid concentrations for skin irritations are in the range of 1%, and some very sensitive individuals may show irritations even at lower levels (Lück, 1976). Considering the average use levels of 0.10–0.30% in food processing, the potential for such irritations in commercial products is minor.

There has been some concern about possible sorbate–nitrite reactions when added to the same substrate (National Academy of Sciences, 1982; Sofos, 1981). Since there is evidence of reactions between nitrite and fatty acids (Benedict, 1980) and because sorbic acid is an unsaturated fatty acid, it may react with nitrite in products such as cured meats. The products of such reactions have the potential to be mutagenic (Sofos, 1989). In general, however, various studies have indicated that: (1) nitrite reacts with sorbic acid to form mutagens optimally at pH 3.5, not at 6.0, which is the average pH of most meat products; (2)

formation of the c-nitroso mutagens requires the presence of excess nitrite; and, (3) mutagen formation is inhibited by such ingredients as ascorbic acid, cysteine, and vegetable juices and is inactivated by heat (Kada, 1974; Khoudokormoff, 1981; Kito et al., 1978; Namiki et al., 1979, 1980, 1981, 1983; Osawa et al., 1979, 1980, 1982, 1986; Osawa and Namiki, 1982; Robach et al., 1980a). Thus, the requirements for the reactions to proceed and the instability of the products make it unlikely for mutagens to be formed at detectable or harmful concentrations in foods treated with nitrite and sorbate (Sofos, 1981, 1989). Walker (1990) concluded that sorbate use, in general, does not appear to create toxicological problems.

It has been regarded that amino-compounds, including lysine and glutamate, may undergo non-enzymatic browning reactions when stored with sorbic acid at a favorable range of a_w and temperature (Quattrucci and Masci, 1992). This reaction was demonstrated in model systems but, to some extent, it may apply also to more complex systems such as food, leading to decreased amino acid availability (Quattrucci and Masci, 1992). Sorbates were found to react easily with proteins, forming high-molecular-weight products which do not seem to result in a nutritional loss, probably because the acid environment of the stomach leads to the breakdown of the sorbic-protein adducts (Quattrucci and Masci, 1992).

Sorbic acid has a conjugated system of double bonds which makes it susceptible to nucleophilic attack, sometimes giving mutagenic products (Ferrand et al., 2000a, b). Under conditions typical of food processing (50–80°C) cyclic derivatives resulting from a double addition reaction between sorbic acid and various amines were analyzed. The formation of new products by addition reactions using ethyl sorbate and various amines led, at 20°C, to linear monoadducts and, at 50°C, to cyclic derivatives resulting from double addition (Ferrand et al., 2000a, b). Mutagenesis studies, involving the Ames test and genotoxicity studies with HeLa cells and on plasmid DNA, in cyclic interaction products, showed that none of the products studied presented mutagenic or genotoxic activities (Ferrand et al., 2000a, b). It should be noted, however, that certain studies have reported inhibition by sorbate of carcinogenic nitrosamine formation in model systems (Chung, 1981; Lathia and Schellhoeh, 1981; Massey et al., 1982; Rao et al., 1982; Tanaka et al., 1978; Yamamoto et al., 1988). Sorbic acid, however, had no effect on nitrosamine formation from aminopyrine and sodium nitrite in animal stomachs (Kawanishi et al., 1981a, b; Sofos, 1989). Potential injury in cell hepatocytes induced by potassium sorbate was prevented by antioxidants (Sugihara et al., 1997).

No synergism in acute toxicity has been detected for combinations of sorbic acid with various other additives, including parabens, benzoate, and propionate (Sofos, 1989). Sorbate, however, did not protect against the toxic effects of other substances (Daoud and Griffin, 1980). The derivatives of sorbic acid are also relatively nontoxic. Overall, sorbates appear to be one of the safest food preservatives available.

4.6 Detection and Analysis

Analytical methods used or tested for qualitative and quantitative detection of sorbates in foods include acidimetry, bromometry, colorimetry, enzymatic, spectrophotometry, polarography, and chromatography. The most widely used methods, however, have been colorimetric (Schmidt, 1960a, 1962) and spectrophotometric (Melnick and Luckman, 1954a), although chromatographic methods have gained acceptance in recent years.

Detection methods require quantitative extraction and separation of sorbic acid from the food material without food ingredient interference (DeLuca et al., 1995; Mandrou et al., 1999; Montano et al., 1995; Sofos, 2000). Extraction methods include acid-steam distillation, selective gas diffusion, and solvent extension using ethyl or petroleum ether, dichloromethane, and isooctane. In some foods, filtration, dialysis, or direct analysis has been used (Sofos, 1989).

Extraction by steam distillation has been used extensively, but it is time-consuming, and the compounds present in the food or generated by decomposition of lipid materials may interfere with colorimetric or spectrophotometric detection of sorbic acid (Harrington et al., 1962; Sofos, 1989). Several modifications have been proposed as useful in avoiding interference and in improving the accuracy of the steam distillation procedure (Luckmann and Melnick, 1955; Sofos, 1989), while combinations of various treatments have also been used to improve extraction and reduce interference (Sofos, 1989). Such combinations have involved extraction under acid conditions from the steam distillate and re-extraction with sodium hydroxide; dialysis and solvent extraction; successive extractions with ether, sodium hydroxide, and methylene chloride; and double distillation and ether extraction (Larsson, 1983; Noda et al., 1973; Puttermans et al., 1984; Sofos, 1989; Tjan and Konter, 1972).

The colorimetric detection of sorbic acid at an absorbance of 532 nm is an official method for quantitative determination in foods and beverages (AOAC Int. method 975.31), cheese (AOAC Int. methods 971.15 and 975.22), and wine (AOAC Int. method 975.10) by the color reaction with α-thiobarbituric acid (AOAC Int., 2000; Roy et al., 1976). The method is simple and usually very specific (Lück, 1980). The spectrophotometric (ultraviolet/UV, absorption) procedure has also been used in many foods, including fruit products, bakery items, wine, cheese, and sausage products (Holley and Millard, 1980; Maxstadt and Karasz; 1972; Stafford, 1976; Wilamowski, 1974). After development by Melnick and Luckman (1954a), the method was modified by Alderton and Lewis (1958); it involves measurement of absorbance at 260nm (250–290 nm). Special extraction and purification steps have been proposed in the literature to reduce interference problems (Sofos, 1989). The method is an official first action procedure for ground beef (AOAC Int. method 980.17) and final action procedure for wines (AOAC Int. method 974.08) and dairy products (AOAC Int. method 974.10) (AOAC Int., 2000).

The literature contains numerous reports of chromatographic methods used or evaluated for the determination of sorbate in food products (Sofos, 1989). They include gas chromatography, high-performance liquid chromatography (HPLC), thin layer chromatography, paper chromatography, and micellar electrokinetic capillary chromatography (MECC) and ion chromatography. A gas chromatographic method is a first action procedure (AOAC Int. method 983.16) for sorbic and benzoic acids in foods (AOAC Int., 2000). Solid extraction has been used as a clean-up procedure for the determination of sorbic acid by liquid chromatography in fruit products (Mandrou et al., 1999). Others, however, have reported that clean-up procedures did not improve determination by liquid chromatography (Benassi and Cecchi, 1998). Aung and Pyell (2016) developed a method combining in-capillary derivatization with separation by capillary electrophoresis; the method uses the rapid derivatization of monosodium glutamate with o-phthalaldehyde (OPA) in the presence of 3-mercaptopropionic acid for non-derivatized sorbic acid. The method is validated with respect to LOD, LOQ, linearity, recovery, and precision and can be applied to real samples such as soy, fish, oyster, and sweet and sour chili sauces after application of appropriate clean-up steps.

Microdialysis was used to extract sorbic acid and benzoic acid from food to be separated and detected by HPLC (Mannino and Cosio, 1996). In addition, HPLC was found to be adequate in obtaining accurate stability data for sorbic acid in creams (de Villiers and Bergh, 2000). A rapid method for the identification and quantitation of sorbic and benzoic acids in beverages and foods by MECC has been reported (Pant and Trenerry, 1995). More recent methods have also been developed using microdialysis using a nylon membrane combined with derivative spectrophotometry to test for sorbic acid in food additives like antioxidants and sweeteners. Such methods can make for faster and inexpensive testing for sorbic acid either as a screening test or for less rigorous applications that do not require follow-up testing. Use of the nylon membrane, 0.2-μm pore size, reduces impurities such as pigments and higher-molecular-weight substances. Reduction of interference was also enhanced by using derivative spectroscopy. Results of this method were compared to HPLC methods and had a very high degree of correlation (Tomoharu Fujiyoshi et al., 2018). Testing has also been done in soft drinks, milk, ketchup sauce, and bread by dispersive liquid-liquid microextraction in combination with HPLC methods to determine usual sorbic acid and benzoic acid consumption in Iran (Farkin Javanmardi et al., 2015).

Ion chromatography has been shown to be in good agreement with HPLC in determinations of sorbic acid in food and pharmaceutical preparations (Chen and Wang, 2001). In addition, ion chromatography was effective in simultaneously determining artificial sweeteners, preservatives, caffeine, theobromine, and theophylline and as such may be a beneficial alternative to conventional HPLC.

A polarographic method employing differential-pulse voltometry with the use of an HMDE was developed for the determination of sorbic acid in fruit juices and drinks. The procedure is simple and specific and is not subject to interference from many substances commonly found in soft drinks such as ethanol and benzoic acid. Moreover, it can be used for the analysis of intensely colored juice samples. Thus, the scope of application of the proposed method for the determination of sorbic acid has been extended to cover juice and drink samples which cannot be analyzed by the AOAC procedure (Fung and Luk,

1990). The use of reversed phase high performance liquid chromatography (RP-HPLC) was found to be a simple, rapid method with high reproducibility for determining sorbic acid and other additives in fruit juices (Zou et al., 2001). More recently HPLC methods have been developed to determine sorbic acid in shortening oil and rice wine. The rice wine method had recovery of spiked samples of 95.1–96.4% and low standard deviation of 2.6–3.8%, and the shortening oil had a 98% recovery of a spiked sample and relative standard deviation of 10.93% (Song Li-Jun et al., 2017; Ye Jing-Wen et al., 2015).

The detection of sorbic acid in beverages using HPLC methods has also been achieved using simple and effective methods that have the additional advantage of a simple mobile phase of acetonitrile–phosphate buffer at pH = 3.5 in an 8:92 ratio, meaning that only 8% of the composition is acetonitrile while still achieving separation for identification and quantification (Biljana Petanovska-Ilievska et al., 2017).

There has also been the development of more environmentally friendly procedures for certain HPLC methods of detecting benzoic and sorbic acid, specifically in soy sauce. The method involves extraction and then back extraction using only 5 mL of diethyl ether in sample preparation and no organic solvents in the preparation of the standard solutions used. The analysis used an internal standard and had excellent accuracy with relative standard deviation of less than 3% with recoveries between 96.1% and 104.3%. Such advantages potentially make the use of such methods valuable for large-scale purposes (Mingzhe Ding et al., 2015). Methods such as this mentioned above have the ability to reduce costs as well as improve safety both for those working in the lab and the environment.

An enzymatic method used to determine sorbic acid based on the spectrophotometric measurement of sorbyl coenzyme A at 300 nm has been developed (Hofer and Jenewein, 2000). The method is based on sorbic acid being converted to sorbyl CoA with acyl CoA synthetase in the presence of coenzyme A and adenosine-5'-triphosphate. The reaction is quantified by irreversibly hydrolyzing pyrophosphate to phosphate in the presence of inorganic pyrophosphatase, where the absorbance is measured at 300 nm and is specific for sorbyl CoA which is proportional to sorbic acid levels in the sample. The method was found to be fast, precise, and reliable, and is thus well-suited for routine determinations especially for high sample throughputs (Hofer and Jenewein, 2000).

Capillary electrophoresis was also applied for detection in citrus juices, wine, and other substrates (Cancalon, 1999; Mercier et al., 1998; Castineira et al., 2000; Dobiasova et al., 2002). Inhibition of microorganisms has also been evaluated as a procedure for the qualitative detection of sorbate (Ellerman, 1977). It is obvious, however, that in addition to sorbate, such inhibition may be caused by a number of other inhibitors.

4.7 Regulatory Status

Several forms of sorbate are allowed for use in a variety of foods throughout the world (Sofos, 1989). This major use may expand in the future, considering their extensive testing, low toxicity, and advantages over other preservatives. In the United States, sorbic acid and potassium sorbate are GRAS, and this status has been reaffirmed (U.S. Department of Agriculture, 1978) by a select committee of the U.S. Food and Drug Administration. In addition, sorbates in the United States are sanctioned in more than 80 food products designated by standards of identity. According to the *Code of Federal Regulations*, when a food preservative is used in a food product, its common name (e.g., potassium sorbate or sorbic acid) should be listed on the product label and its function should be indicated by an explanatory description (e.g., "to maintain freshness," "to extend shelf-life," or "as a preservative"). The use of sorbates may be requested for any food product that allows preservatives.

REFERENCES

Acott, K., Sloan, A. E., and Labuza, T. P. 1976. Evaluation of antimicrobial agents in a microbial challenge study for an intermediate moisture dog food. *J. Food Sci.* 41:541–546.

Ahmad, S. R., Pathak, V., Bhat, Z. F., and Bukhari, S. A. A. 2014. Effect of sorbic acid on the storage quality of Kaladhi: An acid coagulated milk product. *J. Food Sci. Technol.* 51:4040–4046.

Alagoz, S., Turkyilmaz, Tagi, S., and Ozkan, M. 2015. Effects of different sorbic acid and moisture levels on chemical and microbial qualities of sun-dried apricots during storage. *Food Chem.* 174:356–364.

Al-Dagal, M. M., and Bazaraa, W. A. 1999. Extension of shelf life of whole and peeled shrimp with organic acid salts and bifidobacteria. *J. Food Prot.* 62:51–56.

Alderton, G., and Lewis, J. C. 1958. Determination of sorbic acid and its disappearance from pickled brines. *Food Res.* 23:338.

Alli, Z. L. I., and Kermasha, S. 1989. Use of acidification, low temperature, and sorbates for storage of orange juice. *J. Food Sci.* 54:674–678.

Alnoman, M.Udompijitkul, P., Paredes-Sabja, D., and Sarker, M. 2015. The inhibitory effects of sorbate and benzoate against *Clostridium perfringens* type A isolates. *Food Microbiol.* 48:89–98.

Aly, M. E. 1996. Prolongation of the keeping quality of Mozzarella cheese by treatment with sorbate. *Die Nahrung* 40:194–200.

Amano, K., Shibasaki, I., Yokoseki, M., and Kawabata, T. 1968. Preservation of fish sausage with tylosin, furylfuramide and sorbic acid. *Food Technol.* 22:881–885.

Amundson, C. M., Sebranek, J. G., Kraft, A. A., Rust, R. E., Wagner, M. K., and Robach, M. C. 1981a. Effect of belly composition on sorbate-cured bacon. *J. Food Sci.* 47:218–221.

Amundson, C. M., Sebranek, J. G., Kraft, A. A., Rust, R. E., Wagner, M. K., and Robach, M. C. 1981b. Effect of belly handling (chilled, frozen, prerigor) and smoking delay on regular and sorbate-cured bacon. *J. Food Sci.* 47:222–225.

Anderson, T. E., and Costilow, R. N. 1963. Site of inhibition of yeast respiration by sorbic acid. *Bacteriol. Proc.* 63:14.

AOAC Int. 2000. *Official Methods of Analysis*, 17th ed., AOAC International, Gaithersburg, MD.

Aouadhi, C., Mejri, S., and Maaroufi, A. 2015. Inhibitory effects of nisin and potassium sorbate alone or in combination on vegetative cells growth and spore germination of *Bacillus sporothermodurans* in milk. *Food Microbiol.* 46:40–45.

Arya, S. S., and Thakur, B. R. 1988. Degradation products of sorbic acid in aqueous solutions. *Food Chem.* 29:41–49.

Aung, H.-P., and Pyell, U. 2016. In-capillary derivatization with o-phthalaldehyde in the presence of 3-mercaptopropionic acid for the simultaneous determination of monosodium glutamate, benzoic acid, and sorbic acid in food samples via capillary electrophoresis with ultraviolet detection. *J. Chrom. A.* 1449:156–165.

Azukas, J. J., Costilow, R. N., and Sadoff, H. L. 1961. Inhibition of alcoholic fermentation by sorbic acid. *J. Bacteriol.* 81:189–194.

Baldock, J. D., Frank, P. R., Graham, P. P., and Ivey, F. J. 1979. Potassium sorbate as a fungistatic agent in country ham processing. *J. Food Prot.* 42:780–783.

Banks, J. G., Morgan, S., and Stringer, M. F. 1988. Inhibition of heated *Bacillus* spores by combinations of potassium sorbate, sodium benzoate, pH and organic acids. *Lebensm. Wiss. Technol.* 21:250–255.

Barzegar, H., Hossein Azizi, M., Barzegar, M., and Hamidi-Esfahani, Z. 2014. Effect of potassium sorbate on antimicrobial and physical properties of starch-clay nanocomposite films. *Carb. Polymers* 110:26–31.

Bauermann, J. F. 1979. Processing of poultry products with and without sodium nitrite. *Food Technol.* 33:42–43.

Bell, R. G., and De Lacy, K. M. 1987. The efficacy of nisin, sorbic acid and monolaurin as preservatives in pasteurized cured meat products. *Food Microbiol.* 4:277–283.

Bell, T. A., Etchells, J. L., and Borg, A. F. 1959. Influence of sorbic acid on the growth of certain species of bacteria, yeasts and filamentous fungi. *J. Bacteriol.* 77:573–580.

Benassi, M. T., and Cecchi, H. M. 1998. Method development for the simultaneous determination of carboxylic acids, phenolic compounds, and sorbic acid in white wines. *J. Liq. Chromatogr. Rel. Technol.* 21:491–501.

Benedict, R. C. 1980. Biochemical basis for nitrite-inhibition of *Clostridium botulinum* in cured meat. *J. Food Prot.* 43:877–891.

Berry, B. W., and Blumer, T. N. 1981. Sensory, physical, and cooking characteristics of bacon processed with varying levels of sodium nitrite and potassium sorbate. *J. Food Prot.* 46:321–327.

Berry, B. W., Quick, J. A., Douglass, L. W., and Tennent, I. N. 1981. Shelflife characteristics of bacon processed with various levels of sodium nitrite and potassium sorbate. *J. Food Prot.* 46:596–600.

Beuchat, L. R. 1980. Comparison of anti-*Vibrio*, activities of potassium sorbate, sodium benzoate, and glycerol and sucrose esters of fatty acids. *Appl. Environ. Microbiol.* 39:1178–1182.

Beuchat, L. R. 1981a. Effects of potassium sorbate and sodium benzoate on inactivating yeasts heated in broths containing sodium chloride and sucrose. *J. Food Prot.* 44:765–769.

Beuchat, L. R. 1981b. Synergistic effects of potassium sorbate and sodium benzoate on thermal inactivation of yeasts. *J. Food Sci.* 46:771–777.

Beuchat, L. R. 1981c. Combined effects of solutes and food preservatives on rates of inactivation of and colony formation by heated spores and vegetative cells of molds. *Appl. Environ. Microbiol.* 41:472–477.

Beuchat, L. R. 1982. Thermal inactivation of yeasts in fruit juices supplemented with food preservatives and sucrose. *J. Food Sci.* 47:1679–1682.

Bhattacharya, M., and Majumdar, S. K. 1984. Effect of some antifungal compounds on growth of *Penicillium citrinum* and citrinin production. *J. Food Sci. Technol.* 21:38–40.

Bills, S., Restaino, L., and Lenovich, L. M. 1982. Growth response of an osmotolerant sorbate-resistant yeast, *Saccharomyces rouxii*, at different sucrose and sorbate levels. *J. Food Prot.* 45:1120–1124, 1128.

Bizri, J. N., and Wahem, I. A. 1994. Citric acid and antimicrobials affect microbiological stability and quality of tomato juice. *J. Food Sci.* 59:130–134.

Blocher, J. C., and Busta, F. F. 1983. Influence of potassium sorbate and reduced pH on the growth of vegetative cells of four strains of type A and B *Clostridium botulinum. J. Food Sci.* 48:574–575, 580.

Blocher, J. C., and Busta, F. F. 1985. Multiple modes of inhibition of spore germination and outgrowth by reduced pH and sorbate. *J. Appl. Bacteriol.* 59:469–478.

Blocher, J. C., Busta, F. F., and Sofos, J. N. 1982. Influence of potassium sorbate and pH on ten strains of type A and B *Clostridium botulinum. J. Food Sci.* 47:2028–2032.

Bolin, H. R., King, A. D., Jr., and Stafford, A. E. 1980. Sorbic acid loss from high moisture prunes. *J. Food Sci.* 45:1434–435.

Botha, S. J., and Holzapfel, W. H. 1987. Resistance of *Sporolactobacillus* to potassium sorbate and sodium nitrite. *Intl. J. Food Microbiol.* 5:331–336.

Bracey, D., Holyoak, D. D., and Coote, P. J. 1998. Comparison of the inhibitory effect of sorbic acid and amphotericin B on *Saccharomyces cerevisiae*: Is growth inhibition dependent on reduced intracellular pH? *J. Appl. Microbiol.* 85:1056–1066.

Bremner, H. A., and Statham, J. A. 1983. Effect of potassium sorbate on refrigerated storage of vacuum packed scallops. *J. Food Sci.* 48:1042–1047.

Bremner, H. A., Laslett, G. M., and Olley, J. 1978. Taste panel assessment of textural properties of fish minces from Australian species. *J. Food Technol.* 13:307–318.

Bullerman, L. B. 1977. Incidence and control of mycotoxin producing molds in domestic and imported cheeses. *Annu. Nutr. Aliment.* 31:435–446.

Bullerman, L. B. 1983. Effects of potassium sorbate on growth and aflatoxin production by *Aspergillus parasiticus* and *Aspergillus flavus. J. Food Prot.* 46:940.

Bullerman, L. B. 1984. Effects of potassium sorbate on growth and patulin production by *Penicillium roqueforti* and *Penicillium patulum. J. Food Prot.* 47:312–316.

Bullerman, L. B. 1985. Effects of potassium sorbate on growth and ochratoxin production by *Aspergillus ochraceus* and *Penicillium* species. *J. Food Prot.* 48:162.

Bullerman, L. B., and Olivigni, F. J. 1974. Mycotoxin producing-potential of molds isolated from cheddar cheese. *J. Food Sci.* 39:1166–1168.

Burlini, N., Pellegrini, R., Facheris, P., Tortora, P., and Guerritore, A. 1993. Metabolic effects of benzoate and sorbate in the yeast *Saccharomyces cerevisiae* at neutral pH. *Arch. Microbiol.* 159:220–224.

Bushway, A. A., Fickner, N., and Jen, C. W. 1982. Effect of nitrite and sorbate on total number of aerobic microorganisms in chicken white and dark meat patties. *J. Food Sci.* 47:858–860, 863.

Campos, C. A., Rojas, A. M., and Gerschenson, L. N. 1996. Studies of the effect of ethylene diamine tetraacetic acid (EDTA) on sorbic acid degradation. *Food Res. Intl.* 29:259–264.

Cancalon, P. F. 1999. Analytical monitoring of citrus juices by using capillary electrophoresis. *J. AOAC Intl.* 82:95–106.

Carstens, C. K., Salazar, J. K., Bathija, V. M., Narula, S. S., Wang, P., and Tortorello, M. L. 2018. Control of *Listeria monocytogenes* in caramel apples by use of sticks pretreated with potassium sorbate. *J. Food Prot.* 81:1921–1928.

Castineira, A., Pena, R. M., Herrero, C., and Garcia-Martin, S. 2000. Simultaneous determination of organic acids in wine samples by capillary electrophoresis and UV detection: Optimization with five different background electrolytes. *HRC-J. High Res. Chrom.* 23:647–652.

Cerruti, P., Alzamora, S. M., and Chirife, J. 1988. Effect of potassium sorbate and sodium bisulfite on thermal inactivation of *Saccharomyces cerevisiae* in media or lowered water activity. *J. Food Sci.* 53:1911–1912.

Cerruti, P., Alzamora, S. M., and Chirife, J. 1990. A multi-parameter approach to control the growth and *Saccharomyces cerevisiae* in laboratory media. *J. Food Sci.* 55:837–840.

Chaleshtori, F. S., Arian, A., and Chaleshtori, R. S. 2018. Assessment of sodium benzoate and potassium sorbate preservatives in some products in Kashan, Iran with estimation of human health risk. *Food Chem. Toxicol.* 120:634–638.

Chambers, E., IV, Bowers, J. A., Prusa, K., and Craig, J. 1981. Sensory attributes and instron measurements of reduced-nitrite poultry frankfurters with sorbic acid or potassium sorbate. *J. Food Sci.* 47:207–209.

Chang, C.-W., Sebranek, J. G., Walker, H. W., and Galloway, D. E. 1983. Packaging film permeability in conjunction with sodium nitrite, potassium sorbate or lactic acid starter in culture for control of *Clostridium sporogenes* (PA 3679) growth in sliced bologna. *J. Food Sci.* 48:861–864.

Charvalos, E., Tzatzarakis, M., Tsatsakis, A., and Petrikkos, G. 2001. Controlled release of water-soluble polymeric complexes of sorbic acid with antifungal activities. *Appl. Microbiol. Biotech.* 57:770–775.

Chen, Q. C., and Wang, J. 2001. Simultaneous determination of artificial sweeteners, preservatives, caffeine, theobromine and theophylline in food and pharmaceutical preparations by ion chromatography. *J. Chrom. A.* 937:57–64.

Cheng, L., and Piper, P. W. 1994. Weak acid preservatives block the heat shock response and heat-shock-element directed lacZ expression of low pH *Saccharomyces cerevisiae* cultures, an inhibitory action partially relieved by respiratory deficiency. *Microbiology* 140:1085–1096.

Cheng, L., Moghraby, J., and Piper, P. W. 1999. Weak organic acid treatment causes a trehalose accumulation in low-pH cultures of *Saccharomyces cerevisiae*, not displayed by the more preservative-resistant *Zygosaccharomyces bailii. FEMS Microbiol. Lett.* 170:89–95.

Chichester, D. F., and Tanner, F. W. 1972. Antimicrobial food additives. In *Handbook of Food Additives*, edited by T. E. Furia, CRC Press, Boca Raton, FL, pp. 115–184.

Chung, K. C. 1981. Inhibition of nitrosamine formation *in vitro* by sorbic acid. Yongnam Taehakkyo Nonmunjip, Chayon Kwahak Pyon 13:177; *Chem. Abstr.* 94:12081.

Chung, Y. M., and Lee, J. S. 1981. Inhibition of microbial growth in English sole (*Parophrys retulus*). *J. Food Prot.* 44:66–68.

Chung, Y. M., and Lee, J. S. 1982. Potassium sorbate inhibition of microorganisms isolated from seafood. *J. Food Prot.* 45:1310.

Cole, M. B., Franklin, J. G., and Keenan, M. H. J. 1987. Probability of growth of the spoilage yeast *Zygosaccharomyces bailii* in a model fruit drink system. *Food Microbiol.* 4:115–119.

Costilow, R. N., Ferguson, W. E., and Ray, S. 1955. Sorbic acid as a selective agent in cucumber fermentations. I: Effect of sorbic acid on microorganisms associated with cucumber fermentations. *Appl. Microbiol.* 3:341–345.

Costilow, R. N., Coughlin, F. M. Robach, D. L., and Rasheb, H. S. 1956. A study of the acid-forming bacteria from cucumber fermentations in Michigan. *Food Res.* 21:27–33.

Costilow, R. N., Coughlin, F. M., Robins, E. K., and Hsu, W. T. 1957. Sorbic acid as a selective agent in cucumber fermentations. II: Effect of sorbic acid on the yeast and lactic acid fermentations in brined cucumbers. *Appl. Microbiol.* 5:373–379.

Cowles, P. B. 1941. The germicidal action of the hydrogen ion and of the lower fatty acids. *Yale J. Biol Med.* 13:571.

Crowell, E. A., and Guymon, J. F. 1975. Wine constituents arising from sorbic acid addition, and identification of 2-ethoxyhexa-3,5-diene as a source of geranium-like off-odor. *Am. J. Enol. Vitic.* 26:97–102.

Cunningham, F. E. 1979. Shelf-life and quality characteristics of poultry parts dipped in potassium sorbate. *J. Food Sci.* 44:863–864.

Danzinger, M. T., Steinberg, M. P., and Nelson, A. I. 1973. Effect of CO_2, moisture content and sorbate on safe storage for wet corn. *Trans. Am. Soc. Agric. Eng.* 16:679–682.

Daoud, A. H., and Griffin, A. C. 1980. Effect of retinoic acid, butylated hydroxytoluene, selenium and sorbic acid on azo-dye hepatocarcinogenesis. *Cancer Lett.* 9:299–304.

D'Aquino, S., Fadda, A., Barberis, A., Palma, A., Angioni, A., and Schirra, M. 2013. Combined effects of potassium sorbate, hot water and thiabendazole against green mould of citrus fruit and residue levels. *Food Chem.* 141:858–864.

Davidson, P. M., Brekke, C. J., and Branen, A. L. 1981. Antimicrobial activity of butylated hydroxyanisole, tertiary butylhydroquinone, and potassium sorbate in combination. *J. Food Sci.* 46:314–316.

De Nobel, H., Lawrie, L., Brul, S., Klis, F. Davis, M. Allush, H. and Coote, P. 2001. Parallel and comparative analysis of the proteome and transcriptome of sorbic acid-stressed *Saccharomyces cerevisiae. Yeast* 18:1413–1428.

Deak, T., and Beuchat, L. R. 1994. Use of indirect conductimetry to predict the growth of spoilage yeasts, with special consideration of *Zygosaccharomyces bailii. Intl. J. Food Microbiol.* 23:405–417.

Deak, T., and Novak, E. K. 1970. Effect of sorbic acid on the growth of yeasts on various carbohydrates. *Acta Microbiol.* 17:257.

Debevere, J. M., and Voets, J. P. 1972. Influence of some preservatives on the quality of prepacked cod fillets in relation to the oxygen permeability of the film. *J. Appl. Bacteriol.* 35:351–356.

Dejobert, Y., Delaporte, E., Piette, F., and Thomas, P. 2001. Vesicular eczema and systemic contact dermatitis from sorbic acid. *Contact Derm.* 45:291.

DeLuca, C., Passi, S., and Quattrucci, E. 1995. Simultaneous determination of sorbic acid, benzoic acid and parabens in foods: A new gas chromatography-mass spectrophotometry technique adopted in a survey on Italian foods and beverages. *Food Addit. Contam.* 112:1–7.

Demaree, G. E., Sjogren, D. W., McCashland, B. W., and Cosgrove, F. P. 1955. Preliminary studies on the effect of feeding sorbic acid upon the growth reproduction, and cellular metabolism of albino rats. *J. Am. Pharm. Assoc. Sci. Ed.* 44:619–621.

Deuel, H. J., Jr., Alfin-Slater, R. Weil, C. S., and Smyth, H. F., Jr. 1954a. Sorbic acid as a fungistatic agent for foods. I: Harmlessness of sorbic acid as a dietary component. *Food Res.* 19:1–12.

Deuel, H. J., Jr., Calbert, C. E., Anisfeld, L., McKeehan, H., and Blunden, H. D. 1954b. Sorbic acid as a fungistatic agent for foods. II: Metabolism of α,β-unsaturated fatty acids with emphasis on sorbic acid. *Food Res.* 19:13–19.

de Villiers, M. M., and Bergh, J. J. 2000. Comparing HPLC and UV spectrophotometric analysis methods for determining the stability of sorbic acid in nonionic creams containing lactic acid. *Drug Develop. Indus. Phar.* 26:539–547.

Dickens, F., Jones, H. E. H., and Waynforth, H. B. 1968. Further tests on the carcinogenicity of sorbic acid in the rat. *Br. J. Cancer* 22:762–768.

Ding, M., Peng, J., Ma, S., and Zhang. Y. 2015. An environment-friendly procedure for the high performance liquid chromatography determination of benzoic acid and sorbic acid in soy sauce. *Food Chem.* 183:26–29.

Dobiasova, Z., Pazourek, J., and Havel, J. 2002. Simultaneous determination of trans-resveratrol and sorbic acid in wine by capillary zone electrophoresis. *Electrophor*esis 23:263–267.

Doulia, D., Rigas, F., and Katsinis, G. 2000. Multiparametric investigation of mould-free shelf life of bread via factorial design. *Int. J. Food Prot.* 3:363–374.

Draughon, F. A., Sung, S. C., Mount, J. R., and Davidson, P. M. 1982. Effect of the parabens with and without nitrite on *Clostridium botulinum* and toxin production in canned pork slurry. *J. Food Sci.* 47:1635–1637, 1642.

Dudman, W. F. 1963. Sorbic hydroxamic acid, an antifungal agent effective over a wide pH range. *Appl. Microbiol.* 11:362–364.

Edinger, W. D., and Splittstoesser, D. F. 1986a. Production by lactic acid bacteria of sorbic alcohol, the precursor of the geranium odor compound. *Am. J. Enol. Vitic.* 37:34–38.

Edinger, W. D., and Splittstoesser, D. F. 1986b. Sorbate tolerance by lactic acid bacteria associated with grapes and wine. *J. Food Sci.* 51:1077–1078.

Einarsson, H., and Lauzon, H. L. 1995. Biopreservation of brined shrimp (*Pandalus borealis*) by bacteriocins from lactic acid bacteria. *Appl. Environ. Microbiol.* 61:669–676.

Eklund, T. 1980. Inhibition of growth and uptake processes in bacteria by some chemical food preservatives. *J. Appl. Bacteriol.* 48:423–432.

Eklund, T. 1983. The antimicrobial effect of dissociated and undissociated sorbic acid at different pH levels. *J. Appl. Bacteriol.* 54:383–389.

Eklund, T. 1985. The effect of sorbic acid and esters of *p*-hydroxybenzoic acid on the protonmotive force in *Escherichia coli* membrane vesicles. *J. Gen. Microbiol.* 131:7376.

El Halouat, A., and Debevere, J. M. 1996. Influence of modified atmosphere and preservatives on the growth of *Zygosaccharomyces rouxii* isolated from dried fruits. *Intl. J. Food Microbiol.* 33:219–229.

El Halouat, A., Gourama, H., Uyttendaele, M., and Debevere, J. M. 1998. Effects of modified atmosphere packaging and preservatives on the shelf-life of high moisture prunes and raisins. *Intl. J. Food Microbiol.* 41:177–184.

Ellerman, J. A. G. 1977. Assay of sorbate and benzoate by the disc-plate method. *Aust. J. Dairy Technol.* 32:13–14.

Elliott, P. H., and Gray, R. J. H. 1981. *Salmonella* sensitivity in a sorbate/modified atmosphere combination system. *J. Food Prot.* 44:903.

Elliott, P. H., Tomlins, R. I., and Gray, R. J. H. 1982. Inhibition and inactivation of *Staphylococcus aureus* in a sorbate/modified atmosphere combination system. *J. Food Prot.* 45:1112–1116.

Elliott, P. H., Tomlins, R. I., and Gray, R. J. H. 1985. Control of microbial spoilage on fresh poultry using a combination potassium sorbate/carbon dioxide packaging system. *J. Food Sci.* 50:1360–1363.

El-Shenawy, M. A., and Marth, E. H. 1988. Inhibition and inactivation of *Listeria monocytogenes* by sorbic acid. *J. Food Prot.* 51:842–847.

Emard, L. O., and Vaughn, R. H. 1952. Selectivity of sorbic acid media for the catalase negative lactic acid bacteria and clostridia. *J. Bacteriol.* 63:487–494.

Eyles, M. J., and Warth, A. D. 1989. The response of *Gluconobacter oxydans* to sorbic and benzoic acids. *Intl. J. Food Microbiol.* 8:335–342.

Fan, J. J., and Chen, J. H. 1999. Inhibition of aflatoxin-producing fungi by Welsh onion extracts. *J. Food Prot.* 52:414–417.

Fang, Shao W., Li, Chin F., and Shih, D. Y. C. 1994. Antifungal activity of chitosan and its preservative effect on low-sugar candied kumquat. *J. Food Prot.* 57:136–140.

Ferguson, W. E., and Powrie, W. D. 1957. Studies on the preservation of fresh apple juice with sorbic acid. *Appl. Microbiol.* 5:41–43.

Ferrand, C., Marc, F., Fritsch, P., Cassand, P., and de Saint Blanquat, G. 2000a. Chemical and toxicological studies of products resulting from sorbic acid and methylamine interaction in food conditions. *Amino Acids* 18:251–263.

Ferrand, C., Marc, F., Fritsch, P., Cassand, P., and de Saint Blanquat, G. 2000b. Mutagenicity and genotoxicity of sorbic acid-amine reaction products. *Food Add. Contam.* 17:895–901.

Fey, M. S., and Regenstein, J. M. 1982. Extending shelf-life of fresh wet red hake and salmon using carbon dioxide and oxygen modified atmosphere and potassium sorbate ice at 1°C. *J. Food Sci.* 47:1048–1054.

Filya, I., Karabulut, A., Degirmencioglu, T., Canbolat, O., and Kalkan, H. 2001. Improving of conservation and feed value characteristics of citrus pulps. *Turkish J. Vet. Anim. Sci.* 25:939–945.

Finol, M. L., Marth, E. H., and Lindsay, R. C. 1982. Depletion of sorbate from different media during growth of *Penicillium* species. *J. Food Prot.* 45:398–404, 409.

Fletcher, G. C., Murrell, W. G., Statham, J. A., Stewart, B. J., and Bremner, H. A. 1988. Packaging of scallops with sorbate: An assessment of the hazard from *Clostridium botulinum*. *J. Food Sci.* 53:349–352, 358.

Food and Drug Research Laboratories, Inc. 1973. *Scientific Literature Reviews of Generally Recognized as Safe (GRAS) Food Ingredients: Sorbic Acid and Derivatives.* Report prepared for the Food and Drug Administration, Washington, DC.

Food and Drug Research Laboratories, Inc. 1975. Teratologic evaluation of potassium sorbate-sorbistat in mice and rats. U.S. Department of Commerce National Technical Information Service Publication PB-245, Washington, DC.

Frank, P. R. 1977. Potassium sorbate as a fungistatic agent in country ham processing. M.S. thesis, Virginia Polytechnic Institute and State University, Blacksburg, VA.

Freese, E., and Levin, B. C. 1978. Action mechanisms of preservatives and antiseptics. In *Developments in Industrial Microbiology*, edited by L. A. Underkofler, Society for Industrial Microbiology, Washington, DC, pp. 207.

Freese, E., Sheu, C. W., and Galliers, E. 1973. Function of lipophilic acids as antimicrobial food additives. *Nature (London)* 241:321–325.

Fujiyoshi, T., Ikami, T., Kikukawa, Koboyashi, M., Takai, R., Kozaki, D., and Yamamato, A. 2018. Direct quantitation of the preservatives benzoic and sorbic acid in processed foods using derivative spectrophotometry combined with micro dialysis. *Food Chem.* 240:386–390.

Fung, Y.-S., and Luk, S.-F. 1990. Polarographic determination of sorbic acid in fruit juices and soft drinks. *Analyst* 115:1219–1221.

Garcia-Soto, B., Miranda, J. M., Rodriguez-Bernaldo de Quiros, A., Sendon, R., Rodriguez-Martinez, A. V., Barros-Velazquez, J., and Aubourg, S. P. 2015. Effect of biodegradable film (lyophilized alga *Fucus spiralis* and sorbic acid) on quality properties of refrigerated megrim (*Lepidorhombus whiffagonis*). *Intl. J. Food Sci. Technol.* 50:1891–1900.

Gareis, M., Bauer, J., von Montegelas, A., and Gedek, B. 1984. Stimulation of aflatoxin B_1 and T-2 toxin production by sorbic acid. *Appl. Environ. Microbiol.* 47:416–418.

Garza, S., Canela, R., Viñas, I., and Sanchis, V. 1993. Effects of potassium sorbate on growth and penicillic acid production by *Aspergillus ochraceus* and *Penicillium aurantiogriseum*. *Zentralbl. Mikrobiol.* 148:343–350.

Gaunt, I. F., Butterworth, K. R., Hardy, J., and Gangolli, S. D. 1975. Long-term toxicity of sorbic acid in the rat. *Food Cosmet. Toxicol.* 13:31–45.

Geminder, J. J. 1959. Use of potassium and sodium sorbate in extending shelf-life of smoked fish. *Food Technol.* 13:459–461.

Gerschenson, L. N., Alzamora, S. M., and Chirife, J. 1986a. Stability of sorbic acid in model food systems of reduced water activity: Sugar solutions. *J. Food Sci.* 51:1028–1031.

Gerschenson, L. N., Alzamora, S. M., and Chirife, J. 1986b. Technical note: Kinetics of sorbic acid loss during storage of peaches preserved by combined factors. *J. Food Technol.* 21:517–519.

Gerschenson, L. N., Alzamora, S. M., and Chirife, J. 1987. Effect of sodium chloride and glycerol on the stability of sorbic acid solutions of reduced water activity. *Lebensm. Wiss. Technol.* 20:98–99.

Giannakopoulos, A., and Guilbert, S. 1986a. Determination of sorbic acid diffusivity in model food gels. *J. Food Technol.* 21:339–353.

Giannakopoulos, A., and Guilbert, S. 1986b. Sorbic acid diffusivity in relation to the composition of high intermediate moisture model gels and foods. *J. Food Technol.* 21:477–485.

Goettel, M.S. and Duke, G.M. 1996. Effects of sorbic acid and methyl paraben on susceptibility of alfalfa leafcutting bees (*Megachile rotundata*) to chalkbrood (*Ascosphaera aggregata*). *J. Invertebrate Pathol.* 68:94–96.

Golden, D. A., and Beuchat, L. R. 1992. Effects of potassium sorbate on growth patterns, morphology, and heat resistance of *Zygosaccharomyces rouxii* at reduced water activity. *Can. J. Microbiol.* 38:1252–1259.

Golden, D. A., Beuchat, L. R., and Hitchcock, H. L. 1994. Changes in fatty acid composition of various lipid components of *Zygosaccharomyces rouxii* as influenced by solutes, potassium sorbate and incubation temperature. *Intl. J. Food Microbiol.* 21:293–303.

Gooding, C. M. 1945. Process of inhibiting growth of molds. U.S. Patent 2,379,294.

Gooding, C. M., Melnick, D., Lawrence, R. L., and Luckmann, F. H. 1955. Sorbic acid as a fungistatic agent for foods. IX: Physicochemical considerations in using sorbic acid to protect foods. *Food Res.* 20:639–648.

Gourama, H., and Bullerman, L. B. 1988. Effects of potassium sorbate and natamycin on growth and penicillic acid production by *Aspergillus ochraceus*. *J. Food Prot.* 51:139–144.

Gray, R. J. H., Elliot, P. H., and Tomlins, R. I. 1984. Control of two major pathogens on fresh poultry using a combination potassium sorbate/carbon dioxide packaging treatment. *J. Food Sci.* 49:142–145, 179.

Greer, G. G. 1982. Mechanism of beef shelf-life extension by sorbate. *J. Food Prot.* 45:82–83.

Guerrero, S., Alzamora, S. M., and Gerchenson, L. N. 1990. Stability of sorbic acid in aqueous solutions of sodium chloride. *Lebensm. Wiss. Technol.* 23:271–273.

Guerrero, S., Alzamora, S. M., and Gerschenson, L. N. 1994. Development of a shelf-stable banana puree by combined factors: Microbial stability. *J. Food Prot.* 57:902–907.

Hall, M. A., and Maurer, A. J. 1980. Effects of nitrite, ascorbate, and sorbate on *Clostridium botulinum* in turkey frankfurter slurries. *Poultry Sci.* 59:1616–1617.

Hallerbach, C. M., and Potter, N. N. 1981. Effects of nitrite and sorbate on bacterial populations in frankfurters and thuringer cervelat. *J. Food Prot.* 44:341–346.

Hamdan, I. Y., Deane, D. D., and Kunsman, J. E., Jr. 1971. Effect of potassium sorbate on yogurt cultures. *J. Milk Food Technol.* 34:307–311.

Han, J. H., and Floros, J. D. 1998. Simulating diffusion model and determining diffusivity of potassium sorbate through plastics to develop antimicrobial packaging films. *J. Food Process. Preserv.* 22:107–122.

Hansen, J. D., and Appleman, M. D. 1955. The effect of sorbic, propionic and caproic acids on the growth of certain clostridia. *Food Res.* 20:92–96.

Harada, K., Hizuchi, R., and Utsumi, I. 1968. Studies on sorbic acid. Part IV: Inhibition of the respiration in yeast. *Agric. Biol. Chem.* 32:940.

Harrington, W. O., Anne, S. J., and Hills, C. H. 1962. Spectrophotometric determination of sorbic acid in apple cider. *J. Food Sci.* 27:15–19.

Hauser, C., and Wunderlich, J. 2011. Antimicrobial packaging films with a sorbic acid based coating. *Proc. Food Sci.* 1:197–202.

Helmroth, E., Rijk, R., Dekker, M., and Jongen, W. 2002. Predictive modeling of migration from packaging materials into food products for regulatory purposes. *Trends Food Sci. Technol.* 13:102–109.

Hendy, R. J., Hardy, J., Gaunt, I. F., Kiss, I. S., and Butterworth, K. R. 1976. Long-term toxicity studies of sorbic acid in mice. *Food Cosmet. Toxicol.* 14:381–386.

Henriques, M., Quintas, C., and Louriero-Dias, M. 1997. Extrusion of benzoic acid in *Saccharomyces cerevisiae* by an energy-dependent mechanism. *Microbiology* 143:1877–1883.

Hofer, K., and Jenewein, D. 2000. Enzymatic determination of sorbic acid. *Eur. Food Res. Tech.* 211:72–76.

Hoffman, C., Schweitzer, T. R., and Dalby, G. 1944. Fungistatic properties of the fatty acids and possible biochemical significance. *Food Res.* 4:539–545.

Holley, R. A., and Millard, G. E. 1980. Ultraviolet spectrophotometric determination of sorbic acid in dry fermented sausage. *J. Assoc. Off. Analytic. Chem.* 63:1332–1335.

Holyoak, D. C., Stratford, M., McMullin, Z., Cole, M. B., Crimmins, K., Brown, A. J. P., and Coote, P. J. 1996. Activity of the plasma membrane H super (+)-TPase and optimal glycolytic flux are required for rapid adaptation and growth of *Saccharomyces cerevisiae* in the presence of the weak-acid preservative sorbic acid. *Appl. Environ. Microbiol.* 62:3158–3164.

Horwood, J. F., Lloyd, G. T., Ramshaw, E. H., and Stark, W. 1981. An off-flavor associated with the use of the sorbic acid during feta cheese maturation. *Aust J. Diary Technol.* 36:38–40.

Hosseini, S. F., Rezaei, M., Zandi, M., and Farahmandghavi, F. 2016. Development of bioactive fish gelatin/chitosan nanoparticles composite films with antimicrobial properties. *Food Chem.* 194:1266–1274.

Hu, S., Yu, J., Wang, Z., Li, L., Du, Y., Wang, L., and Liu, Y. 2017. Effects of sorbic acid-chitosan microcapsules as antimicrobial agents on the properties of ethylene vinyl alcohol copolymer film for food packaging. *J. Food Sci.* 82:1451–1460.

Huang, E., and Armstrong, J. G. 1970. The effect of microbial inhibitors on the storage life of cottage cheese exposed to ambient temperatures. *Can. Inst. Food Sci. Technol. J.* 3:157–161.

Huhtanen, C. N., and Feinberg, J. 1980. Sorbic acid inhibition of *Clostridium botulinum* in nitrite-free poultry frankfurters. *J. Food Sci.* 45:453–457.

Huhtanen, C., N., Talley, F. B., Feinberg, J., and Phillips, J. G. 1981. Flavor and antibotulinal evaluation of sorbic acid-containing bacon. *J. Food Sci.* 46:1796–1800.

Huhtanen, C. N., Feinberg, J. I., Trenchard, H., and Phillips, J. G. 1983. Acid enhancement of *Clostridium botulinum* inhibition in ham and bacon prepared with potassium sorbate and sorbic acid. *J. Food Prot.* 46:807–810.

Hunter, D. R., and Segel, I. H. 1973. Effect of weak acids on amino acid transport by *Penicillium chrysogenum*: Evidence for a proton or charge gradient as the driving force. *J. Bacteriol.* 113:1184–1192.

Hwang, C., Huang, L., and Juneja, V. 2015. Effect of acidified sorbate solutions on lag-phase durations and growth rates of listeria monocytogenes on meat surfaces. *J. Food Prot.* 6:1154–1160.

Ivey, F. J., and Robach, M. C. 1978. Effect of sorbic acid and sodium nitrite on *Clostridium botulinum* outgrowth and toxin production in canned comminuted pork. *J. Food Sci.* 43:1782–1785.

Ivey, F. J., Shaver, K. J., Christiansen, L. N., and Tompkin, R. B. 1978. Effect of potassium sorbate on toxogenesis by *Clostridium botulinum* in bacon. *J. Food Prot.* 41:621–625.

Javanmardi, F., Nemati, M., Ansarin, M., and Rafie Arefhosseini, S. 2015. Benzoic and sorbic acid in soft drink, milk, ketchup sauce and bread by dispersive liquid-liquid microextraction coupled with HPLC. *Food Addit. Contam. Part B* 8:32–39.

Jideani, V. A., and Wedzicha, B. L. 1994. Reaction of sorbic acid in millet and sorghum doughs: Reaction with thiols. *Food Add. Contam.* 11:539–548.

Jideani, V. A., and Wedzicha, B. L. 1995. ^{14}C-sorbic acid distribution in the aqueous and non-aqueous extracts of cooked millet dough (fura). *Food Add. Contam.* 12:161–166.

Jones, A., and Harper, G. S. 1952. A preliminary study of factors affecting the quality of pickles on the Canadian market. *Food Technol.* 6:304–308.

Juven, B. J. 1976. Bacterial spoilage of citric products of pH lower than 3.5. *J. Milk Food Technol.* 39:819–822.

Kabara, J. J. 1984. Inhibition of *Staphylococcus aureus* in a model agar-meat system by monolaurin: A research note. *J. Food Safety* 6:197–201.

Kada, T. 1974. DNA-damaging products from reaction between sodium nitrite and sorbic acid. *Rep. Natl. Inst. Genet. Jpn.* 24:43–44.

Kadam, P. S., Thomas, G., and Ingle, U. M. 1985. Use of sorbic acid and carboxymethyl cellulose as a surface coating in the preservation of apple jam. *J. Food Sci. Technol.* 22:372–373.

Kaloyereas, S., Crown, R. M., and McCleskey, C. S. 1961. Experiments on preservation with a new ice containing glycol diformate and sorbic acid. *Food Technol.* 15:361–364.

Kaul, A., Singh, J., and Kuila, R. K. 1979. Effect of potassium sorbate on the microbiological quality of butter. *J. Food Prot.* 43:656–657.

Kawanishi, T., Ohno, Y., Takahashi, A., Ishiwata, H., Tanimura, A., Kasuya, Y., and Omori, Y. 1981a. Studies on nitrosamine formation by the interaction between drugs and nitrite. I: Measurement of the amount of nitrosamine formed in rat and guinea pig stomachs. *J. Toxicol.* Sci. 6:261–270.

Kawanishi, T., Ohno, Y., Sunouchi, M., Onoda, K., Takahashi, A., Kasuya, Y., and Omori, Y. 1981b. Studies on nitrosamine formation by the interaction between drugs and nitrite. II: Hepatotoxicity by the simultaneous administration of several drugs and nitrite. *J. Toxicol. Sci.* 6:271.

Kayaci, F., Ertas, Y., and Uyar, T. 2013. Enhanced thermal stability of eugenol by cyclodextrin inclusion complex encapsulated in electrospun polymeric nanofibers. *J. Ag. Food Chem.* 61:8156–8165.

Kemp, J. D., Langlois, B. E., and Fox, J. D. 1983. Effect of potassium sorbate and vacuum packaging on the quality and microflora of dry-cured intact and boneless hams. *J. Food Sci.* 48:1709–1714.

Kemp, J. D., Langlois, B. E., Solomon, M. B., and Fox, J. D. 1979. Quality of boneless dry-cultured ham produced with or without nitrate, netting or potassium sorbate. *J. Food Sci.* 44:914–915.

Khandelwal, G. D., Rimmer, Y. L., and Wedzicha, B. L. 1992. Reaction of sorbic acid in wheat flour doughs: Reaction with thiols. *Food Add. Contam.* 9:493–497.

Khoudokormoff, B. 1981. Evaluation of the mutagenicity of sorbic acid-sodium nitrite reaction products. *Food Cosmet. Toxicol.* 19:405–406.

Kito, Y., Namiki, M., and Tsuji, K. 1978. A new N-nitropyrrole, 1,4-dinitro-2-methylpyrrole, formed by the reaction of sorbic acid with sodium nitrite. *Tetrahedron* 34:505–508.

Kivanç, M. 1992. Fungal contamination of Kaşhar cheese in Turkey. *Die Nahrung.* 36:578–583.

Klindworth, K. J., Davidson, P. M., Brekke, C. J., and Branen, A. L. 1979. Inhibition of *Clostridium perfringens* by butylated hydroxyanisole. *J. Food Sci.* 44:564–567.

Kondaiah, N., Zeuthen, P., and Jul, M. 1985. Effect of chemical dips on unchilled fresh beef inoculated with *E. coli, S. faecalis* and *Cl. perfringens* and stored at 30°C and 20°C. *Meat Sci.* 12:17–30.

Koodie, L., and Dhople, A. M. 2001. Acid tolerance of *Escherichia coli* O157:H7 and its survival in apple juice. *Microbios* 104:167–175.

Kouassi, Y., and Shelef, L. A. 1995a. Listeriolysin O secretion by *Listeria monocytogenes* in the presence of cysteine and sorbate. *Lett. Appl. Microbiol.* 20:295–299.

Kouassi, Y., and Shelef, L. A. 1995b. Listeriolysin O secretion by *Listeria monocytogenes* in broth containing salts of organic acids. *J. Food Prot.* 58:1314–1319.

Kumar, J. K., Sharma, A. K., and Kulkarni, P. R. 2000. Effect of preservation techniques and food additives on staphylococcal thermonuclease. *Die Nahrung.* 44:272–275.

Lahellec, C., Fung, D. Y., and Cunningham, F. E. 1981. Growth effect of sorbate and selected antioxidants on toxigenic strains of *Staphylococcus aureus. J. Food Prot.* 44:531–534.

Larocco, K. A., and Martin, S. E. 1981. Effects of potassium sorbate alone and in combination with sodium chloride on the growth of *Salmonella typhimurium* 7136. *J. Food Sci.* 46:568–570.

Larson, A. E., Johnson, E. A., and Nelson, J. H. 1999. Survival of *Listeria monocytogenes* in commercial cheese brines. *J. Dairy Sci.* 82:1860–1868.

Larsson, B. K. 1983. Gas-liquid chromatographic determination of benzoic acid and sorbic acid in foods: Nordic Committee on Food Analysis collaborative study. *J. Assoc. Off. Analytic. Chem.* 66:775–780.

Lathia, D., and Schellhoeh, B. 1981. Inhibition of nitrosamine formation in vitro in presence of both ascorbic acid and sorbic acid. *Recent Adv. Clin. Nutr.* 1:189–190.

Lennox, J. E., and McElroy, L. J. 1984. Inhibition of growth and patulin synthesis in *Penicillium expansum* by potassium sorbate and sodium propionate in culture. *Appl. Environ. Microbiol.* 48:1031–1033.

Lenovich, L. M., Buchanan, R. L., Worley, N. J., and Restaino, L. 1988. Effect of solute type on sorbate resistance in *Zygosaccharomyces rouxii. J. Food Sci.* 43:914–916.

Liewen, M. B., and Marth, E. H. 1983. Responses of conidia of *Penicillium roqueforti* to sorbic acid. *J. Dairy Sci.* 66:66.

Liewen, M. B., and Marth, E. H. 1984. Inhibition of penicillia and aspergilli by potassium sorbate. *J. Food Prot.* 47:554–556.

Liewen, M. B., and Marth, E. H. 1985a. Growth and inhibition of microorganisms in the presence of sorbic acid: A review. *J. Food Prot.* 48:364–375.

Liewen, M. B., and Marth, E. H. 1985b. Production of mycotoxins by sorbate-resistant molds. *J. Food Prot.* 48:156–157.

Liewen, M. B., and Marth, E. H. 1985c. Growth of sorbate-resistant and -sensitive strains of *Penicillium roqueforti* in the presence of sorbate. *J. Food Prot.* 48:525–529.

Litton Biometrics, Inc. 1974. Mutagenic evaluation of potassium sorbate. U.S. Department of Commerce, National Technical Information Service, Publication PB-245-434, Washington, DC.

Litton Biometrics, Inc. 1977. Mutagenic evaluation of calcium sorbate. U.S. Department of Commerce, National Technical Information Service, Publication PB-266-894, Washington, DC.

Liu, X., Yousef, A. E., and Chism, G. W. 1997. Inactivation of *Escherichia coli* O157:H7 by the combination of organic acids and pulsed electric field. *J. Food Safety* 16:287–299.

Lopez, M., Mazas, M., Gonzalez, I., Gonzalez, J., and Bernardo, A. 1996. Thermal resistance of *Bacillus stearothermophilus* spores in different heating systems containing some approved food additives. *Lett. Appl. Microbiol.* 23:187–191.

Lück, E. 1976. Sorbic acid as a food preservative. *Intl. Flavors Food Addit.* 7:122–124, 127.

Lück, E. 1980. *Antimicrobial Food Additives, Characteristics, Uses, Effects*, Springer-Verlag, Berlin, Germany.

Luckman, F. H., and Melnick, D. 1955. Sorbic acid as a fungistatic agent for foods. X: Spectrophotometric determination of sorbic acid in foods in general. *Food Res.* 20:649–654.

Lund, B. M., George, S. M., and Franklin, J. G. 1987. Inhibition of type A and type B (proteolytic) *Clostridium botulinum* by sorbic acid. *Appl. Environ. Microbiol.* 59:935–941.

Lund, B. M., Graham, A. F., George, S. M., and Brown, D. 1990. The combined effect of incubation temperature, pH and sorbic acid on the probability of growth of non-proteolytic, type B *Clostridium botulinum. J. Appl. Bacteriol.* 69:481–492.

Lusher, P., Denyer, S. P., and Hugo, W. B. 1984. A note on the effect of dilution and temperature on the bactericidal activity of potassium sorbate. *J. Appl. Bacteriol.* 57:179–181.

Lynch, D. J., and Potter, N. N. 1982. Effects of potassium sorbate on normal flora and on *Staphylococcus aureus* added to minced cod. *J. Food Prot.* 45:824–828.

Mackey, B. m., Forestiere, K., and Isaacs, N. 1995. Factors affecting the resistance of *Listeria monocytogenes* to high hydrostatic pressure. *Food Biotech.* 9:1–11.

Mandrou, B., Nolleau, V., Gastaldi, E., and Fabre, H. 1999. Solid phase extraction as a clean up procedure for the liquid chromatographic determination of benzoic and sorbic acids in fruit derived products. *J. Liq. Chromatogr. Rel. Technol.* 21:829–842.

Mannino, S., and Cosio, M. S. 1996. Determination of benzoic and sorbic acids in food by microdialysis sampling coupled with HPLC and UV detection. *Italian J. Food Sci.* 8:311–316.

Marshall, D. L., and Bullerman, L. R. 1986. Effect of sucrose esters in combination with selected mold inhibitors in growth and aflatoxin production by *Aspergillus parasiticus J. Food Prot.* 49:378–382, 388.

Marth, E. H., Capp, C. M., Hasenzahl, L., Jackson, H. W., and Hussang, R. V. 1966. Degradation of potassium sorbate by *Penicillium* species. *J. Dairy Sci.* 49:1197–1205.

Martoadiprawito, W., and Whitaker, J. R. 1963. Potassium sorbate inhibition of yeast alcohol dehydrogenase. *Biochim. Biophys. Acta* 77:536.

Massey, R. C., Forsythe, L., and McWeeny, D. J. 1982. The effects of ascorbic acid and sorbic acid on N-nitrosamine formation in a heterogeneous model system. *J. Sci. Food Agric.* 33:294–298.

Matamoros-Leon, B., Argaiz, A. and Lopez-Malo, A. 1999. Individual and combined effects of vanillin and potassium sorbate on *Penicillium digitatum*, *Penicillium glabrum*, and *Penicillium italicum* growth. *J. Food Prot.* 62:540–542.

Maxstadt, J. J., and Karasz, A. B. 1972. Rapid spectrophotometric method for the determination of sorbic acid in fresh dairy products. *J. Assoc. Off. Analytic. Chem.* 55:7–8.

Mayer, F., and Hillebrandt, J.-O. 1997. Potato pulp: Microbiological characterization, physical modification, and application of this agricultural waste product. *Appl. Microbiol. Biotechnol.* 48:435–440.

McMeekin, T. A., Pennington, P. I., and Thomas, C. J. 1984. Effect of potassium sorbate on the microbiology of vacuum-packed poultry. *J. Food Safety.* 6:261–270.

Melnick, D., and Luckmann, F. H. 1954a. Sorbic acid as a fungistatic agent for foods. III: Spectrophotometric determination of sorbic acid in cheese wrappers. *Food Res.* 19:20–27.

Melnick, D., and Luckmann, F. H. 1954b. Sorbic acid as a fungistatic agent for foods. IV: Migration of sorbic acid from wrapper into cheese. *Food Res.* 19:28–32.

Melnick, D., Luckmann, F. H., and Gooding, C. M. 1954a. Sorbic acid as a fungistatic agent for foods. V: Resistance of sorbic acid in cheese to oxidative deterioration. *Food Res.* 19:33–43.

Melnick, D., Luckmann, F. H., and Gooding, C. M. 1954b. Sorbic acid as a fungistatic agent for foods. VI: Metabolic degradation of sorbic acid in cheese by molds and the mechanism of mold inhibition. *Food Res.* 19:44–58.

Mendonca, A. F., Molins, R. A., Kraft, A. A., and Walker, H. W. 1989. Effects of potassium sorbate, sodium acetate, phosphates and sodium chloride alone or in combination on shelf life of vacuum packaged pork chops. *J. Food Sci.* 54:302–306.

Mercier, J. P., Chaimbault, P., Morin, P., Dreux, M., and Tambute, A. 1998. Identification of phosphoric acids by capillary electrophoresis-ionspray mass spectrometry. *J. Chromatogr. A.* 825:71–80.

Mihyar, G. F., Yamani, M. I., and Al-Sa'Ed, A. K. 1997. Resistance of yeast flora of labneh to potassium sorbate and sodium benzoate. *J. Dairy Sci.* 80:2304–2309.

Miller, S. A., and Brown, W. D. 1984. Effectiveness of chlortetracycline in combination with potassium sorbate or tetrasodium ethylenediamine-tetraacetate for preservation of vacuum packed rockfish fillets. *J. Food Sci.* 49:188–191.

Miller, L. G., and Kaspar, C. W. 1994. *Escherichia coli* O157:H7 acid tolerance and survival in apple cider. *J. Food Prot.* 57:460–464.

Monnet, D., Vidal, D., and Creach, O. 1988. Influence of metabolic and physical factors on production of diacetoxyscirpenol by *Fusarium sambucinum* fuckel. *Appl. Environ. Microbiol.* 54:2167–2173.

Montano, A., Sanchez, A. H., and Rejano, J. 1995. Determination of benzoic and sorbic acids in packaged vegetable products: Comparative evaluation of methods. *Analyst* 120:2483–2487.

Morad, M. M., Branen, A. L., and Brekke, C. J. 1982. Antimicrobial activity of butylated hydroxyanisole and potassium sorbate against natural microflora in raw turkey meat and *Salmonella typhimurium* in cooked turkey meat. *J. Food Prot.* 45:1038–1043.

Myers, B. R., Edmondson, J. E., Anderson, M. E., and Marshall, R. T. 1983. Potassium sorbate and recovery of pectinolytic psychrotrophs from vacuum-packaged pork. *J. Food Prot.* 46:499.

Namiki, M., Osawa, T., and Namiki, K. 1979. Antibacterial effects developed by the reaction of sorbic acid with sodium nitrite, presented at the ACS/CSF Chemical Congress, April 1–6, Honolulu, Hawaii.

Namiki, M., Udaka, S., Osawa, T., Tsuji, K., and Kada, T. 1980. Formation of mutagens by sorbic acid-nitrite reaction: Effects of reaction conditions on biological activities. *Mutat. Res.* 73:21–28.

Namiki, M., Oswa, T., Ishibashi, H., Namiki, K., and Tsuji, K. 1981. Chemical aspects of mutagen formation by sorbic acid-sodium nitrite reaction. *J. Ag. Food Chem.* 29:407–411.

Namiki, M., Oswa, T., Kada, T., Tsuji, K., and Namiki, K. 1983. Formation of C-nitro and C-nitroso mutagens by the reaction of nitrite with sorbic acid and its analogues and their inactivation with food constituents. *Carcinogens Mutagens Environ.* 3:109.

National Academy of Sciences. 1981. *The Health Effects of Nitrate, Nitrite and N-nitroso Compounds*, National Academy Press, Washington, DC.

National Academy of Sciences. 1982. *Alternatives to the Current Use of Nitrite in Foods*, National Academy Press, Washington, DC.

Nelson, K. A., Busta, F. F., Sofos, J. N., and Wagner, M. K. 1983. Effect of polyphosphates in combination with nitrite-sorbate or sorbate on *Clostridium botulinum* growth and toxin production in chicken frankfurter emulsions. *J. Food Prot.* 46:846–850.

Neves, L., Pampulha, M. E., and Loureiro-Dias, M. C. 1994. Resistance of food spoilage yeasts to sorbic acid. *Lett. Appl. Microbiol.* 19:8–11.

Nielsen, P. V., Beuchat, L. R., and Frisvad, J. C. 1989. Growth and fumitremorgin production by *Neosartorya fischeri* as affected by food preservatives and organic acids. *J. Appl. Bacteriol.* 66:197–207.

Noda, K. Kenjo, N., and Takahashi, T. 1973. Determination of sorbic acid in cheese. *J. Food Hyg. Soc. Jpn.* 14:253–257.

Ogundiwin, J. O., Ilori, M. O., Fessehatzion, B., Babalola, G. O., and Olajuyigbe, A. O. 1991. Effect of chemical treatments on the microorganisms associated with malting of sorghum grains and sorghum malt. *J. Appl. Bacteriol.* 71:139–143.

Oka, S. 1960. Transfer of antiseptics to microbes and their toxic effect. I: Accumulation of acid antiseptics in yeast cells. *Bull. Agric. Chem. Soc. Jpn.* 24:59–65.

Oloyede, O. B., and Scholefield, J. 1994. Inhibition of *Bacillus* spores by combinations of heat, potassium sorbate, NaCl and pH. *World J. Microbiol. Biotechnol.* 10:579–589.

Osawa, T., and Namiki, M. 1982. Mutagen formation in the reaction of nitrite with the food components analogous to sorbic acid. *Agric. Biol. Chem.* 46:2299–2304.

Osawa, T., Namiki, M., and Namiki, K. 1982. Mutagen formation in the nitrite-piperic acid reaction. *Agric. Biol. Chem.* 46:3105–3108.

Osawa, T., Kito, Y., Namiki, M., and Tsuji, K. 1979. A new furoxan derivative and its precursors formed by the reaction of sorbic acid with sodium nitrite. *Tetrahedron* 45:4399–4402.

Osawa, T., Ishibashi, H., Namiki, M., and Kada, T. 1980. Desmutagenic actions of ascorbic acid and cysteine on a new pyrrole mutagen formed by the reaction between food additives; sorbic acid and sodium nitrite. *Biochem. Biophys. Res. Commun.* 95:835–841.

Osawa, T., Ishibashi, H., Namiki, M., Kada, T., and Tsuji, K. 1986. Desmutagenic action of food components on mutagens formed by the sorbic acid/nitrite reaction. *Agric. Biol. Chem.* 50:1971–1977.

Ough, C. S., and Ingraham, J. L. 1960. Use of sorbic acid and sulfur dioxide in sweet table wines. *Am. J. Enol. Vitic.* 11:117.

Ozdemir, M., and Floros, J. D. 2003. Film composition effects on diffusion of potassium sorbate through whey protein films. *J. Food Sci.* 68:511–516.

Palou, E., Lopez-Malo, A., Barbosa-Canovas, G. V., Welti-Chanes, J., and Swanson, B. G. 1997. High hydrostatic pressure as a hurdle for *Zygosaccharomyces bailii* inactivation. *J. Food Sci.* 62:855–857.

Pandey, R., Pieper, G. H., Ter Beek, A., Vischer, N. O., Smelt, J. P., Manders, E. M., and Brul, S. 2015. Quantifying the effect of sorbic acid, heat and combination of both on germination and outgrowth of *Bacillus subtilis* spores at single cell resolution. *Food Microbiol.* 52:88–96.

Pant, I., and Trenerry, V. C. 1995. The determination of sorbic acid and benzoic acid in a variety of beverages and foods by micellar electrokinetic capillary chromatography. *Food Chem.* 53:219–236.

Paquette, M. W., Robach, M. C., Sofos, J. N., and Busta, F. F. 1980. Effects of various concentrations of sodium nitrite and potassium sorbate on color and organoleptic qualities of commercially prepared bacon. *J. Food Sci.* 45:1293–1296.

Parish, M. E., and Carroll, D. E. 1988a. Minimum inhibitory concentration studies on antimicrobic combinations against *Saccharomyces cerevisiae* in a model broth system. *J. Food Sci.* 53:237–239.

Parish, M. E., and Carroll, D. E. 1988b. Effects of combined antimicrobial agents on fermentation initiation by *Saccharomyces cerevisiae* in a model broth system. *J. Food Sci.* 53:240–242.

Park, H. S., and Marth, E. H. 1972. Inactivation of *Salmonella typhimurium* by sorbic acid. *J. Milk Food Technol.* 35:532–539.

Park, H. S., Marth, E. H., and Olson, N. F. 1970. Survival of *Salmonella typhimurium* in cold-pack cheese food during refrigerated storage. *J. Milk Food Technol.* 33:383–388.

Patrizi, A., Orlandi, C., Vincenzi, C., and Bardazzi, F. 1999. Allergic contact dermatitis caused by sorbic acid: Rare occurrence. *Am. J. Contact Dermat.* 10:52.

Pederson, M., Albury, N., and Christensen, M. D. 1961. The growth of yeasts in grape juice stored at low temperature. IV: Fungistatic effects of organic acids. *Appl. Microbiol.* 9:162–167.

Perry, G. A., Lawrence, R. L., and Melnick, D. 1964. Extension of poultry shelf-life by processing with sorbic acid. *Food Technol.* 18:891–897.

Petaja, E., Raevuori, M., Poulanne, E., and Hill, P. 1979. The effect of potassium sorbate and sodium nitrite on the organoleptic properties, stability, and growth of *Bacillus cereus* and *Clostridium perfringens* in cooked sausage. *Proc. Eur. Meet. Meat Res. Work* 25:917.

Petanovska-Ilievska, B., Velkoska-Markovska, L., and Jankulovska, M. S. 2017. Development of reverse-phase high-performance liquid chromatography method for simultaneous determination of sodium benzoate and potassium sorbate in beverages. *Acta Chromat.* 29:345–358.

Phillips, G. F., and Mundt, J. O. 1950. Sorbic acid as an inhibitor of scum yeast in cucumber fermentations. *Food Technol.* 4:291–293.

Piccinin, D. M., and Shelef, L. A. 1995. Survival of *Listeria monocytogenes* in cottage cheese. *J. Food Prot.* 58:128–131.

Pierson, M. D., Robach, M. C., Van Tassel, K. R., and Smoot, L. A. 1979. Sodium nitrite and potassium sorbate inhibition of *Clostridium botulinum* as influenced by storage temperature variation. *Presented at the 39th Annual Meeting Institute of Food Technology*, St. Louis, MO, p. 373. (Abstract.)

Piper, P., Mahe, Y., Thompson, S., Pandjaitan, R., Holyoak, C., Egner, R., Muehlbauer, J., Coote, P., and Kuchler, K. 1998. The Pdr12 ABC transporter is required for the development of weak organic acid resistance in yeast. *EMBO J.* 17:4257–4265.

Poerschke, R. E., and Cunningham, F. E. 1985. Influence of potassium sorbate and selected antioxidants on growth of *Salmonella senftenberg*. *J. Food Qual.* 8:113–129.

Prabhu, G. A., Molins, R. A., Kraft, A. A., Sebranek, J. G., and Walker, H. W. 1988. Effect of heat treatment and selected antimicrobials on the shelf-life and safety of cooked, vacuum-packaged, refrigerated pork chops. *J. Food Sci.* 53:1270–1274.

Przybylski, K. S., and Bullerman, L. B. 1980. Influence of sorbic acid on viability and ATP content of conidia of *Aspergillus parasiticus*. *J. Food Sci.* 45:375–376, 385.

Puttermans, M. L., Dryon, L., and Massart, D. L. 1984. Extraction of organic acids by ion-pair formation with tri-*n*-octylamine. V: Simultaneous determination of synthetic dyes, benzoic acid, sorbic acid, and saccharin in soft drinks and lemonade syrups. *J. Assoc. Off. Analytic. Chem.* 67:880–885.

Quattrucci, E., and Masci, V. 1992. Nutritional aspects of food preservatives. *Food Add. Contam.* 9:515–525.

Raevuori, M. 1976. Effect of sorbic acid and potassium sorbate on growth of *Bacillus cereus* and *Bacillus subtilis* in rice filling Karelian pasty. *Eur. J. Appl. Microbiol.* 2:205–213.

Rahn, O., and Conn, J. W. 1944. Effect of increase in acidity on antiseptic efficiency. *Ind. Eng. Chem.* 36:185–187.

Rao, G. S., Osborn, J. C., and Adatia, M. R. 1982. Drug-nitrite interactions in human saliva effects of food constituents on carcinogenic N-nitrosamine formation. *J. Dent. Res.* 61:768–771.

Redl, A., Gontard, N., and Guilbert, S. 1996. Determination of sorbic acid diffusivity in edible wheat gluten and lipid based films. *J. Food Sci.* 61:116–120.

Regenstein, J. M. 1983. The shelf-life extension of haddock in carbon dioxide oxygen atmosphere with and without potassium sorbate. *J. Food Qual.* 5:285–300.

Restaino, L., Komatsu, K. K., and Syracuse, M. J. 1981. Effects of acids on potassium sorbate inhibition of food-related microorganisms in culture media. *J. Food Sci.* 47:134–138.

Restaino, L., Lenovich, L. M., and Bills, S. 1982. Effect of acids and sorbate combinations on the growth of four osmophilic yeasts. *J. Food Prot.* 45:1138–1142.

Restaino, L., Bills, S., Tscherneff, K., and Lenovich, L. M. 1983. Growth characteristics of *Saccharomyces rouxii* isolated from chocolate syrup. *Appl. Environ. Microbiol.* 45:1614–1621.

Rice, K. M., and Pierson, M. D. 1982. Inhibition of *Salmonella* by sodium nitrite and potassium sorbate in frankfurters. *J. Food Sci.* 47:1615–1617.

Robach, M. C. 1979a. Extension of shelf-life of fresh, whole broilers using a potassium sorbate dip. *J. Food Prot.* 42:855–857.

Robach, M. C. 1979b. Effect of processing variables on the outgrowth of *Clostridium sporogenes* PA 3679 spores in comminuted meat cured with sorbic acid and sodium nitrite. *Appl. Environ. Microbiol.* 38:846–849.

Robach, M. C. 1980. Interaction of salt, potassium sorbate and temperature on the outgrowth of *Clostridium sporogenes* PA 3679 spores in a pre-reduced medium. *J. Food Sci.* 45:742–743.

Robach, M. C., and Adam, K. 1980. Preliminary observations on human reactions related to sensory evaluation of cooked, commercial bacon. *J. Food Prot.* 43:933–935, 938.

Robach, M. C., and Hickey, C. S. 1978. Inhibition of *Vibrio parahaemolyticus* by sorbic acid in crab meat and flounder homogenates. *J. Food Prot.* 41:699–702.

Robach, M. C., and Ivey, F. J. 1978. Antimicrobial efficacy of a potassium sorbate dip on freshly processed poultry. *J. Food Prot.* 41:284–288.

Robach, M. C., and Sofos, J. N. 1982. Use of sorbates in meat products, fresh poultry and poultry products: A review. *J. Food Prot.* 45:374–383.

Robach, M. C., and Stateler, C. L. 1980. Inhibition of *Staphylococcus aureus* by potassium sorbate in combination with sodium chloride, tertiary butylhydroquinone, butylated hydroxyanisole or ethylenediamine tetraacetic acid. *J. Food Prot.* 43:208–211.

Robach, M. C., Hickey, C. S., and To, E. C. 1981. Comparison of antimicrobial actions on monolaurin and sorbic acid. *J. Food Safety* 3:89–98.

Robach, M. C., Ivey, F. J., and Hickey, C. S. 1978. System for evaluation of clostridial inhibition in cured meat products. *Appl. Environ. Microbiol.* 36:210–211.

Robach, M. C., DiFate, V. G., Adam, K., and Kier, L. D. 1980a. Evaluation of the mutagenicity of sorbic acid-sodium nitrite reaction products produced in bacon-curing brines. *Food Cosmet. Toxicol.* 18:237–240.

Robach, M. C., Owens, J. L., Paquette, M. W., Sofos, J. N., and Busta, F. F. 1980b. Effects of various concentrations of sodium nitrite and potassium sorbate on nitrosamine formation in commercially prepared bacon. *J. Food Sci.* 45:1280–1284.

Robach, M. C., To, E. C., Meydav, S., and Cook, C. F. 1980c. Effect of sorbates on microbiological growth in cooked turkey products. *J. Food Sci.* 45:638–640.

Roberts, T. A., Gibson, A. M., and Robinson, A. 1982. Factors controlling the growth of *Clostridium botulinum* types A and B in pasteurized, cured meats. III: The effect of potassium sorbate. *J. Food Technol.* 17:307–326.

Roland, J. O., and Beuchat, L. R. 1984. Biomass and patulin production by *Byssochlamys nivea* in apple juice as affected by sorbate, benzoate, SO_2 and temperature. *J. Food Sci.* 49:402–406.

Roland, J. O., Beuchat, L. R., Worthington, R. E., and Hitchcock, H. L. 1984. Effects of sorbate, benzoate, sulfur dioxide and temperature on growth and patulin production by *Byssochlamys nivea* in grape juice. *J. Food Prot.* 47:237–241.

Ronning, I. E., and Frank, H. A. 1987. Growth inhibition of putrefactive anaerobe 3679 caused by stringent-type response induced by protonophoric activity of sorbic acid. *Appl. Environ. Microbiol.* 53:1020–1027.

Ronning, I. E., and Frank, H. A. 1988. Growth response of putrefactive anaerobe 3679 to combinations of potassium sorbate and some common curing ingredients (sucrose, salt, and nitrite), and to noninhibitory levels of sorbic acid. *J. Food Prot.* 51:651.

Ronning, I. E., and Frank, H. A. 1989. Morphological changes in putrefactive anaerobe 3679 (*Clostridium sporogenes*) induced by sorbate, hydrochloric acid, and nitrite. *Can. J. Microbiol.* 35:388–398.

Roy, R. B., Sahn, M., and Conetta, A. 1976. Automated analysis of sorbic acid in food products. *J. Food Sci.* 41:372–374.

Rusul, G., and Marth, E. M. 1987. Growth and aflatoxin production by *Aspergillus parasiticus* NRRL 2999 in the presence of potassium benzoate or potassium sorbate and at different initial pH values. *J. Food Prot.* 50:820–825.

Ryser, E. T., and Marth, E. H. 1988. Survival of *Listeria monocytogenes* in cold-pack cheese food during refrigerated storage. *J. Food Prot.* 51:615–621.

Salmond, C. V., Kroll, R. G., and Booth, I. R. 1984. The effect of food preservatives on pH homeostasis in *Escherichia coli*. *J. Gen Microbiol.* 130:2845–2850.

Sanchis, V., Quilez, M. L., Viladrich, R., Vinas, I., and Canela, R. 1988. Hazelnuts as possible substrate for aflatoxin production. *J. Food Prot.* 51:289–292.

Sayeed, S. A., and Sankaran, R. 1991. Action of sorbic acid on *Staphylococcus* metabolism: A microcalorimetric investigation. *Indian J. Exp. Biol.* 29:628–630.

Schmidt, E. W., Jr., Gould, W. A., and Weiser, H. H. 1969. Chemical preservatives to inhibit the growth of *Staphylococcus aureus* in synthetic cream pies acidified to pH 4.5 to 5.0. *Food Technol.* 23:1197–1220.

Schmidt, H. 1960a. A specific method for the colorimetric determination of sorbic acid (in non-alcoholic foods). *Z. Analytic. Chem.* 178:173–193.

Schmidt, H. 1962. Colorimetric determination of sorbic acid in wine. *Deutsch. Lebensm. Rundschau.* 58:1.

Schmidt, O. C. 1960b. The effect of aureomycin combinations with salts and acids on various kinds of bacteria. *Arch. Lebensmitt. Hyg.* 11:225.

Schroeder, L. L., and Bullerman, L. B. 1985. Potential for development of tolerance by *Penicillium digitatum* and *Penicillium italicum* after repeated exposure to potassium sorbate. *Appl. Environ. Microbiol.* 50:919–923.

Seward, R. A., Deibel, R. H., and Lindsay, R. C. 1982. Effect of potassium sorbates and other antibotulinal agents on germination and outgrowth of *Clostridium botulinum* type E spores in microcultures. *Appl. Environ. Microbiol.* 44:1212–1221.

Sharp, W. F., Jr., Norback, J. P., and Stuiber, D. A. 1986. Using a new measure to define shelf life of fresh whitefish. *J. Food Sci.* 51:936–939.

Shaw, S. J., Bligh, E. C., and Woyewoda, A. D. 1983. Effect of potassium sorbate application on shelf-life of Atlantic cod (*Gadus morhua*). *Can. Inst. Food Sci. Technol. J.* 16:237–241.

Sheneman, J. M., and Costilow, R. N. 1955. Sorbic acid as a preservative for sweet cucumber pickles. *Appl. Microbiol.* 3:186–189.

Sheu, C. W., and Freese, E. 1972. Effects of fatty acids on growth and envelope proteins of *Bacillus subtilis. J. Bacteriol.* 111:516.

Sheu, C. W., and Freese, E. 1973. Lipopolysaccharide layer protection of Gram-negative bacteria against inhibition by long-chain fatty acids. *J. Bacteriol.* 115:869–875.

Sheu, C. W., Konings, W. N., and Freese, E. 1972. Effects of acetate and other short-chain fatty acids on sugar and amino acid uptake of *Bacillus subtilis. J. Bacteriol.* 111:525.

Sheu, C. W., Salomon, D., Simmons, J. L., Sreevalsan, T., and Freese, E. 1975. Inhibitory effect of lipophilic acids and related compounds on bacteria and mammalian cells. *Antimicrob. Agents Chemother.* 7:349–363.

Shtenberg, A. J., and Ignat'ev, A. D. 1970. Toxicological evaluation of some combinations of food preservatives. *Food Cosmet. Toxicol.* 8:369–380.

Skirdal, I. M., and Eklund, T. 1993. Microculture model studies on the effect of sorbic acid on *Penicillium chrysogenum, Cladosporium cladosporioides* and *Ulocladium atrum* at different pH levels. *J. Appl. Bacteriol.* 74:191–195.

Smith, D. P., and Rollin, N. J. 1954a. Sorbic acid as a fungistatic agent for foods. VII: Effectiveness of sorbic acid in protecting cheese. *Food Res.* 19:59–65.

Smith, D. P., and Rollin, N. J. 1954b. Sorbic acid as a fungistatic agent for foods. VIII: Need for efficacy in protecting packaged cheese. *Food Technol.* 8:133–135.

Smoot, L. A., and Pierson, M. D. 1981. Mechanism of sorbate inhibition of *Bacillus cereus* T and *Clostridium botulinum* 62A spore germination. *Appl. Environ. Microbiol.* 42:477–483.

Smyth, H. F., and Carpenter, C. P. 1948. Further experience with the range finding test in the industrial toxicological laboratory. *J. Ind. Hyg. Toxicol.* 30:63–70.

Sofos, J. N. 1981. Nitrite, sorbate and pH interaction in cured meat products. *Proc. Rec. Meat Conf. American Meat Science Association*, Chicago, IL. 34:104–120.

Sofos, J. N. 1986a. Use of phosphates in low-sodium meat products. *Food Technol.* 40:52–69.

Sofos, J. N. 1986b. Antimicrobial activity and functionality of reduced sodium chloride and potassium sorbate in uncured poultry products. *J. Food Sci.* 51:16–19, 23.

Sofos, J. N. 1986c. Influence of sodium chloride, sorbic acid and potassium sorbate on weight losses and gas production in emulsified meat products. *Proc. Eur. Meet. Meat Res. Work* 32:303.

Sofos, J. N. 1989. *Sorbate Food Preservatives*, CRC Press, Boca Raton, FL.

Sofos, J. N. 1992. Sorbic acid, mode of action. In *Encyclopedia of Microbiology*, Vol. 4, edited by J. Lederberg, Academic Press, Inc., San Diego, CA, pp. 43–52.

Sofos, J. N. 1994. Antimicrobial agents. In *Food Additive Toxicology*, edited by J. A. Maga and A. T. Tu, Marcel Dekker, Inc., New York, p. 501.

Sofos, J. N. 2000. Sorbic acid. In *Natural Food Antimicrobial Systems*, edited by A. S. Naidu, CRC Press, Boca Raton, FL, pp. 637–659.

Sofos, J. N., and Busta, F. F. 1980. Alternatives to the use of nitrite as an antibotulinal agent. *Food Technol.* 34:244–251.

Sofos, J. N., and Busta, F. F. 1981. Antimicrobial activity of sorbate. *J. Food Prot.* 44:614–622.

Sofos, J. N., and Busta, F. F. 1982. Chemical food preservatives. In *Principles and Practice of Disinfection, Preservation and Sterilization*, edited by A. D. Russell, W. B. Hugo, and G. A. J. Alyiffe, Blackwell Scientific, Oxford, p. 307.

Sofos, J. N., and Busta, F. F. 1993. Sorbic acid and sorbates. In *Antimicrobials in Food*, edited by P. M. Davidson and A. L. Branen, Marcel Dekker, Inc., New York, pp. 49–94.

Sofos, J. N., Busta, F. F., and Allen, C. E. 1979a. Botulism control by nitrite and sorbate in cured meats: A review. *J. Food Prot.* 42:739–770.

Sofos, J. N., Busta, F. F., Bhothipaksa, K., and Allen, C. E. 1979b. Sodium nitrite and sorbic acid effects on *Clostridium botulinum* toxin formation in chicken frankfurter-type emulsions. *J. Food Sci.* 44:668–672, 675.

Sofos, J. N., Busta, F. F., and Allen, C. E. 1979c. *Clostridium botulinum* control by sodium nitrite and sorbic acid in various meat and soy protein formulations. *J. Food Sci.* 44:1662–1671.

Sofos, J. N., and Busta, F. F., and Allen, C. E. 1979d. Sodium nitrite and sorbic acid effects on *Clostridium botulinum* spore germination and total microbial growth in chicken frankfurter emulsions during temperature abuse. *Appl. Environ. Microbiol.* 37:1103–1109.

Sofos, J. N., Busta, F. F., and Allen, C. E. 1980a. Influence of pH on *Clostridium botulinum* control by sodium nitrite and sorbic acid in chicken emulsions. *J. Food Sci.* 45:7–12.

Sofos, J. N., Busta, F. F., Bhothipaksa, K., Allen, C. E., Robach, M. C., and Paquette, M. W. 1980b. Effects of various concentrations of sodium nitrite and potassium sorbate on *Clostridium botulinum* toxin production in commercially prepared bacon. *J. Food Sci.* 45:1285–1292.

Sofos, J. N., Pierson, M. D., Blocher, J. C., and Busta, F. F. 1986. Mode of action of sorbic acid on bacterial cells and spores. *Intl. J. Food Microbiol.* 3:1–17.

Soljic, I., Vermeulen, A., Nambooze, C., Riviere, A., Samapundo, S., and Devlieghere, F. 2018. Solid fat influences sorbic acid partitioning and enhances the preservation effect on *C. guilliermondii* in biphasic food model systems. *Intl. J. Food Microbiol.* 285:74–80.

Song, L.-J., Zhou, H., Liu C.-W., Tan H.-T., Xiong, L., Li, T.-G., and Zhou, Y.-G. 2017. Determination of benzoic acid, sorbic acid, and saccharin sodium in rice wine by high performance liquid chromatography. *J. Food Safety Quality.* 8:2922–2926.

Splittstoesser, D. F., and Churey, J. J. 1989. Effect of low concentrations of sorbic acid on the heat resistance and viable recovery of *Neosartorya fischeri* ascospores. *J. Food Prot.* 52:821–822.

Splittstoesser, D. F., Queale, D. T., and Mattick, L. R. 1978. Growth of *Saccharomyces bisporus* var. *bisporus*, a yeast resistant to sorbic acid. *Am. J. Enol. Vitic.* 29:272–276.

Splittstoesser, D. F., Lammers, J. M., Downing, D. L., and Churey, J. J. 1989. Heat resistance of *Eurothium herbariorum*, a xerophilic mold. *J. Food Sci.* 54:683–685.

Splittstoesser, D. F., McLellan, M. R., and Churey, J. J. 1996. Heat resistance of *Escherichia coli* O157:H7 in apple juice. *J. Food Prot.* 59:226–229.

Stafford, A. E. 1976. Rapid analysis of potassium sorbate in dried prunes of ultraviolet on colorimetric procedures. *J. Ag. Food Chem.* 24:894–895.

Statham, J. A., and Bremner, H. A. 1983. Effect of potassium sorbate on spoilage of blue grenadier (*Macruronus novaezelandiae*) as assessed by microbiology and sensory profiles. *J. Food Prot.* 46:1084–1091.

Statham, J. A., and McMeekin, T. A. 1988. The effect of potassium sorbate on the structural integrity of *Alteromonas putrefaciens*. *J. Appl. Bacteriol.* 65:469–476.

Statham, J. A., Bremner, H. A., and Quarmby, A. R. 1985. Storage of Morwong (*Nemadactylus macropterus* Bloch and Schneider) in combinations of polyphosphate, potassium sorbate and carbon dioxide at 4°C. *J. Food Sci.* 50:1580–1584, 1587.

Stead, D. 1995. The effect of hydroxycinnamic acids and potassium sorbate on the growth of 11 strains of spoilage yeasts. *J. Appl. Bacteriol.* 78:82–87.

Steels, H., James, S. A., Roberts, I. N., and Stratford, M. 2000. Sorbic acid resistance: The inoculum effect. *Yeast.* 16:1173–1183.

Stratford, M., and Anslow, P. A. 1998. Evidence that sorbic acid does not inhibit yeast as a classic weak acid preservative. *Lett. Appl. Microbiol.* 27:203–206.

Stratford, M. and Anslow, P.A. 1996. Comparison of the inhibitory action on *Saccharomyces cerevisiae* of weak-acid preservatives, uncouplers, and medium-chain fatty acids. *FEMS Microbiol. Lett.* 142:53–58.

Sugihara, N., Shimomichi, K., and Furuno, K. 1997. Cytotoxicity of food preservatives in cultured rat hepatocytes loaded with linolenic acid. *Toxicology* 120:29–36.

Sun, Y., and Oliver, J. D. 1994. Antimicrobial action of some GRAS compounds against *Vibrio vulnificus*. *Food Add. Contam.* 11:549–558.

Tanaka, K., Chung, K. C., Hayatsu, H., and Kada, T. 1978. Inhibition of nitrosamine formation in vitro by sorbic acid. *Food Cosmet. Toxicol.* 16:209–215.

Tellez-Giron, A., Acuff, G. R., Vanderzant, C., Rooney, L. W., and Waniska, R. D. 1988. Microbiological characteristics and shelf life of corn tortillas with and without antimicrobial agents. *J. Food Prot.* 51:945–948.

Thomas, D. J., and Wagner, M. K. 1987. Effect of sodium acid pyrophosphate and/or potassium sorbate on *Staphylococcus aureus* FRI-100 growth and toxin production. *J. Food Sci.* 52:793–794, 800.

Thomas, L. V., Wimpenny, J. W. T., and Davis, J. G. 1993. Effect of three preservatives on the growth of *Bacillus cereus*, Verocytotoxigenic *Escherichia coli* and *Staphylococcus aureus*, on plates with gradients of pH and sodium chloride concentration. *Intl. J. Food Microbiol.* 17:289–301.

Tjan, G. H., and Konter, T. 1972. Thin-layer chromatographic identification of preservatives in food. *J. Assoc. Off. Analytic. Chem.* 55:1223–1225.

To, E. C., and Robach, M. C. 1980a. Potassium sorbate dip as a method of extending shelf-life and inhibiting the growth of *Salmonella* and *Staphylococcus aureus* on fresh, whole broilers. *Poultry Sci.* 59:726–730.

To, E. C., and Robach, M. C. 1980b. Inhibition of potential food poisoning microorganisms by sorbic acid in cooked, uncured, vacuum-packaged turkey products. *J. Food Technol.* 15:543–547.

Tompkin, R. B., Christiansen, L. N., Shaparis, A. B., and Bolin, H. 1974. Effect of potassium sorbate on salmonellae, *Staphylococcus aureus, Clostridium perfringens* and *Clostridium botulinum* in cooked, uncured sausage. *Appl. Microbiol.* 28:262–264.

Tong, C. H., and Draughon, F. A. 1985. Inhibition by antimicrobial food additives of ochratoxin A production by *Aspergillus sulphureus* and *Penicillium viridicatum*. *Appl. Environ. Microbiol.* 49:1407–1411.

Troller, J. A. 1965. Catalase inhibition as a possible mechanism of the fungistatic action of sorbic acid. *Can. J. Microbiol.* 11:611–617.

Troller, J. A., and Olsen, R. A. 1967. Derivatives of sorbic acid as food preservatives. *J. Food Sci.* 32:228–231.

Tsai, W. J., Liewen, M. B., and Bullerman, L. B. 1988. Toxicity and sorbate sensitivity of molds isolated from surplus commodity cheeses. *J. Food Prot.* 51:457–462.

Tsai, W.-Y. J., Shao, K.-P. P., and Bullerman, L. B. 1984. Effects of sorbate and propionate on growth and aflatoxin production by sublethally injured *Aspergillus parasiticus*. *J. Food Sci.* 49:86–90.

Tsay, W. I., and Chou, C. C. 1989. Influence of potassium sorbate on the growth of *Yersinia enterocolitica*. *J. Food Prot.* 52:723–726.

Tuncan, E. U., and Martin, S. E. 1985. Effect of pH, temperature, and potassium sorbate on amino acid uptake in *Salmonella typhimurium* 7136. *Appl. Environ. Microbiol.* 49:505–508.

Tzatzarakis, M. N., Tsatsakis, A. M., Lotter, M. M., Shtilman, M. I., and Vakalounakis, D. J. 2000. Effect of novel water-soluble polymeric forms of sorbic acid against *Fusarium oxysporum f. sp. Radicus-cucumerinum*. *Food Add. Contam.* 17:965–971.

Uljas, H. E., and Ingham, S. C. 1999. Combinations of intervention treatments resulting in 5-$\log_{10}$-unit reductions in numbers of *Escherichia coli* O157:H7 and *Salmonella Typhimurium* DT104 organisms in apple cider. *Appl. Environ. Microbiol.* 65:1924–1929.

Unda, J. R., Molins, R. A., and Walker, H. W. 1990. Microbiological and some physical and chemical changes in vacuum-packaged beef steaks treated with combinations of potassium sorbate, phosphate, sodium chloride and sodium acetate. *J. Food Sci.* 55:323–326.

U. S. Department and Agriculture. 1978. Proposal to affirm potassium sorbate and sorbic acid as GRAS. Fed. Reg. 43:9823.

U. S. Department of Agriculture. 1979a. Food Safety and Quality Service four-plant study to investigate the use of 40 ppm sodium nitrite and 0.26% potassium sorbate in bacon, Final Report July 23.

U. S. Department of Agriculture. 1979b. Final report: Shelf-life, sensory, cooking and physical characteristics of bacon cured with varying levels of sodium nitrite and potassium sorbate, July, 23.

Vareltzis, K., Buck, E. M., and Labbe, R. G. 1984. Effectiveness of a betalain/ potassium sorbate system versus sodium nitrite for color development and control of total aerobes, *Clostridium perfringens* and *Clostridium sporogenes* in chicken frankfurters. *J. Food Prot.* 47:532–536.

Vareltzis, K. P., and Buck, E. M. 1984. Color stability and sensory attributes of chicken frankfurters made with betalains and potassium sorbate versus sodium nitrite. *J. Food Prot.* 47:41–45.

Vaughn, R. H., and Emard, L. O. 1951. Selectivity of sorbic acid media for catalase-negative lactic acid bacteria and obligate, sporulating anaerobes. *Bacteriol. Proc.* 51:38.

Vidyasagar, K., and Arya, S. S. 1983. Stability of sorbic acid in orange squash. *J. Ag. Food Chem.* 31:1262–1264.

Vidyasagar, K., and Arya, S. S. 1984. Degradation of sorbic acid in fruit squashed and fish paste. *J. Food Technol.* 19:447–454.

Vojdani, F., and Torres, J. A. 1990. Potassium sorbate permeability of methylcellulose and hydroxypropyl methylcellulose coatings: Effect of fatty acids. *J. Food Sci.* 55:841–846.

Wagner, M. K., and Busta, F. F. 1983. Effect of sodium acid pyrophosphate in combination with sodium nitrite or sodium nitrite/potassium sorbate on *Clostridium botulinum* growth and toxin production in beef/pork frankfurter emulsions. *J. Food Sci.* 48:990–991, 993.

Wagner, M. K., and Busta, F. F. 1984. Inhibition of *Clostridium botulinum* growth from spore inocula in media containing sodium acid pyrophosphate and potassium sorbate with or without added sodium chloride. *J. Food Sci.* 49:1588–1594.

Wagner, M. K., and Busta, F. F. 1985a. Inhibitory effects of various salts and/or ionic strengths on growth from *Clostridium botulinum* 52A spores or vegetative cells. *J. Food Prot.* 48:421–428.

Wagner, M. K., and Busta, F. F. 1985b. Influence of minimal change in pH on germination of *Clostridium botulinum* 52A in media containing sodium acid pyrophosphate and potassium sorbate. *J. Food Prot.* 48:693–696.

Wagner, M. K., and Gehrke, W. H. 1982. Effect of packaging film on vacuum level on regular and sorbate-cured bacon. *J. Food Sci.* 47:355.

Wagner, M. K., Kraft, A. A., Sebranek, J. G., Rust, R. E., and Amundson, C. M. 1982. Effect of packaging films and vacuum on the microbiology of bacon cured with or without potassium sorbate. *J. Food Prot.* 45:854–858.

Wakil, S. J., and Hubscher, G. 1960. Determination of coenzyme A by sorbyl coenzyme A formation. *J. Biol Chem.* 235:1554–1558.

Walker, R. 1990. Toxicology of sorbic acid and sorbates. *Food Add. Contam.* 7:671–676.

Wang, Q., Wang, J., Ding, W., Zhang, D., Reed, K., and Zhang, B. 2018. Alternatives to carcinogenic preservatives in Chinese Sausage: Sorbic acid-loaded chitosan /tripolyphosphate nanoparticles. *Int. J. Biol. Macro.* 120:28–33.

Warth, A. D. 1977. Mechanism of resistance of *Saccharomyces bailii* to benzoic, sorbic and other weak acids used as food preservatives. *J. Appl. Bacteriol.* 43:215–230.

Warth, A. D. 1985. Resistance of yeast species to benzoic and sorbic acids and to sulfur dioxide. *J. Food Prot.* 48:564–569.

Wedzicha, B. L., and Brook, M. A. 1989. Reaction of sorbic acid with nucleophiles: Preliminary studies. *Food Chem.* 31:29–40.

Wei, Q., Wang, X., Cheng, J.-H., Zeng, G., and Sun, D.-W. 2018. Synthesis and antimicrobial activities of novel sorbic and benzoic acid amide derivatives. *Food Chem.* 268:220–232.

Weng, Y. M., and Chen, M. J. 1997. Sorbic anhydride as antimycotic additive in polyethylene food packaging films. Leben. *Wiss. Tech. J. Food Sci.* 30:485–487.

Whitaker, J. R. 1959. Inhibition of sulfhydryl enzymes with sorbic acid. *Food Res.* 24:37–43.

Wilamowski, G. 1974. Collaborative study of spectrophotometric method for the determination of sorbic acid in fresh dairy products. *J. Assoc. Off. Analytic. Chem.* 57:675–677.

Wind, C. E., and Restaino, L. 1995. Antimicrobial effectiveness of potassium sorbate and sodium benzoate against *Zygosaccharomyces bailii* in a salsa mayonnaise. *J. Food Prot.* 58:1257–1259.

Yamamoto, M., Yamada, T., Yoshihira, K., Tanimura, A., and Tomita, I. 1988. Effect of food components and additives on the formation of nitrosamides. *Food Add. Contam.* 5:289–298.

Yamamura, A., Murai, A., Takamatsu, H., and Watabe, K. 2000. Antimicrobial effect of chemical preservatives on enterohemorrhagic *Escherichia coli* O157:H7. *J. Health Sci.* 46:204–208.

Ye, J.-W., Yan, L.-J., Zheng, X.-H., Guo, C.-H., Xu, D.-M., and Fang, E.-H. 2015. Determinatio of benzoic acid and sorbic acid in shortening oil by high performance liquid chromatography. *J. Food Safety Quality.* 6:2157–2163.

York, G. K., and Vaughn, R. H. 1954. Use of sorbic acid enrichment media for species of *Clostridium*. *J. Bacteriol.* 68:739–744.

York, G. K., and Vaughn, R. H. 1955. Resistance of *Clostridium parabotulinum* to sorbic acid. *Food Res.* 20:60.

York, G. K., and Vaughn, R. H. 1964. Mechanisms in the inhibition of microorganisms by sorbic acid. *J. Bacteriol.* 88:411.

Yousef, A. E., and Marth, E. H. 1983. Incorporation of [^{14}C] acetate into aflatoxin by resting cultures of *Aspergillus parasiticus* in the presence of antifungal agents. *Euro. J. Appl. Microbiol. Biotechnol.* 18:103–108.

Zamora, M. C., and Zaritzky, N. E. 1987a. Potassium sorbate inhibition of microorganisms growing on refrigerated packaged beef. *J. Food Sci.* 53:257–262.

Zamora, M. C., and Zaritzky, N. E. 1987b. Antimicrobial activity of undissociated sorbic acid in vacuum packaged beef. *J. Food Sci.* 52:1449–1454.

Zhao, T., Doyle, M. P., and Besser, R. E. 1993. Fate of enterohemorrhagic *Escherichia coli* O157:H7 in apple cider with and without preservatives. *Appl. Environ. Microbiol.* 59:2526–2530.

Zhuang, R., Beuchat, L. R. Chinnan, M. S., Shewfelt, R. L., and Huang, Y.-W. 1996. Inactivation of *Salmonella montevideo* on tomatoes by applying cellulose-based edible films. *J. Food Prot.* 59:808–812.

Zou, J. H., Chen, W. D., and Shao, J. D. 2001. Simultaneous determination of 8 food additives in fruit juices by reversed phase high performance liquid chromatography. *Chinese J. Analytic. Chem.* 29:1192–1195.

5

Organic Acids

T. M. Taylor and Stephanie X. Doores

CONTENTS

5.1 Introduction

Many parameters govern the survival and growth of microorganisms in food. The acidity, as often indicated by its pH, of a food can affect the type and number of microorganisms, as well as their potential to replicate and thrive, present in a product. All microorganisms have minimum, maximum, and optimum pH levels and ranges for growth; altering the hydrogen ion concentration consequently influences the growth or inhibition of an organism. In general, most foodborne bacteria are more fastidious and prefer to grow at a pH near neutrality (pH 6.5–7.5) with some notable exceptions (e.g., acidophiles such as *Alicyclobacillus* spp.), but they will tolerate a pH range of 4–9. Yeasts are more tolerant of lower pH values than bacteria. Molds have the widest range of acceptable pH; foods with a pH below 3.5 can support the growth of both yeasts and molds. Tolerance of organisms to widely differing pH levels varies naturally and the pH selects the species or group(s) of microorganisms that will predominate in unaltered food products. For example, bacteria primarily spoil proteinaceous foods such as dairy, meat, poultry, and seafoods with a pH range of 5.5–6.5; in many cases acidification of the food through fermentation of acids is a form of spoilage incurred by bacteria. Yeasts and molds more commonly proliferate on fruits and vegetables with naturally lower pH and little buffering capacity, potentially spoiling through acidification but also through other processes and mechanisms.

One effective means of limiting microbial growth is to increase the acidity of a food, thereby creating an unfavorable environment for microbial growth and replication. Adding an acidulant to the food or harnessing microbial fermentation to develop acidity changes the pH of the food. These actions tend to be microbiostatic rather than microbiocidal. Success in limiting the numbers of microorganisms will depend on the species of microorganism(s) and their inherent tolerance to the acidulant (i.e., pre-adapted to acidic condition), the type and concentration of the acidulant, time of exposure, the buffering capacity of the food, and any pre-existing conditions in the food that could enhance inhibition.

The multiple-barrier concept ("hurdle" concept) has gained popularity for use in a variety of foods in recent years. This concept is based on the premise that foods can be preserved using several inhibitors concurrently rather than relying on a single factor. Therefore, adjustments in pH levels can be coupled with changes in temperature, gaseous atmosphere, water activity, and other inhibitory compounds. The use of the multiple-barrier concept results in lowering the concentrations of acidulants or inhibitors. For example, incorporation of an acid into a food can lessen processing demands for heat treatment owing to the lowered heat resistance of microorganisms in foods with increased acidity. Additionally, foods processed by high pressure may require less intense pressurization treatment due to the presence of the acid antimicrobial without compromising product safety and/or quality. The continued presence of acid can effectively inhibit germination and outgrowth of spores that survive the process. Not only would this interaction ensure commercial sterility of the food, but also the decreased processing time would aid in preserving the palatability of the product (Corlett and Brown, 1980). Comprehensive texts describing applications, physical parameters, and chemical analyses can be found in the references by Taylor et al. (2019) and Doores (2002, 2003).

5.2 Effective Use

Various approaches have been used to study the efficiency of antimicrobial effects by organic acids on cells. The inhibitory effect of acids has been compared based on pH, concentration, chain length, type, and degree of branching to inhibit or kill a wide variety of microorganisms. Because some experiments express concentration in percent, normality, or molarity to achieve final pH, it is somewhat difficult to compare acids or make broad statements concerning the choice of the appropriate acid for a particular effect in a category of foods.

Effective use of an acidulant depends heavily upon the processor managing knowledge of the acid dissociation constant(s) (pK_as), the pH at which 50% of the total acid is dissociated when dissolved. The pK_a of most organic acids lies between pH 3 and 5 (Table 5.1). Since the undissociated portion of the molecule is believed primarily responsible for the antimicrobial effect, it would be advantageous to use the acids near these values. This limits the use of acidulants to foods with pH values of less than 5.0. Given that this value lies at the lower limits of growth for many bacteria, organic acids are usually more effective as antimycotic agents. Hoffman et al. (1939) observed that acids with fewer than 7 carbons were more effective at lower pH, and acids with 8-12 carbons were more effective at neutral and alkaline pH values.

In addition to any antimicrobial effects an acid possesses, the choice of an acidulant may depend upon secondary effects. Acids may contribute to the taste and tartness of a product (e.g., acetic acid and vinegars). They can participate in synergistic interactions with antioxidants by chelating metal ions, as well as with other antimicrobials in enhancing solubility of the partnering antimicrobial through pH reduction (e.g., nisin). They can control pectin gel formation, aid in the inversion of sucrose, prevent browning, and protect color. Some can be used as chemical leavening agents, and the salts of the acids are important in regulating the acidity of the foods (Gardner, 1966).

The toxicology and safety of the acids must be considered when selecting an acid for use. Most of the acids appear safe for use in food products. Several animal studies have indicated possible safety hazards; however, many of these studies used high levels of the concentrated acids at low pH values. At the levels and pH values in which these acids are normally used in foods, the same hazards would not be expected. The long-term consumption of these acids, which exist naturally in many foods, would further indicate that hazards are unlikely. In addition, all these acids can be metabolized by the body primarily through lipid oxidation and via the tricarboxylic acid cycle.

TABLE 5.1

Dissociation Constants of Organic Acids in Aqueous Solutions

Acid	pK_1	pK_2	pK_3
Acetic acid	4.75		
Dehydroacetic acid	5.27		
Sodium diacetate	4.75		
Adipic acid	4.43	5.41	
Caprylic acid	4.89		
Citric acid	3.14	4.77	6.39
Fumaric acid	3.03	4.44	
Lactic acid	3.80		
Malic acid	3.40	5.11	
Propionic acid	4.87		
Succinic acid	4.16	5.61	
Tartaric acid	2.98	4.34	

Source: Lide (2002).

5.3 The Acids

Information on the toxicology, antimicrobial properties, application, and regulatory status is presented here for each acid. The lipophilic acids include acetic acid and acetic acid derivatives, propionic acid, and caprylic acid. These acids, as the name implies, are relatively non-polar compounds. They are usually added to foods for their antimicrobial properties, although they could also serve other functions in foods. The carboxylic acids are more polar than the lipophilic acids and are traditionally used in foods for secondary effects rather than for their ability to inhibit microbial growth. Numerous studies included in this chapter compare several organic acids for their antimicrobial effects. The results of these studies are usually presented in the section describing the acid that gave the most antimicrobial effects. In some instances, the information appears separately for each acid.

5.3.1 Acetic Acid

5.3.1.1 Antimicrobial Activity in Foods

Members of the acetic acid bacteria (AA), including *Acetobacter aceti*, and heterofermentative lactic acid bacteria (heterolactics) produce acetic acid as a by-product of metabolism, and as such are more tolerant to this acid than the homofermentative lactic acid bacteria (homolactics), which do not. These organisms are found naturally associated with fermented products, such as pickles, sauerkraut, and vinegar.

Adams and Hall (1988) showed a weakly synergistic inhibitory effect between acetic and lactic acids with *Salmonella* Enteritidis and *Escherichia coli*. Tejero-Sariñena et al. (2012) more recently demonstrated a similar interaction of these fermentation products against the Gram-positive pathogens *Staphylococcus aureus*, *Clostridium difficile*, and *Enterococcus faecalis*, wherein observed antimicrobial effects of mixed fermentates were neutralized when fermentate was adjusted to pH 7.0. Although these acid mixtures might be expected to occur naturally during fermentations by heterolactic bacteria, this apparent synergy has not been demonstrated in other foods. This association could explain the apparent stability that occurs in natural fermentations, such as sauerkraut or sourdough bread (Gagiu et al., 2017). Sauerkraut manufacture is a fermentation process brought about by a natural succession of microbes that inhabit the surface of cabbage leaves. The succession typically begins with growth of coliforms and natural microflora followed by *Leuconostoc mesenteroides* (heterolactic), then by *Lactobacillus plantarum* (homolactic), a more acid-tolerant organism. McDonald et al. (1990) found that growth of *L. mesenteroides* ceased when the internal pH of the cells reached 5.4–5.7, whereas growth stopped in *L. plantarum* at pH 4.6–4.8. *L. plantarum* maintained its pH gradient in the presence of either 160 mM sodium acetate or sodium lactate down to an external pH of 3.0.

Microorganisms vary in their susceptibility to acetic acid. Lower concentrations of acid inhibited *Bacillus cereus* (pH 4.9), *Salmonella* Aertrycke (pH 4.9), and *S. aureus* (pH 5.0) than those needed to inhibit *Saccharomyces cerevisiae* (pH 3.9) and *Aspergillus niger* (pH 4.1) (Levine and Fellers, 1939). *Bacillus* and *Clostridium* spp. and Gram-negative bacteria were more strongly inhibited at pH 6.0 than lactic acid bacteria, Gram-positive bacteria, yeasts, and molds. However, as the pH decreased to 4.0, all of the groups were similarly affected (Woolford, 1975b). Owen (1946) suggested that acetic acid inhibited Gram-positive organisms.

S. aureus was most sensitive to acetic, followed by citric, lactic, malic, tartaric, and hydrochloric acids (Nunheimer and Fabian, 1940). Acetic acid exerted the most inhibition, but had the lowest dissociation constant whereas tartaric acid had one of the highest dissociation constants, yet exerted the weakest inhibitory action. Similar results were noted when various acids were compared at the pH level required for a 90 and 99% reduction in numbers of *S. aureus* in 12 h (Minor and Marth, 1970). Acetic acid was effective at pH 5.2 and 5.0, respectively, followed by lactic, pH 4.9 and 4.6, citric, pH 4.7 and 4.5, and HCl, pH 4.6 and 4.0. Both studies concluded that the antimicrobial effect was due to the undissociated molecule. *S. aureus* preconditioned in acetate buffer (pH 4.4–4.6) at 20°C, shifted to pH 7.0 buffer, and exposed to 40°C became injured, but no injury occurred in cells not exposed to acidic conditions before heating (Smith et al., 1984).

B. cereus was inhibited at pH 5.0 with 0.15 M acetic acid and with 0.33 M lactic acid (Wong and Chen, 1988). A 0.1-M concentration of acetate and lactate completely inhibited multiplication at pH 6.1 and 5.6, respectively, and caused 50% inhibition of spore germination at pH 4.2 and 4.3, respectively. Lues and Theron (2012) reported similar findings from studies on *Listeria monocytogenes* sensitivity to differing organic acids, where lower minimum inhibitory concentration (MIC) requirements for acetic acid were noticed versus lactic, citric, and propionic acids.

Buffered acidulant systems (brain heart infusion (BHI) broth, pH 4.8, 22°C) containing 1.0% acetic acid and 1.0% lactic, malic, tartaric, or citric acid were more effective in reducing numbers of *S. aureus*, *Salmonella* Blockley, *Streptococcus faecalis*, and *E. coli* than unbuffered systems. This finding is important in the manufacture of medium-acid foods containing more than one organic acid, such as mayonnaise (Debevere, 1988).

E. coli O157:H7 is considered more acid-tolerant than other *E. coli*. When *E. coli* O157:H7 was inoculated into tryptic soy broth (TSB) containing acetic, citric, lactic, malic, and tartaric acids growth occurred at pH 5.5 for all acidified media at 10°C except for acetic acid. At 25 and 37°C, *E. coli* O157:H7 grew in all media at pH 5.0 with scant growth in TSB containing acetic acid. Malic acid allowed growth at pH 4.5 and tartaric acid permitted growth at pH 4.0 (Conner and Kotrola, 1995; Conner et al., 1997). In contrast to research detailing stronger antimicrobial activity of acetic acid versus lactic acid, lactic acid exhibited stronger antimicrobial activity against *E. coli* and *Salmonella* isolates by requiring lower minimum bactericidal concentrations (MBC) for both pathogens (Alves de Ávila et al., 2013).

Acid and non-acid food ingredients can interact with acetic acid to influence the degree of inhibitory action. Reduced-calorie mayonnaise (RCM) manufactured with 0.1, 0.3, 0.5, or 0.7% acetic acid in the aqueous phase and adjusted to pH 4.0 with hydrochloric acid was compared to cholesterol-free (egg yolk-free), reduced-calorie mayonnaise (CFM) manufactured with 0.3 and 0.7% acetic acid (Glass and Doyle, 1991). Both formulations were inoculated with approximately 10^8 CFU/mL of an 8-strain mixture of *Salmonella* or a 6-strain cocktail of *L. monocytogenes* and held at 23.9°C. *Salmonella* were inactivated within 48 h in both RCM and CFM formulated with 0.7% acetic acid, with reduction to undetectable levels in 1 and 2 weeks with 0.5 and 0.3% acetic acid formulations, respectively. Destruction of *Salmonella* occurred more quickly in CFM than RCM, likely because of the absence of egg yolks in CFM and the presence of egg white, which has recognized antimicrobial properties. *L. monocytogenes*, although more resistant than *Salmonella*, was reduced by 4.0 $\log_{10}$-cycles in both mayonnaises within 72 h in formulations containing 0.7% acetic acid. *L. monocytogenes* was also reduced to undetectable levels in RCM within 10 days and in CFM within 14 days compared to 2 days for *Salmonella*. It is unlikely that these products would contain such high initial levels of either of these organisms especially with the use of pasteurized egg products; therefore the standard use of 0.7% acetic acid, pH 4.0 in the aqueous phase of CFM and RCM would still produce a microbiologically sound product. A synergistic interaction of 0.5% acetic acid and 3.0% NaCl was reported against *L. monocytogenes* in pickle brine (Bae and Lee, 2015).

A concentration of 0.4–0.8% reduced growth of *Micrococcus* and *Bacillus* spp., some strains of *S. aureus*, Gram-negative aerobic bacteria, and *Enterobacteriaceae* in Japanese fermented soy sauce (Hayashi et al., 1979). An increased toxic effect of acetic acid against *Saccharomyces rouxii* and *Torulopsis versatilis* was demonstrated in brine fermentation of soy sauce as the pH decreased from 5.5 to 3.5 (Noda et al., 1982).

The effect of temperature–acid interactions is noteworthy for the psychrotrophic pathogens *L. monocytogenes* and *Yersinia enterocolitica*. Acetic acid caused greater inactivation of *L. monocytogenes* than lactic or citric acid and increased inhibition as the temperature of incubation decreased from 35 to 13°C; concentrations ≥0.3% inhibited *L. monocytogenes* irrespective of temperature. On the other hand, citric acid caused greater injury than either lactic or acetic acid, and injured organisms survived longer at low temperatures (Ahamad and Marth, 1989; 1990). Acetic acid displayed the greatest antimicrobial action against *L. monocytogenes* when compared to citric, malic, lactic, and hydrochloric acids at equal pH values and at all incubation times and temperatures (Sorrells et al., 1989). When comparisons were based on equal molar concentrations, citric acid was more bactericidal than other acids at 25 and 35°C, but malic acid was more bacteriostatic than other acids at 10°C. The greatest antimicrobial effect occurred at 35°C with the greatest survival at 10°C. The minimum inhibitory pH level was pH 4.4 for malic, citric, and hydrochloric acids, pH 4.4–4.6 for lactic acid, and pH 4.8–5.0 for acetic acid.

Sorrells and Enigl (1990) also demonstrated interactive effects between sodium chloride, acidulants, and temperature. Similar growth inhibition was demonstrated for *Y. enterocolitica* at lower pH levels as the temperature increased. Acetic acid was more effective than lactic or citric acid in producing this effect (Adams et al., 1991; Brocklehurst and Lund, 1990; Karapinar and Gönül, 1992a; Little et al., 1992). Conversely, injury and death rates were not affected by refrigeration temperatures for the mesophile *Salmonella* Bareilly. A 90% reduction of the population required exposure to 0.01 N for 75 min, 0.05 N for 8.5 min, 0.10 N for 5 min, and 0.2 N for 4 min. Acid injury did not appear to involve damage to ribosomes or other nucleic acid material, but recovery required the presence of amino acids or peptones (Blankenship, 1981).

Raw produce is now being subjected to acid rinses in some production practices as a means of reducing high levels of spoilage organisms and foodborne pathogens. Because produce is typically refrigerated with only minimal processing, these foods can be the carriers of psychrotrophic pathogens. Parsley, inoculated with approximately 10^7 CFU/g *Y. enterocolitica*, was dipped in acetic acid solutions (1, 2, or 5%) or vinegar (30, 40 or 50%) for 15 or 30 min. *Y. enterocolitica* was not detectable after exposure to 2 and 5% acetic acid or 40 and 50% vinegar for 30 min (Karapinar and Gönül, 1992b). Application of 0.5 or 1.0% acetic acid to pathogen-inoculated iceberg lettuce produced reductions in *L. monocytogenes* numbers of 2.3 or 2.4 $\log_{10}$ CFU/g, respectively, after 5 min of dipping in the acid-containing washing solution. Substitution of acetic acid with lactic or citric acid at the same concentrations produced greater reductions in pathogen numbers (2.8–3.1 $\log_{10}$ CFU/g) (Akbas and Ölmez, 2007).

Acetic acid applied as a volatile compound (1000 mg/liter of air) for 7 h at 60°C achieved a >3.0 $\log_{10}$-cycle reduction of six *Salmonella* strains inoculated onto alfalfa seeds compared to a 1.9 $\log_{10}$-cycle reduction for non-treated seeds (Weissinger et al., 2001). Exposure at 50°C for 12 h with lower levels (100 and 300 mg/liter of air) reduced the population >1.7 $\log_{10}$ without affecting germination of seeds. Treatment at low temperatures (4 days at 10°C) with 200 and 500 mg/liter of air was effective in reducing *Salmonella* by 2.3 and 5.7 $\log_{10}$-cycles, respectively, but resulted in changes in sensory characteristics. Mung bean seeds inoculated with 3.0 to 5.0 $\log_{10}$-cycles of *S.* Typhimurium, *E. coli* O157:H7, and *L. monocytogenes* were exposed to acetic acid at a concentration of 242 μL/liter of air for 12 h at 45°C. Treatment reduced *S.* Typhimurium and *E. coli* O157:H7 to non-detectable levels, but *L. monocytogenes* was isolated from two of ten samples. In addition, vapor treatment also reduced levels of resident microflora on seed, but did not affect seed germination rates (Delaquis et al., 1999). More recently, use of gaseous acetic acid (4.7 mM) on fenugreek seeds yielded reductions ranging from 3.3 to 4.1 $\log_{10}$ CFU/g of *S.* Enteritidis, whereas *E. coli* O157:H7 was reduced by >5.0 $\log_{10}$ CFU/g by 4.7 mM gaseous acetic acid (Nei et al., 2015). One day exposure to gaseous 8.7% acetic acid reduced *E. coli* O157:H7 and *Salmonella* to non-detectable counts on inoculated alfalfa and radish seeds, without significant compromise of seed germination (Nei et al., 2011).

A variety of organic acid treatments has been used to control microbial loads associated with meat carcasses and products (Dickson and Anderson, 1992; Dorsa, 1997; Dorsa et al., 1997, 1998; Smulders and Greer, 1998). Numerous organic acid combinations, containing from 0.6 to 3.0% acetic acid at pH from 1.5 to 3.0 have been used to rinse meat animal carcasses or meat tissue. Treatments spanned varying amounts of time with concomitant decrease in microbial loads up to 3.0 $\log_{10}$. Depending upon the treatments, there can be adverse sensory effects including discoloration of meat, off odors, and off flavors.

Pork carcasses inoculated with 10^6 CFU/ml *S.* Enteritidis were sprayed with acetic acid at pH 1.5 or 2.0 for 30 or 60 s (Biemuller et al., 1973). Total plate counts were reduced by 4.0 $\log_{10}$ at pH 1.5, a two-fold reduction at pH 2.0, 1.0 $\log_{10}$-cycle at pH 2.5, and no reduction in microbes at pH 3.0. The treatment effectively reduced recovery of *S.* Enteritidis to 1 of 6 samples at pH 1.5 and 11 of 72 for pH 2.0. Although washing of pork carcasses with 1.5% acetic, citric, or lactic acid led to significant decreases in aerobic plate and coliform counts for acetic and citric acid washes after 14 days of storage at 2–4°C, *L. monocytogenes* was detected in 69% of loins and 33% of chops, suggesting that acids had limited effectiveness (Fu et al., 1994). Increasing the concentration to 2% acetic acid or using 200-ppm hypochlorite solutions before vacuum-packaging and storage for 28 days at 4°C significantly lowered aerobic, anaerobic, and lactic acid bacteria counts; however, some discoloration occurred (Cacciarelli et al., 1983). A 3% concentration of acetic acid was quite effective in reducing counts of *Enterobacteriaceae* in vacuum-packaged pork stored for 6 weeks at 2–4°C (Mendonca et al., 1989b). An application of 2%

acetic acid reduced the incidence of *Salmonella* on pork cheek meat in addition to significantly reducing aerobic plate and coliform counts (Frederick et al., 1994). An 18% acetic acid or 12% lactic acid spray significantly reduced bacterial counts on lamb carcasses. At this concentration, the pH level decreased to 2.1 and 1.5, respectively, and bleaching of the carcasses occurred (Ockerman et al., 1974). Mani-López et al. (2012) summarized research on acetic acid sprays and dips usage to decontaminate meat animal carcasses and fabricated products, reporting that 0.1–2.0% acetic acid produced reductions in salmonellae and various meat spoilage microbes between 1.0 and 3.0 $\log_{10}$-cycles, and sometimes enhanced the antimicrobial activity of plant-derived antimicrobials or other physical processing measures. Argentinian researchers reported Shiga toxin-producing *E. coli* (STEC) were reduced on beef carcasses only by 0.5 or 0.7 $\log_{10}$-cycles following application of 2.0% acetic acid by automated or manual application methods, though aerobic bacteria were reduced to a higher degree by acid exposure, without researchers reporting any visual carcass discoloration (Signorini et al., 2018).

As part of recently established guidelines by the U.S. Department of Agriculture Food Safety Inspection Service (FSIS), organic acids can be used in pre-evisceration rinse systems consisting of a water rinse followed by an organic acid rinse. Currently in the U.S., acetic acid is approved by the USDA-FSIS as a spray for dried/fermented meat products, or as a dip-antimicrobial up to 5% for chicken livers, although it's applied extensively with FSIS approval as a component of various peroxyacetic acid mixtures and other mixed acid antimicrobials for treating surfaces of various red meat- and poultry meat-derived products (USDA-FSIS, 2019). Application of a 3% concentration of acetic acid as a spray on beef carcasses was not significantly better than water washes in reducing levels of *E. coli* O157:H7; however a final application of acid washes could provide some residual effect during storage (Brackett et al., 1994; Cutter et al., 1997; Cutter and Siragusa, 1994; Dorsa et al., 1997; Hardin et al., 1995).

Quartey-Papafio et al. (1980) found that a 1% formic and 1% acetic acid combination applied to beef for up to 20 min was effective in reducing a variety of bacterial species. Treatment of beef cubes with 1.2% acetic acid or 0.6% acetic and 0.046% formic acids for 10 s caused discoloration from both treatments, but beef treated with the mixture of acids did not differ in flavor from untreated beef (Bell et al., 1986). Hamby et al. (1987) used intermittent sprays of 1% acetic acid or lactic acids which resulted in significant reductions in aerobic plate count (1.8–4.3 log/cm^2). Meat was sprayed, vacuum-packaged, and stored for 28 days at 2°C in a high-oxygen barrier film. *Lactobacillus* spp. predominated as storage progressed. Application of a 2% lactic or 2% acetic acid spray to beef strip loins and storage at –1°C resulted in significantly lower aerobic and lactic acid bacteria counts over 112 days of storage (Goddard et al., 1996).

Reductions in microbial counts depend upon the water temperature, type of chemicals, and application rate and sequence in the process (Gorman et al., 1995). Anderson et al. (1977) reduced bacterial counts by 99.6% on meat using a 3% concentration of acetic acid before washing. They found that acetic acid was a superior sanitizer compared to hypochlorite and had a greater residual effect. Similar reductions in numbers of *E. coli* O157:H7 and *Salmonella* were reported by other researchers treating beef hides with 10% acetic (55°C) followed by a 30 sec water rinse, although these researchers also reported 10% lactic acid produced greater pathogen reductions on carcass surfaces when applied under similar conditions (Carlson et al., 2008). Further work by Anderson and Marshall (1989) demonstrated that the application of acetic acid at 70°C was most effective as compared to other treatments for sanitizing beef semitendinous muscle inoculated with *E. coli*, *S.* Typhimurium, or manure slurry. The intervention temperature played a greater role in reducing numbers than did the acid concentration. Microbial loads on beef carcasses subjected to machine washing and sanitization with 1.5% acetic acid were reduced to a greater extent when treatments were conducted at 52°C than 14.4°C (Anderson et al., 1987). Incorporation of pulsed power electricity and a 2% acetic acid spray led to improved reduction of inoculated *E. coli* O157:H7 and *S.* Typhimurium on beefsteaks (Tinney et al., 1997). Choi et al. (2009) combined the application of supercritical CO_2 (SC-CO_2) with acetic acid to decontaminate pork cuts from human pathogens. The combination of 3% acetic acid plus SC-CO_2 produced a greater reduction in numbers of *L. monocytogenes*, *Salmonella* Typhimurium, and *E. coli* O157:H7 than SC-CO_2, though combination of lactic acid with SC-CO_2 was equally effective. Similarly, combined application of 1.8% acetic acid with a 15 sec 82°C water spray produced reductions in enteric bacteria and *E. coli* by 2.0 to 2.3 $\log_{10}$ CFU/cm^2 (Eggenberger-Solorzano et al., 2002).

Dickson (1992) contaminated lean and beef tissue surfaces with *S.* Typhimurium followed by treatment with 2% acetic acid. *S.* Typhimurium was reduced by 0.5–0.8 log CFU/cm^2, although this was not significantly different from the controls. *S.* Typhimurium reductions were greater when the bacteria were attached to fatty tissue, possibly because this tissue retained less moisture (20%) than lean tissue (75%), and the reduced water activity could have enhanced the antimicrobial effects. It was noted that the use of acetic acid as a rinse for beef tissue led to sublethal injury of bacterial cells. An increase in organic material, such as rumen fluid, dirt, or manure, resulted in less effective reduction of *S.* Typhimurium. Unda et al. (1991) inoculated beef roasts by injection and on the surface with approximately 10^3 *Clostridium sporogenes* and *L. monocytogenes*. The acetic acid in the brine inhibited anaerobic and aerobic bacteria.

Interventions using acetic acid alone or in combination with other antimicrobials incorporated into poultry processing have been effective in reducing microbial contamination. Poultry scald water containing 0.1% acetic acid at 52°C decreased levels of *S.* Typhimurium and *Campylobacter jejuni* by 0.5–1.5 $\log_{10}$ (Okrend et al., 1986). Addition of 1.0% acetic acid caused instantaneous death. Levels of 0.2–0.5% acetic acid reduced *Enterobacteriaceae* counts by 2.24 $\log_{10}$-cycles (Lillard et al., 1987). Similarly, reductions in numbers of *Salmonella* Kentucky on chicken breast by the combined application of 0.2 mg/mL thymol and 2.0 mg/mL acetic acid was 2.2 $\log_{10}$ CFU/g (Lu and Wu, 2012). Nonetheless, the possible attachment of *Salmonella* to broiler chicken skin required a >4% concentration of acetic acid to reduce pathogen numbers by 2.0 $\log_{10}$-cycles (Tamblyn and Conner, 1997).

Acidification of the chill water at the end of processing delivers a second intervention step to reduce microbial loads and to increase the shelf life of poultry. Adjustment of the chill water to pH 2.5 with acetic acid was the most effective antimicrobial treatment followed by other acids in descending order of effectiveness, adipic, succinic, citric, fumaric, and lactic (Mountney and O'Malley, 1965). Nevertheless, despite the reduction in microbial numbers at this pH, acetic acid caused the skin of the poultry to be hard and leathery. Other studies continued to explore the use of acetic acid in chill water, but changes in pH and concentration of acid were more successful. Immersion of broilers for 10 min in 0.6% acetic solution acid (pH 3.0) did not result in a significant reduction of aerobic plate counts but did significantly reduce levels of *Enterobacteriaceae* by 0.71 log most probable number (MPN)/ ml (Dickens et al., 1994). Expanding on their work, Dickens and Whittemore (1994) exposed broilers to 0.3 and 0.6% acetic acid under the same conditions, with and without the use of air injection to agitate the chill water. Again, aerobic plate counts were unaffected by the treatments, but *Enterobacteriaceae* counts were significantly reduced by 0.86 log MPN/ml for the 0.3% acid and 2.35 log MPN/ml for the 0.6% acid solutions. There was no significant difference in texture or sensory characteristics between the treatments although the skin of the 0.6% acetic acid-treated carcasses was darkened or yellowed (Dickens et al., 1994). Water pockets occurred under the skin of chicken carcasses with air-agitated samples (Dickens and Whittemore, 1994). Conversely, Fabrizio et al. (2002) used 2.0% acetic acid in chiller waters and reported *S.* Typhimurium numbers were reduced by 1.4 $\log_{10}$ CFU/mL compared to control samples, and numbers of the pathogen remained unchanged thereafter over 7 days' storage at 4°C. Coliform bacteria were reduced by 3.0 $\log_{10}$ CFU/mL of chiller waters by acetic acid treatment, and continued to decline over the refrigerated storage (Fabrizio et al., 2002).

Products preserved by a number of antimicrobial compounds use a multiple-barrier approach to preservation. Kurita and Koike (1983) combined 3% ethanol, 0.05% acetic acid, 1 mM perillaldehyde, or 0.5%, 0.05%, and 0.5 mM, respectively, of the same components, with 2% salt to inhibit growth of contaminating microorganisms for over 20 days at 27°C. Rubin (1978) found that acetic and lactic acid acted synergistically to retard growth of *S.* Typhimurium using a model system. Combined application of 5 g/kg sodium dodecyl sulfate (SDS) with 10 g/kg acetic acid produced similar synergism as determined by *S.* Kentucky reductions on chicken skin. Reductions of the pathogen following treatment for 1 and 3 min increased by 2.2 and 4.0 $\log_{10}$-cycles as compared to sequential application of acetic acid followed by SDS (Zaki et al., 2015). Thymol (0.2 mg/mL) plus 5% SDS and 2.0 mg/mL acetic acid produced a similar reduction in *S.* Kentucky numbers on chicken breasts (Lu and Wu, 2012). El-Shenawy and Marth (1989b) investigated the effect of acetic, citric, lactic, and tartaric acid adjusted to pH 5.6 or 5.0 on inhibition of *L. monocytogenes*. Acidified media showed greater inhibitory activity at 35 than at 13°C. Acetic and tartaric acids were more inhibitory than lactic and citric acids.

The acid level could be reduced when used in combination with brines and sugar. Whole pickles preserved with brine at 12°Brix and 0.9% acetic acid did not support growth or toxin formation of *Clostridium botulinum* (Ito et al., 1976). Stroup et al. (1985) found that mushrooms reached an equilibrium pH of 4.6 depending upon the concentration of acid in can brine within 1 day with 1.0% acetic acid compared to 2 days for 0.7% citric acid. For pearl onions, a 0.35% solution required 7 days for equilibration for acetic and citric acid but 1 day for 0.7% acetic and 4 days for the same concentration of citric acid. In egg and pickle brines containing 2.0% acetic acid at pH 4.0, bacterial pathogens were reduced to non-detection during product storage over 10 days. Nevertheless, in contrast to other studies cited here, the addition of sodium benzoate (0.02%) did not reduce the amount of acetic acid needed for pathogen inhibition (Sullivan et al., 2013).

Nisin (an approved bacteriocin; see Chapter 8) coupled with 0.05 and 0.5% concentrations of acetic, lactic, citric acids or glucono-delta lactone were studied using *Bacillus* spores heated at pasteurization temperatures of 65 and 95°C and recovered at 12, 20, and 30°C (Oscroft et al., 1990). Nisin, in combination with organic acids, displayed synergistic effects in increasing the destruction of the organism. It was found that acetic acid was the most effective acid, followed by glucono-delta lactone, lactic, and citric acid when spores were heated at 95°C for 15 min and acetic, lactic, glucono-delta lactone, and citric acids when temperatures were reduced to 65°C for 60 min. *Listeria monocytogenes* on tomato surfaces treated by 0.5% nisin plus either 3.0 or 5.0% acetic acid demonstrated greater pathogen inhibition activity versus nisin plus citric acid at the same concentrations, and both outperformed a 200-ppm chlorine treatment (Oladunjoye et al., 2016). Foods, particularly pre-cooked, chilled, ready-to-eat products, are especially susceptible to spoilage because of indigenous microbial populations. A combination of these parameters can be used to extend the shelf life of these products. For example, frankfurters treated with a combination of bacteriocins and acetic acid reduced *L. monocytogenes* to non-detection over a 35-day storage period, and produced bacteriostatic effects on psychrophilic spoilage microbes on frankfurter surfaces (Castellano et al., 2018). Likewise, vegetative *B. cereus* cells were inhibited from growth on cooked spinach post-processing by the combined application of 0.02% thymol, 0.68% acetic acid, and 0.03% nisin, where the lag time for cells was extended for 3 days at 25°C versus the control (Hyun et al., 2018).

Although acetic acid is generally used as an inhibitor of bacteria and yeasts, it is effective against the black bread molds *A. niger*, and *Rhizopus nigrificans* at pH 3.5. *Aspergillus fumigatus* required 0.4% acetic acid at pH 5.8 and 0.2% acid at pH 5.0 and below to inhibit growth (Kirby et al., 1937). More recently, acetic acid produced by the biopreservation culture *Lactobacillus plantarum* EM inhibited the molds *Aspergillus flavus*, *A. fumigatus*, *Aspergillus ochraceus*, *Aspergillus nidulans*, and *Penicillium expansum* (MIC: 40 AU/mL) (Mun et al., 2019). When used as a surface application, a 1% concentration of acetic acid at pH 4.5 completely inhibited growth and aflatoxin production by *Aspergillus parasiticus*. Lower concentrations of 0.6 or 0.8% partially inhibited growth and decreased toxin formation by 70 and 90%, respectively (Buchanan and Ayres, 1976). Cruess and Irish (1932) showed that apple juice containing 0.8–1.0% acetic acid adjusted to pH 3.5 inhibited *Saccharomyces ellipsoideus* and *Penicillium glaucum*, but a concentration greater than 4% was needed when the pH was at 7.0. Populations of *Pseudomonas aeruginosa* were reduced by 99.999% after exposure to a 1.0% concentration of acetic acid, suggesting its use as a decontamination agent for equipment (Hedberg and Miller, 1969).

Other forms of acetic acid have antimicrobial uses. Peracetic acid is formed by an oxygen molecule bound to the carboxyl carbon atom of acetic acid. Upon contact with organic substrates, it decomposes to yield oxygen and acetic acid. Spore-forming bacteria are killed on surfaces at a concentration of 0.0001 and 40% relative humidity within 10 min (Portner and Hoffman, 1968). A combination of peroxyacetic (64 ppm)/octanoic (53 ppm) acid solution was more effective as an antifungal agent in flume water for vegetables such as celery, cabbage, and potatoes than peroxyacetic (80 ppm) acid alone (Hilgren and Salverda, 2000). This treatment could also be used in recycling of water to reduce levels of yeasts and molds, aerobic bacteria, and coliforms.

5.3.1.2 Application and Regulatory Status

Acetic acid is a monocarboxylic acid with a pungent odor and taste, which limits its use. It is the principal component of vinegars and as such is primarily used for its flavoring abilities. It is highly soluble in

TABLE 5.2
Acceptable Daily Intake for Humans

Acid	Limitations (mg/kg body weight) Unconditional	Conditional	Reference
Acetic acid	Not limited		FAO (1965)
Acetate, Ca^{+2}, Na^{+}, K^{+}	Not limited		FAO (1963, 1973)
Sodium diacetate	0–15		FAO (1973)
Adipic acid		0–5	FAO (1965)
Citric[a]	Not limited		FAO (1966)
Citrate, Ca^{+2}, Na^{+}, K^{+}	Not limited		FAO (1963)
Fumaric acid	0-6		FAO (1974)
Lactic acid	Not limited		FAO (1965)
DL-lactic acid		0–100[b]	FAO (1965)
Lactate, Ca^{+2}, NH_4^{+}, Na^{+}, K^{+}	Not limited		FAO (1973)
Malic acid	Not limited		FAO (1965)
DL-malic acid		0–100[c]	FAO (1966)
Propionic acid	Not limited		FAO (1965)
Propionate, Ca^{+2}, Na^{+}, K^{+}	Not limited		FAO (1973)
Tartaric acid	0–30		FAO (1973)
Tartrate, Na^{+}, K^{+}	0–30		FAO (1973)

[a] Naturally occurring substances; the estimated acceptable daily intakes listed do not include naturally occurring amounts.
[b] Content of D(–) lactic acid.
[c] Content of D(–) malic acid; the malic acid content should not exceed 0.05%.

water. It is used in condiments such as mustard, catsup, salad dressings, and mayonnaise, and is found in pickled products such as sausages and pigs' feet. Because of cost and antimicrobial action, it has been added to infant feeding formulas to replace lactic acid (Seymour et al., 1954). Acetic acid is generally regarded as safe (GRAS) for miscellaneous and general-purpose usage (21 CFR 184.1005). The acceptable daily intakes for humans for acetic acid and its derivatives are listed in Table 5.2. The data are scarce for the use of peracetic acid in food products. It has primarily been used as a surface disinfectant and is used for preventing fruit flies from laying eggs in damaged tomatoes (Ayres et al., 1980).

5.3.2 Acetates

5.3.2.1 Antimicrobial Properties

Several derivatives of acetic acid are currently in use as antimicrobial agents. The sodium and calcium salts are sometimes used in foods and would be expected to have the same antimicrobial properties as acetic acid at the same pH values (Hoffman et al., 1939). The salt form, however, requires different handling and use procedures than the acid. Mendonca et al. (1989a) used a dip of sodium acetate in combination with potassium sorbate (10%) and phosphates (10%) to extend the shelf life of pork chops. Sodium acetate (10%), potassium sorbate (1.5%), sodium lactate (3%), or trisodium citrate were used alone or in combination with *Bifidobacterium breve* (5%) as a dip for fresh camel meat (Al-Sheddy et al., 1999). Sodium acetate extended shelf life to 12 days at 4°C, whereas potassium sorbate increased shelf life to 9 days. *B. breve*, sodium lactate, and trisodium citrate extended shelf life to ≤6 days. A combination of *B. breve* and sodium acetate exhibited an additive effect to extend shelf life still further. Color remained acceptable although a mild acetic acid odor was detected. Utilization of 3% acetic acid or electrolysis-activated potassium acetate (both adjusted to pH 2.2) produced reductions in *S. enterica* and *L. monocytogenes*, while the acid anion demonstrated greater antimicrobial activity against the Gram-positive pathogen during *in vitro* testing (Liato et al., 2017b). Authors suggested the electro-activated acid anions achieved antimicrobial outcomes by pH and oxidative effects of the acid anions as compared to undissociated acid uncoupling of homeostasis regulation.

Sodium acetate (0.5%) was effective in reducing populations of *L. monocytogenes* in vacuum-packaged turkey bologna during storage at 4°C and was similar to 0.26% potassium sorbate or 2% sodium citrate in its efficacy (Wederquist et al., 1994). A combination of 0.25% acetate and 2.5% lactate was effective as an antimicrobial treatment in sliced cervelat sausage over a 4- to 6-week storage life (Blom et al., 1997). Other RTE cured and uncured poultry and meat products were similarly preserved against growth of pathogens including *L. monocytogenes* and *Clostridium perfringens* with acetate-containing antimicrobials (Gupta and Ravishankar, 2005; King et al., 2015; McDonnell et al., 2013).

Sodium acetate (SA) at concentrations of 0.75 and 1% significantly reduced the initial aerobic plate counts of catfish fillets by 0.6–0.7 $\log_{10}$ when stored at 4°C compared to the non-treated fillets and increased shelf life by 6 days. The combination of 0.3–0.7% sodium acetate and 0.4% monopotassium phosphate (MKP) more effectively prolonged shelf life to 12 days compared to SA alone, although MKP alone had no effect. SA alone or SA-MKP-treated fillets were not significantly different in odor from fresh controls up to 9 days, but did appear brownish and watery (Kim et al., 1995b). Sodium acetate in combination with potassium sorbate and *Lactococcus lactis* ssp. *cremoris* culture added to catfish fillets was an effective treatment against the growth of Gram-negative spoilage bacteria at 4°C (Kim and Hearnsberger, 1994). The combination of 0.5% sodium acetate, 0.25% potassium sorbate, and 2.5% lactic cultures or 0.5 to 1.0% sodium acetate alone inhibited growth for more than 6 days at 4°C, and counts increased only by 1.3 or 1.2 logs by day 12. Treatment with sodium acetate alone or in combination with bifidobacteria extended shelf life at 4°C by 3 days (Kim et al., 1995a). Catfish fillets treated with 2% sodium acetate, vacuum-packaged, and held at 4°C for 12 days contained lower population levels of psychrotrophic and anaerobes than untreated samples (Zhuang et al., 1996). Whole and peeled shrimp were immersed for 2 min in solutions containing 10% sodium acetate, 1.5% potassium sorbate, 5% sodium lactate, and 1.5% trisodium citrate with or without *B. breve* culture and stored at 4°C for 9 days (Al-Dagal and Bazaraa, 1999). Psychrotrophic counts from untreated whole shrimp increased to $>10^7$/g, the indicator of spoilage, over 6 days, but bifidobacteria did not grow within 9 days; sodium acetate alone or potassium sorbate with bifidobacteria retarded the growth of psychrotrophs compared to other treatments within 6 days and provided the lowest counts with the potassium sorbate-bifidobacteria mixture. For peeled shrimp, untreated samples reached the spoilage mark within 10 days, but sodium acetate provided better sensory characteristics and more controlled growth of psychrotrophs than potassium sorbate with bifidobacteria. Growth and aflatoxin production of *A. parasiticus* was inhibited by a 1.0% concentration of sodium acetate at pH 4.5. At lower acid concentrations of 0.6 or 0.8%, growth and toxin formation were decreased 70 and 90%, respectively (Buchanan and Ayres, 1976).

Acetates have been tested for fruit and vegetable decontamination, with some success against pathogens inoculated onto commodity surfaces. Gómez-Aldapa et al. (2018) reported 0.5% acetic acid reductions in Gram-negative and -positive bacterial pathogens on romaine lettuce surfaces ranging from 2.0 to 3.5 $\log_{10}$ CFU/g, while an ethyl acetate extract from roselle calyx produced lower pathogen reductions, only 1.8 to 2.2 $\log_{10}$-cycles. Similar differences in the reductions of pathogens were observed on spinach leaves, where acetate extract was out-performed by acetic acid, though the acetate did produce reductions of ~95–99%. On blueberry surfaces, electrolysis-activated potassium acetate (3.0%; pH 2.33) reduced *L. monocytogenes* and *E. coli* O157:H7 by ~1.0 and 1.5 $\log_{10}$ CFU/g, with greater reductions (1.1–2.0 $\log_{10}$ CFU/g) achieved against *Fusarium*, *Botrytis*, and *Alternaria* spp. (Liato et al., 2017a).

5.3.2.2 Application and Regulatory Status

Sodium (21 CFR §184.1721) and calcium (21 CFR §184.1185) acetates are approved as GRAS substances for miscellaneous and general-purpose usage. Sodium acetate can be used as a boiler additive for food-grade steam (Federal Register, 1962). Calcium acetate is classified as a stabilizer when migrating from food packaging materials (21 CFR 181.29).

5.3.3 Sodium Diacetate

5.3.3.1 Antimicrobial Activity in Foods

Sodium diacetate was effective at a 0.5% level in malt syrups. Levels of 0.1–2.0% delayed mold growth in cheese spreads and, when incorporated into butter wrappers, prevented mold growth (Chichester and Tanner, 1972). Sodium diacetate is equally effective in the baking industry where it may prevent the growth of bread mold and rope-forming bacteria, such as *Bacillus subtilis*, while having little effect on *Saccharomyces* spp. (Glabe and Maryanski, 1981). A concentration of 0.05–0.4% at pH 3.5 inhibited *Aspergillus flavus*, *A. fumigatus*, *A. niger*, *Aspergillus glaucus*, *Penicillium expansum*; a concentration of 0.15–0.5% at pH 4.5 inhibited these same strains and *Mucor pusillus* in feeds and silage. Interestingly, recent research has also demonstrated sodium diacetate can effectively reduce *E. coli* in small and large intestines of broiler chickens when supplemented into animal feed at 0.01–0.05%, without substantially acidifying the intestine (Wen et al., 2017).

The minimum inhibitory concentration (MIC) of sodium diacetate was determined for *L. monocytogenes* grown in BHI broth adjusted to pH 5.25, 5.4, 5.55, 6.0, and 6.3 and incubated at 5, 20, and 35°C (Shelef and Addala, 1994). An interactive effect was noted between temperature and concentration of acid in that growth was inhibited with decreasing concentrations of acid and temperatures. The MIC was determined to be 35 mM (pH 5.25) at 35°C, 32 mM (pH 5.4) at 20°C, and 28 mM (pH 5.55) at 5°C. Sodium diacetate was more effective than acetic acid, and the inhibitory effect was attributed to the diacetate moiety rather than pH alone.

Cooked-in-bag ham was formulated with additions of 0.1 or 0.2% sodium diacetate, 2.5 or 3.3% sodium lactate, or 1% buffered sodium citrate (Stekelenburg and Kant-Muermans 2001). *Lactobacillus curvatus* and *L. monocytogenes* were inoculated into the cooked ham products at an initial level of 10^4 and 10^2/g of product, vacuum-packaged, and stored for up to 40 days at 4°C. *L. curvatus* reached an end point of 10^7/g within 2–2.5 weeks in products containing 0.1 and 0.2% sodium diacetate or 1% buffered sodium citrate and within 3 and 5 weeks for products containing 2.5 and 3.3% sodium lactate, respectively. *L. monocytogenes* was inhibited by sodium lactate and 0.2% sodium diacetate. However, in products containing buffered sodium citrate, *L. monocytogenes* increased to approximately 10^6/g within three weeks. The addition of sodium diacetate affected the sensory qualities of the product precluding its use. In ground beef, the addition of 0.3% sodium diacetate was effective in suppressing total aerobic counts. In addition to *L. monocytogenes*, sodium diacetate was also effective against hemorrhagic *E. coli*, *Pseudomonas fluorescens*, *S. enteritidis*, *Shewanella putrefaciens*, and *B. cereus* with no effect against *Pseudomonas fragi*, *Y. enterocolitica*, *Enterococcus faecalis*, *Lactobacillus fermentans*, or *S. aureus* (Shelef and Addala, 1994).

Sodium diacetate at concentrations of 0.1–0.3% and sodium lactate at concentrations of 2–3% were equally effective as antilisterial compounds in meat with little effect on pH and sensory characteristics (Mbandi and Shelef, 2001). *L. monocytogenes* and *S.* Enteritidis were inoculated into a comminuted beef emulsion formulated with 0.1 and 0.2% sodium diacetate, 0.2% sodium acetate, or 1.8 and 2.5% sodium lactate or combinations of the acids and held at 5 or 10°C. The growth of *L. monocytogenes* was not significantly different in the presence of the individual inhibitors compared with control samples after 20 days at 10°C. However, the combination 0.2% sodium diacetate and 2.5% sodium lactate was bacteriostatic to *L. monocytogenes* and bactericidal to *S.* Enteritidis. At lower temperatures, lower concentrations of the antimicrobials were equally as effective. In subsequent work, Mbandi and Shelef (2002) reported the reduction of multiple strains of *L. monocytogenes* and *Salmonella* spp. using 2.5% sodium lactate and 0.2% sodium diacetate alone or in combination in beef bologna, a more complex product, stored aerobically at 5 and 10°C. As expected the combination of acids achieved greater reductions of both organisms at both temperatures compared to individual acids and greatly reduced background spoilage microflora.

A surface application of 1–3 mg/cm^2 (2–6 g/chicken) sodium diacetate powder extended the shelf life of chickens about 4 days when held at 2°C. The powder dissolved in the surface moisture of the carcass to form acetic acid and sodium acetate and produced a surface pH of about 4.8. *Enterobacteriaceae* counts dropped 1000-fold over 6 days at 2°C. Sensory characteristics of roasted chickens were not affected at levels up to 8.0 g per chicken (Moye and Chambers, 1991).

Schlyter et al. (1993b) used the multiple-barrier concept to limit growth of *L. monocytogenes* in turkey mixtures. Slurries of uncured turkey breast, prepared with sodium diacetate (0.1, 0.3, 0.5%), sodium lactate (2.5%), and pediocin (5,000 U/ml) were pasteurized for 10 min at 68°C after which they were inoculated with *L. monocytogenes* at a level of approximately 4.5 log CFU/ml slurry. Mixtures were incubated at 25 or 4°C and sampled up to 7 and 42 days, respectively. Sodium diacetate was determined to be listericidal over the 7-day period at 25°C only at the 0.5% concentration (0.54-log reduction). *L. monocytogenes* in mixtures held at 4°C for 42 days was inhibited at 0.3 and 0.5% concentrations (0.59-log reduction). Lactate and pediocin enhanced the listericidal properties of diacetate, suggesting some type of synergistic activity. Sodium diacetate (0.5%) was listericidal in combination with ALTA™, a fermentation by-product of lactic acid bacteria, producing a 2-log reduction after 7 days at 25°C (Schlyter et al., 1993a).

The growth of *L. monocytogenes* was reduced by 2.6 logs in crabmeat stored at 4°C for 6 days when treated with 2 M sodium diacetate. Although 1 M sodium lactate solution reduced populations within 2 days, populations increased approximately 0.5 logs within 6 days. The application of 4 M sodium acetate reduced the levels of *L. monocytogenes* by 0.8 logs (Degnan et al., 1994).

5.3.3.2 Regulatory Approval

Sodium diacetate is approved as a GRAS substance for miscellaneous and general-purpose usage (21 CFR §184.1754). It is used in cheese spread, malt syrups, butter, and wrapping material. In bread and cake products, use levels range from 0.25 to 0.4% (Glabe and Maryanski, 1981). Its use is approved for USDA-FSIS-inspected products, typically at levels not exceeding 0.25% in the finished product (USDA-FSIS, 2019).

5.3.4 Dehydroacetic Acid

5.3.4.1 Antimicrobial Activity against Foodborne Microorganisms

Dehydroacetic acid (DHA) has one of the highest dissociation constants (Table 5.1) of the acidulants and remains effective at higher pH ranges. This acid is effective against *Enterobacter aerogenes* and *L. plantarum* at levels of 0.3 and 0.1%, respectively, although it is fungistatic at concentrations of 0.005–0.1% (Wolf, 1950). Sodium dehydroacetate was twice as effective as sodium benzoate at pH 5.0 against *S. cerevisiae*, and 25 times more effective against *P. glaucum* and *A. niger* (Banwart, 1989).

Strawberries dipped in or sprayed with 0.25 or 5% sodium dehydroacetate for 60 s were considered more marketable after three days of storage at 20°C than water-dipped berries. Furthermore, treatment with sodium dehydroacetate was effective in reducing respiration rate and delaying ripening, thus extending marketability (Watada, 1971).

The effects of potassium sorbate, sodium *p*-hydroxybenzoate, benzoic acid, and sodium dehydroacetate acid were investigated for three strains of *E. coli* O157:H7 in broth and model meat products. The MIC in broth at pH 6.5 was 4 mg/mL for potassium sorbate, benzoic acid, and sodium dehydroacetate and 16 mg/mL for sodium *p*-hydroxybenzoate. Sorbate was bacteriostatic at half the MIC for *E. coli* in homogenized beef, but not in poultry (Yamamura et al., 2000). Combining the antimicrobial monolaurin with sodium dehydroacetate reduced the MICs of the compounds against *E. coli* and *B. subtilis* in medium, displaying weak synergism as interpreted by fractional inhibitory concentration index (FICI) values (Zhang et al., 2009).

Huang et al. (2010) reported MIC values of sodium dehydroacetate against five foodborne fungal microbes, comparing against sorbic acid and other fatty acid antimicrobials. Compared to sorbic acid, dehydroacetate MICs were either equivalent to or 5–10-fold lower than the corresponding MICs for sorbic acid in microbiological medium adjusted to a_w of 0.85 or 0.90. Tang et al. (2018) reported that mycelial growth of *Geotrichum citri-aurantii* was halted by the application of 0.40 mg/mL sodium dehydroacetate, resulting from membrane permeabilization, declining intracellular ATP content, and increased sodium/potassium ATPase activity in cells, along with leakage of other cytoplasmic components.

5.3.4.2 Application and Regulatory Status

Dehydroacetic acid or its sodium salt is GRAS when used in accordance with good manufacturing practices. It is approved for use as a treatment for cut or peeled squash in amounts equal to or less than 65 ppm remaining in or on prepared squash (21 CFR §172.130). It can also be used as a fungistat in cheese wrappers.

5.3.5 Adipic Acid

5.3.5.1 Antimicrobial Properties

Adipic acid has no antimicrobial properties other than those ascribed to its ability to reduce pH. This acid at the appropriate concentration level would inhibit the growth of any microorganisms susceptible to low pH values. Nevertheless, multiple research studies have reported on antimicrobial activity of adipic acid derivatives and biopolymers constructed using adipic acid against foodborne bacteria and fungal microorganisms. For example, chitosan-derived films containing adipic acid produced clearing zones around Gram-positive bacterial organisms, including *E. faecium* and *S. aureus*, on the surfaces of plating media ranging from 26.8 to 28 mm in diameter. Additionally, numbers of *S.* Typhimurium and *E. coli* O157:H7 cells in liquid medium were decreased by films by ~6.0 $\log_{10}$ CFU/mL compared to chitosan-only films (Medimagh et al., 2016).

5.3.5.2 Application and Regulatory Status

Adipic acid is a non-hygroscopic, saturated dicarboxylic acid having low water solubility. It is four to five times more soluble than fumaric acid at room temperature and has the lowest acidity of any of the food acids. It is used in gelatin desserts, baking powders, powdered fruit beverages, candies, refrigerated rolls, and biscuits because it imparts a smooth, mildly acid flavor. It also serves a function in the canning of vegetables, as a sequestrant in oils, and in improving the melting characteristics of processed cheese (Gardner, 1972). Adipic acid is approved as a GRAS substance when used as a buffer or neutralizing agent not to exceed good manufacturing practices (21 CFR §184.1009). The acceptable daily intake for humans is listed in Table 5.2.

5.3.6 Citric Acid

5.3.6.1 Antimicrobial Properties and Activity in Foods

Inhibition of *L. monocytogenes* was achieved in trypticase soy yeast extract broth (TSBYE) adjusted to pH 5.0 with propionic, pH 4.5 with acetic, lactic, or pH 4.0 with citric and hydrochloric acids (Conner et al., 1990). The effect was temperature-dependent in that survival of *L. monocytogenes* decreased to undetectable levels within 1–3 weeks at 30°C, whereas at 10°C, *L. monocytogenes* survived after 11–12 weeks in TSBYE adjusted with acetic, citric, and propionic acids, and for 6 weeks in media containing lactic acid. Similar pH-temperature relationships were noted by Cole et al. (1990), who determined that the minimum pH values that allowed survival of *L. monocytogenes* after 4 weeks were pH 4.66 at 30°C, 4.36 at 10°C, and 4.19 at 5°C. The minimum pH that permitted growth within 60 days (>100-fold increase in cell numbers) was the same at 30°C, but rose to 4.83 at 10°C and 7.0 at 5°C. These experiments used citrate phosphate buffer in which the citric acid content was calculated at 1.1% at pH 4.4 and 0.7% at pH 6.0. The authors suggested the inhibitory activity might be due to the chelating ability of citric acid. Additional interactive effects were also noted in the presence of salt. Sorrells et al. (1989) reported that on an equal molar basis, citric acid was more inhibitory than lactic acid, followed by acetic acid. Interaction was demonstrated between citric acid and sodium citrate added to BHI broth adjusted to pH 4, 5, 6, and 7, impacting the pathogen's reduction (Buchanan and Golden, 1994). Likewise, Oh et al. (2016) reported that citric acid functioned synergistically with lysozyme to inhibit *L. monocytogenes* in a micro-broth dilution assay, with an MIC of 312.5 mg/liter citric acid alongside 12,500 mg/liter lysozyme.

The inhibitory capacities of hydrochloric (HCL), citric (CIT), acetic (ACE), lactic (LAC), propionic (PRO), or phosphoric (PHO) acids were compared in TSB for *Y. enterocolitica* based on concentration of acid, pH, and degree of dissociation. A comparison of equimolar concentrations identified CIT as the most antimicrobial, followed by HCL, LAC, PHO, PRO, and ACE. Based on pH, the highest inhibitory activity was associated with PRO, then LAC, ACE, CIT, PHO, and HCL. The activity of the undissociated portion of the acid was HCL followed by LAC, PRO, and ACE (Brackett, 1987). Bae and Lee (2015) completed a similar series of experiments, reporting that 0.5% citric acid and 3% NaCl use produced a pH in Luria-Bertani broth of 3.45 and a citric acid content of 0.014 M, a lower undissociated acid content than all other experimental weak organic acids (acetic, lactic, propionic, malic, and tartaric acids, all used at 0.5%). In these experiments, *E. coli* O157:H7 and *S.* Typhimurium cells were more sensitive to citric acid, malic, or tartaric acids combined with salt than they were for lactic or acetic acids.

Citric acid at a concentration of 0.5% (pH 4.5) was more effective than lactic acid (pH 3.4) in inhibiting *Arcobacter butzleri* in a broth system (Phillips, 1999). Sodium citrate was also more effective than 2% sodium lactate, and the combination of 1.5% sodium citrate and 1.5% sodium lactate was equally inhibitory as sodium citrate alone, indicating that citrate was the more effective compound. Citric acid was more inhibitory to thermophilic bacteria than acetic or lactic acids; however, pH was not considered a reliable indicator of preservation effectiveness (Fabian and Wadsworth, 1939; Fabian and Graham, 1953). Citric acid was particularly inhibitory to flat-sour organisms isolated from tomato juice. Little bacteriostatic activity occurred at pH 5.0, but inhibition increased with decreasing pH (Murdock, 1950). When tomato juice was adjusted with citric acid to pH 4.0 and 3.7 and stored for 7 days at 5°C, lower aerobic plate counts of 3.8 and 2.6 log/ml were seen, respectively, compared to counts in unacidified juice (6.22 log/ml); aerobic counts for pH 3.7 juice incubated at 20°C were 6.12 log/ml. Yeast and mold counts were the reverse with counts of 2.09, 2.13, and 1.64 log/ml, respectively; aerobic counts at 20°C for pH 3.7 juice were 4.42 log/ml (Bizri and Wahem, 1994).

The acidification of foods before canning is often used to reduce the thermal process time of foods that are particularly sensitive to changes in sensory qualities, such as texture or appearance. Canned tomatoes acidified with citric, fumaric, or malic acids did not change in physical or chemical attributes during processing compared with unacidified tomatoes (Schoenemann and Lopez, 1973; Schoenemann et al., 1974). Okra, canned in a brine containing acetic, citric, lactic, malic, or tartaric acid to achieve an equilibrium pH of 4.3, was processed for 30 min in boiling water. Acidification impaired the color but enhanced the flavor of canned okra. All acids would be effective antibotulinal agents at that pH level (Nogueira et al., 1997).

Beelman et al. (1989) developed an acid-blanch-chelate (ABC) process designed to reduce mesophilic and thermophilic bacterial spoilage in canned mushrooms. They found that acid blanching in a citric acid (0.05 M) solution buffered to pH 3.5 followed by canning in a solution containing 200 ppm EDTA controlled the outgrowth of spores of *C. sporogenes* PA3679 after a sublethal heating process (Okereke et al., 1990a). By adding citric acid to can brine, this process reduced thermophilic spoilage from 68 to 23.9%, and spoilage was further reduced to 16.8% with the addition of EDTA. Only 2.4% of cans containing mushrooms vacuum hydrated by the ABC process spoiled (Beelman et al., 1989; Okereke et al., 1990b). This process improved quality and increased the microbial stability of canned mushrooms (Okereke and Beelman, 1990). Mushroom extract acidified with citric acid to pH 6.7 (control), 6.22, 5.34, and 4.65 and asparagus with added citric acid or glucono-delta-lactone to pH levels of 5.9, 5.4, 5.1, 4.8, and 4.5 were inoculated with *C. sporogenes* PA 3679 spores. Extracts were then heated at 110, 115, 118, and 121°C. Although the heat resistance of *C. sporogenes* decreased with decreasing pH levels, heat resistance was not affected by higher processing temperatures, leading the authors to conclude that the extracts have a protective effect on *C. sporogenes*, thereby leading to its survival regardless of the change in pH (Ocio et al., 1994; Silla Santos et al., 1992).

Reduction in the thermal process time of canned liver paste was achieved through the addition of 0.14% citric acid (CA), 0.29% CA, 0.31% CA–2.0% trisodium citrate (TCA), or 0.31% CA–2.0% TCA–0.1% potassium sorbate (Houben and Krol, 1991). Liver pastes inoculated with 10^6–10^7 *Bacillus* spores per g, and 10^4 *C. sporogenes* spores per g were heated to an F_0 of 0.05, 0.3, and 0.85. Microbial stability was achieved in liver paste formulated with 0.14% CA (pH 5.69) and processed with an F_0 of 0.3. The addition of potassium sorbate did not enhance product stability. Likewise, *D*-values of *E. coli* O157:H7 were reduced in pureed green beans when acetic, citric, or malic acid was supplemented into the puree. During heating at 65°C, a *D*-value for the pathogen was determined as 0.199 min, whereas the addition of citric acid reduced the pathogen's *D*-value to 0.111 min (Khurana et al., 2006).

Growth curves developed for *B. cereus* strains grown in model vegetable broths were conducted at typical refrigeration and abuse temperatures of 5, 8, 12, and 16°C (Valero et al., 2000; 2003). *B. cereus* grew in zucchini broth (pH 6.5) at 12°C but did not grow in broth acidified to pH 5.0 at temperatures below 16°C. Growth in carrot puree was temperature dependent as *B. cereus* could grow at pH 5.1 and 5.2, provided temperatures were at 16°C, but could grow at 8°C only when the pH was higher, 5.4 and 5.5, indicating a synergistic effect between pH and temperature. Both vegetable broths stimulated germination and growth of spores compared to nutrient broth.

Citric acid dips for foods not only retard spoilage but also act as a chelator of metal ions responsible for enzymatic browning reactions. Cabbage and carrots were pre-treated with the addition of a reduced calorie dressing or with a dip of 0.2 or 1.0 % citric acid for 5 or 30 min. The products were packed with or without modified atmosphere and stored at 4°C for 10–21 days. Samples pre-treated with 1% citric acid displayed significantly lower total numbers, coliforms, and lactic acid bacteria counts than samples dipped with the lower concentration of citric acid. Samples containing the reduced calorie dressing were even lower in microbial numbers than acid-dipped vegetables (Eytan et al., 1992). Notermans et al. (1985) dipped potatoes in a solution of 1% citric acid and 2% ascorbic acid for 2 min before vacuum-packing and cooking. This process inhibited *C. botulinum* type B when held for 70 days at 15°C or 14 days at 20°C. Citric acid dips (0.3 g/kg) for crawfish (*Procambarus clarkii*) stored at 4°C were not effective in retarding the growth of *L. monocytogenes* (Dorsa et al., 1993).

Homemade mayonnaise has been a significant health hazard due to contamination by *Salmonella* from raw eggs (Membre et al., 1997; Xiong et al., 1999). The preparation of these products included adjustment of the pH with acetic acid to 3.4–4.1 with holding temperatures between 18 and 20°C for up to 3 days before consumption. Mayonnaise made with >5% citric acid solution adjusted to pH below 4.05 led to more inactivation of *S.* Enteritidis PT4 at 22 than 5°C, but the shelf life was shortened at this higher temperature. The authors recommended that the final pH should be <3.3 when at least 20 ml of lemon juice (citric acid) per fresh egg yolk was used. If the amount of lemon juice was increased to 20–35 ml per egg yolk, the product should be held at 22°C for over 72 h; when 35 ml lemon juice per egg yolk is added, the time decreased to 48 h before use. Radford and Board (1993) found that acetic acid was more inhibitory than citric acid and recommended that mayonnaise be prepared with vinegar to a pH of 4.1 or less. The addition of garlic or mustard increased the death of *S.* Enteritidis, but salt had a protective effect on this organism.

C. botulinum showed no change in numbers when inoculated in tomato puree, shrimp puree, or shrimp and tomato sauce acidified with citric or acetic acids to pH 4.2 or 4.6. No toxin was detected in any of these foods over an 8-week shelf life (Post et al., 1985). Tsang et al. (1985) found no outgrowth or toxin production from spores of types A and B in TPGY medium acidified to ≤ pH 4.6 and incubated at 35°C; however, type E grew and produced toxin at 26°C in media containing citric acid at pH 4.2. When acetic acid was used, no growth occurred at ≤ pH 5.0.

Citric acid applied to skim milk was more inhibitory to *S.* Typhimurium than lactic and hydrochloric acids (Subramanian and Marth, 1968). *Cronobacter sakazakii* was inhibited in infant formula supplemented with 30.0 mM citric acid, carvacrol (plant-derived antimicrobial) and nisin, wherein reductions of 1.0 to 1.5 $\log_{10}$ CFU/mL were observed in treated formula as compared to controls. As little as 0.3% citric acid lowered the numbers of salmonellae on poultry carcasses, and rinsing with a citric acid solution adjusted to pH 4.0 reduced the numbers of *P. fluorescens* on beef carcasses (Thomson et al., 1967). However, it did not appear to affect attachment (Appl and Marshall, 1984). Similar findings on pathogen-inoculated beef knuckles were reported more recently, wherein the combination of citric, acetic, or lactic acid with tripolyphosphate (TPP) resulted in the three acids performing equivalently in reducing numbers of STEC (O157 and non-O157); researchers reported that reductions in STEC numbers were modest (~1.0 $\log_{10}$-cycle) following pH 3.75 citric acid and TPP application (Byeshalov et al., 2010).

In recent years, multiple researchers have explored the usefulness of citric acid application for pathogen control on surfaces of fresh produce commodities; results of these reports are mixed with respect to pathogen reductions achieved on differing produce item surfaces. Lettuce leaves inoculated with *E.*

coli O157:H7, *S.* Typhimurium, or *L. monocytogenes* were treated with up to 2.0% citric acid; pathogens were moderately reduced but by less than 2.0 $\log_{10}$ CFU/g in all cases (Sagong et al., 2011). A commercial citric acid-containing sanitizer produced similar reductions in numbers of these pathogens (1.6 to 2.2 $\log_{10}$ CFU/g) on lettuce leaves following a 5 min washing application (Jung et al., 2017). Oladunjoye et al. (2016) tested the combination of nisin (5000 IU/mL) and 3 or 5% organic acids (acetic, citric) on tomato surfaces for *L. monocytogenes* decontamination. Citric acid at 3 and 5% when combined with nisin produced greater reductions in *Lm* numbers than those obtained with nisin alone, although acetic acid outperformed citric acid when treated tomatoes were stored at 4, 10, or 25°C for up to 10 days. Treated tomatoes were lighter on the day of treated versus controls, though vitamin C content was not reported to be reduced by treatment. Jalapeño peppers inoculated with *Salmonella* were treated by a lactic + citric acid blend (1.0% application); pathogen numbers were then reduced by 1.1 $\log_{10}$ CFU/g (Adler et al., 2016).

Ascospores of heat-resistant molds are difficult to kill during processing of fruit products, and an increase in processing temperature can lead to adverse changes in quality (Rajashekhara et al., 1998). Survival studies for *Neosartorya fischeri* ascospores were conducted in mango and grape juices and heated at 75, 80, 85, and 90°C. Ascospores were able to withstand more than 6 h of heating at 75°C, 5 h at 80°C, and 3–4 h at 85°C. Thermal resistance studies were conducted in mango juice adjusted to pH 3.5 with citric, lactic, malic, or tartaric acid. Citric acid provided the maximal destruction of ascospores at 85°C. Similar destruction values were seen with potassium sorbate and sodium benzoate. A level of 0.75% for both citric and lactic acids had a slight growth-retarding effect on *A. parasiticus*; a 0.25% concentration greatly reduced toxin production. A level of 0.5% citric or 0.75% lactic acid inhibited growth of *Aspergillus versicolor*, but 0.25% citric acid and 0.5% lactic acid prevented toxin production. A 0.75% concentration of either acid did not affect the growth of *P. expansum* (Reiss, 1976). Other researchers have reported that citric acid application onto animal feedstuffs reduced mycotoxin content, with reductions in deoxynivalenol and aflatoxin B1 (Humer et al., 2016; Rastegar et al., 2017; Shi et al., 2017).

Much as with other organic acids and their acid salts, citrates (e.g., sodium citrate) can also exhibit bacteriostatic activity. Sodium citrate in concentrations of 0.1–4.0% was not as inhibitory to *Streptococcus agalactiae* when added to skim milk or fresh milk as citric acid at concentrations of 1, 2, and 4%. The MIC of 0.8% gave a pH of 4.08–4.12 (Sinha et al., 1968). Concentrations of 12–12.5% sodium citrate were inhibitory to *Salmonella* Anatum and *Salmonella* Oranienburg (Davis and Barnes, 1952). Porto-Fett et al. (2011) formulated a pork scrapple with a 0.64% citrate/diacetate blend and then formulated with other organic acid salt blends prior to inoculating with *L. monocytogenes*. After a 50-day storage period at 4°C, scrapple products formulated with the citrate/diacetate blend bore lower *L. monocytogenes* counts versus scrapple not formulated with citrate/diacetate. For 2% lactate-diacetate-propionate-formulated scrapple not containing citrate/diacetate, the *Lm* count was 7.7 $\log_{10}$ CFU/g, whereas with citrate/diacetate the pathogen load was 4.0 $\log_{10}$ CFU/g. Lactate-diacetate (3%)-formulated scrapple produced a larger difference in *Lm* counts in citrate/diacetate-formulated or non-formulated (7.0 versus 2.6 $\log_{10}$ CFU/g) (Porto-Fett et al., 2011).

S. aureus showed inhibition both with increasing concentrations of citric acid and decreasing pH (Minor and Marth, 1970). Citrate inhibition of *S. aureus* was overcome by adding calcium and magnesium ions, suggesting that citrate was a chelator of ions essential for growth (Rammell, 1962). The amount of sodium citrate had a dual effect on *Lactobacillus casei*. In concentrations of 12–18 μM/ml, sodium citrate stimulated growth, and in concentrations greater than 40 μM/ml, sodium citrate inhibited growth. Branen and Keenan (1970) also suggested that citrate chelation of metal ions inhibited *L. casei*. The metabolic effects of citrate on *Arthrobacter* spp. were two-fold. With *Arthrobacter pascens*, the function of citrate appeared to be one of chelation of metal ions, even in media containing up to 1% citrate. However, with *Arthrobacter simplex*, citrate appeared to inhibit glucose utilization (Imai et al., 1970).

Using the multiple-barrier concept, hard-cooked eggs allowed to equilibrate in 0.5, 0.75, or 1.0% concentrations of citric acid and 0.2% sodium benzoate were held for 30 days at 4°C or in 0.75% citric acid alone for 21 days at 4°C. A concentration of 0.75% citric acid was sufficient to reduce inoculated populations of *S.* Typhimurium, *Y. enterocolitica*, *E. coli*, and *S. aureus* (Fischer et al., 1985). Daly et al. (1973) used a combination of 0.1% citric acid and 0.75% glucono-delta-lactone incorporated into sausage before fermentation. This combination inhibited the growth of *S. aureus* early in the fermentation, and the increased acidity of the meat mixture allowed the fermentation to proceed more quickly. Restaino et al. (1982) used a combination of 0.05% potassium sorbate and citric acid at pH 5.0 to reduce the growth rate of *S. rouxii* and *Saccharomyces bisporus*. Potassium sorbate at 0.2% concentration in combination with citric or lactic acid at pH 5.5 depressed growth of *L. plantarum*, but did not affect *P. aeruginosa* (Restaino et al., 1981).

5.3.6.2 Application and Regulatory Status

Citric acid is a tricarboxylic acid having a pleasant sour taste. It is highly water-soluble and enhances the flavor of citrus-based foods. It is approved for use in ice cream, sherbets and ices, beverages, salad dressings, fruit preserves, and jams and jellies, and it is used as an acidulant in canned vegetables and dairy products. It is a precursor of diacetyl and therefore indirectly improves the flavor and aroma of a variety of cultured dairy products. It can control the pH for optimum gel formation. Citric acid also acts synergistically with antioxidants to prevent rancidity by chelating metal ions (Gardner, 1972). Citric acid is approved as a GRAS substance for miscellaneous and general-purpose usage when used in accordance with good manufacturing practices, in the acid form (21 CFR §184.1033) or as the calcium (21 CFR §184.1195), potassium (21 CFR §184.1625), or sodium salt (21 CFR §184.1751). The acceptable daily intake for humans is listed in Table 5.2.

5.3.7 Fumaric Acid

5.3.7.1 Antimicrobial Properties

Fumaric acid (0.15%) serves as an acidulant (Ough and Kunkee, 1974) and inhibitor of malolactic fermentation in wine (Pilone, 1975). Huhtanen (1983) and coworkers (1985) investigated the effect of various esters of fumaric acid on the growth of *C. botulinum* types A and B in canned bacon. Mono-, dimethyl, and ethyl esters of fumaric acid at concentrations of 0.125, 0.15, or 0.2% showed considerable promise as substitutes for sodium nitrite in bacon. Cans packed with bacon and inoculated with *C. botulinum* swelled in 4–6 days at 30°C, whereas the presence of fumaric acid esters in the bacon prevented swelling for up to 8 weeks. Fumaric acid by itself was only slightly inhibitory. The n-monoalkyl fumarates and maleates esterified with C_{13}–C_{18} alcohols also displayed significant antibotulinal activity (Dymicky et al., 1987). Methyl n-alkyl fumarates demonstrated lower activity, and di-n-alkyl fumarates were almost inactive (Dymicky et al., 1987). Medium chain length alkyl esters had more activity than aliphatic acids (Dymicky and Trenchard, 1982).

The use of fumaric acid to achieve a 5.0 $\log_{10}$-cycle reduction of *E. coli* O157:H7 inoculated into unpreserved apple cider was investigated by Comes and Beelman (2002). The addition of 0.15% fumaric acid coupled with 0.05% sodium benzoate followed by holding at 25°C for 6 h before refrigeration at 4°C for 24 h achieved reduction. These preservatives provided a similar reduction in less than 5 and 2 h when held at 25 and 35°C, respectively. In addition to reduction of *E. coli* O157:H7, aerobic bacteria were reduced to levels below those achieved through pasteurization. Combined application of 0.15% fumaric acid and ultrasound processing (40 kHz, 700 W, up to 5 min) increased the reductions observed in numbers of *E. coli* O157:H7, *S.* Typhimurium, and *L. monocytogenes* in apple juice over those observed for the application of acid or ultrasound processing alone (Park and Ha, 2019).

Fumaric acid was also more effective than lactic or acetic acid when used in meat products (Podolak et al., 1996a, b). Fumaric acid at concentrations of 1.0 and 1.5% applied for 5 s at 55°C to the surface of lean beef inoculated with *L. monocytogenes* and *E. coli* O157:H7 reduced levels by 1.0 and 1.3 $\log_{10}$-cycles, respectively. When used in vacuum-packaged ground beef patties, fumaric acid achieved the greatest reduction of aerobic, psychrotrophic, and coliform bacteria. It was speculated that the antimicrobial properties of fumaric acid were related to a lower pK_a versus lactic or acetic acids. Similar reductions in numbers of *E. coli* O157:H7, *S.* Typhimurium, and *L. monocytogenes* were observed when 0.5% fumaric acid was co-applied to beef surfaces with acidified electrolyzed water treatment, though when the processing temperature was increased from ambient to 60°C observed reductions increased to 2.7–3.1 $\log_{10}$ CFU/g from 1.2 to 1.7 $\log_{10}$ CFU/g (Tango et al., 2014).

Monomethyl and monoethyl (0.2%) and dimethyl and diethyl fumarates (0.05%) or sodium propionate (0.1%) controlled fungal growth in tomato juice. The fumarate esters not only prevented mold growth on bread (Huhtanen et al., 1981; Islam, 1982) but the dimethyl and diethyl esters at a concentration of 20 mg per 2.6 L in the vapor phase were also effective as a mold preventive (Huhtanen and Guy, 1984). Fumaric acid was more fungicidal against *Talaromyces flavus* ascospores than acetic, citric, malic, or tartaric acid based on weight when added to the heating medium. A 0.5% concentration of fumaric acid inactivated ascospores after 20 min of heating at 80°C and lethality increased as the pH decreased from 5.0 to 2.5. However, malic, citric, and tartaric acids at a concentration of 2.5% (pH 2.5–5.0) were not lethal to *T. flavus* ascospores (Beuchat, 1988). Similar effects were noted with *N. fischeri* (Conner and Beuchat, 1987) heated in the presence of 2.0% acid concentration at 82°C.

5.3.7.2 Application and Regulatory Status

Fumaric acid is a non-hygroscopic, unsaturated, dicarboxylic acid having low water solubility. Although this additive has a strong acid taste, it blends with certain flavoring compounds to intensify the aftertaste of a flavor. Fumaric acid is used in fruit drinks, gelatin desserts, pie fillings, biscuit doughs, and wines. It increases the strength of gelatin gels. In addition to supplying acidity to a product, fumaric acid does display some antioxidant properties in fat-containing foods (Gardner, 1972). Fumaric acid and its salts are approved for direct addition to food for human consumption (21 CFR §172.350). The acceptable daily intake for humans is listed in Table 5.2.

5.3.8 Lactic Acid

5.3.8.1 Antimicrobial Properties

Lactic acid has been used extensively for its sensory qualities in addition to its antimicrobial properties in the past, although more recently this acid has been applied extensively as a rinsing agent for beef, pork, and chicken carcasses. The inhibitory capacity of this acid lies in its reduction of pH to levels below which bacteria cannot initiate growth. In fermented foods, lactic acid coupled with other antimicrobial factors excreted by lactic acid-producing microorganisms inhibits competing microorganisms.

Lactic acid was an excellent inhibitor of spore-forming bacteria at pH 5.0, but was ineffective against yeasts and molds (Woolford, 1975a). Lactic acid was about four times more effective than malic, citric, propionic, and acetic in limiting the growth of *Bacillus coagulans*, the organism responsible for flat-sour spoilage in tomato juice (Rice and Pederson, 1954). Lactic acid was more inhibitory to *Mycobacterium tuberculosis* as the pH decreased (Dubos, 1950). Gill and Newton (1982) suggested that the inhibitory effect towards psychrotrophic organisms was related to decrease in pH rather than degree of dissociation. In contrast to these previous findings, short-term exposure (10 min) to acetic, lactic, or citric acid (pH 3; 2.5%) failed to inhibit the germination of *B. cereus* spores determined by plate counts of cells, indicating acidification was not sufficient for pathogen inhibition. The utilization of lactic acid-derived chlorine dioxide, nevertheless, did result in enhanced sporicidal activity, wherein pH 6.0-adjusted organic acids produced greater ClO_2 and thereby reduced viable spores more rapidly to non-detectable counts (Kim et al., 2008). Cold-pack cheese normally prepared with lactic acid was reformulated with acetic acid and inoculated with *S.* Typhimurium. Destruction rates for *S.* Typhimurium were similar for both acids (Park et al., 1970). Additionally, the combination of acids and heat treatment led to a reduction of spore-forming organisms in frankfurters. Frankfurter emulsions incorporated with acids to adjust pH to 5.2 or 4.6 for *Bacillus stearothermophilus* and pH 4.5 and 4.2 for *B. coagulans* were heated to 121°C and 105 or 110°C. Greater inactivation was noted for both organisms at the lower pH level when acetic or lactic acids were used, but not with citric, malic, or hydrochloric acid (Lynch and Potter, 1988).

Lactic acid sprays have been effective in limiting microbial growth on meat carcasses under a variety of storage conditions. A 1.0 to 1.25% lactic acid spray on veal carcasses, followed by vacuum-packaging, lowered microbial counts after storage for 14 days at 2°C, though a 2% concentration led to discoloration of the carcass surface (Smulders and Woolthuis, 1983; Woolthuis and Smulders, 1985). A 2% L-lactic acid treatment at pH 2.3 of veal tongues, combined with vacuum-packaging and storage at 3°C, reduced aerobic mesophiles from 5.6 to 2.7 $\log_{10}$ CFU/cm^2 (Visser et al., 1988). Optimal decontamination was achieved when 1% lactic acid solution at 55°C was sprayed on beef carcasses immediately after dehiding and after evisceration (Prasai et al., 1991).

Similar reductions in microbial counts occurred with a 1.25% lactic acid spray onto beef carcasses followed by a treatment of hot-boned cuts with 2% acid, vacuum-packaging, and storage for 10 days. Acid treatment combined with vacuum-packaging was more effective in prolonging shelf life than vacuum-packaging alone (Smulders and Woolthuis, 1985). *Enterobacteriaceae* contaminated 50% of the untreated samples, which was reduced to 10% after treatment. A study using lactic acid sprays of skinned cow heads found that 1% acid was effective in significantly reducing total counts and extending shelf life by 3 days at 4°C and 1 day at 15 and 20°C (Cudjoe, 1988). Since most other studies chose refrigeration storage temperatures, this study showed that lactic acid was effective at higher temperatures as well.

Hot-boned, sub-primal meat was injected with one of three treatments: 0.3 M calcium chloride solution (CAL, pH 5.1), 0.3 M lactic acid solution (LAC, pH 3.0) or a combination (1:1) CAL/ LAC (COM, pH 2.5) at a volume equal to 10% of the initial sub-primal weight. Significantly higher aerobic plate counts were obtained from hot-boned cuts over 17 days post-mortem compared to cold-boned cuts. Aerobic plate counts of meats injected with CAL were significantly higher than with LAC or COM. Furthermore, COM improved tenderness in muscle cuts, but control cuts retained more desirable flavor profile (Eilers et al., 1994).

Chilled beef rounds inoculated with *E. coli* O157:H7 or *S.* Typhimurium were subjected to an automated wash water system (Castillo et al., 1998; 2001a, b); the system alone produced a 3.0 $\log_{10}$-cycle reduction. Sprays containing 4% lactic acid applied at 55°C for 30 s or at 65°C for 15 or 30 s led to nonrecovery of *E. coli* cells. When sequential treatments were employed using pre-chill and post-chill lactic acid sprays, *E. coli* was reduced by 6.8–7.2 $\log_{10}$-cycles. Microbial counts obtained from ground beef made from the treated beef rounds that were subjected to the multiple sprays were significantly lower than samples receiving a post-chill rinse only, suggesting that lactic acid can exert a continued antibacterial effect during the shelf life of the product. Nevertheless, Tolen et al. (2017) did not report any residual antimicrobial effects of 2% lactic acid (55°C solution temperature) applied to beef trimmings during post-grinding refrigerated storage on beef surfaces for members of the O157 and non-O157 STEC.

Beef trim meat used to make ground beef is of particular concern because of its higher microbial load. Beef trim was fabricated to make predominantly beef trim lean (BTL) and beef trim fat (BTF) products that were inoculated with bovine fecal samples (Kang et al., 2001a, b). The inoculated meat was subjected to sprays of tap water (15–17°C, 65 psi), 2% lactic acid (12–15°C, 30 psi), hot water (30 psi) at 65, 71, 76, 82°C, or hot air at 371, 426, 454, 482, or 510°C. In addition, combinations of the treatments were examined. All products were stored at 4°C and examined for aerobic, psychrotrophic, presumptive lactic acid bacteria, total fecal coliforms, and *E. coli* biotype 1 counts. Greater reductions on all microbial loads were seen in BTF compared to BTL. Lactic acid application as the final step in a multiple-hurdle process led to a residual effect during storage for 7 days at 4°C. Similar work was conducted using lean pork trim (LPT) and fat-covered pork trimmings (FPT) (Castelo et al., 2001). As with beef, the FPT supported lower microbial numbers than LPT, somewhat surprisingly given other research indicating Gram-negative bacteria tended to adhere more strongly to fatty versus lean meat surfaces (Dickson and Koohmaraie, 1989). Similar to multiple other similarly designed studies, Wolf et al. (2012) reported a 4.4% lactic acid applied to beef trimmings by dipping achieved better reductions to numbers of STEC and *Salmonella* (0.5 to 1.4 $\log_{10}$ CFU/g) versus spray-application of lactic acid at the same concentration on trimmings, indicating dipping provided improved meat surface contact and consequent pathogen inhibition.

Acuff et al. (1987) compared a combination of 1% lactic, 2% acetic, 0.25% citric, and 0.1% ascorbic acids adjusted to pH 2.6 with either 1% lactic acid at pH 2.9 or 1% acetic acid at pH 3.3. There was little difference in bactericidal activity between the combination and either acid alone on sub-primal cuts of beef. Dixon et al. (1987) confirmed their findings using beef strip loins stored in polyvinyl chloride (PVC) film for 6 days or high-oxygen barrier film for 28 days. When the temperature changed from 20 to

70°C or lactic and acetic acid concentrations increased individually to 3%, aerobic bacterial counts and *S.* Typhimurium levels decreased approximately 1.0 $\log_{10}$, <1.0 $\log_{10}$ for *Enterobacteriaceae* and 0.5 $\log_{10}$ for *E. coli* (Anderson and Marshall, 1990a, b). The combination of acids performed as well as 1% lactic or acetic acids alone. Lactic acid was most effective against *S.* Typhimurium at 70°C, reducing numbers by 2.0 $\log_{10}$-cycles and *Enterobacteriaceae* by 1.5 $\log_{10}$-cycles (Anderson et al., 1992).

Lactic acid (LA; 2%), low-molecular-weight polylactic acid (LMW-PLA; 2% PLA), and nisin (N; 400 IU/ml) were used alone or in combination in beef to inhibit *L. monocytogenes* (Ariyapitipun et al., 2000). Beef was inoculated with *L. monocytogenes* before vacuum-packaging and storage at 4°C. All treatments significantly reduced *L. monocytogenes*, although there was no significant difference between N, NPLA, LA and PLA. After 42 days of storage the inoculum was reduced to <1.0 $\log_{10}$ after using LA, NLA, and N. Beef inoculated with *E. coli* O157:H7 was subjected to treatments of LMW-PLA, LA, and 200 IU/ml of nisin (N) (Mustapha et al., 2002). After 28 days of storage, significant reductions were seen with all treatments, but no differences were seen between PLA and LA. Nisin use did not enhance the effects of lactic acid against *E. coli* O157:H7. The use of these compounds was effective in reducing psychrotrophs and *Enterobacteriaceae* (Ariyapitipun et al., 1999). Beef cubes were inoculated with *L. monocytogenes* and treated with 2% lactic acid, 40,000 IU/mL nisin, and 3200 units/mL pediocin. These compounds reduced the level of *L. monocytogenes* by 1.7, 1.1, and 0.6 $\log_{10}/6\ cm^2$, respectively (El-Khateib et al., 1993).

A novel approach to shelf-life extension was the use of lactic or acetic acid-impregnated calcium alginate gels applied to the surface of beef tissue (Siragusa and Dickson, 1992, 1993). Although the gels reduced populations of *L. monocytogenes* on beef tissue, non-coated tissue became dehydrated during the 7-day study. Lactic acid treatments alone were only 50% as effective as acid/alginate treatments whereas acetic acid was more effective than the acetic acid/gel treatment. Numbers of *S.* Typhimurium were reduced more by exposure to acetic acid/gel treatment than acetic acid alone.

Lactic acid (2%), acetic acid (0.5%), or sodium lactate (4%) alone or in combination with pulsed electric field or freezing was ineffective in reducing numbers of *E. coli* O157:H7 inoculated into beef trimmings (Bolton et al., 2002). Therefore, these treatments were not deemed usable in the control of *E. coli* O157:H7 in beef burgers. Conversely, combination of 5% lactic acid and electron beam irradiation (0.8 to 0.9 kGy) reduced verotoxin-producing *E. coli* on steaks by >3.0 $\log_{10}$ CFU/g (Li et al., 2015). High-pressure processing and lactic acid application on sliced hams produced synergistic inhibition of both *L. monocytogenes* and *Salmonella* inoculated on ham slices (Aymerich et al., 2005).

The use of lactic acid for reduction of microbial numbers or pathogens has been similarly applied to pork products. Woolthuis et al. (1984) effectively reduced total plate counts, *Enterobacteriaceae*, and lactobacilli by 2.0 to 3.0 $\log_{10}$, by immersing porcine livers in 0.2% lactic acid for 5 min, vacuum-packaging, and storing for 5 days at 3°C. A 1% concentration of lactic acid (pH 2.8, 55°C) had little effect on the aerobic plate counts taken from the surface of pork carcasses. *Salmonella* spp. or *Listeria* spp. were not recovered nor were sensory characteristics affected (Prasai et al., 1992). By increasing the concentration of lactic acid to 2%, numbers of *Salmonella* spp. and *Campylobacter* spp. were reduced immediately and remained lower 24 h after slaughter (Epling et al., 1993). Van Netten and Hust In't Veld (1994) developed a pork skin model system to determine the effect of lactic acid decontamination on microbial counts and changes in predominance of the microflora. Five treatments were delivered to the pork skins: water-treated (control), or exposure at 21°C for 120, 180, 240, or 360 s with 2% lactic acid. Not only did the microbial counts decrease as the time of decontamination increased, but the population of microorganisms shifted. Psychrotrophic Gram-negative bacteria were the most sensitive to lactic acid, followed by mesophilic *Enterobacteriaceae*, psychrotrophic Gram-positive bacteria; lactobacilli and yeasts were the least sensitive.

The effects of 3% lactic acid at 55°C were studied using pork fat and lean tissue artificially inoculated with 10^4–10^5 CFU/cm^2 of *L. monocytogenes*, *Y. enterocolitica*, *Aeromonas hydrophila*, *P. fragi*, or *Brochothrix thermosphacta* (Greer and Dilts, 1995). After inoculation, tissue was dipped for 15 s in water or lactic acid and held for 15 days at 4°C. The spoilage organisms *P. fragi* and *B. thermosphacta* grew on both tissue types after acid treatment, though acid treatment affected growth rates. None of the pathogens grew on lean tissue, and *A. hydrophila* numbers declined by 5.0 $\log_{10}$ within 11 days of storage. Acid treatments of fat tissues led to pathogen reductions to non-detection within 4 days.

Lactic acid dips have also been used successfully in the poultry industry. Total microbial numbers from skin of birds immersed for 15 s at 19°C in 1 or 2% lactic acid at pH 2.2 decreased from 5.2 to 3.7 $\log_{10}$ CFU/g (van der Marel et al., 1988). Levels of psychrotrophs decreased from 3.9 to 2.7 $\log_{10}$ CFU/g and *Enterobacteriaceae* from 3.3 to 2.6 $\log_{10}$ CFU/g. Higher concentrations of the acid did not yield greater decontamination nor did repeated treatments. These treatments inhibited hydrogen sulfide-producing bacteria, such as *Pseudomonas* spp. that contributed to spoilage without affecting sensory quality.

Lactic acid (1%) added to broiler carcass scald water (54°C/2 min) and chill water (0–1.1°C, pH 2.8) reduced inoculated *S.* Typhimurium to almost non-detectable numbers. Lactic acid added to scald water alone had minimal effect on reducing the numbers of contaminated birds. The number of *Salmonella*-positive birds was also reduced as a function of scalding period (Izat et al., 1990a). Lactic acid added to broiler chill water resulted in the development of a brown discoloration. In an effort to eliminate carcass discoloration, reduced levels of lactic acid (0.25%, pH 2.88 or 0.5%, pH 2.62) were combined with propylene glycol (20%) in chill water. *Salmonella* were eliminated from broiler carcasses after a 1-hr exposure; discoloration still occurred and propylene glycol contributed an objectionable flavor (Izat et al., 1990b). Chicken skin subjected to washing treatments using 1% lactic acid led to significant reductions of *Salmonella* spp. and *L. monocytogenes* compared to monosodium phosphate, sodium tripolyphosphate, sodium acid pyrophosphate, or sodium hexametaphosphate. Consumer evaluation of chicken treated with a 0.5% lactic acid/0.5% sodium benzoate dip concluded that sensory qualities were acceptable (Hathcox et al., 1995; Hwang, and Beuchat, 1995). Pre-treatment of broiler carcasses with lactic acid buffer increased shelf life by 6–7 days at 4°C and 5–6 days at 7°C. When treated broilers were packaged using modified atmospheres, shelf life was lengthened to >36 and 35 days at 4 and 7°C, respectively (Sawaya et al., 1995).

Lactic acid was more effective than acetic acid when used as a 10 min rinse for fresh cut vegetables. Although *L. monocytogenes* counts were reduced by 0.5 and 0.2 $\log_{10}$, respectively, other compounds, such as chlorine or chlorine dioxide, produced similar reductions of approximately 1 log (Zhang and Farber, 1996).

The interaction of sodium chloride (0.5–4%) and pH (4.2) was examined for *E. coli* O157:H7 grown in TSB at 37°C (Casey and Condon, 2002). *E. coli* O157:H7 rapidly died at pH 4.2 in media containing lactic, acetic, or formic acids. *E. coli* demonstrated a ten-fold decrease in numbers when grown in medium containing sodium chloride, acidified with lactic acid. Conversely, in NaCl-free acidified medium, *E. coli* decreased 10,000-fold. Sodium chloride offered a degree of protection against the inhibitory activity of lactic acid when the culture was incubated in the medium containing both ingredients, but no protective effect was observed if the sodium chloride was added to the medium 45 min after the inoculation. This protective effect was even more pronounced with acetic acid (100,000-fold). The sodium chloride sparing effect was partially attributed to a reduction in water activity causing the cell size to shrink due to loss of water in the cytoplasm and a concomitant increase of internal pH from 5.23 in cells exposed to acid to 5.79 in acid-free media. *S. aureus* grown at 37 or 46°C in the presence of 1.0 M sodium chloride had marked differences in its sensitivity to acetic and lactic acids (El-Banna and Hurst, 1983). Cells grown at 46°C were more heat resistant and better able to grow in media adjusted with acetic and lactic acids to pH 5.0 than cells grown at 37°C.

El-Gazzar et al. (1987) found that *A. parasiticus* NRRL 2999 grown in a laboratory medium containing 0.5 and 0.75% lactic acid adjusted to pH levels of 3.5 or 4.5 using HCl or NaOH and incubated for 10 days at 28°C produced more aflatoxin B_1 than other cultures after 3 days of incubation. With increasing levels of lactic acid, cultures decreased production of aflatoxin G_1. Luchese and Harrigan (1990) also noted an increase in aflatoxin production adjusted to pH 4.2 using HCl or lactic acid stimulated by substituting half of the carbon content by lactate and a reduction in production at pH 6.8. Aflatoxin production also increased during growth of *A. parasiticus* in association with *L. lactis*.

5.3.8.2 Application and Regulatory Status

Lactic acid is a hygroscopic, syrupy liquid having a moderately strong acid taste. Lactic acid is used in the manufacture of jams, jellies, sherbets, confectionery products, and beverages. It is used to adjust acidity in brines for pickles and olives. Calcium lactate is used as a firming agent for apple slices, to prevent discoloration in fruit, and in baking powders. Lactic acid is approved as a GRAS substance for miscellaneous or general-purpose usage (21 CFR §184.1061). The acceptable daily intake for humans is listed in Table 5.2. It is approved in the U.S. by the USDA-FSIS at up to 5% for carcass surface decontamination purposes/applications, and at up to 2.8% for beef head and tongue decontamination. It may also be combined with other acidulants or applied topically to RTE products during post-lethality handling, or may be used as a component of antimicrobial-bearing commercial biopreservatives (USDA-FSIS, 2019).

5.3.9 Lactates

5.3.9.1 Antimicrobial Properties

The salt forms of lactic acid have been used successfully as antimicrobial agents in various products as they provide a slight salty taste that enhances meat flavor, retains color, contributes to water-holding capacity, improves juiciness, increases meat yields, appears naturally in meat products, and extends the shelf life of products. Chen and Shelef (1992) suggested that, in addition to pH change, lactates also contribute to lowering the a_w of a food, a potential antimicrobial mechanism in addition to that of cytoplasmic accumulation following penetration of the microbe in the undissociated acid (lactic acid) form with subsequent dissociation, summarized by Taylor et al. (2012). Additionally, excellent reviews of the antimicrobial nature of sodium lactate are available from Shelef (1994) and De Wit and Rombouts (1990).

The MIC of sodium lactate was investigated under optimum growth conditions (pH 6.5, 20°C) for a number of bacteria and yeasts isolated from spoiled and unspoiled meat products (Houtsma et al., 1993). MIC values for *Salmonella* ranged from 714 to 982 mM; MICs for *L. monocytogenes* and *L. innocua* ranged from 804 to 982 mM. Recent research placed the MIC of sodium lactate against *L. innocua* isolates at 535 mM (60 mg/mL); when combined with the bacteriocin nisin the MIC for lactate was reduced 20-fold to 3 mg/mL (Verdi et al., 2019). Gram-negative bacteria including *Pseudomonas* and *Yersinia* had comparable lactate sensitivities to *Salmonella*, while *Campylobacter* was more sensitive with an MIC = 179 mM. 200 mM sodium lactate produced a 1.9 $\log_{10}$ CFU/mL reduction in *S.* Enteritidis cells, though when combined with the bacteriocin cerein a 3.0 $\log_{10}$-cycle reduction was achieved in the pathogen's numbers (Lappe et al., 2009). Sodium lactate, in other studies, required >2.0 mg/mL to inhibit *E. coli*, *B. subtilis*, and *S. aureus* in liquid medium; surprisingly, the combination of lactate with monolaurin did not result in enhancement of antimicrobial activity (Zhang et al., 2009). Gram-positive bacteria including lactic acid bacteria, *Carnobacterium*, *Lactococcus*, *Brochothrix*, *Bacillus*, and *Staphylococcus* ranged from 268 for *S. aureus* and *Lactobacillus coryniformis* to 804 mM for *Brochothrix*. Yeasts were very broad in their sensitivity with *Debaryomyces* and *Candida* being the most resistant at 1161 and 1339 mM, respectively (Houtsma et al., 1993).

The effect of sodium lactate on toxin production, spore germination, and heat resistance was tested on proteolytic *C. botulinum* strains in peptone yeast extract medium (Houtsma et al., 1994a). Toxin production occurred within 14 days of incubation at 15°C for cultures grown in 0 and 1% sodium lactate but was delayed in the presence of 2% sodium lactate until day 21; no toxin formation was detected after 49 days of incubation in media containing 3% sodium lactate. At 20°C, toxin formation was noted within 5 days for 0%, 11 days for 1.5%, and 15 days for 2.5%; no growth or toxin formation was detected after 32 days of incubation for media containing 4.0% sodium lactate. At 30°C, toxin production was detected in media containing 4% sodium lactate within 11 days; however, spore germination did not lead to growth after 7 days. Sodium chloride at the same water activity as sodium lactate (0.982) did not inhibit toxin production, indicating that inhibition was not due to a lowering of water activity. Similar findings were reported by others who evaluated the capacity of lactate salts (sodium, potassium, and calcium) to inhibit the germination of *C. perfringens* spores in injected turkey products during cooling. All salts of lactate produced anti-germination effects at 2 and 3%, as compared to control and 1% lactate-treated product, but researchers reported calcium lactate was most effective at limiting spore germination in the product, partially as a result of more lactate from the calcium salt versus other lactate salts based on usage and the pH buffering capacity differences amongst lactate salts (Velugoti et al., 2007).

Lactic acid/sodium lactate and acetic acid/sodium acetate were studied in a BHI broth at pH 7, 6, 5, and 4 using 0.1, 0.5, 1.0, and 2 M concentrations (Buchanan et al., 1993). Survivor curves were developed using *L. monocytogenes* as the target organisms. As expected, the rate of inactivation was dependent upon the acid, its concentration, and the resultant pH. Lactic acid was more effective at higher pH levels whereas acetic acid was more effective at lower pH levels, and there was a correlation with the concentration of dissociated acid. The effect of sodium lactate and sodium chloride on the growth of *L. innocua* was modeled in broth adjusted to pH 5.5, 6.0, 6.5, and 7.0 at 4, 10, 20, and 30°C (Houtsma et al., 1994b). The pH was influential in inhibition by sodium lactate but not by sodium chloride. As expected, inhibition by sodium lactate was directly related to pH and temperature, with lower amounts of sodium lactate needed as pH and temperature dropped. This model was then evaluated for use in a bologna-type sausage system (Houtsma et al., 1996). A response surface model for the probiotic organism *Lactobacillus plantarum* yielded similar findings to those reported above, wherein increasing lactate concentration from 0 to 3% decreased the growth rate of the organism (μ, h^{-1}) to 0.02 from 0.08 at 18°C (Delcanton et al., 2018).

Salts of lactic acid are equally as effective in muscle food systems. A study by Shelef and Potluri (1995) examined the effect of 3% sodium or calcium lactate added to pork liver sausage and various heat treatments. As expected *L. monocytogenes* was inhibited by a combination of 3% lactate and heat sterilization (121°C for 15 min) followed by a water bath heat process (70°C internal temperature) and least affected when lactate was added to sausage without heat treatment. The calcium form was found to be more inhibitory than the sodium form. Calcium lactate was also more effective than sodium lactate in retarding growth of *E. coli* O157:H7 and *S.* Typhimurium when stored for 4 days at 20°C. Others have reported the antimicrobial effects of calcium lactate to be greater than sodium or potassium lactates in treated injected turkey breast products for the inhibition of *Clostridium perfringens* spore germination during product cooling, wherein calcium lactate achieved greater reductions than did other lactate salts in numbers of recovered pathogen cells on plating medium (Velugoti et al., 2007). Fresh pork sausage formulated with 0 to 3% sodium lactate was vacuum-packaged and stored for up to 28 days at 4°C (Brewer et al., 1991; 1995). Using 10^8 CFU/g as a microbial end point, 1% sodium lactate delayed growth for 10 days, while a 2% level extended the time to 24 days. Potassium lactate (4%) and sodium chloride (3%) reduced growth somewhat at 5°C. There did not appear to be a synergistic effect between a 4% concentration of lactate when combined with 140 ppm potassium nitrite. Red color was preserved by using 2–3% sodium lactate. Pork sausages manufactured with 25% fat or 8% fat, with and without 3.0% potassium lactate, showed differences in spoilage rates. Lactates inhibited the growth of psychrotrophic organisms and delayed the pH decline (Bradford et al., 1993). As opposed to this report indicating no apparent synergism between preservatives, others reported that in pork scrapple the combined application of lactate/diacetate blend at 4%, with the use of citrate-diacetate, produced a 7.0 $\log_{10}$ CFU/g difference in *L. monocytogenes* numbers following 50 days of refrigerated storage between the control product (no lactate-diacetate) and the treatment. Researchers also reported use of lactate-diacetate-propionate (2.5%) yielded bacteriostatic effects against the pathogen when coupled to citrate-diacetate usage at pH 5.0, 5.5, and 6.0 (Porto-Fett et al., 2011). Similar results were reported on frankfurters utilizing a 2.1%/0.15% lactate/diacetate blend in frankfurters inoculated with *L. monocytogenes* and then treated with a lactic acid bacteria biopreservative (LactiGuard™) and an antimicrobial supernatant product comprised of fermentates produced by the members of the biopreservative product. The pathogen's growth was kept to less than 1.0 $\log_{10}$ CFU/package during 8 weeks of refrigerated storage (Koo et al., 2012).

Cured meats, such as wieners, rely on sodium nitrite to inhibit growth of *C. botulinum* (the reader is directed to the Chapter 7 on nitrites usage for an extensive discussion of the antimicrobial mechanisms and use of these preservative agents). Retention of nitrite was linked to higher pH levels and was unaffected by the incorporation of 1.2, 2.4, or 3.6% sodium lactate in frankfurters (Kilic et al., 2002). Frankfurters containing 2.0 or 3.0% potassium lactate were inoculated with a five-strain cocktail of *L. monocytogenes* and stored at 4 and 10°C for up to 90 and 60 days, respectively (Porto-Fett et al., 2002). Both concentrations of potassium lactate significantly inhibited or delayed the growth of *L. monocytogenes* during storage at refrigerator and abuse temperatures. Sodium lactate (2%) was also effective in extending the shelf life of lower fat frankfurters to 6 weeks compared to 3 and 4 weeks for low-fat and high-fat control frankfurters (Bloukas et al., 1997). Another study (Bedie et al., 2001) compared the use of 3 or 6% sodium lactate, 0.25 or 0.5% sodium acetate or sodium diacetate as an antilisterial process for frankfurters. Sodium lactate at a 6% concentration or 0.25% sodium diacetate were the most effective treatments throughout the 120-day shelf life conducted at 4°C with minimal change in pH. The growth of *L. monocytogenes* was delayed for 4 and 12 weeks at 7 and 3°C in unsmoked, uncured bratwurst manufactured with 3.4% sodium lactate/0.1% sodium diacetate (Glass et al., 2002). Levels of >3% sodium lactate or 1% each of lactate and diacetate prevented growth of *Listeriae* for 60 days at 4.5°C. Combined use of lactate-diacetate (3.0%) and e-beam application (1.8 or 2.6 kGy) produced similar outcomes on aerobically refrigerated frankfurters, wherein bacteriostatic results were achieved against *L. monocytogenes* and aerobic psychrotrophic bacteria over 4 weeks of storage (Knight et al., 2007).

Lactate influenced the growth of lactic acid bacteria in meat based on the pH level and atmospheric conditions under which post-rigor meat was held (Grau, 1980). At pH 6.1, the amount of L-lactate ordinarily found in muscle was not inhibitory to any of the strains. A 100 mM concentration of sodium lactate buffered to pH 5.5 in a meat system prevented anaerobic growth of *Serratia liquefaciens*, *Y. enterocolitica*, *Enterobacter cloacae*, and *A. hydrophila*, as well as aerobic growth of the aeromonads (Grau, 1981). When raised to 210 mM, sodium lactate permitted aerobic growth of *B. thermosphacta* down to pH 5.5. A 3% concentration of sodium lactate at pH 7.05 and 6.5 reduced surface discoloration in chubs of fresh sausage and extended shelf life from 10 to 20 days. In retail products, sodium lactate caused more rapid surface discoloration (Lamkey et al., 1991). Papadopoulos et al. (1991a; b; c) injected beef top rounds with 0–4% sodium lactate, then cooked and vacuum-packaged the roasts. Although aerobic plate counts for control meat increased from 2.7 to 8.0 $\log_{10}$ CFU/cm^2 after 84 days of storage, the plate counts of treated roasts increased to only 6.5 log CFU/cm^2 in the same period. The microflora changed from predominantly staphylococci and micrococci at day 0 to coryneforms, micrococci, lactobacilli, and yeasts at 42 days, to homo- and heterofermentative lactobacilli and streptococci at 84 days. Lactate increased cooking yields, enhanced flavor at the 1% level by adding a salty taste, but produced a mild throat irritation at a 4% concentration. Although Shelef and Yang (1991) suggested a possible a_w effect for sodium lactate, Papadopoulos et al. (1991b) did not find this effect in their studies.

Shelef and Yang (1991) found that a 5.0% concentration of sodium or potassium lactate delayed the growth of *L. monocytogenes* in TSB, and 4% limited growth of *L. monocytogenes* in comminuted chicken or beef when held at 35, 20, or 5°C. They concluded that the two salts gave comparable results and could be used interchangeably, suggesting that the lactate moiety was the effective component. Lactate contributed to water-holding capacity and increased cooking yields.

Maas et al. (1989) demonstrated that sodium lactate was effective in retarding toxin formation by *C. botulinum* in comminuted raw turkey using levels of 2, 2.5, 3, or 3.5%. The meat was cooked to an internal temperature of 71.1°C, vacuum-packaged, and stored at 27°C. Toxin production was detected within 3 days in turkey without lactate and within 4–5, 4–6, 7, or 7–8 days for each concentration, respectively. Organoleptically, the 2.5% concentration was acceptable while off-flavors developed in meat containing the 3.5% concentration. They also noted that the inhibitory component was the lactate moiety not sodium. Maas et al. (1989) suggested that a possible mechanism for the action of lactate was the inhibition of a major anaerobic energy metabolism pathway necessary for growth. Unda et al. (1991) also found that 2% lactate inhibited the growth of *C. sporogenes*. More recently, a commercial nitrite/lactate/diacetate curing ingredient product demonstrated best antimicrobial activity against *C. perfringens* spores in frankfurters and cured hams stored at 20 or 22°C, with reductions in spores (based on plate counts of germinated spores) to non-detection in frankfurters and by 5.0 $\log_{10}$ CFU/g in hams during storage (Jackson et al., 2011). The conventional blend outperformed other treatments including natural-source pre-converted nitrite products, with or without antimicrobials derived from vinegar and lemon extracts. Researchers reported that treatments bearing the highest levels of residual nitrite post-cooking (notably including the treatment containing lactate in the formulation), regardless of nitrite source, exhibited the best sporostatic outcomes.

The influence of 2, 5, or 10% lactic acid/sodium lactate buffer rinses (pH 3.0) alone or combined with modified atmosphere packaging was examined on the survival of *L. monocytogenes* and shelf life of chicken legs (Zeitoun and Debevere, 1991). Some treated chicken was packaged without further modification, but other samples were placed into packages that were gas flushed with a 90% carbon dioxide/10% oxygen mixture; all were stored at 6°C. Excised skin tissue was analyzed for *Listeriae* for up to 17 days post-treatment. Product was considered spoiled when microbial numbers reached 7–8 log CFU/g. Odor concomitant with growth of psychrotrophic spoilage organisms occurred in untreated samples within 8 days and within 9, 10, and 13 days for the 2, 5, and 10% acid spray, respectively. Modified-atmosphere packaging (MAP) extended storage life for products without acid sprays to 17 days and even longer with sprays. Reduction in the production of hydrogen sulfide and other sulfur-containing compounds resulted from the decreased populations of psychrotrophs. Levels of *L. monocytogenes* increased by only 0.56 log during 17 days of storage for the legs treated with 10% buffer combined with MAP. Sodium lactate (3.3%) was very effective in reducing numbers of *B. thermosphacta* and *E. coli* in a chicken model system after 14 days at 22°C but only delayed the growth of *Lactobacillus alimentarius* by about 2 days (Lemay et al., 2002).

Lactates have been combined with other compounds in multiple-barrier studies. Vacuum-packaged beef was treated with 2% sodium lactate, pediocin (1,400 U/g), nisin (1,400 U/g), Nisaplin™ (500 IU/g) or Microgard™ (2%) (Rozbeh et al., 1993). Of all the treatments sodium lactate was the most effective during an 8-week storage period at 3°C. Beef roasts containing sodium lactate, glycerol monolaurin, or sodium gluconate and inoculated with *L. monocytogenes* and *C. sporogenes* were stored at 2, 7, 10, and 25°C. Sodium lactate was more effective at reducing pathogen numbers at levels up to 3.5% versus sodium gluconate, which did not provide control (Stillmunkes et al., 1993). On the other hand, in studies testing the combination of 0.78 or 1.56% lactate with the antimicrobial protein ovotransferrin (40 mg/mL), while the combination of acid salt and chelating protein exhibited bacteriostatic-type inhibition against *L. monocytogenes* organisms in liquid medium, the acid salt applied by itself did not result in significant inhibition of pathogen growth. The combination of lactate (0.78%) and diacetate (0.52%), likewise, did not inhibit the pathogen's growth compared to the control (Moon et al., 2011).

Harmayani et al. (1991) looked at the interactive effects of sodium alginate and calcium or sodium lactate used in restructured meats on *P. fragi* and *S.* Typhimurium. Ground beef containing sodium alginate at concentrations up to 0.8%, calcium lactate up to 0.3%, or a combination of the two, was inoculated with the two organisms and held at 5°C for 5 days. None of the treatments affected the growth of either organism. Sodium lactate at 3% only minimally affected the growth of *S.* Typhimurium, but did inhibit *P. fragi* at even the 4% level. Raw and cooked ground beef were prepared with 1.8% sodium lactate, 0.1% sodium erythorbate, 1% K-carrageenan, and an alginate binder. The products were inoculated with *L. monocytogenes* and stored aerobically at 4°C. In general, sodium lactate was more effective than the other additives in inhibiting growth of *L. monocytogenes* and total aerobic plate counts (Harmayani et al., 1993). Similar studies used low-fat ground beef patties formulated with carrageenan, encapsulated salt, and hydrolyzed vegetable protein and containing 1, 2, or 3% potassium lactate. Carrageenan had no effect on bacterial loads but the addition of 2 or 3% potassium lactate reduced microbial numbers. Some discoloration and lipid oxidation occurred, thus affecting the sensory qualities of the meat (Egbert et al., 1992).

Comminuted salmon was mixed with sodium lactate, sodium chloride, and sodium nitrite, inoculated with *L. monocytogenes*, vacuum-packaged, and stored at 5 or 10°C. Sodium lactate displayed synergistic activity with nitrite and increasing concentrations of sodium chloride; a 2% lactate and 3% sodium chloride inhibited *L. monocytogenes* for up to 50 days at 5°C. A 3% concentration of lactate in combination with 3% sodium chloride or with 125 ppm sodium nitrite and 3% sodium chloride was effective at 10°C (Pelroy et al., 1994).

5.3.9.2 Application and Regulatory Status

Calcium (21 CFR 184.1207), potassium (21 CFR 184.1639), and sodium lactates (21 CFR 184.1768) are also approved. The acceptable daily intake for humans is listed in Table 5.2.

5.3.10 Malic Acid

5.3.10.1 Antimicrobial Properties

Malic acid is inhibitory to fungi and bacteria, probably as a direct effect of pH manipulation. Nunheimer and Fabian (1940) found that malic acid was inhibitory against *S. aureus* at pH 3.98. At pH 3.3 (37.5 mM) malic acid was reported to produce bacteriostatic effects against *S.* Typhimurium and *L. monocytogenes* in microbiological medium, with bactericidal effects upon increasing the concentration to 75 mM malic acid (Over et al., 2009). Kang and Kang (2017) washed carrots and muskmelons in 2.0% malic acid (pH 2.2), reporting a 1.5 $\log_{10}$-cycle drop in numbers of *S.* Typhimurium on melon surfaces, but a 6.0 $\log_{10}$-cycle decline in pathogen numbers on carrots. Application of the acidulant under conditions of vacuum infusion (21.3 kPa pressure) sped the process of pathogen decrease, but did not increase the degree of pathogen reduction observed.

5.3.10.2 Application and Regulatory Status

Malic acid is a non-hygroscopic, dicarboxylic acid with high water solubility. It has a strong acid flavor, but does not have the same buildup of acid taste as other acids. It is used in sherbets and ices, fruit preserves, jams, jellies, and beverages, primarily for its flavoring and acidification characteristics (Gardner, 1972). Malic acid is approved as a GRAS substance for miscellaneous and general-purpose usage (21 CFR §184.1069). The acceptable daily intake for humans is listed in Table 5.2.

5.3.11 Propionic Acid

5.3.11.1 Antimicrobial Properties

Propionic acid inhibited spore-forming bacteria, especially rope bacteria (*B. subtilis*) at pH 6.0. As the pH decreased to 5.0 or 4.0, the acid inhibited yeasts and molds, but not to the same extent as bacteria (Woolford, 1975b). Chung and Goepfert (1970) compared various organic acids to determine the highest pH level that inhibited growth of *Salmonella* serovars Anatum, Senftenberg, and Tennessee when all other parameters were at their optimum. Propionic acid at pH 5.5 inhibited the growth of *Salmonella* at a higher pH than other acids (pH 5.4, acetic; pH 5.1, adipic; pH 4.6, succinic; pH 4.4, lactic; pH 4.3, fumaric and malic; pH 4.1, tartaric; pH 4.05, citric). Cherrington et al. (1991a) determined that 0.5–0.7 M propionic acid at pH 5.0 was more inhibitory at 60 min compared to formic acid in 3 h for *E. coli* K12 and *Salmonella*. At a 5 mM concentration, the rate of RNA, DNA, protein, lipid, and cell wall synthesis decreased, which led to an increase in cell mass without cell division, and a decrease in viability (Cherrington et al., 1990). Eklund (1985) concluded that propionic acid was not as inhibitory as benzoic acid for *P. aeruginosa*. Small 0.5 $\log_{10}$-cycle reductions in *Salmonella* and *Campylobacter* spp. on chicken frames were obtained by the application of 0.3% propionic acid, while 0.1% peracetic acid produced a 1.4 $\log_{10}$ CFU/g reduction, though reductions were not reported by researchers to differ statistically (Moore et al., 2017). Numbers of aerobic bacteria, unlike pathogens, were not reduced by propionic acid application (Moore et al., 2017).

The salts of propionic acid are also effective antimicrobial agents. Calcium propionate prevented rope formation caused by *B. subtilis* in bread dough at a level of 0.188%, pH 5.8, or at a level of 0.156%, pH 5.6 (O'Leary and Kralovec, 1941). The incorporation of calcium propionate into coatings for RTE products was effective in reducing problems with *L. monocytogenes* (Janes et al., 2002). Zein (Z) film coatings were dissolved in propylene glycol (ZP) or ethanol (ZE) with and without nisin (N; 1,000 IU/g) or calcium propionate (CP; 1%) and coated onto ready-to-eat chicken samples dipped in *L. monocytogenes*. Products were stored for up to 24 days at 4 or 8°C. Non-detectable levels of the pathogen were found in products ZEN, ZPNCP, or ZENCP. Similarly, chitosan films produced to entrap and deliver propionic acid on bread doughs produced complete inhibition of *Candida albicans* and *Penicillium* spp. on plating medium by inhibition zone assay, while *S. aureus* was inhibited (inhibition zone: 1.99 cm) (Rivero et al., 2013).

Sodium propionate at levels of 0.1–5% delayed growth of *S. aureus*, *Sarcina lutea*, *Proteus vulgaris*, *L. plantarum*, *Torula* spp., and *S. ellipsoideus* by as much as 5 days. This substance not only postponed spoilage in fresh figs, syrup, applesauce, cherries, and berries, but also prevented spoilage in vegetables having a more neutral pH, such as lima beans and peas (Wolford and Anderson, 1945). The Gram-negative pathogen *Cronobacter* was inhibited from growth by 0.4% propionate application in milk and microbiological medium; the addition of bacteriocin in milk did not enhance antimicrobial effects as determined by plate counts (Oshima et al., 2012).

Turkey breast meat was formulated with sodium salts of lactate, acetate, pyruvate, citrate, and propionate to a target level of pH 6.0, inoculated with *C. botulinum* and held at 28°C for 0–18 days. At periodic intervals, samples were removed from storage and tested for production of neurotoxin. Toxin production occurred after 2 days for pyruvate, 3 for citrate, 4 for lactate and acetate, and 5 days with a 2% concentration. At a 6% concentration, toxicity occurred after 7 days for pyruvate, 18 for citrate, and greater than 18 days for propionate, acetate, and lactate. Citrate was most effective based on molarity (Miller et al., 1993). In cured deli turkey product, *L. monocytogenes* was inhibited over 12 weeks of refrigerated storage when commercial propionate-containing antimicrobials were used in product formulation to concentrations of 0.3, 0.4, and 0.5% sodium propionate (Glass et al., 2013). On uncured turkey chubs, application of combined sodium propionate and potassium lactate (0.1%), with or without 44 ppm lauric arginate ester, produced antimicrobial effects against *L. monocytogenes* over 90 days' refrigerated storage, yielding under a 2.0 log_{10}-cycle increase in *L. monocytogenes* numbers during storage. Nonetheless, combination of propionate, lactate, and levulinate allowed no growth of *L. monocytogenes* in treated turkey products during the product storage (Porto-Fett et al., 2015).

In laboratory studies examining the effect of sodium propionate on *L. monocytogenes*, the organism was able to grow at up to a 0.3% concentration in tryptose broth adjusted to pH 5.6 and incubated at 4, 13, 21, and 35°C. Reduction of pH to 5.0 minimized growth at 13, 21, and 35°C and prevented growth at 4°C (El-Shenawy and Marth, 1989a). When media containing 0.3% sodium propionate was adjusted to pH 5.6 with acetic, tartaric, citric, or lactic acids in media (El-Shenawy and Marth, 1992), the lag phase of *L. monocytogenes* grown at 13°C was extended by 7, 7, 5, and 4 days, respectively. At pH 5.0 and 13°C, the levels of *L. monocytogenes* were undetectable. Sodium propionate at a 0.3% concentration was also effective at pH 5.0 in reducing numbers of *L. monocytogenes* when used in conjunction with acetic acid in cold-pack cheese food (Ryser and Marth, 1988). Exposure of *L. monocytogenes* to 8% sodium propionate for 60 min caused injury (Buazzi and Marth, 1992). An MIC of 11 mM against *L. monocytogenes* was achieved in cheese production-simulating conditions (Wemmenhove et al., 2016).

Emphasizing the multiple-barrier concept in retarding growth, Golden et al. (1995) examined the combination of sodium propionate and sodium acetate at a number of concentrations, pH levels, and temperatures. A three-strain mixture of *L. monocytogenes* was inoculated into BHI broth containing 0.08% EDTA, 0.02% ascorbic acid, and 0.9% sodium propionate or sodium acetate, adjusted to pH 4.5, 4.0, 3.5, or 3.0 with hydrochloric acid and stored at 28, 19, or 4°C. Inactivation of *L. monocytogenes* was due to incubation temperature, concentration of the organic acid, degree of dissociation of the acid, and pH. Propionic acid was more effective than acetic acid as an antimicrobial, particularly at the higher pH level. The acids contributed to approximately 85% of the antimicrobial activity with 14% contributed by ascorbic acid; little activity was ascribed to EDTA in combination with the acids.

The effect of fat concentration on the efficacy of antimicrobial agents against *L. monocytogenes* was studied in pork liver beaker sausage (Hu and Shelef, 1996). Sausage batter was mixed with 0.2% sodium propionate, 1.8% sodium lactate, and 0.1% sorbic acid or potassium sorbate. Increasing the fat content from 22 to 67% decreased the growth of *L. monocytogenes* by only 1.5 log_{10}-cycles. The antimicrobial activity of sodium propionate was most affected at 4°C by fat concentration with a 1.5 log_{10} reduction of *L. monocytogenes* at 22% fat compared to 2.8 log_{10} reduction at 67% fat; with sodium lactate, 0.5 and 1.8 log_{10} reductions were observed, respectively.

Propionic acids and their salts are primarily inhibitory to molds; however, some species of *Penicillium* grew on media containing 5% propionic acid (Heseltine, 1952b). Propionic acid at a concentration of 0.1%, pH 4.5, reduced growth and aflatoxin formation of *A. flavus*; at 0.2%, no growth occurred. This inhibitory effect was more pronounced with the addition of acid at the time of inoculation, rather than later (Ghosh and Häggblom, 1985). Propionic acid at a concentration of 2435 ppm was more effective than acetic acid in limiting growth of *Fusarium oxysporum* but less effective than sorbic acid and potassium sorbate. Spore germination was inhibited by 1402 ppm (Tzatzarakis et al., 2000).

Concentrations of propionic acid and propionates ranging from 8 to 12% were effective in controlling mold growth on the surface of cheese and butter (Deane and Downs, 1951; Ingle, 1940). A 5% calcium propionate solution acidified to pH 5.5 with lactic acid was as effective as a 10% unacidified solution in preventing surface mold growth on butter (Olson and Macy, 1945). Molds were inhibited by less calcium propionate by weight than sodium propionate. Not only was the final pH of the substrate critical, but also various organisms displayed different tolerances to the compounds (Olson and Macy, 1940).

A novel use of propionic acid to retard spoilage of bread was investigated by Gardner et al. (2001). Adding yeasts extracts previously fermented by *Propionibacterium freundrechii* provided a source of propionic acid in bread formulation. Bread produced with the extracts contained less ethanol and supported less spoilage by molds. Mold spoilage is a significant economic problem in bakery products. Although mold spores are destroyed during baking, post-process contamination from the atmosphere, cooling surfaces, and wrapping materials reintroduce molds. *Eurotium amstelodami*, *herbariorum*, *rubrum*, and *repens* were inoculated by needle into cake analogues prepared with calcium propionate, sodium benzoate, and potassium sorbate at a level of 0.3% by weight. All preservatives were effective at pH 6.0 and a water activity of 0.8–0.85 (Guynot et al., 2002). Subsequent studies (Marin et al., 2002a, b) examined several levels of calcium propionate, potassium sorbate, and sodium benzoate (0.003, 0.03, and 0.3%) at pH 4.5, 6.0, and 7.5 and water activity of 0.8, 0.85, and 0.93 in a model agar system on the retardation of growth of *E. amstelodami*, *herbariorum* and *rubrum*, *A. flavus* and *niger*, and *Penicillium corylophilum*. None of the preservatives was effective at neutral pH. Sub-optimal doses (0.03%) led to enhanced growth of *Aspergillus* and *Penicillium* isolates.

5.3.11.2 Application and Regulatory Status

Propionic acid is a monocarboxylic acid with a slightly pungent, disagreeable odor. Salts of the acid have a slight cheese-like flavor. The acid form is readily miscible with water, and the sodium salt is more soluble than the calcium form. It is normally found in cheese and as a metabolite in the ruminant gastrointestinal tract. Members of the propionibacteria metabolize up to 1% propionic acid in Swiss cheese; these organisms are associated with cheese manufacture and the characteristic Swiss cheese flavor. This naturally formed additive becomes a developed preservative that limits mold growth on Swiss cheese. This additive is also used as a mold inhibitor in cheese foods and spreads. Propionic acid and its salts are used primarily as mold and rope inhibitors in bread. Baking destroys most molds, but surface recontamination can occur during packaging, and growth can be seen under the wrapper during storage. Propionates can be added to bread dough without interfering with leavening, since there is little or no effect on yeast. Sodium propionate is recommended for use in chemically leavened products, because the calcium ion interferes with the leavening action. Calcium propionate is preferred, however, for use in bread and rolls, because the calcium contributes to the enrichment of the product (Chichester and Tanner, 1972).

Propionic acid (21 CFR §184.1081), calcium (21 CFR §184.1221), and sodium propionate (21 CFR §184.1784) salts are approved as GRAS substances for miscellaneous and general-purpose usage. In addition, calcium and sodium propionate are listed as antimycotics when migrating from food-packaging material (21 CFR §181.23). No upper limits are prescribed for use of this additive, except in bread and rolls, which conform to standards of identity. A limit of 0.32% can be used in flour and in white bread and rolls, 0.38% in whole-wheat products, and 0.3% in cheese products (Robach, 1980). The acceptable daily intake for humans is listed in Table 5.2.

5.3.12 Succinic Acid

5.3.12.1 Antimicrobial Properties

Succinic acid successfully lowered microbial loads on poultry carcasses (Mountney and O'Malley, 1965), though a 3 or 5% concentration used at 60°C impaired the appearance of the product (Cox et al., 1974). Thomson et al. (1967) reduced the population of inoculated *S.* Typhimurium by 60% on chicken carcasses after spraying with 1.0% succinic acid. As with other organic acids reviewed in this chapter, recent research efforts focused on the incorporation of succinic acid via edible films. Khalil et al. (2015) added 0.1 to 0.4% succinic acid as a cross-linking agent to corn-derived zein films, or a combination of eugenol and succinic acid at 0.1 to 0.4%, and then subjected films to biophysical and antimicrobial properties analyses. Incorporation of succinic acid into zein films, or succinic acid + eugenol, produced zones of inhibition ranging from 11 to 15 mm against *B. cereus* and *Salmonella*, and smaller inhibition zones against *Y. enterocolitica* and *E. coli*, but did not produce detectable inhibition against Gram-positive bacteria including *L. monocytogenes* and *Clostridium sporogenes*. A similar lack of apparent antimicrobial activity for succinic acid-bearing polyvinyl alcohol films was also reported by researchers testing film activity against *E. coli*, *P. aeruginosa*, and *S. aureus*, wherein inhibition zones did not exceed 1 mm (Suganthi et al., 2018).

5.3.12.2 Application and Regulatory Status

Succinic acid is a non-hygroscopic, dicarboxylic acid having low water solubility. It has a slightly bitter taste and serves as a flavor enhancer. Succinic acid is used to modify the plasticity of bread doughs and in the production of edible fats. It can be incorporated into gelatin deserts and cake flavorings (Gardner, 1972). Succinic acid is approved as a GRAS substance for miscellaneous and general-purpose usage (21 CFR §184.1091). No limit has been set on the acceptable daily intake for humans.

5.3.13 Tartaric Acid

5.3.13.1 Antimicrobial Properties

Any antimicrobial property of tartaric acid is because of its ability to lower the pH. Tartaric acid at 37.5 and 75 mM was applied in combination with 1 to 3 kGy electron beam irradiation for decontaminating *L. monocytogenes* on marinated uncooked or marinated and cooked chicken breast surfaces. Electron beam dose produced a significant effect on pathogen survival, though tartaric acid was not reported to produce any antilisterial effect (Over et al., 2010). The application of 5.0 g/liter tartaric acid, alone or in combination with cinnamon oil, over 10 days of refrigerated storage, resulted in decreased numbers of aerobic bacteria by $\geq$1.0 $\log_{10}$-cycle on surfaces of peeled shrimp. Tartaric acid was lethal to Gram-positive pathogens in medium at 6.25 mg/mL but exhibited increased efficacy against Gram-negative enteric pathogens (*Salmonella enterica*, *Shigella flexneri*), with a reduced minimum lethal concentration of 1.1 mg/mL. When incorporated into carboxymethylcellulose (CMC) films, aerobic bacteria growth was slowed over 10 days of refrigerated storage on raw chicken breasts, wherein a 3.7 $\log_{10}$-cycle difference was observed between CMC-only films and CMC films infused with the organic acid (Lee et al., 2016).

5.3.13.2 Application and Regulatory Status

Tartaric acid, the most soluble of the solid acidulants, has a strong tart taste that enhances grape-like flavors. It is used in fruit jams, jellies and preserves, sherbets, and grape-flavored beverages. The monopotassium salt (cream of tartar) is commonly found in baking powders. Tartaric acid acts synergistically with antioxidants to prevent rancidity, and it prevents discoloration in cheese (Gardner, 1972). Tartaric acid is a GRAS substance for miscellaneous and general-purpose usage in accordance with good manufacturing practices (21 CFR §184.1099). The acceptable daily intake for humans is listed in Table 5.2.

5.4 Mode of Action

The mode of action of organic acids in inhibiting microbial growth appears to be related to the maintenance of acid-base equilibrium, proton donation, and the production of energy by the cells. It is thus essential to understand each of these concepts before consideration of the actual mode of action of the organic acids. Biological and chemical systems depend on an interaction between acid and base systems. The microbial cell normally reflects this equilibrium by attempting to maintain an internal pH near neutrality (Baird-Parker, 1980). Homeostasis is the tendency of a cell to sustain chemical equilibrium despite a fluctuation in the acid-base environment. Through the interaction of a series of chemical mechanisms, this delicate balance is maintained and alteration of this balance causes destruction of the microbial cell. Proteins, nucleic acids, and phospholipids can be structurally altered by pH changes. The availability of metallic ions to the organism is also altered and becomes a function of membrane permeability, because membranes are less permeable to charged molecules than to uncharged molecules. These changes in membrane permeability can exert a dual effect by impairing transport of nutrients into the cell or by causing the leakage of internal metabolites to the outside. Changes due to pH can destroy bacteria, molds, and yeasts, although some microorganisms such as *Acetobacter* can exist at and even require extremes of acidity (Langworthy, 1978).

The terms strong or weak are used to describe acids and reflect the degree with which acids readily donate a proton or dissociate in aqueous solutions. Inorganic acids such as hydrochloric acid almost entirely dissociate in solution, and are thus sometimes referred to as strong acids. Acetic acid and other organic acids only slightly ionize and do not readily give up their proton(s) to water. When an acid enters and ionizes within the cell, the problem becomes one of elimination of the excess protons. This ejection process becomes a fundamental issue in how a cell produces energy. Energy produced through chemical reactions is essential for cellular processes. Synthesis of macromolecules, maintenance of osmotic gradients, and active transport of molecules across the membrane depend on energy generated by the cell in the form of adenosine triphosphate (ATP). In anaerobic microorganisms, the glycolytic pathways inefficiently generate ATP. In aerobic organisms, the electron transport system (ETS) principally generates ATP with oxygen as the terminal electron acceptor (Gould et al., 1983).

Before energy can be produced, materials must be transported into the cell. One means for substrate entry into the cell is active transport, which also allows greater internal concentration of solute than external concentration. To maintain this unequal gradient, energy must be expended. For this reason, active transport is coupled with energy-yielding processes. The energy evolves from the oxidation of the substrates and the respiratory chain. The respiratory chain or ETS transverses the membrane which does not permit H^+ or OH^- ions to penetrate and ejects accumulated protons to the external environment. This process generates a chemical and electric gradient capable of driving metabolic reactions (Dawes and Sutherland, 1992).

Levine and Fellers (1940) demonstrated that acetic acid was more lethal at a higher pH than hydrochloric acid or lactic acid. They concluded that this toxicity was not due to hydrogen ion concentration alone, but seemed to be a function of the undissociated molecule. With acetic acid, lowering the pH increased the inhibitory activity, confirming that the undissociated molecule was the effective inhibitor. Thus, the inhibition by extracellular fatty acids used as antimicrobial agents would increase with decreasing pH, in agreement with pK_a values (Freese et al., 1973).

Sheu and Freese (1972) determined that short-chain fatty acids reversibly reacted with the cell membrane and altered its structure. It was postulated that this interfered with the regeneration of ATP by uncoupling the ETS, or the transport of metabolites into the cell was altered. Further studies by Sheu et al. (1972) indicated that acetate uncoupled the amino acid carrier protein from the ETS and inhibited amino acid transport non-competitively. Serine uptake was inhibited in membrane vesicles of *B. subtilis* when exposed to fatty acids; using the same system, L-leucine and L-malate were shown to uncouple both substrate transport and oxidative phosphorylation from the ETS (Freese et al., 1973).

In a later study, Sheu et al. (1975) recognized that if reducing compounds were no longer available to the cell because of transport inhibition by the fatty acids, oxygen consumption would be reduced. To understand these processes, the bacterial cell was converted to a spheroplast under isotonic conditions and these membrane vesicles were used to study uptake of a substrate against a gradient. Since oxygen consumption was observed in the presence of membrane preparations and an energy source, inhibition transport would not necessarily be linked with ATP generation. The active transport of a molecule would depend upon the proton gradient generated during the oxidation of a substrate. Lipophilic agents, such as the short-chain fatty acids, would shuttle protons through the membrane until the proton motive force had been destroyed and transport thus eliminated. Freese and Levine (1978) further postulated that the most effective antimicrobial agents would be lipophilic enough to attach to microbial membranes yet be soluble in the aqueous phase. This is because they can approach the membrane from the aqueous medium, yet, easily and without requiring energy, penetrate the membrane lipid bilayer. Undissociated acids of short chain length can penetrate the cell more easily because they possess these characteristics.

Not all acids are effective against all microorganisms. *E. coli* and *B. subtilis* are equally inhibited by equal concentrations of compounds containing up to six carbons. However, twice the amount of C_8 (caprylic) acid is needed to inhibit the growth of *E. coli* compared to *B. subtilis*. The difference between the two organisms could be due to the lipopolysaccharide layer that surrounds the Gram-negative *E. coli*, thus acting as a mechanical barrier and preventing passage of the acid into the cell. Another theory proposes that Gram-negative organisms can rapidly metabolize the acidulant and therefore disallow its accumulation within the cell (Freese et al., 1973; Kabara et al., 1972). Furthermore, some acids, lactic and citric, reduce the internal pH of the cell more than acetic acid; however, other changes associated with metabolic or physiological processes may be similarly affected (Ita and Hutkins, 1991).

Hunter and Segal (1973) in studies using *Penicillium chrysogenum* suggest that weak acids at or below their pK_a could discharge the proton gradient and ionize within the cell to acidify the interior. It was postulated that the rate of proton leakage into the cell versus proton ejection would determine the inhibition of the cell (Freese et al., 1973). Eklund (1983) suggested, similarly, that dissociation of an organic acid and its proton cytoplasmically would result in increased acidity of the cell and would induce the cell to attempt to re-establish homeostasis through various means (Brul and Coote, 1999; Hirschfield et al., 2003).

Eklund (1980) has questioned the mode of action of the weak lipophilic acids. He stated that although inhibition of uptake of nutrients contributes to growth inhibition, it does not seem to be the sole cause of the static action. Taylor et al. (2012) reviewed other reported antimicrobial mechanisms of weak organic acids once the acid penetrates the microbial cytoplasm, suggesting multiple effects occur to result in microbial inhibition. Research by Cherrington et al. (1990, 1991b) reported the application of sub-lethal concentrations of organic acids (propionic acid, formic acid) inhibited the uptake of multiple macromolecule building blocks, including leucine, glycerol, and *N*-acetylglucosamine. This group later reported that application of propionic acid or formic acid did not induce membrane damage to cells of *Salmonella* based on lactase extracellular activity assessment (Cherrington et al., 1992), although Alakomi et al. (2000) later reported that *E. coli*, *Salmonella*, and *P. aeruginosa* cells were permeabilized by lactic acid exposure. This indicates that multiple potential complementary antimicrobial mechanisms occur against bacterial cells by organic acid exposure. Similarly, Koczoń (2009) reported the interplay of these factors on yeast cells treated with benzoates.

In summary, the current data suggest that the mode of action of short-chain lipophilic acids requires the destruction of the proton motive force, thereby limiting substrate transport. It is further speculated that acids, which possess both lipoidal and aqueous solubilities, are the most effective antimicrobial agents and that some sort of membrane attachment of the acid is involved. Studies have indicated that cells, when placed in fresh medium devoid of the inhibitor, can reinitiate transport, suggesting a static response rather than a killing effect (Freese et al., 1973). Obviously, there is need for further research on the mode of action of these antimicrobial agents.

5.5 Acid Adaptation and Tolerance

The predictable response of cells to stressful situations is the cornerstone of the multiple-hurdle concept, whereby organisms are subjected to multiple deleterious compounds or processes designed to reduce or eliminate microbial loads. These stresses occur simultaneously, such as the addition of several antimicrobial agents (acid, bacteriocin, preservative) to a food, or sequentially, such as intervention strategies that occur during processing of muscle foods (acid sprays, hot water rinses). In sequential treatments, there may be more of an opportunity for strains to develop resistances to the antimicrobial compound or treatment provided that time of exposure allows for such development.

Cells strive to maintain a balance between competing stresses that occur through exposure to sublethal conditions, namely, acidity, temperature, available water, starvation, and desiccation, among others. Generally, organisms survive and function over a wide range of acidity and alkalinity due to their ability to maintain homeostasis; however, exposure to adverse environments frequently induces a stress response in susceptible microorganisms. Response to such stresses often takes the form of specific stress proteins induced after sublethal exposure. It is hypothesized that these proteins protect against subsequent exposure to the same stress or possibly cross protect against other stresses (Davidson and Harrison, 2002). Adaptation to sublethal stresses allows organisms to adapt to higher levels of the same stress without being killed. For example, exposure to low concentrations of organic acids would cause organisms to become more tolerant to higher concentrations of acids, thus potentially allowing them to survive exposure to commonly used antimicrobial intervention strategies. Depending on the length of time and exposure to mild acid conditions, cells can adapt by means of an acid tolerance response (ATR). Exposure to greater quantities of acid (lower pH levels) causes acid shock, which can lead to cell death. During the shock period, stress-induced proteins are formed, which can be specific to the type of shock conditions.

Adaptive responses to sublethal exposure to acids can often offer cross protection to cells when exposed to other stresses. Cross-protective effects have been demonstrated in *E. coli* between initial exposure to organic acids and subsequent exposure to salt (Garren et al., 1998; Jordan and Davies, 2001), radiation (Buchanan et al., 1999), and heating (Mazzotta, 2001; Ryu and Beuchat, 1998). The adaptive effect has also translated to the survival of microorganisms in acidic foods, such as *E. coli* O157:H7 in salami and apple cider (Leyer et al., 1995), non-O157 STEC in fermented sausage (Montet et al., 2009), *S.* Typhimurium in cheeses (Leyer and Johnson, 1992), and *L. monocytogenes* in fermented dairy products (Gahan et al., 1996). In addition, cell survival is affected by its previous growth environment, including oxygen conditions, growth phase, and temperature (Cheng and Kaspar, 1998). Conversely, pre-treatment with acids increased the sensitivity of *S.* Typhimurium to chlorine and iodine (Leyer and Johnson, 1997) leading to potentially more effective interventions.

Measurement of survival or injury of organisms subjected to stresses is dependent on the methods designed to recover organisms. It is assumed that inhibitory compounds that affect optimal recovery of cells would be avoided. Interestingly, higher survival levels were achieved in acid-injured *E. coli* when dilutions were made in osmotically stable diluents containing sucrose, sorbitol, glucose, or sodium chloride, but not glycerol (Jordan et al., 1999).

Acid adaptation is also seen in Gram-positive organisms, such as *L. monocytogenes* (Kroll and Patchett, 1992). Pre-exposure to mild acid conditions led to increased resistance to higher acid environments such as those encountered in the stomach or macrophage phagosome (O'Driscoll et al., 1996).

Inducement of stress conditions can result in the shortening of processing systems and can optimize the use of multiple hurdles as a preservation process (Brul and Coote, 1999; Brul et al., 2002). Using acid-adapted strains of *Salmonellae* inoculated into lean beef tissue, Dickson and Kunduru (1995) demonstrated that exposure to organic acid rinses did not lead to more resistant organisms. It was further shown (Uyttendaele et al., 2001) that acid-resistant, acid-sensitive, and acid-inducible strains of *E. coli* O157:H7 presented no difference in their survival when inoculated into beef tissue treated with 1 and 2% buffered lactic acid.

Shadbolt et al. (2001) postulated that exposure of *E. coli* to stresses such as low water activity (0.90) or low pH (3.5) could produce a biphasic death phase. This was readily apparent if the first stress was low water activity. If the stresses were in reverse order, pH rapidly inactivated cells such that a biphasic death phase was not observed. The authors hypothesized that disruption of cell homeostasis created a large energy drain, thus sensitizing cells to future environmental stresses.

5.6 Toxicology

5.6.1 Toxicity of Organic Acids Post-Ingestion

Organic acid exposure through various means has been extensively researched and evaluated for the acid species discussed in this chapter, as well as for the acid salts described herein. While consumption of organic acids up to levels not exceeding federal allowances has been generally deemed safe by multiple researchers and regulatory bodies/agencies, exposure to acids in excess or through non-ingestion routes has indicated in some cases potential for toxicity to consumers.

Rats fed 10% acetic acid and cats fed 20% acid developed ulcers in the gastric mucosa (Okabe et al., 1971). Johnson (1968) noted that rat gastric mucosa responded to a 100 mM concentration of acetic acid by releasing histamine into the interstitial fluid. Rats consuming acetic acid in drinking water in concentrations of 0.01–0.2% (0.2 g/kg body weight) did not show any adverse effects. At the 0.5% level, however, rats decreased their food intake level and lost 2.6% of their body weight (Sollman, 1921). Mori (1952) fed rats acetic acid ranging from 10 to 50 cm^3 acid/kg of rice. All rats showed mucosa damage owing to the feeding method, however, it was not possible to determine the actual quantity of acid consumed. Ingestion of acetic acid by humans reportedly caused burned lips (Palmer, 1932), interference with blood coagulation (Fin'ko, 1969), pneumonia (Gerhartz, 1949), renal failure (Paar et al., 1968), gastric erosion, (O'Keane et al., 1971), and death. Nonetheless, recently the European Food Safety Authority (EFSA) Panel on Food Contact Materials, Enzymes and Processing Aids (2018) indicated no ingestion-derived toxicological concern when acetic acid was consumed at levels presumed to occur via the consumption of pork products treated with up to 4% acetic acid in carcass-applied antimicrobial sprays.

The mean lethal dose (LD_{50}) for sodium acetate in mice is presented in Table 5.3. Chickens fed diets supplemented with 5.44% sodium acetate showed decreased appetite, depressed growth rate, and increased mortality. Histological findings showed hypertrophy of the kidneys and ureters. Waterhouse and Scott (1962) attributed the toxic effect to the sodium ion rather than the acid moiety. A 22-year-old woman with sensitivity to acetic anhydride was not allergic to sodium acetate (Weil and Rogers, 1951). No toxicological data were available for calcium acetate (Food and Agricultural Organization, 1963) although it is assumed that this derivative would be assimilated in a manner similar to that for acetic acid. The U.S. Environmental Protection Agency (EPA) (2016) reported sodium diacetate to be an eye irritant, but not a chronic toxicant or mutagen at normal usage/consumption levels.

Adipic acid did not affect female rats fed at levels of up to 40 mg/day in short-term studies (Lang and Bartsch, 1953). However, 800 mg/day dosage depressed growth, produced diarrhea, altered behavior, and caused slight histological aberrations. Longer studies using male and female rats given 400 or 800 mg/day did not affect offspring. However, at the higher dosage levels, weight gains were lower and mortality was higher than in the controls (Lang and Bartsch, 1953). The 90-day feeding trials conducted on rats fed 0.1, 1.0, or 5.0% adipic acid showed no effect at the two lowest levels, but significant growth depression occurred at the 5.0% level even though the food intake level was the same. Sodium adipate at the 5% level decreased the rate of weight gain (Informatics, Inc., 1943). Two-year feeding trials using male rats fed 0.1, 1.0, 3.0, and 5% adipic acid and female rats fed 1.0% acid did not show any abnormalities; however, weight gains for rats fed the two highest levels were significantly lower than those for control rats (Horn et al., 1957).

Short-term studies on rats fed 0.2, 2.4, and 4.8% citric acid in a commercial diet showed lowered weight gains as a result of decreased intake of food and minor blood chemistry abnormalities at the 4.8% level; slight atrophy occurred in regions of the thymus and spleen (Yokotani et al., 1971). Conversely, long-term studies involving rats fed quantities up to 1.2, 3, or 5% citric acid (Bonting and Jansen, 1956; Horn et al., 1957) showed no abnormalities other than lowered weight gains and feed consumption in the 5% group. Cramer et al. (1956) fed rats vitamin D-free diets containing low calcium levels but adequate levels of phosphorus supplemented with 0.02 mol sodium citrate and citric acid. Although there was no effect on weight gain, the urine contained calcium citrate. The authors concluded that citrate had a rachitogenic effect on the test animals. Dalderup (1960) fed rats 1.5, 4.5, or 12 g citric acid per kg of a non-cariogenic diet to show the possible effect of citric acid on the development of teeth. The number of cavities formed did not differ between the control and experimental groups; however, the highest dosage contributed to enamel erosion.

The LD_{50} of lactic acid for test animals varied according to animal species (Table 5.3). Infants died after consuming milk acidified with an unknown quantity of lactic acid. The autopsy revealed hemorrhaging and gangrenous gastritis (Young and Smith, 1944). In two other instances, infants suffered stricture of the esophagus following ingestion of milk acidified with 1 teaspoon of 85% (Pitkin, 1935) or 87.5% lactic acid (Trainer et al., 1945). Premature infants fed acidified milk containing the isomeric forms of lactic acid, D (–) or DL, developed acidosis, weight loss, dehydration, and vomiting. Ballabriga et al. (1970) recommended that the L (+) form should be used in feeding premature infants. The Food and Agriculture Organization of the United Nations (Food and Agriculture Organization, 1973) supported this view. Hamsters fed a cariogenic diet incorporating lactic acid in the drinking water (40 mg/100 ml) or in the feed (45.6 mg/100 g) did not differ in growth rate compared with control groups (Granados et

TABLE 5.3

Reported Acute Toxicity Data (LD_{50}) for Various Acids

Acid	Animal	Route of Administration	LD_{50} (mg/kg body weight)	Reference
Acetic	Mouse	Oral	4960	Spector (1956)
		Intravenous	525	Orö and Wretland (1961)
	Rat	Oral	3310–3530	Spector (1956)
Sodium acetate	Mouse	Intravenous	380	Spector (1956)
Dehydroacetic acid	Rat	Oral	1000	Spencer et al. (1950)
Adipic	Mouse	Oral	1900	Horn et al. (1957)
		Intravenous	680	
		Intraperitoneal	275	
	Rabbit	Oral	2430–4860	Enders (1941)
Caprylic	Mouse	Intravenous	600	Orö and Wretland (1961)
Citric	Mouse	Oral	5040–5790	Yokotani et al. (1971)
		Intravenous	940–960	
		Intravenous	42	Gruber and Halbeisen (1948)
		Intravenous	203	Horn et al. (1957)
		Intraperitoneal	961	Gruber and Halbeisen (1948)
		Subcutaneous	2700	Yokotani et al. (1971)
	Rat	Oral	11,700	
		Intraperitoneal	725	
		Intraperitoneal	884	Gruber and Halbeisen (1948)
		Subcutaneous	5500	Yokotani et al. (1971)
	Rabbit	Intravenous	330	Gruber and Halbeisen (1948)
Sodium citrate	Mouse	Intravenous	44	Gruber and Halbeisen (1948)
		Intraperitoneal	1460	
	Rat	Intraperitoneal	1210	
	Rabbit	Intraperitoneal	338	
Lactic	Rat	Oral	3730	Spector (1956)
	Guinea pig	Oral	1810	
Propionic	Mouse	Intravenous	625	Orö and Wretland (1961)
Calcium propionate	Rat	Oral	3340	Hara et al. (1963)
		Intravenous	580–1020	
Sodium propionate	Rat	Oral	5100	Hara et al. (1963)
		Intravenous	1380–3200	
Tartaric	Mouse	Intravenous	584	Horn et al. (1957)

al., 1949). Similar to work described above focusing on citric acid, enamel decalcification was detected in experimental groups.

In contrast to results reported for citric and lactic acids, feeding trials involving fumaric and malic acids have not identified similar types of significant toxicological effects. Weanling rats fed 0.1, 0.5, 0.8, and 1.2% fumaric acid or rabbits fed sodium fumarate at a concentration of 6.9% (5% fumaric acid) showed no toxic effects (Packman et al., 1963). This contrasted with a previous study wherein at 1.5%, rats showed an increased mortality rate and more atrophy of the liver (Fitzhugh and Nelson, 1947). Two-year studies using rats fed 0.1 and 1.0% fumaric acid and 1.38% sodium fumarate or guinea pigs fed 1.0% fumaric acid did not show any abnormalities. Human feeding trials using a 500 mg/day oral dose for 1 year were also negative. Because of these dosage levels, an LD_{50} of 8000 mg/kg was proposed (Levey et al., 1946). Malic acid fed to rats at levels of 500 and 5000 ppm in the basal diet had no effect; at 50,000 ppm, however, food consumption decreased and growth rate declined (Hazelton Laboratories, 1971a). Dogs fed the same three levels of malic acid showed no effects using the same parameters (Hazelton

Laboratories, 1971b). Reproductive experiments using rats fed 1000 and 10,000 ppm malic acid before mating showed no significant differences from the controls (Hazelton Laboratories, 1970). Consumption of up to 2000 D- or L-malic acid by animals or humans in feed/food was reported to be metabolized safely without detectable toxicological effects (EFSA FEEDAP, 2014).

Orö and Wretland (1961) determined the LD_{50} of propionic acid for mice, and Hara et al. (1963) determined the LD_{50} of calcium and sodium propionate for rats (Table 5.3). Studies involving albino rats fed propionic acid at 50 cm^3/kg of rice for 110 days showed umbilicate or warty lesions on the stomach (Mori, 1953). Additional studies using calcium and sodium propionate fed to mice, rats, and humans showed no toxic effects (Graham et al., 1954; Graham and Grice, 1955; Hara, 1965; Harshbarger, 1942). There is evidence that the sodium salt has some local antihistaminic activity (Heseltine, 1952a).

Feeding studies for succinic acid feeding to rats indicated weight loss in excess feeding regimens (1900 mg/kg body weight) and elevated protein levels in urine, indicating nephrotoxicity, though no histopathology was identified to occur in renal cells (Hartwig and MAK Commission, 2018). Toxicity data in humans were not identified to be available by the commission. Male rabbits consuming a diet containing 7.7% sodium tartrate (5% tartaric acid) did not show any changes in food consumption, growth, mortality, or gross histology compared with controls (Packman et al., 1963). Fitzhugh and Nelson (1947) found similar results in long-term studies using 21-day-old weanling rats fed tartaric acid as 0.1, 0.5, 0.8, and 1.2% of the diet.

5.6.2 Irritation and Inhalational Exposure Toxicological Effects

With regards to its ability to cause irritation, acetic acid has been previously reported to produce symptoms when applied by vapor exposure. Rats exposed to vapor levels of acetic acid showed muscle imbalance at 0.2 or 5 mg/m^3 but not at 0.01 mg/m^3. Levels of 0.48 mg/m^3 altered sensitivity to light, and 0.29 mg/m^3 caused changes in electroencephalograms. Human olfactory thresholds to acetic acid vapor were 0.60 mg/m^3. Takhirov (1969) suggested limits of $\leq$0.06 mg/m^3. Parmeggiani and Sassi (1954) reported that workers exposed for 7–12 years to acetic acid vapors in the production of acetyl cellulose displayed corrosion of the hands, eyes, teeth, pharynx, and lungs with an air concentration of 0.2–0.65 mg/liter. An acute study of vapor exposure to 5 or 10 ppm acetic acid resulted in no effects on study participants on inflammatory markers, pulmonary function, and only a minimal impact on eye-blinking frequency, indicative of mild irritation (Ernstgård et al., 2006). Conversely, men having 7–25 years of exposure displayed blood abnormalities (Cesaro and Granata, 1955). In rare instances acetic acid, as well as citric, lactic, and tartaric acids, caused allergic reactions by producing canker sores (Tuft and Ettelson, 1956), cold sensitivities (Wiseman and Adler, 1956), and epidermal reactions (Weil and Rogers, 1951). Inhalation of adipic acid for 6 h at a rate of 126 μg/liter produced no unusual findings (Gage, 1970).

With respect to peracetic acid toxicity, Ferraris et al. (2005) reported no cytotoxic impacts on trout hepatocytes following treatment with 0.6 to 1.0 mg/L peracetic acid. Heptatocyte viability remained unchanged following PAA application at all doses from the control, while reactive oxygen species formation was also non-differing from the untreated control, indicating no acute toxicity by the disinfectant. Nevertheless, a group of researchers reported that some food process establishment workers could be exposed to significantly higher gaseous PAA than some recommended limits during food processing, at points including sanitizer solution preparation, beverage bottling, and, as might be expected, during facility surface sanitation (Dugheri et al., 2018).

5.6.3 Intravenous Exposure and Toxicological Effects

Acute toxicity data for adipic acid in mice and rabbits indicated varied tolerance levels by different routes of administration (Table 5.3). Intravenous and intraperitoneal injection produced hemorrhaging in the lungs, in addition to acidosis by the former route and intestinal adhesions by the latter route (Horn et al., 1957). Enders (1941) showed similar symptoms in rabbits, which included marked obstruction in the liver and kidneys by the oral route and polyuria by intravenous administration, resulting in the loss of up to 20% of body weight.

Horn et al. (1957) evaluated dosage levels in mice given citric acid intravenously. Deaths resulted from acute acidosis and lung hemorrhages. Acute toxicity data of sodium citrate are shown in Table 5.3. Subcutaneous injections of 320–1200 mg/kg body weight of sodium citrate into dogs decreased blood calcium levels while increasing calcium levels in urine (Gomori and Gulyas, 1944). In another study, 7.7% sodium citrate (5% citric acid) fed to rabbits for 60 days produced no unusual effects (Packman et al., 1963). Acute cytotoxicity in canine bone mesenchymal stem cells by citric acid exposure was reported; the LC_{50} was reported at 0.02 mg/mL, one of the lowest of the feed additives tested by researchers (Ortega et al., 2016). In other research by this group, the LC_{50} for citric acid for canine hepatocyte cells was 1.6 mg/mL following 24 h exposure (Zhang et al., 2015). Acute toxicity data for tartaric acid by the intravenous route for mice indicate an LD_{50} of 485 mg/kg body weight (Table 5.3). Rose's (1924) work on subcutaneous administration of 0.25–1 g to rabbits produced abnormalities in blood chemistry, including increased non-protein nitrogen, sugar, and cholesterol levels. In short-term studies, rats received subcutaneous injections of 0.5 mg succinic acid daily, gradually increasing to 2.0 mg/day at 4 weeks and continuing at this level for 100 days. The test animals did not differ significantly in reproduction rates, hair appearance, tooth eruption, or eye opening from the control group. Chick embryos developed normally when comparable dosages were injected into the air sac (Dye et al., 1944). A commission for determining safe work-related exposures to various chemicals set a maximum value for work-related inhalation exposure to succinic acid at 2.0 mg/m^3.

5.6.4 Mutagenic and Teratogenic Activities of Organic Acids

Teratological studies using mice, rats, and hamsters showed no significant difference from controls at levels of 263, 288, and 205 mg/kg body weight, respectively (Food and Drug Research Laboratory, 1973b). More recent research has focused on the evaluation of toxicity of adipic acid-derived biopolymers and structures. Hydrogels constructed of multiple components, including adipic acid dihydrazide, did not induce any observed systemic toxicity in experimental rodents. Hemo-toxicity markers for hydrogel-treated animals did not differ from control animals, but still functioned to assist in bone repair for treated animals (Peireira et al., 2019). Studies measuring teratogenic activity of tartaric acid in pregnant mice, rats, hamsters, and rabbits demonstrated normal offspring at levels of 274, 181, 225, and 215 mg/kg body weight, respectively (Food and Drug Research Laboratory, 1973a). Workers exposed to tartaric acid dust measuring up to 32 mg/m^3 in a manufacturing plant that produced the acid experienced tooth erosion. The workers also displayed skin eruptions, which disappeared after they left the work site (Barsotti et al., 1954).

5.7 Assay Methods

The various acids can be assayed by relatively simple procedures. Methods for determining acid content in foods are found in the *Official Methods of Analysis of the Association of Analytical Chemists* (Horowitz, 2000) or the *Handbook of Food Analysis* (Nollet, 1996).

Volatile acids can be separated from foods using steam distillation. If only one acid is present, the distillate can be titrated for the particular acid. For a mixture of acids, the distillate is separated on a silicic acid column and the component acids can be identified using enzymatic, chromatographic (gas, paper, thin-layer, or high-performance liquid chromatography (HPLC)), or electrochemical methods. Qualitative or quantitative methods can be used depending upon the degree of sensitivity required. Often derivatives of the acids are required, which adds to sample preparation time (Gomis and Alonso, 1996).

REFERENCES

Acuff, G. R., Vanderzant, C., Savell, J. W., Jones, D. K., Griffin, D. B., and Ehlers, J. G. 1987. Effect of acid decontamination of beef subprimal cuts on the microbiological and sensory characteristics of steaks. *Meat Sci.* 19:217–226.

Adams, M. R., and Hall, C. J. 1988. Growth inhibition of food-borne pathogens by lactic and acetic acids and their mixtures. *Int. J. Food Sci. Technol.* 23:287–292.

Adams, M. R., Little, C. L., and Easter, M. C. 1991. Modelling the effect of pH, acidulant and temperature on the growth rate of *Yersinia enterocolitica*. *J. Appl. Bacteriol.* 71:65–71.

Adler, J. M., Cain-Helfrich, E. D., and Shen, C. 2016. Reductions in natural microbial flora, nonpathogenic *Escherichia coli*, and pathogenic *Salmonella* on jalapeno peppers processed in a commercial antimicrobial cabinet: A pilot plant trial. *J. Food Prot.* 79:1854–1859.

Ahamad, N., and Marth, E. H. 1989. Behavior of *Listeria monocytogenes* at 7, 13, 21, and 35°C in tryptose broth acidified with acetic, citric, or lactic acid. *J. Food Prot.* 52:688–695.

Ahamad, N., and Marth, E. H. 1990. Acid-injury of *Listeria monocytogenes*. *J. Food Prot.* 53:26–29.

Akbas, M. Y., and Ölmez, H. 2007. Inactivation of *Escherichia coli* and *Listeria monocytogenes* on iceberg lettuce by dip wash treatments with organic acids. *Lett. Appl. Microbiol.* 44:619–624.

Alakomi, H.-L., Skyttä, E., Saarela, M., Mattila-Sandholm, T., Latva-Kala, K., and Helander, I. M. 2000. Lactic acid permeabilizes gram-negative bacteria by disrupting the outer membrane. *Appl. Environ. Microbiol.* 66:2001–2005.

Al-Dagal, M. M., and Bazaraa, W. A. 1999. Extension of shelf life of whole and peeled shrimp with organic acid salts and bifidobacteria. *J. Food Prot.* 62:51–56.

Al-Sheddy, I., Al-Dagal, M. M., and Bazaraa, W. A. 1999. Microbial and sensory quality of fresh camel meat treated with organic acid salts and/or *Bifidobacteria*. *J. Food Sci.* 64:336–339.

Alves de Ávila, A. R., Marques, S. C., Piccolli, R. H., and Schwan, R. F. 2013. Sensitivity to organic acids *in vitro* and *in situ* of *Salmonella* spp. and *Escherichia coli* isolated from fresh pork sausages. *J. Food Qual.* 36:155–163.

Anderson, M. E., and Marshall, R. T. 1989. Interaction of concentration and temperature of acetic acid solution in reduction of various species of microorganisms on beef surfaces. *J. Food Prot.* 52:312–315.

Anderson, M. E., and Marshall, R. T. 1990a. Reducing microbial populations on beef tissues: Concentration and temperature of lactic acid. *J. Food Saf.* 10:181–190.

Anderson, M. E., and Marshall, R. T. 1990b. Reducing microbial populations on beef tissues: Concentration and temperature of an acid mixture. *J. Food Sci.* 55:903–905.

Anderson, M. E., Huff, H. E., Naumann, H. D., Marshall, R. T., Damare, J. M., Pratt, M., and Johnston, R. 1987. Evaluation of an automated beef carcass washing and sanitizing system under production conditions. *J. Food. Prot.* 50:562–566.

Anderson, M. E., Marshall, R. T., and Dickson, J. S. 1992. Efficacies of acetic, lactic and two mixed acids in reducing numbers of bacteria on surfaces of lean meat. *J. Food Saf.* 12:139–147.

Anderson, M. E., Marshall, R. T., Stringer, W. C., and Naumann, H. D. 1977. Combined and individual effects of washing and sanitizing on bacterial counts of meat – A model system. *J. Food Prot.* 40:688–690.

Appl, C. P., and Marshall, R. T. 1984. Detachment of *Pseudomonas fluorescens* P26 from beef rinsed in salt and acid solutions. *J. Food Prot.* 47:537–541.

Ariyapitipun, T., Mustapha, A., and Clarke, A. D. 1999. Microbial shelf life determination of vacuum-packaged fresh beef treated with polylactic acid, lactic acid, and nisin solutions. *J. Food Prot.* 62:913–920.

Ariyapitipun, T., Mustapha, A., and Clarke, A. D. 2000. Survival of *Listeria monocytogenes* Scott A on vacuum-packaged raw beef treated with polylactic acid, lactic acid, and nisin. *J. Food Prot.* 63:131–136.

Aymerich, T., Jofre, A., Garriga, M., and Hugas, M. 2005. Inhibition of *Listeria monocytogenes* and *Salmonella* by natural antimicrobials and high hydrostatic pressure in sliced cooked ham. *J. Food Prot.* 68:173–177.

Ayres, J. C., Mundt, J. O., and Sandine, W. E. 1980. *Microbiology of Foods.* W. H. Freeman, San Francisco.

Bae, Y.-M., and Lee, S.-Y. 2015. Combined effects of organic acids and salt depending on type of acids and pathogens in laboratory media and acidified pickle. *J. Appl. Microbiol.* 119:455–464.

Baird-Parker, A. C. 1980. Organic acids. In *Microbial Ecology of Foods*, Vol. 1. *Factors Affecting Life and Death of Microorganisms.* Edited by the International Commission on Microbiological Specifications for Foods. Academic Press, New York, p. 126.

Ballabriga, A., Conde, C., and Gallart-Catala, A. 1970. Metabolic response of prematures to milk formulas with different lactic acid isomers or citric acid. *Helv. Paediatr. Acta* 25:25–34.

Banwart, G. J. 1989. *Basic Food Microbiology.* 2nd ed. Van Nostrand Reinhold, New York.

Barsotti, M., Sassi, C., and Ghetti, G. 1954. Health hazards to workers in a tartaric acid factory. *Med. Lav.* 45:239–243.

Bedie, G. K., Samelis, J., Sofos, J. N., Belk, K. E., Scanga, J. A., and Smith, G. C. 2001. Antimicrobials in the formulation to control *Listeria monocytogenes* postprocessing contamination on frankfurters stored at 4°C in vacuum packages. *J. Food Prot.* 64:1949–1955.

Beelman, R. B., Witowski, M. E., Doores, S., Kilara, A., and Kuhn, G. D. 1989. Acidification process technology to control thermophilic spoilage in canned mushrooms. *J. Food Prot.* 52:178–183.

Bell, M. F., Marshall, R. T., and Anderson, M. E. 1986. Microbiological and sensory tests of beef treated with acetic and formic acids. *J. Food Prot.* 49:207–210.

Beuchat, L. R. 1980. Comparison of anti-*Vibrio* activities of potassium sorbate, sodium benzoate and glycerol and sucrose esters of fatty acids. *Appl. Environ. Microbiol.* 39:1178–1182.

Beuchat, L. R. 1988. Influence of organic acids on heat resistance characteristics of *Talaromyces flavus* ascospores. *Int. J. Food Microbiol.* 6:97–105.

Biemuller, G. W., Carpenter, J. A., and Reynolds, A. E. 1973. Reduction of bacteria on pork carcasses. *J. Food Sci.* 38:261–263.

Bizri, J. N., and Wahem, I. A. 1994. Citric acid and antimicrobials affect microbiological stability and quality of tomato juice. *J. Food Sci.* 59:130–134.

Blankenship, L. C. 1981. Some characteristics of acid injury and recovery of *Salmonella bareilly* in a model system. *J. Food Prot.* 44:73–77.

Blom, H., Nerbrink, E., Dainty, R., Hagtvedt, T., Borch, E., Nissen, H., and Nesbakken, T. 1997. Addition of 2.5% lactate and 0.25% acetate controls growth of *Listeria monocytogenes* in vacuum-packed sensory-acceptable servelat sausage and cooked ham stored at 4°C. *Int. J. Food Microbiol.* 38:71–76.

Bloukas, J. G., Paneras, E. D., and Fournitzis, G. C. 1997. Sodium lactate and protective culture effects on quality characteristics and shelf-life of low-fat frankfurters produced with olive oil. *Meat Sci.* 45:223–238.

Bolton, D. J., Catarame, T., Byrne, C., Sheridan, J. J., McDowell, D. A., and Blair, I. S. 2002. The ineffectiveness of organic acids, freezing and pulse fields to control *Escherichia coli* O157:H7 in beef burgers. *Lett. Appl. Microbiol.* 34:139–143.

Bonting, S. L., and Jansen, B. C. P. 1956. The effect of a prolonged intake of phosphoric acid and citric acid in rats. *Voeding* 17:137–148.

Brackett, R. E. 1987. Effects of various acids on growth and survival of *Yersinia enterocolitica. J. Food Prot.* 50:598–601.

Brackett, R. E., Hao, Y.-Y., and Doyle, M. P. 1994. Ineffectiveness of hot acid sprays to decontaminate *Escherichia coli* O157:H7 on beef. *J. Food Prot.* 57:198–203.

Bradford, D. D., Huffman, D. L., Egbert, W. R., and Mikel, W. B. 1993. Potassium lactate effects on low-fat fresh pork sausage chubs during simulated retail distribution. *J. Food Sci.* 58:1245–1253.

Branen, A. L., and Keenan, T. W. 1970. Growth stimulation of *Lactobacillus casei* by sodium citrate. *J. Dairy Sci.* 53:593–597.

Brewer, M. S., Rostogi, B. K., Argoudelis, L., and Sprouls, G. K. 1995. Sodium lactate/sodium chloride effects on aerobic plate counts and color of aerobically packaged ground pork. *J. Food Sci.* 60:58–62.

Brewer, S. M., McKeith, F., Martin, S. E., Dallmier, A. W., and Meyer, J. 1991. Sodium lactate effects on shelf-life, sensory and physical characteristics of fresh pork sausage. *J. Food Sci.* 56:1176–1178.

Brocklehurst, T. F., and Lund, B. M. 1990. The influence of pH, temperature and organic acids on the initiation of growth of *Yersinia enterocolitica. J. Appl. Bacteriol.* 69:390–397.

Brul, S., and Coote, P. 1999. Preservative agents in foods. *Int. J. Food Microbiol.* 50:1–17.

Brul, S., Coote, P., Oomes, S., Mensonides, F., Hellingwerf, K., Klis, F. 2002. Physiological actions of preservative agents: Prospective of use of modern microbiological techniques in assessing microbial behaviour in food preservation. *Int. J. Food Microbiol.* 79:55–64.

Buazzi, M. M., and Marth, E. H. 1992. Sites of action by propionate on *Listeria monocytogenes. Int. J. Food Microbiol.* 15:109–119.

Buchanan, R. L., and Ayres, J. C. 1976. Effect of sodium acetate on growth and aflatoxin production by *Aspergillus parasiticus* NRRL 2999. *J. Food Sci.* 41:128–132.

Buchanan, R. L., and Golden, M. H. 1994. Interaction of citric acid concentration and pH on the kinetics of *Listeria monocytogenes* inactivation. *J. Food Prot.* 57:567–570.

Buchanan, R. L., Edelson, S. G., and Boyd, G. 1999. Effects of pH and acid resistance on the radiation resistance of Enterohemorrhagic *Escherichia coli. J. Food Prot.* 62:219–228.

Buchanan, R. L., Golden, M. H., and Whiting, R. C. 1993. Differentiation of the effects of pH and lactic or acetic acid concentration on the kinetics of *Listeria monocytogenes* inactivation. *J. Food Prot.* 56:474–484.

Byelashov, O.A., Daskalov, H., Geornaras, I., Kendall, P.A., Belk, K.E., Scanga, J.A., Smith, G.C., Sofos, J.N. 2010. Reduction of *Listeria monocytogenes* on frankfurters treated with lactic acid solutions of various temperatures. *Food Microbiol.* 27: 783–790.

Cacciarelli, M. A., Stringer, W. C., Anderson, M. E., and Naumann, H. D. 1983. Effects of washing and sanitizing on the bacterial flora of vacuum-packaged pork loins. *J. Food Prot.* 46:231–234.

Carlson, B. A., Ruby, J., Smith, G. C., Sofos, J. N., Bellinger, G. R., Warren-Serna, W., Centrella, B., Bowling, R. A., and Belk, K. E. 2008. Comparison of antimicrobial efficacy of multiple beef hide decontamination strategies to reduce levels of *Escherichia coli* O157:H7 and *Salmonella*. *J. Food Prot.* 71:2223–2227.

Casey, P. G., and Condon, S. 2002. Sodium chloride decreases the bacteriocidal effect of acid pH on *Escherichia coli* O157:H7. *Int. J. Food Microbiol.* 76:199–206.

Castelo, M. M., Kang, D.-H., Siragusa, G. R., Koohmaraie, M., and Berry, E. D. 2001. Evaluation of combination treatment processes for the microbial decontamination of pork trim. *J. Food Prot.* 64:335–342.

Castellano, P., Pẽna, N., Ibarreche, M. P., Carduza, F., Soteras, T., and Vignolo, G. 2018. Antilisterial activity of *Lactobacillus* bacteriocins and organic acids on frankfurters. Impact on sensory characteristics. *J. Food Sci. Technol.* 55:689–697.

Castillo, A., Lucia, L. M., Goodson, K. J., Savell, J. W., and Acuff, G. R. 1998. Comparison of water wash, trimming, and combined hot water and lactic acid treatments for reducing bacteria of fecal origin on beef carcasses. *J. Food Prot.* 61:823–828.

Castillo, A., Lucia, L. M., Mercado, I., and Acuff, G. R. 2001b. In-plant evaluation of a lactic acid treatment for reduction of bacteria on chilled beef carcasses. *J. Food Prot.* 64:738–740.

Castillo, A., Lucia, L. M., Roberson, D. B., Stevenson, T. H., Mercado, I., and Acuff, G. R. 2001a. Lactic acid sprays reduce bacterial pathogens on cold beef carcass surfaces and in subsequently produced ground beef. *J. Food Prot.* 64:58–62.

Cesaro, A. N., and Granata, A. 1955. Hematologic cytochemistry in subjects exposed to occupational risk of intoxication due to acetone and to acetic acid. *Folia Med.* 38:957.

Chang, S.-S., Redondo-Solano, M., and Thippareddi, H. 2010. Inactivation of *Escherichia coli* O157:H7 and *Salmonella* spp. on alfalfa seeds by caprylic acid and monocaprylin. *Int. J. Food Microbiol.* 144:141–146.

Chen, N., and Shelef, L. A. 1992. Relationship between water activity, salts of lactic acid, and growth of *Listeria monocytogenes* in a meat model system. *J. Food Prot.* 55:574–578.

Cheng, C.-M., and Kaspar, C. W. 1998. Growth and processing conditions affecting acid tolerance in *Escherichia coli* O157:H7. *Food Microbiol.* 15:157–166.

Cherrington, C. A., Hinton, M., and Chopra, I. 1990. Effect of short-chain organic acids on macromolecular synthesis in *Escherichia coli*. *J. Bacteriol.* 68:69–74.

Cherrington, C. A., Hinton, M., Pearson, G. R., and Chopra, I. 1991a. Inhibition of *Escherichia coli* K12 by short-chain organic acids: Lack of evidence of the SOS response. *J. Appl. Bacteriol.* 70:156–160.

Cherrington, C. A., Hinton, M., Pearson, G. R., and Chopra, I. 1991b. Short-chain organic acids at pH 5.0 kill *Escherichia coli* and *Salmonella* spp. without causing membrane perturbation. *J. Appl. Bacteriol.* 70:161–165.

Cherrington, C.A., Allen, V. and Hinton, M. 1992. The influence of temperature and organic matter on the bactericidal activity of short-chain organic acids on salmonellas. *J. Appl. Bacteriol.* 72:500–503.

Chichester, D. F., and Tanner, F. W. Jr. 1972. Antimicrobial food additives. In *Handbook of Food Additives*. 2nd ed. Edited by T. E. Furia. CRC Press, Cleveland, OH, p. 115.

Choi, Y. M., Kim, O. Y., Kim, K. H., Kim, B. C., and Rhee, M. S. 2009. Combined effect of organic acids and supercritical carbon dioxide treatments against nonpathogenic *Escherichia coli, Listeria monocytogenes, Salmonella typhimurium*, and *E. coli* O157:H7. *Lett. Appl. Microbiol.* 49:510–515.

Chung, K. C., and Goepfert, J. M. 1970. Growth of *Salmonella* at low pH. *J. Food Sci.* 35:326–328.

Code of Federal Regulations. 2001. *Code of Federal Regulations Title 21*. Office of the Federal Register, U.S. Government Printing Office, Washington, DC.

Cole, M. B., Jones, M. V., and Holyoak, C. 1990. The effect of pH, salt concentration and temperature on the survival and growth of *Listeria monocytogenes*. *J. Appl. Bacteriol.* 69:63–72.

Comes, J. E., and Beelman, R. B. 2002. Addition of fumaric acid and sodium benzoate as an alternative method to achieve a 5-log reduction of *Escherichia coli* O157:H7 populations in apple cider. *J. Food Prot.* 65:476–483.

Conner, D. E., and Beuchat, L. R. 1987. Heat resistance of ascospores of *Neosartorya fischeri* as affected by sporulation and heating medium. *Int. J. Food Microbiol.* 4:303–312.

Conner, D. E., and Kotrola, J. S. 1995. Growth and survival of *Escherichia coli* O157:H7 under acidic conditions. *Appl. Environ. Microbiol.* 61:382–385.

Conner, D. E., Kotrola, J. S., Mikel, W. B., and Tamblyn, K. C. 1997. Effects of acetic –lactic acid treatments applied to beef trim on populations of *Escherichia coli* O157:H7 and *Listeria monocytogenes* in ground beef. *J. Food Prot.* 60:1560–1563.

Conner, D. E., Scott, V. N., and Bernard, D. T. 1990. Growth, inhibition, and survival of *Listeria monocytogenes* as affected by acidic conditions. *J. Food Sci.* 53:652–655.

Corlett, D. A., Jr., and Brown, M. H. 1980. pH and acidity. In *Microbial Ecology of Foods*, Vol. 1. Edited by J. H. Silliker. Academic Press, Inc., New York, p. 92.

Cox, N. A., Mercuri, A. J., Juven, B. J., Thomson, J. E., and Chew, V. 1974. Evaluation of succinic acid and heat to improve the microbiological quality of poultry meat. *J. Food Sci.* 39:985–987.

Cramer, J. W., Porrata-Doria, E. I., and Steenbock, H. 1956. A rachitogenic and growth-promoting effect of citrate. *Arch. Biochem. Biophys.* 60:58–63.

Cruess, W. V., and Irish, J. H. 1932. Further observations on the relation of pH value to toxicity of preservatives to microorganisms. *J. Bacteriol.* 23:163–166.

Cudjoe, K. S. 1988. The effect of lactic acid sprays on the keeping qualities of meat during storage. *Int. J. Food Microbiol.* 7:1–7.

Cutter, C. N., and Siragusa, G. R. 1994. Efficacy of organic acids against *Escherichia coli* O157:H7 attached to beef carcass tissue using a pilot scale model carcass washer. *J. Food Prot.* 57:97–103.

Cutter, C. N., Dorsa, W. J., and Siragusa, G. R. 1997. Parameters affecting the efficacy of spray washes against *Escherichia coli* O157:H7 and fecal contamination on beef. *J. Food Prot.* 60:614–618.

Dalderup, L. M. 1960. The effects of citric and phosphoric acids on the teeth. *J. Dent. Res.* 39:420.

Daly, C., LaChance, M., Sandine, W. E., and Elliker, P. R. 1973. Control of *Staphylococcus aureus* in sausage by starter cultures and chemical acidulation. *J. Food Sci.* 38:426–430.

Davidson, P. M. and Harrison, M. A. 2002. Resistance and adaptation to food antimicrobials, sanitizers, and other process controls. *Food Technol.* 56:69–78.

Davis, F., and Barnes, L. A. 1952. Suppression of growth of *Salmonella anatum* and *Salmonella oranienburg* by concentration variation of energy sources in a synthetic basal medium. *J. Bacteriol.* 63:33–38.

Dawes, I. W., and Sutherland, I. W. 1992. *Microbial Physiology*. 2nd ed. Blackwell Scientific Publications, New York.

Deane, D., and Downs, P. A. 1951. Flexible wrappers for cheddar cheese. *J. Dairy Sci.* 34:767–775.

Debevere, J. M. 1988. Effect of buffered acidulant systems on the survival of some food poisoning bacteria in medium acid media. *Food Microbiol.* 5:135–139.

Degnan, A. J., Kaspar, C. W., Otwell, W. S., Tamplin, M. L., and Luchansky, J. B. 1994. Evaluation of lactic acid bacterium fermentation products and food-grade chemicals to control *Listeria monocytogenes* in blue crab (*Callinectes sapidus*) meat. *Appl. Environ. Microbiol.* 60:3198–3203.

Delaquis, P. J., Sholberg, P. L., and Stanich, K. 1999. Disinfection of mung bean seed with gaseous acetic acid. *J. Food Prot.* 62:953–957.

Delcanton, F., Carrasco, E., Pérez-Rodríguez, F., Posada-Izquierdo, G. D., Falcão de Aragão, G. M., and García-Gimeno, R. M. 2018. Modeling the combined effects of temperature, pH, sodium chloride and sodium lactate concentrations on the growth rate of *Lactobacillus plantarum* ATCC 8014. *J. Food Qual.* 2018:1726761.

de Wit, J. C., and Rombouts, F. M. 1990. Antimicrobial activity of sodium lactate. *Food Microbiol.* 7:113–120.

Dickens, J. A., and Whittemore, A. D. 1994. The effect of acetic acid and air injection on appearance, moisture pick-up, microbiological quality, and *Salmonella* incidence on processed poultry carcasses. *Poult. Sci.* 73:582–586.

Dickens, J. A., Lyon, B. G., Whittemore, A. D., and Lyon, C. E. 1994. The effect of an acetic acid dip on carcass appearance, microbiological quality, and cooked breast meat texture and flavor. *Poult. Sci.* 73:576–581.

Dickson, J. S. 1992. Acetic acid action on beef tissue surfaces contaminated with *Salmonella typhimurium*. *J. Food Sci.* 57:297–301.

Dickson, J. S., and Anderson, M. E. 1992. Microbiological decontamination of food animal carcasses by washing and sanitizing systems: A review. *J. Food Prot.* 55:133–140.

Dickson, J. S., and Koohmaraie, M. 1989. Cell surface charge characteristics and their relationship to bacterial attachment to meat surfaces. *Appl. Environ. Microbiol.* 55:832–836.

Dickson, J. S., and Kunduru, M. R. 1995. Resistance of acid-adapted salmonellae to organic acid rinses on beef. *J. Food Prot.* 58:973–976.

Dixon, Z. R., Vanderzant, C., Acuff, G. R., Savell, J. W., and Jones, D. K. 1987. Effect of acid treatment of beef strip loin steaks on microbiological and sensory characteristics. *Int. J. Food Microbiol.* 5:181–186.

Doores, S. 2002. pH control agents and acidulants. In *Food Additives*. 2nd ed. Edited by A. L. Branen, P. M. Davidson, S. Salminen, and J. H. Thorngate III. Marcel Dekker, Inc., New York.

Doores, S. 2003. Acidulants. In *Food Additives Databook*. Edited by J. Smith, and L. H. Hong-Shum. Blackwell Publishing, Oxford, England.

Dorsa, W. J. 1997. New and established carcass decontamination procedures commonly used in the beef-processing industry. *J. Food Prot.* 60:1146–1151.

Dorsa, W. J., Cutter, C. N., and Siragusa, G. R. 1997. Effects of acetic acid, lactic acid and trisodium phosphate on the microflora of refrigerated beef carcass surface tissue inoculated with *Escherichia coli* O157:H7, *Listeria innocua*, and *Clostridium sporogenes*. *J. Food Prot.* 60:619–624.

Dorsa, W. J., Cutter, C. N., and Siragusa, G. R. 1998. Long-term bacterial profile of refrigerated ground beef made from carcass tissue, experimentally contaminated with pathogens and spoilage bacteria after hot water, alkaline, or organic acid washes. *J. Food Prot.* 61:1615–1622.

Dorsa, W. J., Marshall, D. L., and Semien, M. 1993. Effect of potassium sorbate and citric acid sprays on growth of *Listeria monocytogenes* on cooked crawfish (*Procambarus clarkii*) tail meat at 4°C. *Lebensm. Wiss. U. Technol.* 26:480–482.

Dubos, R. J. 1950. The effect of organic acids on mammalian tubercle bacilli. *J. Exp. Med.* 92:319–332.

Dugheri, S., Bonari, A., Pompilio, I., Colpo, M., Montalti, M., Mucci, N., and Arcangeli, G. 2018. Assessment of occupational exposure to gaseous peracetic acid. *Int. J. Occ. Med. Environ. Health.* 31:527–535.

Dye, W. S., Overholse, M. D., and Vinson, C. G. 1944. Injections of certain plant growth substances in rats and chick embryos. *Growth* 8:1–11.

Dymicky, M., and Trenchard, H. 1982. Inhibition of *Clostridium botulinum* 62A by saturated n-aliphatic acids, n-alkyl formates, acetates, propionates and butyrates. *J. Food Prot.* 45:1117–1119.

Dymicky, M., Bencivengo, M., Buchanan, R. L., and Smith, J. L. 1987. Inhibition of *Clostridium botulinum* 62A by fumarates and maleates and relationship of activity to some physicochemical constants. *Appl. Environ. Microbiol.* 53:110–113.

Egbert, W. R., Huffman, D. L., Bradford, D. D., and Jones, W. R. 1992. Properties of low-fat ground beef containing potassium lactate during aerobic refrigerated storage. *J. Food Sci.* 57:1033–1037.

Eggenberger-Solorzano, L., Niebuhr, S. E., Acuff, G. R., and Dickson, J. S. 2002. Hot water and organic acid interventions to control microbiological contamination on hog carcasses during processing. *J. Food Prot.* 65:1248–1252.

Eilers, J. D., Morgan, J. B., Martin, A. M., Miller, R. K., Hale, D. S., Acuff, G. R., and Savell, J. W. 1994. Evaluation of calcium chloride and lactic acid injection on chemical, microbiological and descriptive attributes of mature cow beef. *Meat Sci.* 38:443–451.

Eklund, T. 1980. Inhibition of growth and uptake processes in bacteria by some chemical food preservatives. *J. Appl. Bacteriol.* 48:423–432.

Eklund, T. 1983. The antimicrobial effect of dissociated and undissociated sorbic acid at different pH levels. *J. Appl. Bacteriol.* 54:383–389.

Eklund, T. 1985. Inhibition of microbial growth at different pH levels by benzoic and propionic acids and esters of *p*-hydroxy-benzoic acid. *Int. J. Food Microbiol.* 2:159–167.

El-Banna, A. A., and Hurst, A. 1983. Survival in foods of *Staphylococcus aureus* grown under optimal and stressed conditions and the effect of some food preservatives. *Can. J. Food Microbiol.* 29:297–302.

El-Gazzar, F. E., Rusul, G., and Marth, E. H. 1987. Growth and aflatoxin production by *Aspergillus parasiticus* NRRL 2999 in the presence of lactic acid and at different initial pH values. *J. Food Prot.* 50:940–944.

El-Khateib, T., Yousef, A. E., and Ockerman, H. W. 1993. Inactivation and attachment of *Listeria monocytogenes* on beef muscle treated with lactic acid and selected bacteriocins. *J. Food Prot.* 56:29–33.

Ellebracht, E. A., Castillo, A., Lucia, L. M., Miller, R. K., and Acuff, G. R. 1999. Reduction of pathogens using hot water and lactic acid on beef trimmings. *J. Food Sci.* 64:1094–1099.

El-Shenawy, M. A., and Marth, E. H. 1989a. Behavior of *Listeria monocytogenes* in the presence of sodium propionate. *Int. J. Food Microbiol.* 8:85–94.

El-Shenawy, M. A., and Marth, E. H. 1989b. Inhibition or inactivation of *Listeria monocytogenes* by sodium benzoate together with some organic acids. *J. Food Prot.* 52:771–776.

El-Shenawy, M. A., and Marth, E. H. 1992. Behavior of *Listeria monocytogenes* in the presence of sodium propionate together with food acids. *J. Food Prot.* 55:241–245.

Enders, A. 1941. Physiological compatibility and excretion of dicarboxylic acids. *Arch. Exp. Pathol. Pharmakol.* 197:597–601.

Epling, L. K., Carpenter, J. A., and Blankenship, L. C. 1993. Prevalence of *Campylobacter* spp. and *Salmonella* spp. on pork carcasses and the reduction effected by spraying with lactic acid. *J. Food Prot.* 56:536–537, 540.

Ernstgård, L., Iregren, A., Sjögren, B., and Johanson, G. 2006. Acute effects of exposure to vapours of acetic acid in humans. *Toxicol. Lett.* 165:22–30.

European Food Safety Authority (EFSA) Panel on Additives and Products or Substances used in Animal Feed (FEEDAP). 2014. Scientific opinion on the safety and efficacy of malic acid and a mixture of sodium and calcium malate when used as technological additives for all animal specie. *EFSA J.* 12:3563.

European Food Safety Authority (EFSA) Panel on Food Contact Materials, Enzymes and Processing Aids. 2018. Evaluation of the safety and efficacy of the organic acids lactic and acetic acids to reduce microbiological surface contamination on pork carcasses and pork cuts. *EFSA J.* 16:5482.

Eytan, O., Weinert, I. A. G., and McGill, A. E. J. 1992. Effect of salad dressing and citric acid dip on storage quality of shredded cabbage and carrots packed under modified atmospheres. *Lebensm. Wiss. U. Technol.* 25:445–450.

Fabian, F. W., and Graham, H. T. 1953. Viability of thermophilic bacteria in the presence of varying concentrations of acids, sodium chloride and sugars. *Food Technol.* 7:212–217.

Fabian, F. W., and Wadsworth, C. K. 1939. Experimental work on lactic acid in preserving pickles and pickle products. II. Preserving value of acetic acid and lactic acid in the presence of sucrose. *Food Res.* 4:511–519.

Fabrizio, K. A., Sharma, R. R., Demirci, A., and Cutter, C. N. 2002. Comparison of electrolyzed oxidizing water with various antimicrobial interventions to reduce *Salmonella* speces on poultry. *Poult. Sci.* 81:1595–1605.

Federal Register. 1962. Food additives. Boiler water additives. *Fed. Reg.* 27:6232.

Ferraris, M., Chiesara, E., Radice, S., Giovara, A., Frigerio, S., Fumagalli, R., and Marabini, L. 2005. Study of potential toxic effects on rainbow trout hepatocytes of surface water treated with chlorine or alternative disinfectants. *Chemosphere* 60:65–73.

Fin'ko, L. N. 1969. Blood coagulation changes in acetic acid poisoning. *Klin. Med. (Mosk)* 47:84–87.

Fischer, J. R., Fletcher, D. L., Cox, N. A., and Bailey, J. S. 1985. Microbiological properties of hard-cooked eggs in a citric acid-based preservative solution. *J. Food Prot.* 48:252–256.

Fitzhugh, O. G., and Nelson, A. A. 1947. The comparative chronic toxicities of fumaric, tartaric, oxalic and maleic acids. *J. Am. Pharm. Assoc.* 36:217–219.

Food and Agriculture Organization. 1962. Evaluation of the toxicity of a number of antimicrobials and antioxidants. 6th Report of the Joint Food and Agriculture Organization of the United Nations/World Health Organization Expert Committee on Food Additives. WHO Technical Report Series No. 228. FAO Nutrition Meetings Report Series No. 31.

Food and Agriculture Organization. 1963. Specifications for the identity and purity of food additives and their toxicological evaluation: Emulsifiers, stabilizers, bleaching and maturing agents. 7th Report of the Joint Food and Agriculture Organization of the United Nations/World Health Organization Expert Committee on Food Additives. WHO Technical Report Series No. 281. FAO Nutrition Meetings Report Series No. 35.

Food and Agriculture Organization. 1973. Toxicological evaluation of certain food additives with a review of general principles and of specifications. 17th Report of the Joint Food and Agriculture Organization of the United Nations/World Health Organization WHO Expert Committee on Food Additives. WHO Technical Series No. 539. FAO Nutrition Meetings Report Series No. 53.

Ninth Report of the Joint FAO/WHO Expert Committee on Food Additives. 1966. *Wld. Hlth. Org. Techn. Rep. Ser.*, 1966, 339.

Food and Drug Research Laboratory, Inc. 1973a. *NAS/NRC Questionnaire on Tartaric Acid, Contract 71–75.* Food and Drug Administration, Rockville, MD.

Food and Drug Research Laboratory, Inc. 1973b. *Teratological Evaluation of F D A 71–50 (Adipic Acid) in Mice, Rats and Hamsters.* Food and Drug Administration, Rockville, MD.

Frederick, T. L., Miller, M. F., Thompson, L. D., and Ramsey, C. B. 1994. Microbiological properties of pork cheek meat as affected by acetic acid and temperature. *J. Food Sci.* 59:300–305.

Freese, E., and Levine, B. C. 1978. Action mechanisms of preservatives and antiseptics. In *Developments in Industrial Microbiology*, Vol. 19. Edited by L. A. Underkofler. American Institute of Biological Sciences, Washington, DC, p. 207.

Freese, E., Sheu, C. W., and Galliers, E. 1973. Function of lipophilic acids as antimicrobial food additives. *Nature* 241:321–325.

Fu, A.-H., Sebranek, J. G., and Murano, E. A. 1994. Microbial and quality characteristics of pork cuts from carcasses treated with sanitizing sprays. *J. Food Sci.* 59:306–309.

Gage, J. C. 1970. The subacute inhalation toxicity of 109 industrial chemicals. *Br. J. Ind. Med.* 27:1–18.

Gagiu, V., Israel-Roming, F., Pelinescu, D., Dobre, A. A., Cucu, M. E., Smeu, I., Iorga, E., Cornea, C. P., Stoica, I., and Belc, N. 2017. Synergistic effects of the biological and chemical preservatives in dough fermentations. *Rom. Biotechnol. Lett.* 22:12915–12925.

Gahan, C. G., O'Driscoll, B., and Hill, C. 1996. Acid adaptation of *Listeria monocytogenes* can enhance survival in acidic foods and during milk fermentation. *Appl. Environ. Microbiol.* 62:3128–3132.

Gardner, N., Champagne, C. P., and Gélinas, P. 2001. Effect of yeast extracts containing propionic acid on bread dough fermentation and bread properties. *J. Food Sci.* 67:1855–1858.

Gardner, W. H. 1966. *Food Acidulants*. Allied Chemical Corporation, New York.

Gardner, W. H. 1972. Acidulants in food processing. In *Handbook of Food Additives*. 2nd ed. Edited by T. E. Furia. CRC Press, Cleveland, OH, p. 225.

Garren, D. M., Harrison, M. A., and Russell, S. M. 1998. Acid tolerance and acid shock response of *Escherichia coli* O157:H7 and non-O157:H7 isolates provide cross protection to sodium lactate and sodium chloride. *J. Food Prot.* 61:158–161.

Gerhartz, H. 1949. Changes in the liver in a case of acetic acid poisoning and their significance for the timely measurement of liver regeneration and cirrhotic scar formation. *Virchows Arch. [A]*. 316:456–461.

Ghosh, J., and Häggblom, P. 1985. Effect of sublethal concentrations of propionic or butyric acid on growth and aflatoxin production by *Aspergillus flavus*. *Int. J. Food Microbiol.* 2:323–330.

Gill, C. O., and Newton, K. G. 1982. Effect of lactic acid concentration on growth on meat of Gram-negative psychrotrophs from a meatworks. *Appl. Environ. Microbiol.* 43:284–288.

Glabe, E. F., and Maryanski, J. K. 1981. Sodium diacetate: An effective mold inhibitor. *Cereal Foods World* 26:285–289.

Glass, K. A., and Doyle, M. P. 1991. Fate of *Salmonella* and *Listeria monocytogenes* in commercial, reduced-calorie mayonnaise. *J. Food Prot.* 54:691–695.

Glass, K. A., Granberg, D. A., Smith, A. L., McNamara, A. M., Hardin, M., Mattias, J., Ladwig, K., and Johnson, E. A. 2002. Inhibition of *Listeria monocytogenes* by sodium diacetate and sodium lactate on wieners and cooked bratwurst. *J. Food Prot.* 65:116–123.

Glass, K. A., McDonnell, L. M., VonTayson, R., Wanless, B., and Badvela, M. 2013. Inhibition of *Listeria monocytogenes* by propionic acid-based ingredients in cured deli-style turkey. *J. Food Prot.* 76:2074–2078.

Goddard, B. L., Mikel, W. B., Conner, D. E., and Jones, W. R. 1996. Use of organic acids to improve the chemical, physical, and microbial attributes of beef strip loins stored at -1°C for 112 days. *J. Food Prot.* 59:849–853.

Golden, M. H., Buchanan, R. L., and Whiting, R. C. 1995. Effect of sodium acetate or sodium propionate with EDTA and ascorbic acid on the inactivation of *Listeria monocytogenes*. *J. Food Saf.* 15:53–65.

Gómez-Aldapa, C. A., Rangel-Vargas, E., Torres-Vitela, M. A. R., Villarruel-López, A., Acevedo-Sandoval, O. A., Gordillo-Martínez, A. J., Godínez-Oviedo, A., and Castro-Rosas, J. 2018. Antibacterial activities of *Hibiscus sabdariffa* extracts and chemical sanitizers directly on green leaves contaminated with foodborne pathogens. *J. Food Prot.* 81:209–217.

Gomis, D. B., and Alonso, J. J. M. 1996. Analysis for organic acids. In *Handbook of Food Analysis*, Vol. 1: *Physical Characterization and Nutrient Analysis*. Edited by L. M. L. Nollet. Marcel Dekker, Inc., New York.

Gomori, G., and Gulyas, E. 1944. Effect of parenterally administered citrate on the renal excretion of calcium. *Proc. Soc. Exp. Biol. Med.* 56:226–228.

Gorman, B. M., Sofos, J. N., Morgan, J. B., Schmidt, G. R., and Smith, G. C. 1995. Evaluation of hand-trimming, various sanitizing agents, and hot water spray-washing as decontamination interventions for beef brisket adipose tissue. *J. Food Prot.* 58:899–907.

Gould, G. W., Brown, M. H., and Fletcher, B. C. 1983. Mechanisms of action of food preservation procedures. In *Mechanisms of Food Preservation Procedures*. Edited by T. A. Roberts and F. A. Skinner. Academic Press, New York, p. 67.

Graham, W. D., and Grice, H. C. 1955. Chronic toxicity of bread additives to rats. II. *J. Pharm. Pharmacol.* 7:126–134.

Graham, W. D., Teed, H., and Grice, H. C. 1954. Chronic toxicity of bread additives to rats. *J. Pharm. Pharmacol.* 6:534–545.

Granados, H., Glavind, J., and Dam, H. 1949. Observations on experimental dental caries. III. The effect of dietary lactic acid. *J. Dent. Res.* 28:282–287.

Grau, F. H. 1980. Inhibition of the anaerobic growth of *Brochothrix thermosphacta* by lactic acid. *Appl. Environ. Microbiol.* 40:433–436.

Grau, F. H. 1981. Role of pH, lactate, and anaerobiosis in controlling the growth of some fermentative gram-negative bacteria on beef. *Appl. Environ. Microbiol.* 42:1043–1050.

Greer, G. G., and Dilts, B. D. 1995. Lactic acid inhibition of the growth of spoilage bacteria and cold tolerant pathogens on pork. *Int. J. Food Microbiol.* 25:141–151.

Gruber, C. M., and Halbeisen, W. A. 1948. A study on the comparative toxic effects of citric acid and its sodium salts. *J. Pharmacol. Exp. Ther.* 94:65–67.

Gupta, S., and Ravishankar, S. 2005. A comparison of the antimicrobial activity of garlic, ginger, carrot, and turmeric pastes against *Escherichia coli* O157:H7 in laboratory buffer and ground beef. *Foodborne Pathog. Dis.* 2:330–340.

Guynot, M. E., Ramos, A. J., Sala, D., Sanchis, V., and Marín, S. 2002. Combined effects of weak acid preservatives, pH, and water activity on growth of *Eurotium* species on a sponge cake. *Int. J. Food Microbiol.* 76:39–46.

Hamby, P. L., Savell, J. W., Acuff, G. R., Vanderzant, C., and Cross, H. R. 1987. Spray-chilling and carcass decontamination systems using lactic and acetic acid. *Meat Sci.* 21:1–14.

Hara, S. 1965. Pharmacological and toxic actions of propionates. Examination of general pharmacological actions and toxicity of sodium and calcium propionates. *Chem. Abstr.* 62:977f.

Hara, S., Shibuya, T., Yakazu, K., Kobayashi, T., Takahashi, R., Takeuchi, T., and Tokizaki, K. 1963. Studies on pharmacological and toxic actions of propionates. *Tokyo Med. Coll.* 1–28.

Hardin, M. D., Acuff, G. R., Lucia, L. M., Oman, J. S., and Savell, J. W. 1995. Comparison of methods for decontamination from beef carcass surfaces. *J. Food Prot.* 58:368–374.

Harmayani, E., Sofos, J. N., and Schmidt, G. R. 1991. Effect of sodium lactate, calcium lactate and sodium alginate on bacterial growth and aminopeptidase activity. *J. Food Saf.* 11:269–283.

Harmayani, E., Sofos, J. N., and Schmidt, G. R. 1993. Fate of *Listeria monocytogenes* in raw and cooked ground beef with meat processing additives. *Int. J. Food Microbiol.* 18:223–232.

Harshbarger, K. E. 1942. Report of a study on the toxicity of several food preserving agents. *J. Dairy Sci.* 25:169–174.

Hartwig, A., and MAK Commission. 2018. Succinic acid/butanedioci acid. The MAK collection for occupational health and safety. *MAK-Coll. Occ. Health Saf.* 3:592–605.

Hathcox, A. K., Hwang, C. A., Resurreccion, A. V. A., and Beuchat, L. R. 1995. Consumer evaluation of raw and fried chicken after washing in trisodium phosphate or lactic acid/sodium benzoate solutions. *J. Food Sci.* 60:604–610.

Hayashi, K., Terada, M., Mizunuma, T., and Yokotsuka, T. 1979. Retarding effect of acetic acid on growth of contaminated bacteria during shoyu-koji making process. *J. Food Sci.* 44:359–362.

Hazelton Laboratories. 1970. *Two Generation Reproduction Study-Rats X-5120 (Malic Acid).* Final report submitted to Specialty Chemicals Division. Allied Chemical Corporation, Buffalo, NY.

Hazelton Laboratories. 1971a. *24 Month Dietary Administration-Rats.* Project No. 165–126 (malic acid). Submitted to Specialty Chemicals Division. Allied Chemical Corporation, Buffalo, NY.

Hazelton Laboratories. 1971b. *104 Week Dietary Administration-Dogs. X-5120 (Malic Acid).* Final report submitted to Specialty Chemicals Division. Allied Chemical Corporation, Buffalo, NY.

Hedberg, M., and Miller, J. K. 1969. Effectiveness of acetic acid, betadine, amphyll, polymyxin B, colistin and gentamicin against *Pseudomonas aeruginosa*. *Appl. Microbiol.* 18:854–855.

Heseltine, W. W. 1952a. A note on sodium propionate. *J. Pharm. Pharmacol.* 4:120–122.

Heseltine, W. W. 1952b. Sodium propionate and its derivatives as bacteriostatics and fungistatics. *J. Pharm. Pharmacol.* 4:577–581.

Hilgren, J. D., and Salverda, J. A. 2000. Antimicrobial efficacy of a peroxyacetic/octanoic acid mixture in fresh-cut-vegetable process waters. *J. Food Sci.* 65:1376–1379.

Hirshfield, I. N., Terzulli, S., and O'Byrne, C. 2003. Weak organic acids: A panoply of effects on bacteria. *Sci. Prog.* 86:245–269.

Hoffman, C., Schweitzer, T. R., and Dalby, G. 1939. Fungistatic properties of the fatty acids and possible biochemical significance. *Food Res.* 4:539–543.

Horn, H. J., Holland, E. G., and Hazelton, L. W. 1957. Safety of adipic acid as compared with citric and tartaric acid. *J. Agric. Food Chem.* 5:759–762.

Horowitz, W. (ed.) 2000. *Official Methods of Analysis of AOAC International*, 17th ed. Association of Official Analytical Chemists, Gaithersburg, MD.

Houben, J. H., and Krol, B. 1991. Effect of citric acid and citrate on the shelf stability of canned liver paste processed at reduced F_0 values. *Meat Sci.* 30:185–194.

Houtsma, P. C., de Wit, J. C., and Rombouts, F. M. 1993. Minimum inhibitory concentration (MIC) of sodium lactate for pathogens and spoilage organisms occurring in meat products. *Int. J. Food Microbiol.* 20:247–257.

Houtsma, P. C., Heuvelink, A. Dufrenne, J., and Notermans, S. 1994a. Effect of sodium lactate on toxin production, spore germination and heat resistance of proteolytic *Clostridium botulinum* strains. *J. Food Prot.* 57:327–330.

Houtsma, P. C., Kant-Muermans, M. L., Rombouts, F. M., and Zwietering, M. H. 1996. Model for the combined effects of temperature, pH, and sodium lactate on growth rates of *Listeria innocua* in broth and bologna-type sausages. *Appl. Environ. Microbiol.* 62:1616–1622.

Houtsma, P. C., Kusters, B. J. M., de Wit, J. C., Rombouts, F. M., and Zwietering, M. H. 1994b. Modelling growth rates of *Listeria innocua* as a function of lactate concentration. *Int. J. Food Microbiol.* 24:113–123.

Hu, A. C., and Shelef, L. A. 1996. Influence of fat content and preservatives on the behavior of *Listeria monocytogenes* in beaker sausage. *J. Food Saf.* 16:175–181.

Huang, Y., Wilson, M., Chapman, B., and Hocking, A. D. 2010. Evaluation of the efficacy of four weak acids as antifungal preservatives in low-acid intermediate moisture model food systems. *Food Microbiol.* 27:33–36.

Huhtanen, C. N. 1983. Antibotulinal activity of methyl and ethyl fumarates in comminuted nitrate-free bacon. *J. Food Sci.* 48:1574–1575.

Huhtanen, C. N., and Guy, E. J. 1984. Antifungal properties of esters of alkenoic and alkynoic acids. *J. Food Sci.* 49:281–283.

Huhtanen, C. N., Guy, E. J., and Milnes-McCaffrey, L. 1981. *Antifungal Properties of Fumaric Acid Esters.* Presented at the 41st Annual Meeting of the Institute of Food Technologists, Atlanta, GA, June 7–10.

Huhtanen, C. N., Trenchard, H., and Milnes-McCaffrey, L. 1985. Inhibition of *Clostridium botulinum* in comminuted bacon by short-chain alkynoic and alkenoic acids and esters. *J. Food Prot.* 48:570–573.

Hulankova, R., Borilova, G., and Steinhauserova, I. 2013. Combined antimicrobial effect of oregano essential oil and caprylic acid in minced beef. *Meat Sci.* 95:190–194.

Humer, E., Lucke, A., Harder, H., Metzler-Zebeli, B. U., Böhm, J., and Zebeli, Q. 2016. Effects of citric and lactic acid on the reduction of deoxynivalenol and its derivatives in feeds. *Toxins* 8:285–10.

Hunter, D., and Segel, I. H. 1973. Effect of weak acids on amino acid transport by *Penicillium chrysogenum*: Evidence for a proton or charge gradient as the driving force. *J. Bacteriol.* 113:1184–1192.

Hwang, C.-A., and Beuchat, L. R. 1995. Efficacy of selected chemicals for killing pathogenic and spoilage microorganisms on chicken skin. *J. Food Prot.* 58:19–23.

Hyun, J.-E., Yoon, J.-H., and Lee, S.-Y. 2018. Response surface modeling for the inactivation of *Bacillus cereus* on cooked spinach by natural antimicrobials at various temperatures. *J. Food Saf.* 38:e12484.

Imai, K., Banno, I., and Iijima, T. 1970. Inhibition of bacterial growth by citrate. *J. Gen. Appl. Microbiol.* 16:479–489.

Informatics, Inc. 1943. *The Toxicity of Adipic Acid. Quoted in Monograph on Adipic Acid—NAS/NRC Questionnaire.* Informatics Inc., Rockville, MD.

Ingle, J. D. 1940. Some preliminary observations on the effectiveness of propionates as mold inhibitors on dairy products. *J. Dairy Sci.* 23:509.

Islam, M. N. 1982. Inhibition of mold in bread by dimethyl fumarate. *J. Food Sci.* 47:1710–1712.

Ita, P. S., and Hutkins, R. W. 1991. Intracellular pH and survival of *Listeria monocytogenes* Scott A in tryptic soy broth containing acetic, lactic, citric, and hydrochloric acids. *J. Food Prot.* 54:15–19.

Ito, K., Chen, J. K., Lerke, P. A., Seeger, M. L., and Unverferth, J. A. 1976. Effect of acid and salt concentration in fresh-pack pickles on the growth of *Clostridium botulinum* spores. *Appl. Environ. Microbiol.* 32:121–124.

Izat, A. L., Colberg, M., Thomas, R. A., Adams, M. H., and Driggers, C. D. 1990a. Effects of lactic acid in processing waters on the incidence of salmonellae on broilers. *J. Food Qual.* 13:295–306.

Izat, A. L., Hierholzer, R. E., Kopek, J. M., Adams, M. H., McGinnis, J. P., and Reiber, M. A. 1990b. The use of propylene glycol and/or lactic acid in chill water for reducing salmonellae on broilers. *J. Food Proc. Pres.* 14:369–374.

Janes, M. E., Kooshesh, S., and Johnson, M. G. 2002. Control of *Listeria monocytogenes* on the surface of refrigerated, ready-to-eat- chicken coated with edible zein film coatings containing nisin and/or calcium propionate. *J. Food Sci.* 67:2754–2757.

Johnson, L. R. 1968. Source of the histamine released during damage to the gastric mucosa by acetic acid. *Gastroenterology* 54:8–15.

Joint FAO/WHO Expert Committee on Food Additives, World Health Organization & Food and Agriculture Organization of the United Nations. (1974). Evaluation of certain food additives: eighteenth report of the Joint FAO/WHO Expert Committee on Food Additives, Rome, 4-13 June 1974. World Health Organizati on. https://apps.who.int/iris/handle/10665/41117

Jordan, K. N., and Davies, K. W. 2001. Sodium chloride enhances recovery and growth of acid-stressed *E. coli* O157:H7. *Lett. Appl. Microbiol.* 32:312–315.

Jordan, K. N., Hall, S., and McClure, P. J. 1999. Osmotic stress on dilution of acid injured *Escherichia coli* O157:H7. *Lett Appl. Microbiol.* 28:389–393.

Jung, Y., Jang, H., Guo, M., Gao, J., and Matthews, K. R. 2017. Sanitizer efficacy in preventing cross-contamination of heads of lettuce during retail crisping. *Food Microbiol.* 64:179–185.

Kabara, J. J., Swieczkowski, D. M., Conley, A. J., and Truant, J. P. 1972. Fatty acids and derivatives as antimicrobial agents. *Antimicrob. Agents Chemother.* 2:23–28.

Kang, D.-H., Koohmaraie, M., Dorsa, W. J., and Siragusa, G. R. 2001a. Development of a multiple-step process for the microbial decontamination of beef trim. *J. Food Prot.* 64:63–71.

Kang, D.-H., Koohmaraie, M., and Siragusa, G. R. 2001b. Application of multiple interventions for microbial decontamination of commercial beef trim. *J. Food Prot.* 64:168–171.

Kang, J.-W., and Kang, D.-H. 2017. Antimicrobial efficacy of vacuum impregnation washing with malic acid applied to whole paprika, carrots, king oyster mushrooms and muskmelons. *Food Cont.* 82:126–135.

Karapinar, M., and Gönül, S. A. 1992a. Effects of sodium bicarbonate, vinegar, acetic and citric acids on growth and survival of *Yersinia enterocolitica. Int. J. Food Microbiol.* 16:343–347.

Karapinar, M., and Gönül, S. A. 1992b. Removal of *Yersinia enterocolitica* from fresh parsley by washing with acetic acid or vinegar. *Int. J. Food Microbiol.* 16:261–264.

Khalil, A. A., Deraz, S. F., Elrahman, S. A., and El-Fawal, G. 2015. Enhancement of mechanical properties, microstructure, and antimicrobial activities of zein films cross-linked using succinic anhydride, eugenol, and citric acid. *Prep. Biochem. Biotechnol.* 45:551–567.

Khurana, A., Awuah, G. B., Taylor, B., and Enache, E. 2006. Efficacy of selected acidulants in pureed green beans inoculated with pathogens (*Escherichia coli* O157:H7 and *Listeria monocytogenes*). *J. Food Prot.* 69:1865–1869.

Kilic, B., Cassens, R. G., and Borchert, L. L. 2002. Effect of turkey meat, phosphate, sodium lactate, carrageenan, and konjac on residual nitrite in cured meats. *J. Food Sci.* 67:29–31.

Kim, C. R., and Hearnsberger, J. O. 1994. Gram negative bacteria inhibition by lactic acid culture and food preservatives on catfish fillets during refrigerated storage. *J. Food Sci.* 59:513–516.

Kim, C. R., Hearnsberger, J. O., Vickery, A. P., White, C. H., and Marshall, D. L. 1995a. Sodium acetate and bifidobacteria increase shelf life of refrigerated catfish fillets. *J. Food Sci.* 60:25–27.

Kim, C. R., Hearnsberger, J. O., Vickery, A. P., White, C. H., and Marshall, D. L. 1995b. Extending shelf life of refrigerated catfish fillets using sodium acetate and monopotassium phosphate. *J. Food Prot.* 58:644–647.

Kim, H., Kang, Y., Beuchat, L. R., and Ryu, J.-H. 2008. Production and stability of chlorine dioxide in organic acid solutions as affected by pH, type of acid, and concentration of sodium chlorite, and its effectiveness in inactivating *Bacillus cereus* spores. *Food Microbiol.* 25:964–969.

Kim, S. A., and Rhee, M. S. 2013. Marked synergistic bactericidal effects and mode of action of medium-chain fatty acids in combination with organic acids against *Escherichia coli* O157:H7. *Appl. Environ. Microbiol.* 79:6552–6560.

Kim, S. A., and Rhee, M. S. 2015. Synergistic antimicrobial activity of caprylic acid in combination with citric acid against both *Escherichia coli* O157:H7 and indigenous microflora in carrot juice. *Food Microbiol.* 49:166–172.

King, A. M., Glass, K. A., Milkowski, A. L., and Sindelar, J. J. 2015. Impact of clean-label antimicrobials and nitrite derived from natural sources on the outgrowth of *Clostridium perfringens* during cooling of deli-style turkey breast. *J. Food Prot.* 78:946–953.

Kirby, G. W., Atkin, L., and Frey, C. N. 1937. Further studies on the growth of bread molds as influenced by acidity. *Cereal Chem.* 14:865–878.

Knight, T.D., Castillo, A., Maxim, J., Keeton, J.T., and Miller, R.K., Effectiveness of potassium lactate and sodium diacetate in combination with irradiation to control Listeria monocytogenes on frankfurters. *J. Food Sci.* 72, M26–M30.

Koczoń, P. 2009. Growth inhibition mode of action of selected benzoic acid derivatives against the yeast *Pichia anomala. J. Food Prot.* 72:791–800.

Koo, O.-K., Eggleton, M., O'Bryan, C. A., Crandall, P. G., and Ricke, S. C. 2012. Antimicrobial activity of lactic acid bacteria against *Listeria monocytogenes* on frankfurters formulated with and without lactate/diacetate. *Meat Sci.* 92:533–537.

Kotula, K. L., and Thelappurate, R. 1994. Microbiological and sensory attributes of retail cuts of beef treated with acetic and lactic acid solutions. *J. Food Prot.* 57:665–670.

Kozak, S. M., Margison, K. M., D'Amico, D. J. 2017. Synergistic antimicrobial combinations inhibit and inactivate *Listeria monocytogenes* in neutral and acidic broth systems. *J. Food Prot.* 80:1266–1272.

Kroll, R. G., and Patchett. 1992. Induced acid tolerance in *Listeria monocytogenes. Lett. Appl. Microbiol.* 14:224–227.

Kurita, N., and Koike, S. 1983. Synergistic antimicrobial effect of ethanol, sodium chloride, acetic acid and essential oil components. *Agric. Biol. Chem.* 47:67–75.

Lamkey, J. W., Leak, F. W., Tuley, W. B., Johnson, D. D., and West, R. L. 1991. Assessment of sodium lactate addition to fresh pork sausage. *J. Food Sci.* 56:220–223.

Lang, K., and Bartsch, A. R. 1953. Concerning the metabolism and tolerability of adipic acid. *Biochem. Z.* 323:462–468.

Langworthy, T. A. 1978. Microbial life in extreme pH values. In *Microbial Life in Extreme Environments.* Edited by D. J. Kushner. Academic Press, New York, p. 279.

Lappe, R., Motta, A. S., Sant'Anna, V., and Brandelli, A. 2009. Inhibition of *Salmonella* Enteritidis by cerein 8A, EDTA and sodium lactate. *Int. J. Food Microbiol.* 135:312–316.

Lee, S. E., Han, J. H., and Han, J. 2016. Development and evaluation of apple peel- and carboxymethylcellulose-based biodegradable films with antioxidant and antimicrobial properties. *J. Food Saf.* 36:317–324.

Lemay, M. J., Choquette, J., Delaquis, P. J., Gariépy, C., Rodrigue, N., and Saucier, L. 2002. Antimicrobial effect of natural preservatives in a cooked and acidified chicken meat model. *Int. J. Food Microbiol.* 78:217–226.

Levey, S., Lasichak, A. G., Brimi, R., Orten, J. M., Smyth, C. J., and Smith, A. H. 1946. A study to determine the toxicity of fumaric acid. *J. Am. Pharm. Assoc. Sci. Ed.* 35:298–304.

Levine, A. S., and Fellers, C. R. 1939. The inhibiting effect of acetic acid upon microorganisms in the presence of sodium chloride and sucrose. *J. Bacteriol.* 39:17.

Levine, A. S., and Fellers, C. R. 1940. Action of acetic acid on food spoilage microorganisms. *J. Bacteriol.* 39:499–515.

Leyer, G. J., and Johnson, E. A. 1992. Acid adaptation promotes survival of *Salmonella* spp. in cheese. *Appl. Environ. Microbiol.* 58:2075–2080.

Leyer, G. J., and Johnson, E. A. 1997. Acid adaptation sensitizes *Salmonella typhimurium* to hypochlorous acid. *Appl. Environ. Microbiol.* 63:461–467.

Leyer, G. J., Wang, L. L., and Johnson, E. A. 1995. Acid adaptation of *Escherichia coli* O157:H7 increases survival in acidic foods. *Appl. Environ. Microbiol.* 61:3752–3755.

Li, S., Kundu, D., and Holley, R. A. 2015. Use of lactic acid with electron beam irradiation for control of *Escherichia coli* O157:H7, non-O157 VTEC *E. coli*, and *Salmonella* serovars on fresh and frozen beef. *Food Microbiol.* 46:34–39.

Liato, V., Hammami, R., and Aïder, M. 2017a. Influence of electro-activated solutions of weak organic acid salts on microbial quality and overall appearance of blueberries during storage. *Food Microbiol.* 64:56–64.

Liato, V., Labrie, S., and Aïder, M. 2017b. Study of the antibacterial activity of electro-activated solutions of weak organic acids on *Salmonella enterica, Staphylococcus aureus*, and *Listeria monocytogenes. J. Indust. Microbiol. Biotechnol.* 44:23–33.

Lide, D. R. 2002. *CRC Handbook of Chemistry and Physics*, 83rd ed. CRC Press, Boca Raton, FL.

Lillard, H. S., Blankenship, L. C., Dickens, J. A., Craven, S. E., and Schackelford, S. D. 1987. Effect of acetic acid on the microbiological quality of scalded picked and unpicked broiler carcasses. *J. Food Prot.* 50:112–114.

Little, C. L., Adams, M. R., and Easter, M. C. 1992. The effect of pH, acidulant and temperature on the survival of *Yersinia enterocolitica. Lett. Appl. Microbiol.* 14:148–152.

Lu, Y., and Wu, C. 2012. Reductions of *Salmonella enterica* on chicken breast by thymol, acetic acid, sodium dodecyl sulfate or hydrogen peroxide combinations as compared to chlorine wash. *Int. J. Food Microbiol.* 152:31–34.

Luchese, R. H., and Harrigan, W. F. 1990. Growth of, and aflatoxin production by *Aspergillus parasiticus* when in the presence of either *Lactococcus lactis* or lactic acid and at different initial pH values. *J. Appl. Bacteriol.* 69:512–519.

Lues, J.F.R. and Theron, M.M. 2012. Comparing organic acids and salt derivatives as antimicrobials against selected poultry-borne *Listeria monocytogenes* strains *in vitro. Foodborne Pathog. Dis.* 9:1126–1129.

Lynch, D. J., and Potter, N. N. 1988. Effect of organic acids on thermal inactivation of *Bacillus stearothermophilus* and *Bacillus coagulans* spores in frankfurter emulsion slurry. *J. Food Prot.* 51:475–480.

Maas, M. R., Glass, K. A., and Doyle, M. P. 1989. Sodium lactate delays toxin production by *Clostridium botulinum* in cook-in-bag turkey products. *Appl. Environ. Microbiol.* 55:2226–2229.

Makinen, Y. 1958. The effect of acetic acid and hydrogen peroxide on mitotic frequency in *Allium cepa. Nucleus* 1:131–152.

Mani-López, E., García, H. S., and López-Malo, A. 2012. Organic acids as antimicrobials to control *Salmonella* in meat and poultry products. *Food Res. Int.* 45:713–721.

Manna, G. K., and Mukherjee, P. K. 1966. Spermatocyte chromosome aberrations in two species of grasshoppers at two different ionic activities. *Nucleus* 9:119–131.

Marín, S., Guynot, M. E., Neira, P., Bernadó, Sanchis, V., and Ramos, A. J. 2002a. Risk assessment of the use of sub-optimal levels of weak-acid preservatives in the control of mould growth on bakery products. *Int. J. Food Microbiol.* 79:203–211.

Marín, S., Guynot, M. E., Sanchis, V., Arbonés, J., and Ramos, A. J. 2002b. *Aspergillus flavus, Aspergillus niger*, and *Penicillium corylophilum* spoilage prevention of bakery products by means of weak-acid preservatives. *J. Food Sci.* 67:2271–2277.

Mazzotta, A. S. 2001. Thermal inactivation of stationary-phase and acid-adapted *Escherichia coli* O157:H7, *Salmonella*, and *Listeria monocytogenes* in fruit juices. *J. Food Prot.* 64:315–320.

Mbandi, E., and Shelef, L. A. 2001. Enhanced inhibition of *Listeria monocytogenes* and *Salmonella* Enteritidis in meat by combinations of sodium lactate and diacetate. *J. Food Prot.* 64:640–644.

Mbandi, E., and Shelef, L. A. 2002. Enhanced antimicrobial effects of combination of lactate and diacetate on *Listeria monocytogenes* and *Salmonella* spp. in beef bologna. *Int. J. Food Microbiol.* 76:191–198.

McDonald, L. C., Fleming, H. P., and Hassan, H. M. 1990. Acid tolerance of *Leuconostoc mesenteroides* and *Lactobacillus plantarum. Appl. Environ. Microbiol.* 56:2120–2124.

McDonnell, L., Glass, K. A., and Sindelar, J. J. 2013. Identifying ingredients that delay outgrowth of *Listeria monocytogenes* in natural, organic, and clean-label ready-to-eat meat and poultry products. *J. Food Prot.* 76:1366–1376.

McMahon, C. M. M., Doherty, A. M., Sheridan, J. J., Blair, I. S., McDowell, D. A., and Hegarty, T. 1999. Synergistic effect of heat and sodium lactate on the thermal resistance of *Yersinia enterocolitica* and *Listeria monocytogenes* in minced beef. *Lett. Appl. Microbiol.* 28:340–344.

Medimagh, R., Aloui, H., Jemli, M., Chaabane, H., Belkahla, F., and Khwaldia, K. 2016. Enhanced functional properties of chitosan films cross-linked by biosourced dicarboxylic acids. *Poly. Sci. A.* 58:409–418.

Mellefont, L. A., McMeekin, T. A., and Ross, T. 2003. Performance evaluation of a model describing the effects of temperature, waster activity, pH and lactic acid concentration on the growth of *Escherichia coli. Int. J. Food Microbiol.* 82:45–58.

Membre, J. M., Majchrzak, V., and Jolly, I. 1997. Effects of temperature, pH, glucose, and citric acid on the inactivation of *Salmonella typhimurium* in reduced calorie mayonnaise. *J. Food Prot.* 60:1497–1501.

Mendonca, A. F., Molins, R. A., Kraft, A. A., and Walker, H. W. 1989a. Microbiological, chemical, and physical changes in fresh, vacuum-packaged pork treated with organic acids and salt. *J. Food Sci.* 54:18–21.

Mendonca, A. F., Molins, R. A., Kraft, A. A., and Walker, H. W. 1989b. Effects of potassium sorbate, sodium acetate, phosphates and sodium chloride alone or in combination on shelf life of vacuum-packaged pork chops. *J. Food Sci.* 54:302–306.

Miller, A. J., Call, J. E., and Whiting, R. C. 1993. Comparison of organic acid salts for *Clostridium botulinum* control in an uncured turkey product. *J. Food Prot.* 56:958–962.

Miller, R. K., and Acuff, G. R. 1994. Sodium lactate affects pathogens in cooked beef. *J. Food Sci.* 59:15–19.

Minor, T. E., and Marth, E. H. 1970. Growth of *Staphylococcus aureus* in acidified pasteurized milk. *J. Milk Food Technol.* 33:516–520.

Montet, M. P., Christieans, S., Thevenot, D., Coppet, V., Ganet, S., Delignette Muller, M. L., Dunière, L., Miszczycha, S., and Vernozy-Rozand, C. 2009. Fate of acid-resistant and non-acid resistant Shiga toxin-producing *Escherichia coli* strains in experimentally contaminated French fermented raw meat sausages. *Int. J. Food Microbiol.* 129:264–270.

Moon, S. H., Paik, H.-D., White, S., Daraba, A., Mendonca, A. F., and Ahn, D. U. 2011. Influence of nisin and selected meat additives on the antimicrobial effect of ovotransferrin against *Listeria monocytogenes*. *Poult. Sci.* 90:2584–2591.

Moore, A., Nannapaneni, R., Kiess, A., and Sharma, C. S. 2017. Evaluation of USDA approved antimicrobials on the reduction of *Salmonella* and *Campylobacter* in ground chicken frames and their effect on meat quality. *Poult. Sci.* 96:2385–2392.

Mori, K. 1952. The production of gastric lesions in the rat by acetic acid feeding. *Gan* 43:443–447.

Mori, K. 1953. Production of gastric lesions in the rat by the diet containing fatty acids. *Gan* 44:421–427.

Mountney, G. J., and O'Malley, J. 1965. Acids as poultry meat preservatives. *Poult. Sci.* 44:582–586.

Moye, C. J., and Chambers, A. 1991. An innovative technology for salmonella control and shelf life extension. *Food Austr.* 43:246–249.

Mun, S. Y., Kim, S. K., Woo, E. R., and Chang, H. C. 2019. Purification and characterization of an antimicrobial compound produced by *Lactobacillus plantarum* EM showing both antifungal and antibacterial activities. *LWT – Food Sci. Technol.* 114:108403.

Murdock, D. I. 1950. Inhibitory action of citric acid on tomato juice flat-sour organisms. *Food Res.* 15:107–113.

Mustapha, A., Ariyapitipun, T., and Clarke, A. D. 2002. Survival of *Escherichia coli* O157:H7 on vacuum-packaged raw beef treated with polylactic acid, lactic acid, and nisin. *J. Food Sci.* 67:262–266.

Nakamura, M., and Zangar, M. J. 1968. Effect of fatty acids on *Shigella*. *Proc. Mont. Acad. Sci.* 28:51–57.

Nei, D., Enomoto, K., and Nakamura, N. 2015. A gaseous acid treatment to disinfect fenugreek seeds and black pepper inoculated with pathogenic and spoilage bacteria. *Food Microbiol.* 49:226–230.

Nei, D., Latiful, B. M., Enomoto, K., Inatsu, Y., and Kawamoto, S. 2011. Disinfection of radish and alfalfa seeds inoculated with *Escherichia coli* O157:H7 and *Salmonella* by a gaseous acetic acid treatment. *Foodborne Pathog. Dis.* 8:1089–1094.

Nobmann, P., Smith, A., Dunne, J., Henehan, G., and Bourke, P. 2009. The antimicrobial efficacy and structure activity relationship of novel carbohydrate fatty acid derivatives against *Listeria* spp. and food spoilage microorganisms. *Int. J. Food Microbiol.* 128:440–445.

Noda, F., Hayashi, K., and Mizunuma, T. 1982. Influence of pH on inhibitory activity of acetic acid on osmophilic yeasts used in brine fermentation of soy sauce. *Appl. Environ. Microbiol.* 43:245–246.

Nogueira, J. N., Cantarelli, P. R., Gallo, C. R., Moreno, I. A. M., Matsuura, F. C. A. U., and Tiba, M. A. 1997. Effect of calcium addition and acidification on the quality characteristics of canned okra (*Hibiscus esculentus* L.). *Archivos Latino Americanos de Nutricion* 47:62–65.

Nollet, L. M. L. (ed.). 1996. *Handbook of Food Analysis. Vol. 1 Physical Characterization and Nutrient Analysis.* Marcel Dekker, Inc., New York.

Noordin, W. N. M., Shunmugam, N., Huda, N., and Adzitey, F. 2018. The effects of essential oils and organic acids on microbiological and physicochemical properties of whole shrimps at refrigerated storage. *Curr. Res. Nutr. Food Sci.* 6:273–283.

Notermans, S., and Dufrenne, J. 1981. Effect of glyceryl monolaurate by *Clostridium botulinum* in meat slurry. *J. Food Saf.* 3:83–88.

Notermans, S., Dufrenne, J., and Keybets, M. J. H. 1985. Use of preservatives to delay toxin formation by *Clostridium botulinum* (type B strain okra) in vacuum-packed, cooked potatoes. *J. Food Prot.* 48:851–855.

Nunheimer, T. D., and Fabian, F. W. 1940. Influence of organic acids, sugars, and sodium chloride upon strains of food poisoning staphylococci. *Am. J. Public Health* 30:1040–1049.

Ocio, M. J., Sánchez, T., Fernandez, P. S., Rodrigo, M., and Martínez, A. 1994. Thermal resistance characteristics of PA 3679 in the temperature range of 110–121°C as affected by pH, type of acidulant and substrate. *Int. J. Food Microbiol.* 22:239–247.

Ockerman, H. W., Borton, R. J., Cahill, V. R., Parrett, N. A., and Hoffman, H. D. 1974. Use of acetic and lactic acid to control the quantity of microorganisms on lamb carcasses. *J. Milk Food Technol.* 37:203–204.

O'Driscoll, B., Gahan, C. G., and Hill, C. 1996. Adaptive acid tolerance response in *Listeria monocytogenes*: Isolation of an acid-tolerant mutant which demonstrates increased virulence. *Appl. Environ. Microbiol.* 62:1693–1698.

Okabe, S., Roth, J. L. A., and Pfeiffer, C. J. 1971. Differential healing periods of the acetic acid ulcer model in rats and cats. *Experientia* 27:146–148.

O'Keane, M., Smith, S., and Goldberg, A. 1971. Gastric erosions caused by home-brewed lager. *Lancet* 2:795–797.

Okereke, A., and Beelman, R. B. 1990. Acid-blanching and EDTA effects on yield, quality and microbiological stability of canned mushrooms. *J. Food Sci.* 55:1327–1330.

Okereke, A., Beelman, R. B., and Doores, S. 1990a. Control of spoilage of canned mushrooms inoculated with *Clostridium sporogenes* PA 3679 spores by acid-blanching and EDTA. *J. Food Sci.* 55:1331–1333.

Okereke, A., Beelman, R. B., Doores, S., and Walsh, R. 1990b. Elucidation of the mechanism of the acid-blanch and EDTA process inhibition of *Clostridium sporogenes* PA 36779 spores. *J. Food Sci.* 55:1137–1142.

Okrend, A. J., Johnston, R. W., and Moran, A. B. 1986. Effect of acetic acid on the death rates at 52°C of *Salmonella newport, Salmonella typhimurium* and *Campylobacter jejuni* in poultry scald water. *J. Food Prot.* 49:500–503.

Oladunjoye, A. O., Singh, S., and Ijabadeniyi, O. A. 2016. Biocontrol of *Listeria monocytogenes* ATCC 7644 on fresh-cut tomato (*Lycopersicon esculentum*) using nisin combined with organic acids. *Food Sci. Biotechnol.* 25:1491–1496.

O'Leary, D. K., and Kralovec, R. D. 1941. Development of *B. mesentericus* in bread and control with calcium acid phosphate or calcium propionate. *Cereal Chem.* 18:730–741.

Olson, J. C., and Macy, H. 1940. Propionic acid and its calcium and sodium salts as inhibitors of mold growth. *J. Dairy Sci.* 23:509–510.

Olson, J. C., Jr., and Macy, H. 1945. Propionic acid, sodium propionate and calcium propionate as inhibitors of mold growth. I. Observations on the use of propionate-treated parchment in inhibiting mold growth on the surface of butter. *J. Dairy Sci.* 28:701–710.

Orö, L., and Wretland, A. 1961. Pharmacological effects of fatty acids, triolein and cottonseed oil. *Acta Pharmacol. Toxicol.* 18:141–152.

Ortega, M. T., Jeffery, B., Riviere, J. E., and Monteiro-Riviere, N. A. 2016. Toxicological effects of pet food ingredients on canine bone marrow-derived mesenchymal stem cells and enterocyte-like cells. *J. Appl. Toxicol.* 36:189–198.

Oscroft, C. A., Banks, J. G., and McPhee, S. 1990. Inhibition of thermally-stressed *Bacillus* spores by combinations of nisin, pH and organic acids. *Lebensm. Wiss. U. Technol.* 23:538–544.

Oshima, S., Rea, M. C., Lothe, S., Morgan, S., Begley, M., O'Connor, P. M., Fitzsimmons, A., Kamikado, H., Walton, R., Ross, R. P., and Hill, C. 2012. Efficacy of organic acids, bacteriocins, and the lactoperoxidase system in inhibiting the growth of *Cronobacter spp.* in rehydrated infant formula. *J. Food Prot.* 75:1734–1742.

Osthold, W., Shin, H.-K., Dresel, J., and Leistner, L. 1984. Improving the storage life of carcasses by treating their surfaces with an acid spray. *Fleischwirtschaft* 64:828–830.

Ough, C. S., and Kunkee, R. E. 1974. The effect of fumaric acid on malolactic fermentation in wines from warm areas. *Am. J. Enol. Vitic.* 25:188–190.

Over, K. F., Hettiarachchy, N., Johnson, M. G., and Davis, B. 2009. Effect of organic acids and plant extracts on *Escherichia coli* O157:H7, *Listeria monocytogenes*, and *Salmonella* Typhimurium in broth culture model and chicken meat systems. *J. Food Sci.* 74:M515–M521.

Over, K. F., Hettiarachchy, N. S., Perumalla, A. V. S., Johnson, M. G., Meullenet, J.-F., Dickson, J. S., Holtzbauer, M. J., Niebuhr, S. E., and Davis, B. 2010. Antilisterial activity and consumer acceptance of irradiated chicken breast meat vacuum-infused with grape see and green tea extracts and tartaric acid. *J. Food Sci.* 75:M455–M461.

Owen, C. R. 1946. Acetic acid inhibition of gram-negative bacilli in culture media. *J. Bacteriol.* 52:353–356.

Paar, D., Helmsoth, V., Werner, M., and Bock, K. D. 1968. Haemostatic failure due to consumption of coagulation factors in acute acetic acid poisoning. *Dtsch. Med. Monatsschr.* 13:421.

Packman, E. W., Abbott, D. D., and Harrisson, J. W. E. 1963. Comparative subacute toxicity for rabbits of citric, fumaric and tartaric acids. *Toxicol. Appl. Pharmacol.* 5:163–167.

Palejwala, V. A., Sattur, A. P., and Modi, V. V. 1984. Some factors responsible for the spoilage of mangoes by *Penicillium cyclopium. Food Microbiol.* 1:255–262.

Palmer, A. A. 1932. Two fatal cases of poisoning by acetic acid. *Med. J. Aust.* 1:687–689.

Papadopoulos, L. S., Miller, R. K., Ringer, L. J., and Cross, H. R. 1991a. Sodium lactate effect on sensory characteristics, cooked meat color and chemical composition. *J. Food Sci.* 56:621–626.

Papadopoulos, L. S., Miller, R. K., Acuff, G. R., Vanderzant, C., and Cross, H. R. 1991b. Effect of sodium lactate on microbial and chemical composition of cooked beef during storage. *J. Food Sci.* 56:341–347.

Papadopoulos, L. S., Miller, R. K., Acuff, G. R., Lucia, L. M., Vanderzant, C., and Cross, H. R. 1991c. Consumer and trained sensory comparisons of cooked beef top rounds treated with sodium lactate. *J. Food Sci.* 56:1141–1146.

Park, H. S., Marth, E. H., and Olson, N. F. 1970. Survival of *Salmonella typhimurium* in cold-pack cheese food during refrigerated storage. *J. Milk Food Technol.* 33:383–388.

Park, J.-S., and Ha, J.-W. 2019. Ultrasound treatment combined with fumaric acid for inactivating food-borne pathogens in apple juice and its mechanisms. *Food Microbiol.* 84:103277.

Parmeggiani, L., and Sassi, C. 1954. Injury to health caused by acetic acid in the manufacture of acetylcellulose. *Med. Lav.* 45:319–326.

Peireira, I., Fraga, S., Maltez, L., Requicha, J., Guardão, L., Oliveira, J., Prada, J., Alves, H., Santos, J. D., Teixeira, J. P., Pereira, J. E., Soares, R., and Gama, F. M. 2019. *In vivo* systemic toxicity assessment of an oxidized dextrin-based hydrogel and its effectiveness as a carrier and stabilizer of granular synthetic bone substitutes. *J. Biomed. Mat. Res. A.* 107A:1678–1689.

Pelroy, G. A., Peterson, M. E., Holland, P. J., and Eklund, M. W. 1994. Inhibition of *Listeria monocytogenes* in cold-process(smoked) salmon by sodium lactate. *J. Food Prot.* 57:108–113.

Phillips, C. A. 1999. The effect of citric acid, lactic acid, sodium citrate and sodium lactate, alone and in combination with nisin, on the growth of *Arcobacter butzleri. Lett. Appl. Microbiol.* 29:424–428.

Pilone, G. J. 1975. Control of malo-lactic fermentation in table wines by addition of fumaric acid. In *Lactic Acid Bacteria in Beverages and Food.* Edited by J. G. Carr, C. V. Cutting, and G. C. Whiting. Academic Press, London, p. 121.

Pitkin, C. E. 1935. Lactic acid stricture of esophagus. *Ann. Otol. Rhinol. Laryngol.* 44:842–843.

Podolak, R. K., Zayas, J. F., Kastner, C. L., and Fung, D. Y. C. 1996a. Inhibition of *Listeria monocytogenes* and *Escherichia coli* O157:H7 on beef by application of organic acids. *J. Food Prot.* 59:370–373.

Podolak, R. K., Zayas, J. F., Kastner, C. L., and Fung, D. Y. C. 1996b. Reduction of bacterial populations on vacuum-packaged ground beef patties with fumaric and lactic acids. *J. Food Prot.* 59:1037–1040.

Portner, D. M., and Hoffman, R. K. 1968. Sporicidal effect of peracetic acid vapor. *Appl. Microbiol.* 16:1782.

Porto-Fett, A. C. S., Campano, S. G., Call, J. E., Shoyer, B. A., Yoder, L., Gartner, K., Tufft, L., Oser, A., Lee, J., and Luchansky, J. B. 2011. Validation of food-grade salts of organic acids as ingredients to control *Listeria monocytogenes* on pork scrapple during extended refrigerated storage. *J. Food Prot.* 74:394–402.

Porto-Fett, A. C. S., Campano, S. G., Shoyer, B. A., Israeli, D., Oser, A., and Luchansky, J. B. 2015. Comparative efficacy of potassium levulinate with and without potassium diacetate and potassium propionate versus potassium lactate and sodium diacetate for control of *Listeria monocytogenes* on commercially prepared uncured turkey breast. *J. Food Prot.* 78:927–933.

Porto-Fett, A. C. S., Franco, B. D. G. M., Sant'anna, E. S., Call, J. E., Piva, A., and Luchansky, J. B. 2002. Viability of a five-strain mixture of *Listeria monocytogenes* in vacuum-sealed packages of frankfurters, commercially prepared with and without 2.0 or 3.0% added potassium lactate, during extended storage at 4 and 10°C. *J. Food Prot.* 65:308–315.

Post, L. S., Amoroso, T. L., and Solberg, M. 1985. Inhibition of *Clostridium botulinum* type E in model acidified food systems. *J. Food Sci.* 50:966–968.

Prasai, R. K., Acuff, G. R., Lucia, L. M., Hale, D. S., Savell, J. W., and Morgan, J. B. 1991. Microbiological effects of acid decontamination of beef carcasses at various locations in processing. *J. Food Prot.* 54:868–872.

Prasai, R. K., Acuff, G. R., Lucia, L. M., Morgan, J. B., May, S. G., and Savell, J. W. 1992. Microbiological effects of acid decontamination of pork carcasses at various locations in processing. *Meat Sci.* 32:413–423.

Presser, K. A., Ross, T., and Ratkowsky, D. A. 1998. Modelling the growth limits (growth/no growth interface) of *Escherichia coli* as a function of temperature, pH, lactic acid concentration, and water activity. *Appl. Environ. Microbiol.* 64:1773–1779.

Quartey-Papafio, E. A., Marshall, R. T., and Anderson, M. E. 1980. Short-chain fatty acids as sanitizers for beef. *J. Food Prot.* 43:168–171.

Radford, S. A., and Board, R. G. 1993. Fate of pathogens in home-made mayonnaise and related products. *Food Microbiol.* 10:269–278.

Rajashekhara, E., Suresh, E. R., and Ethiraj, S. 1998. Thermal death rate of ascospores of *Neosartorya fischeri* ATCC 200957 in the presence of organic acids and preservatives in fruit juices. *J. Food Prot.* 61:1358–1362.

Rammell, C. G. 1962. Inhibition by citrate of the growth of coagulase-positive staphylococci. *J. Bacteriol.* 84:1123–1124.

Rastegar, H., Shoeibi, S., Yazdanpanah, H., Amirahmadi, M., Khaneghah, A. M., Campagnollo, F. B., and Sant'Ana, A. S. 2017. Removal of aflatoxin B_1 by roasting with lemon juice and/or citric acid in contaminated pistachio nuts. *Food Cont.* 71:279–284.

Reiss, J. 1976. Prevention of the formation of mycotoxins in whole wheat bread by citric and lactic acid. *Experientia* 32:168–169.

Restaino, L., Komatsu, K., and Syracuse, M. J. 1981. Effects of acids on potassium sorbate inhibition of food-related microorganisms in culture media. *J. Food Sci.* 47:134–143.

Restaino, L., Lenovich, L. M., and Bills, S. 1982. Effect of acids and sorbate combinations on the growth of four osmophilic yeasts. *J. Food Prot.* 45:1138–1142.

Rice, A. C., and Pederson, C. S. 1954. Factors influencing growth of *Bacillus coagulans* in canned tomato juice. II. Acidic constituents of tomato juice and specific organic acids. *Food Res.* 19:124–133.

Rivero, S., Giannuzzi, L., García, M. A., and Pinotti, A. 2013. Controlled delivery of propionic acid from chitosan films for pastry dough conservation. *J. Food Eng.* 116:524–531.

Robach, M. C. 1980. Use of preservatives to control microorganisms in food. *Food Technol.* 34:81–84.

Rose, W. C. 1924. The nephropathic action of the dicarboxylic acids and their derivatives. I. Tartaric, malic and succinic acids. *J. Pharmacol. Exp. Ther.* 24:123–129.

Ross, T., Ratkowsky, D. A., Mellefont, L. A., and McMeekin, T. A. 2003. Modelling the effects of temperature, water activity, pH and lactic acid concentration on the growth of *Escherichia coli*. *Int. J. Food Microbiol.* 82:33–43.

Rozbeh, M., Kalchayanand, N., Field, R. A., Johnson, M. C., and Ray, B. 1993. The influence of biopreservatives on the bacterial level of refrigerated vacuum packaged beef. *J. Food Saf.* 13:99–111.

Rubin, H. E. 1978. Toxicological model for a two-acid system. *Appl. Environ. Microbiol.* 36:623–624.

Ryser, E. T., and Marth, E. H. 1988. Survival of *Listeria monocytogenes* in cold-pack cheese food during refrigerated storage. *J. Food Prot.* 51:615–621.

Ryu, J.-H., and Beuchat, L. R. 1998. Influence of acid tolerance responses on survival, growth, and thermal cross-protection of *Escherichia coli* O157:H7 in acidified media and fruit juices. *Int. J. Food Microbiol.* 45:185–193.

Sagong, H.-G., Lee, S.-Y., Chang, P.-S., Heu, S., Ryu, S., Choi, Y.-J., and Kang, D.-H. 2011. Combined effect of ultrasound and organic acids to reduce *Escherichia coli* O157:H7, *Salmonella* Typhimurium, and *Listeria monocytogenes* on organic fresh lettuce. *Int. J. Food Microbiol.* 145:287–292.

Sawaya, W. N., Elnawawy, A. S., Al-Zenki, S., Al-Otaibi, J., Al-Omirah, H., and Al-Amiri, H. 1995. Storage stability of chicken as affected by map and lactic acid treatment. *J. Food Sci.* 60:611.

Schlyter, J. H., Degnan, A. J., Loeffelholz, J., Glass, K. A., and Luchansky, J. B. 1993a. Evaluation of sodium diacetate and ALTA™ 2341 on viability of *Listeria monocytogenes* in turkey slurries. *J. Food Prot.* 56:808–810.

Schlyter, J. H., Glass, K. A., Loeffelholz, J., Degnan, A. J., and Luchansky, J. B. 1993b. The effects of diacetate with nitrite, lactate, or pediocin on the viability of *Listeria monocytogenes* in turkey slurries. *Int. J. Food Microbiol.* 19:271–281.

Schoenemann, D. R., and Lopez, A. 1973. Heat processing effects on physical and chemical characteristics of acidified canned tomatoes. *J. Food Sci.* 38:195–201.

Schoenemann, D. R., Lopez, A., and Cooler, F. W. 1974. pH and acidic stability during storage of acidified and nonacidified canned tomatoes. *J. Food Sci.* 39:257–259.

Seymour, C. F., Taylor, G., and Welsh, R. C. 1954. Substitution of vinegar for lactic acid as bactericidal agent in infant milk mixtures. *Am. J. Dis. Child.* 88:62–66.

Shadbolt, C., Ross, T., McMeekin, T. A. 2001. Differentiation of the effects of lethal pH and water activity: Food safety implications. *Lett. Appl. Microbiol.* 32:99–102.

Shelef, L. A. 1994. Antimicrobial effects of lactates: A review. *J. Food Prot.* 57:445–450.

Shelef, L. A., and Addala, L. 1994. Inhibition of *Listeria monocytogenes* and other bacteria by sodium diacetate. *J. Food Saf.* 14:103–115.

Shelef, L. A., and Potluri, V. 1995. Behaviour of foodborne pathogens in cooked sausage containing lactates. *Food Microbiol.* 12:221–227.

Shelef, L. A., and Yang, Q. 1991. Growth suppression of *Listeria monocytogenes* by lactates in broth, chicken, and beef. *J. Food Prot.* 54:283–287

Sheu, C. W., and Freese, E. 1972. Effects of fatty acids on growth and envelope proteins of *Bacillus subtilis. J. Bacteriol.* 111:516–524.

Sheu, C. W., Konings, W. N., and Freese, E. 1972. Effects of acetate and other short-chain fatty acids on sugar and amino acid uptake of *Bacillus subtilis. J. Bacteriol.* 111:525–530.

Sheu, C. W., Salomon, D., Simmons, J. L., Sreevalsan, T., and Freese, E. 1975. Inhibitory effects of lipophilic acids and related compounds on bacteria and mammalian cells. *Antimicrob. Agents Chemother.* 7:349–363.

Shi, H., Stroshine, R. L., and Ileleji, K. 2017. Determination of the relative effectiveness of four food additives in degrading aflatoxin in distillers wet grains and condensed distillers solubles. *J. Food Prot.* 80:90–95.

Signorini, M., Costa, M., Teitelbaum, D., Restovich, V., Braseco, H., García, D., Superno, V., Petroli, S., Bruzzone, M., Arduini, V., Vanzini, M., Sucari, A., Suberbie, G., Maricel, T., Rodríguez, R., and Leotta, G. A. 2018. Evaluation of decontamination efficacy of commonly used antimicrobial interventions for beef carcasses against Shiga toxin-producing *Escherichia coli. Meat Sci.* 142:44–51.

Sinha, D. P., Drury, A. R., and Conner, G. H. 1968. The *in vitro* effect of citric acid and sodium citrate on *Streptococcus agalactiae* in milk. *Indian Vet. J.* 45:805–810.

Silla-Santos, M. H., Nuñez Kalasic, H., Casado Goti, A., and Rodrigo Enguidanos, M. 1992. The effect of pH on the thermal resistance of *Clostridium sporogenes* (PA 3679) in asparagus purée acidified with citric acid and glucono-δ-lactone. *Int. J. Food Microbiol.* 16:275–281.

Siragusa, G. R., and Dickson, J. S. 1992. Inhibition of *Listeria monocytogenes* on beef tissue by application of organic acids immobilized in a calcium alginate gel. *J. Food Sci.* 57:293–296.

Siragusa, G. R., and Dickson, J. S. 1993. Inhibition of *Listeria monocytogenes, Salmonella typhimurium* and *Escherichia coli* O157:H7 on beef muscle tissue by lactic or acetic acid contained in calcium alginate gels. *J. Food Saf.* 13:147–158.

Smith, J. L., and Palumbo, S. A. 1980. Inhibition of aerobic and anaerobic growth of *Staphylococcus aureus* in a model sausage system. *J. Food Saf.* 2:221.

Smith, J. L., Benedict, R. C., Buchanan, R. L., Kalinowski, S., and Palumbo, S. A. 1984. pH dependent acetate injury in *Staphylococcus aureus*: Role of temperature. *Lebensm. Wiss. U. Technol.* 17:271–275.

Smulders, F. J., and Greer, G. G. 1998. Integrating microbial decontamination with organic acids in HACCP programmes for muscle foods: Prospects and controversies. *Int. J. Food Microbiol.* 44:149–169.

Smulders, F. J. M., and Woolthuis, C. H. J. 1983. Influence of two levels of hygiene in the microbiological condition of veal as a product of two slaughtering/processing sequences. *J. Food Prot.* 46:1032–1035.

Smulders, F. J. M., and Woolthuis, C. H. J. 1985. Immediate and delayed microbiological effects of lactic acid decontamination of calf carcasses – Influence on conventionally boned versus hot-boned and vacuum-packaged cuts. *J. Food Prot.* 48:838–847.

Sollman, T. 1921. Studies of chronic intoxication on albino rats. III. Acetic and formic acids. *J. Pharmacol. Exp. Ther.* 16:463–474.

Sorrells, K. M., and Enigl, D. C. 1990. Effect of pH, acidulant, sodium chloride, and temperature on the growth of *Listeria monocytogenes. J. Food Saf.* 11:31–37.

Sorrells, K. M., Enigl, D. C., and Hatfield, J. R. 1989. Effect of pH, acidulant, time, and temperature on the growth and survival of *Listeria monocytogenes. J. Food Prot.* 52:571–573.

Spector, W. S. 1956. *Handbook of Toxicology*, Vol. 1. W. B. Saunders, Philadelphia, PA.

Spencer, H. C., Rowe, V. K., and McCollister, D. D. 1950. Dehydro-acetic acid I. Acute and chronic toxicity. *J. Pharmacol. Exp. Ther.* 99:57–68.

Stekelenburg, F. K., and Kant-Muermans, M. L. T. 2001. Effects of sodium lactate and other additives in a cooked ham product on sensory quality and development of a strain of *Lactobacillus curvatus* and *Listeria monocytogenes. Int. J. Food Microbiol.* 66:197–203.

Stillmunkes, A. A., Prabhu, G. A., Sebranek, J. G., and Molins, R. A. 1993. Microbiological safety of cooked beef roasts treated with lactate monolaurin or gluconate. *J. Food Sci.* 58:953–958.

Stroup, W. H., Dickerson, R. W., Jr., and Johnston, M. R. 1985. Acid equilibrium development in mushrooms, pearl onions and cherry peppers. *J. Food Prot.* 48:590.

Suganthi, S., Mohanapriya, S., Raj, V., Kanaga, S., Dhandapani, R., Vignesh, S., and Sundar, J. K. 2018. Tunable physicochemical and bactericidal activity of multicarboxylic-acids-crosslinked polyvinyl alcohol membrane for food packaging applications. *Chem. Select.* 3:11167–11176.

Sullivan, E. K., Manns, D. C., Churey, J. J., Worobo, R. W., and Padilla-Zakour, O. I. 2013. Pickled egg production: Inactivation rate of *Salmonella, Escherichia coli* O157:H7, *Listeria monocytogenes*, and *Staphylococcus aureus* during the acidification step. *J. Food Prot.* 76:1846–1853.

Subramanian, C. S., and Marth, E. H. 1968. Multiplication of *Salmonella typhimurium* in skim milk with and without added hydrochloric, lactic and citric acids. *J. Milk Food Technol.* 31:323–326.

Takhirov, M. T. 1969. Hygienic standards for acetic acid and acetic anhydride in air. *Hyg. Sanit.* 34:122–125.

Tamblyn, K. C., and Conner, D. E. 1997. Bactericidal activity of organic acids against *Salmonella typhimurium* attached to broiler chicken skin. *J. Food Prot.* 60:629–633.

Tang, X., Ouyang, Q., Jing, G., Shao, X., and Tao, N. 2018. Antifungal mechanism of sodium dehydroacetate against *Geotrichum citri-aurantii. World J. Microbiol. Biotechnol.* 34:29.

Tango, C.-N., Mansur, A.-R., Kim, G.-H., and Oh, D.-H. 2014. Synergetic effect of combined fumaric acid and slightly acidic electrolyzed water on the inactivation of food-borne pathogens and extending the shelf life of fresh beef. *J. Appl. Microbiol.* 117:1709–1720.

Taylor, T. M., Joerger, R., Palou, E., López-Malo, A., Avila-Sosa, R., and Cálix-Lara, T. 2012. Alternatives to traditional antimicrobials for organically processed meat and poultry. In *Organic Meat Production and Processing*, 1st ed. Edited by S. C. Ricke, E. J. Van Loo, M. G. Johnson, and C. A. O'Bryan. John Wiley and Sons, Ltd., Ames, IA, pp. 211–238.

Taylor, T. M., Ravishankar, S., Bhargava, K., and Juneja, V. K. 2019. Chemical preservatives and natural food antimicrobials. In *Food Microbiology: Fundamentals and Frontiers*, 5th ed. Edited by M. P. Doyle, F. Diez-Gonzalez, and C. Hill. ASM Press, Washington, DC, pp. 705–731.

Tejero-Sariñena, S., Barlow, J., Costabile, A., Gibson, G. R., and Rowland, I. 2012. *In vitro* evaluation of the antimicrobial activity of a range of probiotics against pathogens: Evidence for the effects of organic acids. *Anaerobe* 18:530–538.

Thomson, J. E., Banwart, G. J., Sanders, D. H., and Mercuri, A. J. 1967. Effect of chlorine, antibiotics, ß-propiolactone, acids, and washing on *Salmonella typhimurium* on eviscerated fryer chickens. *Poult. Sci.* 46:146–151.

Tinney, K. S., Miller, M. F., Ramsey, C. B., Thompson, L. D., and Carr, M. A. 1997. Reduction of microorganisms on beef surfaces with electricity and acetic acid. *J. Food Prot.* 60:625–628.

Tolen, T. N., Ruengvisesh, S., and Taylor, T. M. 2017. Application of surfactant micelle-entrapped eugenol for prevention of growth of the Shiga toxin-producing *Escherichia coli* in ground beef. *Foods* 6:6080069.

Trainer, J. B., Krippaehne, W. W., Hunter, W. C., and Lagozzino, D. A. 1945. Esophageal stenosis due to lactic acid. *Am. J. Dis. Child.* 69:173–175.

Tsang, N., Post, L. S., and Solberg, M. 1985. Growth and toxin production by *Clostridium botulinum* in model acidified systems. *J. Food Sci.* 50:961–965.

Tuft, L., and Ettelson, L. N. 1956. Canker sores from allergy to weak organic acid (citric and acetic). *J. Allergy* 27:536–543.

Tzatzarakis, M., Tsatsakis, A. M., Liakou, A., and Vakalounakis, D. J. 2000. Effect of common food preservatives on mycelial growth and spore germination of *Fusarium oxysporum*. *J. Environ. Sci. Health* B35:527–537.

Unda, J. R., Molins, R. A., and Walker, H. W. 1991. *Clostridium sporogenes* and *Listeria monocytogenes*: Survival and inhibition in microwave-ready beef roasts containing selected antimicrobials. *J. Food Sci.* 56:198–205.

U.S. Department of Agriculture – Food Safety and Inspection Service (USDA-FSIS). 2019. Directive 7120.1: Safe and suitable ingredients used in the production of meat, poultry, and egg products, Revision 52. Available from: https://www.fsis.usda.gov/wps/portal/fsis/topics/regulations/directives/7000-series/safe-suitable-ingredients-related-document (accessed on 11 November 2019).

U.S. Environmental Protection Agency (EPA). 2016. Pesticide reregistration status: Sodium diacetate. Available from: https://archive.epa.gov/pesticides/reregistration/web/html/status.html (accessed on 13 November 2019).

Uyttendaele, M., Jozwik, E., Tutenel, A., De Zutter, L., Uradzinski, J., Pierard, D., and Debevere, J. 2001. Effect of acid resistance of *Escherichia coli* O157:H7 on efficacy of buffered lactic acid to decontaminate chilled beef tissue and effect of modified atmosphere packaging on survival of *Escherichia coli* O157:H7 on red meat. *J. Food Prot.* 64:1661–1666.

Valero, M., Fernández, P. S., and Salmerón, M. C. 2003. Influence of pH and temperature on growth of *Bacillus cereus* in vegetable substrates. *Int. J. Food Microbiol.* 82:71–79.

Valero, M., Leontidis, S., Fernández, P. S., Martínez, A., and Salmerón. 2000. Growth of *Bacillus cereus* in natural and acidified substrates over the temperature range 5-30°C. *Food Microbiol.* 17:605–612.

van der Marel, G. M., van Logtestijn, J. G., and Mossel, D. A. A. 1988. Bacteriological quality of broiler carcasses as affected by in-plant lactic acid decontamination. *Int. J. Food Microbiol.* 6:31–42.

Van Netten, P., and Huis In't Veld, J. H. 1994. The effect of lactic acid decontamination on the microflora on meat. *J. Food Saf.* 14:243–257.

Velugoti, P. R., Bohra, L. K., Juneja, V. K., and Thippareddi, H. 2007. Inhibition of germination and outgrowth of *Clostridium perfringens* spores by lactic acid salts during cooling of injected turkey. *J. Food Prot.* 70:923–929.

Verdi, M. C., Melian, C., Castellano, P., Vignolo, G., and Massani, M. B. 2019. Synergistic antimicrobial effect of lactocin AL705 and nisin combined with organic acid salts against *Listeria innocua* 7 in broth and a hard cheese. *Int. J. Food Sci. Technol.* doi: 10.1111/ijfs.14302.

Visser, I. J. R., Koolmees, P. A., and Bijker, P. G. H. 1988. Microbiological conditions and keeping quality of veal tongues as affected by lactic acid decontamination and vacuum packaging. *J. Food Prot.* 51:208–213.

Watada, A. E. 1971. Postharvest physiology of strawberry fruits treated with sodium dehydroacetate. *Am. Soc. Hort. Sci.* 96:177–179.

Waterhouse, H. N., and Scott, H. M. 1962. Effect of sex, feathering, rate of growth and acetates on chick's need for glycine. *Poult. Sci.* 41:1957–1960.

Weaver, R. A., and Shelef, L. A. 1993. Antilisterial activity of sodium, potassium or calcium lactate in pork liver sausage. *J. Food Saf.* 13:133–146.

Wederquist, H. J., Sofos, J. N., and Schmidt, G. R. 1994. *Listeria monocytogenes* inhibition in refrigerated vacuum packaged turkey bologna by chemical additives. *J. Food Sci.* 59:498–500.

Weil, A. J., and Rogers, H. E. 1951. Allergic reactivity to simple alipathic acids in man. *J. Invest. Dermatol.* 17:227–231.

Weissinger, W. R., and Beuchat, L. R. 2000. Comparison of aqueous chemical treatments to eliminate *Salmonella* on alfalfa seeds. *J. Food Prot.* 63:1475–1482.

Weissinger, W. R., McWatters, K. H., and Beuchat, L. R. 2001. Evaluation of volatile chemical treatments for lethality to *Salmonella* on alfalfa seeds and sprouts. *J. Food Prot.* 64:442–450.

Wemmenhove, E., van Valenberg, H. J. F., Zwietering, M. H., van Hooijdonk, T. C. M., and Well-Bennik, M. H. J. 2016. Minimal inhibitory concentrations of undissociated lactic, acetic, citric, and propionic acid for *Listeria monocytogenes* under conditions relevant to cheese. *Food Microbiol.* 58:63–67.

Wen, A., Bai, X., Shah, A. A., and Hu, H. 2017. Effect of dietary sodium diacetate on growth performance, carcass characteristics, meat quality, intestinal pH and *Escherichia coli* of broilers. *Anim. Prod. Sci.* 58:1323–1328.

Wiseman, R. D., and Adler, D. K. 1956. Acetic acid sensitivity as a cause of cold urticaria. *J. Allergy* 27:50–55.

Wolf, M. J., Miller, M. F., Parks, A. R., Loneragan, G. H., Garmyn, A. J., Thompson, L. D., Echeverry, A., and Brashears, M. M. 2012. Comparing the effectiveness of a lactic acid dip with a lactic acid spray for reducing *Escherichia coli* O157:H7, *Salmonella*, and non-O157 Shiga toxigenic *Escherichia coli* on beef trim and ground beef. *J. Food Prot.* 75:1968–1973.

Wolf, P. A. 1950. Dehydroacetic acid: A new microbiological inhibitor. *Food Technol.* 4:294–297.

Wolford, E. R., and Anderson, A. A. 1945. Propionates control microbial growth in fruits and vegetables. *Food Ind.* 17:622–624, 726, 728, 730, 732, and 734.

Wong, H.-C., and Chen, Y.-L. 1988. Effects of lactic acid bacteria and organic acids on growth and germination of *Bacillus cereus*. *Appl. Environ. Microbiol.* 54:2179–2184.

Woolford, M. K. 1975a. Microbiological screening of food preservatives, cold sterilants and specific antimicrobial agents as potential silage additives. *J. Sci. Food Agric.* 26:226–237.

Woolford, M. K. 1975b. Microbiological screening of the straight chain fatty acids (C_1-C_{12}) as potential silage additives. *J. Sci. Food Agric.* 28:219–228.

Woolthuis, C. H. J., and Smulders, F. J. M. 1985. Microbial decontamination of calf carcasses by lactic acid sprays. *J. Food Prot.* 48:832–837.

Woolthuis, C. H. J., Mossel, D. A. A., Van Logtestijn, J. G., DeKruiff, J. M., and Smulders, F. J. M. 1984. Microbial decontamination of porcine liver with lactic acid and hot water. *J. Food Prot.* 47:220–226.

Wright, L. D., and Skeggs, H. R. 1946. Reversal of sodium propionate inhibition of *Escherichia coli* with ß-alanine. *Arch. Biochem.* 10:383–386.

Xiong, R., Xie, G., and Edmondson, A. S. 1999. The fate of *Salmonella enteritidis* PT4 in home-made mayonnaise prepared with citric acid. *Lett. Appl. Microbiol.* 28:36–40.

Yamamura, A., Murai, A., Takamatsu, H., and Watabe, K. 2000. Antimicrobial effect of chemical preservatives on enterohemorrhagic *Escherichia coli* O157:H7. *J. Health Sci.* 46:204–208.

Yokotani, H., Usui, T., Nakaguchi, T., Kanabayashi, T., Tanda, M., and Aramaki, Y. 1971. Acute and subacute toxicological studies of TAKEDA-citric acid in mice and rats. *J. Takeda Res. Lab.* 30:35.

Young, E. G., and Smith, R. P. 1944. Lactic acid: A corrosive poison. *J. Am. Med. Assoc.* 125:1179–1181.

Zaki, H. M. B. A., Mohamed, H. M. H., and El-Sherif, A. M. A. 2015. Improving the antimicrobial efficacy of organic acids against *Salmonella enterica* attached to chicken skin using SDS with acceptable sensory quality. *LWT – Food Sci. Technol.* 64:558–564.

Zeitoun, A. A. M., and Debevere, J. M. 1990. The effect of treatment with buffered lactic acid in microbial decontamination and on shelf life of poultry. *Int. J. Food Microbiol.* 11:305–311.

Zeitoun, A. A. M., and Debevere, J. M. 1991. Inhibition, survival and growth of *Listeria monocytogenes* on poultry as influenced by buffered lactic acid treatment and modified atmosphere packaging. *Int. J. Food Microbiol.* 14:161–170.

Zhang, H., Wei, H., Cui, Y., Zhao, G., and Feng, F. 2009. Antibacterial interactions of monolaurin with commonly used antimicrobials and food components. *J. Food Sci.*74:M418–M421.

Zhang, L. W., Koci, J., Jeffery, B., Riviere, J. E., and Monteiro-Riviere, N. A. 2015. Safety assessment of potential food ingredients in canine hepatocytes. *Food Chem. Toxicol.* 78:105–115.

Zhang, S., and Farber, J. M. 1996. The effects of various disinfectants against *Listeria monocytogenes* on fresh-cut vegetables. *Food Microbiol.* 13:311–321.

Zhuang, R.-Y., Huang, Y.-W., and Beuchat, L. R. 1996. Quality changes during refrigerated storage of packaged shrimp and catfish fillets treated with sodium acetate, sodium lactate or propyl gallate. *J. Food Sci.* 61:241–244.

6

Sulfur Dioxide and Sulfites

Lilian Were and Leandra Filiaci

CONTENTS

6.1 Overview of Sulfite Use in Food

Sulfites have been used since ancient times by the Romans and Egyptians as an antiseptic, disinfectant, and to sanitize wine vessels. Sulfites are one of the oldest food preservatives and have been used since 1664, when casks that contained sulfur dioxide were filled with cider to prevent spoilage (Roberts and McWeeny 1972). Besides wine, sulfites can also be found in foods such as dried, canned, or frozen fruits and vegetables, canned and dried seafood, gravies, jams, and doughs (Vally, Misso, and Madan 2009, Vally and Misso 2012, Aguilar et al. 2016, USDA 2011). The main food vehicles leading to sulfite exposure for children and teenagers are non-alcoholic beverages, sausages, dried fruit, and nuts in contrast to wine for adults.

Sulfur dioxide can be produced during combustion of hydrogen sulfide (H_2S):

$$2H_2S + 3O_2 \rightarrow 2H_2O + 2SO_2$$

or

oxidizing elemental sulfur:

$$S_8 + 8O_2 \rightarrow 8SO_2 \quad \text{(USDA 2011).}$$

Sulfur dioxide and inorganic salts of sulfurous acid sulfites are used in the food industry as antimicrobials, and bleaching and reducing agents to prevent enzymatic and non-enzymatic browning. The most commonly used sulfite forms in food include sulfur dioxide, sodium and calcium sulfite, sodium and potassium bisulfite, and sodium and potassium metabisulfite.

Although sulfites serve a variety of functions in the food industry, concerns of sulfite toxicity began in the 1970s when reports of adverse reactions to sulfites such as triggering anaphylactic responses, dermatitis, urticaria, flushing, hypotension, abdominal pain, and diarrhea started to emerge (Vally and Misso 2012). Toxicology studies to assess the safety of sulfites have continued to be conducted into recent years (Akdogan et al. 2011, Kocamaz et al. 2012, Fatima Zohra El et al. 2014).

With the negative consumer perceptions of chemical preservatives in foods, added sulfites are included on the "No-No List" for some groceries and restaurants in the USA. However, despite sulfites being controversial preservatives in food and beverages, sulfites are not easy to replace as a food ingredient due to their multiple functions. Alternatives to sulfites are also often more expensive and less effective, and some are not approved for use (Lisanti et al. 2019, Vally and Misso 2012).

This chapter will outline uses of sulfites in food, with an emphasis on sulfites' antimicrobial activity, health concerns, and global regulations on sulfite use. Emerging technologies for sulfite detection, and alternatives to sulfites as an antimicrobial in wine are also outlined.

6.2 Chemical Properties of Sulfites Used in Food

The concentration of free sulfites in food is directly correlated to their preservative activity expressed as equivalents of sulfur dioxide. Depending on the pH, sulfites can exist in various forms: at a low pH (below 1.8), the sulfurous acid form is favored; at an intermediate pH (pH 2–6), the bisulfite ion is favored; while at a higher pH (pH >7), sulfite ions are favored (Vally, Misso, and Madan 2009; Figure 6.1). pH control is extremely important in the effective use of sulfur dioxide as molecular sulfur dioxide is the most effective form for inhibiting yeast. A 10-fold increase in molecular sulfur dioxide occurs when pH is reduced from 4.0 to 3.0 (de Azevedo et al. 2007). The amount of molecular SO_2 (mg/L) in wine can be calculated using free SO_2 mg/L/$(1 + 10^{pH - 1.81})$. A plot of the distribution of the three ionic forms is shown in Figure 6.1.

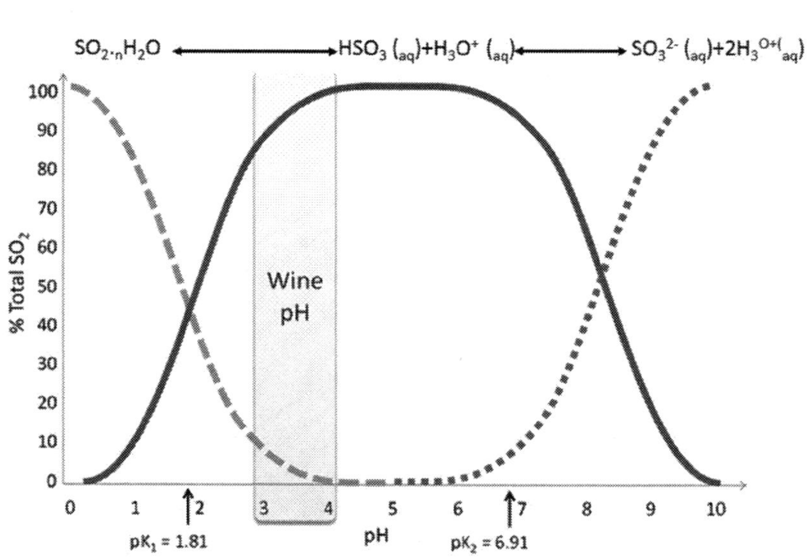

FIGURE 6.1 Sulfite equilibrium as a function of pH. pK_1 is the dissociation constant of the chemical reaction $SO_2 \bullet H_2O \leftrightarrow HSO_3^- + H^+$ and pK_2 that of the reaction $HSO_3^- \leftrightarrow SO_3^{2-} + H^+$. The effective pH range of 3 to 4 in wine is highlighted (Vally, Misso, and Madan 2009, *Nadai* 2013, Cojocaru and Antoce 2012).

It is useful to have the sulfur dioxide in a salt form, as the dry salts are easier to store and are less of a problem to handle than gaseous or liquid sulfur dioxide. The metabisulfite is the anhydride of the acid sulfite:

$$2HSO_3^- \leftrightarrow S_2O_5^{2-} + H_2O$$

Sulfites can exist either as free or bound. Depending on the storage conditions and the food system, sulfites can react with carbonyl groups of reducing sugars (aldehydes, ketones), and proteins. When considering apple juice, the sugars present including glucose, xylose, and L-xylosone bind to sulfites to form covalent hydrosulfonate adducts. In addition to these sugars, sulfites may also bind to arabinose and galacturonic acid from the break-down of pectin, and acetaldehyde, pyruvic acid, and 2-ketoglutaric acid from yeast fermentation (Burroughs and Sparks 1964). The proportion of bound and free sulfites in a food depends on temperature, pH, concentration of sulfites, and the proximate composition of the food (Wedzicha 1992).

During wine fermentation, sulfites may bind to carbonyls formed from yeast fermentation (acetaldehyde, pyruvic acid, and 2-ketoglutaric acid) to a greater extent than sugars and other carbonyls that are naturally present in fruit-based beverages. Jarvis and Lea (2000) found that the binding capacity of sulfites was highest with acetaldehyde at a dissociation constant of 0.002 mM, and lowest with arabinose (dissociation constant of 40 mM). Other compounds found in fermented beverages that can bind to sulfites and their dissociation constants are as follows: 2,5-diketogluconic acid (0.4 mM), L-xylosone (1.4 mM), D-threo-2,5-hexodiulose (3.4 mM), and galacturonic acid 17 mM (Wedzicha, 1992). The yeast strain affects the sulfite-binding capacity in ciders. Yeast strains with higher fermentative capacities in ciders result in higher sulfite binding. Added ingredients also affect the binding capacity of sulfites in ciders (Jarvis and Lea 2000).

Sulfites can also react with anthocyanins and the flavin moiety of flavoenzymes to result in reversible binding (Nikolantonaki, Magiatis, and Waterhouset 2015). The reaction between sulfites and non-carbonyl groups does not significantly change the concentration of sulfites in foods (Wedzicha 1992).

Irreversibly bound sulfites refer to the portion of the chemical that is not recovered when sulfite concentration is analyzed. The main cause of irreversibly bound sulfites in food is from reactions between intermediate compounds formed during non-enzymatic Maillard browning, and between disulfide bonds present in proteins. Irreversibly bound sulfites can also form when polyphenol oxidase (PPO) enzyme reacts with phenols. During this enzymatic browning, monophenols are converted to *o*-diphenols, which then form *o*-quinones that produce dark melanin pigments. In the presence of sulfites, sulfites can react with quinones, forming sulfonated *o*-quinones, which are not further oxidized by PPO (Wedzicha 1992). Bisulfites can also react with quinones and reduce them back to diphenols. The suggested protection of thiols when bisulfites are used in wine is attributed to sulfites more easily reacting with quinones. However, at lower levels of sulfite, thiols and polyphenols can outcompete sulfites and react with quinones (Danilewicz, Seccombe, and Whelan 2008). Sulfites in food can also be converted to sulfate (SO_4^{-2}), due to autooxidation or its reaction with other oxidizing agents in foods. For example, during the dehydration of vegetables, about 10% of sulfite is converted to sulfate (Wedzicha 1992). Cascado et al. (2014) found that after 46 weeks of storage, pickled blanched garlic treated with 100 mg/L sulfites had 70% lower sulfite levels due to its oxidation to sulfate. After 6 months of storage, there was no free sulfite as it was bound to garlic compounds (Cascado et al. 2014).

6.3 Factors Affecting Antimicrobial Activity

It has also been noted that the antimicrobial effects of sulfites are stronger against bacteria than molds, and are least effective against yeasts (Walker 1985, de Oliva-Neto and Yokoya 2001, Faparusi 1969, Rojo-Bezares et al. 2007). Chang, Kim, and Shin (1997) for example found that adding sulfite up to 400 ppm did not affect yeast cells, while a sulfite concentration above 100 ppm decreased the CFU of bacterial strains exponentially. The form of sulfite most effective as an antimicrobial is sulfur dioxide, with 1000 times more effectiveness compared to the hydrogen sulfite ion against *Escherichia coli*. For yeast and *Aspergillus niger*, the antimicrobial activity of sulfur dioxide is 500 and 100 times stronger than the hydrogen sulfite ion, respectively (Wedzicha 1984).

The growth-inhibiting or lethal effects of sulfurous acid are most intense when the acid is in the non-ionized form. The antimicrobial activity of sulfites is pH-dependent as the proportions of sulfur dioxide, bisulfite, and sulfite are influenced by pH. At a low pH, around 3, the active sulfur dioxide form is favored, which is the form with the greatest antibacterial activity (Corte et al. 2012; Wedzicha, 1992). McFeeters (1998), for instance, found that the highest pH in which microbial growth in stored cucumbers was prevented was 3.65 when 300 ppm sodium metabisulfite was added, compared to 100 ppm sodium metabisulfite when the pH was decreased to 3.0. When Faparusi (1969) examined the effect of pH on the preservation activity of metabisulfite against microorganisms (*Leuconostoc mesenteroides, L. plantarum, Sarcina, Acetobacter, Corynebacterium, Streptococcus*, and the yeast *Saccharomyces cerevisiae*) commonly found in palm wine, they found that at pH 2.5, 100 ppm of sodium metabisulfite was sufficient to inhibit bacterial growth but 150 ppm of sodium metabisulfite was needed to suppress yeast growth at pH 3. At pH 4 to 7, yeast growth could not be inhibited at concentrations as high as 2000 ppm sodium metabisulfite. Bacterial growth was however inhibited at 800 ppm at pH 4 but could not be inhibited at pH 4.5 and 7 during the 30-day study. Aforementioned studies suggest that foods with pH above 4 require sulfite levels that may exceed the recommended average daily intake (ADI) for human consumption to inhibit microbial growth.

The concentration of ethanol also affects the inhibition of LAB by sulfites. When 6% ethanol was added to growth media, the MIC_{50} of metabisulfite against *Oenococcus oeni* increased from 50 μg/ml to 100 μg/ml. However, with other LAB, the opposite effect was observed. The presence of 6% ethanol decreased resistance to metabisulfite and doubled the MIC_{50} value when compared to growth media without ethanol (Rojo-Bezares et al. 2007).

6.4 Antimicrobial Activity of Sulfites

Sulfites are broad-spectrum antimicrobials. There are various proposed mechanisms for the antimicrobial activity of sulfites (Figure 6.2). When sulfur dioxide enters bacterial cells, it disrupts protein and enzyme activity leading to antibacterial activity. Only the molecular sulfur dioxide form can enter bacterial cells, and therefore act as an antimicrobial. One mechanism of antimicrobial activity occurs when metal catalyzed sulfites oxidize to produce reactive intermediates. The $SO_4^{\bullet -}$ radicals formed by sulfite oxidation (Chen et al. 2017) can react with components of bacterial cells, leading to oxidation by obstructing hydrogen atoms or electron transfer (Neta and Huie 1985).

Autooxidation of sulfites can destroy tryptophan and allow sulfite to react with other amino acids and nucleic acids. Through these mechanisms, the damage caused by sulfites directly depends on their likelihood of forming radicals, which leads to oxidative damage to cells. Another possible mechanism for the antimicrobial effects of sulfites is the diffusion of sulfurous acid into the cell and disrupting of thiol groups in the structural proteins and enzymes of the cell (Figure 6.2).

The cytoplasmic membrane has a high affinity for reaction with SO_2. One type of activity of sulfite against the yeast cell is its reaction with cellular adenosine triphosphate (ATP) and/or its blocking of the cystine disulfide linkages (Schimz 1980). Hinze, Maier, and Holzer (1981) found reduced ATP activity when SO_2 was added to lactic acid bacteria. Anacleto and Vanuden (1982) suggested that the antimicrobial effects of SO_2 occurred at the surface of the cell. They proposed two receptor sites. One site was directly related to the death process while the other modulated the entropy of activation of the process.

Thiamine pyrophosphate, a required cofactor for many enzymatic reactions, can be destroyed by the action of bisulfite (Williams et al. 1935). Lenz and Holzer (1985) showed that the depletion of thiamine pyrophosphate (TPP) in *Saccharomyces cerevisiae* by SO_2 at levels used for juice and wine preservation caused TPP-dependent enzymes to decrease in activity. Pyruvate decarboxylase and transketolase lost 42% and 87% of their activity, respectively. TPP activity losses were slower than ATP losses, though overall losses were about the same.

Another reaction of significance is that between bisulfite and disulfide bonds:

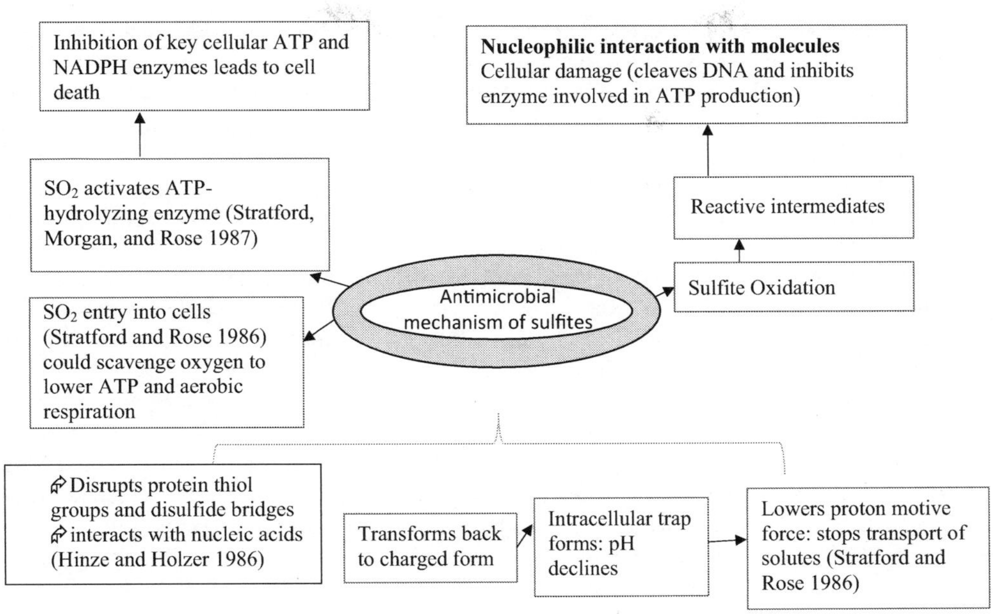

FIGURE 6.2 Antimicrobial mechanism of sulfites.

$$R_1\text{-S-S-}R_2 + HSO_3^- \rightarrow R_1SH + R_2\text{-S-}SO_3^-$$

This reaction can cause conformational changes in enzymes. Both free and bound sulfite have antimicrobial activity, with bound sulfites having lower antimicrobial activity (Osborne and Edwards 2006, Wells and Osborne 2011). The effectiveness of sodium sulfite differs between *in vitro* and food models. Lamas et al. (2016) for instance found that *Salmonella enterica* strains demonstrated a minimum inhibitory concentration (MIC) of 50,000 ppm sodium sulfite while 4 *L. monocytogenes* strains had an MIC of 625 ppm sodium sulfite *in vitro*. The growth inhibition of *S. enterica* and *L. monocytogenes* in minced meat with sodium sulfite was however similar (Lamas et al. 2016).

When considering the antimicrobial activity of sulfur dioxide and its salts, three main groups of microbes are of interest in acidic beverages and fruits. These are as follows: (1) acetic acid-producing and lactic acid-producing bacteria, (2) fermentation and spoilage yeasts, and (3) fruit molds. Sulfites are used in other foods and pharmaceuticals, but their major use as an antimicrobial agent is in beverages and fruits.

6.4.1 Antibacterial Activity

Bacteria common in acidic fruits and beverages are the lactic acid-producing genera *Lactobacillus*, *Leuconostoc*, *Pediococcus*, and *Oenococcus* (Rojo-Bezares et al. 2007). The main bacteria of concern in ethanol fermentation are the proliferation of lactic acid bacteria (LAB). During ethanol fermentation, bacterial contamination can inhibit yeast growth and cause losses in ethanol yield. LAB ferment carbohydrates to lactic acid, resulting in lower ethanol yield and inhibiting yeast fermentation (Chang et al. 1995). *L. mesenteroides* can also cause ropiness in beverages through the formation of dextrans (Montersino et al. 2008).

At low pH, some LAB, e.g. *Oenococcus oeni*, can develop sulfite tolerance attributable to their ability to produce stress proteins and maintain high internal pH of 5.8–6.3 in an environment or a food with a pH in the range of 3.0–5.0 (Guzzo, Jobin, and Divies 1998). At pH 3.5, high sulfite concentrations of 30 mg/L significantly lowered the number of viable cells. This inhibition was lowered after cells were grown in acidic medium of pH 3.5 before exposure to sulfites, as the *O. oeni* grown in acidic conditions had adapted to the low pH and weakened sulfite antimicrobial effects.

Irwin et al. (2017) examined the effects of sodium bisulfite ($NaHSO_3$) and sodium sulfite (Na_2SO_3) on four common beneficial bacterial species: *Lactobacillus* species *casei, plantarum,* and *rhamnosus,* as well as *Streptococcus thermophilus.* The bacteria cultures were exposed for 6 hours to 10–3780 ppm sulfite concentrations. Irwin et al. (2017) observed bactericidal effects in all four species when exposed to 1000–3780 ppm sulfite. Facultative heterofermentative LAB (*L. casei, L. rhamnosus,* and *L. plantarun*) were found to be more susceptible to sulfites than homofermentative LAB (*Streptococcus thermophilus*) due to lower acid production by *S. thermophilus* in BHI growth medium, resulting in lower production of bacteriostatic SO_2 (Irwin et al. 2017). Although more research needs to be done on longer exposure times and in conditions that mimic digestion, Irwin et al. (2017) suggested the possible detrimental effects sulfites may have on beneficial probiotic microbes.

Lactobacillus casei cells were found to be more sensitive to sulfite than *L. fermentum,* with an LC_{50} of sulfite at 130 and 100 ppm, respectively. *D*-values for *L. casei* and *L. fermentum* were 0.79 and 0.96 h in the presence of 134 ppm sulfite, respectively. At 287 ppm sulfite, *D*-values of 0.61 and 0.9 h were observed, respectively (Chang, Kim, and Shin 1997). These researchers observed bactericidal effects when *L. casei* and *L. fermentum* cells were treated with 100 ppm sulfite in the presence of oxygen, although the cells were unaffected by 400 ppm of sulfite treatment in the absence of oxygen. Chang, Kim, Shin, and Lee (1997) hypothesized that the oxidation of sulfite to sulfur trioxide radical in the presence of peroxidase of LAB resulted in death of the LAB. Some LAB can also produce hydrogen peroxide. When H_20_2 reacts with sulfite, sulfur trioxide radicals that are toxic to bacteria may form, resulting in different susceptibilities of bacteria to sulfites depending on H_20_2 production (Irwin et al. 2017).

6.4.2 Activity against Yeasts

The predominant spoilage yeasts in the grape and wine industry belong to the species *Zygosaccharomyces bailii, Schizosaccharomyces pombe, Saccharomycodes ludwigii,* and *Brettanomyces bruxellensis,* among others (Loureiro and Malfeito-Ferreira 2003). The use of sulfur dioxide to deplete the wild yeast in grape juice has been a standard practice dating back many years (Cruess 1912). Yeasts grow best in acidic conditions where competition from bacteria is limited (Pilkington and Rose 1988). The preservative effects of sulfites are strongest at low pH, as sulfites can diffuse into yeast cells, and disrupt cellular processes (Figure 6.2). The pH inside the yeast cells is higher than outside the cell (about pH 6.5), causing a portion of SO_2 entering the cell to be converted to HSO_3^-. When inside the yeast cell, sulfites cannot leave, resulting in an intracellular sulfite trap where the extent of the trap depends on the intracellular pH. The accumulation of sulfites inside the yeast cell causes the intracellular pH to decline, lowering the transmembrane gradient and weakening the proton motive force across the plasma membrane. This causes processes inside the cell, such as the active transport of solutes, that need energy from the proton motive force, to stop (Stratford and Rose 1986). When inside the yeast cell, sulfites can also interact with proteins and nucleic acids and disrupt disulfide bridges, leading to cell death. For example, when *Saccharomyces cerevisiae* was exposed to 5.0 mM sulfites, inhibition of the enzyme glyceraldehyde-3-phosphate dehydrogenase was observed, leading to inhibition of ATP synthesis (Hinze and Holzer 1986).

Corte et al. (2012) exposed *S. cerevisiae* cells to 100 mg/mL of potassium metabisulfite at pH levels 2 to 8; the strongest biocidal effects were observed at pH 3, as to be expected since at this pH a majority of sulfite is present in the SO_2 molecular form. However, at pH 4 to 6, where the ionic form of potassium metabisulfite is high and levels of molecular SO_2 are low, some induced mortality of *S. cerevisiae* cells was still observed. The bisulfite ions could thus be converted to SO_2 before passing through the cell membrane, or could be transported through the membrane (Corte et al. 2012).

The minimum inhibitory concentrations (MICs) of sulfites against yeast species have been extensively studied (Table 6.1). Warth (1985) found that resistance of yeast to sulfur dioxide range from 0.05 M free sulfur dioxide for *Kloeckera apiculata* to 2.8 mM for *Z. bailii* at 25°C under aerobic conditions at pH 3.5 in tryptone yeast extract medium. *Saccharomycodes ludwigii* was nearly as tolerant as *Z. bailii.* Once

TABLE 6.1

Minimum Inhibitory Concentration of Sulfite for Various Yeast Species

Yeast Species	Minimum Inhibitory Concentration (mM free sulfur dioxide)	Reference
Klockera apiculata	0.11	Warth (1985)
Hansenula anomala	0.89	Anaerobic growth prevented at
Sacchraromyces cerevisiae	0.14	pH3.5 and 25°C within 1000 h
Candida krusei	1.2	
S. ludwigii	2.6	
Schizosaccharomyces pombe	1.9	
Zygosaccharomyces bailii	2.9	
Sacchraromyces cerevisiae		(Stratford, Morgan, and Rose 1987)
	1.56 mM free sulfite	
S. ludwigii	>7.8mM	

the inhibition was overcome, the growth rates and cell yields were similar. Sulfur dioxide did not completely inhibit the growth of the yeast strains, possibly due to the yeast strains producing SO_2-binding compounds, such as acetaldehyde, 2-oxoglututarate, and pyruvate, as growth occurred after free SO_2 was depleted (Warth 1985).

S. ludwigii has greater resistance to sulfite than S. cerevisiae (Stratford, Morgan, and Rose 1987). The first few hours of growth of the two yeast strains with sulfites correlated with the loss of free sulfites from the growth media and presence of acetaldehyde, with a higher rate of acetaldehyde production by *S. ludwigii* than by *S. cerevisiae.* Although sulfites can react with several cell components such as carbonyl groups, the antimicrobial mechanism in *S. cerevisiae* at sulfite concentrations of 10–15 mM was attributed to the loss of cellular ATP from inactivation of glyceraldehyde-3-phosphate dehydrogenase (Figure 6.2). Therefore, the difference in sulfite resistance observed between the two yeast strains may be due to difference in the extent to which sulfites access the intracellular parts involved in ATP production (Stratford, Morgan, and Rose 1987). Since transport of sulfite is by free diffusion, the lipid composition of the plasma membrane of yeast cells may affect its resistance to sulfites.

The passage of sulfites into *S. cerevisiae* and *S. bailii* is through free diffusion, with sulfites inhibiting the growth of both yeasts at 3.3 mM, and *S. bailii* being more sensitive to sulfite than *S. cerevisiae* (Pilkington and Rose 1988). Internal pH declined as sulfite accumulated in both yeast strains. When 1 mM sulfite was added to the yeast cultures, there was no effect on growth or ethanol production, but a reduction in free sulfite concentration, and an increase in glycerol and acetaldehyde were observed. In contrast, when 2 mM sulfite was added, the growth of yeast strains decreased as well as the production of ethanol and glycerol. This was also accompanied by a decrease in free sulfite concentration and an increase in acetaldehyde concentration. Therefore, both yeast strains could produce acetaldehyde even when growth was inhibited by 2 mM sulfite (Pilkington and Rose 1988). The difference in production of acetaldehyde in both strains suggests that production of acetaldehyde contributes to the sulfite resistance observed.

Given the importance of yeast in wine fermentation, sulfite resistance which differs by yeast strain is important. The three mechanisms involved in sulfite resistance in yeast are through increasing production of acetaldehyde, regulation of sulfite uptake, and the movement of sulfite in the cell (Casalone et al. 1992). When more acetaldehyde is produced by pyruvate decarboxylation in the fermentation process, the acetaldehyde can combine with sulfites in the cell and therefore lower the amount of free sulfites, reducing cellular stress (Nadai et al. 2016). Sulfite uptake in yeast cells can be regulated by reducing the sulfite entering the cell to hydrogen sulfide, which can be excreted or used in the synthesis of sulfur-containing amino acids. The presence of methionine increases the harmful effects of sulfite by shutting down the genes responsible for turning sulfite into sulfur amino acids. Sulfites can also affect gene expression of MET16, which affects the sulfur-containing amino acid biosynthesis pathway. In strains of yeast that display strong sulfite resistance, expression of MET16 is repressed. In addition to methionine, the presence of adenine can increase sulfite resistance in yeasts (Landete, Ferrer, and Pardo 2005, Aranda et al. 2006), and thus amino acid and nucleic acid metabolism is crucial in affecting sulfite resistance. Lastly, the flow of sulfites in the cell membrane of yeast can be regulated by the membrane pump Ssu1p. The transcription and expression of the SSU1 gene is regulated by the transcription factor FZF1 and is strain dependent. The presence of sulfite can induce the over-expression of the SSU1 gene in some yeast strains, resulting in more resistance (Nardi et al. 2010). Adding an insertion fragment in front of the SSU1 gene can lead to greater sulfite resistance in some yeast strains (Liu et al. 2017). Differences in sulfite resistance between yeast strains can thus be attributed to yeast strains that produce sulfite-binding compounds such as acetaldehyde which prevent the accumulation of free sulfites in the microbes (Pilkington and Rose 1988). In addition to the production of sulfite-binding proteins, sulfite-resistant *S. cerevisiae* strains produce a plasma membrane protein called Ssu1 to efflux sulfite as wine yeast strains lacking the SSU1 gene are more sensitive to sulfites (Garcia-Rios et al. 2019).

Brettanomyces bruxellensis is a yeast of concern in wine-making due to its strong resistance to sulfite (Table 6.2). It has the ability to produce compounds such as 4-ethylphenol and 4-ethylguaiacol that produce off-flavors in wine.

Controlling *B. bruxellensis* is challenging because it is tolerant to sulfites and ethanol, unlike other wine-spoilage microorganisms, and because of its ability to survive in a low-nutrient environment (Guzzon et al. 2018, Smith and Divol 2018, Renouf et al. 2006). *B. bruxellensis* may survive through the primary and malolactic fermentation of the wine, where it has little competition for nutrients and can then metabolize residual sugars, malic acid, acetic acid, and ethanol during wine aging (Renouf, Lonvaud-Funel, and Coulon 2007, Guzzon et al. 2011, Smith and Divol 2018). The resistance to sulfites by *B. bruxellensis* is due to its ability to go into a viable but nonculturable (VNC) phase. Barata et al. (2008) did not observe any VNC state of *B. bruxellensis* when exposed to 0.6–1.0 mg/L SO_2, but several studies have. du Toit, Pretorius, and Lonvaud-Funel (2005) found that adding 0.8 mg/L of SO_2 resulted in cells of *B. bruxellensis* being non-culturable but still viable. Agnolucci et al. (2010) found that the addition of 0.2 mg/L SO_2 resulted in 12–60% culturability of a strain of *B. bruxellensis* while the percent of viable cells was 45–67%. The difference between viable and culturable cells observed suggests that some cells went into a VNC state. In line with these results, Serpaggi et al. (2012) found that after 48 hours of exposure to 0.8 mg/L of SO_2, all cells were no longer culturable, though viability was still high. Edwards and Oswald (2018) found that adding 100 mg/L total SO_2 decreased populations of *B. bruxellensis* to undetectable levels; after 60 days, however, cell growth was observed. Similarly, van Wyk, Farid, and

TABLE 6.2

Effect of Sulfites on *Brettanomyces bruxellensis*

Sulfite Amounts (ppm molecular SO_2)	Key Findings	Reference
50 ppm total SO_2	5 log reduction after 2 months, recovery of *B. bruxellensis* was observed after 10 months where a 1 log reduction similar to untreated wine was observed	(van Wyk, Farid, and Silva 2018)
60–90 ppm total SO_2	*D. bruxellensis* growth increased after declining	(Barata et al. 2008)
100 ppm total SO_2	Cell death observed in 14.5% v/v ethanol Merlot wine potentially attributed to entering a VNC state.	(Edwards and Oswald 2018)
0.5	Strain LO417 isolated from Bordeaux red wine was not viable and culturable after 24 hrs but regrowth was observed after 5 days	(Longin et al. 2016)
0–1	Of 7 strains tested, VNC phase was detected in 5 strains at 0.2 ppm SO_2 and 2 strains at 0.4 ppm	(Agnolucci et al. 2010)
0.8	No culturable cells after 48 hr but viable 9 days later	(Serpaggi et al. 2012)
0.8	No culturable cells, still viable	(du Toit, Pretorius, and Lonvaud-Funel 2005)

Silva (2018) found that cell counts of *B. bruxellensis* were lowered by 5.0 $\log_{10}$-cycles after 2 months of exposure to 50 mg/L free SO_2. However, after 10 months, cell growth increased. Growth recovery after exposure to sulfites occurs sooner when the number of initial cells is higher. Growth recovery is also linked to the cell's ability to decrease cell permeability. This differs between strains of *B. bruxellensis*, which may be the reason for differences observed between studies. Exposure to molecular SO_2 can increase the permeability of *B. bruxellensis* cells (Longin et al. 2016). It has been suggested that lowering temperatures in addition to adding sulfite can lower *B. bruxellensis.* When strain F3 of *B. bruxellensis* was exposed to 0.24 mg/L SO_2, growth was observed at 15°C but not at 10°C. Similarly for strain B1b, growth was observed when exposed to 0.12 mg/L SO_2 at 18°C but not at 15°C (Zuehlke and Edwards 2013).

The control of *B. bruxellensis* involves keeping the winery environment, and any equipment that comes into contact with the wine, as clean as possible. Additionally, the growth of *B. bruxellensis* can be impeded by a combination of free SO_2 concentration of 30 ppm, storage below 15°C, an ethanol concentration of 12–15%, and low residual sugar (Jackson 2008, Guzzon et al. 2018). Finally, *B. bruxellensis* can be removed from wine, prior to bottling, by sterile filtration or by the addition of dimethyl decarbonate/DMDC (Jackson 2008, Renouf, Lonvaud-Funel, and Coulon 2007). The use of DMDC in preserving wine quality is approved in Australia, Europe, and the USA at levels up to 200 ppm. A synergistic effect has been observed when sulfites are added where lower concentrations of each additive can be used.

6.4.3 Activity against Molds

A number of molds can infect fruit during processing, storage, and transit to market. *Botrytis* spp. are likely the most prevalent fungi, but others including *Cladosporium*, *Alternaria*, *Stemphyllium*, *Byssochlamys*, *Penicillium*, *Aspergillus*, *Rhizopus*, and *Uncinula* (powdery mildew) also infect fruit. Whole fruit storage success has depended largely on the use of sulfur dioxide. Fumigation of fruits with sulfur dioxide as a method of preservation during storage has been practiced in California since the 1920s (Lindsey et al. 1989). Table grapes are commonly treated with sulfur dioxide to control growth of gray mold, caused by the fungus *B. cinerea.* Smilanick et al. (1990) found the LD_{50} and LD_{99} of molecular sulfur dioxide for *Botrytis cinerea* spores were 0.1 and 0.2 μg/ml respectively. At pH 3.0, 2.8 μg/ml of SO_2 was sufficient to kill 99% of the spores. At pH 3.5 and 4.0, 9.6 μg/ml and 33.8 μg/ml of SO_2 was needed to kill 99% of spores, respectively. The effect of temperature was also evaluated. At pH 4, exposure of spores to sulfur dioxide at 0, 10, 15, 20, 25, and 32°C for 4 hours resulted in an observed LD_{99} of 886, 343, 300, 57.8, 49.3, and 25. 8 μg/ml sulfur dioxide, respectively, in buffered grape juice (Smilanick et al. 1990). Domingues et al. (2018) reported that "BRS Vitoria" table grapes exposed to SO_2-generating pads presented a lower incidence of gray mold in relation to the control after 50 days of cold storage. Ballinger, Maness, and Nesbitt (1985), using corrugated shipping containers with polyethylene liners and SO_2 generators, stored Euvitis hybrid bunch grapes up to 20 weeks.

Sulfur dioxide effectively eliminates any mold present on fruit surfaces, where SO_2 is used as a fungicide on grapes because of the physical damage resulting from transit, humidity, and temperature. However, to inhibit mold growth inside the fruit, additional sulfur dioxide is needed. Concentrations of 100 ppm sulfur dioxide for 1 hour can eliminate spores of *B. cinerea* on grapes at 32°F (0°C). Alternatively, 200 ppm for 0.5 hours or 25 ppm for 4 hours was effective. It is important to balance the amount of sulfur dioxide needed to inhibit mold growth for various fruits because excessive exposure to sulfur dioxide can damage fruits by bleaching grapes. To get a balance between microbial decay and damage to fruit by sulfur dioxide, Crisosto et al. (2002) suggested filling the boxes with 0.5 lbs of sulfur dioxide (a dose of 260 ppm/hr).

Cantín et al. (2011) tested the effects of adding 25, 50, and 100 (μl/L) of sulfur dioxide to figs and found that adding 25 μl/L sulfur dioxide lowered fig decay by 40% after four days of storage. Decreased mold growth was observed when 100 μl/L of sulfur dioxide was applied to figs at 20°C. This combination was more effective than when sulfur dioxide was applied at 0°C, suggesting fumigation should not happen under refrigeration conditions. At higher concentrations of 50 and 100 (μl/L), SO_2 damaged the fruit by bleaching.

The antifungal effect of sodium bisulfite is dependent on pH and water activity (a_w), with effectiveness increasing when fungi are exposed to lower a_w and pH. The MIC for *Aspergillus flavus* at a_w 0.99 and pH 3.5 was 100 ppm sodium bisulfite, while at pH 4.5 and a_w 0.99, the MIC was higher at 900 ppm on potato dextrose agar. In contrast, the MIC observed at a_w 0.95 and pH 3.5 and 4.5 was 300 and 700 ppm respectively (Lopez-Malo, Alzamora, and Palou 2005). Sulfur dioxide diffuses through membranes and accumulates in the microbial cell by ionization-entrapment. Ionization-entrapment occurs when fungi are exposed to acidic pH. When exposed to acidic pH, molecular sulfur dioxide can diffuse through the plasma membrane where it is then ionized when exposed to the more neutral pH inside the cell. Since the plasma membrane of the cell is hydrophobic, when this happens the sulfite ion is trapped.

6.5 Applications

Dried fruit and fermented beverages are common foods and beverages that expose adults to sulfites (Aguilar et al. 2016), as they are often used as antioxidants in these foods due to their reducing ability (Gunther et al. 1998). Sulfites in wine can prevent oxidation by reaction with hydrogen peroxide as well as their reaction with polyphenols. During the oxidation of polyphenols, hydrogen peroxide is produced. During this reaction, sulfites can scavenge hydrogen peroxide, therefore preventing oxidation of ethanol by hydrogen peroxide. The autooxidation of sulfites in wine, which is another way in which ethanol can be oxidized, is prevented by the presence of polyphenols that scavenge sulfite radicals produced and thus prevent oxidation from sulfite radicals (Danilewicz 2007).

6.5.1 Use in Wine

Sulfur dioxide is used in wine to prevent oxidation, control growth of yeasts, and inhibit growth of spoilage bacteria (Ough and Crowell 1987). Bacterial diversity is lower in sulfite-treated wine than sulfite-free wine during the fermentation process, which can be beneficial in wine-making as a way to limit the growth of unwanted microorganisms. Takahashi et al. (2014) found that sulfites inhibited the growth of *L. mesenteroides* and *Candida* spp. during the wine fermentation process. This is important since some LAB, such as *L. mesenteroides*, can produce undesirable biogenic amines in wine (Landete, Ferrer, and Pardo 2005).

Although sulfites are effective in controlling bacterial contamination that can cause losses in ethanol yield and inhibit yeast growth, sulfites in wine may also be undesirable depending on the type of wine being made. Malolactic fermentation is used to produce wine with less acidity and greater desirable flavors, aromas, and mouthfeel. These wines are processed by adding LAB, which converts malic acid into lactic acid. However, malolactic fermentation is difficult to induce due to the interactions between the yeast present (*Saccharomyces*) and the LAB added. During fermentation, *Saccharomyces* produces sulfites, which then can act as antimicrobials against the LAB. The ratio of endogenous sulfite to added sulfite in wine largely depends on the type of bacterial strain added to it. A "high-sulfite" forming strain of *Saccharomyces* produced 94 mg/L of sulfite during wine fermentation, while a "low-sulfite" forming strain produced 0.5 mg/L (Andorra et al. 2018). Acetaldehyde production of yeast is also strain-specific and correlates with SO_2 production. Strain JN10 of *S. cerevisiae* produced about 50.40 mg/L of SO_2 and 29.00 mg/L acetaldehyde, while strain JN17 produced 9.00 mg/L SO_2 and 9.35 mg/L acetaldehyde.

The molecular basis for the difference in SO_2 production between yeast strains has been attributed to differences in gene expression of the sulfur assimilation pathway, where low-sulfite producer strain overexpressed the MET2 and SKP2 genes controlling the sulfur assimilation pathway (Noble, Sanchez, and Blondin 2015). The extent to which *Saccharomyces* produces sulfur dioxide in wine can also depend on the grape juice composition of the wine as well as nutrient supplementation. Andorra et al. (2018) found that adding commercial amino acids affected production of end-products with thiamin increasing SO_2 production from 90 mg/L in the control to 121 mg/L at 18°C. Higher sugar and nitrogen concentrations increased the amount of sulfur dioxide produced, whereas lower nitrogen concentrations increased hydrogen sulfide production (Andorra et al. 2018).

During wine production, sulfites react with carbonyl compounds, which can lower its antimicrobial activity. In both red and white wine, the main carbonyl compound that binds to sulfite is acetaldehyde (Wells and Osborne 2011), accounting for about 70% carbonyls in white wine and about 55% in red wine (Jackowetz and de Orduna 2013). Wells and Osborne (2011) observed differences in the inhibition of malolactic fermentation in synthetic grape juice between different strains of *S. cerevisiae*. Earliest inhibition occurred after 2 days of growth, other strains inhibited malolactic fermentation after 21 days of growth, while some strains did not inhibit malolactic fermentation at all. The yeast strains that produced the highest concentrations of sulfites inhibited malolactic fermentation, while strains that produced very little sulfite did not. Wells and Osborne (2011) also found that the sulfite produced by *S. cerevisiae* during alcoholic fermentation was almost completely in its bound form, suggesting that bound sulfite inhibited malolactic fermentation, since very little free sulfite was detected. The amount of sulfite and sulfite-binding compounds (acetaldehyde, pyruvic acid, alpha-ketoglutaric acid, and galacturonic acid) produced depends on the yeast strain used in wine making. The bound sulfite detected was mainly acetaldehyde-bound sulfite (Wells and Osborne 2011). Therefore, higher malolactic fermentation would occur when yeast strains that produce lower amounts of sulfite-binding compounds are used.

6.5.2 Use in Beer

Sulfites in beer are most commonly in the bisulfite form due to the pH of beer of 3.8–4.4. The sulfur dioxide concentrations in beer have been estimated to range from <1 to 30 ppm (Ilett 1995). Above 30 ppm, sulfites could result in undesirable flavors in beer. Sulfites in beer are both from the added ingredients (such as fining agents and syrups), as well as from yeast fermentation (Baert et al. 2012). Sulfites in beer production enhance extract yield during grain germination, and act as a bleaching agent during malt and hops kilning, and as a preservative during storage (Ilett 1995, Guido 2016).

The processing conditions and yeast strain determine the amount of sulfites produced from the reduction of sulfates from water and grist material to SO_2 (Ilett 1995, Guido 2016). Sulfite production by yeasts in the beer-making process is affected by the fermentation temperature, the pH of the wort, wort strength, pressure, and wort aeration. Sulfur dioxide production increases at high wort pH, low wort oxygenation, and fermentation temperatures (Ilett 1995). Sulfite production in beer has been suggested to happen in four stages (Figure 6.3).

First, methionine and threonine amino acids present in the wort prevent sulfite excretion by repressing enzymatic activity. During the second stage, the demand for sulfur-containing amino acids for cysteine and methionine biosynthesis prevents a high concentration of sulfite from being formed. During the third stage, yeast growth stops, lowering the demand for sulfur-containing amino acids and allowing sulfite production to occur. In the fourth stage, sulfite excretion stops due to cessation of sulfate reduction (Baert et al. 2012).

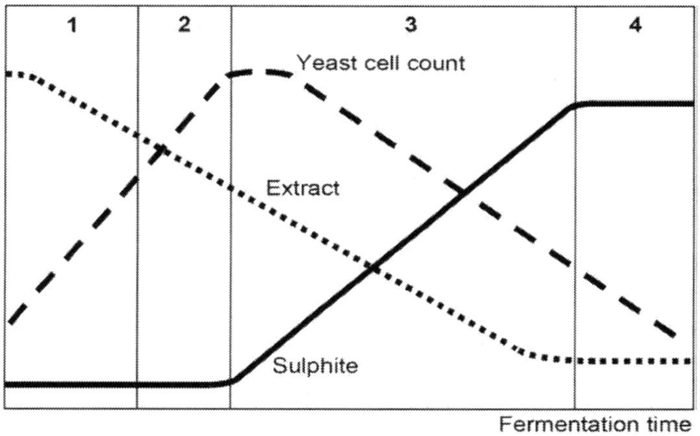

FIGURE 6.3 Sulfite production during yeast fermentation (reprinted with permission from Dufour 1991).

Some strains of *S. cerevisiae* can produce 10–30 mg of SO_2 during the synthesis of cysteine and methionine from sulfate. Yeast strains used for lagers were shown to produce more sulfite (4 mg/L), while yeast strains used for ale produced less than 2 mg/L (Ilett 1995).

In addition to its use as an antimicrobial in beer, sulfite antioxidant capacity controls flavor. Oxidation in beer produces off-flavors due to the formation of carbonyl compounds such as 2-furaldehyde in beer, an indicator of deterioration of beer flavor. Sulfur dioxide's antioxidant capacity can lower the levels of 2-furaldehyde during beer storage (Cameiro et al. 2006). Sulfur dioxide can help mask off-flavors in beer by reacting with carbonyl staling compounds to form adducts that are non-volatile and have radical scavenging ability. Similar to wine, the predominant aldehyde that binds sulfites in beer is acetaldehyde, representing 95% of the aldehydes present in beer (Kaneda et al. 1996, Almeida et al. 2003).

6.5.3 Use in Juices

Fruit and vegetable juices can be preserved by the addition of up to 50 mg sulfites/kg (Codex-Alimentarius 2018, Martins et al. 2011). Sulfites prevent discoloration and deterioration by yeast and mold in juices. Splittstoesser and Mattick (1981) examined the effects of 0, 100, 200, and 300 ppm sulfur dioxide on the storage life of grape juice refrigerated at 5°C. When no sulfur dioxide was added, 85% of the grape juice fermented within 5 weeks in comparison to 13% of samples fermenting in 30 weeks when 100 ppm sulfur dioxide was added. Splittstoesser and Mattick (1981) also found that 200 ppm sulfur dioxide sufficiently inhibited mold growth in most juice samples. In the samples where fermentation occurred, yeast populations were lowered by 54–96% when sulfur dioxide was added. Akhtar et al. (2010) also found increased microbial stability of mango pulp when potassium metabisulfite was added.

The concentration of sulfite needed depends on the sugar content and pH of juice. Less SO_2 would be required in beverages with lower pH. In concentrated juice, sulfur dioxide binds to the reducing sugars present, with increased binding at lower pH (Joslyn and Braverman 1954, Feigenbau and Israelashvili 1949).

6.5.4 Use in Fruit

Examples of dried fruits in which sulfites are used include golden raisins, dried figs, dried apricots, dried peaches, dried apples, dried pineapple, dried papaya, and dried mango. In the USA, about 15% of dried fruit products contain sulfites according to the California Office of Environmental Health and Hazard Assessment (OEHHA 2012). Sulfites are used in dried fruit to preserve its microbial quality, flavor, and color. Dried apricots with added sulfur dioxide for instance had 14–59% less browning than dried apricots without added sulfur dioxide.

Since the pH of dried fruits is usually in the range of 3.2–5.5, sulfites are typically in the bisulfite ion form. The concentration of free sulfite in dried fruits varies, and decreases during storage (OEHHA 2012). Sulfites can be added to dried fruit before and after they are dried. Figs and raisins, for example, can be processed by tray- or sun-drying. When processed by tray drying, sulfur dioxide is applied when the figs are still fresh. In contrast, sun-dried figs are exposed to sulfur dioxide during storage, after they are dried. The method of drying can affect the amount of residual SO_2 on the fruit. When tray dried, sulfur dioxide is applied when fresh, resulting in more surface area for the fruit to absorb the sulfur dioxide, therefore resulting in higher residual levels. Dried fruits such as figs will therefore differ in sulfite residue due to differences in their processing (González-Rompinelli et al. 2013).

The addition of sulfites can also affect the sugar concentration of dried fruits. When considering dried apricots, the addition of sulfites resulted in glucose being the sugar in highest concentrations (260–280 g/kg dry weight), in contrast to sucrose (256 g/kg dry weight) in dried apricots without added sulfites. These changes in sugar composition with added sulfites could be from the breakdown of sucrose to glucose and fructose by invertase, which shows high stability against SO_2 (Hamzaoglu, Turkyilmaz, and Ozkan 2018).

The California OEHHA found that the highest sulfite exposure due to consuming dried fruit was low at 191 µg/day (OEHHA 2012) when compared to red wine, which can have sulfite concentrations of 122 mg/L. The OEHHA concluded that consumption of dried fruit does not pose a risk of exceeding the maximum allowable dose level of sulfites given the low residual free sulfite present in dried fruit. A study done in China on sulfite exposure from dried fruit found that dried fruits had sulfite residues ranging from 1.6 to 9730 mg/kg, with the highest levels found in dried apricot. The 50th percentile of people tested showed exposure levels of 0.001–0.003 mg (kg body weight [bw]/day) for dried fruit, while the 99th percentile showed exposure levels of 819–2620 mg (kg bw/day) (Lou et al. 2017). Similar to the conclusions of the OEHHA in the USA, these researchers found that when considering the 50th and 75th percentiles of the population, exposure levels in China were also below the ADI of 0.7 mg/kg bw.

6.6 Regulatory Status of Sulfites in Food

6.6.1 Uses and Regulatory Limits of Sulfites in the USA

Sulfite use in foods and beverages must comply with global and regional regulations. Sulfur dioxide (SO_2), sodium bisulfite ($NaHSO_3$), potassium bisulfite ($KHSO_3$), sodium metabisulfite ($Na_2S_2O_5$), potassium metabisulfite ($K_2S_2O_5$), and sodium sulfite (Na_2SO_3) are all approved food ingredients in the USA (U.S. Food and Drug Administration 1977). The United States FDA requires the labeling of foods containing sulfites if the concentration is ≥10 ppm total SO_2, declared by the specific name of the ingredient, as stated in 21 CFR §101.100. in the USA (Table 6.3). In California, USA, the OEHHA added SO_2 (CAS number 7446-09-5) to its Proposition 65 list of chemicals known to cause reproductive toxicity in 2011 (OEHHA).

Sulfites are allowed for use as a fumigant during the shipment of grapes (Austin et al. 1997), but SO_2 final residue must still be lower than 10 ppm total SO_2 as per U.S. Environmental Protection Agency (EPA) regulations. Sulfur dioxide is permitted in wine up to 100 ppm. In the USA, the labeling of alcoholic beverages, such as wine and beer, is under the jurisdiction of the Tax and Trade Bureau (TTB) of the U.S. Department of Treasury. According to the TTB, the presence of sulfites must be declared with a label stating "Contains Sulfites" or "Contains (a) sulfiting agent(s)" when in an alcoholic beverage if it is present in 10 ppm or more sulfur dioxide [27 CFR §4.32(e)]. If an alcoholic beverage contains less than 10 ppm, a sulfur dioxide waiver may be submitted to Certificate of Level Approval (COLA) with an official laboratory report from the TTB or a TTB-certified laboratory to waive the labeling requirement, or to say, "contains less than 10 ppm sulfites" (U.S. Department of Treasury, 2018). Similarly, in the EU wine must be labeled with a statement if containing more than 10 mg/L sulfites ([EC] No: 607/2009). Even if no sulfites are added, a "sulfite-free" statement cannot be used as sulfites are produced endogenously during the fermentation process. The maximum amount of sulfites allowed in wine in the EU

TABLE 6.3

Permitted Uses of Sulfites in Food in the USA

Ingredient	Uses	Regulation
Sodium bisulfite ($NaHSO_3$)	Antimicrobial agent, antioxidant, color/coloring adjunct, dough strengthener, flavoring agent/adjuvant, flour treating agent, formulation aid, processing aid, surface-finishing agent, texturizer	21CFR 173.310 21CFR 177.1200 21CFR 182.3739
Potassium metabisulfite ($KHSO_3$)	Antimicrobial agent, antioxidant, color/coloring adjunct, dough strengthener, flour treating agent, oxidizing or reducing agent	21CFR182.3637
Potassium bisulfite ($KHSO_3$)	Antimicrobial agent	21CFR182.3616
Sodium metabisulfite ($Na_2S_2O_5$)	Antimicrobial agent, antioxidant, color/coloring adjunct, dough strengthener, flour treating agent, freezing or cooling agent, direct contact, oxidizing or reducing agent	21CFR173.310 21CFR182.3766
Sodium sulfite (Na_2SO_3)	Antioxidant, color/coloring adjunct, dough strengthener, flour treating agent, formulation aid, oxidizing or reducing agent, pH control agent	21CFR73.85 21CFR 173.310 21CFR 177.1200 21CFR 182.3798

Source: U.S. FDA 1977a,b.

ranges from 150 to 200 mg/L depending on the wine ([EC] No: 607/2009). In Canada, sulfites must also be declared if above 10 ppm.

The FDA banned sulfite use on raw fruits and vegetables in 1986, with the exception of potatoes that are minimally processed for frying. Sulfite use in the USA is also not permitted in foods regarded as good sources of thiamine such as raw meats due to sulfite's ability to cleave the methylene bridge in thiamine, leading to loss of vitamin B1 activity (U.S. Food and Drug Administration 1977). Similar to the FDA regulations, the USDA-FSIS requires labeling in finished meat products containing 10 ppm or higher of sulfites.

In some countries such as New Zealand and Australia, up to 500 mg/kg sulfites are however permitted for use in meat products (WHO 2009). Despite allowance of sulfites in meat, the possibility of fraud with high levels of sulfites in fresh meat products has been detected in Italy (Carrabs et al. 2017). Tomasevic et al. (2018) looked into the effect of the implementation of HACCP on the addition of sulfites to minced meat in Serbia, which is not permitted. They found a total average reduction of 43% sulfur dioxide concentration in minced meat products between 2007 and 2016 after the implementation of HACCP.

6.6.2 Uses and Regulatory Limits of Sulfites outside the USA

Similar to the USA, sulfites present in foods at or above 10 ppm must be labeled in Canada. They can be labeled as a declaration in the list of ingredients or shown as an allergen source and have an added sulfites statement as per the Food and Drug Regulations B.01.008 and B.01.009. In the European Union (EU) sulfites are considered food additives: sulfur dioxide (E 220), sodium sulfite (E 221), sodium bisulfite (E 222), sodium metabisulfite (E 223), potassium metabisulfite (E 224), calcium sulfite (E 226), calcium bisulfite (E 227), and potassium bisulfite (E 228). The safety of food additives in the EU is regularly evaluated by the Scientific Committee of Food and the European Food Safety Authority. Annex II Regulation (EC) No 1333/2008 on food additives has outlined the maximum levels of sulfites permitted in foods in the EU (Table 6.4).

6.7 Alternatives to Sulfites in Food and Beverages

Though the U.S. FDA does not permit sulfite use in fresh produce, sulfites can be used in minimally processed potato products and for processing potato starch. Regulatory limits (Table 6.4), recommendations to lower sulfite use in foods (WHO 2009), and consumer interest have nonetheless resulted in research on reduced-sulfite or sulfite-free products. Using biotechnology, non-browning fresh-cut potatoes that do not need to be preserved with sulfites have been produced (Lacy and Huffman 2016).

The growing interest from some consumers for sulfite-free wines could be due to sensitivities to sulfites and sulfites being part of the allergen lists in Canada and the EU. The addition of SO_2 is not permitted by the USDA in the production of organic wine. Therefore, greater care to minimize spoilage by *B. bruxellensis*, lactic acid-producing bacteria, and wine oxidation is needed by ensuring sanitation in wineries, obtaining a vigorous initial fermentation, and minimizing the introduction of oxygen at all stages of processing (Jackson 2008, Renouf, Lonvaud-Funel, and Coulon 2007). Cravero et al. (2018) and Englezos et al. (2019) demonstrated that ozone fumigation of grapes used to make wine can result in a 2.0 log or greater reduction in *B. bruxellensis* in wines produced from fumigated grapes. However, given that only a small population may cause spoilage, it is still unknown if ozone fumigation could completely prevent *B. bruxellensis* spoilage of organic wine.

When sulfites cannot be eliminated, there has been interest in lowering the levels used. For instance, carbonyl polymers attached to amine functional ion exchange resin (Parish and Babcock 2018) and the use of chloroplast from wheatgrass to oxidize sulfites (Sung-Chyr and Georgiou 2005) have been tested methods to strip sulfites from wine.

Chemical alternatives to using sulfites in wines as an antimicrobial include the addition of lipopeptides, bacteriocins, and phenolic compounds (Santos et al. 2012). Lipoproteins (200 ppm) produced by *Bacillus subtilis* provided greater inhibitory effects against *Aspergillus carbonarius* producing ochratoxin in grape must than the use of 50 ppm SO_2 (Jiang et al. 2017). Nisin can inhibit LAB growth in wine,

TABLE 6.4

Maximum Permitted Levels of Sulfites in Foods in the European Union

Category	Food	Maximum Permitted Level (mg/kg or mg/L)
Peeled/cut vegetables and fruits	Peeled potatoes	50
	Onion, garlic, shallot pulp	300
	Horseradish pulp	800
Fruits and vegetables in vinegar, oil, or brine	(Except olives and golden peppers in brine)	100
	Golden peppers in brine	500
Fruit and vegetable spreads	Jams, jellies, marmalades, and other fruit or vegetable spreads	50
	Jams, jellies, and marmalades made with sulfited fruit	100
Decorations, coatings, fillings (except fruit-based)	Glucose syrup-based confectionary	50
	Toppings (syrup for pancakes, milkshakes, ice cream, etc.)	40
	Fruit fillings for pastries	100
Whole, broken, or flaked grain	Sago and pearl barley	30
Sugars and syrups	Sugars, except glucose syrup	10
	Glucose syrup	20
	Other sugar and syrups	40
	Treacle and molasses	70
Seasonings and condiments	Cinnamon	150
	Citrus juice-based seasoning	200
Vinegar and dilutes acetic acid		170
Mustard	Excluding Dijon mustard	250
	Only Dijon mustard	500
Protein products	Analogues of meat, fish, crustaceans, and cephalopods	200
	Only gelatin	50
Fruit and vegetable juices	Concentrated grape juice for winemaking	2000
	Orange, grapefruit, apple, and pineapple juice	50
	Lime and lemon juice	250
	Grape juice, unfermented	70
Alcoholic beverages	Beer and malt beverages	20–50
	Wine	200
	Mead	260 mg/l
	Distilled alcoholic beverage with whole pears	50
Cider and perry		200
Starches (excluding infant formula), potato, cereal, or starch-based snacks, and processed nuts		50
Dried fruit and vegetables	Dried coconut	50
	Dried mushrooms	100
	Dried ginger	150
	Dried tomatoes	200
	Dried white vegetables	400
	Dried apples and pears	600
	Dried bananas	1000
	Dried apricots, peaches, grapes, prunes, and figs	2000
Meat preparations	Breakfast sausages, burger meat (with min vegetable or cereal content of 4% mixed in)	450
Fresh fruit or vegetables	Only vacuum-packed sweet corn	100
	Only table grapes, fresh lychee, and blueberries	10
Frozen fruit or vegetables	Only while vegetables (mushroom and white pulses)	50
	Frozen or deep-frozen potatoes	100
Processed fish and fishery products	Cooked crustaceans and cephalopods	50
	Cooked crustaceans of *Penaeidae*, *Solenoceridae*, and *Aristaeidae* family over 120 units per kg	270
	Cooked crustaceans of *Penaeidae*, *Solenoceridae*, and *Aristaeidae* family between 80 and 120 units per kg	180
	Cooked crustaceans of *Penaeidae*, *Solenoceridae*, and *Aristaeidae* family up to 80 units per kg	135
	Dried salted fish of the *Gadidae* species	200

Source: Aguilar et al. 2016.

with an MIC_{50} of 0.024 µg/mL against *O. oeni*. When combined with metabisulfite, a synergistic effect was observed as 0.01 µg/mL of nisin added to wine lowered the MIC_{50} of metabisulfite against *O. oeni* from 50 to 25 µg/mL (Rojo-Bezares et al. 2007). Pediocin PA-1 is another bacteriocin with demonstrated antimicrobial activity against *O. oeni*, with an IC_{50} of 19 ng/mL compared to IC_{50} of 312 ng/mL in other LAB species. Similar to the effects with nisin, when 10 ng/mL of pediocin was combined with metabisulfite, a two-fold decrease in the IC_{50} against *O. oeni* was observed (Diez et al. 2012).

Lysozyme has also been extensively studied for use as an antimicrobial in wine to control malolactic fermentation (Santos et al. 2012, Bartowsky et al. 2004). Lysozyme (EC 3.2.1.17) can inhibit the growth of LAB in wine. Although lysozyme is an effective antimicrobial for white wines, it lacks the antioxidant properties of sulfites and could result in browner wines if used alone (Chen et al. 2015). The use of lysozyme as an antimicrobial against LAB in red wine is also not as effective as in white wine due the higher levels of wine phenols in red wine, which bind to lysozyme during the fermentation process (Azzolini et al. 2010). Lysozyme can also form haze in white wine. Reduced sulfite wines have thus been produced, but sulfite alternatives have not been as effective in prolonging the shelf life of wines against oxidation (Lisanti et al. 2019, Santos et al. 2012).

Sulfites as an antimicrobial at lower levels have been investigated individually and in combination with alternative non-synthetic preservatives such as fruit extracts and chitosan in meat. Bañón et al. (2007) found that adding 100 mg SO_2/kg was not enough to delay the growth of total coliforms in raw beef patties. Serrano and Banon (2012) found that 150 mg/kg SO_2 inhibited the growth of *E. coli* and *Salmonella* in pork burgers to levels permitted by the Commission Regulation (EC) No. 2073 (between 500–1000 CFU/g for *E. coli* and total viable counts below 5.6–6.7 $\log_{10}$ CFU/g), but it resulted in a shelf life of only 7 days due to the growth of spoilage organisms. Adding 0.05% chitosan inhibited bacterial growth and extended shelf life to 14 days, while increasing the concentration of SO_2 from 150 to 450 mg/kg (maximum allowable levels) extended microbial shelf life to 21 days. Combining chitosan and sulfite effectively inhibited the growth of pathogenic microorganisms, and was proposed as an option to lower sulfite use in meat. The additive effect when chitosan was used in combination with sulfites was due to chitosan's ability to attach to microbial membranes and alter the membrane permeability, by preventing nutrients from entering the microbe's cell, by inhibiting microbial toxin production, and by binding to microbial DNA (Zheng and Zhu 2003). Serrano and Banon (2012) suggested that chitosan and SO_2 may be effective antimicrobials when added together because chitosan can slow down sulfite breakdown resulting in higher levels of residual SO_2 or by preventing sulfite from irreversible binding to ingredients found in meat. Other combinations tested to lower sulfite levels in meat have been combining sulfites with non-sulfite-reducing agents such as sodium ascorbate and phenolic-rich green tea and grape extracts (Bañón et al. 2007).

6.8 Toxicology and Safety

Sulfites are both endogenous and exogenous to the human body. Endogenous sulfites in the human body are formed from the metabolism of sulfur-containing amino acids methionine and cysteine which oxidize to sulfates by the enzyme sulfite oxidase (SOX) (Tan et al. 2005). Sulfite oxidase in mammalian tissue is found in the liver, heart, and kidneys and may be found in low amounts in the brain, blood, and skeletal muscle. Sulfite oxidase metabolizes sulfites quickly, preventing sulfite build up in the body and ultimately shielding the body from the harmful effects of excess sulfites. A very small portion of the population have SOX deficiency, in which individuals suffer from neurological irregularities and destruction of brain tissue when exposed to sulfites. By 2005, only 21 cases of SOX deficiency had been reported (Garcia-Fuentes et al. 2015, Ramos et al. 2017, Tan et al. 2005).

The European Food Safety Authority (EFSA) and Joint Food and Agriculture and World Health Organization Joint expert Committee on Food Additives (JECFA) have determined an ADI value of up to 0.7 mg SO_2 equivalent/kg bw for humans (Aguilar et al. 2016, WHO 2009). The JECFA suggested that, when possible, alternative preservatives are encouraged to be used in cases where the use of sulfites may lead to high levels of intake. The potential for consumers to exceed the ADI for sulfites exists in certain populations, for instance regular consumers of wine (Irwin et al. 2017). The committee suggested that countries should collect more data on the use of sulfites in foods/beverages to discern whether exposure exceeds the ADI, enforce the national maximum levels, and encourage research on alternative methods of preservation (WHO 2009).

Toxicological studies of sulfites in the lung, liver, and kidney have been carried out in cell cultures (Qin and Meng 2009) and rodents (Table 6.5). Fatima Zohra El et al. (2014) exposed female Wister rats to 0.25%, 1%, and 4% sodium metabisulfite for 90 days. At 4%, sodium bisulfite increased calcium, urea, creatine, and uric acid, as well as decreasing immunoglobin, red blood cells and hemoglobin, and leukocytes. Mahmoud et al. (2015) investigated the toxicity of 200, 500, and 1000 ppm sodium sulfite (Na_2SO_3) for 12 weeks on albino Wister rats, and demonstrated that Na_2SO_3 significantly decreased body weight, red blood cell count, hemoglobin concentration, hematocrit value, white blood cell count, and glucose levels of rats. Additionally, an increase in serum activity of aspartate aminotransferase, alanine aminotransferase, alkaline phosphatase, urea, and creatine was seen in the blood, indicating adverse effects to liver and kidney cells. Lastly, lesions were seen on both liver and kidney tissue. Qu et al. (2017) used high content automated imaging technology to predict human liver toxicity and found that there was a decrease in growth of HepG2 cells as the dose increased, and that sunset yellow and sodium sulfite together lowered cell number, increased membrane permeability, increased oxidative stress, and in high doses caused DNA damage.

Since the brain has low SOX enzyme activity in comparison to the liver, kidney, and heart tissues, it is more susceptible to sulfite toxicity. Akdogan et al. (2011) looked at the effects of sulfite on pyramidal neurons in rats by counting the number of cells in the CA1 and CA2-3 subdivisions of the hippocampus. At 25 mg/kg/day of sodium metabisulfite, a significant decrease in the number of pyramidal neurons in the three subdivisions of the hippocampus was seen. Kocamaz et al. (2012) also investigated the effect of sulfites on pyramidal neurons by looking at CA1 and CA3-2 subdivisions. Since rats have higher levels of SOX enzyme, with some studies estimating 20 times more SOX activity than in the human liver, they used both SOX-deficient and normal rats in their study. These researchers found that sulfites at 70 mg/kg decreased the number of pyramidal neurons in both SOX-deficient and normal rats (Table 6.5).

Adverse reactions from sulfite, including anaphylactic reactions, dermatitis, urticaria, flushing, hypotension, and abdominal pain, have been observed since the 1970s (Vally and Misso 2012). Although the reactions are similar to those of a food allergy, they are not considered a true food allergy as they do not involve a protein. An estimated 3–10% of asthmatics are sulfite-sensitive (Vally, Misso, and Madan 2009). The exact mechanisms that cause these asthmatic reactions from the exposure to sulfites are still unknown.

The first occurrence of anaphylaxis after the ingestion of sodium metabisulfite in salad was reported in 1976 and was followed by a case of SO_2 in orange juice causing asthma in 1977 (Prenner and Stevens 1976, Freedman 1977). Thereafter, further reported incidents related to the ingestion of sulfites were

TABLE 6.5

Select Toxicological Studies of Sulfites (2011–2017)

Form of Sulfite and Concentration	Model	Result	Reference
0 to 4% (w/w) sodium metabisulphite for 7 consecutive days	Male Sprague-Dawley rats	The gastric lesions in rats observed in a 1972 study of dietary sodium metabisulphite for 10–56 days could not be replicated	(Dalefield and Mueller 2016)
Sodium sulfite up to 1000 ppm	Wister rat	Decrease in body weight, red blood cell count, white blood cell count, and hemoglobin concentration	(Mahmoud et al. 2015)
Sodium sulfite, IC_{50} of 0.3 g/L	Cultured human HepG cells	Decrease in HepG cells	(Qu et al. 2017)
Sodium metabisulfite, up to 4%	Wister rat	Decreased immunoglobin, red blood cells, hemoglobin, and leukocytes	(Fatima Zohra El et al. 2014)
Sodium metabisulfite, 25 mg/kg/day	Wister rat	Decrease in pyramidal neurons	(Akdogan et al. 2011)
Sulfite,70 ppm	Wister rat (normal and SOX deficient)	Decrease in pyramidal neurons	(Kocamaz et al. 2012)

TABLE 6.6

U.S. Food and Drug Administration Recalls, Market Withdrawals, and Safety Alerts of Undeclared Sulfites (2016–2018)

Year	Product Type	No. of Products/Companies
2018	Snacks, dried fruit and vegetables (dried bamboo shoot, seedless raisins, licorice slice, white beans, sugar melon candy/wax gourd, lily bulb, snack ginger sliced, dried apricots, dried yellow potatoes, blueberry vanilla, peanut butter and jelly, and apple cinnamon cricket protein bars, dried longya lily)	12
2017	Dried lily flower	1
2016	Dried fruit and vegetables (preserved lemons, sweetened ginger, dried daylily, ginger slices, dry apricots, cauliflower, dried kiwi, and dried mango products)	7

Source: Administration 2018, U.S.FDA 2018.

found through the 1980s. Yang and Purchase (1985) found more than 250 cases of sulfite-related reactions in the USA, including 6 deaths. This led the FDA to prohibit the use of sulfites on fresh fruits and vegetables (except potatoes) in 1986 and to add stricter regulations on sulfites in food. Following these regulatory changes, adverse reactions to sulfites in food continued to be reported, including an outbreak in 1995 of six people having adverse reactions and being hospitalized due to salsa containing high levels of sulfites (Nagy et al. 1995). FDA recall records from 1996 to 1999 showed that 59 recalls for undeclared sulfites were issued and 34 reports of adverse reactions due to undeclared sulfites were received (Timbo et al. 2004). This added up to an average of 10 reports per year between 1996 and1999, a significant decrease from 111 reports per year from 1980 to 1987. This number continues to decrease, with only 20 recalls for undeclared sulfites issued between 2016 and 2018 (Table 6.6). In 2018, there were a total of 12 import alerts in the USA due to undeclared sulfites, with a majority from Mexico (4) and Pakistan (3), and the rest from Afghanistan, China, South Korea, Russia, and Vietnam (FDA 2018).

6.9 Detection and Analysis in Food

Sulfite concentration is commonly determined by the AOAC optimized Monier-Williams (OMW) Official Method 990.28 (AOAC 2000) depending on the matrix composition and sulfite concentration. The OMW method starts by heating and acidifying a sample, converting sulfite to sulfur dioxide. The sulfur dioxide is then oxidized to sulfuric acid by the addition of nitrogen through a reflux condenser. The sulfite content is directly related to the amount of sulfuric acid formed, which is determined by titration with sodium hydroxide. This modified method, though accurate and cheap, is time-consuming and displays low sensitivity below 10 ppm SO_2. The method also produces inaccurate results with certain food matrices, such as *Allium* and *Brassica* vegetables (Perfetti and Diachenko 2003). *Allium* and *Brassica* vegetables can result in false positives when using the OMW method due to the presence of endogenous sulfur compounds which can form SO_2 during analysis (Carlos and de Jager 2016, Addala et al. 2016). As much as 50–200 ppm SO_2 can be detected as a false positive result when analyzing unsulfured dried garlic (Lafeuille, Lefevre, and Achouri 2007). Due to this potential for false positives, the FDA only considers dried garlic sulfited if the values are above 230 ppm using the OMW method. Due to the drawbacks of the OMW method in certain matrices, new detection and analysis methods of sulfites have been investigated (Chen, De Borba, and Rohrer 2016). Modification of the OMW method by increasing pH levels to 2.4–3.4 was shown to successfully prevent alliin breakdown and avoided false positives in dried garlic samples (Lafeuille, Lefevre, and Achouri 2007).

The Monier-Williams official method is time-consuming, with a limit of detection of 10 ppm SO_2 (Carlos and de Jager 2016). Research has thus been done on rapid and more sensitive methods to detect sulfites in food. Various alternative methods for sulfite detection in foods and beverages have been investigated, including electrochemical methods, chromatography, flow injection analysis, and the use of enzyme biosensors (Table 6.7). Due to the relatively high detection limits (10 ppm SO_2) of the OMW

TABLE 6.7

Select Alternative Methods to Monier-Williams Method for Sulfite Detection in Food and Beverages

Method	Limit of Detection (ppm)	Limit of Quantification (ppm)	Spiked Recoveries	Comparison to Modified Monier-Williams	Reference
Enzymatic biosensor to detect sulfites in shrimp	<7	<20	80–110	Faster with higher accuracy and recoveries	(Alonso et al. 2019)
Ion chromatography with conductivity detection	2.7	8.2	85.2–92.1%	Less pre-treatment required, quicker, lower limit of detection, 10 samples in 8 h)	(Iammarino et al. 2010)
Electrochemical method	7.2			Low-cost, easy to implement	(Montes et al. 2012)
Electrochemical sensing using non-modified screen-printed carbon electrodes	0.05	0.17		Low-cost, simple to use, portable, lower limit of detection, quicker	(Ramos et al. 2017)
Liquid chromatography–tandem mass spectrometry	0.5 to 1	0.12–0.75	84–115%	Batched sampling/ extraction method, quicker (30 a day vs 4), lower limit of detection	(Robbins et al. 2015)
Liquid chromatography coupled with plasma mass spectrometry	0.02 SO_3^{2-}	0.06 SO_3^{2-}	78.5–91.6%	Quicker, lower limit of detection	(Ni et al. 2015)
Flow injection analysis	0.043		100%	Quicker (one measurement per minute), lower limit of detection, high reproducibility, little amount of reagent needed; can't be used with viscous foods or large particles	(Martins et al. 2011)
HPLC–IMER	0.01			Lower limit of detection, more specific (due to use of enzyme) with less toxic chemicals used	(Theisen et al. 2010)

method, Iammarino et al. (2010) developed ion chromatography with conductivity detection for the detection of sulfites in meat and shrimp. This method was validated (Regulation 882/2004/EC) as a method for sulfite detection in meats and showed high sensitivity, with a limit of detection (LOD) of 2.7 mg/kg SO_2 and a quantification limit of 8.2 mg/kg SO_2. Chromatographic separation can be done by using an anion-exchange column that is eluted with sodium carbonate and sodium hydroxide. When compared to the official OMW method, ion chromatography with conductivity detection required less pre-treatment, was faster (10 samples could be done in 8 hours) and had increased accuracy and precision.

Montes et al. (2012) found that the OMW method presented high precision but low accuracy, in contrast to most of the literature. They also found low accuracy in the absence of adduct formation with sulfite, as when sulfite was bound to gallic acid accuracy was restored. When the electrochemical method was used to detect sulfites in wine, a range of 10–40 mg of sulfite per liter could be detected, with a limit of detection of 7.2 mg/L. This electrochemical method was thus proposed as an alternative in the industry as it is easy to implement, needs little equipment, and is low-cost.

Robbins et al. (2015) developed a liquid chromatography–tandem mass spectrometry (LC–MS/MS) method for sulfite detection in food. This method derivatizes sulfite into hydroxymethylsulfonate before quantification by LC–MS/MS. The validation studies showed spiked recovery ranging from 84 to 115%. When compared to the OMW method, the LC–MS/MS had an LOD two orders of magnitude below that of OMW at 0.12 and 0.75 ppm SO_2 in white grape juice and dried apricots, respectively. When Carlos and de Jager (2017) conducted a multi-laboratory validation study on LC–MS/MS and also compared it to the OMW method, they found that LC–MS/MS was more sensitive, faster, and could be a reliable sulfite detection method (Carlos and de Jager 2017, Carlos, Treblin, and de Jager 2019, Carlos and de Jager 2016).

Flow injection analysis (FIA) is another alternative method to the OMW method for sulfite detection in food. Similar to the Monier-Williams method, the flow injection method converts free sulfites to sulfuric acid by acidification, though instead of distillation the FIA uses diffusion through a semi-permeable membrane. The sulfuric acid is then collected by a buffered solution, instead of absorption in H_2O_2 solution, and detected amperometrically instead of by titration. These changes result in higher sensitivity (Martins et al. 2011).

6.10 Conclusions

Sulfites are used in the food industry in the form of sulfur dioxide, sodium and calcium sulfite, sodium and potassium bisulfite, and sodium and potassium metabisulfite. Sulfites are effective antimicrobials, antioxidants, and anti-browning agents against enzymatic and non-enzymatic browning in food. Sulfites as a food ingredient have been controversial due to reports of adverse reactions caused by exposure to sulfites. Future studies should be done focusing on how sulfites can interact with other food ingredients such as sunset yellow as this has been reported to increase the toxicity of sulfites. Studies should also continue to evaluate the risk of overconsumption of sulfite in food in different human populations. Continued research should also focus on new detection methods for sulfites, concentrating on increasing sample size, increasing the number of laboratory tests, and testing on more food matrices.

REFERENCES

Addala, R., L. Fong, M. Vasavada, and S. Subramanian. 2016. "Influence of distillation temperature in the determination of added sulfites in dehydrated garlic powders using the modified optimized Monier-Williams method." *Journal of AOAC International* 99 (3):721–724. doi: 10.5740/jaoacint.15-0313.

Agnolucci, M., F. Rea, C. Sbrana, C. Cristani, D. Fracassetti, A. Tirelli, and M. Nuti. 2010. "Sulphur dioxide affects culturability and volatile phenol production by Brettanomyces/*Dekkera bruxellensis*." *International Journal of Food Microbiology* 143 (1–2):76–80. doi: 10.1016/j.ijfoodmicro.2010.07.022.

Aguilar, F., R. Crebelli, A. Di Domenico, B. Dusemund, M. J. Frutos, P. Galtier, D. Gott, U. Gundert-Remy, C. Lambre, J. C. Leblanc, O. Lindtner, P. Moldeus, A. Mortensen, P. Mosesso, D. Parent-Massin, A. Oskarsson, I. Stankovic, I. Waalkens-Berendsen, R. A. Woutersen, M. Wright, M. Younes, and EFSA

Panel Food Additives Nutrient. 2016. "Scientific opinion on the re-evaluation of sulfur dioxide (E 220), sodium sulfite (E 221), sodium bisulfite (E 222), sodium metabisulfite (E 223), potassium metabisulfite (E 224), calcium sulfite (E 226), calcium bisulfite (E 227) and potassium bisulfite (E 228) as food additives." *EFSA Journal* 14 (4). doi: 10.2903/j.efsa.2016.4438.

Akdogan, I., E. Kocamaz, V. Kucukatay, N. G. Yonguc, M. B. Ozdemir, and W. Murk. 2011. "Hippocampal neuron number loss in rats exposed to ingested sulfite." *Toxicology and Industrial Health* 27 (9):771–778. doi: 10.1177/0748233710397418.

Akhtar, S., M. Riaz, A. Ahmad, and A. Nisar. 2010. "Physico-chemical, microbiological and sensory stability of chemically preserved mango pulp." *Pakistan Journal of Botany* 42 (2):853–862.

Almeida, P. J., J. A. Rodrigues, L. F. Guido, J. R. Santos, and A. A. Barros. 2003. "Free sulfur dioxide in beer as the difference between total sulfur dioxide and acetaldehyde: A voltammetric approach." *Journal of the American Society of Brewing Chemists* 61 (4):191–195.

Alonso, S. S., I. Gonzalez, N. Linaza, E. Ramos, M. Fernandez, R. Gonzalez, S. Maza, and A. Jaureguibeitia. 2019. "Validation of the Biofish-300 SUL enzymatic biosensor for the detection of sulfite in crustacean." *Journal of AOAC International* 102 (2):539–556. doi: 10.5740/jaoacint.18-0124.

Anacleto, J., and N. Vanuden. 1982. "Kinetics and activation energetics of death in *Saccharomyces cerevisiae* induced by sulfur dioxide." *Biotechnology and Bioengineering* 24 (11):2477–2486. doi: 10.1002/bit.260241112.

Andorra, I., L. Martin, E. Nart, M. Puxeu, C. Hidalgo, and R. Ferrer-Gallego. 2018. "Effect of grape juice composition and nutrient supplementation on the production of sulfur dioxide and carboxylic compounds by *Saccharomyces cerevisiae*." *Australian Journal of Grape and Wine Research* 24 (2):260–266. doi: 10.1111/ajgw.12325.

AOAC. 2000. *990.28, 17th Edition: Sulfites in Foods, Optimized Monier-Williams Method.*

Aranda, A., E. Jimenez-Marti, H. Orozco, E. Matallana, and M. del Olmo. 2006. "Sulfur and adenine metabolisms are linked, and both modulate sulfite resistance in wine yeast." *Journal of Agricultural and Food Chemistry* 54 (16):5839–5846. doi: 10.1021/jf060851b.

Austin, R. K., W. Clay, S. Phimphivong, J. L. Smilanick, and D. J. Henson. 1997. "Patterns of sulfite residue persistence in seedless grapes during three months of repeated sulfur dioxide fumigations." *American Journal of Enology and Viticulture* 48 (1):121–124.

Azzolini, M., E. Tosi, G. Veneri, and G. Zapparoli. 2010. "Evaluating the efficacy of lysozyme against lactic acid bacteria under different winemaking scenarios." *South African Journal of Enology and Viticulture* 31 (2):99–105.

Baert, J. J., J. De Clippeleer, P. S. Hughes, L. De Cooman, and G. Aerts. 2012. "On the origin of free and bound staling aldehydes in beer." *Journal of Agricultural and Food Chemistry* 60 (46):11449–11472. doi: 10.1021/jf303670z.

Ballinger, W. E., E. P. Maness, and W. B. Nesbitt. 1985. "Sulfur dioxide for long-term low temperature storage of Euvitis hybrid bunch grapes." *Hortscience* 20 (5):916–918.

Bañón, S., P. Díaz, M. Rodríguez, M. D. Garrido, and A. Price. 2007. "Ascorbate, green tea and grape seed extracts increase the shelf life of low sulphite beef patties." *Meat Science* 77:626–633. doi: 10.1016/j.meatsci.2007.05.015.

Barata, A., J. Caldeira, R. Botelheiro, D. Pagliara, M. Malfeito-Ferreira, and V. Loureiro. 2008. "Survival patterns of *Dekkera bruxellensis* in wines and inhibitory effect of sulphur dioxide." *International Journal of Food Microbiology* 121 (2):201–207. doi: 10.1016/j.ijfoodmicro.2007.11.020.

Bartowsky, E. J., P. J. Costello, A. Villa, and P. A. Henschke. 2004. "The chemical and sensorial effects of lysozyme addition to red and white wines over six months' cellar storage." *Australian Journal of Grape and Wine Research* 10 (2):143–150. doi: 10.1111/j.1755-0238.2004.tb00017.x.

Burroughs, L. F., and A. H. Sparks. 1964. "The identification of sulphur dioxide-binding compounds in apple juices and ciders." *Journal of the Science of Food and Agriculture* 15 (3):176.

Cameiro, J. R., L. F. Guido, P. J. Almeida, J. A. Rodrigues, and A. A. Barros. 2006. "The impact of sulphur dioxide and oxygen on the behaviour of 2-furaldehyde in beer: An industrial approach." *International Journal of Food Science and Technology* 41 (5):545–552. doi: 10.1111/j.1365-2621.2005.01104.x.

Cantín, C. M., L. Palou, V. Bremer, T. J. Michailides, and C. H. Crisosto. 2011. "Evaluation of the use of sulfur dioxide to reduce postharvest losses on dark and green figs." *Postharvest Biology and Technology* 59:150–158. doi: 10.1016/j.postharvbio.2010.09.016.

Carlos, K. S., and L. S. de Jager. 2017. "Determination of sulfite in food by liquid chromatography tandem mass spectrometry: Collaborative study." *Journal of AOAC International* 100 (6):1785–1794. doi: 10.5740/jaoacint.17-0033.

Carlos, K. S., M. Treblin, and L. S. de Jager. 2019. "Comparison and optimization of three commercial methods with an LC-MS/MS method for the determination of sulfites in food and beverages." *Food Chemistry* 286:537–540. doi: 10.1016/j.foodchem.2019.02.042.

Carlos, K. S. R., and L. S. de Jager. 2016. "Comparison of multiple methods for the determination of sulphite in Allium and Brassica vegetables." *Food Additives and Contaminants Part A-Chemistry Analysis Control Exposure and Risk Assessment* 33 (10):1509–1517. doi: 10.1080/19440049.2016.1229869.

Carrabs, G., G. Smaldone, L. Carosielli, M. Girasole, M. Iammarino, and E. Chiaravalle. 2017. "Detection of sulfites in fresh meat preparation commercialised at retail in Lazio region." *Italian Journal of Food Safety* 6 (2). doi: 10.4081/ijfs.2017.6482.

Casalone, E., C. M. Colella, S. Daly, E. Gallori, L. Moriani, and M. Polsinelli. 1992. "Mechanism of resistance to sulphite in *Saccharomyces cerevisiae*." *Current Genetics* 22 (6):435–440. doi: 10.1007/bf00326407.

Cascado, F. J., A. H. Sanchez, V. M. Beato, A. De Castro, and A. Montano. 2014. "Effect of sulfites and sorbates on the preservation and color of pickled blanched barlic under different storage conditions." *Journal of Food Processing and Preservation* (3). doi: 10.1111/jfpp.12045.

Chang, I. S., B. H. Kim, and P. K. Shin. 1997. "Use of sulfite and hydrogen peroxide to control bacterial contamination in ethanol fermentation." *Applied and Environmental Microbiology* 63 (1):1–6.

Chang, I. S., B. H. Kim, P. K. Shin, and W. K. Lee. 1995. "Bacterial contamination and its effects on ethanol fermentation." *Journal of Microbiology and Biotechnology* 5 (6):309–314.

Chen, K., S. Y. Han, B. Zhang, M. Li, and W. J. Sheng. 2015. "Development of lysozyme-combined antibacterial system to reduce sulfur dioxide and to stabilize Italian Riesling ice wine during aging process." *Food Science and Nutrition* 3 (5):453–465. doi: 10.1002/fsn3.238.

Chen, L., B. De Borba, and J. Rohrer. 2016. "Determination of total and free sulfite in foods and beverages." https://assets.thermofisher.com/TFS-Assets/CMD/Application-Notes/AN-54-IEX-Sulfite-Food-Beverage-AN70379-EN.pdf (accessed May 30).

Chen, L., M. Tang, C. Chen, M. G. Chen, K. Luo, J. Xu, D. N. Zhou, and F. Wu. 2017. "Efficient bacterial inactivation by transition metal catalyzed auto-oxidation of sulfite." *Environmental Science and Technology* 51 (21):12663–12671. doi: 10.1021/acs.est.7b03705.

Codex-Alimentarius. 2018. *GSFA Online, Food Additive Group Details, Sulfites.*

Cojocaru, G., and O. Antoce. 2012. "Chemical and biochemical mechanisms of preservatives used in wine: A review." *Scientific Papers Series B Horticulture* 56:457–466.

Corte, L., L. Roscini, C. Zadra, L. Antonielli, B. Tancini, A. Magini, C. Emiliani, and G. Cardinali. 2012. "Effect of pH on potassium metabisulphite biocidic activity against yeast and human cell cultures." *Food Chemistry* 134 (3):1327–1336. doi: 10.1016/j.foodchem.2012.03.025.

Cravero, F., V. Englezos, K. Rantsiou, F. Torchio, S. Giacosa, S. R. Segade, V. Gerbi, L. Rolle, and L. Cocolin. 2018. "Control of *Brettanomyces bruxellensis* on wine grapes by post-harvest treatments with electrolyzed water, ozonated water and gaseous ozone." *Innovative Food Science and Emerging Technologies* 47:309–316. doi: 10.1016/j.ifset.2018.03.017.

Crisosto, C. H., L. Palou, D. Garner, and D. A. Armson. 2002. "Concentration by time product and gas penetration after marine container fumigation of table grapes with reduced doses of sulfur dioxide." *HortTechnology*, 12(2): 241–245.

Cruess, W. V. 1912. "The effect of sulfurous acid on fermentation organisms." *Journal of Industrial and Engineering Chemistry-Us* 4:581–585. doi: 10.1021/ie50044a010.

Dalefield, R. R., and U. Mueller. 2016. "Gastric mucosal irritation following oral exposure to sodium metabisulphite: A reproducible effect?" *Regulatory Toxicology and Pharmacology* 80:277–282. doi: 10.1016/j.yrtph.2016.07.005.

Danilewicz, J. C. 2007. "Interaction of sulfur dioxide, polyphenols, and oxygen in a wine-model system: Central role of iron and copper." *American Journal of Enology and Viticulture* 58 (1):53–60.

Danilewicz, J. C., J. T. Seccombe, and J. Whelan. 2008. "Mechanism of interaction of polyphenols, oxygen, and sulfur dioxide in model wine and wine." *American Journal of Enology and Viticulture* 59 (2):128–136.

de Azevedo, L. C., M. M. Reis, L. F. Motta, G. O. da Rocha, L. A. Silva, and J. B. de Andrade. 2007. "Evaluation of the formation and stability of hydroxyalkylsulfonic acids in wines." *Journal of Agricultural and Food Chemistry* 55 (21):8670–8680. doi: 10.1021/jf0709653.

de Oliva-Neto, P., and F. Yokoya. 2001. "Susceptibility of *Saccharomyces cerevisiae* and lactic acid bacteria from the alcohol industry to several antimicrobial compounds." *Brazilian Journal of Microbiology* 32 (1):10–14. doi: 10.1590/s1517-83822001000100003.

Diez, L., B. Rojo-Bezares, M. Zarazaga, J. M. Rodriguez, C. Torres, and F. Ruiz-Larrea. 2012. "Antimicrobial activity of pediocin PA-1 against *Oenococcus oeni* and other wine bacteria." *Food Microbiology* 31 (2):167–172. doi: 10.1016/j.fm.2012.03.006.

Domingues, A. R., S. R. Roberto, S. Ahmed, M. Shahab, O. J. Chaves, C. H. Sumida, and R. T. de Souza. 2018. "Postharvest techniques to prevent the incidence of Botrytis mold of 'BRS Vitoria' seedless grape under cold storage." *Horticulturae* 4 (3). doi: 10.3390/horticulturae4030017.

du Toit, W. J., I. S. Pretorius, and A. Lonvaud-Funel. 2005. "The effect of sulphur dioxide and oxygen on the viability and culturability of a strain of *Acetobacter pasteurianus* and a strain of *Brettanomyces bruxellensis* isolated from wine." *Journal of Applied Microbiology* 98 (4):862–871. doi: 10.1111/j.1365-2672.2004.02549.x.

Dufour, J. P. 1991. "Influence of industrial brewing and fermentation working conditions on beer SO2 level and flavour stability." In *Proceedings of the 23rd Congress of the European Brewery Convention*, 209–216.

Edwards, C. G., and T. A. Oswald. 2018. "Interactive effects between total SO2, ethanol and storage temperature against *Brettanomyces bruxellensis*." *Letters in Applied Microbiology* 66 (1):71–76. doi: 10.1111/lam.12816.

El Kadi, F. Z., A. I. Bénali, B. Mohammed, and S. Belbraouet. 2014. "Effect of sodium metabisulphite on blood metabolic status of wistar rats." *Food and Nutrition Sciences* (15):1529.

Englezos, V., K. Rantsiou, F. Torchio, Matteo Pollon, S. Giacosa, Susana Rio Segade, V. Gerbi, L. Rolle, and L. Cocolin. 2019. "Effectiveness of ozone against different strains of *Brettanomyces bruxellensis* on postharvested winegrapes and impact on wine composition." *American Journal of Enology and Viticulture*. doi: 10.5344/ajev.2019.18058.

Faparusi, S. I. 1969. "Effect of pH on preservation of palm wine by sulfite." *Applied Microbiology* 18 (1):122–123.

FDA. 2018. "Import Alert 99–21." https://www.accessdata.fda.gov/cms_ia/importalert_265.html.

Feigenbau, J., and S. Israelashvili. 1949. "Sulfurous acid in citrus juices." *Industrial and Engineering Chemistry (1923)* 41:797–798.

Freedman, B. J. 1977. "Asthma induced by sulfur-dioxide, benzoate and tartrazine contained in orange drinks." *Clinical Allergy* 7 (5):407–415. doi: 10.1111/j.1365-2222.1977.tb01471.x.

Garcia-Fuentes, A. R., S. Wirtz, E. Vos, and H. Verhagen. 2015. "Short review of sulphites as food additives." *European Journal of Nutrition and Food Safety* 5(2), 113–120 .

Garcia-Rios, E., M. Nuevalos, E. Barrio, S. Puig, and J. M. Guillamon. 2019. "A new chromosomal rearrangement improves the adaptation of wine yeasts to sulfite." *Environmental Microbiology* 21 (5):1771–1781. doi: 10.1111/1462-2920.14586.

González-Rompinelli, E. M., J. J. Rodríguez-Bencomo, A. García-Ruiz, F. Sánchez-Patán, P. J. Martín-Álvarez, B. Bartolomé, and M. V. Moreno-Arribas. 2013. "A winery-scale trial of the use of antimicrobial plant phenolic extracts as preservatives during wine ageing in barrels." *Food Control* 33 (2):440–447. doi: 10.1016/j.foodcont.2013.03.026.

Guido, L. F. 2016. "Sulfites in beer: Reviewing regulation, analysis and role." *Scientia Agricola* 73 (2):189–197. doi: 10.1590/0103-9016-2015-0290.

Gunther, A., T. Konig, W. D. Habicher, and K. Schwetlick. 1998. "Antioxidant action of organic sulfites - II. Esters of sulfurous acid as primary antioxidants." *Polymer Degradation and Stability* 60 (2–3):385–391. doi: 10.1016/s0141-3910(97)00098-0.

Guzzo, J., M. P. Jobin, and C. Divies. 1998. "Increase of sulfite tolerance in *Oenococcus oeni* by means of acidic adaptation." *FEMS Microbiology Letters* 160 (1):43–47. doi: 10.1111/j.1574-6968.1998.tb12888.x.

Guzzon, R., R. Larcher, R. Guarcello, N. Francesca, L. Settanni, and G. Moschetti. 2018. "Spoilage potential of *Brettanomyces bruxellensis* strains isolated from Italian wines." *Food Research International* 105:668–677. doi: 10.1016/j.foodres.2017.11.078.

Guzzon, R., G. Widmann, M. Malacarne, T. Nardin, G. Nicolini, and R. Larcher. 2011. "Survey of the yeast population inside wine barrels and the effects of certain techniques in preventing microbiological spoilage." *European Food Research and Technology* 233 (2):285–291. doi: 10.1007/s00217-011-1523-8.

Hinze, H., and H. Holzer. 1986. "Analysis of the energy-metabolism after incubation of *Saccharomyces cerevisiae* with sulfite or nitrite." *Archives of Microbiology* 145 (1):27–31. doi: 10.1007/bf00413023.

Hinze, H., K. Maier, and H. Holzer. 1981. "Rapid decrease of the adenosine-triphosphate content in *Lactobacillus malii* and *Leuconostoc mesenteroides* after incubation with low concentrations of sulfite." *Zeitschrift Fur Lebensmittel-Untersuchung Und-Forschung* 172 (5):389–392. doi: 10.1007/bf01127670.

Iammarino, M., A. Di Taranto, M. Muscarella, D. Nardiello, C. Palermo, and D. Centonze. 2010. "Development of a new analytical method for the determination of sulfites in fresh meats and shrimps by ion-exchange chromatography with conductivity detection." *Analytica Chimica Acta* 672 (1–2):61–65. doi: 10.1016/j.aca.2010.04.002.

Ilett, O. R. 1995. "Aspects of the analysis, role, and fate of sulphur dioxide in beer: A review." *Technical Quarterly (Master Brewers Association of the Americas) (USA)* 32 (4):213.

Irwin, S. V., P. Fisher, E. Graham, A. Malek, and A. Robidoux. 2017. "Sulfites inhibit the growth of four species of beneficial gut bacteria at concentrations regarded as safe for food." *Plos One* 12 (10). doi: 10.1371/journal.pone.0186629.

Jackowetz, J. N., and R. M. de Orduna. 2013. "Survey of SO2 binding carbonyls in 237 red and white table wines." *Food Control* 32 (2):687–692. doi: 10.1016/j.foodcont.2013.02.001.

Jackson, Ron S. 2008. *Wine Science: Principles and Applications, Food Science and Technology International Series*: Amsterdam/Boston: Elsevier/Academic Press.

Jarvis, B., and A. G. H. Lea. 2000. "Sulphite binding in ciders." *International Journal of Food Science and Technology* 35 (1):113–127. doi: 10.1046/j.1365-2621.2000.00370.x.

Jiang, C., X. Chen, S. Lei, H. Zhao, Y. Liu, and J. Shi. 2017. "Lipopeptides from *Bacillus subtilis* have potential application in the winemaking process: Inhibiting fungal and ochratoxin A contamination and enhancing esters and acids biosynthesis." *Australian Journal of Grape and Wine Research* 23 (3):350–358. doi: 10.1111/ajgw.12289.

Joslyn, M. A., and J. B. S. Braverman. 1954. "The chemistry and technology of the pretreatment and preservation of fruit and vegetable products with sulfur dioxide and sulfites." *Advances in Food Research* 5:97–160. doi: 10.1016/S0065-2628(08)60222-0.

Kaneda, H., M. Takashio, T. Osawa, S. Kawakishi, S. Koshino, and T. Tamaki. 1996. "Analysis of aldehyde-bisulfites in beer by HPLC-fluorescence detection with post-column derivatization." *Journal of Food Science* 61 (1):105–108. doi: 10.1111/j.1365-2621.1996.tb14736.x.

Kocamaz, E., E. Adiguzel, E. R. Buket, G. Gundogdu, and V. Kucukatay. 2012. "Sulfite leads to neuron loss in the hippocampus of both normal and SOX-deficient rats." *Neurochemistry International* 61 (3):341–346. doi: 10.1016/j.neuint.2012.06.010.

Lacy, K., and W. E. Huffman. 2016. "Consumer demand for potato products and willingness-to-pay for low-acrylamide, sulfite-free fresh potatoes and dices: Evidence from lab auctions." *Journal of Agricultural and Resource Economics* 41 (1):116–137.

Lafeuille, J. L., S. Lefevre, and D. Achouri. 2007. "Determination of added sulfites in dried garlic with a modified version of the optimized Monier-Williams method." *Journal of AOAC International* 90 (4):1090–1097.

Lamas, A., J. M. Miranda, B. Vazquez, A. Cepeda, and C. M. Franco. 2016. "An evaluation of alternatives to nitrites and sulfites to inhibit the growth of *Salmonella enterica* and *Listeria monocytogenes* in meat products." *Foods* 5 (4). doi: 10.3390/foods5040074.

Landete, J. M., S. Ferrer, and I. Pardo. 2005. "Which lactic acid bacteria are responsible for histamine production in wine?" *Journal of Applied Microbiology* 99 (3):580–586. doi: 10.1111/j.1365-2672.2005.02633.x.

Lenz, A. G., and H. Holzer. 1985. "Cleavage of thiamine pyrophosphate by sulfite in *Saccharomyces cerevisiae*." *Zeitschrift Fur Lebensmittel-Untersuchung Und-Forschung* 181 (5):417–421. doi: 10.1007/bf01027410.

Lindsey, P. J., S. S. Briggs, K. Moulton, and A. A. Kader. 1989. "Sulfites on grapes: Issues and alternatives." In *Chemical Use in Food Processing and Postharvest Handling: Issues and Alternatives*, edited by Agricultural Issues Center, 5–19. University of California, Davis.

Lisanti, M. T., G. Blaiotta, C. Nioi, and L. Moio. 2019. "Alternative methods to SO2 for microbiological stabilization of wine." *Comprehensive Reviews in Food Science and Food Safety* 18 (2):455–479. doi: 10.1111/1541-4337.12422.

Liu, X. Z., M. Sang, X. A. Zhang, T. K. Zhang, H. Y. Zhang, X. He, S. X. Li, X. D. Sun, and Z. M. Zhang. 2017. "Enhancing expression of SSU1 genes in *Saccharomyces uvarum* leads to an increase in sulfite tolerance and a transcriptome profile change." *FEMS Yeast Research* 17 (3). doi: 10.1093/femsyr/fox023.

Longin, C., C. Degueurce, F. Julliat, M. Guilloux-Benatier, S. Rousseaux, and H. Alexandre. 2016. "Efficiency of population-dependent sulfite against *Brettanomyces bruxellensis* in red wine." *Food Research International* 89:620–630. doi: 10.1016/j.foodres.2016.09.019.

Lopez-Malo, A., S. M. Alzamora, and E. Palou. 2005. "*Aspergillus flavus* growth in the presence of chemical preservatives and naturally occurring antimicrobial compounds." *International Journal of Food Microbiology* 99(2):119–128.

Lou, T. T., W. S. Huang, X. D. Wu, M. M. Wang, L. Y. Zhou, B. Y. Lu, L. F. Zheng, and Y. Z. Hu. 2017. "Monitoring, exposure and risk assessment of sulfur dioxide residues in fresh or dried fruits and vegetables in China." *Food Additives and Contaminants Part A-Chemistry Analysis Control Exposure and Risk Assessment* 34 (6):918–927. doi: 10.1080/19440049.2017.1313458.

Loureiro, V., and M. Malfeito-Ferreira. 2003. "Spoilage yeasts in the wine industry." *International Journal of Food Microbiology* 86 (1–2):23–50. doi: 10.1016/s0168-1605(03)00246-0.

Mahmoud, A. A. T., G. M. Hassan, A. M. S. Hassan, Akma Latif, and M. F. Ramadan. 2015. "Demonstrating adverse effects of a common food additive (sodium sulfite) on biochemical, cytological and histopathological parameters in tissues of albino Wister rats." *European Journal of Integrative Medicine* 7 (3):234–242. doi: 10.1016/j.eujim.2015.03.003.

Martins, P. R., W. D. Popolim, L. A. F. Nagato, E. Takemoto, K. Araki, H. E. Toma, L. Angnes, and M. D. C. Penteado. 2011. "Fast and reliable analyses of sulphite in fruit juices using a supramolecular amperometric detector encompassing in flow gas diffusion unit." *Food Chemistry* 127 (1):249–255. doi: 10.1016/j.foodchem.2010.12.103.

McFeeters, R. F. 1998. "Use and removal of sulfite by conversion to sulfate in the preservation of salt-free cucumbers." *Journal of Food Protection* 61 (7):885–890. doi: 10.4315/0362-028x-61.7.885.

Montersino, S., A. Prieto, R. Munoz, and B. De Las Rivas. 2008. "Evaluation of exopolysaccharide production by leuconostoc mesenteroides strains isolated from wine." *Journal of Food Science* 73 (4):M196–M199. doi: 10.1111/j.1750-3841.2008.00726.x.

Montes, C., J. H. Velez, G. Ramirez, M. Isaacs, R. Arce, and M. J. Aguirre. 2012. "Critical comparison between modified Monier-Williams and electrochemical methods to determine sulfite in aqueous solutions." *Scientific World Journal*. doi: 10.1100/2012/168148.

Nadai, C. 2013. "Microbiological approaches to reduce the sulphite addition in oenology." https://pdfs.semanticscholar.org/c4a9/451d9071188b3c03d2f2abab55c4b2a21d79.pdf.

Nadai, C., L. Treu, S. Campanaro, A. Giacomini, and V. Corich. 2016. "Different mechanisms of resistance modulate sulfite tolerance in wine yeasts." *Applied Microbiology and Biotechnology* 100 (2):797–813. doi: 10.1007/s00253-015-7169-x.

Nagy, S. M., S. S. Teuber, S. M. Loscutoff, and P. J. Murphy. 1995. "Clustered outbreak of adverse reactions to a salsa containing high-levels of sulfites." *Journal of Food Protection* 58 (1):95–97. doi: 10.4315/0362-028x-58.1.95.

Nardi, T., V. Corich, A. Giacomini, and B. Blondin. 2010. "A sulphite-inducible form of the sulphite efflux gene SSU1 in a *Saccharomyces cerevisiae* wine yeast." *Microbiology-Sgm* 156:1686–1696. doi: 10.1099/mic.0.036723-0.

Neta, P., and R. E. Huie. 1985. "Free-radical chemistry of sulfite." *Environmental Health Perspectives* 64:209–217. doi: 10.2307/3430011.

Ni, Z. L., F. B. Tang, Y. H. Liu, D. Y. Shen, and R. H. Mo. 2015. "Rapid determination of low-level sulfite in dry vegetables and fruits by LC-ICP-MS." *Food Science and Technology Research* 21 (1):1–6. doi: 10.3136/fstr.21.1.

Nikolantonaki, M., P. Magiatis, and A. L. Waterhouset. 2015. "Direct analysis of free and sulfite-bound carbonyl compounds in wine by two-dimensional quantitative proton and carbon nuclear magnetic resonance spectroscopy." *Analytical Chemistry* 87 (21):10799–10806. doi: 10.1021/acs.analchem.5b01682.

Noble, J., I. Sanchez, and B. Blondin. 2015. "Identification of new *Saccharomyces cerevisiae* variants of the MET2 and SKP2 genes controlling the sulfur assimilation pathway and the production of undesirable sulfur compounds during alcoholic fermentation." *Microbial Cell Factories* 14. doi: 10.1186/s12934-015-0245-1.

OEHHA. 2012. *Consumption of Sulfur Dioxide in Dried Fruits*. Proposition 65 Interpretive Guideline No. 2012-02. https://oehha.ca.gov/media/downloads/crnr/so2driedfruitig.pdf.

OEHHA. 2020 *Chemicals Known to the State to Cause Cancer of Reproductive Toxicity*. California Environmental Protection Agency, Sacramento, CA. https://oehha.ca.gov/media/downloads/proposition-65//p65list010320.pdf (accessed January 31,2020).

Osborne, J. P., and C. G. Edwards. 2006. "Inhibition of malolactic fermentation by *Saccharomyces* during alcoholic fermentation under low- and high-nitrogen conditions: A study in synthetic media." *Australian Journal of Grape and Wine Research*, 12: 69–78.

Ough, C. S., and E. A. Crowell. 1987. "Use of sulfur-dioxide in winemaking." *Journal of Food Science* 52 (2):386–388. doi: 10.1111/j.1365-2621.1987.tb06620.x.

Parish, W. L. Jr., and M. Babcock. 2018. "Composition, method and apparatus for sulfite removal from beverages." United States Patent Application 20180154348 http://www.freepatentsonline.com/y2018/0154348.html.

Perfetti, G. A., and G. W. Diachenko. 2003. "Determination of sulfite in dried garlic by reversed-phase ion-pairing liquid chromatography with post-column detection." *Journal of AOAC International* 86 (3):544–550.

Pilkington, B. J., and A. H. Rose. 1988. "Reactions of *Saccharomyces cerevisiae* and *Zygosaccharomyces bailii* to sulphite." *Journal of General Microbiology* 134(10):2823–2830.

Prenner, B. M., and J. J. Stevens. 1976. "Anaphylaxis after ingestion of sodium bisulfite." *Annals of Allergy* 37 (3):180–182.

Qin, G., and Z. Meng. 2009. "Effects of sulfur dioxide derivatives on expression of oncogenes and tumor suppressor genes in human bronchial epithelial cells." *Food and Chemical Toxicology* 47 (4):734–744. doi: 10.1016/j.fct.2009.01.005.

Qu, D. F., Y. P. Gu, L. F. Feng, and J. Z. Han. 2017. "High content analysis technology for evaluating the joint toxicity of sunset yellow and sodium sulfite in vitro." *Food Chemistry* 233:135–143. doi: 10.1016/j.foodchem.2017.04.102.

Ramos, R. M., P. F. Brandao, L. M. Goncalves, V. Vyskocil, and J. A. Rodrigues. 2017. "Electrochemical sensing of total sulphites in beer using non-modified screen-printed carbon electrodes." *Journal of the Institute of Brewing* 123 (1):45–48. doi: 10.1002/jib.402.

Renouf, V., M. Falcou, C. Miot-Sertier, M. C. Perello, G. De Revel, and A. Lonvaud-Funel. 2006. "Interactions between *Brettanomyces bruxellensis* and other yeast species during the initial stages of winemaking." *Journal of Applied Microbiology* 100(6):1208–1219. doi: 10.1111/j.1365-2672.2006.02959.x.

Renouf, V., A. Lonvaud-Funel, and J. Coulon. 2007. "The origin of *Brettanomyces bruxellensis* in wines: A review." *OENO One* 41 (3):161–173. doi: 10.20870/oeno-one.2007.41.3.846.

Robbins, K. S., R. Shah, S. MacMahon, and L. S. de Jager. 2015. "Development of a liquid chromatography-tandem mass spectrometry method for the determination of sulfite in food." *Journal of Agricultural and Food Chemistry* 63 (21):5126–5132. doi: 10.1021/jf505525z.

Roberts, A. C., and D. J. McWeeny. 1972. "The uses of sulphur dioxide in the food industry." *Journal of Food Technology* 7 (3):221–238. doi: 10.1080/10871200260293388.

Rojo-Bezares, B., Y. Saenz, M. Zarazaga, C. Torres, and F. Ruiz-Larrea. 2007. "Antimicrobial activity of nisin against *Oenococcus oeni* and other wine bacteria." *International Journal of Food Microbiology* 116 (1):32–36. doi: 10.1016/j.ijfoodmicro.2006.12.020.

Santos, M. C., C. Nunes, J. A. Saraiva, and M. A. Coimbra. 2012. "Chemical and physical methodologies for the replacement/reduction of sulfur dioxide use during winemaking: Review of their potentialities and limitations." *European Food Research and Technology* 234 (1):1–12. doi: 10.1007/s00217-011-1614-6.

Schimz, K. L. 1980. "The effect of sulfite on the yeast *Saccharomyces-cerevisiae*." *Archives of Microbiology* 125 (1–2):89–95. doi: 10.1007/bf00403203.

Serpaggi, V., F. Remize, G. Recorbet, E. Gaudot-Dumas, A. Sequeira-Le Grand, and H. Alexandre. 2012. "Characterization of the "viable but nonculturable" (VBNC) state in the wine spoilage yeast *Brettanomyces*." *Food Microbiology* 30 (2):438–447. doi: 10.1016/j.fm.2011.12.020.

Serrano, R., and S. Banon. 2012. "Reducing SO2 in fresh pork burgers by adding chitosan." *Meat Science* 92 (4):651–658. doi: 10.1016/j.meatsci.2012.06.014.

Smilanick, J. L., P. I. Hartsell, D. Henson, D. C. Fouse, M. Assemi, and C. M. Harris. 1990. "Inhibitory activity of sulfur dioxide on the germination of spores of Botrytis cinerea." *Phytopathology (USA)* 80(2):217.

Smith, B. D., and B. Divol. 2018. "The carbon consumption pattern of the spoilage yeast *Brettanomyces bruxellensis* in synthetic wine-like medium." *Food Microbiology* 73:39–48. doi: 10.1016/j.fm.2017.12.011.

Splittstoesser, D. F., and L. R. Mattick. 1981. "The storage life of refrigerated grape juice containing various levels of sulfur dioxide." *American Journal of Enology and Viticulture* 32(2): 171–173.

Stratford, M., P. Morgan, and A. H. Rose. 1987. "Sulphur dioxide resistance in *Saccharomyces cerevisiae* and *Saccharomycodes ludwigii*." *Journal of General Microbiology* 133:2173–2179.

Stratford, M., and A. H. Rose. 1986. "Transport of sulphur dioxide by *Saccharomyces cerevisiae*." *Journal of General Microbiology* 132:1–6.

Sung-Chyr, L., and G. Georgiou. 2005. "A biocatalyst for removal of sulfite from alcoholic beverages." *Biotechnology and Bioengineering* 89 (1):123–127. doi: 10.1002/bit.20307.

Takahashi, M., T. Ohta, K. Masaki, A. Mizuno, and N. Goto-Yamamoto. 2014. "Evaluation of microbial diversity in sulfite-added and sulfite-free wine by culture-dependent and -independent methods." *Journal of Bioscience and Bioengineering* 117 (5):569–575. doi: 10.1016/j.jbiosc.2013.10.012.

Tan, W. H., F. S. Eichler, S. Hoda, M. S. Lee, H. Baris, C. A. Hanley, E. Grant, K. S. Krishnamoorthy, and V. E. Shih. 2005. "Isolated sulfite oxidase deficiency: A case report with a novel mutation and review of the literature." *Pediatrics* 116 (3):757–766. doi: 10.1542/peds.2004-1897.

Theisen, S., R. Hänsch, L. Kothe, U. Leist, and R. Galensa. 2010. "A fast and sensitive HPLC method for sulfite analysis in food based on a plant sulfite oxidase biosensor." *Biosensors and Bioelectronics* 26:175–181. doi: 10.1016/j.bios.2010.06.009.

Timbo, B., K. M. Koehler, C. Wolyniak, and K. C. Klontz. 2004. "Sulfites - A Food and Drug Administration review of recalls and reported adverse events." *Journal of Food Protection* 67 (8):1806–1811. doi: 10.4315/0362-028x-67.8.1806.

Tomasevic, I., M. Dodevska, M. Simic, S. Raicevic, V. Matovic, and I. Djekic. 2018. "A decade of sulphite control in Serbian meat industry and the effect of HACCP." *Food Additives and Contaminants Part B-Surveillance* 11 (1):49–53. doi: 10.1080/19393210.2017.1403492.

U.S. Food and Drug Administration. 1977a. "21 CFR §182: Substances generally recognized as safe." https://www.ecfr.gov/cgi-bin/text-idx?SID=345be3b47f4d43dbc8c0bcb7150ddc65&mc=true&node=pt21.3.182&rgn=div5 (accessed June 14).

U.S. Food and Drug Administration. 1977b. "Title 21: Food and drugs part 182-substances generally recognized as safe subpart D—Chemical preservatives." https://www.accessdata.fda.gov/scripts/cdrh/cfdocs/cfcfr/CFRSearch.cfm?CFRPart=182 (accessed May 30, 2019).

U.S. Food and Drug Administration. 2018a. "Recalls, market withdrawals, & safety alerts." https://www.fda.gov/Safety/Recalls/default.htm.

U.S. Food and Drug Administration. 2018b. "Recalls, market withdrawals, & safety alerts." https://www.fda.gov/safety/recalls-market-withdrawals-safety-alerts (accessed December 31, 2018).

USDA. 2011. "Technical evaluation report sulfur dioxide-crops." https://www.ams.usda.gov/sites/default/files/media/Sulfur%20dioxide%20smoke%20bombs%20report%202011.pdf (accessed June 25, 2019).

Vally, H., and N. L. A. Misso. 2012. "Adverse reactions to the sulphite additives." *Gastroenterology and Hepatology from Bed to Bench* 5 (1):16–23.

Vally, H., N. L. A. Misso, and V. Madan. 2009. "Clinical effects of sulphite additives." *Clinical and Experimental Allergy* 39 (11):1643–1651. doi: 10.1111/j.1365-2222.2009.03362.x.

van Wyk, S., M. M. Farid, and F. V. M. Silva. 2018. "SO2, high pressure processing and pulsed electric field treatments of red wine: Effect on sensory, Brettanomyces inactivation and other quality parameters during one year storage." *Innovative Food Science and Emerging Technologies* 48:204–211. doi: 10.1016/j.ifset.2018.06.016.

Walker, R. 1985. "Sulphiting agents in foods: Some risk/benefit considerations." *Food Additives and Contaminants* 2 (1):5–24.

Warth, A. D. 1985. "Resistance of yeast species to benzoic and sorbic acids and to sulfur dioxide." *Journal of Food Protection* 48(7):564–569.

Wedzicha, B. L. 1992. "Chemistry of sulphiting agents in food." *Food Additives and Contaminants* 9(5):449–459.

Wells, A., and J. P. Osborne. 2011. "Production of SO2 binding compounds and SO2 by *Saccharomyces* during alcoholic fermentaton and the impact on malolactic fermentation." *South African Journal of Enology and Viticulture* 32 (2):267–279.

WHO. 2009. Technical Report Series 952. *Sixty-Ninth Report of the Joint FAO/WHO Expert Committee on Food Additives.*

Williams, R. R., R. E. Waterman, J. C. Keresztesy, and E. R. Buchman. 1935. "Studies of crystalline vitamin B-1. III. Cleavage of vitamin with sulfite." *Journal of the American Chemical Society* 57 (1):536–537. doi: 10.1021/ja01306a044.

Yang, W. H., and E. C. R. Purchase. 1985. "Adverse reactions to sulfites." *CMAJ: Canadian Medical Association Journal* 133 (9):865.

Zheng, L. Y., and J. A. F. Zhu. 2003. "Study on antimicrobial activity of chitosan with different molecular weights." *Carbohydrate Polymers* 54 (4):527–530. doi: 10.1016/j.carbpol.2003.07.009.

Zuehlke, J. M., and C. G. Edwards. 2013. "Impact of sulfur dioxide and temperature on culturability and viability of *Brettanomyces bruxellensis* in wine." *Journal of Food Protection* 76 (12):2024–2030. doi: 10.4315/0362-028x.jfp-13-243r.

7

Nitrite

R. Bruce Tompkin, Wendy Bedale, Andrew Milkowski, Kathleen Glass, and Jeffrey J. Sindelar

CONTENTS

7.1 Introduction

This review of the antimicrobial properties of nitrite is one among many. It differs, however, in certain respects from other reviews. By generally arranging the information in chronological order, insight is provided into why the research was performed, the selection of the experimental procedure, and the information that was available to influence the researchers' conclusions. Other comprehensive reviews are available that provide additional information and other perspectives (Anderton, 1963; Benedict, 1980; Duncan, 1970; Ingram, 1974; Lechowich et al., 1978; Pivnick, 1980; Sofos et al., 1979a; Holley, 1981; Roberts et al., 1991; Skovgaard, 1992). Binkerd and Kolari (1975) reviewed the history of nitrite and nitrate as curing agents for meats. Cerveny (1980) described many of the changes in the production and marketing of cured meats in the United States since 1900. The status of nitrate and nitrite in food and water until about 1990, particularly in the UK and Europe, is discussed in Hill (1991). Extensive updates on the chemistry and impact of nitrite on the color and flavor of cured meats (Pegg and Shahidi, 2000) and the microbiological, toxicological, and physiological effects of nitrate and nitrite (Archer, 2002) have been developed. Reported outbreaks of botulism attributed to meat and poultry products and an assessment of nitrite for botulinal protection have been summarized (Tompkin, 1980). Eklund (1982) reviewed the potential risk of botulism in fishery products and the role of nitrite as a preservative.

The many chemical reactions that occur when nitrite is added to meat offer clues to the reactions that may occur with microbial cells. For this reason, the reader should consider reactions with heme compounds (Fox, 1966; Bard and Townsend, 1971, Roberts et al., 1991; Pegg and Shahidi, 2000) and other components of meat (Cassens et al., 1979). The reviews by Gray et al. (1981) and Pegg and Shahidi (2000) of the effect of nitrite on the stability of cured meat flavor provide additional possibilities.

The use of nitrite and/or nitrate in foods has been influenced by a variety of factors. Initially, their use was to fulfill the basic need to preserve food. As questions were raised from about 1970 to 2002 over the toxicological safety of nitrite, considerable effort was directed toward verifying the benefits of nitrite relative to control of certain pathogens and quality attributes. At the same time a considerable amount of research was directed toward identifying the risks associated with the use of nitrite in food. This chapter will concentrate on the antimicrobial research. It should be remembered that most of the research during that period was driven by the potential risk to consumers from dietary intake of nitrite and the Delaney Clause that declared additives found to be carcinogenic in any animal test could not be used in food in the United States.

With the new millennium a different perspective has evolved. Current regulations for the use of nitrite and nitrate and modern, controlled processing conditions provide the benefits of food preservation with no measurable risk to consumers. It is tempting to delete a significant quantity of information from this fourth revision of the chapter; however, the information is being retained for its historical value and to provide a foundation for future research when nitrite and nitrate are next called into question.

The discovery in 1969 that nitrosamines can be formed in foods containing nitrite led to a series of events. The time and resources expended to resolve the nitrite issue were enormous and worldwide. At the heart of the issue was whether nitrite would continue to be an approved food additive. If nitrite were not available, then cured meats, as we know them, would disappear from the market. It is a curious thought that if nitrite were disapproved, this chapter would not exist in this book. To better understand this complex issue, the reader is encouraged to read the historical documentary of Cassens (1990). Cassens' text places the research outlined in this chapter into its proper historical perspective relative to the politics and science of the nitrite issue.

Unlike the other antimicrobial agents discussed in this text, there has been a long, controversial history over whether nitrate and nitrite have antimicrobial properties. Until the early 1940s, nitrite was considered to have no value in meat other than for flavor and color development. Nitrate, however, was believed to provide the antimicrobial effect. During the 1940s and early 1950s, data appeared that began to prove the opposite. By the middle 1950s, research had clearly demonstrated that nitrite has significant antimicrobial properties. By then it was also more firmly believed that nitrous acid derived from nitrite in the acid environment of meat is responsible for microbial inhibition. This concept became widely accepted, and the value of nitrate was relegated to serving merely as a reservoir for the generation of nitrite. By the time the nitrosamine controversy arose in 1969 and the early 1970s, the value of nitrate as a preservative had been so degraded that it was almost unanimously agreed in the United States that nitrate could be omitted from most cured meats without loss of antimicrobial protection or preservation.

Clever as the scientific community has been in more clearly defining the antimicrobial properties of nitrite, it remains to be seen whether wisdom has prevailed with the voluntary abandonment of nitrate as an additive to many meat products in the USA. One of the messages from the 1920s is that nitrate serves as a potential reservoir for conversion to nitrite and thus to nitrous acid. Considering that research continues to reveal new interactions among the chemical constituents of cured meat, has the proper research been conducted and interpreted in an unbiased manner to justify the elimination of nitrate from the majority of cured meats? A primary reason for the controversy over the antimicrobial effect of nitrate and nitrite that prevailed into the 1940s was ignorance of the significance of pH. The list of recognized factors influencing the antimicrobial efficacy of nitrite in cured meats continues to grow 50 years later. It is still difficult to design experiments that yield predictable, reproducible results in all instances. We are still unaware of all the significant factors that influence the efficacy of nitrite or, for that matter, the value of nitrate.

Research of the 1920s and 1930s focused on using nitrite to cure meat, solving a problem of sour hams, and dealing with perishable canned hams that were being temperature abused in the marketplace. Most of the research on sour hams involved test tube experiments with culture media above pH 7.0, despite meat having a pH closer to 6.0. This basic flaw in experimental design led to results that misguided and hardened opinions in the industry for 20 years.

7.2 Research History

7.2.1 Before 1950

Tucker (1930) studied the proteolytic activity of a *Clostridium putrificum* isolate from sour ham. Proteolysis was inhibited by 4.0% sodium chloride (salt), 44,000 mg/L (ppm) of sodium nitrate, or 12,000 mg/L of sodium nitrite in a pork infusion broth buffered to pH 7.6 and held at 37°C. He concluded that, of the three, only salt was used in sufficient concentration in commercial hams to prevent proteolytic activity at 37°C. The same results were obtained with *Clostridium putrefaciens* at 23°C. He suggested that nitrite, not nitrate, is responsible for retarding proteolysis by *C. putrificum*.

Tanner and Evans (1933), using *Clostridium botulinum*, *C. putrificum*, and *C. sporogenes*, found that sodium nitrate up to 22,000 mg/L did not inhibit the 12 strains tested in either pork infusion or egg-meat media. At levels between 22,000 and 44,000 mg/L, irregular inhibition occurred among seven botulinum cultures. At 44,000 mg/L, six of the seven cultures were inhibited. The authors concluded that nitrate at concentrations as low as 22,000 mg/L cannot be relied upon to inhibit putrefactive anaerobes if other conditions are favorable for growth.

Tanner and Evans (1934) reported that 5900 mg/L of sodium nitrite inhibited 9 of 12 strains of clostridia in nutrient broth and all 12 strains in dextrose broth. Levels of 3100 mg/L in pork infusion and 3900 mg/L in egg-meat, the highest levels tested, were not inhibitory to any of the strains even though the nitrite was added to the media (pH 7.4) before autoclaving. The authors concluded that reliance cannot be placed upon nitrite alone to prevent spoilage by clostridia. From tests with various blends of salt, nitrite, and nitrate in peptic digest broth (pH 7.4), raw pork, and cooked pork, Evans and Tanner (1934) concluded that "the most effective component in curing mixtures is sodium chloride" and "the sodium nitrite present apparently produced no effect on the organisms." Botulinal toxin was produced in the cooked meat and a few samples of raw meat. Repression of toxin formation in the raw meat was a result of other bacteria that caused a decline in pH to inhibitory levels (pH 4.49–5.3).

Curiously, Tanner and Evans (1933), citing earlier work by MacNeal and Kerr, state that

> potassium nitrate in neutral or alkaline solutions exerted no special restrictive effect on bacterial activity. Under these conditions, it was used as a food. In acid solutions, however, the results were quite different. A marked inhibition was noticed. They said that this effect was incomparably greater than that of salt and was best ascribed to the production of small amounts of nitric acid, and of nitrous acid also, in mixtures containing reducing substances. Potassium nitrate was therefore considered to be particularly effective in restricting acid fermentation of organic substances that are already slightly acid. They further believed that the claim of meat packers that small amounts of nitrate in the pickle produced better preservation of the meat, was borne out by their results. It seemed that nitrate was especially valuable in preventing high degrees of acidity or souring of meat. MacNeal and Kerr stated that the effect of saltpeter was probably due to the oxidizing action of the nitrate ion in the presence of hydrogen ion.

It is surprising that the fundamental importance of pH, so clearly stated and available to researchers at that time, was not pursued.

Perishable (i.e., insufficiently heated to be stable at room temperature) canned chopped spiced ham and luncheon meats containing nitrate, nitrite, and sugar swelled in 1–90 days during temperature abuse at 37°C. Examining 1000 cans representing all manufacturers at the time, the sole cause of the spoilage was found to be nitrate-reducing *Bacillus* species (Jensen et al., 1934). A medium was developed with nitrate, sugar, and cured meat to detect the gas-producing bacteria in raw materials and the plant environment. Nitrite could not replace nitrate. To obtain gas production in the medium, both nitrate and cured meat had to be present. The reason for this specific combination of ingredients remains unknown. Perhaps, the cured meat provides a source of heme as the more recent results suggest (Jacobs et al., 1964).

The opinions expressed by Jensen et al. in 1934 laid the foundation for a new concept. They stated that if nitrate were omitted from the product, the usual fermentative carbon dioxide swells caused by *Bacillus* species would be prevented and conditions for the growth of clostridia and other anaerobes would prevail. Their bold statements that a mixed nitrate-nitrite cure favors most species of aerobes but a nitrite cure "always inhibits fermentation and aids in putrefaction" were to have long-term effects within the meat industry and federal government.

Brooks et al. (1940) described the contemporary method of making bacon in the United Kingdom. After the slaughtering process, pork sides were held at ambient temperature overnight and then moved into refrigeration for 24 h to achieve 5.6°C in the meat. The sides were then immersed in a tank and cured using a pickle solution containing potassium nitrate. Several interesting questions were raised. Does nitrate serve any function other than as a precursor for nitrite? Does nitrate appreciably retard or inhibit the growth of putrefactive anaerobes? What is the mechanism behind nitrite depletion, and what factors influence depletion? Would the more rapid chilling process as practiced in North America adversely affect the quality of English-style bacon? Can bacon acceptable to the English trade be produced by the use of rapidly chilled meat and with nitrite in place of nitrate? Not all the questions could be answered, but the research led to the following general conclusions. Satisfactory bacon can be produced with the use of nitrite. The characteristic cured flavor of bacon is due primarily to the action of nitrite. The conversion of nitrate to nitrite in commercial bacon curing brines is due mainly to the growth of micrococci. The presence of nitrate or microbial action during the curing process is not essential for bacon flavor. Rapid chilling of the meat is not detrimental to quality attributes.

Tarr and coworkers published a series of reports on the possible use of nitrite for preservation of fish. Tarr and Sutherland (1940) concluded that "as far as can be ascertained from the literature, little or nothing is known regarding the possible bacteriostatic (or bactericidal) action of nitrites in meat pickles or in cured meats themselves." Their tests with fish muscle showed that nitrite delayed spoilage. Subsequent tests (Tarr, 1941) demonstrated the importance of pH to the efficacy of nitrite. At pH 7.01, little or no inhibition was observed. At pH 5.7 and 6.0, complete or strong microbial inhibition occurred.

The effect of pH was more firmly documented in a series of tests in nutrient broth inoculated with a variety of aerobes. Later tests involving *C. botulinum* and *C. sporogenes* yielded similar results (Tarr, 1942). Investigations on the mode of action led Tarr to conclude that nitrite is not inhibitory by sole virtue of its toxicity toward aerobic respiratory catalysts (Tarr, 1941) and that the mode of action was still unknown (Tarr, 1942).

While Tarr was demonstrating that nitrite has strong antimicrobial properties, Jensen and Hess (1941) advocated the virtues of nitrate and perpetuated the belief that the sole function of nitrite was for cure color development. They stated the literature shows "unmistakenly that nitrate in the cure exerts a definite inhibitory effect upon bacteria." They postulated the anticlostridial mechanisms to consist of partial conversion of nitrate to hydroxylamine, which inactivates catalase and allows the accumulation of hydrogen peroxide, which destroys anaerobes.

Jensen and Hess stated that nitrite reacts with protein during heating and is destroyed, thus leaving the meat in much the same state as freshly cooked uncured meat. Others have shared this concern (Scott, 1955). Even if nitrite had been uniformly accepted as a critical factor in microbial protection, workers at the time would have had difficulty applying the information to commercial practice. How could reliance be placed upon an unstable preservative that disappears during processing and storing? This question is still valid.

The purpose of Jensen and Hess' work in 1941 was to determine the specific value of nitrate as a bacteriostat and as an inhibitor of bacterial spore germination and toxigenicity and its effect on the anaerobic microbiota of cured meats. Two tests were reported with perishable canned ham. In one, nitrite alone inhibited anaerobic putrefaction during 30 days of abuse at 37.2°C. In the other, fewer cans became putrid in the product with both nitrate and nitrite than in the product with nitrite only. Despite these results, the authors concluded that nitrate is beneficial for preventing the growth of putrefactive anaerobes in abused perishable canned ham. In both tests some cans with nitrate swelled because of the growth of *Bacillus* species, thus serving as a desirable indicator of temperature abuse.

Using the spiced ham medium, they reported earlier, Jensen and Hess found that heat (93°C for 4 h) combined with 5000 mg/L, but not 2000 mg/L, of sodium nitrate, prevented growth of *C. sporogenes* during 4 weeks of incubation at 37.2°C. The authors suggested that the combination of heat, nitrate, nitrite, and salt caused destruction of anaerobic spores at much lower temperatures.

Yesair and Cameron (1942) pursued this idea but concluded that curing salts do not assist in thermal destruction but inhibit outgrowth. Stumbo et al. (1945a) reached the same conclusion from tests in meat processed at 116°C. They also reported (Stumbo et al., 1945b) that nitrite appreciably delayed germination, although salt was the stronger inhibitor. Nitrate, alone or in combination with other ingredients, did not appreciably influence spoilage in tubes processed from F_0 = 1–16 and held for a year at 28°C.

Jensen et al. (1949) also examined the combined effect of heating and curing salts. Greatly increased inhibition occurred in tubes of pork heated in the range of 50–65°C for 30 min. Higher temperatures, longer heating times, or repetitive heating did not increase this effect. Within the levels normally added to canned ham, increasing salt and nitrite caused increased inhibition. However, increasing nitrate did not increase the inhibition of *C. sporogenes* 3679.

Vinton et al. (1947) found the heat resistance of spore crops was not altered by adding nitrite and nitrate to the meat on which the spores were grown, but the condition of the meat was a factor. Spores produced in raw, pasteurized, and sterilized meat were of low, intermediate, and high heat resistance, respectively.

7.2.1.1 Summary for before 1950

A summary of the state of knowledge by 1950 might include the following:

1. The relative roles of nitrate and nitrite as preservatives in cured meats were unclear. Strong opinions were expressed that nitrite alone was ineffective at the levels used in commercial practice.
2. Nitrate was promoted as an inhibitor of anaerobic spoilage and to enhance the swelling of perishable canned cured meat by *Bacillus* species when temperature abused.
3. Nitrite was clearly shown to be an effective antimicrobial agent, especially if the product pH was below 7.0.
4. The disappearance of nitrite during processing and storage was assumed to make it an unreliable preservative.
5. Salt at the levels used at that time was considered the cornerstone for the preservation of cured meats.
6. The concept of a combined effect of heating in the presence of curing salts was proposed. Although thermal destruction was not shown to be enhanced by the presence of curing salts, there was increased inhibition of outgrowth of surviving spores by salt and, to some degree, by nitrite. Nitrate was without effect.
7. An underlying factor in the research before 1940 was the unfavorable experience of 1928 and 1929 with temperature-abused perishable canned hams. Halvorson (1955) later discussed the widespread problem of temperature abuse at the retail level. It is perhaps for this reason that Jensen believed so strongly in the need for nitrate in perishable canned cured meats. During the 1930s some processors incubated perishable canned cured meat at 37°C as an index of "keeping quality." This experience may have led to improvements in meat handling and equipment sanitation.

7.2.2 1950–1960

Research of the 1950s clarified the relative significance of salt, nitrate, and nitrite as preservative agents and expanded this information to other microorganisms.

Steinke and Foster (1951) investigated the effect of packaging liver sausage in an oxygen-impermeable film (Saran) that was gaining acceptance in the industry. Salt was found to be the major factor retarding botulinal outgrowth in temperature-abused products. The basic formulation resulted in a product having a moderately high brine of 5.05–5.37% and a pH range of 6.1–6.5. Adding 200 mg/kg of sodium nitrite was considerably more inhibitory to toxin formation than the addition of 1000 mg/kg of sodium nitrate; however, a combination of both nitrite and nitrate was the most inhibitory.

Bulman and Ayres (1952) found that levels of salt in excess of 4.4% or sodium nitrate in excess of 40,000 mg/kg were required to prevent spoilage from anaerobic spore formers in pork. In the absence of added salt, increased inhibition of PA 3679 was obtained as sodium nitrite was increased from 400 to 800 mg/kg. A mixed cure of salt (3.5%) and nitrite (150–170 mg/kg) was much more effective than either substance alone. Nitrate at the level permitted in meat (1250 mg/kg) was without effect, even in the presence of salt or nitrite. A mixed cure of all three (salt, nitrate, and nitrite), however, yielded the maximum inhibition. The question was raised whether the growth of spoilage organisms is influenced by the depletion of nitrite to non-inhibitory levels.

Henry et al. (1954) reported that at pH 7.5 or above, nitrite enhanced bacterial growth in curing brine. A pH of 5.6–5.8 was optimal for antibacterial efficacy. At pH 5.3 or below, nitrite rapidly disappeared and was ineffective. Nitrite was more inhibitory in the presence of ascorbate.

Castellani and Niven (1955) stated that nitrite was not known to have any practical preservative value against those organisms not inhibited by the high salt content in cured meats. They investigated why *Staphylococcus aureus* is rarely, if ever, found in the interior of cured meats. Upon adding a filter-sterilized solution of nitrite to a broth medium, less nitrite was necessary for inhibition as the pH was decreased from 6.9 to 5.05. The authors postulated that nitrous acid was responsible for the bacteriostatic action of nitrite. Additional tests showed that if the broth medium (pH 6.55) was autoclaved with glucose, a very small amount of added nitrite prevented staphylococcal growth when incubated anaerobically. This effect was reversed by adding sulfhydryl compounds.

Lechowich et al. (1956) examined whether curing salts plus the anaerobiosis of a vacuum package would inhibit the growth of *S. aureus*, as suggested by Castellani and Niven (1955). Tests in broth media and raw pork led them to conclude that *S. aureus* growth can occur in any combination of salt, nitrate, and nitrite that is palatable and permissible. They found *S. aureus* to die rather rapidly in ham-curing pickle, unless protected by meat juices. Also, *S. aureus* was killed in hams when heated to an internal temperature of 58.3°C, the legal requirement for trichina destruction.

Scott (1955) concluded from the literature that since nitrate exhibited relatively poor antimicrobial inhibition and nitrite, although effective, was unstable, the control of salt concentration and resultant water activity was the most reliable bacteriostatic system for cured meats. As late as 1957, this cautious view of the inhibitory effect of nitrite was expressed by Eddy (1957): "Taken in their totality, these observations leave no doubt inhibition by nitrite is at least a possibility."

An international symposium (Ann. Inst. Pasteur, 1955) on semi-preserved meats in hermetically sealed containers included a curious debate. At that time European microbiologists commonly preincubated perishable canned cured meats before microbiological examination. The debated issue was the time and temperature for preenrichment. Today, a similar debate would include the time elapsed since manufacturing.

Halvorson (1955) cited a private communication from L. B. Jensen, who performed a series of tests with spiced ham inoculated with different strains of *C. botulinum*. Jensen concluded that if the brine content were less than 5.0%, the spores could germinate, growth would occur, and toxin would be produced. At 6.0% brine or above, toxin formation was inhibited with a heat treatment as low as $F_0 = 0.27$. Germination was apparently prevented, since viable spores could be recovered. Earlier, Dack (1949) reported toxin inhibition in meat with brine levels above 6.4%. In North America, brine values traditionally have been calculated at % brine = (% salt/% salt + % water) × 100. In the United Kingdom, salt levels may be expressed as the percentage brine as just described, or as the percentage brine or the percentage salt (wt/vol or wt/wt) on the water phase using the equation % brine or % salt = (% salt/% water) × 100.

During the mid-1950s the U.S. Department of Agriculture (USDA) issued a regulation to assure the safety of new perishable canned cured meat products (e.g., 1.5 lb canned ham). Since the procedures traditionally used for large canned hams had a history of safety, these existing products were not affected. The requirements imposed on new products included a brine level of 5.5%, the addition of 1.75% sugar, and 1 ounce of sodium nitrate per 100 pounds meat. This combination of ingredients was derived from the literature and unpublished results from Jensen, who concluded that 1-ounce nitrate was optimal for gas formation by *Bacillus* species in temperature-abused perishable canned luncheon meat.

The USDA concluded that the high brine would retard the growth of *C. botulinum*, and in the presence of nitrate, the growth of *Bacillus* species would occur, causing the cans to swell and forewarn consumers of a potential hazard. A typical formula being used for large canned hams at that time included 0.5% sugar, 1/4-ounce sodium nitrate (156 mg/kg), 1/8-ounce sodium nitrite (78 mg/kg), and a brine level of 3.5%. The brine requirement of 5.5% would have caused consumer complaints of excessive saltiness in the new products. It was also believed within the industry that the higher nitrate level would have enhanced can corrosion and product discoloration owing to reduction of nitrate to nitrite by iron at imperfections in the lacquer coating of the cans. This brought nitrate under renewed attack.

Although shelf-stable canned, cured luncheon meat had been produced for years, the reasons for stability had not been defined. Silliker et al. (1958) used this class of product to show that nitrate played no role in retarding putrid spoilage. As expected, nitrate actively stimulated spoilage by aerobic spore formers. Brine level (0, 3.5, or 5.0%) alone was not responsible for stability. The authors considered the key to stability to be the addition of sodium nitrite (78 mg/kg) and heat injury to the small number of indigenous spores. They concluded that the stability of shelf-stable canned cured meat given less than a botulinal cook was due to the combined effect of nitrite, salt, thermal injury to the spores, and a low indigenous spore level. Since viable spores could be recovered from stable commercial product, the primary effect of nitrite was to prevent germination and/or outgrowth of heat-injured spores. This research and a subsequent review by Silliker (1959) established the future direction for research on nitrite.

Greenberg et al. (1959) subsequently demonstrated botulinal spore outgrowth in perishable canned cured meat having brine levels from 3.0 to 7.125% and abused at 29.4–37.8°C for up to 90 days. Sodium nitrite levels were 22 mg/kg or less at the time the meat was inoculated. At 5.14% brine or less, the product was organoleptically unacceptable – as the cans swelled – and toxin assays were performed. A significant observation was that at brine levels of 6.25, 7.09, and 7.12%, toxin was produced without concomitant organoleptic spoilage. At 8.95% brine, toxin was not produced. There have been no reported attempts to duplicate this effect of brine at levels approaching total inhibition, although there have been several studies under other conditions in which toxin has been detected without obvious organoleptic spoilage.

7.2.2.1 Summary for 1950–1960

As 1960 approached, the following became increasingly clear:

1. Nitrate, per se, had no antimicrobial effect, other than its possible influence on water activity.
2. Nitrate could serve as an electron acceptor, permitting the growth of aerobes, such as micrococci and bacilli. Several new reports of unstable commercially canned cured meats containing nitrate appeared (Hankins et al., 1950; Verhoeven, 1950; Eddy and Ingram, 1956).
3. Although nitrite was recognized as an effective antimicrobial agent, its value as a preservative in perishable cured meats was still in doubt.
4. Nitrite was shown to play a significant role in the stability of shelf-stable canned meat. The system providing stability consisted of the combined effect of nitrite, salt, and thermal injury to the low indigenous level of anaerobic spores. The suspected role of nitrite was to prevent germination or outgrowth of surviving heat-injured spores.
5. Brine content was further shown to influence botulinal outgrowth and toxin formation. At brine levels approaching the inhibitory level, toxin was produced without obvious organoleptic spoilage.
6. *S. aureus* was killed during the heating process given to hams. Anaerobic intolerance to nitrite within the interior of ham, as postulated by test tube experiments, was not confirmed.

7.2.3 1960–1970

After 1960, increased emphasis was directed toward the role of nitrite in the total inhibitory system in cured meats and its effect on thermally injured spores. In the course of this research, the observations of Perigo et al. (1967) opened new possibilities. Research was initiated on irradiation for shelf-stable canned cured meats. The increased use of vacuum packaging raised new questions about the microbiological safety of cured meats. Outbreaks of botulism from temperature-abused vacuum-packed smoked fish during the early 1960s added impetus to this research.

Shank et al. (1962) explored whether nitrite or the oxides of nitrogen were involved in the inhibition of Gram-negative bacteria in cured meats. They concluded that nitrite and nitrate have either neutral or stimulatory effects on bacteria and that nitric oxide has virtually no effect. Nitrous acid was considered primarily responsible for the quantitative as well as qualitative changes that occur in the bacterial microbiota when meat is cured. It was postulated that nitrous acid reacts either with the cell itself or with various constituents of the medium, making them unavailable for subsequent metabolism.

A requirement of hemin for nitrate reduction and growth stimulation under anaerobic conditions for six strains of staphylococci and one strain of *Bacillus subtilis* was shown by Jacobs et al. (1964). Hemin had no effect on nitrate reduction or growth of *Escherichia coli* or *Bacillus polymyxa*. The authors concluded that when oxygen is absent during the growth of some bacteria, heme synthesis is impaired, even if nitrate is present as an alternative electron acceptor. The impaired synthesis of heme prevents the formation of the required cytochrome systems, and nitrate is not reduced anaerobically. Nitrate and oxygen were considered interchangeable as electron acceptors required for heme synthesis in some bacteria, but not in others.

7.2.3.1 Shelf-Stable Cured Meats

Riemann (1963) supplemented existing information on the incidence, heat resistance, inhibition, and outgrowth of spores with unpublished data from the Danish Meat Research Institute. He concluded that most of the spores are killed during the thermal processing of shelf-stable canned cured meat. One reason given was that spores indigenous to meat emulsions have lower heat resistance than spores produced in laboratory media. He estimated the thermal process caused a 4-log reduction of the indigenous spores. The surviving spores are subsequently inhibited by the salt and nitrite. If, however, the initial spore load in the product is too high, the larger number of spores that survive the heat process increases the probability that at least one will be able to grow despite the preservative system. Another factor in spore destruction is that a thermal process of $F_0 = 0.4–0.6$ at the center of the can would yield ten-fold higher values near the surface of the can.

Gould (1964) observed germination, but at a reduced rate, of spores of *Bacillus* species in the presence of less than 300 mg/L of sodium nitrite at pH 6. Immediately after germination, development ceased before lysis or rupture of the spore walls. The inhibited spores did not appear to become fully phase dark. Germination was completely prevented by sodium nitrite levels of 750 mg/kg or higher at pH 6.0. The toxicity of nitrite was three to five times greater at pH 6 than at pH 7.

The thermal process used during the early 1960s by seven US producers of shelf-stable canned cured meat was reported to range from $F_0 = 0.1$ to 0.5, with a median of 0.2 (Greenberg et al., 1965). In a study comparing irradiation with conventional thermal processing, canned chopped ham was inoculated with botulinal spores and given two thermal processes. If the surviving spore counts between the enzyme-inactivating (74°C internal temperature) and the shelf-stable process ($F_0 = 0.203$) were compared, a 4–5-log reduction occurred. Product inoculated with the lowest inoculum level (173 spores/g) had 0.017 viable cells/g remaining after processing to $F_0 = 0.2$. The presence of 25 mg/kg of residual nitrite after processing and a relatively low brine level of 3.36% did not prevent outgrowth, spoilage, and toxin formation from this low level of survivors during storage at 30–37.7°C for 6 months.

Spencer (1966) concluded from the literature that 3.5–4.0% brine, 75–150 mg/kg of added sodium nitrite to provide about 20 mg/kg after processing to $F_0 = 0.1–0.4$, and a low incidence of clostridial spores (less than 1/g) were necessary for the safety and stability of shelf-stable canned cured meats.

The thermal process required to achieve shelf stability of canned hams larger than 1.5 pounds (~0.68 kg) caused excessive purge (e.g., 30%) and poor meat texture. Kueck et al. (1965) developed a new process for the production of 3-pound (1.36 kg) shelf-stable canned hams with a purge of about 17% and acceptable texture and flavor. The process generally consisted of injecting a curing solution and then immersion in a cover pickle for 1–3 days as was typical for that time. A 3-pound (1.36 kg) meat portion was filled into each can, and the product was given a long thermal process at a low temperature (e.g., 2–2 1/2 h in 74°C water) to achieve an internal temperature of at least 71°C. The product was then heated for a short time at a high temperature (e.g., 15 min at 113–118°C). This process caused the surface of the meat in the can to reach 96–107°C. The product was then cooled in tap water and placed into refrigerated storage (e.g., 30 days at 6–8.3°C) before release as a shelf-stable product. They demonstrated that increased stability could be achieved by omitting sodium nitrate and sucrose from the formulation. Increased stability also resulted from improving the canning process to reduce or prevent voids. The success of the process was believed to be the result of a combination of factors, including (1) the destruction of vegetative cells by heating to the initial internal temperature of 71°C, (2) partial destruction of spores by the initial long thermal process but at the same time retaining a high level of residual nitrite, (3) destruction of spores on the surface of the meat by the short but high thermal process, and (4) the germination and autolysis of surviving spores during refrigerated storage of the meat because of the adverse conditions for outgrowth (low temperature and high residual nitrite). The refrigerated storage period was necessary to assure stability. This process was used commercially with the approval of the U.S. Department of Agriculture. The process had to be properly controlled to be successful (Kueck, personal communication).

Schack and Taylor (1966) described a procedure to produce shelf-stable canned, cured meat with improved quality as a result of reduced thermal process. The process involved a unique sequence of adding the curing ingredients. First, the meat was ground or chopped and then deaerated and backflushed with nitrogen to displace free oxygen. Ascorbic acid was then uniformly distributed throughout the meat. A nitric oxide-generating material, such as sodium nitrite, was then added and mixed into the meat while under a positive pressure (e.g., 1 psig). The meat mixture was then filled into cans, sealed, and then thermally processed. The process was considered to greatly increase the amount of nitric oxide available for use as an antibacterial agent. This process was tested extensively in at least one commercial plant, but the process was not dependable and was abandoned.

Roberts and Ingram (1966) heated spores in water from $F_0 = 0.0015$ to 1.5 before placing them onto culture media containing salt, nitrite, and nitrate. Their results confirmed the concept of spore injury and increased sensitivity to salt and nitrite proposed by Silliker et al. (1958). Curing salts interfered with some stage of germination and development of the surviving heated spores at concentrations that were ineffective with unheated spores. The two *Bacillus* species were rendered more sensitive to salt by lower heat treatments than clostridia. These data reemphasized the importance of spore contamination level for the stability of shelf-stable canned cured meats. Nitrate had no effect beyond that on water activity. The effect of nitrite was pH-dependent and led the authors to support the position that undissociated nitrous acid is the inhibitory agent.

Duncan and Foster (1968a) also reported that heated putrefactive anaerobe spores were less tolerant of salt, nitrite, and nitrate than unheated spores. When heated in the presence of these agents, spores tended to be protected by salt and nitrate, but nitrite enhanced heat injury. When heated and cultured in the presence of the agents, sodium nitrate (1000 mg/L) and sodium nitrite (300 mg/L) with a lower (0.5–1%) salt concentration increased the apparent heat resistance of the spores. At the same nitrate and nitrite levels but a higher (2–4%) salt concentration, heat resistance was decreased. Nitrite was strongly inhibitory, especially at pH 6.0, the normal pH value of canned luncheon meat, and appeared to be the chief preservative against putrefactive anaerobic spoilage.

Duncan and Foster (1968b) also reported that nitrite levels up to 600 mg/L at pH 6.0 allowed emergence and elongation of vegetative putrefactive anaerobe cells but blocked cell division. The newly emerged cells then lysed. Even high nitrite levels (i.e., 40,000 mg/L) did not prevent germination or swelling of the spores. Salt levels above 6% prevented germination; levels within the range of brine in shelf-stable canned cured meat (i.e., between 3 and 6%) allowed germination and vegetative cell formation. Sodium nitrate affected neither germination nor outgrowth at levels up to 20,000 mg/L.

Perigo et al. (1967) offered a new possibility for the stability of shelf-stable canned cured meat. When nitrite was heated in a laboratory medium, an unknown substance was formed that was extremely inhibitory to the growth of vegetative cells of *C. sporogenes*. The authors postulated that the unknown substance may be formed from the nitrite that disappears during heating. Thus, in the case of shelf-stable canned cured meat, the level of residual nitrite remaining after processing may be an inappropriate measurement of the inhibitory capacity of the product. For this reason, they selected a medium that enhanced the disappearance of nitrite during autoclaving. Since formation of the inhibitor required substantial heating, they reasoned the inhibitor might play a complementary role in the stability of shelf-stable but not perishable canned cured meats. The stability of perishable cured meats was assigned to undissociated nitrous acid. Perigo and Roberts (1968) subsequently expanded the observation of Perigo et al. (1967) to 30 different clostridial strains. Roberts and Garcia (1973) later reported 9 of 14 strains of *Bacillus* spp. and *Streptococcus durans* to be sensitive to the inhibitor, but not *Streptococcus faecium*, *Streptococcus faecalis*, *Streptococcus faecalis* var. *zymogenes*, or *Salmonella* Typhimurium.

The first attempt to demonstrate the presence of a Perigo-type factor in cured meat was unsuccessful (Johnston et al., 1969). Johnston et al. concluded that the inhibitor produced in media is of little or no consequence for explaining the inhibitory effect of nitrite in commercially produced canned cured meat.

The interrelation of thermal process, salt, nitrite, and spore level in shelf-stable canned cured meat was further substantiated by Pivnick et al. (1969). Ground pork inoculated with 1 botulinal spore/g and processed to $F_0 = 0.6$ became toxic in the absence of salt and nitrite. Meat inoculated with 10^6 spores/g remained nontoxic after the same process if sufficient salt and nitrite were present. This required either a brine level of 6.1% alone, one of 4.6% with 300 mg/kg of sodium nitrite, or intermediate levels of brine and nitrite. The significance of spore level upon the thermal process required for preventing botulinal outgrowth was again demonstrated. Viable botulinal cells were recovered in stable inoculated product 18 months after processing. The authors concluded that the safety of some shelf-stable cured meats is due to the scarcity of botulinal spores in the raw materials. The questions were raised as to what constitutes a safe process and at what point are the products under-processed. Despite the research and experience since the commercialization of these products, to date there has been no agreement on the lower limits of brine, nitrite, and the thermal process required for the safety of shelf-stable cured meats given less than a botulinal cooking.

Four possible mechanisms by which nitrite may inhibit heat-injured spores in shelf-stable canned cured meats were investigated by Labbe and Duncan (1970). Of the four, their results supported only one, the inhibition of outgrowth of heat-injured spores. When testing the Perigo-type factor formed with nitrite levels up to ten times greater than those commercially accepted, the inhibitor did not prevent germination of most of the intact or heat-injured *Clostridium perfringens* spores.

7.2.3.2 Perishable Cured Meats

Thatcher et al. (1962) questioned the safety of vacuum packaging. Growth levels of *S. aureus* were similar in back bacon packaged under nitrogen, vacuum, air, and 5% carbon dioxide in oxygen, and enterotoxin was reportedly produced under all four packaging conditions. The bacon exposed to air or carbon dioxide in oxygen was obviously spoiled. Since the bacon incubated under nitrogen or vacuum was organoleptically acceptable, the authors recommended that vacuum-packaged nonsterile processed foods be stored in the frozen state.

Gough and Alford (1965) showed that spores of *C. perfringens* inoculated into raw hams could survive the thermal process, but no attempt was made to determine whether they could multiply if the hams were subsequently temperature abused. From studies in thioglycolate medium containing salt, nitrate, and nitrite, the authors concluded that the levels used in ham would permit growth. No mention was made of pH, and none of the agents were tested in combination.

Brownlie (1966) demonstrated the combined inhibitory effect of sodium nitrite concentration, pH, and temperature on a *Microbacterium* species in APT broth. At pH 7.0, the presence of nitrite caused very little or no inhibition. At pH 6.0 and below, increasing the amount of nitrite from 25 to 200 mg/L caused progressively greater inhibition. Nitrite was more inhibitory at 0°C than at the other temperatures tested (10 and 25°C). Another 25 strains of microbacteria failed to grow at pH 5.5 in the presence of 200 mg/L of sodium nitrite. Of 25 lactobacilli from cured meat, 21 were able to grow under the same conditions. This would help to explain the dominance of the lactobacilli as spoilage organisms in many perishable cured meats.

Several studies on the growth of *C. botulinum* in vacuum-packaged perishable cured meats were reported during this period. Schmidt and Segner (1964) observed increased delay in the growth of four type E strains in culture media (neutral pH) as the nitrite level increased from 0 to 200 mg/kg. A similar effect occurred with increasing salt from 0 to 4% during incubation at 7.8 or 10°C, but not at 29.4°C. This suggested that salt becomes more inhibitory as storage temperatures are decreased. The four strains failed to produce toxin during 5 weeks at 10°C in sliced bologna (2.7% brine, pH 6.1, 34 mg/kg of residual nitrite). Schmidt and Segner (1964) concluded that factors or a combination of factors provided a much longer shelf life than might be expected from salt content alone.

Christiansen and Foster (1965) found type A botulinal toxin production to occur at the same rate on bologna whether packaged with vacuum or without. The time required for toxin production decreased as the abuse temperature was increased (22, 30, and 37°C). When *S. aureus* was inoculated onto sliced chopped ham, lower maximum population levels occurred in vacuum packages than in nonvacuum packages at 15 and 22°C.

Pivnick and Bird (1965) found the oxygen permeability of packaging films to influence spoilage but not toxigenesis of sliced cured meats inoculated with *C. botulinum*. Their findings that the type of cured meat influenced toxin production were particularly significant. Bologna was relatively resistant to type A toxin production, but not ham. Cooked sliced ham from five producers inoculated with type E spores yielded variable results at 30°C, which was thought to be due to product differences in brine and residual nitrite levels. Toxin was not produced at 10°C or lower. On sliced bologna, type E toxin was not produced, even at 30°C. The combination of brine (3.75%), residual nitrite, a fermentable carbohydrate, and decreasing pH during temperature abuse were the suggested inhibitory factors. Similar results were obtained with jellied corned beef stored at 15 and 25°C. Within 1 week the product pH had decreased from 6.0 to 4.6. In jellied ox tongue, type E toxin was produced at 10–25°C, but not at 5°C. A total of 90% of the samples held at 25°C remained above pH 6.0. Failure to obtain toxin production at 5°C may have been influenced by pH, since 13 of 15 samples had a pH of 5.7 or less after 4 weeks.

The significance of brine level and storage temperature on botulinal toxin formation (types A and B) in vacuum-packaged meats was studied by Pivnick and Barnett (1965). Toxin was produced within 7 days at 30°C in bologna (3.8% brine), but not at 25°C if the brine was 5.1% or higher. A decrease in pH likely occurred during storage and prevented toxin formation in the other variables. In sliced ham, toxin was produced at 20–30°C, but not at 15°C. The rate of toxin formation decreased as the brine level increased from 3.1 to 4.8%. At the highest brine level (4.8%), toxin was produced only at 30°C. In jellied pork tongue, toxin was detected in 2 days at 2.35 and 3.5% brine and in 4 days at 4.4% brine, at 30°C. The effect of brine on type E toxin production was more dramatic. In jellied pork tongue, toxin was detected after 4 days with 2.35% and after 8 days with 3.5% brine at 20°C. Toxin was not produced in product with 4.4% brine. All three products (bologna, ham, and jellied pork tongue) were formulated with commercial levels of nitrate and nitrite. Although not mentioned, as the brine level increased in the bologna and ham, an increase in the level of residual nitrite also occurred.

The effect of nitrite levels on types A and B botulinal toxin production in vacuum-packaged sliced ham (4.1% brine) and jellied pork tongue (3.8–4.3% brine) was studied next (Pivnick et al., 1967). In both products a marked delay in toxin production occurred at 20°C. At each temperature (20, 25, and 30°C), increasing the nitrite level increased the time for toxin production. The authors concluded that heat injury of spores is not necessary for nitrite to prevent toxin production in cured meats.

Warnecke et al. (1967) reported type E botulinal toxin production in ground beef held at 22.2°C for 3–18 days. Toxin was also detected in bacon held at 12.8 and 22.2°C, but this is highly questionable, considering the bacon had a brine level of 5.28%. Toxin was not produced in either bologna or country cured ham.

Blanche-Koelensmid and van Rhee (1968) described a smoke ring sausage produced by stuffing the sausage mixture into casings and smoking at 25–35°C, during which time naturally occurring lactics drop the pH to 5.5–5.8. The raw product was then sold at ambient temperature for cooking at home. The product had been changed to a shelf-stable fermented product precooked to 80°C in an oxygen-impermeable casing. Using juices expressed from the newer sausage product, they determined that type B botulinal outgrowth was inhibited during 4 weeks at 30°C if the juice had a combined pH of 5.8 or less, a brine level of 4.5% or above, and an input nitrite level of 500 mg/kg or more.

An additional factor influencing the apparent significance of nitrite is the growth of competitive microbiota. Kafel and Ayres (1969) demonstrated that the growth of enterococci in culture media and, more importantly, in perishable canned hams can inhibit the growth of *C. botulinum*, *C. perfringens*, *C. sporogenes*, and *Lactobacillus viridescens*.

Riemann (Anon., 1968) emphasized maintaining adequate salt levels to assure the microbiological safety of perishable vacuum-packaged cured meats. He concluded that *C. botulinum* type A seemed to be completely inhibited by 4.5% brine at pH 5.3, 5.5% brine at pH 6.1, and 8.6% brine at pH 6.5. Type E botulinal growth is inhibited by 2% brine at pH 5.2 and 3% brine at all pH values. The growth of *S. aureus* is inhibited by more than 4% brine at pH 5.1, 9% brine at pH 5.3, and more than 10% brine at pH 6.9.

7.2.3.3 Summary for 1960–1970

Progress made during the 1960s toward defining the role of nitrite in cured meats included the following:

1. Evidence was presented that the antimicrobial effect of nitrite is related to the formation of nitrous acid.
2. The system responsible for the shelf stability of canned cured meats given less than a botulinal cooking was studied by several research groups. All of them reached the same general conclusions as those set forth by Silliker et al. (1958).
3. The concept of thermal injury to spores that survive the processing of shelf-stable canned cured meats was confirmed.
4. The process of outgrowth was more sensitive than spore germination to nitrite.
5. A new explanation for the shelf stability of canned cured meat was proposed. Perigo et al. (1967) suggested an inhibitory substance was formed from nitrite, which disappeared during the thermal process.
6. An early test on the effect of vacuum packaging on the safety of perishable cured meat led to the recommendation that they be frozen to prevent the development of microbial health hazards.
7. Certain cured meats are more prone to supporting botulinal growth. Factors influencing growth included brine level, pH, residual nitrite, storage temperature, and inoculum level. Limited data indicated the lowest temperature reported for growth of type E on cured meats to be 10°C. Lücke et al. (1980) would later report growth at 8°C in liver sausage.
8. The significance of decreasing pH on botulinal inhibition in cooked cured meats with sufficient fermentable carbohydrate was reported for the first time.
9. The growth of enterococci in perishable canned hams was reported to inhibit several clostridial species.

7.2.4 1970–1980

From the outset of the 1970s, the formation of nitrosamines in cured meats was an issue of immediate concern. Throughout the decade pressure increased to reduce or eliminate nitrate and/or nitrite from cured meat. The timely report of Perigo et al. in 1967 opened the possibility of a breakthrough on the mechanism of nitrite inhibition. Considerable effort was applied toward isolating and characterizing the Perigo factor, with the goals of greater understanding and its possible use as a preservative to replace nitrite.

Research during this period can be grouped into five categories by the approaches employed. They include tests in (1) test tubes or Petri plates using controlled media and conditions, (2) water slurries of meat processed in bottles, (3) meat products processed in tubes or cans, (4) meat products produced and packaged in pilot plants, and (5) commercially produced meats. Each of the approaches had its favorable and unfavorable features. As in the past, the test tube approach was productive, but not all the results were directly applicable to commercially cured products. The complex, dynamic system in which nitrite plays a role in cured meats has been virtually impossible to duplicate in microbiological media.

7.2.4.1 Shelf-Stable Cured Meat and Perigo Factor

Pivnick et al. (1970) used an aqueous meat mixture without salt or other additives to investigate four possible roles of nitrite in shelf-stable canned cured meats: (1) nitrite induces germination during the heating process and the germinated spores are killed by heat; (2) nitrite potentiates direct killing by heat without the need for prior germination; (3) nitrite increases the germination rate of spores that survive heating; and (4) these rapidly germinating spores are then inhibited by the nitrite remaining after processing and subsequently die before the nitrite has decreased appreciably. Of the four possible roles, nitrite at the level used commercially (200 mg/kg) did not enhance destruction during heating or germination during subsequent storage. The chief value of nitrite appeared to be its ability to prevent growth from spores that survive and germinate following processing.

Ingram and Roberts (1971) heated botulinal spores in water and in water with salt (4%) and sodium nitrite (200 g/ml) before subculturing into media with and without salt (2%) and sodium nitrite (50 g/ml). Salt and nitrite had little effect on the degree of thermal destruction, but their presence in the culture medium markedly reduced the recovery of the heated spores. The data were discussed in terms of applying the D concept to shelf-stable canned cured meat.

Matsuda and Sekiguchi (1971) studied the combined effect of heat (105, 110, and 120°C) and curing salts on the destruction of *Clostridium subterminale* spores. Neither sodium chloride (3%) nor sodium nitrate (0.1%) affected thermal destruction when tested alone. Sodium nitrite alone caused increased destruction. A combination of 3.0% salt, 0.1% sodium nitrate, and 0.01% sodium nitrite caused the greatest destruction when heated. Small differences in the amount of sodium nitrite below 0.01% strongly influenced the results, but not at levels above 0.01%.

The effect of heating spores of PA 3679 at 115.5°C in ground pork containing 2.8% salt, 200 mg/kg sodium nitrite, and 0.5% sodium tripolyphosphate was studied by Nordin et al. (1975). Omitting any one or all three of the additives did not affect thermal destruction, thereby confirming earlier research, as already discussed. The effect of nitrite, salt, pH, and heating at 115.5°C on the outgrowth of surviving spores was also examined by placing the heated tubes at 23°C for up to 1 year. The major factor influencing outgrowth was pH; nitrite and salt levels had lesser, although "clear and undisputable subsidiary effects." The effect of all three factors (pH, nitrite, and salt) appeared to be additive. Outgrowth was determined by blackening of the tubes as a result of ferric citrate having been added to the meat. Since it would later be found that iron concentration can influence the antibotulinal efficacy of nitrite in perishable canned cured meat given a milder heating, did the addition of ferric citrate influence the results? The interaction of iron and nitrite has not been investigated in shelf-stable canned cured meats heated in steam under pressure (e.g., 115°C), as opposed to perishable canned cured meats heated in hot water to a minimum internal temperature of 66°C. Preliminary data, however, exist (Christiansen et al., personal communication), showing a similar iron-nitrite interaction with *C. sporogenes* strain PA 69-82 in perishable canned cured meat as occurs with *C. botulinum*. A similar response with strain 3679 would not be unexpected.

Farkas et al. (1973) studied the effect of adding nitrite to the brine solution for canned Vienna sausages. The sausages were prepared with nitrite-containing curing salt, cooked, and then packed into cans. One of the sausages in each can was inoculated with *C. sporogenes* spores to provide a level of 10^4 spores per can. The cans were then filled with brine (3% salt), sealed, and heated to varying degrees. Sausages packed in brine with 400 mg/kg of sodium nitrite and processed to $F_0 = 0.35$ were as stable as sausages in brine without nitrite and processed to $F_0 = 2.6$. A 1:1 extract was next prepared with 3% saline and ground, cooked sausages, the solids removed, the liquid placed into tubes, inoculated (10^5/ml), overlaid with paraffin oil, and heated ($F_0 = 0.4$). The extract had 2.6% salt, a_w 0.966, and pH 6.4. Between 120 and 200 mg/kg sodium nitrite provided stability through 5 weeks of storage. A significant observation was that residual nitrite did not undergo normal depletion in the extract. After a loss of 15–25% during processing and an initial decline, residual nitrite levels remained above 125 mg/kg throughout 80 days at 30°C.

Johnston and Loynes (1971) obtained Perigo factor formation in the Perigo et al. (1967) medium and reinforced clostridial medium, but not in liver veal or Wynne fluid media. This emphasized the importance of medium selection when studying the antimicrobial activity of nitrite and relating the results to commercially produced meats. Aqueous meat suspensions prepared from raw, normal, and overheated (115°C for 90 min) canned luncheon meat required 500, 150, and 50 mg/kg of sodium nitrite, respectively, for inhibition. The residual nitrite values of the four media and three meat suspensions were about the same after preparation. Considerable differences were noted in their redox potentials. Reducing agents (e.g., ascorbate and cysteine) enhanced the antibotulinal effect of the meat suspensions but did not induce Perigo factor formation.

Ashworth and Spencer (1972) described a model system of ground pork (25 g). The pork was autoclaved before or after adding nitrite, held overnight for nitrite equilibration, and then inoculated with vegetative clostridial cells. The inhibitory level of nitrite was not influenced by pH within the range

5.7–6.3. Reducing agents (e.g., sodium ascorbate, cysteine, and, to some degree, thioglycolate) enhanced the effect of nitrite. It was reported that less nitrite was required for inhibition when it was added before autoclaving. These observations led to the conclusion that a Perigo-type effect was detected in the meat system. The magnitude of the effect was much less than was previously reported in culture media. Perigo et al. (1967) reported equivalent inhibition with 8 mg/kg of nitrite added before heating the medium and 52 mg/kg added after heating the medium. In ground pork, Ashworth and Spencer (1972) reported the equivalent levels to be 60–70 and 90–100 mg/kg. Perigo et al. (1967) based their conclusions on the added level of nitrite; Ashworth and Spencer based their conclusions upon the residual level of nitrite. Actually, more added nitrite was required in the pork if the nitrite was added before (300 mg/kg) rather than after (150 mg/kg) autoclaving.

It was originally suggested (Perigo et al., 1967) that the nitrite that disappears during heating may be involved in the formation of the unknown inhibitor. Ashworth and Spencer (1972) found no relation between the amount of nitrite lost during heating and the inhibition obtained. With excessive heating, higher levels of added nitrite were required for inhibition. The additional nitrite provided comparable residual levels after the various heat treatments, which indicates that residual nitrite influenced the results.

When spores were used, results similar to those just mentioned were obtained during the early stages of temperature abuse. However, with prolonged storage (42 days at 37°C), higher initial residual nitrite levels were required for inhibition. In summary, the conclusion that a Perigo effect was demonstrated must be tempered with the observation that, on the basis of added nitrite, more was required when added before heating the meat. If residual nitrite is a valid means for comparison, the Perigo effect was of low magnitude. Residual nitrite, per se, strongly influenced clostridial inhibition. The data obtained with the spore inoculum should be more relevant to shelf-stable canned cured meats.

Ashworth et al. (1973) extended this research to the milder heating given to large perishable canned hams and obtained similar results. Adding salt (3.5% brine) to the meat had no effect on the inhibitory levels of residual nitrite but appeared to reduce the level of added nitrite required for inhibition. In this case alone, the levels of nitrite required to obtain inhibition were about the same or lower if added before heating than if added after heating. The inhibitory levels of residual nitrite for two heating processes, (1) 80°C for 4 h and (2) heating from 20 to 70°C over a 4 h period, did not differ. Similar residual nitrite levels were required to inhibit vegetative cells (160,000/g) added after heating and spores (20/g) added before or after heating. In all cases, less residual nitrite was required for inhibition when nitrite was added before heating. Inhibition by nitrite diminished gradually over a period of 8 weeks at 37°C. Ashworth et al. concluded that a Perigo-type factor was produced, and it may also play a role in the safety and stability of pasteurized cured meats, such as perishable canned ham.

Johnston et al. (1969) found the inhibitor of Perigo et al. (1967) to be inactivated when added to meat. Pivnick and Chang (1974) considered, however, that although the Perigo inhibitor may not be present in meat, another inhibitory factor may be formed when nitrite is heated with meat. The term Perigo-type factor (PTF) was selected to describe the effect observed in their test. Shelf-stable canned cured luncheon meat was prepared with 0–300 mg/kg of sodium nitrite and processed to $F_0 = 0.4$ The product was held until the highest level of nitrite declined to less than 2 mg/kg. The meat was then inoculated with botulinal spores which had survived heating to $F_0 = 0.4$ in raw meat juice with 4.5% salt and 150 mg/kg of sodium nitrite. The authors found that as the original input nitrite level increased, outgrowth and toxin production decreased. This was considered evidence for an inhibitory factor other than residual nitrite. The amount of PTF formed was related to the amount of original nitrite added to the meat before heating. They concluded that although a PTF was formed, the relative inhibitory effect was quantitatively small. The equation of Pivnick and Petrasovits (1973) was used to estimate the inhibitory effect of the PTF relative to the protection provided by other factors (e.g., thermal destruction of spores and inhibition of the survivors by the salt).

Chang and Akhtar (1974) prepared buffered saline suspensions of canned luncheon meat made with 0, 100, 150, and 200 mg/kg of sodium nitrite and processed to $F_0 = 0.4$. The nitrite level of the suspensions was adjusted after heating to 53.5 or 22.1 mg/kg, deaerated in boiling water, and inoculated. The levels selected (53.5 and 22.1 mg/kg) were the residual nitrite levels after 0 and 2 weeks in product made with 200 mg/kg of nitrite. The depletion rate of the nitrite added to the suspensions was not influenced by the amount of nitrite originally added to the meat. Botulinal inhibition was related to the original amount of nitrite. A second test using canned product held for 4 weeks at 35°C before preparing suspensions with 22.1 mg/kg of sodium nitrite yielded similar results. Decreasing inhibition occurred as the product was held for 0, 2, and 4 weeks before preparing the suspensions. The degree of inhibition was correlated with the amount of nitrite originally added to the product. This correlation agreed with results obtained when canned product was held at 35°C to allow the depletion of residual nitrite and then challenged directly (Chang et al., 1974). It was concluded that a PTF is formed during the heating of canned luncheon meat and that the inhibitor is not stable and loses its activity during storage. Thus, age of product must be considered when investigating the presence or activity of PTF in commercial product or model meat systems.

Roberts and Smart (1974) found that the inhibitor in the medium of Perigo et al. (1967) had a comparable effect on the inhibition of clostridial spores, regardless of the inoculum level or whether the spores were untreated, heat injured, or irradiation injured. There was no evidence of germination (i.e., vegetative cells) in the media with inhibitory levels of nitrite, although the spores remained viable during prolonged incubation. Inoculation of the medium after heating with nitrite and aging for 3 months showed that it was less inhibitory, indicating the Perigo factor is not stable.

Additional studies were reported by 1980 that dealt with the Perigo factor (Mirna and Hofmann, 1969; Wasserman and Huhtanen, 1972; van Roon, 1974; Ashworth et al., 1974; Incze et al., 1974; Mirna and Coretti, 1974; Moran et al., 1975; Hansen and Levin, 1975; Huhtanen and Wasserman, 1975; Asan and Solberg, 1976; Lee et al., 1978; and van Roon, 1979a, b). Van Roon (1979a, b) concluded that Perigo-type inhibitors do not contribute to clostridial inhibition in heated cured meats. Debatable as the formation of a Perigo-type in meat may be, common agreement that nitrite, iron, and sulfhydryl compounds are involved in the formation of the Perigo factor in broth media may be significant.

Riha and Solberg (1975a) studied the formation and effect of a Perigo-type factor on seven strains of *C. perfringens* in a chemically defined medium. They subsequently concluded (Riha and Solberg, 1975b) that inhibition was apparently at the cellular level. They hypothesized that nitrite, in some form, causes inhibition by reacting with enzymes containing functional sulfhydryl groups. Additional data by O'Leary and Solberg (1976) suggested that inhibition of *C. perfringens* may involve an interaction of nitrite, as nitrous acid, with sulfhydryl-containing constituents of the bacterial cell.

7.2.4.2 Perishable Canned Cured Meat and Slurries of Meat

In addition to research on shelf-stable canned cured meats and the composition and role of the Perigo factor in commercial products, considerable effort was devoted to perishable cured meats. Research on perishable canned cured meat or meat slurries is discussed first.

Tjaberg and Kvale (1972) studied the antibotulinal effect of nitrite (0–480 mg/kg) added to canned raw meat, shelf-stable canned meat, and perishable canned meat and held at 4, 22, and 37°C. At 4°C, predictably, toxin was not detected. At 22°C, the rate of swelling ranged from 11 to 48 days across all variables. Neither swelling nor toxin occurred in some treatments having 240 or 480 mg/kg of nitrite. At 37°C, swelling and toxin production ranged from 3 to 17 days, and fewer treatments proved stable. The authors concluded that it was not possible to demonstrate any preservative action of nitrite at the levels (i.e., 200 mg/kg) usually used in meat products.

Perishable canned cured pork was studied by Christiansen et al. (1973). Botulinal outgrowth in product abused at 27°C was influenced by the spore inoculum level and the level of nitrite. Nitrate had a slight but statistically significant inhibitory effect that may have been caused by nitrite generated from nitrate. Nontoxic spoilage occurred at 7°C in uninoculated product made without nitrite. It was stated that botulinal inhibition was related to the level of nitrite added during formulation rather than residual nitrite. This was later reassessed by Christiansen (1980) who concluded that residual nitrite is a significant factor influencing botulinal inhibition.

Baird-Parker and Baillie (1974) studied the growth of a large, diverse collection of botulinal strains in a broth medium and a cooked pork medium. Sodium nitrite and L-ascorbic acid were added as filter-sterilized solutions. The number of strains showing growth in broth was found to decrease with increasing nitrite (50, 100, 150, 200 mg/L), decreasing temperature (25, 20, 15°C), decreasing pH (7.0, 6.5. 6.0, 5.5), increasing salt (1.5, 3.0, 4.5, 6.0% w/v), and decreasing inoculum level (10^6, 10^3, 10^1). Adding L-ascorbic acid (1.0%) markedly increased the effectiveness of nitrite with none of the 38 strains tested showing growth in broth containing 50 mg/L sodium nitrite. Although less effective, a lower concentration of L-ascorbic acid (0.1%) also increased the efficacy of nitrite in broth but had no effect, pro or con, in the cooked pork medium. Furthermore, a preliminary test in bacon led to the conclusion that L-ascorbic acid does not alter the antibotulinal properties of nitrite in bacon.

Grever (1974) examined the stability of emulsions normally used for preparing cooked sausage and liver sausage. The emulsions were filled into cans and heated. Stability was influenced by nitrite content, level of clostridial spores, product pH, and whether the products were treated by a perishable or shelf-stable thermal process. It was suggested that *Bacillus* species are less sensitive to nitrite than clostridia.

Jarvis et al. (1976) found that heating botulinal spores in an aqueous extract of lean pork at 85, 90, or 95°C decreased the number that could grow in a recovery medium containing 2.5% salt. Spores heated at lower temperatures (70 or 80°C) were inhibited by 4.5% but not 3.5% salt. A statistically significant salt–nitrite interaction was observed after heating (70–95°C) only if the recovery medium contained salt levels of 4.5% or above. Spores heated at either 70 or 95°C were more sensitive to nitrite that was autoclaved in a meat slurry or the Perigo medium than to nitrite added after autoclaving. The spores heated at 70 and 95°C were equally sensitive to the autoclaved nitrite. Autoclaving nitrite in the Perigo medium resulted in significantly more inhibition than autoclaving the nitrite in the meat slurry; 20 mg/L of nitrite in the Perigo medium and 210 mg/L in the slurry were equally inhibitory. Jarvis et al. recommended that large-scale tests be used to quantify the many interacting factors influencing the safety of perishable cured meats.

This led to the pork slurry system of Rhodes and Jarvis (1976) and used by Roberts et al. (1976) to study the interactions of temperature of abuse, salt, nitrite, and inoculum level on botulinal growth and toxin production at pH 6.0. Temperature of abuse was the most significant factor influencing the rate of botulinal growth. At 15°C, the greatest inhibition occurred with 1.8 or 3.5% salt (on water phase). At 17.5°C, less inhibition occurred. At 20, 22.5, or 25°C, neither salt level provided much inhibition, even with 300 mg/L of sodium nitrite. At the lower temperatures, a pronounced salt–nitrite interaction occurred. At 20°C, the impact of increasing the nitrite level from 40 to 300 mg/L was minor and became less so as the temperature of abuse increased to 22.5 and 25°C.

The effect of nitrite in perishable canned cured meat was also reported by Tompkin et al. (1977). The rate at which inoculated cans swelled at 27°C was inversely related to the level of nitrite added to the product. The antibotulinal effect of nitrite was considerably greater than that reported for the meat slurries of Roberts et al. (1976). Upon investigation (Tompkin et al., 1978a), one reason was found to be the addition of sodium isoascorbate to the canned meat but not to the meat slurries. Isoascorbate had enhanced the antibotulinal efficacy of nitrite in the canned meat.

Christiansen et al. (1978) concluded that when inoculated perishable canned cured meat is placed at 27°C, a race occurs between death of germinated botulinal cells and depletion of residual nitrite. Botulinal outgrowth was dependent upon the relative levels of residual nitrite and surviving botulinal cells. This agreed with an earlier observation (Tompkin et al., 1978b) involving product held at 4.4 or 10°C for up to half a year before placing at 27°C. The longer the product was held at refrigeration temperatures, the more the residual nitrite levels declined and the less inhibitory the product became when placed at 27°C.

In another study thermal processing within the range of 63–74°C (internal temperature) influenced neither residual nitrite levels nor botulinal inhibition in perishable canned cured meat (Tompkin et al., 1978c). Adding hemoglobin, however, which resulted in a lower level of residual nitrite after processing, decreased botulinal inhibition. It was also shown that the type of meat could influence the degree of botulinal inhibition. Substituting turkey thigh meat, beef round, beef hearts, or pork hearts for fresh pork ham reduced the efficacy of nitrite.

The reason isoascorbate enhanced the antibotulinal effect of nitrite was pursued (Tompkin et al., 1978d). Isoascorbate, ascorbate, cysteine, and ethylenediaminetetraacetic acid (EDTA) were found to share a common function in meat, which later was demonstrated to be the sequestering of iron (Tompkin et al., 1978e). This made it possible to explain the differences in the relative efficacy of nitrite in various meats on the basis of their iron content, with one exception: beef liver, despite its relatively high iron content (49 mg/kg), did not cause a substantial loss in the antibotulinal efficacy of nitrite. In a later unpublished test with pork liver that has a much higher iron content than beef liver (225–243 vs. 49 mg/kg), no botulinal inhibition was observed. Subsequent research more clearly confirmed the relation of available iron to the antibotulinal efficacy of nitrite (Tompkin et al., 1979a).

Suggested mechanisms were proposed for the relation of iron to the antibotulinal efficacy of nitrite (Tompkin, 1978; Tompkin et al., 1978e). One possibility involved the reaction of nitrite with an essential iron-containing compound (e.g., ferredoxin) within the germinated cell. This was based upon the rationale that in clostridia, nonheme iron sulfur proteins (e.g., ferredoxin) are essential for electron transport, energy production, and enzyme activity. Clostridial ferredoxin is believed to contain two independent, single-electron transfer sites consisting of four iron atoms. Each active site is comprised of iron bonded to both cysteine and sulfur (Buchanan and Arnon, 1970). It was assumed that some form of nitrogen oxide–iron reaction could occur with the nonheme iron sulfur proteins as occurs with nitric oxide and the iron in heme compounds (e.g., catalase, peroxidases, cytochromes, cytochrome oxidase, hemoglobin, and myoglobin). On the basis of the data existing in 1978, a simplistic explanation for the inhibition of clostridia consisted of the uptake of undissociated nitrous acid, a nitrogen oxide (e.g., nitric oxide)–iron reaction as in ferredoxin, and interference with energy metabolism in the vegetative cell to prevent outgrowth. Subsequent outgrowth could occur when (1) residual nitrite depletes to non-inhibitory levels, allowing the dissociation of the nitrogen oxide from the iron, as can occur with nondenatured (i.e., nonheated) heme compounds, and/or (2) repair through the formation of new unreacted ferredoxin by the vegetative cell. A third possibility is the germination of dormant spores after residual nitrite has declined to non-inhibitory levels.

Although isoascorbate enhances the antibotulinal effect of nitrite in freshly prepared perishable cured meat that is temperature abused, isoascorbate also reduces the efficacy of nitrite by causing more rapid depletion of residual nitrite (Tompkin et al., 1979b).

Wojton et al. (1978) studied the effect of nitrite on the stability of naturally contaminated perishable canned pork when temperature abused at 37°C. The product was formulated with 2.5% salt and sodium nitrite (0, 50, 100, or 200 mg/kg) and then processed to an internal temperature of 69°C. As the level of nitrite decreased, the rate at which the product spoiled increased. Spoiled cans yielded *C. perfringens*, *C. sporogenes*, *C. bifermentans*, *Bacillus* sp., and enterococci.

The efficacy of nitrite in frankfurters was explored by Sofos et al. (1979b–e). Their test system mainly consisted of extruding a frankfurter emulsion into test tubes, heating to 68.5°C, overlaying with vaspar, cooling, and then storing the tubes at 27°C. This avoided the problem of recontamination inherent in some studies after heating, but as in other model systems receiving a mild heat treatment, a background microbiota of thermodurics existed from the ingredients. When testing a frankfurter emulsion prepared with mechanically deboned chicken meat, low levels of nitrite (20 and 40 mg/kg) did not influence botulinal growth or toxin production (Sofos et al., 1979b). Adding 156 mg/kg of sodium nitrite delayed botulinal toxin production; however, the degree of inhibition was less than that reported by others.

Sofos et al. (1979c) evaluated various soy proteins in their system. The rate of botulinal toxin production in formulations with soy protein was equal to or slower than in an all-meat formulation, in both the presence and absence of nitrite. The type of soy protein influenced the results. Nitrite was ineffective in a formulation of soy isolate and pork fat. The depletion of residual nitrite was faster as the amount of beef was increased in a formulation containing soy. The rate of nitrite depletion was influenced by the type of soy.

Sofos et al. (1979d) compared the efficacy of nitrite (0 and 80 mg/kg) in chicken, beef, and pork in their system. The addition of 80 mg/kg of sodium nitrite was relatively ineffective in chicken and beef. A significant delay in botulinal toxin production occurred in pork. The rate of botulinal toxin production was slower in all soy protein formulations than in all meat formulations. In a meat–soy mixture, the addition of 156 mg/kg of sodium nitrite was very effective in delaying botulinal toxin production.

Sofos et al. (1979e) found the rate of germination of *C. botulinum* spores in a chicken frank formulation to be unaffected by the addition of sodium nitrite (0, 20, 40, and 156 mg/kg). They concluded that the effect of nitrite was upon delaying the development of the cells after germination and before toxin was produced.

7.2.4.3 Vacuum-Packaged and Fermented Meats

The effect of nitrite in vacuum-packaged franks was studied on two occasions. In the first, Hustad et al. (1973) formulated franks with various levels of nitrite and nitrate, 400 mg/kg of sodium ascorbate, about 2.8% fermentable carbohydrate, and 4.52% brine. Adding 50 mg/kg of sodium nitrite delayed botulinal toxin formation for more than 4 weeks during abuse at 27°C, and then only 2 of the 55 franks became toxic. Although product made with 150 mg/kg of sodium nitrite did not become toxic, 450 mg/kg of nitrate gave a variable response. The strong inhibitory effect observed was likely due to the combined effects of nitrite, the relatively high brine, and, more importantly, a fairly rapid decrease in product pH during abuse.

The second test, reported by Bowen and Deibel (1974) and Bowen et al. (1974), examined whether an increase in ascorbate over normal levels would alter the antibotulinal effect of nitrite. The franks were produced in the same facilities and formulated as before (Hustad et al., 1973), except for changes in the ratio of pork to beef, the levels of nitrite (0–150 mg/kg) and ascorbate (9, 105, and 655 mg/kg), and omission of nitrate. The finished franks had a brine of 4.83%. Toxic product occurred with 50 mg/kg of sodium nitrite at all levels of ascorbate. One toxic sample occurred at 100 mg/kg and none at 150 mg/kg of sodium nitrite. It was concluded that ascorbate did not affect the antibotulinal efficacy of nitrite. A later review of the pH data showed that all variables had a pH of 5.0 or less within 14 days, the exception being one variable with 150 mg/kg of nitrite. In that variable the pH was 5.0 or less within 21 days. Assuming a similar pH response in the first test (Hustad et al., 1973), the two studies indicate that the combination of 4.5–4.8% brine and nitrite levels of 50–100 mg/kg delayed botulinal outgrowth until the lactic microbiota decreased the pH to an inhibitory level. Under these circumstances, the conclusions in both studies concerning the inhibitory levels of nitrite, input vs. residual nitrite, and the effect of ascorbate in franks should be restated or reexamined.

The decline in pH caused by a lactic fermentation was also the dominant factor preventing botulinal outgrowth in fermented sausage (Christiansen et al., 1975). In mildly heated (58.3°C) sausages containing dextrose, only those variables without nitrite became toxic at 27°C. The pH of product made with 50, 100, or 150 mg/kg of nitrite declined to 5.08 or lower within 1 week at 27°C, and the products remained nontoxic. Toxin was produced in all variables formulated without dextrose and more quickly in product having 0 or 50 mg/kg of nitrite. Toxin production was slower, and fewer toxic samples occurred in product with 100 or 150 mg/kg nitrite. The pH of the dextrose-free product remained above 5.6.

A request was made by the poultry industry to the U.S. Food and Drug Administration for approval of sodium nitrite as an additive in poultry products. The request included results that demonstrated the efficacy of sodium nitrite in poultry meat. The data included botulinal challenge tests in frankfurters made with chicken meat and ham made with turkey meat. A summary of the data showing the antibotulinal effect of nitrite in the two poultry products appears in Hauschild (1982).

Raevuori (1975) studied the combined effect of sodium nitrite (0, 100, and 200 mg/kg) and sodium erythorbate (0, 500, and 1000 mg/kg) on the growth of *Bacillus cereus* in cooked sausage placed at 20°C for 48 h. Erythorbate alone was not inhibitory. Erythorbate was found to enhance the inhibitory effect of nitrite. A combination of 200 mg/kg of sodium nitrite and 500 mg/kg of sodium erythorbate prevented the growth of the two strains of *B. cereus* tested. Erythorbate caused reductions in the residual nitrite and the redox potential of the product. Raevouri encouraged more widespread use of erythorbate for sausage production in Finland.

Omitting nitrite from commercially produced liver sausage in Finland led to rapid spoilage by gas-producing anaerobes and suggested a potential risk of botulism (Ala-Huikku et al., 1977). Inoculation studies examined the effect of nitrite (0 and 100 mg/kg) and storage temperature (15, 20, and 25°C) on the rate of botulinal toxin formation. After cooking in Saran casings, the sausages had relatively high pH and water activity values of 6.65–6.80 and 0.98, respectively. Product without nitrite became toxic within 3 days at 25°C and within 6 days at 20°C and remained nontoxic through 14 days at 15°C. Product with 100 mg/kg of sodium nitrite remained nontoxic through 14 days at all three temperatures. Non-inoculated product containing nitrite developed a slight unusual taste at 20°C owing to the growth of spore-forming spoilage organisms (e.g., *B. cereus*). The naturally occurring *B. cereus* was evidently more resistant to nitrite than *C. botulinum*.

7.2.4.4 Bacon

Crowther et al. (1976) in the United Kingdom studied the effects of a mixture of nitrite (100 or 200 mg/kg) and nitrate (250 mg/kg), ascorbate (0, 1000, and 2000 mg/kg), and brine level (mild, 2.9–4.9; medium 4.7–6.6%) on botulinal toxin production in vacuum-packaged back bacon. There was a general trend toward reduced residual nitrite levels at the time of inoculation when ascorbate was added. Since brine levels could not be held constant, the interaction of nitrite and ascorbate is difficult to interpret. If all samples having the mild and medium brine were compared, 42.7 and 18.9% became toxic, respectively. If all samples with no mixed cure, mixed cure with 100 mg/kg of nitrite, and mixed cure with 200 mg/kg of nitrite were compared, 67.4, 19.3, and 34.7% became toxic. If all samples containing mixed cure with and without ascorbate were compared, 35.8 and 21.9% became toxic, respectively. A higher percentage of samples were toxic with the addition of 200 mg/kg of nitrite than with 100 mg/kg of nitrite. The addition of ascorbate enhanced the antibotulinal effect of 100 mg/kg but not 200 mg/kg of nitrite. These values raise a question concerning the conclusions that (1) protection was greater if the level of nitrite was increased to 200 mg/kg and (2) sodium ascorbate at a level up to 2000 mg/kg did not reduce the protection afforded by nitrite against *C. botulinum*.

It also was reported that *S. aureus* grew well in the medium-salted bacon, regardless of the level of nitrite or ascorbate. As expected, growth was faster at 25°C than at 15°C. In opened packages higher populations of *S. aureus* developed and enterotoxin A was produced. In vacuum packages, lower populations resulted and enterotoxin was not detected.

Shaw and Harding (1978) studied the effect of nitrate and nitrite on the microbiota of Wiltshire bacon. The predominant bacteria of the bacon after curing consisted of micrococci, *Moraxella* spp., and *Moraxella*-like bacteria. Omitting nitrate led to higher numbers of *Moraxella* spp. in the cured bacon. Bacon that was sliced and vacuum packaged developed a microbiota mainly of micrococci and lactics. Including nitrate in the bacon enhanced the growth of micrococci. The nitrate likely served as an electron acceptor for the growth of micrococci in the vacuum package. Since higher numbers of lactics were present in bacon with the lowest initial nitrite concentration, it was suggested that nitrite could be important in delaying the sour spoilage caused by the growth of lactics.

The USDA conducted a test to verify the antibotulinal effectiveness of a combined nitrite–sorbate mixture in bacon (Johnston, 1979). Three types of bacon were produced in four commercial plants: bacon with no preservative, bacon with 120 mg/kg of sodium nitrite, and bacon with 40 mg/kg of sodium nitrite plus 0.26% potassium sorbate. All three bacons contained 550 mg/kg of sodium ascorbate or sodium isoascorbate. The levels of sodium chloride, phosphate, and fermentable carbohydrate (e.g., sucrose) were left to the discretion of the producing plant. The bacon was shipped to a laboratory, inoculated with 1000 spores/g, and vacuum packaged (100 g per package). The bacon was then incubated at 27°C for up to 56 days. The results demonstrated that the combination of nitrite–sorbate was equal to or better than nitrite alone. There was considerable variation in toxin production in the bacon from the four plants. This was likely due to the interaction of the other components of the product (brine level, fermentable carbohydrate, and growth of a spoilage microbiota). At least on the basis of the botulinal results, the data supported the potential approved use of a nitrite–sorbate blend to produce bacon.

A total of 30 botulinal studies were performed with vacuum-packaged bacon during the 1970s in the United States and Canada. They can be more readily compared, because they were similar in processing and final product. Although the studies were performed for a variety of reasons, it is possible to glean the relative effect of nitrite by examining the nitrite-containing control variables from each test. Of the 30 tests, 10 appear in the technical literature (Christiansen et al., 1974; Bowen and Deibel, 1974; Collins-Thompson et al., 1974; Ivey et al., 1978; Sofos et al., 1980b; Tanaka et al., 1980).

The data indicate the following:

1. The pH of the bacon during abuse significantly influences the potential for botulinal outgrowth. Vacuum packaged bacon prepared with 0.7% sugar (sucrose) or more provides sufficient fermentable carbohydrate that naturally occurring lactics cause a decline in pH to inhibitory levels.
2. Brine levels below 4.0% are not inhibitory to botulinal outgrowth. As the brine level exceeds 4.0%, outgrowth is increasingly delayed. If a lactic fermentation develops in the interim, the combination of relatively higher brine and decreasing pH can prevent botulinal outgrowth.

3. The level of residual nitrite at the time the bacon is abused influences the extent of the delay in botulinal outgrowth. The level of nitrite added to the product is not important, aside from the fact that the amount of added nitrite partially determines the level of residual nitrite.
4. The addition of ascorbate or isoascorbate can act in concert with residual nitrite to retard botulinal outgrowth in freshly produced bacon. However, ascorbate and isoascorbate can also have a negative effect by causing more rapid loss of residual nitrite during processing and storage.
5. Too high a level of botulinal cells when the bacon is abused can overwhelm the inhibitory system. The initial period of delay depends upon the inoculum level.
6. Botulinal outgrowth is influenced by the temperature of abuse. Temperatures below 27 or 30°C increase the delay in outgrowth and the inhibitory effects of the foregoing factors.
7. Since phosphates were present in almost all the studies, it is difficult to assess their effect. Within the levels and types employed, phosphates do not appear to influence botulinal outgrowth in bacon.
8. These factors interact to influence botulinal inhibition.

7.2.4.5 Test with S. aureus *and Other Bacteria*

A cluster of reports appeared during the early 1970s that dealt with the effect of nitrite on *S. aureus*. Collectively, the data along with the aforementioned results of Christiansen and Foster (1965) and Crowther et al. (1976) indicate that the growth of *S. aureus* is reduced and enterotoxin production can be prevented by the environmental conditions existing within vacuum-packaged cured meats. Reasons for the reduced growth and enterotoxin production are evident from the reports, but it remains debatable how much each of the possible factors (e.g., low oxygen level, nitrite, brine, growth of competitive microbiota, and product pH) contributes to this effect.

Nurmi and Turunen (1970) studied the effect of adding nitrite to a previously autoclaved broth medium (pH 6.0). Lactobacilli (78 strains), micrococci and staphylococci (24 strains), and *Pediococcus cerevisiae* (1 strain) were examined for their tolerance to nitrite in the presence and absence of 4.01% salt. Very high levels of nitrite (1000 and 5000 mg/kg) either prevented or severely retarded the growth of all strains. At 200 mg/kg growth was delayed or slower. At 40 mg/kg growth was comparable to that in the control without nitrite; 8 mg/kg enhanced the growth of the lactobacilli and, to a slight extent, the micrococci. Heterofermentative lactobacilli were more tolerant of salt than the homofermentative strains.

Genigeorgis et al. (1969) found the production of type B staphylococcal enterotoxin to occur earlier on ham at 30°C than at 22–24°C and earlier under aerobic than anaerobic conditions. The commercially canned ham used for the test had a brine level of 3.6%, no residual nitrite, and an abnormally low pH of 5.3. Extensive tests were also reported for laboratory-cured ham, but it is unclear whether nitrite level influenced anaerobic enterotoxin production. Toxin was produced in certain variables having residual nitrite levels as high as 145 mg/kg. The probability of toxin production at 10°C reportedly decreased as the level of nitrous acid increased. Nitrous acid levels were calculated from the residual nitrite and pH values as described by Castellani and Niven (1955). Anaerobic toxin production was prevented with 0.6 mg/kg or more nitrous acid and pH values below 5.58. However, since high nitrous acid values (>0.6 mg/kg) were possible only at reduced pH levels (pH 5.6 or less), it is uncertain whether high nitrous acid content, reduced pH, or both prevented anaerobic toxin production. At higher temperatures (22 and 30°C) anaerobic toxin production occurred at pH 5.3 in two samples with negligible nitrite levels. A third sample with a pH 5.34, 22 mg/kg of nitrite, and 0.159 mg/kg of nitrous acid also became toxic.

Buchanan and Solberg (1972) studied the effect of nitrite (0, 200, 500, 1000, and 2000 mg/L), oxygen pressure, and initial pH (6.3, 7.3) on the growth of *S. aureus* in brain–heart infusion broth. At pH 7.3, nitrite was without effect in an aerobic environment and 500 mg/L or less was without effect anaerobically. At pH 6.3 and 200 mg/L, the increase in the adjustment phase was 10 h or less aerobically and about 30 h anaerobically. The authors questioned the hypothesis that nitrite inhibition is dependent upon the conversion of nitrite to nitrous acid. Nitrite was metabolized by *S. aureus* under aerobic but not anaerobic conditions. This suggested either that nitrite was not being used as an electron acceptor or the nitrite reductase system was blocked anaerobically. After considering the effects of changing pH during the growth of *S. aureus* and a review of the literature, Buchanan and Solberg suggested that nitrite inhibits the growth of *S. aureus* by blocking the sulfhydryl sites of either coenzyme A or -lipoic acid, thus preventing the normal metabolism of pyruvate. They also concluded that vacuum packaging should improve the safety of cured meats relative to *S. aureus*.

Of significant importance was a report that the growth of *S. aureus* was found to be restricted to the periphery of commercially fermented sausage (Barber and Deibel, 1972). Understanding why this occurs would be very instructive. One reason given was that *S. aureus* is less acid tolerant under anaerobic conditions; however, future research might demonstrate the involvement of additional factors. Inoculated raw sausage (2.9% salt and 100 mg/kg of sodium nitrite) incubated at 37°C in environments of 20, 15, 10, 5, and 0% oxygen yielded lower *S. aureus* populations as the oxygen level decreased. Although adequate growth occurred within 48 hours, enterotoxin was not detected through 5 days of storage in the absence of oxygen.

Markus and Silverman (1970) reported enterotoxin A production in broth media (pH 6.6–6.8) to be unaffected by sodium nitrite (200 g/ml) or sodium nitrate (1000 g/ml). Perhaps because of the atypically high pH, no synergistic effect occurred between salt, nitrite, and nitrate on either growth or enterotoxin production. Contrasting results were subsequently reported that showed the production of enterotoxin A to decrease as pH decreased, salt increased, and nitrite increased (Tompkin et al., 1973).

Morse and Mah (1973) studied the effect of glucose on enterotoxin B synthesis in a broth medium buffered to an alkaline pH (7.7). Adding glucose caused decreased toxin production as a result of the rapid oxidative decarboxylation of pyruvate. Subjecting the cells to anaerobic shock (sparging with 95% N_2 and 5% CO_2) to inhibit pyruvate decarboxylation caused an increase in the rate and quantity of toxin produced, the percentage of lactic acid as an end product, and, presumably, more available energy. Adding nitrite as an electron acceptor at the time of anaerobic shock prevented increased toxin production, and nitrate was converted to nitrite. Glucose repression of enterotoxin B production was also reported to occur at pH 6.0, but to a lesser degree than at pH 7.7 (Morse and Baldwin, 1973).

Bean and Roberts (1974, 1975) demonstrated the combined effect of pH, salt level, and nitrite on the heat resistance and recovery of *S. aureus*. There was a general pattern of increased heat resistance as salt level (0, 4, and 8%) increased at pH 6, 6.5, and 7.0, but not at pH 5.5 or 5.0. The effect of nitrite on heat resistance was inconsistent. However, when added to the recovery medium, increasing the level of nitrite had the general effect of reducing the recovery of both heat-damaged and undamaged cells. The inhibitory effect of nitrite in the recovery medium increased with increasing salt content, decreasing incubation temperature, and decreasing pH.

In another report on non-spore-forming bacteria, Rowe et al. (1979) reported that nitrite inhibits active transport, oxygen uptake, and oxidative phosphorylation in *Pseudomonas aeruginosa*. The evidence strongly suggested that nitrite exerted its inhibitory effect by oxidizing ferrous ion of an electron carrier(s) (e.g., cytochrome oxidase) to the ferric state. Nitrite did not affect glucose transport by *S. faecalis* or *Streptococcus lactis*, both of which lack cytochromes. They instead rely on glycolysis for generating ATP when grown on glucose and transport glucose via phosphoenolpyruvate-phosphotransferase rather than by active transport.

7.2.4.6 Summary for 1970–1980

7.2.4.6.1 The Perigo Factor in Cured Meats

1. The formation of the Perigo factor in commercially cured meat is highly debatable. Formation of some similar inhibitor(s) (Perigo-type factor) during the thermal processing of shelf-stable canned cured meat appears possible. The data of Pivnick and Chang (1974) are convincing evidence that a small percentage of the total inhibition can occur that is not due to residual nitrite per se.
2. It has been difficult to design tests that prove the presence or absence of an unknown inhibitor that is likely unstable and is apparently formed in low quantity and functions within the midst of a complex system (e.g., salt, residual nitrite, and pH), the sum of which influences stability. If manipulations of thermally processed meat were attempted to extract such an inhibitor, the question could be raised whether the inhibitor would be altered or destroyed.

7.2.4.6.2 *The Perigo Factor in Media*

The data appear conclusive that nitrite, iron, and sulfhydryl groups are involved in the formation of the inhibitor.

7.2.4.6.3 *Perishable Cured Meats*

Tests on a variety of cured meats showed the efficacy of nitrite to be highly variable among the laboratories and the products tested.

The list of recognized factors influencing the antibotulinal efficacy of nitrite continued to grow and included the following:

1. pH of the product during abuse.
2. Brine level.
3. Residual nitrite upon abuse and the rate of depletion during abuse.
4. Level of viable botulinal spores and vegetative cells at the time of abuse.
5. Temperature of abuse.
6. Level of ascorbate or isoascorbate.
7. Level of "available" iron in the product.
8. Type of meat and other formulation ingredients.
9. The thermal process applied to the product.
10. The growth of competitive microbiota.
11. The level and type of phosphate might play a role through their influence on product pH, but this effect was not clearly defined in meats. Whether phosphates can contribute to inhibition by some other mechanism (e.g., sequestering iron) remained unclear.

All the factors are interrelated. Although the thermal process given to perishable cured meats does not appear to cause thermal injury to the spores, long, slow heating times, as used in certain experiments with bacon and large-diameter sausages, can provide the opportunity for germination followed by destruction, and perhaps injury, of the germinated cells as the temperature is increased.

Inhibition seems to be influenced only by the amount of residual nitrite in excess of the asymptotic level on the depletion curve. The low level of residual nitrite that continues to be detected after prolonged storage does not appear to be inhibitory. The disappearance of residual nitrite in cured meat is faster as product pH is decreased and isoascorbate level and storage temperature are increased. Destruction of residual nitrite could occur directly by utilization of nitrite by certain microbes and perhaps indirectly by reaction with degradation products produced during the growth of the microbiota. In addition, certain competitive microbiota are inhibitory to botulinal outgrowth (e.g., enterococci). Perigo-type factors do not appear to play any role in perishable cured meats.

7.2.4.6.4 *Mechanism of Nitrite Inhibition*

1. Inhibition of *C. perfringens* and *S. aureus* was suggested to involve the blocking of sulfhydryl sites within the bacterial cells.
2. Nitrite was reported to inhibit active transport, oxygen uptake, and oxidative phosphorylation in *P. aeruginosa*. The effect appeared to be at the electron carrier level. Glucose transport was not affected in two bacterial species that lack cytochromes.
3. It was suggested that nitrite inhibition of *C. botulinum* may be due to a reaction of nitric oxide with an essential iron-containing compound (e.g., ferredoxin) within the germinated cells.

7.2.5 1980–1990

The National Academy of Sciences (NAS) examined the health effects of dietary nitrate and nitrite and evaluated possible alternatives to the use of nitrite as a preservative in food (Assembly of Life Sciences,

1981, 1982). The investigation was conducted at the request of the Food and Drug Administration and the U.S. Department of Agriculture. The NAS conclusions were then used to determine regulatory policy for the use of nitrite and nitrate in foods. The recommendations are also important because they influenced the intensity and direction of research. Following is a summary of the 11 points in the recommendations:

1. Results of limited experiments suggest that nitrate is neither carcinogenic nor mutagenic.
2. Evidence does not indicate that nitrite acts directly as a carcinogen in animals.
3. Most N-nitroso compounds are carcinogenic in laboratory animals and mutagenic in microbial and mammalian test systems, and some are teratogenic in laboratory animals.
4. The exposure of humans to nitrite and nitrate should be reduced. Exposure to nitrite should be reduced to the extent that protection against botulism is not compromised. Except for its use in dry-cured and fermented products, the use of nitrate should be discontinued in meat and poultry products. Attention should be given to reducing the nitrate content of vegetables and drinking water.
5. Environmental exposure to N-nitroso compounds should be determined, and methods to reduce exposure should be developed. Improved methods are needed to distinguish between free and bound nitrite and whether residual nitrite is a true measure of nitrosating capacity.
6. Sources of human exposure to amines and nitrosamines (e.g., pesticides and drugs) should be modified to reduce exposure. Amino compounds that can be nitrosated *in vivo* should be identified and investigated.
7. Additional research is needed on the metabolism and pharmacokinetics of nitrate in humans. This includes the role of bacteria in the reduction of nitrate to nitrite and N-nitroso compound formation, especially in certain clinical conditions.
8. Ascorbate is now added to bacon to inhibit the formation of nitrosamines. Further research is needed to determine the amount of nitrite that is destroyed in the human stomach and the extent to which nitrosation reactions are modified by various inhibitors. Interactions among inhibitors, catalysts, and other dietary components deserve attention.
9. Further studies are required to determine the mechanisms whereby nitrite controls the outgrowth of *C. botulinum* spores, its action in cured meats, its antioxidant activity, and its effect against spoilage microorganisms and pathogens other than *C. botulinum*.
10. Although it is not possible to estimate the potential morbidity or mortality from *C. botulinum* if nitrite were removed as a curing agent from certain products, for public health and safety it is prudent to consider that certain preserved foods may contain *C. botulinum* and may be abused.
11. In view of the possible but unquantified risk resulting from the use of nitrite as a curing agent, the search for alternatives to the use of nitrite should be continued. No new agent or combination of agents should be used to replace nitrite until the safety of the new agent(s) has been ensured.

If the NAS committee had concluded that nitrite is carcinogenic, the regulatory agencies would have had to disapprove the use of nitrite as a food additive. Such action would have eliminated all cured meat and poultry products from the market.

The momentum established in the previous decade led to new information on the possible mechanisms involved in nitrite inhibition. Although there was continued emphasis on the anticlostridial effect of nitrite, there was also interest in other microorganisms. Research on alternative agents and/or processing methods to replace nitrite as a preservative was actively pursued. Concern for sodium intake in the diet brought new pressure to reduce the salt content of cured meats, further accelerating the trend of the past 40 years. Considering the recent reductions in the amount of nitrite used for curing meats and the voluntary elimination of nitrate from most cured meats in the United States, the impact of further reductions in salt deserved close examination. Another major trend developed during this time. Throughout the decade, the per capita consumption of red meats decreased while the consumption of poultry products increased. In 1983, the U.S. Food and Drug Administration acknowledged that prior sanction existed for the use of nitrite as an additive to poultry meat (Cassens, 1990). Since that time, there has been a very rapid growth in the volume and variety of cured poultry products in the United States.

7.2.5.1 Assessment of the Botulinal Risk in Cured Meats

Hauschild (1982) summarized the existing data from botulinal challenge tests in a wide variety of meat products. He proposed expressing botulinal inhibition as log I/P, where P is the probability of outgrowth and toxigenesis from one spore. This was used to quantitatively compare the data from the various challenge tests. The data provided estimates of the relative degree of botulinal risk for each product. The level of risk differed among the various products. Bacon was considered a low-risk product, whereas turkey ham was a high-risk product. The level of risk could also differ among products of the same type due to differences in their brine concentrations. He concluded that significant reductions in nitrite should not be made for all cured meats. However, selective reductions in nitrite could safely be made provided the levels of brine, carbohydrate, and erythorbate are controlled.

Hauschild and Simonsen (1985) conducted an extensive review of the technology for producing shelf-stable canned cured meats. Their review was to assist in the development of a document for these products. They summarized research, existing government regulations, and commercial practices to arrive at guidelines to assure the microbiological safety of these products. They recommended that the bacterial spore levels in the raw product be rigidly controlled and that 150 mg/kg of sodium nitrite be used to produce products with the brine and thermoprocess values given in Table 7.1.

Reductions in the recommended levels should be based on extensive plant experience and/or experimental work and controls to assure minimal spore levels. Reductions in the level of nitrite can be compensated for by moderate increases in the brine level and/or thermoprocess. This information was reiterated by Hauschild and Simonsen (1986).

7.2.5.2 Botulinal Challenge Tests

7.2.5.2.1 Tests in Meat and Fish Product

Toxin production by a mixture of type E and nonproteolytic type B botulinal strains in liver sausage, fermented sausage, and hams was investigated by Lücke et al. (1980). Liver sausage was prepared with 83 mg/kg of added sodium nitrite, pH 6.2, and a_w 0.98. Toxin was not produced at 5°C. As temperature was increased to 8, 10, and 15°C, toxin production was increasingly more rapid. In fermented sausage, the inhibitory effect of the lactic fermentation and decreasing water activity values during fermentation and subsequent drying prevented botulinal toxicity. Large bone-in hams were inoculated with nonproteolytic botulinal spores before curing in a solution of salt, nitrite, and nitrate at 8°C. Toxin was detected after 9–11 days. As the hams continued through the completion of curing, toxin levels remained high, but the recovery of viable botulinal cells became increasingly difficult. After curing,

TABLE 7.1

Shelf-Stable Canned Cured Meats: Minimum Brine and Thermoprocess with 150 μg/g Sodium Nitrite

Product	% Brine	Thermoprocess (F_0)
Luncheon meat	3–4	1.0–1.5
	4–4.5	1
	4.5–5.0	0.5–1.0
	5–5.5	0.5
Ham and shoulder	3.3	0.3–0.5
	3.7	0.2–0.3
	4	0.1–0.2
Sausages	2.5	1.5

Source: Hauschild and Simonsen (1985).

toxin was detectable 10 cm from the injection site. The hams had a slight off-odor but no apparent putrefaction. Similar results were obtained during the dry curing of hams at 8 or 10°C. They concluded that nonproteolytic, psychrotrophic strains can grow and produce toxin during the early stages of curing before the salt and nitrite levels penetrate and reach inhibitory levels. Additional data on the ham experiments were presented by Lücke et al. (1982). Methods and media were recommended to facilitate the recovery of both the proteolytic and the psychrotrophic nonproteolytic strains of *C. botulinum.* The latter were suspected to be of particular importance as a cause of botulism from raw cured meat products in Europe (Lücke, 1982).

Hauschild et al. (1982) assessed the relative risk of botulism from temperature-abused liver sausage. At brine levels of 3.8–4.2%, nitrite in the range of 0–100 mg/kg had little or no effect on toxinogenesis. Greater protection was obtained with 150 mg/kg. At a higher brine level of 5.44%, nitrite was more effective, and an appreciable delay in toxicity was obtained with 50 and 100 mg/kg of nitrite. Product with low levels of toxin often appeared acceptable. Storage at 8°C until a nearly complete loss of nitrite occurred did not diminish the effect of the nitrite. By increasing the processing temperature and/or prolonging the time of processing, the inhibitory effect of the nitrite was reduced. Another possibility could be offered for this observation. Perhaps the milder thermal process permitted the survival of a competitive microbiota that prevented toxigenesis. By increasing the thermal process, competitive microbiota could have been destroyed, thereby permitting the more heat-resistant *C. botulinum* to dominate and develop toxin. Hauschild and Hilsheimer (1983) examined 138 samples of liver sausage from retail outlets in Canada. The number of botulinal spores was estimated to be about 0.15 spores/kg in the product. The total *C. botulinum* was estimated to be about 0.24/kg.

The use of nitrite in smoked fish in the United States has been restricted. To prevent the growth of *C. botulinum*, a high level of salt (5% brine) has been required. Some consumers find this amount of salt unacceptable. Cuppett et al. (1987) demonstrated that adding sodium nitrite to obtain an initial level of 156 mg/kg could reduce the amount of salt needed for botulinal protection. The tests involved vacuum-packaged smoked whitefish fillets prepared with about 2.3, 4.4, and 6.2% brine. The pH of the fillets decreased during 27°C incubation from about pH 6.4 to as low as 5.2. The rate of pH decrease was slower as the brine level increased.

7.2.5.2.2 Tests in Model Systems

Sofos et al. (1980a) studied the significance of pH upon the efficacy of nitrite and sorbic acid mixtures in a chicken frank emulsion. The test system consisted of a frank emulsion prepared from mechanically deboned chicken meat that was extruded into test tubes, heated, cooled, overlaid with vaspar, and incubated at 27°C. The antibotulinal efficacy of sorbic acid was pH-dependent, with levels of at least below pH 6.0 being required. If a low level of nitrite (40 mg/kg) was added, the pH of the formulation could be increased to pH 6.2 without loss of the benefit of sorbic acid. The nitrite (40 mg/kg) also significantly increased the inhibitory effect of the sorbic acid. The presence of sorbic acid slowed the rate of nitrite depletion.

Wagner and Busta (1983) compared the antibotulinal effect of combinations of sodium nitrite (40 and 120 mg/kg), sodium acid pyrophosphate (0 and 0.4%), and potassium sorbate (0 and 0.26%) in frankfurters. They used the same test system as in Sofos et al. (1979b–e, 1980a). Tubes showing gas or other spoilage were preferentially selected for toxin assay. The pH of the four product variables was between 5.7 and 6.0. The test system avoided the problem of recontamination that occurs when peeling and packaging franks. The combination of 40 mg/kg of sodium nitrite, 0.26% potassium sorbate, and 0.4% sodium acid pyrophosphate caused the greatest delay in toxin production. There appeared to be more botulinal inhibition with the addition of sodium acid pyrophosphate.

Nelson et al. (1983) assessed the antibotulinal effect of three phosphates in the test system just described (Wagner and Busta, 1983). The frankfurters were made from chicken instead of beef and pork. The initial pH for the various products was 5.78–6.19. Polyphosphate increased botulinal inhibition when added to combinations of sorbic acid or potassium sorbate and a low level of nitrite (40 mg/kg). Of the polyphosphates tested, sodium acid pyrophosphate was more effective than sodium hexametaphosphate or sodium tripolyphosphate. The polyphosphates did not influence botulinal inhibition in the absence of nitrite. The combination of nitrite (40 mg/kg), potassium sorbate (0.26%), sodium acid pyrophosphate

(0.4%), and pH control of the emulsion was the most effective to achieve reduced nitrite levels and provide enhanced botulinal inhibition. The pH was important in all treatments tested.

Jarvis et al. (1979) modified their pork slurry to a thicker consistency, more like that of a reformed ham after heating. They found that increasing the heating from $P_{80°C} = 0.65$ to $P_{80°C} = 6.65$, nitrite in the range of 175–300 mg/kg was even more inhibitory to botulinal outgrowth as measured by the percentage of slurries that became toxic. This increased inhibition was not observed with nitrite levels of 75–125 mg/kg. They also found that the antibotulinal efficacy of nitrite could be increased by the addition of phosphate with certain phosphates being more effective than others.

Roberts et al. (1981a, b) evaluated the interactions of salt, nitrite, nitrate, isoascorbate, polyphosphate, and thermal process on botulinal outgrowth in pork slurries of two different pH ranges (pH 5.4–6.36 and 6.27–6.72). Statistically significant reductions in toxin production occurred with increasing salt level, nitrite level, or thermal process; adding isoascorbate or nitrate; or decreasing the abuse temperature. The pH of the pork used for the lower pH slurries ranged from 5.6 to 6.03, except for two outliers of 5.4 and 5.54. The pH of the pork used for the higher pH slurries ranged from 6.27 to 6.56. The addition of polyphosphate caused an increase of about 0.4 units in the lower pH slurries and about 0.1–0.2 units in the higher pH slurries. The increase in pH could explain the decreased inhibition in the lower pH slurries, but the reason for the increased inhibition when adding polyphosphate to the higher pH meat slurries is less evident. The levels of salt found in the slurries were often as much as 0.5% (wt/vol water) below the targeted values. It is unclear whether the target or analytical values for salt were used for statistical analysis although the conclusions were stated on the basis of the target values of 2.5, 3.5, and 4.5% salt (wt/vol water). This may alter the quantitative assessment of the individual and combined effects of the additives, although the trends in the data may not change.

Roberts et al. (1981c) analyzed results from large multifactorial pork slurry tests to arrive at formulas for estimating the probability of toxin formation in their pork slurry system. A separate formula was developed for slurries prepared with low-pH pork (pH 5.5–6.3) and high-pH pork (pH 6.3–6.7).

Gibson et al. (1982) examined the effect of source of the meat upon the anti-botulinal effect of 100 mg/kg of sodium nitrite in a pork slurry system. They found the nitrite to be less effective in slurries prepared with meat from the shoulder than meat from the leg. No systematic difference in inhibition was observed with the meat from three breeds of pig. Considerable variation occurred between the meat from different animals within each breed.

Roberts et al. (1982) examined the interaction of potassium sorbate in their cured pork slurry system. One nitrite level (40 mg/kg) was used throughout. At a given spore inoculum, potassium sorbate significantly decreased toxin production. The effect of sorbate was greater as salt levels increased (2.5–3.5%), as pH levels decreased to below 6.0, and as the storage temperature decreased.

UK legislation was proposed to control the use of nitrite by monitoring the residual nitrite in cured meat products. This led Gibson et al. (1984) to examine the effect of various factors on the rate of nitrite depletion and botulinal outgrowth. They found that the rate of depletion in their pork slurry system was influenced by many intrinsic and extrinsic factors. They concluded that measuring the residual nitrite of a product cannot be used to assess compliance with the nitrite regulations unless the analysis is performed soon after manufacturing the product. As expected, the rate of nitrite depletion was influenced by storage temperature, the addition of nitrate, and the addition of isoascorbate. The significance of microbial growth on nitrite depletion in the slurries during storage was not addressed. A rather high level of isoascorbate (1000 mg/kg) caused nitrite to disappear rapidly and delayed the outgrowth of *C. botulinum*. They concluded that the level of residual nitrite did not influence botulinal outgrowth in their experiments. No suggestion was offered for the increased botulinal inhibition when nitrate was added. The addition of nitrate led to higher residual nitrite levels for a longer period of time.

The primary objective of the Bristol research group was to arrive at a model system to quantitate the effect of the various factors influencing botulinal inhibition in cured meats. Their approach was to conduct large multifactorial experiments using pork slurries and then to analyze the data statistically and develop equations that could be used for predictive modeling. The results are intriguing and leave relatively unexplored the underlying mechanisms influencing the results. For example, why was a high heat treatment (80°C for 7 min followed by additional heating at 70°C for 1 h) more effective for increasing botulinal inhibition? Why should the initial pH of the meat influence the degree of botulinal inhibition?

Szczawinski et al. (1989) prepared ground pork slurries with 0, 50, 100, 156, and 200 mg/kg of sodium nitrite. The slurries were irradiated, then inoculated with botulinal spores, sealed with vaspar, and stored at 30°C for 28 days. Residual nitrite levels after irradiation were inversely related to the irradiation dose. At lower dose levels of 0–9 kGy, there was a slight reduction of residual nitrite and the probability of botulinal outgrowth increased in only one variable (50 mg/kg of sodium nitrite and the lower spore inoculum level). The probability of botulinal outgrowth decreased as the level of added nitrite increased above 50 mg/kg. After higher doses of irradiation (10, 20, 30, 40, and 50 kGy), there was a marked decrease in the residual nitrite, which resulted in rapid botulinal outgrowth.

7.2.5.3 Perigo Factor

Jonsson and Raa (1980) heated cysteine at pH 6 to produce bis-(2-amino-2-carboxyethyl) trisulfide (Bactin) and found it to be bacteriostatic to *Lactobacillus plantarum*. Bactin was not produced when heating cysteine, unless ferric chloride was present and the pH was 6. The bacteriostatic action was reversed by the addition of cysteine. The authors suggested that Bactin formation may be responsible for the inhibition observed in Perigo-type media. They suggested that the bacteriostatic effect is due to facilitated conversion of cysteine to Bactin and a concomitant removal of thio compounds, which otherwise prevent bacteriostatic activity.

In another study of Perigo factor formation, Custer and Hansen (1980) reported that lactoferrin, an iron-binding protein in casein, forms an inhibitor when autoclaved in the presence of nitrite, thioglycolate, and iron in culture media. Transferrin, a similar protein in blood serum, responded similarly. Pretreatment of the lactoferrin and transferrin with a protease was necessary for inhibitor formation. The amount of iron required for inhibitor formation was reportedly similar to the level found in muscle tissue. Higher or lower levels prevented the inhibitory effect. They concluded that lactoferrin and transferrin are precursors of the Perigo factor and that it is likely that nitrite-modified transferrin accounts for some of the antimicrobial effects of nitrite in cured meats. However, the requirements of pretreatment with protease and autoclaving reduce the likelihood that this would occur under natural conditions in meats.

A Perigo-type factor was produced when tryptone was heated with nitrite at 121°C for 20 min (Miwa and Matsuura, 1983). Inhibition was not observed when casein or an acid hydrolysate of tryptone was used. The maximum inhibitory activity was achieved by heating the medium containing 4% tryptone and 0.2% thioglycolate with more than 50 mg/L of sodium nitrite. There was evidence that the factor may be unstable toward oxygen. The factor was inhibitory against *S. aureus*, *B. subtilis*, and *C. botulinum*, but not against *E. coli* or *S. typhimurium*.

Hansen and coworkers published a series of papers devoted to the Perigo factor with the intent of learning the mechanism of nitrite inhibition. *B. cereus* was used throughout the research. They concluded that nitrite inhibits the germination of spores of *B. cereus* by inactivating membrane sulfhydryl groups (Buchman and Hansen, 1987). The relevance of this information to the effect of nitrite on *C. botulinum* and the stability of cured meats that are mildly heated remains to be established.

7.2.5.4 Alternatives to Nitrite for Botulinal Protection

A considerable amount of research has been devoted to finding an acceptable replacement for nitrite. The National Academy of Sciences report (Assembly of Life Sciences, 1982) is the most complete assessment of the possible replacements and the issues involved in finding a replacement. The committee considered the different functions of nitrite as a food additive (e.g., antioxidant, antimicrobial agent, or cured color and flavor development). In addition, a listing of some of the proposed chemical alternatives along with references was later provided by Roberts and Gibson (1986). None of the chemicals investigated has shown promise to date.

Bacon was identified as the one cured meat in which unacceptable levels of nitrosamine could be produced under certain high heat conditions. Thus, special requirements were established by the USDA for the production of bacon and verifying compliance with a nitrosamine limit. When bacon is fried, the high temperature can result in the formation of N-nitroso compounds. It was learned that the addition of ascorbate or isoascorbate (erythorbate) can block nitrosamine formation during the frying process. This led the

USDA to reduce the input level of sodium nitrite to a maximum of 120 mg/kg and to require the addition of 550 mg/kg of either ascorbate or isoascorbate (Code of Federal Regulations, 1990a). Subsequently, two additional options were added to the USDA regulations (Code of Federal Regulations, 1990a). One option is to use a combination of 100 mg/kg of sodium nitrite, 550 mg/kg of sodium ascorbate or isoascorbate, and a USDA-approved partial quality control program to assure compliance. The second option is to manufacture the bacon using a combination of 40–80 mg/kg of sodium nitrite, a readily fermentable carbohydrate (i.e., a minimum of 0.7% sucrose), and a lactic acid-producing bacterial culture (i.e., *Pediococcus acidilactici*). The rationale for this approach is apparent from the summary outlined earlier in this chapter, in the bacon section for 1970–1980. The second option in the USDA regulation is based upon a three-plant study that verified the efficacy of this concept (Tanaka et al., 1985). The research and subsequent regulation ignore the significance of brine level as an important factor that should also be controlled. The process also does not acknowledge the potential value of the natural contaminating bacteria associated with most bacon-processing environments. If the brine level is controlled, an adequate quantity of fermentable carbohydrate is present, and sufficient sodium nitrite is added to produce an acceptable cured product, then the naturally occurring lactic acid-producing bacteria would reduce the pH of the product to inhibitory levels in a high percentage of the packages if the bacon were temperature abused. A key to this approach is to not eliminate the lactics from the environment through an overly aggressive sanitation program.

Nitrite is unique in being able to produce cured meats with the proper attributes of color, oxidative stability, flavor, and microbial stability. A new approach developed in Canada consists of using nitric oxide to cure the heme portion of beef blood. This cured pigment is then further processed, dried, and then microencapsulated to provide stability against degradation by oxygen and light. The cured meat pigment would then be added along with other ingredients (e.g., ascorbate, phosphate, and tertiary butylhydroquinone), to provide oxidative stability. The proposed system also requires the addition of an antimicrobial agent (e.g., sodium hypophosphite). O'Boyle et al. (1990) and Shahidi and Pegg (1992) have summarized the system and provided the status on the various aspects of the research.

Ismaiel and Pierson (1990) demonstrated that origanum oil, alone or in combination with sodium nitrite, could prevent botulinal outgrowth in a broth medium. The two additives acted synergistically. This effect was explainable by their modes of action. The origanum oil affects germination and vegetative growth; sodium nitrite affects outgrowth and vegetative growth. Unfortunately, origanum oil was not as effective when tested in pork. This was thought to be due to the highly soluble oil components being absorbed into the lipid fraction of the meat. At the lowest spore levels tested, there was a consistent trend, although statistically insignificant, toward increasing inhibition with an increasing level of origanum oil in three of the four sets of nitrite levels. Perhaps the results would be more favorable if challenged with a lower botulinal spore level.

7.2.5.5 Mechanism of Nitrite Inhibition

Yarbrough et al. (1980) reported nitrite inhibition of oxygen uptake and proton-dependent proline transport in *E. coli*. Since glucose transport was not affected, this suggested that nitrite was inhibiting glucose catabolism. Nitrite was found to inhibit aldolase, but hexokinase was not affected. The authors concluded that nitrite inhibits bacteria by several mechanisms. First, nitrite interferes with energy conservation by inhibiting oxygen uptake, oxidative phosphorylation, and proton-dependent active transport. Second, nitrite acts as an uncoupler, causing a collapse of the proton gradient. Third, certain metabolic enzymes (e.g., aldolase) are inhibited. Since streptococcal aldolase was sensitive to nitrite but glucose accumulation was unaffected, the authors suggested the high resistance of *S. lactis* and *S. faecalis* is due to the impermeability to nitrite in these bacteria.

Glucose and 2-deoxyglucose transport and accumulation by *S. aureus* are inhibited by nitrite under anaerobic conditions (Simone and Solberg, 1981). The accumulation of glucose decreased in relation to increasing the nitrite level and the time of exposure to nitrite before adding sugar.

Palumbo and Smith (1982) found heat-injured *S. aureus* to undergo repair and grow in a nitrite-containing agar medium at pH 5.6. Recovery of the heat-injured cells was comparable in media with no nitrite and media containing 100 mg/L of sodium nitrite. At levels of 200, 400, and 800 mg/L, the recovery was progressively lower.

Fang et al. (1985) investigated the effect of nitrite on *S. aureus*. The inhibitory effect of nitrite was enhanced by applying stricter anaerobic conditions. Under less rigid conditions of anaerobiosis, *S. aureus* was more resistant to the nitrite and also caused a more rapid depletion of nitrite in the medium. The disappearance of nitrite was faster with *S. aureus*. The mechanism by which *S. aureus* hastens the loss of nitrite was not attributable to nitrite reductase activity, interactions with metabolic products, dismutation due to the reduced pH during growth, or absorption to the cells. Nitrite, nitric oxide, nitrous oxide, and ammonia were not end products of the sodium nitrite disappearance. The increased efficacy of nitrite under strict anaerobic conditions could be a factor affecting the growth of other microbes in cured meat and poultry products.

Kanner and Juven (1980) compared the anticlostridial effect of nitrite and S-nitrosocysteine (RSNO), which is formed during the curing of meats. They found that RSNO was less effective than nitrite for preventing the recovery of heat-shocked *C. sporogenes* spores. When salt was also present in the recovery medium, very little synergistic effect was observed with RSNO and salt, but a high level of synergism was observed with nitrite and salt. This suggested that RSNO is not the major anticlostridial factor in cured meat. The anticlostridial effect of RSNO increased as the pH of the culture medium was lowered from 6.4 to 5.6. On an equimolar basis, RSNO (540 mg/L) was considerably less inhibitory than sodium nitrite (200 mg/L) when tested in inoculated perishable canned turkey at 27°C.

Chumney and Adams (1980) found *C. perfringens* spores to be sensitive to inhibition by salt, nitrite, and a mixture of polymyxin and neomycin after heat injury at 70–100°C. They suggested the increased sensitivity to each agent was caused by the same cellular damage. Recovery of heat-injured spores in a basal medium containing 3% salt or 1% salt with 200 mg/L of nitrite yielded comparable recovery rates. Compared with earlier research, the nitrite was relatively ineffective as an inhibitor. This may have been due to sulfite in the basal medium (tryptone sulfite neomycin agar), since sulfite destroys nitrite (Tompkin et al., 1980).

Collinge et al. (1981) reported that ferric chloride or hemoglobin stimulates the growth of *C. botulinum* (as opposed to reducing the antibotulinal efficacy of nitrite) in canned cured meat. They considered that nitrite inhibits *C. botulinum* by making the iron of the heme groups in meat unavailable for growth. Adding nitrosylated myoglobin to the meat system resulted in a rather inhibitory system compared with adding untreated myoglobin. Also, there was no noticeable relation between botulinal growth rate and residual nitrite level in the meat to which nitrosylmyoglobin had been added.

Meyer (1981) investigated the effect of nitrite and nitric oxide as inhibitors of the nitrogenase enzyme system of *Clostridium pasteurianum*. The enzyme system consists of two proteins, one of which is an iron protein containing nonheme iron and sulfide as a cubic Fe_4S_4 cluster. Nitric oxide was found to rapidly and irreversibly inactivate the iron protein. The results suggested that nitrite is converted to nitric oxide, which then disrupts the iron–sulfur cluster of the protein and thereby inhibits nitrogenase activity.

Woods et al. (1981) found that the addition of nitrite to a suspension of *C. sporogenes* in a glucose medium caused a rapid decrease in intracellular ATP concentration and an accumulation of pyruvate in the medium. The accumulation of pyruvate was due to inhibition of the phosphoroclastic oxidoreductase. This enzyme consists of a single protein molecule containing thiamine pyrophosphate and a single nonheme chromophore. Nitrite did not react directly with ferredoxin, but there was decreased activity in the ferredoxin from cells preincubated with nitrite. The authors concluded that, although nitric oxide can react with ferredoxin *in vitro*, the main site of inhibition within intact cells is the reaction of nitric oxide with the enzyme pyruvate-ferredoxin oxidoreductase. Woods and Wood (1982) next reported that adding sodium nitrite to cell suspensions of *C. botulinum* led to an accumulation of pyruvic acid, suggesting that nitrite inhibits the phosphoroclastic system.

Reddy et al. (1983) employed electron spin resonance to examine the spectra of botulinal cells treated with nitrite and ascorbate. They concluded that botulinal cells contain iron–sulfur proteins that react with nitrite to form iron–nitric oxide complexes, with resultant destruction of the iron–sulfur cluster. With addition of sodium ascorbate, the intensity of the reaction was stronger, thereby verifying observations that ascorbate increases the antibotulinal efficacy of nitrite in meat products. Their results supported the theory that nitrite is inhibitory to *C. botulinum* by inactivating iron–sulfur enzymes (e.g., ferredoxin) that are essential for growth.

Vahabzadeh et al. (1983) found carbon monoxide to be ineffective as a replacement of nitrite for inhibition of *C. botulinum* in pork. Nitric oxide was inhibitory, but this could be attributed to the formation of analytically detectable nitrite in the meat. Adding ferric chloride or myoglobin decreased the antibotulinal efficacy of nitrite. Adding EDTA or denatured nitrosylated myoglobin increased the efficacy of nitrite. The decreased inhibition by nitrite when iron compounds were added appeared to be due to a reduction in residual nitrite levels in the product rather than to the iron serving as nutrient to stimulate botulinal growth. They concluded that the antibotulinal effect of nitrite and related compounds is probably due to inhibition of the iron-containing enzymes within the botulinal cell rather than to the complexing of iron, a nutritional requirement for growth.

The debate over whether clostridial ferredoxin or pyruvate-ferredoxin oxidoreductase is the site of nitrite inhibition for *C. botulinun* was again addressed by Carpenter et al. (1987). Their results led to the conclusion that both are inactivated. On the basis of the research conducted by these researchers and others, it is fairly certain that the mechanism of nitrite involves the inactivation of iron–sulfur proteins, such as ferredoxin and pyruvate oxidoreductase. It appears that the hypothesis proposed earlier is fairly close (Tompkin et al., 1978). One aspect that the more recent data suggest is that the interaction of nitric oxide with iron–sulfur complexes is irreversible.

Research by Payne et al. (1990a, b) examined the effect of nitrite and iron–sulfur–nitrosyl complexes on iron–sulfur clusters in *C. sporogenes* cells and isolated proteins. They were unable to demonstrate a direct action of the compounds on the iron–sulfur proteins, leading them to doubt that the inhibitory effect of nitrite is on preformed iron–sulfur proteins of the bacterial cell. After continued research on the mechanism of compounds associated with the Perigo factor the efficacy of sodium nitroprusside was explored (Joannou et al., 1998). Sodium nitroprusside was found to be a very effective inhibitor of *C. sporogenes*, comparable to the inhibitory properties of the anion of Roussin's black salt. Nitroprusside was used as a model to investigate the inhibitory mechanism of sodium nitroprusside and, more generally, the toxicity of nitrosyl compounds to bacteria. The cell wall surface was the primary point of attack for the nitrosyl complex associated with nitroprusside. The cell walls showed a 90% decrease in thiol content when compared with untreated cells. In addition, blisters at the surface of the cells were observed when viewed under the electron microscope. In the research of Duncan and Foster (1968a) such blister formation was not reported following exposure to sodium nitrite. Additional information and views on the mechanism of nitrite are available in Roberts et al. (1991) and Roberts and Dainty (1991).

7.2.5.6 Miscellaneous Studies in Laboratory Media

Botha and Holzapfel (1987) tested the effect of nitrite on *Sporolactobacillus* isolates from a variety of sources. In a broth medium at pH 6.0, 30–50% less growth was observed after 5 days at 35°C with 100 or 200 mg/L of sodium nitrite compared with no added nitrite.

Gibson and Roberts (1986b) demonstrated the interactions of pH, incubation temperature, sodium chloride, and sodium nitrite on the inhibition of *C. perfringens* and fecal streptococci in broth media. The fecal streptococci proved to be quite resistant to nitrite. A high salt concentration and low incubation temperatures were the more important factors influencing the inhibition of these bacteria. The inhibitory effect of nitrite against *C. perfringens* was enhanced with a decrease in pH, a decrease in temperature of incubation, and/or an increase in the salt concentration. The relative sensitivity of *C. perfringens* to 3–4% salt could explain why there have been few outbreaks in *C. perfringens* food-borne illness involving cured meats. Most cured meats have traditionally had brine levels of 3% or greater.

The effect of nitrite on the growth of *Listeria monocytogenes* was examined in trypticase soy broth by Shahamat et al. (1980). Sodium nitrite was added to the broth before autoclaving. At the pH values (6.5, 5.5, and 5.0) found in cured meats, some reduction in growth was observed with the addition of 200 mg/L of sodium nitrite. However, the data more strongly demonstrate that *L. monocytogenes* is quite resistant to the inhibitory effect of nitrite, even though the nitrite had been autoclaved with the broth.

Juntilla et al. (1989) found the rate of death of *L. monocytogenes* to be similar in fermented sausage manufactured with different levels of salt (3.0 and 3.5%) and sodium nitrite (50, 120, and 200 mg/kg). The final pH of the products was about pH 4.6. It can be concluded that factors (e.g., decreasing pH and a_w) influenced the rate of death.

Buchanan et al. (1989) tested the effect of nitrite on *L. monocytogenes* in a broth medium at pH 7.5 and 6.0. The incubation temperature (5, 19, 28, and 37°C) had the strongest effect on growth rate and lag-phase duration. Consistent with previous reports over the past 60 years, nitrite was not an effective inhibitor at pH 7.5. At pH 6.0, which is more typical of cured meat and poultry products, nitrite was inhibitory. When comparing 0, 50, 100, and 200 mg/L of sodium nitrite, the degree of inhibition was most strongly enhanced by reducing the incubation temperature, to a lesser extent by anaerobiosis, and even less by increasing the salt level (0.5 and 4.5%). The results of Glass and Doyle (1989) and Schmidt and Kaya (1990) indicate that the level of nitrite typically added to cured meats is insufficient to significantly inhibit *L. monocytogenes* growth in processed meat and poultry products at refrigeration temperature.

7.2.5.7 Spoilage of Cured Meats

Zeuthen (1980) studied the effect of meat pH on the rate of microbial growth on sliced ham. Fresh pork was divided into two groups: low-pH meat (5.5–5.7) and high-pH meat (6.3–6.5). The hams were boned and pumped with a pickle solution containing sodium tripolyphosphate. The meat was then used for processing as canned hams. After cooking and chilling, the canned hams were opened, sliced, vacuum packaged, and placed at 5°C. The lower pH meat resulted in a ham with a pH of about 6.0 and a residual nitrite level of 38 mg/kg after processing. The higher pH meat resulted in a ham with a pH of 6.35 and a residual nitrite level of 90 mg/kg. Both products had a brine level of approximately 4.0%. During the 7–8 weeks of storage at 5°C, the rate of microbial growth was considerably slower in the sliced ham prepared from the lower pH meat.

Collins-Thompson and Lopez (1981) demonstrated that nitrite depletion rate can be increased by certain lactic acid bacteria that are typically responsible for the spoilage of many cured meats. The loss of residual nitrite in bologna as a result of the action of the lactic acid bacteria was estimated as 30%. The authors questioned the role of nitrite during cured meat spoilage, since there is evidence demonstrating that nitrite can either limit or enhance the population of lactics.

Competitive inhibition can be a factor affecting the spoilage bacteria of cured meats. Collins-Thompson and Lopez (1982) reported the inhibition to be species-dependent. The growth of *Brochothrix thermosphacta* was restricted when grown on sliced vacuum-packaged bologna with *Lactobacillus brevis* or *L. plantarum* but not with *L. viridescens* or *Leuconostoc mesenteroides*. Inhibitory species caused the pH of the bologna to decrease to 5.3 within 2 weeks at 5°C. Noninhibitory strains decreased the pH of the bologna to only 5.5 within 4 weeks. The authors suggested that inhibitory strains produced specific antimicrobial substances other than hydrogen peroxide or lactic acid. The data of Grau (1980), however, demonstrate that inhibition can be dependent upon the concentration of undissociated lactic acid and the degree of anaerobiosis existing at the site of growth.

Under certain circumstances, the level of analytically detectable residual nitrite can influence microbial inhibition. Thus, the rate of nitrite depletion in cured meats can be important. Collins-Thompson and Lopez (1982) and Dodds and Collins-Thompson (1984) further demonstrated that lactic acid bacteria, particularly homofermentative lactobacilli, can accelerate nitrite depletion. *L. viridescens* was among the most active species. The depletion of nitrite was attributed to enzymatic activity (i.e., nitrite reductase) rather than lowering the pH of the medium. Subsequent tests demonstrated that a strain of *L. lactis* was able to reduce nitrite to N_2O (Dodds and Collins-Thompson, 1985).

Collins-Thompson and Thomson (1986) found two strains of *L. plantarum*, commonly involved in the spoilage of cured meats, which were resistant to the addition of 50 mg/L of sodium nitrite if grown aerobically in APT broth. Both cultures were sensitive to the inhibitory effect of nitrite under anaerobic but not under aerobic conditions. When grown anaerobically, both strains accumulated manganese in their cells. The significance of these observations was discussed, and some interesting questions were posed. For example, it was suggested that in trace metal metabolism, nitrite might stimulate the uptake and transport of ions (e.g., manganese).

By 1988, it was established that nitrite reduction could occur in some strains of *Lactobacillus plantarum*, *L. acidophilus*, *L. lactis*, *L. viridescens*, *L. leichmannii*, and *L. buchneri* (Wolf and Hammes, 1988). The nitrite reductases generally are heme-containing enzymes; however, the lactic acid bacteria were considered incapable of synthesizing heme compounds. Following the clue provided by Whittenbury

(1964) that some strains of lactobacilli produce heme-containing catalase when grown in the presence of blood or hematin, Wolf and Hammes (1988) found all the strains of *L. plantarum* tested were able to reduce nitrite when hematin was provided to cell suspensions. In addition, all the strains of *L. plantarum* and *L. sake* tested were able to produce catalase when provided with hematin. Protoporphyrin, an iron-free compound, did not activate nitrite reductase or catalase activity, whereas other iron-containing porphyrins (i.e., hemoglobin, myoglobin, hematin) did activate the enzymes.

Nielsen (1983a) studied the effect of sodium nitrite (0, 100, and 200 mg/kg) on the spoilage bacteria of a sliced vacuum-packed bologna-type sausage. *B. thermosphacta*, Enterobacteriaceae, and *Moraxella* and *Moraxella*-like organisms were increasingly inhibited by increasing the nitrite concentration and/or decreasing the temperature of storage (2, 5, 10, and 20°C). Gram-positive cocci, yeasts, and lactic acid bacteria were inhibited to a limited extent. This led to the lactic acid bacteria becoming the dominant spoilage bacteria in the vacuum-packed product. Gram-negative bacteria often constituted the major microbiota of the product without nitrite. Also, as the storage temperature increased, the Gram-negative bacteria proliferated in product with nitrite.

Nielsen (1983b) subsequently studied the effect of sodium nitrite (0, 100, and 200 mg/kg) on the spoilage bacteria of cooked pork loin. The formulated raw meat was stuffed into a casing, cooked, chilled, sliced, and then vacuum packed. Nitrite inhibited the growth of *B. thermosphacta* and Enterobacteriaceae at all temperatures of storage (2, 5, and 10°C). Again, the lactic acid bacteria, being insensitive to the presence of nitrite, dominated the spoilage bacteria in product with nitrite. It was concluded that nitrite has a significant effect in extending the keeping quality of vacuum-packed cured meat when used in combination with low storage temperatures and/or the use of packaging films with a low oxygen permeability.

Silla and Simonsen (1985) studied the rate and type of spoilage of four cured meat products packaged under vacuum and in modified atmospheres. Lactobacilli were the dominant microbiota in all four products at the time of spoilage. High levels of *B. thermosphacta* may also have significantly influenced spoilage of two of the products. A correlation was found between the nitrite concentration at the time of slicing and packaging and the time for spoilage to occur.

Chyr et al. (1980) found the most important factors influencing the rate of microbial spoilage of liver sausage to be the degree of cooking and the level of nitrite. A marginal thermal process (i.e., 63–68°C) led to the growth of enterococci during storage at 7°C. Increasing the level of sodium nitrite (0, 50, 100, and 150 mg/kg) retarded microbial growth. Omitting nitrite resulted in a perfumy odor that was associated with the growth of enterococci. The level of surviving enterococci decreased as the processing temperature increased (63, 68, and 74°C). Using good-quality raw materials, 156 mg/kg of sodium nitrite, and processing to 68°C resulted in a shelf life of more than 16 weeks at 5°C. The spoilage of liver sausage was shown to be due to *S. faecalis* (Chyr et al., 1981). The spoilage microbiota of the liver sausage prepared without nitrite also included *Pediococcus pentosaceus*. The pediococcus isolate was not as heat-resistant as the enterococcus. The enterococcus survived 65°C for 5 min or 60°C for 60 min with only 10^3 cells/ml broth. An industry survey found that liver sausage was commonly cooked to an internal temperature of 65°C.

Asplund et al. (1988) reported that the growth of *B. cereus* in liver sausage has been a problem in Finland, particularly in the summer. If the product is temperature abused, *B. cereus* can grow in the product even with the normal curing ingredients of salt, phosphate, and sodium nitrite (120 mg/kg). Additional additives were tested to reduce this risk. The growth of *B. cereus* was delayed or prevented by including a combination of erythorbate, gluconolactone (glucono delta lactone, GDL), and citric acid in the product. Adding GDL was considered the most important. Product with GDL had a pH value of 5.7–5.8, about 0.2–0.3 units lower than with the other variables tested.

Kafel and Jozwik (1987) investigated the spoilage rates of commercially produced products from six meat processing plants in Poland. Polish regulations require the heating of perishable canned meats to a minimum internal temperature of 68.8°C. The regulations also require that product from each "lot" be incubated at 37°C for 3 days to assess the relative stability and quality of the product. From among 4322 cans of 8 different classes of perishable canned cured meats, 980 cans, or 23%, swelled. When spoilage first occurred in each lot, additional cans (8290) were removed from storage at around 8°C and incubated at 37°C for 3 days. The time elapsed was normally 7–10 days after production. The spoilage rate of this second group of cans was 4%. The authors assumed that the lower spoilage rate was due to the death of spores that had been heat injured and then subsequently exposed to the effect of salt and nitrite. These data reflect the influence of a mixture of naturally occurring spores in fairly large cans of commercial product.

The effect of nitrite on the aerobic spoilage microbiota of raw chicken patties was studied by Bushway et al. (1982). The ground meat was aerobically packaged and stored at 4–5°C. Nitrite was considerably more effective in white meat than in dark meat. With the addition of sodium chloride (2.5%), even a low level of nitrite (40 mg/kg) caused some inhibition in white meat but not in dark meat.

Bushway and Jen (1984) found lower residual nitrite levels in chicken white meat than in dark meat, whether cooked or raw. This could be influenced by the fact that white meat is lower in pH and iron content than dark meat. Bushway and Serreze (1984) found that adding iron to raw white meat caused an increase in the rate of aerobic microbial growth. If nitrite (130 mg/kg) was also added, the effect of iron on aerobic microbial growth depended upon the amount of iron added. Low levels of iron (5–15 mg/kg) reduced the efficacy of nitrite. On the other hand, high levels of iron (40–160 mg/kg) increased the efficacy of nitrite. The reason for this effect on the aerobic spoilage bacteria was not known.

Bushway and Serreze (1984) also studied the effect of adding EDTA to raw dark chicken meat. Adding EDTA alone had no significant impact on the rate of aerobic microbial growth. Adding EDTA enhanced the efficacy of nitrite and caused an increase in the lag phase for aerobic microbial growth. Perhaps some adjustment occurred in the microbial population. The authors also investigated the effect of the inherent pH of the meat on the efficacy of nitrite. The pH of the white and dark meat was 5.6 and 6.4, respectively. The antimicrobial effect of nitrite was blocked when the pH of the white meat was adjusted to 6.4. Likewise, the efficacy of nitrite increased when the pH of dark meat was adjusted to 5.6. This led to the conclusion that the inherent pH of the meat is the major factor determining the effectiveness of nitrite on the rate of aerobic spoilage in chicken meat. The authors concluded that the differences in iron content between the dark and white meat may contribute in a small way to the differences observed. A test was not conducted in which the pH of the two meats was adjusted and iron was added along with the nitrite.

7.2.5.8 Summary for 1980–1990

1. The NAS concluded that the evidence available does not indicate that nitrite can act directly as a carcinogen in animals. Furthermore, the NAS was not able to conclude that nitrate can act as either a carcinogen or mutagen. These conclusions meant that neither nitrite nor nitrate would be banned as food additives.
2. A means to quantitatively compare the data from various challenge tests based on the probability of outgrowth and toxigenesis from one spore was proposed. This provided a basis for assessing the risk associated with different cured meats.
3. Tests demonstrated that nonproteolytic, psychrotrophic strains of *C. botulinum* can multiply and produce toxin in dry-cured hams during the early stages before the salt and nitrite levels penetrate and reach inhibitory levels. This could explain the high number of cases that have occurred during certain periods in Europe (Tompkin, 1980).
4. Extensive studies in meat and meat slurries demonstrated the interactions of nitrite, pH, isoascorbate, phosphate, and thermal process on botulinal outgrowth. This led to the development of a model system for predicting botulinal outgrowth in food, a concept that has been expanded to include other pathogens and has gained acceptance throughout the world.
5. Some research continued on the Perigo factor but no evidence was generated to link the Perigo factor to the antimicrobial effect that has been demonstrated in perishable, cured products that receive no heat treatment or are cooked to final internal temperatures in the range of 60 to 75°C.
6. Alternatives to nitrite continued to be investigated but none have been adopted commercially.
7. The USDA adopted a regulation for bacon that requires a maximum of 120 mg/kg sodium nitrite and the addition of 550 mg/kg sodium ascorbate or isoascorbate. Other available options have not received commercial acceptance.
8. The mechanism of nitrite inhibition differs in different bacterial species.

7.2.6 1990–2002

C. perfringens had not been a concern in cured meat and poultry products as evidenced by the relative lack of research on this pathogen during the preceding 40 years. This probably reflects the effect of the higher salt levels and nitrite found in cured products compared with noncured products. Despite this evidence, the USDA in 1988 initiated a series of increasingly restrictive policies on the rate of chilling for perishable cured meat and poultry products manufactured under USDA inspection. This is a case where the epidemiologic data indicate a negligible public health concern for cured meats but the evidence from challenge studies and predictive modeling suggests otherwise. This situation is a reminder of Morris Ingram's frustration with the increase in research on nitrite's role in botulinal inhibition in the 1970s. At the time he stated, "What we need at the present time, in my opinion, is not more inoculated pack experiments but a rationale for interpreting them" (Ingram, 1974). The debate over the risk of *C. perfringens* in commercially manufactured cured meat and poultry products has continued into the new millennium and serves as an example of the need for a risk assessment to clarify appropriate risk management policies.

The potential impact of various blood fractions as ingredients in sausage was investigated (Miller and Menichillo, 1991). A beef sausage formulated with 156 mg/kg sodium nitrite and an average iron level of 20±5 mg/kg was used as the control. Sausage formulated without sodium nitrite became toxic within 1 week compared with 3 weeks at 27°C when nitrite was present. Adding hemoglobin, red blood cells, or whole blood reduced the time for botulinal outgrowth and toxin formation compared with the control with nitrite. Adding dried plasma, however, delayed the time for toxin to be detected. In subsequent tests the effect of various decolorized blood fractions was evaluated. Again, the rate of toxin formation was inversely related to the level of iron and the addition of plasma delayed the time for toxigenesis compared to the control sausage (Miller et al., 1993).

Since 1990 there has been increased interest in control of *Listeria monocytogenes* in ready-to-eat foods. Buchanan and Phillips (1990) studied the growth of *L. monocytogenes* in tryptose phosphate buffer under a wide variety of conditions (salt, pH, temperature, aerobic/anaerobic conditions) and found sodium nitrite to interact with combinations of the factors tested. The database was incorporated into the USDA Pathogen Modeling Program.

McClure et al. (1991) using an automated turbidimetric system with multi-welled plates containing trypticase soy broth found the efficacy of sodium nitrite was temperature- and pH-dependent. At 20°C and below no visible growth occurred even with 50 mg/L sodium nitrite and a pH of 5.3. At pH values of 6.0 sodium nitrite had little effect in delaying the time to detect visible growth except at the highest level tested (200 mg/L) and a temperature of 15°C or below. At pH 6.0 and 5°C no growth was observed with any of the levels of sodium nitrite evaluated (50, 100, 200, 400 mg/L). In the absence of nitrite very slow growth did occur. Autoclaving the sodium nitrite in the medium did not increase the degree of inhibition against *L. monocytogenes* to the degree that had been observed with clostridia (see Perigo Factor). McClure et al. (1997) subsequently investigated the effects of salt, pH, temperature, and sodium nitrite in laboratory medium and a predictive model for the growth of *L. monocytogenes.*

An extensive series of experiments led to the conclusion that non-thermal inactivation of *L. monocytogenes* by sodium nitrite is pH-dependent and related to the concentration of undissociated nitrous acid (Buchanan and Golden, 1995; Buchanan et al. 1997).

Duffey et al. (1994) inoculated a variety of vacuum-packaged cooked sliced meats with *L. monocytogenes* and found the lag time increased and the rate of growth decreased at 0 and 5°C with the addition of sodium nitrite (0 to 315 mg/kg). Three M of residual undissociated nitrite doubled the time taken for a 3-$\log_{10}$ increase in numbers. As in the case of earlier research with *C. botulinum*, the effectiveness of sodium nitrite was significantly increased by the addition of sodium ascorbate.

Farber et al. (1995) examined the effect of sodium nitrite (0 and 200 mg/kg), salt (1, 3%), sodium erythorbate (0, 550 mg/kg), and temperature of storage (4, 10°C) on the rate of growth of *L. monocytogenes* in liver pate. Only temperature proved to be a significant factor. Perhaps, the high iron content of liver pate negated the effect of sodium nitrite and sodium erythorbate in this product as had been reported earlier for *C. botulinum.* Liver pate has been implicated in outbreaks of listeriosis in the UK, Australia, and the US.

Qvist and Bernbom (2000) manufactured bologna-style sausage with 60 and 150 mg/kg sodium nitrite and found neither level retarded the growth of *L. monocytogenes* in the sliced vacuum-packaged product when stored at 4 or 8°C.

Farkas and Andrássy (1992a) described a test system to evaluate the combined effects of formulation, heat, inoculum level, and other factors for inactivating spores and preventing the outgrowth of survivors. Using *C. sporogenes* (9×10^5/ml) in reinforced clostridial medium they found that broth adjusted to pH 5.8 and containing 3.5% sodium chloride and 50 or 200 mg/L sodium nitrite inhibited growth during 125 days storage at 37°C. If the broth had been adjusted to pH 6.7, a combination of heat and sodium nitrite was necessary to prevent survival and growth ($F_0 = 0.2$ with 200 mg/L, $F_0 = 0.7$ with 50 mg/L). Applying the procedures described by Hauschild (1982) they estimated the various formulations and thermal processes provided the equivalent of 4–7-$\log_{10}$ destruction of *C. sporogenes*. The model test system was then used to evaluate various combinations of heat, irradiation, and additives to achieve shelf stability of canned luncheon meat (Farkas, Andrássy and Horti, 1992), although sodium nitrite concentration was not varied. The procedures described show promise for estimating and comparing the combined effects of multiple treatments on pathogen survival and growth.

The effect of salt and nitrite on the growth of 24 lactics isolated from vacuum-packed cooked sausage was evaluated by Korkeala et al. (1992). Low levels of salt (1–2% enhanced growth but levels above 3% were inhibitory. *Leuconostoc* isolates seemed more sensitive to the effects of sodium nitrite and salt in comparison to the homofermentative lactobacilli. Sameshima et al. (1997) examined the effect of nitrite on three lactics in vacuum-packed cooked sausage made with and without nitrite. Adding 200 mg/kg sodium nitrite to the sausage increased the time required to multiply from 10 cfu/g to 10^5 cfu/g by 9, 4, and 15 days for *L. viridescens*, *L. mesenteroides*, and *Enterococcus faecalis*, respectively.

7.2.6.1 Fermented Meats and Dry-Cured Hams

Although removed from most cured meat products in the USA during the 1970s, nitrate continued to be used in the manufacture of fermented meats and dry-cured hams throughout much of Europe and to a lesser degree in the USA. In regions of Northern Europe, sausages made with nitrite and having pH 5.0 and a sour, smoked flavor can be found. These sausages are produced at temperatures below those generally used in the USA but higher than the temperatures used to manufacture traditional raw fermented dry sausage. Since about 1990, some manufacturers in the Mediterranean countries have been shifting away from the more traditional, low-temperature processing that involves long holding times and reduction of nitrate by micrococci or staphylococci to yield nitrite and other nitrogen oxides that are necessary for cure color development. The cocci also play an important role in flavor and aroma development. In France, Italy, Spain, Hungary, and the Balkan countries these traditional raw fermented sausages have had higher pH values (e.g., >5.8) and flavors that are more balanced, complex, and non-acidic. The trend toward using starter cultures and slightly higher fermentation temperatures to reduce processing times is accompanied by certain negative features. Specifically, the rapid production of acid renders the cocci inactive and, thus, incapable of reducing nitrate or influencing flavor development. In Spain, loss of the traditional flavor and increased acidity has led to a decrease in consumer acceptance and declining sales. Several reviews, published since 1987, describe the procedures used and changes occurring in Europe (Incze, 1987, 1989 and 1992; Nagy et al., 1989; Nychas and Arkoudelos, 1990; Lücke, 1994; Flores and Bermell, 1996; Flores, 1997; Fernandez et al., 2000). It is curious that very little information has been published during that period from researchers in North America. The basic parameters of raw fermented sausage processing in Europe are summarized from Incze (1992) in Table 7.2.

Several studies pertaining to the manufacture of fermented meats will be summarized. In pursuit of more effective starter cultures for use in manufacturing dry sausage, Wolf et al. (1990) conducted further investigations on nitrite reductase activity among the lactic acid bacteria. Tests on 70 strains of lactic acid bacteria using cell suspensions incubated anaerobically led to two groups. One group, consisting of *L. plantarum*, *L. pentosus*, and *Pediococcus pentosaceus*, was hematin-dependent for nitrite reduction and yielded ammonia as the sole product. The other group was able to reduce nitrite in the absence of hematin and yielded nitric oxide and N_2O. Subsequent research found the lactic acid bacteria could be separated into heme-dependent and non-heme-dependent catalase producers (Wolf et al., 1991). Tests by

TABLE 7.2

Basics of Raw Fermented Sausage Processing in Europe

Processing Characteristics	High-Acid Sausage	Low-Acid Sausage
Fermentation	22–25°C	<12°C
Drying/ageing	15–18°C	14–16°C
Final pH	<5.3	~6.0
Added carbohydrate	0.3–0.7%	None
Added nitrate/nitrite	Nitrite and, perhaps, nitrate	Nitrate
Added starter culture(s)	Yes	Not necessary
Product characteristics	Salty, strongly acidic flavor; softer, rubbery, or spreadable texture	Salty, round, rich, non-acidic flavor; firm texture

Source: Incze (1992).

Mares et al. (1994) found that sodium nitrite (80 mg/kg) reduced the catalase activity of a strain of *L. plantarum* but not *P. acidilactici* or *L. pentosus*. Scannell et al. (2001) found the growth of two strains of *Lactococcus lactis* and production of the bacteriocin, lacticin 3147, was inversely related to the added level of sodium nitrite (20, 50, and 100 mg/kg) in salami.

In Europe, nitrate has traditionally been used for dry sausages and hams requiring long ripening times. In these processes, micrococci convert the nitrate to nitrite, which is responsible for the reddening and other effects. In modern processes, the rate of fermentation is faster and micrococci become inactive due to an accumulation of acid. In studies involving faster fermentation, over 50% of the initial nitrite disappearance was attributed to conversion to nitrate. Color stability was mainly affected by the presence of residual added ascorbate (Alley et al., 1992).

Neubauer and Götz (1996) summarized nitrate utilization as being either assimilation or respiration. In assimilation, which may occur aerobically or anaerobically, nitrate is reduced to ammonia that is then incorporated into the cell and used as a source of nitrogen. In respiration, nitrate is used as an alternative electron acceptor when oxygen is not present. Two main forms have been described for respiration and, in both, nitrate reduction is coupled to the generation of a proton motive force that is directly used as a source of energy or transformed into ATP. Members of Enterobacteriaceae reduce nitrate to nitrite, which is either then excreted or further reduced to ammonia. Obligately respiring bacteria (e.g., pseudomonads) reduce nitrate to gaseous oxides of nitric oxide and nitrous oxide, which may be further reduced to nitrogen. Certain bacteria, such as strict aerobes, are able to perform assimilatory nitrate reduction but cannot use nitrate as an electron acceptor. Other bacteria, such as *Salmonella* Typhimurium, do not assimilate nitrogen through nitrate reduction during aerobic growth but grow anaerobically with nitrate as the terminal acceptor. In contrast, certain other bacteria use both assimilatory and respiratory nitrate reduction.

Neubauer and Götz (1996) investigated the nitrate reducing system of *Staphylococcus carnosus*, which is used commercially as a starter culture for fermented sausages and dry-cured ham. They found *S. carnosus* reduces nitrate to ammonia in two steps. First nitrate is taken up, reduced to nitrite, and excreted. Second, after nitrate is depleted the accumulated nitrite is taken up and reduced to ammonia. Synthesis of the nitrate and nitrite reductases is inhibited by oxygen and induced to greater or lesser degrees by nitrate or nitrite, respectively. A study of cell suspensions from seven strains of staphylococci isolated from dry sausage (six) and cheese (one) found all but one (*S. warneri*) was able to synthesize nitrate reductase (Talon et al., 1999). Adding nitrate to the growth medium favored the synthesis of catalase, particularly in *S. carnosus*. *S. warneri* had previously been reported to have no nitrate reductase activity and a low catalase activity and when inoculated into sausage caused the product to become rancid (Berdagué et al., 1993).

A new class of non-fermented raw dried sausages of small diameter that has gained popularity in countries such as Spain and Italy is made with nitrate and/or nitrite, but nitrate alone being the more common (Sanz et al., 1998). A study comparing sausages made with only nitrate or nitrite and a 26-day process at temperatures that did not exceed 15°C found the aroma and flavor to be more acceptable when the product was made with nitrate.

Parma and San Daniele hams made in Italy, Karst ham made in Yugoslavia, and Savoie ham made in France involve long ripening times and are made without any added nitrate or nitrite (Leistner, 1986). Morita and co-workers investigated the red pigment found in Parma ham and its formation. Morita et al. (1994) found that *S. carnosus* and *S. caseolyticus* isolated from fresh pork were able to convert metmyoglobin to red myoglobin derivatives. Subsequently, Morita et al. (1996) demonstrated that the red pigment found in Parma ham was much more stable and different from nitrosomyoglobin, the pigment found in hams made with nitrate and/or nitrite. Microbiological examination of 471 isolates from Parma ham found the majority was able to convert metmyoglobin to a red myoglobin derivative. Of the ten strains selected for further study all were staphylococci (*S. epidermidis*, *S. warneri*, and *S. lentus*). Tests with two strains of *S. lentus* in a sausage mix without nitrite or nitrate found both to produce a pigment in the meat that had the same spectral pattern as the pigment found in Parma ham. Morita et al. (1998) then evaluated a strain of *S. xylosus* and found it to produce a red myoglobin derivative in the same meat system without any added nitrate or nitrite. Spectral analysis and electron spin resonance studies determined the pigment was nitrosomyoglobin, the red pigment that is associated with products made with nitrate or nitrite. The mechanism for the formation of the pigment in the absence of nitrate and nitrite was not determined.

7.2.6.2 Alternatives to the Use of Nitrate or Nitrite

Concern for the potential toxicological effects of nitrite and nitrate led to continued research to find an alternative. Among the options considered was the use of a red pigment producing mold used to ferment rice, *Monascus purpureus* (Leistner et al., 1991; Fink-Gremmels et al., 1991; Park et al., 1999; Tseng et al., 2000). In the Far East, particularly in China, rice is fermented with molds of the genus *Monascus* to form a pleasant red color. For more than 1500 years the rice has been dried, ground, and, in this form (called Angkak), added to a variety of foods (e.g., pork, poultry, fish, tofu, rice wine) to improve color, flavor, and keeping quality (Leistner et al., 1991).

7.2.6.3 Cheese

Nitrate has been added during the manufacture of certain European cheeses for over 150 years (Pivnick, 1980). Although not permitted in the United States, Canadian regulations permit this practice (Assembly of Life Sciences, 1981). This chapter will not review the extensive literature on the use of nitrate to prevent cheese defects (e.g., late fermentation and gassiness due to *Clostridium tyrobutyricum* or *Clostridium butyricum*). An introduction to the literature on this subject can be obtained through references cited in Langsrud and Reinbold (1974), Assembly of Life Sciences (1981), and others (Mayenobe et al., 1983; Galesloot and Hassing, 1983; Gilles and Fryer, 1984; Ariga et al., 1984; Gouda and El-Zayat, 1988; Klijn et al., 1995).

7.2.6.4 Seafood

The comprehensive review by Jarvis (1950) provides very few examples for the use of nitrate (i.e., saltpeter) in the curing of fishery products prior to 1950 throughout the world and virtually no mention is made of its value as a preservative. Knøchel and Huss (1984b) state that nitrate traditionally has been added to the salt used in producing sugar-salted fish in Scandinavia. In Denmark, for example, the allowable level has been 500 mg potassium nitrate per kg fish to provide about 300 mg/kg nitrate. They conducted an extensive study in which 0, 25, or 50 g potassium nitrate was added to each of 200 barrels containing 90–100 kg fish, 16 kg salt, and 6 kg sucrose (Knøchel and Huss, 1984a, b). The 600 barrels were topped off with brine solution, sealed, and evaluated over 18 months' storage at 4–6°C. Adding nitrate caused the herring to be redder in color but no differences were noted for flavor or texture. As the amount of sodium nitrate increased, fewer barrels were spoiled with sour, sweet/sour, fruity, or putrid off-odors and

-flavors. The incidence of spoilage was 13, 8, and 2 barrels, respectively, with 0, 25, and 50 g of sodium nitrate. The reduction in spoilage was not considered to be due to microbial inhibition by nitrate or its reaction products (nitrite, nitric oxide, nitrous acid) since a high percentage of the isolates from all three variables were able to reduce nitrate. Nitrate was found to delay the depletion of trimethylamineoxide (TMAO). Nitrate and TMAO serve as alternative electron acceptors in barrel-cured herring and when in sufficient concentration delay growth of the strict anaerobes that are responsible for the most common type of spoilage. The reader may recall that during the 1950s nitrate was found to be an electron acceptor in certain cured meat products.

Two microbial hazards of concern in ready-to-eat fish products are *C. botulinum* and *L. monocytogenes* (Gram, 2001a, b). Cuppett et al. (1987) found that toxin production by *C. botulinum* type E (10^3 spores/g) at 27°C was inhibited for 35 days in smoked whitefish containing 4.4% salt in the water phase. When sodium nitrite (156 mg/kg) and sodium ascorbate (550 mg/kg) were added, toxin production was inhibited for 56 days. The additives provided no additional protection when the salt level in the fish was reduced to about 2.3%.

Hyytiä et al. (1997) found the addition of sodium nitrite delayed or prevented toxin production by *C. botulinum* type E in vacuum-packed cold-smoked trout having 3.4% salt in the water phase. In one study with 166 mg/kg sodium nitrite toxin was detected at 4 weeks at 4°C with a high inoculum level (20,000 cfu/kg) and after 4 weeks with a low inoculum (800 cfu/kg). In a second study with 109 mg/kg sodium nitrite and lower inoculum levels (4000 and 140 cfu/kg) toxin was not detected through 6 weeks at 4 or 8°C; whereas with salt alone toxin was detected in 3–4 weeks. Adding potassium nitrate at 347 or 686 mg/kg was equal to or more effective than sodium nitrite. Analysis demonstrated that nitrite was produced during storage. The addition of nitrite or nitrate increased the time the fish was considered acceptable by 3 or 4 weeks at 4°C.

With the availability of lower-cost farm-raised salmon the international market for cold-smoked salmon has continued to increase. In 1992, more than 32,000 tons shipped internationally from 28 exporting countries to 52 importing countries (Duffes, 1999). In many countries there is no legal requirement for the minimum amount of salt, although a minimum of 3% salt in the water phase is strongly recommended (Huss et al., 1995). A level of 3% salt in the water phase will yield perfectly acceptable lightly preserved fish products (cold-smoked fish, gravad fish, brined shrimp), and with refrigeration this combination is effective in inhibiting growth of *C. botulinum* (Huss, 1997). In France with a production of 12,380 tons of cold-smoked salmon in 1997, the salt in the water phase can range from 2 to 4.6% (Duffes, 1999). Considering the lack of regulations and apparent variability in salt concentration in large quantities of cold-smoked salmon distributed domestically and internationally, it would appear the risk of botulism is lower than the research has suggested.

Two options are available in the event commercially manufactured product is implicated in botulism. First, a minimum concentration for salt in the water phase can be established. Controlling the salt in the water phase and temperature should be adequate but, if necessary, sodium nitrite could be used as an additional control measure. Until there is evidence of a need, it is highly unlikely that sodium nitrite will be adopted as an additive in smoked fish in Europe. In response to outbreaks of botulism in the US in the early 1960s, when less information was available about commercial practices and from research, federal regulations were adopted requiring vacuum-packed, smoked fish to contain at least 3.5% salt in the water phase or 3.0% salt if combined with 200 mg/kg sodium nitrite (Gram, 2001a). If packaged aerobically, a lower salt content is permitted (i.e., 2.5% salt in the water phase), however, questions remain about "aerobic packaging" and its value for controlling botulinal outgrowth. The addition of sodium nitrite is limited to smoked salmon, sablefish, shad, and chub.

Following a review of the literature a panel of experts reached the following consensus for its report to the U.S. Food and Drug Administration (Gram, 2001a):

- Psychrotrophic *C. botulinum* can be expected to occur in smoked fish although the numbers, if present, are low.
- Toxin production can be controlled with a combination of a moderate level of salt (3.5% in the water phase) and cold storage (<4.4°C) for at least 4 weeks.

Furthermore, that combination of salt and temperature, allowing for short periods of elevated temperatures up to 10°C, will prevent toxin production in cold-smoked fish for several weeks beyond its sensory shelf life when packaged under reduced oxygen conditions.

The panel was reluctant to recommend expanded use of sodium nitrite despite evidence showing that psychrotrophic *C. botulinum* would be further inhibited, particularly in combination with controlled levels of salt (Pelroy et al., 1982; Cuppett et al., 1987; Hyytiä et al., 1997). The panel expressed concern for the potential carcinogenic effects of nitrosamines, but the panel did not consider the more recent risk assessments demonstrating the safety of dietary nitrite (see section on toxicology). *N*-nitroso compounds were not detected in two studies in which nitrite had been added to smoked fish (Pelroy et al., 1982; Cuppett et al., 1987).

Surveys conducted before the year 2000 indicate *L. monocytogenes* is common in smoked fish. This may reflect survival or contamination depending on the conditions of processing. Following its review of the literature the panel of experts concluded that it is not possible to consistently produce cold-smoked salmon that is free of *L. monocytogenes* (Gram, 2001a). It is possible, however, with strict adherence to GMPs to produce cold-smoked salmon with low levels, often less than 1 cfu/g and at a low prevalence. Such levels would ensure the number does not increase above 100 cfu/g when the product is held at 5°C and eaten within 3 to 4 weeks. Due to the relatively high tolerance of *L. monocytogenes* to salt and the adverse effect on flavor, increasing the salt concentration above the normal range for smoked fish is an unacceptable option for control. Peterson et al. (1993) found the rate of growth to be similar at 5°C for salt concentrations of 3 to 5% in the water phase. In subsequent research, adding 125 mg/kg sodium nitrite resulted in a slower rate of growth in vacuum-packed salmon having 3% salt in the water phase (Pelroy et al., 1994a). Sodium lactate (2%, w/w) was much more effective and prevented growth through 50 days at 5°C when used in combination with 3% salt. Pelroy et al. (1994b) found the addition of sodium nitrite (190–200 mg/kg) to be very effective in slowing the growth *L. monocytogenes* in salmon containing 3% salt in the water phase. This combination was more effective when a low inoculum level (10 cfu/g) was used as a challenge. Previous studies had found levels of 0.5 to 11.7 cfu/g in positive samples of cold-smoked fish from 5 plants. The initial level of contamination is an important factor influencing growth (Guyer and Jemmi, 1991; Peterson et al., 1993; Pelroy et al., 1994b; Rørvik, 2000).

7.2.6.5 Effect of Nitrite on Enteric Pathogens

Blanche Koelensmid and van Rhee (1974) observed the growth of three of four *Salmonella* strains at 20°C in ham jelly containing 500 mg/kg of sodium nitrite and 5.6% salt. At 5°C nitrite accelerated the death of all four strains.

Meers and Goode (1965) reported the rapid growth of *Salmonella* Typhi in inoculated canned corned beef stored at room temperature and at 37°C. After reaching 10^7/g, there was a gradual decline; however, viable cells remained even after 6 months of storage. Cans inoculated with *E. coli* remained flat; cans inoculated with *C. perfringens* swelled within 2 days at 37°C. The authors implied that, since *S. typhi* has a nitrase, it utilizes the nitrate in the product in place of oxygen as an alternative hydrogen acceptor; however, tests were not performed to show whether the presence of nitrate influenced growth, nor were analytical data presented for the product tested. Kittaka and Greenberg (personal communication) obtained similar results in corned beef inoculated with three different *S. typhi* strains, including an isolate from the 1964 Aberdeen outbreak involving canned corned beef. Product from two sources, one domestic and one imported, was used for testing. The domestic product contained 4.13% brine, 14 mg/kg of sodium nitrite, and 720 mg/kg of sodium nitrate at the time of opening and inoculation; the imported product had 4.26% brine, 15 mg/kg of sodium nitrite, and 45 mg/kg of sodium nitrate. Growth occurred in both products. A lower maximum population (10^6 vs. 10^8/g) and slower die-off occurred in the imported product. Viable *S. typhi* remained after 20 months at 80–100°F.

Goepfert and Chung (1970) inoculated the surface of sliced bologna and liver cheese with *S.* Typhimurium or *Salmonella* Anatum. In both products, *S.* Typhimurium declined during the first 2 weeks at 5°C and then remained stable through 6 weeks. *S.* Typhimurium grew in both vacuum-packaged products when held at room temperature. *S.* Anatum grew in bologna but not in liver cheese. The growth of *S.* Typhimurium occurred more rapidly and to higher levels when the packages were opened, rewrapped in Saran, and held at room temperature. Dehydration prevented growth if the packages were opened and left exposed at room temperature. Chemical data were not provided for either luncheon meat. It is reasonable to assume the products were recently produced.

Davidson and Webb (1973) also studied the growth of *S. typhimurium* in vacuum-packaged luncheon meats. Survival of the inoculum in product held at 7°C varied with the product and inoculum level. Mock chicken supported the best growth at both temperatures tested (24 and 37°C). Very erratic growth occurred on bologna at 18°C, but a steady increase in numbers occurred at 24°C. Similar results were obtained on ham. The inocula died in wieners at both 18 and 24°C, although not before some initial growth at 24°C. At the end of storage, the pH of mock chicken and bologna was about 6.0, that of the ham was near 7.0, and the pH for the wieners was 4.7–4.8. No other chemical data were provided.

Leistner et al. (1973) inoculated vacuum-packaged luncheon meat with a mixture of *Salmonella* strains and held the product at 8°C for 22 days. The rate of growth or death was influenced by the sodium nitrite content (0, 50, 75, and 100 mg/kg) and the level of inoculum (100 or 100,000 CFU/g). A similar response occurred with enteropathogenic *E. coli*, but a mixture of *Enterobacter*, *Klebsiella*, and *Hafnia* strains readily multiplied at all nitrite levels. Fermented sausage (teewurst) was prepared with four sodium nitrite levels (0, 65, 98, and 130 mg/kg) and two *Salmonella* inoculum levels (100 and 100,000). Growth occurred in the sausage containing 0 and 65 mg/kg of nitrite. Death occurred after 3 days in the sausage with the lower inoculum and 130 mg/kg of nitrite. The growth of *Klebsiella* and *Enterobacter* sp. on vacuum-packaged luncheon meats at 3°C was reported earlier by Hechelmann et al. (1974). Both genera were prevalent in commercially produced luncheon meats and fermented sausages.

The resistance of *Salmonella* to nitrite in culture media was demonstrated by Roberts and Garcia (1973). Neither of two strains of *S.* Typhimurium were inhibited at pH 6.0 by 160 mg/L of sodium nitrite, whether the nitrite was added before or after autoclaving the medium of Perigo et al. (1967). Castellani and Niven (1955) reported *S.* Typhimurium to be resistant to nitrite (2000–4000 mg/L) when added as a filter-sterilized solution to a broth medium having a higher pH of 6.6–6.7.

The trend in the United Kingdom toward using less salt for processing ham was studied by Akman and Park (1974). A low-brine ham was prepared with 1.8–2.2% salt and 60 mg/kg of potassium nitrite and a finished brine level of about 2.8%. A high-brine ham had a brine level of about 6%; nitrite values were not given. In the low-brine ham at 22°C, four of five strains of *Salmonella* grew readily. *Salmonella* Enteritidis was able to grow at 17°C but not at 10°C. In the high-brine ham held at 22°C, viable counts of the five strains remained near the inoculum level. Two strains of *S.* Enteritidis were then used to study the effect of nitrite and salt upon growth in nutrient broth at 22°C. At a low pH of 5.0, 5% salt, and 100 mg/kg of sodium nitrite, growth was reduced by only about 30%. At a high pH of 7.0 and 5% salt, nitrite was without effect. The authors did not state whether the nitrite was added before or after autoclaving the medium. It was concluded that cooked ham is a more likely source of salmonellosis if prepared with a low salt content.

Stiles and Ng (1979) studied ham and chopped ham obtained fresh from two processors. The products were inoculated with five "enteropathogens" (*C. perfringens*, *E. coli*, *S.* Typhimurium, *S. aureus*, and *B. cereus*), and then vacuum packaged. The moisture content of the hams was 70%; that of the chopped hams was 70 and 59%. Although salt and nitrite values were not reported, pH values were followed through storage. The mean pH for the ham during storage decreased to no lower than 5.58. The pH of the chopped ham with 70% and 59% moisture decreased to 5.47 and 5.78, respectively. All the enteropathogens grew in one or more of the products within 24 h at 30°C. The primary factors influencing growth were the temperature of storage and growth of competitive microbiota. It is not possible to assess the influence of nitrite from the data, other than to say that fresh, commercially produced sliced ham and chopped ham did support the growth of the pathogens studied, including *Salmonella*, given the proper circumstances.

Stiles et al. (1979) tested the same bacteria on sliced bologna from two processors. Again, no data were provided for brine or nitrite levels, but pH changes were noted. Growth of *C. perfringens* did not occur under any condition. The other four species were able to grow under certain conditions. Aside from temperature of storage, the primary factor influencing growth was product pH. The product from one supplier remained above pH 6.0, whereas the other frequently decreased to below pH 5.5. This probably reflects the quantity of fermentable carbohydrate added to the product and/or the nature of the competitive microbiota.

Page and Solberg (1980) found *S.* Typhimurium to utilize nitrate, nitrite, and ammonia as the sole nitrogen source in a chemically defined minimal medium. Nitrite and ammonia were assimilated at comparable rates. Under anaerobic conditions, the fastest generation times occurred in media containing both nitrate and ammonia. *S.* Typhimurium utilized nitrite as the sole nitrogen source at concentrations as high as 400 mg/L. At concentrations of 500 mg/L and above, complete inhibition occurred.

Page et al. (1985) identified a nitrite reductase in the cytoplasmic fraction of *S.* Typhimurium when grown anaerobically with nitrite as the sole nitrogen source. The reductase required NADH and was most active at pH 8.0. When grown anaerobically in a glucose-limited minimal medium with peptone and nitrite, *S.* Typhimurium had shorter generation times and increased cell yields compared with nitrite-free cultures. The presence of nitrite had no effect on aerobically grown cultures.

Gibson and Roberts (1986a) examined the interaction of salt (1–10% wt/vol), pH (5.6, 6.2, and 6.8), and sodium nitrite (0–400 mg/L) in brain–heart infusion broth and incubation at 10–35°C on the growth of *Salmonella* and *E. coli*. The results demonstrate that these bacteria are quite resistant to the effect of salt and nitrite at the levels most commonly present in such foods as perishable cured meats and smoked fish. Thomas et al. (1991), using a gradient gel plate system, found *S.* Typhimurium to be relatively resistant to sodium nitrite across the ranges of salt and pH values found in nonfermented and acidulated foods. The nitrite resistance of *E. coli* was also apparent in tests involving two-dimensional gradient plates incubated at 20, 27, and 37°C (McClure and Roberts, 1987).

Hughes and McDermott (1989) used a procedure similar to that of Gibson and Roberts (1986a) to determine the additional effect of eight different phosphates of *E. coli*. The phosphates generally delayed the outgrowth of *E. coli*. When examined after 10 weeks of incubation, the effect of added phosphate was relatively minor, except when the cultures were incubated at the lowest temperature tested (10°C).

Rice and Pierson (1982) tested the effect of sodium nitrite (0, 50, and 156 mg/kg) in frankfurters on a mixture of two strains of *Salmonella*. The franks were inoculated on the surface, vacuum packaged, and incubated at 15 or 27°C for 21 days. Nitrite was only inhibitory with the highest level of nitrite and when the franks were incubated at 15°C.

Collins-Thompson et al. (1982) were able to isolate Gram-negative bacteria from commercial packages of vacuum-packaged bologna only after storage for 7–8 weeks at 5°C. Isolates of *Enterobacter* and *Serratia* species were tolerant to nitrite, being able to utilize nitrite in broth culture under anaerobic conditions at 5°C.

The rate of death of *Salmonella* Senftenberg in fermented sausage was fastest with 200 mg/kg of sodium nitrite and next fastest with 100 mg/kg of sodium nitrite (Puolanne, 1977). The death rate was slower with potassium nitrate (150 or 300 mg/kg) or a combination of nitrate and a low level of nitrite (75 mg/kg of potassium nitrate and 50 mg/kg of sodium nitrite). He recommended that a starter culture and/or gluconolactone be added to provide optimum safety.

Collins-Thompson et al. (1984) prepared fermented sausages with sodium nitrite (0, 50, 100, 150, and 200 mg/kg) and three strains of *Salmonella*. Growth did not occur during fermentation or drying. The presence of nitrite did not influence the survival of Salmonellae. Similar results were observed with *S. aureus* and *C. sporogenes*.

Schillinger and Lücke (1989) measured *Salmonella* growth and/or survival in mettwurst, a cured raw spreadable sausage. Factors influencing *Salmonella* growth in mettwurst that was cured and stored at 20°C include the amount of nitrite and salt, the inherent pH of the meat (normal pH 5.6–5.8 vs. 6.0–6.2), the addition of sugar, and the presence of lactic acid-producing bacteria. This research was primarily concerned with preventing the growth of *Salmonella*. None of the variables tested would likely assure the destruction of *Salmonella* in this product category.

S. Typhimurium numbers decreased in BHI broth adjusted to pH 4.8 and held at 24°C independent of sodium nitrite concentration. At pH 5.8 numbers increased during 72 hours in the presence of 0, 50, and 100 mg/kg sodium nitrite but growth was delayed with 200 mg/kg (Turantaş and Ünlütürk, 1991). In a traditional Turkish dry sausage that is fermented and dried at temperatures not exceeding 24°C the death rate of *S.* Typhimurium was not affected by the addition of 200 mg/kg sodium nitrite (Turantaş and Ünlütürk, 1993).

The data for *Salmonella* can be summarized as follows. *S.* Typhimurium has a nitrate reductase to reduce nitrate to nitrite and can assimilate inorganic nitrogen from nitrate, nitrite, or ammonia under anaerobic conditions. The data indicate that the nitrite content of cured meats can influence the growth and survival of *Salmonella* and *E. coli*. That cured meats have been implicated in outbreaks of salmonellosis supports the available research that *Salmonella* can survive, if not multiply, in a variety of commercially cured products. Preventing salmonellosis from cured meats is best accomplished by means other than relying on the level of nitrite in the product, particularly since salmonellae are destroyed during the manufacturing process for most cured meats. This is not true for the clostridia.

Inhibition of *E. coli* O157:H7 in a broth system containing sodium nitrite was strongly pH-dependent (i.e., pH 5.5) and enhanced by lowering the incubation temperature (Buchanan and Bagi, 1994). In tryptic soy broth 4 strains of *E. coli* O157:H7 at 37°C were inhibited by 200 mg/L sodium nitrite at pH 5.0 but even 1 g/L sodium nitrite was ineffective at higher pH values (Tsai and Chou, 1996). In ground pork held at 5 or 25°C sodium nitrite at 70–150 mg/kg did not affect the growth or survival of *E. coli* O157:H7 (Yu and Chou, 1996a). The effect of nitrite in Chinese-style sausage during drying at 50–60°C for different time periods (2.5–6 hours) was next studied (Yu and Chou, 1997). This sausage is not fermented or cooked and relies on relatively rapid dehydration to achieve microbial stability. They found some increased death of *E. coli* O157:H7 with the addition of curing agents and drying at time–temperature combinations that appear to cause stress of the inoculum. Adding sodium nitrite did not increase death of *E. coli* O157:H7 when lower temperature and/or shorter time combinations (e.g., 55°C for 4 h) were used. Riordan et al. (1998) found *E. coli* O157:H7 numbers decreased less than 1-$\log_{10}$ in pepperoni formulated with 2.5% salt, 100 mg/kg sodium nitrite, fermented to pH 4.8, and then dried for 7 days. In Lebanon bologna, a fermented non-cooked beef sausage, the addition of sodium nitrite (0, 78, and 156 mg/kg) had no effect on the death rate of *E. coli* O157:H7 (Chikthimmah et al., 2001).

Inhibition of *Shigella flexneri* in broth incubated aerobically was significantly influenced by the combined effect of nitrite concentration, salt level, pH, and temperature. The inhibitory effect of nitrite was much greater at pH 5.5, the lowest pH tested and with increasing salt levels. Inhibition was much greater at 28°C than at 37°C (Zaika et al., 1991). Another study involving anaerobic conditions yielded similar trends (Zaika et al., 1994).

Inhibition of *Y. enterocolitica* in broth incubated aerobically was significantly influenced by the combined effect of nitrite concentration, salt level, pH, and temperature (Bhaduri et al., 1994 and 1995). The effect of sodium nitrite (0, 50, 80, 100, and 120 mg/kg) on the growth and survival of *Y. enterocolitica* O:3 during the manufacture of raw dry fermented sausage was examined by Asplund et al. (1993). The inoculum decreased from 1.7×10^5/g to nondetectable levels by 35 days in the sausages formulated with 80 mg/kg nitrite. The highest survival involved a starter culture that produced the least amount of acid and yielded a relatively high final pH of 5.5. Raccach and Henningsen (1997) reported sodium nitrite and salt to act synergistically in the reduction of *Y. enterocolitica* when tested in a meat system.

Inhibition of *Aeromonas* in broth has been found to be highly dependent on nitrite concentration, pH, and incubation temperature (Chou and Tsai, 1996; Pin et al., 1996).

7.2.6.6 Effect of Nitrite on Yeasts and Molds

A wide variety of yeasts and molds have been isolated from commercially cured meats, but there is no published information that nitrite or nitrate inhibits their growth. The lack of information on the significance of nitrite and nitrate stems from the general acceptance that other factors largely determine the potential for yeast and mold spoilage of cured meats. These factors include water activity, the availability of oxygen, the inhibitory effect of smoke, storage temperature, and, where permitted, the use of preservatives (e.g., sorbate). In addition, yeasts and molds are of little or no public health importance. Except for a limited number of products to which nitrite and nitrate might be added, they are not major causes of spoilage.

The effect of nitrite on aflatoxin production by *Aspergillus parasiticus* in fresh pork sausage was examined by Obioha et al. (1983). During storage at 5°C for 28 days, no aflatoxin production or mold growth occurred. At 26 and 37°C, the product spoiled before aflatoxin reached detectable levels. Despite this observation, the results in pork sausage plus a test with media led to the conclusion that high levels of nitrite (e.g., 156–200 mg/kg) initially restricted the profuse growth of mold. As residual nitrite decreased to less inhibitory levels, the remaining nitrite enhanced growth and aflatoxin production.

7.2.6.7 Green Discoloration in Cured Meats

Niven et al. (1949) described the characteristics of *Lactobacillus* and *Leuconostoc* strains isolated from commercially cured meats with greenish discoloration. All were salt-tolerant and capable of growth at low temperatures (5°C). Evidence was presented indicating that the discoloration results from a reaction of hydrogen peroxide with the cured meat pigment. Niven et al. (1954) demonstrated that the heterofermentative lactobacilli responsible for greening on the surface of cured meats were more sensitive to heat than those developing in the center of sausages. Increased heat resistance could be acquired by repetitive exposure to thermal processing. The name *L. viridescens* was given to these bacteria (Niven and Evans, 1957). The mechanism of green discoloration was further elucidated by Shank and Lundquist (1963).

Nitrite burn is another form of green discoloration that is caused by a combination of excessive levels of nitrite and reduced pH (Deibel and Evans, 1957). High levels of nitrite can result from the conversion of nitrate to nitrite by micrococci near the periphery of sausage or by *Staphylococcus* sp. within the sausage (Bacus and Deibel, 1972). Control of these bacteria lies in proper handling of cooked product to be reworked into succeeding lots of meat. These bacteria must not be given the opportunity to develop a population of increased heat resistance whereby the process will no longer be adequate to assure their destruction. The heat resistance of these bacteria has been reported (Niven et al., 1954; Iwata et al., 1965; Gardner, 1967; Vrchlabsky and Leistner, 1971; Milbourne, 1983). Unpublished research by Shaparis and Christiansen has shown D values in minced ham to be $D_{63°C}$ = 30.5, $D_{65.5°C}$ = 12.5–15, $D_{68°C}$ = 5.5–6.5, and $D_{71°C}$ = 1–2 min (Tompkin, 1986). Additional information on factors affecting green discoloration by *L. viridescens* and other lactics has been developed by Grant and McCurdy (1986) and Grant et al (1988).

7.2.6.8 Summary for 1990–2002

1. The most significant development during this period was the growing body of data that dietary nitrate and nitrite are not carcinogenic (see Toxicology).
2. Naturally occurring nitrate in vegetables accounts for over 85% of the dietary intake of nitrate. Conversion of nitrate in the saliva is the major source of nitrite in humans.
3. Endogenously formed nitric oxide and oxygen radicals can play an important role in human health.
4. Nitrite can contribute to the inhibition of *L. monocytogenes* under certain circumstances, but not all.
5. A system was developed to evaluate the combined effects of formulation, heat, inoculum level and other factors for controlling mesophilic clostridia.
6. The role of nitrate and nitrite in the manufacture of traditional and more rapidly produced fermented meats was clarified. The more rapidly produced products closely resemble those that have been produced in the US over the past 30–40 years.
7. The cured ham appearance of traditional European dry-cured hams manufactured without the addition of nitrate or nitrite was found to be due to the formation of stable pigments by naturally occurring staphylococci.
8. The use of nitrate or nitrite as additives in cheese and seafood products was summarized.
9. Nitrite is of relatively little value for controlling Gram-negative enteric pathogens in commercially prepared foods.

7.2.7 2002–2019

7.2.7.1 Mechanisms of Nitrite Antimicrobial Activity

In the years since 2002, more details regarding how nitrite inhibits microbial activities have been pieced together, although a complete understanding remains elusive.

While the central roles of nitric oxide and reactive nitrogen species in nitrite's antimicrobial activities have been understood for some time, recent progress in the identification of the molecular targets of nitric oxide and other reactive nitrogen species has been made. Nitric oxide reacts with many molecules, including metals (e.g., iron, zinc), organic radicals, cysteine, and tyrosine, and can react differently depending upon bacterial species and even strain (Meng et al., 2018; Richardson et al., 2011; Vázquez-Torres and Bäumler, 2016). Reaction with nitric oxide can disable the biological function of heme proteins, iron–sulfur proteins, protein thiols, and other redox-active centers, resulting in the disruption of oxygen transport and electron transfer reactions needed for cellular metabolism, thereby inhibiting bacterial growth (Reddy et al., 1983; Yarbrough et al., 1980). Reactive nitrogen species derived from NO also react via N-nitrosylation, S-nitrosylation, and disulfide formation with various molecules in the bacterial cell to inhibit bacterial growth (Hammes, 2012). Reactive nitrogen species (RNS) can chemically damage DNA and inhibit its repair (Schairer et al., 2012). The reactivity of NO and RNS with so many bacterial metabolic targets helps explain why nitrite (as its precursor) is such a broad-spectrum antimicrobial agent, showing activity against many different bacterial species (Richardson et al., 2011).

The importance of ascorbate and erythorbate in enhancing nitrite's antimicrobial activity has been more completely appreciated in recent years. As discussed earlier, ascorbate and erythorbate have been historically viewed as "cure accelerators" because their antioxidant activity hastens the conversion of nitrite to nitric oxide and shortens the time required to cure meat. The utility of these compounds to greatly reduce nitrosamine formation was also recognized and adopted by the meat industry and regulators (Cassens, 1990; Gray, 1976). Other work has suggested that the synergistic antimicrobial activity noted between nitrite and ascorbate (or erythorbate) is related to its ability to chelate iron, a necessary cofactor in the recovery of nitrite damage (Tompkin et al., 1978a). An enhancement of nitrite's inhibition of bacterial growth has been noted previously for ascorbate and erythorbate in many studies (for example, Henry, 1954; Ravouri, 1975; Tompkin 1978a). The movement towards using lower levels of nitrite (discussed below) in meat products has therefore led researchers to question "how low is too low" when reducing nitrite levels, with the corollary: "can we augment lower nitrite levels by supplying ascorbate?" In two studies covering a wide range of nitrite and erythorbate concentrations, the ability of defined levels of erythorbate to effectively compensate for lower nitrite levels in inhibiting the growth of *C. perfringens* in cured deli meats or ham was demonstrated (King et al., 2015a; Osterbauer et al., 2017). These studies led the USDA FSIS to assert that a lower concentration of nitrite (75 mg/kg) together with 500 mg/kg ascorbate represents an acceptable way of meeting the stabilization requirements when cooling or hot-holding fully or partially heat-treated meat and poultry products (USDA Food Safety and Inspection Service, 2018b).

The ability of ascorbate to reduce nitrite to nitric oxide and other short-lived nitrogen species means that residual nitrite will be depleted more rapidly in the presence of the antioxidant. The long-standing discussion over the relative importance of ingoing vs. residual nitrite has not yet been resolved. Several studies suggested that residual nitrite levels were important in controlling *C. botulinum* and *C. perfringens* outgrowth from germinating spores in cured meats, rather than inhibiting germination (Tompkin et al., 1978b; Christiansen et al., 1973; Duncan and Foster, 1968b; Taormina and Dorsa, 2004). However, EFSA concluded that ingoing nitrite rather than residual nitrite is most important in terms of inhibitory effect on microorganisms in fermented meat products (EFSA, 2003). A 2012 study found no correlation between residual nitrite or nitrate and *L. monocytogenes* growth in alternatively cured (using celery or other vegetable extract to generate nitrite) frankfurters and hams nor in *C. perfringens* growth in alternatively cured frankfurters (Sullivan et al., 2012b), but lower residual nitrate and nitrite was associated with greater *C. perfringens* growth in alternatively cured ham. More recently, both ingoing and residual nitrite have been proposed to play a role in the antimicrobial activity of nitrite in controlling *L. monocytogenes* growth in ready-to-eat meats (King et al., 2016). Because residual nitrite appears to be important in nitrosamine formation during frying of cured bacon (Gray, 1976), various strategies have been described to decrease residual nitrite levels in processed meat and poultry products (Sebranek and Bacus, 2007; Viuda-Martos et al., 2009). However, it remains unclear how strategies to reduce residual nitrite could impact food safety or quality.

Recent studies regarding how bacteria respond to nitrate, nitrite, and nitric oxide (obtained not only from food but also generated within the bodies of humans and animals) have provided insight at the molecular level as to how nitrite inhibits bacterial growth but also regarding how bacteria use nitrite and nitrate. Nitric oxide upregulates numerous proteins in bacteria which help them defend against the agent, including the flavohemoglobins and other globins under aerobic conditions (Bowman et al., 2011), while under anaerobic conditions, cytochrome C nitrite reductase and flavorubredoxin protect against nitric oxide (Mills et al., 2008). Other more global responses to nitrosative stress have been studied and reviewed in detail (Bowman et al., 2011; Cole, 2018; Muller-Herbst et al., 2016).

While most of this work (done largely *in vitro*) has been discussed with respect to nitric oxide and related compounds derived from the host, similar molecular responses of bacteria to these compounds obtained from the diet are expected. One study specifically examined the ability of resistance of *Salmonella* strains with mutations in genes encoding proteins involved in nitrosative stress responses (flavohemoglobin, cytochrome C nitrite reductase, and flavorubredoxin discussed above) to 150 mg/kg sodium nitrite within sausage (Mühlig et al., 2014). These mutants showed enhanced sensitivity to nitrite *in vitro* when compared to a wild-type strain. Surprisingly, within the sausage, none of the single mutants showed increased sensitivity to nitrite, perhaps because of the complex chemical environment within the sausage or because other multiple and potentially unknown additional systems also play a role in protecting bacteria against nitrosative damage (Mühlig et al., 2014).

Another study found that nitric oxide and reactive nitrogen species interfere with *Salmonella*'s response to acid, potentially adding to the host's innate gastric defenses against foodborne pathogens (Bourret et al., 2008). Reactive nitrogen species were demonstrated to block transcription and translation of signaling components needed within *Salmonella* to turn on its acid-tolerance response, although the timing and therefore the *in vivo* relevance of these effects are not clear.

Exposure of *S. enterica* serovar Typhimurium to 150 mg/L sodium nitrite at pH 5.5 resulted in intracellular acidification of the bacteria, accompanied by the induction of genes involved with nitrosative and other stress responses (Muhlig et al., 2014). Other genes involved in translation, transcription, and replication are down-regulated, resulting in slower growth of the bacteria (Muhlig et al., 2014). In the Gram-positive organism *L. monocytogenes*, acidified nitrite also resulted in profound transcriptional changes, including down-regulation of most genes involved in translation, DNA replication and repair, and cell wall/membrane biogenesis (Muller-Herbst et al., 2016).

At least one Gram-negative organism, *Shewanella oneidensis*, uses cytochrome C molecules in its first line of defense against nitric oxide (Meng et al., 2018); however, nitrite still appears to inhibit growth even in the absence of cytochrome C. *S. oneidensis* is unusual in that it is resistant to nitric oxide but sensitive to nitrite, which suggests that nitrite may work against slightly different targets than nitric oxide, at least in some bacterial species (Meng et al., 2018).

7.2.7.2 Antibotulinum Activity of Nitrite

Research on the behavior of *C. botulinum* in foods was severely curtailed after 2002, when the US government restricted possession, use, and transfer of the agent to protect against bioterrorism (Federal Select Agent Program., 2017). Hence, little work has been published since that time, with most new work in this area focusing on psychrotrophic *C. botulinum* rather than mesophilic strains or comparing the effect of nitrite derived from synthetic vs. naturally occurring sources (discussed below). In one study with nonproteolytic strains, Finnish researchers reported that uncured (no nitrite or nitrate added) sausages and ham stored at 8°C supported toxin production within 3 weeks, but no toxin was detected in any of the treatments stored at 8°C for 5 weeks when supplemented with 75 mg/kg sodium nitrite (Keto-Timonen et al., 2012). The effect of purified sodium nitrite was compared to nitrite added via cultured celery powder in frankfurters (Wanless et al., 2010). As expected, the addition of 156 mg/kg sodium nitrite and 550 mg/kg sodium erythorbate inhibited toxin production by *C. botulinum* for the duration of the 12-week storage at 10°C, but toxin was detected at 8 weeks in the treatment with 40 mg/kg nitrite and at 14 days for the treatment with no nitrite added.

Nitrite has also been explored to inhibit botulinum toxin production in the event of temperature abuse of other refrigerated foods. For example, no botulinum toxin was detected in Dijon pork (pH 6.0) supplemented with 80 mg/kg nitrite (derived from cultured celery juice powder) stored at 15°C for 8 weeks, but toxin was detected at 5 weeks in controls without nitrite (Golden et al., 2017a).

7.2.7.3 Nitrite Effects on *Clostridium perfringens*

The USDA's Pathogen Reduction and HACCP regulation finalized in 1996 set into motion procedures that require industry to demonstrate that their manufacturing procedures would consistently produce safe products. To assist in this process FSIS issued guidelines in 1999, one for cooking (Appendix A) and one for cooling (Appendix B) meat and poultry products (USDA Food Safety and Inspection Service, 1999). Furthermore, FSIS regulations require that processes used to manufacture meat and poultry products be subjected to a hazard analysis. When this is done for cooling perishable cooked meat and poultry products, the potential growth of *C. perfringens* is normally identified as a biological hazard of concern. The HACCP team must then consider options for preventing the multiplication of *C. perfringens* during cooling to levels that could cause illness. FSIS regulations also require that the conditions for controlling *C perfringens* growth during cooling be validated and proven effective. This can be done using a validated predictive model and by microbial challenge tests.

The ability of *C. perfringens* to cause foodborne disease has been known since the 1940s when the first outbreaks attributed to this pathogen were documented (Doyle, 2002). *C. perfringens* outbreaks have typically been associated with improper cooling of uncured cooked meat and poultry products in homes, restaurants, institutions, or at catering events. Commercially cured meats with at least 100 mg/kg added sodium nitrite have not been associated with *C. perfringens* outbreaks, although two *C. perfringens* outbreaks attributed to corned beef were reported in 1993 (Centers for Disease Control and Prevention, 1994).

Foodborne disease outbreaks linking this pathogen to improperly cooled RTE meat and poultry produced in federally inspected establishments are also virtually unknown (Kalinowski et al., 2003; Grass et al., 2013; USDA Food Safety and Inspection Service, 2017). Nevertheless, the potential for *C. perfringens* to grow in RTE meats after cooking in federally inspected establishments became an issue when the USDA FSIS issued proposed performance standards for certain meat and poultry products in 1999 (USDA Food Safety and Inspection Service, 1999), which were later expanded to include all RTE and partially cooked meat and poultry products (USDA Food Safety and Inspection Service, 2001).

The performance standards require processors to ensure that no more than 1 log CFU/g growth of *C. perfringens* occurs during cooling after cooking. This value was chosen to ensure that *C. perfringens* levels high enough to cause disease (6 to 7 log CFU/g food) would not be present at the time of consumption and was based on FSIS microbiological data that found some meat samples to contain 4 log CFU/g presumptive *C. perfringens* spores (McNamara and Lattuada, 1998; USDA Food Safety and Inspection Service, 2001; Golden et al., 2009). Other studies, however, suggested that these data overestimated the populations of heat-stable *C. perfringens* spores in meats (Golden et al., 2009; Taormina, 2018; Taormina et al., 2003). As a result, the 1-log growth limit for *C. perfringens* is considered by many to be overly conservative.

To assist processors in identifying safe processing conditions to limit *C. perfringens* growth, USDA issued specific guidance (Appendix B) for cooling different products (USDA Food Safety and Inspection Service, 1999). The guidelines stated that cured products (with a minimum of 100 mg/kg ingoing sodium nitrite) should be cooled from 54.4°C to 26.7°C within 5 hours and then from 26.7°C to 7.2°C within 10 hours, whereas products without nitrite are required to cool to within 1.5 and 5 hours, respectively. The use of a cure accelerator or salt level was not mentioned in the initial guidelines.

In response to the restrictive cooling limits identified in the 1999 USDA stabilization performance standards, a significant amount of work has been undertaken to understand better the ability of *C. perfringens* to grow in meat and poultry products, including products containing nitrite and other formulation components. These studies not only identified and refined alternative cooling programs that satisfy the requirement for *C. perfringens* control but also highlighted the effects of nitrite and the synergistic impact of cure accelerator (e.g., erythorbate or ascorbate), salt, pH, and moisture.

In the later updates, Appendix B specified curing with at least 100 mg/kg ingoing sodium nitrite (either from a purified or naturally occurring source) and 250 mg/kg sodium erythorbate or ascorbate, but allowed lower sodium nitrite levels (≥40 mg/kg sodium nitrite or its equivalent) as long as a higher brine (≥6%) concentration was used. The option of a 20-hour cooling period for fully cooked products with 120 mg/kg sodium nitrite and 3.5% brine condition was deleted. In 2017, manufacturers were given an option to use processes resulting in up to a 2 log CFU/g increase in *C. perfringens* if they demonstrated low baseline *C. perfringens* spore levels (e.g., 2 log spores/g) in their ingredient blends before cooking (USDA Food Safety and Inspection Service, 2017). Further revisions to Appendix B are anticipated pending additional research currently underway (Taormina, 2018).

In support of the efficacy of nitrite as a growth inhibitor, a 2003 study observed that populations of *C. perfringens* did not increase more than 1 log over the inoculated level in a variety of cured meat products (bologna, chunked ham, whole-muscle ham) during various chilling conditions (ranging from 4 to 10 hours cooling from 54.4 to 26.7°C; 4.5 to 20 hours from 26.7 to 7.2°C) (Taormina et al., 2003). The authors measured pH (range 5.9 to 6.3), water activity (range 0.965 to 0.985), and residual nitrite levels in the raw meat batter (range 103 to 140 mg/kg), but did not report levels of erythorbate, salt, or moisture (Taormina et al., 2003). A comparison of *C. perfringens* growth in uncured vs. cured turkey (156 mg/kg ingoing sodium nitrite; use of cure accelerator was not reported) found that during 6 hours of cooling between 54.4 and 7.2°C, *C. perfringens* growth was inhibited to <1 log in cured turkey, while the pathogen increased >2 logs in uncured turkey (Kalinowski et al., 2003). The combination of nitrite and 1 or 2% salt extended the lag phase of *C. perfringens* for 3 hours in cured meat stored at 43.3°C, but no growth was observed in cured product with 3% salt when stored at this severely abusive temperature for 24 hours. Similar results were obtained in another study in which *C. perfringens* levels in cooked cured ground pork decreased during the initial phase of cooling (from 54.4 to 26.7°C) followed by a slight increase during the last phase of cooling (Marquez-Gonzalez et al., 2012). However, no overall increase in *C. perfringens* numbers was observed over a 20-hour extended cooling period (from 54.4 to 7.2°C).

The levels of nitrite required to inhibit *C. perfringens* depend on salt (in moisture phase), acidity, and temperature. Nitrite present during initial meat processing steps (~156 mg/kg) inhibits the germination and outgrowth of *C. perfringens* spores (Juneja et al., 2006), although other reports suggest that higher levels of 300 mg/kg might be needed during severe temperature abuse (Crouch and Golden, 2005; Garde et al., 2014). Redondo-Solano et al. (2013) found that when the concentration of nitrite increased in ham (0 to 200 mg/kg) with or without sodium erythorbate, there was an increasing inhibitory effect on spore germination and outgrowth following cooking (i.e., heat shocking of the spores) and cooling from 54.4 to 7.2°C in 15 hours. Another study assessed the effects of various concentrations of nitrite with erythorbate against *C. perfringens. C. perfringens* growth in ham (75% moisture, pH 6.15, 2.2% salt) was limited to <1-log growth during extended cooling regimens of up to 25 hours with ≥150 mg/kg sodium nitrite and 250 mg/kg erythorbate (Osterbauer et al., 2017). The study also confirmed that higher pH levels (e.g., 6.3) and lower salt concentrations (e.g., 1.8%) allowed *C. perfringens* growth under the same cooling conditions.

These studies and others investigated the effect that reducing (i.e., antioxidative) agents such as erythorbate or its isomer, ascorbate, have on nitrite's inhibition of *C. perfringens.* In one report, 547 mg/kg erythorbate decreased the inhibitory effect of low nitrite concentrations (≤100 mg/kg) against *C. perfringens* in ham (Redondo-Solano et al., 2013). It was suggested this could possibly be due to an oxidative effect arising from residual oxygen present in the early meat processing steps which was blocked by the antioxidant effects of erythorbate (Redondo-Solano et al., 2013), although other explanations (such as the sequestering effect of iron in the meat) are possible. In a more recent study, ascorbate enhanced the activity of nitrite (either purified sodium nitrite or from a natural occurring source) against *C. perfringens* in deli-style turkey breast (King et al., 2015a). Subsequently, USDA FSIS stated that they will not object to a combination of 75 mg/kg nitrite and 500 mg/kg ascorbate from natural occurring sources to limit the growth of *C. perfringens* to ≤1 log CFU/g when fully cooked products are cooled between 54.6 and 26.7°C for no more than 5 hours and then between 26.7 and 7.2°C for no more than 10 hours (15 hours total cooling time) (USDA Food Safety and Inspection Service, 2018b).

Although the above recommendation was initially based on data from plant-based ("naturally occurring") sources of nitrite and ascorbate, these results extend to purified (synthetic) sources of nitrite and ascorbate as well. As will be discussed below, a number of studies have demonstrated that equivalent inhibition of pathogen growth occurs when equivalent levels of nitrite from natural vs. synthetic sources are used in foods. While alternatively cured (utilizing a natural occurring source of curing ingredients) vs. conventionally cured (utilizing purified sources) products did not demonstrate equivalent protection against *C. perfringens* growth during storage in several early studies of naturally (i.e. alternatively) cured products (Jackson et al., 2011b; Jackson et al., 2011a), the authors attribute the difference to variable concentrations of nitrite that were produced in the natural (i.e. alternative) curing process. More recently, synthetic and naturally occurring sources of nitrite were shown to be equivalent when used to control for *C. perfringens* during cooling of deli-style turkey meat (King et al., 2015a).

Consumer and regulatory demands for reduced levels of nitrite in foods (discussed below) have prompted researchers to augment the ability of reduced nitrite levels to control *C. perfringens* by using additional antimicrobial ingredients or processes. By itself, a naturally occurring source of nitrite at 50 mg/kg permitted a >4 log CFU/g increase in *C. perfringens* levels during 15-hour cooling of deli-turkey, but when combined with a 0.7% dried vinegar or 2.0% lemon–vinegar blend, *C. perfringens* levels were held to only a ~1.5 log CFU/g increase above initial levels during the same cooling schedule (King et al., 2015b). In mortadella sausage, the addition of winter savory essential oil with 100 mg/kg sodium nitrite resulted in greater *C. perfringens* reductions during storage at 25°C over 30 days than when nitrite was used alone (de Oliveira et al., 2010). The addition of other essential oils together with reduced ingoing nitrite was more effective than reduced nitrite alone in the control of *C. perfringens* growth in mortadella during 20 days of storage at 25°C (Dias et al., 2015); the combination of essential oils with nitrite was associated with increased residual nitrite levels. Essential oils in combination with sodium nitrite have also shown promise for protection against *C. perfringens* during storage of vacuum-packaged cooked beef fillets for 30 days at room temperature (Bakhtiary et al., 2018).

7.2.7.4 Antilisterial Effects of Nitrite

Although nitrite by itself is less effective against *Listeria monocytogenes* than it is against anaerobic spore-forming bacteria, it has emerged in as a useful hurdle in controlling the pathogen, particularly in ready-to-eat (RTE) meats and poultry.

L. monocytogenes can be found in the environment, can survive and grow at refrigeration temperatures, and is able to form sanitation-resistant biofilms: all properties which make it a potential inhabitant in food manufacturing environments. Despite aggressive sanitation and monitoring programs, *L. monocytogenes* can sometimes contaminate RTE foods after a lethality step prior to packaging. Listerial outbreaks have been associated with RTE meat and poultry products, including cured products (USDA Food Safety and Inspection Service, 2014). In addition, the death rate for listeriosis is unusually high, about 20%. As a result, the USDA's FSIS has established a "zero tolerance" for *L. monocytogenes* in RTE products, and requires manufacturing establishments to follow one of several alternatives (including the use of antimicrobial agents) to control the pathogen in post-lethality exposed RTE food products (USDA Food Safety and Inspection Service, 2014). In Europe, regulations promulgated in 2005 and amended in 2007 established a tolerance level of 100 CFU/g *L. monocytogenes* in RTE foods during shelf life (Dalzini et al., 2014b).

Nitrite is significantly less efficacious against *L. monocytogenes* than it is against *C. botulinum in vitro*. At a pH 7, the minimum inhibitory concentration (MIC) of sodium nitrite against *L. monocytogenes* (incubated for 18 hours in Mueller-Hinton broth) was 3500 mg/kg (Kashani et al., 2012). In contrast, the MIC for *C. botulinum* 62A at the same pH (in TPGY medium for 48 hours) was 6 to 8 mg/kg (Cui et al., 2010). However, at a lower pH (5.5), the MIC for sodium nitrite against *L. monocytogenes* was 100 mg/kg (Kashani et al., 2012). A more recent study reported MIC values between 125 and 300 mg/kg for various food isolates of *L. monocytogenes* in Mueller-Hinton broth (pH not reported, but typically ~7.3) at 37°C for 24 hours (Lamas et al., 2016). These *in vitro* results suggest that in foods at lower pH levels or in combination with other antimicrobial hurdles, nitrite could be part of an effective *L. monocytogenes* control program.

Several recent studies tested nitrite's ability to control *L. monocytogenes* in dry fermented sausages, which have other properties such as low water activity, reduced pH, and potential bacteriocins to help control this pathogen. In Spanish *salchichón*-type dry fermented sausages, a mixture of 150 mg/kg sodium nitrate and 150 mg/kg sodium nitrite reduced *L. innocua* (a surrogate for *L. monocytogenes* which was inoculated in the meat batter) levels by 2 logs relative to a nitrate- and nitrite-free control during 30 days of ripening (Hospital et al., 2012). This level of nitrate plus nitrite is the highest allowed by European Union regulations. Smaller reductions in *L. innocua* levels were obtained when lower concentrations of nitrate and nitrite were used in the formulation, which suggests potential food safety concerns might arise if the use of curing salts is restricted (Hospital et al., 2012).

Another study also tested various levels of nitrate and nitrite in French dry fermented sausages (batter pH 5.9, a_w 0.97; pH 5.05 at end of fermentation) for their ability to control *L. monocytogenes* (inoculated in the batter) after 34 days fermentation/drying (24 to 13°C) and 56 days storage at 18°C. The addition of nitrate and nitrite at ≥80 mg/kg each prevented the growth of *L. monocytogenes* during fermentation/drying (relative to a 250 mg/kg nitrate/0 mg/kg nitrite control, which allowed about 1 log of growth). In addition, the presence of 80 mg/kg nitrite and nitrate together maintained at least a 1- to 2-log lower level of *L. monocytogenes* than the control during storage, during which time levels in all formulations decreased to <1 log CFU/g (Christieans et al., 2018).

The effects of nitrite on the complex interaction between various microorganisms and *L. monocytogenes* were evaluated in *linguiça*, a Portuguese traditional dry-fermented sausage produced at various factories (Gonzales-Barron et al., 2015). When nitrite was added during maceration, the pH of the sausage did not drop as quickly. This allowed *L. monocytogenes* to grow in the nitrite- and nitrate-containing sausages. However, the pathogen was inactivated during smoking and ripening, and the authors attributed some of this inactivation to the presence of nitrite.

Another study examined the survival of *L. monocytogenes* applied to slices of a low-fat salami (pH 5.0, a_w = 0.945, 3% sodium chloride at beginning of shelf life) formulated with nitrite and nitrate at 150 mg/kg each. Pathogen numbers declined from a starting level of 1.44 to 1.70 log CFU during storage (8 to 12°C), with no detectable *L. monocytogenes* at 60 days of storage (Dalzini et al., 2014b).

In dry-cured hams, the addition of sodium nitrite and sodium nitrate to a salt- and sugar-cure mix did not affect the reduction in *L. monocytogenes* levels observed during 120 (Reynolds et al., 2001) or 234 days' curing (Portocarrero et al., 2002) at ~28-29°C. In a more recent study in dry-cured ham, the addition of nitrite to the cure reduced *Listeria innocua* by 1.5 log CFU/g at the end of resting in the *semimembranosus* but not *biceps femoris* tissue, possibly because of better penetration of the nitrite into the more external muscles such as the *semimembranosus* (Hospital et al., 2017). In whole brined hams, using a significantly faster process (<1 day vs. >15 days in the original process from pork loins until vacuum packaging) resulted in greater residual nitrite levels which were associated with a >2 log CFU/g decrease in *L. monocytogenes* levels in the ham over 35 days of storage at 10°C (Omori et al., 2010). Products from the conventional vs. the new method had similar salt concentrations, pH, and water activities. However, the faster process was also associated with higher *L. monocytogenes* levels at the beginning of storage and during the first 28 days of storage.

In addition to accelerating lethality of *L. monocytogenes* in dry-cured meats, nitrite enhances the inhibition of *L. monocytogenes* by other ingredients during refrigerated storage in high moisture, ready-to-eat meat products (Glass et al., 2002; Legan et al., 2004; Seman et al., 2002; Seman et al., 2008). Even at low levels (30 mg/kg), nitrite extended the lag phase of *L. monocytogenes* in turkey bologna supplemented with lactate and diacetate; when used at 120 mg/kg nitrite along with 1.6% lactate and 0.12% diacetate, *L. monocytogenes* did not grow for 15 weeks' storage at 4°C (Glass et al., 2008). Similarly, 150 mg/kg sodium nitrite plus 1% sodium lactate in sliced turkey *bresaola* prevented *L. monocytogenes* growth for 90 days when stored at 5 to 8°C (Dalzini et al., 2014a). In sliced restructured hams, sodium nitrite (at 156 mg/kg with 550 mg/kg sodium erythorbate plus lactate-diacetate) prevented growth of *L. monocytogenes* during 35 days of storage at 4°C (Sullivan et al., 2012a). Naturally occurring nitrite alone (at up to 200 mg/kg) allowed *L. monocytogenes* growth, but the addition of other naturally occurring antimicrobials resulted in *L. monocytogenes* levels similar to the conventionally cured control (Sullivan et al., 2012a; Horsch et al., 2014; Golden et al., 2014).

Nitrite has also shown efficacy against *L. monocytogenes* growth during storage in short-ripened or fresh spreadable raw sausages (Birzele et al., 2005; Muller-Herbst et al., 2016).

7.2.7.5 Nitrite Controversy and Conflicting Messages

In the last two decades, several high-profile national and international reports related to the safety of nitrite, nitrate, and processed meats have been published, receiving significant attention in the popular media and among consumers.

In 2001, the National Toxicology Program, a US interagency government group, issued a technical report on the toxicology and carcinogenesis of sodium nitrite in rats and mice (National Toxicology, 2001). No evidence of carcinogenic activity was observed in rats or male mice consuming drinking water containing 3000 mg/kg sodium nitrite (the highest level tested) for two years. In female mice, equivocal evidence of carcinogenic activity was reported based on a trend in increased incidence of certain forestomach lesions (squamous cell papillomas and carcinomas).

Several years later, in 2006, the International Agency for Research on Cancer (IARC) convened a working group of international experts to evaluate all pertinent epidemiological studies, cancer bioassays in experimental animals, and mechanistic, exposure, and other data associated with the cancer hazards of ingesting nitrate and nitrite. Their report, released in 2010, concluded that "ingested nitrate or nitrite under conditions that result in endogenous nitrosation is probably carcinogenic to humans Group 2A)." The report also concluded "there is limited evidence in humans for the carcinogenicity of nitrite in food. Nitrite in food is associated with an increased incidence of stomach cancer" (International Agency for Research on Cancer, 2010).

Just after the 2006 IARC meeting, the World Cancer Research Fund released a report containing ten universal guidelines for healthy nutrition, which included "limit intake of red meat and avoid processed meats" (World Cancer Research Fund and American Institute for Cancer Research, 2007). The report cited nitrite's ability to form carcinogenic N-nitroso compounds within meat or in the body as rationale for this recommendation (Demeyer et al., 2008). Several years later, the American Cancer Society issued a similar but slightly muted recommendation to "limit consumption of processed meat and red meat" (Kushi et al., 2012).

In 2015, another IARC working group evaluated red and processed meats and concluded that sufficient evidence existed for processed meat to be classified as a Group 1 carcinogen in humans (International Agency for Research on Cancer and World Health Organization, 2015; Bouvard et al., 2015). The full report of this group was recently published (International Agency for Research on Cancer, 2018). The report cites the potential for meat curing and smoking to form carcinogens, including N-nitroso compounds and polycyclic aromatic hydrocarbons, in making their decision.

Commentary related to the IARC conclusions on nitrite and processed meat has been made (Bryan et al., 2012; Mills et al., 2017). Much of the evidence used by IARC consisted of epidemiological studies, which among other limitations, can typically only demonstrate an association, not a cause (Farmer, 2007; Milkowksi, 2010; Boffetta, 2010). In addition, the magnitude of the carcinogenic effect noted by the IARC for processed meats was small (Science Media Centre, 2015). Residual nitrite levels present in food are now much lower than levels in the past; for example, residual nitrite in frankfurters dropped from 55 mg/kg in 1970 to 7 mg/kg in 2009 (A. Milkowski, personal communication), and residual nitrite levels in processed meats remain low (<30 mg/kg) in recent US, German, and Korean surveys (Honikel, 2008; Nunez De Gonzalez et al., 2012; Lee et al., 2018). Among meat products, only bacon is routinely cooked at temperatures where nitrosamines could potentially form, and cognizant of this, the USDA set ingoing nitrite level limits lower for bacon than for other cured meats and additionally requires ascorbate or erythorbate be present to prevent nitrosamine formation.

Other authorities have concluded that the levels of nitrate and nitrite consumed by most individuals is low enough to not cause safety concerns. In 2017, the European Food Safety Authority's Panel on Food Additives reviewed updated information and concluded that their existing limits for nitrite and nitrate added to foods (3.7 mg/kg bw/day for nitrate, 0.07 mg/kg bw/day for nitrite) were still safe (EFSA Panel on Food Additives and Nutrient Sources Added to Foods, 2017; EFSA Panel on Food Additives and Nutrient Sources Added to Food, 2017). The group acknowledged the recent IARC conclusion classifying processed meat as a carcinogenic hazard, but stated that the level of nitrite exposure from consumption of processed meats would be within safe limits for nearly all populations.

Meanwhile, the health benefits of nitric oxide and its generation from dietary nitrate via nitrite were clarified (Bondonno et al., 2016). The fact that nitrate and nitrite from dietary sources can be converted by the body to nitric oxide coupled with increasing evidence of the critical role of nitric oxide in various physiological processes, especially cardiovascular and metabolic, is changing how both researchers and consumers view nitrate and nitrite (Lundberg et al., 2018; McNally et al., 2015; Milkowski et al., 2010). Nitrate-rich foods such as beet juice have become very popular among consumers as performance-enhancing products or for cardiovascular health benefits (Van der Avoort et al., 2018). Somewhat ironically, arguments have even been made that nitrate or nitrite could prevent cancer, and certainly many nitrate-rich vegetables have been associated with decreased risks of certain cancers (Mills et al., 2017).

7.2.7.6 Changing Consumer Demands

The migration of consumers toward foods with fewer ingredients and which lack synthetic chemicals (including antimicrobials like nitrite) went mainstream after the enormously influential publication of Michael Pollan's 2008 book *In Defense of Food: An Eater's Manifesto* (Pollan, 2008). The coincidence of this movement with international public health organizations' advice to avoid or limit nitrite-containing processed meats re-awakened consumers' latent fears about the use of purified nitrite in meat products. In the subsequent rush by food manufacturers to respond and sell foods with "clean labels," meat processors began exploring ways to minimize the use of nitrite on cured meat labels without compromising product safety or quality.

Removing nitrite and nitrate from cured meats such as ham suffered from a major problem: consumers did not like the taste (Froehlich et al., 1983). As no suitable replacements existed, meat scientists turned to a natural source of nitrite. Vegetable juice powders, rich in nitrate, can be treated with bacterial cultures to form nitrite, and this natural source of nitrite was shown capable of producing products with color and taste similar to conventionally cured counterparts (Sindelar et al., 2007a; Sindelar et al., 2007b; Posthuma et al., 2018). The first generation of "naturally cured" or "alternatively cured" products relied upon the reduction of nitrate to nitrite to be performed within the meat product, which resulted in inconsistent nitrite levels (Jackson et al., 2011b). However, converting the nitrate in the vegetable powder to nitrite before adding to meat resulted in "pre-converted" natural nitrite at higher and more consistent levels. Numerous studies in recent years have shown that as long as nitrite (added in formulation) and pH levels are comparable, purified nitrite and naturally occurring nitrite are similarly effective in control of unwanted microorganisms in food (Table 7.3).

Alternatively cured products such as bacon are now widely available in the US and in Canada. Regulatory acceptance of the antimicrobial activity of the vegetable-derived nitrite in these products is growing, with the USDA recognizing nitrite derived from naturally occurring sources as safe and suitable as an antimicrobial when used in combination with ascorbate (USDA Food Safety and Inspection Service, 2018a, 2017). The use of naturally occurring sources of nitrite is even being considered as an antimicrobial in foods beyond meats, including prepared meals whose storage at abusive temperatures represents a potential risk from *C. botulinum* (Golden et al., 2017b). However, as the legal definition of "natural" for food labeling is refined and consumers become more aware that alternatively cured products do, in fact, contain nitrite, a potential exists for pushback against alternatively cured products (Shanker and Mulvany, 2019).

New naturally occurring sources of nitrate or nitrite are continually being tested, including young radish (Ko et al., 2017), parsley (Riel et al., 2017), beetroot (Sucu and Turp, 2018), and more. Other novel ways of using nitrite with foods which might be perceived by consumers as "clean" are being developed, including cold plasma treatment (which generates nitrite within the meat (Lee et al., 2017; Jung et al., 2017b; Jung et al., 2017a; Misra and Jo, 2017)) and nitrite-containing packaging films (Narvaez-Bravo et al., 2017). Other strategies that are being tested include the use of reduced levels of nitrite in combination with other naturally occurring antimicrobials, including carotenoprotein from blue crab shells (Hamdi

TABLE 7.3

Studies Demonstrating Equivalent Antimicrobial Activity between Purified and Natural Nitrite

Microorganism	Food Product	Reference
Clostridium perfringens	Deli-style turkey breast	King et al., 2015a
Listeria monocytogenes	Ham	Horsch et al., 2014
Listeria monocytogenes	Ham	Sullivan et al., 2012a
Listeria monocytogenes	Deli-style turkey breast	Golden et al., 2014
Spoilage organisms	Turkey bologna	Djeri and Williams, 2014

et al., 2018), lysozyme (Herath et al., 2015), probiotic strains (Ben Slima et al., 2017), and plant essential oils (de Oliveira et al., 2011). Reduced nitrite in levels concert with physical antimicrobial processes such as thermal treatments (Lavieri et al., 2015) and high-pressure processing have also shown antimicrobial efficacy (Omer et al., 2010; Duranton et al., 2012; Myers et al., 2013).

Whether it is purified or from naturally occurring sources, nitrite is expected to continue playing an important role in protecting and preserving meat products in the future. While more than half of US respondents to a 2015 survey reported they are reducing their consumption of processed meats (Neff et al., 2018), frankfurters and sausages are expected to remain popular consumer choices, with their production in the US continuing to grow through at least 2023 (IBIS World, 2019). Nitrite may be used less frequently now to preserve certain types of meat products (for example, refrigerated canned hams) than at the time this chapter was last revised (Tompkin, 2005). However, nitrite remains an important agent against *C. perfringens*, *C. botulinum*, and *L. monocytogenes* in many different meat products, and even allows a slight reduction in thermal processing shelf-stable canned meat products like SPAM® for shelf stability, serving a function beyond color and flavor (Pivnick et al., 1969; Duncan and Foster, 1969).

7.2.7.7 Predictive Models of Microbial Growth or Inactivation That Incorporate Nitrite

Predictive models have been developed in the past several decades that include nitrite concentration as a variable in predicting microbial growth and inactivation. Models developed in laboratory media or in specific foods can be found online or published in scientific literature; these models are widely used by food manufacturers and regulators for product development and to assess cooling deviations.

7.2.7.7.1 Established Modeling Programs Incorporating Nitrite as a Variable

ComBase is a long-standing, ongoing collaboration between the University of Tasmania and the USDA Agricultural Research Service (ARS), which has developed several online tools for quantitative food microbiology (ComBase, 2019; Baranyi and Tamplin, 2004). The service collects data from researchers to update and revise the models on a regular basis. Among the models which include nitrite as a variable is the ComBase Perfringens Predictor which can be used to predict *C. perfringens* growth; however, at the time of this writing, this model does not currently incorporate the positive impact of erythorbate on nitrite's ability to control *C. perfringens* (King et al., 2015a) and, therefore, overpredicts growth of this pathogen in a traditionally cured product. However, the Perfringens Predictor for uncured meats delivers useful predictions (Mohr et al., 2015).

The Pathogen Modeling Program (PMP) developed by the USDA ARS in the early 1990s has been frequently updated (USDA Agricultural Research Service, Undated). PMP includes online models for pathogen growth, survival, thermal inactivation, and cooling. Nitrite is included in models for *C. perfringens* growth inhibition for select meat systems; nitrite is included in the formulations for cooked cured ham and fermented sausage to evaluate the survival of *E. coli* O157:H7 and *L. monocytogenes* but nitrite levels are not adjustable. Like the ComBase model, these models may need further validation to confirm accuracy.

The Corbion® Listeria Control Model is an online tool that predicts *Listeria monocytogenes* growth in foods for proprietary organic acid-based ingredients alone, or in combination with nitrite (Corbion, 2019). The model generates predictions as a "best fit" line for most probable growth, as well as a line representing 95% confidence intervals.

Danish Meat Research Institute (DMRI) provides free predictive models useful for assessing pathogen levels (especially *Listeria monocytogenes*) in meat products (Danish Meat Research Institute, 2019).

Food Spoilage and Safety Predictor (FSSP) is a software program that can be downloaded for free that includes a variety of models, including a generic model to predict growth of spoilage organisms and pathogens in meat and seafood using nitrite, organic acids, salt, pH, and temperature (Technical University of Denmark, Undated).

7.2.7.7.2 Models Described in Primary Literature Incorporating Nitrite as a Variable

A selection of more recent published experimentally derived models (some of which may have been incorporated in the established models discussion in the previous section) in nitrite-containing foods or that include nitrite as a variable are described in this section.

Selected Gram-positive pathogen growth and related models that include nitrite as a variable or were developed in nitrite-containing foods are shown in Table 7.4. Selected growth and related models for Gram-negative pathogens that include nitrite as a variable or were developed in nitrite-containing foods are shown in Table 7.5. Selected growth and related models for spoilage organisms that include nitrite as a variable or were developed in nitrite-containing foods are shown in Table 7.6.

TABLE 7.4

Selected Gram-Positive Predictive Models Which Incorporate Nitrite

Pathogen	Type of Model	Food or Broth Tested	Reference
Bacillus cereus	Aerobic growth	Brain heart infusion (Singh et al.) broth	Benedict et al. (1993)
Clostridium botulinum (psychrotrophic strains)	Growth	Pasteurized RTE meat products	Gunvig et al. (2013)
Clostridium perfringens	Growth during cooling	Cooked beef	Juneja et al. (2013)
Clostridium perfringens	Growth during cooling	Cooked cured pork	Juneja et al. (2006)
Listeria monocytogenes	Growth	Tryptone soya broth (TSB)	McClure et al. (1997)
Listeria monocytogenes	Growth	Seafood	Mejlholm and Dalgaard (2007)
Listeria monocytogenes	Growth	Cooked sausages with other natural antimicrobials	Ghabraie et al. (2016)
Listeria monocytogenes	Growth	Cooked cured pork sausage	King et al. (2016)
Listeria monocytogenes	Growth	Low-salt, low-nitrite frankfurters	Gwak et al. (2015)
Listeria monocytogenes	Growth	Ham model system	Dussault et al. (2016)
Listeria monocytogenes	Reduction with HHP	Simulated meat medium	Possas et al. (2018)
Listeria monocytogenes	Reduction during fermentation and maturation	Fermented and matured sausages	Gunvig et al. (2016)
Listeria monocytogenes	Growth	Ready-to eat cured and uncured meat products	Legan et al. (2004)
Staphylococcus aureus	Growth and toxin production	Cooked pork sausage	Lin et al. (2018)
Staphylococcus aureus	Growth	Nutrient broth with various salt and nitrite combinations	Lee et al. (2016b)
Staphylococcus aureus	Growth and toxin production	Heated and fermented meat products	Gunvig et al. (2018)
Staphylococcus aureus	Lag time and growth	Tryptone soya broth (TSB)	Zurera-Cosano et al. (2004)

TABLE 7.5

Selected Gram-Negative Predictive Models Which Reference Nitrite

Pathogen	Type of Model	Food or Broth Tested	Reference
E. coli O157:H7	Growth	Brain heart infusion broth	Buchanan and Bagi (1994)
Shiga-toxin production E. coli	Reduction during fermentation and maturation	Fermented and matured sausages	Gunvig et al. (2016)
Salmonella spp.	Reduction during fermentation and maturation	Fermented and matured sausages	Gunvig et al. (2016)
Salmonella spp.	Growth	Low-salt, low-nitrite frankfurters	Gwak et al. (2016)
Salmonella Typhimurium DT104	Growth	Chicken frankfurters	Oscar (2008)
Shigella flexneri	Growth	Brain heart infusion broth	Zaika et al. (1998)
Yersinia enterocolitica	Growth	Nutrient broth	Divya and Varadaraj (2015)
Yersinia enterocolitica	Growth	Brain heart infusion broth	Bhaduri et al. (1994)

TABLE 7.6

Selected Spoilage Organism Predictive Models Which Reference Nitrite

Organism	Type of Model	Food or Broth Tested	Reference
Enterococcus spp.	Growth	DeMan, Rogosa, and Sharpe broth	(Lee et al., 2016a)
Lactobacillus spp.	Growth	Low-salt, low-nitrite frankfurters	(Lee et al., 2015)
Lactobacillus spp. (psychrotolerant)	Growth	Seafood and meats	(Mejlholm and Dalgaard, 2013)
Leuconostoc mesenteroides	Growth	Tryptone soya broth	(Zurera-Cosano et al., 2006; Garcia-Gimeno et al., 2005)
Pseudomonas spp.	Growth initiation	Processed meats (frankfurters, bacon)	(Jo et al., 2014)
Natural microbiota	Growth	Salted and cured meat products	(Kajak and Kolozyn-Krajewska, 2006)

7.2.7.8 Summary for 2002–2019

1. The expansion of research in the response of pathogens (foodborne and others) to nitric oxide and reactive nitrogen species has generated considerable knowledge of genes, proteins, and pathways used by bacteria to protect against nitrosative damage. A large number of molecular targets for nitric oxide and reactive nitrogen species have been identified, which helps explain the broad antimicrobial specific of nitrite as it is a precursor for these chemicals.
2. The role of ascorbate and erythorbate in enhancing nitrite's antimicrobial activity has been better defined.
3. The USDA's Appendix B guidances, first issued in 1999 and considered overly conservative because they were based on overestimates of the levels of *C. perfringens* spores likely present in meat, led to considerable work to better understand the growth of the pathogen in meat products and to identify conditions which satisfy the requirement for less than 1 log CFU/g growth of the pathogen during cooling or holding. As a result, the ability of nitrite to control *C. perfringens* growth under many different conditions in a variety of different meat and poultry products has been well-tested. In contrast, bioterrorism concerns and increased oversight of the use of select agents led to a decline in research related to *Clostridium botulinum*.
4. The realization that *Listeria monocytogenes* could be a significant and serious post-lethality contaminant of processed meat and poultry products led to work demonstrating that nitrite did have significant antilisterial activity with commercial utility, especially at lower pHs or in combination with other antimicrobials.
5. IARC's conclusions that ingested nitrate or nitrite under conditions that result in endogenous nitrosation are probably carcinogenic and that processed meats are carcinogenic, despite the conclusions of other authorities that the levels of nitrate and nitrite consumed by most people should not cause safety concerns, were key factors which put nitrate and nitrite back into the public spotlight as a food additive of potential concern.
6. IARC's pronouncements against nitrite and the general movement of consumers away from foods with synthetic ingredients has pushed food processors to limit the use of purified nitrite in processed foods (for example, using "natural nitrite" derived from microbial reduction of nitrate obtained from celery juice powder).
7. The antimicrobial equivalency of purified vs. naturally occurring nitrite has been established under a variety of different conditions for several different pathogens.
8. Increased reliance of food processors on predictive models to determine whether pathogens and spoilage organisms are likely to grow under processing conditions has led to many of the models including nitrite levels as a key variable.

7.3 Regulations Governing the Use of Nitrite and Nitrate

7.3.1 Historical Perspective on Regulations

The regulations limiting the use of nitrite and nitrate in cured meats vary widely among countries and have continued to change. The regulations were summarized in two comprehensive surveys (Meester, 1974; Braathen, 1979). The status of nitrite and nitrate in cured meats and cheese in the United Kingdom was reviewed in 1978 (Ministry of Agriculture, Fisheries, and Food, 1978). Germany reduced the permitted level of nitrite in nitrite-containing curing salt by 20% (Leistner, 1981; Gerhardt, 1995). The most recent summary of the regulations in different countries appears in Cassens (1990).

Nitrite and nitrate levels for curing meats in the United States were recommended by a USDA Expert Panel on Nitrites and Nitrosamines (Food Safety and Quality Service, 1978). The current USDA regulations controlling the use of nitrite and nitrate in meat and poultry products are specified in 9 CFR 317.17 and 424.21–23 of the federal regulations (Code of Federal Regulations, 1990a). Procedures for calculating the amount of nitrite and nitrate that can be added to curing solutions and product were specified in the USDA calculation handbook (Food Safety and Inspection Service, 1986). The use of nitrite and nitrate in the processing of fish products is specified in 21 CFR 172.160, 172.170, 172.175, and 172.177 (Code of Federal Regulations, 1990b).

A unique feature of the US meat and poultry regulations is that nitrite and nitrate are added on the basis of the meat block weight in the product formulation. Other countries limit the amount of nitrite and nitrate on the basis of the total weight of the product formulation. The net effect of the US regulations is that less nitrite is added to many products than might be assumed with a regulation that permits 156 mg/kg of sodium nitrite. Examples of the effect of this regulation on the maximum allowable level of nitrite in several sausage and loaf products are listed in Table 7.7. As the percentage of extenders and other nonmeat ingredients is increased in the products, the level of added nitrite is decreased. Also listed is the actual amount of nitrite that could be added to the total formulation of the products if the permitted level of nitrite was reduced from 156 to 100 mg/kg, as proposed by the Expert Panel on Nitrites and Nitrosamines (Food Safety and Quality Service, 1978). It is also common to generate a certain amount of rework during the manufacture of sausage products. An example of rework is the round ends of sausages that are trimmed off before slicing the product. This material is normally used as an ingredient at a level of about 5% in the manufacture of succeeding lots of the same product. Since it has already been cured, the weight of this meat is subtracted from the fresh meat portion of the formulation and the amount of nitrite added is reduced accordingly. This further reduces the nitrite level in the product formulation and, depending upon the circumstances, leads to lower residual nitrite levels after processing.

7.3.2 Nitrite and Nitrate Regulations in the United States

7.3.2.1 US Regulations for Formulation of Cured Meat and Poultry Products

US requirements for nitrite and nitrate used in cured meat products include both minimum and maximum input requirements which vary depending upon the type of product (USDA Food Safety and Inspection Service, 1995).

Either sodium or potassium salts of nitrite are permitted. Except for bacon, the same amount (in terms of ppm or mg/kg) of sodium or potassium nitrite salts is allowed in most products despite their differing total nitrite content (USDA Food Safety and Inspection Service, 1995), which means that for the same ingoing level, potassium nitrite contains less nitrite than sodium nitrite.

7.3.2.1.1 Minimum Levels of Nitrite

A minimum of 120 mg/kg (ppm) of ingoing sodium or potassium nitrite is required in all cured "keep refrigerated" products (including bacon) unless additional controls are in place. No minimum ingoing nitrite level exists for shelf-stable cured products (USDA Food Safety and Inspection Service, 1995). If nitrite is from a natural source, a minimum of 75 mg/kg nitrite with a minimum of 500 mg/kg ascorbate from a natural source are accepted in meat and poultry products, as is a minimum of

TABLE 7.7

Examples of Sodium Nitrite Levels Added to Cured Meats in the United States

Product	$NaNO_2$ (mg/kg) Added on Basis of Meat Content	Actual $NaNO_2$ (mg/kg) in Total Formulation	Actual $NaNO_2$ (mg/kg) in Total Formulation if Input $NaNO_2$ Is Reduced to 100 g/g in the Meat
Bologna	156	121	78
Bologna with NFDM3	156	117	75
Bologna (imitation)	156	105	67
Franks (meat) 1	156	115	74
Franks (meat) 2	156	107	69
Franks with NFDM	156	114	73
Liver sausage 1	156	152	97
Liver sausage 2	156	140	90
Cured beef loaf	156	143	92
Blood pudding	156	130	83
Pickle and pimento loaf	156	91	58
Olive loaf	156	94	60
Family loaf	78	50	50
Veal loaf	78	47	47

[a] NFDM, nonfat dry milk.

Source: Based on formulations in Pearson and Tauber (1984).

100 mg/kg nitrite with a minimum of 250 mg/kg of ascorbate (USDA Food Safety and Inspection Service, 2018c).

7.3.2.1.2 Maximum Levels of Nitrite

Maximum levels of nitrite allowed by USDA FSIS in meat and poultry products are dependent upon the type of product and the way that it is cured. For immersion cured, massaged, or pumped products, sodium or potassium nitrite is allowed up to 200 mg/kg, while for comminuted products such as frankfurters, a lower level (156 mg/kg) is the maximum level allowed. Dry-cured products are permitted up to 625 mg/kg of sodium or potassium nitrite.

Due to its lower toxicity, higher maximum levels of nitrate are acceptable in cured meat and poultry products: 700 mg/kg sodium or potassium nitrate are allowed in immersion cured, massaged or pumped products, while 1718 mg/kg and 2187 mg/kg of sodium or potassium nitrate are allowed in comminuted and dry-cured products, respectively (USDA Food Safety and Inspection Service, 1995). If both nitrite and nitrate are used, each can be added up to the maximum individual limits, but the finished product cannot contain more than 200 mg/kg of analytically measured nitrite (USDA Food Safety and Inspection Service, 1995; Sebranek and Bacus, 2007).

Because consumers may fry bacon at high temperatures, increasing the risk of nitrosamine formation, the US regulations (9 CFR Part 424.22 and 9 CFR Part 424.21) specify lower maximum levels of nitrite than for other cured products and also require the presence of a cure accelerator (sodium ascorbate or sodium erythorbate) for pumped or massaged bacon (USDA Food Safety and Inspection Service, 1995). Immersion cured or pumped and/or massaged bacon (rind-off) are allowed a maximum of 120 mg/kg ingoing sodium nitrite or 148 mg/kg ingoing potassium nitrite. Dry-cured bacon (rind-off) is allowed up to 200 mg/kg ingoing sodium nitrite and 246 mg/kg ingoing potassium nitrite. Lower levels of nitrite are permitted under certain conditions which have been demonstrated to provide additional protection against *C. botulinum* (USDA Food Safety Inspection Service, 1995).

Per 9 CFR Part 424.23, nitrate is prohibited in bacon because it could serve as an additional source of nitrite and lead to nitrosamine formation. However, natural nitrate (from celery powder, for example) is allowed in uncured bacon products (to be discussed further below).

7.3.2.2 US Guidelines for Stabilization (Cooling) of Cooked Cured Meat and Poultry Products

The USDA is in the process of revising its guidelines ("Appendix B") for cooling and "hot holding" of heat-treated meat and poultry products (USDA Food Safety and Inspection Service, 2017). Although additional modifications are anticipated, the current guidelines (USDA Food Safety and Inspection Service, 2019) include additional cooling strategies for meat and poultry products that contain nitrite in order to limit the growth of *C. perfringens* and prevent *C. botulinum* growth. These additional cooling strategies apply to products made with nitrite from either purified or natural sources as long as certain ingoing concentration requirements are met.

7.3.2.3 US Requirements for Natural Sources of Nitrate or Pre-Converted Nitrite in Meat and Poultry Products

Naturally occurring sources of nitrate or pre-converted nitrite are allowed to be used as antimicrobials in meat and poultry products (Directive 7120.1) or as flavorings, but the USDA doesn't recognize them as "curing agents," per 9 CFR 424.21(c) (USDA Food Safety and Inspection Service, 2018a). This means that products such as bacon, frankfurters, or corned beef, which are normally cured with nitrate or nitrite, must be labeled "uncured" if they are made without purified nitrate or nitrite, even if they contain nitrate or nitrite from naturally occurring sources (9 CFR Part 319.2).

7.3.2.4 US Requirements Regarding Other Uses of Nitrate and Nitrite in Food Products

Various other uses of nitrite or nitrate in US food products are permitted. Nitrite is allowed to be incorporated in one side of a food packaging film used for vacuum packaging raw red meat (as a color enhancer) at 113 mg/m^2 of film (FSIS Directive 7120.1). Nitrate and nitrite are also permitted for use as a food additive in cured fish at no more than 500 mg/kg sodium nitrate and 200 mg/kg sodium nitrite in the finished product (200 ppm potassium nitrate is allowed in processing cod roe (21 CFR Part 172.175)).

The US currently has no limits currently on the use of cultured celery powder and other natural sources of nitrate and nitrite in food products, and food processors are beginning to explore the use of "natural" nitrite in foods other than meats in order to provide protection against *C. botulinum* as well as spore-formers which can reduce product quality (Golden et al., 2017b).

Several uses of nitrate or nitrite in food products are prohibited:

- In the US, foods labeled "natural" or "organic" cannot contain added purified nitrate or nitrite (Sebranek and Bacus, 2007).
- Nitrate and nitrites are not allowed in baby, junior, or toddler foods (9 CFR Part 424.21).
- Nitrate and nitrite are not permitted for use in cheese in the US (Genualdi et al., 2018).

7.3.2.5 US Regulations on Nitrite and Nitrate in Water

The U.S. Environmental Protection Agency has set a maximum contaminant level (MCL) of 10 mg/L for nitrate and 1 mg/L for nitrite in drinking water (U.S. Environmental Protection Agency, 2018). Similarly, bottled drinking water in the US is allowed to contain 10 mg/L nitrate (as nitrogen), 1 mg/L nitrite (as nitrogen), and total nitrate and nitrite: 10 mg/L (as nitrogen) (21 CFR Part 165.110).

7.3.3 Canadian Regulations Regarding Nitrite and Nitrate in Foods

Canadian regulations involving the use of nitrite or nitrate in foods are similar but slightly different from US regulations.

As in the US, both the sodium and potassium salts of nitrite and nitrate may be used in cured meat products (Canadian Food Inspection Agency, 2019)

In cured meats other than side bacon, including comminuted meats (which are limited to 156 mg/kg nitrite in the US), the maximum input level of sodium nitrite is 200 mg/kg. In side bacon, the limit is 120 mg/kg sodium nitrite. Sodium nitrate is allowed in slow cured meat products at an input level of 200 mg/kg, which may be used in addition to nitrite (Government of Canada, 2018).

Unlike the current US regulations, "alternatively cured" products made with nitrite or nitrate from natural sources are considered cured and do not need to be labeled "uncured" (Canadian Food Inspection Agency, 2019). Pre-converted nitrites (from cultured celery juice powder or other approved products) may be used in curing as an alternative source of nitrite, but such products must achieve a minimum ingoing level of 100 mg/kg while not exceeding the 200 mg/kg maximum limit (Canadian Food Inspection Agency, 2019). If using non-converted nitrate from a naturally occurring source (celery juice powder, for example), the operator must ensure that the amount of nitrate used will achieve a nitrite level (following conversion in the product) of between 100 and 200 mg/kg. Nitrate (not pre-converted to nitrite) from naturally occurring sources cannot be used for poultry products or cured meat products made by pumping or injection procedures (such as brine-injected hams) (Canadian Food Inspection Agency, 2019).

The Canadian Food Inspection Agency specifies schedules for cooling cooked meat products to prevent pathogen growth which are dependent upon nitrite concentration, water activity, and brine concentration in finished product (Canadian Food Inspection Agency, 2018).

7.3.4 European Union Regulations Regarding Nitrite and Nitrate in Foods

As in the US, the following nitrite and nitrate compounds are permitted as food additives in the European Union (EU):

- E 249: Potassium nitrite
- E 250: Sodium nitrite
- E 251: Sodium nitrate
- E 252: Potassium nitrate

In contrast to the US, whose regulations specify ingoing nitrite and nitrate levels that should be used (USDA Food Safety and Inspection Service, 1995), the EU still regulates residual nitrite or nitrate in some foods (mostly for traditional cured products), but ingoing or added nitrite or nitrate levels are now specified for some food products and are becoming more commonly used (Adler-Nissen et al., 2014; European Union, 2006).

The levels of nitrate and nitrite allowed in foods (including cheese, raw and processed meats, and processed fish) in the EU are described in the food additives directive 2006/52/EC (European Union, 2006) and implemented in the regulations EC1129/2011 (updated from regulation 1333/2008/EC) (European Union, 2011b). Nitrite and nitrate regulations are generally harmonized across the EU, but Denmark has different (lower) requirements than the EU (Adler-Nissen et al., 2014; Food Chain Evaluation Consortium of the European Commission, 2016).

In most meat products, 150 mg/kg nitrate plus 150 mg/kg ingoing nitrite is allowed, while in some traditional cured meat products, the EU allows up to 300 mg/kg ingoing nitrate (when no nitrite is added) or up to 180 mg/kg ingoing nitrite. There is some recent movement to potentially reduce these levels (Christieans et al., 2018; Food Chain Evaluation Consortium of the European Commission, 2016).

In contrast to the US, purified nitrite is allowed in organic meat products in Europe, although at lower concentrations than for non-organic foods. Per (EC) 780/2006, sodium (or potassium) nitrite can be applied at an ingoing concentration of 80 mg/kg with a maximum residual concentration in the product of 50 mg/kg (Hammes, 2012).

In other foods, the EU permits nitrate at 150 mg/kg to be used in milk that will be used to make hard, semi-hard, and semi-soft cheeses in order to control clostridia that can cause late-blowing cheese defects. Nitrate at 500 mg/kg is also allowed in some types of pickled fish (European Union, 2006).

The EU does not allow nitrites or nitrates from naturally occurring sources to be used in foods if they are being used for preservation or coloration; the nitrite or nitrate would be considered a food additive which is not currently approved as a preservative and also does not meet purity criteria (Cantwell and Elliott, 2017; Food Safety Authority of Ireland, 2019). In Europe, therefore, "alternatively cured" products that use celery juice or other naturally occurring source of nitrate (pre-converted to nitrite or not) are not currently marketed.

7.3.5 Regulations in Other Regions Regarding Nitrite and Nitrate Use in Foods

The Australia New Zealand Food Standards Code limits sodium or potassium nitrite to 125 mg/kg (125 ppm) in cured, dried, and slow dried cured meat. Commercially sterile and canned cured meat are limited to 50 mg/kg potassium nitrite. Potassium and sodium nitrate are allowed in slow dried cured meats or fermented, uncooked processed comminuted meat products at 500 mg/kg (Australian Government, 2018). Nitrate (potassium or sodium salts) is allowed in cheese at 50 mg/kg (calculated as nitrate ion).

A number of other countries have strict limits on residual nitrite levels allowed in foods. Both Korea and Japan limit residual nitrite to not more than 70 mg/kg in meat products, and the Codex Committee on Food Additives specifies maximal residual levels of 80 mg/kg nitrite in cured ham and other meat products (Lee et al., 2018; Hong et al., 2016).

While the EU limits the maximum levels of nitrate in certain vegetables, cereal products, and baby foods (European Union, 2011a), at least one country, China, limits the concentration of nitrite that can be present in fresh vegetables, with a limit of 4 mg/kg (4 ppm) in fresh vegetables and 20 mg/kg (20 ppm) in fermented vegetables to 20 mg/kg (Wang and Shao, 2018).

7.4 Assay Method

Since the levels of nitrite and nitrate permitted in foods are established by regulation, the method of assay applicable to each country should be employed. Within the US the methods of assay appear in the Official Methods of Analysis of the Association of Official Analytical Chemists (AOAC international, 2002). International agreement has been reached on the specifications and analytical methods for potassium nitrate and sodium nitrite (JECFA, 1995).

Nitrate and nitrite can often be analyzed by similar methods by quantitatively reducing nitrate to nitrite. Sample preparation for nitrite determination is critical because of the reactivity of nitrite. This reactivity, however, also allows it to form derivatives with many compounds which then allow its quantitation. The Griess reaction, which takes advantage of nitrite being the only compound which can form a diazo dye, has been used for over a century in the quantitation of nitrite (Fox and Suhre, 1985). Analytical methods currently used to detect and quantitate nitrite and nitrate, including spectrophotometric (such as the Griess reaction), chemiluminescent, electrochemical, chromatographic, capillary electrophoresis, spectrofluorimetric, and electrochemiluminescent methods, have been reviewed (Merino et al., 2017; Wang et al., 2017). New research methods for assessing nitrite and nitrate levels in foods continue to be developed. In recent years, a voltammetric electronic tongue has been developed that can measure nitrite and nitrate in brines or minced meat quickly and without destroying the sample (Campos et al., 2010).

7.5 Toxicology

Toxicities which have been associated with nitrite ingestion include acute toxicity from brief exposures to high nitrite doses (methemoglobinemia, where nitrite oxidizes the iron within hemoglobin to an oxidation state that cannot bind oxygen) and potential chronic toxicities resulting from long-term administration of lower nitrite doses (Umbreit, 2007; Agency for Toxic Substances and Disease Registry, 2016). In most cases, the largest exposures to dietary nitrate and nitrite are from those compounds naturally present in vegetables and water and not from cured meat products (Milkowski et al., 2010). Acute toxic reactions from natural sources of nitrite or nitrate are rare, although young infants have suffered methemoglobinemia after being given formula made with water containing high levels (>20 mg/L) nitrate (Knobeloch et al., 2000). Acute toxic reactions to nitrite can occur when purified nitrite is misused in foods (Cvetkovic et al., 2019; Wang et al., 2013).

The acceptable daily intakes (Campos et al., 2010) for potassium nitrate and sodium nitrite have been reported to be 0–3.7 and 0–0.06 mg/kg body weight, respectively (JECFA, 1995). The values are expressed as nitrate and nitrite ions, respectively, and apply to all sources of intake for adults but do not apply to infants below the age of 3 months.

For potassium nitrate, estimates of the lethal dose have ranged from 4 to 30 g (about 70–500 mg/kg body weight) while 30–60 g of sodium nitrate have been given for 2 months as an acidifying diuretic without manifestation of adverse effects. A realistic estimate of a lethal dose for adults is about 20 g nitrate ion or 330 mg nitrate ion/kg body weight (Gangolli et al., 1994).

Two years of rat studies on nitrate showed no evidence of carcinogenicity. Data on reproductive toxicity for nitrate were limited, and it appears that nitrate per se is not genotoxic. The no-adverse-effect dose level was estimated to be 2500 mg sodium nitrate/kg body weight/day (Gangolli et al., 1994). A 2015 long-term (17-month) dietary nitrate study in mice (at a human equivalent of 350 mg/day) found no obvious adverse long-term adverse effects of nitrate, but did find decreased plasma insulin and IL-10 levels accompanied by improved insulin tolerance and survival (Hezel et al., 2015).

For sodium nitrite, estimates for the lethal dose have ranged from 2 to 9 g (about 33–250 mg/kg body weight) with the lower doses applying to infants who are more sensitive to methemoglobinemia, the criterion most commonly used for nitrite toxicity. The normal range of methemoglobin levels is 0.5–2.0%, the uppermost level being found in infants and pregnant women. Cyanosis, as evidence of toxicity, occurs at methemoglobin levels in excess of 10%. Using this criterion, estimates for the toxic dose range from 1 to 8.3 mg sodium nitrite/kg body weight (Gangolli et al., 1994).

Concern for the formation of N-nitroso compounds (NOCs) in foods began with the discovery that adding sodium nitrite to preserve raw herring for fish meal production resulted in malignant liver disease in cattle and sheep. In 1963 it was proposed that nitrosodimethylamine formed during the herring meal process was the toxic agent. For humans the greatest exposure to NDMA is from tobacco smoke (Koppang, 1998). Investigations for NDMA in foods led to cured meats, especially bacon when prepared and cooked under certain conditions, and beer from exposing the malt to nitrogen oxides. Controls adopted by the food industry over the past 20 years have led to reductions in the risk of exposure to nitrosamines (Lijinsky, 1999). For example, Cassens (1997) reported that residual nitrite levels in cured meats at the retail level in the US had decreased over the past two decades by approximately 80%. In addition, higher residual ascorbate levels (mean value of 209 mg/kg) in the cured meat products would reduce the risk of nitrosation reaction products being formed.

An extensive review was conducted during the development of two reports by the National Academy of Sciences (Assembly of Life Sciences, 1981, 1982). Those reports provide the most complete assessment of the safety of nitrite as a food additive for that period of time. Cassens (1995) has provided an update of the nitrite controversy, including a review of epidemiologic studies seeking to establish a link between intake of nitrate and nitrite from food and water and cancer, particularly among young children. Murphy et al. (1998) concluded that the increase in the incidence of brain and nervous system cancer observed among both children and adults between 1970 and 1990 was not consistent with the decrease in the consumption of cured meats and marked reductions in residual nitrite content during that period. Information presented at a Ceres Forum leads to the overall conclusion that available information either does not or is inadequate to support a link between dietary intake of nitrite and NOCs and brain and nervous system cancers in children and adults (Ceres Forum, 1997).

Sodium nitrite is now viewed in an entirely different light than during 1970–1990 (CAST, 1997). Assessments conducted during the past decade have led to the conclusion that nitrite, per se, is not carcinogenic. Several countries that monitor their foods for nitrate, nitrite, and NOCs have concluded that dietary nitrite and NOCs represent a relatively small contribution to the total body burden of these compounds (Cassens, 1995). Even the American Cancer Society has concluded that "nitrites in foods are not a significant cause of cancer among Americans" (American Cancer Society, 1996).

More recently, an extensive review of the literature and testimony from experts led the California Developmental and Reproductive Toxicant Identification Committee to vote on June 2, 2000, not to list sodium nitrite as a developmental toxicant under the state's Proposition 65 law. This was followed by the final report from the National Toxicology Program on May, 2001, summarizing studies conducted over a two-year period on the carcinogenesis and genetic toxicology of sodium nitrite in drinking water. As discussed previously, the report concluded that there was no evidence of carcinogenic activity for sodium nitrite in rats and mice exposed to 750, 1500, or 3000 g/ml. There was equivocal evidence of carcinogenic activity in the forestomach of the female mice tested (National Toxicology Program, 2001). This same report also discusses several shorter chronic toxicology studies (13–14 weeks) in mice and rats at doses up to 5000 mg/L sodium nitrite in drinking water. At the highest doses, clinical and pathological findings were noted in the sodium nitrite groups relative to control groups, and methemoglobin concentrations were elevated in all exposed groups in the rat study. A battery of genotoxicity tests all demonstrated negative results.

In humans, meta-analyses have found varying results with respect to nitrite, nitrate, and stomach cancer. A review of nitrate and nitrite and its association with stomach cancer funded by the American Meat Institute concluded that prospective epidemiological studies do not consistently indicate a relationship between ingested nitrate, nitrite, or N-nitrosamines and stomach cancer (Bryan et al., 2012). A more recent meta-analysis of studies of dietary nitrate, nitrite, and nitrosamine intake concluded that high consumption of nitrites and nitrosamines could increase gastric cancer risk, but available data were not considered sufficient to make firm conclusions (Song et al., 2015). A 2019 meta-analysis found that a high or moderate nitrite intake increased the risk of gastric cancer, while a high or modest nitrate intake reduced the risk of gastric cancer (Zhang et al., 2019).

For other cancers, a meta-analysis of epidemiological studies looked for associations between dietary nitrate and nitrite intake and specific cancer types across 62 studies and >60,000 cancers; no significant associations were observed between dietary nitrate or nitrite and breast, bladder, colorectal, esophageal, renal cell, non-Hodgkin lymphoma, ovarian, and pancreatic cancers, with only modest associations with adult glioma and thyroid cancer (Xie et al., 2016).

An assessment of NOCs by Hotchkiss et al. (1992) concluded that there was only indirect evidence that some portion of human cancer risk is related to NOC exposure and that tobacco was the major source. Endogenous formation of NOCs is influenced by dietary intake and a variety of other factors.

Gangolli et al. (1994) conducted a comprehensive risk assessment on nitrate, nitrite, and NOC and determined that vegetables provide over 85% of the average daily human dietary intake of nitrate. Endogenous synthesis of nitrate is an additional important source of nitrate. A review of animal toxicologic studies, human effects and epidemiologic surveys led to the conclusion that there is no firm scientific evidence to recommend drastic reductions beyond the average levels of nitrate encountered in vegetables grown following good agricultural practice.

Gangolli et al. (1994) concluded that dietary sources of nitrite and NOCs contribute relatively small amounts to the body burden. The major source of these compounds in the body is derived through metabolism of ingested nitrate. In humans, the saliva is the major site for formation of nitrite.

Dykhuizen et al. (1996) reported that ingested nitrate is absorbed from the gastrointestinal tract into the bloodstream and concentrated in the saliva, increasing in concentration up to ten times the level found in plasma. Salivary nitrate is converted to nitrite by nitrate-reducing microorganisms in the mouth. While this source of nitrite would increase the risk of NOC formation when introduced into the acid conditions of the stomach, the acidified nitrite also has antimicrobial activity that coincides with the formation of nitric oxide. Expelled stomach air contains a high concentration of the antimicrobial gaseous nitric oxide that is enhanced by dietary nitrate intake. As might be expected from other information in this chapter, *in vitro* tests demonstrated the bactericidal effect of acidified nitrite against several enteric pathogens (Dykuizen et al., 1996). Patients with infective gastroenteritis have increased plasma nitrate levels compared with healthy control (Dykhuizen et al., 1995).

A review of the pros and cons of nitrate and nitrite as food additives in light of the information available by 2001 and again by 2015 has been provided by Archer (2002) and Habermeyer et al. (2015).

Nitric oxide is produced by three different nitric oxide synthases (Bogdan, 2001; Chen et al., 2002). There is a growing body of evidence showing that endogenously produced nitric oxide is an important component of the natural defense system to infectious disease in humans (Karupiah et al., 2000; Shiloh and Nathan, 2000; Cherayil and Antos, 2001; Bogdan, 2001; Alam et al., 2002). Nitric oxide is involved in the pathogenesis and control of infectious diseases, tumors, autoimmune processes, and chronic degenerative diseases (Bogdan, 2001). More recent research indicates that nitric oxide plays many diverse roles in infection and immunity. It is now clear that inducible nitric oxide synthase (iNOS)

> is detrimental in some infectious disease processes and that it helps to counteract excessive immune reactions, protects to some degree against autoimmunity and functions as an intra- and intercellular signaling molecule shaping the immune response. In addition, nNOS and eNOS (neuronal and endothelial NOS) are now known to participate in important immunological processes such as apoptosis, cell adhesion, autoimmunity and perhaps antimicrobial defense. We have also begun to learn about the possible role of NO in thymic education. (Bogdan, 2001)

Endogenously formed nitric oxide also plays a role in neurotransmission, blood clotting, and blood pressure control (CAST, 1997). The enzyme, nitric oxide synthase, catalyzes the oxidation of L-arginine and l-citrulline to release nitric oxide that can be converted to nitrite and nitrate and subsequently excreted (CAST, 1997).

Some of the most important effects that nitric oxide exerts in the vascular wall are potentially vasoprotective because these effects maintain important physiological functions such as vasodilation, anticoagulation, leukocyte adhesion, smooth muscle proliferation, and antioxidative capacity (Gewaltig and Kojda, 2002). Impairment of endothelium-dependent nitric oxide activity is believed to be due to increased production of reactive oxygen species, particularly superoxide. This has been found in atherosclerosis, hypertension, diabetes, hypercholesterolemia, heart failure, and cigarette smoking (Gewaltig and Kojda, 2002; Kurowska, 2002). Inhaled nitric oxide therapy shows promise for improving oxygenation and reducing mortality in preterm infants with pulmonary illness (Srisuparp et al., 2002). On the other hand, the mostly beneficial effects of nitrate and nitrite on blood vessel relaxation and vasodilation have been associated at times with headaches, and occupational workers chronically exposed to nitrate dusts have suffered increased mortality from strokes and heart disease (Agency for Toxic Substances and Disease Registry, 2016).

The vasodilation potential for nitrite and related compounds has led to their use (and abuse) by athletes and others as performance enhancement agents. Anonymous inquiries into an online drug information database developed for athletes showed a large increase in queries related to vasodilators, specifically those in the nitric/nitrate group, prior to the 2008 Beijing Olympics (Petróczi and Naughton, 2010). Caution against the uncontrolled use of nitrate and nitrite salts by athletes has been raised, and at least one anecdotal report of methemoglobinemia associated with unintended use of nitrite by an athlete has been described (Lundberg et al., 2011).

Evidence suggests that nitric oxide and oxygen radicals (e.g., superoxide) are generated in response to various infectious diseases (Akaika and Maeda, 2000). Nitric oxide biosynthesis, particularly through action of an inducible nitric oxide synthase, is beneficial in the case of bacterial and protozoal infections. Generation of nitric oxide in response to viral infections such as influenza virus and certain other neurotropic viruses, however, may have an overall detrimental effect upon the host. Excessive amounts of nitric oxide produced for long periods of time allows the generation of peroxynitrite which can cause oxidative tissue injury through oxidation and nitration reactions of various biomolecules. Nitric oxide also appears to adversely affect the host's immune response (Akaika and Maeda, 2000). On the positive side, recent studies demonstrate that nitric oxide is a potent inhibitor of both rhinovirus-induced cytokine production and viral replication, factors considered of importance in asthma (Sanders, 1999). It has been speculated that nitric oxide plays an important anti-inflammatory and antiviral role in rhinovirus infections and that nitric oxide donors may represent a novel therapeutic approach for the treatment of colds and their related complications (Sanders, 1999).

The effects of dietary nitrate and nitrite on human microbiota are only beginning to be understood (Lundberg et al., 2018; Koch et al., 2017). Many normal inhabitants of the human gut microbiota are sensitive to nitrite *in vitro* at levels relevant to those which might be found in the stomach after eating a nitrite-containing meal (Hrncirova et al., 2019). Shifts in the gut microbiota of mice were observed following long-term deficiency of nitrate and nitrite, although the relevance of these changes is not yet known (Kina-Tanada et al., 2017). In the acidic environment of the stomach, nitrite may help protect against enteric pathogens; however, host-derived nitrate may also facilitate the growth of some pathogens such as *Salmonella* by serving as an alternate electron acceptor for these organisms (but not all bacteria) within a competitive microbial environment (Vázquez-Torres and Bäumler, 2016; Winter et al., 2013). Nitrite- or nitrate-induced changes in the oral microbiome have been postulated to cause both blood pressure benefits and more frequent migraine headaches (Tribble et al., 2019; Ormesher et al., 2018; Gonzalez et al., 2016), while nitrite produced by commensal skin bacteria has been proposed to play a role in maintaining a healthy skin microbiome (Whitlock and Feelisch, 2009).

Research over the past 40 years has clarified many aspects of the role of nitric oxide in physiology and pathology. The research has demonstrated that nitric oxide is a double-edged sword mediator in that it can exert a beneficial or detrimental effect depending on the physiopathological context. At the time that previous editions of this chapter were published, this dual nature may have impaired the development of new drugs that would take advantage of the beneficial properties of nitric oxide (Cirino et al., 2002). However, in 2019 at the time of this writing, the clinical trials registry clinicaltrials.gov shows more than 100 clinical trials currently underway in which nitric oxide is being tested as a therapeutic agent for a wide variety of conditions, with additional studies testing oral sodium nitrite or nitrate as interventions for many different diseases.

REFERENCES

Adler-Nissen, J., Ekgreen, M. H., and Risum, J. 2014. Practical use of nitrite and basis for dosage in the manufacture of meat products. National Food Institute, Technical University of Denmark. https://www.food.dtu.dk/-/media/Institutter/Foedevareinstituttet/Publikationer/Pub-2014/Report-on-the-use-of-nitrite-2014.ashx?la=da&hash=532221A86C395DFC8D727C8459A4128EE9D7A739 (accessed February 6, 2019).

Agency for Toxic Substances and Disease Registry. 2016. Nitrate/nitrite toxicity: What are the health effects from exposure to nitrates and nitrites? Centers for Disease Control. https://www.atsdr.cdc.gov/csem/csem.asp?csem=28&po=10 (accessed May 28, 2019).

Akaike, T., and Maeda, H. 2000. Nitric oxide and virus infection. *Immunology* 101:300.

Akman, M., and Park, R. W. A. 1974. The growth of *Salmonellas* on cooked cured pork. *J. Hyg.* 72:369.

Ala-Huikka, K., Nurmi, E., Pajulahti, H., and Raevuori, M. 1977. Effect of nitrite, storage temperature and time on *Clostridium botulinum* type A toxin formation in liver sausage. *Eur. J. Appl. Microbiol.* 4:145.

Alam, M. S., Akaika, T., Okamoto, S., Kubota, T., Yoshitake, J., Sawa, T., Miyamoto, Y., Tamura, F., and Maeda, H. 2002. Role of nitric oxide in host defense in murine salmonellosis as a function of its antibacterial and antiapoptotic activities. *Infect. Immun.* 70:3130.

Alley, G., Cours, D., and Demeyers, D. 1992. Effect of nitrate, nitrite and ascorbate on colour and colour stability of dry, fermented sausage prepared using "back slopping". *Meat Sci.* 32:279.

American Cancer Society. 1996. Guidelines on diet, nutrition, and cancer prevention: Reducing the risk of cancer with healthy food choices and physical activity. The American Cancer Society 1996 Dietary Guidelines Advisory Committee, American Cancer Society, Atlanta, GA.

Anderton, J. I. 1963. Pathogenic organisms in relation to pasteurized cured meats. Scientific and Technical Surveys No. 40. British Food Manufacturing Industries Research Association, Leatherhead, Surrey.

Ann. Inst. Pasteur. 1955. An approach to the standardization of the bacteriological examination of semi-preserved canned meats. *Ann. Inst. Pasteur (Lille)* 7:265.

Anon. 1968. Cured meats need built-in safety factor. *West. Meat Ind.* 14:15.

Archer, D. L. 2002. Evidence that ingested nitrite and nitrate are beneficial to health. *J. Food Prot.* 65:872.

Ariga, H., Ohtsuka, T., Hattori, S. Ohnishi, T., and Sukegawa, K. 1984. Reduction of nitrate and nitrite in cheese and nitrate reducibility of *Penicillium candidum* as fungal starter. *Nippon Shokuhin Kogyo Gakkaishi* 31:710.

Asan, T., and Solberg, M. 1976. Inhibition of *Clostridium perfringens* by heated combinations of nitrite, sulfur, and ferrous or ferric ions. *Appl. Environ. Microbiol.* 31:49.

Ashworth, J., and Spencer, R. 1972. The Perigo effect in pork. *J. Food Technol.* 7:111.

Ashworth, J., Didcock, A., Hargreaves, L. L., Jarvis, B., Walters, C. L., and Larkworthy, L. F. 1974. Chemical and microbiological comparisons of inhibitors derived thermally from nitrite with iron thionitrosyl (Roussin black salt). *J. Gen. Microbiol.* 84:403.

Ashworth, J., Hargreaves, L. L., and Jarvis, B. 1973. The production of an antimicrobial effect in pork heated with sodium nitrite under simulated commercial pasteurization conditions. *J. Food Technol.* 8:477.

Asplund, K., Nurmi, E., Hill, P., and Him, J. 1988. The inhibition of the growth of *Bacillus cereus* in liver sausage. *Int. J. Food Microbiol.* 7:349.

Asplund, K. et al. 1993. Survival of *Yersinia enterocolitica* in fermented sausages manufactured with different levels of nitrite and different starter cultures. *Journal of Food Protection* 56(8): 710–712.

Assembly of Life Sciences. 1981. The health effects of nitrate, nitrite, and N-nitroso compounds. Part 1 of 2-part study by the committee on nitrite and alternative curing agents in food. National Research Council, National Academy Press, Washington, DC.

Assembly of Life Sciences. 1982. Alternatives to the current use of nitrite in foods. Part 2 of a 2-part study by the committee on nitrite and alternative curing agents in food. National Research Council, National Academy Press, Washington, DC.

Association of Analytical Communities International. 2002. Official Methods of Analysis of AOAC International, Association of Analytical Communities International, Gaithersburg, MD. Available at: http://www.aoac.org/index.htm.

Australian Government. 2018. Schedule 15: Substances that may be used as food additives. https://www.legislation.gov.au/Details/F2018C00915 (accessed May 11, 2019).

Bacus, J. N., and Deibel, R. H. 1972. Nitrite burn in fermented sausage. *Appl. Microbiol.* 24:405.

Baird-Parker, A. C., and Baillie, M. A. H. 1974. The inhibition of *Clostridium botulinum* by nitrite and sodium chloride. In *Proceedings of the International Symposium on Nitrite in Meat Products*, edited by B. J. Tinbergen and B. Krol, Pudoc, Wageningen, The Netherlands, p. 77.

Bakhtiary, F., Sayevand, H. R., Khaneghah, A. M., Haslberger, A. G., and Hosseini, H. 2018. Antibacterial efficacy of essential oils and sodium nitrite in vacuum processed beef fillet. *Appl. Food Biotechnol.* 5:1.

Baranyi, J., and Tamplin, M. L. 2004. ComBase: A common database on microbial responses to food environments. *J. Food Prot.* 67:1967.

Barber, L. E., and Deibel, R. H. 1972. Effect of pH and oxygen tension on staphylococcal growth and enterotoxin formation in fermented sausage. *Appl. Microbiol.* 24:891.

Bard, J., and Townsend, W. E. 1971. Meat curing. In *The Science of Meat and Meat Products*, edited by J. F. Price and B. S. Schweigert, W. H. Freeman, San Francisco, p. 452.

Bean, P. G., and Roberts, T. A. 1974. The effect of pH, sodium chloride, and sodium nitrite on heat resistance of *Staphylococcus aureus* and growth of damaged cells in laboratory media. *Proc. Int. Cong. Food Sci. Technol.* 3:93.

Bean, P. G., and Roberts, T. A. 1975. Effect of sodium chloride and sodium nitrite on the heat resistance of *Staphylococcus aureus* NCTC 10652 in buffer and meat macerate. *J. Food Technol.* 10:327.

Ben Slima, S., Ktari, N., Trabelsi, I., Triki, M., Feki-Tounsi, M., Moussa, H., Makni, I., Herrero, A., Jimenez-Colmenero, F., Perez, C. R. C., and Ben Salah, R. 2017. Effect of partial replacement of nitrite with a novel probiotic *Lactobacillus plantarum* TN8 on color, physico-chemical, texture and microbiological properties of beef sausages. *LWT-Food Sci. Technol.* 86:219.

Benedict, R. C. 1980. Biochemical basis for nitrite-inhibition of *Clostridium botulinum* in cured meat. *J. Food Prot.* 43:877.

Benedict, R. C., Partridge, T., Wells, D., and Buchanan, R. 1993. *Bacillus cereus*: Aerobic growth kinetics. *J. Food Prot.* 56:211.

Berdagué, J. L., Monteil, P., Montel, M. C., and Talon, R. 1993. Effects of starter cultures on the formation of flavour compounds in dry sausage. *Meat Sci.* 35:275.

Bhaduri, S., Buchanan, R. L., and Phillips, J. G. 1995. Expanded response surface model for predicting the effects of temperature, pH, sodium chloride content and sodium nitrite concentration on the growth rate of *Yersinia enterocolitica*. *J. Appl. Bacteriol.* 79:163.

Bhaduri, S., Turner-Jones, C. O., Buchanan, R. L., and Phillips, J. G. 1994. Response surface model of the effect of pH, sodium chloride and sodium nitrite on growth of *Yersinia enterocolitica* at low temperatures. *Int. J. Food Microbiol.* 23:333.

Binkerd, E. F., and Kolari, O. E. 1975. The history and use of nitrate and nitrite in the curing of meat. *Food Cosmet. Toxicol.* 13:655.

Birzele, B., Djordjevic, S., and Kramer, J. 2005. A study of the role of different nitrite concentrations on human pathogenic bacteria in fresh spreadable ham and onion sausage. *Food Contr.* 16:695.

Blanche Koelensmid, W. A. A., and van Rhee, R. 1968. Intrinsic factors in meat products counteracting botulinogenic conditions. *Antonie Van Leeuwenhoek* 34:287.

Blanche Koelensmid, W. A. A., and van Rhee, R. 1974. *Salmonella* in meat products. *Ann. Inst. Pasteur (Lille)* 15:85.

Boffetta, P. 2010. Causation in the presence of weak associations. *Crit. Rev. Food Sci. Nutr.* 50:13.

Bogdan, C. 2001. Nitric oxide and the immune response. *Nat. Immunol.* 2:907.

Bondonno, C. P., Croft, K. D., and Hodgson, J. M. 2016. Dietary nitrate, nitric oxide, and cardiovascular health. *Crit. Rev. Food Sci. Nutr.* 56:2036.

Botha, S. J., and Holzapfel, W. H. 1987. Resistance of *Sporolactobacillus* to potassium sorbate and sodium nitrite. *Int. J. Food Microbiol.* 5:331.

Bourret, T. J., Porwollik, S., McClelland, M., Zhao, R., Greco, T., Ischiropoulos, H., and Vázquez-Torres, A. 2008. Nitric oxide antagonizes the acid tolerance response that protects *Salmonella* against innate gastric defenses. *PLoS One* 3:e1833.

Bouvard, V., Loomis, D., Guyton, K. Z., Grosse, Y., El Ghissassi, F., Benbrahim-Tallaa, L., Guha, N., Mattock, H., and Straif, K. 2015. Carcinogenicity of consumption of red and processed meat. *Lancet Oncol.* 16:1599.

Bowen, V. G., and Deibel, R. H. 1974. Effects of nitrite and ascorbate on botulinal toxin formation in wieners and bacon. In *Proceedings of the Meat Industry Research Conference*, American Meat Institute, Chicago, IL, p. 63.

Bowen, V. G., Cerveny, J. G., and Deibel, R. H. 1974. Effect of sorbate and sodium nitrite on toxin formation of *Clostridium botulinum* in wieners. *Appl. Microbiol.* 17:605.

Bowman, L. A. H., McLean, S., Poole, R. K., and Fukuto, J. M. 2011. The diversity of microbial responses to nitric oxide and agents of nitrosative stress: Close cousins but not identical twins. *Adv. Microb. Physiol.* 59:135.

Braathen, O. 1979. Update: Nitrite free meat products. In *Proceedings of the 32nd Annual Reciprocal Meat Conference of the American Meat Science Association*, Brookings, South Dakota, National Livestock and Meat Board, Chicago, IL, p. 143.

Brooks, J., Haines, R. B., Moran, T., and Pace, J. 1940. The function of nitrate, nitrite and bacteria in the curing of bacon and hams. Department of Scientific and Industrial Research. Food Investigation Special Report No. 49. His Majesty's Stationery Office, London.

Brownlie, L. E. 1966. Effect of some environmental factors on psychrophilic microbacteria. *J. Appl. Bacteriol.* 29:447.

Bryan, N. S., Alexander, D. D., Coughlin, J. R., Milkowski, A. L., and Boffetta, P. 2012. Ingested nitrate and nitrite and stomach cancer risk: An updated review. *Food Chem. Toxicol.* 50:3646.

Buchanan, B. B., and Arnon, D. I. 1970. Ferredoxins: Chemistry and function in photosynthesis, nitrogen fixation, and fermentative metabolism. *Adv. Enzymol.* 33:119.

Buchanan, R. L., and Bagi, L. K. 1994. Expansion of response surface models for the growth of *Escherichia coli* O157:H7 to include sodium nitrite as a variable. *Int. J. Food Microbiol.* 23:317.

Buchanan, R. L., and Golden, M. H. 1995. Model for the non-thermal inactivation of *Listeria monocytogenes* in a reduced oxygen environment. *Food Microbiol.* 12:203.

Buchanan, R. L., and Phillips, J. G. 1990. Response surface model for predicting the effects of temperature, pH, sodium chloride content, sodium nitrite concentration and atmosphere on the growth of *Listeria monocytogenes*. *J. Food Prot.* 53:370.

Buchanan, R. L., and Solberg, M. 1972. Interaction of sodium nitrite, oxygen and pH on growth of *Staphylococcus aureus*. *J. Food Sci.* 37:81.

Buchanan, R. L. Golden, M. H., and Phillips, J. G. 1997. Expanded models for the non-thermal inactivation of *Listeria monocytogenes*. *J. Appl. Microbiol.* 82:567.

Buchanan, R. L., Stahl, H. G., and Whiting, R. C. 1989. Effects and interactions of temperature, pH, atmosphere, sodium chloride, and sodium nitrite on the growth of *Listeria monocytogenes*. *J. Food Prot.* 52:844.

Buchman, G. W., and Hansen, J. N. 1987. Modification of membrane sulfhydryl groups in bacteriostatic action of nitrite. *Appl. Environ. Microbiol.* 53:79.

Bulman, C., and Ayres, J. C. 1952. Preservative effect of various concentrations of curing salts in comminuted pork. *Food Technol.* 6:255.

Bushway, A. A., and Jen, K.-W. C. 1984. Residual nitrite concentration and total plate counts in white and dark chicken patties. *J. Food Prot.* 47:119.

Bushway, A. A., and Serreze, D. V. 1984. Difference in the PH and iron content of chicken white and dark meat that affect the action of nitrite against aerobes. *Can. Inst. Food Sci. Technol. J.* 17:214.

Bushway, A. A., Picker, N., and Jen, C. W. 1982. Effect of nitrite and sorbate on total number of aerobic microorganisms in chicken white and dark meat patties. *J. Food Sci.* 47:858.

Campos, I., Masot, R., Alcañiz, M., Gil, L., Soto, J., Vivancos, J. L., García-Breijo, E., Labrador, R. H., Barat, J. M., and Martínez-Mañez, R. 2010. Accurate concentration determination of anions nitrate, nitrite and chloride in minced meat using a voltammetric electronic tongue. *Sensor Actuator B Chem.* 149:71.

Canadian Food Inspection Agency. 2018. Preventive control recommendations for cooling heat processed meat products. http://www.inspection.gc.ca/food/requirements/preventive-controls-food-businesses/meat/cooling-heat-processed/eng/1522948253869/1522948254134 (accessed May 11, 2019).

Canadian Food Inspection Agency. 2019. Preventive control recommendations on the use of nitrites in the curing of meat products. http://inspection.gc.ca/food/requirements/preventive-controls-food-businesses/meat/nitrites/eng/1522949763138/1522949763434#a32 (accessed May 11, 2019).

Cantwell, M., and Elliott, C. 2017. Nitrates, nitrites and nitrosamines from processed meat intake and colorectal cancer risk. *J. Clin. Nutr. Diet.* 3:27.

Carpenter, C. E., Reddy, D. S. A, and Cornforth, D. P. 1987. Inactivation of clostridial ferredoxin and pyruvate-ferredoxin oxidoreductase by sodium nitrite. *Appl. Environ. Microbiol.* 53:549.

Cassens, R. G. 1990. *Nitrate-Cured Meat: A Food Safety Issue in Perspective.* Trumbull, CT: Food & Nutrition Press, Inc.

Cassens, R. G. 1995. Use of sodium nitrite in cured meats today. *Food Technol.* 49:72.

Cassens, R. G. 1997. Residual nitrite in cured meat. *Food Technol.* 51:53.

Cassens, R. G., Greaser, M. L., Ito, T., and Lee, M. 1979. Reactions of nitrite in meat. *Food Technol.* 33:46.

CAST (Council for Agricultural Science and Technology). 1997. Examination of dietary recommendations for salt-cured, smoked, and nitrite-preserved foods. Issue Paper No. 8, November, Council for Agricultural Science and Technology, Ames, IA.

Castellani, A. G., and Niven, C. F. 1955. Factors affecting the bacteriostatic action of sodium nitrite. *Appl. Microbiol.* 3:15.

Centers for Disease Control and Prevention. 1994. *Clostridium perfringens* gastroenteritis associated with corned beef served at St. Patrick's Day meals--Ohio and Virginia, 1993. *Morb. Mortal. Wkly. Rep.* 43:137.

Ceres Forum. 1997. Nitrite as a food additive: State of the science. A Ceres Forum organized by Georgetown University's Center for Food and Nutrition Policy, December 10–11, Washington, DC.

Cerveny, J. G. 1980. Effects of changes in the production and marketing of cured meats on the risk of botulism. *Food Technol.* 34:240.

Chang, P. C., and Akhtar, S. M. 1974. The Perigo effect in luncheon meat. *Can. Inst. Food Sci. Technol. J.* 7:117.

Chang, P. C., Akhtar, S. M., Burke, T., and Pivnick, H. 1974. Effect of sodium nitrite on *Clostridium botulinum* in canned luncheon meat: Evidence for a Perigo-type factor in the absence of nitrite. *Can. Int. Food Sci. Technol. J.* 7:209.

Chen, A. F., Ren, J., and Miao, C. Y. 2002. Nitric oxide synthase gene therapy for cardiovascular disease. *Jpn. J. Pharmacol.* 89:327.

Cherayil, B. J., and Antos, D. 2001. Inducible nitric oxide synthase and *Salmonella* infection. *Microbes Infect.* 3:771.

Chikthimmah, N., Anantheswaran, R. C., Roberts, R. F., Mills, E. W., and Knabel, S. J. 2001. Influence of sodium chloride on growth of lactic acid bacteria and subsequent destruction of *Escherichia coli* O157:H7 during processing of Lebanon bologna. *J. Food Prot.* 64:1145.

Chou, C. C., and Tsai, H-J. 1996. Antimicrobial activity of sodium nitrite against *Aeromonas hydrophila* at various pH and temperatures. *Food Sci., Taiwan* 23:159.

Christiansen, L. N. 1980. Factors influencing botulinal inhibition by nitrite. *Food Technol.* 34:237.

Christiansen, L. N., and Foster, E. M. 1965. Effect of vacuum packaging on growth of *Clostridium botulinum* and *Staphylococcus aureus* in cured meats. *Appl. Microbiol.* 13:1023.

Christiansen, L. N., Johnston, R. W., Kautter, D. A., Howard, J. W., and Aunan, W. J. 1973. Effect of nitrite and nitrate on toxin production by *Clostridium botulinum* and on nitrosamine formation in perishable canned comminuted cured meat. *Appl. Microbiol.* 25:357.

Christiansen, L. N., Tompkin, R. B., and Shaparis, A. B. 1978. Fate of *Clostridium botulinum* in perishable canned cured meat at abuse temperature. *J. Food Prot.* 41:354.

Christiansen, L. N., Tompkin, R. B., Shaparis, A. B., Johnston, R. W., and Kautter, D. A. 1975. Effect of sodium nitrite and nitrate on *Clostridium botulinum* growth and toxin production in a summer style sausage. *J. Food Sci.* 40:488.

Christiansen, L. N., Tompkin, R. B., Shaparis, A. B., Kueper, T. V., Johnston, R. W., Kautter, D. A., and Kolari, O. J. 1974. Effect of sodium nitrite on toxin production by *Clostridium botulinum* in bacon. *Appl. Microbiol.* 27:733.

Christieans, S., Picgirard, L., Parafita, E., Lebert, A., and Gregori, T. 2018. Impact of reducing nitrate/nitrite levels on the behavior of *Salmonella* Typhimurium and *Listeria monocytogenes* in French dry fermented sausages. *Meat Sci.* 137:160.

Chry, C.-Y., Walker, H. W., and Sebranek, J. G. 1980. Influence of raw ingredients, nitrite levels, and cooking temperatures on the microbiological quality of braunschweiger. *J. Food Sci.* 45:1732.

Chry, C.-Y., Walker, H. W., and Sebranek, J. G. 1981. Bacteria associated with spoilage of braunschweiger. *J. Food Sci.* 46:468.

Chumney, R. K., and Adams, D. M. 1980. Relationship between the increased sensitivity of heat injured *Clostridium perfringens* spores to surface active antibiotics and to sodium chloride and sodium nitrite. *J. Appl. Bacteriol.* 49:55.

Cirino, G., Santagada, V., and Caliendo, G. 2002. The nitric oxide related therapeutic phenomenon: A challenging task. *Curr. Pharm. Des.* 8:233.

Code of Federal Regulations. 1990a. Animals and animal products. 9CFR 317.17 and 424.21-23. U.S. Government Printing Office, Washington, DC. Available at: http://www.gpo.gov/nara/cfr/index.html.

Code of Federal Regulations. 1990b. Food and drugs. 21CFR 172.160; 172.170; 172.175; and 172.177. U.S. Government Printing Office, Washington, DC. Available at: http://www.gpo.gov/nara/cfr/index.html.

Cole, J. A. 2018. Anaerobic bacterial response to nitrosative stress. *Adv. Microb. Physiol.* 72:193.

Collinge, S. K., Mahoney, A. W., Cornforth, D. P., and Post, F. J. 1981. Effect of nitrosylation of myoglobin on gas production by *Clostridium botulinum* in pork sausage product. Paper Presented at the 41st Annual Meeting of the Institute of Food Technology, Atlanta, Georgia, June 7–10.

Collins-Thompson, D. L., and Lopez, G. R. 1981. Depletion of sodium nitrite by lactic acid bacteria isolated from vacuum-packed bologna. *J. Food Prot.* 44:593.

Collins-Thompson, D. L., and Lopez, G. R. 1982. Control of *Brochothrix thermosphacta* by *Lactobacillus* species in vacuum-packed bologna. *Can. Inst. Food Sci. Technol. J.* 15:307.

Collins-Thompson, D. L., and Thomson, I. Q. 1986. Changes in manganese content in *Lactobacillus plantarum* during inhibition with sodium nitrite. *J. Food Prot.* 49:602.

Collins-Thompson, D. L., Chang, P. C., Davidson, C. M., Larmond, E., and Pivnick, H. 1974. Effect of nitrite and storage temperature on the organoleptic quality and toxinogenesis by *Clostridium botulinum* in vacuum-packaged side bacon. *J. Food Sci.* 39:607.

Collins-Thompson, D. L., Krusky, B., Usborne, W. R., and Hauschild, A. H. W. 1984. The effect of nitrite on the growth of pathogens during manufacture of dry and semi-dry sausage. *Can. Inst. Food Sci. Technol. J.* 17:102.

Collins-Thompson, D. L., Wood, D., and Jones, M. 1982. Factors controlling the growth of Gram-negative bacteria in vacuum-packed bologna. *J. Food Prot.* 45:305.

ComBase. 2019. ComBase: A web resource for quantitative and predictive food microbiology. https://www.combase.cc/index.php/en/ (accessed April 9, 2019).

Corbion. 2019. Corbion *Listeria* control model. https://clcm.corbion.com/ (accessed April 9, 2019).

Crouch, E., and Golden, N. J. 2005. A risk assessment for *Clostridium perfringens* in ready-to-eat and partially cooked meat and poultry products. https://www.fsis.usda.gov/shared/PDF/CPerfringens_Risk_Assess_Sep2005.pdf?redirecthttp=true (accessed May 13, 2019).

Crowther, J. S., Holbrook, R., Baird-Parker, A. C., and Austin, B. L. 1976. Role of nitrite and ascorbate in the microbiological safety of vacuum-packed sliced bacon. In *Proceedings of the 2nd International Symposium on Nitrite in Meat Products*, edited by B. J. Tinbergen and B. Krol, Pudoc, Wageningen, The Netherlands, p. 13.

Cui, H., Gabriel, A. A., and Nakano, H. 2010. Antimicrobial efficacies of plant extracts and sodium nitrite against *Clostridium botulinum*. *Food Contr.* 21:1030.

Cuppett, S. L., Gray, J. I., Pestka, J. J., Booren, A. M., Price, J. F., and Kuth, C. L. 1987. Effect of salt level and nitrite on toxin production by *Clostridium botulinum* type E spores in smoked great lakes whitefish. *J. Food Prot.* 50:212.

Custer, M. C., and Hansen, J. N. 1980. Reaction of nitrite with lactoferrin or transfer-rin forms an inhibitor of bacterial spore outgrowth. Paper presented at the 80th Annual Meeting of the American Society for Microbiology, Miami Beach, Florida, May 11–16.

Cvetkovic, D., Zivkovic, V., Lukic, V., and Nikolic, S. 2019. Sodium nitrite food poisoning in one family. *Forensic Sci. Med. Pathol.* 15:102.

Dack, G. M. 1949. Botulism. In *Food Poisoning*, University of Chicago Press, Chicago, IL, p. 65.

Dalzini, E., Cosciani-Cunico, E., D'Amico, S., Sfameni, C., Bertasi, B., Losio, M. N., Serraino, A., and Daminelli, P. 2014a. Growth potential of *Listeria monocytogenes* in sliced turkey bresaola packed in modified atmosphere. *Ital. J. Food Saf.* 3:2231.

Dalzini, E., Cosciani-Cunico, E., Pavoni, E., Bertasi, B., Daminelli, P., Finazzi, G., Losio, M. N., and Varisco, G. 2014b. Study of growth potential of *Listeria monocytogenes* in low fat salami: An innovative Italian meat product. *Ital. J. Food Saf.* 3:2112.

Danish Meat Research Institute. 2019. Predictive models for meat. http://dmripredict.dk/ (accessed April 9, 2019).

Davidson, C. M., and Webb, G. 1973. The behavior of *Salmonella* in vacuum-packaged cooked cured meat products. *Can. Inst. Food Sci. Technol. J.* 6:41.

de Oliveira, T., de Araujo Soares, R., Ramos, E. M., das Gracas Cardoso, M., Alves, E., and Piccoli, R. H. 2010. Antimicrobial activity of *Satureja montana* L. essential oil against *Clostridium perfringens* type A inoculated in mortadella-type sausages formulated with different levels of sodium nitrite. *Int. J. Food Microbiol.* 144:546.

de Oliveira, T. L. C., Soares, R. D., Ramos, E. M., Cardoso, M. D., Alves, E., and Piccoli, R. H. 2011. Antimicrobial activity of *Satureja montana* L. essential oil against *Clostridium perfringens* type A inoculated in mortadella-type sausages formulated with different levels of sodium nitrite. *Int. J. Food Microbiol.* 144:546.

Deibel, R. H., and Evans, J. B. 1957. "Nitrite burn" in cured meat products— particularly in fermented sausages. Bulletin No. 32, American Meat Institute Foundation, Chicago, IL.

Demeyer, D., Honikel, K., and De Smet, S. 2008. The World Cancer Research Fund report 2007: A challenge for the meat processing industry. *Meat Sci.* 80:953.

Dias, N. A. A., Rodrigues, L. T. D., Palhares, P. C., Ramos, E. M., and Piccoli, R. H. 2015. Antimicrobial activity of essential oils on *Clostridium perfringens* type A inoculated in mortadella. *J. Food Saf.* 35:466.

Divya, K. H., and Varadaraj, M. C. 2015. Growth kinetics of a native toxigenic isolate of *Yersinia enterocolitica* CFR 2301 under the influence of incubation temperature, pH, sodium chloride and sodium nitrite. *J. Food Sci. Tech. Mys.* 52:7014.

Djeri, N., and Williams, S. K. 2014. Celery juice powder used as nitrite substitute in sliced vacuum-packaged turkey bologna stored at 4°C for 10 weeks under retail display light. *J. Food Qual.* 37:361.

Dodds, K. L., and Collins-Thompson, D. L. 1984. Incidence of nitrite depleting lactic acid bacteria in cured meats and in meat starter cultures. *J. Food Prot.* 47:7.

Dodds, K. L., and Collins-Thompson, D. L. 1985. Production of N_2O and CO_2 during the reduction of NO_2^- by *Lactobacillus lactis* TS4. *J. Food Prot.* 50:1550.

Doyle, M. E. 2002. Survival and growth of *Clostridium perfringens* during the cooling step of thermal processing of meat products. Food Research Institute. https://fri.wisc.edu/files/Briefs_File/cperfsurvivgrow.pdf (accessed May 13, 2019).

Duffes, F. 1999. Improving the control of *Listeria monocytogenes* in cold smoked salmon. *Trends Food Sci. Technol.* 10:211.

Duffy, L. L., Vanderlinde, P. B., and Grau, F. H. 1994. Growth of *Listeria monocytogenes* on vacuum-packed cooked meats: Effects of pH, a_w, nitrite and ascorbate. *Int. J. Food Microbiol.* 23:377.

Duncan, C. L. 1970. Arrest of growth from spores in semipreserved foods. *J. Appl. Bacteriol.* 33:60.

Duncan, C. L. and E. M. Foster.1968. Role of curing agents in preservation of shelf-stable canned meat products. *Applied Microbiology* 16(2): 401–405.

Duncan, C. L., and Foster, E. M. 1968a. Role of curing agents in the preservation of shelf-stable canned meat products. *Appl. Microbiol.* 16:401.

Duncan, C. L., and Foster, E. M. 1968b. Effect of sodium nitrite, sodium chloride, and sodium nitrate on germination and outgrowth of anaerobic spores. *Appl. Microbiol.* 16:406.

Duranton, F., Guillou, S., Simonin, H., Cheret, R., and de Lamballerie, M. 2012. Combined use of high pressure and salt or sodium nitrite to control the growth of endogenous microflora in raw pork meat. *Innov. Food Sci. Emerg. Technol.* 16:373.

Dussault, D., Vu, K. D., and Lacroix, M. 2016. Development of a model describing the inhibitory effect of selected preservatives on the growth of *Listeria monocytogenes* in a meat model system. *Food Microbiol.* 53:115.

Dykhuizen, R. S., Copland, M., Smith, C. C., Douglas, G., and Benjamin, N. 1995. Plasma nitrate concentration and urinary nitrate excretion in patients with gastroenteritis. *J. Infect.* 31:73.

Dykhuizen, R. S., Fraser, R., Duncan, C., Smith, C. C., Golden, M., Benjamin, N., and Leifert, C. 1996. Antimicrobial effect of acidified nitrite on gut pathogens: Importance of dietary nitrate in host defense. *Antimicrob. Agents Chemother.* 40:1422.

Eddy, B. P. 1957. The inhibitory action of nitrite. In *Proceedings of the 2nd International Symposium on Food Microbiology*, Her Majesty's Stationery Office, London.

Eddy, B. P., and Ingram, M. 1956. A salt-tolerant denitrifying *Bacillus* strain which "blows" canned bacon. *J. Appl. Bacteriol.* 19:62.

EFSA. 2003. Opinion of the scientific panel on biological hazards related to the effects of nitrites/nitrates on the microbiological safety of meat products. *EFSA J.* 2:14.

EFSA Panel on Food Additives and Nutrient Sources Added to Food. 2017. Re-evaluation of sodium nitrate (E 251) and potassium nitrate (E 252) as food additives. *EFSA J.* 15:123.

EFSA Panel on Food Additives and Nutrient Sources Added to Foods. 2017. Re-evaluation of potassium nitrite (E 249) and sodium nitrite (E 250) as food additives. *EFSA J.* 15:157.

Eklund, M. W. 1982. Significance of *Clostridium botulinum* in fishery products preserved short of sterilization. *Food Technol.* 36(12):307.

European Union. 2006. Directive 2006/52/EC of the European Parliament and of the Council of 5 July 2006 amending Directive 95/2/EC on food additives other than colours and sweeteners and Directive 94/35/EC on sweeteners for use in foodstuffs. Publication Office of the European Union. https://publications.europa.eu/en/publication-detail/-/publication/c5234caa-cfee-4e9e-8479-b174798b0b20/language-en (accessed February 6, 2019).

European Union. 2011a. Commission Regulation (EC) No. 1258/2011 amending Regulation (EC) No. 1881/2006 as regards maximum levels for nitrates in foodstuffs. http://eur-lex.europa.eu/legal-content/EN/TXT/PDF/?uri=CELEX:32011R1258&from=EN (accessed May 11, 2019).

European Union. 2011b. Commission Regulation (EU) No 1129/2011 of 11 November 2011 amending Annex II to Regulation (EC) No 1333/2008 of the European Parliament and of the Council by establishing a Union list of food additives. https://eur-lex.europa.eu/LexUriServ/LexUriServ.do?uri=OJ:L:2011:295:0001:0177:en:PDF (accessed February 6, 2018).

Evans, F. L., and Tanner, F. W. 1934. The effect of meat curing solutions on anaerobic bacteria. IV. The effect of mixed curing solutions. *Zentralbl. Bakteriol. Parasitenkd. Infektionskr. Hyg. Abt* 2 91:135.

Fang, C.-S., Post, L. S., and Solberg, M. 1985. Antimicrobial effect and disappearance of sodium nitrite in *Staphylococcus aureus* cultures. *J. Food Sci.* 50:1412.

Farber, J. M., McKellar, R. C., and Ross, W. H. 1995. Modelling the effects of various parameters on the growth of *Listeria monocytogenes* on liver pate. *Food Microbiol.* 12:447.

Farkas, J., and Andrássey, É. 1992a. Combined effects of physical treatments and sporostatic factors on *Clostridium sporogenes* spores. I. Combined effects of heat treatment, nitrite, reduced a_w and reduced pH in an anaerobic nutrient medium. *Acta. Alimentaria* 21:39.

Farkas, J., and Andrássey, É. 1992b. Combined effects of physical treatments and sporostatic factors on *Clostridium sporogenes* spores. II. Combined effects of gamma radiation, heat treatment, reduced a_w and reduced pH in canned luncheon meat. *Acta. Alimentaria* 21:49.

Farkas, J., Incze, K., and Zukal, E. 1973. Effect of nitrite on the microbiological stability of canned Vienna sausages preserved by mild heat treatment or combinations of heat and irradiation. *Acta Aliment. Acad. Sci. Hung.* 2:361.

Farkas, J, et al. 1992. Combined effects of physical treatments and sporostatic factors on clostridium-sporogenes spores: Combined effects of gamma-radiation, heat-treatment, reduced aw and reduced pH in canned luncheon meat. *Acta Alimentaria* 21(1): 49–66.

Farmer, R. 2007. The problems with some epidemiological studies. *Maturitas* 57:11.

Federal Select Agent Program. 2017. Federal select agent program overview. https://www.selectagents.gov/index.html (accessed June 3, 2019).

Fernández, M., Ordóñez, J. A., Bruna, J. M., Herranz, B., and Hoz, L. 2000. Accelerated ripening of dry fermented sausages. *Trends Food Sci. Technol.* 11:201.

Fink-Gremmels, J., Dresel, J., and Leistner, L. 1991. Use of monascus extracts as an alternative to nitrite in meat products. *Fleischwirtschaft* 71:1184.

Flores, J. 1997. Mediterranean vs Northern European meat products. Processing technologies and main differences. *Food Chem.* 59:505.

Flores, J., and Bermell, S. 1996. Dry-cured sausages. Factors influencing souring and their consequences. *Fleischwirtschaft* 76:163.

Food Chain Evaluation Consortium of the European Commission. 2016. Study on the monitoring of the implementation of Directive 2006/52/EC as regards the use of nitrites by industry in different categories of meat products. https://www.fecic.es/img/galeria/file/BUTLLETi%20INTERNACIONAL/ARXIUS%20BUTLLETI%20INTERNACIONAL/setmana%205/05.pdf (accessed May 9, 2019).

Food Safety and Inspection Service. 1986. Processing inspectors' calculations handbook. U.S. Department of Agriculture. Food Safety and Inspection Service Program Training Division, Denton, TX.

Food Safety and Quality Service. 1978. Final report on nitrites and nitrosamines. Report to the secretary of agriculture by the expert panel on nitrites and nitrosamines. U.S. Department of Agriculture, Washington, DC.

Food Safety Authority of Ireland. 2019. Use and removal of nitrite in meat products. https://www.fsai.ie/faq/use_and_removal_of_nitrite.html#ingredient_sources_nitrate (accessed May 10, 2019).

Fox, J. B. 1966. The chemistry of meat pigments. *Agric. Food Chem.* 14:207.

Fox, J. B., and Suhre, F. B. 1985. The determination of nitrite: A critical review. *Crit. Rev. Anal. Chem* 15:283.

Froehlich, D. A., Gullett, E. A., and Usborne, W. R. 1983. Effect of nitrite and salt on the color, flavor and overall acceptability of ham. *J. Food Sci.* 48:152.

Galesloot, T. E., and Hassing, F. 1983. Effect of nitrate and mixtures of these salts on the growth of coliform bacteria. Results of model experiments related to gas defects in cheese. *Neth. Milk Dairy J.* 37:1.

Gangolli, S. D., Brandt, P. A., Feron, V. J., Janzowsky, C., Koeman, J. H., Speijers, G. J. A., Spiegelhalder, B., Walker, R., and Wishnok, J. S. 1994. Assessment: Nitrate, nitrite and *N*-nitosos compounds. *Eur. J. Pharmacol., Environ. Toxicol. Pharmacol. Section* 292:1.

Garcia-Gimeno, R. M., Hervas-Martinez, C., Rodriguez-Perez, R., and Zurera-Cosano, G. 2005. Modelling the growth of *Leuconostoc mesenteroides* by artificial neural networks. *Int. J. Food Microbiol.* 105:317.

Garde, S., Gomez-Torres, N., Hernandez, M., and Avila, M. 2014. Susceptibility of *Clostridium perfringens* to antimicrobials produced by lactic acid bacteria: Reuterin and nisin. *Food Contr.* 44:22.

Gardner, G. A. 1967. Discoloration in cooked ham. *Proc. Biochem.* 2:49.

Genigeorgis, C., Riemann, H., and Sadler, W. W. 1969. Production of enterotoxin-B in cured meats. *J. Food Sci.* 34:62.

Genualdi, S., Jeong, N., and DeJager, L. 2018. Determination of endogenous concentrations of nitrites and nitrates in different types of cheese in the United States: Method development and validation using ion chromatography. *Food Addit. Contam. Part A Chem. Anal. Control Expo. Risk Assess.* 35:614.

Gerhardt, U. 1995. Preserving agents in meat products – A real necessity? *Fleischerei* 10:VI–IX.

Gewaltig, M. T., and Kojda, G. 2002. Vasoprotection by nitric oxide: Mechanisms and therapeutic potential. *Cardiovasc. Res.* 55:250.

Ghabraie, M., Vu, K. D., Huq, T., Khan, A., and Lacroix, M. 2016. Antilisterial effects of antibacterial formulations containing essential oils, nisin, nitrite and organic acid salts in a sausage model. *J. Food Sci. Tech. Mys.* 53:2625.

Gibson, A. M., and Roberts, T. A. 1986a. The effect of pH, water activity, sodium nitrite and storage temperature on the growth of enteropathogenic *Escherichia coli* and *Salmonella* in a laboratory medium. *Int. J. Food Microbiol.* 3:183.

Gibson, A. M., and Roberts, T. A. 1986b. The effect of pH, sodium chloride, sodium nitrite, and storage temperature on the growth of *Clostridium perfringens* and faecal streptococci in laboratory media. *Int. J. Food Microbiol.* 3:195.

Gibson, A. M., Roberts, T. A., and Robinson, A. 1982. Factors controlling the growth of *Clostridium botulinum* types A and B in pasteurized cured meats. IV. The effect of pig breed, cut and batch of pork. *J. Food Technol.* 17:471.

Gibson, A. M., Roberts, T. A., and Robinson, A. 1984. Factors controlling the growth of *Clostridium botulinum* types A and B in pasteurized cured meats. VI. Nitrite monitoring during storage of pasteurized pork slurries. *J. Food Technol.* 19:29.

Gilles, J., and Fryer, T. F. 1984. The effect of storage temperature on "late blowing" in Gouda cheese. *New Zealand J. Sci. Technol.* 19:83.

Glass, K., McDonnell, L. M., Sawyer, C., and Claus, J. 2008. Minimum nitrite levels required to control *Listeria monocytogenes* on ready-to-eat poultry products manufactured with lactate and diacetate. American Meat Institute Foundation. https://meatpoultryfoundation.org/namif/wp-content/uploads/05-226.pdf (accessed January 24, 2019).

Glass, K. A., and Doyle, M. P. 1989. Fate of *Listeria monocytogenes* in processed meat products during refrigerated storage. *Appl. Environ. Microbiol.* 55:1565.

Glass, K. A., Granberg, D. A., Smith, A. L., McNamara, A. M., Hardin, M., Mattias, J., Ladwig, K., and Johnson, E. A. 2002. Inhibition of *Listeria monocytogenes* by sodium diacetate and sodium lactate on wieners and cooked bratwurst. *J. Food Prot.* 65:116.

Goepfert, J. M., and Chung, K. C. 1970. Behavior of salmonellae in sliced luncheon meats. *Appl. Microbiol.* 19:190.

Golden, M. C., McDonnell, L. M., Sheehan, V., Sindelar, J. J., and Glass, K. A. 2014. Inhibition of *Listeria monocytogenes* in deli-style turkey breast formulated with cultured celery powder and/or cultured sugar-vinegar blend during storage at 4 degrees C. *J. Food Prot.* 77:1787.

Golden, M. C., Wanless, B. J., David, J. R., Lineback, D. S., Talley, R. J., Kottapalli, B., and Glass, K. A. 2017a. Effect of equilibrated pH and indigenous spoilage microorganisms on the inhibition of proteolytic *Clostridium botulinum* toxin production in experimental meals under temperature abuse. *J. Food Prot.* 80:1252.

Golden, M. C., Wanless, B. J., David, J. R. D., Kottapalli, B., Lineback, D. S., Talley, R. J., and Glass, K. A. 2017b. Effect of cultured celery juice, temperature, and product composition on the inhibition of proteolytic *Clostridium botulinum* toxin production. *J. Food Prot.* 80:1259.

Golden, N. J., Crouch, E. A., Latimer, H., Kadry, A. R., and Kause, J. 2009. Risk assessment for *Clostridium perfringens* in ready-to-eat and partially cooked meat and poultry products. *J. Food Prot.* 72:1376.

Gonzales-Barron, U., Cadavez, V., Pereira, A. P., Gomes, A., Araujo, J. P., Saavedra, M. J., Estevinho, L., Butler, F., Pires, P., and Dias, T. 2015. Relating physicochemical and microbiological safety indicators during processing of linguica, a Portuguese traditional dry-fermented sausage. *Food Res. Int.* 78:50.

Gonzalez, A., Hyde, E., Sangwan, N., Gilbert, J. A., Viirre, E., and Knight, R. 2016. Migraines are correlated with higher levels of nitrate-, nitrite-, and nitric oxide-reducing oral microbes in the American Gut Project cohort. *mSystems* 1:e00105.

Gouda, A., and El-Zayat, A. I. 1988. Effect of some preservatives on the quality of processed cheese spreads. *Egypt. J. Food Sci.* 16:135.

Gough, B. J., and Alford, J. A. 1965. Effect of curing agents on the growth and survival of food-poisoning strains of *Clostridium perfringens. J. Food Sci.* 30:1025.

Gould, G. W. 1964. Effect of food preservatives on the growth of bacteria from spores. In *Microbial Inhibitors in Food*, edited by N. Molin, Almqvist and Wiksell, Stockholm, p. 17.

Government of Canada. 2018. List of permitted preservatives. https://www.canada.ca/en/health-canada/services/food-nutrition/food-safety/food-additives/lists-permitted/11-preservatives.html (accessed February 4, 2019).

Gram, L. 2001a. Potential hazards in cold smoked fish: *Clostridium botulinum. J. Food Sci. (Suppl.)* 66:S1082.

Gram, L. 2001b. Potential hazards in cold smoked fish: *Listeria monocytogenes. J. Food Sci. (Suppl.)* 66:S1072.

Grant, G. F., and McCurdy, A. R. 1986. Effect of curing agents on the growth of *Lactobacillus viridescens* in model meat systems. *Can. Inst. Food Sci. Technol. J.* 19:78.

Grant, G. F., McCurdy, A. R., and Osborne, A. D. 1988. Bacterial greening in cured meats: A review. *Can. Inst. Food Sci. Technol. J.* 21:50.

Grass, J. E., Gould, L. H., and Mahon, B. E. 2013. Epidemiology of foodborne disease outbreaks caused by *Clostridium perfringens*, United States, 1998–2010. *Foodborne Pathog. Dis.* 10:131.

Grau, F. H. 1980. Inhibition of the anaerobic growth of *Brochothrix thermosphacta* by lactic acid. *Appl. Environ. Microbiol.* 40:433.

Gray, J. I. 1976. N-nitrosamines and their precursors in bacon: A review. *J. Milk Food Technol.* 39:686.

Gray, J. I., MacDonald, B., Pearson, A. M., and Morton, I. D. 1981. Role of nitrite in cured meat flavor: A review. *J. Food Prot.* 44:302.

Greenberg, R. A., Bladel, B. O., and Zingelmann, W. J. 1965. Radiation injury of *Clostridium botulinum* spores in cured meat. *Appl. Microbiol.* 13:743.

Greenberg, R. A., Silliker, J. H., and Fatta, L. D. 1959. The influence of sodium chloride on toxin production and organoleptic breakdown in perishable cured meat inoculated with *Clostridium botulinum. Food. Technol.* 13:509.

Grever, A. B. G. 1974. Minimum nitrite concentrations for inhibition of clostridia in cooked meat products. In *Proceedings of the International Symposium on Nitrite in Meat Products*, edited by B. J. Tinbergen and B. Krol, Pudoc, Wageningen, The Netherlands, p. 103.

Gunvig, A., Andresen, M. S., Jacobsen, T., and Borggaard, C. 2018. Staphtox predictor – A dynamic mathematical model to predict formation of *Staphylococcus* enterotoxin during heating and fermentation of meat products. *Int. J. Food Microbiol.* 285:81.

Gunvig, A., Borggaard, C., Hansen, F., Hansen, T. B., and Aabo, S. 2016. ConFerm – A tool to predict the reduction of pathogens during the production of fermented and matured sausages. *Food Contr.* 67:9.

Gunvig, A., Hansen, F., and Borggaard, C. 2013. A mathematical model for predicting growth/no-growth of psychrotrophic *C. botulinum* in meat products with five variables. *Food Contr.* 29:309.

Guyer, S., and Jemmi, T. 1991. Behavior of *Listeria monocytogenes* during fabrication and storage of experimentally contaminated smoked salmon. *Appl. Environ. Microbiol.* 57:1523.

Gwak, E., Lee, H., Lee, S., Oh, M. H., Park, B. Y., Ha, J., Lee, J., Kim, S., and Yoon, Y. 2016. Evaluation of *Salmonella* growth at low concentrations of $NaNO_2$ and NaCl in processed meat products using probabilistic model. *Asian Australas. J. Anim. Sci.* 29:1013.

Gwak, E., Oh, M. H., Park, B. Y., Lee, H., Lee, S., Ha, J., Lee, J., Kim, S., Choi, K. H., and Yoon, Y. 2015. Probabilistic models to predict *Listeria monocytogenes* growth at low concentrations of $NaNO_2$ and NaCl in frankfurters. *Korean J. Food Sci. Anim. Resour.* 35:815.

Habermeyer, M., Roth, A., Guth, S., Diel, P., Engel, K. H., Epe, B., Furst, P., Heinz, V., Humpf, H. U., Joost, H. G., Knorr, D., de Kok, T., Kulling, S., Lampen, A., Marko, D., Rechkemmer, G., Rietjens, I., Stadler, R. H., Vieths, S., Vogel, R., Steinberg, P., and Eisenbrand, G. 2015. Nitrate and nitrite in the diet: How to assess their benefit and risk for human health. *Mol. Nutr. Food Res.* 59:106.

Halvorson, H. O. 1955. Factors determining survival and development of *Clostridium botulinum* in semi-preserved meats. *Ann. Inst. Pasteur (Lille)* 7:53.

Hamdi, M., Nasri, R., Dridi, N., Moussa, H., Ashour, L., and Nasri, M. 2018. Improvement of the quality and the shelf life of reduced-nitrites turkey meat sausages incorporated with carotenoproteins from blue crabs shells. *Food Contr.* 91:148.

Hammes, W. P. 2012. Metabolism of nitrate in fermented meats: The characteristic feature of a specific group of fermented foods. *Food Microbiol.* 29:151.

Hankins, O. G., Sulzbacher, W. L., Kauffman, W. R., and Mayo, M. E. 1950. Factors affecting the keeping quality of bacon. *Food Technol.* 4:33.

Hansen, J. N., and Levin, R. A. 1975. Effect of some inhibitors derived from nitrite on macromolecular synthesis in *Bacillus cereus. Appl. Microbiol.* 30:862.

Hauschild, A. H. W. 1982. Assessment of botulism hazards from cured meat products. *Food Technol.* 36:95.

Hauschild, A. H. W., and Hilsheimer, R. 1983. Incidence of *Clostridium botulinum* in commercial liver sausage. *J. Food Prot.* 46:242.

Hauschild, A. H. W., and Simonsen, B. 1985. Safety of shelf-stable canned cured meats. *J. Food Prot.* 48:997.

Hauschild, A. H. W., and Simonsen, B. 1986. Safety assessment for shelf-stable canned cured meats—An unconventional approach. *Food Technol.* 40:155.

Hauschild, A. H. W., Hilsheimer, R., Jarvis, G., and Raymond, D. P. 1982. Contribution of nitrite to the control of *Clostridium botulinum* in liver sausage. *J. Food Prot.* 45:500.

Hechelmann, H., Bern, Z., Uchida, K., and Leistner, L. 1974. Vorkommen des tribus Klebsielleae bei kuhlgelagertem fleisch und fleischwaren. *Fleischwirtschaft* 54:1515.

Henry, M., Joubert, L., and Goret, P. 1954. Mecanisme biochimique de Faction du nitrite dans la conservation des viandes. Conditions physicochimiques favorables a son action bacteriostatique. *C. R. Soc. Biol.* 148:819.

Henry, M. et al. 1954. Mecanisme biochimique de laction du nitrite dans la conservation des viandes – conditions physicochimiques favorables a son action bacteriostatique. *Comptes Rendus des Seances de la Societe de Biologie et de Ses Filiales* 148(9–10): 819–821.

Herath, I., Priyanath, J. J. M., Ahn, D. U., and Abeyrathne, E. 2015. Use of lysozyme from chicken egg white as a nitrite replacer in an Italian-type chicken sausage. *Funct. Food Health Dis.* 5:320.

Hezel, M. P., Liu, M., Schiffer, T. A., Larsen, F. J., Checa, A., Wheelock, C. E., Carlstrom, M., Lundberg, J. O., and Weitzberg, E. 2015. Effects of long-term dietary nitrate supplementation in mice. *Redox Biol.* 5:234.

Hill, M. J. 1991. *Nitrates and Nitrites in Food and Water.* Ellis Horwood, New York, NY.

Holley, R. A. 1981. Review of the potential hazard from botulism in cured meats. *Can. Inst. Food Sci. Technol. J.* 14:183.

Hong, S. H., Kim, H. S., and Yoon, K. S. 2016. Survival and risk comparison of *Campylobacter jejuni* on various processed meat products. *Int. J. Environ. Res. Public Health* 13:14.

Honikel, K. O. 2008. The use and control of nitrate and nitrite for the processing of meat products. *Meat Sci.* 78:68.

Horsch, A. M., Sebranek, J. G., Dickson, J. S., Niebuhr, S. E., Larson, E. M., Lavieri, N. A., Ruther, B. L., and Wilson, L. A. 2014. The effect of pH and nitrite concentration on the antimicrobial impact of celery juice concentrate compared with conventional sodium nitrite on *Listeria monocytogenes. Meat Sci.* 96:400.

Hospital, X. F., Hierro, E., Arnau, J., Carballo, J., Aguirre, J. S., Gratacos-Cubarsi, M., and Fernandez, M. 2017. Effect of nitrate and nitrite on *Listeria* and selected spoilage bacteria inoculated in dry-cured ham. *Food Res. Int.* 101:82.

Hospital, X. F., Hierro, E., and Fernandez, M. 2012. Survival of *Listeria innocua* in dry fermented sausages and changes in the typical microbiota and volatile profile as affected by the concentration of nitrate and nitrite. *Int. J. Food Microbiol.* 153:395.

Hotchkiss, J. H., Helser, M. A., Maragos, C. M., and Weng, Y. M. 1992. Nitrate, nitrite and *N*-nitroso compounds: Food safety and biological implications. In *ACS Symposium Series: Food Safety Assessment*, edited by J. W. Finley, S. F. Robinson, and D. J. Armstrong, American Chemical Society, Washington, DC, pp. 400–418.

Hrncirova, L., Hudcovic, T., Sukova, E., Machova, V., Trckova, E., Krejsek, J., and Hrncir, T. 2019. Human gut microbes are susceptible to antimicrobial food additives in vitro. *Folia Microbiol.* doi: 10.1007/s12223-018-00674-z.

Hughes, A. H., and McDermott, J. C. 1989. The effect of phosphate, sodium chloride, sodium nitrite, storage temperature and pH on the growth of enteropathogenic *Escherichia coli* in a laboratory medium. *Int. J. Food Microbiol.* 9:215.

Huhtanen, C. N., and Wasserman, A. E. 1975. Effect of added sodium nitrite on the formation of clostridial inhibitors. *Appl. Microbiol.* 30:768.

Huss, H. H. 1997. Control of indigenous bacteria in seafood. *Food Contr.* 8:91–98.

Huss, H. H., Ben Embarek, P. K., and Jeppesen, V. F. 1995. Control of biological hazards in cold smoked salmon production. *Food Contr.* 6:335.

Hustad, G. O., Cerveny, J. G., Trenk, H., Deibel, R. H., Kautter, D. A., Fazio, T., Johnston, R. W., and Kolari, O. E. 1973. Effect of sodium nitrite and sodium nitrate on botulinal growth production and nitrosamine formation in wieners. *Appl. Microbiol.* 26:22.

Hyytiä, E., Eerola, S., Hielm, S., and Korkeala, H. 1997. Sodium nitrite and potassium nitrate in control of nonproteolytic *Clostridium botulinum* outgrowth and toxigenesis in vacuum packed cold smoked rainbow trout. *Int. J. Food Microbiol.* 37:63.

IBIS World. 2019. Hot dog and sausage production: Industry outlook. https://www.ibisworld.com/united-states/market-research-reports/hot-dog-sausage-production-industry/.

Incze, K. 1987. The technology and microbiology of Hungarian salami. Tradition and current status. *Fleischwirtschaft* 67:445.

Incze, K. 1989. Ripening and storage of Hungarian salami. Chemical and organoleptic changes. *Fleischwirtschaft* 69:587.

Incze, K. 1992. Raw fermented and dried meat products. *Fleischwirtschaft* 72:58.

Incze, K., Farkas, J., Mihalyi, V., and Zukal, E. 1974. Antibacterial effect of cysteine-nitrosothiol and possible precursors thereof. *Appl. Microbiol.* 27:202.

Ingram, M. 1974. The microbiological effects of nitrite. In *Proceedings of the International Symposium on Nitrite in Meat Products*, edited by B. J. Tinbergen and B. Krol, Pudoc, Wageningen, The Netherlands, p. 63.

Ingram, M., and Roberts, T. A. 1971. Application of the "D-Concept" to heat treatments involving curing salts. *J. Food Technol.* 6:21.

International Agency for Research on Cancer. 2010. Ingested nitrate and nitrite, and cyanobacterial peptide toxins. *IARC Mongr. Eval. Carcinog. Risks Hum.* https://monographs.iarc.fr/iarc-monographs-on-the-evaluation-of-carcinogenic-risks-to-humans-27/.

International Agency for Research on Cancer. 2018. Red meat and processed meat. *IARC Mongr. Eval. Carcinog. Risks Hum.* https://monographs.iarc.fr/iarc-monographs-on-the-evaluation-of-carcinogenic-risks-to-humans-2/.

International Agency for Research on Cancer, and World Health Organization. 2015. IARC Monographs evaluate consumption of red meat and processed meat. https://www.iarc.fr/wp-content/uploads/2018/07/pr240_E.pdf (accessed April 15, 2019).

Ismaiel, A. A., and Pierson, M. D. 1990. Effect of sodium nitrite and origanum oil on growth and toxin production of *Clostridium botulinum* in TYG broth and ground pork. *J. Food Prot.* 53:958.

Ivey, F. J., Shaver, K. J., Christiansen, L. N., and Tompkin, R. B. 1978. Effect of potassium sorbate on toxinogenesis by *Clostridium botulinum* in bacon. *J. Food Prot.* 41:621.

Iwata, J., Tomishima, M., and Sakamoto, H. 1965. Microbiological studies on the greening of cured meat products. I. Taxonomic characteristics of lactic acid bacteria which cause the greening. *Jpn. Soc. Sci. Fish.* 31:527.

Jackson, A. L., Kulchaiyawat, C., Sullivan, G. A., Sebranek, J. G., and Dickson, J. S. 2011a. Use of natural ingredients to control growth of *Clostridium perfringens* in naturally cured frankfurters and hams. *J. Food Prot.* 74:417.

Jackson, A. L., Sullivan, G. A., Kulchaiyawat, C., Sebranek, J. G., and Dickson, J. S. 2011b. Survival and growth of *Clostridium perfringens* in commercial no-nitrate-or-nitrite-added (natural and organic) frankfurters, hams, and bacon. *J. Food Prot.* 74:410.

Jacobs, N. J., Heady, R. E., Jacobs, J. M., Chan, K., and Deibel, R. H. 1964. Effect of hemin and oxygen tension on growth and nitrate reduction by bacteria. *J. Bacteriol.* 87:1406.

Jarvis, B., Rhodes, A. C., King, S. E., and Patel, M. 1976. Sensitization of heat-damaged spores of *Clostridium botulinum*, type B to sodium chloride and sodium nitrite. *J. Food Technol.* 11:41.

Jarvis, B., Rhodes, A. C., and Patel, M. 1979. Microbiological safety of pasteurized cured meats: Inhibition of *Clostridium botulinum* by curing salts and other additives. In *Food Microbiology and Technology*, edited by B. Jarvis, J. H. B. Christian, and H. D. Michener. Medicina Viva Servizio Congressi S.r.l., Parma, Italy, p. 251.

Jarvis, N. D. 1950. Curing of fishery products. Research Report No. 18, Fish and Wildlife Service, United States Department of the Interior, Washington, DC.

JECFA (Joint FAO/WHO Expert Committee on Food Additives). 1995. Potassium nitrate, sodium nitrite. Summary of Evaluations Performed by the Joint FAO/WHO Expert Committee on Food Additives (JECFA 1956–2001), Internet edition, ILSI Press. International Life Sciences Institute, Washington, DC. Both evaluations are available at: http://jecfa.ilsi.org/index.htm.

Jensen, L. B., and Hess, W. R. 1941. A study of the effects of sodium nitrate on bacteria in meat. *Canner* 92:82.

Jensen, L. B., Wood, I. H., and Jansen, C. A. 1934. Swelling in canned chopped hams. *Ind. Eng. Chem.* 26:1118.

Jensen, M., Madsen, J. T., and Ronnenberg, P. 1949. Kogning af dasekinker. I. Kog-ningens haemmende betydning overfor bombagefremkaldende bakterier. *Konserves* 7:105.

Jo, H., Park, B., Oh, M., Gwak, E., Lee, H., Lee, S., and Yoon, Y. 2014. Probabilistic models to predict the growth initiation time for *Pseudomonas* spp. in processed meats formulated with NaCl and $NaNO_2$. *Korean J. Food Sci. Anim. Resour.* 34:736.

Joannou, C. L., Cui, X-Y., Rogers, N., Vielotte, N., Martinez, C. L. T., Vugman, N. V., Hughes, M. N., and Cammack, R. 1998. Characterization of the bactericidal effects of sodium nitroprusside and other pentacyanonitrosyl complexes on the food spoilage bacterium *Clostridium sporogenes*. *Appl. Environ. Microbiol.* 64:3195.

Johnston, M. A., and Loynes, R. 1971. Inhibition of *Clostridium botulinum* by sodium nitrite as affected by bacteriological media and meat suspensions. *Can. Inst. Food Technol. J.* 4:179.

Johnston, M. A., Pivnick, H., and Samson, J. M. 1969. Inhibition of *Clostridium botulinum* by sodium nitrite in a bacteriological medium and in meat. *Can. Inst. Food Technol. J.* 2:52.

Johnston, R. W. 1979. Microbiological results from the four plant commercial bacon study with nitrite, nitrite/sorbate and control variables. Report submitted to D. L.

Jonsson, S., and Raa, J. 1980. Conversion of cystine or cysteine to the antibacterial compound bis-(2-amino-2-carboxyethyl) trisulfide (bactin) enhanced by ferric chloride and sodium nitrite. *J. Food Sci.* 45:1641.

Juneja, V. K., Huang, L. H., and Thippareddi, H. H. 2006. Predictive model for growth of *Clostridium perfringens* in cooked cured pork. *Int. J. Food Microbiol.* 110:85.

Juneja, V. K., Marks, H., Mohr, T., and Thippareddi, H. H. 2013. Predictive model for growth of *Clostridium perfringens* during cooling of cooked beef supplemented with NaCl, sodium nitrite and sodium pyrophosphate. *J. Food Proc. Technol.* 4:275.

Jung, S., Lee, C. W., Lee, J., Yong, H. I., Yum, S. J., Jeong, H. G., and Jo, C. 2017a. Increase in nitrite content and functionality of ethanolic extracts of *Perilla frutescens* following treatment with atmospheric pressure plasma. *Food Chem.* 237:191.

Jung, S., Lee, J., Lim, Y., Choe, W., Yong, H. I., and Jo, C. 2017b. Direct infusion of nitrite into meat batter by atmospheric pressure plasma treatment. *Innov. Food Sci. Emerg. Technol.* 39:113.

Juntilla, J., Hirn, J., Hill, P., and Nurmi, E. 1989. Effect of different levels of nitrite and nitrate on the survival of *Listeria monocytogenes* during the manufacture of fermented sausage. *J. Food Prot.* 52:158.

Kafel, S., and Ayres, J. C. 1969. The antagonism of enterococci on other bacteria in canned hams. *J. Appl. Bacteriol.* 32:217.

Kafel, S., and Jozwik, E. 1987. Effect of a short cold storage on frequency of spoilage in pasteurized (perishable) canned meat products subjected to the incubation test. *J. Food Prot.* 50:56.

Kajak, K., and Kolozyn-Krajewska, D. 2006. Construction of predictive models of growth of microorganisms in salted and cured meat products. *Innov. Food Sci. Emerg. Technol.* 7:152.

Kalinowski, R. M., Tompkin, R. B., Bodnaruk, P. W., and Pruett, W. P. 2003. Impact of cooking, cooling, and subsequent refrigeration on the growth or survival of *Clostridium perfringens* in cooked meat and poultry products. *J. Food Prot.* 66:1227.

Kanner, J., and Juven, B. J. 1980. S-nitrosocysteine as an antioxidant, color-developing and anticlostridial agent in comminuted turkey meat. *J. Food Sci.* 45:1105.

Karupiah, G., Hunt, N. H., King, N. J., and Chaudri, G. 2000. NADPH oxidase, Nrampl and nitic oxide synthase in the host antimicrobial response. *Rev. Immunogenet.* 2:387.

Kashani, H. H., Nikzad, H., Mobaseri, S., and Hoseini, E. S. 2012. Synergism effect of nisin peptide in reducing chemical preservatives in food industry. *Life Sci. J.* 9:496.

Keto-Timonen, R., Lindstrom, M., Puolanne, E., Niemisto, M., and Korkeala, H. 2012. Inhibition of toxigenesis of group II (nonproteolytic) *Clostridium botulinum* type b in meat products by using a reduced level of nitrite. *J. Food Prot.* 75:1346.

Kina-Tanada, M., Sakanashi, M., Tanimoto, A., Kaname, T., Matsuzaki, T., Noguchi, K., Uchida, T., Nakasone, J., Kozuka, C., Ishida, M., Kubota, H., Taira, Y., Totsuka, Y., Kina, S.-i., Sunakawa, H., Omura, J., Satoh, K., Shimokawa, H., Yanagihara, N., Maeda, S., Ohya, Y., Matsushita, M., Masuzaki, H., Arasaki, A., and Tsutsui, M. 2017. Long-term dietary nitrite and nitrate deficiency causes the metabolic syndrome, endothelial dysfunction and cardiovascular death in mice. *Diabetologia* 60:1138.

King, A. M., Glass, K. A., Milkowski, A. L., Seman, D. L., and Sindelar, J. J. 2016. Modeling the impact of ingoing sodium nitrite, sodium ascorbate, and residual nitrite concentrations on growth parameters of *Listeria monocytogenes* in cooked, cured pork sausage. *J. Food Prot.* 79:184.

King, A. M., Glass, K. A., Milkowski, A. L., and Sindelar, J. J. 2015a. Comparison of the effect of curing ingredients derived from purified and natural sources on inhibition of *Clostridium perfringens* outgrowth during cooling of dell-style turkey breast. *J. Food Prot.* 78:1527.

King, A. M., Glass, K. A., Milkowski, A. L., and Sindelar, J. J. 2015b. Impact of clean-label antimicrobials and nitrite derived from natural sources on the outgrowth of *Clostridium perfringens* during cooling of deli-style turkey breast. *J. Food Prot.* 78:946.

Klijn, N., Nieuwenhof, F. F. J., Hoolwerf, J. D., van der Waals, C. B., and Weerkamp, A. H. 1995. Identification of *Clostridium tyrobutyricum* as the causative agent of late blowing in cheese by species-specific PCR amplification. *Appl. Environ. Microbiol.* 61:2919.

Knobeloch, L., Salna, B., Hogan, A., Postle, J., and Anderson, H. 2000. Blue babies and nitrate-contaminated well water. *Environmental Health Perspectives* 108:675.

Knøchel, S., and Huss, H. H. 1984a. Ripening and spoilage of sugar salted herring with and without nitrate. I. Microbiological and related chemical changes. *J. Food Technol.* 19:203.

Knøchel, S., and Huss, H. H. 1984b. Ripening and spoilage of sugar salted herring with and without nitrate. II. Effect of nitrate. *J. Food Technol.* 19:215.

Ko, Y. M., Park, J. H., and Yoon, K. S. 2017. Nitrite formation from vegetable sources and its use as a preservative in cooked sausage. *J. Sci. Food Agr.* 97:1774.

Koch, C. D., Gladwin, M. T., Freeman, B. A., Lundberg, J. O., Weitzberg, E., and Morris, A. 2017. Enterosalivary nitrate metabolism and the microbiome: Intersection of microbial metabolism, nitric oxide and diet in cardiac and pulmonary vascular health. *Free Radic. Biol. Med.* 105:48.

Koppang, N. 1998. Nitrosodimethylamine as a toxic and carcinogenic substance in herring meal. *Norsk Vetrinaertidsskrift* 110:493.

Korkeala, H., Alanko., T., and Tiusanen, T. 1992. Effect of sodium nitrite and sodium chloride on growth of lactic acid bacteria. *Acta. Vet. Scand.* 33:27.

Kueck, D. R., Adams, A. T., and Froning, L. L. 1965. Process for canned meat. U.S. Patent 3,192,053. Patented June 29, 1965. U.S. Patent Office, Washington, DC.

Kurowska, E. M. 2002. Nitric oxide therapies in vascular diseases. *Curr. Pharm. Des.* 8:155.

Kushi, L. H., Doyle, C., McCullough, M., Rock, C. L., Demark-Wahnefried, W., Bandera, E. V., Gapstur, S., Patel, A. V., Andrews, K., Gansler, T., and Amer Canc, S. 2012. American Cancer Society guidelines on nutrition and physical activity for cancer prevention reducing the risk of cancer with healthy food choices and physical activity. *CA-Cancer J. Clin.* 62:30.

Labbe, R. G., and Duncan, C. L. 1970. Growth from spores of *Clostridium perfringens* in the presence of sodium nitrite. *Appl. Microbiol.* 19:353.

Lamas, A., Miranda, J. M., Vazquez, B., Cepeda, A., and Franco, C. M. 2016. An evaluation of alternatives to nitrites and sulfites to inhibit the growth of *Salmonella enterica* and *Listeria monocytogenes* in meat products. *Foods* 5:11.

Langsrud, T., and Reinbold, G. W. 1974. Flavor development and microbiology of Swiss cheese – A review. IV. Defects. *J. Milk Food Technol.* 37:26.

Lavieri, N. A., Sebranek, J. G., Cordray, J. C., Dickson, J. S., Horsch, A. M., Jung, S., Manu, D. K., Mendonca, A. F., and Stecher, B. B. 2015. Control of *Listeria monocytogenes* on alternatively cured ready-to-eat ham using natural antimicrobial ingredients in combination with post-lethality interventions. *J. Food Proc. Technol.* 6:493.

Lechowich, R. V., Brown, W. L., Deibel, R. H., and Somers, I. I. 1978. The role of nitrite in the production of canned cured meat products. *Food Technol.* 32:45.

Lechowich, R. V., Evans, J. B., and Niven, C. F. 1956. Effect of curing ingredients and procedures on the survival and growth of staphylococci in and on cured meats. *Appl. Microbiol.* 4:360.

Lee, H., Gwak, E., Lee, S., Kim, S., Lee, J., Ha, J., Oh, M. H., Park, B. Y., Yang, M., Choi, K. H., and Yoon, Y. 2016a. Model to predict growth/no growth interfaces of *Enterococcus* as a function of NaCl and $NaNO_2$. *J. Food Saf.* 36:537.

Lee, J., Gwak, E., Ha, J., Kim, S., Lee, S., Lee, H., Oh, M. H., Park, B. Y., Oh, N. S., Choi, K. H., and Yoon, Y. 2016b. Mathematical model for predicting the growth probability of *Staphylococcus aureus* in combinations of NaCl and $NaNO_2$ under aerobic or evacuated storage conditions. *Korean J. Food Sci. An.* 36:752.

Lee, J., Lee, C. W., Yong, H. I., Lee, H. J., Jo, C., and Jung, S. 2017. Use of atmospheric pressure cold plasma for meat industry. *Korean J. Food Sci. An.* 37:477.

Lee, S., Lee, H., Kim, S., Lee, J., Ha, J., Choi, Y., Oh, H., Choi, K. H., and Yoon, Y. 2018. Microbiological safety of processed meat products formulated with low nitrite concentration – A review. *Asian Australas. J. Anim. Sci.* 31:1073.

Lee, S., Lee, H., Kim, S., Lee, J., Ha, J., Gwak, E., Oh, M. H., Park, B. Y., Kim, J. S., Choi, K. H., and Yoon, Y. 2015. Probabilistic models to describe the effect of $NaNO_2$ in combination with NaCl on the growth inhibition of *Lactobacillus* in frankfurters. *Meat Sci.* 110:302.

Lee, S. H., Cassens, R. G., and Sugiyama, H. 1978. Factors affecting inhibition of *Clostridium botulinum* in canned meats. *J. Food Sci.* 43:1371.

Legan, J. D., Seman, D. L., Milkowski, A. L., Hirschey, J. A., and Vandeven, M. H. 2004. Modeling the growth boundary of *Listeria monocytogenes* in ready-to-eat cooked meat products as a function of the product salt, moisture, potassium lactate, and sodium diacetate concentrations. *J. Food Prot.* 67:2195.

Leistner, L. 1981. New nitrite regulation of West Germany. *Fleischwirtschaft* 61:252.

Leistner, L. 1986. An overall look at raw hams. *Fleischwirtschaft* 66:496, 498, 503, 504, 506, 508, 510, 567.

Leistner, L., Fink-Gremmels, J., and Dresel, J. 1991. Monascus extract – A possible alternative to nitrite in meats. In *Proceedings of 37th International Congress of Meat Science and Technology* 3:1252–1256, September 1–6, Kulmbach, Germany.

Leistner, L., Hechelmann, H., and Uchida, K. 1973. Hemmung von Enterobacteriaceae, einschliesslich salmonellen, in fleischerzeugnissen durch nitrit. In *Proceedings of the 19th of the European Meeting Meat Research Workers*, Paris, France, p. 1541.

Lijinsky, W. 1999. *N*-nitroso compounds in the diet. *Mutation Res.* 443:129.

Lin, L., Hu, J. Y., Wu, Y., Chen, M., Ou, J., and Yan, W. L. 2018. Assessment of the inhibitory effects of sodium nitrite, nisin, potassium sorbate, and sodium lactate on *Staphylococcus aureus* growth and *staphylococcal* enterotoxin A production in cooked pork sausage using a predictive growth model. *Food Sci. Human Wellness* 7:83.

Lücke, F.-K. 1982. The enrichment and isolation of *Clostridium botulinum* from raw ham. *Arch. Lebensmittelhyg.* 33:1.

Lücke, F.-K. 1994. Fermented meat products. *Food Res. Int.* 27:299.

Lücke, F.-K., Hechelmann, H., and Leistner, L. 1980. The relevance to meat products of psychrotrophic strains of *Clostridium botulinum.* In *Psychrotrophic Microorganisms in Spoilage and Pathogenicity*, edited by T. A. Roberts, G. Hobbs, J. H. B. Christian, and N. Skovgaard, Academic Press, New York, p. 491.

Lücke, F.-K., Hechelmann, H., and Leistner, L. 1982. Botulism after consumption of raw hams. *Fleischwirtschaft* 62:203.

Lundberg, J. O., Carlström, M., and Weitzberg, E. 2018. Metabolic effects of dietary nitrate in health and disease. *Cell Metab.* 28:9.

Lundberg, J. O., Larsen, F. J., and Weitzberg, E. 2011. Supplementation with nitrate and nitrite salts in exercise: A word of caution. *J. Appl. Physiol.* 111:616.

Mares, A., Neyts, K., and Debevere, J. 1994. Influence of pH, salt and nitrite on the heme-dependent catalase activity of lactic acid bacteria. *Int. J. Food Microbiol.* 24:191.

Markus, Z. H., and Silverman, G. J. 1970. Factors affecting the secretion of staphylococcal enterotoxin A. *Appl. Microbiol.* 20:492.

Marquez-Gonzalez, M., Cabrera-Diaz, E., Hardin, M. D., Harris, K. B., Lucia, L. M., and Castillo, A. 2012. Survival and germination of *Clostridium perfringens* spores during heating and cooling of ground pork. *J. Food Prot.* 75:682.

Matsuda, N., and Sekiguchi, M. 1971. Effect of curing agents on the heat tolerance of putrefactive anaerobic bacterial spores. *J. Food Hyg. Soc. Jpn.* 12:384.

Mayenobe, D., Didienne, R., and Pradel, G. 1983. Characterization of the late blowing in St Nectaire cheese and some pressed cheese. *Le Lait* 63:15.

McClure, P. J., and Roberts, T. A. 1987. The effect of incubation time and temperature on growth of *Escherchia coli* on gradient plates containing sodium chloride and sodium nitrite. *J. Appl. Bacteriol.* 63:401.

McClure, P. J., Beaumont, A. L., Sutherland, J. P., and Roberts, T. A. 1997. Predictive modelling of growth of *Listeria monocytogenes*: The effects on growth of NaCl, pH, storage temperature and $NaNO_2$. *Int. J. Food Microbiol.* 34:221.

McClure, P. J., Kelly, T. M., and Roberts, T. A. 1991. The effects of temperature, pH, sodium chloride and sodium nitrite on the growth of *Listeria monocytogenes. Int. J. Food Microbiol.* 14:77.

McNally, B., Griffin, J. L., and Roberts, L. D. 2015. Dietary inorganic nitrate: From villain to hero in metabolic disease? *Mol. Nutr. Food Res.* 60:67.

McNamara, A. M., and Lattuada, C. P. 1998. Chapter 13: Examination of meat and poultry products for *Clostridium perfringens.* In *USDA/FSIS Microbiology Laboratory Guidebook*, 3rd edition. https://www.fsis.usda.gov/wps/wcm/connect/87a04da8-5c60-4983-84a5-8b4d0b24db87/Mlgchp13.pdf?MOD=AJPERES

Meers, P. D., and Goode, D. 1965. *Salmonella typhi* in corned beef. *Lancet* 1:426.

Meester, J. 1974. Nitrate and nitrite allowances in meat products. In *Proceedings of the International Symposium on Nitrite in Meat Products*, edited by B. J. Tinbergen and B. Krol, Pudoc, Wageningen, The Netherlands, p. 265.

Mejlholm, O., and Dalgaard, P. 2007. Modeling and predicting the growth boundary of *Listeria monocytogenes* in lightly preserved seafood. *J. Food Prot.* 70:70.

Mejlholm, O., and Dalgaard, P. 2013. Development and validation of an extensive growth and growth boundary model for psychrotolerant *Lactobacillus* spp. in seafood and meat products. *Int. J. Food Microbiol.* 167:244.

Meng, Q., Sun, Y. J., and Gao, H. C. 2018. Cytochromes C constitute a layer of protection against nitric oxide but not nitrite. *Appl. Environ. Microbiol.* 84:13.

Merino, L., Örnemark, U., and Toldrá, F. 2017. Analysis of nitrite and nitrate in foods: Overview of chemical, regulatory and analytical aspects. *Adv. Food Nutr. Res.* 81:65.

Meyer, J. 1981. Comparison of carbon monoxide, nitric oxide, and nitrite as inhibitors of nitrogenase from *Clostridium pasteurianum. Arch. Biochem. Biophys.* 210:246.

Ministry of Agriculture, Fisheries, and Food. 1978. Food additives and contaminants committee. Report on the review of nitrites and nitrates in cured meats and cheese, London.

Milkowksi, A. L. 2010. Nutritional epidemiology methods and interpretation criteria. American Meat Science Association. https://meatscience.org/docs/default-source/publications-resources/rmc/2010/amsa-rmc-2010---andrew-milkowski.pdf?sfvrsn=259dbab3_0 (accessed April 16, 2019).

Milbourne, K. 1983. Thermal tolerance of *Lactobacillus viridescens* in ham. *Meat Science* 9(2): 113–119.

Milkowski, A., Garg, H. K., Coughlin, J. R., and Bryan, N. S. 2010. Nutritional epidemiology in the context of nitric oxide biology: A risk-benefit evaluation for dietary nitrite and nitrate. *Nitric Oxide-Biol. Ch.* 22:110.

Miller, A. J. amd Menichillo, D. A. 1991. Blood fraction effects on the antibotulinal efficacy of nitrite in model beef sausages. *J. Food Sci.* 56:1158.

Miller, A. J., Murphey, C. A., Call, J. E., Konieczny, P., and Uchman, W. 1993. Enhanced botulinal toxin development in beef sausages containing decolourized red blood cell fractions. *Lett. Appl. Microbiol.* 16:80.

Mills, C. E., Khatri, J., Maskell, P., Odongerel, C., and Webb, A. J. 2017. It is rocket science – Why dietary nitrate is hard to 'beet'! Part II: Further mechanisms and therapeutic potential of the nitrate-nitrite-NO pathway. *Br. J. Pharmacol.* 83:140.

Mills, P. C., Rowley, G., Spiro, S., Hinton, J. C. D., and Richardson, D. J. 2008. A combination of cytochrome c nitrite reductase (NrfA) and flavorubredoxin (NorV) protects *Salmonella enterica* serovar Typhimurium against killing by NO in anoxic environments. *Microbiology* 154:1218.

Mima, A., and Coretti, K. 1974. Uber den verbleib von nitrit in fleischwaren. II. Mitteilung. Untersuchungen iiber chemische und bakteriostatische eigenschaften verschiedener reaktionsprodukte des nitrite. *Fleischwirtschaft* 54:507.

Mirna, A., and Hofmann, K. 1969. Uber den verbleib von nitrit in fleischwaren. I. Mitteilung. Umsatz von nitrit mit sulfhydrylverbindungen. *Fleischwirtschaft* 49:1361.

Misra, N. N., and Jo, C. 2017. Applications of cold plasma technology for microbiological safety in meat industry. *Trends Food Sci. Tech.* 64:74.

Miwa, M., and Matsuura, S. 1983. Formation of antibacterial factor by heating system containing nitrite and tryptone. *Agric. Biol. Chem.* 47:687.

Mohr, T. B., Juneja, V. K., Thippareddi, H. H., Schaffner, D. W., Bronstein, P. A., Silverman, M., and Cook, L. V. 2015. Assessing the performance of *Clostridium perfringens* cooling models for cooked, uncured meat and poultry products. *J. Food Prot.* 78:1512.

Moran, D. M., Tannenbaum, S. R., and Archer, M. C. 1975. Inhibitor of *Clostridium perfringens* formed by heating sodium nitrite in a chemically denned medium. *Appl. Microbiol.* 30:838.

Morita, H., Niu, J., Sakata, R., and Nagata, Y. 1996. Red pigment of Parma ham and bacterial influence on its formation. *J. Food Sci.* 61:1021.

Morita, H., Sakata, R., and Nagata, Y. 1998. Nitric oxide complex of iron (II) myoglobin converted from metmyoglobin by *Staphylococcus xylosus. J. Food Sci.* 63:352.

Morita, H., Sakata, R., Sonoki, S., and Nagata, Y. 1994. Metmyoglobin conversion to red myoglobin derivatives and citrate utilization by bacteria obtained from meat products and pickles for curing. *Anim. Sci. Technol. (Jpn).* 65:1026.

Morse, S. A., and Baldwin, J. N. 1973. Factors affecting the regulation of staphylococcal enterotoxin B. *Infect. Immun.* 7:839.

Morse, S. A., and Mah, R. A. 1973. Regulation of staphylococcal enterotoxin B: Effect of anaerobic shock. *Appl. Microbiol.* 25:553.

Muhlig, A., Behr, J., Scherer, S., and Muller-Herbst, S. 2014. Stress response of *Salmonella enterica* serovar Typhimurium to acidified nitrite. *Appl. Environ. Microbiol.* 80:6373.

Mühlig, A., Kabisch, J., Pichner, R., Scherer, S., and Müller-Herbst, S. 2014. Contribution of the NO-detoxifying enzymes HmpA, NorV and NrfA to nitrosative stress protection of *Salmonella* Typhimurium in raw sausages. *Food Microbiol.* 42:26.

Muller-Herbst, S., Wustner, S., Kabisch, J., Pichner, R., and Scherer, S. 2016. Acidified nitrite inhibits proliferation of *Listeria monocytogenes*: Transcriptional analysis of a preservation method. *Int. J. Food Microbiol.* 226:33.

Murphy, R. S., Sadler, C. J., and Blot, W. J. 1998. Trends in cured meat consumption in relation to childhood and adult brain cancer in the United States. *Food Contr.* 9:299.

Myers, K., Cannon, J., Montoya, D., Dickson, J., Lonergan, S., and Sebranek, J. 2013. Effects of high hydrostatic pressure and varying concentrations of sodium nitrite from traditional and vegetable-based sources on the growth of *Listeria monocytogenes* on ready-to-eat (RTE) sliced ham. *Meat Sci.* 94:69.

Nagy, A., Mihályi, V., and Incze, K. 1989. Ripening and storage of Hungarian salami: Chemical and organoleptic changes. *Fleischwirtschaft* 69:587.

Narvaez-Bravo, C., Rodas-Gonzalez, A., Ding, C. M., Lopez-Campos, O., Galbraith, J., Larsen, I. L., Ye, J., Siegel, D., and Aalhus, J. L. 2017. Effects of novel nitrite packaging film on the bacterial growth of bison strip-loin steaks. *J. Food Process. Pres.* 41:7.

National Toxicology Program. 2001. Toxicology and carcinogenesis studies of sodium nitrite (CAS NO. 7632-00–0) in F344/N rats and B6C3F1 mice (drinking water studies). NIH Publication No. 01-3954:7.

Neff, R. A., Edwards, D., Palmer, A., Ramsing, R., Righter, A., and Wolfson, J. 2018. Reducing meat consumption in the USA: A nationally representative survey of attitudes and behaviours. *Public Health Nutr.* 21:1835.

Nelson, K. A., Busta, F. F., Sofos, J. N., and Wagner, M. K. 1983. Effect of polyphosphates in combination with nitrite-sorbate or sorbate on *Clostridium botulinum* growth and toxin production in chicken frankfurter emulsions. *J. Food Prot.* 46:846.

Neubauer, H., and Götz, F. 1996. Physiology and interaction of nitrate and nitrite reduction in *Staphylococcus carnosus*. *J. Bacteriol.* 178:2005.

Nielsen, H.-J. S. 1983a. Composition of bacterial flora in sliced vacuum packed bologna-type sausage as influenced by nitrite. *J. Food. Technol.* 18:371.

Nielsen, H.-J. S. 1983b. Influence of nitrite addition and gas permeability of packaging film on the microflora in a sliced vacuum-packed whole meat product under refrigerated storage. *J. Food Technol.* 18:573.

Niven, C. F., and Evans, J. B. 1957. *Lactobacillus viridescens* nov. spec., a heterofermentative species that produces a green discoloration of cured meat pigments. *J. Bacteriol.* 73:758.

Niven, C. F., Buettner, L. G., and Evans, J. B. 1954. Thermal tolerance studies on the heterofermentative lactobacilli that cause greening of cured meat products. *Appl. Microbiol.* 2:26.

Niven, C. F., Castellani, A. G., and Allanson, V. 1949. A study of the lactic acid bacteria that cause surface discolorations of sausages. *J. Bacteriol.* 58:633.

Nordin, H. R., Burke, T., Webb, G., Rubin, L. J., and van Binnendyke, D. 1975. Effect of pH, salt and nitrite in heat processed meat on destruction and out-growth of PA3679. *Can. Inst. Food Sci. Technol. J.* 8:58.

Nunez De Gonzalez, M. T., Osburn, W. N., Hardin, M. D., Longnecker, M., Garg, H. K., Bryan, N. S., and Keeton, J. T. 2012. Survey of residual nitrite and nitrate in conventional and organic/natural/uncured/indirectly cured meats available at retail in the United States. *J. Agric. Food Chem.* 60:3981.

Nurmi, E. V., and Turunen, S. O. 1970. The effect of nitrite on the growth curve of lactobacilli and micrococci. In *Proceedings of the 16th European Meeting of Meat Research Workers*, Vol. 1, Bristol, England, p. 413.

Nychas, G. J. E and Arkoudelos, J. S. 1990. *Staphylococci*: Their role in fermented sausages. *J. Appl. Bacteriol.* 19:167S–188S.

Obioha, W. I., Stahr, H. M., and Kraft, A. A. 1983. Effect of sodium nitrite on aflatoxin production in pork sausage at different temperatures and the effect of nitrite on growth of *Aspergillus parasiticus* in culture. *J. Agric. Food Chem.* 31:1039.

O'Boyle, A. R., Rubin, L. J., Diosady, L. L., Aladin-Kassam, N., Comer, F., and Brightwell, W. 1990. A nitrite-free curing system and its application to the production of wieners. *Food Technol.* 44(5):88.

O'Leary, V., and Solberg, M. 1976. Effect of sodium nitrite inhibition on intracellular thiol groups and on the activity of certain glycolytic enzymes in *Clostridium perfringens. Appl. Environ. Microbiol.* 31:208.

Omer, M. K., Alvseike, O., Holck, A., Axelsson, L., Prieto, M., Skjerve, E., and Heir, E. 2010. Application of high pressure processing to reduce verotoxigenic *E. coli* in two types of dry-fermented sausage. *Meat Sci.* 86:1005.

Omori, Y., Sakikubo, T., Nakane, M., Fuchu, H., Miake, K., Kodama, Y., Sugiyama, M., and Nishikawa, Y. 2010. Fates of foodborne pathogens in raw hams manufactured rapidly using a new patented method. *J. Food Prot.* 73:1803.

Ormesher, L., Myers, J. E., Chmiel, C., Wareing, M., Greenwood, S. L., Tropea, T., Lundberg, J. O., Weitzberg, E., Nihlen, C., Sibley, C. P., Johnstone, E. D., and Cottrell, E. C. 2018. Effects of dietary nitrate supplementation, from beetroot juice, on blood pressure in hypertensive pregnant women: A randomised, double-blind, placebo-controlled feasibility trial. *Nitric Oxide-Biol. Ch.* 80:37.

Oscar, T. P. 2008. Development and validation of a stochastic model for predicting the growth of *Salmonella typhimurium* DT104 from a low initial density on chicken frankfurters with native microflora. *J. Food Prot.* 71:1135.

Osterbauer, K. J., King, A. M., Seman, D. L., Milkowksi, A. L., Glass, K. A., and Sindelar, J. J. 2017. Effects of nitrite and erythorbate on *Clostridium perfringens* growth during extended cooling of cured ham. *J. Food Prot.* 80:1697.

Page, G. V., and Solberg, M. 1980. Nitrogen assimilation by *Salmonella typhimurium* in a chemically defined minimal medium containing nitrate, nitrite or ammonia. *J. Food Soc.* 45:75.

Page, G. V., Solberg, M., and Carman, G. M. 1985. Nitrite reductase in *Salmonella typhimurium. J. Food Safety* 7:1.

Palumbo, S. A., and Smith, J. L. 1982. Repair of heat-injured *Staphylococcus aureus* 196E on food substrates and additives and at different temperatures. *J. Food Prot.* 45:458.

Park, S. Y., Mah, J. H., Choi, Y. I. Kim, D. H., and Hwang, H. J. 1999. Optimization of red pigmentation and effect of the metabolites produced by Monascus strains on microbial inhibition and colorization in processed ham. *Korean J. Appl. Microbiol. Biotechnol.* 27:172.

Payne, M. J., Glidewell, C., and Cammack, R. 1990a. Interactions of iron-thiol-nitrosyl compounds with the phosphoroclastic system of *Clostridium sporogenes. J. Gen. Microbiol.* 136:2077.

Payne, M. J., Woods, L. F., Gibbs, P., and Cammack, R. 1990b. Electron paramagnetic resonance spectroscopic investigation of the inhibition of the phosphoroclastic system of *Clostridium sporogenes* by nitrite. *J. Gen. Microbiol.* 136:2067.

Pearson, A. M., and Tauber, F. W. 1984. *Processed Meats*, AVI Publishing, Westport, CT, p. 211.

Pegg, R. B., and Shahidi, F. 2000. *Nitrite Curing of Meat. The N-Nitrosamine Problem and Nitrite Alternatives.* Food and Nutrition Press, Trumbull, CT.

Pelroy, G. A., Eklund, M. W., Paranjpye, R. N., Suzuki, E. M., and Peterson, M. E. 1982. Inhibition of *Clostridium botulinum* types A and E toxin formation by sodium nitrite and sodium chloride in hot-process (smoked) salmon. *J. Food Prot.* 45:833.

Pelroy, G. A., Peterson, M. E., Holland, P. J., and Eklund, M. W. 1994a. Inhibition of *Listeria monocytogenes* in cold process (smoked) salmon by sodium lactate. *J. Food Prot.* 57:108.

Pelroy, G. A., Peterson, M. E., Paranjpye, R., Almond, J., and Eklund, M. 1994b. Inhibition of *Listeria monocytogenes* in cold process (smoked) salmon by sodium nitrite and packaging method. *J. Food Prot.* 57:114.

Perigo, J. A., and Roberts, T. A. 1968. Inhibition of clostridia by nitrite. *J. Food Technol.* 3:91.

Perigo, J. A., Whiting, E., and Bashford, T. E. 1967. Observations on the inhibition of vegetative cells of *Clostridium sporogenes* by nitrite which has been autoclaved in a laboratory medium, discussed in the context of sublethally processed cured meats. *J. Food Technol.* 2:377.

Peterson, M. A., Pelroy, G. A., Paranjpye, R., Poysky, F. T., Almond, J. S., and Eklund, M. W. 1993. Parameters for control of *Listeria monocytogenes* in smoked fishery products: Sodium chloride and packaging method. *J. Food Prot.* 56:938.

Petróczi, A., and Naughton, D. P. 2010. Potentially fatal new trend in performance enhancement: A cautionary note on nitrite. *J. Int. Soc. Sport Nutr.* 7:25.

Pin, C. Benito, Y., Carcia, M. L., Selgas, D., Tormo, J., and Casas, C. 1996. Influence of temperature, pH, sodium chloride and sodium nitrite on the growth of clinical and food motile *Aeromonas* spp. strains. *Arch. Lebensmittelhygiene* 47:50.

Pivnick, H. 1980. Curing salts and related materials. In *Microbial Ecology of Foods*, Academic Press, New York, p. 136.

Pivnick, H., and Barnett, H. 1965. Effect of salt and temperature on toxinogenesis by *Clostridium botulinum* in perishable cooked meats vacuum-packed air-impermeable plastic pouches. *Food Technol.* 19:1164.

Pivnick, H., and Bird, H. 1965. Toxinogenesis by *Clostridium botulinum* types A and E m perishable cooked meats vacuum packed in plastic pouches. *Food Technol.* 19:1156.

Pivnick, H., and Chang, P.-C. 1974. Perigo effect in pork. In *Proceedings of the International Symposium on Nitrite in Meat Products*, edited by B. J. Tinbergen and B. Krol, Pudoc, Wageningen, The Netherlands, p. 111.

Pivnick, H., and Petrosovits, A. 1973. A rationale for the safety of canned shelf stable cured meat: Protection = destruction + inhibition. In *Proceedings of the 19th European Meeting of Meat Research Workers*, Paris, France, p. 1086.

Pivnick, H., Barnett, H. W., Nordin, H. R., and Rubin, L. J. 1969. Factors affecting the safety of canned, cured shelf-stable luncheon meat inoculated with *Clostridium botulinum*. *Can. Inst. Food Technol. J.* 2:141.

Pivnick, H., Johnston, M. A., Thacker, C., and Loynes, R. 1970. Effect of nitrite on destruction and germination of *Clostridium botulinum* and putrefactive anaerobes 3679 and 3679h in meat and in buffer. *Can. Inst. Food Technol. J.* 3:103.

Pivnick, H., Rubin, L. J., Barnett, H. W., Nordin, H. R., Ferguson, P. A., and Perrin, C. H. 1967. Effect of sodium nitrite and temperature on toxinogenesis by *Clostridium botulinum* in perishable cooked meats vacuum-packed in air-impermeable plastic pouches. *Food Technol.* 21:100.

Pollan, M. 2008. *In Defense of Food: An Eater's Manifesto.* Penguin Press.

Portocarrero, S. M., Newman, M., and Mikel, B. 2002. Reduction of *Listeria monocytogenes*, *Salmonella* spp., and *Escherichia coli* O157:H7 during processing of country-cured hams. *J. Food Sci.* 67:1892.

Possas, A., Perez-Rodriguez, F., Valero, A., Rincon, F., and Garcia-Gimeno, R. M. 2018. Mathematical approach for the *Listeria monocytogenes* inactivation during high hydrostatic pressure processing of a simulated meat medium. *Innov. Food Sci. Emerg. Technol.* 47:271.

Posthuma, J. A., Rasmussen, F. D., and Sullivan, G. A. 2018. Effects of nitrite source, reducing compounds, and holding time on cured color development in a cured meat model system. *LWT-Food Sci. Tech.* 95:47.

Puolanne, E. 1977. The effect of lowered addition of nitrite and nitrate on the properties of dry sausage. *J. Sci. Agric. Soc. Finland* 49:1.

Qvist, S., and Bernbom, N. 2000. Effect of sodium lactate, sodium nitrite, pH and temperature on growth of *Listeria monocytogenes*. Paper Presented at the Second NSF International Conference on Food Safety, October 11–13, Savannah, GA.

Raccach, M., and Henningsen, E. C. 1997. The effect of chloride salts on *Yersinia enterocolitica* in meat. *Food Microbiol.* 14:431.

Raevuori, M. 1975. Effect of nitrite and erythorbate on growth of *Bacillus cereus* in cooked sausage and in laboratory media. *Zentralbl. Bakteriol. Hyg. I. Abt. Orig. B.* 161:280.

Reddy, D., Lancaster, J. R., and Cornforth, D. P. 1983. Nitrite inhibition of *Costridium botulinum*: Electron-spin resonance detection of iron nitric-oxide complexes. *Science* 221:769.

Redondo-Solano, M., Valenzuela-Martinez, C., Cassada, D. A., Snow, D. D., Juneja, V. K., Burson, D. E., and Thippareddi, H. 2013. Effect of meat ingredients (sodium nitrite and erythorbate) and processing (vacuum storage and packaging atmosphere) on germination and outgrowth of *Clostridium perfringens* spores in ham during abusive cooling. *Food Microbiol.* 35:108.

Reynolds, A. E., Harrison, M. A., Rose-Morrow, R., and Lyon, C. E. 2001. Validation of dry cured ham process for control of pathogens. *J. Food Sci.* 66:1373.

Rhodes, A. C., and Jarvis, B. 1976. A pork slurry system for studying inhibition of *Clostridium botulinum* by curing salts. *J. Food Technol.* 11:13.

Rice, K. M., and Pierson, M. D. 1982. Inhibition of *Salmonella* by sodium nitrite and potassium sorbate in frankfurters. *J. Food Sci.* 47:1615.

Richardson, A. R., Payne, E. C., Younger, N., Karlinsey, J. E., Thomas, V. C., Becker, L. A., Navarre, W. W., Castor, M. E., Libby, S. J., and Fang, F. C. 2011. Multiple targets of nitric oxide in the tricarboxylic acid cycle of *Salmonella enterica* serovar Typhimurium. *Cell Host Microbe* 10:33.

Riel, G., Boulaaba, A., Popp, J., and Klein, G. 2017. Effects of parsley extract powder as an alternative for the direct addition of sodium nitrite in the production of mortadella-type sausages impact on microbiological, physicochemical and sensory aspects. *Meat Sci.* 131:166.

Riemann, H. 1963. Safe heat processing of canned cured meats with regard to bacterial spores. *Food Technol.* 17:39.

Riha, W. E., and Solberg, M. 1975a. *Clostridium perfringens* inhibition by sodium nitrite as a function of pH, inoculum size and heat. *J. Food Sci.* 40:439.

Riha, W. E., and Solberg, M. 1975b. *Clostridium perfringens* growth in a nitrite containing defined medium sterilized by heat or filtration. *J. Food Sci.* 40:443.

Riordan, D. C., Duffy, G., Sheridan, J., Eblen, B. S., Whiting, R. C., Blair, I. S., and McDowell, D. A. 1998. Survival of *Escherichia coli* O157:H7 during the manufacture of pepperoni. *J. Food Prot.* 61:146.

Roberts, T. A., and Dainty, R. H. 1991. Nitrite and nitrate as food additives: Rationale and mode of action. In *Nitrates and Nitrites in Food and Water*, edited by M. J. Hill, Ellis Horwood, New York, NY.

Roberts, T. A., and Garcia, C. E. 1973. A note on the resistance of *Bacillus* spp., faecal streptococci and *Salmonella typhimurium* to an inhibitor of *Clostridium* spp. formed by heating sodium nitrite. *J. Food Technol.* 8:463.

Roberts, T. A., and Gibson, A. M. 1986. Chemical methods for controlling *Clostridium botulinum* in processed meats. *Food Technol.* 40(4):164.

Roberts, T. A., and Ingram, M. 1966. Effect of sodium chloride, potassium nitrate and sodium nitrite on the recovery of heated bacterial spores. *J. Food Technol.* 1:147.

Roberts, T. A., and Smart, J. L. 1974. Inhibition of spores of *Clostridium* spp. by sodium nitrite. *J. Appl. Bacteriol.* 37:261.

Roberts, T. A., Gibson, A. M., and Robinson, A. 1981a. Factors controlling the growth of *Clostridium botulinum* types A and B in pasteurized, cured meats. I. Growth in pork slurries prepared from "low" pH meat (pH ranges 5.5–6.3). *J. Food Technol.* 16:239.

Roberts, T. A., Gibson, A. M., and Robinson, A. 1981b. Factors controlling the growth of *Clostridium botulinum* types A and B in pasteurized, cured meats. II. Growth in pork slurries prepared from "high" pH meat (range 6.3–6.8). *J. Food Technol.* 16:267.

Roberts, T. A., Gibson, A. M., and Robinson, A. 1981c. Prediction of toxin production by *Clostridium botulinum* in pasteurized pork slurry. *J. Food Technol.* 16:337.

Roberts, T. A., Gibson, A. M., and Robinson, A. 1982. Factors controlling the growth of *Clostridium botulinum* types A and B in pasteurized, cured meats. III. The effect of potassium sorbate. *J. Food Technol.* 17:307.

Roberts, T. A., Jarvis, B., and Rhodes, A. C. 1976. Inhibition of *Clostridium botulinum* by curing salts in pasteurized pork slurry. *J. Food Technol.* 11:25.

Roberts, T. R., Woods, L. F. J., Payne, M. J., and Cammack, R. 1991. Nitrite. In *Food Preservatives*, edited by N. J. Russell and G. W. Gould, Blackie and Son, Ltd., Glasgow.

Rørvik, L. M. 2000. *Listeria monocytogenes* in the smoked fish industry. *Int. J. Food Microbiol.* 62:183–190.

Rowe, J. J., Yarbrough, J. M., Rake, J. B., and Eagan, R. G. 1979. Nitrite inhibition of aerobic bacteria. *Curr. Microbiol.* 2:51.

Sameshima, T., Takeshita, K., Miki, T., Arihara, K., Itoh, M., and Kondo, Y. 1997. Effect of sodium nitrite and sodium lactate on the growth of lactic acid spoilage bacteria isolated from cured meat products. *Japanese J. Food Microbiol.* 13:159.

Sanders, S. P. 1999. Asthma, viruses, and nitric oxide. *J. Experimental Biol. Med.* 220:123.

Sanz, Y., Vila, R., Toldrá, F., and Flores, J. 1998. Effect of nitrate and nitrite curing salts on microbial changes and sensory quality of non-fermented sausages. *Int. J. Food Microbiol.* 42:213.

Scannell, A. G. M., Arendt, E. K., Hill, C., Ross, R. P., and Schwarz, G. 2001. Effect of nitrite on a bacteriocinogenic *Lactococcus lactis* transconjugant in fermented sausage. *Eur. Food Res. Technol.* 213:48.

Schack, W. R., and Taylor, R. E. 1966. Extending shelf life of cured meats. U.S. Patent 3,258,345. Patented June 28, 1966. U.S. Patent Office, Washington, DC.

Schairer, D. O., Chouake, J. S., Nosanchuk, J. D., and Friedman, A. J. 2012. The potential of nitric oxide releasing therapies as antimicrobial agents. *Virulence* 3:271.

Schillinger, U., and Lücke, F.-K. 1989. Inhibiting salmonellae growth in fresh, spread-able Mettwurst (dry sausage eaten relatively fresh) made with sugar. *Fleischwirtschaft* 69:879.

Schmidt, C. F., and Segner, W. P. 1964. The bacteriology of type E *Clostridium botulinum*. In *Proceedings of the Meat Industry Research Conference*. American Meat Institute Foundation, Chicago, IL, p. 13.

Schmidt, U., and Kaya, M. 1990. Behaviour of *Listeria monocytogenes* in vacuum-packed sliced frankfurter-type sausage. *Fleischwirtschaft* 70:236.

Science Media Centre. 2015. Expert reaction to IARC classification of processed meat as "carcinogenic to humans" and red meat as "probably carcinogenic to humans". http://www.sciencemediacentre.org/xpert-reaction-to-iarc-classification-of-processed-meat-as-carcinogenic-to-humans-group-1-and-red-meat-as-probably-carcinogenic-to-humans-group-2a/ (accessed April 16, 2019).

Scott, W. J. 1955. Factors in canned ham controlling *Cl. botulinum* and *Staph. aureus*. *Ann. Inst. Pasteur (Lille)* 7:68.

Sebranek, J. G., and Bacus, J. N. 2007. Cured meat products without direct addition of nitrate or nitrite: What are the issues? *Meat Sci.* 77:136.

Seman, D. L., Borger, A. C., Meyer, J. D., Hall, P. A., and Milkowski, A. L. 2002. Modeling the growth of *Listeria monocytogenes* in cured ready-to-eat processed meat products by manipulation of sodium chloride, sodium diacetate, potassium lactate, and product moisture content. *J. Food Prot.* 65:651.

Seman, D. L., Quickert, S. C., Borger, A. C., and Meyer, J. D. 2008. Inhibition of *Listeria monocytogenes* growth in cured ready-to-eat meat products by use of sodium benzoate and sodium diacetate. *J. Food Prot.* 71:1386.

Shahamat, M., Seaman, A., and Woodbine, M. 1980. Influence of sodium chloride, pH and temperature on the inhibitory activity of sodium nitrite on *Listeria monocytogenes*. In *Microbial Growth and Survival in Extremes of Environment*, edited by G. W. Gould and J. E. L. Corry, Academic Press, New York, p. 227.

Shahidi, F., and Pegg, R. B. 1992. Nitrite-free meat curing systems: Update and review. *Food Chem.* 43:185.

Shank, J. L., and Lundquist, B. R. 1963. The effect of packaging conditions on the bacteriology, color, and flavor of table-ready meats. *Food Technol.* 17:83.

Shank, J. L., Silliker, J. H., and Harper, R. H. 1962. The effect of nitric oxide on bacteria. *Appl. Microbiol.* 10:185.

Shanker, D., and Mulvany, L. 2019. That 'natural' bacon you bought may contain nitrates. *Los Angeles Times*. https://www.latimes.com/business/la-fi-hormel-natural-bacon-nitrates-antibiotics-preservatives-20190410-story.html (accessed April 18, 2019).

Shaw, B. G., and Harding, C. D. 1978. The effect of nitrate and nitrite on the microbial flora of Wiltshire bacon after maturation and vacuum-packed storage. *J. Appl. Bacteriol.* 45:39.

Shiloh, M. U., and Nathan, C. F. 2000. Reactive nitrogen intermediates and the pathogenesis of *Salmonella* and mycobacteria. *Curr. Opin. Microbiol.* 3:35.

Silla, H., and Simonsen, B. 1985. Shelf life of cured, cooked and sliced meat products. I. Influence of composition, vacuum packaging and modified atmospheres. *Fleischwirtschaft* 65:66.

Silliker, J. H. 1959. The effect of curing salts on bacterial spores. In *Proceedings of the Meat Industry Research Conference*, American Meat Institute Foundation, Chicago, IL, p. 51.

Silliker, J. H., Greenberg, R. A., and Schack, W. R. 1958. Effect of individual curing ingredients on the shelf stability of canned comminuted meats. *Food Technol.* 12:551.

Simone, A. J., and Solberg, M. 1981. Effect of nitrite on anaerobic glucose transport. Paper Presented at the 41st Annual Meeting of the Institute of Food Technology, Atlanta, Georgia, June 7–10.

Sindelar, J. J., Cordray, J. C., Olson, D. G., Sebranek, J. G., and Love, J. A. 2007a. Investigating quality attributes and consumer acceptance of uncured, no-nitrate/nitrite-added commercial hams, bacons, and frankfurters. *J. Food Sci.* 72:S551.

Sindelar, J. J., Cordray, J. C., Sebranek, J. G., Love, J. A., and Ahn, D. U. 2007b. Effects of varying levels of vegetable juice powder and incubation time on color, residual nitrate and nitrite, pigment, pH, and trained sensory attributes of ready-to-eat uncured ham. *J. Food Sci.* 72:S388.

Skovgaard, N. 1992. Microbiological aspects and technological need: Technological needs for nitrates and nitrites. *Food Addit. Contamin.* 9:391.

Sofos, J. N., Busta, F. F., and Allen, C. E. 1979a. Botulism control by nitrite and sorbate in cured meats: A review. *J. Food Prot.* 42:739.

Sofos, J. N., Busta, F. F., Bhothipaska, K., and Allen, C. E. 1979b. Sodium nitrite and sorbic acid effects on *Clostridium botulinum* toxin formation in chicken frankfurter-type emulsions. *J. Food Sci.* 44:668.

Sofos, J. N., Busta, F. F., and Allen, C. E. 1979c. Effects of sodium nitrite on *Clostridium botulinum* toxin production in frankfurter emulsions formulated with meat and soy proteins. *J. Food Sci.* 44:1267.

Sofos, J. N., Busta, F. F., and Allen, C. E. 1979d. *Clostridium botulinum* control by sodium nitrite and sorbic acid in various meat and soy protein formulations. *J. Food Sci.* 44:1662.

Sofos, J. N., Busta, F. P., and Allen, C. E. 1979e. Sodium nitrite and sorbic acid effects on *Clostridium botulinum* spore germination and total microbial growth in chicken frankfurter emulsions during temperature abuse. *Appl. Environ. Microbiol.* 37:1103.

Sofos, J. N., Busta, F. F., and Allen, C. E. 1980a. Influence of pH on *Clostridium botulinum* control by sodium nitrite and sorbic acid in chicken emulsions. *J. Food Sci.* 45:7.

Sofos, J. N., Busta, F. F., Bhothipaksa, K., Allen, C. E., Robach, M. C., and Paquette, M. W. 1980b. Effects of various concentrations of sodium nitrite and potassium sor-bate on *Clostridium botulinum* toxin production in commercially prepared bacon. *J. Food Sci.* 45:1285.

Song, P., Wu, L., and Guan, W. 2015. Dietary mitrates, nitrites, and nitrosamines intake and the risk of gastric cancer: A meta-analysis. *Nutrients* 7:9872.

Spencer, R. 1966. Processing factors affecting stability and safety of non-sterile canned cured meats. *Food Manuf.* 41:39.

Srisuparp, P., Heitschmidt, M., and Schreiber, M. D. 2002. Inhaled nitric oxide therapy in premature infants with mild to moderate respiratory distress syndrome. *J. Med. Assoc. Thai.* 85 Supplement 2:S469.

Steinke, P. K. W., and Foster, E. M. 1951. Botulinum toxin formation in liver sausage. *Food Res.* 16:477.

Stiles, M. E., and Ng, L.-K. 1979. Fate of pathogens inoculated onto vacuum-packaged sliced hams to simulate contamination during packaging. *J. Food Prot.* 42:464.

Stiles, M. E., Ng, L.-K., and Paradis, D. C. 1979. Survival of enteropathogenic bacteria on artificially contaminated bologna. *Can. Inst. Food. Sci. Technol. J.* 12:128.

Stumbo, C. R., Gross, C. E., and Vinton, C. 1945a. Bacteriological studies relating to thermal processing of canned meats. II. Influence of a putrefactive anaerobic bacterium in meat. *Food Res.* 10:283.

Stumbo, C. R., Gross, C. E., and Vinton, C. 1945b. Bacteriological studies relating to thermal processing of canned meats. III. Influence of meat-curing agents upon growth of a putrefactive anaerobic bacterium in meat. *Food Res.* 10:293.

Sucu, C., and Turp, G. Y. 2018. The investigation of the use of beetroot powder in Turkish fermented beef sausage (sucuk) as nitrite alternative. *Meat Sci.* 140:158.

Sullivan, G. A., Jackson-Davis, A. L., Niebuhr, S. E., Xi, Y., Schrader, K. D., Sebranek, J. G., and Dickson, J. S. 2012a. Inhibition of *Listeria monocytogenes* using natural antimicrobials in no-nitrate-or-nitrite-added ham. *J. Food Prot.* 75:1071.

Sullivan, G. A., Jackson-Davis, A. L., Schrader, K. D., Xi, Y., Kulchaiyawat, C., Sebranek, J. G., and Dickson, J. S. 2012b. Survey of naturally and conventionally cured commercial frankfurters, ham, and bacon for physio-chemical characteristics that affect bacterial growth. *Meat Sci.* 92:808.

Szczawinski, J., Szczawinska, M., and Szulc, M. 1989. Effect of irradiation on antibotulinal efficacy of nitrite. *J. Food Sci.* 54:1313.

Talon, R., Walter, D., Chartier, S., Barrière, C., and Montel, M. C. 1999. Effect of nitrate and incubation conditions on the production of catalase and nitrate reductase by *staphylococci*. *Int. J. Food Microbiol.* 52:47–56.

Tanaka, N., Meske, L., Doyle, M. P., Traisman, E., Thayer, D. W., and Johnston, R. W. 1985. Plant trials of bacon made with lactic acid bacteria, sucrose and lowered sodium nitrite. *J. Food Prot.* 48:679.

Tanaka, N., Traisman, E., Lee, M. H., Cassens, R. G., and Foster, E. M. 1980. Inhibition of botulinum toxin formation in bacon by acid development. *J. Food Prot.* 43:450.

Tanner, F. W., and Evans, F. L. 1933. Effect of meat curing solutions on anaerobic bacteria. II. Sodium nitrate. *Zentrabl. Baketeriol. Parasitenkd. Infektionskr. Hyg.* 88(2):48.

Tanner, F. W., and Evans, F. L. 1934. Effect of meat curing solutions on anaerobic bacteria. III. Sodium nitrite. *Zentralbl. Bakteriol. Parasitenkd. Infektionskr. Hyg.* 91(2):1.

Taormina, P. 2018. Revised USDA cooking and cooling compliance guidelines: Impact on validation and process deviation. *Food Safety Magazine.* https://www.foodsafetymagazine.com/magazine-archivel/december-2018january-2019/revised-usda-cooking-and-cooling-compliance-guidelines-impact-on-validation-and-process-deviation/ (accessed February 13, 2019).

Taormina, P. J., and Dorsa, W. J. 2004. Growth potential of *Clostridium perfringens* during cooling of cooked meats. *J. Food Prot.* 67:1537.

Taormina, P. J., Bartholomew, G. W., and Dorsa, W. J. 2003. Incidence of *Clostridium perfringens* in commercially produced cured raw meat product mixtures and behavior in cooked products during chilling and refrigerated storage. *J. Food Prot.* 66:72.

Tarr, H. L. A. 1941. The action of nitrites on bacteria. *J. Fish. Res. Board Can.* 5:265.

Tarr, H. L. A. 1942. The action of nitrites on bacteria: Further experiments. *J. Fish Res. Board Can.* 6:74.

Tarr, H. L. A.; and Sutherland, P. A. 1940. The comparative value of preservatives for fresh fillets. *J. Fish. Res. Board Can.* 5:148.

Technical University of Denmark. Undated. Food spoilage and safety predictor. http://fssp.food.dtu.dk/default.aspx (accessed April 9, 2019).

Thatcher, F. S., Robinson, J., and Erdman, I. 1962. The "vacuum pack" method of packaging foods in relation to the formation of the botulinum and staphylococcal toxins. *J. Appl. Bacteriol.* 25:120.

Thomas, L. V., Wimpenny, J. W. T., and Peters, A. C. 1991. An investigation of the effects of four variables on the growth of *Salmonella typhimurium* using two types of gradient gel plates. *Int. J. Food Microbiol.* 14:261.

Tjaberg, T. B., and Kvale, O. 1972. The effect of sodium nitrite on germination and growth of spores from *Clostridium botulinum* type B in a meat product. In *Proceedings of the 18th Meeting of Meat Research Workers*, Guelph, ON, p. 119.

Tompkin, R. B. 1978. The role and mechanism of the inhibition of *C. botulinum* by nitrite—is a replacement available? In *Proceedings of the 31st Annual Reciprocal Meat Conference of American Meat Science Association*, Storrs, CT, p. 135.

Tompkin, R. B. 1980. Botulism from meat and poultry products—a historical perspective. *Food Technol.* 34:229.

Tompkin, R. B. 1986. Microbiology of ready-to-eat meat products. In *Advances in Meat Research, Volume 2, Meat and Poultry Microbiology*, edited by A. M. Pearson and T. R. Dutson, AVI Publishing, Westport, CT, p. 89.

Tompkin, R. B. 2005. Nitrite. *Antimicrobials in Food, Third Ed.* 143:169.

Tompkin, R. B., Ambrosino, J. M., and Stozek, S. K. 1973. Effect of pH, sodium chloride, and sodium nitrite on enterotoxin A production. *Appl. Microbiol.* 26:833.

Tompkin, R. B., Christiansen, L. N., and Shaparis, A. B. 1977. Variation in inhibition of *C. botulinum* by nitrite in perishable canned comminuted cured meat. *J. Food Sci.* 42:1046.

Tompkin, R. B., Christiansen, L. N., and Shaparis, A. B. 1978. Enhancing nitrite inhibition of *Clostridium botulinum* with isoascorbate in perishable canned cured meat. *Appl. Environ. Microbiol.* 35:59.

Tompkin, R. B., Christiansen, L. N., and Shaparis, A. B. 1978a. Effect of iron on botulinal inhibition in perishable canned cured meat. *J. Food Technol.* 13:521.

Tompkin, R. B., Christiansen, L. N., and Shaparis, A. B. 1978b. Effect of prior refrigeration on botulinal outgrowth in perishable canned cured meat when temperature abused. *Appl. Environ. Microbiol.* 35:863.

Tompkin, R. B., Christiansen, L. N., and Shaparis, A. B. 1978c. Causes of variation in botulinal inhibition in perishable canned cured meat. *Appl. Environ. Microbiol.* 35:886.

Tompkin, R. B., Christiansen, L. N., and Shaparis, A. B. 1978d. Antibotulinal role of isoascorbate in cured meat. *J. Food Sci.* 43:1368.

Tompkin, R. B., Christiansen, L. N., and Shaparis, A. B. 1978e. The effect of iron on botulinal inhibition in perishable canned cured meat. *J. Food Technol.* 13:521.

Tompkin, R. B., Christiansen, L. N., and Shaparis, A. B. 1979a. Iron and the antibotulinal efficacy of nitrite. *Appl. Environ. Microbiol.* 37:351.

Tompkin, R. B., Christiansen, L. N., and Shaparis, A. B. 1979b. Isoascorbate level and botulinal inhibition in perishable canned cured meat. *J. Food Sci.* 44:1147.

Tompkin, R. B., Christiansen, L. N., and Shaparis, A. B. 1980. Antibotulinal efficacy of sulfur dioxide in meat. *Appl. Environ. Microbiol.* 39:1096.

Tribble, G. D., Angelov, N., Weltman, R., Wang, B. Y., Eswaran, S. V., Gay, I. C., Parthasarathy, K., Dao, D. H. V., Richardson, K. N., Ismail, N. M., Sharina, I. G., Hyde, E. R., Ajami, N. J., Petrosino, J. F., and Bryan, N. S. 2019. Frequency of tongue cleaning impacts the human tongue microbiome composition and enterosalivary circulation of nitrate. *Front. Cell Infect. Mi.* 9:16.

Tsai, S. H., and Chou, C. C. 1996. Injury, inhibition and inactivation of *Escherichia coli* O157:H7 by potassium sorbate and sodium nitrite as affected by pH and temperature. *J. Sci. Food Agric.* 71:10.

Tseng, Y. Y.., Chen, M. T., and Lin, C. F. 2000. Growth, pigment production and protease activity of *Monascus purpureus* as affected by salt, sodium nitrite, polyphosphate and various sugars. *J. Appl. Microbiol.* 88:31.

Tucker, W. H. 1930. *Studies on Clostridium putrificum and Clostridium putrefaciens.* Institute of American Meat Packers, Chicago, IL.

Turantaş, F., and A. Ünlütürk. 1991. The effect of sodium nitrite and pH on the growth of *Salmonella typhimurium* and *Staphylococcus aureus. Chem. Mikrobiol. Technol. Lebensmittel* 13:167.

Turantaş, F., and A. Ünlütürk. 1993. The effect of nitrite, garlic and starter culture on the survival of *Salmonella typhimurium* in Turkish soudjuk. *J. Sci. Food Agric.* 61:95.

U.S. Environmental Protection Agency. 2018. National primary drinking water regulations. https://www.epa.gov/ground-water-and-drinking-water/national-primary-drinking-water-regulations#Inorganic (accessed February 1, 2019).

Umbreit, J. 2007. Methemoglobin—It's not just blue: A concise review. *Am. J. Hematol.* 82:134.

USDA Agricultural Research Service. Undated. Pathogen modeling program. https://pmp.errc.ars.usda.gov/tutorial.aspx (accessed April 9, 2019).

USDA Food Safety and Inspection Service. 1995. Processing inspectors' calculations handbook. https://www.fsis.usda.gov/wps/wcm/connect/7d364131-137e-4da3-905b-fa240974a5a9/7620-3.pdf?MOD=AJPERES (accessed July 17, 2017).

USDA Food Safety and Inspection Service. 1995. Processing inspectors' calculations handbook. http://www.fsis.usda.gov/OPPDE/rdad/FSISDirectives/7620-3.pdf (accessed November 5, 2015).

USDA Food Safety and Inspection Service. 1999. Performance standards for the production of certain meat and poultry products: Final rule. *Feder. Regis.* 64:732.

USDA Food Safety and Inspection Service. 2001. Performance standards for the production of processed meat and poultry products: Proposed rule. *Feder. Regist.* 66:12589.

USDA Food Safety and Inspection Service. 2014. FSIS compliance guideline: Controlling *Listeria monocytogenes* in post-lethality exposed ready-to-eat meat and poultry products. http://www.fsis.usda.gov/wps/wcm/connect/d3373299-50e6-47d6-a577-e74a1e549fde/Controlling-Lm-RTE-Guideline.pdf?MOD=AJPERES (accessed May 11, 2016).

USDA Food Safety and Inspection Service. 2017. FSIS compliance guideline for stabilization (cooling and hot-holding) of fully and partially heat-treated RTE and NRTE meat and poultry products produced by small and very small establishments and revised Appendix B. https://www.fsis.usda.gov/wps/wcm/connect/9ac49aba-46bc-443c-856b-59a3f51b924f/Compliance-Guideline-Stabilization-Appendix-B.pdf?MOD=AJPERES (accessed June 21, 2017).

USDA Food Safety and Inspection Service. 2018a. Part 1 of 3: Use of celery powder and other natural sources of nitrite as curing agents, antimicrobials or flavorings. https://askfsis.custhelp.com/app/answers/detail/a_id/2029/kw/nitrite (accessed April 18, 2019).

USDA Food Safety and Inspection Service. 2018b. Part 2 of 3: Revised Appendix B: Stabilization option 3 for products containing natural sources of nitrite and natural sources of ascorbate or ascorbic acid. https://askfsis.custhelp.com/app/answers/detail/a_id/2030/kw/nitrite/related/1 (accessed May 21, 2019).

USDA Food Safety and Inspection Service. 2018c. Safe and suitable ingredients used in the production of meat, poultry, and egg products, rev. 48. https://www.fsis.usda.gov/wps/wcm/connect/bab10e09-aefa-483b-8be8-809a1f051d4c/7120.1.pdf?MOD=AJPERES (accessed January 16, 2019).

USDA Food Safety and Inspection Service. 2019. FSIS notice: Extension of delayed implementation of verification of revised Appendix A and B. https://www.fsis.usda.gov/wps/wcm/connect/05853dc9-3551-401f-914d-c5940243f8cc/01-19.pdf?MOD=AJPERES (accessed February 4, 2019).

Vahabzadeh, F. et al. 1983. Evaluation of iron-binding compounds as inhibitors of gas and toxin production by clostridium-botulinum in ground pork. *Journal of Food Science* 48(5): 1445–1451.

Van der Avoort, C. M. T., Van Loon, L. J. C., Hopman, M. T. E., and Verdijk, L. B. 2018. Increasing vegetable intake to obtain the health promoting and ergogenic effects of dietary nitrate. *European Journal of Clinical Nutrition* 72:1485.

Van Roon, P. S. 1974. Inhibitors in cooked meat products. In *Proceedings of the International Symposium on Nitrite in Meat Products*, edited by B. J. Tinbergen and B. Krol, Pudoc, Wageningen, The Netherlands, p. 117.

Van Roon, P. S. 1979a. Effect of reaction products from nitrite on *Clostridium sporogenes* in heated cured meats. *Fleischwirtschaft* 59:542.

Van Roon, P. S. 1979b. *Effect of Reaction Products from Nitrite on Clostridium sporogenes.* Drukkerij Elinkwij BV, Utrecht.

Vázquez-Torres, A., and Bäumler, A. J. 2016. Nitrate, nitrite and nitric oxide reductases: From the last universal common ancestor to modern bacterial pathogens. *Curr. Opin. Microbiol.* 29:1.

Verhoeven, W. 1950. On a spore-forming bacterium causing swelling of cans containing cured hams. *Antonie Van Leeuwenhoek* 16:269.

Vinton, C., Martin, S., and Gross, C. E. 1947. Bacteriological studies relating to thermal processing of canned meats. VII. Effect of substrate upon thermal resistance of spores. *Food Res.* 12:173.

Viuda-Martos, M., Fernandez-Lopez, J., Sayas-Barbera, E., Sendra, E., Navarro, C., and Perez-Alvarez, J. A. 2009. Citrus co-products as technological strategy to reduce residual nitrite content in meat products. *J. Food Sci.* 74:R93.

Vrchlabsky, J., and Leistner, L. 1971. The heat resistance of lactobacilli at different aw values. *Fleischwirtschaft* 54:1368.

Wagner, M. K., and Busta, F. F. 1983. Effect of sodium acid pyrophosphate in combination with sodium nitrite or sodium nitrite/potassium sorbate on *Clostridium botulinum* growth and toxin production in beef/pork frankfurter emulsions. *J. Food Sci.* 48:990.

Wang, Z. X. and Y. Y. Shao. 2018. Effects of microbial diversity on nitrite concentration in pao cai, a naturally fermented cabbage product from China. *Food Microbiology* 72: 185–192.

Wang, Q. H., Yu, L. J., Liu, Y., Lin, L., Lu, R. G., Zhu, J. P., He, L., and Lu, Z. L. 2017. Methods for the detection and determination of nitrite and nitrate: A review. *Talanta* 165:709.

Wang, R., Teng, C. G., Zhang, N., Zhang, J., and Conway, G. 2013. A family cluster of nitrite poisoning, Suzhou City, Jiangsu Province, China, 2013. *Western Pac. Surveill. Response J.* 4:33.

Wanless, B. J., McDonnell, L. M., and Glass, K. A. 2010. Effect of nitrite and natural antimicrobials on *Clostridium botulinum* toxigenesis in model frankfurter and ham products. *Institute of Food Technologists Annual Meeting Abstracts.*

Warnecke, M. O., Carpenter, J. A., and Saffle, R. L. 1967. A study of *Clostridium botulinum* type E, in vacuum packaged meat. *Food Technol.* 21:433.

Wasserman, A. E., and Huhtanen, C. N. 1972. Nitrosamines and the inhibition of clostridia in medium heated with sodium nitrite. *J. Food Sci.* 37:785.

Whitlock, D., and Feelisch, M. 2009. Soil bacteria, nitrite and the skin. In *The Hygiene Hypothesis and Darwinian Medicine*, edited by G. A. W. Rook, Birkhäuser, Basel.

Whittenbury, R. 1964. Hydrogen peroxide formation and catalase activity in the lactic acid bacteria. *J. Gen Microbiol.* 35:13–26.

Winter, S. E., Winter, M. G., Xavier, M. N., Thiennimitr, P., Poon, V., Keestra, A. M., Laughlin, R. C., Gomez, G., Wu, J., Lawhon, S. D., Popova, I. E., Parikh, S. J., Adams, L. G., Tsolis, R. M., Stewart, V. J., and Baumler, A. J. 2013. Host-derived nitrate boosts growth of *E. coli* in the inflamed gut. *Science* 339:708.

Wojton, B., Kossakowska, A., Ryglewicz, Z., and Moczybroda, J. 1978. Effect of sodium nitrite on the stability of pasteurized canned meat. *Bull. Vet. Inst. Pulawy* 22:38.

Wolf, G., and Hammes, W. P. 1988. Effect of hematin on the activities of nitrite reductase and catalase in lactobacilli. *Arch. Microbiol.* 149:220.

Wolf, G., Arendt, E. K., Pfähler, U., and Hammes, W. P. 1990. Heme-dependent and heme-independent nitrite reduction by lactic acid bacteria results in different N-containing products. *Int. J. Food Microbiol.* 10:323.

Wolf, G., Strahl, A., Meisel, J., and Hammes, W. P. 1991. Heme-dependent catalase activity of lactobacilli. *Int. J. Food Microbiol.* 12:133.

Woods, L. F. J., and Wood, J. M. 1982. A note on the effect of nitrite inhibition on the metabolism of *Clostridium botulinum. J. Appl. Bacteriol.* 52:109.

Woods, L. F. J., Wood, J. M., and Gibbs, P. A. 1981. The involvement of nitric oxide in the inhibition of the phosphoroclastic system in *Clostridium sporogenes* by sodium nitrite. *J. Gen. Microbiol.* 125:399.

World Cancer Research Fund, and American Institute for Cancer Research. 2007. Food, nutrition, physical activity, and the prevention of cancer: A global perspective. http://discovery.ucl.ac.uk/4841/1/4841.pdf (accessed April 15, 2019).

Xie, L., Mo, M., Jia, H. X., Liang, F., Yuan, J., and Zhu, J. 2016. Association between dietary nitrate and nitrite intake and site-specific cancer risk: Evidence from observational studies. *Oncotarget* 7:56915.

Yarbrough, J. M., Rake, J. B., and Eagon, R. G. 1980. Bacterial inhibitory effects of nitrite: Inhibition of active transport, but not of group translocation, and of intracellular enzymes. *Appl. Environ. Microbiol.* 39:831.

Yesair, J., and Cameron, E. J. 1942. Inhibitive effect of curing agents on anaerobic spores. *Canner* 94:89.

Yu, C. H., and Chou, C. C. 1996a. Effect of curing salts on the growth of hemorrhagic *Escherichia coli* O157:H7 in ground pork at different temperatures. *J. Chinese Agric. Chem. Soc.* 34:165.

Yu, C. H., and Chou, C. C. 1996b. Fate of *Escherichia coli* O157:H7 in Chinese-style sausage during the drying step of the manufacturing process as affected by the drying condition and curing agent. *J. Sci. Food Agric.* 74:551.

Yu, C. F. and C. C. Chou. 1997. Fate of *Escherichia coli* O157:H7 in Chinese-style sausage during the drying step of the manufacturing process as affected by the drying condition and curing agent. *Journal of the Science of Food and Agriculture.* 74(4): 551–556.

Zaika, L. L., Kim, A. H., and Ford, L. 1991. Effect of sodium nitrite on growth of *Shigella flexneri*. *J. Food Prot.* 54:424.

Zaika, L. L., Moulden, E., Weimer, L., Phillips, J. G., and Buchanan, R. L. 1994. Model for the combined effects of temperature, initial pH, sodium chloride and sodium nitrite concentrations on anaerobic growth of *Shigella flexneri*. *Int. J. Food Microbiol.* 23:345.

Zaika, L. L., Phillips, J. G., Fanelli, J. S., and Scullen, O. J. 1998. Revised model for aerobic growth of *Shigella flexneri* to extend the validity of predictions at temperatures between 10 and 19 degrees C. *Int. J. Food Microbiol.* 41:9.

Zeuthen, P. 1980. Some observations on the bacterial growth in sliced, vacuum-packed ham in relation to pH of the product. In *Proceedings of 26th European Meeting of Meat Research Workers*, Volume 2, American Meat Science Association National Live Stock and Meat Board, Chicago, IL, p. 265.

Zhang, F. X., Miao, Y., Ruan, J. G., Meng, S. P., Dong, J. D., Yin, H., Huang, Y., Chen, F. R., Wang, Z. C., and Lai, Y. F. 2019. Association between nitrite and nitrate intake and risk of gastric cancer: A systematic review and meta-analysis. *Med. Sci. Monit.* 25:1788.

Zurera-Cosano, G., Castillejo-Rodriguez, A. M., Garcia-Gimeno, R. M., and Rincon-Leon, F. 2004. Performance of response surface and Davey model for prediction of *Staphylococcus aureus* growth parameters under different experimental conditions. *J. Food Prot.* 67:1138.

Zurera-Cosano, G., Garcia-Gimeno, R. M., Rodriguez-Perez, R., and Hervas-Martinez, C. 2006. Performance of response surface model for prediction of *Leuconostoc mesenteroides* growth parameters under different experimental conditions. *Food Contr.* 17:429.

8

Nisin

Noushin Eghbal, Nour-Eddine Chihib, and Adem Gharsallaoui

CONTENTS

8.1 Introduction

The bacteriocin nisin belongs to class-Ia bacteriocin known as lantibiotics and is the most characterized and commercially used antimicrobial peptide (Ross et al., 2002). Nisin is a heat-stable and cationic low-molecular-weight peptide produced by the bacterial dairy starter *Lactococcus lactis* subsp. *lactis*. Generally, application of nisin is carried out for inhibiting spore germination and growth of pathogens spoiling the surface of food products. The growth of a wide spectrum of Gram-positive bacteria, including the pathogens *Listeria* and *Staphylococcus* spp., and spore-forming bacteria such as *Bacillus* and *Clostridium* can be prevented by nisin (Silva et al., 2018). This bacteriocin was commercialized for the first time in the 1950s for the prevention of late blowing in cheese caused by gas-producing *Clostridium* spp. (García et al., 2010; Galvez et al., 2008). Nisin is produced naturally in raw milk and dairy products mainly by *Lactococcus lactis*, but complex nutrition conditions are required for its production by the bacterium, resulting in increasing its production costs (Gharsallaoui, et al., 2016a). Thus, commercial nisin preparations containing 2.5% of pure nisin are standardized to 10^6 IU/g with NaCl and denatured milk proteins. One gram of pure nisin has the activity of 40×10^6 IU, so 1 μg of pure nisin has a biological activity of 40 IU (Müller-Auffermann et al., 2015).

Gram-negative bacteria, yeasts, and molds are resistant toward nisin, and such microorganisms should be killed by other preservation techniques like pasteurization treatments and chemical antimicrobials. Some Gram-positive bacteria develop resistance to nisin through various mechanisms which were reviewed by Zhou et al. (2017). Regarding the resistance of Gram-negative bacteria like *Escherichia coli* and *Salmonella* spp. to nisin, chelators like ethylenediaminetetraacetic acid (EDTA), heat treatment, and freezing can be combined with the bacteriocin to disrupt the lipopolysaccharide (LPS) layer and increase the interactions between nisin and the cytoplasmic membrane (Belfiore et al., 2007). Moreover, thermally injured spores are more sensitive to nisin, and extended shelf life can be achieved with a combination of nisin and a reduced heat treatment.

Nisin was approved for use in food in 1969 by the Joint Food and Agriculture Organization (FAO)/World Health Organization Committee on Food Additives and has been approved as generally recognized as safe (GRAS) by the US Food and Drug Administration (FDA) since 1988 (FDA, 1988; Favaro et al., 2015). It is the only bacteriocin allowed to be used in food as a preservative (E234) without having adverse effects on human health and the "FAO/WHO Codex Committee on Milk and Milk Products" has permitted the use of nisin with a concentration of 12.5 mg per kilogram (Ross et al., 2002).

Before its development as a food preservative there were attempts to use nisin as a medical drug, but these attempts failed due to nisin's lack of activity against Gram-negative bacteria and its rapid digestion and breakdown within the body. Generally antibiotics are secondary metabolites. Nisin is a primary metabolite produced by a process involving ribosomal transcription and translation. Nisin is not used therapeutically in human or veterinary medicine nor as an animal feed additive or for growth promotion. Various studies have shown that nisin will not contribute to antibiotic drug resistance. For instance, passaging of bacteria in media containing sub-lethal concentrations of nisin did not alter the sensitivity of the test organisms to therapeutic antibiotics and other chemotherapeutic drugs (Hossack et al., 1983). We recommend that nisin (and other bacteriocins) should be considered as antimicrobial peptides and not referred to as antibiotics.

8.2 Structure

Several factors including molecular weight, biological activity, and post-translational modifications are used for the classification of bacteriocins produced by LAB. Another classification criterion is the antibacterial activity of the bacteriocin against *Listeria monocytogenes* due to the crucial problems of listeriosis disease. Generally, bacteriocins are classified into three classes: class I (lantibiotics), class II (small, heat-stable, and non-lantibiotics peptides), and class III (large and heat-sensitive proteins). Lantibiotics are the most important bacteriocins containing an unusual amino acid named lanthionine (Lan). Nisin is a widely investigated lantibiotic (Gharsallaoui et al., 2016a). It is made up of 34 amino acids with a molar mass of 3500 Da displaying cationic and hydrophobic characteristics (de Aruaz et al., 2009). There are several unusual amino acids in the structure of nisin such as dehydroalanine (Dha), dehydrobutyrine (Dhb), aminobutyric acid (Aba), and β-methyllanthionine (Punyauppa-path et al., 2015).

Many nisin variants with different antimicrobial activities have been studied such as nisin A, nisin Z, nisin Q, nisin F, nisin U, and nisin U2. Nisin A, Z, Q, and F are produced by *Lactococcus lactis* while the other nisin variants are produced by *Streptococcus* spp. (Punyauppa-path et al., 2015). Nisin A, Z, and Q are the main variants (Figure 8.1) and possess different biological activities. However, nisin A and Z are the most active peptides of these which are often marketed. These two variants are differentiated

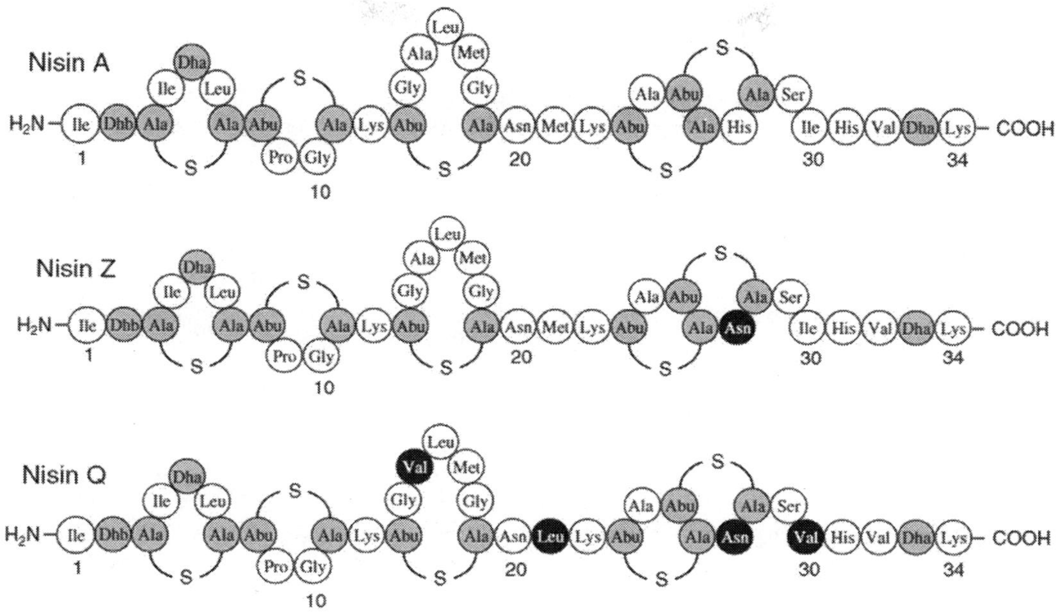

FIGURE 8.1 Main natural nisin variants. The black-filled residues indicate the substituted residues as compared with nisin A. The grey-filled residues indicate unusual amino acids. Ala-S-Ala, lanthionine; Abu-S-Ala, 3-methyllanthionine; S, the sulfur atom of the thioether linkage. (Adapted from: Fukao et al., 2008.)

by one amino acid at position 27 that is histidine for nisin A and asparagine for nisin Z. This difference in the structure between the two nisin variants does not have any effect on their biological activity or their thermal stability, but nisin Z has higher diffusion properties and solubility at neutral pH than nisin A (Gharsallaoui et al., 2016a; de Aruaz et al., 2009).

8.3 Physicochemical Properties

The aqueous solubility and structural stability of nisin are pH-dependent and affect its antimicrobial efficacy. Nisin shows higher solubility and stability under acidic conditions up to 12 wt% solubility at pH 2.5 and 4 wt% at pH 5.0. At higher pH values and near to neutrality, the solubility of nisin reaches zero. The net charge of nisin is pH-dependent (isoelectric point ~8–9). The formation of multimers *via* inter-molecular interactions results in structural modifications which occur at a pH range higher than the isoelectric point (Gharsallaoui et al., 2016a). Reduction in the biological activity of nisin upon increasing pH can be attributed to the occurrence of irreversible modifications of the nisin molecular structure. Nisin has a flexible, three-dimensional structure which is determined by its internal thio-ether rings. The absence of aromatic amino acids in the nisin structure is the reason for its having no absorbance at 280 nm unlike other proteins.

Nisin is amphiphilic in character; the N-terminal contains several hydrophobic residues whereas the C-terminal is more hydrophilic. Nisin has been shown to have emulsifying activity in a comparative study with Tween-80 (a common nonionic food emulsifier) and β-casein (a food emulsion stabilizer), but this activity was highly dependent on pH and nisin concentration (Bani-Jaber et al., 2000). Nisin has no significant taste and imparts no negative taste to foods in which it is used. In triangle taste trials conducted by Danisco, volunteers could not detect 200 mg/l Nisaplin® in mineral water (equivalent to 5 μg pure nisin/g).

Heating the antimicrobial peptide can contribute to its inactivation while mild and low temperatures can preserve the nisin stability. According to Vessoni et al. (2005), it is possible to preserve the nisin activity during autoclaving at 121°C for 15 min at pH 2.0 indicating high thermal stability of the bacteriocin at lower pH values.

8.4 Antimicrobial Activity

The mechanism of nisin action against target bacteria (Figure 8.2) includes adsorption on the cell surface causing the disintegration of the cytoplasmic membrane structure. Positively charged nisin can be adsorbed on the target cell by establishing electrostatic interactions with membrane phospholipids having negative charges. Mutagenesis of nisin's charged amino acids demonstrated the importance of the positive charges in the carboxyl terminal region for membrane binding. When the anionic lipid content of the membranes was varied, it was found that the amount of nisin binding increased dramatically at levels above 40% anionic lipid content (Demel et al., 1996; Breukink et al., 1997; van Kraaij et al., 1997). This indicated that nisin needed a relatively large amount of anionic lipids for efficient binding (Breukink and de Kruijff, 1999). The nature and content of the cell membrane phospholipids play an important role in determining the sensitivity of different bacterial strains to nisin.

The hydrophobicity of the N-terminal extremity of nisin acts subsequently to allow the insertion of nisin in the lipid cell membrane, leading to its permeabilization and dissipating the proton motive force. This inhibits the uptake of amino acids and promotes rapid efflux of small metabolites, ions, or cytoplasmic solutes such as amino acids and nucleotides (Jung and Sahl, 1991; Abee, 1995; Chihib et al., 1999a, 1999b). It is described in one of the several nisin action modes that lipid II, a precursor molecule in the synthesis of peptidoglycan, can establish a complex with nisin. By insertion of the formed complex into the cytoplasmic membrane, pores can be formed, leading to the efflux of essential cellular compounds and thus death of the bacteria (Gharsallaoui et al., 2016a). Cells containing low levels of lipid II were shown to be less sensitive to nisin (Brotz et al., 1998). Nisin mutant peptides have since been identified that were unable to form pores but retained considerable antimicrobial activity by blocking the incorporation of lipid II into peptidoglycan, thus inhibiting cell wall synthesis (Wiedemann et al., 2001). This demonstrated that nisin has two potentially cidal mechanisms.

Budde and Jakobsen (2000) examined real-time interactions between nisin and cells on a solid surface. The method was based on fluorescence ratio-imaging microscopy and measurements of changes in the intracellular pH of carboxyfluorescein succinimidyl ester-stained cells during exposure to nisin. In the absence of nisin the intracellular pH of the cells was almost constant at pH 8.0, and independent of the external pH that ranged from pH 5–9. In the presence of nisin, dissipation of the pH gradient was observed, which was dependent on nisin content and time. In this system, immobilized cells were significantly less sensitive to nisin than planktonic cells.

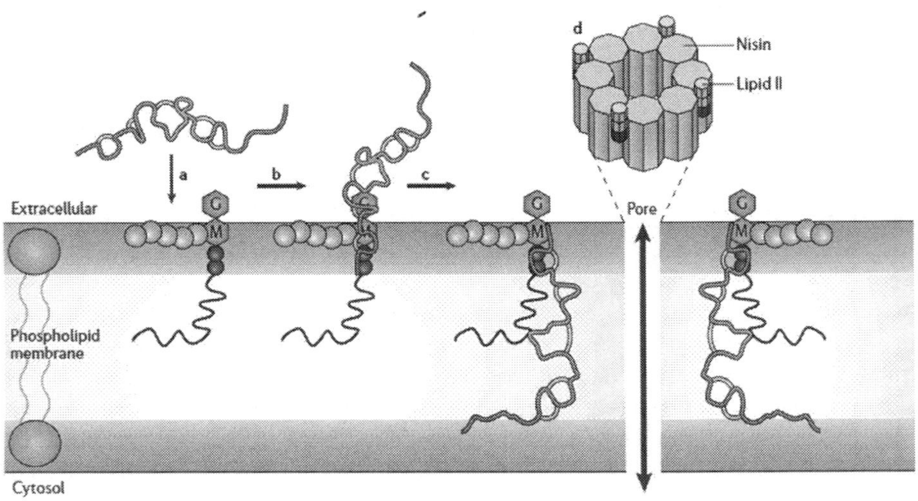

FIGURE 8.2 A schematic representation of the mechanism of action of nisin (from Breukink & de Kruijff, 2006). First, nisin reaches the bacterial plasma membrane (**a**), where it binds to Lipid II *via* two of its amino-terminal rings (**b**). This is then followed by pore formation (**c**), which involves a stable transmembrane orientation of nisin. During or after assembly of four 1:1 (nisin: Lipid II) complexes, four additional nisin molecules are recruited to form the pore complex (**d**).

Pathogenic bacteria such as *Listeria monocytogenes* and *Clostridium botulinum* are known to be nisin-sensitive. Nisin has a strong bactericidal activity against Gram-positive bacteria (*Streptococcus* spp., *Staphylococcus* spp., and *Listeria* spp.), especially spore-forming ones. Nisin-sensitive microorganisms relevant to food are shown in Table 8.1. Vegetative forms and the outgrowth of spores of spore-forming bacteria can be inhibited by nisin. Although it is speculated that fermented foods are pathogen-free products, survival and growth of *Listeria* spp. phenomena have been reported in fermented foods prepared from contaminated raw materials (Bastos and Ceotto, 2011). However, it should be mentioned that some Gram-positive bacteria can inactivate nisin through synthesizing an enzyme named nisinase (Ray, 1992).

Heat treatments usually eliminate all vegetative cells but surviving spore-forming bacteria can spoil foods. Being active against *Bacillus* and *Clostridium* bacterial cells and their heat-resistant spores, nisin can be used as a preservative in pasteurized or heat-treated foods that are not fully sterilized. Spores of significant food spoilage and pathogenic bacteria such as *Bacillus cereus* and *Clostridium botulinum* were found to have higher sensitivity to nisin than vegetative cells especially after heat treatment. Inhibition of spore germination by nisin is associated with pore formation in the cell membranes in the early germination stages (Bastos and Ceotto, 2011; Bierbaum and Sahl, 2009). The outcome of nisin activity can also vary from achieving total kill of the cells to a growth inhibitory effect (Delves-Broughton et al., 1996).

TABLE 8.1

Gram-Positive Species Sensitive to Nisin

Genus	Species	Description
Alicyclobacillus	*acidoterrestris*	Heat-resistant spore former; growth at pH 2.5–6, 25–60°C; spoilage organism of fresh/pasteurized fruit juice stored at ambient temperature
Bacillus	*brevis, cereus, coagulans, licheniformis, macerans, megaterium, pumilis, subtilis, sporothermodurans, stearothermophilus*	Heat-resistant aerobic and facultative anaerobic spore formers; includes psychrotrophs, thermophiles, acid-resistant, UHT-resistant, spoilage organisms (cause of "rope" in bread and flat sour spoilage), and food-poisoning pathogens
Brochothrix	*thermosphacta*	Heat-sensitive spoilage organisms of meat; growth between 0 and 30°C. Very nisin-sensitive
Clostridium	*bifermentans, botulinum, butyricum, histolyticum, pasteurianum, perfringens, putrificum, sordelli, sporogenes, tertium, thermosaccharolyticum, tyrobutyricum*	Heat-resistant spore-forming obligate anaerobes; cause spoilage (blowing, etc.) and food-poisoning
Desulfotomaculum	*nigrificans*	Heat-resistant spore-forming thermophiles; obligate anaerobes; cause blackening of canned food
Enterococcus	*faecalis, faecium*	Aerobes/anaerobes; varied nisin sensitivity; cause spoilage
Lactobacillus	spp., *bulgaricus, brevis, buchneri, casei, curvatus, helveticus, fermentum, lactis, plantarum*	Cause spoilage of acid products, salad dressings, wine, beer, cider, cured meat products; can grow at low pH; some used for cheese production
Leuconostoc	*oenos, mesenteroides*	Aerobes/anaerobes
Listeria	*innocua, monocytogenes*	Aerobes/anaerobes; varied nisin sensitivity; cause spoilage of wine and vegetables, producing slime
Micrococcus	*luteus, varians*	Aerobes; *M. luteus* is indicator strain for the bioassay.
Pediococcus	spp. *damnosus, pentosaceus*	Aerobes/anaerobes; cause spoilage of wine and beer.
Sporolactobacillus	*inulinus*	Aerobe/anaerobe; spore forming; growth at low pH
Staphylococcus	*aureus*	Aerobe/anaerobe; varied sensitivity; causes food poisoning

(Derived from Thomas et al., 2000)

Individual application of nisin does not have any effect on the yeasts, molds, or Gram-negative bacteria (possessing impermeable outer membrane). In fact, the lipopolysaccharides (LPS) of their outer layer make Gram-negative bacteria resistant to nisin by creating a barrier against nisin action on the cytoplasmic wall. Destabilization of LPS can take place as a result of chelating agents like EDTA confining the divalent magnesium and calcium ions of the layer. Thus, transportation of the antimicrobial peptide through the disrupted LPS layer can form pores in the cytoplasmic membrane contributing to the loss of intercellular components (Millette et al., 2004). Khan et al. (2015) demonstrated also that the antimicrobial activity of nisin against Gram-negative bacteria such as *Escherichia coli* and *Salmonella* Typhimurium was increased in combination with EDTA as a chelating agent.

8.4.1 Factors Affecting the Antimicrobial Activity in Foods

The outcome of nisin activity within a food system depends on numerous factors such as pH and temperature, inactivation, interactions with food ingredients, the natural microbiota of the food, and heterogeneous distribution of the bacteriocin in the food matrix. Nisin antimicrobial activity can be enhanced by combining the use of this antimicrobial peptide with various methods like heating, freezing, and high-pressure processing as well as other preservatives including plant essential oils and chelating agents, such as EDTA (Prudêncio et al., 2016). Table 8.2 gives some examples of nisin minimal inhibitory concentrations (MIC) for common food-contaminating bacteria that can help in using the appropriate nisin concentration in foods.

8.4.1.1 Food Composition

Nisin works better in liquid or homogeneous foods compared to solid or heterogeneous products because the bacteriocin can be better and more evenly distributed throughout the food matrix of the former. Due to nisin's hydrophobic nature, fat in food may hinder its uniform distribution and render it unavailable for activity. According to Deegan et al. (2006), the interference of nisin with meat ingredients like phospholipids restricts its activity and makes this bacteriocin less effective in meat products than in dairy ones. The presence of certain food additives can significantly decrease the antimicrobial activity of nisin used as a preservative. For example, nisin is degraded in the presence of sodium metabisulphite (used as an antioxidant, bleach, and antimicrobial) and titanium dioxide (used as a whitener) (de Aruaz et al., 2009).

TABLE 8.2

Examples of Nisin Minimal Inhibitory Concentration (MIC) for Common Food-Contaminating Bacteria

Bacteria	MIC (IU/mL)	Conditions
Bacillus cereus	16	BHI medium, 21°C
Lactobacillus brevis	16	BHI medium, 21°C
Lactobacillus plantarum	16	BHI medium, 21°C
Brochothrix thermosphacta	0.4	BHI medium, 21°C
Pediococcus acidilactici	0.4	BHI medium, 21°C
Listeria innocua	32	BHI medium, 21°C
Listeria monocytogenes	16	BHI medium, 21°C
Listeria monocytogenes	32	BHI medium, 30°C
Bacillus cereus	32	Milk, 21°C
Listeria monocytogenes	16	Milk, 21°C
Listeria monocytogenes	48	Milk, 30°C
Lactococcus lactis	3.2	-----
Micrococcus luteus	2.8	-----
Listeria monocytogenes	156	Culture media

(Gharsallaoui et al., 2016b)

8.4.1.2 Temperature

Loss during storage was first observed by Hirsch (1951) in Swiss-type cheese made with a nisin-producing starter culture. The decline of nisin levels during storage is temperature-dependent. Levels remain relatively stable during refrigerated storage, but loss is faster at ambient temperatures (Delves-Broughton, 1990). Higher nisin levels (12.5–15.5 μg/g) are needed for products stored at high ambient temperatures for long periods. However, nisin does offer the possibility of avoiding storing products including processed cheese at chill temperatures. This can be important when cold storage distribution chains cannot be guaranteed or when refrigerated storage is not possible.

Abee et al. (1994) found that the rate of nisin Z-induced K^+ efflux from listerial cells grown at 30°C was reduced when the temperature was lowered. This was attributed to a decrease in membrane fluidity caused by the temperature reduction. *Listeria* cells can adapt to growth at low temperatures by increasing the fluidity of the membrane by increasing the proportion of short and/or branched fatty acyl chains of the lipids (Russell et al., 1995, Chihib et al., 2003). This concurs with reported nisin Z efficacy against foodborne bacteria grown in laboratory medium, low-fat milk, and at high and low temperatures (Abee et al., 1995). Many foods are stored at low temperatures. It seems from this evidence that if bacteria are capable of growth at these low temperatures, their membranes will have adapted and will be sensitive to nisin action.

8.4.1.3 pH

The solubility and stability of nisin solutions decrease as the pH increases. Nisin is often added to acidic foods, but it remains relatively active in a pH range up to 8 (Liu and Hansen, 1990). Rayman et al. (1981) showed that the effectiveness of nisin decreased with increasing pH, and Scott and Taylor (198la, 1981b) reported similar results.

8.4.1.4 Proteolytic Enzymes

In foods that are not heat-treated, or that have been minimally processed, nisin may be degraded during storage by proteolytic enzymes derived from microbial, plant, or animal cells within the food. The biological activity of nisin in the cheese-making process can be hampered due to proteolysis. Some proteolytic enzymes including pancreatin, α-chymotrypsin, and ficin can inactivate nisin by damaging its peptide chain. However, the antibacterial activity of nisin did not change in the presence of enzymes like trypsin, pepsin, and carboxypeptidase (Chollet et al., 2008). Nisinase, a specific enzyme that targets nisin, has been reported from a few strains exhibiting reduced nisin sensitivity. The enzyme was isolated from *Bacillus* and proved to be dehydropeptide reductase (Ray, 1992).

8.4.1.5 Process

Any treatment, such as sub-lethal heat, hydrostatic pressure, or freezing, which disrupts the outer membrane may render Gram-negative cells sensitive to nisin. Rapid chilling in the presence of nisin caused a dose-dependent reduction in the populations of several Gram-negative bacteria, despite the fact that appreciable injury to the outer membrane was not detected (Boziaris and Adams, 2000). This was attributed to a transient susceptibility caused by phase changes in the lipids associated with the outer membrane, which were rapidly reversed when the cells returned to higher temperatures. During the melting process used in processed cheese manufacture as much as 20% nisin will be lost, and nisin addition levels need to be adjusted to compensate for this. The degree of nisin loss depends on the extent of heat treatment, the period of time the cheese is kept molten, and the pH of the product. Nisin loss is considerable if a UHT process is involved; up to 40% may be lost. However, much less nisin will subsequently be needed to control the reduced number of heat-damaged spores that can survive this treatment.

Novel non-thermal pasteurization methods such as high hydrostatic pressure and,pulsed electric fields are also being investigated in combination with nisin. For example, a pulsed electric field has been shown to enhance the cidal action of nisin against *Bacillus cereus* and *Micrococcus luteus* (Dutreux et al., 2000;

Pohl et al. 2000). In contrast, an *E. coli* study by Terebiznik et al. (2000) reported nisin inactivation by pulsed electric fields and that nisin was protected by the presence of bacterial cells. Kalchayanand et al. (1994) found that ultra-high pressure and electroporation injured cells of *L. monocytogenes*, *E. coli*, and *Salmonella* Typhimurium such that they became more sensitive to nisin. Hauben et al. (1996) showed that ultra-high pressure sensitized *E. coli* to nisin and postulated that this was due to disruption of the outer membrane. Masschalck et al. (2000; 2001) later confirmed this effect. A synergy between nisin and ultra-high pressure against *Bacillus* and *Clostridium* spores has been reported (Roberts and Hoover, 1996; Stewart et al., 2000). A combination of nisin and sublethal ultra-high pressure was synergistic against a yeast strain (ter Steeg et al., 1999).

8.4.1.6 Synergy with Other Preservatives

Certain compounds are synergistic with nisin. These include essential oils, organic acids, bacteriocins, plant extracts, lysozyme, chelating agents, lactoperoxidase, non-ionic and amphoteric surfactants, and emulsifiers such as monoglycerides (particularly monolaurin) and sucrose fatty acid esters (Gharsallaoui et al., 2016a). Protective cultures producing other bacteriocins *in situ* may also enhance nisin activity (Hanlin et al., 1993; Schllinger, 1999). Future new preservative systems are likely to use combinations of agents in order to enhance the preservative effect and increase the antimicrobial spectrum of nisin. This type of combined approach has already proved successful. For example, a combination of nisin with chelating agents like disodium EDTA was effective against Gram-negative bacteria such as *Salmonella* Typhimurium and *E. coli* O157:H7 (Fang and Tsai, 2003; Prudêncio et al., 2016).

8.4.1.7 Bacterial Load

The effectiveness of nisin decreases with the size of the inoculum (Scott and Taylor, 1981b; Rayman et al., 1981). Ramseier (1960) published a study illustrating this point with spores of *Clostridium butyricum*. He found a close relation between the nisin concentrations required to inhibit outgrowth and spore load. Increasing the spore load ten-fold requires an increased nisin concentration of about half a logarithmic cycle expressed in international units (Ramseier, 1960). Prevention of the outgrowth of *C. botulinum* requires 10–100 times more nisin than that needed to prevent the outgrowth of *C. butyricum*.

8.5 Applications as Food Preservative

Nisin is applied for food preservation by several strategies including inoculating food with a nisin-producing strain, utilizing a fermented product containing a nisin-producing strain as an ingredient, incorporating purified or semi-purified nisin as a natural preservative, and immobilizing nisin in polymers, beads, and gels, or encapsulating this bacteriocin for controlling its release and protecting it from degradation by proteolytic enzymes. The suitable method of nisin application is determined according to the food composition as well as intrinsic and environmental factors during production, storage, and distribution (Gharsallaoui et al., 2016a).

Nisin can be used in a wide range of liquid or solid foods, cans, or packages, stored at ambient or chill temperatures. Its usage is mainly in heat-processed foods to prevent spoilage by spore-forming bacteria, but it is also used to control spoilage by heat-sensitive lactic acid bacteria and *Brochthrix thermosphacta*. It is also used as part of the safety measures to control pathogens such as *B. cereus*, *L. monocytogenes*, and *C. botulinum*. There are several good reasons to use nisin as a preservative in food. For example, it can extend the shelf life of both chilled and ambient stored food. It can protect chilled foods from temperature abuse. It can enable thermal processing treatments to be reduced (if this is legally permissible). This has the consequence of reducing the negative effects of heat treatments on product quality as well as reducing manufacturing costs. Distribution costs can also be reduced since the nisin use may eliminate the need for chilled transport.

The concentration of added nisin depends on the type of food, heat processing, pH, used packaging, storage conditions, bacterial load, and required shelf life. Nisin can preserve its activity in a pH range up to 8 and is used frequently for extending the shelf life of aged cheese, pasteurized milk, canned soups, fruit juices, and fresh-cut fruits and vegetables. This bacteriocin can be used as a part of hurdle technology like in canned foods where nisin is added after thermal processing for damaging heat-resistant spores of thermophilic bacteria.

8.5.1 Dairy Products

8.5.1.1 General Considerations

The main use of nisin still remains in pasteurized processed cheese and associated products such as spreads and slices. The anaerobic environment, high moisture (ranging from a minimum of 44% in block cheese to 60–65% in spreads, sauces, or dips), and acid pH usually found in processed cheese favor particularly the growth of some *Clostridium* species. Facultative anaerobic *Bacillus* may also cause problems. Spoilage is observed as blowing, off-odors, and liquefaction. Incidence of *C. botulinum* is rare but is possible, particularly when temperature abuse occurs (Collins-Thompson and Woods, 1992).

Somers and Taylor (1987) found that water activity and the levels of sodium chloride and phosphate substantially affect the minimum concentration of nisin for the inhibition of *C. botulinum* types A and B. Tanaka et al. (1986) found that the pH of the pasteurized cheese spread, as well as the levels of both sodium chloride and phosphate, affected the required amount of nisin needed for the antibotulinal effect. Indeed, higher nisin levels are required in formulations with high moisture content and low concentrations of sodium chloride and phosphate. The bioload of contaminant spores and the duration and temperature of shelf life also need to be considered when deciding nisin addition levels.

It was rapidly realized after the early work of Hirsch et al. (1951) that a major problem with nisin use in cheese was its ability to inhibit the bacterial starter cultures used for ripening and the production of cheese characteristics (Kooy, 1952). To overcome this problem, starter cultures were trained to grow in the presence of nisin, but although this was achieved it proved impractical (Lipinska, 1977). Some developments in gene technology have revived interest in this application. Nisin is encoded as a transposon, and therefore can be moved from cell to cell relatively easily (Rauch and de Vos, 1992). A nisin-producing starter culture for Gouda cheese was developed by a Dutch group (Hugenholtz et al., 1995). An improvement from earlier attempts was to use lactose-negative strains as donor cells, enabling the selection of transconjugants for the analysis of good cheese sensory characteristics. Cheese was prepared with different nisin levels by using nisin-producing and nisin-immune strains in different ratios as starters. The nisin produced within the cheese protected against growth of *C. tyrobutyricum* and *Staphylococcus aureus* during ripening. A nisin-producing transconjugant *L. lactis* subsp. *cremoris* has also been used as a starter culture in a pH-controlled bulk Cheddar cheese starter culture system. A high level of 17.5 μg/g nisin was produced (Roberts et al., 1992; Yezzi et al., 1993). Such nisin-containing cheese has been used to make processed cheese spreads, which showed better protection against *C. sporogenes* in inoculated trials (Roberts and Zottola, 1993).

Nisin-producing strains should have some technological properties to be applicable in cheese production including being good acidifiers, having acid phosphatase and phosphohydrolase activities, as well as tolerating salt. Recently, isolated nisin producing *Lactococcus lactis* strains from goat and cow milk, fermented wheat, and buckwheat samples were characterized and applied for controlling the growth of *Listeria monocytogenes* in fresh cheese (Kondrotiene et al., 2018). The strains encoding novel nisin variant GL03 were not appropriate for usage in the dairy food products due to being resistant to tetracycline and the production of harmful β-glucosidase.

8.5.1.2 Main Uses for the Preservation of Dairy Products

Nisin distribution is affected by the presence of various ingredients and production steps especially when milk is curdled into curd and whey. Nisin (at 400 and 500 IU/mL concentrations) showed an antibacterial activity against *S. aureus* when it is present in curd leading to an increase of the cheese shelf life. In fact, the high moisture, low salt content, and pH close to 6.0 of Minas Frescal cheese are the main parameters that enhance *S. aureus* growth during storage (Felicio et al., 2015). In another study, nisin at 600 IU/g concentration was used along with bovicin in Minas Frescal cheese containing an initial *S. aureus* count of 10^4 CFU/g. Only 1 log cycle reduction was obtained until 15 days of storage at 4°C after which bacteria grew again (Pimentel-Filho et al., 2014).

Nisin has been shown to be effective against *L. monocytogenes* in surface mold-ripened soft cheeses. Maisnier-Patin et al. (2001) prepared Camembert cheeses with milk containing 1, 3 or 5 log CFU *Listeria monocytogenes*. A nisin-producing starter of *Lactococcus lactis* subsp *lactis* was used in the cheese to inhibit L monocytogenes. A maximum of approximately 700 IU/g nisin was detected at 9 h followed by a slow decrease in the first 24 h and more rapid decrease during ripening. *L. monocytogenes* was reduced by 3.3 log compared to the initial level in cheese milk. Inhibition continued until the end of the second week of ripening but thereafter, regrowth occurred. Ultimately, cheese with the nisin-producing starter had a 2.4 log lower number of *L. monocytogenes* than the control after 6 weeks.

Nisin, at concentrations of 2.50–6.24 µg/g, was shown to be extremely effective in preventing the growth of relevant *Clostridium* spp. inoculated in processed cheese and cheese spreads (with or without ham) incubated at 37°C (Delves-Broughton and Gasson, 1994). This effectiveness was also reported by Plocková et al. (1996) who found that similar nisin levels inhibited *Clostridium* and *Bacillus* in processed cheese during a 3-month storage period. Nisin at 4.0–5.1 µg/g also reduced spoilage microflora in processed cheese analogues incorporating fat substitutes over a 4-month storage at 5°C (Muir et al., 1999).

Soft white cheese such as ricotta, paneer, queso fresco, and fresh Hispanic-style cheeses are not made using starter cultures, and nisin can therefore be used in these cheeses without causing problems with their production. Davies et al. (1997) investigated the efficacy of nisin added directly to the milk used to prepare ricotta cheese by direct acidification. The growth of inoculated *L. monocytogenes* was inhibited during storage at 6–8°C for over 8 weeks. Control cheese not containing nisin contained unsafe levels of the pathogen after 2 weeks. Nisin bound preferentially to the curd with minimum losses in the whey. Similar success has been reported in queso fresco cheese, with nisin produced *in situ* in a pre-fermentation step with a *L. lactis* subsp. *lactis* producer strains, or by salting nisin onto the finished curd (Degnan et al., 1996). Nisin, added at a level of 50 µg/g to cottage cheese after production, caused a 3-log reduction of inoculated *L. monocytogenes* in 3 days at 20°C. Controls showed only a 1 log reduction after 7 days (Ferreira and Lun, 1996).

The incorporation of nisin at 1000 AU/mL concentration into both skim milk and whole milk decreased the initial counts of *Enterococcus faecalis*, *Enterococcus italicus*, *Enterococcus mundtii*, *Bacillus cereus*, *Bacillus thuringiensis*, *Lactobacillus paracasei*, *Pseudomonas fluorescens*, and *Acinetobacter* spp. under detectable levels during incubation at 8°C for 8–20 h (Mitra et al., 2011). Flavored milk such as chocolate milk can have higher numbers of contaminating spores derived from the flavoring ingredients. Nisin at 2.5 µg/mL was effective in controlling spoilage of such milk inoculated with spores of the mesophile, *Clostridium* PA3679, and thermophile, *Geobacillus* (formerly *Bacillus*) *stearothermophilus*, prior to a heat treatment at an F_0 value of 3.2. Nisin at 1.0 µg/mL also prevented the spoilage of chocolate milk in a large-scale trial for 3 weeks at 55°C and 6 months at ambient temperature (Heinemann et al., 1965).

In dairy desserts and milks, nisin can be added to a small quantity of milk, mixed into the bulk milk, filled, and processed. As with all nisin applications, efficacy is best in homogeneous products. Nisin at 14 mg/g has been shown to reduce numbers of *Listeria* in ice cream during the manufacturing process and frozen storage at –18°C for 3 months (Dean and Zottola, 1996). This was observed in both 1% and 3% fat ice cream and is noteworthy because nisin activity occurred at such low temperatures. Nissen at al. (2001) studied the germination and growth of both inoculated and naturally occurring *Bacillus* strains in heat-treated cream with and without nisin. In heat-treated cream (90°C for 15 min), bacterial growth was mainly due to *B pumilus*, *B. licheniformis*, and other *Bacillus* spp. *B. cereus* was also isolated. Bacterial growth during storage of the cream at 8 or 10°C was completely inhibited by low concentrations of nisin. It was concluded that the safety of cream, for instance in "ready-to-eat" products, could be improved by the use of nisin.

Despite the problems associated with nisin use in yoghurt (inhibition of starter bacteria and consequent delay of fermentation), several studies have reported good results. Use of nisin at 0.625–1.25 µg/g was recommended by Bayoumi (1991). This produced acceptable yoghurt after 5.5 h that had good organoleptic and textural characteristics and a shelf life of 19 days at 10°C or 33 days at 6°C. Prevention

TABLE 8.3

Examples of Nisin Uses for Dairy Products Preservation

Food	Nisin Concentration	Target Microorganisms	Main Observations
UHT skim milk	≤2.5 mg/L	*L. monocytogenes*	After an initial effectiveness, a regrowth of nisin-resistant cells was observed.
Fresh, pasteurized, and/or homogenized whole and skim milk	≤12.5 mg/L	*Listeria innocua, L. monocytogenes*	Antilisterial activity lost or reduced when whole milk was homogenized.
Traditional Greek whey cheese	≤12.5 mg/kg	*L. monocytogenes*	Direct nisin addition to whey prevented the increase of *L. monocytogenes* population for more than 30 days when cheeses were stored at 4°C under vacuum. However, addition of nisin reversed the dominant flora from Gram-positive (lactic acid bacteria) to Gram-negative bacteria during refrigerated storage.
Skim milk	≤0.5 mg/L	*Streptococcus thermophilus*	Exposure to subminimal inhibitory concentrations of nisin induced resistance to nisin (possibly caused by changes in *S. thermophilus* cell wall).
Processed cheese products	2.5–12.5 mg/kg	*Clostridium sporogenes, C. butyricum, C. tyrobutyricum* or *C. botulinum*	Nisin was effective in delaying or preventing growth and subsequent toxin production by inoculated spores of *C. botulinum* types A and B.
Ewe's milk Greek soft acid-curd cheese	3.75 mg/kg	Lactobacilli, lactococci, and yeasts	• Nisin addition induced a significant and expected decrease of lactococci, lactobacilli and an unexpected decrease of yeasts. • Nisin did not affect sensory quality and extended shelf life based on this criterion.
Swiss-type (Emmental) cheese	1 g/L in cheese-water (2:1, w:w) slurry	*Kocuria rhizophila*	Nisin antimicrobial activity reduction can be ascribed to its adsorption on fat globules and proteins but not to its hydrolysis by proteolytic enzymes.
Reconstituted powdered infant milk formula	≤40 mg/L	*Cronobacter* spp.	Nisin was not effective likely because they are Gram-negative bacteria and because of the presence of fat globules and proteins.
Minas Serro cheese	≤12.5 mg/L (nisin addition to milk before enzymatic coagulation)	*Staphylococcus aureus*	• Up to 2 log cycles reduction in *S. aureus* count from the 7th day of ripening. • Decrease of the ripening index.

(Gharsallaoui et al., 2016a)

of over-acidity can be achieved by mixing nisin into stirred yoghurt when the required fermentation is completed (Thomas and Delves-Broughton, 2001). More examples concerning nisin uses for dairy products preservation are given in Table 8.3.

8.5.1.3 Uses in Combination with Other Treatments

Nisin resistance has emerged in several species of bacteria like *S. aureus* (Blake et al., 2011) which is an opportunistic pathogen leading to human infections and spoiling food (Argudin et al., 2010). Hindering the generation of resistant mutants and achieving synergy among various active agents are two main reasons for the application of an antimicrobial combination treatment with the purpose of increasing activity against a broad spectrum of microorganisms. Combined active substances display synergy as a result of attacking different targets in a metabolic pathway (Shi et al., 2017a). In recent years, the synergistic antimicrobial activity of nisin with other natural active compounds like cinnamaldehyde, thymol, and ρ-Anisaldehyde has been investigated against *S. aureus* and *L. monocytogenes* in milk (Shi et al., 2017a; b; Chen and Zhong, 2017). Nisin can protect against temperature abuse and extend shelf life. It is possible to decrease the concentration of required nisin in dairy products like whole milk by mixing with other antimicrobial substances having antilisterial activity. Chen and Zhong (2017) showed that the minimum inhibitory concentration (MIC) of nisin (125 IU/mL), thymol (0.25 mg/mL), and lactobionic acid, a natural chelating agent, (10 mg/mL) for *Listeria*, were significantly lowered upon combination (31.25 IU/mL, 0.0625 mg/mL, and 1.25 mg/mL, respectively). Ternary combination of nisin (250 IU/mL), thymol (2 mg/mL), and lactobionic acid (10 mg/mL) could totally prevent the growth of *L. monocytogenes* to below the detection limit in 2% reduced-fat milk within 72 h of incubation at 21°C. The same ternary mixture, but with 500 IU/mL nisin, reduced approximately 2 log CFU/mL of pathogen in whole milk after 4 h of incubation at 21°C.

A study by Wirjantoro et al. (2001) demonstrated the effectiveness of a reduced-heat treatment (117°C for 2 sec) and nisin (1.88 and 3.75 μg/mL). The milk had an extended shelf life at refrigeration temperatures, and its quality was considerably better than milk subjected to UHT processing of 142°C for 2 sec.

8.5.2 Meat Products

There are many reported investigations of nisin use as a decontaminant in meat and meat products. However, interaction with phospholipids, low solubility, and partial inactivation by endogenous meat enzymes restrict successful application of this bacteriocin in various meat products (Figueiredo and Almeida, 2017). Thus, when nisin is used for the preservation of meat products, it is often associated with other preservatives. The main examples concerning nisin uses for meat products preservation are given in Table 8.4.

Nisin was included in a study to investigate control of *L. monocytogenes* in refrigerated and frozen ground beef (Park et al., 2001). Nisin achieved a greater reduction of bacterial numbers in refrigerated beef compared to lacticin, with an even better reduction occurring in frozen beef. Govaris et al. (2010) assessed the antimicrobial activity of oregano essential oil at 0.6 or 0.9%, nisin at 500 or 1000 IU/g, and their combination against *Salmonella* Enteritidis in minced sheep meat during storage at 4 or 10°C for 12 days. Binary usage of nisin with essential oil showed stronger antibacterial activity against the pathogen

TABLE 8.4

Examples of Nisin Uses for Meat Products Preservation

Food	Nisin Concentration	Target Microorganisms	Main Observations
Raw buffalo minced meat	10 or 20 mg/kg	*Listeria monocytogenes*	*L. monocytogenes* growth inhibition more pronounced at 4°C than at 37°C
Round steak	Beef meat cubes dipped for 10 min in 125 mg/L nisin solutions	*Brochothrix thermosphacta*	Once vacuum packed, *B. thermosphacta* growth was inhibited for more than 25 days at 4°C
Cooked sausage	11.25 mg/kg	No antimicrobial activity assay	More than 68% of initially added nisin still detected after 28 days storage at 6°C
Sous vide cooked (11 min, 97°C) seasoned beef	2.5 or 12.5 mg/kg	*Bacillus cereus*, *Clostridium perfringens*	Increase of shelf life at 4°C and/or increased stability in case of temperature abuse (simulated by a storage at 25°C)
Meat emulsions (fresh lean beef, 20% w/w bovine fat, 2.5% w/w NaCl)	0.25 to 2.5 mg/kg	*Listeria monocytogenes*	Time to increase initial *L. monocytenes* population from 10^4 UFC/g to 10^7 UFC/g at 20°C increased from 1 to 7 days
Natural sausage casings	Dipping for 8 days in 50 mg/L nisin solutions	*Clostridium sporogenes*	90% reduction of *C. sporogenes* spores compared with a control

(Gharsallaoui et al., 2016a)

compared to the samples treated with each active agent. Both concentrations of nisin were not adequate in the prevention of bacteria growth, indicating the importance of synergism between bacteriocin and essential oil for increasing food shelf life. Similarly, the *L. monocytogenes* count in minced beef meat treated with thyme essential oil at 0.6% combined with nisin at 1000 IU/g was decreased below the official limit of the European Union (2 log CFU/g) during storage at 4°C for 12 days (Solomakos et al., 2008). Bacteriophage is a bacterial virus that is used as a biocontrol agent for decreasing *Salmonella* spp. populations in foods like meat and poultry products (Guenther et al., 2012; Hungaro et al., 2013). Regarding the narrow antimicrobial spectrum of phages, prolonging meat shelf life can be obtained by using various natural antimicrobial agents like nisin. Total viable counts, total volatile basic nitrogen, and oxidation (TBARS test) of the fresh chilled pork decreased significantly upon treating with a combination of nisin (5000 IU/g), bacteriophage (9 log PFU/g), and potassium sorbate (2 mg/g) during the storage period (21 days at 4°C) (Wang et al., 2017).

Several studies have focused on nisin use as a carcass disinfectant together with chelating agents, a combination known to act against Gram-negative bacteria as well as Gram-positive bacteria *in vitro* as mentioned above. Regrettably the promising results observed in buffer systems did not translate successfully to raw meat treatment, either by spraying or dipping. An example of this was the work of Cutter and Siragusa (1995a, 1995b) who demonstrated good cidal effects against *E. coli* O157 and *S.* Typhimurium in a buffer system using a nisin/chelator combination. However, this treatment was much less effective when used as a treatment of lean beef tissue. In another study, nisin at 2.5 µg/mL achieved an effective kill of *L. monocytogenes* when used in scald water at 52°C for 3 min, but the effect was much less when the bacteria were located on the turkey skin (Mahadeo and Tahini, 1994). An optimum treatment of nisin (100 mg/mL) with a chelating agent (EDTA, citric acid) and a surfactant (Tween 20) was described by Shefet et al. (1995) for use on chicken drumsticks against *S.* Typhimurium and for extension of shelf life.

Cutter and Siragusa (1994) examined the decontamination of raw meat surfaces using a pilot-scale model carcass washer and achieved a reduction in total bacterial count, and, in a further study, a reduction in *Listeria innocua*, *Carnobacterium divergens*, and *Brochothrix thermosphacta* was observed (Cutter and Siragusa, 1996b). This group also examined the effect of nisin incorporation into a surface coating of calcium alginate, which resulted in better reduction and control of *Br. thermosphacta* on beef surfaces and better nisin retention (Cutter and Siragusa, 1996a). Chung et al. (1989) investigated the effect of nisin added to raw meat inoculated with meat-associated bacteria. Nisin activity, when extracted from the meat, decreased by 95% during storage for 4 days at ambient temperature and by 70% at 5°C. The growth of the inoculated Gram-positive bacteria (*L. monocytogenes* and *S. aureus*) was delayed but that of the Gram-negative bacteria was not.

Pawar et al. (2000) examined the effect of nisin (10 and 20 mg/g) in combination with 2% NaCl incorporated into raw buffalo meat mince. The growth of inoculated *L. monocytogenes* was significantly inhibited compared to samples without nisin. There have also been investigations into the feasibility of replacing sulfur dioxide with a nisin/organic acid combination in raw pork sausages. Scannell et al. (1997) reported a combination of sodium lactate and nisin that reduced the total bacterial count. Luo and Zhu (2000) showed that nisin had a significant effect on the preservation of fresh beef slices, and that this was synergistic with sodium lactate.

In cooked meat products, nitrite must be added for development of the desired pink color. Some prevention of growth of *C. botulinum* and *C. perfringens* was reported, but in general results were inconsistent and extension of shelf life was not reliable. It was suggested that this could be due to binding of nisin to meat particles, uneven distribution, and poor solubility (de Vuyst and Vandamme, 1994; Henning et al., 1986). Better results can be obtained from nisin use in cooked, cured, chilled sausages and pâté products. This may partly be due to the fact that nisin can be homogeneously distributed throughout the matrix of these food systems. The formulation of cooked cured meat products can vary considerably in their composition and levels of fat, moisture, NaCl, nitrite, and phosphate. In a chilled, vacuum-packed bologna sausage product, nisin was more effective against lactic acid bacterial growth in formulations with higher percentages of lean meat (Davies et al., 1999). The type of phosphate was also found to influence both nisin efficacy and bacterial growth. Optimum efficacy was reported for a formulation containing 25% fat and diphosphate as an emulsifier. Significant shelf life extension was obtained with 6.25–25 µg/g nisin.

Increasing shelf life of sausages might be achieved through the addition of mixed nisin with essential oils as a natural process for preventing the growth of both Gram-positive and Gram-negative bacteria (Mathenjwa et al., 2012). Treating fresh sausage with 20 mg/kg nisin + 125 µL/kg garlic essential oil + 62.5 µL/kg allyl isothiocyanate could significantly decrease *E. coli* O157H7 and spoilage lactic acid bacteria counts, and the physicochemical properties of sausage were not changed (Araújo et al., 2018). One of the most common foodborne pathogens related to the cooked meat products is known to be *S. aureus*. However, Lin et al. (2018) found that nisin, sodium nitrite, and potassium sorbate could not inhibit the growth of *S. aureus* and had no effect on the prevention of staphylococcal enterotoxin A production in cooked pork sausage during storage at 37°C for 36 h. The synergistic and individual antibacterial activity of nisin, bacteriophage P100, and sodium lactate have been evaluated against *L. monocytogenes* in ready-to-eat sliced pork ham (Figueiredo and Almeida, 2017). Treating sliced pork ham with a mixture of bacteriophage P100 and nisin decreased the number of viable *L. monocytogenes* cells by 3 log cycles at 72 h of storage at 6–8°C which was more effective than nisin alone (1.67 log cycle reduction). These authors demonstrated that bacteriophage P100 showed the highest efficiency in reduction of the counts of tested *Listeria* strains to undetectable levels during 72 h of storage among the individual antimicrobial treatments.

A combination of lysozyme, nisin, and EDTA was investigated by Gill and Holley (2000a) on ham and bologna. The treatment reduced counts of lactic acid bacteria and *B. thermosphacta*. Inhibition of *E. coli* O157 was achieved for 4 weeks in ham sausages. The next step was to explore this combination applied as a surface coating to the sausages using gelatin (Gill and Holley, 2000b). This had an immediate cidal effect against Gram-positive bacteria and inhibited growth during the 4-week storage period. The gel also had a bactericidal effect on *S.* Typhimurium during storage. Kassaify et al. (2000) have shown good nisin uptake in natural and collagen casings; this could be an alternative approach to the preservation of cooked sausages.

The use of nisin in meat is mainly focused on the presence of glutathione, a molecule capable of inactivating nisin by a glutathione S-transferase catalyzed reaction (Rose et al., 1999). This glutathione inactivation is lower in cooked meat due to the loss of free sulfhydryl groups, which catalyze the reaction between glutathione and proteins, during the heating process (Stergiou et al., 2006). Nisin can also be inactivated by proteolytic enzymes generally found in fresh meat (Rose et al., 1999). In addition to glutathione and the presence of proteolytic enzymes, nisin can easily interact with the meat fats which can reduce its antimicrobial efficacy.

8.5.3 Seafood Products

The psychrotrophic pathogen *L. monocytogenes* represents a major microbiological risk in fresh and lightly preserved seafood products. These products are often not heated before consumption and some are transported at ambient temperatures. The addition of nisin to vacuum-packed raw shrimps stored at 4°C decreased the surface population of *L. monocytogenes*, *Salmonella*, and native microflora (Wan Norhana et al., 2012). The combination of nisin with preservation methods including radio frequency or moderate heat treatment was successful in preventing growth of *L. innocua*, *L. monocytogenes*, and total mesophilic microorganisms in caviar sturgeon (*Acipenser transmontanus*) and Pacific salmon (*Oncorhynchus keta*) without changing sensory properties (Al-Holy et al., 2005). Saelao et al. (2018) reported that inoculation of shrimp (*kung-som*) with nisin Z producing *Lactococcus lactis* KTH0-1S as a starter culture significantly decreased *Enterobacteriaceae*, *Staphylococcus aureus*, and total viable count in comparison with non-treated shrimp. Also, tyramine (harmful biogenic amine) accumulation was lowered by 100 mg/kg in shrimp treated with the nisin-producing starter.

Nisin (100–150 µg/g) and sodium lactate synergistically inhibited listeria when injected into rainbow trout before cold smoking (Nykanen et al., 2000). These antimicrobial treatments did not adversely affect the sensory quality of the fish, and they remained acceptable to a sensory panel for one week more than untreated samples. Combined chitosan, nisin, and sodium lactate reduced the growth of *Listeria innocua*, *Shewanella putrefaciens*, and psychrophilic bacteria isolated from fish (Schelegueda et al., 2016). Encapsulation is one the promising techniques used for protecting the biological activity of nisin from harmful environmental parameters including pH, temperature, and food ingredients (Maresca et al., 2016). Using chitosan for the production of nisin delivery systems has many advantages such as being nontoxic, biodegradable, and presenting wide antimicrobial spectrum against Gram-positive and Gram-negative bacteria (Ming et al., 2010). Recently, treatment of small yellow croaker (*Pseudosciaena polyactis*) with nisin-loaded chitosan microcapsules resulted in prolonged shelf life of fish during storage by 6 to 9 days compared to the use of chitosan or nisin individually based on total viable count (TVC) assay (Wu et al., 2017).

Canned frozen lobster is usually heat processed, but the treatment cannot be greater than 60°C for 5 min, otherwise product shrinking occurs. This only reduces counts of *L. monocytogenes* by 2 logs, unless nisin is added to the brine at a concentration of 25 µg/mL. This achieves a 5–6 log reduction (Budu-Amoaka et al., 1999). The synergy between heat and nisin against *L. monocytogenes* was confirmed (Modi et al., 2000). Washing of blue crab (*Callinectes sapidus*) with 2.5 or 5 µg/mL nisin reduced listerial counts by 2 and 3 logs respectively (Degnan et al., 1994).

8.5.4 Fruits and Vegetables

Heat-resistant endospores can survive and turn into vegetative cells as soon as the physicochemical conditions become favorable in heat-processed food (Naim et al., 2008). Botulinum neurotoxins are produced by *Clostridium botulinum* which is a proteolytic, anaerobic, and heat-resistant spore forming Gram-positive bacillus (Diao et al., 2014) playing a major role in spoilage of canned food at home and during storage. Hurdle technology including electro-activated aqueous solutions (EAS) of potassium acetate and potassium citrate, nisin (250, 500, 750, and 1000 IU/mL), and moderate heat processing (95, 105, and 115°C for 5, 15, and 30 min) was studied against the growth of *Clostridium sporogenes* PA 3679 spores in green beans puree and whole green beans. An inactivation efficacy of 5.9–6.1 log CFU/mL was obtained in green beans puree due to the synergistic effect of the hurdle principle. Higher spore inactivation (6.5 log CFU/mL) occurred for experiments done with whole green beans (El Jaam et al., 2017). Recently, it was shown that addition of nisin A to minimally processed vegetables such as lettuce could lower and postpone the growth of *L. monocytogenes* pathogen without affecting sensory properties for 2–5 days of storage (McManamon et al., 2019). Treatment of fresh-cut carrots with combined 1% chitosan solution containing 64 µg/mL nisin and 250 µg/mL ε-polylysine along with 1% lactic acid solution (v/v) led to significant reduction of microorganisms such as yeast and mold, *Staphylococcus aureus*, *Pseudomonas* spp., and total coliform counts while enhancement of phenylalanine activity and total phenol content was achieved after 9 days of storage at 4°C (Song et al., 2017). Nisin at a concentration of 5000 IU/mL was able to inhibit the growth of *L. monocytogenes* ATCC 7644 on fresh-cut tomato during 10 days of storage at refrigerator temperatures. However, the efficacy of nisin in the prevention of *Listeria* growth in tomato increased by combining the bacteriocin with an organic salt, sodium citrate (Oladunjoye et al., 2016). According to Xu et al. (2007) the shelf life of fresh-cut ready-to-eat vegetables like cucumber and lettuce could be extended by applying a mixture of nisin along with citric acid and grapefruit seed extract. This mixture resulted in growth inhibition of three *Salmonella* spp. strains and three *L. monocytogenes* strains. It should be mentioned that nisin is usually used in combination with various chemical preservatives or physical treatments for the conservation of fruits and vegetables.

8.5.5 Beverages

The resistance of unwanted bacteria at high pH values is known to be the major challenge in prolonging the shelf life of carrot juice (pH 6.4 to 6.8), a nutritious beverage containing high amounts of beta-carotene and antioxidants. Pokhrel et al. (2019) used nisin (25 and 50 ppm) in combination with high pressure (200–500 MPa) and mild treatments (35 and 50°C) for the inactivation of *Listeria innocua* and *Escherichia coli* in carrot juice. Incorporation of 25 ppm nisin in juice processed at 500 MPa at 20°C for 2 min led to more reduction of both microorganisms (7 log CFU/mL) than the same sample without added nisin (4 to 5 log CFU/mL reduction). According to the obtained results, synergism between nisin, high pressure, and heat could inhibit the growth of bacteria in food more efficiently.

Cantaloupe is one of the sources for human pathogen outbreak (Martiñon et al., 2014) that can be treated by natural antimicrobial agents as a novel approach for minimally processed food. Nisin and thymol were stabilized in cantaloupe juice through the formation of an oil-in-water emulsion using starch octenyl succinate (Sarkar et al., 2017). Preservation of nisin and thymol by emulsion during storage at 4°C resulted in better inhibition of *L. monocytogenes* V7 and *Salmonella enterica* serovar Typhimurium growth than non-emulsion juice formulation. Nisin can have synergistic activity with thymol against the outgrowth of Gram-negative bacteria. One of the most consumed fruit juices is apple juice which can lose its freshness and quality due to discoloration and harmful microbial growth. The number of pathogens in apple fruit juice should be reduced by 5 log cycles along with keeping qualities including nutritional value, taste, and color according to USFDA standards. Microbiologically safe but undesirable brown-colored juice with an unacceptable taste can be produced by using traditional thermal pasteurization (Arroyo et al., 2012). However, incorporating nisin in acidic apple juice (pH ~ 4.0) is suitable for inhibiting bacteria growth due to being highly active at lower pH. Besides, combining nisin with other antimicrobials or with ultrasound technology showed a synergy in inactivating various microorganisms. In fact, ultrasound is considered as an alternative to heat processing, and the microorganisms can be damaged through both mechanical and chemical effects (Erkaya et al., 2015; Ferrario et al., 2015). Liao et al. (2018) demonstrated that nisin did not have any effect on yeast and molds present in fresh apple juice but it showed antibacterial activity against aerobic bacteria. They found that processing apple juice with thermosonication and 100 ppm nisin at 52°C for 30 min prolonged shelf life to 15 days at 8°C with the preservation of 89% of ascorbic acid.

Some properties make products prone to higher contamination risk including low acidity, low alcohol, low bitterness, or low content of carbon dioxide, and beverages mixed with beer containing sugar. Using nisin with acids, pasteurization, filtration, storing at low temperatures, and aseptic filling are known to be useful in the brewing industry through reducing the time and temperature of processing and diminishing the number of deteriorated sterilized products (Müller-Auffermann et al., 2015). The growth of *Pectinatus frisingensis* (a Gram-negative anaerobic beer-spoilage bacteria) was inhibited by the combination of added nisin (over 400 IU/mL), low pH (pH 4.0), and mild heat treatment (50°C) (Chihib et al., 1999b).

8.5.6 Other Food Products

Whole liquid egg is pasteurized to ensure the destruction of *Salmonella*, but it is not possible to use a more severe heat treatment as this will coagulate the egg. Heat-resistant and spore-forming Gram-positive bacteria may survive. Growth of these organisms is effectively controlled by nisin (Delves-Broughton et al., 1992). Significant shelf life extension as well as control of psychrotrophic *B. cereus* can be achieved using nisin at 2.5–5.0 µg/g. A non-thermal pasteurization process has been described by Calderon-Miranda et al. (1999) using nisin with a pulsed electric field treatment. A synergistic combination of nisin with high hydrostatic pressure achieving enhanced kill of *L. innocua* and *E. coli* in liquid egg has also been reported (Ponce et al., 1998).

Work in Danisco laboratories has shown that nisin at 1–2.5 µg/mL controls spoilage and extends the shelf life of UHT and pasteurized soy milk. The growth of *G. stearothermophilus* spores was prevented by 2.5 µg/mL nisin in soy milk for 7 days at 55°C (Thomas and Delves-Broughton, 2001). UHT treatment can produce a bitter taste in the milk, and nisin addition would enable a reduction in the heat process, resulting in a better product with good shelf life (Thomas and Delves-Broughton, 2001). Coconut milk/water can become discolored by UHT processing, but low levels of nisin can protect against spoilage if the heat process is reduced. The use of titanium dioxide as a whitening agent should be avoided in these products if nisin is used. Nisin as a preservative in other soybean products including tofu (non-fermented) and the fermented Japanese product miso has also been investigated (Kato et al., 2001; Schillinger et al., 2001; Thomas and Delves-Broughton, 2001). Fang et al. (1997) reported successful inhibition by nisin of *S. aureus* and *B. cereus* in a dried bean curd product. Nisin together with nisin-resistant lactic acid bacterial starter cultures has also been investigated in fermented vegetable products including fermented cabbage, sauerkraut, and kimchi (Harris et al., 1992; Breidt et al., 1993; Oh et al., 1994).

There is potential for nisin use as a preservative in potato cakes, a high-moisture product stored at ambient temperatures. The associated spoilage and safety problems are the same as described above for crumpets. Potato cakes have a moisture content of approximately 56%, even higher than crumpets, and examination of such products showed high levels of *Bacillus* (Thomas and Delves-Broughton, 2001). Ham pizza stored in modified atmosphere packaging under 20% carbon dioxide spoils as a result of the growth of lactic acid bacteria and yeasts. Cabo et al. (2001) used a rotatable factorial design to investigate the effects of nisin (as Nisaplin®) and carbon dioxide on the stability of ham pizza during storage. The combination of carbon dioxide and nisin significantly increased the shelf life of ham pizza by the inhibitory action of the gas against the yeasts and the nisin against the spoilage bacteria. Optimum spoilage prevention was achieved using highly enriched carbon dioxide-atmospheres with 100 mg/kg Nisaplin® (2.5 μg/g nisin) added to the top of the pizza and mixed in the tomato paste.

8.5.7 Food Packaging Containing Nisin

8.5.7.1 General Considerations

Bacteria growth and the biological activity of nisin depend on various parameters including food composition (carbohydrates, proteins, lipids, etc.), and physicochemical (water activity, pH, temperature, oxygen partial pressure, ionic strength, etc.) and biological (enzymes, etc.) properties. The solubility and antimicrobial activity of nisin as well as the growth of acidophilic bacteria increase in foods having low pH values. Applying free nisin into a food matrix is restricted due to the fast depletion (degradation by proteases and diffusing within food), and producing an appropriate delivery system like polymer films seems to be indispensable for enhancing the biological activity of this bacteriocin. Determining the amount of nisin released from antimicrobial films onto the surface of the food matrix plays an important role in predicting its antimicrobial activity. Long-term packaged food preservation might be achieved by the slow release of the bacteriocin from the film which would remain active at concentrations required for the inhibition of bacterial growth.

The main difference between edible film and edible coating can be recognized upon their application on the food products. Edible coatings create a thin layer on the product immediately after the application, but edible films as stand-alone wrapping polymers are placed separately on the product. As a result, edible coatings and films exhibit different properties despite being produced from the same gelling agents. Film-forming biopolymers belong generally to the class of hydrocolloids (polysaccharides, proteins, and composites). The application of proteins in film formation can improve its mechanical stability, lipids can lower water permeability, and polysaccharides provide a barrier against the transmission of gas. The incorporation of both hydrocolloids and lipids in film-forming solutions produces composites encompassing the advantages of both biopolymers (Parreidt et al., 2018). Cellulose ethers, chitosan, starch, and alginate are among the most studied polysaccharides for the production of active packaging films containing nisin. Generally, homogeneous and transparent polysaccharide films show high water permeability and moderate mechanical properties.

The nature of the polymer molecule and the added substances in film-forming solutions like plasticizers and cross-linking agents determines the functional properties of prepared packaging films. Migration of nisin into the food matrix by diffusion and partitioning should be taken into account in the elaboration of packaging food films with incorporated nisin. It has been reported that nisin is distributed in the amorphous regions of the polymer in most studied cases. However, saturation of amorphous areas can happen upon the addition of high nisin concentrations, and it can diffuse into the crystalline regions (Gharsallaoui et al., 2016b).

In contact with food, prepared films should be able to release the incorporated nisin from the polymer matrix and diffuse into food (Karam et al., 2013; Gharsallaoui et al., 2016b). The release kinetics of antimicrobial agents like nisin from packaging material into food systems or food simulants can be described by Fick's second law. In fact, the capacity and efficiency of packaging polymers containing nisin might be improved by modeling its diffusion from polymer matrix into the food system. It was reported that increasing nisin concentration (3000, 5000, and 10000 IU) resulted in increased diffusion coefficients (k) and consequently faster release of nisin from polyvinylalcohol–*Alyssum homolocarpum* seed gum composite films (Marvdashti et al., 2019).

TABLE 8.5

Polymer Packaging Containing Nisin

Polymer	Concentration	Food/Simulant
Soy protein isolate, corn zein films	----	Culture media
Polylactic acid	400 IU/mL of PLA solution (2%)	Raw bee
Soy protein isolate or whey protein isolate	6000 IU/g of dry film	Culture media
Soy protein	40,000 IU/g of dry film	Turkey bologna
Corn zein	70,000 IU/g of dry film	Water
Cast corn zein	81,000 IU/g of dry film	Water
Cast wheat gluten	110,000 IU/g of dry film	Water
Heat-pressed corn zein	43,000 IU/g of dry film	Water
Heat-pressed wheat gluten	49,000 IU/g of dry film	Water
Methylcellulose	2500–10,000 IU/cm^2	Culture media
HPMC	2500-10,000 IU/cm^2	Culture media
Chitosan	51,000 IU/g of film	Culture media
Glucomannan-chitosan	42,000 IU/g of film	Culture media
Tapioca starch	881–2204 IU/cm^2	Tryptic soy broths
Polylactic acid	250,000 IU/g or 1600 IU/cm^2 of film	Orange juice
LDPE	2000 IU/cm^2	Cold-smoked salmon
Soy protein isolate	10,000 IU/g	Culture media
Chitosan-coated plastic film	500 IU/cm^2	Cold-smoked salmon
Pectin/polylactic acid composite films	500 IU/mg of films	BHI media, orange juice, liquid egg white
PBAT	1000–5000 IU/cm^2	Tryptic soy broth
Sodium caseinate	1000 IU/cm^2	Cheese

(Gharsallaoui et al., 2016b)

The polymeric food packaging materials are classified as biodegradable (mainly polysaccharides and proteins) and non-biodegradable (mainly synthetic petrol derivatives) food packaging films according to the characteristics of used polymers. Table 8.5 gives some examples of polymer-based packaging containing different nisin concentrations, and more details about nisin incorporation in films and their uses will be reviewed in the next section.

8.5.7.2 Non-Biodegradable Films

Various types of non-biodegradable polymers have been applied in the production of food packaging films containing nisin such as polyvinylchloride (PVC), polyvinylalcohol (PVOH), polypropylene (PP), polyethylene (PE), polylactic acid (PLA), and polybutylene adipate terephthalate (PBAT). Synthetic polymer-based films possess excellent barrier and mechanical properties. The inclusion of nisin in PVOH films exhibited high antilisterial activity in fermented sausages during storage at refrigerator temperatures for 90 days (Marcos et al., 2013). The thermal stability of nisin in PLA and PLA/polyethylene glycol (PEG) blends elaborated by solvent cast technique has been studied at 90–180°C up to 48 h by Holcapkova et al. (2018). Preservation of nisin antimicrobial activity (~70%) was achieved in PLA film even after heat treatment at 160°C for 15 min. Nisin degradation was observed in PLA/PEG blend films by increasing the temperature (>120°C). PEG can interact with the nisin molecules in their molten state because of its very low melting temperature (T_m). Non-toxic dicarboxylic acid, glutaric acid was used as a cross-linking agent in the formation of PVOH film with imbued nisin (Hrabalikova et al., 2016). Glutaric acid presented synergistic activity with the antimicrobial peptide nisin against *S. aureus* and *E. coli*. Stable films with long-lasting antimicrobial activity were obtained at higher crosslinking degrees

(40% or more). In the most recent study, Pattanayaiying et al. (2019) produced thermoplastic starch/PBAT (40/60) film through coating gelatin (15% v/v) containing lauric arginate (0.8 mg/cm^2) alone or in combination with nisin Z (69.4 AU/cm^2). Blend films with added nisin and lauric arginate exhibited excellent inhibitory activity against the growth of inoculated *S.* Typhimurium and *V. parahaemolyticus* on chilled and frozen sea food (bigeye snapper and tiger prawn slices) during storage at 4°C for 28 days and –20°C for 90 days.

Environmental pollution is one of the pivotal reasons for reducing the use of the films based on synthetic polymers. Biodegradable antimicrobial packaging films are more favorable in the long-term application as a result of being reproducible and environmentally friendly. Packaging films based on natural polymers present good properties including strength, flexibility, stiffness, and a barrier to oxygen and moisture (Huang et al., 2019). The addition of lipids in films or modifying the polymer structure with crosslinking agents have been suggested for reducing the high permeability of biopolymer films to water vapor (Gharsallaoui et al., 2016b). Also, the application of high hydrostatic pressure (HHP) treatment improved the structural and physical properties of nisin-soy-protein-isolate (SPI) films due to partial unfolding of protein molecules (Wei et al., 2018).

8.5.7.3 Cellulose-Based Films

Plant cellulose is the most abundant natural polymer that is environmentally friendly and biodegradable (Huang et al., 2019). Flexibility, resistance to fats and oils, and strong mechanical properties are the advantages of cellulose derivatives (methylcellulose (MC), hydroxypropylcellulose (HPC), hydroxypropylmethylcellulose (HPMC), and carboxymethyl cellulose (CMC)) films. However, applications of cellulose films can be restricted due to the relatively high cost and water sensitivity. Imran et al. (2010) evaluated the effect of commercial nisin (Nisaplin) and glycerol (plasticizer) content on the physicochemical and antimicrobial properties of HPMC films elaborated through casting method (Figure 8.3). The thickness, water sorption, and transparency of HPMC films were greatly influenced because of the components present in Nisaplin including denatured proteins and salt. Active cellulose films containing nisin exhibited antimicrobial activity against *Listeria* strains without being affected by glycerol. A simple Schiff base reaction has been used for anchoring nisin onto oxidized cellulose to form long-term antimicrobial active food packaging for the first time by Wu et al. (2018). Nisin-anchored cellulose films were produced as a result of establishing a grafting reaction between the amino groups of nisin and the aldehyde groups of oxidized cellulose. These films presented excellent physical properties including high thermal stability, reduced oxygen permeability, and acceptable transparency (>90% transmittance through 300–800 nm). Good antimicrobial activity against *Alicyclobacillus acidoterrestris* DSM 3922T (spoilage bacteria of some juices) was observed through disk inhibition zone assay even after three months of storage in a glass dryer.

Bacterial cellulose generated by *Gluconacetobacter xylinus* K3 was used for producing polysaccharide-based films containing nisin (Nguyen et al., 2008). Covering frankfurters with the produced active cellulosic films decreased the population of *L. monocytogenes* by 2 log CFU/g after 14 days of storage.

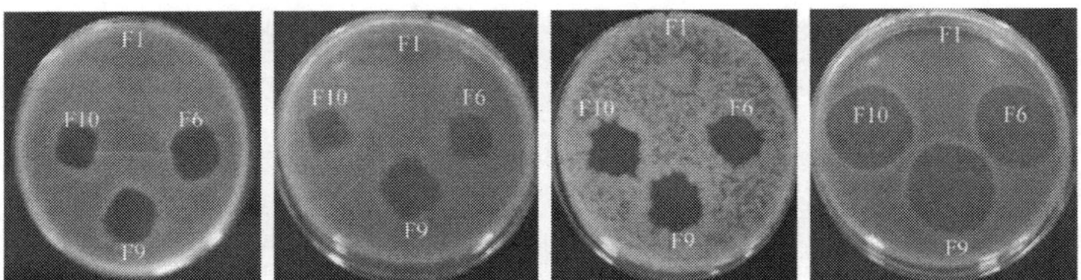

FIGURE 8.3 Inhibition zone of nisin-based active films against bacteria of food origin. (**F1** = HPMC film, **F6** = HPMC+ 10^4 IU Nisaplin®, **F9** = HPMC+ 30% glycerol + 10^4 IU Nisaplin®, **F10** = HPMC+ 50% glycerol + 10^4 IU Nisaplin®). (Nisaplin® is a commercial preparation of nisin) (Imran et al., 2010).

8.5.7.4 Chitosan-Based Films

The second most abundant polysaccharide is chitosan, a natural polymer having a linear structure made up of β-(1-4)-2-acetamide-2-deoxi-D-glucan units which is a deacetylated derivative of chitin. Chitosan films possess good characteristics including high flexibility and high resistance to manipulation (Remedio et al., 2019). Chitosan films with incorporated nisin at 60,000 IU/g of polymer did not have any biological activity against Gram-negative bacteria including *E. coli*, *S.* Typhimurium, and *Shigella flexneri* while displaying high inhibition zone against Gram-positive *S. aureus*. The addition of 0.5% v/v eugenol oil into chitosan films containing nisin contributed to significant inhibition of both Gram-positive (such as *S. aureus*, *L. monocytogenes*, and *Bacillus cereus*) and Gram-negative pathogens (Singh et al., 2015). In addition, in another study, Cé et al. (2012) found that the inclusion of nisin in chitosan films was effective against *S. aureus*, *L. monocytogenes*, and *B. cereus* with no activity against Gram-negative bacteria. The authors showed also that the incorporation of nisin with peptide P34 into chitosan films as part of hurdle technology increased the antimicrobial activity against *E. coli*. Covering minimally processed pear samples with the formed active chitosan films could significantly decrease total counts during 7 days of storage. In another study, coating fresh pork loins with 2% chitosan solution containing 0.2% nisin and 0.2% gallic acid in combination with high-oxygen modified atmosphere packaging (80% O_2 + 20% CO_2) was shown to better protect meat color, decrease protein and lipid oxidation, and show synergistic bacterial effect (the lowest total viable count value of 4.67 log CFU/g) compared with samples treated with chitosan solution alone and chitosan coating with added gallic acid during storage at 2 ± 1°C for 20 days (Cao et al., 2019).

8.5.7.5 Alginate-Based Films

Alginates, naturally occurring polysaccharides, are obtained from different brown algae genera like *Laminaria hyperborean* and some bacterial strains including *Azotobacter vinelandii* and *Pseudomonas aeruginosa*. Alginate is composed of linear binary copolymers of β-D-mannuronic acid (M) and α-L-guluronic acid (G) linked through 1–4 glycosidic bonds. Immobilized nisin at concentrations of 500 and 1000 IU/mL in hydrophobic and water vapor-resistant modified alginate films was able to decrease *Staphylococcus aureus* counts by 0.91 and 1.86 CFU/cm^2, respectively, on sliced beef after 7 days of storage at 4°C (Millette et al., 2007). The incorporation of nisin with other preservatives including cinnamon and rosemary essential oils into sodium alginate coating displayed stronger antimicrobial activity than individual treatments against tested microorganisms (*Enterobacteriaceae* spp., *Pseudomonas* spp., lactic acid bacteria, psychrotrophic bacteria, and yeast and molds) on chicken meat fillets stored in a refrigerator for 15 days (Raeisi et al., 2016). According to the obtained results, alginate coating containing just nisin had lower efficacy than essential oils in the prevention of increasing total viable count values.

Composite films have shown to have better functionalities than films produced from one biopolymer (Farris et al., 2009). In this context, Chandrasekar et al. (2016) evaluated the release kinetics of nisin from chitosan/alginate polyelectric complex films as a function of increasing alginate concentration (33%, 50%, and 66%) into an aqueous medium. Complexations between alginate and chitosan as well as between alginate and nisin led to a reduction in the total amount of released nisin by increasing alginate concentration.

8.5.7.6 Protein-Based Films

Various molecular characteristics including flexibility, conformation, molecular weight, charge, and thermal stability directly affect the ability of proteins to form films and coatings. The process of protein-based film formation consists of denaturing the protein through heat treatment, subsequently casting the solution, and evaporating the solvent. The gas barrier and mechanical properties of films made from proteins are stronger than those of polysaccharide-based films due to the structure of proteins composed of different amino acids. The major problems in the application of proteins for the production of food packaging films are associated with their high water vapor permeability and moisture sensitivity (Gharsalloui et al., 2016b). Batpho et al. (2017) used a vacuum impregnation method (pressure: 680 mmHg) to produce collagen casing containing 10,000 ppm nisin. The shelf life of the sausage increased at least 90 days at 4°C and 49 days at 10°C after the application of nisin-impregnated collagen casing. Inoculated *L. monocytogenes* counts were decreased by 1.1 log CFU/g during storage at 4°C. In another study, it was reported that sodium caseinate film with incorporated nisin at a concentration of 1000 IU/cm^2 was able to reduce 1.1 log CFU/g of *Listeria innocua* population in surface-inoculated cheese samples after 7 days of storage at 4°C (Cao-Hoang et al., 2010). Assessment of nisin antimicrobial activity incorporated in caseinate film against the growth of a tested *Listeria* strain within in-depth inoculated cheese revealed it to be dependent on the distance from the surface of food in contact with active film to the food's matrix. Rossi-Márquez et al. (2009) used a casting method for the formation of whey protein isolate films containing nisin plasticized with sorbitol. Covering ham with nisin-incorporated protein film at concentration of 473 IU/cm^2 was able to lower *Brochothrix thermosphacta* by 4 log after storing for 8 days at 4°C. According to the authors, increasing the temperature from 5 to 10°C did not have any effect on the diffusivity of nisin, while decreasing the pH could enhance bacteriocin diffusion due to weak interactions between positively charged nisin and whey protein under acidic conditions.

The type of protein and the film-forming method including casting or heat pressing significantly influence the antibacterial activity of nisin added to the film. For example, wheat gluten film was more successful than corn zein film in the preservation of nisin activity (15.8% and 12.1% of the original activity, respectively). Also, the obtained results showed that the biological activity of nisin was retained better by applying the casting method (Dawson et al., 2003). Recently, it has been demonstrated that corn zein-based edible coatings containing nisin significantly reduced *L. monocytogenes* counts in both refrigerated and frozen fish fillets (hybrid striped bass) in both polyvinyl chloride (PVC) and vacuum packing treatments (Hager et al., 2019). The incorporation of lemongrass essential oil (8%) in zein coating displayed intermediate activity against the growth of *Listeria* strains in fish fillets compared to nisin.

8.6 Regulatory Status

Several bacteriocins produced by LAB have been purified, characterized, and studied for their antibacterial properties. For example, pediocin (Class IIa) possesses a confirmed antilisterial activity and could have several potential applications in food industries. However, nisin is the only bacteriocin whose use is currently permitted in food. In 1969 the Joint Food and Agriculture Organization/World Health Organization (FAO/WHO) Expert Committee on Food Additives gave nisin international acceptance as a food preservative. The FAO/WHO Expert Committee recommended that use of nisin as a food preservative "be considered acceptable, the unconditional Average Daily Intake being 0–33,000 units/kg of body weight." The FAO/WHO Codex Committee on milk and milk products recommended standards for acceptance by national governments. For processed cheese, nisin was included as a food additive at 12.5 mg pure nisin per kilogram product.

This peptide was added to the list of food additives under the European number E234 (EEC, 1983). The authorization of the use of this bacteriocin has many reasons: (i) the peptide is easily degradable with intestinal proteases, (ii) it presents no risk to human health, and (iii) it does not alter the organoleptic and sensory properties of foods. The legislation concerning the maximum nisin amount used differs from one country to another: for example, if nisin can be added to the cheese without any limit in the United Kingdom, Australia, and France, the amount added should not exceed 100 IU/g in Belgium, 500 IU/g in Argentina, and 10,000 IU/g in the USA (Cleveland et al., 2001). More data concerning the maximum authorized nisin levels are given in Table 8.6. Finally, it is important to highlight that nisin cannot be regarded as a "natural" preservative when used in concentrations higher than those naturally found in foods fermented with nisin-producing strains.

TABLE 8.6

Examples of World-Wide Maximum Levels of Use of Nisin

Country	Food in Which Nisin Is Permitted	Maximum Level (IU/g)
Argentina	Processed cheese	500
Australia	Cheese, processed cheese, canned tomatoes	No limit
Belgium	Cheese	100
EU	E234, labeled as "natural preservative"	May also vary according to product state and member
France	Processed cheese	No limit
Italy	Cheese	500
Mexico	Nisin is a permitted additive	500
Netherlands	Factory cheese, processed cheese, cheese powder	800
Russia	Dietetic processed cheese, canned vegetables	8000
UK	Cheese, canned foods, clotted cream	No limit
US	Pasteurized processed cheese spreads	10,000

(Adapted from Cleveland et al., 2001)

REFERENCES

Abee, T. 1995. Pore-forming bacteriocins of Gram-positive bacteria and self-protection mechanisms of producer organisms. *FEMS Microbiol. Lett.* 129:1–10.

Abee, T., Rombouts, F. M., Hugenholtz, J., Guihard, G., and Letellier, L. 1994. Mode of action of nisin Z against *Listeria monocytogenes* Scott A grown at high and low temperatures. *Appl. Environ. Microbiol.* 60:1962–1968.

Al-Holy, M., Lin, M., and Rasco, B. 2005. Destruction of *Listeria monocytogenes* in sturgeon (*Acipenser transmontanus*) caviar by a combination of nisin with chemical antimicrobials or moderate heat. *J. Food Prot.* 68:512–520.

Araújo, M. K., Gumiela, A. M., Bordin, K., Luciano, F. B., and Macedo, R. E. F. de. 2018. Combination of garlic essential oil, allyl isothiocyanate, and nisin Z as bio-preservatives in fresh sausage. *Meat Sci.* 143:177–183.

Argudin, M. A., Mendoza, M. C., and Rodicio, M. R. 2010. Food poisoning and Staphylococcus aureus enterotoxins. *Toxins (Basel).* 2:1751–1773.

Arroyo, C., Cebrián, G., Pagán, R., and Condón, S. 2012. Synergistic combination of heat and ultrasonic waves under pressure for *Cronobacter sakazakii* inactivation in apple juice. *Food Control* 25:342–348.

Bani-Jaber, A., McGuire, J., Ayres, J. W., and Daeschel, M. A. 2000. Efficacy of the antimicrobial peptide nisin in emulsifying oil in water. *J. Food Sci.* 65:502–506.

Bastos, M. C. F., and Ceotto, H. 2011. Bacterial antimicrobial peptides and food preservation. In *Natural Antimicrobials in Food Safety and Quality*, edited by M. Ray, and M. Chikindas, pp. 62–76. Oxfordshire:CAB International.

Batpho, K., Boonsupthip, W., and Rachtanapun, C. 2017. Antimicrobial activity of collagen casing impregnated with nisin against foodborne microorganisms associated with ready-to-eat sausage. *Food Control* 73:1342–1352.

Bayoumi, S. 1991. Nisin and yoghurt manufacture. *Chemie Mikrobiologie Technologie Lebensmittel* 13:65–69.

Belfiore, C., Castellano, P., and Vignolo, G. 2007. Reduction of *Escherichia coli* population following treatment with bacteriocins from lactic acid bacteria and chelators. *Food Microbiol.* 24:223–229.

Bierbaum, G., and Sahl, H.-G. 2009. Lantibiotics: Mode of action, biosynthesis and bioengineering. *Curr. Pharm. Biotechnol.* 10:2–18.

Blake, K. L., Randall, C. P., and Neill, A. J. 2011. In vitro studies indicate a high resistance potential for the lantibiotic nisin in *Staphylococcus aureus* and define a genetic basis for nisin resistance. *Antimicr. Agents Chemother.* 55:2362–2368.

Boziaris, I. S., and Adams, M. R. 2000. Transient sensitivity to nisin in cold-shocked Gram-negatives. *Lett. Appl. Microbiol.* 31:233–237.

Breidt, F., Crowley, K. A., and Fleming, H. P. 1993. Isolation and characterization of nisin-resistant *Leuconostoc mesenteroides* for use in cabbage fermentations. *Appl. Environ. Microbiol.* 59:3778–3783.

Breukink, E., and de Kruijff, B. 1999. The lantibiotic nisin, a special case or not? *Biochim. Biophys. Acta* 1462:223–234.

Breukink, E., and de Kruijff, B. 2006. Lipid II as a target for antibiotics. *Nat. Rev. Drug Discov.* 5:321–323.

Breukink, E., van Karaaij, C., Demel, R. A., Siezen, R. J., Kuipers, O. P., and de Kruijff, B. 1997. The C-terminal region of nisin is responsible for the initial interaction of nisin with the target membrane. *Biochemistry* 36:6968–6976.

Brotz, H., Josten, M., Wiedmann, I., Schneider, U., Gotz, F., Bierbaum, G., and Sahl, H.-G. 1998. Role of peptidoglycan precursors in the formation of pores by nisin, epidermin and other lantibiotics. *Mol. Microbiol.* 30:317–327.

Budde, B. B., and Jakobsen, M. 2000. Real-time measurements of the interaction between single cells of *Listeria monocytogenes* and nisin on a solid surface. *Appl. Environ. Microbiol.* 66:3586–3591.

Budu-Amoaka, E., Ablett, R. F., Harris, J., and Delves-Broughton, J. 1999. Combined effect of nisin and moderate heat on the destruction of *Listeria monocytogenes* in cold-pack lobster meat. *J. Food Prot.* 62:46–50.

Cabo, M. L., Pastoriza, L., Bernandez, M., and Herrera, J. J. R. 2001. Effectiveness of carbon dioxide and Nisaplin on increasing shelf-life of fresh pizza. *Food Microbiol.* 18:489–498.

Calderon-Miranda, M. L., Barbosa-Canovas, G. V., and Swanson, B. G. 1999. Inactivation of *Listeria innocua* in liquid whole egg by pulsed electric fields and nisin. *Int. J. Food Microbiol.* 51:7–17.

Cao, Y., Warner, R. D., and Fang, Z. 2019. Effect of chitosan/nisin/gallic acid coating on preservation of pork loin in high oxygen modified atmosphere packaging. *Food Control* 101:9–16.

Cao-Hoang, L., Chaine, A., Grégoire, L., and Waché, Y. 2010. Potential of nisin-incorporated sodium caseinate films to control *Listeria* in artificially contaminated cheese. *Food Microbiol.* 27:940–944.

Cé, N., Noreña, C. P. Z., and Brandelli, A. 2012. Antimicrobial activity of chitosan films containing nisin, peptide P34, and natamycin. *CyTA - J. Food.* 10:21–26.

Chandrasekar, V., Coupland, J. N., and Anantheswaran, R. C. 2016. Release kinetics of nisin from chitosan–alginate complex films. *J. Food Sci.* 81:E2503–E2510.

Chen, H., and Zhong, Q. 2017. Lactobionic acid enhances the synergistic effect of nisin and thymol against *Listeria monocytogenes* Scott A in tryptic soy broth and milk. *Int. J. Food Microbiol.* 260:36–41.

Chihib, N. E., Crepin, T., Delattre, G., and Tholozan, J. L. 1999a. Involvement of cell envelope in nisin resistance of *Pectinatus frisingensis*, a Gram-negative, strictly anaerobic beer-spoilage bacterium. *FEMS Microbiol. Lett.* 177:167–175.

Chihib, N. E., Monnerat, L., Membre, J. M., and Tholozan, J.-L. 1999b. Nisin, temperature and pH effects on growth and viability of *Pectinatus frisingensi*, A Gram-negative, strictly anaerobic beer spoilage bacterium. *J. Appl. Microbiol.* 87:438–446.

Chihib, N. E., Ribeiro da Silva, M., Delattre, G., Laroche, M., and Federighi, M. 2003. Different cellular fatty acid pattern behaviours of two strains of *Listeria monocytogenes* Scott A and CNL 895807 under different temperature and salinity conditions. *FEMS Microbiol. Lett.* 218:155–160.

Chollet, E., Sebti, I., Martial-Gros, A., and Degraeve, P. 2008. Nisin preliminary study as a potential preservative for sliced ripened cheese: NaCl, fat and enzymes influence on nisin concentration and its antimicrobial activity. *Food Control* 19:982–989.

Chung, K. T., Dickson, J. S., and Grouse, J. D. 1989. Effects of nisin on growth of bacteria attached to meat. *Appl. Environ. Microbiol.* 55:1329.

Cleveland, J. E., Montville, T.J., Nes, I.F. and and Chikindas, M. L. 2001. Bacteriocins: safe, natural antimicrobials for food preservation *Intl. J. Food Microbiol.* 71:1–20.

Collins-Thompson, D. L., and Woods, D. S. 1992. Control in dairy products. In *Clostridium botulinum, Ecology and Control in Foods*, edited by A. H. W. Hauschild and K. L. Dodds, pp. 261–277. New York, NY: Marcel Dekker.

Cutter, C. N., and Siragusa, G. R. 1994. Decontamination of beef carcass tissue with nisin using a pilot scale carcass washer. *Food Microbiol.* 11:481–489.

Cutter, C. N., and Siragusa, G. R. 1995a. Population reduction of Gram-negative pathogens following treatments with nisin and chelators under various conditions. *J. Food Prot.* 58:977–983.

Cutter, C. N., and Siragusa, G. R. 1995b. Treatments with nisin and chelators to reduce *Salmonella* and *Escherichia coli* on beef. *J. Food Prot.* 58:1028–1030.

Cutter, C. N., and Siragusa, G. R. 1996a. Reduction of *Brochothrix thermosphacta* on beef surfaces following immobilization of nisin in calcium alginate gels. *Lett. Appl. Microbiol.* 23:9–12.

Cutter, C. N., and Siragusa, G. R. 1996b. Reductions of *Listeria innocua* and *Brochothrix thermosphacta* on beef following nisin spray treatments and vacuum packaging. *Food Microbiol.* 13:23–33.

Davies, E. A., Bevis, H. E., and Delves-Broughton, J. 1997. The use of the bacteriocin, nisin, as a preservative in ricotta-type cheeses to control the food-borne pathogen *Listeria monocytogenes. Lett. Appl. Microbiol.* 24:343–346.

Davies, E. A., Milne, C. F., Bevis, H. E., Potter, R. W., Harris, J. M., Williams, G. C., Thomas, L. V., and Delves-Broughton, J. 1999. Effective use of nisin to control lactic acid bacterial spoilage in vacuum-packed Bologna–type sausage. *J. Food Prot.* 62:1004–1010.

Dawson, P. L., Hirt, D. E., Rieck, J. R., Acton, J. C., and Sotthibandhu, A. 2003. Nisin release from films is affected by both protein type and filmforming method. *Food Res. Int.* 36:959–968.

de Arauz, L. J., Jozala, A. F., Mazzola, P. G., and Vessoni Penna, T. C. 2009. Nisin biotechnological production and application: A review. *Trends Food Sci. Technol.* 20:146–154.

De Vuyst, L., and Vandamme, E. J. 1994. Nisin, a lantibiotic produced by *Lactococcus lactis* subsp. *lactis*: Properties, biosynthesis, fermentation and applications. In *Bacteriocins of Lactic Acid Bacteria. Microbiology, Genetics and Applications*, edited by L. de Vuyst and E. J. Vandamme, pp. 151–221. London: Blackie Academic & Professional.

Dean, J. P., and Zottola, E. A. 1996. Use of nisin in ice cream and effect on survival of *Listeria monocytogenes. J. Food Prot.* 59:476–480.

Deegan, L. H., Cotter, P. D., Hill, C., & Ross, P. 2006. Bacteriocins: Biological tools for bio-preservation and shelf-life extension. *Int. Dairy J.* 16:1058–1071.

Degnan, A. J., Farkye, N. Y., Johnson, M. E., and Luchansky, J. B. 1996. *Use of Nisin to Control Listeria monocytogenes in Queso Fresco Cheese.* Paper presented at the 83rd Annual Meeting of IAMFES, Seattle, WA, USA, 30 June–3 July.

Degnan, A. J., Kaspar, C. W., Otwell, W. S., Tamplin, M. L., and Luchansky, J. B. 1994. Evaluation of lactic acid bacterium fermentation products and food grade chemicals to control *Listeria monocytogenes* in blue crab (*Callinectes sapidus*) meat. *Appl. Environ. Microbiol.* 60:3198–3203.

Delves-Broughton, J. 1990. Nisin and its uses as a food preservative. *Food Technol.* 44:100, 102, 104, 106, 108, 111–112 and 117.13–17.

Delves-Broughton, J., and Gasson, M. J. 1994. Nisin. In *Natural Antimicrobial Systems and Food Preservation*, edited by V. M. Dillon and R. G. Board, pp. 99–131. Wallingford, UK: CAB International.

Delves-Broughton, J., Blackburn, P., Evans, R. J., and Hugenholtz, J. 1996. Applications of the bacteriocin nisin. *Antonie van Leuwenhoek* 69:193–202.

Delves-Broughton, J., Williams, G. C., and Wilkinson, S. 1992. The use of the bacteriocin, nisin, as a preservative in pasteurized liquid whole egg. *Lett. Appl. Microbiol.* 15:133–136.

Demel, R. A., Peelen, T., Siezen, R. J., de Kruijf, B., and Kuipers, O. P. 1996. Nisin Z, a mutant nisin Z and lacticin 481 interactions with anionic lipids correlate with antimicrobial activity. A monolayer study. *Euro. J. Biochem.* 235:267–274.

Diao, M. M., André, S., Membré, J.–M. 2014. Meta-analysis of D-values of proteolytic *Clostridium botulinum* and its surrogate strain *Clostridium sporogenes* PA 3679, *Int. J. Food Microbiol.* 174:23–30.

Dutreux, N., Notemans, S., Gongora-Nieto, M. M., Barbosa-Canovas, G. V., and Swanson, B. G. 2000. Effects of combined exposure of *Micrococcus luteus* to nisin and pulsed electric fields. *Int. J. Food Microbiol.* 60:147–152.

EEC. 1983. EEC Commission Directive 83/463/EEC.

El Jaam, O., Fliss, I., and Aïder, M. 2017. Effect of electro-activated aqueous solutions, nisin and moderate heat treatment on the inactivation of *Clostridium sporogenes* PA 3679 spores in green bean puree and whole green beans. *Anaerobe* 47:173–182.

Erkaya, T., Baslar, M., Sengul, M., Ertugay, M. F. 2015. Effect of thermosonication on physicochemical, microbiological and sensorial characteristics of ayran during storage. *Ultrason. Sonochem.* 23:406–412.

Fang, T. J., and Lin, C. -Y., and Chen, H. L. 1997. Inhibition of *Staphylococcus aureus* and *Bacillus cereus* on a vegetarian food treated with nisin combined with potassium sorbate or sodium benzoate. *J. Food Safety* 17:69–87.

Fang, T. J., and Tsai, H.-C. 2003. Growth patterns of *Escherichia coli* O157:H7 in ground beef treated with nisin, chelators, organic acids and their combinations immobilized in calcium alginate gels. *Food Microbiol.* 20:243–253.

Farris, S., Schaich, K. M., Liu, L., Piergiovanni, L., and Yam, K. L. 2009. Development of polyion-complex hydrogels as an alternative approach for the production of bio-based polymers for food packaging applications: A review. *Trends Food Sci Technol.* 20:316–32.

Favaro, L., Penna, A. L. B., and Todorov, S. D. 2015. Bacteriocinogenic LAB from cheeses–application in biopreservation? *Trends Food Sci Technol.* 41:37–48.

FDA. 1988. Nisin preparation: Affirmation of GRAS status as a direct human food ingredient. *Fed Reg.* 53:11247–11251.

Felicio, B. A., Pinto, M. S., Oliveira, F. S., Lempk, M. W., Pires, A. C. S., and Lelis, C. A. 2015. Effects of nisin on *Staphylococcus aureus* count and physicochemical properties of Minas Frescal cheese. *J. Dairy Sci.* 98:4364–4369.

Ferrario, M., Alzamora, S. M., Guerrero, S. 2015. Study of the inactivation of spoilage microorganisms in apple juice by pulsed light and ultrasound. *Food Microbiol.* 46:635–642.

Ferreira, M. A. S. S., and Lun, B. M. 1996. The effect of nisin on *Listeria monocytogenes* in culture medium and long-life cottage cheese. *Lett. Appl. Microbiol.* 22:432–438.

Figueiredo, A. C. L., and Almeida, R. C. C. 2017. Antibacterial efficacy of nisin, bacteriophage P100 and sodium lactate against *Listeria monocytogenes* in ready-to-eat sliced pork ham. *Braz. J. Microbiol.* 48:724–729.

Fukao, M., Obita, T., Yoneyama, F., Kohda, D., Zendo, T., Nakayama, J., and Sonomoto, K. 2008. Complete covalent structure of nisin Q, new natural nisin variant, containing post-translationally modified amino acids. *Biosci. Biotech. Bioch.* 72:1750–1755.

Gálvez, A., López, R. L., Abriouel, H., Valdivia, E., and Ben Omar, N. 2008. Application of bacteriocins in the control of foodborne pathogenic and spoilage bacteria. *Crit. Rev. Biotechnol.* 28:125–152.

García, P., Rodríguez, L., Rodríguez, A., and Martínez, B. 2010. Food biopreservation: Promising strategies using bacteriocins, bacteriophages and endolysins. *Trends Food Sci. Technol.* 21:373–382.

Gharsallaoui, A., Oulahal, N., Joly, C., and Degraeve, P. 2016a. Nisin as a food preservative: Part 1: Physicochemical properties, antimicrobial activity, and main Uses. *Crit. Rev. Food Sci. Nutr.* 56:1262–1274.

Gharsallaoui, A., Joly, C., Oulahal, N., and Degraeve, P. 2016b. Nisin as a food preservative: Part 2: Antimicrobial polymer materials containing nisin. *Crit. Rev. Food Sci. Nutr.* 56:1275–1289.

Gill, A. O., and Holley, R. A. 2000a. Inhibition of bacterial growth on ham and bologna by lysozyme, nisin and EDTA. *Food Res. Int.* 33:83–90.

Gill, A. O., and Holley, R. A. 2000b. Surface application of lysozyme, nisin, and EDTA to inhibit spoilage and pathogenic bacteria on ham and bologna. *J. Food Prot.* 63:1338–1346.

Govaris, A., Solomakos, N., Pexara, A., and Chatzopoulou, P. S. 2010. The antimicrobial effect of oregano essential oil, nisin and their combination against *Salmonella enteritidis* in minced sheep meat during refrigerated storage. *Int. J. Food Microbiol.* 137:175–180.

Guenther, S., Herzig, O., Fieseler, L., Klumpp, J., & Loessner, M. J. 2012. Biocontrol of *Salmonella* Typhimurium in RTE foods with the virulent bacteriophage FO1-E2. *Int. J. Food Microbiol.* 154:66–72.

Hager, J. V., Rawles, S. D., Xiong, Y. L., Newman, M. C., and Webster, C. D. 2019. Edible corn-zein-based coating incorporated with nisin or lemongrass essential oil inhibits *Listeria monocytogenes* on cultured hybrid striped bass, *Morone chrysops* × *Morone saxatilis*, fillets during refrigerated and frozen storage. *J. World Aquacult. Soc.* 50:204–218.

Hanlin, M. B., Kalchayanand, N., Ray, P., and Ray, B. 1993. Bacteriocins of lactic acid bacteria in combination have greater antibacterial activity. *J. Food Prot.* 56:252–255.

Harris, L. J., Fleming, H. P., and Klaenhammer, T. R. 1992. Characterization of two nisin producing *Lactococcus lactis* subsp *lactis* strains isolated from a commercial sauerkraut fermentation. *Appl. Environ. Microbiol.* 58:1477–1487.

Hauben, K. J. A., Wuytack, E. C., Soontjens, G. C. F., and Michiels, C. W. 1996. High pressure transient sensitization of *Escherichia coli* to lysozyme and nisin by disruption of outer membrane permeability. *J. Food Prot.* 59:350–355.

Heinemann, B., Voris, L., and Stumbo, C. R. 1965. Use of nisin in processing food products. *Food Technol.* 19:592–596.

Henning, S., Metz, R., and Hammes, W. P. 1986. New aspects for the application of nisin to foods based on its mode of action. *Int. J. Food Microbiol.* 3:13–141.

Hirsch, A. 1951. Growth and nisin production of a strain of *Streptococcus lactis*. *J. Gen. Microbiol.* 5:208.

Hirsch, A., Grinsted, E., Chapman, H. R., and Mattick, A. T. R. 1951. A note on the inhibition of an anaerobic sporeformer in Swiss-type cheese by a nisin producing streptococcus. *J. Dairy Res.* 18:205–207.

Holcapkova, P., Hurajova, A., Bazant, P., Pummerova, M., and Sedlarik, V. 2018. Thermal stability of bacteriocin nisin in polylactide-based films. *Poly. Degrad. Stab.* 158:31–39.

Hossack, D. J. N., Bird, M. C., and Fowler, G. G. 1983. The effect of nisin on the sensitivity of microorganisms to antibiotics and other chemotherapeutic agents. In *Antimicrobials and Agriculture*, edited by M. Woodbine, pp. 425–433. London: Butterworths.

Hrabalikova, M., Holcapkova, P., Suly, P., and Sedlarik, V. 2016. Immobilization of bacteriocin nisin into a poly(vinyl alcohol) polymer matrix crosslinked with nontoxic dicarboxylic acid. *J. Appl. Polym. Sci.* 133:43674.

Huang, T., Qian, Y., Wei, J., and Zhou, C. 2019. Polymeric antimicrobial food packaging and its applications. *Polymers* 11:560.

Hugentholtz, J., Twigt, M., Stomp, M., and Smith, M. R. 1995. *Development of Nisin-Producing Starters for Gouda Cheese Manufacture.* Paper presented at International Dairy Lactic Acid Bacteria Conference, Palmerston North, New Zealand, 19–23 February 1995.

Hungaro, H. M., Mendonça, R. C. S., Gouvêa, D. M., Vanetti, M. C. D., and Pinto, C. L. D. O. 2013. Use of bacteriophages to reduce *Salmonella*, in chicken skin in comparison with chemical agents. *Food Res. Int.* 52:75–81.

Imran, M., El-Fahmy, S., Revol-Junelles, A. M., and Desobry, S. 2010. Cellulose derivative based active coatings: Effects of nisin and plasticizer on physico-chemical and antimicrobial properties of hydroxypropyl methylcellulose films. *Carbohydr. Polym.* 81:219–225.

Jung, G., and Sahl, H.-G. 1991. *Nisin and Novel Lantibiotics.* Leiden, The Netherlands: Escom.

Kalchayanand, N., Sikes, T., Dunne, C. P., and Ray, B. 1994. High hydrostatic pressure and electroporation have increased bactericidal efficiency in combination with bacteriocins. *Appl. Environ. Microbiol.* 60:4174–4177.

Karam, L., Jama, C., Dhulster, P., and Chihib, N. E. 2013. Study of surface interactions between peptides, materials and bacteria for setting up antimicrobial surfaces and active food packaging. *J. Mater. Environ Sci.* 4:798–821.

Kassaify, Z. G., Thomas, L. V., Potter, R. W., Rosenthal, A. J., and Delves-Broughton, J. 2000. The potential use of nisin in sausage casings. In *Revista Latinoamerican de Microbiologia. Proceedings of the XV Congreso Latinoamericano de Microbiologia and XXX1 Congreso Nacional de Microbiologia* 42:Supplement. 9–13 April, Merida, Mexico.

Kato, T., Inuzuka, L., Kondo, M., and Matsuda, T. 2001. Growth of nisin-producing lactococci in cooked rice supplemented with soybean extract and its application to inhibition of *Bacillus subtilis* in rice miso. *Biosci. Biotech. Biochem.* 65:330–337.

Khan, A., Vu, K. D., Riedl, B., and Lacroix, M. 2015. Optimization of the antimicrobial activity of nisin, Na-EDTA and pH against Gram-negative and Gram-positive bacteria. *LWT - Food Sci. Technol.* 61:124–129.

Kondrotiene, K., Kasnauskyte, N., Serniene, L., Gölz, G., Alter, T., Kaskoniene, V., Maruska, A. S., and Malakauskas, M. 2018. Characterization and application of newly isolated nisin producing *Lactococcus lactis* strains for control of *Listeria monocytogenes* growth in fresh cheese. *LWT* 87:507–514.

Kooy, J. S. 1952. Strains of *Lactobacillus plantarum* which destroy the antibiotic made by *Streptococcus lactis*. *Neth. Milk Dairy J.* 6:223.

Liao, H., Jiang, L., Cheng, Y., Liao, X., and Zhang, R. 2018. Application of nisin-assisted thermosonication processing for preservation and quality retention of fresh apple juice. *Ultrasonics Sonochemistry* 42:244–249.

Lin, L., Hu, J. Y., Wu, Y., Chen, M., Ou, J., and Yan, W. L. 2018. Assessment of the inhibitory effects of sodium nitrite, nisin, potassium sorbate, and sodium lactate on *Staphylococcus aureus* growth and staphylococcal enterotoxin A production in cooked pork sausage using a predictive growth model. *Food Science and Human Wellness* 7:83–90.

Lipinska, E. 1977. Nisin and its applications. In *Antibiotics and Antibiosis in Agriculture*, edited by M. Woodbine, pp. 103–130. London: Butterworths.

Liu, W., and Hansen, J. N. 1990. Some chemical and physical properties of nisin, a small-protein antibiotic produced by *Lactococcus lactis*. *Appl. Environ. Microbiol.* 56:2551–2558.

Luo, X., and Zhu, Y. 2000. Application of nisin to preservation of refrigerated beef. *Food Sci., China* 21:53–57. (English summary).

Mahadeo, M., and Tatini, S. R. 1994. The potential use of nisin to control *Listeria monocytogenes* in poultry. *Lett. Appl. Microbiol.* 18:323–326.

Maisnier-Patin, S., Deschamps, N. Tatini, S.R. and Richard, J. 1992. Inhibition of *Listeria monocytogenes* in Camembert cheese made with a nisin-producing starter. *Le Lait* 72:249-263.

Marcos, B., Aymerich, T., Garriga, M., and Arnau, J. 2013. Active packaging containing nisin and high pressure processing as post-processing listericidal treatments for convenience fermented sausages. *Food Control* 30:325–330.

Maresca, D., Prisco, A. D., Storia, A. L., Cirillo, T., Esposito, F., and Mauriello, G. 2016. Microencapsulation of nisin in alginate beads by vibrating technology: Preliminary investigation. *LWT - Food Sci. Technol.* 66:436–443.

Martiñon, M. E., Moreira, R. G., Castell-Perez, M. E., and Gomes, C. 2014. Development of a multilayered antimicrobial edible coating for shelf-life extension of fresh-cut cantaloupe (*Cucumis melo* L.) stored at 4°C. *LWT - Food Sci. Technol.* 56:341–350.

Marvdashti, L. M., Yavarmanesh, M., and Koocheki, A. 2019. Controlled release of nisin from polyvinyl alcohol - *Alyssum homolocarpum* seed gum composite films: Nisin kinetics. *Food Biosci.* 28:133–139.

Masschalck, B., Garcia-Graells, C., van Haver, E., and Michiels, C. W. 2000. Inactivation of high pressure resistant *Escherichia coli* by lysozyme and nisin under high pressure. *Innov. Food Sci. Emerg. Technol.* 1:39–47.

Masschalck, B., Houdt, R., and van Michiels, C. W. 2001. High pressure increases bactericidal activity and spectrum of lactoferrin, lactoferricin and nisin. *Int. J. Food Microbiol.* 64:325–332.

Mathenjwa, S. A., Hugo, C. J., Bothma, C., and Hugo, A. 2012. Effect of alternative preservatives on the microbial quality, lipid stability and sensory of boerewors. *Meat Sci.* 91:165–172.

McManamon, O., Kaupper, T., Scollard, J., and Schmalenberger, A. 2019. Nisin application delays growth of *Listeria monocytogenes* on fresh-cut iceberg lettuce in modified atmosphere packaging, while the bacterial community structure changes within one week of storage. *Postharvest Biol. Technol.* 147:185–195.

Millette, M., Smoragiewicz, W., and Lacroix, M. 2004. Antimicrobial potential of immobilized *Lactococcus lactis* subsp. *lactis* ATCC 11454 against selected bacteria. *J. Food Prot.* 67:1184–1189.

Millette, M., Le Tien, C., Smoragiewicz, W., and Lacroix, M. 2007. Inhibition of *Staphylococcus aureus* on beef by nisin-containing modified alginate films and beads. *Food Control* 18:878–884.

Ming, K., Xi, G. C., Ke, X., and Park, H. J. 2010. Antimicrobial properties of chitosan and mode of action: A state of the art review. *Int. J. Food Microbiol.* 144:51–63.

Mitra, S., Mukhopadhyay, B. C., and Biswas, S. R. 2011. Potential application of the nisin Z preparation of *Lactococcus lactis* W8 in preservation of milk. *Lett. Appl. Microbiol.* 53:98–105.

Modi, K. D., Chimindas, M. L., and Montville, T. J. 2000. Sensitivity of nisin-resistant *Listeria monocytogenes* to heat and the synergistic action of heat and nisin. *Lett. Appl. Microbiol.* 30:249–253.

Muir, D. D., Tamime, A. Y., Shenan, M. E., and Dawood, A. H. 1999. Processed cheese analogues incorporating fat-substitutes. I. Composition, microbiological quality and flavour changes during storage at 5°C. *Lebensm. –Wiss. u-Technol.* 32:41–49.

Müller-Auffermann, K., Grijalva, F., Jacob, F., and Hutzler, M. 2015. Nisin and its usage in breweries: A review and discussion. *J. Inst. Brew.* 121:309–319.

Naim, F., Zareifard, M. R., Zhu, S., Huizing, R. H., Grabowski, S., and Marcotte, M. 2008. Combined effects of heat, nisin and acidification on the inactivation of *Clostridium sporogenes* spores in carrot-alginate particles: From kinetics to process validation. *Food Microbiol.* 25:936–941.

Nguyen, V., Gidley, M., and Dykes, G. 2008. Potential of a nisin-containing bacterial cellulose film to inhibit *Listeria monocytogenes* on processed meats. *Food Microbiol.* 25:471–478.

Nissen, H., Holo, H., Axelsson, L., and Blom, H. 2001. Characterization and growth of *Bacillus* spp. in heat-treated cream with and without nisin. *J. Appl. Microbiol.* 90:530–534.

Nykanen, A., Weckman, K., and Lapvetelainen, A. 2000. Synergistic inhibition of *Listeria monocytogenes* on cold-smoked rainbow trout by nisin and sodium lactate. *Int. J. Food Microbiol.* 61:63–72.

Oh, H. I., Leon, M. J., and Lee, W. J. 1994. Use of nisin to enhance storage stability of Kimchi. *36C-12 IFT Annual Meeting Technical Program, June 25–29.* Abstracts.

Oladunjoye, A. O., Singh, S., and Ijabadeniyi, O. A. 2016. Inactivation of *Listeria monocytogenes* ATCC 7644 on fresh-cut tomato using nisin in combinations with organic salts. *Braz. J. Microbiol.* 47:757–763.

Park, H. J., Lee, N.-K., Choi, J.-O., Ha, J.-U., and Paik, H.-D. 2001. Control of *Listeria monocytogenes* in ground beef by lactococcal bacteriocins. *Food Sci. Technol.* 10:199–203.

Parreidt, T. S., Müller, K., and Schmid, M. 2018. Alginate-based edible films and coatings for food packaging applications. *Foods* 7:170.

Pattanayaiying, R., Sane, A., Photjanataree, P., and Cutter, C. N. 2019. Thermoplastic starch/polybutylene adipate terephthalate film coated with gelatin containing nisin Z and lauric arginate for control of foodborne pathogens associated with chilled and frozen seafood. *Int. J. Food Microbiol.* 290:59–67.

Pawar, D. D., Malik, S. V. S., Bhilegaonkar, K. N., and Barbuddhe, S. B. 2000. Survivability of pathogenic *Listeria monocytogenes* against nisin and its combination with sodium chloride in raw buffalo meat mince. *J. Food Sci. Technol.* 38:138–141.

Pimentel-Filho, N. D. J., Martins, M. C. D. F., Nogueira, G. B., Mantovani, H. C., and Vanetti, M. C. D. 2014. Bovicin HC5 and nisin reduce *Staphylococcus aureus* adhesion to polystyrene and change the hydrophobicity profile and Gibbs free energy of adhesion. *Int. J. Food Microbiol.* 190:1–8.

Plocková, M., Štěpánek, M., Demnerová, K., Čurda, L., and Šviráková, E. 1996. Effect of nisin for improvement in shelf-life and quality of processed cheese. *Adv. Food Sci. (CMTL)* 18:78–83.

Pohl, I. E., Mastwijk, H. C., Bartels, P. V., and Smid, E. J. 2000. Pulsed-electric field treatment enhances the bactericidal action of nisin against *Bacillus cereus*. *Appl. Environ. Microbiol.* 66:428–430.

Pokhrel, P. R., Toniazzo, T., Boulet, C., Oner, M. E., Sablani, S. S., Tang, J., and Barbosa-Cánovas, G. V. 2019. Inactivation of *Listeria innocua* and *Escherichia coli* in carrot juice by combining high pressure processing, nisin, and mild thermal treatments. *Innov. Food Sci. Emerg. Technol.* 54:93–102.

Ponce, E., Pla, R., Sendra, J., Guamis, B., and Mor-Mur, M. 1998. Combined effect of nisin and high hydrostatic pressure on destruction of *Listeria innocua* and *Escherichia coli* in liquid whole egg. *Int. J. Food Microbiol.* 43:15–19.

Prudêncio, C. V., Mantovani, H. C., Cecon, P. R., Prieto, M., and Vanetti, M. C. D. 2016. Temperature and pH influence the susceptibility of *Salmonella* Typhimurium to nisin combined with EDTA. *Food Control* 61:248–253.

Punyauppa-path, S., Phumkhachorn, P., and Rattanachaikunsopon, P. 2015. Nisin: Production and mechanism of antimicrobial action. *Int. J. Curr. Res. Rev.* 7:47–53.

Raeisi, M., Tabaraei, A., Hashemi, M., and Behnampour, N. 2016. Effect of sodium alginate coating incorporated with nisin, *Cinnamomum zeylanicum*, and rosemary essential oils on microbial quality of chicken meat and fate of *Listeria monocytogenes* during refrigeration. *Int. J. Food Microbiol.* 238:139–145.

Ramseier, H. R. 1960. The action of nisin on *Clostridium butyricum*. *Arch. Mikrobiol.* 37:57–94.

Rauch, P. J. G., and de Vos, W. M. 1992. Characterization of the novel nisin-sucrose conjugative transposon Tn 5276 and its insertion in *Lactococcus lactis*. *J. Bacteriol.* 174:1280–1287.

Ray, B. 1992. Nisin of *Lactococcus lactis* subsp. *lactis* as a food preservative. In *Food Biopreservatives of Microbiol Origins*, edited by B. Ray and M. A. Daeschel, pp. 207–264. Boca Raton, FL: CRC Press.

Rayman, M. K., Aris, B., and Hurst, A. 1981. Nisin: A possible alternative or adjunct to nitrite in the preservation of meats. *Appl. Environ. Microbiol.* 41:375–380.

Remedio, L. N., Silva dos Santos, J. W., Vieira Maciel, V. B., Yoshida, C. M. P., and Aparecida de Carvalho, R. 2019. Characterization of active chitosan films as a vehicle of potassium sorbate or nisin antimicrobial agents. *Food Hydrocoll.* 87:830–838.

Roberts, C. M., and Hoover, D. G. 1996. Sensitivity of *Bacillus coagulans* spores to combinations of high hydrostatic pressure, heat, acidity and nisin. *J. Appl. Bacteriol.* 84:363–368.

Roberts, R. F., and Zottola, E. A. 1993. Shelf life of pasteurized process cheese spreads made from Cheddar cheese manufactured with a nisin-producing starter culture. *J. Dairy Sci.* 76:1829–1836.

Roberts, R. F., Zottola, E. A., and McKay, L. L. 1992. Use of a nisin producing starter culture suitable for Cheddar cheese manufacture. *J. Dairy Sci.* 75:2353–2363.

Rose, N. L., Sporns, P., and McMullen, L. M. 1999. Detection of bacteriocins by matrix-assisted laser/desorption/ionization time of flight mass spectrometry. *Appl. Environ. Microbiol.* 65:2238–2242.

Ross, R., Morgan, S., and Hill, C. 2002. Preservation and fermentation: Past, present and future. *Int. J. Food Microbiol.* 79:3–16.

Rossi-Márquez, G., Han, J. H., García-Almendárez, B., Castaño-Tostado, E., and Regalado-González, C. 2009. Effect of temperature, pH and film thickness on nisin release from antimicrobial whey protein isolate edible films. *J. Sci. Food Agric.* 89:2492–2497.

Russell, N. J., Evans, R. I., ter Steeg, P. F., Hellemons, J., Verheul, A., and Abee, T. 1995. Membranes as a target for stress adaptation. *Int. J. Food Microbiol.* 28:255–263.

Saelao, S., Maneerat, S., Thongruck, K., Watthanasakphuban, N., Wiriyagulopas, S., Chobert, J.-M., and Haertlé, T. 2018. Reduction of tyramine accumulation in Thai fermented shrimp (kung-som) by nisin Z-producing *Lactococcus lactis* KTH0-1S as starter culture. *Food Control* 90:249–258.

Sarkar, P., Bhunia, A. K., and Yao, Y. 2017. Impact of starch-based emulsions on the antibacterial efficacies of nisin and thymol in cantaloupe juice. *Food Chem.* 217:155–162.

Scannel, A. G. M., Hill, C., Buckley, D. J., and Arendt, E. K. 1997. Determination of the influence of organic acids and nisin on shelf life and microbiological safety aspects of fresh pork sausage. *J. Appl. Microbiol.* 83:407–412.

Schelegueda, L. I., Zalazar, A. L., Gliemmo, M. F., and Campos, C. A. 2016. Inhibitory effect and cell damage on bacterial flora of fish caused by chitosan, nisin and sodium lactate. *Int. J. Biol. Macromol.* 83:396–402.

Schillinger, U. 1999. Protective cultures improve the efficacy of nisin. In *European Research Towards Safer and Better Food. Technological Methods to Improve Food Quality. Proceedings of the 3rd Karlsuhe Symposium, Karlsruhe, October, 1998*, pp. 253–261.

Schillinger, U., Becker, B., Vignolo, G., and Holzapfel, W. H. 2001. Efficacy of nisin in combination with protective cultures against *Listeria monocytogenes* Scott A in tofu. *Int. J. Food Microbiol.* 71:159–168.

Scott, V. N., and Taylor, S. L. 1981a. Effect of nisin on the outgrowth of *Clostridium botulinum* spores. *J. Food Sci.* 46:117–120, 126.

Scott, V. N., and Taylor, S. L. 1981b. Temperature, pH and spore load effects on the ability of nisin to prevent outgrowth of *Clostridium botulinum* spores. *J. Food Sci.* 46:121–126.

Shefet, S. M., Sheldon, B. W., and Klaenhammer, T. R. 1995. Efficacy of optimized nisin-based treatments to inhibit *Salmonella typhimurium* and extend shelf life of broiler carcasses. *J. Food Prot.* 58:1077–1082.

Shi, C., Zhang, X., Zhao, X., Meng, R., Liu, Z., Chen, X., and Guo, N. 2017a. Synergistic interactions of nisin in combination with cinnamaldehyde against *Staphylococcus aureus* in pasteurized milk. *Food Control* 71:10–16.

Shi, C., Zhao, X., Meng, R., Liu, Z., Zhang, G., and Guo, N. 2017b. Synergistic antimicrobial effects of nisin and p-anisaldehyde on *Staphylococcus aureus* in pasteurized milk. *LWT* 84:222–230.

Silva, C. C. G., Silva, S. P. M., and Ribeiro, S. C. 2018. Application of bacteriocins and protective cultures in dairy food preservation. *Front. Microbiol.* 9:594.

Singh, T. P., Chatli, M. K., and Sahoo, J. 2015. Antimicrobial activity of chitosan based bioactive film incorporated with nisin and eugenol. *J. Vet. Pub. Hlth.* 13:59–62.

Solomakos, N., Govaris, A., Koidis, P., and Botsoglou, N. 2008. The antimicrobial effect of thyme essential oil, nisin, and their combination against *Listeria monocytogenes* in minced beef during refrigerated storage. *Food Microbiol.* 25:120–127.

Somers, E. E., and Taylor, S. L. 1987. Antibotulinal effectiveness of nisin in pasteurized processed cheese spreads. *J. Food Prot.* 50:842.

Song, Z., Li, F., Guan, H., Xu, Y., Fu, Q., and Li, D. 2017. Combination of nisin and ε-polylysine with chitosan coating inhibits the white blush of fresh-cut carrots. *Food Control* 74:34–44.

Stergiou, V. A., Thomas, L. V., and Adams, M. R. 2006. Interactions of nisin with glutathione in a model protein system and meat. *J. Food Prot.* 69:951–956.

Stewart, C. M., Dunne, P., Sikes, A., and Hover, D. G. 2000. Sensitivity of spores of *Bacillus subtilis* and *Clostridium sporogenes* PA 3679 to combinations of high hydrostatic pressure and other processing parameters. *Innov. Food Sci. Emerging Technol.* 1:49–56.

Tanaka, N., Traisman, E., Plantingua, P., Finn, L., Flora, W., Meske, L., and Guggisberg, J. 1986. Evaluation of factors involved in antibotulinal properties of pasteurized processed cheese spreads. *J. Food Prot.* 49:526–531.

Ter Steeg, P. F., Hellemons, J. C., and Kok, A. E. 1999. Synergistic actions of nisin, sublethal ultrahigh pressure, and reduced temperature on bacteria and yeast. *Appl. Environ. Microbiol.* 65:4148–4154.

Terebiznik, M. R., Jagus, R. J., Cerrutti, P., de Huergo, M. S., and Pilosoe, A. M. R. 2000. Combined effect of nisin and pulsed electric fields on the inactivation of *Escherichia coli*. *J. Food Prot.* 63:741–746.

Thomas, L. V., and Delves-Broughton, J. 2001. New advances in the application of the food preservative nisin. *Res. Adv. Food Sci.* 2:11–22.

Thomas, L. V., Clarkson, M. R., and Delves-Broughton, J. 2000. Nisin. In *Natural Food Antimicrobial Systems*, edited by A. S. Naidu. CRC Press. Boca Raton, FL, pp. 463–524.

Van Kraaij, C., Breukink, E., Rollema, H. S., Siezen, R., Demel, R. A., de Kruijff, B., and Kuipers, O. P. 1997. Influence of charge differences in the C-terminal part of nisin on antimicrobial activity and signalling capacity. *Euro. J. Biochem.* 247:114–120.

Vessoni Penna, T. C., Jozala, A. F., Novaes, L. C. L., Pessoa Jr., A., and Cholewa, O. 2005. Production of nisin by *Lactococcus lactis* in media with skimmed milk. *Appl. Biochem. Biotechnol.* 121–124:619–637.

Wan Norhana, M. N., Poole, S. E., Deeth, H. C., and Dykes, G. A. 2012. Effects of nisin, EDTA and salts of organic acids on *Listeria monocytogenes*, *Salmonella* and native microflora on fresh vacuum packaged shrimps stored at 4°C. *Food Microbiol.* 31:43–50.

Wang, C., Yang, J., Zhu, X., Lu, Y., Xue, Y., and Lu, Z. 2017. Effects of *Salmonella* bacteriophage, nisin and potassium sorbate and their combination on safety and shelf life of fresh chilled pork. *Food Control* 73:869–877.

Wei, J., Zhang, Z., Cai, Q., and Peng, B. 2018. Effects of high hydrostatic pressure on structural and physical properties of nisin-SPI film. *Int. J. Biol. Macromol.* 111:976–982.

Wiedemann, I., Breukink, E., van Kraaij, C., Kuipers, O. P., Bierbaum, G., de Kruijff, B., and Sahl, H.-G. 2001. Specific binding of nisin to the peptidoglycan precursor lipid II combines pore formation and inhibition of cell wall biosynthesis for potent antibiotic activity. *J. Biol. Chem.* 276:1772–1779.

Wirjantoro, T. I., Lewis, M. J., Grandison, A. S., Williams, G. C., and Delves-Broughton, J. 2001. The effect of nisin on the keeping quality of reduced heat-treated milks. *J. Food Prot.* 64:213–219.

Wu, H., Teng, C., Liu, B., Tian, H., and Wang, J. 2018. Characterization and long term antimicrobial activity of the nisin anchored cellulose films. *Int. J. Biol. Macromol.* 113:487–493.

Wu, T., Wu, C., Fang, Z., Ma, X., Chen, S., and Hu, Y. 2017. Effect of chitosan microcapsules loaded with nisin on the preservation of small yellow croaker. *Food Control* 79:317–324.

Xu, W., Qu, W., Huang, K., Guo, F., Yang, J., Zhao, H., and Luo, Y. 2007. Antibacterial effect of grapefruit seed extract on food-borne pathogens and its application in the preservation of minimally processed vegetables. *Postharvest Biol. Technol.* 45:126–133.

Yezzi, T. L., Ajao, A. B., and Zottola, E. A. 1993. Increased nisin in Cheddar-type cheese prepared with pH control of the bulk starter culture system. *J. Dairy Sci.* 76:2827–2831.

Zhou, H., Fang, J., Tian, Y., and Lu, X. Y. 2014. Mechanisms of nisin resistance in Gram-positive bacteria. *Ann. Microbiol.* 64:413–420.

9

Natamycin

P. Michael Davidson and Craig Doan

CONTENTS

9.1 Introduction

Natamycin (CAS 7681-93-8) is a compound used to delay or prevent the spoilage of certain foods and beverages by yeast or mold growth. Natamycin was first discovered in 1955 in a culture filtrate of *Streptomyces natalensis* isolated from a soil sample in the Natal Province of South Africa, and it is from this region that the name is derived. Prior to the current naming, the compound was called "pimaricin" (named after the nearby town of Pietermaritzberg). Although this name can be found in earlier publications, the World Health Organization (WHO) no longer accepts it. Another antimicrobial compound isolated from a culture of *Streptomyces chattanoogensis* in 1959 was called "Tennecetin" (Burns and Holtman, 1959); however, it was later discovered to be identical to natamycin (Brik, 1981). Natamycin preparations are available commercially under the trade names Natamax® (DuPont) and Delvo®Cid (DSM), both of which contain approximately 50% natamycin blended with lactose. Commercial preparations are made by the fermentation in dextrose-based media by selected *Streptomyces* strains. Dried natamycin recovered from the fermentation broth is white to cream-colored, has little or no odor or taste, and in the crystalline form is very stable. The optimization of nutrients in the fermentation media for natamycin production by *Streptomyces natalensis* in submerged batch culture was described by Farid et al. (2000).

Natamycin is particularly useful in foods to prevent the growth of fungi. Fungal growth is favored on foods that have low pH, moisture, and water activity and are stored at low temperatures. Some fungi can produce mycotoxins that are carcinogenic, teratogenic, hemorrhagic, or dermatitic (Davis and Diener, 1987). Natamycin can be used in human and veterinary medicine for the topical treatment of fungal infections, but such use has now been mainly superseded by other antimycotic preparations.

Natamycin has been the subject of several reviews (Raab, 1972; Brik, 1981; Stark, 1999; Thomas and Delves-Broughton, 2001; Stark and Tan, 2003; Delves-Broughton and Weber, 2011).

9.2 Physical and Chemical Properties

Natamycin is a tetraene polyene macrolide with a molecular weight of 665.7 Daltons. The full chemical name is 22-[(3-amino-3,6-dideoxy-b-D-manno pyranosol)oxy]-1,3,26 trihydroxy-12-methyl-10-oxo-6, 11,28-trioxiatricyclo (22.3.1.0) octacosa-8,14,16,18,20-pentanene-25-carboxylic acid. The empirical formula is $C_{33}H_{47}NO_{13}$ and the full structure is shown in Figure 9.1.

In its common crystalline form, natamycin exists as a trihydrate. As a dry powder it can be stored for several years with minimal loss of activity. Factors affecting the stability are extreme pH value, light, oxidants, chlorine, and heavy metals. Like many polyene macrolides, natamycin is amphoteric, possessing one acidic and one basic group. Natamycin has low solubility in water (approximately 40 µg/ml). This low solubility is an advantage for the surface treatment of food, since it ensures that the preservative remains on the surface of food where it is needed, instead of migrating into the food. Solubility ranges in other solvents include: (µg/ml) methanol, 1200; *n*-butanol, 50–120; glycerol, 3000; and ethanol, 100 (Brik, 1981; Raab, 1972). The pH of a 2% suspension of Natamax™ or Delvocid® is 5–7.5.

Raab (1972) reports in detail on the effect of pH on the stability of natamycin solutions in phosphate buffers over the pH range 2 to 10. At pH values of 4.5 and 9 natamycin shows good stability at 37°C over a 72-hour period whereas at extreme pH values of 2 and 10, stability is significantly reduced. Stability at 100°C is optimum at pH 6 and 7 and less optimum at pH levels higher and below. Raab (1967) reported the isoelectric point of natamycin as 6.5.

The biosynthesis of natamycin was first described by Aparicio et al. (1999) who identified and cloned a large *Streptomyces natalensis* gene cluster involved in the biosynthesis of the polyketide backbone that formed the 26-membered tetraene macrolide ring of natamycin. Natamycin is synthesized by polyketide synthases that sequentially assemble acyl precursors in a mechanism that was described as resembling fatty acid biosynthesis (Aparicio et al. 2003). A complete description of molecular biosynthesis of natamycin by *S. natalensis* was described in detail by Aparicio et al. (2016).

9.3 Antimicrobial Activity

Natamycin is active against nearly all yeasts and molds but has no effect on bacteria or viruses. Natamycin has been shown to be inhibitory to some protozoa *in vitro* (Aparicio et al., 2016). Most molds are inhibited at concentrations of 0.5 to 6 µg/ml, but some species require 10–25 µg/ml (Table 9.1).

Most yeasts are inhibited at concentrations from 1.0 to 5.0 µg/ml. (Table 9.2).

Marin et al. (2017) studied the influence of water activity adjusted with either NaCl or glycerol on the antifungal activity of natamycin against the cheese spoilage molds, *Aspergillus varians*, *Mucor racemosus*, *Penicillium solitum*, *P. chrysogenum*, and *P. roqueforti*. Natamycin at 10 µg/ml and a water activity

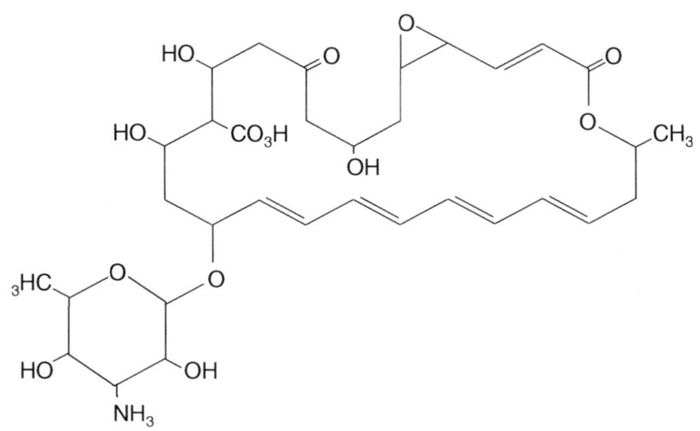

FIGURE 9.1 Structure of natamycin.

TABLE 9.1

Sensitivity of Selected Molds to Natamycin

	MIC[a] (μg/ml)
Byssochlamys fulva 040021	0.1–1.25
Penicillium candidum S66	
P. chrysogenum S138	
P. commune ABC118	
P. cyclopium S124	
P. nalgiovense S125	
Aspergillus chevalieri 4298	0.1–2.5
A. clavatus	
A. nidulans	
A. ochraceus 4069	
Cladosporium cladosporioides	
Gloeosporium album	
Penicillium chrysogenum	
P. islandicum	
P. verrucolosum var. *cyclopium*	
Sclerotinia fructicola	
Botrytis cinerea	1–25
Aspergillus niger CBS733.88	2.5
A. versicolor 108959	
B. nivea 163642	
Fusarium solani S200	
P. roqueforti S44	
Absidia sp.	4.0–8.0
Acremomium sclerotigenum	
Alternaria sp.	
Aspergillus flavus (Strains CBS 3005, BB 67, Madagascar, Port Lamy)	
A. niger	
A. versicolor	
Mucor mucedo	
Penicillium digitatum	
P. expansum	
P. notatum 4640	
P. nigricans	
P. viridicatum Westling	
Scopulariopsis asperula	
Aspergillus oryzae	10
Fusarium sp.	
Geotrichum candidum	
Penicillium roqueforti var. *punctatum* 6018	
Rhizopus oryzae 4758	
P. discolor (Strains 547.95, 549.95, 551.95)	>40

[a] Minimum inhibitory concentration (defined as no growth after 5 d at 25°C. Inoculum of ca. 10^4 spores in center of agar plate).

Source: adapted from Klis et al. (1959), de Boer and Stolk-Horsthuis (1977), Delves-Broughton et al. (2005).

of 0.996 inhibited the growth rate of all the molds at 20°C but to different extents. It was least effective against *P. solitum*. Reducing the water activity to 0.99, 0.97, 0.95, and 0.93 again had different effects depending on mold species and type of solute. In some cases, reducing water activity increased antifungal activity but, surprisingly, with either solute water activity in the range of a_w = 0.93–0.97 actually stimulated the growth of *A. varians* and *P. roqueforti*. In a related study, Zalazar et al. (2019) determined the effect of NaCl and xanthan gum on the antifungal effect of natamycin against *Zygosaccharomyces*

TABLE 9.2

Sensitivity of Selected Yeasts to Natamycin

	MIC[a] (μg/ml)
Brettanomyces bruxellensis	1.0–2.5
Candida albicans	
C. krusei H66	
C. pseudotropicalis H3	
C. valida H74	
C. vini	
Debaryomyces hansenii H42	
Dekkera bruxellensis (Strains CBS2796, CBS4459, CBS6055)	
Hanseniasporum uvarum CBS5074	
Hansenula polymorpha	
Pichia membranaefaciens H67	
Rhodotorula mucilaginosa CBS8161	
Saccharomyces bayanus IO18-2007	
S. carlsbergensis CRA6413	
S. cerevisiae (Strains ATCC9763, CRA124, H78, 8021)	
S. cerevisiae var. *ellipsoideus*	
S. exiguus	
S. ludwigii 0339	
Torulopsis candida	
Zygosaccharomyces bailii	
Z. bailii CRA229	
Z. rouxii CBS1640	
Candida guilliermondii	3.0–10.0
C. kefyr H2	
C. paralopsilosis NCYC458	
C. utilis H41	
Kloeckera apiculata	
Kluyveromyces lactis H17	
Rhodotorula gracilis	
Saccharomyces cerevisiae	
S. cerevisiae var. *paradoxus H103*	
S. exiguus Rees CBS1514	
S. florentinus H79	
S. unisporus H104	
S. sake 0305	
Torulopsis lactis-condensi	
Torulaspora rosei	
Zygosaccharomyces rouxii 0562	
Z. barkerii	

[a] Minimum inhibitory concentration (defined as inhibition of growth for 14 d at 25°C. Inoculum level at ca. 10^3 cfu/ml).

Source: adapted from Klis et al. (1959), Delves-Broughton et al. (2005).

bailii growth in a model acid sauce system. Natamycin alone was required at 12 μg/ml to inhibit growth of the yeast. Similarly to Marin et al., NaCl at 3.0 or 6.0% had an antagonistic effect on inhibition by natamycin. Additionally, 0.25 or 0.5% xanthan gum had an antagonistic effect on natamycin activity against *Z. bailii.* The results of these studies indicate the importance of evaluating all of the potential interacting factors when applying natamycin as an antifungal agent in foods. Several researchers have shown natamycin is capable of inhibiting mycotoxin production. Ray and Bullerman (1982) reported that 10 μg/ml natamycin inhibited aflatoxin B_1 production of *Aspergillus flavus* by 62.0% and eliminated ochratoxin production by *A. ochraceus.* The same level of natamycin inhibited penicillic acid production by *Penicillium cyclopium* by 98.8% and eliminated patulin production of *P. patulum.* The inhibitory effect of natamycin was reported to be greater against mycotoxin production than against mycelial growth. Gourama and Bullerman (1988) studied the effect of natamycin on growth and mycotoxin production (penicillic acid) by *Aspergillus ochraceus* OL24 in yeast extract sucrose (YES) medium and olive paste. Natamycin at 20 µg/ml delayed onset of growth of *A. ochraceus* by 2, 4, and 16 days in YES broth compared to the control at 15, 25, and 35°C, respectively. Sporulation of the microorganism was completely inhibited at 15°C by 20 µg/ml natamycin and was delayed 6 days at 25°C and 13 days at 35°C. In terms of mycelial weight increase, 20 µg/ml natamycin was effective in reducing growth by 80–100% at all three temperatures. In addition, penicillic acid production by *A. ochraceus* at all temperatures was nearly totally inhibited by 20 µg/ml. Even at 10 µg/ml natamycin, penicillic acid production was inhibited 20–90%. Significantly these data were in contrast to those found for potassium sorbate which, at sublethal concentrations, caused stimulation of penicillic acid production by *A. ochraceus*. In olive paste, natamycin was added at 85, 175, and 350 µg/g. Growth initiation was delayed 1–4 days by the 3 levels of natamycin at 25°C. At 85, 175, and 350 µg/g in olive paste, penicillic acid production was inhibited 70, 90, and 96%, respectively. The authors concluded that natamycin could provide protection against fungal growth and mycotoxin formation in olives. Similarly Shahani et al. (1977) showed that the mycotoxin produced by the molds *A. flavus*, *A. ochraceus*, *P. cyclopium*, and *P. patulum* was inhibited at levels of natamycin that did not inhibit their growth. At the lowest level of 1.0 µg/ml natamycin the growth of the molds was inhibited by 0.3 to 23% whereas their toxin was inhibited by 25 to 97%. Comparative studies were made to determine effects of commonly used antimycotic agents, sorbate, propionate, and benzoate, and they were found to be far less effective. Rusul and Marth (1988) showed that natamycin at concentrations up to 20 µg/ml in glucose yeast extract medium could delay growth and aflatoxin production by *A. parasiticus* and that effectiveness could be enhanced by combining it with other additives or by using other conditions such as low temperature and pH. Medina et al. (2017) evaluated the influence of natamycin on ochratoxin production by *Aspergillus carbonarius* under different environmental conditions. The researchers used fresh red grape extract medium at 0.98, 0.96, and 0.94 a_w and 15–25°C, and 0 to 20 ng/ml. They found that the growth of all strains at all water activities and temperatures was completely inhibited by 20 ng/ml natamycin. Significant inhibition occurred at 5 and 10 ng/ml at 15°C and 10 ng/ml at 20°C. Total inhibition of OTA production was demonstrated at 5 ng/ml natamycin at 15°C and 0.96 or 0.98 a_w and 10 ng/ml natamycin at 25°C and 0.98 a_w. At 20°C, total inhibition of OTA production was achieved only at 0.94 at 5 ng/ml. The authors suggest that natamycin at 50–100 ng/ml could inhibit growth and OTA production of *A. carbonarius* over a range of environmental conditions and that natamycin might be used to prevent OTA contamination in wine and during the drying and production of vine fruits.

Mann and Beuchat (2008) evaluated natamycin along with potassium sorbate and sodium benzoate for inhibition of the molds *Penicillium brevicompactum*, *Penicillium roqueforti*, *Paecilomyces variotii*, *Aspergillus niger*, and *Cephaloascus fragrans*, all of which are capable of producing 1,3-pentadiene, a volatile compound with an unpleasant hydrocarbon-like odor, in beverages. The compounds were tested in a model agar media containing Parmesan cheese (PRM agar) at pH 5.5, and growth was monitored by increased colony diameter. All the molds grew at 21°C on PRM agar containing up to 3500 µg/ml of potassium sorbate or 3000 µg/ml calcium propionate. At 20 µg/ml natamycin, growth was delayed by approximately 14, 20, and 20 days for *P. brevicompactum*, *P. roqueforti*, and *C. fragrans*, respectively, while *P. variotii* and *A. niger* growth was inhibited but not stopped by the same concentration. Combinations of potassium sorbate (250–1000 µg/ml), calcium propionate (250–1000 µg/ml), and/or natamycin (10–18 µg/ml) greatly inhibited or prevented growth of molds on PRM agar, which, with a pH of 6.2, were well outside the range of pH effectiveness of the organic acids. Natamycin was tested against *Saccharomyces cerevisiae*, *Zygosaccharomyces rouxii*, and *Yarrowia lipolytica* in model systems and on cheese surfaces by Resa et al. (2014). In a model liquid food system (16% whey protein concentrate adjusted to pH 5.5), 50 µg/ml natamycin was required to reduce viable yeast and delay, but not prevent recovery, of *S. cerevisiae* and *Y. lipolytica*. In contrast, *Z. rouxii* was only slightly inhibited by the same concentration of natamycin. Further, the researchers prepared inoculated cheese and coated it with either a tapioca starch-based film containing 9.25 mg natamycin/dm^2 or a 50 µg/ml natamycin spray. Neither the film nor the spray completely prevented growth of any of the three yeasts but the film was as or more effective than the spray.

9.4 Mode of Action

The target of natamycin inhibition in fungi is ergosterol, the primary sterol in fungal cell membranes. Aparacio et al. (2016) stated that the amphiphilic character of natamycin allowed it interact with ergosterol via hydrophobic interactions between the lipophilic regions of the two molecules. Ergosterol has many functions in the fungal cell membrane and has been correlated with endocytosis, exocytosis, vacuole fusion, polarity, and morphogenesis (Aparacio et al., 2016). While some polyene macrolides cause permeabilization of the cell membrane following binding with ergosterol, natamycin does not (Welscher et al., 2008). Instead, Welscher et al. (2012) found that the antifungal activity of natamycin was the result of inhibition of ergosterol-dependent transport of glucose and amino acid transport across the cell membrane.

Since ergosterol is absent from the cell membranes of viruses and bacteria these microorganisms are resistant to natamycin (Hamilton-Miller, 1974). Ziogas et al. (1983) studied 17 natamycin mutants of *A. nidulans* which contained either no ergosterol or reduced levels compared to the wild type. Ergosterol-deficient mutants were most resistant to natamycin (approximately 14–16 µg/ml) compared to the wild type (around 2 µg/ml) but grew at much slower rates. Some wild-type fungi, presumably because the ergosterol content of their membranes is low, have reduced sensitivity to natamycin. Examples are *P. discolour*, *Verticillium cinnabarinum*, and *Botrytis cinerea*.

De Boer and Stolk-Horsthuis (1977) investigated the potential for development of resistance to natamycin among fungi. They found no evidence of resistant fungi in cheese warehouses in which natamycin had been used for various periods of time, up to several years. They also attempted to induce tolerance in 26 strains of fungi by transferring each culture 25–31 times in media containing concentrations of natamycin equal to and higher than the minimum inhibitory concentration (MIC; defined as the lowest concentration which prevented growth after 4 days and 25°C). The MIC, following multiple transfers, increased in only 8 of 26 strains and only by a maximum of 4 µg/ml. They concluded that the lack of resistance was due to the strongly fungicidal activity of the compound, along with its environmental instability. Stark (1999) attributes the lack of reported natamycin resistance to be possibly due to the fact that natamycin occurs as micelles in solution. Thus when a cell comes into contact with the preservative it encounters a high and probably lethal dose of the antimycotic. More recently, Seekstra et al. (2016) subjected 20 species of yeasts and molds representing the genera *Saccharomyces*, *Candida*, *Rhodotorula*, *Trichosporon*, *Geotrichum*, *Fusarium*, *Aspergillus*, *Cladophialophora*, *Neosartorya*, *Penicillium Mucor*, *Verticillium*, *Colletotrichum*, and *Trichoderma* to increasing concentrations of natamycin starting with 10% of the minimum inhibitory concentration (MIC) up to 250% of the MIC over for a prolonged time. Prior exposure caused an increase in the MIC in 13 of the 20 strains although the increases were small, with the exceptions of *Aspergillus ochraceus*, *Fusarium oxysporum*, *Trichosporon asahii*, and *Colletotrichum musae* which had two- to three-fold increases. The strains which demonstrated the greatest MIC increase were then exposed to the polyene antifungal agents amphotericin B and nystatin. Only one, *A. ochraceus*, had increased resistance to natamycin and also had increased resistance to amphotericin B. While there was some increased resistance, the authors concluded that fungi would likely not encounter such multiple exposure conditions in the processing or handling of foods.

9.5 Food Applications

Natamycin is particularly useful for food products produced using bacterial fermentations since bacteria have no ergosterol and are thus immune to inhibition by the compound. Natamycin is generally applied as a surface application, particularly for treatment of the surfaces of hard cheese and dry or ripened sausages such as salami. Because natamycin is crystalline and hydrophobic, it does not readily penetrate but remains at the surface where mold spoilage generally occurs in these products. Other applications such as yoghurt, wine, and fruit juice incorporate the natamycin into the food matrix. Table 9.3 outlines various applications of natamycin, method of application, and recommended dosage. The majority of

TABLE 9.3

Applications of Natamycin, with Suggested Dosage Levels and Methods of Application

Food Application	Suggested Dosage, mg/kg	Method
Hard/semi-hard cheese	1250–2000	Surface treatment by spray or immersion
	500	Direct addition to coating or film
Grated cheese	15–20	Surface treatment by spray or direct addition
Meat products: dry sausage	1250–2000	Surface treatment by spray or immersion
Yoghurt	5–10	Direct addition to yoghurt mix
Bakery products	1250–2000	Surface treatment by spray
Tomato purée/paste	7.5	Direct addition
Olives	7.5	Direct addition
Fruit juice, malt beverage	2.5–10	Direct addition
Wine	30–40	Direct addition to stop fermentation
	3–10	Added prior to bottling to prevent secondary fermentation

Sources: Thomas and Delves-Broughton (2001), Delves-Broughton et al. (2005), Delves-Broughton and Weber (2011).

research on natamycin has focused on application to dairy products or fermented sausages either directly or via films/coatings.

9.5.1 Cheese

The surface growth of mold on cheese can be a major factor limiting the shelf life of cheeses. As well as being unsightly there is a danger of mycotoxin production by the mold. Many cheeses are ripened or matured for several months in ripening rooms at temperatures in the range of 10–12°C, and during this period they are very susceptible to mold colonization and surface growth. There are numerous ways of applying natamycin preparations. As a surface treatment natamycin can be applied by spray, painting, immersion, or by adding to a plastic coating then dipping or painting the cheese. Often many such applications are given over a period of time to ensure complete protection. When applied in a polyvinyl acetate (PVA) coating for ripening the coating can be easily removed prior to retail sale so the consumer receives a clean yet mature cheese. When natamycin is added by immersion it is recommended that, to prevent bacterial growth, 10% salt is added to the solution. Natamycin when applied as dipping solutions or sprays will tend to settle out unless stirred or agitated. Modified formulations are available that contain food-grade thickening agents that both prevent settling of the natamycin and improve adhesion to the surface of the product (Stark, 1999). Similarly blends are available that are combined with anticaking agents such as powdered cellulose that can be used in the production of shredded cheese. A method of applying natamycin to the surface of cheese by electrostatic coating was proposed by Elayedath and Barringer (2001).

Natamycin is approved by the U.S. Food and Drug Administration (FDA) for use on any standardized cheese for which antimycotics are allowed. In the European Union (EU) the maximum level of natamycin permitted on the surface of hard, semi-hard, and semi-soft cheese is 1 mg/dm^2 surface, and natamycin should not be present at a depth greater than 5 mm. Numerous studies have confirmed the efficacy of natamycin as a surface treatment of cheeses, often comparing it to sorbate. Pugazhenti et al. (1999) compared natamycin and potassium sorbate control of *P. citrinum*, a mold which produces the mycotoxin citrinin. Isolates from Swiss cheese samples were inoculated onto the cheese surface. Cheese samples were either left untreated or surface treated with 5 and 10 µg/g natamycin, or 500 and 1000 µg/g potassium sorbate. The cheeses were stored at 25°C for 21 days and then analyzed for counts of the fungus. Counts were highest in the control cheeses and lowest in the 10 µg/g natamycin-treated cheese. The lower level of natamycin tested (5 µg/g) was more effective than the highest level of potassium sorbate. A study by de Ruig and van den Berg (1985) demonstrated further advantages of natamycin over sorbate for the control of mold growth on cheese surfaces. They reported that sorbate, unlike natamycin, caused adverse effects on both the interior and surface of the cheese. Sensory evaluation found that sorbate lowered the quality of the cheese, particularly under the rind, and this was most obvious for young cheese. Inner portions of the cheese were also adversely affected by the migration of the sorbate. Natamycin had no adverse effect on the rind or the flavor of the cheese. Engel et al. (1983), using HPLC, showed that natamycin does not penetrate far into the cheese and therefore has less effect on the taste of the cheese and remains predominantly on the surface where mold contamination occurs and needs to be prevented. The depth of penetration of natamycin into the cheese was found to depend on the initial natamycin concentration, storage time, and cheese type. Experiments were carried out at 15°C.

Nilson et al. (1975) determined the effect of natamycin in comparison with mycostatin, on the shelf life of cottage cheese stored at 4.4, 10.0, or 15.6°C. The compounds were added through curd wash water or in the cheese dressing. The cottage cheese was inoculated with *A. niger* or *Saccharomyces cerevisiae* or was uninoculated. Natamycin (100 μg/ml) added in the wash water was effective in increasing the days to spoilage of uninoculated cottage cheese by 13.6, 7.7, and 6.3 days over the control when stored at 4.4, 10.0, and 15.6°C, respectively. Cottage cheese inoculated with *A. niger* stored at the same temperatures had increased days to spoilage of 12.7, 6.0, and 4.3 days, and samples inoculated with *S. cerevisiae* had increased shelf life of 10.3, 6.3, and 3.7 days, respectively. Adding natamycin to the cottage cheese dressing was even more effective in extending shelf life. At 4.4, 10.0, and 15.6°C, the inoculated and uninoculated cottage cheese had increased days to spoilage ranges of 20.4–26.7, 9.7–12.3, and 2.6–5.0, respectively. Natamycin was found to be slightly more effective than mycostatin. *A. niger* was found to be the most sensitive of the two microorganisms used in the inoculation studies. Lück and Cheeseman (1978) found that 500 or 1000 μg/ml concentration of natamycin delayed mold growth on cheese for up to 6 months but did not prevent it completely. Neviani et al. (1981) treated Italian Caciotta cheese by immersion in 2000 μg/ml natamycin and found it effective in controlling mold growth, with no effect on the population of desired ripening surface bacteria that produce the red color of Caciotta cheese. Verma et al. (1988) evaluated the effectiveness of natamycin against sorbic acid, benzoic acid, and nystatin in inoculated and uninoculated butter and cheese. All four preservatives were effective in reducing fungal growth on uninoculated butter and cheese samples stored 30 days at 7°C. Maximum effect was shown by the antifungal agent nystatin, followed by natamycin. Lodi et al. (1989) found that natamycin was effective in preserving seven types of Italian cheeses with no detrimental effect on ripening. *Aspergillus niger* spores were inhibited for 30 days on semi-hard cheese in a methyl cellulose and wheat gluten film at 5 mg natamycin per 10 g of film (Ture, et al. 2011). However, *Penicillium roqueforti* was not controlled at 20 mg natamycin per 10 g of film in the same study. Natamycin has been shown to be effective against "thread" mold *Phoma glomerata*, which causes infections of hard cheeses (Fente-Sampayo et al. 1995; Basilico et al. 2001). Morris and Castberg (1980) used a 1000 μg/ml natamycin dip in a 10–15% brine to control surface mold of blue cheese without affecting the desired internal mold growth. Kallinteri et al. (2013) evaluated the use of nisin, natamycin, and/or their combination as antimicrobial treatments to improve the shelf life of Galotyri cheese (a curd cheese made from milk and yogurt). Samples were treated with nisin 100–200 IU/g and/or natamycin 0.01–0.02% w/w and their combinations. Yeast populations in cheese samples with natamycin alone or natamycin plus nisin were suppressed throughout storage. The observed shelf life of Galotyri, based on overall acceptability data, was longest with natamycin containing cheese samples at >28 days compared to the control of 14–15 days.

Natamycin is used to treat shredded cheese particularly in the USA (Berry, 1999). A natamycin suspension of 1500 μg/ ml, finely sprayed onto 1 ton of shredded cheese using a tumbler and a commercial spray system, will provide protection against yeast and mold spoilage in modified atmosphere or vacuum-packed packs of shredded cheese that lose their modified atmosphere or vacuum due to pack damage.

9.5.2 Meat

Natamycin is often used in Europe to treat the surface of dry fermented sausages, and has also been used to treat the surface of cooked hams. EU regulations for meat are similar to those for cheese. The maximum level of natamycin allowed on the surface of dry cured sausages is 1 mg/square decimeter surface, with none present at a depth greater than 5 mm.

Holley (1986) compared the efficacy of sorbate and natamycin on the surface mold and ripening of Italian dry salami. Although complete control was not achieved, a dip of 2000 µg/ml immediately after stuffing the casings helped minimize the development of surface mold. Natamycin spray (2 × 1000 µg/ml) was as good or better than 2500 µg/ml sorbate. However, the sorbate treatments proved unsatisfactory because of the antibacterial activity of sorbate which inhibited the bacteria responsible for the salami-ripening process. Baldini et al. (1979) sprayed Mortadella sausages, salami, and raw hams with different aqueous solutions of natamycin (1000–5000 µg/ml) during maturation of the meats at the beginning of storage. For all the products, this treatment effectively prevented surface mold growth when examined in normal conditions of preparation and storage. The preservative did not cause any change to the typical microflora of the salami (the desired growth that contributes to the flavor), nor did it generate off-flavors in the ham. For most samples natamycin was undetectable at the end of the ageing and storage period.

Cattaneo et al. (1978) also investigated the protective effects of natamycin spraying and immersion of similar meat products (salami, cured beef) and found the preservative gave good protection against mold growth. Protection was achieved by immersing the sausage casings in the preservative. Natamycin content was higher in natural compared to synthetic casings. Moerman (1972) carried out similar studies with sausages (Snijworst), applying the natamycin either during brining, by dipping or spraying with 50–1000 µg/g before or after smoking or preliminary soaking of the casings (natural, protein, or fiber). Sausages were stored for 5 weeks, and inspected for mold formation weekly. With the natural casings, little or no mold growth was observed after 4 weeks, regardless of the method of application. With the protein-based casings, similar results were obtained except with the presoaked casings, where considerable mold growth occurred. With the cellulose-based casings, results were only satisfactory on spraying with 1000 µg/ml natamycin. For practical purposes, spraying with 1000–2000 µg/ml was recommended. De Boer et al. (1979) showed that molds and yeasts had comparable natamycin sensitivities irrespective of the natamycin content of the sausages from which they were isolated. This further confirms that development of resistance to natamycin is not a problem.

9.5.3 Fruits and Fruit Juices

Natamycin has been shown to be an effective preservative in both unpasteurized and pasteurized fruit juices preventing the growth of yeasts and molds. Thomas and Delves-Broughton (2001) investigated the efficacy of natamycin in preventing visible spoilage (surface growth or gas production) of a wide range of yeasts and molds in apple, orange, and pineapple juices. Incubation was at 25°C for the organisms except for the *Byssochlamys* spp.; these were incubated at their optimum temperature of 37°C. *Byssochlamys* spp. are often associated with heat-processed foods and beverages as they produce spores that have high heat resistance that can survive the pasteurization processes typically used in fruit juice processing (Pitt and Hocking, 1999). Results for pineapple juice are shown in Table 9.4.

More or less similar results were achieved in apple and orange juice. Unpublished data indicate that pasteurization of orange fruit juice at 80°C for 10 minutes results in 70% retention of natamycin (Delves-Broughton et al., 2005).

Shirk and Clark (1963) found that a natamycin addition of 20 µg/g had an immediate reduction effect on yeast and mold counts in orange juice inoculated with *S. cerevisiae* as well as samples which were not inoculated. After 1 week of storage at 2.5–4°C, yeast and mold counts were undetectable and the sample remained unspoiled for the 8-week duration of the test. Control samples not containing natamycin were spoiled within 1 week. A second study then compared the effectiveness of natamycin to sorbic acid. In this study juice inoculated with natural contaminants spoiled after 1 week of storage, whereas levels as low as 1.25 µg/g of natamycin prevented spoilage during the 8 weeks. In comparison, levels as high as 1000 µg/g of sorbic acid were needed to retard yeast growth. The sorbic acid imparted an unpalatable flavor to the fruit juice; natamycin did not.

He et al. (2019) determined the effect of natamycin on the postharvest disease-causing molds *Botrytis cinerea*, which causes gray mold disease, and *Penicillium expansum*, which causes blue mold disease. Natamycin completely inhibited growth of *B. cinerea* at 2 µg/ml in fungal media and *P. expansum* at 3 µg/ml. On grapes inoculated with *B. cinerea*, applied at 100 µg/ml, natamycin completely inhibited

TABLE 9.4

Natamycin Control of Yeast and Mold Growth in Pineapple Juice

	Days until Visible Spoilage at Each Natamycin Concentration (ppm)						
Organism	**0**	**1.25**	**2.5**	**5**	**10**	**20**	**40**
Debaryomyces hansenii H42	3	72	>72	>72	>72	>72	>72
Pichia membranaefaciens H67	3	11	>72	>72	>72	>72	>72
Saccharomyces carlsbergensis ATCC6413	2	8	8	8	>63	>63	>63
Saccharomyces cerevisiae ATCC9763	2	9	>63	>63	>63	>63	>63
S. cerevisiae H78	2	16	16	35	35	>72	>72
S. cerevisiae CRA124	2	14	29	35	35	>72	>72
S. cerevisiae var. *paradoxus* H103	2	7	7	7	>72	>72	>72
Zygosaccharomyces bailii CRA229	3	8	18	66	66	>72	>72
Penicillium roqueforti S44	2	3	3	6	61	>61	>61
Byssochlamys fulva 040021	6	6	8	>63	>63	>63	>63
B. nivea 163642	6	6	6	13	13	17	23
B. fulva TSD 0.0796	6	6	6	>63	>63	>63	>63

Source: Delves-Broughton et al. (2005).

gray mold disease for 4 days at 25°C. On jujube fruit inoculated with *P. expansum*, 200 μg/ml natamycin prevented blue mold disease.

9.5.4 Miscellaneous Applications

Aside from the above applications, natamycin has potential applications in other food products; however these have much more limited regulatory approval. Promising applications of natamycin include yogurt and bakery products. Thomas and Delves-Broughton (2001) indicate that a level of 5–10 μg/g natamycin is effective in preventing the growth of yeasts in yogurt. Such use of natamycin up to 10 mg/kg is approved in South Africa (Anonymous, 2017). There is no detrimental effect on the yogurt starter cultures, and the natamycin can be added prior to the addition of the starter cultures, or with the fruit preparation after the fermentation in the manufacture of fruit-flavored yogurts. Snyder et al. (2016) studied the antifungal effect of natamycin on the spoilage mold *Mucor circinelloides* in yogurt. Natamycin was added to yogurt at 0, 5, 10, 15, and 20 μg/ml to determine its ability to inhibit *M. circinelloides* growth and carbon dioxide production over 30 days at 4°C, 15°C, and 25°C. The addition of at least 15 μg/ml natamycin inhibited the hyphal phase growth and carbon dioxide production by the yeast-like phase of *M. circinelloides* at all temperatures compared to the untreated control.

Natamycin has also been found to have potential as a spray on bakery products. The spraying of tortillas with natamycin to achieve surface concentrations of 1 and 5 μg/cm² resulted in increased protection against mold growth. Ticha (1975) found that a level of 100 μg/g natamycin effectively controlled the growth of five molds including *A. flavus* isolated from bakery products. Tsiraki et al. (2018) determined the antifungal activity of natamycin with chitosan on vacuum-packaged phyllo pastry. Chitosan (1.5%, w/v) and natamycin (10 μg/ml, w/v) were tested alone and in combination on uninoculated phyllo pastry. The combination of chitosan and natamycin caused 1 to 3 log CFU/g reductions in total viable counts, yeasts and molds, psychrotrophic bacteria, lactic acid bacteria, Enterobacteriaceae, and enterococci after 18 days of storage. The authors concluded that a combination of chitosan and natamycin was effective in delaying spoilage of phyllo pastry while maintaining acceptable sensory characteristics.

Another potential application is olives. Contamination of these fermented products with toxigenic molds presents a potential health risk as well as an economic problem, especially in Mediterranean countries where olives are popular. In a study investigating an implicated mold, *A. ochraceus*, Gourama and Bullerman (1988) found that natamycin was effective at controlling fungal growth and mycotoxin production on olives, and recommended this as a practical solution to the problem. In a more recent study, natamycin was able to inhibit the growth of yeast in black Conservolea olive fermentation and, at the same time, improve the natural fermentation by fermentative bacteria (Hondrodimou et al. 2011). Additionally, natamycin inhibited surface mold on the fermentation brine resulting in less spoilage compared to the natural process.

Natamycin is used in the wine industry in South Africa for the production of sweet wines. Natamycin is added before the fermentation is complete in order to arrest the fermentation of the yeast; natamycin is also used as a preservative to prevent yeast or mold spoilage. Gerbaux et al. (2000) showed that addition of 40 μg/ml was effective in controlling *Brettanomyces* yeast infections of the French Pinot noir red wine.

Natamycin was tested with an oxygen treatment on the microbial and enzymatic qualities of button mushrooms (*Agaricus bisporus*) (Jiang, 2012). Mushrooms were dipped in a 333 μg/ml solution of natamycin for 10 min at 20°C with or without pure oxygen and stored for 16 days at 4°C. A combination of natamycin and oxygen reduced counts of yeasts and molds as well as improving the physical characteristics of the mushrooms compared to either natamycin or oxygen alone.

9.5.5 Incorporation into Packaging Films and Coatings for Foods

In addition to direct application as an ingredient in food products, natamycin has been extensively studied for its use in packaging films and in coatings for food products. The purpose of these applications is generally to extend the shelf life of foods by gradual release over time. One such application is the use of chitosan, an antifungal agent on its own, as the basis of the film with natamycin incorporated into the film. Da Silva et al. (2013) developed alginate and alginate/chitosan films with natamycin incorporated and tested them against *Debaromyces hansenii*, *Penicillium commune*, and *Penicillium roqueforti* in an agar diffusion assay. Natamycin was inhibitory at as low as 0.005 g/g of alginate films and 0.01 g/g of alginate/chitosan films against all fungi tested. Sodium caseinate and low-methoxyl pectin were utilized to prepare films containing 45 μg/ml natamycin that were tested against *Penicillium chrysogenum* and *Aspergillus brasiliensis* in an agar diffusion assay (Eghbal et al., 2019). Films containing natamycin were more effective in inhibiting the growth of *P. chrysogenum* than *A. brasiliensis* in diffusion tests and reduced the growth of *P. chrysogenum* inoculated on a cured meat model.

de Oliveira et al. (2007) were some of the first researchers to evaluate natamycin films to inhibit fungi on cheese. They prepared cellulosic films containing up to 4% natamycin and used them to cover Gorgonzola cheese during ripening. Cheeses wrapped in natamycin-containing film demonstrated significant reduction in surface mold compared to controls. Resa et al. (2016) used a tapioca starch-based film with natamycin and nisin incorporated against mixed cultures of *Saccharomyces cerevisiae* and *Listeria innocua* on the surface of Port Salut cheese. The films contained 9.25 mg natamycin/dm^2 of film and were shown to prevent the growth of natural contaminating yeasts and molds on the cheese stored at 7°C for over 20 days. Additionally, no growth of *S. cerevisiae* was detected in cheese inoculated with a mixed culture of the two test microorganisms. Mehyar et al. (2018) prepared coatings containing chitosan and natamycin and used them on brined Halloumi cheese in 5%, 10%, or 15% (w/v) NaCl. The coatings increased shelf life by ≥5 days than the control in 5% and 10% NaCl and reduced psychrotrophs, anaerobes, lactic acid bacteria, and yeasts and molds. Coatings did not adversely affect the sensory properties of cheese. Several other studies have reported on the efficacy of natamycin-containing film coatings on cheeses (Moatsou et al., 2015; Fajardo et al., 2010; Pintado et al., 2010; Hanusová et al., 2010).

9.6 Regulatory Status

In the United States, the Code of Federal Regulations (2019) 21.172.155 states that "Natamycin (CAS Reg. No. 7681-93-8) … may be applied on cheese, as an antimycotic, in amounts not to exceed 20 milligrams per kilogram (20 parts per million)." In the European Union, natamycin (E235) is approved for external treatment of uncut hard, semi-hard, and semi-soft ripened cheeses at a maximum limit of 1 mg/kg on surface of dried cured sausage (not present at a depth of 5 mm) and heat-treated and non-heated treated meats at the limit of 1 mg/dm^2 on the surface (not present at a depth of 5 mm) (European Commission,

2019). Internationally, natamycin is approved for use on cheeses in most countries worldwide, in some countries for specific types of meats, and in selected countries for use on baked products, fruit juices, fish products, yogurt, and wines (Anonymous, 2019). The acceptable daily intake (ADI) allowed for natamycin specified by the FAO/WHO Expert Committee on Food Additives (JECFA, 2002) is 0.3 mg/kg body weight/day. As natamycin is mainly used for surface treatments the ADI is not reached, even by consuming large quantities of treated products. The current ADI was reviewed again in 2009 (EFSA, 2009) and confirmed to be safe based upon use and current knowledge.

9.7 Toxicology

A review of the toxicological data on natamycin was published by EFSA (2009). The intravenous route is the path by which polyene macrolide antimicrobials are most toxic, and oral administration is least toxic (Table 9.5). There is apparently no absorption of up to 500 mg/day natamycin from the human intestinal tract after 7 days' administration (Brik, 1981).

Levinskas et al. (1966) carried out a study to determine the acute and chronic toxicity of natamycin. The single oral dose LD_{50} for natamycin in the male rat was found to be 2.73 g/kg (1.99–3.73 g/kg) and 4.67 g/kg (3.0–7.23 g/kg) for female rats. The oral LD_{50} for fasted male albino rabbits was 1.42 g/kg (0.46–4.39 g/kg). The single dermal dose LD_{50} was estimated at >1.25 g/kg. For rats, no signs of toxicity occurred after large single doses, and no gross lesions could be detected related to natamycin ingestion. Rabbits which died had congested and hemorrhagic gastric mucosa. In a 3-month feeding study with rats, Levinskas et al. (1966) found that animals fed 8000-ppm natamycin had body weights which averaged 54–67% of the control group (no natamycin). At 2000 ppm, animal weight averaged 85% of the control group. Natamycin had no apparent effect on body organs nor did it produce any lesions. Oral administration of natamycin to rats for 2 years at up to 1000 ppm did not have an effect on survival of the animals (Levinskas et al., 1966). The diet containing natamycin was judged to have had no adverse effects on food utilization, reproductive performance, neoplasms, or other lesions above that encountered in controls. Dogs were fed diets containing 125, 250, or 500 ppm natamycin for 2 years (Levinskas et al., 1966). Oral administration of 500 ppm resulted in a slight decrease in body weight. No significant hematologic abnormalities nor significant lesions could be attributed to the consumption of natamycin (Levinskas et al., 1966). Nausea, vomiting, and diarrhea in humans given natamycin orally in doses exceeding 1000 mg/day (Brik, 1981) and vomiting and diarrhea in dogs fed 5000 ppm natamycin (Levinskas et al., 1966) have been reported.

9.8 Assay

Shirk et al. (1962) first developed a bioassay for natamycin using *S. cerevisiae* as an indicator organism. The standard curve was linear between 0.5 and 15 µg/ml. Natamycin has a strong UV-visible light absorption spectrum with a characteristic multipeak pattern (Aparicio et al., 2016). Ultraviolet spectra

TABLE 9.5

Acute Toxicity of Various Polyene Antimicrobials by Different Routes in Mice or Rats

	LD_{50} (mg/kg)			
	Intravenous	**Subcutaneous**	**Intraperitoneal**	**Oral**
Amphotericin B	4–6.6	-	28.0	>8000
Candicidin	-	160–280	2.1–7	90–400
Natamycin	5–10	5000[a]	250[a]	1500 [a]
Nystatin	3	24	8–14	>3500

[a] Rats.

Source: adapted from Hamilton-Miller (1973).

maxima for natamycin are 281, 291, 304, and 319 nm. Using this property, De Ruig et al. (1987) developed a method for the detection of natamycin in cheese and cheese rind by extraction with methanol and ultraviolet spectrometric detection or liquid chromatographic separation followed by UV detection. In a subsequent collaborative study, De Ruig (1987) reported that both methods gave reasonable to good results for levels of natamycin down to 15 mg/kg (0.9 mg/dm^2 surface area). Riedl et al. (1984) also reported using methanol extraction of natamycin on cheese and ultraviolet spectrometry to detect concentrations of 2.5 mg/kg natamycin in cheese. Fletouris et al. (1995) developed a rapid spectrophotometric method for assay of nisin in cheese and cheese rind. Quantitation and detection limits were estimated at 0.5 and 0.25 µg/g respectively.

High-performance liquid chromatographic (HPLC) assay became the preferred method of assay, and methods for natamycin were developed by a number of researchers (Engel et al., 1983; Tuinstra and Traag, 1982; Guldborg, 1984). Luf and Brandl (1986) extracted cheese samples with methanol and subjected extracts to gradient HPLC with ultraviolet detection in a mixture of acetonitrile/acetate buffer (pH 4.4). Using this method, they were able to detect 0.1 µg/ml natamycin which corresponded to 0.5 ppm natamycin at the cheese surface. Maruyama et al. (1988) extracted surface and inner area samples of cheese with acetonitrile/phosphate buffer and, using reversed-phase HPLC, the limit of detection for natamycin was 0.2 µg/g cheese. An enzyme immunoassay for the detection of natamycin in cheese was described by Maertlbauer et al. (1990). The assay involved a methanol extraction of the cheese followed by detection of natamycin using an anti-natamycin antibody from rabbits. The detection limit was 0.005 mg/dm^2. Capitán-Vallvey et al. (2000) describe rapid ultraviolet spectrophotometric and liquid chromatographic methods for the detection of natamycin in lactoserum matrix. The methods involve protein precipitation followed by centrifugation and no clean-up stage is necessary. Detection and quantitation limits were 0.07 and 0.23 µg/g respectively for the UV spectrophotometric method, and 0.1 and 0.32 µg/g for the liquid chromatographic method. The current standardized method required by many regulatory agencies for assay of natamycin in cheese is ISO 9233 (International Dairy Federation Standard 140) (ISO, 2018). It is in two parts, one using a molecular absorption spectrometric method for natamycin in cheese rind and the second an HPLC method.

REFERENCES

Anonymous. 2017. Regulation additives: Regulations—Preservatives and antioxidants (R965/1977). Department of Health, Republic S. Africa. http://www.health.gov.za/index.php/shortcodes/2015-03-29-10-42-47/2015-04-30-09-10-23/2015-04-30-09-11-35/category/182-regulations-addictives

Anonymous. 2019. Use of natamycin by country. https://www.natamycinvgp.com/use-of-natamycin-by-countries

Aparicio, J.F., Colina, A., Ceballos, E. and Martin, J.F. 1999. The biosynthetic gene cluster for the 26-membered ring polyene macrolide pimaricin. *J. Biol. Chem.* 274:10133.

Aparicio, J.F., Caffrey, P., Gil, J.A. and Zotchev, S.B. 2003. Polyene antibiotic biosynthesis gene clusters. *Appl. Microbiol. Biotechnol.* 61:179–88.

Aparicio, J.F., Barreales, E.G., Payero, T.D., Vicente, C.M., de Pedro, A. and Santos-Aberturas, J. 2016. Biotechnological production and application of the antibiotic pimaricin: Biosynthesis and its regulation. *Appl. Microbiol. Biotechnol.* 100:61–78.

Baldini, P., Palmia, F., Racznski, R.G. and Campanini, M. 1979. Use of pimaricin in preventing development of molds on Italian meat products. *Industrie Conserve* 54:305.

Basílico, J.C., de Basílico, M.Z., Chiericatti, C. and Vindero, C.G. 2001. Characterization and control of thread mould in cheese. *Lett. Appl. Microb.* 32:419.

Berry, D. 1999. Natamycin for shredded cheese. *Dairy Foods* 100:45.

Brik, H. 1981. Natamycin. p. 513. In *Analytical Profiles of Drug Substances*, (ed.) Flory, K. Academic Press, New York.

Burns, J. and Holtman, D.F. 1959. Tennecetin: A new fungal antibiotic. *Antibiot, Chemother.* 9:398.

Capitán-Vallvey, L.F., Checa-Moreno, R. and Navas, N. 2000. Rapid ultraviolet spectrophotometric and liquid chromatographic methods for the determination of natamycin in lactoserum matrix. *J. Assoc. Off. Anal. Chem.* 83:802.

Cattaneo, P., D'Aubert, S. and Righetti, A. 1978. Attività antifungina della pimaracina in salumi crudi stagionati. *Industrie Alimentari* 17:658.

Code of Federal Regulations. 2019. *Title 21: Food and Drugs, Part 172—Food Additives Permitted For Direct Addition to Food for Human Consumption*, Subpart B—Food Preservatives. §172.155 Natamycin (pimaricin)

Da Silva, M.A., Iamanaka, B.T., Taniwaki, M.H. and Kieckbusch, T.G. 2013. Evaluation of the antimicrobial potential of alginate and alginate/chitosan films containing potassium sorbate and natamycin. *Packaging Technol. Sci.* 26:479–492.

Davis, N.D. and Diener, U.L. 1987. Mycotoxins. p. 517. In *Food and Beverage Mycology*, (ed.) Beuchat, L.R. Van Nostrand Reinhold, New York.

De Boer, E. and Stolk-Horsthuis, M. 1977. Sensitivity to natamycin (pimaricin) of fungi isolated in cheese warehouses. *J. Food Prot.* 40:533.

De Boer, E., Labots, H., Stolk-Horsthuis, M. and Visser, J.N. 1979. Sensitivity to natamycin of fungi in factories producing dry sausage. *Fleishwirtsh.* 59:1868.

Delves-Broughton, J., Thomas, L.V., Doan, C.H. and Davidson, P.M. 2005. Natamycin. pp. 275–88. In *Antimicrobials in Food*, 3rd ed., (eds) Davidson, P.M., Sofos, J.N. and Branen, A.L.. CRC Press, Boca Raton, FL.

Delves-Broughton, J. and Weber, G. 2011. Nisin, natamycin and other commercial fermentates used in food biopreservation. pp. 63–100. In *Protective Cultures, Antimicrobial Metabolites and Bacteriophages for Food and Beverage Biopreservation*, (ed.) Lacroix, C. Woodhead Publ, Cambridge.

De Oliveira T.M., Soares, N.F., Pereira, R.M. and Fraga, K.F. 2007. Development and evaluation of antimicrobial natamycin-incorporated film in gorgonzola cheese conservation. *Packag. Technol. Sci.* 20:47–153.

De Ruig, W.G. and Van den Berg, G. 1985. Influence of the fungicides sorbate and natamycin in cheese coatings on the quality of the cheese. *Neth. Milk. Dairy* 39:165.

De Ruig, W.G. 1987. Determination of natamycin in cheese and cheese rind: Interlaboratory collaborative study. *J. Assoc. Off. Anal. Chem.* 70:949.

De Ruig, W.G., van Oostrom, J.J. and Leenheer, K. 1987. Spectrometric and liquid chromatographic determination of natamycin in cheese and cheese rind. *J. Assoc. Off. Anal. Chem.* 70:944.

Eghbal, N., Dumas, E., Yarmand, M.S., Mousavi, M.E., Oulahal, N. and Gharsallaoui, A. 2019. Antimicrobial films based on pectin and sodium caseinate for the release of antifungal natamycin. *J. Food Proc. Preserv.* 43:e13953.

Elayedath, S. and Barringer, S.A. 2001. Electrostatic coating of shredded cheese with anticaking and antimycotic agents. Abstract from: Institute of Food Technologists Annual Conference 2001.

Engel, G., Hertel, K. and Teuber, M. 1983. Detection and degradation of natamycin (pimaricin) on the surface of cheese. *Milchwissenschaft* 38:145.

European Commission. 2019. Database of food additives—Annex II of Regulation (EC) No 1333/2008. https://webgate.ec.europa.eu/foods_system/main/index.cfm?event=substance.view&identifier=73

European Food Safety Authority (EFSA). 2009. Scientific opinion on the use of natamycin (E 235) as a food additive. *EFSA J.* 7:1412.

Fajardo, P., Martins, J.T., Fucinos, C., Pastrana, L., Teixeira, J.A. and Vicente, A.A. 2010. Evaluation of a chitosan-based edible film as carrier of natamycin to improve the storability of Saloio cheese. *J. Food Eng.* 101:349–356.

Farid, M.A., El-Enhasy, H., El-Diwany, A.E and El Sayed, E.-S.A. 2000. Optimization of the cultivation medium for natamycin production by *Streptomyces natalensis*. *J. Basic Microbiol.* 40:157.

Fente-Sampayo, C.A., Vazquez-Belda, B., Franco-Abuin, C., Quinto-Fernandez, E., Rodrriquez-Otero, J.L. and Cepeda-Saez, A. 1995. Distribution of fungal genera in cheese and dairies. Sensitivity of potassium sorbate and natamycin. *Arch. F.Lebensmittelhyg.* 46:49.

Fletouris, D.J., Botosglou, N.A. and Mantis, A.J. 1995. Rapid spectrophotometric method for analysing natamycin in cheese and cheese rind. *J. Assoc. Off. Anal. Chem.* 78:1024.

Gerbaux, V., Jeudy, S., and Monamy, C. 2000. Etude des phenols volatils dans les vins de Pinot noir en Bourgogne. *Bull. de L'O.I.V.* 73:835.

Gourama, H. and Bullerman, L.B. 1988. Effects of potassium sorbate and natamycin on growth and penicillic acid production by *Aspergillus ochraceus*. *J. Food Prot.* 51:139.

Guldborg, M. 1984. Pimaricin on cheese for the Danish home market. Statens Levensmiddelinstitut, No. 90, Søberg, Denmark. Dairy Sci. Abst. 46:90.

Hamilton-Miller, J.M.T. 1973. Chemistry and biology of the polyene macrolide antibiotics. *Bact. Rev.* 37:166.

Hamilton-Miller, J.M.T. 1974. Fungal sterols and the mode of action of the polyene antibiotics. *Adv. Appl. Microbiol.* 17:109.

Hanusová, K., Stastna, M., Votavova, L., Klaudisova, K., Dobias, J., Voldrich, M. and Marek, M. 2010. Polymer films releasing nisin and/or natamycin from polyvinyldichloride lacquer coating: Nisin and natamycin migration, efficiency in cheese packaging. *J. Food Eng.* 99:491–496.

He, C, Zhang, Z., Li, B., Xu, Y and Tian, S. 2019. Effect of natamycin on *Botrytis cinerea* and *Penicillium expansum*—Postharvest pathogens of grape berries and jujube fruit. *Postharvest Biol. Technol.* 151:134–141.

Holley, R.A. 1986. Effect of sorbate and pimaricin on surface mould and ripening of Italian dry salami. *Lebens.-Wiss, u.- Technol.* 19:59–65.

Hondrodimou, Y., Kourkoutas, Y. and Panagou, E.Z. 2011. Efficacy of natamycin to control fungal growth in natural black olive fermentation. *Food Microbiol.* 28:621–627.

International Standards Organization (ISO). 2018. Methods 9233–1 (https://www.iso.org/standard/74628.html) and 9233-2 https://www.iso.org/standard/74629.html).

Joint FAO/WHO Expert Committee on Food Additives. 2002. Preservative: Natamycin. pp. 49–78. In *Safety Evaluation of Certain Food Additives and Contaminants.* WHO Food Additives Series 48. International Programme on Chemical Safety, World Health Organization.

Jiang, T. 2012. Effect of natamycin in combination with pure oxygen treatment on postharvest quality and selected enzyme activities of button mushrooms (*Agaricus bisporus*). *J. Ag. Food Chem.* 60:2562–2568.

Kallinteri, L.D., Kostoula, O.K. and Savvaidis, I.N. 2013. Efficacy of nisin and/or natamycin to improve the shelf-life of Galotyri cheese. *Food Microbiol.* 36:176–181.

Klis, J.B., Witter, L.D. and Ordal, Z.J. 1959. The effect of several antifungal antibiotics on the growth of common food spoilage fungi. *Food Technol.* 13:124.

Levinskas, G.J., Ribelin, W.E. and Shaffer, C.B. 1966. Acute and chronic toxicity of pimaricin. Toxicol. Appl. Pharmacol. 8:97.

Lodi, R. Todesco, R. and Bozzetti, V. 1989. New applications of natamycin with different types of Italian cheese. *Microbiol., Aliments, Nutr.* 7:81.

Lück, H. and Cheeseman, C.E. 1978. Mould growth on cheese as influenced by pimaricin or sorbate treatments. *S. Afr. J. Dairy Technol.* 10:143.

Luf, W. and Brandl, E. 1986. Simultaneous determination of sorbic acid and natamycin in and on cheese surfaces by means of high performance liquid chromatography (HPLC). *Meijeritieteellinen Aikakauskirja* 44:56. [*Dairy Sci. Abst.* 1987. 49:460.]

Maertlbauer, E., Ali, H. Dietrich, R. and Terplan, G. 1990. Enzyme immunoassay for the detection of natamycin in cheese rind. *Arch. Lebensmittelhyg.* 41:112. [*Chem. Abst.* 114:553.]

Mann, D.A. and Beuchat, L.R. 2008. Combinations of antimycotics to inhibit the growth of molds capable of producing 1,3-pentadiene. *Food Microbiol.* 25:144–153.

Marin, P., Ginés, C., Kochaki, P. and Jurado, M. 2017. Effects of water activity on the performance of potassium sorbate and natamycin as preservatives against cheese spoilage moulds. *Irish J. Ag. Food Res.* 56:85–92.

Maruyama, T., Watanabe, Y., Tobano, H., Matsuura, N. and Asai, Y. 1988. An HPLC method for determination of natamycin in natural cheese. *J. Food Hyg. Soc. Japan* 29:413.

Medina, Á., Jiménez, M., Mateo, R. and Magan, N. 2017. Efficacy of natamycin for control of growth and ochratoxin A production by *Aspergillus carbonarius* strains under different environmental conditions. *J. Appl. Microbiol.* 103:2234–2239.

Mehyar, G.F., Al Nabulsi, A.A., Saleh, M., Olaimat, A.N. and Holley, R.A. 2018. Effects of chitosan coating containing lysozyme or natamycin on shelf-life, microbial quality, and sensory properties of Halloumi cheese brined in normal and reduced salt solutions. *J. Food Proc. Preserv.* 42:e13324.

Moatsou, G., Moschopoulou, E., Beka, A. and Tsermoula, P. 2015. Effect of natamycin-containing coating on the evolution of biochemical and microbiological parameters during the ripening and storage of ovine hard-Gruyère-type cheese. *Intl. Dairy J.* 50:1–8.

Moerman, P.C. 1972. Mould control in meat products using pimaricin. *Voedingsmiddelentechnologie* 3:261–264.

Morris, H.A. and Castberg, H.B. 1980. Control of surface growth of blue cheese using pimaricin. *Cultured Products J.* May:21.

Neviani, E., Emaldi, G.C. and Carini, S. 1981. L'impiegio di pimaracina come antifungino sulle croste dei formaggi: tecnologia e microflora di suprficie. *Lait* VI:335.

Nilson, K.M., Shahani, K.M., Vakil, J.R. and Kilara, A. 1975. Pimaricin and mycostatin for retarding cottage cheese spoilage. *J. Dairy Sci.* 58:668.

Pintado, C.M.B.S., Ferreira, M.A.S.S. and Sousa, I. 2010. Control of pathogenic and spoilage microorganisms from cheese surface by whey protein films containing malic acid, nisin and natamycin. *Food Control* 21:240–246.

Pitt, J.I. and Hocking, A.D. 1999. *Fungi and Food Spoilage*, 2nd ed. Aspen Publishers, Inc., New York.

Pugazhenti, T.R., Dhanalakshmi, B., Narasimhan, Shibu, A.V. and Madhan, S. 1999. Effect of anti-mycotic agents on *Penicillium citrinum* in cheese. *Indian Vet. J.* 76:537.

Raab, W. 1967. Pimaricin ein Antibiotikum gegen Pilze und Trichomonaden. *Arzneimittel-Forschung* 17:538.

Raab, W. 1972. *Natamycin (Pimaricin). Its Properties and Possibilities in Medicine.* Georg Thieme Publishers, Stuttgart.

Ray, L.L. and Bullerman, L.B. 1982. Preventing growth of potentially toxic moulds using antifungal agents. *J. Food Prot.* 45:953.

Resa, C.P.O., Jagus, R.J. and Gerschenson, L.N. 2014. Natamycin efficiency for controlling yeast growth in models systems and on cheese surfaces. *Food Control* 35:101–108.

Resa, C.P.O., Gerschenson, L.N. and Jagus, R.J. 2016. Starch edible film supporting natamycin and nisin for improving microbiological stability of refrigerated Argentinian Port Salut cheese. *Food Control* 59:737–752.

Riedl, R., Luf, W. and Brandl, E. 1984. Spectrophotometric estimation of natamycin in cheese. *Zeitschrift Lebens.-Unters. und -Forschung* 179:394. [*Dairy Sci. Abst.* 1986. 48:263.]

Rusul, G. and Marth, E.H. 1988. Growth and mycotoxin production of *Aspergillus parasiticus* in a medium at different pH values and with and without pimaricin. *Z. Lebensm Unters Forsch* 187:436.

Seekstra, H., Verkennis, A.E.E., Jacobs, R. Dekker, A., Stark, J. and Dijksterhuis, J. 2016. Fungal strains and the development of tolerance to natamycin. *Intl. J. Food Microbiol.* 238:15–22.

Shahani, K.M., Bullerman, L.B., Evans, T.A. and Arnold, R.G. 1977. Prevention of toxin mould growth in cheese by pimaricin. *Archives de L' Institut Pastuer de Tunis.* 54:511.

Shirk, R.J., Whitelhall, A.R. and Clark, W.L. 1962. The bioassay of pimaricin and its binding effect in orange juice. *J. Food Sci.* 27:605.

Shirk, R.J. and Clark, W.L. 1963. The effect of pimaricin in retarding the spoilage of fresh orange juice. *Food Technol.* 17:1062.

Snyder, A.B., Churey, J.J. and Worobo, R.W. 2016. Characterization and control of *Mucor circinelloides* spoilage in yogurt. *Intl. J. Food Microbiol.* 228:14–21.

Stark, J. 1999. Permitted presevatives—natamycin. p. 1780. In *Encyclopaedia of Food Microbiology*, Academic Press, Cambridge, MA.

Stark, J. and Tan, H.S. 2003. Natamycin. In *Food Preservatives*, (ed.) Russell, N.J. and Gould, G.W. Springer, Boston, MA.

Thomas, L.V. and Delves-Broughton, J. 2001. Applications of the natural food preservative natamycin. *Res. Adv. Food Sci.* 2:1.

Ticha, J. 1975. A new fungicide, pimaricin, and its application in the baking industry. *Mlynsko-Pekarensky Prumysl* 21:225. [*Food Sci. Technol. Abst.* 8:163.]

Tsiraki, M.I., El-Obeid, T., Yehia, H.M., Karam, L. and Savvaidis, I.N. 2018. Effects of chitosan and natamycin on vacuum-packaged phyllo: A pastry product. *J. Food Prot.* 81:1982–1987.

Tuinstra, L.G.M.T. and Traag, W.A. 1982. Liquid chromatographic determination of natamycin in cheese at residue levels. *J. Assoc. Off. Anal. Chem.* 65:820.

Ture, H., Eroglu, E., Ozen, B. and Soyer, F. 2011. Effect of biopolymers containing natamycin against *Aspergillus niger* and *Penicillium roquefortii* on fresh kashar cheese. *Intl. J. Food Sci. Technol.* 46:154–160.

Verma, H.S., Yadav, J.S. and Neelakantan, S. 1988. Preservative effect of selected antifungal agents on butter and cheese. *Asian J. Dairy Res.* 7:34.

Welscher, Y.M.T., Ten Napel, H.H., Balague, M.M., Souza, C.M., Riezman, H., de Kruijff, B. and Breukink, E. 2008. Natamycin blocks fungal growth by binding specifically to ergosterol without permeabilizing the membrane. *J. Biolog. Chem.* 283:6393–6401.

Welscher, Y.M.T., van Leeuwen, M.R., de Kruijff, B., Dijksterhuis, J. and Breukink, E. 2012. Polyene antibiotic that inhibits membrane transport proteins. *Proc. Natl. Acad. Sci. USA* 109:11156–11159.

Zalazar, A.L., Gliemmo, M.F., Soria, M. and Campos, C.A. 2019. Modelling growth/no growth interface of *Zygosaccharomyces bailii* in simulated acid sauces as a function of natamycin, xanthan gum and sodium chloride concentrations. *Food Res. Intl.* 116:916–924.

Ziogas, B.N., Sisler, H.D. and Lusby, W.R. 1983. Sterol content and other characteristics of pimaricin-resistant mutants of *Aspergillus nidulans. Pesticide Biochem. Physiol.* 20:320.

10

Lauric Arginate Ethyl Ester

Cangliang Shen and T. Matthew Taylor

CONTENTS

10.1 Chemistry Structure and Characteristics

Ethyl lauroyl arginate (LAE), as shown in Figure 10.1 (Kawamura and Whitehouse, 2008), composed of lauric acid, L-arginine hydrochloride, and ethanol, is synthesized through L-arginine monohydrochloride esterification with ethanol using thionyl chloride as an esterification agent (patented by the Spanish company LAMIRSA, European Patent No.1294678, Kawamura and Whitehouse, 2008). Its active component is the hydrochloride salt of an N-fatty acyl-substituted amino acid ethyl ester, referred as ethyl-N^{α}-lauroyl-L-arginate hydrochloride.

The chemical name of LAE is ethyl-N^{α}-dodecanoyl-l-arginate hydrochloride (CAS number: 60372-77-2), and it is white hygroscopic powder with molecular weight of 421.02 g/mole. The compound melts at 50.5–58.0°C and decomposes above 107°C (FDA, 2005). LAE has good water solubility (>247 g/kg) because of its polar characteristics and a low oil–water equilibrium partition coefficient ($K_{ow} < 0.1$), which can be concentrated in the water phase of products (Asker et al., 2009; FDA, 2005). In addition to being soluble in aqueous solutions, LAE is also soluble in other organic chemical solutions including ethanol, propylene glycol, and glycerol, with pH values ranging from 3.0 to 5.0 (1% solution) (Kawamura and Whitehouse, 2008). LAE can be stably stored in a closed container protected from light at 25°C for more than 2 years. However, LAE can be degraded under severe conditions over extended time periods, including very low pH (<1.5) and high temperature (more than 5 h at 100°C and 1 h at 121°C) (FDA, 2005). The degradation products, such as lauric acid, ethyl laurate, arginine, and arginine ethyl ester, are the same nontoxic compounds as impurities or residuals in LAE. It is well-known that LAE interacts with anionic components through electrostatic interactions. It can precipitate from solution at pH >4.5 and at high ionic strength (Asker et al., 2009; FDA, 2005), and it can be hydrolyzed by nitrites or proteins composed of alkali amino acids (Kawamura and Whitehouse, 2008).

LAE, at the concentration ranging from 50 to 200 ppm, was tested effectively to inhibit bacterial growth (i.e., bacteriostatic), including *Staphylococcus carnosus*, *Listeria innocua*, *Listeria*

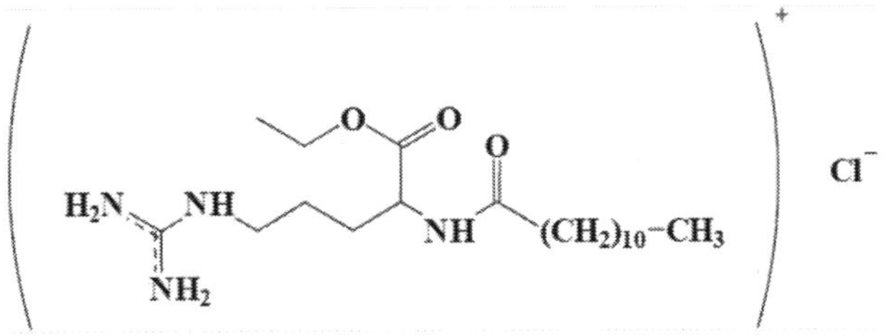

FIGURE 10.1 Chemical structure of ethyl-N^{α}-lauroyl-L-arginate hydrochloride.

monocytogenes, generic *Escherichia coli* K12, *E. coli* O157:H7, *Pseudomonas fluorescens*, *Salmonella*, and *Campylobacter* spp. in food systems (Ma et al., 2013; Manrique et al., 2017). LAE also exerts fungistatic impacts against yeasts such as *Candida utilis*, *Saccharomyces cerevisiae*, and *Torulopsis pinus*, and molds such as *Aspergillus niger*, *Cladosporium cladosporioides*, and *Penicillium chrysogenum*, with MIC values ranging from 4 to 120 ppm on solid or liquid medium (Higueras et al., 2013).

10.2 Antimicrobial Activity in Microbiological Media

Brandt et al. (2010) evaluated the inhibitory effect on *L. monocytogenes* by LAE alone or in combination with nisin in broth using a checkerboard assay to determine MIC, MBC, optimal inhibitory combinations (OICs), optimal bactericidal combinations (OBCs), and fractional inhibitory concentrations (FICs) to classify the types of interactions of antimicrobials with regards to their inhibition of the pathogen, identifying interactions as synergistic (FIC < 1.00), additive (FIC = 1.00), or antagonistic (FIC > 1.00). The MIC and MBC of LAE were 12.50 ppm, the same for all four tested *L. monocytogenes* strains. The OIC and OBC values of nisin with LAE ranged from 0.78 to 3.13 and 1.56 to 6.25, respectively. The FIC values of nisin with LAE ranged from 0.63 to 1.00, suggesting the combination produced additive-type inhibition. Benli et al. (2011) reported the efficacy of LAE (100 or 200 mg/liter) alone or sprayed sequentially with ε-poly-lysine (EPL) and acidic calcium sulfate (ACS) to reduce *Salmonella* on a membrane filter model system. Results showed spraying of LAE 200 ppm alone for 20 or 60 s reduced *Salmonella* by 3.5 log CFU/ml on filter papers, while greater reductions of 4.2 and 5.5 log CFU/ml were achieved with the combination of LAE 200 ppm with ACS 20 ppm and LAE 200 ppm with EPL 300 ppm, respectively.

Soni et al. (2012) investigated the anti-listerial efficacy of LAE (200 ppm) in a broth model during storage at 4 and 30°C. LAE alone (200 ppm) reduced the pathogen in broth by 3.0 log CFU/ml after 4 h storage, and the reduction increased to 9.0 log CFU/ml after 24 h storage at 4 and 30°C. Ma et al. (2013) determined the MIC and MBC of LAE against *L. monocytogenes*, *Escherichia coli* O157:H7, and *Salmonella* Enteritidis in liquid medium alone or with beef extract (2 or 5%) and soluble potato starch (2–10%). These researchers also investigated the growth of the pathogens in 2% reduced milk during 48 h storage at 21°C. Results showed that the MIC and MBC for LAE against *L. monocytogenes*, *E. coli* O157:H7, and *S.* Enteritidis were 11.8, 11.8, and 23.5 ppm, and 23.5, 11.8, and 23.5 ppm, respectively. Beef extract showed no effect on the MIC and MBC against *L. monocytogenes*, though 2–5% potato starch increased MICs to 93.8 to 187.5 ppm and MBCs to 187.5 to 375 ppm. LAE at the concentration of 375 and 750 ppm reduced the population of *L. monocytogenes* in 2% milk by 1.0 and 6.2 log CFU/ml, respectively. The combination of LAE with essential oils cinnamon, thymol, and eugenol showed synergism against *L. monocytogenes*, but antagonism against *E. coli* O157:H7 and *S.* Enteritidis in milks during the storage period.

Oladunjoye et al. (2013) tested the antimicrobial efficacy of LAE (25 to 200 ppm) against *Salmonella* in a broth medium stored at 4, 22, and 45°C. Results showed that at 4°C, 200 and 100 ppm of LAE reduced the pathogen by 6.0 and 3.6 log CFU/ml, respectively. Conversely, 25 and 50 ppm treatment had no effect on the pathogen survivals. At 22 and 45°C, 50 to 200 ppm of LAE all achieved ~6.0 log CFU/ml reduction which reduced the pathogen levels below the detection limit (1.0 log CFU/ml). Higueras et al. (2013) reported that MIC values of LAE were 8 and 16 ppm for Gram-positive (*L. monocytogenes* and *S. aureus*) and Gram-negative bacteria (*E. coli, P. putida*, and *S. enterica*) on solid or in liquid medium, respectively. MIC values of LAE for yeasts (*Candida utilis, Saccharomyces cerevisiae*, and *Torulopsis pinus*) were 4–16 ppm in liquid medium and 32–120 ppm on a solid medium. MIC values of LAE for molds (*Aspergillus niger, Cladosporium cladosporioides, Penicillium chrysogenum*) ranged from 24 to 120 ppm on solid medium.

Nair et al. (2014) evaluated the efficacy of LAE against *Campylobacter jejuni* subsp. *jejuni* in broth. Results showed that 50–200 ppm of LAE reduced the pathogen by >6.0 log CFU/ml in broth after storage at 4°C for 2 h. Kang et al. (2014) evaluated the antilisterial efficacy of LAE in brain heart infusion broth (BHI) during storage at 7°C over 60 days. Results showed that 10 ppm LAE showed promising bactericidal effects in broth yielding 3.7 log CFU/ml initial reduction (Nr), extending the lag phase time (λ) up to 22.2 days, reducing the growth rate (μ_{max}) to 0.1 log CFU/ml/day, and lowering the maximum cell density (N_{max}) to 6.4 log CFU/ml, than the controls ($\lambda = 3.5$ days, $\mu_{max} = 0.4$ log CFU/ml/day, $N_{max} = 8.7$ log CFU/ml).

Pattanayaiying et al. (2014) studied the antimicrobial effect of LAE alone and in combination with nisin Z against *E. coli* O157:H7, *L. monocytogenes*, and *Brochothrix thermosphacta* and its related antimicrobial mechanism by SEM and TEM. Results showed that 2,000 ppm LAE alone reduced the growth of *E. coli* O157:H7, *L. monocytogenes*, and *B. thermosphacta* by 4.5, 7.0, and 7.0 log CFU/ml in liquid medium. Treatment of LAE alone to bacteria caused cellular morphology changes, cytoplasm coagulation, and low-level leakage of potassium and phosphate ions. Combination of LAE (2,000 ppm) with nisin Z (320 AU/ml) reduced all three tested bacteria by approximately 7.0 log CFU/ml, which caused severe changes of cytoplasmic membrane and cell lysis, and significant potassium (16.62 to 50.35 mg/L) and phosphate ion leakage (97.96 to 122.66 mg/L) among three tested bacteria cells.

Shen et al. (2015) evaluated the behavior of acid stress-adapted *L. monocytogenes* exposure to LAE in liquid growth medium, milk, and carrot juice during storage at 37, 22, and 4°C for up to 24h. Acid-adapted cells introduced the cross-protection to LAE by exhibiting ~2.0 log CFU/ml greater survival than non-adapted cells in TSBYE with 50 ppm of LAE. Eight hundred (800) or 700 ppm LAE reduced the non-adapted cells below detection limits in milk and carrot juice after 60 min, whereas the acid-adapted cells showed about 2.0 log CFU/ml survival after lethal LAE inactivation. They also reported that the cross-protection against LAE in *L. monocytogenes* was reversible within 1 h at 4°C in the absence of sub-lethal acid stress.

Kozak et al. (2017) determined the MIC and MBC of LAE alone or in binary combination with hydrogen peroxide and ε-polylysine against *L. monocytogenes* in bacterial medium adjusted to pH 7.4 and 5.5. The lowest MIC of LAE in BHI broth was 10 ppm and 6.25 ppm at pH of 5.5 and 7.4, respectively. Similarly, the lowest MBC of LAE in BHI broth was 20 ppm at pH of 5.5 and increased to 25 ppm at pH of 7.4. The combination of LAE with ε-polylysine demonstrated the most effective bactericides at pH 7.4 with the fractional bactericidal concentration of 0.292, which indicated promising future applications as the combination of these two compounds showed low use concentrations (<10 ppm) and no significant change of pH in broth.

Manrique et al. (2017) tested the efficacy of sequential applications of eugenol and LAE against *S. carnosus, L. innocua, E. coli* K12, and *P. fluorescens*. Applications of antimicrobials were at half of their minimum lethal concentrations (MLC) or sequentially at t = 0 h and t = 3, 4, 6, or 8 h. The most effective treatments were eugenol and LAE when applied within 3–4 h and eugenol first applied, reducing *S. carnosus* and *P. fluorescens* by 2.3–2.6 and 6.8 log CFU/ml, respectively. However, no sequence effects were observed for *L. innocua*, and sequential treatments proved to be ineffective against *E. coli* K12 (Manrique et al., 2017).

10.3 Antimicrobial Mechanism

The antimicrobial mechanism of LAE has been reported in several studies, using scanning electron microscopy (SEM) and transmission electron microscopy (TEM), flow cytometry, and ion efflux technology to investigate the integrity of cell wall and membrane components of treated microorganisms.

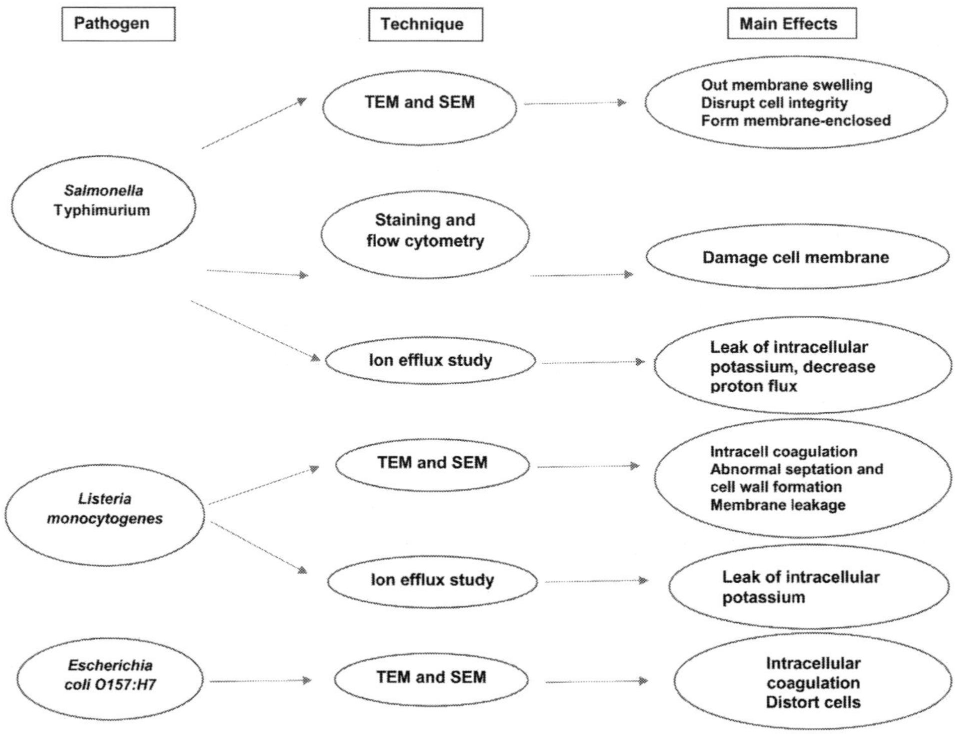

FIGURE 10.2 Antimicrobial mechanisms of LAE to inactivate bacterial foodborne pathogens.

Results of these previous studies concluded that LAE, as a cationic surfactant, primarily attacks bacterial cell wall, further damages microbial membranes, produces alteration of membrane electrical potential, and changes membrane permeability, which results in leakage of cytoplasmic material and eventually kills bacterial cells (Rodríguez et al., 2004). Ion efflux analysis reported by Rodríguez et al. (2004) and Pattanayaiying et al. (2014) also observed that treatment of LAE to bacteria causes cellular morphology change, cytoplasm coagulation, low-level leakage of potassium and phosphate ions, and decreased proton flux. The summary of antimicrobial mechanism of LAE to inhibit microorganisms (Rodríguez et al., 2004, Becerril et al., 2013, Pattanayaiying et al., 2014) is shown in the Figure 10.2.

10.4 Application in Food Systems

10.4.1 Application Alone or with Other Chemicals on Raw Meat and Poultry Products

Benli et al. (2011) reported the efficacy of LAE (100 or 200 mg/liter) alone or sprayed sequentially with ε-poly-lysine (EPL) and acidic calcium sulfate (ACS) to reduce *Salmonella* on skins of raw chicken carcasses. The sequential spraying of LAE 200 (20 s) followed by ACS 30 (20 s) reduced *Salmonella* on chicken carcasses by 2.2 log CFU/ml, which was the most effective combination among the eight tested treatments. Sharma et al. (2013) studied the antimicrobial efficacy of 200 and 400 ppm LAE against *Salmonella* and mesophilic microorganisms in fresh, skinless, and boneless chicken breast fillets. Results showed that 400 ppm of LAE produced reductions in numbers of *Salmonella* ranging from 0.7 to 1.0 log CFU/g over 7 days of storage at 4°C; 200 ppm treatment generated a reduction of 0.7 log CFU/g after 0, 1, and 7 days as compared to controls. They also found that LAE did not generate any effect for the mesophilic microorganism population, pH, and color of chicken fillet samples during the storage period.

Oladunjoye et al. (2013) reported testing 2,000 to 5,000 ppm LAE alone or in combination with 0.1 to 5% of carvacrol against *Salmonella* in turkey samples with 1, 7, and 15% fat content. Applying 2,000 to

5,000 ppm of LAE on the surface of ground turkey reduced the pathogen by 2 to 5 log CFU/g. LAE at 2,000 ppm reduced the pathogen by 4.0, 1.5, and 1.5 log CFU/g in 1, 7, and 15% fat-containing samples, respectively, indicating no specific relationship between fat content and LAE for *Salmonella* reduction. Results also showed that the combination of 2,000 ppm of LAE with 1% carvacrol showed a synergistic effect by reaching 4.0 log CFU/g reduction of the pathogen in 7% fat ground turkey samples. Moore et al. (2017) examined the efficacy of 0.1% LAE against *Salmonella* and *Campylobacter jejuni* in ground chicken frames during storage at 4°C for 24 h. Results showed that 0.1% LAE dip treatment reduced *Salmonella* by 0.9 and 0.7 log CFU/g on days 0 and 1. The same treatment reduced *C. jejuni* by ~1.0 log CFU/g by day 1. LAE treatment did not affect the pH and colors of the ground chicken frames. Nair et al. (2014) determined the efficacy of LAE against *Campylobacter jejuni* on chicken breast fillets. Results showed that LAE at 200 and 400 ppm significantly reduced the pathogen on chicken breast fillets compared to the controls, with the maximum reduction of 1.5 log CFU/g in the 400-ppm treatment during storage at 4°C for 7 days. The 400 ppm of LAE decreased psychrotrophic bacteria by 2.3 log CFU/g compared to the controls at day 0, though this inhibition did not extend after 3 days of storage. The LAE treatment did not generate any reduction effects on mesophilic organisms.

Pattanayaiying et al. (2014) investigated the antimicrobial efficacy of LAE with nisin Z (produced from *Lactococcus lactis* subsp. *lactis* isolated from catfish gut) alone or in combination for inactivating foodborne pathogens on fresh and processed muscle products. Films containing 2% pf LAE alone reduced *Salmonella* Typhimurium and Enteritidis counts on raw turkey breast slices by 2.5 to 4.5 log CFU/cm^2 during storage at 4°C up to 28 days. The reductions increased to 3.5 to 5.1 log CFU/cm^2 when LAE were combined with nisin Z (320 AU/ml). Films manufactured with LAE and nisin Z reduced *S. aureus* and *L. monocytogenes* on ham surfaces by 5.5 and 5.6 log CFU/cm^2, respectively, during storage at 4°C up to 28 days. The same treatment also reduced *E. coli* O157:H7, O111, and O26 by >4.0 log CFU/cm^2 on raw beef slices during storage at 4°C up to 28 days.

Magrinya et al. (2015) investigated the impact of fat content on the antimicrobial activity of LAE against natural microbiota in minced meat. Minced meat containing 0, 15, and 50 wt% of fat were added with 0.5, 1, 1.5, 2.0, and 2.5 mg/g LAE. Results showed that LAE at concentrations >0.5 mg/g inhibited the growth of mesophilic microorganisms, coliforms, and lactic acid bacteria (LAB), with the highest antimicrobial activity being against LAB with 80% equaling a total inhibition. The inhibition significantly increased with the presence of 50 wt% fat versus 0 and 15 wt% samples.

Sukamaran et al. (2015) evaluated the efficacy of LAE (25 or 200 ppm) alone or in combination with lytic bacteriophage (8.0 and 9.0 log PFU/ml) in reducing *Salmonella* spp. *in vitro* and on chicken breast fillets and chicken skins. Results showed 25 ppm LAE reduced *S.* Typhimurium, *S.* Heidelberg, and *S.* Enteritidis by 2.3, 2.4, and 3.5 log CFU/ml at 2 h of storage, respectively; reductions increased to 3.8, 2.6, and 3.4 log CFU/ml after 24 h storage for the *in vitro* studies. Application of bacteriophage with LAE (25 ppm) reduced *Salmonella* spp. up to 5 log CFU/ml *in vitro* at 4°C. Surface application of 200 ppm of LAE alone reduced *Salmonella* by 0.7–0.8 log CFU/g on chicken breast fillets as compared to controls during 7 days' storage. Immersion of chicken skin into 200 ppm LAE generated 0.5–0.6 log CFU/cm^2 reduction, which increased to 0.8–1.0 log CFU/cm^2 after sequential application with phage treatment. Yang et al. (2017) evaluated the inhibitory effect of LAE alone or in combination with bacteriophage on *L. monocytogenes* in chicken breast tissue during storage at 4°C for 1 to 3 days. LAE alone with concentrations at 100, 200, and 400 ppm produced lower reductions of the Gram-positive pathogen versus those reported against *Salmonella* serovars by Sukamaran et al. (2015), with pathogen loads being reduced by 0.3 to 1.0 log CFU/g at day 0, 0.5 to 1.2 log CFU/g at day 1, 0.2 to 1.0 log CFU/g at day 2, and 0.1 to 0.8 at day 3, with the best reductions produced from the 400 ppm treatment. The combination of LAE with bacteriophage (8.0 log PFU/g) increased pathogen reductions to 0.8 to 1.5, 1.2 to 2.0, 0.6 to 1.4, and 0.4 to 1.1 log CFU/g at day 0, 1, 2, and 3, respectively. They also found no significant differences in organoleptic properties (surface color, pH, TBARS value, or sensory scores) of chicken breast tissues treated by LAE versus controls over the experimental period.

Fisher et al. (2016) evaluated the efficacy of a new antimicrobial solution (AMS) composed of 5% chitosan, 2% acetic, lactic, and levulinic acids, and 4% LAE to inactivate *E. coli* O157:H7, non-O157 STEC, *Salmonella*, and *L. monocytogenes* on beef top round steaks. Results showed that reductions of >1.5 log CFU/cm^2 and a 3.5 log CFU/cm^2 were achieved in AMS-treated beef samples (after 6 h of marination) with 2.5 and 4.5 log CFU/cm^2 inoculation levels, respectively. The antimicrobial efficacy of diluted AMS was affected by marination time, and pathogen and inoculation levels. Better anti-*E. coli* O157:H7 results were observed in AMS diluted to 1:5 and 1:10 after 6 and 24 h of marination with samples inoculated with 2.5 log CFU/cm^2 of *E. coli* O157:H7. However, AMS 1:10 showed better anti-*E. coli* O157:H7 results then the 1:5 solution with an inoculation level of 4.5 log CFU/cm^2. They also found that AMS, regardless of concentrations, exhibited better antimicrobial efficacy against *L. monocytogenes* than *E. coli* O157:H7, non-O157 STEC members, and *Salmonella*.

10.4.2 Application in Ready-to-Eat Meat Products

Luchansky et al. (2005) evaluated the effectiveness of "Sprayed Lethality in Container" methods with 5 or 10% LAE to inactivate *L. monocytogenes* on vacuum-packaged "table brown" ham during storage at 4°C for 24 h. Applying 2, 4, 6, and 8 ml of 5% LAE in shrink-wrap bags reduced pathogen counts by 3.3, 6.5, 5.6, and 6.5 log CFU/ham. The 10% LAE treatments generated reductions of 6.5 log CFU/ham for all tested volumes on ham surfaces. Further shelf life studies (4°C, 60 days) with 3.0 or 7.0 log CFU inoculum levels on hams showed that adding 4, 6, and 8 ml of 5% LAE into shrink-wrap bags reduced the pathogen on hams by 5.1, 5.4, and 5.5 log CFU/ham, respectively, within 24 h storage, but the pathogen recovered by approximately 2.0–5.0 log CFU/ham from day 1 to 60. For hams inoculated with 3.0 log CFU *L. monocytogenes*, 4, 6, and 8 ml of 5% LAE reduced the pathogen to below the detection threshold (1.5 log CFU/ham) within 24 h of storage. After 60 days at 4°C, the pathogen counts increased by 2.0 log CFU/ham for the 4- and 6-ml treatments but remained below the assay detection limit for the 8-ml treatment. Porto-Fett et al. (2010) studied the viability of *L. monocytogenes* on frankfurters without or with potassium lactate (PL, 0.68 or 1.36%) and sodium diacetate (SD, 1:7 ratio, 0.097 or 0.19%) treated with 22 or 44 ppm of LAE using the "Sprayed Lethality in Container" method. Results showed that regardless of the presence of PL and SD, LAE reduced the pathogen counts by 1.8 log CFU/package within 2 h, although pathogen counts later increased to approximately 7.0 log CFU/package after 120 days. LAE reduced the pathogen counts by 2.0 log CFU/package on samples formulated with PL and SD within 2 h, and it remained unchanged during the rest of storage time period.

Taormina and Dorsa (2009a) evaluated the short-term bactericidal efficacy of LAE alone or with an antimicrobial liquid smoke extract against *L. monocytogenes* inoculated onto vacuum-packaged frankfurters. Results showed that applying 2 or 3 ml of 5,000 ppm LAE reduced *L. monocytogenes* by 1.3 to 1.4 log CFU/package on day 0. Reductions of 1.8 to 2.2 log were achieved on frankfurters treated with 3.0 to 4.5 ml of 5,000 ppm LAE after 48 h storage at 4.4°C. Spraying 2.5 ml of 5,000 or 8,000 ppm of LAE achieved additional reductions of *L. monocytogenes* by 1.6 to 1.8 log CFU/package (dip-inoculated) and by 1.4 to 2.0 log CFU/package compared to the water treatment. However, the combination of LAE and smoke flavor did not increase the reductions compared to the LAE alone. The same group evaluated the antilisterial efficacy of LAE on various surfaces of large-diameter ham (Taormina and Dorsa, 2009b). LAE solution application reduced *L. monocytogenes* by 5.0 log-cycles within 180 min at 4.4 and 23°C *in vitro*. Automatic spraying of 15 and 29 ml of 9,090 ppm LAE achieved a 2.0 log CFU/cm^2 reduction on six types of ham surfaces after storage at 4.4°C for 48 h. Spraying 22 ml LAE (9,090 ppm) on a large surface ham (1,129 cm^2) also produced a 2.0 log CFU/cm^2 reduction in counts of the organism under the same storage conditions.

Terjung et al. (2014) evaluated the antimicrobial activity of LAE applied as an aqueous solution (2,000–10,000 ppm), oil-in-water emulsion (2,000–4,000 ppm), and solid lipid particles (2,000–4,000 ppm) on "Lyoner style" sausage slices stored at 6°C for 24 days. Results showed that applying aqueous LAE (2,000 and 2,500 ppm) completely inhibited the growth of *L. innocua* and *Lactobacillus curvatus* on sausage surfaces, but no impact was observed on the survival of generic *E. coli* C600. However, the use of LAE in emulsion and solid lipid particles showed reduced antimicrobial efficacy, with *L. innocua* and *Lac. curvatus* growing from 1.2 to 4–6 log CFU/g by the end of storage period (24 days). Lavieri et al. (2014) investigated the efficacy of LAE as a post-lethality intervention to inactivate *L. monocytogenes* on alternatively cured frankfurters during vacuum-packaged storage at 4°C for 98 days. A 10% LAE treatment was dispensed into bags containing frankfurter links, delivering 43.16 mg/kg antimicrobial on each product. Results showed that LAE reduced the pathogen by 2.3 log CFU/g compared to untreated controls. However, LAE-treated samples supported the pathogen's growth after 28 days with no difference in survival compared to the control by the end of the storage period. These results indicated that LAE did not inhibit growth of *L. monocytogenes* during an extended storage time period. Luchansky et al. (2016) monitored the survival of *L. monocytogenes* on casing-cooked ham slices, prepared with and without commercial antimicrobials (potassium lactate, sodium diacetate, buffered vinegar, potassium lactate), and subsequently surface-treated with LAE (44 ppm) during refrigeration and/or frozen storage. Results showed that LAE-treated hams without antimicrobials support the growth of *L. monocytogenes* by ~4.0 log CFU/slice, and the pathogen decreased by 0.9 to 1.9 log CFU/slice when LAE and antimicrobials were combined. During 6 months of frozen storage, applied LAE reduced *L. monocytogenes* by 4.2 to 5.2 log CFU/slice, greater reductions than those not treated with LAE (2.0 to 3.5 log CFU/slice).

Terjung et al. (2016) evaluated the antimicrobial efficacy of LAE as an ingredient in emulsion type sausages and reconstructed hams against *L. innocua*. Results showed that LAE (200 ppm) inhibited the growth of the pathogen surrogate on emulsion-type sausages less than 1 day and retarded the growth on reconstructed ham after 1 day. They concluded that results were comparable to most previous studies, in that the application as a matrix of formulation in this study was comparable to the surface application in most other ones. The incorporation of LAE with proteins and salts in meat products may prevent the inclusion of LAE into bacterial cell membrane, potentially reducing its antimicrobial efficacy. Stopforth et al. (2010) investigated the antilisterial effects of a lactate–diacetate blend (1.68% potassium lactate and 0.12% sodium diacetate, PLSD) in the formulation of cooked cured ham with a surface treatment of 0.07% LAE during vacuum storage at 4°C for 90 days. Use of LAE alone generated the initial reduction of *L. monocytogenes* by ~1.0 log CFU/g and inhibited the pathogen growth for 8 days. The combination of PLSD and LAE produced an initial reduction of 0.7 log CFU/g and completely inhibited the pathogen's growth during 90 days of storage.

10.4.3 Application in Other Food Products

Soni et al. (2012) investigated the antilisterial efficacy of LAE (200 ppm) alone or in combination with bacteriophage (P100) in queso fresco-style cheese (QFC) during storage at 4 and 30°C. 200 ppm LAE reduced the pathogen by 2.0 log CFU/g in QFC samples after 1 day of storage at 4°C, but was followed by subsequent growth. The combination of P100 and 200 ppm LAE reduced the initial population of the pathogen by 2.0 to 4.0 log CFU/cm^2 and kept the pathogen survival to a reduced degree (<4.7 log CFU/cm^2) for 28 days at 4°C.

Kang et al. (2014) evaluated the antilisterial efficacy of LAE on cold-smoked salmon during storage at 7°C over 60 days. Results showed that 200 ppm of LAE did not produce bactericidal activity against *L. monocytogenes* in salmon samples, with Nr = 0.2 log CFU/ml, $\lambda = 2.5$ days, $\mu_{max} = 0.19$, and $N_{max} = 8.4$, which were similar to those from the control (Nr = 0.0 log CFU/ml, $\lambda = 3.2$ days, $\mu_{max} = 0.18$, and $N_{max} = 8.6$).

Chen et al. (2015) investigated the antimicrobial efficacy of LAE against *Salmonella* Tennessee in a glycerol-sucrose model, and peanut paste (<5 and 50% fat) during storage at 25°C for 3 to 5 days. Results showed that 100 and 200 ppm of LAE reduced the pathogen to below the study's detection limit (0.0 log CFU/g) by day 2 for water activity (a_w) ranging from 0.3 to 1.0. LAE did not show any antimicrobial effect in low-fat peanut paste at $a_w = 1.0$, although inactivation was enhanced with reducing water activity from 1.0 to 0.3. LAE at 2,500 and 5,000 ppm produced a 4.0 to 5.0 log reduction in low-fat peanut paste in 5 days compared with the control at $a_w = 0.3$. However, reduction in *Salmonella* numbers declined to 1.5 log CFU/g when the same concentration of LAE was applied in high-fat peanut paste.

Frelka and Harris (2015) evaluated the effects of antimicrobial sprays of 200 ppm of PAA with 2% LAE to inactivate aerobic bacteria, *E. coli*, and coliforms (ECC) from in-shell California walnuts. LAE spray bars [69 kPa (10 psi liquid pressure and 140 kPa (20 psi air pressure)] were positioned behind PAA spray bars to coat walnuts with the LAE product. Results showed that the 200 ppm PAA spray after hulling with a subsequent 2% LAE spray reduced APC and ECC by less than 1.0 log CFU/nut, which was not different from the water control. Frelka et al. (2016) evaluated the efficacy of PAA in combination with LAE to reduce aerobic plate counts (APCs) and ECCs on in-shell walnuts during ambient and controlled-temperature storage. The initial APCs on PAA-LAE-sprayed walnuts were 6.3 log CFU/

nut, decreasing by 1.4 log CFU/nut during ambient storage; no significant decline was observed during controlled-temperature storage. Initial ECCs on PAA-LAE-treated walnuts were 5.3 log CFU/nut, which decreased by 1.3 log CFU/nut in the first 34 days of ambient temperature storage. However, the final ECC populations of PAA-LAE-treated samples (3.6 log CFU/nut) were similar to the PAA-only-treated samples (3.9 log CFU/nut).

10.4.4 Application within Packaging Material

Jiang et al. (2011) evaluated the antilisterial efficacy of four polysaccharide-based edible coatings incorporating potassium lactate (PL) with sodium diacetate (SD) and LAE on roasted turkey during storage at 4°C for 8 weeks without or with pre-frozen storage at 4°C for 4 weeks. The PL+SD and LAE treatment reduced *L. monocytogenes* by 2.1–2.7, 1.7–2.4, 1.0–2.2, and 1.6–3.6 log CFU/cm^2 on turkey after storage at 4°C for 1, 2, 4, and 8 weeks. After frozen storage at 4°C for 4 weeks, the reductions of the PL with SD and LAE treatment were 0–0.2, 2–2.3, 1.6–2.4, 1.1–2.3, and 1.6–2.3 log CFU/cm^2 after storage at 4°C for 0, 2, 4, 6, and 8 weeks, respectively. Jin et al. (2013) evaluated the efficacy of antimicrobial coatings using chitosan with LAE to inactivate *Salmonella* Enteritidis on shell eggs. Results showed that including 0.1, 0.5, and 1.0% LAE into chitosan coatings reduced *Salmonella* by 0.5, 2.5, and 4.7 log CFU/ml as compared to non-treated controls. Nevertheless, adding 0.1% LAE into chitosan coatings did not increase ($P \geq 0.05$) reductions to the pathogen compared to the chitosan alone. Storage studies indicated that chitosan coatings infused with 0.5% LAE reduced the pathogen to 3.9, 3.3, 2.2, and <0.5 log CFU/ml after 1, 3, 7, and 14 days, respectively. The same coatings with 1.0% LAE reduced the pathogen level by 5.6 and >6.3 log CFU/ml after 1 and 3 days, respectively.

Muriel-Galet et al. (2012) developed an active packaging antimicrobial film containing EVOH-29 and EVOH-44 copolymers with LAE (0.25 to 10%) and evaluated their antimicrobial activities in an infant formula milk (4°C, 6 days). EVOH-29 and EVOH-44 films incorporating 5 and 10% LAE completely inhibited the growth of the three tested bacteria. For EVOH-29 films, 1% LAE completely inhibited *L. monocytogenes*, reduced the growth of *E. coli* (ATCC 25922) by 4.0 log CFU, and reduced *Salmonella* by 5.3 log CFU. Use of 0.25% LAE in films reduced pathogen populations by 2.2 to 3.1 log CFU. For EVOH-44 films, 1% and 0.25% LAE reduced pathogen counts by 3.0 to 4.2 and 1.4 to 2.5 log CFU, respectively. They also found that EVOH-29 and EVOH-44 with 10% LAE produced a 4.0 log CFU reduction of *L. monocytogenes* in infant formula after 2 days of storage. EVOH-29 films with 5 and 10% LAE reduced the salmonellae by 3.7 to 4.0 log CFU; 5 and 10% LAE in EVOH-44 reduced the pathogen by 1.0 to 3.3 log CFU in the same infant formula.

The survival of the natural microbial population, and *L. innocua*, on whole raw shrimp following treatment with ozonated water (OW), chitosan with 5% LAE (CC3), and cryogenic freezing (CF, –75°C, 2 min, Cryo-Test chamber with liquid nitrogen vapor), alone or in combination, was studied by Guo et al. (2013a). Results showed that CC3, CC3+OW, and OW+CC3+CF reduced psychrotrophic bacteria by 2.0–3.0, 2.8, and 2.3 log CFU/g, respectively. CC3 decreased *L. innocua* by 3.0–4.0 log CFU/g on shrimp, while reductions increased to 4.0–5.0 log CFU/g when applying OW+CC3 and OW+CC3+CF treatments. In a subsequent study, Guo et al. (2013b) studied the efficacy of an antimicrobial coating consisting of chitosan with LAE (50 ppm), with or without cryogenic freezing, on the survival of *L. innocua* on frozen RTE shrimp. Results showed that chitosan with LAE reduced *L. innocua* by 4.1 log CFU/g and 0.6 log CFU/g from the initial populations of 7.8 and 4.9 log CFU/g, respectively. However, their results indicated that cryogenic freezing in combination with chitosan plus LAE did not achieve synergistic effects for *L. innocua* (only additional 0.3 log CFU/g reduction).

Higueras et al. (2013) evaluated the antimicrobial activities of cast chitosan films made with 1, 5, or 10% LAE and 20% glycerol. Chitosan films bearing 1% LAE achieved 1.0 to 1.5 log-cycle reductions of all tested bacterial growth. Tested Gram-negative bacteria were reduced by 2.5 and 4.8 log-cycles by CS-5% and CS-10% LAE films, respectively. Greater reductions of 4.0 and 6.0 log-cycles for Gram-positive bacteria were achieved by CS-5% and CS-10% LAE films. The presence of LAE in films completely inhibited the growth of tested yeast populations. CS-5% LAE film reached >4.0 log-cycles reduction for mesophilic bacteria, psychrophiles, and *Pseudomonas* spp., and reductions of 1.8–3.4 for LAB on chicken breast fillets after 2 to 8 days' storage.

Guo et al. (2014) evaluated the efficacy of antimicrobial coating with chitosan and LAE against *L. innocua* on RTE deli turkey meat with or without prior flash pasteurization in a pilot plant setting. Results showed that antimicrobial coating incorporating 1.94 mg/cm^2 chitosan and 0.39 mg/cm^2 LAE (PLA pouch) reduced the pathogen by 4.6 log CFU/cm^2, and that reductions increased to 5.2 log CFU/cm^2 with prior steam flash pasteurization (FP). The combination of FP and PLA also decreased *Salmonella* Typhimurium by approximately 3.3 log CFU/cm^2 on the same product. Wang et al. (2015) evaluated the antilisterial efficacy of novel coating solutions containing 5% chitosan, 2% each of acetic, lactic, and levulinic acids, and 200 ppm of LAE (stock solution) on inoculated roast beef stored at 4°C for 30 days. Results showed that the stock coating solutions, 1:5- and 1:10-diluted solutions, reduced *L. monocytogenes* by 0.8, 0.6, and 0.5 log CFU/ml, respectively, after 20 min of contact time. For beef samples, no growth of the pathogen was seen on stock solutions, and 1.6 and 4.6 log CFU/cm^2 increases were seen for the 1:5 and 1:10 diluted solutions.

Ma et al. (2016a) investigated the antimicrobial properties of chitosan films with LAE and EDTA against *L. monocytogenes*, *E. coli* O157:H7, and *Salmonella enterica* in liquid medium. Results showed that film discs impregnated with 0.1 g/100 g LAE alone created inhibitory zones of 20.1, 16.3, and 14.9 mm against *L. monocytogenes*, *E. coli* O157:H7, and *Salmonella enterica*, respectively, after 48 h incubation. Addition of 0.25 g/100 g EDTA into chitosan with LAE increased the inhibitory zone to 16.1, 18.0, and 22.5 mm against *L. monocytogenes*, *E. coli* O157:H7, and *Salmonella enterica*, respectively. Similarly, film discs containing 0.2 g/100 g LAE and 0.25 g/100 g EDTA achieved increased inhibitory zone diameters when tested against *L. monocytogenes* compared to film discs with LAE alone after 24 (23.3 versus 20.6 mm) or 48 h incubation (22.8 versus 20.3 mm). However, no differences in inhibitory zones were found for *E. coli* O157:H7 and *Salmonella enterica* regardless of adding EDTA into LAE chitosan films. In a related study, Ma et al. (2016b) reported that adding 500 mg/L EDTA into 5 mg/L LAE and 200 mg/L cinnamon oil increased the reductions of *E. coli* O157:H7, *Salmonella enterica*, and *L. monocytogenes* by approximately 4.0, 5.0, and 1.0 log CFU/ml, respectively, compared to the double or single treatment. The results from atomic force microscopy analysis indicated that LAE caused the aggregation of DNA molecules. EDTA increased the permeability of Gram-negative bacteria outer membranes to assist the penetration of LAE. The same group, Ma et al. (2016c), also investigated pathogen survival on cantaloupe-coating chitosan films containing 0.1% each of LAE, cinnamon oil, and EDTA. Results showed the combination immediately reduced *E. coli* O157 and *L. monocytogenes* on cantaloupes by >3.0 log CFU/cm^2 and reduced *Salmonella enterica* (>2.2 log CFU/cm^2) to below detection limits during a 14-day storage period. The same treatment also reduced yeast and mold populations by >3.8 log CFU/cm^2 compared to untreated cantaloupes.

Hilbig et al. (2016) investigated the antimicrobial activities of nano-emulsions of LAE with Tween 80 and cinnamon bark essential oil (CBO). Results showed that the MIC values of LAE alone in TSB were 11, 6, and 6 ppm for *Salmonella* Enteritidis, *E. coli* O157:H7, and *L. monocytogenes*, respectively, which increased to 12, 7, and 8 ppm after nano-emulsion of LAE with Tween 80 and CBO. All LAE (750 ppm) containing nano-emulsion treatments completely inactivated *L. monocytogenes* in 2% reduced fat milk below the detection limit (1 log CFU/ml). However, CBO was more effective than LAE for inactivating *Salmonella* Enteritidis and *E. coli* O157:H7, possibly a result of negatively charged lipopolysaccharides on the outer membranes of Gram-negative bacteria binding with LAE, reducing its antimicrobial activities compared to Gram-positive *L. monocytogenes*.

Guo et al. (2017) evaluated the antimicrobial activities of edible coatings with 0.5% Corn-Bio-fiber gum (C-BFG), 1% chitosan, and incorporated with 1–4% LAE against foodborne pathogens in broth, RTE deli meat, and strawberries. Results showed that films with 1–4% LAE reduced *L. innocua* in TSB to below 1.0 log CFU/ml after 1 day, and reduced the same pathogen from 5.0 to 3.0 log CFU/cm^2 at day 0 and maintained this level to day 5. The LAE film reduced the populations of *E. coli* O157:H7 and *Salmonella* in TSB from 8.5 log CFU/ml to below the detection limit for 5 days. The LAE film reduced the populations of *E. coli* O157:H7 and *Salmonella* on strawberries from 6 to 4.5 log-cycles at day 1 and maintained the same level through day 5.

Moreno et al. (2017) evaluated the antilisterial efficacy of biodegradable active films composed by glycerol-plasticized-oxidized corn starch (OS) and bovine gelatin (BG) blends (1:1 mass ratio) with ethyl lauroyl arginate (LAE) (1.3 g LAE/100g polymer) *in vitro* test and on marinated salmon. Results showed that the incorporation of LAE into the films led to a total bactericidal effect after 5 h (>6.6 log CFU/ml reduction) and 24 h incubation (>8.6 log CFU/ml reduction) *in vitro*. During the first 25 storage days, the LAE presence in the films did not produce significant differences in the *L. innocua* counts. However, at the end of storage time (45 days), salmon coated with the OS:BG:LAE film achieved a reduction of 1.0 log CFU/g in *L. innocua* compared to the control. Pattanayaiying et al. (2019) also evaluated the antimicrobial activities of thermoplastic starch/polybutylene adipate terephthalate (TPS/PBAT; 40/60) film produced by coating gelatin (15% v/v) containing LAE (0.8 mg/cm^2), alone or combination with nisin Z (69.4 AU/cm^2), to inactivate *Vibrio parahaemolyticus* and *Salmonella* Typhimurium on bigeye snapper (*Lutjanus lineolatus*) and tiger prawn (*Penaeus monodon*) slices during 28 days' storage at 4°C, and 90 days' storage at −20°C. LAE-Gelatin-TPS/PBAT film reduced *S.* Typhimurium on bigeye snapper and tiger prawn slices (stored at 4°C) by 3.2 log CFU/g after 28 days and 7.0 log CFU/g after 21 days, respectively. Nisin-LAE-Gelatin-TPS/PBAT film reduced *S.* Typhimurium on bigeye snapper and tiger prawn slices by 3.5 log CFU/g after 28 days and 7.0 log CFU/g after 14 days at 4°C, respectively. The LAE-Gelatin-TPS/PBAT and Nisin-LAE-Gelatin-TPS/PBAT films reduced *V. parahaemolyticus* on chilled bigeye snapper slices approximately 2.6 and 4.2 log CFU/g, respectively, after storage for 28 days at 4°C. Both films reduced *V. parahaemolyticus* on chilled tiger prawn slices approximately 7.1 log CFU/g after 28 days at 4°C. The LAE-Gelatin-TPS/PBAT and Nisin-LAE-Gelatin-TPS/PBAT films also reduced *S.* Typhimurium on bigeye snapper and tiger prawn slices by 5.8 and 5.6 log CFU/g, respectively, after 60 days at −20°C. *V. parahaemolyticus* was reduced by 5.8 log CFU/g on frozen bigeye snapper and tiger prawn slices after treatment with Nisin-LAE-Gelatin-TPS/PBAT film after 14 and 21 days, respectively. The LAE-Gelatin-TPS/PBAT film reduced *V. parahaemolyticus* by 5.8 log CFU/g on both frozen seafood slices after 28 days.

10.5 Regulatory Status

LAE is a chemically synthetic cationic surfactant, approved to be used as an antimicrobial during food product manufacturing processes by the U.S. Food and Drug Administration (FDA) in 2005 (FDA, 2005) and approved as a food preservative by the European Food Safety Authority (EFSA) in 2007 (EFSA, 2007). In September 2005, the FDA issued a Letter of No Objection regarding a submission that LAE is generally recognized as safe (GRAS; Notice No. GRN 000164; FDA, 2005) for use as an antimicrobial at levels up to 200 ppm of ethyl-N-lauroyl-L-arginate HCl in a wide range of food products as listed in Table 10.1. Since December 2007, the U.S. Department of Agriculture (USDA) issued a letter considering the use of ethyl lauroyl arginate to be consistent with the FDA's definition of a processing aid that is listed in Title 21 of the U.S. Code of Federal Regulations (CFR), §101.100(a)(3)(ii) on fresh cuts of meat and poultry products. Since July 2008, the USDA Food Safety and Inspection Service (FSIS) Directive 7120.1 approved the use of LAE as an antimicrobial on ready-to-eat (RTE) meat products up to 44 ppm applied inside packaging and at no more than 53 ppm by weight of the final products (USDA-FSIS, 2019). LAE has also been approved to be used on ground poultry and ground beef (not to exceed 200 ppm by weight in the finished product) since October 2011 (USDA-FSIS, 2019).

10.6 Evaluation of Acute Toxicity

Ruckman et al. (2004) reported low acute toxicity of LAE following the completion of multiple assays determining oral, ocular, and dermal exposure results. High-concentration acute exposure (2,000 mg LAE/kg) produced scabbing, edema, and other dermal irritation symptoms in test animals. Oral gavage of 2,000 mg LAE/kg in rats produced no fatalities in any test animals, with piloerection and intensified salivation resolving in affected animals within 4 days of exposure. These researchers reported no mutagenic activity against *S.* Typhimurium cells dosed with LAE at non-cytotoxic levels, and described LAE as not bearing mutagenic activity. Metabolism evaluation indicates the compound is excreted in urine via amino acid and fatty acid breakdown pathways (Ruckman et al., 2004). In April 2007, the European

TABLE 10.1

Food Categories Allowed to Be Applied with Ethyl-Nα-lauroyl-L-arginate Hydrochloride (≤200 ppm)

Food Categories	Notes/Limitations
Beverages and beverage bases	Excludes dairy products, soft drinks, and alcoholic beverages, ≤100 ppm
Carbonated beverages	≤100 ppm
Cheeses	Includes curd and whey cheeses, cream, natural, grating, processed, spread, dip, and miscellaneous cheeses
Chewing gums	Includes all forms
Tea	Includes regular, decaffeinated, and instant types
Condiments and relishes	Includes plain seasoning sauces and spreads, olives, pickles, and relishes, but not spices or herbs
Egg dishes	Includes egg roll, egg foo young, egg salad, and frozen multicourse egg meals, but not fresh eggs
Margarine and margarine-like table spreads	Includes mayonnaise, and spoonable and pourable dressings for salads
Fish products	Includes all prepared main dishes, salads, appetizers, frozen multicourse meals, and spreads containing fish, shellfish, and other aquatic animals, but not fresh fish
Fresh fish	Only fresh and frozen fish, shellfish, and other aquatic animals
Fresh meats	Only fresh or home-frozen beef or veal, pork, lamb or mutton, and home-prepared fresh meat-containing dishes, salads, appetizers, or sandwich spreads made therefrom
Fresh poultry	Only fresh or home-frozen poultry and game birds and home-prepared fresh poultry-containing dishes, salads, appetizers, or sandwich spreads made therefrom
Pie fillings	
Gravies and sauces	Includes all meat sauces and gravies, and tomato, milk, buttery, and specialty sauces
Hard candies	Includes dietetic hard candies, cough drops, mints and dietetic mints
Meat products	Includes all meats and meat-containing dishes, salads, appetizers, frozen multicourse meat meals, and sandwich ingredients prepared by commercial processing or using commercially processed meats with home preparation
Poultry products	Includes all poultry and poultry-containing dishes, salads, appetizers, frozen multicourse poultry meals, and sandwich ingredients prepared by commercial processing or using commercially processed poultry with home preparation
Processed fruits and fruit juices	Excludes apple juice, includes commercially processed juices and juice punches, concentrates, dilutions, and drink substitutes made therefrom, and dried fruits; also including strained fruits and fruit juices excluding apple juice as baby or toddler foods
Processed vegetables and vegetable juices	Includes potato salads, raw vegetables, vegetable juices and blends, and tomato sauces
Soups and soup mixes	Includes commercially prepared meat, fish, poultry, vegetable, and combination soups and soup mixes

Food Safety Authority (EFSA) issued the opinion of the Scientific Committee on ethyl lauroyl arginate as a new food preservative for use in a range of food categories, which indicated that the acceptable daily intake (ADI) is 0–0.5 mg/kg body weight (bw). In June 2008, the Joint FAO/WHO Expert Committee on Food Additives (JECFA) considered LAE as a food additive and allocated an ADI of 0–4 mg/kg bw. The difference in the ADIs established by EFSA and JECFA is due to a difference in the interpretation of hematology data obtained in animal toxicity studies (EFSA, 2007).

10.7 Chemical Assays

LAE is commonly subjected to antimicrobial activity assays via *in vitro* testing, or applied via sprays or dips to differing food products inoculated with microbial pathogens or pathogen indicators. LAE can be infused into edible and inedible films for subsequent application in and onto food products. Its water solubility also facilitates its quantification via chromatographic methods. Quantification of LAE content can be made by the development of standard curve with peak retention identification via HPLC (JECFA, 2008). Bonnaud et al. (2010), following the mixing of LAE with differing polysaccharides, determined the mixing capacity and thermodynamic properties shifts of mixtures via isothermal titration calorimetry (ITC). As a function of polysaccharide charge characteristics and LAE addition, system entropy increased during sample processing and mixing. Hawkins et al. (2009) described the metabolism of ^{13}C- and ^{14}C-labeled LAE in simulated gastric fluid and in human clinical trials to determine its pharmacokinetic properties and breakdown/excretion. Metabolized LAE was analyzed from blood plasma samples via LC–MS/MS procedures to determine the antimicrobial's breakdown and metabolites.

10.8 Conclusion

LAE is a broadly active antimicrobial preservative used for inhibition of the growth of both bacterial and fungal microorganisms causing foodborne disease and/or food spoilage. Its applications range from plant-derived to various animal-derived foods, with multiple federal regulatory agencies in the U.S. giving approval for its use in meat/poultry foods and other mixed-ingredient foods. The FAO has recognized its relative safety as indicated by low acute toxicity, and multiple researchers have indicated its metabolism into non-toxic compounds in the body. It can be applied directly via spray or dip, or through incorporation into food packaging films. Chromatographic assays have been published detailing its quantification from foods and bodily fluids to determine its stability and metabolism. It has successfully been applied for food protection on its own as well as in combination with other traditional and/or natural food preservatives/antimicrobials, delivered through aqueous solution or in various emulsion structures (e.g., micelles) to food products and food product surfaces.

REFERENCES

Asker, D., Weiss, J., & McClements, D. J. (2009). Analysis of the interactions of a cationic surfactant (Lauric Arginate) with an anionic biopolymer (pectin): Isothermal titration calorimetry, light scattering, and microelectrophoresis. *Langmuir, 25*, 116–122.

Becerril, R., Manso, S., Nerin, C., & Gómez-Lus, R. (2013). Antimicrobial activity of Lauroyl Arginate Ethyl (LAE), against selected food-borne bacteria. *Food Control, 32*, 404–408.

Benli, H., Sanchez-Plata, M. X., & Keeton, J. T. (2011). Efficacy of ε-polylysine, lauric arginate, or acidic calcium sulfate applied sequentially for salmonella reduction on membrane filters and chicken carcasses. *Journal of Food Protection, 74*, 743–750.

Bonnaud, M., Weiss, J., & McClements, D. J. (2010). Interaction of a food-grade cationic surfactant (lauric arginate) with food-grade biopolymers (pectin, carrageenan, xanthan, alginate, dextran, and chitosan). *Journal of Agricultural and Food Chemistry, 58*, 9770–9777.

Brandt, A. L., Castillo, A., Harris, K. B., Keeton, J. T., Hardin, M. D., & Taylor, T. M. (2010). Inhibition of Listeria monocytogenes by food antimicrobials applied singly and in combination. *Journal of Food Science, 75*, M557–M563.

Chen, W., Golden, D. A., Critzer, F. J., & Davidson, P. M. (2015). Antimicrobial activity of cinnamaldehyde, carvacrol, and lauric arginate against Salmonella Tennessee in a glycerol-sucrose model and peanut paste at different fat concentrations. *Journal of Food Protection, 78*, 1488–1495.

European Food Safety Authority (EFSA). (2007). Opinion of the Scientific Panel on food additives, flavourings, processing aids and materials in contact with food on a request from the Commission related to an application on the use of ethyl lauroyl arginate as a food additive. *EFSA Journal, 511*, 1–27.

Fisher, K. D., Bratcher, C. L., Jin, T. Z., Bilgili, S. F., Owsley, W. F., & Wang, L. (2016). Evaluation of a novel antimicrobial solution and its potential for control Escherichia coli O157:H7, non-O157:H7 shiga toxin-producing *E. coli*, *Salmonella* spp., and *Listeria monocytogenes* on beef. *Food Control, 64*, 196–201. doi: 10.1016/j.foodcont.2015.12.007

Frelka, J.C., and Harris, L.J. 2015a. Evaluation of microbial loads and the effects of antimicrobial sprays in postharvest handling of California walnuts. *Food Microbiol.* 48:133–142.

Frelka, J.C. Davidson,G.R., Harris, L.J. (2016) Changes in aerobic plate and *Escherichia coli*: Coliform counts and in populations of inoculated foodborne pathogens on inshell walnuts during storage. *J. Food. Prot.* 79 (7): 1143–1153. https://doi.org/10.4315/0362-028X.JFP-15-553

Guo, M., Jin, T. Z., Scullen, O. J., & Sommers, C. H. (2013a). Effects of antimicrobial coatings and cryogenic freezing on survival and growth of Listeria innocuaon frozen ready-to-eat shrimp during thawing. *Journal of Food Science*, *78*, M1195–M1200.

Guo, M., Jin, T. Z., Yang, R., Antenucci, R., Mills, B., Cassidy, J., … Sommers, C. H. (2013b). Inactivation of natural microflora and inoculated *Listeria innocua* on whole raw shrimp by ozonated water, antimicrobial coatings, and cryogenic freezing. *Food Control*, *34*, 24–30.

Guo, M., Jin, T. Z., Wang, L., Scullen, O. J., & Sommers, C. H. (2014). Antimicrobial films and coatings for inactivation of *Listeria innocua* on ready-to-eat deli turkey meat. *Food Control*, *40*, 64–70.

Guo, M., Yadav, M. P., & Jin, T. Z. (2017). Antimicrobial edible coatings and films from micro-emulsions and their food applications. *International Journal of Food Microbiology*, *263*, 9–16.

Hawkins, D. R., Rocabayera, X., Ruckman, S., Segret, R, & Shaw, D. (2009). Metabolism and pharmacokinetics of ethyl N^{α}-lauroyl-L-arginate hydrochloride in human volunteers. *Food and Chemical Toxicology*, 47, 2711–2715.

Higueras, L., López-Carballo, G., Hernández-Muñoz, P., Gavara, R., & Rollini, M. (2013). Development of a novel antimicrobial film based on chitosan with LAE (ethyl-Nα-dodecanoyl-l-arginate) and its application to fresh chicken. *International Journal of Food Microbiology*, *165*, 339–345.

Hilbig, J., Ma, Q., Davidson, P. M., Weiss, J., & Zhong, Q. (2016). Physical and antimicrobial properties of cinnamon bark oil co-nanoemulsified by lauric arginate and Tween 80. *International Journal of Food Microbiology*, *233*, 52–59.

Jiang, Z., Neetoo, H., & Chen, H. (2011). Efficacy of freezing, frozen storage and edible antimicrobial coatings used in combination for control of *Listeria monocytogenes* on roasted turkey stored at chiller temperatures. *Food Microbiology*, *28*, 1394–1401.

Jin, T. Z., Gurtler, J. B., & Li, S.-Q. (2013). Development of Antimicrobial Coatings for Improving the Microbiological Safety and Quality of Shell Eggs. *Journal of Food Protection*, *76*, 779–785.

Joint FAO/WHO Expert Committee on Food Additives (JECFA). (2008). Compendium of food additive specifications. In 69th meeting 2008. Food and Agriculture Organization, Rome, Italy.

Kang, J., Stasiewicz, M. J., Murray, D., Boor, K. J., Wiedmann, M., & Bergholz, T. M. (2014). Optimization of combinations of bactericidal and bacteriostatic treatments to control *Listeria monocytogenes* on cold-smoked salmon. *International Journal of Food Microbiology*, *179*, 1–9.

Kawamura, Y., & Whitehouse, B. (2008). Ethyl Lauroyl Arginate—Chemical and technical assessment. In 69th JEFCA 5: FAO JEFCA Monographs.

Kozak, S. M., Margison, K. M., & Damico, D. J. (2017). Synergistic Antimicrobial Combinations Inhibit and Inactivate *Listeria monocytogenes* in Neutral and Acidic Broth Systems. *Journal of Food Protection*, *80*, 1266–1272.

Lavieri, N. A., Sebranek, J. G., Brehm-Stecher, B. F., Cordray, J. C., Dickson, J. S., Horsch, A. M., & Mendonca, A. F. (2014). Investigating the control of *Listeria monocytogenes* on alternatively-cured frankfurters using natural antimicrobial ingredients or post-lethality interventions. *Meat Science*, *97*, 568–574.

Luchansky, J., Call, J., Hristova, B., Rumery, L., Yoder, L., & Oser, A. (2005). Viability of Listeria monocytogenes on commercially-prepared hams surface treated with acidic calcium sulfate and lauric arginate and stored at 4°C. *Meat Science*, *71*, 92–99.

Luchansky, J. B., Campano, S. G., Shoyer, B. A., & Porto-Fett, A. C. S. (2016). Viability of Listeria monocytogenes on boneless, water-added hams, commercially prepared with and without food-grade chemicals, during Extended Storage at 4 and/or –2.2°C. *Journal of Food Protection*, *79*, 613–619.

Ma, Q., Davidson, P. M., & Zhong, Q. (2013). Antimicrobial properties of lauric arginate alone or in combination with essential oils in tryptic soy broth and 2% reduced fat milk. *International Journal of Food Microbiology*, *166*, 77–84.

Ma, Q., Zhang, Y., & Zhong, Q. (2016a). Physical and antimicrobial properties of chitosan films incorporated with lauric arginate, cinnamon oil, and ethylenediaminetetraacetate. *LWT Food Science and Technology*, *65*, 173–179.

Ma, Q., Davidson, P. M., Critzer, F., & Zhong, Q. (2016b). Antimicrobial activities of lauric arginate and cinnamon oil combination against foodborne pathogens: Improvement by ethylenediaminetetraacetate and possible mechanisms. *LWT Food Science and Technology*, *72*, 9–18.

Ma, Q., Zhang, Y., Critzer, F., Davidson, P. M., & Zhong, Q. (2016c). Quality attributes and microbial survival on whole cantaloupes with antimicrobial coatings containing chitosan, lauric arginate, cinnamon oil and ethylenediaminetetraacetic acid. *International Journal of Food Microbiology, 235*, 103–108.

Magrinyà, N., Terjung, N., Loeffler, M., Gibis, M., Bou, R., & Weiss, J. (2015). Influence of fat addition on the antimicrobial activity of sodium lactate, lauric arginate and methylparaben in minced meat. *International Journal of Food Microbiology, 215*, 86–94.

Manrique, Y., Gibis, A., Schmidt, H. and Weiss, J. (2017). Influence of application sequence and timing of eugenol and lauric arginate (LAE) on survival of spoilage organisms. *Food Microbiol.* 64:210–218.

Moore, A., Nannapaneni, R., Kiess, A., & Sharma, C. S. (2017). Evaluation of USDA approved antimicrobials on the reduction of *Salmonella* and *Campylobacter* in ground chicken frames and their effect on meat quality. *Poultry Science, 96*, 2385–2392.

Moreno, O., Gil, À., Atarés, L., & Chiralt, A. (2017). Active starch-gelatin films for shelf-life extension of marinated salmon. *LWT Food Science and Technology, 84*, 189–195.

Muriel-Galet, V., López-Carballo, G., Gavara, R., & Hernández-Muñoz, P. (2012). Antimicrobial food packaging film based on the release of LAE from EVOH. *International Journal of Food Microbiology, 157*, 239–244.

Nair, D. V. T., Nannapaneni, R., Kiess, A., Mahmoud, B., & Sharma, C. S. (2014). Antimicrobial efficacy of lauric arginate against *Campylobacter jejuni* and spoilage organisms on chicken breast fillets. *Poultry Science, 93*, 2636–2640.

Oladunjoye, A., Soni, K. A., Nannapaneni, R., Schilling, M. W., Silva, J. L., Mikel, B., … Sharma, C. S. (2013). Synergistic activity between lauric arginate and carvacrol in reducing *Salmonella* in ground turkey. *Poultry Science, 92*, 1357–1365.

Pattanayaiying, R., H-Kittikun, A., & Cutter, C. N. (2014). Effect of lauric arginate, nisin Z, and a combination against several food-related bacteria. *International Journal of Food Microbiology, 188*, 135–146.

Pattanayaiying, R., H-Kittikun, A., & Cutter, C. N. (2015). Incorporation of nisin Z and lauric arginate into pullulan films to inhibit foodborne pathogens associated with fresh and ready-to-eat muscle foods. *International Journal of Food Microbiology, 207*, 77–82.

Pattanayaiying, R., Sane, A., Photjanataree, P., & Cutter, C. N. (2019). Thermoplastic starch/polybutylene adipate terephthalate film coated with gelatin containing nisin Z and lauric arginate for control of foodborne pathogens associated with chilled and frozen seafood. *International Journal of Food Microbiology, 290*, 59–67.

Porto-Fett, A., Campano, S., Smith, J., Oser, A., Shoyer, B., Call, J., & Luchansky, J. (2010). Control of *Listeria monocytogenes* on commercially-produced frankfurters prepared with and without potassium lactate and sodium diacetate and surface treated with lauric arginate using the Sprayed Lethality in Container (SLIC®) delivery method. *Meat Science, 85*, 312–318.

Rodríguez, E., Seguer, J., Rocabayera, X., & Manresa, A. (2004). Cellular effects of monohydrochloride of L-arginine, N-lauroyl ethylester (LAE) on exposure to *Salmonella typhimurium* and *Staphylococcus aureus*. *Journal of Applied Microbiology, 96*, 903–912.

Ruckman, S. A., Rocabayera, X., Borzelleca, J. F., & Sandusky, C. B. (2004). Toxicological and metabolic investigations of the safety of N-alpha-lauroylL-arginine ethyl ester monohydrochloride (LAE). *Food and Chemical Toxicology, 42*, 245–259.

Sharma, C. S., Ates, A., Joseph, P., Nannapaneni, R., & Kiess, A. (2013). Reduction of Salmonella in skinless chicken breast fillets by lauric arginate surface application. *Poultry Science, 92*, 1419–1424.

Shen, Q., Soni, K. A., & Nannapaneni, R. (2015). Stability of sublethal acid stress adaptation and induced cross protection against lauric arginate in Listeria monocytogenes. *International Journal of Food Microbiology, 203*, 49–54.

Soni, K. A., Desai, M., Oladunjoye, A., Skrobot, F., & Nannapaneni, R. (2012). Reduction of Listeria monocytogenes in queso fresco cheese by a combination of listericidal and listeriostatic GRAS antimicrobials. *International Journal of Food Microbiology, 155*, 82–88.

Stopforth, J. D., Visser, D., Zumbrink, R., Dijk, L. V., & Bontenbal, E. W. (2010). Control of Listeria monocytogenes on cooked cured ham by formulation with a lactate-diacetate blend and surface treatment with lauric arginate. *Journal of Food Protection, 73*, 552–555.

Sukumaran, A. T., Nannapaneni, R., Kiess, A., & Sharma, C. S. (2015). Reduction of Salmonella on chicken meat and chicken skin by combined or sequential application of lytic bacteriophage with chemical antimicrobials. *International Journal of Food Microbiology, 207*, 8–15.

Taormina, P. J., & Dorsa, W. J. (2009a). Short-term bactericidal efficacy of lauric arginate against *Listeria monocytogenes* present on the surface of frankfurters. *Journal of Food Protection*, *72*(6), 1216–1224.

Taormina, P. J., & Dorsa, W. J. (2009b). Inactivation of Listeria monocytogenes on hams shortly after vacuum packaging by spray application of lauric arginate. *Journal of Food Protection*, *72*, 2517–2523.

Terjung, N., Monville, C., Loeffler, M., Gibis, M., Hinrichs, J., & Weiss, J. (2014). Impact of lauric arginate application form on its antimicrobial activity on the surface of a model meat product. *Journal of Food Science*, *79*, M2056–M2065.

Terjung, N., Holzwarth, S., Loeffler, M., Gibis, M., Herrmann, K., Hinrichs, J., & Weiss, J. (2016). Antimicrobial efficacy of a spice ferment in emulsion type sausages and restructured ham. *Food Control*, *59*, 139–147.

U.S. Department of Agriculture, Food Safety and Inspection Service. (2019). Safe and suitable ingredients used in the production of meat, poultry, and egg products. Directive 7120.1, Rev. 42. Food Safety and Inspection Service, Washington, DC. https://www.fsis.usda.gov/wps/wcm/connect/bab10e09-aefa-483b-8be8-809a1f051d4c/7120.1.pdf?MOD=AJPERES

U.S. Food and Drug Administration. (2005). Agency response letter GRAS Notice No. GRN 000164. https://wayback.archiveit.org/7993/20171031021544/https://www.fda.gov/Food/IngredientsPackagingLabeling/GRAS/NoticeInventory/ucm154571.htm

Wang, L., Zhao, L., Yuan, J., & Jin, T. Z. (2015). Application of a novel antimicrobial coating on roast beef for inactivation and inhibition of *Listeria monocytogenes* during storage. *International Journal of Food Microbiology*, *211*, 66–72.

Yang, S., Sadekuzzaman, M., & Ha, S.-D. (2017). Treatment with lauric arginate ethyl ester and commercial bacteriophage, alone or in combination, inhibits *Listeria monocytogenes* in chicken breast tissue. *Food Control*, *78*, 57–63.

11

Medium-Chain Fatty Acids (>C8) and Monoesters

P. Michael Davidson, Jon J. Kabara and Douglas L. Marshall

CONTENTS

11.1 Introduction and History

Fatty acids and their corresponding esters are a large group of chemicals found in nature and are considered to have little or no toxicity to humans. Early records suggest that fatty acids have a long history of antimicrobial activity. Additionally, soaps (salts of fatty acids) have long been used as cleaning and disinfecting agents. Levey (1958) stated that ancient Mesopotamians obtained sodium carbonate (washing soda) from partially burning plants. The most important use of this compound was in the manufacture of a type of soap. In Ancient Babylon, during the reign of Nabonidus (556–539 BCE), a typical recipe for soap was given as a mixture of ash, cypress oil, and sesame oil which was used "washing the stones for the servant girls." The Gauls and Germanic tribes used soap before the Christian era, and soap came to the Romans before the year 79 CE (Common Era). From the Romans, the Greeks learned the value of soap.

In more recent history, the earliest mention of soap as an antifungal agent can be found in a paper by Clark (1899). Another early worker, Reichenback (1908), studied the action of soaps on *Escherichia coli* and found that the potassium salts of myristate, palmitate, and stearate were inhibitory against this bacterium, but oleate, erucate (ω9-C22:1), and linoleate were inactive. Much of the work carried out on the germicidal activity of soaps can be found in the literature between 1920 and 1940 (Bayliss, 1936; Kodicek, 1949; Nieman, 1954; Kabara et al., 1977). Despite early findings that fatty acids had antimicrobial activity, their use in the food industry was limited most likely because other compounds with better properties were discovered such as sorbates, benzoates, and short-chain organic acids. Despite the lack of interest in food applications, it is obvious that some fatty acids have played an important historical role in food preservation. Following is a review of the antimicrobial activity of fatty acids and their corresponding esters.

11.2 Antimicrobial Activity of Fatty Acids

Fatty acids are saturated or unsaturated carbon chains attached to a carboxyl group which makes them amphiphilic in nature. The aqueous solubility of fatty acids is inversely related to the length of the carbon chain. The common saturated and unsaturated medium-chain fatty acids greater than eight carbons are shown in Table 11.1.

11.2.1 Antibacterial Activity

In early research, Walker (1924) found that using either sodium or potassium salt to make soaps had little effect on their antimicrobial activity. The lower members of this series, butyric (C4:0), valeric (C5:0), and caproic (C6:0), had little or no germicidal effect against a *Streptococcus pneumoniae* (pneumococcus), *Streptococcus* sp., *Salmonella*, or *Staphylococcus aureus*. The most effective antimicrobial fatty acid salts against pneumococcus and *Streptococcus* were laurate (C12:0) and the unsaturated fatty acids oleate (C18:1), linoleate (C18:2), and linolenate (C18:3). *Salmonella* was inhibited only by saturated fatty acid salts and at much higher concentrations while, in this study, *S. aureus* was not inhibited by any of the fatty acid salts. In contrast, Hettche (1934) found that oleate, linoleate, and linolenate were lethal to *Staphylococcus*, but not to *E. coli*.

Bayliss (1936) did a comprehensive study on the relation between the structure of fatty acid salts and their antimicrobial effectiveness against the Gram-positive bacteria, *Streptococcus pneumoniae* and *Lactococcus lactis*. A comparison of the pneumococcidal properties of laurate, myristate, palmitate, and stearate showed that the laurate had the maximum antimicrobial activity. In some cases, the presence of unsaturated bonds in the molecule significantly altered activity against pneumococci. For example, sodium

TABLE 11.1

Most Common Food-Related Saturated and Unsaturated Fatty Acids

Acid Name	Chemical Structure	Carbons:Double Bonds
Caprylic	$CH_3(CH_2)_6COOH$	8:0
Capric	$CH_3(CH_2)_8COOH$	10:0
Lauric	$CH_3(CH_2)_{10}COOH$	12:0
Myristic	$CH_3(CH_2)_{12}COOH$	14:0
Palmitic	$CH_3(CH_2)_{14}COOH$	16:0
Stearic	$CH_3(CH_2)_{16}COOH$	18:0
Arachidic	$CH_3(CH_2)_{18}COOH$	20:0
Palmitoleic	$CH_3(CH_2)_5CH{=}CH(CH_2)_7COOH$	16:1 (cis-Δ9)
Oleic	$CH_3(CH_2)_7CH{=}CH(CH_2)_7COOH$	18:1 (cis-Δ9)
Elaidic	$CH_3(CH_2)_7CH{=}CH(CH_2)_7COOH$	18:1 (trans-Δ9)
Vaccenic	$CH_3(CH_2)_5CH{=}CH(CH_2)_9COOH$	18:1 (trans-Δ11)
Linoleic	$CH_3(CH_2)_4CH{=}CHCH_2CH{=}CH(CH_2)_7COOH$	18:2 (cis,cis-Δ9,Δ12)
Linolenic	$CH_3CH_2CH{=}CHCH_2CH{=}CHCH_2CH{=}CH(CH_2)_7COOH$	18:3 (cis,cis,cis-Δ9,Δ12,Δ15
Arachidonic	$CH_3(CH_2)_4CH{=}CHCH_2CH{=}CHCH_2CH{=}CHCH_{2\text{-}}$ $CH{=}CH(CH_2)_3COOH$	20:4 (cis,cis,cis,cis-Δ5Δ8,Δ11,Δ14)

oleate was >100 times more effective than sodium stearate (18:0) in inactivating pneumococci. The addition of more double bonds, as in linoleate or linolenate salts, caused no change in this property. Sodium α- and β-elaeostearates (C18:3), although isomeric with sodium linolenate, were relatively ineffective against pneumococcus. Thus the three poly-unsaturated fatty acid salts had pneumococcidal properties considerably weaker than those of sodium oleate. Ricinelaidic acid salt ((9E,12R)-12-hydroxyoctadec-9-enoic acid), a geometric isomer of sodium ricinoleate that has physical properties much like stearate and palmitate, was somewhat less effective than ricinoleate. Ricinstearolate (12-hydroxy-9-octadecynoic acid salt), containing a triple bond and with physical properties very similar to those of ricinoleate, is even less active against bacteria. Sodium undecylenate (C11:1) required a relatively high concentration to destroy pneumococcus. Lactococci are generally relatively resistant to chemical disinfection. However, the saturated fatty acid salts, myristate and laurate, exhibited lethality to *Lactococcus lactis* (Bayliss, 1936). Sodium oleate was capable of killing *L. lactis* at concentrations equal to that of sodium laurate and myristate. The addition of a second double bond, sodium linoleate, enhanced this lethality considerably. Further addition of a third double bond, however, conferred no greater ability to kill lactococci. Kodicek (1949) tested a number of fatty acids on *Lactobacillus casei* and found that in the saturated series lauric acid depressed growth which was affected by cholesterol. Inhibition by other fatty acids, including the more active and branched-chain α-methyl lauric acid, could be reversed by the addition of cholesterol. Similarly to Bayliss (1936), Kodicek (1949) found that of the unsaturated series, activity was limited to *cis* isomers rather than *trans* isomers. Nieman (1954) reviewed the antimicrobial activity of unsaturated fatty acids against Gram-positive bacteria. In contrast to Bayliss, he concluded that the inhibitory effects of unsaturated fatty acids increased as the number of double bonds in the molecule increased. Thus, the inhibitory effect of linoleic acid was far greater than that of oleic acid (Fuller and Moore, 1967). However, the minimum inhibitory concentration of arachidonic acid (C20:4) was about the same as that of linoleic acid, whereas the minimum inhibitory concentration of linolenic acid was somewhat greater than that of linoleic acid. Several meat spoilage bacteria were tested for their sensitivity to saturated and unsaturated fatty acids (C12 to C18) (Ouattara et al., 1997). *Brochothrix thermosphacta*, *Pseudomonas fluorescens*, and *Serratia liquefaciens* were not inhibited by any of the fatty acids but *Carnobacterium piscicola*, *Lactobacillus curvatus*, and *Lactobacillus sake* were inhibited to the greatest extent by lauric and palmitoleic acids and less so by linoleic and linolenic acids. *Carnobacterium piscicola* was the most susceptible to inhibition among the bacteria tested. Thus, based on these studies, there are significant differences in the susceptibility of the Gram-positive bacteria, most likely due to the species tested as well as the conditions and methods utilized.

Gram-negative bacteria have been generally thought to have greater resistance to fatty acids because of the outer cell membrane (Branen et al., 1980). Altieri et al. (2009) evaluated the antimicrobial activity of lauric, myristic, and palmitic acid against *E. coli* O157:H7, *Salmonella* spp., and *Y. enterocolitica*. Lauric acid was shown to initially inhibit (as measured by "inhibition index") *E. coli* O157:H7 at 20 and 30 mg/l but no inhibition was detected after that time. Surprisingly, higher concentrations of 40 and 50 mg/l showed no inhibitory effect. Measuring growth by turbidity increase over time, 20 or 50 mg/l lauric acid had little or no effect on the growth rate of *E. coli* O157:H7. Similarly, myristic had a high inhibitory index at 20 mg/l but only for the initial sampling time of 10 h, and the same concentration had no effect on growth rate. Twenty mg/l palmitic acid demonstrated a high inhibition index for the up to 48 h, but higher concentrations had no effect. *E. coli* O157:H7 lag phase and growth rate were slower with 20 mg/l palmitic acid. *Salmonella* demonstrated a significantly high inhibition index in the presence of 50 mg/l of lauric acid for up to 48 h. *Yersinia enterocolitica* was more susceptible to lauric acid and at 30–50 mg/l demonstrated significantly a high inhibition index over a 100-h incubation period. However, a growth study was only carried out at 20 mg/l for *Y. enterocolitica* which demonstrated a very slight increase in lag phase only. Myristic and palmitic acids did not significantly inhibit either *Salmonella* or *Y. enterocolitica*. Skřivanová et al. (2006) determined the antimicrobial activity of fatty acids (C2:0–C18:0, C18:1, C18:2), citric, succinic, fumaric, malic, and lactic acid against multiple strains of *E. coli*, *Salmonella*, and *Clostridium perfringens*. Antimicrobial activity was determined by minimum inhibitory concentration (MIC). Among the fatty acids, caprylic (C8:0) was the only acid that inhibited all cultures with MICs of 1000–3000 mg/l. Strains of *E. coli* also were inhibited by capric (C10:0) acid at 5000 mg/l. Strains of *C. perfringens* were inhibited by medium-chain fatty acids (C8:0

to C14:0), oleic acid (C18:1), and one strain by linoleic acid (C18:2). The lowest MICs for *C. perfringens* were those of lauric (C12:0) and myristic (C14:0) acid at 100–200 mg/l. All strains were resistant to succinic, fumaric, malic, and lactic acid alone at 5000 mg/l. Activity of caprylic acid was further demonstrated against other Gram-negative bacteria. The MIC of caprylic acid was 100 mg/l in microbiological media for *Vibrio parahaemolyticus* (Beuchat, 1980). Caprylic acid at 0.3% inhibited *E. coli* when grown in a chemically defined medium and was more effective than 1.5% acetic acid or 1.8% propionic acid. Caprylic acid was also more inhibitory to *Shigella* spp. than either acetic or propionic acids (Nakamura and Zangar, 1968). *E. coli* O157:H7 cells exposed to 1.0 mM caprylic acid and acetic or lactic acid produced significant reductions in numbers of the pathogen in medium compared with controls, producing synergistic inhibition of the pathogen (Kim and Rhee, 2013). Woolford (1975) found that, at pH 6.0, caprylic acid inhibited bacteria and fungi at a lower concentration than either acetic or propionic acid but, at pH 5.0, the inhibitory concentrations were not dramatically lower. Conversely, the MIC for caprylic acid exhibited a drop from 3250 mg/l to 50 mg/l µg/ml against *L. monocytogenes* when the medium was acidified to pH 5.5 from pH 7.4 (Kozak et al., 2017). The MIC of caprylic acid was lower against *Listeria* spp. than Gram-negative bacteria and extended the lag phase of *Listeria monocytogenes* growth (Nobmann et al., 2009).

Interest in using caprylic acid as an antimicrobial for treating meat products has increased both because of its effectiveness and status as "natural." Burnett et al. (2006) evaluated the antimicrobial activity of caprylic (octanoic) acid-based treatment against *L. monocytogenes* in ready-to-eat (RTE) meat and poultry products. RTE products were inoculated with *L. monocytogenes* and an acidified 1% octanoic acid treatment was applied to the products prior to packaging and vacuum sealing. The octanoic acid treatment of RTE products reduced *L. monocytogenes* numbers up to 1.46–3.34 log CFU per sample depending on product and holding conditions. Similarly, Lavieri et al. (2014) studied post-lethality interventions for control of *L. monocytogenes* growth and recovery on an RTE ham product. Among the antimicrobials evaluated was octanoic acid and post-packaging thermal treatment. Octanoic acid reduced *L. monocytogenes* population compared to the control after 1 day of storage by 1.73 log CFU/g. Octanoic acid however had no effect on recovery and growth of *L. monocytogenes* during storage. Using less caprylic acid (0.5%) than previous studies along with 0.2% oregano essential oil applied to beef, Hulankova et al. (2013) found no reduction in *L. monocytogenes*, but a 2.0 log decline in psychrotrophic bacteria over 10 days of storage at 3°C did occur. The United States Department of Agriculture Food Safety and Inspection Service (USDA FSIS, 2019) regulations allow the use of octanoic acid as a secondary direct food additive with the following limitations, (1) an aqueous solution of sodium octanoate or octanoic acid and either glycerin and/or propylene glycol and/or a polysorbate surface active agent (quantity sufficient to achieve the intended technical effect of octanoic acid emulsification) adjusted to a final solution pH of 1.5 to 4.0 using sodium hydroxide, potassium hydroxide, or an acceptable GRAS acid for various non-standardized RTE meat and poultry products and standardized meat and poultry products that permit the use of any safe and suitable antimicrobial agent. Applied to the surface of the product at a rate not to exceed 400 µg/ml octanoic acid by weight of the finished food product, or (2) an aqueous solution of sodium octanoate, potassium octanoate, or octanoic acid and either glycerin and/or propylene glycol and/or a Polysorbate surface active agent (quantity sufficient to achieve the intended technical effect of octanoic acid emulsification) adjusted to a final solution pH of 1.5 to 6.0 using sodium hydroxide, potassium hydroxide, or an acceptable GRAS acid for fresh meat primals and subprimals and cuts. Applied to the surface of the product at a rate not to exceed 400 µg/ml octanoic acid by weight of the final product. Octanoic acid is also allowed to be used with peroxyacetic acid (21 CFR 173.370) on meat and poultry carcasses, parts, trim, and organs.

Lourenco et al. (2017) studied antimicrobial treatments that could be used for queso fresco, a Hispanic non-fermented cheese highly susceptible to contamination with *L. monocytogenes*. Antimicrobials included caprylic acid and nisin. Caprylic acid (2.9 g/kg) and nisin (0.5 g/kg) alone caused >5 and >3 log CFU/g reductions compared to the control, respectively. Mixtures of nisin and caprylic acid consistently reduced viable *L. monocytogenes* by 6 log CFU/g. Samples treated with nisin and caprylic acid (0.5 and 0.7 g/kg, respectively) were evaluated in a consumer sensory panel. Panelists slightly preferred the control over the treated samples, but all samples were on average rated between "slightly liking" and "moderately liking."

Kim and Rhee (2015) reported synergistic inhibition of *E. coli* O157:H7 cells in carrot juice when 720 mg/l caprylic acid was combined with 5.0 mM citric acid. Research investigating the antimicrobial activity of caprylic acid on alfalfa seeds indicated *E. coli* O157:H7 cells were reduced to non-detection 12 h post-treatment with 75 mM antimicrobial; similar reductions were observed for *Salmonella* cells, though seed germination rates declined by 12% (Chang et al., 2010). The antimicrobial effects of caprylic acid, citric acid, and vanillin on the inactivation of *Cronobacter sakazakii* and *Salmonella enterica* serovar Typhimurium were examined in reconstituted infant formula (Choi et al., 2013). The samples were treated with the compounds alone or a combination of caprylic and citric acid or caprylic acid and vanillin. The combined treatments had synergistic antimicrobial activity compared with the sum of the effects of each individual treatment. For example, there was little reduction in viable *C. sakazakii* with 20 mM caprylic or 30 mM citric acid for 5 min at 40°C, whereas the population was reduced to undetectable levels (> 7.3 log CFU/ml) after exposure to a combination at the same concentrations and time.

As with other antimicrobials, the activity of fatty acid salts is influenced by environmental and processing factors. Churchward et al. (2018) stated that, in addition to the bacterial species investigated, the exposure time, temperature, pH, and constituents of the culture medium may all influence apparent susceptibility. Cowles (1941) showed that at low pH, fatty acids have bactericidal action and that their activity increases with chain length. As fatty acids are organic acids and are more undissociated and thus more effective at lower pH, it is not surprising that decreasing pH would increase activity. Other physical parameters may also influence apparent susceptibility. For example, after exposure to sublethal heat stress, cells of *E. coli* became more sensitive to medium-chain fatty acids (Tsuchido and Takano, 1988).

The antimicrobial activity of unsaturated fatty acids on bacteria *in vitro* helped account for the known biocidal activity of some biological materials. For example, Humfeld (1947) found an antimicrobial fatty acid in wheat bran, and McKee et al. (1947) found antibacterial lipids produced by the ciliated protozoa, *Tetrahymena geleii*. Subsequent findings have documented numerous other antimicrobial lipids in natural products. For example, the medium-chain fatty acids (lauric acid and myristic acid) produced by blue vein cheese molds have been shown to inhibit growth of *L. monocytogenes* (Kinderlerer et al., 1996). Another antimicrobial fatty acid that was active against Gram-positive bacteria was extracted from moss (Borel et al., 1993). The compound dicranin (Z,Z,Z-octadeca-6-yne-9,12,15 trienoic acid) was active against several *Bacillus* spp., *S. aureus*, and *Enterococcus faecalis*.

11.2.2 Antifungal Activity

As systemic antifungal agents, the fatty acids have generally been shown to have a low order of effectiveness. This may be because they are readily metabolized by the fungi through the fatty acid pathways. In general, they may be esterified to form glycerides and/or may be degraded to small fragments by β-oxidation. Despite the limitations, varying levels of antifungal activity of fatty acids have been reported by several researchers (Kiesel, 1913; Kitajima and Kawamura, 1931; Tetsuomoto, 1933a, b; Wyss et al., 1945; Thornton, 1963; Kodicek and Worden, 1945). Antifungal activity of fatty acids has been found to be dependent on chain length, concentration, and pH of the medium. As with bacteria, fatty acids increase in antifungal activity with decreasing pH provided that the low pH does not make the compound insoluble. The change in activity with hydrogen ion concentration was shown to be much greater for the short-chain acids. Several materials also can act as protective agents against the antimicrobial action of the fatty acids. These antagonistic substances include serum albumin, starch, cholesterol, lecithin, saponin, and charcoal (Nieman, 1954).

Kiesel (1913) at the Pasteur Institute studied the characteristics of the antimycotic action of fatty acids and determined that: (1) activity of saturated fatty acids increased as the number of carbon atoms in the fatty acid chain increased up to 11 carbon atoms, (2) branched-chain fatty acids were less active than those with straight chains and an equal number of carbon atoms, and (3) substitution of hydrogen by hydroxyl decreased activity. Japanese workers (Kitajima and Kawamura, 1931; Tetsuomoto, 1933a, b) confirmed Keisel's work but found that, with a saprophytic wood-rotting fungi, the optimum activity was with the 12-carbon acid. Unsaturated acids were found to be more active than the corresponding saturated acids, and dicarboxylic acids were almost without activity. Other efforts provided further confirmatory evidence using different conditions and different microorganisms (Kirby et al., 1937; Peck and Rosenfeld, 1938; Peck et al., 1939; Baechler, 1939; Hoffman et al., 1939, 1940; Rigler and Greathouse, 1940). Keeney (1943, 1944) reported fungicidal activity for the long-chain acids but much less or no lethality by short-chain fatty acids. Data showed that the optimum chain length was species-dependent and was affected by the solubility of the fatty acids in question. Thus, for *Aspergillus niger* the optimum chain length was 11 carbon atoms, but for *Trichoderma interdigitale*, the 13-carbon acids were most active. *Trichoderma purpureum* was more sensitive and was inhibited by 14-carbon acids, thus suggesting that longer chain length compounds failed to show antimycotic effects because of their low solubility.

11.2.3 Effect of Chemical Modifications on Antimicrobial Activity

Because fatty acids are metabolized by fungi, 2-fluoro fatty acids, which do not undergo β-oxidation, were compared to their non-fluorinated analogs for antifungal activity (Gershon and Parmegiani, 1967). 2-fluoro fatty acids up to a chain length of 20 carbon atoms were tested against *Aspergillus niger, Trichoderma viride, Myrothecium verrucaria*, and *Trichophyton mentagrophytes* (Gershon and Parmegiani, 1967). Both fluorinated and non-fluorinated compounds were about equally active, except that the non-fluorinated fatty acids showed maximal activity at chain lengths of 4 to 10 carbons, whereas the 2-fluoro fatty acids were most active at chain lengths of 8 to 14 carbons. Since the pK_a values of the 2-fluoro fatty acids are generally two units lower than those of the non-fluorinated fatty acids, it appeared that the pK_a did not play an important role in antifungal activity.

Eggerth (1929a) found that α-bromo fatty acids were usually more effective antimicrobials than unsubstituted fatty acids against *Streptococcus, Staphylococcus aureus, Corynebacterium diptheriae, Micrococcus*, and *Vibrio cholerae*. All of the Gram-negative bacteria tested were most susceptible to salts of fatty acids with 12 or 14 carbon atoms while Gram-positives were more susceptible to 16 or 18 carbon atoms.

Similarly, a hydroxyl group in the α position of fatty acid salts was shown to increase the antimicrobial action of saturated fatty acid soaps of greater than 16 carbons (Eggerth, 1929b). In contrast, the hydroxylated salts, gluconate and trihydroxystearate, seem to lack any ability to kill pneumococci under the test conditions employed (Bayliss, 1936). Stearate is unique in that addition of a 4-hydroxyl group restores the lethality against bacteria at fairly low concentrations. Hydroxylation, in the case of the ricinoleate, enhanced the bactericidal activity against streptococci, whereas there was a decrease in activity against pneumococcus.

The replacement of the ethylenic linkage in fatty acids by an acetylenic linkage caused a marked decrease in germicidal activity (Bayliss, 1936). Sodium abietate ($C_{20}H_{29}NaO_2$), an organic compound found in tree resin, had destructive action on pneumococcus at 1/10 the concentration necessary for phenol. The same was true for chaulmoograte, esters of chaulmoogra oil, which was once used as a chemotherapeutic agent in the treatment of leprosy (Hansen's disease) (Bayliss, 1936). The introduction of a 6-phenyl group on an 11-carbon acid decreased antimicrobial activity markedly, regardless of the degree of unsaturation of the fatty acid chain. Substitutions in the phenyl ring with an amino or hydroxyl group on the fatty acid chain resulted in activity comparable to that of the corresponding straight-chain acids. No changes made to the arrangement of the carbon atoms brought about greater activity than that displayed by the straight-chain acids (Bayliss, 1936).

With an equivalent number of carbon atoms, branching of the carbon chain decreases activity. The effect of branching is less, however, if only those carbon atoms in the longest chain are considered. Fatty acids substituted with aldehyde, acetate, ethyl ester, amide, or substituted amide have considerable activity but are less active than the corresponding acid. Alcohols have high activity, but owing to their limited solubility, the effectiveness of long-chain alcohols can be demonstrated only on more sensitive microorganisms (Kabara et al., 1972). Covalent bonding with palmitic, myristic, or stearic acids to the antibacterial enzyme lysozyme was found to increase the spectrum of activity of the hydrolytic protein to include both Gram-positive (native protein) and Gram-negative bacteria (Ibrahim et al., 1991; 1993).

11.2.4 Structure–Function Relationship

There is consensus among many microbiologists that the antimicrobial effects of fatty acids is due to the undissociated molecule (protonated acid), not the anion (Winslow and Lochridge, 1906; Reid, 1932; Hoffman et al., 1939; Levine and Fellers, 1940; Albert, 1960). Thus, their activity should be markedly affected by pH, since this determines the degree of dissociation of the acid. Indeed, this fact has been demonstrated in that more rapid killing occurs at lower pH (Meynell, 1955; Prince, 1959). An example of this is shown for *C. perfringens*, where the minimum inhibitory concentrations (MIC) of selected fatty acids are shown at pH 6.5 and 7.5 (Table 11.2).

An increase from pH 6.5 to 7.5 increased the MIC of the caprylic and capric but decreased the MIC of the lauric and myristic. The MIC of the unsaturated fatty acids were unaffected by a change in pH in the medium. Of the fatty acids tested, those with the lowest MIC were the polyunsaturated acids, such as linoleic, linolenic, and arachidonic acids. However, the MIC of linoleic acid, with two double bonds, was somewhat less than that of linolenic acid, with three double bonds.

In an attempt to bring some order to the vast amount of literature on structure–function activity, a series of studies was carried out with very pure (99%) fatty acids and derivatives. The results in Table 11.3 and elsewhere (Kabara et al., 1972, 1973; Conley and Kabara, 1973; Kabara, 1982) are the basis for a number of generalizations that help to explain most of the data in the literature:

1. Fatty acids have less effect on Gram-negative bacteria than Gram-positive bacteria.
2. The most active chain length for saturated fatty acids generally is C12; the most active monounsaturated fatty acid is C16:1, and C18:2 is the most active polyunsaturated fatty acid.
3. The position and number of double bonds are more important to long-chain (>C12) than shorter-chain (<C12) fatty acids.
4. The *cis* form is active; the *trans* isomer is inactive.
5. Acetylenic derivatives, compared to ethylenic fatty acids, are more active against fungi.
6. Yeasts are affected by fatty acids with chain lengths of C10 to C12.
7. Fatty acids esterified to monohydric alcohol are inactive; esterification to polyols increases activity.

TABLE 11.2

Minimum Inhibitory Concentrations of Fatty Acids against *Clostridium perfringens (welchii)*[a]

	Minimum Inhibitory Concentration (mg/100 ml)	
Fatty Acid	**pH 6.5**	**pH 7.5**
Caprylic	721	3610
Capric	172	862
Lauric	1000	200
Myristic	2280	457
Palmitic	2560	2560
Stearic	2850	2850
Oleic	283	283
Linoleic	5.61	5.61
Linolenic	27.90	27.90
Arachidonic	6.09	6.09
Erucic (C22:1)	339	339
Methyl linoleate	146	733
Methyl linolenate	293	1460

[a] Fatty acids and methyl esters in 0.1 M phosphate buffer (pH 6.5 or 7.5) containing 0.1% (wt/vol) Tween 80.

Source: Tetsumoto (1933b).

TABLE 11.3

Minimum Inhibitory concentrations[a] of Saturated and Unsaturated Fatty Acids[a]

	Microorganism[b]				
Fatty Acid	**Pn**	**SA**	**Snon-A**	**Ca**	**Sta**
Capric	1.45	1.45	2.9	2.9	2.9
Lauric	0.062	0.124	0.249	2.49	2.49
Myristic	0.218	0.547	2.18	4.37	4.37
Myristoleic	0.110	0.110	0.110	0.552	0.441
Palmitic	0.48	3.9	3.9	NI	NI
Palmitoleic	0.024	0.098	0.049	0.491	0.983
Stearic	NI	NI	NI	NI	NI
Oleic	NI	1.77	NI	NI	NI
Elaidic	NI	NI	NI	NI	NI
Linoleic	0.044	0.089	0.089	0.455	NI
Linolenic	0.179	0.35	0.35	NI	1.79
Linolelaidic	NI	NI	NI	NI	NI
Arachidonic	NI	NI	NI	NI	NI

[a] Results are given in mM.
[b] Pn – Pneumococci, SA – *Streptococcus* group A, Snon-A – *Streptococcus* β-hemolytic non-A, Ca – *Candida albicans*, Sta – *Staphylococcus aureus.*
[c] NI, not inhibitory at the concentrations tested (1.0 mg/ml or 3–6.0 mM).

11.3 Fatty Acid Monoesters

11.3.1 Sucrose Monoesters

Kato and Arima (1971) first reported that the sucrose ester of lauric acid was active against the Gram-negative bacterium, *E. coli*, while Conley and Kabara (1973) and Shibasaki and Kato (1979) indicated that these and other esters were primarily active only against Gram-positive bacteria and fungi. The latter has been confirmed by other researchers (Nobmann et al., 2009; Skřivanová et al., 2014). In contrast to data generated for glycerides, the diester of sucrose, rather than the monoester, was more active against several Gram-positive bacteria. Sucrose dicaprylate possessed the highest activity of the sucrose esters but was still less active than monolaurin (Kato and Shibasaki, 1975a, b; Beuchat, 1980). The monoesters of glycerol and the diesters of sucrose not only have higher antimicrobial activity than their corresponding free fatty acids but also compare favorably in activity with commonly used antimicrobials, such as parabens, sorbic acid, and dehydroacetic acid (Tables 11.4 and 11.5).

TABLE 11.4

Comparison of Antifungal Activities of Fatty Acid Esters with Some Commonly Used Preservatives

	Minimum Inhibitory Concentration (μg/ml)[a]		
Food Additive	***Aspergillus niger***	***Candida utilis***	***Saccharomyces cerevisiae***
Monocaprin	123	123	123
Monolaurin	137	69	137
Butyl-p-hydroxybenzoate	200	200	200
Sodium lauryl sulfate	100	400	100
Sorbic acid	1000	1000	1000
Dehydroacetic acid	100	200	200

[a] By agar dilution method.
Source: Kato and Shibasaki (1975b).

TABLE 11.5

Comparison of Antibacterial Activities of Fatty Acid Esters with Some Commonly Used Preservatives

	Minimum Inhibitory Concentration (μg/ml)[a]		
Food Additive	***Bacillus subtilis***	***Bacillus cereus***	***Staphylococcus aureus***
Sucrose dicaprylin	74	74	148
Monocaprin	123	123	123
Monolaurin	17	17	17
Butyl-p-hydroxybenzoate	400	200	200
Sodium lauryl sulfate	100	100	50
Sorbic acid	4000	4000	4000

[a] By agar dilution method.
Source: Kato and Shibasaki (1975b).

The antimicrobial activity of sucrose esters of medium- to long-chain fatty acids was studied against two microorganisms involved in the spoilage of salad dressings, *Zygosaccharomyces bailii* and *Lactobacillus fructivorans* (Yang et al., 2003). Sucrose monoesters were most inhibitory when the esterified fatty acid was myristic or palmitic acid. The least inhibition was generally observed with sucrose oleate. In salad dressing, 1% sucrose monoesters of lauric, myristic, or palmitic acid significantly inhibited the growth of *Z. bailii* and *L. fructivorans*, and were comparable with or more effective than 0.1% sodium benzoate. Sucrose monostearate was not inhibitory to either *Z. bailii* or *L. fructivorans* in salad dressing. Skřivanová et al. (2014) tested the sucrose monoesters of capric and lauric acid at 100 to 5000 mg/l against *E. coli* and *C. perfringens*. Sucrose monocaprate decreased viable *E. coli* at 100 mg/l by 1–2 logs after 18 h up to 4–5 logs at 5000 mg/l. Sucrose monolaurate had little inhibitory effect on *E.* coli under those same conditions (i.e., less than 1 log reduction at 5000 mg/l). The sucrose monoesters were more effective against *C. perfringens*. Sucrose monocaprate reduced viable *C. perfringens* by 1.5–2.5 logs at 100 mg/l and to non-detectable levels (>7 log reduction) at 2000–3000 mg/l. Sucrose monolaurate was even more effective, reducing the viable cell population by >7 logs at the lowest concentrations tested of 100 mg/l.

Marshall and Bullerman (1986) evaluated the antimicrobial effect of sucrose esterified with a mixture of palmitic and stearic acids in combination with cinnamon, potassium sorbate, or calcium propionate on growth of and aflatoxin production by *Aspergillus parasiticus*. Only a combination of 0.3% propionate or sorbate and 1% sucrose ester resulted in a synergistic effect on inhibition of growth, while aflatoxin production remained relatively unaffected. Sucrose palmitate combined with the bacteriocin nisin showed synergistic activity against Gram-positive bacteria but not Gram-negative bacteria (Thomas et al., 1998). Low levels of the combined compounds inhibited *L. monocytogenes*, *B. cereus*, *Lactobacillus plantarum*, and *S. aureus*. Sucrose laurate (1%) combined with high pressure (392 mPa) for 15 minutes at 45°C provided up to a 5 log reduction of *Bacillus subtilis* in milk, *B. cereus* in beef, *Bacillus coagulans* and *Alicyclobacillus* sp. in tomato juice, and *Alicyclobacillus* sp. in apple juice (Shearer et al., 2000). Enhanced inhibition of *E. coli* O157:H7 was achieved by 500 μg/ml sucrose monolaurate combined with 300 μg/ml EDTA compared to either compound used alone (Hathcox and Beuchat, 1996). Sucrose monolaurate enhanced thermal inactivation of *E. coli* O157:H7 when used alone and was even greater when combined with EDTA. Beef slices that were contaminated with the pathogen and surface dipped in 1 to 3% sucrose laurate had significantly reduced populations compared to water-dipped controls. Guava fruits treated with atomized 1% sucrose esters had reduced decay after 1 to 3 weeks of storage compared to controls (Combrink et al., 1990). Others have shown similar effects. When 200 μg/ml sucrose esters were combined with 200 μg/ml tiabendazol ($C_{10}H_7N_3S$; a compound used to control mold, blight, and other fungal diseases in fruits), *Penicillium* citrus rot mold was completely controlled on lemons (Bellone et al., 1993). A review of the antimicrobial properties of sucrose esters is available (Marshall and Bullerman, 1994). Sucrose monolaurate and sucrose monocaprate were tested alone and in combination with EDTA, propionic acid, or citric acid for *Yersinia enterocolitica* on eggshells (Favier et al., 2009). *Y. enterocolitica* was decreased by 3.2 and 3.0 logs with 1000 mg/l sucrose monolaurate plus 0.1% citric acid or 1000 mg/l sucrose monolaurate plus 600 mg/l EDTA. A review of the antimicrobial properties of sucrose esters is available (Marshall and Bullerman, 1994).

Shao et al. (2018) studied the antibacterial activity and mechanism of action of sucrose monolaurate. Sucrose monolaurate had varying antibacterial activity depending upon the test microorganism. Minimum inhibitory and bactericidal concentrations were 2.5 and 20 mM for *L. monocytogenes*, 2.5 and 20 mM for *B. subtilis*, and 10 and 40 mM for *S. aureus*, respectively. No significant inhibition was observed for *E. coli* at 80 mM. The researchers reported that the mechanism of bacterial inactivation against Gram-positive bacteria was primarily through damage to the cell membrane integrity.

11.3.2 Polyglycerol Esters

Another series of food-grade esters was derived from a group of compounds called polyglycerol esters. These linear polymers of glycerol were discovered and advocated for food use by Babayan et al. (1964). As with sucrose esters, the medium-chain fatty acids appeared to be the most active (Conley and Kabara, 1973). It was of interest that, regardless of whether the polyol was tri-, hexa-, or decapolyglycerol, the fatty acid moiety seemed to determine overall biocidal activity. Generally, as the polyol became bulkier, the spectrum of biocidal activity became narrower. Other polyol esters have been tested, and except for special applications, none have proven to be more useful than monolaurin (Table 11.6).

11.3.3 Glycerol Monoesters

While a great number of fatty acids could be used to produce glycerol monoesters, it has been shown that those made with lauric acid are often the most effective antimicrobials even in the presence of food (Beuchat, 1980; Shibasaki, 1982; Wang and Johnson, 1992; 1997). Shibasaki (1982) stated that "lauric acid (esters) appear to have the greatest overall antimicrobial activity … [and] appear to have the highest potential for use in foods and cosmetics." However, as is true of other lipophilic preservatives, antimicrobial activity is influenced by test microorganism, the presence of starchy and proteinaceous material (Kato and Shibasaki, 1975b; Bal'a and Marshall, 1996b), proper homogeneous mixing (Cotton and Marshall, 1997), proper pH range (Bal'a and Marshall, 1996a), and effective acidulants such as acetic, citric, or lactic acids (Oh and Marshall, 1994). The purity of the monoglyceride is apparently also very important. Many monoglyceride preparations have a number of impurities, particularly notable quantities of di- and tri-esters. It was found that monoglycerides less than 90% monoester were not effective antimicrobials, which is why Kabara (1978) coined the name Lauricidin® to emphasize its high monoester content.

Prior to research on monolaurin, Kabara et al. (1972) showed that esterification of fatty acids with a monohydric alcohol (e.g., methanol and ethanol) resulted in esters that were inactive. However, when an α-hydroxy fatty acid was esterified this was not true (Kabara et al., 1972). From these initial results it was concluded that some hydrophilic group was necessary for biological activity and that the esterification of a polyhydric alcohol would yield active derivatives (Conley and Kabara, 1973). Thus, it was shown that there existed a need for a free single or multiple hydroxyl groups before biological activity of the ester

TABLE 11.6

Minimum Inhibitory Concentrations of Various Polyglycerol Esters (mM) against *Corynebacterium* sp.

Fatty Acid Moiety	Triglycerol	Hexaglycerol	Decaglycerol
Acetate	>3.55[a]	>1.98[a]	>1.25[a]
Butyrate	>3.55[a]	>1.98[a]	>1.21[a]
Caproate	2.96	>1.79[a]	>1.17[a]
Caprylate	1.37	0.85	1.13
Caprate	0.32	0.39	1.10
Laurate	0.24	0.38	0.53

[a] No inhibition, maximum concentration tested.

Source: Conley and Kabara (1973).

could be achieved. One of the more common polyhydric alcohols, glycerol, was esterified and found to be more active than the corresponding fatty acids (Table 11.7). Details of these findings have been published (Conley and Kabara, 1973; Kabara et al., 1977; Kabara, 1982), and confirmed by others (Kato and Shibasaki, 1975b; Shibasaki and Kato, 1979; Sands et al., 1979; Beuchat, 1980).

As a general statement, the fatty acid used to esterify the polyol determines the antimicrobial potency of the ester. The structure–function relation for saturated and unsaturated esters follows the activity of their respective fatty acids, as reviewed earlier. This means that lauric acid (C12) and palmitoleic acid (C16:1) form the most active saturated and unsaturated derivatives, respectively (Conley and Kabara, 1973; Sands et al., 1979, Wang and Johnson, 1992); however caprylic (C8) also has been shown to have antimicrobial activity.

11.3.3.1 Antibacterial Activity

Ababouch et al. (1994) conducted experiments on the ability of fatty acids and monolaurin to inhibit spores and vegetative cells of *Bacillus cereus*. They demonstrated that linolenic acid and monolaurin were more effective against cell growth from spores than lauric, linoleic, oleic, and stearic acids. Vegetative cell growth was inhibited similarly by monolaurin, linolenic, and oleic acids. Monolaurin (20–30 mg/l), linoleic acid (50–90 mg/l), and linolenic acid (10–15 mg/l) inhibited both spore germination and vegetative cell outgrowth. They also found that starch neutralized the inhibitory activity of the fatty acids with the exception of lauric acid. Spores of other foodborne spoilage (*Clostridium sporogenes*) and pathogenic (*B. cereus*, *Clostridium botulinum*) bacteria were also generally more sensitive to monoglycerides than their respective vegetative cells (Chaibi et al., 1996a). The general order of activity from high to low was monolinolenin, monolaurin, monomyristin, and monolinolein. Monopalmitin, monostearin, and monoolein were much less effective against the spore-formers. The concentrations of monoglycerides needed to inhibit spore germination and outgrowth were higher than the concentrations needed to inhibit cell multiplication. The main target of inhibition appeared to be spore germination and cell growth rather than spore outgrowth (Chaibi et al., 1996b). Monolaurin and monolinolenin were shown to reduce *B. cereus* heat resistance (Chaibi et al., 1998), which may be useful when lower heating temperatures are desired for commercial sterilization of foods. It is important to note that the manner in which monolaurin is incorporated into test systems will affect its activity (Cotton and Marshall, 1997).

Either monolaurin or lauric acid at 0.5% was able to eliminate *B. subtilis* from a water-in-oil emulsion stored at 23°C (Plockova et al., 1999b). The same treatments also reduced the growth of *E. coli* in the emulsion. The Gram-positive bacterium *Helicobacter pylori*, involved in causing peptic ulcers, was sensitive to several medium-chain monoglycerides (Petschow et al., 1996). The greatest activity was seen with C10 to C14 monoglycerides, with 1 mM levels resulting in a 4 log reduction after 1-hour exposure. Lauric acid was the only free fatty acid demonstrating activity against this pathogen. *S. aureus* production of β lactamase, toxic shock toxin 1, and other exoproteins can be inhibited by monolaurin (Schlievert et al., 1992; Projan et al., 1994). These inhibitory effects by monolaurin were presumed to be due to interference with membrane-associated signal transduction resulting in transcriptional blockage of promoters (Ruzin and Novick, 2000).

Early work by Kabara (1979a) indicated that the monoesters alone were generally inactive against Gram-negative bacteria. The reason given for this was that Gram-negative bacteria were afforded protection by the outer cell membrane (Branen et al., 1980). This was supported by Skřivanová et al. (2006) who found that monolaurin had no inhibitory effect against strains of *E. coli* or *Salmonella* at 5000 mg/l. However, Kato and Shibasaki (1975a, 1976) reported that Gram-negative bacteria could be made susceptible in the presence of chelating acids (citric and polyphosphoric acid). They suggested that chelating acids released lipopolysaccharide from the outer membrane, increasing the susceptibility of the cells. This finding was supported by Branen and Davidson (2004) who found that the chelator EDTA synergistically enhanced the activity of monolaurin against two strains of enterohemorrhagic *E. coli* strains. However *Salmonella* Enteritidis and *P. fluorescens* strains were resistant to all antimicrobial combinations.

The presence of chelators is not always a prerequisite for inhibitory action of monoglycerides as Beuchat (1980) showed with *V. parahaemolyticus*. He compared the effects of glycerol and sucrose fatty

TABLE 11.7

Minimum Inhibitory Concentrations (µg/ml) for Fatty Acids and Their Corresponding Monoglycerides

Microorganism[a]	Undecanoic Acid	10-Undecenoic Acid	10-Undecenoyl Monoglyceride	10-Undecynoyl Monoglyceride	11-Dodecenoic Acid	Dodecanoic Acid	Dodecanoyl Monoglyceride	12-Tridecenoic Acid	Tridecanoic Acid	Tridecanoyl Monoglyceride
Enterococcus faecalis	NI[b]	NI	500	500	NI	500	NI	1000	NI	NI
Streptococcus pyogenes	125	1000	125	125	250	62	8	125	1000	62
Staphylococcus aureus	1000	1000	500	500	NI[b]	500	250	1000	NI	NI
*Corynebacterium*sp.	31	31	62	62	125	31	16	31	NI	NI
Nocardia asteroides	62	125	125	62	62	62	16	125	1000	125
Candida albicans	1000	1000	250	100	1000	1000	500	1000	NI	NI
Saccharomyces cerevisiae	500	500	250	100	500	1000	250	500	1000	NI

[a] *Escherichia coli* and *Pseudomonas aeruginosa* were not affected.
[b] NI, minimum inhibitory concentration greater than 1000 µg/ml.
Source: Kabara et al. (1977).

acid esters, benzoate, and sorbic acid and its potassium salt against *Vibrio parahaemolyticus*. Results indicated that the C12 monoglyceride (monolaurin) was more active than lower (C8 or C10) or higher (C14) chain length derivatives. Also, the low minimum inhibitory concentration of monolaurin (5 mg/l) showed it to be more effective on a concentration basis than sodium benzoate (300 mg/l) or sorbic acid (70 mg/l). Sucrose caprate and sucrose caprylate were also found to be effective against *V. parahaemolyticus* but at higher concentrations than monolaurin. Altieri et al. (2009) evaluated the antimicrobial activity of monolaurin, monomyristin, and monopalmitin acid against *E. coli* O157:H7, *Salmonella* spp., and *Y. enterocolitica*. Measuring growth by turbidity increase over time, 50 mg/l monolaurin completely prevented growth of *E. coli* O157:H7 for up to 100 h. Neither monomyristin nor monopalmitin at 20 mg/l had a significant effect on the growth rate of *E. coli* O157:H7. *Salmonella* demonstrated a moderately high inhibition index in the presence of 30–50 mg/l of monolaurin for up to 100 h but only an increased lag phase at 50 mg/l during a growth study. *Yersinia enterocolitica* demonstrated a dose response to monolaurin with 50 mg/l causing 90–100% inhibition for up to 100 h and 20 mg/l slightly increasing lag phase and reducing growth rate in a growth study. Neither monomyristin nor monopalmitin significantly inhibited *Salmonella* or *Y. enterocolitica*.

Marounek et al. (2012) tested sorbic, caprylic, capric, and lauric acids, monosorbin, monocaprylin, monocaprin, monolaurin, and sucrose monocaprate against the neonatal pathogens *Cronobacter sakazakii* and *Cronobacter malonaticus*. Both *Cronobacter* strains were susceptible to caprylic acid, monocaprylin, and, to a lesser extent, sorbic acid. Capric acid, lauric acid, monosorbin, monocaprin, monolaurin, and sucrose caprate exhibited no inhibitory activity. Caprylic acid and monocaprylin treatment (2 mg/ml) of *C. sakazakii* reduced the number of viable cells by 5 logs in 30 min. In contrast, Al-Holy et al. (2010) tested monolaurin and copper (II) against five strains of *Cronobacter* inoculated into reconstituted infant formula. They found that a sublethal concentration of monolaurin (2000 mg/l) with 50 mg/l of copper (II) resulted in a complete elimination of the *Cronobacter*. Monolaurin alone exhibited a slight inhibitory activity against *Cronobacter*.

Zhang et al. (2019) studied the antimicrobial activity of monolaurin compared to sodium benzoate and potassium sorbate against the foodborne pathogens *E. coli*, *Salmonella* Typhimurium, *S. aureus*, *L. monocytogenes*, *Candida albicans*, and *Cryptococcus neoformans*. The minimum inhibitory concentrations (MIC) of monolaurin against the six pathogens were 0.16, 0.16, 0.08, 0.04, 0.64, and 1.25 mg/ml, and the minimum bactericidal concentrations (MBC) were 0.64, 1.25, 0.32, 0.16, 5.0, and 5.0 mg/ml, respectively. The MICs and MBCs were both lower than sodium benzoate (6.25–25 mg/ml and 50 to >100 mg/ml, respectively) or potassium sorbate (1.6–25 mg/ml and 25 to >100 mg/ml, respectively). While the MIC of benzoate and sorbate decreased as pH decreased, the MIC of monolaurin remained stable at pH 3 to 9.

The objective of a study by Nair et al. (2004) was to determine the antibacterial effect of caprylic acid and monocaprylin on *L. monocytogenes* and *E. coli* O157:H7 in whole milk. At 37°C, 0.86% capric acid or 1.1% monocaprylin reduced the viable population of both pathogens by approximately 5.0 log CFU/ml in 6 h. At the same concentration, monocaprylin and capric acid decreased viable *L. monocytogenes* and *E. coli* O157:H7 by > 5.0 log CFU/ml and to undetectable levels at 24 h at 8°C and 48 h and 4°C, respectively.

11.3.3.2 Antifungal Activity

Altieri et al. (2008, 2009) determined the antifungal activity of lauric, myristic, and palmitic acids and their monoesters against *Aspergillus niger*, *A. terreus*, *Emericella nidulans*, *Fusarium oxysporum*, *F. avenaceum*, *Penicillium glabrum*, and *P. italicum*. Inhibition was evaluated by monitoring colony diameter, radial growth rate, lag time, and minimum detection time. Monolaurin at the highest concentrations (30 and 40 mg/l) completely inhibited the radial growth of *Penicillium*, *Emericella*, and *Aspergillus* within 30 days of storage and was more effective than lauric acid. In contrast, myristic and palmitic acids were more inhibitory than their corresponding monoesters. The effectiveness of the active compounds seemed to be strain- and genus-dependent. Palmitic acid was the most effective chemical against *Aspergillus*, whereas *Penicillium* were strongly inhibited by myristic acid. Additionally, *Aspergillus* sp. were more susceptible to fatty acids than were *Penicillium* sp. A low concentration of lauric acid (20 mg/l) was required to completely inhibit *F. oxysporum* for 30 days, while monolaurin had little effect on growth. In contrast, both monolaurin and lauric acid (20 mg/l) completely inhibited growth of *F. avenaceum* for 30 days. Palmitic and myristic acids and their monoglycerides had moderate effectiveness against *Fusarium* sp. but inhibition was reversible. Chipley et al. (1981) studied the antifungal and antiaflatoxin effects of monolaurin. In their study, monolaurin alone, sorbic acid alone, and a commercial mixture of monolaurin and sorbic acid (Lauribic™; ML+sorbic) were tested for aflatoxin inhibition using the aflatoxigenic molds *Aspergillus flavus* and *Aspergillus parasiticus*. Sorbic acid levels of 1000 mg/l were required for inhibition of aflatoxin formation, but monolaurin and ML + sorbic were active at 750 mg/l. Mycelial growth inhibition also was significant, with controls reaching an average of 952 mg mycelia per flask while mycelia in the presence of sorbic acid, monolaurin, and ML + sorbic reached only 540, 582, and 480 mg per flask, respectively. Mansour et al. (1996) found that monolaurin was a less effective inhibitor of nontoxigenic aspergilli and penicillia than were eugenol or isoeugenol. However, monolaurin (100 to 200 mg/l) was capable of significantly reducing mycelial dry weight and aflatoxin B1 and G1 in growth media. Lisker and Paster (1982) carried out extensive studies with commercial monolaurin and monolaurin plus other antimicrobials against 16 different molds with different cell wall structures. The results are presented in Tables 11.8 and 11.9. The combination of monolaurin and sorbic acid proved to be more active than either compound alone. The most effective combination was a commercial mixture of monolaurin, tert-butylhydroxyanisole (BHA), and ethylenediaminetetraacetic acid (EDTA). Plockova et al. (1999a) tested lauric acid and its glycerol esters against several food spoilage molds including *Alternaria*, *Cladosporium*, *Fusarium*, *Geotrichum*, and *Penicillium*. The order of effectiveness in decreasing order was monolaurin, dilaurin, trilaurin, and lauric acid.

11.3.3.3 Antiviral Activity

Some fatty acids and monoglycerides have demonstrated activity against enveloped viruses such as cytomegalovirus, herpes simplex virus, rhinoviruses, and human immunodeficiency virus (HIV) (Clarke and May, 2000; Hierholzer and Kabara, 1982; Hilmarsson et al., 2005; Isaacs et al., 1986, 1990; 1994;

TABLE 11.8

Antifungal Activity (Percentage of Mycelial Growth Inhibition) of 0.1% Monolaurin and Monolaurin Mixtures Compared with 0.1% Sorbic and Propionic Acids

Fungi	Monolaurin (Lauricidin®)	Monolaurin + BHA + EDTA (Lauricidin® plus F)	Monolaurin + Sorbic Acid (Lauribic™)	Sorbic Acid	Propionic Acid
Aspergillus niger	40.3	100	100	100	97.6
Aspergillus ochraceus	21.1	100	100	73.5	84.1
Asperglllus flavus	28.6	100	100	83.5	79.9
Candida albicans	17.5	100	100	100	11.7
Fusarium graminearum	20.1	100	100	100	100
Mucor circinelloides	30.5	100	94.3	100	78.7
Penicillium digitatum	39.3	100	100	100	100
Penicillium patulum	18.9	100	100	75.2	65.7
Phytophthora citrophthora	41.2	100	100	100	100
Rhizoctonia solani	54.9	100	100	98.2	100
Pythium elongatum	10.0	100	100	100	100
Rhizopus stolonifer	10.1	100	100	100	100
Rhodotorula mucilaginosa	98.0	100	100	100	100
Saccharomyces cerevisiae	26.5	100	100	100	23.2
Sclerotium rolfsii	63.3	100	100	100	100
Sporobolomyces sp.	100.0	100	100	100	100

Source: Lisker and Paster (1982).

TABLE 11.9

Antifungal Activity (Percentage of Mycelial Growth Inhibition) of 0.5% Monolaurin and 0.05% Monolaurin Mixtures

Fungi	Monolaurin (Lauricidin®)	Monolaurin + Sorbic Acid (Lauribic™)	Monolaurin + BHA + EDTA (Lauricidin® plus F)
Aspergillus flavus	73.6	44.8	85.5
Aspergillus niger	71.7	48.3	89.4
Aspergillus ochraceus	53.4	40.4	90.5
Candida albicans	0	100	0
Fusarium graminearum	57.5	100	82.5
Mucor circinelloides	56.4	48.5	100
Penicillium digitatum	76.1	100	86.7
Penicillium patulum	34.3	24.6	85.1
Phytophthora citrophthora	70.5	100	100
Rhizoctonia solani	61.7	100	100
Rhizopus stolonifer	29.8	81.0	100
Rhodotorula mucilaginosa	9.2	100	95.7
Saccharomyces cerevisiae	56.5	100	100
Sclerotium rolfsii	65.6	100	100
Sporobolomyces sp.	79.9	100	100

Source: Lisker and Paster (1982).

Orloff et al., 1993; Thormar et al., 1987; Welsh et al., 1979). Exposure to 500 mg/l monolaurin completely inhibited growth of cytomegalovirus in culture (Clarke and May, 2000).

11.3.3.4 Combinations with Other Antimicrobials

Kato and Shibasaki (1976) described the usefulness of acidulants in increasing the antimicrobial spectrum and activity of monoglycerides, particularly monolaurin. Citric, lactic, malic, and polyphosphoric acids were found to enhance the antibacterial effect of monoglycerides and sucrose esters against *E. coli*. These combinations also were effective against other Gram-negative bacteria, such as *Serratia marcescens*, *Proteus vulgaris*, and *Salmonella* Typhimurium. These same workers reported the combined use of monolaurin, organic acids, and EDTA (Tsuchido et al., 1981). The enhanced bactericidal activity of monolaurin by acid or EDTA was due to the increased uptake of the monoglyceride into the cells. Individually, neither monolaurin nor citrate had any effect of their own on *E. coli*. Yuda et al. (1977) reported that organic acids, such as fumaric acid, when combined with lauric esters of propylene glycol, glycerol, and polysorbate 80 (sorbitan) helped preserve starch-containing foods.

As discussed above, a mixture of monolaurin with various compounds including sorbic acid, BHA, and/or EDTA enhanced the antimicrobial activity (Kabara, 1978, 1979b, 1980). This was further demonstrated by a combined treatment of 0.005% monolaurin and 1.0 or 1.5% lactic acid that was able to inactivate (>8 log reduction) *E. coli* O157:H7 within 15 minutes at 40°C (Venkitanarayanan et al., 1999). A 5 log reduction was observed at 22°C using the same combination.

The bacteriocin nisin was found to be more effective in the presence of monolaurin (Blackburn et al., 1993). Nisin at concentrations of 0.2 mg/l was not active in complex milk medium. Monolaurin (1%) also was not active. However, a combination of the two agents reportedly had synergistic antimicrobial activity against *Streptococcus agalactiae*. Tkarskyy and Marshall (2008) tested combinations of lactic acid, monolaurin, and nisin against *L. monocytogenes*. They determined that a combination of all three was synergistically inhibitory compared to the compounds used alone. They studied the influence of the presence of the compounds on the cell membrane lipids. Monolaurin exposure caused incorporation of lauric acid into the cell membrane, and monolaurin plus lactic acid increased the incorporation further. Nisin had no influence. An increase in lauric acid in the membrane could lead to decreased fluidity and a more highly ordered membrane. The authors stated that this could disrupt membrane function by affecting signal transduction due to blockage of promoters, uncoupling of energy systems, altered respiration, and altered amino acid uptake. The authors further postulated that lactic acid-enhanced monolaurin C12:0 incorporation into the cell membrane changed membrane fluidity which resulted in increased nisin activity. Zhang et al. (2009) using a microtiter plate assay showed that the minimum inhibitory concentrations for monolaurin were 25 µg/mL against *E. coli*, 12.5 µg/mL against *S. aureus*, and 30 µg/mL against *Bacillus subtilis*. They then evaluated interaction of monolaurin with nisin, sodium dehydroacetate, EDTA, calcium propionate, and sodium lactate. Monolaurin and nisin acted synergistically against the test microorganisms while monolaurin plus sodium dehydroacetate or EDTA were synergistic against *E. coli* and *B. subtilis* but not *S. aureus*. Monolaurin combined with calcium propionate or sodium lactate showed no synergistic effects against any test microorganism. Branen and Davidson (2004) used a microtiter plate assay to systematically assess the interaction between EDTA or lactoferrin (chelators) with monolaurin against strains of *L. monocytogenes*, enterohemorrhagic *E. coli*, *Salmonella* Enteritidis, and *Pseudomonas fluorescens*. EDTA synergistically enhanced the activity of monolaurin against both enterohemorrhagic *E. coli* strains. Also, low concentrations of EDTA acted additively with monolaurin against *L. monocytogenes*. Lactoferrin in combination with monolaurin inhibited growth of *E. coli* O157:H7 but not *E. coli* O104:H21. In 2% fat milk, monolaurin alone at 1000 mg/l caused a 4 log reduction of *E. coli* O157:H7 but had no effect on *E. coli* O104:H21. Presence of EDTA up to 5000 mg/l did not enhance monolaurin inhibition of *E. coli* O157:H7 but did increase log reduction of *E. coli* O104:H21 up to 2 logs in milk. *Salmonella* Enteritidis and *P. fluorescens* strains were consistently more resistant to all antimicrobial combinations. It was suggested that the resistance of the latter Gram-negative bacteria may be due to differences in the outer membrane and/or LPS structure.

A treatment combination of alkaline shock (pH 10.5), high salt stress (10% NaCl), and 100 µg/ml monolaurin or lauric acid was able to cause a 5 to 8 log reduction in populations of *L. monocytogenes*, *Pseudomonas fragi*, and *P. fluorescens* (Vasseur et al., 2001). A unique approach for studying various combinations of factors on monolaurin activity was described by Bal'a and Marshall (1996a). Their approach used the double gradient plate technique to analyze the combined interaction of pH, salt, monolaurin, and temperature on growth of *L. monocytogenes*. Results of their work are shown in Figure 11.1A–D, where monolaurin activity is clearly seen to increase with high salt and low pH. A clear graphic display of growth domains of the bacterium was produced showing the combined effect of the four variables (Figure 11.2).

Oh and Marshall (1992, 1993a) showed that low pH improved monolaurin efficacy against *L. monocytogenes*. Razavi-Rohani and Griffiths (1996) further demonstrated that combinations of monolaurin with EDTA, low pH, high salt, and either lysozyme or BHA increased activity against a series of Gram-positive and Gram-negative bacteria. Monolaurin activity against Gram-negative bacteria was enhanced when combined with EDTA and low pH and/or high NaCl levels (Razavi-Rohani and Griffiths, 1994). Citrate, another cation chelator, was not an effective adjunct to enhance monolaurin activity against Gram negatives.

The antibacterial activity of tarragon, pennyroyal, and spearmint essential oils (EOs) against *L. monocytogenes* in culture medium and fresh white cheese was evaluated alone, and in combination with monolaurin (Hassan et al., 2014). The combination of monolaurin with EOs had an additive effect. Addition of tarragon EO plus monolaurin produced the greatest inhibition. No bacteria were detected at combination of 0.6% tarragon EO with 400 µg/ml monolaurin after 7 days of storage. A problem associated with the use of essential oils as antimicrobials in foods is that they contribute strong flavors which may not be compatible with the food product. Yamazaki et al. (2004) found that diglycerol monolaurate and nisin in combination with the essential oil (EO) components, carvacrol, thymol, and eugenol, significantly enhanced inhibition of *L. monocytogenes* allowing for a reduction in EO dosage, thereby resulting in the reduction of undesirable flavors.

Blaszyk and Holley (1998) found enhanced activity of monolaurin when combined with either eugenol or sodium citrate against a number of meat spoilage and pathogenic bacteria. Lactic acid bacteria and *E. coli* O157:H7 were most resistant and *L. monocytogenes* and *B. thermosphacta* were most sensitive to control.

Monolaurin was tested in combination with the naturally occurring enzyme system, the lactoperoxidase system (lactoperoxidase enzyme, thiocyanate, and hydrogen peroxide), against *E. coli* O157:H7 and

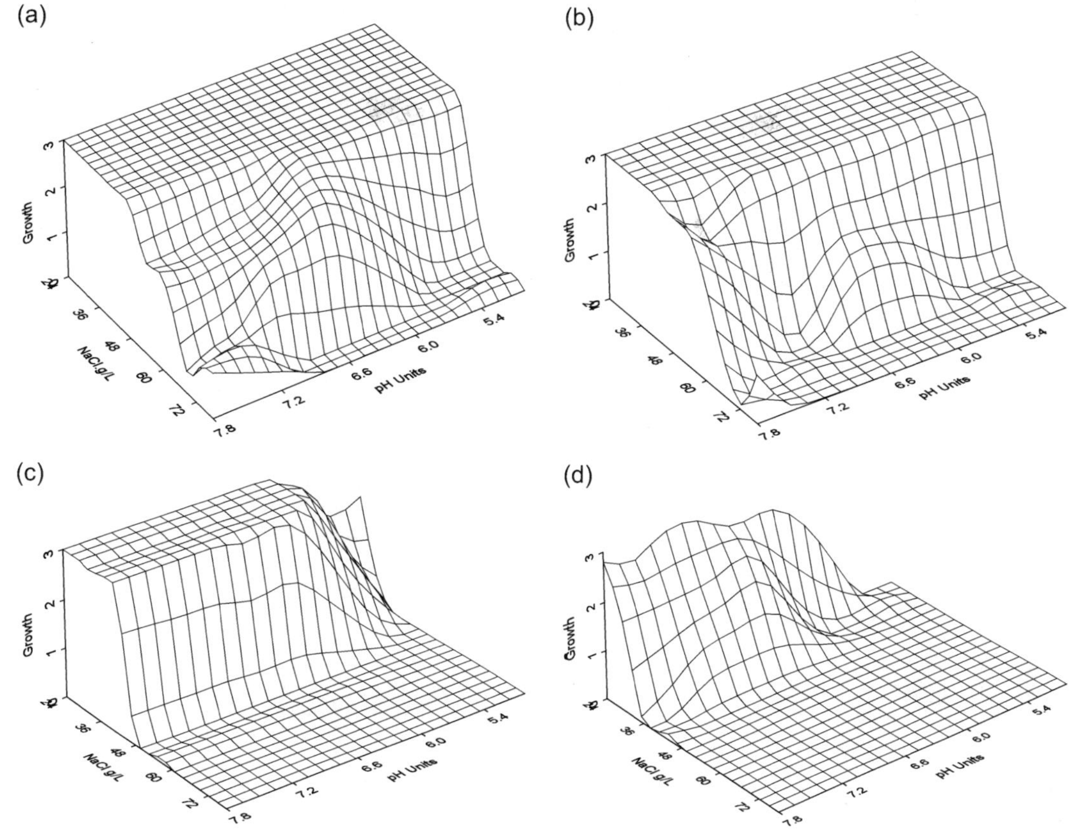

FIGURE 11.1 A–D. Growth profile of *L. monocytogenes* on two-dimensional salt and pH gradient plates incubated at 35°C for 48 h in the presence of (A) 0 μg, (B) 2 μg, (C) 4 μg, or (D) 8 μg of monolaurin per ml (from Bal'a and Marshall, 1996a).

S. aureus (McLay et al., 2002). The lactoperoxidase system (LPS) is considered to have greatest activity against Gram-negative bacteria while monolaurin is generally considered to have greatest activity against Gram-positive bacteria. An LPS–monolaurin system with LPS in the range 5–200 mg/kg and monolaurin in the range of 50–1000 μg/ml inhibited growth of *E. coli* O157:H7 and *S. aureus*. Growth of *S. aureus* was inhibited more strongly by LPS–monolaurin in microbiological medium than in milk and more strongly in milk than ground beef. A similar pattern was observed for *E. coli* O157:H7 except that there was no enhanced inhibition by LPS–monolaurin over LPS alone in ground beef.

Monocaprin has been shown to enhance the activity of ethanol as a bactericide (Furuta et al., 1984). The effect was reported to be synergistic. Oh and Marshall (1993b) found similar enhanced activity of ethanol and monolaurin against *L. monocytogenes.*

11.3.3.5 Combination with Heat and Cold

The use of monolaurin for enhancing inactivation by heat and cold has been studied by several research groups. Fatty acids and their esters were shown to have an effect on the thermal destruction of *E. coli* and *P. aeruginosa* (Kato and Shibasaki, 1975c). The enhancing effect by monolaurin was at a very low concentration of 100 μg/ml. Thus, lower heat treatments (improved food quality) may be possible using monolaurin in combination with other hurdles. Annamalai, et al. (2003) studied the effect temperatures up to 50°C in combination with monolaurin and an antimicrobial peptide, PR-26, on *E. coli* O157:H7 in apple cider. *E. coli* O157:H7 was inoculated into apple cider (pH 3.53) containing 0.01% PR-26 or 0.02% monolaurin or a combination of both. The apple cider was heated at 40, 45, or 50°C for 30 s, 5 min, 10

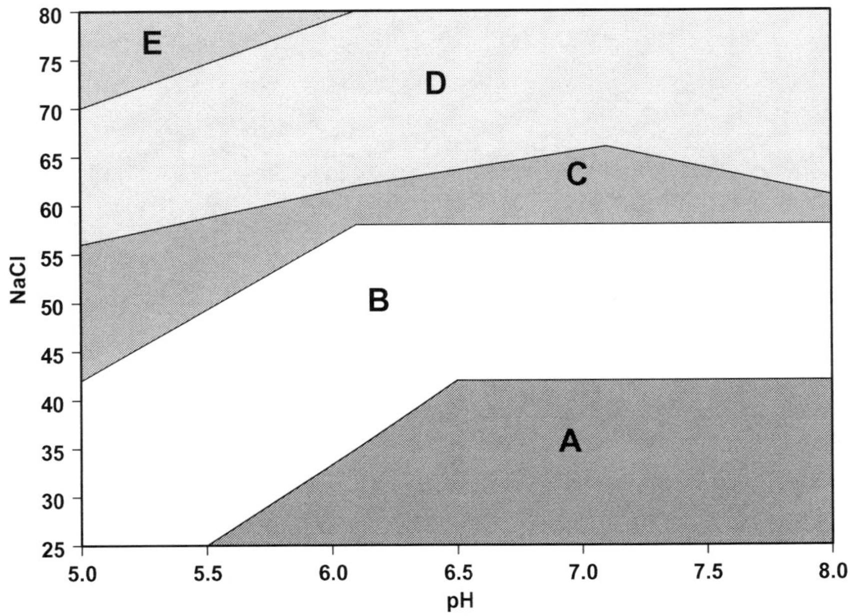

FIGURE 11.2 Effect of NaCl (g/L) and pH on growth domains of *Listeria monocytogenes* with monolaurin concentrations of (A) 8 μg/ml, (A+B) 4 μg/ml, (A+B+C) 2 μg/ml, (A+B+C+D) 0 μg/ml, and (E) area of no growth at 0 μg/ml (from Bal'a and Marshall, 1996a).

min, and 15 min. At each of the specified time intervals, the surviving population of *E. coli* O157:H7 was determined. At all the temperatures, there was no effect of the presence of PR-26. However, the treatments containing monolaurin alone or combination with PR-26 resulted in 5 log reductions of *E. coli* O157:H7. The thermodynamic characteristics of bactericidal action of monolaurin and heat were studied by Tsuchido et al. (1981), who found that the bactericidal effect of increasing concentrations of monolaurin would be more effective in low-temperature long-time treatments than high-temperature short-time processes.

The usefulness of monolaurin in the heat inactivation of spores was substantiated by Kimsey et al. (1981), who showed enhanced thermal inactivation of *Geobacillus stearothermophilus* 1518 in the presence of 18 mM monolaurin. The D value (time required to reduce the number of viable spores by 90%) at 120°C was reduced by 49.8%, while the D value for *B. subtilis* was reduced by 47.9%. The effectiveness of various saturated monoglycerides on *G. stearothermophilus* spores is shown in Table 11.10. Monocaprin and monolaurin were the most active of the series C8 to C16. Monolaurin was also tested in 10% nonfat dry milk and in the fluid portion of cream-style corn. In these foods

TABLE 11.10

Effectiveness of Various Saturated Monoglycerides against *Geobacillus stearothermophilus* Spores at 120°C

Spores Heated in 18 mM	Number of C in Chain	Average D_{120C} (s)	Range
Control	–	265	236–294
Monocaprylin	8	195	180–210
Monocaprin	10	88	84–91
Monolaurin	12	119	117–120
Monomyristin	14	221	210–228
Monopalmitin	16	244	234–248

Source: Kimsey et al. (1981).

the $D_{120°C}$ time was lowered by approximately 50%. It was concluded that monolaurin could be used as a food additive (emulsifier) to reduce the heat treatment required to achieve commercial sterility of foods. Naruki et al. (1983) confirmed earlier studies by demonstrating that fatty acids and monoglycerides inhibited *B. subtilis*. Fatty acids were less active than their corresponding monoglyceride (Table 11.11).

In addition to enhancement of thermal death by monolaurin, Takano et al. (1979) showed that freezing significantly injured cells of Gram-negative bacteria exposed to monolaurin. Of the various nonionic derivatives tested, monolaurin was the most active in reducing viability (Table 11.12). Carbohydrate fatty acid esters were shown to enhance damage in the growth phase rather than during the freezing process. Freezing with these agents was also found to be lethal to Gram-positive bacteria.

TABLE 11.11

Effect of Lauric Acid (LA) and Its Monoglyceride Monolaurin (ML) on the Loss of Heat Resistance of Spores During Germination[a]

	Viability (%) after Heat Treatment of Spores		
Exposure Time (min)	**None**	**LA (0.2 mM)**	**ML (0.05 mM)**
0	100	100	100
30	11.5	59	55
60	6.4	35	33

[a] Spores exposed to fatty acid, then heat treated at 85°C for 30 min; viability expressed as percentage of initial spore count (approximately 7 log CFU/ml).

TABLE 11.12

Effect of Surfactants and Fatty Acid Monoesters on Survival of *Escherichia coli* Following Frozen Storage

Chemical Agent	**Surviving Fraction after Frozen Storage[a]**
Nonionic surfactants (100 μg/ml)	
None	0.1
Tween 20	0.001
Tween 40	0.03
Tween 60	0.02
Tween 80	0.02
Span 20	0.000001
Span 60	0.04
Span 80	0.02
Span 85	0.01
Monoglycerides (10 μg/ml)	
Monocaprylin	0.03
Monocaprin	0.0001
Monolaurin	0.000001
Monomyristin	0.00005
Monopalmitin	0.0004
Monostearin	0.03
Monoolein	0.0001

[a] Surviving fraction of unfrozen controls was > 0.1.
Source: Takano et al. (1979).

11.4 Food Applications

Monoglyceride antimicrobial activity can be neutralized by a variety of polymers, including fat and starch. Glass and Johnson (2004) did an extensive study on the influence of milk fat and soybean oil on the antimicrobial activity of fatty acids and monolaurin against *C. botulinum* growth and toxin production. The compounds evaluated included (all at 500 mg/l) monolaurin, caprylic, capric, lauric, myristic, palmitic, stearic, oleic, and linoleic acids. They found that the antibotulinal effects of monolaurin and the free fatty acids were significantly reduced or eliminated in treatments supplemented with 20% fat. Zhang et al. (2009) showed that the antibacterial effectiveness of monolaurin was reduced by fat or starch while the monolaurin activity remained unchanged in the presence of protein.

The amount of monolaurin that can be added to a food is likely restricted more by taste than by governmental regulation. For most applications, 500 to 750 μg/ml of monolaurin is effective and remains undetected by most sensory panelists (Kabara, unpublished data). Higher application levels are possible if the bitter or soapy taste can be masked.

11.4.1 Dairy Products

For dairy applications, a 35% increase in cottage cheese shelf life (5 to 10 days longer) was obtained using 500 μg/ml monolaurin (Bautista et al., 1993). The shelf life increase was accomplished due to a 100-fold reduction in the number of spoilage bacteria in the monolaurin-containing product. However, sensory analysis demonstrated that trained panelists could detect 250 μg/ml monolaurin in cottage cheese as a notable chemical off flavor, which would limit its use in this application.

Kappa-conjugated linoleic acid and monolaurin were found to be effective antimicrobials against *L. monocytogenes* in broth (Wang and Johnson, 1992). Monolaurin was effective to inactivate *L. monocytogenes* in skim milk but not whole milk, presumably due to higher fat content in the latter food. K-conjugated linoleic acid was bacteriostatic, preventing growth of *L. monocytogenes* in whole and skim milk at 4°C. A combined treatment of 250 μg/ml monolaurin combined with 100 IU/ml nisin was able to eliminate 7 log populations of *B. cereus*, *B. coagulans*, *B. subtilis*, and *Bacillus licheniformis* in milk (Mansour et al., 1999; Mansour and Milliere, 2001). Italian Stracchino cheese surface treated with 200 μg monolaurin/cm^2 prevented growth of *L. monocytogenes* during storage at 5°C for 12 days (Stecchini et al., 1996). As reported by Duxbury (1986), Kabara and Brady (Michigan State University) found that control cheese samples and those containing 0.05 or 0.10% monolaurin had a shelf life of only 5–8 days. Lactic acid (0.2%) alone extended the shelf life to 25 days, but the addition of 0.2% monolaurin plus lactic acid increased the time to 120 days. This combination of monolaurin and acidulant was also successfully used in an onion-cucumber powdered snack dip.

Mohamad et al. (2018) developed an antimicrobial cellulose-chitosan (CC) film containing monolaurin (0.5 and 1%) using the sol-gel method. It was analyzed for physical characteristics and antimicrobial properties against *L. monocytogenes in vitro* and on ultrafiltered cheese during storage at 4°C for 14 days. The results of antimicrobial activity revealed that the addition of monolaurin significantly ($P < 0.05$) increased inhibition by the CC film. The addition of 0.5 and 1% monolaurin into CC films caused a 2.3–2.4 log reduction in *L. monocytogenes* on cheese after 14 days.

An effective teat dip preparation using monolaurin, saturated fatty acids, and lactic acid has been described (Boddie and Nickerson, 1992). The treatment reduced mammary infections by *S. aureus* and *Streptococcus agalactiae* 40 to 80%.

11.4.2 Meat Products

Clostridium sporogenes and *L. monocytogenes* inactivation by monolaurin during cooking was studied by Unda et al. (1991). Beef roasts were pumped with brine containing phosphates and acetic acid, sodium lactate, potassium sorbate, or monolaurin. The meat was cooked in the bag once or twice to 62.8°C. In roasts recooked after being inoculated externally, monolaurin and sodium lactate were very effective in inhibiting bacteria during temperature abuse. These results suggest that perhaps combinations of monolaurin and sodium lactate may offer potential control of clostridia and listeriae in "microwave-ready" beef products.

Smith and Palumbo (1980) investigated the inhibition of aerobic and anaerobic growth of *S. aureus* in a model sausage system. They compared potassium sorbate, sorbic acid, and monolaurin. Whereas anaerobic growth of *S. aureus* was inhibited by 2500 μg/ml sorbate and monolaurin, twice as much was required to inhibit aerobic growth. To enhance activity it was necessary to acidify the system with 20 mg lactic acid. Under these same acidulating conditions, sorbic acid and commercial ML + sorbic acid inhibited both aerobic and anaerobic growth equally well at 500 and 750 μg/ml, respectively. In these experiments, the amount of lactic acid added, not the pH, was critical. The addition of 10 to 20 mEq lactic acid resulted in a small change in pH value (0.4 units) and no change in microbial titer. The addition of 25 and 30 mEq acids resulted in a pH decline of 0.6 units to pH 5.7 but a drop in the preserved meat emulsion of 4 to 5 logs from control levels. Thus, when sorbic acid is added to such a test system, one is adding not only a preservative but also an acidulating agent. When comparisons between sorbic acid and monolaurin are made, this important factor should be recognized.

Using the same model sausage system and controlling pH, Kabara (1984) showed that monolaurin was slightly more active than potassium sorbate. Commercial ML + sorbic acid gave greater inhibition than either monolaurin or sorbic acid alone. Inhibition was greater against anaerobic than aerobic growth. In the latter instance, high (>10,000 μg/ml) concentrations of sorbate and lower (5000 μg/ml) amounts of monolaurin were required for a good reduction in the bacterial population. ML + sorbic acid was effective at 500 μg/ml (Figure 11.3).

The utility of monolaurin in meat products was further evaluated by Notermans and Dufrenne (1981). Monolaurin (Lauricidin®), when used in the proportion of 5 g/kg meat slurry (pH 6.0 to 6.2), completely inhibited toxin production by *C. botulinum* types A, B, and E. The inhibitory effect of monolaurin was not improved by the addition of lactic acid to the meat slurry until the pH was lowered to less than 5.2. Under these conditions, potassium sorbate inhibited toxin production of type B at 32 g/kg (3.2%). Since monolaurin was six times more active than potassium sorbate, Notermans and Dufrenne (1981) recommended it as a possible alternative to nitrite for the preservation of meat products. Since nitrite is necessary for the color and taste of meat products, the suggestion to lower rather than eliminate nitrite was indicated. Hall and Maurer (1986) tested monolaurin and propylene glycol as inhibitors of *C. botulinum*

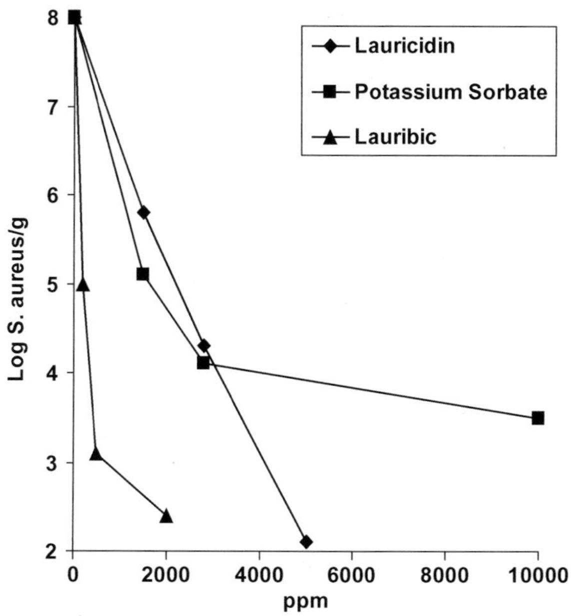

FIGURE 11.3 Effect of Monolaurin (Lauricidin®), potassium sorbate or Molaurin plus sorbic acid (Lauribic™) on survival of *Staphylococcus aureus* in a model sausage after 3 days at 35°C (from Kabara, 1984).

in turkey frankfurter slurries as possible replacements for nitrite as a preservative, or so that lower nitrite levels could be used. They reported that inhibition of *C. botulinum* was achieved; however, it was not clear from the report what monolaurin concentration was used. Monolaurin at 2.5% was shown to inhibit *C. sporogenes* and *L. monocytogenes* in cooked beef roasts (Stillmunkes et al., 1993). Monolaurin activity was less than sodium lactate, particularly in roasts that were temperature abused. In addition, monolaurin caused significant purge during storage of the roasts, which is a quality defect.

A wider application of monolaurin in meat products was reported by Baker et al. (1982). They added monolaurin to deboned chicken meat, minced fish, and chicken sausage. Spoilage for these meat systems was evident in 5, 8, and 9 days, respectively. The addition of 0.2% citric and 250 µg/ml monolaurin increased the shelf life of deboned chicken meat by 7 days, minced fish by 5 days, and chicken sausage by 8 days.

The action of Lauri-lac™ (a mixture of monolaurin and lactic acid) on the multiplication of *Salmonella* Typhimurium in minced beef was examined at 0–2°C, 30°C, and 37°C by Catsaras et al. (1987). The effects at lower temperatures and concentrations between 0.1 and 0.5% were negligible. At 37°C and 0.5%, the original inoculum was reduced from 5.4 log CFU/g to less than 2 logs in the time period examined. This confirms similar studies that suggest that monoglyceride preparations are more effective under abusive temperature conditions.

Nisin, sorbic acid, monolaurin, and commercial ML + sorbic acid were studied by Bell and DeLacy (1987). Determinations were made of their ability to prevent *B. licheniformis* growth in a model cooked luncheon meat product. This product was made from a 3:1 mixture of lean beef and beef fat and contained 2.5% salt, 0.3% sodium tripolyphosphate, and 200 µg/ml sodium nitrite. Various concentrations of the test preservatives were added to batches of the raw meat emulsion inoculated with 4.3 log *B. licheniformis* spores/g. Emulsions were cooked to a pasteurization value of 36 (reference temperature 70°C), chilled overnight, and then stored at 25°C. Minimum concentrations of the test preservatives necessary to prevent product spoilage within 10 days during abusive storage were 500 IU nisin/g, 0.125% sorbic acid, 0.5% monolaurin, and 0.25% commercial ML + sorbic acid. These concentrations also were effective in controlling the growth of natural *Enterococcus* spp. spoilage microflora in mildly cooked products. The minimum effective concentration (0.125%) of sorbic acid was made more effective by monolaurin. Thus, combinations of sorbic acid and monolaurin may provide additional protection beyond that produced by a given concentration of sorbic acid used alone. The researchers did note a soapy odor in luncheon meat treated with 500 µg/ml monolaurin.

Zein films with 1% monolaurin and 2 or 3% *Zataria multiflora* Boiss. (a thyme-like plant of the Lamiaceae family) essential oil (ZEO) were prepared by casting methods, and their antimicrobial effects against *Listeria monocytogenes* and *E. coli* O157: H7 *in vitro* and in minced beef meat were determined (Moradi et al., 2016). Antimicrobial activity was significantly increased by adding ZEO and monolaurin but the extent of antimicrobial activity was dependent upon concentration and type of bacteria. Overall, *L. monocytogenes* was more susceptible than *E. coli*. In minced meat, the growth *of L. monocytogenes* was significantly inhibited by monolaurin- and/or ZEO-containing films during 3 days of storage at 4°C, whereas the growth of *E. coli* was only inhibited by films containing ZEO.

Saturated monoglycerides (C6 to C12) inhibit mutagenesis by N-nitroso-dimethylamine in cultures of *E. coli* (Japanese Patent, 1985). Thus, a food additive consisting of monoglyceride (C8 to C12), glycerin, and gelatin inhibited nitrosamine mutagenesis. This finding may be of importance to cured meat processors desiring to control nitrosamine formation in their products.

11.4.3 Seafood

Monolaurin and fumaric acid added to kamaboko (fish paste) at levels of 0.12 and 0.05% were effective preservatives (Katagiri and Yamamoto, 1978). These investigators also designed a combination that included a 0.5% monocaprin–0.2% monolaurin mixture, lactic acid, glycine, and calcium lactate. This mix was added to kamaboko at 0.8% and preserved the product for 20 days at 30°C. The control product spoiled in 6 to 10 days. Ogawa (1980) used one or more salts of acetic, citric, tartaric, malic, and lactic acids to potentiate the effect of C8:0 to C12:0 monoglycerides. Their use improved the flavor and aroma of frozen cooked foods.

A treatment combination of modified atmosphere packaging, monolaurin, and lactic acid was able to reduce growth of *L. monocytogenes* in crawfish tail meat (Oh and Marshall, 1995a). Similarly, monolaurin combined with lactic acid was able to reduce the population of *L. monocytogenes* on catfish fillets (Verhaegh et al., 1996). Pre-harvest and post-harvest control of the seafood-borne pathogen *Streptococcus iniae* may be achieved using a combination of low to moderate pH, salt, and monolaurin (Trotter and Marshall, 2003). Biofilms of *L. monocytogenes* on stainless steel were eliminated using combinations of monolaurin with moderate heat (Oh and Marshall, 1995b) or acetic acid (Oh and Marshall, 1996).

11.4.4 Other Foods

Kato (1981) made a direct comparison of fatty acids, monoglycerides, and sucrose fatty acid esters in soy sauce. He found that capric acid and monolaurin were the most active preservatives in soy sauce that contained osmophilic yeast. In a series of sucrose esters, sucrose monolaurate was more active than sucrose monocaprate. The antimicrobial activity of monolaurin compared favorably with that of sodium lauryl sulfate and was superior to butyl-p-hydroxybenzoate, which was approved as an antimicrobial additive for soy sauce. Shibasaki et al. (1979) added several different glycerol and sucrose esters to tofu (soybean protein curd) and placed the mixture in a casing kept at 30°C for 5 days. Coliform, total, and thermostable bacterial counts were less than 300 per gram for the product. Controls produced by conventional methods showed levels of 3.9 log, 5.3 log, and 5.5 log CFU/g after storage, respectively.

Robach et al. (1981) compared the activity of monolaurin and sorbic acid in laboratory media, cottage cheese, and a pork homogenate. In general they found sorbic acid more active than monolaurin, but they reported that monolaurin under similar conditions was more effective than sorbic acid against the growth of *Salmonella* Typhimurium. Contrary to the results of others, Robach et al. (1981) found no advantage to the use of monolaurin alone or monolaurin–sorbic acid combinations in their systems.

Wang and Johnson (1997) evaluated C:10:0 and 12:0 monoglycerides against *L. monocytogenes* in microbiological media and in several foods. Monolaurin was inhibitory to the bacterium at 50 mg/l in microbiological media and the presence of phenolic antioxidants, acetic acid, lactic acid, and potassium sorbate all potentiated monolaurin activity. Monoglycerides were inhibitory to *L. monocytogenes* in certain foods including beef frank slurries (pH 5.0 and 5.5) and seafood salad (pH 4.9) at 4°C, but were less active at 12 than at 4°C. Monoglycerides were less inhibitory to *L. monocytogenes* in other foods tested including turkey frank slurries (pH 5.5), imitation crabmeat, cooked shrimp, summer sausage, yogurt, cottage cheese, and Camembert cheese. Combinations of monocaprin and monolaurin increased antilisterial activity in beef and turkey frank slurries. The degree of inhibition by the compounds was influenced by pH.

11.5 Mechanism of Action

Fatty acids and their esters are heterogeneous in their structures as well as their physico-chemical characteristics. Thus, mechanisms of action against various microorganisms are likely non-specific. In one sense, that is beneficial because the development of mutations to a single mode of action causing resistance is not likely. There exists a number of microbial possible targets for fatty acids and monoglycerides. One commonality among fatty acids and monoglycerides is that they are amphiphilic. Churchward et al. (2018) suggested that high concentrations of these compounds may act as a detergent and cause solubilization of the lipids in the cytoplasmic membrane. However, multiple studies have found evidence that lower but still inhibitory concentrations probably involve the disruption of the cytoplasmic membrane which is more than a simple insertion (Churchward et al., 2018). Early studies focused on the inhibition by fatty acids on oxygen and membrane-associated amino acid transport (Galbraith and Miller, 1973; Gale and Folkes, 1967). As these both involve bacterial cytoplasmic membrane and associated enzymes, this suggested that the mechanism of action was indeed a type of disruption of the structure.

Hyldgaard et al. (2012) studied the effect of monocaprylin on *E. coli*, *Staphylococcus xylosus*, and *Zygosaccharomyces bailii*. The effect of monocaprylin on cell structure and membrane integrity, as well as its interaction with model membranes, were evaluated with atomic force microscopy and propidium iodide staining. Changes in cell structure were visible and showed membrane disruption and increased permeability, indicating the membrane as a site of action. It was concluded that monocaprylin destabilizes membranes by increasing membrane fluidity and that cytoplasmic membrane phospholipid composition, resultant fluidity, and membrane curvature all play a role in susceptibility. Zhang et al. (2019) studied the mechanism of action of monolaurin by detecting plasma membrane permeability, membrane surface damage, and the impact of cell cycle and intracellular biomacromolecules synthesis. Monolaurin was found to increase cell membrane permeability, destroy cell membrane integrity, arrest the cell cycle, and inhibit biomacromolecule synthesis. They suggested that monolaurin inhibited bacteria and yeast by action on the cell membrane and intracellular inhibition. These results are in agreement with Kim and Rhee (2013) who used flow cytometry and microscopic analysis of *E. coli* O157:H7 treated with 24 mg/l, 2.4 mg/l, or 22 mg/l of monocaprylin, monocaprin, or monolaurin. They detected membrane disruption with the compounds; however it was found to be reversible in that cells were capable of recovery to form colonies.

REFERENCES

Ababouch, L.H., Bouqartacha, F. and Busta, F.F. 1994. Inhibition of *Bacillus cereus* spores and vegetative cells by fatty acids and glycerol monododecanoate. *Food Microbiol.* 11:327–336.

Albert, A. 1960. *Selective Toxicity*, 2nd ed. Methuen, London, 233 p.

Al-Holy, M.A., Castro, L.F. and Al-Qadiri, H.M. 2010. Inactivation of *Cronobacter* spp. (*Enterobacter sakazakii*) in infant formula using lactic acid, copper sulfate and Monolaurin. *Lett. Appl. Microbiol.* 50:246–251.

Altieri, C., Cardillo, D., Bevilacqua, A. and Sinigaglia, M. 2007. Inhibition of *Aspergillus* spp. and *Penicillium* spp. by fatty acids and their monoglycerides. *J. Food Prot.* 70:1206–1212.

Altieri, C., Bevilacqua, A., Cardillo, D. and Sinigaglia, M. 2008. Antifungal activity of fatty acids and their monoglycerides against *Fusarium* spp. in a laboratory medium. *Intl. J. Food Sci. Technol.* 44: 242–245.

Altieri, C., Bevilacqua, A., Cardillo, D. and Sinigaglia, M. 2009. Effectiveness of fatty acids and their monoglycerides against gram-negative pathogens. *Intl. J. Food Sci. Technol.* 44:359–366.

Annamalai, T., Vasudevan, P., Marek, P., Hoagland, T. and Venkitanarayanan, K. 2003. Heat inactivation of *Escherichia coli* O157:H7 in apple cider in the presence of glycerol monolaurate and a synthetic antimicrobial peptide, PR-26. *J. Food Proc. Preserv.* 27:37–49.

Babayan, V.K., Kaufman, T.G., Lehjman, H. and Tkaczuk, R.J. 1964. Some uses and applications of polyglycerol esters in cosmetics and pharmaceutical preparations. *J. Soc. Cosmet. Chem.* 15:473–483.

Baechler, R.H. 1939. Toxicity of normal aliphatic alcohols, acids, and sodium salts. *Proc. Am. Wood Preserv. Assoc.*, 364–372.

Baker, R.C., Kline, D., Poon, W. and Vadehra, D.H. 1982. Antimicrobial properties of Lauricidin in mechanically deboned chicken and minced fish. *J. Food Safety* 4:177–184.

Bal'a, M.F.A. and Marshall, D.L. 1996a. Use of double gradient plates to study combined effects of salt, pH, monolaurin, and temperature on *Listeria monocytogenes*. *J. Food Prot.* 59:601–607.

Bal'a, M.F.A. and Marshall, D.L. 1996b. Testing matrix, inoculum size, and incubation temperature affect monolaurin activity against *Listeria monocytogenes*. *Food Microbiol.* 13:467–473.

Bautista, D.A., Durisin, M.D., Razavi-Rohani, S.M, Hill, A.R. and Griffiths, M.W. 1993. Extending the shelf-life of cottage cheese using monolaurin. *Food Res. Intl.* 26:203–208.

Bayliss, M. 1936. Effect of the chemical constitution of soaps upon their germicidal properties. *J. Bacteriol.* 31:489–504.

Bell, R.G. and DeLacy, M. 1987. The efficacy of nisin, sorbic acid and monolaurin as preservatives in pasteurized cured meat products. *Food Microbiol.* 4:277–283.

Bellone, C.H., de Farolfi, H.V., Jaime, M.A., de Kristal, N.I. and de Torres, C.R. 1993. Activity fungitoxic of combinations of tiabendazol and sucrose fatty acid esters on harvested citrus fruit. *Rev. Agron. N.O. Argent.* XXVII (1–4):5–25.

Beuchat, L.R. 1980. Comparison of anti-*Vibrio* activities of potassium sorbate, sodium benzoate and glycerol and sucrose esters of fatty acids. *Appl. Environ. Microbiol.* 39:1178–1182.

Blackburn, P., Polak, J., Gusik, S.A. and Rubino, S.D. 1993. Nisin composition for use as enhanced broad range bactericides. US Patent 5,217,950.

Blaszyk, M. and Holley, R.A. 1998. Interaction of monolaurin, eugenol and sodium citrate on growth of common meat spoilage and pathogenic organisms. *Int. J. Food Microbiol.* 39:175–183.

Boddie, R.L. and Nickerson, S.C. 1992. Evaluation of postmilking teat germicides containing Lauricidin, saturated fatty acids, and lactic acid. *J. Dairy Sci.* 75:1725–1730.

Borel, C., Welti, D.H., Fernandez, I. and Colmenares, M. 1993. Dicranin, an antimicrobial and 15-lipoxygenase inhibitor from the moss *Dicranum scoparium*. *J. Natural Products* 56:1071–1077.

Branen, A.L., Davidson, P.M. and Katz, B. 1980. Antimicrobial properties of lipids and phenolic antioxidants. *Food Technol.* 34:42–53, 63.

Branen, J.K. and Davidson, P.M. 2004. Enhancement of nisin, lysozyme, and monolaurin antimicrobial activities by ethylenediaminetetraacetic acid and lactoferrin. *Intl. J. Food Microbiol.* 90:63–74.

Burnett, S.L., Chopskie, J.H., Podtburg, T.C., Gutzmann, T.A., Gilbreth, S.E. and Bodnaruk, P.W. 2006. Use of octanoic acid as a postlethality treatment to reduce *Listeria monocytogenes* on ready-to-eat meat and poultry products. *J. Food Prot.* 70:392–398.

Catsaras, M., Danjoy, J.P. and Seynave, R. 1987. Temperature and action of Lauri-Lac on multiplication of *Salmonella* in minced meat. *Bull. Acad. Vet. Fr.* 60:359.

Chaibi, A., Ababouch, L.H. and Busta, F.F. 1996a. Inhibition of bacterial spores and vegetative cells by glycerides. *J. Food Prot.* 59:716–722.

Chaibi, A., Ababouch, L.H. and Busta, F.F. 1996b. Inhibition by monoglycerides of L-alanine-triggered *Bacillus cereus* and *Clostridium botulinum* spore germination and outgrowth. *J. Food Prot.* 59:832–837.

Chaibi, A., Ababouch, L.H., Ghouila, M.R. and Busta, F.F. 1998. Effect of monoglycerides on the thermal inactivation kinetics of *Bacillus cereus* F4165/75 spores. *Food Microbiol.* 15:527–537.

Chang, S.-S., Redondo-Solano, M. and Thippareddi, H. 2010. Inactivation of *Escherichia coli* O157:H7 and *Salmonella* spp. on alfalfa seeds by caprylic acid and monocaprylin. *Intl. J. Food Microbiol.* 144:141–146.

Chipley, J.R., Story, L.D., Todd, P.T. and Kabara, J.J. 1981. Inhibition of *Aspergillus* growth and aflatoxin release by sorbic acid and derivatives of fatty acids. *J. Food Safety* 3:352–357.

Choi, M.J., Kim, S.A., Lee, N.Y. and Rhee, M.S. 2013. New decontamination method based on caprylic acid in combination with citric acid or vanillin for eliminating *Cronobacter sakazakii* and *Salmonella enterica* serovar Typhimurium in reconstituted infant formula. *Intl. J. Food Microbiol.* 166:499–507.

Churchward, C.P., Alany, R.G. and Snyder, L.A.S. 2018. Alternative antimicrobials: The properties of fatty acids and monoglycerides. *Crit. Rev. Microbiol.* 44:561–570.

Clark, J.R. 1899. On the toxic effect of deleterious agents on the germination and development of certain filamentous fungi. *Bot. Gaz.* 28:289–327.

Clarke, N.M. and May, J.T. 2000. Effect of antimicrobial factors in human milk on rhinoviruses and milk-borne cytomegalovirus in vitro. *J. Med. Microbiol.* 49:719–723.

Combrink, J.C., de Kock, S.L. and van Eeden, C.J. 1990. Effect of post-harvest treatment and packaging on the keeping quality of fresh guava fruit. *Acta Horticult.* 275:639–645.

Conley, A.J. and Kabara, J.J. 1973. Antimicrobial action of esters of polyhydric alcohols. *Antimicrob. Agents Chemother.* 4:501–506.

Cotton, L.N. and Marshall, D.L. 1997. Monolaurin preparation method affects activity against vegetative cells of *Bacillus cereus*. *Food Sci. Technol.* 30:830–833.

Cowles, C. 1941. The germicidal action of the hydrogen ion and of the lower fatty acids. *J. Biol. Med.* 13:571.

Duxbury, D.D. 1986. Combination emulsifier/acidulant extends cheese sauce shelf life. *Food Proc.* 46:38–39.

Eggerth, A.H. 1929a. Germicidal and hemolytic action of a-bromo soaps. *J. Exp. Med.* 49:53–62.

Eggerth, A.H. 1929b. Germicidal and action of a-hydroxy soaps. *J. Exp. Med.* 50:299–262.

Favier, G.I., Escudero, M.E. and de Guzman, A.M.S. 2009. Inhibitory effects of sucrose fatty acid esters on survival of *Yersinia enterocolitica* on eggshell surface and use of blue lake as indicator of bacterial penetration into eggs. *J. Appl. Microbiol.* 106: 774–783.

Fuller, R. and Moore, J.H. 1967. The inhibition of the growth of *Clostridium welchii* by lipids isolated from the contents of the small intestine of the pig. *J. Gen. Microbiol.* 46:23–41.

Furuta, T., Komori, K. and Ueda, A. 1984. Combined effects of various food additives on the bactericidal activity of ethanol against *Escherichia coli* and *Staphylococcus aureus*. *BokinBobai* (Jpn.) 12:65–72.

Galbraith, H. and Miller, T.B. 1973. Effect of long-chain fatty acids on bacterial respiration and amino acid uptake. *J. Appl. Bacteriol.* 36:659–675.

Gale, E.F. and Folkes, E.S. 1967. The effect of lipids on the accumulation of certain amino acids by *Staphylococcus aureus*. *Biochim. Biophys. Acta* 144:461–466.

Gershon, H. and Parmegiani, R. 1967. Organic fluorine compounds. II: Synthesis and antifungal properties of 1-fluoro fatty acids. *J. Med. Chem.* 10:186–188.

Glass, K.A. and Johnson, E.A. 2004. Antagonistic effect of fat on the antibotulinal activity of food preservatives and fatty acids. *Food Microbiol.* 21:675–682.

Hall, M.A. and Maurer, A.J. 1986. Spice extracts, Lauricidin and propylene glycol as inhibitors of *Clostridium botulinum* in turkey frankfurter slurries. *Poultry Sci.* 65:1167–1171.

Hassan, H., Razavi-Rohani, S.M. and Gandomi, H. 2014. Combination effect of essential oils of some herbs with monolaurin on growth and survival of *Listeria monocytogenes* in culture media and cheese. *J. Food Proc. Preserv.* 38: 304–310.

Hathcox, A.K. and Beuchat, L.R. 1996. Inhibitory effects of sucrose fatty acid esters, alone and in combination with ethylenediaminetetraacetic acid and other organic acids, on viability of *Escherichia coli* O157:H7. *Food Microbiol.* 13:213–225.

Hettche, H.O. 1934. The nature of bactericidal and hemolytic constituents of lipid from *Pyocyaneus*. *Z. Immun.* 83:804–805.

Hierholzer, J.C. and Kabara, J.J. 1982. In vitro effects of monolaurin compounds on enveloped RNA and DNA viruses. *J. Food Safety* 4:1–12.

Hilmarsson, H., Kristmundsdóttir, T. and Thormar, H. 2005. Virucidal activities of medium- and long-chain fatty alcohols, fatty acids and monoglycerides against herpes simplex virus types 1 and 2: comparison at different pH levels. *APMIS* 113:58–65.

Hoffman, C., Schweitzer, T.R. and Dalby, G. 1939. Fungistatic properties of the fatty acids and possible biochemical significance. *Food Res.* 4:539–545.

Hoffman, C., Schweitzer, T.R. and Dalby, G. 1940. A comparison of the fungistatic properties of enanthaldehyde and enonthic acid as related to their possible anticarcinogenic effect. *Am. J. Cancer* 38:569–570.

Hulankova, R., Borilova, G. and Steinhauserova, I. 2013. Combined antimicrobial effect of oregano essential oil and caprylic acid in minced beef. *Meat Sci.* 95:190–194.

Humfeld, H. 1947. Antibiotic activity of the fatty acid-like constituents of wheat bran. *J. Bacteriol.* 54:513–517.

Hyldgaard, M., Sutherland, D.S., Sundh, M., Mygind, T. and Meyer, R.L. 2012. Antimicrobial mechanism of monocaprylate. *Appl Environ Microbiol.* 78:2957–2965.

Ibrahim, H.R., Kato, A. and Kobayashi, K. 1991. Antimicrobial effects of lysozyme against Gram-negative bacteria due to covalent binding of palmitic acid. *J. Agric. Food Chem.* 39:2077–2082.

Ibrahim, H.R., Kobayashi, K. and Kato, A. 1993. Length of hydrocarbon chain and antimicrobial action to Gram-negative bacteria of fatty acylated lysozyme. *J. Agric. Food Chem.* 41:1164–1168.

Isaacs, C.E., Thormar, H. and Pessolano, T. 1986. Membrane-disruptive effect of human milk: Inactivation of enveloped viruses. *J. Infect. Dis* 154:966–971.

Isaacs, C.E., Kashyap, S., Heird, W.C. and Thormar, H. 1990. Antiviral and antibacterial lipids in human milk and infant formula feeds. *Arch. Dis. Childhood* 65:861–864.

Isaacs, C.E., Kim, K.S. and Thormar, H. 1994. Inactivation of enveloped viruses in human bodily fluids by purified lipids. *Ann. New York Acad. Sci.* 724:457–464.

Japanese Patent. 1985. Food mutagen-inhibiting monoglycerides additives. Japan Kokai Tokkyo Koho JP 60/84219 A2 (85/84219), May 13, 1985.

Kabara, J.J. 1978. Synergistic microbiocidal composition and methods. U.S. Patent 4,067,997, January 10, 1978.

Kabara, J.J. 1979a. Fatty acids and derivatives as antimicrobial agents: A review. In *Pharmacological Effects of Lipids*, J.J. Kabara (Ed.), American Oil Chemists Society, Champaign, IL, AOCS Monograph No. 5, pp. 1–14.

Kabara, J.J. 1979b. Multi-functional food-grade preservatives in cosmetics. *Drug Cosmet. Ind.* 125:60, 62, 64, 66, 68, 70, 75, 76, 140, 142, 144, 145.

Kabara, J.J. 1980. GRAS antimicrobial agents for cosmetic products. *J. Cosmet. Chem.* 31:1–10.

Kabara, J.J. 1982. A new preservative system for food. *J. Food Safety* 4:13–25.

Kabara, J.J. 1984. Inhibition of *Staphylococcus aureus* in a model agar-meat system by monolaurin. *J. Food Safety* 6:197–201.

Kabara, J.J., Swieczkowski, D.M., Conley, A.J. and Truant, J.P. 1972. Fatty acids and derivatives as antimicrobial agents. *Antimicrob. Agents Chemother.* 2:23–28.

Kabara, J.J., Conley, A.J., Swieczkowski, D.J., Ismail, I.A., Lie Ken Jie, M. and Gunstone, F.D. 1973. Antimicrobial action of isomeric fatty acids on group A *Streptococcus*. *J. Med. Chem.* 16:1060–1063.

Kabara, J.J., Vrable, R. and Lie Ken Jie, M. 1977. Antimicrobial lipids: Natural and synthetic fatty acids and monoglycerides. *Lipids* 9:753–759.

Katagiri, K. and Yamamoto, H. 1978. Food preservatives. Japan Kokai Tokkyo Koho JP 53/148528 (78/148528), December 15, 1978.

Kato, N. 1981. Antimicrobial activity of fatty acids and their esters against film-forming yeast in soy sauce. *J. Food Safety* 3:121–126.

Kato, A. and Arima, K. 1971. Inhibitory effect of sucrose ester of lauric acid on the growth of *E. coli. Biochem. Biophys. Res. Commun.* 42:596–601.

Kato, A. and Shibasaki, I. 1975a. Combined effect of different drugs on the antibacterial activity of fatty acids and their esters. *J. Antibacterial Antifung. Agents (Jpn.)* 8:355–361.

Kato, N. and Shibasaki, I. 1975b. Comparison of antimicrobial activities of fatty acids and their esters. *J. Ferment. Technol.* 53:793–801.

Kato, N. and Shibasaki, I. 1975c. Enhancing effect of fatty acids and their esters on the thermal destruction of *E. coli* and *Pseudomonas aeruginosa. J. Ferment. Technol.* 53:802–807.

Kato, N. and Shibasaki, I. 1976. Combined effect of citric and polyphosphoric acid on the anti-bacterial activity of monoglycerides. *J. Antibacterial. Antifung. Agents* 4:254–261.

Keeney, E.L. 1943. Fungistatic and fungicidal effect of sodium propionate on common pathogens. *Bull. Johns Hopkins Hosp.* 73:379–390.

Keeney, E.L. 1944. New perspectives for the treatment of fungous infections. In vitro and in vivo experiments with fatty acid salts, penicillin and sodium sulfathiazole. *Clin. Invest.* 23:929.

Kiesel, A. 1913. The action of different acids and acid salts upon the development of *Aspergillus niger. Ann. Inst. Pasteur* 27:193–196.

Kim, S.A. and Rhee, M.S. 2013. Marked synergistic bactericidal effects and mode of action of medium-chain fatty acids in combination with organic acids against *Escherichia coli* O157:H7. *Appl. Environ. Microbiol.* 79:6552–6560.

Kim, S.A. and Rhee, M.S. 2015. Synergistic antimicrobial activity of caprylic acid in combination with citric acid against both *Escherichia coli* O157: H7 and indigenous microflora in carrot juice. *Food Microbiol.* 49:166–172.

Kimsey, M., Adams, H.R. and Kabara, J.J. 1981. Increased inactivation of bacterial spores in the presence of monolauryl-glyceride. *J. Food Safety* 3:69–82.

Kinderlerer, J.L., Matthais, H.E. and Finner, P. 1996. Effect of medium-chain fatty acids in mould ripened cheese on the growth of *Listeria monocytogenes. J. Dairy Res.* 63:593–606.

Kirby, G.W., Atkin, L. and Frey, C.N. 1937. Growth of bread molds is influenced by acidity. *Cereal Chem.* 14:865–878.

Kitajima, K. and Kawamura, J. 1931. Antiseptic action of higher fatty acids against wood-attacking fungi. *Bull. Imp. Forestry Exp. Sta. Jpn.* 31:108–113.

Kodicek, E. 1949. The antagonism of unsaturated fatty acids on Gram-positive bacteria. *Soc. Exp. Biol. Symp.* 3:946–960.

Kodicek, E. and Worden, A.N. 1945. Effect of unsaturated fatty acids on *Lactobacillus helveticus* and other Gram-positive microorganisms. *Biochem. J.* 39:78–85.

Kozak, S.M., Margison, K.M. and D'amico, D.J. 2017. Synergistic antimicrobial combinations inhibit and inactivate *Listeria monocytogenes* in neutral and acidic broth systems. *J. Food Prot.* 80:1266–1272.

Lavieri, N.A., Sebranek, J.G., Brehm-Stecher, B.F., Cordray, J.C., Dickson, J.S., Horsch, A.M., Jung, S., Larson, E.M., Manu, D.K. and Mendonça, A.F. 2014. Investigating the control of *Listeria monocytogenes* on a ready-to-eat ham product using natural antimicrobial ingredients and postlethality interventions. *Foodborne Pathog Dis.*11:462–467.

Levey, M. 1958. Gypsum, salt and soda in ancient Mesoptamian chemical technology. *J. History of Sci. Soc.* 49:336–342.

Levine, A.S. and Fellers, C.R. 1940. Action of acetic acid on food-spoilage microorganisms. *J. Bacteriol.* 39:499–514.

Lisker, N. and Paster, N. 1982. Antifungal activity of monolaurin and related compounds. *J. Food Safety* 4:27–34.

Lourenco, A., Kamnetz, M.B., Gadotti, C., Diez-Gonzalez, F. 2017. Antimicrobial treatments to control *Listeria monocytogenes* in queso fresco. *Food Microbiol.* 64:47–55.

Mansour, M. and Milliere, J.B. 2001. An inhibitory synergistic effect of a nisin-monolaurin combination on *Bacillus* sp. vegetative cells in milk. *Food Microbiol.* 18:87–94.

Mansour, N., Yousef, A.E. and Kime, J.G. 1996. Inhibition of surface growth of toxigenic and nontoxigenic aspergilli and penicillia by eugenol, isoeugenol, and monolaurin. *J. Food Safety* 16:219–229.

Mansour, M., Amri, D., Bouttefroy, A., Linder, M. and Milliere, J.B. 1999. Inhibition of *Bacillus licheniformis* spore growth in milk by nisin, monolaurin, and pH combinations. *J. Appl. Microbiol.* 86:311–324.

Marounek M., Putthana V., Benada O. and Lukešova D. 2012. Antimicrobial activities of medium chain fatty acids and monoacylglycerols on *Cronobacter sakazakii* DBM 3157T and *Cronobacter malonaticus* DBM 3148. *Czech J. Food Sci.* 30:573–580.

Marshall, D.L. and Bullerman, L.B. 1986. Effect of sucrose esters in combination with selected mold inhibitors on growth and aflatoxin production by *Aspergillus parasiticus. J. Food Prot.* 49:378–382.

Marshall, D.L. and Bullerman, L.B. 1994. Antimicrobial properties of sucrose fatty acid esters. In *Carbohydrate Polyesters as Fat Substitutes*, C.C. Akoh and B.G. Swanson (Eds.), Marcel Dekker, Inc., New York, Chapter 8, pp. 149–167.

McKee, C.M., Butcher, J.D., Groope, V. and Moore, M. 1947. Antibacterial lipids from *Tetrahymena geleii. Proc. Soc. Exp. Biol. Med.* 65:326–332.

McLay, J.C., Kennedy, M.J., O'Rourke, A.-L., Elliot, R.M. and Simmonds, R.S. 2002. Inhibition of bacterial foodborne pathogens by the lactoperoxidase system in combination with monolaurin. *Intl. J. Food Microbiol.* 73:1–9.

Meynell, G.G. 1955. Some factors affecting the resistance of mice to oral infection by *Salmonella. Proc. R. Soc. Med.* 48:916.

Mohamad L., Hossein, T., Moradi, M., Forough, M., Divsalar, E. and Kuswandi, B. 2018. Nanostructured chitosan/monolaurin film: preparation, characterization and antimicrobial activity against *Listeria monocytogenes* on ultrafiltered white cheese. *LWT Food Sci. Technol.* 92:576–583.

Moradi, M., Tajik, H., Razavi Rohani, S.M. and Mahmoudian, A. 2016. Antioxidant and antimicrobial effects of zein edible film impregnated with *Zataria multiflora* Boiss. Essential oil and monolaurin. *LWT Food Sci. Technol.* 72:37–43.

Nair, M.K.M., Vasudevan, P., Hoagland, T. and Venkitanarayanan, K. 2004. Inactivation of *Escherichia coli* O157: H7 and *Listeria monocytogenes* in milk by caprylic acid and monocaprylin. *Food Microbiol.* 21:611–616.

Nakamura, M. and Zangar, M.J. 1968. Effect of fatty acids on *Shigella. Proc. Montana Acad. Sci.* 28:51.

Naruki, S., Tsuchido, T. and Shibasaki, I. 1983. Inhibition of germination and outgrowth of *Bacillus subtilis* var. niger spores by dodecanoic acid and its monoglyceride. *Ann. Rep. Int. Center Coop. Res. Dev. Micro. Eng. (Jpn.)* 161–171.

Nieman, C. 1954. Influence of trace amounts of fatty acids on the growth of microorganisms. *Bacteriol. Rev.* 18:147–163.

Nobmann, P., Smith, A., Dunne, J., Henehan, G. and Bourke, P. 2009. The antimicrobial efficacy and structure activity relationship of novel carbohydrate fatty acid derivatives against *Listeria* spp. and food spoilage microorganisms. *Intl. J. Food Microbiol.* 128:440–445.

Notermans, S. and Dufrenne, J. 1981. The effect of glyceryl monolaurate on toxin production by *Clostridium botulinum* in meat slurry. *J. Food Safety* 3:83–88.

Ogawa, H. 1980. Quality improvement of frozen cooked foods. Japan Tokkyo Koho JP 55/2269 (80/2269), January 16, 1980.

Oh, D.H. and Marshall, D.L. 1992. Effect of pH on the minimum inhibitory concentration of monolaurin against *Listeria monocytogenes. J. Food Prot.* 55:449–450.

Oh, D.H. and Marshall, D.L. 1993a. Influence of temperature, pH, and glycerol monolaurate on growth and survival of *Listeria monocytogenes. J. Food Prot.* 56:744–749.

Oh, D.H. and Marshall, D.L. 1993b. Antimicrobial activity of ethanol, glycerol monolaurate or lactic acid against *Listeria monocytogenes. Intl. J. Food Microbiol.* 20:239–246.

Oh, D.H. and Marshall, D.L. 1994. Enhanced inhibition of *Listeria monocytogenes* by glycerol monolaurate with organic acids. *J. Food Sci.* 59:1258–1261.

Oh, D.H. and Marshall, D.L. 1995a. Influence of packaging method, lactic acid and monolaurin on *Listeria monocytogenes* in crawfish tailmeat homogenate. *Food Microbiol.* 12:159–163.

Oh, D.H. and Marshall, D.L. 1995b. Destruction of *Listeria monocytogenes* biofilms on stainless steel using monolaurin and heat. *J. Food Prot.* 58:251–255.

Oh, D.H. and Marshall, D.L. 1996. Monolaurin and acetic acid inactivation of *Listeria monocytogenes* attached to stainless steel. *J. Food Prot.* 59:249–252.

Orloff, S.L., Wallingford, J.C. and McDougal, J.S. 1993. Inactivation of human immunodeficiency virus type 1 in human milk: Effects of intrinsic factors in human milk and of pasteurization. *J. Human Lact.* 9:13–17.

Ouattara, B., Simard, R.E., Holley, R.A., Piette, J.P. and Begin, A. 1997. Antibacterial activity of selected fatty acids and essential oils against six meat spoilage organisms. *Intl. J. Food Microbiol.* 37:155–162.

Peck, S.M. and Rosenfeld, H. 1938. Effects of H-ion concentration, fatty acids, and vitamin C on the growth of fungi. *Invest. Dermatol.* 1:237–265.

Peck, S.M., Rosenfeld, H., Leifer, W. and Bierman, W. 1939. Sweat as a fungicide-use of constituents of sweat in the therapy of fungous infections. *Arch. Dermatol. Syphilol.* 39:126–148.

Petschow, B.W., Bateman, R.P. and Ford, L.L. 1996. Susceptibility of *Helicobacter pylori* to bactericidal properties of medium-chain monoglycerides and free fatty acids. *Antimicrob. Agents Chemother.* 40:302–306.

Plockova, M., Rihakova, Z. and Filip, V. 1999a. Inhibition of food moulds by lauric acid derivatives. *Adv. Food Sci.* 21:44–47.

Plockova, M., Filip, V., Kukackova, O., Smidrkal, J. and Rihakova, Z. 1999b. Antimicrobial effect of monolauroylglycerol and lauric acid in a model emulsion system. *Czech J. Food Sci.* 17:49–54.

Prince, H.N. 1959. Effect of pH on the antifungal activity of undecylenic acid and its calcium salt. *J. Bacteriol.* 78:788–791.

Projan, S.J., Brown-Skrobot, S., Schlievert, P.M., Vandenesch, F. and Novick, R.P. 1994. Glycerol monolaurate inhibits the production of β-lactamase, toxic shock syndrome toxin-1 and other staphylococcal exoproteins by interfering with signal transduction. *J. Bacteriol.* 176:4204–4209.

Razavi-Rohani, S.M. and Griffiths, M.W. 1994. The effect of mono and polyglycerol laurate on spoilage and pathogenic bacteria associated with foods. *J. Food Safety* 14:131–151.

Razavi-Rohani, S.M. and Griffiths, M.W. 1996. Inhibition of spoilage and pathogenic bacteria associated with foods by combinations of antimicrobial agents. *J. Food Safety* 16:87–104.

Reichenback, H. 1908. The effect of soaps on E. coli. *A. Hyg. Infektionskr.* 59:296.

Reid, J.D. 1932. The disinfectant action of certain organic acids. *Am. J. Hyg.* 16:540–556.

Rigler, N.E. and Greathouse, G.A. 1940. The chemistry of resistance of plants to *Phymatotri chum* root. VI: Fungicidal properties of fatty acids. *Am. J. Bot.* 27:701–704.

Robach, M.C., Hickey, C. and To, E.C. 1981. Comparison of antimicrobial actions of monolaurin and sorbic acid. *J. Food Safety* 3:89–98.

Ruzin, A. and Novick, R.P. 2000. Equivalence of lauric acid and glycerol monolaurate as inhibitors of signal transduction in *Staphylococcus aureus*. *J. Bacteriol.* 182:2668–2671.

Sands, J.A., Auperin, D.A., Landin, Ph.D., Reinhardt, A. and Cadden, S.P. 1979. Antiviral effects of fatty acids and derivatives. In *Pharmacological Effect of Lipids*, J.J. Kabara (Ed.), American Oil Chemists Society, Champaign, IL, Monograph No. 5, p. 75.

Schlievert, P.M., Deringer, J.R., Kim, M.H., Projan, S.J. and Novick, R.P. 1992. Effect of glycerol monolaurate on bacterial growth and toxin function. *Antimicrob. Agents Chemother.* 36:626–631.

Shao, S.-Y., Shi, Y.-G., Wu, Y., Bian, L.-Q., Zhu, Y.-J., Huang, X.-Y., Pan, Y., Yeng, L.-Y. and Zhang, R.-R. 2018. Lipase-catalyzed synthesis of sucrose monolaurate and its antibacterial property and mode of action against four pathogenic bacteria. *Molecules* 23:1118.

Shearer, A.E.H., Dunne, C.P., Sikes, A. and Hoover, D.G. 2000. Bacterial spore inhibition and inactivation in foods by pressure, chemical preservatives, and mild heat. *J. Food Prot.* 63:1503–1510.

Shibasaki, I. 1982. Food preservation with nontraditional antimicrobial agents. *J. Food Safety* 4:35–58.

Shibasaki, L. and Kato, N. 1979. Combined effects on antibacterial activity of fatty acids and their esters against gram-negative bacteria. In *Pharmacological Effect of Lipids*, J.J. Kabara (Ed.), American Oil Chemists Society, Champaign, IL, Monograph No. 5, pp. 15–24.

Shibasaki, L, Yuta, K., Matsuura, H. and Iwasaki, T. 1979. Food preservative composition. *Japan Tokkyo Koho* JP 54/34061 (79/34061), October 24, 1979.

Skřivanová, E., Marounek, M., Benda, V. and Brezina, P. 2006. Susceptibility of *Escherichia coli*, *Salmonella* sp. and *Clostridium perfringens* to organic acids and monolaurin. *Veterinarni Medicina* 51:81–88.

Skřivanová, E., Pražáková, Š., Benada, O., Hovorková, P. and Marounek, M. 2014. Susceptibility of *Escherichia coli* and *Clostridium perfringens* to sucrose monoesters of capric and lauric acid. *Czech J. Anim. Sci.* 59:374–380.

Smith, J.L. and Palumbo, S.A. 1980. Inhibition of aerobic and anaerobic growth of *Staphylococcus aureus* in a model sausage system. *J. Food Safety* 2:221–233.

Stecchini, M.L., Di Luch, R., Bortolussi, G. and Del Torre, M. 1996. Evaluation of lactic acid and monolaurin to control *Listeria monocytogenes* on Stracchino cheese. *Food Microbiol.* 13:483–488.

Stillmunkes, A.A., Prabhu, G.A., Sebranek, J.G. and Molins, R.A. 1993. Microbiological safety of cooked beef roasts treated with lactate, monolaurin or gluconate. *J. Food Sci.* 58:953–958.

Takano, M., Simbol, A.B., Yasin, M. and Shibasaki, I. 1979. Bactericidal effect of freezing with chemical agents. *J. Food Sci.* 44:112–115.

Tetsumoto, S. 1933a. Sterilizing action of acids. II: Sterilizing action of saturated monobasic fatty acids. *J. Agric. Chem. Soc. Jpn.* 9:388–397.

Tetsumoto, S. 1933b. Sterilizing action of acids. III: Sterilizing action of saturated monobasic fatty acids. *J. Agric. Chem. Soc. Jpn.* 9:563–567.

Thomas, L.V., Davies, E.A., Delves-Broughton, J. and Wimpenny, J.W.T. 1998. Synergist effect of sucrose fatty acid esters on nisin inhibition of Gram-positive bacteria. *J. Appl. Microbiol.* 85:1013–1022.

Thormar, H., Isaacs, E.C., Brown, H.R., Barshatzky, M.R. and Pessolano, T. 1987. Inactivation of enveloped viruses and killing of cells by fatty acids and monoglycerides. *Antimicrob. Agents Chemother.* 31:27–31.

Thornton, R.H. 1963. Antifungal activity of fatty acids to *Pithomyces chartatum* and other fungi. *N.Z. J. Agric. Res.* 6:469–483.

Tokarskyy, O. and Marshall, D.L. 2008. Mechanism of synergistic inhibition of *Listeria monocytogenes* growth by lactic acid, monolaurin, and nisin. *Appl. Environ. Microbiol.* 74:7126–7129.

Trotter, T.M. and Marshall, D.L. 2003. Influence of pH and NaCl on monolaurin inactivation of *Streptococcus iniae*. *Food Microbiol.* 20:187–192.

Tsuchido, T. and Takano, M. 1988. Sensitization by heat treatment of *Escherichia coli* K-12 cells to hydrophobic antibacterial compounds. *Antimicrob. Agents Chemother.* 32:1680–1683.

Tsuchido, T., Saeki, T. and Shibasaki, I. 1981. Death kinetics of *Escherichia coli* in a combined treatment of heat and monolaurin. *J. Food Safety* 3:57–68.

USDA FSIS. 2019. FSIS Directive. Safe and suitable ingredients used in the production of meat, poultry, and egg products. 7120.1 Revision 15. USDA Food Safety and Inspection Service, Washington, DC.

Unda, J.R., Molins, R.A. and Walker, H.W. 1991. *Clostridium sporogenes* and *Listeria monocytogenes*: Survival and inhibition in microwave-ready beef roasts containing selected antimicrobials. *J. Food Sci.* 56:198–205, 219.

Vasseur, C., Rigaud, N., Hebraud, M. and Labadie, J. 2001. Combined effects of NaCl, NaOH, and biocides (monolaurin or lauric acid) on inactivation of *Listeria monocytogenes* and *Pseudomonas* spp. *J. Food Prot.* 64:1442–1445.

Verhaegh, E.G.A., Marshall, D.L. and Oh, D.H. 1996. Effect of monolaurin and lactic acid on *Listeria monocytogenes* attached to catfish fillets. *Intl. J. Food Microbiol.* 29:403–410.

Venkitanarayanan, K.S., Zhao, T. and Doyle, M.P. 1999. Inactivation of *Escherichia coli* O157:H7 by combinations of GRAS chemicals and temperatures. *Food Microbiol.* 16:75–82.

Walker, J.E. 1924. Germicidal properties of chemically pure soaps. *J. Infect. Dis.* 35:557–566.

Wang, L.L. and Johnson, E.A. 1992. Inhibition of *Listeria monocytogenes* by fatty acids and monoglycerides. *Appl. Environ. Microbiol.* 58:624–629.

Wang, L.L. and Johnson, E.A. 1997. Control of *Listeria monocytogenes* by monoglycerides in foods. *J. Food Prot.* 60:131–138.

Welsh, J.K., Arsenakis, M., Coelen, R.J. and May, J.T. 1979. Effect of antiviral lipids, heat, freezing on the activity of viruses in human milk. *J. Infect. Dis.* 140:322–328.

Winslow, C.E.A. and Lochridge, E.E. 1906. The toxic effect of certain acids upon typhoid and colon bacilli in relation to the degree of their dissociation. *J. Infect. Dis* 3:547–571.

Woolford, M.K. 1975. Microbiological screening of the straight chain fatty acids (C1-C12) as potential silage additives. *J. Sci. Food Agric.* 28:219.

Wyss, O., Ludwig, B.J. and Joiner, R.R. 1945. Fungistatic and fungicidal action of fatty acids and related compounds. *Arch. Biochem.* 7:415–425.

Yamazaki, K., Yamamoto, T., Kawai, Y. and Inoue, N. 2004. Enhancement of antilisterial activity of essential oil constituents by nisin and diglycerol fatty acid ester. *Food Microbiol.* 21:283–289.

Yang, C.-M. Luedecke, L.O., Swanson, B.G. and Davidson, P.M. 2003. Inhibition of microorganisms in salad dressing by sucrose and methylglucose fatty acid monoesters. *J. Food Proc. Preserv.* 27:285–298.

Yuda, M., Matsuura, H., Takeuchi, A. and Iwasaki, T. 1977. Preservatives for starch-containing foods. Japan Kokai JP 52/21317 (77/21317), February 17, 1977.

Zhang, H., Wei, H., Cui, Y. Zhao, G. and Feng, F. 2009. Antibacterial interactions of monolaurin with commonly used antimicrobials and food components. *J. Food Sci.* 74:M418–M421.

Zhang, S., Xiong, J., Lou, W., Ning, Z., Zhang, D. and Yang, J. 2019. Antimicrobial activity and action mechanism of triglycerol monolaurate on common foodborne pathogens. *Food Control* 98:113–119.

12

Parabens

P. Michael Davidson

CONTENTS

The methyl and propyl esters of p-hydroxybenzoic acid, are allowed to be directly added to foods as antimicrobials in many countries. The following sections will review various characteristics of these compounds along with the ethyl, butyl, and heptyl esters which are approved for use in foods by selected countries.

12.1 Chemical and Physical Properties

Parabens have the general structure shown in Figure 12.1. The molecular weights of the various esters are as follows: methyl (Chemical Abstracts Service (C.A.S.) 99-76-3), 152.15; ethyl (C.A.S. 120-47-8), 166.18; propyl (C.A.S. 94-13-3), 180.21; butyl (C.A.S. 94-26-8), 194.23; and heptyl, 236.21. Solubility data for these compounds are shown in Table 12.1. Water solubility is inversely related to alkyl chain length.

Parabens are stable in air and are resistant to cold and heat, including steam sterilization. Aalto et al. (1953) detected no hydrolysis in solutions of parabens buffered to pH 3.0 and 6.0 and heated at 120°C for 30 min. At pH 8.0, 6% hydrolysis was detected under the same conditions. Solutions of the parabens buffered to pH 3.0, 6.0, and 8.0 remained unchanged during storage at 25°C for 6 weeks (Aalto et al., 1953). The compounds are all odorless, except for methyl paraben, which reportedly has a faint characteristic odor (Aalto et al., 1953, Chichester and Tanner, 1972).

12.2 Antimicrobial Activity

The first reports on the antimicrobial activity of the parabens came from Sabalitschka and co-workers in the early 1920s (Prindle, 1983). Esterification of the carboxyl group of benzoic acid allows the molecule to remain undissociated up to pH 8.5 versus the pK_a of benzoic acid which is 4.20. Thus, while the pH optimum for antimicrobial activity of benzoic acid is 2.5–4.0, parabens are effective at pH 3–8 (Aalto et al., 1953; Chichester and Tanner, 1972). Propyl paraben has been demonstrated to be two to eight times

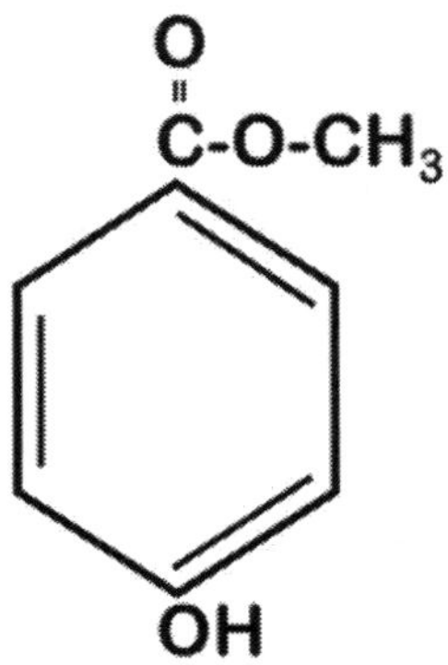

Methyl

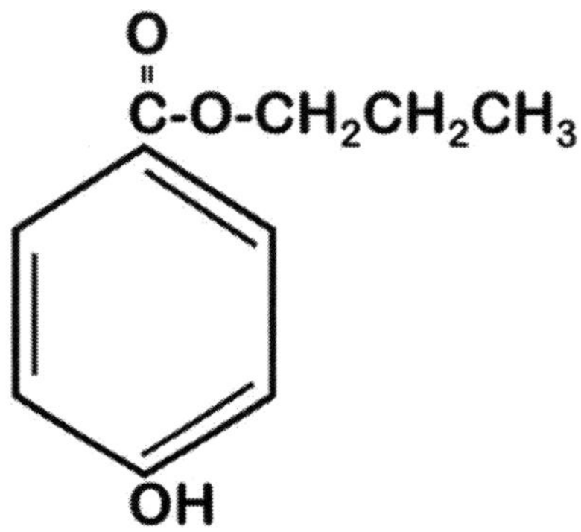

Propyl

FIGURE 12.1 Structure of the methyl and propyl esters of *p*-hydroxybenzoic acid.

TABLE 12.1
Solubility of the Parabens in Various Solvents

		Solubility (g/100 g)				
Solvent	**Temperature**[a]	**Methyl**	**Ethyl**	**Propyl**	**Butyl**	**Heptyl**
Water	25°C	0.25	0.17	0.05	0.02	1.5 mg
	10°C	0.20	0.07	0.025	0.005	-[b]
	80°C	2.0	0.86	0.30	0.15	-
Ethanol	25°C	52.0	70.0	95.0	210.0	-
	50% (25°C)	18.0	-	18.0	-	-
	10% (25°C)	0.5	-	0.1	-	-
Propylene glycol	25°C	22.0	25.0	26.0	110.0	-
	50% (25°C)	2.7	-	0.9	-	-
	10% (25°C)	0.3	-	0.06	-	-
Olive oil	25°C	2.9	3.0	5.2	9.9	-
Peanut oil	25°C	0.5	1.0	1.4	5.0	-

[a] Solubilities were determined at the temperatures specified; all solvents were pure, except where percentages are shown, which indicates percentage in water.
[b] Not reported.
Sources: Aalto et al. (1953); Lück and Jager (1997).

more effective at inhibiting bacterial growth than sodium benzoate or sorbate at pH 6.8–7.0 (Aalto et al., 1953; Jurd et al., 1971). Additionally, Eklund (1985a) showed that, while the antimicrobial activity of parabens was related to pH, this dependence was not related to dissociation of the compounds.

12.2.1 Bacteria

The antimicrobial activity of parabens has been evaluated against a wide variety of Gram-negative and Gram-positive food-related bacteria (Table 12.2). It should be noted that many inhibition studies were done with different strains of bacteria, incubation conditions (pH, time, temperature), media, assay techniques, and data analyses. Because of these differences, it is difficult to compare results of different studies, except in relative terms.

From the results of the minimum inhibitory concentration studies shown in Table 12.2, it can be seen that, as the alkyl chain length of the parabens increases, inhibitory activity generally increases. Increasing

TABLE 12.2

Concentration Ranges of Esters of p-Hydroxybenzoic Acid Necessary for Total Growth Inhibition[a] of Growth of Various Bacteria (pH, Incubation Temperature, and Time Vary)

	Concentration (μg/ml)				
Microorganism	**Methyl**	**Ethyl**	**Propyl**	**Butyl**	**Heptyl**
Gram-positive					
Bacillus cereus	1000–2000	830–1000	125–400	63–400	12
Bacillus megaterium	1000	-	320	100	-
Bacillus subtilis	1980–2130	1000–1330	250–450	63–115	-
Clostridium botulinum	1000–1200	800–1000	200–400	200	-
Clostridium perfringens	500 (3:1 methyl and propyl)				
Lactococcus lactis	-	-	400	-	12
Listeria monocytogenes	1430–1600	-	512	-	-
Micrococcus sp.	-	60–110	10–100	-	-
Micrococcus (Sarcina) lutea	4000	1000	400–500	125	12
Staphylococcus aureus	1670–4000	1000–2500	350–540	120–200	12
Streptococcus faecalis	-	130	40	-	-
Gram-negative					
Aeromonas hydrophila	550	-	100	-	-
Enterobacter aerogenes	2000	1000	1000	4000	-
Escherichia coli	1200–2000	1000–2000	400–1000	1000	-
Klebsiella pneumoniae	1000	500	250	125	-
Pseudomonas aeruginosa	4000	1000–4000	8000	8000	-
Pseudomonas fluorescens	1310	-	670	-	-
Pseudomonas fragi	-	-	4000	-	-
Pseudomonas putida	450	-	-	-	-
Pseudomonas stutzeri	500–750	400–500	250–300	100	-
Salmonella	2000	1000	1000	1000	-
Salmonella Typhimurium	-	-	180–300	-	-
Vibrio parahaemolyticus	-	-	50–100		
Yersinia enterocolitica	350	-	-	-	-

Sources: Aalto et al. (1953), Bargiota et al. (1987), Dymicky and Huhtanen (1979), Eklund et al. (1981), Eklund (1985a), Jurd et al. (1971), Juneja and Davidson (1993), Kato and Shibasaki (1975), Klindworth et al. (1979), Lee (1973), Lewis and Jurd (1972), Lück and Jager (1997), Moir and Eyles (1992), Moustafa and Collins (1969), Payne et al. (1989), Pierson et al. (1980), Reddy et al. (1982), Robach and Pierson (1978), Sokol (1952), Tattawasart et al. (1999), Venugopal et al. (1984).

[a] As defined by the authors of the references cited.

activity with decreasing polarity is more evident against Gram-positive than against Gram-negative bacteria. Gram-positive bacteria are generally more susceptible to non-polar phenolic compounds than Gram-negative bacteria. Both Eklund (1980) and Freese et al. (1973) stated that Gram-negative bacteria were, most likely, resistant to the parabens owing to a screening effect by the cell wall lipopolysaccharide layer. Fukahori et al. (1996) studied the relationship between uptake and the antimicrobial activity of the methyl, ethyl, propyl, and butyl esters of *p*-hydroxybenzoic acid using *Escherichia coli*. They reported that the uptake of the parabens was logarithmically proportional to the alkyl chain length from methyl to butyl. However, free energy change measurements indicated that the transfer of the compounds also involved hydrophilic interactions. In addition, they found that the concentration of parabens necessary for antimicrobial activity decreased in logarithmic relationship to alkyl chain length. They concluded that the antibacterial activity of parabens is dependent upon alkyl chain length both for uptake and for concentration at the cellular target.

In addition to total inhibition achieved in studies shown in Table 12.2, others have reported variable results with partial or no inhibition. For example, Martin et al. (1972) found no inhibition of *Alcaligenes viscolactis* in skim milk with up to 600 µg/ml propyl paraben. Moustafa and Collins (1969) showed that, while 4000 µg/ml propyl paraben inhibited the growth of *Pseudomonas fragi*, 2000 µg/ml actually stimulated growth. At 200 µg/ml, propyl paraben inhibited protease secretion by *Aeromonas hydrophila* (Venugopal et al., 1984). Ahmedy et al. (1999) found no difference in susceptibility of pathogenic and non-pathogenic strains of *Yersinia* to methyl paraben despite the fact that the same strains showed differences in resistance to certain antibiotics and cationic biocides.

Darwish and Bloomfield (1997) evaluated the effect of co-solvents ethanol, propylene glycol, and glycerol on the activity of methyl and propyl parabens against *Staphylococcus aureus* and *Pseudomonas aeruginosa*. Interestingly, the antimicrobial activity of the parabens increased with increasing hydrophobicity of the co-solvent, being greatest with the most hydrophobic co-solvent, ethanol. In contrast, the uptake of parabens was apparently determined by the hydrophilicity of the co-solvent, with glycerol causing the greatest uptake by the cells. Uptake however, was not correlated with inhibition. It was concluded that inhibition was a combination of the action of the solvent and the paraben on the integrity of the outer (*P. aeruginosa*) and cytoplasmic (both genera) membranes.

Moir and Eyles (1992) compared the effectiveness of methyl paraben and potassium sorbate on the growth of four psychrotrophic foodborne bacteria, *A. hydrophila*, *Listeria monocytogenes*, *Pseudomonas putida*, and *Yersinia enterocolitica*. At pH 5, little difference was found between minimum inhibitory concentrations (MIC) of methyl paraben and potassium sorbate at 5°C or 30°C. At pH 6 however, methyl paraben was effective at a lower concentration than potassium sorbate for all pathogens except *A. hydrophila*, where the two were equal. Little or no adaptation was found to occur when cells were exposed to sub-inhibitory concentrations of antimicrobials. At 5°C in the presence of 1000 µg/ml methyl paraben, *A. hydrophila* survived for 1–2 days, *P. putida* and *Y. enterocolitica* for 1–2 weeks, and *L. monocytogenes* for over 4 months. Injury occurred with *A. hydrophila* and *L. monocytogenes* but was variable for *Y. enterocolitica* and did not occur for *P. putida*. Razavilar and Genigeorgis (1998) studied the influence of temperature, time, and inoculum on the efficacy of methyl paraben against growth of *Listeria monocytogenes*, *L. innocua*, *L. ivanovii*, and *L. seeligeri*. Methyl paraben at 1000 µg/ml allowed growth of all species of *Listeria* at pH 6.0–6.2 in brain heart infusion (BHI) broth at 20 and 30°C. In contrast, no growth occurred with any species at 4 or 8°C and 1000 µg/ml methyl paraben. At 1500 µg/ml and 20°C, methyl paraben had inhibitory effects (e.g., increased lag time, decreased final growth level) on the *Listeria* species but did not completely inhibit growth of any species except *L. ivanovii*.

Fyfe et al. (1998) evaluated the antimicrobial activity of 1000 µg/ml methyl paraben or benzoic acid with plant oil extracts (fennel, anise, or basil) against *Listeria monocytogenes* and *Salmonella* Enteritidis. Methyl paraben, by itself, did not inhibit the growth of either microorganism under the test conditions. However, when combined with 0.2% anise, fennel, or basil oil, *L. monocytogenes* was reduced by 5.1, 5.7, and >8 logs compared to the control after 24 h, respectively. The combinations were even more effective against *Salmonella* Enteritidis, reducing viable cells by >8 logs compared to the control with all combinations at 24 and 48 h. All combinations containing methyl paraben were more inhibitory than those containing benzoic acid. This was not surprising since the tests were done in a microbiological medium with a pH around 7.0. Er et al. (2014) investigated the inhibitory concentrations of methyl and propyl parabens, among others, against *Salmonella* spp. isolated from chicken as both planktonic and biofilm cells. The MIC (as defined as no growth in 24 h) for the planktonic cells ranged from 60 to 70 µg/ml for methyl and 50–120 for propyl. Biofilm cells were significantly more resistant than the planktonic cells, requiring up to around 2000 µg/ml for methyl and propyl (although one strain required ca. 8000 µg/ml for propyl paraben at pH 6.0). They demonstrated the reduced susceptibility of *Salmonella* in biofilms.

Propyl paraben was evaluated against *Listeria monocytogenes* Scott A in 10% chicken and hot dog suspensions (Dje et al., 1989). The effectiveness of propyl paraben was dependent upon the product. In chicken, *L. monocytogenes* was inhibited 99.9% compared to the control after 24 h incubation at 35°C. In contrast, little inhibition of *L. monocytogenes* growth was demonstrated by propyl paraben in the hot dog suspension. The difference in effectiveness was theorized to be due to a higher fat content in the hot dogs. Dje et al. (1990) evaluated the effect of 1000 µg/ml propyl paraben and 1000 µg/ml methyl:propyl paraben in salt brine (13%) on *L. monocytogenes* suspended in the brine or inoculated on hot dogs which were dipped in brine. Propyl paraben alone had little or no effect on *L. monocytogenes* survival in salt brine at 4°C or on the surface of hot dogs which were dipped in brine for 5 min and incubated at 24°C. In contrast, 1000 µg/ml methyl:propyl paraben caused a 2–3 log decrease in viable *L. monocytogenes* in the salt brine at 4°C. On hot dogs, the antimicrobial combination was less effective, delaying growth for around 4 h at 24°C of one of two strains tested. In a similar study by Blom et al. (1997), propyl paraben added to vacuum-packaged sliced ham or sausage inoculated with *Listeria monocytogenes* and stored at 4 or 9°C for 5 weeks was ineffective in controlling the microorganism. The efficacy of a blend of benzalkonium chloride, acetic acid, and methyl paraben was evaluated as a wash to reduce *Listeria monocytogenes*, *Salmonella*, and *Escherichia coli* O157:H7 on fresh-cut romaine and iceberg lettuce (Sevart et al., 2017). While up to a 3% solution reduced the numbers of test microorganisms in wash water, it did not significantly reduce cells on the lettuce itself.

Robach and Pierson (1978) investigated the effect of methyl and propyl paraben on toxin production of *Clostridium botulinum* NCTC 2021. At 100 µg/ml of methyl and 100 µg/ml of propyl paraben, toxin formation was prevented while 1200 µg/ml methyl and 200 µg/ml propyl were necessary for growth inhibition. Reddy and Pierson (1982) and Reddy et al. (1982) determined the effect of methyl, ethyl, propyl, and butyl parabens on the growth and toxin production of ten *Clostridium botulinum* strains (5 Type A, 5 Type B). In TYG medium at pH 7.0 and 37°C, 1000 µg/ml methyl paraben blocked growth and toxin formation for only 1 day. Ethyl and propyl paraben, at the same concentration, prevented growth and toxin production for the maximum incubation time of 7 d. Butyl paraben, as might be expected, was most effective and prevented growth and toxin production for 7 d at 200 µg/ml. Reddy and Pierson (1982) also evaluated ethyl, propyl, and butyl parabens in TYG medium in the presence of 0.05 M phosphate buffer at pH 7.0 and 6.0. Ethyl paraben prevented growth and toxin production by *C. botulinum* at 37°C for 7 d at 1000 µg/ml (pH 7.0) and 800 µg/ml (pH 6.0). Propyl paraben at 800 µg/ml and 400 µg/ml, and butyl paraben at 200 µg/ml and 100 µg/ml, were similarly effective against *C. botulinum* at pH 7.0 and 6.0, respectively. Draughon et al. (1982) demonstrated effective growth inhibition of *C. botulinum* by 1000 µg/ml of all esters of the parabens. Ethyl paraben at 1000 µg/ml was also effective in inhibiting toxin formation in canned comminuted pork. However, inhibition of *C. botulinum* by the parabens in many food systems has been reported to be much less than in laboratory media (Sofos and Busta, 1980).

Little research on the activity of the n-heptyl ester in foods has been published. Chan et al. (1975) did show that this compound was very effective in inhibiting bacteria involved in the malolactic fermentation of wines.

12.2.2 Fungi

The antifungal effectiveness of the parabens has been evaluated against several food-related fungi (Table 12.3). In comparison to bacteria, fungi are much more susceptible to parabens. As with bacteria, inhibition of fungi increases as the alkyl chain length of the parabens increases.

Thompson (1994) evaluated butyl, propyl, ethyl, and methyl parabens, alone and in combination, against strains of mycotoxigenic *Aspergillus*, *Penicillium*, and *Fusarium*. The most effective parabens were the propyl and butyl esters with minimum inhibitory concentrations of 1.0–2.0 mM in potato dextrose agar. Combinations of the various parabens were reported to have synergistic activity against the

TABLE 12.3

Concentration Ranges of Esters of p-Hydroxybenzoic Acid Necessary for Total Growth Inhibition[a] of Various Fungi (pH, Incubation Temperature, and Time Vary)

	Concentration (μg/ml)				
Fungi	**Methyl**	**Ethyl**	**Propyl**	**Butyl**	**Heptyl**
Alternaria sp.	-	-	100	-	50–100
Aspergillus flavus	-	-	200	-	-
Aspergillus niger	1000	400–500	200–250	125–200	-
Byssochlamys fulva	-	-	200	-	-
Candida albicans	1000	500–1000	125–250	125	-
Debaryomyces hansenii	-	400	-	-	-
Penicillium digitatum	500	250	63	<32	-
Penicillium chrysogenum	500	250	125–200	63	-
Rhizopus nigricans	500	250	125	63	-
Saccharomyces bayanus	930	-	220	-	-
Saccharomyces cerevisiae	1000	500	125–200	32–200	25–100
Torula utilis	-	-	200	-	25
Torulaspora delbrueckii	-	700	-	-	-
Zygosaccharomyces bailii	-	900	-	-	-
Zygosaccharomyces bisporus	-	400	-	-	-
Zygosaccharomyces rouxii	-	700	-	-	-

Sources: Aalto et al. (1953), Jermini and Schmidt-Lorenz (1987), Jurd et al. (1971), Kato and Shibasaki (1975), Lewis and Jurd (1972), Marwan and Nagel (1986).

[a] As defined by the authors of the references cited.

mold species. Nesci et al. (2003) determined that 180 μg/ml propyl paraben partially inhibited conidial germination of *Aspergillus flavus* at a_w of 0.982 and totally inhibited aflatoxin B_1 production. Similarly, Passone et al. (2005) showed that propyl paraben at 1800 μg/ml significantly extended the germination and lag phase of several aflatoxigenic *Aspergillus* section *Flavi* in peanut extract meal agar adjusted to various a_w from 0.937 to 0.982. Additionally, 180 μg/ml propyl paraben, which had inhibited but not stopped growth of mold, significantly reduced aflatoxin B_1 production by several toxigenic strains of *A. flavus* and *A. parasiticus*. The inhibitory effect of propyl paraben added to peanut kernals on ochratoxin A production by *Aspergillus* section *Nigri* was studied by Barbaeris et al. (2010). They found that at 18°C, 1800 μg/ml propyl paraben significantly increased the lag phase of two strains of *A. carbonarius* and *A. niger* and 3600 μg/ml extended the lag phase of the strains beyond the maximum time measured (estimated at ca. 7 days). At 20°C, 3600 μg/ml reduced but did not eliminate growth. Propyl paraben was capable of eliminating ochratoxin A production by both *A. carbonarius* and *A. niger* at 18°C and a_w from 0.93 to 0.98. However, raising the temperature to 25°C resulted in greater ochratoxin A in the presence of propyl paraben than the control with both strains at all but 0.93 a_w. Torres et al. (2003) tested propyl paraben as a potential inhibitor of growth and toxin production by *Fusarium* sp. on maize. Propyl paraben at 500 μg/ml significantly increased the lag phase and decreased the growth rate of *F. verticillioides* and *F. proliferatum* at a_w of 0.95, 0.98, and 0.995. In addition, 500 μg/ml propyl paraben was effective in totally inhibiting production of fumonisin toxin production by both species at 0.95 a_w. At a_w of 0.98 and 0.995, fumonisin production was reduced 94–98% and 20–30%, respectively, by 500 μg/ml propyl paraben. Sodium methyl, ethyl, or propyl paraben were all found to be effective at controlling black spot on cherry tomatoes after 6 days at 20°C when added at 2.0% in a hydroxypropyl methylcellulose-lipid edible coating on cherry tomatoes inoculated with *Alternaria alternata* (Fagundes et al., 2013).

Jermini and Schmidt-Lorenz (1987) evaluated ethyl paraben against osmotolerant yeasts at various water activities and pH levels. They found that the concentration of ethyl paraben necessary for inhibition was a function of initial number of yeast cells present. At 600 μg/ml ethyl paraben, the time to initiate growth was approximately 15, 12, 5, and 2–3 days for 10^2, 10^3, 10^4, and 10^5 cells at a_w 0.900 and pH 4.8. Inhibition by parabens was also a function of pH, with more acid environments requiring less ethyl paraben for inhibition. Little effect of water activity or type of humectant was observed. Of the genera tested, *Zygosaccharomyces bailii* was most resistant, requiring 900 μg/ml at 25°C, a_w 0.900 and pH 4.8. Other yeasts evaluated included *Torulaspora delbrueckii*, *Z. rouxii*, *Z. bisporus*, and *Debaryomyces hansenii* with minimum inhibitory concentrations of 700 μg/ml, 700 μg/ml, 400 μg/ml, and 400 μg/ml, respectively. They concluded that the concentration of ethyl paraben required to preserve a product from the effect of osmotolerant yeast for 30 days at 25°C and a_w 0.795–0.980 was 900 μg/ml or 400 μg/ml at a pH of 4.8 or pH 4.0, respectively. Murata et al. (2019) investigated the enhancement of parabens with sulforaphane, a cruciferous vegetable-derived isothiocyanate, against *Saccharomyces cerevisiae*. Sulforaphane was effective in decreasing the growth inhibitory concentrations of ethyl-, propyl-, butyl-, and methylparaben against *S. cerevisiae*. The effect of methylparaben-sulforphane combination caused cell membrane damage and leakage of internal substances. Additionally, sulforaphane decreased the minimum concentration of methyl and ethylparaben to inhibit growth of *Candida albicans*, *Aspergillus niger*, and *Escherichia coli*.

12.3 Mechanism of Action

While an exact cause and effect relation for the mode of action of parabens has not been established, various studies have shown that these compounds may be most active at the cytoplasmic membrane. Leakage of intracellular compounds could indicate disruption of the cytoplasmic membrane. Furr and Russell (1972) detected leakage of intracellular RNA by *Serratia marcescens* in the presence of the parabens. The amount of leakage was proportional to the alkyl chain length of the paraben. Freese et al. (1973) found that the parabens inhibited serine uptake as well as the oxidation of α-glycerol phosphate and NADH in membrane vesicles of *Bacillus subtilis*. They concluded that the parabens were capable of inhibiting both membrane transport and the electron transport system. Bredin et al. (2005) observed increased potassium efflux in *E. coli* cells following incubation with propyl paraben, reporting that K^+ efflux rates were similar to those observed following application of other antimicrobials known to destabilize cation regulation processes such as colicin A and polymyxin B. Eklund (1980) did a similar study using *E. coli*, *B. subtilis*, and *P. aeruginosa*. He determined the uptake of alanine by whole cells, and alanine, serine, phenylalanine, and glucose by vesicles. Parabens generally caused a decrease in amino acid uptake, but not in glucose uptake. Eklund (1980) postulated that since the parabens are known to cause leakage of cellular contents, they are capable of neutralizing chemical and electrical forces which establish a normal membrane gradient. In continued work with *E. coli*, Eklund (1985b) found that parabens eliminated the ΔpH of the cytoplasmic membrane of the organism. In contrast, the compounds did not significantly affect the ΔΨ (membrane potential) component of the proton motive force. He concluded that neutralization of the proton motive force and subsequent transport inhibition could not be the only mechanism of inhibition for the parabens.

Oka (1960) suggested that parabens act on yeast by absorbing on the solid phase of cells rather than in the cell fluid or lipid layers. This conclusion was reached even though there was a direct relation between the antimicrobial dissolved in the lipid phase and the minimum concentration necessary to inhibit the yeast. Bargiota et al. (1987) examined the relationship between lipid composition of *S. aureus* and resistance to parabens. Differences were found for total lipid, phospholipids, and fatty acids between *S. aureus* strains which were relatively resistant and sensitive to parabens. The paraben-resistant strain was shown to have a higher percentage total lipid, higher relative percentage of phosphatidyl glycerol, and decreased cyclopropane fatty acids than sensitive strains. It was suggested that these changes could influence membrane fluidity and therefore adsorption of the parabens to the membrane. Juneja and Davidson (1993) altered the lipid composition of *L. monocytogenes* by growth in the presence of added fatty acids (C14:0, C18:0 or C18:1). Growth of *L. monocytogenes* in the presence of exogenously added C14:0 or C18:0 fatty acids increased the resistance of the cells to TBHQ and parabens. However, growth in the presence of C18:1 led to increased sensitivity to the antimicrobial agents. Results indicated that, for *L.*

monocytogenes, a correlation existed between lipid composition of the cell membrane and susceptibility to antimicrobial compounds.

Rico-Munoz et al. (1987) investigated the effect of BHA, BHT, TBHQ, propyl gallate, p-coumaric acid, ferulic acid, caffeic acid, methyl, and propyl paraben on the membrane-bound ATPase of two strains of *S. aureus*. Only BHA was found to significantly stimulate the activity of the enzyme. Propyl gallate, TBHQ, p-coumaric, ferulic, and caffeic acids caused inhibition of ATPase activity. There was no effect on the enzyme with BHT or the parabens. The authors concluded that phenolic compounds probably do not have the same mechanism of action and there may be several targets which lead to inhibition of microorganisms by these compounds.

Resistance of bacteria to the parabens was studied by Valkova et al. (2002). They reported that a strain of *Enterobacter cloacae* produced an esterase that was responsible for the hydrolysis of parabens. The chromosomal gene *prbA* encoded for the production of the esterase, PrbA (Valkova et al., 2003). Valkova et al. (2002) further demonstrated that 41 strains of *Enterobacter gergoviae*, *E. aerogenes*, *E. coli*, *Pseudomonas agglomerans*, *P. aeruginosa*, and *Burkholderia cepacia* also were resistant to methyl paraben but that only two strains of *E. gergoviae* degraded methyl paraben utilizing a similar mechanism. While these microorganisms were isolated primarily from cosmetics, mineral supplements, and as clinical isolates, they illustrate the potential for development of resistance by target foodborne microorganisms. Additional research has investigated membrane-associated channels and the interactions of these bacterial systems with parabens as a method of explaining the sensitivity of Gram-negative bacteria to parabens. Nguyen et al. (2005) determined that the addition of propyl paraben, and to a lesser extent ethyl paraben, to *E. coli* cells containing a mechanosensitive channel functioning to maintain osmotic balance induced spontaneous opening of the channel in non-osmotically stressful environments, resulting in fatal leakage of cytoplasmic contents.

12.4 Applications

To take advantage of their respective solubility and increased activity, methyl and propyl parabens are normally used in a combination of 2–3:1 (methyl:propyl). Chichester and Tanner (1972) recommended initial testing in foods with 0.05% of a 2:1 methyl:propyl paraben combination. In high-fat foods, they recommended a 0.1% methyl:propyl combination.

The compounds may be incorporated into foods by dissolving in water, ethanol, propylene glycol, or the food product itself. To make an aqueous solution, the water may be at room temperature; however hot water (70–82°C) is recommended (Chichester and Tanner, 1972). The compounds may also be dry blended with other water-soluble components before addition to the food product. Parabens may be dissolved in ethanol or propylene glycol to make a 10–20% stock solution. According to Lück and Jager (1997), parabens have a perceptible taste at use concentrations; however, other sources refute this (Aalto et al., 1953).

Parabens have been suggested for use in a variety of foods (Table 12.4). However, the compounds do not seem to be used widely in foods. In addition to products in Table 12.4, other products reported to have been tested with parabens include margarine, butter, ices, confections, soy sauce, maple syrup, and meats (Chichester and Tanner, 1972). Foods in United States markets with propyl paraben include baked goods (cookies, muffins, cake), snack bars and trail mixes, tortillas, and candies. However, because of the some of the controversial concerns about parabens (see toxicological section) use of these compounds has apparently declined.

With the development of antimicrobial packaging, the parabens have been studied for their potential incorporation in polymeric films. Dobias et al. (2000) incorporated ethyl and propyl parabens into a low-density polyethylene film at 5 and 10 mg/kg. The parabens were found to migrate into a model water and olive oil food system. Chung et al. (2001a, b) studied the release and inhibitory activity of propyl paraben from a styrene-acrylate copolymer coating. They demonstrated that the propyl paraben not only released from the polymer coating but was capable of inhibiting the growth of *Saccharomyces cerevisiae* better than direct addition of the compound.

TABLE 12.4

Potential Applications of the Parabens in Food Products

Product(s)	Ester(s)	Concentration
Bakery products[a] (cakes, crusts, pastries, toppings, fillings)	Methyl:propyl (3:1)	0.03–0.06%
Beverages (soft drinks)	Methyl:propyl (2:1)	0.03–0.05%
Beer	n-Heptyl; butyl	12 µg/ml; 0.01%
Fish (marinated, smoked, jellied)	Methyl, propyl	0.03–0.06%
Flavor extracts	Methyl, propyl	0.05–0.1%
Fruit products	Methyl:propyl (2:1) + sodium benzoate	0.05%
Gelatin	Methyl (or combination)	0.05–0.1%
Jams, jellies, preserves (artificially sweetened)	Methyl:propyl (2:1)	0.07%
Malt extracts	Methyl:propyl (2:1) Propyl	0.05% 0.04%
Olives	Methyl, propyl + sodium benzoate	0.1%
Pickles	Methyl:propyl (2:1)	0.1%
Salad dressings	Methyl:propyl (2:1)	0.1%
Sorbitol	Methyl:propyl (2:1)	0.07%
Syrups	Methyl Propyl	0.07% 0.02%
Tomato concentrate	Methyl (with sulfur dioxide and benzoate)	100 µg/ml
Wines	Methyl:propyl (2:1)	0.1%

[a] Not for use in yeast-leavened products.

Sources: Chichester and Tanner (1972), Dan et al. (1996), Sethi (1994).

12.5 Regulatory Status

In the United States, the methyl (21 CFR 184.1490) and propyl (21 CFR 184.1670) esters of p-hydroxybenzoic acid are generally recognized as safe (GRAS) at a maximum concentration of 0.1% each (Code of Federal Regulations, 2019). When used in combination, the total may not exceed 0.1%. Both methyl and propyl paraben are permitted as antimycotic agents in food packaging material (21 CFR 181.23). The n-heptyl (21 CFR 172.145) ester is also approved for use in fermented malt beverages (beers) at a maximum of 12 µg/ml and in certain noncarbonated soft drinks and fruit-based beverages at a maximum of 20 µg/ml. In Canada, methyl and propyl paraben are allowed for use in jams, juices, jellies, marinated cold processed meat or fish, pickles, relish, tomato products, and unstandardized foods at a maximum of 1000 ppm (Government of Canada, 2019). Japanese regulations allow the addition of the ethyl, propyl, butyl, isopropyl, and isobutyl esters on the rinds of fruit (≤0.012 g/kg (as p-hydroxybenzoic acid (PHBA))), fruit sauce (≤0.2 g/kg (as PHBA)), non-alcoholic beverage syrup (≤0.1 g/kg (as PHBA)), soy sauce (≤0.25 g/kg (as PHBA)), and vinegar (≤0.1 g/kg (as PHBA)) (Japan External Trade Association, 2011). In the European Union, the methyl (E 218) and ethyl (E 214) and their sodium salts (E219 and E215, respectively) are permitted for use in some foods. Those include up to 500 ppm in table-top sweeteners; 1000 ppm in jelly coatings of meat products, heat-treated pate, coated processed nuts, and potato, cereal, flour, or starch-based snacks; and 1500 ppm in some confectionary, coatings and fillings, and surface dried meat products (Anonymous, 2019). Similarly in Russia, the methyl and ethyl esters are allowed in jelly coatings of meat products, breakfast cereal snacks, sugar confectionery, filled chocolate, and the surface of cured meat products (USDA Foreign Agricultural Service Global Agricultural Information Network, 2019). Chinese regulations also allow only the methyl and ethyl esters on the surface of fresh fruits and vegetables, jams, jellies, marmalades, fillings and topping syrups, vinegar, soy sauce, and certain beverages (USDA Foreign Agricultural Service Global Agricultural Information Network, 2015).

12.6 Toxicology

Parabens have been safely used for many decades as preservatives in the food, drug, and personal care and cosmetic industries. They are derived from p-hydroxybenzoic acid (PHBA) that occurs naturally in some plants, including many fruits and vegetables (Tomaszewski, 1960). For example, Baardseth and Russwurm (1978) found that the Scandinavian cloudberry (*Rubus chamaemorus*) contained benzoic acid, sorbic acid, salicylic acid, PHBA, and methyl and propyl parabens. They hypothesized that the compounds were antimicrobial agents synthesized by the cloudberry plant to protect from microbial challenges. Likewise, PHBA is also found naturally in humans as a key component of the ubiquinone (also known as coenzyme Q (CoQ)) biosynthetic pathway, which is evolutionarily conserved across eukaryotic cells (Wang et al., 2015). Liao et al. (2013) measured the concentrations of methyl, ethyl, propyl, butyl, and benzyl parabens in 267 various foods purchased at retail. The mean concentrations of the methyl and propyl esters for all foods were 5.83 ng/g (5.83 ppt) and 1.51 ng/g (1.51 ppt), respectively. The authors did not specify the types of foods tested other than that they were packaged in some way and represented various food categories including beverages, dairy products, fats and oils, fish and shellfish, grains, meat, and fruits and vegetables. The authors calculated the daily intake of all paraben esters for infants, children, and adults as 940, 470, and 307 ng/kg bw/day, respectively. As parabens are not normally added to most of the foods analyzed and there was no indication if they were listed on the ingredient labels, it must be concluded that most of these are naturally occurring parabens.

The methyl and propyl parabens used in the food and cosmetics industries are identical to those found in nature. The human body metabolizes them into PHBA and eliminates them. Parabens are absorbed from the gastrointestinal tract and the ester linkage is hydrolyzed in the liver and kidney (Jones et al., 1956). The resulting PHBA is excreted in the urine unchanged or as *p*-hydroxyhippuric acid, glucuronic acid esters, or sulfates (Lück and Jager, 1997). A majority of the metabolites of the parabens are excreted within 6 and 24 h following intravenous and oral doses, respectively (Jones et al., 1956).

Detailed reviews on the toxicological aspects of methyl and propyl paraben were published by Soni et al. (2001, 2002). The acute toxicity of the parabens has been determined to be of a low order. Matthews et al. (1956) found the oral LD_{50} values in mice for methyl and propyl paraben to be >8000 mg/kg body weight. The sodium salts of methyl, ethyl, propyl, and butyl paraben elicited oral LD_{50} values of 2000, 2500, 3700, and 950 mg/kg, respectively. The chief toxic effect in dogs was acute myocardial depression with hypotension, but the effects were transient and non-cumulative.

In subchronic testing, 500 mg/kg of methyl paraben caused no ill effects to rabbits over a 6-day period, while 3000 mg was toxic to the animals (Lück and Jager, 1997). Lück and Jager (1997) also reported that propyl paraben showed similar subchronic toxicity responses. Prindle (1983) reported that feeding 2–20 mg/kg per day of the "lower" esters of parabens to rabbits, guinea pigs, or rats caused no harmful effects after 120 days. Rats fed up to 60 mg/kg per day for 30 days also showed no effects.

For chronic toxicity testing, white rats were fed diets containing 2% (0.9–1.2 g/kg per day) and 8% (5.5–5.9 g/kg per day) each of methyl and propyl paraben (Matthews et al., 1956). After 96 weeks, the animals at the 2% level had no depressed weight gains or any histological changes in internal organs. At 8% of the diet however, a mild growth retardation was observed. The same researchers found that mongrel dogs could tolerate daily doses of 1 g/kg of the methyl and propyl esters for 1 year with no ill effects. Tissue samples from these animals were normal.

Matthews et al. (1956) reported that none of the esters produced skin irritation in humans at concentrations of 5%. Epstein (1968) however, reported that parabens in foods were associated with dermatitis. Contact dermatitis involving parabens has been reported, although it is associated with topical use (Reitschel and Fowler, 2001). The concentrations required to elicit such a response are generally high and no mechanism is known for the sensitivity (Soni et al., 2001). Similarly, allergic-type reactions have been reported with parabens but evidence of allergenicity of the compounds is lacking (Soni et al., 2001).

In the early 2000s, the parabens family of preservatives came under intense public scrutiny because of concerns over their potential estrogenicity or endocrine disruption capabilities and due to possible links of paraben-containing cosmetics to breast cancer (Darbre et al., 2004). While this issue attracted a great deal of public and global media attention, no objective evidence has demonstrated any significant risk from current cosmetic or food usage of parabens, and various scientific and regulatory bodies have consistently reported that parabens do not cause cancer (CIR, 2008; US FDA, 2018).

Like many other food ingredients, parabens have safe dietary intake amounts that have been established by scientific and government regulatory agencies (U.S. *Code of Federal Regulations*, 2019; 21 CFR 184.1490, 21 CFR 184.1670). Bio-monitoring studies on levels of parabens found in human urine reflect the cumulative exposures across all product types, including foods, cosmetics, and drugs (Calafat et al., 2010). These monitoring studies on levels of parabens provide physicians and public health officials with reference values that facilitate ongoing research on exposure and health effects.

While parabens have been proven safe over decades of research and scrutiny, consumers have shown a growing trend and desire for products that are "natural," "naturally derived," or "nature-inspired." This trend has provoked a strong push by the foods and cosmetics industries to search for natural ingredients that have similar antimicrobial efficacy and can ultimately replace traditional preservatives like parabens. The irony behind this public perception driving the consumer trend away from traditional preservatives is that PHBA is the building block of parabens and is found in nature as a truly natural preservative (Baardseth and Russwurm, 1978).

12.7 Assay

Several methods are available for the qualitative or quantitative determination of the parabens. According to the Joint FAO/WHO Expert Committee on Food Additives (2000), 2 g of a dried sample to be assayed for parabens should be weighed to the nearest mg and transferred to a flask. Forty milliliters of 1 N sodium hydroxide are added and the sides of the flask rinsed with water. The flask is covered with a watch glass, and the solution boiled gently for 1 h and cooled. Five drops of bromothymol blue TS are added and the mixture titrated with 1 N sulfuric acid, comparing the color with a buffer solution (pH 6.5) containing the same proportion of indicator. Blank determinations with the reagents are performed to make any necessary corrections. Each ml of 1 N sodium hydroxide is equivalent to 152.2 mg of (methyl) $C_8H_8O_3$, 166.18 mg of (ethyl) $C_9H_{10}O_3$, or 180.2 mg of (propyl) $C_{10}H_{12}O_3$.

With the increased interest in the toxicological aspects of parabens has come with a plethora of newly developed methods for analysis. These involve more sophisticated extraction procedures and automated analytical techniques. In an article describing the development of a method for screening of foods using high-performance liquid chromatography (HPLC) and photodiode array detection, Maher et al. (2018) reviewed some of the methods that had been developed to analyze for parabens in foods. Those methods include gas chromatographic analysis following derivatization flame ionization detector (Jain et al., 2013), capillary electrophoresis (Alshana et al., 2015; Bottoli et al., 2011), HPLC with UV detection (Saad et al., 2005; Yang et al., 2014) and mass spectrometry detection (Cao et al., 2013; Yin et al., 2018; Molognoni et al., 2018; Song et al., 2017), and ultraperformance liquid chromatography with an electrochemical detector (Chuto et al., 2013), among others.

REFERENCES

Aalto, R.R., Firman, M.C., and Rigler, N.E. 1953. p-Hydroxybenzoic acid esters as preservatives. I. Uses, antibacterial and antifungal studies, properties and determination. *J. Am. Pharm. Assoc.* 42:449.

Ahmedy, A.M., Hussein, M.M.M., El-Falaha, B.M.A., and Megid, R.M.A. 1999. Antimicrobial susceptibility and surface hydrophobicity of *Yersinia* species in relation to genetic elements. *Al-Azhar J. Microbiol.* 45:125.

Alshana, U., Ertaş, N., and Göğer, N.G. 2015. Determination of parabens in human milk and other food samples by capillary electrophoresis after dispersive liquid–liquid microextraction with back-extraction. *Food Chem* 181:1–8.

Anonymous. 2019. Eur-Lex Access to European Union Law. https://eur-lex.europa.eu/eli/reg/2008/1333/2019-06-18

Baardseth, P., and Russwurm, H. 1978. Content of some organic acids in cloudberry (*Rubus chamaemorus*). *Food Chem.* 3:43.

Barberis, C.L., Astoreca, A.L., Dalcero, A.M., and Magnoli, C.E. 2010. Food-grade antioxidants and antimicrobials to control growth and ochratoxin a production by *Aspergillus* Section Nigri on peanut kernels. *J. Food Prot.* 73:1493–1501.

Bargiota, E.E., Rico-Munoz, E., and Davidson, P.M. 1987. Lethal effect of methyl and propyl parabens as related to *Staphylococcus aureus* lipid composition. *Intl. J. Food Microbiol.* 4:257.

Blom, H., Nerbrink, E., Dainty, R., Hagtvedt, T., Borch, E., Nissen, H., and Nesbakken, T. 1997. Addition of 2.5% lactate and 0.25% acetate controls growth of *Listeria monocytogenes* in vacuum-packed, sensory acceptable servelat sausage and cooked ham stored at 4°C. *Intl. J. Food Microbiol.* 38:71–76.

Bottoli, C.B.G., Gutiérrez-Ponce, M.D.S., Aguiar, V.S., and de Aquino W.M. 2011. Determination of parabens in sweeteners by capillary electrochromatography *Braz. J. Pharm. Sci.* 47: 779–785.

Bredin, J., Davin-Regli, A., and Pages, J.-M. 2005. Propyl paraben induces potassium efflux in *Escherichia coli. J. Antimicrob. Chemother.* 55:1013–1015.

Calafat, A.M., Ye, X. Wong, L-Y., Bishop, A.M. and Needham, L.L. 2010. Urinary concentrations of four parabens in the U.S. Population: NHANES 2005–2006. *Environ. Health Perspectives* 118:679–685.

Cao, S., Liu, Z., Zhang, L., Xi, C., Li, X., Wang, G., Yuana, R., and Mu, Z. 2013. Development of an HPLC–MS/MS method for the simultaneous analysis of six kinds of parabens. *Food Anal. Methods* 5:1016–1023.

Chan, L., Weaver, R., and Ough, C.S. 1975. Microbial inhibition caused by p-hydroxybenzoate esters in wines. *Am. J. Enol. Vitic.* 26:201.

Chichester, F., and Tanner, F.W. 1972. Antimicrobial food additives. In *Handbook of Food Additives*, 2nd Ed., edited by T.E. Furia. CRC Press, Cleveland, OH, p. 115.

Chung, D., Papadakis, S.E., and Yam, K.L. 2001a. Release of propyl paraben from a polymer coating into water and food simulating solvents for antimicrobial packaging applications. *J. Food Proc. Preserv.* 25:71.

Chung, D., Chikindas, M.L., and Yam, K.L. 2001b. Inhibition of *Saccharomyces cerevisiae* by slow release of propyl paraben from a polymer coating. *J. Food Prot.* 64:1420.

Chuto, M., Chaiyo, S., Siangproh, W., and Chailapakul, O. 2013. Rapid separation and highly determination of paraben species by ultra-performance liquid chromatography—electrochemical detection. *Detection* 1:21–29.

Code of Federal Regulations. 2019. *Title 21: Food and Drugs, Parts 170–199.* Off. Federal Regulations, National Archives Records Serv., General Service Admin., Washington, DC. Accessed at https://www.ecfr.gov/cgi-bin/text-idx?gp=&SID=d8a31c956015980bae96c405675b1226&mc=true&tpl=/ecfrbrowse/Title21/21tab_02.tpl

Cosmetic Ingredient Review Expert Panel (CIR). 2008. Final amended report on the safety assessment of methylparaben, ethylparaben, propylparaben, isopropylparaben, butylparaben, isobutylparaben, and benzylparaben as used in cosmetic products. *Intl. J. Toxicol.* 27 (Suppl 4):1–82.

Dan, V., Raducan, A., Stoicescu, A., and Nicolau, A. 1996. Effect of esterparabens upon biotechnological processes in brewing. *Roum. Biotechnol. Lett.* 1:181.

Darbre, P.D., Aljarrah, A., Miller, W.R., Coldham, N.G., Sauer, M.J., and Pope, G.S. 2004. Concentrations of parabens in human breast tumours. *J. Appl. Tox.* 24:5.

Darwish, R.M., and Bloomfield, S.F. 1997. Effect of ethanol, propylene glycol and glycerol on the interaction of methyl and propyl *p*-hydroxybenzoate with *Staphylococcus aureus* and *Pseudomonas aeruginosa. Intl. J. Pharma.* 147:51.

Dje, Y., Payne, K.D., and Davidson, P.M. 1989. Inhibition of *Listeria monocytogenes* Scott A by propyl paraben in non-fat milk and chicken and hot dog suspensions. *Ann. Mtg. Inst. Food Technol.*, Chicago, IL.

Dje, Y., Davidson, P.M., and Rico, E. 1990. Effect of antimicrobials against *Listeria monocytogenes* on hot dogs and in brine. *Ann. Mtg. Inst. Food Technol.*, Anaheim, CA.

Dobias, J., Chudackova, K., Voldrich, M., and Marek, M. 2000. Properties of polyethylene films with incorporated benzoic anhydride and/or ethyl and propyl esters of 4-hydroxybenzoic acid and their suitability for food packaging. *Food Addit. Contam.* 17:1047.

Draughon, F.A., Sung, S.C., Mount, J.R., and Davidson, P.M. 1982. Effect of the parabens with and without nitrite on *Clostridium botulinum* toxin production in canned pork slurry. *J. Food Sci.* 47:1635.

Dymicky, M., and Huhtanen, C.N. 1979. Inhibition of *Clostridium botulinum* by p-hydroxybenzoic acid n-alkyl esters. *Antimicrob. Agents Chemother.* 15:798.

Eklund, T. 1980. Inhibition of growth and uptake processes in bacteria by some chemical food preservatives. *J. Appl. Bact.* 48:423.

Eklund, T., Nes, I.F., and Skjelkväle, R. 1981. Control of *Salmonella* at different temperatures by propyl paraben and butylated hydroxyanisole. In *Psychrotrophic Microorganisms in Spoilage and Pathogenicity*, edited by T.A. Roberts, G. Hobbs, J.H.B. Christian and N. Skovgaard. Academic Press, London, UK.

Eklund, T. 1985a. Inhibition of microbial growth at different pH levels by benzoic and propionic acids and esters of p-hydroxybenzoic acid. *Intl. J. Food Microbiol.* 2:159.

Eklund, T. 1985b. The effect of sorbic acid and esters of p-hydroxybenzoic acid on the protonmotive force in *Escherichia coli* membrane vesicles. *J. Gen. Microbiol.* 131:73.

Epstein, S. 1968. Paraben sensitivity: Subtle trouble. *Ann. Alergy* 26:185.

Er, B., Demirhan, B., Onurdag, F., Özgacar, S., and Öktem, A. 2014. Antimicrobial and antibiofilm effects of selected food preservatives against *Salmonella* spp. isolated from chicken samples. *Poultry Sci.* 93:695–701.

Fagundes, C., Perez-Gago, M.B., Monteiro, A.R., and Palou, L. 2013. Antifungal activity of food additives in vitro and as ingredients of hydroxypropyl methylcellulose-lipid edible coatings against *Botrytis cinerea* and *Alternaria alternata* on cherry tomato fruit. *Intl. J. Food Microbiol.* 166:391–398.

FAO/WHO. 2000. Joint FAO/WHO Expert Committee on Food Additives. FAO Food and Nutrition Papers 55th Meeting.

Freese, E., Sheu, C.W., and Galliers, E. 1973. Function of lipophilic acids as antimicrobial food additives. *Nature* 241:321.

Fukahori, M., Akatsu, S., Sato, H., and Yotsuyanagi, T. 1996. Relationship between uptake of p-hydroxybenzoic acid esters by *Escherichia coli* and antibacterial activity. *Chem. & Pharm. Bull.* 44:1567.

Furr, J.R., and Russell, A.D. 1972. Some factors influencing the activity of esters of p-hydroxybenzoic acid against *Serratia marcescens. Microbios* 5:189.

Fyfe, L., Armstrong, F., and Stewart, J. 1998. Inhibition of *Listeria monocytogenes* and *Salmonella enteritidis* by combinations of plant oils and derivatives of benzoic acid: The development of synergistic antimicrobial combinations. *Intl. J. Antimicrob. Agents* 9:195.

Government of Canada. 2019. List of Permitted Preservatives (Lists of Permitted Food Additives), Accessed 5 May 2019 at https://www.canada.ca/en/health-canada/services/food-nutrition/food-safety/food-additives/lists-permitted.html

Jain, R., Mudiam, M.K.R., Chauhan, A., Ratnasekhar, Ch., Murthy, R.C., and Khan, H.A. 2013. Simultaneous derivatisation and preconcentration of parabens in food and other matrices by isobutyl chloroformate and dispersive liquid–liquid microextraction followed by gas chromatographic analysis. *Food Chem.* 141:436–443.

Japan External Trade Association. 2011. Specifications and Standards for Foods, Food Additives, etc. Under the Food Sanitation Act (Abstract) 2010. Accessed at https://www.jetro.go.jp/ext_images/en/reports/regulations/pdf/foodext2010e.pdf

Jermini, M.F.G., and Schmidt-Lorenz, W. 1987. Activity of Na-benzoate and ethyl-paraben against osmotolerant yeasts at different water activity values. *J. Food Prot.* 50:920–927.

Jones, P.S., Thigpen, D., Morrison, J.L., and Richardson, A.P. 1956. p-Hydroxybenzoic acid esters as preservatives. III. The physiological disposition of p-hydroxybenzoic acid and its esters. *J. Am. Pharm. Assoc.* 45:268.

Juneja, V.K., and P.M. Davidson. 1993. Influence of altered fatty acid composition on resistance of *Listeria monocytogenes* to antimicrobials. *J. Food Prot.* 56:302–305.

Jurd, L., King, A.D., Mihara, K., and Stanely, W.L. 1971. Antimicrobial properties of natural phenols and related compounds. I. Obtusastyrene. *Appl. Microbiol.* 21:507.

Kato, A., and Shibasaki, I. 1975. Combined effect of different drugs on the antibacterial activity of fatty acids and their esters. *J. Antibact. Antifung. Agents (Japan)* 8:355.

Klindworth, K.J., Davidson, P.M., Brekke, C.J., and Branen, A.L. 1979. Inhibition of *Clostridium perfringens* by butylated hydroxyanisole. *J. Food Sci.* 44:564.

Lee, J.S. 1973. What seafood processors should know about *Vibrio parahaemolyticus. J. Milk Food Technol.* 36:405.

Lewis, J.C., and Jurd, L. 1972. Sporostatic action of cinnamylphenols and related compounds on *Bacillus megaterium*. In *Spores V*, edited by H.O. Halvorson, R. Hewson, and L.L. Campbell. *Amer. Soc. Microbiol.*, Washington, DC.

Liao, C. Liu, F., and Kurunthachalam, K. 2013. Occurrence of and dietary exposure to parabens in foodstuffs from the United States. *Environ. Sci. & Technol.* 47:3918–3925.

Lück, E., and Jager, M. 1997. *Antimicrobial Food Additives. Characteristics, Uses, Effects*, 2nd Ed. Springer-Verlag, Berlin.

Maher, H.M., Alzoman, N.Z., Almeshal, M.A., Alotaibi, H.A., Alotaibi, N.N., and Al-Showiman, H. 2018. Quantitative screening of parabens in ready-to-eat foodstuffs available in the Saudi market using high performance liquid chromatography with photodiode array detection. *Arabian J. Chem.* 13:2897–2911. https://doi.org/10.1016/j.arabjc.2018.07.019

Martin, J.H., Chung, K.S.S., and Ogrosky, L. 1972. Inhibition of growth of *Alcaligenes viscolactis* by some common food preservatives. *J. Dairy Sci.* 55:1179.

Marwan, A.G., and Nagel, C.W. 1986. Quantitative determination of infinite inhibition concentrations of antimicrobial agents. *Appl. Env. Microbiol.* 51:559.

Matthews, C., Davidson, J., Bauer, E., Morrison, J.L., and Richardson, A.P. 1956. p-Hydroxybenzoic acid esters as preservatives. II. Acute and chronic toxicity in dogs, rats, and mice. *J. Am. Pharm. Assoc.* 45:260.

Moir, C.J., and Eyles, M.J. 1992. Inhibition, injury and inactivation of four psychrotrophic foodborne bacteria by the preservatives methyl p-hydroxybenzoate and potassium sorbate. *J. Food Prot.* 55:360.

Molognoni, L., Daguer, H., de Sá Ploêncio, L.A., and De Dea Lindner, J. 2018. A multi-purpose tool for food inspection: Simultaneous determination of various classes of preservatives and biogenic amines in meat and fish products by LC-MS. *Talanta* 178:1053–1066.

Moustafa, H.H., and Collins, E.B. 1969. Effects of selected food additives on growth of *Pseudomonas fragi*. *J. Dairy Sci.* 52:335.

Murata, W., Yamaguchi, Y., Fujita, K.-I., Yamauchi, K., Tanaka, T., and Ogita, A. 2019. Enhancement of paraben-fungicidal activity by sulforaphane, a cruciferous vegetable-derived isothiocyanate, via membrane structural damage in *Saccharomyces cerevisiae*. *Lett. Appl. Microbiol.* 69:doi.org/10.1111/lam.13230

Nesci, A., Rodriguez, M., and Etcheverry, M. 2003. Control of *Aspergillus* growth and aflatoxin production using antioxidants at different conditions of water activity and pH. *J. Appl. Microbiol.* 95:279.

Nguyen, T., Clare, B., Guo, W., and Martinac, B. 2005. The effects of parabens on the mechanosensitive channels of *E. coli*. *Eur. Biophys. J.* 34:389–395.

Oka, S. 1960. Studies on the transfer of antiseptics to microbes and their toxic effect. Part III. Adsorption of esters of p-hydroxybenzoic acid on yeast cell and their toxic effect. *Bull. Agric. Chem. Soc. Jpn.* 24:412.

Passone, M.A., Resnik, S.L., and Etcheverry, M. 2005. In vitro effect of phenolic antioxidants on germination, growth and aflatoxin B_1 accumulation by peanut *Aspergillus* section Flavi. *J. Appl. Microbiol.* 99:682–691.

Payne, K.D., Rico-Munoz, E., and Davidson, P.M. 1989. The antimicrobial activity of phenolic compounds against *Listeria monocytogenes* and their effectiveness in a model milk system. *J. Food Prot.* 52:151–153.

Pierson, M.D., Smoot, L.A., and Vantassell, K.R. 1980. Inhibition of *Salmonella typhimurium* and *Staphylococcus aureus* by butylated hydroxyanisole and the propyl ester of p-hydroxybenzoic acid. *J. Food Prot.* 43:191.

Prindle, R.F. 1983. Phenolic compounds. In *Disinfection, Sterilization, and Preservation*, 3rd Ed., edited by S.S. Block. Lea and Febiger, Philadelphia, p. 197.

Razavilar, V., and Genigeorgis, C. 1998. Prediction of *Listeria* spp. growth as affected by various levels of chemicals, pH, temperature and storage time in a model broth system. *Intl. J. Food Microbiol.* 40:149.

Reddy, N.R., and Pierson, M.D. 1982. Influence of pH and phosphate buffer on inhibition of *Clostridium botulinum* by antioxidants and related phenolic compounds. *J. Food Prot.* 45:925.

Reddy, N.R., Pierson, M.D., and Lechowich, R.V. 1982. Inhibition of *Clostridium botulinum* by antioxidants, phenols and related compounds. *Appl. Environ. Microbiol.* 43:835.

Reitschel, R.L., and Fowler, J.F. 2001. *Fisher's Contact Dermatitis*, 5th Ed. Lippincott, Williams and Wilkins, Philadelphia.

Rico-Munoz, E., Bargiota, E.E., and Davidson, P.M. 1987. Effect of selected phenolic compounds on the membrane-bound adenosine triphosphatase of *Staphylococcus aureus*. *Food Microbiol.* 4:239.

Robach, M.C., and Pierson, M.D. 1978. Influence of para-hydroxybenzoic acid esters on the growth and toxin production of *Clostridium botulinum* 10755A. *J. Food Sci.* 43:787.

Saad, B., Bari, M.F., Saleh, M.I., Ahmad, K., and Talib, M.K. 2005. Simultaneous determination of preservatives (benzoic acid, sorbic acid, methylparaben and propylparaben) in foodstuffs using high-performance liquid chromatography. *J. Chromatogr. A* 1073:393–397.

Sethi, V. 1994. Efficacy of various preservatives for preserving whole tomato concentrate. *Ind. Food Packer* 48:11.

Sevart, N., Gragg, S., Krug, M., Jenott, J., and Phebus, R. 2017. A Blend of benzalkonium chloride, acetic acid, and methyl paraben effectively reduces *Escherichia coli* O157:H7, *Salmonella* spp., and *Listeria monocytogenes* in produce wash water, but lacks consistent efficacy on romaine and iceberg lettuce. International Association of Food Protection P3-60, Tampa, FL.

Sofos, J.N., and Busta, F.F. 1980. Alternatives to the use of nitrite as an antibotulinal agent. *Food Technol.* 34:244.

Sokol, H. 1952. Recent developments in the preservation of pharmaceuticals. *Drug Stand.* 20:89.

Song, S., Zhang, Z., Zou, N., Chen, R., Han, L., Pan, C., and Sapozhnikova, Y. 2017. Determination of six paraben residues in fresh-cut vegetables using QuEChERS with multi-walled carbon nanotubes and high-performance liquid chromatography–tandem mass spectrometry. *Food Anal. Methods* 10:3972–3979.

Soni, M.G., Burdock, G.A., Taylor, S.L., and Greenberg, N.A. 2001. Safety assessment of propyl paraben: A review of the published literature. *Food & Chem. Toxicol.* 39:513.

Soni, M.G., Taylor, S.L., Greenberg, N.A., and Burdock, G.A. 2002. Evaluation of the health aspects of methyl paraben: A review of the published literature. *Food & Chem. Toxicol.* 40:1335.

Tattawasart, U., Maillard, J.-Y., Furr, J.R., and Russell, A.D. 1999. Comparative responses of *Pseudomonas stutzeri* and *Pseudomonas aeruginosa* to antibacterial agents. *J. Appl. Microbiol.* 87:323.

Thompson, D.P. 1994. Minimum inhibitory concentration of esters of p-hydroxybenzoic acid (paraben) combinations against toxigenic fungi. *J. Food Prot.* 57:133–135.

Tomaszewski, M. 1960. The occurrence of p-hydroxybenzoic acid and some other simple phenols in vascular plants. *Bull. Acad. Pol. Sci.* 8:61.

Torres, A.M., Ramirez, M.L., Arroyo, M. Chulze, S.N., and Magan, N. 2003. Potential use of antioxidants for control of growth and fumonisin production by *Fusarium verticillioides* and *Fusarium proliferatum* on whole maize grain. *Intl. J. Food Microbiol.* 83:319.

US Food and Drug Administration. 2018. Parabens. Accessed at http://www.fda.gov/cosmetics/productsingredients/ingredients/ucm128042.htm

USDA Foreign Agricultural Service Global Agricultural Information Network. 2015. Chinese Standards for Food Additives - GB2760-2015. https://gain.fas.usda.gov/Recent%20GAIN%20Publications/Standard%20for%20Food%20Additive%20Use%20-%20GB2760-2015_Beijing_China%20-%20Peoples%20Republic%20of_4-28-2015.pdf

USDA Foreign Agricultural Service Global Agricultural Information Network. 2019. Russian Federation Food and Agricultural Import Regulations and Standards Report RS1838. Accessed at http://agriexchange.apeda.gov.in/IR_Standards/Import_Regulation/FoodandAgriculturalImportRegulationsandStandardsReportMoscowRussianFederation2132019.pdf

Valkova, N., Lepine, F., Bollet, C., Dupont, M., and Villemur, R. 2002. *prbA*, a gene coding for an esterase hydrolyzing parabens in *Enterobacter cloacae* and *Enterobacter gergoviae* strains. *J. Bacteriol.* 184:5011.

Valkova, N., Lepine, F., Labrie, L., Dupont, M., and Beaudet, R. 2003. Purification and characterization of PrbA, a new esterase from *Enterobacter cloacae* hydrolyzing the esters of 4-hydroxybenzoic acid (parabens). *J. Biol. Chem.* 278:12779.

Venugopal, V., Pansare, A.C., and Lewis, N.F. 1984. Inhibitory effect of food preservatives on protease secretion by *Aeromonas hydrophila. J. Food Sci.* 49:1078.

Wang, Y., Oxer, D., and Hekimi, S. 2015. Mitochondrial function and lifespan of mice with controlled ubiquinone biosynthesis. *Nat Commun.* 6:6393. doi: 10.1038/ncomms7393

Yang, J., Li, Y., Gong, W., Wang, C., Liu, B., and Sun, C. 2014. Simultaneous determination of six parabens in foods by matrix liquid-phase dispersion extraction combined with high-performance liquid chromatography. *Food Anal. Methods* 7: 1693–1702.

Yin, Q., Zhu, Y., and Yang, Y. 2018. Dispersive liquid–liquid microextraction followed by magnetic solid-phase extraction for determination of four parabens in beverage samples by ultra-performance liquid chromatography tandem mass spectrometry *Food Anal. Methods* 11:797–807.

13

Dimethyl Dicarbonate and Diethyl Dicarbonate

Randy W. Worobo, Rebecca M. Cheng, and Cornelius S. Ough

CONTENTS

13.1 Introduction

Diethyl dicarbonate (DEDC), also known as diethyl pyrocarbonate (DEPC), has been mentioned in the Russian literature as a cause for effervescence in sparkling wine (Parfentjev and Kovalenko, 1951a, b; Merzhanian, 1951). Kozenko (1952), in further discussing DEDC, mentioned that in 1933 a Russian investigator expressed the opinion that the neutral esters of carbonic acid were involved in sparkling wine effervescence. Although these works were erroneous in their conclusions, they stimulated interest in these types of compounds.

Boehm and Mehta (1938) first isolated and properly identified the alkyl esters of pyrocarbonic acid. Bayer AG first introduced DEDC into the market for trial tests about 1959, at which time they also applied for a U.S. patent (Bernhard et al., 1959). Hennig (1959) reported on the effectiveness of DEDC as a fungicide in wine and also reported that it caused no off-aromas or off-flavors. Before the introduction into the market, Bayer AG tested a number of analogs of DEDC. The choice of DEDC over the equally effective fungicide dimethyl dicarbonate (DMDC), also known as dimethyl pyrocarbonate, was made, according to Genth (1972), because of the innocuous hydrolysis products, ethanol and carbon dioxide. However, a 1972 ban on the use of DEDC in the United States resulted in greater focus on DMDC.

13.2 Chemistry

13.2.1 Description

DEDC is a colorless liquid with a faint fruit-like odor. The boiling point at 760 mm Hg is 155°C, at which temperature it decomposes. The density of DEDC is 1.12 g/cm^3 at 20°C, the refractive index is 1.397 at 25°C, and the viscosity is 1.97 cP at 20°C. It is only slightly soluble in water, with a solubility of about 0.6 g per 100 g at 18°C. The compound also hydrolyzes rapidly in water. In ethanol, solubility is greater (50 g per 100 g) and hydrolysis is slower. DEDC is a lacrimator and causes skin irritation.

DMDC is a colorless, fruity smelling liquid, and it has a melting point of 15.2°C, a boiling point of 123–149°C, with decomposition, and a density of 1.26. It is also only slightly soluble in water and more soluble in organic solvents. Skin and eye contact should be avoided.

13.2.2 Ethyl Carbamate Formation

One source of this compound is the reaction of carbamoyl phosphate with ethanol. This ubiquitous compound, of the ornithine cycle, is found in all living cells. The reaction is:

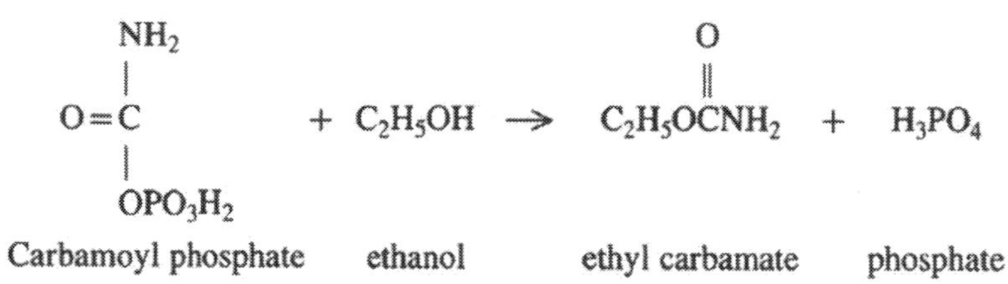

(13.1)

It has been conclusively shown by Ough et al. (1988b, 1990) and Monteiro et al. (1989) that fermenting yeasts metabolize arginine and excrete urea into the medium. This is the primary source in wine.

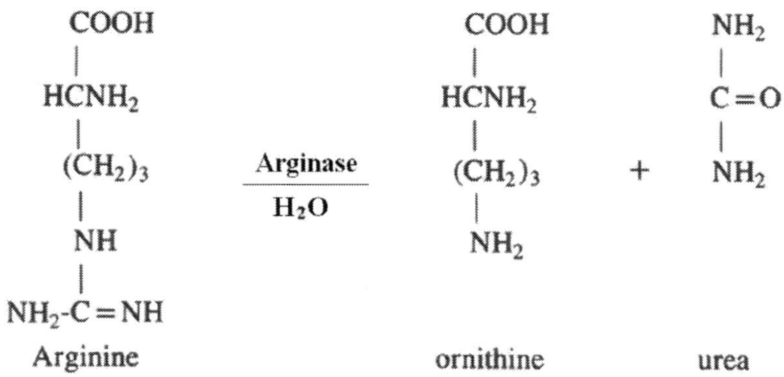

(13.2)

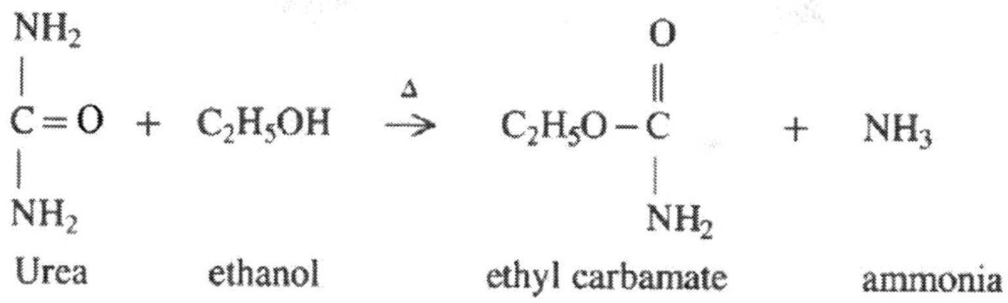

(13.3)

13.2.3 Synthesis

The synthesis of dialkyl dicarbonate esters was described by Boehm and Mehta (1938). Their method involved the heating of the chlorocarbonic ethyl esters of emetine (6', 7', 10, 11 - tetromethoxyemetan), an alkaloid, with dilute potassium hydroxide. DEDC could be formed, as well as some other dialkyl dicarbonates, but DMDC could not be recovered. Kovalenko (1952) devised a more direct method and was able to synthesize DMDC. The reaction used was:

$$CH_3OCOOCl + NaOCOOCH_3 \rightarrow CH_3OCOOCOOCH_3 + NaCl \quad (13.4)$$

13.2.4 Reactions

The diethyl and dimethyl dicarbonates are extremely reactive substances. Some of the reactions that can take place are as follows:

$$\text{R-dicarbonate} = \text{DMDC or DEDC} \quad (13.5)$$

(R = either two methyl groups or two ethyl groups).

$$R_1\text{-dicarbonate} + H_2O \rightarrow 2R_1OH + 2CO_2 \quad (13.6a)$$

$$R_1\text{-dicarbonate} + R_2OH \rightarrow R_1OCOOR_2 + R_1OH + CO_2 \quad (13.6b)$$

(R_2 = any alkyl or aromatic group).

$$R_1\text{-dicarbonate} + R_2NH_2 \rightarrow R_1OCONHR_2 + R_1OH + CO_2 \quad (13.6c)$$

(R_2 = alkyl group, alkyl-C-–, aromatic, or H).

$$R_1\text{-dicarbonate} + R_2COOH \rightarrow R_1OCOOCOOR_2 + R_1OH + CO_2 \quad (13.6d)$$

(R_2 = any alkyl or aromatic group).

$$R_1\text{-dicarbonate} + R_2SH \rightarrow R_1OCOSR_2 + R_1OH + CO_2 \quad (13.6e)$$

(R_2 = any alkyl or aromatic group).

The hydrolysis reaction (6a) was recognized by Kovalenko (1952). The alcoholysis (6b) was documented by Kielhöfer and Würdig (1963a). Larrouquère (1963) reacted alkyl and aromatic amines with DEDC to produce reaction (6c). Thoukis et al. (1962) reported reactions of DEDC with $-NH_2$ groups and noted carbethyoxy compounds could be formed with amino acids. The mixed esters formed by reaction of DEDC with the desired alkyl acid (6d) were indicated by Thoma and Rinke (1959). Mühlrád et al.

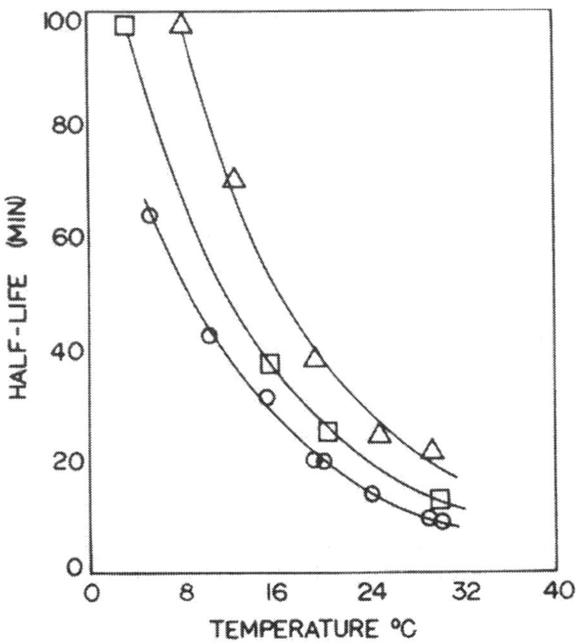

FIGURE 13.1 Hydrolysis of DEDC and DMDC at different temperatures. Half-life refers to the time for half of the DEDC or DMDC to hydrolyze: DMDC in water (circles); DMDC in 14.6% vol/vol ethanol (squares); DEDC in water (triangles).

(1967) stated that the sulfydryl groups of cysteine were carbethoxylated by DEPC (6e). Larrouquère (1965) also investigated the reaction of DEDC with thiol groups. Similar reactions [(4) through (6)] for DMDC have been reported by Peterson and Ough (1979), Ough and Langbehn (1976), and Stafford and Ough (1976).

Other reactions of DEDC that have been investigated include those with phenols and phenol glucosides (Paulus and Lorke, 1967), those with vitamins (Fischer, 1970), and those with malic acid (Schelenz and Fischer, 1971). The report of Duhm et al. (1966) indicated that a large range of food chemicals reacted with DEDC.

The hydrolysis rate of DEDC was considered by Pauli and Genth (1966) and further by Schelenz and Fischer (1970). The kinetics of hydrolysis of DMDC have been reported by Peterson (1978) and Genth (1979). Figure 13.1 gives the half-life of DMDC and DEDC in water and that of DMDC in 14% vol/vol ethanol at a variety of temperatures.

13.3 Antimicrobial Activity

After the initial publications of the effectiveness of DEDC against wine yeast, a number of investigators further examined its usefulness as a bactericide and fungicide. Because the dicarbonates rapidly hydrolyze to the corresponding alcohol (methyl or ethyl) and carbon dioxide, the challenge is to destroy target microorganisms quickly, before hydrolysis is completed. As such, the dicarbonates cannot be relied upon for long-term protection against recontamination or later outgrowth of surviving organisms.

13.3.1 Yeast

Reviews by Pauli and Genth (1966) and Genth (1964) included numerous reports on the fungicidal action of DEDC on yeasts. Their fungicidal data plus those of Alimukhamedova (1975), Grospicova et al. (1969), and Todor et al. (1967) are shown in Table 13.1. Many of the differences in antifungal effectiveness of DEDC are due to the conditions under which each test was made. Test conditions as well as tolerance differences between genera, species, or even strains could account for most of the variation.

TABLE 13.1
Fungicidal Activity of Diethyl Dicarbonate[a]

Yeast	Starting Cell Concentration per mL	DEDC Concentration (mg/L) Sufficient to Be Fungicidal (100% kill)
Saccharomyces acidifaciens	5.05×10^2	200
S. apiculatus	1.0×10^5	60–90
S. cerevisiae	5.0×10^2–1.0×10^4	30–80
S. cerevisiae	5.0×10^4	50
S. cerevisiae	1.50×10^2	50–100
S. cerevisiae	5.00×10^2	50
S. cerevisiae	4.0×10^3	500–1000
S. cerevisiae	1.0×10^3	500
S. cerevisiae	3.6×10^6	50
S. cerevisiae	1.4×10^4	12
S. cerevisiae var. *ellipsoides* Burgundy	5.0×10^4	100
S. cerevisiae var. *ellipsoides* Champagne	6.0×10^5	100
S. carlsbergensis	4.0×10^3	250–1000
S. chevalieri	5.6×10^2	30
S. globosus	8.0×10^2	30
S. heterogenicus	5.2×10^2	30
S. ludwigii	5.0×10^4	25
S. oviformis	5.0×10^4	50
S. pastorianus	4.0×10^5	100
S. pastorianus	4.0×10^2	100
S. pastorianus	4.0×10^5	100
S. rouxii	1.0×10^5	100–500
S. uvarum	5.0×10^4	25
S. uvarum	2.0×10^3	200
Schizosaccharomyces pombe	5.0×10^2	400
Brettanomyces bruxellensis	1.0×10^5	250
B. vini	1.0×10^4	25
B. claussenii	5.0×10^2	200
B. species	5.0×10^4	25
B. species	4.6×10^3	100
Pichia alcoholophila	1.0×10^3	100
P. farinosa	4.0×10^5	100
P. farinosa	4.0×10^3	100
P. membranefaciens	5.0×10^2	25
P. membranefaciens	2.0×10^4	14
Torula utilis	1.0×10^5	250
Torulopsis candida	1.0×10^5	100
T. colliculosa	5.6×10^2	30
T. colliculosa	5.0×10^2	200
T. versatilis	5.0×10^2	200
Rhodotorula glutinosa	5.6×10^2	30
R. mucilaginosa	1.0×10^5	200
R. rubra	2.0×10^3	200
R. rubra	5.0×10^3	300
Candida krusei	1.0×10^5	60–200

(*Continued*)

TABLE 13.1 (CONTINUED)
Fungicidal Activity of Diethyl Dicarbonate[a]

Yeast	Starting Cell Concentration per mL	DEDC Concentration (mg/L) Sufficient to Be Fungicidal (100% kill)
C. krusei	5.0×10^2	200
C. krusei	5.0×10^4	200
C. lipolytica	1.0×10^5	90–500
C. mycoderma	5.0×10^2	200
C. mycoderma	2.5×10^3	100
C. parapsilosis	5.0×10^2	200
C. pseudotropicalis	5.0×10^2	200
C. pulcherima	5.0×10^2	200
Cryptococcus diffluens	5.0×10^4	75
Kloeckera apiculata	1.0×10^4	200
Debaryomyces kloeckeri	5.0×10^2	200
Hanseniaspora apiculta	1.0×10^4	25
Hansenula anomala	4.0×10^3	250–500
H. anomala	5.0×10^2	200
H. anomala	2.0×10^3	100
Fabospora macedoniensis	4.0×10^2	30
Zygosaccharomyces eupagicus	6.3×10^2	30
Zygosaccharomyces sp.	1.0×10^5	80

[a] Reports of various tests done under varied environmental conditions.

Genth (1979) reported the effectiveness of DMDC as a fungicide when used in alcohol-free beverages (Table 13.2). Daudt and Ough (1980) reported its use as a sterilant against yeast in wine (Table 13.3). DMDC has proven to be equal or superior to DEDC as a fungicide in both instances.

Pauli and Genth (1966) showed that, for every ten-fold increase in cell count, the amount of DEDC required for fungicidal action was doubled. Daudt and Ough (1980) showed that 50 mg/L of DMDC effectively sterilized a wine with 10% vol/vol ethanol containing 2×10^4 cells. Viable cells remained, however, if only 25 mg/L of DMDC was used, with as few as 50 cells/mL in the wine initially.

The rate of kill is proportional to the concentration of DEDC (Ough and Ingraham, 1961). Over 2 hours were required for complete kill of 50 yeast cells/mL with the addition of 40 mg/L of DEDC; less than half an hour was required with 120 mg/L of DEDC. Similar results were found using DMDC (Daudt and Ough, 1980). A delay in the lethal effect at lower DEDC or DMDC concentrations could be due to the capability of the yeast cells to reproduce for short periods before permanent damage and death of the organism.

Ough et al. (1988c) investigated the interaction of sulfur dioxide and DMDC. They found additive effects between the two inhibitors when testing a very resistant yeast. For a lethal dose of DMDC for this yeast, a much larger than normal amount was required.

13.3.2 Bacteria

A comparison of the antibacterial effectiveness of DEDC on a number of different genera has been summarized in Table 13.4 (Pauli and Genth, 1966). Similar tests using DMDC (Genth, 1979, 1980) are shown in Table 13.5. Results on the antimicrobial activity of these compounds indicated that they had approximately equivalent bactericidal activity (Genth, 1979, 1980).

Murata (1974) found that 1300 mg/L of DEDC was capable of destroying a *Lactobacillus caseii* bacteriophage. Ough et al. (1988c) found *Oenococcus oeni* relatively insensitive to DMDC at normal wine treatment levels. It had a slight additive effect on the bacteria when used with sulfur dioxide.

TABLE 13.2

Effective Fungicidal Concentration of DMDC in Alcohol-Free Beverages [a]

Yeast	DMDC Concentration (mg/L) Sufficient to Be Fungicidal (100% kill)
Saccharomyces carlsbergensis	100
S. carlsbergensis	70
S. diastaticus	200
S. oviformis	100
S. bailii	120
S. cerevisiae	30
S. uvarum	20
S. pastorianus	100
S. apiculatus	70
S. globosus	40
Zygosaccharomyces priorianus	90
Rhodotorula mucilaginosa	50
R. glutinosa	30
R. rubra	200
Candida krusei	200
Pichia membranefaciens	30
P. farinosa	100
Torulopis candida	100
T. versatilis	100
T. stallata	75
Torula utilis	250
Endomyces lactis	70
Kloeckera apiculata	30
Hansenula anomala	50

[a] The cell counts were about 500 cells/mL, temperature 28°C, and pH between 2.8 and 4.7.

13.3.3 Molds

For activity against molds, higher levels of DEDC were required than for either yeast or bacteria. Table 13.6 shows the results summarized by Pauli and Genth (1966). Genth (1979) reported that 500 spores per milliliter of *Penicillium glaucum*, *Byssochlamys fulva*, *Botrytis cinerea*, *Mucor racemosus*, and *Fusarium oxysporum* required 200, 100, 100, 500, and 100 mg/L of DMDC, respectively, for effective killing. Vegetative cells and mold conidia are relatively sensitive to both dicarbonates. However, heat-resistant structures such as ascospores of *B. fulva* are quite resistant and, thus, likely to survive exposure to up to 200 ppm DMDC (Splittstoesser and Wilkinson, 1973; Van der Riet et al., 1989).

13.3.4 Physical and Chemical Effects on Activity

It has been recognized that DEDC activity could be altered by some of the physical or chemical variables existing during handling of products. Such variables are temperature, pH, chemical makeup of product, and the amount of other reactive substances in the treated material.

13.3.4.1 Temperature

The hydrolysis rate of DEDC or DMDC is dependent on temperature—the higher the temperature the more rapid the rate. The half-life values for both DEDC and DMDC are given in Figure 13.1. Despite the

TABLE 13.3

Antimicrobial Effectiveness of DMDC in a Standard Wine Containing 10% Ethanol

	DMDC Added (mg/L)					
Microorganism	**0**	**6.25**	**12.5**	**25**	**50**	**100**
Saccharomyces cerevisiae Montrachet	58[a]	16	0	0	0	0
S. bayanus Champagne	2500	419	729	0	0	0
Schizosaccharomyces pombe	42	14	4	0	0	0
Saccharomyces cerevisiae Steinberg	96	50	43	0	0	0
S. fermantati Flor[b]	203	53	9	0	0	0
Brettanomyces spp.	10,000	2000	100	0	0	0
Saccharomyces cerevisiae Distillers	78	0	0	0	0	0
S. cerevisiae Geisenhiem	88	0	0	0	0	0
S. cerevisiae Burgundy	2000	310	0	0	0	0
S. cerevisiae Rankine 350	89	64	38	0	0	0
S. cerevisiae Eperney Rankine 729	2500	2000	844	0	0	0
S. cerevisiae Tokay[b]	887	129	144	0	0	0
Champagne Epernay (Geisenheim Reinzucthefen)	5000	3000	350	0	0	0
Saccharomyces bayanus (formerly *oviformis*)	17	19	0[c]	0	0	0
S. oviformis (S. bayanus) Inst. Pasteur	214	346	200	11	0	0
Schizosaccharomyces pombe Moscovin[b]	2200	529	2000	4	0	0
Geisenheim 1949 (Geisenheim Reinzucthefen)	1200	800	640	5	0	0
Dekkera intermedia FS&T 71-12	2,000	600	70	0	0	0
Rhodotorula rubra	1,000	800	720	650	0	0

[a] Values are the number of viable cells remaining in wine after 38 h.

[b] Yeasts did not grow well in wines with 10% ethanol; wines of 8% ethanol were used instead. *S. fermentati* Flor needs to stay at least 4 days in the incubator (at 37°C) to grow on the plates; *S. cerevisiae* Tokay and *S. pombe* Moscovin need 3 days. *S. fermentati* Flor also grew better in 8% ethanol wines, but the data presented are for 10% ethanol wines.

[c] Although nothing grew on the plates, growth developed later in the bottles, indicating some viable cells remaining.

more rapid hydrolysis at higher temperatures, Turtura (1966) found that increased temperature enhanced the effectiveness of DEDC as a sterilant. He noted that DEDC was twice as effective at 27°C as at 0°C. Grospicova et al. (1969) also studied the effect of treatment temperature and found a two-fold increase in effectiveness from 6 to 20°C on two *Lactobacillus* species. Shibasaki et al. (1969) noted the temperature coefficient for molds in distilled water and in phosphate buffer was 2.5 and 2.3, respectively, but the temperature coefficient for the microbiocidal activity was 5 and 10 for the same two solutions. Their work with *Penicillium thomii* showed that at pH 4.0, about a four-fold increase in the antiseptic effectiveness of DEDC was evident as the temperature was increased from 10 to 20°C, and another ten-fold increase occurred at 20–30°C.

Splittstoesser and Wilkinson (1973) reported a 100-fold increase in effectiveness of DEDC against *S. cerevisiae* and *Lactobacillus plantarum* by raising the temperature from 20 to 40°C. Goto et al. (1970) found no temperature effect; however, they used only 250 mg/L of DEDC against high numbers of yeast cells. At the higher temperature (35°C), the rapid growth rate could have overcome the sterilization effects in the rich medium used. They also found no effect of ethanol or sugars on the kill rates (Goto et al., 1970).

13.3.4.2 pH

Reaction of R–NH groups with DEDC and DMDC (one of the proposed modes of action) is dependent on the degree of ionization of the –NH group. In the pH range 3.0–4.0, this relation is demonstrated by the

TABLE 13.4

Antibacterial Effectiveness of Diethyl Dicarbonate[a]

Bacteria	Starting Cell Concentration per mL	DEDC Concentration (mg/L) Sufficient to Be Bactericidal (100% kill)
Acetobacter pasteurianus	5.0×10^2	80
A. xylinum	6.0×10^3	300
Bacterium coli	1.0×10^5	400
Escherichia coli	1.0×10^5	1500
Bacterium aceticum	5.0×10^4	500
Staphylococcus aureus	1.0×10^3	70–100
Lactobacillus arabinosus	1.0×10^6	250
L. helveticus	1.0×10^7	100–300
L. pastorianus	6.0×10^2	300
Lactobacillus buchneri	4.8×10^2	30
Bacillus subtilis	4.0×10^9	3000
Salmonella Typhimurium	1.0×10^5	2000
Bacillus mediosporum	4.8×10^2	30
B. megatherium	4.8×10^2	30
Salmonella Thompson	1.0×10^5	2000
Pseudomonas cocciformis	8.0×10^2	30
Pseudomonas, pigmented	1.0×10^5	200
Pseudomonas, unpigmented	1.0×10^5	500
P. erithrogloem	7.2×10^2	30
P. flavum	4.8×10^2	30
P. mildenbergii	6.4×10^2	30
Micrococcus annulatus	4.8×10^2	30
M. lacteus	6.4×10^2	30
M. lacteus	6.4×10^2	30
Sarcina flavescens	8.0×10^2	30

[a] Determined under various environmental conditions.

reaction of DEDC and DMDC with ammonia to form urethanes (Figure 13.2). As the ammonia becomes ionized at the lower pH, the reaction is lessened (Ough, 1976b; Ough and Langbehn, 1976).

13.3.4.3 Ethanol and Other Constituents

The rate of hydrolysis of DEDC or DMDC can be significantly decreased by increasing the ethanol concentration. Other water-soluble organic solvents also have similar effect. The reactivity of DEDC or DMDC is also such that any proteinaceous material suspended in the solution reacts and decreases the antimicrobial effectiveness of the treatment. Porter and Ough (1982) found that ethanol increased the effectiveness of DMDC and that 20°C was the optimum use temperature.

13.4 Mechanism of Action

The inactivation of microorganisms by DEDC or DMDC appears strongly related to the inactivation of the enzymes of the organism. Protein modification, through reaction of nucleophilic groups, such as imidazoles, amines, or thiols, can readily occur with the dicarbonate (Osterman-Golkar et al., 1974; Ehrenberg et al., 1976). The point of attack is one of the central carbon atoms. The residual portion of the dicarbonate is unstable and decomposes rapidly. Reactions with protein imidazole (7) and amine

TABLE 13.5

Antibacterial Effectiveness of Dimethyl Dicarbonate[a]

Bacteria	Starting Cell Concentration per mL	DEDC Concentration (mg/L) Sufficient to Be Bactericidal (100% kill)
Acetobacter pasteurianus	5.0×10^2	80
Escherichia coli	5.0×10^2	400
Pseudomonas aeruginosa	5.0×10^2	100
Staphylococcus aureus	5.0×10^2	100
Lactobacillus buchneri	5.0×10^2	30
Lactobacillus pastorianus	5.0×10^2	300
Lactobacillus brevis	5.0×10^2	200
Pediococcus cerevisiae	5.0×10^2	300

[a] Temperature 28°C and pH between 2.8 and 4.7 in an artificial medium.

TABLE 13.6

Fungicidal (Mold) Effectiveness of Diethyl Dicarbonate

Mold	Starting Spore Concentration per mL	DEDC Concentration (mg/L) Sufficient to Be Fungicidal (100% kill)
Aspergillus flavus	5.0×10^2	1000
A. niger	1.0×10^4	150
A. niger	2.9×10^2	750
A. terreus	3.0×10^2	250–400
Penicillium digitatum	3.0×10^2	120–250
P. expansum	1.0×10^2	100
P. glaucum	3.0×10^2	250
P. italicum	3.0×10^2	100
P. leteum	3.0×10^2	300
Mycoderma species	4.0×10^3	500–100
Trichoderma viride	1.0×10^3	500
Botrytis cinerea	1.0×10^3	100
Oidium lactis	3.0×10^2	300–700
Fusarium oxysporum	1.0×10^3	100
Fusarium orthoceras	1.0×10^3	100
Neurospora sitophila	3.0×10^2	50–200
Mucor racemosus	3.0×10^2	300–700
Rhizophus nigricans	3.0×10^2	300
Pullularia pullulans	3.0×10^2	100–200
Cladosporium herbarum	5.0×10^2	100
Paecilomyces spp.	1.3×10^2	1000

[a] Tested under various environmental conditions.

groups are of particular importance owing to the ease and rapidity with which they occur (Means and Feeney, 1971).

(13.7)

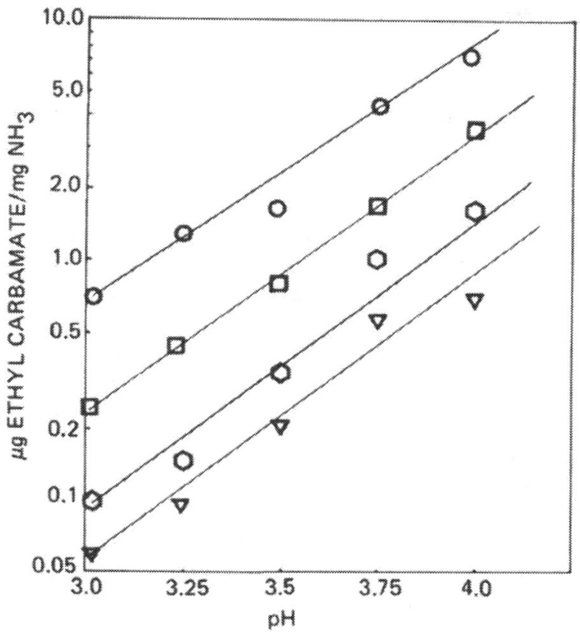

FIGURE 13.2 Influence of ammonia, pH, and DEDC on the amounts of ethyl urethane formed, mg/L of DEDC: 50 (triangles); 100 (hexagons); 200 (squares); 400 (circles).

Mühlrád et al. (1967) investigated the reactions of DEDC with a number of amino acids. The reaction with histidine is similar to that of the imidazole reaction, i.e., a reaction with the imino nitrogen. Holbrook and Ingram (1973) found that the enzyme lactate dehydrogenase was completely inactivated by reaction of DEDC with the histidyl groups of the enzyme. Many other reports (Huc et al., 1971; Dann and Briton, 1974; Choong et al., 1977) with other enzymes have shown that DEDC reacts primarily with the histidyl moeity to form carbethoxy-histidine. Enzyme inhibition results from active site blocking and conformational changes. For example, the histidyl group in alcohol dehydrogenase has been proposed (Ringold, 1966) to be involved in the oxidation-reduction process. Blocking of either nitrogen in the histidine ring would inactivate the enzyme. DEDC has been used routinely to inhibit nuclease in the extradition of nucleic acids. Any single key enzyme inactivated could eventually cause the death of an organism. A report by Ehrenberg et al. (1976) reviews the many reactions that can take place with DEDC.

13.5 Applications

Pauli (1984) reviewed the nature of food additives and their possible hazardous reactions. DEDC was one of concern. The usefulness of DMDC as a beverage preservative, its antimicrobial activity, and some of its chemistry were reviewed by Thoukis (1983). Because of its regulatory status, DMDC is approved only for use in various beverages. Furthermore, its application to beverages requires use of special equipment for metering and mixing DMDC.

13.5.1 Fruit Juices

Martienssen (1961) found DEDC could be used to stabilize fruit juice and prevent fermentation for several days. The authors suggested that syrups with over 50% sugar would delay the hydrolysis of the ester. Mehlitz and Gierschner (1964), who investigated DEDC sterilization failures, noted that possible mold contamination points, such as corks, had to be sterilized before treatment of juices with DEDC. Wucherpfenning (1966) pointed out that the DEDC reacts with many components in the juices and therefore the juice can no longer be considered "natural." Attempts to use DEDC to prevent malolactic fermentation in apple juice have met with little success (Hara and Otsuka, 1967). Genth (1969) has reviewed the use of DEDC for cold sterilization of fruit juices.

Fisher and Golden (1998) reported that *Escherichia coli* O157:H7 populations were reduced from about 7.0 log CFU/mL to undetectable levels within three days in apple cider containing 250 ppm DMDC and stored at 4°C; increasing the storage temperature to 25°C enhanced DMDC effectiveness, and *E. coli* O157:H7 was undetectable within two days of storage. Basaran-Akgul et al. (2009) treated 8 different apple ciders inoculated with different strains of *E. coli* O157:H7 and found that 250 ppm of DMDC was able to achieve greater than a 5.0-log reduction after 6 hours at room temperature. DMDC has also been found to be effective at reducing *Salmonella enterica* in orange juice at concentrations of 172 and 200 ppm, resulting in a 5-log reduction after 24 hours when held at 4°C (Cheng et al., 2018).

DMDC has also been used with other non-thermal processes, such as high-pressure processing, ultraviolet (UV) radiation, or additional preservatives to increase DMDC effectiveness. Combination of DMDC with sodium benzoate or potassium sorbate offers enhanced protection by providing a secondary barrier against surviving spoilage organisms (Golden, 2002; Worobo, 2002). *E. coli* O157:H7 populations were reduced from 7.0 log CFU/mL to undetectable levels at 4°C within 72 hours in apple cider and 48 hours in orange juice containing 250 ppm DMDC (Lakins, 2002). A combination of 250 ppm DMDC and 450 ppm sodium benzoate rendered *E. coli* O157:H7 undetectable within 48 hours in apple cider and 24 hours in orange juice. Lakins (2002) reported that inoculated *Salmonella* (7.0 log CFU/mL initial) was undetectable in apple cider containing 250 ppm DMDC within 48 hours; the same result was achieved within 24 hours in cider containing a combination of 250 ppm DMDC and 450 ppm sodium benzoate.

Williams et al. (2006) reported that a combination of ozone, DMDC, and 24-hour storage at 4°C was effective at reducing populations of *E. coli* O157:H7 in apple cider by 5.7 log CFU/mL and *Salmonella* spp. in apple cider and orange juice by 5.8 and 5.6 log CFU/mL, respectively.

DMDC has also been used in conjunction with non-thermal processes, such as high pressures or UV radiation to achieve microbial inactivation. Combination treatments of 400 MPa with 62.5 ppm DMDC and 550 MPa with 125 ppm DMDC were able to achieve greater than 5.0 log reductions of *Salmonella* Agona in orange juice and *E. coli* O157:H7 in apple juice, respectively (Whitney et al., 2008). After 24-hour storage at 4°C, *E. coli* O157:H7 was undetectable in apple juice treated at 550 MPa with both 62.5 and 125 ppm DMDC. Gouma et al. (2015) combined DMDC with UV radiation and mild heat (55°C) to inactivate *E. coli* in apple juice. Combination UV and heat treatments using 25 mg/L, 50 mg/L, and 75 mg/L of DMDC resulted in a greater than 6.0-log reduction after 3.58 min, with the addition of DMDC reducing UV dose and processing time (Gouma et al., 2015).

Worobo (2002) reported that strain variation and differences in apple varietals used in cider-making affect the sensitivity of *E. coli* O157:H7 in apple cider. Using 250 ppm DMDC, *E. coli* O157:H7 D-values ranged from a low of 0.24 hours in apple made from McIntosh apples to a high of 1.02 hours in cider made from Red Delicious apples.

B. fulva ascospores were resistant to DMDC up to 1000 mg/L concentration in apple juice, while only 25–75 mg/L was required to kill the vegetative cells. Increased effectiveness was noted at higher treatment temperatures (Van der Riet et al., 1989).

13.5.2 Grape Juice and Wines

The effectiveness of DEDC as a sterilant, particularly in grape juice and wines, has been reviewed by Rankine (1964), Pauli and Genth (1966), Pauli (1969), Genth (1971), Barnick (1973), and Ough (1978). Beuchat (1976) investigated the ability of DEDC to inhibit the growth of *Byssochlamys nivea* ascospores in grape juice. This microorganism is known to produce the toxin patulin. It took 500 mg/L of DEDC to inhibit growth of the spores.

The initial work outside the Bayer AG laboratories was reported by Hennig (1959, 1960) and indicated that DEDC was a satisfactory inhibitor and had active cidal properties. He indicated that the inhibitory effect required about one-third the amount of DEDC, as did the killing effect. Hennig (1961) also developed the atomizing pump system that came into general use for application of the compound. Mayer and Luthi (1960) also did some of the initial testing on wine in Switzerland and suggested 50–100 mg/L of DEDC as an effective sterilizing level. Ough and Ingraham (1961) confirmed that 50–100 mg/L of DEDC was a safe sterilizing level in wine and also noted no identifiable sensory problems with DEDC. A sensory threshold for DEDC was found at about 600 mg/L (Van Zyl, 1962; Blouin and Barthe, 1963). Van Zyl (1962) also noted the synergistic effect of DEDC with sulfur dioxide.

Italian studies by Fontana and Colagrande (1964) indicated that levels of 150–200 mg/L DEDC were required for effective sterilization. This was further verified for both wine and grape juice by Garoglio and Stella (1964) and Minarik (1964) in Czechoslavakia, Adams (1965) in Canada, Kalugina (1965) in Russia, Todor et al. (1967) in Romania, Ivanov et al. (1967) in Bulgaria, Prillinger (1964) in Austria, and Rankine and Pilone (1974) in Australia. All these researchers tested DEDC in a practical manner, and most determined it to be a very effective antimicrobial agent for use in bottling wine or grape juice in conjunction with the use of good standard sanitary practices. These are only a representative group of the published reports.

Relatively large amounts of ethyl urethane (up to 600 mg/L) were found in a number of commercial Japanese sakes (Ough, 1978). No diethyl carbonate was found in these samples, indicating that no ethanol was present when the DEDC was used. It appears that this method was used commercially in Japan to sterilize the koji before the alcoholic fermentation (Hara et al., 1970). Further studies (Ough et al., 1988a) showed that heating of ethyl alcohol and urea in water solutions resulted in the formation of ethyl carbamate. In wine, Ough et al. (1988b) found that urea when formed during fermentation could remain in the wine. If the wine was heated ethyl carbamate formed. Freshly made sake has no more urea than any other similar ferment; however, in the processing of sake it is heated. This results in significant amounts of ethyl carbamate.

DMDC has been reviewed as an agent for inhibiting yeasts and fermentation in wine (Delfini et al., 2002; Costa et al., 2008; Zuehlke et al., 2015). Porter and Ough (1982) investigated the use of DMDC as an alternative sterilant to DEDC for alcoholic beverages, especially in wine. DMDC at 100 mg/L in 8% v/v ethanol at 30°C effectively reduced *S. cerevisiae* Montrachet to undetectable levels after 20 minutes. For the same concentration of DMDC in 10% v/v ethanol at 30°C, the same result was achieved within 10 minutes, suggesting DMDC is more effective at higher ethanol concentrations. Divol et al. (2005) found that DMDC used in concentrations of 200 mg/L was more effective than SO_2 at stopping fermentation in pure cultures of *S. cerevisiae*, *Candida stellata*, and *Zygosaccharomyces bailii* and in botrytized must.

Fermentative spoilage of grape juice at 21°C was prevented by addition of 0.8 mM DMDC when juice was inoculated with *S. cerevisiae* Montrachet at levels of 2 or 200 CFU/mL (Terrell at al., 1993). These researchers also noted that DMDC was generally more effective than sulfur dioxide and sorbic acid, and that the effectiveness of DMDC, but not the other chemicals, was enhanced when the storage temperature was increased to 31°C. Similarly, Morris et al. (1996) demonstrated that 0.8 mM DMDC effectively prevented spoilage of grape juice held at 31°C, while sorbic acid, potassium metabisulfite, and combinations of these two delayed but did not prevent spoilage.

13.5.3 Soft Drinks

DEDC appears to have some value in inhibiting growth of osmophilic yeasts in natural fruit juice soft drinks (Pozsonyi, 1972). A maximum of 300 mg/L of DEDC was suggested for use in sterilizing the drink against yeast or bacterial growth (Pozsonyi, 1972). Molds were not found to be sufficiently killed or inhibited. Pátkay et al. (1973) investigated the use of DEDC with various diluted and carbonated commercial syrups and found that a level of 200 mg/L of DEDC was sufficient to protect the drinks.

DMDC has been suggested for commercial use in Germany as a sterilant for artificial drinks (Genth, 1980). The advantage of using DMDC is that no reactions occur with sugar, sugar alcohols, or artificial sweeteners, such as saccharin or cyclamate.

13.5.4 Foods

Rash (2003) reported that soaking cantaloupes for 3 minutes in a 10,000-ppm solution of DMDC reduced the population of *Salmonella* from 5.01 log cfu/cm^2 to undetectable levels; *Salmonella* populations were detected only by enrichment after a 3-minute treatment with 5000 ppm DMDC. Mount et al. (1999)

demonstrated that shelf life of fresh salsa could be extended by several weeks by a combination of 100 ppm DMDC and 1% potassium sorbate. They reported that the aerobic plate count and yeast/mold count of fresh salsa decreased from 6 and 2 log CFU/g, respectively, to undetectable levels within 8 days of storage at 4°C.

Chen et al. (2013) dipped Chinese cabbage parts in a 200 mg/L DMDC solution and found that the solution reduced total aerobic plate counts and total yeast and mold count of the leaf and stalk by at least 4.0 log CFU/g and 3.0 log CFU/g, respectively. The DMDC solution was also able to inhibit polyphenol oxidase and peroxidase activity, which is linked to enzymatic browning reactions, thus preserving the quality of Chinese cabbage (Chen et al., 2013). Wang et al. (2013) performed similar analyses using a 200 mg/L DMDC solution as a disinfectant for fresh cut carrots, reporting that total aerobic plate count and total yeast and mold counts were reduced by more than 3.0 log CFU/g.

Molin et al. (1963) summarized the possible uses of DEDC in foods. They found that the growth of *B. cinerea* on strawberries was delayed by dipping the fruit into 100 and 1000 mg/L solutions of DEDC. No adverse sensory effects of the DEDC were noted. In testing apple sauce, they noted that up to 1000 mg/L was required to inhibit mold growth and was not very effective in preventing fermentation of apple sauce.

The storage life of freshly slaughtered chicken carcasses dipped into a solution of DEDC by Schmidt-Lorentz (1962) was extended by 65%. Surface bacteria showed a 10- to 20-fold reduction.

Hara and Otsuka (1966a) studied the use of DEDC to sterilize koji. They determined the most satisfactory approach was to mix the koji in water and then add DEDC. The wild yeast and some bacteria present were killed by 100–200 mg/L, but the lactic bacteria and *Aspergillus oryzae* required more than 600 mg/L of DEDC.

13.5.5 Beer

Mönch (1961) appears to have been the first to test DEDC in beer sterilization. His studies indicated levels between 250 and 1000 mg/L were required for effective action against spoilage by yeast. He noted a taste threshold in beer of 1000 mg/L of added DEDC. Molin et al. (1963) indicated that 0.01% DEDC doubled the shelf life of a beer. A system was suggested by Kozulis et al. (1971) in which DEDC was used in conjunction with *p*-hydroxybenzoate esters for beer sterilization.

13.6 Regulatory Status

DEDC may not be used for treatments involving food or beverages in the United States. In 1972, based upon a report by Löfroth and Gejvall (1971), the U.S. Food and Drug Administration (FDA) banned the use of DEDC (FDA, 1972). Until that time, it was legally used in a number of beverages. The first approved use for DMDC was for addition to wines as a yeast inhibitor at a level not to exceed 200 ppm (FDA, 1988). The FDA, in 2001, amended the food additive regulations to provide for a more descriptive term, "microbial control agent," in place of "inhibitor of yeast," for the safe use of DMDC based upon a food additive petition filed by Bayer Corp. (FDA, 2001a). DMDC presently is approved for use as a direct food additive to be used as a microbial control agent in certain beverages in which the microbial population has been reduced to 500 microorganisms per mL or less by current good manufacturing practices. Beverages in which DMDC is approved for use are: (1) wine, dealcoholized wine, and low-alcohol wine, at a limit of 200 ppm; (2) ready-to-drink teas, at a limit of 250 ppm; (3) carbonated or non-carbonated, non-juice-containing ($\leq$1% juice), flavored or unflavored beverages containing added electrolytes (5–20 meq/L Na^+ and 3–7 meq/L K^+), at a limit of 250 ppm; and (4) carbonated, dilute beverages containing juice, fruit flavor, or both, in which the juice content does not exceed 50%, at a limit of 250 ppm (FDA, 2001a).

In 2000, Bayer Corp. filed an effective notification of food-contact substance (FCN 0035) with the FDA, thereby allowing for use of DMDC as a microbial control agent in non-carbonated juice beverages containing up to and including 100% juice (FDA, 2000). A food-contact substance is defined as one intended for use as a component of material used in manufacturing, packaging, transporting, or holding food where the substance is not intended to have a technical effect on the food (FDA, 2002). Limitations for use of DMDC as a food-contact substance are similar to those specified for its use as a food additive, i.e., the microbial load of the juice must be reduced by current technologies (e.g., heat) prior to the addition of DMDC at a limit of 250 ppm (FDA, 2003). DMDC breaks down into carbon dioxide and methanol at very low concentrations, making it an incidental additive, a processing aid that breaks down into constituents that are either normally present in the food or found at insignificant levels (FDA, 2001b). Incidental additives are exempt from label reporting. DMDC is marketed and sold for application to foods under the registered trademark, Velcorin®.

The basis for the ban on DEDC was that a small amount of ethyl carbamate, a carcinogen, was formed by the reaction of ammonia with the DEDC. Despite extensive proof of the error of that report (Fischer, 1972; Ough, 1976a), the ban was upheld, based on the Delaney Clause of the Food Additive Amendment. Interestingly, *all* fermented products tested in one study, including yogurt, beer, and soy sauce, had small but detectable amounts of ethyl carbamate (Ough, 1976a). As it turns out, the amount of ethyl carbamate added to the food chain by the use of DEDC would be minor. The natural formation of precursors in fermented foods and beverages and in drugs using alcohol as either a carrier or as an antiseptic far exceeds any formed from the use of DEDC. Nevertheless, this has not altered the FDA's position on this compound.

13.7 Toxicology

Since the dicarbonates are expected to be decomposed before consumption, the reaction products are of major concern when considering their use in foods. Direct consumption or contact with the material, however, must not be ignored.

13.7.1 Foods

Since DMDC is hydrolyzed to carbon dioxide and methanol almost immediately after addition to beverages, the FDA determined that there would be virtually no exposure of consumers to the additive itself when used within the limits of 200 ppm in beverages (FDA, 1988). Methanol is the principal hydrolysis product of concern resulting from addition of DMDC to wine. Theoretically, complete hydrolysis of DMDC would yield two moles of methanol from one mole of DMDC added to wine. Using a worst-case scenario, the FDA determined that consumption of wine containing 200 ppm DMDC at a rate of 232 grams per person per day would result in a daily intake of not more than 22 milligrams per person per day, well within safe limits (FDA, 1987). An adult human can metabolize up to 1500 milligrams of methanol per hour with no adverse symptoms or effects (Lehman, 1963). A fairly extensive report of the safe use of DMDC in wine, including reference to breakdown and reaction products, is described in the FDA response to the original DMDC food additive petition (FDA, 1988).

Hecht (1961) found no significant toxic effects in rats consuming DEDC-treated grape juice for a 59-day period. This observation was confirmed by Bornmann and Loeser (1961). Although DEDC-treated milk caused some weight loss, no other symptoms were detected by these workers.

Bornmann and Loeser (1966) did long-range studies using the reaction product of DEDC and ethanol (diethyl carbonate, DEC). No harmful effects were found when DEC was given to rats at 0.3% in drinking water for 100 weeks or by force feeding it to dogs at 600 mg/kg per day for 3 months.

Sharratt et al. (1972) repeated the short-term tests using wine, beer, orange juice, and blackcurrant juice. Treatments with DEDC caused no adverse effects from the reaction products on the rats being tested.

Studies by Lang et al. (1966) showed that most of the reaction products were hydrolyzed enzymatically to the original compound plus carbon dioxide and ethanol. The exception was the carbethyoxy ascorbic acid. This product hydrolyzed into the original ascorbic acid, dehydroascorbic acid, diketogluconic acid, and furfural after a few hours. Paulus and Lorke (1967) tested these products in short-term studies on rats and found no adverse effects. Wolf et al. (1969) found no toxic effects in five human volunteers, who over a 40-day period ingested 250 mL wine per day that was treated with 150 mg/L of DEDC. A single treatment of 1 liter of the treated wine caused no toxic symptoms in the group. Zaitsev et al. (1970) suggested that DEDC not be used in foods because of its toxicity. Their fear was that the compound would not completely hydrolyze.

Santini et al. (1985) tested the mutagenicity of DEDC on two strains of *Salmonella* Typhimurium (His^-) and on *Bacillus subtilis* 170 (Trp^-, Ura^-) and found no reversion to prototrophy. They also found no effect of ethyl carbamate on the mutagenicity of *B. subtilis*. There was some transformation reduction.

The Joint Food and Agriculture Organization/World Health Organization Expert Committee on Food Additives (JECFA) (1972) studied the use of DEDC and concluded beverages over pH 4.0 with a significant content of ammonia, amino acids, or proteins (for example, milk) should not be treated with DEDC. Treatment of beer and fruit juices is not technically justified, nor should wine be treated with DEDC. The acceptable level of treatment for soft drinks, carbonated or not, is 250 mg/L. Under no circumstances should the human daily intake of ethyl carbamate exceed 100 μμg/day.

Short-term toxicological studies reported by Genth (1980) indicated the acute toxicity for rats of DMDC was 330–900 mg/kg body weight. The Ames test for mutagenicity was negative. No adverse effects were found after feeding fruit juice and alcoholic beverages treated with 4000 mg/L of DMDC for 3 months. The amounts of methanol produced were well below toxicological levels. Methyl urethane, another reaction product, was reported not to be a carcinogen (Pound, 1967). The JECFA (1991) has also reviewed studies regarding short- and long-term DMDC toxicity studies in rats and dogs, where no evidence of toxic effects was found from consumption of DMDC-treated beverages. The JECFA has ruled that since the concentrations of methanol resulting from DMDC use are similar to naturally occurring amounts in fruit and alcoholic beverages, DMDC is acceptable to use as a cold sterilization agent up to a concentration of 250 mg/L and in accordance with good manufacturing practices (GMPs).

13.7.2 Direct Exposure

The median lethal dose (LD_{50}) values of DEDC for mice, rats, cats, rabbits, and dogs were reported to be 1558, 850, 100–250, 500–750, and greater than 500 mg/kg body weight, respectively, by the oral route (Joint FAO/WHO Expert Committee on Food Additives, 1972). Tests with rabbits, guinea pigs, rats, and mice (Hecht, 1961) showed 1 mg/L exposure in air caused chronic respiratory symptoms and 10 mg/L was lethal. It is a strong skin and eye irritant.

13.7.3 Indirect Exposure

The carcinogenicity of ethyl carbamate has been well-documented in animals (Allen et al., 1986; Miller and Miller, 1983; Mirvish, 1968; Schmähl et al., 1977; Woo, 1983).

Uzvolgyi et al. (1983) found that DEDC and ammonia incubated *in vitro* in rat gastric juice and then administered to mice orally caused lung cancer in the mice. If DEDC and ammonia were administered by probe *in vivo* to mice, the same effect was found. Later Uzvolgyi (1986) reported contrary results when testing lactating mice and their offspring.

13.8 Analysis

Both DEDC and DMDC are very reactive, and if any water or other proton donor substance is present, they break down autocatalytically. Two types of analyses are available: one to measure the pure or nearly pure substance and the other to measure the residual amount of a specified reactant in the treated material and calculate the amount of dicarbonate added. The first is used primarily to determine the purity of the dicarbonate or to study the kinetics of its reaction. The other method is used at some later date to determine the amounts added originally.

13.8.1 Direct

The dicarbonates, either diethyl or dimethyl, can be determined. DEDC was measured with a relative standard deviation of ±0.3% by Cuzner et al. (1971). A measured amount of dicarbonate was combined with an excess but known amount of morpholine. The morpholine reacted with the dicarbonate.

The remaining morpholine was titrated with standardized methanolic HCI solution to a pH end point of 4.0. Hara and Otsuka (1966b) tested three other amines besides morpholine and determined that *n*-butylamine and isobutylamine gave the sharpest end point. They extracted dicarbonate from sake with organic solvents before analysis. Moncelsi (1970) has suggested a spectrophotometric method using 4-aminoantipyrine, which requires an extraction step.

Peterson (1978) used the method suggested by Berger (1975) for his studies on hydrolysis rate. This spectrophotometric method lends itself to multi-sample analysis. The reactions involved in this method are as follows:

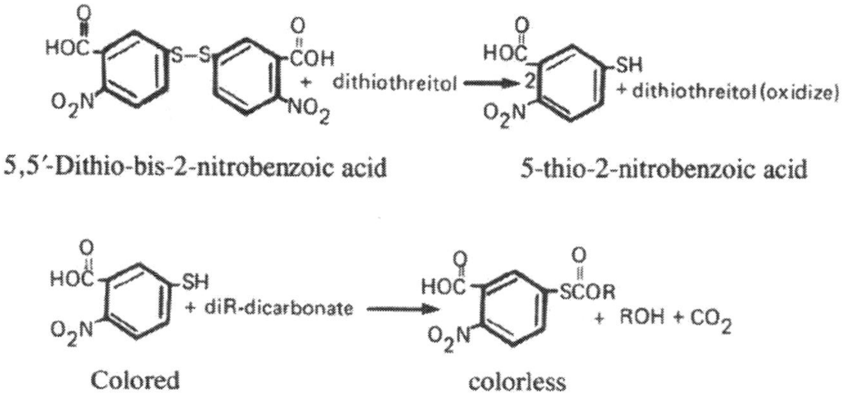

(13.8)

The colored compound 5-thio-2 nitrobenzoic acid is destroyed by the dicarbonate. Pseudo-first-order kinetics describe the reaction. The rate of the color disappearance is proportional to the initial amount of dicarbonate present.

13.8.2 Indirect

The reactions of DEDC with alcohols were postulated earlier. Kielhöfer and Würdig (1963a, b) were first to use the formation of diethyl carbonate to measure the original amount of DEDC added to an alcoholic beverage. They found no DEC in untreated wine. The wines treated with DEDC formed DEC in amounts proportional to the amount of ethanol present in the wine. The method consisted of extraction of the wine with carbon disulfide or pentane and determining the DEC by gas chromatography. The amounts found are in the range of 5–10 mg/L of DEC in most wines treated with the usual amount of DEDC. Prillinger (1964) verified the earlier work. Garschagen (1967) found that DEC could be measured by gas chromatography directly without extraction. A very sensitive headspace method that could detect DEC to 0.1 mg/L was suggested by Kunitake (1969).

Wunderlich (1972) reported a collaborative study for the Association of Official Analytical Chemists. The method involved extracting the DEC and determining the amount with quantitative gas chromatography. McCalla et al. (1977) questioned the use of the 15% Carbowax 20 M column used for the separations. They found an unknown peak at the same retention time as DEC. Using headspace sampling and chromatography coupled to a mass spectrometer, Van Lierop and Nootenboom (1979) were able to measure as little as 1 g/L of DEC. The specificity of measurement was enhanced when the mass spectrometer was selectively tuned to direct M/E 63 and 91, the main fragments of DEC. The official method (Horwitz, 1980) is extraction with carbon disulfide, separation on the gas chromatographic column, and comparison of the DEC peak area with nonalcoholic foods or beverages.

The DMDC reaction product with ethanol is ethyl methyl carbonate (EMC). Stafford and Ough (1976) reported a method to detect the amount of EMC formed and to relate it to the amount of DEDC added to an alcoholic beverage. They used a gas chromatographic method similar to that of Wunderlich (1972) for DEC measurement but used an internal standard (DEC) to allow more accurate determination. Bandion et al. (1979) reported a similar determination using glass capillary columns for better separation.

REFERENCES

Adams, A. M. 1965. The diethyl ester of pyrocarbonic acid as an antimicrobial agent. Rep. Hort. Exp. Sta. Prod. Lab., Vineland Sta., ON, Canada, pp. 138–145.

Alimukhamedova, O. S. 1975. DPK (diethyl pyrocarbonate), an effective antiseptic (transl.). *Uzb. Biol. Zh.* 19:13.

Allen, J. W., Stoner, G. D., Pereira, M. A., Backer, L. C., Sharief, Y., Hatch, G. G., Campbell, J. A., Stead, A. G., and Nesnow, S. 1986. Tumorigenesis and genotoxicity of ethyl carbamate and vinyl carbamate in rodent cells. *Cancer Res.* 47:4911.

Bandion, F., Valenta, M., and Fankel, R. 1979. Nachweis von Methyläthylcarbonat in Wein. *Mitt. Rebe Wein Kosterneuburg* 29:195.

Barnick, J. W. F. K. 1973. Diëthyldicarbonaat. *Chem. Technol.* 28:253.

Basaran-Akgul, N., Churey, J. J., Basaran, P., and Worobo, R. W. 2009. Inactivation of different strains of *Escherichia coli* O157:H7 in various apple ciders treated with dimethyl dicarbonate (DMDC) and sulfur dioxide (SO_2) as an alternative method. *Food Microbiol.* 26:8–15.

Berger, S. L. 1975. Diethylpyrocarbonate: An examination of its properties in buffered solutions with a new assay technique. *Anal. Biochem.* 67:428.

Bernhard, H., Thoma, W., and Genth, H. 1959. Preserving carbohydrate-containing perishable material. U.S. Patent No. 2,910,400, October 27.

Beuchat, L. R. 1976. Effectiveness of various food preservatives in controlling the out-growth of *Byssochlamys nivea* ascospores. *Mycopathologia* 59:175.

Blouin, J., and Barthe, J. C. 1963. Étude du pyrocarbonate d'ethyle. *Vigne Vin* 119:113.

Boehm, T., and Mehta, D. 1938. Ester der Pyrokohlensäure. *Berichte* 17:1797.

Bornmann, G., and Loeser, A. 1961. Toxikologisch Studie über den Pyrokohlensäurediäthylester. *Arch. Toxicol.* 19:69.

Bornmann, G., and Loeser, A. 1966. Untersuchungen zur Toxikologie von Diäthylcarbonat. *Arch. Toxicol.* 22:98.

Chen, Y., Wang, H., Xu, Y., Wu, J., and Xiao, G. 2013. Effect of treatment with dimethyl dicarbonate on microorganisms and quality of Chinese cabbage. *Postharvest Biol Technol.* 76:139–144.

Cheng, R. M., Churey, J. J., and Worobo, R. W. 2018. Inactivation of Salmonella enterica and spoilage microorganisms in orange juice treated with dimethyl dicarbonate (DMDC). *Intl. J. Food Microbiol.* 285:152–157.

Choong, Y. S., Sheperd, M. G., and Sullivan, P. A. 1977. Modification of lactate oxidase with diethyl pyrocarbonate. *Biochem, J.* 165:385.

Costa, A., Barata, A., Malfeito-Ferreira, M., and Loureiro, V. 2008. Evaluation of the inhibitory effect of dimethyl dicarbonate (DMDC) against wine microorganisms. *Food Microbiol.* 25:422–427.

Cuzner, J., Bayne, P. D., and Rehberger, A. J. 1971. Analytical methods for the application of diethyl dicarbonate to beer. *Am. Soc. Brew. Chem. Proc.* 1971:116.

Dann, L. G., and Britton, H. G. 1974. The reaction of pyrocarbonate with pyruvate kinase. *Biochem, J.* 137:405.

Daudt, C. E., and Ough, C. S. 1980. Action of dimethyl dicarbonate on various yeasts. *Am. J. Enol. Vitic.* 31:21.

Delfini, C., Gaia, P., Schellino, R., Strano, M., Pagliara, A., and Ambrò, S. 2002. Fermentability of grape must after inhibition with dimethyl dicarbonate (DMDC). *J. Agric. Food Chem.* 50:5605–5611.

Divol, B., Strehaiano, P., and Lonvaud-Funel, A. 2005. Effectiveness of dimethyldicarbonate to stop alcoholic fermentation in wine. *Food Microbiol.* 22:169–178.

Duhm, B., Maul, W., Medenwald, H., Patzschke, K., and Wegner, L. A. 1966. Zur Kenntnis des Pyrokohlensäurediäthylesters. II. Radioaktive Untersuchungen zur Klärung von Reaktionen mit Getränkebestandteilen. *Z. Lebensm. Unters. Forsch.* 132:200.

Ehrenberg, L., Fedorscsak, I., and Solymosy, F. 1976. Diethyl pyrocarbonate in nucleic acid research. *Prog. Nucleic Acid Res. Mol. Biol.* 16:189.

FDA.1972. Food additives: Diethyl pyrocarbonate. *Fed. Reg.* 37:15426.

FDA. 1987. Memorandum dated January 14, 1987, Regulatory Food Chemistry Branch to GRAS Review Branch, "Dimethyl Dicarbonate (DMDC) in Wine. Submission of September 5, 1986; Exposure Estimate for Methyl Carbamate (MC) and Methanol in Wine."

FDA. 1988. Direct food additives; food additives permitted for direct addition to food for human consumption; dimethyl dicarbonate. *Fed. Reg.* 53(204):41325.

FDA. 2000. Bayer Co.; Filing of Food Additive Petition; Amendment; Withdrawal in Part. *Fed. Reg.* 65(188): 58091.

FDA. 2001a. Food additives permitted for direct addition to food for human consumption. Dimethyl dicarbonate. *Fed. Reg.* 66(45):13653.

FDA. 2001b. Food Labeling: Exemptions from Food Labeling Requirements. 21 CFR 101.100.

FDA. 2002. Food additives: Food contact substance notification system. *Fed. Reg.* 67(98):35724.

FDA. 2003. Inventory of effective premarket notifications for food contact substances. College Park, MD: U.S. Food and Drug Administration. Available from: www.cfsan.fda.gov/~dms/opa-fcn.html. Accessed July, 2003.

Fischer, E. 1970. Untersuchung mit ^{14}C-markierten Pyrokohlensäurediäthylester. II. Reaktionen mit Vitaminen und anderren Lebensmittelbestandteilen. *Z. Lebesm. Unters. Forsch.* 144:262.

Fischer, E. 1972. Über die Bildung von Carbaminsäureäthylester. (Urethan) in Getränken nach Behandlung mit Pyrokohlensäurediäthylester. *Z. Lebensm. Unters. Forsch.* 148:221.

Fisher, T. L., and Golden, D. A. 1998. Survival of *Escherichia coli* O157:H7 in apple cider as affected by dimethyl dicarbonate, sodium bisulfite, and sodium benzoate. *J. Food Sci.* 63:904–906.

Fontana, P., and Colagrande, O. 1964. Recherches sur l'emploi de l'ether diéthylpyrocarbonique pour la stabilisation des vins doux. *Ind. Aliment. Agric.* 81:197.

Garoglio, P. G., and Stella, C. 1964. Ricerche sull'impiego in enologia dell'estere dietilico dell'acido pirocarbonico (DEPC) allo stato puro. *Riv. Vitic. Enol.* 17:422.

Garschagen, H. 1967. Über die Bestimmung von Diäthylcarbonat in mit Pyrokohlensäurediäthylester behandelten weinen. *Weinberg Keller* 14:131.

Genth, H. 1964. On the action of diethylpyrocarbonate on microorganisms. In *Fourth International Symposium on Food Microbiology*, edited by N. Molin, SIK, Göteborg, Sweden, p. 77.

Genth, H. 1969. Gyümölcslevek elöallitása hideg úton. *Elelmez. Ipar* 23:181.

Genth, H. 1971. Diäthyldicarbonat, ein im Getränk zerfallender Zusatzstoff. *Gordian* 71:255.

Genth, H. 1972. Personal communication.

Genth, H. 1979. Dimethyldicarbonate-ein neuer Verschwindestoff für alkoholfreie fruchtsafthaltige Erfrischungsegetränke. *Erfrischungsgetränke Mineralwasser Z.* 13:6p.

Genth, H. 1980. Dimethyldicarbonat ein neuer Verschwindestoff. *Braueri J.* 6:129.

Golden, D. A. 2002. Personal communication.

Goto, S., Yamakawa, Y., and Yokotsuka, I. 1970. Sterilization of food and wine by diethylpyrocarbonate. I. Eradication of microorganisms by diethylpyrocarbonate. *Yamanashi Daigaku Hakko Kenkyusho Kenkyu Hokoku* 14:23.

Gouma, M., Gayán, E., Raso, J., Condón, S., and Álvarez, I. 2015. Influence of dimethyl dicarbonate on the resistance of *Escherichia coli* to a combined UV-Heat treatment in apple juice. *Frontiers Microbiol* 6:501. https://doi.org/10.3389/fmicb.2015.00501

Grospicova, A., Kyzlink, V., and Uhlířová, L. 1969. Einfluss des Pyrokohlkensäurediäthylesters auf die entwicklung eininger Mikroben. *Vys. Sk. Che. Technol. Praze Protaviny* E25:49.

Hara, S., and Otsuka, K. 1966a. Fermentation control of sake mash. IV: Sterilization of koji. *Hakko Kogaku Zasshi* 44:824.

Hara, S., and Otsuka, K. 1966b. Determination and decomposition of diethylpyrocarbonate. I: Determination with isobutylamine and decomposition in various solutions. *Shokuhin Eiseigaku Zasshi* 7:497.

Hara, S., and Otsuka, K. 1967. Utilization of diethylpyrocarbonate in sake and wine making. II: Utilization of diethylpyrocarbonate on malo-lactic fermentation in wine and cider making. *Hakko Kogaku Zasshi* 45:137.

Hara, S., Murakami, H., Katagiri, K., and Oba, T. 1970. Influence of subculture age and inoculum size on hiochi bacteria on antimicrobial activities of various preservatives against hiochi bacteria. *J. Soc. Brew Jpn.* 65:1007.

Hecht, G. 1961. Zur Toxikologie des Pyrokohlensäurediäthylesters. *Z. Lebensm. Unters. Forsch.* 114:292.

Hennig, K. 1959. Pyrokohlensäure-Diäthylester, ein neues gärhemmendes Mittel. *Dtsch. Lebensm. Rundshau* 55:297.

Hennig, K. 1960. Der Pyrokohlensäurediäthylester ein neues, rückstandloses, gärhemmendes Mittel. *Weinberg Keller* 7:351.

Hennig, K. 1961. Die technische Anwendung des Pyrokohlensäurediäthylesters ein der Kellerei. *Weinberg Keller* 8:215.

Holbrook, J. J., and Ingram, V. A. 1973. Ionic properties of an essential histidine residue in pig heart lactate dehydrogenase. *Biochem. J.* 131:729.

Horwitz, W. 1980. *Official Methods of Analysis of the Association of Official Analytical Chemists*, 13th ed., Association of Official Analytical Chemists, Washington, DC, pp. 192–193.

Huc, C., Olomucki, A., Thi-Lan, L., Pho, D. B., and van Thoai, N. 1971. Essential histidyl residues of octopine dehydrogenase. *Eur. J. Biochem.* 21:161.

Ivanov, T., Tonchev, T., and Bambalov, G. 1967. Diethyl ester of pyrocarbonic acid (DEPA) for use in the biological stability of wine with residual sugar. *Lozar. Vinar.* 16(14):38.

Joint FAO/WHO Expert Committee on Food Additives. 1972. Evaluation of mercury, lead, cadmium and the food additives amaranth, diethylpyrocarbonate and octyl gallate. WHO Food Additives Series No. 4, World Health Organization, Geneva, p. 67.

Joint FAO/WHO Expert Committee on Food Additives, World Health Organization (eds.). 1991. Evaluation of certain food additives and contaminants: Thirty-seventh report of the Joint FAO/WHO Expert Committee on Food Additives. World Health Organization, Geneva, pp. 23–24.

Kalugina, G. I. 1965. Pyref-diethyl ester of pyrocarbonic acid and its application in the wine making industry. In *Diethyl Pyrocarbonic Ester (Pyref). A New Conservant in Wine Making*, edited by K. Moldavenyaske, Academy of Sciences of Moldova, Kishinev, p. 18.

Kielhöfer, E., and Würdig, G. 1963a. Nachweis und Bestimmung von Diäthylcarbonat und Pyrokohlensäurediäthylester im Wein and Schaumwein. *Dtsch. Lebensm. Rundshau* 59:197.

Kielhöfer, E., and Würdig, G. 1963b. Nachweis und Bestimmung eines Zusatzes von Pyrokohlensäurediäthylester zum Wein. Dtsch. *Lebensm. Rundshau* 59:224.

Kovalenko, V. I. 1952. Preparation of dimethyl pyrocarbonate. *Zh. Obschch. Khim.* 22:154.

Kozenko, E. M. 1952. About the content of stable ester of carbonic acid. *Vinodel. Vinograd. SSSR* 12(4):25.

Kozulis, J. A., Bayne, P. D., and Cuzner, J. 1971. New technique for cold sterilization of beer. *Am. Soc. Brew. Chem. Proc.* 1971:105.

Kunitake. N. 1969. High-sensitive determination of diethylcarbonate-a decomposition by-product of diethyl pyrocarbonate added to alcoholic beverages. *Bull. Brew. Sci.* 15:1.

Lakins, C. 2002. Inactivation of *Escherichia coli* O157:H7 and *Salmonella* spp. in apple cider and orange juice by chemical preservatives and ozone. M.S. thesis, University of Tennessee, Knoxville.

Lang, K., Fingerhut, M., Krug, E., Reimold, W., and Pauli, O. 1966. Über enzymatische Spaltung und stoffwechsel von Umsetzungsprodukten des Pyrokohlensäurediäthylesters mit Bestandteilen von Lebensmitteln. *Z. Ernaehrungswiss.* 6:219.

Larrouquère, J. 1963. Réaction et dosage du pyrocarbonate d'ethyle. *Bull. Soc. Chim. Fr.* 1963:1026.

Larrouquère, J. 1965. Réaction du pyrocarbonate d'ethyle avec les thiols. *Bull. Soc. Chim. Fr.* 1965:382.

Lehman, A. J. 1963. February 12, 1963 letter in Food Additive Petition No.0A0043 (FAP 0A0043).

Löfroth, G., and Gejvall, T. 1971. Diethyl pyrocarbonate formation of urethane in treated beverages. *Science* 174:1248.

Martienssen, E. 1961. Pyrokohlensäurediäthylester als technisches Hilfsmittel bei der Herstellung keimarmer Fruchsäftgetränke. *Fruschsaft Ind.* 6:201.

Mayer, K., and Luthi, H. 1960. Versuche mit Pyrokohlensäurediäthylester, einem neuen Getränkekonservierungsmittel. *Mitt. Geb. Lebensm. Unters. Hyg.* 51:132.

McCalla, M., Barr, G. R., and Kipp, W. H. 1977. Anomalous results in the determination of diethylcarbonate in wine. *J. Assoc. Off. Anal. Chem.* 60:237.

Means, G. E., and Feeney, R. E. 1971. *Chemical Modification of Proteins*, Holden-Day, San Francisco, p. 81.

Mehlitz, A., and Gierschner, K. 1964. Weitere Erfahrungen mit Pyrokohlensäurediäthylester (PKE) bei der Haltbarmachung von Fruchtsäften und restzuckerhaltigen Fruchtweinen. *Ind. Obst. Gemueseverwert.* 49:1.

Merzhanian, A. A. 1951. The behavior of diethylester of pyrocarbonic acid in sparkling and aerated wines. *Vinodel. Vinograd. SSSR* 11(3):19.

Miller, J. A., and Miller, E. C. 1983. The metabolic activation of nucleic acid adducts of naturally occurring carcinogens: Recent results with ethyl carbamate and spice flavors safrole and estragole. *Br. J. Cancer* 48:1.

Minarik, E. 1964. Tokay choice wine stabilization against biological turbidities. *Kvasny Prum.* 10:131.

Mirvish, S. S. 1968. The carcinogenic action and metabolism of urethan and N-hydroxyurethan. In *Advances in Cancer Research*, Vol. 11, edited by A. Haddow and S. Weinhouse, Academic Press, New York, pp. 1–42.

Molin N., Satmark, L., and Thorell, M. 1963. Pyrocarbonic acid diethyl ester as a potential food preservative. *Food Technol.* 17:797.

Moncelsi, E. 1970. Determinazione spettrofotometrica del dietile pircocarbonato con 4-amminoantipirina. *Chem. Ind. (Milan)* 52:637.

Mönch, G. 1961. Pyrokohlens ure-Di thylester, ein neues Konservierungsmittel f r Bier? *Brawissenshaft* 14:257.

Monteiro, F. F., Trousdale, E. K., and Bisson, L. F. 1989. Ethyl carbamate formation in wine: Use of radioactively labeled precursor to demonstrate the involvement of urea. *Am. J. Enol. Vitic.* 40:1.

Morris, J. R., Main, G., and Threlfall, R. 1996. Fermentations: Problems, solutions, and prevention. *Vitic. Enol. Sci.* 51(3):210.

Mount, J. R., Kasrai, S., and Draughon, F. A. 1999. Dimethyl dicarbonate and potassium sorbate effects on microbial growth in fresh salsa. *Ann. Mtg. Inst. Food Technol.*, Chicago, IL, July 24–28, No.22D-9, p. 49.

Mühlrád, A., Hegyi, G., and Tóth, G. 1967. Effect of diethylprocarbonate on proteins. I. Reaction of diethylpyrocarbonate with amino acids. *Acta Biochim. Biophys. Acad. Sci. Hung.* 2:19.

Murata, A. 1974. Growth inhibition of *Lactobacillus casei* phage J1 by diethyl dicarbonate. *J. Gen. Appl. Microbiol.* 20:71.

Osterman-Golkar, A., Ehrenberg, L., and Solymosy, F. 1974. Reaction of diethyl pyrocarbonate with nucleophiles. *Acta Chem. Scand.* 28B:215.

Ough, C. S. 1976a. Ehthylcarbonate in fermented beverages and foods. I: Naturally occurring ethylcarbamate. *J. Agric. Food Chem.* 24:323.

Ough, C. S. 1976b. Ethylcarbamate in feremented beverages and foods. II: Possible formation of ethylcarbamate from diethyl dicarbonate addition to wine. *J. Agric. Food Chem.* 24:328.

Ough, C. S. 1978. A Davis scientist reviews and excellent fungicide, DEDC. *Wine Vines* 59(4): 30.

Ough, C. S., and Ingraham, J. L. 1961. The diethylester of pyrocarbonic acid as a bottled-wine sterilizing agent. *Am. J. Enol. Vitic.* 12:149.

Ough, C. S., and Langbehn, L. 1976. Measurement of methylcarbamate formed by the addition of dimethyl dicarbonate to model solutions and to wines. *J. Agric. Food Chem.* 24:428.

Ough, C. S., Crowell, E. A., and Gutlove, B. R. 1988a. Carbamyl compound reactions with ethanol. *Am. J. Enol. Vitic.* 39:239.

Ough, C. S. Crowell, E. A., and Mooney, L. A. 1988b. Ethyl carbamate precursors during grape juice fermentation. I. Addition of amino acids, urea and ammonia. *Am. J. Enol. Vitic.* 39:243.

Ough, C. S., Kunkee, R. E., Vilas, M. R., Bordeu, E., and Huang, M. C. 1988c. The interaction of sulfur dioxide, pH and dimethyl dicarbonate on the growth of *Saccharomyces cerevisiae* Montrachet and *Leuconostoc oenos*. *Am. J. Enol. Vitic.* 39:279.

Ough, C. S., Stevens, D., Sendovski, T., Huang, Z., and An, D. 1990. Factors contributing to urea formation in commercially fermented wines. *Am. J. Enol. Vitic.* 41:68.

Parfentjev, L. N. and Kovaklenko, V. I. 1951a. The possible participation of pyrocarbon esters in forming champagne properties in sparkling wines. *Vinodel. Vinograd. SSSR* 11(3):16.

Parfentjev, L. N., and Kovaklenko, V. I. 1951b. The theory of champagne fermentation. *Vinodel. Vinograd. SSSR* 113:17.

Pátkay, G., Hergár , E., and Gyns, L. 1973. Szénsavas, Üdititalok tartósitása piroszénsavdietiléeszterrel. *K. Paprikaipar.* 38:95.

Pauli, O., and Genth, H. 1966. Diethyl pyrocarbonate. I: Properties, effects, and analysis. *Zeitschrift fuer Lebensmittel-Untersuchung und -Forschung* 132:216–227.

Pauli, O. 1969. Le pyrocarbonate d'ethyle, un nouveau produit pour la sterilisation des boissons. Spectre d'efficacit , utilisation, analyse, toxicologie et legislation. *Anal. Falsif. Expert. Chim.* 61:377.

Pauli, G. H. 1984. Chemistry of food additives: Direct and indirect effects. *J. Chem. Educ.* 61:332.

Paulus, W., and Lorke, D. 1967. Diethyl pyrocarbonate. III. Preparation and toxicological testing of the representative reaction products of diethyl pyrocarbonate. *Zeitschrift fuer Lebensmittel-Untersuchung und -Forschung* 132:325–333.

Peterson T., and Ough, C. S. 1979. Dimethyldicarbonate reaction with higher alcohols. *Am. J. Enol. Vitic.* 30:119.

Peterson, T. W. 1978. Reaction of dimethyldicarbonate with higher alcohols and water, M.S. thesis, University of California, Davis.

Porter, L. J., and Ough, C. S. 1982. The effects of ethanol, temperature and dimethyl dicarbonate on viability of *Saccharomyces cerevisiae* Montrachet No. 522 in Wine. *Am. J. Enol. Vitic.* 33:222.

Pound, A. W. 1967. Initiation of skin tumours in mice by homologous and N-substituted derivatives of ethylcarbamate. *Aust. J. Exp. Biol. Med. Sci.* 45:507.

Pozsonyi, F. 1972. A dietilpirokarbon t alkalmaz s nak el yer az dit ital gy rt sban. *Szeszipar* 20:28.

Prillinger, F. 1964. ber den Nachweis und die Bestimmung von Di thylkarbonat in mit Baycovin behandelten Weinen. *Mitt. Rebe Wein Klosterneuburg 14A*:29.

Rankine, B. C. 1964. Di-ethyl pyro-carbonate. *Int. Bottler Packer*, December, p. 68.

Rankine, B. C., and Pilone, D. A. 1974. Yeast spoilage in bottle table wine and its prevention. *Aust. Wine Brew. Spirit Rev.* 92(11):36, 38, 40.

Rash, V. A. 2003. Physical and chemical treatments for control of *Salmonella* on cantaloupe rinds. M.S. thesis. University of Tennessee, Knoxville.

Ringold, H. J. 1966. Proposed catalytic role for histidine in pyridine nucleotide-linked alcohol dehydrogenases. *Nature* 210:535.

Santini, P., Moretton, J., and D'Aquino, M. 1985. Detection of the genetic toxicity of diethyl pyrocarbonate using bacterial systems. *Rev. Latinoam. Microbiol.* 27:157 [CA 103(19):159244z].

Schelenz, R., and Fischer, E. 1970. Untersuchungen mit ^{14}C-markeitem Pyrokohlens ure-Di thylester. III. Zur hydrolyse and Alkoholyse. *Z. Lebensm. Unters. Forsch.* 145:279.

Schelenz, R., and Fischer, E. 1971. Untersuchung mit ^{14}C-markiertiem Pyrokohlens ure-Di thylester. IV. Reaktion mit dl-Apfels ure. *Z. Lebensm. Unters.* Forsch. 147:145.

Schmahl, D., Port, R., and Wahrendorf, J. 1977. A dose response study of urethane carcinogenesis in rats and mice. *Intl. J. Cancer* 19:77.

Schmidt-Lorenz, W. 1962. Uber die Anwendlbarkeit von Pyrokohlens ure-di thylester als Konzervierungsmittel f r Schlachtgefl gel. *Z. Lebensm. Unters. Forsch.* 117:231.

Sharratt, M., Gaunt, I. F., Grasso, P., Kiss, I. S. Hooson, J., Wright, M. G., and Ganolli, S. D. 1972. Short-term toxicity studies on diethyl pyrocarbonate-treated beverages in the rat. *Food Cosmet. Toxicol.* 10:743.

Shibasaki, I., Matsumoto, K., and Horie, H. 1969. Factors affecting the microbicidal activity of diethylpyrocarbonate. *Nippon Shokuhin Kogyo Gakkai-Shi* 16:405.

Splittstoesser, D. F., and Wilkinson, M. 1973. Some factors affecting the activity of diethylpyrocarbonate as a sterilant. *Appl. Microbiol.* 25:853.

Stafford, P. A., and Ough, C. S. 1976. Formation of methanol and ethylmethyl carbonate by dimethyl dicarbonate in wine and model solutions. *Am. J. Enol. Vitic.* 27:7.

Terrell, F. R., Morris, J. R., Johnson, M. G., Gbur, E. E., and Makus, D. J. 1993. Yeast inhibition in grape juice containing sulfur dioxide, sorbic acid, and dimethyldicarbonate. *J. Food Sci.* 58(5):1132.

Thoma, W., and Rinke, H. 1959. Synthesen mit Pyrokohlens ure-estern. *Liebigs Ann. Chem.* 624:30.

Thoukis, G. 1983. Dimethyl dicarbonate as a beverage conservative. In *Instrumental Analysis of Foods*, edited by G. Charalamvous, Academic Press, New York, pp. 455–462.

Thoukis, G. R. Bouthilet, J., Ueda, M., and Caputi, A., Jr. 1962. Fate of diethyl pyrocarbonate in wine. *Am. J. Enol. Vitic.* 13:105.

Todor, T., Trifon, I., and Georgi, B. 1967. Effect of pyrocarbonic acid diethyl ester on some microorganisms in grape juice. Lozar. *Vinar* 15(3):36.

Turtura, G. S. 1966. Ricerche sul potere microbicida dell'estere dietilico dell'acido pirocarbonico. *Ric. Sci.* 36:638.

Uzvolgyi, E., Bojan, F., and Arany, I. 1983. In vitro and in vivo formation of carcinogens from diethyl pyrocarbonate in the presence of ammonia. *Magy. Onkol.* 27:165 [CA 99(25):211314n].

Uzvolgyi, E. 1986. Another possibility of development of carcinogenic substances from noncarcinogenic precursors occurring in foods. *Egeszsegtudomany* 30(1):32 [CA 105(3):2321w].

Van der Riet, W. B., Botha, A., and Pinches, S. E. 1989. The effect of dimethyldicarbonate on the vegetative growth and ascospores of *Byssochlamys fulva* suspended in apple juice and strawberry nectar. *Intl. J. Food Microbiol.* 8:95.

Van Lierop, B. H., and Nootenboom, H. 1979. Gas-liquid chromatographic-mass fragmentographic determination of low levels of diethylcarbonate in beverages. *J. Assoc. Off. Anal. Chem.* 62:253.

Van Zyl, J. A. 1962. The microbiology of South African wine making II. The preservation of musts and wines with pyrocarbonic acid diethyl ester. *S. Afr. J. Agric. Sci.* 5:293.

Wang, C., Chen, Y., Xu, Y., Wu, J., Xiao, G., Zhang, Y., and Liu, Z. 2013. Effect of dimethyl dicarbonate as disinfectant on the quality of fresh-cut carrot (*Daucus carota* L.): Dimethyl dicarbonate used on fresh-cut carrot. *J. Food Process Preserv.* 37:751–758.

Whitney, B. M., Williams, R. C., Eifert, J., and Marcy, J. 2008. High pressures in combination with antimicrobials to reduce *Escherichia coli* O157:H7 and *Salmonella* Agona in apple juice and orange juice. *J. Food Prot.* 71:820–824.

Williams, R. C., Sumner, S. S., and Golden, D. A. 2006. Inactivation of *Escherichia coli* O157:H7 and *Salmonella* in apple cider and orange juice treated with combinations of ozone, dimethyl dicarbonate, and hydrogen peroxide. *J Food Sci* 70:M197–M201.

Wolf, A., Hrivnak, D., and Machovcova. 1969. Beitrag zur Toxicit t des Dikohlens uredi thylesters. *Z. Lebensm. Unters. Forsch.* 139:287.

Woo, Y. 1983. Carcinogenicity, mutagenicity and teratogenicity of carbamates, thiocarbamates and related compounds: An overview of structure-activity relationships and environmental concerns. *Environ. Sci. Health* C1:97.

Worobo, R. W. 2002. Personal communication.

Wucherpfenning, K. 1966. Moeglichkeiten des Einsatzes von Pyrokohlens uredi thylester zur haltbarmachung von Fruschs ften und seive Beurteilung hinsichtlich de Erhaltung der Naturreinkeit. *Int. Fruchs ftunion Wiss. Tech. Komm.* 7:49.

Wunderlich, H. 1972. Collaborative study of a gas-liquid chromatographic method for the determination of diethylcarbonate in wine. *J. Assoc. Off. Anal. Chem.* 55:557.

Zaitsev, A. N., Rakhamanina, N. L., and Dyubyuk, N. E. 1970. Hygenic study of PIREF and its possible use in the food industry. *Vopr. Pitan.* 29:14.

Zuehlke, J. M., Glawe, D. A., and Edwards, C. G. 2015. Efficacy of dimethyl dicarbonate against yeasts associated with Washington State grapes and wines: Use of DMDC in winemaking. *J. Food Process Preserv.* 39:1016–1026.

14

Lysozyme

T. Matthew Taylor, Eric A. Johnson, and Ann E. Larson

CONTENTS

14.1 Background and General Properties of Lysozyme

Lysozyme was discovered in 1921 by Alexander Fleming (1881–1955), who described it as a "remarkable bacteriolytic principle" (Fleming, 1922). His subsequent discovery of penicillin occurred in the late 1920s. Fleming's lysozyme research emerged from his demonstration that chemical antiseptics were ineffective in treating infections. He showed that his own nasal mucus inhibited the growth of a *Micrococcus* spp., a fortuitous discovery since micrococci are among the most sensitive organisms to lysozyme. He first thought that the inhibitory factor was a bacteriophage, but later showed that it was an enzyme that lysed bacterial cells (Fleming, 1922). He established that lysozyme was an endogenous (innate) antimicrobial in the body, and his work supported the concept that an effective way to treat infections was to enhance the body's own immune responses. Fleming found that lysozyme was present in nasal mucus and tears, and that hen egg white had a particularly high level of the protein. It is noteworthy that Laschenko (1909) and Rettger and Sperry (1912) showed that egg white was capable of causing lysis of certain strains of bacteria and spores. It was Fleming, however, who actually isolated the lytic agent and showed that it occurred in human secretions.

Lysozyme was the first enzyme whose primary amino acid sequence was determined, and was also the first enzyme whose structure was determined by X-ray crystallography (Blake et al., 1965; Phillips, 1966). It has an ellipsoid structure with dimensions of approximately of $4.5 \times 3.0 \times 3.0$ nm. Its catalytic activity was demonstrated in 1966 by John Rupley, who showed that lysozyme cleaved purified oligosaccharides of N-acetylglucosamine containing a β-(1-4) glycosidic bond (Rupley, 1967). Oligosaccharides with more than three sugar units were hydrolyzed, and the hexasaccharide was shown to be the optimum length substrate. Several families of lysozyme have been found, but they all share the characteristic property of cleaving a β-glycosidic bond between the C-1 of N-acetylmuramic acid and the C-4 on N-acetylglucosamine of bacterial peptidogylcan.

Although traditionally associated with eggs of birds, especially those of domestic hens, lysozyme is widespread in nature and is found in many sources including certain vegetables, insects, plants, and fungi (Masschalck et al., 2001; Tranter, 1994; Jolles and Jolles, 1984). Lysozyme is present in human colostrum (Mathur et al., 1990) and mammalian tissues and fluids such as milk, saliva, mucus, blood, and tears. It is also present in high concentrations in macrophages, leukocytes, monocytes, and neutrophilic granulocytes. Lysozyme has important roles in the immune response of organisms in reaction to infections and inflammation (Osserman et al., 1974; Sava, 1996). The major families of naturally occurring lysozyme that differ markedly in amino acid primary structure and secondary structure: "C" for chicken or classical, "G" for goose, bacterial lysozyme (autolysins), phage lysozyme, and plant lysozyme (Jolles, 1996), but only the C enzyme from the hen egg whites is currently permitted for use in food preservation. Approximately 3.5% of the total protein content in hen egg white is lysozyme (Alderton et al., 1945; Sofos et al., 1998; Yamamoto et al., 1997). Lysozyme is present at concentrations of about 0.1, 0.13, and 0.25 μg/ml in milk from sheep, cows, and goats, respectively (Chandan et al., 1968), while human milk contains 0.4 μg/ml lysozyme (Reiter, 1978). Lysozyme is considered to be one of the most important factors of nonspecific immunity in human breast milk (Hennart et al., 1991). It also occurs at much higher concentrations in colostrum than in milk, and may have positive attributes on the intestinal flora of nursing infants (Barbara and Pellegrini, 1976). Lysozyme has been proposed for use in various clinical applications, including antibacterial, antiviral, and anti-inflammatory treatments in humans and animal species (Sava, 1996; Proctor and Cunningham, 1988). As described in detail in the following sections, lysozyme is also used to control microbial growth in foods such as cheese and wine, and has potential uses as a preservative in other food systems. Lysozyme serves as a good model of an ideal food preservative in many respects, since it is an innate component of the human immune system and thus would be expected to have low toxicity, it is an enzyme that acts catalytically and can be used at low concentrations in foods, it is specific for bacterial peptidoglycan and does not react with human tissue, and it has certain desirable resistance properties to heat, reduced pH, and other intrinsic and extrinsic factors of foods.

14.2 Properties

14.2.1 Chromatographic Elution Characteristics

Early researchers described various methods for the purification and assay of lysozyme from egg white (Alderton et al., 1945; Meyer et al., 1936). Purified commercial lysozyme for food use is currently produced from hen egg whites in high yield by cation exchange resins. Its high affinity to cation resins is due to the basic character of the enzyme, its isoelectric point of 10.5–11, and its monomeric structure and molecular weight of ~14,400 Da (Cunningham et al., 1991). Lysozyme is eluted from the resins by NaCl, and is recovered as the hydrochloride salt. In addition to removal of lysozyme from egg albumin, the process also removes more than 90% of the avidin from eggs. The extracted egg whites are not nutritionally or functionally changed, and are approved for food uses. Lysozyme is generally used as the hydrochloride salt, which is freely soluble in water and buffers. Lysozyme hydrochloride appears as a white, fine powder and has a slightly sweet taste. A 2% lysozyme hydrochloride solution has a pH of about 3.3. It is insoluble in most organic salts and concentrated salt solutions, but regains activity when transferred back to an aqueous solution. Depending on the pH conditions, lysozyme can polymerize (Sophianopoulos and Van Holde, 1964). Basic properties of lysozyme are described in Table 14.1.

14.2.2 Basic Enzymatic and Hydrolytic Properties

Lysozyme (E.C. 3.2.1.17) is classified as a mucopeptide *N*-acetylmuramyl hydrolase by the Commission on Enzymes (Connor, 1993) and is commonly referred to as muramidase (Proctor and Cunningham, 1988). The primary amino acid sequence and structure as determined by X-ray crystallography of hen egg-white lysozyme (HEWL) has been described in detail (Blake et al., 1965; Phillips, 1966; Jolles et al., 1963; Rees and Offord, 1972). Egg-white lysozyme is a single polypeptide chain of 129 amino acids cross-linked by four disulfide bridges (Jolles et al., 1963; Jolles, 1996; Phillips, 1966) with a molecular weight of approximately 14,400 Da (Barbara and Pellegrini, 1976; Jolles and Jolles, 1984). Enzymatic activity of the molecule is lost if at least two of the disulfide bridges are not intact, or if all disulfide bonds are reduced (Proctor and Cunningham, 1988; Wang and Shelef, 1992). Lysozyme has a hydrophobic core with hydrophilic amino acid side chains toward the surface, giving the molecule stability (Proctor and Cunningham, 1988).

The lytic activity of lysozyme has traditionally been measured by observing lysis of UV-killed and lyophilized *Micrococcus luteus* (*M. lysodeikticus*) cells based on the early observations of Fleming, and this assay has been optimized by Shugar (Fleming, 1922; Parry et al., 1969; Shugar, 1952). Numerous variations for assay of lysozyme by turbidimetry have been reported (detailed in Proctor and Cunningham, 1988). One unit of lysozyme is defined as the smallest quantity leading to complete lysis of *Micrococcus luteus* in a serial dilution test (Alderton et al., 1945). Specifically, one Shugar unit is that quantity of enzyme in 1 ml of a suspension of *Micrococcus luteus* inactivated cells at pH 7.0, with an initial absorbance of 0.750 at 450 nm in a pathlength of 1 cm, which causes the absorbance to decrease at a rate of 0.001 per minute. The maximal specific activity of lysozyme at 25°C using *M. luteus* as the substrate generally is ~50,000 units/mg, but can be much less (~20,000 units/mg) depending on the

TABLE 14.1

Properties and Specifications of Food-Grade Lysozyme Hydrochloride for Use in the United States[a]

a. GRAS substance: Egg-white lysozyme (CAS Reg. No. 9001-62-2) obtained by extraction of egg whites.
b. Enzyme property: Peptidioglycan N-acetylmuramoylhydrolase (EC No. 3.2.1.17) that catalyzes the hydrolysis of peptidoglycan in the cell walls of certain bacteria. The ingredient is used as an enzyme as defined in §170(o)(9) in 21 CFR §184.
c. Primary food application: Cheeses as defined in 21 CFR, §51, Sec. 170.3 (n) (5), in accordance with §184.1(b)(3). The primary enzyme target is germinating spores of *Clostridium tyrobutyricum*, an organism responsible for late blowing of certain cheese varieties.
d. Use level: Levels not to exceed current good manufacturing practices in cheese.
e. Label requirement: Bulk and packaged foods that contain cheese manufactured using egg-white lysozyme shall include the usual name "egg-white lysozyme" on their ingredient labels.
f. Commercial ingredient: Food-grade lysozyme hydrochloride.
g. Food use properties: Freely soluble in water; pH 3.3; electrophoretically pure.
h. Limit specifications: Satisfies limit tests and regulations for heavy metals, arsenic, and microbial limit tests. According to IDF (1987) salmonellae absent in 25 g; *Staphylococcus aureus*, *Pseudomonas aeruginosa*, *E. coli*, sulfite-reducing clostridia absent in 1 g. Coliforms max 30/g.

[a] Sources include: International Dairy Federation 1987; U.S. FDA 1998; www.fordras.com; www.inovatech.com. For regulatory information for other countries, the manufacturer is advised to contact the major manufacturers of lysozyme, whose websites are listed above.

lysozyme preparation. From a practical perspective, it is often desirable to use a wavelength of 650 nm to reduce absorption by colored substances in food extracts. Certain compounds can inhibit the lysozyme assay including surfactants such as sodium dodecyl sulfate (SDS), iodine, and fatty acids and alcohols of C12 or higher (Smith and Stroker, 1949). Several alternate methods of assay have been developed including chromatographic separations, agar plate lysis, radioimmunoassays, and fluorescence systems, but these methods are generally only used for special purposes and the *M. luteus* lysis method is the standard (Proctor and Cunningham, 1988; Jolles and Jolles, 1984). A sensitive reverse-phase HPLC detection method for lysozyme was recently developed for use in milk and cheese samples (Pellegrino and Tirelli, 2000).

The optimum pH for lysis of *Micrococcus lysodeikticus in vitro* for egg-white lysozyme is ~6.6 (Barbara and Pellegrini, 1976). Enzymatic activity is observed over the pH range of 3.5–7.0 (Wang and Shelef, 1992), although lower activity has been observed at pH values <3.5 or >7.0 (Smolelis and Hartsell, 1952; Yang and Cunningham, 1993). The optimum pH for activity is strongly dependent on salt concentration (Chang and Carr, 1971; Davies et al., 1969). At pH values <5.3, reaction products can inhibit the enzyme (Barbara and Pellegrini, 1976). Lysozyme is active *in vitro* from 1°C to near-boiling temperature (Johnson, 1994).

The principal natural substrate of lysozyme is the peptidoglycan of bacteria cell walls, in which the polysaccharide is comprised of alternating β-(1-4) linked N-acetyl glucosamine and N-acetyl-muramic acid residues (Salton, 1957; Sharon, 1967). Lysozyme targets the peptidoglycan layer of bacterial cell walls, hydrolyzing the β(1-4) linkage between N-acetylmuramic acid and N-acetylglucosamine, ultimately resulting in cell lysis (Proctor and Cunningham, 1988; Pakkanen and Aalto, 1997; Salton, 1957). Lysozyme also has weaker enzymatic activity on chitin, such as occurs in the bud scars of certain yeasts and in the cell walls of filamentous fungi. Lysozyme is most active on certain Gram-positive bacteria, particularly *Micrococcus luteus* (*M. lysodeikticus*) which is used for assay of activity, and is also highly active on certain clostridia, bacilli, and other Gram-positive bacteria (Table 14.2). Lysozyme activity varies widely among Gram-positive bacteria. For example, in its application for prevention of late blowing by spores of *C. tyrobutyricum* (International Dairy Federation, 1987), the lactic acid starter cultures are generally not affected (Bester and Lombard, 1990; Wasserfall, 1978). Lysozyme is inactive against certain Gram-positive bacteria such as *Staphylococcus aureus* in which the amino group on the peptidoglycan is modified by chemical groups, or on certain species containing large capsules. Lysozyme has weak or negligible activity on Gram-negative bacteria, probably due mainly to its inability to penetrate the outer membrane. However, being strongly cationic, lysozyme also reacts with negatively charged components including nucleic acids, certain proteins, and lipopolysaccharide with some loss of activity (Cunningham and Lineweaver, 1967; Giangiacomo et al., 1992; Ohno and Morrison, 1989). Under most conditions encountered in foods, the lipopolysaccharide layer in the outer membrane of Gram-negative bacteria offers protection from lysozyme through a combination of negative charge and cross bridging with divalent cations (Hancock, 1984; Nikaido, 2001; Nikaido and Vaara, 1985). However, disruption of the outer membrane by physical (e.g., heating, shearing) or chemical treatments (e.g., EDTA, bacteriocin) can increase the sensitivity of Gram-negative bacteria and Gram-positive bacteria to lysozyme (Chai et al., 2016; Conte et al., 2011). A "rough" mutant of *Salmonella enterica* Typhimurium with a defect in the LPS component of the cell wall was inhibited by lysozyme without EDTA treatment, unlike "smooth" strains, which required the presence of EDTA for inhibition (Sanderson et al., 1974). Due to its basic character, lysozyme can cause flocculation of various bacteria or toxins (Barbara and Pellegrini, 1976).

14.2.3 Common Properties amongst Differing Lysozyme Types

Lysozymes from different organisms have the same substrate specificity and catalytic activity (U.S. FDA, 1998). However, depending on the evolutionary lineage of lysozymes (e.g., types C, G, or phage), the biochemical properties are different. The amino acid content of bovine milk lysozyme is different than that of human milk lysozyme or egg-white lysozyme (Eitenmiller et al., 1976). The optimum pH for activity of bovine milk lysozyme is 7.9 (Chandan et al., 1965). Bovine milk lysozyme contains 154

TABLE 14.2

Antimicrobial Spectrum of Egg White Lysozyme against Foodborne Pathogenic and Spoilage Organisms

Organisms
Organisms Strongly Lysed or Inhibited
Paenibacillus coagulans
Geobacillus stearothermophilus
Clostridium thermosaccharolyticum
Clostridium tyrobutyricum
Micrococcus spp.
Sarcina spp.
Organisms Moderately Inhibited or Showing Marked Strain Sensitivity
Bacillus cereus
Brucella spp.
Campylobacter jejuni subsp. *jejuni*
Clostridium botulinum serotypes types A, B, and E
Listeria monocytogenes
Enterococcus faecalis
Lactobacillus spp.
Moraxella spp.
Pseudomonas aeruginosa
Yersinia enterocolitica
Yeasts – *Candida albicans*, *Cryptococcus neoformans*, select others
Organisms Usually Not Lysed or Inhibited
Aeromonas hydrophila
Brocothrix thermosphacta
Clostridium butyricum
Clostridium perfringens
Escherichia coli O157:H7
Klebsiella pneumoniae
Lactococcus spp.
Leuconostoc spp.
Salmonella enterica serotype Typhimurium
Shewanella putrefaciens
Shigella spp.
Staphylococcus aureus
Streptococcus thermophilus and other starter streptococci
Vibrio cholerae

The data included in this table are collected from a variety of sources including primary sources: Barbara and Pellegrini 1976; Cunningham et al., 1991; Hughey and Johnson 1987; Salton 1957; Wooley and Blue 1974; and other papers cited in the text of this chapter. It should be noted that many of the resistance organisms can be inhibited or lysed when certain physical or chemical treatments are used in conjunction with lysosyzme. Also, these data are mostly from *in vitro* experiments and do not represent the activity of lysozyme against food-related organisms in food systems. These aspects are discussed in the text.

amino acids and has a molecular weight around 18,000 (Eitenmiller et al., 1976). Bovine milk lysozyme is more heat-stable than egg-white lysozyme at pH 4, but less heat-stable at pH 7 or 9 (Eitenmiller et al., 1976). Lysozymes found in plants often exhibit weak chitinase activity to protect plant tissues by hydrolyzing the β-(1-4) linkages binding N-acetylglucosamine molecules of chitin found in certain fungi (Jolles et al., 1963; Jolles and Jolles, 1984). Some lysozymes also exhibit esterase activity (Tranter, 1994; Jolles and Jolles, 1984).

14.3 Uses and Stability in Foods

Lysozyme has several desirable biochemical properties for use as a food preservative.

Lysozyme is generally stable under a variety of conditions, including those typically found in many minimally processed foods. Lysozyme in Camembert cheese was detected for several weeks (Hughey et al., 1989). Lysozyme is stable in certain cheeses for months to years since it effectively prevents late-blowing (incidentally the application for which it was originally granted GRAS status by the U.S. FDA), which commences during long-term aging of the cheese (International Dairy Federation, 1987). Lysozyme (10–20 μg/ml) was stable in sweetened sake (mirin) containing less than 20% alcohol for 20 weeks at 37°C (Yashitake and Shinichiro, 1977), for 1 year at room temperature (Uchida et al., 1972), and during pasteurization at 65°C for 1–5 hours (Uchida et al., 1972; Yashitake and Shinichiro, 1977). Lysozyme activity can be inhibited by certain surfactants, fatty acids of ≥C12, certain ions, and lipopolysaccharide (LPS). During processing, it can also react with proteins, probably mainly forming mixed disulfides with itself or other proteins or by ionic interactions. Lysozyme is stable in many acidic solutions (Wang and Shelef, 1991), but has limited stability under alkaline conditions (Alderton et al., 1945). Lysozyme can be frozen (Sofos et al., 1998) and is stable in dry storage at 5°C for several years (Wilkinson and Dorrington, 1975). Lysozyme was more stable in stored, spray-dried, whole-egg powder stored at 0°C than 37°C, although lytic activity was still present after 3 months even at the higher temperature (Hartsell, 1948). When considered for a food application, the stability of lysozyme should be tested in the food system.

14.3.1 Thermal Stability in Solution and in Food Products during Heat-Processing

Lysozyme is among the most heat-stable enzymes, and can resist boiling for 1–2 min at pH 4–5, but it is denatured by heat at higher pHs (Ahern and Klibanov, 1985; Sofos et al., 1998). Irreversible denaturation at 100°C occurred much slower at pH 4 than at pH 8 (Ahern and Klibanov, 1985). Recent research has confirmed these results in systems adjusted to pH 3 and 5 wherein egg-white lysozyme, natively folded or in crystalline form, was shown to have very similar Fourier transform-infrared (FTIR spectra (Galm et al., 2015). Heating for 5 minutes at 100°C at pH 9.0 resulted in loss of most activity (Meyer et al., 1936). Lysozyme is usually not affected by temperatures up to 55°C, although stability during heating varies with the food system, pH, and buffer/salt composition and concentration(s) (Esquembre et al., 2013; Wang and Shelef, 1992). The heat stability of lysozyme decreased above pH 7.0 (Smolelis and Hartsell, 1952, Wang and Shelef, 1991) with much greater stability at pH 7.0 than pH 9.0 in egg white (Cunningham and Lineweaver, 1965), probably due to the reduction of at least one disulfide bond by the sulfhydryl group of ovalbumin (Cunningham and Lineweaver, 1967). At pH 6.2, egg-white lysozyme lost only 5% of lytic activity after heating at 80°C for 30 minutes, and lost 25% of activity after heating at 100°C for 20 minutes (Smolelis and Hartsell, 1952). Heat stability between 85 and 95°C was highest at approximately pH 5.5 (Beychok and Warner, 1959).

Inactivation of lysozyme occurs during heating in certain foods. In skim milk, activity decreased at temperatures above 60°C (Weaver and Kroger, 1978). Nonetheless, sufficient activity remains after pasteurization of milk to prevent late blowing of cheeses by *Clostridium tyrobutyricum*. At high temperatures, it is likely that lysozyme can bind to and become inactivated by proteins in certain foods. The heat-stability of lysozyme is dependent on food type (Gould, 2002). No loss of lysozyme activity occurred in egg white heated at 60°C for 60 minutes at pH 5 (Sandow, 1926). Enzymatic activity of lysozyme remained stable in camel, cow, and buffalo milk samples heated to 65°C for 30 minutes, although activity was reduced 56–81% and 94–100% in milk samples heated for 30 minutes at 85°C or 100°C, respectively, potentially through the impacts of partial unfolding and loss of enzyme tertiary structure at higher heating temperatures (Elagamy, 2000; Esposito et al., 2009). Extremely high temperatures such as those used in roasting can inactivate lysozyme. After dry heating at 150°C for 20 minutes, polymerization through disulfide linkages was observed, and both polymerization and degradation were noted after heating at 180°C for 20 minutes (Hayase et al., 1975). However, some enzymatic activity remained after treatment at both temperatures (Hayase et al., 1975). At a higher temperature (250°C), decomposition of almost all amino acids in the polypeptide occurred (Fujimaki et al., 1972). The development of stable formulations of lysozyme for use in foods that receive a heat treatment would increase the utility of the enzyme. Sugars and polyols can protect lysozyme against heat denaturation, possibly due to the stabilizing effect of the compounds on hydrophobic interactions (Back et al., 1979). For instance, sucrose increased the thermal stability of lysozyme, especially with a decrease in pH from 6.0 to 3.2 (Yashitake and Shinichiro, 1977).

14.3.2 Lysozyme Stability to Non-Thermal Processing and Food Preservatives

Non-thermal treatments used in food processing can have a deleterious effect on lysozyme. Pulsed electric field treatment of lysozyme in buffer resulted in a 15–60% loss of enzymatic activity compared to untreated controls (Ho et al., 1997). Inactivation of lysozyme by ultraviolet radiation follows first-order kinetics (Shugar, 1952). Egg-white lysozyme, bovine milk lysozyme, and human milk lysozyme were sensitive to >10 krad gamma-irradiation at ambient temperature in phosphate buffer at pH 6.2, although inactivation was concentration-dependent (Eitenmiller et al., 1971). Inactivation of lysozymes by gamma-irradiation was due in part to hydroxyl radicals (Eitenmiller et al., 1971). Kume et al. (1973) observed a 37% inactivation of lysozyme activity in aqueous solution by 0.35 Mrad gamma-irradiation, although only slight inactivation of lysozyme occurred in irradiated egg white treated under the same conditions, possibly due to protection by high protein levels in the egg white. Results of another study showed that both purified lysozyme in buffer and endogenous lysozyme in egg white were not significantly affected by 0.195 kGray gamma-irradiation at 60°C (Schaffner et al., 1989). Gamma-irradiation at 20 kGy polymerized lysozyme through covalent bonds, especially in the presence of peroxidizing lipids and at higher water activity levels (Kanner and Karel, 1976).

Lysozyme activity can be increased by physical processes used in many food preservation systems. Freezing and thawing of *Escherichia coli* caused sensitivity to lysozyme (Kohn, 1960; Ray et al., 1984), although the organism regained resistance after incubation at 37°C or exposure to calcium (Ray et al., 1984). Transmission electron microscopy demonstrated protoplast formation and cell wall damage in unfrozen or frozen/thawed *L. monocytogenes* treated with lysozyme (El-Kest and Marth, 1992). The sensitization of Gram-negative cells to lysozyme by high pressure is dependent on factors such as pressure, temperature, pH, medium, growth stage, and bacterial species (Masschalck et al., 2001). High pressure has been shown to sensitize *E. coli* to lysozyme (Garcia-Graells et al., 1999; Masschalck et al., 2000), although the sensitization is transient (Hauben et al., 1996; Masschalck et al., 2001). In addition, four of six Gram-negative bacteria tested were sensitive to lysozyme during treatment by high hydrostatic pressure (Masschalck et al., 2001). EDTA increased the activity of lysozyme against *E. coli* during high-pressure treatment (Hauben et al., 1996). However, exposure of *E. coli* to lysozyme or lysozyme-EDTA immediately after high-pressure treatment did not decrease the number of viable cells (Hauben et al., 1996). Pulses of high hydrostatic pressure may sensitize some bacterial strains more effectively to lysozyme than treatment with continuous pressure (Masschalck et al., 2001). Sublethal injury by high pressure might be a useful hurdle along with lysozyme or lysozyme-EDTA in certain foods. Lower pressures required to inactivate certain bacteria in the presence of lysozyme (Masschalck et al., 2001) may lead to new combination treatments for the non-thermal preservation of foods. Similarly, pulsed electric field (PEF) was enhanced in its anti-*S. aureus* efficacy in skim milk when milk was first treated with 5.0 IU/mL nisin and 3000 IU/mL lysozyme in milk adjusted to pH 5.0, yielding a 4.8 $\log_{10}$-cycle reduction as compared to other treatments. PEF application time was also a key factor in determining *S. aureus* reductions observed (Sobrino-López and Martín-Belloso, 2008).

Lysozyme retains activity in the presence of many common components and compounds inherently present in or added to foods. Activity of lysozyme was not affected by the presence of certain antimicrobial compounds used in foods, including sodium nitrite, ethanol, butylated hydroxyanisole, calcium propionate, potassium sorbate, and propyl paraben (Yang and Cunningham, 1993). Potassium dichromate had no effect on lysozyme activity in mastitic bovine milk (Weaver and Kroger, 1978). Some organic solvents used in foods, such as ethanol, do not denature lysozyme (Sofos et al., 1998).

Certain chemicals which may be present in foods can have a detrimental effect on lysozyme activity. Lysozyme can bind to certain food components, leading to a loss in activity (Cunningham and

Lineweaver, 1967; Sofos et al., 1998). Lysozyme may be less heat-stable in proteinaceous foods than *in vitro*, due to the formation of mixed disulfide-linked molecules or other interactions with proteins or other constituents in the food (Johnson, 1994). Purified lysozyme in buffer at pH 6.2 was inactivated by egg yolk (Cunningham and Cotterill, 1971), possibly due to electrostatic interaction between lysozyme and substances in the yolk (Proctor and Cunningham, 1988). Activity was lost in the presence of high concentrations of lactic acid, acetic acid, and chlorine (Yang and Cunningham, 1993). Lysozyme can also be inhibited by substances such as alkyl sulfates, aliphatic long chain alcohols, and fatty acids and their sodium salts (Smith and Stroker, 1949). Polysaccharides with carboxylic acid and sulfonic acid groups, such as pectin and alginate, also reduced lysozyme activity, although the addition of 1–6% salt minimized inactivation by these compounds (Yashitake and Shinichiro, 1977). High concentrations of polyvalent and divalent cations such as cobalt, manganese, mercury, and copper can inactivate lysozyme, as well as low concentrations of iodine (Proctor and Cunningham, 1988).

Salt level plays a critical role in the enzymatic activity of lysozyme. Non-specific activation occurs at low salt concentrations (Tranter, 1994; Chang and Carr, 1971), although lysozyme activity is inhibited at high salt concentrations greater than 0.05–0.1 M (Chang and Carr, 1971; Davies et al., 1969; Smolelis and Hartsell, 1952), at least partially through loss of enzyme solubility and formation of crystalline structures and/or precipitation from aqueous solution (Galm et al., 2015). The addition of 3 mM NaCl improved the lysis of rapidly growing sensitive organisms by lysozyme (Vakil et al., 1969). Inhibition of lysozyme activity in high ionic strength solutions was stronger at alkaline pH values (Davies et al., 1969). The presence of sodium nitrate (180 mg/L) or sodium chloride (270 mg/L) did not inhibit the activity of lysozyme (450 mg/L) against several Gram-positive and Gram-negative bacteria under non-growth-restricted conditions (Gill and Holley, 2002).

14.4 Antimicrobial Spectrum of Activity

Lysozyme exhibits antimicrobial activity against vegetative cells of a wide variety of organisms, including numerous foodborne pathogens and spoilage organisms. Egg-white lysozyme is generally most active against certain Gram-positive organisms, and ineffective against dormant bacterial spores and Gram-negative bacteria. Antimicrobial activity of the enzyme can vary widely within groups of similar organisms, and its lytic activity is dependent on growth medium, test conditions, age of cells, presence of interfering substances, and many other parameters.

14.4.1 Gram-Positive and Gram-Negative Bacteria Sensitivity to Lysozyme

Gram-positive bacteria vary markedly in their sensitivity to lysozyme depending on the strain, species, and conditions of microbe handling and testing. Spoilage of foods by Gram-positive organisms including the heterofermentative LAB, staphylococci, *Brochothrix*, and certain other Gram-positives is a major problem in the food industry, but these groups of organisms are mostly resistant to the enzyme. Species of the Gram-positive genera *Micrococcus* and *Sarcina* are generally more sensitive to lysozyme than are *Lactobacillus* and *Bacillus* species (Connor, 1993). Ashton et al. (1975) demonstrated that the prevention of growth of *Geobacillus. stearothermophilus* and *Paenibacillus coagulans* by egg albumin was due to the presence of lysozyme. Human lysozyme inhibited *S. aureus* (Maga et al., 1998), but the egg white enzyme is generally inactive by itself against *S. aureus*. Lysozyme susceptibility among species of Gram-positive bacteria may decrease by derivatization of the amino groups and may vary due to differences in the relative proportions of N-acetylamino sugars in the cell walls (Salton and Pavlik, 1960). Specifically, higher proportions of β-1-6 and β-1-3 linkages between N-acetylmuramic acid and N-acetylglucosamine are found in the peptidoglycan of some Gram-positive bacterial species, which are more resistant to lysozyme than β-1-4 linkages. These differences in cell surface structure contribute to lysozyme susceptibility amongst the Gram-positives (Vakil et al., 1969). Bacterial spores can be sensitized to lysozyme by strong reducing agents or oxidizing agents to rupture disulfide bonds (Gould and Hitchins, 1963), but these methods are not compatible with food use (Gould, 2002). Recent research, nevertheless, has demonstrated spores of the gastrointestinal pathogen *Clostridium difficile* to be sensitive to combined application of egg-white lysozyme and nisin in synergistic fashion (Chai et al., 2016).

Gram-negative bacteria are generally less sensitive than Gram-positive bacteria to lysozyme due mainly to protection of the cell wall by the outer membrane. Resistant organisms include the important pathogens *Salmonella* spp. and *E. coli* O157:H7. However, certain physiological conditions, physical treatments, or chemicals that disrupt or alter the outer membrane can enhance the activity of lysozyme against Gram-negatives. At pH 3.5, most Gram-negative bacteria tested in one study exhibited some sensitivity to lysis by lysozyme, especially *Salmonella* and *Brucella* species (Peterson and Hartsell, 1955). Ibrahim (1998) reported enhanced antimicrobial activity of heat-denatured lysozyme against Gram-negative bacteria. In another study, human and bovine milk lysozymes displayed activity against five Gram-negative bacteria, including some strains of Gram-negative bacteria that are resistant to egg-white lysozyme (Vakil et al., 1969), demonstrating variability in antimicrobial activity between lysozymes isolated from different sources. The susceptibility of various species of food-related organisms, summarized in Table 14.2, represents the authors' assessment of the available literature. However, as discussed throughout this chapter, resistant organisms can often be made sensitive by physical treatments or chemical agents that disrupt outer membrane layers and capsules, allowing penetration of lysozyme and consequent lytic activity. Thus, in evaluating the potential for lysozyme function as a preservative in foods, a systems-based approach is often desirable whereby combinations of physical treatment or chemical agents are used in combination with lysozyme.

In recent years, researchers have explored opportunities for increasing the spectrum of antimicrobial activity of lysozyme against Gram-negative genera through multiple means, including thermal or chemical modification of lysozyme, or incorporating lysozyme into some form of carrier system. For example, a lysozyme–dextran conjugate produced a consistent 2.0 $\log_{10}$-cycle decrease in *E. coli* numbers in a buffer compared to native, non-modified lysozyme when tested at 100, 250, or 400 µg/mL. When applied in cheese curd aged 39 days at 4°C, *E. coli* numbers treated by the dextran–lysozyme (400 µg/mL lysozyme) conjugate were >3.0 $\log_{10}$ CFU/g less than non-treated controls at day 39, and ~2.0 $\log_{10}$ CFU/g less than numbers of the same organism treated by native lysozyme (Amiri et al., 2008). Similar findings have been reported for lysozyme conjugated with polysaccharides such as xanthan and guar gum, wherein conjugates demonstrated enhanced antimicrobial activity against Gram-negative bacteria, including *E. coli* and *Salmonella*, versus native egg-white lysozyme (Hamdani et al., 2018; Hashemi et al., 2014). Thermo-chemical modified lysozyme (egg white), treated with 1 or 2% hydrogen peroxide and heating at 70°C produced significant improvement in the inhibition of *Pseudomonas fluorescens* versus non-modified lysozyme during 24 h of incubation at 37°C (Cegielska-Radziejewska et al., 2010). However, populations of the organism recovered to non-differing numbers at 48 h of incubation. Exposure to conjugate egg-white lysozyme and the antibacterial phenolic triclosan caused a 6.0 to almost 10.0 $\log_{10}$ CFU/mL decrease in *Pseudomonas aeruginosa*, *Salmonella* Typhimurium, *Klebsiella pneumoniae*, or *E. coli* depending upon the organism tested (Hoq et al., 2008).

14.4.2 Lysozyme Sensitivity as a Function of Microbial Physiological Status

Sensitivity of organisms to lysozyme may be dependent on physiological state. Most studies examining the effectiveness of lysozyme have tested cells in nutrient-deficient environments. Cell wall synthesis in rapidly growing cultures may exceed the rate of degradation by lysozyme (Hughey and Johnson, 1987). Lysozyme–EDTA combinations were less effective against rapidly growing Gram-negative organisms in broth media than in buffer (Gill and Holley, 2002). The antimicrobial effect of lysozyme and nisin against lactic acid bacteria was sometimes enhanced when the growth media was diluted (Chung and Hancock, 2000). Inhibition of growth of *L. monocytogenes* in culture media by lysozyme at 5°C and 25°C increased when the pH was lowered from 7.2 to 5.5, although lysis of non-growing cells in buffer by lysozyme was not increased by a reduction in pH (Johansen et al., 1994). In that study, lower pH may have inhibited growth of the organism enough to allow the rate of hydrolysis by lysozyme to exceed the growth rate (Johansen et al., 1994). Heat-treated cells of *Yersinia enterocolitica* grown at 4°C were resistant to lysozyme, but heat-treated cells grown at 37°C were sensitive (Pagan et al., 1999). However, the activity of lysozyme against Gram-positive organisms was enhanced by EDTA under non-growth-restricted conditions (Gill and Holley, 2002). During evaluation of the antimicrobial efficacy of lysozyme and lysozyme–antimicrobial combinations, testing of both growing and non-growing cells may be advantageous.

14.5 Activity against Food-Related Microorganisms

In the following section, selected studies of the effects of lysozyme on food-related organisms are described. *Salmonella* Senftenberg was not detected on inoculated turkey drumsticks treated with 0.1% lysozyme for 3 hours at 22°C, or 0.05% lysozyme for 1 minute at 63°C (Teotia and Miller, 1975). The heat resistances of *Salmonella* serovars Enteritidis and Typhimurium, *E. coli* O157:H7, and *S. aureus* were significantly reduced in buffer in the presence of eggshell membranes containing lysozyme (Poland and Sheldon, 2001). In this system, components of the albumin may have enhanced the activity of enzyme against these organisms.

14.5.1 Antimicrobial Activity of Lysozyme against Gram-Positive Rods in Foods

The antimicrobial activity of lysozyme has been extensively tested against *L. monocytogenes*, a Gram-positive pathogen of great concern to the food industry. Lysozyme (>2 mg/ml) prevented growth of *L. monocytogenes* in culture media at pH 7.0, and 2 mg/ml had listericidal effects at pH 9.0 (Wang and Shelef, 1991). Lysozyme was bactericidal to *L. monocytogenes* in milk, mozzarella, taleggio, and grana cheeses, frankfurters, and on the surface of chickens (Dell'Acqua et al., 1989). Poor survival of the organism in egg white (Sofos et al., 1998; Erickson and Jenkins, 1992) is probably due to the presence of lysozyme. Lysozyme was bactericidal to *L. monocytogenes* in egg white-containing cholesterol-free mayonnaise (Erickson and Jenkins, 1991) and may be partially responsible for the rapid inactivation of *E. coli* O157:H7 in commercial mayonnaise made with whole eggs (Raghubeer et al., 1995). Pretreatment of raw cod-fish fillets with 3 mg/ml lysozyme delayed growth of *L. monocytogenes* at abuse temperatures, and was listericidal at refrigeration temperatures, although it did not affect growth of natural microflora in the product (Wang and Shelef, 1992). Egg-white lysozyme alone (100 ppm) killed or prevented growth of *L. monocytogenes* in several foods, particularly vegetables (Hughey et al., 1989). Lysozyme in Camembert cheese initially was bactericidal to *L. monocytogenes*, although the organism eventually grew in the cheese after 35 days (Hughey et al., 1989). *Listeria monocytogenes* cells grown at low temperatures were more sensitive to lysozyme than cells grown at ambient temperature (Johnson, 1994) or 37°C (Smith et al., 1991), indicating lysozyme may be useful for controlling the organism in refrigerated foods. Milk pasteurization in combination with added lysozyme may eliminate *L. monocytogenes* in certain cheeses prepared from milk contaminated with the organism. Numbers of heat-stressed *L. monocytogenes* significantly decreased in Cheddar and Camembert-type cheeses containing lysozyme (Johnson, 1994).

Conversely, growth of *Listeria monocytogenes* was not inhibited on frankfurters formulated with 0.01% lysozyme (Bedie et al., 2001). Mangalassary et al. (2008) treated turkey bologna surfaces with 0.5 U lysozyme/cm^2 prior to a thermal intervention, which reduced numbers of *L. monocytogenes* on the deli meat surfaces to non-detectable counts during subsequent refrigeration storage. Lysozyme had limited effectiveness against *L. monocytogenes* in pork sausage (Hughey et al., 1989) and pork, beef, or turkey frankfurters (Johnson, 1994). *L. monocytogenes* was resistant to hen egg-white lysozyme in milk (Hughey et al., 1989; Kihm et al., 1994), but sensitive to the enzyme in media or buffer, or in mineralized milk during heating (Kihm et al., 1994). Mineral or mineral-associated components in milk may protect the organisms from inactivation by lysozyme and heat in milk, probably by increasing the stability of the cell surface (Kihm et al., 1994).

Egg-white lysozyme was antimicrobial to *C. botulinum* in foods such as turkey, pork sausage, salmon, asparagus, potatoes, tomatoes, and mushrooms (Johnson, 1994; Sofos et al., 1998). Lysozyme delayed botulinum toxin production in bratwurst and salmon, as well as cod suspensions (Johnson, 1994). Lysis of *C. botulinum* by lysozyme in culture medium was enhanced at lower temperatures and pH (Johnson, 1994).

The effect of lysozyme against clostridial species has been well-documented. Spores of *C. tyrobutyricum* survive milk pasteurization and cause "late blowing" of cheeses such as Edam and Gouda during ripening, which results in cracks or fissures in the finished cheeses. Lysozyme lyses *C. tyrobutyricum* cells as they outgrow from germinated spores and can prevent blowing of cheeses through lactate utilization by the organism (Carini and Lodi, 1982). Late blowing by sporeformers in Edam cheese prepared with low-quality milk (i.e., high sore load) was inhibited by 500 units/ml lysozyme (Wasserfall et al., 1976). Growth of *C. tyrobutyricum* in Italian cheeses was also inhibited by 50 ppm lysozyme or non-toxic lysozyme salts added to butter or cheese (Ferrari and Dell'Acqua, 1979). Vegetative cells of *C. tyrobutyricum* can develop resistance to lysozyme (Wasserfall and Teuber, 1979). However, lysozyme prevented the growth of the organism in Edam cheese, even when spores from lysozyme-resistant vegetative cells were used to contaminate the cheese milk (Wasserfall and Prokopek, 1978).

Lysozyme inhibition of LAB members, including many starter cultures used in foods as well as spoilage organisms, is variable among species. Lysozyme in milk is usually more inhibitory to pathogenic and spoilage organisms than LAB (Cunningham et al., 1991). In culture media, lysozyme had no effect on the growth of Gouda cheese starter cultures (Bester and Lombard, 1990). Another study demonstrated inhibition of one *Streptococcus* starter culture from semi-hard cheese (Wasserfall, 1978). Spoilage lactobacilli were inhibited by 20 ppm lysozyme in sweetened sake (Uchida et al., 1972). *Lactobacillus delbrueckii* subsp. *lactis* was sensitive to $\geq$0.5 mg/l lysozyme in culture media, though susceptibility of the organism to the bacteriocin acidocin was reduced in the presence of lysozyme (Chumchalova et al., 1998). The application of lysozyme did not affect growth of LAB or yeasts on ready-to-use pear cubes, demonstrating that the action of lysozyme may be affected by substrate (Pittia et al., 1999).

14.5.2 Antifungal Activity against Foodborne Yeasts and Spoilage Bacteria in Foods

Chitinase activity exhibited by lysozyme is probably responsible for the weak activity of lysozyme against several pathogenic and spoilage yeasts, including *Candida albicans*, *Cryptococcus neoformans*, *Saccharomyces cerevisiae*, and *Zygosaccharomyces rouxii* (Johnson, 1994). Lysozyme inhibits growth of spoilage lactic acid bacteria, and thermo-modified lysozyme was shown by researchers to inhibit wine-spoiling acetic acid bacteria (AAB) members, including *Acetobacter aceti* and *Gluconobacter oxydans* at contents ranging from 0.01 to 0.09 mg/mL following thermo-modification (Carillo et al., 2014; Gerbaux et al., 1997).

14.6 Effect on Heat Resistance of Bacterial Spores

The safety of minimally processed, extended shelf-life foods, including sous-vide and cook-chill products, is usually due to a minimal heat treatment to achieve a 6-log reduction in heat-sensitive nonproteolytic *Clostridium botulinum* spores, followed by refrigeration below 10°C (Gould, 1999). However, heat treatment guidelines at 65–95°C do not usually take into consideration the effect of lysozyme on spore heat resistance (Fernandez and Peck, 1999). Lysozyme, which is naturally present in many foods, is relatively heat-resistant and may be active in foods after mild heat processes such as pasteurization (Peck et al., 1992).

At concentrations as low as 0.1 µg/ml (Peck et al., 1992) lysozyme increases the apparent heat resistance of non-proteolytic *Clostridium botulinum* spores (Alderton et al., 1974; Fernandez and Peck, 1999). Increased heat resistance of proteolytic type A *C. botulinum* spores has been shown, although the effect was minimal compared to the effect on type E nonproteolytic spores (Alderton et al., 1974). Lysozyme also increased the recovery of heat-treated spores of *C. perfringens* (Adams, 1974), and possibly *Bacillus cereus* (Blocher and Busta, 1982), but not *Bacillus thuringiensis* (Faille et al., 1999). Approximately 0.1–20% of nonproteolytic *C. botulinum* spores are permeable to lysozyme, resulting in a biphasic survival curve upon heating (Peck et al., 1993). Lysozyme in the recovery medium is thought to compensate for a heat-damaged germination system (Broda et al., 1998) in lysozyme-permeable heat-sensitive bacterial spores (Peck et al., 1992), probably by degrading the spore cortex, allowing core hydration, and inducing spore germination (Gould, 1989).

Vegetable juice extracts with measurable lytic activity when added to the recovery medium increased the measurable heat resistance of nonproteolytic *C. botulinum* (Stringer and Peck, 1996). In other research, heating in a meat medium resulted in a 6-log reduction in numbers of viable nonproteolytic *C. botulinum* in the absence of lysozyme, but not with lysozyme present (Peck and Fernandez, 1995). It is unknown whether the unusually high measured heat resistance of type E *C. botulinum* in Dungeness crab meat (Peterson et al., 1997) was due to the presence of lysozyme in the product. A combination of lysozyme with high-pressure processing increased the lethality against *B. cereus* spores in milk during post-process aging and cheese manufacture (López-Pedemonte et al., 2003). Polypeptides produced from egg-white lysozyme also were reported to reduce *Bacillus subtilis* spores to non-detectable counts at levels of 10 up to 100 μg/mL (Abdou et al., 2007).

Predictive models for the inactivation of nonproteolytic *C. botulinum* in minimally processed refrigerated foods in the presence of lysozyme were developed by Fernandez and Peck (1999). Because inadequate heat treatments may lead to subsequent growth of the organism during refrigerated storage, additional research was needed to verify the ability of currently used heat treatments to adequately kill nonproteolytic *C. botulinum* in such foods containing lysozyme or similar lytic enzymes. Stochastic modeling of *B. cereus* growth inhibition in liquid medium at 16 or 25°C when supplemented with nisin (0.13 μM) and 0.66 mM lysozyme produced significant extensions in the lag phase of vegetative cells of the pathogen as compared to controls ranging anywhere from 3.25 to 13 h (Antolinos et al., 2011).

14.7 Enhancement of Activity by Other Chemical Agents

The outer membrane of most Gram-negative bacteria protects the cell wall from lysozyme and may limit the effectiveness of lysozyme in food applications (Gill and Holley, 2002; Davidson and Harrison, 2002). The antimicrobial effect of lysozyme, particularly against Gram-negative bacteria, can often be enhanced by the presence of specific chemicals or physical treatments (Table 14.3). The sensitivity of

TABLE 14.3

Physical Treatments and Chemical Agents Reported to Increase the Activity of Lysozyme against Selected Organisms

Physical Treatments
Heat
Alkaline pH
Osmotic shock
Freeze-thawing
High pressure
Chemical Treatments
EDTA, DTPA, other divalent metal chelators
Trisodium phosphate
Cysteine, proline
Conalbumin
Lactoferrin, other transferrins
Lysolecithin, polyethylene glycol, other surfactants.
Glycine
Poly-L-lysine
Nisin, possibly other bacteriocins
Hydrogen peroxide
Ascorbic acid
Lactic acid
Trypsin, proteinase K
Lipase
Carvacrol, nerolidol, certain other terpenoids

Gram-negative species to lysozyme is increased by chelators, certain antibiotics, amino acids, alkaline pH, osmotic shock, drying, and freeze-thawing (Tranter, 1994; Proctor and Cunningham, 1988), as well as hydrogen peroxide and ascorbic acid (Miller, 1969). Lysozyme activity is also enhanced by food additives including EDTA, butyl paraben, and tripolyphosphate, as well as naturally occurring antimicrobials such as conalbumin, lactoferrin, nisin, and transferrin (Johnson, 1994). Other food components promoting lysozyme activity include lactic acid, trypsin, proteinase K, and lipase (Sofos et al., 1998).

14.7.1 Enhancement of Lysozyme by Chelators against Foodborne Bacteria

Many chemical agents can permeabilize the outer membrane of Gram-negative bacteria, as classically shown by the metal chelator EDTA. Early studies demonstrated that Versene (EDTA) could increase the lytic activity of lysozyme against certain Gram-negative bacteria (Repaske, 1958). Chelators such as EDTA may remove stabilizing divalent cations (e.g., Mg^{+2}, Ca^{+2}) from lipopolysaccharide in the outer membrane, increasing permeability of the cells to antimicrobial agents, including lysozyme (Hancock, 1984; Lam et al., 2014).

The combination EDTA-Tris-lysozyme was bactericidal to eight of ten Gram-negative bacterial pathogens tested in buffer, and two of six Gram-positive bacteria tested, including *L. monocytogenes* (Wooley and Blue, 1974). The presence of EDTA increased the activity of lysozyme against *C. botulinum* and *L. monocytogenes* (Hughey and Johnson, 1987) as well as *B. subtilis* and *P. aeruginosa* (Vakil et al., 1969). Lysis of coliforms implicated in bovine mastitis by bovine sera was markedly increased upon addition of lysozyme and EDTA, although lysozyme alone did not lyse the cells (Carroll, 1979). Growth of predominantly Gram-negative shrimp microflora in nutrient broth was slightly delayed by 150 μg/ml lysozyme, but strongly inhibited by $\geq$50 μg/ml lysozyme in the presence of 0.02% EDTA (Chander and Lewis, 1980). Lysozyme alone was shown to inhibit only 2 of 12 fungal species tested, although all but 1 species were inhibited by a lysozyme–EDTA combination (Razavi-Rohani and Griffiths, 1999).

The potentiation of lysozyme activity by EDTA has also been demonstrated in food systems, and research has continued until the present. Somewhat contrary to others' findings, Ko et al. (2008) did not report the addition of EDTA to be beneficial in the inhibition of *L. monocytogenes* growth in cooked hams, though a combination of ovotransferrin + lysozyme did produce a reduction in populations of the pathogen over post-lethality storage. Similar results were reported for the inhibition of *E. coli* O157:H7 populations on pork chops, though the blend of ovotransferrin with both EDTA and lysozyme proved best at reducing pathogen growth, with reductions of approximately 8.0 $\log_{10}$-cycles as compared to the use of only EDTA and lysozyme (Ko et al., 2009). Similarly, *Pseudomonas* numbers on pre-cooked chicken treated by 1.5% w/w of EDTA and lysozyme, stored under vacuum, were 3.0 $\log_{10}$-cycle lower compared to control samples not treated with antimicrobials and stored aerobically (Ntzimani et al., 2010). Ostrich meat patty counts of aerobic and lactic acid bacteria stored refrigerated aerobically or under vacuum were significantly reduced by the application of EDTA (20 mM), lysozyme, and nisin (250 ppm each), with lactic bacteria being reduced to a greater extent versus controls as compared to total aerobic loads (Mastromatteo et al., 2010). Bevilacqua et al. (2010) also reported 1.0 g/L lysozyme and 100 mM EDTA produced rapid inhibition of *Hafnia alvei* in medium. Most recently, pseudomonads were reported to be inhibited via application of silver-based nanoparticles bearing lysozyme and EDTA when applied in burrata-style cheese over 12 days' storage under modified atmosphere storage (Costa et al., 2017).

The effectiveness of lysozyme against *C. botulinum* can be increased using chelators such as EDTA, DTPA, and cysteine (Johnson, 1994). EDTA, nisin, butyl-paraben, and tripolyphosphate increased the delay of botulinal toxin production by lysozyme in bratwurst, salmon, and cod suspensions (Johnson, 1994). Toxin production by *Clostridium botulinum* was also delayed in turkey suspensions, potatoes, asparagus, and salmon by the addition of lysozyme, and the delay was increased in the presence of EDTA (Johnson and Dell'Acqua, 1995). Nonetheless, some food components, notably fat, have been reported to hamper/lower the antibotulinal effects of antimicrobial food preservatives, including lysozyme and EDTA (Glass and Johnson, 2004).

14.7.2 Lysozyme Use in Combination with Antimicrobial Polypeptides and Bacteriocins

Combinations of lysozyme and nisin, a food preservative effective against Gram-positive bacteria, are effective against numerous pathogenic and spoilage bacteria. Nisin concentrations of 100–500 units/g are usually sufficient to increase the antimicrobial effectiveness of lysozyme (Tranter, 1994). The increased efficacy of lysozyme–nisin combinations may be due to increased membrane damage, cell lysis, or the inhibition of energy-dependent processes that repair lysozyme/nisin damage to the cell (Chung and Hancock, 2000). The presence of lysozyme increased the depolarization of the cytoplasmic membranes of *S. aureus* by nisin (Chung and Hancock, 2000). Severe cell damage in the form of surface disruption was observed by scanning electron microscopy in *Lactobacillus sakei* treated with lysozyme–nisin combinations (Chung and Hancock, 2000).

Lysozyme and nisin in combination was more effective than lysozyme alone against many LAB in culture media (Chung and Hancock, 2000), and can be bacteriostatic or bactericidal to *L. monocytogenes* under certain conditions (Monticello, 1990). Growth of *L. monocytogenes* was inhibited on sliced pork bologna by the use of a dip containing nisin and lysozyme, although lysozyme alone had no effect (Kain et al., 2001). Lysozyme–nisin combinations were generally more effective than either lysozyme or nisin alone to control the growth of the Gram-positive meat-spoilage bacteria *Brochothrix thermosphacta* and *Carnobacterium* spp. in culture media, meat juice extract, and on pork tissue (Nattress et al., 2001). The addition of 500 mg/kg lysozyme–nisin (1:3 mixture) and 500 mg/kg EDTA to ham or bologna sausage batter before cooking resulted in growth inhibition at 8°C of *B. thermosphacta*, *L. monocytogenes*, *Lactobacillus curvatus*, *Lactobacillus mesenteroides* on sausage, and *E. coli* O157:H7, *L. mesenteroides*, *B. thermosphacta*, and *L. curvatus* on ham (Gill and Holley, 2000a). The same combination was ineffective against *L. sakei*, *Salmonella* Typhimurium, *Serratia grimesii*, and *Shewanella putrefaciens* on both products (Gill and Holley, 2000a). Combinations of lysozyme, nisin, and hop acids were bactericidal to *L. monocytogenes*, *B. cereus*, and *B. subtilis* in culture media, and to *L. monocytogenes* on hot dogs or cooked hams when applied as a dip or spray, respectively (King and Ming, 2002). However, without nisin, hop resins did not increase the effectiveness of lysozyme against *E. coli* (Fukao et al., 2000). When added as a third preservative factor, lysozyme increased the bactericidal activity of nisin and carvacrol against *B. cereus* and *L. monocytogenes* in buffer (Pol and Smid, 1999). The bactericidal activity of pulsed electric field (PEF) against *Salmonella* Typhimurium in orange juice was increased in the presence of lysozyme or a combination of lysozyme and nisin (Liang et al., 2002).

While nisin–lysozyme combinations are effective against many Gram-positive bacteria, lactoferrin may enhance the effectiveness of lysozyme against Gram-negative bacteria. Lactoferrin, a protein found in milk, appears to damage the outer membrane of Gram-negative bacteria, causing the release of LPS molecules and increasing bacterial susceptibility to lysozyme by increasing penetration of lysozyme through the outer membrane (Ellison and Giehl, 1991; Yamauchi et al., 1993). Lactoferrin has been shown to enhance the antimicrobial activity of lysozyme against *E. coli* (Yamauchi et al., 1993, Suzuki et al., 1989). The combination of apo-lactoferrin and lysozyme delayed growth of *L. monocytogenes* in UHT milk, but was not as effective against the organism as an EDTA–lysozyme combination (Payne et al., 1994). Similar delays in growth of *L. innocua* were reported in microbiological medium for a combination of lysozyme and lactoferrin delivered via carboxy methylcellulose (CMC)-derived films. The lag phase for untreated (control) cells was 1.86 h while for lysozyme alone it was 5.81 h, and for lysozyme + lactoferrin it was 6.5 h (Barbiroli et al., 2012). Egg-white lysozyme (0.08 mg/ml) increased the antimicrobial effect of bovine lactoferrin against *Salmonella* Enteritidis in buffer, though lysozyme alone was not bactericidal to the organism (Facon and Skura, 1996). The bactericidal effect of lysozyme–lactoferrin combinations is dose-dependent, is blocked by iron saturation of lactoferrin, and is inhibited by high calcium levels (Ellison and Giehl, 1991).

Numerous other substances commonly found in foods have been shown to affect lysozyme activity. Glycine–lysozyme combinations are effective against Gram-positive and Gram-negative bacteria, and are used commercially in Japan in certain foods (Proctor and Cunningham, 1988). Lysozyme at 100 μg/g delayed botulinum toxin production by in aqueous turkey slurries, especially in the presence of cysteine, proline, and sodium lactate (Johnson, 1994). *Pseudomonas aeruginosa*, but not *E. coli* O157:H7 or *Salmonella* Typhimurium, was sensitized to lysozyme by lactic acid or HCl disruption of the outer membrane (Alakomi et al., 2000). Oh et al. (2016) similarly reported synergistic inhibition of *L. monocytogenes* in liquid microbiological medium by the combined application of lysozyme with acetic, citric, lactic, and malic acid.

14.8 Enhancement by Physical Processes

Lysozyme activity can be increased by physical processes used in many food preservation systems. Freezing and thawing of *Escherichia coli* causes sensitivity to lysozyme (Kohn, 1960; Ray et al., 1984), although the organism regained resistance after incubation at 37°C or exposure to calcium (Ray et al., 1984). Transmission electron microscopy demonstrated protoplast formation and cell wall damage in unfrozen or frozen/thawed *L. monocytogenes* treated with lysozyme (El-Kest and Marth, 1992).

The sensitization of Gram-negative cells to lysozyme by high pressure is dependent on factors such as pressure, temperature, pH, medium, growth stage, and bacterial species (Masschalck et al., 2001). High pressure has been shown to sensitize *E. coli* to lysozyme (Garcia-Graells et al., 1999; Masschalck et al., 2000), although the sensitization is transient (Hauben et al., 1996; Masschalck et al., 2001). In addition, four of six Gram-negative bacteria tested were sensitive to lysozyme during treatment by high hydrostatic pressure (Masschalck et al., 2001). EDTA increased the activity of lysozyme against *E. coli* during high-pressure treatment (Hauben et al., 1996). However, exposure of *E. coli* to lysozyme or lysozyme–EDTA immediately after high-pressure treatment did not decrease the number of viable cells (Hauben et al., 1996). Pulses of high hydrostatic pressure may sensitize some bacterial strains more effectively to lysozyme than treatment with continuous pressure (Masschalck et al., 2001). Sublethal injury by high pressure might be a useful hurdle along with lysozyme or lysozyme–EDTA in certain foods. Lower pressures required to inactivate certain bacteria in the presence of lysozyme (Masschalck et al., 2001) may lead to new combination treatments for the non-thermal preservation of foods. Similarly, pulsed electric field (PEF) was enhanced in its anti-*S. aureus* efficacy in skim milk when milk was first treated with 5.0 IU/mL nisin and 3000 IU/mL lysozyme in milk adjusted to pH 5.0, yielding a 4.8 $\log_{10}$-cycle reduction as compared to other treatments. PEF application time was also a key factor in determining *S. aureus* reductions observed (Sobrino-López and Martín-Belloso, 2008).

14.9 Food Applications

Egg-white lysozyme is desirable as a food preservative because of its ease of purification in economically feasible quantities from egg white, low toxicity, specific activity against target bacteria and fungi, low effective usage levels, and low impact on sensory qualities of foods (Sofos et al., 1998). Lysozyme is stable under many conditions typically found in foods, and is highly soluble in aqueous environments (Razavi-Rohani and Griffiths, 1996b). Lysozyme is typically used in foods at 20–400 ppm (Gould, 2002), and commercial preparations for food use are generally available in powdered or granular form. Possible limitations of lysozyme include cost, inactivation by endogenous food components, potential for allergenic response by some consumers, and limited antimicrobial spectrum (Holzapfel et al., 1995).

14.9.1 Functionality in Ready-to-Eat Further Processed Foods

As described, lysozyme is currently used in foods such as cheese, frankfurters, cooked meat, and poultry products (Davidson and Harrison, 2002; Pina-Pérez et al., 2015; Taylor et al., 2019). A patented process in the United Kingdom applies to the addition of lysozyme to butter or milk used to make Italian cheeses in order to control undesirable organisms such as *C. tyrobutyricum* (Ferrari and Dell'Acqua, 1979). Lysozyme, used in combination with standard preservatives, reduced microbial spoilage of cooked, salami, and Vienna sausages (Akashi, 1969, 1970, 1971). Microbial lytic enzymes may also have applications as food preservatives. An N-acetylmuramidase purified from *Streptomyces rutgerensis* extended the shelf life of adzuki bean paste (Hayashi et al., 1989).

Lysozyme may be used in lieu of nitrate to prevent late blowing in hard and semi-hard cheeses such as Gouda, Edam, Provolone, and Emmental in the U.S. as well as numerous other countries (International Dairy Federation, 1987). Late blowing is a ripening defect resulting from hydrogen and carbon dioxide gas production during the fermentation of lactate by *C. tyrobutyricum*, a heat-resistant spore-former commonly found in milk. Lysozyme is usually added to milk at 20–35 ppm, resulting in levels in cheese ranging from 200 to 400 ppm, which does not generally inhibit starter cultures (Sofos et al., 1998) or affect physical or organoleptic properties of the cheese, though its presence must be labeled in order to prevent potential food allergenicity-related food safety concerns in cheese consumers (EC Rule 2003/89/EC).

Numerous Japanese patents cover the use of lysozyme as a preservative in foods such as fresh vegetables, meats, seafood such as oysters and shrimp, tofu, cheeses, butter, potato salad, noodles, custard, and infant milk formula (Tranter, 1994; Cunningham et al., 1991). Lysozyme has been investigated extensively as a preservative in vegetable, meat, and seafood products in Japan (Tranter, 1994), and has shown effectiveness as a preservative in products such as kimchi (fermented cabbage), rice sushi, noodles, and creamed custard (Cunningham et al., 1991). Other Japanese patents apply to the use of lysozyme applied as a coating on fruits, vegetables, meats, and seafoods (Cunningham et al., 1991; Sofos et al., 1998). Lysozyme used in a gelatin dip extended the shelf life of a type of seafood product made from lizard-fish (Akashi and Oono, 1972).

Lysozyme has been used in Japan to extend the shelf life of wine and sake (Sofos et al., 1998, Tranter, 1994), and is utilized with increasing frequency to improve the quality of wines produced in a number of other countries. The addition of 250 mg/L lysozyme after malolactic fermentation increased the microbiological stability of red wine (Gerbaux et al., 1997). Such inhibition of spoilage bacteria in wine by lysozyme may contribute to the use of lower sulfite levels in some products.

Numerous patents have been issued in the United States for lysozyme applications in food systems. These include patents for listericidal treatments utilizing lysozyme in dairy or meat products (Dell'Acqua et al., 1989), lysozyme and EDTA on vegetables (Johnson et al., 1991), and combinations of hop acids, nisin, and/or lysozyme as ingredients in or applied to the surface of solid foods (King and Ming, 2002). Another patent covers the use of lysozyme or nontoxic lysozyme salts along with chelators such as EDTA to control growth of *Clostridium botulinum* in animal or vegetable foods (Johnson and Dell'Acqua, 1995). Lysozyme added to infant formula or milk encourages the growth of *Lactobacillus* spp. (Schwimmer, 1981) and *Bifidobacterium bifidus* (Sawada et al., 1967; Nishihava and Isoda, 1967) in infant intestines, which may prevent growth of harmful bacteria. Daeschel and Kennedy (2006) reported a method to reduce lysozyme binding by wine tannins via tannin binding agents, thereby retaining greater antimicrobial activity of lysozyme against spoilage bacteria.

14.9.2 Lysozyme Utility on Animal Carcass and Derived Product Surfaces

Lysozyme may also be effective in immersion or spray methods to decontaminate animal carcasses. Gram-negative organisms, including *Salmonella*, can become sensitive to lysozyme after a sudden decrease in osmolarity, possibly increasing the antimicrobial effectiveness of lysozyme used in the decontamination of poultry (Chatzolopou et al., 1993).

Combinations of lysozyme with other antimicrobials may be useful when applied to the surface of meat products, either through a dip, application of an edible gel, or on packaging film (Gill and Holley, 2000a). The use of lysozyme–antimicrobial combinations as a surface application in packaging components would prevent inactivation of the enzyme during cooking (Gill and Holley, 2000b). A lysozyme–nisin gel coating (1:3 mixture; 25.5 mg/ml) with 25.5 mg/ml EDTA applied to the surface of cooked ham and bologna was bactericidal to *B. thermosphacta*, *L. sakei*, *Leuconostoc mesenteroides*, *L. monocytogenes*, and *Salmonella* Typhimurium, and inhibited growth of the organisms during storage at 8°C for 4 weeks (Gill and Holley, 2000b). Nisin and lysozyme incorporated together into whey-protein isolate film was antimicrobial to *Salmonella enterica* serotype Typhimurium colonies on agar plates (Rodrigues et al., 2002). Lysozyme incorporated into biodegradable packaging films was bactericidal to *Lactobacillus plantarum* and, in combination with EDTA, *E. coli* (Padgett et al., 1998). Incorporation of lysozyme into packaging films has been reported to allow direct contact of antimicrobials with pathogens present on the surface of foods (Barbiroli et al., 2012; Wang et al., 2018).

14.9.3 Lysozyme Applications for Food Contact Surfaces Disinfection

The use of lysozyme on food contact surfaces may prevent colonization by pathogens such as *L. monocytogenes* as well as other biofilm-forming bacteria (Bower et al., 1998). Hen egg-white lysozyme forms multilayers on hydrophobic metal surfaces (Schmidt et al., 1990). However, some loss of antimicrobial activity by adsorbed lysozyme has been demonstrated (Schmidt et al., 1990). Lysozyme treatment of stainless steel resulted in the attachment of increased numbers of vegetative *Bacillus* species, possibly due to removal of extracellular polysaccharides by the enzyme (Parkar et al., 2001). On the other hand, Sheffield et al. (2012) reported that 25–50 μg/mL lysozyme produced near-complete removal of *K. pneumoniae* biofilm material from steel coupons after a 1-h treatment, compared to only ~60% removal by dextranase at the same content under similar exposure conditions.

14.10 Non-Enzymatic Antimicrobial Activity of Lysozyme and Lysozyme-Peptides

The antimicrobial activity of lysozyme resulting from its enzymatic muramidase activity on bacterial cell walls has been studied extensively. However, lysozyme has also been reported to have antimicrobial activity unrelated to its enzymatic activity (Chung and Hancock, 2000; Ibrahim, 1998; Ibrahim et al., 2001).

14.10.1 Bactericidal Characteristics of Lysozyme

Cationic proteins often possess bactericidal properties (Pelligrini et al., 1992). The basic nature of the lysozyme molecule itself may result in antimicrobial activity against some organisms (Mayes and Takeballi, 1983; Salton, 1957). For instance, *S. aureus* will grow in egg white at pH 7.2, but not at pH 8.0, possibly due to negative charges on the cell leading to agglutination by positively charged molecules such as lysozyme (Ng and Garibaldi, 1975). Treatment of *E. coli* with egg-white lysozyme resulted in disintegration of the cell cytoplasm visible by electron microscopy, and lysozyme was detected within the cytoplasm of frozen cells by immunogold labeling (Pelligrini et al., 1992).

Bactericidal properties of denatured or partially denatured lysozyme against Gram-negative and Gram-positive bacteria have been shown, with increased sensitivity of *E. coli* to heat-denatured lysozyme compared to native lysozyme (Ibrahim et al., 1996a, 1996b, 2001). Pelligrini et al. (1992) found that denaturation of lysozyme by dithiothreitol (DTT) to reduce disulfide bonds did not lead to a loss of antimicrobial activity against Gram-negative and Gram-positive bacteria, suggesting that the bactericidal efficacy of lysozyme is due not only to lytic muramidase activity, but also to cationic and hydrophobic properties. Native lysozyme and lysozyme denatured by DTT were equally bactericidal to oral streptococci (Liable and Germaine, 1985). The microbiocidal activity of T4 lysozyme may be due to membrane disruption by small peptide sequences rather than enzymatic cell wall degradation (During et al., 1999). Heat-denatured bacteriophage T4 lysozyme exhibited antimicrobial activity but no enzymatic activity (During et al., 1999).

14.10.2 Lysozyme-Derived Antimicrobial Peptides

Antimicrobial cationic peptides with no enzymatic activity have been isolated from lysozyme (Pelligrini et al., 1997; During et al., 1999). Peptide digests of lysozyme without enzyme activity were strongly bactericidal to *E. coli* and *S. aureus*, with damage to cell membranes of both species evident by scanning electron microscopy (Mine et al., 2004). A 15-amino-acid sequence (amino acids 98–112) isolated from egg-white lysozyme and an identical synthesized peptide both exhibited antimicrobial activity without catalytic muramidase activity (Pelligrini et al., 1997). However, the peptide sequences were weaker against Gram-negative bacteria than parent lysozyme and were ineffective against some organisms that were lysed by the parent lysozyme, such as *P. aeruginosa*, *M. luteus*, and *S. aureus* (Pelligrini et al., 1997). Memarpoor-Yazdi et al. (2012) reported peptides, the products of enzymatic breakdown of HEWL, exhibited both antioxidant and antimicrobial activity against both Gram-positive and -negative bacterial organisms. Carrillo et al. (2018a, b) reported that heat-denatured hydrolysates exhibited greater antibacterial spectrum than native HEWL, producing significant structural damage in cell walls of *Staphylococcus carnosus* and *E. coli*, producing reductions in numbers of both pathogens at 0.5 to 2.0 mg/mL application. These researchers also previously reported inhibition of spoilage bacteria within the LAB and AAB groups relevant to wine spoilage (Carillo et al., 2014).

14.11 Recombinant Lysozymes for Use as Food Antimicrobials

For several years, researchers have been interested in expressing recombinant (cloned) human lysozyme, egg-white lysozyme, or lysozymes from other sources. Recombinant lysozyme from humans or other sources may have reduced allergenicity compared to egg-white lysozyme. Directed mutations of the cloned lysozyme gene could result in lysozymes with enhanced resistance to intrinsic and extrinsic factors in foods, increased specificity for target molecules, and other desirable features. Lysozyme from human, bovine, and microbial sources has been expressed in organisms that are often compatible for use in foods (Sofos et al., 1998). For instance, active egg-white lysozyme was expressed by *Saccharomyces cerevisiae* (Oberto and Davison, 1985) and *Aspergillus niger* (Archer et al., 1990). Inactive egg-white lysozyme but active T4 and bacteriophage lysozymes were produced by transformed *Lactococcus lactis* subsp. *lactis* strains (van de Guchte et al., 1992). Human lysozyme has been expressed during growth in whey by the yeast *Kluyveromyces lactis* after transformation with the human lysozyme, potentially leading to use of the yeast as a lysozyme-producing starter culture in cheeses prone to late blowing caused by clostridial fermentation (Maullu et al., 1999). Similarly, human lysozyme secreted into the milk of transgenic mice retained its lytic activity, leading to potential applications in the dairy industry of lysozyme-containing milk produced by transgenic animals (Maga et al., 1995; 1998). Wu et al. (2015) demonstrated production of human lysozyme from transgenic chicken eggs, wherein nearly equivalent antibacterial activity was generated from the recombinant human lysozyme as compared to commercial sources of both hen and human lysozymes as determined from disc diffusion assay. Additionally, the pH and thermo-stability of the recombinant lysozyme were non-differing from commercial lysozymes as well.

Genetic engineering resulted in a modified T4 lysozyme with a new disulfide bond, which increased thermostability by stabilizing the tertiary structure (Genentech and Genencor, 1985). Such modifications may increase applications of lysozyme in cheese-making and other food applications (Proctor and Cunningham, 1988). A hyperstable chicken lysozyme was isolated by directed mutations and expression (Shih and Kirsch, 1995). Increased numbers of recombinant enzymes with advantages over native are being introduced into the food industry, but their safety must be carefully evaluated prior to their use in commerce (Pariza and Johnson, 2001). Modification of pig-derived lysozyme with a short (six) multi-peptide add-on resulted in inhibition of *E. coli*, *K. pneumoniae*, and *Salmonella* Enteritidis, where for native pig lysozyme no antibacterial activity was detectable (Zhu et al., 2017).

14.12 Regulatory Status and Toxicology

Egg whites are well-known to be allergenic to sensitive individuals, particularly children (Poulsen et al., 2001), and lysozyme may constitute one of the allergens in egg albumin. However, the most allergenic proteins in eggs appear to be (in descending order) conalbumin (ovotransferrin), ovomucoid, ovalbumin, and lysozyme. Allergic reactions produced by egg-white lysozyme in animals and humans were fewer than reactions to other egg proteins such as ovalbumin and albumin (Bianchi, 1982), which have long been used as food ingredients. The acute toxicity of lysozyme in humans is unknown (Lück and Jager, 1997), although rodents and rabbits tolerated high intravenous doses (Barbara and Pellegrini, 1976; Bianchi, 1982). Several studies have indicated that lysozyme from egg white has negligible acute, subacute, and chronic toxicity in animals (Barbara and Pellegrini, 1976). In addition, lysozyme has a long history of safe consumption as an endogenous food ingredient in eggs, milk, and other foods. The presence of added egg-white lysozyme in foods at permitted levels is generally not considered to be a health concern.

Lysozyme was previously considered as generally recognized as safe (GRAS) by the United States Food and Drug Administration for use in certain cheeses according to the tentative final rule (U.S. FDA, 1998). However, because of the potential for allergenicity of lysozyme, the FDA has tentatively concluded that bulk and packaged foods containing lysozyme must be labeled to indicate presence of "egg-white lysozyme" (U.S. FDA, 1998). This is a condition of lysozyme use for GRAS status in such treated cheese. It is permitted for use in cheese at levels in accordance with good manufacturing practices. In addition, the FAO-WHO Joint Expert Committee on Food Additives concluded that the low additional intake of lysozyme in cheese made from treated milk was not a hazard to consumer health (JECFA, 1993). Use of hen egg-white lysozyme has been approved in many countries to control the growth of spoilage organisms in foods (Wang and Shelef, 1992). Lysozyme was approved as a preservative (E1105) under the European Additives Directive to prevent late blowing in ripened cheese. Lysozyme is also approved for use in cheese in Austria, Australia, Belgium, Canada, Denmark, Finland, France, Germany, Italy, the United Kingdom, and Spain, but as mentioned previously its use is to be labeled for purposes of food safety for consumers allergic to egg. The Organisation Internationale du Vin (OIV) approved the use of lysozyme in wine in 1997 to control undesirable bacterial growth, and its use in wines is awaiting approval in other countries. Applications for uses in other foods are being evaluated. The United States Department of Agriculture recently issued an acceptability determination regarding the use of a mixture of hops beta acids, egg-white lysozyme, and cultured skim milk in salad dressings used as a preservative in refrigerated meat and poultry deli salads (USDA, 2002). Interestingly, lysozyme has been removed in the current version of the U.S. Code of Federal Regulations, Title 21 Section 184, available at www.ecfr.gov, where it was previously listed alongside other GRAS compounds, though authors are not aware of a reversal of GRAS status notice being promulgated by the FDA.

14.13 Summary and Perspectives

In response to increased consumer perception and demand for more nutritious foods, the food industry is introducing foods into the marketplace that receive minimal processing and contain reduced levels of traditional preservatives such as salt, sugar, and acids, and few or no chemical preservatives, increasingly in favor of natural and/or "clean label" ingredients. To attain adequate shelf life for quality and to prevent microbial foodborne disease, there is significant interest by the industry in the use of natural preservation systems. A myriad of naturally occurring antimicrobials are being considered to develop safety systems usually based a multiple-hurdle system to inactivate or prevent the growth of pathogenic or spoilage organisms in foods (Sofos et al., 1998). In designing these systems, it is important to consider the interaction and multiple functions of antimicrobials as pioneered in the writings of Reiter (1978) and others. Naturally occurring antimicrobials, in conjunction with GMPs usage, can enhance the safety and quality of minimally processed refrigerated foods (Johnson, 1994). Egg-white lysozyme is a good example of a naturally occurring enzyme that has been employed by the industry to maintain product quality and reduce the incidence of spoilage.

Combinations of natural antimicrobials and mild preservation techniques (i.e., hurdle processing) are an attractive approach to control pathogens and spoilage organisms in many foods. The addition of lysozyme to certain foods before heat processing may reduce the thermal requirements necessary to inactivate spores of some strains of thermophilic spore-forming spoilage bacteria that are particularly sensitive to lysozyme, such as *G. stearothermophilus* and *Clostridium thermosaccharolyticum* (Johnson, 1994).

Lysozyme could play a key role in the development of antimicrobial systems for use in foods. Lysozyme is relatively inexpensive, readily available, and is stable under a wide variety of conditions. Lysozyme and lysozyme–antimicrobial combinations have many current and potential uses in the food industry to control pathogenic and spoilage organisms. Activity of lysozyme occurs over a wide range of temperatures and pHs. It is among the most stable of enzymes in foods and resists many food-processing procedures.

These properties may lead to numerous applications for lysozyme minimally processed foods. The use of lysozyme in these foods for the control of pathogens and spoilage organisms may result in products with increased safety, enhanced quality, and long shelf life. Future research may consider the interaction of lysozyme with other antimicrobials in minimally processed or alternatively processed food products, the incorporation and delivery of lysozyme to foods via encapsulation technologies or edible films, and the potential utility of other forms of lysozyme (e.g., goose egg lysozyme, insect-derived lysozymes) or chemi-modified lysozyme derivatives as innovative antimicrobials for food safety protection and spoilage prevention.

REFERENCES

Abdou, A. M., Higashiguchi, S., Aboueleinin, A. M., Kim, M., and Ibrahim, H. R. 2007. Antimicrobial peptides derived from hen egg lysozyme with inhibitory effect against *Bacillus* species. *Food Cont.* 18:173–178.

Adams, D. M. 1974. Requirements for and sensitivity to lysozyme by *Clostridium perfringens* spores heated at ultra-high temperatures. *Appl. Microbiol.* 27:797–801.

Ahern, T. J., and Klibanov, A. M. 1985. The mechanism of irreversible enzyme inactivation at 100°C. *Science* 228:1280–1284.

Akashi, A. 1969. Preservative effect of egg white lysozyme added to cooked sausage. *Nippon Chikusan Gakkaih* 40:243–348.

Akashi, A. 1970. Preservative effect of egg-white lysozyme on salami sausage. *Nippon Chikusan Gakkaih* 41:143–150.

Akashi, A. 1971. Preservative effect of egg-white lysozyme on vienna sausage. *Nippon Chikusan Gakkaih* 42:289–295.

Akashi, A., and Oono, A. 1972. The preservative effect of egg-white lysozyme to non-packaged Kamaboko. *Nippon Nogeikagaku Kaishi* 46:177–183.

Alakomi, H.-L., Skytta, E., Saarela, M., Mattila-Sandholm, T., Latva-Kala, K., and Helander, I. M. 2000. Lactic acid permeabilizes gram-negative bacteria by disrupting the outer membrane. *Appl. Environ. Microbiol.* 66:2001–2005.

Alderton, G., Chen, J. K., and Ito, K. A. 1974. Effect of lysozyme on the recovery of heated *Clostridium botulinum* spores. *Appl. Microbiol.* 27:613–615.

Alderton, G., Ward, W. H., and Fevold, H. L. 1945. Isolation of lysozyme from egg white. *J. Biol. Chem.* 157:43–58.

Amiri, S., Ramezani, R., and Aminlari, M. 2008. Antibacterial activity of dextran-conjugated lysozyme against *Escherichia coli* and *Staphylococcus aureus* in cheese curd. *J. Food Prot.* 71:411–415.

Antolinos, V., Muñoz, M., Ros-Chumilla, M., Aznar, A., Periago, P. M., and Fernández, P. S. 2011. Combined effect of lysozyme and nisin at different incubation temperature and mild heat treatment on the probability of time to growth of *Bacillus cereus*. *Food Microbiol.* 28:305–310.

Archer, D. B., Jeenes, D. J., MacKenzie, D. A., et al. 1990. Hen egg white lysozyme expressed in, and secreted from, *Aspergillus niger* is correctly folded and processed. *Bio/Technol.* 8:741–745.

Ashton, D. H., Wilson, A. A., and Evancho, G. M. 1975. Identification of a component of crystalline egg albumin bactericidal for thermophilic aerobic sporeformers. *Appl. Microbiol.* 30:821–824.

Back, J. F., Oakenfull, D., and Smith, M. B. 1979. Increased thermal stability of proteins in the presence of sugars and polyols. *Biochemistry.* 18:5191–5196.

Barbara L. and Pellegrini R. 1976. Fleming's Lysozyme: Biological significance and therapeutic applications. Torino, Italy: *Minerva medica.* Monograph.

Barbiroli, A., Bonomi, F., Capretti, G., et al. 2012. Antimicrobial activity of lysozyme and lactoferrin incorporated in cellulose-based food packaging. *Food Cont.* 26:387–392.

Bedie, G. K., Samelis, J., Sofos, J. N., Belk, K. E., Scanga, J. A., and Smith, G. C. Antimicrobials in the formulation to control post-processing contamination by *Listeria monocytogenes* on frankfurters stored at 4°C in vacuum. Abstract of poster presented at 2001 Institute of Food Technologists annual meeting, New Orleans, LA.

Bester, B. H., and Lombard, S. H. 1990. Influence of lysozyme on selected bacteria associated with Gouda cheese. *J. Food Prot.* 53:306–311.

Bevilacqua, A., Sinigaglia, M., and Corbo, M. R. 2010. Use of lysozyme and Na_2-ethylenediaminetetraacetic acid for the inhibition of *Hafnia alvei* of dairy origin: individuation of the minimal inhibitory concentrations. *J. Food Proc. Preserv.* 34:904–914.

Beychok, S., and Warner, R. C. 1959. Denaturation and electrophoretic behavior of lysozyme. *J. Am. Chem. Soc.* 81:1892–1897.

Bianchi, C. 1982. Antigenic properties of hen egg white lysozyme (Fleming's lysozyme) and notes on its acute/sub-acute toxicity. *Curr. Therap. Res.* 31:494–505.

Blake, C. C. F., Koenig, D. F., Mair, G. A., North, A. C. T., Phillips, D. C., and Sarma, V. R. 1965. Structure of hen egg-white lysozyme. *Nature* 4986:757–761.

Blocher, J. C., and Busta, F. F. 1982. Constituents of media for recovery of sporeformers from food. *Arch. Lebensm.* 33:138–142.

Bower, C. K., Daeschel, M. A., and McGuire, J. 1998. Protein antimicrobial barriers to bacterial adhesion. *J. Dairy Sci.* 81:2771–2778.

Broda, D. M., De Lacy, K. M., and Bell, R. G. 1998. Efficacy of heat and ethanol spore treatments for the isolation of psychrotrophic *Clostridium* spp. associated with the spoilage of chilled vacuum-packed meats. *Int. J. Food Microbiol.* 39:61–68.

Carini, S., and Lodi, R. 1982. Inhibition of germination of clostridial species by lysozyme. *Ind. Latte.* 18:35–48.

Carneiro de Melo, A. M. S., Cassar, C. A., and Miles, R. J. 1998. Trisodium phosphate increases sensitivity of gram-negative bacteria to lysozyme and nisin. *J. Food Prot.* 61:839–844.

Carillo, W., Lucio, A., Gaibor, J., Morales, D., and Vasquez, G. 2018a. Isolation of antibacterial hydrolysates from hen egg white lysozyme and identification of antibacterial peptides. *J. Med. Food* 21:808–818.

Carillo, W., and Ramos, M. 2018b. Identification of antimicrobial peptides of native and heated hen egg white lysozyme. *J. Med. Food* 21:915–926.

Carrillo W., García-Ruiz, A., and Moreno-Arribas, M. V. 2014. Antibacterial activity of hen egg white lysozyme modified by heat and enzymatic treatments against oenological lactic acid bacteria and acetic acid bacteria. *J. Food Prot.* 77:1732–1739.

Carroll, E. J. 1979. The role of lysozyme in killing and lysis of coliform bacteria in the bovine animal. 1. Serum and milk concentrations of lysozyme and susceptibility of coliform strains to its action. *Vet. Microbiol.* 4:61–72.

Cegielska-Radziejewska, R., Lesnierowski, G., Szablewski, T., and Kijowski, J. 2010. Physico-chemical properties and antibacterial activity of modified egg white—lysozyme. *Eur. Food Res. Technol.* 231, 959–964. https://doi.org/10.1007/s00217-010-1347-y

Chai, C., Lee, K.-S., Imm, G.-S., Kim, Y. S., and Oh, S.-W. 2016. Inactivation of *Clostridium difficile* spore outgrowth by synergistic effects of nisin and lysozyme. *Can. J. Microbiol.* 63:638–643.

Chandan, R. C., Parry, R. M., and Shahani, K. M. 1965. Purification and some properties of bovine milk lysozyme. *Biochim. Biophys. Acta* 110:389–398.

Chandan, R. C., Parry, R. M., and Shahani, K. M. 1968. Lysozyme, lipase, and ribonuclease in milk of various species. *J. Dairy Sci.* 51:606–607.

Chander, R., and Lewis, N. F. 1980. Effect of lysozyme and sodium EDTA on shrimp microflora. *Eur. J. Appl. Microbiol. Biotechnol.* 10:253–258.

Chang, K. Y., and Carr, C. W. 1971. Studies on the structure and function of lysozyme. I: The effect of pH and cation concentration on lysozyme activity. *Biochim. Biophys. Acta* 229:496–503.

Chatzolopou, A., Miles, R. J., Anagnostopoulos, G., inventors and assignees. 1993. Destruction of gram negative bacteria. International Patent Application WO 93/00822.

Chumchalova, J., Josephsen, J., and Plockova, M. 1998. The antimicrobial activity of acidocin CH5 in MRS broth and milk with added NaCl, $NaNO_2$ and lysozyme. *Int. J. Food Microbiol.* 43:33–38.

Chung, W., and Hancock, R. E. W. 2000. Action of lysozyme and nisin mixtures against lactic acid bacteria. *Int. J. Food Microbiol.* 60:25–32.

Connor, D. E. 1993. Naturally occurring compounds. In: Davidson, P. M. and Branen, A. L., eds. *Antimicrobials in Foods.* 2nd ed. New York: Marcel Dekker. pp. 441–468.

Conte, A., Brescia, I., and del Nobile, M. A. 2011. Lysozyme/EDTA disodium salt and modified-atmosphere packaging to prolong the shelf life of burrata cheese. *J. Dairy Sci.* 94:5289–5297.

Costa, C., Lucera, A., Conte, A., Zambrini, A. V., and del Nobile, M. A. 2017. Technological strategies to preserve burrata cheese quality. *Coatings* 7:97 (10 pages).

Cunningham, F. E., and Cotterill, O. J. 1971. The influence of yolk on egg white lysozyme. *Poult. Sci*, 50:1013–1016.

Cunningham, F. E., and Lineweaver, H. 1965. Stabilization of egg-white proteins to pasteurizing temperatures above 60°C. *Food Technol.* 19:136–141.

Cunningham, F. E., and Lineweaver, H. 1967. Inactivation of lysozyme by native ovalbumin. *Poult. Sci.* 46:1471–1477.

Cunningham, F. E., Proctor, V. A., and Goetsch, S. J. 1991. Egg-white lysozyme as a food preservative: An overview. *W. Poult. Sci. J* 47:141–163.

Daeschel, M. A., and Kennedy, J. E. 2006. Enhancing the effective concentration of lysozyme in wine. U.S. Patent 11/993,789.

Davidson, P. M., and Harrison, M. A. 2002. Resistance and adaptation to food antimicrobials, sanitizers, and other process controls. *Food Technol.* 56:69–78.

Davies, R. C., Neuberger, A., and Wilson, B. M. 1969. The dependence of lysozyme activity on pH and ionic strength. *Biochim. Biophys. Acta* 178:294–305.

Dell'Acqua, E., Bruzzese, T., and Van Den Huevel, H. H., inventors; Soc. Prod. Ant., assignee. 1989. Process of obtaining foods free of *Listeria* bacteria. U.S. Patent 4,810,508.

During, K., Porsch, P., Mahn, A., Brinkmann, O., and Gieffers, W. 1999. The non-enzymatic microbicidal activity of lysozymes. *FEBS Lett.* 449:93–100.

Eitenmiller, R. R., Barnhart, H. M., and Shahani, K. M. 1971. Effect of gamma-irradiation on bovine and human milk lysozymes. *J. Food Sci.* 36:1127–1129.

Eitenmiller, R. R., Friend, B. A., and Shahani, K. M. 1976. Relationship between composition and stability of bovine milk lysozyme. *J. Dairy Sci.* 59:834–839.

Elagamy, E. I. 2000. Effect of heat treatment on camel milk proteins with respect to antimicrobial factors: A comparison with cows' and buffalo milk proteins. *Food Chem.* 68:227–232.

El-Kest, S. E., and Marth, E. H. 1992. Transmission electron microscopy of unfrozen and frozen/thawed cells of *Listeria monocytogenes* treated with lipase and lysozyme. *J. Food Prot.* 55:687–696.

Ellison, R. T., and Giehl, T. J. 1991. Killing of gram-negative bacteria by lactoferrin and lysozyme. *J. Clin. Invest.* 88:1080–1091.

Erickson, J. P., and Jenkins, P. 1991. Comparative *Salmonella* spp. and *Listeria monocytogenes* inactivation rates in four commercial mayonnaise products. *J. Food Prot.* 54:913–916.

Erickson, J. P., and Jenkins, P. 1992. Behavior of psychrotrophic pathogens *Listeria monocytogenes*, *Yersinia enterocolitica*, and *Aeromonas hydrophila* in commercially pasteurized eggs held at 2, 6.7, and 12.8°C. *J. Food Prot.* 55:8–12.

Esposito, A., Comez, L., Cinelli, S., Scarponi, F., and Onori, G. 2009. Influence of glycerol on the structure and thermal stability of lysozyme: dynamic light scattering and circular dichroism study. *J. Phys. Chem. B* 113:16420–16424.

Esquembre, R., Sanz, J. M., Wall, J. G., del Monte, F., Mateo, C. R., and Ferrer, M. L. 2013. Thermal unfolding and refolding of lysozyme in deep eutectic solvents and their aqueous dilutions. *Phys. Chem. Chem. Physics* 15:11248–11256.

Facon, M. J., and Skura, B. J. 1996. Antibacterial activity of lactoferricin, lysozyme and EDTA against *Salmonella enteritidis*. *Int. Dairy J.* 6:303–313.

Faille, C., Fontaine, F., and Membre, J. M. 1999. Factors influencing recovery of heat-injured *Bacillus thuringiensis* spores: Statistical approach. *J. Food Sci.* 64:363–366.

Fernandez, P. S., and Peck, M. W. 1999. A predictive model that describes the effect of prolonged heating at 70 to 90°C and subsequent incubation at refrigeration temperatures on growth from spores and toxigenesis by nonproteolytic *Clostridium botulinum* in the presence of lysozyme. *Appl. Environ. Microbiol.* 65:3449–3457.

Ferrari, L., and Dell'Acqua, E., inventors and assignees. 1979. Process for the production of dairy products. U.K. patent 2014032.

Fleming, A. 1922. On a remarkable bacteriolytic element found in tissues and secretions. *Proc. Roy. Soc. Ser. B.* 93:306–317.

Fujimaki, M., Kato, H., and Hayase, F. 1972. Studies on roasting changes of proteins. Part I: Changes of casein and lysozyme during roasting. *Agric. Biol. Chem.* 36:416–425.

Fukao, T., Sawada, H., and Ohta, Y. 2000. Combined effect of hop resins and sodium hexametaphosphate against certain strains of *Escherichia coli*. *J. Food Prot.* 63:735–740.

Galm, L., Morgenstern, J., and Hubbuch, J. 2015. Manipulation of lysozyme phase behavior by additives as a function of conformational stability. *Int. J. Pharm.* 494:370–380.

Garcia-Graells, C., Masschalck, B., and Michiels, C. W. 1999. Inactivation of *Escherichia coli* in milk by high-hydrostatic-pressure treatment in combination with antimicrobial peptides. *J. Food Prot.* 62:1248–1254.

Genentech, Genencor. 1985. Cover focus: T4 lysozyme. Food Technol. 39:2.

Gerbaux, V., Villa, A., Monamy, C., and Bertrand, A. 1997. Use of lysozyme to inhibit malolactic fermentation and to stabilize wine after malolactic fermentation. *Am. J. Enol. Vitic.* 48:49–54.

Giangiacomo, R., Nigro, F., Messina, G., and Cattaneo, T. M. 1992. Lysozyme: Just an additive or a technological aid as well? *Food Addit. Contam.* 9:427–433.

Gill, A. O., and Holley, R. A. 2000a. Inhibition of bacterial growth on ham and bologna by lysozyme, nisin and EDTA. *Food Res. Int.* 33:83–90.

Gill, A. O., and Holley, R. A. 2000b. Surface application of lysozyme, nisin, and EDTA to inhibit spoilage and pathogenic bacteria on ham and bologna. *J. Food Prot.* 63:1338–1346.

Gill, A. O., and Holley, R. A. 2002. Interactive inhibition of meat spoilage and pathogenic bacteria by lysozyme, nisin and EDTA in the presence of nitrite and sodium chloride at 24°C. *Int. J. Food Microbiol.* 80:251–259.

Glass, K. A., and Johnson, E. A. 2004. Antagonistic effect of fat on the antibotulinal activity of food preservatives and fatty acids. *Food Microbiol.* 21:675–682.

Gould, G. W. 1989. Heat induced injury and inactivation. In: Gould, G. W., ed. *Mechanisms of Action of Food Preservation Procedures.* London: Elsevier. pp. 11–42.

Gould, G. W. 1999. Sous vide foods: Conclusions of an ECFF botulinum working party. *Food Cont.* 10:47–51.

Gould, G. W. 2002. Control with naturally occurring antimicrobial systems including bacteriolytic enzymes. In: Juneja, V. K. and Sofos, J. N., eds. *Control of Foodborne Microorganisms.* New York: Marcel Dekker. pp. 281–302.

Gould, G. W., and Hitchins, A. D. 1963. Sensitization of bacterial spores to lysozyme and to hydrogen peroxide with agents which rupture disulphide bonds. *J. Gen. Microbiol.* 33:413–423.

Hamdani, A. M., Wani, I. A., Bhat, N. A., and Siddiqi, R. A. 2018. Effect of guar gum conjugation on functional, antioxidant and antimicrobial activity of egg white lysozyme. *Food Chem.* 240:1201–1209.

Hancock, R. E. W. 1984. Alterations in outer membrane permeability. *Ann. Rev. Microbiol.* 38:237–264.

Hartsell, S. E. 1948. Lysozyme activity of rehydrated, spray-dried, whole-egg powder. *Food Res.* 13:136–142.

Hashemi, M. M., Aminlari, M., and Moosavinasab, M. 2014. Preparation of and studies on the functional properties and bactericidal activity of the lysozyme-xanthan gum conjugate. *LWT Food Sci. Technol.* 57:594–602.

Hauben, K. J. A., Wuytack, E. Y., Soontjens, C. C. F., and Michiels, C. W. 1996. High-pressure transient sensitization of *Escherichia coli* to lysozyme and nisin by disruption of outer-membrane permeability. *J. Food Prot.* 59:350–355.

Hayase, F., Kato, H., and Fujimaki, M. 1975. Chemical investigation on the changes in lysozyme during heating. *Agric. Biol. Chem.* 39:1255–1262.

Hayashi, K., Kasumi, T., Kubo, N., Haraguchi, K., and Tsumura, T. 1989. Effects of N-acetylmuramidase from *Streptomyces rutgersensis* H-46 as a food preservative. *Agric. Biol. Chem.* 53:3173–3177.

Hennart, P. F., Brasseur, D. J., Delogne-Desnoeck, J. B., Dramaix, M. M., and Robyn, C. E. 1991. Lysozyme, lactoferrin and secretory immunoglobulin A content in breast milk: Influence of duration of lactation, nutrition status, prolactin status, and parity of mother. *Am. J. Clin. Nutr.* 53:32–39.

Ho, S. Y., Mittal, G. S., and Cross, J. D. 1997. Effects of high field electric pulses on the activity of selected enzymes. *J. Food Eng.* 31:69–84.

Holzapfel, W. H., Geisen, R., and Schillinger, U. 1995. Biological preservation of foods with reference to protective cultures, bacteriocins and food-grade enzymes. *Int. J. Food Microbiol.* 24:343–362.

Hoq, M. I., Mitsuno, K., Tsujino, Y., Aoki, T., and Ibrahim, H. R. 2008. Triclosan-lysozyme complex as novel antimicrobial macromolecule: a new potential of lysozyme as phenolic drug-targeting molecule. *Int. J. Biol. Macromol.* 42:468–477.

Hughey, V. L., and Johnson, E. A. 1987. Antimicrobial activity of lysozyme against bacteria involved in food spoilage and food-borne disease. *Appl. Environ. Microbiol.* 53:2165–2170.

Hughey, V. L., Wilger, P. A., and Johnson, E. A. 1989. Antibacterial activity of hen egg white lysozyme against *Listeria monocytogenes* Scott A in foods. *Appl. Environ. Microbiol.* 55:63–638.

Ibrahim, H. R. 1998. On the novel catalytically-independent antimicrobial function of hen egg-white lysozyme: A conformation-dependent activity. *Nahrung* 42:187–193.

Ibrahim, H. R., Higashiguchi, S., Juneja, L. R., Kim, M., and Yamamoto, T. 1996a. A structural phase of heat-denatured lysozyme with novel antimicrobial action. *J. Agric. Food Chem.* 44:1416–1423.

Ibrahim, H. R., Higashiguchi, S., Koketsu, M., et al. 1996b. Partially unfolded lysozyme at neutral pH agglutinates and kills gram-negative and gram-positive bacteria through membrane damage mechanism. *J. Agric. Food Chem.* 44:3799–3806.

Ibrahim, H. R., Matsuzaki, T., and Aoki, T. 2001. Genetic evidence that antibacterial activity of lysozyme is independent of its catalytic function. *FEBS Lett.* 506:27–32.

International Dairy Federation. 1987. The use of lysozyme in the prevention of late blowing in cheese. Bull. No. 216: 16 p. The International Dairy Federation, Brussels.

Johansen, C., Gram, L., and Meyer, A. S. 1994. The combined inhibitory effect of lysozyme and low pH on growth of *Listeria monocytogenes. J. Food Prot.* 57:561–566.

Johnson, E. A. 1994. Egg white lysozyme as a preservative for use in foods. In: Sim, J. S. and Nakai, S., eds. *Egg Uses and Processing Technologies: New Developments.* Wallingford: CAB International. pp. 177–191.

Johnson, E. A., Dell'Acqua, E., inventors; Solchem Italiana SPA, assignee. 1995. Composition active against botulism. U.S. Patent 5,393,545.

Johnson, E. A., Dell'Acqua, E., and Ferrari, L., inventors; Soc. Prod. Ant., assignee. 1991. Treating vegetables with lysozyme and EDTA to kill *Listeria.* U.S. Patent 5,019,411.

Joint FAO/WHO Expert Committee on Food Additives (JECFA). 1993. Food Additive Series 30. Geneva: World Health Organization. pp. 25–31.

Jolles, J., Jauregui-Adell, J., Bernier, I., and Jolles, P. 1963. The chemical structure of hen's egg-white lysozyme: The detailed study. *Biochim. Biophys. Acta* 78:668–689.

Jolles, P., and Jolles, J. 1984. What's new in lysozyme research? *Mol. Cell Biochem.* 63:165–189.

Jolles, P. (ed.). 1996. *Lysozymes: Model Enzymes in Biochemistry and Biology.* Basel; Boston: Birkhäuser Verlag.

Kain, M. L., Samelis, J., Sofos, J. N., Belk, K. E., Scanga, J. A., and Smith, G. C. 2001. Evaluation of antimicrobials as post-processing solutions against *Listeria monocytogenes* on sliced pork bologna stored at 4°C in vacuum packages. Abstract of poster presented at Institute of Food Technologists annual meeting, New Orleans, LA.

Kanner, J., and Karel, M. 1976. Changes in lysozyme due to reactions with peroxidizing methyl linoleate in a dehydrated model system. *J. Agric. Food Chem.* 24:468–472.

Kihm, D. J., Leyer, G. J., An, G. H., and Johnson, E. A. 1994. Sensitization of heat-treated *Listeria monocytogenes* to added lysozyme in milk. *Appl. Environ. Microbiol.* 60:3854–3861.

King, W., and Ming, X., inventors; Rhodia, Inc., assignee. 2002, September 17. Antibacterial composition for control of gram positive bacteria in food applications. U.S. Patent 6,451,365.

Ko, K. Y., Mendonca, A. F., and Ahn, D. U. 2008. Effect of ethylenediaminetetraacetate and lysozyme on antimicrobial activity of ovotransferrin against *Listeria monocytogenes. Poult. Sci.* 87:1649–1658.

Ko, K. Y., Mendonca, A. F., Ismail, H., and Ahn, D. U. 2009. Ethylenediaminetetraacetate and lysozyme improves antimicrobial activities of ovotransferrin against *Escherichia coli* O157:H7. *Poult. Sci.* 88:406–414.

Kohn, A. 1960. Lysis of frozen and thawed cells of *Escherichia coli* by lysozyme, and their conversion into sphaeroplasts. *J. Bacteriol.* 79:697–706.

Kume, T., Sato, Y., and Umemoto, Y. 1973. Effect of irradiation on egg white lysozyme in aqueous solution. *Nippon Nogeikagaku Kaishi* 47:549–555.

Lam, N. H., Ma, Z., and Ha, B.-Y. 2014. Electrostatic modification of the lipopolysaccharide layer: Competing effects of divalent cations and polycationic or polyanionic molecules. *Soft Matter* 10:7528–7544.

Laschenko, P. 1909. The effect of egg albumin in killing spores and in retarding development. *Z. Hyg.* 64: 419.

Liable, N. F., and Germaine, G. R. 1985. Bactericidal activity of human lysozyme, muramidase-inactive lysozyme, and cationic polypeptides against *Streptococcus sanguis* and *Streptococcus faecalis*: Inhibition by chitin oligosaccharides. *Infect. Immun.* 48:720–728.

Liang, Z., Mittal, G. S., and Griffiths, M. W. 2002. Inactivation of *Salmonella* Typhimurium in orange juice containing antimicrobial agents by pulsed electric field. *J. Food Prot.* 65:1081–1087.

López-Pedemonte, T. J., Roig-Sagués, A. X., Trujillo, A. J., Capellas, M., and Guamis, B. 2003. Inactivation of spores of *Bacillus cereus* in cheese by high hydrostatic pressure with the addition of nisin or lysozyme. *J. Dairy Sci.* 86:3075–3081.

Luck, E., and Jager, M. 1997. *Antimicrobial Food Additives*. 2nd ed. Berlin: Springer. 260 p.

Maga, E. A., Anderson, G. B., Cullor, J. S., Smith, W., and Murray, J. D. 1998. Antimicrobial properties of human lysozyme transgenic mouse milk. *J. Food Prot.* 61:52–56.

Maga, E. A., Anderson, G. B., and Murray, J. D. 1995. The effect of mammary gland expression of human lysozyme on the properties of milk from transgenic mice. *J. Dairy Sci.* 78:2645–2652.

Mangalassary, S., Han, I., Rieck, J., Acton, J., and Dawson, P. 2008. Effect of combining nisin and/or lysozyme with in-package pasteurization for control of *Listeria monocytogenes* in ready-to-eat turkey bologna during refrigerated storage. *Food Microbiol.* 25:866–870.

Masschalck, B., Garcia-Graells, C., Van Haver, E., and Michiels, C. 2000. Inactivation of high pressure resistant *Escherichia coli* by lysozyme and nisin under high pressure. *Innov. Food Sci. Emerg. Technol.* 1:39–47.

Masschalck, B., Van Houdt, R., Van Haver, E. G. R., and Michiels, C. W. 2001. Inactivation of gram-negative bacteria by lysozyme, denatured lysozyme, and lysozyme-derived peptides under high hydrostatic pressure. *Appl. Environ. Microbiol.* 67:339–344.

Mastromatteo, M., Lucera, A., Sinigaglia, M., and Corbo, M. R. 2010. Synergistic antimicrobial activity of lysozyme, nisin, and EDTA against *Listeria monocytogenes* in ostrich meat patties. *J. Food Sci.* 75: M422–M429.

Mathur, N. B., Dwarkadas, A. M., Sharma, V. K., Saha, K., and Jain, N. 1990. Anti-infective factors in preterm human colostrum. *Acta Paediatr. Scand.* 79:1039–1044.

Maullu, C., Lampis, G., Basile, T., Ingianni, A., Rossolini, G. M., and Pompei, R. 1999. Production of lysozyme-enriched biomass from cheese industry by-products. *J. Appl. Microbiol.* 86:182–186.

Mayes, F. J., and Takeballi, M. A. 1983. Microbial contamination of the hen's egg: A review. *J. Food Prot.* 46:1092–1098.

Memarpoor-Yazdi, M., Asoodeh, A., and Chamani, J. 2012. A novel antioxidant and antimicrobial peptide from hen egg white lysozyme hydrolysates. *J. Funct. Foods* 4:278–286.

Meyer, K., Thompson, R., Palmer, J. W., and Khorazo, D. 1936. The purification and properties of lysozyme. *J. Biol. Chem.* 113:303–309.

Miller, T. E. 1969. Killing and lysis of gram-negative bacteria through the synergistic effect of hydrogen peroxide, ascorbic acid and lysozyme. *J. Bacteriol.* 98:949–955.

Mine, Y., Ma, F., and Lauriau, S. 2004. Antimicrobial peptides released by enzymatic hydrolysis of hen egg white lysozyme. *J. Agric. Food Chem.* 52:1088–1094.

Monticello, D. J., inventor; Miles, Inc., assignee. 1990 June 27. Control of microbial growth with nisin/lysozyme formulations. European patent EP0374823.

Nattress, F. M., Yost, C. K., and Baker, L. P. 2001. Evaluation of the ability of lysozyme and nisin to control meat spoilage bacteria. *Int. J. Food Microbiol.* 70:111–119.

Ng, H., and Garibaldi, J. A. 1975. Death of *Staphylococcus aureus* in liquid whole egg near pH 8. *Appl. Microbiol.* 29:782–786.

Nikaido, H. 2001. Preventing drug access to targets: Cell surface permeability barriers and active efflux in bacteria. *Seminars Cell Develop. Biol.* 12:215–223.

Nikaido, H., and Vaara, M. 1985. Molecular basis of bacterial outer membrane permeability. *Microbiol. Rev.* 49:1–32.

Nishihava, K., and Isoda, K. 1967. Fecal bacterial flora in the use of lysozyme treated dried milk. *Acta Paed. Jap.* 71:95–102.

Ntzimani, A. G., Giatrakou, V. I., and Savvaidis, I. N. 2010. Combined natural antimicrobial treatments (EDTA, lysozyme, rosemary and oregano) on semi cooked coated chicken meat stored in vacuum packages at 4°C: Microbiological and sensory evaluation. *Innov. Food Sci. Emerg. Technol.* 11:187–196.

Oberto, J., and Davison, J. 1985. Expression of chicken egg white lysozyme by *Saccharmyces cervisiae*. *Gene* 40:57–65.

Oh, M., Lee, J., and Kim, M. 2016. Synergistic antilisterial effects of mixtures of lysozyme and organic acids. *J. Food Prot.* 79:2184–2189.

Ohno, N., and Morrison, D. C. 1989. Lipopolysaccharide interaction with lysozyme. Binding of lipopolysaccharide to lysozyme and inhibition of lysozyme enzymatic activity. *J. Biol. Chem.* 264:4434–4441.

Osserman, E. F., Canfield, R. E., and Beychok, S. (eds.). 1974. *Lysozyme*. New York: Academic Press.

Padgett, T., Han, I. Y., and Dawson, P. L. 1998. Incorporation of food-grade antimicrobial compounds into biodegradable packaging films. *J. Food Prot.* 61:1330–1335.

Pagan, R., Manas, P., Raso, J., and Trepat, F. J. S. 1999. Heat resistance of *Yersinia enterocolitica* grown at different temperatures and heated in different media. *Int. J. Food Microbiol.* 47:59–66.

Pakkanen, R., and Aalto, J. 1997. Growth factors and antimicrobial factors of bovine colostrum. *Int. Dairy J.* 7:285–297.

Pariza, M. W., and Johnson, E. A. 2001. Evaluating the safety of microbial enzyme preparations used in food processing: Update for a new century. *Regul. Toxicol. Pharm.* 33:173–186.

Parkar, S.G., Flint, S.H., Palmer, J.S., and Brooks, J.D. 2001. Factors influencing attachment of thermophilic bacilli to stainless steel. *J. Appl. Microbiol.* 90:901–908. https://doi.org/10.1046/j.1365-2672.2001.01323.x

Parry, R. M., Chandan, R. C., and Shahini, K. M. 1969. A rapid and sensitive assay of muramidase. *Proc. Soc. Exp. Biol. Med.* 119:384–388.

Payne, D. K., Oliver, S. P., and Davidson, P. M. 1994. Comparison of EDTA and apo-lactoferrin with lysozyme on the growth of foodborne pathogenic and spoilage bacteria. *J. Food Prot.* 57:62–65.

Peck, M. W., Fairbairn, D. A., and Lund, B. M. 1992. The effect of recovery medium on the estimated heat-inactivation of spores of nonproteolytic *Clostridium botulinum*. *Lett. Appl. Microbiol.* 15:146–151.

Peck, M. W., Fairbairn, D. A., and Lund, B. M. 1993. Heat-resistance of spores of non-proteolytic *Clostridium botulinum* estimated on medium containing lysozyme. *Lett. Appl. Microbiol.* 16:126–131.

Peck, M. W., and Fernandez, P. S. 1995. Effect of lysozyme concentration, heating at 90°C, and then incubation at chilled temperatures on growth from spores of non-proteolytic *Clostridium botulinum*. *Lett. Appl. Microbiol.* 21:50–54.

Pellegrino, L., and Tirelli, A. 2000. A sensitive HPLC method to detect hen's egg white lysozyme in milk and dairy products. *Int. Dairy J.* 10:435–442.

Pelligrini, A., Thomas, U., Bramaz, N., Klauser, S., Hunziker, P., and von Fellenberg, R. 1997. Identification and isolation of a bactericidal domain in chicken egg white lysozyme. *J. Appl. Microbiol.* 82:372–378.

Pelligrini, A., Thomas, U., von Fellenberg, R., and Wild, P. 1992. Bactericidal activities of lysozyme and aprotinin against gram-negative and gram-positive bacteria related to their basic character. *J. Appl. Bacteriol.* 72:180–187.

Peterson, M. E., Pelroy, G. A., Poysky, F. T., et al. 1997. Heat pasteurization process for inactivation of non-proteolytic types of *Clostridium botulinum* in picked Dungeness crabmeat. *J. Food Prot.* 60:924–928.

Peterson, R. G., and Hartsell, S. E. 1955. The lysozyme spectrum of the gram-negative bacteria. *J. Infect. Dis.* 96:75–81.

Phillips, D. C. 1966. The three-dimensional structure of an enzyme molecule. *Sci. Amer.* 215:78–90.

Pina-Pérez, M. C., Rodrigo, D., and Martinez, A. 2015. Using natural antimicrobials to enhance the safety and quality of milk. In Taylor, T. M., ed. *Handbook of Natural Antimicrobials for Food Safety and Quality*, Chap. 15. New York, NY: Woodhead Publishing, Ltd.

Pittia, P., Nicoli, M. C., Comi, G., and Massini, R. 1999. Shelf-life extension of fresh-like ready-to-use pear cubes. *J. Sci. Food Agric.* 79:955–960.

Pol, I. E., and Smid, E. J. 1999. Combined action of nisin and carvacrol on *Bacillus cereus* and *Listeria monocytogenes*. *Lett. Appl. Microbiol.* 29:166–170.

Poland, A. L., and Sheldon, B. W. 2001. Altering the thermal resistance of foodborne bacterial pathogens with an eggshell membrane waste by-product. *J. Food Prot.* 64:486–492.

Poulsen, L. K., Hansen, T. K., Nørgaard, A., Vestergaard, H., Stahl, S. K., and Bindslev-Jensen, C. 2001. Allergens from fish and egg. *Allergy* 56 (Suppl. 67): 39–42.

Proctor, V. A., and Cunningham, F. E. 1988. The chemistry of lysozyme and its use as a food preservative and a pharmaceutical. *CRC Crit. Rev. Food Sci. Nutr.* 26:359–395.

Raghubeer, E. V., Ke, J. S., Campbell, M. L., and Meyer, R. S. 1995. Fate of *Escherichia coli* O157:H7 and other coliforms in commercial mayonnaise and refrigerated salad dressing. *J. Food Prot.* 58:13–18.

Ray, B., Johnson, C., and Wanismail, B. 1984. Factors influencing lysis of frozen *Escherichia coli* cells by lysozyme. *Cryo-Lett.* 5:183–190.

Razavi-Rohani, S. M., and Griffiths, M. W. 1996. Inhibition of spoilage and pathogenic bacteria associated with foods by combinations of antimicrobial agents. *J. Food Safety* 16:87–104.

Razavi-Rohani, S. M., and Griffiths, M. W. 1999. The antifungal activity of butylated hydroxyanizole and lysozyme. *J. Food Safety* 19:97–108.

Rees, A. R., and Offord, R. E. 1972. X-ray crystallography and the amino acid sequence of lysozyme. *Biochem. J.* 130:965–968.

Reiter, B. 1978. Review of the progress of dairy science: Antimicrobial systems in milk. *J. Dairy Res.* 45:131–147.

Repaske, R. 1958. Lysis of gram-negative organisms and the role of versene. *Biochim. Biophys. Acta* 30:225–232.

Rettger, L. O., and Sperry, J. A. 1912. The antiseptic and bactericidal properties of egg white. *J. Med. Res.* 26:55.

Rodrigues, E. T., Han, J. H., and Holley, R. A. 2002. Optimized antimicrobial edible whey protein films against spoilage and pathogenic bacteria. Abstract of poster presented at Institute of Food Technologists Annual Meeting, Anaheim, CA.

Rupley, J. A. 1967. The binding and cleavage of lysozyme by N-acetylglucosamine oligosaccharides. *Proc. Roy. Soc. B.* 167: 416–428.

Salton, M. R. J. 1957. The properties of lysozyme and its action on microorganisms. *Bacteriol. Rev.* 21:82–99.

Salton, M. R. J., and Pavlik, J. G. 1960. Studies of the bacterial cell wall. VI: Wall composition and sensitivity to lysozyme. *Biochim. Biophys. Acta* 39:398–407.

Sanderson, K. E., MacAlister, T., Costerton, J. W., and Cheng, K. J. 1974. Permeability of lipopolysaccharide-deficient (rough) mutants of *Salmonella typhimurium* to antibiotics, lysozyme, and other agents. *Can. J. Microbiol.* 20:1135–1145.

Sandow, A. 1926. The anti-bacterial activity of egg-white. *Proc. Soc. Exp. Biol. Med.* 24:172.

Sava, G. 1996. Pharmacological aspects and therapeutic applications of lysozymes. In: Jolles, P., ed. *Lysozymes: Model Enzymes in Biochemistry and Biology*. Basel: Birkhauser Verlag. pp. 433–449.

Sawada, J., Misaki, T., Yamagishi, M., and Kitahara, T., inventors; Taisho Pharm. Co. Ltd., assignee. 1967 August 29. Process of treating milk with muramidase. U.S. Patent 3,338,719.

Schaffner, D. F., Hamdy, M. K., Toledom R. T., and Tift, M. L. 1989. *Salmonella* inactivation in liquid whole egg by thermoradiation. *J. Food Sci.* 54:902–905.

Schmidt, C. F., Zimmermann, R. M., and Gaub, H. E. 1990. Multilayer adsorption of lysozyme on a hydrophobic substrate. *Biophys. J.* 57:577–588.

Schwimmer, S. 1981. *Source Book of Food Enzymology*. Westport, CT: AVI. 662 p.

Scott, V. N. 1989. Interaction of factors to control microbial spoilage of refrigerated foods. *J. Food Prot.* 52:431–435.

Sharon, N. 1967. The chemical structure of lysozyme substrates and their cleavage by the enzyme. *Proc. Roy. Soc. B.* 167:402–415.

Sheffield, C. L., Crippen, T. L., Poole, T. L., and Beier, R. C. 2012. Destruction of single species biofilms of *Escherichia coli* or *Klebsiella pneumoniae* subsp. *Pneumoniae* by dextranase, lactoferrin, and lysozyme. *Int. Microbiol.* 15:183–187.

Shih, P., and Kirsch, J. F. 1995. Design and structural analysis of an engineered thermostable chicken lysozyme. *Protein Sci.* 4:2063–2072.

Shugar, D. 1952. The measurement of lysozyme activity and the ultraviolet inactivation of lysozyme. *Biochim. Biophys. Acta* 8:302–309.

Smith, G. N., and Stroker, C. 1949. Inhibition of crystalline lysozyme. *Arch. Biochem.* 21:383–394.

Smith, J. L., McColgan, C., and Marmer, B. S. 1991. Growth temperature and action of lysozyme on *Listeria monocytogenes*. *J. Food Sci.* 56:1101, 1103.

Smolelis, A. N., and Hartsell, S. E. 1952. Factors affecting the lytic activity of lysozyme. *J. Bacteriol.* 63:665–674.

Sobrino-López, A., and Martín-Belloso, O. 2008. Enhancing the lethal effect of high-intensity pulsed electric field in milk by antimicrobial compounds as combined hurdles. *J. Dairy Sci.* 91:1759–1768.

Sofos, J. N., Beuchat, L. R., Davidson, P. M., and Johnson, E. A. 1998. Naturally occurring antimicrobials in foods. Council for Agricultural Science and Technology, Task Force Report No. 132, Ames, IA. 103 p.

Sophianopoulos, A. J., and Van Holde, K. E. 1964. Physical studies of muramidase (lysozyme). II. pH-dependent dimerization. *J. Biol. Chem.* 239: 2516–2524.

Stringer, S. C., and Peck, M. W. 1996. Vegetable juice aids the recovery of heated spores of non-proteolytic *Clostridium botulinum. Lett. Appl. Microbiol.* 23:407–411.

Suzuki, T., Yamauchi, K., Kawase, K., Tomita, M., Kiosawa, I., and Okonogi, S. 1989. Collaborative bacteriostatic activity of bovine lactoferrin with lysozyme against *Escherichia coli* O111. *Agric. Biol. Chem.* 53:1705–1706.

Taylor, T. M., Ravishankar, S., Bhargava, K., and Juneja, V. K. 2019. Chemical preservatives and natural food antimicrobials. In Doyle, M. P., Diez-Gonzalez, F., and Hill, C. eds. *Food Microbiology: Fundamentals and Frontiers*. 5th ed., Chap. 27. Washington, DC: ASM Press.

Teotia, J. S., and Miller, B. F. 1975. Destruction of salmonellae on poultry meat with lysozyme, EDTA, x-ray, microwave and chlorine. *Poult. Sci.* 54:1388–1395.

Tranter, H. S. 1994. Lysozyme, ovotransferrin and avidin. In: Dillon, V. M. and Board, R. G., eds. *Natural Antimicrobial Systems and Food Preservation.* Wallingford: CAB International. pp. 65–97.

Uchida, M., Yodomura, S., and Nagahama, G. 1972. Studies on lactobacilli isolated from mirin liquor. V: Antiseptic effects of egg white lysozyme on the growth of lactobacilli in mirin liquor. *J. Ferm. Technol.* 50:292–297.

United States Department of Agriculture Food Safety and Inspection Service (USDA-FSIS). 2019. Directive 7120.1 Rev. 52 Available from www.fsis.usda.gov.

United States Department of Agriculture – Food Safety and Inspection Service (USDA-FSIS). 2019. Directive 7120.1, Rev. 52: Safe and suitable ingredients used in the production of meat, poultry, and egg products. Accessed: April 15, 2020. Available at: https://www.fsis.usda.gov/wps/wcm/connect/bab10e09-aefa-483b-8be8-809a1f051d4c/7120.1.pdf?MOD=AJPERES

United States Food and Drug Administration (FDA). 1998. Direct food substances affirmed as GRAS: Egg white lysozyme. Tentative final rule. *Fed. Regist.* 63:12421–12426.

United States Food and Drug Administration (FDA). 2001. Agency response letter: GRAS notice No. GRN 000064. Washington, DC: Food and Drug Administration, Center for Food Safety and Applied Nutrition. Available from www.fda.gov.

Vakil, J. R., Chandan, R. C., Parry, R. M., and Shahani, K. M. 1969. Susceptibility of several microorganisms to milk lysozymes. *J. Dairy Sci.* 52:1192–1197.

van de Guchte, M., van der Wal, F. J., Kok, J., and Venema, G. 1992. Lysozyme expression in *Lactococcus lactis. Appl. Microbiol. Biotechnol.* 37:216–224.

Wang, C., and Shelef, L. A. 1991. Factors contributing to antilisterial effects of raw egg albumen. *J. Food Sci.* 56:1251–1254.

Wang, C., and Shelef, L. A. 1992. Behavior of *Listeria monocytogenes* and the spoilage microflora in fresh cod fish treated with lysozyme and EDTA. *Food Microbiol.* 9:207–213.

Wang, D., Lv, R. L., Ma, X. B., et al. 2018. Lysozyme immobilization on the calcium alginate film under sonication: Development of an antimicrobial film. *Food Hydrocoll.* 83:1–8.

Wasserfall, F. 1978. Basis of lysozyme action on the technology of semihard cheese. II: Influence of lysozyme on starter cultures and coliforms. *Milchwissenschaft* 33:91–96.

Wasserfall, F., and Prokopek, D. 1978. Trials on cheese ripening. VI: Prevention of late blowing of semi-hard cheese provoked by spores of *Clostridium tyrobutyricum* by means of lysozyme. *Milchwissenschaft* 33:288–291.

Wasserfall, F., and Teuber, M. 1979. Action of egg white lysozyme on *Clostridium tyrobutyricum. Appl. Environ. Microbiol.* 38:197–199.

Wasserfall, F., Voss, E., and Prokopek, D. 1976. Studies on cheese ripening. V: The use of lysozyme instead of nitrate to inhibit late fermentation of cheese. *Kiel Milchwirtsch Forschung* 28:3–16.

Weaver, G. L., and Kroger, M. 1978. Lysozyme activity of high-leucocyte-count milk and the effect of heat and potassium dichromate on lysozyme activity. *J. Dairy Sci.* 61:1089–1092.

Wilkinson, B. R., and Dorrington, R. E. 1975. Lysozyme (muramidase) from waste egg white. *Proc. Biochem.* 10:24–25.

Wooley, R. E., and Blue, J. L. 1974. In-vitro effect of EDTA-tris-lysozyme solutions on selected pathogenic bacteria. *J. Med. Microbiol.* 8:189–193.

Wu, H., Cao, D., Liu, T., Zhao, J., Hu, X., and Li, N. 2015. Purification and characterization of recombinant human lysozyme from eggs of transgenic chickens. *PLoS One* 10:e0146032.

Yamamoto, T., Juneja, L. R., Hatta, H., and Kim, M. (eds.). 1997. *Hen Eggs: Their Basic and Applied Science.* Boca Raton: CRC Press.

Yamauchi, K., Toimita, M., Giehl, T. J., and Ellison, R. T. 1993. Antibacterial activity of lactoferrin and a pepsin-derived lactoferrin peptide fragment. *Infect. Immun.* 61:719–728.

Yang, T. S., and Cunningham, F. E. 1993. Stability of egg white lysozyme in combination with other antimicrobial substances. *J. Food Prot.* 56:153–156.

Yashitake, S., and Shinichiro, A. 1977. Use of egg white lysozyme in the food industry. *New Food Ind.* 19:17–23.

Zhu, D., Cai, G., Li, X., Lu, J., and Zhang, L. 2017. Enhancing the antimicrobial activity of *Sus scrofa* lysozyme by N-terminal fusion of a sextuple unique homologous peptide. *J. Biotechnol.* 243:61–68.

15

Bacteriocins and Their Applications in Foods

Pushpinder Kaur Litt, Dallas G. Hoover and Haiqiang Chen

CONTENTS

15.1 Introduction

Interest in the development of bacteriocins as food preservatives and antimicrobial agents has robustly continued into the twenty-first century. A substantial portion of this effort has occurred in food technology laboratories working with bacteriocins produced by lactic acid bacteria (LAB). In the United States, the authorization by the U.S. Food and Drug Administration (FDA) in 1988 for the use of nisin as a food additive in pasteurized processed cheese spreads opened the door for the potential application of other bacteriocins as preservatives in foods. Covered earlier in this book (Chapter 8), nisin is an antibiotic-like, heat-stable peptide produced by *Lactococcus lactis* subsp. *lactis* that, among bacteriocins produced by LAB, is the most potent bacteriocin known. It is effective against an array of Gram-positive bacteria, most notably vegetative types and spores that are involved in foodborne illness and food spoilage. The effectiveness of nisin as a processing aid has been demonstrated in many products worldwide for over 50 years. An important contributory factor for the sustained interest in bacteriocins is the great concern that has grown in the food industry regarding *Listeria monocytogenes*, especially in American processed foods where there is zero tolerance for the pathogen. *L. monocytogenes* is a Gram-positive, psychrotrophic bacterium that is relatively common in raw foods and stubbornly resists normal means of control and elimination. Its human foodborne syndrome can sometimes be fatal, and the infective dose is highly variable among human subpopulations. The relevance is that nisin and other bacteriocins produced by some strains of LAB are specifically antagonistic against *L. monocytogenes*. Therefore, with the legal precedent set in the U.S. for use of nisin as an additive in foods, the potential exists for application of other bacteriocins as effective inhibitive agents against *L. monocytogenes* and other important Gram-positive pathogens to enhance product safety. Recently, the nano-engineered antimicrobial peptides of nisin have shown to be effective against Gram-negative bacteria (Vukomanović et al., 2017). The production of bacteriocins from lactic cultures potentially shortens the regulatory process because most lactic acid bacteria have generally regarded as safe (GRAS) status, having been consumed in large numbers in acidified fermented foods by people for thousands of years with no ill effects. In this regard, bacteriocins from LAB have been described as "natural" inhibitors. This chapter focuses upon review of bacteriocins from LAB exclusive of nisin. Using current taxonomy, LAB associated with foods include members of the genera *Lactococcus*, *Lactobacillus*, *Pediococcus*, *Leuconostoc*, *Carnobacterium*, *Enterococcus*, *Oenococcus*, *Streptococcus*, *Tetragenococcus*, *Vagococcus*, and *Weissella* (Stiles and Holzapfel, 1997). We have not attempted to inclusively present the vast volume of information that now exists in this area. We have primarily addressed bacteriocins from the first six LAB genera listed above with inclusion of some specific examples of work on bacteriocins from *Propionibacterium* and *Bifidobacterium*, given the importance of these two genera in foods.

15.2 Classification of Bacteriocins

The prototype bacteriocins were the colicins. First discovered by Gratia in 1925, "principe V" was produced by one strain of *Escherichia coli* against another culture of *E. coli*. The term "colicine" was coined by Gratia and Fredericq (1946); "bacteriocine" was used by Jacob et al. (1953) as a general term for highly specific antibacterial proteins. The term colicin now implies a bactericidal protein produced by varieties of *E. coli* and closely related Enterobacteriaceae (Cascales et al., 2007; Konisky, 1982).

The original meaning of the term bacteriocin was thus greatly influenced by characteristics common to colicins. That is, bacteriocins were originally defined as bactericidal proteins characterized by lethal biosynthesis, a very narrow range of activity, and adsorption to specific cell envelope receptors (Jacob et al., 1953). A later amendment was the association of bacteriocin biosynthesis with plasmids. The description has expanded to recognize the differences between colicins and bacteriocins produced by Gram-positive bacteria (Yang et al., 2014). Bacteriocins from Gram-positive bacteria usually do not possess a specific receptor for adsorption, are commonly of lower molecular weight than colicins, can have different modes of killing, a wider range of effect, different modes of release and cell transport, and possess leader sequences cleaved during maturation (Reeves, 2012). Today, bactericidal peptides or proteins produced by bacteria are labeled as bacteriocins. Normally, to demonstrate the proteinaceous nature of a newly characterized bacteriocin, sensitivity to proteolytic enzymes such as trypsin, α-chymotrypsin, and pepsin is an expected demonstration. Usage in food preservation usually requires estimation of its heat resistance given the widespread use of thermal processing in food production.

Bacteriocins are widely found throughout the bacterial world. These compounds are assumed to provide producing bacterial cells with a selective advantage over other bacteria. Synthesis of bacteriocins is found across all major groups of eubacteria and archaebacteria (Riley, 1998). There have been many valuable monographs written over the years that review colicins, bacteriocins, bacteriocins from LAB, and applications of specific bacteriocins (Yang et al., 2014; Chikindas et al., 2018; Field et al., 2018; Juturu and Wu, 2018; Alvarez- Sieiro et al., 2016).

Normally, bacteriocins from LAB are cationic, and hydrophobic or amphiphilic molecules composed of 20 to 60 amino acid residues (Nes and Holo, 2000). Bacteriocins produced by Gram-positive bacteria are commonly classified into three classes (Rea et al., 2011; Cintas et al., 2001; Klaenhammer, 1993).

Class I contains bacteriocins termed lantibiotics (from *lan*thionine-containing an*tibiotic*). They are small (<5 kDa) peptides containing the unusual amino acids lanthionine (Lan), α-methyllanthionine (MeLan), dehydroalanine, and dehydrobutyrine. Class I is further subdivided into Type A and Type

B lantibiotics based on their chemical structures and their antimicrobial activities (Cooper, Li, and van der Donk, 2010; Guder et al., 2000). Type A lantibiotics are elongated peptides with a net positive charge. They exert their activity through the formation of pores in bacterial membranes. Some examples of type A lantibiotics are nisin, lactocin S produced by *Lactobacillus sake*, epidermin produced by *Staphylococcus epidermidis* and *Streptococcus mutans*, gallidermin from *S. epidermidis*, and lacticin 481 from *L. lactis* (Gharsallaoui et al., 2016; Draper et al., 2015; Kuipers, Rink, and Moll, 2011). Type B lantibiotics are smaller globular peptides, which have a negative or no net charge. Their antimicrobial activity is related to the inhibition of specific enzymes. Some examples are mersacidin from *Bacillus* ssp., cinnamycin from *Streptomyces cinnamoneus*, ancovenin and duramycin from *Streptomyces* and *Streptoverticillium* sp., and actagardine from *Actinoplanes liguriae* (Repka et al., 2017).

Class II bacteriocins are the largest group that have been characterized. They contain small (<10 kDa) heat-stable, non-lanthionine containing, unmodified peptides. They can be further divided into three subgroups. Class IIa includes pediocin-like peptides having an N-terminal consensus sequence -Tyr-Gly-Asn-Gly-Val-Xaa-Cys. They have attracted much of the attention due to their anti-*Listeria* activity (Cui et al., 2012; Ennahar et al., 2000). Some examples are pediocin PA-1/AcH, sakacins A and P, leucocin A-UAL 187, mesentericin Y105, enterocin A, divercin V41, and lactococcin MMFII (Cui et al., 2012). Class IIb contains bacteriocins requiring two different peptides for activity. Some examples are lactococcins G and M, lactacin F, plantaricins A, S, EF, and JK (Cui et al., 2012). Class IIc contains the remaining peptides of the class including *sec*-dependent secreted bacteriocins. Some examples are acidocin B, carnobacteriocin A (Worobo et al., 1994), divergicin A, and enterocins P and B (Cui et al., 2012; Cotter et al., 2005).

Class III bacteriocins are large (>30 kDa) heat-labile proteins that are not as well characterized, e.g., helveticins J and V-1829 (Gálvez et al., 2007). A fourth class consisting of complex bacteriocins that require carbohydrate or lipid moieties for activity has also been suggested by Klaenhammer (1993); however, bacteriocins in this class have not been characterized adequately at the biochemical level such that the definition of this class still requires additional information (Preciado et al., 2016; Veskovic-Moracanin et al., 2014; McAuliffe et al., 2001; Jimenez-Diaz et al., 1995).

15.3 Genetics, Biosynthesis, and Mode of Action

15.3.1 Organization of Gene Clusters

The genetic determinants involved in the production of class I and II bacteriocins are by far the most studied due to the success of nisin and the potential applications of the bacteriocins in these two classes. Bacteriocins are ribosomally synthesized and the genes encoding bacteriocin production and immunity are usually organized in operon clusters (Singh et al., 2015; Cooper, Li, and van der Donk, 2010; Field et al., 2007; McAuliffe et al., 2001; Sahl and Bierbaum, 1998). For linear unmodified bacteriocins, which include the plantaricins, carnobacteriocins, and sakacins, it appears that specific inducing peptides or peptide pheromones stimulate the synthesis of bacteriocins whose genes are usually located nearby on the same gene cluster. Bacteriocin gene clusters can be located on the chromosome, as in the case of subtilin and mersacidin, or plasmids, as in the case of divergicin A and sakacin A, or transposons, as in the case of nisin and lacticin 481 (McAuliffe et al., 2001; Dufour et al., 2000).

The genetic analysis of the biosynthesis of several lantibiotics, epidermin, nisin, subtilin, lacticin 481, and mersacidin (Kuipers et al., 2011; Field et al., 2007; Altena et al., 2000), has revealed that the genes involved in biosynthesis of lantibiotics generally include genes coding for the prepeptide (LanA – the abbreviation lan refers to homologous genes of different lantibiotic gene clusters), enzymes responsible for modification reactions (LanB,C/LanM), processing proteases responsible for removal of the leader peptide (LanP), ABC (ATP-binding cassette), superfamily transport proteins involved in peptide translocation (LanT), regulatory proteins (LanR, K), and proteins involved in producer self-protection (immunity) (LanI, FEG).

The genetics of many class II bacteriocins, such as lactococcins A, B, and M, pediocin PA-1/AcH, and plantaricin A, have been studied (Belguesmia et al., 2011; Nissen-Meyer et al., 2010; Deegan et al.,

2006; Field et al., 2018). Genes encoding the biosynthesis of class II bacteriocins share many similarities in their genetic organizations, consisting of a structural gene coding for precursor peptide, followed immediately by a dedicated immunity gene and genes for a dedicated ABC-transporter and an accessory protein. In some cases, regulatory genes have been found. The accessory proteins are essential for the export of class II bacteriocins. No counterparts of these accessory proteins in lantibiotics have been found (Sablon et al., 2000). For more elaborate reviews on the genetics of lantibiotics and nonlantibiotics see Field et al., (2018), Juturu and Wu (2018), Kuipers et al., (2011), Field et al. (2007), Sablon et al. (2000), and McAuliffe et al. (2001).

15.3.2 Biosynthetic Pathway

All bacteriocins are synthesized as a biologically inactive prepeptide carrying an N-terminal leader peptide that is attached to the C-terminal propeptide. For lantibiotics, the serine, threonine, and cysteine residues in their propeptide parts undergo extensive post-translational modification to form Lan/MeLan. The biosynthetic pathway of lantibiotics follows a general scheme (as shown in Figure 15.1) formation of prepeptide, modification reactions, proteolytic cleavage of the leader peptide, and the translocation of the modified prepeptide or mature propeptide across the cytoplasmic membrane. The cleavage of the leader peptide may take place prior to, during, or after export from the cell. Based on the biosynthetic pathway, two groups of genetic organization of lantibiotics can be identified (Sahl and Bierbaum, 1998; McAuliffe et al., 2001). In the producers of the group I lantibiotics, as in the case of nisin, epidermin, subtilin, and Pep5, the dehydration reaction is presumably catalyzed by the LanB enzyme, while LanC is involved in the thioether formation. The modified prepeptide is processed by a serine protease LanP and translocated through the ABC-transporter LanT. In contrast, lantibiotics of group II, as in the case of cytolysin, lacticin 481, and mersacidin, are very likely modified by a single LanM enzyme (McAuliffe et al., 2001) and processing takes place concomitantly with transport by LanT(P). Lactocin S is the exception to this classification. It is modified by a single LanM enzyme and processing takes place prior to export, and lactocin S may therefore represent a new group (Skaugen et al., 1997).

With the exception of class IIc bacteriocins, which are produced with a typical N-terminal signal sequence of the *sec*-type and processed and secreted through the general secretory pathway (Leer et al.,

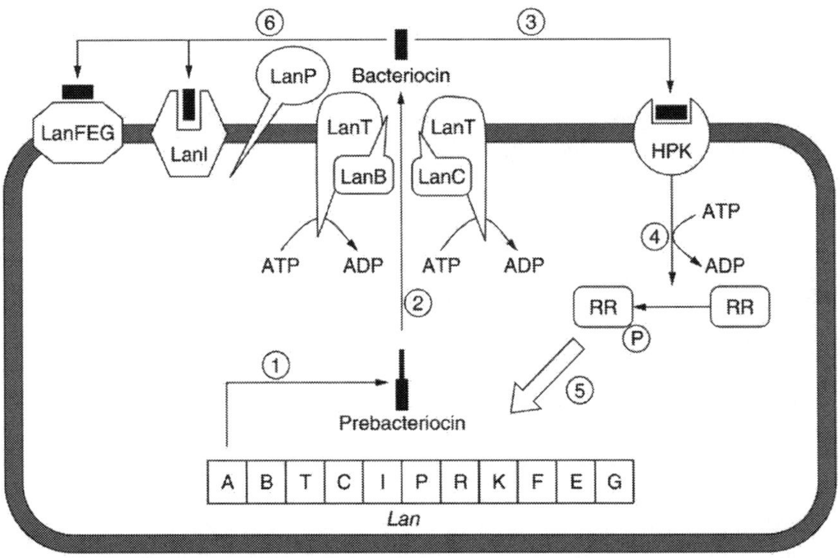

FIGURE 15.1 A schematic diagram of the biosynthesis of lantibiotics. 1: Formation of prebacteriocin. 2: The prebacteriocin is modified by LanB and LanC, translocated through a dedicated ABC-transporter LanT, and processed by LanP, resulting in the release of mature bacteriocin. 3: Histidine protein kinase (HPK) senses the presence of bacteriocin and autophosphorylates. 4: The phosphoryl group (P) is subsequently transferred to the response regulator (RR). 5: RR activates transcription of the regulated genes. 6: Producer immunity.

1995; Worobo et al., 1995), class II bacteriocins are synthesized as a prepeptide containing a conserved N-terminal leader and a characteristic double-glycine proteolytic processing site. Unlike the lantibiotics, they do not undergo extensive post-translational modification. Following the formation of prepeptide, the prepeptide is processed to remove the leader peptide concomitant with export from the cell through a dedicated ABC-transporter and its accessory protein (Nes et al., 1996; Ennahar et al., 2000). The biosynthetic pathway of class II bacteriocins is shown in Figure 15.2.

Several functions of the leader peptides have been proposed. They may serve as a recognition site which directs the prepeptide towards the maturation and transport proteins, protect the producer strain by keeping the lantibiotic in an inactive state while it is inside the producer, and interact with the propeptide domain to ensure a suitable conformation essential for enzyme–substrate interaction (Repka et al., 2017; Nissen-Meyer et al., 2010; McAuliffe et al., 2001).

15.3.3 Post-Translational Modification, Activation, and Transport

A two-step post-translational modification reaction of a pre-lantibiotic leading to formation of Lan/MeLan was first proposed by Ingram (1969, 1970). In the first step, the hydroxyl amino acids serine and threonine are dehydrated to yield 2,3-didehydroalanine or 2,3-didehydrobutyrine, respectively. Some dehydrated amino acids do not contain cysteine residues and remain as such in the mature peptide; others undergo an intramolecular Michael addition reaction that involves the thiol groups of neighboring cysteine residues and the double bonds of the didehydroamino acids, resulting in the formation of thioether bridges.

Following the modification reactions, the modified pre-lantibiotics undergo proteolytic processing to release the leader peptide that leads to activation of the lantibiotic. For group I lantibiotics, the removal of the leader peptide is performed by a serine protease LanP, and depending on the location of LanP this can take place before or after the peptide is exported from the producing cell via a dedicated ABC-transporter LanT. For example, the proteases LanP of epicidin 280 and Pep5 (Sahl and Bierbaum, 1998) are located intracellularly and proteolytic processing takes place within the cell. In contrast, the proteases of nisin (van der Meer et al., 1993) and epidermin (Geissler et al., 1996), which are located extracellularly, activate the lantibiotics only after export by the ABC transporter. The ABC-transporter contains 500–600

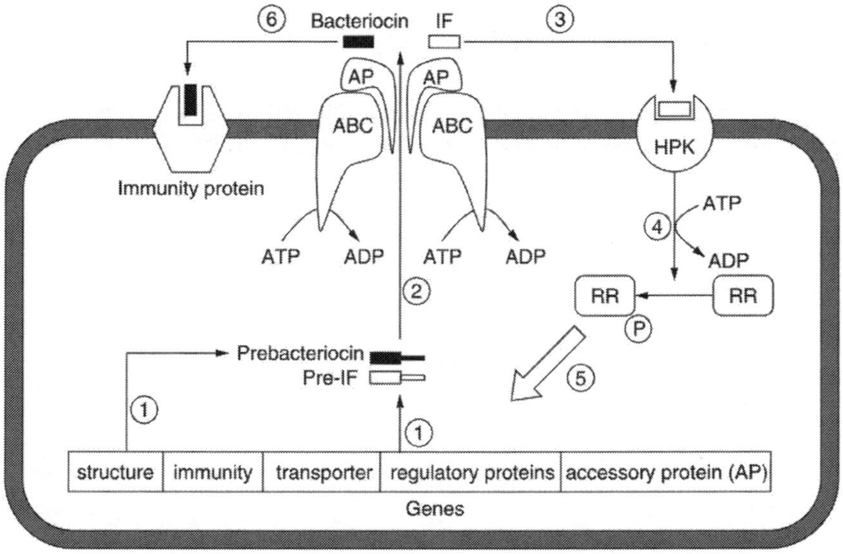

FIGURE 15.2 A schematic diagram of the biosynthesis of class II bacteriocins. 1: Formation of prebacteriocin and prepeptide of induction factor (IF). 2: The prebacteriocin and pre-IF are processed and translocated by the ABC-transporter, resulting in the release of mature bacteriocin and IF. 3: Histidine protein kinase (HPK) senses the presence of IF and autophosphorylates. 4: The phosphoryl group (P) is subsequently transferred to the response regulator (RR). 5: RR activates transcription of the regulated genes. 6: Producer immunity.

amino acids and is characterized by two membrane-associated domains. The N-terminal domain consists of six membrane-spanning helices that can recognize the substrate and form its pathway across the membrane, while the cytoplasmic C-terminal domain contains two ATP-binding domains with the conserved ATP-binding or Walker motif. ATP hydrolysis, which likely occurs at the ATP-binding domains, provides the energy for the export process (Fath and Kolter, 1993; McAuliffe et al., 2001). The LanB and LanC enzymes, together with LanT transporter, probably form a multimeric membrane-associated complex (Siegers et al., 1996; Kiesau et al., 1997). For group II lantibiotics, which possess a conserved double-glycine cleavage site, proteolytic processing takes place concomitantly with export through a hybrid ABC transporter. This unique ABC transporter possesses an N-terminal protease domain of approximately 150 amino acid residues that cleaves the double-glycine leader (Nes et al., 1996; Sablon et al., 2000). This is exemplified in Figure 15.3.

The leader peptides of class IIa and b share considerable similarities with those of group II lantibiotics and contain the characteristic double-glycine cleavage site (Nes et al., 1996; Ennahar et al., 2000). The conservation of the cleavage site strongly suggests that the mechanism of processing and translocation of class IIa and b bacteriocins is very similar to that of the group II lantibiotics. Class IIc bacteriocins are processed by a signal peptidase during translocation across the cytoplasmic membrane.

15.3.4 Regulation of Biosynthesis

Regulation of the biosynthesis of lantibiotic and nonlantibiotics is usually achieved through well-known two-component regulatory systems. These regulatory systems consist of two signal-producing proteins, a membrane-bound histidine protein kinase (HPK), and a cytoplasmic response regulator (RR) (Stock et al., 1989; Parkinson, 1993; Nes et al., 1996). In this signal transduction pathway, HPK autophosphorylates the conserved histidine residue in its intracellular domain when it senses a certain concentration of bacteriocin in the environment. The phosphoryl group is subsequently transferred to the conserved aspartic acid residue on the RR receiver domain, and the resulting intramolecular change triggers the response regulator to activate the transcription of the regulated genes. These regulated genes include the structural gene, the export genes, the immunity genes, and in some cases the regulatory genes themselves (Kuipers et al., 2011).

The bacteriocin molecule itself may act as an external signal, as in the case of nisin and subtilin, autoregulating its own biosynthesis via signal transduction (Guder et al., 2000; Kuipers et al., 1995). In contrast, most class II bacteriocins produce a bacteriocin-like peptide with no antimicrobial activity and use it as an induction factor (IF) to activate the transcription of the regulated genes. The IF is a small, heat-stable, cationic, and hydrophobic peptide that is first synthesized as prepeptide with a double-glycine leader sequence. A dedicated ABC transporter specifically cleaves the leader peptide of IF concomitant with export of the mature peptide from the cell. The secreted IP acts as an external signal that triggers the transcriptions of the genes involved in bacteriocin production as shown in Figure 15.2 (Repka et al., 2017; Ennahar et al., 2000; Nes et al., 1996).

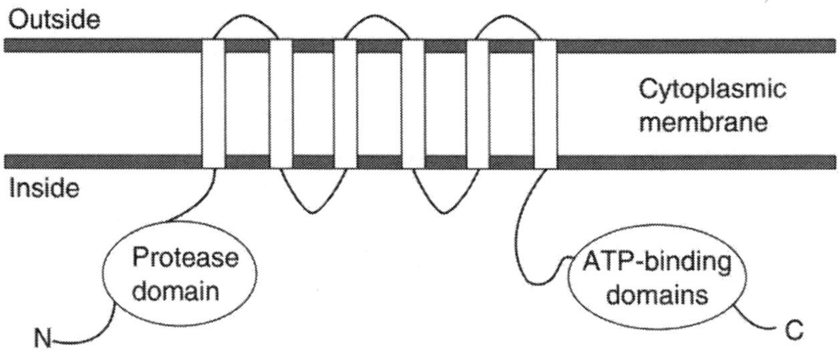

FIGURE 15.3 ABC-transporter with an N-terminal protease domain.

15.3.5 Producer Immunity

The antimicrobial activity of bacteriocin forces the producers to develop a protection system against their own bacteriocin (Abee et al., 1995). Two systems of lantibiotic immunity have been identified. Protection can be mediated by immunity proteins, LanI, and dedicated ABC transport proteins, LanFEG, which can be encoded on multiple open reading frames (Clemens et al., 2018;Field et al., 2007; McAuliffe et al., 2001). These two immunity systems work synergistically to protect the producing cells from their own bacteriocin (Klein and Entian, 1994). LanI, which is most likely attached to the outside of the cytoplasmic membrane, probably confers immunity to the producer cells by preventing pore formation by the bacteriocin of the producer. The LanFEG could act by transporting bacteriocin molecules that have inserted into the membrane back to the surrounding medium, thus keeping the concentration of the bacteriocin in the membrane under a critical level.

The immunity gene in class II bacteriocins usually codes a dedicated protein that is loosely interacted with the cytoplasmic membrane. The immunity protein, which is cationic and ranges in size from 51 to 254 amino acids, provides total immunity against the bacteriocin of the producer. It is speculated that the interaction of the immunity protein with the membrane protects the producer against its bacteriocin (Field et al., 2018; Cui et al., 2012; Nissen-Meyer et al., 2010).

15.3.6 Mode of Action

Most bacteriocins of lactic acid bacteria interact with anionic lipids that predominate in the cytoplasmic membrane of Gram-positive bacteria, leading to pore formation (Abee et al., 1995; Moll et al., 1999). Conductivity and stability of pores induced by class I lantibiotics may be heightened by docking molecules (lipid II, the peptidoglycan precursor), while in the case of class II bacteriocins receptors in the target membrane apparently act to determine specificity (Venema et al., 1995a, b). Class I bacteriocins may induce pore formation according to a wedge-like model, and class II bacteriocins may function by creating barrel stave-like pores or by creating a carpet mechanism whereby peptides orient parallel to the membrane surface and interfere with membrane structure (Moll et al., 1999).

Nisin is likened to a surface-active cationic detergent, with adsorption to the bacterial cell envelope the necessary first step of membrane disruption, followed by the inactivation of sulfhydryl groups. Bruno et al. (1992) showed nisin to completely dissipate the basal proton-motive force of *L. monocytogenes* Scott A. Other strains of *L. monocytogenes* were found to be equally sensitive to nisin and exhibit similar disruption to the pH gradient and membrane potential of the cells. Other class I bacteriocins of lactic acid bacteria behave in a similar manner. Carnocin UI49 acts upon the cytoplasmic membrane in a manner very similar to nisin (Stoffels et al., 1994). Ion permeability or channel formation in cytoplasmic membranes caused by bacteriocins of *L. acidophilus* was demonstrated by Palmeri et al. (1999) using artificial lipid membranes. Increases in membrane conductance were detected, and channels with elementary conductance of 68–70 pS were measured by applying external voltages of different polarity.

In earlier research, Upreti and Hinsdill (1975) found lactocin 27, produced by *L. helveticus* LP27 and described as a bacteriocin whose effect was bacteriostatic, adsorbed equally well to sensitive and resistant cells terminating protein synthesis; however, there were substantial effects on DNA synthesis, RNA synthesis, or levels of ATP. With the lactostrepcin, Las 5, Zajdel et al. (1985) found the bacteriocin to immediately blocks syntheses of DNA, RNA, and protein, but these responses are probably a secondary reaction to severe membrane disruption and loss of intracellular constituents. For Las 5, both sensitive and insensitive cells adsorbed the bacteriocin equally well. Protoplasts of sensitive strains were not affected by Las 5, indicating that cell wall receptors were required for the action of the bacteriocin.

In 1986, Andersson demonstrated resistant Gram-negative cells of *E. coli*, *Erwinia carotovora*, *P. aeruginosa*, and *S. marcescens* could be made sensitive to the bacteriocin of a strain of *L. plantarum* by transforming the Gram-negative cells to spheroplasts (Andersson, 1986). It is now established that inactivation of Gram-negative pathogens with bacteriocins produced from Gram-positive bacteria normally requires the use of chelating agents (i.e., EDTA) that function to diminish the barrier properties provided by the outer LPS membrane of Gram-negative bacteria (Prudêncio et al., 2015; Helander et al., 1997; Scanell et al., 1997). Additional stress to Gram-negative bacteria as in a hurdle approach also can potentially enhance the effectiveness of a bacteriocin. Ganzle et al. (1999a) demonstrated that curvacin A and nisin in combination with low pH, greater than 5% NaCl, or propyl paraben also leads to an increased sensitivity of *Salmonella enterica* and *E. coli* towards these bacteriocins. So it would appear that combination packages of bacteriocins, processes, and other additives can significantly broaden the target spectrum of some LAB bacteriocins into the Gram-negative segment.

15.4 Production and Modeling

To produce bacteriocins on a commercial basis, optimization of yield is desirable. In a study of nisin, pediocin AcH from *P. acidilactici*, leuconocin Lcm1 by *Leuconostoc carnosum*, and sakacin A by *L. sake* Lb 706, Yang and Ray (1994) found that in general growth medium the key factors were maintaining optimum pH and supplementing specific nutrients for each strain or species. Essentially, the conditions favored high cell density and resulted in a comparable bacteriocin yield, indicative of a primary metabolite. To further enhance yield and purity of LAB bacteriocins from industrial fermentations and lower production costs, Coventry et al. (1996) demonstrated that nisin, pediocin PO2, brevicin 286, and piscicolin 126 could be significantly free of contaminating proteins (92–99%) as compared to original culture supernatants and ammonium sulfate preparations by using food-grade diatomite calcium silicate and several desorbing agents.

An important part of the successful development for bacteriocin use rests on the infusion of biotechnological techniques, such as strain improvement and protein engineering, as well as optimization of production methods to lower costs. Following this path, Carolissen-Mackay et al. (1997) reviewed protocols for the purification of bacteriocins from lactic acid bacteria. In order to lower production costs, improvements in the speed and yield of purification methods for bacteriocins from culture broth continue (Guyonnet et al., 2000).

To optimize the application of bacteriocins, models have been devised to predict the commercial effectiveness of bacteriocins based on information gathered in test systems. Using traditional agar plates, Blom et al. (1997) designed a method for simultaneous examination of the effect of intrinsic factors (pH, concentration of indicator cells, agar, soy oil, and sodium chloride concentration) on the diffusion and efficacy of bacteriocins. They tested sakacins A and P, piscicolin 61, pediocin PA-1, and nisin, and found that each of the bacteriocins had a unique intrinsic factor effect profile, although pH and the concentration of indicator cells similarly affected all of the bacteriocins. Except for use of a commercial preparation of nisin, the other bacteriocins were produced using *L. sake* Lb790, modified to produce each of the remaining bacteriocins. Parente et al. (1998) used logistical regression analysis of categorial data to generate predictive models for the probability of survival for *L. monocytogenes* versus leucocin F10 and nisin in TSB with 0.6% yeast extract. Nisin had a significant effect on the probability of survival, but supplementation with leucocin F10 was necessary to eliminate *L. monocytogenes*. Lower pH values in the broth decreased survival, while NaCl and EDTA demonstrated a limited effect. In a somewhat different approach, Pleasants et al. (2001) modeled the growth of *L. monocytogenes* L70 in response to *L. sake* 706, a producer of sakacin A. In MRS broth it was found that the *L. sake* strain grew much faster than *L. monocytogenes* L70 at reduced pH. Excretion of sakacin A by its producer rapidly reduced the presence of *L. monocytogenes* in mixed culture. It was deemed that the model could accurately predict growth and interaction of the two bacteria in the test medium. Improvements in the production of bacteriocins continue to be addressed for potential commercialization (Guerra, 2012).

15.5 Resistance

As has been well-established in microbial ecology, the presence of an antimicrobial substance will eventually select for resistant varieties of bacteria in a given environment. As with therapeutic antibiotics, bacteriocin-resistant mutants have arisen. Gravesen et al. (2002) examined the responses of a number of strains of *L. monocytogenes* to pediocin PA-1 and nisin, and found a wide range of resistances to the two bacteriocins occurring naturally. The influence of environmental stress (reduced pH, low temperature, and the presence of sodium chloride) was apparently bacteriocin-specific. That is, these stresses did not influence the frequency of resistance to pediocin PA-1, but the frequency of nisin resistance was significantly reduced. Also in contrast, the stability of the phenotype of nisin resistance varied substantially, while resistance to the pediocin was stable with ongoing growth of *L. monocytogenes*. Fitness costs, as measured by reduced growth rate in pediocin-resistant mutants, were demonstrable but nisin-resistant mutants showed limited growth rate reductions. Lastly, it was found that the bacteriocin-resistant mutants of *L. monocytogenes* were not more sensitive to the applied environmental stresses than wild-type strains; in a model sausage system growth differences were minimal.

Horn et al. (1998, 1999) moved the genes of pediocin PA-1 into nisin-producing *L. lactis* FI5876. As a result both pediocin PA-1 and nisin A were synthesized concurrently in the strain. Production levels of pediocin PA-1 in *L. lactis* FI5876 were at the same levels as found in the parent, *P. acidilactici* 347. Since pediocin PA-1 and nisin A are unrelated bacteriocins, the use of *L. lactis* FI5876 as a starter culture in fermented dairy products should prevent the emergence of any bacteriocin-resistant isolates of *L. monocytogenes* since the frequency for emergence of a strain resistant to both peptides should be low.

15.6 Examples of Activity Spectra and Biochemical Properties

15.6.1 *Lactococcus*

The inhibition of commercial cheese starter cultures by similar dairy bacteria has been known since the early 1900s (Rogers and Whittier, 1928; Whitehead and Riddet, 1933). It was demonstrated that these effects were caused by inhibitory proteins (Mattick and Hirsch, 1947; Whitehead, 1933). At that time, the two best-characterized proteins causing this phenomenon in lactococci were nisin and diplococcin. Nisin is the recognized standard for bacteriocins that other potential applicants for commercialization hope to follow. On the other hand, diplococcin (produced by some strains of *Lactococcus lactis* subsp. *cremoris*) lacks the effectiveness of nisin because of extreme instability when purified, an activity spectrum limited to other cultures of *L. lactis* (diplococcin is ineffective against sporeformers), and a relatively high degree of sensitivity to trypsin, pronase, and α-chymotrypsin (Oxford, 1944; Davey, 1981; Davey and Pearce, 1982). Diplococcin has a higher molecular weight than nisin, 5300 versus 3500 daltons (Davey and Richardson, 1981; Gross and Morell, 1967), and lacks the unusual sulfur-containing amino acids (lanthionine and α-methyllanthionine) that nisin possesses (Davey and Pearce, 1980; Gross et al., 1969). Diplococcin also does not contain dehydroalanine or dehydrobutyrine (Davey and Pearce, 1982). In many ways, nisin has greater similarities to subtilin, a bacteriocin produced by *Bacillus subtilis*, than to diplococcin.

Apparently the majority of *L. lactis* subsp. *cremoris* strains do not synthesize diplococcin, as indicated in a study by Davey and Richardson (1981) in which only 1 of 150 strains was positive for the bacteriocin. Nonetheless, the capability of a culture to produce diplococcin is a significant concern to cheesemakers because diplococcin-positive strains rapidly predominate over other lactococci used in multiple-strain starter cultures (Davey and Pearce, 1980; Klaenhammer, 1988). Diplococcin-negative variants are desirable in such a situation; Davey and Pearce (1980) developed diplococcin-negative strains of *Streptococcus cremoris* 346 (now *L. lactis* subsp. *cremoris* 346) whose fermentative performance for cheesemaking was unaffected by loss of this unwanted characteristic. As with most other bacteriocins, diplococcin is normally associated with plasmid DNA. Davey (1984) found diplococcin biosynthesis mediated by a 54-megadalton (MD) conjugative plasmid in *S. cremoris* 346.

Lactostrepcins are bacteriocins produced by non-nisin-producing *L. lactis* subsp. *lactis* strains, almost all *Lactococcus lactis* subsp. *diacetylactis* strains, and some *L. lactis* subsp. *cremoris* strains (Kozak et al., 1978). The lactostrepcins function optimally only at acid pH (<5.0) and have a molecular weight of 10,000 or more. Although heat-stable (121°C for 10 min), they are especially susceptible to most proteolytic enzymes, as well as phospholipases A and C and lipases (Bardowski et al., 1979; Dobrzanski et al., 1982). According to the system of Bradley (1967), they were first categorized as group I bacteriocins.

Five types of lactostrepcins have been differentiated, primarily based upon the range of bactericidal activity (Dobrzanski et al., 1982). Lactostrepcins are stabilized by enriched, complex growth media, such as milk, and have shown antagonism not only against lactococci but streptococci of groups A, C, and G, *Bacillus cereus*, and a few varieties of *Leuconostoc* and *Lactobacillus* (Zajdel and Dobrzanski, 1983).

Probably the most studied of the lactostrepcins has been lactrostrepcin 5 (Las 5) as produced by *S. cremoris* 202. Like diplococcin, Las 5 contains no unusual amino acids, nor a lipid component like other lactostrepcins. Its molecular weight has been estimated as 20,000 by Laemli and Favre (1973) and 6000 by Davey and Richardson (1981). Dobrzanski et al. (1982) searched for association of Las 5 synthesis with one of the plasmids of *S. cremoris* 202 but were unable to establish linkage. A similar attempt was made to link plasmids of *Streptococcus lactis* 71 with its bacteriocins, lactostrepcins 3 and 4, but the results were also inconclusive.

Bacteriocin typing with lactostrepcins has been investigated (Bardowski and Kozak, 1981). Lactostrepcin typing of lactococci appears possible since many of the strains are sensitive to at least one lactostrepcin. Seven different lactostrepcin types of lactococci can be determined, although lactostrepcin sensitivity does not correspond to taxonomical classification of these bacteria. Of potential benefit would be the correlation of industrially important lactococci with lactostrepcin type (Dobrzanski et al., 1982).

Lactococci are capable of producing other bacteriocins besides nisin, diplococcin, and the lactostrepcins. Geis et al. (1983) conducted a survey of bacteriocin activity encompassing 280 commercial isolates of lactococci. In comparing antagonism between broth extracts and direct agar plating, they found 14 positives in broth and 5 positives on agar for 54 strains of *S. lactis*, 36 positives in broth and 1 positive on agar for 93 cultures of *Streptococcus lactis* subsp. *diacetylactis*, and 15 positives in broth and 10 positives on agar for 133 strains of *S. cremoris*. Clearly there was a difference in expression of inhibiting activity between liquid and solid systems. Geis et al. (1983) defined eight bacteriocin types based on sensitivity and immunity to other bacteriocin-positive cultures. Although some of the antagonistic reactions very closely resembled nisin or diplococcin, others did not; none were characteristic of the lactostrepcins. One group of non-nisin-producing strains was inhibitive versus species of *Clostridium*.

Lacticin 3147, a class II bacteriocin, is produced by a lactococcal strain originally isolated from kefir; it is a broad-spectrum bacteriocin consisting of two individual peptides genetically residing on a self-transmissable plasmid that can be easily moved to other strains of lactococci used as cheese starter cultures (Ross et al., 2000). In cottage cheese, a 3-log reduction in CFU/g of *L. monocytogenes* was consistently shown using a lacticin 3147-producing culture of *L. lactis* subsp. *lactis*, while a control non-lacticin 3147-producing strain demonstrated no effect on *L. monocytogenes* (McAuliffe et al., 1999). On the surface of mold-ripened cheese a lacticin 3147-producing *L. lactis* subsp. *lactis* again reduced *L. monocytogenes* by 3 log CFU/g, but this time as compared to a nisin-producing *Lactococcus* (Ross et al., 2000). Lacticin 3147 acts by selectively dissipating the membrane potential (McAuliffe et al., 1998).

Another antibacterial peptide produced by lactococci is bacteriocin S50. It has a relatively narrow spectrum of activity. Studied by Kojic et al. (1991), this bacteriocin is produced by *L. lactis* subsp. *diacetylactis* S50. It is active only against other *Lactococcus* species and is heat-stable (100°C for 60 min) and active in the pH range 2–11.

Not all investigations of antimicrobial action by dairy streptococci have involved the determination of specific bacteriocin effect. Pulusani et al. (1979) examined the activity of skim milks fermented by *Streptococcus thermophilus* as well as *Lactobacillus bulgaricus* and *Lactobacillus acidophilus*. There was strong inhibition of *Pseudomonas fluorescens*, *Pseudomonas aeruginosa*, and *B. subtilis*, but bacteriocin activity could not be inferred. Reddy and Ranganathan (1983) measured the presence of antimicrobial substances in the cell-free filtrates from 14 strains of *S. lactis* subsp. *diacetylactis*. Subculturing resulted in mutant derivatives with inhibitory action that exceeded the parent activity. Greater inhibition was shown against Gram-negative bacteria than Gram-positive bacteria, suggesting the action of organic acids. With the exception of *Geotrichum candidum*, molds were not affected.

In *Lactococcus*, bacteriocin production has been used as a marker in plasmid transfer studies and in other genetic procedures. Neve et al. (1984) characterized bacteriocin plasmids from strains of *S. lactis*, *S. lactis* subsp. *diacetylactis*, and *S. cremoris*. Conjugal transfers were done into a plasmid-free recipient strain of *S. lactis* subsp. *diacetylactis*. The activity spectra of these bacteriocins included strains of lactococci, lactobacilli, pediococci, *Streptococcus faecalis*, *Streptococcus sanguis*, *Leuconostoc* spp., *B. subtilis*, and clostridia.

A bacteriocin produced by *S. lactis* subsp. *diacetylactis* WM_4, has been studied by Scherwitz et al. (1983). Bacteriocin production was conjugally transferred into *S. lactis* and found associated with an 88-MD plasmid. The bacteriocin from *S. lactis* subsp. *diacetylactis* WM_4 was active against a variety of lactococci. Cloning of restriction enzyme fragments of the 88-MD plasmid into a recombinant plasmid expressed in *S. lactis* resulted in greater inhibition versus the indicator strain of *S. diacetylactis*; this is an excellent example of enhanced production of a bacteriocin in the development of commercial dairy strains (Scherwitz-Harmon and McKay, 1987). Further work with WM_4 (Stoddard et al., 1992) involved the characterization of the lactococcin A gene cluster by DNA restriction fragment deletion, subcloning, and nucleotide sequence analysis. This molecular analysis suggests the existence of a signal sequence-independent secretory pathway in the lactococci that is similar to the secretion and processing system in *B. subtilis*.

van Belkum et al. (1992) cloned two bacteriocin genes from a *L. lactis* subsp. *cremoris* plasmid that specified both bacteriocin production and immunity. This culture, 9B4, was originally described by Geis et al. (1983). Two regions on the plasmid were identified that had different specified inhibitory activities toward *L. lactis* indicator strains; however, hybridization experiments revealed a high level of homology between the two bacteriocin gene regions. The authors noted the desirability for the isolation of bacteriocin production and immunity genes for use as food-grade vectors that are selectively retained within the plasmid-bearing population.

Further study of the two bacteriocin operons of the plasmid from *L. lactis* subsp. *cremoris* 9B4 by van Belkum et al. (1991) resulted in the sequencing and analysis by deletion and frameshift mutation of the regions responsible for bacteriocin production and immunity. It was found that the nucleotide sequences of both fragments upstream of the first detectable open reading frames and the first 20 base pairs of the first open reading frame of both bacteriocin operons appeared identical. From this finding, movement of the operons by transposition-like events was suggested because each fragment encoding for bacteriocin synthesis was flanked by similar nucleotide sequences.

As studies continue, the genus *Lactococcus* continues to supply bacteriocins active against *L. monocytogenes* and other Gram-positive pathogens. Isolated from raw pork, lacticin FS92 is another example; it contains 32 amino acids, is heat-stable, and as a food preservative appears desirable as resistance to it from strains of *L. monocytogenes* is low (Mao et al., 2001). Unlike pediocin PA-1, curvaticin FS47, and lacticin FS56, resistant mutants of *L. monocytogenes* remained sensitive to lacticin FS92. However, strain-to-strain variations are great whether considering resistance of target strains or effectiveness of bacteriocins. For example, Castellano et al. (2001) found susceptibility of strains of *L. monocytogenes*, *L. innocua*, and *L. seeligeri* to be strain-dependent at each pH that was examined in response to lactocin 705, enterocin CRL35, and nisin.

In a study using MRS agar to isolate lactic bacteria isolated from raw milks (ewe, goat, and cow) and survey bacteriocin production, the majority of the bacteriocin producers were lactococci (Rodriguez et al., 2000). Nisin-producing LAB were the most frequently encountered. Almost as abundant as the nisin producers were lactococci synthesizing lacticin 481 and enterococci producing AS-48. Using PCR techniques, nisin was found in 39 of the 67 bacteriocin-producing lactococci with lacticin 481 found from 23 isolates. Antilisterial activity was well-represented as well as activity against a typical range of Gram-positive bacteria normally used in bacteriocin studies. This work demonstrates the relatively high incidence of bacteriocin-producing bacteria occurring in milk, primarily lactococci. The lactococci, regardless of origin, continue to be investigated for applications in foods and other products (Balciunas et al., 2013). For example, various novel bacteriocins from *Lactococcus* have been identified, such as a nisin variant, nisin Q, a two-peptide bacteriocin, lactococcin Q, a leaderless bacteriocin, lacticin Q, and a circular bacteriocin, lactocyclicin Q (Zendo, 2013).

15.6.2 *Pediococcus*

The pediococci share many characteristics with the lactococci, including the ability to produce a range of interesting bacteriocins. It was recognized in cucumber fermentations using *P. pentosaceus* and *Lactobacillus plantarum* as starter cultures that the pediococci inhibited the growth of lactobacilli (Etchells et al., 1964). Since *L. plantarum* is as resistant to acids, if not more so than *P.*

pentosaceus, it was assumed that other antagonistic substances besides organic acids were active. Further investigation by Fleming et al. (1975) resulted in the isolation and characterization of *P. pentosaceus* FBB61 and L7230. These cultures showed similar inhibitory activity against a number of Gram-positive bacteria (most notably the pathogens *Staphylococcus aureus* and *B. cereus*), but were not active versus Gram-negative bacteria, yeasts, or each other. Catalase was included in their assays to obviate hydrogen peroxide effects. The bacteriocin of *P. pentosaceus* FBB61 was characterized by Rueckert in 1979 (M.S. thesis, Michigan State University, East Lansing). It was bactericidal, heat-stable (100°C for 60 min), sensitive to pronase, and nondialyzable. Daeschel and Klaenhammer (1985) further examined the antagonistic activity and host cell immunity of strains FBB61 and L7230. The bacteriocin produced by these cultures, designated pediocin A, was effective against other pediococci, *Lactobacillus brevis* and *L. plantarum*, the nisin-producing ATCC 11454 *S. lactis*, four strains of *S. aureus*, and cultures of clostridia representing *C. botulinum*, *Clostridium perfringens*, and *Clostridium sporogenes*. The bacteriocin was associated with a plasmid of approximately 13.6 MD; however, as noted by Klaenhammer (1988) further study of this bacteriocin was hindered by the low levels produced and difficulties in purification, which curbed detailed biochemical characterization.

Graham and McKay (1985) found the apparent biosynthesis of a bacteriocin in *Pediococcus cerevisiae* FBB-63 linked to a plasmid of approximately 10.5 MD. Gonzalez and Kunka (1987) correlated production of bacteriocin activity, designated pediocin PA-1, with a 6.2-MD plasmid in *P. acidilactici* PAC 1.0. Antagonism was demonstrated against other pediococci, some lactobacilli, and a strain of *Leuconostoc mesenteroides* subsp. *dextranicum*. No activity was found against lactococci. Partially purified pediocin PA-1 was ineffective versus micrococci, staphylococci, and some lactobacilli. Pediocin PA-1 was proteinaceous and bactericidal in action and sensitive to protease, papain, pepsin, and α-chymotrypsin. It was heat-stable (100°C for 10 min) and, as determined by gel filtration, had a molecular weight of approximately 16,500.

Marugg et al. (1992) reported the cloning, expression, and nucleotide sequence of the genes involved in the synthesis of pediocin PA-l. The bacteriocin was associated with a 9.4-kbp plasmid. The genes were cloned and expressed in *E. coli*. Deletion analysis demonstrated that a 5.6-kbp fragment from the plasmid was necessary for production of the bacteriocin. Four clustered open reading frames were found in the 5.6-kbp fragment. The first encoded for the 62 amino acid precursor of pediocin PA-1, with amino acid residues 19–62 corresponding to the purified bacteriocin. Although functions could not be determined for the next two downstream open reading frames, the last gene specified a large protein of 724 amino acids required for pediocin PA-1 production in *E. coli* that probably corresponded to an ATP-dependent transport protein. Comparison of the amino acid sequence of pediocin PA-1 with leucocin A-UAL 187 (Hastings et al., 1991), lactacin F (Muriana and Klaenhammer, 1991a), lactococcin M (van Belkum et al., 1991), lactococcin A (Holo et al., 1991; van Belkum et al., 1991), and lactococcin B (van Belkum et al., 1992) showed few similarities, except for the N-terminal leaders of 18–24 amino acids of the precursor proteins, particularly near the cleavage site. This site appears to be uniquely conserved in bacteriocin precursors from lactic acid bacteria, suggesting the requirement for separate processing enzyme(s) in the maturation of small heat-stable bacteriocins synthesized by lactic acid bacteria.

Hoover et al. (1988) examined bacteriocin activity associated with a plasmid of approximately 5.5 MD detectable in *P. acidilactici* P02, B5627, and PC and *P. pentosaceus* MC-03. Strain MC-03, isolated from pepperoni, was speciated primarily on the basis of growth temperature characteristics. Besides antagonism against other pediococci, *S. aureus*, *B. cereus*, *S. faecalis*, and *L. mesenteroides*, four of five strains of *L. monocytogenes* were also inhibited, as shown by agar plate diffusion assays. *P. acidilactici* PAC 1.0 shares a common heritage with *P. acidilactici* B5627 (Carlos Gonzalez, personal communication), and as found by rDNA typing, strains P02, PC, and B5627 are essentially the same genetically (Susan Harlander, personal communication). Most of the commercial starter cultures of *Pediococcus* used to manufacture fermented meat products in the U.S. have the capability of producing a bacteriocin with the potential to inhibit several important Gram-positive foodborne pathogens.

Hoover et al. (1989) surveyed 37 cultures of pediococci for antagonistic effects against 8 strains of *L. monocytogenes* using an agar plate diffusion overlay method. There was an observable or positive response in 28% of the matchings, indicating that a bacteriocin effect of these lactic acid bacteria versus

L. monocytogenes may not be limited to a few industrial starter cultures. Indeed, five strains of pediococci that were isolated from the gastrointestinal tract of humans inhibited all the *L. monocytogenes* cultures. These strains of pediococci were not only of interest because of their overall effectiveness against the *L. monocytogenes* cultures, but also because of their site of isolation. Members of the genus *Pediococcus* are not considered pathogenic; however, these strains were isolated from surgical treatment of human abdominal disease (William Riebel, personal communication). Although these pediococci were probably opportunistic in their presence and may not have contributed to the disease syndrome, their presence in such an environment was nonetheless unexpected.

Hoover et al. (1989) also more closely examined the inhibition of *L. monocytogenes* 19113 by dialyzed growth supernatant of *P. acidilactici* P02. Besides ruling out the inhibitive effects of lactic acid, the emergence of *L. monocytogenes* growth over time was also evident using a liquid assay. Apparently, prolonged exposure of the sensitive bacteria to the bacteriocin either resulted in the selection of a resistant subpopulation of *Listeria*, or the bacteriocin was bound or made unavailable so that the bacteriocin-sensitive population could then reestablish itself.

Pucci et al. (1988) prepared a dried powder of growth extract from *P. acidilactici* PAC 1.0 amended with 10% nonfat dry milk. Against *L. monocytogenes* LM01 they found the PA-1 powder effective over a pH range of 5.5–7.0 at both 4 and 32°C. PA-1 powder acted to reduce growth or eliminate an inoculum of approximately 10^2–10^3 CFU/ml of *L. monocytogenes* in cottage cheese, half-and-half cream, and cheddar cheese sauce for up to 2 weeks at 4°C. Lysis of the *L. monocytogenes* was evident.

Pediocin AcH was characterized by Bhunia et al. (1987, 1988), and optimal conditions for its production were established (Biswas et al., 1991). Synthesized by *P. acidilactici* H, pediocin AcH was approximately 2700 daltons, sensitive to proteolytic enzymes (trypsin, papain, ficin, and chymotrypsin and proteases IV, XIV, XXIV, and K), heat-resistant (121°C/15 min), resistant to a range of organic solvents, and effective over a pH range of 2.5–9.0. A rather wide range of activity was noted. Although only single strains of some genera were examined by Bhunia et al. (1988), inhibition by pediocin AcH of Gram-negative bacteria was reported (*Aeromonas hydrophila* and *Pseudomonas putida*). Inhibition of *A. hydrophila* was also found with culture derivatives of *P. pentosaceus* FBB61/L7230 (Thomas Montville, personal communication). This is of interest because bacteriocins produced by Gram-positive bacteriocin commonly have no effect against Gram-negative bacteria.

A 7.4-MD plasmid harboring the genes for activity and immunity of pediocin AcH from *P. acidilactici* H was conjugated into another strain of *P. acidilactici* by Ray et al. (1989a). For *P. acidilactici* strains E, F, and H, bacteriocin activity was associated with plasmids of 7.4–8.0 MD (Ray et al., 1989b). Overall, the strains contained a total of two to three plasmids ranging from 7.4 to 40.2 MD in size. Association of carbohydrate fermentation and antibiotic resistance could not be made with any of the detectable plasmids. The authors noted that mutanolysin was more effective than the more often used lysozyme for lysing cells of pediococci.

Given the use of pediococci as a meat starter culture by the food industry, the effectiveness of bacteriocin-producing pediococci against *L. monocytogenes* has been investigated in meat systems. Berry et al. (1990) evaluated *Pediococcus* spp. strains JD1-23 and MP1-08 versus *L. monocytogenes* Scott A in summer sausage chubs employing two different spice blends. Inoculated at an initial level of about 10^6 CFU/g, *L. monocytogenes* Scott A was reduced approximately 2 log cycles using a bacteriocin-positive culture and approximately 1 log cycle using a bacteriocin-negative isolate; however, recovery of Scott A using McBride *Listeria* agar was intermittent and did not indicate that the bacteriocin was effective for inactivation of Scott A that survived the smoking process given the sausages. In a subsequent study (Berry et al., 1991), bacteriocin-producing *P. acidilactici* were found effective against *L. monocytogenes* in frankfurters.

Nielsen et al. (1990) examined strain Scott A in a fresh meat system. The bacteriocin producer was *P. acidilactici* from a commercial culture identified as Lactacel 110. Radiation-sterilized fresh lean beef muscle cut into small pieces represented the model food system. Their approach was to evaluate the bacteriocin present in a filter-sterilized growth extract against *L. monocytogenes* on the meat surface. Cut meat pieces were exposed to various buffer solutions containing bacteriocin and growth extract and strain Scott A in alternating sequences. Using this protocol Nielsen and coworkers demonstrated a reduction of attached *L. monocytogenes* by 0.5–2.2 log cycles within 2 min of treatment with the bacteriocin.

Meat pieces treated initially with the bacteriocin showed 1.0–2.5 log cycle decreases in attached bacteria. The bacteriocin was functional on meat surfaces after a month of refrigerated storage.

Harris et al. (1989) examined 15 strains representing the genera *Lactobacillus*, *Pediococcus*, *Lactococcus*, and *Leuconostoc* for inhibition against 8 strains of *L. monocytogenes*. From these 15 lactic cultures, only cell-free supernatants from *Lactobacillus* sp. UAL11, *P. acidilactici* PAC 1.0, and *Leuconostoc* sp. UAL14 inhibited all 8 strains of *L. monocytogenes* as measured by deferred antagonism and well diffusion methods. This inhibition was prevented by the addition of proteolytic enzymes. All cultures of *L. monocytogenes* that were examined showed an identical response to the bacteriocins.

Foegeding et al. (1992) showed that pediocin production by meat starter cultures enhanced safety against growth of *L. monocytogenes* during both the fermentation and drying portions of dry-fermented sausage manufacture. Use of a bacteriocin-producing strain of *P. acidilactici* P02 effectively inactivated a mixture of five strains of *L. monocytogenes* (inoculated to a level of approximately 10^{-5} CFU/g) when the pH at the end of the fermentation segment of the process was less than 4.9.

Further effort to optimize pediocin AcH as a food preservative was attended by Motlagh et al. (1992b). Sterile ground beef, sausage mix, cottage cheese, ice cream, and reconstituted dry milk food systems were investigated using *Listeria ivanovii* and *L. monocytogenes* as target organisms. They found that the greatest bactericidal effect of pediocin AcH occurred within 1 h and was not significantly affected by food type, the *Listeria* strains that were examined varied in sensitivity to the bacteriocin, and the surviving cells of *Listeria* were capable of replicating in the presence of residual AcH during storage at 4 and 10°C.

In an effort to improve the effectiveness of pediocin AcH against *L. monocytogenes*, Degnan and Luchansky (1992) used liposome encapsulation of the bacteriocin. In heated slurries of beef tallow and beef muscle, encapsulation of pediocin AcH within phosphatidylcholine liposomes improved bacteriocin activity nearly 30% compared to slurries containing free pediocin.

The pediocin family of bacteriocins are not just contained in *Pediococcus*, or even the lactic acid bacteria (pediocins have been found in *Leuconostoc*, *Lactobacillus*, and *Carnobacterium*; Atrih et al., 2001). Coagulin, a peptide of 44 amino acids produced by *Bacillus cereus*, differs from pediocins AcH and PA-1 by only 1 amino acid in the C terminus (LeMarrec et al., 2000). The 3.5-kb operon of four genes for coagulin is plasmid-encoded. An isolate of *Lactobacillus plantarum* WHE 92, isolated from Muenster cheese, was found to produce a bacteriocin identical to pediocin AcH produced from *P. acidilactici* H (Ennahar et al., 1996). WHE 92 was identified at the DNA level. It was found that pediocin AcH was produced more effectively in *L. plantarum* in the pH range of 5.0–6.0 than in *P. acidilactici* H. The authors noted that since dairy products are normally higher than pH 5.0, *L. plantarum* WHE 92 offers a more effective means of antagonism against *L. monocytogenes*. WHE 92 grew very well in cheeses. Clearly, the same bacteriocin in LAB can be expressed across a range of different genera.

Pediocin binding to target membranes has been shown to be related to electrostatic interactions by Chen et al. (1997a, b). Pediocin PA-1 can function in the absence of a protein receptor. Positive patches of amino acid residues of pediocin PA-1 bound to negatively charged phospholipid head groups in the target membrane. Low pH strengthened interaction with the membrane by increasing the net charge of the bacteriocin; low ionic strength (i.e., low salt) enhances pediocin activity as high salt concentrations weaken the electrostatic interactions. Pediocin N5p, produced by a wine isolate of *P. pentosaceus*, nonspecifically absorbed to both Gram-positive and Gram-negative bacteria whether or not the cells were sensitive or resistant to it; however, for sensitive cells higher values up to 30% imply the presence of lethal receptors (Manca de Nadra et al., 1998). Cell death occurred without lysis and the cations of magnesium and manganese were found to improve pediocin binding by 80–100%. In the case of pediocin N5p, a lipid moiety plays a key role in antagonism as suggested by the inactivation of N5p by lipase and chloroform.

PCR random mutagenesis was used to create amino acid substitution mutants of pediocin AcH (Miller et al., 1998). Seventeen mutations of AcH were isolated having 14 of the 44 amino acids substituted. Most of the mutations were inactive or demonstrated reduced antagonism. One mutant peptide showed a 2.8-fold higher activity against the indicator strain (*L. plantarum* NCDO955). The critical amino acids necessary for activity of the pediocins were a few basic amino acids and nonpolar amino acids in the hydrophobic C-terminal region. Structural models were proposed. For pediocin PA-1, there is a loss of activity upon storage at refrigeration or ambient temperatures. Johnsen et al. (2000) engineered increased stability for the peptide by replacing a methionine residue (Met 31) with alanine, isoleucine, or leucine for protection from oxidation.

15.6.3 *Lactobacillus*

The lactobacilli are perhaps the most heterogeneous genus of lactic acid bacteria, and within such diversity the potential exists for equally disparate antagonistic peptides. A not uncommon trait of bacteriocins from *Lactobacillus* is a limited or narrow range of inhibitory activity, usually within the Lactobacillaceae. Another feature encountered in earlier studies involving the lactobacilli was an inability to link plasmid production with plasmid-encoded genes (Barefoot and Klaenhammer, 1983; McCormick and Savage, 1983; Joerger and Klaenhammer, 1986); however, Muriana and Klaenhammer (1987) were able to document plasmid involvement in *L. acidophilus* 88. Also, some isolates of lactobacilli are strong producers of hydrogen peroxide (Wheater et al., 1952). Along with lactic acid, these compounds can interfere when screening for authentic bacteriocin activity in lactobacilli.

Early surveys of strains of lactobacilli for bacteriocin activity were conducted by De Klerk and coworkers (De Klerk and Coetzee, 1961, 1967; De Klerk and Smit, 1967) and Upreti and Hinsdill (1973, 1975). Barefoot and Klaenhammer (1983) detected bacteriocin production in 42 of 52 strains of *L. acidophilus* examined. Bacteriocin producers have been well-represented in the genus *Lactobacillus*. De Klerk and Coetzee (1961) originally screened the supernatants of 189 strains of lactobacilli for lysogeny; however, for a small percentage of strains the antagonism was not caused by phage. Follow-up work (De Klerk and Coetzee, 1967) with 125 strains of *Lactobacillus fermenti* found 20% of these cultures to be bacteriocin-positive versus strains of *L. fermenti* and *L. acidophilus* when screened on agar.

The broad-spectrum antibiotic lactocidin was first discovered in *L. acidophilus* (Vincent et al., 1959); however, as pointed out by Klaenhammer (1988), a significant portion of the early work involved antagonistic substances that were not characterized very well. The activity spectrum of lactocidin included both Gram-negative and Gram-positive bacteria, but lactocidin has not been demonstrated in lactobacilli since its original citation.

Other antibiotic substances demonstrating a wide spectrum of activity have been detected in *Lactobacillus*. Two bacteriocins produced by strains of *L. acidophilus*, acidolin (Hamdan and Mikolajcik, 1974) and acidophilin (Shahani et al., 1977), have wide spectra of activity. In the case of an antibiotic substance from *L. bulgaricus*, a negative biuret test indicated the purified antagonistic compound was not a polypeptide (Abdel-Bar and Harris, 1984; Abdel-Bar et al., 1987). Ultraviolet absorption and mass spectra revealed an aromatic moiety with a molecular weight of approximately 700. This substance was active against *Pseudomonas fragi*. The compound examined by Abdel-Bar and coworkers has characteristics similar to reuterin, a low-molecular-weight, nonproteinaceous substance produced by *Lactobacillus reuteri* (Axelsson et al., 1988; unpublished manuscript, North Carolina State University, Raleigh, in Daeschel, 1989). Besides such Gram-positive varieties as *Clostridium*, *Staphylococcus*, and *Listeria*, reuterin was also active versus the Gram-negative enteric bacteria *Salmonella* and *Shigella*, as well as *Candida* (yeast) and *Trypanosoma* (protozoan). Such an extremely wide range of effectiveness usually does not represent bacteriocin activity, although Kabuki et al. (1997) studied a bacteriocin produced by *L. reuteri* LA6, reutericin 6. The 2.7-kDa peptide lacked lanthionine residues and featured 67% hydrophobic and polar neutral amino acids; a narrow spectrum of activity was suggested. However, reutericyclin isolated from *L. reuteri* LTH2584 is not a bacteriocin but a negatively charged, highly hydrophobic tetrameric acid derivative with a molecular weight of 349 (Ganzle et al., 2000). Reutericyclin was effective against Gram-positive bacteria, but not Gram-negative bacteria. Resistant to proteolytic digestion, reuticyclin is about 20 times less potent than the bacteriocins, curvacin A and sakacin P. For example, these two bacteriocins could inhibit strains of *L. sakei* at levels of 0.003 mg/L while against its most sensitive indicator strain (*E. faecalis*) the level of inhibition by reutericyclin was about 0.05 mg/L.

Gilliland and Speck (1977) surveyed the antagonistic activity of *L. acidophilus* toward intestinal and foodborne pathogens that included *S. aureus*, enterotoxigenic *E. coli*, *Salmonella* Typhimurium, and *C. perfringens*. The presence or extent of bacteriocin activity was not specifically addressed, but it was estimated that antagonism was not directly related to acids produced and that hydrogen peroxide was only partially responsible for the inhibitory effect since it was evident under anaerobic conditions in a prereduced medium.

Evidence for a 5400-dalton peptide with a broad range of activity was described by Mehta et al. (1983). Although the method for measurement of antagonistic effect was not clear, *Salmonella* Typhosa, *P. aeruginosa*, and *E. coli*, as well as *S. aureus* and several bacilli, were inhibited by this trypsin- and chymotrypsin-sensitive bacteriocin produced by *L. acidophilus* AC.

Bulgarican was named by Reddy et al. (1983). Growth of *L. bulgaricus* DDS 14 at 45°C for 48 h in milk medium with methanol-acetone extraction and silica gel chromatography of the supernatant resulted in inhibition of both Gram-positive and Gram-negative bacteria, including *B. subtilis*, *E. coli*, *Proteus vulgaris*, *Sarcina lutea*, *S. aureus*, *P. aeruginosa*, *P. fluorescens*, and *Serratia marcescens*. There was no demonstratable antifungal activity. The best effect was at a pH of less than 4.5, with pH 2.2 optimum.

As noted earlier, many bacteriocins from lactobacilli inhibit only other lactobacilli; examples include lactacins B and F. These proteins resemble each other and are produced by *L. acidophilus*. Lactacin B as produced by *L. acidophilus* N2 was first characterized by Barefoot and Klaenhammer (1983). The protein formed aggregates of approximately 100,000 daltons. Purification showed the protein to actually be of 6000–6500 daltons (Barefoot and Klaenhammer, 1984). It was active against *L. bulgaricus*, *Lactobacillus leichmannii*, *Lactobacillus helveticus*, and *Lactobacillus lactis* and was bactericidal, sensitive to proteases, and heat-stable (100°C for 1 h). Nonspecific adsorption was demonstrated by the adherence of lactacin B to sensitive and insensitive cells. Production of lactacin B appears to be chromosomally linked because *L. acidophilus* N2 does not contain plasmid DNA.

Lactacin F is produced by *L. acidophilus* 88 (Barefoot and Klaenhammer, 1983). Besides the lactacin B indicators, lactacin F also inhibits *Lactobacillus fermentum* 1750 and *S. faecalis* 19433 (Muriana and Klaenhammer, 1987). Lactacin F is heat-stable (121°C for 15 min) and sensitive to trypsin, proteinase K, ficin, and subtilisin, and its production is pH-dependent. As noted earlier, lactacin F is a plasmid-linked property of *L. acidophilus* 88 (Muriana and Klaenhammer, 1987). Bacteriocin production and immunity are associated with resident plasmids of 4 and 27 MD. Genetic analysis of transconjugants acquiring these phenotypes suggested the involvement of transient plasmid determinants capable of chromosomal integration. That is, the occurrence of detectable plasmids or the "plasmid state" may be the result of a defective or incomplete mobilization event.

Muriana and Klaenhammer (1991a) purified lactacin F. Electron microscopy demonstrated micelle-like globular particles of the native protein. Gel filtration of the bacteriocin suggested a large macromolecular complex with a molecular weight of 180,000; further purification and sodium dodecyl sulfate-polyacrylamide gel electrophoresis identified a 2500-dalton peptide that showed a 474-fold increase in activity; however, a protein as large as 6200 daltons or 56 amino acids was revealed by amino acid composition analysis. A segment of 25 N-terminal amino acid residues was determined by sequence analyses that aided in the genetic cloning of the lactacin F structural gene (Muriana and Klaenhammer, 1991b).

Raccach et al. (1989) described the inhibition of *L. monocytogenes* by *L. acidophilus* strains NU-A and 88 (the latter strain producing lactacin F; Muriana and Klaenhammer, 1987). Inhibition was demonstrated in milk, in which the presence of *L. acidophilus* prevented the growth of *L. monocytogenes* to the levels it would normally reach in pure culture. The lactobacilli lowered the pH of the milk to 4.7 within 24 h indicating that acid was active in antibiosis, but it was shown that *L. acidophilus* 88 was bactericidal against *L. monocytogenes* OH in the agar test system.

In an examination of the antibacterial activity of *L. acidophilus* for intended use as a dietary adjunct, Fernandes et al. (1988) used bile salts added to MRS broth to simulate concentrations in the human intestinal tract. Nutrient content also significantly affected the antimicrobial activity of *L. acidophilus*.

Ferreira and Gilliland (1988) presented an example of the disadvantage of having a bacteriocin producer in a mixed-strain starter culture. When *L. acidophilus* NCFM, a mixed-strain culture, was grown at pH 6.0, a slow-growing, bacteriocin-producing variety was able to predominate by killing the other strain(s) in the culture. When a multiple starter culture is used (e.g., probiotics, fermented foods and beverages), it should be examined during its formulation for a lack of strain compatibility because of the production of a bacteriocin by one member that eliminates other strains.

Lactocin 27 and helveticin J are both produced by strains of *L. helveticus*. Lactocin 27, as produced by *L. helveticus* LP27, was originally described as a protein-lipopolysaccharide complex of over 200,000

Da (Upreti and Hinsdill, 1973), but refinement of technique defined a glycoprotein of approximately 12,400 Da (Upreti and Hinsdill, 1975). The formation of aggregates by bacteriocins produced by lactic acid bacteria appears to be a rather common occurrence.

The bacteriocin produced by *L. fermenti* 466 appears to be similar to lactocin 27 (De Klerk and Smit, 1967; Upreti and Hinsdill, 1973). Lactocin 27 is digested by trypsin and pronase, and its stability at high temperatures is exceptional. There is no demonstratable loss of activity after a 1-h exposure to 100°C (Upreti and Hinsdill, 1973, 1975). Biochemically, the bacteriocin from *L. fermenti* 466 also appears as a heat-resistant macromolecular lipopolysaccharide-protein complex. Loss of activity results with removal of the carbohydrate component. It is also inactivated by exposure to trypsin and pepsin, but is resistant to urea, lysozyme, and heat (96°C for 30 min).

Helveticin J also has a limited range of antagonistic activity, inhibiting some strains of *L. helveticus*, *L. bulgaricus*, and *L. lactis* (Joerger and Klaenhammer, 1986). It is sensitive to pronase, trypsin, pepsin, ficin, proteinase K, and subtilisin, as well as heat (100°C for 30 min). In culture supernatants helveticin J forms aggregates of over 300,000 daltons, but with purification a molecular weight of 37,000 was identified that retained functionality. Extensive genetic investigation led Joerger and Klaenhammer (1986) to conclude that, like lactocin 27, the genes for production and immunity of helveticin J reside on the bacterial chromosome.

L. plantarum is an industrially important variety of lactobacilli, especially in the manufacture of pickles and olives. Lactolin was the first documentation of an antagonistic protein produced by *L. plantarum* (Kodama, 1952). It was originally distinguished in that it differed from nisin and diplococcin. Work on bacteriocins from *L. plantarum* languished from 1952 until Andersson (1986) published the inhibition of *S. aureus* and spheroplasts of Gram-negative bacteria by a bacteriocin from *L. plantarum* 83. After this study, research activity increased in this area (Klaenhammer, 1988).

West and Warner (1988) evaluated bacteriocin production in six strains each of *L. plantarum* and *L. mesenteroides* and found bacteriocin activity in *L. plantarum* 1193. This bacteriocin was sensitive to proteolytic enzymes, and its activity was reduced by lipase and α-amylase. It was antagonistic toward two other strains of *L. plantarum*, *L. mesenteroides* 8015 and *Pediococcus damnosus* 1832, but ineffective versus single strains of *Leuconostoc cremoris*, *Lactobacillus casei*, *L. fermentum*, and *L. acidophilus*. The bacteriocin was not isolatable from liquid culture. The authors stated that a bacteriocin effective against clostridia would be most valuable in a *L. plantarum* starter culture used to make silage.

Plantaricin A, as produced by *L. plantarum* C-11, was reported by Daeschel et al. (1990). Their study showed a bacteriocin that was heat-stable (100°C for 30 min) and active over a pH range of 4.0–6.5 and had an approximate molecular weight of less than 8000 as based on retention in dialysis membranes. The spectrum of activity was limited to those lactic acid bacteria that *L. plantarum* C-11 would normally encounter in its function of fermenting cucumbers, that is, other *L. plantarum*, *P. pentosaceus*, and *Leuconostoc paramesenteroides* populations. A similar range of inhibition was found by Rammelsberg and Radler (1990) with lactobacilli isolated from plants or fermenting material. For the 15% of isolates that were found to be bacteriocin-positive, the range of effectiveness was primarily *Lactobacillus*, *Pediococcus*, and *Leuconostoc*.

Dry sausage was the source of *L. plantarum* UG1, producer of plantaricin UG1 (Enan et al., 1996). The thermostable peptide has a molecular weight of 3 and 10 KDa and amylolytic enzymes reduce its activity. Plantaricin UG1 is chromosomally encoded in strain UG1. Whether killed or not, all Gram-positive bacteria absorbed the bacteriocin, while Gram-negative bacteria did not absorb it. Plantaricin UG1 inhibited more than LAB as is the case with plantaricin A, *L. monocytogenes*, *B. cereus*, *C. perfringens*, and *Clostridium sporogenes* were inhibited.

Plantaricin C19 is a pediocin-like bacteriocin produced by *L. plantarum* C19 originally isolated from cucumbers (Atrih et al., 2001). With a weight of 3845.3 Da (36 amino acids) as determined by mass spectroscopy, the peptide is antilisterial with a preponderance of hydrophobic and basic amino acids. Its absorption to cells of target *Listeria* was diminished by the presence of NaCl, KCl, $MgSO_4$, and $CaCl_2$ possibly due to competitive ion adsorption on the cell surface by the salts.

Plantaricin LP84 from *L. plantarum* NCIM 2084 was found to be bactericidal and lytic against both Gram-positive bacteria (*B. cereus*, *S. aureus*, and lactic acid bacteria) and Gram-negative bacteria (*E.*

coli and *P. aeruginosa*; Suma et al., 1998). It is a heat-stable peptide of 1 to 5 kDa; its ability to inactivate these two types of Gram-negative bacteria is quite interesting as a broad-spectrum preservative.

A lantibiotic, plantaricin C, was isolated from a dairy strain of *L. plantarum*. In its 2-part, 27-amino acid sequence, Turner et al. (1999) found 1 dehydroalanine, 1 lanthionine, and 3 α-methyl-lanthionine residues. The first part of the molecule (six amino acids) confers a strong positive charge, while the larger second part contains a lanthionine bridge to stabilize a globular structure to the bulk of the molecule that allows insertion of the bacteriocin into the target membrane opening pores to cause leakage of cytoplasm from the bacterial cell.

Strains of *L. sake* have been found to possess bacteriocin activity. Schillinger and Lucke (1989) surveyed 221 strains of lactobacilli isolated from meats and found 19 strains of *L. sake*, as well as 3 strains of *L. plantarum* and 1 strain of *Lactobacillus curvatus*, to inhibit other lactobacilli. Evaluation of the antimicrobial spectra of the supernatants suggested that the bacteriocins were not identical. A bacteriocin from *L. sake* Lb 706, termed sakacin A, was effective against *L. monocytogenes* strains 8732 and 17a. Another four strains of *L. sake* and one strain of *L. plantarum* proved antilisterial as well. Synthesis and immunity of sakacin A was linked to an 18-MDa plasmid. In mixed culture, it was found that once the bacteriocin-positive strain reached maximal cell density, cells of the bacteriocin-sensitive strain were killed.

Lactocin S was identified by Mortvedt and Nes (1990). The bacteriocin is produced by *L. sake* L45 isolated from a dry, sausage fermentation. Lactocin S was described as moderately heat-stable, sensitive to proteases, and active against members in the genera *Lactobacillus*, *Pediococcus*, and *Leuconostoc*. It has approximately 33 amino acid residues (Mortvedt et al., 1991). Initial investigation of the plasmid biology in *L. sake* L45 revealed instability of bacteriocin production and immunity (Mortvedt and Nes, 1990). Two plasmids of approximately 50 and 34 kb were associated with these traits. It was found that isolates of *L. sake* L45 that were bacteriocin-negative (Bac-) and negative for immunity to lactocins (Imm-) had lost the 50-kb plasmid; however, restriction enzyme analyses of the 50-kb plasmids from Bac+Imm+ and Bac-Imm+ isolates revealed distinct differences suggestive of a rearrangement of DNA bases between the plasmids.

In a study examining the effectiveness of *L. sakei* CTC 494 as a potential starter culture for sausage fermentation, Leroy and de Vuyst (1999) found that sodium chloride and sodium nitrite in the meat mixture interfered with production of sakacin K by the lactobacilli. With *L. sakei* CTC494 and sakacin K, Hugas et al. (1998) demonstrated an antagonistic effect to different extents against *L. innocua* CTC1014 in meats over a range of 4 to 30°C, initial pH range of 5.5 to 6.5 packaged with oxygen-permeable film, under vacuum and under a modified atmosphere (80% oxygen, 20% carbon dioxide). Vacuum-packaged poultry breasts and cooked pork, and modified atmosphere-packaged raw minced pork demonstrated the greatest degree of inhibition of *L. innocua*. An immediate bactericidal effect was evident in every product and atmosphere evaluated when sakacin K was added.

Ganzle et al. (1999b) used a dynamic model of the stomach and small intestine to evaluate the antagonistic effect of bacteriocin-producing *L. curvatus* LTH 1174 on *E. coli* LTH 1600 and *L. innocua* DSM20649. The model, kept at 37°C, consisted of four serial compartments simulating the stomach, duodenum, jejunum, and ileum; it could be disinfected with steam, and the flexible walls allowed contractions duplicating peristaltic movements. Results indicated that the ability to produce bacteriocins was unlikely to contribute to colonization of the human intestine with LAB starter cultures. It was felt that due to the low bile tolerance of meat lactobacilli, it was unlikely that bacteriocin production was a significant factor in killing *E. coli* or *L. innocua* in the gut; however, bacteriocin-producing lactobacilli that are bile-resistant may be useful in controlling pathogens in the small intestine.

A system for heterologous expression using a bacteriocin-negative strain (Lb790) of *L. sake* was developed by Axelsson et al. (1998). In it, two plasmids introduced into Lb790 allowed various bacteriocins (sakacin P, pediocin PA-1, and piscicolin 61) to be produced at levels equal to or exceeding levels obtained with the corresponding wild-type cultures. Genes necessary for transcriptional activation of the sakacin A promoter, and export and processing of bacteriocin precursors are contained on the first plasmid. The second plasmid has the sakacin A promoter fused to the structural and immunity genes for the bacteriocin.

Two-peptide bacteriocins are more frequently encountered. For example, Cuozzo et al. (2000) investigated lactocin 705, a class IIb bacteriocin from *L. casei* CRL 705. For activity, lactocin 705 was found to require the presence of both peptides, 705α and 705β, each containing 33 amino acids and encoded on a 35-kb plasmid. Similar DNA base sequences with other class II bacteriocins were found regarding immunity determinants, and maturation and transport proteins. Garneau et al. (2002) discussed the classification, structure, production, regulation, biological activity, and potential applications of two-peptide bacteriocins.

Prior to the discovery of two forms of lactocin 705, Vignolo et al. (1998) evaluated the effectiveness with which lactocin 705 inhibited *L. monocytogenes* when contained in a meat slurry. It was found that ingredients (sodium chloride and sodium nitrite, ascorbic acid, alginate, and sodium lactate) used in the meat cure protected *L. monocytogenes* from the effects of lactocin 705; sodium lactate neutralized the effects of lactocin 705 to the greatest extent.

15.6.4 *Carnobacterium*

Once members of *Lactobacillus*, *Carnobacterium* is comprised of non-aciduric lactobacilli producing L(+) lactic acid from glucose (Collins et al., 1987). In the past, the carnobacteria were sometimes referred to as the meat lactobacilli. It includes what were formerly species of nonaciduric *Lactobacillus piscicola* and *Lactobacillus divergens* and other nonaciduric lactobacilli from poultry and vacuum-packaged meat. Bacteriocins produced by this genus appear quite diverse although common types are distributed within the group. Other bacteriocin-producing members of the lactic acid bacteria are commonly isolated from various animal sources consumed as meat (da Costa et al., 2019).

Ahn and Stiles (1990) examined bacteriocin activity produced by *Carnobacterium piscicola* LV 17. Activity was found against closely related lactic acid bacteria, *Enterococcus* spp. and a strain of *L. monocytogenes*, but not Gram-negative bacteria. Cell death was dependent on contact of the bacteriocin with the cell membrane of the target organism. They found bacteriocin production and resistance to the bacteriocin residing on plasmids of 40 and 49 MD. It was postulated that two bacteriocins arc produced, since the inhibitor substances of mutant strains containing either the 40- or 49-MD plasmids have different antimicrobial spectra.

Carnocin UI49 is a bacteriocin from a strain of *Carnobacterium* isolated from fish. Stoffels et al. (1992) purified and characterized UI49. It is a protein with a molecular weight of 4635 as determined by mass spectrometry. Amino acid analysis suggests that carnocin UI49 belongs to the lantibiotic class of bacteriocins. Apparently, U149 is distinct from the bacteriocins produced by *C. piscicola* LV17, which do not contain lanthiobiotic-like amino acids.

C. piscicola JG126 was isolated from spoiled ham and found to produce piscicolin 126 that is effective against *L. monocytogenes* and other Gram-positive bacteria (Jack et al., 1996). The peptide has a molecular weight of 4417 (44 amino acids), and sequence analysis showed it a member of class IIa with close resemblance to pediocin PA-1 and sakacin P. In trials in deviled ham paste, piscicolin 126 inhibited *L. monocytogenes* for 14 days at 10°C, proving itself (final concentration 2048 AU/g) superior to two commercial preservatives, ALTA™2341 (final concentration 1% or 2048 AU/g) and Nisaplin™ (400 IU/g). The ingredient declaration of ALTA™2341 (Quest International Bioproducts) states it is a fermentation by-product of a lactic acid bacterium that includes corn syrup solids, hydrolyzed yeast, and hydrolyzed vegetable protein. Antibacterial components are not specified.

By comparing DNA sequences in data banks and confirming results with PCR, Herbin et al. (1997) found carnocin CP51 from *C. piscicola* CP5 to share homologies with carnobacteriocin BM1, while carnocin CP52 and carnobacteriocin B2, both produced by *C. piscicola* LV17, were similar. Strains CP5 and LV17 were isolated from different foods: *C. piscicola* CP5 from soft cheese and *C. piscicola* LV17 from vacuum-packed meat.

Blom et al. (2001) examined carnocin H, a class II bacteriocin produced by *Carnobacterium* sp. 377. Carnocin H contained approximately 75 amino acids. Increasing levels of pH, indicator bacteria, sodium chloride, agar, and soybean oil impeded antagonism by the peptide. Produced during the late stationary growth phase, carnocin H inhibited the nine strains of *L. monocytogenes* examined. Its inhibitory spectrum differed from that of *C. piscicola* LV61 and *C. divergens* 750.

Zhang et al. (2019) conducted a comparative analysis of bacteriocins from *Carnobacterium* isolated from meat in conjunction with organic acids. The species examined were *C. maltaromaticum* and *C.*

divergens. Formic and acetic acids were the principal antibacterial acids demonstrating effectiveness with several bacteriocins from the carnobacteria.

15.6.5 *Leuconostoc* and Other Lactic Acid-Associated Bacteria

Leuconostoc is a genus of heterofermentative lactic organisms with important commercial applications in the manufacture of cultured dairy and vegetable products. Usually the inhibition of other organisms by *Leuconostoc* is attributed to the excretion of acetic and lactic acids by its metabolizing cells (Devoyod and Poullain, 1988). Orberg and Sandine (1984) found six strains of *Leuconostoc* to produce substances with antibacterial activity versus other leuconostocs and *Streptococcus lactis* strains. These effects were measured using the agar plate methods of Kekessy and Piguet (1970). One strain, *Leuconostoc* spp. P0184, inhibited *S. cremoris* U134 but not the nisin producer *S. lactis* ATCC 11454. In subsequent work, Tsai and Sandine (1987) demonstrated the capability of *Leuconostoc dextranicum* to produce nisin. A 17.5-MDa plasmid from *S. lactis* 7962 that encoded for nisin biosynthesis and sucrose fermentation was transferred by conjugation into *L. dextranicum* 181. Not only was nisin production detected in a transconjugant, but there was significant overproduction of the bacteriocin (enough for the transconjugant to be termed a supernisin producer). Such a phenomenon is of obvious interest in the optimization of fermentations for commercial bacteriocin production.

Branen et al. (1975) investigated the inhibition of pseudomonads by culture supernatants of *Leuconostoc citrovorum* and *Streptococcus diacetylactis*. Organic acids were believed to play a more significant role in the antagonism of *Pseudomonas* by *L. citrovorum* than for *S. diacetylactis*. From growth of *S. diacetylactis* a small-molecular-weight, heat-stable protein was isolatable that possessed activity against *Pseudomonas* at pH 6.0.

An amylase-sensitive bacteriocin, leuconocin S, produced by a strain of *L. paramesenteroides* isolated from retail lamb, was characterized by Lewus et al. (1992). Leuconocin S is active against *L. monocytogenes*, *S. aureus*, *Yersinia enterocolitica*, and some strains of *C. botulinum*. Leuconocin S is a glycoprotein sensitive to α-amylase, trypsin, α-chymotrypsin, protease, and proteinase K but not to lipase or treatment at 60°C for 30 min. Apparently, the bacteriocin forms aggregates. Molecular weights of 2000 and 10,000 were determined from sodium dodecyl sulfate-polyacrylamide gels (SDS-PAGE) of the supernatants from culture broths.

L. mesenteroides TA33a was found to produce three bacteriocins, each with a different spectrum of inhibition (Papathanasopoulos et al., 1997). After purification of the bacteriocins, it was discovered that leucocin A-TA33a was identical to leucocin A-UAL 187 (Fleury et al., 1996). Leucocin B-TA33a (3466 Da) appeared similar to mesenterocin 52B (Revol-Junelles and Lefebvre, 1996) and dextranicin 24 (Revol-Junelles et al., 1996), and leucocin C-TA33a was a novel compound of 4598 Da. A-T33a and C-T33a both were antilisterial. Papathanasopoulos et al. (1997) conducted a survey of other strains of *Leuconostoc* and *Weissella* for the occurrence of the three different bacteriocins and found them relatively widespread; strains of *L. carnosum*, *W. paramesenteroides*, and *L. gelidum* produced the three bacteriocins.

In an effort to optimize bacteriocin production during batch fermentation, Krier et al. (1998) used *L. mesenteroides* subsp. *mesenteroides* FR52 for the synthesis of mesenteroicins 52A and 52B to find that syntheses for both bacteriocins were stimulated by slow growth rates of the culture. The optimal temperature and pH for mesenterocin 52A were 20°C and pH 5.5 while for mesenterocin 52B optimal parameters were 25°C and 5.0. As a result, strain FR52 could be grown under different conditions to favor one bacteriocin over the other. Additionally, different production kinetics were exhibited by the two bacteriocins.

Leucocin H is another two-peptide bacteriocin (Blom et al., 1999). Synthesized by *Leuconostoc* sp. MF215B, the two peptides called leucocin Hα and Hβ both contain 40 amino acids and inhibit *L. monocytogenes*, *B. cereus*, and *C. perfringens* among other Gram-positive bacteria Blom et al. (1999) found the two peptides to share little sequence homology with each other, and no sequence homology could be found with any other bacteriocin either. It was noted that the additional of soy oil to agar had no influence on the effectiveness of leucocin H.

Callewaert et al. (2000) tested *Enterococcus faecium* CCM 4231 and *E. faecium* RZS C13 as starter cultures in Spanish-style dry sausage fermentation against nonpathogenic *Listeria*. They found the bacteriocins produced by both strains to be inactive against lactic acid bacteria but effective against *L. innocua* and *L. ivanovii*. Results indicated that the enterococci were effective against *Listeria* spp., decreasing their counts after 1 week of fermentation by 1 to 3 log CFU/g. The sensory quality of the sausages was judged normal.

Enterocin CRL35 from *E. faecium* CRL35 has effectiveness against herpes viruses. Wachsman et al. (1999) demonstrated that this antibacterial polypeptide (3.5 KDa), inhibitive of *L. plantarum* and Gram-positive pathogens, showed a 2-log reduction of viable thymidine-kinase positive and deficient strains of herpes simplex virus type 1 and 2 in Vero and BHK-21 cells. Viral multiplication was inhibited in a dose-dependent manner. The antiviral effect of enterocin CRL35 did not appear to be due to direct inactivation, suggesting a different mechanism from human defensins for antiviral effect. The bacteriocin may affect the viral reduction of host protein synthesis, thus impairing virus replication. Kassaa et al. (2014) reviewed the potential of lactic acid bacteria and their bacteriocins as antiviral agents.

Additional bacteriocins from *Enterococcus* include enterocin P, enterocin B, and enterocins L50A and L50B. Enterocin P from *E. faecium* is strongly antilisterial, pediocin-like in sequence, stable to 121°C for 15 min and also effective versus *C. botulinum* (Cintas et al., 1997). Enterocin B, from *E. faecium* T136, acts synergistically with enterocin A and closely resembles carnobacteriocin A (Casaus et al., 1997). As noted by others, combinations of bacteriocins, in this case enterocins A and B, were highly bactericidal as compared to individual bacteriocins. For enterocins L50A and L50B from *E. faecium* L50, a similar synergism was demonstrated against target Gram-positive bacteria, although L50A and L50B did not resemble the pediocin family of bacteriocins, but rather staphylococcal cytolysins having a hemolytic activity (Cintas et al., 1998).

The use of bacteriocin-producing enterococci in cheese has been shown to inhibit histamine formation in Manchego cheese. In separate cheese-making experiments, Joosten and Nunez (1996) demonstrated that inclusion of two bacteriocin-producing enterococci or a nisin-producing strain of *L. lactis* was effective in preventing the formation of histamines in the cheese by histamine-producing *Lactobacillus buchneri*. In control cheeses, levels of histamine reached 200 mg per kg of cheese after 4 months of ripening. The production of pediocin PA-1, bavaricin A, lactococcin A, and a bacteriocin from *E. faecalis* was found to be ineffective in preventing histamine production.

Du et al. (2017) also targeted *L. monocytogenes*. These workers characterized two bacteriocins from *Enterococcus durans* 152 with comparison to nisin usage. They found the enterocins effective, rating their anti-listerial activity superior to nisin at 500 ppm in refrigerated ham.

Another group of food-grade organisms that are sometimes classified with the traditional lactic acid bacteria are the bifidobacteria. These anaerobic, Gram-positive rods are known for their fastidious nature. The benefit derived from their presence in the human intestinal tract is also well-established (Hughes and Hoover, 1991). One of the major benefits is the reduction of enteric pathogens in the intestinal tract by bifidobacteria. The bifidobacteria excrete high levels of acetate, lactate, and some formate from their metabolism of fermentable sugars, and this is the primary mechanism of inhibiting detrimental gut organisms. Of course, the production of bacteriocins is another ecological weapon that many acidulating bacteria use. Jao et al. (1978) suggested that the antibacterial activity of *Bifidobacterium bifidum* may be caused by the production of antibacterial substances in addition to acids. Anand et al. (1985) detected potential bacteriocin activity in strains of *B. bifidum*. Cultures of both Gram-positive and Gram-negative bacteria were inhibited by growth supernatants, but the effect was absent above pH 4.5, and fractionated metabolites singly or in combination did not show antagonistic activity. Meghrous et al. (1990) partially characterized a bacteriocin from an unspeciated strain of *Bifidobacterium*. Susceptible to trypsin, pepsin, and pronase and functional between pH 2.0 and 10.0, the activity of the bacteriocin remained after 121°C for 15 min against lactococci, *Streptococcus salivarius* subsp. *thermophilus*, *L. acidophilus*, *Clostridium tyrobutyricum*, and *Clostridium butyricum*. No Gram-negative varieties were inhibited. In later work, Yildirim and Johnson (1998) screened for bacteriocin production in MRS broth from growth of five strains of *B. bifidum*, and found only one strain (NCFB 1454) capable of producing a detectable bacteriocin. It was a proteinaceous compound of 3.3 kDa with resistance to exposure to 121°C for 15 min as well as pH extremes of 2.0 and 10.0. Designated bifidocin B, it was active against most strains from species representing *Listeria*, *Bacillus*, *Enterococcus*, *Lactobacillus*, *Leuconostoc*, and *Pediococcus*.

Propionibacterium freudenreichii subsp. *shermanii* is used to produce a preservative ingredient called MicroGARD™. It is produced by the pasteurization of propionibacteria-fermented skim milk and is estimated to be used in about one-third of all cottage cheese made in the United States because of its ability to inhibit psychrotrophic spoilage bacteria (Daeschel, 1989). Its effectiveness is directed against most Gram-negative bacteria and some fungi, but not Gram-positive bacteria. Although propionic acid, diacetyl, acetic acid, and lactic acid are present, the most active component of MicroGARD is a heat-stable peptide of about 700 Da that is inactivated by proteolytic enzymes (N. Al-Zoreky, M.S. thesis, Oregon State University, Corvallis, 1988).

Evidence of a bacteriocin produced by *Propionibacterium thoenii* P127 was shown by Lyon and Glatz (1991). The bacteriocin, called propionicin PLG-1, was found to be heat-labile (85°C for 30 min), sensitive to several proteolytic enzymes, stable at pH 3.0–9.0 (greatest activity is at pH 7.0), and effective against a wide range of Gram-positive and Gram-negative bacteria, as well as some yeasts and molds. As measured by well-diffusion assays, activity was discovered versus all propionibacteria, except for strains of *P. freudenreichii* subsp. *freudenreichii* and *P. freudenreichii* subsp. *shermanii*. Lactic acid bacteria were inhibited (lactococci, lactobacilli, and pediococci), but representative strains of clostridia, bacilli, *S. aureus*, and *S. faecalis* were not. For Gram-negative bacteria, pseudomonads were significantly inhibited, but cultures of salmonellae, *A. hydrophila*, and *Y. enterocolitica* were not. Fungistatic rather than fungicidal activity was found against some *Saccharomyces* strains and *Candida lipolytica*. *Aspergillus wentii* and seven other fungi were inhibited. Distinct from MicroGARD, propionicin PLG-1 demonstrated two molecular weights, indicative of different protein aggregate forms. Gel-filtration studies determined complexes of approximately 10,000 Da and of more than 150,000 Da, with further purification revealing activity specific to the 10,000-Da molecule. The authors suggested a parallel relationship of propionicin PLG-1 with lactacin B as described by Barefoot and Klaenhammer (1983), in that both bacteriocins were first isolated as large aggregates that required SDS-PAGE purification into their monomeric forms. The need was also stated for a defined growth medium free of contaminating proteins that associate with bacteriocins to hinder production and purification. Detection and production of propionicin PLG-1 was improved by Hsieh et al. (1996) who increased the sensitivity of a standard well-diffusion assay. The assay was optimized by the use of wells of specific dimensions into a solid medium containing 2.5% agar, 0.85% NaCl, and 0.1% Tween 80. The use of 0.1% Tween 80 in the buffer used for dialysis and resuspension of precipitated protein also improved purification of the bacteriocin. The preferred medium for growth of *P. thoenii* P127 for production of PLG-1 was 3:1 beet molasses:corn steep liquor whose yield was five times higher than sodium lactate broth.

Jenseniin G is a heat-stable bacteriocin produced by *Propionibacterium jensenii* P126. Grinstead and Barefoot (1992) found *P. jensenii* P126 inhibited closely related varieties of dairy propionibacteria, as well as species of *Lactobacillus* and *Lactococcus*, under anaerobic conditions. The bacteriocin was inactivated by pronase E, proteinase K, and type 14 protease but stable to freezing, cold storage (4°C for 3 days), and heat treatment (100°C for 15 min). The stability of jenseniin G differs from the lability of propionicin PLG-1 to 85°C. The lack of detectable plasmids in *P. jensenii* P126 suggests a chromosomal location for the gene(s) of jenseniin G.

Faye et al. (2000) characterized propionicin T1 at the molecular level. It was isolated from two different strains of *P. thoenii*, and showed no sequence similarity to other bacteriocins. Propionicin T1 inhibited most of the propionibacteria tested as well as a strain of *L. sake*. Evidence was lacking that T1 possesses a wider target range. The mature form of the bacteriocin had 65 amino acids. It was noted that both producers of propionicin T1 were sensitive to it as well, suggesting that the immunity factors may not be constitutively expressed, but instead be regulated to production of T1. Cells early in the growth cycle seemed sensitive to the bacteriocin that is normally produced in the late logarithmic stage.

A recent book by Vinderola et al. (2019) includes an overview of bacteriocin production by lactic acid bacteria and other acidulating bacteria. The content updates production, ecology, taxonomic status, and genomics. Normally, bacteriocin activity is screened using conventional agar plate methods, and initial characterization of bacteriocins follows a systematic approach used for decades. Kanauchi (2018) overviews the procedures commonly used in screening and characterization.

15.7 Applications in the Food Industry

Bacteriocin application in foods can be categorized as partially purified bacteriocins (e.g., Nisaplin®, containing 2.5% nisin), crude bacteriocin fermentate (e.g., MicroGARD® series), and bacteriocin-producing protective cultures (Makhal et al., 2015; Saucier and Champagne, 2005; Aymerich et al., 2008). Over the past several decades, several groups have studied bacteriocins and their applications in controlling foodborne pathogens in a variety of food products (Drider and Rebuffat, 2011). Work continues on improving bacteriocin performance in various food environments and appropriate food-grade delivery systems. Studies have been conducted to develop bacteriocin derivatives with increased resistance to proteolytic enzymes, enhanced antimicrobial activity, and a broadened inhibitory spectrum (O'Shea et al., 2010; Field et al., 2012; Mills et al., 2017).

15.7.1 Dairy Products

Several bacteriocin applications have been aimed at inactivation of foodborne pathogens or spoilage bacteria in milk and dairy products. Nisin is widely used in dairy products to prevent post-processing contamination by such foodborne pathogens as *Listeria monocytogenes* and *Clostridium botulinum*, and to inhibit *Clostridium tyrobutyricum* to avoid gas-blowing defects in cheeses (Sobrino-Lopez et al., 2009; Thomas and Delves-Broughton, 2005). Immobilized nisin in the packaging (cellulose-based packaging paper) of cheese slices has been utilized to reduce the populations of *Listeria innocua* (commonly used surrogate for *L. monocytogenes*), and *Staphylococcus aureus* by approximately 2 logs CFU/ on cheese slices (Scannell et al., 2000). It has also been tested in various other pasteurized dairy products, including flavored or canned evaporated milks, and chilled desserts.

In recent years nisin applications have also been tested in combination with other technologies such as pulsed electric fields (PEF) and high-pressure processing (HPP) in multi-hurdle approaches (Black, Kelly, and Fitzgerald, 2005; Sobrino-Lopez et al., 2009). Increased reductions of *L. monocytogenes*, *S. aureus*, *B. cereus*, and *E. coli* were observed in skim milk, and whey or simulated milk ultrafiltrate medium during storage when treated with nisin and in combination with PEF (Sobrino-Lopez et al., 2009). In another study, the application of nisin along with HPP in cheese showed increased reductions of *L. innocua*, *Pseudomonas fluorescens*, *E. coli*, and endospores of *Bacillus cereus* (Gálvez et al., 2007; Galvez et al., 2008).

Many other bacteriocins have been isolated and tested for their effectiveness against foodborne pathogens and spoilage bacteria in dairy products and other foods. Examples include lacticin 3147 (a lactococcal bacteriocin), and enterocins CRL35, CCM 4231, and AS-48 (enterococcal bacteriocins) to control *L. monocytogenes*, *S. aureus*, and *B. cereus* in dairy products (Morgan et al., 2001; Giraffa, 1995; Foulquié Moreno et al., 2006; Galvez et al., 2008). Another bacteriocin which has been used in commercial preparations (e.g., Alta™ products) is pediocin PA-1/AcH which can effectively reduce *L. monocytogenes* in half-and-half cream, cottage cheese, and cheese sauces (Rodríguez et al., 2002). Propionicins from propionibacteria are another group of bacteriocins of interest in dairy products (Holo et al., 2002). The long-established MicroGARD® composite produced using *P. freudenreichii* ssp. *shermanii* relies on effectives from an antimicrobial peptide. MicroGARD® has been approved by the FDA to control *Listeria*, *Salmonella*, *E. coli*, yeasts, and molds in a wide variety of food products. MicroGARD® 100 is commonly used in yogurt, sour cream, cottage or other cheeses, dairy desserts, and chocolate confectionery applications. (Elsser-Gravesen and Elsser-Gravesen, 2013; von Staszewski and Jagus, 2008). Moreover, MicroGARD® 200 was also reported to control the growth of certain fungi and *Salmonella* spp., *Pseudomonas* spp., and *Yersinia* spp. in sauces, cottage cheese, salad dressing, pasta, baked goods, and fruit applications (Prospect, 2019b).

Makhal et al., (2015) evaluated the effects of MicroGARD® on the physico-chemical, microbiological, and organoleptic properties of directly acidified cottage cheese during 4-day refrigerated storage. MicroGARD® 400 consists of *Propionibacterium freudenreichii* subsp. *shermanii* ATCC 9616 culture fermented in skim milk, It was tested at levels of 0.20, 0.35, and 0.50% and found to effectively minimize the growth of psychrotrophic bacteria, and yeasts and molds (below 1 log CFU/g) up to 16 days of the storage at 0.50% while reducing the pH of the cheese. This change in pH enhanced the antimicrobial activity of the MicroGARD® 400 and caused protein hydrolysis (breakdown of protein molecule into smaller peptides). Moreover, compared to other concentrations, MicroGARD® 400 at 0.50% improved the flavor and aesthetic quality of the product, extending its shelf-life up to 26 days with refrigeration.

Mills et al. (2017) evaluated the efficacies of single- and double-bacteriocin-producing *Lactococcus lactis* capable of producing the class I bacteriocins nisin A and/or lacticin 3147 in terms of starter culture performance and inhibition of *Listeria*. They generated a single-bacteriocin producing commercial starter culture of *L. lactis* CSK2775 by inserting conjugative plasmid pMRC01, carrying either lacticin genes, or the conjugative transposon Tn5276 with nisin genes. The authors also evaluated the effect of production of both bacteriocins by generating a single *L. lactis* strain incorporating both lacticin- and nisin-encoding genes. Coproduction of both bacteriocins was evaluated by superimposing pMRC01 into the nisin transconjugant. Overall, results showed that in laboratory-scale cheeses, individual or combinations of bacteriocins (class I) effectively inactivated *L. innocua* populations; however, *Lactobacillus plantarum* (class II bacteriocin producer) along with double-bacteriocin producers were most effective in reducing *L. innocua* populations in cheese, where *Listeria* counts were reduced to undetectable levels by week 4. These results suggest that a double-bacteriocin producer along with *L. plantarum* as a protective culture can be utilized in the dairy industry to control the growth of pathogenic bacteria.

15.7.2 Meat Products

Bacteriocins alone or in combination with other hurdles have been assessed for the preservation of meats as well as the decontamination of carcasses. To enhance shelf-life and avoid cross-contamination, raw meat is usually chilled, followed by modified atmospheric packaging (MAP), vacuum-packaging, or active-packaging with O_2 scavengers or CO_2 systems (Véronique, 2008; McMillin, 2008). Additional combinations include the incorporation of low-dose irradiation, HPP, and UV irradiation (Aymerich et al., 2008). These treatments in combination have shown that the incorporation of bacteriocins improves product safety and extends shelf-life with some (~2 logs reduction) effects on the initial meat microbiota. Woraprayote et al. (2016) surveyed various applications of bacteriocins in meat and meat products, especially those incorporating hurdle technology. Studies have revealed reductions in *L. monocytogenes*, *E. coli* O157:H7, and *Brochothrix thermosphacta* (a spoilage bacterium) populations in meat products treated with nisin on the surface followed by lysozyme, organic acids, chelators, vacuum-packaging, or MAP treatment (Stergiou et al., 2006; Thomas and Delves-Broughton, 2005). Pediocin PA-1/Ach treatment of raw meat reduced *L. monocytogenes* and *C. perfringens* populations and inhibited the growth of *B. thermosphacta* (Nieto-Lozano et al., 2006; Rodríguez et al., 2002). Various other bacteriocins have also been tested against spoilage or pathogenic bacteria in raw meats or poultry, such as bifidocins, pentocins, sakacins, lactococcins, lactocins, and carnobacteriocins with variable results (Galvez et al., 2008; Aymerich et al., 2008). Bacteriocins have also been tested in combination with plant-derived essential oils to improve the inactivation of *S.* Enteritidis and *L. monocytogenes* in minced meat products (Solomakos et al., 2008; Govaris et al., 2010).

Another approach has also been evaluated which involves bacteriocin surface treatment along with refrigeration, irradiation, and HPP, after packaging. It has been shown that the application of combinations of nisin (5000 AU/ml), and nisin + lysozyme (2 mg nisin + 10 mg lysozyme/ml) along with in-package heat treatments (65°C for 32 sec) reduced *L. monocytogenes* populations to undetectable levels for 2–3 weeks of storage of turkey bologna (Mangalassary et al., 2008). In another study, frankfurters inoculated with *L. monocytogenes* and treated with pediocin (6000 AU/ml) followed by vacuum packaging and in-package heat treatments (71°C, 81°C, or 96°C) for 30, 60, or 120 sec, were stored at 4°C, 10°C, and 25°C for up to 12 weeks. Results showed an overall reduction of 1.4 to 3.0 log CFU/g in *L. monocytogenes* populations during storage (Chen et al., 2004a). In ready-to-eat (RTE) sausages (20-mm diameter and ~53-mm length) *L. innocua* populations were reduced (1.4 logs CFU/sausage) during refrigerated storage after a 2-minute dip treatment with Nisaplin™ followed by 9 pulses of pulsed light per cm^2 exposure (Uesugi and Moraru, 2009). The authors suggested that to enhance the safety of RTE meat products, pulsed light alone or in combination with bacteriocin treatment can be applied at a post-processing step to reduce contamination. Studies have shown that low-dose irradiation in combination with pediocin-treated hot-dogs reduced *L. monocytogenes* populations by 5.0 logs throughout storage (Chen et al., 2004b). Another study revealed that vacuum-packaged RTE turkey meat with active (pectin–nisin) film increased *L. monocytogenes* inactivation (5.1 logs CFU/g to undetectable levels) by the end of the storage (Jin et al., 2009). Moreover, the application of bacteriocin-containing packing films along with HPP treatment can effectively reduce *L. monocytogenes* cell counts from about 6 logs CFU/g to undetectable levels in RTE meat products after 60 days of storage at 4°C (Aymerich et al., 2008). Enterocin-containing packaging film in combination with HPP reduced counts of *L. monocytogenes* and limited recovery of injured cells during temperature abuse of cooked ham (Marcos et al., 2008). A combination treatment of nisin addition and HPP to sliced cooked ham with nisin and HPP reduced *Salmonella* populations to undetectable levels during refrigerated storage. One can assume that combination treatments with other antimicrobials or food-processing technologies could potentiate bacteriocin biocontrol capabilities.

Studies using fermented meats have shown reductions of populations of *S. aureus* and *L. monocytogenes* using leucocins, nisin, and the enterocins AS-48 and CCM 4231 A and B bacteriocin-producing pediococci and enterococci (Chen and Hoover, 2003; Amezquita and Brashears 2002; Aymerich et al., 2008; Galvez et al., 2008; Rodríguez et al., 2002). Moreover, starter cultures producing pediocin PA-1 usually do not antagonize desirable natural meat bacteria, such as *Staphylococcus carnosus* and lactobacilli, while inhibiting pathogenic types (Gonzalez and Kunka, 1987). Nisin-positive lactococcal isolates recovered from fermented pork sausages (*nham*, usually consumed uncooked in parts of Thailand), could be utilized as starter cultures in meat fermentations. This fermented sausage is usually produced at household levels under unhygienic conditions and has been associated with product inconsistency, spoilage, and pathogen contamination (Noonpakdee et al., 2003). Sparo et al. (2008) used craft dry-fermented sausages composed of beef, pork, and bacon (in ratio of 80:15:5) inoculated with *Enterococcus faecalis* CECT7121 (enterocin MR99 producer) to reduce *S. aureus*, Enterobacteriaceae, and other Gram-positive cocci to undetectable levels (1.5–2 logs CFU/g reductions) at the end of a drying process on day 21.

15.7.3 Fruits and Vegetables

The application of bacteriocins such as pediocin, enterocin AS-48, or nisin in whole or processed fruits and vegetables has been shown to inactivate *B. cereus*, *L. monocytogenes*, *Salmonella*, and *E. coli*, during storage (Silveira et al., 2008; Randazzo et al., 2009; Abriouel et al., 2010; Galvez et al. 2008; Ukuku et al., 2005). In general, these results revealed that bacteriocin alone can significantly reduce (2 logs CFU/g) Gram-positive pathogens as compared to Gram-negatives. Also, Randazzo et al. (2009) found that bacteriocin treatment was effective only in combination with other antimicrobials, such as phenolic compounds, and inorganic or organic acids. Studies have also evaluated the effectiveness of bacteriocin-producing live cultures against foodborne pathogens on fresh produce surfaces (Trias, et al., 2008a; Trias et al., 2008b; Galvez et al., 2008). However, the efficacy of these cultures significantly depends on ecological factors such as the capability to grow and produce bacteriocins on produce surfaces while competing with resident microbiota.

Another approach has been proposed in which nisin and enterocins are added to fruit juices and drinks to control spoilage bacteria such as endospore-forming *Alicyclobacillus acidoterrestris* and thermophilic *Geobacillus stearothermophilus*. In low-acid juices or drinks such as sports and energy beverages, where foodborne pathogens can proliferate if the pH increases during storage, enterocins (1 μg/ml) have been reported to inactivate *L. monocytogenes*, *S. aureus*, and *B. cereus* populations from 4–6 logs to below the detection limit of 10 CFU/ml after 24 hours of treatment (Viedma et al., 2009). Moreover, in fresh fruit juices, combined treatments of nisin and enterocin AS-48, and PEF increased antibacterial effectiveness by 2.27 logs CFU/ml and reduced the risks of survivor proliferation with product storage (Liang et al., 2002; Mosqueda-Melgar et al., 2008).

Addition of the bacteriocin AS-48 at 16 μg/ml to boiled rice can prevent outgrowth of *B. cereus* endospore after a 1-minute treatment at 90°C; the treatment also avoided enterotoxin production between temperatures 6 to 37°C (Grande et al., 2006). Studies have shown that the application of nisin, or pediocin, alone or in-combination with thermal treatments, can effectively reduce *B. cereus* populations to undetectable levels in ready-to-eat products (potato, mushrooms, soups, purees) for up to 30 days at 6°C (Cabo et al., 2009; Galvez et al., 2008). The incorporation of bacteriocins in low-acid canned fruits and vegetables such as canned peaches syrup, tomato paste, and canned pineapple juice, along with heat treatment (80–95°C for 5 minutes) can significantly increase *B. coagulans* endospore deactivation (2.2 logs) (Lucas et al., 2006).

15.8 Summation

Screening LAB isolates with bacteriocin-like activity against *L. monocytogenes* and *S. aureus* from various food sources such as milk has been very popular as the search for valuable new peptides draws continual interest (Silva et al., 2018; Li, 2018; Castilho et al., 2019). Although the production of bacteriocins by pathogens effective against other pathogens is well-documented, it is doubtful that, even with confirmed safety and purity, bacteriocins such as monocins of *L. monocytogenes*, boticins of *C. botulinum*, or staphylococcins of *S. aureus* would be approved for food use. The use of these peptides or proteins would bear the stigma and concern related to their site of production.

One possible concern is that when a bacteriocin is used as a food additive, it may far exceed the concentration to which a consumer is normally exposed by ingestion of fermented foods containing bacteriocin-producing cultures. Such high levels may trigger an allergic or toxicological response, although this is believed highly unlikely. If bacteriocins of LAB are added at levels equivalent to that found in cultured foods, then they should be safe for consumption; however, toxicological testing is required and costs can be daunting.

Adding to the issue of cost is the required levels of bacteriocin necessary in the food to demonstrate effectiveness over the shelf-life of the product. Efficacy testing continues over a range of different, individual products; examples include the optimization of enterocin CCM 4231 versus *L. monocytogenes* in dry fermented salami (Laukova et al., 1999), the development of a lacticin 3147-enriched whey powder against *L. monocytogenes* and *S. aureus* in dairy-related products (Morgan et al., 1999), and improved control of *L. monocytogenes* with semipurified bacteriocins from *Carnobacterium* species in vacuum-packed, refrigerated cold-smoked salmon (Duffes et al., 1999). Studies such as these indicate that necessary bacteriocin concentrations can be quite high and issues of stability and consistency of effectiveness in the product remain, as do problems of low yield in the production of purified or semipurified preparations. Fermentation optimization is necessary to lower costs. As a result of the inherent difficulties with purified bacteriocin preparations, the use of bacteriocin-producing cultures of lactic acid bacteria in food fermentations or the incorporation of food ingredients containing bacteriocin-containing growth extracts (e.g., MicroGARD® and ALTA™2341) are alternative approaches that will continue to be employed in those situations and products where these procedures or approaches have merit. The modification of starter cultures used in commercial fermentations with genes for bacteriocin production and immunity can be expected to continue.

Elucidation of signaling mechanisms for the synthesis of bacteriocins along with a better understanding of the mode of action of bacteriocins should lead to expanded applications for these peptides. Clearly similar research thrusts in these and related areas will steadily evolve to increase our knowledge base and provide answers to address the questions limiting the use of bacteriocins. Additionally, it can be assumed that nisin and other bacteriocins will continue to be incorporated and developed into hurdle concept methods and processes of food preservation for individual products. The use of bacteriocins of LAB to specifically inhibit pathogens such as *L. monocytogenes* and *S. aureus* can expect attention in order to improve food safety. It may be possible to target Gram-negative bacteria in foods by inserting genes from bacteriocins from Gram-negative bacteria into fermentative Gram-positive bacteria.

Protein engineering of bacteriocins can be expected to persist and possibly thrive. For example, Rollema et al. (1995) improved the solubility and stability of nisin Z by replacing Asn-27 or His-31 with lysine; Johnsen et al. (2000) improved the stability of pediocin PA-1 at 4°C and room temperature by replacing Met-31 with alanine, isoleucine, or leucine. Examples such as these indicate solutions to some of the problems related to bacteriocin application.

REFERENCES

Abdel-Bar, N. M. and Harris, N. D. 1984. Inhibitory effect of *Lactobacillus bulgaricus* on psychrotrophic bacteria in associative cultures and in refrigerated foods. *J. Food Prot.* 47:61.

Abdel-Bar, N. M., Harris, N. D. and Rill, R. L. 1987. Purification and properties of an antimicrobial substance produced by *Lactobacillus bulgaricus. J. Food Sci.* 52:411.

Abee, T., Krockel, L. and Hill, C. 1995. Bacteriocins: Modes of action and potentials in food preservation and control of food poisoning. *Intl. J. Food Microbiol.* 28:169.

Abriouel, H., Lucas, R., Omar, N. B., Valdivia, E. and Gálvez, A. 2010. Potential applications of the cyclic peptide enterocin AS-48 in the preservation of vegetable foods and beverages. *Probiotics Antimicrob. Prot.* 2:77–89.

Ahn, C. and Stiles, M. E. 1990. Plasmid-associated bacteriocin production by a strain of *Carnobacterium piscicola* from meat. *Appl. Environ. Microbiol.* 56:2503.

Al Kassaa, I., Hober, D., Hamze, M., Chihib, N. E. and Didier, D. 2014. Antiviral potential of lactic acid bacteria and their bacteriocins. *Probiotics Antimicrob. Prot.* 6:177.

Altena, K., Guder, A., Cramer, C. and Bierbaum, G. 2000. Biosynthesis of the lantibiotic Mersacidin: Organization of a Type B lantibiotic gene cluster. *Appl. Environ. Microbiol.* 66:2565–2571.

Alvarez-Sieiro, P., Montalbán-López, M., Mu, D. and Kuipers, O. P. 2016. Bacteriocins of lactic acid bacteria: Extending the family. *Appl. Microbiol. Biotech.* 100:2939.

Amezquita, A. and Brashears, M. M. 2002. Competitive inhibition of *Listeria monocytogenes* in ready-to-eat meat products by lactic acid bacteria. *J. Food Prot.* 65:316–325.

Anand, S. K., Srinivasan, R. A. and Rao, L. K. 1985. Antibacterial activity associated with *Bifidobacterium bifidum. Cultured Dairy Prod. J.* 20:21.

Andersson, R. 1986. Inhibition of *Staphylococcus aureus* and spheroplasts of Gram-negative bacteria by an antagonistic compound produced by a strain of *Lactobacillus plantarum. Intl. J. Food Microbiol.* 3:149–160.

Atrih, A., Rekhif, N., Moir, A. J. G., Lebrihi, A. and Lefebvre, G. 2001. Mode of action, purification and amino acid sequence of plantaricin C19, an anti-*Listeria* bacteriocin produced by *Lactobacillus plantarum. Intl. J. Food Microbiol.* 68:93.

Axelsson, L.T., Ahrné, S.E.I., Andersson, M.C. and Ståhl, S.R. 1988. Identification and cloning of a plasmid-encoded erythromycin resistance determinant from *Lactobacillus reuteri. Plasmid* 20:1717–174.

Axelsson, L., Katla, T., Bjornslett, M., Eijsink, V. G. H. and Holck, A. 1998. A system for heterologous expression of bacteriocins in *Lactobacillus sake. FEMS Microbiol. Lett.* 168:137.

Aymerich, T., Picouet, P. A. and Monfort, J. M. 2008. Decontamination technologies for meat products. *Meat Sci.* 78:114–129.

Balciunas, E. M., Martinez, F. A. C., Todorov, S. D., de Melo Franco, B. D. G., Converti, A. and de Souza Oliveira, R. P. 2013. Novel biotechnological applications of bacteriocins: A review. *Food Control* 32:134.

Bardowski, J., Kozak, W. and Dobrzanski, W. T. 1979. Further characterization of lactostrepcins-acid bacteriocins of lactic streptococci. *Acta Microbiol. Pol.* 28:93.

Bardowski, J. and Kozak, W. 1981. Preliminary attempts at bacteriocin typing of lactic streptococci. *Acta Microbiol. Pol.* 30:27.

Barefoot, S. F. and Klaenhammer, T. R. 1983. Detection and activity of lactacin B, a bacteriocin produced by *Lactobacillus acidophilus. Appl. Environ. Microbiol.* 45:1808.

Barefoot, S. F. and Klaenhammer, T. R. 1984. Purification and characterization of the *Lactobacillus acidophilus* bacteriocin lactacin B. *Antimicrob. Agents Chemother.* 26:328.

Belguesmia, Y., Naghmouchi, K., Chihib, N.-E. and Drider, D. 2011. Class IIa bacteriocins: Current knowledge and perspectives. In *Prokaryotic Antimicrobial Peptides*, pp. 171–195. Springer, New York, NY.

Berry, E. D., Liewen, M. B., Mandigo, R. W. and Hutkins, R. W. 1990. Inhibition of *Listeria monocytogenes* by bacteriocin-producing *Pediococcus* during the manufacture of fermented semidry sausage. *J. Food Prot.* 53:194.

Berry, E. D., Hutkins, R. W. and Mandigo, R. W. 1991. The use of bacteriocin-producing *Pedicoccus acidilactici* to control post processing *Listeria monocytogenes* contamination of frankfurters. *J. Food Prot.* 54:681.

Bhunia, A. K., Johnson, M. C. and Ray, B. 1987. Direct detection of an antimicrobial peptide of *Pediococcus acidilactici* in sodium dodecyl sulfate polyacrylamide gel electrophoresis. *J. Industrial Microbiol.* 2:319.

Bhunia, A. K., Johnson, M. C. and Ray, B. 1988. Purification, characterization and antimicrobial spectrum of a bacteriocin produced by *Pediococcus acidilactici*. *J. Appl. Bacteriol.* 65:261.

Biswas, S. R., Ray, P., Johnson, M. C. and Ray, B. 1991. Influence of growth conditions on the production of a bacteriocin, pediocin AcH, by *Pediococcus acidilactici* H. *Appl. Environ. Microbiol.* 57:1265.

Black, E. P., Kelly, A. L. and Fitzgerald, G. F. 2005. The combined effect of high pressure and nisin on inactivation of microorganisms in milk. *Innov. Food Sci. Emerg. Technol.* 6:286–292.

Blom, H., Katla, T., Hagen, B. F. and Axelsson, L. 1997. A model assay to demonstrate how intrinsic factors affect diffusion of bacteriocins. *Intl. J. Food Microbiol.* 38:103.

Blom, H., Katla, T., Holck, A., Sletten, K., Axelsson, L. and Holo, H. 1999. Characterization, production, and purification of leucocin H, a two-peptide bacteriocin from *Leuconostoc* MF215B. *Curr. Microbiol.* 39:43.

Blom, H., Katla, T., Nissen, H. and Holo, H. 2001. Characterization, production and purification of carnocin H, a bacteriocin produced by *Carnobacterium* 377. *Curr. Microbiol.* 43:227.

Bradley, D. E. 1967. Ultrastructure of bacteriophages and bacteriocins. *Bacteriological Review* 31:230.

Branen, A. L., Go, H. C. and Genske, R. P. 1975. Purification and properties of antimicrobial substances provided by *Streptococcus diacetylactis* and *Leuconostoc citrovorum*. *J. Food Sci.* 40:446.

Bruno, S. E., Kaiser, A. and Montville, T. J. 1992. Depletion of proton motive force by nisin in *Listeria monocytogenes* cells. *Appl. Environ. Microbiol.* 58:2255.

Cabo, M. L., Torres, B., Herrera, J. J. R., Bernardez, M.and Pastoriza, L. 2009. Application of nisin and pediocin against resistance and germination of *Bacillus* spores in sous vide products. *J. Food Prot.* 72:515–523.

Callewaert, R., Hugas, M. and De Vuyst, L. 2000. Competitiveness and bacteriocin production of *Enterococcus* in the production of Spanish-style dry fermented sausages. *Intl. J. Food Microbiol.* 57:33.

Carolissen-Mackay, V., Arendse, G. and Hastings, J. W. 1997. Purification of bacteriocins from lactic acid bacteria: Problems and pointers. *Intl. J. Food Microbiol.* 34:1.

Casaus, P., Nilsen, T., Cintas, L. M., Nes, I. F., Hernandez, P. E. and Holo, H. 1997. Enterocin B, a new bacteriocin from *Enterococcus faecium* T136 which can act synergistically with enterocin A. Microbiol. 143:2287.

Cascales, E., Buchanan, S. K., Duché, D., Kleanthous, C., Lloubes, R., Postle, K., Riley, M., Slatin, S. and Cavard, D. 2007. Colicin biology. *Microbiol. Molecular Biol. Rev.* 71:158–229.

Castellano, P., Farias, M. E., Holzapfel, W. and Vignolo, G. 2001. Sensitivity variations of *Listeria* strains to the bacteriocins, lactocin 705, enterocin CRL35 and nisin. *Biotech. Lett.* 23:605.

Castilho, N. P. A., Colombo, M., de Oliveira, L. L., Todorov, S. D. and Nero, L. A. 2019. *Lactobacillus curvatus* UFV-NPAC1 and other lactic acid bacteria isolated from calabresa, a fermented meat product, present high bacteriocinogenic activity against *Listeria monocytogenes*. *BMC Microbiol.* 19:63.

Chen, C. M., Sebranek, J. G., Dickson, J. S. and Mendonca, A. F. 2004a. Combining pediocin (ALTA 2341) with post-packaging thermal pasteurization for control of *Listeria monocytogenes* on frankfurters. *J. Food Prot.* 67:1855–1865.

Chen, C. M., Sebranek, J. G., Dickson, J. S. and Mendonca, A. F. 2004b. Combining pediocin with post-packaging irradiation for control of *Listeria monocytogenes* on frankfurters. *J. Food Prot.* 67:1866–1875.

Chen, H. and Hoover, D. G. 2003. Bacteriocins and their food applications. *Comp. Rev. Food Sci. Food Safety* 2:82–100.

Chen, Y., Ludescher, R. D. and Montville, T. J. 1997a. Electrostatic interactions, but not the UGNGV consensus motif, govern the binding of pediocin PA-1 and its fragments to phospholipid vesicles. *Appl. Environ. Microbiol.* 63:4770.

Chen, Y., Shapira, R., Eisenstein, M. and Montville, T. J. 1997b. Functional characterization of pediocin PA-1 binding to liposomes in the absence of a protein receptor and its relationship to a predicted tertiary structure. *Appl. Environ. Microbiol.* 63:524.

Chikindas, M. L., Weeks, R., Drider, D., Chistyakov, V. A. and Dicks, L. M. T. 2018. Functions and emerging applications of bacteriocins. *Curr. Opinion in Biotech.* 49:23–28.

Cintas, L. M., Casaus, P., Håvarstein, L. S., Hernández, P. E. and Nes, I. F. 1997. Biochemical and genetic characterization of enterocin P, a novel *sec* dependent bacteriocin from *Enterococcus faecium* P13 with a broad antimicrobial spectrum. *Appl. Environ. Microbiol.* 63:4321.

Cintas, L.M., P. Casaus, Holo, H., Hernandez, P.E., Nes, I.F. and Håvarstein, L.S. 1998. Enterocins L50A and L50B, two novel bacteriocins from *Enterococcus faecium* L50, are related to Staphylococcal hemolysins, *J. Bacteriol.* 180:1988–1994.

Cintas, L. M., Herranz, C., Hernandez, P. E., Casaus, M. P. and Nes, I. F. 2001. Review: Bacteriocins of lactic acid bacteria. *Food Sci. & Technology Intl.* 7:281.

Clemens, R., Zaschke-Kriesche, J., Khosa, S. and Smits, S. H. J. 2018. Insight into two ABC transporter families involved in lantibiotic resistance. *Front. Molecular BioSci.* 4:91.

Collins, M. D., Farrow, J. A. E., Phillips, B. A., Ferusu, S. and Jones, D. 1987. Classification of *Lactobacillus divergens*, *Lactobacillus piscicola* and some catalase-negative, asporogenous, rod-shaped bacteria from poultry in a new genus, *Carnobacterium*. *Intl. J. System. Evol. Microbiol.* 37:310.

Cooper, L. E., Li, B. and van der Donk, W. A. 2010. Biosynthesis and mode of action of lantibiotics. In *Comprehensive Natural Products II: Chemistry and Biology*, pp. 217–256. Elsevier Ltd.

Cotter, P. D., Hill, C. and Ross, R. P. 2005. Food Microbiology. Bacteriocins: Developing innate immunity for food. *Nature Rev. Microbiol.* 3:777.

Coventry, M. J., Gordon, J. B., Alexander, M., Hickey, M. W. and Wan, J. 1996. A food-grade process for isolation and partial purification of bacteriocins of lactic acid bacteria that uses diatomite calcium silicate. *Appl. Environ. Microbiol.* 62:1764.

Cui, Y., Zhang, C., Wang, Y., Shi, J., Zhang, L., Ding, Z., Qu, X. and Cui, H. 2012. Class IIa bacteriocins: diversity and new developments. *Intl. J. Molec. Sci.* 13:16668–16707.

Cuozzo, S. A., Sesma, F., Palacios, J. M., de Ruiz-Holgado, A. P. and Raya, R. R. 2000. Identification and nucleotide sequence of genes involved in the synthesis of lactocin 705, a two-peptide bacteriocin from *Lactobacillus casei* SRL 705. *FEMS Microbiol. Lett.* 185:157.

da Costa, R. J., Flávia, L., Voloski, S., Mondadori, R. G., Duval, E. H. and Fiorentini, Â. M. 2019. Preservation of meat products with bacteriocins produced by lactic acid bacteria isolated from meat. *J. Food Quality*. DOI 10.1155/2019/4726510.

Daeschel, M. A. and Klaenhammer, T. R. 1985. Association of a 13.6-megadaton plasmid in *Pediococcus pentosaceus* with bacteriocin activity. *Appl. Environ. Microbiol.* 50:1538.

Daeschel, M. A. 1989. Antimicrobial substances from lactic acid bacteria for use as food preservatives. *Food Technol.* 43:1641.

Daeschel, M. A., McKenney, M. C. and McDonald, L. C. 1990. Bacteriocidal activity of *Lactobacillus plantarum* C-11. *Food Microbiol.* 7:91.

Davey, G. P. 1981. Mode of action of diplococcin, a bacteriocin from *Streptococcus cremoris* 346. *N.Z. J. Dairy Sci. and Technol.* 16:187.

Davey, G. P. 1984. Plasmid associated with diplococcin production in *Streptococcus cremoris*. *Appl. Environ. Microbiol.* 48:895.

Davey, G. P. and Pearce, L. E. 1980. The use of *Streptococcus cremoris* strains cured of diplococcin production as cheese starters. *N.Z. J. Dairy Sci. Technol.* 15:51.

Davey, G. P. and Richardson, B. C. 1981. Purification and some properties of diplococcin from *Streptococcus cremoris* 346. *Appl. Environ. Microbiol.* 41:84.

Davey, G. P. and Pearce, L. E. 1982. Production of diplococcin by *Streptococcus cremoris* and its transfer to nonproducing group N streptococci. In *Microbiology 1982*, edited by D. Schlessinger, p. 221. American Society for Microbiology, Washington, DC.

De Klerk, H. C. and Coetzee, J. N. 1961. Antibiosis among lactobacilli. *Nature* 192:340.

De Klerk, H. C. and Coetzee, J. N. 1967. Bacteriocinogeny in *Lactobacillus fermenti*. *Nature* 214:609.

De Klerk, H. C. and Smit, J. A. 1967. Properties of a *Lactobacillus fermenti* bacteriocin. *J. General Microbiol.* 48:309.

Deegan, L. H., Cotter, P. D., Hill, C. and Ross, P. 2006. Bacteriocins: Biological tools for bio-preservation and shelf-life extension. *Intl. Dairy J.* 16:1058–1071.

Degnan, A. J. and Luchansky, J. B. 1992. Influence of beef tallow and muscle on the antilisterial activity of pediocin AcH and liposome-encapsulated pediocin AcH. *J. Food Prot.* 55:552.

Devoyod, J. J. and Poullain, F. 1988. The Leuconostocs. Properties: Their role in dairy technology. *Milk* 68:249.

Dobrzanski, W. T., Bardowski, J., Kozak, W. and Zajdel, J. 1982. Lactostrepcins: Bacteriocins of lactic streptococci. In *Microbiology*, edited by D. Schlessinger, p. 225. American Society for Microbiology, Washington, DC.

Draper, L. A., Cotter, P. D., Hill, C. and Ross, R. P. 2015. Lantibiotic resistance. *Microbiol. Molec. Biology Rev.* 79:171–191.

Drider, D. and Rebuffat, S., eds. 2011. *Prokaryotic Antimicrobial Peptides: From Genes to Applications*. Springer Science and Business Media.

Du, L., Liu, F., Zhao, P., Zhao, T. and Doyle, M. P. 2017. Characterization of *Enterococcus durans* 152 bacterocins and their inhibition of *Listeria monocytogenes* in ham. *Food Microbiol.* 68:97.

Duffes, F., Corre, C., Leroi, F., Dousset, X. and Boyaval, P. 1999. Inhibition of *Listeria monocytogenes* by in situ produced and semipurified bacteriocins of *Carnobacterium* spp: On vacuum-packed, refrigerated cold-smoked salmon. *J. Food Prot.* 62:1394.

Dufour, A., Rince, A., Uguen, P. and LePennec, J. P. 2000. IS1675, a novel lactococcal insertion element, forms a transposon-like structure including the lacticin 481 lantibiotic operon. *J. Bacteriol.* 182:5600.

Elsser-Gravesen, D. and Elsser-Gravesen, A. 2013. Biopreservatives. In *Biotechnology of Food and Feed Additives*, pp. 29–49. Springer.

Enan, G., El-Essawy, A., Uyttendaele, M. and Debevere, J. 1996. Antibacterial activity of *Lactobacillus plantarum* UG1 isolated from dry sausage: Characterization, production and bactericidal action of plantaricin UG1. *Intl. J. Food Microbiol.* 30:189.

Ennahar, S., Aoude-Werner, D., Sorokine, O., Van Dorsselaer, A., Bringel, F., Hubert, J. and Hasselmann, C. 1996. Production of pediocin AcH by *Lactobacillus plantarum* WHE 92 isolated from cheese. *Appl. Environ. Microbiol.* 62:4381.

Ennahar, S., Sashihara, T., Sonomoto, K. and Ishizaki, A., 2000. Class IIa bacteriocins: Biosynthesis, structure and activity. *FEMS Microbiol. Rev.* 24:85.

Etchells, J. L., Costilow, R. N., Anderson, T. E. and Bell, T. A. 1964. Pure culture fermentation of brined cucumbers. *Appl. Environ. Microbiol.* 12:523–535.

Fath, M. J. and Kolter, R. 1993. ABC transporters: Bacterial exporters. *Microbiol. Review* 57:995.

Faye, T., Langsrud, T., Nes, I. F. and Holo, H. 2000. Biochemical and genetic characterization of propionicin T1, a new bacteriocin from *Propionibacterium thoenii. Appl. Environ. Microbiol.* 66:4230.

Fernandez, C. F., Shahani, K. M. and Amer, M. A. 1988. Effect of nutrient media and bile salts on growth and antimicrobial activity of *Lactobacillus acidophilus. J. Dairy Sci.* 71:3222.

Ferreira, C. L. and Gilliland, S. E. 1988. Bacteriocin involved in premature death of *Lactobacillus acidophilus* NCFM during growth at pH 6. *J. Dairy Sci.* 71:306.

Field, D., Cotter, P., Hill, C. and Ross, R. P. 2007. Bacteriocins biosynthesis, structure, and function. In *Bacteriocins Current Research and Applications*, edited by M. A. Riley and O. Gillor, pp. 5–43. Horizon Biosciences, Norfolk.

Field, D., Begley, M., O'Connor, P. M., Daly, K. M., Hugenholtz, F., Cotter, P. D., Hill, C. and Ross, R. P. 2012. Bioengineered nisin A derivatives with enhanced activity against both Gram positive and Gram negative pathogens. *PloS ONE* 7:e46884.

Field, D., Ross, R. P. and Hill, C. 2018. Developing bacteriocins of lactic acid bacteria into next generation biopreservatives. *Curr. Opinion in Food Sci.* 20:1–6.

Fleming, H. P., Etchells, J. L. and Costilow, R. N. 1975. Microbial inhibition by an isolate of *Pediococcus* from cucumber brines. *Appl. Microbiol.* 30:1040.

Fleury, Y., Dayem, M. A., Montagne, J. J., Chaboisseau, E., Le Caer, J. P., Nicolas, P. and Delfour, A. 1996. Covalent structure, synthesis, and structure-function studies of mesentericin Y105[37], a defensive peptide from gram-positive bacteria *Leuconostoc mesenteroides. J. Biol. Chem.* 271:14421.

Foegeding, P. M., Thomas, A. B., Pilkington, D. H. and Klaenhammer, T. R. 1992. Enhanced control of *Listeria monocytogenes* by in situ-produced pediocin during dry fermented sausage production. *Appl. Environ. Microbiol.* 58:884.

Foulquié Moreno, M.R., Sarantinopoulos, P., Tsakalidou, E. and De Vuyst, L. 2006. The role and application of enterococci in food and health. *Intl. J. Food Microbiol.* 106:1–24.

Gálvez, A., Abriouel, H., Lucas López, R. and Omar, N. B. 2007. Bacteriocin-based strategies for food biopreservation. *Intl. J. Food Microbiol.* 120:51–70.

Galvez, A., Lucas López, R., Abriouel, H., Valdivia, E. and Omar, N. B. 2008. Application of bacteriocins in the control of foodborne pathogenic and spoilage bacteria. *Crit. Rev. Biotech.* 28:125–152.

Ganzle, M. G., Hertel, C. and Hammes, W. P. 1999a. Resistance of *Escherichia coli* and *Salmonella* against nisin and curvacin A. *Intl. J. Food Microbiol.* 48:37.

Ganzle, M. G., Hertel, C., van der Vossen, J. M. B. M. and Hammes, W. P. 1999b. Effect of bacteriocin-producing lactobacilli on the survival of *Escherichia coli* and *Listeria* in a dynamic model of the stomach and small intestine. *Intl. J. Food Microbiol.* 48:21.

Ganzle, M. G., Holtzel, A., Walter, J., Jung, G. and Hammes, W. P. 2000. Characterization of reutericyclin produced by *Lactobacillus reuteri* LTH2584. *Appl. Environ. Microbiol.* 66:4325.

Garneau, S., Martin, N. I. and Vederas, J. C. 2002. Two-peptide bacteriocins produced by lactic acid bacteria. *Biochimie* 84:577.

Geis, A., Singh, J. and Teuber, M. 1983. Potential of lactic streptococi to produce bacteriocin. *Appl. Environ. Microbiol.* 45:205.

Geissler, S., Gőtz, F. and Kupke, T. 1996. Serine protease EpiP from *Staphylococcus epidermidis* catalyzes the processing of the epidermin precursor peptide. *J. Bacteriol.* 178:284-288.

Gharsallaoui, A., Oulahal, N., Joly, C. and Degraeve, P. 2016. Nisin as a food preservative. Part 1: Physicochemical properties, antimicrobial activity, and main uses. *Crit. Rev. Food Sci. Nutr.* 56:1262–1274.

Gilliland, S. E. and Speck, M. L. 1977. Antagonistic action of *Lactobacillus acidophilus* toward intestinal and foodborne pathogens in associative culture. *J. Food Prot.* 40:820.

Giraffa, G. 1995. Enterococcal bacteriocins: their potential as anti-*Listeria* factors in dairy technology. *Food Microbiol.* 12:291–299.

Gonzalez, C. F. and Kunka, B. S. 1987. Plasmid-associated bacteriocin production and sucrose fermentation in *Pediococcus acidilactici. Appl. Environ. Microbiol.* 53:2534.

Govaris, A., Solomakos, N., Pexara, A. and. Chatzopoulou, P. S. 2010. The antimicrobial effect of oregano essential oil, nisin and their combination against *Salmonella* Enteritidis in minced sheep meat during refrigerated storage. *Intl. J. Food Microbiol.* 137:175–180.

Graham, D. C. and McKay, L. L. 1985. Plasmid DNA in strains of *Pediococcus cerevisiae* and *Pediococcus pentosaceus. Appl. Environ. Microbiol.* 50:532.

Grande, M. J., Lucas, R., Abriouel, H., Valdivia, E., Omar, N. B., Maqueda, M., Martínez-Bueno, M., Martínez-Cañamero, M. and Gálvez, A. 2006. Inhibition of toxicogenic *Bacillus cereus* in rice-based foods by enterocin AS-48. *Intl. J. Food Microbiol.* 106:185–194.

Gratia, A. and Fredericq, P. 1946. *C. R. Seanc. Soc. Biol.* 140:1032, in Mayr-Harting et al. 1972.

Gravesen, A., Axelsen, A. M. J., da Silva, J. M., Hansen, T. B. and Knochel, S. 2002. Frequency of bacteriocin resistance development and associated fitness costs in *Listeria monocytogenes. Appl. Environ. Microbiol.* 68:756.

Grinstead, D. A. and Barefoot, S. F. 1992. Jenseniin G, a heat-stable bacteriocin produced by *Propionibacterium jensenii* P126. *Appl. Environ. Microbiol.* 58:215.

Gross, E. and Morell, J. L. 1967. The presence of dehydroalanine in the antibiotic nisin and its relationship to activity. *J. the American Chemical Society* 89:2791.

Gross, E., Morell, J. L. and Craig, L. C. 1969. Dehydroalanyllysine: Identical COOH-terminal structures in the peptide antibiotics nisin and subtilin. *Proceedings of the National Academy of Sci.* 62:952.

Guder, A. Wiedemann, I. and Sahl, H.-G. 2000. Posttranslationally modified bacteriocins—the lantibiotics. *Biopolymers* 55:62–73.

Guerra, N. R. 2012. A review on some chemical engineering and microbiological aspects considered in the production of highly concentrated probiotic cultures and bacteriocins by lactococci and lactobacilli. *ISRN Chemical Engineering* 2012:1.

Guyonnet, D., Fremaux, C., Cenatiempo, Y. and Berjeaud, J. M. 2000. Method for rapid purification of class IIa bacteriocins and comparison of their activities. *Appl. Environ. Microbiol.* 66:1744.

Hamdan, J. Y. and Mikolajcik, E. M. 1974. Acidolin: An antibiotic produced by *Lactobacillus acidophilus. J. Antibiotics* 27:631.

Harris, L. J., Daeschel, M. A., Stiles, M. E. and Klaenhammer, T. R. 1989. Antimicrobial activity of lactic acid bacteria against *Listeria monocytogenes. J. Food Prot.* 52:384.

Hastings, J.W., Sailer, M., Johnson, K., Roy, K.L., Vederas, J.C. and Stiles, M.E. 1991. Characterization of leucocin A-UAL 187 and cloning of the bacteriocin gene from *Leuconostoc gelidum. J. Bacteriol.* 173:7491–7500.

Helander, I. M., von Wright, A. and Mattila-Sandholm, T.-M. 1997. Potential of lactic acid bacteria and novel antimicrobials against Gram-negative bacteria. *Trends Food Sci. Technol.* 8:146.

Herbin, S., Mathieu, F., Brule, F., Branlant, C., Lefebvre, G. and Lebrihi, A. 1997. Characteristics and genetic determinants of bacteriocin activities produced by *Carnobacterium piscicola* CP5 isolated from cheese. *Curr. Microbiol.* 35:319.

Holo, H., Faye, T., Brede, D. A., Nilsen, T., Ødegård, I., Langsrud, T., Brendehaug, J. and Nes, I. F. 2002. Bacteriocins of propionic acid bacteria. *Le Lait* 82:59–68.

Hoover, D. G., Walsh, P. M., Kolaetis, K. M. and Daly, M. M. 1988. A bacteriocin produced by *Pediococcus* species associated with a 5.5-megadalton plasmid. *J. Food Prot.* 51:29.

Hoover, D. G., Dishart, K. J. and Hermes, M. A. 1989. Antagonistic effect of *Pediococcus* spp. against *Listeria monocytogenes. Food Biotech.* 3:183.

Horn, N., Martinez, M. I., Martinez, J. M., Hernandez, P. E., Gasson, M. J., Rodriguez, J. M. and Dodd, H. M. 1998. Production of pediocin PA-1 by *Lactococcus lactis* using the lactococcin A secretory apparatus. *Appl. Environ. Microbiol.* 64:818.

Horn, N., Martinez, M. I., Martinez, J. M., Hernandez, P. E., Gasson, M. J., Rodriguez, J. M. and Dodd, H. M. 1999. Enhanced production of pediocin PA-1 and coproduction of nisin and pediocin PA-1 by *Lactococcus lactis. Appl. Environ. Microbiol.* 65:4443.

Hsieh, H., Paik, H. and Glatz, B. A. 1996. Improvement of detection and production of propionicin PLG-1, a bacteriocin produced by *Propionibacterium thoenii. J. Food Prot.* 59:734.

Hugas, M. 1998. Bacteriocinogenic lactic acid bacteria for the biopreservation of meat and meat products. *Meat Sci.* 49:S139.

Hughes, D. B. and Hoover, D. G. 1991. Bifidobacteria: Their potential for use in American dairy products. *Food Technolnology.* 45:74.

Ingram, L. C. 1969. Synthesis of the antibiotic nisin: formation of lanthionine and ß-methyl-lanthionine. *Biochim. Biophys. Acta* 184:216.

Ingram, L. C. 1970. A ribosomal mechanism of synthesis for peptides related to nisin. *Biochim. Biophys. Acta* 224:263.

Jack, R. W., Wan, J., Gordon, J., Harmark, K., Davidson, B. E., Hillier, A. J., Wettenhall, R. E. H., Hickey, M. W. and Coventry, M. J. 1996. Characterization of the chemical and antimicrobial properties of piscicolin 126, a bacteriocin produced by *Carnobacterium piscicola* JG126. *Appl. Environ. Microbiol.* 62:2897.

Jacob, F., Lwoff, A., Siminovitch, L. and Wallman, E. 1953. Definition de quelques termes relatifs a la Pysogenie. *Annales de l'Institut Pasteur* 84:222.

Jao, Y. C., Mikolajcik, E. M. and Hansen, P. M. T. 1978. Growth of *Bifidobacterium bifidum* var. *pennsylvanicus* in laboratory media supplemented with amino sugars and spent broth from *Escherichia coli. J. Food Sci.* 43:1257.

Jimenez-Diaz, R., Ruiz-Barba, J. L., Cathcart, D. P., Holo, H., Nes, I. F., Sletten, K. H. and Warner, P. J. 1995. Purification and partial amino acid sequence of plantaricin S, a bacteriocin produced by *Lactobacillus plantarum* LPCO10, the activity of which depends on the complementary action of two peptides. *Appl. Environ. Microbiol.* 61:4459.

Jin, T., Liu, L., Sommers, C. H., Boyd, G. and Zhang, H. 2009. Radiation sensitization and post-irradiation proliferation of *Listeria monocytogenes* on ready-to-eat deli meat in the presence of pectin-nisin films. *J. Food Prot.* 72:644–649.

Joerger, M. C. and Klaenhammer, T. R. 1986. Characterization and purification of helveticin J and evidence for a chromosomally determined bacteriocin produced by *Lactobacillus helveticus* 481. *J. Bacteriol.* 167:439.

Johnsen, L., Fimland, G., Eijsink, V. and Nissen-Meyer, J. 2000. Engineering increased stability in the antimicrobial peptide pediocin PA-1. *Appl. Environ. Microbiol.* 66:4798.

Joosten, H. M. L. and Nunez, M. 1996. Prevention of histamine formation in cheese by bacteriocin-producing lactic acid bacteria. *Appl. Environ. Microbiol.* 62:1178.

Juturu, V. and Wu, J. C. 2018. Microbial production of bacteriocins: Latest research development and applications. *Biotech. Advances* 36:2187–2200.

Kabuki, T., Saito, T., Kawai, Y., Uemura, J. and Itoh, T. 1997. Production, purification and characterization of reutercin 6, a bacteriocin with lytic activity produced by *Lactobacillus reuteri* LA6. *Intl. J. Food Microbiol.* 34:145.

Kanauchi, M. 2018. *Lactic Acid Bacteria: Methods and Protocols.* Humana Press, New York, NY.

Kekessy, D. A. and Piguet, J. D. 1970. New method for detecting bacteriocin production. *Appl. Microbiol.* 20:282.

Kiesau, P., Eikmanns, U., Gutowski-Eckel, Z., Weber, S., Hammelmann, M. and Entian, K.-D. 1997. Evidence for a multimeric subtilin synthetase complex. *J. Bacteriol.* 179:1475.

Klaenhammer, T.R. 1988. Bacteriocins of lactic acid bacteria. *Biochemie* 70:337349.

Klaenhammer, T. R. 1993. Genetics of bacteriocins produced by lactic acid bacteria. *FEMS Microbiol. Review* 12:39.

Klein, C. and Entian, K.D. 1994. Genes involved in self-protection against the lantibiotic subtilin produced by *Bacillus subtilis* ATCC 6633. *Appl. Environ. Microbiol.* 60:2793–2801.

Kodama, R. 1952. Studies on lactic acid bacteria. 1. Lactolin, a new antibiotic substance produced by lactic acid bacteria. *J. Antibiotics* 5:72.

Kojic, M., Svircevic, J., Banina, A. and Topisirovic, L. 1991. Bacteriocin-producing strain of *Lactococcus lactis* subsp. *diacetylactis* S50. *Appl. Environ. Microbiol.* 57:1835.

Konisky, J. 1982. Colicins and other bacteriocins with established modes of action. *Ann. Rev. Microbiol.* 36:125–144.

Kozak, W., Bardowski, J. and Dobrzanski, W. T. 1978. Lactostrepcins-acid bacteriocins produced by lactic streptococci. *J. Dairy Research* 45:247.

Krier, F., Revol–Junelles, A. M. and Germain, P. 1998. Influence of temperature and pH on production of two bacteriocins by *Leuconostoc mesenteroides* subsp. *mesenteroides* FR52 during batch fermentation. *Appl. Microbiol. Biotech.* 50:359.

Kuipers, A., Rink, R. and Moll, G. N. 2011. Genetics, biosynthesis, structure, and mode of action of lantibiotics. In *Prokaryotic Antimicrobial Peptides*, pp. 147–169. Springer, New York, NY.

Kuipers, O. P., Beerthuyzen, M. M., de Ruyter, P. G. G. A., Luesink, E. J. and de Vos, W. M. 1995. Autoregulation of nisin biosynthesis in *Lactococcus lactis* by signal transduction. *J. Biol. Chem.* 270:27299–27304.

Laemli, U. K. and Favre, M. 1973. Maturation of the head of bacteriophage T_4. I. DNA packing events. *J. Molec. Biology* 80:575.

Laukova, A., Czikkova, S., Laczkova, S. and Turek, P. 1999. Use of enterocin CCM 4231 to control *Listeria monocytogenes* in experimentally contaminated dry fermented Hornad salami. *Intl. J. Food Microbiol.* 52:115.

Leer, R.J. van der Vossen, J.M.B.M., van Giezen, M., van Noort J.M. and Pouwels, P.H. 1995. Genetic analysis of acidocin B, a novel bacteriocin produced by *Lactobacillus acidophilus*. Microbiology 141:1629–1635.

LeMarrec, C., Hyronimus, B., Bressollier, P., Verneuil, B. and Urdaci, M. C. 2000. Biochemical and genetic characterization of coagulin, a new antilisterial bacteriocin in the pediocin family of bacteriocins, produced by *Bacillus coagulans* I_4. *Appl. Environ. Microbiol.* 66:5213.

Leroy, F. and de Vuyst, L. 1999. The presence of salt and a curing agent reduces bacteriocin production by *Lactobacillus sakei* CTC 494, a potential starter culture for sausage fermentation. *Appl. Environ. Microbiol.* 65:5350.

Lewus, C. B., Sun, S. and Montville, T. J. 1992. Production of an amylase-sensitive bacteriocin by an atypical *Leuconostoc paramesenteroides. Appl. Environ. Microbiol.* 58:143.

Li, Pinglan. 2018. Purification and characterization of plantaricin LPL-1, a novel class IIa bacteriocin produced by *Lactobacillus plantarum* LPL-1 isolated from fermented fish. *Frontiers in Microbiol.* 9:2276.

Liang, Z., Mittal, G. S. and Griffiths, M. W. 2002. Inactivation of *Salmonella* Typhimurium in orange juice containing antimicrobial agents by pulsed electric field. *J. Food Prot.* 65:1081–1087.

Lucas, R., Grande, M. J., Abriouel, H., Maqueda, M., Omar, N. B., Valdivia, E., Martínez-Cañamero, M. and Gálvez, A. 2006. Application of the broad-spectrum bacteriocin enterocin AS-48 to inhibit *Bacillus coagulans* in canned fruit and vegetable foods. *Food Chem. Toxicol.* 44:1774–781.

Lyon, W. J. and Glatz, B. A. 1991. Partial purification and characterization of a bacteriocin produced by *Propionibacterium thoenii. Appl. Environ. Microbiol.* 57:701.

Makhal, S., Kanawjia, S. K. and, Giri, A. 2015. Effect of microGARD on keeping quality of direct acidified Cottage cheese. *J. Food Sci. Technol.* 52:936–943.

Manca de Nadra, M. C., Sandino de Lamelas, D. and Strasser de Saad, A. M. 1998. Pediocin N5p from *Pediococcus pentosaceus*: Adsorption on bacterial strains. *Intl. J. Food Microbiol.* 39:79.

Mangalassary, S., Han, I., Rieck, J., Acton, J. and Dawson, P. 2008. Effect of combining nisin and/or lysozyme with in-package pasteurization for control of *Listeria monocytogenes* in ready-to-eat turkey bologna during refrigerated storage. *Food Microbiol.* 25:866–870.

Mao, Y., Muriana, P. M. and Cousin, M. A. 2001. Purification and transpositional inactivation of lacticin FS92, a broad-spectrum bacteriocin produced by *Lactococcus lactis* FS92. *Food Microbiol.* 18:165.

Marcos, B., Aymerich, T., Monfort, J.M. and Garriga, M. 2008. High-pressure processing and antimicrobial biodegradable packaging to control *Listeria monocytogenes* during storage of cooked ham. *Food Microbiol.* 25:177–182.

Marugg, J.D., Gonzalez, C.F., Kunka, B.S. Ledeboer, A.M., Pucci, M.J., Toonen, M.Y., Walker, S.A., Zoetmulder, L.C. and Vandenbergh, P.A. 1992. Cloning, expression, and nucleotide sequence of genes involved in production of pediocin PA-1, and bacteriocin from *Pediococcus acidilactici* PAC1.0. *Appl. Environ. Microbiol.* 58:2360–2367.

Mattick, A. T. R. and Hirsch, A. 1947. Further observation on an inhibitor (nisin) from lactic streptococci. *Lancet* 2:5.

McAuliffe, O., Ryan, M. P., Ross, R. P., Hill, C., Breeuwer, P. and Abee, T. 1998. Lacticin 3147, a broad-spectrum bacteriocin which selectively dissipates the membrane potential. *Appl. Environ. Microbiol.* 64:439.

McAuliffe, O., Hill, C. and Ross, R. P. 1999. Inhibition of *Listeria monocytogenes* in cottage cheese manufactured with a lacticin 3147-producing starter culture. *J. Appl. Microbiol.* 86:251.

McAuliffe, O. Ross, R.P. and Hill, C. 2001. Lantibiotics: structure, biosynthesis and mode of action. *FEMS Microbiol. Rev.* 25:285–308.

McCormick, E. L. and Savage, D. C. 1983. Characterization of *Lactobacillus* sp. strain 100–37 from the murine gastrointestinal tract: Ecology, plasmid content and antagonistic activity toward *Clostridium rhamnosum* H1. *Appl. Environ. Microbiol.* 46:1103.

McMillin, K. W. 2008. Where is MAP going? A review and future potential of modified atmosphere packaging for meat. *Meat Sci.* 80:43–65.

Meghrous, J., Euloge, P., Junelles, A. M., Ballongue, J. and Petitdemanage, H. 1990. Screening of *Bifidobacterium* strains for bacteriocin production. *Biotech. Lett.* 12:575.

Mehta, A. M., Patel, K. A. and Dave, P. J. 1983. Isolation and purification of an inhibitory protein from *Lactobacillus acidophilus* AC. *J. Microbiol.* 37:37.

Miller, K. W., Schamber, R., Osmanagaoglu, O. and Ray, B. 1998. Isolation and characterization of pediocin AcH chimeric protein mutants with altered bactericidal activity. *Appl. Environ. Microbiol.* 64:1997.

Mills, S., Griffin, C., O'Connor, P. M., Serrano, L. M., Meijer, W. C., Hill, C. and Ross, R. P. 2017. A multibacteriocin cheese starter system, comprising nisin and lacticin 3147 in *Lactococcus lactis*, in combination with plantaricin from *Lactobacillus plantarum*. *Appl. Environ. Microbiol.* 83:e00799–17.

Moll, G.N., Konings, W.N. and Driessen, A.J.M. 1999. Bacteriocins: mechanism of membrane insertion and pore formation. *Antonie van Leeuwenhoek* 76:185–198.

Morgan, S. M., Galvin, M., Kelly, J., Ross, R. P. and Hill, C. 1999. Development of a lacticin 3147-enriched whey powder with inhibitory activity against foodborne pathogens. *J. Food Prot.* 62:1011.

Morgan, S.M., Galvin, M., Ross, R.P. and Hill, C. 2008. Evaluation of a spray-dried lacticin 3147 powder for the control of *Listeria monocytogenes* and *Bacillus cereus* in a range of food systems. *Lett. Appl. Microbiol.* 33:387–391.

Mortvedt, C.I. and Nes, I.F. 1990. Plasmid-associated bacteriocin production by a *Lactobacillus sake* strain. *J. General Microbiol.* 136:1601–1607.

Mortvedt, C. I., Nissen-Meyer, J., Sletten, K. and Nes, I. F. 1991. Purification and amino acid sequence of lactacin S, a bacteriocin produced by *Lactobacillus sake* 1045. *Appl. Environ. Microbiol.* 57:1829.

Mosqueda-Melgar, J., Elez-Martinez, P., Raybaudi-Massilia, R. M. and Martin-Belloso, O. 2008. Effects of pulsed electric fields on pathogenic microorganisms of major concern in fluid foods: A review. *Crit. Rev. Food Sci. Nutr.* 48:747–759.

Motlagh, A.M., Holla, S., Johnson, M.C., Ray, B. and Field, R.A. 1992. Inhibition of *Listeria* spp. in sterile food systems by pediocin AcH, a bacteriocin produced by *Pediococcus acidilactici* H. *J. Food Prot.* 55:337–343.

Muriana, P. M. and Klaenhammer, T. R. 1987. Conjugal transfer of plasmid-encoded determinants for bacteriocin production and immunity in *Lactobacillus acidophilus* 88. *Appl. Environ. Microbiol.* 53:553.

Muriana, P. M. and Klaenhammer, T. R. 1991a. Purification and partial characterization of lactacin F, a bacteriocin produced by *Lactobacillus acidophilus* 11088. *Appl. Environ. Microbiol.* 57:114.

Muriana, P. M. and Klaenhammer, T. R. 1991b. Cloning, phenotypic expression, and DNA sequence of the gene for lactacin F, an antimicrobial peptide produced by *Lactobacillus* spp. *J. Bacteriol.* 173:1779.

Nes, I.F., Diep, D.B., Havarstein, L.S., Brurberg, M.B., Eijsink, V. and Holo, H. 1996. Biosynthesis of bacteriocins in lactic acid bacteria. *Antonie van Leeuwenhoek Intl. J. General Molec. Microbiol.* 70:113–128.

Nes, I.F. and Holo, H. 2000. Class II antimicrobial peptides from lactic acid bacteria. *Biopolymers* 55:50–61.

Neve, H., Geis, A. and Teuber, M. 1984. Conjugal transfer and characterization of bacteriocin plasmids in group N (lactic acid) streptococci. *J. Bacteriol.* 157:833.

Nielsen, J. W., Dickson, J. S. and Crouse, J. D. 1990. Use of a bacteriocin produced by *Pediococcus acidilactici* to inhibit *Listeria monocytogenes* associated with fresh meat. *Appl. Environ. Microbiol.* 56:2142.

Nieto-Lozano, J. C., Reguera-Useros, J. I., Peláez-Martínez, M. C. and de la Torre, A. H. 2006. Effect of a bacteriocin produced by *Pediococcus acidilactici* against *Listeria monocytogenes* and *Clostridium perfringens* on Spanish raw meat. *Meat Sci.* 72:57–61.

Nissen-Meyer, J., Oppegård, C., Rogne, P., Haugen, H. S. and Kristiansen, P. E. 2010. Structure and mode-of-action of the two-peptide (class-IIb) bacteriocins. *Probiotics Antimicrob. Prot.* 2:52–60.

Noonpakdee, W., Santivarangkna, C., Jumriangrit, P., Sonomoto, K. and Panyim, S. 2003. Isolation of nisin-producing *Lactococcus lactis* WNC 20 strain from *nham*, a traditional Thai fermented sausage. *Intl. J. Food Microbiol.* 81:137–145.

Orberg, P. K. and Sandine, W. E. 1984. Common occurrence of plasmid DNA and vancomycin resistance in *Leuconostoc* spp. *Appl. Environ. Microbiol.* 48:1129.

O'Shea, E. F., O'Connor, P. M., Cotter, P. D., Ross, R. P. and Hill, C. 2010. Synthesis of trypsin-resistant variants of the *Listeria*-active bacteriocin salivaricin P. *Appl. Environ. Microbiol.* 76:5356–5362.

Oxford, A. E. 1944. Diplococcin, an antibacterial protein elaborated by certain milk streptococci. *Biochem. J.* 38:178.

Palmeri, A., Pepe, I. M., Rolandi, R., Pagani, S. and Morelli, A. 1999. Ion permeability induced by bacteriocins of *Lactobacillus acidophilus* M247 on artificial membranes. *Materials Sci. and Engineering: C* 8:539–542.

Papathanasopoulos, M. A., Krier, F., Revol-Junelles, A. M., Lefebvre, G., Le Caer, J. P., von Holy, A. and Hastings, J. W. 1997. Multiple Bacteriocin Production by *Leuconostoc mesenteroides*TA33a and Other *Leuconostoc*/*Weissella* Strains. *Curr. Microbiol.* 35:331–335.

Parente, E., Giglio, M. A., Ricciardi, A. and Clementi, F. 1998. The combined effect of nisin, leucocin F10, pH, NaCl and EDTA on the survival of *Listeria monocytogenes* in broth. *Intl. J. Food Microbiol.* 40:65.

Parkinson, J. S. 1993. Signal transduction schemes of bacteria. *Cell* 73:857.

Pleasants, A. B., Soboleva, T. K., Dykes, G. A., Jones, R. J. and Filippov, A. E. 2001. Modeling of the growth of populations of *Listeria monocytogenes* and a bacteriocin-producing strain of *Lactobacillus* in pure and mixed cultures. *Food Microbiol.* 18:605.

Preciado, G. M., Michel, M. M., Villarreal-Morales, S. L., Flores-Gallegos, A. C., Aguirre-Joya, J., Morlett-Chávez, J., Aguilar, C. N. and Rodriguez-Hererra, R. 2016. Bacteriocins and its use for multidrug-resistant bacteria control. In *Antibiotic Resistance—Mechanisms and New Antimicrobial Approaches*, pp. 329–345. Academic Press. https://doi.org/10.1016/B978-0-12-803642-6.00016-2

Prospect. 2019. Available at https://www.ulprospector.com/en/na/Food/Detail/3882/397046/MicroGARD-200. Accessed on 8/11/2019.

Prudêncio, C. V., Santos, M. T. D. and Vanetti, M. C. D. 2015. Strategies for the use of bacteriocins in Gram-negative bacteria: relevance in food microbiology. *J. Food Sci. and Technology* 52:5408–5417.

Pucci, M. J., Vedamuthu, E. R., Kunka, B. S. and Vandenburgh, P. A. 1988. Inhibition of *Listeria monocytogenes* by using bacteriocin PA-I produced by *Pediococcus acidilactici* PAC 1.0. *Appl. Environ. Microbiol.* 54:2349.

Pulusani, S. R., Rao, D. R. and Sunki, G. R. 1979. Antimicrobial activity of lactic cultures: Partial purification and characterization of antimicrobial compound(s) produced by *Streptococcus thermophilus*. *J. Food Sci.* 44:575.

Raccach, M., McGrath, R. and Daftarian, H. 1989. Antibiosis of some lactic acid bacteria including *Lactobacillus acidophilus* toward *Listeria monocytogenes*. *Intl. J. Food Microbiol.* 9:25.

Rammelsberg, M. and Radler, F. 1990. Antibacterial polypeptides of *Lactobacillus* species. *J. Appl. Bacteriol.* 69:177.

Randazzo, C. L., Pitino, I., Scifò, G. O. and Caggia, C. 2009. Biopreservation of minimally processed iceberg lettuces using a bacteriocin produced by *Lactococcus lactis* wild strain. *Food Control* 20:756–763.

Ray, S. K., Kim, W. J., Johnson, M. C. and Ray, B. 1989a. Conjugal transfer of a plasmid encoding bacteriocin production and immunity in *Pediococcus acidilactici* H. *J. Appl. Bacteriol.* 66:393.

Ray, S. K., Johnson, M. C. and Ray, B. 1989b. Bacteriocin plasmids of *Pediococcus acidilactici*. *J. Microbiol.* 4:163.

Rea, M. C., Ross, R. P., Cotter, P. D. and Hill, C. 2011. Classification of bacteriocins from Gram-positive bacteria. In *Prokaryotic Antimicrobial Peptides*, pp. 29–53. Springer, New York, NY.

Reddy, G. V., Shahani, K. M., Friend, B. A. and Chandan, R. C. 1983. Natural antibiotic activity of *Lactobacillus acidophilus* and *bulgaricus*.111: Production and partial purification of bulgarican from *Lactobacillus bulgaricus*. *Cult. Dairy Prod. J.* 18:15.

Reddy, N. S. and Ranganathan, B. 1983. Preliminary studies on antimicrobial activity of *Streptococcus lactis* subsp. *diacetylactis*. *J. Food Prot.* 46:222.

Reeves, P. 2012. *The Bacteriocins*, Vol. 11. Springer Science & Business Media.

Repka, L. M., Chekan, J. R., Nair, S. K. and Donk, W. A. V. D. 2017. Mechanistic understanding of lanthipeptide biosynthetic enzymes. *Chemical Reviews* 117:5457–5520.

Revol-Junelles, A. M. and Lefebvre, G. 1996. Purification and N-terminal amino acid sequence of dextranicin 24, a bacteriocin of *Leuconostoc* sp. *Curr. Microbiol.* 33:136.

Revol-Junelles, A. M., Mathis, R., Krier, F., Fleury, Y., Delfour, A. and Lefebvre, G. 1996. *Leuconostoc mesenteroides* subsp. *mesenteroidies* FR52 synthesizes two distinct bacteriocins. *Lett. Appl. Microbiol.* 23:120.

Riley, M. A. 1998. Molecular mechanisms of bacteriocin evolution. *Annual Review of Genetics* 32:255.

Rodriguez, E., Gonzalez, B., Gaya, P., Nunez, M. and Medina, M. 2000. Diversity of bacteriocins produced by lactic acid bacteria isolated from raw milk. *Intl. Dairy J.* 10:7.

Rodríguez, J. M., Martínez, M. I. and Kok, J. 2002. Pediocin PA-1, a wide-spectrum bacteriocin from lactic acid bacteria. *Crit. Rev. Food Sci. Nutr.* 42:91–121.

Rogers, L. A. and Whittier, E. O. 1928. Limiting factors in lactic fermentation. *J. Bacteriol.* 16:211.

Rollema, H. S., Kuipers, O. P., Both, P., de Vos, W. M. and Siezen, R. J. 1995. Improvement of solubility and stability of the antimicrobial peptide nisin by protein engineering. *Appl. Environ. Microbiol.* 61:2873.

Ross, R. P., Stanton, C., Hill, C., Fitzgerald, G. F. and Coffey, A. 2000. Novel cultures for cheese improvement. *Trends Food Sci. & Technol.* 11:96.

Sablon, E., Contreras, B. and Vandamme, E. 2000. Antimicrobial peptides of Lactic Acid Bacteria: mode of action, genetics and biosynthesis. *Adv. Biochem. Engr./Biotech.* 68:21.

Sahl, H.-G. and Bierbaum, G. 1998. Lantibiotics: biosynthesis and biological activities of uniquely modified peptides from Gram-positive bacteria. *Ann. Rev. Microbiol.* 52:41–79.

Saucier, L. and Champagne, C. P. 2005. Immobilised-cell technology and meat processing. In *Applications of Cell Immobilization Biotech*, pp. 337–353. Springer.

Scanell, A. G. M., Hill, C., Buckley, D. J. and Arendt, E. K. 1997. Determination of the influence of organic acids and nisin on shelf-life and microbiological safety aspects of fresh pork. *J. Appl. Microbiol.* 83:407.

Scannell, A. G. M., Hill, C., Ross, R. P., Marx, S., Hartmeier, W. and Arendt, E. K. 2000. Development of bioactive food packaging materials using immobilized bacteriocins Lacticin 3147 and Nisaplin®. *Intl. J. Food Microbiol.* 60:241–249.

Scherwitz, K. M., Baldwin, K. A. and McKay, L. L. 1983. Plasmid linkage of a bacteriocin-like substance in *Streptococcus lactis* subsp. *diacetylactis* strain WM4: Transferability to *Streptococcus lactis*. *Appl. Environ. Microbiol.* 45:1506.

Scherwitz-Harmon, K. and McKay, L. L. 1987. Restriction enzyme analysis of lactose and bacteriocin plasmids from *Streptococcus lactis* subsp. *diacetylactis* WM, and cloning of BclI fragments coding for bacteriocin production. *Appl. Environ. Microbiol.* 53:1171.

Schillinger, U. and Lucke, F. K. 1989. Antibacterial activity of *Lactobacillus sake* isolated from meat. *Appl. Environ. Microbiol.* 55:1901.

Shahani, K. M., Vakil, J. R. and Kilara, A. 1977. Natural antibiotic activity of *Lactobacillus acidophilus* and *bulgaricus*. Z. Isolation of acidophilin from *L. acidophilus*. *Cult. Dairy Prod. J.* 12:8.

Siegers, K., Heinzmann, S. and Entian, K.-D. 1996. Biosynthesis of lantibiotic nisin: post-translational modification of its prepeptide occurs at a mutimeric membrane-associated lanthionine synthetase complex. *J. Biol. Chem.* 271:12294.

Silva, C. C. G., Silva, S. P. M. and Ribeiro, S. C. 2018. Application of bacteriocins and protective cultures in dairy food preservation. *Front. Microbiol.* 9:594.

Silveira, A. C., Conesa, A., Aguayo, E. and Artes, F. 2008. Alternative sanitizers to chlorine for use on fresh-cut "Galia" (*Cucumis melo* var. *catalupensis*) melon. *J. Food Sci.* 73:M405–M411.

Singh, N. P., Tiwari, A., Bansal, A., Thakur, S., Sharma, G. and Gabrani, R. 2015. Genome level analysis of bacteriocins of lactic acid bacteria. *Comput. Biol. Chem.* 56:1.

Skaugen, M., Abildgaard, C. I. M. and Nes, I. F. 1997. Organization and expression of a gene cluster involved in the biosynthesis of the lantibiotic lactocin S. *Molec. Genet. Genomics* 253:674.

Sobrino-Lopez, A., Viedma-Martínez, P., Abriouel, H., Valdivia, E., Gálvez, A. and Martin-Belloso, O. 2009. The effect of adding antimicrobial peptides to milk inoculated with *Staphylococcus aureus* and processed by high-intensity pulsed-electric field. *J. Dairy Sci.* 92:2514–2523.

Solomakos, N., Govaris, A., Koidis, P. and Botsoglou, N. 2008. The antimicrobial effect of thyme essential oil, nisin, and their combination against *Listeria monocytogenes* in minced beef during refrigerated storage. *Food Microbiol.* 25:120–127.

Sparo, M., Nuñez, G. G., Castro, M., Calcagno, M. L., García Allende, M. A., Ceci, M., Najle, R. and Manghi, M. 2008. Characteristics of an environmental strain, *Enterococcus faecalis* CECT7121, and its effects as additive on craft dry-fermented sausages. *Food Microbiol.* 25:607–615.

Stergiou, V. A., Thomas, L. V. and Adams, M. R. 2006. Interactions of nisin with glutathione in a model protein system and meat. *J. Food Prot.* 69:951–956.

Stiles, M. E. and Holzapfel, W. H. 1997. Lactic acid bacteria of foods and their current taxonomy. *Intl. J. Food Microbiol.* 36:1.

Stock, J. B., Ninfa, A. J. and Stock, A. M. 1989. Protein phosphorylation and regulation of adaptive responses in bacteria. *Microbiol. Review* 53:450.

Stoddard, G. W., Petzel, J. P., van Belkum, M. J., Kok, J. and McKay, L. L. 1992. Molecular analyses of the lactococcin a gene cluster from *Lactococcus lactis* subsp. *lactis* biovar *diacetylactis* WM_4. *Appl. Environ. Microbiol.* 58:1952.

Stoffels, G., Nissen-Meyer, J., Gudmundsdottir, A., Sletten, K., Holo, H. and Nes, I. F. 1992. Purification and characterization of a new bacteriocin isolated from a *Carnobacterium* sp. *Appl. Environ. Microbiol.* 58:1417.

Stoffels, G., Gudmundsdottir, A. and Abee, T. 1994. Membrane-associated proteins encoded by the nisin gene cluster may function as a receptor for the lantibiotic carnocin UI49. *Microbiol.* 140:1443.

Suma, K., Misra, M. C. and Varadaraj, M. C. 1998. Plantaricin LP84, a broad spectrum heat-stable bacteriocin of *Lactobacillus plantarum* NCIM 2084 produced in a simple glucose broth medium. *Intl. J. Food Microbiol.* 40:17.

Thomas, L. V. and Delves-Broughton, J. 2005. Nisin. In: Davidson, P.M., Sofos, J.N., Branen, A.L. (Eds.), *Antimicrobials in Foods*, 3rd ed. Taylor and Francis Group, Boca Raton, pp. 237–274.

Trias, R., Badosa, E., Montesinos, E. and Bañeras, L. 2008a. Bioprotective *Leuconostoc* strains against *Listeria monocytogenes* in fresh fruits and vegetables. *Intl. J. Food Microbiol.* 127:91–98.

Trias, R., Bañeras, L., Badosa, E. and Montesinos, E. 2008b. Bioprotection of Golden Delicious apples and Iceberg lettuce against foodborne bacterial pathogens by lactic acid bacteria. *Intl. J. Food Microbiol.* 123:50–60.

Tsai, H.-J. and Sandine, W. E. 1987. Conjugal transfer of nisin plasmid gene from *Streptococcus lactis* 7962 to *Leuconostoc dextranicum* 181. *Appl. Environ. Microbiol.* 53:352.

Turner, D. L., Brennan, L., Meyer, H. E., Lohaus, C., Siethoff, C., Costa, H. S., Gonzalez, B., Santos, H. and Suarez, J. E. 1999. Solution structure of plantaricin C, a novel lantibiotic. *European J. Biochem.* 264:833.

Uesugi, A. R. and Moraru, C. I. 2009. Reduction of *Listeria* on ready-to-eat sausages after exposure to a combination of pulsed light and nisin. *J. Food Prot.* 72:347–353.

Ukuku, D. O., Bari, M. L., Kawamoto, S. and Isshiki, K. 2005. Use of hydrogen peroxide in combination with nisin, sodium lactate and citric acid for reducing transfer of bacterial pathogens from whole melon surfaces to fresh-cut pieces. *Intl. J. Food Microbiol.* 104:225–233.

Upreti, G. C. and Hinsdill, R. D. 1973. Isolation and characterization of a bacteriocin from a homofermentative *Lactobacillus. Antimicrob. Agents Chemother.* 4:487.

Upreti, G. C. and Hinsdill, R. D. 1975. Production and mode of action of lactocin 27: Bacteriocin from homofermentative *Lactobacillus. Antimicrob. Agents Chemother.* 7:145.

van Belkum, M.J., Hayema, B.J., Jeeninga, R.E., Kok, J. and Venema, G. 1991. Organization and nucleotide sequences of two lactococcal bacteriocin operons. *Appl. Environ. Microbiol.* 57:492–498.

van Belkum, M.J. Kok, J. and Venema, G. 1992. Cloning, sequencing, and expression in *Escherichia coli* of lcnB, a third bacteriocin determinant from the lactococcal bacteriocin plasmid p9B4-6. *Appl. Environ. Microbiol.* 58:572–577.

van der Meer, J. R., Polman, J., Beerthuyzen, M. M., Siezen, R. J., Kuipers, O. P. and de Vos, W. M. 1993. Characterization of the *Lactococcus lactis* nisin A operon genes *nisP*, encoding a subtilisin-like serine protease involved in precursor processing, and *nisR*, encoding a regulatory protein involved in nisin biosynthesis. *J. Bacteriol.* 175:2578.

Venema, K., Kok, J., Marugg, J.D., Toonen, M.Y., Ledeboer, A.M., Venema, G. and Chikindas, M.L. 1995a. Functional-analysis of the pediocin operon of *Pediococcus acidilactici* PAC1.0/PedB is the immunity protein and PedD is the precursor processing enzyme. *Molec. Microbiol.* 17:515–522.

Venema, K., Venema, G. and Kok, J. 1995b. Lactococcins: Mode of action, immunity and secretion. *Int. Dairy J.* 5:815–832.

Véronique, C. 2008. Bioactive packaging technologies for extended shelf life of meat-based products. *Meat Sci.* 78:90–103.

Vesković-Moračanin Slavica, M., Đukić Dragutin, A. and Memiši Nurgin, R. 2014. Bacteriocins produced by lactic acid bacteria: A review. *Acta Periodica Technologica* 45:271. DOI:10.2298/APT1445271V.

Viedma, P. M., Abriouel, H., Omar, N. B., López, R. L., Valdivia, E. and Galvez, A. 2009. Antibacterial protection by enterocin AS-48 in sport and energy drinks with less acidic pH values. *J. Food Prot.* 72:881–884.

Vignolo, G., Fadda, S., de Kairuz, M. N., de Holgado, A. P. and Oliver, G. 1998. Effects of curing additives on the control of *Listeria monocytogenes* by lactocin 705 in meat slurry. *Food Microbiol.* 15:259.

Vincent, J. G., Veomett, R. C. and Riley, R. F. 1959. Antibacterial activity associated with *Lactobacillus acidophilus. J. Bacteriol.* 78:477.

Vinderola, G., Ouwehand, A., Salminen, S. and von Wright, A. 2019. *Lactic Acid Bacteria: Microbiological and Functional Aspects.* CRC Press/Taylor & Francis, Boca Raton, FL.

von Staszewski, M. and Jagus, R. J. 2008. Natural antimicrobials: Effect of Microgard™ and nisin against *Listeria innocua* in liquid cheese whey. *Intl. Dairy J.* 18:255–259.

Vukomanović, M., Žunič, V., Kunej, Š., Jančar, B., Jeverica, S. and Suvorov, D. 2017. Nano-engineering the antimicrobial spectrum of lantibiotics: activity of nisin against gram negative bacteria. *Scientific Rept.* 7: 4324.

Wachsman, M. B., Farias, M. E., Takeda, E., Sesma, F., Holgado, A. P., de Torres, R. A. and Coto, C. E. 1999. Antiviral activity of enterocin CRL35 against herpesviruses. *Intl. J. Antimicrob. Agents* 12:293.

West, C. A. and Warner, P. J. 1988. Plantacin B, a bacteriocin produced by *Lactobacillus plantarum* NCDO 1193. *FEMS Microbiol. Lett.* 49:163.

Wheater, D. M., Hirsch, A. and Mattick, A. T. R. 1952. Possible identity of "lactobacillin" with hydrogen peroxide produced by lactobacilli. *Nature* 170:623.

Whitehead, H. R. 1933. A substance inhibiting bacterial growth, produced by certain strains of lactic streptococci. *Biochem. J.* 27:1793.

Whitehead, H. R. and Riddet, W. 1933. Slow development of acidity in cheese manufacture: Investigation of a typical case of "non-acid" milk. *N. Z. J. Agric. Res.* 46:225.

Woraprayote, W., Malila, Y., Sorapukdee, S., Swetwiwathana, A., Benjakul, S. and Visessanguan, W. 2016. Bacteriocins from lactic acid bacteria and their applications in meat and meat products. *Meat Sci.* 120:118.

Worobo, R. W., Henkel, T., Sailer, M., Roy, K. L., Vederas, J. C. and Stiles, M. E. 1994. Characteristics and genetic determinant of a hydrophobic peptide bacteriocin, carnobacteriocin A, produced by *Carnobacterium piscicola* LV17A. *Microbiology* 140:517–526.

Worobo, R. W., VanBelkum, M. J., Sailer, M., Roy, K. L., Vederas, J. C. and Stiles, M. E. 1995. A signal peptide secretion-dependent bacteriocin from *Carnobacterium divergens. J. Bacteriol.* 177:3143.

Yang, R. and Ray, B. 1994. Factors influencing production of bacteriocins by lactic acid bacteria. *Food Microbiol.* 11:281.

Yang, S.-C., Lin, C.-H., Sung, C. T. and Fang, J.-Y. 2014. Antibacterial activities of bacteriocins: application in foods and pharmaceuticals. *Front. Microbiol.* 5:241.

Yildirim, Z. and Johnson, M. G. 1998. Characterization and antimicrobial spectrum of bifidocin B, a bacteriocin produced by *Bifidobacterium bifidum* NCFB 1454. *J. Food Prot.* 61:47.

Zajdel, J. K. and Dobrzanski, W. T. 1983. Isolation and preliminary characterization of *Streptococcus cremoris* strain 202 bacteriocin. *Acta Microbiologica Polonica.* 32:119.

Zajdel, J. K., Ceglowski, P. and Dobrzanski, W. T. 1985. Mechanism of action of lactostrepcin 5, a bacteriocin produced by *Streptococcus cremoris* 202. *Appl. Environ. Microbiol.* 49:969.

Zendo, T. 2013. Screening and characterization of novel bacteriocins from lactic acid bacteria. *BioSci. Biotech. Biochem.* 77:893.

Zhang, P., Ganzle, M. and Yang, X. 2019. Comparative analysis of *Carnobacterium* spp. from meat reveals complementary antibacterial effects between bacteriocins and organic acids. *Appl. Environ. Microbol.* DOI:10.1128/AEM.01227-19.

16

Bacteriophages

Pushpinder Kaur Litt, Kalmia E. Kniel, and Manan Sharma

CONTENTS

16.1 History of Bacteriophages

Bacteriophages are among the most abundant biological particles on earth and exert enormous influences on the biosphere. Their total number on earth has been estimated to be in the range of 10^{30} to 10^{32} (Comeau et al. 2008, Brussow and Kutter 2005). They are found in the same varied environments as their bacterial hosts – their numbers are estimated to be 10^7 per milliliter in coastal waters, 10^7 per gram in soil, and 10^9 per gram in marine sediments (Brussow and Kutter 2005). Phages are viruses that infect bacteria; their self-replication depends on access to a bacterial host. Phages kill between 4–50% of the bacteria produced every day, and are a driver of global geochemical cycles and a reservoir of the greatest genetic diversity on earth (Suttle 2005). Phages are consumed daily by humans via the various foods they eat and water they drink. They have been isolated from diverse sources of food, ranging from chicken and pork sausages, ground beef, freshwater and saltwater fish, raw skim milk, cheese, various deli meats, mushrooms, lettuce, refrigerated biscuit dough, and frozen chicken pot pies and the food production environment (Litt, Saha and Jaroni 2018a, Litt and Jaroni 2017a, Greer 2005, Kennedy et al. 1984, Kennedy et al. 1986).

It is surprising that they were not recognized for almost 40 years after the beginning of serious bacteriological work in European and American laboratories around 1880. Two independent observations were made by Frederick W. Twort (1915) and Felix d'Herelle (1917) who described glassy transformations of micrococci colonies and antagonistic bacterial microbes, respectively (Summers 1999, Monk et al. 2010). However, phages were first formally documented in 1896 by Ernest Hankin, a British bacteriologist who reported observing a filterable antibacterial activity against *Vibrio cholerae* in the rivers Ganges and Jamuna in India (Sulakvelidze et al. 2001). These observations suggested the presence of an unidentified entity which passed through filters, was heat liable, and limited the spread of cholera, thus cholera epidemics (Sulakvelidze et al. 2001). Two years later the Russian bacteriologist Gamalea documented similar observations against *Bacillus subtilis*, although these observations were not further investigated until the work of Fredrick Twort, a medically trained English bacteriologist, suggested that the observed phenomenon was due to the presence of a virus (Duckworth 1976). Two years later phages were officially discovered by Felix d'Herelle, a French-Canadian self-trained microbiologist (Keen 2012). d'Herelle's discovery came from the observation of small, clear plaques on an agar medium containing stool samples from dysentery patients (Summers 2001). It was d'Herelle who came up with the name bacteriophages, derived from 'bacteria' and the Greek 'phagein', which means 'to eat'. He also introduced the term plaque to describe the circular area of clearing caused by infection of a single phage on double-layered agar plates. Soon after their discovery, d'Herelle observed that phage titers increased in stool samples from dysentery patients. This finding encouraged scientists almost immediately to explore the therapeutic potential of these bacterial viruses. The first report of the therapeutic use of phage was documented in a note by Bruynoghe and Maisin (1921), who noticed reduction in swelling and pain after the injection of Staphylococcal phage in the local region of cutaneous boils. Although many early phage therapy trials were reported to be successful and many of the major pharmaceutical companies sold phage preparations, there were failures too. Various factors, including the success of broad-spectrum antibiotics, an inadequate understanding of phage biology and imperfections in the diagnostic bacteriology techniques available at that time, and issues in quality control of the production of bacteriophage products (Summer 2001), contributed to the failure of early phage therapy studies and decline of interest in phage therapy in the Western countries. At the same time, phage therapy for human infections continued to be offered in the Eliava Institute (Tbilisi, Georgia), and later also by others, like the Hirszfeld Institute of Immunology and Experimental Therapy in Poland (Deresinski 2009). The study of phages played a central role in some of the most significant discoveries in the biological sciences, from the identification of DNA as the genetic material, to the deciphering of the genetic code and the phenomenon of restriction and modification, to the development of molecular recombinant technology and more recently CRISPR technology.

16.2 The Diversity of Bacteriophages: Phage Morphology Forms the Basis for Classification

Bacteriophages can be defined as obligatory intracellular bacterial parasites which lack an independent metabolism. The genome sizes of phages vary enormously, from a few thousand base pairs up to 498 kilobases. Phage G is the largest phage sequenced to date (Guttman et al. 2005). The size of the phage G genome approximates that of an average bacterium, but it lacks genes for essential cellular machinery like ribosomes, emphasizing the purely parasitic nature of these organisms. A bacteriophage particle or virion consists of a single- or double-stranded (ss or ds) DNA or RNA molecule, encapsulated inside a protein or lipoprotein coat. The International Committee on the Taxonomy of Viruses (ICTV) used virion morphology and nucleic acid composition as a basis for the classification of phages into 13 families. Over 95% of all phages described in the literature belong to the order of the *Caudovirales* or tailed dsDNA phages. The two main families comprising the *Caudovirales* are distinguished by their very distinct tail morphologies: 60% of the characterized phages are *Siphoviridae*, with long, flexible tails; 25% are *Myoviridae*, with double-layered, contractile tails (Fauquet et al. 2005) (Figure 16.1). Polyhedral, filamentous, and pleomorphic phages represent only 3–4% of the studied phages and belong to ten families, some of which are very small (Ackermann 2007), such as *Tectiviridae* phages or tail-less dsDNA phages which contain double capsid, pseudo tail, and a lipid layer around (Figure 16.2). Compared to tailed phages, these are oddly distributed and found to be associated with γ proteobacteria such as enterobacteria and their relatives, and also occur in evolutionarily distant hosts as *Bacillus* and *Thermus* bacteria (Ackermann 2007). Virions with a tail are typically composed of proteins and dsDNA in 1:1 ratio. In all dsDNA phages, the highly condensed chromosome represents 20–50% of its mass, packed within the virion head (Earnshaw and Harrison 1977). It is important for the chromosome to maintain a highly condensed form in the mature capsid in order to inject its DNA to infecting host cell. All *Caudovirales* have heads with icosahedral (20 sides/12 vertices) symmetry or elongated derivatives thereof with known triangulation numbers of T = 4, 7, 13, 16, and 52, assembled from many copies of one

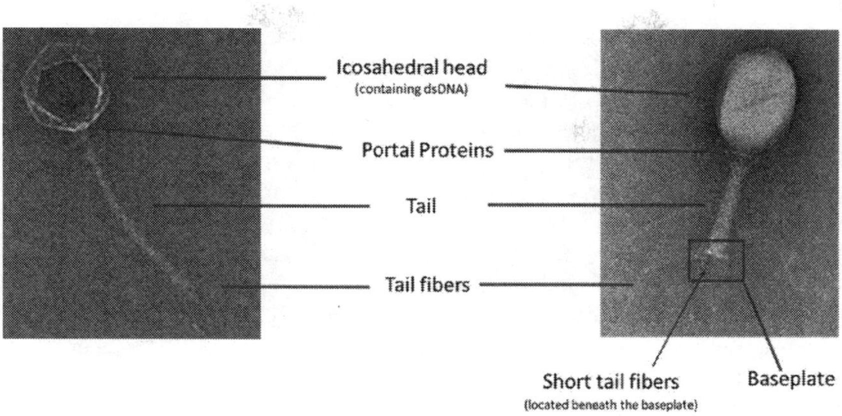

FIGURE 16.1 Basic morphology of the two families of the *Caudovirales*. From left to right, electron micrographs of a *Siphoviridae* (long, non-contractile tail), and *Myoviridae* (long, contractile tail) are shown. See text for more details. (Adapted from Leiman et al. (2004) and Serwer et al. (2008).)

or two major head proteins. The size of the phage head correlates with the genome size being packaged, and can vary in diameter between 45 and 100 nm. The corners of the icosahedral are generally made up of pentamers of the capsid proteins, while the rest of each side is made up of hexamers of similar protein (Guttman et al. 2005).

The phage proteins that connect the head and tail structures (portal protein) play a significant role in several steps of the infection cycle. This connector structure is found in a unique vertex of the phage head and controls the entry of the DNA and assembly of the tail to the immature head during the morphogenesis of new virion particles. During infection, it undergoes a conformation change permitting the phage DNA to exit the capsid towards the bacterial cell (Valpuesta et al. 2000). The tail facilitates attachment of the phage particle to the host bacterial cell and serves as the conduit for genome injection. Tail shafts have three- or six-fold symmetry, and are composed of helical subunits varying from 3 to 825 nm in length. Tail shafts usually have base plates, spikes, or terminal fibers at the distal end. These spikes may contain enzymes – polysaccharide depolymerases (like glycanases) – which allow the phage more intimate contact with the host cell (Hughes et al. 1998, Lindberg 1997) by degrading portions of the protective exopolysaccharride layer of the bacterial cell. The phage tails can be contractile, as found in the *Myoviridae* (e.g., phage T4), or non-contractile, as found in the *Siphoviridae* (e.g., phage λ). Long-tailed phages possess a large gene (usually, >2 kbp) encoding a tape measure protein, which is responsible for precisely determining tail length. The contractile tails are often complex structures, for example, the tail of T4 is composed of 22 different proteins (Mesyanzhinov et al. 2002), whereas the non-contractile tails

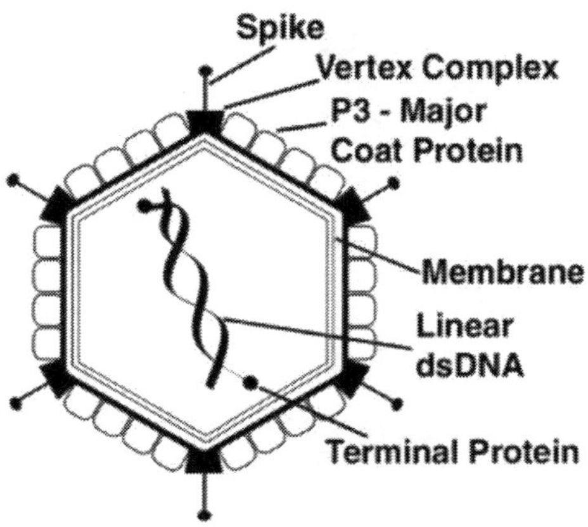

FIGURE 16.2 Basic morphology of the *Tectiviridae* (no tail) is shown. See text for more details. (Adapted from San Martın et al. (2001).)

are primarily composed of multiple copies of one protein, known as the major tail protein; hence, the overall shape of the non-contractile tail is maintained, during infection. In the case of contractile tails, a central tail tube protein (TTP) is surrounded by a tail sheath protein. Therefore, upon infection, the tail sheath contracts and the TTP penetrates the outer cell membrane and cell wall (Leiman et al. 2004).

The tail-less *Tectiviridae* (e.g., PRD1) virion has an icosahedral protein capsid with an underlying membrane that encloses its genome, which is locked with cementing proteins, and spike complexes forming vertices (Rydman et al. 1999). A unique vertex of the phage particle contains the DNA-packaging machinery, anchored into the viral membrane through association with integral membrane proteins. It has been reported that the entire mass of PRD1 consists of 70% protein, 15% lipid, and 15% DNA (Bamford et al. 1995). The phage bilayer membrane contains integral proteins and lytic enzymes, associated with its periphery, and plays a crucial role in the infection process (Rydman and Bamford 2000, 2002, Grahn et al. 2002). The viral membrane undergoes structural rearrangements and transforms into a tubular structure protruding from one vertex, which might penetrate the host cell wall to form a channel through which the phage genome enters the host cell (Daugelavicius et al. 1997, Grahn et al. 2002).

16.3 Phage Lifecycle

Phages have two lifecycles, the lytic and the lysogenic lifecycles (Figure 16.3). Any phage lifecycle begins with the complex process of attachment (Letellier, Bertin et al. 2011). This process starts with initial binding of the phage particle to the host cell, through recognition of specific receptors located on the host surface (Marsh and Helenius 2006). The interactions that occur between the phage receptor binding protein and the host cell surface trigger a signaling cascade required for viral entry into the host cell (Marsh and Helenius 2006). Studies have indicated that in tailed phages such as T2, initial attachment is followed by contraction of the viral tail, bringing the phage capsid containing genetic material closer to the host cell before injecting it into the host cell (Cota-Robles and Coffman 1963). The genome then travels into the host cell, leaving an empty phage particle attached to the host surface (Bertin et al. 2011). This attachment process appears to be general across phage families however; the mechanisms by which genomes are internalized within hosts differ between phages (Rakhuba et al. 2010).

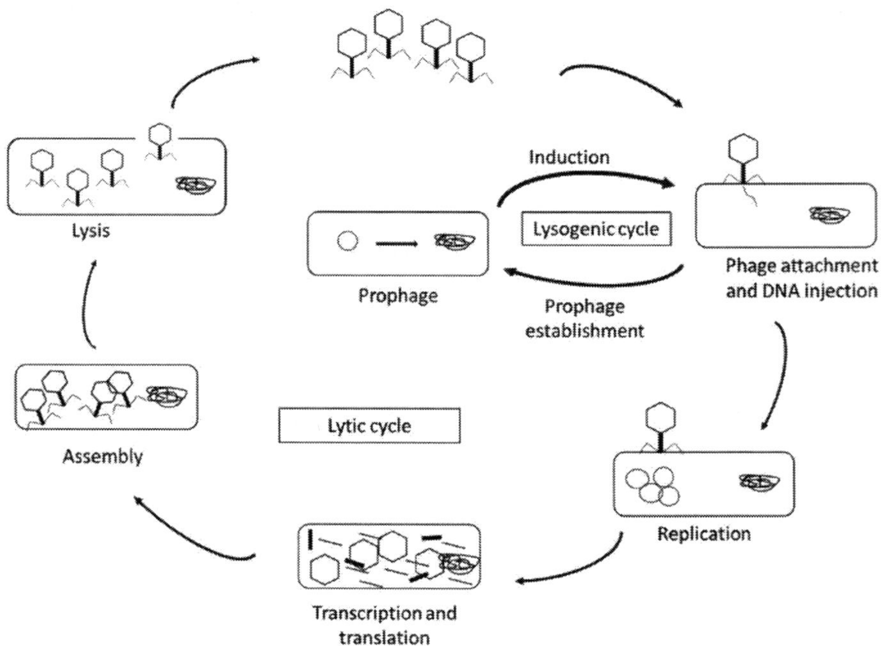

FIGURE 16.3 Lifecycle of phage. See text for more details. (Adapted from Salmond et al. (2015).)

16.3.1 Lytic Pathway

Activation of the lytic pathway results in a number of processes which lead to the development and accumulation of newly synthesized viral particles (Echols 1972). Following viral synthesis, rapid cell lysis occurs, resulting in the release of progeny phage (Echols 1972). The lytic cycle begins when a phage comes into contact with a bacterial cell through random motion (Hanlon 2007). This contact is followed by initial attachment or adsorption to bacterial surface receptor sites, which include proteins, teichoic acid, and peptidoglycan (Marsh and Helenius 2006). The phage adsorption rate depends upon the concentration of bacteria and phage present in the medium (Puck et al. 1951). However, the addition of divalent ions has been shown to promote phage attachment onto the bacterial surface due the high strength of the electrostatic forces between both cells (Puck et al. 1951). The initial attachment is reversible and later becomes irreversible and is followed by the transfer of the viral genome into the host bacteria (Maura and Debarbieux 2011). A viral genome is transferred to the host through many routes, the most common being the contraction of the tail through a pore formed on the host cell surface (Hanlon 2007). Once inside the host cell the viral genome hijacks host cell machinery and metabolic processes to replicate and synthesize components required for viral assembly (Echols 1972, Hanlon 2007). Synthesis involves the development of proteins required for encapsulating viral genetic material, head, and tail structures (Echols 1972). Once viral components have been synthesized and assembled, bacteriophages are released through cell lysis which occurs when lysins are released, disrupting the cell membrane and resulting in the degradation of bacterial peptidoglycan (Matsuzaki et al. 2005) and structural rupture of the cell.

16.3.2 Lysogenic Pathway

Phages replicating through the lysogenic pathway are known as temperate (Hanlon 2007). In this form of replication the lytic functions of the virus are turned 'off' and the host cell survives (Echols 1972). The lysogenic lifecycle is similar to the lytic cycle during the initial contact, attachment, and transfer of phage genetic material, but at the end, the phage genome is either integrated into host cell DNA or exists episomally (Matsuzaki et al. 2005). The process of integration occurs through site-specific recombination at specific regions of both the host and phage genomes (Matsuzaki et al. 2005). Bacterial cells that have integrated a phage genome are known as lysogenic (Hanlon 2007). As the host cell genome replicates, the phage genome also replicates, and each daughter cell receives a copy; these sequences are termed prophages (Hanlon 2007). Temperate phages remain successful by preventing other phages from infecting the host cell, by repressing the genes responsible for lytic replication, or by altering the host cell wall receptor sites (Labrie and Moineau 2007, Labrie et al. 2010). Although prophages can replicate through the lysogenic cycle for many generations, exposure of the host cell to stressful conditions such as UV irradiation, mutagenic agents, and osmotic and nutrient stress can activate the lytic cycle, leading to cell lysis and release of progeny phage (Hanlon 2007, Rice and Bayles 2008). Low numbers of progeny phage particles are often released from temperate phage replication compared to lytic phages because the majority of phage produced are from a small percentage of the stressed host cells (Lwoff 1953). For this reason temperate phage are not considered suitable for biocontrol purposes (Matsuzaki et al. 2005). In addition to the lytic and lysogenic lifecycles, an alternate lifecycle, pseudolysogeny, has been described (Ripp and Miller 1998).

Pseudolysogeny is defined as an unstable state whereby the phage genome co-exists with the bacterial host for extended periods of time (Ripp and Miller 1998). In the pseudolysogenic state the phage genome does not actively replicate through the lytic or lysogenic pathways; instead it enters a dormant state and is maintained within the host cell (Ripp and Miller 1998). Phages have been observed to enter the pseudolysogenic state when nutrients are limited for the host, and then when environmental conditions improve, the phage begins to actively replicate through either the lytic or lysogenic cycle (Ripp and Miller 1998). Some phages may also kill bacterial cells through a phenomenon known as lysis from without (Abedon 2011).

Lysis from without (LO) occurs when a high number of phages simultaneously infect a single bacterial cell, causing destruction of the bacterial cell through destabilization of the cell membrane integrity

(Monk et al. 2010, Abedon 2011). This process results in the rapid death of bacterial cells and does not require active phage replication (Abedon 2011). LO requires a high multiplicity of infection (MOI), which is the ratio of bacteriophage particles to bacterial cells, in order to have multiple phages to attach and lyse a single bacterial cell. LO is specifically relevant to the use of bacteriophages in food because many commodities and products are stored or refrigerated at temperatures that are not conducive to bacterial growth and therefore the lytic cycle cannot be propagated, making LO an efficient bacteriophage-mediated mechanism of pathogen reduction.

16.4 Phage Application within the Food Industry

16.4.1 Fresh Produce

Lytic bacteriophage application to fresh produce to reduce pathogenic bacterial contamination may be of particular interest due to some of their unique qualities. Phages are highly target- (species) specific; therefore, phage treatments can be directed towards a particular pathogen without disrupting the non-pathogenic microbial community (microbiome) present in/on produce commodity. Recent studies have shown that as a biocontrol agent, phage can effectively reduce bacterial foodborne pathogens on surfaces of whole and fresh-cut commodities, but the reduction is dependent upon how the phage is applied and how commodities are stored after phage application. Previous studies include applications of phages in produce to control pathogens using a cocktail of phages to ensure effectiveness against different strains of species, and to ensure that the development of bacteriophage-insensitive mutants (BIMs) are minimized. Phage cocktail treatments are typically targeted against specific pathogens (Sharma et al. 2013, Litt et al. 2018b). In most cases when applying phages to the surface of produce commodities, LO seems to be the mechanism by which pathogens are reduced (Sharma et al. 2013).

Evaluating the effectiveness of bacteriophages on produce commodities has involved spray or immersion application methods on whole or fresh-cut commodities. A combination of three *E. coli* O157:H7-specific lytic bacteriophages (termed ECP-100) was sprayed at a level of 5.9 log plaque-forming units (PFU)/cm^2 onto fresh-cut lettuce which had been inoculated with *E. coli* O157:H7 at a level of 2.6 log colony-forming units (CFU)/cm^2, and stored at 4°C for up to 2 days.[11] *E. coli* O157:H7 populations on the fresh-cut lettuce treated with control (phosphate buffered saline) were significantly ($P < 0.05$) higher than those on lettuce receiving ECP-100 spray (Sharma et al. 2009), indicating the effectiveness of the phage treatment. Furthermore, the bactericidal effect due to the application of ECP-100 occurred very quickly after spraying: the initial application of ECP-100 reduced *E. coli* O157:H7 counts by 1.92 log CFU/cm^2 on day 0 within 30 min. Spraying the same ECP-100 bacteriophage mixture at ca. 7 log PFU/ml on spinach inoculated with ca. 4 log CFU *E. coli* O157:H7 /g resulted in reductions of 100% after 24 h and 120 h, and 99% by 168 h when stored at 10°C (Abuladze et al. 2008). In another study, a mixture of eight lytic bacteriophages (BEC 8) specific to *E. coli* O157:H7 were deposited at a titer of 10^6 PFU/leaf on fresh-cut baby romaine lettuce or baby spinach leaves and inoculated with either 10^4, 10^5, or 10^6 CFU *E. coli* O157:H7/leaf and stored at either 4, 8, 23, or 37°C. The results from these experiments also showed that as the multiplicity of infection (MOI) and incubation temperature increased, inactivation of the pathogen on the surface of the leaf increased (Viazis et al. 2011a).

Lytic bacteriophages have also been evaluated under minimal processing and commercial storage conditions used for leafy greens. The immersion of fresh-cut lettuce in solutions of bacteriophages to simulate wash water used for fresh-cut leafy greens did provide protection against cross-contamination with *E. coli* O157:H7, but required a longer storage time (24 hours) compared to spraying to show differences between control and phage treatments (Ferguson et al. 2013). After immersion treatment, populations were reduced to below the detectable limits after 3 days of storage at 4°C. The lack of immediate reduction by immersion is most likely due to the lower number of phages distributed to the surface of lettuce by immersion compared to spraying. The treatment of fresh-cut lettuce with a combination of *E. coli* O157-specific bacteriophage (spray treatment) and 50 ppm sodium hypochlorite (immersion treatment) reduced *E. coli* O157:H7 populations on fresh-cut iceberg lettuce more than either bacteriophage or hypochlorite treatment alone. Bacteriophages specific to *E. coli* O157:H7 were also effective on spinach packaged under modified atmosphere packaging (MAP) and stored at 4°C. MAP is commonly used to preserve the quality of fresh-cut leafy greens sold at retail. Bacteriophage treatment significantly reduced populations of *E. coli* O157:H7 after 24 h on spinach, green leaf, and romaine lettuce, respectively, compared to untreated controls under MAP conditions (Boyacioglu et al. 2013). These findings describing the effectiveness of lytic bacteriophages on leafy greens after immersion, under modified atmosphere packaging, and on fresh-cut products show that phages can be integrated into current leafy green processing and packaging conditions.

Lytic bacteriophages have been effective in reducing pathogenic bacterial populations on several types of melons. A cocktail of four lytic bacteriophages (SCPLX-1), specific to *Salmonella* Enteritidis, was applied to inoculated fresh-cut honey dew melons through a spot treatment (5×10^6 PFU/spot) and reduced *Salmonella* by approximately 3.5 log CFU when stored at 5 and 10°C after 3 days when compared to untreated controls; however, the bacteriophage treatment was less effective at a storage temperature of 20°C (Leverentz et al. 2001). Bacteriophages specific to *Listeria monocytogenes* were more effective in reducing the pathogen on fresh-cut honeydew melon stored at 10°C for 7 days compared to treatments of nisin (a bacteriocin) or water alone; a combination of bacteriophages specific to *L. monocytogenes* and nisin reduced populations of the pathogen more effectively than nisin alone (Leverentz et al. 2003). This work showed that the activity of bacteriophages occurs almost instantaneously once applied to inoculated flesh of melons, and that bacteriophage activity can be enhanced at lower temperatures, which limit growth and potential regrowth of the pathogen. Lytic phages specific to *Salmonella* spp. were applied to reduce *Salmonella* Newport on whole and fresh-cut cucumbers (Sharma et al. 2017). Spray application of 10^{10} PFU/ml *Salmonella*-specific phages to whole mini cucumbers inoculated with *S.* Newport (5 log CFU/cucumber) resulted in significantly lower *S.* Newport levels compared to non-phage treated controls at both 10°C or 22°C storage temperatures; however, lytic phages applied to whole, inoculated cucumbers were not able to prevent the transfer of *S.* Newport to cut pieces of cucumbers.

The effectiveness of lytic bacteriophages against foodborne pathogens on fresh-cut tomatoes and tomato plants has been variable. Initial *E. coli* O157:H7 populations of ca. 3 log CFU were reduced by ca. 1 log CFU after 168 h on fresh-cut tomatoes when spray-treated with an ECP-100, a mixture of lytic phages specific to *E. coli* O157:H7 (Abuladze et al. 2008). A cocktail of five lytic bacteriophages specific to *Salmonella* spp. were applied to blossoms of tomato plants at a level of 6 log PFU/ml, and also treated with 6 log CFU/ml of *Salmonella* Javiana. Tomato fruits had a significantly lower incidence of internalized *S.* Javiana compared to fruits which were inoculated with *S.* Javiana and no bacteriophage (Ye et al. 2009). These experiments also showed that a combination of a *Salmonella*-specific lytic bacteriophages, combined with potential bacterial biocontrol (*E. absuriae*), was more effective at reducing *S.* Javiana compared to the bacteriophage mixture alone.

Bacteriophages specific to *Salmonella* spp. have been applied to seeds and sprouts to limit contamination with the pathogen. A combination of two bacteriophages, one with specific lytic ability against *S.* Montevideo, and one with specificity against both *S.* Typhimurium and *S.* Enteritidis, applied at a level of 6.7 log PFU/ml, reduced *Salmonella* populations by 1.50 log CFU/ml from an initial population of ca. 7 log CFU/ml in soak-water of broccoli seeds compared to non-bacteriophage-treated seeds (Pao et al. 2004). A cocktail of six lytic bacteriophages specific to *Salmonella* spp., applied at a concentration of 6 log PFU/ml in combination with *E. absuriae* (a bacterial biocontrol agent), reduced *Salmonella* spp. populations on inoculated mung bean sprouts by ca. 6 log CFU/ml compared to sprouts which did not receive any bacteriophage or biocontrol treatments (Ye et al. 2010). Interestingly, the combination of *Salmonella*-specific lytic phages and *E. absuriae* reduced *Salmonella* spp. populations on mung bean sprouts more than either bacteriophages or *E. absuriae* alone. Although different phage cocktails and *Salmonell*a serovars were used on mung bean sprouts, these results differ from those observed with tomatoes which showed the bacteriophage mixture contributing a minimal antimicrobial effect in combination with *E. absuriae.* These contrasting results with the same bacteriophage mixture indicate that lytic phages may be more effective on some types of produce commodities compared to others.

A wide array of products, like chocolate milk, mozzarella cheese brine, hot dogs, sliced turkey meat, smoked salmon, seafood, sliced cabbage, and lettuce leaves, have all been treated with phage cocktails (Guenther et al. 2009) to reduce bacterial pathogen loads. Foods were dip-inoculated (immersed) in an inoculum of *L. monocytogenes* (10^3 CFU/g), and then a phage cocktail specific to *L. monocytogenes* was applied (10^6–10^8 PFU/g), followed by storage at 6°C for 6 days. In general, phages were more effective at reducing *L. monocytogenes* in liquid foods (chocolate milk, cheese brine); *L. monocytogenes* populations rapidly fell by 3 log CFU/g. In other foods, pathogen populations were reduced up to 5 log CFU/g, compared to the non-phage treated control. In the case of cabbage, the *L. monocytogenes* population was reduced by approximately 4 log CFU/g and on lettuce by 2.6 log CFU/g. Overall, treatment with a high phage concentration (10^8 PFU/g) was more effective than treatment with a lower concentration. Phages stability study results showed that an anti-listeria effect was maintained during storage in foods of animal origin, whereas phages titers fell to ineffective levels on leafy greens. These results suggest that phage effectiveness may be affected not only by the food type but also by the other non-pathogenic microflora present in those foods.

16.4.2 Meat and Dairy Food Products

Phage applications in the dairy industry have been long studied for their ability to inactivate microorganisms, but also pose a unique threat to the cultures necessary for cheese production (Fernández et al. 2017, Gutiérrez et al. 2019). Our focus in this chapter remains on pathogen inactivation. One study investigated the phage activity against *Salmonella* Enteritidis on Cheddar cheeses produced from raw and pasteurized milk (Modi et al. 2001). Raw milk and pasteurized milk were inoculated with a luminescent strain of *S.* Enteritidis (10^4 CFU/ml) and 10^8 PFU/ml of a *Salmonella*-specific phage. Raw and pasteurized milk samples were processed into Cheddar cheese following standard procedures. Populations of *S.* Enteritidis decreased by 1–2 logs CFU/g in cheese made from both raw and pasteurized milk cheeses compared to the control (no phage treatment), and *S.* Enteritidis survived for shorter durations in phage-treated cheese compared to control.

The application of phages to control pathogenic bacteria has been tested on dairy and meat products during both pre- and post-harvest applications. Lytic phages specific to *Salmonella* Typhimurium and *Campylobacter jejuni* were applied to cooked and raw beef stored at refrigeration (5°C) and room temperature (24°C) (Bigwood et al. 2008). Raw and cooked meat samples (2×2 cm^2) were spot-inoculated (10^2 or 10^4 CFU/cm^2) with either *Salmonella* Typhimurium or *C. jejuni*, followed by respective phage application (10 or 10^4 MOI), and storage at 5°C or 24°C for 8 days. Samples stored at 5°C and 24°C showed significant reductions of 2–2.3 log CFU/cm^2 and 4.7–6 logs CFU/cm^2 of *S.* Typhimurium populations, respectively, when compared to the control after 24 hours of incubation in raw and cooked meat at high MOI. The *C. jejuni* population was reduced between 1.5–2.4 logs CFU/cm^2 at 5°C after 24 hours on cooked and raw meats after phage treatment at high MOI. At 24°C, *C. jejuni* populations were reduced significantly by 2.2 logs CFU/cm^2 in raw meat after 24 hours and by >4 logs CFU/cm^2 in cooked meat after 3 hours when compared to the control. An increase in *C. jejuni* populations was observed after 24 h at 24°C, which it was suggested could be due to moisture loss and restricted phage movement in the complex meat surface.

Another study evaluated lytic bacteriophage application to eliminate *E. coli* O157:H7 on beef steak. The steak meat was cut into equal pieces (18 pieces) and spot-inoculated with a rifampin-resistant strain of *E. coli* O157:H7 (P1432) at 10^3 CFU/ml concentration and incubated at 37°C for 1 hour (O'Flynn et al. 2004). Afterwards, 1 ml of a phage cocktail at a titer of 10^8 PFU/ml was spot-inoculated onto the meat surfaces (9 pieces) and incubated for 1 h. In seven of the nine samples no *E. coli* O157:H7 was recorded after enrichment, and the other two pieces had less than 10 CFU/ml of viable *E. coli* O157:H7. Another study also showed that single phage (DT6), applied at a high MOI (10^3–10^4 MOI) to beef cuts (1 cm^2), reduced non-O157 and O157 STEC populations by 0.33–0.77 log CFU/cm^2 after 3 h, and 0.47–1.15 log CFU /cm^2, respectively, and these reductions were significantly greater than control (non-bacteriophage) treatments (Tomat et al. 2013).

16.4.3 Food-Contact Surfaces

Food-contact surfaces vary within food production and retail environments and are often areas of contamination and present risk of cross-contamination between food products. Lytic bacteriophages have been assessed for their effectiveness on several food-contact surfaces (Litt et al. 2017b, Viazis et al.

2011a). The effect of a bacteriophage cocktail (BEC8) to control *E. coli* O157:H7 on food-contact surfaces was assessed (Viazis et al. 2011). Sterile stainless steel (SS), ceramic tile (CT), and high-density polyethylene (HDPE) were inoculated with *E. coli* O157:H7 (10^6, 10^5, 10^4 CFU/chip). Inoculated chips were treated with controls and BEC8 (10^6 PFU/chip) and incubated at 4, 12, 23, and 37°C and the surviving pathogen population was enumerated. In general, *E. coli* O157:H7 cells in low inoculum on chips were reduced by ca. 4 log CFU/chip quickly (10–60 min) at temperatures other than 4°C. *E. coli* O157:H7 populations (10^4, 10^6 CFU/chip) were reduced significantly compared to controls only after 24 hours at 37°C by ca. 5, 2.5, and 3 log CFU/chip on SS, CT, and HDPE, respectively (Viazis et al. 2011).

Another study determined the effectiveness of a lytic bacteriophage (KH1) against *E. coli* O157:H7 attached to sterile stainless-steel coupons or which had formed biofilms on these coupons (Sharma et al. 2005). Sterile stainless-steel coupons were inoculated with *E. coli* O157:H7 (7–8 log CFU/coupon) and incubated at 22°C for 24 h, to encourage attachment with minimal biofilm formation, or 96 hours to encourage biofilm formation, before treatment with a single lytic phage specific to *E. coli* O157:H7 (7.7 logs PFU/ml) for 4 days (biofilm) or 1 day (attached cells). Phage treatment reduced attached bacterial cells by 1.2 logs CFU/coupon, after day 1, and maintained those levels of inactivation up to 4 days. In biofilms treated with KH1, *E. coli* O157:H7 ATCC 43895 populations did not differ significantly compared to populations in biofilms treated with controls (water), when treated with phage. By comparison, alkaline cleaners reduced both *E. coli* O157:H7 strain populations by 5–6 logs CFU per coupon. Overall, phage treatment was more effective against bacterial cells attached to stainless-steel coupon surfaces compared to cells in the biofilms (Sharma et al. 2005).

Reduction of non-O157 and O157 shiga-toxigenic *E. coli* (STEC) strains on stainless-steel surfaces was also achieved by applying a two-phage cocktail on inoculated stainless-steel coupons at low (10^4) or high (10^6) MOI values. Both non-O157 and O157 STEC populations on stainless-steel surfaces were reduced more at 37°C (ca. 2–8 log CFU reductions) than at 4°C (ca. 2–4 log CFU reductions), and, in general, reductions were greater after 24 hours than after 1 or 3 hours (Tomat et al. 2014).

16.4.4 Pre-Harvest Control

Some researchers have considered the application of bacteriophages in agricultural and food production systems to reduce pathogen contamination of foods (Atterbury et al. 2007). This may include the application of phages in crop or animal production, in agricultural waste management equipment, or oral administration in live food animals. A study was conducted to evaluate the efficacy of phage therapy to reduce *C. jejuni* colonization in broiler chickens (Carrillo et al. 2005). The 25-day-old broiler chickens were experimentally colonized with *C. jejuni* HPC5 and GIIC8 and treated with phages specific to *C. jejuni* (CP8 and CP34) through oral administration at different dosages. Phage-treated *C. jejuni*-colonized birds showed a reduction of *C. jejuni* populations ranging between 0.5 and 5 logs CFU/g of cecal contents compared to untreated controls after 5-day period. Reductions varied on the basis of the phage combination, the dose of phage given to the bird, and the time elapsed after administration. However, bacteriophage-insensitive mutant (BIM) *C. jejuni* cells were recovered from phage-treated chickens at a frequency of less than 4%. These BIM *C. jejuni* were compromised in their ability to subsequently colonize experimental chickens and rapidly reverted to a phage-sensitive phenotype *in vivo*. These results show the importance of using a cocktail of multiple phages specific to the target pathogen in the food animal (Carrillo et al. 2005).

Another study looked at the application of phage at varying multiplicity of infection (MOI) in the treatment of dairy manure compost comprised of cow manure and sawdust, wasted feed, old hay, and vegetable wastes (Heringa et al. 2010). Compost (600 g) was spread and dried on trays in a biosafety hood at room temperature for 1 day. *Salmonella* Typhimurium strain 8243 (ca. 10^7 CFU/ml) was sprayed on the surface of compost and stored overnight at room temperature. A mixture of five lytic bacteriophages effective against several *Salmonella* serovars was sprayed at MOIs of 1, 10, and 50 onto 150 g of inoculated compost, mixed, and stored at room temperature (25°C) for 24 hours. Results showed significant reductions (~2.5 logs CFU/g) in *S.* Typhimurium populations compared to untreated controls following phage treatment at all MOI levels evaluated and at all compost moisture content levels (30, 40, 45, and 50%). Populations of *S.* Typhimurium were decreased by phage treatment within the first 4 h at all MOI levels evaluated, and these levels were maintained over 24 hours (Heringa et al. 2010). These results suggest that phage treatment can be applied during the compost process, even at higher moisture contents, to reduce foodborne pathogen counts.

16.4.5 Bacteriophage Insensitive Mutants

Bacterial cells can become resistant to lytic phage infection by preventing phage adsorption via modifying the surface phage receptor structure, or by hindering the access of the phage to the receptor through the production of an excess of the extracellular matrix, or even by producing competitive inhibitors (Buckling and Rainey 2002). Phage-driven bacterial evolution increases its chances to adapt and survive in altered environmental conditions, ultimately becoming beneficial to bacteria (Pal et al. 2007). However, the coevolution of phage and bacteria in the natural environment has been shown, where phage-susceptible and -resistant bacterial populations intertwine with parent phages and evolved phages (mutant phages) that regain infectivity against the resistant bacteria (Buckling and Rainey 2002). Several studies have evaluated bacterial pathogens for BIM development after phage treatment in foods. BIM mutants of *E. coli* O157:H7 developed when meat was treated with individual lytic or a cocktail of lytic phages (O'Flynn et al. 2004), but most BIM colonies remained susceptible to at least of the one of lytic phages used in the cocktail. Another study showed that of eight presumptive BIM *E. coli* O157:H7 colonies isolated after phage treatment of beef samples, only two (25%) showed resistance to a single O157-specific phage infection (Tomat et al. 2013). A similar study examining *S.* Newport inactivation by lytic phages on whole cucumbers showed a relatively low rate (0.7%, 1/127 isolates) of BIM development (Sharma et al. 2017). The development of BIM may also be based on the storage temperature of foods treated with lytic bacteriophages, with higher temperatures potentially promoting higher growth rates and mutation rates of bacterial cells. However, BIM development for lytic phage applications to foods is unlikely, due to the presence of low levels of the pathogen in the environment, the use of multi-phage preparations in commercial lytic bacteriophage preparations which have different targets, and the lack/slow growth of most bacterial pathogens at refrigeration or storage temperatures that slow the development of BIM mutant populations.

16.5 Where Does Phage Technology Stand Today?

Several commercial phage products have been approved by regulatory agencies in the United States as food additives or to enhance food safety. The U.S. Department of Agriculture's Food Safety and Inspection Service (USDA FSIS) has recorded of 12 entries for use of phage products which are listed as safe and suitable antimicrobials and could be utilized as processing aids in food production systems (FDA 2019). Most of these phage-based products do not have labeling requirements, except in cases where the product requires labeling of any additives. Phage products can be applied at any point including live animals/poultry pre-slaughter, poultry carcasses, raw poultry, beef parts and trims, and ready-to-eat (RTE) meat and poultry. The Food and Drug Administration (FDA) has issued ten notices as generally recognized as safe (GRAS) for phage-based food safety products with 'no questions' notices and has not contested the manufacturer's claim of GRAS (FDA 2019).

Many companies have entered and are continuing to enter the bacteriophage production space. A few are mentioned here, and this discussion, showing the variety of materials approved and available, is not intended to be a complete discussion of all available technologies as this is a large area of research and development. A Dutch company, Micreos, PhageGuard, produces multiple FDA-approved phage preparations against foodborne pathogens, such as Listex P100 or PhageGuard Listex against *L. monocytogenes*, Salmonelex or PhageGuard S targeting *Salmonella* spp., and PhageGuard E for *E. coli* O157:H7 in raw and RTE foods (PhageGuard 2019a). PhageGuard Listex and S are approved for clean-label processing in the US, European Union (EU), Canada, Australia, New Zealand, Switzerland, and Israel (PhageGuard 2019a, 2019b). Intralytix is a United States-based biotech company in Baltimore, MD, producing phage-based products including ListShield targeting *L. monocytogenes*, EcoShield for *E. coli* O157: H7, SalmoFresh against *Salmonella* spp., and ShigaShield targeting *Shigella* spp. SalmoFresh has received GRAS status from the FDA and Health Canada, and is approved to be used as a processing aid in fresh and processed fruits and vegetables against *Salmonella* (Intralytix 2019a). Moreover, the National Food Service, Ministry of Health in Tel Aviv, Israel, approved the use of all FDA-approved phage preparations for similar applications on food (Intralytics 2019b). Another US phage company, Passport, has developed a phage product, Finalyse, to control *E. coli* O157:H7 and other STEC at preharvest level including animal hides and processing plants (Passport 2019).

16.6 Conclusion

Bacteriophages have numerous applications for food safety, and can effectively reduce foodborne pathogens on various foods. Several commercial phage-based products have been approved for use in the food industry in a number of countries. Commercial phage products can be utilized for the reduction of specific foodborne pathogens at various timepoints in food production systems, such as in animal feed, sprayed on animal hides, animal contact surfaces or carcasses, and on raw poultry and beef trim, ready-to-eat (RTE) meat and poultry, dairy products, dip or spray wash produce, whole and fresh-cut fruits and vegetables, in fruit juices, or for surface-treatment of processing facilities. Lytic bacteriophage treatments seem to be most effective against bacterial foodborne pathogens when applied to foods in a way that delivers a high titer of phages to a large area, and then foods are stored at a temperature which prevents or slows the growth of foodborne pathogens after bacteriophage treatment. Bacteriophages can be an additional tool in a multi-hurdle approach to minimize the hazard of foodborne pathogens in foods with minimal processing and decrease the risk of bacterial infections to consumers, especially in fresh-cut produce (Abedon 2016, Mahony et al. 2011, Hagens and Loessner 2010). More understanding about the interactions of lytic bacteriophages with the overall microbiomes of foods can only improve the vast potential of lytic phages in foods to reduce the risk of foodborne illness.

REFERENCES

Abedon, S. T. 2011. Lysis from without. *Bacteriophage* 1:46–49.

Abedon, S. T. 2016. Phage therapy dosing: The problem(s) with multiplicity of infection (MOI). *Bacteriophage* 6:e1220348.

Abuladze, T., Li, M., Menetrez, Y. M., Dean, T., Senecal, A. and Sulakvelidze, A. 2008. Bacteriophages reduce experimental contamination of of hard surfaces, tomato, spinach, broccoli, and ground beef with *Escherichia coli* O157:H7. *Appl. Environ. Microbiol.* 74:6230–6238.

Ackermann, H. W. 2007. 5500 Phages examined in the electron microscope. *Arch. Virology* 152:227–243.

Atterbury, R. J., Van Bergen, M. A. P., Ortiz, F., Lovell, M. A., Harris, J. A., De Boer, A., Wagenaar, J. A., Allen, V. M. and Barrow, P. A. 2007. Bacteriophage therapy to reduce *Salmonella* colonization of broiler chickens. *Appl. Environ. Microbiol.* 73:4543–4549.

Bamford, D. H., Caldentey, J. and Bamford, J. K. H. 1995. Bacteriophage PRD1: A broad host range dsDNA tectivirus with an internal membrane. *Adv. Virus Res.* 45:281–319.

Bertin, A., de Frutos, M. and Letellier, L. 2011. Bacteriophage–host interactions leading to genome internalization. *Curr. Opinion Microbiol.* 14:492–496.

Bigwood, T., Hudson, J. A., Billington, C., Carey-Smith, G. V. and Heinemann, J. A. 2008. Phage inactivation of foodborne pathogens on cooked and raw meat. *Food Microbiol.* 25:400–406.

Boyacioglu, O., Goktepe, I., Sharma, M. and Sulakvelidze, A. 2013. Biocontrol of *Escherichia coli* O157:H7 on fresh-cut leafy greens: Using a bacteriophage cocktail in combination with modified atmosphere packaging. *Bacteriophage* 3:e24620.

Brussow, H. and Kutter, E. 2005. Genomics and evolution of tailed phages. In *Bacteriophages: Biology and Applications*. E. Kutter and A. Sulakvelidze (eds), 91–128. Washington DC: CRC Press.

Bruynoghe, R. A. J. M. and Maisin, J. 1921. Essais de thérapeutique au moyen du bacteriophage. *Compt Rend Soc Biol.* 85:1120–1121.

Buckling, A. and Rainey, P. B. 2002. Antagonistic coevolution between a bacterium and a bacteriophage. *Proc. Royal Soc. London. Ser. B: Biol. Sci.* 269:931–936.

Carrillo, C. L., Atterbury, R. J., El-Shibiny, A., Connerton, P. L., Dillon, E., Scott, A. and Connerton, I. F. 2005. Bacteriophage therapy to reduce *Campylobacter jejuni* colonization of broiler chickens. *Appl. Environ. Microbiol.* 71:6554–6563.

Comeau, A. M., Hatfull, G. F., Krisch, H. M., Lindell, D., Mann, N. H. and Prangishvili, D. 2008. Exploring the prokaryotic virosphere. *Res. Microbiol.* 159:306–313.

Cota-Robles, E. H. and Coffman, M. D. 1963. Fine structure of the bacteriophage attachment process. *J. Bacteriol.* 86:266–273.

d'Herelle, F. 1917. Sur un microbe invisible antagoniste des bacilles dysentériques. *Compt Rend Soc Biol.* 165:373–375.

Daugelavicius, R., Bamford, J. K. and Bamford, D. H. 1997. Changes in host cell energetics in response to bacteriophage PRD1 DNA entry. *J. Bacteriol.* 179:5203–5210.

Deresinski, S. 2009. Bacteriophage therapy: Exploiting smaller fleas. *Clin. Infect. Dis.* 48:1096–1101.

Duckworth, D. H. 1976. Who discovered bacteriophage? *Bacteriol. Rev.* 40:793.

Earnshaw, W. C. and Harrison, S. C. 1977. DNA arrangement in isometric phage heads. *Nature* 268:598–602.

Echols, H. 1972. Developmental pathways for the temperate phage: Lysis vs lysogeny. *Ann. Rev. Genetics* 6:157–190.

Fauquet, C. M., Mayo, M. A., Maniloff, J., Desselberger, U. and Ball, L. A. 2005. Virus Taxonomy: *VIIIth Report of the International Committee on Taxonomy of Viruses.* London: Academic Press.

FDA. 2019. GRAS Notice. https://www.accessdata.fda.gov/scripts/fdcc/?set=GRASNotices&sort=GRN_No&order=DESC&startrow=1&type=basic&search=Phage

Ferguson, S., Roberts, C., Handy, E. and Sharma, M. 2013. Lytic bacteriophages reduce *Escherichia coli* O157:H7 on fresh-cut lettuce introduced through cross-contamination. *Bacteriophage* 3:e24323.

Fernández, L., Escobedo, S., Gutiérrez, D., Portilla, S., Martínez, B., García, P. and Rodríguez, A. 2017. Bacteriophages in the dairy environment: From enemies to allies. *Antibiotics* 6:27.

Grahn, A. M., Daugelavičius, R. and Bamford, D. H. 2002. Sequential model of phage PRD1 DNA delivery: Active involvement of the viral membrane. *Molec. Microbiol.* 46:1199–1209.

Greer, G. G. 2005. Bacteriophage control of foodborne bacteria. *J. Food Prot.* 68:1102–1111.

Guenther, S., Huwyler, D., Richard, S. and Loessner, M. J. 2009. Virulent bacteriophage for efficient biocontrol of *Listeria monocytogenes* in ready-to-eat foods. *Appl. Environ. Microbiol.* 75:93–100.

Gutiérrez, D., Fernández, L., Rodríguez, A. and García, P. 2019. Role of bacteriophages in the implementation of a sustainable dairy chain. *Frontiers Microbiol.* 10:12.

Guttman, B., Raya, R. and Kutter, E. 2005. Basic phage biology. *Bacteriophages: Biol. Applic.* 4.

Hagens, S. and Loessner, M. J. 2010. Bacteriophage for biocontrol of foodborne pathogens: Calculations and considerations. *Curr. Pharm. Biotechno.* 11:58–68.

Hanlon, G. W. 2007. Bacteriophages: An appraisal of their role in the treatment of bacterial infections. *Intl. J. Antimicrob. Agents* 30:118–128.

Heringa, S. D., Kim, J. K., Jiang, X., Doyle, M. P. and Erickson, M. C. 2010. Use of a mixture of bacteriophages for biological control of *Salmonella enterica* strains in compost. *Appl. Environ. Microbiol.* 76:5327–5332.

Hughes, K. A., Sutherland, I. W. and Jones, M. V. 1998. Biofilm susceptibility to bacteriophage attack: The role of phage-borne polysaccharide depolymerase. *Microbiol.* 144:3039–3047.

Intralytix. 2019a. Food Safety. http://www.intralytix.com/index.php?page=food

Intralytix. 2019b. Health Canada approved the use of SalmoFresh as a processing aid for various foods in Canada. http://www.intralytix.com/index.php?page=prod&id=3

Keen, E. C. 2012. Felix d'Herelle and our microbial future. *Future Microbiol.* 7:1337–1339.

Kennedy, J. E., Oblinger, J. L. and Bitton, G. 1984. Recovery of coliphages from chicken, pork sausage and deli meats. *J. Food Prot.* 47:623–626.

Kennedy, J. E., Wei, C. I. and Oblinger, J. L. 1986. Characterization of coliphages recovered from foods according to temperature of infectivity. *J. Food Prot.* 49:952–954.

Labrie, S. J. and Moineau, S. 2007. Abortive infection mechanisms and prophage sequences significantly influence the genetic makeup of emerging lytic lactococcal phages. *J. Bacteriol.* 189:1482–1487.

Labrie, S. J., Samson, J. E. and Moineau, S. 2010. Bacteriophage resistance mechanisms. *Nature Rev. Microbiol.* 8:317–327.

Leiman, P. G., Chipman, P. R., Kostyuchenko, V. A., Mesyanzhinov, V. V. and Rossmann, M. G. 2004. Three-dimensional rearrangement of proteins in the tail of bacteriophage T4 on infection of its host. *Cell* 118:419–429.

Leverentz, B., Conway, W. S., Alavidze, Z., Janisiewicz, W. J., Fuchs, Y., Camp, M. J., Chighladze, E. and Sulakvelidze, A. 2001. Examination of bacteriophage as a biocontrol method of *Salmonella* on fresh cut fruit: A model study. *J. Food Prot.* 64:1116–1121.

Leverentz, B., Conway, W. S., Camp, M. J., Janisiewicz, W. J., Abuladze, T., Yang, M., Saftner, R. and Sulakvelidze, A. 2003. Biocontrol of *Listeria monocytogenes* on fresh-cut produce by treatment with lytic bacteriophages and a bacteriocin. *Appl. Environ. Microbiol.* 69:4519–4526.

Lindberg, A. A. 1997. Bacterial surface carbohydrates and bacteriophage adsorption. In *Surface Carbohydrates of the Prokaryotic Cell.* I. W. Sutherland (ed.), 289–356. London: Academic Press.

Litt, P. K. and Jaroni, D. 2017a. Isolation and physio-morphological characterization of *Escherichia coli* O157:H7-infecting bacteriophages recovered from beef cattle operations. *Intl. J. Microbiol.* 2017. https://doi.org/10.1155/2017/7013236

Litt, P. K., Kountoupis, T., Lamichhane, P. and Jaroni, D. 2017b. Disruption of shiga-toxigenic *E. coli* biofilms *in-vitro* and on food contact surfaces using bacteriophages. *Proceedings of Annual Meeting of International Association of Food Protection.*

Litt, P. K., Saha, J. and Jaroni, D. 2018a. Isolation and characterization of bacteriophages targeting non-O157 shiga-toxigenic *Escherichia coli. J. Food Prot.* 81:785–794.

Litt, P. K., Jadeja, R., Kakani, R., Saha, J., Kountoupis, T. and Jaroni, D. 2018b. Biocontrol of shiga-toxigenic *Escherichia coli* on fresh produce using bacteriophages. *Proceedings of Annual Meeting of International Association of Food Protection.*

Lwoff, A. 1953. Lysogeny. *Bacteriological Rev.* 17:269.

Mahony, J., McAuliffe, O., Ross, R. P. and Sinderen, D. V. 2011. Bacteriophages as biocontrol agents of food pathogens. *Curr. Opinion Biotechnol.* 22:157–163.

Marsh, M. and Helenius, A. 2006. Virus entry: Open sesame. *Cell* 124:729–740.

Matsuzaki, S., Rashel, M., Uchiyama, J., Sakurai, S., Ujihara, T., Kuroda, M., Imai, S., Ikeuchi, M., Tani, T. and Fujieda, M. 2005. Bacteriophage therapy: A revitalized therapy against bacterial infectious diseases. *J. Infec. Chemother.* 11:211–219.

Maura, D. and Debarbieux, L. 2011. Bacteriophages as twenty-first century antibacterial tools for food and medicine. *Appl. Microbiol. Biotechnol.* 90:851–859.

Mesyanzhinov, V. V., Robben, J., Grymonprez, B., Kostyuchenko, V. A., Bourkaltseva, M. V., Sykilinda, N. N., Krylov, V. N. and Volckaert. G. 2002. The genome of bacteriophage φKZ of *Pseudomonas aeruginosa. J. Molec. Biol.* 317:1–19.

Modi, R., Hirvi, Y., Hill, A. and Griffiths, M. W. 2001. Effect of phage on survival of *Salmonella enteritidis* during manufacture and storage of cheddar cheese made from raw and pasteurized milk. *J. Food Prot.* 64:927–933.

Monk, A. B., Rees, C. D., Barrow, P., Hagens, S. and Harper, D. R. 2010. Bacteriophage applications: Where are we now? *Lett. Appl. Microbiol.* 51:363–369.

O'Flynn, G., Ross, R. P., Fitzgerald, G. F. and Coffey, A. 2004. Evaluation of a cocktail of three bacteriophages for biocontrol of *Escherichia coli* O157:H7. *Appl. Environ. Microbiol.* 70:3417–3424.

Pal, C., Maciá, M. D., Oliver, A., Schachar, I. and Buckling, A. 2007. Coevolution with viruses drives the evolution of bacterial mutation rates. *Nature* 450:1079.

Pao, S., Randolph, S. P., Westbrook, E. W. and Shen, H. 2004. Use of bacteriophages to control *Salmonella* in experimentally contaminated sprout seeds. *J. Food Sci.* 69:M127–M129.

Passport. 2019. Finalyse. http://www.passportfoodsafety.com/finalyse

PhageGuard. 2019a. Don't give *Listeria* a chance with PhageGuard Listex. https://phageguard.com/listeria-solution/

PhageGuard. 2019b. Don't give *Salmonella* a chance with PhageGuard S. https://phageguard.com/salmonella-solution/

Puck, T. T., Garen, A. and Cline, J. 1951. The mechanism of virus attachment to host cells: I. The role of ions in the primary reaction. *J. Exper. Med.* 93:65.

Rakhuba, D. V., Kolomiets, E., Dey, E. S. and Novik, G. I. 2010. Bacteriophage receptors, mechanisms of phage adsorption and penetration into host cell. *Polish J. Microbiol.* 59:145–155.

Rice, K. C. and Bayles, K. W. 2008. Molecular control of bacterial death and lysis. *Microbiol. Molec. Biol. Rev*, 72:85–109.

Ripp, S. and Miller, R. V. 1998. Dynamics of the pseudolysogenic response in slowly growing cells of *Pseudomonas aeruginosa. Microbiol.* 144:2225–2232.

Rydman, P. S. and Bamford, D. H. 2000. Bacteriophage PRD1 DNA entry uses a viral membrane-associated transglycosylase activity. *Molec. Microbiol.* 37:356–363.

Rydman, P. S. and Bamford, D. H. 2002. The lytic enzyme of bacteriophage PRD1 is associated with the viral membrane. *J. Bacteriol.* 184:104–110.

Rydman, P. S., Caldentey, J., Butcher, S. J., Fuller, S. D., Rutten, T. and Bamford, D. H. 1999. Bacteriophage PRD1 contains a labile receptor-binding structure at each vertex. *J. Molec. Biol.* 291:575–587.

Salmond, G. P. C. and Fineran, P. C. 2015. A century of the phage: Past, present and future. *Nature Reviews Microbiol.* 13:777–786.

San Martin, C., Burnett, R. M., de Haas, F., Heinkel, R., Rutten, T., Fuller, S. D., Butcher, S. J. and Bamford, D. H. 2001. Combined EM/X-ray imaging yields a quasi-atomic model of the adenovirus-related bacteriophage PRD1 and shows key capsid and membrane interactions. *Structure* 9:917–930.

Serwer, P., Wright, E. T., Hakala, K. W. and Weintraub, S. T. 2008. Evidence for bacteriophage T7 tail extension during DNA injection. *BMC Res. Notes* 1:36.

Sharma, M. 2013. Lytic bacteriophages: Potential interventions against enteric bacterial pathogens on produce. *Bacteriophage* 3:e25518.

Sharma, M., Dashiell, G., Handy, E. T., East, C. L., Reynnells, R., White, C., Nyarko, E., Micallef, S., Hashem, F. and Millner, P. D. 2017. Survival of *Salmonella* Newport on whole and fresh-cut cucumbers treated with lytic bacteriophages. *J. Food Prot.* 80:668–673.

Sharma, M., Patel, J. R., Conway, W. S., Ferguson, S. and Sulakvelidze, A. 2009. Effectiveness of bacteriophages in reducing *Escherichia coli* O157:H7 on fresh-cut cantaloupes and lettuce. *J. Food Prot.* 72:1481–1485.

Sharma, M., Ryu, J. H. and Beuchat, L. R. 2005. Inactivation of *Escherichia coli* O157:H7 in biofilm on stainless steel by treatment with an alkaline cleaner and a bacteriophage. *J. Appl. Microbiol.* 99:449–459.

Sulakvelidze, A., Alavidze, Z. and Morris, J. G. 2001. Bacteriophage therapy. *Antimicrob. Agents Chemother.* 45:649–659.

Summers, W. C. 1999. *Felix de'Herelle and the Origins of Molecular Biology*. New Haven: Yale University Press.

Summers, W. C. 2001. Bacteriophage therapy. *Ann. Rev. Microbiol.* 55:437–451.

Suttle, C. A. 2005. Viruses in the sea. *Nature* 437:356–361.

Tomat, D., Migliore, L., Aquili, V., Quiberoni, A. and Balagué, C. 2013. Phage biocontrol of enteropathogenic and shiga toxin-producing *Escherichia coli* in meat products. *Frontiers Cell. Infect. Microbiol.* 3:20.

Tomat, D., Quiberoni, A., Mercanti, D. and Balagué, C. 2014. Hard surfaces decontamination of enteropathogenic and Shiga toxin-producing *Escherichia coli* using bacteriophages. *Food Res. Intl.* 57:123–129.

Twort, F. W. 1915. An investigation on the nature of ultra-microscopic viruses. *The Lancet* 186:1241–1243.

Valpuesta, J. M., Sousa, N., Barthelemy, I., Fernández, J. J., Fujisawa, H., Ibarra, B. and Carrascosa, J. L. 2000. Structural analysis of the bacteriophage T3 head-to-tail connector. *J. Structural Biol.* 131:146–155.

Viazis, S., Akhtar, M., Feirtag, J. and Diez-Gonzalez, F. 2011a. Reduction of *Escherichia coli* O157:H7 viability on hard surfaces by treatment with a bacteriophage mixture. *Intl. J. Food Microbiol.* 145:37–42.

Viazis, S., Mastura, A., Feirtag, J. and Diez-Gonzalez, F. 2011b. Reduction of *Escherichia coli* O157:H7 viability on leafy green vegetables by treatment with a bacteriophage mixture and trans-cinnaldehyde. *Food Microbiol.* 28:149–157.

Ye, J., Kostrzynska, M., Dunfield, K. and Warriner, K. 2009. Evaluation of a biocontrol preparation consisting of *Enterobacter asburiae* JX1 and a lytic bacteriophage cocktail to suppress the growth of *Salmonella* Javiana associated with tomatoes. *J. Food Prot.* 72:2284–2292.

Ye, J., Kostrzynska, M., Dunfield, K. and Warriner, K. 2010. Control of *Salmonella* on sprouting mung bean and alfalfa seeds by using a biocontrol preparation based on antagonistic bacteria and lytic bacteriophages. *J. Food Prot.* 73:9–17.

17

Naturally Occurring Compounds – Plant Sources

Aurelio Lopez-Malo, Stella M. Alzamora, María J. Paris, Leonor Lastra-Vargas, María Bernarda Coronel, Paula L. Gómez, and Enrique Palou

CONTENTS

17.1 Introduction

Food market trends are changing, consumers more frequently demand high-quality foods with fresh-like attributes (Alzamora et al., 2016; Gould, 1995a, 1995b, 1996), and consequently less extreme treatments and/or additives are being required. To satisfy consumer demands, adjustments or reductions in conventionally used preservation techniques must be accomplished. Gould (2002) identified some food characteristics that must be attained in response to consumer demands; most of them occur within the minimal processing concept (Berdejo et al., 2019). To satisfy market requirements, the safety and quality of foods have to be based on substantial improvements in traditional preservation methods (Aguilar-González et al., 2015; Lorenzo-Leal et al., 2019a, 2019b; Mani-López et al., 2018; Reyes-Jurado et al., 2019b).

Safe food may have different meanings to different people involved in the food chain; consumers for instance, relate a risk-free food as a safe one, and associate increased risk with the increased use of added substances such as antimicrobials, synthetic additives, and higher levels of sodium or fat, among others (Reyes-Jurado et al., 2019b). Scientists, public health officers, and international organizations define a safe food as one that provides maximum nutrition and quality while posing a minimal hazard to public health, and expect any risks that are present to be minimal (Shank and Carson, 1992).

Consequences of quality loss caused by microorganisms include consumer hazards, due to the presence of pathogenic organisms or microbial toxins, as well as economic losses due to spoilage (Davidson, 2001). Inactivation, growth delay, or growth prevention of spoilage and pathogenic microorganisms are the main objectives of food preservation. Several factors influence microbial growth and survival; the appropriate modification and/or application of these factors is the base of preservation technologies. Food preservation technologies protect foods from the effects of microorganisms and inherent deterioration. Major food preservation technologies can be classified as those that act mainly by preventing or slowing down microbial growth (low temperature, reduced water activity, acidification, fermentation, modified atmosphere packaging, addition of antimicrobials, and compartmentalization in water-in-oil emulsions, among others), those that act by inactivating microorganisms (heat pasteurization and sterilization, microwave heating, ionizing radiation, high pressure, pulsed electric fields, and high-frequency ultrasound, among others), and those that prevent or minimize the entry of microorganisms into food or remove them (aseptic handling or packaging, centrifugation, and filtration, among many others) (Gould, 2002). In addition, techniques in combination, based on the "hurdle technology" concept (Alzamora et al., 2016; Leistner, 2000), may act by inhibiting or inactivating microorganisms, depending on the combination of hurdles applied to achieve food preservation (Berdejo et al., 2019; López-Malo et al., 2000).

The use of chemical agents exhibiting antimicrobial activity (by inhibiting and/or reducing microbial growth or even by inactivating undesirable microorganisms) is one of the oldest and most traditional food preservation techniques. Antimicrobial agents are chemical compounds added to, or present in foods that retard microbial growth or cause microbial death. The antimicrobial activities of several plants used today as seasoning agents in foods and beverages have been recognized for centuries. The early Egyptians utilized plant extracts (spices and oils) as antimicrobials for the preservation of food as well as for embalming. Although ancient civilizations recognized the antiseptic or antimicrobial potential of many plant extracts, it was not until the eighteenth century that scientific information established the preservative effects of several plants.

Antimicrobial agents are somewhat arbitrarily classified as traditional and naturally occurring (Davidson, 2001). Traditional antimicrobials are those that have been used for a long time, approved by many countries as antimicrobials in foods, or are produced by synthetic means or are inorganic. Antimicrobial agents may be either synthetic compounds intentionally added to foods or naturally occurring, biologically derived substances (the so-called naturally occurring antimicrobials), which may be used commercially as additives for food preservation besides exhibiting antimicrobial properties in the biological systems in which they are originally found (Sofos et al., 1998). However, as Davidson (2001) stated, ironically synthetic antimicrobial agents are found in nature, such as acetic, benzoic, and sorbic acids, among many others.

Concerns about the use of antimicrobial agents in food products have been discussed for decades. Both the increasing demand for reduced-additive (including antimicrobial agents) and more "natural" foods, and the increasing demand for greater convenience have promoted the search for alternative antimicrobial agents or combinations to be used by the food industry (López-Malo et al., 2000). In this search, a wide range of natural systems from animals, plants, and microorganisms are being studied (Burt, 2004; Mani-López et al., 2018; Reyes-Jurado et al., 2019b). However, mainly economic aspects originating in the strict requirements to obtain approval and efforts to get the product onto the market restrict the spectrum of new chemical compounds that can help in the preservation of foods. The application of chemical antimicrobial agents in food preservation is regulated in the United States of America by the Food and Drug Administration (FDA) and in other countries by appropriate corresponding authorities. The Food and Agriculture Organization (FAO) and the World Health Organization (WHO) of the United Nations, testing and recommending usage and safety of chemicals in foods and acceptable daily intakes (ADI), regulate chemical antimicrobials internationally. In the US, chemicals in foods are examined according to the Food Additives Amendment of the Food, Drug and Cosmetic Act, specifying the procedures and conditions required for a chemical food additive to be approved. These obstacles have originated the search for emerging preservatives by examining compounds already utilized by the food industry, perhaps with other purposes, but with potential as antimicrobials, approved and not toxic in the levels used, many of them classified as generally recognized as safe (GRAS). Within these compounds are, for example, the so-called "green chemicals" present in plants that are utilized as flavor ingredients.

Although many natural systems have potential to be used as antimicrobials, this chapter will focus mainly on natural antimicrobials from plants and their possible application in foods.

17.2 Sources of Natural Antimicrobials from Plants

Plants, herbs, and spices as well as their derived essential oils and isolated compounds contain a large number of substances that are known to inhibit various metabolic activities of bacteria, yeast, and molds, although many of them are not yet completely exploited. The antimicrobial compounds in plant materials are commonly contained in the essential oil fraction of leaves (rosemary, sage), flowers and flower buds (clove), bulbs (garlic, onion), rhizomes (asafoetida), fruit (pepper, cardamom), or other parts of the plant (Nychas, 1995). Table 17.1 presents a list of some of the most highly recognized plants, herbs, and spices that have been reported as sources of natural antimicrobials. These compounds may be lethal to microbial cells or they may simply inhibit the production of a metabolite (e.g. mycotoxins) (Davidson, 2001).

Zaika (1988) reviewed the literature reporting the antimicrobial activity of many spices and classified their inhibitory activities as strong, medium, or weak. According to this ranking, cinnamon and

TABLE 17.1

Selected Plants and Their Major Antimicrobial Compounds

Plant (*Scientific Name*)	Major Components (in Descending Order)
Allspice (*Pimenta dioica*)	Eugenol, methyl ether cineol
Basil (*Ocimum basilicum*)	D-linalool, methyl chavicol, eugenol, cineol, geraniol
Black pepper (*Pipper nigrum*)	Monoterpenes, sesquiterpenes
Bay (*Laurus nobilis*)	Cineol, l-linalool, eugenol, geraniol
Caraway seed (*Carum carvi*)	Carvone, limonene
Celery seed (*Apium graveolens*)	D-limonene
Cinnamon (*Cinnamomum zeylanicum*)	Cinnamic aldehyde, l-linalool, p-cymene, eugenol
Clove (*Syzygium aromaticum*)	Eugenol, caryophyllene
Coriander (*Coriandum sativum*)	D-linalol, d-α-pinene, β-pinene
Cumin (*Cuminum cyminum*)	Cuminaldehyde, p-cymene
Fennel (*Foeniculum vulgare*)	Anethole
Garlic (*Allium sativum*)	Diallyl disulfide, diethyl sulfide, diallyl trisulfide, allicin
Lemongrass (*Cymbopogon citratus*) citral	Geraniol
Marjoram (*Origanum majorana*)	Linalool, cineol, methyl chavicol, eugenol, terpinineol
Mustard (*Brassica hirta, B. juncea, B. nigra*)	Allyl-isothiocyanate
Onion (*Allium cepa*)	D-n-propyl disulfide, methyl-n-propyl disulfide
Oregano (*Origanum vulgare*)	Thymol, carvacrol, α-pinene, p-cymene
Parsley (*Petroselinum crispum*)	A-pinene, fenol-eter-apiol
Rosemary (*Rosmarinus officinalis*)	Borneol, cineol, camphor, α-pinene, bornyl acetate
Sage (*Salvia officinalis*)	Thujone, cineol, borneol, thymol, eugenol
Tarragon (*Artemisia dracunculus*)	Methyl chavicol, anethole
Thyme (*Thymus vulgaris*)	Thymol, carvacrol, l-linalool, geraniol, p-cymene
Vanilla (*Vanilla planifolia, V. pompona, V. tahilensis*)	Vanillin, vanillic, p-hydroxybenzoic, p-coumaric acids

clove were listed as exhibiting a strong inhibitory effect while allspice was classified with a medium inhibitory effect. Conner and Beuchat (1984a,b) screened 32 essential oils from plant sources for inhibitory effects on 13 food spoilage and industrial yeasts and identified cinnamon, allspice, and clove among the most inhibitory tested. Antimicrobial activity of cinnamon, allspice, and cloves is attributed to eugenol (2-methoxy-4-allyl phenol) and cinnamic aldehyde, which are major constituents of the volatile oils of these spices. Cinnamon contains 0.5 to 1.0% volatile oil, which contains 65 to 75% cinnamic aldehyde and 8% eugenol. Allspice contains up to 4.5% volatile oil, of which 80% is eugenol. Clove buds have an average essential oil content of 17% that is 93 to 95% eugenol (Bullerman et al.; 1977; Farrell, 1990).

Major components with antimicrobial activity found in plants, herbs, and spices are phenolic compounds, terpenes (also called isoprenoids), terpenoids, aliphatic alcohols, aldehydes, ketones, acids, and isoflavonoids. Certain plant-based proteins and peptides, as well as some alkaloids, have also been reported as antimicrobials against foodborne bacteria, yeast, and molds (Savoia, 2012). As a rule, it has been reported that the antimicrobial activity of essential oils depends on the chemical structure of their components and on their concentration. Chemical structures of selected antimicrobial compounds from plant origin are presented in Figure 17.1.

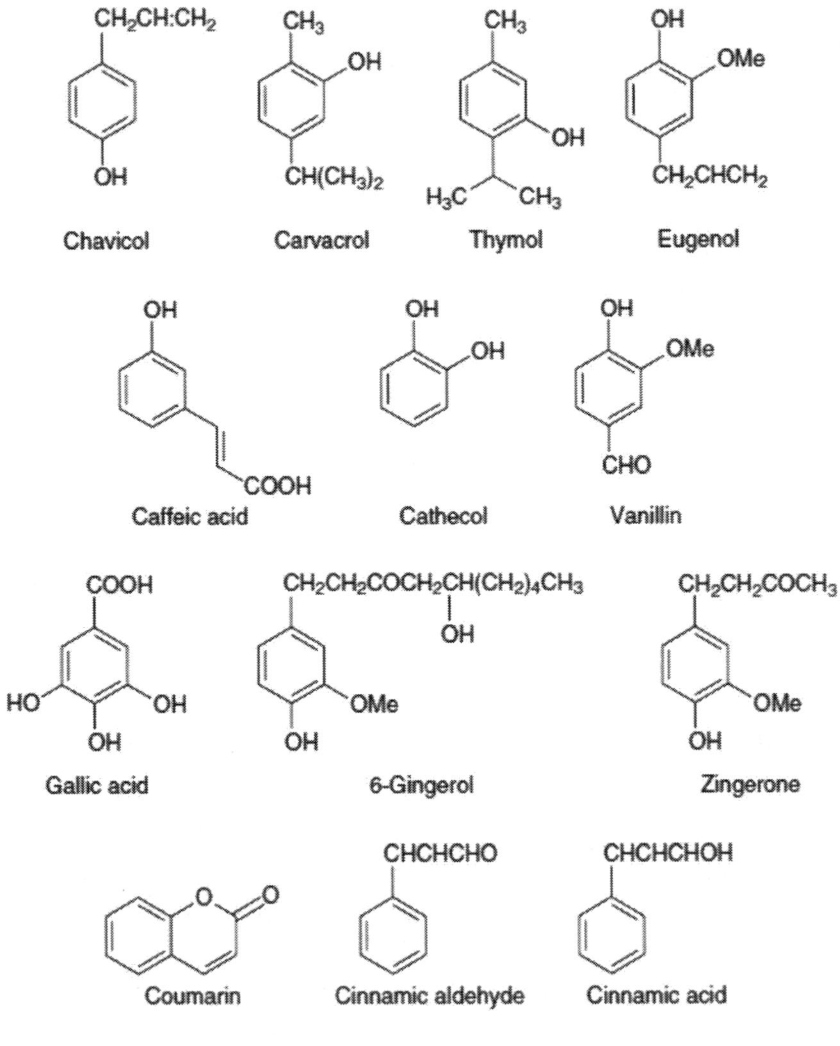

FIGURE 17.1 Selected naturally occurring plant compounds with antimicrobial activity.

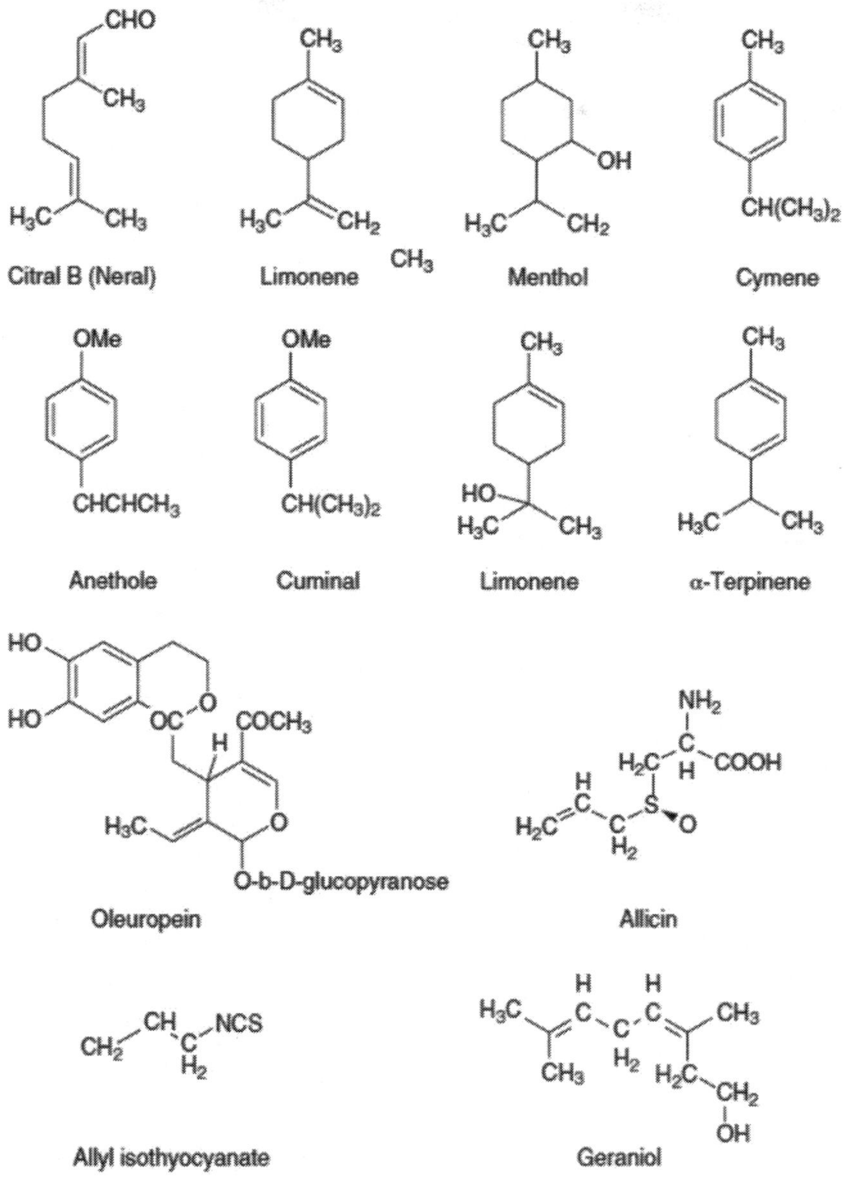

FIGURE 17.1 Continued.

17.2.1 Natural Phenolic Compounds

Derivatives of phenol, called phenolics, make up the second largest group of nutraceuticals after terpenoids and contain a phenol molecule with one or more hydroxyl groups (Gutierrez-del-Rio et al., 2018). This chemical alteration of phenol may increase its antimicrobial activity. Phenolic compounds have been used as antimicrobial agents since the early use of phenol as a sanitizer in 1867 (Davidson, 1993). As food antimicrobials, phenolic compounds can be classified, following Davidson (1993), as those currently approved (parabens), those approved for other uses (antioxidants), and those found in nature (polyphenolics, phenol). Naturally occurring phenolic compounds are widespread in plants and may be found in a great variety of food systems, and as phenol derivatives they may have antimicrobial activity (Chan et al., 2018; Gutierrez-del-Rio et al., 2018). These naturally occurring phenols and phenolic compounds may be classified into the following groups: simple phenols and phenolic acids (e.g. p-cresol, 3-ethylphenol, vanillic, gallic, ellagic, hydroquinone), hydroxycinnamic acid derivatives (e.g. p-coumaric, caffeic, ferulic, sinapic), flavonoids (e.g. catechins, proanthocyanins, anthocyanidins and flavons, flavonols and their glycosides), and "tannins" (e.g. plant polymeric phenolics with the ability to precipitate protein from aqueous solutions).

These groups of phenolic compounds share the ability to inhibit microorganisms; therefore, they may have a common mode of action. Phenolics exert antimicrobial activity by injuring lipid-containing membranes, which results in changes in membrane structure and functionality, and in leakage of cellular contents (Freitas and Cattelan, 2018). They are extensively found in essential oils, as well as in many different plant extracts, and there are innumerous studies demonstrating their antimicrobial activity against fungi and bacteria. It is recognized that Gram-positive bacteria are generally more sensitive to these compounds, compared to Gram-negative ones.

Flavonoids, a subgroup of phenolic compounds, can be found in fruits and vegetables, and their ability to work as antimicrobials is due to their capacity to form complexes with extracellular and soluble proteins, as well as with the bacterial membrane (Savoia, 2012). In a study developed by Yakoub et al. (2018), an ethanolic extract, rich in phenolic acids and flavonoid compounds, was obtained from *Tossa jute*. Phenolics acids such as quinic acid, 4-O-caffeoylquinic acid, 3,4-di-O-caffeoyquinic acid, protocatchuic acid, and caffeic acid, as well as flavonoids such as quercetin and cirsiliol, were among the major components detected. It was observed that increasing concentrations of phenolic extracts in the medium improved *Micrococcus luteus*, *Bacillus cereus*, *Escherichia coli*, *Salmonella enterica*, *Salmonella typhi*, and *Enterobacter* sp. inhibition (Yakoub et al., 2018); previous work had already suggested the efficacy of *Tossa jute* extract against bacteria and fungi. Vitamin P is another flavonoid that has been suggested as a potential antimicrobial; its addition to Chinese sausage resulted in *Staphylococcus* count reductions, although it allowed slight growth of certain beneficial and harmful bacteria (Tang et al., 2019).

Unlike essential oils and their components that, as will be described later, are in part lipophilic, anthocyanins have the particularity of being water-soluble. This characteristic makes them interesting, given the aqueous nature of the majority of fresh food. Anthocyanins can be a potential choice of antimicrobials for direct application in food, while essential oils with considerable volatility are interesting for application in their vapor phase. Anthocyanins from Chinese wild blueberry extract destroyed the cell membrane of different pathogens, including *Listeria monocytogenes*, *Staphylococcus aureus*, *Staphylococcus* Enteritidis, and *Vibrio parahaemolyticus*; membrane destruction caused nucleic acid and protein leakage through the membrane (Sun et al., 2018). Sun et al. (2018) reported that anthocyanins could pass through the membrane and lower the activity of bacterial enzymes such as alkaline phosphatase, adenosine triphosphatase, and superoxide dismutase. Besides, they suggested that anthocyanins also weakened cellular respiration which, by affecting cellular energy supply, led to cell death. Apart from anthocyanins, both phenols and flavonoids with antimicrobial activity are found in Chinese wild blueberry extract (Sun et al., 2018). Epigallocatechin-3-gallate (EGCG) from blueberry ash fruit and macadamia skin extracts was also effective against pathogenic and spoilage bacteria, as well as against fungi, when incorporated in active pea starch-guar gum films; in this case, among the microorganisms studied, *Salmonella* Typhimurium and *Rhizopus* sp. were the most resistant to active films with the polyphenols, and in general, fungi resisted better than bacteria (Saberi et al., 2017).

Grape seed is widely known for its polyphenol content. Grape seed extract is rich in cathechins, epicatechins, gallic acid, and proanthocyanidins; thus, it was used to mitigate *L. monocytogenes* growing in doner kebabs. The use of grape seed extract made the pathogen less resistant to heat and decreased the time needed for 4-log inactivation; the higher the concentration, the less time was needed for 4-log inactivation, and the less resistant the pathogen became. One percent grape seed extract was enough to cause immediate lethality to *L. monocytogenes* in doner kebabs at 65°C (Haskaraca et al., 2019). In a study ran by Zambrano et al. (2019), black grape extracts, also rich in phenolic compounds, showed antimicrobial activity against *Bacillus subtilis* and *B. cereus*. Pterostilbene, a polyphenol present in Xinjiang wine grape, had antifungal and antibacterial activity, damaging *S. aureus* and *E. coli* morphology, making cells irregular and wrinkled, and also causing cell lysis with nucleic acid and protein leakage; up-regulation of oxidative stress genes and down-regulation of genes involved in cell wall synthesis were also caused by the polyphenol (Ren et al., 2019). In another study, gallic acid grafting onto chitosan significantly inhibited *E. coli* and *S. aureus* growth, disrupting the cell membrane, which caused the release of cytoplasmic content and increased relative conductivity; cell morphology was also changed to an irregular shape, and cells gained a tendency to aggregate. The gallic acid–chitosan complex was able to cross the membrane and disrupt DNA synthesis in the nucleus (Li et al., 2019). Chitosan is, *per se*, recognized for its antimicrobial activity. However, the addition of gallic acid made the film more effective and more water-soluble, and made reactions with bacteria easier, compared to chitosan films without gallic acid. The higher the concentration of gallic acid, the more effect was observed. Gallic acid and chitosan were also loaded in glucomannan film, with remarkable antimicrobial activity against *S. aureus* and *E. coli* O157:H7 (Wu et al., 2019). Fang et al. (2018) developed a chitosan edible coating with gallic acid for fresh pork covering and, as in the last study mentioned, gallic acid improved chitosan film antimicrobial activity.

Vanillin, a phenolic compound present in vanilla pods (Figure 17.1), also has antifungal activity (Cerrutti and Alzamora, 1996; López-Malo et al., 1995, 1997, 1998). Another phenolic compound present in plants with antimicrobial activity is oleuropein, obtained from green olive extracts (Beuchat and Golden, 1989; Nychas, 1995). The antimicrobial activity of naturally occurring phenols and phenolic compounds from olives, tea, and coffee has also been examined. Green and black tea, as well as coffee, are, in effect, known to be rich in polyphenols. Nychas (1995) reported that phenolic extracts of black and green tea and/or coffee could be bactericidal or bacteriostatic against *Campylobacter jejuni*, *Campylobacter coli*, *Streptococcus mutans*, *Vibrio cholerae*, *Staphylococcus aureus*, *S. epiderimidis*, *Plesiomonas shigelloides*, *Salmonella typhi*, *S.* Typhimurium, *S.* Enteritidis, *Shigella flexneri*, and *S. dysenteriae.* Moreover, *Pseudomonas fragi*, *Lactobacillus plantarum*, *S. aureus*, *S. carnosus*, *Enterococcus faecalis*, and *Bacillus cereus* were inhibited by ethyl acetate extracts from olives while *B. subtilis*, *B. cereus*, and *S. aureus* were inhibited by oleuropein. Coatings containing green tea extract, rich in polyphenols such as catechins, were used to protect strawberries and raspberries against viral attack (Falcó et al., 2019). Green tea extract owes its properties mainly to epigallocatechin-3-gallate (EGCG), the main catechin in green tea (Falcó et al., 2019). On the other hand, coffee waste containing gallic, chlorogenic, p-coumaric, sinapic acids, and caffeine inhibited *Fusarium verticillioides*, *Fusarium* sp., and *Colletotrichum gloeosporioides* when added to gellan gum films (Mirón-Mérida et al., 2019).

Other extracts rich in phenolic compounds were obtained from *Cynara cardunculus*, an edible thistle; aqueous, ethanolic, and methanolic extracts of the leaves, containing chlorogenic acid, cynarin, luteolin-7-O-rutinoside, and cymaroside, had antimicrobial activity against foodborne Gram-negative bacteria, although they were not as effective against Gram-positive bacteria (Scavo et al., 2019). Also containing epicatechin, *Hamelia patens* extracts obtained by maceration, percolation, or Soxhlet extraction were successfully utilized as antimicrobials against *S. aureus*, *E. coli*, *Salmonella* Paratyphi, and *S.* Typhi (Paz et al., 2018).

Pomegranate peel powder has a variety of polyphenols, including anthocyanins, gallotannins, ellagitannins, gallagyl esters, hydroxybenzoic acids, hydroxy cinnamic acids, and dihydroflavonol (Sharayei et al., 2019). There is evidence of *E. coli* O104:H4 inoculated in ground chicken becoming more sensitive to heat in the presence of the pomegranate powder extract (Juneja et al., 2016). Films prepared with fish gelatin and incorporated with pomegranate peel powder exhibited antibacterial activity against *S. aureus*, *L. monocytogenes*, and *E. coli*, with a more pronounced effect on Gram-positive bacteria rather than in *E. coli* (Hanani et al., 2019). When phenolic extracts from pomegranate peel, oregano, and clove were tested against Gram-positive and Gram-negative bacteria, the former were in general more sensitive to the extracts, as has been already mentioned. Foodborne pathogens such as *Shigella flexneri* and *S. aureus* were strongly inhibited by the extracts (Chan et al., 2018). In the same study, over-acidification was reported as one of the modes of action of phenolic acids, and as an outcome, lactic acid bacteria (LAB) were more resistant to the extracts compared to other foodborne bacteria, since they are naturally prepared to resist acidic environments. LAB were able to detoxify the phenolic acids through their metabolism, not being therefore inhibited by polyphenol-rich extracts (Chan et al., 2018). Pomegranate peel extract obtained by ultrasound-assisted extraction was also effective against *Aspergillus niger*; when pomegranate peel (3000 mg/L) and potassium sorbate (1000 mg/L) antifungal activities were compared, they were not significantly different in efficiency (Sharayei et al., 2019).

Tannins, mainly catechin, catechin-3-gallate, gallocatechin, and gallocatechin-3-gallate, are present in persimmon fruit (Besada and Salvador, 2011). Persimmon tannins' antibacterial activity is expressed through cell membrane hyperpolarization, which decreases membrane potential, leakage of intracellular ATP content, and cell protein defective synthesis and loss, as well as cell wall and membrane damage or partial disintegration. In fact, persimmon tannins were active against methicillin-resistant *S. aureus* isolated from pork, suppressing cell proliferation, changing cell morphology and destroying its integrity, decreasing membrane potential and intracellular APT concentration, and damaging the cell wall and membrane (Liu et al., 2019).

Gingeron, zingerone, and capsaicin have been also found to be sporostatic for *B. subtilis* (Al-Khayat and Blank, 1985). When studying the relationship between structure and inhibitory actions, Katayama and Nagai (1960) assayed 32 pure phenol compounds. They found that 0.05% eugenol, carvacrol, isoborneol, thymol, vanillin, or salicyldehyde in agar were inhibitory against *B. subtilis*, *S.* Enteritidis, *P. aeruginosa*, *P. morganii*, and *E. coli*, concluding that the presence of a hydroxyl group enhanced the antimicrobial activity. López-Malo et al. (1995) demonstrated that vanillin concentration and the type of agar significantly ($p < 0.05$) affected the radial growth rate of *Aspergilli*. Also, differences among mold response were reported; the most resistant mold to the conditions assayed was *A. niger*, followed by *A. parasiticus*, *A. flavus*, and *A. ochraceus*. The presence of 1000 ppm of vanillin inhibited *A. ochraceus* growth for more than 2 months at 25°C in potato dextrose agar (PDA), while the growth of *A. niger*, *A. flavus*, and *A. parasiticus* was inhibited by 1500 ppm; results obtained *in vitro* were corroborated for *A. flavus* and *A. parasiticus* in apple-, banana-, mango-, papaya-, and pineapple-based agars and for *A. ochraceus* and *A. niger* in papaya-, pineapple-, and apple-containing agars. On the other hand, inhibitory concentrations of vanillin for *A. niger* and *A. ochraceus* in mango- and banana-based agars were greater than those found in PDA. Resnik et al. (1996) analyzed the effect of the concentration of vanillin on the growth rate and aflatoxin accumulation of *A. parasiticus*; the growth rate decreased abruptly in the presence of 250 ppm of vanillin, whereas 1500 ppm of vanillin inhibited mold growth during at least 37 days of storage at 28°C. However, 500 ppm of vanillin enhanced aflatoxins B1 and G1 accumulation, the toxin levels exceeding those of the control.

Plant extracts, which may contain active compounds such as polyphenols or essential oils, can be used as a whole in microbial activity mitigation. Ginger, thyme, and mostly pomegranate and clove ethanolic extracts were effective against *B. cereus*, *S. aureus*, *E. coli*, and *S. typhi* (Mostafa et al., 2018). On the other hand, rosemary extract was effective against *B. subtilis*, mainly when nano-encapsulated in chitosan (Lee et al., 2019). Bullerman (1974) observed the inhibitory effect of cinnamon on an aflatoxin-producing mold, and reported that a 1% to 2% level of ground cinnamon in broth allowed some growth of *A. parasiticus* but eliminated approximately 99% of the production of aflatoxins.

17.2.2 Essential Oils

Essential oils are one of the most studied groups of natural antimicrobials from plant sources. They contain a variety of compounds of different chemical classes, although natural phenolics have been reported to be the main antimicrobial compounds. Due to their low toxicity and easy degradation, they are considered environmentally friendly (Jafri et al., 2019), and as a matter of fact, most of the essential oils studied for their antimicrobial activity come from herbs or spices utilized as food (Freitas and Cattelan, 2018).

Essential oils are volatile liquids, constituted by both polar and non-polar low-molecular-weight compounds, whose antimicrobial activity relies mostly on one or a few major components found in relatively high concentrations. Nevertheless, their effect on microbial cells can be due to synergy among different components (Gutierrez-del-Rio et al., 2018; Mani-López et al., 2018). Sometimes, their antimicrobial activity can be greater than conventional antimicrobials; such was the case of clove essential oil encapsulated in sodium alginate particles which showed stronger activity *in situ* against *S. aureus* compared to nitrite (Radunz et al., 2019).

In general, the lipophilic and hydrophobic nature of an essential oil allows its passage through cell membranes, affecting the conformation of the polysaccharides, fatty acids, and phospholipids, and interacting with membrane proteins; as consequence, the leakage of ATP and other cellular content, as well as movement of proton flux outward from the cell, cause cell damage and death. The mitochondrial membrane is also damaged by essential oil components (Gutierrez-del-Rio et al., 2018; Mostafa et al., 2018; Dutra et al., 2019).

The active compounds present in essential oils mainly belong to terpenes (monoterpenes, diterpenes, and sesquiterpenes) and terpenoids (monoterpenoids), although they can also be aldehydes, phenols, and methoxy derivatives (Gutierrez-del-Rio et al., 2018; Baldim et al., 2019; Jafri et al., 2019). Terpenes are hydrocarbons formed by jointed isoprene units and are the most diverse type of secondary metabolites with more than 55,000 constituents (Jafri et al., 2019). Thymol and carvacrol are, so far, considered the most active monoterpenes (Jafri et al., 2019) and together with p-cymene are the major volatile components of oregano, thyme, and savory, and likely account for their antimicrobial activity (Beuchat, 1976). The essential oil of oregano contains up to 50% thymol; thyme has 43% thymol and 36% p-cymene, while savory has 30–45% carvacrol and 30% p-cymene (Farag et al., 1989). On the other hand, terpenoids are terpene derivatives with additional oxygen or other elements, and highly active compounds with considerable antimicrobial activities; examples of terpenoids found in essential oils are linalool, linalyl acetate, citronellal, piperitone, menthol, and geraniol (Jafri et al., 2019).

It has been reported that thymol, cinnamic aldehyde, eugenol, and other essential oil components have a wide antimicrobial spectrum (Wilkins and Board, 1989; Beuchat and Golden, 1989; Davidson, 1993). Thymol is the major component of thyme essential oil, as already suggested, and apart from interacting with periplasmatic protein and enzymes, it also induces the overexpression of outer-membrane proteins, causing unfolding of outer-membrane proteins and changing the membrane permeability. There is a loss of membrane potential, K^+ leaking, and ATP losses (Jafri et al., 2019). Carvacrol, the main component of oregano essential oil, is believed to increase membrane fluidity and permeability due to the formation of channels in the membrane, allowing ion passage through the membrane (H^+ inward and K^+ outward), and leaking of nucleic acid and other intracellular materials (Dutra et al., 2019). It also affects ATP production and might even interfere in the bacterial quorum sensing signaling mechanism (D'Amato et al., 2018; Gutierrez-del-Rio et al., 2018; Dutra et al., 2019). Indeed, carvacrol was effective against *E. coli* strains O157, O26, O45, O103, O111, O121, O145, and O104, affecting their membrane permeability and also intracellular ATP levels (Stratakos et al., 2018).

It has been reported that carvacrol and thymol had lesser inhibitory effects on mold growth than extracts from oregano and clove, suggesting that the antimicrobial activity may be due to several compounds from the extracts. These findings show that the phenolic extracts from plants may contain not only phenolics but also other compounds that could possess antimicrobial activities, and that, in effect, essential oil components can work synergistically. Paster et al. (1990), evaluating the antimycotic capacity of essential oil of oregano and clove against three strains of *Aspergillus*, indicated that the studied molds differed in their sensibility to tested extracts and found that *A. flavus* was sensitive to the essential oil of oregano. Zaika et al. (1983) found that little if any inhibition of growth and acid production by *Lactobacillus plantarum* and *Pediococcus acidilactici* was noted in the presence of 40 ppm oregano oil, whereas levels >200 ppm were bactericidal to both organisms; in addition, they reported that bacteria exposed to sub-lethal concentrations of oregano oil were able to overcome the inhibition and to develop resistance to the toxic effect of oregano oil or oregano. In a recent study, oregano essential oil was effective against bacteria of the *Alicyclobacillus* genera (Dutra et al., 2019).

Apart from oregano and thyme, other essential oils such as cinnamon, clove, and rosemary essential oils are well-known as being very effective against microorganisms (Xu et al., 2018; Stojiljkovic et al., 2018). Cinnamon essential oil's main component is cinnamaldehyde (or cinnamic aldehyde), a phenylpropene (Jafri et al., 2019; Chuesiang et al., 2019b), and was capable of affecting *Alternaria alternata* cell membrane integrity and permeability while changing the morphology of its hyphae (Xu et al., 2018). It is suggested that in fungi, cinnamaldehyde also inhibits cell division and disturbs the function of enzymes, including those involved in cell wall construction (Jafri et al., 2019). Cinnamaldehyde also has the ability to interfere with the activities of cytokinases, disrupting bacteria cell membranes and amino acid decarboxylases (D'Amato et al., 2018; Ji et al., 2019). Bullerman et al. (1977) demonstrated that the essential oil of cinnamon at a concentration of 200 ppm was inhibitory to growth and subsequent toxin production by *A. parasiticus* and that cinnamic aldehyde, the major component of cinnamon oil, was effectively inhibitory at a level of 150 ppm; however, they did not discard other minor constituents that may also have antimicrobial activity. In nanoemulsions formed by phase inversion temperature, cinnamon essential oil was able to disrupt the cell wall, leading to the expulsion of internal cellular material of *E. coli*, *S.* Typhimurium, *S. aureus*, and *V. parahaemolyticus* (Chuesiang et al., 2019a). Likewise, corn starch film with 2% cinnamon essential oil was effective against *S. aureus*, *E. coli*, and *S. enterica* for more than 120 days (Ke et al., 2019).

Eugenol, a phenylpropene, can also be found in cinnamon essential oil (Ji et al., 2019), and it is known to affect the yeast cell membrane and wall, causing leakage of cellular content (Ji et al., 2019). Eugenol is also one of the main components of clove essential oil, together with carvacrol, thymol, and cinnamaldehyde (Chaieb et al., 2007). Eugenol has also exhibited antimicrobial activity, which is due to its capacity to permeate through the cell membrane causing K^+ and ATP leakage, preventing cell enzyme activity, and interacting with proteins (D'Amato et al., 2018; Jafri et al., 2019). Eugenol has been reported by Al-Khayat and Blank (1985) as one of the most effective natural antimicrobials from plant origins acting as a sporostatic agent: significant reductions of viable *Bacillus subtilis* spores were obtained when exposed to 0.1–1.0 % eugenol for 8 days at 37°C. Clove essential oil had a bactericidal effect against *S. aureus*, *E. coli*, *L. monocytogenes*, and *S.* Typhimurium (Radunz et al., 2019).

Rosemary essential oil is another effective antimicrobial and its main components are 1,8-cineole (33.08–38.50%) and camphor (13.55–18.13%) (Hcini et al., 2013; Stojiljkovic et al., 2018). The oil inhibited *L. monocytogenes*, *E. coli*, and *S. enterica* growth, and had an additive effect against Gram-positive and Gram-negative bacteria in combination with clove oil (Stojiljkovic et al., 2018).

Ji et al. (2019) evaluated the essential oils vapors of 97 essential oils, among which garlic, cinnamon bark, mountain pepper, citronella, thyme, oregano, and spearmint displayed considerable activities against *Penicillium corylophilum*. The synergistic activity of cinnamon bark, citronella, and mountain pepper was also useful in the inhibition of *P. corylophilum* in beef jerky (Ji et al., 2019).

α-terpineol and terpene-4-ol, the main components of tea tree essential oil, interfered with *A. niger* cell membrane permeability, hyphae morphology and mycelium growth, and spore morphology and germination, as well as with metabolic pathways (An et al., 2019). The main components, as well as the oil itself, showed strong antimicrobial activities against *A. niger*, *in vitro* and in grapes (An et al., 2019).

Geraniol, nerol, and citronelol are examples of non-phenolic essential oil components with reported antimicrobial activity; Mahmoud (1994) found that 1000 ppm of geraniol, nerol, or citronelol inhibited *Aspergillus flavus* growth in nutritive broth (pH 5.5) during 15 days of incubation at 28°C, the MIC being 500 ppm for the three alcohols. López-Malo and Argaiz (1999) evaluated the effects of pH (6.5, 5.5, 4.5, or 3.5) and citral (3,7-dimethyl-2, 6-octadienal) concentration (0, 500, 1000, 1500, or 2000 ppm) on the growth of *A. flavus*, *A. parasiticus*, *Penicillium digitatum*, and *P. italicum* in PDA adjusted with sucrose to a_w 0.97; the radial growth rate of molds increased as the pH increased and citral concentration decreased. Conversely, reducing pH and increasing citral concentration increased mold germination times; inhibitory citral concentration depended on pH and differed among molds. *P. digitatum* and *P. italicum* were inhibited with 500 ppm citral at pH 3.5 and 6.5, whereas 1000 ppm was required to inhibit *A. flavus* and *A. parasiticus*, while at pH 4.5 and 5.5 inhibitory concentrations varied from 1000 ppm for *Penicillium* to 2000 ppm for *Aspergillus*.

The evaluation of essential oil antimicrobial activity against foodborne microorganisms is no longer confined to traditional *in vitro* methods; they have been tested in vapor phase, in foods, in food-packaging models, combined with other technologies such as high-pressure and modified atmosphere, or even utilized in nets for crop protection against pests (Loziene et al., 2018; Black-Solis et al., 2019; Park et al., 2019; Reyes-Jurado et al., 2019b). Essential oil incorporation in carrier polymers has been widely considered recently, aiming at their holding and preservation, controlled release, improved activity, or to diminish the effect of their strong odors (Baldim et al., 2019; Mazarei and Rafati, 2019; Radunz et al., 2019). Studies have been implemented in which the antimicrobial activity of encapsulated or entrapped essential oil has been tested, reporting liposomes (Hammoud et al., 2019), starch films or bionanocomposites (Garrido-Miranda et al., 2018; Ke et al., 2019), and micro- and nanoemulsions (Mazarei and Rafati, 2019), among other means to entrap the essential oil. On this basis, lipid nanosystems were used for the encapsulation of *Lippia sidoides* essential oil, which were effective against *Candida albicans* (Baldim et al., 2019), while encapsulation using randomly methylated β cyclodextrin (RAMEB) improved thyme essential oil antimicrobial activity against *Schizosaccharomyces pombe*, *S. aureus*, and *E. coli* by 2 to 4 fold (Das et al., 2019). Nanoemulsions of carvacrol and *Satureja khuzestancia* essential oils showed antimicrobial activity against *S. enterica* and *S. aureus* (Mazarei and Rafati, 2019) while solid lipid micro-particles loaded with cinnamon oleoresin were effective against *Candida pseudointermedia* and *Penicillium paneum* (Procopio et al., 2018). Other nanostructures that were useful against fungi were zinc oxide nanoparticles, which combined with carvacrol showed bactericidal activity against *Campylobacter jejuni*; when carvacrol was used on its own, the effect was merely bacteriostatic while carvacrol induced cell membrane damage, and zinc oxide nanoparticles induced cell leakage (Windiasti et al., 2019).

Regarding essential oil-loaded films with antimicrobial activity, corn-starch-based film containing cinnamon essential oil was effective against *S. aureus*, *E. coli*, and *S. enterica* (Ke et al., 2019). Tests have also been performed in foods, with successful results; coatings prepared with quinoa protein and chitosan, and filled with thymol nanoparticles, were able to reduce *Botrytis cinerea* growth in tomato (Robledo et al., 2018). In another study, the use of chitosan/montmorillonite bio-nanocomposites with rosemary essential oil for fresh poultry meat packaging reduced *Bacillus cereus* and *Salmonella enterica* counts in the meat (Souza et al., 2019). Likewise, carvacrol and thymol were effective in reducing microbial growth in marinated fresh chicken stored in vacuum packaging for 21 days at 4°C (Karam et al., 2019).

Although some of the above-mentioned tests prove the efficacy of essential oils in protecting foods against foodborne microorganisms, essential oil efficacy in foods may differ from their efficacy *in vitro*. Essential oil antimicrobial activity is affected by different factors such as pH, temperature, and water activity, with increasing stability and effectiveness at low temperatures and pHs (Mani-López et al., 2018). Food composition may, indeed, increase or decrease essential oil activity (Freitas and Cattelan, 2018) and hinder or favor microbial growth. When microorganisms grow in food, they may find a friendlier environment, with more nutrients available, compared to growing medium, which help them fight the antimicrobial damage (Freitas and Cattelan, 2018). Besides, essential oil antimicrobial activity can decrease due to interaction with protein, fat, or enzymes present in a particular food, and because of such interactions, results are difficult to compare to other foods (Guitiérrez-del-Rio., 2018). If, in foods, higher essential oil concentrations are needed to inhibit microbial growth compared to growing media, food sensory properties might, in some cases, be affected (Goni et al., 2009). Requena et al. (2019) found carvacrol, eugenol, oregano, and clove essential oils more effective against *E. coli* and *Listeria innocua in vitro*, compared to their inoculation in foods such as cheese, chicken breast, pumpkin, and melon. The essential oils and their components had different effect on bacterial growth depending on the tested food, which highlighted the importance of food composition in foodborne microorganism nutrition and survival (Requena et al., 2019).

Other spices and oils have also been shown to have some antimicrobial activity; Marth (1966) reviewed studies that reported antimicrobial activity by spice oils and extracts of laurel, peppers (chilies), coriander, anise, carvone, peppermint, caraway, cardamom, cumin, fennel, celery, dill, and mustard. Many of these preparations seemed to be active against only one microbial strain, and some were even active in one study but not in another. The normal inhibitory concentrations used in these studies were very high, indicating little activity. Other spices, including rosemary, sage, and turmeric, also possess antimicrobial activities. Curcumin, found in turmeric, for instance, was used for *Aspergillus flavus* spore inactivation (Temba et al., 2016).

Hydrosols, the aqueous phase obtained from plant-containing essential oil hydro-distillation, hold small amounts of dispersed essential oil and can be used as antimicrobials, both *in vitro* and in foods, against bacteria and fungi. The essential oil volatile components present in the hydrosol, mainly phenolics, will affect the bacterial membrane structure and function through similar mechanisms known for pure essential oils (D'Amato et al., 2018).

17.2.3 Phytoalexins

Phytoalexins are normally low-molecular-weight compounds (Ahuja et al., 2012; Kuc and Rush, 1985) elicited *de novo* by a plant when exposed to a hostile biotic or abiotic agent (Luis et al., 1997; Ejike et al., 2013). This occurs at the infection site, as a way to surpass the aggression. Biotic elicitation can be caused by bacteria, fungi, viruses, nematodes, and insects (Kuc and Rush, 1985), while abiotic elicitation is normally triggered by certain environmental condition or compounds such as ethylene and fungicides (Kuc and Rush, 1985). Their effectiveness against plant pathogens such as fungi and bacteria (Kuc and

Rush, 1985; Ahuja et al., 2012) suggests their potential as foodborne pathogen antimicrobials, and it is, in fact, known that they are broad-spectrum antimicrobial agents.

More than 200 phytoalexin compounds have been isolated from more than 20 families of plants, normally belonging to different classes of compounds such as terpenes, terpenoinds, and isoflavonoids and, to a lesser extent, proteins (chitinases, thionins, zeamatins, thaumatins, etc.) (Kuc and Rush, 1985; Sofos et al., 1998; González-Lamothe et al., 2009; Savoia, 2012). Their action against pathogens is thought to be multisite, and so affects different physiological and biochemical pathways. Phytoalexins with terpene, terpenoid, and isoflavonoid structures are believed to disrupt the cell membrane, a statement that could be supported by the phytoalexins' lipophilic nature (Guest, 2017). Indeed, it was suggested that the mechanism through which glyceollins, a class of phytoalexins, express their antibacterial activity is by affecting cell membranes (Arora and Strange, 1991). On the other hand, there have been suggestions of dead causative oxidative phosphorylation in fungal cells due to phytoalexins, for which free phenolic groups seem to impart an important contribution (Guest, 2017). As a matter of fact, phytoalexins are mainly active against fungi, although they also have an effect on bacteria (Sofos et al., 1998; Smid and Gorris, 1999; González-Lamothe et al., 2009). There are reports of phytoalexin development in different food-based plants due to fungal attack, which is related, in a broad way, to phytoalexin antifungal capacity. Tsibulins 1d and 2d accumulated in onion bulbs, preferentially in necrotic tissue rather than in fresh tissue, as triggered by *B. cinerea* (Dmitriev et al., 1990). Likewise, Arora and Strange (1991) reported the development of phytoalexin in lettuce, as a response to the threat of *B. cinerea*, *Bremia lactuacae*, and *Pseudomonas syringae*. Three years later Bennett et al. (1994) concluded that lettucenin A, found in lettuce, had antimicrobial activity against *B. cinerea*, *B. lactuacae*, and *P. syringae*. Indeed, a correlation was found between lettucenin A accumulation and the inhibition of microbial growth, and it was also suggested that lettucenin A concentrations obtained *in vivo* could inhibit bacteria and fungi *in vitro* (Bennett et al., 1994). Phytoalexins have also been identified in food products such as banana leaves and fruits, triggered by kanamycin or *Mycosphaerella fijiensis* (Arora and Strange, 1991).

Phytoalexins have also been found in legumes and cereals triggered either by fungal activity or by other factors. Phytoalexins occur in peanuts as a result of elicitation with resveratrol (Arora and Strange, 1991), and in sorghum the flavonoid 3-deoxyanthocyanidin can be found. In corn the diterpernoid kauralexin and the sesquiterpenoid zealexins are triggered after infection with fungi (Dmitriev et al., 1990). Phytoalexins are also produced in rice, normally diterpenoids, triggered by substances such as chitin oligosaccharides, proteins such as xylanase, fungus, or even UV light (Dmitriev et al., 1990). Phaseolin A occurs in bean pods, and it is triggered by monilicolin A from *Monilinia fructicola*, while glyceollins and glycinol are found in soybean cotyledons because of elicitation by *Phytophthora megasperma* cell wall glucan (Kuc and Rush, 1985). Finally, gluceollins, phytoalexins originated in soybean, showed antifungal activity against *Diaporthe phaseolorum*, *Macrophomina phaseolina*, *Phytophthora sojae*, *Sclerotinia sclerotiorum*, *Cercospora sojina*, *Phialophora gregata*, and *Rhizoctonia solani* and antibacterial activity against *P. syringae*, *Xanthomonas campestris*, *Bacillus licheniformis*, *B. subtilis*, *S. aureus*, and *Corynebacterium flaccumfaciens* (Arora and Strange, 1991). In many cases, the high concentrations needed to exert antimicrobial action in food matrices may explain the scarcity of examples of their actual use in foods (Sofos et al., 1998).

17.2.4 Phytoanticipins

Phytoanticipins, compounds belonging to terpenoid derivatives, diterpenes, sesquiterpenes, flavonoid derivatives, and phenylpropanoid phenols, can also possess antimicrobial activity (Pedras and Yaya, 2015). Unlike phytoalexins, phytoanticipins are already present in plant tissue before it experiences unfavorable conditions. However, under hostile conditions, phytoanticipin concentration goes up (Ejike et al., 2013).

Soledade et al. (1997) mentioned sinalexin as a phytoanticipin produced in white mustard, elicited by *Alternaria brassicae* (Soledade et al., 1997), that prevents white mustard from being attacked by *Alternaria* blackspot. This disease is caused by *Alternaria brassicae*, which is a hint of protection of some sort by the phytoanticipin (Soledade et al., 1997).

Allicin, a phytoanticipin present in *Allium* genus plants, was described as a colorless oil, extremely pungent, that characterized the principal odor and taste of garlic and onion. It was reported that allicin in concentrations of 1:85,000 in broth was bactericidal to a wide variety of Gram-negative and Gram-positive organisms. Investigations have shown that extracts from *Allium* bulbs inhibit the growth and respiration of pathogenic fungi and bacteria. Aqueous extracts from fresh garlic bulbs at levels of 3, 5, and 10% inhibited the growth of *B. cereus* on nutrient agar plates by 31.3, 58.2, and 100%, respectively (Saleem and Al-Delaimy, 1982). Similarly, allicin had a bactericidal effect against *E. coli*, proportional to the concentration of allicin used, and was effective against *M. luteus* in vapor phase. Other bacteria that were inhibited by allicin were *Agrobacterium tumefaciens*, *Erwinia carotovora*, *P. syringae*, and *Xanthomonas campestris*; some molds such as *Alternaria brassicicola*, *B. cinerea*, *Magnaporthe grisea*, and *Plectosphaerella cucumerina* were inhibited as well (Curtis et al., 2004).

Many foodborne pathogenic bacteria are sensitive to onion and garlic extracts; *S. aureus*, *B. cereus*, *Clostridium botulinum*, *S.* Typhimurium, and *E. coli* have been adversely affected by garlic extracts (Beuchat and Golden, 1989). The antimicrobial effects reported for garlic and onion were attributed to the allicin concentrations and other sulfur compounds present in their essential oils, which might interfere with microorganism cell protein synthesis (Putnik et al., 2018). Mantis et al. (1978) studied the effect of garlic extracts on *S. aureus* in culture media; they reported that a 5% garlic extract concentration had a germicidal effect on *S. aureus*, whereas concentrations of garlic extracts equal to or greater than 2% had a clear inhibitory effect, while concentrations less than 1% were not considered inhibitory. An investigation of the effect of garlic and onion oils on toxin production by *C. botulinum* in meat slurry indicated that these oils, when used in the proportion of 1500 μg/g meat slurry, inhibited toxin production by *C. botulinum* type A; however, the inhibition was incomplete, and toxin production by *C. botulinum* type B and type E was not inhibited (DeWit et al., 1979). Yin and Tsao (1999) evaluated the antifungal activity of *Allium* plant (including garlic) extracts against *Aspergillus niger*, *A. flavus*, and *A. fumigatus* and found that garlic was effective as a fungicidal in concentrations between 35 and 104 ppm.

17.2.5 Antimicrobial Peptides (AMPs)

Although not as well-known as phenolic compounds, plant proteins and peptides can also present antibiotic and antifungal activity, and be attractive for food applications. AMPs lack toxicity to mammalian cells (Fliss et al., 2015; Nordström and Malmsten, 2017); they are thermostable and specific to certain targets (Ben Said et al., 2019); these might be appealing characteristics. They normally have a molecular weight between 2 to 10 kDa, variable amino acid sequences and structure motifs (Fliss et al., 2015), are high in cysteine residues (4 to 12), and form disulfide bonds (Lipkin et al., 2005).

AMPs are constituents of the plant immune defense mechanism, and as such can be isolated from different plant organs such as seeds, roots, stems, leaves, and flowers (Ben Said et al., 2019; Fliss et al., 2015; Tang et al., 2018). As part of plant defense, the specimen has increased resistance against the pathogen; this was observed in tomato plants expressing alpha-helical cationic peptide, which better resisted bacterial wilt caused by *Ralstonia solanacearum* (Morais et al., 2019).

AMPs' antimicrobial activity is based on their amphipathic structure, allowing attachment to both lipidic and phospholipidic groups, and the formation of ion channels in the membrane; these might destabilize and disintegrate the membrane, with irreversible damage to the cell (Cowan, 1999; González et al., 2017; Ben Said et al., 2019). They can also affect the formation of the cytoplasmatic membrane septum and cell wall or attach to outer-membrane AMP receptors, allowing the membrane to be permeable to metabolites (Ben Said et al., 2019; Tang et al., 2018). Some might instead interfere with intracellular functions, such as DNA replication and nucleic acid and protein synthesis; cellular enzymatic activity might also be altered, affecting microbial growth (Ben Said et al., 2019; Tang et al., 2018). When a proteolytic fraction from *Vasconcellea cundinamarcencis* was used against *B. cinerea*, the mold endured conidia germination and adhesion inhibition, as well as germ tube elongation; this mold also suffered from changes in the membrane integrity, and mycelium cells became hyper-sensible to cell wall-perturbing agents (Torres-Ossandón et al., 2019). *B. cinerea* was also inhibited by an AMP isolated from amaranth (*Amaranthus retroflexus* L.) seeds, along with other fungi such as *Fusarium culmorum*, *Helminthosporium sativum*, and *Alternaria consortiale*; the AMP, although not effective against

Gram-negative bacteria, was effective against Gram-positive ones (Lipkin et al., 2005). As for Gram-negative bacteria, a 10 kDa AMP isolated from bitter melon seed extract was effective against *E. coli*, *S. typhi*, and *Pseudomonas aeruginosa* (Jabeen and Khanum, 2017).

As natural antimicrobials regarded for food applications, AMP immobilization in suitable matrices has been suggested, aiming at improving their effectiveness (Ben Said et al., 2019; González et al., 2017). AMP isolation, however, is a challenging task that relies on elaborated techniques to be accomplished (Tang et al., 2018; Jabeen and Khanum, 2017).

17.2.6 Other Sources of Plant Antimicrobial Agents

Organic acids are also known to hold antimicrobial properties, and their use in food preservation is not a recent practice. Organic acids such as citric, succinic, malic, and tartaric can be found in fresh fruits and vegetables. Others, such as lactic and propionic acids, can be obtained during fruit or vegetable fermentation (Gould, 2002).

Another group of compounds that might sometimes present antimicrobial activity are alkaloids; those of *Gymnema montanum* and *Tabernaemontana catharinensis* extracts have antimicrobial properties and the ability to change membrane permeability (Savoia, 2012).

Certain medium chain fatty acids such as caprylic, capric, and lauric acid are known as natural antimicrobials and were utilized in combination with essential oils for *E. coli* O157:H7 inhibition (Kim and Rhee, 2016). Likewise, a water-soluble polysaccharide labeled PLP1 was isolated from root barks of *Periploca laevigata* and exhibited antimicrobial activity against Gram-positive and Gram-negative bacteria; the polysaccharide was effective against *E. coli* but rather ineffective against *S.* Typhimurium (Hajji et al., 2019).

Interestingly, Maillard reaction products obtained from the reaction of protein hydrolysate from sunflower meals, xylose, and cysteine (pH 7.4; heated at 120°C for 2 h) showed antimicrobial effect against *S. aureus* and *E. coli*; antimicrobial activity was attributed to melanoidins, which were suggested to destruct cell membrane through chelation of the stabilizing cation Mg^{2+} (Habinshuti et al., 2019).

17.3 Testing the Efficacy of Antimicrobials

Essential oils of many plants possess useful biological and therapeutic activities and are extensively utilized for the preparation of pharmacologic drugs. They are commercially recovered from plant materials primarily by steam distillation, solvent extraction, or pressing, and their use in the food industry is influenced by the nature of their constituents. The antimicrobial activity of essential oils and extracts from plants, herbs, and spices depends on the extraction method but also on the initial quantity of essential oil in the plant. Within the same spice or plant, the levels of constituents and therefore active antimicrobial groups can substantially vary. Also, the geographic zone of cultivation may influence extract composition. The antimicrobial effectiveness of essential oils varies depending on the harvesting time within the year (Quesada et al., 2016). Therefore, there is a necessity to establish methods to fix or standardize essential oil purity or concentration of active components. Knowledge of the qualitative and quantitative composition of essential oils will allow valuable data to be obtained in systematic biological studies of plants as antimicrobial agents. Differences in the antimicrobial effectiveness of plant extract compounds from different sources, extraction methods, and geographic zones can be diminished if the active antimicrobial compounds from extracts or essential oils are identified. Gas chromatography is an interesting tool to measure the qualitative and quantitative composition of essential oils and its use has been widely reported in the literature (Velázquez-Nuñez et al., 2013; Hernández-Hernández et al., 2019; Requena et al., 2019). The adoption of plant or spice essential oils as alternatives to other preservatives will depend on antimicrobial uniformity and/or on the analytic methods available to normalize their antimicrobial potency (López-Malo et al., 2000). The concentration of antimicrobial compounds varies depending on the type of plant; among the richest are clove, nutmeg, laurel, and mustard, while the major constituents in most cases are phenolic compounds (Turgis et al., 2008; Aguila-González et al., 2015;

Gómez-Heincke et al., 2016). Many publications about the antimicrobial activity of extracts, oils, spices, and herbs can be found in the food-related scientific literature. However, it is difficult to make quantitative comparisons of their effects, at least partially, because of the great variety of methods utilized to evaluate antimicrobial efficiency (Reyes-Jurado et al., 2019b; Freitas and Cattelan, 2018; Ncama et al., 2019). Parish and Davidson (1993) have critically discussed the numerous tests that have been used as well as the factors that are important to consider in determining the effectiveness of natural antimicrobials in foods. The screening and isolation of antimicrobial agents from plants require a multidisciplinary approach (Davidson and Naidu, 2000); *in vitro* or *in vivo* screening bioassays are useful determining factors for the successful isolation of active compounds. Solvent selection for extraction including solvent ability to extract components is an important factor determining operation efficiency. Dry plants could be extracted with a variety of solvents and sometimes sequentially from low to high polarity. Polar solvents such as ethyl acetate or methanol are often used. Ethyl acetate theoretically only extracts by leaching the sample while alcoholic solvents presumably rupture cell structures (membranes), extracting also intracellular materials. For fresh plant materials (with high water content) a solvent mixture of dichloro-methane-methanol gave better results during extraction. Methanol separation and sample partition followed by ethyl acetate and butanol extraction help to separate lipophilic compounds from water-soluble materials.

17.3.1 *In Vitro* Testing

In vitro or explanatory (endpoint and descriptive methods) and applied (inhibition curves and endpoint methods) tests are the most utilized methods for antimicrobial efficiency evaluation of essential oils of plants and spices. A great number of these studies have been accomplished *in vitro*, and fewer have been accomplished in foods. Different methods, with different principles, are applied for the *in vitro* study of essential oils effectivity, namely dilution and diffusion methods. The former consists of diluting essential oil into the growing medium, while in the latter essential oil is loaded into a filter paper which then is added to the surface of a culture medium (Freitas and Cattelan, 2018). Ncama et al. (2019) distinguished between qualitative and quantitative methods. Some qualitative methods described are disc- or agar-diffusion, cylinder method, and thin-layer chromatography. In the first method a microbial culture is inoculated in agar, and a filter paper, dipped in the plant extract, is placed on the agar. In the second method, a specific volume of antimicrobial agent is placed into a cylinder containing the inoculated agar and mycelium growth is observed. In the third method a thin-layer chromatography plate with antimicrobial agents is dipped into broth containing the fungi and incubated in a humid atmosphere at a suitable temperature for fungal growth. The zones where inhibition occur can be visualized using reagents such as tetrazolium salts. A quantitative method described by Ncama et al. (2019) is known as the mycelia plug method and consists of a modified agar dilution method, in which agar is enriched with the antimicrobial agent at a specific concentration; inhibition is quantified through comparison between radial mycelia growth observed in medium treated with the antimicrobial agent and in the untreated control. Within the evaluation methods in model systems, descriptive (inhibition curves) methods have been commonly utilized for bacteria and yeasts, while endpoint tests have been used for bacteria, yeasts, and molds. Zaika (1988) concluded that many factors affect the antimicrobial activity of spices, extracts, and essential oils and mentioned that several aspects must be considered and reported; the observed inhibition depends on the evaluation method. Microorganisms differ in their resistance toward spices or herbs. Additionally, food components can increase (by the presence of acids, humectants, antimicrobials, etc.) or reduce (by partitioning of active components into the lipid phase, etc.) the antimicrobial capacity.

17.3.2 Vapor Phase

Many of the studies regarding the application of essential oils against foodborne microorganisms entail the direct use of the oils in the medium or food. This practice carries some disadvantages when one considers the real application of essential oils in food products for human consumption. The most notorious drawback of direct application of essential oil in food is the effect on food sensory properties, which is not as pronounced when essential oils are used in the vapor phase, since essential oils are not in direct contact with food. On the other hand, when essential oils are used in the vapor phase, lower concentrations are needed to guarantee the same effectiveness, which also contributes to lowering the effect on sensory properties (Velázquez-Nuñez et al., 2013; Lorenzo-Leal et al., 2019). Some studies have, in fact, suggested that essential oils are more efficient in vapor phase compared to direct application in foods, and authors such as Aguilar-González et al. (2015), Songsamoe et al. (2017), and Chen et al. (2018b) provided recent results on essential oil effectiveness against bacteria and molds when applied in vapor phase.

Essential oil antimicrobial components are mainly volatiles which supports their application in vapor phase for potentially improved results (Nedorostova et al., 2009). When essential oils are used in the vapor phase, volatiles act by destabilizing microbial cell membranes (Goni et al., 2009). Besides, essential oil antimicrobial components are known to be in the majority lipophilic, highly hydrophobic, and presenting low water solubility; therefore, due to their lack of affinity with water present in most food products, they form micelles in aqueous solutions, which hinders their activity as antimicrobials (Reyes-Jurado et al., 2019b). In direct application, essential oil activity is mainly due to the action of hydrophilic substances, which are normally not as effective as the hydrophobic ones. All of the abovementioned explains why essential oil application in vapor phase could be an alternative to reduce their impact on preserved food flavor (Lorenzo-Leal et al., 2019), and to reduce irritation while allowing the utilization of lower concentrations of an essential oil for the same antimicrobial activity (Reyes-Jurado et al., 2019b). Reyes-Jurado et al. (2019b) point out that since essential oils contain different volatile compounds with different volatilities, components will be released with different releasing times. If on one hand the antimicrobial activity of essential oils in vapor phase is directly related to their volatility, on the other hand in the direct addition of an essential oil to the medium, diffusivity and solubility are important for the distribution of the active compounds (Reyes-Jurado et al., 2019b). These factors might also suggest that the choice between applying essential oils in their vapor phase or directly to the medium depends on the food product to be preserved, the target microorganisms, and the solution sought. The vapor phase seems to be more suitable for microbial growth inhibition on food surfaces rather than in the inner parts of the food. Indeed, antimicrobial essential oil components with rather relevant volatilities can be well-distributed on the surface of growing medium or foods, and are effective against molds, which are very common on food surfaces. Reyes-Jurado et al. (2019b) emphasize the especial importance of using essential oils in vapor phase for mold growth inhibition due to easier airborne conidia inhibition, which are difficult to inactivate with the direct addition of antimicrobials. Another important application of essential oils in vapor phase consists in their incorporation into food package material for controlled release to the package headspace containing foods. Poly(vinyl alcohol) (PVA) films were incorporated with clove oil (1, 3, 5, 7, or 9% w/w) and used to cover minced *Trichiurus haumela* meat, without direct contact with the food; total viable counts in the fish meat preserved by the film lowered slightly with increasing essential oil content, and shelf life was extended for two days when films with at least 5% w/w essential oil were used (Chen et al., 2018b). Azadbakht et al. (2018) prepared chitosan films with *Eucalyptus globulus* essential oil, which reduced *Listeria monocytogenes* counts on sliced sausages, stored at 23°C. Although films with *E. globulus* essential oil showed increasing antibacterial activity with increasing essential oil concentration, the films were more efficient in direct contact with liquid medium than in vapor phase (Azadbakht et al., 2018). In another project, *Michelia alba* essential oil was used incorporated into an absorbent material based on the roots of water hyacinth for the inhibition of *Aspergillus flavus* in brown rice (Songsamoe et al., 2017).

Several methods have been utilized to analyze essential oil effectiveness in vapor phase such as the inverted Petri dish method. In this method, a paper disc is impregnated with essential oil and introduced in the small headspace left by the Petri dish. Other methods comprise the use of a glass or plastic box in which the essential oil and agar (or food) inoculated with microorganism are introduced and kept apart, allowing contact of the essential oil vapor with the microorganisms (Reyes-Jurado et al., 2019b). In a study run by Tyagi and Malik (2011), *E. globulus* essential oil was tested in the vapor phase using the disc volatilization method and in direct addition using agar dilution and well diffusion methods; significantly higher antimicrobial activity was observed in vapor phase against Gram-positive (*Bacilus subtilis* and *Staphylococcus aureus*) and Gram-negative (*Escherichia coli*, *Pseudomonas aeruginosa*, and *Pseudomonas fluorescens*) bacteria, molds (*Penicillium digitatum*, *Aspergillus flavus*, *Aspergillus niger*, *Mucor* spp., *Rhizopus nigricans*, and *Fusarium oxysporum*), and yeasts (*Candida albicans* and *Saccharomyces cerevisiae*). Increasing essential oil concentration in the vapor phase increased microorganism inhibition zones, and these were higher in trials in the vapor phase compared to the ones with direct essential oil addition; microbial viability was also affected with 100% viability reductions for *B. subtilis* and *C. albicans* while 76.6, 72.3, and 82.9% viability reductions for *P. aeruginosa*, *P. fluorescens*, and *E. coli*, respectively, were observed after 12 h exposure. Ncama et al. (2019) cited a method consisting of putting the extract inside a well, and covering it with microtitration plates, on top of each of which the inoculated product is seated. Exposing the extracts to higher temperatures promoted evaporation, which allowed the vapors to be in contact with the product and avoid fungal spore or mycelial growth (Ncama et al., 2019).

Additionally, time-saving methods have been suggested, using systems in which several microorganisms and essential oils can be tested at the same time. Kloucek et al. (2012) proposed a modified traditional disc volatilization method aiming at reducing labor and material needed, by using a four-section Petri dish (90 mm diameter), each with 5 mL of agar. Three compartments were inoculated while one was left un-inoculated for contamination control; a filter paper disc (85 mm diameter) was evenly impregnated with tested EO and laid on the walls dividing the compartments, allowing a 2 mm distance between the filter paper and growing media. Agar was poured into the lid as well, to function as a seal for the Petri dish. Once the Petri dish was closed with the agar-containing lid, it was incubated at selected conditions.

There are several studies in the literature which report the synergistic effect of different essential oil components in the vapor phase; Ji et al. (2019) used an experimental set-up consisting of a modified checkerboard array to test the effect of combinations of garlic, cinnamon bark, may chang, citronella, oregano, and spearmint essential oils against *Penicillium coryphilumon.* A polycarbonate vial upper well containing agar inoculated with *P. corylophilum* was put into contact with a lower well containing a paper disc impregnated with different combinations of essential oils that eventually vaporized. After sealing the junction between the upper and lower wells with Parafilm, this system was incubated at 25°C for 120 hours, after which it was visualized for mold growth. Beef jerky inoculated with *P. corylophilum* was maintained in airtight container in the presence of essential oil to allow contact between essential oil vapor and food. Among the different combinations of essential oil tested, cinnamon bark combined with citronella essential oils' vapors showed the highest synergistic antifungal activity against *P. corylophilum*, and in beef jerky this combination allowed a decrease in microbial population. The combination of cinnamon bark, citronella, and may chang essential oils was also very effective in vapor phase. In both cases, synergy between 1,8-cineole or eugenol with linalool was suggested. When each essential oil was tested separately, garlic essential oil showed the highest antimicrobial activity. Also using a checkerboard methodology, Netopilova et al. (2018) combined carvacrol and thymol in vapor phase, which generally generated a strong additive antimicrobial effect against 12 *Staphylococcus aureus* strains.

When clove and cinnamon essential oils were combined in vapor phase, an antagonist effect was observed for *E. coli*, while against *L. monocytogenes*, *B. cereus*, and *Yersinia enterocolitica* the combination had a synergetic effect (Goni et al., 2009); the activity in vapor phase was better than with direct contact for the microorganisms mentioned, although this was not the case for *P. aeruginosa*. Solid-phase microextraction analysis identified eugenol as the major component found in the headspace; 1,8-cineole and camphor had a synergistic effect with eugenol when present in the headspace (Goni et al., 2009). According to the authors, the synergistic effect can be related to the sequential inhibition of certain biochemical pathways, the inhibition of protective enzymes, and the enhancing of antimicrobial uptake by the cell wall. On the other hand, an antagonist effect can be generated due to a decrease in components' water solubility, making them less available to act on microorganisms (Goni et al., 2009). As previously mentioned, the effect of the application of *M. alba* essential oil in the vapor phase was tested against *A. flavus* on brown rice by Songsamoe et al. (2017), inhibiting mold spore germination and mycelium growth, and increasing brown rice shelf life up to four times; these authors also mentioned the advantage of essential oil components' synergy, by pointing out the increasing antimicrobial effect of linalool, the major component of *M. alba* essential oil, by using only a small amount of caryophyllene, the second major component. Reyes-Jurado et al. (2019a) evaluated the effect of thyme, Mexican oregano, and mustard essential oils in vapor phase, against foodborne Gram-positive and -negative bacteria, as well as against fungi; the three studied oils were rather effective against tested fungi, among which are *Aspergillus fumigatus*, *A. niger*, *Candida*, and *Penicillium expansum*. They observed different sensitivities of the tested microorganisms to the essential oils, mustard vapor being the most effective in general. Oregano essential oil was not effective against the bacteria studied, including *E. coli*, *P. aeruginosa*, *Salmonella* Typhimurium, *B. subtilis*, and *S. aureus*. There also was a synergetic effect with the combination of thyme and Mexican oregano essential oils when these were used in proportions of either 25:25% or 25:50%. Aguilar-González et al. (2015) tested clove and mustard essential oils' antimicrobial efficacy in vapor phase, by incorporating the essential oil in filter paper; tests were run against *Botrytis cinerea in vitro* and in strawberries. The two essential oils had synergistic effect, with lower minimal inhibitory concentrations compared to when the essential oils were used separately. Among a wide variety of essential oils tested by Hyun et al. (2015), thyme, oregano, and lemongrass essential oils showed strong activities in vapor phase against *E. coli* O157:H7, *Salmonella* Typhimurium, *L. monocytogenes*, *S. aureus*, and *B. cereus*, using a disc volatilization method. Lemongrass essential oil was effective in maintaining a reduced count of spoilage and pathogenic bacteria in cabbage stored at modified atmosphere with 100% CO_2; SEM images of *Candida albicans* revealed shrunk and deformed surfaces in cells exposed to lemongrass essential oil in vapor phase, contrary to smooth surfaces in cells that did not undergo treatment. Although when the fungus was treated with direct application of lemongrass essential oil at its minimal inhibitory concentration the surface was also damaged, the harm was more pronounced in vapor phase. In general, lemongrass vapor was effective in a much lower concentration compared to the essential oil used directly, probably due to its better capacity to penetrate the exopolymeric matrix of *Candida* biofilms.

While searching through the information available in the literature, it does not go unnoticed the number of studies regarding essential oil antimicrobial effectivity in vapor phase that have been performed using food products. An example is the application of garlic extract in zein film for the inhibition of *P. expansum* in bread (Heras-Mozos et al., 2019). Another example was the elaboration of polyvinyl alcohol microcapsules with oregano essential oil for microbial growth control in iceberg lettuce; a polypropylene-coated paper sachet was used to hold the spray-dried microcapsules, and these reduced the bacteria, mold, and yeast count in iceberg lettuce stored at 20°C and 85% RH (Chang et al., 2017).

As has been widely observed throughout effectiveness studies in vapor phase, microorganism survival and growth differ depending on whether inoculation is made in growing media or foods. In the aforementioned study by Ji et al. (2019), *P. corylophilum* growth was easier in beef jerky than in agar DG18 due to beef's lower carbohydrate content, which does not favor fungal growth. Lorenzo-Leal et al. (2019) found that allspice essential oil applied in vapor phase was more effective against *L. monocytogenes* and *S.* Typhimurium inoculated in alfalfa seeds, than in culture media. Phillips et al. (2012) also observed that microbial growth varied between culture medium and food; in culture medium, mycelial growth inhibition was observed, and spore germination was decreased for *Penicillium chrysogenum*, *A. niger*, and *Alternaria alternata*, using Citri-V™® (50:50 mix of orange:bergamot essential oils) in vapor phase. However, when Citri-V™® antifungal effectiveness against *A. alternata* was tested in tomatoes, there was no mold inhibition. In milling grain, on the other hand, the essential oils combination decreased *A. niger* and *P. chrysogenum* conidiophore growth (Phillips et al., 2012).

17.4 Mechanisms of Action

The preceding reports indicate that some spice extracts (essential oils) have a broad spectrum of biological effects, whereas other extracts may be specific toward certain groups of microorganisms, such as Gram-positive or Gram-negative bacteria, or only bacteria, not yeasts or molds. Some essential oils, plant extracts, and oleoresins influence certain biochemical and/or metabolic functions, such as respiration or production of toxins or acids, indicating that the active components in various oils and oleoresins may have different specificity with regard to target sites, on or in microbial cells.

The possible modes of action of phenolic compounds have been reported in different reviews; Swamy, Akhtar, and Sinniah (2016) explained the proposed mechanisms of action of essential oils against bacterial and fungal pathogens, while Basak and Guha (2018) and Grande-Tovar et al. (2018) focused on the effects against fungi involved in fruit decay. Friedman (2017) and Marinelli, Di Stefano, and Cacciatore (2018) report findings related to the mode of action of cinnamaldehyde and carvacrol on bacteria respectively. A good summary of the experimental approaches used to identify target sites and modes of action of essential oils can be found in Hyldgaard et al. (2012). Table 17.2 gives an overview of the experimental approaches more recently taken in elucidating the mode of action of different plant antimicrobial components.

The effect of phenolic compounds on microbial growth and toxin production could be the result of the ability of phenolic compounds to alter microbial cell permeability, permitting the loss of macromolecules from the interior. They could also interact with membrane proteins causing a disruption to the membrane structure and thus the functionality of embedded proteins (Reyes-Jurado et al., 2019b). Juven et al. (1994) found that increasing thyme essential oil, thymol, or carvacrol concentration was not reflected in a direct relationship with antimicrobial effects; however, they reported that after exceeding a certain (critical) concentration, a rapid and drastic reduction in viable cells of *Salmonella* Typhimurium was observed. Phenolic compounds could sensitize cellular membranes and when sites were saturated, a serious damage and a rapid collapse of cytoplasmatic membrane integrity could be presented, with the consequent loss of cytoplasmatic constituents. Ruiz-Barba et al. (1990), using scanning electron microscopy, observed that cells without treatment were smooth compared with those treated with phenols for 24 h, which appeared rugged and with irregular surfaces. Phenolic compounds could also denaturalize enzymes responsible for spore germination or interfere with amino acids necessary in germination processes (Nychas, 1995). So far, the mechanism of action proposed for the different plant antimicrobials tested have to do with one or several of the following aspects: induction of apoptosis, loss of membrane permeability, changes in DNA content, genetic expression, cell components, morphology and metabolism, and interaction with DNA.

Ren et al. (2019) studied the effect of pterostilbene, a plant polyphenol, derived from Xinjiang wine grape on *Staphylococcus aureus* and *Escherichia coli*. The authors concluded that the antibacterial effect is explained by induction of bacterial cell wall and cell membrane damage and an interruption of cellular processes. These were evidenced by an increase detection of reactive oxygen species (ROS), a factor associated with apoptosis and an indicator of oxidative stress. These in turn explained the membrane depolarization, measured by an increase in membrane potential, and damage, as revealed by transmission electron microscopy (TEM) photographs in which cell lysis and leakage of intracellular substances could be observed. DNA content was also measured and a significant decrease in treated bacteria was observed. Leakage of intracellular nucleic acids and proteins revealed by significant changes in optical density values measured at 260 and 280 nm showed that pterostilbene increased cell membrane permeability with dose–effect relationships. Evident morphology changes were observed by scanning electron microscopy (SEM), in contrast to untreated cells presenting smooth and regular structures, *S. aureus* treated cells were described as irregular, shriveled, and adhered to each other while *E. coli* treated cells were found incomplete and with uneven shapes. Analysis on the expression of the genes TocL, DinF, Pal, MtgA, and NagA by qRT-PCR revealed changes at the molecular level. DinF was found to be significantly up-regulated which was explained as associated with oxidative stress. Pal, MtgA, and Nag were found to be down-regulated, and this was related to an inhibition of cell wall synthesis processes. TolC, the gene coding for tolerant colicin protein, an important component of the efflux pump of Gram-negative bacteria, was found to be up-regulated only in *E. coli*. Similar mechanisms involving the loss of membrane permeability, induction of apoptosis, and changes in morphology have been reported for different plant components against a variety of microorganisms (Table 17.2).

The effects of clove oil on *Listeria monocytogenes* were studied by Cui et al. (2018); TEM analysis revealed a significantly deformed cell membrane and an incomplete and destroyed phospholipid bilayer. Evidence of damage to membrane integrity was found as a decrease in intracellular protein and ATP content as well as the detection of nucleic acid outside the cells. A decrease in DNA cell content was also found; this, and the decrease in cellular protein content, are explained by the authors both in terms of damage to membrane integrity as well as suppression of nucleic acid synthesis with a resultant negative effect on gene expression and protein synthesis. After treatment with clove oil, the activity of three types of ATPase (Na^+K^+-ATP, Ca^{2+}-ATP, and Mg^{2+}-ATP), as well as of the enzymes β-galactosidase and

TABLE 17.2

Experimental Approaches Recently Made in Elucidating the Mode of Action of Different Plant Antimicrobial Components and the Targeted Microorganisms

Mode of Action	Experiment	Agent	Targeted Microorganisms	References
Induction of apoptosis	Cell autolysis by measurement in absorbance changes at 600nm	*Inula helenium* L. root essential oil	*Staphylococcus aureus*	Stojanović-Radić et al. (2012)
	Reactive oxygen species (ROS) detection	Pterostilbene	*S. aureus* and *Escherichia coli*	Ren et al. (2019)
	DNA fragmentation analysis by agarose gel electrophoresis	Phytic acid (2%), *Syzygium cumini* methanolic seed extract (2%), and sodium chloride (0.5%)	*Bacillus subtilis*	Yadav et al. (2018)
Loss of membrane permeability	Changes in optical density (OD) at 260 nm indicating leakage of 260 nm absorbing compounds such as nucleic acids	Pterostilbene	*S. aureus* and *E. coli*	Ren et al. (2019)
		Pink pepper essential oil	Gram-positive bacteria	Dannenberg et al. (2019)
		Anthocyanins	*L. monocytogenes*, *S. aureus*, *S.* Enteritidis, and *V. parahaemolyticus*	Sun et al. (2018)
		Zataria multiflora Boiss. essential oil	*Lactobacillus curvatus*	Ziaee et al. (2018)
		Carvacrol	O157 and non-O157 Shiga-toxin producing *E. coli*	Stratakos et al. (2018)
		Acantholippia seriphioides, *Aloysia polystachya*, *Buddleja globose*, *Lippia turbinate*, *Minthostachys mollis*, *Schinus mole*, and *Solidago chilensis* essential oils	*Paenibacillus larvae*	Pellegrini et al. (2017)
		Helichrysum italicum oil	*E. coli* and *S. aureus*	Cui et al. (2015a)
		Salvia sclarea essential oil	*E. coli* and *S. aureus*	Cui et al. (2015c)
		Enteromorpha linza L. essential oil	*E. coli* ATCC 43890	Patra et al. (2015)
		Chlorogenic acid	*S. aureus*	Li et al. (2014)
		Taxus cuspidata leaf essential oil	*B. cereus* ATCC 13061 and *E. coli* ATCC 43889	Bajpai et al. (2014)
		Eleutherococcus senticosus seed essential oil	*B. cereus* ATCC 13061 and *E. coli* O157:H7 ATCC 43889	Bajpai et al. (2013a)
		Taxus cuspidata stem essential oil	*B. cereus* ATCC 13061 and *E. coli* ATCC 43889	Bajpai et al. (2013b)
		Allium sativum essential oil	*B. cereus* ATCC 13061 and *E. coli* ATCC 43889	Sharma et al. (2013)

(*Continued*)

TABLE 17.2 (CONTINUED)

Experimental Approaches Recently Made in Elucidating the Mode of Action of Different Plant Antimicrobial Components and the Targeted Microorganisms

Mode of Action	Experiment	Agent	Targeted Microorganisms	References
		BIOLL+®, a commercial citrus fruit extract	*Brachyspira hyodysenteriae*, *S. enterica*, and *E. coli*	Álvarez-Ordoñez et al. (2013)
		Inula helenium L. root essential oil	*S. aureus* ATCC 6538	Stojanović-Radić et al. (2012)
		Inula graveolens and *Santolina corsica* essential oils	*S. aureus*	Guinoiseau et al. (2010)
		Cinnamaldehyde	*Alternatia alternata*	Xu et al. (2018)
		Mint oil	*Botrytis cinerea*	Xueuan et al. (2018)
		Tea tree oil	*Botrytis cinerea*	Shao et al. (2013a,b)
	Protein leakage by measuring changes in OD at 280 nm or at OD 595 after staining with Coomassie Brilliant Blue	Pterostilbene	*S. aureus* and *E. coli*	Ren et al. (2019)
		Artemisia argyi essential oil	*S. aureus*	Xiang et al. (2018)
		Anthocyanins	*L. monocytogenes*, *S. aureus*, *S.* Enteritidis, and *V. parahaemolyticus*	Sun et al. (2018)
		Allium sativum extracts	*Erwinia carotovora*	Chen et al. (2018a)
		Carvacrol	O157 and non-O157 Shiga-toxin producing *E. coli*	Stratakos et al. (2018)
		Macleaya cordata R. leaf essential oil	*Ralstonia solanacearum*	Li et al. (2015)
		Mint oil	*Botrytis cinerea*	Xueuan et al. (2018)
	Measurement of intracellular protein content	Clove oil	*L. monocytogenes*	Cui et al. (2018)
	Leakage of phosphate, potassium, and other ions	*Enteromorpha linza* L. essential oil	*E. coli* ATCC 43890	Patra et al. (2015)
		Taxus cuspidata leaf essential oil	*B. cereus* ATCC 13061 and *E. coli* ATCC 43889	Bajpai et al. (2014)
		Eleutherococcus senticosus seed essential oil	*B. cereus* ATCC 13061 and *E. coli* O157:H7 ATCC 43889	Bajpai et al. (2013a)
		Methyl gallate isolated from *Acacia farnesiana*	*Vibrio cholerae*	Sánchez et al. (2013)

(Continued)

TABLE 17.2 (CONTINUED)

Experimental Approaches Recently Made in Elucidating the Mode of Action of Different Plant Antimicrobial Components and the Targeted Microorganisms

Mode of Action	Experiment	Agent	Targeted Microorganisms	References
		Taxus cuspidata stem essential oil	*B. cereus* ATCC 13061 and *E. coli* ATCC 43889	Bajpai et al. (2013b)
		Allium sativum essential oil	*B. cereus* ATCC 13061 and *E. coli* ATCC 43889	Sharma et al. (2013)
		Cinnamomum verum bark essential oil	*S. aureus* and *P. aeruginosa*	Bouhdid et al. (2010)
		Cistus ladanifer L. essential oil	*Aspergillus flavus*	Upadhyay et al. (2018)
		BIOLL+®, a commercial citrus fruit extract	*Brachyspira hyodysenteriae*, *S. enterica*, and *E. coli*	Álvarez-Ordoñez et al. (2013)
		Inula helenium L. root essential oil	*S. aureus* ATCC 6538	Stojanović-Radić et al. (2012)
	Measuring leakage of reducing sugars	*Macleaya cordata* R. leaf essential oil	*Ralstonia solanacearum*	Li et al. (2015)
	Changes in electrical conductivity	Pink pepper essential oil	Gram-positive bacteria	Dannenberg et al. (2019)
		Allium sativum extracts	*Erwinia carotovora*	Chen et al. (2018a)
		Carvacrol	O157 and non-O157 Shiga-toxin producing *E. coli*	Stratakos et al. (2018)
		Macleaya cordata R. leaf essential oil	*Ralstonia solanacearum*	Li et al. (2015)
		Enteromorpha linza L. essential oil	*E. coli* ATCC 43890	Patra et al. (2015)
		Cinnamomum cassia bark essential oil	*S. aureus* and *E. coli*	Huang et al. (2014)
		Methyl gallate isolated from *Acacia farnesiana*	*Vibrio cholerae*	Sánchez et al. (2013)
		α-terpineol and terpene-4-alcohol	*Aspergillus ochraceus*	Kong et al. (2019)
		α-terpineol and terpene-4-ol	*Aspergillus niger*	An et al. (2019)
		Cinnamaldehyde	*Alternaria alternata*	Xu et al. (2018)
		Tea tree oi, terpinen-4-ol and 1,8-cineole	*Botrytis cinerea*	Yu et al. (2015)
		Tea tree oil	*Botrytis cinerea*	Shao et al. (2013a,b)
		Zataria multiflora Boiss. essential oil	*Lactobacillus curvatus*	Ziaee et al. (2018)
	Measuring dye intake into permeable cells	Phytic acid (2%), *Syzygium cumini* methanolic seed extract (2%), and sodium chloride (0.5%)	*Bacillus subtilis*	Yadav et al. (2018)
		Pterostilbene	*S. aureus* and *E. coli*	Ren et al. (2019)
		Artemisia argyi essential oil	*S. aureus*	Xiang et al. (2018)

(*Continued*)

TABLE 17.2 (CONTINUED)

Experimental Approaches Recently Made in Elucidating the Mode of Action of Different Plant Antimicrobial Components and the Targeted Microorganisms

Mode of Action	Experiment	Agent	Targeted Microorganisms	References
		Acantholippia seriphioides, *Aloysia polystachya*, *Buddleja globose*, *Lippia turbinate*, *Minthostachys mollis*, *Schinus mole*, and *Solidago chilensis* essential oils	*Paenibacillus larvae*	Pellegrini et al. (2017)
		BIOLL+®, a commercial citrus fruit extract	*Brachyspira hyodysenteriae*, *S. enterica*, and *E. coli*	Álvarez-Ordoñez et al. (2013)
		Cinnamomum verum bark essential oil	*S. aureus* and *P. aeruginosa*	Bouhdid et al. (2010)
		(*E*)-2-Hexenal	*Aspergillus flavus*	Ma et al. (2019)
		Cinnamaldehyde	*Alternaria alternata*	Xu et al. (2018)
		Tea tree oi, terpinen-4-ol and 1,8-cineole	*Botrytis cinerea*	Yu et al. (2015)
		Pterostilbene	*S. aureus* and *E. coli*	Ren et al. (2019)
		Artemisia argyi essential oil	*S. aureus*	Xiang et al. (2018)
		Chlorogenic acid	*S. aureus*	Li et al. (2014)
	Measuring alkaline phosphatase (AKP), superoxide dismutase (SOD), and/or adenosine triphosphatase (ATPase) levels	Anthocyanins	*L. monocytogenes*, *S. aureus*, *S.* Enteritidis, and *V. parahaemolyticus*	Sun et al. (2018)
		Clove oil	*L. monocytogenes*	Cui et al. (2018)
		Tea tree oil	*Botrytis cinerea*	Shao et al. (2013a,b)
	Intra-extracellular ATP	Carvacrol	O157 and non-O157 Shiga-toxin producing *E. coli*	Stratakos et al. (2018)
		Helichrysum italicum oil	*E. coli* and *S. aureus*	Cui et al. (2015a)
		Salvia sclarea essential oil	*E. coli* and *S. aureus*	Cui et al. (2015c, 2016)
		Chlorogenic acid	*S. aureus*	Li et al. (2014)
		Taxus cuspidata leaf essential oil	*B. cereus* ATCC 13061 and *E. coli* ATCC 43889	Bajpai et al. (2014)
		Methyl gallate isolated from *Acacia farnesiana*	*Vibrio cholerae*	Sánchez et al. (2013)
		Taxus cuspidata stem essential oil	*B. cereus* ATCC 13061 and *E. coli* ATCC 43889	Bajpai et al. (2013b)

(*Continued*)

TABLE 17.2 (CONTINUED)

Experimental Approaches Recently Made in Elucidating the Mode of Action of Different Plant Antimicrobial Components and the Targeted Microorganisms

Mode of Action	Experiment	Agent	Targeted Microorganisms	References
		Allium sativum essential oil	*B. cereus* ATCC 13061 and *E. coli* ATCC 43889	Sharma et al. (2013)
	Cytoplasmic pH	Methyl gallate isolated from *Acacia farnesiana*	*Vibrio cholerae*	Arnon-Rips et al. (2019)
	Fatty acid composition	Cinnamon oil	*Shewanella putrefaciens*	Lyu et al. (2017)
		Cinnamaldehyde	*Alternaria alternata*	Xu et al. (2018)
		Cistus ladanifer L. essential oil	*Aspergillus flavus*	Upadhyay et al. (2018)
Changes in DNA content	Measurement of 4´, 6-diamidino-2-phenylindole (DAPI) fluorescence. DAPI can penetrate bacterial cells and bind to DNA and RNA	Clove oil	*L. monocytogenes*	Cui et al. (2018)
	Random amplification of polymorphic bacterial DNA and resolution by gel electrophoresis	Ultrasonicated spinach leaf extracts	*S. aureus* ATCC 29213 and *E. coli* OH17. H7 ATCC 25922	Altemimi et al. (2017)
Interaction with DNA	Ultraviolet absorption spectroscopy	Eugenol	pBR322 DNA	Cui et al. (2018)
Changes in morphology	Scanning electron microscopy (SEM) and/or transmission electron microscopy (TEM) analysis	Pterostilbene	*S. aureus* and *E. coli*	Ren et al. (2019)
		Clove oil	*L. monocytogenes*	Cui et al. (2018)
		Helichrysum italicum oil	*E. coli* and *S. aureus*	Cui et al. (2015a)
		Cinnamon oil nanoemulsions	*E. coli*, *S.* Typhimurium, and *V. parahaemolyticus*	Chuesiang et al. (2019)
		Artemisia argyi essential oil	*S. aureus*	Xiang et al. (2018)
		Phytic acid (2%), *Syzygium cumini* methanolic seed extract (2%), and sodium chloride (0.5%)	*Bacillus subtilis*	Yadav et al. (2018)
		Zataria multiflora Boiss. essential oil	*Lactobacillus curvatus*	Ziaee et al. (2018)
		Anthocyanins	*L. monocytogenes*, *S. aureus*, *S.* Enteritidis, and *V. parahaemolyticus*	Sun et al. (2018)
		Piper peltatum and *Piper marginatum* extracts	*Alicyclobacillus acidoterrestris*	De Pascoli et al. (2018)

(Continued)

TABLE 17.2 (CONTINUED)

Experimental Approaches Recently Made in Elucidating the Mode of Action of Different Plant Antimicrobial Components and the Targeted Microorganisms

Mode of Action	Experiment	Agent	Targeted Microorganisms	References
		Allium sativum extracts	*Erwinia carotovora*	Chen et al. (2018a)
		Ultrasonicated spinach leaf extracts	*S. aureus* ATCC 29213 and *E. coli* OH17. H7 ATCC 25922	Altemimi et al. (2017)
		Benzyl isothiocyanate	*E. coli*	Clemente et al. (2017)
		Salvia sclarea essential oil	*E. coli* and *S. aureus*	Cui et al. (2015b)
		Macleaya cordata R. leaf essential oil	*Ralstonia solanacearum*	Li and Yu (2015)
		Enteromorpha linza L. essential oil	*E. coli* ATCC 43890	Patra et al. (2015)
		Chlorogenic acid	*S. aureus*	Li et al. (2014)
		Cinnamomum cassia bark essential oil	*S. aureus* and *E. coli*	Huang et al. (2014)
		Taxus cuspidata leaf essential oil	*B. cereus* ATCC 13061 and *E. coli* ATCC 43889	Bajpai et al. (2014)
		Eleutherococcus senticosus seed essential oil	*B. cereus* ATCC 13061 and *E. coli* O157:H7 ATCC 43889	Bajpai et al. (2013a)
		Taxus cuspidata stem essential oil	*B. cereus* ATCC 13061 and *E. coli* ATCC 43889	Bajpai et al. (2013b)
		Allium sativum essential oil	*B. cereus* ATCC 13061 and *E. coli* ATCC 43889	Sharma et al. (2013)
		Inula helenium L. root essential oil	*S. aureus* ATCC 6538	Stojanović-Radić et al. (2012)
		Cold-pressed Valencia orange oil	Methicillin-resistant *S. aureus*	Muthaiyan et al. (2012)
		Inula graveolens and *Santolina corsica* essential oils	*S. aureus*	Guinoiseau et al. (2010)
		Cinnamomum verum bark essential oil	*S. aureus* and *P. aeruginosa*	Bouhdid et al. (2010)
		α-terpineol and terpene-4-alcohol	*Aspergillus ochraceus*	Kong et al. (2019)
		(*E*)-2-Hexenal	*Aspergillus flavus*	Ma et al. (2019)
		α-terpineol and terpene-4-ol	*Aspergillus niger*	An et al. (2019)
		Cinnamaldehyde	*Alternaria alternata*	Xu et al. (2018)
		Mint oil	*Botrytis cinerea*	Xueuan et al. (2018)
		Thymus vulgaris essential oil and thymol	*Alternatia alternata*	Perina et al. (2015)

(Continued)

TABLE 17.2 (CONTINUED)

Experimental Approaches Recently Made in Elucidating the Mode of Action of Different Plant Antimicrobial Components and the Targeted Microorganisms

Mode of Action	Experiment	Agent	Targeted Microorganisms	References
		Tea tree oil, terpinen-4-ol, and 1,8-cineole	*Botrytis cinerea*	Yu et al. (2015)
		Tea tree oil	*Botrytis cinerea*	Shao et al. (2013)
		Lemon grass essential oil	*Candida albicans*	Tyagi and Malik (2010)
Differential genetic expression	qRT-pCR, whole-genome DNA microarray approach, single-gene expression assessment	Pterostilbene	*S. aureus* and *E. coli*	Ren et al. (2019)
		Thyme essential oil, carvacrol, 2-(*E*)-hexenal and citral	*E. coli*	Siroli et al. (2018)
		Methyl gallate isolated from *Acacia farnesiana*	*Vibrio cholerae*	Sánchez et al. (2013)
		Cinnamon and lavender oils	*S. aureus* ATCC 29213	Herman et al. (2013)
		Cold-pressed Valencia orange oil	Methicillin-resistant *S. aureus*	Muthaiyan et al. (2012)
		Zataria multiflora Boiss, *Rosmarinus officinalis* L., and *Mentha longifolia* L. essential oils	*S. aureus* ATCC 29213	Azizkhani et al. (2012)
Metabolic changes	Several	α-terpineol and terpene-4-alcohol	*Aspergillus ochraceus*	Kong et al. (2019)
		α-terpineol and terpene-4-ol	*Aspergillus niger*	An et al. (2019)
	Specific enzyme inhibition	(*E*)-2-Hexenal	*Aspergillus flavus*	Ma et al. (2019)
	Tricarboxylic acid (TCA) pathway	Clove oil	*L. monocytogenes*	Cui et al. (2018)
Changes in cell components	Raman spectroscopy	Benzyl isothiocyanate	*E. coli*	Clemente et al. (2017)
		Antimicrobial packaging containing benzyl Isothiocyanate	*Aspergillus ochraceus*	Clemente et al. (2016)
	FT-IR spectroscope	BIOLL+®, a commercial citrus fruit extract	*Brachyspira hyodysenteriae*, *S. enterica*, and *E. coli*	Álvarez-Ordoñez et al. (2013)
	Electron paramagnetic resonance spectroscopy	Oregano essential oil	*L. monocytogenes* ATCC 7644	Serio et al. (2010)
	Changes in membrane protein profile by microfluid chip technology	Cinnamon bark and clove oil	*P. aeruginosa* ATCC 27853 and *P. aeruginosa* ATCC 27857	Felso et al. (2013)

alkaline phosphatase was significantly reduced. The study (Cui et al., 2018) also reported that clove oil could inhibit the physiological activity of *L. monocytogenes* by affecting metabolites and key regulatory enzymes of the tricarboxylic acid (TCA) pathway; the activity of the enzymes isocitrate dehydrogenase and α-ketoglutarate dehydrogenase decreased while the activity of citrate synthase increased or remained unchanged in treated cells. Intermediate compounds of the TCA cycle (citric, succinic, and L-malic acids) were significantly reduced after treatment; finally, eugenol, a major component of clove oil, was found to change the structure of DNA through formation of a eugenol-DNA chimera.

An et al. (2019) studied the effect that tea tree oil and its components have on *Aspergillus niger*. An increase in relative electric conductivity, indicative of membrane permeability destruction, was reported as follows, from most to least damaging: α-terpineol > terpene-4-ol > tea tree oil > 3-carene. Amongst the morphological changes in external and internal structures described are a badly twisted, broken, wrinkling, and coarse mycelium with a flat-trip shape and absence of cytoplasmic content when cells were treated with α-terpineol; treatment with α-terpineol, terpene-4-ol, and tea tree oil clearly damaged the cell walls and structural clumps gathered from organelles, and abnormal vacuolations of varying size were observed. The damage to membranes was not restricted to the cytoplasmic membrane but included organelle membranes as well; metabolic changes reflected in a significant alteration of four metabolic pathways were also revealed. Changes in the histidine degradation pathway resulted in the formation of lower amounts of 10-formyltetrahydrofolate and L-glutamate, a detrimental consequence to the metabolism of *A. niger*. Levels of L-aspartate and oxaloacetate from the asparagine degradation pathway and 2-keto-3deoxy-6-phosphogluconate, glyceraldehyde-3-phosphate, and pyruvate from the 2-keto-3deoxy-6-phosphogluconate cleavage pathway were also lowered; these led to lower availability of oxaloacetate and pyruvate to enter the TCA cycle and lower amounts of glyceraldehyde-3-phosphate entering the glycolytic pathway and participating in energy metabolism. All of these had an effect on compromising fungal life processes (An et al., 2019).

Raman micro-spectroscopy is a spectroscopic technique that allows the researcher to obtain information on the chemical nature of the sample studied while information on the molecular composition, orientation, symmetry, and structure can be acquired (Clemente, Aznar and Nerín, 2016). Clemente, Aznar, and Nerín (2016) and Clemente et al. (2017) used this approach to investigate the effect of antimicrobial packaging containing benzyl isothiocyanate, an active agent found in plants of the mustard family (*Brassicaceae*), against *Aspergillus ochraceus* and *E. coli*, respectively. For *A. ochraceus*, it was concluded that the isothiocyanate group of benzyl isothiocyanate reacts with saccharides, amino acids, proteins, and lipids; possibly by the action of the highly electrophilic central carbon atom which can react readily with oxygen-, sulfur-, or nitrogen-centered nucleophiles, as reported by others (Zhang and Talalay, 1994). This in turn has impacts on cellular functions such as respiration, metabolism, and cell cycle; the effect was observed independently of the type of contact, liquid or vapor, although when spores were treated in vapor phase, some of them presented a reduced effect. In the case of *E. coli*, despite no cellular damage being observed, benzyl isothiocyanate accumulated in the cells and caused alterations to the previously mentioned essential cell components, saccharides, amino acids, proteins, and lipids (Clemente et al., 2016; Clemente et al., 2017).

The major antimicrobial constituent of garlic and onion is allicin, along with several other sulfur-containing compounds. Cavallito and Bailey (1944) were the first to isolate the major antimicrobial component from garlic bulbs by steam distillation of ethanol extracts. It was noted that allicin was extremely pungent and characterized the principal odor and taste of garlic and onion. The high antimicrobial activities of garlic and onion extracts are probably due to a high content of allicin and other sulfides in the essential oils of garlic and onion. Wills (1956) observed that the inhibition of sulfhydryl enzymes was associated with the presence of the –SO–S– grouping, not the –SO–, –S–S–, or –S– groups. Barone and Tansey (1977) reported that allicin disrupted microbial cell metabolism primarily by inactivation of –SH proteins by the oxidation of thiols to disulfides; competitive inhibition of the activity of sulfhydryl components, such as cysteine and glutathione, occurred by binding with them and/or by noncompetitive inhibition of enzyme functions by oxidation of the binding to –SH groups at allosteric sites. Chen et al. (2018a) reported that garlic extracts could alter cell membrane permeability and destroy the structural integrity of cell membranes of *Erwinia carotovora*, which they found by observing changes in membrane conductivity and leakage of proteins, respectively. SEM images showed different degrees of cell rupture and adherence in a time- and concentration-dependent manner; the authors concluded that the mechanism by which garlic extracts affected the microorganisms was by disruption of the cytoplasmic membrane.

Siroli et al. (2018) studied the effect that different essential oils and their components had on *E. coli* K12 cells. Cells were exposed to sub-lethal concentrations of carvacrol, 2-(E)-hexenal, citral, and thyme essential oil for 1 hour and the composition of fatty acids and transcriptional response were analyzed; for all treatments, an increase in unsaturated and cyclic fatty acids was observed, where the greatest increase in unsaturated fatty acids was observed when cells were exposed to carvacrol. Exposure to citral, 2-(E)-hexenal, and thyme essential oil lead to an increase in mean chain length while citral treatment also caused an increase in branched fatty acids; authors reported that the increase in unsaturated fatty acids is related to bacterial response to different stresses – temperature, oxidative, and ethanol stresses or high pressures – and that the increase in chain length and cyclic fatty acids is then a response to counteract the fluidizing effect of unsaturated fatty acids. These findings were in agreement with what was observed in the transcriptomic analyses where the up-regulation of genes involved in fatty acid biosynthesis was evident. Despite clear differences between the numbers of genes being significantly up- or down-regulated with each treatment, the effects observed were similar for all antimicrobials tested; overall results show that most of the affected genes are those involved in energy metabolism, protection against oxidative stress, purine/pyrimidine metabolism, fatty acid and phospholipid metabolism, and protein synthesis (Siroli et al., 2018).

17.5 Factors Affecting Activity

17.5.1 Plant Source Variation

Several factors have a significant role in the composition/content of naturally occurring antimicrobial compounds: botanical source (species, variety, and genotype; part(s) of the plant, geographic origin, sampling time, and season and environmental conditions), growth and harvest conditions (wild or cultivated, soil composition and texture, good agricultural practices, site of collection, climate, time of harvest, stage of growth, storage conditions, and pre- and post-harvest phytosanitary treatments), steps of preparation, and extraction methods (Mukherjee et al., 2017).

Essential oils are concentrated hydrophobic liquids containing more than 200 compounds; the volatile fraction (90–95 % w/w) contains monoterpene and sesquiterpene hydrocarbons and their oxygenated derivatives along with aliphatic aldehydes, alcohols, and esters, while the nonvolatile residue contains hydrocarbons, fatty acids, sterols, carotenoids, and flavonoids (Thakur et al., 2017). Their antimicrobial effect varies greatly due to differences in their chemical properties (Burt 2004). In a recent study, Semeniuc et al. (2018) compared and classified parsley, lovage, basil, and thyme essential oils based on their chemical composition, total phenolic content (TPC), and antibacterial activities against *Salmonella* Enteritidis and *Listeria monocytogenes*. Although all tested essential oils showed inhibitory and bactericidal effect on both bacteria, thyme essential oil showed the highest level of TPC and the stronger antibacterial activity; parsley essential oil was characterized by a low TPC and weak antibacterial activity against *S.* Enteritidis; lovage essential oil by a low TPC and moderate antibacterial activity; and basil essential oil by a low TPC and weak bacterial activity against *L. monocytogenes*.

Moreover, plant bioactive compounds are secondary metabolites, which protect plants from predators and pathogens on the basis of their toxicity and repellency to herbivores and microbes, increasing plant's ability to survive and overcome local challenges. These metabolites are often produced after growth, and their compositions vary due to genetic and environmental factors that influence genetic expression (Azmir et al., 2013; Thakur et al., 2017); some of them are preformed and some of them are induced by infection and adverse environmental factors (Taiz and Seiger 2002).

As reviewed by Talhaoui et al. (2015), abiotic and biotic factors qualitatively and quantitatively affected the phenolic compounds composition in olive leaves. Oleuropein, the main phenolic compound of olive leaves, increased in response to water or cold stress while salinity stress associated with high sunlight enhanced the biosynthesis of other phenolic compounds, particularly flavonoids. Tyrosol, catechin, and oleuropein increased in leaves and other parts of the olive tree after fertilization with urea, nitrogen, copper, manganese, and zinc; changes in levels of phenolic compounds had been found to be involved in defensive reactions of olive leaves against fungi and bacteria (Talhaoui et al., 2015).

Torres et al. (2018) demonstrated that mycorrhizal symbiosis induced the accumulation of flavonols and anthocyanins in the leaves of Tempranillo grapevines under stressful temperature conditions, although clonal diversity affected the secondary metabolism. The effect of the fertilizer regime (balanced or over-supply of macronutrients), locations (different soil type), cultivar, polyphenol content, and profile of polyphenols and their interactions in globe artichoke were examined by Lombardo et al. (2015); both the cultivar and the soil type modulate the plant's response to fertilizer while the balanced fertilizer regime markedly favored the accumulation of polyphenols.

The age of American cranberry plants significantly influenced secondary metabolites synthesis (total soluble phenolic compounds, flavonoids, catechins, procyanidins, and anthocyanins) (Berezina et al. 2017); catechins and flavonoids had the same accumulation dynamics, gradually decreasing their content as plants reached reproductive age. Phenolic compounds' content in the berries was lower than in the leaves, the only exception being anthocyanins. This natural and induced variability in chemical composition and content of bioactive compounds dramatically influences their antimicrobial properties; consequently, all botanicals and their derived products and extracts should be properly characterized and standardized through analytical techniques, such as GC, HPLC–NMR, HPLC–MS, GC–MS, and CE–MS.

17.5.2 Extraction Methods

Bioactive components, including essential oils, proteins, terpenoids, flavonoids, and phenols, among others, can be obtained from one or more plant parts: leaves, flowers, peels, barks, seeds, roots, fruits, rhizomes, and gums and oleoresin exudations. The composition of the essential oils from different parts of the plant may vary widely (Burt 2004). Vegetable matrices are very complex, and the type of structure in which the bioactive is located depends on the plant type and is plant family-specific. For instance, essential oils may occur in specialized secretary structures, modified parenchyma cells, resin canals, oil tubes, or gum canals (Dalla Nora and Borges, 2017). Thus, many techniques are usually performed before or during the extraction to allow a better recovery or quality of the bioactive compounds (Gallego et al., 2019).

A variety of methods are commercially used or have been proposed for the extraction and isolation of plant natural compounds; extensive studies on the subject have been compiled in many reviews (Azmir et al., 2013; Thakur et al., 2017; Gallego et al., 2019; Zhang et al., 2018b; Selvamuthukumaran and Shi, 2017). Extraction is the first step to separate the desired natural compounds from the plant matrix and can be performed by using conventional and non-conventional methods. Classical extraction techniques (solvent extraction, maceration, and hydro-distillation) are the most widely utilized methods on a commercial basis and are mainly based on the extracting power of different solvents and the application of heat and/or mixing. Major disadvantages of these methods are long extraction times, requirement of large volumes of solvents, low extraction selectivity, and damage to thermo-labile compounds. Techniques that involve high temperatures may cause the loss of volatile compounds due to long extraction times, oxidation and degradation of unsaturated or esterified compounds by thermal or hydrolytic reactions, and yield losses. To overcome these limitations, new techniques that require lower organic solvent amounts and energy consumption and shorter extraction times, and exhibit higher selectivity have been developed; some of these non-conventional techniques are super-critical fluid extraction, pressurized liquid extraction, microwave-assisted extraction, ultrasound-assisted extraction, enzyme-assisted extraction, and pulsed electric field-assisted extraction; these techniques may be scalable and can be coupled to other extraction processes (Gallego et al., 2019). Many of these advanced extraction techniques comply with green chemistry principles (Chemat et al., 2012).

Extracts are complex and usually contain a variety of natural components, and further separation could be required. Separation of extract components can be performed by methods based on the difference in molecular size (membrane filtration, gel filtration chromatography) or on the ionic strength (ion-exchange chromatography) as well as using other modern separation techniques such as molecular distillation, supercritical fluid chromatography, preparative gas chromatography, molecular imprinted technology, and multi-dimensional chromatographic separation (Zhang et al., 2018b). Extraction techniques may affect the physicochemical properties of essential oils and alter the natural proportion of the chemical constituents as in their original state. Process parameters, such as extraction solvent and temperature, can dramatically change the composition of the extracts. The extraction conditions may be modulated to change the selectivity according to the target compound. In addition, the coupling of processes has been employed to make the extraction more efficient (Gallego et al., 2019). Thus, there is a wide array of possibilities going from the extraction method or the combination of extraction methods to the selection of process variables to obtain selected bioactive compounds. Accordingly, the chemical composition of extracts can be affected in a different way and so their antimicrobial activities (Azis et al., 2018; Elyemni et al., 2019; Tao et al., 2014).

17.5.3 Interaction with Food Matrix

Most of natural antimicrobials demonstrated excellent activities against microorganisms in model media and beverages but failed to be as effective when used in complex foods. The antimicrobial may require up to 1000 times the *in vitro* dose (nutrient broth, planktonic media) to be equally effective *in vivo* situations (complex foods), overcoming the low thresholds of these compounds. The antimicrobial efficacy was found to be a function of food ingredients and structure; active compounds tend to interact with food components such as proteins, fats, sugars, and salts, and/or be adsorbed onto the multiple interfaces of foods. Hence, only part of the total compound amount is available to perform the antimicrobial action. High fat concentration was found to have an adverse influence on the antimicrobial action of various essential oils in hotdogs, milk, and other foods (Singh et al., 2003; Cava et al., 2007; Canillac and Mourey, 2004). Regarding the effect of carbohydrates and starch, the reports are somehow contradictory, although the negative impact on antimicrobial effects would also occur at high concentrations (Devlieghere et al., 2004; Ofman et al., 2004; Gutiérrez et al., 2008). Tassou et al. (2000) and Min et al. (2010) reported a negative impact of high protein contents.

Weiss et al. (2015) proposed a useful "structural" explanation on why preservatives lose activity in foods; in contrast with planktonic media where structure is homogeneous, foods are in generally highly structured and may simultaneously contain various phases with different physical properties and chemical compositions. Two types of physical interactions between antimicrobials and food ingredients may occur: electrostatic and hydrophobic ones. Antimicrobial compounds carry a certain charge depending on pH, temperature, and ionic strength and can electrostatically interact with proteins, forming soluble or insoluble complexes. Hence, antimicrobials, bound to the proteins, are no longer homogeneously distributed in the food and are not available for acting on the microorganisms. Hydrophobic interactions arise between permanent or induced dipoles and promote the association between polar molecules, that tend to form cluster-like structures, while non-polar molecules tend to form a separate phase where hydrophobic molecules associate with other hydrophobic ones. For example, phenolic compounds associated to proteins and non-polar molecules may partition into lipid phases, hindering a direct interaction with microorganisms and rendering antimicrobials less effective. The interaction of antimicrobials with food will result always in a loss of antimicrobial activity excepting when both the antimicrobials and the microorganisms are located in the same substructures. Hence, considering the difficulty in predicting the interaction between antimicrobials and components in complex foods, each food needs to be individually assessed (Weiss et al., 2015).

17.6 Increasing the Efficacy of Natural Antimicrobials from Plants

Besides their high variability in composition and concentration of active substances, low water solubility, high volatility, limited stability against chemical or physical degradation, uncontrolled release, interaction with food matrix/components, and strong organoleptic properties limit the application of essential oils and other bioactive compounds in food systems. Some strategies employed/suggested to partially overcome these deficiencies are the application of essential oils in gaseous (vapor) phase, the use of the antimicrobials in combination with other factors (the so-called "hurdle technology"), and the employment of novel delivery systems.

17.6.1 Application of Essential Oils in Gaseous Phase

Many bioactive compounds of essential oils are volatile, a property that makes them useful as possible antimicrobials for stored foods (Tzortzakis, 2010). Antimicrobial activity depends on the composition and volatility of the compounds. Due to the hydrophobic nature of essential oils, water is not a good solvent for these substances, and a higher effectivity at relatively lower concentrations than in liquid phase could be expected. Treatments in gaseous phase exhibit also antimicrobial activity over larger areas than those in liquid phase. As the essential oils are not directly added to foods, they may have a relative minor effect on sensory attributes. Various studies about the application of plant volatile compounds have been carried out in recent years.

Tyagi and Malik (2011) compared the effectiveness of *E. globulus* essential oil against 14 food-spoilage bacteria and molds in liquid and vapor phases. The significant high antimicrobial activity detected in the vapor phase was attributed to the higher percentage of monoterpene hydrocarbons in the vapor than in the liquid phase (54.7 *vs* 44.5 %). Mexican oregano, thyme, and mustard essential oils in vapor phase were tested against various bacterial and fungal species; evaluated microorganisms exhibited different sensitivity to tested oils, the bacterial strains being the most resistant to thyme and oregano essential oils (Reyes-Jurado et al., 2019a). The growth of *Botrytis cinerea* and *Rhizopus stolonifer* conidia in artificially inoculated strawberries was significantly reduced by tee tree oil vapor at 0.9 g/L, and treated berries maintained a fresher quality than untreated strawberries during storage (20°C, 3 days); tee tree oil induced fruit defense responses as shown by the induction of H_2O_2 levels and several defense-related enzymes in the first period of incubation (Shao et al., 2013a,b). Allyl isothiocyanate (AITC, 0.81–1.41 µg/ml, gas phase) was evaluated for preserving mung bean sprouts and fresh-cut iceberg lettuce; although AITC indeed showed microbicidal or microstatic activity against the native flora of both fresh produce, color and respiratory activity were severely affected (Kramer et al., 2018). Exposure of radish sprouts inoculated with *Listeria monocytogenes* (6.3 log CFU/g) to oregano, thyme, or cinnamon bark essential oils gases (78–625 µL/L; 24 h, 30°C) resulted in a very small decrease in counts; even at the highest concentration, the microorganism was not completely inactivated. This lower antimicrobial activity of essential oils compared to the effects obtained in a laboratory medium was attributed to: (a) a reaction of the antimicrobial compounds of the essential oils with organic matter on the surface of the sprouts, and (b) structural protection by the produce; *L. monocytogenes* may be embedded in the stomata and crevices of the sprout and/or internalized into the open tissue of the sprout (Lee et al., 2018).

Interesting, many studies reported a synergistic effect when two or more essential oils were applied in combination in gaseous phase. The use of different essential oils, which develop synergistic interactions at low concentrations, could minimize their sensory impact. Tunc et al. (2007) studied the effect of three aroma compounds (carvacrol, cinnamaldehyde, and allyl isothiocyanate), ethanol, and SO_2, individually and in binary combinations, on the growth of *Penicillium notatum*, simulating an active packaging and the effect of a modified atmosphere enriched with antimicrobial compounds; tested aroma compounds were more effective than ethanol and SO_2; six combinations were synergistic, enabling the reduction of the concentration of each antimicrobial and reducing their organoleptic impact. Combinations of clove and mustard essential oils demonstrated a synergistic interaction *in vitro* and *in vivo* (strawberry) to inhibit *Botrytis cinerea*, a ubiquitous plant pathogen of fruit; moreover, the antimicrobial effect was greater *in vivo* than *in vitro* probably due to the low pH of the berries (Aguilar-González et al., 2015). In other study, the antifungal activities of the essential oils of 97 plants against *Penicillium corylophilum*, a moderately xerophilic mold, were tested in a laboratory medium and on beef jerky, an intermediate-moisture food; the combination of cinnamon bark, citronella, and may chang essential oil vapors had synergistic activity in inhibiting the growth of *P. corylophilum* on beef jerky (Ji et al., 2019).

The antimicrobial activity of volatile plant compounds makes them extremely attractive, in particular because of their low toxicity at low concentrations, renewability, and biodegradability, for application as gaseous treatments to control postharvest diseases in an eco-chemical approach ("biofumigation").

Studies about the effect of different plant essential oils on disease development caused by fungi during storage of some commercially important horticultural products (papaya, tomato, kiwifruit, apple, avocado, strawberry, blueberry, and cucumber) were reviewed by Mari et al. (2016). In general, higher concentrations of essential oils provided *in vitro* a lower survival of the fungus. Besides the structural modification of hyphal morphology by partitioning the lipid layer of the cell membrane and the change in cell membrane permeability, some essential oils would act as a signaling compound to stimulate plant defense mechanisms.

17.6.2 Combination with Other Antimicrobials and/or Preservation Factors

The addition of plant bioactive compounds at concentrations needed to control the microorganisms may adversely affect the food flavor. Moreover, their antimicrobial activity is usually moderate when applied as the only preservation factor. The use of combinations of bioactive compounds with other antimicrobials or other preservation factors ("hurdle technology") as a strategy to overcome these limitations has been a point of intensive research. Overall, multiple disturbances of microbial homeostasis have been used/suggested in different arrangements: (a) using a plant antimicrobial with other stressors simultaneously to prevent the growth of spoilage and pathogenic microorganisms, and (b) using one or more stressors (one may be a plant antimicrobial) simultaneously or in sequence to inactivate/injure target microorganisms, and then, in sequential mode, a plant antimicrobial compound alone or with more stressors to prevent the survival/proliferation of remaining refractory or sub-lethally damaged cells (Alzamora et al., 2016). Hundreds of papers in this topic have been published in recent years (Burt, 2004; Berdejo et al., 2019; Calo et al., 2015; Coutinho de Oliveira et al., 2015). Although the effect of combinations strongly depends on the type and amount of microorganisms, the doses of the stressors, and the food matrix, some general points can be addressed to facilitate the selection of hurdles or stressors:

- Many essential oils showed synergistic antimicrobial activity when applied simultaneously in a combined approach, or when used with traditional synthetic food preservatives, probably because of the effect on different cell targets.
- Many combinations of essential oils with non-thermal technologies or mild heat resulted in synergistic lethal effects associated with modifications of microbial cell envelopes, which facilitated the access of the antimicrobial to the target and/or cell membrane disruption and loss of membrane potential.
- Synergism or additivity was observed due to the microbicidal effect of the non-thermal or other physical stressors and the subsequent microstatic effect of the natural antimicrobial during storage.

17.6.3 Delivery Systems

Over the last years, many studies have been carried out on the development and potential application of delivery systems of active components in the food, pharmaceutical, and cosmetic industries. Results have been compiled in several excellent review articles and book chapters (Fu et al., 2016; Prakash et al., 2018a,b; Weiss et al., 2009; Castro-Rosas et al., 2017; Donsi and Ferrari, 2016). Many potential benefits have been claimed for these systems as compared with the direct addition of antimicrobial compounds: enhanced long-term chemical and physical stability (impaired interaction with food ingredients, lower losses of volatile compounds), improved solubility in aqueous phase and thus better bioavailability of active compounds, decreased impact on food sensory properties, release control, dispersion uniformity, and rapid penetration of bioactive compounds into the target sites (Fu et al., 2016; Pavoni et al., 2019).

Overall, these delivery systems are still relatively new technologies and only limited information is available. Most of the studies have been in culture medium models, and there are very few studies in food matrices; they did not consider the interactions between the delivery system and the food matrix, their sensory impact, the technological aspects for the incorporation of antimicrobial structures, and the toxicological aspects of formulated nanostructures (Fu et al., 2016; Brandelli and Taylor, 2015). Other points to be addressed are the economics of the nanoscale system, the consideration of food-grade ingredients for their formulation, and the physical stability of the delivery systems throughout the shelf life of the product (Weiss et al., 2009). However, interest in this subject is growing worldwide, and it is expected that within the next few years encapsulated antimicrobials will be available at the commercial level. Delivery systems can be classified as: (a) emulsion-based, (b) nanosized carrier-based, and (c) packaging films or coating-based (Fu et al., 2016).

17.6.3.1 Emulsions

Colloidal delivery systems based on microemulsions, nanoemulsions, and Pickering emulsions are increasingly being investigated to encapsulate, protect, and deliver lipophilic bioactive components in the food and pharmaceutical industries (McClements, 2012). Conventional emulsions, thermodynamically unstable, consist of liquid-liquid dispersions of two immiscible compounds in which one is dispersed in the form of small droplets (0.1–20 μm) to another one, and an emulsifier that forms an interfacial layer that improves the stability of the system. Pickering emulsions are conventional emulsions stabilized by solid particles, which gives them greater resistance to destabilization mechanisms, while nanoemulsions are conventional emulsions with mean droplet size ranging from 20 to 200 nm. In contrast, microemulsions are thermodynamically stable, and consist of isotropic liquids formed by mixing oil, water, and surfactant (emulsifier) together; they are characterized by the small size of the disperse phase ($d < 100$ nm). Due to their exceptional solubility properties, microemulsions are able to deliver both lipophilic and hydrophilic compounds (Pavoni et al., 2019).

The small size of the droplets ($d < 200$ nm) in micro- and nanoemulsions results in a number of potential benefits for certain applications: long-term stability, high optical clarity, and enhanced transport of active molecules through biological membranes. Only a limited number of studies addressed the use of micro- and nanoemulsions in real food systems, i.e. by direct mixing with a liquid food (milk and fruit juices), washing the food surface with the antimicrobial aqueous dispersions (fresh lettuce, spinach leaves, and pears), infusion in porous food matrices (pork sausages and zucchini), and coating with a biopolymeric layer containing the antimicrobial dispersion (apple pieces, broccoli florets, plums, green beans, rucula leaves, and sliced bread) (Donsi and Ferrari, 2016; Fu et al., 2016). Regarding Pickering emulsions, they have been used for the protection and release of various bioactive compounds, including flavonoids and thyme oil (Fu et al., 2016).

The antimicrobial efficacy of nanoemulsions is strongly influenced by essential oil components, tested microbial strain, emulsion formulation, and food matrix. Emulsion formulation affects the transport of essential oils to the microbial cell membrane and their interaction with the molecular sites of the membrane. In some cases, encapsulation in nanoemulsions inhibited the antimicrobial activity (Donsi and Ferrari, 2016; Terjung et al., 2012; Nielsen et al., 2017).

17.6.3.2 Nanosized Carriers/Coatings

Nanoparticles (particle size less than 1 μm) which could be used as coating material/carrier agents for the plant bioactive compounds include, among others, liposomes, polymer nanoparticles, and solid-liquid particles (Brandelli and Taylor, 2015; Fu et al., 2016; Prakash et al., 2018a,b; Weiss et al., 2009). Liposomes are nano-vesicles with an internal aqueous core rounded by a self-assembled coat of amphiphilic lipids and can be used for the encapsulation and delivery of hydrophilic, hydrophobic, and amphiphilic molecules. Polymer nanoparticles refer to polymer nanospheres and nanocapsules, where bioactive molecules may be adsorbed at the surface or encapsulated within the particle in the case of nanospheres, or confined to an aqueous core and surrounded by the polymeric shell around them in the case of nanocapsules. Solid-liquid particles have a relatively rigid core consisting of hydrophobic lipids that are solid at room temperatures, surrounded by a monolayer of phospholipids.

Some examples of nanoparticle delivery systems are the incorporation of essential oils of *Artemisia arborescens* L. and *Zataria multiflora* in solid-liquid particles, *Oreganum dictamnus* L. in phosphatidylcholine liposomes, *Atractylodes macrocephala* L. in phosphatidylcholine and cholesterol liposomes, and lime in nanocapsules (Prakash et al., 2018a,b); in all these cases, enhanced antimicrobial activities compared to the free essential oil form were demonstrated.

17.6.3.3 Packaging Films/Coatings

Antimicrobials derived from plants can be added directly to the food or incorporated into active packaging systems. The incorporation of antimicrobial agents in active packaging has been extensively investigated to control their diffusion and maintain them in adequate concentration on food surfaces. An active packaging (AP) system can be classified into two main types: (a) non-migratory AP, which contains antimicrobial agents that show their activity without true migration into foods, and (b) active-releasing packaging, which allows the controlled migration of antimicrobial agents in direct contact with the food surface (non-volatile migration), or the release of volatile antimicrobials which act in the headspace around the food surface and in the food itself after absorption (volatile migration) (Khaneghah et al., 2018a,b).

Numerous methods have been investigated to develop antimicrobial packaging systems (Figure 17.2). Among them are: (a) the addition of volatile antimicrobial compounds in a sachet or pad in order to released them within the packaging headspace, (b) the incorporation of antimicrobial compounds (volatile and non-volatile) into the structure of package polymers, (c) the coating or absorption of the antimicrobial compounds onto the packaging, (d) the use of polymers (e.g. chitosan) with antimicrobial properties, and (e) the addition of antimicrobial compounds into edible coatings.

The incorporation of antimicrobial agents into synthetic (e.g. petroleum-based polymers) and bio-based polymers (e.g. bio-polyethylene, alginate, cellulose, and polylactic acid (PLA) among others) represent good strategies for increasing the shelf life of packed food products (Khaneghah et al., 2018b; Scaffaro et al., 2018b) The antimicrobial efficiency depends on the release rate of the antimicrobial

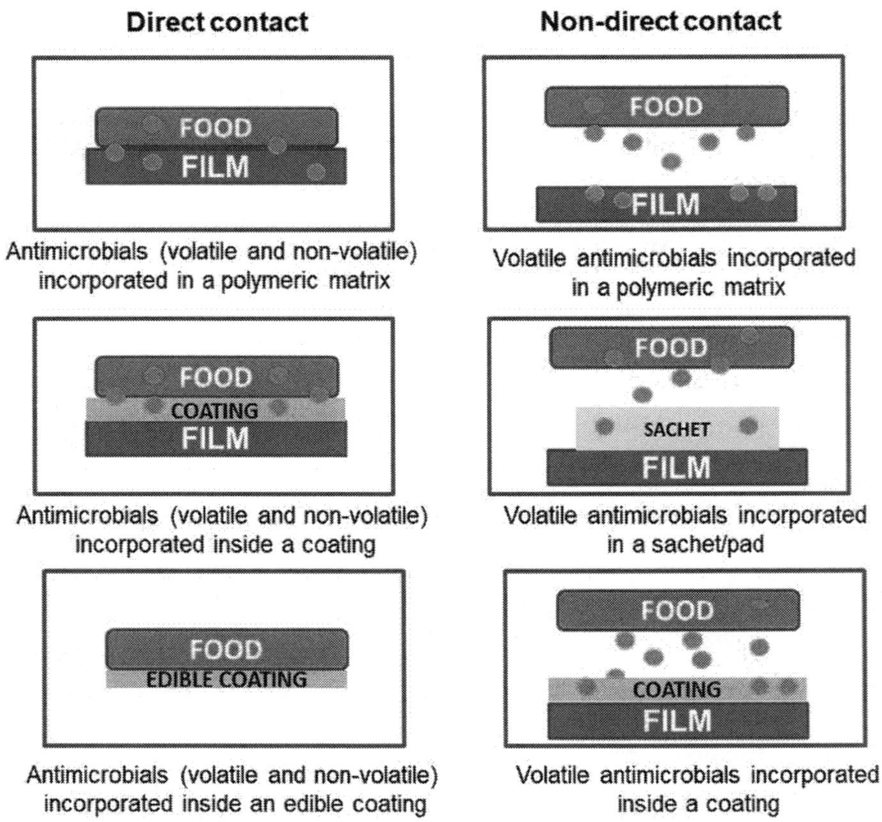

FIGURE 17.2 Some methods evaluated for the release of antimicrobials compounds in active packaging.

agents from the packaging. The most critical factors for controlling the release rate and their transfer into the foodstuff are the method of incorporation, the nature of the polymeric matrix, the mechanism of release, food properties, temperature, and environmental relative humidity (Wu et al., 2018).

Methods commonly utilized for the incorporation of antimicrobial agents into polymeric materials include thermomechanical processes, such as melt extrusion, or a casting procedure, which involves the dissolution of the polymer and the antimicrobial substance into an appropriate solvent, followed by the evaporation of the solvent. For thermal-sensitive antimicrobials such as essential oils and organic acids, coating has emerged as a useful and alternative incorporation system. Coating consists of the application of at least one layer of fluid containing the antimicrobial substance onto the surface of a polymeric matrix. Another technique utilized is the covalent immobilization to the polymeric material of antimicrobial agents containing reactive functional groups, such as antimicrobial enzymes, peptides, polyamines, organic acids, etc.; some advantages of this method are the reduction of the amount of the antimicrobial added to the polymer and the minimization of antimicrobial activity losses (Khaneghah et al., 2018a,b).

There is an increasing tendency of exploring the application of bio-based packaging due to environmental concerns. In this context, polylactic acid-based antimicrobial materials received considerable attention due to the good mechanical properties, renewability, biodegradability, and biocompatibility of PLA (Llorens et al., 2015; Scaffaro et al., 2018a). Sacaffaro et al. (2018b) carried out an extensive review of the incorporation of essential oils and other antimicrobial compounds of natural origin to PLA by different techniques; they highlighted the need to conduct more *in vivo* studies to further assess the effectiveness of the PLA-based antimicrobial systems.

Chitosan, a natural polysaccharide, has also potential applications in active food packaging systems due to its excellent film-forming ability and its intrinsic antimicrobial properties. Chitosan is soluble in diluted aqueous acidic solutions due to protonation of their amino groups; this cationic character confers its antimicrobial properties and its ability to carry and slow-release functional ingredients. Chitosan-based films have been evaluated in the form of pure chitosan films or combined with other biopolymers or synthetic polymers (Wang et al., 2018a). The inclusion of gallic acid into chitosan-based films significantly increased antioxidant activity, antimicrobial activity, and tensile strength of the film, and decreased water vapor and oxygen permeability (Wu et al., 2016a; Xie et al., 2014). Chitosan film enriched with cinnamon oil also showed good antimicrobial activity (Wang et al., 2011). Other tested oils, such as olive oil, rosemary essential oil, and oregano essential oil, also improved the mechanical and barrier properties, and the antioxidant activity of chitosan films (Jian et al., 2012; Pelissari et al., 2009).

The addition of natural antimicrobial compounds into polymeric edible coatings produced with polysaccharides, proteins, or lipids obtained from renewable agricultural resources and/or food-processing wastes has also been investigated to control the diffusion of these compounds into different foodstuffs. Edible coatings produced by other biopolymers such as chitosan have also been extensively studied (Aloui and Khwaldia, 2016; Irkin and Esmer, 2015). These coatings consist of thin layers of biopolymers that are applied on food surfaces by different mechanical procedures, such as spraying, brushing, dipping, or by new techniques such as layer-by-layer assembly (Dhall, 2013; Poverenov et al., 2014; Bilbao-Sainz et al., 2018). Several studies have been reported on the use of biopolymer-based coatings carrying different plant-derived antimicrobial compounds to preserve the quality and increase the safety of many fresh and minimally processed fruit and meat products (Fernández- Pan et al., 2014; Aloui et al., 2014; Aloui and Khwaldia, 2016; Duran et al., 2016; Shakila et al., 2016; Valdés et al., 2017). However, for practical applications some shortcomings have been reported such as the weak adhesion of coating materials to the hydrophilic surface of the food, degradation of the antimicrobial agent, or its rapid desorption through coating materials (Aloui and Khwaldia, 2016). In recent years, with the advances in the field of nanotechnology, the incorporation of nanoemulsions containing antimicrobial compounds into edible coating solutions has been investigated as an effective strategy to improve their distribution and enhance their adhesion to the surface of solid foods (Kim et al., 2014; Salvia-Trujillo et al., 2015; Wu et al., 2016b; Acevedo- Fani et al., 2017). Another innovative and promising approach is the use of antimicrobial multilayer coatings which rely on incorporating antimicrobial compounds into multilayered/nanolaminate systems formed using the electrostatic deposition technique; these multilayer coatings are able to act as efficient antimicrobial delivery systems, due to their inner and barrier layers that could control the diffusion rate of natural antimicrobial compounds embedded in the matrix layer (Mantilla et al., 2013; Sipahi et al., 2013; Moreira et al., 2014; Aloui and Khwaldia, 2016).

17.7 Toxicity of Natural Antimicrobials and Its Evaluation and Regulatory Aspects

Studies to determine the dose to elicit the antimicrobial effect are necessary for each food. Once the dose is selected, the impact on organoleptic characteristics must be assessed. It is also important to evaluate whether such dose will be toxic or not to the consumers.

Variability in the composition of botanical products affects their safety assessment. Plant extracts contain more than one component, from which one or more will be the component(s) of interest (due to their benefit), but others might involve an adverse effect (Schilter et al., 2003). Unlike the internationally accepted approaches for the safety evaluation of food additives or food contaminants (WHO, 1987), in which *in vitro* or *in vivo* studies with animals lead to the determination of health-based guidance values of intake and consider large uncertainty factors, a similar approach in the case of botanical products may not be appropriate (Schilter et al., 2003). Tests in the case of botanical products would involve high amounts of food to be administered to the animals in order to obtain uncertainty factors that are comparable to those obtained through the administration of pure compounds (Schilter et al., 2003). A guidance for the safety assessment of botanical products used in food and food supplements, consistent with the existing food regulations at that time, was proposed by an expert group of the Natural Toxin Task Force of the European Branch of the International Life Sciences Institute (ILSI Europe); they considered the framework of risk assessment as a basis for the safety evaluation of these products (Schilter et al., 2003). The document highlights the importance of standardization of the products, considers data regarding history of use, use of the product for different applications, intended use, dietary exposure, dietary consequences, and refers to hazard identification and characterization, risk characterization, and post-launch monitoring, and includes a decision tree as a guidance tool for the determination of the information that needs to be considered for the safety evaluation of these products. Other publications complement the criteria for safety assessment, such as the decision tree proposed by Walker (2004) as well as examples of assessments by van den Berg et al. (2011) and Speijers et al. (2010).

Plant extracts/essential oils intended to be utilized in foods are generally regulated as food additives, food flavorings, or classified as "generally recognized as safe" (GRAS) substances, after fulfilling the requirements for each case. Specific databases of safe compounds, permitted concentrations, and permitted food uses are provided by different regional authorities (Malhotra, Keshwani, and Kharkwal, 2015). In the European Union (EU), the procedure for authorization of a flavoring substance is common to the one established for food additives and enzymes (European Commission, 2008a, 2008b). Annex I of Regulation 1334/2008 (European Commission, 2008b) lists the flavoring substances that can be utilized in foods. It is suggested to refer to its amendments in EUR-Lex for updates (European Union, 2019), and the Food Flavorings Database is also available (DG SANTE, 2019; European Commission, 2019). In the case of active packaging, all components not available in the positive lists of the regulations and not recognized as a food additive must be below 10 μg per kg of food or food simulant (Chibane et al., 2018), according to the EU regulation for food-contact materials (European Commission, 2009, 2011). The United States Food and Drug Administration (FDA) refers to food additives and GRAS substances in the Code of Federal Regulations, Title 21 (U.S. Food and Drug Administration, 2018). There is also the "Substances Added to Food" inventory (U.S. Food and Drug Administration, 2019), which includes ingredients regulated by the FDA, such as food additives, flavoring substances, and GRAS substances, as well as substances formerly utilized (prohibited substances, delisted color additives, and no longer GRAS substances); the inventory is only a partial list of food ingredients, and inclusion of information from non-FDA entities does not indicate an FDA approval or evaluation of this use.

Even though some plant extracts/essential oils/botanical products constituents have been approved for their use in foodstuffs and classified in one of the categories mentioned, some intoxications reported with botanical products were due to contamination with other plant species or through misidentification of species (Schilter et al., 2003). Other cases of toxic effects were reported, such as allergic contact dermatitis when used frequently as in the case of aromatherapy (Carson and Riley, 2001; Bleasel, Tate, and Rademaker, 2002; Schaller and Korting, 1995; Trattner, David, and Lazarov, 2008). Cytotoxicity of *Pimenta dioica* and *Rosmarinus officinalis* essential oils was observed when evaluated *ex vivo* by Lorenzo-Leal et al. (2019) using the model of human-derived macrophage THP-1 cells. An *in vivo* experiment, however, showed a different result: a wound treatment of diabetic mice showed recovery when the animals were treated with *Rosmarinus officinalis* essential oil (Abu-Al-Basal, 2010). Wright et al. (2019) reported that Kakadu plum (*Terminalia ferdinandiana*) fruit and leaf extracts did not have any toxic effects against nauplii stage *Artemia franciscana* crustaceans but recommended that further toxicity studies in human cell lines be performed before the extracts are used commercially. Using a similar toxicity assessment using brine shrimp (*Artemia salina*) larvae, Kayode et al. (2018) found that *Adansonia digitata* leaf and stem-bark essential oils were non-toxic.

Carvacrol showed a concentration-dependent toxic effect *in vitro* (Llana-Ruiz-Cabello et al., 2014), despite being registered as flavoring by the EU, FDA, and the Joint Food and Agriculture Organization/World Health Organization (FAO/WHO); therefore, potential risks associated with the use of essential oils constituents should be continuously monitored (Donsì and Ferrari, 2016). In addition to the *in vitro* studies of the antimicrobial effect and the *in vitro/in vivo* toxicity evaluation, further research is needed regarding the interaction of the botanical products with the food matrix, the delivery systems of their active compounds, and their stability throughout shelf life. These data and the standardization of the composition of botanical products will be convenient for their safe use in foods at doses needed to produce the desired antimicrobial effect.

17.8 Application in Food and Sensory Analysis

For many years, many of the published reports regarding the application of phenolic compounds (antioxidants and constituent of extracts and essential oils) as antimicrobials had been accomplished in model and laboratory systems while recently more studies have been carried out in real foods. The growing consumer demand for safe and natural products has led to the increase in publications focused on the study of plant extracts, essential oils, and their components, as antimicrobials in real food systems including different green leaves; fresh and processed tomatoes; a variety of whole and sliced fruits and vegetables; fruit juices and jams; nuts, seeds, and beans; flours and bread; fresh, cooked, and processed meats; cheeses, milk, and puddings; salsas and salad dressings. Table 17.3 displays some of the studies published between 2015 and 2019.

The essential oils of spices and plants, as well as their major components, are more effective in microbiological media than when evaluated in real foods (Gutierrez, Barry-Ryan and Bourke, 2009). In most cases the inhibitory concentrations found in model systems increase significantly when evaluated with the same microorganisms in actual foods and, in consequence, few of the applications of phenolic antioxidants as antimicrobials have been successful (Kabara and Eklund, 1991). This reduction in the effectiveness observed *in vivo* represents an important limitation to the use of essential oils and phenolic antioxidants as antimicrobial agents in foods (Juven et al., 1994). Ribes et al. (2016) reported a difference between the effectiveness of oil-in-water emulsions of clove and cinnamon leaf essential oils against *Aspergillus flavus*, *A. niger*, and *Penicillium expansum* when tested in strawberry jams and in agar media; the time to observe the effectiveness of the essential oil was significantly longer in jams. Authors attributed this behavior to the difference in matrix structure and water contents and the possible antagonistic interactions with ingredients present in the jam. Yuan and Yuk (2018) found that 32 mg/mL of *Syzygium antisepticum* plant extract was able to inhibit and retard staphylococcal growth at 10°C and ambient temperature, respectively, in cooked chicken; MIC values obtained *in vitro* for the extract against the same *S. aureus* strains were reported by the authors as 0.125 mg/mL, 256 times less than the MICs needed in the food.

Interactions among phenolic groups and proteins, lipids, and aldehydes could explain, at least partially, the reduction of the antimicrobial effect of essential oils where major constituents are phenols.

TABLE 17.3

Recent Studies on the Application of Plant Essential Oils and Extracts as Antimicrobials in Real Food Systems

Food Evaluated	Target Microorganisms	Tested Agent(s)	References
Fruits, Vegetables, and Roots			
Lamb's lettuce	Mesophilic aerobic bacteria, yeasts, and lactic acid bacteria	Oregano and thyme essential oils	Siroli et al. (2015)
Cherry tomatoes	*Aspergillus flavus*, *A. oryzae*, *A. niger*, and *Alternaia alternata*	Perillaldehyde, the main constituent of *Perilla frutescens* essential oil	Tian et al. (2015)
	Botrytis cinerea	Quinoa protein-chitosan edible film containing thymol nanoemulsion	Robledo et al. (2018)
Mature green tomato	Rotting rate of tomatoes	Encapsulated allyl isothiocyanate	Wu et al. (2015)
Tomatoes	*Fusarium solani* and *Rhizopus stolonifer*	Combination of thymol and salicylic acid	Kong et al. (2016)
	Total microbial count	Essential oil from bark and leaf of *Adansonia digitata*	Kayode et al. (2018)
Romaine lettuce leaves	E. coli O157:H7	Ultrasound assisted with oregano essential oil	Millan-Sango et al. (2015)
Iceberg lettuce and chard	Mixed inoculum *of E. coli*, *L. monocytogenes*, and *S. enterica*	Essential oils of *Origanum vulgare* L. and *Rosmarinus officinalis* L. alone and in combination	Barbosa et al. (2016)
Fresh-cut red chard	*L. monocytogenes* and *Salmonella* Typhimurium	Cinnamon bark oil emulsion washing and ultraviolet-C radiation	Park et al. (2018)
Lettuce	*E. coli* O157:H7	Cold nitrogen plasma and clove oil alone and in combination	Cui, Ma, and Lin (2016)
	Salmonella enterica	Ultrasound treatment combined with oregano and thyme essential oils	Millan-Sango et al. (2016)
	Pseudomonas fluorescens indole-3-acetic acid-dependent type III secretion system	Caryophyllene-rich white and black pepper essential oils	Myszka et al. (2017)
	E. coli O157:H7	Combination of thymol-eugenol and carvacrol-eugenol rinse	Yuan et al. (2019)
Fresh cabbage	Total mesophilic microorganisms	Thyme, oregano, and lemongrass essential oil in vapor phase	Hyun et al. (2015)
Shredded cabbages	*E. coli* and *Pichia pastoris*	Carvacrol nanoemulsion and acidic electrolyzed water	Sow et al. (2017)
Kale leaves	*E. coli* O157:H7 and *L. monocytogenes*	Cinnamon leaf essential oil emulsions	Kang et al. (2019)
Fresh-cut cauliflower	*L. monocytogenes*, *E. coli* O157:H7, and total yeasts and molds	UV-C, gamma irradiation combined with oregano or lemongrass essential oil in formulation with citrus extract and lactic acid	Tawema et al. (2016)
Watermelon juice	*S.* Typhimurium, *S. aureus*, and *E. coli*	Trans-cinnamaldehyde nanoemulsions	Jo et al. (2015)
Guavas	Total mesophilic bacteria, mold and yeast, and thermotolerant coliforms	Edible chitosan-cassava starch coatings with Lippia gracilis (LGRA106 and LGRA107 genotypes) essential oil	De Aquino et al. (2015)
Garlic crops	White rot (*Sclerotium cepivorum* Berk)	Essential oils from *Tagetes minuta* L., *Tagetes filifolia* L., *Origanum vulgare* L. *Origanum x majoricum*, and *Laurus nobilis* L. and their binary combinations with iprodione	Camiletti et al. (2016)

(Continued)

TABLE 17.3 (CONTINUED)

Recent Studies on the Application of Plant Essential Oils and Extracts as Antimicrobials in Real Food Systems

Food Evaluated	Target Microorganisms	Tested Agent(s)	References
Licorice	Aflatoxin B1 production by *A. flavus*	*Litsea cubeba* essential oil	Li et al. (2016)
Golden Delicious apples	*Penicillium expansum*	Clove, cinnamon, lemongrass, and oregano essential oils combined with warm air flow	Frankova et al. (2016)
Apples and pears slices	*Aspergillus caronarius*	Sodium alginate-cinnamon essential oil coatings	Kapetanakou et al. (2018)
Organic cantaloupes	*E. coli* O157:H7, *S. enterica*, and *L. monocytogenes*	Organic thyme oil emulsion	Zhang et al. (2016)
Dragon fruit (*Hylocereus undatus*)	*Aletrnaria alternata*	Cinnamon (*Cinnamomun zeylanicum*) and clove (*Eugenia caryophyllus*) essential oil	Castro et al. (2017)
Fresh-cut carrot slices	Total aerobic mesophilic and psychrophilic bacteria, Enterobacteriaceae, lactic acid bacteria, and yeast and molds	Carvacrol-loaded chitosan-tripolyphosphate nanoparticles	Martínez-Hernández et al. (2017)
Orange and pomegranate juices	*Saccharomyces cerevisiae*	Thermo-ultrasound and cinnamon leaf essential oil	Sánchez-Rubio et al. (2016)
Cantaloupe juice	*L. monocytogenes* and *Salmonella enterica* serovar Typhimurium	Starch octenyl succinate stabilized emulsions with nisin and/or thymol	Sarkar et al. (2017)
Raw apple juice	*A. flavus* and *P. expansum*	Betel leaf (*Piper betle* L.) essential oil	Basak and Guha (2017)
Reconstituted orange juice	*Alicyclobacillus acidoterrestris*	*Piper peltatum* and *Piper marginatum*	de Pascoli et al. (2018)
Table grapes	Yeast-mold growth	Alginate preharvest spray and postharvest coating with and without vanillin	Takama and Korel (2017)
Grape berries (*Vitis vinifera* L. x V. labruscana Bailey)	*S.* Typhimurium, mesophilic aerobes, yeasts and molds	Lemongrass oil droplet size in chitosan emulsion as coating	Oh et al. (2017)
Mangaba fruit	Total aerobic mesophilic bacteria, yeast and mold, and *Bacillus cereus*	*Myrcia oata* Cambeseedes essential oils	Frazão et al. (2017)
Peach (*Prunus persica* L. Batsch) slices	Mesophiles, *Pseudomonas* spp., Enterobacteriaceae, and yeasts and fungi	*Posidonia oceanica* and green tea extract	Piva et al. (2017)
Fresh-cut melons	Total microbial count and mold and yeast counts	Citral-based edible coatings	
Strawberries	*Colletotrichum acutatum*	Carboxymethylcellulose coating with *Lippia sidoides* essential oil	Oliveira et al. (2019)
	Fungal decay percentage and total yeast and mold counts	Alginate-limonene edible coatings	Dhital et al. (2018)
Strawberry jams	*Aspergillus flavus*, *A. niger*, and *Penicillium expansum*	Clove and cinnamon leaf essential oils in oil-in-water emulsions	Ribes et al. (2016)
	A. flavus, *P. expansum*, *Zygosaccharomyces rouxii*, and *Z. bailii*	Cinnamon bark-xanthan gum emulsions	Ribes et al. (2017)
	Aspergillus niger	Oil-in-water emulsions containing cinnamon bark oil, zinc gluconate and transferulic acid	Ribes et al. (2018)

(Continued)

TABLE 17.3 (CONTINUED)

Recent Studies on the Application of Plant Essential Oils and Extracts as Antimicrobials in Real Food Systems

Food Evaluated	Target Microorganisms	Tested Agent(s)	References
Tomato paste	*Salmonella enterica* serovar Typhimurium	Carvacrol and potassium sorbate	Batista et al. (2019)
Cereals, Grains, and Nuts			
Young coconut liquid endosperm	*E. coli* O157:H7	1500 ppm malic acid and 75 ppm nisin	Garbriel and Estilo (2015)
Peanut, cashew, almond, walnut pistachio, and hazelnut	Aflatoxins produced by *Aspergillus parasiticus*	Isothiocyanates generated in oriental and yellow mustard flours	Hontanaya et al. (2015)
Stored peanut seeds	*A. flavus*	(E)-2-hexenal	Ma et al. (2017)
Almonds	*Salmonella* Enteriridis PT30 and *S.* Tennessee K4643	*Cinnamonum cassia* oil	Tsai et al. (2017)
Tofu	*S. aureus*	Liposome-encapsulated clove oil	Cui et al. (2015)
Fresh green beans (*Phaseolus vulgaris* L.)	*E. coli* and *S.* Typhimurium	Carvacrol incorporated in modified chitosan coating, gamma irradiation, and modified atmosphere packaging alone or in combinations	Severino et al. (2015)
Clover (*Trifolium resupinatum*) sprouts	Total aerobic bacteria, total coliforms, and molds and yeasts	Thyme (*Thymus vulgaris*) essential oil	Shirvani et al. (2016)
Wheat flour	Mycotoxin production by *Aspergillus parasiticus* and *Fusarum poae*	Gaseous allyl isothiocyanate	Nazareth et al. (2016)
Loaf bread	*A. parasiticus* growth and aflatoxin reduction	Allyl, benzyl, and phenyl isothiocyanates	Saladino et al. (2017)
Bread	Mold spoilage	Lavender and melissa waste	Vasileva et al. (2018)
Par-baked wheat and sourdough bread	*Aspergillus niger* and *Penicillium paneum*	Thyme essential oil	Debonne et al. (2018)
Cake	Aerobic mesophilic bacteria, yeast and molds, and *Salmonella*	Microencapsulated and free thyme (*Thymus vulgaris*) essential oil	Gonçalves et al. (2017)
Valencia rice grain	*Alternaria alternate*, *Bipolaris oryzae*, *Fusarium graminearum*, *F. equiseti*, and *F. verticillioides*	Clove (*Syzygium aromaticum* L. Merr. & Perry) essential oil	Santamarina et al. (2016)
Corn kernels	*A. parasiticus*, *Fusarium tricinctum*, *F. verticilloides*, *A. alternata*, and *Gibberela zeae*	Gaseous allyl isothiocyanate	Tracz et al. (2017)
Maize kernels	*Fusarium graminearum* growth and deoxynivalenol and zearalenone production	*Cananga odorata* essential oil	Kalagatur et al. (2018)
Oats	*Aspergillus flavus*, *A. parasiticus*, and *A. clavatus*	Vapor phase essential oil of cinnamon leaves, thyme, clove, oregano, and ginger	Božik et al. (2017)
Meat and Seafood			
Ground beef	Methicillin-resistant *Staphylococcus aureus*	Yerba mate (*Ilex paraguarensis*) aqueous extracts	Burris et al. (2015)
	Listeria monocytogenes	Clave (*Syzygium aromaticum*) and cinnamon (*Cinnamomus cassia*) essential oils	Khaleque et al. (2016)

(*Continued*)

TABLE 17.3 (CONTINUED)

Recent Studies on the Application of Plant Essential Oils and Extracts as Antimicrobials in Real Food Systems

Food Evaluated	Target Microorganisms	Tested Agent(s)	References
Ground chicken	*Salmonella* spp.	Gallic acid, eugenol, and heating	López-Romero et al. (2018)
Beef meatballs	*Staphylococcus aureus*, *Listeria monocytogenes*, *Salmonella* Enteritidis, and *Campylobacter jejuni*	*Thymus vulgaris*, *Origanum vulgare*, *Cinnamomum zeylanicum*, *Rosmarinus officinalis*, and *Salvia officinalis*	Pesavento et al. (2015)
Cooked sausage	Total aerobic mesophilic bacteria, yeasts and molds, *E. coli*, *Clostridium* spp., and total Enterobacteriaceae count	Nutmeg (*Myristica frangrans*) essential oil	Šojić et al. (2015)
Dry fermented sausage-chouriço	*Salmonella* spp., *L. monocytogenes*, *S. aureus*, and lactic acid bacteria	Bay, garlic, nutmeg, oregano, rosemary, and thyme essential oils	García-Díez et al. (2016)
Portuguese sausages paínhos and alheiras	Total viable aerobic bacteria, *E. coli*, *Staphylococcus* spp., *Salmonella* spp., and *Listeria monocytogenes*	Whey protein coating containing *Origanum virens* essential oil	Catarino et al. (2017)
Turkish fermented sausages (sucuk)	Surface mold, total mesophilic bacteria, Micrococcaceae, Enterobacteriaceae, yeasts and molds, and lactic acid bacteria	Chitosan coatings with thyme or rosemary essential oils	Soncu et al. (2018)
Sausage	*Salmonella* Typhimurium, *Listeria monocytogenes*, *Staphylococcus aureus*, *Shigella flexneri*, and *Escherichia coli*	Potato starch-based coating with *Hibiscus sabdariffa* acetone and methanol extracts	Cruz-Gálvez et al. (2018)
	Listeria innocua	Hydrogels made of turmeric residue and gelatin or cassava starch and gelatin with purified curcumin and UV-A light treatment	Tosati et al. (2018)
Chinese sausages	Fungus community assessed by high-throughput sequencing	Flos Sophorae (*Sophora Japonica* L. flower buds) or rutin	Tang et al. (2019)
Doner kebabs	*Listeria monocytogenes*	Grapefruit seed extract	Haskaraca et al. (2019)
Sheep burgers	Total viable counts, *Pseudomonas* spp., Enterobacteriaceae, and lactic acid bacteria	Oregano extract	Fernandes et al. (2016)
Low/sodium sliced vacuum/packed turkey breast ham	Lactic acid bacteria, total psychrotrophic bacteria, *Listeria innocua*, thermo-tolerant coliforms, *Salmonella* sp., and *Staphylococcus* sp.	Carvacrol and high-pressure processing	de Oliveira et al. (2015)
Raw chicken fillets	*Listeria monocytogenes*, psychrotrophic and aerobic mesophilic bacteria, lactic acid bacteria, *Pseudomonas* spp., Enterobacteriaceae, and molds and yeasts	Sodium alginate, galbanum oleoresin gum, and *Ziziphora persica* essential oil-based edible coatings	Hamedi et al. (2017)
Fresh poultry meat	Total mesophilic aerobic bacteria, psychotropic aerobic bacteria, and total coliforms	Chitosan/montmorillonite nanocomposites incorporated with rosemary essential oil	Lauriano de Souza et al. (2019)

(*Continued*)

TABLE 17.3 (CONTINUED)

Recent Studies on the Application of Plant Essential Oils and Extracts as Antimicrobials in Real Food Systems

Food Evaluated	Target Microorganisms	Tested Agent(s)	References
Marinated chicken meat	Total viable count, lactic acid bacteria, *Brochothrix thermosphacta*, *Pseudomonas* spp., total coliforms, *Escherichia coli*, yeasts and molds	Thymol and carvacrol combined with air storage and vacuum packaging	Karam et al. (2019)
Beef meat	Psychrotrophic bacteria, *Brochothrix thermosphacta*, *Pseudomonas* spp., and Enterobacteriaceae	Rosemary essential oil and modified atmosphere packaging	Sirocchi et al. (2017)
Refrigerated beef steaks	Total viable counts, *Pseudomonas* spp., lactic acid bacteria, Enterobacteriaceae, and yeasts and molds	Chitosan edible coating with Zataria multiflora essential oil and hydroalcoholic extract of sumac in modified atmosphere packaging	Langroodi et al. (2018)
Pork loins	Total viable counts	Gallic acid/chitosan coating	Fang et al. (2018)
Black wildebeest biceps femoris muscles	Total viable counts, lactic acid bacteria, and coliform counts	Oregano essential oil	Shange et al. (2019)
Lamb meat	Total viable counts, lactic acid bacteria, and *Pseudomonas* spp.	Chitosan coatings with free or nano-encapsulated *Satureja khuzestanica* essential oil	Pabast et al. (2018)
Ostrich meat	Total viable counts, lactic acid bacteria, *Staphylococcus aureus*, and *Escherichia coli*	Kefiran/waterborne polyurethane film incorporated with *Zataria mltiflora* and *Rosmarinus officinalis* essential oils	Hedayati Rad et al. (2018)
Cooked chicken	*Staphylococcus aureus* and methicillin-resistant *S. aureus*	*Syzygium antisepticum* extract	Yuan et al. (2019)
Raw and thermally processed chicken meat	Aerobic and anaerobic mesophilic bacteria, fecal coliforms, lactic acid bacteria, *Staphylococcus aureus*, *Salmonella* spp., *Escherchia coli*, and *Salmonella* Enteritidis	Basil and/or rosemary essential oils	Stojanović-Radić et al. (2018)
Reconstructed chicken product	Mesophilic and psychrotrophic aerobic counts	Chitosan films containing peanut skin and pink pepper residue	Serrano-León et al. (2018)
Ready to cook pork chops	Total viable counts, lactic acid bacteria, Enterobacteriaceae, and *Pseudomonas* spp.	Chitosan edible coatings with bamboo vinegar	Zhang et al. (2018a)
Organic beef jerky	*Penicillium corylophilum*	Cinnamon bark, citronella, and may chang essential oils and their combinations in vapor phase	Ji et al. (2019)
Minced trout fillets	*L. monocytogenes*, total mesophilic and psychrotrophic bacteria, *Pseudomonas* spp., *P. fluorescens*, *Shewanella putrefaciens*, lactic acid bacteria, and Enterobacteriaceae	Chitosan-gelatin film incorporated with grape seed extract and *Ziziphora clinopodioides* essential oil	Kakaei and Shahbazi (2016)
Rainbow trout (*Oncorhynchus mykiss*)	Total viable count, Enterobacteriaceae counts, lactic acid bacteria, H2S-producing bacteria	Carrageenan-based edible coatings with lemon essential oil	Volpe et al. (2015)
	Total viable count, lactic acid bacteria, and *Pseudomonas* spp.	Carboxymethyl cellulose coatings incorporated with *Zataria multiflora* Boiss. essential oil and grape seed extract	Raeisi et al. (2015)

(Continued)

TABLE 17.3 (CONTINUED)

Recent Studies on the Application of Plant Essential Oils and Extracts as Antimicrobials in Real Food Systems

Food Evaluated	Target Microorganisms	Tested Agent(s)	References
Semi-fried coated rainbow trout (*Oncorhynchus mykiss*)	Total viable count, psychrophilic bacterial counts, and *Pseudomonas* spp.	Shallot (*Allium ascalonicum* L.) fruit and ajwain (*Trachyspermum ammi* (L.) Sprague) seed extracts	Raeisi et al. (2016)
Abalone (*Haliotis discus* hannai Ino)	Total viable count, *Pseudomonas* spp., H2S-producing bacteria, Enterobacteriaceae, and lactic acid bacteria	Sodium alginate coating with bamboo leaf extract or rosemary extract	Hao et al. (2017)
Bream (*Megalobrama ambycephala*)	Total viable count, psychrophilic bacteria, lactic acid bacteria, *Enterobacteriaceae*, *Pseudomonas* spp., and H2S-producing bacteria	Pectin-based coatings enriched with clove essential oil	Nisar et al. (2019)
Grass carp (*Ctenopharyngodon idellus*)	*Pseudomonas* spp. and *Aeromonas* spp.	Cinnamon bark oil	Huang et al. (2017)
Sea bass (*Dicentrarchus labrax*)	Total colony counts, total aerobic counts, and psychrotrophic counts	Superchilling storage-ice glazing and clove essential oil	He et al. (2019)
Black sea bream (*Acanthopagrus butcheri* Munro) fillets	*Shewanella* spp.	Kakadu plum (*Terminalia ferdinandiana*) extracts	Wright et al. (2019)
Atlantic salmon (*Salmo salar*)	Mesophilic and psychrophilic bacteria and *Pseudomonas* spp.	Chitosan films with grape seed extract and carvacrol microcapsules	Alves et al. (2018)
Salted sardines	Total microbial counts, mesophilic rod lactic acid bacteria, Enterobacteriaceae, staphylococci and halophilic bacteria	Lemon essential oil micro-emulsions	Alfonzo et al. (2017)
Deepwater pink shrimp (*Parapenaeus longirostris* Lucas 1846)	Total viable counts, psychrotrophic bacterial counts, and Enterobacteriaceae	Edible gelatin coating enriched with orange (*Citrus sinensis* (L.) Osbeck) leaves essential oil	Alparslan et al. (2016)
Dairy			
Cheese	*Listeria monocytogenes*	Pink pepper tree (*Schinus tereinthifolius* Raddi) essential oil	Dannenberg et al. (2016)
Low-fat cut cheese	*Staphylococcus aureus*, psychrophilic bacteria, and yeasts and molds	Sodium alginate-mandarin fiber-oregano essential oil edible coatings	Artiga-Artigas et al. (2017)
Milk	*Staphylococcus aureus*, *Bacillus licheniformis*, and *Enterococcus hirae* Farrow and Collins	Free and nanoemulsified thyme essential oil	Jemaa et al. (2017)
Skim milk	*Pseudomonas putida* and *Staphylococcus* spp.	Odorless citrus extract	Bevilacqua et al. (2016)
Pasteurized milk	Foodborne isolated *Staphylococcus aureus*	Nisin and cinnamaldehyde	Shi et al. (2017)
Vanilla cream pudding	Five strain mixture of *Listeria monocytogenes*	Cinnamon extract and storage temperature	Lianou et al. (2018)
Seasonings			
Salsa (tomatoes, jalapeño peppers, onions, garlic, and distilled vinegar)	*E. coli* O157:H7, *S.* Typhimurium, and *L. monocytogenes*	Carvacrol and/or ohmic heating treatments	Kim and Kang (2017)
Mayonnaise-based salad dressing	*S. pombe* and *Z. bailii*	Carvacrol, eugenol, and trans-cinnamaldehyde	Monu et al. (2016)

Juven et al. (1972) showed that oleuropein antimicrobial activity could be reduced by the addition of triptone and/or yeast extract to the culture medium. Tassou and Nychas (1994) demonstrated that inoculum size, oleuropein concentration, and pH influenced significantly *S. aureus* growth and lag time while proving that the efficiency of phenolic compound antimicrobial action was reduced in foods with relatively low protein content; the principal cause of antimicrobial activity loss could be probably the solubilization of these compounds in the lipidic phase of the medium, reducing its availability to act as an antimicrobial. Aureli et al. (1992), evaluating the antimicrobial capacity of thyme against *L. monocytogenes* growth in model systems and in a real food, found that essential oil antilisteric efficiency decreased when used *in vivo* (ground pork meat) in comparison with the behavior in laboratory media (solid and liquid).

Spencer et al. (1988) reported that the interaction or complex formation between phenols and proteins depends partially on protein characteristics, pH, and on the phenolic group contained by the molecule. This interaction takes place by hydrogen bonds between phenolic groups and peptides and also by hydrophobic interactions. The interactions of aldehydes with proteins have been extensively studied, since protein addition to aldehyde solutions can decrease the effective concentration of these groups (Cha and Ho, 1988; Hansen and Heinis, 1991; Montgomery and Day, 1965). Citral (lemon flavor component) concentration was reduced almost 100% when 5% of casein or soy protein isolate was added to aqueous solutions; 68% initial vanillin concentration, measured by HPLC, was lost after 26 hours in drinks containing aspartame (Tateo et al., 1988).

The presence of fats and/or proteins in food could also provide a protective layer for bacteria while the physical structure of foods may reduce the diffusion of active compounds to target microorganisms (Perricone et al., 2015). Carvalho et al. (2015) observed that the effect of thyme essential oil against *S. aureus* and *L. monocytogenes* was reduced in a semi-solid cheese model in comparison with a cheese-based broth; the effect was attributed in part to the limited diffusion of the EO in the solid media. Burris, Higginbotham, and Stewart (2015) observed that higher concentrations of yerba mate extracts were needed to inactivate methicillin-resistant *S. aureus* in ground beef with high fat content.

The effectiveness of plant-derived antimicrobials in a certain food matrix is different for different bacterial strains; Bajpai et al. (2019) studied the effect of the biflavonoid ametoflavone, isolated from *Nandina domestica*, against *S. aureus* and *E. coli* inoculated in two different food models: minced chicken breast meat and apple juice after 10 days of storage at 4°C. It was found that MIC levels of ametoflavone (62.5 and 125 μg/mL for *S. aureus* and *E. coli*, respectively) were more effective against *E. coli* in apple juice but more effective against *S. aureus* in chicken. Stojanović-Radić et al. (2018) observed that basil essential oil has a higher effect against *Salmonella* when it is applied in fresh chicken meat, while its combination with rosemary essential oil or rosemary essential oil alone are more efficient in the drier, thermally processed meat.

The considerably higher levels of essential oils or plant extracts needed in food products to reach the desired antimicrobial effect will most likely affect sensory characteristics of the product and potentially turn it unacceptable. Thus, it is important to consider which essential oil/food combinations will be of liking to the consumer (Van Haute et al., 2016). Only four (thyme, holybasil, cassia, and clove) out of nine tested essential oils were acceptable to consumers at the required concentration to have an antimicrobial effect in chicken sausages (Sharma et al., 2017). Frankova et al. (2016) studied the effect of oregano, clove, and cinnamon oils in vapor phase in combination with warm airflow as treatment for apples inoculated with *P. expansum*; the highest tested concentration of oregano essential oil (16 μL/L) was the most effective treatment after 21 days of storage. That same concentration of oregano essential oil was shown to negatively affect the organoleptic properties of treated apples, in contrast with treatments with other studied essential oils or lower concentrations of oregano which were satisfactorily evaluated; thus, it was concluded that the most effective concentration of oregano essential oil cannot be utilized against the growth of phytopathogenic fungi on apples (Frankova et al., 2016). In orange and pomegranate juices, concentrations of cinnamon leaf essential oil above 0.04 mg/mL were found to have negative changes in aroma and flavor which made them unacceptable for consumers; the accepted concentration was 0.02 mg/mL which, in combination with ultrasonic treatment (24 kHz; 105 μm; 33.31 W/mL 30 min) at 50°C, resulted in 2.52 and 2.81 log reductions of *S. cerevisiae* in orange and pomegranate juice, respectively (Sánchez-Rubio et al., 2016).

The use of natural antimicrobials in combination with other environmental stress factors not only can enhance their antimicrobial properties but also makes possible the development of products that consumers are demanding by reducing the amounts of synthetic or natural antimicrobials needed to assure microbial stability. The combined effect of carvacrol at acceptable sensory levels (200 ppm) with high-pressure processing (600 MPa/180s/25°C) was able to significantly extend the shelf life of low-sodium sliced vacuum-packed turkey breast ham (de Oliveira et al., 2015). It has been demonstrated that vanillin (4-hydroxy-3-methoxy-benzaldehyde) inhibited microbial growth in laboratory media and fruit purées stored at 25–27°C (López-Malo et al., 2000). Promising results have been obtained by Cerrutti et al. (1997) and Cerrutti and Alzamora (1996) in strawberry (pH 3.4, a_w 0.95) and apple purées (pH 3.5, $a_w \cong$ 0.99) preserved by combined factors (hurdle technology); strawberry purée with 3000 ppm vanillin and reduced a_w inhibited the native and inoculated flora (*Saccharomyces cerevisiae, Zygosaccharomyces rouxii, Z. bailii, Schizosaccharomyces pombe, Pichia membranaefaciens, Botrytis* sp., *Byssochlamys fulva, Bacillus coagulans*, and *Lactobacillus delbrueckii*) for at least 60 days of storage at 25°C, while in apple purée with 2000 ppm vanillin, a germicidal effect was observed on inoculated *S. cerevisiae, Z. rouxii*, and *D. hansenii*. Cerrutti and Alzamora (1996) also reported that 3000 ppm of vanillin in a banana purée with pH 3.4 and a_w 0.98 was inhibitory for *S. cerevisiae, Z. rouxii*, and *D. hansenii*. Castañón et al. (1999) evaluated the effects of vanillin (1000 or 3000 ppm) or potassium sorbate (1000 ppm) addition on the microbial stability during storage at 15, 25, or 35°C of banana purée preserved by combined methods (a_w 0.97, pH 3.4); native flora (standard plate, yeasts and molds) counts during storage at different temperatures in the control purées (without antimicrobial addition), and in those containing 1000 ppm vanillin, demonstrated that a reduced storage temperature (15°C) was not enough to detain or delay the spoilage of the purée; after six days of storage at 15°C, native yeasts and molds reached counts of 10^4 CFU/g (about 3 log cycles greater than the initial count), and these purées were at this time sensorially unacceptable (odor and textural changes accompanying the microbial spoilage). The addition of 1000 ppm of vanillin increased the lag phase up to $\cong$ 16 days at 15°C, and the time to detect the microbial spoilage was extended to around 21 days. In the presence of 3000 ppm of vanillin or 1000 ppm potassium sorbate no microbial growth (< 10 CFU/g) was detected after 6 days and up to 60 days of storage (15, 25, or 35°C). Results obtained by Cerrutti and Alzamora (1996), Cerrutti et al. (1997), and Castañón et al. (1999) demonstrated that the addition of vanillin in combination with a slight reduction of a_w and pH may be a promising method for fruit purée preservation and confirmed the antimicrobial properties of vanillin; they found that the protein and fat contents of the fruit partially determined the vanillin concentration necessary to obtain a sound product. The relatively high protein (1.2%) and fat (0.3%) contents in banana, compared with those of other fruits, explain the necessity of a higher vanillin concentration to obtain the same antimicrobial effects as in the other studied fruits.

Castañón et al. (1999) reported results of sensory evaluation of banana purées containing 3000 ppm of vanillin or 1000 ppm of potassium sorbate; mean scores corresponded to products with a good overall acceptability with scores around 6 (like slightly). Studied purées were significantly different ($p < 0.05$) in odor (the one with vanillin was preferred) and flavor (the one with potassium sorbate was better) whereas there was not significant difference ($p < 0.05$) in color and overall acceptability. The flavoring characteristics of vanillin are well-accepted and have demonstrated compatibility with many fruits in concentrations up to 3000 ppm (Cerrutti and Alzamora, 1996). The microbial stability of mango juice (pH 4.9) supplemented with extracts of ginger (main antimicrobial compounds: zingerone, gingerol, and shogaol) and nutmeg (main antimicrobial compounds: myristicin and sabinene) was investigated during 3 months at room temperature by Ejechi et al. (1998); the combination of heating (15 min at 55°C) and 4% v/v of each spice extract inhibited microbial growth (yeasts and non-spore-forming bacteria) and created products with acceptable taste. Surface disinfection of tomatoes using cinnamic aldehyde was studied by Smid et al. (1996); whole tomatoes were dipped for 30 min in a solution containing 13 mM cinnamic aldehyde and then stored at 18°C in sealed plastic bags. The combination of the treatment with the natural phenolic compound and packaging under modified atmosphere reduced spoilage-associated fungi and bacteria on the tomatoes' surface, increasing their shelf life up to 11 days. Batista et al. (2019) observed that the synergistic interaction of carvacrol and potassium sorbate in tomato paste eliminated *Salmonella* spp. at the third day of storage with no effects on physicochemical properties and a sensory acceptance index of 67% from a nine-point hedonic scale. A less severe thermal process could be applied to young coconut liquid endosperm to preserve sensory attributes while inhibiting *E. coli* O157:H7 growth if combined with malic acid and nisin treatments (Gabriel and Estilo, 2015).

Strategies to reduce the amount of essential oil needed to ensure the microbiological safety of food products include the encapsulation of essential oils into emulsions or their utilization as components of active packaging systems instead of using them as ingredients in the food products. This in turn diminishes the unacceptable organoleptic changes in food that accompany elevated essential oil concentrations (Hyldgaard, Mygind and Meyer, 2012). Gonçalves et al. (2017) confirmed that the encapsulation of thyme essential oil decreased its minimal inhibitory concentration (MIC) values when tested *in vitro*, reasoning that the protective micro-environment that the particle wall provides retards the volatilization of the bioactive compounds, thus rendering the system more efficient to inhibit microbial growth. This functional barrier generated between the core and wall material helps to maintain the biological function of the core by avoiding chemical and physical reactions (Bakry et al., 2016). Martínez-Hernandez, Amodio, and Colelli (2017) reported that, unlike sliced carrots treated with a carvacrol solution, treatments with carvacrol-loaded chitosan-tripolyphosphate nanoparticles at the same concentration exhibited the lowest aroma score related to carvacrol off-flavors; encapsulated carvacrol was also effective in controlling microbial growth throughout 13 days of storage. Lee et al. (2019) found that encapsulation of rosemary extract in chitosan and γ-poly glutamic acid nanoparticles was able to improve its dispersibility in barley tea and enhanced its antimicrobial activity against *B. subtilis*. Chang et al. (2017) used sachets containing polyvinyl alcohol-oregano essential oil microcapsules to be utilized in iceberg lettuce containers as active packaging to control *Dickeya chrysanthemi*, yeasts and molds, and mesophilic aerobic bacteria; it was observed that the sachets allowed effective antimicrobial activity of oregano essential oil under high humidity and temperature conditions.

Other important aspects to consider from the use of essential oils and plant extracts as antimicrobials in food systems are the effects that food processing techniques such as cooking or pressure could have on the antimicrobial activity of the compounds (Nikmaram et al., 2018). In this sense, Wang et al. (2018b) developed β-cyclodextrin and porous starch microcapsules for the encapsulation of clove oil; their microcapsules could endure 30 min of boiling and still present inhibition of mold spores on meat products.

Given that food-contact surfaces are an important source of food contamination when not properly cleaned (Rouger, Tresse and Zagorec, 2017), as well as biofilm tolerance to chemical biocides and concerns about the environmental toxicity of residues from classical synthetic sanitizers (Bridier et al. 2015), recent studies have focused on the use of compounds extracted from aromatic plants; Rodrigues et al. (2017) and Tapia-Rodriguez et al. (2017) applied oregano essential oil and carvacrol to prevent *S. aureus* biofilm formation as well as biofilm formation and quorum sensing of *P. aeruginosa*, respectively. Cui et al. (2016) used salvia oil nanoliposomes to control *S. aureus* biofilm formation on milk containers.

Plant bioactive compounds with antimicrobial properties are also being studied for their application as feed supplements for farm animals as antibiotic substitutes intended to improve animal health (Guil-Guerrero et al., 2016), over the concern that antibiotic use in livestock has contributed to the emergence and transmission to humans of antibiotic-resistant bacteria. Careful consideration has to be given since compounds like carvacrol, oregano, and thyme essential oils have a broad-spectrum antimicrobial activity and, while potent against pathogens, they can also have an effect on beneficial ruminant bacteria (Benchaar et al., 2008). Of interest are the studies reviewed by Hassoun and Çoban (2017) since the effect of the use of essential oils and their components as fish dietary additives extended the delay of microbial growth in fish meat *post-mortem*.

17.9 Final Remarks

The use of spices, herbs, plants, essential oils, and related phenolic compounds is limited due to the high minimal inhibitory concentrations needed in actual foods with high protein and/or fat contents, which impart undesirable flavor and/or odor. These undesirable effects can be minimized if the natural compound is utilized in combination with other environmental stress factors in the frame of the "hurdle technology." Table 17.4 presents some general guidelines to be followed for selecting plant-derived

TABLE 17.4

Factors That Influence Plant-Derived Antimicrobial Selection and Effectiveness

Antimicrobial Characteristics	Toxicological and Legal Aspects, Solubility, Sensory Impact, Cost
Possible synergistic, antagonist, or additive interaction effects with other antimicrobial factors (composition of food)	Moisture content, fat and protein content, water activity and pH, presence of other inhibitors (acids, salts, smoke, antimicrobials), interactions with food matrix and other food additives
Initial contamination level	Sanitary conditions of ingredients and raw materials, sanitary conditions of equipment, processing conditions, type of potential growing microorganisms
Handling and distribution	Length of storage, temperature of storage, packaging

antimicrobials. In this way, considering the consumers' interest in more "natural" foods, the potential for applications in minimally processed fruit and vegetables appears to be good (Alzamora et al., 1995, 2016). However, for a wider and more rational use of these natural compounds, some points should be addressed:

- The extraction methodology utilized to obtain the essential oil or extract.
- The response of key microorganisms to the multi-target preservation system *in vitro* and then evaluation of its efficacy *in vivo.*
- The form in which the plant, herb, spice, essential oil, or extract will be incorporated in the foods without adversely affecting sensory, nutritional, and safety characteristics, and without increasing significantly the formulation, processing, or marketing costs of the minimally processed product to which they are added.

17.10 Acknowledgments

Authors López-Malo, Paris, Lastra-Vargas, and Palou gratefully acknowledge financial support from the National Council for Science and Technology (CONACyT) of Mexico (Project CB-2016-01-283636) and Universidad de las Américas Puebla (UDLAP Projects 2409 and 3555). Authors Alzamora, Coronel, and Gómez thankfully acknowledge financial support from the National Council of Scientific and Technical Research (CONICET) of Argentina and Universidad de Buenos Aires.

REFERENCES

Abu-Al-Basal, M.A. 2010. Healing potential of *Rosmarinus officinalis* L. on full-thickness excision cutaneous wounds in alloxan-induced-diabetic BALB/c mice. *J. Ethnopharmacol.* 131:443–450.

Acevedo-Fani, A., Soliva-Fortuny, R. and Martín-Belloso, O. 2017. Nanoemulsions as edible coatings. *Curr. Opinion Food Sci.* 15:43–49.

Aguilar-González, A.E., Palou, E. and López-Malo, A. 2015. Antifungal activity of essential oils of clove (*Syzygium aromaticum*) and/or mustar (*Brassica nigra*) in vapor phase against gray mold (*Botrytis cinerea*) in strawberries. *Innov. Food Sci. Emerging Technol.* 32:181–185.

Ahuja, I., Kissen, R. and Bones, A.M. 2012. Phytoalexins in defense against pathogens. *Trends Plant Sci.* 17:73–90.

Alfonzo, A., Martorana, A., Guarrasi, V., Barbera, M., Gaglio, R., Santulli, A., Settanni, L., Galati, A., Moschetti, G. and Francesca, N. 2017. Effect of the lemon essential oils on the safety and sensory quality of salted sardines (*Sardina pilchardus* Walbaum 1792). *Food Control* 73:1265–1274.

Al-Khayat, M.A. and Blank, G. 1985. Phenolic spice components sporostatic to *Bacillus subtilis. J. Food Sci.* 50:971–974.

Aloui, H. and Khwaldia, K. 2016. Natural antimicrobial edible coatings for microbial safety and food quality enhancement. *Comp. Rev. Food Sci. Food Safety* 15:1080–1103.

Aloui, H., Khwaldia, K., Sanchez-González, L., Muneret, L., Jeandel, C., Hamdi, M. and Desobry, S. 2014. Alginate coatings containing grapefruit essential oil or grapefruit seed extract for grapes preservation. *Intl. J. Food Sci. Technol.* 49:952–959.

Alparslan, Y., Yapıcı, H.H., Metin, C., Baygar, T., Günlü, A. and Baygar, T. 2016. Quality assessment of shrimps preserved with orange leaf essential oil incorporated gelatin. *LWT Food Sci. Technol.* 72:457–466.

Altemimi, A., Lakhssassi, N., Abu-Ghazaleh, A. and Lightfoot, D.A. 2017. Evaluation of the antimicrobial activities of ultrasonicated spinach leaf extracts using RAPD markers and electron microscopy. *Arch. Microbiol.* 199:1417–1429.

Álvarez-Ordóñez, A., Carvajal, A., Arguello, H., Martínez-Lobo, F.J., Naharro, G. and Rubio, P. 2013. Antibacterial activity and mode of action of a commercial citrus fruit extract. *J. Appl. Microbiol.* 115:50–60.

Alves, V.L.C.D., Rico, B.P.M., Cruz, R.M.S., Vicente, A.A., Khmelinskii, I. and Vieira, M.C. 2018. Preparation and characterization of a chitosan film with grape seed extract-carvacrol microcapsules and its effect on the shelf-life of refrigerated Salmon (*Salmo salar*). *LWT* 89:525–534.

Alzamora, S.M., Cerrutti, P., Guerrero, S. and López-Malo, A. 1995. Minimally processed fruits by combined methods. In Barbosa-Cánovas, G. and Welti, J., eds. *Food Preservation by Moisture Control: Fundamentals and Applications*, pp. 463–491. Technomic Publishing Co. Inc., Lancaster.

Alzamora, S.M., Guerrero, N.S., Raffellini, S., Ferrario, M. and Schenk, M. 2016. Hurdle technology in fruit processing. In Pareek, S. ed. *Fresh-Cut Fruits and Vegetables: Technology, Physiology and Safety*, pp. 101–138. CRC Press, Boca Raton, FL.

An, P., Yang, X., Yu, J., Qi, J., Ren, X. and Kong, Q. 2019. α-terpineol and terpene-4-ol, the critical components of tea tree oil, exert antifungal activities in vitro and in vivo against *Aspergillus niger* in grapes by inducing morphous damage and metabolic changes of fungus. *Food Control* 98:42–53.

Arnon-Rips, H., Porat, R. and Poverenov, E. 2019. Enhancement of agricultural produce quality and storability using citral-based edible coatings; the valuable effect of nano-emulsification in a solid-state delivery on fresh-cut melons model. *Food Chem.* 277:205–212.

Arora, M.K. and Strange, R.N. 1991. Phytoalexin accumulation in groundnuts in response to wounding. *Plant Sci.* 78:157–163.

Artiga-Artigas, M., Acevedo-Fani, A. and Martín-Belloso, O. 2017. Improving the shelf life of low-fat cut cheese using nanoemulsion-based edible coatings containing oregano essential oil and mandarin fiber. *Food Control* 76:1–12.

Aureli, P., Constantini, A., and Zolea, S. 1992. Antimicrobial activity of some plant essential oils against *Listeria monocytogenes*. *J. Food Prot.* 55:344–348.

Azadbakht, E., Maghsoudlou, Y., Khomiri, M. and Kashiri, M. 2018. Development and structural characterization of chitosan films containing *Eucalyptus globulus* essential oil: Potential as an antimicrobial carrier for packaging of sliced sausage. *Food Pkg. Shelflife* 17:65–72.

Azis, Z.A.A., Ahmad, A., Setapar, S.H.M., Siti, H.M., Karakucuk, A., Azim, M.M., Lokhat, D., Rafatullah, M., Ganash, M., Kamal, M.A. and Ashraf, G.M. 2018. Essential oils: Extraction techniques, pharmaceutical and therapeutic potential: A review. *Curr. Drug Metabol.* 19:1–11.

Azizkhani, M., Akhondzadeh Basti, A., Misaghi, A. and Tooryan, F. 2012. Effects of *Zataria multiflora* Boiss., *Rosmarinus officinalis* L. and *Mentha longifolia* L. essential oils on growth and gene expression of enterotoxins C and E in *Staphylococcus aureus* ATCC 29213. *J. Food Safety* 32:508–516.

Azmir, J., Zaidul, I.S.M., Rahman, M.M., Sharif, K.M., Mohamed, A., Sahena, F., Jahurul, M.H.A., Ghafoor, K., Norulaini, N.A.N. and Omar, A.K.M. 2013. Technique for the extraction of bioactives compounds from plant materials: A review. *J. Food Engr.* 117:426–436.

Bajpai, V.K., Park, I., Lee, J., Shukla, S., Nile, S.H., Chun, H.S., Khan, I., Oh, S.Y., Lee, H., Huh, Y.S., Na, M. and Han, Y. 2019. Antioxidant and antimicrobial efficacy of a biflavonoid, amentoflavone from *Nandina domestica* in vitro and in minced chicken meat and apple juice food models. *Food Chem.* 271:239–247.

Bajpai, V.K., Sharma, A. and Baek, K.H. 2013a. Antibacterial mode of action of seed essential oil of Eleutherococcus senticosus against foodborne pathogens. *Intl. J. Food Sci. Technol.* 48:2300–2305.

Bajpai, V.K., Sharma, A. and Baek, K.H. 2013b. Antibacterial mechanism of action of *Taxus cuspidata* stem essential oil against selected foodborne pathogens. *J. Food Safety* 33:348–359.

Bajpai, V.K., Sharma, A., Moon, B. and Baek, K.H. 2014. Chemical composition analysis and antibacterial mode of action of *Taxus Cuspidata* leaf essential oil against foodborne pathogens. *J. Food Safety* 34:9–20.

Bakry, A.M., Abbas, S., Ali, B., Majeed, H., Abouelwafa, M.Y., Mousa, A. and Liang, L. 2016. Microencapsulation of oils: A comprehensive review of benefits, techniques and applications. *Comp. Rev. Food Sci. Food Safety* 15:143–182.

Baldim, I., Tonani, L., von Zeska Kress, M.R. and Oliveira, W.P. 2019. *Lippia sidoides* essential oil encapsulated in lipid nanosystem as an anti-*Candida* agent. *Industrial Crops Prod.* 127:73–81.

Barbosa, I.M., Medeiros, J.A.C., Rimá de Oliveira, K.A., Gomes-Neto, N.J., Tavares, J.F., Magnani, M. and de Souza, E.L. 2016. Efficacy of the combined application of oregano and rosemary essential oils for the control of *Escherichia coli*, *Listeria monocytogenes* and *Salmonella* Enteritidis in leafy vegetables. *Food Control* 59:468–477.

Barone, F.E. and Tansey, M.R. 1977. Isolation, purification, identification, synthesis and kinetics of activity of the anticandidal component of *Allium sativum* and a hypothesis for its mode of action. *Mycologia* 69:793–799.

Basak, S. and Guha, P. 2017. Use of predictive model to describe sporicidal and cell viability efficacy of betel leaf (*Piper betle* L.) essential oil on *Aspergillus flavus* and *Penicillium expansum* and its antifungal activity in raw apple juice. *LWT* 80:510–516.

Basak, S. and Guha, P. 2018. A review on antifungal activity and mode of action of essential oils and their delivery as nano-sized oil droplets in food system. *J. Food Sci. Technol.* 55:4701–4710.

Batista, A.F.P., dos Santos, A.R., da Silva, A.F., Trevisan, D.A.C., Ribeiro, L.H., Campanerut-Sá, P.A.Z., Filho, B.A.A., Junior, M.M. and Mikcha, J.M.G. 2019. Inhibition of *Salmonella enterica* serovar Typhimurium by combined carvacrol and potassium sorbate in vitro and in tomato paste. *LWT* 100:92–98.

Ben Said, L., Fliss, I., Offret, C. and Beaulieu, L. 2019. Antimicrobial peptides: The new generation of food additives. *Enc. Food Chem.* P. 576–582.

Benchaar, C., Calsamiglia, S., Chaves, A.V., Fraser, G.R., Colombatto, D., McAllister, T.A. and Beauchemin, K.A. 2008. A review of plant-derived essential oils in ruminant nutrition and production. *Anim. Feed Sci. Technol.* 145:209–228.

Bennett, M.H., Gallagher, M.D.S., Bestwick, C.S., Rossiter, J.T. and Mansfield, J.W. 1994. The phytoalexin response of lettuce to challenge by *Botrytis cinerea*, *Bremia lactucae* and *Pseudomonas syringae* pv. phaseolicola. *Physiol. Molec. Plant Pathol.* 44:321–333.

Berdejo, D., Pagán, E., García-Gonzalo, D. and Pagán, R. 2019. Exploiting the synergism among physical and chemical processes for improving food safety. *Curr. Opinion Food Sci.* 30:14–20.

Berezina, E.V., Brilfina, A.A. and Veselov, A.P. 2017. Content of phenolic compounds, ascorbic acid, and photosynthetic pigments in *Vaccinium macrocarpon* Ait. Dependent on seasonal plant development stages and age (the example of introduction in Russia). *Sci. Hortic.* 218:139–146.

Besada, C. and Salvador, A. 2011. Tannins of persimmon fruit: Artificial removal of astringency. *Tannins: Types, Foods Containing and Nutrition*. 217–230.

Beuchat, L.R. 1976. Sensitivity of *Vibrio parahaemolyticus* to spices and organic acids. *J. Food Sci.* 41:899–902.

Beuchat, L.R. and Golden, D.A. 1989. Antimicrobials occurring naturally in foods. *Food Technol.* 43:134–142.

Bevilacqua, A., Speranza, B., Iorio, M.C., Loi, M., Sinigaglia, M. and Corbo, M.R. 2016. US-inactivation of foodborne bacteria: Screening in distilled water and combination with citrus extract in skim milk. *LWT Food Sci. Technol.* 70:135–141.

Bilbao-Sainz, C., Chiou, B.S., Punotai, K., Olson, D., Williams, T., Wood, D., Rodov, V., Poverenov, E. and Mc.Hugh, T. 2018. Layer-by-layer alginate and fungal chitosan based edible coating applied to fruit bars. *J. Food Sci.* 83:1880–1887.

Black-Solis, J., Ventura-Aguilar, R.I., Correa-Pacheco, Z., Corona-Rangel, M.L. and Bautista-Baños, S. 2019. Preharvest use of biodegradable polyester nets added with cinnamon essential oil and the effect on the storage life of tomatoes and the development of *Alternaria alternata*. *Sci. Hortic.* 245:65–73.

Bleasel, N., Tate, B. and Rademaker, M. 2002. Allergic contact dermatitis following exposure to essential oils. *Australasian J. Dermatol.* 43:211–213.

Bouhdid, S., Abrini, J., Amensour, M., Zhiri, A., Espuny, M.J. and Manresa, A. 2010. Functional and ultrastructural changes in *Pseudomonas aeruginosa* and *Staphylococcus aureus* cells induced by *Cinnamomum verum* essential oil. *J. Appl. Microbiol.* 109(4):1139–1149.

Božik, M., Císarová, M., Tančinová, D., Kouřimská, L., Hleba, L. and Klouček, P. 2017. Selected essential oil vapours inhibit growth of *Aspergillus* spp. in oats with improved consumer acceptability. *Industrial Crops Prod.* 98:146–152.

Brandelli, A. and Taylor, T.M. 2015. Nanostructured and nanoencapsulated natural antimicrobials for use in food products. In *Handbook of Natural Antimicrobials for Food Safety and Quality*, ed. T.M. Taylor, 229–257. Elsevier Ltd., Cambridge.

Bridier, A., Sanchez/Vizuete, P., Guilbaud, M., Piard, J.-C., Naïtali, M. and Briandet, R. 2015. Biofil-associated persistence of food-borne pathogens. *Food Microbiol.* 45(B):167–178.

Bullerman, L.B. 1974. Inhibition of aflatoxin production by cinnamon. *J. Food Sci.* 39:1163–1167.

Bullerman, L.B., Lieu, F.Y. and Seier, S.A. 1977. Inhibition of growth and aflatoxin production by cinnamon and clove oils: Cinnamic aldehyde and eugenol. *J. Food Sci.* 42:1107–1109.

Burris, K.P., Higginbotham, K.L. and Stewart, C.N. 2015. Aqueous extracts of yerba mate as bactericidal agents against methicillin-resistant *Staphylococcus aureus* in a microbiological medium and ground beef mixtures. *Food Control* 50:748–753.

Burt, S. 2004. Essential oils: Their antibacterial properties and potential applications in foods: A review. *Intl. J. Food Microbiol.* 94:223–253.

Calo, J.R., Crandall, P.G., O´Bryan, C.A. and Ricke, S.C. 2015. Essential oils as antimicrobials in food systems: A review. *Food Control* 54:111–119.

Camiletti, B.X., Asensio, C.M., Gadban, L.C., Pecci, M.P.G., Conles, M.Y. and Lucini, E.I. 2016. Essential oils and their combinations with iprodione fungicide as potential antifungal agents against withe rot (*Sclerotium cepivorum* Berk) in garlic (*Allium sativum* L.) crops. *Industrial Crops Prod.* 85:117–124.

Canillac, N. and Mourey, A. 2004. Effect of several environmental factors on the anti-*Listeria monocytogenes* activity of essential oil of *Picea excelsa*. *Intl. J. Microbiol.* 92:95–103.

Carson, C.F. and Riley, T.V. 2001. Safety, efficacy and provenance of tea tree (*Melaleuca alternifolia*) oil. *Contact Dermatitis* 45:65–67.

Carvalho, R.I., de Souza, G.T., Honório, V.G., de Sousa, J.P., da Conceiçã, M.L., Maganani, M. and de Souza, E.L. 2015. Comparative inhibitory effects of *Thymus vulgaris* L. essential oil against *Staphylococcus aureus*, *Listeria monocytogenes* and mesophilic starter co-culture in cheese-mimicking models. *Food Microbiol.* 52:59–65.

Castañón, X., Argaiz, A. and López-Malo, A. 1999. Effect of storage temperature on the microbial and color stability of banana purées prepared with the addition of vanillin or potassium sorbate. *Food Sci.Technol. Intl.* 5:56–60.

Castro, J.C., Endo, E.H., de Souza, M.R., Zanqueta, E.B., Polonio, J.C., Pamphile, J.A., Ueda-Nakamura, T., Nakamura, C.V., Filho, B.P.D. and Filho, B.A.A. 2017. Bioactivity of essential oils in the control of *Alternaria alternata* in dragon fruit (*Hylocereus undatus* Haw.). *Industrial Crops Prod.* 97:101–109.

Castro-Rosas, J., Ferreira-Grosso, C.R., Gómez-Aldapa, C.A., Rangel-Vargas, E., Rodríguez-Marin, M.L., Guzmán-Ortiz, F.A. and Falfan-Cortes, R.N. 2017. Recent advances in microencapsulation of natural sources of antimicrobial compounds used in food: A review. *Food Res. Intl.* 102:575–587.

Catarino, M.D., Alves-Silva, J.M., Fernandes, R.P., Gonçalves, M.J., Salgueiro, L.R., Henriques, M.F. and Cardoso, S.M. 2017. Development and performance of whey protein active coatings with *Origanum virens* essential oils in the quality and shelf life improvement of processed meat products. *Food Control* 80:273–280.

Cava, R., Novak, E., Taboada, A. and Marín-Iniesta, A. 2007. Antimicrobial activity of clove and cinnamon essential oils against *Listeria monocytogenes* in pasteurized milk. *J. Food Prot.* 70:2757–2763.

Cavallito, C.J. and Bailey, J.H. 1944. Allicin, the antibacterial principle of *Allium sativum*. I: Isolation, physical properties and antibacterial action. *J. Amer. Chem. Soc.* 66:1950–1960.

Cerrutti, P. and Alzamora, S.M. 1996. Inhibitory effects of vanillin on some food spoilage yeasts in laboratory media and fruit purées. *Intl. J. Food Microbiol.* 29:379–386.

Cerrutti, P., Alzamora, S.M. and Vidales, S.L. 1997. Vanillin as an antimicrobial for producing shelf-stable strawberry purée. *J. Food Sci.* 62:608–610.

Cha, A.S. and Ho, C.T. 1988. Studies of the interaction between aspartame and flavor vanillin by high performance liquid chromatography. *J. Food Sci.* 53:562–564.

Chaieb, K., Hajlaoui, H., Zmantar, T., Kahla-Nakbi, A.B., Rouabhia, M., Mahdouani, K. and Bakhrouf, A. 2007. The chemical composition and biological activity of clove essential oil, Eugenia caryophyllata (*Syzigium aromaticum* L. Myrtaceae): A short review. *Phytotherapy Res.* 21:501–506.

Chan, C.L., Gan, R.Y., Shah, N.P. and Corke, H. 2018. Polyphenols from selected dietary spices and medicinal herbs differentially affect common food-borne pathogenic bacteria and lactic acid bacteria. *Food Control.* 92:437–443.

Chang, Y., Choi, I., Cho, A.R. and Han, J. 2017. Reduction of *Dickeya chrysanthemi* on fresh-cut iceberg lettuce using antimicrobial sachet containing microencapsulated oregano essential oil. *LWT Food Sci. Technol.* 82:361–368.

Chemat, F., Aber Vian, M. and Cravotto, G. 2012. Green extraction of natural products: Concept and principles. *Intl. J. Molec. Sci.* 13:8615–8627.

Chen, C., Liu, C.H., Cai, J., Zhang, W., Qi, W.L., Wang, Z., Liu, Z. and Yang, Y. 2018a. Broad-spectrum antimicrobial activity, chemical composition and mechanism of action of garlic (*Allium sativum*) extracts. *Food Control* 86:117–125.

Chen, C., Xu, Z., Ma, Y., Liu, J., Zhang, Q., Tang, Z., Fu, K., Yang, F. and Xie, J. 2018b. Properties, vapour-phase antimicrobial and antioxidant activities of active poly (vinyl alcohol) packaging films incorporated with clove oil. *Food Control.* 88:105–112.

Chibane, L.B., Degraeve, P., Ferhout, H., Bouajila, J. and Oulahal, N. 2018. Plant antimicrobial polyphenols as potential natural food preservatives. *J. Sci. Food Agric.* 99:1457–1474.

Chuesiang, P., Siripatrawan, U., Sanguandeekul, R., Mcclements, D.J. and Mclandsborough, L. 2019a. Antimicrobial activity of PIT-fabricated cinnamon oil nanoemulsions: Effect of surfactant concentration on morphology of foodborne pathogens. *Food Control* 98:405–411.

Chuesiang, P., Siripatrawan, U., Sanguandeekul, R., Yang, J.S., McClements, D.J. and McLandsborough, L. 2019b. Antimicrobial activity and chemical stability of cinnamon oil in oil-in-water nanoemulsions fabricated using the phase inversion temperature method. *LWT* 110:190–196.

Clemente, I., Aznar, M. and Nerín, C. 2016. Raman Imaging Spectroscopy as a tool to investigate the cell damage on *Aspergillus ochraceus* caused by an antimicrobial packaging containing benzyl isothiocyanate. *Analytical Chem.* 88:4772–4779.

Clemente, I., Aznar, M., Salafranca, J. and Nerín, C. 2017. Raman spectroscopy, electronic microscopy and SPME-GC-MS to elucidate the mode of action of a new antimicrobial food packaging material. *Anal. Bioanal. Chem.* 409:1037–1048.

Conner, D.E. and Beuchat, L.R. 1984a. Effects of essential oils from plants on growth of food spoilage yeasts. *J. Food Sci.* 49:429–434.

Conner, D.E. and Beuchat, L.R. 1984b. Sensitivity of heat-stressed yeasts to essential oils of plants. *Appl Environ Microbiol.* 47:229–233.

Coutinho de Oliveira, T.L., Ramos, A.L.S., Mendes Ramos, E., Piccoli, R.H. and Cristianini, M. 2015. Natural antimicrobials as additional hurdles to preservation of foods by high pressure processing. *Trends Food Sci. Technol.* 45:60–85.

Cowan, M.M. 1999. Plant products as antimicrobial agents. *Clin. Microbiol. Rev.* 12:564–582.

Cruz-Gálvez, A.M., Castro-Rosas, J., Rodríguez-Marín, M.L., Cadena-Ramírez, A., Tellez-Jurado, A., Tovar-Jiménez, X., Chavez-Urbiola, E.A., Abreu-Corona, A. and Gómez-Aldapa, C.A. 2018. Antimicrobial activity and physicochemical characterization of a potato starch-based film containing acetonic and methanolic extracts of *Hibiscus sabdariffa* for use in sausage. *LWT* 93:300–305.

Cui, H., Ma, C. and Lin, L. 2016. Synergetic antibacterial efficacy of cold nitrogen plasma and clove oil against *Escherichia coli* O157:H7 biofilms on lettuce. *Food Control* 66:8–16.

Cui, H., Zhang, C., Li, C. and Lin, L. 2018. Antimicrobial mechanism of clove oil on *Listeria monocytogenes. Food Control* 94:140–146.

Cui, H., Zhao, C. and Lin, L. 2015a. The specific antibacterial activity of liposome-encapsulated clove oil and its application in tofu. *Food Control* 56:128–134. .

Cui, H., Zhao, C. and Lin, L. 2015b. Antibacterial activity of *Helichrysum italicum* oil on vegetables and its mechanism of action. *J. Food Proc. Preserv.* 39:2663–2672.

Cui, H., Zhang, X., Zhou, H., Zhao, C. and Lin, L. 2015c. Antimicrobial activity and mechanisms of *Salvia sclarea* essential oil. *Botanical Stud.* 56:1–8.

Cui, H., Zhou, H. and Lin, L. 2016. The specific antibacterial effect of the Salvia oil nanoliposomes against *Staphylococcus aureus* biofilms on milk container. *Food Control* 61:92–98.

Curtis, H., Noll, U., Stormann, J. and Slusarenko, A.J. 2004. Broad-spectrum activity of the volatile phytoanticipin allicin in extracts of garlic (*Allium sativum* L.) against plant pathogenic bacteria, fungi and Oomycetes. *Physiol. Molec. Plant Pathol.* 65:79–89.

Dhall, R.K. 2013. Advances in edible coatings for fresh fruits and vegetables: A review. *Crit. Rev. Food Sci. Nutr.* 53:435–450.

Dalla Nora, F.M. and Borges, C.D. 2017. Ultrasound pretreatment as an alternative to improve essential oils extraction. *Ciencia Rural, Santa María* 47:1–9.

D'Amato, S., Serio, A., López, C.C. and Paparella, A. 2018. Hydrosols: Biological activity and potential as antimicrobials for food applications. *Food Control* 86:126–137.

Dannenberg, G. da S., Funck, G.D., Silva, W.P. da, and Fiorentini, Â.M. 2019. Essential oil from pink pepper (*Schinus terebinthifolius* Raddi): Chemical composition, antibacterial activity and mechanism of action. *Food Control* 95:115–120.

Dannenberg, G.S., Funck, G.D., Mattei, F.J., da Silva, W.P. and Fiorentini, Â.M. 2016. Antimicrobial and antioxidant activity of essential oil from pink pepper tree (*Schinus terebinthifolius* Raddi) in vitro and in cheese experimentally contaminated with *Listeria monocytogenes*. *Innov. Food Sci. Emerg. Technol.* 36:120–127.

Das, S., Gazdag, Z., Szente, L., Meggyes, M., Horváth, G., Lemli, B., Kunsági-Máté, S., Kuzma, M. and Kőszegi, T. 2019. Antioxidant and antimicrobial properties of randomly methylated β cyclodextrin–captured essential oils. *Food Chem.* 278:305–313.

Davidson, P.M. 1993. Parabens and phenolic compounds. In Davidson, P.M. and Branen, A.L, eds. *Antimicrobials in Foods*, pp. 263–305, 2nd ed. Marcel Dekker, Inc., New York.

Davidson, P.M. 2001. Chemical preservatives and naturally antimicrobial compounds. In Doyle, M.P., Beuchat, L.R. and Montville, T.J., eds. *Food Microbiology: Fundamentals and Frontiers*, pp. 593–628, 2nd ed.. ASM Press, Washington DC.

Davidson, P.M. and Naidu, A.S. 2000. Phyto-phenols. In Naidu, A.S., ed. *Natural Food Antimicrobial Systems*, pp. 265–294. CRC Press, Boca Raton, FL.

de Aquino, A.B., Blank, A.F. and de Aquino Santana, L.C.L. 2015. Impact of edible chitosan–cassava starch coatings enriched with *Lippia gracilis* Schauer genotype mixtures on the shelf life of guavas (*Psidium guajava* L.) during storage at room temperature. *Food Chem.* 171:108–116.

de Oliveira, T.L.C., Leite Junior, B.R.C., Ramos, A.L.S., Ramos, E.M., Piccoli, R.H. and Cristianini, M. 2015. Phenolic carvacrol as a natural additive to improve the preservative effects of high pressure processing of low-sodium sliced vacuum-packed turkey breast ham. *LWT Food Sci. Technol.* 64:1297–1308.

De Pascoli, I.C., dos Anjos, M.M., da Silva, A.A., Lorenzetti, F.B., Cortez, D.A.G., Mikcha, J.M.G., Nakamura, T.U., Nakamura, C.V. and de Abreu Filho, B.A. 2018. Piperaceae extracts for controlling *Alicyclobacillus acidoterrestris* growth in commercial orange juice. *Industrial Crops Prod.* 116:224–230.

Debonne, E., De Leyn, I., Verwaeren, J., Moens, S., Devlieghere, F., Eeckhout, M. and Van Bockstaele, F. 2018. The influence of natural oils of blackcurrant, black cumin seed, thyme and wheat germ on dough and bread technological and microbiological quality. *LWT* 93:212–219.

Devlieghere, F., Vermeulen, A. and Debevere, J. 2004. Chitosan: Antimicrobial activity, interaction with food components and applicability as a coating on fruit and vegetables. *Food Microbiol.* 21:703–714.

DeWit, J.C., Notermans, S., Gorin, N. and Kampelmacher, E.H. 1979. Effect of garlic oil or onion oil on toxic production by *Clostridium botulinum* in meat slurry. *J. Food Prot.* 42:222–227.

DG SANTE. 2019. Food Flavourings. https://webgate.ec.europa.eu/foods_system/main/?sector=FFL&auth=SANCAS (accessed June 17, 2019).

Dhital, R., Mora, N.B., Watson, D.G., Kohli, P. and Choudhary, R. 2018. Efficacy of limonene nano coatings on post-harvest shelf life of strawberries. *LWT* 97:124–134.

Dmitriev, A.P., Tverskoy, L.A., Kozlovsky, A.G. and Grodzinsky, D.M. 1990. Phytoalexins from onion and their role in disease resistance. *Physiol. Molec. Plant Pathol.* 37:235–244.

Donsì, F. and Ferrari, G. 2016. Essential oil nanoemulsions as antimicrobial agents in food. *J. Biotechnol.* 233:106–120.

Duran, M., Aday, M.S., Zorba, N.N.D., Temizkan, R., Buyükcan, M.B. and Caner, C. 2016. Potential of antimicrobial active packaging containing natamycin, nisin, pomegranate and grape seed extract in chitosan coating to extend shelf-life of fresh strawberry. *Food Bioprod. Proc.* 98:354–363.

Dutra, T.V., Castro, J.C., Menezes, J.L., Ramos, T.R., do Prado, I.N., Junior, M.M., Mikcha, J.M.G. and Filho, B.A.A. 2019. Bioactivity of oregano (*Origanum vulgare*) essential oil against *Alicyclobacillus* spp. *Industrial Crops Prod.* 129:345–349.

Ejechi, B.O., Sousey, J.A. and Akpomedaye, D.E. 1998. Microbial stability of mango (*Mangífera indica* L.) juice preserved by combined applications of mild heat and extracts of two tropical spices. *J. Food Prot.* 61:725–727.

Ejike, C.E., Gong, M. and Udenigwe, C.C. 2013. Phytoalexins from the Poaceae: Biosynthesis, function and prospects in food preservation. *Food Res. Intl.* 52:167–177.

Elyemni, M., Louaste, B., Nechad, I., Elkamli, T., Bouia, A., Taleb, M., Chaouch, M. and Eloutassi, N. 2019. Extraction of essential oils of *Rosmarinus officinalis* L. by two different methods: Hydrodistillation and microwave assisted hydrodistillation. *Sci. World J.* 2019:1–6.

European Commission. 2019. EU Lists of Flavourings. https://ec.europa.eu/food/safety/food_improvement_agents/flavourings/eu_lists_flavourings_en (accessed June 17, 2019).

European Commission. 2008a. Regulation (EC) No 1331/2008 of the European Parliament and of the Council of 16 December 2008 establishing a common authorisation procedure for food additives, food enzymes and food flavourings. *Official J. of the European Union* L 354.

European Commission. 2008b. Regulation (EC) No 1334/2008 on flavourings and certain food ingredients with flavouring properties for use in and on foods and amending Council Regulation (EEC) nº 1601/91, Regulations (EC) No 2232/96 and (EC) No 110/2008 and Directive 2000/13/EC. *Official J. of the European Union* L 354.

European Commission. 2009. Commission Regulation (EC) No 450/2009 of 29 May 2009 on active and intelligent materials and articles intended to come into contact with food. *Official J. of the European Union* L 135.

European Commission. 2011. Commission Regulation (EU) nº 10/2011 of 14 January 2011 on plastic materials and articles intended to come into contact with food. *Official J. of the European Union* L 12.

European Union. 2019. EUR-Lex. Access to European Union Law. https://eur-lex.europa.eu/homepage.html (accessed June 17, 2019).

Falcó, I., Flores-Meraz, P.L., Randazzo, W., Sánchez, G., López-Rubio, A. and Fabra, M.J. 2019. Antiviral activity of alginate-oleic acid based coatings incorporating green tea extract on strawberries and raspberries. *Food Hydrocoll.* 87:611–618.

Fang, Z., Lin, D., Warner, R.D. and Ha, M. 2018. Effect of gallic acid/chitosan coating on fresh pork quality in modified atmosphere packaging. *Food Chem.* 260:90–96.

Farag, R.S., Dawm, Z.Y., Hewedi, F.M. and El-Baroty, G.S.A. 1989. Antimicrobial activity of some Egyptian spice essential oils. *J. Food Prot.* 52:665–667.

Farrell, K.T. 1990. *Spices, Condiments and Seasonings*, 2nd ed. New York: Van Nostrand Reinhold.

Felso, P., Horváth, G., Bencsik, T., Godányi, R., Lemberkovics, É., Böszörményi, A., Böddi, K., Takátsy, A., Molnár, P. and Kocsis, B. 2013. Detection of the antibacterial effect of essential oils on outer membrane proteins of *Pseudomonas aeruginosa* by lab-on-a-chip and MALDI-TOF/MS. *Flavour Fragrance J.* 28:367–372.

Fernandes, R.P.P., Trindade, M.A., Lorenzo, J.M., Munekata, P.E.S. and de Melo, M.P. 2016. Effects of oregano extract on oxidative, microbiological and sensory stability of sheep burgers packed in modified atmosphere. *Food Control* 63:65–75.

Fernández-Pan, I., Carrión-Granda, X. and Maté, J.I. 2014. Antimicrobial efficiency of edible coatings on the preservation of chicken breast fillets. *Food Control* 36:69–75.

Fliss, I., Offret, C. and Beaulieu, L. 2015. *Antimicrobial Peptides: The New Generation of Food Additives.* Elsevier.

Frankova, A., Smid, J., Bernardos, A., Finkousova, A., Marsik, P., Novotny, D., Legarová, V., Pulkrabek, J. and Kloucek, P. 2016. The antifungal activity of essential oils in combination with warm air flow against postharvest phytopathogenic fungi in apples. *Food Control* 68:62–68.

Frazão, G.G.S., Blank, A.F. and de Aquino Santana, L.C.L. 2017. Optimisation of edible chitosan coatings formulations incorporating *Myrcia ovata* Cambessedes essential oil with antimicrobial potential against foodborne bacteria and natural microflora of mangaba fruits. *LWT Food Sci. Technol.* 79:1–10.

Freitas, I.R. and Cattelan, M.G. 2018. Antimicrobial and antioxidant properties of essential oils in food systems: An overview. In *Microbial Contamination and Food Degradation*, pp. 443–470. Academic Press, Cambridge, MA.

Friedman, M. 2017. Chemistry, antimicrobial mechanisms, and antibiotic activities of cinnamaldehyde against pathogenic bacteria in animal feeds and human foods. *J. Agric. Food Chem.* 65:10406–10423.

Fu, Y., Sarkar, P., Bhunia, A.K. and Yao, Y. Delivery systems of antimicrobial compounds to food. 2016. *Trends Food Sci. Technol.* 57:165–177.

Gabriel, A.A. and Estilo, E.E.C. 2015. Influences of malic acid and nisin supplementations on the decimal reduction times of *Escherichia coli* O157:H7 in mildly-heated young coconut liquid endosperm. *Food Control* 50:645–651.

Gallego, R., Bueno, M. and Herrero, M. 2019. Sub- and supercritical fluid extraction of bioactive compounds from plants, food-by-products, seaweeds and microalgae: An update. *Trends Anal. Chem.* https://doi.org/110.1016/j.trac.2019.04.030.

García-Díez, J., Alheiro, J., Pinto, A.L., Soares, L., Falco, V., Fraqueza, M.J. and Patarata, L. 2016. Behaviour of food-borne pathogens on dry cured sausage manufactured with herbs and spices essential oils and their sensorial acceptability. *Food Control* 59:262–270.

Garrido-Miranda, K.A., Rivas, B.L., Pérez-Rivera, M.A., Sanfuentes, E.A. and Peña-Farfal, C. 2018. Antioxidant and antifungal effects of eugenol incorporated in bionanocomposites of poly (3-hydroxybutyrate)-thermoplastic starch. *LWT* 98:260–267.

Gómez-Heincke, D., Martínez, I., Partal, P., Guerrero, A. and Gallegos, C. 2016. Development of antimicrobial active packaging materials based on gluten proteins. *J. Sci. Food Agric.* 96:3432–3438.

Gonçalves, N.D., Pena, F. de L., Sartoratto, A., Derlamelina, C., Duarte, M.C.T., Antunes, A.E.C. and Prata, A.S. 2017. Encapsulated thyme (*Thymus vulgaris*) essential oil used as natural preservative in bakery product. *Food Res. Intl.* 96:154–160.

Goni, P., López, P., Sánchez, C., Gómez-Lus, R., Becerril, R. and Nerín, C. 2009. Antimicrobial activity in the vapour phase of a combination of cinnamon and clove essential oils. *Food Chem.* 116:982–989.

González, I., Oliver-Ortega, H., Tarrés, Q., Delgado-Aguilar, M., Mutjé, P. and Andreu, D. 2017. Immobilization of antimicrobial peptides onto cellulose nanopaper. *Intl. J. Biol. Macromolec.* 105:741–748.

González-Lamothe, R., Mitchell, G., Gattuso, M., Diarra, M., Malouin, F. and Bouarab, K. 2009. Plant antimicrobial agents and their effects on plant and human pathogens. *Intl. J. Molec. Sci.* 10:3400–3419.

Gould, G.W. 1995a. Overview. In *New Methods of Food Preservation.* Blackie Academic & Professional, Glasgow.

Gould, G.W. 1995b. The microbe as a high pressure target. In Ledward, D.A., Johnston, D.E., Earnshaw, R.G., Hasting, A.P.M., eds. *High Pressure Processing of Foods.* Nottingham University Press, Nottingham, UK.

Gould, G.W. 1996. Industry perspectives on the use of natural antimicrobials and inhibitors for food applications. *J. Food Prot.* 59:82–86.

Gould, G.W. 2002. Control with naturally occurring antimicrobial systems including bacteriolytic enzymes. In Juneja, V.K., Sofos, J.N., eds. *Control of Foodborne Microorganisms*, pp. 281–302. Marcel Dekker, Inc., New York.

Grande-Tovar, C.D., Chaves-Lopez, C., Serio, A., Rossi, C. and Paparella, A. 2018. Chitosan coatings enriched with essential oils: Effects on fungi involve in fruit decay and mechanisms of action. *Trends Food Sci. Technol.* 78:61–71.

Guest, D.I. 2017. *Phytoalexins, Natural Plant Protection.* Elsevier.

Guil-Guerrero, J.L., Ramos, L., Moreno, C., Zúñiga-Paredes, J.C., Carlosama-Yapez, M. and Ruales, P. 2016. Antimicrobial acivity of plant-food by-products: A review focusing on the tropics. *Livestock Sci.* 189:32–49.

Guinoiseau, E., Luciani, A., Rossi, P.G., Quilichini, Y., Ternengo, S., Bradesi, P. and Berti, L. 2010. Cellular effects induced by *Inula graveolens* and *Santolina corsica* essential oils on *Staphylococcus aureus. Euro. J. Clinical Microbiol. Infect. Dis.* 29:873–879.

Gutiérrez, J., Barry-Ryan, C. and Bourke, P. 2008. The antimicrobial efficacy of plant essential oil combinations and interactions with food ingredients. *Intl. J. Food Microbiol.* 124:91–7.

Gutierrez, J., Barry-Ryan, C. and Bourke, P. 2009. Antimicrobial activity of plant essential oils using food model media: Efficacy, synergistic potential and interactions with food components. *Food Microbiol.* 26:142–150.

Gutierrez-del-Rio, I., Fernández, J. and Lombo, F. 2018. Plant nutraceuticals as antimicrobial agents in food preservation: Terpenoids, polyphenols and thiols. *Intl. J. Antimicrob. Agents* 52:309–315.

Habinshuti, I., Chen, X., Yu, J., Mukeshimana, O., Duhoranimana, E., Karangwa, E., Muhoza, B., Zhang, M., Xia, S. and Zhang, X. 2019. Antimicrobial, antioxidant and sensory properties of Maillard reaction products (MRPs) derived from sunflower, soybean and corn meal hydrolysates. *LWT* 101:694–702.

Hajji, M., Hamdi, M., Sellimi, S., Ksouda, G., Laouer, H., Li, S. and Nasri, M. 2019. Structural characterization, antioxidant and antibacterial activities of a novel polysaccharide from *Periploca laevigata* root barks. *Carbohydr. Polym.* 206:380–388.

Hamedi, H., Kargozari, M., Shotorbani, P.M., Mogadam, N.B. and Fahimdanesh, M. 2017. A novel bioactive edible coating based on sodium alginate and galbanum gum incorporated with essential oil of *Ziziphora persica*: The antioxidant and antimicrobial activity, and application in food model. *Food Hydrocoll.* 72:35–46.

Hammoud, Z., Gharib, R., Fourmentin, S., Elaissari, A. and Greige-Gerges, H. 2019. New findings on the incorporation of essential oil components into liposomes composed of lipoid S100 and cholesterol. *Intl. J. Pharmaceut.* 561:161–170.

Hanani, Z.N., Yee, F.C. and Nor-Khaizura, M.A.R. 2019. Effect of pomegranate (*Punica granatum* L.) peel powder on the antioxidant and antimicrobial properties of fish gelatin films as active packaging. *Food Hydrocoll.* 89:253–259.

Hansen, A.P. and Heinis, J.J. 1991. Decrease of vanillin flavor perception in the presence of casein and whey proteins. *J. Dairy Sci.* 74:2936–2940.

Hao, R., Liu, Y., Sun, L., Xia, L., Jia, H., Li, Q. and Pan, J. 2017. Sodium alginate coating with plant extract affected microbial communities, biogenic amine formation and quality properties of abalone (*Haliotis discus hannai* Ino) during chill storage. *LWT Food Sci. Technol.* 81:1–9.

Haskaraca, G., Juneja, V.K., Mukhopadhyay, S. and Kolsarici, N. 2019. The effects of grapefruit seed extract on the thermal inactivation of *Listeria monocytogenes* in sous-vide processed döner kebabs. *Food Control* 95:71–76.

Hassoun, A. and Çoban, Ö.E. 2017. Essential oils for antimicrobial and antioxidant applications in fish and other seafood products. *Trends Food Sci. Technol.* 68:26–36.

Hcini, K., Sotomayor, J.A., Jordan, M.J. and Bouzid, S. 2013. Chemical composition of the essential oil of rosemary (*Rosmarinus officinalis* L.) of Tunisian origin. *Asian J. Chem.* 25:2601–2603.

He, Q., Gong, B., He, J. and Xiao, K. 2019. A novel superchilling storage-ice glazing (SS-IG) approach using anti-oxidative and antimicrobial essential oil (EO) for freshness-keeping of sea bass (*Dicentrarchus labrax*). *Aquaculture* 500:243–249.

Hedayati Rad, F., Sharifan, A. and Asadi, G. 2018. Physicochemical and antimicrobial properties of kefiran /waterborne polyurethane film incorporated with essential oils on refrigerated ostrich meat. *LWT* 97:794–801.

Heras-Mozos, R., Muriel-Galet, V., López-Carballo, G., Catalá, R., Hernández-Muñoz, P. and Gavara, R. 2019. Development and optimization of antifungal packaging for sliced pan loaf based on garlic as active agent and bread aroma as aroma corrector. *Intl. J. Food Microbiol.* 290:42–48.

Herman, A., Bochenek, J. and Herman, A.P. 2013. Effect of cinnamon and lavender oils on FtsZ gene expression in the *Staphylococcus aureus* ATCC 29213. *Appl. Biochem. Microbiol.* 49:481–484.

Hernández-Hernández, E., Castillo-Hernández, G., González-Gutiérrez, C.J., Silva-Dávila, A.J., Gracida-Rodríguez, J.N., García-Almendárez, B.E., Di Pierro, P., Vázquez-Landaverde, P. and Regalado-González, C. 2019. Microbiological and physicochemical properties of meat coated with microencapsulated Mexican oregano (*Lippia graveolens* kunth) and basil (*Ocimum basilicum* l.) essential oils mixture. *Coatings* 9:414.

Hontanaya, C. Meca, G., Luciano, F.B., Mañes, J. and Font, G. 2015. Inhibition of aflatoxin B1, B2, G1 and G2 production by *Aspergillus parasiticus* in nuts using yellow and oriental mustard flours. *Food Control* 47:154–160.

Huang, D.F., Xu, J.-G., Liu, J.-X., Zhang, H. and Hu, Q.P. 2014. Chemical constituents, antibacterial activity and mechanism of action of the essential oil from *Cinnamomum cassia* bark against four food-related bacteria. *Microbiology* 83:357–365.

Huang, Z., Liu, X., Jia, S. and Luo, Y. 2017. Antimicrobial effects of cinnamon bark oil on microbial composition and quality of grass carp (*Ctenopharyngodon idellus*) fillets during chilled storage. *Food Control* 82:316–324.

Hyldgaard, M., Mygind, T. and Meyer, R.L. 2012. Essential oils in food preservation: Mode of action, synergies, and interactions with food matrix components. *Frontiers Microbiol.* 3:1–24.

Hyun, J.-E., Bae, Y.-M. Yoon, J.-H. and Lee, S.-Y. 2015. Preservative effectiveness of essential oils in vapor phase combined with modified atmosphere packaging against spoilage bacteria on fresh cabbage. *Food Control* 51:307–313.

Irkin, R. and Esmer, O.K. 2015. Novel Food packaging systems with natural antimicrobial agents. *J. Food Sci. Technol.* 52:6095–6111.

Jabeen, U. and Khanum, A. 2017. Isolation and characterization of potential food preservative peptide from *Momordica charantia* L. *Arabian J. of Chem.* 10:S3982–S3989.

Jafri, H., Ansari, F.A. and Ahmad, I. 2019. Prospects of essential oils in controlling pathogenic biofilm. In *New Look to Phytomedicine*, pp. 203–236. Academic Press, Cambridge, MA.

Jemaa, M.B., Falleh, H., Neves, M.A., Isoda, H., Nakajima, M. and Ksouri, R. 2017. Quality preservation of deliberately contaminated milk using thyme free and nanoemulsified essential oils. *Food Chem.* 217:726–734.

Ji, H., Kim, H., Beuchat, L.R. and Ryu, J.H. 2019. Synergistic antimicrobial activities of essential oil vapours against *Penicillium corylophilum* on a laboratory medium and bbef jerky. *Intl. J. Food Microbiol.* 291:104–110.

Jiang, T., Feng, L. and Zheng, X. 2012. Effect of chitosan coating enriched with thyme oil on postharvest quality and shelf life of shiitake mushroom (*Lentinus edodes*). *J. Agric. Food Chem.* 60:188–196.

Jo, Y.-J., Chun, J.-Y., Kwon, Y.-J., Min, S.-G., Hong, G.-P. and Choi, M.-J. 2015. Physical and antimicrobial properties of trans-cinnamaldehyde nanoemulsions in water melon juice. *LWT Food Sci. Technol.* 60(1):444–451.

Juneja, V.K., Cadavez, V., Gonzales-Barron, U., Mukhopadhyay, S. and Friedman, M. 2016. Effect of pomegranate powder on the heat inactivation of *Escherichia coli* O104:H4 in ground chicken. *Food Control* 70:26–34.

Juven, B., Henis, Y. and Jacoby, B. 1972. Studies on the mechanism of the antimicrobial action of oleuropein. *J. Appl. Bacteriol.* 35:559–567.

Juven, B.J., Kanner, J., Schved, F. and Weisslowicz, H. 1994. Factors that interact with the antibacterial action of thyme essential oil and its active constituents. *J. Appl. Bacteriol.* 76:626–631.

Kabara, J.J. and Eklund, T. 1991. Organic acids and esters. In Russel, N.J. and Gould, G.W., eds. *Food Preservatives*, pp. 44–71. Blackie & Son Ltd., Glasgow.

Kakaei, S. and Shahbazi, Y. 2016. Effect of chitosan-gelatin film incorporated with ethanolic red grape seed extract and Ziziphora clinopodioides essential oil on survival of *Listeria monocytogenes* and chemical, microbial and sensory properties of minced trout fillet. *LWT Food Sci. Technol.* 72:432–438.

Kalagatur, N.K., Mudili, V., Kamasani, J.R. and Siddaiah, C. 2018. Discrete and combined effects of Ylang-Ylang (*Cananga odorata*) essential oil and gamma irradiation on growth and mycotoxins production by Fusarium graminearum in maize. *Food Control* 94:276–283.

Kang, J.-H., Park, S.-J., Park, J.-B. and Song, K.B. 2019. Surfactant type affects the washing effect of cinnamon leaf essential oil emulsion on kale leaves. *Food Chem.* 271:122–128.

Kapetanakou, A.E., Nestora, S., Evageliou, V. and Skandamis, P.N. 2018. Sodium alginate–cinnamon essential oil coated apples and pears: Variability of *Aspergillus carbonarius* growth and ochratoxin A production. *Food Res. Intl.* 119:876–885.

Karam, L., Roustom, R., Abiad, M.G., El-Obeid, T. and Savvaidis, I.N. 2019. Combined effects of thymol, carvacrol and packaging on the shelf-life of marinated chicken. *Intl. J. Food Microbiol.* 291:42–47.

Katayama, T. and Nagai, I. 1960. Chemical significance of the volatile components of spices in the food preservative view point. VI: Structure and antibacterial activity of terpenes. *Bull. Japan Soc. Sci. Fisheries* 26:29–32.

Kayode, R.M., Azubuike, C.U., Laba, S.A., Dauda, A.O., Balogun, M.A. and Ajala, S.A. 2018. Chemical composition and anti-microbial activities of the essential oil of *Adansonia digitata* stem-bark and leaf on post-harvest control of tomato spoilage. *LWT* 93:58–63.

Ke, J., Xiao, L., Yu, G., Wu, H., Shen, G. and Zhang, Z. 2019. The study of diffusion kinetics of cinnamaldehyde from corn starch-based film into food simulant and physical properties of antibacterial polymer film. *Intl. J. Biol. Macromolec.* 125:642–650.

Khaleque, M.A., Keya, C.A., Hasan, K.N., Hoque, M.M., Inatsu, Y. and Bari, M.L. 2016. Use of cloves and cinnamon essential oil to inactivate *Listeria monocytogenes* in ground beef at freezing and refrigeration temperatures. *LWT* 74:219–223.

Khaneghah, A.M., Hashemi, S.M.B, Es, I., Fracassetti, D. and Limbo, S. 2018a. Efficacy of antimicrobial agents for food contact applications: Biological activity, incorporation into packaging, and assessment methods: A review. *J. Food Prot.* 81:1142–1156.

Khaneghah, A.M., Hashemi, S.M.B, and Limbo, S. 2018b. Antimicrobial agents and packaging systems in antimicrobial active food packaging: An overview of approaches and interactions. *Food Bioprod. Proc.* 3:1–19.

Kim, I.H., Oh, Y.A., Lee, H., Song, K.B. and Min, S.C. 2014. Grape berry coatings of lemongrass oil-incorporating nanoemulsion. *LWT Food Sci. Technol.* 58:1–10.

Kim, S.A. and Rhee, M.S. 2016. Highly enhanced bactericidal effects of medium chain fatty acids (caprylic, capric, and lauric acid) combined with edible plant essential oils (carvacrol, eugenol, β-resorcylic acid, trans-cinnamaldehyde, thymol, and vanillin) against *Escherichia coli* O157:H7. *Food Control.* 60:447–454.

Kim, S.-S. and Kang, D.-H. 2017. Synergistic effect of carvacrol and ohmic heating for inactivation of *E. coli* O157:H7, *S.* Typhimurium, L. *monocytogenes*, and MS-2 bacteriophage in salsa. *Food Control* 73:300–305.

Kloucek, P., Smid, J., Frankova, A., Kokoska, L., Valterova, I. and Pavela, R. 2012. Fast screening method for assessment of antimicrobial activity of essential oils in vapor phase. *Food Res. Intl.* 47:161–165.

Kong, J., Xie, Y.-F., Guo, Y.-H., Cheng, Y.-L., Qian, H. and Yao, W.-R. 2016. Biocontrol of postharvest fungal decay of tomatoes with a combination of thymol and salicylic acid screening from 11 natural agents. *LWT Food Sci. Technol.* 72:215–222.

Kong, Q., Zhang, L., An, P., Qi, J., Yu, X., Lu, J. and Ren, X. 2019. Antifungal mechanisms of α-terpineol and terpene-4-alcohol as the critical components of *Melaleuca alternifolia* oil in the inhibition of rot disease caused by *Aspergillus ochraceus* in postharvest grapes. *J. Appl. Microbiol.* 126:1161–1174.

Kramer, B., Wunderlich, J. and Muranyi, P. 2018. Impact of volatile allyl isothiacianate on fresh produce. *Food Packag. Shelf Life* 16:220–224.

Kuc, J. and Rush, J.S. 1985. Phytoalexins. *Arch. Biochem. Biophys.* 236:455–472.

Langroodi, A.M., Tajik, H., Mehdizadeh, T., Moradi, M., Kia, E.M. and Mahmoudian, A. 2018. Effects of sumac extract dipping and chitosan coating enriched with *Zataria multiflora* Boiss oil on the shelf-life of meat in modified atmosphere packaging. *LWT* 98:372–380.

Lauriano de Souza, V.G., Pires, J.R.A., Vieira, E.T., Coelhoso, I.M., Duarte, M.P. and Fernando, A.L. 2019. Activity of chitosan-montmorillonite bionanocomposites incorporated with rosemary essential oil: From in vitro assays to application in fresh poultry meat. *Food Hydrocoll.* 89:241–252.

Lee, G., Kim, Y., Kim, H., Beuchat, L.R. and Ryu, J.H. 2018. Antimicrobial activities of gaseous essential oils against *Listeria monocytogenes* on a laboratory medium and radish sprouts. *Intl. J. Food Microbiol.* 265:49–54.

Lee, K.H., Lee, J.-S., Kim, E.S. and Lee, H.G. 2019. Preparation, characterization, and food application of rosemary extract-loaded antimicrobial nanoparticle dispersions. *LWT* 101:138–144.

Leistner, L. 2000. Hurdle technology in the design of minimally processed foods. In Alzamora, S., Tapia, M.S. and López-Malo, A., eds. *Minimally Processed Fruits and Vegetables. Fundamentals Aspects and Applications*, pp. 13–28. Aspen Publishers, Inc., Gaithersburg.

Li, C.M. and Yu, J.P. 2015. Chemical composition, antimicrobial activity and mechanism of action of essential oil from the leaves of *Macleaya cordata* (Willd.) R. *Brit. J. Food Safety* 35:227–236.

Li, G., Wang, X., Xu, Y., Zhang, B. and Xia, X. 2014. Antimicrobial effect and mode of action of chlorogenic acid on *Staphylococcus aureus. Euro. Food Res. Technol.* 238:589–596.

Li, K., Guan, G., Zhu, J., Wu, H. and Sun, Q. 2019. Antibacterial activity and mechanism of a laccase-catalyzed chitosan–gallic acid derivative against *Escherichia coli* and *Staphylococcus aureus. Food Control.* 96:234–243.

Li, Y., Kong, W., Li, M., Liu, H., Zhao, X., Yang, S. and Yang, M. 2016. *Litsea cubeba* essential oil as the potential natural fumigant: Inhibition of *Aspergillus flavus* and AFB1 production in licorice. *Industrial Crops Prod.* 80:186–193.

Lianou, A., Moschonas, G., Nychas, G.-J.E. and Panagou, E.Z. 2018. Growth of *Listeria monocytogenes* in pasteurized vanilla cream pudding as affected by storage temperature and the presence of cinnamon extract. *Food Res. Intl.* 106:1114–1122.

Lipkin, A., Anisimova, V., Nikonorova, A., Babakov, A., Krause, E., Bienert, M., Grishin, E. and Egorov, T. 2005. An antimicrobial peptide Ar-AMP from amaranth (Amaranthus retroflexus L.) seeds. *Phytochem.* 66:2426–2431.

Liu, M., Yang, K., Wang, J., Zhang, J., Qi, Y., Wei, X. and Fan, M. 2019. Young astringent persimmon tannin inhibits methicillin-resistant Staphylococcus aureus isolated from pork. *LWT* 100:48–55.

Llana-Ruiz-Cabello, M., Maisanaba, S., Puerto, M., Prieto, A.I., Pichardo, S., Jos, Á. and A.M. Cameán. 2014. Evaluation of the mutagenicity and genotoxic potential of carvacrol and thymol using the Ames *Salmonella* test and alkaline, Endo III- and FPG-modified comet assays with the human cell line Caco-2. *Food Chem. Toxicol.* 72:122–128.

Llorens, E., del Valle, L.J. and Puiggalí, J. 2015. Electrospun scaffolds of polylactide with a different enantiomeric content and loaded with anti-inflammatory and antibacterial drugs. *Macromolec. Res.* 23:636–648.

Lombardo, S., Pandino, G. and Mauromicales, G. 2015. The nutraceutical response of two globe artichoke cultivars to contrasting NPK fertilizer regimes. *Food Res. Intl.* 76:852–859.

López-Malo, A., Alzamora, S.M. and Argaiz, A. 1995. Effect of natural vanillin on germination time and radial growth of moulds in fruit-based agar systems. *Food Microbiol.* 12:213–219.

López-Malo, A., Alzamora, S.M. and Argaiz, A. 1997. Effect of vanillin concentration, pH and incubation temperature on *Aspergillus flavus, A. niger, A. ochraceus* and *A. parasiticus* growth. *Food Microbiol.* 14:117–124.

López-Malo, A., Alzamora, S.M. and Argaiz, A. 1998. Vanillin and pH synergistic effects on mold growth. *J. Food Sci.* 63:143–146.

López-Malo, A., Alzamora, S.M. and Guerrero, S. 2000. Natural antimicrobials from plants. In Alzamora S.M., Tapia, M.S. and López-Malo, A., eds. *Minimally Processed Fruits and Vegetables: Fundamentals Aspects and Applications*, pp. 237–264. Aspen Publishers, Inc., Gaithersburg.

López-Malo, A. and Argaiz, A. 1999. Citral and pH synergistic effects on mold growth. Paper N° 37D-15, presented at IFT Annual Meeting. Chicago, July 24–28, 1999.

López-Romero, J.C., Valenzuela-Melendres, M., Juneja, V.K., García-Dávila, J., Camou, J.P., Peña-Ramos, A. and González-Ríos, H. 2018. Effects and interactions of gallic acid, eugenol and temperature on thermal inactivation of *Salmonella* spp. in ground chicken. *Food Res. Intl.* 103:289–294.

Lorenzo-Leal, A.C., Palou, E. and López-Malo, A. 2019a. Evaluation of the efficiency of allspice, thyme and rosemary essential oils on two foodborne pathogens in in-vitro and on alfalfa seeds, and their effect on sensory characteristics of the sprouts. *Intl. J. Food Microbiol.* 295:19–24.

Lorenzo-Leal, A.C., Palou, E., López-Malo, A. and Bach, H. 2019b. Antimicrobial, cytotoxic, and anti-inflammatory activities of *Pimenta dioica* and *Rosmarinus officinalis* essential oils. *BioMed Res. Intl.*, 1–8. doi:10.1155/2019/1639726.

Loziene, K., Svediene, J., Paskevicius, A., Raudoniene, V., Sytar, O. and Kosyan, A. 2018. Influence of plant origin natural α-pinene with different enantiomeric composition on bacteria, yeasts and fungi. *Fitoterapia* 127:20–24.

Luis, J.G., Lahlou, E.H. andrés, L.S., Echeverri, F. and Fletcher, W.Q. 1997. Phenylphenalenonic phytoanticipins: New acenaphtylene and dimeric phenylphenalenones from the resistant Musa selected hybrid SH-3481. *Tetrahedron* 53:8249–8256.

Lyu, F., Gao, F., Wei, Q. and Liu, L. 2017. Changes of membrane fatty acids and proteins of *Shewanella putrefaciens* treated with cinnamon oil and gamma irradiation. *Biores. Bioproc.* 4:1–10.

Ma, W., Zhao, L. and Xie, Y. 2017. Inhibitory effect of (E)-2-hexenal as a potential natural fumigant on *Aspergillus flavus* in stored peanut seeds. *Industrial Crops Prod.* 107:206–210.

Ma, W., Zhao, L., Zhao, W. and Xie, Y. 2019. (E)-2-Hexenal, as a potential natural antifungal compound, inhibits *Aspergillus flavus* spore germination by disrupting mitochondrial energy metabolism. *J. Agric. Food Chem.* 67(4):1138–1145.

Mahmoud, A.L.E. 1994. Antifungal action and antiaflatoxigenic properties of some essential oil constituents. *Lett. Appl. Microbiol.* 19:110–113.

Malhotra, B., Keshwani, A. and Kharkwal, H. 2015. Antimicrobial food packaging: Potential and pitfalls. *Front. Microbiol.* 6:611. doi: 10.3389/fmicb.2015.00611

Mani-López, E., Palou, E. and López-Malo, A. 2018. Biopreservatives as agents to prevent food spoilage. In *Microbial Contamination and Food Degradation*, pp. 235–270. Academic Press, Cambridge, MA.

Mantilla, N., Castell-Perez, M.E., Gomes, C. and Moreira, R.G. 2013. Multilayered antimicrobial edible coating and its effect on quality and shelf-life of fresh-cut pineapple (*Ananas comosus*). *LWT Food Sci. Technol.* 51:37–43.

Mantis, A.J., Karaioannoglou, P.G., Spanos, G.P. and Panetsos, A.G. 1978. The effect of garlic extract on food poisoning bacteria in culture media. I: *Staphylococcus aureus. Lebensm. Wiss. Technol.* 11:26–31.

Mari, M., Bautista-Baños, S. and Sivakumar, D. 2016. Decay control in the postharvest system: Role of microbial and plant volatile organic compounds. *Postharvest Biol. Technol.* 122:70–81.

Marinelli, L., Di Stefano, A. and Cacciatore, I. 2018. Carvacrol and its derivatives as antibacterial agents. *Phytochem. Rev.* 17:903–921.

Marth, E.H. 1966. Antibiotics in foods--naturally occurring, developed, and added. *Residue Rev.* 12:65-161.

Martínez-Hernández, G.B., Amodio, M.L. and Colelli, G. 2017. Carvacrol-loaded chitosan nanoparticles maintain quality of fresh-cut carrots. *Innov. Food Sci. Emerg. Technol.* 41:56–63.

Mazarei, Z. and Rafati, H. 2019. Nanoemulsification of *Satureja khuzestanica* essential oil and pure carvacrol; comparison of physicochemical properties and antimicrobial activity against food pathogens. *LWT* 100:328–334.

McClements, D.J. 2012. Nanoemulsions versus microemulsions: Terminology, differences, and similarities. *Soft Matter* 8:1719–1729.

Millan-Sango, D., Garroni, E., Farrugia, C., Van Impe, J.F.M. and Valdramidis, V.P. 2016. Determination of the efficacy of ultrasound combined with essential oils on the decontamination of Salmonella inoculated lettuce leaves. *LWT* 73:80–87.

Millan-Sango, D., McElhatton, A. and Valdramidis, V.P. 2015. Determination of the efficacy of ultrasound in combination with essential oil of oregano for the decontamination of *Escherichia coli* on inoculated lettuce leaves. *Food Res. Intl.* 67:145–154.

Min, K.J., Kwon, K.J. and Yoon, K.S. 2010. Effect of various antimicrobials on the growth kinetics of foodborne pathogens in ready-to-eat pyeonyuk (cooked and pressed pork). *Food Sci.Biotechnol.* 19:99–106.

Mirón-Mérida, V.A., Yáñez-Fernández, J., Montañez-Barragán, B. and Huerta, B.E.B. 2019. Valorization of coffee parchment waste (*Coffea arabica*) as a source of caffeine and phenolic compounds in antifungal gellan gum films. *LWT* 101:167–174.

Montgomery, M.W. and Day, E.A. 1965. Aldehyde-amine condensation reaction: A possible fate of carbonyls in foods. *J. Food Sci.* 31:829–832.

Monu, E.A., Techathuvanan, C., Wallis, A., Critzer, F.J. and Davidson, P.M. 2016. Plant essential oils and components on growth of spoilage yeasts in microbiological media and a model salad dressing. *Food Control* 65:73–77.

Morais, T.P., Zaini, P.A., Chakraborty, S., Gouran, H., Carvalho, C.P., Almeida-Souza, H.O. and Nascimento, R. 2019. The plant-based chimeric antimicrobial protein SlP14a-PPC20 protects tomato against bacterial wilt disease caused by *Ralstonia solanacearum*. *Plant Sci.* 280:197–205.

Moreira, S.P., de Carvalho, W.M., Alexandrino, A.C., de Paula, H.C.B., Rodrigues, M.C.P, de Figueiredo, R.W., Maia, G.A., de Figueiredo, E.M.A.T. and Brasil, I.M. 2014. Freshness retention of minimally processed melon using different packages and multilayered edible coating containing microencapsulated essential oil. *Intl. J. Food Sci. Technol.* 49:2192–2203.

Mostafa, A.A., Al-Askar, A.A., Almaary, K.S., Dawoud, T.M., Sholkamy, E.N. and Bakri, M.M. 2018. Antimicrobial activity of some plant extracts against bacterial strains causing food poisoning diseases. *Saudi J. Biol. Sci.* 25:361–366.

Mukherjee, P.K., Harwansh, R.K., Bahadur, S., Duraipandiyan, V. and Al-Dhabi, N.A. 2017. Factors to consider in development of nutraceuticals and dietary supplements. In S. Badal, and Delgoda, R., eds. *Pharmacognosy: Fundamental Applications and Strategies*, pp. 653–661. Academic Press, London, UK.

Muthaiyan, A., Martin, E.M., Natesan, S., Crandall, P.G., Wilkinson, B.J. and Ricke, S.C. 2012. Antimicrobial effect and mode of action of terpeneless cold-pressed Valencia orange essential oil on methicillin-resistant *Staphylococcus aureus*. *J. Appl. Microbiol.* 112:1020–1033.

Myszka, K., Schmidt, M.T., Majcher, M., Juzwa, W. and Czaczyk, K. 2017. β-Caryophyllene-rich pepper essential oils suppress spoilage activity of *Pseudomonas fluorescens* KM06 in fresh-cut lettuce. *LWT Food Sci. Technol.* 83:118–126.

Nazareth, T.M., Bordin, K., Manyes, L., Meca, G., Mañes, J. and Luciano, F.B. 2016. Gaseous allyl isothiocyanate to inhibit the production of aflatoxins, beauvericin and enniatins by *Aspergillus parasiticus* and *Fusarium poae* in wheat flour. *Food Control* 62:317–321.

Ncama, K., Mditshwa, A., Tesfay, S.Z., Mbili, N.C. and Magwaza, L.S. 2019. Topical procedures adopted in testing and application of plant-based extracts as bio-fungicides in controlling postharvest decay of fresh produce. *Crop Prot.* 115:142–151.

Nedorostova, L., Kloucek, P., Kokoska, L., Stolcova, M. and Pulkrabek, J. 2009. Antimicrobial properties of selected essential oils in vapour phase against foodborne bacteria. *Food Control* 20:157–160.

Netopilova, M., Houdkova, M., Rondevaldova, J., Kmet, V. and Kokoska, L. 2018. Evaluation of in vitro growth-inhibitory effect of carvacrol and thymol combination against *Staphylococcus aureus* in liquid and vapour phase using new broth volatilization chequerboard method. *Fitoterapia.* 129:185–190.

Nielsen, C.K., Kjems, J., Mygind, T., Snabe, T., Schwarz, K., Serfert, Y. and Meyer, R.L. 2017. Antimicrobial effect of emulsion-encapsulated isoeugenol against biofilms of food pathogens and spoilage bacteria. *Intl. J. Food Microbiol.* 242:7–12.

Nikmaram, N., Budaraju, S., Barba, F.J., Lorenzo, J.M., Cox, R.B., Mallikarjunan, K. and Roohinejad, S. 2018. Application of plant extracts to improve the shelf-life, nutritional and health-related properties of ready-to-eat meat products. *Meat Sci.* 145:245–255.

Nisar, T., Yang, X., Alim, A., Iqbal, M., Wang, Z.-C. and Guo, Y. 2019. Physicochemical responses and microbiological changes of bream (*Megalobrama ambycephala*) to pectin based coatings enriched with clove essential oil during refrigeration. *Intl. J. Biol. Macromolec.* 124:1156–1166.

Nordstrom, R. and Malmsten, M. 2017. Delivery systems for antimicrobial peptides. *Adv. Colloid Interface Sci.* 242:17–34.

Nychas, G.J.E. 1995. Natural antimicrobials from plants. In Gould, G.W., ed. *New Methods of Food Preservation*, pp. 58–89. Blackie Academic and Professional, Glasgow.

Ofman, M.H., Campos, C.A. and Gerschenson, L.N. 2004. Effect of preservatives on the functional properties of tapioca starch: Analysis of interactions. *LWT Food Sci. Technol.* 37:355–361.

Oh, Y.A., Oh, Y.J., Song, A.Y., Won, J.S., Song, K.B. and Min, S.C. 2017. Comparison of effectiveness of edible coatings using emulsions containing lemongrass oil of different size droplets on grape berry safety and preservation. *LWT* 75:742–750.

Oliveira, J., Gloria, E.M., Parisi, M.C.M., Baggio, J.S., Silva, P.P.M., Ambrosio, C.M.S. and Spoto, M.H.F. 2019. Antifungal activity of essential oils associated with carboxymethylcellulose against *Colletotrichum acutatum* in strawberries. *Sci. Hortic.* 243:261–267.

Pabast, M., Shariatifar, N., Beikzadeh, S. and Jahed, G. 2018. Effects of chitosan coatings incorporating with free or nano-encapsulated *Satureja* plant essential oil on quality characteristics of lamb meat. *Food Control* 91:185–192.

Parish, M.E. and Davidson, P.M. 1993. Methods of evaluation. In Davidson, P.M., Branen, A.L., eds. *Antimicrobials in Foods*, pp. 597–615. Marcel Dekker, Inc., New York.

Park, J.-B., Kang, J.-H. and Song, K.B. 2018. Combined treatment of cinnamon bark oil emulsion washing and ultraviolet-C irradiation improves microbial safety of fresh-cut red chard. *LWT* 93:109–115.

Park, J.B., Kang, J.H. and Song, K.B. 2019. Clove bud essential oil emulsion containing benzethonium chloride inactivates *Salmonella* Typhimurium and *Listeria monocytogenes* on fresh-cut pak choi during modified atmosphere storage. *Food Control.* 100:17–23.

Paster, N., Juven, B.J., Shaaya, E., Menasherov, M., Nitzan, R., Weisslowicz, H. and Ravid, U. 1990. Inhibitory effect of oregano and thyme essential oils on moulds and foodborne bacteria. *Lett Appl Microbiol.* 11:33–37.

Patra, J.K., Das, G. and Baek, K.H. 2015. Antibacterial mechanism of the action of *Enteromorpha linza* L. essential oil against *Escherichia coli* and *Salmonella* Typhimurium. *Bot. Stud.* 56:1–9.

Pavoni, L., Maggi, F., Mancianti, F., Nardoni, S., Ebani, V.V., Cespi, M., Bonacucina, G. and Palmieri, G.F. 2019. Microemulsions: An effective encapsulation tool to enhance the antimicrobuial activity of selected EOs. *J. Drug Delivery Sci. Technol.* 53. doi:https://doi.org/10.1016/j.jddst.2019.05.050.

Paz, J.E.W., Contreras, C.R., Munguía, A.R., Aguilar, C.N. and Inungaray, M.L.C. 2018. Phenolic content and antibacterial activity of extracts of *Hamelia patens* obtained by different extraction methods. *Brazilian J. Microbiol.* 49:656–661.

Pedras, M.S.C and Yaya, E.E. 2015. Plant chemical defenses: are all constitutive antimicrobial metabolites phytoanticipins? *Nat. Prod. Commun.* 10(1):1934578X1501000142.

Pelissari, F.M., Grossmann, M.V., Yamashita, F. and Pineda, E.A. 2009. Antimicrobial, mechanical, and barrier properties of cassava starch chitosan films incorporated with oregano essential oil. *J. Agric. Food Chem.* 57:7499–7504.

Pellegrini, M.C., Alonso-Salces, R.M., Umpierrez, M.L., Rossini, C. and Fuselli, S.R. 2017. Chemical composition, antimicrobial activity, and mode of action of essential oils against Paenibacillus larvae, etiological agent of American foulbrood on *Apis mellifera*. Chem. and Biodiver. 14:1–18. https://onlinelibrary.wiley.com/doi/epdf/10.1002/cbdv.201600382.

Perina, F.J., Amaral, D.C., Fernandes, R.S., Labory, C.R.G., Teixeira, G.A. and Alves, E. 2015. *Thymus vulgaris* essential oil and thymol against *Alternaria alternata* (Fr.) Keissler: Effects on growth, viability, early infection and cellular mode of action. *Pest Management Sci.* 71:1371–1378.

Perricone, M., Arace, E., Corbo, M.R., Sinigaglia, M. and Bevilacqua, A. 2015. Bioactivity of essential oils: A review on their interaction with food components. *Frontiers Microbiol.* 6:1–7.

Pesavento, G., Calonico, C., Bilia, A.R., Barnabei, M., Calesini, F., Addona, R., Mencarelli, L., Carmagnini, L., Di Martino, M.C. and Lo Nostro, A. 2015. Antibacterial activity of Oregano, Rosmarinus and Thymus essential oils against *Staphylococcus aureus* and *Listeria monocytogenes* in beef meatballs. *Food Control* 54:188–199.

Phillips, C.A., Laird, K. and Allen, S.C. 2012. The use of Citri-V™®—An antimicrobial citrus essential oil vapour for the control of *Penicillium chrysogenum*, *Aspergillus niger* and *Alternaria alternata* in vitro and on food. *Food Res. Intl.* 47:310–314.

Piva, G., Fracassetti, D., Tirelli, A., Mascheroni, E., Musatti, A., Inglese, P., Piergiovanni, L. and Rollini, M. 2017. Evaluation of the antioxidant/antimicrobial performance of *Posidonia oceanica* in comparison with three commercial natural extracts and as a treatment on fresh-cut peaches (*Prunus persica* Batsch). *Postharvest Biol. Technol.* 124:54–61.

Poverenov, E., Zaitsev, Y., Arnon, H., Granit, R., Alkalai-Tuvia, S., Perzelan, Y., Weinberg, T. and Fallik, E. 2014. Effects of a composite chitosan–gelatin edible coating on postharvest quality and storability of red bell peppers. *Postharvest Biol. Technol.* 96:106–109.

Prakash, A., Baskaran, R., Paramasivam, N. and Vadivel, V. 2018a. Essential oil based nanoemulsions to improve the microbial quality of minimally processed fruits and vegetables: A review. *Food Res. Intl.* 111:509–523.

Prakash, A., Kujur, A., Yadav, A., Kumar, A., Singh, P.P. and Dubey, N.K. 2018b. Nanoencapsulation: An efficient technology to boost the antimicrobial potential of plant essential oils in food system. *Food Control* 89:1–11.

Procopio, F.R., Oriani, V.B., Paulino, B.N., do Prado-Silva, L., Pastore, G.M., Sant'Ana, A.S. and Hubinger, M.D. 2018. Solid lipid microparticles loaded with cinnamon oleoresin: Characterization, stability and antimicrobial activity. *Food Res. Intl.* 113:351–361.

Putnik, P., Gabrić, D., Roohinejad, S., Barba, F.J., Granato, D., Mallikarjunan, K., Lorenzo, J.M. and Kovačević, D.B. 2018. An overview of organosulfur compounds from *Allium* spp.: From processing and preservation to evaluation of their bioavailability, antimicrobial, and anti-inflammatory properties. *Food Chem.* 15:680–691.

Quesada, J., Sendra, E., Navarro, C. and Sayas-Barberá, E. 2016. Antimicrobial active packaging including chitosan films with *Thymus vulgaris* L. essential oil for ready-to-eat meat. *Foods* 5:57.

Radunz, M., da Trindade, M.L.M., Camargo, T.M., Radünz, A.L., Borges, C.D., Gandra, E.A. and Helbig, E. 2019. Antimicrobial and antioxidant activity of unencapsulated and encapsulated clove (*Syzygium aromaticum*, L.) essential oil. *Food Chem.* 276:180–186.

Raeisi, M., Tajik, H., Aliakbarlu, J., Mirhosseini, S.H. and Hosseini, S.M.H. 2015. Effect of carboxymethyl cellulose-based coatings incorporated with *Zataria multiflora* Boiss. essential oil and grape seed extract on the shelf life of rainbow trout fillets. *LWT Food Sci. Technol.* 64:898–904.

Raeisi, S., Sharifi-Rad, M., Quek, S.Y., Shabanpour, B. and Sharifi-Rad, J. 2016. Evaluation of antioxidant and antimicrobial effects of shallot (*Allium ascalonicum* L.) fruit and ajwain (*Trachyspermum ammi* (L.) Sprague) seed extracts in semi-fried coated rainbow trout (*Oncorhynchus mykiss*) fillets for shelf-life extension. *LWT Food Sci. Technol.* 65:112–121.

Ren, X., An, P., Zhai, X., Wang, S. and Kong, Q. 2019. The antibacterial mechanism of pterostilbene derived from xinjiang wine grape: A novel apoptosis inducer in *Staphyloccocus aureus* and *Escherichia coli. LWT* 101:100–106.

Requena, R., Vargas, M. and Chiralt, A. 2019. Eugenol and carvacrol migration from PHBV films and antibacterial action in different food matrices. *Food Chem.* 277:38–45.

Resnik, S.L., Pacin, A.M., Alvarez, G. and Alzamora, S.M. 1996. Effect of vanillin concentration on aflatoxins accumulation. Presented at the IX International Symposium on Mycotoxins and Phycotoxins, May 27-31, 1996, Rome, Italy.

Reyes-Jurado, F., Cervantes-Rincón, T., Bach, H. and López-Malo, A. 2019a. Antimicrobial activity of Mexican oregano (*Lippia berlandieri*), thyme (*Thymus vulgaris*) and mustard (*Brassica nigra*) essential oils in gaseous phase. *Industrial Crops Prod.* 131:90–95.

Reyes-Jurado, F., Navarro-Cruz, A.R., Ochoa-Velasco, C.E., Palou, E., López-Malo, A. and Ávila-Sosa, R. 2019b. Essential oils in vapor phase as alternative antimicrobials: A review. *Crit. Rev. Food Sci. Nutri.* 18:1–10.

Ribes, S., Fuentes, A., Talens, P. and Barat, J.M. 2016. Use of oil-in-water emulsions to control fungal deterioration of strawberry jams. *Food Chem.* 211:92–99.

Ribes, S., Fuentes, A., Talens, P. and Barat, J.M. 2017. Application of cinnamon bark emulsions to protect strawberry jam from fungi. *LWT Food Sci. Technol.* 78:265–272.

Ribes, S., Fuentes, A., Talens, P. and Barat, J.M. 2018. Combination of different antifungal agents in oil-in-water emulsions to control strawberry jam spoilage. *Food Chem.* 239:704–711.

Robledo, N., Vera, P., López, L., Yazdani-Pedram, M., Tapia, C. and Abugoch, L. 2018. Thymol nanoemulsions incorporated in quinoa protein/chitosan edible films; antifungal effect in cherry tomatoes. *Food Chem.* 246:211–219.

Rodrigues, J.B. dos S., de Carvalho, R.J., de Souza, N.T., Oliveira, K. de S., Franco, O.L., Schaffner, D., de Souza, E.L. and Magnani, M. 2017. Effects of oregano essential oil and carvacrol on biofilms of *Staphylococcus aureus* from food-contact surfaces. *Food Control* 73:1237–1246.

Rouger, A., Tresse, O. and Zagorec, M. 2017. Bacterial contaminants of poultry meat: Sources, species and dynamics. *Microorganisms* 5:1–6.

Ruiz-Barba, J.L., Rios-Sanchez, R.M., Fedriani-Triso, C., Olias, J.M., Rios, J.L. and Jimenez-Diaz, R. 1990. Bactericidal effect of phenolic compounds from green olives against *L. plantarum* system. *Appl Microbiol.* 13:199–205.

Saberi, B., Chockchaisawasdee, S., Golding, J.B., Scarlett, C.J. and Stathopoulos, C.E. 2017. Characterization of pea starch-guar gum biocomposite edible films enriched by natural antimicrobial agents for active food packaging. *Food Bioprod. Proc.* 105:51–63.

Saladino, F., Quiles, J.M., Luciano, F.B., Mañes, J., Fernández-Franzón, M. and Meca, G. 2017. Shelf life improvement of the loaf bread using allyl, phenyl and benzyl isothiocyanates against *Aspergillus parasiticus*. *LWT Food Sci. Technol.* 78:208–214.

Saleem, Z.M. and Al-Delaimy, K.S. 1982. Inhibition of *Bacillus cereus* by garlic extracts. *J. Food Prot.* 45:1007–1011.

Salvia-Trujillo, L., Rojas-Grau, A., Soliva-Fortuny, R. and Martín-Belloso, O. 2015. Use of antimicrobial nanoemulsions as edible coatings: Impact on safety and quality attributes of fresh-cut Fuji apples. *Food Hydrocoll.* 43:547–556.

Sánchez, E., Heredia, N., Camacho-Corona, M. del R. and García, S. 2013. Isolation, characterization and mode of antimicrobial action against *Vibrio cholerae* of methyl gallate isolated from *Acacia farnesiana*. *J. Appl. Microbiol.* 115:1307–1316.

Sánchez-Rubio, M., Taboada-Rodríguez, A., Cava-Roda, R., López-Gómez, A. and Marín-Iniesta, F. 2016. Combined use of thermo-ultrasound and cinnamon leaf essential oil to inactivate *Saccharomyces cerevisiae* in natural orange and pomegranate juices. *LWT* 73:140–146.

Santamarina, M.P., Roselló, J., Giménez, S. and Blázquez, M.A. 2016. Commercial *Laurus nobilis* L. and *Syzygium aromaticum* L. Merr. & Perry essential oils against post-harvest phytopathogenic fungi on rice. *LWT Food Sci. Technol.* 65:325–332.

Sarkar, P., Bhunia, K. and Yao, Y. 2017. Impact of starch-based emulsions on the antibacterial efficacies of nisin and thymol in cantaloupe juice. *Food Chem.* 217:155–162.

Savoia, D. 2012. Plant-derived antimicrobial compounds: Alternatives to antibiotics. *Future Microbiol.* 7:979–990.

Scaffaro, R., Lopresti, F., D'Arrigo, M., Marino, A. and Nostro, A. 2018a. Efficacy of poly(lactic acid)/carvacrol electrospun membranes against *Staphylococcus aureus* and *Candida albicans* in single and mixed cultures. *Appl. Microbiol. Biotechnol.* 102:4171–4181.

Scaffaro, R., Lopresti, F., Marino, A. and Nostro, A. 2018b. Antimicrobial additives for poly(lactic acid) materials and their applications: Current state and perpectives. *Appl. Microbiol. Biotechnol.* 102:7739–7756.

Scavo, A., Pandino, G., Restuccia, C., Parafati, L., Cirvilleri, G. and Mauromicale, G. 2019. Antimicrobial activity of cultivated cardoon (*Cynara cardunculus* L. var. altilis DC.) leaf extracts against bacterial species of agricultural and food interest. *Industrial Crops Prod.* 129:206–211.

Schaller, M.S. and Korting, H.C. 1995. Allergic airborne contact dermatitis from essential oils used in aromatherapy. *Clinical Exper. Dermatol.* 20:143–145.

Schilter, B., Andersson, C., Anton, R., Constable, A., Kleiner, J., O'Brien, J., Renwick, A.G., Korver, O., Smit, F. and Walker, R. 2003. Guidance for the safety assessment of botanicals and botanical preparations for use in food and food supplements. *Food Chem. Toxicol.* 41:1625–1649.

Selvamuthukumaran, M. and Shi, J. 2017. Recent advances in extraction of antioxidants from plant by-products processing industries. *Food Quality Safety* 1:61–81.

Semeniuc, C.a., Soaciu, M.I., Socaci, S.A., Muresan, V., Fogarasi, M., and Rotar, A.M. 2018. Chemometric comparison and classification of plant essential oils extracted from plants belonging to *Apiaceae* and *Lamiaceae* families based on their chemical composition and biological activities. *Molecules* 23:2261–2274.

Serio, A., Chiarini, M., Tettamanti, E. and Paparella, A. 2010. Electronic paramagnetic resonance investigation of the activity of *Origanum vulgare* L. essential oil on the *Listeria monocytogenes* membrane. *Lett. Appl. Microbiol.* 51:149–157.

Serrano-León, J.S., Bergamaschi, K.B., Yoshida, C.M.P., Saldaña, E., Selani, M.M., Rios-Mera, J.D., Alencar, S.M. and Contreras-Castillo, C.J. 2018. Chitosan active films containing agro-industrial residue extracts for shelf life extension of chicken restructured product. *Food Res. Intl.* 108:93–100.

Severino, R., Ferrari, G., Dang Vu, K., Donsì, F., Salmieri, S. and Lacroix, M. 2015. Antimicrobial effects of modified chitosan based coating containing nanoemulsion of essential oils, modified atmosphere packaging and gamma irradiation against *Escherichia coli* O157:H7 and *Salmonella* Typhimurium on green beans. *Food Control* 50:215–222.

Shakila, R.J., Jeevithan, E., Arumugam, V. and Jeyasekaran, G. 2016. Suitability of antimicrobial grouper bone gelatin films as edible coatings for vacuum-packaged fish steaks. *J. Aquatic Food Prod. Technol.* 25:724–734.

Shange, N., Makasi, T., Gouws, P. and Hoffman, L.C. 2019. Preservation of previously frozen black wildebeest meat (*Connochaetes gnou*) using oregano (*Oreganum vulgare*) essential oil. *Meat Sci.* 148:88–95.

Shank, F.R. and Carson, K.L. 1992. What is safe food? In Finley, J.W., Robinson, S.F. and Armstrong, D.J., eds. *Food Safety Assessment*, pp. 26–34. American Chemical Society, ACS Symposium Series 484.

Shao, X., Cheng, S., Wang, H., Yu, D. and Mungai, C. 2013a. The possible mechanism of antifungal action of tea tree oil on *Botrytis cinerea. J. Appl. Microbiol.* 114:1642–1649.

Shao, X., Wang, H., Xu, F. and Cheng, S. 2013b. Effects and possible mechanisms of tea tree oil vapor treatment on the main disease in postharvest strawberry fruit. *Postharvest Biol. Technol.* 77:94–101.

Sharayei, P., Azarpazhooh, E., Zomorodi, S. and Ramaswamy, H.S. 2019. Ultrasound assisted extraction of bioactive compounds from pomegranate (*Punica granatum* L.) peel. *LWT* 101:342–350.

Sharma, A., Bajpai, V.K. and Baek, K.H. 2013. Determination of antibacterial mode of action of *Allium sativum* essential oil against foodborne pathogens using membrane permeability and surface characteristic parameters. *J. Food Safety* 33:197–208.

Sharma, H., Mendiratta, S.K., Agrawal, R.K., Gurunathan, K., Kumar, S. and Singh, T.P. 2017. Use of various essential oils as bio preservatives and their effect on the quality of vacuum packaged fresh chicken sausages under frozen conditions. *LWT Food Sci. Technol.* 81:118–127.

Shi, C., Zhang, X., Zhao, X., Meng, R., Liu, Z., Chen, X. and Guo, N. 2017. Synergistic interactions of nisin in combination with cinnamaldehyde against *Staphylococcus aureus* in pasteurized milk. *Food Control* 71:10–16.

Shirvani, A., Goli, S.A.H., Shahedi, M. and Soleimanian-Zad, S. 2016. Changes in nutritional value and application of thyme (*Thymus vulgaris*) essential oil on microbial and organoleptic markers of Persian clover (*Trifolium resupinatum*) sprouts. *LWT Food Sci. Technol.* 67:14–21.

Singh, A., Singh, R.K., Bhunia, A.K. and Singh, N. 2003. Efficacy of plant essential oils as antimicrobial agentes against *Listeria monocytogenes* in hotdogs. *LWT Food Sci. Technol.* 36:787–794.

Sipahi, R.E., Perez, M.E.C., Moreira, R.G., Gomes, C. and Castillo, A. 2013. Improved multilayered antimicrobial alginate-based edible coating extends the shelf life of fresh-cut watermelon (*Citrullus lanatus*). *LWT Food Sci. Technol.* 51:9–15.

Sirocchi, V., Devlieghere, F., Peelman, N., Sagratini, G., Maggi, F., Vittori, S. and Ragaert, P. 2017. Effect of *Rosmarinus officinalis* L. essential oil combined with different packaging conditions to extend the shelf life of refrigerated beef meat. *Food Chem.* 221:1069–1076.

Siroli, L., Braschi, G., de Jong, A., Kok, J., Patrignani, F. and Lanciotti, R. 2018. Transcriptomic approach and membrane fatty acid analysis to study the response mechanisms of *Escherichia coli* to thyme essential oil, carvacrol, 2-(E)-hexanal and citral exposure. *J. Appl. Microbiol.* 125(5):1308–1320.

Siroli, L., Patrignani, F., Serrazanetti, D.I., Tappi, S., Rocculi, P., Gardini, F. and Lanciotti, R. 2015. Natural antimicrobials to prolong the shelf-life of minimally processed lamb's lettuce. *Postharvest Biol. Technol.* 103:35–44.

Smid, E.J. and Gorris, L.G.M. 1999. Natural antimicrobials for food preservation. In Rahman, M.S., ed. *Handbook of Food Preservation*, pp. 285–308. Marcel Dekker, Inc., New York.

Smid, E.J., Hendriks, L., Boerrigter, H.A.M. and Gorris, L.G.M. 1996. Surface disinfection of tomatoes using the natural plant compound trans-cinnamaldehyde. *Postharvest Biol. Technol.* 8:343–348.

Sofos, J.N., Beuchat, L.R., Davidson, P.M. and Johnson, E.A. 1998. Naturally occurring antimicrobials in food. Task Force Report No 132. Council for Agricultural Science and Technology, Ames, IA.

Šojić, B., Tomović, V., Kocić-Tanackov, S., Škaljac, S., Ikonić, P., Džinić, N., Živković, N., Jokanović, M., Tasić, T. and Kravić, S. 2015. Effect of nutmeg (*Myristica fragrans*) essential oil on the oxidative and microbial stability of cooked sausage during refrigerated storage. *Food Control* 54:282–286.

Soledade, M., Pedras, C. and Smith, K.C. 1997. Sinalexin, a phytoalexin from white mustard elicited by destruxin B and *Alternaria brassicae*. *Phytochem.* 46:833–837.

Soncu, E.D., Arslan, B., Ertürk, D., Küçükkaya, S., Özdemir, N. and Soyer, A. 2018. Microbiological, physicochemical and sensory characteristics of Turkish fermented sausages (sucuk) coated with chitosan-essential oils. *LWT* 97:198–204.

Songsamoe, S., Matan, N. and Matan, N. 2017. Antifungal activity of *Michelia alba* oil in the vapor phase and the synergistic effect of major essential oil components against *Aspergillus flavus* on brown rice. *Food Control* 77:150–157.

Souza, V.G.L., Pires, J.R., Vieira, E.T., Coelhoso, I.M., Duarte, M.P. and Fernando, A.L. 2019. Activity of chitosan-montmorillonite bionanocomposites incorporated with rosemary essential oil: From in vitro assays to application in fresh poultry meat. *Food Hydrocoll.* 89:241–252.

Sow, L.C., Tirtawinata, F., Yang, H., Shao, Q. and Wang, S. 2017. Carvacrol nanoemulsion combined with acid electrolysed water to inactivate bacteria, yeast in vitro and native microflora on shredded cabbages. *Food Control* 76:88–95.

Speijers, G., Bottex, B., Dusemund, B., Lugasi, A., Toth, J., Amberg-Müller, J., Galli, C., Silano, V. and Rietjens, I.M.C.M. 2010. Safety assessment of botanicals and botanical preparations used as ingredients in food supplements: Testing an European Food Safety Authority-tiered approach. *Risk Assessment of Phytochemicals in Food: Novel Approaches* 54:175–85.

Spencer, C.M., Cai, Y., Martin, R., Gaffney, S.H., Goulding, P.N., Magnolato, D., Lilley, T.H. and Haslam, E. 1988. Polyphenol complexation: Some thoughts and observations. *Phytochem.* 27:2397–2409.

Stojanović-Radić, Z., Čomić, L., Radulović, N., Blagojević, P., Denić, M., Miltojević, A., Rajković, J. and Mihajilov-Krstev, T. 2012. Antistaphylococcal activity of *Inula helenium* L. root essential oil: Eudesmane sesquiterpene lactones induce cell membrane damage. *Euro. J. Clinical Microbiol. Infect. Dis.* 31:1015–1025.

Stojanović-Radić, Z., Pejčić, M., Joković, N., Jokanović, M., Ivić, M., Šojić, B., Škaljac, S., Stojanović, P. and Mihajilov-Krstev, T. 2018. Inhibition of *Salmonella* Enteritidis growth and storage stability in chicken meat treated with basil and rosemary essential oils alone or in combination. *Food Control* 90:332–343.

Stojiljkovic, J., Trajchev, M., Nakov, D. and Petrovska, M. 2018. Antibacterial activities of rosemary essential oils and their components against pathogenic bacteria. *Adv. Cytology Pathol.* 3:93–96.

Stratakos, A.C., Sima, F., Ward, P., Linton, M., Kelly, C., Pinkerton, L., Stef, L., Pet, I. and Corcionivoschi, N. 2018. The in vitro effect of carvacrol, a food additive, on the pathogenicity of O157 and non-O157 Shiga-toxin producing *Escherichia coli*. *Food Control* 84:290–296.

Sun, X.H., Zhou, T.T., Wei, C.H., Lan, W.Q., Zhao, Y., Pan, Y.J. and Wu, V.C. 2018. Antibacterial effect and mechanism of anthocyanin rich Chinese wild blueberry extract on various foodborne pathogens. *Food Control.* 94:155–161.

Swamy, M.K., Akhtar, M.S. and Sinniah, U.R. 2016. Antimicrobial properties of plant essential oils against human pathogens and their mode of action: An updated review. *Evidence-Based Complem. Alter. Med.* 2016:1–21. https://www.hindawi.com/J.s/ecam/2016/3012462/abs/.

Taiz, L. and Seiger, E. 2002. *Plant Physiology*, 3rd ed. Sinauer Associates, Sunderland, MA.

Takama, D.K. and Korel, F. 2017. Impact of preharvest and postharvest alginate treatments enriched with vanillin on postharvest decay, biochemical properties, quality and sensory attributes of table grapes. *Food Chem.* 221:187–195.

Talhaoui, N., Taamalli, A., Gómez-Caravaca, A.M., Fernández-Gutiérrez, A. and Segura-carretero, A. 2015. Phenolic compounds in olive leaves: Analytical determination, biotic and abiotic influence, and health benefits. *Food Res. Intl.* 77:92–108.

Tang, R., Luo, J., Wang, W., Liu, D., Wang, G. and Guo, X. 2019. Rutin's natural source Flos Sophorae as potential antioxidant and improver of fungal community in Chinese sausages. *LWT* 101:435–443.

Tang, S.S., Prodhan, Z.H., Biswas, S.K., Le, C.F. and Sekaran, S.D. 2018. Antimicrobial peptides from different plant sources: Isolation, characterisation, and purification. *Phytochem.* 154:94–105.

Tao, N., Hia, L. and Zhou, H. 2014. Anti-fungal activity of *Citrus reticulate* Blanco essential oils against *Penicillium italicum* and *Penicillium digitatum*. *Food Chem.* 153:265–271.

Tapia-Rodriguez, M.R., Hernandez-Mendoza, A., Gonzalez-Aguilar, G.A., Martinez-Tellez, M.A., Martins, C.M. and Ayala-Zavala, J.F. 2017. Carvacrol as potential quorum sensing inhibitor of *Pseudomonas aeruginosa* and biofilm production on stainless steel surfaces. *Food Control* 75:255–261.

Tassou, C., Koutsoumanis, K. and Nychas, G. 2000. Inhibition of *Salmonella enteritidis* and *Staphylococcus aureus* in nutrient broth by mint essential oil. *Food Res. Intl.* 33:273–280.

Tassou, C.C. and Nychas, G.J.E. 1994. Inhibition of *Staphylococcus aureus* by olive phenolics in broth and in a model food system *J. Food Prot.* 57:120–124.

Tateo, F., Triangeli, L., Panna, E., Berte, F. and Verderio, E. 1988. Stability and reactivity of aspartame in cola-type drinks. In Charambous, G., ed. *Frontiers of Flavor.* Elsevier Science, Amsterdam.

Tawema, P., Han, J., Dang Vu, K., Salmieri, S. and Lacroix, M. 2016. Antimicrobial effects of combined UV-C or gamma radiation with natural antimicrobial formulations against *Listeria monocytogenes*, *Escherichia coli* O157:H7, and total yeasts/molds in fresh cut cauliflower. *LWT Food Sci. Technol.* 65:451–456.

Temba, B.A., Fletcher, M.T., Fox, G.P., Harvey, J.J. and Sultanbawa, Y. 2016. Inactivation of *Aspergillus flavus* spores by curcumin-mediated photosensitization. *Food Control.* 59:708–713.

Terjung, N., Loffler, M., Gibis, M., Hinrichs, J., Silva, J. and Weiss, J. 2012. Influence of droplet size on the efficacy of oil-in-water emulsions loaded with phenolic antimicrobials. *Food Function* 3:290–301.

Thakur, P., Kapila, P., Sharma, R.B., Agarwal, S. and Vashist, H. 2017. Natural essential oils from extraction to encapsulation. *World J. Pharm. Pharmaceut. Sci.* 6:488–518.

Tian, Zeng, X., Lü, A., Zhu, A., Peng, X. and Wang, Y. 2015. Perillaldehyde, a potential preservative agent in foods: Assessment of antifungal activity against microbial spoilage of cherry tomatoes. *LWT Food Sci. Technol.* 60:63–70.

Torres, N., Antolín, M.C., Garmendia, I. and Goicochea, N. 2018. Nutritional properties of Tempranillo grapevine leaves are affected by clonal diversity, mycorrhizal symbiosis and air temperature regime. *Plant Physiol. BioChem.* 130:542–554.

Torres-Ossandón, M.J., Vega-Gálvez, A., Salas, C.E., Rubio, J., Silva-Moreno, E. and Castillo, L. 2019. Antifungal activity of proteolytic fraction (P1G10) from (*Vasconcellea cundinamarcensis*) latex inhibit cell growth and cell wall integrity in *Botrytis cinerea*. *Intl. J. Food Microbiol.* 289:7–16.

Tosati, J.V., de Oliveira, E.F., Oliveira, J.V., Nitin, N. and Monteiro, A.R. 2018. Light-activated antimicrobial activity of turmeric residue edible coatings against cross-contamination of *Listeria innocua* on sausages. *Food Control* 84:177–185.

Tracz, B.L., Bordin, K., de Melo Nazareth, T., Costa, L.B., de Macedo, R.E.F., Meca, G. and Luciano, F.B. 2017. Assessment of allyl isothiocyanate as a fumigant to avoid mycotoxin production during corn storage. *LWT* 75:692–696.

Trattner, A., David, M. and Lazarov, A. 2008. Occupational contact dermatitis due to essential oils. *Contact Dermatitis* 58:282–284.

Tsai, H.-C., Sheng, L. and Zhu, M.-J. 2017. Antimicrobial efficacy of cinnamon oil against *Salmonella* in almond based matrices. *Food Control* 80:170–175.

Tunc, S., Chollet, E., Chalier, P., Preziosi-Belloy, L. and Gontard, N. 2007. Combined effect of volatile antimicrobial agents on the growth of *Penicillium notatum*. *Intl. J. Food Microbiol.* 113:263–270.

Turgis, M., Han, J., Borsa, J. and Lacroix, M. 2008. Combined effect of natural essential oils, modified atmosphere packaging, and gamma radiation on the microbial growth on ground beef. *J. Food Prot.* 71:1237–1243.

Tyagi, A.K. and Malik, A. 2010. Liquid and vapour-phase antifungal activities of selected essential oils against *Candida albicans*: Microscopic observations and chemical characterization of *Cymbopogon citratus*. *BMC Compl. Altern. Med.* 10:1–11.

Tyagi, A.K. and Malik, A. 2011. Antimicrobial potential and chemical composition of *Eucalyptus globulus* oil in liquid and vapour phase against food spoilage microorganisms. *Food Chem.* 126:228–235.

Tzortzakis, N.G. 2010. Ethanol, vinegar and *Origanum vulgare* oil vapour suppress the development of anthracnose rot in tomato fruit. *Intl. J. Food Microbiol.* 142:14–18.

Upadhyay, N., Singh, V.K., Dwivedy, A.K., Das, S., Chaudhari, A.K. and Dubey, N.K. 2018. *Cistus ladanifer* L. essential oil as a plant based preservative against molds infesting oil seeds, aflatoxin B1 secretion, oxidative deterioration and methylglyoxal biosynthesis. *LWT* 92:395–403.

U.S. Food and Drug Administration. 2018. CFR: Code of Federal Regulations Title 21. https://www.accessdata.fda.gov/scripts/cdrh/cfdocs/cfcfr/CFRSearch.cfm (accessed June 17, 2019).

U.S. Food and Drug Administration. 2019. Substances Added to Food (Formerly EAFUS). https://www.accessdata.fda.gov/scripts/fdcc/?set=FoodSubstances (accessed June 17, 2019).

Valdés, A. Ramos, M., Beltran, A., Jimenez, A. and Garrigos, M.C. 2017. State of the art of antimicrobial edible coating for food packaging applications. *Coating* 7:1–23.

van den Berg, S.J.P.L., Serra-Majem, L., Coppens, P. and Rietjens, I.M.C.M. 2011. Safety assessment of plant food supplements (PFS). *Food Function* 2:760–768.

Van Haute, S., Raes, K., Van der Meeren, P. and Sampers, I. 2016. The effect of cinnamon, oregano and thyme essential oils in marinade on the microbial shelf life of fish and meat products. *Food Control* 68:30–39.

Vasileva, I., Denkova, R., Chochkov, R., Teneva, D., Denkova, Z., Dessev, T., Denev, P. and Slavov, A. 2018. Effect of lavender (*Lavandula angustifolia*) and melissa (*Melissa Officinalis*) waste on quality and shelf life of bread. *Food Chem.* 253:13–21.

Velázquez-Nuñez, M.J., Avila-Sosa, R., Palou, E. and López-Malo, A. 2013. Antifungal activity of orange (*Citrus sinensis* var. Valencia) peel essential oil applied by direct addition or vapor contact. *Food Control* 31:1–4.

Volpe, M.G., Siano, F., Paolucci, M., Sacco, A., Sorrentino, A., Malinconico, M. and Varricchio, E. 2015. Active edible coating effectiveness in shelf-life enhancement of trout (*Oncorhynchus mykiss*) fillets. *LWT Food Sci. Technol.* 60:615–622.

Walker, R. 2004. Criteria for risk assessment of botanical food supplements. *Toxicology Lett.* 149:187–195.

Wang, H., Qian, I. and Ding, F. 2018a. Emerging chitosan-based films for food packaging applications. *J. Agric. Food Chem.* 66:395–413.

Wang, L., Liu, F., Jiang, Y., Chai, Z., Li, P., Cheng, Y., Jing, H. and Leng, X. 2011. Synergistic antimicrobial activities of natural essential oils with chitosan films. *J. Agric. Food Chem.* 59:12411–12419.

Wang, Y.-F., Jia, J.-X., Tian, Y.-Q., Shu, X., Ren, X.-J., Guan, Y. and Yan, Z.-Y. 2018b. Antifungal effects of clove oil microcapsule on meat products. *LWT* 89:604–609.

Weiss, J., Gaysinsky, S., Davidson, P.M. and McClements, J. 2009. Nanostructured encapsulation systems: Food antimicrobials. In Barbosa-Canovas, G., Mortimer, A., Lineback, D., Spiess, W., Buckle, K. and Colonna, P., eds. *Global Issues in Food Science and Technology*, pp. 425–479. Academic Press, New York.

Weiss, J., Loeffler, M. and Terjung, N. 2015. The antimicrobial paradox: Why preservatives lose activity in foods. *Curr. Opinion Food Sci.* 4:69–75.

WHO (World Health Organization). 1987. Principles for the safety assessment of food additives and contaminants in food. IPCS Environmental Health Criteria 70. Geneva.

Wilkins, K.M. and Board, R.G. 1989. Natural antimicrobial systems. In Gould, G.W., ed. *Mechanisms of Action of Food Preservation Procedures*, pp. 285–362. Elsevier Science Publishers Ltd., New York.

Wills, E.D. 1956. Enzyme inhibition by allicin, the active principle of garlic. *Biochemistry* 63:514–520.

Windiasti, G., Feng, J., Ma, L., Hu, Y., Hakeem, M.J., Amoako, K., Delaquis, P. and Lu, X. 2019. Investigating the synergistic antimicrobial effect of carvacrol and zinc oxide nanoparticles against *Campylobacter jejuni*. *Food Control* 96:39–46.

Wright, M.H., Shalom, J., Matthews, B., Greene, A.C. and Cock, I.E. 2019. *Terminalia ferdinandiana* Exell: Extracts inhibit *Shewanella* spp. growth and prevent fish spoilage. *Food Microbiol.* 78:114–122.

Wu, C., Li, Y., Du, Y., Wang, L., Tong, C., Hu, Y., Pang, J. and Yan, Z. 2019. Preparation and characterization of konjac glucomannan-based bionanocomposite film for active food packaging. *Food Hydrocoll.* 89:682–690.

Wu, C., Tian, J., Li, S., Wu, T., Hu, Y., Chen, S., Sugawara, T. and Ye, X. 2016a. Structural properties of films and rheology of film-forming solutions of chitosan gallate for food packaging. *Carbohyd. Polym.* 146:10–19.

Wu, C., Wang, L., Hu, Y., Chen, S., Liu, D. and Ye, X. 2016b. Edible coating from citrus essential oil-loaded nanoemulsions: Physicochemical characterization and preservation performance. *RSC Advances* 6:892–900.

Wu, H., Xue, N., Hou, C., Feng, J. and Zhang, X. 2015. Microcapsule preparation of allyl isothiocyanate and its application on mature green tomato preservation. *Food Chem.* 175:344–349.

Wu, Y.M., Wang, Z.W., Hu, C.Y. and Nerín, C. 2018. Influence factors on release of antimicrobials from antimicrobial packaging materials. *Crit. Rev. Food Sci. Nutr.* 58:1108–1121.

Xiang, F., Bai, J., Tan, X., Chen, T., Yang, W. and He, F. 2018. Antimicrobial activities and mechanism of the essential oil from *Artemisia argyi* Levl. et Van. var. argyi cv. Qiai. *Industrial Crops Prod.* 125:582–587.

Xie, M., Hu, B., Wang, Y. and Zeng, X. 2014. Grafting of gallic acid onto chitosan enhances antioxidant activities and alters rheological properties of the copolymer. *J. Agric. Food Chem.* 62:9128–9136.

Xu, L., Tao, N., Yang, W. and Jing, G. 2018. Cinnamaldehyde damaged the cell membrane of *Alternaria alternata* and induced the degradation of mycotoxins in vivo. *Industrial Crops Prod.* 112:427–433.

Xueuan, R., Dandan, S., Zhuo, L. and Qingjun, K. 2018. Effect of mint oil against Botrytis cinerea on table grapes and its possible mechanism of action. *Euro. J. Plant Pathol.* 151:321–328.

Yadav, A.K., Sirohi, P., Saraswat, S., Rani, M., Singh, M.P., Srivastava, S. and Singh, N.K. 2018. Inhibitory mechanism on combination of phytic acid with methanolic seed extract of *Syzygium cumini* and sodium chloride over *Bacillus subtilis. Curr. Microbiol.* 75:849–856.

Yakoub, A.R.B., Abdehedi, O., Jridi, M., Elfalleh, W., Nasri, M. and Ferchichi, A. 2018. Flavonoids, phenols, antioxidant, and antimicrobial activities in various extracts from Tossa jute leave (*Corchorus olitorus* l.). *Industrial Crops Prod.* 118:206–213.

Yin, M.-C. and Tsao, S.-M. 1999. Inhibitory effect of seven *Allium* plants upon three *Aspergillus* species. *Intl. J. Food Microbiol.* 49:49–56.

Yu, D., Wang, J., Shao, X., Xu, F. and Wang, H. 2015. Antifungal modes of action of tea tree oil and its two characteristic components against *Botrytis cinerea. J. Appl. Microbiol.* 119:1253–1262.

Yuan, W., Teo, C.H.M. and Yuk, H.-G. 2019. Combined antibacterial activities of essential oil compounds *against Escherichia coli* O157:H7 and their application potential on fresh-cut lettuce. *Food Control* 96:112–118.

Yuan, W. and Yuk, H.G. 2018. Antimicrobial efficacy of *Syzygium antisepticum* plant extract against *Staphylococcus aureus* and methicillin-resistant *S. aureus* and its application potential with cooked chicken. *Food Microbiol.* 72:176–184.

Zaika, L.L. 1988. Spices and herbs: Their antimicrobial activity and its determination. *J. Food Safety* 9:97–118.

Zaika, L.L., Kissinger, J.C. and Wasserman, E. 1983. Inhibition of lactic bacteria by herbs. *J. Food Sci.* 48:1455–1459.

Zambrano, C., Kerekes, E.B., Kotogán, A., Papp, T., Vágvolgyi, C., Krisch, J. and Takó, M. 2019. Antimicrobial activity of grape, apple and pitahaya residue extracts after carbohydrase treatment against food-related bacteria. *LWT* 100:416–425.

Zhang, H., He, P., Kang, H. and Li, X. 2018a. Antioxidant and antimicrobial effects of edible coating based on chitosan and bamboo vinegar in ready to cook pork chops. *LWT* 93:470–476.

Zhang, Q.W., Lin, L.G. and Ye, W.C. 2018b. Techniques for extraction and isolation of natural products: A comprehensive review. *Chinese Med.* 13:1–26.

Zhang, Y., Ma, Q., Critzer, F., Davidson, P.M. and Zhong, Q. 2016. Organic thyme oil emulsion as an alternative washing solution to enhance the microbial safety of organic cantaloupes. *Food Control* 67:31–38.

Zhang, Y. and Talalay, P. 1994. Anticarcinogenic activities of organic isothiocyanates: Chemistry and mechanisms. *Cancer Res.* 54(7 Supplement) Supplement:1976s–1981s.

Ziaee, E., Razmjooei, M., Shad, E. and Eskandari, M.H. 2018. Antibacterial mechanisms of *Zataria multiflora* Boiss. essential oil against *Lactobacillus curvatus. LWT* 87:406–412.

18

Naturally Occurring Compounds – Animal Sources

Jarret Stopforth and Travis Kudron

CONTENTS

18.1 Introduction

Natural antimicrobials include agents found in plants, microbes, insects, and animals (Naidu, 2000; Sofos et al., 1998). This chapter focuses on the natural antimicrobials that exist in animals where they evolved as host defense mechanisms and may exhibit antimicrobial activity in foods as natural ingredients or be extracted and used as additives in other foods. In general, animal-derived antimicrobials have been known for many years and, thus, there has been much interest in the isolation and purification of these compounds to be used as food antimicrobials or as adjuncts to other antimicrobials.

The discussion in this chapter targets mainly those antimicrobials isolated from milk, especially of bovine origin, and from poultry eggs, since these are the major food products produced by animals for human consumption that possess antimicrobial activity devised by nature for protection and immune development in the neonate. The antimicrobials isolated from these products are generally broad-spectrum agents providing protection to the neonate against bacteria, fungi, parasites, and viruses, and in some cases are the only means of transferring protective factors to the unborn or newly born young (e.g., bovine colostrum is the only source of immune factors including antibodies from mother to newborn) (Floris et al., 2003). Maternal transfer of immune factors and antimicrobial substances provides the offspring with acquired immunity, leaving them relatively immunocompetent against diseases. Substances with antimicrobial properties from the mother, such as lactolipids in bovine milk, are not strictly immune factors and may possess both nutritional and protective qualities.

Major challenges faced by the industry include difficulty in the production of these compounds in large enough quantities to be effective in a food and the incorporation of these compounds in foods while minimizing undesirable interactions, undesirable effects, or subsequent inactivation of their desirable effect. It should also be noted that the use of these compounds as antimicrobials in foods may not have the same effectiveness as if the compound were a natural ingredient in food. This is especially true in milk where the total antimicrobial activity is not necessarily a function of individual antimicrobials but probably the additive or synergistic effect of multiple antimicrobials (e.g., lactoferrin, lactoperoxidase, lactoglobulins). Despite the challenge of obtaining effective activity of antimicrobial compounds in a food matrix, there has been commercial application of various compounds isolated from animal products including the fortification of dairy products with milk antibodies, the fortification of dairy products including milk with substrates to facilitate lactoperoxidase production, and the use of monoacylglycerols from milk as biopreservatives in many foods. Furthermore, there are many proposed applications of antimicrobial compounds in foods or consumables including: (i) passive immunization with milk- and egg-derived antibodies; (ii) the addition of transferrins, lactoperoxidase, and immunoglobulins in oral healthcare products; and (iii) the use of lactolipids, immunoglobulins, and transferrins as ingredients in infant formulas. Products derived from animal sources, in particular eggs and milk, are usually characterized as generally recognized as safe (GRAS) compounds; however, these products do need to be evaluated for potential allergenic properties in sensitive individuals. The following sections provide a summary of the most important naturally occurring antimicrobial compounds found in animal tissues and in particular in bovine milk and poultry eggs.

18.2 Lactoperoxidase

Lactoperoxidase (LP), a hemoprotein present in milk, tears, and saliva (Tenovuo and Pruitt, 1984), is the most abundant enzyme in bovine milk (Reiter, 1985a). The peroxidase activity associated with bovine milk was first demonstrated by Arnold in 1881 and the protein, termed lactoperoxidase, was isolated by Theorell and Akeson in 1943 (Naidu, 2000). The lactoperoxidase–thiocyanate–hydrogen peroxide interaction constitutes what is referred to as the LP system, wherein hydrogen peroxide serves as a substrate for LP in oxidizing thiocyanate (SCN^-) and iodide ions, resulting in the generation of highly reactive oxidizing agents (Thomas, 1985; Naidu, 2003). The association of LP with the inhibition of microbial growth was first demonstrated by Wright and Trammer (1958), while characterization of the complete LP system including enzymes and substrates occurred later (Jago and Morrison, 1962; Reiter et al., 1963). Additionally, the LP system possesses hexokinase and glyceraldehyde-3-phosphate dehydrogenase activities, which may contribute to the antimicrobial action of the system (Carlsson et al., 1983). The LP system has the ability to inhibit bacteria, fungi, parasites, and viruses and is, thus, considered a broad-spectrum natural antimicrobial contributing to protecting the gut of weaning calves from enteric pathogens, protecting the mammary gland from disease, and indeed preserving milk (Bjorck et al., 1979; Reiter and Bramley, 1975; Reiter et al., 1980).

18.2.1 Molecular Properties

18.2.1.1 Occurrence and Biosynthesis

LP is synthesized and secreted by ductal epithelial cells of the mammary gland and other exocrine glands (Parkos, 1997). The compound constitutes approximately 1% (10–30 µg/ml) of the whey proteins in the milk (Reiter, 1985a, b) and levels may be influenced by feeding practices, udder irritation, and estrogen levels (Janota-Bassalik et al., 1977; Kern et al., 1963; Kiermeier and Kuhlmann, 1972). The level of LP in bovine milk is about 20 times higher than that of human milk and changes constantly during the postpartum. Thiocyanate, which is required for the antimicrobial activity of the LP system, may be present in significant amounts in saliva, milk, and airway secretions, while hydrogen peroxide may be generated by microbial flora, usually bacteria, resident in sites such as the oral cavity, respiratory tract, or mammary gland or *in situ* by epithelial cells in these regions (Reiter and Perraudin, 1991). In bovine milk, the initial concentration of LP in colostrum is low, increasing to a peak at 4–5 days postpartum, after which it declines to a level considered relatively high and remains unchanged at that level during lactation (Kiermeier and Kuhlmann, 1972). In order to combat infections, the concentrations of LP and SCN^- increase in milk from infected bovine udders as compared with normal, healthy udders (Janota-Bassalik et al., 1977).

18.2.1.2 Chemistry and Structure

It has been determined that bovine LP is comprised of a single peptide chain, with eight disulfide bonds contributing to the rigidity of the molecule (Sievers, 1980). The single polypeptide chain contains 612 amino acid residues with a molecular weight of about 80 kDa (Cals et al., 1991; Paul and Ohlson, 1985). The polypeptide chain contains 15 half-cystine residues and 4 or 5 potential *N*-glycosylation sites, and carbohydrate moieties account for almost 10% of the molecular weight (Cals et al., 1991). LP is a heme-containing enzyme that shares 50–70% amino acid sequence homology (particularly among the active-site related residues) with myelo-, thyro-, and eosinophil peroxidases (Dull et al., 1990), all of which have a protoporyphyrin IX catalytic center (Fenna et al., 1995; Sievers, 1979). The covalent linkages between heme and the protein are well-represented in the myeloperoxidase X-ray crystal structure (Fenna et al., 1995; Zeng and Fenna, 1992); however, little work has been done concerning the crystallographic data for LP. Although the details regarding the heme-binding site in LP have not been established, the heme structure has been studied in terms of its electron transfer mechanisms (Nakamura et al., 1986) as the heme moiety is essential for the development of the oxidation-reduction reaction associated with LP activity. The presence of an odd number of half-cystines supports the theory that a heme thiol is released from this enzyme by a reducing agent and suggests that the heme is bound *via* disulfide links to the peptide chain (Naidu, 2000), indicating that there are no free thiol groups present in the enzyme molecule (Ekstrand, 1994). The iron content of LP is 0.07% which corresponds to one iron atom per LP molecule as part of the heme group (Kussendrager and van Hooijdonk, 2000). The molecular conformation of LP is thought to be stabilized by the strong binding of a calcium ion (Kussendrager and van Hooijdonk, 2000). Although earlier research (Sievers, 1980) revealed that leucin is found on the N-terminus of the LP polypeptide chain, Watanabe et al. (2000) found that different preparations of natural LP may have different N-terminal amino acid residues. This heterogeneity may be a result of variation in terms of isolation methods (i.e., disk-electrophoresis and ion-exchange chromatography); however, the finding suggests that the change in N-terminus structure does not affect the conformation of LP. Other findings of this research using circular dichroism (CD) revealed that the secondary structure (α-helix) of LP is influenced by peptide bonds and that the protein is rich in β-sheet structure (Watanabe et al., 2000).

18.2.1.3 Stability

It was reported that the LP system stored in airtight containers lost only 35% of the initial thiocyanate concentration during 18 months and that the system was strong enough to kill 10^6 CFU/ml of 4 test organisms. When the LP system was stored in the presence of air it lost thiocyanate activity after 7 d but after 516 days was still able to kill inocula of 10^6 CFU/ml *Pseudomonas aeruginosa*, *Staphylococcus aureus*, and *Candida albicans* and *Escherichia coli* within 2 h, 4 h, and 1 week, respectively (Bosch et al., 2000). It has been shown that during pasteurization, whole milk loses about 75% of its LP activity, while the purified LP was rendered unstable after 15 min of exposure (Wutrich et al., 1964). Research (Herandez et al., 1990) indicated that heat denaturation of LP in milk, whey, permeate, and buffer started at about 70°C (close to the temperature at which the native structure of the enzyme unfolds to follow first-order kinetics) and that the calcium ion concentration influenced the heat sensitivity of LP. The heat stability of LP is lower under acidic (pH 5.3) conditions and may be related to the release of calcium from the molecule (de Wit and van Hooijdonk, 1996). LP is deactivated during storage at pH 3 with partial denaturation at <pH 4, while there is no deactivation of the enzyme at values of up to pH 10 (de Wit and van Hooijdonk, 1996; Paul and Ohlsson, 1985; Herandez et al., 1990; Watanabe et al., 2000). The optimum pH for the LP catalyzed reaction lies between 5 and 6, and, based on this, it was concluded that determination of LP should be conducted in these ranges (Bardsley, 1985). LP has been found to be quite resistant to proteolytic enzymes including trypsin and thermolysin, although chymotrypsin was found to inactivate LP relatively slowly (Paul and Ohlsson, 1985; Sievers, 1979). Gothefors and Marklund (1975) found that LP is not inactivated by the gastric juice of an infant (pH 5) but that pepsin at pH 2.5 inactivated LP. The compound appears to be sensitive to light when in the presence of riboflavin (Herandez et al., 1990), while adsorption of LP to certain surfaces may cause a significant decrease in activity (Paul and Ohlsson, 1985).

18.2.2 Antimicrobial Activity

18.2.2.1 Mode of Action

LP is an enzyme with a primary function to oxidize thiocyanate and some halides (I^- and Br^-) at the expense of H_2O_2 to generate products which kill or inhibit the growth of many species of microorganisms

(Bosch et al., 2000; Kussendrager and van Hooijdonk, 2000). The reaction mechanisms involved in the generation of oxidative products is systematic. The first step is the initiation of the resting LP (Fe^{3+}) to its ground state by H_2O_2 followed by propagation reactions including the conversion of LP from its ground state to the compound I state (the compound that oxidizes SCN^- and halides) (de Wit and van Hooijdonk, 1996). At low SCN^- and halide concentrations, compound I reacts with any one-electron donors that are present to form compound II, which is continually reduced to the ground state. With an excess of H_2O_2, compound II may react to form compound III. This leads to a ferrylperoxidase adduct and irreversible inactivation of LP. The oxidation of SCN^- is catalyzed by LP and leads to temporary oxidation products which are responsible for antimicrobial activity (de Wit and van Hooijdonk, 1996). With the oxidation of SCN^-, the generation of $OSCN^-$ (hypothiocyanate) and HOSCN (hypothiocyanous acid) is in equilibrium and at the pH of maximal LP activity (pH 5.3), they exist in equal quantities. Both oxidized forms are antimicrobial, although there is evidence that hypothiocyanous acid is of stronger activity (Thomas, 1985). The oxidation of sulfhydryl (SH) groups of microbial proteins by $OSCN^-$ (hypothiocyanate) and HOSCN (hypothiocyanous acid) is considered to be the key to the antimicrobial action of the LP-system (Aune and Thomas, 1978); however, these forms do not appear to oxidize the SH groups of milk proteins (de Wit and van Hooijdonk, 1996). The structural damage to microbial cytoplasmic membranes through oxidation of SH-groups causes leakage of potassium ions, amino acids, and peptides into the medium as well as inhibition of the uptake of glucose, amino acids, purines, and pyrimidines, and subsequent synthesis of proteins, RNA, and DNA (Reiter and Perraudin, 1991).

18.2.2.2 Specificity

Mammalian cells are not affected by antimicrobial products of the LP system and in fact it is suggested that LP protects human cells against the toxic effects of H_2O_2 (Reiter and Perraudin, 1991). Although there are contrasting views, it appears that LP adsorption to the microbial cell surface is necessary to elicit antimicrobial effects (Tenovuo and Knuuttila, 1977). The cytoplasmic membrane and/or the cytoplasm are considered to be the major targets for antimicrobial products generated by the LP system (Naidu, 2000). Accessibility of products generated by the LP system to the interior of the microbial cell may be limited by the cell wall and membrane although they are not entirely excluded (Naidu, 2000). The cytoplasmic membrane of bacteria and cell walls of Gram-positive bacteria may be altered by antimicrobial products from the LP system (Marshall and Reiter, 1980). Some of the alterations that may occur at the biological barriers protecting bacteria include damage to the inner membrane resulting in extensive leakage of amino acids, hydrogen ions, and other molecules essential for bacterial survival (Law and John, 1981), inhibition of amino acid transport, and inhibition of glucose and oxygen transport and uptake (Hamon and Klebanoff, 1973; Slowey et al., 1968). Other antimicrobial effects from the products of the LP system include the inhibition of several glycolytic enzymes (e.g., hexokinase, aldolase, pyruvate kinase, etc.). The antimicrobial products of the LP system have a varying effect on different groups of microorganisms. The products from the LP system could exhibit static and/or cidal activity on bacteria, fungi, viruses, and parasites, and the molecular mechanism(s) of inhibition (oxidative killing, blockage of glycolytic pathways, or cytopathic effects) would be dependent on the type of organism, type of electron donor present, temperature, pH, and other factors (Naidu, 2000). Gram-negative, catalase-positive bacteria (pseudomonads, coliforms, salmonellae, shigellae, etc.) are not only inhibited but killed depending on pH, temperature, incubation time, cell density, etc. of the substrate (Reiter and Perraudin, 1991; de Wit and van Hooijdonk, 1996). Gram-positive, catalase-negative bacteria (streptococci, lactobacilli, etc.) are generally inhibited but not necessarily killed by the LP system, and the difference in their sensitivity may be explained by the differences in biological barriers (Reiter and Perraudin, 1991; de Wit and van Hooijdonk, 1996). Products of the LP system appear to cause more extensive damage to the inner membrane of Gram-negative than Gram-positive bacteria (Marshall and Reiter, 1980). Gram-negative bacteria appear to be more sensitive to inactivation by hypothiocyanate than Gram-positive cells, especially at low temperatures, and are also more sensitive to pH-dependent killing (Naidu, 2000; Pruitt and Reiter, 1985). The antifungal effects of the LP system have been demonstrated against *Candida tropicalis*, *C. albicans*, *Saccharomyces cerevisiae*, *Aspergillus niger*, *Rhodotorula rubra*, *Mucor rouxii*, *Byssochlamys fulva*, and species of *Geotrichum* (Hamon and Klebanoff, 1973; Lehrer, 1969; Popper and Knorr, 1997). Additionally, Doyle and Marth (1978) observed the ability of the LP system to degrade aflatoxin in the presence of salt. Antiviral properties of the LP system have also been displayed against RNA (poliovirus) and DNA (vaccinia) viruses (Belding et al., 1970). Courtois et al. (1990) found the LP system to be effective against HSV1 virus and the activity to be time-dependent. Inhibition of human immunodeficiency virus has also been demonstrated (Pourtois et al., 1990; Yamaguchi et al., 1993).

18.2.3 Applications in Food

The utilization of the LP system as a natural biopreservative in food, in particular dairy products, has increased significantly since the implementation of industrial processes for LP isolation from milk and whey (Kussendrager and van Hooijdonk, 2000). Milk may be cooled and stored for long periods of time prior to distribution and as such is susceptible to psychrotrophic spoilage, especially by pseudomonads surviving pasteurization (Naidu, 2000). Another and more severe problem originates in developing countries where milk may be unpasteurized and even uncooled prior to distribution and as such may spoil faster and even allow pathogenic organisms to proliferate. Introducing substrates for the production of antimicrobials from the LP system would be a method by which the milk may be protected from contaminating organisms. A two-enzyme coupled system of β-galactosidase and glucose oxidase was developed by Bjorck (1976) to activate the LP system and improve the storage of raw milk. Activation of the LP system by the addition of hypothiocyanate and H_2O_2 resulted in considerable reduction of the bacterial flora in the milk and prevented the growth of psychrotrophic bacteria for up to 5 days without altering the physico-chemical properties of the milk (Bjorck, 1978). An alternative to preserving milk with LP may be achieved by combining the activation of the LP system with acidification and subsequent souring of the milk (Kangumba et al., 1997; Wray and McLaren, 1987). This method was also found effective against pathogenic microorganisms, especially *Listeria monocytogenes* (Kamau et al., 1990; Kangumba et al., 1997; Rodriguez et al., 1997). Boussouel et al. (2000) concluded that the addition of LP in combination with nisin as a stepwise addition after the maximum inhibitory effect of nisin was reached inhibited *L. monocytogenes* in milk. The addition of LP to starter cultures for yogurt production has also been studied (Nakada et al., 1996), and the results indicated that the acid production was inhibited but the culture was not killed and as such a new type of yogurt was produced with extended shelf-life. A review of potential and actual application of LP-systems in dairy products by de Wit and van Hooijdonk (1996) indicated that raw milk could be preserved for 4 days at 4°C and the shelf-life extended for 3 days at 10°C; the shelf-life of pasteurized milk could be extended to 21 days at 10°C; the shelf-life of cheese milk could be extended to 8 days at 4–7°C; and the acidity of yogurt could be suppressed for 14 days at 20°C.

A combination of chitosan coating and LP was evaluated as an edible coating for extending the shelf-life of fish. The chitosan alone maintained samples of rainbow trout without LP and was even more efficient in shelf-life extensions than LP during cold storage. The indication was that the combination of chitosan and LP may prove more effective than either alone in extending shelf-life for chill-stored fish (Jasour et al., 2015).

Another major application of LP-systems has been described for oral-care products (Pellico and Montgomery, 1989), and indeed toothpaste (Biotene™) containing complete LP-systems ($LP/SCN^-/H_2O_2$) has been on the market for a number of years (Kussendrager and van Hooijdonk, 2000). Johansen et al. (1997) evaluated the effects of LP on model biofilms of several oral bacteria and found that the LP combined with oxidoreductases was bactericidal to the biofilms and when combined with polysaccharide-hydrolyzing enzymes the mixture was able to not only kill but remove the biofilms. Work done with ground beef revealed that LP-systems controlled pathogenic contamination and as such have the potential for incorporation into ground beef products as a "natural preservative" (Kennedy et al., 2000). McLay et al. (2002) confirmed that LP-systems may be considered as natural antimicrobials for incorporation into food systems including ground beef. Structural determination of lactoperoxidase in buffalo and goat milk *via* the use of RP-UHPLC-MS/MS showed that the lactoperoxidase from these sources is very similar to that of lactoperoxidase from bovine. A common glycosylation feature was identified between the three sources, and this finding may be useful in looking at the positive role these glycans may play as functional food ingredients and understanding the glycan role in lactoperoxidase function (Kumar at el., 2018). Detection of lactoperoxidase has also been explored in the form of a rapid detection test strip for raw milk and dairy products (Che et al., 2015). A lower detection limit of 1–2 mg/L was achieved. Such test strips can be useful for rapid testing of lactoperoxidase in the field or an industrial setting.

Lactoperoxidase can also be heat denatured and rendered inactive. Only 5.2% of enzyme activity was preserved after five minutes at 70°C. This denaturation happened due to a narrowing of the channel connecting the protein surface to the binding pocket (Aprodu et al., 2014). High hydrostatic pressure treatment in combination with the lactoperoxidase system showed effective antimicrobial action against *S.* Enteritidis and *E. coli* O157:H7 on cured beef carpaccio. This combination may be used to reduce the intensity of high-pressure treatments that may lower the quality of meat products (Bravo et al., 2014).

18.3 Transferrins

18.3.1 Lactoferrin, Lactoferricin B, and Activated Lactoferrin

Lactoferrin (LF), also termed lactotransferrin, or lactosiderophilin (Nagasawa et al., 1972), is an iron-binding, bioactive glycoprotein of the transferrin family that contributes to the control of iron in biological fluids. The transferrin family is subdivided into two branches: the melanotransferrins and the soluble glycoproteins that apart from lactoferrin also include two avian transferrins: the serum transferrin and the ovotransferrin (Baker and Lindley, 1992; Thakurta et al., 2003). The functional properties, including antimicrobial activity, of LF are mainly associated with its ability to bind two Fe^{3+} in combination with two CO_3^{2-} ions. It is found mainly in milk, the mucosal surfaces (e.g., intestinal epithelial cells), and exocrine secretions of mammals such as saliva, tears, and seminal fluid, as well as in the secondary granules (vesicles) of polymorphonuclear neutrophils or lymphocytes (Levy, 1996; Levay and Viljoen, 1995; Lonerdal and Lyer, 1995; Naidu, 2000; van der Strate et al., 2001). Human and porcine milk contain significantly higher concentrations of LF (almost ten times) than mature bovine milk (Masson and Herremans, 1971; Playford et al., 2000; Reiter, 1978a; Sanchez et al., 1988); however, the highest levels of LF exist in bovine colostrum, whereas increases may also be observed in mature milk after mastitic infection (Reiter and Oram, 1967; Mandel and Ellison, 1985; Hagiwara et al., 2003; Playford et al., 2000). LF has an important role in many physiological pathways, such as the adsorption of metal ions in the intestinal tract (Davidson and Lönnerdal, 1986; Davidson et al., 1990, 1994; Naidu, 2000), assistance in digestion and utilization of micronutrients and macronutrients from the milk (Lonnerdal, 2003), suppression of myelopoiesis (Broxmeyer et al., 1987), protection of the intestinal flora in young animals against enteropathogenic bacteria, protection from mastitis infection in non-lactating udders, and opsonic activity and immunoregulatory function (i.e., contribution to the pre-immune innate defense of mammals) (Ekstrand, 1994; Naidu, 2000, 2002). LF possesses antioxidant properties and has a broad antimicrobial spectrum including antibacterial, antifungal, antiprotozoal, antiviral, and antitumor properties (Naidu, 2000).

Lactoferricin (LFcin) is an active peptide which is derived from peptic hydrolysis of human and bovine LF close to the N-terminus (Jones et al., 1994; Tomita et al., 1994). It exhibits marked antimicrobial effects and, in a majority of studies, is more efficacious than LF. Activated LF (ALF) is an immobilized form of LF on edible substrates and was developed as a deterrent to pathogenic bacteria that may be present on meat surfaces (Naidu et al., 2003). Specifically, ALF is presented to act as a blocking agent, which interferes with microbial adhesion/colonization, detaches live or dead microorganisms from biological surfaces, inhibits microbial growth/multiplication, and neutralizes the activity of endotoxins (Naidu, 2002). It is considered a generally recognized as safe (GRAS) compound by the Food and Drug Administration (21 CFR.170.36[f]), since it is a derivative of the naturally occurring LF in milk, which in turn is isolated from cheese-whey and skim milk of bovine origin. The use of ALF on fresh beef was approved (October of 2001) by the U.S. Department of Agriculture to prevent bacterial contamination during processing (Naidu et al., 2003).

18.3.1.1 Molecular Properties

18.3.1.1.1 Occurrence and Biosynthesis

LF is present at 20–200 mg/l and >2000 mg/l in bovine and human milk, respectively, as well as two major reservoirs of mammals, including a circulatory pool in the secondary (specific) granules of polymorphonuclear neutrophils at 15 to 30 μg/10^7 cells, depending on the age of the individual (Baggliolini et al., 1970; Baveye et al., 1999; Bennet and Kokocinski, 1978; Levay and Viljoen, 1995; Naidu, 2000), and a stationary pool on the mucosal surfaces (van Hooijdonk et al., 2000; Levy, 1996; Levay and Viljoen, 1995; Lonerdal and Lyer, 1995). The presence of LF in mucosal surfaces originates from epithelial cells or sub-mucosal glands that excrete LF by the influence of the parasympathetic nervous system (Naidu, 2000; Raphael et al., 1989; Testa, 2002). Examples of mucosal surfaces where LF can be found are lactating breast tissues, gastric tissues, and duodenal epithelial cells (Mason and Taylor, 1978). LF accounts for 11.5% of the total secretory proteins excreted by bronchial glands (Harbitz et al., 1984) and 25% of the total tear protein produced by lachrymal glands (Kijlstra et al., 1983). In addition, concentrations of 0.2 to 1.0 mg/ml are found in seminal plasma (Witchman et al., 1989) and have been found in the sperm-coating antigen (Ashorn et al., 1986).

The highest levels of LF (5–7g/l) are observed in colostrum and gradually decrease by almost sevenfold in mature milk during lactation (Hennart et al., 1991; Hirai et al., 1990; Levay and Viljoen, 1995; Lonnerdal and Iyer, 1995; Montagne et al., 1998; Playford et al., 2000). In plasma it is normally found at low levels (approximately 0.2–1.6 μg/ml) (Levay and Viljoen, 1995; Steijns and van Hooijdonk, 2000; van der Strate et al., 1999). Milk from mastitic cows contains higher levels of LF than that from healthy ones. In addition, the concentration of LF is higher when infection is caused by *Staphylococcus aureus* and streptococci species rather than when caused by coagulase-negative staphylococci or *Corynebacterium bovis* (Hagiwara et al., 2003; Kai et al., 2002), indicating the antimicrobial role of the compound and its potential to prevent disease.

Increase of LF in plasma may occur during acute-phase host responses, such as infection, inflammation, or toxic shock (Klasing, 1984; Baynes et al., 1986; Naidu, 2000; Qadri et al., 2002). van der Strate et al. (1999) established a linear correlation between concentrations of LF in plasma and counts of neutrophils. Elevated levels of LF in plasma are a primary indication of an inflammatory state, such as septicemia and endotoxemia (Gutteberg et al., 1988) and are commonly followed by an increase of LF in serum, thus making LF levels a reliable diagnostic tool for pathological conditions (Qadri et al., 2002). Increases of LF both in plasma and serum also correlate with increases in C-reactive protein during meningococcal septicemia (Gutteberg et al., 1984), cystic fibrosis protein during cystic fibrosis (Barthe et al., 1989), cholecystokinin secretion during pancreatitis (Dite et al., 1989), or lysozyme during localized juvenile periodontitis (Friedman et al., 1983). The iron-chelating properties of LF are an additional reason for increased clinical levels of LF (e.g., during rheumatoid arthritis) (Blake et al., 1981; Bukhardt and Schwingel, 1986) as well as in various pathogenetic malignancies (especially carcinomas) when iron uptake by neoplastic cells is necessary (Barresi and Tuccari, 1987).

18.3.1.1.2 Physico-Chemistry

LF is a single polypeptide chain with a molecular weight of 75 to 80 kDa and consists of approximately 690 amino acid residues (Baker et al., 2000; Metz-Boutigue et al., 1984; Powel and Ogden, 1990; Rey et al., 1990). The amino acid composition of LF includes 16–18 different amino acids (of which 8 are basic); the frequency and types of amino acids depend on the mammalian species (Naidu, 2000).

The reported isoelectric point (pI value) for LF varies with the measuring method. According to electrophoretic, chromatofocusing, and Rotafors methods, the pI of LF is around 8.0 (Groves, 1960; Shimazaki et al., 1993; Szuchet-Drechin and Johnson, 1965), although it may range from 5.5 to 10.0 (Naidu, 2000). In general, LF is considered a basic protein (with an expected pI higher than 7.0), because its N-terminal region contains multiple arginine (arg) and lysine (lys) residues that make this region extremely basic (Vogel et al., 2002). The heat resistance (known to be stable at 90°C for 60 min) and heat-induced enthalpy of LF change depends on its iron-binding status and the pH of the substrate. The iron-saturated form of LF (i.e., holo-form [87 ± 3°C]) is more stable than the non-saturated form (i.e., apo-form [67 ± 3°C]) (Naidu, 2000; Paulsson et al., 1993; Reiter and Oram, 1967; Rossi et al., 2002).

Neutral and basic pH enhances thermal denaturation of LF, whereas LF is more heat-stable at low pH, especially around 4.0 (Saito et al., 1994). The holo-form of LF is more resistant to proteolysis than the apo-form (Brock et al., 1976). Moreover, binding of Ca^{2+} *via* the carboxylate groups of the protein surface increases the chemical and thermal stability of bovine LF (Rossi et al., 2002). Specifically, the presence of Ca^{2+} causes an approximate 9°C increase in the denaturation temperature of both holo- and apo-forms of LF (Rossi et al., 2002).

18.3.1.1.3 Structure

LF is characterized by an amphipathic structure with strong cationic properties on the N-terminus (Bellamy et al., 1992). The three-dimensional structure of human and bovine LF have been determined to a 2.8 Å resolution (Naidu, 2000; Thakurta et al., 2003), and a high degree of homology of this protein has been observed between species. LF is folded into similarly sized globular and homologous terminal N- and C-lobes, which are stabilized by intrachain disulfide bonds and linked by an α-helix that provides flexibility (Baker et al., 1987; Baveye et al., 1999; Naidu, 2000; Vorland, 1999). The symmetry of the lobes derives from gene duplication, and they are further divided into two similar domains of about 160 amino acids (Testa, 2002). Each lobe contains glycosylation sites (the number varies among mammalian species) in homologous position and has the capacity to bind one Fe^{3+} with high affinity (Kd = 10^{20} M^{-1}) together with one CO_3^{2-} or HCO_3^- ion that is held electrostatically to an arginyl side group (Naidu, 2000). Each iron atom is coordinated to four protein ligands in accordance with the carbonate or bicarbonate ion. Certain areas within the molecular structure serve as potential cavities for binding not only of carbonate but also of larger anions. Specifically, the HCO_3^- occupies a pocket between the iron and two positively charged groups and serves to neutralize this positive charge, thus facilitating binding of iron ions. Using X-ray structure analysis it has been concluded that the multi-domain nature and the flexibility of LF allow for binding of various metals with no significant changes in the total structure (Naidu, 2000). Apart from the sites for iron absorption, metal-binding sites for other cations, such as Cu(II), Cr(III), Mn(III), and Co(III), are available within the structure of LF similarly to serum transferrin (Ainscough et al., 1980; Hirose, 2000; Naidu, 2000). The anion-binding properties of LF are affected by the type of metal ion that is already bound. Anions that have been investigated for their potential to bind LF in the presence of Fe^{2+} or Cu^{2+} include carbonate, citrate, oxalate, and hybrid carbonate–oxalate complexes (Brodie et al., 1994). The role of the citrate/bicarbonate ratio is very crucial for the apo- to holo-form transition of LF (Nonnecke and Smith, 1984), that controls the iron-binding process and, hence, the biological activity of LF.

LFcin is produced by peptic hydrolysis (also mentioned as cleavage degradation) of bovine or human LF (Bellamy et al., 1992; Tomita et al., 1994) with different proteases. Bovine LFcin (commonly termed "lactoferricin B" [LfcinB]), was found to consist of 25 amino acid corresponding to the sequence of residues 17 to 41 close to the N-terminal of the molecule (Pierce et al., 1991; Tomita et al., 1994; Wakabayashi et al., 1992). Human LFcin (commonly termed "lactoferricin H" [LfcinH]) was found to consist of 47 amino acid residues, corresponding to the sequence of residues 1 to 47 at the N-terminal of the human LF (Bellamy et al., 1992; Metz-Boutigue et al., 1984). Both LFcinB and H have an almost circular structure consisting of a loop of 18 amino acid residues that links two linear subfragments, and have molecular weights of 3126 and 5558, respectively (Bellamy et al., 1992; Metz-Boutigue et al., 1984; Pierce et al., 1991). The linkage of subfragments occurs through a disulfide bond between the two terminus cysteine residues of the loop (Bellamy et al., 1992).

18.3.1.2 Antimicrobial Activity

18.3.1.2.1 Mode of Action

The diversity of substrates in which LF is present, its coexistence with other physiological substances, and the regulatory role of LF as acute-phase reactants explain its multifunctionality and its wide spectrum of antimicrobial activity (Naidu, 2000; van Hooijdonk et al., 2000). It is considered that the antimicrobial activity of LF is generally dependent on its protein conformation (structural characteristics and spatial orientation) and substrate conditions (Naidu and Arnold, 1997). In particular, two basic biochemical properties of LF contribute to its antimicrobial effect and involvement in the host defense: the extremely powerful iron-binding capability, and its strong interaction with other molecules and surfaces. The amphipathic structure and the net positive charge, especially on the strongly cationic N-terminus region of LF, are considered deterministic for its interaction with microbial membranes (Bellamy et al., 1992). Binding of LF occurs *via* specific membrane receptors that exist for example in the leukemic lines, the activated T-lymphocytes, the monocytes-macrophages, the brush-border membranes, the parenchymal liver cells, the breast epithelial cells, platelets, and neuronal cells (Testa, 2002). Specific molecules that interact with LF are glycosaminoglycans of epithelial milieu (e.g., heparin sulfate in mucins), collagens, fibronectins, and DNA of mammalian cells (Naidu, 2002; Nichols et al., 1990; Wu et al., 1995).

As previously mentioned, LF is a broad-spectrum antimicrobial, and modes of action appear to be common for different types of microorganisms. Among the studies conducted to elucidate the mechanisms of antimicrobial activity, most have been performed with bacteria and viruses although work has been done with fungi and protozoa. The following section introduces potential mechanisms by which LF may inhibit microorganisms, specifically bacteria and viruses. LF may be bacteriostatic or bactericidal, and iron deprivation is the most likely underlying mechanism for the direct bacteriostatic effect of LF (Naidu, 2000) as demonstrated by *in vitro* studies. However, bacteria such as *Escherichia coli*, *Moraxella catarrhalis*, *Neisseria* spp., and *Vibrio* spp. have potential defensive mechanisms to counteract iron depletion, mainly by the formation of siderophores that mediate iron uptake. Alternatively non-siderophore-mediated iron uptake by bacteria may occur either through outer-membrane protein receptors (*Neisseria meningitis*) that recognize the complex of LF-iron and internalize the chelated iron cation (Mickelsen et al., 1982; Naidu, 2000; Schryvers and Morris, 1988; Tranter, 1994), or by the production of reductases as is the case with *L. monocytogenes* (Cowart and Foster, 1985). Moreover, the bacteriostatic activity of LF may be reduced by factors that are irrelevant to microbial resistance, such as inappropriate citrate/bicarbonate ratios, given that bicarbonate is essential for iron chelating, whereas citrate competes with LF for iron and, hence, makes the iron available for growth by the organisms (Reiter, 1978b).

The bactericidal activity of LF is iron-independent and relates to direct binding of positively charged LF on negatively charged microbial outer membranes, which results in dispersion of lipopolysaccharides (or negatively charged fatty acids), increase in membrane permeability, and eventually cell death (Arnold et al., 1980; Baveye et al., 1999; Caccavo et al., 2002; Ellison et al., 1988; Naidu and Arnold, 1997; Pellegrini 2003; Rossi et al., 2002; Yamauchi et al., 1993). Ellison et al. (1990) suggested that Ca^{2+} and Mg^{2+} are the major cations that modulate the LF-induced damage in Gram-negative bacteria, and this conclusion was confirmed by Rossi et al. (2002). The binding of LF to microbial cell surfaces, however, is enhanced by bacterial receptors that exist in a variety of Gram-positive bacteria (e.g., lipoteichoic or teichoic acids, surface layer proteins, peptidoglycan components, heat shock proteins [*Helicobacter pylori*]) and Gram-negative bacteria (e.g., lipid A, whereas the pore-forming outer membrane proteins [porins] are a very common binding target of LF) (Amini et al., 1996; Baveye et al., 1999; Caccavo et al., 2002; Dhaenens et al., 1997; Erdei et al., 1994; Naidu, 2000; Naidu and Arnold, 1994; Naidu et al., 1993; Sallmann et al., 1999; Vorland et al., 1999c). Likewise, antibiotic potentiation has also been demonstrated *in vitro* as a result of increase in the permeability of outer membranes (Naidu and Arnold, 1994; Vorland et al., 1999c; Wakabayashi et al., 2002). An additional antimicrobial function that stems from the interaction of LF with the outer membrane of Gram-negative bacteria is the inhibition of microbial attachment to sub-epithelial matrix proteins, or alternatively the detachment of bacteria from mucosal surfaces (Kawasaki et al., 2000; Naidu and Bidlack, 1998). The blocking of cell-surface attachment factors, such as fimbriae and other adhesins, and inhibition of the colonization-factor antigens synthesis are associated with this antimicrobial mechanism (Naidu and Bidlack, 1998; Naidu, 2002). Moreover, LF binds on many tissue surfaces with higher affinity than cell-surface anchors of pathogens (Naidu, 2002). Bellamy et al. (1993) demonstrated an additional mechanism derived from membrane disruption (similar to that of polymyxin B) related to the inhibition of proline uptake. Optimal binding (dose-dependent) to membranes and biocidal effect was obtained at pH 6.0 and pH 7.5 for *E. coli* and *Bacillus subtilis*, respectively. The interaction of LF and LFcin with microbial membranes has also been found to account for their fungicidal activity (Bellamy et al., 1993; Wakabayashi et al., 1996).

Lactoferrin has been studied *in vivo* and *in vitro* against *E. coli* O157:H7. The mechanisms used against *E. coli* O157:H7 and others are bacteriostatic, bactericidal, and anti-adhesion effects. By use of these three mechanisms lactoferrin has the potential to decrease colonization pressure on farms and in the long run potentially decrease *E. coli* O157:H7 illness in humans. There have been no reports of bacteria developing lactoferrin resistance (Rybarczyk et al., 2017). Another interesting aspect of bovine lactoferrin is that it has been shown to be safe for preterm neonates while holding promise to reduce major morbidities in preterm infants (Embleton et al., 2013).

Many new techniques for food preservation are being researched and developed. Techniques such as HHP, irradiation, ohmic heating, cold plasma, and ultrasound are some of the newest in the last 10 years that have been worked on, yet not much information is available on the effect of a variety of compounds such as milk proteins and enzymes such as lactoferrin. As lactoferrin is susceptible to degradation by heat treatment some of these new techniques may be useful in preserving lactoferrins' useful properties. Also lactoferrin may be able to enhance the safety and efficiency of the new technologies as well in addition to its existing biological function (Franco et al., 2018). More research is needed with these new technologies as food matrix interactions with lactoferrin can affect its properties.

A more detailed investigation of the mode of action of both LF and its hydrolysates, oriented towards the effect of LF on liposomes, electrochemical potential ($\Delta\Psi$), and pH gradient of Gram-negative pathogens, has become available in the literature. Many studies aim to fully elucidate all the potential underlying mechanisms of antimicrobial activity, as well as to detect synergism phenomena between LF and other compounds. It was shown that 0.1 to 6.4 µM of iron-free LF increased permeability of the outer and inner membrane of *E. coli*, reduced by 50% or dissipated the electrical potential and the pH gradient, and also caused selective ion permeabilization on liposomes of *E. coli* (Aguilera et al., 1999, 2003). Different amino sequences of LFcinB were found to enter the cytoplasm of *S. aureus* and *E. coli* within 15 min of exposure, whereas lower quantities of LFcinB were also found in the cell wall (Haukland et al., 2001). Penetration and attachment to the cell wall was proven to be time- and concentration-dependent. Ulvatne et al. (2001) illustrated that 30 to 100 µg/ml of LFcinB, corresponding to concentrations below the minimum inhibitory concentrations (MIC) or minimum bactericidal concentrations (MBC), caused morphological changes of cells and depolarized cytoplasmic membrane and destabilized liposomes, causing leakage and fusion of liposome contents, whereas no cell lysis was observed. These indications emphasize that membrane destabilization is the major mechanism for the antimicrobial activity of LFcinB.

Investigation of the synergism between LFcinB and antibiotics has provided useful indications about potential clinical applications as well as for elucidation of the mode of action of LFcinB and/or common antibiotics. The antimicrobial effect of antibiotics, which are usually excluded by the outer membrane of Gram-negative bacteria (e.g., erythromycin, vancomycin, and penicillin) may be enhanced by LFcin which destabilized bacterial membranes and facilitated their penetration (Diarra et al., 2002; Vorland et al., 1999b). Recent studies have further elucidated the mode by which LFcinB destabilizes bacterial membranes. Hossain et al. (2019) present the importance that rapid permeabilization of bacterial plasma membranes has in cessation of the cells based on cell length during interaction with LFcinB. The researchers further found that LFcinB-induced local rupture of lipid bilayers and rapid permeabilization of Gram-negative bacterial cell membranes is significantly affected by membrane potential.

Other antibiotics, such as polymyxins which act on membranes, may compete with LFcin with a moderated antimicrobial effect. Similarly, the prevention of entry of ribosome-targeting aminoglycosides (gentamycin) into cells due to disturbance in the respiratory chain of the cytoplasmic membrane caused by LFcinB may also result in a reduction of the antimicrobial effect of antibiotics (Vorland et al., 1999a). For the above reasons, LFcinB acted synergistically with penicillin against *S. aureus* and with erythromycin against *E. coli*, whereas antagonism was evident between gentamicin and LFcinB against *S. aureus* (Vorland et al., 1999a). In another study, minocyclin and other compounds (e.g., acids, alcohols, and acylglycerols) enhanced the antimicrobial activity of LFcinB against antibiotic-resistant strains of *S. aureus* (Wakabayashi et al., 2002). The effect of sequence of application of LFcin and antibiotics was studied *in vitro* by Haukland and Vorland (2001), who suggested that LFcinB had no overwhelming effect when applied at the post-antibiotic stage against *E. coli* and *S. aureus*. Microbial resistance to LF is a potential response, and a possible resistance mechanism for *E. coli* and *S. aureus* involves the production of proteases that decompose LFcin (Ulvatne et al., 2002). Additionally, the presence of Mg^{2+} ions in the growth environment of cells prior to or during exposure to LFcin may also offer protection against subsequent exposure to LFcin by altering the properties of the outer membrane (Masschalk et al., 2003).

With respect to antiviral activity of LF, according to Naidu and Bidlack (1998) and van der Strate et al. (2001), the potential mechanisms for antiviral activity are the following: (i) prevention of viral infection by direct binding of LF to virus particles (e.g., envelope proteins); (ii) interference with virus docking into cells by binding of LF either to sulfate proteoglycans (HSPGs), or to viral receptors of the host cells that are used by the virus for intracellular entry; (iii) inhibition of virus proliferation; (iv) intracellular activity of LF likely associated with interference with antigen synthesis of virus (e.g., during infection by rotavirus); and (v) other indirect mechanisms, such as cell cytopathy, regulatory function of LF on myelopoiesis, and cytopathy during viral infection (e.g., friend virus complex, inhibition of viral hemagglutination, like human influenza virus). Factors such as the stage of infection (i.e., duration of infection before addition of LF) (Vorland, 1999), other synergistically acting compounds, such as zidovudine (Viani et al., 1999), and the saturation of LF with different metals (e.g., Zn^{2+}, Mn^{2+}, Fe^{3+}) are crucial for the antiviral outcome of LF (van der Strate et al., 2001).

Finally, LF is important in immunoregulation and specifically in the first line of the host defense system, which is an interaction of neutrophils, lymphocytes, macrophages, and their secretory products. Synergism with antibodies, or the activation of a complex series of reactions, is associated with the contribution of LF to the overall protective immune response after infection or inflammation (Levay and Viljoen, 1995; Lonnerdal and Iyer, 1995; Sanchez et al., 1992; Machnicki et al., 1993; Ward et al., 2002; Zagulski et al., 1989). Specifically, when contamination with a pathogen occurs, polymorphonuclear neutrophils capture the invader (phagocytosis) and specific granules release low iron-saturated LF (6–8%) into the blood, which in turn creates an hypoferremic environment (depletion of iron), thus preventing the pathogen from acquiring sufficient iron for growth (Miyauchi et al., 1998; Sanchez et al., 1992; Ward et al., 2002). The affinity of LF with lipopolysaccharides (LPS) is also beneficial, since binding of LF to lipid A, the toxin moiety of LPS of Gram-negative pathogens, prevents the toxic shock induced by interaction of this lipid with monocytes/macrophages (Baveye et al., 1999; Caccavo et al., 2002). A review of mechanisms related to the modulation and amplification of the inflammatory process by LF that was presented by Baveye et al. (1999) listed the following mechanisms: (i) enhancement of phagocytosis by preventing the deactivation of complement factor C3; (ii) inhibition of extracellularly formed hydroxy-radicals by sequestration of free iron; (iii) enhancement of polymorphonucleocytes (PMNs) recruitment to inflammatory sites; (iv) regulation of the proliferation and differentiation of immune cells (suppression of myelopoiesis) by suppressing the production of interleukins IL-1 and IL-6 by monocytes; (v) inhibition of platelet aggregation; and (vi) promotion of the recruitment and activation of immune cells in inflammatory sites by blocking the release of other cytokines (e.g., the tumor necrosis factor alpha-TNF-α) (Crouch et al., 1992; Machnicki et al., 1993; Mattsby-Baltzer et al., 1996; Shinoda et al., 1996; van Hooijdonk et al., 2000; Vorland, 1999; Wagstrom et al., 2000 ; Zagulski et al., 1989). A number of studies reviewed by Ward et al. (2002) illustrated that inhibition of TNF-α by LF may protect human and mice from allergen-induced skin inflammation. Moreover, it is shown that LFcin and some of its acylated derivatives inhibit the lipid peroxidation in liposomes (Wakabayashi et al., 1999a). In one of the latest reviews on LF, Wakabayashi et al. (2003) used terms such as immunosuppression and immunostimulation to describe the above immunoregulatory activities of LF and LFcin or its derivatives with shorter amino acid sequences. Immunostimulation, and specifically the stimulation of phagocytic and cytotoxic properties of macrophages, was also termed the opsonic effect (Naidu, 2000). A list of major binding targets of these peptides, including LPS, heparin, DNA, glycosaminoglycan, etc., and some more specific mechanisms, such as the induction of apoptosis in monocytic and mycloid leukemic cells, or the activation of kinase CK2, commonly representative of different LFcin residues (Maekawa et al., 2002; Roy et al., 2002; Yang et al., 2002).

ALF as an immobilized derivative of LF possesses identical mechanisms for antimicrobial activity, summarized into three major categories: (i) blocking of microbial adhesion; (ii) bacterial detachment; and (iii) microbial growth inhibition (Naidu, 2002). The interaction of ALF with the outer membranes of bacteria, interference with adhesin/fimbrial synthesis or colonization factors, and competition with bacteria in adhesion on tissue-matrix components are the properties responsible for the first two of the above mechanisms. Iron deprivation with further consequences on ATP synthesis and cellular multiplication accounts for the growth inhibitory mechanism, and the interaction of ALF with nucleic acid is likely associated with its antiviral activity.

18.3.1.2.2 Specificity

LF is known to possess a wide antimicrobial spectrum including Gram-positive and Gram-negative bacteria, such as *B. subtilis*, *E. coli* O157:H7, *E. coli* O111, *Helicobacter pylori*, *Klebsiella pneumoniae*, *L. monocytogenes*, *Micrococcus flavus*, *P. aeruginosa*, *S. aureus*, *Proteus mirabilis*, and *S.* Typhimurium (Bellamy et al., 1993; Naidu, 2000; Viejo-Diaz et al., 2003; Wakabayashi et al., 2003), yeast of the genus *Candida* spp. (Kuipers et al., 2002; Samaranayake et al., 1997; Ueta et al., 2001; Wakabayashi et al., 1996, 1998), fungi, such as *Rhodotorula rubra* (Andersson et al., 2000), *Penicillium* spp. (Liceaga-Gesualdo et al., 2001), *Trichophyton* spp. (Wakabayashi et al., 2000), RNA and DNA, enveloped or nonenveloped viruses (Naidu, 2000; van der Strate, 2001; Vorland, 1999), and protozoa, including *Toxoplasma gondii* (Isamida et al., 1998), *Giardia lamblia* (Turchany et al., 1995), and *Tritrichomonas foetus* (Grab et al., 2001).

For determination of the MIC and MBC of bovine and human LF as well as their hydrolysates (i.e., LFcinB and LFcinH) the standard microdilution method of Vorland et al. (1999b) in 1% peptone water has been used by the majority of researchers. In general, hydrolysates of LF (i.e., LFcins) and especially LFcinB are more inhibitory than the original LF (Naidu, 2000). It is important to note that LFcin of human origin is more active than LFcin of bovine, murine, and caprine origin (Vorland et al., 1998). According to Bellamy et al. (1992), who performed a screening on MIC of bovine and human LF as well as the corresponding LFcins against *E. coli* O111, the MICs of human and bovine LF were 2000 µg/ml and 3000 µg/ml (37µM), respectively; whereas the MICs of LFcinH was 100 µg/ml (ca. ten-fold lower than that of LF), and that of LFcinB was 6 µg/ml (almost 1000-fold lower than LF). In the same review, the MIC of LFcinB was found lower than the MIC of bovine LF against a variety of other microorganisms, including *K. pneumoniae*, *L. monocytogenes*, *P. aeruginosa*, and *S. aureus*. In a more recent study, Kimura et al. (2000) reported that the MIC of Korean goat LF against *E. coli* O111 was 5000 µg/ml (i.e., higher than that of bovine LF but lower than the MIC of sheep and horse LFs) (Lee et al., 1997). The results of this study also agree with the observed differences between the antimicrobial activity of bovine LF and its hydrolysate, since the MIC of Korean goat LFcin was only 100 µg/ml (i.e., similar to that of LFcinH and 50-fold lower than that of Korean goat LF). Ulvatne et al. (2001) demonstrated that *P. mirabilis* was more resistant to LFcinB than *E. coli*.

In a study of Recio and Visser (2000) the antibacterial effects of the apo- and holo-forms of bovine, ovine, and caprine LF were comparatively evaluated in a medium containing 1% peptone and 1% glucose, against *E. coli* and *M. flavus*. Of the apo-lactoferrins, caprine LFcin had the highest effectiveness against both microorganisms, followed by bovine and then ovine LF. Holo-forms of LF were almost ineffective. Griffiths et al. (2003) comparatively investigated the effect of the apo-form and the holo-form bovine and human LF against *E. coli* and *S.* Typhimurium in coculture with probiotic bacteria, such as bifidobacteria. A conclusion of high importance for the protective role of intestinal flora was that the apo-form of LF retarded the growth of enteric pathogens without affecting the growth of probiotic bacteria. This conclusion is beneficial for the potential application of LF in foods compared to antibiotics, given the deleterious effect of the latter to the intestinal flora of humans.

The antimicrobial effect of several LFcinB analogs (i.e., peptides with shorter amino acid sequences) has also been considered by many research groups with varied results related to their effectiveness compared to the original LFcinB (Chapple et al., 1998; Chen et al., 2003; Hoek et al., 1997; Kang et al., 1996; Odell et al., 1996; Rekdal et al., 1999; Strøm et al., 2002a, b; Ueta et al., 2001; Wimley and White, 2000). Schibli et al. (1999) reported that the six-residue center of LFcinB is responsible for the antimicrobial activity of the whole molecule. Usually, the examined analogues are 11- or 15-residue LFcin derivatives synthesized by the incorporation of certain amino acids on residues 6 and 8 (for addition of tryptophan), 5 and 9 (for addition of arginine), or other residues (e.g., lysine) of LFcin (Chen et al., 2003; Haug and Svendsen, 2001; Kang et al., 1996; Strøm et al., 2000, 2002a, b). The antimicrobial activity of these analogues depends on the type of incorporated amino acids, the terminus of LFcin where addition of amino acids is made (N-terminal analogues show higher variability on antimicrobial effectiveness than C-terminus analogues), as well as the charge and the lipophilicity of the resulting analogue (Strøm et al., 2000, 2002a). Tryptophane derivatives seem to be the most effective analogues, probably due to the high membrane affinity of this amino acid (Haug and Svendsen, 2001; Strøm et al., 2002b; Wimley and White, 2000).

Wakabayashi et al. (1999b) demonstrated that N-acylated and D-enantiomer derivatives of LFcin possess increased antibacterial and antifungal activity as indicated by the higher MICs of LFcinB. The MICs of these derivatives ranged from 3 to 12 μg/ml against bacteria, such as *E. coli*, *P. aeruginosa*, *S. aureus*, and the fungus *Trichophyton mentagrophytes*, whereas significantly higher MICs, ranging from 25 to 100 μg/ml were observed for *C. albicans*. Vorland et al. (1999a) made a comparative test on the MIC of LFcin and its peptides synthesized by L- and D-amino acids, on spheroplasts of *E. coli*, *P. mirabilis*, and protoplasts of *S. aureus* as well as on the above microorganisms with intact cell walls. The D-enantiomer of LFcin was more active than the L-enantiomer. LFcin (MIC 30 μg/ml and MBC 80 μg/ml) and its D-enantiomer (MIC and MBC equal to 30 μg/ml) had the same effect on *S. aureus*, but the D-enantiomer with MIC and MBC equal to 5 μg/ml was far more effective than LFcin against *E. coli*. None of the tested peptides was effective against *P. mirabilis*. Spheroplasts and protoplasts were more sensitive than their counterparts, suggesting that the cell wall has a protective role against LFcin. Lower temperatures (22 and 28°C) sensitized bacteria to all tested peptides in comparison with 37°C, and similar results for the effect of the D-amino acid counterpart of LFcin were reported by Ulvatne and Vorland (2001).

The antifungal properties of LF and LFcins have also been investigated. Wakabayashi et al. (1996) reviewed the antifungal spectrum and fungicidal mechanism of LFcinB. *Candida* species, due to their clinical significance as the causative agents of many types of candidosis, are the most extensively used organisms in studies for the investigation of the antifungal effect of LF and LFcins. Indeed, the antifungal effect of LF, LFcin, and LFcin residues alone (Xu et al., 1999) or in combination with other compounds (e.g., azole antifungal agents) (Wakabayashi et al., 1996, 1998), or in the form of sodium alginate tablets (Kuipers et al., 2002), against *C. albicans* (Ueta et al., 2001), and other species (e.g., *C. glabrata* or *C. krusei*) has been well-established (Samaranayake et al., 1997). An inhibitory effect has also been reported for other fungi (Wakabayashi et al., 2003), including *Rhodotorula rubra* (Andersson et al., 2000), spores of *Penicillium* sp. (Liceaga et al., 2001), and *Trichophyton* spp. (Wakabayashi et al., 2000).

Additionally, LF and its hydrolysates have been shown *in vivo* (in mice) to inhibit protozoa, such as *Toxoplasma gondii* (Isamida et al., 1998), *Giardia lamblia* (Turchany et al., 1995), and tumor metastasis (Yang et al., 2002; Yoo et al., 1997). Some protozoa, however, such as *Tritrichomonas foetus* are reported to sequester the iron necessary for survival from the host (cattle) LF (Grab et al., 2001). Moreover, LF has antiviral activity against a wide range of human and animal RNA- and DNA-viruses, enveloped or not (Naidu, 2000; van der Strate, 2001; Vorland, 1999), such as hepatitis C virus (Ikeda et al., 1998, 2000), cytomegalovirus (Andersen et al., 2001), rotavirus (Superti et al., 1997), friend virus (Vorland, 1999), poliovirus (Marchetti et al., 1999), respiratory syncytial virus (Grover et al., 1997), human immunodeficiency virus (Swart et al., 1996; 1998; Vorland, 1999), human influenza virus (Kawasaki et al., 1993), spleen focus forming virus (Hangoc et al., 1987), feline immunodeficiency virus (Sato et al., 1996), and herpes simplex virus (Hammer et al., 2000; Hasegawa et al., 1994). In contrast, LFcin does not always demonstrate antiviral activity (Ikeda et al., 2000).

18.3.1.3 Applications in Foods

LF is available in ready-to-use form such as liquid or spray-dried powder. In the past two decades, its metal-chelating property has been the primary claim for the application of LF in several infant food formulas in South-East Asian markets (Satue-Gracia et al., 2000). Nandi et al. (2002) successfully induced expression of LF in transgenic rice grains at levels up to 0.5% by linking a synthetic LF gene to rice glutelin 1 promoter for future application in infant formulas. The pharmaceutical applications of LF and its hydrolysates LFcinB and LFcinH are also well-established (Clare et al., 2000, 2003; Yamamoto et al., 2003), as well as their potential benefits in curing infected fish and seafood (Gallardo-Cigarroa et al., 2000; Kakuta, 2000; Koshio et al., 2000). Oral administration of bovine LF is reported to enhance the survival of rainbow trout, red sea bream, and goldfish upon infection with *Aeromonas salmonicida*, *Cryptocaryon irritans*, and *Ichthyopthirius multifiliis*, respectively (Kakuta, 2000). However, the application of these compounds in foods is still limited so far due to the requirement of high doses to obtain preservative effect. For instance, the susceptibility of LF to pH, elevated levels of calcium or phosphates, excess of cations (especially iron), and improper citrate/bicarbonate ratios are factors with the potential to decrease the activity of LF (Naidu, 2002). The antimicrobial activity may also be reversed by trypsin, ferrous sulfate, magnesium sulfate, and hematin (Reiter and Oram, 1967). In addition, phenomena related to the risk of denaturation, structural alterations, or even charge-induced aggregation during the isolation of LF pose further limitations in the application of lactoferrin in foods. Therefore, research has been focused on the discovery of alternatives to overcome the limitations in application of LF. Potential alternatives include: (i) activation of LF under conditions that protect its structure and minimize the negative impact of milieu conditions, thus resulting in activated lactoferrin; and, (ii) LF digestion derivatives (i.e., LFs and their analogues with smaller amino acid residues) in an effort to evaluate the potential of compounds, which are released naturally as a result of proteolysis of LF and LFcin, respectively.

Investigation of the antimicrobial properties of LF and LFcins in foods is still in its infancy, with a few published studies available. A common conclusion drawn from these studies is that the activity of LF is reduced due to the excess of metal cation that saturates the peptide. Thus, the presence of other compounds with chelating properties, such as EDTA, or the dilution of cations concentrations are necessary to enhance LF activity (Branen and Davidson, 2000; Chantaysakorn and Richter, 2000; Masschalck et al., 2003). Alternatively, the combination of LFcin with other emerging technologies, such as high hydrostatic pressure, has been proven promising, especially when prevention of saturation of LFcin with ions is not feasible (Masschalck et al., 2001, 2003).

LFcinB was primarily tested in ground beef at concentrations of 50 or 100 μg/ml, where it was found to cause a maximum of 2 $\log_{10}$ CFU/g reduction at 4 or 10°C (Venkitanarayanan et al., 1999). The positive contribution of EDTA on the antimicrobial activity of LFcin was demonstrated by Branen and Davidson (2000) against *E. coli* and *L. monocytogenes* in TSB and in a medium containing 1% peptone, 0.05% yeast extract, and 1% glucose. LFcinB at concentrations of 1600 μg/ml was unable to inhibit the growth of *E. coli* at 37°C, whereas the addition of 100 or 400 μg/ml EDTA, depending on the strain of *E. coli*, totally prevented growth. Similar results were obtained against *L. monocytogenes*; however, lower concentrations of LFcin were necessary for inhibition due to the higher sensitivity of *L. monocytogenes* compared to *E. coli*. However, there seems to be a dispute on whether EDTA enhances LF activity in foods. Murdock and Matthews (2002) found no antimicrobial effect by the combination of the two compounds in UHT milk acidified to pH 4.0 and stored at 37°C, even at higher concentrations of LFcin and EDTA (i.e., 4000 μg/ml and 10 mg/ml, respectively). With respect to the increase of LFcin activity as a result of reduced cation concentration, Chantaysakorn and Richter (2000) found that 5000 and 10,000 μg/ml of LFcin had no effect on *E. coli* in carrot juice, but significantly retarded growth when added in a filtrate of carrot juice through 500 or 10,000 Da molecular weight rejection membranes and totally inhibited growth in dialysate of the filtrate. Considering the decrease in cation concentration in the filtrate and further reduced cation concentration in the dialysate of carrot juice, there is a likely negative correlation between antimicrobial activity and presence of metal cations.

Emerging preservation technologies, such as high hydrostatic pressure, when combined with LF have been proven promising in the effort to overcome the limitations in its application (Masschalck et al., 2001, 2003). High pressure (155 to 400 mPa) was reported to enhance the bactericidal activity of bovine LF and LFcinB against *E. coli*, *Salmonella* Enteritidis, *S.* Typhimurium, *Shigella sonnei*, *Shigella flexneri*, *P. fluorescens*, and *S. aureus* in potassium phosphate buffer at 20°C (Masschalck et al., 2001). High pressure up to 300 mPa as a single treatment caused 1 to 2 $\log_{10}$ reductions, whereas when applied in combination with antimicrobial peptides approximately 2 additional log reductions were sustained for all microorganisms. LFcinB caused consistently higher reductions than bovine LF, apart from the combination with 400 mPa, where bovine LF indicated higher reductions than LFcinB. The same group found that a combination of 100 or 270 mPa for 15 min with 20 μg/ml of LFcin in phosphate buffer increased reduction *S.* Typhimurium and *P. aeruginosa* by 1 to 2 log and 3 to 5 log (depending on strain) in comparison with the single application of high pressure and LFcinB, respectively. Recently Federico et al.

(2015) and Baruzzi et al. (2015) demonstrated the significant antimicrobial effect of LFcinB to reduce pseudomonads on ready-to-eat vegetables.

ALF is a novel LF derivative that has gained increasing interest as an alternative of LF, due to the limitations in application of the latter in foods. However, since ALF technology was established (Naidu, 2001), there is still limited information on the efficacy of this compound in foods. In a review by Naidu (2002), preliminary research data related to the detachment efficacy and bacteriostatic effect of ALF against *E. coli* in broth, beefsteaks, and fresh beef are shown. ALF and LF at a concentration of 1% were comparatively evaluated for their efficacy to detach collagen-bound *E. coli* and delay its growth. The ALF had a 2.7 log higher detachment efficacy than LF and caused 17.4 h more stasis on *E. coli* than LF according to impedimetric data in tryptic soy broth. Accordingly, MIC of ALF and LF was 62 μg/ml and >1000 μg/ml, respectively. Similar results demonstrating the better bacteriostatic performance of ALF compared to LF were obtained against 4 log contamination of *E. coli* on beef steaks. ALF was also tested in combination with current meat decontamination intervention strategies (cold/hot water and/or organic acids) for its efficacy to detach *E. coli* from fresh beef. Specifically, a set of sequential decontamination treatments (i.e., cold water for 10 s, hot water at 82°C for 30 s, and 2% lactic acid for 10 s) was applied alone or followed by spraying of 1% solution of ALF for 10 sec to detach an approximate *E. coli* population of 7 logs. The regular sanitizing assembly alone resulted in 72.2% detachment of *E. coli* per gram of beef tissue, whereas additional spraying with ALF increased the detachment efficacy to almost 100%/g of beef tissue. ALF was deposited on the meat surface as a fine mist created by electrostatic or high-pressure liquid spray-nozzles in a flow adjusted by digital controllers. ALF has also demonstrated activity against other pathogenic microorganisms, such as *L. monocytogenes*, *Salmonella* spp., and some meat spoilage organisms including *Pseudomonas* spp. and *Klebsiella* spp. Sensory evaluation of strip loins treated with ALF, stored under vacuum below 3.3°C and periodically exposed to retail display within a period of 35 days, showed that ALF might extend retail display life by 1.7–2.5 days. This conclusion was based on evaluation of lean and fat color, percentage discoloration, and overall appearance of strip loin samples treated with ALF compared to nontreated samples.

Furthermore, Ransom et al. (2003) comparatively evaluated the effect of single or sequential dipping into solutions of ALF, LF, and lactic acid (2%) on the survival and growth of *E. coli* O157:H7, *L. monocytogenes*, and *S.* Typhimurium, inoculated pre- or post-dipping, on vacuum-packaged bologna at 10°C for 33 days, as well as on beef plates, beef carcasses adipose, and beef lean tissues stored aerobically at 12°C for 2 to 29 days. It was shown that dipping beef bologna slices into solutions of ALF after inoculation enhanced the inactivation of *E. coli* O157:H7, and inhibited the growth of *L. monocytogenes* compared to untreated slices after 33 days of vacuum-packaged storage at 10°C. Treatment with ALF before inoculation delayed the growth of *S.* Typhimurium, whereas post-inoculation treatment prevented growth during storage; however, dipping into LF did not affect the growth of S. Typhimurium and *L. monocytogenes* compared with samples dipped in water. Of the treatments evaluated, lactic acid was the most effective treatment in inhibiting or reducing pathogen populations on all products, followed by ALF applied either pre- or post-inoculation, exclusively on bologna, while LF did not appear to affect survival, or growth of the bacterial populations. The same study also showed that dipping into ALF followed by dipping into 2% lactic acid was the most effective treatment for decontaminating inoculated beef carcass adipose tissue.

18.3.1.4 Safety and Tolerance

Bovine LF as a naturally occurring compound in milk has been consumed for years at levels ranging from 50 to 75 mg/day by humans of all ages (children, teens, and adults). The adverse effects of LF on iron absorption, modulation of microflora, and prevention of infection have been investigated with human studies, whereas animal studies involved acute toxicity (Nishimura, 1991), oral toxicity (Nishimura, 1997, 2000), and Ames assay (Kawai and Tanaka, 1997). No adverse effects on iron absorption, modulation of microflora, and prevention of infection are reported for daily intake of LF ranging from 0.3 to 1.0 g/kg/day for 11 days to 5 months for infants and 1.7 mg/kg/day to 60 mg/kg/day for a single-dose study for a total of 8 weeks for adults. Likewise, no adverse effects were evident in animal studies.

The regulatory frames of LF and ALF are based on their natural occurrence in milk and consequently in beef tissue, since both of them are consumed by humans. Initially, LF was characterized as GRAS by the U.S. Food and Drug Administration (USFDA) (GRAS proposal), and DMV International (De Melkindustrie Veghel International) suggested that its use is considered safe in sports and functional foods at a level of 100 mg/product serving, estimating that the daily uptake of LF from that use is 1.0 g per person per day (GRN 000077). In a previous GRAS notice (GRN 000067) National Beef suggested the use of bovine milk lactoferrin as a GRAS compound in uncooked beef (carcasses, subprimals, and finished cuts of beef) as a component of an antimicrobial spray at concentrations up to 2%. The estimated daily uptake of LF from this use was 4.1 mg per person per day. However, allergenic and immunological data related to both GRAS notices emphasized the necessity of stating the source of LF (e.g., cow milk) in the ingredient statement of the products that contain this compound in order to prevent consumption of these products by people who are allergic to milk or its constituents. In a GRAS Notice (GRN 000130), ALF Ventures (the company that commercially produces and markets ALF) suggested that milk-derived LF can be considered GRAS as a component of an antimicrobial spray for dressed beef, which will subsequently be rinsed to reduce the residues of milk-derived LF, even for people who are allergic to milk, and, hence, disclaiming its presence in ingredients statements is no longer necessary.

Milk-derived ALF was also considered GRAS (21 CFR.170.36[f]) and permitted at levels of 65.2 mg/kg of beef (Naidu et al., 2003) and, accordingly, in October of 2001, the USDA approved the use of ALF on fresh beef. However, the regulatory status of ALF and LF in other countries is not clear yet (Naidu, 2002). The European Union has issued a directive (83/417/EEC) for the use of proteins derived from milk, but LF is not currently included. Therefore, permitted use of LF in foods may only result from the consideration of LF as a milk protein (provision 79/112/EEC). Asian countries, such as Japan, South Korea, and Taiwan, have listed LF among other natural compounds in the List of Existing Food Additives; however, its use (in Taiwan) is limited to some specific nutritional foods under the condition "only for supplementing foods with an insufficient nutritional content and may be used in appropriate amounts according to the actual requirements" (Naidu, 2002).

18.3.2 Ovotransferrin

Ovotransferrin (OTF, also called conalbumin) is an iron-binding monomeric glycoprotein that constitutes at least 10–12% of the total egg white solids (Beuchat and Golden, 1989; Parkinson, 1966). There are many similarities between OTF and LF and thus the following section focuses only on those characteristics unique to OTF. The isolation and purification of OTF may be accomplished using solvent fractionation and chromatographic methods (ion-exchange chromatography or metal-affinity chromatography). Structural variability in terms of molecular size, amino acid composition, and visible/ultraviolet (UV) absorption spectra of OTF is evident among different birds and/or strains (Clark et al., 1963; Itoh et al., 1979; Lush, 1961; Osuga and Feeney, 1968). Many similarities exist between all transferrin family proteins, including OTF (i.e., a total number of 680 to 700 amino acid residues subdivided in two halves, one C-lobe and one N-lobe with 35 to 40% internal homology) (Testa, 2002). Moreover, the reported homology between OTF and human LF and serum transferrin is 49% and 51%, respectively. OTF is specified by the same gene as serotransferrin, but differs from the serotransferrin of the same species only in its glycan part (Testa, 2002).

Like LF, OTF reversibly binds two Fe^{3+} ions per molecule concomitantly with two bicarbonate anions. Its high affinity for iron (~10^{30} M^{-1}) renders the latter unavailable for bacteria and, hence, inhibits their growth (Tranter and Board, 1984; Valenti et al., 1983a). Therefore, the stoichiometric balance of iron in the environment is essential to maintain the effectiveness of OTF (Conner, 1993; Valenti et al., 1983b). It is well-known that saturation of OTF with iron reduces its effectiveness against many Gram-negative bacteria in hen egg albumen; however, OTF remains effective against Gram-positive bacteria, including lysozyme-resistant strains, at 30 or 39.5°C regardless of the presence or absence of iron (Tranter and Board, 1984). The antimicrobial activity of OTF is highly dependent on milieu conditions and the target organism. Alkaline pH and elevated temperature close to the physiological temperature of birds (~40°C) enhance the antimicrobial activity of OTF (Tranter and Board, 1984).

OTF is mainly considered to have bacteriostatic activity, although there is evidence of a biocidal effect independent of iron depletion, against a wide range of bacteria, such as *E. coli*, *Klebsiella* spp., *Proteus* spp., *Pseudomonas* spp., and *S. aureus* (Valenti et al., 1982, 1983a, 1984). It has been reported to extend the lag phase and reduce the growth rate of many Gram-positive and Gram-negative microorganisms, with the latter being less sensitive. Of the Gram-positive bacteria, *Bacillus* sp. and micrococci are the most sensitive groups (Board, 1969). An inhibitory effect against some yeasts species, such as *Candida* sp., has also been observed (Valenti et al., 1983a, 1985). Despite the negative impact of iron-saturation on the antimicrobial activity of OTF, the complex of OTF with other metal cations is reported to increase its antimicrobial effectiveness. Among the complexes of OTF with metal cations tested *in vitro*, the OTF-Zn^{2+} complex indicated the highest antimicrobial effectiveness (Valenti et al., 1983b).

OTF is heat-sensitive and 80% of the activity is lost with heating to 70–79°C for 3 min (Banwart, 1979), or 60°C for 5 min (Tranter, 1994). However, the presence of ions, such as phosphate or citrate at pH values above 6.0 (Nakamura and Omori, 1979), or saturation with iron in the presence of an anion, increases the resistance of OTF to heat and proteolytic action and disulfide reduction by thiols (Banwart, 1979; Williams et al., 1985).

18.4 Immunoglobulins

Immunoglobulins (Ig), or antibodies, are a complex, heterogeneous mixture of glycoproteins that are produced by plasma cells (lymphocytes or immunocytes) and are present on the B-cell membrane or secreted by plasma cells (Goldsby et al. 2000). Ig are the effectors of the immune system responsible for binding specific antigen molecules which are foreign to a host system to enable an immune response and clearance of the foreign substances and any associated harmful effects. Ig are able to recognize, bind, and occasionally neutralize bacteria, viruses, polysaccharides, nucleotides, peptides, and proteins (Bostwick et al., 2000). Due to the diversity and great multitude of foreign molecules or antigens (Ag) in nature, there is a need for a large number of Ig to recognize and counter all the different antigens that may exist. The great diversity of Ig molecules originates from subtle structural differences in their antigen combining sites, or variable regions, accounting for unique antigen binding specificities (Goldsby et al., 2000). The structural differences in the regions other than the antigen combining sites, constant regions, are related to the different effector functions mediated by antibodies, like complement activation or binding to one or more of the Ig receptors expressed on monocytes and granulocytes (Goldsby et al., 2000; Wilson and Stanfield, 1994). The main functions of Ig are thus, (i) Ag binding – Ig binding specifically to one or more closely related antigens *via* the antigenic determinant (Stanfield and Wilson, 1995); and (ii) effector functions – the Ig has no direct biological effect on the Ag and secondary "effector functions" are sequestered to aid in the clearance of the Ag and are usually preceded by Ag binding. Effector functions may include (i) complement fixation – a group of serum proteins that participate in an enzymatic cascade that generates the cytolytic membrane-attack-complex and results in lysis of cells through the release of biologically active molecules; and (ii) binding to various cell types such as phagocytic cells, lymphocytes, platelets, mast cells, and basophils that have receptors that bind Ig and activate the cells to perform a clearance function (Stanfield and Wilson, 1995). Those functions associated with lacto- and ovoglobulins mainly complement activation and Ig-augmented activity including agglutination, opsonization, adherence-blocking, and neutralization to aid in the clearance of microorganisms.

18.4.1 Lactoglobulins

Colostrum, as a source for transporting immune factors including Ig from mother to newborn, has been known for more than 100 years. While colostrum remains an important source, in humans, rabbits, and rodents, passive immunity is conferred mainly through the transfer of immune factors *via* the uterus. In contrast, for livestock (pigs, horses, sheep, and cattle), colostrum serves as the major source of immunity (Bostwick et al., 2000; Butler, 1994; Hilpert et al., 1987; Reddy et al., 1988; Stephan et al., 1990; Yolken et al., 1985). The use of commercially produced and purified Ig as antimicrobials for the prevention and treatment of microbial diseases in humans was proposed by Weiner et al. (1999), and use in food systems for preservation has been suggested (Korhonen et al., 1998; Pellegrini, 2003; Zommara et al., 2002). The limiting factors for the use of Ig from animal products are the cost of purification and the need for hyperimmunization to achieve broader specificity (Bostwick et al., 2000). Ig may be derived from a number of commercial sources, including colostrum and milk from different livestock, eggs, and cell culture; however, the most practical sources are bovine milk and colostrum followed by avian eggs (Bostwick et al., 2000).

18.4.1.1 Molecular Properties

18.4.1.1.1 Occurrence

Ig may be characterized according to their existence in the host, either as serum fractions or as secretory Ig, and since this chapter discusses antimicrobials in food originating from animal sources the focus will be on secretory Ig from lacteal secretions (milk and colostrum) and avian eggs. There are five general classes of Ig, namely IgG, IgA, IgM, IgE, and IgD, and another specific class, IgY, which is exclusively found in eggs (Bostwick et al., 2000). The Ig found in bovine and colostral whey include: (i) IgA; (ii) IgG with two major subclasses IgG_1 (comprising approximately 75% of IgG in milk), IgG_2, and IgG fragments; (iii) IgM; (iv) J-chain or component; and (v) the free secretory component (Butler, 1994). The nomenclature of Ig is based on their immunological cross-reaction with reference proteins of human origin and was proposed by the World Health Organization (Butler, 1983; Butler et al., 1971). The protein content of bovine colostrum and milk is higher than that from humans, with Ig levels in bovine colostrum higher than that of humans and, conversely, higher Ig levels in human milk than found in bovine milk (Butler, 1994; Hanson et al., 1993; Kulkarni and Pimpale, 1989). IgG accounts for the majority of the Ig fraction in colostrum and milk (ca. 75–80%) followed by IgM and IgA at similar levels (Butler, 1994; Hanson et al., 1993; Kulkarni and Pimpale, 1989).

18.4.1.1.1.1 Structure All Ig molecules are symmetrical glycoproteins of approximately 180 kD that are monomers or polymers of four polypeptide chains comprising two identical non-glycosylated light chains (L) (each approximately 20 kD) and two identical glycosylated (with mannoses and N-acetylglucosamine) heavy chains (H) (each approximately 50–70 kD) linked together with disulfide bonds (Butler, 1974; Lascelles, 1979; Vasilov and Ploegh, 1982). The H and L chains of each molecule have a constant (C) and a variable (V) region. The V regions are composed of approximately 100 amino acids near the N-terminal while the C region makes up the remainder of the molecule toward the C-terminal (Butler, 1974; Edelman, 1973; Lascelles, 1979). The variable N-terminal regions of both the H and L chains are considered to be important in the specificity of the Ig to which it can bind, whereas the complement fixation, membrane transport, species-specific, and class-specific Ag determinants are related to the C region of the H chains (Butler, 1974; Lascelles, 1979). The primary difference between the Ig molecules found in bovine colostrum and milk is in the structure of the heavy chains and especially in the V regions (Bostwick et al., 2000).

18.4.1.1.2 Biosynthesis

The genes coding for the H and L chains for the production of an Ig molecule are located on different chromosomes and there are separate genes coding for the V and C regions (Bostwick et al., 2000). The genes for the V and C regions on the same chromosome have discrete coding segments or exons separated by noncoding segments or introns, and the mRNA produced from these genes undergoes a series of splices to remove the introns which moves the V and C regions closer together and results in a functional H or L chain (Tonegawa et al., 1976). After synthesis the H and L chains are glycosylated mainly with mannose oligosaccharide (some linked to N-acetylgalactosamine) at asparagines primarily, as well as serine and threonine (Kemp et al., 1983). The assembly of the H and L chains into a complete molecule occurs when cysteine molecules form disulfide bonds, through either the formation and crosslinking of two H chains or the combination of an H and L chain (Buxbaum et al., 1971; Parkhouse, 1971). The Ig molecules are synthesized in a class of lymphocytes known as beta cells (B-cells) which may differentiate into plasma cells when activated and secrete Ig (Bostwick et al., 2000; Goldsby et al., 2000). The process of generating mucosal Ig that is found in bovine colostrum and milk begins with the uptake of Ag by M-cells of the Peyer's patches in the intestine or lamina propria of the bronchi (Goldsby et al., 2000). The Ag is presented to B-cells in lymphoid tissue where it is bound preferentially by membrane-bound IgD on the B-cells, activating the B-cells and causing them to secrete IgM, which later switches to IgG (Goldsby et al., 2000). The primed B-cells may then proliferate and differentiate into plasma cells and be directed *via* lactation hormones to the mammary gland where they produce mucosal Ig in colostrum and milk (Goldsby et al., 2000; Swain et al., 1999).

18.4.1.1.3 Stability

The stability of Ig molecules during processing is affected by thermal treatment (Dominguez et al., 1997; Li-Chan et al., 1995; Lindstrom et al., 1994), and although Ig is heat-sensitive, a substantial proportion of the Ig activity will survive pasteurization (Chen and Chang, 1998). The use of high-temperature/short-time (HTST) pasteurization techniques results in only 10–30% loss of Ig activity, while ultrahigh temperature (UHT) processing and evaporation destroy almost all the specific immune activity (Kummer et al., 1992; Li-Chan et al., 1995). Rapid inactivation of Ig starts at temperatures exceeding 65°C, and it has been demonstrated that at 81°C most of the antiviral activity is lost (Mainer et al., 1999). Anema (2000) indicated that thermal denaturation of Ig is retarded with increasing levels of milk solid concentration. The specific antimicrobial activity of Ig appears to not be affected by storage temperature, retaining its activity up to 12 months storage at 4, 20, and 37°C (Husu et al., 1993).

18.4.1.2 Antimicrobial Activity

18.4.1.2.1 Mode of Action

Bovine colostrum and milk may contain many naturally occurring antimicrobial substances including the antibody-complement system and complementary antigen-antibody binding activity (Pakkanen and Aalto, 1997; Regester et al., 1997; Reiter, 1985a). The antibody-complement system is considered to be one of the major antimicrobial activities in colostrum (Mueller et al., 1983). The antibodies or Ig absorbed from colostrum after birth in combination with the complement system have a crucial role in conferring passive immunity to a newborn calf (Butler, 1986; Staak, 1992). The Ig-mediated antimicrobial activity of complement is well-established in colostrum where all the complement components are present, while in milk the activity is not always demonstrated due to the fact that not all components are present (Eckblad et al., 1981; Husu et al., 1993; Korhonen et al., 1995; Reiter and Brock, 1975). The complement system consists of over 20 different proteins involved in an enzymatic cascade that may be activated by Ag–Ig interactions (classical pathway), by certain carbohydrates (lectin pathway), or by surfaces not protected by natural inhibitors (alternative pathway) (Korhonen et al., 2000). Complement may kill microorganisms, clear immune complexes, or induce and enhance antibody responses through lytic functions of the membrane-attack complex generated by the cascading enzymes (Fearon, 1998). The activation of the classical pathway involves binding of the first component of complement to Ag–Ig interactions or directly to microbes, while in the absence of Ig, complement activation occurs through the lectin pathway with lectins bound to pathogen surfaces or through the alternative pathway, involving the presence of bacterial cell membrane components such as lipopolysaccharides (Holmskov and Jensenius, 1996; Turner, 1996). Although Gram-positive and some Gram-negative bacteria are resistant to complement, most Gram-negative bacteria are sensitive to the lytic actions of complement and both may become inactivated due to opsonization and subsequent opsonophagocytosis (Rautemaa and Meri, 1999).

Aside from complement activation, the major Ig-augmented antimicrobial activity occurs through the highly specific, reversible, noncovalent Ag–Ig interactions responsible for immobilizing target Ag as well as the sequestration of other components of the immune system to remove the Ag (Bostwick et al., 2000). The mechanisms that may be used by Ig to target an Ag include agglutination, opsonization, adherence-blockade, toxin and virus neutralization, or stasis and cidal activity. Agglutination is brought about by the flexibility in Ig molecules, allowing cells with common surface Ag to be cross-linked to one another, and usually involves flagella and pili and, to a lesser extent, the cell membrane. Opsonization or covering of target Ag by Ig molecules may serve to signal effector immune cells to assist in targeting, enhancing phagocytic activity, and even destroying foreign bodies. Ig may bind to specific cell surface components on microorganisms and form an adherence-blockade, preventing the microbe from binding to its host cell surface receptor. Although Ig are rarely cidal, they do tend to display stasis more often by binding to microorganisms and causing alterations or disturbances within the cells by impairing growth, multiplication, and other cellular processes until other immune effectors may aid in killing or removal of the Ag. Additionally, Ig may serve to neutralize toxins and viruses by binding to various portions of the cells or molecules and thereby preventing receptor-mediated internalization of the Ag molecules.

18.4.1.2.2 Specificity

Ig molecules are one of the most effective antimicrobial agents in their ability to target a wide spectrum of pathogenic agents including bacteria, viruses, fungi, protozoa, toxins, and other proteinaceous or polysaccharide molecules due to the polyclonal nature of their composition. The two major effects on pathogenic agents involve blockage of attachment and invasiveness or immobilization followed by the destruction or removal of Ag regardless of its nature and origin.

18.4.1.3 Applications in Food

Commercial whey or colostral Ig has been used for many years as a feed supplement for farm animals, mainly newborns, to combat contagious diseases, and have been proven useful particularly against diarrheal diseases (Korhonen et al., 2000). Commercial products containing milk Ig have been developed and marketed for human use in the prevention and/or treatment of microbial disease (Ebina, 1992; Reddy et al., 1988). The main sources of Ig for commercial production are bovine milk and colostrum due to their availability and safety as compared with serum-derived analogues (Korhonen et al., 2000). Hyperimmune milk or whey derived from the immunization of cows with specific antibodies has been produced and found to contain higher levels of Ig than that of non-immunized cows (Casswall et al., 1998; Cordle et al., 1991; Facon et al., 1995; Greenberg and Cello, 1996; Kelly et al., 1996; Tacket et al., 1992) and has indeed been marketed in Asian countries for many years (Casswall et al., 1998, 2000; Hilpert et al., 1987). The extent of the increase in specific activity of Ig appears to be dependent on the specific organism and agent used for vaccination (Li-Chan et al., 1995; Facon et al., 1995; Tomita et al., 1995). Li-Chan et al. (1995) found increased ELISA activity for only one of five bacterial Ags in immunized cows in comparison to the levels of activity in milk from commercial dairies. Tomita et al. (1995) found that although cows vaccinated with a lipopolysaccharide-protein conjugate derived from *E. coli* J5 enhanced serum antibody titer to the organism, it did not enhance whey IgG titers. Several studies have suggested that feeding hyperimmune milk increases Ig against the immunizing bacteria and may reduce disease (Boedeker et al., 1987; Chernokhvostova et al., 1990; Murosaki et al., 1991). In early studies, Stephan et al. (1990) found that Ig preparation from the first colostrum of multiple cows displayed antimicrobial activity against a number of microorganisms and when tested *in situ* in a human host conferred Ig against *Cryptosporidium* (Shield et al., 1993). Ormrod and Miller (1991) reported that Ig from the milk of dairy cows immunized with a multivariate bacterial vaccine had anti-inflammatory activity in the rat hind-paw edema assay. Kobayashi et al. (1991) and Ishida et al. (1992) showed that hyperimmune milk conferred a higher survival rate for mice given lethal doses of irradiation as compared to control mice not receiving the milk. Although bovine colostrum and milk appear to be the ideal sources of Ig for human use, they originate from a foreign species and, as such, are limited to application only against oral and gastrointestinal pathogens or for topical applications in humans (Korhonen et al., 2000).

The immunoglobulin β-lactoglobulin in combination with hypericin was effective at reducing *S. aureus* by up to 5 logs. The complex of hypericin with β-lactoglobulin, a protein carrier that has full compatibility with dairy processes, indicates that it may be used as a disinfectant for food manufacturing and handling (Rodríguez-Amigo et al., 2015).

18.4.2 Ovoglobulins

Much like mammalian species that produce Ig in serum and lactations, domestic avian species such as chickens, turkeys, and ducks produce Ig in serum and in their eggs (Rose et al., 1974). The Ig in the avian serum is transferred to the yolks of eggs to provide the offspring with acquired immunity to avian diseases and other Ag, leaving the newly hatched chick relatively immunocompetent (Rose and Orlans, 1981). Although avian sera contains three main Ig (IgG, IgM, and IgA), the Ig found in egg yolk is referred to as IgY (Leslie and Clem, 1969) due to the different structure and immunological properties as compared with mammalian IgG (Akerstrom et al., 1985; Higgins, 1975; Jensenius et al., 1981; Kobayashi and Hirai, 1980). In comparison with mammalian IgG, IgY is much larger (Kobayashi and Hirai, 1980), more acidic, less rigid, and does not fix complement (Higgins, 1975).

18.4.2.1 Molecular Properties

18.4.2.1.1 Occurrence and Biosynthesis

Avian blood contains at least three distinguishable kinds of Ig, including IgG, IgM, and IgA, of which IgG comprises approximately 75% of the total Ig (Leslie and Martin, 1973). IgG (known as IgY in egg yolk) is transferred from the maternal serum to the egg yolk and subsequently to the circulation of the chick through the endoderm of the yolk sac, while IgM and IgA are secreted into the ripening egg follicle and incorporated into the egg *via* the oviduct (Locken and Roth, 1983; Patterson et al., 1962). Subsequently, the transfer of Ig to the embryonic gut occurs when amniotic fluid is swallowed and thereafter may provide passive immunity to the newly hatched chick (Losch et al., 1986). Hens lay on average of 240 eggs per annum, and the associated production of IgY totals 24 g, which is relatively high in comparison with certain mammals such as rabbits, mice, and goats (Hatta et al., 1997; Leslie and Clem, 1969). In a study comparing two species of chickens, it was determined that although the ratio of yolk weight to egg white was similar in both types of chickens and the total content of IgY in the yolk was relatively constant during 18 weeks, the level was influenced by hen species, egg weight, and egg production per day (Li et al., 1998).

18.4.2.1.2 Chemistry and Structure

IgY, much like the Ig, found in serum consists of two heavy (H) chains and two light (L) chains and has a molecular weight of approximately 180 kDa (Parvari et al., 1988). Unlike the H chains of mammalian IgG which have three constant (C) regions and a hinge region, IgY has four C regions and no hinge (Parvari et al., 1988). The content of β-sheet structure in the constant region of IgY is lower than that of mammalian IgG, and the flexibility of the region corresponding to the hinge region in IgG was less than that of IgG (Ohta et al., 1991; Shimizu et al., 1992). A lack of disulfide linkage in the IgY L-chain, lower flexibility in the hinge region, and other structural properties (molecular size, intramolecular bonding, and domain conformation) may all influence the lower molecular stability of IgY as compared with IgG (Shimizu et al., 1992).

18.4.2.1.3 Stability

The stability of IgY was compared to that of mammalian IgG when exposed to adverse pH and temperature environments (Hatta et al., 1993; Otani et al., 1991; Shimizu et al., 1988, 1992, 1993). In doing so, it was determined that above 70°C, IgY was more sensitive to the effects of heat than IgG, and the maximum temperature of denaturation for IgY was 73.9°C as compared with 77°C for IgG (Hatta et al., 1993). Shimizu et al. (1988, 1992, 1993) found that heating for 15 min at 65°C or higher, especially at temperatures above 75°C, reduced the activity of IgY. Otani et al. (1991) reported that at pH 2 and 3, IgY activity was more sensitive to the acidic conditions than IgG. Similarly, it was found that the activity of IgY was decreased at pH values below 3.5 and almost lost at values of 3.0, while at the opposite extreme, alkaline conditions up to pH 11 did not affect activity whereas incubation at pH 12 and above resulted in a significant decrease (Shimizu et al., 1988, 1992, 1993). It is well-established that although IgY is sensitive to pepsin digestion, it is relatively resistant to digestion by trypsin or chymotrypsin and that overall it is more sensitive to digestion by all three enzymes in comparison to IgG (Hatta et al., 1993; Otani et al., 1991; Shimizu et al., 1993).

18.4.2.2 Antimicrobial Activity

18.4.2.2.1 Mode of Action

The essential mechanism for Ig to clear foreign material including pathogenic microbes is related to the specific binding to complementary Ag. The Ag–Ig interaction is often referred to as the "lock-and-key" mechanism for the specific noncovalent interactions including hydrogen bonds, hydrophobic bonds, coulombic interactions, and van der Waals forces (Sim et al., 2000). The target Ag may be composed of amino acids, nucleic acids, carbohydrates, or lipids, and it is the strength of association and complementarity which determines the overall binding strength (Sim et al., 2000). IgY has mostly been documented for its antiviral and antibacterial activity achieved by the exclusion of these pathogenic agents from infection of host cells. In order for infection to develop in a host, the etiologic agent needs to form an association with the host cells either by contact or by subsequent internalization. Viruses express surface receptor molecules specific for the host cell membrane to initiate and assist in internalization of the infective material and, thus, the potential for preventing this is the ability of Ig to specifically bind to the viral receptor and block infection, a process termed viral neutralization (Goldsby et al., 2000). Bacterial pathogens are similar to viruses in that most bacteria need to form an adherence with the host cell before the infection process is initiated and the adhesion to the host cell usually triggers invasive mechanisms of the bacteria or toxin production, which may be internalized to cause infection. It was demonstrated that anti-fimbriae Ig prevented the attachment of bacteria to piglet intestinal epithelia by blocking the mucosal receptor, interfering with binding to mucins, or neutralizing the colonization factor (Imberechts et al., 1997; Jin et al., 1998; Wanke et al., 1990).

18.4.2.2.2 Specificity

IgY has been shown to be specific against infectious pathogens of bacterial or viral origin. Specific IgY activity against pathogenic agents has been elucidated from *in vitro* studies in hens immunized with pathogens and through passive immunization of animals with specific IgY (Sim et al., 2000). Antibacterial effects involving the use of specific IgY *in vivo* or *in vitro* have included use of *E. coli*, *Edwardsiella tarda*, *P. aeruginosa*, *Salmonella* spp., *Streptococcus mutans*, and *S. aureus* (Akita et al., 1998; Hamada et al., 1991; Hatta et al., 1994, 1997; Ikemori et al., 1992; Jin et al., 1998; Ozpinar et al., 1996; O'Farrelly et al., 1992; Wiedemann et al., 1991; Yokoyama et al., 1992; Yoshiko et al., 1996). Antiviral effects involving the use of specific IgY *in vivo* or *in vitro* have included the use of rotavirus, coronavirus, and IBD-virus (Ebina, 1996; Eterradossi et al., 1997; Ikemori et al., 1997; Kuroki et al., 1993, 1994, 1997).

18.4.2.3 Applications in Food

IgY may find application as a microstatic agent in preventing the growth of pathogenic bacteria using specific polyclonal Ig to effectively reduce or neutralize bacterial proliferation in food products and especially meat, thereby reducing the food safety risk associated with these products (Sim et al., 2000). It was suggested that IgY could be used as an ingredient for foods or even mouthwashes to prevent the colonization of invading microorganisms (Hatta et al., 1997). Furthermore, IgY may be used as a food adjuvant to control bacterial growth and prevent attachment of microorganisms to the intestinal epithelium (Sim et al., 2000). Akita and Nakai (1993) proposed the use of IgY in infants under 6 months as the reduced acidity of infant gastric acid would allow the molecule to maintain its protective effect against peptic digestion. However, numerous studies have documented the role of colostrum and milk in conferring protection to newborns (Jason et al., 1984), and others examined the potential of nonmaternal Ig for passive immunization of the GI tract (Hatta et al., 1997; Yokoyama et al., 1992). Passive immunization may be the most valuable application of Ig and involves the introduction of pathogen-specific IgY into a host to combat infectious diseases (Hatta et al., 1997). However, for passive immunization to be implemented, large amounts of Ig would need to be administered to an individual, and this may be achieved through commercial-scale production from eggs laid by hens (Hatta et al., 1997).

18.5 Avidin

Avidin is a 66-kDa positively charged glycoprotein isolated from various avian egg whites and egg jelly of invertebrates (Green 1964; Gyorgy et al., 1942). Avidin received much attention due to its anti-nutritive effects (Green, 1975); however, subsequent interest has revolved around its ability to bind up to four biotin molecules and form stable complexes (Green, 1964, 1975). The interaction between avidin and biotin is the strongest known protein–ligand binding found in nature (Boas, 1927). The discovery of an antibacterial molecule, streptavidin, produced by *Streptomyces* spp. (a 60-kDa avidin analogue) yielded a similar primary structure compared to avidin and indeed confirmed early suspicions that avidin possesses antimicrobial properties (Bayer and Wilchek, 1990).

18.5.1 Molecular Properties

18.5.1.1 Occurrence and Biosynthesis

Avidin is a minor protein in avian albumen accounting for only 0.05% of the egg white (Mine, 2000). Avidin production occurs in the goblet cells of the epithelium of the oviduct exclusively in laying hens, indicating that the production is regulated by functioning of the ovaries and specifically the hormone progesterone, and several steroids (Elo and Korpela, 1984; Mine, 2000).

18.5.1.2 Chemistry and Structure

Avidin is a basic glycoprotein consisting of four identical subunits each with an approximate molecular weight of 16 kDa (approximately 66 kDa for the molecule), and has a *pI* of approximately 10 (Green, 1975; Woolley and Longsworth, 1942). Each subunit consists of 128 amino acid residues and a single intramolecular disulfide bond between cystine residues at positions 4 and 83 (DeLange and Huang, 1971). Each subunit is organized in an eight-stranded antiparallel orthogonal β-barrel with extended loop regions providing the biotin-binding pocket (Livnah et al., 1993; Pugliese et al., 1994). Avidin binds one biotin molecule to each of the four comprising subunits, specifically in the highly complementary polar, protein-core pocket, and in the absence of biotin the site may be partially occupied by a water molecule (Gitlin et al., 1988; Pugliese et al., 1994). The structure of avidin reveals the existence of several surface-exposed lysine and arginine residues, which may contribute to the basic nature of the molecule (Hendrickson et al., 1989). Avidin possesses a carbohydrate moiety (about 10% of the total molecular weight) that is comprised of a single oligosaccharide chain with four or five mannose units and three N-acetylglucosamine residues linked to *Asp-17* of each polypeptide subunit (Kett et al., 2003).

18.5.1.3 Stability

Avidin may be resistant to treatment with iodine applied at neutral pH, acetylation of the amino groups, and esterification of the carboxyl groups; however, inactivation of the molecule may result from oxidation with H_2O_2 in the presence of Fe^{2+} or treatment with formaldehyde in the presence of alanine or hydroxylamine at 50°C (Fraenkel-Conrat and Fraenkel-Conrat, 1952). Although avidin is relatively stable over a wide range of pH and temperature in regards to its biotin-binding ability, all four tryptophans associated with the molecule are rapidly oxidized at pH 4.0 and biotin-binding activity is lost when at least two are destroyed (Green, 1963). Avidin may also lose stability at low ionic strength, with 0.1M HCl, 0.1 M sodium-dodecyl-sulfate, and 6 M guanidine HCl resulting in dissociation of avidin subunits and loss of biotin-binding ability (Green, 1963). In addition, it has been shown that avidin may be denatured at temperatures exceeding 70°C; however, the avidin–biotin complex offers stabilization to the molecule up to 100°C (Gitlin et al., 1988; Pugliese et al., 1994).

18.5.2 Antimicrobial Activity

18.5.2.1 Mode of Action

Although the antimicrobial activity of avidin has not been established, it has been suggested that the compound is involved in antimicrobial responses based on the production of streptavidin by *Streptomyces* in the derivation of an antibiotic system and supported by initiation of avidin production at the site of tissue injury in chickens (Elo and Korpela, 1984). It is proposed (Elo et al., 1978; Korpela, 1984) that avidin production and its secretion by macrophages is induced during inflammation and cellular damage and as such may constitute a host-defense factor for bacterial and viral infection. Studies (Miller and Tauig, 1964) indicating increased amounts of avidin in chicken tissues after intraperitoneal and intravenous administration of *E. coli* support the view that avidin is directed toward combating microbial infection. Korpela (1984) revealed that bacterial binding of avidin was independent of the saturation of its biotin-binding site and in *E. coli* the receptor was the porin protein of the outer membrane. Porins extend through the outer membrane of Gram-negative bacteria and are indeed among the most abundant proteins of the cell, constituting an integral part of the structure and function, and since no porins or similar analogues exist in Gram-positive bacteria it is assumed that porin is the only avidin-binding component of the cell envelope (Lugtenberg and van Alphen, 1983). It has been hypothesized (Campbell et al., 1972; Eakin et al., 1941; Herts, 1946) that due to the high affinity avidin has for biotin, it may function as an antimicrobial by rendering it unavailable to biotin-requiring microbes, and indeed it has been shown to inhibit the *in vitro* growth of biotin-requiring yeasts and bacteria (Green, 1975). Furthermore, binding of extracellular/intracellular biotin may even render it unavailable to the animal cell, thereby resulting in decreased activity of biotin enzymes and altered cellular metabolism and growth (Messmer and Young, 1977; Miller and Tauig, 1964; Wood and Barden, 1977).

18.5.2.2 Specificity

Although there is very little documentation of the inhibitory effects of avidin on pathogenic bacteria, Korpela (1984) reported on the *in vitro* binding of avidin to various Gram-negative and -positive bacteria. The Gram-negative bacteria found to bind avidin were *E. coli*, *K. pneumoniae*, *P. aeruginosa*, and *Serratia marcescens*, while the Gram-positive bacteria that bound avidin were *S. aureus* and *S. epidermis*.

18.5.3 Applications in Food

Since little is known about the antimicrobial activity of avidin, there are no current applications of this glycoprotein in food as an antimicrobial. However, avidin is used in the avidin-biotin system as a diagnostic tool in immunoassays, and there is potential for the application of avidin as an antimicrobial compound in foods (Mine, 2000).

18.6 Lactolipids

Lipids, especially fatty acids, isolated from milk and other dairy products possess both nutritional and protective qualities (Kabara, 1978). Detail on the antimicrobial activity of lipids and their esters is covered in Chapter 11 (Medium-Chain Fatty Acids (>C8) and Monoesters). Lipids were identified to possess antimicrobial activity in the mid nineteenth century when it was evidenced that fatty acids generated during the ripening of cheese prevented neurotoxin production by *C. botulinum* (Grecz et al., 1959). In neonates, the antimicrobial activity conferred by milk lipids is due to the release of antimicrobial fatty acids and monoglycerides from milk triglycerides present in fat globules, which constitute 98% of milk fat (Hamosh, 1991, 1995; Jensen, 1995). Other lipids of animal origin that have provided evidence of antimicrobial activity include epidermis-derived skin lipids shown to inactivate *S. aureus* (Bibel et al., 1989; Miller et al., 1988), free fatty acids at mucosal surfaces shown to inactivate pneumococci (Coonrod, 1987), and porcine intestinal lipids shown to inactivate *Clostridium perfringens* (Fuller and Moore, 1967). Lipids may serve to inhibit the establishment, multiplication, and proliferation of pathogenic microorganisms in a host, and their effects may be enhanced by interaction with other antimicrobial factors present in milk (e.g., lactoferrin, lactoperoxidase, lactoglobulins) (Mandel and Ellison, 1985; Watanabe et al., 1984).

18.6.1 Molecular Properties

18.6.1.1 Occurrence and Biosynthesis

Fatty acid biosynthesis occurs after a meal when the body is energy rich (Drackley, 2000). The events after consumption involve the generation of adenosine triphosphate (ATP) by glycolysis, generation of NADPH (the reduced state of nicotinamide adenine dinucleotide phosphate) by the pentose phosphate pathway, and the storage of glucose as glycogen. Any excess glucose is converted to fatty acids and stored as triacylglycerols (Drackley, 2000). The majority of fatty acid biosynthesis occurs in the cytosol of the liver, although there is limited biosynthesis in other tissues, including adipose tissue, the brain, and mammary glands (Drackley, 2000; Jensen, 1996). The antimicrobial activity in milk is contained mainly in long-chain unsaturated fatty acids and the medium-chain saturated fatty acids (Lampe and Isaacs, 2000). The presence of fatty acids in milk originates from the diet, the mobilization of stored fatty acids, and synthesis in the mammary glands (Jensen, 1996). The diet of lactating animals affects the proportion of fatty acids present in the milk, as a high-energy meal will increase production of medium-chain fatty acids and thereby promote the antimicrobial properties of the milk (Francois et al., 1998; Spear et al., 1992).

18.6.1.2 Chemistry and Structure

Fatty acid molecules are best defined by a description of the length of the carbon chain (number of carbon molecules) and the number of double bonds present and their exact positioning (Drackley, 2000). Common fatty acids are straight-chain compounds and usually possess an even number of carbon atoms (Drackley, 2000). Fatty acid chain-lengths are classed into three categories based on their length, namely short-chain fatty acids with a chain length between 2 and 4, medium-chain fatty acids with between 6 and 10 molecules, and long-chain fatty acids commonly with 12 to 24 carbon molecules but potentially up to 80 (Garrett and Grisham, 1999). The simplest fatty acids have no unsaturated linkages and cannot be modified by hydrogenation or halogenation; these are referred to as saturated fatty acids (Garrett and Grisham, 1999). The presence of double bonds between carbon molecules defines the chains as unsaturated and specifically monounsaturated if only one double bond is present and polyunsaturated if two or more double bonds are present (Garrett and Grisham, 1999). Alternatively, fatty acids in animals may possess branched chains or contain a variety of other functional groups including acetylenic bonds, epoxy-, hydroxy-, or keto-groups, and even ring structures (Garrett and Grisham, 1999). Monoglycerides or monoacylglycerols are fatty acid monoesters of glycerol and are found in very low amounts in cell extracts as an intermediate product in the degradation of triacylglycerides or diacylglycerides during lipolysis (Garrett and Grisham, 1999). Monoglycerides are the most polar components of simple lipids since they possess only one hydrocarbon chain and two alcohol groups and, thus, need careful manipulation to prevent their loss in hydrophilic solutions and on chromatographic columns (Ledoux et al., 2000; Precht et al., 2001). Furthermore, monoglycerides possess detergent properties and as such they easily form micelles in water solutions (Garrett and Grisham, 1999).

18.6.1.3 Stability

The presence of proteins, especially albumin, may reduce the antimicrobial activity of fatty acids through specific and nonspecific binding (Shibasaki and Kato, 1978). The antimicrobial activity of specifically unsaturated fatty acids may also be reduced by the presence of other surface-active agents like cholesterol (Ammon, 1985; Kabara, 1978). Since the antimicrobial activity of short-chain fatty acids is due to the undissociated form rather than the anionic form, their activity is highly pH-dependent, with activity decreasing as the degree of dissociation increases (Kabara, 1978; Lundblad and Seng, 1991). The effect pH has on antimicrobial activity is observed mainly with short- and medium-chain fatty acids where the minimum inhibitory concentration of short-chain fatty acids increases while that of medium-chain acids decreases with corresponding increases in pH (Kabara, 1978). The activity of long-chain unsaturated fatty acids as well as monoglyceride esters and ethers of short-chain fatty acids appear to be unaffected by pH alterations (Kabara, 1978).

18.6.2 Antimicrobial Activity

18.6.2.1 Mode of Action

The antimicrobial products produced by the hydrolysis of milk triglycerides are fatty acids and monoglycerides (Thormar et al., 1987). Milk lipids are mainly antiviral, and it has been shown that short- and long-chain saturated fatty acids have minimal antiviral activity, while medium-chain saturated and long-chain unsaturated fatty acids are strongly antiviral (Thormar et al., 1987). Monoglycerides of fatty acids produced by hydrolysis of milk lipids are also effective at inactivating enveloped viruses at five to ten times lower concentrations than their corresponding fatty acid (Thormar et al., 1987). In addition, the antiviral activity of fatty acids and monoglycerides is additive (Isaacs and Thormar, 1990). Isaacs and Thormar (1990) demonstrated that the virus envelope is the target for lipid-dependent viral inactivation. Furthermore, research (Noseda et al., 1989; Verdonck and van Heugten, 1997) indicates that antimicrobial lipids destabilize the membrane of cells. The antimicrobial activity of lipids through membrane destabilization has also been demonstrated against fungi and bacteria (Kabara, 1978; Shibasaki and Kato, 1978). The mechanisms by which lipids target microorganisms may be through disruption of the bacterial cell wall or membrane or envelope in viruses, blockage of receptor–ligand interactions, inhibition of intracellular replication, or by inhibition of an intracellular target (Lampe and Isaacs, 2000).

18.6.2.2 Specificity

The antimicrobial activity of milk lipids has been demonstrated on a variety of enveloped viruses including herpes simplex virus, influenza virus, respiratory syncytial virus, measles virus, vesicular stomatitis virus, visna virus, mouse mammary tumor virus, dengue virus types 1–4, cytomegalovirus, Semliki forest virus, Japanese B encephalitis virus, and human immunodeficiency virus (Lampe and Isaacs, 2000). Milk lipids have also been shown to inactivate Gram-positive bacteria including *B. cereus*, *B. subtilis*, *C. botulinum*, *Corynebacterium* spp., *L. monocytogenes*, *Micrococcus* spp., *Pneumococcus* spp., *S. aureus*, *S. epidermis*, *Streptococcus* spp., and Gram-negative bacteria including *C. trachomatis*, *E. coli*, *N. gonorrhoeae*, *P. aeruginosa*, and *S.* Enteriditis (Hernell et al., 1986; Isaacs et al., 1990; Kato and Shibasaki, 1975; Lampe et al., 1998; Qu et al., 1996; Reiner et al., 1986; Rohrer et al., 1986; Rabe et al., 1997; Wang and Johnson, 1992, 1997; Wang et al., 1993). In addition, milk lipids display antimicrobial activity against fungi including *Aspergillus niger* and *Trichoderma viride*, yeasts, including *Alternaria* spp., *C. albicans*, *C. utilis*, *Cladosporium* spp., *Kluveromyces marxianus*, and *S. cerevisiae*, and the protozoal pathogen *G. lamblia* (Isaacs et al., 1990; Kato and Shibasaki, 1975; Marshall and Bullerman, 1986; Sofos et al., 1998).

18.6.3 Applications in Food

The antimicrobial activity of lipids has been used in food preservation for decades (Grecz et al., 1959; Shibasaki and Kato, 1978). The majority of the lipids isolated from animal products such as milk are considered as GRAS chemicals and can as such find application as antimicrobials in food products (Lampe and Isaacs, 2000). Monoacylglycerols have increased the shelf-life of various foods including soy sauce, miso, sausage, Worcestershire sauce, sponge cake, and noodles (Shibasaki, 1982; Sofos et al., 1998). In addition to these foods, the lauric acid ester of monoacylglycerol has shown antimicrobial potential in seafood salad, Camembert cheese, and various flesh foods including deboned chicken meat, minced fish, refrigerated beef roasts, and turkey frankfurter slurries (Baker et al., 1985; Hall and Mauer,

1986; Unda et al., 1991; Wang and Johnson, 1997). Monoacylglycerols may also be combined with other antimicrobials (e.g., nisin) to increase their effectiveness in foods (Sofos et al., 1998). Considering that monoacylglycerols lower the heat resistance of certain bacteria and fungi, they may find application in decreasing the heat-treatment requirements of some foods (Kimsey et al., 1981; Sofos et al., 1998). A proposed application of antimicrobial lipids has been the use in infant formulas to provide protection after hydrolysis of the triglycerides into antimicrobial fatty acids and monoglycerides in the gastrointestinal tract following consumption (Isaacs et al., 1995).

18.7 Defensins

Defensins are a group of antimicrobial peptides folded in a characteristic β-sheet and a framework of six disulfide-linked cysteines (Ganz, 2003; Selsted et al., 1985). Defensins are widely distributed in nature and indeed in mammalian epithelial cells and leukocytes, often at high concentrations, and have a broad spectrum of antimicrobial activity (Ganz, 2003; Selsted et al., 1985; Tiwari et al., 2009). Some of the key defensins conferring antimicrobial activity are protamine and magainin (Humblot et al., 2009). Protamine and magainin have been shown to be effective against bacteria and fungi (Humblot et al., 2009).

18.7.1 Molecular Properties

18.7.1.1 Occurrence and Biosynthesis

Defensin peptides have been found in all mammals examined, in chickens, and in turkeys, and are abundant in cells and tissues active in host defense against microorganisms (Brockus et al., 1998; Harwig et al., 1994; Zhao et al., 2001). The highest (>10 mg/ml) concentration of defensins is usually in the granules or storage organelles of leukocytes (Ganz, 2003). Upon ingestion of microorganisms by leukocytes into phagocytic vacuoles, the vacuoles fuse with the granules and the granule material is delivered onto the target organism (Ganz, 2003). Another site containing high (>10 mg/ml) concentrations of defensins are the Paneth cells, specialized cells of the small intestine, that contain secretory granules which are released into intestinal pits or crypts (Ganz, 2003; Harder et al., 1997). Other cells containing defensins at lower concentrations (10–100 μg/ml) include barrier and secretory epithelial cells, which may produce defensins constitutively or only upon infection, and to a lesser extent cells of the immune system such as monocytes, macrophages, and lymphocytes (Agerberth et al., 2000; Harder et al., 1997; Ryan et al., 1998).

18.7.1.2 Chemistry and Structure

There are two main defensin subfamilies, α- and β-defensins, that differ in peptide segment length between the six cysteine residues and the cysteine pairs connected by disulfide bonds (Hill et al., 1991; Hoover et al., 2000; Sawai et al., 2001; Zhang et al., 1992; Zimmermann et al., 1995). Defensins exist as a triple-strand β-sheet that have the distinctive fold brought about by the disulfide bonds between cysteine residues (Hoover et al., 2000; Sawai et al., 2001). The amino acid sequence and composition of defensins are highly variable, however, the cysteine framework is conserved in each defensin subfamily (Ganz, 2003). Most α- and β-defensins possess clusters of positively charged amino acids, although their distribution in the molecule is variable (Ganz, 2003). Subcellular storage organelles such as leukocytes and Paneth cells are rich in negatively charged glycosaminoglycans (Parmley et al., 1986). The majority of α- and β-defensins from leukocytes and Paneth cells contain arginine as the main cationic amino acid, while β-defensins secreted from epithelial cells contain similar amounts of arginine and lysine (Fromm et al., 1995; Kostoulas et al., 1997). The general synthesis of α-defensins occurs through the encoding of tripartite pre-propeptide sequences to produce an amino acid precursor consisting of 90–100 amino acids with an amino (N)-terminal signal sequence (approximately 19 amino acids), an anionic pro-piece (approximately 45 amino acids), and a carboxy (C)-terminal mature cationic defensin (approximately 30 amino acids) (Ganz, 2003). The synthesis of β-defensin precursors is similar, consisting of a signal sequence at the N-terminus, a short or absent pro-piece, and the mature defensin peptide at the C-terminus; however, it lacks the anionic pro-piece which is prominent in the α-defensin precursor (Ganz, 2003).

18.7.2 Antimicrobial Activity

Defensins are mainly antibacterial and antimycotic, especially at low ionic strength conditions, low concentrations of divalent cations, plasma proteins, and other interfering substances (Ganz, 2003; Lehrer and Ganz, 2002; Selsted et al., 1993). Under these optimal conditions, defensins may be active at very low (1–10 μg/ml) concentrations (Ganz, 2003). Defensins have also been shown to be effective against some enveloped viruses (Daher et al., 1986). The main mechanism responsible for antimicrobial activity is attributed to the permeabilization of target membranes, and subsequent cell leakage, inhibition of RNA, DNA, and protein synthesis, and decreased cellular viability (Ganz, 2003).

18.7.3 Applications in Food

Although defensins are antimicrobial peptides that may be isolated and purified from animals, there is, other than the antimicrobial activities as a natural ingredient in raw food products, no known application of the peptides as an additive in foods.

18.8 Chitosan

There is a significant amount of literature referencing the use of animal-derived peptides with antimicrobial properties; however there also are a number of polysaccharides and lipids from animals that also show antimicrobial effects. Chitosan [(1-4)-2-amino-2-deoxy-β-D-glucan] comprises a series of polymers with differing ratios of glucosamine and N-acetylglucosamine. Chitosan is a marine polysaccharide derived from alkaline-hydrolysis of the chitin in the exoskeleton of crustaceans and arthropods. It is also a component of fungal cell walls. Chitosan is polycationic and has been shown to have antifungal (Ben-Shalom et al., 2003) and antibacterial (Fernandes et al., 2008) properties. Chitosan inhibits the growth of foodborne bacteria and fungi including lactic acid bacteria, *Bacillus cereus*, *Staphylococcus aureus*, *Listeria monocytogenes*, *Salmonella* Typhimurium, *Escherichia coli*, *Shigella dysenteriae*, *Byssochlamys* spp., and *Zygosaccharomyces bailii* (Yang et al., 2005; Chung et al., 2011; Davidson et al., 2013). Minimum inhibitory concentrations for bacteria and yeasts vary widely depending on the molecular weight of the polymer, degree of acetylation, pH, temperature, and interfering compounds (e.g., proteins and lipids).

The antimicrobial mode of action of chitosan is not completely understood, although the polycationic nature of the polymer, the spacing between moieties, and the hydrophilicity have all been identified as having significant impact on the overall activity of the molecule and as contributors to the mechanism of microbial inhibition (Kong et al. 2010). Even though the mode of action is not completely understood, the two suggested mechanisms are (i) the binding of cationic chitosan to sialic acid in phospholipids, consequently restraining the movement of microbiological substances (Chatterjee et al., 2005), and (ii) penetration of oligomeric chitosan into the cells of microorganisms and prevention of the growth of cells by preventing the transformation of DNA into RNA (Sashiwa and Aiba, 2004).

The use of chitosan alone in food safety is somewhat limited due to its low solubility at neutral and alkaline pH and interference of the food matrix with the ionic interactions required to inhibit the bacterial cell (Hu and Gänzle, 2019). However it has been extensively studied for its use as a biodegradable film having antimicrobial properties (Sundaram et al., 2016, Torlak and Sert, 2013. The chitosan matrix can be formed into films, gels, fibers, beads, and nanoparticles (Dutta et al., 2012). Edible films cast from chitosans also have been infused with other antimicrobials to enhance the inhibition of microorganisms and improve applicability. For example, incorporating lysozyme in a chitosan film resulted in significant inhibition of *Escherichia* coli O157:H7 and *Listeria monocytogenes*, and lowered numbers of both pathogens in broth. Chitosan films with added essential oil of oregano reduced viable *L. monocytogenes* and *E. coli* O157:H7 on inoculated bologna samples stored for 5 d at 10°C (Zivanovic et al. 2005). Park et al. (2005) found that chitosan-based coatings infused with potassium sorbate applied to strawberries produced significant reductions in *Rhizopus* and *Cladosporium* spp. as compared with controls. A chitosan coating with 0.1% lauric arginate, 0.1% EDTA, and 1% cinnamon oil was effective for inactivating >3 log CFU/cm^2 *Escherichia coli* O157:H7 and *Listeria monocytogenes* immediately after coating and reduced *Salmonella enterica* to below the detection limit on cantaloupes during a 14-day storage (Ma et al., 2016). This coating also reduced total molds and yeasts to the detection limit. Chitosan with nisin incorporated has been shown to have antimicrobial activity against several Gram-positive pathogenic bacteria (see Chapter 8, "Nisin").

Chitosan can potentiate the efficacy of other limiting technologies such as high pressure and heat, thereby creating multiple hurdles (Kumar et al., 2009). Food quality can also be improved by chitosan outside of antimicrobial activity in the retardation of lipid oxidation, melanosis, and plant metabolism even if antimicrobial activity is limited (Hu and Gänzle, 2018).

18.9 Other Antimicrobials of Animal Origin

18.9.1 Pleurocidin

As discussed in this chapter, there are a number of antimicrobial peptides of animal origin that can act quickly to destroy the cellular lipid bilayer membranes even of fast-growing microorganisms such as Gram-negative and Gram-positive bacteria (Tiwari et al., 2009) while also exhibiting antifungal and antiviral activities (Aires et al., 2009). Indeed pleurocidin has been shown to effectively combat a number of bacteria and fungi including *E. coli* O157:H7, *L. monocytogenes*, *P. expansum*, *S. cerevisiae*, and *V. parahemolyticus* (Burrowes et al., 2004). Pleurocidin is found in the skin mucus membrane of the winter flounder (*Pleuronectes americanus*).

18.9.2 Casocidin

Much like the antimicrobial peptides (i.e., lactoperoxidase, lactoferricin B) mentioned earlier in this chapter, casocidin is another antimicrobial peptide found in milk. Casocidin is produced by the hydrolysis of S2-casein *via* the enzyme chymosin found in rennet. Antibacterial activity has been demonstrated against *Staphylococcus* spp., *Sarcina* spp., *B. subtilis*, *Diplococcus pneumoniae*, and *Streptococcus pyogenes* (Szwajkowska et al., 2011). Peptides such as casein A and casein B have been shown to inhibit *Cronobacter sakazakii* which has possible market application in milk-based formulations for neonates (Norberg et al., 2011). Furthermore, peptides from S2-casein, S1-casein, and k-casein have demonstrated antibacterial effects against *E. coli* and *B. subtilis* (Elbarbary et al., 2012).

18.9.3 Lysozyme (Consult Chapter 14, "Lysozyme" for More Information)

As discussed previously, lactoferrin, lactoferricin B, lactoglobulins, and lysozyme together have demonstrated synergistic effects against bacteria and fungi. Lysozyme, however, when used alone has demonstrated strong bacteriolytic effects against food-spoilage organisms (Potter et al., 2005). Lysozyme targets primarily Gram-positive bacteria as the cell wall consists of peptidoglycans (Tiwari et al., 2009), while Gram-negative bacteria have demonstrated some resistance due to their outer membrane that acts as a physical barrier. The use of surfactants and chelating agents such as EDTA has proven to assist lysozyme in antibacterial effects targeting Gram-negative bacteria due to their action disruption of the outer membrane (Branen and Davidson, 2004).

18.9.4 Lipids

There is very little work done on elucidating the antimicrobial properties of animal-derived lipids. Research has shown that lipids present in milk have mild antibacterial and antifungal effects (Shin et al., 2007). The combination of purified monoglycerides and fatty acids, especially those found in milk, has demonstrated almost twice the antibacterial effect than when applied alone (Isaacs, 2001). Specific animal lipids that have shown promise as antibacterial agents are eicosapentaenoic and docosahexaenoic acid (Shin et al., 2007).

REFERENCES

Agerberth, B., Charo, J., Werr, J., Olsson, B., Idali, F., Lindbom, L., Kiessling, R., Jornvall, H., Wigzell, H., and Gudmundsson, G.H. 2000. The human antimicrobial and chemotactic peptides LL-37 and alpha-defensins are expressed by specific lymphocyte and monocyte populations. *Blood* 96:3086–3093.

Aguilera, O., Ostolaza, H., Quirós, L.M., and Fierro, J.F. 1999. Permeabilizing action of an antimicrobial lactoferricin-derived peptide on bacterial and artificial membranes. *FEBS Lett.* 462:273–277.

Aguilera, O., Quirós, L.M., and Fierro, J.F. 2003. Transferrins selectively cause ion efflux through bacterial and artificial membranes. *FEBS Lett.* 548:5–10.

Ainscough, E.W., Brodie, A.M., Plowman, J.E., Bloor, S.J., Loehr, J.S., and Loehr, T.M. 1980. Studies on human lactoferrin by electron paramagnetic resonance, fluorescence, and resonance *Raman spectroscopy. Biochem.* 19:4072–4079.

Aires, A., Mota, V., Saavedra, M., Rosa, E., and Bennett, R. 2009. The antimicrobial effects of glucosinolates and their respective enzymatic hydrolysis products on bacteria isolated from the human intestinal tract. *J. Appl. Microbiol.* 106:2086–2095.

Akerstrom, G., Brodin, T., Reis, K., and Borg, L. 1985. Protein G: A powerful tool for binding and detection of monoclonal and polyclonal antibodies. *J. Immunol.* 135:2589–2592.

Akita, E.M., and Nakai, S. 1993. Comparison of four purification methods for the production of immunoglobulins from eggs laid by hens immunized with an enterotoxigenic *E.coli* strain. *J. Immunol. Methods* 160:207–214.

Akita, E.M., Li-Chan, E.C.Y., and Nakai, S. 1998. Neutralization of enterotoxigenic *Escherichia coli* heat-labile toxin by chicken egg yolk immunoglobulin Y and its antigen-binding fragments. *Food Agri. Immunol.* 10:161–172.

Amini, H.-R., Ascencio, F., Ruiz-Bustos, E., Romero, M.J., and Wadström, T. 1996. Cryptic domains of a 60 kDa heat shock protein of *Helicobacter pylori* bound to bovine lactoferrin. *FEMS Immunol. Med. Microbiol.* 16:247–255.

Ammon, H.V. 1985. Effects of fatty acids on intestinal transport. In *The pharmacological Effect of Lipids – II*, edited by J.J. Kabara, The American Oil Chemists Society, Champaign, IL, pp. 173–181.

Andersen, J.H., Osbakk, S.A., Vorland, L.H., Traavik, T., and Gutteberg, T.J. 2001. Lactoferrin and cyclic lactoferrin inhibit the entry of human cytomegalovirus into human fibroblasts. *Antiviral Res.* 51:141–149.

Andersson, Y., Lindquist, S., Lagerqvist, C., and Hernell, O. 2000. Lactoferrin is responsible for the fungistatic effect of human milk. *Early Human Dev.* 59:95–105.

Anema, S.G. 2000. Effect of milk concentration on the irreversible thermal denaturation and disulfide aggregation of beta-lactoglobulin. *J. Agric. Food Chem.* 48:4168–4175.

Aprodu, I., Stănciuc, N., Dumitrascu, L., Râpeanu, G., and Stanciu, S. 2014. Investigations towards understanding the thermal denaturation of lactoperoxidase. *Intl. Dairy J.* 38:47–54.

Arnold, R.R., Brewer, M., and Gauthier, J.J. 1980. Bactericidal activity of human lactoferrin: Sensitivity of a variety of microorganisms. *Infect. Immun.* 28:893–898.

Ashorn, R.G., Eskola, J. Tuohimaa, P.J., and Krohn, K.J. 1986. Effect of progesterone-inducible proteins human lactoferrin and avidin on lymphocyte proliferation. *Hum. Reprod.* 1:149–151.

Aune, T.M., and Thomas, E.L. 1978. Oxidation of protein sulfhydryls by products of peroxidase-catalyzed oxidation of thiocyanate ion. *Biochemistry* 17:1005–1010.

Baggiolini, M., De Duve, C., Masson, P.L., and Heremans, J.F. 1970. Association of lactoferrin with specific granules in rabbit heterophil leukocytes. *J. Exp. Med.* 131:559–570.

Baker, E.N., Rumball, S.V., and Andersson, B.F. 1987. Transferrins: Insight into structure and function from studies on lactoferrin. *Trends Biochem. Sci.* 12:350–353.

Baker, E.N., and Lindley, P.F. 1992. New perspectives on the structure and function of transferrins. *J. Inorg. Biochem.* 47:147.

Baker, H.M., Anderson, B.F., Kidd, R.D., Shewry, S.C., and Baker, E.N. 2000. Lactoferrin three-dimensional structure: A framework for interpreting function. In *Lactoferrin: Structure, Function and*

Applications, edited by K.I. Shimazaki, et al., Elsevier Science B.V., Amsterdam, The Netherlands, pp. 3–15.

Baker, R.C., Poon, W., Kline, D., and Vadehra, D.V. 1985. Antimicrobial properties of lauricidin in mechanically deboned chicken, minced fish and chicken sausage. In *The Pharmacological Effect of Lipids – II*, edited by J.J. Kabara, American Oil and Chemical Society, Champaign, IL, pp. 65–70.

Banwart, G.J. 1979. *Basic Food Microbiology*, AVI, Publishing, Westport, CT.

Bardsley, W.G. 1985. Steady-state kinetics of lactoperoxidase-catalyzed reactions. In *The Lactoperoxidase System*, edited by K. Pruitt and J.O. Tenovuo, Marcel Dekker, New York, NY, pp. 55–87.

Barresi, G., and Tuccari, G. 1987. Iron-binding proteins in thyroid tumours: An immunocytochemical study. *Pathol. Res. Pract.* 182:344–351.

Barthe, C., Galabert, C., Guy-Crotte, O., and Figarella, C. 1989. Plasma and serum lactoferrin levels in cystic fibrosis: Relationship with the presence of cystic fibrosis protein. *Clin. Chim. Acta* 181:183–188.

Baruzzi, F., Pinto, L., Quintieri, L., Carito, A., Calabrese, N., and Caputo, L. 2015. Efficacy of lactoferricin B in controlling ready-to-eat vegetable spoilage caused by *Pseudomonas spp. Intl. J. Food Microbiol.* 215:179–186.

Baveye, S., Elass, E., Mazurier, J., Spik, G., and Legrand, D. 1999. Lactoferrin: A multifunctional glycoprotein involved in the modulation of the inflammatory process. *Clin. Chem. Lab. Med.* 37:281–286.

Bayer, E.A., and Wilchek, M. 1990. Avidin column as a highly efficient and stable alternative for immobilization of ligands for affinity chromatography. *J. Molec. Recog.* 3:102–107.

Baynes, R.D., Bezwoda, W.R., Khan, Q., and Mansoor, N. 1986. Plasma lactoferrin content: Differential effect of steroid administration and infective illnesses: Lack of effect of ambient temperature at which specimens are collected. *Scand. J. Haemat.* 37:353–359.

Belding, M.E., Klebanoff, S.J., and Ray, G.C. 1970. Peroxidase-mediated virucidal systems. *Science* 167:195–196.

Bellamy, W.R., Takase, M., Yamauchi, K., Wakabayashi, H., Kawase, K., and Tomita, M. 1992. Identification of the bactericidal domain of lactoferrin. *Biochim. Biophys. Acta* 1121:130–136.

Bellamy, W.R., Wakabayashi, H., Takase, M., Kawase, K., Shimamura, S., and Tomita, M. 1993. Role of cell-binding in the antibacterial mechanism of lactoferricin B. *J. Appl. Bacteriol.* 75:478–484.

Bennett, R.M., and Kokocinski, T. 1978. Lactoferrin content of peripheral blood cells. *Brit. J. Haematol.* 39:509–521.

Ben-Shalom, N., Ardi, R., Pinto, R., and Aki, C., and Fallik, E. 2003. Controlling gray mould caused by *Botrytis cinerea* in cucumber plants by means of chitosan. *Crop Prot.* 22:285–290.

Beuchat, L.R., and Golden, D.A. 1989. Antimicrobials occuring naturally in foods. *Food Technol.* 43:134–142.

Bibel, D.J., Miller, S.J., Brown, B.E., Pandey, B.B., Elias, P.M., Shinefield, H.M., and Aly, R. 1989. Antimicrobial activity of stratum corneum lipids from normal and essential fatty acid-deficient mice. *J. Invest. Dermatol.* 92:632–638.

Bjorck, L. 1976. An immobilized two-enzyme system for the activation of the lactoperoxidase antibacterial system in milk. *Biotechnol. Bioeng.* 18:1463–1472.

Bjorck, L. 1978. Antibacterial effect of the lactoperoxidase system on psychotrophic bacteria in milk. *J. Dairy Res.* 45:109–118.

Bjorck, L., Claesson, O., and Schulthess, W. 1979. The lactoperoxidase/thiocyanate/hydrogen peroxide system as a temporary preservative for raw milk in developing countries. *Milchwissenschaft* 34:726–729.

Blake, D.R., Hall, N.D., and Bacon, P.A. 1981. The importance of iron in rheumatoid disease. *Lancet.* 21:1142–1143.

Board, R.G. 1969. The microbiology of the hen's egg. *Adv. App. Microbiol.* 11:245–281.

Boas, M.A. 1927. The effect of desiccation upon the nutritive properties of egg white. *Biochem.* 21:712–724.

Boedeker, E.C., Cheney, C.P., and Cantey, J.R. 1987. Inhibition of enteropathogenic *Escherichia coli* (strain RDEC-1) adherence to rabbit intestinal brush borders by milk immune secretory immunoglobulin A. *Adv. Exp. Med. Biol.* 216:919–930.

Bosch, E.H., Van Doorne, H., and De Vries, S. 2000. The lactoperoxidase system: The influence of iodide and the chemical and antimicrobial stability over the period of about 18 months. *J. Appl. Microbiol.* 89:215–224.

Bostwick, E.F., Steijns, J., and Braun, S. 2000. Lactoglobulins. In *Natural Food Antimicrobial Systems*, edited by A.S. Naidu, CRC Press, Boca Raton, FL, pp. 133–158.

Boussouel, N., Mathieu, F., Revol-Junelles, A.M., and Milliere, J.B. 2000. Effects of combinations of lactoperoxidase system and nisin on the behavior of Listeria monocytogenes ATCC 15313 in skim milk. *Intl. J. Food Microbiol.* 61:169–175.

Branen, J.K., and Davidson, P.M. 2000. Activity of hydrolysed lactoferrin against foodborne pathogenic bacteria in growth media: The effect of EDTA. *Lett. Appl. Microbiol.* 30:233–237.

Branen, J.K., and Davidson, P.M. 2004. Enhancement of nisin, lysozyme, and monolaurin antimicrobial activities by ethylenediaminetetraacetic acid and lactoferrin. *Intl. J. Food Microbiol.* 90:63–74.

Bravo, D., de Alba, M., and Medina, M. 2014. Combined treatment of high-pressure with the lactoperoxidase system or lactoferrin on the inactivation of *Listeria monocytogenes*, *Salmonella* Enteritidis and *Escherichia coli O157:H7* in beef carpaccio. *Food Microbiol.* 41:27–32.

Brock, J.H., Arzabe, F., Lampreave, F., and Piñeiro, A. 1976. The effect of trypsin on bovine transferrin and lactoferrin. *Biochim. Biophys. Acta* 446:214–225.

Brockus, C.W., Jackwood, M.W., and Hamon, B.G. 1998. Characterization of β-defensin prepropetide mRNA from chicken and turkey bone marrow. *Anim. Genet.* 29:283–289.

Brodie, A.M., Ainscough, E.W., Baker, E.N., Baker, H.M., Shongwe, M.S., and Smith, C.A. 1994. Synergism and substitution in the lactoferrins. In *Lactoferrin: Structure and Function*, edited by T.W. Hutchens et al., Plenum Press, New York, pp. 33–44.

Broxmeyer, H.E., Williams, D.E., Hangoc, G., Cooper, S., Gentile, P., Shen, R.N., Ralph, P., Gillis, S., and Bicknell, D.C. 1987. The opposing action in vivo on murine myelopoiesis of purified preparations of lactoferrin and the colony stimulating factors. *Blood Cells* 13:31–48.

Bukhhardt, H., and Schwingel, M. 1986. Oxygen radicals as effectors of cartilage destruction. *Arthritis Rheum.* 30:57–63.

Burrowes, O.J., Hadjicharalambous, C., Diamond, G., and Lee, T.-C. 2004. Evaluation of antimicrobial spectrum and cytotoxic activity of pleurocidin for food applications. *J. Food Sci.* 69:FMS66–FMS71.

Butler, J.E., Winter, A.J., and Wagner, G.G. 1971. Symposium: Bovine immune system. *J. Dairy Sci.* 54:1309–1314.

Butler, J.E. 1974. Immunoglobulins of the mammary secretions. In *Lactation, a Comprehensive Treatise*, edited by B.L. Larson and V.L. Smith, Academic Press, New York, NY, pp. 217–225.

Butler, J.E. 1983. Bovine immunoglobulins: An augmented review. *Vet. Immunol. Immunopath.* 4:43–152.

Butler, J.E. 1986. Biochemistry and biology of ruminant immunoglobulins. *Prog. Vet. Microbiol. Immunol.* 2:1–53.

Butler, J.E. 1994. Passive immunity and Ig diversity. In *Indigenous Antimicrobial Agents in Milk: Recent Developments*, International Dairy Federation Press, Brussels, pp. 14–47.

Buxbaum, J., Zolla, S., Scharff, M., and Franklin, E. 1971. Syntheses and assembly of Ig by malignant human plasmacytes and lymphocytes. *J. Exp. Med.* 133:1118.

Caccavo, D., Pellegrino, N.M., Altamura, M., Rigon, A., Amati, L., Amoroso, A., and Jirillo, E. 2002. Antimicrobial and immunoregulatory functions of lactoferrin and its potential therapeutic application. *J. Endotoxin Res.* 8:403–417.

Cals, M.M., Mailliart, P., Brignon, G., Anglade, P., and Dumas, B.R. 1991. Primary structure of bovine lactoperoxidase, a fourth member of a mammalian heme peroxidase family. *Euro. J. Biochem.* 198:733–739.

Campbell, A., del Campillo-Campbell, A., and Chang, R. 1972. A mutant of *Escherichia coli* that requires high concentration of biotin. *Proc. Natl. Acad. Sci. U.S.A.* 69:676–680.

Carlsson, J., Iwami, Y., and Yamada, T. 1983. Hydrogen peroxide excretion by oral streptococci and effect of lactoperoxidase-thiocyanate-hydrogen peroxide. *Infect. Immun.* 40:70–80.

Casswall, T.H., Sarker, S.A., Albert, M.J., Fuchs, G.J., Bergstrom, M., Bjorck, L., and Hammerstrom, M. 1998. Treatment of *Helicobacter pylori* infection in infants in rural Bangladesh with oral Ig from hyperimmune bovine colostrums. *Aliment Pharmacol. Therapeutics* 12:563–568.

Casswall, T.H., Sarker, S.A., Faruque, S.M., Weintraub, A., Albert, M.J., Fuchs, G.J., Alam, N.H., Dahlstrom, A.K., Link, H., Brussow, H., and Hammarstrom, L. 2000. Treatment of enterotoxigenic and enteropathogenic Escherichia coli-induced diarrhoea in children with bovine immunoglobulin milk concentrate from hyperimmunized cows: A double-blind, placebo-controlled, clinical trial. *Scand. J. Gastroenterol.* 35:711–718.

Chantaysakorn, P., and Richter, R.L. 2000. Antimicrobial properties of pepsin-digested lactoferrin added to carrot juice and filtrates of carrot juice. *J. Food Prot.* 63:376–380.

Chapple, D.S., Mason D.J., Joannou, C.L., Odell, E.W., Gant, V., and Evans, R.W. 1998. Structure-function relationship of antibacterial synthetic peptides homologous to a helical surface region on human lactoferrin against *Escherichia coli* serotype O111. *Infect. Immun.* 66:2434–2440.

Chatterjee, S., Adhya, M., Guha, A., and Chatterjee, B. 2005. Chitosan from *Mucor rouxii*: Production and physico-chemicul characterization. *Process Biochem.* 40:395–400.

Che, H., Tian, B., Bai, L., Cheng, L., Liu, L., Zhang, X., Jiang, Z., and Xu, X. 2015. Development of a test strip for rapid detection of lactoperoxidase in raw milk. *J. Zhejiang Univ. Sci. B.* 16:672–679.

Chen, C.-C., and Chang, H.-M. 1998. Effect of thermal protectants on the stability of bovine milk Ig G. *J. Agric. Food Chem.* 46:3570–3576.

Chen, P.-W., Shyu, C.-L., and Mao, F.C. 2003. Antibacterial activity of short hydrophobic and basic-rich peptides. *Am. J. Vet. Res.* 64:1008–1092.

Chernokhvostova, E.V., Lyubinskaya, M.M., Belkin, Z.P., and Levenson, V.I. 1990. Protective milk O antibodies induced in guinea pigs by parenteral *Shigella* ribosomal vaccine. *Intl. Arch. Allergy Appl. Immunol.* 92:265–267.

Chung, Y.-C., Yeh, J.-Y., and Tsai, C.-F. 2011. Antibacterial characteristics and activity of water-soluble chitosan derivatives prepared by the Maillard reaction. *Molecules* 16:8504–8514.

Clare, D.A., and Swaisgood, H.E. 2000. Bioactive milk peptides: A prospectus. *J. Dairy Sci.* 83:1187–1195.

Clare, D.A., Catignani, G.L., and Swaisgood, H.E. 2003. Biodefense properties of milk: The role of antimicrobial proteins and peptides. *Curr. Pharm. Des.* 9:1239–1255.

Clark, J.R., Osuga, D.T., and Feeney, R.E. 1963. Comparison of avian egg white conalbumins. *J. Biol. Chem.* 238:3621–3631.

Conner, D.E. 1993. Naturally occurring compounds. In *Antimicrobials in Foods: Second Edition, Revised and Expanded*, edited by P.M. Davidson and A.L. Branen, Marcel Dekker, Inc., New York, NY, pp. 441–468.

Coonrod, J.D. 1987. Role of surfactant-free fatty acids in antimicrobial defenses. *Euro. J. Respir. Dis.* 71:209–214.

Cordle, C.T., Schaller, J.P., Winshi, T.R., Candler, E.L., Hilty, M.D., Smith, K.L., Saif, L.J., Kohler, E.M., and Krakowka, S. 1991. Passive immune protection from diarrhea caused by rotavirus or *E. coli*: Animal model to demonstrate and quantitate efficacy. *Adv. Exp. Med. Biol.* 310:317–327.

Courtois, P., van Beers, D., de Foor, M., Mandelbaum, I.M., and Poutois, M. 1990. Abolition of herpes simplex cytopathic effect after treatment with peroxidase generated hypothiocyanite. *J. Biol. Buccale* 18:71–74.

Cowart, R.E., and Foster, B.G. 1985. Differential effects of iron on the growth of *L. monocytogenes*: Minimum requirements and mechanism of acquisition. *J. Inf. Dis.* 151:721–730.

Crouch, S.P., Slater, K.J., and Fletcher, J. 1992. Regulation of cytokine release from mononuclear cells by the iron-binding protein lactoferrin. *Blood* 80:235–240.

Daher, K.A., Selsted, M.E., and Lehrer, R.I. 1986. Direct inactivation of viruses by human granulocyte defensins. *J. Virol.* 60:1068–1074.

Davidson, L.A., and Lönnerdal, B. 1986. Isolation and characterization of rhesus monkey lactoferrin. *Pediatr. Res.* 20:197–201.

Davidson, L.A., Litov, R.E., and Lönnerdal, B. 1990. Iron retention from lactoferrin-supplemented formulas in infant rhesus monkeys. *Pediatr. Res.* 27:176–180.

Davidson, L., Kastenmayer, P., Yuen, M., Lönnerdal, B., and Hurrell, R.F. 1994. Influence of lactoferrin on iron absorption from human milk in infants. *Pediatr. Res.* 35:117–124.

Davidson, P.M., Critzer, F.J., and Taylor, T.M. 2013. Naturally occurring antimicrobials for minimally processed foods. *Ann. Rev. Food Sci. Technol.* 4:163–190.

Delange, R., and Huang, T.S. 1971. Egg white avidin 3. Sequence of the 78-residue middle cyanogen bromide peptides: Complete amino acid sequence of the protein subunit. *J. Biol. Chem.* 246:698–709.

De Wit, J.N., and van Hooijdonk, A.C.M. 1996. Structure, functions, and applications of lactoperoxidase in natural antimicrobial systems. *Netherlands Milk Dairy J.* 50:227–244.

Dhaenens, L., Szczebara, F., and Husson, M.O. 1997. Identification, characterization, and immunogenicity of the lactoferrin-binding protein from *Helicobacter pylori*. *Infect. Immun.* 65:514–518.

Diarra, M.S., Petitclerc, D., and Lacasse, P. 2002. Effect of lactoferrin in combination with penicillin on the morphology and the physiology of *Staphylococcus aureus* isolated from bovine mastitis. *J. Dairy Sci.* 85:1141–1149.

Dite, P., Bartova, M., and Skaunic, V. 1989. Assessment of lactoferrin in the diagnosis of pancreatic disease. *Vnitrni. Lek.* 35:132–136.

Dominguez, E., Perez, M.D., and Calvo, M. 1997. Effect of heat treatment on the antigen-binding activity of anti-peroxidase immunoglobulins in bovine colostrums. *J. Dairy Sci.* 80:3182–3187.

Doyle, M.P., and Marth, E.H. 1978. Degradation of aflotoxin by lactoperoxidase. *Z. Lebensm. Unters. Forsch.* 166:271–273.

Drackley, J.K. 2000. Lipid metabolism. In *Farm Animal Metabolism and Nutrition*, edited by J.P.F. D'Mello, CABI Publishing, Wallingford, Oxfordshire, UK.

Dull, T.J., Uyeda, C., Strosberg, A.D., Nedwin, G., and Seilhamer, J.J. 1990. Molecular, cloning of cDNAs encoding bovine and human lactoperoxidase. *DNA Cell. Biol.* 9:499–509.

Dutta, J., Tripathi, S., and Dutta, P.K. 2012. Progress in antimicrobial activities of chitin, chitosan and its oligosaccharides: A systematic study needs for food applications. *Food Sci. Technol. Intl.* 18:3–34.

Eakin, R.E., Snell, E.E., and Williams, R.J. 1941. The concentration and assay of avidin, the injury-producing protein in raw egg white. *J. Biol. Chem.* 140:535–543.

Ebina, T. 1992. Passive immunizations of suckling mice and infants with bovine colostrum containing antibodies to human rotavirus. *J. Med. Virol.* 38:117–123.

Ebina, T. 1996. Prophylaxis of rotavirus gastroenteritis using immunoglobulin. *Arch. Virol.* 12:217–223.

Eckblad, W.P., Hendrix, K.M., and Olson, D.P. 1981. Total complement hemolytic activity of colostral whey and sera from dairy cows. *Cornell Vet.* 71:54–58.

Edelman, G.M. 1973. Antibody structure and molecular immunology. *Science* 180:830–840.

Ekstrand, B. 1994. Lactoperoxidase and lactoferrin. In *Natural Antimicrobial Systems and Food Preservation*, edited by V.D. Dillon and R.G. Board, CAB Intentional, Wallingford, UK, pp. 15–63.

Elbarbary, H.A., Abdou, A.M., Nakamura, Y., Park, E.Y. Mohamed, H.A. and Sato, K. 2012. Identification of novel antibacterial peptides isolated from a commercially available casein hydrolysate by autofocusing technique. *BioFactors* 38:309–315.

Ellison, R.T. III, Giehl, T.J., and Laforce, F.M. 1988. Damage of the outer membrane of enteric gram-negative bacteria by lactoferrin and transferrin. *Infect. Immun.* 56:2774–2781.

Ellison, R.T, III., LaForce, F.M., Giehl, T.J., Boose, D.S., and Dunn, B.E. 1990. Lactoferrin and transferrin damage of the gram-negative outer membrane is modulated by Ca^{2+} and Mg^{2+}. *J. Gen. Microbiol.* 136:1437–1446.

Elo, H.A., Kulomaa, M., and Tuohimaa, P. 1978. Avidin induction by tissue injury and inflammation in male and female chickens. *Comp. Biochem. Physiol.* 62:237–240.

Elo, H.A., and Korpela, J. 1984. The occurrence and production of avidin: A new conception of the high-affinity biotin-binding protein. *Comp. Biochem. Physiol. B.* 78:15–20.

Embleton, N., Berrington, J., McGuire, W., Stewart, C., Cummings, S. 2013. Lactoferrin: Antimicrobial activity and therapeutic potential. *Semin. Fetal. Neonatal Med.* 18:143–149.

Erdei, J., Forsgren, A., and Naidu, A.S. 1994. Lactoferrin binds to porins OmpF and OmpC in *Escherichia coli*. *Infect. Immun.* 62:1236–1240.

Eterradossi, N., Toquin, D., Addassi, H., Rivallan, G., Cotte, J.P., and Guttet, M. 1997. Passive protection of specific pathogen free chicks against infectious bursal disease by *In-ovo* injection of semi-purified egg yolk antiviral immunoglobulins. *J. Vet. Med.* B 44:371–383.

Facon, M., Skura, B.J., and Nakai, S. 1995. Antibodies to a colonization factor of human enterotoxic *Escherichia coli* in cow's milk and colostrum. *Food Res. Intl.* 28:387–391.

Fearon, D.T. 1998. The complement system and adaptive immunity. *Seminars Immunol.* 10:355–361.

Federico, B., Pinto, L., Quintieri, L., Carito, A., Calabrese, N. and Caputo, L. 2015. Efficacy of lactoferricin B in controlling ready-to-eat vegetable spoilage caused by *Pseudomonas* spp. *Intl. J. Food Microbiol.* 215:179–186.

Fenna, R., Zeng, J., and Davey, C. 1995. Structure of the green heme in myeloperoxidase. *Arch. Biochem. Biophys.* 316:653–656.

Fernandes, J.C., Tavaria, F.K., Soares, J.C., Ramos, O.S., Joao Monteiro, M., Pintado, M.E., and Xavier Malcata, F. 2008. Antimicrobial effects of chitosans and chitooligosaccharides, upon *Staphylococcus aureus* and *Escherichia coli*, in food model systems. *Food Microbiol.* 25:922–928.

Floris, R., Recio, I., Berkhout, B., and Visser, S. 2003. Antibacterial and antiviral effects of milk proteins and derivatives thereof. *Curr. Pharm. Des.* 9:1257–1275.

Fraenkel-Conrat, J., and Fraenkel-Conrat, H. 1952. Metabolic fate of biotin and of avidin-biotin complex upon parenteral administration. *Biochim. Biophys. Acta.* 8:66–70.

Franco, I., Pérez M., Conesa, C., Calvo, M., and Sanchez, L. 2018. Effect of technological treatments on bovine lactoferrin: An overview. *Food Res. Intl.* 106:173–182.

Francois, C.A., Connor, S.L., Wander, R.C., and Connor, W.E. 1998. Acute effects of dietary fatty acids on the fatty acids of human milk. *Am. J. Clin. Nutr.* 67:301–308.

Friedman, S.A., Mandel, I.D., and Herrera, M.S. 1983. Lysozyme and lactoferrin quantitation in the crevicular fluid. *J. Periodontol.* 54:347–350.

Fromm, J.H., Hileman, R.E., Caldwell, E.E., Weder, J.M., and Linhardt, R.J. 1995. Differences in the interaction of heparin with arginine and lysine and the importance of these basic amino acids in the binding of heparin to acidic fibroblast factor. *Arch. Biochem. Biophys.* 323:279–287.

Fuller, R., and Moore, J.H. 1967. The inhibition of the growth of *Clostridium welchii* by lipids isolated from the contents of the small intestine of the pig. *J. Gen. Microbiol.* 46:23–41.

Gallardo-Cigarroa, F.J., Koshio, S., Hayasawa, H., and Oshida, K. 2000. Effect of bovine lactoferrin on performance of Japanese flounder *Paralichthys olivaceus.* In *Lactoferrin: Structure, Function and Applications*, edited by K.I. Shimazaki, et al., Elsevier Science B.V., Amsterdam, The Netherlands, pp. 443–450.

Ganz, T. 2003. Defensins: Antimicrobial peptides of innate immunity. *Nat. Rev. Immunol.* 3:710–720.

Garrett, R.H., and Grisham, C.M. 1999. Molecular components of cells – lipids. In *Biochemistry*, 2nd ed., Saunders College Publishing, San Antonio, TX, p. 138.

Gitlin, G., Bayer, E.D., and Wilchek, M. 1988. Studies on the biotin-binding site of avidin. *Biochem. J.* 250:291–294.

Goldsby, R.A., Kindt, T.J., and Osborne, B.A. 2000. *Kuby Immunology*, 4th ed., W.H. Freeman and Company, New York, NY, pp. 83–115.

Gothefors, L., and Marklund, S. 1975. Lactoperoxidase activity in human milk and in saliva of newborn infants. *Infect. Immun.* 11:1210–1215.

Grab, D.J., Lonsdale-Eccles, J.D., Oli, M.W., and Corbeil, Lynette B. 2001. Lactoferrin-binding proteins of *Tritrichomonas foetus. J. Parasitol.* 87:1064–1070.

Grecz, N., Wagenaar, R.O., and Dack, G.M. 1959. Relation of fatty acids to the inhibition of *Clostridium botulinum* in aged surface ripened cheese. *Appl. Microbiol.* 7:228–234.

Green, N.M. 1963. Avidin 3: The nature of the biotin-binding site. *Biochem. J.* 89:599–609.

Green, N.M. 1964. The molecular weight of avidin. *Biochem. J.* 92:16–17.

Green, N.M. 1975. Avidin. *Adv. Protein Chem.* 29:85–133.

Greenberg, P.D., and Cello, J.P. 1996. Treatment of severe diarrhea caused by the *Crytosporidium parvum* with oral bovine Ig concentrate in patients with AIDS. *J. AIDS Human Retrovirol.* 13:348–354.

Griffiths, E.A., Duffy, L.C., Schanbacher, F.L., Dryja, D., Leavens, A., Neiswander, R.L., Qiao, H., DiRienzo, D., and Ogra, P. 2003. *In vitro* growth responses of bifidobacteria and enteropathogens to bovine and human lactoferrin. *Dig. Dis. Sci.* 48:1324–1332.

Groves, M. 1960. The isolation of a red protein from milk. *J. Am. Chem. Soc.* 82:3345–3350.

Grover, M., Giouzeppos, O., Schnagl, R.D., and May, J.T. 1997. Effect of human milk prostoglandins and lactoferrin on respiratory syncytial virus and rotavirus. *Acta Paed.* 86:315–316.

Gutteberg, T.J., Haneberg, B., and J$_{\emptyset}$rgensen, T. 1984. The latency of serum acute phase proteins in meningococcal septicemia, with special emphasis on lactoferrin. *Clin. Chem. Acta* 136:173–178.

Gutteberg, T.J., Røkke, O., Andersen, O., and Jørgensen, T. 1988. Lactoferrin as an indicator of septicemia and endotoxaemia in pigs. *Scand. J. Infect. Dis.* 20:659–666.

Gyorgy, P. Rose, C.S. and Tomarelli, R. 1942. Investigations on the stability of avidin. *J. Biol. Chem.* 144:169–173.

Hagiwara, S.-I., Kawai, K., Anri, A., and Nagahata, H. 2003. Lactoferrin concentrations in milk from normal and subclinical mastitic cows. *J. Vet. Med Sci.* 65:319–323.

Hall, M.A., and Maurer, A.J. 1986. Spice extracts, lauricin, and propylene glycol as inhibitors of *Clostridium botulinum* in turkey frankfurter slurries. *Poult. Sci.* 65:145–151.

Hamada, S., Horikoshi, T., Minami, T., Kawabata, S., Hiraoka, J., Fujiwara, T., and Ooshima, T. 1991. Oral passive immunization against dental caries in rats by use of hen egg yolk antibodies specific for cell-associated glycosyltransferase of *Streptococcus mutans. Infect. Immun.* 59:4161–4167.

Hammer, J., Haalheim, H., and Gutteberg, T.J. 2000. Bovine lactoferrin is more efficient than bovine lactoferricin in inhibiting HSV-I/-II replication *in vitro*. In *Lactoferrin: Structure, Functions and Applications*, edited by K. Shimazaki, Elsevier Science, Amsterdam, pp. 239–243.

Hamon, C.B., and Klebanoff, S.J. 1973. A peroxidase-mediated, *Streptococcus mitis*-dependent antimicrobial system in saliva. *J. Exp. Med.* 137:438–450.

Hamosh, M. 1991. Lipid metabolism. In *Neonatal Nutrition and Metabolism*, edited by W.W. Hay, Jr., Mosby Year Book, St. Louis, MO, pp. 122–142.

Hamosh, M. 1995. Lipid metabolism in pediatric nutrition. *Ped. Clinics N. Amer.* 42:839–859.

Hangoc, G., Lu, L., Oliff, A., Gillis, S., Hu, W., Bicknell, D.C., Williams, D., and Broxmeyer, H.E. 1987. Modulation of friend virus infectivity *in vivo* by administration of purified preparations of human lactoferrin and recombinant murine interleukin-3 to mice. *Leukemia* 1:762–764.

Hanson, L.A., Ashraf, R.N., Hahn-Zoric, M., Carlsson, B., Herias, V., Weidermann, U., Dahlgren, U., Motas, C., Mattsby-Baltzer, I., Gonzales-Cossio, T., Cruz, J.R., Karlberg, J., Lindblad, B.S., and Jalil, F. 1993. Breast milk: Roles in neonatal host defense. In *Immunophysiology of the Gut*, pp. 248–249.

Harbitz, O., Jensen, A.O., and Smidsrod, O. 1984. Lysozyme and lactoferrin in sputum from patients with chronic obstructive lung disease. *Euro. J. Res. Dis.* 65:512–520.

Harder, J., Siebert, R., Zhang, Y., Matthiesen, P., Christophers, E., Schlegelberger, B., Schroder, J.M. 1997. Mapping of the gene encoding human beta-defensin-2 (DEFB2) to chromosome region 8p22-p23.1. *Genomics* 46:472–475.

Harwig, S.S., Swiderek, K.M., Kokryakov, V.N., Tan, L., Lee, T.D., Panyutich, E.A., Aleshina, G.M., Shamova, O.V., and Lehrer, R.I. 1994. Gallinacins: Cysteine-rich antimicrobial peptides of chicken leukocytes. *FEBS Lett.* 342:281–285.

Hasegawa, K., Motsuchi, W., Tanaka, S., and Dosako, S. 1994. Inhibition with lactoferrin of in vitro infection with human herpes virus. *Jap. J. Med. Sci. Biol.* 47:73–85.

Hatta, H., Tsuda, K., Akachi, S., Kim, M., Yamamoto, T., and Ebina, T. 1993. Oral passive immunization effect of anti-human rotavirus IgY and its behavior against proteolytic enzymes. *Biosci. Biotech. Biochem.* 57:1077–1081.

Hatta, H., Mabe, K., Kim, M., Yamamoto, T., Gutierrez, M.A., and Miyazaki, T. 1994. Prevention of fish disease using egg yolk antibody. In *Egg Uses and Processing Technologies-New Developments*, edited by J.S. Sim and S. Nakai, CAB International, p. 241.

Hatta, H., Ozeki, M., and Tsuda, K. 1997. Egg yolk antibody IgY and its application. In *Hen Eggs: Their Basic and Applied Science*, edited by T. Yamamoto, L.R. Juneja, H. Hatta, and M. Kim, CRC Press, Boca Raton, FL, pp. 151–178.

Haug, B.E., and Svendsen, J.S. 2001. The role of tryptophan in the antibacterial activity of a 15-residue bovine lactoferricin peptide. *J. Peptide Sci.* 7:190–196.

Haukland, H.H., and Vorland, L.H. 2001. Post-antibiotic effect of the antimicrobial peptide lactoferricin on *Escherichia coli* and *Staphylococcus aureus*. *J. Antimicrob. Chemother.* 48:569–571.

Haukland, H.H., Ulvatne, H., Sandvik, K., and Vorland, L.H. 2001. The antimicrobial peptides lactoferricin B and magainin 2 cross over the bacterial cytoplasmic membrane and reside in the cytoplasm. *FEBS Lett.* 508:389–393.

Hendrickson, W.A., Pähler, A., Smith, J.L., Satow, Y., Merrit, E.A., and Phizackerley, R.P. 1989. Crystal structure of core streptavidin determined from multiwavelength anomalous diffraction of synchrotron radiation. *Proc. Natl. Acad. Sci. U.S.A.* 86:2190–2194.

Hennart, P.F., Brasseur, D.J., Delogne Desnoeck, J.B., Dramaix, M.M., and Robyn, C.E. 1991. Lysozyme, lactoferrin and secretory immunoglobulin A content in breast milk: Influence of duration of lactation, nutrition status and parity of mother. *Am. J. Clin. Nutr.* 53:32–39.

Herandez, C.M.M., van Markwijk, B.W., and Vreeman, H.J. 1990. Isolation and properties of lactoperoxidase from bovine milk. *Neth. Milk and Dairy J.* 44:213–231.

Hernell, O., Ward, H., Blackberg, L., and Pereira, M.E.A. 1986. Killing of *Giardia lamblia* by human milk lipases: An effect mediated by lipolysis of milk lipids. *J. Infect. Dis.* 153:715–720.

Herts, R. 1946. Biotin and the avidin-biotin complex. *Physiol. Rev.* 26:479–494.

Higgins, D.A. 1975. Physical and chemical properties of fowl immunoglobulins. *Vet. Bull.* 45:139–154.

Hill, C.P., Yee, J., Selsted, M.E., and Eisenberg, D. 1991. Crystal structure of defensin HNP-3, an amphiphilic dimer: Mechanisms of membrane permeabilization. *Science* 251:1481–1485.

Hilpert, H., Brussow, H., Mietens, C., Sidoti, J., Lerner, L., and Werschau, H. 1987. Use of bovine milk concentrate containing antibody to rotavirus to treat rotavirus gastroenteritis in infants. *J. Infect. Dis.* 156:158–166.

Hirai, Y., Kawakata, N., Satoh, K., Ikeda, Y., Hisayasu, S., Orimo, H., and Yoshino, Y. 1990. Concentration of lactoferrin and iron in human milk at different stages of lactation. *J. Nutr. Sci. Vitaminol.* 36:531–544.

Hirose, M. 2000. The structural mechanism for iron uptake and release by transferrins. *Biosci. Biotechnol. Biochem.* 64:1328–1336.

Hoek, K.S., Milne, J.M., Grieve, P.A., Dionusius, D.A., and Smith, R. 1997. Antibacterial activity of bovine lactoferrin-derived peptides. *Antimicrob. Agents Chemother.* 41:54–59.

Holmskov, U., and Jensenius, J.C. 1996. Two bovine collectins: Conglutinin and CL-43. In *Collectins and Innate Immunity*, edited by R.A.B. Ezekowitz, K. Sastry, and K.B.M. Reid, Springer, Heidelberg, pp. 51–71.

Hoover, D.M., Rajashankar, K.R., Blumenthal, R., Puri, A., Oppenheim, J.J., Chertov, O., and Lubkowski, J. 2000. The structure of human beta-defensin-2 shows evidence of higher order oligomerization. *J. Biol. Chem.* 275:32911–32918.

Hossain, F., Rahman Moghal, M., Islam, Z., Moniruzzaman, M., and Yamazaki, M. 2019. Membrane potential is vital for rapid permeabilization of plasma membranes and lipid bilayers by the antimicrobial peptide lactoferrin *British J. Biol. Chem.* 294:10449–10462.

Hu, Z., and Gänzle, M.G. 2019. Challenges and opportunities related to the use of chitosan as a food preservative. *J. Appl. Microbiol.* 126:1318–1331.

Humblodt, V., Yala, J.-F., Thebault, P., Boukerma, K., Hequet, A., Berjeaud, J.-M., and Pradier, C.M. 2009. The antibacterial activity of magainin I immobilized onto mixed thiols self-assembled monolayers. *Biomaterials* 30:3503–3512.

Husu, J., Syvaoja, E., Ahola-Luttila, H., Kalsta, H., Sivela, S., and Kosunen, T.U. 1993. Production of hyperimmune bovine colostrums against *Campylobacter jejuni*. *J. Appl. Bacteriol.* 74:564–569.

Ikeda, M., Sugiyama, K., Tanaka, T., Tanaka, K., Sekihara, H., Shimotohno, K., and Kato, N. 1998. Lactoferrin markedly inhibits hepatitis C virus infection in cultured human hepatocytes. *Biochem. Biophys. Res. Comm.* 245:549–553.

Ikeda, M., Nozaki, A., Sugiyama, K., Tanaka, T., Naganuma, A., Tanaka, K., Sekihara, H., Shimotohno, K., Saito, M., and Kato, N. 2000. Characterization of antiviral activity of lactoferrin against hepatitis C virus infection in human cultured cells. *Virus Res.* 66:51–63.

Ikemori, Y., Kuroki, M., Peralta, R.C., Yokoyama, H., and Kodama, Y. 1992. Protection of neonatal calves against fatal enteric colibacillosis by administration of egg yolk powder from hens immunized with K99-piliated enterotoxigenic Escherichia coli. *Am. J. Vet. Res.* 53:2005–2008.

Ikemori, Y., Ohta, M., Umeda, K., Icalto, F.C., Kuroki, M., Yokoyama, H., and Kodama, Y. 1997. Passive protection of neonatal calves against bovine coronavirus-induced diarrhea by administration of egg yolk or colostrums antibody powder. *Vet. Microbiol.* 58:105–111.

Imberechts, H., Deperez, P., Van Driessche, E., and Pohl, P. 1997. Chicken egg yolk antibodies against F18ab fimbriae of *Escherichia coli* inhibit shedding of F18 positive *E. coli* by experimentally infected pigs. *Vet. Microbiol.* 54:329–341.

Isaacs, C.E., and Thormar, H. 1990. Human milk lipids inactivate enveloped viruses. In *Breastfeeding, Nutrition, Infection and Infant Growth in Developed and Emerging Countries*, edited by S.A. Atkinson, L.A. Hanson, and R.K. Chandra, ARTS Biomedical Publishers and Distributors, Newfoundland, Canada, pp. 161–174.

Isaacs, C.E., Kashyap, S., Heird, W.C., and Thormar, H. 1990. Antiviral and antibacterial lipids in human milk and infant formula feeds. *Arch. Dis. Child.* 65:861–864.

Isaacs, C.E., Litov, R.E., and Thormar, H. 1995. Antimicrobial activity of lipids added to human milk, infant formula, and bovine milk. *J. Nutr. Biochem.* 6:362–366.

Isaacs, C.E. 2001. The antimicrobial function of milk lipids. In *Advances in Nutritional Research, Vol. 10, Immunological Properties of Milk*, edited by B. Woodward, H.H. Draper, Kluwer Academic/Plenum Publishers, New York, NY, pp. 271–285.

Isamida, T., Tanaka, T., Omata, Y., Yamauchi, K., Shimazaki, K., and Saito, A. 1998. Protective effect of lactoferricin against *Toxoplasma gondii* infection in mice. *J. Vet. Med Sci.* 60:241–244.

Ishida, A., Yoshikai, Y., Murosaki, S., Kubo, C., Hidaka, Y., and Nomoto, K. 1992. Consumption of milk from cows immunized with intestinal bacteria influences age-related changes in immune competence in mice. *J. Nutr.* 122:1875–1883.

Itoh, T., Sugawara, H., and Adachi, S. 1979. Comparative chemical studies on the quail (*Coturnix coturnix japonica*) egg ovotransferrin. *Compar. Biochem. Physiol.* 62B:41–44.

Jago, G.R., and Morrison, M. 1962. Anti-streptococcal activity of lactoperoxidase. *Proc. Soc. Exp. Biol. Med.* 111:585–588.

Janota-Bassalik, L., Zajac, M., Pietraszek, A., and Piotrowska, E. 1977. Bacteriostatic activity in cow's milk from udders infected with *Streptococcus agalactiae* and *Staphylococcus aureus*. *Acta Microbiol. Pol.* 26:413–419.

Jason, J.M., Nieburg, P., and Marks, J.S. 1984. Mortality and infectious disease associated with infant-feeding practices in developing countries. *Pediatrics* 74:702–727.

Jasour, M.S., Ehsani, A., Mehryar, L., and Naghibi, S.S. 2015. Chitosan coating incorporated with the lactoperoxidase system: An active edible coating for fish preservation. *J. Sci. Food Agric.* 95:1373–1378.

Jensen, R.G. 1995. Comments on the extraction of fat from human milk for analysis of contaminants. *Chemosphere* 31:4197–4200.

Jensen, R.G. 1996. The lipids in human milk. *Prog. Lipid Res.* 35:53–92.

Jensenius, J.C., Andersen, I., Hau, J., Crone, M., and Koch, C. 1981. Eggs: Conveniently packaged antibodies. Methods for purification of yolk IgG. *J. Immunol. Methods* 46:63–68.

Jin, L.Z., Baidoo, S.K., Marquardt, R.R., and Frohlich, A.A. 1998. *In vitro* inhibition of adhesion of enterotoxigenic *Esherichia coli* K88 to piglet intestinal mucus by egg-yolk antibodies. *FEMS Immunol. Med. Microbiol.* 21:313–321.

Johansen, C., Falholt, P., and Gram, L. 1997. Enzymatic removal and disinfection of bacterial biofilms. *Appl. Environ. Microbiol.* 63:3724–3728.

Jones, E.M., Smart, A., Bloomberg, G., Burgess, L., and Millar, M.R. 1994. Lactoferricin, a new antimicrobial peptide. *J. Appl. Bacteriol.* 77:208–214.

Kabara, J.J. 1978. Fatty acids and derivatives as antimicrobial agents: A review. In *The Pharmacological Effect of Lipids –1*, edited by J.J. Kabara, American Oil Chemists Society, Champaign, IL, pp. 1–14.

Kai, K., Komine, K., Komine, Y., Kuroishi, T., Kozutsumi, T., Kobayashi, J., Ohta, M., Kitamura, H., and Kumagai, K. 2002. Lactoferrin stimulates a *Staphylococcus aureus* killing activity of bovine phagocytes in the mammary gland. *Microbiol. Immunol.* 46:187–194.

Kakuta, I. 2000. Lactoferrin improves physiological conditions of fish held under deteriorating states. In *Lactoferrin: Structure, Function and Applications*, edited by K.I. Shimazaki, et al., Elsevier Science B.V., Amsterdam, The Netherlands, pp. 429–441.

Kamau, D.N., Doores, S., and Pruitt, K.M. 1990. Enhanced thermal destruction of *Listeria monocytogenes* and *Staphylococcus aureus* by the lactoperoxidase system. *Appl. Environ. Microbiol.* 56:2711–2716.

Kang, J.H., Lee, M.K., Kim, K.L., and Hahm, K.-S. 1996. Structure-biological activity relationships of 11-residue highly basic peptide segment of bovine lactoferrin. *Intl. J. Peptide Protein Res.* 48:357–363.

Kangumba, J.G., Venter, E.H., and Coetzer, J.A. 1997. The effect of activation of the lactoperoxidase system and souring on certain potential human pathogens in cows' milk. *J. S. African Vet. Assoc.* 68:130–136.

Kato, N., and Shibasaki, I. 1975. Enhancing effect of fatty acids and their esters on the thermal destruction of *E. coli* and *Pseudomonas aeruginosa*. *J. Ferment. Technol.* 53:802–807.

Kawai, A., and Tanaka, K. 1997. *Reverse Mutation Test of Monl-01 Using Bacteria*. Gotemba Laboratory, Bozo Research Center Inc., Setagaya-ku, Tokyo, Japan.

Kawasaki, Y., Isoda, H., Shinmoto, H., Tanimoto, M., Dosako, S., Idota, T., and Nakajima, I. 1993. Inhibition by kappa-casein glycomacropeptide and lactoferrin of influenza virus hemagglutination. *Biosci. Biotechnol. Biochem.* 57:1214–1215.

Kawasaki, Y., Tazume, S., Shimizu, K., Matsuzawa, H., Dosako, S., Isoda, H., Tsukiji, M., Fujimura, R., Muranaka, Y., and Isihida, H. 2000. Inhibitory effects of bovine lactoferrin on the adherence of enterotoxigenic Escherichia coli to host cells. *Biosci. Biotechnol. Biochem.* 64:348–354.

Kelly, C.P., Pothoulakis, C., Vavva, F., Castagliuolo, I., Bostwick, E.F., O'Keane, J.C., Keates, S., and J.T. 1996. Anti-*Clostridium difficile* bovine Ig concentrate inhibits cytotoxicity and enterotoxicity of *C. difficile* toxins. *Antimicrob. Agents Chemother.* 40:373–379.

Kemp, D.J., Morahan, G., Cowman, A.F., and Harris, A.W. 1983. Production of RNA for secreted Ig μ chains does not require transcriptional termination 5′ to the μm exons. *Nature* 301:84.

Kennedy, M., O'Rourke, A.L., McLay, J., Simmonds, R. 2000. Use of a ground beef model to assess the effect of the lactoperoxidase system on the growth of *Escherichia coli* 0157:H7, *Listeria monocytogenes* and *Staphylococcus aureus* in red meat. *Intl. J. Food Microbiol.* 57:147–158.

Kern, V.R., Wildbrett, G., and Kiermeier, F. 1963. Dependence of peroxidase activity in milk on the sexual cycle in cows. *Z. Naturforsch.* 18:1082–1084.

Kett, W.C., Osmond, R.I.W., Moe, L., Skett, S.E., Kinnear, B.F., and Coombe, D.R. 2003. Avidin is a heparin-binding protein: Affinity, specificity, and structural analysis. *Biochim. Biophys. Acta* 1620:225–234.

Kiermeier, F., and Kuhlmann, H. 1972. Lactoperoxidase activity in human and in cows' milk: Comparative studies. *Munch. Med. Wochenshr.* 114:2144–2146.

Kijlstra, A.S., Jeurissen, H.M., and Koning, K.M. 1983. Lactoferrin levels in normal human tears. *Brit. J. Opthalmol.* 67:199–202.

Kimsey, H.R., Adams, D.M., and Kabara, J.J. 1981. Increased inactivation of bacterial spores at high temperatures in the presence of monoglycerides. *J. Food Safety* 3:69–82.

Kimura, M., Nam, M.-S., Ohkouchi, Y., Kumura, H., Shimazaki, K.-I., and Yu, D.-Y. 2000. Antimicrobial peptide of Korean native goat lactoferrin and identification of the part essential for this activity. *Biochem. Biophys. Res. Communic.* 268:333–336.

Klasing, K.C. 1984. Effect of inflammatory agents and interleukin 1 on iron and zinc metabolism. *Am. J. Physiol.* 247:R901–R904.

Kobayashi, K., and Hirai, H. 1980. Studies on subunit components of chicken polymeric immunoglobulins. *J. Immunol.* 124:1695–1704.

Kobayashi, T., Ohmori, T., Yanai, M., Kawanishi, G., Mitsuyama, M., and Nomoto, K. 1991. Protective Effect of orally administering immune milk on endogenous infection in X-Irradiated mice. *Agric. Biol. Chem.* 55:2265–2272.

Kong, M., Chen, X.G., Xing, K., Park, H.J. 2010. Antimicrobial properties of chitosan and mode of action: A state of the art review. *Intl. J. Food Microbiol.* 144:51–63.

Korhonen, H., Syvaoja, E.-L., Ahola-Luttila, H., Sivela, S., Kopola, S., Husu, J., and Kosunen, T.U. 1995. Bactericidal effect of bovine normal and immune serum, colostrum and milk against *Helicobacter pylori*. *J. Appl. Bacteriol.* 78:655–662.

Korhonen, H., Syvaoja, E.-L., Vasara, E., Kosunen, T., and Marnila, P. 1998. Pharmaceutical composition, comprising complement proteins, for the treatment of *Helicobacter* infections and a method for the preparation of the composition. Patent application PCT No. W098/00150.

Korhonen, H., Marnila, P., and Gill, H.S. 2000. Milk immunoglobulins and complement factors. *Brit. J. Nutr.* 84 Supplement 1:S75–S80.

Korpela, J. 1984. Avidin, a high affinity biotin-binding protein as a tool and subject of biological research. *Med. Biol.* 62:5–26.

Koshio, S., Yokoyama, S., Ishikawa, M., Hayasawa, H., and Oshida, K. 2000. Effect of lactoferrin on growth, survival and tolerance to stress of kuruma prawn *Penaeus japonicus*. In *Lactoferrin: Structure, Function and Applications*, edited by K.I. Shimazaki, et al., Elsevier Science B.V., Amsterdam, The Netherlands, pp. 451–456.

Kostoulas, G., Honer, D., Naggi, A., Casu, B., and Baci, A. 1997. Electrostatic interactions between human leukocyte elastase and sulfated glycosaminoglycans: Physiological implications. *Biol. Chem.* 378:1481–1489.

Kuipers, M.E., Heegsma, J., Bakker, H.I., Meijer, D.K., Swart, P.J., Frijlink, E.W., Eissens, A.C., de Vries-Hospers, H.G., and van den Berg, J.J. 2002. Design and fungicidal activity of mucoadhesive lactoferrin table for the treatment of oropharyngeal candidosis. *Drug Delivery* 9:31–38.

Kulkarni, P.K. and Pimpale, N.V. 1989. Colostrum – A review. *Indian J. Dairy Sci.* 42:216–224.

Kumar, G., Reddy, P.M., and Kottekad, S. 2018. Comparative site-specific N-glycosylation analysis of lactoperoxidase from buffalo and goat milk using RP-UHPLC-MS/MS reveals a distinct glycan pattern. *J. Agric. Food Chem.* 66:11492–11499.

Kumar, S., Thippareddi, H., Subbiah, J., Zivanovic, S., Davidson, P.M., and Harte, F. 2009. Inactivation of *Escherichia coli* K-12 in apple juice using combination of high-pressure homogenization and chitosan. *J. Food Sci.* 74:M8–M14.

Kummer, A., Kitts, D.D., Li-Chan, E., Losso, J.N., Skkura, B.J., and Nakai, S. 1992. Quatification of Bovine IgG in milk using enzyme-linked immunosorbent assay. *Food Agri. Immunol.* 4:93–102.

Kuroki, M., Ikemori, Y., Yokoyama, H., Peralta, R.C., Icalto, F.C., and Kodama, Y. 1993. Passive protection against bovine rotavirus-induced diarrhea in murine model by specific immunoglobulins from chicken egg yolk. *Vet. Microbiol.* 37:135–146.

Kuroki, M., Ohta, M., Ikemori, Peralta, R.C., Yokoyama, H., and Kodama, Y. 1994. Passive protection against bovine rotavirus in calves by specific immunoglobulins from chicken egg yolk. *Arch. Virol.* 138:143–148.

Kuroki, M., Ohta, M., Ikemori, Y., Icalto, F.C., Kobayashi, C., Yokoyama, H., and Kodama, Y. 1997. Field evaluation of chicken egg yolkimmunoglobulins specific for bovine rotavirus in neonatal calves. *Arch. Virol.* 142:843–851.

Kussendrager, K.D., and van Hooijdonk, A.C. 2000. Lactoperoxidase:physico-chemical properties, occurrence, mechanism of action and applications. *Brit. J. Nutr.* 84 Supplement 1:S19–25.

Lampe, M.F., Ballweber, L.M., Isaacs, C.E., Patton, D.L., and Stamm, W.E. 1998. Inhibition of *Chlamydia trachomatis* by novel antimicrobial lipids adapted from human breast milk. *Antimicrob. Agents Chemother.* 42:1239–1244.

Lampe, M.F., and Isaacs, C.E. 2000. In *Natural Food Antimicrobial Systems*, edited by A.S. Naidu, CRC Press, Boca Raton, FL, pp. 159–183.

Lascelles, A.K. 1979. The immune system on the ruminant mammary gland and its role in the control of mastitis. *J. Dairy Sci.* 62:154–167.

Law, B.A., and John, P. 1981. Effect of lactoperoxidase bactericidal system on the formation of the electrochemical proton gradient in *Escherichia coli. FEMS Microbiol. Lett.* 10:67–70.

Ledoux, M., Laloux, L., and Wolff, R.L. 2000. Analytical methods for the determination of tans-C18 fatty acid isomers in milk fat: A review. *Analysis* 28:402–412.

Lee, T.H., Shimazaki, K., Yu, S.L., Nam, M.S., Kim, S.J., Lee, K.K., and Yu, D.Y. 1997. Polymorphic sequence of Korean native goat lactoferrin exhibiting greater antibacterial activity. *Anim. Genet.* 28:367–369.

Lehrer, R.I. 1969. Antifungal effects of peroxidase systems. *J. Bacteriol.* 99:361–365.

Lehrer, R.I., and Ganz, T. 2002. Defensins of vertebrate animals. *Curr. Opin. Immunol.* 14:96–102.

Leslie, G.A., and Clem, L.W. 1969. Phylogeny of immunoglobulin structure and function. III: Immunoglobulins of the chicken. *J. Exp. Med.* 130:1337–1352.

Leslie, G.A., and Martin, L.N. 1973. Studies on the secretory immunologic system of fowl. III: Serum and secretory IgA of the chicken. *J. Immunol.* 110:1–9.

Levay, P.F., and Viljoen, M. 1995. Lactoferrin: A general review. *Haematologica* 80:252–267.

Levy, O. 1996. Antibiotic proteins of polymorphonuclear leukocytes. *Euro. J. Haematol.* 56:263–277.

Li, X., Nakano, T., Chae, H.S., Sunwoo, H.H., and Sim, J.S. 1998. Production of chicken egg yolk antibody (IgY) against bovine proteoglycan. *Can. J. Anim. Sci.* 78:287–291.

Li-Chan, E., Kummer, A., Losso, J.N., Kitts, D.D., and Nakai, S. 1995. Stability of bovine Igs to thermal treatment and processing. *Food Res. Intl.* 28:9–16.

Liceaga-Gesualdo, A., Li-Chan, E.C.Y., and Skura, B.J. 2001. Antimicrobial effect of lactoferrin digest on spores of a *Penicillium* sp. isolated from bottled water. *Food Res. Intl.* 34:501–506.

Lindstrom, P., Paulsson, M., Nylander, T., Elofsson, U., and Lindmark-Mansson, H. 1994. The effect of heat treatment on bovine immunoglobulins. *Milchwissenschaft* 49:67–70.

Livnah, O., Bayer, E.A., Wilchek, M., and Sussman, J.L. 1993. *Proc. Natl. Acad. Sci. U.S.A.* 90:5076–5080.

Locken, M.R., and Roth, R.E. 1983. Analysis of maternal IgG subpopulations which are transported into the chicken oocyte. *Immunol.* 49:21–28.

Lönnerdal, B., and Iyer, S. 1995. Lactoferrin: Molecular structure and biological function. *Annual Rev. Nutr.* 15:93–110.

Lönnerdal, B. 2003. Nutritional and physiologic significance of human milk proteins. *Am. J. Clin. Nutr.* 77:1537S–1543S.

Losch, Y., Schranner, I., Wanke, R., and Jurgens, L. 1986. The chicken egg, and antibody source. *J. Vet. Med* B 33:609–619.

Lugtenberg, B., and Van Alphen, L. 1983. Molecular architecture and functions of the outer membrane of *Escherichia coli* and other Gram-negative bacteria. *Biochim. Biophys. Acta.* 737:51–115.

Lundblad, J.L., and Seng, R. 1991. Inactivation of lipid-enveloped viruses in proteins with caprylate. *Vox. Sang.* 60:75–81.

Lush, I.E. 1961. Genetic polymorphisms in the egg albumen proteins of the domestic fowl. *Nature* 189:981–984.

Ma, Q., Zhang, Y., Critzer, F., Davidson, P.M., and Zhong, Q. 2016. Quality attributes and microbial survival on whole cantaloupes with antimicrobial coatings containing chitosan, lauric arginate, cinnamon oil and ethylenediaminetetraacetic acid. *Intl. J. Food Microbiol.* 235:103–108.

Machnicki, M., Zimecki, M., and Zagulski, T. 1993. Lactoferrin regulates the release of tumor necrosis factor-alpha and interleukin-6 in vivo. *Intl. Exp. Path.* 74:433–439.

Maekawa, T., Fujihara, M., and Ohtsuki, K. 2002. Characterization of human lactoferricin as a potent protein kinase CK2 activator regulated by A-kinase *in vitro*. *Biol. Pharm. Bull.* 25:118–121.

Mainer, G., Domingues, E., Randrup, M., Sanchez, L., and Calvo, M. 1999. Effect of heat treatment on anti-rotavirus activity of bovine immune milk. *J. Dairy Res.* 66:131–137.

Mandel, I.D., and Ellison, S.A. 1985. The biological significance of the nonimmunoglobulin defense factors. In *The Lactoperoxidase System Chemistry and Biological Significance*, edited by K.M. Pruitt and J.O. Tenovuo, Marcel Dekker, New York, NY, pp. 1–14.

Marchetti, M., Superti, F., Ammendolia, M.G., Rossi, P., Valenti, P., and Seganti, L. 1999. Inhibition of poliovirus type 1 infection by iron-, manganese- and zinc-saturated lactoferrin. *Med. Microbiol. Immunol.* 187:199–204.

Marshall, D.L., and Bullerman, L.B. 1986. Antimicrobial activity of fatty acid ester emulsifiers. *J. Food Sci.* 51:468–470.

Marshall, V.M., and Reiter, B. 1980. Comparison of the antibacterial activity of the hypothiocyanite anion towards *Streptococcus lactis* and *Escherichia coli*. *J. Gen. Microbiol.* 120:513–516.

Mason, D.Y., and Taylor, C.R. 1978. Distribution of transferrin, ferritin, and lactoferrin in human tissues. *J. Clin. Pathol.* 31:316–327.

Masschalck, B., van Houdt, R., and Michiels, C.W. 2001. High pressure increases bactericidal activity and spectrum of lactoferrin, lactoferricin and nisin. *Intl. J. Food Microbiol.* 64:325–332.

Masschalck, B., Deckers, D., and Michiels, C.W. 2003. Sensitization of outer-membrane mutants of *Salmonella* Typhimurium and *Pseudomonas aeruginosa* to antimicrobial peptides under high pressure. *J. Food Prot.* 66:1360–1367.

Masson, P.L., and Heremans, J.F. 1971. Lactoferrin in milk from different species. *Comp. Biochem. Physiol.* 39:119–129.

Mattsby-Baltzer, I., Roseanu, A., Motas, C., Elverfors, J., Engberg, I., and Hanson, L.A. 1996. Lactoferrin or a fragment thereof inhibits the endotoxin-induced interleukin-6 response in human monocytic cells. *Pediatr. Res.* 40:257–262.

Messmer, T.O., and Young, D.V. 1977. The effect of biotin and fatty acids on SV3T3 cell growth in the presence of normal calf serum. *J. Cell Physiol.* 90:265–270.

Metz-Boutigue, M.-H., Jollès, J., Mazurier, J., Schoentgen, F., Legrand, D., Spik, G., Montreuil, J., and Jollès, P. 1984. Human lactotransferrin: Amino acid sequence and structural comparisons with other transferrins. *Euro. J. Biochem.* 145:659–676.

McLay, J.C., Kennedy, M.J., O'Rourke, A.L., Elliot, R.M., and Simmonds, R.S. 2002. Inhibition of bacterial foodborne pathogens by the lactoperoxidase system in combination with monolaurin. *Intl. J. Food Microbiol.* 73:1–9.

Mickelsen, P.A., Blackman, E., and Sparling, F.P. 1982. Ability of *Neisseria gonnorhoeae*, *Neisseria meningitidis* and commensal *Neisseria* species to obtain iron from lactoferrin. *Inf. Immun.* 35:915–920.

Miller, A.K., and Tauig, F. 1964. Biotin-binding by parenterally-administered streptavidin or avidin. *Biochem. Biophys. Res. Commun.* 14:210–214.

Miller, S.J., Aly, R., Shinefield, H.R., and Elias, P.M. 1988. In vitro and in vivo antistaphylococcal activity of human stratum corneum lipids. *Arch. Dermatol.* 124:209–215.

Mine, Y. 2000. Avidin. In *Natural Food Antimicrobial Systems*, edited by A.S. Naidu, CRC Press, Boca Raton, FL, pp. 253–261.

Miyauchi, H., Hashimoto, S.-I., Nakajima, M., Shinoda, I., Fukuwatari, Y., and Hayasawa, H. 1998. Bovine lactoferrin stimulates the phagocytic activity of human neutrophils: Identification of its active domain. *Cell. Immunol.* 187:34–37.

Montagne, P., Cuilliere, M.L., Mole, C., Bene, M.C., and Faure, G. 1998. Microparticle-enhanced nephelometric immunoassay of lysozyme in milk and other human body fluids. *Clin. Chem.* 44:1610–1615.

Mueller, R., Boothy, J.T., Carroll, E.J., and Panico, L. 1983. Changes of complement values in calves during the first month of life. *Am. J. Vet. Res.* 44:747–750.

Murdock, C.A., and Matthews, K.R. 2002. Antibacterial activity of pepsin-digested lactoferrin on foodborne pathogens in buffered broth systems and ultra-high temperature milk with EDTA. *J. Appl. Microbiol.* 93:850856.

Murosaki, S., Yoshikai, Y., Kubo, C., Ishida, A., Matsuzaki, G., Sato, T., Endo, K., and Nomoto, K. 1991. Influence of intake of skim milk from cows immunized with intestinal bacterial antigens on onset of renal disease in (NZB x NZW) F1 mice fed ad libitum or restricted in energy intake. *J. Nutr.* 121:1860–1868.

Nagasawa, T., Kiyosawa, I., and Kuwahara, K. 1972. Amounts of lactoferrin in human colostrum and milk. *J. Dairy Sci.* 55:1651–1659.

Naidu, A.S., and Arnold, R.R. 1994. Lactoferrin interaction with salmonellae potentiates antibiotic susceptibility *in vitro. Diagn. Microbiol. Infect. Dis.* 20:69–75.

Naidu, A.S., and Arnold, R.R. 1997. Influence of lactoferrin on host-microbe interactions. In *Lactoferrin: Interactions and Biological Functions*, edited by T.W. Hutchens, and B. Lönerdal, Humana Press, Ottawa, NJ, pp. 259–275.

Naidu, A.S., and Bidlack, W.R. 1998. Milk lactoferrin—Natural microbial blocking agent (MBA) for food safety. *Environ. Nutr. Interact.* 2:35–50

Naidu, A.S. 2000. Lactoperoxidase. In *Natural Food Antimicrobial Systems*, edited by A.S. Naidu, CRC Press, Boca Raton, FL, pp. 103–132.

Naidu, A.S. 2001. Immobilized lactoferrin antimicrobial agents and the use. U.S. patent 6,172,040 B1.

Naidu, A.S. 2002. Activated lactoferrin: A new approach to meat safety. *Food Technol.* 56:40–45.

Naidu, A.S. 2003. Antimicrobials from animals. In *Natural Antimicrobials for the Minimal Processing of Foods*, edited by S. Roller, CRC Press, New York, NY, pp. 133–158.

Naidu, A.S., Tulpinski, J., Gustilo, K., Nimmagudda, R., and Morgan, J.B. 2003. Activated lactoferrin part 2: Natural antimicrobial for food safety. *Agrofood* Industry Hi-Tech May/June: 27–31.

Naidu, S.S., Svensson, U., Kishore, A.R., and Naidu, A.S. 1993. Relationship between antibacterial activity and porin binding of lactoferrin in *Escherichia coli* and *Salmonella typhimurium. Antimicrob. Agents Chemother.* 37:240–245.

Nakada, M., Dasako, S., Hirano, R., Oooka, M., and Nakajima, I. 1996. Lactoperoxidase suppresses acid production in yoghurt during storage under refrigeration. *Intl. Dairy J.* 6:33–42.

Nakamura, R., and Omori, I. 1979. Protective effect of some anions on the heat denaturation of ovotransferrin. *Agric. Biol. Chem.* 43:2393–2394.

Nakamura, M., Yamazaki, I., Ohtaki, S., and Nakamura, S. 1986. Characterization of one- and two-electron oxidations of glutathione coupled with lactoperoxidase and thyroid peroxidase reactions. *J. Biol. Chem.* 261:13923–13927.

Nandi, S., Suzuki, Y.A., Huang, J., Yalda, D., Pham, P., Wu, L., Bartley, G., Huang, N., and Lönnerdal, B. 2002. Expression of human lactoferrin in transgenic rice grains for the application in infant formula. *Plant Sci.* 163:713–722.

Nichols, B.L., McKee, K.S., and Huebers, H.A. 1990. Iron is not required in the lactoferrin stimulation of thymidine incorporation into the DNA of rat crypt enterocytes. *Pediatr. Res.* 27:525–528.

Nishimura, N. 1991. *Single Dose Oral Toxicity Study of monl-01 in Rats.* Gotemba Laboratory, Bozo Research Center Inc., Setagaya-ku, Tokyo, Japan.

Nishimura, N. 1997. Four-week oral repeated dose toxicity study of monl-01 in rats. Gotemba Laboratory. Bozo Research Center Inc. Setagaya-ku, Tokyo, Japan.

Nishimura, N. 2000. Thirteen-week oral repeated dose toxicity study of monl-01 in rats. *Food Chem. Tox.* 38:489–498.

Nonnecke, B.J., and Smith, K.L. 1984. Inhibition of mastitic bacteria by bovine milk apo-lactoferrin evaluated by *in vitro* microassay of bacterial growth. *J. Dairy Sci.* 67:606–613.

Norberg, S., O'Connor, P.M., Stanton, C., Ross, R.P., Hill, C., Fitzgerald, G.F. and Cotter, P.D. 2011. Altering the Composition of caseicins A and B as a means of determining the contribution of specific residues to antimicrobial activity. *Appl. Environ. Microbiol.* 77: 2496–2501.

Noseda, A., White, J.G., Godwin, P.L., Jerome, W.G., and Modest, E.J. 1989. Membrane damage in leukemic cells induced by ether and ester lipids: An electron microscopic study. *Exper. Mol. Pathol.* 50:69–83.

Odell, E.W., Sarra, R., Foxworthy, M., Chapple, D.S., and Evans, R.W. 1996. Antibacterial activity of peptides homologous to a loop region in human lactoferrin. *FEBS Lett.* 382:175–178.

O'Farrelly, C., Branton, D., and Wanke, C.A. 1992. Oral ingestion of egg yolk immunoglobulin from hens immunized with an enterotoxigenic Escherichia coli strain prevents diarrhea in rabbits challenged with the same strain. *Infect. Immun.* 60:2593–2597.

Ohta, M., Hamako, J., Yamamoto, S., Hatta, H., Kim, M., Yamamoto, T., Oka, S., Mizuochi, T., and Matuura, F. 1991. Structures of asparagines-linked oligosaccharides from hen egg-yolk antibody (IgY). Occurrence of unusual glucosylated oligomannose type oligosaccharides in a mature glycoprotein. *Glycoconjugate J.* 8:400–413.

Ormrod, D.J., and Miller, T.E. 1991. Assessment of antiinflammatory agents using 125I-labeled human serum albumin to quantify footpad edema volume in the rat. *Pharm Res.* 8:1270–1273.

Osuga, D.T., and Feeney, R.E. 1968. Biochemistry of the egg white proteins of the Ratite group. *Arch. Biochem. Biophys.* 124:560–574.

Otani, H., Matsumoto, K., Saeki, A., and Hosono, A. 1991. Comparative studies on properties of hen egg yolk IgY and rabbit serum IgG antibodies. *Lebensm. Wiss. Technol.* 24:152–158.

Ozpinar, H., Erhard, M.H., Aytug, N., Ozpinar, A., Baklaci, C., Karamuptuoglu, S., Hofmann, A., and Losch, U. 1996. Dose-dependent effects of specific egg-yolk antibodies on diarrhea of newborn calves. *Preventive Vet. Med.* 27:67–73.

Park, S.-I., Stan, S.D., Daeschel, M.A., and Zhao, Y. 2005. Antifungal coatings on fresh strawberries (*Fragaria* x *ananassa*) to control mold growth during cold storage. *J. Food Sci.* 70:M202–7.

Pakkanen, R., and Aalto, J. 1997. Growth factors and microbial factors of bovine colostrum. *Intl. Dairy J.* 7:285–297.

Parkhouse, R.M.E. 1971. IgM biosynthesis: Production of intermediates and excess of light chain in mouse myelome MOPC-104E. *Biochem. J.* 123:635.

Parkinson, T.L. 1966. The chemical composition of eggs. *J. Sci. Food Agric.* 17:101–111.

Parkos, C.A. 1997. Molecular events in neutrophil transepithelial migration. *Bioassays* 19:865–873.

Parmley, R.T., Doran, T., Boyd, R.L., and Gibert, C. 1986. Unmasking and redistribution of lysosomal sulfated glycoconjugates in phagocytic polymorphonuclear leukocytes. *J. Histochem. Cytochem.* 34:1701–1707.

Parvari, R., Avivi, A., Lentner, F., Ziv, E., Tel-Or, S., Burstein, Y., and Schechter, I. 1988. Chicken immunoglobulin γ-heavy chains: Limited V_H gene repertoire, combinatorial diversification by D gene segments and evolution of the heavy chain locus. *EMBO J.* 7:739–744.

Patterson, R., Younger, J.S., Weigle, W.O., and Dixon, F.J. 1962. The metabolism of serum proteins in the hen and chick and secretions of serum proteins by the ovary of the hen. *J. Physiol.* 45:501–513.

Paul, K.G., and Ohlsson, P.I. 1985. The chemical structure of lactoperoxidase. In *The Lactoperoxidase System*, edited by K. Pruitt and J.O. Tenovuo, Marcel Dekker, New York, NY, pp. 15–29.

Paulsson, M.A., Svensson, U., Kishore, A.R., and Naidu, A.S. 1993. Thermal behavior of bovine lactoferrin in water and its relation to bacterial interaction and antibacterial activity. *J. Dairy Sci.* 76:3711–3720.

Pellegrini, A. 2003. Antimicrobial peptides from food proteins. *Curr. Pharm. Des.* 9:1225–1238.

Pellico, M.A., and Montgomery, R.E. 1989. Di-enzymatic dentrifice. European Patent Specification EP0133736 B1.

Pierce, A., Colavizza, D., Benaissa, M., Maes, P., Tartar, A., Montreuil, J., and Spik, G. 1991. Molecular cloning and sequence analysis of bovine lactotransferrin. *Euro. J. Biochem.* 196:177–184.

Playford, R.J., Macdonald, C.E., and Johnson, W.S. 2000. Colostrum and milk-derived peptide growth factors for the treatment of gastrointestinal disorders. *Am. J. Clin. Nutr.* 72:5–14.

Popper, L., and Knorr, D. 1997. Inactivation of yeast and filamentous fungi by the lactoperoxidase-hydrogen peroxide-thiocyanate-system. *Nahrung* 41:29–33.

Potter, R., Hansen, L.T., and Gill, T.A. 2005. Inhibition of foodborne bacteria by native and modified protamine: Importance of electrostatic interactions. *Intl. J. Food Microbiol.* 103:23–34.

Pourtois, M., Binet, C., Van Tieghem, N., Courtois, P., Vendenabbeele, A., and Thiry, L. 1990. Inhibition of HIV infectivity by lactoperoxidase-produced hypothiocyanite. *J. Biol. Buccale* 18:251–253.

Powell, M.J., and Ogden, J.E. 1990. Nucleotide sequence of human lactoferrin cDNA. *Nucl. Acids Res.* 18:4013.

Precht, D., Molkentin, J., Destaillats, F., and Wolff, R.L. 2001. Comparative studies on individual isomeric 18:1 acids in cow, goat, and ewe milk fats by low-temperature high-resolution capillary gas-liquid chromatography. *Lipids* 36:827–832.

Pruitt, K.M., and Reiter, B. 1985. Biochemistry of peroxidase system—Antimicrobial effects. In *The Peroxidase System—Chemistry and Biological Significance*, edited by K.M. Pruitt and J.O. Tenovuo, Marcel Dekker, New York, NY, pp. 143–177.

Pugliese, L., Malcovait, M., Coda, A., and Bolognesi, M. 1994. Crystal structure of apoavidin from hen egg-white. *J. Mol. Biol.* 235:42–46.

Qadri, F., Raqib, R., Ahmed, F., Rahman, T., Wenneras, C., Das, S.K., Alam, N.H., Mathan, M.M., and Svennerholm, A.-M. 2002. Increased levels of inflammatory mediators in children and adults infected with *Vibrio cholerae* O1 and O139. *Clin. Diagn. Lab. Immunol.* 9:221–229.

Qu, X.D., Harwig, S.S., Oren, A.M., Shafer, W.M., and Lehrer, R.I. 1996. Susceptibility of *Neisseria gonorrhoeae* to protegrins. *Infect. Immun.* 64:1240–1245.

Rabe, L.K., Coleman, M.S., Hillier, S.L., and Isaacs, C. 1997. The in vitro activity of a microbicide, octylglycerol, against *Neisseria gonorrhoeae, Trichomonas vaginalis,* and vaginal microflora. International Congress of Sexually Transmitted Diseases, Seville, Spain.

Ransom, J.R., Sofos, J.N., Belk, K.E., Scanga, J.A., Hossner, K.L., Childs, K.D., Yemm, R.S., and Smith, G.C. 2003. Susceptibility of *Escherichia coli* O157:H7, *Salmonella* Typhimurium, and *Listeria monocytogenes,* inoculated onto beef tissues, steaks and RTE products, to lactic acid, lactoferrin, and activated lactoferrin. pp. 1–41. Final Report to National Cattlemen's Beef Association, Centennial, CO. Center for Red Meat Safety, Colorado State University, Fort Collins, CO.

Raphael, G.D., Jeney, E.V., Baranuik, J.N., Kim, I., Meridith, S.D., and Kaliner, M.A. 1989. Pathophysiology of rhinitis: Lactoferrin and lysozyme in nasal secretion. *J. Clin. Invest.* 84:1528–1535.

Rautemaa, R., and Meri, S. 1999. Complement resistance mechanisms of bacteria. *Microbes Infect.* 1:785–794.

Recio, I., and Visser, S. 2000. Antibacterial and binding characteristics of bovine, ovine and caprine lactoferrins: A comparative study. *Intl. Dairy J.* 10:597–605.

Reddy, N.R., Roth, S.M., Eigel, W.N., and Pierson, M.D. 1988. Foods and food ingredients for prevention of diarrheal disease in children in developing countries. *J. Food Prot.* 51:66–75.

Regester, G.O., Smithers, G.W., Mitchell, I.R., McIntosh, G.H., and Dionysis, G.H. 1997. Bioactive factors in milk: Natural and induced. In *Milk Composition, Production and Technology,* edited by D.J.W. Welch, S.R. Davis, A.I. Popay, and C.G. Prosser, CAB International, Wallingford, UK, pp. 119–132.

Reiner, D.S., Wang, C.-S., and Gillin, F.D. 1986. Human milk kills *Giardia lamblia* by generating toxic lipolytic products. *J. Infect. Dis.* 154:825–832.

Reiter, B., Pickering, A., Oram, J.D., and Pope, G.S. 1963. Peroxidase-thiocyanate inhibition of streptococci in raw milk. *J. Gen. Microbiol.* 33:12.

Reiter, B., and Oram, J.D. 1967. Bacterial inhibitors in milk and other biological fluids. *Nature* 216:328–332.

Reiter, B., and Bramley, A.J. 1975. Document 85. In *Seminar on Mastitis Control,* edited by F.H. Dodd, T.K. Griffin, and R.G. Kingwill, IDF, Reading, pp. 210–215.

Reiter, B., and Brock, J.H. 1975. Inhibition of *Escherichia coli* by bovine colostrum and post-colostral milk I. Complement-mediated bactericidal activity of antibodies to a serum susceptible strain of *E.coli* of the stereotype O 111. *Immunology* 28:71–82.

Reiter, B., Marshall, V.M., and Philips, S.M. 1980.The antibiotic activity of the lactoperoxidase-thiocyanate-hydrogen peroxide system in calf abomasum. *Res. Vet. Sci.* 28:116, 122.

Reiter, B., and Perraudin, J.P. 1991. Lactoperoxidase: Biological functions. In *Peroxidases in Chemistry and Biology,* vol. 1, edited by J. Everse, K.E. Everse, and M.B. Grisham, CRC Press, Boca Raton, FL, pp. 143–180.

Reiter, B. 1978a. Review of nonspecific antimicrobial factors in colostrum. *Ann. Rech. Vet.* 9:205–224.

Reiter, B. 1978b. Review of the progress of dairy science: Antimicrobial systems in milk. *J. Dairy Res.* 45:131–147.

Reiter, B. 1985a. Protective proteins in milk—Biological significance and exploitation. *IDF Bulletin* 191:1–35.

Reiter, B. 1985b. The lactoperoxidase system of bovine milk. In *The Peroxidase System—Chemistry and Biological Significance,* edited by K.M. Pruitt and J.O. Tenovuo, Marcel Dekker, New York, NY, pp. 123–141.

Rekdal, Ø., Andersen, J., Vorlund, L.H., and Svendsen, J.S. 1999. Construction and synthesis of lactoferrin derivatives with enhanced antibacterial activity. *J. Peptide Sci.* 5:32–45.

Rey, M.W., Woloshuk, S.L., deBoer, H.A., and Pieper, F.R. 1990. Complete nucleotide sequence of human mammary gland lactoferrin. *Nucl. Acids Res.* 18:5288.

Rodriguez, E., Tomillo, J., Nunes, M., and Medina, M. 1997. Combined effect of bacteriocin-producing lactic acid bacteria and lactoperoxidase system activation on *Listeria monocytogenes* in refrigerated raw milk. *J. Appl. Microbiol.* 83:389–395.

Rodríguez-Amigo, B., Delcanale, P., Rotger, G., Juárez-Jiménez, J., Abbruzzetti, S., Summer, A., Luque, F., Nonell, S., and Viappiani, C. 2015. The complex of hypericin with β-lactoglobulin has antimicrobial activity with potential applications in dairy use. *J. Dairy Sci.* 98:89–94.

Rohrer, L., Winterhalter, K.H., Eckert, J., and Kiohler, P. 1986. Killing of *Giardia lamblia* by human milk is mediated by unsaturated fatty acids. *Antimicrob. Agents Chemother.* 30:254–257.

Rose, M.E., Orlans, E., and Buttress, N. 1974. Immunoglobulin classes in the hen's egg: Their segregation in yolk and white. *Euro. J. Immunol.* 4:521–523.

Rose, M.E., and Orlans, E. 1981. Immunoglobulins in the egg, embryo and young chick. *Devel. Comp. Immunol.* 5:15–20, 371–375.

Rossi, P., Giansanti, F., Boffi, A., Ajello, M., Valenti, P., Chiancone, E., and Antonini, G. 2002. Ca^{2+} binding to bovine lactoferrin enhances protein stability and influences the release of bacterial lipopolysaccharide. *Biochem. Cell Biol.* 80:41–48.

Roy, M.K., Kuwabara, Y., Hara, K., Watanabe, Y., and Tamai, Y. 2002. Peptides from the N-terminal end of bovine lactoferrin induce apoptosis in human leukemic (HL-60) cells. *J. Dairy Sci.* 85:2065–2074.

Ryan, L.K., Rhodes, J., Bhat, M., and Diamond, G. 1998. Expression of beta-defensin genes in bovine alveolar macrophages. *Infect. Immun.* 66:878–881.

Rybarczyk, J., Kieckens, E., Vanrompay, D., and Cox, E. 2017. In vitro and in vivo studies on the antimicrobial effect of lactoferrin against *Escherichia coli* O157:H7. *Vet. Microbiol.* 202:23–28.

Saito, H., Takase, M., Tamura, Y., Shimamura, S., and Tomita, M. 1994. Physicochemical and antibacterial properties of lactoferrin and its hydrolysate produced by heat treatment at acidic pH. In *Lactoferrin: Structure and Function*, edited by T.W. Hutchens et al., Plenum Press, New York, NY, pp. 219–226.

Sallmann, F.R., Baveye-Descamps, S., Pattus, F., Salmon, V., Branza, N., Spik, G., and Legrand, D. 1999. Porins OmpC and PhoE of *Escherichia coli* as specific cell-surface targets of human lactoferrin—Binding characteristics and biological effects. *J. Biol. Chem.* 274:16107–16114.

Samaranayake, Y.H., Samaranayake, L.P., Wu, P.C., and So, M. 1997. The antifungal effect of lactoferrin and lysozyme on *Candida krusei* and *Candida albicans*. *APMIS* 105:875–883.

Sanchez, L., Aranda, P., Perez, M.D., and Calvo, M. 1988. Concentration of lactoferrin and transferrin throughout lactation in cow's colostrum and milk. *Biol. Chem. Hoppe Seyler.* 369:1005–1008.

Sanchez, L., Calvo, M., and Brock, H. 1992. Biological role of lactoferrin. *Arch. Dis. Child.* 67:657–661.

Sashiwa, H., and Aiba, S.I. 2004. Chemically modified chitin and chitosan as biomaterials. *Progress Polymer Sci.* 29:887–908.

Sato, R., Inanami, O., Tanaka, Y., Takase, M., and Naito, Y. 1996. Oral administration of bovine lactoferrin for treatment of intractable stomatitis in feline immunodeficiency virus (FIV)-positive and FIV-negative cats. *Am. J. Vet. Res.* 57:1443–1446.

Satué-Gracia, M.T., Frankel, E.N., Rangavajhyala, N., and German, J.B. 2000. Lactoferrin in infant formulas: Effect on oxidation. *J. Agric. Food Chem.* 48:4984–4990.

Sawai, M.V., Jia, H.P., Liu, L., Aseyev, V., Wiencek, J.M., McCray, P.B. Jr., Ganz, T., Kearney, W.R., and Tack, B.F. 2001. The NMR structure of human beta-defensin-2 reveals a novel alpha-helical segment. *Biochem.* 40:3810–3816.

Schibli, D.J., Hwang, P.M., and Vogel, H.J. 1999. The structure of the antimicrobial active center of lactoferricin B bound to sodium dodecyl sulfate micelles. *FEBS Lett.* 446:213–217.

Schryvers, A.B., and Morris, L.J. 1988. Identification and characterization of the human lactoferrin-binding protein *Neisseria meningitidis*. *Inf. Immun.* 56:1144–1149.

Selsted, M.E., Harwig, S.S., Ganz, T., Schilling, J.W., and Lehrer, R.I. 1985. Primary structures of three human neutrophil defensins. *J. Clin. Invest.* 76:1436–1439.

Selsted, M.E., Tang, Y.Q., Morris, W.L., McGuire, P.A., Novotny, M.J., Smith, W., Henschen, A.H., and Cullor, J.S. 1993. Purification, primary structures, and antibacterial activities of beta-defensins, a new family of antimicrobial peptides from bovine neutrophils. *J. Biol Chem.* 268:6641–6648.

Shibasaki, I., and Kato, N. 1978. Combined effects on antibacterial activity of fatty acids and their esters against gram-negative bacteria. In *The Pharmacological Effects of Lipids*, edited by J.J. Kabara, The American Oil Chemists Society, St. Louis, MO, pp. 15–24.

Shibasaki, I. 1982. Food preservation with non-traditional antimicrobial agents. *J. Food Safety* 4:35–58.

Shield, J., Melville, C., Novelli, V., Anderson, G., Scheimberg, I., Gibb, D., and Milla, P. 1993. Bovine colostrum Ig concentrate for cryptosporidiosis in AIDS. *Arch. Dis. Child* 69:451–453.

Shimazaki, K., Kawaguchi, A., Sato, T., Ueda, Y., Tommimura, T., and Shimamura, S. 1993. Analysis of human and bovine milk lactoferrins by rotafor and chromatofocusing. *Intl. J. Biochem.* 25:1653–1658.

Shimizu, M., Fitzsimmons, R.C., and Nakai, S. 1988. Anti-*E.coli* immunoglobulin Y isolated from egg yolk of immunized chickens as a potential food ingredient. *J. Food Sci.* 53:1360–1366.

Shimizu, M., Nagashima, H., Sano, K., Hashimoto, K., Ozeki, M., Tsuda, K., and Hatta, H. 1992. Molecular stability of chicken and rabbit immunoglobulin G. *Biosci. Biotech. Biochem.* 56:270–274.

Shimizu, M., Nagashima, H., and Hashimoto, K. 1993. Comparative studies on molecular stability of immunoglobulin G from different species. *Comp. Biochem. Physiol.* 106B:255–261.

Shin, S.Y., Bajpai, V.K., Kim, H.R., and Kang, S.C. 2007. Antibacterial activity of bio-converted eicosapentaenoic (EPA) and docosahexaenoic acid (DHA) against foodborne pathogenic bacteria. *Intl. J. Food Microbiol.* 113:233–236.

Shinoda, I., Takase, M., Fukuwatari, Y., Shimamura, S., Köller, M., and König, W. 1996. Effects of lactoferrin and lactoferricin on the release of interleukin 8 from human polymorphonuclear leukocytes. *Biosci. Biotech. Biochem.* 60:521–523.

Sievers, G. 1979. The prosthetic group of milk lactoperoxidase is protoheme IX. *Biochim. Biophys. Acta* 579:181–190.

Sievers, G. 1980. Structure of milk lactoperoxidase is protoheme IX. *Biochim. Biophys. Acta* 579:181–190.

Sim, J.S., Sunwoo, H.H., and Lee, E.N. 2000. Ovoglobulin IgY. In *Natural Food Antimicrobial Systems*, edited by A.S. Naidu, CRC Press, Boca Raton, FL, pp. 227–252.

Slowey, R.R., Eidelman, S., and Klebanoff, S.J. 1968. Antibacterial activity of the purified peroxidase from human parotid saliva. *J. Bacteriol.* 96:575–579.

Sofos, J.N., Beuchat, L.R., Davidson, P.M., and Johnson, E.A. 1998. Naturally occurring antimicrobials in food, Council for Agricultural Science and Technology, Task Force Report, No. 132.

Spear, M.L., Bitman, J., Hamosh, M., Wood, D.L., Gavula, D., and Hamosh, P. 1992. Human mammary gland function at the onset of lactation: Medium-chain fatty acid synthesis. *Lipids* 27:908–911.

Staak, C. 1992. Bovine colostrum and protection of young animals. *Berliner und Munchener Tierarztliche Wochenschrift* 105:219–224.

Stanfield, R.L., and Wilson, I.A. 1995. Protein-peptide interactions. *Curr. Opin. Struct. Biol.* 5:103–113.

Steijns, J.M., and van Hooijdonk, A.C.M. 2000. Occurrence, structure, biochemical properties and technological characteristics of lactoferrin. *Brit. J. Nutr.* 84 Supplement 1:S11–S17.

Stephan, W., Dichtelmüller, H., and Lissner, R. 1990. Antibodies from colostrums in oral immunotherapy. *J. Clin. Chem. Clin. Biochem.* 28:19–23.

Strøm, M.B., Rekdal, Ø., and Svenden, J.S. 2000. Antibacterial activity of 15-residue lactoferricin derivatives. *J. Peptide Res.* 56:265–274.

Strøm, M.B., Haug, B.E., Rekdal, Ø., Skar, M.L., Stensen, W., and Svenden, J.S. 2002a. Important structural features of 15-residue lactoferricin derivatives and methods for improvement of antimicrobial activity. *Biochem. Cell Biol.* 80:65–74.

Strøm, M.B., Rekdal, Ø., and Svenden, J.S. 2002b. Antimicrobial activity of short arginine- and tryptophan-rich peptides. *J. Peptide Sci.* 8:431–437.

Sundaram, J., Pant, J., Goudie, M.J., Mani, S., and Handa, H. 2016. Antimicrobial and physicochemical characterization of biodegradable, nitric oxide-releasing nanocellulose-chitosan packaging membranes. *J. Agric. Food Chem.* 64:5260–5266.

Superti, F., Ammendolia, M.G., Valenti, P., and Seganti, L. 1997. Antirotaviral activity of milk proteins: Lactoferrin prevents rotavirus infection in the enterocyte-like cell line HT-29. *Med. Microbiol. Immunol.* 186:83–91.

Swain, L.S., Hu, H., and Huston, G. 1999. Class II-independent generation of CD4 memory of T cells form effectors. *Science* 286:1381–1383.

Swart, P.J., Kuipers, M.E., Smit, C., Pauwels, R., deBethune, M.P., de Clercq, E., Meijer, D.K., and Huisman, J.G. 1996. Antiviral effects of milk proteins: Acylation results in polyanionic compounds with potent activity against human immunodeficiency virus types 1 and 2 in vitro. *AIDS Res. Human Retroviruses.* 12:769–775.

Swart, P.J., Kuipers, E.M., Smit, C., van der Strate, B.W., Harmsen, M.C., and Meijer, D.K. 1998. Lactoferrin. Antiviral activity of lactoferrin. *Adv. Exp. Med. Biol.* 443:205–213.

Szuchet-Derechin, S., and Johnson, P. 1965. The 'Albumin' fraction of bovine milk-II: The isolation and physicochemical properties of the red proteins'. *Euro. Polymer. J.* 1:283–291.

Szwajkowska, M., Wolanciuk, A., Barlowska, J., Krol, J., and Litwinczuk, Z. 2011. Bovine milk proteins as the source of bioactive peptides isolated from a commercially available casein hydrolysate by autofocusing technique. *Biofactors.* 38:309–315.

Tacket, C.O., Binion, S.B., Bostwick, E., Losonsky, G., Roy, M.J., and Edelman, R. 1992. Efficacy of bovine milk Ig concentrate in preventing illness after *Shigella flexneri* challenge. *Am. J. Trop. Med. Hyg.* 47:276–283.

Tenovuo, J., and Knuuttila, M.L. 1977. Antibacterial effect of salivary peroxidases on a cariogenic strain of *Streptococcus mutans*. *J. Dent. Res.* 56:1608–1613.

Tenovuo, J., and Pruitt, K.M. 1984. Relationship of the human salivary peroxidase system to oral health. *J. Oral Pathol.* 13:573–584.

Testa, U. 2002. Lactoferrin. In *Proteins of Iron Metabolism*, edited by U. Testa, CRC Press, Boca Raton, FL, pp. 143–222.

Thakurta, P.G., Choudhury, D., Dasgupta, R., and Dattagupta, J.K. 2003. Structure of diferric hen serum transferring at 2.8Å resolution. *Acta. Cryst.* D 59:1773–1781.

Thomas, E.L. 1985. Products of lactoperoxidase-catalyzed oxidation of thiocyanate and halides. In *The Lactoperoxidase System, Chemistry and Biological Significance*, edited by K.M. Pruitt and T.J. Tenovuo, Marcel Dekker, New York, NY, pp. 31–53.

Thormar, H., Isaacs, C.E., Brown, H.R., Barshatzky, M.R., and Pessolano, T. 1987. Inactivation of enveloped viruses and killing of cells by fatty acids and monoglycerides. *Antimicrob. Agents Chemother.* 31:27–31.

Tiwari, B., Valdramidis, V., O'Donnell, C., Muthukumarappan, K., Cullen, P., and Bourke, P. 2009. Application of natural antimicrobials for food preservation. *J. Agric. Food Chem.* 57:5987–6000.

Tomita, G.M., Todhunter, D.A., Hogan, J.S., and Smith, K.L. 1995. Antigenic crossreactivity and lipopolysaccharide neutralization properties of bovine immunoglobulin G. *J. Dairy Sci.* 78:2745–2752.

Tomita, M., Takase, M., Wakabayashi, H., and Bellamy, W. 1994. Antimicrobial peptides of lactoferrin. In *Lactoferrin: Structure and Function*, edited by T.W. Hutchens et al., Plenum Press, New York, NY, pp. 209–218.

Tonegawa, S., Hozumi, N., Matthyssens, G., and Schuller, R. 1976. Somatic changes in the content and context of Ig genes. *Cold Springs Harb. Symp. Quant. Biol.* XLI:877.

Torlak, E., and Sert, D. 2013. Antibacterial effectiveness of chitosan-propolis coated polypropylene films against foodborne pathogens. *Intl. J. Biol. Macromolec.* 60:52–55.

Tranter, H.S., and Board, R.G. 1984. The influence of incubation temperature and pH on the antimicrobial properties of hen egg albumen. *J. Appl. Bacteriol.* 56:53–61.

Tranter, H.S. 1994. Lysozyme, ovotransferrin and avidin. In *Natural Antimicrobial Systems and Food Preservation*, edited by V.D. Dillon and R.G. Board, CAB Intentional, Wallingford, UK, pp. 15–63.

Turchany, J.M., Aley, S.B., and Gillin, F.D. 1995. Giardicidal activity of lactoferrin and N-terminal peptides. *Infect. Immun.* 63:4550–4552.

Turner, W.M. 1996. Mannose-binding lectin—The pluripotent molecule of the innate immune system. *Immun. Today* 17:532–540.

Ueta, E., Tanida, T., and Osaki, T. 2001. A novel bovine lactoferrin peptide, FKCRRWQWRM, suppresses *Candida* cell growth and activates neutrophils. *J. Pept. Res.* 57:240–249.

Ulvatne, H., Haukland, H.H., Olsvik, Ø., and Vorland, L.H. 2001. Lactoferricin B causes depolarization of the cytoplasmic membrane of *Escherichia coli* ATCC 25922 and fusion of negatively charged liposomes. *FEBS Lett.* 492:62–65.

Ulvatne, H., and Vorland, L.H. 2001. Bactericidal kinetics of 3 lactoferricins against *Staphylococcus aureus* and *Escherichia coli*. *Scand. J. Infect. Dis.* 33:507–511.

Ulvatne, H., Haukland, H.H., Samuelsen, Ø., Krämer, M., and Vorland, L.H. 2002. Proteases in *Escherichia coli* and *Staphylococcus aureus* confer reduced susceptibility to lactoferricin B. *J. Antimicrob. Chemother.* 50:461–467.

Unda, J.R., Molins, R.A., and Walker, H.W. 1991. *Clostridium sporogenes* and *Listeria monocytogenes*: Survival and inhibition in microwave-ready beef roasts containing selected antimicrobials. *J. Food Sci.* 56:198–205.

Valenti, P., Antonini, G., Rossi Fanelli, M.R., Orsi, N., and Antonini, E. 1982. Antibacterial activity of matrix-bound ovotransferrin. *Antimicrob.Agents Chemother.* 21:840–841.

Valenti, P., Antonini, G., Von Hunolstein, C., Visca, P., Orsi, N., and Antonini, E. 1983a. Studies on the antimicrobial activity of ovotransferrin. *Intl. J. Tiss. Reac.* 1:97–105.

Valenti, P., Visca, P., Von-Hunolstein, C., Antonini, G., Creo, C., and Orsi, N. 1983b. Importance of the presence of metals in the antibacterial activity of ovotransferrin. *Ann. Ist. Super Sanita* 18:471–472.

Valenti, P., Antonini, G., Von-Hunolstein, C., Visca, P., Orsi, N., and Antonini, E. 1984. Studies of the antimicrobial activity of ovotransferrin. *J. Appl. Bacteriol.* 56:53–61.

Valenti, P., Visca, P., Antonini, G., and Orsi, N. 1985. Antifungal activity of ovotransferrin towards genus *Candida*. *Mycopathologia* 89:169–175.

van der Strate, B.W.A., Harmsen, M.C., The, T.H., Sprenger, H.G., De Vries, H., Eikelboom, M.C., Kuipers, M.E., Meijer, D.K.F., and Swart, P.J. 1999. Plasma lactoferrin levels are decreased in end-stage AIDS patients. *Viral Immunol.* 12:197–203.

van der Strate, B.W.A., Harmsen, M.C., Schäfer, P., Swart, P.J., The, T.H., Jahn, G., Speer, C.P., Meijer, D.K.F., and Hamprecht, K. 2001. Viral load in breast milk correlates with transmission of human cytomegalovirus to preterm neonates, but lactoferrin concentrations do not. *Clin. Diagn. Lab. Immunol.* 8:818–821.

Van Hooijdonk, A.C.M., Kussendrager, K.D., and Steijns, J.M. 2000. *In vivo* antimicrobial and antiviral activity of components in bovine milk and colostrum involved in non-specific defense. *Brit. J. Nutr.* 84 Supplement 1:S127–S134.

Vasilov, R.G., and Ploegh, H.L. 1982. Biosynthesis of murine immunoglobulin D: Heterogeneity of glycosylation. *Euro. J. Immunol.* 12:804–813.

Venkitanarayanan, K.S., Zhao, T., and Doyle, M.P. 1999. Antibacterial effect of lactoferricin B on *Escherichia coli* O157:H7 in ground beef. *J. Food Prot.* 62:747–750.

Verdonck, L.F., and van Heugten, H.G. 1997. Ether lipids are effective cytotoxic drugs against multidrug-resistant acute leukemia cells and can act by the induction of apoptosis. *Leukemia Res.* 21:37–43.

Viani, R.M., Gutteberg, T.J., Lathey, J.L., and Spector, S.A. 1999. Lactoferrin inhibits HIV-1 replication *in vitro* and exhibits synergy when combined with zidovuline. *AIDS* 13:1273–1274.

Viejo-Diaz, M., Andrés, M.T., Pérez-Gil, J., Sánchez, M., and Fierro, J.F. 2003. Potassium efflux induced by a new lactoferrin-derived peptide mimicking the effect of native human lactoferrin on the bacterial cytoplasmic membrane. *Biochem.* 68:217–227.

Vogel, H.J., Schibli, D.J., Jing, W., Lohmeier-Vogel, E.M., Epand, R.F., and Epand, R.M. 2002. Towards a structure-function analysis of bovine lactoferricin and related tryptophan- and arginine-containing peptides. *Biochem. Cell Biol.* 80:49–63.

Vorland, L.H., Ulvatne, H., Andersen, J., Haukland, H.H., Rekdal, Ø., Svendsen, J.S., and Gutteberg, T.J. 1998. Lactoferricin of bovine origin is more active than lactoferricins of human, murine and caprine origin. *Scand. J. Infect. Dis.* 30:513–517.

Vorland, L.H. 1999. Lactoferrin: A multifunctional glycoprotein. *APMIS* 107:971–981.

Vorland, L.H., Osbakk, S.A., Perstølen, T., Ulvatne, H., Rekdal, Ø., Svendsen, J.S., and Gutteberg, T.J. 1999a. Interference of the antimicrobial peptide lactoferricin B with the action of various antibiotics against *Escherichia coli* and *Staphylococcus aureus. Scand. J. Infect. Dis.* 31:173–177.

Vorland, L.H., Ulvatne, H., Andersen, J., Haukland, H.H., Rekdal, Ø., Svendsen, J.S., and Gutteberg, T.J. 1999b. Antibacterial effects of lactoferricin B. *Scand. J. Infect. Dis.* 31:179–184.

Vorland, L.H., Ulvatne, H., Rekdal, Ø., and Svendsen, J.S. 1999c. Initial binding sites of antimicrobial peptides in *Staphylococcus aureus* and *Escherichia coli. Scand. J. Infect. Dis.* 31:467–473.

Wagstrom, E.A., Yoon, K.-J., and Zimmerman, J.J. 2000. Immune components in porcine mammary secretions. *Viral Immunol.* 13:383–397.

Wakabayashi, H., Bellamy, W., Takase, M., and Tomita, M. 1992. Inactivation of *Listeria monocytogenes* by lactoferricin, a potent antimicrobial peptide derived from cow's milk. *J. Food.Prot.* 55:238–240.

Wakabayashi, H., Hiratani, T., Uchida, K., and Yamaguchi, H. 1996. Antifungal spectrum and fungicidal mechanism of an N-terminal peptide of bovine lactoferrin. *J. Infect. Chemother.* 1:185–189.

Wakabayashi, H., Abe, S., Teraguchi, S., Hayasawa, H., and Yamaguchi, H. 1998. Inhibition of hyphal growth of azole-resistant strains of Candida albicans by triazole antifungal agents in the presence of lactoferrin-related compounds. *Antimicrob. Agents Chemother.* 42:1587–1591.

Wakabayashi, H., Matsumoto, H., Hashimoto, K., Teraguchi, S., Takase, M., and Hayasawa, H. 1999a. Inhibition of iron/ascorbate-induced lipid peroxidation by an N-terminal peptide of bovine lactoferrin and its acylated derivatives. *Biosci. Biotechnol. Biochem.* 63:955–957.

Wakabayashi, H., Matsumoto, H., Hashimoto, K., Teraguchi, S., Takase, M., and Hayasawa, H. 1999b. N-acylated and D enantiomer derivatives of a nonamer core peptide of lactoferricin B showing improved antimicrobial activity. *Antimicrob. Agents Chemother.* 43:1267–1269.

Wakabayashi, H., Uchida, K., Yamauchi, K., Teraguchi, S., Hayasawa, H., and Yamaguchi, H. 2000. Lactoferrin given in food facilitates dermatophytosis cure in guinea pig models. *J. Antimicrob. Chemother.* 46:595–601.

Wakabayashi, H., Teraguchi, S., and Tamura, Y. 2002. Increased *Staphylococcus*-killing activity of an antimicrobial peptide, lactoferricin B, with minocycline and monoacylglycerol. *Biosci. Biotechnol. Biochem.* 66:2161–2167.

Wakabayashi, H., Takase, M., and Tomita, M. 2003. Lactoferricin derived from milk protein lactoferrin. *Curr. Pharm. Des.* 9:1277–1287.

Wang, L.-L., and Johnson, E.A. 1992. Inhibition of *Listeria monocytogenes* by fatty acids and monoglycerides. *Appl. Environ. Microbiol.* 58:624–629.

Wang, L.-L., Yang, B.K., Parkin, K.L., and Johnson, E.A. 1993. Inhibition of *Listeria monocytogenes* by monoacylglycerols synthesized from coconut oil and milkfat by lipase-catalyzed glycerolysis. *J. Agric. Food Chem.* 41:1000–1005.

Wang, L.-L., and Johnson, E.A. 1997. Control of *Listeria monocytogenes* by monoglycerides in foods. *J. Food Prot.* 60:131–138.

Wanke, C.A., Cronan, S., Goss, C., Chadee, K., and Guerrant, R.L. 1990. Characterization of binding of *Escherichia coli* strains which are enteropathogens to small-bowel mucin. *Infect. Immun.* 58:794–800.

Ward, P.P., Uribe-Luna, S., and Conneely, O.M. 2002. Lactoferrin and host defense. *Biochem. Cell Biol.* 80:95–102.

Watanabe, S., Murata, S., Kumura, H., Nakamura, S., Bollen, A., Moguilevsky, N., and Shimazaki, K.-I. 2000. Bovine lactoperoxidase and its recombinant: Comparison of structure and some biochemical properties. *Biochem. Biophys. Res. Commun.* 274:756–761.

Watanabe, T., Nagura, H., Watanabe, K., and Brown, W.R. 1984.The binding of human milk lactoferrin to immunoglobulin A. *FEBS Lett.* 168:203–207.

Weiner, C., Pan, Q., Hurtif, M., Boren, T., Bostwick, E., and Hammerstrom, L. 1999. Passive immunity against human pathogens using bovine antibodies. *Clin. Exp. Immunol.* 166:193–205.

Wiedemann, V., Linckh, E., Kuhlmann, R., Schmidt, P., and Losch, U. 1991. Chicken egg antibodies for prophylaxis and therapy of infectious intestinal diseases. *J. Vet. Med.* B 38:283–291.

Williams, J., Moreton, K., and Goodnearl, A.D.J. 1985. Selective reduction of a disulphide bridge in hen ovotransferrin. *Biochem. J.* 228:661–665.

Wilson, I.A., and Stanfield, R.L. 1994. Antibody-antigen interactions: New structures and new conformational changes. *Curr. Opin. Struct. Biol.* 4:857–867.

Wimley, W.C., and White, S.H. 2000. Determining the membrane topography of peptides by fluorescence quenching. *Biochemistry* 39:161–170.

Witchmann, L., Vaalasti, A., Vaalasti, T., and Tuohimaa, P. 1989. Localization of lactoferrin in the male reproductive tract. *Intl. J. Androl.* 12:179–186.

Wood, H.G., and Barden, R.E. 1977. Biotin enzyme. *Rev. Biochem.* 46:385–413.

Woolley, D.W., and Longsworth, L.G. 1942. Isolation of an antibiotin factor from egg white. *J. Biol. Chem.* 142:285–290.

Wray, C., and McLaren, I. 1987. A note on the effect of the lactoperoxidase systems on salmonellas *in vitro* and *in vivo*. *J. Appl. Bacteriol.* 62:115–118.

Wright, R.C., and Trammer, J. 1958. Factors influencing the activity of cheese starters: The role of milk peroxidase. *J. Dairy Res.* 25:104–118.

Wu, H.-F., Monroe, D.M., and Church, F.C. 1995. Characterization of the glycosaminoglycan-binding region of lactoferrin. *Arch. Biochem. Biophys.* 317:85–92.

Wutrich, S., Richterich, R., and Hostettler, H. 1964. *Zeitschrift Lebensmittel-Untersuch. Forsh* 124:345.

Xu, Y.Y., Samaranayake, Y.H., Samaranayake, L.P., and Nikawa, H. 1999. *In vitro* susceptibility of *Candida* species to lactoferrin. *Med. Mycolog.* 37:35–41.

Yamaguchi, Y., Semmel, M., Stanislawski, L., Strosberg, A.D., and Stanislawski, M. 1993. Virucidal effects of glucose oxidase and peroxidase or their protein conjugates on human immunodeficiency virus type 1. *Antimicrob. Agents Chemother.* 37:26–31.

Yamamoto, N., Ejiri, M., and Mizuno, S. 2003. Biogenic peptides and their potential use. *Curr. Pharm. Res.* 9:1345–1355.

Yamauchi, K., Tomita, M., Giehl, T.J., and Ellison, R.T. III. 1993. Antibacterial activity of lactoferrin and a pepsin-derived lactoferrin peptide fragment. *Infect. Immun.* 61:719–728.

Yang, N., Stensen, W., Svendsen, J.S., and Rekdal, Ø. 2002. Enhanced antitumor activity and selectivity of lactoferrin-derived peptides. *J. Peptide Res.* 60:187–197.

Yang, T.-C., Chou, C.-C., and Li, C.-F. 2005. Antibacterial activity of alkylated disaccharide chitosan derivatives. *Intl. J. Food Microbiol.* 97:237–245.

Yokoyama, H., Peralta, R.C., Diaz, R., Sendo, S., Ikemori, Y., and Kodama, Y. 1992. Passive protective effect of chicken egg yolk immunoglobulins against experimental entertoxigenic *Escherichia coli*. *Infect. Immun.* 60:998–1007.

Yolken, R.H., Losonsky, G.A., Vonderfect, S., Leister, F., and Wee, S.-B. 1985. Antibody to human rotavirus in cow's milk. *N. Eng. J. Med.* 312:605–610.

Yoo, Y.-C., Watanabe, S., Watanabe, R., Hata, K., Shimazaki, K.-I., and Azuma, I. 1997. Bovine lactoferrin and lactoferricin, a peptide derived from bovine lactoferrin, inhibit tumor metastasis in mice. *Jpn. J. Cancer Res.* 88:184–190.

Yoshiko, S.K., Shibata, K., Yun, S.S., Yukiko, H.K., Yamaguchi, K., and Kumagai, S. 1996. Immune functions of immunoglobulin Y isolated from egg yolk of hens immunized with various inceftious bacteria. *Biosci. Biotech. Biochem.* 60:886–888.

Zagulski, T., Lipinski, P., Zagulska, A., Broniek, S., and Jarzabek, Z. 1989. Lactoferrin can protect mice against a lethal dose of *Escherichia coli* in experimental infection *in vivo*. *Brit. J. Exp. Pathol.* 70:697–704.

Zeng, J., and Fenna, R.E. 1992. X-Ray crystal structure of canine myeloperoxidase at 3 Å resolution. *J. Mol. Biol.* 226:185–207.

Zhang, X.L., Selsted, M.E., and Pardi, A. 1992. NMR studies of defensin antimicrobial peptides. 1. Resonance assignment and secondary structure determination of rabbit NP-2 and human HNP-1. *Biochem.* 31:11348–11356.

Zhao, C., Nguyen, T., Liu, L., Sacco, R.E., Brogden, K.A., and Lehrer, R.I. 2001. Gallinacin-3, an inducible epithelial beta-defensin in the chicken. *Infect. Immun.* 69:2684–2691.

Zimmermann, G.R., Legault, P., Selsted, M.E., and Pardi, A. 1995. Solution structure of bovine neutrophil beta-defensin-12: The peptide fold of the beta-defensins is identical to that of the classical defensins. *Biochem.* 34:13663–13671.

Zivanovic, S., Chi, S., Draughon, F.A. 2005. Antimicrobial activity of chitosan films enriched with essential oils. *J. Food Sci.* 70:M45–51.

Zommara, M.A., Toubo, H., and Imaizumi, K. 2002. Supplementing bovine milk immunoglobulin G prevents rats fed on a vitamin E-deficient diet from developing peroxidation stress. *Ann. Nutr. Metab.* 46:97–102.

19

Use of Antimicrobials as Processing Aids in Food Processing

Emefa A. Monu and Jairus R. D. David

CONTENTS

19.1 Importance

Food-processing aids are a large category of ingredients that are approved for use in food processing operations to facilitate the batching and production of products, such as plating agents, antifoaming and anticaking agents, acidifiers, and antimicrobials – e.g., lactic acid and chlorine, etc. Approved antimicrobial processing aids play a very important role in ensuring food microbiological quality and safety in the upstream production and pre-processing of food. Food safety is built on a series of processing techniques and applied compounds that each decrease the microbial load of the targeted food. There may not be one that completely eliminates resident microorganisms (thermal processing, irradiation, and certain "non-thermal" processes are the only techniques which can truly accomplish this), but together they act to reduce the microbial load at the front end so that the safety and quality are not dependent solely on the final kill step applied by the processor. Several processes are validated for a 5D or 6D reduction (5–6 log reduction) of the target microorganisms, so if the microbial load of the incoming product is higher than that, there will still be quality and safety issues in the final food product. Most of the treated end products are intermediate or starting products for manufacturing a range of other products that may or may not be ready to eat (RTE); additional preparation and cooking may be needed at the point of consumption.

Many antimicrobial interventions function more effectively when applied to lower microbial loads. As they are often applied during the initial and intermediate portions of processing and subsequent steps may remove or dilute the antimicrobials, it is thus important that processing aids are fast-acting. It is also important that an antimicrobial has low toxicity, even in concentrated form, as this can affect the employees who work with the substance, and may also require that extensive pretreatment is done to any waste water that contains the compound before discharge into the environment, as directed by the US Environmental Protection Agency (EPA). Therefore, there are several characteristics processors look for in an antimicrobial processing aid:

1) Fast-acting
2) Degrades over time so that residual concentrations are not present in the final product
3) Effective at low concentrations so that there will not be significant amounts in the final product
4) Does not pose a health hazard to employees
5) Low toxicity to consumers
6) Cost-effective

For each category of food product, there are common types of antimicrobials used. For example, for beef carcass washes, lactic acid is most commonly used, whereas for poultry products chlorine is traditionally used. Peracetic acid has rapidly become popular for application at various stages of primary processing including sprays, rinses, and chill water.

Other chapters of this book have discussed compounds and microorganisms that are added to foods or food-contact surfaces for the sole purpose of their antimicrobial functionality. This chapter will discuss processing aids used as antimicrobials in food processing operations. The role of antimicrobial processing aids will be illustrated using two diverse unit operations with specific objectives: animal post-evisceration decontamination, and decontamination of fresh produce and fresh-cut produce. In both of these unit operations processing aids are intended and used for the reduction of the microbial bioburden rather than total elimination. This level of microbial reduction may be adequate for spoilage control and assurance of shelf-life quality, but may or may not be adequate for the control of foodborne human pathogens and assuring food safety. Processing aids are a small and flexible hurdle and to be effective should always be used in conjunction with all other recommended best practices of Good Agricultural Practices (GAPs), Current Good Manufacturing Practices (cGMPs), and Hazard Analysis Critical Control Point (HACCP) throughout the supply chain; animal husbandry, slaughtering, and handling and storage; and agronomy, harvest, and post-harvest handling and storage.

19.2 Regulatory Perspectives

The definition and context of application of processing aids is defined in 21 CFR 101.100(a)(3) (Food and Drug Administration, 2018). The three contextual categories of processing aids are:

(a) substances that are added during the processing of a food but are removed in some manner from the food before it is packaged in its finished form;
(b) substances that are added to a food during processing, are converted into constituents normally present in the food, and do not significantly increase the amount of the constituents naturally found in the food; or
(c) substances that are added to a food for their technical or functional effect in the processing but are present in the finished food at insignificant levels and do not have any technical or functional effect in that food.

(Food and Drug Administration, 2018)

The definition illustrates that it is not only the inherent nature of the substance, but its contextual contribution to the final food product that determines whether it is an additive or a processing aid. For example, if sulfur dioxide is used in the apple peeling and slicing stage of the production of an apple pie to reduce the browning of the apples, it would be a processing aid rather than an additive. It is not present in significant amounts in the final product (most of it will be driven off during baking), has no sensory or functional effect in the final product, and was only used as an aid to the manufacturing of the product (Saltmarsh and Insall, 2013). If, however, that same sulfur dioxide was added as an antimicrobial agent to wine, it would be considered an additive rather than a processing aid.

Processing aids, by definition, do not need to be labeled along with other ingredients on the package and for the United States Department of Agriculture (USDA) label review and approval process. The use of any antimicrobial ingredient that does not comply with the three contexts described above as per 21 CFR 101.100(a)(3) should be considered as a direct or indirect food additive and should be declared explicitly in the ingredient listing (see Chapter 1, "Food Antimicrobials—An Introduction," Table 1.3).

Although the inclusion of processing aids on product labels is not legally required in most countries, many companies still disclose the inclusion of these substances on their product label voluntarily. This practice is due to the needs and concerns of customers and consumers of the affected company. For example, maltodextrin is a processing aid commonly used in plating liquid antimicrobials to convert them into powder form when spray drying, and does not need to be labeled in the ingredient listing. However, many companies have chosen to list maltodextrin as an ingredient for increased transparency and to build trust with consumers.

Due to the way they are defined, processing aids do not need to be included on food labels, which is why they often get overlooked when discussing food antimicrobials. The above definition was determined by the Food and Drug Administration (FDA), which regulates most of the U.S. food supply. The U.S. Department of Agriculture (USDA), which regulates meat and eggs, does not have its own definition listed in its regulations; for the purpose of determining whether a substance is a processing aid, the FDA definition is used. The United States is not the only country to look at processing aids in this manner. Mexico defines a processing aid as

> substance or material, excluding instruments, utensils and additives, that is not consumed as a food ingredient by itself and is used intentionally in the production of raw materials, foods or their ingredients to achieve a technological function during the treatment or processing and that can lead to the unintentional presence of residues or derivatives in the final product

(Magnuson et al., 2013). The Codex Alimentarius Commission, a joint international food standards program between the Food and Agriculture Organization and the World Health Organization, defines processing aids as

> any substance or material, not including apparatus or utensils, and not consumed as a food ingredient by itself, intentionally used in the processing of raw materials, foods or its ingredients, to fulfil a certain technological purpose during treatment or processing and which may result in the non-intentional but unavoidable presence of residues or derivatives in the final product.

(Codex Secretariat, 2018). The European Union (European Commission, 2008) and Australia (Food Standards Australia New Zealand, 2008) also use similar definitions. However, some countries such as Canada have no regulatory definition of the term processing aid, although Health Canada does have a working definition of the term that is similar to that used by Codex and does not require them to be included on the ingredient label (Canadian Food Inspection Agency, 2018). Countries also differ in how they classify specific antimicrobials. For example, in Canada ozone gas can be applied as a processing aid to poultry meat carcasses (up to 10 ppm) and poultry parts (up to 200 ppm) (Health Canada Bureau of Chemical Safety Food Directorate, 2018). However, in the U.S., such application is considered a secondary direct food additive by the USDA (US Department of Agriculture Food Safety Inspection Service, 2019).

19.3 Types of Processing Aids

Processing aid substances include cleaners and sanitizers, water-treatment chemicals, catalysts, exchange resins, desiccating agents, filtering agents, washing and peeling agents, emulsifying agents, and head space/packaging gases. They are often applied in meat and poultry processing, fruit and vegetable processing, water treatment, and included in a variety of further processed foods and beverages. Seafood products are unique in that they serve as both a source and an application of processing aids. Enzymes harvested from the wastes of fish processing (such as protease inhibitors) can then be used in seafood and other types of products. Although cleaners and sanitizers have antimicrobial properties, they will not be discussed here in detail as they are not applied to the food.

Various processing aids have purposes in multiple commodities, for example, acidified sodium chlorite is used in poultry, red meat, seafood, and fresh produce. Examples of some processing aids, their approved usage levels in the United States, and applications are shown in Table 19.1.

19.4 Processing Aids in Meat and Poultry Products

19.4.1 Foodborne Illness and Microorganisms of Concern in Meat and Poultry

Meat is a perishable product. There are several bacteria that can cause spoilage of meat products and reduce the quality even at refrigeration temperatures. These bacteria may be naturally present on the carcass, or may be introduced through contact with knives, equipment, and surfaces in the processing environment. Improper employee handling techniques may also increase the number of bacteria on meat products. There are many bacteria that are commonly found on fresh meats and poultry. Those that cause spoilage include slime-inducing *Pseudomonas*, *Lactobacillus*, *Enterococcus*, and *Brochothrix*; *Shewanella*, which can cause greening; and *Clostridium*, which can cause bone taints (Nychas et al., 2008; Sohaib et al., 2016). Although fresh meat is usually not RTE and proper cooking times and temperatures will eliminate most pathogenic bacteria, if these microorganisms are present in fresh meat they may still pose a threat to consumer health due to improper cooking temperatures, improper handling in the home leading to cross contamination, and improper storage temperatures (Mataragas et al., 2008; Ravishankar et al., 2010). There are several microbial hazards associated with raw meat and poultry. These include *Salmonella*, *Campylobacter*, *Clostridium perfringens*, and Shiga toxin-producing *Escherichia coli* (Borch and Arinder, 2002; Mead, 2004). Prevalence rates for these organisms vary by country and product. For example, an Ethiopian study from two beef-processing plants found *E. coli* O157:H7 in 0.54% of skin and internal carcass swabs and in 0.8% of samples from retail stores (Abdissa et al., 2017). Similarly, analysis of carcasses at a vertically integrated cattle operation in Mexico showed *E. coli* O157:H7 in 0.4% of carcasses post-processing (Narvaez-Bravo et al., 2013). In the EU, from 2007 to 2009, fresh meat was found to have a prevalence of 0.3–2.3% and 0.1–0.7% of STEC and O157 *E. coli*, respectively (Takkinen et al., 2011). In the U.S., prevalence rates of STEC in post-processing beef carcasses have been shown as 10% (Arthur et al., 2002). In the U.S., the USDA has set performance standards (Food Safety and Inspection Service, 2016) for the presence of viable *Salmonella* and *Campylobacter* in comminuted (ground) and whole carcasses of chicken and turkey as well as chicken parts (Table 19.2) and has established *E. coli* O157:H7 and six other strains of Shiga toxin-producing *E.coli* (STEC) as adulterants in ground beef, with zero tolerance for any presence of these *E.coli* in these types of products (Food Safety and Inspection Service, 2011).

In combination with including antimicrobials in the finished product and utilizing processing techniques such as thermal processing, irradiation, acidification, and drying to destroy these pathogens, it is important to control both spoilage and pathogenic microorganisms through the primary processing chain. In order to keep pathogens to acceptable levels in the final product, it is necessary to employ several techniques. To ensure this, the USDA Food Safety and Inspection Service (FSIS) established requirements for meat and poultry facilities to implement a Hazard Analysis and Critical Control Point plan. This plan involves the development of a "system of preventive controls designed to improve the safety of their products" (Food Safety and Inspection Service, 1996). Processing aids play a vital role

TABLE 19.1

Some Select Examples of Approved Antimicrobial Applications of Processing Aids in Various Foods in the United States

Substance	Intended Use	Amount
Aqueous solution of citric and hydrochloric acids adjusted to a pH of 1.0 to 2.0	Poultry carcasses, parts, trim, and organs	An aqueous solution of citric and hydrochloric acids adjusted to a pH of 1.0 to 2.0 applied as a spray or dip with a minimum contact time of 2 to 5 seconds, pH measured prior to application
Aqueous solution of peroxyacetic acid, hydrogen peroxide, acetic acid, and 1-hydroxyethylidene-1, 1-diphosphonic acid (HEDP)	Poultry scald tanks	Level of peroxyacetic acid will not exceed 220 ppm, hydrogen peroxide will not exceed 110 ppm, and 1-hydroxyethylidene-1, 1-diphosphonic acid (HEDP) will not exceed 13 ppm
Blend of lactic acid, polysorbate 80, and xanthan gum	Processing of beef heads and carcasses	Not to exceed 5% lactic acid solution, not to exceed 0.07% polysorbate 80, and not to exceed 0.05% xanthan gum; applied as a spray; exposure time 5–30 seconds, pressure 20–60 psi, temperature 18–55°C
Bacteriophage preparation (*Salmonella*-targeted)	Ready-to-eat (RTE) poultry products prior to slicing and on raw poultry, including carcasses and parts applied as a spray	Bacteriophage preparation (*Salmonella*-targeted) applied as a spray at 10^6 to 10^7 plaque-forming units (pfu) per gram of food product
Combination of natural source of nitrite and natural source of ascorbate	Meat or poultry product (including ground, formed, or whole muscle meat) that will be heat-treated and processed to be NRTE or RTE	For use as a component in the product formulation at (1) a rate of a minimum 75 ppm of nitrite from natural sources and minimum 500 ppm of ascorbate from natural sources or (2) a rate of a minimum 100 ppm of nitrite from natural sources and minimum 250 ppm of ascorbate from natural sources by weight of the finished food product
Lauramide arginine ethyl ester (LAE) dissolved in ethanol or water	Fresh cuts of beef and pork	Applied and dried to the inside of packaging at a concentration not to exceed 105 ppm LAE by weight of finished food product
Colicin protein preparation	Nine recombinant proteins intended for use singly or in combination as an antimicrobial spray on meat products	Colicin protein preparation applied as a spray at a rate of 1–10 mg/kg
Acidified sodium chlorite	Spray or dip for fresh produce	500–1200 ppm at pH 2.3–2.9
Potassium acid tartrate	pH lowering in wine and other beverages	Not specified
Sodium pentachlorophenate	Preservative for ammonium alginate, a processing aid in the manufacture of polyvinyl chloride emulsion polymers for use as articles or components of articles that contact food	At temperatures not exceeding room temperature, $\leq$0.5 % w/w of ammonium alginate solids

Source: USDA (U.S. Department of Agriculture Food Safety Inspection Service, 2019), and 21 CFR (Food and Drug Administration, 2018a, 2018c)

in these systems and, as seen in Table 19.3, can be applied anywhere from the fresh carcass to the final in-package product.

19.4.2 Animal Post-Evisceration Carcass Decontamination

Meat and poultry products go through several stages; first-processing (also known as primary processing) involves the conversion of the live animals to carcasses and parts such as roasts, wings, organ meats,

TABLE 19.2

Performance Standards for Raw Poultry in Processing Plants in the United States

	Maximum % Positive (Performance Standard)	
Product	***Salmonella***	***Campylobacter***
Broiler chicken carcass	9.8 (5 out of 51)	15.7 (8 out of 51)
Turkey carcass	7.1 (4 out of 56)	5.4 (3 out of 56)
Comminuted chicken	25 (13 out of 52)	1.9 (1 out of 52)
Comminuted turkey	13.5 (7 out of 52)	1.9 (1 out of 52)
Chicken parts	15.4 (8 out of 52)	7.7 (4 out of 52)

Source: FSIS (Food Safety and Inspection Service, 2016).

etc. Further processing converts the carcasses and parts into other products such as frozen fried chicken, burgers, deli meats, canned soups, etc. (Fletcher, 2004). Processing aids are one strategy applied during meat and poultry processing to achieve pathogen reduction. After evisceration and before further processing, they can be applied as carcass washes, or be added to water used in chilling or scalding of the carcass.

19.4.2.1 Carcass Washes

One application of processing aids in meat and poultry first-processing facilities is as antimicrobials in the carcass washes applied post-slaughter. Water is used to remove fecal matter, hair/feathers and microorganisms, and in carcass chilling and is therefore a prime vehicle for antimicrobial compounds. In the U.S. poultry industry, it has been estimated that 26 L of water is used per bird (Northcutt and Jones, 2004), which is a significant amount considering that many plants produce over 1 million birds per week. Beef carcasses are sprayed, which is always applied from the top of the carcass to the bottom. Other red meat carcasses such as goat, pork, veal, and lamb are washed in a similar manner, but application times are shorter due to the smaller sizes of the carcasses. The beef carcass is first washed with warm or hot water (usually for 2 minutes) followed by a 5-minute drip. This drip time is necessary to allow the excess film of water to be removed from the carcass surface, which would impede the interaction of the antimicrobial applied in the following step with any bacteria on the carcass surface. The antimicrobial rinse is applied for about a minute and can be made up of different compounds. Organic acids are usually used as antimicrobial agents during carcass washing. The most common antimicrobial rinse for beef carcasses is a 2% lactic acid solution, which has been shown to result in a 3.5, 4.7, and 5.0 log reduction of *E. coli* O157:H7, *Salmonella* Typhimurium, and *Campylobacter* spp., respectively, on beef carcasses (Flowers, 2006). Acetic acid is also commonly used at 2%, and commercial mixes such as Fresh Bloom™ that include several organic acids are also available. However, there are several mixtures of antimicrobials that are approved as safe and suitable ingredients by the Food Safety Inspection Service (FSIS) for use as sprays or dips on meat carcasses (Table 19.3). This list is a living document that is constantly being updated as companies or researchers apply for approval. One commercially available example is Amplon, which contains sulfuric acid and sodium sulfate and is approved for use as a processing aid as a spray, dip, or wash for poultry parts and carcasses and has exhibited a 0.8–1.2 log reduction of *Salmonella* on inoculated chicken wings (Kim et al., 2017; Scott et al., 2015).

Poultry carcass washing takes place at several points of first-processing. The outside of the bird is rinsed after de-feathering, and there is often an inside wash as well where nozzles are inserted into the abdominal cavity. Another wash can be applied just prior to the chilling of the carcass (Barbut, 2016). Inside-outside bird washers (IOBW) and dip tanks are then used both before and after the chilling step to create multiple hurdles (Singh et al., 2015). As in red meat production, organic acids can be used in poultry carcass washes. One study showed that wash sprays containing >5% lactic acid significantly reduced *Salmonella* counts on chicken thighs with or without skin (Ramirez-Hernandez et al., 2017). Similarly, a dip treatment of acidified lactic acid (ALA), containing 0.5% lactic acid, reduced *S.* Heidelberg and *Campylobacter jejuni* in ground chicken frames (Moore et al., 2017).

TABLE 19.3

Some Select Examples of Processing Aids Deemed Safe and Suitable for Use in Meat and Poultry Products by the USDA

Substance	Intended Use	Amount
Aqueous potassium hydroxide solution	Hide-on carcass wash in spray cabinet	Wash solution used at a final concentration of 0.01–0.40 % w/w
Aqueous solution of citric and hydrochloric acids adjusted to a pH of 0.5 to 2.0	Meat carcasses, parts, trim, and organs	Adjusted to a pH of 1.0 to 2.0 applied as a spray or dip for a contact time of 2 to 5 seconds
Aqueous mixture of peroxyacetic acid (PAA), hydrogen peroxide (HP), acetic acid, sulfuric acid (optional), and 1-hydroxyethylidene-1,1-diphosphonic acid (HEDP), catalyzed with sulfuric acid	(1) In process water used for washing, rinsing, or cooling whole or cut meat including carcasses, parts, trim, and organs; (2) in process water or ice for washing, rinsing, storing, or cooling of processed and pre-formed meat products	(1) PAA not to exceed 1800 ppm, HP not to exceed 600 ppm, and HEDP not to exceed 22.5 ppm; (2) Not to exceed 495 ppm PAA, 165 ppm HP, and 14 ppm HEDP
Blend of citric acid, phosphoric acid, and hydrochloric acid	Poultry carcasses	Citric acid 1.87%, phosphoric acid 1.72%, and hydrochloric acid 0.8% applied as a spray with a minimum contact time of 1 to 2 seconds and allowed to drip from the carcasses for 30 seconds
Aqueous solution of acidic calcium sulfate and lactic acid	Continuous spray or dip on raw poultry carcasses, parts, giblets, and ground poultry	Acidic calcium sulfate sufficient for purpose; lactic acid not to exceed 5.0 % and 55°C
Acetic acid	Chicken livers	Immersion dip at a concentration of up to 5% and not to exceed 2 minutes
Aqueous solution of citric acid and hydrochloric acid	Processed and comminuted red meat products in an enclosed mixing, grinding, and/or blending system.	Solution adjusted to a pH of 0.5–2.0
Aqueous solution of citric acid and hydrochloric acid	Permeable and impermeable casings of meat and poultry products applied as a spray, dip, or immersion to casings prior to opening, removal, or slicing operations	Solution adjusted to a pH <2.5
Aqueous mixture PAA, HP, acetic acid, and HEDP	(1) For use in: brines, sauces, and marinades applied either on the surface or injected into processed or unprocessed, cooked or uncooked, whole or cut poultry or parts and pieces; (2) surface sauces and marinades applied on processed and preformed meat and poultry products	Aqueous mixture not exceeding 50 ppm PAA, 33 ppm HP, 3.3 ppm HEDP
Lauramide arginine ethyl ester (LAE) dissolved in either ethanol or water	Packaged fresh cuts of beef and pork	Applied and dried to the inside of packaging at a concentration not to exceed 105 ppm LAE by weight of finished food product
Blend of citric acid and sorbic acid in a 2:1 ratio	Reduction of the microbial load of purge trapped inside soaker pads in packages of raw whole muscle cuts of meat and poultry	Blend of citric acid and sorbic acid in a 2:1 ratio incorporated into soaker pads at a level not to exceed 1 to 3 grams per pad

Source: USDA (U.S. Department of Agriculture Food Safety Inspection Service, 2019).

Although most carcass washes are applied post-evisceration, there are some processing aids that are approved for use through direct application to the hides or feathers of live animals while still in holding pens prior to slaughter, including several bacteriophage mixtures (US Department of Agriculture Food Safety Inspection Service, 2019).

19.4.2.2 Processing Aids in Scald and Chilling Water

In meat and poultry first-processing an abundance of water is used not only for the cleaning of the carcass, but also to aid in the removal of hair and feathers through the scalding process, and cooling through spray or immersion chilling of the carcass before refrigeration/freezing. Antimicrobials can be added to the water used at these points as processing aids to reduce microbial numbers. This is especially important in immersion chilling, as it can lead to cross-contamination as several carcasses come in contact with the same water and each other during the process, although this can be mitigated by applying a countercurrent flow of water near the exit of the chiller so that carcasses are exposed to cleaner water as they progress through the process (Singh et al., 2015). Chlorine is the most commonly used antimicrobial in meat and poultry processing water in the U.S., although several countries in the EU do not allow for its use due to concerns over its toxicity (namely the production of carcinogens); trihalomethanes (THMs) are formed when chlorine reacts with any organic material in water (Cords et al., 2005). THMs have been identified as carcinogens, and therefore their presence in any food product poses a risk to human health. In water, added chlorine is converted to hypochlorous acid hypochlorite, and the ratio is dependent upon the pH of wash water. At pH of 6.0, roughly 97% of the chlorine is hypochlorous acid, and at pH 9.0, 97% is hypochlorite ion. The antimicrobial activity is due largely to hypochlorous rather than to hypochlorite. Chlorine works most effectively in the pH range of 5–7. Peroxyacetic acid (PAA) is an equilibrium mixture of the peroxy compound, hydrogen peroxide, and acetic acid. The primary mode of action of peroxyacetic acid is oxidation. PAA has a stronger oxidizing potential than chlorine, chlorine dioxide, chlorous acid, and hydrogen peroxide, but less than ozone. It has become a widely used antimicrobial in poultry processing at various stages of rinsing, chilling, and scalding, and several applications are also approved for use in red meat processing. Table 19.1 shows one of the approved applications of PAA in combination with hydrogen peroxide and 1-hydroxyethylidene-1, 1-diphosphonic acid (HEDP). In this mixture, hydrogen peroxide reacts with acetic acid to form peroxyacetic acid. This results in an equilibrium of peroxyacetic acid, acetic acid, hydrogen peroxide, and HEDP. HEDP is a necessary part of the mixture to ensure the solution is stable, as peroxy compounds are unstable (Mattia et al., 2006). It is important to note that not all commonly used approved applications of PAA and other antimicrobial compounds are considered processing aids. This is usually related to the level of the compound that is used; higher concentrations would be considered direct additives, secondary direct additives, or indirect food additives/food-contact substances (FCS) depending on the stage of processing at which they are applied.

19.4.2.3 Post-Chill Processing Aids Added for Antimicrobial Activity

In poultry processing, antimicrobials may also be applied post-chilling. These could take the form of sprays or dips that are applied to the carcass after air or water chilling. For example, several mixtures of PAA with hydrogen peroxide and HEDP are approved for this use.

In addition to incorporation directly into or onto the product, processing aids may be included in the packaging. For example, compounds such as lauric arginate (LAE) may be added to the packaging film that comes in direct contact with the chicken parts (Table 19.3). LAE (which is described in detail in Chapter 10, "Lauric Arginate Ethyl Ester") is a surfactant derived from lauric acid and arginine. As a surfactant, it disintegrates bacterial cell membranes at low concentrations, thereby increasing cell permeability and causing cell death (Becerril et al., 2013). Laboratory studies have shown LAE to be effective against several pathogenic and spoilage bacteria, and when incorporated into polylactic acid film at concentrations of 0.07%, 0.14%, and 0.28% (w/w) successfully decreased *L. monocytogenes* and *Salmonella* in sliced ham from 6 log to ca. 4 log CFU/g and maintained cell count at this level over 7 days of refrigerated storage while films without LAE rose to cell counts of ca. 8 log CFU/g (Theinsathid et al., 2012).

Absorbent pads are used in the packaging of raw poultry products to absorb exudates from the meat during refrigerated storage. The liquid that is absorbed into the pads contains spoilage and potentially pathogenic microorganisms. The addition of antimicrobials to the pads can help maintain the microbial quality and safety of the product (Barbut, 2015; Otoni et al., 2016). Antimicrobials incorporated into these pads in order to reduce bacterial growth may be classified as processing aids, such as the citric acid and ascorbic acid blend described in Table 19.3.

19.4.3 Addition of Processing Aids During Further Processing of Meat and Poultry Products

19.4.3.1 Non-RTE Meat and Poultry Products

In addition to its use as a wash or in cooling water, PAA can be added during further processing of meat and poultry. It can be added in combination with other compounds to brines, sauces, or marinades that are injected into the meat or applied to the surface (Table 19.3). Note that the maximum concentrations approved for use in these applications (50 ppm PAA combined with 33 ppm hydrogen peroxide, 3.3 ppm HEDP acetic acids) are much lower than those used in poultry scald tanks (220 ppm PAA with 110 ppm hydrogen peroxide, and 13 ppm HEDP). This relates to the FDA definition of a processing aid. In the case of the scald water, the antimicrobials are "substances that are added during the processing of a food but are removed in some manner from the food before it is packaged in its finished form." When the antimicrobials are applied as marinades or sauces, they are not removed and will still be present in the final product; therefore we must ensure that they are "added to a food for their technical or functional effect in the processing but are present in the finished food at insignificant levels and do not have any technical or functional effect in that food." These marinades may be used on products that are sold raw or partially cooked and are therefore sold as NRTE, or they may be used in products that will be subsequently be sufficiently treated during processing to be sold as RTE.

19.4.3.2 RTE Meat and Poultry Products

Natural antimicrobials are more frequently used in RTE meat and poultry products as consumer demand for natural ingredients grows. Some of these can be classified as processing aids. For example, a combination of a natural source of nitrite and natural source of ascorbate is approved as a processing aid for use in meat or poultry products (see Table 19.1). Natural sources of nitrite may include celery extract or celery powder (Sebranek et al., 2012).

Biopreservatives, including bacteriocins and bacteriophage, are also categorized as natural antimicrobials. The mechanism of action of biopreservatives will not be discussed here as they will be covered in detail in Chapters 15, "Bacteriocins and Their Applications in Foods," and 16, "Bacteriophages." Bacteriocins are small, antimicrobial peptides that are produced by bacteria. They can be applied to foods in several ways. The first is to add the purified bacteriocin itself to the food product or packaging material. The second is to add a crude extract of a fermentate that includes the bacteriocin, but in a less concentrated form. The third is to add a fermentate produced by adding the bacteriocin-producing organism to a growth medium that includes not only the bacteriocin, but several other compounds, some of which also have antimicrobial activity. Finally, the bacteriocin-producing culture itself can be added to the food product, allowing for the production of the bacteriocin and possibly other antimicrobials in the food throughout its shelf life. Several of these options can be used as processing aids in foods. In general, bacteriocin-producing cultures are regarded as processing aids in foods. It is important to keep in mind the definition of processing aids when determining whether an antimicrobial is considered a processing aid rather than an ingredient, namely it should be added during processing, and is either removed, leaves insignificant or nonfunctional residues, or is converted into normal food constituents (Food and Drug Administration, 2018). Through this reasoning, several bacteriocins have been approved as processing aids for use in meat and poultry products after submissions by producing companies. For example, in 2015 colicins were granted GRAS status by FDA for controlling pathogenic *E. coli* in produce based on a submission by the company Nomad. An approved application of these bacteriocins as a processing aid was added to the USDA list of safe and suitable ingredients for use in meat, poultry, and egg products (see Table 19.1) (Hahn-Löbmann et al., 2019; US Department of Agriculture Food Safety Inspection Service, 2019). In contrast, there is a mixture of the bacteriocin-producing *Carnobacterium maltaromaticum*, sodium acetate, and sodium diacetate which is also on the USDA approved list, but is considered an additive that must be included in the ingredient label, as is the case with several other mixtures of bacteriocin-producing lactic acid bacteria that are listed (US Department of Agriculture Food Safety Inspection Service, 2019).

Bacteriophages are viruses which infect and consequently destroy target bacteria; each bacteriophage is specific to a particular species or sometimes even strain of bacteria. Commercial preparations of bacteriophage are utilized as processing aids against pathogens such as *Salmonella*, *E. coli*, and *Listeria* in deli meat and other RTE meat (see Table 19.1) as well as various other foods including raw poultry and beef carcasses, processed seafood (for example smoked salmon or battered fish fillets), cheese, and fresh produce. As with bacteriocins and their various preparations, bacteriophage can be applied to a food by spraying, dipping, direct incorporation into the product, or spraying into the individual bags as the product is being packaged, and are used in other foods in addition to meat and poultry. For example, Listex P100® is used as a processing aid to control the growth of *Listeria* on fruits and vegetables in several ways (Oliveira et al., 2014). Firstly, it can be used as a final rinse before produce is packaged. It can also be sprayed on frozen vegetables post-blanching before the freezer tunnel to decrease the risk of *Listeria* growth in the moist conditions of the first portion of the freezer tunnel.

19.4.4 The Role of Acidifiers in Microbial Reduction of Meat and Poultry Products

Acidifiers can increase the efficacy of antimicrobials in foods even if they do not exert a substantial antimicrobial effect on their own. Acidic conditions can act as an additional hurdle, causing stress to the microorganisms and thereby increasing the effect of other added antimicrobials (David et al., 2013). In addition, some antimicrobials are more effective at lower pH, such as organic acids and sodium hypochlorite. Organic acids (such as lactic acid, acetic acid, and propionic acid) and the salts of these compounds are weak which means they dissociate depending on the pH (the undissociated form is required to cross the cell membrane and enter the cytoplasm, where it then dissociates and inhibits the target microorganisms). Therefore, lowering the pH of the solution to below the pK_a of the acid will increase the antimicrobial effect (Mani-López et al., 2012). Sodium hypochlorite is more effective at a pH below 5.5, as there is a greater concentration of hypochlorous acid at those conditions (Lillard, 1979). Acidifiers can be used at any point of the production stream. Table 19.4 lists several compounds and blends that although approved for use as acidifiers by the USDA rather than antimicrobials, still aid in the reduction of the microbial load of meat and poultry products at various stages of processing.

TABLE 19.4

Some Select Examples of Acidifiers Approved for Use as Processing Aids in Meat and Poultry by the USDA

Substance	Intended Use	Amount
Combination of sulfuric acid, ammonium sulfate, copper sulfate, and water	Adjust the pH in meat and poultry processing water, including processing water used as a spray or dip	Combination of sulfuric acid, ammonium sulfate, copper sulfate, and water, sufficient for purpose
Aqueous solution of citric and hydrochloric acids	pH control agent in water used in poultry processing	Solution of citric and hydrochloric acids, sufficient for purpose
Aqueous solution of sulfuric acid, citric acid, and phosphoric acid	Adjust the pH of peroxyacetic acid (PAA) for use on poultry carcasses as a spray or dip	Blend of sulfuric (35%), citric (1%), and phosphoric acid (1%) solution that is injected into a diluted water stream of PAA (≤100 ppm), hydrogen peroxide, acetic acid, and HEDP – to lower the pH of the PAA water stream from ca. 4.5 to <2.5
Aqueous solution of sulfuric acid, citric acid, and phosphoric acid	Adjust the pH in poultry chiller water and the processing water in meat and poultry plants	Solution of sulfuric acid, citric acid, and phosphoric acid, sufficient for purpose
Aqueous solution of hydrochloric acid, phosphoric acid, and lactic acid	pH control agent on raw and ready-to-eat (RTE) meat products and in water used in poultry processing	Hydrochloric acid and phosphoric acid – sufficient for purpose; lactic acid not to exceed 5.0%

Source: USDA (U.S. Department of Agriculture Food Safety Inspection Service, 2019).

19.4.5 Summary

There are a number of pathogenic microorganisms that are associated with red meat and poultry products. The processing of these products involves multiple steps and intervention methods that can be applied at various stages throughout production to reduce the microbial load, from carcass washing to incorporation into RTE products. Water in particular is used in several stages of first-processing and provides an efficient vehicle for antimicrobial application. Processing aids are a vital tool for these intervention strategies, and the USDA offers a large number of antimicrobial options for processors including organic acids, biopreservatives, and surfactants, along with information on acceptable levels of use and context (Food and Drug Administration, 2018).

19.5 Decontamination of Fresh and Fresh-Cut Produce

19.5.1 Fresh Produce – Benefits and Risks

The consumption of fresh produce (vegetables and fruits) is seen as a healthy choice by consumers. Several health and wellness initiatives that have been implemented to reduce obesity and other complications arising from unhealthy diets have effectively promoted the health benefits of fresh produce and have popularized the consumption of fresh vegetables and fruits. It has been estimated that over 1 billion bags of salad vegetables are sold in the U.S. each year. Also, the increased consumer demand has resulted in increased imports of fresh produce. Sales of pre-cut salad mixes were at $1.94 billion in year 2001 and increased to $2.11 billion in 2003 (Hodge, 2003). In restaurants alone, fresh salads/vegetable comprise 9% of the main menu (Sloan, 2003).

Unlike fresh (raw) animal carcass, fresh produce is considered ready-to-eat (RTE) as it is traditionally eaten without further processing by the end user/consumer. However, health benefits from consumption of fresh RTE produce or fresh-cut produce and associated microbiological food safety are mutually exclusive, i.e., they are independent of one another. The increase in consumption of RTE fruits and vegetables calls for increased focus on minimizing microbial food safety hazards.

Contamination of fruits and vegetables by human pathogens can occur anywhere in the farm-to-table continuum including contamination of seed stocks and during production, harvesting, postharvest handling, storage, processing, transport distribution, retail display, and/or preparation at home or in restaurants and food service. Produce contamination with human pathogens cannot be completely disinfected by washing or rinsing the product in an aqueous solution, and low sporadic levels of human pathogens can be found on produce.

The microbial food safety of fresh produce is a complex issue. Fresh produce comes with its own microbial food safety problems for upstream farmers, intermediate suppliers, processors, retailers, and consumers. It is well-known that fresh produce will encounter potential safety hazards somewhere along the supply chain. Fruit and vegetable producers are responsible for maintaining hygiene in their products. Food hygiene can be minimized but will always exist as long as produce is grown outdoors in the fields (as compared to produce grown in green houses and indoor hydroponics). Therefore, to date, a preventative approach to the contamination of fresh fruits and vegetables by the use of GAPs, cGMPs, and HACCP has proven to be the most effective means of ensuring produce food safety.

19.5.2 Fresh Produce – Foodborne Illness and Microorganisms of Concern

Fruits and vegetables are highly perishable and are easily spoiled by a number of factors including the growth of spoilage microorganisms, naturally occurring enzymes present in the fresh produce, chemical reactions, and structural changes. The quality and safety of fresh fruits and vegetables very much depend on their microbiological flora and the storage conditions.

Gram-negative bacteria dominate the microflora associated with most vegetables, whereas mold and weakly fermentative yeasts often comprise the majority of the microflora on raw fruits with the acidic pH of the fruit tissue, below 4.0. The majority of the microorganisms associated with product tissues contribute to product spoilage. The presence of human pathogens in fresh RTE produce is a public health concern, as evidenced by several contaminated produce multi-state recalls during the past decade (Harris et al., 2003). Harris et al. (2003) extensively reviewed outbreaks associated with fresh produce and reported that the most common human pathogens associated with produce foodborne illness outbreaks can be categorized as follows:

- Soil-associated pathogenic bacteria (*Clostridium botulinum*, *Listeria monocytogenes*)
- Feces-associated pathogenic bacteria (*Salmonella* spp., *Shigella* spp., *E. coli* O157:H7, and others)
- Pathogenic parasites (*Cryptosporidium*, *Cyclospora*)
- Pathogenic viruses (hepatitis A, enterovirus, Norwalk-like viruses)

Many of these pathogens are spread via a human or animal to food to human transmission route.

Some of the commonly incriminated fresh produce vegetables and fruits including raw unpasteurized fruit juices are sprouts, cucumber, lettuce, vegetable salad, apple cider, apple juice, raspberries, radish sprouts, iceberg lettuce, watercress, vegetables including carrots, strawberries, tossed salad, coleslaw and onions, watermelon, cantaloupes, beansprouts, alfalfa sprouts, root vegetables and dried seaweed, mixed salad, salad crops and vegetables, broccoli florets, cauliflower florets, cut celery, shredded cabbage, and cut melon (Mitchell, 2003).

According to Mitchell (2003), from 1996 to 2006, 72 foodborne illness outbreaks in the U.S. were associated with the consumption of fresh produce. CDC data also indicate that produce-related outbreaks have become larger, involving more individuals and increasing in frequency (Sivapalasingam et al., 2004). From 1998 to 2016, there have been 30–60 reported outbreaks linked to fresh produce annually (sickening 900–3000 people per year), and 20–40% of these outbreaks were caused by leafy greens specifically, corresponding to 100–1100 illnesses annually (Congressional Research Service, 2019).

Of the 72 produce related outbreaks from 1996 to 2006, 25% (18 outbreaks) implicated fresh-cut produce. Processing fresh produce for use in ready-to-eat (RTE) foods (e.g., thinly sliced celery in wet salads, e.g., chicken or turkey or tuna salads) or shredding of lettuce or cabbage increase the risk of bacterial growth and contamination by breaking the natural exterior barrier of the produce and introducing surface bacteria into the nutrient-rich and high-water-activity interior.

19.5.3 Washing and Sanitizing Treatments for Fruits and Vegetables

Most freshly harvested fruits and vegetables are washed by the grower, packer, or processor to remove soil, plant debris, pesticide residues, and microorganisms from the commodity surface.

GAPs are intended for reducing risk due to contamination during the growing and harvesting of produce. If produce contamination occurs post-harvest and contamination sources cannot be eliminated, the washing and sanitizing of produce and equipment become the first line of defense. Fresh produce and fresh-cut produce processors usually rely on wash water sanitizers to reduce microbial counts in order to maintain quality and extend shelf life. Also, it is common for producers to recycle wash water for conservation. However, reusing wash water increases the risk of spreading contamination as it acts as a vector for the transmission of pathogens from the contaminated to uncontaminated produce.

The regulations of processing aids that are used to reduce the microbial load of fresh fruits and vegetables are complex and, in some areas, uncertain. In U.S., the wash water processing aids used for fresh-cut produce are regulated by the FDA as a secondary direct food additive, unless they are considered to be generally recognized as safe (GRAS). If the product is a raw agricultural commodity (RAC) that is washed in a processing facility producing fresh-cut produce, both the Environmental Protection Agency (EPA) and the FDA have regulatory jurisdiction and the processing aid products must be registered as pesticides with the EPA. A list of wash water processing aid sanitizers approved by FDA is referenced in the Code of Federal Regulations 21CFR Parts 173.315 and 178.1010 (Food and Drug Administration, 2003a, 2003b).

The main effect of sanitizing treatments for washing fresh produce is to reduce the microbial load and keep process water free from re-contamination rather than having a persistent preservative effect in the finished product (RTE fresh produce and fresh-cut produce). Chlorine and peroxyacetic acid are approved and very commonly used washing agents and are discussed below (Herdt and Feng, 2009). While these treatments may reduce the levels of natural microflora on the surface of fruits and vegetables, they cannot guarantee the total elimination of spoilage microorganisms and human pathogens. The choice of treatment for sanitizing or preserving fruits and vegetables depends on the type of commodity, the processing conditions, and the shelf life of the product.

19.5.3.1 Chlorine

Chlorine is the most widely used sanitizing agent in wash, spray, and flume waters in the fresh produce industry. It may be added to wash water as chlorine gas, or more commonly as sodium or calcium hypochlorite. The U.S. FDA and Drug Administration (FDA) specifies a use level for washing fruits and vegetables not to exceed 0.2% chlorine when followed by a potable water rinse. Commonly, 50–200 ppm calcium hypochlorite is used for most commodities, at a pH of <8.0 and with a contact time of 1–2 minutes for raw fruits and vegetables. However, as much as 2000 to 20,000 ppm calcium hypochlorite may be used to sanitize alfalfa seeds intended for sprout production, for inactivation of pathogenic bacteria present on seed surfaces.

Chlorine is highly effective for inactivating planktonic cells of bacteria, yeasts, molds, and viruses, although spore forms of bacteria and molds are more resistant. Chlorine is less effective for inactivating bacteria attached to produce surfaces or embedded within the product. The efficacy of chlorine as measured by microbial reduction is no greater than 2 logs. While such reductions can greatly reduce spoilage, they are not adequate to ensure safety in the event of gross contamination with human pathogens. Major advantages of chlorine are its broad-spectrum antimicrobial efficacy, ease of application, and low cost. However, inactivation of free chlorine as a result of contact with the surfaces and tissues of produce, as well as with soil and other field debris, is an important limiting factor in the overall effectiveness. Regular and periodic monitoring of the level of free chlorine and pH is critical in chlorinated waters used in the fresh produce industry.

Advantages of chlorine:

- Chlorine is "broad-spectrum antimicrobial," effective against a wide variety of bacteria, fungi, and viruses, including bacteriophage.
- Antimicrobial efficacy is dependent upon use-solution pH.
- It is effective at fairly low temperatures and is not temperature-sensitive as other common sanitizers.
- It is relatively inexpensive.
- It does not foam.

Disadvantages of chlorine:

- Chlorine is corrosive to many metals.
- It is unstable and dissipates rapidly from solution.
- It loses activity rapidly in the presence of organic materials, light, air, and metals.
- Liquid chlorine deteriorates during storage, and stability decreases with increasing temperature (Grab, 2001).

19.5.3.2 Peroxyacetic Acid (PAA)

Peroxyacetic acid is approved for fresh produce and fresh-cut produce as an antimicrobial process water additive at a maximum concentration of 80 ppm. Like chlorine, microbial reduction for PAA is limited to less than 2 logs, given the regulatory maximum concentration of 80 ppm approved for the decontamination of fresh produce and fresh-cut produce.

The advantages of PAA over other oxidizing treatments are its higher tolerance to organic materials and soil loading in recycled water, and its broad spectrum efficacy; it also does not need pH control. Peroxyacetic acid has environmentally friendly decomposition by-products (i.e., oxygen, acetic acid, octanoic acid, and water).

Advantages of peroxyacetic acid:

- As an oxidizer and acid-type sanitizer, it combines acid-type rinse in one step.
- It leaves no residues.
- It is tolerant to organic matter.
- It is generally non-corrosive.
- It has broad spectrum of activity.
- Wider range of pH up to 7.5.
- Readily biodegradable and breaks down into water, oxygen, and acetic acid.

Disadvantages of peroxyacetic acid:

- Loss of activity in the presence of some metals and high organic loads.
- Corrosive to yellow metals as well as mild and galvanized steel (Grab, 2001).

19.5.4 Summary

The consumption of RTE fresh vegetables and fruits is seen as a healthy choice by consumers. Both fresh vegetables and fruits are known to carry spoilage microorganisms, and on rare occasions human pathogens. The health benefits from consumption of fresh and fresh-cut fruits and vegetables and associated microbiological food safety are independent of one another. The microbiological food safety of fresh produce is a complex issue. Contamination of vegetables and fruits by human pathogens can occur anywhere in the farm-to-table continuum. The prevention of microbial contamination at all points of the food chain from upstream production to the final consumer is preferred over washing and the application of disinfectant after contamination occurs. Also, the efficacy of washing with sanitizers in reducing human pathogens and other microorganism on fresh produce surfaces is limited to about 1 to 2 logs. This reduction may be good for microbial spoilage control and shelf life assurance, but is not adequate to ensure food safety. A multiple prevention-intervention approach throughout the supply chain must be considered to enhance the safety of fresh and fresh-cut produce, built upon sound, well-documented, and verifiable GAPs, cGMPs, and HACCP. Incremental improvements in washing efficacy with approved conventional sanitizers such as chlorine and peroxyacetic acid can be obtained through buffering, addition of surfactants, wetting agents, temperature elevation, full immersion (as compared to topical spray), and mechanical agitation during washing. Microbial reductions in the range of >2–5 logs may be possible with the use of emerging new technologies such as surface pasteurization or irradiation subject to minimal negative impact on total quality and freshness of treated fresh produce, regulatory approval, consumer acceptance, and messaging.

REFERENCES

Abdissa, R., Haile, W., Fite, A. T., Beyi, A. F., Agga, G. E., Edao, B. M., Tadesse, F., Korsa, M. G., Beyene, T., Beyene, T. J., De Zutter, L., Cox, E., and Goddeeris, B. M. 2017. Prevalence of *Escherichia coli* O157:H7 in beef cattle at slaughter and beef carcasses at retail shops in Ethiopia. *BMC Infectious Dis.* 17:277.

Arthur, T. M., Barkocy-Gallagher, G. A., Rivera-Betancourt, M., and Koohmaraie, M. 2002. Prevalence and characterization of non-O157 Shiga toxin-producing *Escherichia coli* on carcasses in commercial beef cattle processing plants. *Appl. Environ. Microbiol.* 68:4847–4852.

Barbut, S. 2015. The science of poultry and meat processing [Self-published online book]. Available at: https://atrium.lib.uoguelph.ca/xmlui/handle/10214/9300.

Barbut, S. 2016. Primary processing of poultry. pp. 97–124. In *Poultry Products Processing*, CRC Press, Boca Raton, FL.

Becerril, R., Manso, S., Nerin, C., and Gómez-Lus, R. 2013. Antimicrobial activity of lauroyl arginate ethyl (LAE), against selected food-borne bacteria. *Food Control* 32:404–408.

Borch, E., and Arinder, P. 2002. Bacteriological safety issues in red meat and ready-to-eat meat products, as well as control measures. *Meat Sci.* 62:381–390.

Canadian Food Inspection Agency. 2018. Processing aids. Available at: http://www.inspection.gc.ca/food/requirements/labelling/-f-for-industry/former-food-additives/eng/1511448488183/1511448488947?chap=9.

Codex Secretariat. 2018. Basic texts and definitions. p. 24. In *Procedural Manual of the Codex Alimentarius Commission 26th Edition*. FAO, Rome, Italy.

Congressional Research Service. 2019. Foodborne illnesses and outbreaks from fresh produce. Available at: https://crsreports.congress.gov/product/pdf/IF/IF11092.

Cords, B. R., Burnett, S. L., Hilgren, J., Finley, M., and Magnuson, J. 2005. Sanitizers: Halogens, surface-active agents, and peroxides. pp. 512–577. In P. M. Davidson, J. N. Sofos, and A. L. Branen (eds.), *Antimicrobials In Food*, CRC Press, Boca Raton, FL.

David, J. R. D., Steenson, L. R., and Davidson, P. M. 2013. Expectations and applications of natural antimicrobials to foods: A guidance document for users, suppliers. research and development and regulatory agencies. *Food Prot. Trends* 33:238–247.

European Commission. 2008. Regulation (EC) No 1333/2008 of the European Parliament and of the Council of 16 December 2008 on food additives. *Official Journal of the European Union*. 354:16–33.

Fletcher, D. 2004. Further processing of poultry. pp. 108–134. In *Poultry Meat Processing and Quality*, Elsevier.

Flowers, S. L. 2006. PhD Thesis: Identification and validation of antimicrobial interventions for red meat carcasses processed in very small meat establishment. Pennsylvania State University.

Food and Drug Administration. 2003a. 21CFR Part 173: Chemicals used in washing to assist in the peeling of fruits and vegetables. Available at: http://www.cfsan.fda.gov/-Ird/FCF173.html.

Food and Drug Administration. 2003b. 21 CFR Section 178.1010: Sanitizing solutions. Available at: http://www.cfsan.fda.gov/-Ird/FCF173.html.

Food and Drug Administration. 2018a. CFR-code of federal regulations title 21 Part 173. Available at: https://www.accessdata.fda.gov/scripts/cdrh/cfdocs/cfcfr/CFRSearch.cfm?fr=173.325.

Food and Drug Administration. 2018b. Code of federal regulations title 21 Part 101. Available at: https://www.accessdata.fda.gov/scripts/cdrh/cfdocs/cfcfr/cfrsearch.cfm?fr=101.100.

Food and Drug Administration. 2018c. Code of federal regulations title 21 Part 178. Available at: https://www.accessdata.fda.gov/scripts/cdrh/cfdocs/cfcfr/CFRSearch.cfm?CFRPart=178.

Food Safety and Inspection Service. 1996. Pathogen reduction; Hazard Analysis and Critical Control Point (HACCP) systems. *Federal Register.* 61:38806–38989.

Food Safety and Inspection Service. 2011. Shiga toxin-producing *Escherichia coli* in certain raw beef products. *Federal Register.* 76:58157–51865.

Food Safety and Inspection Service. 2016. New performance standards for *Salmonella* and *Campylobacter* in Not-Ready-to-Eat comminuted chicken and turkey products and raw chicken parts and changes to related agency verification procedures: response to comments and announcement of implementation schedule. *Federal Register.* 81:7285.

Food Standards Australia New Zealand. 2008. Australia New Zealand food standards code standard 1.3.3. Available at: http://www.foodstandards.gov.au/code/Documents/1.3.3%20Processing%20aids%20v157.pdf.

Grab, L. 2001. Methods of testing sanitizers and bacteriostatic substances. In S. S. Block (ed.), *Disinfection, Sterilization, and Preservation*, Lippincott Williams & Wilkins, Philadelphia, PA.

Hahn-Löbmann, S., Stephan, A., Schulz, S., Schneider, T., Shaverskyi, A., Tusé, D., Giritch, A., and Gleba, Y. 2019. Colicins and salmocins–new classes of plant-made non-antibiotic food antibacterials. *Frontiers Plant Sci.* 10:437.

Harris, L., Farber, J., Beuchat, L., Parish, M., Suslow, T., Garrett, E., and Busta, F. 2003. Outbreaks associated with fresh produce: incidence, growth, and survival of pathogens in fresh and fresh-cut produce. *Comp. Rev. Food Sci. Food Safety* 2:78–141.

Health Canada Bureau of Chemical Safety Food Directorate. 2018. Antimicrobial food processing aid uses for which Health Canada has Expressed No Objection.

Herdt, J., and Feng, H. 2009. Aqueous antimicrobial treatments to improve fresh and fresh-cut produce safety. *Microbial Safety of Fresh Produce*. 169.

Hodge, K. 2003. Salads still hot after all these years. *Fresh Cut Magazine* July: 22–24.

Kim, S. A., Park, S. H., Lee, S. I., Owens, C. M., and Ricke, S. C. 2017. Assessment of chicken carcass microbiome responses during processing in the presence of commercial antimicrobials using a next generation sequencing approach. *Sci. Rep.* 7:43354.

Lillard, H. 1979. Levels of chlorine and chlorine dioxide of equivalent bactericidal effect in poultry processing water. *J. Food Sci.* 44:1594–1597.

Magnuson, B., Munro, I., Abbot, P., Baldwin, N., Lopez-Garcia, R., Ly, K., McGirr, L., Roberts, A., and Socolovsky, S. 2013. Review of the regulation and safety assessment of food substances in various countries and jurisdictions. *Food Addit. Contam.: Part A* 30:1147–1220.

Mani-López, E., García, H., and López-Malo, A. 2012. Organic acids as antimicrobials to control *Salmonella* in meat and poultry products. *Food Res. Intl.* 45:713–721.

Mataragas, M., Skandamis, P., and Drosinos, E. 2008. Risk profiles of pork and poultry meat and risk ratings of various pathogen/product combinations. *Intl. J. Food Microbiol.* 126:1–12.

Mattia, A., Merker, R., Choudhuri, S., DiNovi, M., and Walker, R. 2006. Safety Evaluation of certain food additives: Peroxyacid antimicrobial solutions containing 1-hydroxyethylidene-1, 1-diphosphonic acid (HEDP) In. *WHO Food Additives Series*. 54:87.

Mead, G. 2004. Microbial hazards in production and processing. pp. 232–257. In G. Mead (ed.), *Poultry Meat Processing and Quality*, Elsevier.

Mitchell, M. 2003. The evaluation of vegetable treatment in maintaining the safety of produce used in manufacturing. *RFA (Refrigerated Food Association) Technical Report*. 2003.

Moore, A., Nannapaneni, R., Kiess, A., and Sharma, C. S. 2017. Evaluation of USDA approved antimicrobials on the reduction of *Salmonella* and *Campylobacter* in ground chicken frames and their effect on meat quality. *Poultry Sci.* 96:2385–2392.

Narvaez-Bravo, C., Miller, M. F., Jackson, T., Jackson, S., Rodas-Gonzalez, A., Pond, K., Echeverry, A., and Brashears, M. M. 2013. *Salmonella* and *Escherichia coli* O157:H7 prevalence in cattle and on carcasses in a vertically integrated feedlot and harvest plant in Mexico. *J. Food Prot.* 76:786–795.

Northcutt, J., and Jones, D. 2004. A survey of water use and common industry practices in commercial broiler processing facilities. *J. Appl. Poultry Res.* 13:48–54.

Nychas, G.-J. E., Skandamis, P. N., Tassou, C. C., and Koutsoumanis, K. P. 2008. Meat spoilage during distribution. *Meat Sci.* 78:77–89.

Oliveira, M., Viñas, I., Colàs, P., Anguera, M., Usall, J., and Abadias, M. 2014. Effectiveness of a bacteriophage in reducing *Listeria monocytogenes* on fresh-cut fruits and fruit juices. *Food Microbiol.* 38:137–142.

Otoni, C. G., Espitia, P. J., Avena-Bustillos, R. J., and McHugh, T. H. 2016. Trends in antimicrobial food packaging systems: Emitting sachets and absorbent pads. *Food Res. Intl.* 83:60–73.

Ramirez-Hernandez, A., Brashears, M. M., and Sanchez-Plata, M. X. 2017. Efficacy of lactic acid, lactic acid–acetic acid blends, and peracetic acid to reduce *Salmonella* on chicken parts under simulated commercial processing conditions. *J. Food Prot.* 81:17–24.

Ravishankar, S., Zhu, L., and Jaroni, D. 2010. Assessing the cross contamination and transfer rates of Salmonella enterica from chicken to lettuce under different food-handling scenarios. *Food Microbiol.* 27:791.

Saltmarsh, M., and Insall, L. 2013. Food additives and why they are used. pp. 1–13. In M. Saltmarsh (ed.), *Essential Guide to Food* Additives, Royal Society of Chemistry, Cambridge, UK.

Scott, B. R., Yang, X., Geornaras, I., Delmore, R. J., Woerner, D. R., Reagan, J. O., Morgan, J. B., and Belk, K. E. 2015. Antimicrobial efficacy of a sulfuric acid and sodium sulfate blend, peroxyacetic acid, and cetylpyridinium chloride against *Salmonella* on inoculated chicken wings. *J. Food Prot.* 78:1967–1972.

Sebranek, J., Jackson-Davis, A., Myers, K., and Lavieri, N. 2012. Beyond celery and starter culture: Advances in natural/organic curing processes in the United States. *Meat Sci.* 92:267–273.

Singh, M., Smith, J., and Bailey, M. 2015. Using natural antimicrobials to enhance the safety and quality of poultry. pp. 375–401. In T. M. Taylor (ed.), *Handbook of Natural Antimicrobials for Food Safety and Quality*, Woodhead Publishing, Cambridge, UK.

Sivapalasingam, S., Friedman, C. R., Cohen, L., and Tauxe, R. V. 2004. Fresh produce: a growing cause of outbreaks of foodborne illness in the United States, 1973 through 1997. *J. Food Prot.* 67:2342–2353.

Sloan, A. E. 2003. Going for the green: Super salads-consumer trends. *Food Technol.* 57:18–19.

Sohaib, M., Anjum, F. M., Arshad, M. S., and Rahman, U. U. 2016. Postharvest intervention technologies for safety enhancement of meat and meat based products; a critical review. *J. Food Sci. Technol.* 53:19–30.

Takkinen, J., Struelens, M., Niskanen, T., Makela, P., Rizzi, V., Caprioli, A., and Scheutz, F. 2011. Shiga toxin/verotoxin-producing *Escherichia coli* in humans, food and animals in the EU/EEA, with special reference to the German outbreak strain STEC O104, ECDC & EFSA, Stockholm, Sweden.

Theinsathid, P., Visessanguan, W., Kruenate, J., Kingcha, Y., and Keeratipibul, S. 2012. Antimicrobial activity of lauric arginate-coated polylactic acid films against *Listeria monocytogenes* and *Salmonella* Typhimurium on cooked sliced ham. *J. Food Sci.* 77:M142–M149.

US Department of Agriculture Food Safety Inspection Service. 2019. FSIS Directive 7120.1. Safe and suitable ingredients used in the production of meat, poultry, and egg products. Available at: https://www.fsis.usda.gov/wps/portal/fsis/topics/regulations/directives/7000-series/safe-suitable-ingredients-related-document.

20

Antimicrobial Delivery Systems

Adriano Brandelli, Cicero C. Pola, and Carmen L. Gomes

CONTENTS

20.1 Introduction

Microorganisms that cause economic losses associated with food spoilage, and the spread of microbial pathogens along the food chain, constitute a topic of utmost concern in the food sector. Hence, the guarantee of food quality and safety is a constant challenge to the food industry during either conventional food manufacturing operations or the development of novel foodstuffs and food-processing technologies. Despite innovative preservation technologies such as high hydrostatic pressure, ohmic heating, irradiation, and pulsed electric field that have been proposed and in some cases adopted by members of the food industry, thermal processing measures (e.g., pasteurization, retort-processing, hot-fill) have remained in widespread use as a major food stabilization method. In addition, chemical preservatives have been extensively used to control microbial growth in some food products. Furthermore, consumer interest in food quality and the demand for food products without extensive processing and/or chemical additives has continuously grown during the last decades. In this regard, the use of natural antimicrobial substances has gained importance as a food preservation technology. Some antimicrobial substances are already naturally present in the food, generally causing no toxic or undesirable effect on the consumer. Also, several natural antimicrobials have been characterized as safe to humans and could be incorporated as preservatives into food formulations. Some natural antimicrobials show effective inhibition of important food pathogens and spoilage microorganisms. However, undesirable interactions with food components and antimicrobial inactivation, once incorporated in the food matrix, have limited the application of some promising natural antimicrobials in food products. Consequently, encapsulation stands out as a means to overcome these hurdles. Different methods of encapsulation can be applied to produce delivery systems for antimicrobials into food products. Herein, we discuss in more detail the need for delivery systems, the different delivery systems available, current applications, and relevant properties to consider when using these systems in food products.

20.2 The Need for Delivery Systems

Bioactive compounds such as antimicrobials, antioxidants, and enzymes can be protected against harsh processing conditions by encapsulation, which may also improve their controlled-release behavior during storage, and consequently their functionality. Thus, different delivery systems for bioactive compounds, including antimicrobial substances, have been developed to improve microbiological food quality and safety (Blanco-Padilla et al., 2014; Brandelli & Taylor, 2015). Several antimicrobial delivery systems are based on the encapsulation of the active substance into nanostructures or microstructures, which can be incorporated in the food matrix or into packaging materials (Brandelli, Brum, & dos Santos, 2017a; Dasgupta et al., 2015). Nanoencapsulation offers a potential solution to protect the antimicrobial substance, increasing its stability and application efficiency.

Nanotechnology deals with the manipulation of materials, devices, and other structures in which at least one dimension is in the order of 1 to 100 nm (NNI, 2014). Thus, nanotechnology can be defined as a science, engineering, and technology conducted at the nanoscale, in general, when the material has a dimension below 100 nm. However, the nanoscale can be extended for up to 1000 nm in the pharmaceutical and biomedical fields, provided these materials have different physical, chemical, and/or biological properties to those presented by the same material at bulk state. A critical point is that the nanomaterial must perform its function at the nanoscale (Lee, Yun, & Park, 2015).

A diverse array of materials has been described as suitable for use in biological and food systems for the development of nano- and micro-scaled delivery technologies. The best delivery system will depend on the nature of the antimicrobial compound and the desired functionality in the candidate food product. Lipids, for example, can be used in formulations to develop nano-sized micelles, reverse micelles, liposomes, or solid lipid nanoparticles, delivering hydrophobic and/or hydrophilic payloads to the food product. Antimicrobial compounds can also be associated with metallic nanoparticles coated with polymeric films. Nanoparticles loading antimicrobials have been developed with both synthetic and natural biocompatible polymers. Among synthetic bio-based polymers, poly(ε-caprolactone) and poly(L-lactic acid) are among the most widely used, while natural polymers, such as polysaccharides like chitosan and alginate, and proteins such as gelatin and casein, have been used to develop nanoparticles (Lopes & Brandelli, 2018; Ranjan et al., 2014). In addition, some proteins and peptides can undergo self-assembly resulting in structures such as nanospheres or nanotubes (Brandelli, 2012).

Several nanostructures can be used to load antimicrobial compounds and later on applied into food systems (Figure 20.1). These nanostructures can be formed with different morphologies such as nanoparticles, nanofibers, nanotubes, and nanocomposite films that are formed by the incorporation of nanostructures into a polymeric matrix. Next, useful encapsulating materials and their combination to produce nanostructures, with emphasis on antimicrobial delivery systems, and their main properties and uses are presented.

20.3 Emulsion-Based Delivery Systems

Emulsions can be prepared in the form of oil drops in water (O/W) or water drops in oil (W/O), but in both cases, the drops need to be stabilized to avoid coalescence, the formation of large oil droplets in O/W emulsions stemming from rupture of lamellae of smaller droplets in close proximity. The stabilization

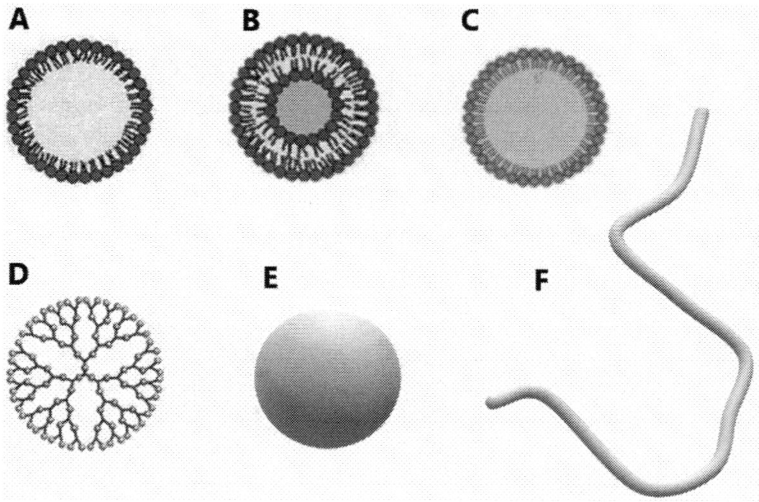

FIGURE 20.1 Schematic representation of the structural characteristics of more commonly used nanostructures in antimicrobial delivery. (A) Nano-emulsions, (B) nanoliposomes, (C) solid-lipid nanoparticles, (D) dendrimers, (E) polymeric nanoparticles, and (F) nanofibers.

of emulsion droplets is usually achieved by the addition of amphiphilic molecules such as emulsifiers or surfactants, which decrease the interfacial tension between the phases, increasing the steric hindrances and/or the electrostatic repulsion between the droplets. The term nano-emulsion denotes an emulsion that is either an O/W or W/O dispersion stabilized by an interfacial film of surfactant molecule, yielding droplet sizes ranging from 20 to 600 nm (Lovelyn & Attama, 2011). Two significant advantages of nano-emulsions when compared with conventional emulsions are the considerably smaller amount of surfactant required for their formation and the visual transparent aspect due to the small size of the droplets.

20.3.1 Preparation Methods

Generally, formulations used for nano-emulsions include the active substance, the additive, and the emulsifier. Nano-emulsion preparation can be achieved by high-energy and low-energy emulsification methods (Calligaris et al., 2016; Maali & Mosavian, 2013). The high-energy methods comprise high-pressure homogenization, high-energy stirring, ultrasonic emulsification, microfluidization, and membrane emulsification. The low-energy methods include phase inversion temperature, emulsion inversion point, and spontaneous emulsification. Meanwhile, the combination of high-energy and low-energy emulsification techniques makes possible the production of reverse nano-emulsions in a highly viscous system (Koroleva & Yurtov, 2012). Food-grade nano-emulsions can be prepared using the methods mentioned above.

The utilization of a high-pressure homogenizer/piston homogenizer is required for the preparation of nano-emulsions by the high-pressure homogenization method. High-pressure homogenizers force liquid samples through very narrow openings, achieving drop disruption via shear forces, pressure differentials, and cavitation. This technique often results in nano-emulsions of extremely small particle size (up to 10 nm) (Gupta et al., 2019). Ultrasonic emulsification is very efficient in reducing droplet size. In ultrasonic emulsification, the energy is supplied by a sonicator probe, which contains a piezoelectric quartz crystal that can expand and contract in response to an alternating electric voltage. As the tip of sonicator contacts the liquid, mechanical vibration is produced and cavitation occurs. Cavitation is the formation and collapse of vapor cavities in liquid (Bisten & Schuchmann, 2016). Thus, ultrasound can be directly used to produce emulsion; it is mainly used in laboratories where emulsion droplet size as low as 0.2 μm can be obtained. Microfluidization is a patented mixing technology, which makes use of a device called a microfluidizer. This device uses high pressure that forces the active substance through the interaction chamber, resulting in a very fine particle of sub-micron dimension. The process is repeated several times until the desired particle size to produce a uniform nano-emulsion is obtained (Jafari, He, & Bhandari, 2007).

Other methods of nano-emulsion formulation have been reported effective for the development of antimicrobial encapsulation systems. The phase inversion temperature method involves the change in phase by applying a higher temperature to a microemulsion. The spontaneous emulsification method involves three sequential steps. Initially, a homogeneous organic solution consisting of oil and a lipophilic surfactant is prepared in water-miscible solvent and hydrophilic surfactant. Then, the organic phase is injected into an aqueous phase under continuous magnetic stirring forming an O/W emulsion. Finally, the organic phase is removed by evaporation under reduced pressure (Fernandez, André, Rieger, & Kühnle, 2004).

The emulsification process is strongly influenced by the preparation method, resulting in very different droplet size distribution, which is strictly associated to the product stability. The phase inversion process leads to the formation of finely dispersed droplets in a continuous phase. Phase inversion refers to a phenomenon that occurs when an agitated O/W emulsion reverts to a W/O and vice versa. Emulsification via phase inversion has been largely used in the manufacture of pharmaceutical products (e.g., vesicles for drug delivery), foodstuff, and detergents (Preziosi, Perazzo, Caserta, Tomaiuolo, & Guido, 2013). A phenomenon always accompanying phase inversion is the spontaneous change in the surfactant arrangements at the oil–water interface.

Some important factors should be considered during nano-emulsion preparation. The surfactant must be carefully selected to guarantee that an ultra-low interfacial tension may be achieved, which is a primary requirement to produce nano-emulsions. In addition, surfactant concentration and bio-physical characteristics, such as critical micelle concentration, are crucial to stabilize the nano-droplets and produce the nano-emulsion.

20.3.2 Applications

Emulsion-based strategies have been used to incorporate plant-derived essential oils, and their components, into antimicrobial formulations. Emulsions and nano-emulsions containing clove, thyme, lemongrass, or palmarosa essential oils were characterized and their antimicrobial activity against *Escherichia coli* was evaluated *in vitro*. Sequential processing of blends by high-shear homogenization and microfluidization resulted in nano-emulsions with reduced droplet size. The type of essential oil used in the formulation significantly influenced the droplet size, size distribution, and viscosity of coarse emulsions. In comparison with conventional emulsions, nano-emulsions showed the strongest antimicrobial activity (Salvia-Trujillo, Rojas-Graü, Soliva-Fortuny, & Martín-Belloso, 2015).

The phenolic compound quercetin has been associated with multiple bio-active effects, including antioxidant and antimicrobial activities. Water-soluble formulations of quercetin have been developed by a pressurized ethyl acetate-water emulsion technique. Quercetin formulations prepared with the natural amphiphilic surfactant lecithin yielded the best results, with encapsulation efficiencies around 76% and droplet sizes in the range of nanometers. The solubility of emulsified quercetin in aqueous suspensions of emulsified quercetin was 315 times more soluble when compared with pure quercetin in water (Gonçalves et al., 2015). Curcumin is also difficult to incorporate into aqueous food systems because it is highly unstable and hydrophobic. For this reason, curcumin was encapsulated inside carrier oil in an emulsion form, reaching an encapsulation efficiency of 90%; the most stable preparation maintained its initial properties for up to 27 days of storage at 25°C (Sari et al., 2015). Reverse micelles prepared with vegetable oils were used to encapsulate the antimicrobial peptide nisin; antimicrobial activity was thereafter evaluated in lettuce leaves and minced meat. The nanostructured nisin presented an inhibitory effect on *Staphylococcus aureus* and *Listeria monocytogenes* cells in foods, and a synergistic antimicrobial effect between nisin and some of the nanocarrier ingredients was observed (Chatzidaki et al., 2018).

The essential oils of black cumin were encapsulated into a nano-emulsion system prepared with modified waxy maize starch, and the resulting antimicrobial activity was evaluated. The nano-emulsion containing the essential oil produced a lower minimum inhibitory concentration (MIC) value against *L. monocytogenes* as compared with the free essential oil (Sharif et al., 2017). In another study, nano-emulsions of clove bud and oregano essential oils were incorporated into water-soluble methylcellulose films and their antimicrobial activity was tested in sliced bread. The antimicrobial films containing emulsified essential oils reduced the development of yeasts and molds in sliced bread during 15 days (Otoni, Pontes, Medeiros, & Soares, 2014). The antilisterial activity of thymol nano-emulsions was evaluated in model milk samples, including skimmed milk, reduced-fat milk, and whole milk. In all cases, the nano-emulsions were more effective in reducing the viable counts of *L. monocytogenes* (Xue, Davidson, & Zhong, 2017). These studies suggest that nano-emulsions can be a good strategy to produce stable delivery systems for hydrophobic antimicrobials.

20.4 Lipid-Based Delivery Systems

Lipid-based delivery systems have demonstrated significant usefulness as drug delivery platforms in biomedical applications. Lipid-based delivery systems range from simple oil solutions to complex mixtures of lipids, surfactants, and co-surfactants. Two important lipid-based delivery systems widely studied as antimicrobial carriers are liposomes and solid lipid nanoparticles.

Liposomes are solid colloidal structures with an aqueous liquid core, formed by self-assembly of amphiphilic molecules, frequently phospholipids, in aqueous solution. The presence of both aqueous and lipid phases allows the entrapment of hydrophilic, hydrophobic, or amphiphilic molecules. Consequently, nanoliposomes have great potential for the delivery of antimicrobials, enzymes, vitamins, flavor, and bioactive peptides (Mozafari, Johnson, Hatziantoniou, & Demetzos, 2008). Liposomes are often prepared from phospholipids and are generally classified according to size and/or lamellar structure as follows: multilamellar vesicles (MLVs > 400 nm), large unilamellar vesicles (LUVs, 80 nm–1 μm), small unilamellar vesicles (SUVs, 20–80 nm), and giant unilamellar vesicles and multivesicular vesicles (GUVs and MVVs > 1 μm) (Jesorka & Orwar, 2008).

Solid lipid nanoparticles (SLNs) are another platform that can be useful for the distribution of antimicrobials. These delivery systems are generally in the range of 50 to 1000 nm, formed by solid lipids at room temperature and a surfactant for emulsification. The materials used in the formulations are solid lipids such as fatty acids (palmitic, decanoic, behenic acids), triglycerides (trilaurine, tripalmitin, trimiristine), sterols (cholesterol), monoglycerides (glyceryl monostearate), and waxes (cetyl palmitate). Various types of surfactants are used for emulsification, including lecithin, phosphatidylcholine, cholate, glycolate, and poloxamer 188 (Zhang, Pornpattananangkul, Hu, & Huang, 2010).

Nanostructured lipid carriers (NLCs) have been introduced as the next generation of the SLNs to overcome the potential limitations of SLNs. They are distinct from SLNs by the composition of the solid matrix, as the lipid phase in NLCs contains both solid and liquid lipids at ambient temperature. In fact, NLCs are modified cohorts of SLNs that present a mixture of solid and liquid phase (oil) resulting in an unstructured matrix, which improves the stability and loading capacity (Ganesan & Narayanasamy, 2017; Naseri, Valizadeh, & Zakeri-Milani, 2015). The general types of SLNs and NLCs are summarized in Figure 20.2.

Solid lipid nanoparticles have several unique characteristics that make them attractive as antimicrobial delivery systems. The excipients of SLNs are occlusive, meaning that a thin film is rapidly formed with the application on a surface, reducing the evaporation of water and retaining moisture from the surface. This property promotes an enhancement of the active substance penetration into the substrate because of the components of SLNs (Ghasemiyeh & Mohammadi-Samani, 2018). The production of SLNs is relatively simple and allows reproducible production on a large scale, without the need for organic solvents.

20.4.1 Preparation Methods

Most preparation methods of lipid-based delivery systems result in heterogeneous mixtures of lipid vesicles, including MLVs and LUVs. In general, more homogeneous mixtures can be obtained by using heat or the application of shear forces such as sonication or extrusion. The challenge is to achieve the formation of particles with adequate size, low dispersion, and good encapsulation efficiency (Patil & Jadhav, 2014).

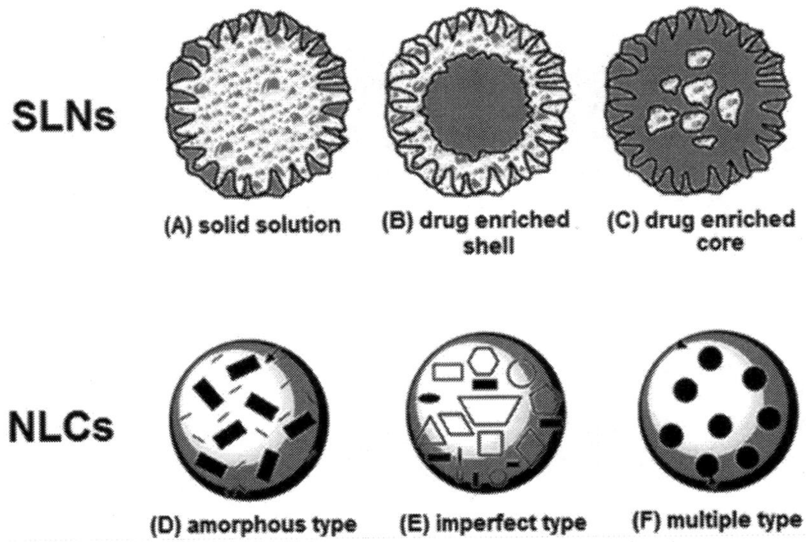

FIGURE 20.2 Types of solid-lipid nanoparticles (SLNs) and nanostructured lipid carriers (NLCs). Reproduced from Ganesan and Narayanasamy (2017), with permission from Elsevier.

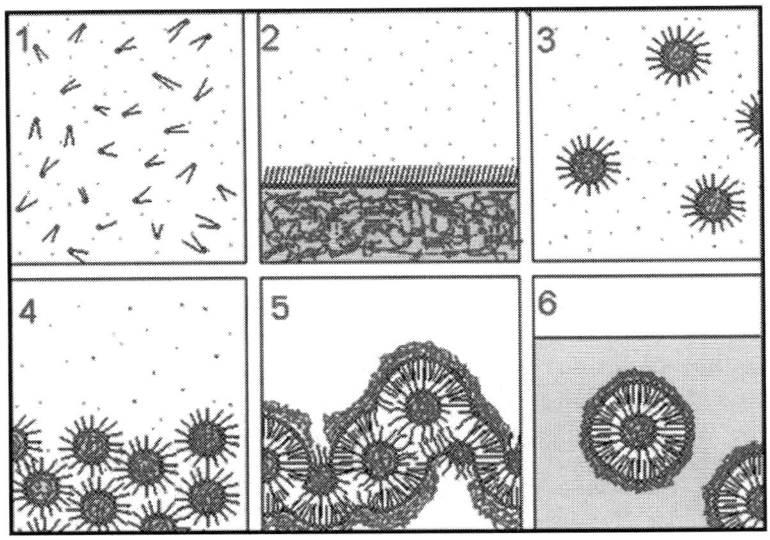

FIGURE 20.3 Liposome formation by the reversed-phase method. (1) Phosphatidylcholine is dissolved in an organic solvent. (2) A portion of aqueous solution containing chitosan is added. (3) An emulsion of reverse micelles containing chitosan is obtained under sonication. (4) The organic solvent is evaporated. (5) The organogel is obtained. (6) Vesicles containing chitosan are formed after water addition and shaking. Reproduced from Mertins, Lionzo, Micheletto, Pohlmann, and Silveira (2009), with permission from Elsevier.

The most used methods for antimicrobial encapsulation into liposomes are the film hydration method and the reverse micelle method. The film hydration method starts with a solution of the phospholipid in an organic solvent and after solvent evaporation, a thin lipid film is formed in the wall of the flask. An aqueous solution containing the antimicrobial substance is added under agitation, which often results in MLVs. Ultrasound, heat, or extrusion is then applied to obtain LUVs and then SUVs (Malheiros, Daroit, & Brandelli, 2010a).

On the other hand, the reverse micelle method includes the phospholipid dissolution in the organic solvent and the addition of the aqueous phase with the active substance (Figure 20.3). A W/O emulsion

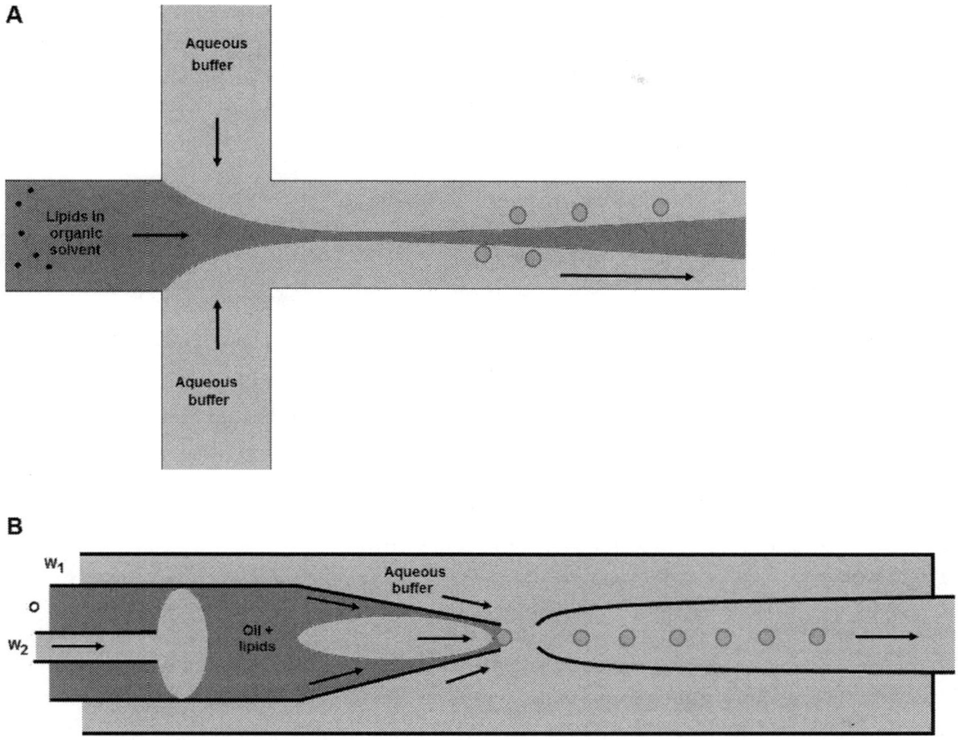

FIGURE 20.4 Production of nanometric liposomes using microfluidic approaches. (A) Schematic representation of microfluidic device for the production of liposomes of controlled size by hydrodynamic focusing. (B) Schematic representation of the microfluidic drop system for microfluidic fabrication of water-in-oil-in-water double emulsion drops with ultrathin shells.

is formed with the phospholipid at the interface. Next, the mixture is ultrasonicated and the reverse micelles are formed, where water droplets are coated by the phospholipid. After evaporation, many micelles collapse and a high-viscosity organogel is formed. In the final stage, the liposomes are formed by adding water and stirring (Mertins, Sebben, Pohlmann, & da Silveira, 2005).

More recently, the hydrodynamic focusing approach in microfluidic devices has been developed and applied for the production of liposomes (Patil & Jadhav, 2014). The formation of the liposome is carried out by the introduction of a channel containing the phospholipids dispersed in alcohol, generally ethanol or isopropanol, through a central channel. Then, the constriction of that channel occurs by two perpendicular water lines (Figure 20.4A). The hydrodynamic focus of the central channel causes controlled diffusion of the alcohol to the aqueous phase and vice versa. This results in a change of solubility and self-organization of the phospholipids into bilayers and finally into SUVs (Jahn, Vreeland, DeVoe, Locascio, & Gaitan, 2007; Zizzari et al., 2017). The production of liposomes by microfluidic drop systems has also been described (Figure 20.4B). Microfluidic devices are used to generate W/O/W (water-in-oil-in-water) double emulsions of sub-micron size. Phospholipids are solubilized in partially water-soluble organic solvents to accelerate the solvent evaporation process. With the evaporation of the intermediate phase, giant liposomes are formed (Trantidou, Friddin, Salehi-Reyhani, Ces, & Elani, 2018).

Typical methods of SLN preparation include high-pressure homogenization, atomization, mixing with high shear force, and ultrasound (Naseri et al., 2015). Spray drying (atomization) is an economical alternative to freeze-drying that is used to convert an aqueous SLN dispersion into a powdered product. This method causes particle aggregation due to the high temperature, shear forces, and partial melting of the particle during the process. Lipids with melting point(s) higher than 70°C are recommended for spray drying. The greatest advantage of high-speed stirring and sonication in SLN production is related to the easy access to the required equipment, which is common in many laboratories (Ganesan & Narayanasamy, 2017). The problem of these methods is the generation of broader particle size distributions, ranging into the micrometer range. Such variability results in physical instability, including particle aggregation upon storage. Potential metal contamination due to sonication is also a significant problem in ultrasound-based methods. To achieve stable formulations, high-speed stirring and ultrasound should be used in combination and performed at high temperatures (Zhang et al., 2010).

20.4.2 Applications

Liposomes have been employed for a long time as a drug delivery system, efficiently carrying antimicrobial substances (Brandelli, Pinilla, & Lopes, 2017c). The liposome encapsulation of antimicrobials intended for food use, such as nisin, lysozyme, curcumin, and essential oils, among others, has been described as well (Lopes & Brandelli, 2018). Nisin and other food-grade antimicrobials showing a broad antimicrobial activity may have impaired effectiveness in food systems due to undesirable interaction with some food components or enzyme inactivation (Malheiros et al., 2010a). Liposomes encapsulating natural antimicrobials have been shown to be active against food pathogens. For example, nisin has been successfully incorporated into liposomal systems prepared from different phospholipids, soy lecithin, and marine lecithin (Imran et al., 2015; Taylor, Gaysinsky, Davidson, Bruce, & Weiss, 2007). In addition, the bacteriocin pediocin PA-1/AcH, belonging to the Class II bacteriocins known as "antilisterial" peptides, was incorporated into phosphatidylcholine liposomes. The nanostructured pediocin showed antimicrobial activity against diverse *Listeria* spp. (de Mello et al., 2013).

Some examples of liposomes as antimicrobial carriers in food systems were previously described. Phosphatidylcholine/phosphatidylglycerol liposomes encapsulating nisin and EDTA were tested in liquid medium, resulting in a significant extension of the lag phases of *L. monocytogenes* and *E. coli* O157:H7. The bacteriostatic inhibition of these pathogens for at least 48 h was reported (Taylor, Bruce, Weiss, & Davidson, 2008). In another study, the lag phase of *L. monocytogenes* Scott A incubated at 5 or 20°C in fluid milk was significantly increased with the addition of encapsulated nisin (Schmidt, Holub, Sturino, & Taylor, 2009). Similar inhibition of *L. monocytogenes* was observed in whole or skimmed fluid milk at 7 and 30°C after the addition of liposome-encapsulated nisin (Malheiros, Daroit, da Silveira, & Brandelli, 2010b), garlic extract (Pinilla, Noreña, & Brandelli, 2017), or nisin/lysozyme mixtures (Lopes, Barreto Pinilla, & Brandelli, 2019). Phosphatidylcholine liposomes containing nisin were also able to control *L. monocytogenes* in cheese (Malheiros, Sant'Anna, Barbosa, Brandelli, & Franco, 2012).

Solid lipid nanoparticles (SLNs) are ideal for topical applications and commonly used in the administration of drugs that are extremely insoluble in water, to combat superficial fungal infections (Gupta & Vyas, 2012). In addition to topical applications, SLNs are also used as capsules for oral administration, for example, to combat infections of the gastrointestinal tract by *Pseudomonas* spp. This utility can be extended to deliver poorly water-soluble antimicrobials, such as essential oils, in food systems. SLNs are interesting systems for the delivery of hydrophobic bioactive compounds that can be prepared using food-grade lipids. In this regard, nisin-loaded SLNs were produced with Inwitor 900 (glyceryl monostearate) by three cycles of high-pressure homogenization and using poloxamer 188 and sodium deoxycholate as surfactant and co-surfactant, respectively (Prombutara, Kulwatthanasal, Supaka, Sramala, & Chareonpornwattana, 2012). These nanostructures promoted the controlled release of nisin, inhibiting *L. monocytogenes* for up to 20 days during *in vitro* release studies.

20.5 Polymer-Based Delivery Systems

Biocompatible and biodegradable polymers have been widely used for the controlled release of antimicrobial compounds. Such delivery systems can be developed in the form of nanoparticles or composite films and coatings. A variety of biodegradable polymers has been used to prepare delivery systems, including synthetic polymers such as poly(lactic acid) (PLA), poly(lactic-co-glycolic acid) (PLGA), poly(ε-caprolactone) (PCL), and natural polymers including proteins such as gelatin, casein, and whey proteins, or polysaccharides such as cellulose, starch, alginate, and chitosan (Kumari, Yadav, & Yadav, 2010).

Polymeric nanoparticles have some distinctive characteristics. They have structural stability and well-defined properties (such as size and zeta potential), which are achieved by precise control over the polymer size, surfactant, and solvent used during their synthesis. In addition, the surface of polymeric nanoparticles generally contains functional groups that can be modified with the addition of a drug or a ligand. This modification allows the development of targeting strategies by the attachment of specific affinity ligands such as antibodies and lectins at the particle's surface.

Dendrimers are nanoparticles defined as highly ordered globular branched macromolecules (Boas & Heegaard, 2004). The structure of a dendrimer consists of three regions: the nucleus (or focus), the dendritic structures (highly branched), and the external surface that contains functional groups. The first synthetic methods for branched amines were described in 1978, then the patents of highly branched poly-L-lysine dendrimers appeared, and in 1984 the synthesis of the first family of polyamidoamine (PAMAM) appeared (Abbasi et al., 2014). Dendrimers have several unique properties that produce nanoparticles for the diffusion of antimicrobials. The highly branched dendrimers provide an enormous surface to size ratio and, therefore, allow a high reactivity with microorganisms *in vivo*. The drug can be loaded in the hydrophobic structure or by electrostatic interaction with a multivalent hydrophilic surface. In addition, the dendrimer can be antimicrobial by itself, for example, when quaternary ammonium compounds are used as terminal functional groups, which are known for their capacity to disrupt bacterial membranes. PAMAM dendrimers are the most studied for the supply of antimicrobials, and some studies show that the antifungal and antibacterial activities of some existing chemotherapeutics can be improved by dendrimer encapsulation (Kalomiraki, Thermos, & Chaniotakis, 2015).

A polymeric film or coating can be defined as a thin, continuous layer of polymeric material formed or placed on food or a food component. Besides acting as protective barriers, films can be used as carriers of antimicrobial compounds, thus enhancing the shelf-life of the food product and potentially promoting health benefits (Brandelli et al., 2017a). Biopolymer-based films and coatings usually exhibit lower moisture barriers when compared to petroleum-based films due to their hydrophilicity. To improve moisture barrier characteristics, hydrophobic compounds such as essential oils can be included in the formulations prepared from hydrophilic materials, not only increasing the resistance to moisture, but also providing them with antimicrobial properties. Antimicrobial films and coatings made from polysaccharides have been demonstrated to improve the physicochemical and microbiological quality of fresh-cut fruits and vegetables (Cé, Noreña, & Brandelli, 2012; Ramos, Fernandes, Silva, Pintado, & Malcata, 2012).

Nanofiber mats have also been developed as antimicrobial delivery systems. Nanofibers are produced from polymers specifically treated to form filaments with a diameter in the nanometer range. Different materials can be used to produce nanofibers, including both synthetic and natural polymers. Nanofibers have great potential for the delivery of antimicrobials because they have a very high surface to volume ratio (Wei, 2012). Antimicrobial loading onto the nanofiber can occur by encapsulation during the electrospinning process or by simple physical adsorption onto the nanofiber surface, generally given by hydrogen and van der Waals bonds. Another possibility is the adsorption of nanoparticles loaded with the antimicrobial on the surface of the fibers. A layer-by-layer assembly method allows a few nanometers' coverage of nanofiber surface by the deposition of polyelectrolytes. Charged polysaccharides such as heparin or chitosan can be used (Brandelli, 2012).

20.5.1 Preparation Methods

Polymeric nanoparticles can be prepared using mainly two methods that consist of self-assembly and emulsification. Self-assembled nanoparticles formed by copolymers are organized into hydrophobic and hydrophilic segments. In general, the hydrophobic segment forms the core containing the drug and the hydrophilic segment provides protection for the core against degradation. The general preparation method is by solvent displacement. In that process, the encapsulating polymer and the drug payload are dissolved in a water-miscible solvent, such as acetonitrile or acetone, and then added to an aqueous suspension. As the solvent evaporates, the copolymers undergo a nanoprecipitation that traps the active substance (Vauthier & Bouchemal, 2009).

Another technique of polymeric nanoparticles preparation is emulsification-polymerization, which forms nano-capsules. In this method, the monomer is dissolved in the polymerization medium in the presence of a surfactant, polymerization initiators are added, and the drug can be adsorbed to the nanocapsules during the polymerization or covalently conjugated after displacement of the solvent. As the solvent evaporates, the copolymers undergo a nanoprecipitation that traps the active substance (Zhang et al., 2010). Other methods for preparation of polymeric nanoparticles are interfacial polymerization, salting-out, coacervation, solvent diffusion, and self-assembling macromolecules (Vauthier & Bouchemal, 2009; Yadav, Kumari, & Yadav, 2011).

Dendrimers can be chemically synthesized by two different methods, either divergent or convergent (Boas & Heegaard, 2004). In the divergent approach, the dendritic structure is synthesized from the core as the initial point and built up generation by generation. The alternative convergent method starts from the surface and terminates at the core, where the dendrimer segments are coupled together. In the convergent approach, only a small number of reactive sites are functionalized in each step, giving a smaller number of possible side-reactions per step (Figure 20.5).

Polymer-based films containing antimicrobials for food packaging applications can be produced by the casting method or by thermal processing. The production of thin polymer films by casting methods has been employed for a long time. This technique involves the distribution of a film-forming solution onto a uniform surface, followed by the solvent evaporation, and finally the formed film is peeled off the support surface. Casting methods have been widely employed to produce biopolymer-based antimicrobial packaging films, mostly using starch, chitosan, carboxymethyl cellulose, and gelatin as polymer matrices (Siemann, 2005). The production of packaging films by thermal processing, such as extrusion, melt blending, and injection molding, has been extensively described and can also be applied to incorporate antimicrobials into thermoplastic polymers. The thermal stability of the antimicrobial compound and compatibility with the polymer matrix are critical aspects to be considered in thermal processing. Since melting temperatures must be applied, these thermal processes use often use temperatures above 140°C, and therefore conventional petroleum-based thermoplastics, such as low-density polyethylene (LDPE), polypropylene (PP), and poly-lactic acid (PLA), are employed (Gavara, 2015). Both casting and thermal processes have been used to incorporate natural antimicrobials such as the bacteriocin nisin into packaging films (Imran et al., 2012; Zehetmeyer et al., 2016).

Several methods are described for the production of ultra-fine fibers, but the electrospinning method is the most effective and straightforward for producing nanofibers in large quantities. In this method, one electrode is located in the solution containing the polymer and the other is linked to a collector, which is generally a metal tube or plate. The charge forms a Taylor cone at the tip of the needle and with the acceleration of the polymer solution the solvent evaporates and forms the nanofiber (Wei, 2012). A

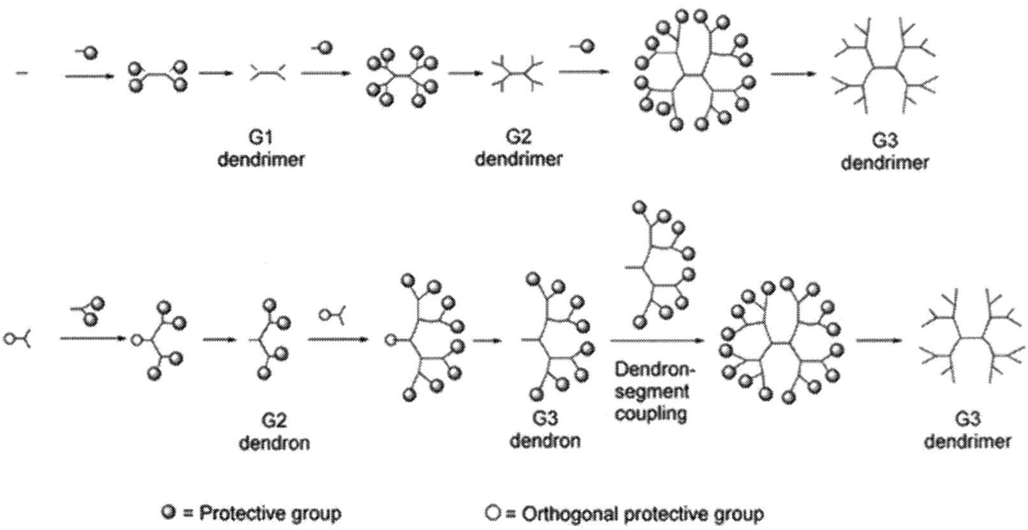

FIGURE 20.5 Schematic representation of dendrimer synthesis pathways. Top: divergent strategy. Bottom: convergent strategy. Reproduced from Boas and Heegaard (2004), with permission from Royal Society of Chemistry.

schematic representation of the electrospinning apparatus is shown in Figure 20.6. Coaxial electrospinning is used to produce nanofibers with a core-shell structure. Nanofibers produced by coaxial electrospinning that have been previously demonstrated include polyethylene oxide (shell) and polydodecyl thiophene (core).

Nanofibers formed by nano-emulsification are prepared from an aqueous phase containing the hydrophilic polymer mixed into an organic phase containing the hydrophobic polymer that makes the coating. This nano-emulsion is then subjected to electrospinning to produce the nanofibers (Xu et al., 2006). Poly-ethylene glycol-poly lactic acid (PEG-PLA) copolymer has been used previously in the fabrication of nanofibers via the nano-emulsion method (Heunis & Dicks, 2010). Another method is the co-electrospinning, where the active substances are present in the polymer suspension. An example is the conjugation of the antimicrobial peptide $(Ser\text{-}Glu\text{-}Glu)_3$ with polyethylene oxide. During the electrospinning, the peptide is polarized with orientation to the nanofiber surface (Yoo, Kim, & Park, 2009).

The quality and characteristics of the nanofiber depend on several factors, such as manufacture temperature, viscosity, surface tension of the suspension, and the intensity of the electric field. Thicker fibers are obtained by increasing the polymer concentration, for example, 4% polyvinylpyrrolidone (PVP) produces fibers of 20 nm and 8% PVP produces 50 nm nanofibers, while 10% PVP results in 300 nm. Also, the morphology of the nanofibers can change according to the concentration of the polymer. For instance, PLA nanofibers prepared from 1 or 3% suspensions can result in a smooth morphology or a bead-on-string shape, respectively (Heunis & Dicks, 2010).

Nanofibers are amenable to surface modification methods and consequently broadening their applications, including targeted drug delivery, also including delivery to food-contaminating microorganisms. Several methods to modify synthetic polymers are used to improve the diffusion capacity of drugs from nanofibers, such as wet chemical methods (Brandelli, Lopes, & Boelter, 2017b). For example, the NaOH treatment of PLA nanofibers increases the ability to bind calcium (for bone-regeneration purposes). The reaction of diamines with polyester nanofibers can increase drug-loading capacity through the adhesion of the charged surface. Plasma treatment is also a technique that has been used. For example, treatment with oxygen or air plasma can produce carboxyl and amino groups on their surfaces. The treatment of PCL nanofibers with argon or air plasma has been used to increase the number of carboxyl groups and shown to increase cell adhesion and proliferation (Yoo et al., 2009). The introduction of functional groups on the surface of nanofibers can also be done by graft polymerization (copolymer) (Yoo et al., 2009). In general, UV or plasma treatment is used to generate free radicals for polymerization.

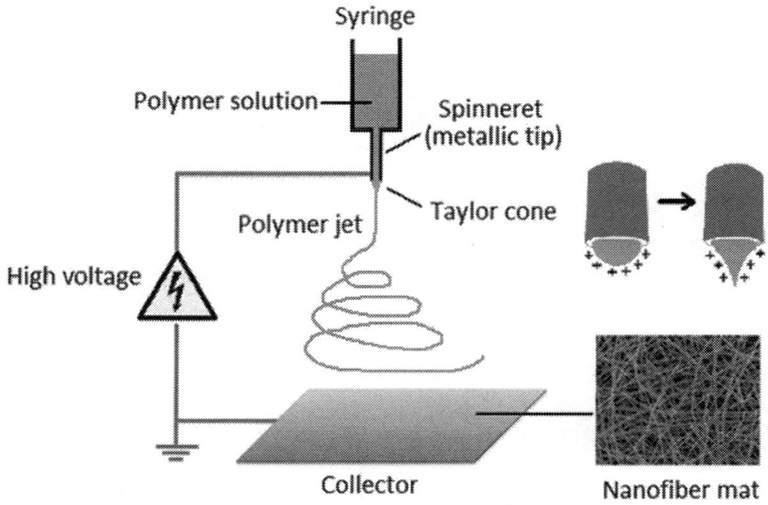

FIGURE 20.6 Schematic representation of an electrospinning device showing its component and the resulting nanofiber mat (representative SEM image) collected after the process. Reproduced from Brandelli, Lopes, and Boelter (2017b), with permission from Elsevier.

20.5.2 Applications

Polymeric nanoparticles encapsulating natural antimicrobials as delivery systems have been widely studied. The great interest in polymeric nanoparticles is associated with their advantages not only regarding their capacity to encapsulate a vast selection of hydrophobic and hydrophilic molecules, but also due to their ability to release them in a controlled or triggered fashion. Polymeric particles for delivery systems consist of a shell formed by the polymer chain, surrounding the entrapped core containing the bioactive compound, enhancing the stability of the latter (Borel & Sabliov, 2014). Furthermore, there is a variety of synthetic and natural polymeric materials available that are considered biodegradable, biocompatible, and non-toxic, which have been used to encapsulate different natural antimicrobials, including bacteriocins, enzymes, essential oils, and phenolic compounds. These delivery systems have been shown to improve antimicrobial activity of the entrapped compound against both Gram-positive and Gram-negative pathogens including *L. monocytogenes*, *E. coli*, and *Salmonella enterica* (Hill & Gomes, 2014; Hill, Taylor, & Gomes, 2013b; Ozdemir et al., 2018; Pereira et al., 2018; Pola et al., 2019; Silva, Hill, Figueiredo, & Gomes, 2014). Consequently, there is a plethora of polymeric material and antimicrobial compound combinations available to develop antimicrobial delivery systems for specific applications. Therefore, polymeric nanoparticle-based delivery systems development is expected to continue expanding in the future.

Polymer-based antimicrobial films and coatings have been extensively studied in the last decades, and several examples of applications have been described in the literature. These include the utilization of either conventional or natural polymers, and effective films have been produced by the incorporation of antimicrobials such as essential oils, phenolic compounds, bacteriocins, and lysozyme among others (Azeredo, 2013; Brandelli et al., 2017b; Brasil, Gomes, Puerta-Gomez, Castell-Perez, & Moreira, 2012; Mantilla, Castell-Perez, Gomes, & Moreira, 2013; Martiñon, Moreira, Castell-Perez, & Gomes, 2014; Rhim, Park, & Ha, 2013; Sipahi, Castell-Perez, Moreira, Gomes, & Castillo, 2013; Vasile, 2018).

Packaging materials having controlled-release properties for multiple active compounds, showing both antimicrobial and antioxidant properties, have been developed. Active films prepared with a blend of zein with fatty acids showed lower initial release rates for model bioactive compounds, namely (+)-catechin and lysozyme, than the zein-derived control films. Both catechin and lysozyme release rates were affected by modification of the fatty acid chain length, while the number of double bonds in the fatty acid influenced only the catechin release rate. The analysis of film morphology indicated that the controlled-release properties of the blend films was mostly related to the microspheres formed within their matrix and encapsulation of bioactive compounds (Arcan & Yemenicioğlu, 2014).

Cinnamaldehyde and cinnamon bark extract have been recognized by their strong antimicrobial activity against diverse foodborne pathogens, such as *S. aureus*, *E. coli*, *Bacillus cereus*, and *P. aeruginosa* (Gomes, Moreira, & Castell-Perez, 2011; Hill, Gomes, & Taylor, 2013a; Hill et al., 2013b; Loquercio, Castell-Perez, Gomes, & Moreira, 2015; Siddiqua, Anusha, Ashwini, & Negi, 2015). Different concentrations of this natural antimicrobial compound were incorporated in sorbitan monooleate/chitosan films. The obtained films ranged from 145 to 345 nm in thickness, and these structured chitosan/cinnamaldehyde materials exhibited the ability to control bacterial colonization on a variety of surfaces, including food process equipment surfaces. NMR analysis suggested that cinnamaldehyde was mostly physically incorporated into the hydrophilic chitosan films. Spin-coating can be used to incorporate and deliver high loads of cinnamaldehyde from ultrathin chitosan films (Rieger, Eagan, & Schiffman, 2015). These natural bioactive films show potential for use as bioactive materials in food packaging.

20.6 Inorganic Particles-Based Delivery Systems

Inorganic particles have been extensively investigated over the past few decades. Several metallic and metal oxide nanoparticles are recognized by their antimicrobial activities, but growing evidence indicates the potential of inorganic nanoparticles as effective carriers for multiple types of antimicrobial substances. The potential of silver nanoparticles as an important antimicrobial agent has been demonstrated by several studies, which showed their usefulness to combat different infectious diseases and can be used as novel nanomedicine (Brandelli, 2012; Rai, Yadav, & Gade, 2009). In addition, other metallic nanoparticles, including noble metal nanoparticles and metal oxide nanoparticles, have been also investigated for their antimicrobial potential and effectiveness as antimicrobial drug carriers (Rai, Ingle, Gupta, & Brandelli, 2015).

Inorganic clay nanoparticles are layered aluminum phyllosilicates containing neutral or negatively charged layers. Nanoclays are often employed in the manufacture of nanocomposite materials, functioning as reinforcement materials in polymeric films and improving their functional properties. Some active molecules with specific dimensions can be entrapped into montmorillonite nanosheets or halloysite nanotubes (Unalan, Cerri, Marcuzzo, Cozzolino, & Farris, 2014).

Aquasomes are described as a class of drug carriers based on three-layer organized nanostructures (Umashankar, Sachdeva, & Gulati, 2010). A ceramic core has the surface non-covalently modified by a carbohydrate, which is exposed to the adsorption of a hydrophilic drug. Disaccharides, such as trehalose and cellobiose, have been used for their protective properties against desiccation and to help maintain the structural integrity of the drug (Banerjee & Sen, 2018).

20.6.1 Preparation Methods

The preparation of inorganic nanoparticles is completed by either chemical, physical, or biological synthesis. Biological synthesis has gained space because it is associated with green synthesis in which plant extracts or microorganisms are used and, consequently, less toxic processes with minimum release of hazardous materials into the environment. The chemical reduction of metal ions is the most common and easy method for the preparation of the metallic nanoparticles. The chemical transformation of the metal ions into metal nanoparticles can be achieved using wet chemical synthesis, a photochemical process, by employing liquid crystal, polymer templates, or solution-based methodologies (Chouhan, 2018; Lee & Jun, 2019). Usually, the chemical synthesis process of metal nanoparticles in solution employs the following three main components: (a) metal precursors (often inorganic salts like $AgNO_3$ AgCl, $AgClO_4$), (b) reducing agents (e.g., $NaBH_4$, N,N-dimethylformamide, N_2H_4, formaldehyde), and (c) stabilizing/capping agents (e.g., citrate, 2-mercapto ethanol, polyols). The different reducing power of different reductants may have a significant influence in determining the final shape of nanoparticles (Chouhan, 2018).

The synthesis of metallic nanoparticles by green methods is carried out with a salt that has the metal and the plant extract or microorganism (Singh, Kim, Zhang, & Yang, 2016). In general, these methods generate heterogeneous nanoparticles with low yield. The optimization of parameters is done by the control of time, mixing rate, temperature, pH, and aeration. After optimization, stable and homogeneous nanoparticles with good performance are often obtained. By manipulating the parameters of the process, it is possible to control of shape and morphology of the nanoparticles (Figure 20.7). For example, pear extract results in triangular and hexagonal AuNPs at alkaline pH, but these do not form under acidic pH conditions (Ghodake, Deshpande, Lee, & Jin, 2010).

There are numerous synthesis methods reported that allow precise control of nanoparticle size (Singh et al., 2016), specially for the synthesis of Ag and AuNPs. For example, a fine control of the size of the nanoparticles was achieved through the control of pH, temperature, and concentration of Ag nitrate, citrate, and borohydride (Figure 20.8). The synthesis was carried out in two stages, the first reduction at 60°C and the second one at 90°C. With the adjustment of synthesis parameters homogeneous AgNPs ranging from 5 to 100 nm were achieved (Agnihotri, Mukherji, & Mukherji, 2014).

20.6.2 Applications

The potential applications of metal nanoparticles have been described in many areas, including the biomedical, pharmaceutical, agriculture, environmental, and chemical fields. They can be used as drug carriers, imaging devices, diagnostics, cosmetics, optical devices, and sensors. Silver nanoparticles are known for their antimicrobial activity and have been used in applications to eliminate microorganisms in air filters, water, and medicine. The Ag^+ ions are effective at millimolar concentrations while the Ag nanoparticles in the nanomolar range, showing that they are much more effective antimicrobials (Rai et al., 2009). In one antimicrobial application study, a combination of Ag nanoparticles and lysozyme

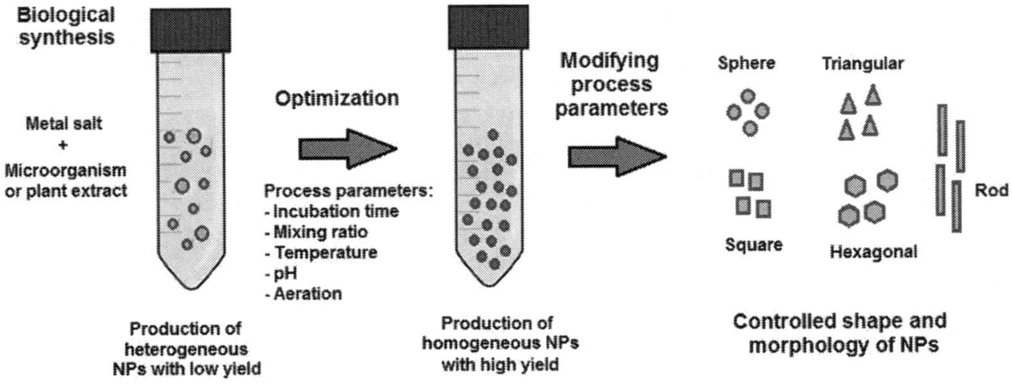

FIGURE 20.7 Different approaches available for the biological synthesis of metal nanoparticles.

was produced as a colloidal suspension to cover medical instrumentation. These nanocomposite materials showed antimicrobial activity against various bacteria, reducing the viability of *S. aureus*, *Bacillus anthracis*, *Klebsiella pneumoniae*, and *Acinetobacter baylyi* by ≥ 1.5 $\log_{10}$-cycles in up to 3 h (Eby,

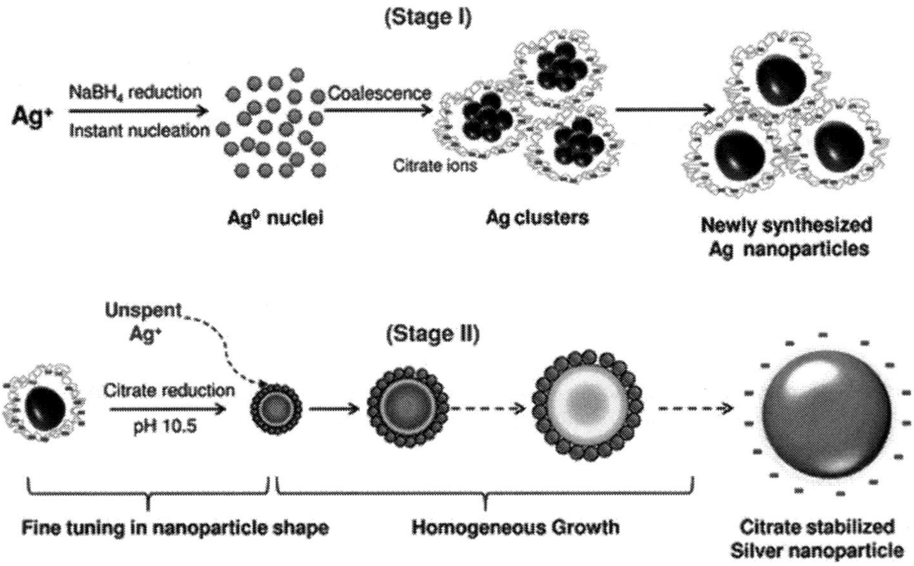

FIGURE 20.8 Schematic representation of size-controlled silver nanoparticles synthesized by employing the co-reduction approach. Reproduced from Agnihotri et al. (2014), under the terms of the Creative Commons Attribution 3.0 Unported License (http://creativecommons.org/licences/by/3.0/).

Luckarift, & Johnson, 2009). The modification of silver and other metallic nanoparticles with biopolymers may facilitate their use for different applications such as catalysis, optics, drug delivery, and antimicrobial activity. In particular, improved solubility and stability can be useful for antimicrobial delivery systems (Rai et al., 2015).

Gold nanoparticles (NPs) are recognized for their antimicrobial activity and as a drug carrier and diagnostic tools. Additionally, they are characterized by a high electron density and strong optical absorption (Daniel & Astruc, 2004). The development of gold NPs decorated with biomolecules has been conducted by surface modification. Peptides with cellular internalization signaling such as TAT (transactivator of transcription, rich in Lys and Arg) or NLS (nuclear localization sequence) have been adsorbed with success onto gold NPs. The interaction of biomolecules occurs by chemosorption, electrostatic interactions, and hydrophobic interactions. Gold NPs have a very high affinity for thiol groups, with covalent bond-type energy of approximately 200 kJ/mol (Sperling & Parak, 2010). Magnetic and metal oxide nanoparticles are also relevant as delivery systems and sensing tools. These nanoparticles have been used for targeted drug delivery and diagnostic methods. Magnetic NPs can be visualized by MRI, conducted by a magnetic field, and heated by a magnetic field for drug release (Arruebo, Fernández-Pacheco, Ibarra, & Santamaría, 2007).

Inorganic nanoparticles based on clay minerals or silica not only have been extensively employed as nanofillers in composite packaging, but also these materials could be used as carriers for antimicrobial substances. Montomorillonite and halloysite nanoclays have been successfully evaluated as carriers for the bacteriocins nisin and pediocin (Meira, Zehetmeyer, Werner, & Brandelli, 2017). In addition, nisin physically adsorbed or covalently attached to silica nanoparticles showed antimicrobial activity against *S. aureus* (Behzadi et al., 2018).

20.7 Important Properties for Antimicrobial Delivery Systems and Methods of Characterization

20.7.1 Antimicrobial Loading and Encapsulation Efficiency

The entrapment efficiency provides an indication of the amount of antimicrobial drug that is successfully entrapped/adsorbed into the nanoparticles. The loading capacity helps one to find the effective drug content of nanoparticles after their separation from the medium. The encapsulation efficiency and the actual loading capacity expressed in percentage are calculated according to Equations 20.1 and 20.2, respectively.

$$\text{Encapsulation Efficiency}\,(\%) = \frac{\text{Drug added} - \text{Free (untrapped) drug}}{\text{Drug added}} \tag{20.1}$$

$$\text{Loading capacity}\,(\%) = \frac{\text{Entrapped drug weight}}{\text{Nanoparticles weight}} \times 100 \tag{20.2}$$

For example, if the encapsulation efficiency is 40%, it means that 40% of the drug included in the formulation is entrapped into the nanoparticles. If the loading capacity is 40%, it means that 40% of the nanoparticles weight is composed of the drug, i.e., each 1 mg nanoparticles contains 0.4 mg drug. Consequently, the selection of an encapsulation method and the encapsulant materials are usually driven to achieve a high encapsulation efficiency in order to minimize antimicrobial drug and encapsulant materials wastage, while loading capacity can still be low as it is dependent on the molecular weight of the encapsulant materials (Tao, Hill, Peng, & Gomes, 2014).

The encapsulation efficiency of the delivery systems depends upon different factors associated with the process parameters of the preparation method (e.g., stirring rate, ultrasound energy, temperature, rate of solvent removal, drying process type) and the formulation, such as concentration of the components, solubility of polymer in solvent, drug–polymer interaction, and miscibility of components, among others (Hill et al., 2013b; Loquercio et al., 2015; Oliveira, Angonese, Ferreira, & Gomes, 2017a; Oliveira, Mezzomo, Gomes, & Ferreira, 2017b; Ozdemir et al., 2018; Pereira et al., 2015; Pereira et al., 2018; Pola et al., 2019). These parameters have been mostly studied for polymeric nanoparticles (Vauthier & Bouchemal, 2009). A low encapsulation efficiency has been related with slow solidification of particles, which can be associated with a high solubility of the polymer in the organic solvent, low solubility of the organic solvent in water, low concentration of the polymer, high dispersed phase to continuous phase (DP/CP) ratio, and/or slow solvent removal rate. By contrast, a high encapsulation efficiency is often associated with fast solidification of particles, related with low solubility of the polymer in the organic solvent, high solubility of the organic solvent in water, high concentration of the polymer, low DP/CP ratio, and fast solvent removal rate (Jyothi et al., 2010).

20.7.2 Size, Zeta Potential, and Morphology

The particle size is a key property of any delivery system since it is related to the surface to volume ratio and may influence the release kinetics of the payload release of the active compound. Several methods have been used for the determination of particle size in nanotechnology, including light scattering techniques and microscopy methods (Fairhurst & Weiner, 2018).

Particle size determination by dynamic light scattering (DLS) has been widely used. This method is based on measuring light scattering intensity fluctuations, which are time-related and occur around a mean intensity value, being caused by the particles that are moving in the suspension under the influence of Brownian motion. The intensity fluctuation recorded is temporally correlated with its own delayed value, resulting in a decaying intensity autocorrelation function from which the translational diffusion coefficient can be determined (Stetefeld, McKenna, & Patel, 2016). With the optical properties of the particles, equivalent volume- and number-weighted distributions, a hydrodynamic equivalent spherical diameter is obtained. As DLS is a widely applied method, commercial DLS instruments are relatively accessible at laboratories. This apparatus is equipped with a red light-emitting He–Ne laser, and the scattered light is collected by a photodiode detector located at a backward scattering angle of 173° (Ruf, 1993).

In particle-tracking analysis (PTA), also known as nanoparticle-tracking analysis (NTA), individual nanoparticles that are moving under Brownian motion in suspension are tracked. The sample is irradiated by a laser and a video of the light scattered by the particles is recorded by a light-sensitive camera through a magnifying objective. From this video, the Brownian motion of the particles is reconstructed and the translational diffusion constant, and lastly, the size is inferred from the particle paths. In order to distinguish the individual particles, the sample must be diluted to ensure the mean particle distance is larger than the diffraction limit of the microscope. Thus, the measurements are performed at a much higher dilution (around 1500×) than other methods (Filipe, Hawe, & Jiskoot, 2010; Gross, Sayle, Karow, Bakowsky, & Garidel, 2016).

The centrifugal liquid sedimentation (CLS) method measures the sedimentation time of nanoparticles under increased gravitational forces and uses the sedimentation time to estimate the supposed Stokes particle diameter. Before the measurements, a transparent rotating disc is partially filled with a sequence of liquids of decreasing density, creating a density gradient. Then, a few microliters of nanoparticle suspension are injected into the center of the rotating disc. If the nanoparticles have a higher density than the density gradient, they will sediment moving radially to the outer edge of the disc. The sedimentation time is monitored by a detector near the outer edge, which records the loss of light intensity of a laser beam passing through the disc (Langevin et al., 2018).

Small-angle X-ray scattering (SAXS) is a method that estimates the angular distribution of an X-ray beam scattered off the suspended particles in the forward direction under small angles. The scattering contrast is caused by electron density differences in the sample. The scattering data is given by the scattering intensity, $I(q)$, as a function of the momentum transfer by the Equation 20.3:

$$q(\theta)=\frac{4\pi}{\lambda}\sin\theta \tag{20.3}$$

where λ is the wavelength of the X-ray beam and θ is half of the scattering angle. In the case of monodispersed particle suspensions, the scattering curve $I(q)$ shows noticeable fluctuations, which depend on the particle diameter and can be evaluated by fitting the scattered intensity with a mathematic model. Consequently, this analysis can provide the size distribution of nanoparticle suspensions (Sakurai, 2017).

Information on the aggregation and physical state of the nanoparticles is relevant to define the most appropriate method for particle size determination. If the particles are suspended in a stable medium and have not formed aggregates, either DLS or transmission electron microscopy (TEM) would provide good single particle size information. If the nanoparticles exist as a dry powder and formed aggregates, then a reasonable single particle size measurement may be obtained either by TEM or by calculating an average particle diameter using a measured Brunauer–Emmett–Teller (BET) surface area (Fairhurst & Weiner, 2018). It is important to mention that the eventual difference between calculated BET surface area value and TEM results is due to a small amount of surface area being uncounted due to the particle aggregates. In the case of measuring aggregate particle size, a combination of microscopy and light scattering methods, either dynamic or static light scattering, would be appropriate. Since a significant amount of energy is required to disperse the aggregate nanoparticle materials, single particle size is not obtainable using a DLS technique (Ponce, Mejía-Rosales, & José-Yacamán, 2012).

Surface charge of the nanoparticles plays a key role in the preparation of a stable particle suspension as well as on the selection of substrates and surfaces for particle distribution (Guerrini, Alvarez-Puebla, & Pazos-Perez, 2018). Surface electrical properties of the particle and the substrate can be expressed by their zeta potential values. Thus, the measurement of zeta potential has become critical for delivery system research. However, the interpretation of nanoparticle electrophoretic mobility, especially that of ligand-coated particles, can be an intricate task. Despite the inherent complexity of the data, key concepts from colloidal science can help to refine valuable information from electrophoretic analysis. Zeta potential can be influenced by the ionic strength and pH value of the bulk solution, and by the functional groups associated with the particle surface (Clogston & Patri, 2011).

Besides the characterization of particle size, microscopy techniques may provide useful information about particle structure and morphology. Microscopic methods, both TEM and scanning electron microscopy (SEM), are possibly the most popular for nanoparticle imaging analysis. The similarities in operation and data output may suggest that these techniques are often treated as interchangeable, but in fact the applications to which they are best suited differ considerably, as do their respective magnification ranges and image resolution capabilities. Electron microscopy works by bombarding a sample with a stream of electrons and monitoring the resulting transmission (TEM) or scattering (SEM) effects (Egerton, 2005; Yao & Wang, 2005). These electrons are detected and converted into magnified images. This information is often processed by an image analysis software, generating particle size data for individual particles, size distributions for the entire dispersion, and several shape and morphological parameters (Woehrle, Hutchison, Özkar, & Finke, 2006). The main difference in data output between SEM and TEM is the way in which the images are resolved. While SEM produces three-dimensional images of particles, TEM produces two-dimensional images at higher magnifications that require further interpretation. Despite that images rendered by TEM are two-dimensional, this technique is capable of delivering much higher resolution (approximately three orders of magnitude higher). Surface and shape analysis can be better assessed by SEM, particularly in applications such as quality control of colloidal nanoparticles, or for evaluating surfaces and structural features of nano-sized powdered materials. As TEM can provide a higher resolution, it may be better suited for particle size analysis in the nano- to sub-nanoscale region (Yao & Wang, 2005).

Nanoparticle analysis can also be performed using atomic force microscopy (AFM), which is one of the most popular scanning probe microscopy methods. The operating principle of AFM is based on a functionally sharp tip (probe) placed at the end of a flexible cantilever beam that is brought into physical contact with the sample surface. The cantilever deflects in proportion to the force of interaction. A piezoelectric transducer allows positioning and scanning via the probe in three dimensions over the sample with very precise movements; a response system detects the interaction of the probe with the sample. Scanning across the surface, the probe follows the bumps and grooves formed by the atoms on the surface. A topography of the surface can be generated by monitoring the deflections of the flexible cantilever beam (Henry, 2005). The size of isolated nanoparticles of spherical shape deposited on a perfectly flat substrate can be determined from the AFM image by measuring the nanoparticle image height. This number is not biased by probe-sample convolution effects and can give precise nanoparticle size results. However, particle size analysis is not so straightforward if the nanoparticles are deposited on rough or curved surfaces, and therefore it may be difficult to implement particle analysis in AFM image-processing software. Another effect strongly influencing the AFM analysis of nanoparticles is particle agglomeration and self-ordering on the substrate. In addition, the interaction of nanoparticles with surfaces and the AFM probe can be evaluated through probe tip functionalization (Ong & Sokolov, 2007). Chemically functionalized tips gain selectivity and sensitivity to perform interactions at the molecular level. Besides, the magnitude of these interactions can be analyzed to allow the detailed mapping of surface chemistry at nanoscale resolution (Fritz, 2008; García & Pérez, 2002).

20.7.3 Controlled-/Triggered-Release Properties

Controlled release of antimicrobial substances can be achieved by their incorporation either in dissolved or dispersed form of the above-discussed delivery systems. The systems can be customized according to the desired final release profile and/or targeted organism(s) to deliver the antimicrobial at a predictable rate, in pulsatile manner, in the form of extended, triggered, and target-responsive forms.

The investigation of novel antimicrobial formulations and delivery systems must achieve the essential understanding of the drug release mechanism by applying transport- and time-dependent kinetic equations. Several products based on controlled release are available, mostly for clinical applications, but limited explanation of their releasing mechanisms is provided. Mechanisms of controlled drug release can be roughly classified based on the release of active agents from delivery systems, namely: diffusion, degradation, swelling followed by diffusion, and active efflux (Gomes et al., 2011; Hill et al., 2013a; Oliveira et al., 2017a).

However, *ad libitum* antimicrobial-releasing systems often exhibit a burst release followed by a long slow release with sub-lethal concentrations, which could lead to antimicrobial resistance in bacterial strains. To circumvent this, delivery systems that respond to an external stimulus such as pH, temperature, magnetic field, light, and electric pulses have been designed. For example, stimuli-responsive polymers could provide methods to trigger antimicrobial release when foods are stored at a temperature that promotes microbial growth, or spoilage occurs, providing a shift in pH that raises the likelihood of pathogen growth. Antimicrobials would be released when they are needed the most, rather than being metabolized or degraded before they are in contact with pathogenic bacteria. Dual-stimuli responsive delivery systems have been demonstrated that combine pH- and temperature-responsive polymers via crosslinking, for hydrophobic antimicrobial compounds delivery, improving their antimicrobial effects (Hill et al., 2013a). In this study, for the same pH and temperature conditions, the chitosan-co-PNIPAAM nanoparticles were significantly more potent bacterial inhibitors against both pathogens (*S. enterica* and *L. monocytogenes*) and also exhibited a faster cinnamon bark extract release over time as well as slightly higher entrapment efficiency (Hill et al., 2013a). A recent study developed and optimized pH-responsive nanoparticles based on PLGA-chitosan for triggered release of antimicrobials (Pola et al., 2019). Optimized nanoparticles characterization indicated a satisfactory TCIN encapsulation ($33.20 \pm 0.85\%$), spherical shape, pH-responsive controlled release, with faster release in the presence of CHIT at low pH, and enhanced antimicrobial activity against both pathogens (Pola et al., 2019) (Figure 20.9).

One attractive type of such delivery systems responds to the presence of adhering bacteria to release antimicrobial content. Recently, a series of such "self-defensive", bacterially triggered coatings has been developed, responding to bacterial presence only when and where needed (Cado et al., 2013; Komnatnyy, Chiang, Tolker-Nielsen, Givskov, & Nielsen, 2014; Pavlukhina et al., 2014; Zhuk et al., 2014). Both enzymes (Cado et al., 2013; Francesko et al., 2016), and acids (Andersen & von Meyenburg, 1980; Smith, 1991) excreted by bacteria have been used as triggers for antimicrobial release to combat adhering bacteria. In all cases, the pH-responsive "activation" of the delivery systems was clearly demonstrated when the coating was exposed to differing strains of *S. aureus* or *E. coli*, both known to acidify the medium in which they grow as a result of secretion of lactic and acetic acid, respectively (Andersen & von Meyenburg, 1980; Smith, 1991). Albright et al. (2017) recently demonstrated pH- and otherwise bacterially triggered polymer delivery systems that were directly linked to bacterial presence and pH activation through localized acidification. In this study, localized interfacial acidification induced by adhering bacteria was demonstrated and correlated with pH-triggered release of gentamicin and polymyxin B, by making use of antibiotic-loaded poly(metacrylic acid) (PMAA) hydrogel coatings (Albright et al., 2017).

All of these mechanisms employ the physical transformation of constituents of the system when they are placed into a biological environment. Although chemically driven drug delivery systems are feasible, they involve chemical modifications with active agents and carrier vehicles that need regulatory approval. In addition, adequate toxicology and safety evaluation including clinical trials are required before approval for final application. Therefore, simple delivery systems with approved active compounds and excipients are often utilized in the preparation of the controlled antimicrobial delivery systems used for food and biomedical applications.

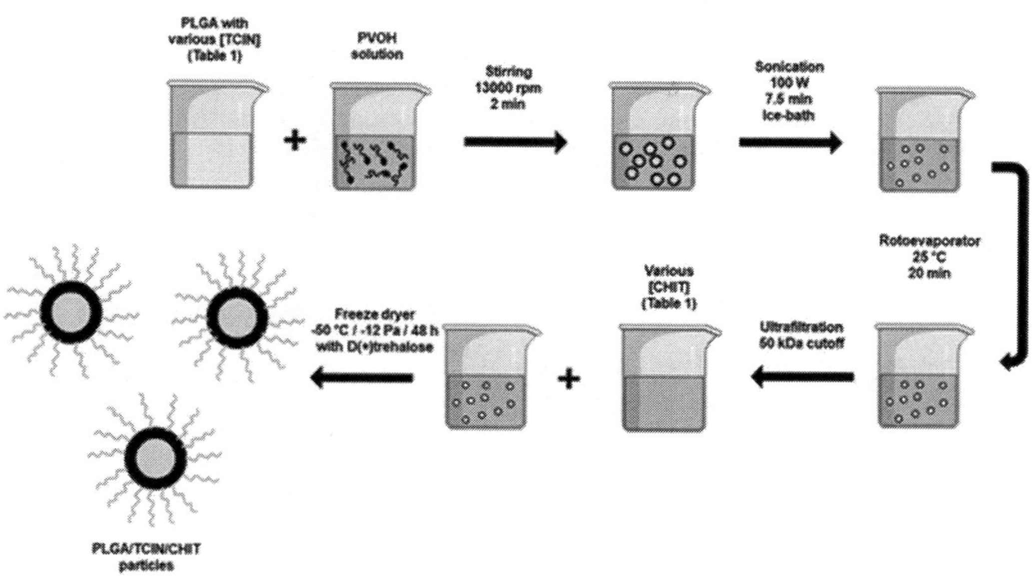

FIGURE 20.9 Schematic diagram of the PLGA/trans-cinnamaldehyde (TCIN)/chitosan (CHIT) nanoparticles development with pH-responsive behavior for the triggered release of antimicrobials. Reproduced from Pola et al. (2019), with permission from Elsevier.

The widespread development of controlled-release systems intended for numerous applications makes it quite complicated to classify and describe each system with its respective mechanism. Release kinetics for the majority of the antimicrobial delivery systems currently investigated can be explained by diffusion or empirical models (Benita, 1996). These systems can be classified as: (a) monolithic devices where the antimicrobial agent is dispersed in a polymer matrix and its release is controlled by diffusion along the matrix, and (b) reservoir systems (membrane-controlled devices), where the antimicrobial agent is contained in a core that is surrounded by a thin polymer membrane. Release to the surrounding environment occurs by diffusion through the rate-controlling membrane. Monolithic devices and reservoir systems have recently been employed and delivery systems based on this scheme have proven their efficacy in model foods (Huynh & Lee, 2015). There are also chemically controlled release approaches that include matrix erosion, combined erosion with diffusion, desorption of adsorbed drug, and drug covalently attached to a polymer. Moreover, externally controlled devices are also available that include magnetic microspheres, antibody recognition, and electrically powered microelectromechanical systems (MEMs) that are controlled via wireless communication to release the drug (Benita, 1996; Nisar, Afzulpurkar, Mahaisavariya, & Tuantranont, 2008; Razzacki, Thwar, Yang, Ugaz, & Burns, 2004; Tsai & Sue, 2007). Chemically and externally controlled-release approaches have been widely studied and applied in the biomedical field, while less explored in food safety applications, which could be associated with the inherent cost of these approaches.

20.7.4 Antimicrobial Delivery System Efficacy Evaluation

The initial *in vitro* confirmation of antimicrobial activity for an antimicrobial delivery system is often carried out by agar diffusion or broth dilution methods based on visualization of inhibition zones of microbial growth and optical density changes over time during incubation, respectively (Brandelli & Taylor, 2015). The main advantages of agar diffusion assays over other methods are the low cost, simplicity, facilitating the testing of large numbers of microorganisms and antimicrobial agents, and the easy interpretation of results (Balouiri, Sadiki, & Ibnsouda, 2016). The use of standard methods is recommended for a quantitative measurement of the inhibitory activity of antimicrobial agents. Many approved guidelines for dilution antimicrobial susceptibility testing of microorganisms are available, and the most recognized standards are supplied by the Clinical and Laboratory Standards Institute (CLSI, 2018) and the European Committee on Antimicrobial Susceptibility Testing (EUCAST, 2003).

Dilution methods are most appropriate for the determination of MIC values, since they offer the possibility to estimate the concentration of the tested antimicrobial agent in the agar (agar dilution) or broth medium (macro- or micro-dilution). The MIC is defined as the lowest concentration of an antimicrobial agent that completely inhibits growth of the microorganism in tubes or microdilution wells as detected by the absence of visible turbidity. The determination of minimum bactericidal concentration (MBC) or minimum fungicidal concentration (MFC) is the most common method for the estimation of bactericidal or fungicidal activity. The MBC is defined as the lowest concentration of antimicrobial agent needed to kill 99.9% of the final inoculum after incubation for 24 h under standardized conditions (Balouiri et al., 2016).

Time-kill testing is most suitable for determining the bactericidal effects of an antimicrobial agent. This test is a robust methodology for obtaining information about the dynamic interaction between the antimicrobial substance and the targeted microbial organism(s). Time-kill testing provides useful information about time- or concentration-dependent antimicrobial effects (Tsuji, Yang, Forrest, Kelchlin, & Smith, 2008). This assay is more laborious than MBC testing and is generally performed in research settings when a new antimicrobial is being evaluated, to confirm unusual MBC results, or as an effort to explain treatment failure when bactericidal action is mandatory (Brandelli & Taylor, 2015).

Methods based on colorimetric or fluorescent probes, associated with active microbial metabolism, are also used to test antimicrobial susceptibility. The tetrazolium/formazan (2,3,5-triphenyltetrazolium chloride, TTC) couple is a particular redox system acting as proton acceptor or oxidant. This redox system has been widely employed in different sectors of the biological sciences. In the presence of microorganisms, TTC is reduced to red formazan, which is directly proportional to the amount of viable active cells. Thus, the TTC method is considered as a comparatively fast method for evaluating the antibacterial activity of antimicrobial agents. In TTC assay, less than 12 h was required to reach susceptibility results and fewer bacterial counts are required (Moussa, Tayel, Al-Hassan, & Farouk, 2013). Similar tetrazolium salts such as MTT and XTT are widely used to estimate cell viability and proliferation in toxicity assays of nanoparticles. However, this method can be influenced by some conditions such as acidic pH and nanomaterials. Some nanoparticles can interfere in the results by inducing the formation of reactive oxygen species, which can lead to inaccurate evaluation (Wang, Yu, & Wickliffe, 2011).

Adenosine triphosphate (ATP) is a key energy-storing molecule in all living cells, and it is present in relatively constant amounts in a cell. Consequently, ATP quantification has been used to estimate the microbial population in a sample, and the ATP bioluminescence assay was developed to measure the ATP produced by bacteria or fungi. This assay is based on the luciferase-catalyzed reaction of luciferin and ATP, thus quantifying existing ATP, indicating the presence of metabolically active cells (Ivancic et al., 2008). In the presence of ATP, luciferase converts D-luciferin to oxyluciferin that generates light. The amount of the emitted light is measured by a luminometer and expressed as relative light unit (RLU), which can be converted into RLU per mole of ATP. Thus, a linear relationship between cell viability and measured luminescence can be obtained.

20.7.5 Comparison among Delivery Systems – Advantages/Disadvantages

A comparison among different delivery systems is important in order to provide information and assist the selection of antimicrobial delivery systems, and consequently for these systems to fully penetrate the food safety market and be used in industry (Hill et al., 2017). Table 20.1 shows a summary of advantages and disadvantages based on the important performance characteristics for delivery systems aforementioned, namely: encapsulation efficiency, loading capacity, size and size distribution, and controlled-/triggered-release ability, and challenges such as storage stability. Some important differences among delivery systems that are not summarized in Table 20.1 are cost, ease of application, i.e., embedded in packaging or as a food additive, and toxicity (these parameters vary too considerably to generalize in a table format). Each of these is critical for the adoption of a delivery system by the food industry.

TABLE 20.1

Summary of Advantages and Disadvantages Based on Key Performance Characteristics of Delivery Systems*

Delivery System	Advantages	Disadvantages
Nano-emulsion	• Versatile: hydrophilic, hydrophobic, amphiphilic molecules • Controlled release • Visually transparent • 20 to 500 nm diameter • Superior stability compared to emulsions • High surface area – improved delivery • Solubility of hydrophobic antimicrobial compounds • Reduced toxicity of encapsulated molecule • Transdermal, ocular, and parenteral administration • Ease of production process	• Stabilizers and surfactants are needed • Low kinetic/thermodynamic stability • Susceptibility to hydrolysis • Use of organic solvents • Organic solvents residues • Toxicity issues • Weak physical stability • Low resistance to changes in temperature, pH, or ionic strength
Liposomes	• Versatile: hydrophilic, hydrophobic, amphiphilic molecules • Can be tailored to site-specific delivery • Controlled release • Low toxicity • Reduce toxicity of encapsulated materials • Biocompatibility	• Polydispersed size distribution • High-cost production • Low loading capacity • Fast release rates • Low storage stability • Susceptible to oxidation • Limited solubility
Solid lipid nanoparticles	• Compatible with drugs that are extremely insoluble in water • Hydrophilic and lipophilic drugs • Ideal for topical and oral applications • Biocompatible • Organic-solvent free	• Limited loading capacity • Low entrapment efficiency • Premature release • Instability during storage • Potential toxicity for parenteral application
Polymer-based Dendrimers	• High level of control over size (usually ranging from 1.1 nm to 9 nm) • Hydrophilic and hydrophobic drugs • Controllable surface functionality • High surface/size ratio • Intracellular delivery capability • High reactivity • Can be modified to acquire targeted drug delivery • Charged end groups – polyvalence • Intrinsic antimicrobial activity	• Potential toxicity • Cytotoxicity • Hemolysis • Promote cell lysis • Limited efficiency with large molecules • Functionalization is required to reduced toxicity and increase entrapment efficiency

(Continued)

TABLE 20.1 (CONTINUED)

Summary of Advantages and Disadvantages Based on Key Performance Characteristics of Delivery Systems*

Delivery System	Advantages	Disadvantages
Nanofibers	• Use of both synthetic and natural polymers • High surface area • High porosity • Morphology can be controlled by the concentration of polymer • Amendable to surface modification methods • High mechanical strength • Cost-effective • Allows application of intrinsic antimicrobial polymers	• Depends on temperature, viscosity, surface tension, intensity of the electric field • Small particles can be released during degradation • Release of toxic byproducts during degradation • Limited application
Nanoparticles	• Great capacity of encapsulation of vast diversity of hydrophobic and hydrophilic molecules • High entrapment efficiency • Controlled-release and triggered-response capability • Long-lasting release • Cost-effective • Particle configuration increases the stability of the bioactive compound • Use of intrinsic antimicrobial polymers • Several production methods	• Use of organic solvents • Burst release • Polydispersed size distribution • Aggregation/flocculation problems • Susceptible to pH and temperature changes • Require purification steps • Batch-to-batch variation
Inorganic particles	• Intrinsic antimicrobial activity • Tunable to controlled release and triggered response • Control of nanoparticle size (usually between 5 and 100 nm) • High surface area • Easily functionalized • Stable (AuNP)	• Potential toxicity • Limited colloidal stability (AgNP) • Require photoactivation (TiO_2NP) • Limited drug loading • Require purification steps • Use of strong chemical reagents • Non-biodegradable

* Compiled from the following references: Abbasi et al., 2014; Borel & Sabliov, 2014; Claire et al., 2015; Kesharwani, Jain, & Jain, 2014; Kriegel, Kit, McClements, & Weiss, 2010; Kumari, Yadav, & Yadav, 2010; Liu & Fréchet, 1999; Lovelyn & Attama, 2011; Malheiros, Daroit, & Brandelli, 2010a; Miller, Wang, Benicewicz, & Decho, 2015; Mozafari, Johnson, Hatziantoniou, & Demetzos, 2008; Otoni, Pontes, Medeiros, & Soares, 2014; Rai, Ingle, Gupta, & Brandelli, 2015; Sherje, Jadhav, Dravyakar, & Kadam, 2018; Simões et al., 2017; Svenson, 2009; Weber, Zimmer, & Pardeike, 2014; Wei, 2012; Zhang, Pornpattananangkul, Hu, & Huang, 2010.

20.8 Conclusions and Future Trends

Delivery systems such as nano-emulsions, nanoliposomes, solid-liquid polymeric and inorganic nanoparticles, and nanofibers have been used to enhance the antimicrobial efficacy and utilization of natural antimicrobials in foods. Their ability to control-/trigger-release the antimicrobial compound and to reduce the interactions of the antimicrobial payload with food components in the food matrix provides great promise for food safety applications and their use in the replacement of synthetic preservatives. While there are many materials that can be used for the preparation of delivery systems in food matrices, the choice of an appropriate encapsulant and preparation method is driven by cost, availability of materials, and performance characteristics (including encapsulation efficiency, loading capacity, size and size distribution, controlled-/triggered-release ability, and storage stability). Future research should focus on developing delivery systems that are based on all-natural components (i.e., encapsulant and antimicrobial compounds) and also on exploring different preparation methods to optimize the antimicrobial activity of active compounds over the entire food product shelf-life. In addition, the discovery of new natural antimicrobial compounds and their combination that could be applied in foods and also understanding of their interaction with the food matrix would be valuable. Furthermore, the development of triggered mechanisms of antimicrobial release in food systems would enable enhancement of antimicrobial efficacy. This approach enables antimicrobial release-on-demand based on the presence of targeted microorganism(s) in a highly localized way, consequently reducing unwanted interaction between the antimicrobial and the food matrix, which dilutes antimicrobial concentration, and which may subsequently induce antimicrobial adaptation by the microbe. Finally, the development of effective antimicrobial delivery systems in combination with non-thermal and thermal food processing such as high pressure, plasma treatment, microwave, irradiation, high-temperature/short-time pasteurization, among others, to promote a synergistic response for microbial safety protection would benefit the food industry. There are many different needs in the food chain associated with food safety and food security. The scientific community continues to contribute and develop a large library of antimicrobial delivery systems that have the potential to be used to ensure food safety. These delivery systems must be beneficial to consumers and the food industry, while complying with regulatory agencies' requirements in order to be adopted. It is time to apply these different delivery system technologies to the benefit of society.

REFERENCES

Abbasi, E., Aval, S. F., Akbarzadeh, A., Milani, M., Nasrabadi, H. T., Joo, S. W., Hanifehpour, Y., Nejati-Koshki, K., and Pashaei-Asl, R. (2014). Dendrimers: Synthesis, applications, and properties. *Nanoscale Research Letters*, 9:247–247.

Agnihotri, S., Mukherji, S., and Mukherji, S. (2014). Size-controlled silver nanoparticles synthesized over the range 5–100 nm using the same protocol and their antibacterial efficacy. *RSC Advances*, 4:3974–3983.

Albright, V., Zhuk, I., Wang, Y., Selin, V., van de Belt-Gritter, B., Busscher, H. J., van der Mei, H. C., and Sukhishvili, S. A. (2017). Self-defensive antibiotic-loaded layer-by-layer coatings: Imaging of localized bacterial acidification and pH-triggering of antibiotic release. *Acta Biomaterialia*, 61:66–74.

Andersen, K. B., and von Meyenburg, K. (1980). Are growth rates of Escherichia coli in batch cultures limited by respiration? *Journal of bacteriology*, 144:114–123.

Arcan, I., and Yemenicioğlu, A. (2014). Controlled release properties of zein–fatty acid blend films for multiple bioactive compounds. *Journal of Agricultural and Food Chemistry*, 62:8238–8246.

Arruebo, M., Fernández-Pacheco, R., Ibarra, M. R., and Santamaría, J. (2007). Magnetic nanoparticles for drug delivery. *Nano Today*, 2:22–32.

Azeredo, H. M. C. (2013). Antimicrobial nanostructures in food packaging. *Trends in Food Science & Technology*, 30:56–69.

Balouiri, M., Sadiki, M., and Ibnsouda, S. K. (2016). Methods for in vitro evaluating antimicrobial activity: A review. *Journal of Pharmaceutical Analysis*, 6:71–79.

Banerjee, S., and Sen, K. K. (2018). Aquasomes: A novel nanoparticulate drug carrier. *Journal of Drug Delivery Science and Technology*, 43:446–452.

Behzadi, F., Darouie, S., Alavi, S. M., Shariati, P., Singh, G., Dolatshahi-Pirouz, A., and Arpanaei, A. (2018). Stability and antimicrobial activity of nisin-loaded mesoporous silica nanoparticles: A game-changer in the war against maleficent microbes. *Journal of Agricultural and Food Chemistry*, 66:4233–4243.

Benita, S. (1996). *Microencapsulation: Methods and Industrial Applications*, New York: Taylor & Francis.

Bisten, A., and Schuchmann, P. H. (2016). Optical measuring methods for the investigation of high-pressure homogenisation. *Processes*, 4.

Blanco-Padilla, A., Soto, K. M., Hernández-Iturriaga, M., and Mendoza, S. (2014). Food antimicrobials nanocarriers. *The Scientific World Journal*, 2014:11.

Boas, U., and Heegaard, P. M. H. (2004). Dendrimers in drug research. *Chemical Society Reviews*, 33:43–63.

Borel, T., and Sabliov, C. M. (2014). Nanodelivery of bioactive components for food applications: Types of delivery systems, properties, and their effect on ADME profiles and toxicity of nanoparticles. *Annual Review of Food Science and Technology*, 5:197–213.

Brandelli, A. (2012). Nanostructures as promising tools for delivery of antimicrobial peptides. *Mini-Reviews in Medicinal Chemistry*, 12:731–741.

Brandelli, A., Brum, L. F. W., and dos Santos, J. H. Z. (2017a). Nanostructured bioactive compounds for ecological food packaging. *Environmental Chemistry Letters*, 15:193–204.

Brandelli, A., Lopes, N. A., and Boelter, J. F. (2017b). Food applications of nanostructured antimicrobials. In A. M. Grumezescu (Ed.), *Food Preservation* (pp. 35–74). London: Academic Press.

Brandelli, A., Pinilla, C. M. B., and Lopes, N. A. (2017c). Nanoliposomes as a plataform for delivery of antimicrobials In M. Rai and C. A. Santos (Eds.), *Nanotechnology Applied to Pharmaceutical Technology* (pp. 55–90). Cham: Springer.

Brandelli, A., and Taylor, T. M. (2015). Nanostructured and nanoencapsulated natural antimicrobials for use in food products. In T. M. Taylor (Ed.), *Handbook of Natural Antimicrobials for Food Safety and Quality* (pp. 229–257). Oxford: Woodhead Publishing.

Brasil, I. M., Gomes, C., Puerta-Gomez, A., Castell-Perez, M. E., and Moreira, R. G. (2012). Polysaccharide-based multilayered antimicrobial edible coating enhances quality of fresh-cut papaya. *LWT - Food Science and Technology*, 47:39–45.

Cado, G., Aslam, R., Séon, L., Garnier, T., Fabre, R., Parat, A., Chassepot, A., Voegel, J.-C., Senger, B., Schneider, F., Frère, Y., Jierry, L., Schaaf, P., Kerdjoudj, H., Metz-Boutigue, M.-H., and Boulmedais, F. (2013). Self-defensive biomaterial coating against bacteria and yeasts: Polysaccharide multilayer film with embedded antimicrobial peptide. *Advanced Functional Materials*, 23:4801–4809.

Calligaris, S., Plazzotta, S., Bot, F., Grasselli, S., Malchiodi, A., and Anese, M. (2016). Nanoemulsion preparation by combining high pressure homogenization and high power ultrasound at low energy densities. *Food Research International*, 83:25–30.

Cé, N., Noreña, C. P. Z., and Brandelli, A. (2012). Antimicrobial activity of chitosan films containing nisin, peptide P34, and natamycin. *CyTA-Journal of Food*, 10:21–26.

Chatzidaki, M. D., Papadimitriou, K., Alexandraki, V., Balkiza, F., Georgalaki, M., Papadimitriou, V., Tsakalidou, E., and Xenakis, A. (2018). Reverse micelles as nanocarriers of nisin against foodborne pathogens. *Food Chemistry*, 255:97–103.

Chouhan, N. (2018). Silver nanoparticles: Synthesis, characterization and applications. In K. Maaz (Ed.), *Silver Nanoparticles: Fabrication, Characterization and Applications* (pp. 21–57). London: IntechOpen.

Claire, M., Wan Li, L., Abhishek, G., Mohd Cairul Iqbal Mohd, A., Iza, R., Stephen, T. B., Prem, R., and Ken, K. (2015). Strategies for antimicrobial drug delivery to biofilm. *Current Pharmaceutical Design*, 21:43–66.

Clogston, J. D., and Patri, A. K. (2011). Zeta potential measurement. In S. E. McNeil (Ed.), *Characterization of Nanoparticles Intended for Drug Delivery* (pp. 63–70). Totowa, NJ: Humana Press.

CLSI. (2018). *Methods for Dilution Antimicrobial Susceptibility Tests for Bacteria that Grow Aerobically CLSI Standard M07.* Wayne, PA: Clinical and Laboratory Standards Institute.

Daniel, M. C., and Astruc, D. (2004). Gold nanoparticles: Assembly, supramolecular chemistry, quantum-size-related properties, and applications toward biology catalysis, and nanotechnology. *Chemical Reviews*, 104:293–346.

Dasgupta, N., Ranjan, S., Mundekkad, D., Ramalingam, C., Shanker, R., and Kumar, A. (2015). Nanotechnology in agro-food: From field to plate. *Food Research International*, 69:381–400.

de Mello, M. B., da Silva Malheiros, P., Brandelli, A., Pesce da Silveira, N., Jantzen, M. M., and de Souza da Motta, A. (2013). Characterization and antilisterial effect of phosphatidylcholine nanovesicles containing the antimicrobial peptide pediocin. *Probiotics and Antimicrobial Proteins*, 5:43–50.

Eby, D. M., Luckarift, H. R., and Johnson, G. R. (2009). Hybrid antimicrobial enzyme and silver nanoparticle coatings for medical instruments. *ACS Applied Materials & Interfaces*, 1:1553–1560.

Egerton, R. F. (2005). *Prhysical Principles of Electron Microscopy*. New York: Springer.

EUCAST. (2003). Determination of minimum inhibitory concentrations (MICs) of antibacterial agents by broth dilution. *Clinical Microbiology and Infection*, 9:ix–xv. doi:10.1046/j.1469-0691.2003.00790.x

Fairhurst, D., and Weiner, B. (2018). *A Guide to Determination of Particle Size—Making an Effective and Reliable Measurement*. Available from: https://www.brookhaveninstruments.com/wp-content/uploads/2019/11/guide-determination-of-particle-size-brookhaven-instruments.pdf. Accessed on April 20, 2020.

Fernandez, P., André, V., Rieger, J., and Kühnle, A. (2004). Nano-emulsion formation by emulsion phase inversion. *Colloids and Surfaces A: Physicochemical and Engineering Aspects*, 251:53–58.

Filipe, V., Hawe, A., and Jiskoot, W. (2010). Critical evaluation of Nanoparticle Tracking Analysis (NTA) by NanoSight for the measurement of nanoparticles and protein aggregates. *Pharmaceutical Research*, 27:796–810.

Francesko, A., Fernandes, M. M., Ivanova, K., Amorim, S., Reis, R. L., Pashkuleva, I., Mendoza, E., Pfeifer, A., Heinze, T., and Tzanov, T. (2016). Bacteria-responsive multilayer coatings comprising polycationic nanospheres for bacteria biofilm prevention on urinary catheters. *Acta Biomaterialia*, 33:203–212.

Fritz, J. (2008). Cantilever biosensors. *Analyst*, 133:855–863.

Ganesan, P., and Narayanasamy, D. (2017). Lipid nanoparticles: Different preparation techniques, characterization, hurdles, and strategies for the production of solid lipid nanoparticles and nanostructured lipid carriers for oral drug delivery. *Sustainable Chemistry and Pharmacy*, 6:37–56.

García, R., and Pérez, R. (2002). Dynamic atomic force microscopy methods. *Surface Science Reports*, 47:197–301.

Gavara, R. (2015). *Practical Guide to Antimicrobial Food Packaging*. Shawbury: Smithers Rapra Technology.

Ghasemiyeh, P., and Mohammadi-Samani, S. (2018). Solid lipid nanoparticles and nanostructured lipid carriers as novel drug delivery systems: Applications, advantages and disadvantages. *Research in Pharmaceutical Sciences*, 13:288–303.

Ghodake, G. S., Deshpande, N. G., Lee, Y. P., and Jin, E. S. (2010). Pear fruit extract-assisted room-temperature biosynthesis of gold nanoplates. *Colloids and Surfaces B: Biointerfaces*, 75:584–589.

Gomes, C., Moreira, R. G., and Castell-Perez, E. (2011). Poly (DL-lactide-co-glycolide) (PLGA) nanoparticles with entrappedtrans-cinnamaldehyde and eugenol for antimicrobial delivery applications. *Journal of Food Science*, 76:N16–N24.

Gonçalves, V. S. S., Rodríguez-Rojo, S., De Paz, E., Mato, C., Martín, Á., and Cocero, M. J. (2015). Production of water soluble quercetin formulations by pressurized ethyl acetate-in-water emulsion technique using natural origin surfactants. *Food Hydrocolloids*, 51:295–304.

Gross, J., Sayle, S., Karow, A. R., Bakowsky, U., and Garidel, P. (2016). Nanoparticle tracking analysis of particle size and concentration detection in suspensions of polymer and protein samples: Influence of experimental and data evaluation parameters. *European Journal of Pharmaceutics and Biopharmaceutics*, 104:30–41.

Guerrini, L., Alvarez-Puebla, A. R., and Pazos-Perez, N. (2018). Surface modifications of nanoparticles for stability in biological fluids. *Materials*, 11.

Gupta, P. K., Bhandari, N., Shah, H. N., Khanchandani, V., Keerthana, R., Nagarajan, V., and Hiremath, L. (2019). An update on nanoemulsions using nanosized liquid in liquid colloidal systams. In K. S. Koh and V. L. Wong (Eds.), *Nanoemulsions: Properties, Fabrication and Applications*. London: IntechOpen.

Gupta, S. H., and Vyas, S. P. (2012). Development, characterization and in vivo assessment of effective lipidic nanoparticles for dermal delivery of fluconazole against cutaneous candidiasis. *Chemistry and Physics of Lipids*, 165:454–461.

Henry, C. R. (2005). Morphology of supported nanoparticles. *Progress in Surface Science*, 80:92–116.

Heunis, T. D. J., and Dicks, L. M. T. (2010). Nanofibers offer alternative ways to the treatment of skin infections. *Journal of Biomedicine & Biotechnology*, 2010:510682.

Hill, L. E., Gomes, C., and Taylor, T. M. (2013a). Characterization of beta-cyclodextrin inclusion complexes containing essential oils (trans-cinnamaldehyde, eugenol, cinnamon bark, and clove bud extracts) for antimicrobial delivery applications. *LWT - Food Science and Technology*, 51:86–93.

Hill, L. E., Taylor, T. M., and Gomes, C. (2013b). Antimicrobial efficacy of poly (DL-lactide-co-glycolide) (PLGA) nanoparticles with entrapped cinnamon bark extract against Listeria monocytogenes and Salmonella typhimurium. *Journal of Food Science*, 78:N626–N632.

Hill, L. E., and Gomes, C. L. (2014). Optimization of synthesis process of thermally-responsive poly-n-isopropylacrylamide nanoparticles for controlled release of antimicrobial hydrophobic compounds. *Materials Research Express*, 1:045404.

Hill, L. E., Oliveira, D. A., Hills, K., Giacobassi, C., Johnson, J., Summerlin, H., Taylor, T. M., and Gomes, C. L. (2017). A comparative study of natural antimicrobial delivery systems for microbial safety and quality of fresh-cut lettuce. *Journal of Food Science*, 82:1132–1141.

Huynh, C. T., and Lee, D. S. (2015). Controlled release. In S. Kobayashi and K. Müller (Eds.), *Encyclopedia of Polymeric Nanomaterials* (pp. 2672). Berlin: Springer.

Imran, M., Revol-Junelles, A.-M., Paris, C., Guedon, E., Linder, M., and Desobry, S. (2015). Liposomal nanodelivery systems using soy and marine lecithin to encapsulate food biopreservative nisin. *LWT - Food Science and Technology*, 62:341–349.

Imran, M., Revol-Junelles, A.-M., René, N., Jamshidian, M., Akhtar, M. J., Arab-Tehrany, E., Jacquot, M., and Desobry, S. (2012). Microstructure and physico-chemical evaluation of nano-emulsion-based antimicrobial peptides embedded in bioactive packaging films. *Food Hydrocolloids*, 29:407–419.

Ivancic, V., Mastali, M., Percy, N., Gornbein, J., Babbitt, J. T., Li, Y., Landaw, E. M., Bruckner, D. A., Churchill, B. M., and Haake, D. A. (2008). Rapid antimicrobial susceptibility determination of uropathogens in clinical urine specimens by use of ATP bioluminescence. *Journal of Clinical Microbiology*, 46:1213–1219.

Jafari, S. M., He, Y., and Bhandari, B. (2007). Optimization of nano-emulsions production by microfluidization. *European Food Research and Technology*, 225:733–741.

Jahn, A., Vreeland, W. N., DeVoe, D. L., Locascio, L. E., and Gaitan, M. (2007). Microfluidic directed formation of liposomes of controlled size. *Langmuir*, 23:6289–6293.

Jesorka, A., and Orwar, O. (2008). Liposomes: Technologies and analytical applications. *Annual Review of Analytical Chemistry*, 1:801–832.

Jyothi, N. V. N., Prasanna, P. M., Sakarkar, S. N., Prabha, K. S., Ramaiah, P. S., and Srawan, G. Y. (2010). Microencapsulation techniques, factors influencing encapsulation efficiency. *Journal of Microencapsulation*, 27:187–197.

Kalomiraki, M., Thermos, K., and Chaniotakis, N. A. (2015). Dendrimers as tunable vectors of drug delivery systems and biomedical and ocular applications. *International Journal of Nanomedicine*, 11:1–12.

Kesharwani, P., Jain, K., and Jain, N. K. (2014). Dendrimer as nanocarrier for drug delivery. *Progress in Polymer Science*, 39:268–307.

Komnatnyy, V. V., Chiang, W.-C., Tolker-Nielsen, T., Givskov, M., and Nielsen, T. E. (2014). Bacteria-triggered release of antimicrobial agents. *Angewandte Chemie International Edition*, 53:439–441.

Koroleva, M. Y., and Yurtov, E. V. (2012). Nanoemulsions: The properties, methods of preparation and promising applications. *Russian Chemical Reviews*, 81:21–43.

Kriegel, C., Kit, K. M., McClements, D. J., and Weiss, J. (2010). Nanofibers as carrier systems for antimicrobial microemulsions. II: Release characteristics and antimicrobial activity. *Journal of Applied Polymer Science*, 118:2859–2868.

Kumari, A., Yadav, S. K., and Yadav, S. C. (2010). Biodegradable polymeric nanoparticles based drug delivery systems. *Colloids and Surfaces B: Biointerfaces*, 75:1–18.

Langevin, D., Lozano, O., Salvati, A., Kestens, V., Monopoli, M., Raspaud, E., Mariot, S., Salonen, A., Thomas, S., Driessen, M., Haase, A., Nelissen, I., Smisdom, N., Pompa, P. P., Maiorano, G., Puntes, V., Puchowicz, D., Stępnik, M., Suárez, G., Riediker, M., Benetti, F., Mičetić, I., Venturini, M., Kreyling, W. G., van der Zande, M., Bouwmeester, H., Milani, S., Rädler, J. O., Mülhopt, S., Lynch, I., and Dawson, K. (2018). Inter-laboratory comparison of nanoparticle size measurements using dynamic light scattering and differential centrifugal sedimentation. *NanoImpact*, 10:97–107.

Lee, B. K., Yun, Y. H., and Park, K. (2015). Smart nanoparticles for drug delivery: Boundaries and opportunities. *Chemical Engineering Science*, 125:158–164.

Lee, S. M., and Jun, B.-H. (2019). Silver nanoparticles: Synthesis and application for nanomedicine. *International Journal of Molecular Sciences*, 20:865.

Liu, M., and Fréchet, J. M. J. (1999). Designing dendrimers for drug delivery. *Pharmaceutical Science & Technology Today*, 2:393–401.

Lopes, N. A., Barreto Pinilla, C. M., and Brandelli, A. (2019). Antimicrobial activity of lysozyme-nisin co-encapsulated in liposomes coated with polysaccharides. *Food Hydrocolloids*, 93:1–9.

Lopes, N. A., and Brandelli, A. (2018). Nanostructures for delivery of natural antimicrobials in food. *Critical Reviews in Food Science and Nutrition*, 58:2202–2212.

Loquercio, A., Castell-Perez, E., Gomes, C., and Moreira, R. G. (2015). Preparation of chitosan-alginate nanoparticles for trans-cinnamaldehyde entrapment. *Journal of Food Science*, 80:N2305–N2315.

Lovelyn, C., and Attama, A. A. (2011). Current state of nanoemulsions in drug delivery. *Journal of Biomaterials Nanobiotechnology*, 2:626–639.

Maali, A., and Mosavian, M. T. H. (2013). Preparation and application of nanoemulsions in the last decade (2000–2010). *Journal of Dispersion Science and Technology*, 34:92–105.

Malheiros, P. S., Daroit, D. J., and Brandelli, A. (2010a). Food applications of liposome-encapsulated antimicrobial peptides. *Trends in Food Science & Technology*, 21:284–292.

Malheiros, P. S., Daroit, D. J., da Silveira, N. P., and Brandelli, A. (2010b). Effect of nanovesicle-encapsulated nisin on growth of *Listeria monocytogenes* in milk. *Food Microbiology*, 27:175–178.

Malheiros, P. S., Sant'Anna, V., Barbosa, M. S., Brandelli, A., and Franco, B. D. G. d. M. (2012). Effect of liposome-encapsulated nisin and bacteriocin-like substance P34 on Listeria monocytogenes growth in Minas frescal cheese. *International Journal of Food Microbiology*, 156:272–277.

Mantilla, N., Castell-Perez, M. E., Gomes, C., and Moreira, R. G. (2013). Multilayered antimicrobial edible coating and its effect on quality and shelf-life of fresh-cut pineapple (Ananas comosus). *LWT - Food Science and Technology*, 51:37–43.

Martiñon, M. E., Moreira, R. G., Castell-Perez, M. E., and Gomes, C. (2014). Development of a multilayered antimicrobial edible coating for shelf-life extension of fresh-cut cantaloupe (Cucumis melo L.) stored at 4 °C. *LWT - Food Science and Technology*, 56:341–350.

Meira, S. M. M., Zehetmeyer, G., Werner, J. O., and Brandelli, A. (2017). A novel active packaging material based on starch-halloysite nanocomposites incorporating antimicrobial peptides. *Food Hydrocolloids*, 63:561–570.

Mertins, O., Sebben, M., Pohlmann, A. R., and da Silveira, N. P. (2005). Production of soybean phosphatidylcholine–chitosan nanovesicles by reverse phase evaporation: A step by step study. *Chemistry and Physics of Lipids*, 138:29–37.

Mertins, O.et al.. (2009). Chitosan effect on the mesophase behavior of phosphatidylcholine supramolecular systems. *Materials Science and Engineering*: C. 29. 463–469. 10.1016/j.msec.2008.08.038.

Miller, K. P., Wang, L., Benicewicz, B. C., and Decho, A. W. (2015). Inorganic nanoparticles engineered to attack bacteria. *Chemical Society Reviews*, 44:7787–7807.

Moussa, S. H., Tayel, A. A., Al-Hassan, A. A., and Farouk, A. (2013). Tetrazolium/formazan test as an efficient method to determine fungal chitosan antimicrobial activity. *Journal of Mycology*, 2013:7.

Mozafari, M. R., Johnson, C., Hatziantoniou, S., and Demetzos, C. (2008). Nanoliposomes and their applications in food nanotechnology. *Journal of Liposime Research*, 18:309–327.

Naseri, N., Valizadeh, H., and Zakeri-Milani, P. (2015). Solid lipid nanoparticles and nanostructured lipid carriers: Structure, preparation and application. *Advanced Pharmaceutical Bulletin*, 5:305–313.

Nisar, A., Afzulpurkar, N., Mahaisavariya, B., and Tuantranont, A. (2008). MEMS-based micropumps in drug delivery and biomedical applications. *Sensors and Actuators B: Chemical*, 130:917–942.

NNI. (2014). *National Nanotechnology Initiative*. National Nanotechnology Initiative Strategic Plan, Executive Office of the President of the United States.

Oliveira, D. A., Angonese, M., Ferreira, S. R. S., and Gomes, C. L. (2017a). Nanoencapsulation of passion fruit by-products extracts for enhanced antimicrobial activity. *Food and Bioproducts Processing*, 104:137–146.

Oliveira, D. A., Mezzomo, N., Gomes, C., and Ferreira, S. R. S. (2017b). Encapsulation of passion fruit seed oil by means of supercritical antisolvent process. *The Journal of Supercritical Fluids*, 129:96–105.

Ong, Q. K., and Sokolov, I. (2007). Attachment of nanoparticles to the AFM tips for direct measurements of interaction between a single nanoparticle and surfaces. *Journal of Colloid and Interface Science*, 310:385–390.

Otoni, C. G., Pontes, S. F. O., Medeiros, E. A. A., and Soares, N. d. F. F. (2014). Edible films from methylcellulose and nanoemulsions of clove bud (Syzygium aromaticum) and oregano (Origanum vulgare) essential oils as shelf life extenders for sliced bread. *Journal of Agricultural and Food Chemistry*, 62:5214–5219.

Ozdemir, N., Pola, C. C., Teixeira, B. N., Hill, L. E., Bayrak, A., and Gomes, C. L. (2018). Preparation of black pepper oleoresin inclusion complexes based on beta-cyclodextrin for antioxidant and antimicrobial delivery applications using kneading and freeze drying methods: A comparative study. *LWT*, 91:439–445.

Patil, Y. P., and Jadhav, S. (2014). Novel methods for liposome preparation. *Chemistry and Physics of Lipids*, 177:8–18.

Pavlukhina, S., Zhuk, I., Mentbayeva, A., Rautenberg, E., Chang, W., Yu, X., van de Belt-Gritter, B., Busscher, H. J., van der Mei, H. C., and Sukhishvili, S. A. (2014). Small-molecule-hosting nanocomposite films with multiple bacteria-triggered responses. *NPG Asia Materials*, 6:e121–e121.

Pereira, M. C., Hill, L. E., Zambiazi, R. C., Mertens-Talcott, S., Talcott, S., and Gomes, C. L. (2015). Nanoencapsulation of hydrophobic phytochemicals using poly (dl-lactide-co-glycolide) (PLGA) for antioxidant and antimicrobial delivery applications: Guabiroba fruit (Campomanesia xanthocarpa O. Berg) study. *LWT - Food Science and Technology*, 63:100–107.

Pereira, M. C., Oliveira, D. A., Hill, L. E., Zambiazi, R. C., Borges, C. D., Vizzotto, M., Mertens-Talcott, S., Talcott, S., and Gomes, C. L. (2018). Effect of nanoencapsulation using PLGA on antioxidant and antimicrobial activities of guabiroba fruit phenolic extract. *Food Chemistry*, 240:396–404.

Pinilla, C. M. B., Noreña, C. P. Z., and Brandelli, A. (2017). Development and characterization of phosphatidylcholine nanovesicles, containing garlic extract, with antilisterial activity in milk. *Food Chemistry*, 220:470–476.

Pola, C. C., Moraes, A. R. F., Medeiros, E. A. A., Teófilo, R. F., Soares, N. F. F., and Gomes, C. L. (2019). Development and optimization of pH-responsive PLGA-chitosan nanoparticles for triggered release of antimicrobials. *Food Chemistry*, 295:671–679.

Ponce, A., Mejía-Rosales, S., and José-Yacamán, M. (2012). Scanning transmission electron microscopy methods for the analysis of nanoparticles. In M. Soloviev (Ed.), *Nanoparticles in Biology and Medicine: Methods and Protocols* (pp. 453–471). Totowa, NJ: Humana Press.

Preziosi, V., Perazzo, A., Caserta, S., Tomaiuolo, G., and Guido, S. (2013). *Phase Inversion Emulsification* (Vol. 32).

Prombutara, P., Kulwatthanasal, Y., Supaka, N., Sramala, I., and Chareonpornwattana, S. (2012). Production of nisin-loaded solid lipid nanoparticles for sustained antimicrobial activity. *Food Control*, 24:184–190.

Rai, M., Ingle, A. P., Gupta, I., and Brandelli, A. (2015). Bioactivity of noble metal nanoparticles decorated with biopolymers and their application in drug delivery. *International Journal of Pharmaceutics*, 496:159–172.

Rai, M., Yadav, A., and Gade, A. (2009). Silver nanoparticles as a new generation of antimicrobials. *Biotechnology Advances*, 27:76–83.

Ramos, Ó. L., Fernandes, J. C., Silva, S. I., Pintado, M. E., and Malcata, F. X. (2012). Edible films and coatings from whey proteins: A review on formulation, and on mechanical and bioactive properties. *Critical Reviews in Food Science and Nutrition*, 52:533–552.

Ranjan, S., Dasgupta, N., Chakraborty, A. R., Melvin Samuel, S., Ramalingam, C., Shanker, R., and Kumar, A. (2014). Nanoscience and nanotechnologies in food industries: opportunities and research trends. *Journal of Nanoparticle Research*, 16:2464.

Razzacki, S. Z., Thwar, P. K., Yang, M., Ugaz, V. M., and Burns, M. A. (2004). Integrated microsystems for controlled drug delivery. *Advanced Drug Delivery Reviews*, 56:185–198.

Rhim, J.-W., Park, H.-M., and Ha, C.-S. (2013). Bio-nanocomposites for food packaging applications. *Progress in Polymer Science*, 38:1629–1652.

Rieger, K. A., Eagan, N. M., and Schiffman, J. D. (2015). Encapsulation of cinnamaldehyde into nanostructured chitosan films. *Journal of Applied Polymer Science*, 132.

Ruf, H. (1993). Data accuracy and resolution in particle sizing by dynamic light scattering. *Advances in Colloid and Interface Science*, 46:333–342.

Sakurai, S. (2017). SAXS evaluation of size distribution for nanoparticles. In A. E. Ares (Ed.), *X-Ray Scattering*. London: IntechOpen.

Salvia-Trujillo, L., Rojas-Graü, A., Soliva-Fortuny, R., and Martín-Belloso, O. (2015). Physicochemical characterization and antimicrobial activity of food-grade emulsions and nanoemulsions incorporating essential oils. *Food Hydrocolloids*, 43:547–556.

Sari, T. P., Mann, B., Kumar, R., Singh, R. R. B., Sharma, R., Bhardwaj, M., and Athira, S. (2015). Preparation and characterization of nanoemulsion encapsulating curcumin. *Food Hydrocolloids*, 43:540–546.

Schmidt, S. E., Holub, G., Sturino, J. M., and Taylor, T. M. (2009). Suppression of listeria monocytogenes Scott A in fluid milk by free and liposome-entrapped nisin. *Probiotics and Antimicrobial Proteins*, 1:152–158.

Sharif, H. R., Abbas, S., Majeed, H., Safdar, W., Shamoon, M., Khan, M. A., Shoaib, M., Raza, H., and Haider, J. (2017). Formulation, characterization and antimicrobial properties of black cumin essential oil nanoemulsions stabilized by OSA starch. *Journal of Food Science and Technology*, 54:3358–3365.

Sherje, A. P., Jadhav, M., Dravyakar, B. R., and Kadam, D. (2018). Dendrimers: A versatile nanocarrier for drug delivery and targeting. *International Journal of Pharmaceutics*, 548:707–720.

Siddiqua, S., Anusha, B. A., Ashwini, L. S., and Negi, P. S. (2015). Antibacterial activity of cinnamaldehyde and clove oil: Effect on selected foodborne pathogens in model food systems and watermelon juice. *Journal of Food Science and Technology*, 52:5834–5841.

Siemann, U. (2005). Solvent cast technology—A versatile tool for thin film production. In N. Stribeck and B. Smarsly (Eds.), *Scattering Methods and the Properties of Polymer Materials* (Vol. 130, pp. 1–14). Berlin: Springer.

Silva, L. M., Hill, L. E., Figueiredo, E., and Gomes, C. L. (2014). Delivery of phytochemicals of tropical fruit by-products using poly (dl-lactide-co-glycolide) (PLGA) nanoparticles: Synthesis, characterization, and antimicrobial activity. *Food Chemistry*, 165:362–370.

Simões, L. S., Madalena, D. A., Pinheiro, A. C., Teixeira, J. A., Vicente, A. A., and Ramos, Ó. L. (2017). Micro- and nano bio-based delivery systems for food applications: In vitro behavior. *Advances in Colloid and Interface Science*, 243:23–45.

Singh, P., Kim, Y.-J., Zhang, D., and Yang, D.-C. (2016). Biological synthesis of nanoparticles from plants and microorganisms. *Trends in Biotechnology*, 34:588–599. doi:10.1016/j.tibtech.2016.02.006

Sipahi, R. E., Castell-Perez, M. E., Moreira, R. G., Gomes, C., and Castillo, A. (2013). Improved multilayered antimicrobial alginate-based edible coating extends the shelf life of fresh-cut watermelon (Citrullus lanatus). *LWT - Food Science and Technology*, 51:9–15.

Smith, S. M. (1991). D-lactic acid production as a monitor of the effectiveness of antimicrobial agents. *Antimicrobial Agents and Chemotherapy*, 35:237.

Sperling, R. A., and Parak, W. J. (2010). Surface modification, functionalization and bioconjugation of colloidal inorganic nanoparticles. *Philosophical Transactions of the Royal Society A: Mathematical, Physical and Engineering Sciences*, 368:1333–1383.

Stetefeld, J., McKenna, S. A., and Patel, T. R. (2016). Dynamic light scattering: A practical guide and applications in biomedical sciences. *Biophysical Reviews*, 8:409–427.

Svenson, S. (2009). Dendrimers as versatile platform in drug delivery applications. *European Journal of Pharmaceutics and Biopharmaceutics*, 71:445–462.

Tao, F., Hill, L. E., Peng, Y., and Gomes, C. L. (2014). Synthesis and characterization of β-cyclodextrin inclusion complexes of thymol and thyme oil for antimicrobial delivery applications. *LWT - Food Science and Technology*, 59:247–255.

Taylor, T. M., Bruce, B. D., Weiss, J., and Davidson, P. M. (2008). *Listeria monocytogenes* and *Escherichia coli* O157:H7 inhibition *in vitro* by liposome-encapsulated nisin and ethylene diaminetetraacetic acid. *Journal of Food Safety*, 28:183–197.

Taylor, T. M., Gaysinsky, S., Davidson, P. M., Bruce, B. D., and Weiss, J. (2007). Characterization of Antimicrobial-bearing Liposomes by ζ-Potential, Vesicle Size, and Encapsulation Efficiency. *Food Biophysics*, 2:1–9.

Trantidou, T., Friddin, M. S., Salehi-Reyhani, A., Ces, O., and Elani, Y. (2018). Droplet microfluidics for the construction of compartmentalised model membranes. *Lab on a Chip*, 18:2488–2509.

Tsai, N.-C., and Sue, C.-Y. (2007). Review of MEMS-based drug delivery and dosing systems. *Sensors and Actuators A: Physical*, 134:555–564.

Tsuji, B. T., Yang, J. C., Forrest, A., Kelchlin, P. A., and Smith, P. F. (2008). In vitro pharmacodynamics of novel rifamycin ABI-0043 against Staphylococcus aureus. *Journal of Antimicrobial Chemotherapy*, 62:156–160.

Umashankar, M. S., Sachdeva, R. K., and Gulati, M. (2010). Aquasomes: A promising carrier for peptides and protein delivery. *Nanomedicine: Nanotechnology, Biology and Medicine*, 6:419–426.

Unalan, I. U., Cerri, G., Marcuzzo, E., Cozzolino, C. A., and Farris, S. (2014). Nanocomposite films and coatings using inorganic nanobuilding blocks (NBB): Current applications and future opportunities in the food packaging sector. *RSC Advances*, 4:29393–29428.

Vasile, C. (2018). Polymeric nanocomposites and nanocoatings for fFood packaging: A review. *Materials (Basel, Switzerland)*, 11:1834.

Vauthier, C., and Bouchemal, K. (2009). Methods for the preparation and manufacture of polymeric nanoparticles. *Pharmaceutical Research*, 26:1025–1058.

Wang, S., Yu, H., and Wickliffe, J. K. (2011). Limitation of the MTT and XTT assays for measuring cell viability due to superoxide formation induced by nano-scale TiO2. *Toxicology in Vitro*, 25:2147–2151.

Weber, S., Zimmer, A., and Pardeike, J. (2014). Solid Lipid Nanoparticles (SLN) and Nanostructured Lipid Carriers (NLC) for pulmonary application: A review of the state of the art. *European Journal of Pharmaceutics and Biopharmaceutics*, 86:7–22.

Wei, Q. (2012). *Functional Nanofibers and their Applications*. Oxford: Woodhead Publishing.

Woehrle, G., Hutchison, J. E., Özkar, S., and Finke, R. G. (2006). Analysis of nanoparticle transmission electron microscopy data using a public domain image-processing program, image. *Turkish Journal of Chemistry*, 30:1–13.

Xu, X., Zhuang, X., Chen, X., Wang, X., Yang, L., and Jing, X. (2006). Preparation of core-sheath composite nanofibers by emulsion electrospinning. *Macromolecular Rapid Communications*, 27:1637–1642.

Xue, J., Davidson, P. M., and Zhong, Q. (2017). Inhibition of Escherichia coli O157:H7 and Listeria monocytognes growth in milk and cantaloupe juice by thymol nanoemulsions prepared with gelatin and lecithin. *Food Control*, 73:1499–1506.

Yadav, S. C., Kumari, A., and Yadav, R. (2011). Development of peptide and protein nanotherapeutics by nanoencapsulation and nanobioconjugation. *Peptides*, 32:173–187.

Yao, N., and Wang, Z. L. (2005). *Handbook of Microcopy for Nanotechnology*. New York: Kluwer Academic Publishers.

Yoo, H. S., Kim, T. G., and Park, T. G. (2009). Surface-functionalized electrospun nanofibers for tissue engineering and drug delivery. *Advanced Drug Delivery Reviews*, 61:1033–1042.

Zehetmeyer, G., Meira, S. M. M., Scheibel, J. M., de Oliveira, R. V. B., Brandelli, A., and Soares, R. M. D. (2016). Influence of melt processing on biodegradable nisin-PBAT films intended for active food packaging applications. *Journal of Applied Polymer Science,* 133.

Zhang, L., Pornpattananangkul, D., Hu, C. M. J., and Huang, C. M. (2010). Development of nanoparticles for antimicrobial drug delivery. *Current Medicinal Chemistry*, 17:585–594.

Zhuk, I., Jariwala, F., Attygalle, A. B., Wu, Y., Libera, M. R., and Sukhishvili, S. A. (2014). Self-defensive layer-by-layer films with bacteria-triggered antibiotic release. *ACS Nano*, 8:7733–7745.

Zizzari, A., Bianco, M., Carbone, L., Perrone, E., Amato, F., Maruccio, G., Rendina, F., and Arima, V. (2017). Continuous-flow production of injectable liposomes via a microfluidic approach. *Materials (Basel, Switzerland)*, 10:1411.

21

Hurdle Technology – or Is It? Multifactorial Food Preservation for the Twenty-First Century

J. David Legan and Jairus R. D. David

CONTENTS

21.1 Introduction

For 30 years or more, consumers have demanded foods that are less severely processed, free from "chemical preservatives" (represented in Europe as absence of "E" numbers), formulated with "natural" ingredients to be more natural and healthier, and still convenient. While it is not always possible to deliver on those expectations, even attempting to do so requires detailed knowledge of food and ingredient technology and a sound understanding of food preservation.

Food preservation is conceptually very simple: we must prevent microbes from growing to the point where they make the food unsafe or unpalatable. In principle, we can do this by killing any microbes and preventing them from re-contaminating the food or by reducing their growth rate to the point that they are not able to achieve harmful concentrations during the time for which we want the food to be palatable. At the same time, we must prevent chemistry and physics from reducing the food quality to an unacceptable extent. We have many more tools available for controlling the microbiology than we do for controlling the chemistry and physics, and in this chapter, we will explore, with examples, the benefits of combining multiple tools. But first, let us briefly review how food science arrived at the idea of "hurdle technology."

Traditional methods of food preservation, including drying, salting, smoking, fermenting to create acid or alcohol, etc., presumably "evolved" through an extended process of empirical observation and modification before the existence of microorganisms was even known. "Modern" methods of food preservation began with Nicolas Appert's invention of the extended preservation of food by heating it in a sealed glass jar (Garcia and Adrian, 2009) and developed into the technology that today we recognize as canning. Later, Louis Pasteur's discovery that wine could be prevented from spoilage by mild heating (Vallery-Radot, 1919). This provided an example of a milder preservation process that would come to be recognized by the name "pasteurization." The advent of mechanical refrigeration brought us reliable chilling and freezing, and the power of chemistry also brought synthetic organic acid preservatives at much lower cost than the same compounds produced naturally. As these technologies advanced through scientific research, the emphasis was naturally focused on each single innovation. Only a few of these technologies, such as thermal-sterilization, thermal-pasteurization, and irradiation (and more recently high-pressure process (HPP)-pasteurization), act by *inactivating* microorganisms. The other processes such as chilling, freezing, lowering a_w, lowering pH, modified-atmosphere packaging, and the addition of antimicrobial preservatives act via *inhibiting* microorganisms to varying degrees. And as a final part of the preservation continuum, package integrity and hermetic seal restrict the access of microorganisms into the product.

It is now well-known that use of these inhibitory factors in combination even at sub-lethal levels, and in lieu of extreme use of any single treatment, leads to improvements in desirable safety, quality, and nutritional attributes. This knowledge is critical to creating "premium" versions of existing products and developing new products, while, simultaneously, optimizing for microbial stability and food safety. This realization that multiple factors contribute to the preservation of a single food, and the association of that realization with the description of each individual factor as a "hurdle" can be attributed to Leistner and Rödel (1976) at a symposium on Intermediate Moisture Foods (IMF) in Weybridge, UK. These authors stated that

> The inhibition of micro-organisms in IMF does not solely depend on the a_w, but the pH, Eh, t-value, F-value, preservative and the competitive microflora might be of importance too. How many of these "hurdles" and at what level they are needed for a sufficient inhibition of microorganisms during the desired storage period ... depends not only on the types but also much on the number of organisms present.

The hurdle concept has been reviewed many times, from multiple angles (e.g. Khan et al., 2017; Leistner, 2000; Leistner and Gorris, 1995; Rahman, 2015; Singh and Shalini, 2016) as well as in earlier editions of this book, and the reader can find extensive information on multifactorial preservation systems in the research literature.

Over time, the literature has more and more come to describe preservation based on multiple factors as "hurdle technology" (Leistner and Gorris, 1995; Chawla and Chander, 2004; Gupta et al., 2012; García-García et al., 2015; Moreira et al., 2017). As a pair of industry practitioners, the authors would argue that each "hurdle" may be the embodiment of a different technology and that the true power of the hurdle concept is in its call to understand all of the elements contributing to the preservation of a particular food, their relative contributions, and any interactions between them. This understanding can help us to develop preservation systems with the minimum impact on food quality for a particular amount of "preserving power." The products developed using this principle may often be considered "minimally-processed products," and hitherto have been considered "hurdle technology products." In this review we have chosen to explicitly identify them as products of multifactorial preservation to recognize that they depend on a collection of technologies, and to make the connection with modern design principles, such as the use of multifactorial models, more intuitive.

21.2 Benefits of the Hurdle Concept

The first benefit of the hurdle concept is simply that it requires us to understand all of the different factors contributing to the preservation of a particular product. When we do that, we realize that multiple

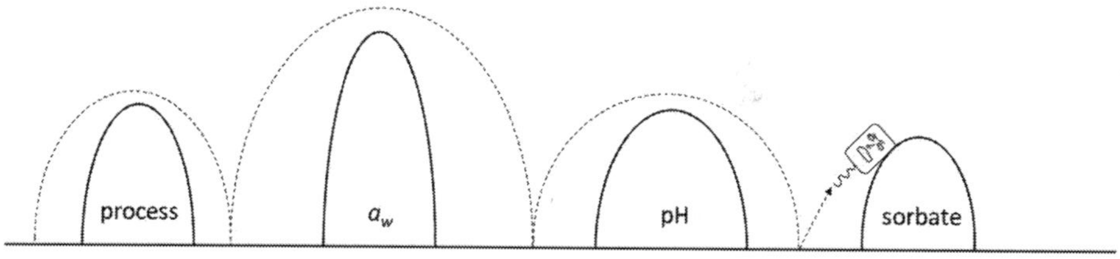

FIGURE 21.1 "Traditional" illustration of the hurdles concept. The figure could be extended with additional hurdles or other hurdles could be substituted for those illustrated. (Modified from Scott, 1989.)

factors are involved in the preservation of most foods, even when we think primarily of a single factor as responsible for preservation. For example, with commercially sterile low-acid canned foods processed by heat we think of the thermal death profile as being the preserving factor, but preventing recontamination, and the climate of the area of distribution are also relevant. For example, for tropical distribution the intensity of the thermal process must be increased to eliminate thermophilic spores that might have been present in the raw materials.

The second benefit is that once we understand the factors, we can determine the relative contribution of some or all of them, and design a preservation system optimized to provide the preserving power that we need with the lowest overall impact on the quality of the food. In principle, more barriers, each at "milder" intensity, can give the required safety and/or preservation with better perceived quality, particularly if there are any truly synergistic interactions between any of the factors.

21.3 Drawbacks of the Hurdle Concept

The hurdle concept is not so much defined as, usually, illustrated in a "boiled eggs" diagram (Figure 21.1). This diagram illustrates that different factors (hurdles) may provide different amounts of "preserving power" shown by the height and/or width of the hurdle. It also indicates that the effort involved in overcoming the "preserving power" of one hurdle weakens the microbe so that it has less capacity to overcome the next hurdle. Eventually the microbe reaches a hurdle that it has been too weakened to overcome, and at that point preservation has been achieved. Many people, the authors included, do not find this model particularly intuitive or illustrative, possibly because we are familiar with the notion of hurdle racing, where athletes compete to see who can be the fastest to traverse a course obstructed by a number of hurdles. All hurdles are the same size and separated by a constant distance. The race winner is the one that traverses the course the fastest, and it is rare to see a runner fail to complete the course. Thus, there is immediate confusion when comparing our mental concept of a hurdle race (Figure 21.2) with the classical "hurdle technology" illustration (Figure 21.1). It may also in large part be because Figure 21.1 is unable to convey the possible interaction between a "gentler" inactivation process and multiple inhibitory factors, and combinations which may act additively and/or synergistically.

21.4 Alternative "Multifactorial" Metaphors

21.4.1 The Swiss Cheese Model

In other areas of risk management, different models are used to illustrate the concept of adding multiple barriers to improve security. The Swiss cheese model of risk management and accident causation (Reason, 1990; Reason et al., 2001) (Figure 21.3) is used in many areas, for example health care (Perneger, 2005), aviation (Shorrock et al., 2006), and railroad safety (Underwood and

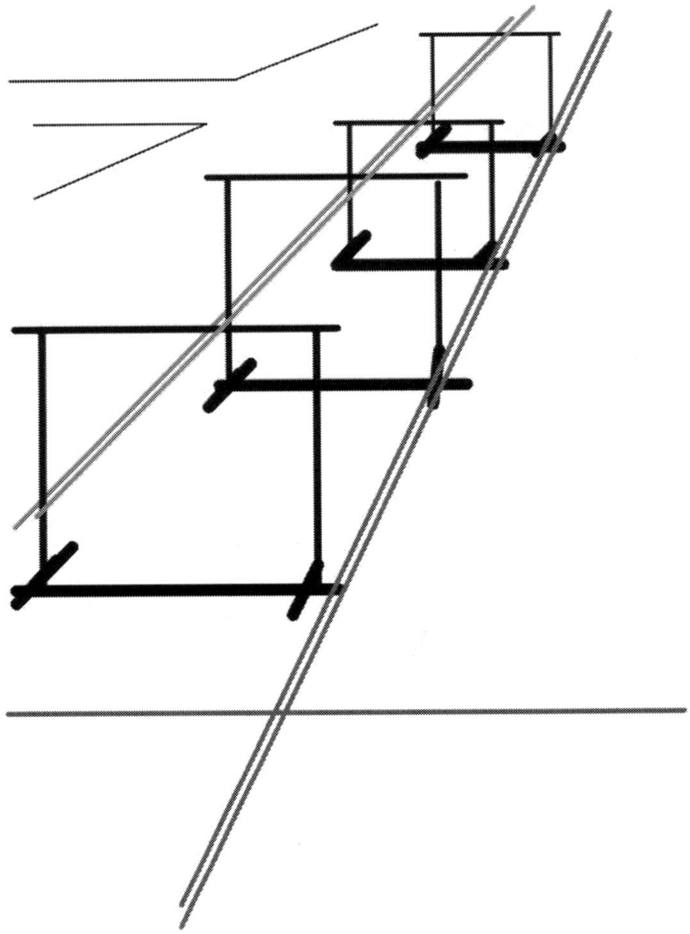

FIGURE 21.2 Racing hurdles showing uniform height and separation: The question is not whether an athlete will clear them, but how fast?

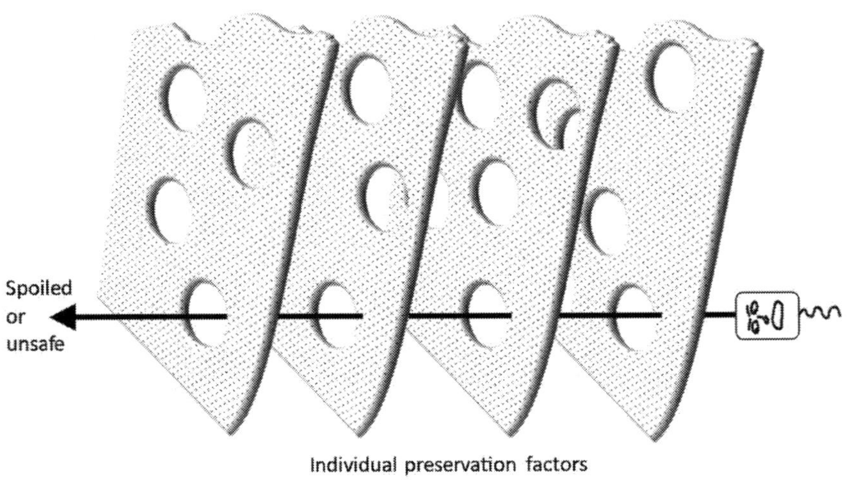

FIGURE 21.3 The Swiss cheese model of accident causation. Each preservation factor, represented as a slice of Swiss cheese, is imperfect but product stability is secure provided that each barrier has a different failure mode. If there is a single failure mode for each barrier (the holes align) then the product will spoil or become unsafe. (Adapted from Reason et al., 2001.)

Waterson, 2014). In this food-centered metaphor each preservation factor would be a slice of Swiss cheese. The presence of holes would indicate that the barrier does not control all risks. For example, one slice might represent a pasteurization step. A hole would indicate that the barrier is not complete against spore-formers. The next slice might indicate a reduced pH that restricts the growth of dangerous spore-formers but permits the growth of *Alicyclobacillus*, providing a spoilage risk. The next slice might represent purchasing specifications to ensure sourcing only of raw materials free from *Alicyclobacillus* spores, and so on. By gradually reducing the likelihood that multiple holes line up as additional barriers are added, the concept illustrates the benefits of multiple barriers, even if each may be incomplete by itself. Failure of the "final" barrier would clearly cause system failure, but note that failure of an earlier barrier may open a different path through the slices leading to failure but by a different mode.

21.4.2 The Pole Vault: Hurdles Upended

The Swiss cheese model has considerable appeal, not least because a food is itself central to the metaphor. However, as we write, the field of food science has invested over 40 years (and how many tens of thousands of lectures?) in the hurdle concept. In an attempt to make the established metaphor more intuitive, we conceive it more as a high-jump or pole vault (Figure 21.4). Each new preservation factor or barrier (B1, B2, B3, etc.), causes "the bar to be raised." The factors may contribute unequal additional heights but eventually enough factors accumulate such that the barrier cannot be overcome. During the competition, as the bar height is raised, the number of athletes able to clear the bar is progressively reduced. Think of this as progressively reducing the number of available failure modes. As the "preservation pressure" of additional factors increases, so too the number and types of organisms able to overcome the barrier and grow become smaller. At the extremes of preservation factors just before all growth is prevented there is usually only a single species able to overcome the pressure – the Olympic gold medalists of preservation resistance – e.g. as a_w reduces, the last mold able to grow is *Xeromyces bisporus*, the most xerotolerant mold known.

There are multiple potential microbial responses to consider once the bar is raised high enough to prevent the growth of any microbe (Figure 21.4). That organism may be injured to the point where it dies such that we do not need to consider it again, or it may be simply injured, able to persist in the food for either a short time, or a long time, until conditions permit its recovery, leading to failure of the preservation system.

In presenting the Legan and David "Pole Vault" Model we aim to modify the established metaphor in a way that makes it more intuitive and more flexible; perhaps merely addressing our own mental challenges but, we hope, providing a service to food scientists and food microbiologists in general. However, we would like to keep the focus on the individual factors or barriers, since this is the level at which the system can be modeled and modified. We also want to emphasize that successful preservation increasingly relies on explicitly multifactorial preservation systems.

21.5 Risks

In any preservation project, we must understand whether we are primarily concerned with mitigating safety risks, spoilage risks, or both. A thorough understanding of the intrinsic factors such as a_w, pH, etc., helps us to determine the nature of the risk that we are managing, and detailed knowledge of the relationships between them allows us to understand how secure our preserving effect is in the event of typical variability in the magnitude of the individual factors. Developments in statistical modeling techniques and the ready availability of modeling software have allowed rapid progress in quantifying the impact of different factors on microbial growth within a food.

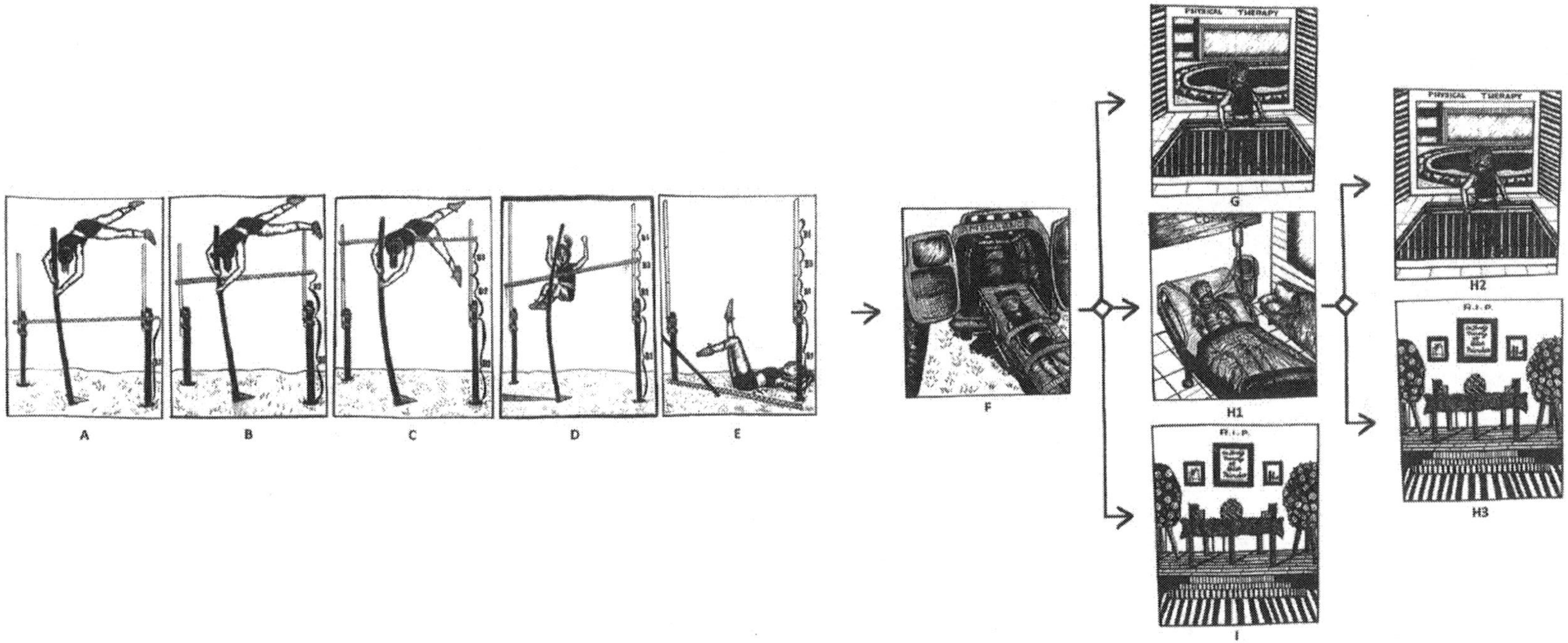

FIGURE 21.4 The Legan and David Pole Vault Model of preservation systems. Barriers (B1, B2, B3, and B4 alongside the poles) may be unequal in effects. Initially with only one factor, the bar can be easily cleared (A). As factors are added, it becomes progressively harder to overcome them (B, C), building to a point where the combined barrier (bar height) is too great to be overcome (D), and the ability of microorganisms to grow is prevented (E). Following failure to grow in the presence of multiple preservation factors, initially the microbes are injured (F). They might quickly recover (G), though this would suggest that the multifactorial preservation system is not yet robust enough. Alternatively, they might quickly die (I) or enter an extended period of quiescence and/or recovery (H1) from which they ultimately recover (H2), limiting product shelf life to an acceptable degree, or die (H3).

21.6 Microbial Modeling to Quantify Factors and Control Risk

21.6.1 Pasteurized Processed Cheese Spread and Control of *Clostridium botulinum*

Tanaka et al. (1986) explored factors affecting toxin production in pasteurized processed cheese spreads. In a series of experiments, they determined conditions permitting toxin formation by *Clostridium botulinum* over 42 weeks' incubation at 30°C. They used a quadratic logistic regression model (not presented in their paper) to determine the boundary between toxin formation and non-toxic conditions and presented seven figures showing the boundary at different moisture contents. We present those boundaries compiled into a single contour plot in Figure 21.5 which compiles four factors: concentrations of NaCl and Na_2HPO_4, pH, and moisture content, all with three further factors fixed: time to end point, incubation temperature, and a pasteurization heat process. This work informed the safe formulation of pasteurized processed cheese spreads with multifactorial preservation for *C. botulinum* control and product safety assurance for many years. However, consumer tastes began to change, and lower fat formulations were demanded, and that created a need to review the basis of product safety.

Many formulations of reduced-fat pasteurized processed cheese products were studied by Glass and Johnson (2004), including even more factors for the control of toxin formation by *C. botulinum*. Beginning with pasteurized processed cheese formulated using full-fat Cheddar, 30% reduced-fat Cheddar, or skim milk cheese and standardized to 59% moisture, pH 5.75, with 2.8% or 3.2% total salts, and 15% to 19% fat, they observed that toxin formation was delayed by 2 days in product formulated with skim milk cheese with 3.1% total salts. Next these authors compared the effect of fat levels and additional ingredients. They found that at similar fat levels (15.1–19.1%) the delay in toxin formation in skim milk pasteurized processed cheese was not statistically significant. However, when fat levels were < 1% in

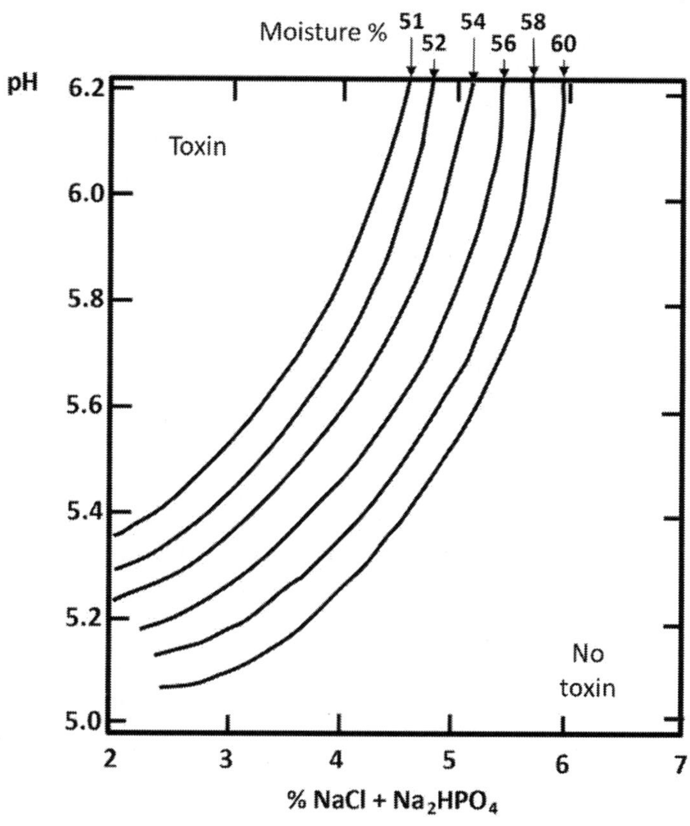

FIGURE 21.5 Model from Tanaka et al. (1986) redrawn as a contour plot. Contours represent the combinations of % NaCl + Na_2HPO_4 and pH that lead to toxin production in 42 weeks at 30. °C at different moisture contents.

skim milk pasteurized processed cheese, 13% in reduced-fat cheese, and 24% fat in full-fat cheese, toxin formation in the skim milk pasteurized processed cheese was delayed by 2 weeks relative to the other 2 formulations. Sodium lactate added at 1.5% significantly delayed toxin formation at all fat levels, though the delay increased as fat content decreased. Enzyme-modified cheese used as a flavor enhancer (1.5% addition) and monolaurin (0.05%) both delayed toxin formation in skim milk pasteurized processed cheese but not in the other formulations. So, this study identified three new formulation hurdles, adding to those previously noted by Tanaka et al. (1986), and identified some complex relationships between them, most notably that added lactate and monolaurin were more effective at low fat concentrations.

Consumers again began to demand product changes, this time looking for lower sodium products in response to publicity about the relationship between sodium consumption and blood pressure and new US dietary guidelines recommending a maximum intake of 2300 mg of sodium per day (U.S. Department of Health and Human Services and U.S. Department of Agriculture, 2015). That created a new need to review the basis of product safety, this time with reduced-sodium formulations. A new model was created under the leadership of Glass et al. (2017). This determined probability and time to toxin formation over 56 weeks in a central composite design of 80 treatments, exploring the influence of seven factors: pH, moisture, Na_2HPO_4, NaCl, sorbate, fat, and fraction of replacement of sodium by potassium. The effects of pH, sorbic acid, moisture, Na_2HPO_4, and NaCl on toxin formation were all highly significant ($P < 0.0001$). Fat content and replacement of sodium by potassium were also significant, but less so ($P = 0.0168$ and $P = 0.0526$, respectively). Highly significant ($P < 0.0001$) pairwise interactions of sorbate with pH, and sorbate with moisture were identified along with a highly significant quadratic term for $(Na_2HPO_4)^2$. Several other interactions were also identified, albeit with lower statistical significance ($P = 0.0041–0.0526$). These interactions demonstrate that, in this case, the overall effect of the several inhibitory factors was more than simply additive.

Reviewing the output from all of the models (Table 21.1) we see that there are 10 factors whose effect on toxin formation is quantified (11 if we count NaCl + Na_2PO_4 as 1 factor, following Tanaka et al., 1986). In addition, incubation temperature was fixed, but would be expected to play a role, and end-point time to toxin was either fixed, or measured as a response variable, rather than being considered a factor contributing to safety. Moreover, the whole preservation system also relies on thermal pasteurization and packaging to protect the product from recontamination. Thus, we have a *minimum* of 13 identified factors contributing to the safety of these products (10 from Table 21.1, plus pasteurization, packaging, and storage temperature), and all for a "simple cheese spread."

TABLE 21.1

Quantified Factors Controlling Toxin Formation by *Clostridium botulinum* in Pasteurized Processed Cheese Spreads Identified through Modeling and Designed Experimentation

Tanaka et al. (1986)	Glass et al. (2017)	Glass and Johnson (2004)
pH	pH	
Moisture	Moisture	
Na_2HPO_4 + NaCl		
	Na_2HPO_4	
	NaCl	
	Fat	Fat
		Sodium lactate
		Monolaurin
		Enzyme modified cheese (EMC)
	Fraction of replacement of Na by K	
	Potassium sorbate	

These factors were determined using a fixed incubation temperature and defined or measured end times, and all rely on the product having been thermally pasteurized and protected from recontamination by packaging.

21.6.2 Ready-to-Eat Meat Products and *Listeria* Control

Another class of products where modeling has been very influential in establishing safe formulation is ready-to-eat (RTE) cooked meats. Following the emergence of *Listeria monocytogenes* as a significant pathogen in the 1980s, and several large and costly recalls in the late 1990s, *Listeria* became the single largest cause of class I recalls (those with a reasonable probability that the hazard will cause adverse health consequences or death; Gorton and Stasiewicz, 2017). The meat-processing industry was determined to find improved ways to control *Listeria* and invested heavily in plant conditions and sanitation to reduce the risk of *Listeria* contamination, plus modified preservation systems to inhibit the growth of any *Listeria* that did manage to contaminate the product.

Seman et al. (2002) used a central composite response surface design to investigate the effects of added sodium chloride (0.8 to 3.6%), sodium diacetate (0 to 0.2%), potassium lactate syrup (60% [wt/wt]; 0.25 to 9.25%), and finished-product moisture (45.5 to 83.5%) on the growth rate of *L. monocytogenes* in cured ready-to-eat (RTE) meat products stored at 4°C. Increasing concentrations of sodium diacetate ($P < 0.11$) and potassium lactate ($P < 0.001$) both gave significant reductions in the growth rates of *L. monocytogenes*. Increased finished-product moisture ($P < 0.11$) significantly increased growth rates, and sodium chloride had no significant effect. Thus three significant factors controlling the growth of *Listeria* were identified.

Working with the same experimental data, Legan et al. (2004) first examined all growth curves and determined the time needed for 1 log of growth to occur, following the recommendations of the expert food microbiology panel of the Institute of Food Technologists (2003). This "time to growth" was used as the response variable in a boundary modeling approach with the same four continuous variables: added sodium chloride, sodium diacetate, potassium lactate syrup, and finished-product moisture. In addition, data generated (but not reported) by Seman et al. (2002) with the design repeated for ready-to-eat uncured meat products were included, giving a fifth, categorical, variable for cure status. Products were stored at 4°C. The results were modeled using a generalized regression approach. All five main effects were statistically significant: added sodium chloride, potassium lactate, and sodium diacetate increased time to growth. Increased finished-product moisture decreased time to growth, and "curing" increased time to growth. In addition, six two-factor interactions, and two quadratic terms were also significant, with all interactions excluding moisture further increasing time to growth, but those including moisture reducing it. The overall effect on the growth and no-growth space is shown in Figure 21.6 for cured products, and Figure 21.7 for uncured products. Excluded from both of these models was a temperature variable, but we know from long experience (and other models, e.g. Buchanan and Phillips, 1990; Buchanan et al., 1989) that increasing temperature increases the growth rate of *L. monocytogenes*. Products were vacuum-sealed to minimize exposure to oxygen and protected from recontamination by a laminated flexible package. Also excluded from the models was the lethality treatment applied to the meat during cooking, which limits the initial level of *Listeria* nominally to zero unless there is some failure of control between cooking and packaging. Hence, we can consider that there are nine factors at play here, of which eight protect the product from the growth of *Listeria*.

Continuing the quest for improved and lower-cost *Listeria* control factors, Seman et al. (2008) built on the work of Glass et al. (2006, 2007), the insights on diacetate effectiveness from Legan et al. (2004), and their own unpublished observations, and designed a modeling study to investigate the combined effect of sodium benzoate (0.08 to 0.25%) and sodium diacetate (0.05 to 0.15%), with salt (0.8 to 2%) and finished-product moisture content (55 to 75%) on the growth of *L. monocytogenes* in RTE cured meat using time to 1 log of growth as the endpoint. Again, products were cooked (but inoculated for experimental purposes), vacuum-sealed in a laminated flexible package, and stored at 4°C. All main effects were significant except product moisture, which was significant when included in the two- and three-way interactions ($P < 0.05$). Sodium benzoate was more effective (lengthening time to growth) when used with increasing concentrations of sodium diacetate and salt and decreasing finished-product moisture.

The reader is invited to read the ingredients list the next time they buy a package of RTE meat. If any of the ingredients listed above are seen, they will know that one of these models was influential in ensuring the safety of the product in their hands. In the event that they find a claim such as "contains no added nitrates or nitrites except those naturally occurring in added vegetable ingredient" they will know that

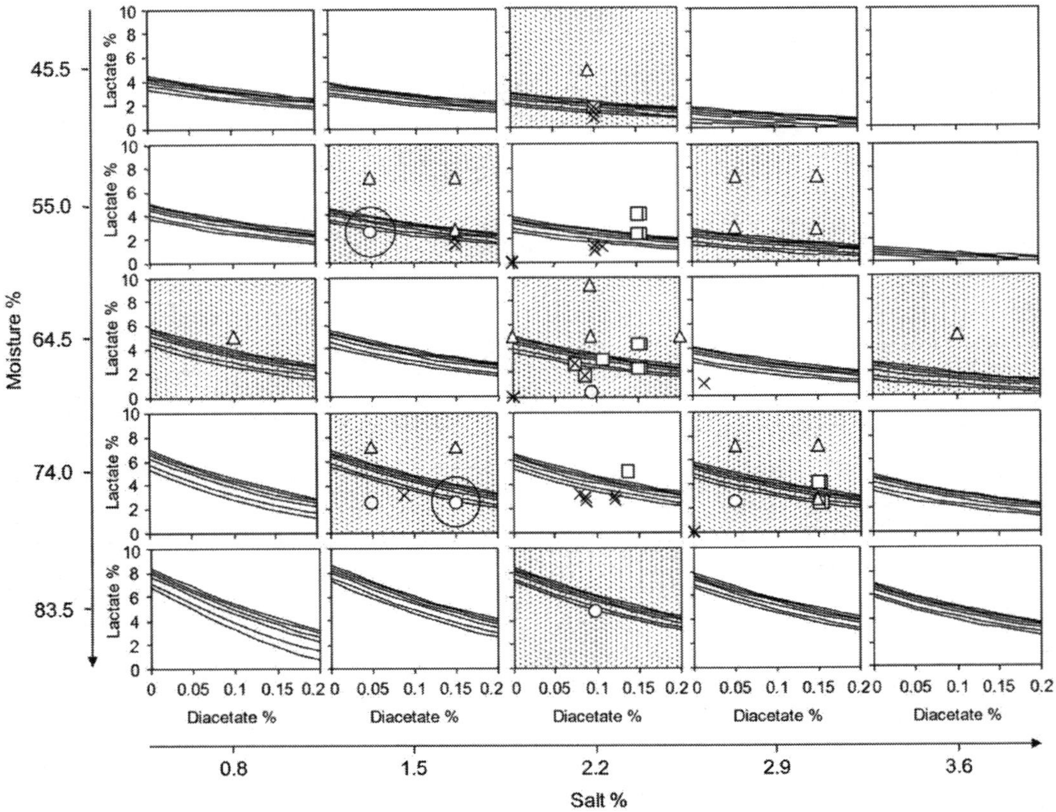

FIGURE 21.6 Compilation of contour plots showing predicted weeks to growth against salt, moisture, sodium diacetate, and potassium lactate syrup concentrations for cured model products. Contours are plotted from 18 weeks in the lower left position rising at 6-week intervals to 48 weeks in the upper-right position in each individual plot. Shaded plots indicate salt and moisture concentrations from the model design treatments. Points within the individual graphs show associated diacetate and lactate syrup concentration where 1 log or more of growth was observed in the modeling experiments (○), where less than 1 log of growth was observed in the modeling experiments (△), where 1 log or more of growth was observed in the validation data (×), and where less than 1 log of growth was observed in the validation data (□). In these plots, the positions of validation data points are approximate, since the salt concentration, moisture content, or both were rounded to allow the points to show within this matrix. Where possible, rounding was in the direction of a reasonable decrease in growth potential so that points were more likely to show in the no-growth space, whether or not growth was observed. Two circled points show where more than 1 log but 2 logs or less of growth was seen in the modeling data. (Republished with permission of International Association for Food Protection, from Legan et al., 2004; permission conveyed through Copyright Clearance Center, Inc.)

the product was "cured" by a process involving conversion of the naturally high nitrate content of celery or another vegetable using a microbial culture and substituting the output for the sodium nitrite used in a "traditional" cure. This process is described in Husgen et al. (2008). Accompanying such a claim, they are likely to notice "cultured sugar" or "cultured corn sugar" and possibly vinegar in the ingredients list. This culture process produces lactic acid (neutralized to lactate), and if the vinegar is included in the fermentation it is simultaneously neutralized to diacetate, which means that the models above, or one of their successors are still helping to ensure safety.

21.7 Multifactorial Preservation in Action

Severely processed products have a large margin of safety. However, when using multifactorial preservation, the applied process must be defined and rigorously controlled with full regulatory compliance and scientific understanding of the factors controlling product safety and stability. Globally this

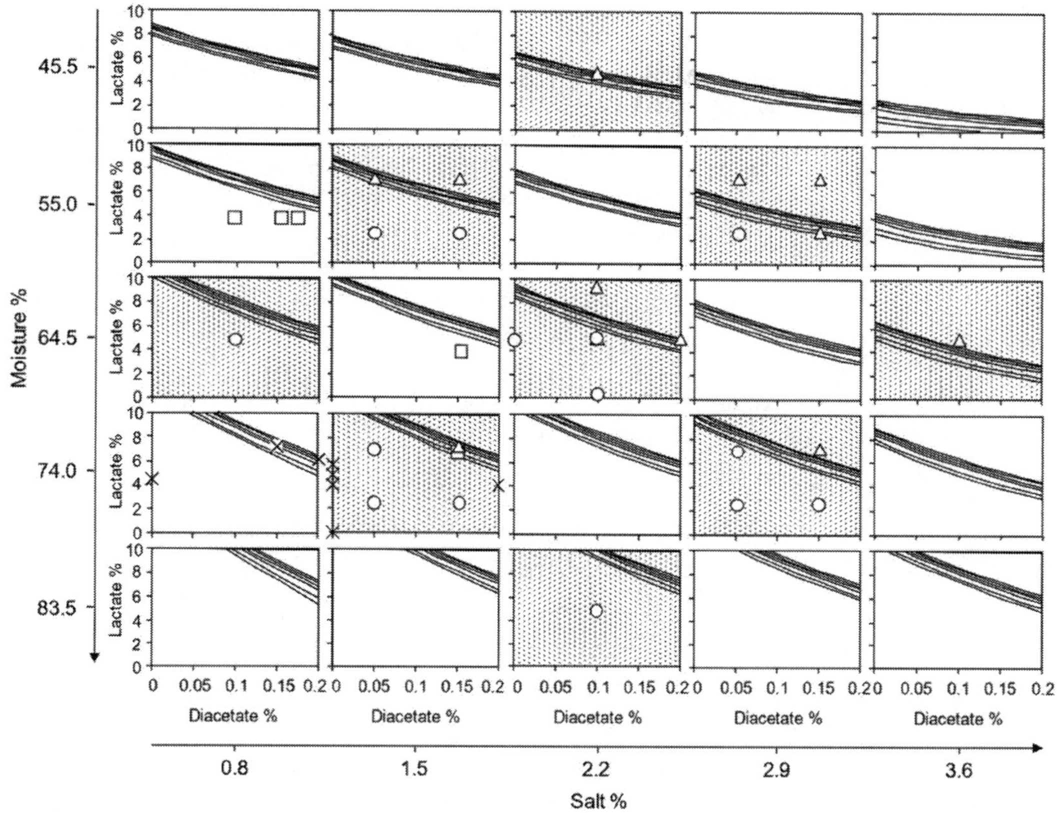

FIGURE 21.7 Compilation of contour plots showing predicted weeks to growth against salt, moisture, sodium diacetate, and potassium lactate syrup concentrations for uncured model products. Contours are plotted from 18 weeks in the lower left position rising at 6-week intervals to 48 weeks in the upper right position in each individual plot. Shaded plots indicate salt and moisture concentrations from the model design treatments. Points within the individual graphs show associated diacetate and lactate syrup concentration where 1 log or more of growth was observed in the modeling experiments (◯), where less than 1 log of growth was observed in the modeling experiments (△), where 1 log or more of growth was observed in the validation data (×), and where less than 1 log of growth was observed in the validation data (◻). In these plots, the positions of validation data points are approximate, since the salt concentration, moisture content, or both were rounded to allow the points to show within this matrix. Where possible, rounding was in the direction of a reasonable decrease in growth potential so that points were more likely to show in the no-growth space, whether or not growth was observed. (Republished with permission of International Association for Food Protection, from Legan et al., 2004; permission conveyed through Copyright Clearance Center, Inc.)

would involve compliance with local food law and the principles of risk management built into the Hazard Analysis and Critical Control Point (HACCP) system (Codex Alimentarius Commission, 1997). Within the US the additional bio-terrorism and deliberate adulteration/food fraud provisions included in the Hazard Analysis and Risk Based Preventive Controls (HARPC) provisions of the Food Safety Modernization Act (2011; FSMA) apply.

Paradoxically, almost all products in the marketplace already rely on multiple factors to provide for their safety and preservation. Most products are:

- Processed thermally or non-thermally as needed to *inactivate* microorganisms.
- Formulated to lower a_w and/or pH, and include antimicrobial preservatives to *inhibit* microbial growth.
- Packaged to *protect from recontamination*, and sometimes with modified atmosphere to *inhibit* microbial growth.
- Distributed with controlled temperature when this is needed to *inhibit* microbial growth.

Hence it is rare to find a product at retail that is only processed by a single severe process. For example, even the "most processed" products, shelf-stable canned foods [low acid canned foods (LACF)] are thermally processed to "commercial sterility" (to *inactivate* microorganisms including mesophilic spores) in a double-seamed hermetically sealed container (to *protect from post-process recontamination*). Acidified canned foods receive a gentler thermal process (to *inactivate vegetative microorganisms*), along with lowered pH (to *inhibit spore germination*), and hermetic seal (to *protect from recontamination*).

While designing and developing multifactorial products, co-optimization of quality and safety attributes is based on the design and precise delivery of:

1. The inactivation process
2. Inhibitory factors
3. Post-process protection including packaging and environmental controls
4. Additional factors listed in the HACCP program including critical control points (CCPs) and/or control points listed in production directives and GMPs

It is important to emphasize that the factors or barriers apply to specific product/processing/packaging situations. When product design relies on modeling, any extrapolation beyond the model design space should be validated with microbial challenge studies and life cycle stability assessment.

21.7.1 Market Examples of Multifactorial Products

Discussed below are examples of minimally processed products commercially available in the market place to demonstrate the use of multifactorial preservation to deliver products that are simultaneously optimized for food safety, stability, and quality. Outlined for each product are the design elements described above. Note that important information about commercial products might not appear in the scientific literature, but rather be disclosed through patents or patent applications, and there might be several applications or granted patents pertaining to a single product covering different aspects of manufacture, preservation, etc.

21.7.1.1 Co-Extruded Cereal Bar with Fruit Core

This is an intermediate-moisture food (IMF) with a finished product a_w of 0.65–0.67, intended to be stored and distributed at ambient temperature. The product consists of two co-extruded physical compartments with gradients of pH and a_w. The outside is a cereal-based dough with near-neutral pH and low a_w after baking. This surrounds a fruit core that has low pH and high a_w. After extrusion, the assembly is baked, cooled, and packaged. The expected shelf life of compliant mold-free products is 9–12 months at ambient. Defective products have a reduced shelf life of about 1–2 months and fail through the development of visible mold growth on the product surface.

21.7.1.1.1 Inactivation Process

Baking of coextruded raw product, to provide the required product texture and flavor characteristics, simultaneously inactivates non-spore forming vegetative pathogens and molds.

21.7.1.1.2 Inhibition Parameters

The control of Brix and total soluble solids (TSS) of the formulated dough and the fruit core, along with control of the baking and cooling conditions, leads to the final product a_w. The use of antimycotic preservative in both dough and fruit core boosts mold control, and the antimicrobial activity of the preservative is optimized through control of pH.

21.7.1.1.3 Production Directives

- HACCP: cooling to 49.1–54.4°C (120–130°F); not to exceed 57.2°C (135°F) at exit prior to packaging

- Finished product release criteria: a_w of 0.65 to 0.67 and not to exceed 0.70
- Regulatory compliance: environmental monitoring of cooling zones and the air handling system, and proper package integrity and heat seal

21.7.1.1.4 Shelf-Life Failure

Significant reduction in finished product ambient shelf life from 9–12 months to 1–2 months occurs due to the growth of xerophilic molds such as *Aspergillus* species, *Wallemia sebi*, and some *Penicillium* spp. on the product surface. Note that many food practitioners would recognize the *Aspergillus* colony appearance on low-a_w product under their *Eurotium* teleomorph names. These names are no longer taxonomically valid (Houbraken et al., 2014: Hubka et al., 2013). This spoilage occurs because recontamination with aerial mold spores on exposed product can occur during post-bake cooling. If these re-contaminated products are packaged warm, condensation of moisture on the inside of the packaging material may be re-absorbed at the product surface, raising the a_w to a level favorable for the surface growth of spoilage xerophilic molds.

21.7.1.2 Optimization of "Non-C. botulinum*" Thermophilic Cook with Use of Nisin in Canned Foods*

The D value at 121°C for spores of *Clostridium botulinum* is 0.1 to 0.3 minutes. In the canning industry, a "12D process" is used to eliminate *C. botulinum* spores which is approximately 3 minutes at 121°C (12 × 0.3) and is commonly known as a "Bot Cook." However, sometimes foods thermally processed to inactivate spore-forming pathogens exhibit high spoilage rates as a result of the growth of non-pathogenic thermophilic spore-forming bacteria. This most often occurs when canned foods are stored at high temperatures (38–52°C; 100–125°F) in tropical or desert areas where thermophilic aerobes and anaerobes can grow. In the food industry, this is commonly known as economic spoilage because there is no health hazard. The D value at 121°C for the causative spore-formers, *Geobacillus stearothermophilus* (Flat Sour) and *C. thermosaccharolyticum*, is about 4 minutes, and thus a 5D process for these extremely heat-resistant spores would result in an F_0 of 20 minutes (David et al., 2013) or more than 6 times longer than the "Bot Cook." If such a thermal process were utilized to inactivate these thermophilic spore-formers, a product of poor quality would result due to the many heat-induced physical and chemical changes.

For the above reasons, nisin has been extensively used as a natural preservative in low-acid canned foods (LACF). By using nisin it is possible to control thermophilic spoilage when the cans are stored under warm conditions for prolonged periods, or to allow a reduction in thermal processing with an F_0 of 5 minutes without the risk of thermophilic spoilage occurring. This reduction in total thermal process leads to improved product quality because of reduced thermal damage to the texture, taste, and appearance of certain vegetables, an increased throughput in the canning operation, and reduced energy consumption in the operation of retorts.

High-acid foods such as canned tomatoes and tomato juice with pH below 4.5 can be processed with a gentler thermal process, as *C. botulinum* is not the target microorganism. However, spoilage can occur due to acid-tolerant, Gram-positive spore-formers such as *Clostridium pasteurianum* and *Bacillus macerans*, and nisin has been used to control the growth of such bacteria in these acidic foods (Abee and Delves-Broughton, 2003).

21.7.1.2.1 Inactivation Process

- Thermal sterilization optimized for reduced "thermophilic cook," but fully adequate for "12D Bot cook"
- Adequate cooling of cans below 32–35°C (90–95°F) before palletizing and stacking to prevent "stack burn" in the warehouse, and potential for spore germination

21.7.1.2.2 Inhibition Process

- Nisin is used to reduce the severity of the "thermophilic cook" and to conserve product quality.

- Package integrity via double seam and hermetic seal.
- Reduced pH control and use of nisin for processing tomatoes and tomato juice.

21.7.1.2.3 Production Directive and Ingredient Specifications

- Incoming ingredients specification for spores – mesophilic, thermophilic, and aerobic and anaerobic
- HACCP and regulatory compliance

21.7.1.3 Bottled Water

In the United States, bottled water is regulated by the U.S. Food and Drug Administration (FDA), which defines bottled water as "water that is intended for human consumption and that is sealed in bottles or other containers with no added ingredients except that it may optionally contain safe and suitable antimicrobial agents." As an option and within limitations, fluoride may also be added (Robin, and Feng, 2015).

21.7.1.3.1 Inactivation and Inhibition Process

The FDA allows disinfection of all water types, typically by ozonation and/or ultraviolet light.

21.7.1.3.2 QA, GMP, and Production Directives

- Water prior to ozonation must meet all microbiological, physical, chemical, and radiological quality standards for bottled water.
- Post-ozonation, the residual ozone level must not exceed 0.4 mg/liter of bottled water.
- Weekly coliform analysis of bottled water and source water; not more than 1 coliform per 100 mL.
- If coliform is detected in source water or bottled water, the manufacturer must determine whether any of the coliform bacteria are *E. coli*. Bottled water containing *E. coli* is considered adulterated. Source water containing *E. coli* must be treated for the eradication of *E. coli*.
- Microbiological testing of containers and closures should be done once every 3 months analyzing for heterotrophic plate count (HPC) and coliforms using 4 different containers and closures. All of the samples must be free of coliforms, and not more than one of the four samples may have HPC in excess of one colony per ml capacity of the container or one colony per square centimeter of surface area of closure.

21.7.1.4 Refrigerated Liquid Egg Containing Vegetable Blend

Liquid whole egg pools the contents of multiple individual eggs, raising the risk of *Salmonella* contamination of the bulk liquid egg. For this reason, conditions for the pasteurization of liquid egg are defined in the regulations (e.g. 9 CFR 590.570; Egg Products Regulations, 1993). These specify both the temperature and time that must be met, but also allow for processes of equivalent lethality. Refrigerated liquid eggs containing vegetable blend are made by adding mixed vegetables into previously pasteurized liquid egg purchased from an upstream supplier. The blend must then be reprocessed so as to maintain safety and deliver a 90-day refrigerated shelf life while minimizing damage to the functional properties of the egg proteins. One method of preparing such a product was disclosed by Rapp (1989).

21.7.1.4.1 Inactivation Process

The post-blending thermal process is optimized for protecting egg proteins while destroying *Salmonella* and spoilage microbes. Incoming liquid egg, previously pasteurized by the ingredient vendor, is shipped refrigerated in stainless totes. The apparently redundant "double pasteurization" is for compliance and for ensuring food safety integrity through the supply chain.

21.7.1.4.2 Inhibition Parameters

- Liquid whole egg has a slightly alkaline pH, typically around 7.2–7.9.
- The natural antimicrobial nisin is added to control the germination of spores of *Bacillus cereus* and the growth of vegetative cells of *Micrococcus*, both tolerant of the mild thermal process intended for *Salmonella* destruction.
- Final product is filled in a cleanroom using a filler equipped and configured for extended shelf-life (ESL) operations. Packaging web and cap/pour fitment are sanitized prior to filling.
- Finished product storage and handling refrigerated below 3.3°C (38°F).

21.7.1.4.3 QA, Ingredient Specifications, and Production Directives

- Incoming pasteurized refrigerated eggs – temperature history and certificate of analysis (COA)
- COA and microbiological specification for vegetable slurry, especially psychrotrophic and mesophilic spores and vegetative cells
- Order of addition of nisin: nisin added while batching vegetable slurry stream prior to comingling with liquid eggs followed by final pasteurization
- Full compliance with USDA, FSMA, HACCP, and GMPs

21.7.1.5 High-Pressure Processed Products

High-pressure pasteurization (HPP) became commercially viable for food processing in the late 1990s, though the early experiments showing a lethal effect on vegetative bacterial cells were done in the late 1800s (Hite 1899). The major advantage of HPP is its ability to reduce or eliminate vegetative cells while maintaining a fresher taste and appearance when applied to suitable substrates. Additionally, HPP, like its thermal predecessor, does not kill spores, so additional physical or chemical factors are needed to protect from spore germination and outgrowth. Typically, these factors include refrigeration and ingredients applied during product formulation. Many different types of foods treated by HPP are available in grocery stores around the world, for example:

- High-pressure pasteurized jams were introduced into the Japanese market in 1990 (Balasubramaniam et al., 2008).
- One of the earliest US commercial HPP products was extended shelf-life guacamole: a preparation of avocado, introduced to the market in 1997 (Balasubramaniam et al., 2008).
- "Natural" RTE meats are a significant market segment in the US (Balasubramaniam et al., 2008).

21.7.1.5.1 Inactivation Process

In the context of multi-factorial preservation, HPP is unique in that it offers a non-thermal pasteurization in which the lethal principle (pressure) is transmitted instantaneously to all parts of a food product. For solid foods, this treatment is applied once the product has been sealed inside a package with at least some flexible component and the package is submerged in water or some other pressure-transmitting fluid. Lethality is a function of applied pressure (commercially commonly 600 MPa) and hold time (typically around 3 min). The lethality of HPP to *L. monocytogenes* in RTE meats was demonstrated by Youart et al. (2010), showing that 600 MPa for 3 min is more than enough to achieve the 2 log post-lethality treatment qualifying for reduced sampling under Alternatives 1 or 2 of the *Listeria* rule (U.S. Department of Agriculture Food Safety and Inspection Service, 2006).

21.7.1.5.2 Inhibition Parameters

- HPP Jams: pH and a_w are important factors preventing the outgrowth of any surviving spores.
- HPP guacamole: low pH and refrigeration prevent spore germination.

- Natural RTE meats: usually have "natural" variants of the lactates, diacetates, nitrites, etc., addressed in Section 21.6.2, above. These ingredients are produced by the fermentation of e.g. corn-derived sugars or bio-conversion of nitrates naturally present in some vegetable ingredients, particularly celery.
- In all cases the sealed, flexible package protects from recontamination after processing until the package is opened. At that point maintenance of refrigeration and limits on time to consumption are needed.

21.7.1.5.3 Ingredient Specifications and Production Directives

- Purchasing specifications including raw material microbial limits.
- Formulation design and control to achieve critical values of a_w, pH, lactate, nitrite, etc.
- Critical limits on pressure and process time.
- HACCP plan including control of celery allergen when RTE meats receive HPP treatment in the same equipment as other products, e.g. at a co-manufacturing facility.
- Temperature control.

21.8 Future Development of Novel Opportunities

As consumers continue to demand ever more "natural" and "fresh-like" food products, industry innovators will do everything possible to satisfy that demand. As we have just seen, multifactorial preservation systems provide opportunities for effective preservation with maximum benefit to total product quality. However, these systems can be very complex. That means that in future we can expect modeling to play a more prominent role in optimizing the effectiveness of the factors that we already know about. The existence of public modeling portals such as the USDA Pathogen Modeling Program (https://pmp.errc.ars.usda.gov/PMPOnline.aspx), and databases with both models and validation data such as ComBase (www.combase.cc/index.php/en/) are very helpful for guiding the design of more detailed modeling studies. As the number of variables increases, modeling becomes more challenging and expensive, and when we consider that some of the preservation systems discussed earlier have 10 to 13 identifiable factors, we can see that we may be pushing the limits of affordability for the food industry.

Since the early 2000s, "Big Data" has been having large effects on consumer research and product design in a wide range of industries (Blackburn et al., 2017), and we might expect that it will be influential in a complex space like multifactorial food preservation. Of course, many factors besides microbial growth can affect food quality, but to this point quality is not typically considered by microbiological models. However, models of changes in food quality have now been developing for some time (Boekel, 2008). Combining microbial models with quality models will give us a greater ability to make an early assessment of the shelf life of newly developed products. One example of including multiple microbiological and food characteristics in a food safety model was presented by Halder et al. (2011). These authors describe an integrated software package that combines a physics-based process model with kinetic models of chemical and microbiological processes and a database with the composition and properties of many different foods.

Then we encounter the reality that there have been few, if any, new preservation ingredients for foods developed in the past 50 years because of the costs involved in the research, and more recently because the US self-GRAS process has come in for some criticism. Also, concerns have been raised about the solvent extraction processes that can be involved in separating an antimicrobial from a plant source. A potentially interesting approach originating in the field of drug discovery arises from the recognition that microbial resistance to antimicrobial compounds is often connected with the ability of the cell to either exclude them, or pump them out. This active exclusion is driven by a number of multi-drug resistance (MDR) efflux pumps (Marquez, 2005). Inhibiting these efflux pumps can potentiate the activity of natural antimicrobials by two or more orders of magnitude (Tegos et al., 2002). Some MDR pump inhibitors are known (Garvey and Piddock, 2008; Stermitz et al., 2000), and in the pharmaceutical world this has led to identified approaches for overcoming antimicrobial resistance (Pagès and Amaral, 2009; Tegos et al., 2011). These approaches could be loosely described as looking for synergies of effect between unrelated materials. In the world of food science there are many published papers describing synergies between antimicrobials. Most are small studies (Ye et al., 2013; Guttierez et al., 2009; Nazer et al., 2005), and very few investigate the taste or flavor impact of the natural antimicrobials used.

To change the trajectory of food science research into natural antimicrobials (particularly those derived from plant materials that might provide for "consumer-friendly" labeling), and enable the next generation of multifactorial preservation, we could adopt the high-throughput screening approaches (De La Fuente et al., 2006; Blondelle and Lohner, 2010) successfully used in the pharmaceutical industry for drug discovery, but must be aware that the path from novel compound, whether natural or synthetic, to market is long, complex, and expensive, encompassing as it does technical, regulatory, and consumer considerations.

Another area to be very aware of is the dramatic progress in gene-sequencing technologies. According to the National Human Genome Research Institute (2019) the cost of sequencing a human genome approximately tracked Moore's law in electronics (Moore 1965) which predicted a doubling in circuit complexity per year at the same minimum component cost. After 2008, when a switch from "Sanger sequencing" to "next-generation sequencing" took place, the cost per human genome decreased by three orders of magnitude more than Moore's law's prediction. The practical consequence of this cost reduction is that sequencing has become accessible for many activities that would have been unaffordable in 2008, and analysis of changes in the microbiome of a food product over time is a realistic way to track the progression of the spoilage microflora. Benson et al. (2014) showed how changes in the microbiome of chilled, fresh pork sausage over time were altered by incorporation of a lactate-diacetate preservative. Both with and without lactate-diacetate the ultimately dominant bacterial population was dominated by *Lactobacillus graminis*, but the population switch to this organism was much faster in the presence of lactate-diacetate and the peak population was lower. This correlated with a reduction in sensory changes over time. This improved ability to measure changes in the microbiome of foods as reviewed by Cao et al. (2017) and to understand the microbiome of food production environments (Bokulich and Mills, 2013) should lead to significantly deeper understanding of the mechanisms of food spoilage. Combining this understanding with advances in methods for modifying microbial population changes could give us the ultimate in multifactorial preservation systems.

21.9 Concluding Remarks

The food industry's understanding of multifactorial preservation systems has come a long way over the past 40 years. This progress has partly enabled, and partly been in response to, consumers' demands for fresher, less processed, more natural products that retain all of the convenience and taste that they have come to expect. As this is a long-running trend, it seems unlikely to reverse any time soon, if ever. Product design using clean-label ingredients combined with multifactorial preservation is the best way that we know of to continue satisfying these consumer demands. We expect this field will continue to develop over the next 40 years and to incorporate at least some of the novel opportunities emerging from universities, research institutes, research consortia, and transferring from other industries. The best chance for success in developing "big ideas" is for all stakeholders in food safety and preservation to work collaboratively in the discovery and feasibility stages and reserve competition for applications and consumer product development.

21.10 Acknowledgments

The authors wish to thank Heather Legan of BP America Inc. for discussions on risk management and the "Swiss cheese" model. Morgan Fields and Blake David were instrumental in developing the art work and drawing for "Pole Vaulting and Injured Patient Status" shown in Figure 21.4. Their patience and diligence in translating the authors' concept through several iterations are acknowledged with gratitude.

21.11 Disclaimer

References to commercial products and trade names are made with the understanding that no discrimination and/or no endorsement by the authors or the organizations that they are involved with are implied.

REFERENCES

9 CFR 590.570 Pasteurization of liquid eggs. https://ecfr.io/Title-09/pt9.2.590#se9.2.590_1570. Accessed 05 October 2019.

Abee, T., and Delves-Broughton, J. 2003. Bacteriocins—Nisin. In *Food Preservatives*. Second Edition. Edited by Russell, N.J., and Gould, G.W. 146–178. Kluwer Academic/Plenum Publishers, New York.

Balasubramaniam, V.M., Farkas, D., and Turek, E.J. 2008. Preserving foods through high pressure processing. *Food Technol.* 11:33–38.

Benson, A.K., David, J.R.D., Gilbreth, S.E., Smith, G., Nietfeldt, J., Legge, R., Kim, J., Sinha, R., Duncan, C.E., Ma, J., and Singh, I. 2014. Microbial successions are associated with changes in chemical profiles of a model refrigerated fresh pork sausage during an 80-day shelf life study. *Appl. Environ. Microbiol.* 80:5178–5194.

Blackburn, M., Alexander, J., Legan, J.D., and Klabjan, D. 2017. Big data and the future of R&D management. *Res. Technol. Manage.* 60:43–51.

Blondelle, S.E., and Lohner, K. 2010. Optimization and high-throughput screening of antimicrobial peptides. *Curr. Pharma. Design* 16:3204–3211.

Boekel, M.A.J.S. Van 2008. Kinetic modeling of food quality: A critical review. *Comp. Reviews Food Sci. and Food Safety* 7:144–158.

Bokulich, N.A., and Mills, D.A. 2013. Facility-specific "house" microbiome drives microbial landscapes of artisan cheesemaking plants. *Appl. Environ. Microbiol.* 7:5214–5223.

Buchanan, R.L., and Phillips, J.G. 1990. Response surface model for predicting the effects of temperature, pH, sodium chloride, sodium nitrite concentration and atmosphere on the growth of *Listeria monocytogenes*. *J. Food Prot.* 53:370–376.

Buchanan, R.L., Stahl, H.G., and Whiting, R.C. 1989. Effects and interactions of the temperature, pH, atmosphere, sodium chloride and sodium nitrite on the growth of *Listeria monocytogenes*. *J. Food Prot.* 52:844–851.

Cao, Y., Fanning, S., Proos, S., Kieran Jordan, K., and Shabarinath Srikumar, S. 2017. A review on the applications of next generation sequencing technologies as applied to food-related microbiome studies. *Front. Microbiol.* 8 article 1829. doi: 10.3389/fmicb.2017.01829

Chawla, S.P., and Chander, R. 2004. Microbiological safety of shelf-stable meat products prepared by employing hurdle technology. *Food Control* 15:559–563.

Codex Alimentarius Commission 1997. Hazard analysis and critical control point (HACCP) system and guidelines for its application. Annex to General *principles of food hygiene* CAC/RCP 1–1969, *Rev.* 3 (1997), 15–23. See http://www.fao.org/3/Y1579E/y1579e03.htm#bm3. Accessed 07 October 2019.

David, J.R.D. 2013. Thermal processing and optimization. In *Handbook of Aseptic Processing and Packaging*. Edited by David, J.R.D., Graves, R.H., and Szemplenski, T. 167–186. CRC Press, Taylor and Francis Group, Boca Raton, Florida.

De La Fuente, R., Sonawane, N.D., Arumainayagam, D., and Verkman, A.S. 2006. Small molecules with antimicrobial activity against *E. coli* and *P. aeruginosa* identified by high-throughput screening. *British J. Pharmacol.* 149:551–559.

Egg Products Regulations. 1993. UK Statutory Instruments 1993 No. 1520. http://www.legislation.gov.uk/uksi/1993/1520/schedule/2/made. Accessed 05 October 2019.

Food Safety Modernization Act. 2011. Public Law 111–353—Jan. 4, 2011 124 Stat. 3885 Government Printing Office. See https://www.govinfo.gov/content/pkg/PLAW-111publ353/pdf/PLAW-111publ353.pdf. Accessed 07 October 2019.

Garcia, R., and Adrian, J. 2009. Nicolas appert: Inventor and manufacturer. *Food Rev. Intl.* 25:115–125.

García-García, R., Escobedo-Avellaneda, Z., Tejada-Ortigoza, V., Martín-Belloso, O., Valdez-Fragoso, A., and Welti-Chanes, J. 2015. Hurdle technology applied to prickly pear beverages for inhibiting *Saccharomyces cerevisiae* and *Escherichia coli*. *Letters in Appl. Microbiol.* 60:558–564.

Garvey, M.I., and Piddock, L.J.V. 2008. The efflux pump inhibitor reserpine selects multidrug resistant *Streptococcus pneumonia* strains that overexpress the ABC transporters PatA and PatB. *Antimicrob. Agents. Chemother.* 52:1677–1685.

Glass, K., Claus, J., McDonnell, L., Rassel, R., and Zierke, K. 2006. Controlling *Listeria monocytogenes* on ready-to-eat meat and poultry products using food-approved antimicrobials, benzoate, propionate and sorbate. American Meat Institute Foundation final report, June 2006. http://meatpoultryfoundation.org/namif/wp-content/uploads/04-221.pdf. Accessed 28 August 2019.

Glass, K., Preston, D., and Veesenmeyer, J. 2007. Inhibition of Listeria monocytogenes in turkey and pork-beef bologna by combinations of sorbate, benzoate and propionate. *J. Food Prot.* 70:214–217.

Glass, K.A., and Johnson, E.A. 2004. Factors that contribute to the botulinal safety of reduced-fat and fat-free process cheese products. *J. Food Prot.* 67:1687–1693.

Glass, K.A., Mu, M., Levine, B., and Rossi, F. 2017. Inhibition of *Clostridium botulinum* in model reduced-sodium pasteurized prepared cheese products. *J. Food Prot.* 80:1478–1488.

Gorton, A., and Stasiewicz, M.J. 2017. Twenty-two years of U.S. meat and poultry product recalls: Implications for food safety and food waste. *J. Food Prot.* 80:674–684.

Gupta, S., Chatterjee, S., Vaishnav, J., Kumar, V., Variyar, P.S., and Sharma, A. 2012. Hurdle technology for shelf stable minimally processed French beans (*Phaseolus vulgaris*): A response surface methodology approach. *LWT Food Sci. Tech.* 48:182–189.

Gutierrez, J., Barry-Ryan, C., and Bourke, P. 2009. Antimicrobial activity of plant essential oils using food model media: Efficacy, synergistic potential and interactions with food components. *Food Microbiol.* 26:142–150.

Halder, A., Dhall, A. Datta, A.K., Black, D.G., Davidson, P.M., Li, J., and Zivanovic, S. 2011. A user-friendly general-purpose predictive software package for food safety. *J. Food Engr.* 104: 173–185.

Hite, B.H. 1899. The effects of pressure in the preservation of milk. Morgantown. *Bulletin of theWest Virginia Agricultural Experimental Station* 58:15–35.

Houbraken, J., de Vries, R.P., and Samson, R.A. 2014. Modern taxonomy of biotechnologically important *Aspergillus* and *Penicillium* Species. *Advances. Appl. Microbiol.* 86:199–249.

Hubka, V., Kolarik, M., Kubatova, A., and Peterson, S.W. 2013. Taxonomic revision of the genus *Eurotium* and transfer of species to *Aspergillus. Mycologia* 105:912–937.

Husgen, A., Bauman, K., McKlem, L., Papinaho, P., and Jones, B. 2008. Method and composition for preparing cured meat products. U.S. patent application US 2008/0305213 A1.

Institute of Food Technologists. 2003. Evaluation and definition of potentially hazardous foods, chapter VI Microbiological challenge testing, Vol. 2, 2003—Compr. *Rev. Food Sci. Food Safety*, 2: 46–50. https://www.fda.gov/media/103613/download. Accessed 28 August 2019.

Khan, I., Tango, N.F., Miskeen, S., Lee, B.H., and Oh, D.-H. 2017. Hurdle technology: A novel approach for enhanced food quality and safety: a review. *Food Control* 73:1426–1444.

Legan, J.D., Seman, D.L., Milkowski, A.L., Hirschey, J.A., and Vandeven, M.H. 2004 Modeling the growth boundary of *Listeria monocytogenes* in ready-to-eat cooked meat products as a function of product salt, moisture, potassium lactate and sodium diacetate. *J. Food Prot.* 67:2195–2205.

Leistner, L. 2000. Basic aspects of food preservation by hurdle technology. *Intl. J. Food Microbiol.* 55:181–186.

Leistner, L., and Gorris, L.G.M. 1995. Food preservation by Hurdle Technology. *Trends Food Sci. Technol.* 6:41–46.

Leistner, L., and Rödel, W. 1976. The stability of intermediate moisture foods with respect to micro-organisms. In *Intermediate Moisture Foods*. Edited by Davis, R., Birch, G.G., and Parker, K.J. 120–137. Applied Science Publishers, London.

Marquez, B. 2005. Bacterial efflux systems and efflux pumps inhibitors. *Biochimie* 87:1137–1147.

Moore, G.E. 1965. Cramming more components onto integrated circuits. *Electronics* 38:114–117.

Moreira, M.R., Álvarez, M.V., Martín-Belloso, O., and Soliva-Fortuny, R. 2017. Effects of pulsed light treatments and pectin edible coatings on the quality of fresh-cut apples: A hurdle technology approach. *J. Sci. Food Agric.* 97:261–268.

National Human Genome Research Institute (2019). DNA sequencing costs: Data. https://www.genome.gov/about-genomics/fact-sheets/DNA-Sequencing-Costs-Data. Accessed 10 October 2019.

Nazer, A.I., Kobilinsky, A., Tholozan, J.-L., and Dubois-Brissonnet, F. 2005. Combinations of food antimicrobials at low levels to inhibit the growth of *Salmonella* sv. Typhimurium: A synergistic effect? *Food Microbiol.* 22:391–398.

Pagès, J-M., and Amaral, L. 2009. Mechanisms of drug efflux and strategies to combat them: Challenging the efflux pump of Gram-negative bacteria. *Biochemica et Biphysica Acta* 1794:826–833.

Perneger, T.V. 2005. The Swiss cheese model of safety incidents: Are there holes in the metaphor? *BMC Health Serv. Res.* 5:71–78.

Rahman, M.S. 2015. Hurdle technology in food preservation. In *Minimally Processed Foods Technologies for Safety, Quality and Convenience.* Edited by Siddiqui, M.W. and Rahman, M.S. 17–33. Springer International Publishing, Switzerland.

Rapp, H. 1989. Process of treating vegetables for use in a vegetable omelet mix. US Patent 4,910,036.

Reason, J. 1990. The contribution of latent human failures to the breakdown of complex systems. *Philos. Trans. R. Soc. Lond. B, Biol. Sci.* 327:475–484.

Reason, J.T., Carthey, J., and de Laval, M.R. 2001. Diagnosing "vulnerable system syndrome": An essential prerequisite to effective risk management. *Qual. Health Care*, 10 (Suppl II): ii21–ii25.

Robin, L.P., and Feng, P. 2015. Bottled water. In *Compendium of Methods for the Microbiological Examination of Foods.* Fifth Edition. Edited by Salfinger, Y., and Tortorello, M.L. 791–796. APHA Press, Washington, DC.

Scott, V.N. 1989. Interaction of Factors to Control Microbial Spoilage of Refrigerated Foods. *J. Food Prot.* 52:431–435.

Seman, D.L., Borger, A.C., Meyer, J.D., Hall, P.A., and Milkowski, A.L. 2002. Modeling the growth of *Listeria monocytogenes* in cured ready-to-eat processed meat products by manipulation of sodium chloride, sodium diacetate, potassium lactate and product moisture content. *J. Food Prot.* 65:651–658.

Seman, D.L., Quickert, S.C., Borger, A.C., and Meyer, J.D. 2008. Inhibition of *Listeria monocytogenes* growth in cured ready-to-eat meat products by use of sodium benzoate and sodium diacetate. *J. Food Prot.* 71:1386–1392.

Shorrock, S.T., Young, M., and Faulkner, J. 2006. Seeking and finding organizational accident causes: Comments on the Swiss cheese model. *Aviation Safety Letter* Issue 2. Transport Canada, Ottawa, ON.

Singh, S., and Shalini, R. 2016. Effect of Hurdle Technology in food preservation: A review. *Critical Reviews Food Sci. Nutrition*, 56:641–649. Doi: 0.1080/10408398.2012.761594

Stermitz, F.R., Lorenz, P., Tawara, J.N., Zenewicz, L.A., and Lewis, K. 2000. Synergy in a medicinal plant: antimicrobial action of berberine potentiated by 5′-methoxyhydnocarpin, a multidrug pump inhibitor. *Proc. Natl. Acad. Sci. USA* 97:1433–1437.

Tanaka, N., Traisman, E., Planinga, P., Finn, L., Flom, W., Meske, L., and Guiggisberg, J. 1986. Evaluation of factors involved in antibotulinal properties of pasteurized process cheese spreads. *J. Food Prot.* 49:526–531.

Tegos, G., Stermitz, F.R., Lomovskaya, O., and Lewis, K. 2002. Multidrug pump inhibitors uncover remarkable ability of plant antimicrobials. *Antimicrob. Agents and Chemother.* 46:3133–3141.

Tegos, G.P., Haynes, M., Stouse, J.J., Khan, M.Md.T., Bologa, C.G., Oprea, T.I., and Sklar, L.A. 2011. Microbial efflux pump inhibition: tactics and strategies. *Curr. Pharma. Design* 17:1291–1302.

Underwood, P., and Waterson, P. 2014. Systems thinking, the Swiss Cheese Model and accident analysis: A comparative systemic analysis of the Grayrigg train derailment using the ATSB, AcciMap and STAMP models. *Accid. Anal. Prev.* 68:75–94.

U.S. Department of Agriculture Food Safety and Inspection Service. 2006. Compliance guidelines to control *Listeria monocytogenes* in post-lethality exposed ready-to-eat meat and poultry products. Available at: http://www.fsis.usda.gov/oppde/rdad/FRPubs/97-013F/LM_Rule_Compliance_Guidelines_May_2006.pdf. Accessed 02 September 2019.

U.S. Department of Health and Human Services and U.S. Department of Agriculture. 2015. 2015–2020 Dietary guidelines for Americans. 8th edition. Available at: https://health.gov/dietaryguidelines/2015/guidelines/chapter-1/key-recommendations/. Accessed 21 August 2019.

Vallery-Radot, R. 1919. *The Life of Pasteur.* Translated by Devonshire, R.L. 113. London: Constable & Company. See https://archive.org/stream/in.ernet.dli.2015.173907/2015.173907.The-Life-Of-Pasteur#page/n135/mode/2up. Accessed 26 September 2019.

Ye, H., Shen, S., Xu, J., Lin, S., Yuan, Y., and Jones, G.S. 2013. Synergistic interactions of cinnamaldehyde in combination with carvacrol against food-borne bacteria. *Food Control* 34:619–623.

Youart, A.M., Huang, Y., Stewart, C.M., Kalinowski, R.M., and Legan, J.D. 2010. Modeling time to inactivation of *Listeria monocytogenes* in response to high pressure, sodium chloride, and sodium lactate. *J. Food Prot.* 73:1793–1802.

22

Applications of Antimicrobials to Foods – A Food Industry Perspective

Jairus R. D. David and Peter Taormina

CONTENTS

22.1 Introduction

The objective of this chapter is to outline the important aspects of the application of antimicrobials to foods, including selection of the antimicrobial, determination of target microorganisms in a specific food matrix, efficacy testing against target microorganisms *in vitro* and in food matrices, and issues that must be addressed in the commercial application of the antimicrobial and product launch. Expectations of the performance of antimicrobials must be realistic and should include other considerations, such as effect on sensory and quality attributes of the food, cost and cost-in-use of the antimicrobial, and regulatory and labeling considerations, in addition to efficacy against target microorganisms in the food matrix. The "idea-to-launch" business framework and governance is recommended for pairing of a potential antimicrobial with a complex food matrix, along with clearly defined objectives, inputs, outputs, and technical success and business decision criteria.

At the outset, it is vitally important to remember that the use of antimicrobials in food is meant to add to the robustness of existing safety and quality programs and production directives, and not to correct or mask poor practices (Davidson, et al., 2005). In fact, existing antimicrobials are not efficacious enough to overcome marginal or poor microbiological quality of a food. Thus, the effective use of antimicrobials in food begins with the existence of sound prerequisite programs, such as sanitation and Good Manufacturing Practices (GMPs), and appropriate food safety controls and process controls. It may be argued that prerequisite programs, and in fact any measures used to enhance the safety and quality of foods, can be viewed as hurdles and be included in a hurdle concept plan. While this is not the classical view when describing hurdle technology (Leistner, 2000), it may be practical to think in this way when setting up food protection programs in the manufacturing environment. Thus, any protective measure, including programs such as sanitation, GMPs, production directives, and even HACCP or food safety preventive controls, can be included in the hurdle concept. Similarly, antimicrobial systems do not preclude the need for downstream measures for food protection such as adequate temperature control during distribution, storage, display, and appropriate handling and preparation and use at the point of consumption.

22.2 Research and Development Perspectives

Many companies have an idea of what microorganism(s) should be the target for antimicrobial control but not what type(s) of antimicrobial compound(s) may be useful against the target microorganism(s). Thus, the first step in selecting an antimicrobial is to determine its efficacy. Although many studies on the activity of antimicrobials have been published, it may be necessary to establish efficacy *de novo*. Because there are no standard methods for determining efficacy, researchers have generally used methods used by clinical microbiologists, such as agar diffusion assays, microbroth dilution assays, and the "time-kill" curves (Davidson and Parish, 1989). The appropriate methods for determining the activity of antimicrobials for food use are described in detail in Chapter 2 ("Methods for Activity Assay and Evaluation of Results"). Because many antimicrobials are hydrophobic or at least bear some degree of hydrophobicity, the commonly used agar diffusion assay that relies on consistent and rapid diffusion of compounds in the polar agar gel, may yield inaccurate results. Dilution assays are more appropriate for testing food antimicrobials. Reports in the literature on the efficacy of compounds tested with the agar diffusion method might be considered suspect unless these compounds have been highly purified and standardized, such as with nisin (Ripoche et al., 2006).

In the evaluation of antimicrobials for potential use in foods, the suggested steps include *in vitro* testing in media to determine endpoints and dynamic inhibition, followed by application to foods and challenge studies. The endpoint assays, generally broth or agar dilution assays, involve adding the compound to a microbiological medium, adding the test microorganisms, and incubating for a specific time under conditions conducive to the microbes' growth. This type of assay generates a "minimum inhibitory concentration (MIC)," or the concentration that prevents growth of the microorganism, as measured by a lack of turbidity (in broth) or colony formation on agar. A "minimum lethal concentration (MLC)" may be determined in the broth dilution assay by transferring media from tubes or wells where no growth occurred to fresh media. If no growth occurs in the fresh media, the assumption is that the microorganism was inactivated and thus that the concentration results in a 99.9% (3 log) reduction in microbial numbers. Obviously, both the MIC and MLC depend highly upon environmental growth conditions (e.g., pH, temperature) and the initial number of microorganisms.

Following an endpoint assay to determine appropriate concentrations, one can determine the influence of the compound on dynamic growth by incubating the target microorganism in a microbiological broth medium and taking repeated samples over time to determine the changes in numbers of survivors. The plot of survivors over time is sometimes referred to a "time-kill" curve, a term used in clinical microbiology. From this type of assay, it can be determined what type of inhibition the test compound causes over time. The type may manifest itself in a number of different ways (Figure 22.1). Compared with the control, concentrations of an antimicrobial that are at the MIC may prevent an increase in the final cell number, delay and prolong lag phase, suppress growth for a time and then allow recovery, or cause lethality indicated by a reduction in cell numbers. A success criterion for further evaluation of an antimicrobial in such a test would likely be (a) an increase in lag phase, or (b) slower/delayed growth rate in exponential or logarithmic phase, or (c) lower final numbers of microorganisms in the stationary phase, or (d) some type of inactivation with no regrowth (David et al., 2013). One point to remember in these types of assays is that an antimicrobial neutralizer should be used in the medium/diluent being used for enumeration of survivors so as to avoid obtaining any false positive results.

Before investing in elaborate and expensive challenge studies in actual food matrices, it is customary to assess the efficacy of promising antimicrobials in simple food systems. These studies may be done in culture tubes, using commercially sterile shelf-stable apple juice or ultra-high temperature (UHT)-sterilized shelf-stable 2% fat milk (Figure 22.2, Phase-2 Technical Success Criteria). These simple food models can be used to evaluate the effect of the food, including pH and binding of the antimicrobial by fat or protein. Obviously, antimicrobials that are bound or inactivated during thermal processing are not available to act against target microorganisms. Generally, one can expect the effect of juice to be similar to that of microbiological media because of the lack of protein and fat. In milk, there generally will be a dramatic drop in the activity of because of the high pH and binding by protein and fat. The purpose of these tests is to get an idea of what concentrations might be efficacious in the food product of interest. The next logical progression is to evaluate the antimicrobial in the actual food matrix of interest for testing the expected/desired shelf life and simulating production, processing, and packaging conditions present at the manufacturing plant. In some cases, commercially available antimicrobial products would have already undergone such testing in model systems against certain microorganisms of concern, and usage ranges may already be provided by the developer/vendor of the

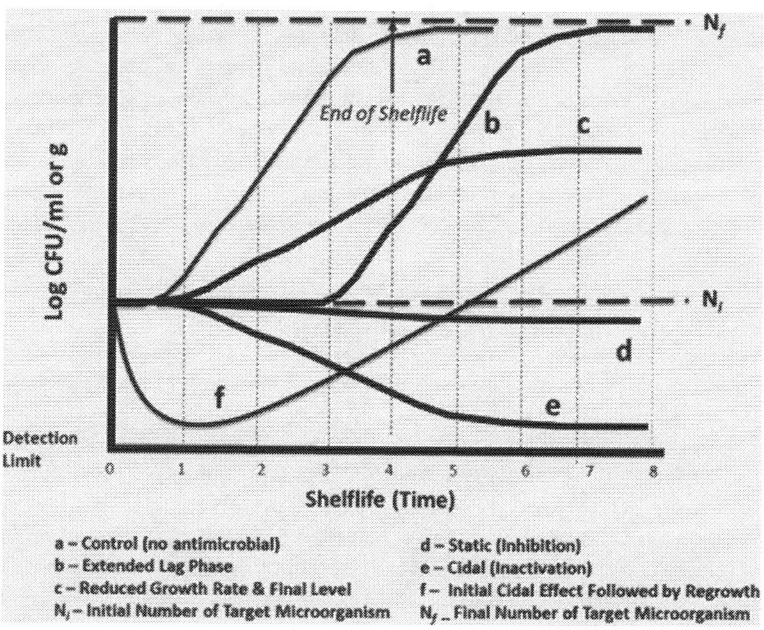

FIGURE 22.1 Diagrammatic representation of the impact of antimicrobials use on microorganism growth in a model media system (David et al., 2013): (a) control without antimicrobial; (b) extended lag phase; (c) reduced growth rate and final level; (d) static [inhibition]; (e) cidal [inactivation]; (f) initial cidal effect followed by regrowth; N_i: initial number of target microorganism; N_f: final number of target microorganism.

antimicrobial product. A product developer or microbiologist may decide if such data are suitable for the purpose of skipping the juice/milk model step, thereby saving time. However, often, when there are specific microorganisms of concern, such as a specific type of spoilage microorganisms, this may not be possible.

22.2.1 Combination Studies

Antimicrobials when used alone are often generally not effective enough or may have a negative effect on food properties at concentrations needed to be effective when used alone. Thus, it is often desirable to use them in combinations of other antimicrobials or with other functional ingredients such as acidulants and emulsifiers. When combinations of antimicrobials are elevated, three outcomes are possible. A combination may be "additive," i.e., the effect of the combined treatments is equivalent to the sum of the effects of the treatments acting independently. The two components can be "antagonistic" toward one another, actually resulting in a reduced efficacy of the combined treatments compared with their use independently. This might result, for example, from a chemical reaction between components to form a new, non-inhibitory, compound. The most desirable outcome is termed "synergistic," in which the activity of the combination is enhanced compared with the sum of individual treatments. Measuring synergism *in vitro* is most easily done with a microtiter "checkerboard" assay and by calculating a fractional inhibitory concentration (FIC), defined as the concentration of each antimicrobial in combination (see Chapter 2, "Methods for Activity Assay and Evaluation of Results") which produces inhibition of growth expressed as a fraction of the concentration that inhibits growth when the antimicrobial is used alone (Branen and Davidson, 2004; Davidson and Parish, 1989), or a fractional lethal concentration (FLC), defined as the concentration of each antimicrobial in the combination that produces lethality, expressed as a fraction of the concentration that is lethal when the antimicrobial is used alone (Techathuvanan et al., 2014). In foods, it is more difficult to determine synergistic activity, although a modified checkerboard assay is possible.

Antimicrobials may be used with physical preservation processes to enhance the effectiveness of the process or as a safeguard for post-process contamination. Corbo et al. (2009) described potential process interactions as: (a) partial inactivation of the microorganism by the preservation process, followed by continued inhibition or inactivation by the antimicrobial during storage; (b) enhancement of the process inactivation of the microorganism by the antimicrobial or vice versa; or (c) totally independent effects. Such physical processes might include heat (pasteurization) or non-thermal processes, such as high hydrostatic pressure processing (HPP), high-pressure homogenization, or pulsed electric fields. Several studies have demonstrated that antimicrobials enhance the effectiveness of physical preservation processes in inactivating target microorganisms (Black et al., 2005; Calderon-Miranda et al., 1999; Liang et al., 2002; Mosqueda-Meglar et al., 2012; Nakimbugwe et al., 2006; Nguyen and Mittal, 2007; Pathanibul et al., 2009).

22.2.2 Standardization of Efficacy Determination

One of the major needs in the arena of food antimicrobials, both natural and traditional, is the adoption of standard methods for the determination of efficacy. While medically important antimicrobials (i.e., antibiotics) and sanitizers have regulatory guidelines on efficacy evaluation, no such guidelines exist for food antimicrobials. In fact, there are no governmental standards concerning the efficacy of most commercially available antimicrobial food preservatives used as antimicrobials, with the exception of nisin and lysozyme. Thus, many commercial antimicrobial food preservatives, such as sorbate or benzoate, have not been evaluated for their intended purpose, i.e., inhibition or inactivation of microorganism. While this may not be a large problem if one is attempting to extend shelf life, it certainly is important if the compound is being used to control pathogenic microorganisms. Recommendations for the use of standard methods were called for over 20 years ago (Davidson and Parish, 1989) but to date, there has been no regulatory adoption. There are however regulatory requirements in the U.S. concerning the identification of preventive controls as part of a food safety plan, and if food antimicrobials are deemed as process preventive controls against pathogens for instance, a scientific validation of that control would be required. This could be a published study with similar parameters to the food in question or a challenge study in the actual food system.

22.3 Considerations for Commercial Application of Antimicrobials in Food

Attempts at pairing a specific food matrix in need of a secondary barrier for food protection with a potential antimicrobial is very rarely a linear or straightforward exercise. In the food industry, several competing factors need to be reviewed and co-optimized to meet predetermined technical success criteria and business decision criteria, as illustrated in Figure 22.2. Key factors that must be considered include: (a) efficacy against target microorganisms in the food matrix during processing; (b) business case and justification; (c) cost-in-use; (d) sensory effects; (e) storage; (f) end use by consumers; (g) regulatory and labeling considerations; and (h) sustainable supply (David, 2015a). To achieve the goal of successful application of an antimicrobial, certain "technical success criteria" must be established up front for managing business expectations, cost structure, implementation at the manufacturing plant, and product launch.

Figure 22.2 is a modified Stage® gate business process based on the "idea-to-launch" framework for product innovation and reducing time-to-market (Cooper and Edgett, 2012). The proposed framework is for the systematic pairing of a potential antimicrobial system with a food matrix, with clearly defined objectives, inputs, outputs, and success criteria for each of the three phases: PHASE-1 – Discovery (Proof of Concept), PHASE-2 – Technology Development, PHASE-3 –Technology Transfer (Scale-up and Commercialization).

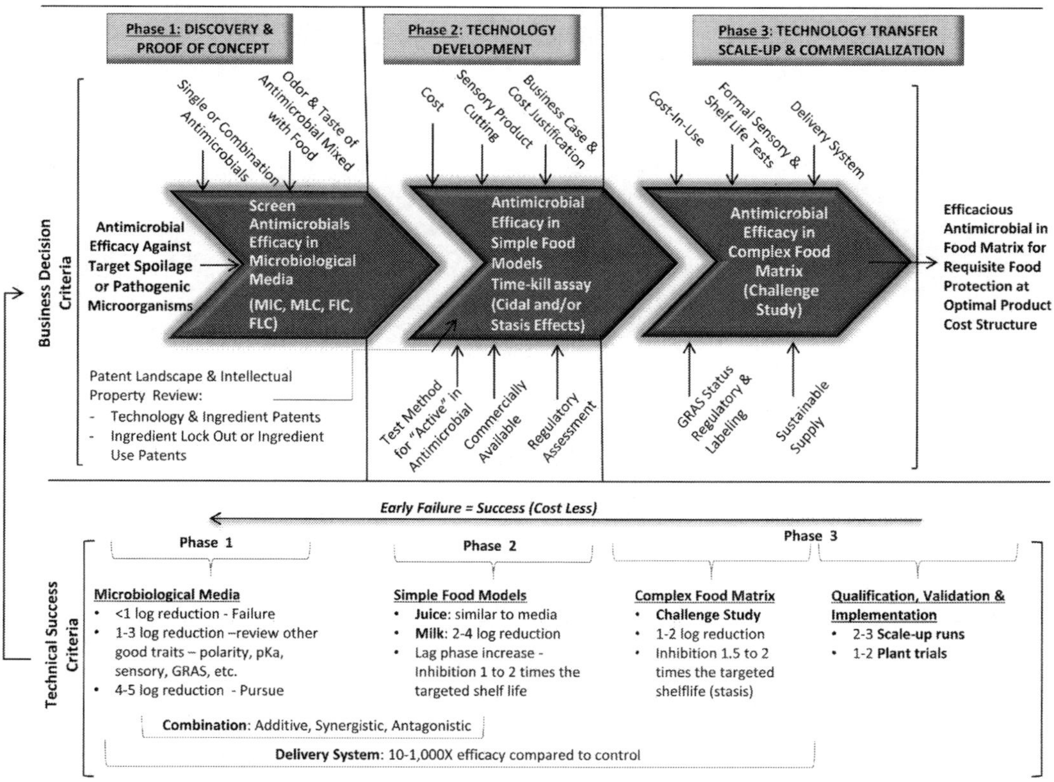

FIGURE 22.2 A modified Stage Gate® business process based on the "idea-to-launch" framework, governance, and decision criteria for pairing a food product with a potential antimicrobial system for requisite food protection at an acceptable product cost structure (David et al., 2013; David, 2015b).

22.3.1 Phase 1 – Discovery or Proof of Concept

This phase consists of high-throughput screening of promising antimicrobials against target microorganisms via appropriate assays to determine MIC and MLC. Antimicrobials differ in their ability to inhibit or inactivate vegetative cells and spores of Gram-positive bacteria, Gram-negative bacteria, and yeasts and molds. As previously stated, the first step in choosing an antimicrobial is to correctly identify and characterize target spoilage and/or pathogenic microorganisms from food. In addition, one should have a good understanding of factory microbial ecology, including vectors, incoming bio-burden load in ingredients, and data trends from the environmental monitoring program. No single antimicrobial can control all types of bacteria, yeasts, and molds in all food matrices. Lower dose concentrations for MIC and MLC are indicative of higher efficacy. Also, an order of magnitude of reduction in microbial numbers relative to initial inoculum level at time zero can be approximated. Thus, for example, successful candidate antimicrobials causing a 4- to 5-log reduction would be moved to the next phase of technology development. It is customary to review those antimicrobials with a score of 1- to 3-log reduction for other good traits as well, such as polarity, pK_a, sensory effects, GRAS status, etc. Even though most antimicrobials come with vendor-generated technical information and MIC and log reduction values, it is prudent for the user to re-check the MIC and MLC under desired environmental conditions of pH and temperature and against microorganisms isolated from product recall or spoiled product or factory-specific environmental microbiome (Benson et al., 2014; David, 2015a). Vendor MIC or MLC data may look promising, but may not have undergone peer-review. Scientists evaluating food antimicrobials data should investigate the methods used in the development of such data, such as whether or not stressed cells were resuscitated using appropriate methods (conservative approach) or samples were plated only on selective and differential media.

A quick test for antimicrobial impact on the odor and taste of the target product is essential. Usually, three levels of antimicrobials (MIC, below MIC, above MIC) are mixed with finished product to assess the concentration of the subject antimicrobial. Because the finished product is the basis for this quick test, it does not account for the impact of processing conditions on final product sensory characteristics or efficacy of the antimicrobial. Combination systems with other antimicrobials or other intrinsic or extrinsic hurdles may also help lower the use and dosage of individual antimicrobials for minimizing negative sensory impact and optimizing cost-in-use. In-company food technologists may also prepare multiple pilot batches to assess differing concentrations of antimicrobial on the resulting finished product's sensory characteristics. Such evaluations can range from the highly analytical to less so, and such evaluation of antimicrobial impacts on finished product characteristics is not necessarily limited to only the largest food manufacturing companies.

22.3.2 Phase 2 – Technology Development

This phase is where the bulk of the investment (resources and cross-functional teams), testing, and assessment work are staged and completed to facilitate making the "go/no-go" technical recommendations and business decision. Presented below are several work streams in a linear manner, but not rank-prioritized. Almost all of the below-described work streams must be reviewed simultaneously to evaluate early failure, not to proceed or to further resource to expedite results.

22.3.2.1 Efficacy

Interestingly, in the product development and product commercialization process, efficacy and validation are often the last consideration. Efficacy is often assumed, based upon prior experience, textbooks, or "taking an ingredient company's word for it." As such, technical personnel can be left scrambling to meet deadlines that well surpass the time needed to perform proper research and validation. Nonetheless, the demonstration of efficacy in product is essential to having a viable, wholesome, safe product in the marketplace that will not cause illness or spoil prematurely. For product renovation, efficacy studies should include both chemical antimicrobials currently used in formulation and potential replacement clean-label ingredients to assess the equivalency of efficacy in food matrix. As a rule of thumb, the efficacy of natural not-purified natural clean-label ingredients will be slightly lower than traditional purified chemical ingredients.

22.3.2.2 Sensory Impact

Antimicrobials may leave a residual odor or taste that can potentially impact finished product sensory and quality attributes. Organic acids acetic and lactic acids are widely used in many different food systems and are used topically on some food but impart a noticeable flavor. Plant-based extracts from citrus or spice must match the flavor profile of the food product. Oregano, thyme, or rosemary extracts are sometimes used, but only in savory-type foods that mesh with those seasonings. Fermentates are byproducts of lactic acid bacterial fermentation that contain mixtures of organic acids and bacteriocins and are sold as dry blends on a maltodextrin carrier or in a syrup form. The pungency and acidity of fermentates may preclude the usage level and desired efficacy, especially if propionic acid is prevalent in the blend. These relatively obvious impacts should enable product developers to rule out certain antimicrobials as options early in the process.

It is important to do a formal cutting of food product with antimicrobial concentrations equal to and less than MIC values established from Phase 1 work, to quickly evaluate any negative sensory impact. A more formal sensory cutting such as a consumer liking test (CLT) is required during Phase-3 prototyping and scale-up work streams (Figure 22.2). It may be important to develop a "tool box" of antimicrobials to appropriately pair with different food matrices, e.g., savory, dessert (sweet), and neutral. For example, sweet essential oils may be more appropriate for use in puddings and dessert pies, whereas essential oils from coriander, lemongrass, or mustards may be more compatible with savory sauces, meat and vegetable entrees, and pot pies. Combinations of antimicrobials need to be developed that may be more efficacious (additive or synergistic) and cause less impact on the sensory properties of the food. Taste and total quality in the final saleable product are not negotiable; antimicrobial candidates that perform well with respect to pathogen or spoilage microbe inhibition but which yield unpalatable flavor/odor outcomes will not be useful preservatives.

22.3.2.3 Cost-in-Use

Cost-in-use is one of the business critical and sensitive constraints that should be evaluated early on for the determination of increased cost per unit of finished packaged product. The cost-in-use analysis should also include the cost of inert carriers (e.g., maltodextrin, vegetable oil, glycerol, NaCl, etc.), enablers (i.e., emulsifiers, surfactants, and wetting agents), and delivery systems for carrying the antimicrobial to target microorganisms in the food matrix. It is important to understand and develop specifications for incoming microbial loads (including bacterial spores) for the above ingredients.

Often, natural and clean-label antimicrobials are more expensive than traditional chemical preservatives, and cost can be higher by a factor of ten or more. The vendor-provided cost per pound list price for Phase-1 successful antimicrobials helps one to assess whether the product in question can absorb upcharge per case of finished product, and thus to make a reasonable business case with marketing and other key stakeholders. The rule of thumb is that the cost of antimicrobials should be less than or equal to $0.015 to $0.025 per pound of finished packaged product. Higher costs of $0.03 to $0.07 per pound of premium-type or innovation product can be justified for antimicrobials meeting or exceeding many or all of the success criteria shown in Figure 22.2 (see Chapter 1, "Food Antimicrobials—An Introduction," Table 1.7).

It is important to recognize that the structure-function of the active agent is identically the same in both traditional chemical antimicrobials and clean-label natural formats (e.g., nitrite in sodium nitrite and nitrite in cultured celery juice powder; propionate in calcium propionate and cultured wheat starch). Also, traditional chemical antimicrobial preservatives have regulatory approval and a long history of use in commercial products (sorbates, benzoates, nitrites, propionates, etc.). However, some consumers prefer and demand premium foods that are not severely processed and made with clean-label natural ingredients including antimicrobial preservatives, despite the potential for increased cost of such products at retail over other foods made without such ingredients. As a result, manufacturers, retailers, and food service operators are responding to the clean-label movement with new and improved products (David, 2015b; see Chapter 1, "Food Antimicrobials—An Introduction," Tables 1.4 and 1.5).

In contrast to foods made with natural and/or clean-label antimicrobials, for many products, the goal is simply to have a fit-for-purpose product that appeals to cost-conscious customers, and high-volume retailers. In such cases, these consumers are not choosing based upon claims such as "no preservatives," but instead are more concerned with price and budget management. For the processor, such products can be produced in high volume and with a relatively long shelf life with the use of traditional preservatives, such as a sodium benzoate and potassium sorbate blend (sometimes with sodium hexametaphosphate (SHMP) and/or EDTA (ethylenediaminetetraacetic acid) in sauces, condiments, and beverages, or such as a sodium lactate and sodium diacetate blend in ready-to-eat meat products).

22.3.2.4 Capital Expenditure

Capital expenditure may also need to be factored in, such as a need for the installation of special mixing and dosing units at batching or at filler operations. For example, special mixing and dosing units are needed for the application of natamycin and dimethyl decarbonate (DMDC). Precise delivery of the approved antimicrobial into food is critical for compliance with rules on the regulatory agency-approved dose at the point of application. The regulatory agency-approved maximum limit is based on the concentration of antimicrobial at the point of application and not on residual activity in downstream finished product during or at the end of shelf life.

22.3.2.5 Minimum Order Quantity (MOQ)

For antimicrobials imported from other countries, it is important for a procurement department to negotiate minimum order quantity (MOQ) with ingredient suppliers. The standard threshold is about 10,000-pound per shipment in 50 to 100-pound sub-container units. Often MOQ may greatly exceed annual usage of ingredient needed for producing and stocking finished product inventory, impacting working capital. Scheduling, stock keeping unit (SKU) volumes, ingredient inventory, ingredient shelf life, and market price will determine and justify accepting large volume MOQ from vendors.

22.3.2.6 Patent Landscape

Patent landscape and intellectual property (IP) review are in order at this stage to ensure that there is room for creating unique and compelling consumer solutions and innovation, and to create substantial competitive advantage via IP and patents. It is important to ensure that the application of an antimicrobial or combination of antimicrobials in target categories of products is not protected by either technology or ingredient patents, or by ingredient-lock out or by ingredient-use patents. For example, there may be restrictions on the use of a lactate-diacetate system in certain types of fresh meat products or restriction on the use of nisin plus natamycin for the control of spoilage due to lactic acid bacteria in salad dressings. Due diligence is the key so as not to infringe on any domestic or international published patents or patent applications. It is possible to use ingredients protected by patents by way of paying royalties directly to the inventors or entity, or indirectly paid to the ingredient supplier as a part of the ingredient price.

22.3.2.7 Regulatory Assessment

Regulatory assessment is also done at this stage to ensure that there are no "red flags" with regard to use, human safety, and toxicology of both parent antimicrobial and breakdown chemical compounds in buffers or food models. In addition, one needs to be cognizant of constraints due to regulatory limits set for most additives and GRAS antimicrobials (FDA, 2012; USDA-FSIS, 2012). At this phase, an assessment should be made of USDA and FDA boundaries for the labeling of antimicrobials in any subject food. Similar assessment should be made for antimicrobial use in foods manufactured in the U.S. that are destined for export to other countries.

22.3.2.8 Challenge Studies

Challenge studies are done in a food matrix with the antimicrobials at the pre-process, in-process, or post-process stages at storage and abuse conditions, and in packaging over 1× and 1.5× the normal shelf life of the product. The National Advisory Committee on Microbiological Criteria for Foods (NACMCF) (Anon., 2009) has developed guidelines for conducting challenge studies with regard to pathogen inhibition and inactivation that should be adapted when developing and executing challenge studies.

For some meat products, it is important to conduct separate independent microbial challenge studies targeting specific pathogens – *Listeria monocytogenes*, and spores of toxigenic clostridia (*Clostridium botulinum*, *Clostridium perfringens*). Each challenge test will have specific success criteria, method of inoculation (surface versus in-product, or both), incubation and sampling protocols, total test duration, and cost. In addition, it is customary to test uninoculated treatment samples for indicator microorganisms to monitor the growth of natural microflora which may undermine product shelf life and affect sensory and quality attributes (Golden et al., 2017). For refrigerated foods, it is customary to test for both psychrotrophic spores and psychrotrophic vegetative cells (Benson et al., 2014; David, 2015a; Bozkurt et al., 2016).

22.3.2.9 Sensory and Shelf Life Study

For proper change management, scale-up, and directional next steps, sensory and shelf life studies are done with a formulation containing preservative made on fully scalable pilot plant-scale equipment and/or small batches of product made in the production plant. A consumer liking test (CLT) sensory testing of baseline control and treatment samples is done using a nine-point hedonic scale with preference scores and overall liking scores, and overall preference at 90% confidence interval (CI). The decision to move forward is based on parity for sensory and quality attributes – color, flavor, texture, mouthfeel, and other product-specific traits. Also, formulations may need to be adjusted for salt and sodium levels while using antimicrobials diluted with salt or sea salt (nisin, celery juice powder, etc.) for optimizing taste, label claims, and compliance.

There are some additional considerations when using calcium propionate in products baked with fruit puree. For example, there can be syneresis of fruit core in coextruded cereal bar during baking due to interaction between calcium in calcium propionate and sodium in alginate in the fruit base (David, 2015b). Also the use of propionate at higher levels for mold-free extended shelf life may contribute to bitterness in baked goods, necessitating the use of other ingredients such as bitterness maskers. Bitterness maskers can be labeled as a natural flavor, but with significant upcharge per case of product.

22.3.2.10 The Production Process

There may be several aspects to the production process to consider when selecting antimicrobials. Physical processes like in-container pasteurization can reduce or eliminate the need for antimicrobials. Heat treatments can render antimicrobials less (or, in some cases, more) efficacious. The configuration and surface area of the food can influence the selection of antimicrobials: smooth-surface products are more conducive to topical antimicrobial treatments. The degree to which the surface area of food components may be exposed to antimicrobial ingredients by being chopped, ground, blended, comminuted, or emulsified into foods is a critical factor in selection. The physical aspects of the packaging format can preclude the effective use of certain antimicrobials. For example, many fresh meats are now sprayed or dipped with processing aids like lauric arginate or octanoic/citric acid combinations that can cause a reduction in *Salmonella* or *E. coli* O157:H7. The efficacy of these antimicrobials depends not only on concentration and volume applied but also upon the application method (e.g., dip versus spray), contact time, and, for finished products, the capillary action necessary to distribute the antimicrobial across the product surface.

22.3.2.11 Occupational Safety and Health Act (OSHA)

The human safety of operators in the warehouse and batching stations is important while handling powder, liquid, or frozen formats of antimicrobial preservatives. The requirements for handling and batching provided by vendors in the safety document sheet (SDS) should be duly followed with appropriate use of personal protective equipment (PPE) and proper ventilation. Also, disposal of out-of-date ingredients should be done per vendor recommendations. Additional precautions should be taken by operators while handling certain antimicrobials that may be considered allergenic or contain allergenic inert ingredients/stabilizers. For example, antimicrobials derived from celery or mustard are considered allergenic in Canada and EU countries. Other routine precautions of line isolation, allergen wash-down procedures, and labeling and SKU palletizing should be followed through per company production directives.

22.3.3 Phase 3 – Technology Transfer (Scale-up and Commercialization)

A determination of efficacy against target microorganisms in the food matrix of choice is a prerequisite for the further investment of resources. A desired success criterion needs to be formulated for these efficacy tests, as shown in Figure 22.2. Additionally, FSMA mandates scientific validation occur prior to actual commercial processing for process preventive controls, or within 90 days of startup, so this would be the opportune time for in-plant validations in commercial plants and/or pilot plants.

22.3.3.1 Food Product Matrix

The efficacy of antimicrobials in foods is most easily determined and interpreted in carbohydrate-based beverages, followed by bakery products, fruits, vegetables and produce, dairy products, and meat, poultry, and seafood products (Davidson et al., 2013). Pilot-scale production tests or small batches in full-scale processing plants can be as simple as adding the preserving ingredient and collecting samples for evaluation. However, it is more likely that adding antimicrobials will require a great deal of forethought and planning as it influences other attributes of the food system that require a modification of the process. In either case, it is at this stage that the incorporation of the antimicrobial into or onto the food must result in adequate coverage and/or distribution and interaction with target microorganisms. A key objective of scaling and commercializing antimicrobial technologies into full-scale food production is confirming that the characteristics hold true, that the product impact is as expected at the time of processing and during shelf life, and that the naturally occurring microorganisms in the food system and processing environment are also affected similarly to laboratory strains.

22.3.3.2 Beverages

Carbohydrate-based beverages such as fruit juices and soft drinks present the fewest challenges associated with incorporating antimicrobials. They are not dissimilar to microbiological media; these systems have relatively low pH and very low to non-existent protein and lipid contents, and they provide a homogenous liquid system for the dispersion of antimicrobials. The only hurdle for application in this type of food would be undesirable interactions of the antimicrobial with simple and complex carbohydrates and minerals (e.g., the chelator sodium hexametaphophate works well in unfortified fruit beverages, but not in calcium-fortified versions). Protection against the microbiological spoilage of beverages can be achieved using antimicrobials and/or processing techniques such as hot-filling, tunnel pasteurization, ultra-high temperature treatment (UHT), or pasteurization followed by aseptic packaging, and/or pasteurization followed by chilling the beverage. Generally, beverages with a pH < 4.6 can be preserved with antimicrobials, heat-processed, and filled into packages such that the product is not re-contaminated. For example, process techniques such as cold-filling with antimicrobials or pasteurization followed by cold-filling may be used to preserve this type of beverage. In a similar manner, this same beverage may be processed using non-preserved techniques such as hot-filling, tunnel pasteurization, pasteurization followed by aseptic filling, or even requiring the beverage to be chilled (i.e., under refrigeration following the pasteurization step). Beverages having a pH ≥ 4.6 must be processed such that spores are destroyed using ultra-high temperatures followed by aseptic filling into packages or retorting sealed packages of product.

One of the most common preservation systems for acidic, shelf-stable carbonated and non-carbonated soft drinks relies on weak acid preservatives (e.g., benzoic and/or sorbic acid). Benzoic and sorbic acids (and salts thereof) effectively inhibit yeast, bacteria, and molds with only a few exceptions. Weak acids in beverages exist in equilibrium between their dissociated and undissociated forms which are dependent upon the dissociation constant of the acid (pK_a) and the beverage pH (Table 22.1). The pK_a for benzoic acid is 4.19 and the pK_a of sorbic acid is 4.76. A beverage pH below the pK_a of the particular acid shifts the equilibrium towards the undissociated form (i.e., benzoic acid versus benzoate). The undissociated form is more efficacious against microorganisms; therefore, weak acid preservatives are most effective in the low pH range. The preservation properties of weak acids may be enhanced by the addition of chelating compounds to the beverage. For example, common chelating compounds added to beverages include calcium disodium ethylenediaminetetraacetic acid (EDTA) or one or more of the polyphosphates, such as sodium hexametaphosphate (SHMP). In high-nutrient non-carbonated products, such as those beverages containing juice, vitamins, and/or minerals, the weak acids are more likely to exert inhibition if used in conjunction with preservative enhancers.

Weak acid preservation systems, however, have limitations. Genetic adaptation and subsequent tolerance by microorganisms may be one of the biggest concerns. Certain yeasts, such as *Zygosaccharomyces bailli*, *Zygosaccharomyces bisporus*, *Candida krusei*, and *Saccharomyces cerevisiae*, have specific genes that enable them to resist the weak acid preservatives and grow, despite their presence and regardless of the co-presence of EDTA or SHMP (Piper, 2001; Pitt and Hocking, 2009). Some bacteria, such as *Gluconobacter* spp., are also thought to be preservative-resistant. The levels of weak acids necessary to overcome this resistance have been shown to be far beyond regulatory limits on use levels. Spoilage of preserved teas, juice-containing beverages, and carbonated beverages is commonly due to preservative-resistant yeasts. Weak acids are also known to impart a sensory impact in the form of throat or mouth burn when used at levels needed to prevent spoilage. Although there are certain shelf-stable beverages where this may be acceptable, often this sensory perception is considered negative. In addition, non-government organizations and also some international government agencies have raised concerns regarding the use of weak acid preservatives in beverages and foods. Finally, some consumers take note of ingredients and certain niche markets prefer preservative-free products. Natural antimicrobials may be alternatives to traditional ways of preserving beverages. Antimicrobials that could be labeled as natural could also eliminate hot-fill requirements for unpreserved shelf-stable preservative-free beverages. Thus, it would be desirable to provide a natural antimicrobial and/or an antimicrobial system that inhibits the growth of microorganisms.

Various antimicrobials are capable of maintaining shelf-stable carbonated sodas either by killing yeast and bacteria at the point of production or by the inhibition of yeast and bacteria growth in the product at ambient temperature. Challenge testing is recommended to validate microbial control and shelf-stability. Sodium benzoate added at about 300 ppm alone is sometimes sufficient for the preservation of carbonated sodas with pH 3.5 or less. At lower pH such as 2.5, lower sodium benzoate concentration such as 175 ppm may be adequate. As stated above however, certain yeasts and some bacteria can be resistant to weak acid preservatives. Typically, this does not happen in carbonated sodas, but it is possible. In such cases, the addition of potassium sorbate and SHMP or potassium sorbate and EDTA can help enhance the antimicrobial properties of benzoate and suppress even preservative-resistant yeasts and bacteria. These factors would influence choice of use and usage level.

TABLE 22.1

Recommended pH Range for Some Commonly Used Antimicrobials

Antimicrobial Agent	Recommended pH Range
Benzoic acid	2.5–4.0
Sorbic acid	3.0–6.5
Propionic acid	2.5–5.0
Acetic acid	3.0–5.0
Parabens	3.0–9.0
Sulfites	2.5–5.0
Nitrites	5.0–5.5

TABLE 22.2

Efficacious Combinations of Lauric Arginate with Cinnamic Acid at Different pH Levels

pH	Cinnamic Acid (ppm)	Lauric Arginate (ppm)
3.0	40	2.5
3.5	50	2.5
4.0	50	10

(Adapted from Dai et al., 2010.)

Natural preservation may be a viable option. Cinnamic acid is a natural preservative that has been researched as a preservative against yeast and bacteria in acidic beverages such as teas and juice-containing beverages. Efficacy varies by pH; lower concentrations are needed to preserve against yeast at lower pH. Typically, 60 ppm would be the minimum needed to inhibit yeasts. Similarly, lauric arginate ester (LAE) is a C-12 fatty acid-based molecule that can be used in conjunction with cinnamic acid to preserve beverages. Efficacious combinations of LAE with cinnamic acid at different pH levels are given in Table 22.2.

In some applications, the use of dimethyl dicarbonate (DMDC) (see Chapter 13, "Dimethyl Dicarbonate and Diethyl Dicarbonate") can be an effective alternative to treating packaged beverages at the point of filling and sealing because it is slightly soluble in water, extremely reactive, and will quickly undergo hydrolysis reactions with organic matter (e.g., microbial cells). The remaining byproducts are alcohol and carbon dioxide; thus, there are typically no labeling requirements when used in accordance with current regulatory agency-approved levels, and it is considered a processing aid. Hours after use, DMDC is not detectable in the product. The challenge is to destroy target microorganisms quickly before the hydrolysis is completed. DMDC was first approved for addition to wines as a yeast inhibitor at a level not to exceed 200 ppm. DMDC can be used in carbonated, dilute beverages containing juice, fruit flavor, or both, in which the juice content does not exceed 50%, at a limit of 250 ppm. In order to ensure efficacy, DMDC is supplied to each individual bottle after filling and immediately prior to capping. Therefore, dosing equipment is needed in order to ensure each bottle receives the proper dose and that caps are applied immediately so that product cannot be recontaminated after DMDC has dissipated. This can be a significant up-front expense.

Molds like *Penicillium* can become problematic if the hygienic conditions of the plant and equipment are not maintained. Most molds will be inhibited by sodium benzoate and/or potassium sorbate, but certain *Penicillium* spp. can become tolerant to sorbic acid if allowed to persist in the environment or equipment. Once these resistant molds reach the product in a package, they will potentially overcome the preservative and produce a colony or a mass within the package.

22.3.3.3 Bakery Products

Bakery products may require antimicrobials to inhibit the growth of spoilage organisms and increase shelf life. These products have relatively low protein content, and surface application would inhibit spoilage fungi. Negative aspects of bakery products as to antimicrobial application are a neutral pH and non-homogeneity of the food system. Calcium propionate is commonly used to prevent mold growth on shelf-stable bakery products. If bakery products contain fillings or sauces, the utilization of weak acid preservatives can prevent microbial spoilage for several days at ambient temperature. However, scientific support for bacterial pathogen control would also be needed for ambient-stored products with a pH and a_w combination that falls within the potentially hazardous foods ranges. Some breads intended for convenience store (c-store) distribution or distribution to warm climates may include calcium propionate, sorbic acid, and natamycin to protect against spoilage by molds.

22.3.3.4 Fresh and Minimally Processed Fruits and Vegetables

The application of antimicrobials to minimally processed fruits should be similar to application to fruit juices. Fruits generally have low lipid and protein content, which decreases the opportunity for interactions. They generally also have lower pH and lower a_w in their edible flesh, rind, and/or skin than vegetables, which may stress the target microorganisms. The only negatives to application on fruits are potential non-homogeneity as to composition and topography, as well as contamination sites of fruits. Those with higher pH (e.g., cantaloupe) may show decreased efficacy of antimicrobials, since low pH would not be working in concert with the antimicrobials to inhibit target microorganisms. Cut fruits intended for several hours or days of shelf life, and used at food service establishments, are preserved by use of a solution of citric acid, calcium chloride, sodium benzoate, and potassium sorbate to prevent not only microbial spoilage but also enzymatic browning through polyphenoloxidase (PPO) denaturation.

Vegetables and fresh produce generally have low lipid and protein content but have high pH and a_w and are highly non-homogenous as to composition and site of contamination. Produce can be treated with antimicrobials to reduce pathogen counts. Cut produce is treated with antimicrobials such as chlorine or peroxyacetic acid (PAA) via flume water or by spray application (see Chapter 19, "Use of Antimicrobials as Processing Aids in Food Processing"). Chorine that has been stabilized by a phosphoric acid and propylene glycol mixture has been shown to enhance the efficacy against *Salmonella* and *E. coli* O157:H7 in fresh-cut lettuce processing (Nou et al., 2011).

22.3.3.5 Dairy Products

Dairy products are very difficult systems in which to achieve activity of antimicrobials because these products have high pH, high a_w, and high levels of protein, fat, and divalent cations as interfering compounds, all of which can interact to reduce the activity of antimicrobials. As a result, antimicrobials have been applied to dairy products with varying success. Nisin has efficacy against *C. botulinum* and other clostridia in cheese for spoilage prevention during ripening. Nisin may also be used in refrigerated dairy products like sour cream. Natamycin is commonly used as a mold-inhibitor on sliced and shredded cheese.

22.3.3.6 Meat and Poultry Products

Meat and poultry products may be the commodities with the highest number of inherent properties that will limit effective application of antimicrobials. These include a non-homogenous substrate surface (particularly for meat animal carcasses but also for derived further processed food products), neutral pH, and high concentrations of lipids and proteins. In addition to commodity effects, products formulated with gums, phosphates, titanium dioxide, or other additives may tend to bind or inactivate antimicrobials, leading to diminished efficacy and much higher cost-in-use. Developing appropriate antimicrobial delivery systems (see Chapter 20, "Delivery Systems"), such as encapsulation (e.g., emulsions, nanoparticles) can reduce their interactions with food components and increase interaction with the target microorganisms and allow for controlled release of the antimicrobial during the shelf life of the food (Davidson et al., 2013).

22.3.3.6.1 Fresh Meats and Poultry Products

Fresh meats and poultry products could be formulated with various ingredients in order to sustain fresh, desirable organoleptic properties, and restrict microbial growth. Many raw meats and poultry items are enhanced for moisture, tenderness, and yield with solutions containing sodium phosphates and sodium chloride. These are not preservatives per se, but they do impart a change (e.g., buffering) to the muscle, and the process of introducing these solutions into the muscle can result in the translocation of bacteria from exterior surfaces to internal meat. Whole-muscle products that are distributed and displayed refrigerated may be moisture-enhanced by needled injection or vacuum tumbling to incorporate preserving ingredients and antioxidants as well as salt, phosphates, and, in some products, flavorings. Sodium lactate has been used for this purpose in moisture-enhanced pork (Banks et al., 1998; Brewer et al., 1995). Antioxidants like rosemary extract can be incorporated into the brine solution to delay the oxidation of lipids. Topical treatments to fresh meat and poultry show promise in reducing bacterial pathogens and may be selected to reduce risk. The application of 200 to 400 ppm LAE to chicken filets caused a 0.7 to 1.0 log reduction of *Salmonella* (Sharma et al. 2013a), but no significant reduction of *Salmonella* occurred in ground chicken treated with 200 ppm LAE (Sharma et al., 2013b). Other treatments with octanoic acid, levulinic acid, and caprylic acid, either individually or in combination with other antimicrobials, have similar outcomes against pathogens in meat and poultry.

22.3.3.6.2 Raw Ground Products

Raw, ground products could be combined with antimicrobials like acidified sodium chlorite, as a means to cause a reduction of pathogens like *E. coli* O157:H7 (Visvalingam and Holley, 2018). Some novel uses of plant-derived antimicrobials like blueberry extract (Das et al., 2017) and combinations of lemon juice and essential oils derived from herbs (Chung et al., 2018) have had moderate success for pathogen reduction in fresh meats at the levels below sensory impact.

22.3.3.6.3 Cooked, Ready-to-Eat Meats

Cooked, ready-to-eat (RTE) meats are examples of substrates in which antimicrobials have inhibitory activity against multiple pathogens and spoilage microorganisms simultaneously. It has been well-established that the organic salts such as potassium and sodium lactate and sodium diacetate are inhibitory to *L. monocytogenes* in RTE meat (Mbandi and Shelef, 2001). There are three main mechanisms of action of organic salts against the pathogen: (a) lowering of the a_w of the meat (Chen and Shelef, 1992); (b) organic salts in the undissociated acid form pass across the bacterial cell membrane and disrupt pH balance; and (c) feedback inhibition occurs when lactate and acetate force a shift of *L. monocytogenes* toward fermentative production of acetoin and away from aerobic respiration and the production of lactate and acetate (Stasiewicz et al., 2011). Further, it has been shown that diacetate, lactate, and product moisture content significantly impacted the growth rate of the pathogen, whereas the influence of sodium chloride was not statistically significant (Legan et al., 2004; Seman et al., 2002). Studies in ground pork showed that pH, a_w, and organic salt were significant factors in the growth of *L. monocytogenes* (Vandeven, 2004; Zuliani et al., 2007).

The MICs of sodium lactate and sodium chloride against five strains of *Staphylococcus aureus* were researched (Houtsma et al., 1993). The MICs for sodium lactate ranged from 268 to 625 mM (or about 2.5 to 5.8%), while the MIC for NaCl was 1873 mM (about 10%). Other researchers showed that 5% sodium lactate in a broth solution extended the lag phase, diminished the overall growth rate, and reduced the stationary phase ultimate cell concentration of *S. aureus* compared to broth at equal pH and a_w with 2.6% NaCl (de Wit and Rombouts, 1990). Further, *S. aureus* was inhibited on ham formulated with 1.8% sodium lactate when held at 1 or 6°C (Jofré et al., 2008). Various organic salts are antimicrobial to *Clostridium perfringens* in meat and poultry. Calcium lactate at 1.0% or potassium lactate and sodium lactate at ≥2.0% controlled *C. perfringens* germination and outgrowth to <1 log CFU/g in boneless pork loins injected with minimal levels of salt (0.85% NaCl), potato starch (0.25%), and potassium tetra pyrophosphate (0.2%) at a 12% pump rate (amount of marinate incorporated into the meat, w/w) (Reddy et al., 2007). Calcium lactate lowered pH but did not change a_w, while potassium lactate and sodium lactate did not change pH but lowered a_w. Model roast beef (≥0.85% NaCl, ≥0.25% potato starch, and ≥0.2% potassium tetra phosphate) formulated with sodium and potassium lactate (≥1.5% PURASAL), sodium lactate and sodium diacetate (≥1.5% OptiForm), buffered sodium citrate (BSC, ≥0.75% Ional), and BSC with diacetate (≥1.00% Ional Plus) resulted in <1 log CFU/g increase following 18 and 21 h cooling rate from 130 to 45°F (Juneja and Thippareddi, 2004a). Turkey breast containing NaCl, potato starch, and potassium tetra pyrophosphate at 0.85, 0.25, and 0.20%, and formulated with sodium lactate (1.0, 2.0, 3.0, or 4.0%), sodium acetate (1.0 or 2.0%), buffered sodium citrate (Ional, 1.0%), or buffered sodium citrate plus sodium diacetate (Ional Plus, 1.0%) resulted in <1 log CFU/g increase following 15h cooling (Juneja and Thippareddi, 2004b). Higher concentrations of lactate and acetate were necessary to inhibit the pathogen during 18 h of cooling. Ional at 1.0% resulted in a <1 log CFU/g increase during 15, 18, and 21 h cooling. Organic salts inhibit multiple pathogens in meat products and therefore are good choices for safety and shelf life. For clean-label options, these can also be added in the form of byproducts of lactic acid bacterial fermentation.

22.3.3.6.4 Curing Agents for Meat and Poultry

As covered in Chapter 7 ("Nitrite"), the direct addition of sodium nitrite to cured meat and poultry products is completed to produce a consistent, safe product that prevents the growth of *Clostridium botulinum* (Tompkin, 1993). The combination of nitrite and salt is also inhibitory to the foodborne bacterial pathogen *Clostridium perfringens* (Taormina and Dorsa, 2004; Zaika, 2003). Also, the presence of nitrite enhances the suppression of growth of *Listeria monocytogenes* in ready-to-eat meats formulated with organic salts such as lactate and diacetate (Legan, et al., 2004; McClure et al., 1997; Seman et al., 2002; Shelef and Yang, 1991; Whiting and Masana, 1994). The USDA requires a minimum of 120 ppm ingoing sodium nitrite for cured products that are refrigerated.

22.3.3.6.5 Alternatively Cured Meat Products

Cured products like ham, frankfurters, and bacon are popular and in demand. However, some consumers have negative perceptions about sodium nitrite and/or the potential formation of nitrosamines in cured meats. Therefore, there is strong consumer and customer interest in products that are "natural," "uncured," or have "no nitrites or nitrates added," and yet have that characteristic cured color, flavor, and eating experience. USDA has required that uncured versions of "typical" cured meats be "labeled with such standard name when immediately preceded with the term 'Uncured' in the same size and style of lettering as the rest of such standard name" (USDA FSIS, 2006). "Natural" processed meats must not "contain any artificial flavor or flavoring, color ingredient, or chemical preservative" (USDA FSIS, 2005). "Organic" products are defined and controlled by the Organic Foods Production Act (1990), which is established by the National Organic Program Standards, and nitrite and nitrate are not permitted in such products. Thus, nitrite or nitrate, by themselves, cannot be directly added to any natural or organic processed meat product. Further, any "typically" cured meat product that does not have added nitrite or nitrate must be labeled "uncured."

"Uncured" or alternatively cured meats are made to appear cured and do have characteristic cure color and flavor but without the direct addition of sodium nitrite. Flavorings or spices such as celery juice or powder and sea salt act as natural sources of nitrate (Taormina, 2014). The naturally occurring nitrate is converted into nitrite by the use of starter cultures composed of nitrate reductase-positive and coagulase-negative cocci, such as *Kocuria varians*, *Staphylococcus xylosus*, and *Staphylococcus carnosus*. The process includes an incubation period of about 1 h at about 107°F to enable nitrate reduction before typical cook cycles. The industry has largely adopted a faster process utilizing pre-converted nitrite, which eliminates the need for a starter culture and incubation step for meat blends. Chemically, these alternatively cured meat products have similar total quality attributes to cured meats produced by the direct addition of sodium nitrite. Additional stability can be achieved by the addition of powder of Swiss chard or acerola cherry. These are natural sources of ascorbic acid, which serves as a cure accelerator as well as an antioxidant.

Alternatively cured meats can have variable residual levels of nitrite compared to cured meats made by the direct addition of sodium nitrite and sodium erythorbate. However, it is difficult to interpret residual nitrite survey data because only about 50% of the nitrite originally included in the product formulation can be detected by chemical analysis after the heating process is completed, and the subsequent loss of nitrite continues during product storage (such as in distribution and retail) (Cassens et al., 1979; Sauter et al., 1977). A research group at Iowa State University found significantly different levels of residual nitrite in frankfurters, ham, and bacon purchased in Midwest groceries (Jackson et al., 2011a). However, they stated, "The only meaningful conclusion that can be drawn from the commercial product residual nitrite and nitrate data is that residual nitrite and nitrate were indeed present in all naturally cured and conventionally cured products." A retail survey of 470 conventionally cured meat products and organic/natural/uncured/indirectly cured meat products from six major cities in the United States showed that residual nitrite levels were not different between these two categories (Nunez de Gonzalez et al., 2012). Levels of residual nitrite were also comparable to a prior retail survey in the 1990s (Cassens, 1997). The survey did reveal that some products varied in nitrate concentration.

Given the chemical similarities between conventionally cured and alternatively cured meat products, the behavior of foodborne bacterial pathogens of concern is also expected to be similar. Naturally cured ham, frankfurter, and bacon products can permit more growth of *C. perfringens* at abusive temperatures than their conventionally cured counterparts (Jackson et al., 2011b). Additional work showed that naturally cured hams permitted about 2-log more growth of *C. perfringens* growth at 68–71.6°F than conventionally cured products (Jackson et al., 2011a). Naturally cured frankfurters allowed about 1-log more growth than conventionally cured. The authors concluded that natural antimicrobials can contribute to suppressing the growth of this pathogen. Similarly, other researchers (King et al., 2015a) found that 1.0% fruit extract and a 1.0% cultured sugar–vinegar blend were effective at controlling *C. perfringens* in alternatively cured deli-style turkey breast, whether or not cultured celery juice powder (2.25% sodium nitrite) was included. Without celery juice powder, 0.7% dried vinegar and a 2.0% lemon–vinegar blend allowed for 2.0- and 2.5-log increases, respectively, and, 1.5-log increases in turkey with celery juice powder, all of which were in excess of USDA-FSIS mandatory limits on growth for stabilization compliance. Also, it was shown that equivalent concentrations of nitrite (regardless of the source) provided similar inhibition of *C. perfringens* during chilling (Taormina and Dorsa, 2004), and that ascorbate (added as a cure accelerant) enhanced the antimicrobial effect of nitrite on *C. perfringens* at concentrations commonly used in alternatively cured meats (King et al., 2015b).

Researchers from the Food Research Institute at the University of Wisconsin-Madison inoculated *C. botulinum* into conventionally cured ham with 142 ppm residual nitrite, no-nitrite ham, and three different naturally cured hams with 36–38 ppm residual nitrite after cooking, measuring time to toxin formation at refrigeration and abusive temperatures (Wanless et al., 2010). At 39.2°F, toxin was not formed in any of the samples. At 50°F, toxin formed in no-nitrite ham within 14 days, and in one of the three naturally cured hams in 21 days, but two naturally-cured hams and the conventionally cured control did not support toxin formation. At 71.6°F, the no-nitrite ham had toxin within 3 days, two naturally cured hams had toxin at 5 days, one naturally cured ham had toxin at 21 days, while the control did not permit toxin formation for the entire 28-day incubation period. In frankfurters, the no-nitrite version permitted toxin formation at 39.2°F within 63 days, but all other versions did not. At 50°F, cured frankfurters and two of three naturally cured frankfurters did not allow toxin formation for the 84-day duration, while one naturally cured frank had toxin at 56 days and the no-nitrite version had toxin at 14 days. At 71.6°F, one naturally cured frank did not allow toxin formation at all, but two naturally cured franks had toxin at 5 days. The cured control franks allowed toxin formation at 14 days. These data demonstrate a similar inhibitory effect in naturally cured products, albeit not as robust as conventional curing agents prepared by the direct addition of sodium nitrite.

The ingoing amounts of nitrate and nitrite and residual amounts in products are important but do not tell the whole story with respect to pathogen control in meat products. Sodium ascorbate, commonly referred to as a cure accelerator, increases the rate of some of these chemical reactions to expedite the development of cured color during processing. As a result of its function to reduce nitrite to nitric oxide, the use of ascorbate also results in a more rapid depletion of residual nitrite during the curing process (Hustad et al., 1973). Although residual nitrite concentrations deplete during processing and storage, regardless of the presence of ascorbate, the loss of added nitrite can be more pronounced when ascorbate is included (Brown et al., 1974; Perez-Rodriguez et al., 1996). Tompkin et al. (1978) demonstrated an additional layer of prevention against botulinal toxin formation when ascorbate was added to cured meats and proposed that the effect was likely due to the ability of ascorbate to chelate some essential cation required for recovery from nitrite-induced injury to *C. botulinum*. Duffy et al. (1994) demonstrated that 420 ppm of ascorbate used with up to 315 ppm of nitrite significantly improved nitrite's ability to influence *L. monocytogenes* lag time, reporting a difference of 25.4 versus 9.9 days of lag time in cured treatments with and without ascorbate, respectively. Although conventionally cured meats commonly contain both a cure accelerator (e.g., ascorbate or erythorbate) and 100 ppm of sodium nitrite, alternatively cured meats often are made with natural sources of nitrite, which provide 100 ppm of ingoing sodium nitrite, but without a curing accelerator such as ascorbate. Based on the work of King et al. (2015a, b), the presence of ascorbate in alternatively cured meats in conjunction with nitrite from natural sources is important in restricting the growth of *C. perfringens*. At least 50 ppm pre-converted nitrite and 250 ppm ascorbate from cherry powder is necessary to restrict *C. perfringens* growth to less than 1-log during 15 h cooling of ground turkey breast.

22.3.3.7 Sauces and Condiments

Sauces and condiments are traditionally acidified foods with pH less than 4.6. These products will typically contain vinegar (acetic acid) and/or citric or phosphoric acids. They are often preserved with weak acids such as benzoic and sorbic acid, and may be supplemented or boosted by the addition of EDTA. Natural preservation of these products can be achieved with nisin for the control of *Bacillus* (Poi et al., 2001) and other Gram-positive bacteria and natamycin for the control of yeast and mold.

22.4 Conclusion

In conclusion, pairing an antimicrobial with a food matrix is not a simple exercise. Due diligence and governance are in order while adding or removing a traditional chemical antimicrobial and replacing it with an equivalent natural and/or clean-label antimicrobial. Because antimicrobials are small but smart and nimble hurdles, expectations of their role, performance, and context must be realistic. To underscore, the use of antimicrobials in food is meant to add to the robustness of existing safety programs, quality assurance, and appropriate food safety controls and process controls, and not to correct or mask poor production practices. Unacceptable cost-in-use, negative impact on the quality and sensory attributes of the final product, and inadequate efficacy in the food matrix are the primary reasons for failure when attempts are made to couple a food with an antimicrobial system.

22.5 Acknowledgments

The core and framework for this chapter was based on a guidance document that was published in *Food Protection Trends* (J. R. D. David, L. Steenson, and P.M. Davidson. 2013, Vol 33, No. 4, pp. 238–247). We would like to acknowledge Dr. P. Michael Davidson for his pioneering contributions to the advances in antimicrobials and food safety, and to our professional development in the field covered. Also, we acknowledge Brennan David for designing Figure 22.2 on the "idea to launch" business framework and governance.

Disclaimer

References to commercial products and trade names are made with the understanding that no discrimination and/or endorsement by the authors or the organization they are involved with are implied.

REFERENCES

Anonymous. 2009. Parameters for inoculated pack/challenge study protocol. National Committee on the Microbiological Specifications for Foods. https://www.fsis.usda.gov/wps/wcm/connect/3b52f9c0-0585-4c0a-abf2- b4fc89a9668c/NACMCF_Inoculated_Pack_2009F.pdf?MOD=AJPERES

Banks, W.T., Wang, C., and Brewer, M.S. 1998. Sodium lactate/sodium tripolyphosphate combination effects on aerobic plate counts, pH and color of fresh pork longissimus muscle. *Meat Sci.* 50:499–504.

Benson, A.K., David, J.R.D., Gilbreth, S.E., Smith, G., Nietfeldt, J., Legge, R., Kim, J., Sinha, R., Duncan, C.E., ma, J., and Singh, I. 2014. Microbiome successions are associated with changes in chemical profiles of a model refrigerated fresh pork sausage during an 80-day shelf life study. *Appl. Environ. Microbiol.* 80:5178–5194.

Black, E.P., Kelly, A.L., and Fitzgerald, G.F. 2005. The combined effect of high pressure and nisin on inactivation of microorganisms in milk. *Innov. Food Sci. Emerging Tech.* 6:286–292.

Bozkurt, H., David, J.R.D., Talley, R., Lineback, S., and Davidson, P.M. 2016. Thermal inactivation kinetics of *Sporolactobacillus nakayamae* spores, a spoilage bacterium isolated from a model mashed potato-scallion mixture. *J. Food Prot.* 79:1482–1489.

Branen, J.K., and Davidson, P.M. 2004. Enhancement of nisin, lysozyme, and monolaurin antimicrobial activities by ethylenediaminetetraacetic acid and lactoferrin. *Intl. J. Food Microbiol.* 90:63–74.

Brewer, M.S., Rastogi, B.K., Rgoudelis, l., and Sprouls, G.K. 1995. Sodium lactate/sodium chloride effects on aerobic plate counts and color of aerobically packaged ground pork. *J. Food Sci.* 60:58–62.

Brown, C.L., Hedrick, H.B., and Bailey, M.E. 1974. Characteristics of cured ham as influenced by levels of sodium nitrite and sodium ascorbate. *J. Food Sci.* 39: 977–979.

Calderon-Miranda, M.L., Barbosa-Canovas, G.V., and Swanson, B.G. 1999. Inactivation of listeria innocua in liquid whole egg by pulsed electric fields and nisin. *Intl. J. Food Microbiol.* 51:7–17.

Cassens, R.G. 1997. Composition and safety of cured meats in the USA. *Food Chem.* 59:561–566.

Cassens, R.G., Greaser, M.L., Ito, T., and Lee, M. 1979. Reactions of nitrite in meat. *Food Tech* 33:46–57.
Chen, N., and Shelef, L.A. 1992. Relationship between water activity, salts of lactic acid, and growth of Listeria monocytogenes in a meat model system. *J. Food Prot.* 55:574–578.
Chung, D., Cho, T.J., and Rhee, M.S., 2018. Citrus fruit extracts with carvacrol and thymol eliminated 7-log acid-adapted *Escherichia coli* O157:H7, *Salmonella typhimurium*, and *Listeria monocytogenes*: A potential of effective natural antibacterial agents. *Food Res. Intl.* 107:578–588.
Corbo, M.R., Bevilacqua, A., Campaniello, D., D'Amato, D., Speranza, B., and Sinigaglia, M. 2009. Prolonging microbial shelf life of foods through the use of natural compounds and non-thermal approaches: A review. *Intl. J. Food Sci. Tech.* 44:223–241.
Cooper, R.G., and Edgett, S.J. 2012. Best practices in the idea-to-launch process and its governance. *Res. Technol. Manag.* 55:43–54.
David, J.R.D., Steenson, L.R., and Davidson, P.M. 2013. Expectations and applications of natural antimicrobials to foods: A guidance document for users, suppliers, research and development, and regulatory agencies. *Food Prot. Trends* 33:241–250.
David, J.R.D. 2015a. How industry uses big data: Metagenomics and beyond. 119th AFDO Annual Educational Conference, Indianapolis, Indiana, June 22, 2015. Session: Whole Genome Sequencing. Session Co-Chairs: Peter Salsbury and Dr. Eric Brown, FDA CFSAN.
David, J.R.D. 2015b. Application of natural antimicrobials in food: Food industry user perspective. IFT Annual Meetings, Session 008: Future of Food Preservatives: Emerging Natural and Potential Alternatives. McCormick Place, Chicago, IL, July 12, 2015.
Dai, Y., Normand, M.D., Weiss, J., and Peleg, P., 2010. Modeling the efficacy of triplet antimicrobial combinations: Yeast suppression by lauric arginate, cinnamic acid, and sodium benzoate or potassium sorbate as a case study. *J. Food Prot.* 73:515–523.
Das, Q., Islam, M.R., Marcone, M.F., Warriner, K., and Diarra, M.S. 2017. Potential of berry extracts to control foodborne pathogens. *Food Control* 73:650–662.
Davidson, P.M., and Parish, M.E. 1989. Methods for testing the efficacy of food antimicrobials. *Food Tech.* 43:148–155.
Davidson, P.M., Critzer, F.J., and Taylor, T.M. 2013. Naturally-occurring antimicrobials for minimally processed foods. *Annual Rev. Food Sci. Tech.* 4:9.1–9.28.
Davidson, P.M., Sofos, J.N., and Branen, A.L. 2005. *Antimicrobials in Food*, 3rd ed. Taylor & Francis, New York.
De Wit, J.C., and Rombouts, F.M., 1990. Antimicrobial activity of sodium lactate. *Food Microbiol.* 7:113–120.
Duffy, L.L., Vanderlinde, P.B., and Grau, F.H. 1994. Growth of *Listeria monocytogenes* on vacuum-packed cooked meats: Effects of pH, aw, nitrite and ascorbate. *Intl. J. Food Microbiol.* 23:377–390.
FDA. 2012. *Code of Federal Regulations: Title 21.* US Government Printing Office, Washington DC.
Golden, M.C., Wanless, B.J., David, J.R.D., Lineback, S., Talley, R.J., Kottapalli, B., and Glass, K.A. 2017. Effect of equilibrated pH and indigenous spoilage microorganisms on the inhibition of proteolytic *Clostridium botulinum* toxin production in experimental meals under temperature abuse. *J. Food Prot.* 80:1252–1258.
Houtsma, P.C., De Wit, J.C., and Rombouts, F.M. 1993. Minimum inhibitory concentration (MIC) of sodium lactate for pathogens and spoilage organisms occurring in meat products. *Intl. J. Food Microbiol.* 20:247–257.
Hustad, G.O., Cerveny, J.G., Trenk, H., Deibel, R.H., Kautter, D.A., Fazio, T., Johnston, R.W., and Kolari., O.E. 1973. Effect of sodium nitrite and sodium nitrate on botulinal toxin production and nitrosamine formation in wieners. *Appl. Microbiol.* 26:22–26.
Jackson, A.L., Sullivan, G.A., Kulchaiyawat, C., Sebranek, J.G., and Dickson, J.S. 2011a. Survival and growth of *Clostridium perfringens* in commercial no-nitrate-or-nitrite-added (natural and organic) frankfurters, hams, and bacon. *J. Food Prot.* 74:410–416.
Jackson, A.L., Sullivan, G.A., Kulchaiyawat, C., Sebranek, J.G., and Dickson, J.S. 2011b. Use of natural ingredients to control growth of *Clostridium perfringens* in naturally cured frankfurters and hams. *J. Food Prot.* 74:417–424.
Jay, J.M. 2000. *Modern Food Microbiology*, 6th ed. Aspen, Gaithersburg, MD.
Jofré, A., Garriga, M., and Aymerich, T. 2008. Inhibition of Salmonella sp. Listeria monocytogenes and Staphylococcus aureus in cooked ham by combining antimicrobials, high hydrostatic pressure and refrigeration. *Meat Sci.* 78:53–59.
Juneja, V.K., and Thippareddi, H. 2004a. Control of *Clostridium perfringens* in a model roast beef by salts of organic acids during chilling. *J. Food Safety* 24:95–108.

Juneja, V.K., and Thippareddi, H. 2004b. Inhibitory effects of organic acid salts on growth of *Clostridium perfringens* from spore inocula during chilling of marinated ground turkey breast. *Intl. J. Food Microbiol.* 93:155–163.

King, A.M., Glass, K.A., Milkowski, A.L., and Sindelar, J.J. 2015a. Impact of clean-label antimicrobials and nitrite derived from natural sources on the outgrowth of *Clostridium perfringens* during cooling of deli-style turkey breast. *J. Food Prot.* 78:946–953.

King, A.M., Glass, K.A., Milkowski, A.L., and Sindelar, J.J. 2015b. Comparison of the effect of curing ingredients derived from purified and natural sources on inhibition of *Clostridium perfringens* outgrowth during cooling of deli-style turkey breast. *J. Food Prot.* 78:1527–1535.

Legan, J.D., Seman, D.L., Milkowski, A.L., Hirschey, J.A., and Vandeven, M.H. 2004. Modeling the growth boundary of *Listeria monocytogenes* in ready-to-eat cooked meat products as a function of the product salt, moisture, potassium lactate, and sodium diacetate concentrations. *J. Food Prot.* 67:2195–2204.

Leistner, L. 2000. Basic aspects of food preservation by hurdle technology. *Intl. J. Food Microbiol.* 55:181–186.

Liang, Z.W., Mittal, G.S., and Griffiths, M.W. 2002. Inactivation of *Salmonella typhimurium* in orange juice containing antimicrobial agents by pulsed electric field. *J. Food Prot.* 65:1081–1087.

Mbandi, E., and Shelef, L.A. 2001. Enhanced inhibition of *Listeria monocytogenes* and *Salmonella enteritidis* in meat by combinations of sodium lactate and diacetate. *J. Food Prot.* 64:640–644.

McClure, P.J., Beaumont, A.L., Sutherland, J.P., and Roberts, T.A. 1997. Predictive modelling of growth of *Listeria monocytogenes* the effects on growth of NaCl, pH, storage temperature and NaNO2. *Intl. J. Food Microbiol.* 34:221–232.

Mosqueda-Melgar, J., Raybaudi-Massilia, R.M., and Martin-Belloso, O. 2012. Microbiological shelf life and sensory evaluation of fruit juices treated by high-intensity pulsed electric fields and antimicrobials. *Food Bioprod. Proc.* 90:205–214.

Nakimbugwe, D., Masschalck, B., Anim, G., and Michiels, G.W. 2006. Inactivation of gram-negative bacteria in milk and banana juice by hen egg white and lambda lysozyme under high hydrostatic pressure. *Intl. J. Food Microbiol.* 112:19–25.

Nguyen, P., and Mittal, G.S. 2007. Inactivation of naturally occurring microorganisms in tomato juice using pulsed electric field (PEF) with and without antimicrobials. *Chem. Engr. Proc.* 46:360–365.

Nou, X., Luo, Y., Hollar, L., Yang, Y., Feng, H., Millner, P., and Shelton, D., 2011. Chlorine stabilizer T-128 enhances efficacy of chlorine against cross-contamination by *E. coli* O157:H7 and Salmonella in fresh-cut lettuce processing. *J. Food Sci.* 76:M218–M224.

Nũnez De González, M.T., Osburn, W.N., Hardin, M.D., Longnecker, M., Garg, H.K., Bryan, N.S., and Keeton, J.T. 2012. Survey of residual nitrite and nitrate in conventional and organic/natural/uncured/indirectly cured meats available at retail in the United States. *J. Agric. Food Chem.* 60:3981–3990.

Organic Foods Production Act. 1990. Electronic Code of Federal Regulations, Part 205: National Organic Program. https://www.ecfr.gov/cgi-bin/text-idx?c=ecfr&sid=6f623e1de5457587ccdfec12bc34ed1c&rgn=div5&view=text&node=7:3.1.1.9.32&idno=7#7:3.1.1.9.32.7.354

Pathanibul, P., Taylor, M.T., Davidson, P.M., and Harte, F. 2009. Inactivation *of Escherichia coli* and *Listeria innocua* in apple and carrot juices using high pressure homogenization and nisin. *Intl. J. Food Microbiol.*, 129:316–320.

Pérez-Rodríguez, M.L., Bosch-Bosch, N., and Garciá-Mata, M. 1996. Monitoring nitrite and nitrate residues in frankfurters during processing and storage. *Meat Sci.* 44:65–73.

Piper, P. 2001. Weak acid adaptation: The stress response that confers yeasts with resistance to organic acid food preservatives. *Microbiology* 147:2635–2642.

Pitt, J.I., and Hocking, A.D. 2009. *Fungi and Food Spoilage*, 3rd ed. Springer, New York.

Pol, I.E., Mastwijk, H.C., Slump, R.A., Popa, M.E., and Smid, E.J. 2001. Influence of food matrix on inactivation of *Bacillus cereus* by combinations of nisin, pulsed electric field treatment, and carvacrol. *J. Food Prot.* 64:1012–1018.

Reddy Velugoti, P., Rajagopal, L., Juneja, V., and Thippareddi, H. 2007. Use of calcium, potassium, and sodium lactates to control germination and outgrowth of *Clostridium perfringens* spores during chilling of injected pork. *Food Microbiol.* 24:687–694.

Ripoche, A.C., Chollet, E., Peyrol, E., and Sebti, I. 2006. Evaluation of nisin diffusion in a polysaccharide gel: Influence of agarose and fatty acid content. *Innov. Food Sci. Emerging Tech.* 7:107–111.

Sauter, E.A., Kemp, J.D., and Langlois, B.E. 1977. Effect of nitrite and erythorbate on recovery of *Clostridium perfringens* spores in cured pork. *J. Food Sci.* 42:1678–1679.

Seman, D.L., Borger, A.C., Meyer, J.D., Hall, P.A., Milkowski, A.L. 2002. Modeling the growth of Listeria monocytogenes in cured ready-to-eat processed meat products by manipulation of sodium chloride, sodium diacetate, potassium lactate, and product moisture content. *J. Food Prot.* 65:651–658.

Sharma, C.S., Ates, A., Joseph, P., Nannapaneni, R., and Kiess, A. 2013a. Reduction of Salmonella in skinless chicken breast fillets by lauric arginate surface application. *Poultry Sci.* 92:1419–1424.

Sharma, C.S., Ates, A., Joseph, P., Soni, K.A., Schilling, M.W., and Kiess, A. 2013b. Evaluation of antimicrobial effects of lauric arginate on reduction of S salmonella spp. in ground chicken. *International J. Food Sci. Tech.* 48:1410–1415.

Shelef, L., and Yang, Q. 1991. Growth suppression of *Listeria monocytogenes* by lactates in broth, chicken, and beef. *J. Food Prot.* 54:283–287.

Stasiewicz, M.J., Wiedmann, M., and Bergholz, T.M. 2011. The transcriptional response of *Listeria monocytogenes* during adaptation to growth on lactate and diacetate includes synergistic changes that increase fermentative acetoin production. *Appl. Environ. Microbiol.* 77:5294–5306.

Taormina, P.J. 2014. Meat and poultry: Curing of meat, pp. 501–507. In *Encyclopedia of Food Microbiology*, 2nd ed., M.L. Tortorello (ed.). Academic Press, Oxford.

Taormina, P.J., and Sofos, J.N. 2014. Low-water activity meat products, pp. 127–164. In *The Microbiological Safety of Low Water Activity Foods and Spices*, Joshua B. Gurtler, Michael P. Doyle, and Jeffrey L. Kornacki (eds.). Springer, New York.

Taormina, P.J., and Dorsa, W.J. 2004. Growth potential of *Clostridium perfringens* during cooling of cooked meats. *J. Food Prot.* 67:1537–1547.

Techathuvanan, C., Reyes, F., David, J.R.D., and Davidson, P.M. 2014. Efficacy of commercial natural antimicrobials alone and in combination against pathogenic and spoilage microorganisms. *J. Food Prot.* 77:269–275.

Tompkin, R.B. 1993. Nitrite, pp. 191–262. In *Antimicrobials in Foods*, P.M. Davidson, and A.L. Branen (eds.). Marcel Dekker, Inc., New York, NY.

Tompkin, R.B., Christiansen, L.N., and Shaparis, A.B. 1978. Enhancing nitrite inhibition of *Clostridium botulinum* with isoascorbate in perishable canned cured meat. *Appl. Environ. Microbiol.* 35:59–61.

United States Department of Agriculture-Food Safety and Inspection Service. 2005. Food Standards and Labeling Policy Book. In Office of Policy Program and Employee Development (ed.), vol. 2009. United States Department of Agriculture, Food Safety and Inspection Service.

United States Department of Agriculture-Food Safety and Inspection Service.2006. Code of Federal Regulations. Animals and animal products. Vol. 9 CFR 317.17, 9 CFR 319.2 and 9 CFR 424.21-23. Washington, DC.

United States Department of Agriculture-Food Safety and Inspection Service. 2012. Safe and suitable ingredients used in the production of meat, poultry, and egg products - FSIS Directive 7120.1, Revision 11.. Food Safety and Inspection Service, United States Department of Agriculture. Washington, DC.

Vandeven, M.H. 2004. Modeling the growth boundary of *Listeria monocytogenes* in ready-to-eat cooked meat products as a function of the product salt, moisture, potassium lactate, and sodium diacetate concentrations. *J. Food Prot.* 67:2195–2204.

Visvalingam, J., and Holley, R.A., 2018. Evaluation of chlorine dioxide, acidified sodium chlorite and peroxyacetic acid for control of *Escherichia coli* O157:H7 in beef patties from treated beef trim. *Food Res. Intl.* 103:295–300.

Wanless, B.J., McDonnell, L.M., Sebranek, J.G., and Glass, K.A. 2010. Effect of nitrite and natural antimicrobials on *Clostridium botulinum* toxigenesis in model frankfurter and ham products. *Institute of Food Technologists Annual Meeting.*

Whiting, R.C., and Masana, M.O. 1994. *Listeria monocytogenes* survival model validated in simulated uncooked-fermented meat products for effects of nitrite and pH. *J. Food Sci.* 59:760–762.

Zaika, L.L. 2003. Influence of NaCl content and cooling rate on outgrowth of *Clostridium perfringens* spores in cooked ham and beef. *J. Food Prot.* 66:1599–1603.

Zuliani, V., Lebert, I., Augustin, J.C., Garry, P., Vendeuvre, J.L., and Lebert, A. 2007. Modelling the behaviour of *Listeria monocytogenes* in ground and pork as a function of pH, water activity, nature and concentration of organic acid salts. *J. Appl. Microbiol.* 103:536–550.

23

Mechanisms of Action, Resistance, and Stress Adaptation

M. Hyldgaard

CONTENTS

23.1 Introduction

Numerous compounds prepared synthetically or isolated from natural sources possess antimicrobial activity, of which some have been used for nearly 100 years as food antimicrobials to extend the shelf-life of food or to ensure food remains safe for ingestion (Davidson and Harrison, 2002). With increased consumer demand for natural, healthy, and minimally processed food, increased efforts have been put into investigating naturally derived compounds (e.g., from microorganisms, plants, or animals) for their potential as food antimicrobials that could substitute for traditional antimicrobials to ensure foods maintain organoleptic quality for longer and remain safe (Juneja et al., 2012; Lucera et al., 2012; Davidson et al., 2013). Thus, the antimicrobials described in this chapter can be found either as naturally occurring in the raw material used to prepare the food or added specifically to protect the food. The application of antimicrobials into food requires detailed knowledge about their stability, *in vitro* and *in situ* spectrum and efficacy, interaction with food matrix components, and mechanism of action (Davidson et al., 2013), of which the latter will be the focal point of this chapter.

Mechanism of action knowledge describes the cause-effect of antimicrobials and can be utilized to understand why some microorganisms are more susceptible to a specific antimicrobial than others, to describe possible resistance, inhibition, or kill mechanisms, or to help design and optimize antimicrobial combinations. Considering the multitude of chemical structures of natural antimicrobials and available cellular targets, many natural antimicrobials may possess more than one mechanism of action (Wiedemann et al., 2001, 2006; Burt 2004; Hyldgaard et al., 2012a; Lou et al., 2012), attack a broad range of targets (Rabinkov et al., 1998; Miron et al., 2000; Kong et al., 2010; Müller et al., 2016a), or show concentration-dependent effects (Raafat et al., 2008; Pisoschi et al., 2018). Hence, describing individual compounds' interaction with cellular targets will be too large a topic to cover in this chapter. Instead, this chapter will focus on describing possible targets within microorganisms and discuss how naturally derived antimicrobials interact with them, provide an overview and guide of experiments that can be used to elucidate the mechanism of actions of antimicrobials, and lastly describe possible resistance mechanisms and adaptation responses of microorganisms to antimicrobials. The emphasis in this chapter will be on Gram-negative and Gram-positive bacteria, whereas fungi and spores are touched briefly.

23.2 Mechanism of Antimicrobial Action on Potential Target Sites of Vegetative Bacteria

The molecular components and cell wall structural elements of bacteria, spores, and fungi are fundamentally different from each other (Figure 23.1), thus implying that unique potential target sites exist on or within each specific type of microorganism (Table 23.1). Notice that this interaction not only depends on the target site (Münch and Sahl, 2015), but also the structural and physicochemical properties of the antimicrobial under scrutiny (Shima et al., 1984). Furthermore, it should be kept in mind that antimicrobials often have more than one target site and that the outcome depends on the applied concentration, with the severity following the order: activation of adaptation response (sub-inhibitory), inhibition of growth/metabolism, and inactivation.

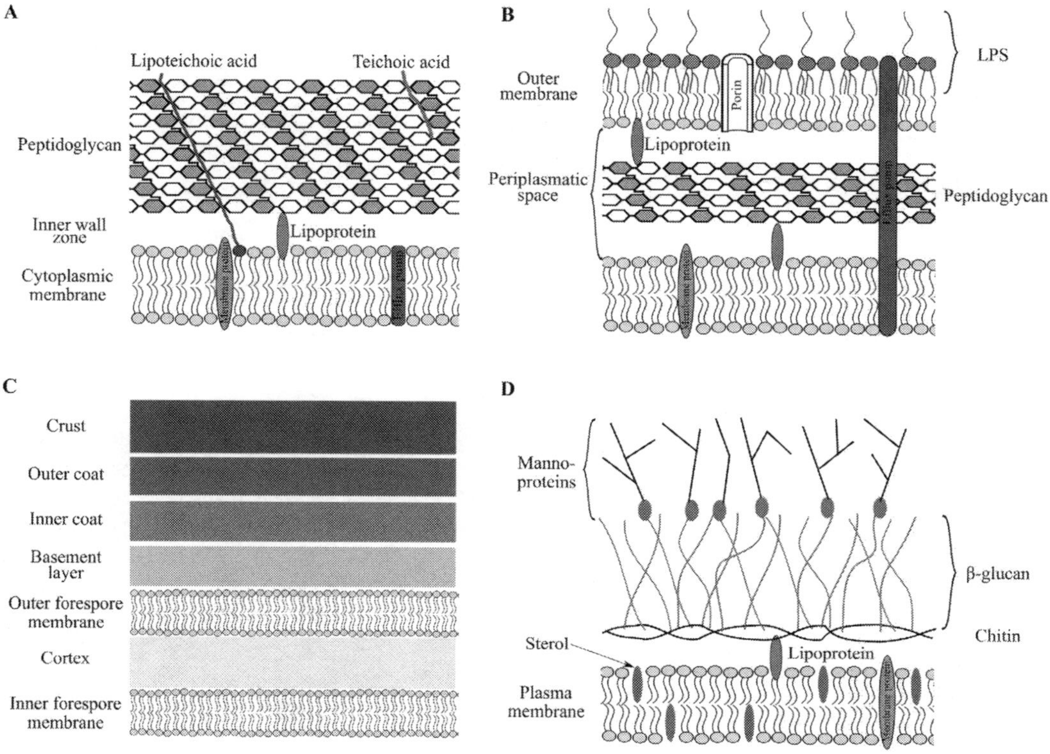

FIGURE 23.1 Representation of cellular structures in Gram-positive bacteria (A), Gram-negative bacteria (B), spores (C), and fungi (D). Abbreviations: extracellular polymeric substances (EPS), lipopolysaccharide (LPS). Figures 23.1A, B, and D were re-drawn from Brown et al. (2015) with permission (License Number 4590070933976). Adapted by permission from Macmillan Publishers Ltd.: [Springer Nature] (Through the wall: Extracellular vesicles in Gram-positive bacteria, mycobacteria and fungi (*Nature Rev. Microbiol.* 2015,13,620–3), Copyright (2015)). Figure 23.1C was re-drawn from Mckenney et al. (2013) with permission (License Number 4590100281936). Adapted by permission from Macmillan Publishers Ltd.: [Springer Nature] (The *Bacillus subtilis* endospore: assembly and functions of the multilayered coat (*Nature Rev. Microbiol.* 2013,11,33–44), Copyright (2013)).

In general, cellular target sites found on or within Gram-positive and Gram-negative bacteria are the outer membrane, the cell wall, the cytoplasmic membrane, membrane-associated proteins, or substrates of the cytoplasm such as nucleic acids, proteins, and enzymes (Table 23.1). In the following, potential cellular target sites will be described and this description will form the basis for discussing how naturally derived antimicrobials interact with them.

23.2.1 Antimicrobial Interactions with the Outer Membrane of Gram-Negative Bacteria

One of the first target sites an antimicrobial will meet in Gram-negative bacteria is the outer membrane. Structurally, the outer membrane is an asymmetric membrane composed of phospholipids on the internal leaflet, and lipopolysaccharides (LPS) on the external leaflet (Figure 23.1B). LPS molecules consist of three subunits, lipid A, an inner- and outer-core oligosaccharide, and a highly variable O-antigen polysaccharide. A typical structure of lipid A is found in *Escherichia coli* and consists of two β-1',6-linked glucosamine disaccharides with adjoining phosphate groups at positions 1 and 4' (Nikaido 2003). The lipid A moiety is anchored to the hydrophobic core of the outer membrane by a high number of saturated fatty acid chains that pack tightly and decrease the outer-membrane fluidity to a gel-like state (Nikaido 1994, 2003). The inner-core oligosaccharide is composed of 3-deoxy-D-manno-oct-2-ulosonic acid (Kdo) and L-glycero-D-manno-heptose (Hep) residues that can be substituted with anionic phosphate groups (Yethon et al., 2000; Wang and Quinn, 2010). Phosphate groups on the lipid A moiety and the

TABLE 23.1

Potential Targets and Mechanisms against Different Microorganisms

Cell Compartment	Potential Target Sites	Type of Microorganism	Potential Mechanism
Cell wall	Outer membrane	Gram-negative bacteria	Porin pathway Hydrophobic pathway Self-promoted uptake (chelator and chelator-independent)
	Peptidoglycan	Gram-negative and Gram-positive bacteria	Degradation of peptidoglycan Diffusion through the peptidoglycan layer
	Mannoproteins, glucan, and chitin	Yeast and mold	Disruption of cell wall
	Crust, outer and inner coat, basement layer, cortex	Spores	Disruption of spore coat
Membrane	Cytoplasmic/plasma membrane	Gram-negative and Gram-positive bacteria, spores, yeast, and mold	Dissipation of proton motive force Affect membrane integrity/permeabilization Inhibit membrane-embedded proteins/enzymes
	Plasma membrane	Yeast and mold	Binding to ergosterol
Cytoplasm	Nucleic acids, proteins, enzymes	Gram-negative and Gram-positive bacteria, spores, yeast, and mold	Affect the function of enzymes and proteins Bind to or degrade nucleic acid components Challenge cytoplasm homeostasis

inner-core oligosaccharide make the outer membrane hydrophilic, and allow cross-linking of adjacent LPS molecules by ionic bridges with divalent cations like Ca^{2+} and Mg^{2+}, ensuring strong lateral interactions between LPS molecules (Nikaido, 2003). The outer membrane is known to limit the entry of many naturally derived antimicrobials, thus making them inefficient against Gram-negative bacteria.

In general, three pathways are known to facilitate the entry of antimicrobials through the outer membrane, and these are the porin pathway, the bilayer pathway, or by self-promoted uptake (Poole, 2002).

23.2.1.1 The Porin Pathway

Associated to the outer membrane are non-specific water-filled hydrophilic channels called porins, which exclude hydrophobic compounds and allow an influx of nutrients and hydrophilic compounds below 600–800 Daltons in size (Benz, 1985; Nikaido, 1992, 2003). The hydrophilic antimicrobial molecule hypothiocyanite ($OSCN^-$), formed by lactoperoxidase, showed reduced efficiency against mutant strains of *E. coli* MG1655 with decreased numbers of porins compared with wildtype, thus, hypothesized to enter the outer membrane through the porin pathway (De Spiegeleer et al., 2005).

Although some hydrophilic antimicrobials might enter Gram-negative bacteria through porins, the porins themselves can also be targeted. For instance, epigallocatechin gallate (EGCg), a polyphenol flavonoid isolated from green tea, has been suggested to interact with and bind to arginine residues within the inner surface of the pore channel of the porin protein OmpG through hydrogen bonding, and thereby inhibit the passive influx of nutrients (Nakayama et al., 2013).

Many antimicrobial agents are hydrophobic or large hydrophilic compounds, which limits their influx through porins, and, therefore, they rely on the hydrophobic (bilayer) pathway or self-promoted uptake.

23.2.1.2 The Hydrophobic Pathway

The bilayer pathway is dependent upon passive diffusion of molecules through the membrane, and even though the outer membrane functions as an efficient permeability barrier that decreases the diffusion rate of hydrophobic compounds more than a hundred-fold, it still allows their influx (Nikaido, 1989,

2003). This means that small hydrophobic compounds such as essential oil constituents (Hyldgaard et al., 2012a), in theory, should be able to diffuse across the outer membrane at decreased rates. Active efflux pumps might counteract the passive influx of hydrophobic compounds, but they too can be targeted (Poole, 2002; Stavri et al., 2007). For instance, geraniol from the plant *Helichrysum italicum* inhibited efflux pump function and thereby potentiated the activity of β-lactam, quinolones, and chloramphenicol up to 16-fold against *Enterobacter aerogenes*, *E. coli*, and *Pseudomonas aeruginosa* (Lorenzi et al., 2009).

23.2.1.3 Self-Promoted Uptake Pathway

The self-promoted uptake of cationic antimicrobials into Gram-negative cells has been proposed to occur through a three-step mechanism: displacement of divalent cations from outer membrane, ionic interactions with phosphate groups and acidic sugars in the LPS, and passage through the membrane (Hancock and Chapple, 1999; Masschalck and Michiels, 2003).

Outer membranes treated with polycationic or antimicrobials with chelator properties might decrease LPS order or cause the release of LPS from the external leaflet, thereby generating phospholipid bilayer patches in the outer membrane that increase the influx rate of hydrophobic antimicrobials (Vaara, 1992; Helander et al., 1997). Polycationic chitosan has been proposed to compete with divalent cations for binding to anionic groups in LPS, and thereby disrupt the outer membrane integrity (Helander et al., 2001; Kong et al., 2010). The polycationic biopolymer ε-poly-L-lysine (ε-PL) causes removal of the outer membrane of *E. coli* by competing with divalent cations for binding to the phosphate groups Hep I and Hep II in the inner core of LPS (Shima et al., 1984; Ye et al., 2013; Hyldgaard et al., 2014). The well-known iron-chelator lactoferrin, which usually inhibits other microorganisms by scavenging iron (Farnaud and Evans, 2003), is also proposed to act on Gram-negative bacteria through an iron-chelator independent mechanism. In this context, lactoferrin interacts with phosphates in lipid A, rigidifies fatty acid chains of LPS (Ellison 3rd and Giehl, 1991; Brandenburg et al., 2001), and releases LPS by its ability to chelate Ca^{2+} (Rossi et al., 2002). Additional compounds with chelating properties include ethylenediaminetetraacetic acid (EDTA), trisodium phosphate (TSP), polyethyleneimine, and the organic acids lactic acid and citric acid (Alakomi et al., 2000; Helander and Mattila-Sandholm, 2000; Masschalck and Michiels, 2003).

23.2.1.3.1 Chelator-Independent Self-Promoted Uptake

Hydrophobic antimicrobials utilize a less defined chelator-independent effect for their self-promoted uptake. Chelation-independent removal of LPS has been proposed for essential oil constituents, like *p*-coumaric acid (Lou et al., 2012), thymol, and carvacrol (Helander et al., 1998), yet the mechanism leading to LPS removal is unknown (Hyldgaard et al., 2012a). For instance, increasing the concentration of $MgCl_2$ did not reverse the outer membrane permeabilization and LPS-releasing properties that thymol and carvacrol had on *E. coli* O157:H7 and *Salmonella* Typhimurium, thus suggesting a chelation-independent mechanism (Helander et al., 1998).

23.2.2 The Peptidoglycan Layer as a Potential Target Site

The peptidoglycan layer is one of the first possible target sites an antimicrobial encounters in Gram-positive bacteria, or after gaining access through the outer membrane of Gram-negative bacteria (Figure 23.1). Gram-positive bacteria have a 20–80 nm thick peptidoglycan layer and a thin periplasmic space positioned between the peptidoglycan and the cytoplasmic membrane (Cabeen and Jacobs-Wagner, 2005; Zuber et al., 2006). The peptidoglycan layer in Gram-negative bacteria is 1–7 nm thin and situated in the periplasmic space between the cytoplasmic and outer membrane (Cabeen and Jacobs-Wagner, 2005). The peptidoglycan layer provides structural integrity and mechanical resistance to intracellular turgor pressure (Vollmer et al., 2008). Compromising the integrity of peptidoglycan will cause lysis of bacterial cells.

Peptidoglycan encapsulates the cytoplasmic membrane in a three-dimensional mesh-like layer of linear glycan strands of repeating disaccharide subunits composed of *N*-acetylglucosamine (Glc*N*A*c*) and

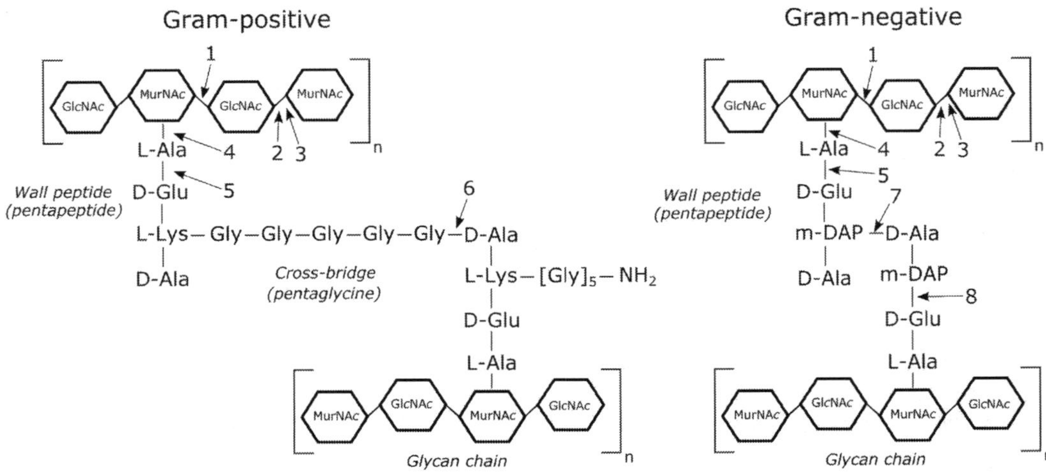

FIGURE 23.2 Known cleavage sites for bacteriophage endolysins in Gram-positive and Gram-negative bacteria according to enzymatic activity: (1) N-acetyl-β-D-glucosaminidase, (2) N-acetyl-β-D-muramidase, (3) lytic transglycosylase, (4) N-acetylmuramoyl-L-alanine amidase, (5) L-alanoyl-D-glutamate endopeptidase, (6) D-alanyl-glycyl endopeptidase, (7) D-alanine-D-meso-DAP endopeptidase, and (8) D-glutamyl-m-DAP endopeptidase. N-acetylglucosamine (GlcNA*c*), N-acetylmuramic acid (MurNA*c*), meso-diaminopimelic acid (m-DAP), *L*-alanine (L-Ala), D-glucosamine (D-Glu), Glycine (Gly). Figures were re-drawn from Oliveira et al. (2012) with permission (License Number 4590121389373). Adapted from *Trends in Food Science and Technology*, Vol 28, Issue 2, Oliveira, H. Azeredo, J. Lavigne, R. Kluskens, L. D. Bacteriophage endolysins as a response to emerging foodborne pathogens, pp. 103–115, Copyright (2012), with permission from Elsevier.

N-acetylmuramic acid (MurNA*c*) (Cabeen and Jacobs-Wagner, 2005). Short interpeptide bridges cross-link chains of MurNA*c* and GlcNA*c*, with the type of interpeptide bridge depending on the amino acid residue in position three of the pentapeptide (Figure 23.2). The meso-2,6-diaminopimelic acid type (m-DAP-type) interpeptide bridge is conserved in Gram-negative bacteria, whereas the L-lysine type (Lys-type) found in Gram-positive bacteria is more diverse in terms of length and composition (Vollmer et al., 2008; Nelson et al., 2012; Oliveira et al., 2012). Long anionic polysaccharides of glycerol and ribitol with phosphate (teichoic acids) or carboxyl (teichuronic acids) groups are covalently linked to MurNA*c* and embedded in the peptidoglycan, whereas lipoteichoic acids are anchored in the outer leaflet of the cytoplasmic membrane (Fischer, 1994; Schäffer and Messner, 2005). Gram-negative bacteria do not have secondary polymers like teichoic acids present in their peptidoglycan meshwork, whereas these polymers provide the peptidoglycan layer of Gram-positive bacteria with negative charges (Cabeen and Jacobs-Wagner, 2005; Malanovic and Lohner, 2016).

23.2.2.1 Peptidoglycan Degrading Compounds

Lysozyme and endolysins are naturally derived antimicrobials for potential use in food, having the peptidoglycan as the target site.

Lysozyme catalyzes the hydrolyzation of β-(1,4) linkages between MurNA*c* and GlcNA*c* (Figure 23.2, position 2) (Masschalck and Michiels, 2003). Besides this well-known mechanism, a non-enzymatic activity has also been proposed for human lysozyme, suggesting an electrostatic binding of cationic lysozyme to negatively charged teichoic and lipoteichoic acids that stimulate autolysis activity in Gram-positive bacteria (Laible and Germaine, 1985). Autolysins are peptidoglycan hydrolases that under certain conditions or stresses break down peptidoglycan at points of cell wall growth (Masschalck et al., 2002). Interestingly, a non-lytic mechanism of bactericidal action has been proposed for hen egg-white lysozyme against *Staphylococcus aureus* (Ibrahim et al., 2001) and *Listeria innocua* that induce cytoplasmic membrane permeabilization without peptidoglycan lysis (Masschalck et al., 2002).

Endolysins are bacteriophage-derived murein hydrolases usually expressed in host cells to achieve cell lysis after bacteriophages have infected and replicated within the host. Exogenously added endolysins alone or in combination with other antimicrobials, such as the bacteriocin nisin or the essential oil constituent carvacrol, eliminate or prevent growth of the foodborne pathogens *Listeria monocytogenes* and *S. aureus* in queso fresco (Ibarra-Sánchez et al., 2018), milk, ham (Chang et al., 2017a), and lean beef (Chang et al., 2017b).

Endolysins consists of cell wall binding domains (CBDs) for target specificity and enzymatic catalytic domains (ECDs) for peptidoglycan degradation activity. Both the CBDs and ECDs give rise to genus-, species-, strain-, and serovar-specific targeting and defined peptidoglycan degradation patterns (Schmelcher et al., 2010; Oliveira et al., 2013).

CBDs recognize and bind to various ligands (structural features provided by carbohydrates or proteins) on the surface of bacteria with equilibrium association constants in the same range as antibodies (Loessner et al., 2002; Oliveira et al., 2013). For instance, the CBDs of *Listeria* endolysins Ply118, Ply500 (Loessner et al., 2002), and PlyP35 (Eugster et al., 2011) rely on serovar- and strain-specific carbohydrate ligands, whereas the CBD of the Cpl-1 endolysin from pneumococcal phage Cp-1 possesses choline-binding repeats for anchoring to choline moieties in the pneumococcal wall teichoic acids, and the CBD of endolysin Lyb5 from *Lactobacillus fermentum* bacteriophage contains LysM domains that bind to GlcNA*c* in peptidoglycan (Eugster et al., 2011; Oliveira et al., 2013). Ligands can be site-specific and not equally distributed on the cell surface, for instance, the CBD of Ply500 covered *L. monocytogenes* 1042 (serovar 4b) cells homogenously, whereas Ply118 preferentially bound to the septal and polar regions of cells (Loessner et al., 2002). CBDs confer strong binding affinities for their ligand, ensuring endolysins stay attached to the peptidoglycan (Loessner et al., 2002). Thus, most Gram-positive endolysins are single-use antimicrobials.

Most Gram-negative endolysins do not confer CBDs and are therefore often composed of only a single catalytic domain (Nelson et al., 2012; Oliveira et al., 2012). The absence of CBDs in Gram-negative endolysins is believed to arise from the inability of Gram-negative endolysins to lyse subsequent cells because the outer membrane shields the underlying peptidoglycan. Thus, Gram-negative endolysins have no evolutionary pressure to acquire CBDs (Oliveira et al., 2012). Still, a few Gram-negative endolysins with CBDs have been observed like *Pseudomonas* phage endolysins KZ144 and EL188 (Briers et al., 2007, 2009).

Four general enzymatic activities of endolysin ECDs are known (Oliveira et al., 2012, 2013): (i) transglycosidases cleave glycosidic at GlcNA*c* (Figure 23.2, position 1) or MurNA*c* (Figure 23.2, position 3), (ii) muramidases (lysozymes) cleave the β-(1,4) linkages between MurNA*c* and GlcNA*c* (Figure 23.2, position 2), (iii) amidases separate the glycan polymer from the short stem peptide (Figure 23.2, position 4), and (iv) carboxy/endopeptidases cleave peptide bonds between two amino acids in the cross-linking peptide stem (Figure 23.2, position 5 and 6) and interpeptide bridges (Figure 23.2, position 7 and 8) (Baker et al., 2006; Drulis-Kawa et al., 2012; Nelson et al., 2012; Oliveira et al., 2013). Endolysins with amidase activity have a broader lytic spectrum compared with other classes of endolysins because the linkage between MurNA*c* and *L*-alanine is conserved (Navarre et al., 1999).

Even though endolysins typically display only one type of enzymatic activity, bifunctional lysins do exist. For instance, the *S. aureus* phage φ11 endolysin has two independent muralytic activities in the form of endopeptidase and amidase (Navarre et al., 1999). For endolysins with two ECDs, both catalytic sites should, in theory, be equally active, but in all cases, it has been shown that the N-terminal ECD encodes the domain with the greatest activity (Oliveira et al., 2013). Because of the differences in substrate (ligand) recognition and enzymatic reactions of endolysins, their activity is narrow and related to the bacterial species from which they derive. However, endolysins with broad lytic activity are known, e.g., PlyV12 from the Enterococcal bacteriophage phi1, active against 14 *Enterococcus faecalis* and *Enterococcus faecium* strains, *Streptococcus pyogenes*, and *S. aureus* (Yoong et al., 2004).

The class IIa bacteriocin Sakacin A, produced by *Lactobacillus sakei* (DSMZ 6333), primarily inactivates *Listeria ivanoii* by dissipation of the proton motive force but also shows proteolytic activity that slowly degrades the peptidoglycan into 1–2.5 kDa fragments (Trinetta et al., 2012). The cleavage site was proposed to be the bond formed by MurNA*c* and L-alanine (Figure 23.2, position 4) (Trinetta et al., 2012).

23.2.2.2 Diffusion through the Peptidoglycan

Uncharged hydrophilic globular molecules up to 25 kDa can diffuse through the peptidoglycan mesh of *B. cereus* and *E. coli* (Demchick and Koch, 1996). In general, further investigations are needed to understand the role of peptidoglycan and secondary polymers in relation to attraction, interaction, adsorption, and diffusion of antimicrobial peptides (Malanovic and Lohner, 2016) and charged or hydrophobic antimicrobials. Antimicrobials not specifically targeting peptidoglycan or secondary polymers might instead interact with targets at the cytoplasmic membrane.

23.2.3 Potential Targets and Mechanisms Associated with Cytoplasmic Membranes

The cytoplasmic membrane situated below the peptidoglycan layer is the primary target site for many antimicrobials, often termed membrane-active antimicrobials (Figure 23.1). Membrane-active antimicrobials might affect more than one target, and the mechanism may differ widely between various compounds. Disruptive effects on membrane integrity, stability, or function of membrane-embedded proteins/enzymes by antimicrobials can result in microbial inhibition or inactivation.

Membrane bilayers are entities made of amphiphilic phospholipids, with hydrophobic fatty acid tails and hydrophilic headgroups, forced together by hydrophobic interactions, van der Waals attractions, and hydrogen bonds between adjacent phospholipids (Seddon et al., 2009). The cytoplasmic membrane is an effective semi-permeable barrier that regulates the transfer of material in and out of the cell interior using membrane-spanning transport proteins and enzymes to sustain the proton-motive force (PMF) across the membrane required for survival (Murínová and Dercová, 2014).

The fatty acid tail moiety of phospholipids is essential for maintaining membrane fluidity under different conditions. The membrane fluidity depends on the average melting temperature of fatty acids in the membrane, which in turn depends on fatty acid chain length, saturated or unsaturated acyl chains, or the proportion of other acyl chain modifications like branched or cyclic chains (Russell, 1984). For instance, a saturated fatty acid with 16 carbon atoms (C16:0) has a melting temperature of 62.2°C, which declines to 1.2°C when introducing a *cis*-double bond at carbon 9 (C16:1Δ9*c*) (Knothe and Dunn, 2009).

The types of phospholipid headgroups found in most bacteria are the zwitterionic (neutral) phosphatidylethanolamine (PE), and the negatively charged phosphatidylglycerol (PG) and cardiolipin (CL) having one or two anionic phosphate groups, respectively (Epand and Epand, 2011). The headgroup composition of the cytoplasmic membrane of Gram-negative bacteria consists of 5–20% CL, 5–21% PG, and 60–80% PE, whereas the majority of lipid headgroup species found in Gram-positive bacteria belongs to the anionic phospholipids with 0–40% PE, 40–70% PG, and 15–50% CL (Epand and Epand, 2009). Other types of phospholipids also exist, such as phosphatidylcholine, lyso-PE (PE with one acyl chain), and lysyl-PG and lysyl-CL (positively charged lysine esters of PG and CL) (den Kamp et al., 1969; Okabe et al., 1980; Fischer and Leopold, 1999; Sohlenkamp et al., 2007, 2015). All bacterial membranes have more than 15% of anionic lipid headgroups, which together with the LPS or secondary polymers provide bacteria with an overall anionic surface charge (Epand and Epand, 2009).

The discussed variation in phospholipid headgroups and their fatty acids' chain length, structure, and saturation can result in cytoplasmic membranes composed of more than a hundred different phospholipids (Dowhan, 1997).

Membrane-active naturally derived antimicrobials interact with the cytoplasmic membrane by three mechanisms: dissipation of PMF, permeabilization of membranes, and inhibition of membrane-associated proteins/enzymes.

23.2.3.1 Dissipation of Proton Motive Force

The cytoplasmic membrane functions as an insulator and barrier for the movement of ions, and therefore requires proteins and enzymes for the movement of ions to generate energy by oxidative phosphorylation through the electron transport chain and chemiosmosis. The chemiosmotic model describes how the electron transport chain pumps out protons generating a pH gradient (ΔpH) across the membrane, leaving the inside alkaline (low [H^+]) (Mitchell, 1961). This proton gradient favors the influx of protons through adenosine triphosphate (ATP) synthase, which generates ATP from adenosine diphosphate (ADP) and inorganic phosphate (P_i) (Mitchell, 1961). The PMF is a measure of potential energy stored as gradients of pH (ΔpH) and membrane potential (Δψ) across the cytoplasmic membrane (Krulwich et al., 2011). The transmembrane electrical potential (Δψ) is a measure of the concentration difference of ions (Na^+, K^+, Cl^-) on opposite sides of the cytoplasmic membrane, with the inner being negatively charged with respect to the outside (Krulwich et al., 2011).

Antimicrobials that dissipate the PMF might affect the ΔpH and Δψ of the cell, thus affecting cell homeostasis and inhibiting ATP generation used as energy for various cellular processes like active transport. Examples of naturally derived food antimicrobials that dissipate the PMF are weak organic acids and hop plant constituents.

Weak organic acids like lactic, acetic, propionic, butyric, or benzoic acid are either added to food products as acidulants or produced as metabolic byproducts by lactic acid bacteria (e.g., in cheese), acetobacteria (e.g., in vinegar), or *Propionibacterium* (e.g., in cheese) during fermentation (Ricke, 2003; Garmiene et al., 2010; Reis et al., 2012; Khan et al., 2017a). Weak organic acids do not entirely dissociate in water but exist in an environmental pH-dependent equilibrium between a dissociated or non-dissociated form that also depends on the pK_a value (pH where 50% ionization occurs) of the weak organic acid (Sofos et al., 1986; Brul and Coote, 1999). When the pH is above the pK_a value of an acid, the antimicrobial activity decreases rapidly (Sofos et al., 1986). The upper pH limits for the effectiveness of propionate and benzoate are pH 5.5, and pH 4.5, respectively (Sofos et al., 1986).

Traditionally, non-dissociated organic acids diffuse across the cytoplasmic membrane to the cytoplasm, thus going from a proton-rich to proton-poor environment where the weak acids dissociate into a proton and the ionized form of the acid. Influx and dissociation of weak organic acid continue until an equilibrium has settled in accordance to ΔpH, resulting in an accumulation of anions and protons that stress the intracellular pH homeostasis and inhibit various metabolic reactions (Brul and Coote, 1999; Pisoschi et al., 2018). Eklund (1983, 1985) showed that acidification of the cell interior (dissipation of ΔpH) could not wholly explain how sorbic acid inhibits bacteria close to neutral pH, indicating that the dissociated acid anion plays an inhibitory role at neutral pH levels. Also, at low environmental pH, the theory that organic acid anions work as pH uncouplers could not explain the presence of acid-resistant bacteria (Russell, 1992). Acid-tolerant *Streptococcus bovis* exposed to acetic acid at declining extracellular pH maintained a relatively constant pH gradient across the membrane by decreasing its intracellular pH to prevent intracellular accumulation of acetate anions, indicating that the accumulation of weak acid anions has a toxic impact on cells rather than acting as uncouplers (Russell, 1991, 1992). Thus, both the proton and anion contribute to the inhibitory activity of weak acids.

The iso-α-acids *cis*- and *trans*-isohumulone derived from flowers of the hop plant *Humulus lupulus* L. provide bitterness and protection against Gram-positive spoilage bacteria in beer through their function as weak acids with ionophoric properties (Simpson, 1993). Ionophores like *trans*-isohumulone trap protons outside the cytoplasmic membrane, diffuse across the cytoplasmic membrane, and dissociate inside cells because of the increased internal pH (Simpson, 1993). The inhibitory ionophoric property of undissociated *trans*-isohumulone requires the presence of monovalent cations, and the ionophoric activity increases when decreasing the pH of the external medium (Simpson and Smith, 1992). Influx of protons results in reduced intracellular pH, thus the dissipation of ΔpH with little influence on Δψ (Simpson, 1993). It has been proposed that the undissociated *trans*-isohumulone acts as an ionophore, internalizing in cells due to electroneutral influx, internal dissociation, and efflux of an isohumulone complex with divalent cations like Mn^{2+}, resulting in an electroneutral H^+/Mn^{2+} exchange of cations (Simpson, 1993; Sakamoto and Konings, 2003). But model membrane investigations could not verify the electroneutral H^+/Mn^{2+} exchanger hypothesis, and instead showed that iso-α-acids belong to proton ionophores of class I/II capable of transporting protons within a wide range of pH (Behr and Vogel, 2009). In extension to this, cyclic voltammetry investigations on model membranes indicated that transmembrane redox reactions driven by iso-α-acids and their manganese complexes caused intracellular oxidative stress at low pH (Behr and Vogel, 2010; Schurr et al., 2015).

Lastly, it is well-known that the PMF in energized cells attracts bacteriocins and promotes their interaction with the cytoplasmic membrane, for example as seen for lacticin 3147 (McAuliffe et al., 1998), plantaricin MG (Gong et al., 2010), and nisin (Garcerá et al., 1993).

Identifying the PMF as the primary target for antimicrobials is hampered by the fact that many antimicrobials permeabilize the cytoplasmic membrane, causing leakage of intracellular constituents that result in PMF dissipation.

23.2.3.2 Membrane Permeabilization

Interactions between antimicrobials and membranes can result in a myriad of different membrane damages that affect its permeability properties. The severity of membrane integrity loss can be inferred from the size of molecules able to leak from the cell interior, with the smallest being ions like K^+ or H^+ (Ultee et al., 1999), followed by inorganic phosphates (P_i) (McAuliffe et al., 1998), amino acids (Ruhr and Sahl., 1985), ATP, and nucleic acids (Russell, 2005; Sun et al., 2017). Efflux of ions indicates an overall loss of the permeability barrier function of membranes and can be related to bacteriostasis but is not necessarily indicative of cell inactivation (Russell, 2005).

Antimicrobial agents that act on membranes cause depolarization or increased permeability by altering bulk membrane properties through non-specific interactions, or by targeting ligands in the membrane (Epand et al., 2016).

23.2.3.2.1 Membrane Permeabilization through Alteration of Bulk Properties

Cells need to maintain an optimal membrane structure and fluidity to support proper cell functions, such as diffusion, energy production, and function of embedded proteins (Russell, 1984; Dowhan, 1997; Mykytczuk et al., 2007). Hydrophobic antimicrobials, like some essential oil constituents belonging to the groups terpenes, terpenoids, or phenylpropenes, alter membrane fluidity, curvature, thickness, and structure-function of membranes by partitioning into and accumulating in the hydrophobic core or at the polar headgroup region of phospholipids in membranes (Turina et al., 2006; Mykytczuk et al., 2007; Witzke et al., 2010; Hyldgaard et al., 2012a, 2015).

A correlation exists between the hydrophobicity of molecules and their partitioning into membranes, and at the same time, there exists a correlation between hydrophobicity and membrane toxicity (Weber et al., 1996). The logarithm to the *n*-octanol-water partition coefficient ($\log(P_{OW})$) of a compound is a measure for its hydrophobicity. Compounds with low $\log(P_{OW})$ values solubilize in aqueous solution but require higher concentrations to interact with membranes, and *vice versa* (Weber et al., 1996). Molecules that partition into membranes and affect the viability of cells typically have $\log(P_{OW})$ values below 4–5 (Weber et al., 1996). Essential oil constituents belonging to the group of terpenoids have $\log(P_{OW})$ values in the range of 1.8 to 4.5 (Griffin et al., 1999). Ethanol with a $\log(P_{OW})$ of –0.28 requires several percentages in aqueous solution to affect membranes, whereas the essential oil constituents carvacrol and *p*-cymene with $\log(P_{OW})$ of 3.49–3.64 and 4.10 inhibit the growth of *B. cereus* at 0.75 mM (0.11‰) and above 10 mM (>1.34‰), respectively (Weber et al., 1996; Griffin et al., 1999; Ultee et al., 2002).

Integration of hydrophobic compounds into the membrane affects lipid ordering and membrane melting temperature (T_m), and influences membrane stability, function, and permeability (Weber et al., 1996). Membrane-active antimicrobials can affect lipid membrane organization as substitutional or interstitial impurities depending on their influence on T_m and enthalpy difference (ΔH) (Cristani et al., 2007). Many essential oil constituents act as substitutional impurities that substitute position with lipids in the membrane, thereby affecting both T_m and ΔH and partly perturbing the membrane (Cristani et al., 2007; Gharib et al., 2018). The group of compounds acting as interstitial impurities, like eucalyptol and pulegone, intercalate among acyl chains of lipids and thereby cause variation of T_m but not ΔH (Cristani et al., 2007; Gharib et al., 2018). It should be noted that fluidity changes do not always influence membrane permeability, for instance, *p*-cymene acts as a substitutional impurity that increases membrane fluidity, expands membranes and affects PMF, yet it is generally regarded as having no effect on membrane permeability (Ultee et al., 2002; Cristani et al., 2007; Hyldgaard et al., 2012a).

Despite the *p*-cymene example above, changes to the membrane fluidity usually affect the membrane permeability barrier properties to non-ionic molecules from the cytoplasm. According to the solubility-diffusion model of permeation, the permeability barrier decreases with decreasing fluidity, and *vice versa* (Lande, 1995). Although the solubility-diffusion model cannot explain the permeation of charged molecules from the cytoplasm, *in silico* modeling indicates that leakage of ions requires the involvement of transient pores or ion-induced defects (Vorobyov et al., 2014; Shinoda, 2016).

Hydrophobic compounds partitioning into membranes can also cause membrane expansion or affect surface curvature (Turina et al., 2006). For instance, the monoterpenes α-pinene and β-pinene partition into the hydrophobic core of membranes and induce membrane expansion and stiffening, whereas the alcohol monoterpenoid menthol position at the polar lipid headgroup region and induce a membrane curvature increase that potentially could lead to membrane budding (Witzke et al., 2010; Zunino et al., 2011). Molecular dynamics simulations indicate that the monoterpene limonene and the monoterpenoid perillyl alcohol localize in the hydrophobic core of membranes and increase the order and thickness of membranes (Witzke et al., 2010). Carnosic acid and carnosol are phenolic diterpenes from rosemary extract proposed to accumulate at the polar headgroup region of phospholipids and the hydrophobic core of membranes, respectively (Pérez-Fons et al., 2010). Both phenolic diterpenes rigidify membranes, whereas carnosol also promotes the formation of phospholipid/diterpene domains with altered lipid structure in a concentration-dependent manner (Pérez-Fons et al., 2006).

23.2.3.2.2 Ligand-Mediated Membrane Permeabilization

Naturally derived antimicrobials that interact with specific targets in membranes are often of proteinaceous origin, positively charged, and rely on electrostatic interactions with the anionic cell surface. Membrane-associated ligands such as anionic phospholipids or specific docking molecules can facilitate membrane permeabilization by antimicrobials.

Electrostatic interaction is the molecular interaction that has the farthest reach, and thus plays an essential role for positively charged antimicrobials having anionic targets. The molecular interaction potential w is a function of the distance r by which two molecules are separated and is proportional to $1/r$ (Suriyarak et al., 2014). In comparison, then van der Waals and steric interactions are short-ranged interactions proportional to $1/r^6$ (Suriyarak et al., 2014).

Electrostatic interactions attract positively charged chitosan, ε-PL, or peptides like bacteriocins to bind to anionic phospholipids or lipoteichoic acids in membranes to exert their mechanisms. The polycationic chitosan interacts with anionic surface components such as LPS and lipoteichoic acids causing extensive alterations to the cell surface, leading to leakage of intracellular substances in an unknown mechanism of action (Raafat et al., 2008; Ganan et al., 2009; Ma et al., 2017). Anionic lipoteichoic acids are anchored in the cytoplasmic membrane of Gram-positive bacteria and could provide a molecular linkage for chitosan to destabilize membranes and cause leakage of K^+ as no membrane disruption or pore formation occurred (Raafat et al., 2008; Kong et al., 2010). The hydrodynamic radius of chitosan in solution is 24.59 nm, and it would theoretically not be able to diffuse through the peptidoglycan layer (Raafat et al., 2008). However, using atomic force microscopy (AFM) for indentation experiments, it was shown that chitosan decreased the cell wall stiffness in *Staphylococcus aureus* and *E. coli* (Eaton et al., 2008), allowing the diffusion of chitosan to the cell membrane for electrostatic interaction with anionic lipids causing slight membrane permeabilization (Kong et al., 2010). Liu et al. (2004) proposed that chitosan acetate interacts with phosphoryl groups in anionic lipids to permeabilize outer and cytoplasmic membranes, evidenced by leakage of nucleic acid from *E. coli* and *S. aureus*. Recently, Hu and Gänzle (2018) proposed that differences in charge distribution on cell surfaces could give species and strain-specific differences in tolerance to chitosan.

The concept of lateral organization of anionic phospholipids into lipid domains can partly explain species- and strain-specific differences in tolerance to cationic antimicrobials (Epand and Epand, 2009). Polycationic compounds induce the formation of lipid domains by the clustering of anionic phospholipids away from zwitterionic phospholipids (e.g., PE), which cause sudden but transient phase boundary defects to occur that allow leakage of cytoplasmic content and depolarization of the membrane (Epand et al., 2008, 2009, 2011, 2016). Bacterial membranes composed predominantly of anionic phospholipids would be less susceptible to cationic antimicrobials because of insufficient zwitterionic phospholipids to form separate domains (Epand and Epand, 2009). Thus, variations in phospholipid headgroup composition in different species or strains of bacteria could explain differences in their susceptibility to polycationic compounds.

Antimicrobial peptides often undergo structural rearrangements when interacting with the membrane, and either cause lipid organization or induce membrane permeabilization through various types of mechanisms known as wedge-, barrel-, toroidal-, and carpet-mechanisms (Driessen et al., 1995; Moll et al., 1999; Yoneyama et al., 2009; Hyldgaard et al., 2014). At high concentrations, the bacteriocin nisin has been proposed to form lipid II-independent wedge-like pores. Nisin molecules bind parallel to the membrane surface and form wedge-like pores by bending their C-terminal part into the membrane to generate local strain that forms a transmembrane pore (Driessen et al., 1995; Moll et al., 1999). Class II bacteriocins bundle as α-helical peptides and produce pores through barrel-stave or carpet mechanisms (Moll et al., 1999). In the barrel-stave mechanism, peptides form α-helical structures when interacting with the membrane surface, and then peptides insert perpendicular to the membrane surface with their hydrophobic part facing the hydrophobic core of the membrane, and the hydrophilic part of the peptide forming the pore lining (Moll et al., 1999). For example, the amino acid sequence of the class IIb bacteriocin Enterocin 1071A and 1071B predicts the formation of pores through the barrel-stave mechanism (Balla et al., 2000). Lacticin Q (class IId) forms huge toroidal pores by displacing phospholipid headgroups with its hydrophobic residues to enforce a positive curvature folding of the membrane, thus creating short-lived pores lined with phospholipid headgroups and bacteriocin molecules that concomitantly close with Lacticin Q translocated to the cytoplasm (Yoneyama et al., 2009). Toroidal-like pore formation has also been observed for the bacteriocins AS-48 (class IIc), which later was proposed to resemble a leaky slit model (Sánchez-Barrena et al., 2003; Maqueda et al., 2004; Cruz et al., 2013). ε-PL confers a carpet-like mechanism, where the polycationic antimicrobial interacts with the anionic headgroup region of phospholipids, and after reaching a threshold concentration, the membrane permeabilizes in a detergent-like manner (Hyldgaard et al., 2014).

Even though electrostatic interactions might facilitate the initial interaction between bacteriocins and membrane surfaces, the membrane pore-forming capabilities are enhanced when bacteriocins and docking molecules form a complex.

Several bacteriocins that belong to class I (lantibiotics) and some class II bacteriocins target docking molecules to enhance their membrane permeabilization potency (Cotter, 2014). For instance, the presence of lipid II in membrane vesicles increased the membrane-permeating capability of nisin by three orders of magnitude compared with vesicles without lipid II (Wiedemann et al., 2001). Recently, the number of docking molecules has expanded and now constitutes the mannose phosphotransferase system targeted by pediocin PA1 (class IIa), lipid II interacting with most lantibiotics and some class II bacteriocin, maltose ABC transporter targeted by garvicin ML (class IIc), Zn-dependent metallopeptidase targeted by LsbB (class IId), and undecaprenyl pyrophosphate phosphatase targeted by Lactococcin G (Cotter, 2014).

Lipid II is a unique non-protein target proposed to sequester at septum-localized anionic phospholipid patches in bacterial cells, where it plays a vital role in the peptidoglycan biosynthesis pathway (Müller et al., 2017). Lipid II is known as the Achilles' heel of bacterial cell wall biosynthesis for four reasons: (i) lipid II molecules are highly conserved among bacteria, (ii) bacterial cells have a finite number of lipid II molecules available for driving biosynthesis of peptidoglycan, which requires the cell to recycle lipid II (Kramer et al., 2004; Van Heijenoort, 2007), (iii) lipid II most likely partitions into domains with anionic phospholipids and enzymes like MurG (Chugunov et al., 2013; Müller et al., 2016b, 2017), and (iv) lipid II could play an essential role in bacterial signal transduction pathways (Hardt et al., 2017).

The N-terminal thioether rings A and B in nisin form a cage that forms hydrogen bonds to the pyrophosphate moiety of lipid II, thus exhibiting a dual mechanism of action as this interaction inhibits the biosynthesis of peptidoglycan and induces pore formation (Wiedemann et al., 2001; Hsu et al., 2004). Lipid II stabilizes the transfer of the C-terminal of nisin through the membrane and facilitates the assembly of large pore-forming complexes consisting of eight nisin molecules and four lipid II molecules (Hasper et al., 2004; Hsu et al., 2004). Nisin and other lantibiotics that exhibit a pyrophosphate cage form complexes with lipid II, and cause segregation and abnormal distribution of lipid II domains (Hasper et al., 2006). Lipid II–nisin domains could be of biological relevance because these domains might induce permeabilization of model membranes through curvature effects (Scherer et al., 2015).

Interestingly, the ability of lantibiotics to form a pore in membranes depends on their peptide length and the membrane thickness. Bacteriocins that resemble nisin (34 amino acids) inactivate cells by forming pores, whereas shorter lantibiotics like gallidermin (22 amino acids) and epidermin cannot form transmembrane pores unless the membrane consists of short-chain phospholipids having 14 carbon atoms or fewer, yet they remain highly antimicrobial (Bonelli et al., 2006; Hasper et al., 2006; Müller et al., 2017).

23.2.3.3 Inhibition of Membrane-Embedded Proteins/Enzymes

Antimicrobials can directly or indirectly inhibit or inactivate cells by affecting the function of membrane-embedded proteins and enzymes (Figure 23.1) including ATPases, transport proteins, transferases, oxidoreductases, and signal transducers (Sikkema et al., 1995).

Antimicrobials can inactivate cells by the disruption of energy generation (ATPase) or storage (ATP synthase). An iron-independent mechanism of lactoferrin has been hypothesized to target and block H^+-ATPase, thus inactivating *Lactococcus lactis* by intracellular accumulation of H^+, and *P. aeruginosa* by blocking reentry of H^+, causing uncoupling of ATP synthase phosphorylation of ADP and accumulation of H^+ in the periplasmic space (Andrés and Fierro 2010). Issa et al. (2018) investigated the efficacy of structurally related essential oil constituents to inhibit *E. coli* membrane-bound F_1F_0 ATP synthase and showed that 4 out of 13 tested terpenes such as *p*-cymene did not affect ATP synthase activity, whereas α-terpinene and (-)-β-pinene caused 87.5% inhibition of ATP synthase activity. They also confirmed that four phenylpropenes inhibited ATP synthase activity up to 50% as observed for isoeugenol (Issa et al., 2018).

Transport proteins and transferases are vital for cell growth and the movement of essential ions and molecules across membranes. Two-dimensional electrophoresis results from *B. subtilis* treated with or without the polyphenol EGCg showed removal of cytoplasmic protein spots in the presence of EGCg, thus indicating EGCg inhibits growth by binding to transport proteins such as oligopeptide ABC transporter, glucose phosphotransferase (PTS) system transporter protein, and phosphate ABC transporter (Nakayama et al., 2015). In other instances, membrane-embedded proteins or enzymes might act as receptors, such as the mannose phosphotransferase system, proposed to act as a receptor for the bacteriocin lactococcin Z that has bactericidal activity towards *L. lactis* IL1403 without the dissipation of membrane potential or pore-formation (Daba et al., 2018).

Changes in membrane curvature, fluidity, thickness, hydration, or sequestering of phospholipid headgroups induced by antimicrobials interacting with membranes could indirectly affect the function, complex-formation, or activity of integral membrane proteins (Sikkema et al., 1995; Lee, 2004). Membrane-embedded proteins need an optimum hydrophobic thickness of membranes to match their transmembrane domain-length. Otherwise, the proteins deform to match the surrounding membrane thickness with a concomitant loss of activity (Lee, 2004). Thus, antimicrobials that affect bulk membrane properties also disturb the activity of membrane-embedded proteins/enzymes (Sikkema et al., 1995). The primary causative of cell inactivation can be especially challenging to discern in cases where compounds are proposed both to interact with membranes and to affect the function of membrane-embedded proteins and enzymes, like many essential oil constituents (Hyldgaard et al., 2012a). Gill and Holley (2006a) showed that although eugenol, carvacrol, and cinnamaldehyde inhibited the ATPase activity of *E. coli* and *L. monocytogenes* membrane suspensions, it required concentrations in the range required for membrane disruption (Gill and Holley, 2006b), thus indicating membrane disruption as the primary lethal event and ATPase inhibition as a secondary cause of cell inactivation.

23.2.4 Antimicrobial Mechanisms on Cytoplasmic Targets

Antimicrobials that gain access to the cytoplasm and disturb healthy cell metabolism or cause cell inactivation might inhibit enzymes and proteins, interact with nucleic acids, or modulate the intracellular environment.

23.2.4.1 Inhibition of Intracellular Enzymes and Proteins

Aldehydes (R-CHO) are reactive compounds characterized by having a highly electronegative oxygen atom and an electrophile carbonyl group that readily reacts with nucleophiles, thus enabling cross-linking covalently with DNA and proteins through amine groups (Feron et al., 1991). Vanillin, citral, and cinnamaldehyde are well-known antimicrobial aldehydes found as essential oil constituents (Hyldgaard et al., 2012a). The extensively studied cinnamaldehyde has three specific mechanisms at increasing concentrations: at low concentrations, cinnamaldehyde inhibits cytokinesis and various enzymes, at higher yet sub-lethal concentrations it inhibits ATPase, and at lethal concentrations, cinnamaldehyde might perturb membranes (Hyldgaard et al., 2012a). Cinnamaldehyde at sub-inhibitory concentrations inhibits cell division by interacting with the T7 loop of FtsZ protein to prevent GTPase-dependent polymerization of the cell division Z-ring structure, resulting in the formation of filamentous structures of *E. coli* O157:H7 and *B. cereus* (Kwon et al., 2003; Cabeen and Jacobs-Wagner, 2005; Domadia et al., 2007). Eight phenylpropanoids like *p*-coumaric acid and cinnamic acid inhibited filament assembly of purified FtsZ *in vitro* and *B. subtilis* 168 cells in a concentration-dependent manner. However, their primary mechanism of action likely remains cell wall damage (Hemaiswarya et al., 2011). Cinnamaldehyde and the phenylpropene eugenol inhibit the activity of intracellular enzymes such as histidine-, lysine-, and ornithine-decarboxylases (Wendakoon and Sakaguchi, 1995).

Allicin isolated from garlic has a reactive thiosulfinate group (R_1-S(O)-S-R_2) and readily diffuses across membranes ($Log(P_{OW}) = 1.35$) to interact non-specifically with SH groups in intracellular enzymes (Rabinkov et al., 1998; Miron et al., 2000; Leontiev et al., 2018). The reactive (-S(O)-S-) group of allicin induces thiol-stress through *S*-allylmercapto modification of protein cysteines, reversed by thiol-containing components like glutathione (Rabinkov et al., 1998). Allicin partly inhibits DNA and protein synthesis and inhibits RNA synthesis with 90% in *S.* Typhimurium, indicating that allicin primarily inhibits RNA polymerases (Feldberg et al., 1988). Recently, Müller et al. (2016a) proposed that allicin's mechanism of action might involve a combination of a decrease in intracellular glutathione levels, inhibition of metabolic enzymes through *S*-allylmercapto modification of cysteine, protein aggregation, and reduction of protein synthesis in *E. coli* cells.

Allyl isothiocyanate (AITC) isolated from mustard oil might readily diffuse into cells as it did not cause cell wall damage or ATP leakage from *L. monocytogenes* (Ahn et al., 2001), and the cell membrane remained intact in *E. coli* (Nowicki et al., 2016). Contrary to those results, AITC induced metabolite leakage from *E. coli* and *Salmonella* Montevideo (Lin et al., 2000). AITC has a reactive electrophile carbon group (–N=C=S) suggested to react with oxygen-, sulfur-, or nitrogen-centered nucleophiles found in glutathione, amino acids, oligopeptides, proteins, and water (Kawakishi and Namiki, 1982; Kawakishi and Kaneko, 1985, 1987; Cejpek et al., 2000; Luciano et al., 2008). The reactive electrophile carbon group of AITC enables oxidative cleavage of cysteine disulfide bridges in proteins (Kawakishi and Namiki, 1982), and the formation of adducts with free ε-amino groups of lysine and arginine residues in proteins to give ε-allylthiourea derivatives (Kawakishi and Kaneko, 1987). Other examples of proposed intracellular mechanisms of AITC and other isothiocyanates include protein aggregation and misfolding, induction of ROS and amino-acid deprivation, and inhibition of enzyme activity, quorum sensing, and RNA and DNA synthesis (Dufour et al., 2013, 2015; Nowicki et al., 2016).

23.2.4.2 Targeting Nucleic Acids

Interactions between nucleic acids and antimicrobials inhibit cellular functions related to replication, transcription, and gene expression. The dual mechanism of action of the essential oil constituent *p*-coumaric acid disrupts bacterial cell membranes, and interacts with bacterial genomic DNA (Lou et al., 2012). The planar aromatic group of *p*-coumaric acid is hypothesized to localize in the hydrophobic environment of DNA or intercalate between adjacent base pairs of DNA, evidenced by the release of DNA-intercalated ethidium bromide (EB) into solution (Lou et al., 2012).

The proposed intracellular mechanism of the polycationic ε-PL involves binding to DNA and displacement of EB intercalated in DNA, either because of competition for binding sites or DNA compaction that impairs intercalation of EB as observed for α-poly-L-lysine (Dubois et al., 2013; Liu et al., 2015).

23.2.4.3 Modulating the Intracellular Environment

The cytoplasm is tightly regulated to maintain an optimum environment for driving biophysical processes. Thus, antimicrobials that affect cell homeostasis would also impact on the metabolic activity and survival of a microorganism.

Reactive oxygen species (ROS) such as superoxide anions ($O_2^{\bullet-}$), hydrogen peroxide (H_2O_2), and hydroxyl radicals ($OH^{\bullet}$) are natural by-products of aerobic respiration that damage cellular components like nucleic acids, lipids, and proteins (Cabiscol et al., 2000; Ezraty et al., 2017). Some antimicrobials induce oxidative stress, e.g., in *E. coli* treated with carvacrol (Khan et al., 2017b), honey containing hydrogen peroxide (Brudzynski et al., 2012), or curcumin exposed to light (Shlar et al., 2017), in *E. coli* O157:H7 treated with ε-PL (Ye et al., 2013), and in *Pseudomonas fluorescens* treated with EGCg (Liu et al., 2017). Lipid peroxidation of polyunsaturated fatty acids by ROS decreases membrane fluidity and affects membrane properties (Cabiscol et al., 2000). DNA damage by ROS includes single- and double-strand breaks, adducts of base and groups, cross-linking to other molecules, and blocking of replication (Cabiscol et al., 2000). ROS covalently modifies arginine, lysine, proline, threonine, histidine, and especially cysteine and methionine residues in proteins, thereby destabilizing and inactivating proteins (Ezraty et al., 2017).

Active transport in eukaryotic cells involves cytoskeletal filaments and protein motors (Golding and Cox, 2006), whereas molecular transport in bacteria relies on diffusion within a crowded cytoplasm with physical properties resembling a glass-like liquid (Parry et al., 2014). The diffusion rate of cellular constituents in the bacterial cytoplasm declines with increasing size (Parry et al., 2014). Interestingly, results from tracking the intracellular movement of green fluorescent protein (GFP)-tagged crescentin indicated that metabolic activity fluidizes the cytoplasm and enhances the motion of GFP-crescentin (Parry et al., 2014). Furthermore, the motion of GFP-crescentin decreased in starving cells, indicating that the fluidity of dormant cells might be different from that of cells with an active metabolism (Parry et al., 2014). Antimicrobials that influence intracellular pH, dissipate PMF, or cause ROS generation could modulate the metabolic activity of cells by affecting their cytoplasmic fluidity, yet this would need more investigation.

23.3 Antimicrobial Target Sites and Action Mechanisms in Bacterial Spores

Bacterial spore-formers mainly belong to the Gram-positive bacterial genera *Bacillus* and *Clostridium*, both of particular concern to the ready-to-eat and minimally processed food industry because of their occurrence and persistence to resist thermal processing (Postollec et al., 2012; Egan et al., 2016). Sporulation involves the formation of endospores within their mother cells to adopt a metabolic inactive, dormant phenotype, able to resist extreme conditions such as high pressure, extreme heat or cold, antimicrobials, and UV radiation (Russell, 2003a; Mckenney et al., 2013). The low susceptibility of spores to antimicrobial treatments might be due to the nature and composition of the spore coat, outer forespore membrane, cortex, and inner forespore membrane, restricting access of antimicrobials to the dehydrated core (Figure 23.1C) (Setlow, 2014a). The core contains a high content of dipicolinic acid, α/β-type small acid-soluble spore proteins (SASPs), divalent cations, DNA, proteins, and ribosomes (Russell, 2003a; Mckenney et al., 2013; Setlow 2014a, b). Dipicolinic acid chelates predominantly with Ca^{2+} and plays a role in maintaining spores in a dormant state, and, together with SASPs, they protect DNA from damage (Setlow, 2014a). Cells can remain in the dormant state for years, but still sense if conditions become favorable for them to germinate and resume vegetative growth (Mckenney et al., 2013; Setlow, 2014b). Stage I in germination consists of the release of monovalent cations, whereas the release of Ca^{2+}-dipicolinic acid triggers stage II, followed by degradation of the cortex and rehydration of the germ cell (Egan et al., 2016). Spores lose the majority of their resistance properties when germinating, rendering them susceptible to antimicrobial and food processing treatments (Russell, 2003a; Egan et al., 2016).

In contrast to the detailed descriptions of mechanisms leading to the inhibition or inactivation of vegetative cells, it is somewhat unknown what governs naturally derived antimicrobials to act sporicidal, or inhibit germination or outgrowth. Various herb extracts mixed with lactic acid reduced *B. subtilis* spore counts with more than 4 log in a mechanism proposed to act by disruption of the spore coat, followed by penetration into the cell interior to dissipate PMF and decrease pH to levels that denature acid-sensitive proteins and DNA (Cho and Chung, 2017). Thymol shows sporicidal activity and the ability to inhibit germination of *Alicyclobacillus acidoterrestris* spores by interacting with the hydrophobic spore surface structures, resulting in a damaged spore structure that leaks nucleic acids and proteins (Cai et al., 2019). Many bacteriocins inhibit or inactivate vegetative spore-formers, whereas only a small number of bacteriocins show activity towards spores in the germination phase (Egan et al., 2016). For instance, nisin required germinated spores of *Bacillus anthracis* to express lipid II to inhibit spore outgrowth through a mechanism involving membrane pore formation and prevention of membrane potential establishment (Gut et al., 2011).

23.4 Antimicrobial Target Sites and Action Mechanisms in Fungi

Yeast and mold have a cytoplasm compartmentalized with intracellular organelles, a plasma membrane, and a cell wall composed of a conserved alkali-insoluble inner core of β-(1,3)-glucan and chitin linked to a heterogeneous outer layer of mannoproteins by β-(1,6)-glucan (Figure 23.1D) (Gow et al., 2017). The mannoprotein layer of the cell wall maintains the structure and rigidity of cells, and for exponential growing cells the wall porosity allows molecules up to 5.8 nm in hydrodynamic radius (~400,000 Dalton) to permeate the wall (De nobel and Barnett, 1991). Phosphate groups in the wall mannan and carboxyl groups in the wall proteins render the cell wall of fungi negatively charged (De nobel and Barnett, 1991). The plasma membrane forms a relatively impermeable barrier for hydrophilic compounds and is embedded with proteins for the transport of solutes, signal transduction, cytoskeleton anchoring, energy generation, and enzymatic complexes for outer-wall polysaccharides (Van der Rest et al., 1995; Gow et al., 2017). The major lipid classes in the plasma membrane found in, e.g., *Saccharomyces cerevisiae* and *Pichia pastoris* consist of glycerophospholipids, sphingolipids, and sterols, and the composition is distinct from the organelle membranes (Van der Rest et al., 1995; Grillitsch et al., 2014). Specialized membranes accommodating organelle function surround the intracellular compartmentalized organelles mitochondria, endoplasmic reticulum and Golgi apparatus, nucleus, and vacuoles (Van der Rest et al., 1995). Sterols are known to decrease the fluidity of membranes, with ergosterol being the primary sterol found in fungi membranes at increased occurrence in plasma membrane than in organelle membranes (Zinser et al., 1993; Van der Rest et al., 1995; Grillitsch et al., 2014). The primary target of many compounds with activity against fungi is the plasma membrane, triggering oxidative stress, or interacting with membrane-associated ergosterol.

23.4.1 Mechanisms of Cell Wall and Membrane-Targeting Antifungals

The primary target of chitosan is the fungal plasma membrane, where chitosan electrostatically interacts with anionic phospholipids and induces membrane permeabilization (Palma-Guerrero et al., 2008, 2010). Fungi with increased amounts of polyunsaturated fatty acids have increased sensitivity towards chitosan, indicating that membrane fluidity plays a critical factor for membrane permeabilization (Palma-Guerrero et al., 2010). Fungi can decrease their membrane fluidity to reduce binding of chitosan by increasing ergosterol biosynthesis (Palma-Guerrero et al., 2010), or induce transcription of, e.g., the *ARL1* gene coding for a GTPase protein regulating membrane traffic (Jaime et al., 2012). Interestingly, studies showed that the tolerance of fungi towards chitosan depended upon cell wall content of β-glucan, indicating that the cell wall could be a physical barrier for chitosan to reach the membrane (Aranda-Martinez et al., 2016).

S. cerevisiae cells treated with ε-PL induce biosynthesis of cell wall components β-1,3-glucan, chitin, and mannosylphosphate, probably because of the cell wall-disruptive effect of ε-PL (Tan et al., 2018). At fungistatic concentrations, ε-PL interacts electrostatically with and affects plasma membrane fluidity, structure, composition, and permeability, which impact on the transportation of essential growth factors, inhibit the central carbon metabolism, and induce apoptosis by stimulating the accumulation of ROS (Bo et al., 2014, 2016). At fungicidal concentrations, ε-PL forms membrane pores according to the carpet model, leading to cell death (Bo et al., 2014). ε-PL inhibited mycelial growth, spore germination, and germ tube elongation of the gray mold *Botrytis cinerea*, in mechanisms involving the accumulation of ROS, damaged membrane integrity, and when applied *in situ* to jujube fruit it induced expression of the plant defense response to pathogens (Li et al., 2019).

Essential oil constituents are known to exert antifungal activity involving the plasma membrane, but their mechanisms are not well-known and proposed to involve multiple other targets as well. For instance, thymol and eugenol caused lysis of proliferating *S. cerevisiae* (Bennis et al., 2004), and carvacrol and thymol showed antifungal activity towards 111 *Candida* isolates by impairment of ergosterol biosynthesis and disruption of membrane integrity (Ahmad et al., 2011). Cinnamaldehyde act as a non-competing inhibitor of β-(1,3)-glucan synthase and chitin synthase 1 in *S. cerevisiae*, causing cell wall biosynthesis inhibition (Bang et al., 2000). Wang et al. (2018) showed that *Penicillium expansum* treated with a combination of cinnamaldehyde and citral resulted in oxidative stress and many membrane damages/changes, causing leakage of proteins.

The fungal plasma membrane lipid ergosterol is unique to fungi, and a frequent target for antifungal agents such as natamycin, thymol, carvacrol (Ahmad et al., 2011), and cinnamaldehyde in combination with citral (Wang et al., 2018). The polyene macrolide natamycin, also known as pimaricin, is produced by *Streptomyces natalensis* and known to inhibit most fungi specifically by binding to ergosterol, thereby blocking fungal growth and impairing membrane fusion with no effect on membrane integrity (Te Welscher et al., 2008, 2010).

23.4.2 Mechanisms of Intracellular Targeting Antifungals

In contrast to the nonspecific membrane targeting described above, Rao et al. (2010) proposed that thymol, carvacrol, and eugenol exerted fungicidal activity towards *S. cerevisiae* that involved cellular loss of metabolic activity and Ca^{2+} influx bursts. Carvacrol furthermore caused efflux of protons from the vacuolar lumen, thus disrupting the ion homeostasis in yeast (Rao et al., 2010). Transcriptional profiling of *S. cerevisiae* treated with carvacrol revealed a mechanism that resembled Ca^{2+} stress, and inhibition of target of rapamycin (TOR) signaling pathways involved in nutrient sensing (Rao et al., 2010) or regulation of growth and proliferation (Mazu et al., 2016).

The thiosulfinate compound allicin traverses fungal membranes to inactivate intracellular thiol-containing enzymes and proteins, and was found to target the glutathione pool and glutathione metabolism of *S. cerevisiae* (Gruhlke et al., 2010, 2017; Leontiev et al., 2018), thus shifting the cellular redox-potential to a more oxidized state that potentially induces apoptosis (Gruhlke et al., 2010). Acetic acid inhibited *S. cerevisiae* strains by intracellular acidification resulting in a pH drop from 6.7 to 4.7 at inhibitory concentrations similar to the weak-acid inhibition mechanism (Stratford et al., 2013). Interestingly, sorbic acid did not acidify the intracellular environment at inhibitory concentrations, but instead inhibited the plasma H^+-ATPase proton pump, probably because sorbic acid partitioned into the membrane and thereby influenced membrane fluidity and pump function (Stratford et al., 2013).

23.5 Mechanism of Action Investigations

In this section, an array of the most common experimental approaches to identify both the site of action and the mechanism of action will be discussed. Lastly, a guided overview to elucidating the mechanism of action of an antimicrobial is presented.

23.5.1 Identifying the Target Site

The site of action refers to the part of the cell that antimicrobials interact with, for instance, the cell wall, cell membrane, or constituents of the cytoplasm.

23.5.1.1 Initial Antimicrobial Activity Assessment

When working with antimicrobials having an unknown mechanism, the first step is to assess the *in vitro* inhibitory or killing efficacy using broth turbidity measurements to identify minimum inhibitory concentration (MIC) or agar-based colony counting methods to evaluate the minimum bactericidal concentration (MBC) (Davidson and Parish, 1989). Another evaluation of antimicrobial efficacy is termed the minimum duration for killing (MDK), explained in the section "Resistance and Stress Adaptation." Kinetic growth patterns of target strains exposed to sub-inhibitory concentrations of an antimicrobial might reveal the potential target site if the growth response compares to that of antibiotics with known target sites and mechanisms of action. For instance, Schneider et al. (2010) showed that the kinetic growth response of *Bacillus subtilis* in the presence of plectasin, a fungal defensin, equaled that induced by cell wall-interfering compounds like vancomycin, penicillin, and bacitracin. Further experiments indeed demonstrated that plectasin formed a complex with lipid II to inhibit peptidoglycan synthesis.

23.5.1.2 Cell Wall as the Target Site

Cellular lysis is the most extreme cell wall damage induced by antimicrobials that target the peptidoglycan, cause gross membrane damages, or activate autolysins. The progression of cellular lysis of a dense cell culture in broth after the addition of, e.g., various endolysins can be visualized by a light microscope (Loessner, 2005), or monitored as decreasing optical density measured at 600 nm (OD_{600}) over time in a lysis assay (Mayer et al., 2008; Chang et al., 2017a).

Visualization techniques such as scanning electron microscopy (SEM) or atomic force microscopy (AFM) can image the external part of the cell wall to evaluate the extent of antimicrobial-induced cell wall damage by comparing the cell morphology of untreated and treated cells (Lv et al., 2011; Hyldgaard et al., 2014, 2015; Tan et al., 2018). AFM utilizes a cantilever with or without a sharp tip, bead, cell, or chemically modified tip to image topography, or map and quantify biological samples' physical, chemical, or biological properties at the sub-nanometer scale in air or under physiological conditions without the need for chemical sample fixation (Kasas et al., 2013; Longo and Kasas, 2014; Alsteens et al., 2017). AFM imaging of cells under physiological conditions requires surface immobilization techniques like adhesive mussel proteins to sustain the lateral forces imposed by the cantilever during imaging without causing detachment of cells from the substrate or affecting cell viability and morphology (Meyer et al., 2010). Transmission electron microscopy (TEM) can visualize ultra-thin cross sections of microorganisms, revealing exterior and interior structures like carbohydrates, nucleic acids, or lipids depending on the type of fixatives and sectioning staining used (Maunsbach and Afzelius, 1999a, b). The main drawback of AFM, SEM, and TEM is that cellular structures must be identified according to morphology unless it is possible to label biomolecules for fluorescence microscopy (Hyldgaard et al., 2012a; Alsteens et al., 2017). During AFM, SEM, or TEM imaging one must be attentive if the observed cellular structures are real, sampling artifacts, or induced by antimicrobial exposure. Severe cell wall damage, not leading to lysis, can be visualized as leakage of intracellular constituents or missing parts of the cell wall structure, e.g., the outer membrane (Gao et al., 2011; Hyldgaard et al., 2014). Cases where cells are inhibited or inactivated but the cell wall remains intact, or the cell wall has wrinkles, indentations, blebs, or is collapsed, indicate that the primary target might be intracellular or at the cell membrane (Kwon et al., 2003; Bouhdid et al., 2009, 2010; Hyldgaard et al., 2012b, 2014, 2015; Shen et al., 2015).

23.5.1.3 Cell Membrane as the Target Site

Membrane-active antimicrobials might cause dissipation of PMF, membrane permeabilization, or affect the function of membrane-embedded proteins. PMF dissipation results in disruption of ion flow across membranes and can be measured as the efflux of intracellular ions like K^+, H^+, Ca^{2+}, and inorganic phosphate (McAuliffe et al., 1998; Lambert et al., 2001; Bouhdid et al., 2010; Clementi et al., 2014). Fluorescent pH probes can detect changes in intracellular pH like 2',7'-Bis-(2-Carboxyethyl)-5-(and-6)-carboxyfluorescein (BCECF) or carboxyfluorescein diacetate succinimidyl ester (cFDA-SE), whereas the cell-permeant acetoxymethyl ester version of the ion-sensitive fluorescent dye PBFI can detect intracellular K^+ levels, Fura-2 measures intracellular Ca^{2+} levels, and the fluorescent dyes Bis-(1,3-dibutylbarbituric acid) trimethine oxonol ($DiBAC_4(3)$) and 3,3'-dipropylthiadicarbocyanine iodide ($DiSC_3(5)$) can measure changes in $\Delta\psi$ (McAuliffe et al., 1998; Krulwich et al., 2011; Clementi et al., 2014). Ion efflux is not always indicative of a complete loss of membrane function and collapse of the PMF, but can be indicative of cells entering a reversible viable but non-cultivable (VNC) state as a survival strategy (Bouhdid et al., 2010).

Strategies for measuring membrane permeabilization leading to cell inactivation can broadly be divided into membrane efflux and influx types of investigations. Efflux types of investigations evaluate the ability of specific ions or molecules with known size or polarity to leak out from membrane compartments. The most straightforward experiment able to evaluate the membrane integrity is to measure increases in optical density at 260 nm wavelength (OD_{260}) of the supernatant, indicating the release of intracellular constituents such as nucleic acids (Carson et al., 2002; Bajpai et al., 2013). The efflux of ions, amino acids, ATP, and enzymes depends on the size of membrane defects. For instance, the diameter of bacteriocin-induced pores determined by current measurements across membranes showed that lacticin 3147 forms pores with a diameter of 0.6 nm, allowing leakage of H^+ and K^+, but not ATP (McAuliffe et al., 1998; Wiedemann et al., 2006). The same type of experiment showed that nisin formed pores with a diameter of 2–2.5 nm, allowing leakage of molecules the size of ions, amino acids, and ATP (Ruhr and Sahl., 1985; Wiedemann et al., 2004). Fluorescence-labeled dextran encapsulated in liposomes can be used to estimate the pore size diameter of membrane defects as demonstrated for the bacteriocin lacticin Q, which produced huge toroidal membrane-pores with a diameter of 4.6 nm, allowing leakage of small proteins (Yoneyama et al., 2009). The efflux of membrane impermeable fluorescent stains from cells is indicative of membrane damage leading to cell death. One example is the nonpolar carboxyfluorescein diacetate (cFDA) stain that enters the cytoplasm unaffected, where esterases cleave it to polar carboxyfluorescein (cF) unable to cross intact membranes (Nguefack et al., 2004). Membrane vesicles with encapsulated calcein at self-quenching concentrations can be used to monitor the membrane-permeabilizing efficiency of antimicrobials by inducing efflux of calcein into solvent where it will be diluted and start to fluoresce (Hyldgaard et al., 2015). Influx type of investigations indicative of cell death by membrane permeabilization typically monitor uptake of large polar nucleic acid binding stains such as EB (Gill and Holley, 2006b; Khan et al., 2017b) and propidium iodide (PI) (Paparella et al., 2008; Hyldgaard et al., 2012b). In case no effects are observed on membrane functionality, it can be inferred that the antimicrobial target site is intracellular.

23.5.1.4 Intracellular Target Site

Intracellular targets can be nucleic acids, proteins, and enzymes, or essential cellular processes involved in biosynthesis or energy generation. A scintillation assay with radiolabeled nucleotides or amino acids can be conducted to assess changes in protein, RNA, and DNA synthesis (Feldberg et al., 1988; Schneider et al., 2010). Antimicrobial interactions with proteins can be evaluated using SDS polyacrylamide gel electrophoresis (SDS-PAGE) of lysate from treated and untreated cells. For instance, ε-PL-treated *S. aureus* resulted in the disappearance of two low-molecular-weight protein bands (10.5 kDa and 14.4 kDa) and the simultaneous emergence of two heavyweight bands (36 kDa and 110 kDa) compared with untreated cells, indicating that ε-PL interfered with protein synthesis and caused protein leakage and aggregation (Li et al., 2014). General inhibition of intracellular enzyme activity can be detected using the nonpolar fluorescein diacetate (FDA) stain, which readily crosses membranes and is cleaved by active esterases to the polar membrane-impermeable fluorescent molecule fluorescein (Hyldgaard et al., 2015). The inhibition of amino acid decarboxylase activity to produce amines can be detected by analyzing solutions consisting of cell lysate with added enzyme substrates in the presence or absence of an antimicrobial by high-performance liquid chromatography (HPLC) (Wendakoon and Sakaguchi, 1995).

23.5.1.5 Biosynthesis or Metabolic Pathways as Targets

Application of genomic, transcriptomic, proteomic, and metabolomic techniques are useful for identifying one or more known or unknown target sites such as cellular constituents, cell structures, or metabolic processes (dos Santos et al., 2016).

Differential transcription of mRNA in the absence or presence of a sub-lethal antimicrobial concentration can be detected using DNA microarrays or metatranscriptomic RNA sequencing (RNA-Seq) (dos Santos et al., 2016). Upregulated expression of genes associated with metabolic or biosynthesis pathways in the presence of an antimicrobial can point to the adaptation stress responses of the cell trying to compensate for affected cellular structures or processes (Siroli et al., 2018).

Proteomics enables the assessment of all proteins expressed in microorganisms at a given time to identify proteins differentially expressed during exposure to an antimicrobial, useful for identifying targets and antimicrobial–ligand binding, and revealing antimicrobial mechanisms (dos Santos et al., 2016; Rempe et al., 2017). Two-dimensional gel electrophoresis (2D electrophoresis) separates proteins according to isoelectric point and molecular mass, so differently expressed proteins can be located as protein spots that disappear or have different intensity when comparing gels from untreated and antimicrobial treated cell lysates, and identified by mass spectrometry after enzymatic digestion of protein spots of interest (Nakayama et al., 2013, 2015). For instance, allicin induced up-regulation of proteins in *E. coli* associated with heat shock and oxidative stress responses, and at the same time showed that many proteins had a shift in their isoelectric point, indicating a direct modification of those proteins by allicin (Müller et al., 2016a). Proteomics evaluation has also been used to assess cellular responses of *E. coli* in the presence of curcumin (Shlar et al., 2017), *E. coli* and *B. subtilis* exposed to EGCg (Nakayama et al., 2013, 2015), and *Helicobacter pylori* treated with allitridi (Liu et al., 2010).

Metabolomics can identify if antimicrobials target specific metabolic functions, thus affecting the end products of cellular processes providing a metabolic signature determined by nuclear magnetic resonance (NMR), gas chromatography–mass spectrometry (GC–MS), and liquid chromatography–mass spectrometry (LC–MS) combined with multivariate analysis (dos Santos et al., 2016). Metabolomic analysis utilizing GC–MS data indicated that ε-PL disrupted the membrane function of *S. cerevisiae*, which indirectly led to inhibition of the central carbon metabolism (Bo et al., 2014).

23.5.2 Identifying the Mechanism of Action

After identifying the potential target site, the interactions leading to cell inhibition or inactivation can be assessed to elucidate the mechanism of action that yields elaborate knowledge about the intermolecular interactions or molecular mechanisms leading to cell inhibition or inactivation. The mechanism of action investigations can take multiple directions depending on the identified target site, which means that this section only covers a fraction of possible experiments.

23.5.2.1 Elucidating Membrane-Destabilizing Action Mechanisms

Antimicrobials identified to target membranes can affect membrane permeability, potential, bulk properties, or function of membrane-embedded proteins/enzymes. Model membrane studies conducted with native membranes from cells or purified lipids with or without membrane-associated proteins/enzymes/ligands provide a controllable framework for mimicking and evaluating how antimicrobials interact with membranes without interference from other cell components. Depending on the type of question to answer, membranes of different composition and morphology can be used such as supported lipid bilayers (SLBs) on a supporting surface, or multi- or unilamellar vesicles of different sizes in suspension, for instance, >1 μm (giant unilamellar vesicles, GUVs), 100–1000 nm (large unilamellar vesicles, LUVs), and ~20–100 nm (small unilamellar vesicles, SUVs). Fluorophore-encapsulated LUVs prepared with a membrane-associated docking molecule like lipid II can be used to investigate if an antimicrobial requires the presence of a specific ligand to permeabilize membranes (Bakhtiary et al., 2017). Fluorescence-labeled GUVs enable visualization of the membrane-destabilizing effects of antimicrobials by confocal laser scanning microscopy (CLSM) (Hyldgaard et al., 2015). Changes in bulk membrane properties, such as fluidity or phase transitions, can be assessed using DPH or Laurdan-stained LUVs combined with fluorescence anisotropy and spectrofluorometry, respectively (Nielsen and Otzen, 2010; Hyldgaard et al., 2015). AFM allows visualization of the direct effect of an antimicrobial on integrity, topography, and mechanical properties of SLBs, depending on the chosen imaging mode (Hyldgaard et al., 2012b; Morandat et al., 2013). The interaction between antimicrobials and membranes can be followed in real-time by monitoring changes in SLB mass and viscoelastic properties using quartz crystal microbalance with dissipation (QCM-D). QCM-D has been used to assess how peptides (Wang et al., 2014), fatty acid derivatives (Hyldgaard et al., 2012b), and essential oil constituents (Hyldgaard et al., 2015) affect membranes. Although model membrane studies reveal essential information about an antimicrobial's mechanism of action, it is a simplified system, and therefore does not consider the influence of other cellular structures, extracellular environment, and cellular stress adaptation mechanisms.

23.5.2.2 Intracellular Action Mechanisms

The influence of antimicrobials on intracellular targets can be challenging to assess, but a couple of examples will be discussed briefly. Mutant strains with GFP fusions to intracellular proteins can be used to assess how an antimicrobial affects the localization and function of proteins involved in regulating the cytoskeleton shape, cell division formation and regulation, chromosome compaction, and biosynthesis of RNA, DNA, and proteins. Using this approach with *B. subtilis* mutants indicated that daptomycin delocalizes cell division proteins FtsZ, FtsA, DivIVA, and MinD, and cell shape proteins MreB, MreC, and MreD (Müller et al., 2016b). Antimicrobials that target DNA can be evaluated *in vitro* using a DNA competitive binding assay, in which displacement of highly fluorescent EB intercalated in DNA result in quenching of EB's fluorescent signal (Lou et al., 2012; Liu et al., 2015). DNA fragmentation induced by antimicrobials can be visualized in yeast cells by fluorescence microscopy using DAPI staining (Tan et al., 2018).

23.5.2.3 Molecular Mechanism of Action

Random transposon mutagenesis is performed to prepare a mutant library of strains with altered tolerance towards an antimicrobial. Comparing the whole genome sequence of the wildtype and mutant strains with altered tolerance, the gene product that the antimicrobial interacts with can be directly identified (De Spiegeleer et al., 2005; Shapira and Mimran, 2007; Bociek et al., 2015). For instance, *E. coli* mutants with increased sensitivity towards thymol had mutated *qseC* and *rfaQ* genes coding for a sensor kinase protein QseC that is part of the quorum sensing system, and a glycosyltransferase involved in LPS core biosynthesis, respectively, indicating these gene products protected against thymol (Shapira and Mimran, 2007).

The molecular interaction between an antimicrobial and its target protein or the membrane can be modeled using structural information gathered by nuclear magnetic resonance (NMR) or protein crystallography, and combined with isothermal titration calorimetry (ITC) to consider thermodynamic parameters for the interaction. NMR enables elucidation of the three-dimensional structure of peptides in solution (Martin-Visscher et al., 2009) or in contact with membranes (Ouellet et al., 2010; Vad et al., 2010a), and the assessment of how an antimicrobial interacts with target molecules (Schneider et al., 2010) or changes lipid order in membranes (Vad et al., 2010b). Protein crystallography provides atomic-scale structural information useful for studying the molecular structure of proteins and enabling the modeling of molecular interactions between an antimicrobial and its target. Target proteins (like ATPase) dissolved in solution can form crystals when the solution becomes supersaturated, inducing proteins to pack in repeating arrays held together by non-covalent interactions (Gledhill et al., 2007). Soaking protein crystals of ATPase with antimicrobial allows diffusion into the protein crystal to its interaction site to infer their molecular interaction, previously done with resveratrol, quercetin, and piceatannol (Gledhill et al., 2007). Modeling a high-resolution structure of the circular bacteriocin AS-48 indicated that structural transitions occur upon membrane binding (Sánchez-Barrena et al., 2003).

Molecular dynamics simulations allow predictions about how an antimicrobial positions in membranes, and how the interaction affects membrane stability during treatment. Molecular dynamics simulations have been used to investigate the interaction of terpenes or bacteriocins with model lipid bilayers in atomic details, and ITC data can be included to provide thermodynamic parameters to the simulation (Sengupta et al., 2008; Witzke et al., 2010). For instance, simulations of the interaction between the essential oil constituent perillyl alcohol and membranes indicated that the hydroxyl-group of perillyl alcohol interacts with lipid phosphates near the aqueous interface of the glycerol backbone of phospholipids, and anchors its nonpolar region in the hydrophobic part of the membrane (Witzke et al., 2010). Furthermore, perillyl alcohol insertion into the hydrophobic core causes ordering of the lipid tails and thickening of the membrane by 1.7 Å, which could alter the function of membrane-embedded proteins (Witzke et al., 2010).

23.5.3 Guide for Conducting a Mechanism of Action Study

The following is not intended as an exhaustive description of every type of investigation that leads to elucidation of the mechanism of action, but rather a guided example of how such a study could be performed utilizing a defined set of experiments (Table 23.2 and Figure 23.3).

The first step in evaluating the mechanism of action of an antimicrobial is to determine its MIC and MBC against target strain(s), or its MDK (explained under the section "Resistance and Stress Adaptation") (Table 23.2, Figure 23.3). Knowing the antimicrobial efficiency enables subsequent experiments to be conducted at concentrations known to elicit a particular cellular response under comparable experimental conditions, for instance, at sub-inhibitory or sub-lethal concentrations, or whole integers of MIC or MBC.

The second step is to identify the target site (Table 23.2, Figure 23.3). Five sets of investigations potentially allow the target site to be identified. First, the potential target site could be identified by comparing growth kinetic measurements for a cell culture treated with either the antimicrobial of interest or antibiotics with known target sites (Schneider et al., 2010). Second, a cell lysis assay will identify if the cell wall is the primary target (Table 23.2, Figure 23.3). Third, imaging with AFM, TEM, or SEM enables visualization and evaluation of the severity of cellular damage. Observing some or no damage to the cell wall means that the target site might be the cytoplasmic cell membrane or the intracellular space. Fourth, cytoplasmic membrane damage leading to cell death can be assessed by monitoring the influx of a membrane impermeable nucleic acid stain like PI, visualized using fluorescence microscopy or quantified by flow cytometry (Hyldgaard et al., 2012b, 2015). No or a low level of challenged membrane integrity indicates that the target site probably is intracellular. Fifth, the inhibition of intracellular enzymes like esterases is relatively easy to assess by evaluating their ability to cleave the nonpolar cFDA stain (Nguefack et al., 2004).

The third step is to investigate the mechanism of action of the antimicrobial with its target site (Table 23.2, Figure 23.3). If for instance, the antimicrobial permeabilizes membranes, then the permeabilization mechanism can be investigated using model membranes such as SLBs and LUVs (Hyldgaard et al., 2012b, 2014, 2015). AFM and QCM-D allow the investigation of how antimicrobials affect SLB topography and mechanical and viscoelastic properties. If the membrane topography is not affected, but mechanical and viscoelastic properties indicate that the membrane fluidity is affected, then the influence of the antimicrobial on membrane fluidity can be monitored using LUVs stained with DPH and measured using fluorescence anisotropy (Hyldgaard et al., 2015).

Lastly, further investigations into the molecular mechanism or intermolecular interactions of antimicrobials require advanced techniques such as NMR and protein crystallography to model interactions or perform molecular dynamics simulation (Table 23.2, Figure 23.3).

The presented approach for elucidating the mechanism of action mainly focused on target-directed investigations, where each experiment answers a specific research question, whereas systems biology methods like transcriptomics, proteomics, and metabolomics are un-directed experiments that provide a broad overview of possible targets (Rempe et al., 2017). The advantage of using systems biology is that these methods can reveal known and unknown mechanisms of action (Rempe et al., 2017).

Complementary experiments that cross-validate results provide confidence in proposed hypothesizes for an antimicrobial's mechanism of action. Notably, the potential mechanisms identified by un-directed experimentation need to be cross-validated using target-directed investigations.

23.6 Resistance and Stress Adaptation

Microbial stress adaptation mechanisms and the potential emergence of microbial resistance to antimicrobials used in food or sanitation, and their potential role in the development of cross-resistance to antibiotics are of significant concern.

Studies examining possible stress adaptation or the development of microbial resistance to subsequent exposures of natural food antimicrobials often rely on *in vitro* observations (Davidson and Harrison, 2002; de Freire Bastos et al., 2015; Leite de Souza 2016). The majority of studies examining

TABLE 23.2

Overview of Experimental Techniques to Investigate the Mechanism of Action of Antimicrobials

Target Site	Mechanism of Action	Assessment Type/Experimental Techniques*	References
Cell wall	Cellular lysis	*Lysis activity*: cell lysis detected as a decrease in OD_{600} over time or visualized using light microscopy.	(Loessner, 2005, Mayer et al., 2008)
Cell wall	Morphological changes	*Scanning electron microscope*: cell surface visualization technique that produces a topographical image of a sample by scanning a focused beam of electrons across a surface coated with conducting heavy-metals.	(Bo et al., 2014, Suriyarak et al., 2014, 2018)
		Transmission electron microscope: visualize cross-sections of cells, revealing damages to the interior or exterior cellular structures depending on the type of fixative and sectioning staining used.	(Maunsbach and Afzelius, 1999a, b, Bo et al., 2016)
		Atomic force microscopy: cell surface technique that visualizes topography or measures mechanical and functional details of biological samples under physiological conditions or in air without the need for chemical sample fixation.	(Meyer et al., 2010, Hyldgaard et al., 2012b, 2014, 2015)
Cell wall	Autolysin-induced cell wall degradation	*Antimicrobials induce autolysin activity*: involvement of autolysins in antimicrobial-induced cell lysis can be assessed by adding autolysin inhibitors (e.g., Co^{2+} and sodium dodecyl sulfate), or by isolation of supernatant from antimicrobial-treated cells transferred to a dense culture of heat-killed cells. A subsequent decrease in OD_{600} of the heat-killed culture indicates the presence of autolysins.	(Martinez-Cuesta et al., 2000, Masschalck et al., 2002)
		Activity of autolysin affected by antimicrobial: influence of antimicrobial treatment on autolysin activity, monitored by OD_{600} reduction of suspensions of cell wall fragments to which autolysins are added in the presence or absence of the antimicrobial.	(Martinez-Cuesta et al., 2000)
Cell wall	Cell wall synthesis	*Inhibition of fungal cell wall synthesizing enzymes*: inhibition of β-(1,3)-glucan synthase and chitin synthase in *S. cerevisiae* can be monitored using radiolabeled cell wall substrates UDP[^{14}C]-Glu and UDP-[U-^{14}C]-GlcNAc, respectively. Incorporation kinetics of radiolabeled substrates can be determined using a liquid scintillation counter.	(Bang et al., 2000)
		Inhibition of peptidoglycan biosynthesis: isolated cytoplasmic membranes with enzymes necessary for peptidoglycan biosynthesis incubated with antimicrobial and radiolabeled substrates for lipid II synthesis (e.g., [^{3}H]-labeled C_{55}P, [^{14}C]-GlcNAc, and [^{14}C]-glycine) allow assessment of which step in peptidoglycan biosynthesis an antimicrobial inhibits by thin-layer chromatography and scintillation counting of stained and excised bands.	(Schneider et al., 2010)
Cell wall	Cell wall integrity	*Identification of cleavage site in peptidoglycan*: peptidoglycan breakdown products and cleavage position can be determined using matrix-assisted laser desorption-ionization with time-of-flight (MALDI–TOF) mass spectroscopy followed by post-source decay fragment ion spectra analysis.	(Trinetta et al., 2012)
		Identification of cleavage site in stem peptide and cross bridge: cleavage site of synthetic peptides mimicking cellular stem peptide and cross bridge of peptidoglycan can be determined by capillary electrophoresis and mass spectrometry.	(Baker et al., 2006)

(*Continued*)

TABLE 23.2 (CONTINUED)

Overview of Experimental Techniques to Investigate the Mechanism of Action of Antimicrobials

Target Site	Mechanism of Action	Assessment Type/Experimental Techniques*	References
Membrane	Outer membrane interactions	*NPN uptake*: outer membrane damages that affect permeability barrier properties permit cellular uptake of the hydrophobic fluorescent probe 1-*N*-phenyl naphthylamine (NPN), monitored by fluorescence spectroscopy (excitation at 350 nm, emission at 420 nm).	(Ellison 3rd and Giehl, 1991, Helander et al., 1998, 2001, Briers et al., 2008)
		Release of LPS: release of LPS and phospholipids can be detected by SDS-PAGE of proteinase K-treated cell-free supernatant from treated and untreated cells. Capillary GC of cell-free supernatants can detect LPS-specific fatty acids.	(Helander et al., 1998, 2001)
		Identify LPS interaction site: the interaction of an antimicrobial with specific LPS moieties in mutant strains having truncated or altered LPS can be evaluated by a combination of MIC and cell surface charge (zeta potential) measurements.	(Helander et al., 2001, Hyldgaard et al., 2014)
		Outer-membrane protein composition: changes to the protein composition of isolated outer membrane can be quantified by microfluid chip technology, which labels and separate proteins according to mass followed by laser-induced detection.	(Horváth et al., 2009)
Membrane	Membrane integrity	*β-galactosidase leakage*: challenged membrane permeability can be monitored by the release of cytoplasmic β-galactosidase, enabling degradation of the membrane-impermeable substrate O-Nitrophenyl-β-D-Galactopyranoside (ONPG) into o-nitrophenol detected spectrophotometrically at 420 nm.	(Shen et al., 2015)
		Uptake of nucleic acid stain: upon membrane permeabilization, the membrane-impermeable nucleic acid stains EB, SYTOX green, and PI enter cells, bind to nucleic acid, and become fluorescent.	(Hyldgaard et al., 2012b, Jaime et al., 2012, Bakhtiary et al., 2017, Khan et al., 2017b)
		Leakage of intracellular stains: nonpolar cFDA is readily taken up by cells and cleaved by esterases to the polar fluorescent cF retained in cells with an intact cytoplasmic membrane. Thus, cFDA also detects metabolic activity of cells.	(Paparella et al., 2008, McBee et al., 2017)
		Leakage of nucleic acid: cell lysis or cellular leakage of nucleic acids and cell content from antimicrobial treated cell suspension can be monitored spectrophotometrically at 260 nm.	(Carson et al., 2002, Bajpai et al., 2013, Khan et al., 2017b)
		Leakage of ATP: extra and intracellular ATP levels assessed before or after antimicrobial treatment of energized or ATP-depleted cells can be detected by quantifying bioluminescence output from a luciferase activity assay.	(McAuliffe et al., 1998, Gill and Holley, 2004)
		Leakage of H^+: intracellular pH and the transmembrane pH gradient can be measured by loading cells with the pH-sensitive fluorescent probes BCECF or cFDA-SE. Esterases hydrolyze cFDA-SE into the carboxyfluorescein succinimidyl ester (cFSE). The intracellular pH can be determined from the ratio of the fluorescent cFSE emission signal detected at 525 nm after excitation at 490 nm (pH sensitive) and 440 nm (pH insensitive). The transmembrane pH gradient is then calculated as $pH_{intracellular} - pH_{outside}$.	(Breeuwer et al., 1996, McAuliffe et al., 1998, Fisher and Phillips, 2009, Gong et al., 2010)

(Continued)

TABLE 23.2 (CONTINUED)

Overview of Experimental Techniques to Investigate the Mechanism of Action of Antimicrobials

Target Site	Mechanism of Action	Assessment Type/Experimental Techniques*	References
		Leakage of K^+: intracellular and extracellular K^+ content can be determined using an atomic absorption spectrophotometer, photometrically with a Kalium/Potassium kit, or by ion-selective electrodes.	(Lambert et al., 2001, Bouhdid et al., 2010, Gong et al., 2010, Clementi et al., 2014)
		Leakage of Ca^2: changes in intracellular Ca^{2+} content can be determined using the ion-sensitive fluorescent Fura-2 stain.	(Clementi et al., 2014)
		Leakage of inorganic phosphate: leakage of inorganic phosphate can be determined using an inorganic phosphate assay kit, using ion-selective electrodes, or by the formation of a colored complex with molybdate and the malachite green dye and measured using a spectrophotometer at 630 nm.	(Geladopoulos et al., 1991, Lambert et al., 2001, Gong et al., 2010)
		Assessment of electrostatic interactions and ligands in model membrane vesicle permeabilization: the membrane-permeabilizing efficacy of antimicrobials can be evaluated using model membrane vesicles with encapsulated calcein or carboxy-fluorescein. Vesicles can be prepared with purified lipids or lipids extracted from cells, and with or without ligands added. Evaluation of electrostatic interaction can be conducted using different lipid compositions of zwitterionic and negatively charged phospholipids.	(Wiedemann et al., 2001, Hyldgaard et al., 2015, Bakhtiary et al., 2017)
		Membrane binding interactions: isothermal titration calorimetry (ITC) measures the binding interaction of antimicrobials to membranes in terms of binding enthalpies and binding constants. Differences in the interaction of antimicrobials with neutral or negatively charged model membrane vesicles can be determined.	(Hunter et al., 2005)
Membrane	Cytosolic pH in yeast	*Evaluate cytosolic pH homeostasis*: *S. cerevisiae* BY4742 transformed with a plasmid containing the pHluorin gene codes for a pH-sensitive GFP derivative that localizes in the cytoplasm, allowing cytosolic pH determination.	(Brett et al., 2005, Rao et al., 2010)
Membrane	Vacuolar pH	*Effect on vacuolar pH*: the vacuolar pH can be monitored as a fluorescent signal from BCECF AM (excited at 450 and 485 nm, emission at 520 nm).	(Brett et al., 2005, Rao et al., 2010)
Membrane	Disruption of intracellular Ca^{2+} levels in yeast	*Disturb intracellular Ca^{2+} levels*: the luminescent aequorin, transformed into yeast and expressed, emits light when complexing with the cofactor coelenterazine and binds to cytosolic Ca^{2+}. The signal can be quantified by measuring luminescence.	(Rao et al., 2010)
Membrane	Dissipation of membrane potential	*Depolarized membrane stain*: depolarized membranes can be stained using $DiBAC_4(3)$, which is a lipid-soluble anionic probe with a high voltage affinity.	(Bouhdid et al., 2010, Clementi et al., 2014)
		Transmembrane potential stain: changes in ion gradients across membranes can be evaluated by spectrofluorometry using the membrane potential stains Rhodamine 123 and $DiSC_3(5)$.	(Trinetta et al., 2012, Müller et al., 2016b, Khan et al., 2017b)

(Continued)

TABLE 23.2 (CONTINUED)

Overview of Experimental Techniques to Investigate the Mechanism of Action of Antimicrobials

Target Site	Mechanism of Action	Assessment Type/Experimental Techniques*	References
Membrane	Changes in sterol content of yeast	*Effect on sterol biosynthesis*: antimicrobial-induced biosynthesis of sterol can be assessed using GC–MS with cholesterol as an internal standard. The sterol content can also be evaluated spectrophotometrically by scanning an intracellular heptane extract between 240 and 300 nm.	(Ahmad et al., 2011, Bo et al., 2016)
Membrane	Membrane proteins and enzymes	*Inhibition of ATPase activity*: specific ATPase activity of isolated cellular membranes is determined by adding ATP to antimicrobial treated membranes, leaving it to react for a while, then terminating the reaction by the addition of trichloroacetic acid. The ATP content is then quantified by bioluminescence output from luciferase assay.	(Gill and Holley, 2006a)
		Inhibition of ATP synthase: the inner membrane-bound ATP synthase can be isolated and utilized in a phosphate quantification assay to measure the hydrolysis reaction of ATP to ADP and P_i in the presence and absence of antimicrobials.	(Issa et al., 2018)
		Screening for bacteriocin docking molecules: sensitive strains repeatedly exposed to bacteriocin may induce mutations that confer resistance. Sensitive and resistant strains can be screened for the presence of putative genes of docking molecules, which can be amplified and sequenced to identify missense, nonsense, point deletions, or frameshift mutations, implying knockdown of functional genes of docking molecule.	(Daba et al., 2018)
Membrane	Membrane properties affected	*Membrane melting temperature, fluidity, and phase separation*: Fourier-transform infrared spectrometry can determine membrane ordering and lipid phase transitions, whereas GC can detect differences in the fatty acid composition of membrane lipids extracted from untreated and treated cells. Differential scanning calorimetry (DSC) allows evaluation of the impact of antimicrobials on the melting temperature and lipid phase changes of model membrane vesicles. Small-angle X-ray diffraction and ^{31}P NMR can detect lipid phase changes.	(Ultee et al., 2000, Hunter et al., 2005, Trombetta et al., 2005, Pérez-Fons et al., 2006, Cristani et al., 2007)
		Viscoelastic properties of membranes: changes in mass and viscoelastic properties of SLBs exposed to antimicrobials can be monitored in real-time by QCM-D.	(Hyldgaard et al., 2012b, 2015)
		Bilayer to micelle transition (solubilization): fluorescence of membrane situated octadecyl rhodamine β quenches at high concentrations (1–10% molar). Bilayer-to-micelle transition results in increased fluorescence, indicating solubilizing effects of an antimicrobial.	(Pérez-Fons et al., 2010)
		Membrane fluidity: fluorescence anisotropy of DPH probe situated in lipid membranes allows the influence of an antimicrobial on lipid ordering and bilayer fluidity to be measured.	(Trevors 2003, Hyldgaard et al., 2015, Bo et al., 2016)
		Membrane phase transition: the fluorescent probe Laurdan situates in membranes and shows specific generalized polarization values depending on the gel and liquid-crystalline phases of phospholipids.	(Pérez-Fons et al., 2010)
Membrane	Membrane destabilization	*Visualization of interaction with membranes*: AFM visualizes membrane damage and measures changes in mechanical properties of SLBs. Real-time interactions between antimicrobials and GUVs stained with membrane-bound Alexa 488 dextran dye and intravesicular Alexa Flour 633 hydrazide are visualized using CLSM.	(Morandat et al., 2013, Hyldgaard et al., 2014, 2015)

(Continued)

TABLE 23.2 (CONTINUED)

Overview of Experimental Techniques to Investigate the Mechanism of Action of Antimicrobials

Target Site	Mechanism of Action	Assessment Type/Experimental Techniques*	References
Intracellular	Inhibition of intracellular enzymes	*Inhibition of intracellular enzymes activity*: changes in intracellular enzyme activity such as amino acid decarboxylases, acetyl-CoA synthases (yeast), acetate kinase, phosphotransacetylase, alcohol dehydrogenases, unspecific esterases, and thioredoxin reductase can be investigated using HPLC, bioluminescence, flow cytometry, and spectrophotometry. Liquid scintillation counter enables detection of [^{3}H]-acetyl-CoA incorporation during fatty acid biosynthesis.	(Focke et al., 1990, Wendakoon and Sakaguchi, 1995, Rabinkov et al., 1998, Luciano and Holley, 2009, Hyldgaard et al., 2015)
Intracellular	Cell division inhibition	*Visualization of inhibited FtsZ polymerization*: light microscopy visualizes if cells treated with an antimicrobial are unable to separate from each other after division, indicating an inhibited cell division apparatus (FtsZ). TEM visualizes the inhibition of protofilamentation of purified FtsZ, whereas CLSM visualizes changes in the spatial distribution of Z-ring distribution in cells using GFP-tagged FtsZ.	(Rai et al., 2008, Müller et al., 2016b)
		Inhibition of FtsZ polymerization: inhibition of GTP-dependent FtsZ polymerization into the cell division Z-ring structure can be determined by assaying polymerization of purified FtsZ in the presence of GTP measured in a fluorometer using 90° angle light scattering at 350 nm of excitation and emission wavelength. The malachite green colorimetric assay quantifies the release of inorganic phosphate from hydrolysis of GTP during polymerization of purified FtsZ.	(Geladopoulos et al., 1991, Mukherjee and Lutkenhaus, 1999, Domadia et al., 2007, Hemaiswarya et al., 2011)
Intracellular	Respiratory activity	*Reductive respiratory activity*: 5-cyano-2,3-ditoyl tetrazolium chloride (CTC) penetrate membranes and is reduced by the respiratory electron transport chain to a fluorescent formazan salt, which accumulates in cells with an active metabolism. Cells with active respiration can then be detected using flow cytometry or fluorescence microscopy.	(Bouhdid et al., 2009, 2010, McBee et al., 2017)
		Oxygen electrodes: Oxygen electrodes detect changes in cellular oxygen consumption of untreated or treated cells.	(Cox et al., 1998)
Intracellular	Metabolic activity in yeast	*Assess metabolic activity of yeast*: The membrane-permeable FUN-1 green fluorescence probe binds to nucleic acids. Cells with an inactive metabolism will fluoresce green, whereas cells with an active metabolism convert the stain into a red intravacuolar spindle-like structure. Assessed using fluorescence microscopy or plate reader.	(Millard et al., 1997, Rao et al., 2010)
Intracellular	Delocalization of intracellular proteins	*Effect on secondary FtsZ structure*: circular dichroism measures the secondary structure of purified FtsZ (α-helix and β-strand) in the presence or absence of an antimicrobial.	(Rai et al., 2008)
		Localize GFP-tagged proteins: fluorescence microscopy visualizes the intracellular distribution of *B. subtilis* expressed GFP-tagged proteins involved in DNA, RNA, and protein synthesis, cell division, cell shape-determination, and peripheral membrane proteins.	(Müller et al., 2016b)
Intracellular	Reactive oxygen species	*Quantify ROS affected cells*: intracellular superoxide ($O_2{\bullet}^-$)-induced ROS activates the DNA-binding capability of CellROX Green dye that starts to fluoresce. Dye-activated cells can be quantified using flow cytometry. Not able to detect H_2O_2-induced ROS.	(McBee et al., 2017)

(Continued)

TABLE 23.2 (CONTINUED)

Overview of Experimental Techniques to Investigate the Mechanism of Action of Antimicrobials

Target Site	Mechanism of Action	Assessment Type/Experimental Techniques*	References
		Visualize ROS-induced cells: fluorescence microscopy visualizes antimicrobial-induced oxidation stress by the accumulation of ROS in cells that together with esterases convert the cell permeant 5(6)-carboxy-2',7'-dichlorofluorescein diacetate (H_2DCFDA) to the highly fluorescent 2',7'-dichlorofluorescein (DCF).	(Khan et al., 2017b)
Intracellular	DNA interactions	*Evaluate DNA intercalation activity*: EB intercalates between adjacent DNA base pairs in purified DNA, emitting a fluorescent signal at 550–750 nm when excited at 535 nm. Antimicrobial that displaces intercalated EB from DNA quenches the fluorescent signal.	(Lou et al., 2012, Liu et al., 2015)
		Assess DNA fragmentation activity: the membrane permeable nucleic acid stain DAPI can be used to visualize how antimicrobial treatment affects distribution or causes fragmentation of DNA (chromatin) in fungi.	(Tan et al., 2018)
Systems biology	General biosynthesis	*Evaluate the effect on biosynthesis*: peptidoglycan, protein, DNA, or RNA biosynthesis can be monitored by following incorporation of radiolabeled glucosamine, isoleucine, thymidine, or uridine in cells using a scintillation assay, respectively.	(Feldberg et al., 1988, Schneider et al., 2010)
Systems biology	Affect transcriptional response	*DNA microarray*: hybridization of cDNA with chip surface-attached probes coding for specific gene sequences of a microorganism. cDNA from untreated and treated cells is labeled with different fluorophores. Thus, comparison of fluorescent signal intensities between treatments can be performed.	(Visvalingam et al., 2013, Siroli et al., 2018)
		RNA-Seq: RNA-Seq monitor and quantify global transcriptomic changes between untreated and treated cells.	(Liu et al., 2017)
Systems biology	Affect proteome	*2D electrophoresis for protein target identification*: differences in spot intensity or position in stained 2D electrophoresis gels indicate changes in the proteomic profile between untreated and treated cells. Antimicrobials that directly target a protein might form a complex that limits their migration in the gel. Changes in spot intensities indicate up- or down-regulation of proteins due to antimicrobial treatment. Differentially expressed proteins can be identified by excising gel spots of interest, followed by enzymatic digestion and analyzed by mass spectrometry.	(Nakayama et al., 2013, 2015)
		Identify newly synthesized proteins: newly synthesized proteins can be identified by adding L-[^{35}S] methionine to the cell culture before stopping protein biosynthesis with chloramphenicol followed by 2D electrophoresis. The newly synthesized proteins can be detected and quantified by densitometry as radioactive-labeled spots on a phosphor screen.	(Müller et al., 2016a)
		SDS-PAGE: changes in protein biosynthesis can be detected by excising differentially Coomassie blue-stained bands in an SDS-PAGE gel, followed by identification using mass spectrometry. Western blot analysis can then confirm the observed changes in levels of the identified protein(s).	(Burt et al., 2007)

(Continued)

TABLE 23.2 (CONTINUED)

Overview of Experimental Techniques to Investigate the Mechanism of Action of Antimicrobials

Target Site	Mechanism of Action	Assessment Type/Experimental Techniques*	References
		Protein aggregation assay: proteins from soluble and insoluble moieties of crude cell extract are separated on an SDS-PAGE gel. An increasing amount of protein in the insoluble fraction at increasing concentrations of the antimicrobial indicates that the antimicrobial causes protein aggregation in a concentration-dependent manner.	(Müller et al., 2016a)
Systems biology	Affect metabolite production	*Metabolomics*: metabolic pathways inhibited by an antimicrobial can be determined using NMR, GC–MS, or LC–MS combined with multivariate statistical analysis.	(Bo et al., 2014, dos Santos et al., 2016)
In silico	Molecular mechanism	*Modeling molecular interactions*: molecular interactions between the antimicrobial and the binding pocket of the target molecule can be modeled *in silico* by using structural information gathered from protein crystallography or NMR. Thermodynamic parameters of binding events can be measured using ITC.	(Sánchez-Barrena et al., 2003, Hsu et al., 2004, Domadia et al., 2007, Hemaiswarya et al., 2011, Park et al., 2014, Bakhtiary et al., 2017)
		Simulation of molecular interactions: *in silico* simulations can be used to calculate membrane binding energies and identify the spatial position of small molecules and peptides in lipid bilayers.	(Sengupta et al., 2008, Witzke et al., 2010, Lomize et al., 2011)
Systems biology	Molecular mechanism	*Heterologously expressing docking molecule*: whole-genome sequencing of bacteriocin-sensitive and bacteriocin-resistant mutants can be done to identify the gene coding for the bacteriocin-receptor. Regaining the bacteriocin-sensitive phenotype by heterologously expressing the wildtype gene coding for the docking molecule in a bacteriocin-resistant mutant strain verifies the receptor function of the identified gene.	(Kjos et al., 2014)
		Transposon mutagenesis: random transposon-insertion mutations can induce gene mutations that alter the cell susceptibility to an antimicrobial. Affected genes can be identified by cloning and sequencing the flanking transposon-chromosome of the sensitive or resistant mutants. Complementing sensitive/resistant mutants with a plasmid containing the wildtype gene should restore the wildtype phenotypical susceptibility to the antimicrobial.	(De Spiegeleer et al., 2005, Shapira and Mimran, 2007, Van Hoang et al., 2011, Bociek et al., 2015)

* **AFM** (atomic force microscope), **BCECF AM** (2′,7′-Bis-(2-carboxyethyl)-5-(and-6)-carboxyfluorescein, acetoxymethyl ester), **C_{55}P** (bactoprenol), **cF** (carboxyfluorescein), **cFDA** (carboxyfluorescein diacetate), **cFDA-SE** (carboxyfluorescein diacetate succinimidyl ester), **CLSM** (confocal laser scanning microscope), **DAPI** (4′,6-diamidino-2-phenylindole), **DiBAC$_4$(3)** (bis(1,3-dibutylbarbituric acid) trimethine oxonol), **DiSC$_3$(5)** (3,3′-Dipropylthiadicarbocyanine iodide), **DPH** (1,6-diphenylhexatriene), **EB** (ethidium bromide), **GC–MS** (gas chromatography – mass spectrometry), **GFP** (green fluorescence protein), **GlcNAc** (*N*-acetyl-glucosamine), **Glu** (glucose), **GUV** (giant unilamellar vesicles), **LC–MS** (liquid chromatography – mass spectrometry), **LPS** (lipopolysaccharide), **LUV** (large unilamellar vesicles), **MIC** (minimum inhibitory concentration), **NMR** (nuclear magnetic resonance), **OD** (optical density), **PI** (propidium iodide), **QCM-D** (quartz crystal microbalance with dissipation), **ROS** (reactive oxygen species), **SDS-PAGE** (sodium dodecyl sulfate-polyacrylamide gel electrophoresis), **SLB** (supported lipid bilayer), **UDP** (uridine diphosphate).

(Adapted from Hyldgaard et al., 2012a)

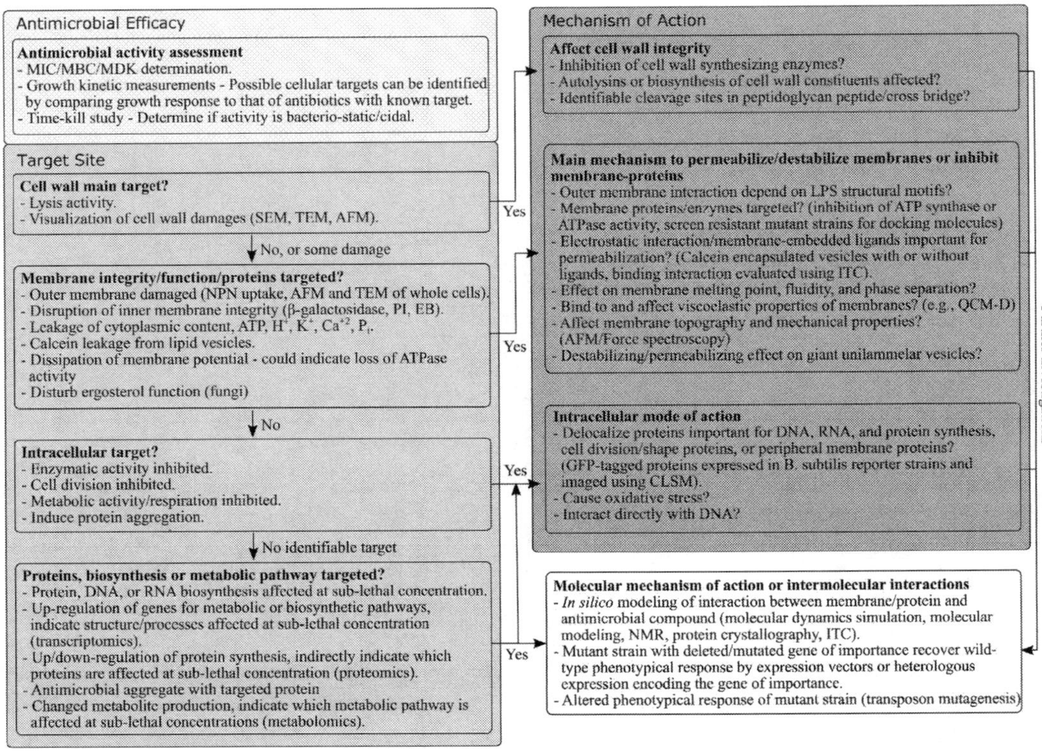

FIGURE 23.3 Pathway representation of potential experimental approaches to investigate the mechanism of action of antimicrobials. The steps follow the order, determine the antimicrobial efficacy, identify the target site, and elucidate the mechanism of action. After elucidating the mechanism of action, further investigations can be conducted to explain interactions at the molecular level. Pathways and questions are meant as guidelines for the investigator, and experiments to answer specific questions can be found in Table 23.2.

microorganisms exposed to sublethal concentrations of essential oils or their constituents did not result in decreased susceptibility to other antimicrobials, or increased antibiotic resistance, revealing that *in vitro* adaptational stresses likely do not impose a significant issue (Leite de Souza, 2016). Fadli et al. (2014) found that the exposure of *E. coli* to sublethal concentrations of *Thymus maroccanus* essential oil, carvacrol, and thymol induced a more than four-fold increase in original MIC by increasing the expression of membrane-embedded AcrAB-TolC efflux pump homologous and decreased expression of outer-membrane porins OmpF and OmpC, indicating increased efflux and decreased influx.

Stress-induced responses or the development of antimicrobial resistance reported *in vitro* still has to be verified to occur in food products. The missing proof for resistance development in food can be regarded positively, but concern remains for four reasons: (i) the accelerated emergence of resistance to antibiotics used in healthcare; (ii) increased reliance on antimicrobials and sanitizers as the primary strategy to control the outgrowth of pathogens in foods; (iii) evidence indicates that specific stresses may induce microbial tolerance to food antimicrobials, sanitizers, and other preservation techniques (Davidson and Harrison, 2002), and (iv) evidence that microbial stress adaptation may induce increased virulence and invasiveness (Humphrey et al., 1996; Gahan and Hill, 1999; Samelis et al., 2003; Leite de Souza, 2016).

23.6.1 Definition of Terms

Microbial tolerance is a transient and normally a non-heritable phenotype, whose adaptational response to physicochemical or environmental stressing conditions ensures the survival of microorganisms, with limited effect on the ability of an antimicrobial to interact with its target (Russell, 2005; Leite de Souza, 2016). Persister cells express a dormant phenotypical behavior found in a subpopulation of a community and are highly tolerant towards antimicrobials (Brauner et al., 2016).

On the contrary, then antimicrobial resistance is a heritable trait that can be intrinsic or acquired, providing a reduced uptake as a result of changes to the permeability barrier function, genetic mutations, cellular efflux, and enzymatic degradation (Davidson and Harrison, 2002; Russell 2003b; Pendleton and Davidson, 2017). Acquired resistance involves horizontal gene transfer of plasmids, transposons, or other genetic elements and may code for genes that offer resistance to various classes of antibiotics or biocides (Pendleton and Davidson, 2017).

Apparent resistance depends on assay or application conditions such as decreased bioavailability because of interactions between antimicrobial and food matrix components (Davidson and Harrison, 2002). If the antimicrobial efficiency decreases with increasing inoculum size, there might be an inoculum effect. The inoculum effect provides apparent resistance at a high cell load because of (i) too few antimicrobial molecules compared with the number of target sites; (ii) a higher chance for the presence of a resistant strain in the initial inoculum; (iii) some cells might slow down growth rate to increase bacterial tolerance; or (iv) the presence of persister cells (Steels et al., 2000; Li et al., 2017).

Friedman (2015) and Brauner et al. (2016) proposed a new set of definitions for resistance, tolerance, and persistence that give a framework for classifying the responses of microorganisms to antibiotics, which might apply to food antimicrobials as well. They proposed that tolerance can be measured as the minimum duration for killing 99% of the population (MDK_{99}, Figure 23.4) (Brauner et al., 2016). MDK_{99} is quantified from time-kill curves and based on the concept that tolerant strains require longer exposure time to be killed compared to susceptible ones, and defined as the duration of treatment required to kill a given population at concentrations exceeding the MIC (Brauner et al., 2016). Resistance describes the inherited capability of strains to proliferate at higher antimicrobial concentrations than a susceptible strain, defined as having greater MIC values (Brauner et al., 2016). Tolerant strains have an MIC comparable to that of a susceptible strain, but a markedly higher MDK_{99} (Figure 23.4). Two forms of tolerance can be identified by comparing the MDK_{99} response of antimicrobial treated cells inoculated from stationary or exponential growing cultures, which group into "tolerance by slow growth" if the MDK is high for both cultures, or "tolerance by lag" if the MDK is high only for the stationary culture (Brauner et al., 2016). "Tolerance by slow growth" is strain-related (inherited) or because of poor growth conditions or stresses such as antimicrobials that cause cells to reduce or arrest growth. The activation of stress response mechanisms, such as the production of efflux pumps, further enhances the protective effect already conferred by slow growth. "Tolerance by lag" is a transient phenotypical trait

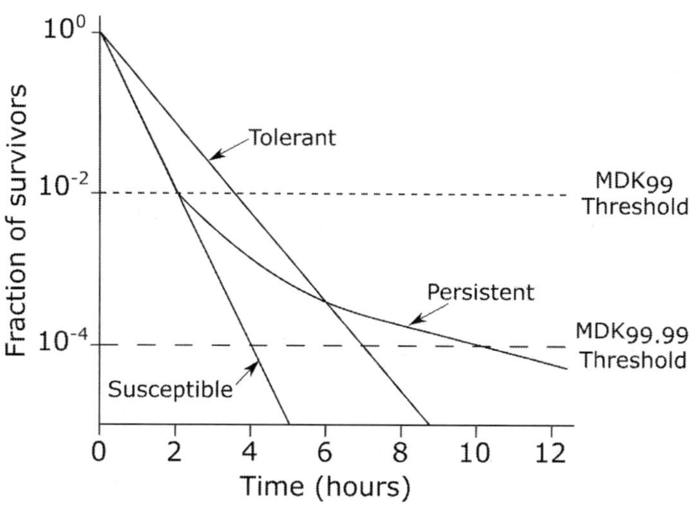

FIGURE 23.4 Schematic definition of minimum duration for killing (MDK) terminology for susceptibility, tolerance, and persistence of microorganisms to antimicrobial treatment. The MIC for tolerant and persistent strains are equal to that of the susceptible strain, but their MDK varies. The timescale is chosen for illustration purpose only. Figure was re-drawn from Brauner et al. (2016) with permission (License Number 4590100416171). Adapted by permission from Macmillan Publishers Ltd.: [Springer Nature] (Distinguishing between resistance, tolerance and persistence to antibiotic treatment (*Nature Rev. Microbiol.* 2016,14,320–30), Copyright (2016)).

that occurs when treatment is shorter than the duration of growth arrest, with an MDK of many hours or days (Brauner et al., 2016). Less than 1% of cells in a population are persisters with MIC and MDK_{99} values similar to those of susceptible cells, but a markedly longer $MDK_{99.99}$ (Figure 23.4). Persisters are grouped into "dose-dependent persistence" if a resistance mechanism is transiently present in a subpopulation, and "time-dependent persistence" if a sub-population either has longer lag-phase or slower growth (Brauner et al., 2016).

23.6.2 Intrinsic Resistance and Stress Adaptation Mechanisms

23.6.2.1 LPS Modifications

Phosphate groups in positions 1 and 4' of the lipid A moiety can be removed or modified by the addition of phosphoethanolamine or 4-amino-4-deoxy-α-L-arabinose, thus protecting bacteria from cationic peptides (Yan et al., 2007; Wang and Quinn, 2010). LPS-attached acyl chains can be hydroxylated, removed, or added to strengthen the integrity of the outer-membrane permeability barrier (Needham and Trent, 2013). For instance, the addition of an extra palmitic fatty acid to the β2-position of lipid A results in seven acyl chains, and thus an increased membrane rigidity that prevents the insertion of cationic peptides (Wang and Quinn, 2010). Ca^{2+} and Mg^{2+} can diminish charge repulsion between adjacent LPS molecules, thus ensuring tighter packing of LPS that increases the LPS permeability barrier properties (Murínová and Dercová, 2014). The O-antigen polysaccharide of LPS affects bacterial surface properties such as hydrophobicity and adhesion-to-surface abilities, but at the same time protects against hydrophobic antimicrobials (Caroff and Karibian, 2003; Murínová and Dercová, 2014).

23.6.2.2 Peptidoglycan

Some bacteria have developed resistance towards lysozyme hydrolysis of peptidoglycan by *N*-deacetylation, *N*-glycolylation, or *O*-acetylation modifications of GlcNAc and MurNAc (Vollmer, 2008; Bernard et al., 2011). Possible alterations to the cell wall of Gram-positive bacteria proposed to induce nisin-resistance include a thickened cell wall, increased cell wall hydrophobicity, or D-alanylation of teichoic acids that change the charge of peptidoglycan to become positive and thereby minimize electrostatic attraction (Zhou et al., 2014). Bacteria can also modify the cell wall hydrophobicity to become more hydrophilic and limit their interaction with hydrophobic compounds (Sikkema et al., 1995).

23.6.2.3 Cytoplasmic Membrane Modifications

The exposure of cells to sub-inhibitory concentrations of membrane-active compounds might trigger a stress adaptation response that modifies the membrane fatty acid composition to maintain membrane structure and fluidity. This response is termed homeoviscous adaptation and covers changes to the saturated/unsaturated fatty acid ratio, *cis/trans* isomerization, branching position, fatty acid chain length, and specific release of free fatty acids (Sinensky, 1974; Di Pasqua et al., 2006; Siroli et al., 2015). Increasing the ratio of long saturated acyl chains or changing unsaturated acyl chains from having *cis* to *trans* configuration make membranes more rigid because of increased membrane ordering, whereas increasing the number of cyclic or branched fatty acids increases fluidity because of less ordered membrane (Russell, 1984; Ultee et al., 2000; Mykytczuk et al., 2007; Murínová and Dercová, 2014). For instance, prevention of nisin insertion into membranes had been proposed to involve reduced transcription of the *fab* operon, which decreases the levels of proteins involved in the saturation of fatty acid chains and results in a more fluid membrane (Kramer et al., 2006). The presence of carotenoids in membranes decreases membrane fluidity and counteracts the membrane fluidifying effects of, for instance, oleic acid (Chamberlain et al., 1991). *S. cerevisiae* can increase the content of ergosterol and the ratio of short fatty acids and unsaturated fatty acids as an adaptive response to counteract the physical-chemical changes in membranes caused by antimicrobials (Bo et al., 2016). Ergosterols are rigid hydrophobic molecules with a polar hydroxyl group essential for maintaining yeast membrane fluidity and thickness. Ergosterol is also a critical factor in stress resistance against hyperosmotic perturbation (Dupont et al., 2011), and plays a role in substrate transport across the cell membrane, maintaining the activity of membrane-bound enzymes and cell viability (Bo et al., 2016).

Bacterial cells adapt to membrane perturbations by changing the ratio of the bilayer and non-bilayer forming phospholipids (e.g., PE, PG) in response to environmental stresses (Murínová and Dercová, 2014). For instance, cells that increase their ratio of cardiolipin (CL) because of environmental stresses have been proposed to enhance structural integrity, or support stress-related increases in energy transduction (Prossnigg et al., 2010). Gram-positive bacteria like *S. aureus* can increase the amount of lysyl-PG or lysyl-CL to reduce their membrane's negative charge, thus protecting against electrostatic interactions with cationic antimicrobials (Epand and Epand, 2011).

23.6.2.4 Efflux Pumps

Efflux pumps reduce the accumulation of antimicrobials inside bacteria by extruding structurally unrelated harmful compounds from the cytoplasm or the cytoplasmic membrane to the exterior. This process can be passive or active (Nikaido, 1994, 1996; Murínová and Dercová, 2014). For example, nisin-resistant *L. lactis* IL-1403 cells express ten-fold greater levels of ATP-binding cassette transporters that might remove nisin from the cell envelope without affecting the efficacy of other lipid II binding peptides (Kramer et al., 2006). In *S. cerevisiae* the ATP-binding cassette transporter pdr12 pumps out anions of weak-organic acids from the cytosol to the extracellular environment (Holyoak et al., 1999).

23.6.2.5 Proteome

The outer membrane has protein channels that allow access of small external molecules into the cell, e.g., porins are non-specific, whereas other channels might be highly specific (Murínová and Dercová, 2014). Porins can be modified or removed by bacteria to limit the access of hazardous compounds (Nikaido, 2003).

The activity and function of membrane-proteins are affected by changes in membrane fluidity, thickness, phospholipid headgroup, and fatty acid composition. Thus, adaptation to some organic solvents results in an increased ratio of proteins to phospholipids to compensate for lost protein activity (Sikkema et al., 1995; Murínová and Dercová, 2014).

Antimicrobials might induce damage in the intracellular proteome, which a cell can adapt to in three ways: (i) activation of sigma factors RpoS or RpoH upregulate gene transcription of proteins involved in stress response such as DnaK and GroEL (Dodd and Aldsworth, 2002; Di Pasqua et al., 2010; Müller et al., 2016a). Chaperone GroEL and its cofactor GroES facilitate *de novo* folding of newly synthesized proteins but can also assist in restoring activity to damaged proteins (Burt et al., 2007; Di Pasqua et al., 2010). (ii) Enzymes can reverse certain forms of protein damage, such as *cis-trans* proline isomerization, that impede proper protein folding (Murínová and Dercová, 2014). Enzymes also play a role in repairing oxidatively damaged cysteine or methionine residues in proteins, for instance, thioredoxins are used to repair oxidized cysteine residues in proteins (Cabiscol et al., 2000; Ezraty et al., 2017). (iii) If protein damage is beyond repair, then the affected proteins undergo proteolysis (Murínová and Dercová, 2014).

Oxidative stress triggers an adaptive stress response in *E. coli* cells, thus upregulating oxidative stress-related genes for superoxide dismutase and enhanced antioxidative defense mechanisms involving increased activities of catalases and peroxidases (Ye et al., 2013). For instance, oxidatively stressed *E. coli* cells activate the bacterial transcription factor OxyR through oxidation of two cysteine residues, resulting in the formation of a disulfide bond and conformational change, enabling OxyR to bind to DNA and assist RNA polymerase in initiating transcription of antioxidant genes encoding peroxiredoxins and catalases (Ezraty et al., 2017). ROS-damaged DNA stimulates increased DNA repair mechanisms by upregulation of the SOS response gene *recA* to cleave the transcriptional repressor of DNA damage repair genes, with a cumulative decrease in expression of the same repressor (Ye et al., 2013).

Sub-lethal concentrations of allicin induce transcription of the Oxidative Stress Induced 1 (*OSI1*) gene, which has a promoter with putative binding sites for the redox-dependent Yap1p transcription factor known to regulate expression of *S. cerevisiae* oxidative stress-response genes involved in antioxidative protection, glutathione metabolism, and detoxification (Gruhlke et al., 2017). Allicin oxidizes Yap1p that then accumulates in the nucleus to activate gene expression of oxidative stress response genes, providing *S. cerevisiae* with increased tolerance towards allicin and antioxidative measures to circumvent oxidative stress-elicited apoptosis (Gruhlke et al., 2010, 2017).

23.6.2.6 The Stringent Response

Starvation or other environmental stresses trigger the stringent response that regulates many cellular processes, such as biofilm formation, quorum sensing, adaptation processes, and virulence (Potrykus and Cashel, 2008; Nowicki et al., 2014, 2016). In *E. coli*, different kinds of environmental stresses lead to synthesis and accumulation of the alarmone guanosine penta/tetraphosphate ((p)ppGpp) mediated by RelA and SpoT proteins. RelA is a ribosome-associated enzyme that activates (p)ppGpp synthesis under amino acid limiting conditions, whereas SpoT has a weak synthase and strong hydrolase activity (Potrykus and Cashel, 2008). The stringent response inhibits processes related to growth such as transcription by RNA polymerase to limit energy consumption under unfavorable conditions (Potrykus and Cashel, 2008). The (p)ppGpp mediates increased transcription of *rpoS* genes involved in providing stationary phase cells with increased resistance to several stress factors such as the presence of ROS, or heat and osmotic stresses (Dodd and Aldsworth, 2002).

23.6.2.7 Biofilm

Bacteria predominantly grow in a multispecies sessile community, known as a biofilm, where cells are attached to a substrate, an interface, or each other (Giaouris et al., 2014). Biofilms are imperative in the food industry environment as they may create a persistent contamination source of pathogenic or spoilage microorganisms on food-processing equipment, thus facilitating cross-contamination of food products when cells detach from the surface-associated biofilm (Shi and Zhu, 2009). Bacteria in a biofilm are embedded within a matrix of extracellular polymeric substances (EPSs) composed of extracellular DNA (eDNA), colonic acid, exopolysaccharides, alginate, and polyglutamate (Giaouris et al., 2014). Biofilms shield microorganisms from harmful environmental conditions, sequester valuable enzymes and nutrients, and facilitate cell-to-cell communication, horizontal gene transfer for the acquisition of new phenotypic traits, and recycling of lysed cell components (Okshevsky and Meyer, 2013). The protective properties of a biofilm towards antimicrobials include (i) diminished transport through the biofilm matrix, (ii) easy transfer of genes between cells encoding resistance mechanisms, (iii) different physiological states of bacteria that are less susceptible to antimicrobials (e.g., dormant or persistent bacteria), (iv) variation in local microenvironments (e.g., different pH), and (v) masking of the negative charge of bacterial cells because eDNA sequester divalent cations (Okshevsky and Meyer, 2013; Giaouris et al., 2014). These numerous protective properties provide sessile state bacteria in a biofilm with up to 1000-fold increased tolerance towards some antimicrobials compared with planktonic cells (Ceri et al., 1999).

23.6.2.8 Antimicrobial Inactivation

Enzymes produced by microorganisms can degrade or modify antimicrobials to become less toxic towards the cell (Shimoni et al., 2000; Fitzgerald et al., 2003; Murínová and Dercová, 2014; Zhou et al., 2014). For example, hydrophobic compounds that generally accumulate in membranes can be modified with a hydroxyl-group to minimize their interaction with membranes (Murínová and Dercová, 2014). The periplasmic space contains potent inhibitory proteins like Inhibitory of Vertebrate lysozyme (Ivy) protein against C-type lysozyme, speculated to situate in the periplasmatic space of *E. coli* to protect it from its natural vertebrate host lysozyme as an additional line of defense (Monchois et al., 2001).

Other intrinsic resistance and stress adaptation mechanisms involve modifications of the receptor or active site of the target to avoid the attack, commonly observed for selective antimicrobials such as antibiotics (Hemaiswarya et al., 2008; Pendleton and Davidson, 2017).

23.6.3 Acquired Resistance Mechanisms

Horizontal gene transfer involves the uptake of foreign nucleic acids, such as naked DNA, or mobile elements like plasmids, transposons, or integrons, followed by their expression within the recipient cell.

Plasmids are self-replicating DNA molecules within cells separated from a chromosome or nuclear DNA. Plasmids are not necessary for the survival of bacteria, but often they carry genes that allow host cells to survive in harsh environments, or provide host strains with greater competitiveness than other organisms (Cui et al., 2015). Some plasmids have been associated with important functions in lactic acid bacteria species, including (i) hydrolysis of proteins; (ii) carbohydrate, amino acid, and citrate metabolism; (iii) production of bacteriocin, exopolysaccharide, and pigments; (iv) resistance to bacteriophages, antibiotics, and other stress responses; and (v) DNA restriction-modification systems (Cui et al., 2015).

The three common horizontal gene transfer mechanisms to share genetic material between closely or distantly related bacterial cells are transformation, transduction, and conjugation (Van Reenen and Dicks, 2011). Transformation and transduction allow the sharing of genetic material between bacterial cells without the involvement of cell-to-cell contact.

Transformation is the genetic alteration of cells involving uptake, incorporation, and expression of exogenous genetic material from the surroundings (Van Reenen and Dicks, 2011). Exogenous genetic material in the form of whole plasmids or segments of genomic DNA usually comes from dead bacteria lysing and releasing their genetic contents into the environment. The transformation process requires the recipient cell to be in a state of competence (time-limited response to environmental conditions such as starvation), and in some cases even cells regarded as non-competent (like *L. lactis*) harbor genes for competence, but these species may require unusual conditions for competence development (Johnsborg et al., 2007; Mell and Redfield, 2014). Naturally competent bacteria actively pull DNA fragments from the environment from any source into their cytoplasm, whereas some bacteria like *Neisseriaceae* show a species-specific preference for DNA uptake (Mell and Redfield, 2014). These DNA fragments provide nucleotide building blocks to the cell, and if the fragment has high enough similarity with the chromosomal DNA, they undergo homologous recombination, in a process known as natural transformation (Mell and Redfield, 2014).

Transduction is a process where foreign DNA is transferred into a cell by bacteriophages or viral vectors from another cell, and the foreign DNA is included in the recipient's chromosome due to recombination during replication (Van Reenen and Dicks, 2011).

Conjugation is a transfer of genetic material (plasmid or transposon) between bacteria via direct cell-to-cell contact or by a bridge-like connection between two living cells (Rauch and De Vos, 1992; Van Reenen and Dicks, 2011). Conjugation requires cell-to-cell contact, usually via a pilus or pore that forms a channel to allow plasmid exchange (Huddleston, 2014). Conjugation of plasmids depends on the presence and type of plasmids as some are self-transmissible, which means they harbor genes for the transfer machinery (Huddleston, 2014). On the other hand, the non-conjugative plasmids are mobilizable if they can be transferred through conjugation when a helper self-transmissible plasmid is present (Huddleston, 2014). It has been proposed that conjugate plasmids (>30 kb) do not have as significant an impact in horizontal gene transfer in nature as do mobilizable plasmids (small plasmids of <15 kb carrying a mobilization region) (Cui et al., 2015). Still, the frequency of transfer and compatibility between individuals is donor- and recipient-dependent.

Plasmids are common in *Lactococcus* spp., with most strains carrying 4–7 plasmids ranging in size from 3 to 80 kb. Many industrially important traits are plasmid encoded, including lactose catabolism, proteinase production, bacteriocin production and immunity, and bacteriophage resistance. Examples of bacteria expressing plasmid-encoded bacteriocins and bacteriocin immunity genes are many and include, but are not limited to, *Lactobacillus plantarum* 423 expressing plantaricin 423, *L. lactis* spp. expressing, e.g., nisin, lacticin 481, and lacticin 3147, *Pediococcus* strains expressing pediocin PA-1/AcH, *Enterococcus* spp. expressing enterocin 1071A and B, bacteriocin 51, and lactococcin (Cui et al., 2015). Three commercially interesting cases of bacteriocins encoded on mobile DNA elements together with their immunity genes are nisin, encoded on a transposon and produced industrially by *L. lactis* subsp. *cremoris* (Rauch and De Vos, 1992), pediocin PA-1/AcH, an anti-listerial class IIa bacteriocin, produced by several *Pediococcus* strains and by *L. plantarum* strains (Cui et al., 2015), and salivaricin A and salivaricin B encoded on a megaplasmid in *Streptococcus salivarius* K12 (Patras et al., 2015).

REFERENCES

Ahmad, A., Khan, A., Akhtar, F., Yousuf, S., Xess, I., Khan, L., Manzoor, N. 2011. Fungicidal activity of thymol and carvacrol by disrupting ergosterol biosynthesis and membrane integrity against *Candida*. *Eur. J. Clin. Microbiol. Infect. Dis.* 30:41–50.

Ahn, E. S., Kim, Y. S., Shin, D. H. 2001. Observation of bactericidal effect of allyl isothiocyanate on *Listeria monocytogenes*. *Food Sci. Biotech.* 10:31–35.

Alakomi, H. L., Skytta, E., Saarela, M., Mattila-Sandholm, T., Latva-Kala, K., Helander, I. M. 2000. Lactic acid permeabilizes gram-negative bacteria by disrupting the outer membrane. *Appl. Environ. Microbiol.* 66:2001–2005.

Alsteens, D., Gaub, H. E., Newton, R., Pfreundschuh, M., Gerber, C., Müller, D. J. 2017. Atomic force microscopy-based characterization and design of biointerfaces. *Nature Rev. Mater.* 2:1–16.

Andrés, M. T., Fierro, J. F. 2010. Antimicrobial mechanism of action of transferrins: Selective inhibition of H+-ATPase. *Antimicrob. Agents Chemother.* 54:4335–4342.

Aranda-Martinez, A., Lopez-Moya, F., Lopez-Llorca, L. V. 2016. Cell wall composition plays a key role on sensitivity of filamentous fungi to chitosan. *J. Basic Microbiol.* 56:1059–1070.

Bajpai, V. K., Sharma, A., Baek, K. H. 2013. Antibacterial mechanism of action of *Taxus cuspidata* stem essential oil against selected foodborne pathogens. *J. Food Safety* 1–12.

Baker, J. R., Liu, C., Dong, S., Pritchard, D. G. 2006. Endopeptidase and glycosidase activities of the bacteriophage B30 lysin. *Appl. Environ. Microbiol.* 72:6825–6828.

Bakhtiary, A., Cochrane, S. A., Mercier, P., McKay, R. T., Miskolzie, M., Sit, C. S., Vederas, J. C. 2017. Insights into the mechanism of action of the two-peptide lantibiotic Lacticin 3147. *J. Am. Chem. Soc.* 139:17803–17810.

Balla, E., Dicks, L. M. T., Du Toit, M., Van Der Merwe, M. J., Holzapfel, W. H. 2000. Characterization and cloning of the genes encoding Enterocin 1071A and Enterocin 1071B, two antimicrobial peptides produced by *Enterococcus faecalis* BFE 1071. *Appl. Environ. Microbiol.* 66:1298–1304.

Bang, K. H., Lee, D. W., Park, H. M., Rhee, Y. H. 2000. Inhibition of fungal cell wall synthesizing enzymes by *trans*-cinnamaldehyde. *BioSci., Biotech. Biochem.* 64:1061–1063.

Behr, J., Vogel, R. F. 2009. Mechanisms of hop inhibition: Hop ionophores. *J. Ag. Fd. Chem.* 57:6074–6081.

Behr, J., Vogel, R. F. 2010. Mechanisms of hop inhibition include the transmembrane redox reaction. *Appl. Environ. Microbiol.* 76:142–149.

Bennis, S., Chami, F., Chami, N., Bouchikhi, T., Remmal, A. 2004. Surface alteration of *Saccharomyces cerevisiae* induced by thymol and eugenol. *Lett. Appl. Microbiol.* 38:454–458.

Benz, R. 1985. Porin from bacterial and mitochondrial outer membrane. *Crit. Rev. Biochem.* 19:145–190.

Bernard, E., Rolain, T., Courtin, P., Guillot, A., Langella, P., Hols, P., Chapot-Chartier, M. P. 2011. Characterization of *O*-acetylation of *N*-acetylglucosamine: A novel structural variation of bacterial peptidoglycan. *J. Biological Chem.* 286:23950–23958.

Bo, T., Liu, M., Zhong, C., Zhang, Q., Su, Q. zhi Z., Tan, Z. lei L., Han, P. pei P., Jia, S. ru R. 2014. Metabolomic analysis of antimicrobial mechanisms of ε-poly-L-Lysine on *Saccharomyces cerevisiae*. *J. Ag. Fd. Chem.* 62:4454–4465.

Bo, T., Han, P. P., Su, Q. Z., Fu, P., Guo, F. Z., Zheng, Z. X., Tan, Z. L., Zhong, C., Jia, S. R. 2016. Antimicrobial ε-poly-L-lysine induced changes in cell membrane compositions and properties of Saccharomyces cerevisiae. *Food Control* 61:123–134.

Bociek, K., Ferluga, S., Mardirossian, M., Benincasa, M., Tossi, A., Gennaro, R., Scocchi, M. 2015. Lipopolysaccharide phosphorylation by the WaaY kinase affects the susceptibility of *Escherichia coli* to the human antimicrobial peptide LL-37. *J. Biol. Chem.* 290:19933–19941.

Bonelli, R. R., Schneider, T., Sahl, H. G., Wiedemann, I. 2006. Insights into in vivo activities of lantibiotics from gallidermin and epidermin mode-of-action studies. *Antimicrob. Agents Chemother.* 50:1449–1457.

Bouhdid, S., Abrini, J., Zhiri, A., Espuny, M. J., Manresa, A. 2009. Investigation of functional and morphological changes in *Pseudomonas aeruginosa* and *Staphylococcus aureus* cells induced by *Origanum compactum* essential oil. *J. Appl. Microbiol.* 106:1558–1568.

Bouhdid, S., Abrini, J., Amensour, M., Zhiri, A., Espuny, M. J., Manresa, A. 2010. Functional and ultrastructural changes in *Pseudomonas aeruginosa* and *Staphylococcus aureus* cells induced by *Cinnamomum verum* essential oil. *J. Appl. Microbiol.* 109:1139–1149.

Brandenburg, K., Jürgens, G., Fukuoka, S., Koch, M. H. J., Borstel, F., Biophysik, A. 2001. Biophysical characterization of lipopolysaccharide and Lipid A inactivation by lactoferrin. *Biol. Chem.* 382:1215–1225.

Brauner, A., Fridman, O., Gefen, O., Balaban, N. Q. 2016. Distinguishing between resistance, tolerance and persistence to antibiotic treatment. *Nature Rev. Microbiol.* 14:320–330.

Breeuwer, P., Drocourt, J. louis, Rombouts, F. M. 1996. A novel method for continuous determination of the intracellular pH in Bacteria with the internally conjugated fluorescent probe 5 (and 6-)-Carboxyfluorescein Succinimidyl Ester. *Appl. Environ. Micrbiol.* 62:178–183.

Brett, C. L., Tukaye, D. N., Mukherjee, S., Rao, R. 2005. The yeast endosomal Na+(K+)/H+ exchanger Nhx1 regulates cellular pH to control vesicle trafficking. *Mol. Biol. Cell* 16:1396–1405.

Briers, Y., Volckaert, G., Cornelissen, A., Lagaert, S., Michiels, C. W., Hertveldt, K., Lavigne, R. 2007. Muralytic activity and modular structure of the endolysins of *Pseudomonas aeruginosa* bacteriophages φKZ and EL. *Molec. Microbiol.* 65:1334–1344.

Briers, Y., Cornelissen, A., Aertsen, A., Hertveldt, K., Michiels, C. W., Volckaert, G., Lavigne, R. 2008. Analysis of outer membrane permeability of *Pseudomonas aeruginosa* and bactericidal activity of endolysins KZ144 and EL188 under high hydrostatic pressure. *FEMS Microbiol. Letters* 280:113–119.

Briers, Y., Schmelcher, M., Loessner, M. J., Hendrix, J., Engelborghs, Y., Volckaert, G., Lavigne, R. 2009. The high-affinity peptidoglycan binding domain of *Pseudomonas* phage endolysin KZ144. *Biochem. Biophys. Res. Communic.* 383:187–191.

Brown, L., Wolf, J. M., Prados-Rosales, R., Casadevall, A. 2015. Through the wall: Extracellular vesicles in Gram-positive bacteria, mycobacteria and fungi. *Nature Rev. Microbiol.* 13:620–630.

Brudzynski, K., Abubaker, K., Miotto, D. 2012. Unraveling a mechanism of honey antibacterial action: Polyphenol/H2O2-induced oxidative effect on bacterial cell growth and on DNA degradation. *Food Chem.* 133:329–336.

Brul, S., Coote, P. 1999. Preservative agents in foods: Mode of action and microbial resistance mechanisms. *Int. J. Food Microbiol.* 50:1–17.

Burt, S. 2004. Essential oils: Their antibacterial properties and potential applications in foods: A review. *Int. J. Food Microbiol.* 94:223–253.

Burt, S. A., van der Zee, R., Koets, A. P., de Graaff, A. M., Van Knapen, F., Gaastra, W., Haagsman, H. P., Veldhuizen, E. J. A. 2007. Carvacrol induces heat shock protein 60 and inhibits synthesis of flagellin in *Escherichia coli* O157:H7. *Appl. Environ. Microbiol.* 73:4484–4490.

Cabeen, M. T., Jacobs-Wagner, C. 2005. Bacterial cell shape. *Nature Rev. Microbiol.* 3:601–610.

Cabiscol, E., Tamarit, J., Ros, J. 2000. Oxidative stress in bacteria and protein damage by reactive oxygen species. *Intl. Microbiol.* 3:3–8.

Cai, R., Zhang, M., Cui, L., Yuan, Y., Yang, Y., Wang, Z., Yue, T. 2019. Antibacterial activity and mechanism of thymol against *Alicyclobacillus acidoterrestris* vegetative cells and spores. *LWT Food Sci. Technol.* 105:377–384.

Caroff, M., Karibian, D. 2003. Structure of bacterial lipopolysaccharides. *Carbohydr. Res.* 338:2431–2447.

Carson, C. F., Mee, B. J., Riley, T. V, 2002. Mechanism of action of *Melaleuca alternifolia* (Tea Tree) Oil on *Staphylococcus aureus* Determined by time-kill, lysis, leakage, and salt tolerance assays and electron microscopy. *Antimicrob. Agents Chemother.* 46:1914–1920.

Cejpek, K., Valusek, J., Velisek, J. 2000. Reactions of allyl isothiocyanate with alanine, glycine, and several peptides in model systems. *J. Ag. Fd. Chem.* 48:3560–3565.

Ceri, H., Olson, M. E., Stremick, C., Read, R. R., Morck, D., Buret, A. 1999. The Calgary Biofilm Device: New technology for rapid determination of antibiotic susceptibilities of bacterial biofilms. *J. Clin. Microbiol.* 37:1771–1776.

Chamberlain, N. R., Mehrtens, B. G., Xiong, Z., Kapral, F. A., Boardman, J. L., Rearick, J. I. 1991. Correlation of carotenoid production, decreased membrane fluidity, and resistance to oleic acid killing in *Staphylococcus aureus* 18Z. *Infect. Immun.* 59:4332–4337.

Chang, Y., Kim, M., Ryu, S. 2017a. Characterization of a novel endolysin LysSA11 and its utility as a potent biocontrol agent against *Staphylococcus aureus* on food and utensils. *Food Microbiol.* 68:112–120.

Chang, Y., Yoon, H., Kang, D. H., Chang, P. S., Ryu, S. 2017b. Endolysin LysSA97 is synergistic with carvacrol in controlling *Staphylococcus aureus* in foods. *Int. J. Food Microbiol.* 244:19–26.

Cho, W. Il, Chung, M. S. 2017. Antimicrobial effect of a combination of herb extract and organic acid against *Bacillus subtilis* spores. *Food Sci. Biotech.* 26:1423–1428.

Chugunov, A., Pyrkova, D., Nolde, D., Polyansky, A., Pentkovsky, V., Efremov, R. 2013. Lipid-II forms potential ‘landing terrain’ for lantibiotics in simulated bacterial membrane. *Sci. Rep.* 3:1–11.

Clementi, E. A., Marks, L. R., Roche-Håkansson, H., Håkansson, A. P. 2014. Monitoring changes in membrane polarity, membrane integrity, and intracellular ion concentrations in *Streptococcus pneumoniae* using fluorescent dyes. *J. Vis. Exp.* e51008:1–8.

Cotter, P. D. 2014. An 'Upp'-turn in bacteriocin receptor identification. *Molec. Microbiol.* 92:1159–1163.

Cox, S. D., Gustafson, J. E., Mann, C. M., Markham, J. L., Liew, Y. C., Hartland, R. P., Bell, H. C., Warmington, J. R., Wyllie, S. G. 1998. Tea tree oil causes K+ leakage and inhibits respiration in *Escherichia coli. Lett. Appl. Microbiol.* 26:355–358.

Cristani, M., D'Arrigo, M., Mandalari, G., Castelli, F., Sarpietro, M. G., Micieli, D., Venuti, V., Bisignano, G., Saija, A., Trombetta, D., Rombetta, D. O. T. 2007. Interaction of four monoterpenes contained in essential oils with model membranes: Implications for their antibacterial activity. *J. Ag. Fd. Chem.* 55:6300–6308.

Cruz, V. L., Ramos, J., Melo, M. N., Martinez-salazar, J. 2013. Bacteriocin AS-48 binding to model membranes and pore formation as revealed by coarse-grained simulations. *Biochim. Biophys. Acta - Biomembranes* 1828:2524–2531.

Cui, Y., Hu, T., Qu, X., Zhang, L., Ding, Z., Dong, A. 2015. Plasmids from food lactic acid bacteria: Diversity, similarity, and new developments. *Intl. J. Mol. Sci.* 16:13172–13202.

Daba, G. M., Ishibashi, N., Gong, X., Taki, H., Yamashiro, K., Lim, Y. Y., Zendo, T., Sonomoto, K. 2018. Characterisation of the action mechanism of a *Lactococcus*-specific bacteriocin, lactococcin Z. *J. BioSci. Bioeng.* 126:603–610.

Davidson, P. M., Parish, M. E. 1989. Methods for testing the efficacy of food antimicrobials. *Food Technol.* 43:148–155.

Davidson, P. M., Harrison, M. A. 2002. Resistance and adaptation to food antimicrobials, sanitizers, and other process controls. *Food Technol.* 56:69–78.

Davidson, P. M., Critzer, F. J., Matthew Taylor, T. 2013. Naturally occurring antimicrobials for minimally processed foods. *Annu. Rev. Food Sci. Technol.* 4:163–190.

de Freire Bastos, M. D. C., Varella Coelho, M. L., da Silva Santos, O. C. 2015. Resistance to bacteriocins produced by gram-positive bacteria. *Microbiol. (United Kingdom)* 161:683–700.

De nobel, J. G., Barnett, J. A. 1991. Passage of molecules through yeast-cell walls: A brief essay-review. *Yeast* 7:313.

De Spiegeleer, P., Sermon, J., Vanoirbeek, K., Aertsen, A., Michiels, C. W. 2005. Role of porins in sensitivity of *Escherichia coli* to antibacterial activity of the Lactoperoxidase enzyme system. *Appl. Environ. Microbiol.* 71:3512–3518.

Demchick, P., Koch, A. L. 1996. The permeability of the wall fabric of *Escherichia coli* and *Bacillus subtilis. J. Bacteriol.* 178:768–773.

den Kamp, J. A., Redai, I., van Deenen, L. L. 1969. Phospholipid composition of *Bacillus subtilis. J. Bacteriol.* 99:298–303.

Di Pasqua, R., Hoskins, N., Betts, G., Mauriello, G. 2006. Changes in membrane fatty acids composition of microbial cells induced by addiction of thymol, carvacrol, limonene, cinnamaldehyde, and eugenol in the growing media. *J. Ag. Fd. Chem.* 54:2745–2749.

Di Pasqua, R., Mamone, G., Ferranti, P., Ercolini, D., Mauriello, G. 2010. Changes in the proteome of *Salmonella enterica* serovar Thompson as stress adaptation to sublethal concentrations of thymol. *Proteomics* 10:1040–1049.

Dodd, C. E. R., Aldsworth, T. G. 2002. The importance of RpoS in the survival of bacteria through food processing. *Int. J. Food Microbiol.* 74:189–194.

Domadia, P., Swarup, S., Bhunia, A., Sivaraman, J., Dasgupta, D. 2007. Inhibition of bacterial cell division protein FtsZ by cinnamaldehyde. *Biochem. Pharmacol.* 74:831–840.

dos Santos, B. S., da Silva, L. C. N., da Silva, T. D., Rodrigues, J. F. S., Grisotto, M. A. G., Correia, M. T. do. S., Napoleão, T. H., da Silva, M. V., Paiva, P. M. G. 2016. Application of omics technologies for evaluation of antibacterial mechanisms of action of plant-derived products. *Front. Microbiol.* 7:1–13.

Dowhan, W. 1997. Molecular basis for membrane phospholipid diversity: Why are there so many lipids? *Annu. Rev. Biochem.* 66:199–232.

Driessen, A. J. M. M., van den Hooven, H. W., Kuiper, W., van de Kamp, M., Sahl, H. G. G., Konings, R. N. H. H., Konings, W. N. 1995. Mechanistic studies of lantibiotic-induced permeabilization of phospholipid vesicles. *Biochem.* 34:1606–1614.

Drulis-Kawa, Z., Majkowska-Skrobek, G., Maciejewska, B., Delattre, A. S., Lavigne, R. 2012. Learning from bacteriophages: Advantages and limitations of phage and phage-encoded protein applications. *Curr. Protein Pept. Sci.* 13:699–722.

Dubois, A. V, Midoux, P., Gras, D., Si-Tahar, M., Bréa, D., Attucci, S., Khelloufi, M. K., Ramphal, R., Diot, P., Gauthier, F. 2013. Poly-L-lysine compacts DNA, kills bacteria, and improves protease inhibition in cystic fibrosis sputum. *Am. J. Respir. Crit. Care Med*, 188:703–709.

Dufour, V., Stahl, M., Rosenfeld, E., Stintzi, A., Baysse, C. 2013. Insights into the mode of action of benzyl isothiocyanate on *Campylobacter jejuni*. *Appl. Environ. Microbiol.* 79:6958–6968.

Dufour, V., Stahl, M., Baysse, C. 2015. The antibacterial properties of isothiocyanates. *Microbiol. (United Kingdom)* 161:229–243.

Dupont, S., Beney, L., Ferreira, T., Gervais, P. 2011. Nature of sterols affects plasma membrane behavior and yeast survival during dehydration. *Biochim. Biophys. Acta - Biomembranes* 1808:1520–1528.

Eaton, P., Fernandes, J. C., Pereira, E., Pintado, M. E., Xavier Malcata, F. 2008. Atomic force microscopy study of the antibacterial effects of chitosans on *Escherichia coli* and *Staphylococcus aureus*. *Ultramicroscopy* 108:1128–1134.

Egan, K., Field, D., Rea, M. C., Ross, R. P., Hill, C., Cotter, P. D. 2016. Bacteriocins: Novel solutions to age old spore-related problems? *Front. Microbiol.* 7:1–21.

Eklund, T. 1983. The antimicrobial effect of dissociated and undissociated sorbic acid at different pH levels. *J. Appl. Bacteriol.* 54:383–389.

Eklund, T. 1985. The effect of sorbic acid and esters of p-hydroxybenzoic acid on the protonmotive force in *Escherichia coli* membrane vesicles. *J. General Microbiol.* 131:73–76.

Ellison 3rd, R. T., Giehl, T. J. 1991. Killing of gram-negative bacteria by lactoferrin and lysozyme. *J. Clin. Invest.* 88:1080–1091.

Epand, R. M., Rotem, S., Mor, A., Berno, B., Epand, R. F. 2008. Bacterial membranes as predictors of antimicrobial potency. *J. Am.Chem. Soc.* 130:14346–14352.

Epand, R. M., Epand, R. F. 2009. Lipid domains in bacterial membranes and the action of antimicrobial agents. *Biochim. Biophys. Acta - Biomembranes* 1788:289–294.

Epand, R. M., Epand, R. F. 2011. Bacterial membrane lipids in the action of antimicrobial agents. *J. Peptide Sci.* 17:298–305.

Epand, R. M., Walker, C., Epand, R. F., Magarvey, N. A. 2016. Molecular mechanisms of membrane targeting antibiotics. *Biochim. Biophys. Acta - Biomembranes* 1858:980–987.

Eugster, M. R., Haug, M. C., Huwiler, S. G., Loessner, M. J. 2011. The cell wall binding domain of *Listeria* bacteriophage endolysin PlyP35 recognizes terminal GlcNAc residues in cell wall teichoic acid. *Molec. Microbiol.* 81:1419–1432.

Ezraty, B., Gennaris, A., Barras, F., Collet, J. F. 2017. Oxidative stress, protein damage and repair in bacteria. *Nature Rev. Microbiol.* 15:385–396.

Fadli, M., Chevalier, J., Hassani, L., Mezrioui, N. E., Pagès, J. M. 2014. Natural extracts stimulate membrane-associated mechanisms of resistance in Gram-negative bacteria. *Lett. Appl. Microbiol.* 58:472–477.

Farnaud, S., Evans, R. W. 2003. Lactoferrin—A multifunctional protein with antimicrobial properties. *Mol. Immunol* 40:395–405.

Feldberg, R. S., Chang, S. C., Kotik, A. N., Nadler, M., Neuwirth, Z., Sundstrom, D. C., Thompson, N. H. 1988. In vitro mechanism of inhibition of bacterial cell growth by allicin. *Antimicrob. Agents Chemother.* 32:1763–1768.

Feron, V. J., Til, H. P., De Vrijer, F., Woutersen, R. A., Cassee, F. R., Van Bladeren, P. J. 1991. Aldehydes: Occurrence, carcinogenic potential, mechanism of action and risk assessment. *Mutat. Res.* 259:363–385.

Fischer, W. 1994. Lipoteichoic acid and lipids in the membrane of *Staphylococcus aureus*. *Med. Microbiol. Immunol.* 183:61–76.

Fischer, W., Leopold, K. 1999. Polar lipids of four listeria species containing L-lysylcardiolipin, a novel lipid structure, and other unique phospholipids. *Intl. J. Syst. Bacteriol.* 49:653–662.

Fisher, K., Phillips, C. 2009. The mechanism of action of a citrus oil blend against *Enterococcus faecium* and *Enterococcus faecalis*. *J. Appl. Microbiol.* 106:1343–1349.

Fitzgerald, D. J., Stratford, M., Narbad, A. 2003. Analysis of the inhibition of food spoilage yeasts by vanillin. *Int. J. Food Microbiol.* 86:113–122.

Focke, M., Feld, A., Lichtenthaler, H. K. 1990. Allicin, a naturally occurring antibiotic from garlic, specifically inhibits acetyl-CoA synthetase. *FEBS Letters* 261:106–108.

Friedman, M. 2015. Antibiotic-Resistant bacteria: Prevalence in food and inactivation by food-compatible compounds and plant extracts. *J. Ag. Fd. Chem.* 63:3805–3822.

Gahan, C. G. M., Hill, C. 1999. The relationship between acid stress responses and virulence in *Salmonella typhimurium* and *Listeria monocytogenes*. *Int. J. Food Microbiol.* 50:93–100.

Ganan, M., Carrascosa, V., Martínez-Rodríguez, A. J. 2009. Antimicrobial activity of chitosan against *Campylobacter* spp. and other microorganisms and its mechanism of action. *J. Food Prot.* 72:1735–1738.

Gao, C., Tian, C., Lu, Y., Xu, J., Luo, J., Guo, X. 2011. Essential oil composition and antimicrobial activity of *Sphallerocarpus gracilis* seeds against selected food-related bacteria. *Food Control* 22:517–522.

Garcerá, M. J. G., Elferink, M. G. L., Driessen, A. J. M., Konings, W. N. 1993. *In vitro* pore-forming activity of the lantibiotic nisin: Role of protonmotive force and lipid composition. *Eur. J. Biochem.* 212:417–422.

Garmiene, G., Salomskiene, J., Jasutiene, I., Macioniene, I., Miliauskiene, I. 2010. Production of benzoic acid by lactic acid bacteria from *Lactobacillus*, *Lactococcus* and *Streptococcus* genera in milk. *Milchwissenschaft* 65:295–298.

Geladopoulos, T. P., Sotiroudis, T. G., Evangelopoulos, A. E. 1991. A malachite green cCalorimetric assay for protein phosphatase activity. *Analytical Biochem.* 192:112–116.

Gharib, R., Auezova, L., Charcosset, C., Greige-Gerges, H. 2018. Effect of a series of essential oil molecules on DPPC membrane fluidity: A biophysical study. *J. Iran. Chem. Soc.* 15:75–84.

Giaouris, E., Heir, E., Hébraud, M., Chorianopoulos, N., Langsrud, S., Møretrø, T., Habimana, O., Desvaux, M., Renier, S., Nychas, G. J. J. 2014. Attachment and biofilm formation by foodborne bacteria in meat processing environments: Causes, implications, role of bacterial interactions and control by alternative novel methods. *Meat Sci.* 97:298–309.

Gill, A. O., Holley, R. A. 2004. Mechanisms of bactericidal action of cinnamaldehyde against *Listeria monocytogenes* and of eugenol against *L. monocytogenes* and *Lactobacillus sakei*. *Appl. Environ. Microbiol.* 70:5750–5755.

Gill, A. O., Holley, R. A. 2006a. Inhibition of membrane bound ATPases of *Escherichia coli* and *Listeria monocytogenes* by plant oil aromatics. *Int. J. Food Microbiol.* 111:170–174.

Gill, A. O., Holley, R. A. 2006b. Disruption of *Escherichia coli*, *Listeria monocytogenes* and *Lactobacillus sakei* cellular membranes by plant oil aromatics. *Int. J. Food Microbiol.* 108:1–9.

Gledhill, J. R., Montgomery, M. G., Leslie, A. G. W., Walker, J. E. 2007. Mechanism of inhibition of bovine F1-ATPase by resveratrol and related polyphenols. *Proc. Natl. Acad. Sci. U.S.A.* 104:13632–13637.

Golding, I., Cox, E. C. 2006. Physical nature of bacterial cytoplasm. *Phys. Rev. Lett.* 96:14–17.

Gong, H. sheng, Meng, X. chen, Wang, H. 2010. Mode of action of plantaricin MG, a bacteriocin active against *Salmonella typhimurium*. *J. Basic Microbiol.* 50 Supplement 1:S37–S45.

Gow, N. A. R., Latge, J. P., Munro, C. A. 2017. The fungal cell wall: Structure, biosynthesis, and function. *Microbiol. Spectrum* 5:1–25.

Griffin, S., Wyllie, S. G., Markham, J. 1999. Determination of octanol–water partition coefficient for terpenoids using reversed-phase high-performance liquid chromatography. *J. Chromatogr. A* 864:221–228.

Grillitsch, K., Tarazona, P., Klug, L., Wriessnegger, T., Zellnig, G., Leitner, E., Feussner, I., Daum, G. 2014. Isolation and characterization of the plasma membrane from the yeast *Pichia pastoris*. *Biochim. Biophys. Acta - Biomembranes* 1838:1889–1897.

Gruhlke, M. C. H., Portz, D., Stitz, M., Anwar, A., Schneider, T., Jacob, C., Schlaich, N. L., Slusarenko, A. J. 2010. Allicin disrupts the cell's electrochemical potential and induces apoptosis in yeast. *Free Radic. Biol. Med.* 49:1916–1924.

Gruhlke, M. C. H., Schlembach, I., Leontiev, R., Uebachs, A., Gollwitzer, P. U. G., Weiss, A., Delaunay, A., Toledano, M., Slusarenko, A. J. 2017. Yap1p, the central regulator of the *S. cerevisiae* oxidative stress response, is activated by allicin, a natural oxidant and defence substance of garlic. *Free Radic. Biol. Med.* 108:793–802.

Gut, I. M., Blanke, S. R., Van Der Donk, W. A. 2011. Mechanism of inhibition of *Bacillus anthracis* spore outgrowth by the lantibiotic nisin. *ACS Chem. Biol.* 6:744–752.

Hancock, R. E. W., Chapple, D. S. 1999. Peptide Antibiotics. *Antimicrob. Agents Chemother.* 43:1317–1323.

Hardt, P., Engels, I., Rausch, M., Gajdiss, M., Ulm, H., Sass, P., Ohlsen, K., Sahl, H. G., Bierbaum, G., Schneider, T., Grein, F. 2017. The cell wall precursor lipid II acts as a molecular signal for the Ser/Thr kinase PknB of *Staphylococcus aureus*. *Int. J. Med. Microbiol.* 307:1–10.

Hasper, H. E., De Kruijff, B., Breukink, E. 2004. Assembly and stability of nisin-Lipid II Pores. *Biochem.* 43:11567–11575.

Hasper, H. E., Kramer, N. E., Smith, J. L., Hillman, J. D., Zachariah, C., Kuipers, O. P., de Kruijff, B., Breukink, E. 2006. Peptides that target Lipid II. *Science* 1501:2005–2006.

Helander, I. M., Von Wright, A., Mattila-Sandholm, T. M. 1997. Potential of lactic acid bacteria and novel antimicrobials against Gram-negative bacteria. *Trends Food Sci. Technol.* 8:146–150.

Helander, I. M., Alakomi, H. L. L., Latva-Kala, K., Mattila-sandholm, T., Pol, I., Smid, E. J., Gorris, L. G. M., Von Wright, A., Wright, A. Von, 1998. Characterization of the action of selected essential oil components on Gram-Negative bacteria. *J. Ag. Fd. Chem.* 8561:3590–3595.

Helander, I. M., Mattila-Sandholm, T. 2000. Permeability barrier of the Gram-negative bacterial outer membrane with special reference to nisin. *Int. J. Food Microbiol.* 60:153–161.

Helander, I. M., Nurmiaho-Lassila, E. L., Ahvenainen, R., Rhoades, J., Roller, S. 2001. Chitosan disrupts the barrier properties of the outer membrane of Gram-negative bacteria. *Int. J. Food Microbiol.* 71:235–244.

Hemaiswarya, S., Kruthiventi, A. K., Doble, M. 2008. Synergism between natural products and antibiotics against infectious diseases. *Phytomedicine* 15:639–652.

Hemaiswarya, S., Soudaminikkutty, R., Narasumani, M. L., Doble, M. 2011. Phenylpropanoids inhibit protofilament formation of *Escherichia coli* cell division protein FtsZ. *J. Med. Microbiol.* 60:1317–1325.

Holyoak, C. D., Bracey, D., Piper, P. W., Kuchler, K., Coote, P. J., Piper, P., Mahe, Y., Thompson, S., Pandjaitan, R., Holyoak, C., Egner, R., Muhlbauer, M., Coote, P., Kuchler, K. 1999. The *Saccharomyces cerevisiae* weak-acid-inducible ABC Transporter Pdr12 transports fluorescein and preservative anions from the cytosol by an energy-dependent mechanism. *J. Bacteriol.* 181:4644–4652.

Horváth, G., Kovács, K., Kocsis, B., Kustos, I. 2009. Effect of thyme (*Thymus vulgaris* L.) essential oil and its main constituents on the outer membrane protein composition of *erwinia* strains studied with microfluid chip technology. *Chromatographia* 70:1645–1650.

Hsu, S. T. D., Breukink, E., Tischenko, E., Lutters, M. A. G., De Kruijff, B., Kaptein, R., Bonvin, A. M. J. J., Van Nuland, N. A. J. 2004. The nisin-lipid II complex reveals a pyrophosphate cage that provides a blueprint for novel antibiotics. *Nat. Struct. Mol. Biol.* 11:963–967.

Hu, Z., Gänzle, M. G. 2018. Challenges and opportunities related to the use of chitosan as a food preservative. *J. Appl. Microbiol.* 307:1–14.

Huddleston, J. R. 2014. Horizontal gene transfer in the human gastrointestinal tract: Potential spread of antibiotic resistance genes. *Infect. Drug Resist.* 7:167–176.

Humphrey, T. J., Williams, A., McAlpine, K., Lever, M. S., Guard-Petter, J., Cox, J. M. 1996. Isolates of *Salmonella enterica* Enteritidis PT4 with enhanced heat and acid tolerance are more virulent in mice and more invasive in chickens. *Epidemiol. Infect.* 117:79–88.

Hunter, H. N., Jing, W., Schibli, D. J., Trinh, T., Park, I. Y., Kim, S. C., Vogel, H. J. 2005. The interactions of antimicrobial peptides derived from lysozyme with model membrane systems. *Biochim. Biophys. Acta - Biomembranes* 1668:175–189.

Hyldgaard, M., Mygind, T., Meyer, R. L. 2012a. Essential oils in food preservation: Mode of action, synergies, and interactions with food matrix components. *Front. Microbiol.* 3:1–24.

Hyldgaard, M., Sutherland, D. S., Sundh, M., Mygind, T., Meyer, R. L. 2012b. Antimicrobial mechanism of monocaprylate. *Appl. Environ. Micrbiol.* 78:2957–2965.

Hyldgaard, M., Mygind, T., Vad, B. S., Stenvang, M., Otzen, D. E., Meyer, R. L. 2014. The antimicrobial mechanism of action of epsilon-poly-L-lysine. *Appl. Environ. Microbiol.* 80:7758–7770.

Hyldgaard, M., Mygind, T., Piotrowska, R., Foss, M., Meyer, R. L. 2015. Isoeugenol has a non-disruptive detergent-like mechanism of action. *Front. Microbiol.* 6:1–14.

Ibarra-Sánchez, L. A., Van Tassell, M. L., Miller, M. J. 2018. Antimicrobial behavior of phage endolysin PlyP100 and its synergy with nisin to control *Listeria monocytogenes* in Queso Fresco. *Food Microbiol.* 72:128–134.

Ibrahim, H. R., Matsuzaki, T., Aoki, T. 2001. Genetic evidence that antibacterial activity of lysozyme is independent of its catalytic function. *FEBS Letters* 506:27–32.

Issa, D., Najjar, A., Greige-Gerges, H., Nehme, H. 2018. Screening of Some Essential Oil Constituents as potential inhibitors of the ATP synthase of *Escherichia coli. J. Food Sci.* 0:1–9.

Jaime, M. D. L. A., Lopez-Llorca, L. V, Conesa, A., Lee, A. Y., Proctor, M., Heisler, L. E., Gebbia, M., Giaever, G., Westwood, J. T., Nislow, C. 2012. Identification of yeast genes that confer resistance to chitosan oligosaccharide (COS) using chemogenomics. *BMC Genomics* 13:1–26.

Johnsborg, O., Eldholm, V., Håvarstein, L. S. 2007. Natural genetic transformation: Prevalence, mechanisms and function. *Res. Microbiol.* 158:767–778.

Juneja, V. K., Dwivedi, H. P., Yan, X. 2012. Novel natural food antimicrobials. *Annu. Rev. Food Sci. Technol.* 3:381–403.

Kasas, S., Longo, G., Dietler, G. 2013. Mechanical properties of biological specimens explored by atomic force microscopy. *J. Phys. D Appl. Phys.* 46:133001 (12 p.).

Kawakishi, S., Namiki, M. 1982. Oxidative cleavage of the disulfide bond of cystine by allyl isothiocyanate. *J. Ag. Fd. Chem.* 30:618–620.

Kawakishi, S., Kaneko, T. 1985. Interaction of oxidized glutathione with allyl isothiocyanate 24:715–718.

Kawakishi, S., Kaneko, T., Protein, J. M. I. 1987. Interaction of proteins with allyl isothiocyanate. *J. Ag. Fd. Chem.* 35:85–88.

Khan, I., Tango, C. N., Miskeen, S., Lee, B. H., Oh, D. H. 2017a. Hurdle technology: A novel approach for enhanced food quality and safety: A review. *Food Control* 73:1426–1444.

Khan, I., Bahuguna, A., Kumar, P., Bajpai, V. K., Kang, S. C. 2017b. Antimicrobial potential of carvacrol against uropathogenic *Escherichia coli* via membrane disruption, depolarization, and reactive oxygen species generation. *Front. Microbiol.* 8:1–9.

Kjos, M., Oppegård, C., Diep, D. B., Nes, I. F., Veening, J. W., Nissen-Meyer, J., Kristensen, T. 2014. Sensitivity to the two-peptide bacteriocin lactococcin G is dependent on UppP, an enzyme involved in cell-wall synthesis. *Molec. Microbiol.* 92:1177–1187.

Knothe, G., Dunn, R. 2009. A comprehensive evaluation of the melting points of fatty acids and esters determined by differential scanning calorimetry. *J. Am. Oil Chem. Soc.* 86:843–856.

Kong, M., Guang, X., Xing, K., Jin, H., Chen, X. G., Xing, K., Park, H. J. 2010. Antimicrobial properties of chitosan and mode of action: A state of the art review. *Int. J. Food Microbiol.* 144:51–63.

Kramer, N. E., Smid, E. J., Kok, J., De Kruijff, B., Kuipers, O. P., Breukink, E. 2004. Resistance of Gram-positive bacteria to nisin is not determined by Lipid II levels. *FEMS Microbiol. Letters* 239:157–161.

Kramer, N. E., Van Hijum, S. A. F. T., Knol, J., Kok, J., Kuipers, O. P. 2006. Transcriptome analysis reveals mechanisms by which *Lactococcus lactis* acquires nisin resistance. *Antimicrob. Agents Chemother.* 50:1753–1761.

Krulwich, T. A., Sachs, G., Padan, E. 2011. Molecular aspects of bacterial pH sensing and homeostasis. *Nat. Rev. Microbiol* 9:330–343.

Kwon, J. A., Yu, C. B., Park, H. D. 2003. Bacteriocidal effects and inhibition of cell separation of cinnamic aldehyde on *Bacillus cereus*. *Lett. Appl. Microbiol.* 37:61–65.

Laible, N. J., Germaine, G. R. 1985. Bactericidal Activity of human lysozyme, muramidase-inactive lysozyme, and cationic polypeptides against *Streptococcus sanguis* and *Streptococcus faecalis*: Inhibition by chitin oligosaccharides. *Infect. Immun.* 48:720–728.

Lambert, R. J. W., Skandamis, P. N., Coote, P. J., Nychas, G. E. J. E. 2001. A study of the minimum inhibitory concentration and mode of action of oregano essential oil, thymol and carvacrol. *J. Appl. Microbiol.* 91:453–462.

Lande, M. B. 1995. The relationship between membrane fluidity and permeabilities to water, solutes, ammonia, and protons. *J. Gen. Physiol.* 106:67–84.

Lee, A. G. 2004. How lipids affect the activities of integral membrane proteins. *Biochim. Biophys. Acta - Biomembranes* 1666:62–87.

Leite de Souza, E. 2016. The effects of sublethal doses of essential oils and their constituents on antimicrobial susceptibility and antibiotic resistance among food-related bacteria: A review. *Trends Food Sci. Technol.* 56:1–12.

Leontiev, R., Hohaus, N., Jacob, C., Gruhlke, M. C. H., Slusarenko, A. J. 2018. A comparison of the antibacterial and antifungal activities of thiosulfinate analogues of allicin. *Sci. Rep.* 8:1–19.

Li, H., He, C., Li, G., Zhang, Z., Li, B., Tian, S. 2019. The modes of action of epsilon-polylysine (ε-PL) against *Botrytis cinerea* in jujube fruit. *Postharvest Biol. Technol.* 147:1–9.

Li, J., Xie, S., Ahmed, S., Wang, F., Gu, Y., Zhang, C., Chai, X., Wu, Y., Cai, J., Cheng, G. 2017. Antimicrobial activity and resistance: Influencing factors. *Front. Pharmacol.* 8:1–11.

Li, Y. Q., Han, Q., Feng, J. L., Tian, W. L., Mo, H. Z. 2014. Antibacterial characteristics and mechanisms of ε-poly-lysine against *Escherichia coli* and *Staphylococcus aureus*. *Food Control* 43:22–27.

Lin, C. M., Preston Iii, J. F., Wei, C. I. 2000. Antibacterial mechanism of allyl isothiocyanate. *J. Food Prot.* 63:727–734.

Liu, H., Du, Y., Wang, X., Sun, L. 2004. Chitosan kills bacteria through cell membrane damage. *Int. J. Food Microbiol.* 95:147–155.

Liu, H., Pei, H., Han, Z., Feng, G., Li, D. 2015. The antimicrobial effects and synergistic antibacterial mechanism of the combination of ε-polylysine and nisin against *Bacillus subtilis*. *Food Control* 47:444–450.

Liu, S., Sun, Y., Li, W., Yu, H., Li, X., Liu, Z., Zeng, J., Zhou, Y., Chen, C., Jia, J. 2010. The antibacterial mode of action of allitridi for its potential use as a therapeutic agent against *Helicobacter pylori* infection. *FEMS Microbiol. Letters* 303:183–189.

Liu, X., Shen, B., Du, P., Wang, N., Wang, J., Li, J., Sun, A. 2017. Transcriptomic analysis of the response of *Pseudomonas fluorescens* to epigallocatechin gallate by RNA-seq. *PLoS ONE* 12:1–19.

Loessner, M. J., Kramer, K., Ebel, F., Scherer, S. 2002. C-terminal domains of *Listeria monocytogenes* bacteriophage murein hydrolases determine specific recognition and high-affinity binding to bacterial cell wall carbohydrates. *Molec. Microbiol.* 44:335–349.

Loessner, M. J. 2005. Bacteriophage endolysins: Current state of research and applications. *Curr. Opin. Microbiol.* 8:480–487.

Lomize, A. L., Pogozheva, I. D., Mosberg, H. I. 2011. Anisotropic solvent model of the lipid bilayer. 2. Energetics of insertion of small molecules, peptides, and proteins in membranes. *J. Chem. Inf. Model.* 51:930–946.

Longo, G., Kasas, S. 2014. Effects of antibacterial agents and drugs monitored by atomic force microscopy. *WIREs Nanomed NanoBiotechnol.* 6:230–244.

Lorenzi, V., Muselli, A., François, A., Berti, L., Pagès, J. M. M., Amaral, L., Bolla, J. M. 2009. Geraniol restores antibiotic activities against multidrug-resistant isolates from gram-negative species. *Antimicrob. Agents Chemother.* 53:2209–2211.

Lou, Z., Wang, H., Rao, S., Sun, J., Ma, C., Li, J. 2012. *p*-Coumaric acid kills bacteria through dual damage mechanisms. *Food Control* 25:550–554.

Lucera, A., Costa, C., Conte, A., Del Nobile, M. A. 2012. Food applications of natural antimicrobial compounds. *Front. Microbiol.* 3:287.

Luciano, F. B., Hosseinian, F. S., Beta, T., Holley, R. A. 2008. Effect of free-SH containing compounds on allyl isothiocyanate antimicrobial activity against *Escherichia coli* O157:H7. *J. Food Sci.* 73:M214–M220.

Luciano, F. B., Holley, R. A. 2009. Enzymatic inhibition by allyl isothiocyanate and factors affecting its antimicrobial action against *Escherichia coli* O157:H7. *Int. J. Food Microbiol.* 131:240–245.

Lv, F., Liang, H., Yuan, Q., Li, C. 2011. In vitro antimicrobial effects and mechanism of action of selected plant essential oil combinations against four food-related microorganisms. *Food Research Intl.* 44:3057–3064.

Ma, Z., Garrido-Maestu, A., Jeong, K. C. 2017. Application, mode of action, and in vivo activity of chitosan and its micro- and nanoparticles as antimicrobial agents: A review. *Carbohydr. Polym.* 176:257–265.

Malanovic, N., Lohner, K. 2016. Gram-positive bacterial cell envelopes: The impact on the activity of antimicrobial peptides. *Biochim. Biophys. Acta - Biomembranes* 1858:936–946.

Maqueda, M., Gálvez, A., Bueno, M. M., Sanchez-Barrena, M. J., González, C., Albert, A., Rico, M., Valdivia, E. 2004. Peptide AS-48: Prototype of a new class of cyclic bacteriocins. *Curr. Protein Pept. Sci.* 5:399–416.

Martin-Visscher, L. A., Gong, X., Duszyk, M., Vederas, J. C. 2009. The three-dimensional structure of carnocyclin A reveals that many circular bacteriocins share a common structural motif. *J. Biol. Chem.* 284:28674–28681.

Martinez-Cuesta, M. C., Kok, J., Herranz, E., Pelaez, C., Requena, T., Buist, G. 2000. Requirement of autolytic activity for bacteriocin-induced lysis. *Appl. Environ. Microbiol.* 66:3174–3179.

Masschalck, B., Deckers, D., Michiels, C. W. 2002. Lytic and nonlytic mechanism of inactivation of gram-positive bacteria by lysozyme under atmospheric and high hydrostatic pressure. *J. Food Prot.* 65:1916–1923.

Masschalck, B., Michiels, C. W. 2003. Antimicrobial properties of lysozyme in relation to foodborne vegetative bacteria. *Crit. Rev. Microbiol.* 29:191–214.

Maunsbach, A. B., Afzelius, B. A. 1999a. Section staining. In: Afzelius, A. B. M. A. (ed.), *Biomedical Electron Microscopy.* Academic Press, San Diego, pp. 207–228.

Maunsbach, A. B., Afzelius, B. A. 1999b. Fixatives. In: Afzelius, A. B. M. A. (ed.), *Biomedical Electron Microscopy.* Academic Press, San Diego, pp. 31–58.

Mayer, M. J., Narbad, A., Gasson, M. J. 2008. Molecular characterization of a *Clostridium difficile* bacteriophage and its cloned biologically active endolysin. *J. Bacteriol.* 190:6734–6740.

Mazu, T. K., Bricker, B. A., Flores-Rozas, H., Ablordeppey, S. Y. 2016. The mechanistic targets of antifungal agents: An overview. *Mini Rev Med. Chem.* 16:555–578.

McAuliffe, O., Ryan, M. P., Ross, R. P., Hill, C., Breeuwer, P., Abee, T. 1998. Lacticin 3147, a broad-spectrum bacteriocin which selectively dissipates the membrane potential. *Appl. Environ. Microbiol.* 64:439–445.

McBee, M. E., Chionh, Y. H., Sharaf, M. L., Ho, P., Cai, M. W. L., Dedon, P. C. 2017. Production of superoxide in bacteria is stress- and cell state-dependent: A gating-optimized flow cytometry method that minimizes ROS measurement artifacts with fluorescent dyes. *Front. Microbiol.* 8:1–17.

Mckenney, P. T., Driks, A., Eichenberger, P. 2013. The *Bacillus subtilis* endospore: Assembly and functions of the multilayered coat. *Nature Rev. Microbiol.* 11:33–44.

Mell, J. C., Redfield, R. J. 2014. Natural competence and the evolution of DNA uptake specificity. *J. Bacteriol.* 196:1471–1483.

Meyer, R. L., Zhou, X., Tang, L., Arpanaei, A., Kingshott, P., Besenbacher, F. 2010. Immobilisation of living bacteria for AFM imaging under physiological conditions. *Ultramicroscopy* 110:1349–1357.

Millard, P. J., Roth, B. L., Thi, H. P. T., Yue, S. T., Haugland, R. P. 1997. Development of the FUN-1 family of fluorescent probes for vacuole labeling and viability testing of yeasts. *Appl. Environ. Microbiol.* 63:2897–2905.

Miron, T., Rabinkov, A., Mirelman, D., Wilchek, M., Weiner, L. 2000. The mode of action of allicin: Its ready permeability through phospholipid membranes may contribute to its biological activity. *Biochim. Biophys. Acta - Biomembranes* 1463:20–30.

Mitchell, P. 1961. Coupling of Phosphorylation to electron and hydrogen transfer by a chemi-osmotic type of mechanism. *Nature* 191:144.

Moll, G. N., Konings, W. N., Driessen, A. J. 1999. Bacteriocins: Mechanism of membrane insertion and pore formation. *Antonie Van Leeuwenhoek* 76:185–198.

Monchois, V., Abergel, C., Sturgis, J., Jeudy, S., Claverie, J. 2001. *Escerichia coli ykfE* ORFan gene encodes a potent inhibitor of C-type lysozyme. *J. Biol. Chem.* 276:18437–18441.

Morandat, S., Azouzi, S., Beauvais, E., Mastouri, A., El Kirat, K. 2013. Atomic force microscopy of model lipid membranes. *Anal. Bioanal. Chem.* 405:1445–1461.

Mukherjee, A., Lutkenhaus, J. 1999. Analysis of FtsZ assembly by light scattering and determination of the role of divalent metal cations. *J. Bacteriol.* 181:823–832.

Müller, A., Eller, J., Albrecht, F., Prochnow, P., Kuhlmann, K., Bandow, J. E., Slusarenko, A. J., Leichert, L. I. O. 2016a. Allicin induces thiol stress in bacteria through S-allylmercapto modification of protein cysteines. *J. Biol. Chem.* 291:11477–11490.

Müller, A., Wenzel, M., Strahl, H., Grein, F., Saaki, T. N. V., Kohl, B., Siersma, T., Bandow, J. E., Sahl, H. G., Schneider, T., Hamoen, L. W. 2016b. Daptomycin inhibits cell envelope synthesis by interfering with fluid membrane microdomains. *Proc. Natl. Acad. Sci. U.S.A.* 113:E7077–E7086.

Müller, A., Klöckner, A., Schneider, T. 2017. Targeting a cell wall biosynthesis hot spot. *Nat. Prod. Rep.* 34:909–932.

Münch, D., Sahl, H. G. 2015. Structural variations of the cell wall precursor lipid II in Gram-positive bacteria - Impact on binding and efficacy of antimicrobial peptides. *Biochim. Biophys. Acta - Biomembranes* 1848:3062–3071.

Murínová, S., Dercová, K. 2014. Response mechanisms of bacterial degraders to environmental contaminants on the level of cell walls and cytoplasmic membrane. *Intl. J. of Microbiol.* 2014:16.

Mykytczuk, N. C. S., Trevors, J. T., Leduc, L. G., Ferroni, G. D. 2007. Fluorescence polarization in studies of bacterial cytoplasmic membrane fluidity under environmental stress. *Prog. Biophys. Mol. Biol.* 95:60–82.

Nakayama, M., Shimatani, K., Ozawa, T., Shigemune, N., Tsugukuni, T., Tomiyama, D., Kurahachi, M., Nonaka, A., Miyamoto, T. 2013. A study of the antibacterial mechanism of catechins: Isolation and identification of *Escherichia coli* cell surface proteins that interact with epigallocatechin gallate. *Food Control* 33:433–439.

Nakayama, M., Shimatani, K., Ozawa, T., Shigemune, N., Tomiyama, D., Yui, K., Katsuki, M., Ikeda, K., Nonaka, A., Miyamoto, T. 2015. Mechanism for the antibacterial action of epigallocatechin gallate (EGCg) on *Bacillus subtilis. BioSci., Biotech. Biochem.* 79:845–854.

Navarre, W. W., Ton-That, H., Faull, K. F., Schneewind, O. 1999. Multiple enzymatic activities of the murein hydrolase from Staphylococcal Phage φ11: Identification of a D-alanyl-glycine endopeptidase Activity. *J. Biol. Chem.* 274:15847–15856.

Needham, B. D., Trent, M. S. 2013. Fortifying the barrier: The impact of lipid A remodelling on bacterial pathogenesis. *Nat. Rev. Microbiol.* 11:467–481.

Nelson, D. C., Schmelcher, M., Rodriguez-Rubio, L., Klumpp, J., Pritchard, D. G., Dong, S., Donovan, D. M. 2012. Endolysins as antimicrobials. *Adv. Virus Res.* 83:299–365.

Nguefack, J., Budde, B. B., Jakobsen, M. 2004. Five essential oils from aromatic plants of Cameroon: Their antibacterial activity and ability to permeabilize the cytoplasmic membrane of *Listeria innocua* examined by flow cytometry. *Lett. Appl. Microbiol.* 39:395–400.

Nielsen, S. B., Otzen, D. E. 2010. Impact of the antimicrobial peptide Novicidin on membrane structure and integrity. *J. Colloid and Interface Sci.* 345:248–256.

Nikaido, H. 1989. Outer membrane barrier as a mechanism of antimicrobial resistance. *Antimicrob. Agents Chemother.* 33:1831–1836.

Nikaido, H. 1992. Porins and specific channels of bacterial outer membranes. *Molec. Microbiol.* 6:435–442.

Nikaido, H. 1994. Prevention of drug access to bacterial targets: Permeability barriers and active efflux. *Science* 264:382–388.

Nikaido, H. 1996. Multidrug efflux pumps of Gram-negative bacteria. *J. Bacteriol.* 178:5853–5859.

Nikaido, H. 2003. Molecular basis of bacterial outer membrane permeability revisited. *Microbiol. Mol. Biol. Rev.* 67:593–656.

Nowicki, D., Maciąg-Dorszyńska, M., Kobiela, W., Herman-Antosiewicz, A., Węgrzyn, A., Szalewska-Pałasz, A., Węgrzyn, G. 2014. Phenethyl isothiocyanate inhibits Shiga Toxin production in Enterohemorrhagic Escherichia coli by stringent response induction. *Antimicrob. Agents Chemother.* 58:2304–2315.

Nowicki, D., Rodzik, O., Herman-Antosiewicz, A., Szalewska-Pałasz, A. 2016. Isothiocyanates as effective agents against enterohemorrhagic Escherichia coli: Insight to the mode of action. *Sci. Rep.* 6:1–12.

Okabe, A., Hirai, Y., Hayashi, H., Kanemasa, Y. 1980. Alteration in phospholipid composition of *Staphylococcus aureus* during formation of autoplast. *Biochim. Biophys. Acta* 617:28–35.

Okshevsky, M., Meyer, R. L. 2013. The role of extracellular DNA in the establishment, maintenance and perpetuation of bacterial biofilms. *Crit. Rev. Microbiol.* 41:341–352.

Oliveira, H., Azeredo, J., Lavigne, R., Kluskens, L. D. 2012. Bacteriophage endolysins as a response to emerging foodborne pathogens. *Trends Food Sci. Technol.* 28:103–115.

Oliveira, H., Melo, L. D. R., Santos, S. B., Nóbrega, F. L., Ferreira, E. C., Cerca, N., Azeredo, J., Kluskens, L. D. 2013. Molecular Aspects and comparative genomics of bacteriophage endolysins. *J. Virology* 87:4558–4570.

Ouellet, M., Voyer, N., Auger, M. 2010. Membrane interactions and dynamics of a 21-mer cytotoxic peptide: A solid-state NMR study. *Biochim. Biophys. Acta - Biomembranes* 1798:235–243.

Palma-Guerrero, J., Jansson, H. B., Salinas, J., Lopez-Llorca, L. V. 2008. Effect of chitosan on hyphal growth and spore germination of plant pathogenic and biocontrol fungi. *J. Appl. Microbiol.* 104:541–553.

Palma-Guerrero, J., Lopez-Jimenez, J. A., Pérez-Berná, A. J., Huang, I. C., Jansson, H. B., Salinas, J., Villalaín, J., Read, N. D., Lopez-Llorca, L. V. 2010. Membrane fluidity determines sensitivity of filamentous fungi to chitosan. *Molec. Microbiol.* 75:1021–1032.

Paparella, A., Taccogna, L., Aguzzi, I., Chaves-López, C., Serio, A., Marsilio, F., Suzzi, G. 2008. Flow cytometric assessment of the antimicrobial activity of essential oils against *Listeria monocytogenes. Food Control* 19:1174–1182.

Park, Y., Lim, J. A., Kong, M., Ryu, S., Rhee, S. 2014. Structure of bacteriophage SPN1S endolysin reveals an unusual two-module fold for the peptidoglycan lytic and binding activity. *Molec. Microbiol.* 92:316–325.

Parry, B. R., Surovtsev, I. V., Cabeen, M. T., O'Hern, C. S., Dufresne, E. R., Jacobs-Wagner, C. 2014. The bacterial cytoplasm has glass-like properties and is fluidized by metabolic activity. *Cell* 156:183–194.

Patras, K. A., Wescombe, P. A., Rösler, B., Hale, J. D., Tagg, J. R., Doran, K. S. 2015. *Streptococcus salivarius* K12 limits Group B Streptococcus vaginal colonization. *Infect. Immun.* 83:3438–3444.

Pendleton, S., Davidson, P. M. 2017. Microbial Resistance to antimicrobials. In: Juneja, V. K. H. P. Dwivedi and J. N. Sofos (eds.), *Microbial Control and Food Preservation*. Springer, New York, pp. 173–198.

Pérez-Fons, L., Aranda, F. J., Guillén, J., Villalaín, J., Micol, V. 2006. Rosemary (*Rosmarinus officinalis*) diterpenes affect lipid polymorphism and fluidity in phospholipid membranes. *Arch. Biochem. Biophys.* 453:224–236.

Pérez-Fons, L., Garzón, M. T., Micol, V. 2010. Relationship between the antioxidant capacity and effect of rosemary (*Rosmarinus officinalis* L.) polyphenols on membrane phospholipid order. *J. Ag. Fd. Chem.* 58:161–171.

Pisoschi, A. M., Pop, A., Georgescu, C., Turcuş, V., Olah, N. K., Mathe, E. 2018. An overview of natural antimicrobials role in food. *European J. of Medicinal Chem.* 143:922–935.

Poole, K. 2002. Outer membranes and efflux: The path to multidrug resistance in Gram-negative bacteria. *Curr. Pharm. Biotech.* 3:77–98.

Postollec, F., Mathot, A. gabrielle G., Bernard, M., Divanac, M. laure, Pavan, S., Sohier, D., Divanac'h, M. L., Pavan, S., Sohier, D. 2012. Tracking spore-forming bacteria in food: From natural biodiversity to selection by processes. *Int. J. Food Microbiol.* 158:1–8.

Potrykus, K., Cashel, M. 2008. *(p)ppGpp*: Still Magical? *Annu. Rev. Microbiol.* 62:35–51.

Prossnigg, F., Hickel, A., Pabst, G., Lohner, K. 2010. Packing behaviour of two predominant anionic phospholipids of bacterial cytoplasmic membranes. *Biophys. Chem.* 150:129–135.

Raafat, D., Bargen, K. Von, Haas, A., Sahl, H. georg G., Von Bargen, K., Haas, A., Sahl, H. georg G. 2008. Insights into the mode of action of chitosan as an antibacterial compound. *Appl. Environ. Microbiol.* 74:3764–3773.

Rabinkov, A., Miron, T., Konstantinovski, L., Wilchek, M., Mirelman, D., Weiner, L. 1998. The mode of action of allicin: Trapping of radicals and interaction with thiol containing proteins. *Biochim. Biophys. Acta - General Subjects* 1379:233–244.

Rai, D., Singh, J. K., Roy, N., Panda, D. 2008. Curcumin inhibits FtsZ assembly: An attractive mechanism for its antibacterial activity. *Biochemical J.* 410:147–155.

Rao, A., Zhang, Y., Muend, S., Rao, R. 2010. Mechanism of antifungal activity of terpenoid phenols resembles calcium stress and inhibition of the TOR Pathway. *Antimicrob. Agents Chemother.* 54:5062–5069.

Rauch, P. J. G., De Vos, W. M. 1992. Characterization of the novel nisin-sucrose conjugative transposon Tn*5276* and its insertion in *Lactococcus lactis*. *J. Bacteriol.* 174:1280–1287.

Reis, J. A., Paula, A. T., Casarotti, S. N., Penna, A. L. B. 2012. Lactic acid bacteria antimicrobial compounds: characteristics and applications. *Food Eng. Rev.* 4:124–140.

Rempe, C. S., Burris, K. P., Lenaghan, S. C., Stewart, C. N. 2017. The potential of systems biology to discover antibacterial mechanisms of plant phenolics. *Front. Microbiol.* 8:422.

Ricke, S. C. 2003. Perspectives in the use of organic acids and short chain fatty acids as microbials. *Poultry Sci.* 82:632–639.

Rossi, P., Giansanti, F., Boffi, A., Ajello, M., Valenti, P., Chiancone, E., Antonini, G. 2002. Ca^{2+} binding to bovine lactoferrin enhances protein stability and influences the release of bacterial lipopolysaccharide. *Biochem. Cell Biology* 80:41–48.

Ruhr, E., Sahl, H. G. 1985. Mode of action of the peptide antibiotic nisin and influence on the membrane potential of whole cells and on cytoplasmic and artificial membrane vesicles. *Antimicrob. Agents Chemother.* 27:841–845.

Russell, A. D. 2003a. Similarities and differences in the responses of microorganisms to biocides. *J. Antimic. Chemother.* 52:750–763.

Russell, A. D. 2003b. Biocide use and antibiotic resistance: The relevance of laboratory findings to clinical and environmental situations. *Lancet Infect. Dis.* 3:794–803.

Russell, A. D. 2005. Mechanisms of action, resistance, and stress adaptation. In: Davidson, P. M. J. N. Sofos and A. L. Branen (eds.), *Antimicrobials in Food.* Third edition. Taylor & Francis, pp. 633–657.

Russell, J. B. 1991. Resistance of *Streptococcus bovis* to acetic acid at low pH: Relationship between intracellular pH and anion accumulation. *Appl. Environ. Microbiol.* 57:255–259.

Russell, J. B. 1992. Another explanation for the toxicity of fermentation acids at low pH: Anion accumulation versus uncoupling. *J. Appl. Bacteriol.* 73:363–370.

Russell, N. J. 1984. Mechanisms of thermal adaptation in bacteria: Blueprints for survival. *Trends Biochem. Sci.* 9:108–112.

Sakamoto, K., Konings, W. N. 2003. Beer spoilage bacteria and hop resistance. *Int. J. Food Microbiol.* 89:105–124.

Samelis, J., Ikeda, J. S., Sofos, J. N. 2003. Evaluation of the pH-dependent, stationary-phase acid tolerance in *Listeria monocytogenes* and *Salmonella Typhimurium* DT104 induced by culturing in media with 1% glucose: A comparative study with *Escherichia coli* O157:H7. *J. Appl. Microbiol.* 95:563–575.

Sánchez-Barrena, M. J., Martínez-Ripoll, M., Gálvez, A., Valdivia, E., Maqueda, M., Cruz, V., Albert, A., Martı, M., Maqueda, M., Cruz, V., Albert, A. 2003. Structure of Bacteriocin AS-48: From soluble state to membrane bound state. *J. Molec. Biology* 334:541–549.

Schäffer, C., Messner, P. 2005. The structure of secondary cell wall polymers: How Gram-positive bacteria stick their cell walls together. *Microbiol.* 151:643–651.

Scherer, K. M., Spille, J. H., Sahl, H. G., Grein, F., Kubitscheck, U. 2015. The lantibiotic nisin induces Lipid II aggregation, causing membrane instability and vesicle budding. *Biophysical J.* 108:1114–1124.

Schmelcher, M., Shabarova, T., Eugster, M. R., Eichenseher, F., Tchang, V. S., Banz, M., Loessner, M. J. 2010. Rapid multiplex detection and differentiation of *Listeria* cells by use of fluorescent phage endolysin cell wall binding domains. *Appl. Environ. Microbiol.* 76:5745–5756.

Schneider, T., Kruse, T., Wimmer, R., Wiedemann, I., Sass, V., Pag, U., Jansen, A., Nielsen, A. K., Mygind, P. H., Raventós, D. S., Ravn, B., Bonvin, A. M. J. J., Maria, L. De, Neve, S., Ravn, B., Bonvin, A. M. J. J., De Maria, L., Andersen, A. S., Gammelgaard, L. K., et al., 2010. Plectasin, a fungal defensin, targets the bacterial cell wall precursor Lipid II. *Science* 328:1168–1172.

Schurr, B. C., Hahne, H., Kuster, B., Behr, J., Vogel, R. F. 2015. Molecular mechanisms behind the antimicrobial activity of hop iso-α-acids in *Lactobacillus brevis. Food Microbiol.* 46:553–563.

Seddon, A. M., Casey, D., Law, R. V, Gee, A., Templer, R. H., Ces, O. 2009. Drug interactions with lipid membranes. *Chem. Soc. Rev.* 38:2509–2519.

Sengupta, D., Leontiadou, H., Mark, A. E., Marrink, S. J. 2008. Toroidal pores formed by antimicrobial peptides show significant disorder. *Biochim. Biophys. Acta - Biomembranes* 1778:2308–2317.

Setlow, P. 2014a. Spore resistance properties. *Microbiol. Spectrum* 2:1–14.

Setlow, P. 2014b. Germination of spores of *Bacillus* species: What we know and do not know. *J. Bacteriol.* 196:1297–1305.

Shapira, R., Mimran, E. 2007. Isolation and characterization of *Escherichia coli* mutants exhibiting altered response to thymol. *Microb. Drug Resist.* 13:157–165.

Shen, S., Zhang, T., Yuan, Y., Lin, S., Xu, J., Ye, H. 2015. Effects of cinnamaldehyde on *Escherichia coli* and *Staphylococcus aureus* membrane. *Food Control* 47:196–202.

Shi, X., Zhu, X. 2009. Biofilm formation and food safety in food industries. *Trends Food Sci. Technol.* 20:407–413.

Shima, S., Matsuoka, H., Iwamoto, T., Sakai, H. 1984. Antimicrobial action of ε-poly-L-lysine. *J. Antibiot.* 37:1449–1455.

Shimoni, E., Ravid, U., Shoham, Y. 2000. Isolation of a *Bacillus* sp. capable of transforming isoeugenol to vanillin. *J. Biotechnol.* 78:1–9.

Shinoda, W. 2016. Permeability across lipid membranes. *Biochim. Biophys. Acta - Biomembranes* 1858:2254–2265.

Shlar, I., Droby, S., Rodov, V. 2017. Modes of antibacterial action of curcumin under dark and light conditions: A toxicoproteomics approach. *J. Proteom.* 160:8–20.

Sikkema, J. A. N., Jan, A. M., Poolman, B., De Bont, J. A. M., Poolman, B. 1995. Mechanisms of membrane toxicity of hydrocarbons. *Microbiol. Rev.* 59:201–222.

Simpson, W. J., Smith, A. R. W. 1992. Factors affecting antibacterial activity of hop compounds and their derivatives. *J. Appl. Bacteriol.* 72:327–334.

Simpson, W. J. 1993. Ionophoric action of *trans*-isohumulone on *Lactobacillus brevis. J. General Microbiol.* 139:1041–1045.

Sinensky, M. 1974. Homeoviscous adaptation: A homeostatic process that regulates the viscosity of membrane lipids in *Escherichia coli. Proc. Natl. Acad. Sci. U.S.A.* 71:522–525.

Siroli, L., Patrignani, F., Gardini, F., Lanciotti, R. 2015. Effects of sub-lethal concentrations of thyme and oregano essential oils, carvacrol, thymol, citral and *trans*-2-hexenal on membrane fatty acid composition and volatile molecule profile of *Listeria monocytogenes*, *Escherichia coli* and *Salmonella enteritidis. Food Chem.* 182:185–192.

Siroli, L., Braschi, G., de Jong, A., Kok, J., Patrignani, F., Lanciotti, R. 2018. Transcriptomic approach and membrane fatty acid analysis to study the response mechanisms of *Escherichia coli* to thyme essential oil, carvacrol, 2-(E)-hexanal and citral exposure. *J. Appl. Microbiol.* 125:1308–1320.

Sofos, J. N., Pierson, M. D., Blocher, J. C., Busta, F. F. 1986. Mode of action of sorbic acid on bacterial cells and spores. *Int. J. Food Microbiol.* 3:1–17.

Sohlenkamp, C., Galindo-Lagunas, K. A., Guan, Z., Vinuesa, P., Robinson, S., Thomas-Oates, J., Raetz, C. R. H., Geiger, O. 2007. The lipid lysyl-phosphatidylglycerol is present in membranes of *Rhizobium tropici* CIAT899 and confers increased resistance to Polymyxin B under acidic growth conditions. *Molec. Plant-Microbe Interact.* 20:1421–1430.

Sohlenkamp, C., Geiger, O. 2015. Bacterial membrane lipids: Diversity in structures and pathways. *FEMS Microbiol. Rev.* 40:133–159.

Stavri, M., Piddock, L. J. V, Gibbons, S. 2007. Bacterial efflux pump inhibitors from natural sources. *J. Antimic. Chemother.* 59:1247–1260.

Steels, H., James, S. A., Roberts, I. N., Stratford, M. 2000. Sorbic acid resistance: The inoculum effect. *Yeast* 16:1173–1183.

Stratford, M., Nebe-von-Caron, G., Steels, H., Novodvorska, M., Ueckert, J., Archer, D. B. 2013. Weak-acid preservatives: pH and proton movements in the yeast *Saccharomyces cerevisiae*. *Int. J. Food Microbiol.* 161:164–171.

Sun, Z., Li, P., Liu, F., Bian, H., Wang, D., Wang, X., Zou, Y., Sun, C., Xu, W. 2017. Synergistic antibacterial mechanism of the *Lactobacillus crispatus* surface layer protein and nisin on *Staphylococcus saprophyticus*. *Sci. Rep.* 7:1–12.

Suriyarak, S., Gibis, M., Schmidt, H., Villeneuve, P. 2014. Antimicrobial mechanism and activity of dodecyl rosmarinate against *Staphylococcus carnosus* LTH1502 as influenced by addition of salt and change in pH. *J. Food Prot.* 77:444–452.

Suriyarak, S., Schmidt, H., Villeneuve, P., Weiss, J. 2018. Morphological and dose-dependent study on the effect of methyl, hexyl, and dodecyl rosmarinate on *Staphylococcus carnosus* LTH1502: Use of the Weibull Model. *J. Food Prot.* 81:598–605.

Tan, Z., Bo, T., Guo, F., Cui, J., Jia, S. 2018. Effects of ε-Poly-L-lysine on the cell wall of *Saccharomyces cerevisiae* and its involved antimicrobial mechanism. *Intl. J. Biol. Macromol.* 118:2230–2236.

Te Welscher, Y. M., Ten Napel, H. H., Balagué, M. M., Souza, C. M., Riezman, H., De Kruijff, B., Breukink, E. 2008. Natamycin blocks fungal growth by binding specifically to ergosterol without permeabilizing the membrane. *J. Biol. Chem.* 283:6393–6401.

Te Welscher, Y. M., Jones, L., Van Leeuwen, M. R., Dijksterhuis, J., De Kruijff, B., Eitzen, G., Breukink, E. 2010. Natamycin inhibits vacuole fusion at the priming phase via a specific interaction with ergosterol. *Antimicrob. Agents Chemother.* 54:2618–2625.

Trevors, J. T. 2003. Fluorescent probes for bacterial cytoplasmic membrane research. *J. Biochem. Biophys. Methods* 57:87–103.

Trinetta, V., Morleo, A., Sessa, F., Iametti, S., Bonomi, F., Ferranti, P. 2012. Purified sakacin A shows a dual mechanism of action against *Listeria* spp: Proton motive force dissipation and cell wall breakdown. *FEMS Microbiol. Letters* 334:143–149.

Trombetta, D., Castelli, F., Sarpietro, M. G., Venuti, V., Cristani, M., Daniele, C., Saija, A., Mazzanti, G., Bisignano, G. 2005. Mechanisms of antibacterial action of three monoterpenes. *Antimicrob. Agents Chemother.* 49:2474–2478.

Turina, A. d V, Nolan, M. V, Zygadlo, J. A., Perillo, M. A. 2006. Natural terpenes: Self-assembly and membrane partitioning. *Biophys. Chem.* 122:101–113.

Ultee, A., Kets, E. P. W., Smid, E. J. 1999. Mechanisms of action of carvacrol on the food-borne pathogen *Bacillus cereus*. *Appl. Environ. Microbiol.* 65:4606–4610.

Ultee, A., Kets, E. P. W., Alberda, M., Hoekstra, F. A., Smid, E. J. 2000. Adaptation of the food-borne pathogen *Bacillus cereus* to carvacrol. *Archives of Microbiol.* 174:233–238.

Ultee, A., Bennik, M. H. J., Moezelaar, R. 2002. The phenolic hydroxyl group of carvacrol is essential for action against the food-borne pathogen *Bacillus cereus*. *Appl. Environ. Microbiol.* 68:1561–1568.

Vaara, M. 1992. Agents that increase the permeability of the outer membrane. *Microbiol. Rev.* 56:395–411.

Vad, B., Thomsen, L. A., Bertelsen, K., Franzmann, M., Pedersen, J. M., Nielsen, S. B., Vosegaard, T., Valnickova, Z., Skrydstrup, T., Enghild, J. J., Wimmer, R., Nielsen, N. C., Otzen, D. E. 2010a. Divorcing folding from function: How acylation affects the membrane-perturbing properties of an antimicrobial peptide. *Biochim. Biophys. Acta - Proteins and Proteomics* 1804:806–820.

Vad, B. S., Bertelsen, K., Johansen, C. H., Pedersen, J. M., Skrydstrup, T., Nielsen, N. C., Otzen, D. E. 2010b. Pardaxin permeabilizes vesicles more efficiently by pore formation than by disruption. *Biophysical J.* 98:576–585.

Van der Rest, M. E., Kamminga, A. H., Nakano, A., Anraku, Y., Poolman, B., Konings, W. N. 1995. The plasma membrane of *Saccharomyces cerevisiae*: Structure, function, and biogenesis. *Microbiol. Rev.* 59:304–322.

Van Heijenoort, J. 2007. Lipid intermediates in the biosynthesis of bacterial peptidoglycan. *Microbiol. Mol. Biol. Rev.* 71:620–635.

Van Hoang, K., Stern, N. J., Saxton, A. M., Xu, F., Zeng, X., Lin, J. 2011. Prevalence, development, and molecular mechanisms of bacteriocin resistance in *Campylobacter*. *Appl. Environ. Microbiol.* 77:2309–2316.

Van Reenen, C. A., Dicks, L. M. T. 2011. Horizontal gene transfer amongst probiotic lactic acid bacteria and other intestinal microbiota: What are the possibilities? A review. *Archives of Microbiol.* 193:157–168.

Visvalingam, J., Hernandez-Doria, J. D., Holley, R. A. 2013. Examination of the Genome-wide transcriptional response of *Escherichia coli* O157:H7 to cinnamaldehyde exposure. *Appl. Environ. Microbiol.* 79:942–950.

Vollmer, W. 2008. Structural variation in the glycan strands of bacterial peptidoglycan. *FEMS Microbiol. Rev.* 32:287–306.

Vollmer, W., Blanot, D., De Pedro, M. A. 2008. Peptidoglycan structure and architecture. *FEMS Microbiol. Rev.* 32:149–167.

Vorobyov, I., Olson, T. E., Kim, J. H., Koeppe, R. E., Andersen, O. S., Allen, T. W. 2014. Ion-induced defect permeation of lipid membranes. *Biophysical J.* 106:586–597.

Wang, K. F., Nagarajan, R., Camesano, T. A. 2014. Antimicrobial peptide alamethicin insertion into lipid bilayer: A QCM-D exploration. *Colloids Surf. B Biointerfaces* 116:472–481.

Wang, X., Quinn, P. J. 2010. Lipopolysaccharide: Biosynthetic pathway and structure modification. *Prog. Lipid Res.* 49:97–107.

Wang, Y., Feng, K., Yang, H., Yuan, Y., Yue, T. 2018. Antifungal mechanism of cinnamaldehyde and citral combination against: *Penicillium expansum* based on FT-IR fingerprint, plasma membrane, oxidative stress and volatile profile. *RSC Advances* 8:5806–5815.

Weber, F. J., De Bont, J. A. M., Bont, J. A. M. De, 1996. Adaptation mechanisms of microorganisms to the toxic effects of organic solvents on membranes. *Biochim. Biophys. Acta - Reviews on Biomembranes* 1286:225–245.

Wendakoon, C. N., Sakaguchi, M. 1995. Inhibition of amino acid decarboxylase activity of *Enterobacter aerogenes* by active components in spices. *J. Food Prot.* 58:280–283.

Wiedemann, I., Breukink, E., Van Kraaij, C., Kuipers, O. P., Bierbaum, G., De Kruijff, B., Sahl, H. G. 2001. Specific binding of nisin to the peptidoglycan precursor lipid II combines pore formation and inhibition of cell wall biosynthesis for potent antibiotic activity. *J. Biol. Chem.* 276:1772–1779.

Wiedemann, I., Benz, R., Sahl, H. G. 2004. Lipid II-Mediated pore formation by the peptide antibiotic nisin: a black lipid membrane study. *J. Bacteriol.* 186:3259–3261.

Wiedemann, I., Böttiger, T., Bonelli, R. R., Wiese, A., Hagge, S. O., Gutsmann, T., Seydel, U., Deegan, L., Hill, C., Ross, P., Sahl, H. G. 2006. The mode of action of the lantibiotic lacticin 3147 - A complex mechanism involving specific interaction of two peptides and the cell wall precursor lipid II. *Molec. Microbiol.* 61:285–296.

Witzke, S., Duelund, L., Kongsted, J., Petersen, M., Mouritsen, O. G., Khandelia, H. 2010. Inclusion of terpenoid plant extracts in lipid bilayers investigated by molecular dynamics simulations. *J. Phys. Chem. B* 114:15825–15831.

Yan, A., Guan, Z., Raetz, C. R. H. 2007. An undecaprenyl phosphate-aminoarabinose flippase required for polymyxin resistance in *Escherichia coli*. *J. Biol. Chem.* 282:36077–36089.

Ye, R., Xu, H. H., Wan, C., Peng, S., Wang, L., Xu, H. H., Aguilar, Z. P., Xiong, Y., Zeng, Z., Wei, H. 2013. Antibacterial activity and mechanism of action of ε-poly-L-lysine. *Biochem. Biophys. Res. Communic.* 439:148–153.

Yethon, J. A., Vinogradov, E., Perry, M. B., Whitfield, C. 2000. Mutation of the lipopolysaccharide core glycosyltransferase encoded by *waaG* destabilizes the outer membrane of *Escherichia coli* by interfering with core phosphorylation. *J. Bacteriol.* 182:5620–5623.

Yoneyama, F., Imura, Y., Ohno, K., Zendo, T., Nakayama, J., Matsuzaki, K., Sonomoto, K. 2009. Peptide-lipid huge toroidal pore, a new antimicrobial mechanism mediated by a lactococcal bacteriocin, lacticin Q. *Antimicrob. Agents Chemother.* 53:3211–3217.

Yoong, P., Schuch, R., Nelson, D., Fischetti, V. A. 2004. Identification of a broadly active phage lytic enzyme with lethal activity against antibiotic-resistant *Enterococcus faecalis* and *Enterococcus faecium*. *J. Bacteriol.* 186:4808–4812.

Zhou, H., Fang, J., Tian, Y., Lu, X. 2014. Mechanisms of nisin resistance in Gram-positive bacteria. *Ann. Microbiol.* 64:413–420.

Zinser, E., Paltauf, F., Daum, G. 1993. Sterol composition of yeast organelle membranes and subcellular distribution of enzymes involved in sterol metabolism. *J. Bacteriol.* 175:2853–2858.

Zuber, B., Haenni, M., Ribeiro, T., Minnig, K., Lopes, F., Moreillon, P., Dubochet, J. 2006. Granular Layer in the periplasmic space of gram-positive bacteria and fine structures of *Enterococcus gallinarum* and *Streptococcus gordonii* septa revealed by cryo-electron microscopy of vitreous sections. *J. Bacteriol.* 188:6652–6660.

Zunino, M. P., Turina, A. V, Zygadlo, J. A., Perillo, M. A. 2011. Stereoselective effects of monoterpenes on the microviscosity and curvature of model membranes assessed by DPH steady-state fluorescence anisotropy and light scattering analysis. *Chirality* 23:867–877.

Index

Made in the USA
Monee, IL
13 September 2026

62978964R10453